The Radio Amateur's Handbook

By the HEADQUARTERS STAFF of the AMERICAN RADIO RELAY LEAGUE

Newington, CT USA 06111

Editor
George Woodward, W1RN

Assistant Editor
George Collins, KC1V

Contributors
Gerald Hall, K1TD
Wes Hayward, W7ZOI
Gerald Hull, VE1CER
Edward Kalin, K1RT
John Lawson, K5IRK
John Lindholm, W1XX
Dennis Lusis, W1LJ
Bob Shriner, WAØUZO
Hal Steinman, K1FHN
Jay Rusgrove, W1VD
Edward Wetherhold, W3NQN

1982

Fifty-Ninth Edition

Foreword

The foreword to each annual edition of the Radio Amateur's Handbook is generally pretty much cut and dried. After one has written the foreword year after year, it becomes each year a little more difficult to be original, to say something different. And so one lapses into a sameness which seems never to be challenged by the readers. Could it be the foreword is never read?

This is the 59th edition of the Handbook, a title which has been published by ARRL since 1925. Its first edition was published when this writer was about five years old, and was nine years away from his first amateur license. His first personal copy, in 1934, was a prized possession which helped him immeasurably during his beginning as a radio amateur. In those early days, never in his wildest dreams did he imagine that someday he might play an active role in the production of this respected title. But that privilege came, and now it is about to go. Before the next edition is printed, the challenge of managing ARRL publications will have been transferred to some other person.

There have through the years been many changes in the identities of those responsible for the Handbook, but the quality and the usefulness of the Handbook have remained consistently high. It is a book which has, since its beginning, sold some six million copies. It is a book which has gotten thicker and heavier with each passing year, as the field of knowledge which it had to cover grew and grew. It is a book that has been constantly revised to meet the changing technology of amateur radio. It is a book which is widely used as a reference manual not only by individual amateurs but by many in professional and government laboratories.

Again this year the Radio Amateur's Handbook has been extensively revised, in an effort to maintain its wide appeal. We hope, and think, that it will serve you well.

Richard L. Baldwin, W1RU
General Manager

Newington, CT
November 1981

Schematic Symbols Used in Circuit Diagrams

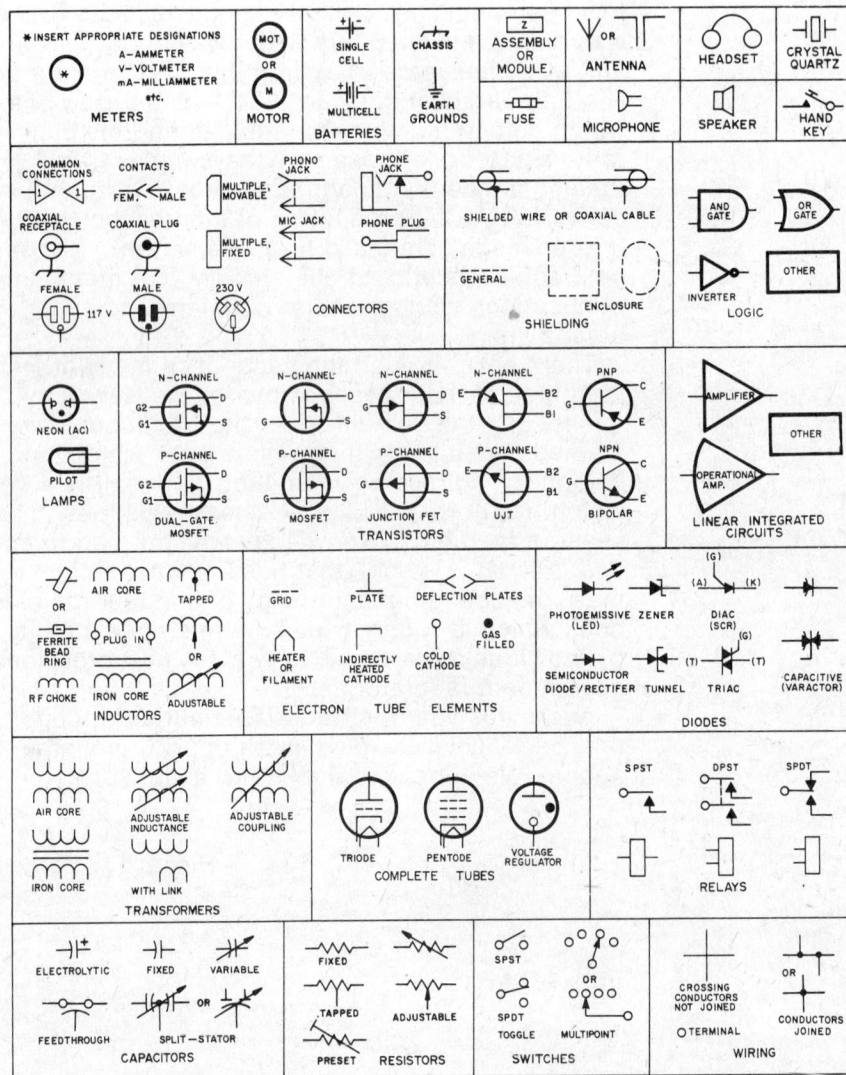

Contents

The Amateur's Code

ONE

The Amateur is Considerate . . . He never knowingly uses the air in such a way as to lessen the pleasure of others.

TWO

The Amateur is Loyal . . . He offers his loyalty, encouragement and support to his fellow radio amateurs, his local club and to the American Radio Relay League, through which Amateur Radio is represented.

THREE

The Amateur is Progressive . . . He keeps his station abreast of science. It is well-built and efficient. His operating practice is above reproach.

FOUR

The Amateur is Friendly . . . Slow and patient sending when requested, friendly advice and counsel to the beginner, kindly assistance, cooperation and consideration for the interests of others; these are marks of the amateur spirit.

FIVE

The Amateur is Balanced . . . Radio is his hobby. He never allows it to interfere with any of the duties he owes to his home, his job, his school, or his community.

SIX

The Amateur is Patriotic . . . His knowledge and his station are always ready for the service of his country and his community.

— PAUL M. SEGAL

Chapter 1

Amateur Radio

For many years, the name Hiram Percy Maxim was synonymous with Amateur Radio. The cofounder and first President of ARRL was the first person to be inducted into the ARRL Hall of Fame. Aside from his pioneering work with radio, Mr. Maxim, who held the call sign W1AW, contributed to the development of the automobile and the movies.

Amateur Radio. You've heard of it. You probably know that Amateur Radio operators are also called "hams." (Nobody knows quite why!) But who are these people and what do they do?

Every minute of every hour of every day, 365 days a year, radio amateurs all over the world communicate with each other. It's a way of discovering new friends while experimenting with different and exciting new ways to advance the art of their hobby. Ham radio is a global fraternity of people with common and yet widely varying interests, able to exchange ideas and learn more about each other with each new on-the-air contact. Because of this Amateur Radio has the ability to enhance international relations as does no other hobby. How else is it possible to talk to an engineer involved in a space program, a Tokyo businessman, a U.S. legislator, a Manhattan store owner, a camper in a Canadian national park, the head of state of a Mediterranean-area country, a student at a high school radio club in Wyoming, or a sailor on board a ship in the middle of the Pacific? And all without leaving your home! Only with Amateur Radio — that's how!

The way communication is accomplished is just as interesting as the people you get to "meet." Signals can be sent around the world using reflective layers of the earth's ionosphere or beamed from point to point from mountaintops by relay stations. Orbiting satellites that hams built are used to achieve communication. Still other hams bounce their signals off the moon! Possibilities are almost unlimited. Not only do radio amateurs use international Morse code and voice for communication, but they also use radioteletype, facsimile and various forms of television. Some hams even have computers hooked up to their equipment. As new techniques and modes of communication are developed, hams continue their long tradition of being among the first to use them.

What's in the future? Digital voice-encoding techniques? Three-dimensional TV? One can only guess. But if there is ever such a thing as a Star Trek transporter unit, hams will probably have them!

Once radio amateurs make sure that their gear does work, they look for things to do with the equipment and special skills they possess. Public service is a very large and integral part of the whole Amateur Radio Service. Hams continue this tradition by becoming involved and sponsoring various activities in their community.

Field Day, just one of many public service-type activities, is an annual event occurring every June when amateurs take their equipment into the great outdoors (using electricity generated at the operation site) and test it for use in case of disaster. Not only do they test their equipment, but they also make a contest out of the exercise and try to contact as many other hams operating emergency-type stations as possible (along with "ordinary" types). Often they make Field Day a club social event while they are operating.

Traffic nets (networks) meet on the airwaves on a schedule for the purpose of handling routine messages for people all over the country and in other countries where such third-party traffic is permitted. By doing so, amateurs stay in practice for handling messages should any real emergency or disaster occur which would require operating skill to move messages efficiently. Nets also meet because the members often have common interests: similar jobs, interests in different languages, different hobbies (yes, some people have hobbies other than ham radio!), and a whole barrelful of other reasons. It is often a way to improve one's knowledge and to share experiences with other amateurs for the good of all involved.

DX (distance) contests are popular and awards are actively sought by many amateurs. This armchair travel is one of the more alluring activities of amateur radio. There are awards for Worked All States (WAS), Worked All Provinces (WAVE), Worked All Continents (WAC), Worked 100 Countries (DXCC), and many others.

Mobile operation (especially on the very high frequencies) holds a special attraction to many hams. It's always fun to keep in touch with ham friends over the local repeater (devices which receive your signal and retransmit it for better coverage of the area) or finding new friends on other frequencies while driving across the country. Mobile units are often the vital link in emergency communications, too, since they are usually first on the scene of an accident or disaster.

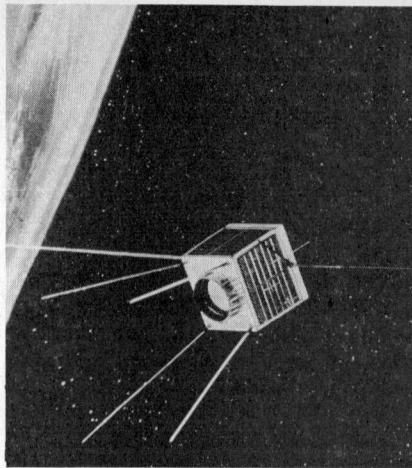

Hurtling through the airless reaches of space, OSCAR 8 is a relay station for amateurs around the world. It provides reliable vhf and uhf communications, and is the focus of a wide range of experiments that advance the radio art. (*Artwork courtesy W6TUF*)

The OSCAR (*O*rbiting *S*atellite *C*arrying *A*mateur *R*adio) program is a relatively new challenge for the Amateur Radio fraternity. Built by hams from many countries around the world, these ingenious devices hitch rides as secondary payloads on space shots for commercial and government communications or weather satellites. OSCAR satellites receive signals from the ground on one frequency and convert those signals to another frequency to be sent back down to earth. Vhf (very high frequency) and uhf (ultra-high frequency) signals normally do not have a range much greater than the horizon, but when beamed to these satellites, a vhf/uhf signal's effective range is greatly increased to make global communication a possibility. These OSCAR satellites also send back telemetry signals either in Morse or radioteleprinter (RTTY) code, constantly giving information on the condition of equipment aboard the satellite.

Self-reliance has always been a trademark of the radio amateur. This is often best displayed by the many hams who design and build their own equipment. Many others prefer to build their equipment from kits. The main point is that hams want to know how their equipment functions, what to do with it and how to fix it if a malfunction should occur. Repair shops aren't always open during hurricanes or floods and they aren't always out in the middle of the Amazon jungle, either. Hams often come up with variations on a circuit design in common use so that they may perform a special function, or a ham may bring out a totally original electronic design, all in the interest of advancing the radio art.

Radio Clubs

Amateur Radio clubs often provide social as well as operational and technical activities. The fun provided by Amateur Radio is greatly enhanced when hams get together so they can "eyeball" (see) each other. It's a good supplement to talking to each other over the radio. The swapping of tales (and sometimes equipment), and a general feeling of high spirits add a bit of spice to club meetings along with technical matters on the agenda. Clubs offer many people their first contact with Amateur Radio by setting up displays in shopping centers and at such events as county fairs, Scout jamborees and parades.

Nearly half of all U.S. amateurs belong to a radio club. And nearly every amateur radio club is affiliated with the ARRL. Club affiliation is available to most organized Amateur Radio groups. The benefits are many: Use of films and slide shows for club classes and meetings, rebate on ARRL membership dues, special publication offers, and other services. Complete information on the requirements and privileges of affiliation is available from the Club and Training Department, ARRL Hq., Newington, CT 06111.

Getting Started in Amateur Radio?

"All of this sounds very interesting and seems to be a lot of fun, but just how do I go about getting into this hobby? Don't you almost need a degree in electronics to pass the test and get a license?"

Nothing could be further from the truth. Although you are required to have a license to operate a station, it only takes a minimal amount of study and effort on your part to pass the basic, entry-grade exam and get on the air.

"But what about the code? Don't I have to know code to get a license?" Yes, you do. International agreements require Amateur Radio operators to have the ability to communicate in international Morse code. But the speed at which you are required to receive it is relatively low so you should have no difficulty. Many grade-school students have passed their tests and each month hundreds of people from 8 to 80 join the ever-growing number of Amateur Radio operators around the world.

Concerning the written exam: To get a license you need to know some basic electrical and radio principles and regulations governing the class of license applied for. The ARRL's basic beginner package, *Tune in the World with Ham Radio*, is available for $8.50 from local radio stores or by mail from ARRL.

Finding Help

One of the first obstacles for a person seriously interested in Amateur Radio is finding a local amateur to provide assistance. This volunteer amateur is called an "Elmer." A nearby ham can help a newcomer with technical advice, putting up and testing antennas, advice on

Whether it's trekking to the North Pole or flying high in the sky, where hams go Amateur Radio often goes as well. Japanese explorer Naomi Uemura, JG1QFW, used Amateur Radio for backup emergency communications during his solo adventure to the North Pole in 1978. Fred Hyde, KØLIS, was one of four crew members on the DaVinci Trans America Balloon, which set a long-distance flight record for balloonists in the continental U.S. before crash-landing in Ohio because of a severe storm. Amateur Radio kept the crew in touch with hams on the ground. (balloon photo courtesy WØHSK)

buying that first radio or just some needed encouragement. Also, nearly all would-be amateurs attend an Amateur Radio class for code, regulations and electronic theory instruction. Where do you find this assistance? The ARRL Club and Training Department helps the prospective amateur in every possible way. It coordinates the work of more than 5000 volunteer Amateur Radio instructors thoughout the United States and Canada and provides a large variety of audio-visual aids and refers inquiries on Amateur Radio to one of the 5000 instructors. If you are looking for an Amateur Radio class or advice on how to get started, write the ARRL Club and Training Department for the name and address of the nearest Elmer.

Looking Back

How did Amateur Radio become the almost unlimited hobby it is today? The beginnings are slightly obscure, but electrical experimenters around the turn of the century, inspired by the experiments of Marconi and others of the time, began duplicating those experiments and attempted to communicate among themselves. There were no regulatory agencies at that time and much interference was caused by these "amateur" experimenters to other stations until governments the world over stepped in and established licensing, laws and regulations to control the problems involved in this new technology. "Amateur" experimenter stations were then restricted to the "useless" wavelengths of 200 meters and below. Amateurs suddenly found that they could achieve communication over

longer distances than commercial stations on the longer wavelengths. Even so, signals often had to be relayed by intermediate amateur stations to get a message to the proper destination. Because of this, the American Radio Relay League was organized to establish routes of Amateur Radio communication and serve the public interest through Amateur Radio. But the dream of eventual transcontinental and even transoceanic Amateur Radio contact burned hot in the minds of Radio Amateur experimenters.

World War I broke out and Amateur Radio, still in its infancy, was ordered out of existence until further notice. Many former Amateur Radio operators joined the armed sevices and served with distinction as radio operators, finding their skills to be much needed.

After the close of the "War to End All Wars," Amateur Radio was still banned by law; yet there were many hundreds of formerly licensed amateurs just itching to "get back on the air." The government had tasted supreme authority over the radio services and was half inclined to keep it. Hiram Percy Maxim, one of the founders of the American Radio Relay League, called the pre-war League's officers together and then contacted all the old members who could be found in an attempt to re-establish Amateur Radio. Maxim traveled to Washington, DC and after considerable effort (and untold red tape) Amateur Radio was opened up again on October 1, 1919.

Experiments on shorter wavelengths were then begun with encouraging results. It was found that as the wavelength

dropped (i.e., frequency increased) greater distances were achieved. The commercial stations were not about to miss out on this opportunity. They moved their stations to the new shorter wavelengths while the battle raged over who had the right to transmit in this new area. Usually, it turned out to be the station with the stronger signal, able to blot out everyone else.

National and international conferences were called in the twenties to straighten out the tangle of wavelength allocations. Through the efforts of ARRL offficials, amateurs obtained frequencies on various bands similar to what we have today: 160 through 6 meters. When the amateur operators moved to 20 meters, the dream of coast-to-coast and transoceanic communication without a relay station was finally realized. (A more detailed history of the early days of Amateur Radio is contained in the ARRL publication *Two Hundred Meters and Down* by Clinton B. DeSoto.)

Public Service

Amateur Radio is a grand and glorious hobby, but this fact alone would hardly merit the wholehearted support given it by nearly all the world's governments at international conferences. There are other reasons. One of these is a thorough appreciation of the value of amateurs as sources of skilled radio personnel in time of war. Another asset is best described as "public service."

The "public service" record of the amateur is a brilliant tribute to his work. These activities can be roughly divided into two classes, expeditions and emergencies. Amateur cooperation with expeditions began in 1923, when a League member, Don Mix, ITS, accompanied MacMillan to the Arctic on the schooner *Bowdoin* with an amateur station. Amateurs in Canada and the U.S. provided the home contacts. The success of this venture was so outstanding that other explorers followed suit. During subsequent years Amateur Radio assisted perhaps 200 voyages and expeditions, the several explorations of the Antarctic being perhaps the best known. And this kind of work is not all in the distant past, either: In 1978 Japanese explorer Naomi Uemura, JG1QFW, became the first person to trek to the North Pole alone. Amateur Radio, through member stationsof the National Capitol DX Association and the Polar Amateur Radio Club, VE8RCS, at Alert, NWT, Canada, provided important backup communications.

Sometimes Mother Nature goes on a rampage — with earthquakes such as those in Alaska in 1964, Peru in 1970, California in 1971, Guatemala in 1976 and Italy in 1980; floods like those in Big Thompson Canyon, Colorado, in 1976, Kentucky, Virginia, West Virginia, and Johnstown, Pennsylvania in 1977, Jackson, Miss. in 1979; the big forest fires of California, particularly in 1977; tornadoes, hurricanes and typhoons, most

Table 1

Canadian Amateur Bands

Band (limitations)	Frequency (MHz)	Emissions
80 meters	3.500-3.725	A1, F1
(1, 3, 4, 5)	3.725	A1, A3, F3
40 meters	7.000-7.050	A1, F1
(1, 3, 4, 5)	7.050-7.100	A1, A3, F1
	7.100-7.150	A1, F1
	7.150-7.300	A1, A3, F3
20 meters	14.000-14.100	A1, F1
(1, 3, 4, 5)	14.100-14.350	A1, A3, F3
15 meters	21.000-21.100	A1, F1
(1, 3, 4, 5)	21.000-21.450	A1, A3, F3
10 meters	28.000-28.100	A1, F1
(2, 3, 4, 5)	28.100-29.700	A1, A3, F3
6 meters	50.000-50.050	A1
(3, 4)	50.050-51.000	A1, A2, A3, F1, F2, F3
	51.000-54.000	AØ, A1, A2, A3, A4, F1, F2, F3, F4
2 meters	144.000-144.100	A1
(3,4)	144.100-145.500	AØ, A1, A2, A3, A4, F1, F2, F3, F4
(3, 4, 7)	144.500-145.800	PØ, P1, AØ, A1, A2, A3, A4, F1, F2, F3, F4
(3,4)	145.800-148.000	AØ, A1, A2, A3, A4, F1, F2, F3, F4
(3,4)	220.000-220.100	AØ, A1, A2, A3, A4, F1, F2, F3, F4
(9, 10, 13, 15)	220.000-220.500	
(9, 10, 13, 15)	220.500-221.000	
(10, 13, 14, 15)	221.000-223.000	
(9, 12, 13, 15)	223.000-223.500	
(3, 4)	223.500-225.000	AØ, A1, A2, A3, A4, F1, F2, F3, F4
(4, 6)	430.000-433.000	AØ, A1, A2, A3, A4, A5, F1, F2, F3, F4, F5
(12, 13, 14, 15)	433.000-434.000	
(3, 4, 8)	434.000-434.500	PØ, P1, P2, P3, AØ, A1, A2, A3, A4, A5, F1, F2, F3, F4, F5
(4, 6)	434.500-450.000	AØ, A1, A2, A3, A4, A5, F1, F2, F3, F4, F5
	902.000-928.000	A3, F3
	1215.000-1300.000	AØ, A1, A2, A3, A4, A5, F1, F2, F3, F4, F5
	2300.000-2450.000	AØ, A1, A2, A3, A4, A5, F1, F2, F3, F4, F5, PØ, P1, P2, P3, P4, P5, P9
	3300.000-3500.000	AØ, A1, A2, A3, A4, A5, F1, F2, F3, F4, F5, PØ, P1, P2, P3, P4, P5, P9
	5650.000-5925.000	AØ, A1, A2, A3, A4, A5, F1, F2, F3, F4, F5, PØ, P1, P2, P3, P4, P5, P9
	10000.000-10500.000	AØ, A1, A2, A3, A4, A5, F1, F2, F3, F4, F5, PØ, P1, P2, P3, P4, P5, P9
(14, 15)	24000.000-24010.000	
	24010.000-24250.000	AØ, A1, A2, A3, A4, A5, F1, F2, F3, F4, F5, PØ, P1, P2, P3, P4, P5, P9

Limitations

1) Phone privileges are restricted to holders of advanced Amateur Radio Operators Certificates, and of Commercial Certificates.

2) Phone privileges are restricted as in footnote 1, and to holders of Amateur Radio Operators Certificates, whose certificates have been endorsed for operation on phone in these bands.

3) Amplitude modulation (A2, A3, A4) shall not exceed ± 3 kHz (6A3).

4) Frequency modulation (F2, F3, F4) shall not produce a carrier deviation exceeding ± 3 kHz, (6F3) except that in the 52.54 MHz and 144.1-148 MHz bands and higher the carrier deviation shall not exceed ± 15 kHz (30F3).

5) Slow scan television (A5), permitted by special authorization, shall not exceed a bandwidth greater than that occupied by a normal single-sideband voice transmission.

6) Television (A5), permitted by special authorization, shall employ a system of standard interlace and scanning with a bandwidth of not more than 4 MHz.

7) Pulse modulation with any mode of transmission shall not produce signals of a bandwidth exceeding 15 kHz.

8) Pulse modulation with any mode of transmission shall not produce signals of a bandwidth exceeding 30 kHz.

9) Any mode may be used.

10) Packet transmissions shall not produce signals exceeding 10 kHz.

11) Packet transmissions shall not produce signals exceeding 25 kHz.

12) Packet transmissions shall not produce signals exceeding 100 kHz.

13) Licensees performing an Amateur Experimental Service may use such modulation techniques or types of emission for packet transmission as they may select by experimentation on conditions that they do not exceed the bandwidths established in 10, 11 and 12.

14) Only packet transmissions shall be used.

15) Final rf output power used for packet transmissions shall not exceed 100 watts peak power and 10 watts average power.

Operation in frequency band 1.800-2.000 MHz shall be limited to the area as indicated in the following table and shall be limited to the indicated maximum dc power input to the anode of the final radio frequency stage of the transmitter during day and night hours respectively; for the purpose of this table "day" means the hours between sunrise and sunset, and "night" means the hours between sunset and sunrise. A1, A3 and F3 emissions are permitted.

	A	B	C	D	E	F	G	H
British Columbia	3¹	3	3	1	0	0	0	0
Alberta	3¹	3	3	3	1	0	0	1
Saskatchewan	3¹	3	3	3	3	1	1	3
Manitoba	3¹	2	2	2	2	2	2	3¹
Ontario North of 50° N.	3	1	1	1	1	0	0	2
Ontario South of 50° N.	3¹	2	1	0	0	0	0	1
Province of Quebec North of 52° N.	1	0	0	1	1	0	0	2
Province of Quebec South of 52° N.	3	2	1	0	0	0	0	0
New Brunswick	3	2	1	0	0	0	0	0
Nova Scotia	3	2	1	0	0	0	0	0
Prince Edward Island	3	2	1	0	0	0	0	0
Newfoundland (Island)	3	1	1	0	0	0	0	0
Newfoundland (Labrador)	2	0	0	0	0	0	0	0
Yukon Territory	3¹	3	1	0	0	0	0	0
District of MacKenzie	3¹	3	3	3	1	0	0	1
District of Keewatin	3	1	1	3	2	0	0	2
District of Franklin	0	0	0	0	1	0	0	1

The power levels 500 day/100 night may be increased to 1000 day/200 night when authorized by a Radio Inspector of the Department of Communications.

Frequency Band

A 1.800-1.825 MHz E 1.900-1.925 MHz
B 1.825-1.850 MHz F 1.925-1.950 MHz
C 1.850-1.875 MHz G 1.950-1.975 MHz
D 1.875-1.900 MHz H 1.975-2.000 MHz

Power Level — Watts

0 — Operation not permitted
1 — 25 night   125 day
2 — 50 night   250 day
3 — 100 night   500 day

anywhere, any year, and the blizzards of 1979 and 1980. When disaster strikes, amateurs are ready, with equipment not needing power from the electric company, to carry on communications for police, fire departments, and relief organizations. The ability of radio amateurs to help the public in emergencies is one big reason Amateur Radio has survived and prospered.

Technical Developments

Amateurs started the hobby with spark-gap transmitters, which took up great hunks of frequency space. Then they moved on to tubes when these devices came along. Much later, transistors were utilized; now integrated circuits are a part of the everyday hardware in the Amateur Radio shack. This is because the amateur is constantly in the forefront of technical progress. His incessant curiosity and eagerness to try anything new are two reasons. Another is that ever-growing Amateur Radio continually overcrowds its frequency assignments, spurring amateurs to the development and adoption of new techniques to permit the accommodation of more stations.

Amateurs have come up with ideas in their shacks while at home and then taken them to industry with surprising results. During World War II, thousands of skilled amateurs contributed their knowledge to the development of secret radio devices, both in government and private laboratories. Equally as important, the prewar technical progress by amateurs provided the keystone for the development of modern military communications equipment.

In the fifties, the Air Force was faced with converting its long range communications from Morse to voice; jet bombers had no room for skilled radio

Table 2
U.S. Amateur Radio Frequency Allocations

Frequency Band	Emissions	Limitations	Frequency Band	Emissions	Limitations	Frequency Band	Emissions	Limitations
kHz								
1800-1900	A1, A3		28.000-29.700	A1			A4, A5, Fø, F1, F2, F3, F4, F5, P	5, 8
1900-2000	A1, A3	1, 2	28.000-28.500	F1		3300-3500	Aø, A1, A2, A3, A4, A5, Fø, F1, F2, F3, F4, F5, P	5, 12
3500-4000	A1		28.500-29.700	A3, F3, A5, F5		5650-5925	Aø, A1, A2, A3, A4, A5, Fø, F1, F2, F3, F4, F5, P	5, 9
3500-3775	F1		50.000-54.000	A1				
3775-3890	A5, F5		50.100-54.000	A2, A3, A4, A5 F1, F2, F3, F5				
3775-4000	A3, F3	4	51.000-54.000	Aø		*GHz*		
4383.8	A3J/A3A	13	144-148	A1		10.0-10.5	Aø, A1, A2, A3, A4, A5, Fø, F1, F2, F3, F4, F5	5
7000-7300	A1	3, 4	144.100-148.000	Aø, A2, A3, A4 A5, Fø, F1, F2, F3, F5		24.0-24.25	Aø, A1, A2, A3, A4, A5, Fø, F1, F2, F3, F4, F5	5,10
7000-7150	F1	3, 4				48-50, 71-76	Aø, A1, A2, A3, A4, A5, Fø, F1, F2, F3, F4, F5, P	
7075-7100	A3, F3	11	220-225	Aø, A1, A2, A3, A4, A5, Fø, F1, F2, F3, F4, F5	5	165-170, 240-250		
7150-7225	A5, F5	3, 4	420-450	Aø, A1, A2, A3, A4, A5, Fø, F1, F2, F3, F4, F5	5,7	Above 300	Aø, A1, A2, A3, A4, A5, Fø, F1, F2, F3, F4, F5, P	
7150-7300	A3, F3	3, 4	1215-1300	Aø, A1, A2, A3, A4, A5, Fø, F1, F2, F3, F4, F5	5			
14000-14350	A1		2300-2450	Aø, A1, A2, A3,				
14000-14200	F1							
14200-14275	A5, F5							
14200-14350	A3, F3							
MHz								
21.000-21.450	A1							
21.250-21.350	A5, F5							
21.250-21.450	A3, F3							

Limitations

1,2) There are certain power and operating restrictions at 1900-2000 kHz. Contact ARRL Hq., or see July 1981 *QST*, p. 10, for details.

3) Where, in adjacent regions or subregions, a band of frequencies is allocated to different services of the same category, the basic principle is the equality of right to operate. Accordingly, the stations of each service in one region or subregion must operate so as not to cause harmful interference to services in the other regions or subregions (No. 117, the Radio Regulations, Geneva, 1959).

4) 3900-4000 kHz and 7100-7300 kHz are not available in the following U.S. possessions: Baker, Canton, Enderbury, Guam, Howland, Jarvis, the Northern Mariana Islands, Palmyra, American Samoa and Wake Islands.

5) Amateur stations shall not cause interference to the Government radio-location service.

6) (Reserved)

7) In the following areas dc plate input power to the final transmitter stage shall not exceed 50 watts, except when authorized by the appropriate Commission engineer in charge and the appropriate military area frequency coordinator.

i) Those portions of Texas and New Mexico bounded by latitude 33°24′ N., 33°53′ N., and longitude 105°40′ W. and 106°40′ W.

ii) The state of Florida, including the Key West area and the areas enclosed within circles of 200-mile radius centered at 28°21′ N, 80°43′ W. and 30°30′ N., 86°30′ W.

iii) The state of Arizona.

iv) Those portions of California and Nevada south of latitude 37°10′ N. and the area within a 200-mile radius of 34°09′ N., 119°11′ W.

8) No protection in the band 2400-2450 MHz is afforded from interference due to the operation of industrial, scientific and medical devices on 2450 MHz.

9) No protection in the band 5725-5875 is afforded from interference due to the operation of industrial scientific and medical devices on 5800 MHz.

10) No protection in the band 24.00-24.25 GHz is afforded from interference due to the operation of industrial, scientific and medical devices on 24.125 GHz.

11) The use of A3 and F3 in this band is limited to amateur radio stations located outside Region 2.

12) Amateur stations shall not cause interference to the Fixed-Satellite Service operating in the band 3400-3500 MHz.

13) The frequency 4383.8 kHz, maximum power 150 watts, may be used by any station authorized under this part of communications with any other station authorized in the state of Alaska for emergency communications. No airborne operations will be permitted on this frequency. Additionally, all stations operating on this frequency must be located in or within 50 nautical miles of the state of Alaska.

14) All amateur frequency bands above 29.5 MHz are available for repeater operation except 50.0-52.0 MHz, 144.0-144.5 MHz, 145.5-146.0 MHz, 220.0-220.5 MHz, 431.0-433.0 MHz, and 435.0-438.0 MHz. Both the input (receiving) and output (transmitting) frequencies of a station in repeater operation shall be frequencies available for repeater operation.

15) All amateur frequency bands above 220.5 MHz, except 431-433 MHz, and 435-438 MHz, are available for auxiliary operation.

NOTE
The types of emission referred to in the amateur rules are as follows:
Type Aø — Steady, unmodulated pure carrier.
Type A1 — Telegraphy on pure continuous waves.
Type A2 — Amplitude tone-modulated telegraphy.
Type A3 — A-m telephony including single and double sideband, with full, reduced or suppressed carrier.
Type A4 — Facsimile.
Type A5 — Television.
Type Fø — Steady, unmodulated pure carrier.
Type F1 — Carrier-shift telegraphy.
Type F2 — Audio frequency-shift telegraphy.
Type F3 — Frequency- or phase-modulated telephony.
Type F4 — Fm facsimile.
Type F5 — Fm television.
Type P — Pulse emissions.

operators. At the time, amateurs had been using single sideband for about a decade, and were communicating by voice at great distances with both homemade and commercially built equipment. Generals LeMay and Griswold, both radio amateurs, hatched an experiment in which ham equipment was used to keep in touch with Strategic Air Command headquarters in Omaha, Nebraska, from an airplane travelling around the world. The system worked well; the equipment needed only slight modification to meet Air Force needs, and the expense and time of normal research and development procedures was saved.

Many youngsters build an early interest in Amateur Radio into a career. Later, as professionals, they may run into ideas which they try out in ham radio. A good example is the OSCAR series of satellites, initially put together by amateurs who worked in the aerospace industry, and launched as secondary payloads with other space shots. At this writing eight Amateur Radio satellites have been launched. OSCARs 7 and 8, portions of which were built by amateurs of several different countries, are currently in space relaying the signals of amateurs. OSCARs 7 and 8 can be heard on almost any 29-MHz receiver. Development of third-generation Phase III satellites proceeds under the guidance of The Radio Amateur Satellite Corporation (AMSAT) with the assistance of Project OSCAR, Inc., the original nonprofit group, both affiliated with ARRL. The Phase III program was temporarily set back in May of 1980 when the first satellite in the series, Phase III-A, ended up in the Atlantic Ocean because of a malfunctioning launch vehicle. It is a credit to its builders that up to the mo-

ment of its demise, the satellite functioned flawlessly. The thousands of hours of experience that went into the design and construction of Phase III-A are now being used as the starting point for its successor, Phase III-B. The new Phase III satellites being built by AMSAT will continue the OSCAR program as older spacecraft are taken out of service. Write ARRL for more information.

The American Radio Relay League

Since its establishment in 1914 by Hiram Percy Maxim and Clarence Tuska, the American Radio Relay League has been and is today not only the spokesman for Amateur Radio in the U.S. and Canada, but also the largest amateur organization in the world. It is strictly of, by and for amateurs, is noncommercial and has no stockholders. The members of the League are the owners of the ARRL and *QST*, the monthly journal of Amateur Radio published by the League.

The League is pledged to promote interest in two-way amateur communication and experimentation. It is interested in the relaying of messages by Amateur Radio. It is concerned with the advancement of the radio art. It stands for the maintenance of fraternalism and a high standard of conduct. It represents the amateur in legislative matters.

One of the League's principal purposes is to keep amateur activities so well conducted that the amateur will continue to justify his existence. Amateur Radio offers its followers countless pleasures and unending satisfaction. It also calls for the shouldering of responsibilities — the maintenance of high standards, a cooperative loyalty to the traditions of Amateur Radio, a dedication to its ideals

and principles — so that the institution of Amateur Radio may continue to operate "in the public interest, convenience and necessity."

In addition to publishing *QST*, the ARRL maintains an active Amateur Radio station, W1AW, which conducts code practice and sends bulletins of interest to amateurs the world over. ARRL sponsors an Intruder Watch Program so that unauthorized use of the amateur radio frequencies may be detected and appropriate action taken. At the Headquarters of the League in Newington, Connecticut, is a well-equipped laboratory to assist staff members in preparation of technical material for *QST* and *The Radio Amateur's Handbook*. Among its other activities, the League maintains a Communications Department concerned with the operating activities of League members. A large field organization is headed by a section communications manager in each of the League's 73 sections. There are appointments for qualified members in various fields, as outlined in chapter 22. Special activities and contests promote operating skill. A special place is reserved each month in *QST* for amateur news from every section.

The ARRL publishes a library of information on Amateur Radio. *Tune in the World with Ham Radio* is written for the person without previous contact with Amateur Radio. It is designed to assist the prospective amateur to get into the hobby in the shortest possible time. *Tune in the World* comes complete with a code instruction and practice tape. For the person seeking the General class or higher license, there are the *License Manual* and the *ARRL Code Kit*. The ARRL also

---

	PAUSE BETWEEN ELEMENTS (ONE UNIT PULSE)	PAUSE BETWEEN CHARACTERS (THREE UNIT PULSES)	PAUSE BETWEEN WORDS (SEVEN UNIT PULSES)

DIDAHDIT DIDAH DAHDIDIT DIDIT DAHDAHDAH

TIME →	DOT LENGTH (ONE UNIT PULSE)	DASH LENGTH (THREE UNIT PULSES)			

A	didah	• —	I	didit	••	S	dididit	•••
B	dahdididit	— •••	J	didahdahdah	• — — —	T	dah	—
C	dahdidahdit	— • — •	K	dahdidah	— • —	U	dididah	•• —
D	dahdidit	— ••	L	didahdidit	• — ••	V	didididah	••• —
E	dit	•	M	dahdah	— —	W	didahdah	• — —
F	dididahdit	•• — •	N	dahit	— •	X	dahdididah	— •• —
G	dahdahdit	— — •	O	dahdahdah	— — —	Y	dahdidahdah	— • — —
H	didididit	••••	P	didahdahdit	• — — •	Z	dahdahdidit	— — ••
			Q	dahdahdidah	— — • —			
			R	didahdit	• — •			

1	didahdahdahdah	• — — — —	4	didididah	•••• —	8	dahdahdahdidit	— — — ••
2	dididahdahdah	•• — — —	5	dididididit	•••••	9	dahdahdahdahdit	— — — — •
3	didididahdah	••• — —	6	dahdidididit	— ••••	0	dahdahdahdahdah	— — — — —
			7	dahdahdidididit	— — •••			

Period: didahdidahdidah.   • — • — • —
Comma: dahdahdididahdah.   — — •• — —
Question mark: dididahdahdidit.   •• — — ••
Error: dididididididit.   ••••••••

Double dash: dahdidididah.   — ••• —
Colon: dahdahdahdididit.   — — — •••
Semicolon: dahdidahdidahdit.   — • — • — •
Parenthesis: dahdidahdahdidah.   — • — — • —
Fraction bar: dahdidididahdit.   — •••• — ...

Wait: didahdididit.   • — •••
End of message: didahdidahdit.   • — • — •
Invitation to transmit: dahdidah.   — • —
End of work: dididahdidahdah.   ••• — • —

A code practice oscillator that can be constructed from commonly available parts.

Assembly photo of the code practice oscillator. The loudspeaker is cemented to the front panel, as is the transformer. A single screw and a dab of cement support the circuit board, and the battery is secured to the panel by an elastic band anchored to convenient points.

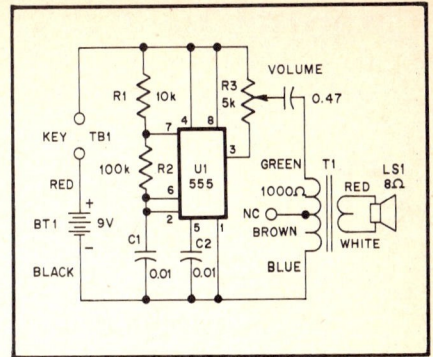

Fig. 1 — Schematic diagram of the code practice oscillator, a useful accessory for newcomers to Amateur Radio.
BT1 — 9-V battery, RS 23-553, 23-583 or equiv.
C1, C2 — Capacitor, 0.01 μF, RS 272-131, 272-1051, 272-1065.
C3 — Capacitor, 0.47 μF, RS 272-1054, 272-1071.
LS1 — 2-in. loudspeaker, 8 Ω, RS 40-245.
R1 — Resistor, 10 kΩ, ±5%, 1/4 W, RS 271-1335.
R2 — Resistor, 100 kΩ, ±5%, 1/4 W, RS 271-1347.
R3 — Potentiometer, 5 kΩ, RS 271-1714, 271-1720.
T1 — Audio output transformer, 1 kΩ CT to 8 Ω, RS 273-1380.
TB1 — Terminal strip, RS 274-663.
U1 — IC timer, type 555, RS 276-1723.
Battery connector clip — RS 270-325.

publishes a series of question and answer manuals for each class of license. All are available from either your local radio store or the ARRL.

Once you have studied, taken the test and have received your license, you will find that there is no other thrill quite the same as Amateur Radio. You may decide to operate on the lowest amateur band, 160 meters. Or you may prefer to operate in the gigahertz bands (billions of cycles per second), where the entire future of communications may lie. Whatever your interest, you are sure to find it in Amateur Radio.

The International Morse Code

Knowledge of and some proficiency with the Morse code is a requirement for all amateur license classes in the United States. The alphanumeric characters and common punctuation marks and prosigns are reproduced symbolically on the previous page. The three fundamental elements of the code are the *dot, dash* and *space*. Dots and dashes are short and long signals, respectively, referenced to a constant time interval called the *unit pulse*. A dot and space each occupy one unit pulse, while a dash occupies three unit pulses. Each dot or dash is separated from others within a character by one unit pulse. The pause between characters is three unit pulses, and the pause between words is seven unit pulses. The chart shows the proper timing for the message "amateur radio" diagrammed with the Morse code.

Morse code communication can be accomplished by means of flashing lights, punched paper tape and even a signal flag, but in radio the code takes the form of *sound*. Short elements make the characteristic sound *dit*, while long elements sound like *dah*. When properly drawn, the dot and dash symbols convey complete information about a code character, but when interpreted orally, the letter "L" becomes "didahdidit" rather than "dot dash dot dot." In the chart, the Morse code is rendered phonetically as well as symbolically.

A Code Practice Oscillator

The simple oscillator described here is a good introduction to electronic construction as well as a useful tool for learning to send the Morse code. Radio Shack stock numbers are given for all components to simplify the parts-acquisition problem for the neophyte. However, the $15 outlay can be dramatically cut through intelligent substitution and liberal scrounging from a ham's "junk box."

The schematic diagram and photographs show the unit in detail, but exact duplication is not necessary. Construction techniques are presented in Chapter 17, and the schematic symbols are printed near the front of this volume. The project can easily be assembled in an evening. This code practice oscillator is intended for beginners using hand keys. Because the circuit is keyed in the positive battery line, the oscillator may not be compatible with some electronic keyers.

Chapter 2

Electrical Laws and Circuits

Some of the manifestations of electricity and magnetism are familiar to everyone. The effects of static electricity on a dry, wintry day, an attraction by the magnetic north pole to a compass needle, and the propagation and reception of radio waves are just a few examples. Less easily recognized as being electrical in nature perhaps, the radiation of light and even radiant heat from a stove are governed by the same physical laws that describe a signal from a TV station or an amateur transmitter. The ability to transmit electrical energy through space without any reliance on matter that might be in that space (such as in a vacuum) or the creation of a disturbance in space that can produce a force are topics that are classified under the study of *electromagnetic fields*. Knowledge of the properties and definitions of fields is important in understanding such devices as transmission lines, antennas, and circuit-construction practices such as shielding.

Once a field problem is solved, it is often possible to use the results over and over again for other purposes. The field solution can be used to derive numerical formulas for such entities as *resistance, inductance* and *capacitance* or the latter quantities can be determined experimentally. These elements, then, form the building blocks for more complex configurations called *networks* or *circuits*. Since there is no need to describe the physical appearance of the individual elements, a pictorial representation is often used and it is called a *schematic diagram*. However, each element must be assigned a numerical value, otherwise the schematic diagram is incomplete. If the numerical values associated with the sources of energy (such as batteries or generators) are also known, it is then possible to determine the power trans-

ferred from one part of the circuit to another element by finding the numerical values of entities called *voltage* and *current*.

Finally, there is the consideration of the fundamental properties of the matter that makes up the various circuit elements or devices. It is believed that all matter is made up of complex structures called *atoms* which in turn are composed of more or less unchangeable particles called *electrons, protons* and *neutrons*. Construction of an atom will determine the chemical and electrical properties of matter composed of like atoms. The periodic table of chemical elements is a classification of such atoms. Electrons play an important role in both the chemical and electrical properties of matter and elements where some of the electrons are relatively free to move about. These materials are called *conductors*. On the other hand, elements where all of the electrons are tightly bonded in the atomic structure are called *insulators*. Metals such as copper, aluminum, and silver are very good conductors while glass, plastics, and rubber are good insulators.

Although electrons play the principal role in the properties of both insulators and conductors, it is possible to construct matter with an apparent charge of opposite sign to that of the electron. Actually, the electron is still the charge carrier but it is the physical absence of an electron location that moves. However, it is convenient to consider that an actual charge carrier is present and it has been labeled a *hole*. Materials in which the motion of electrons and holes determine the electrical characteristics are called *semiconductors*. Transistors, integrated circuits and similar *solid-state devices* are made up from semiconductors. While

there are materials that fall in between the classifications of conductor and insulator, and might be labeled as semiconductors, the latter term is applied exclusively to materials where the motion of electrons and holes is important.

Electrostatic Field and Potentials

All electrical quantities can be expressed in the fundamental dimensions of *time, force* and *length*. In addition, the quantity or dimension of *charge* is also required. The metric system of units (SI — *Systeme International d'Unites*) is almost exclusively used now to specify such quantities, and the reader is urged to become familiar with this system. In the metric system, the basic unit of charge is the *coulomb*. The smallest known charge is that of the electron which is -1.6×10^{-19} coulombs. (The proton has the same numerical charge except the sign is positive.)

The concept of electrical charge is analogous to that of mass. It is the mass of an object that determines the force of gravitational attraction between the object and another one. A similar phenomenon occurs with two charged objects. If the charges can be considered to exist at points in space, the force of attraction (or repulsion if the charges have like signs) is given by the formula

$$F = \frac{Q_1 Q_2}{4 \pi \varepsilon r^2}$$

where Q_1 is the numerical value of one charge, Q_2 is the other charge value, r is the distance in meters, ε is the *permittivity* of the medium surrounding the charges, and F is the force in newtons. In the case of free space of a vacuum, ε has a value of 8.854×10^{-12} · and is the permittivity of

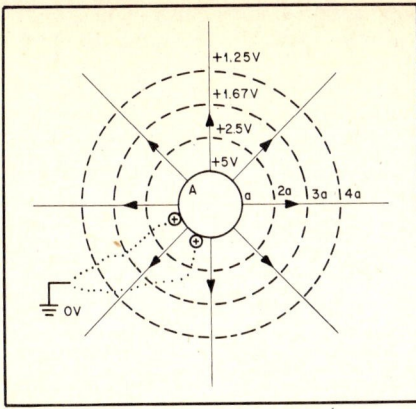

Fig. 1 — Field (solid lines) and potential (dotted lines) lines surrounding a charged sphere.

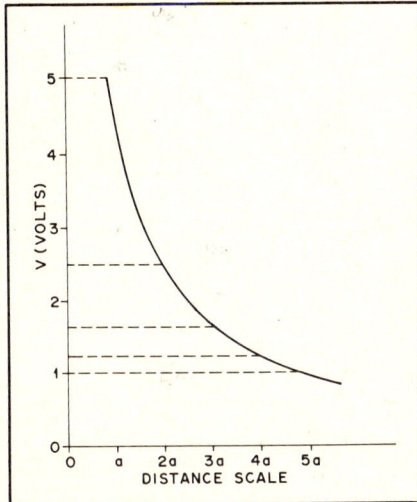

Fig. 2 — Variation of potential with distance for the charged sphere of Fig. 1.

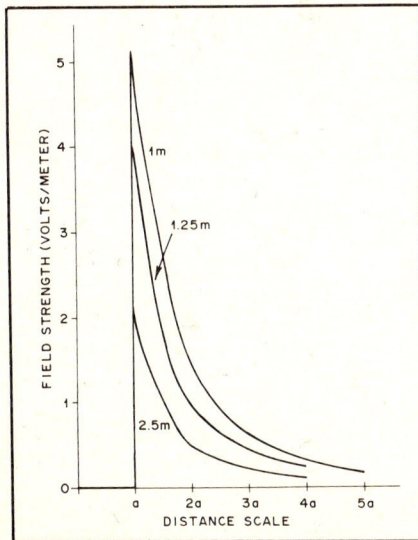

Fig. 3 — Variation of field strength with distance around a sphere charged to 5 volts for spheres of different radii.

free space. The product of relative permittivity and ε_o (the permittivity of free space) gives the permittivity for a condition where matter is present. Permittivity is also called the *dielectric constant* and relative dielectric constants for plastics such as polyethylene and Teflon are 2.26 and 2.1, respectively. (The relative dielectric constant is also important in transmission-line theory. The reciprocal of the square root of the dielectric constant of the material used to separate the conductors in a transmission line gives the *velocity factor* of the line. The effect of velocity factor will be treated in later chapters.)

If instead of just two charges, a number of charged objects are present, the force on any one member is likely to be a complicated function of the positions and magnitudes of the other charges. Consequently, the concept of *electric-field strength* is a useful one to introduce. The field strength or *field intensity* is defined as the force on a given charge (concentrated at a point) divided by the numerical value of the charge. Thus, if a force of 1 newton existed on a test charge of 2 coulombs, the field intensity would be 0.5 newtons/coulomb.

Whenever a force exists on an object, it will require an expenditure of energy to move the object against that force. In some instances, the mechanical energy may be recovered (such as in a compressed spring) or it may be converted to another form of energy (such as heat produced by friction). As is the case for electric-field intensity, it is convenient to express energy ÷ charge as the *potential* or voltage of a charged object at a point. For instance, if it took the expenditure of 5 newton-meters (5 joules) to move a charge of 2 coulombs from a point of zero energy to a given point, the voltage or potential at that point would be 2.5 joules/coulomb. Because of the frequency of problems of this type, the dimension of joules/coulombs is given a special designation and one joule/coulomb is defined as 1 *volt*. Notice that if the voltage is divided by length (meters), the dimensions of field intensity will be obtained and a field strength of one newton/coulomb is also defined as one volt/meter. The relationship between field intensity and potential is illustrated by the following example shown in Fig. 1.

A conducting sphere receives a charge until its surface is at a potential of 5 volts. As charges are placed on the surface of a conductor, they tend to spread out into a uniform distribution. Consequently, it will require the same amount of energy to bring a given amount of charge from a point of zero reference to any point on the sphere. The outside of the sphere is then said to constitute an *equipotential* surface. Also, the amount of energy expended will be independent of the path traveled to get to the surface. For instance, it will require 5

joules of energy to bring a charge of 1 coulomb from a point of 0 voltage to any point on the sphere (as indicated by the dotted lines in Fig. 1). The *direction* of the force on a charged particle *at the surface* of the surface of the sphere must be perpendicular to the surface. This is because charges are able to flow about freely *on* the conductor but not *off* it. A force with a direction other than a right angle to the surface will have a component that is parallel to the conductor and will cause the charges to move about. Eventually, an equilibrium condition will be reached and any initial field component parallel to the surface will be zero. This motion of charge under the influence of an electric field is a very important concept in electricity. The *rate* at which charge flows past a reference point is defined as the current. A rate of 1 coulomb per second is defined as 1 *ampere*.

Because of the symmetry involved, the direction of the electric force and electric field can be represented by the solid straight lines in Fig. 1. The arrows indicate the direction of the force on a positive charge. At points away from the sphere, less energy will be required to bring up a test charge from zero reference. Consequently, a series of concentric spherical shells indicated by the dashed lines will define the equipotential surfaces around the sphere. From mathematical considerations (which will not be discussed here), it can be shown that the potential will vary as the inverse of the distance from the center of the sphere. This relationship is indicated by the numbers in Fig. 1 and by the graph in Fig. 2.

While the electric field gives the direction and magnitude of a force on a charged object, it is also equal to the negative *slope* numerical value of the curve in Fig. 2. The slope of a curve is the rate of change of some variable with distance and in this case, the variable is the potential. This is why the electric field is sometimes called the *potential gradient* (gradient being equivalent to slope). In the case of a curve that varies as the inverse of the distance, the slope at any point is proportional to the inverse of the distance squared.

An examination of Fig. 1 would indicate that the potential variation is only dependent upon the shape of the conductor and not its actual physical size. That is, once the value of the radius a of the sphere in Fig. 1 is specified, the potential at any other point a given distance from the sphere is also known. Thus, Fig. 1 can be used for any number of spheres with different radii. When it is changed by a certain percentage, all the other values would change by the same percentage too. However, the amount of charge required to produce a given voltage, or voltage change, does depend upon the size of the conductor, its shape, and its position in relation to other conductors and insulators. For a given conductor configuration, the voltage is related to the required charge by the formula

$$V = \frac{Q}{C}$$

where the entity C is defined as the capacitance. Capacitance will be discussed in more detail in a later section.

Since the electric-field intensity is related to the change in potential with distance, like potential, the manner in which it changes will be unaffected by the absolute physical size of the conductor configuration. However, the exact numerical value at any point *does* depend on the dimensions of the configuration. This is illustrated in Fig. 3 for spheres with different radii. Note that for larger radii, the numerical value of the field strength at the surface of the sphere (distance equal to a) is less than it is for smaller radii. This effect is important in the design of transmission lines and *capacitors*. (A capacitor is a device for storing charge. In older terminology, it was sometimes called a *condenser*.) Even though the same voltage is applied across the terminals of a transmission line or capacitor, the field strength between the conductors is going to be higher for configurations of small physical size than it is for larger ones. If the field strength becomes too high, the insulating material (including air) can "break down." On the other hand, the effect can be used to advantage in spark gaps used to protect equipment connected to an antenna which is subject to atmospheric electricity. The spark-gap conductors or electrodes are filed to sharp points. Because the needlepoints appear as conductors of very small radii, the field strength is going to be higher for the same applied potential than it would be for blunt electrodes (Fig. 4). This means the separation can be greater and the effect of the spark gap on normal circuit operation will not be as pronounced. However, a blunt electrode such as a sphere is often

used on the tip of a whip antenna in order to lower the field strength under transmitting conditions.

An examination of Fig. 3 reveals that the field strength is zero for distances less than a which includes points *inside* the sphere. The implication here is that the effect of fields and charges cannot penetrate the conducting surface and disturb conditions inside the enclosure. The conducting sphere is said to form an *electrostatic shield* around the contents of the enclosure. However, the converse is not true. That is, charges inside the sphere will cause or induce a field on the outside surface. This is why it is very important that enclosures designed to confine the effects of charges be connected to a point of zero potential. Such a point is often called a *ground*.

Fields and Currents

In the last section, the motion of charged particles in the presence of an electric field was mentioned in connection with charges placed on a conducting sphere and the concept of current was introduced. It was assumed that charges could move around unimpeded on the surface of the sphere. In the case of actual conductors, this is not true. The charges appear to bump into atoms as they move through the conductor under the influence of the electric field. This effect depends upon the kind of material used. Silver is a conductor with the least amount of opposition to the movement of charge while carbon and certain alloys of iron are rather poor conductors of charge flow. A measure of how easily charge can flow through a conductor is defined as the conductivity and is denoted by σ.

The *current density* J, in a conductor is the rate of charge flow or current through a given cross-sectional area. It is related to the electric field and conductivity by the formula

$$J = \sigma E$$

In general, the conductivity and electric field will not be constant over a large

cross-sectional area, but for an important theoretical case this is assumed to be true (Fig. 5).

A cylinder of a material with conductivity σ is inserted between two end caps of infinite conductivity. The end caps are connected to a voltage source such as a battery or generator. (A battery consists of a number of cells that convert chemical energy to electrical energy and a generator converts mechanical energy of motion to electrical energy.) The electric field is also considered to be constant along the length, l, of the cylinder and, as a consequence, the slope of the potential variation along the cylinder will also be a constant. This is indicated by the dashed lines in Fig. 5. Since the electric field is constant, the current density will also be constant. Therefore, the total current entering the end caps will just be the product of the current density and the cross-sectional area. The value of the electric field will be the quotient of the total voltage and the length of the cylinder. Combining the foregoing results and introducing two new entities gives the following set of equations:

$$J = \sigma\left(\frac{V}{l}\right) \text{ since } J = \sigma E \text{ and } E = \frac{V}{l}$$

$$I = J(A) = \frac{\sigma A V}{l}$$

$$\rho = \frac{1}{\sigma} \text{ and } V = I\left(\frac{\rho l}{A}\right)$$

$$R = \frac{\rho l}{A} \text{ and } V = IR$$

where ρ = the *resistivity* of a conducting material, R = the resistance. The final equation is a very basic one in circuit theory and is called Ohm's Law. Configurations similar to the one shown in Fig. 5 are very common ones in electrical circuits and are called *resistors*.

It will be shown in a later section that the power dissipated in a resistor is equal to the product of the resistance and the square of the current. Quite often resistance is an undesirable effect (such as in a wire carrying current from one location to another one) and must be reduced as much as possible. This can be accomplished by using a conductor with a low resistivity such as silver (or copper which is close to silver in resistivity, but is not as expensive) with a large cross-sectional area and as short a length as possible. The current-carrying capability decreases as the diameter of a conductor size gets smaller.

Potential Drop and Electromotive Force

The application of the relations between fields, potential, and similar concepts to the physical configuration shown in Fig. 5 permitted the derivation of the formula that eliminated further consideration of the field problem. The idea of an electrical energy source was also introduced. A similar analysis involving

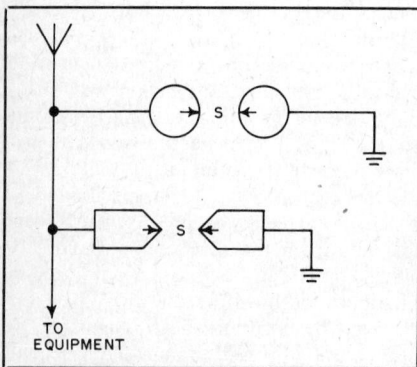

Fig. 4 — Spark gaps with sharp points break down at lower voltages than ones with blunt surfaces even though the separation is the same.

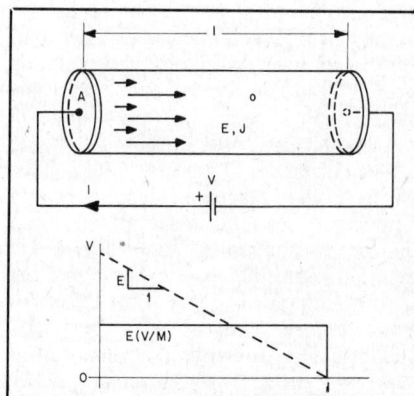

Fig. 5 — Potential and field strength along a current-carrying conductor.

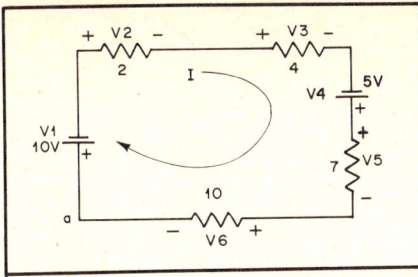

Fig. 6 — A series circuit illustrating the effects of emf and potential drop.

mechanics and field theory would be required to determine the characteristics of an electrical generator and an application of chemistry would be involved in designing a chemical cell. However, it will be assumed that this problem has been solved and that the energy source can be replaced with a symbol such as that used in Fig. 5.

The term *electromotive force* (emf) is applied to describe a source of electrical energy, and potential drop (or voltage drop) is used for a device that consumes electrical energy. A combination of sources and resistances (or other elements) that are connected in some way is called a network or circuit. It is evident that the energy consumed in a network must be equal to the energy produced. Applying this principle to the circuit shown in Fig. 6 gives an important extension of Ohm's Law.

In Fig. 6, a number of sources and resistances are connected in tandem or in *series* to form a *circuit loop*. It is desired to determine the current I. The current can be assumed to be flowing in either a clockwise or counterclockwise direction. If the assumption is not correct, the sign of the current will be negative when the network equations are solved and the direction can be corrected accordingly. In order to solve the problem, it is necessary to find the sum of the emfs (which is proportional to the energy produced) and to equate this sum to the sum of the potential drops (which is proportional to

the energy consumed). Assuming the current is flowing in a clockwise direction, the first element encountered at point a is an emf, V1, but it appears to be connected "backward." Therefore, it receives a minus sign. The next source is V4, and it appears as a voltage rise so it is considered positive. Since the current flow in all the resistors is in the same direction, all the potential drops have the same sign. The potential drop is the product of the current in amperes and the resistance in *ohms*. The *sums* for the emfs and potential drops and the resulting current are given by

$$\text{Sum of emf} = V_1 + V_4 = -10 + 5$$
$$= -5 \text{ volts}$$

$$\text{Sum of pot. drops} = V_2 + V_3 + V_5$$
$$+ V_6 = I(2 + 4 + 7 + 10) = 23I$$

$$I = \frac{-5}{23} = -0.217 \text{ ampere}$$

Because the sign of the current is negative, it is actually flowing in a counterclockwise direction. The physical significance of this phenomenon is that one source is being "charged." For instance, the circuit in Fig. 6 might represent a direct current (dc) generator and a battery.

Charge Polarity and Electron Flow

The "+" and "−" symbols assigned to electromotive forces and potential drops are important in that they define the *polarity* of voltage and *direction* of current flow. These plus and minus (representing positive and negative) symbols were first used in the 18th century by Benjamin Franklin to describe two types of electric charge. Charged atoms (called *ions*) having more than the usual number of electrons have a negative charge, and those having less than the usual number are said to be positively charged. If a polarizing force is applied to some matter and then removed, the atoms will tend to revert to their natural states. This means that atoms deficient in electrons will attract the needed particles from those

atoms having excess electrons. Thus, the transfer of electrons is from negative ions to positive ions. When the path for electrons is an electrical circuit, the current flows from negative to positive around the potential generator.

The direction of electron flow is important in applications of thermionic and semiconductor devices. The cathode of a vacuum tube is heated so that it will boil off electrons. Current will flow in the tube if and only if the anode is biased positive with respect to the cathode. This is known as the Edison effect. A more familiar example of the polarity of current flow can be seen in automotive engines: When the center electrode of a spark plug is made negative with respect to the shell, the voltage required to fire the gap and initiate current flow is significantly reduced as a result of the elevated temperature of the center.

Most modern circuits employ a chassis or ground plane or bus as a common conductor. This practice reduces the wiring or printed circuitry required and simplifies the schematic diagram. When the negative terminal of the power source is connected to this "ground" system, electrons flow from the negative terminal through the ground system and through the circuit elements to the positive terminal. While this is certainly a correct description of the action, it is more convenient to think of the common conductor as the *return* leg for all circuits. To accommodate this reasoning, electrical engineers have adopted a positive-to-negative convention. This convention is adhered to in most of the technical literature. The arrows in semiconductor schematic symbols point in the direction of conventional current and *away* from actual electron flow.

In discussing network elements having one terminal connected to the "common," "ground" or "return" leg of the circuit, engineers use the terms "source" and "sink" to describe the current flow. A device is a *current source* if current flows *away* from the ungrounded terminal and a *current sink* if current flows *into* the ungrounded terminal.

Resistance and Conductance

Given two conductors of the same size and shape, but of different materials, the amount of current that will flow when a given emf is applied will be found to vary with what is called the *resistance* of the material. The lower the resistance, the greater the current for a given value of emf.

Resistance is measured in *ohms* (Ω). A circuit has a resistance of 1 ohm when an applied emf of 1 volt causes a current of 1 ampere to flow. The *resistivity* of a material is the resistance, in ohms, of a cube of the material measuring one centimeter on each edge. One of the best conductors is copper, and it is frequently convenient, in making resistance calculations, to compare the resistance of the material under consideration with that of a copper conductor of the same size and shape. Table 1 gives the ratio of the resistivity of various conductors to that of copper.

The longer the path through which the current flows, the higher the resistance of that conductor. For direct current and low-frequency alternating currents (up to a few thousand cycles per second) the resistance is inversely proportional to the

Table 1
Relative Resistivity of Metals

Materials	Resistivity Compared to Copper
Aluminum (pure)	1.6
Brass	3.7-4.9
Cadmium	4.4
Chromium	1.8
Copper (hard-drawn)	1.03
Copper (annealed)	1.00
Gold	1.4
Iron (pure)	5.68
Lead	12.8
Nickel	5.1
Phosphor Bronze	2.8-5.4
Silver	0.94
Steel	7.6-12.7
Tin	6.7
Zinc	3.4

cross-sectional area of the path the current must travel; that is, given two conductors of the same material and having the same length, but differing in cross-sectional area, the one with the larger area will have the lower resistance.

Resistance of Wires

The problem of determining the resistance of a round wire of given diameter and length — or its opposite, finding a suitable size and length of wire to supply a desired amount of resistance — can be easily solved with the help of the copper wire table given in a later chapter. This table gives the resistance, in ohms per thousand feet, of each standard wire size.

Example: Suppose a resistance of 3.5 ohms is needed and some no. 28 wire is on hand. The wire table in chapter 17 shows that no. 28 has a resistance of 66.17 ohms per thousand feet. Since the desired resistance is 3.5 ohms, the length of wire required will be

$$\frac{3.5}{66.17} \times 1000 = 52.89 \text{ feet.}$$

Or, suppose that the resistance of the wire in the circuit must not exceed 0.05 ohm and that the length of wire required for making the connections totals 14 feet. Then

$$\frac{14}{1000} \times R = 0.05 \text{ ohm}$$

where R is the maximum allowable resistance in ohms per thousand feet. Rearranging the formula gives

$$R = \frac{0.05 \times 1000}{14} \ 3.57 \text{ ohms/1000 ft.}$$

Reference to the wire table shows that no. 15 is the smallest size having a resistance less than this value.

When the wire is not copper, the resistance values given in the wire table should be multiplied by the ratios given in Table 1 to obtain the resistance.

Example: If the wire in the first example were nickel instead of copper, the length required for 3.5 ohms would be

$$\frac{3.5}{66.17 \times 5.1} \times 1000 = 10.37 \text{ feet}$$

Temperature Effects

The resistance of a conductor changes with its temperature. Although it is seldom necessary to consider temperature in making resistance calculations for amateur work, it is well to know that the resistance of practically all metallic conductors increases with increasing temperature. Carbon, however, acts in the opposite way; its resistance *decreases* when its temperature rises. The temperature effect is important when it is

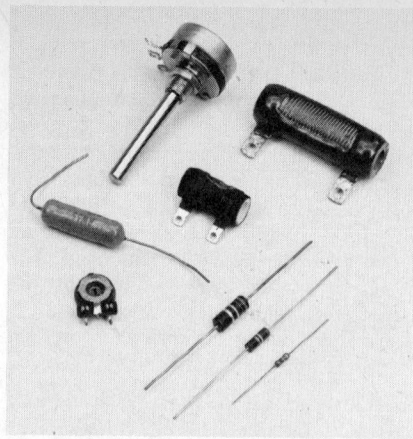

Fig. 7 — Examples of various resistors. In the foreground are 1/4-, 1/2- and 1-watt composition resistors. The three larger cylindrical components at the center are wirewound power resistors. The remaining two parts are variable resistors, pc-board mount at the lower left and panel mount at the upper center.

necessary to maintain a constant resistance under all conditions. Special materials that have little or no change in resistance over a wide temperature range are used in that case.

Resistors

A "package" of resistance made up into a single unit is called a *resistor*. Resistors having the same resistance value may be considerably different in size and construction (Fig. 7). The flow of current through resistance causes the conductor to become heated; the higher the resistance and the larger the current, the greater the amount of heat developed. Resistors intended for carrying large currents must be physically large so the heat can be radiated quickly to the surrounding air. If the resistor does not get rid of the heat quickly it may reach a temperature that will cause it to melt or burn.

Skin Effect

The resistance of a conductor is not the same for alternating current as it is for direct current. When the current is alternating there are internal effects that tend to force the current to flow mostly in the outer parts of the conductor. This decreases the effective cross-sectional area of the conductor, with the result that the resistance increases.

For low audio frequencies the increase in resistance is unimportant, but at radio frequencies this skin effect is so great that practically all the current flow is confined within a few thousandths of an inch of the conductor surface. The rf resistance is consequently many times the dc resistance, and increases with increasing frequency. In the rf range a conductor of thin tubing will have just as low resistance as a solid conductor of the same diameter, because

material not close to the surface carries practically no current.

Conductance

The reciprocal of resistance (1/R) is *conductance*. It is usually represented by the symbol G. A circuit having high conductance has low resistance, and vice versa. In radio work the term is used chiefly in connection with electron-tube and field-effect-transistor characteristics. The unit of conductance is the *mho*, a symbol for which is ℧. Recently, this unit has been renamed the siemens (abbreviated S), although mho is more useful as a functionally descriptive term. A resistance of 1 ohm has a conductance of 1 mho or 1 siemens, a resistance of 1000 ohms has a conductance of 0.001 mho or 0.001 siemens, and so on. A unit frequently used in connection with electron devices is the *micromho*, or one-millionth of a mho. It is the conductance of a one-million-ohm resistance.

Ohm's Law

The simplest form of electric circuit is a battery with a resistance connected to its terminals, as shown by the symbols in Fig. 8. A complete circuit must have an unbroken path so current can flow out of the battery, through the apparatus connected to it, and back into the battery. The circuit is broken, or open, if a connection is removed at any point. A switch is a device for making and breaking connections and thereby closing or opening the circuit, either allowing current to flow or preventing it from flowing.

The values of current, voltage and resistance in a circuit are by no means independent of each other. The relationship between them is known as *Ohm's Law*. It can be stated as follows: The current flowing in a circuit is directly proportional to the applied emf and inversely proportional to the resistance. Expressed as an equation, it is

$$I\text{(amperes)} = \frac{E \text{ (volts)}}{R \text{ (ohms)}}$$

The equation above gives the value of current when the voltage and resistance are known. It may be transposed so that each of the three quantities may be found when the other two are known:

$$E = IR$$

(that is, the voltage acting is equal to the current in amperes multiplied by the resistance in ohms) and

$$R = \frac{E}{I}$$

(or, the resistance of the circuit is equal to the applied voltage divided by the current).

All three forms of the equation are used almost constantly in radio work. It must

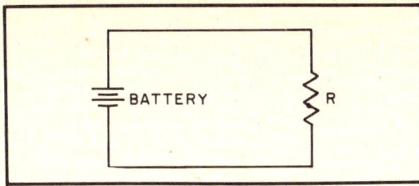

Fig. 8 — A simple circuit consisting of a battery and resistor.

be remembered that the quantities are in *volts, ohms* and *amperes;* other units cannot be used in the equations without first being converted. For example, if the current is in milliamperes it must be changed to the equivalent fraction of an ampere before the value can be substituted in the equations.

Table 2 shows how to convert between the various units in common use. The prefixes attached to the basic-unit name indicate the nature of the unit. These prefixes are

pico — one-trillionth (abbreviated p)
nano — one-billionth (abbreviated n)
micro — one-millionth (abbreviated μ)
milli — one-thousandth (abbreviated m)
kilo — one thousand (abbreviated k)
mega — one million (abbreviated M)
giga — one billion (abbreviated G)

For example, 1 microvolt is one-millionth of a volt, and 1 megohm is 1,000,000

Table 2
Conversion Factors for Fractional and Multiple Units

Change From	To	Divide By	Multiply By
Units	picounits		10^{12}
	nanounits		10^{9}
	microunits		10^{6}
	milliunits		10^{3}
	kilounits	10^{3}	
	megaunits	10^{6}	
	gigaunits	10^{9}	
Picounits	nanounits		10^{3}
	microunits		10^{6}
	milliunits		10^{9}
	units		10^{12}
Nanounits	picounits		10^{3}
	microunits	10^{3}	
	milliunits	10^{6}	
	units	10^{9}	
Microunits	picounits		10^{6}
	nanounits		10^{3}
	milliunits	10^{3}	
	units	10^{6}	
Milliunits	picounits		10^{9}
	nanounits		10^{6}
	microunits		10^{3}
	units	10^{3}	
Kilounits	units	10^{3}	
	megaunits	10^{3}	
	gigaunits	10^{6}	
Megaunits	units		10^{6}
	kilounits	10^{3}	
	gigaunits	10^{3}	
Gigaunits	units		10^{9}
	kilounits		10^{6}
	megaunits		10^{3}

ohms. There are therefore 1,000,000 microvolts in one volt, and 0.000001 megohm in 1 ohm.

The following examples illustrate the use of Ohm's Law:

The current flowing in a resistance of 20,000 ohms is 150 milliamperes. What is the voltage? Since the voltage is to be found, the equation to use is E = IR. The current must first be converted from milliamperes to amperes, and reference to the table shows that to do so it is necessary to divide by 1000. Therefore,

$$E = \frac{150}{1000} \times 20,000 = 3000 \text{ volts}$$

When a voltage of 150 is applied to a circuit, the current is measured at 2.5 amperes. What is the resistance of the circuit? In this case R is the unknown, so

$$R = \frac{E}{I} = \frac{150}{2.5} = 60 \text{ ohms}$$

No conversion was necessary because the voltage and current were given in volts and amperes.

How much current will flow if 250 volts is applied to a 5000-ohm resistor? Since I is unknown

$$I = \frac{E}{R} = \frac{250}{5000} = 0.05 \text{ ampere}$$

Milliampere units would be more convenient for the current, and 0.05 ampere $\times$ 1000 = 50 milliamperes.

Series and Parallel Resistances

Very few actual electric circuits are as simple as the illustration in the preceding section. Commonly, resistances are found connected in a variety of ways. The two fundamental methods of connecting resistances are shown in Fig. 9. In the upper drawing, the current flows from the source of emf (in the direction shown by the arrow, let us say) down through the first resistance, R1, then through the second, R2, and then back to the source. These resistors are connected in series. The current everywhere in the circuit has the same value.

In the lower drawing, the current flows to the common connection point at the top of the two resistors and then divides, one part of it flowing through R1 and the other through R2. At the lower connection point these two currents again combine; the total is the same as the current that flowed into the upper common connection. In this case the two resistors are connected in parallel.

Resistors in Series

When a circuit has a number of resistances connected in series, the total resistance of the circuit is the sum of the individual resistances. If these are numbered R1, R2, R3, and so on, then R (total) = R1 + R2 + R3 + R4 +...where

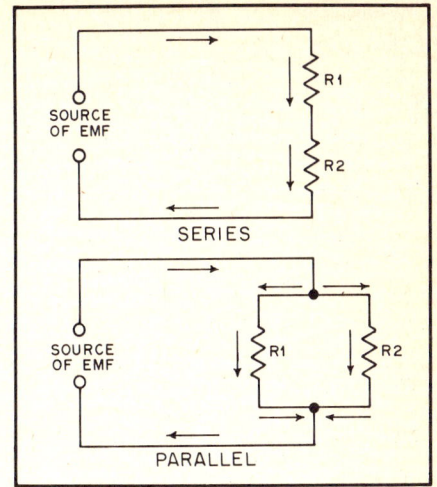

Fig. 9 — Resistors connected in series and in parallel.

the dots indicate that as many resistors as necessary may be added.

Example: Suppose that three resistors are connected to a source of emf as shown in Fig. 10. The emf is 250 volts. R1 is 5000 ohms, R2 is 20,000 ohms, and R3 is 8000 ohms. The total resistance is then

R = R1 + R2 + R3
= 5000 + 20,000 + 8000
= 33,000 ohms

The current flowing in the circuit is then

$$I = \frac{E}{R} = \frac{250}{33,000} = 0.00757 \text{ ampere}$$

$$= 7.57 \text{ mA.}$$

(We need not carry calculations beyond three significant figures, and often two will suffice because the accuracy of measurements is seldom better than a few percent.)

Voltage Drop: Kirchhoff's First Law

Ohm's Law applies to *any part* of a circuit as well as to the whole circuit. Although the current is the same in all three of the resistances in the example, the total voltage divides among them. The voltage appearing across each resistor (the voltage drop) can be found from Ohm's Law.

Example: If the voltage across R1 (Fig. 10) is called E1, that across R2 is called

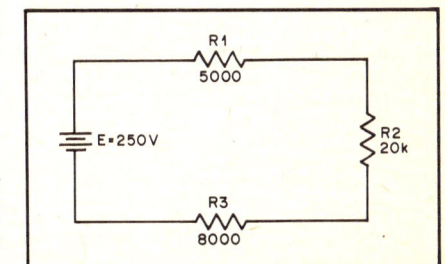

Fig. 10 — An example of resistors in series. The solution of the circuit is worked out in the text.

E2, and that across R3 is called E3, then

E1 = IR1 = 0.00757 × 5000 = 37.9 volts
E2 = IR2 = 0.00757 × 20,000 = 151.4 volts
E3 = IR3 = 0.00757 × 8000 = 60.6 volts

The applied voltage must equal the sum of the individual voltage drops (Kirchhoff's voltage law):

$$E = E1 + E2 + E3$$
$$= 37.9 + 151.4 + 60.6$$
$$= 249.9 \text{ volts}$$

The answer would have been more nearly exact if the current had been calculated to more decimal places, but as explained above a very high order of accuracy is not necessary.

In problems such as this considerable time and trouble can be saved, when the current is small enough to be expressed in milliamperes, if the resistance is expressed in kilohms rather than ohms. When resistance in kilohms is substituted directly in Ohm's Law the current will be milliamperes if the emf is in volts.

Resistors in Parallel: Kirchhoff's Second Law

In a circuit with resistances in parallel, the total resistance is *less* than that of the *lowest* value of resistance present. This is because the total current is always greater than the current in any individual resistor. The formula for finding the total resistance of resistances in parallel is

$$R = \cfrac{1}{\cfrac{1}{R1} + \cfrac{1}{R2} + \cfrac{1}{R3} + \cfrac{1}{R4} + \cdots}$$

where the dots again indicate that any number of resistors can be combined by the same method. For only two resistances in parallel (a very common case), the formula becomes

$$R = \frac{R1 \times R2}{R1 + R2}$$

Example: If a 500-ohm resistor is paralleled with one of 1200 ohms, the total resistance is

$$R = \frac{R1 \, R2}{R1 + R2} = \frac{500 \times 1200}{500 + 1200}$$

$$= \frac{600,000}{1700}$$

$$= 353 \text{ ohms}$$

It is probably easier to solve practical problems by a different method than the "reciprocal of reciprocals" formula. Suppose the three resistors of the previous example are connected in parallel as shown in Fig. 11. The same emf, 250 volts, is applied to all three of the resistors. The current in each can be found from Ohm's Law as shown below, I1 being the current through R1, I2 the current through R2 and I3 the current through R3.

For convenience, the resistance will be expressed in kilohms so the current will be in milliamperes.

$$I1 = \frac{E}{R1} = \frac{250}{5} = 50 \text{ mA}$$

$$I2 = \frac{E}{R2} = \frac{250}{20} = 12.5 \text{ mA}$$

$$I3 = \frac{E}{R3} = \frac{250}{8} = 31.25 \text{ mA}$$

The total current is

$$I = I1 + I2 + I3$$
$$= 50 + 12.5 + 31.25$$
$$= 93.75 \text{ mA}$$

The above example illustrates Kirchhoff's current law: *The current flowing into a node or branching point is equal to the sum of the individual branch currents.* The total resistance of the circuit is therefore

$$R = \frac{E}{I} = \frac{250}{93.75} = 2.66 \text{ kilohms}$$

$$= 2660 \text{ ohms}$$

Resistors in Series-Parallel: Thevinin's Theorum

An actual circuit may have resistances both in parallel and in series. To illustrate, we use the same three resistances again, but now connected as in Fig. 12. The method of solving a circuit such as Fig. 12 is as follows: Consider R2 and R3 in parallel as though they formed a single resistor. Find their equivalent resistance. Then this resistance in series with R1 forms a simple series circuit, as shown at the right in Fig. 12. An example of the arithmetic is given under the illustration.

Using the same principles, and staying within the practical limits, a value for R2 can be computed that will provide a given voltage drop across R3 or a given current through R1. Simple algebra is required.

Example: The first step is to find the equivalent resistance of R2 and R3. From the formula for two resistances in parallel,

$$R_{eq.} = \frac{R2 \times R3}{R2 + R3} = \frac{20 \times 8}{20 + 8} = \frac{160}{28}$$

$$= 5.71 \text{ k}\Omega$$

The total resistance in the circuit is then

$$R = R1 + R_{eq.} = 5 \text{ k}\Omega + 5.71 \text{ k}\Omega$$

$$= 10.71 \text{ k}\Omega$$

The current is

$$I = \frac{E}{R} = \frac{250}{10.71} = 23.3 \text{ mA}$$

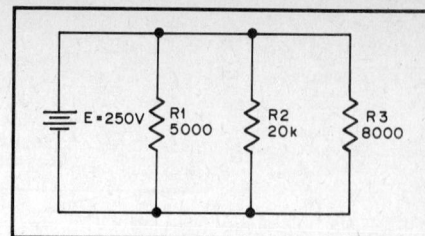

Fig. 11 — An example of resistors in parallel. The solution is worked out in the text.

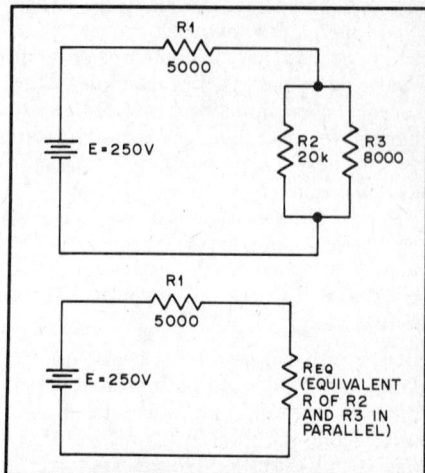

Fig. 12 — An example of resistors in series-parallel. The equivalent circuit is below. The solution is worked out in the text.

The voltage drops across R1 and R_{eq} are

E1 = I × R1 = 23.3 × 5 = 117 volts
E2 = I × R_{eq} = 23.3 × 5.71 = 133 volts

with sufficient accuracy. These total 250 volts, thus checking the calculations so far, because the sum of the voltage drops must equal the applied voltage. Since E2 appears across both R2 and R3

$$I2 = \frac{E2}{R2} = \frac{133}{20} = 6.65 \text{ mA}$$

$$I3 = \frac{E2}{R3} = \frac{133}{8} = 16.6 \text{ mA}$$

where
I2 = current through R2
I3 = current through R3

The total is 23.25 mA, which checks closely enough with 23.3 mA, the current through the whole circuit.

A useful tool for simplifying electrical networks is *Thevinin's Theorum*, which states that any two-terminal network of resistors and voltage sources can be replaced by a single voltage source and a series resistor. This transformation expedites the calculation of the current in a parallel branch. To apply Thevinin's Theorum to the series-parallel circuit of Fig. 12, first remove R3 and calculate the potential developed across R2. Now replace the battery with its internal resistance (a short circuit). This places R1 and R2 in parallel. To find the current in R3 in the original circuit, add the value of R3 to the parallel combination of R1 and R2, and divide the result into the R2 potential derived earlier. The arithmetic proceeds this way: Neglecting R3, the circuit resistance is 25 kΩ, which draws 10 mA from the 250-V battery. This current develops 200 V across R2. The Thevinin equivalent circuit feeding R3 is now the parallel combination of R1 and R2, or 4 kΩ, in series with a 200-V potential source. Installing R3, the total circuit resistance becomes 12 kΩ, which draws 16.6 mA. This value agrees with I3 as calculated by the previous method.

Power and Energy

Power — the rate of doing work — is equal to voltage multiplied by current. The unit of electrical power, called the *watt*, is equal to 1 volt multiplied by 1 ampere. The equation for power therefore is

$$P = EI$$

where

P = power in watts
E = emf in volts
I = current in amperes

Common fractional and multiple units for power are the *milliwatt*, one one-thousandth of a watt, and the *kilowatt*, or 1000 watts:

Example: The plate voltage on a transmitting vacuum tube is 2000 volts and the plate current is 350 milliamperes. (The current must be changed to amperes before substitution in the formula, and so is 0.35 ampere.) Then

$$P = EI = 2000 \times 0.35 = 700 \text{ watts}$$

By substituting the Ohm's Law equivalent for E and I, the following formulas are obtained for power:

$$P = \frac{E^2}{R}$$

$$P = I^2R$$

These formulas are useful in power calculations when the resistance and either the current or voltage (but not both) are known.

Example: How much power will be used up in a 4000-ohm resistor if the potential applied to it is 200 volts? From the equation

$$P = \frac{E^2}{R} = \frac{(200)^2}{4000} = \frac{40,000}{4000}$$

$$= 10 \text{ watts}$$

Or, suppose a current of 20 milliamperes flows through a 300-ohm resistor. Then

$$P = I^2R = (0.02)^2 \times 300$$
$$= 0.0004 \times 300$$
$$= 0.12 \text{ watt}$$

Note that the current was changed from milliamperes to amperes before substitution in the formula.

Electrical power in a resistance is turned into heat. The greater the power the more rapidly the heat is generated. Resistors for radio work are made in many sizes, the smallest being rated to "dissipate" (or carry safely) about 1/10 watt. The largest resistors commonly used in amateur equipment will dissipate about 100 watts.

When electrical energy is converted into mechanical energy, and vice versa, the following relationship holds: 1 horsepower = 746 watts. This formula assumes lossless transformation; the matter of practical efficiency is taken up shortly.

Generalized Definition of Resistance

Electrical power is not always turned into heat. The power used in running a motor, for example, is converted to mechanical motion. The power supplied to a radio transmitter is largely converted into radio waves. Power applied to a loudspeaker is changed into sound waves. But in every case of this kind the power is completely "used up" — it cannot be recovered. Also, for proper operation of the device the power must be supplied at a definite ratio of voltage to current. Both these features are characteristics of resistance, so it can be said that any device that dissipates power has a definite value of "resistance." This concept of resistance as something that absorbs power at a definite voltage/current ratio is very useful, since it permits substituting a simple resistance for the *load* or power-consuming part of the device receiving power, often with considerable simplification of calculations. Of course, every electrical device has some resistance of its own in the more narrow sense, so a part of the power supplied to it is dissipated in that resistance and hence appears as heat even though the major part of the power may be converted to another form.

Efficiency

In devices such as motors and vacuum tubes, the object is to obtain power in some other form than heat. Therefore power used in heating is considered to be a loss, because it is not the *useful* power. The *efficiency* of a device is the useful power output (in its converted form) divided by the power input to the device. In a vacuum-tube transmitter, for example, the object is to convert power from a dc source into ac power at some radio frequency. The ratio of the rf power output to the dc input is the efficiency of the tube. That is,

$$\text{Eff.} = \frac{Po}{Pi}$$

where

Eff. = efficiency (as a decimal)
Po = power output (watts)
Pi = power input (watts)

Example: If the dc input to the tube is 100 watts, and the rf power output is 60 watts, the efficiency is

$$\text{Eff.} = \frac{Po}{Pi} = \frac{60}{100} = 0.6$$

Efficiency is usually expressed as a percentage; that is, it tells what percent of the input power will be available as useful output. The efficiency in the above example is 60 percent.

Suppose a mobile transmitter has an rf power output of 100 W at an efficiency of 52 percent at 13.8 V. The vehicular alternator system charges the battery at a 5-A rate at this voltage. Assuming an alternator efficiency of 68 percent, how much horsepower must the engine produce to operate the transmitter and charge the battery? *Solution:* To charge the battery, the alternator must produce 13.8 V × 5 A = 69 W. The transmitter dc input power is 100 W ÷ 0.52 = 192.3 W. Therefore, the total electrical power required from the alternator is 192.3 + 69 = 261.3 W. The engine load then is

$$\frac{261.3}{746 \times 0.68} = 0.515 \text{ hp.}$$

Energy

In residences, the power company's bill is for electrical energy, not for power. What you pay for is the *work* that electricity does for you, not the *rate* at which that work is done. Electrical work is equal to power multiplied by time; the common unit is the *watt-hour*, which means that a power of 1 watt has been used for one hour. That is,

$$W = PT$$

where

W = energy in watt-hours
P = power in watts
T = time in hours

Other energy units are the *kilowatt-hour* and the *watt-second (joule)*. These units should be self-explanatory.

Energy units are seldom used in amateur practice, but it is obvious that a small amount of power used for a long time can eventually result in a "power" bill that is just as large as though a large amount of power had been used for a very short time.

Capacitance

Suppose two flat metal plates are placed close to each other (but not touching) and are connected to a battery through a switch, as shown in Fig. 13. At the instant the switch is closed, electrons will be attracted from the upper plate to the positive terminal of the battery, and the same number will be repelled into the lower plate from the negative battery terminal. Enough electrons move into one plate and out of the other to make the emf between them the same as the emf of the battery.

If the switch is opened after the plates have been *charged* in this way, the top plate is left with a deficiency of electrons and the bottom plate with an excess. The plates remain charged despite the fact that the battery no longer is connected. However, if a wire is touched between the two plates (*short-circuiting* them) the excess electrons on the bottom plate will flow through the wire to the upper plate, thus restoring electrical neutrality. The plates have then been *discharged.*

The two plates constitute an electrical *capacitor;* a capacitor possesses the property of storing electricity. (The energy actually is stored in the electric field between the plates.) During the time the electrons are moving — that is, while the capacitor is being charged or discharged— a current is flowing in the circuit even though the circuit is "broken" by the gap between the capacitor plates. However, the current flows only during the time of charge and discharge, and this time is usually very short. There can be no continuous flow of direct current "through" a capacitor, but an alternating current can pass through easily if the frequency is high enough.

The charge or quantity of electricity that can be placed on a capacitor is proportional to the applied voltage and to the capacitance of the capacitor. The larger the plate area and the smaller the spacing between the plate the greater the capacitance. The capacitance also depends upon the kind of insulating material between the plates; it is smallest with air insulation, but substitution of other insulating materials for air may increase the capacitance many times. The ratio of the capacitance with some material other than air between the plates, to the capacitance of the same capacitor with air insulation, is called the *dielectric constant* of that particular insulating material. The material itself is called a *dielectric.* The dielectric constants of a number of materials commonly used as dielectrics in capacitors are given in Table 3. If a sheet of polystyrene is substituted for air between the plates of a capacitor, for example, the capacitance will be increased

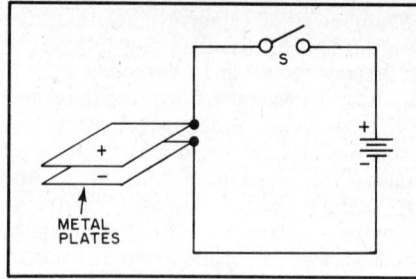

Fig. 13 — A simple capacitor.

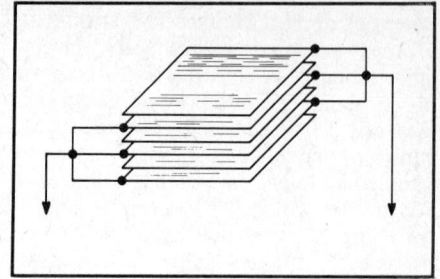

Fig. 14 — A multiple-plate capacitor. Alternate plates are connected together.

Table 3
Dielectric Constants and Breakdown Voltages

Material	Dielectric Constant*	Puncture Voltage**
Air	1.0	240
Alsimag 196	5.7	240
Bakelite	4.4-5.4	300
Bakelite, mica-filled	4.7	325-375
Cellulose acetate	3.3-3.9	250-600
Fiber	5-7.5	150-180
Formica	4.6-4.9	450
Glass, window	7.6-8	200-250
Glass, Pyrex	4.8	335
Mica, ruby	5.4	3800-5600
Mycalex	7.4	250
Paper, Royalgrey	3.0	200
Plexiglas	2.8	990
Polyethylene	2.3	1200
Polystyrene	2.6	500-700
Porcelain	5.1-5.9	40-100
Quartz, fused	3.8	1000
Steatite, low-loss	5.8	150-315
Teflon	2.1	1000-2000

* At 1 MHz ** In volts per mil (0.001 inch)

2.6 times.

Units

The fundamental unit of capacitance is the farad, but this unit is much too large for practical work. Capacitance is usually measured in *microfarads* (abbreviated μF) or *picofarads (pF).* The microfarad is one-millionth of a farad, and the picofarad (formerly micromicrofarad) is one-millionth of a microfarad. Capacitors nearly always have more than two plates, the alternate plates being connected together to form two sets as shown in Fig. 14. This makes it possible to attain a fairly large capacitance in a small space, since several plates of smaller individual area can be stacked to form the equivalent of a single large plate of the same total area. Also, all plates, except the two on the ends, are exposed to plates of the other group on *both sides,* and so are twice as effective in increasing the capacitance.

The formula for calculating capacitance is

$$C = 0.224 \frac{KA}{d} (n - 1)$$

where C = capacitance in pF
K = dielectric constant of material between plates
A = area of one side of one plate in square inches
d = separation of plate surfaces in inches
n = number of plates

If the plates in one group do not have the same area as the plates in the other, use the area of the *smaller* plates.

Capacitors in Radio

The types of capacitors used in radio work differ considerably in physical size, construction, and capacitance. Some representative types are shown in the photograph (Fig. 15). In *variable* capacitors (almost always constructed with air for the dielectric) one set of plates is made movable with respect to the other set so that the capacitance can be varied. Fixed capacitors — that is, assemblies having a single, nonadjustable value of capacitance — also can be made with metal plates and with air as the dielectric, but usually are constructed from plates of metal foil with a thin solid or liquid dielectric sandwiched in between, so that a relatively large capacitance can be secured in a small unit. The solid dielectrics commonly used are mica, paper and special ceramics. An example of a liquid dielectric is mineral oil. The *electrolytic* capacitor uses aluminum-foil plates with a semiliquid conducting chemical compound between them; the actual dielectric is a very thin film of insulating material that forms on one set of plates through electrochemical action when a dc voltage is applied to the capacitor. The capacitance obtained with a given plate area in an electrolytic capacitor is very large, compared with capacitors having other dielectrics, because the film is so thin — much less than any thickness that is practicable with a solid dielectric.

The use of electrolytic and oil-filled capacitors is confined to power-supply filtering and audio-bypass applications. Mica and ceramic capacitors are used throughout the frequency range from

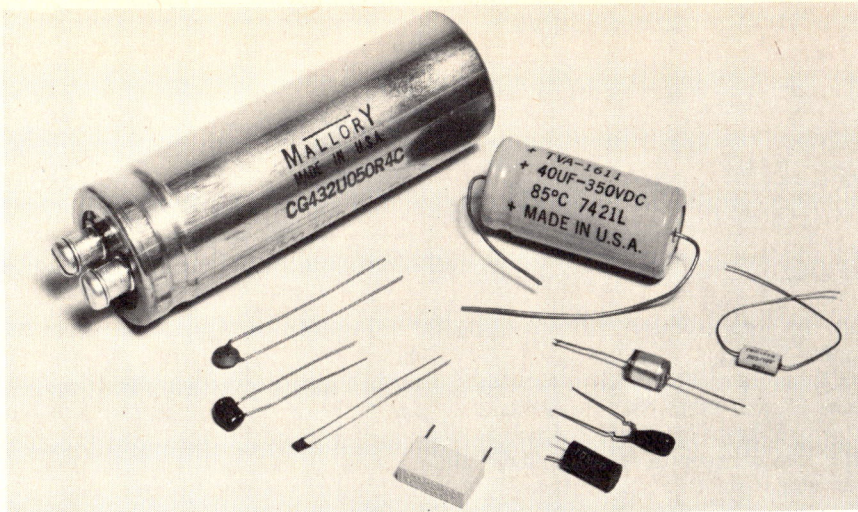

(A)

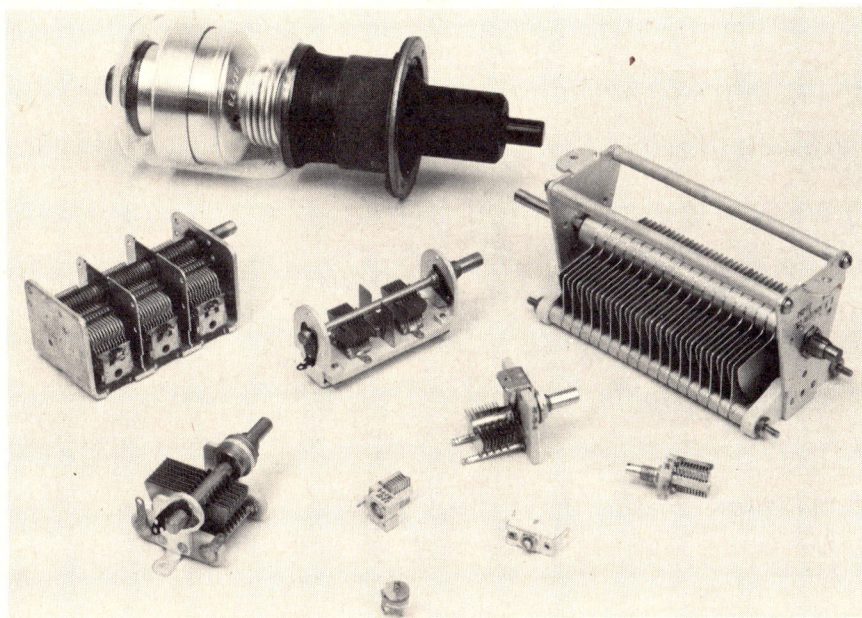

(B)

Fig. 15 — Fixed-value capacitors are seen at A. A large computer-grade unit is at the upper left. The 40-μF unit is an electrolytic capacitor. The smaller pieces are silver-mica, disk-ceramic, tantalum, polystyrene and ceramic chip capacitors. The small black unit (cylindrical) is a pc-board-mount electrolytic. Variable capacitors are shown at B. A vacuum variable is at the upper left.

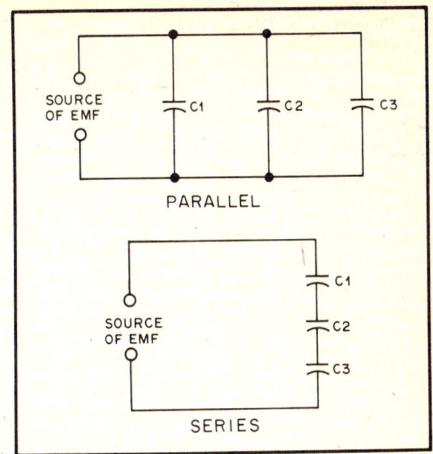

Fig. 16 — Capacitors in parallel and in series.

same circuit meaning as with resistances. When a number of capacitors are connected in parallel, as in Fig. 16, the total capacitance of the group is equal to the sum of the individual capacitances, so

$$C_{total} = C1 + C2 + C3 + C4 + \ldots$$

However, if two or more capacitors are connected in series, as in the second drawing, the total capacitance is less than that of the smallest capacitor in the group. The rule for finding the capacitance of a number of series-connected capacitors is the same as that for finding the resistance of a number of *parallel*-connected resistors. That is,

$$C_{total} = \cfrac{1}{\dfrac{1}{C1} + \dfrac{1}{C2} + \dfrac{1}{C3} + \dfrac{1}{C4} + \ldots}$$

and, for only two capacitors in series,

$$C_{total} = \frac{C1 \times C2}{C1 + C2}$$

The same units must be used throughout; that is, all capacitances must be expressed in either μF or pF; both kinds of units cannot be used in the same equation.

Capacitors are connected in parallel to obtain a larger total capacitance than is available in one unit. The largest voltage that can be applied safely to a group of capacitors in parallel is the voltage that can be applied safely to the one having the *lowest* voltage rating.

When capacitors are connected in series, the applied voltage is divided up among them, and the situation is much the same as when resistors are in series and there is a voltage drop across each. However, the voltage that appears across each capacitor of a group connected in series is in *inverse* proportion to its capacitance, as compared with the capacitance of the whole group.

Example: Three capacitors having capaci-

audio to several hundred megahertz.

Voltage Breakdown

When a high voltage is applied to the plates of a capacitor, a considerable force is exerted on the electrons and nuclei of the dielectric. Because the dielectric is an insulator the electrons do not become detached from atoms the way they do in conductors. However, if the force is great enough the dielectric will "break down;" usually it will puncture and may char (if it is solid) and permit current to flow. The breakdown voltage depends upon the kind and thickness of the dielectric, as shown in Table 3. It is not directly proportional to the thickness; that is, doubling the thickness does not quite double the breakdown voltage. If the dielectric is air or any other gas, breakdown is evidenced by a spark or arc between the plates, but if

the voltage is removed the arc ceases and the capacitor is ready for use again. Breakdown will occur at a lower voltage between pointed or sharp-edged surfaces than between rounded and polished surfaces; consequently, the breakdown voltage between metal plates of given spacing in air can be increased by buffing the edges of the plates.

Since the dielectric must be thick to withstand high voltages, and since the thicker the dielectric the smaller the capacitance for a given plate area, a high-voltage capacitor must have more plate area than a low-voltage one of the same capacitance. High-voltage, high-capacitance capacitors are physically large.

Capacitors in Series and Parallel

The terms "parallel" and "series" when used with reference to capacitors have the

tances of 1, 2 and 4 μF, respectively, are connected in series as shown in Fig 17. The total capacitance is

$$C = \frac{1}{\dfrac{1}{C1} + \dfrac{1}{C2} + \dfrac{1}{C3}} = \frac{1}{\dfrac{1}{1} + \dfrac{1}{2} + \dfrac{1}{4}}$$

$$= \frac{1}{\dfrac{7}{4}} = \frac{4}{7} = 0.571 \ \mu F$$

The voltage across each capacitor is proportional to the *total* capacitance divided by the capacitance of the capacitor in question, so the voltage across C1 is

$$E1 = \frac{0.571}{1} \times 2000 = 1142 \text{ volts}$$

Similarly, the voltages across C2 and C3 are

$$E2 = \frac{0.571}{2} \times 2000 = 571 \text{ volts}$$

$$E3 = \frac{0.571}{4} \times 2000 = 286 \text{ volts}$$

totaling approximately 2000 volts, the applied voltage.

Capacitors are frequently connected in series to enable the group to withstand a larger voltage (at the expense of decreased total capacitance) than any individual capacitor is rated to stand. However, as

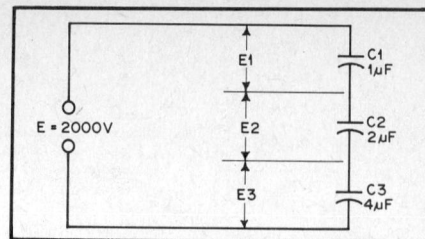

Fig. 17 — An example of capacitors connected in series. The solution to this arrangement is worked out in the text.

shown by the previous example, the applied voltage does not divide equally among the capacitors (except when all the capacitances are the same) so care must be taken to see that the voltage rating of no capacitor in the group is exceeded.

Inductance

It is possible to show that the flow of current through a conductor is accompanied by magnetic effects; a compass needle brought near the conductor, for example, will be deflected from its normal north-south position. The current, in other words, sets up a magnetic field.

The transfer of energy to the magnetic field represents work done by the source of emf. Power is required for doing work, and since power is equal to current multiplied by voltage, there must be a voltage drop in the circuit during the time in which energy is being stored in the field. This voltage "drop" (which has nothing to do with the voltage drop in any resistance in the circuit) is the result of an opposing voltage "induced" in the circuit while the field is building up to its final value. When the field becomes constant the *induced emf* or *back emf* disappears, since no further energy is being stored.

Since the induced emf opposes the emf of the source, it tends to prevent the current from rising rapidly when the circuit is closed. The amplitude of the induced emf is proportional to the rate at which the current is changing and to a constant associated with the circuit itself, called the *inductance* of the circuit.

Inductance depends on the physical characteristics of the conductor. If the conductor is formed into a coil, for example, its inductance is increased. A coil of many turns will have more inductance than one of few turns, if both coils are otherwise physically similar. Also, if a coil is placed around an iron core its inductance will be greater than it was without the magnetic core.

The polarity of an induced emf is always such as to oppose any change in

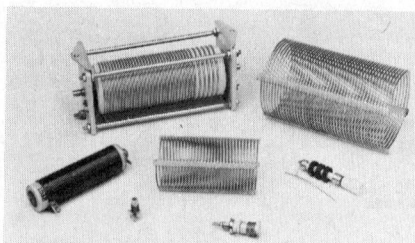

Fig. 18 — Assorted inductors. A rotary (continuously variable) coil is at the upper left. Slug-tuned inductors are visible in the lower foreground. An rf choke (three pi windings) is seen at the lower right.

the current in the circuit. This means that when the current in the circuit is increasing, work is being done against the induced emf by storing energy in the magnetic field. If the current in the circuit tends to decrease, the stored energy of the field returns to the circuit, and thus adds to the energy being supplied by the source of emf. This tends to keep the current flowing even though the applied emf may be decreasing or be removed entirely.

The unit of inductance is the *henry*. Values of inductance used in radio equipment vary over a wide range. Inductance of several henrys is required in power-supply circuits (see chapter on power supplies), and to obtain such values of inductance it is necessary to use coils of many turns wound on iron cores. In radio-frequency circuits, the inductance values used will be measured in *millihenrys* (a mH, one one-thousandth of a henry) at low frequencies, and in microhenrys (μH, one one-millionth of a henry) at medium frequencies and higher. Although coils for radio frequencies may be

wound on special iron cores (ordinary iron is not suitable), most rf coils made and used by amateurs are of the "air-core" type; that is, wound on an insulating support consisting of nonmagnetic material (Fig. 18).

Every conductor has inductance, even though the conductor is not formed into a coil. The inductance of a short length of straight wire is small, but it may not be negligible because if the current through it changes its intensity rapidly enough the induced voltage may be appreciable. This will be the case in even a few inches of wire when an alternating current having a frequency of the order of 100 MHz, or higher is flowing. However, at much lower frequencies the inductance of the same wire could be ignored because the induced voltage would be negligibly small.

Calculating Inductance

The approximate inductance of single-layer air-core coils may be calculated from the simplified formula

$$L \ (\mu H) = \frac{a^2 n^2}{9a + 10b}$$

where

L = inductance in microhenrys
a = coil radius in inches
b = coil length in inches
n = number of turns

The notation is explained in Fig. 19. This formula is a close approximation for coils having a length equal to or greater than 0.8a.

Example: Assume a coil having 48 turns wound 32 turns per inch and a diameter of 3/4 inch. This a = 0.75/2 = 0.375, b = 48/32 = 1.5, and n = 48. Substituting,

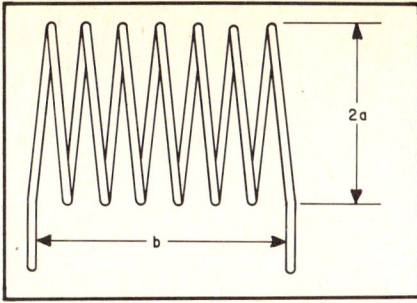

Fig. 19 — Coil dimensions used in the inductance formula. The wire diameter does not enter into the formula. The spacing has been exaggerated in this illustration for clarity. The formula is for closewound coils.

$$L = \frac{.375 \times .375 \times 48 \times 48}{(9 \times .375) + (10 \times 1.5)} = 17.6 \ \mu H$$

To calculate the number of turns of a single-layer coil for a required value of inductance,

$$n = \frac{\sqrt{L(9a + 10b)}}{a}$$

Example: Suppose an inductance of 10 μH is required. The form on which the coil is to be wound has a diameter of one inch and is long enough to accommodate a coil of 1-1/4 inches. Then a = 0.5, b = 1.25, and L = 10. Substituting,

$$n = \frac{\sqrt{10 \ (4.5 + 12.5)}}{0.5} = \frac{\sqrt{170}}{0.5}$$

$$= 26.1 \ \text{turns}$$

A 26-turn coil would be close enough in practical work. Since the coil will be 1.25 inches long, the number of turns per inch will be 26.1/1.25 = 20.8. Consulting the wire table, we find that no. 17 enameled wire (or anything smaller) can be used. The proper inductance is obtained by winding the required number of turns on the form and then adjusting the spacing between the turns to make a uniformly spaced coil 1.25 inches long.

Inductance Charts

Most inductance formulas lose accuracy when applied to small coils (such as are used in vhf work and in low-pass filters built for reducing harmonic interference to television) because the conductor thickness is no longer negligible in comparison with the size of the coil. Fig. 20 shows the measured inductance of vhf coils, and may be used as a basis for circuit design. Two curves are given: curve A is for coils wound to an inside diameter of 1/2 inch; curve B is for coils of 3/4 inch inside diameter. In both curves the wire size is no. 12, winding pitch eight turns to the inch (1/8 inch center-to-center turn spacing). The inductance values given include leads 1/2 inch long.

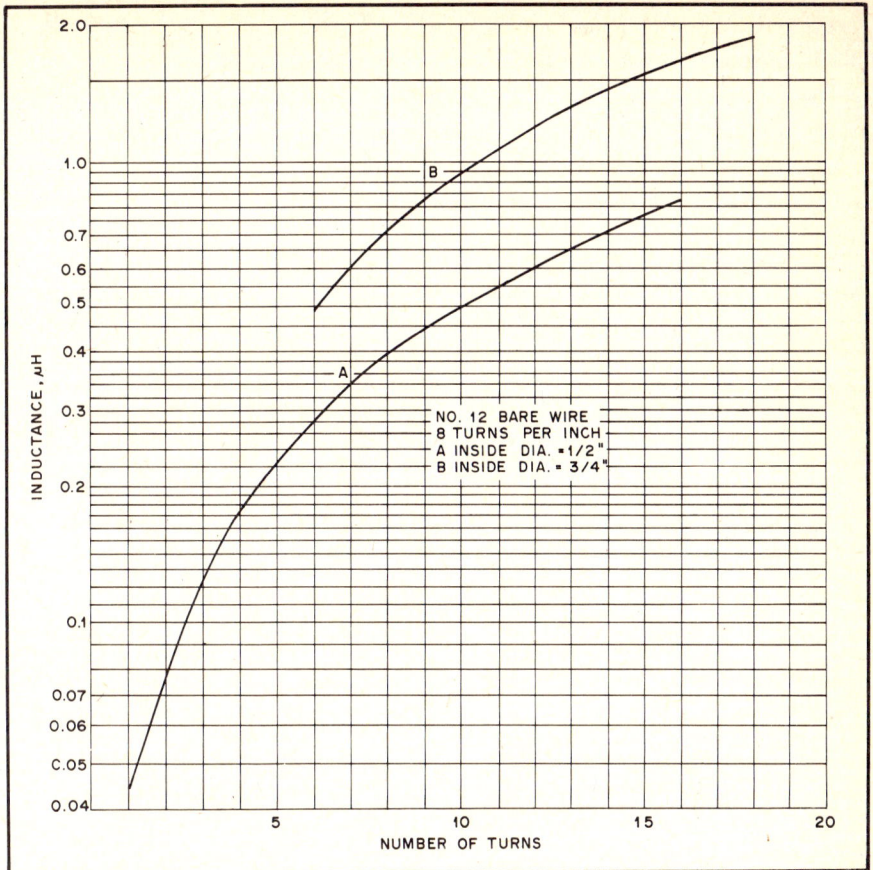

Fig. 20 — Measured inductance of coils wound with no. 12 bare wire, eight turns to the inch. The values include half-inch leads. Inches × 25.4 = mm.

Machine-wound coils with the diameters and turns per inch given in Tables 4 and 5 are available in many radio stores, under the trade names of ''B&W Miniductor,'' ''Air-dux'' and ''Polycoil.'' Figs. 21 and 22 are used with Tables 4 and 5.

While forming a wire into a solenoid increases its inductance, this procedure also introduces *distributed capacitance*. Since each turn is at a slightly different (ac) potential, each pair of turns forms a parasitic capacitor. At some frequency the effective capacitance will have a reactance equal to that of the inductance, and the inductor will show self-resonance. (Reactance and resonance are treated in the section on alternating current.) Above the self-resonant frequency, a coil takes on the reactive properties of a capacitor instead of an inductor. The behavior of a coil with respect to frequency is illustrated in Fig. 23.

Sometimes it is useful to know the inductance of a straight wire, such as a component lead. A straight, round, nonmagnetic wire in free space has an inductance approximated by the formula

$$L = 0.0002b \left[\left(\ell n \ \frac{2b}{a} \right) - 0.75 \right]$$

Table 4

Machine-Wound Coil Specifications

Coil Dia, Inches	No. of Turns Per Inch	Inductance in μH
1-1/4	4	2.75
	6	6.3
	8	11.2
	10	17.5
	16	42.5
1-1/2	4	3.9
	6	8.8
	8	15.6
	10	24.5
	16	63
1-3/4	4	5.2
	6	11.8
	8	21
	10	33
	16	85
2	4	6.6
	6	15
	8	26.5
	10	42
	16	108
2-1/2	4	6.6
	6	23
	8	41
	10	64
3	4	14
	6	31.5
	8	56
	10	89

Inches × 25.4 = mm.

Table 5

Machine-Wound Coil Specifications

Coil Dia, Inches	No. of Turns Per Inch	Inductance in μH
1/2 (A)	4	0.18
	6	0.40
	8	0.72
	10	1.12
	16	2.9
	32	12
5/8 (A)	4	0.28
	6	0.62
	8	1.1
	10	1.7
	16	4.4
	32	18
3/4 (B)	4	0.6
	6	1.35
	8	2.4
	10	3.8
	16	9.9
	32	40
1 (B)	4	1.0
	6	2.3
	8	4.2
	10	6.6
	16	16.9
	32	68

Inches × 25.4 = mm.

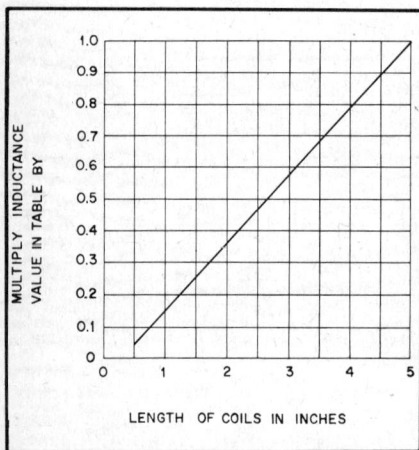

Fig. 21 — Factor to be applied to the inductance of coils listed in Table 4 for coil lengths up to five inches.

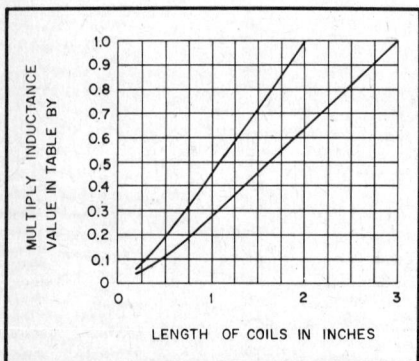

Fig. 22 — Factor to be applied to the inductance of coils listed in Table 5, as a function of coil length. Use curve A for coils marked A, and curve B for coils marked B.

or

$$L = 0.0002b \left[(2.303 \log_{10} \frac{2b}{a}) - 0.75 \right]$$

where

L = inductance in μH
a = wire radius in mm
b = wire length in mm

If the dimensions are expressed in inches the length coefficient (outside the brackets) becomes 0.00508. These formulas are valid for low frequencies; the skin effect reduces the inductance at vhf and above. As the frequency approaches infinity, the constant within the brackets approaches unity. As a practical matter, the skin effect won't reduce the inductance by more than a few percent.

As an example, let a = 2 mm and b = 100 mm. Most pocket calculators can compute either natural or common logarithms. Using the natural logarithm function, the problem is formulated as follows:

$$L = 0.0002(100) \left[(ln \frac{2(100)}{2}) - 0.75 \right]$$

$$= 0.02 \left[(ln\ 100) - 0.75 \right]$$
$$= 0.02\ (4.606 - 0.75)$$
$$= (0.02)\ (3.855) = 0.077\ μH$$

Fig. 24 is a graph of the inductance for wires of various radii as a function of length.

A vhf or uhf tank circuit can be fabricated from a wire parallel to a ground plane, with one end grounded. A formula for the inductance of such an arrangement is

$$L = 0.0004605b \left\{ \log_{10} \left[\frac{2h}{a} \left(\frac{b + \sqrt{b^2 + a^2}}{b + \sqrt{b^2 + 4h^2}} \right) \right] \right\}$$
$$+ 0.0002 \left(\sqrt{b^2 + 4h^2} - \sqrt{b^2 + a^2} + \frac{b}{4} - 2h + a \right)$$

where

L = inductance in μH
a = wire radius in mm
b = wire length parallel to ground plane in mm
h = wire height above ground plane in mm

If the dimensions are in inches, the numerical coefficients become 0.0117 for the first term and 0.00508 for the second term.

Suppose it is desired to find the inductance of a wire 100 mm long and 2 mm in radius, suspended 40 mm above a ground plane. (The inductance is measured between the free end and the ground plane,

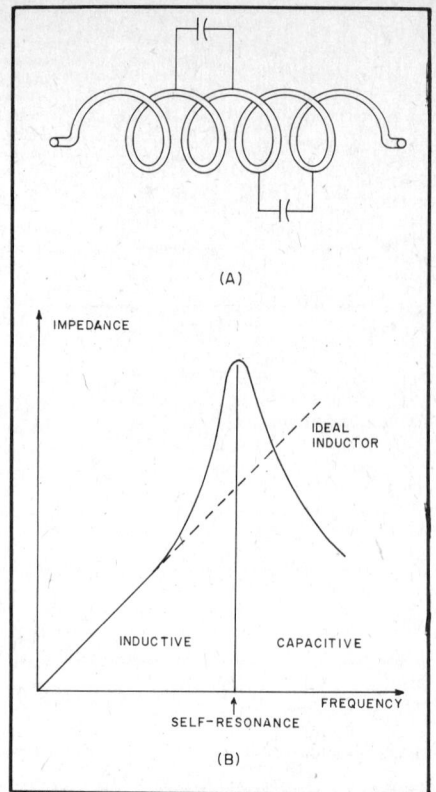

Fig. 23 — The proximity of the turns on a solenoid forms parasitic capacitors, as sketched in A. The net effect of these capacitors is called the distributed capacitance, and causes the coil to exhibit a self-resonance, illustrated in B.

and the formula includes the inductance of the 40-mm grounding link.) A person skilled in the use of a sophisticated calculator could produce the answer with only a few key strokes, but to demonstrate the use of the formula, begin by evaluating these quantities:

$$b + \sqrt{b^2 + a^2} = 100 + 100.02 = 200.02$$

$$b + \sqrt{b^2 + 4h^2} = 100 + 128.06 = 228.06$$

$$\frac{2h}{a} = 40$$

$$\frac{b}{4} = 25$$

Substituting these figures into the formula yields:

$$L = 0.0004605(100) \left\{ \log_{10} \left[40 \left(\frac{200.02}{228.06} \right) \right] \right\}$$

$$+ 0.0002\ (128.06 - 100.02 + 25 - 80 + 2) = 0.066\ μH.$$

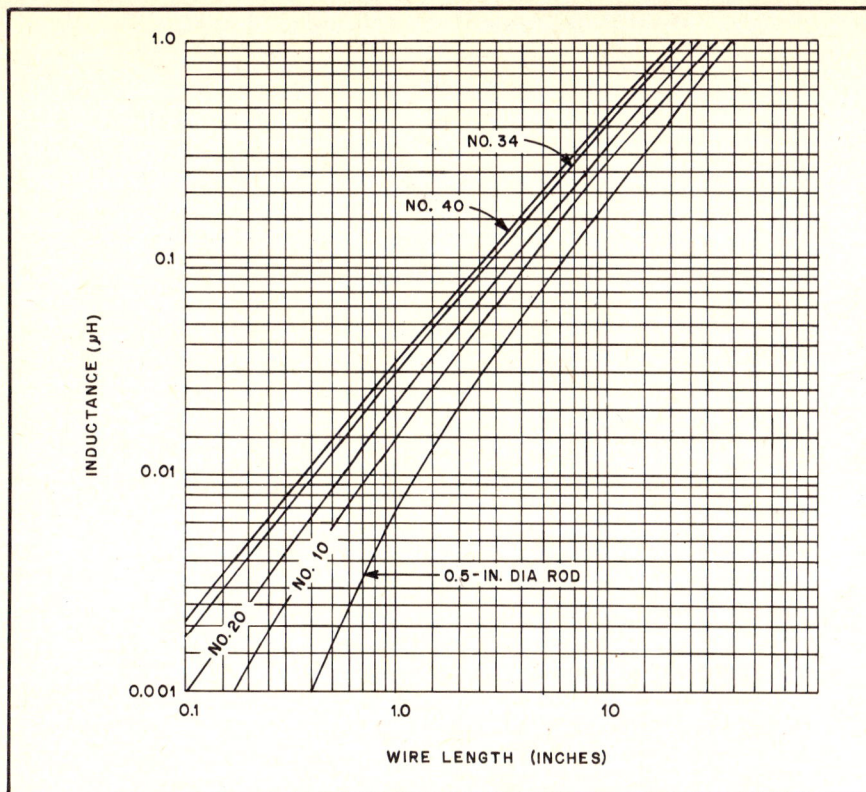

Fig. 24 — Inductance of various conductor sizes when arranged as straight members.

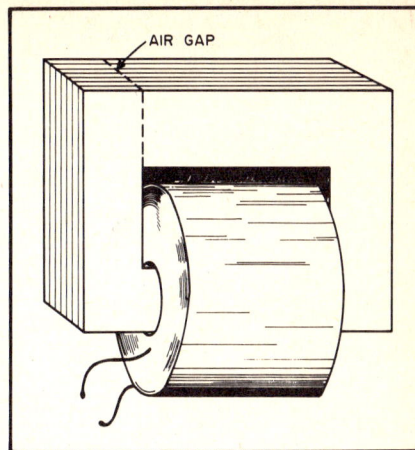

Fig. 25 — Typical construction of an iron-core inductor. The small air gap prevents magnetic saturation of the iron and thus maintains the inductance at high currents.

These straight-wire equations cannot be simply solved for length as a function of desired inductance and given radius, but the proper length can be determined quickly with the aid of a pocket calculator. The technique is to estimate the required length and plug that estimate into the formula to see if it produces the proper inductance. A few iterations will yield a length that is as close as the accuracy of the formula will permit.

Iron-Core Coils: Permeability

Suppose that the coil in Fig. 25 is wound on an iron core having a cross-sectional area of 2 square inches. When a certain current is sent through the coil it is found that there are 80,000 lines of force in the core. Since the area is two square inches, the flux density is 40,000 lines per square inch. Now suppose that the iron core is removed and the same current is maintained in the coil, and that the flux density without the iron core is found to be 50 lines per square inch. The ratio of the flux density with the given core material to the flux density (with the same coil and same current) with an air core is called the *permeability* of the material. In this case the permeability of the iron is 40,000/50 = 800. The inductance of the coil is increased 800 times by inserting the iron core since, other things being equal, the inductance will be proportional to the magnetic flux through the coil.

The permeability of a magnetic material varies with the flux density. At low flux densities (or with an air core) increasing the current through the coil will cause a proportionate increase in flux, but at very high flux densities, increasing the current may cause no appreciable change in the flux. When this is so, the iron is said to be saturated. Saturation causes a rapid decrease in permeability, because it decreases the ratio of flux lines to those obtainable with the same current and an air core. Obviously, the inductance of an iron-core inductor is highly dependent upon the current flowing in the coil. In an air-core coil, the inductance is independent of current because air does not saturate.

Iron core coils such as the one sketched in Fig. 25 are used chiefly in power-supply equipment. They usually have direct current flowing through the winding, and the variation in inductance with current is usually undesirable. It may be overcome by keeping the flux density below the saturation point of the iron. This is done by opening the core so that there is a small "air gap," as indicated by the dashed lines. The magnetic "resistance" introduced by such a gap is so large — even though the gap is only a small fraction of an inch — compared with that of the iron that the gap, rather than the iron, controls the flux density. This reduces the inductance, but makes it practically constant regardless of the value of the current.

For radio-frequency work, the losses in iron cores can be reduced to a satisfactory figure by grinding the iron into a powder and then mixing it with a "binder" of insulating material in such a way that the individual iron particles are insulated from each other. By this means cores can be made that will function satisfactorily even through the vhf range — that is, at frequencies up to perhaps 100 MHz. Because a large part of the magnetic path is through a nonmagnetic material, the permeability of the iron is low compared with the values obtained at power-supply frequencies. The core is usually in the form of a "slug" or cylinder that fits inside the insulating form on which the coil is wound. Despite the fact that with this construction the major portion of the magnetic path for the flux is in air, the slug is quite effective in increasing the coil inductance. By pushing the slug in and out of the coil, the inductance can be varied over a considerable range.

Eddy Currents and Hysteresis

When alternating current flows through a coil wound on an iron core an emf will be induced, as previously explained, and since iron is a conductor a current will flow in the core. Such currents (called *eddy currents*) represent a waste of power because they flow through the resistance of the iron and thus cause heating. Eddy-current losses can be reduced by *laminating* the core; that is, by cutting it into thin strips. These strips or *laminations* must be insulated from each other by painting them with some insulating material such as varnish or shellac.

There is also another type of energy loss: The iron tends to resist any change in its magnetic state, so a rapidly-changing current such as ac is forced continually to supply energy to the iron to overcome this "inertia." Losses of this sort are called *hysteresis* losses.

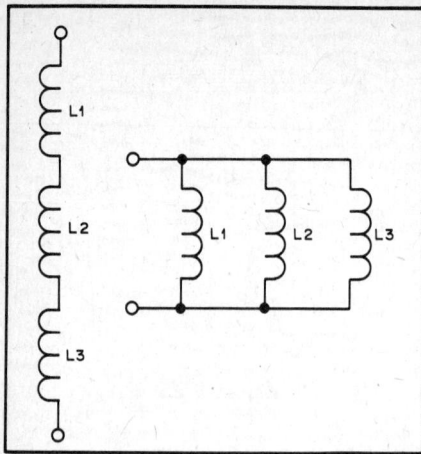

Fig. 26 — Inductances in series and parallel.

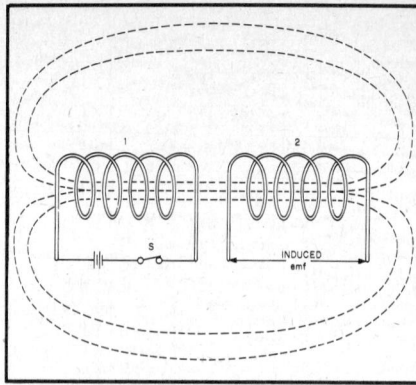

Fig. 27 — Mutual inductance. When the switch, S, is closed current flows through coil no. 1, setting up a magnetic field that induces an emf in the turns of coil no. 2.

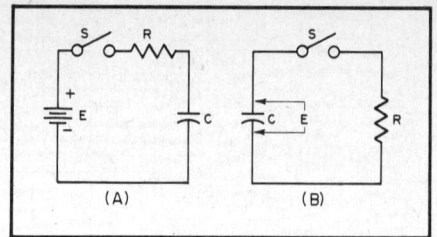

Fig. 28 — Illustrating the time constant of an RC circuit.

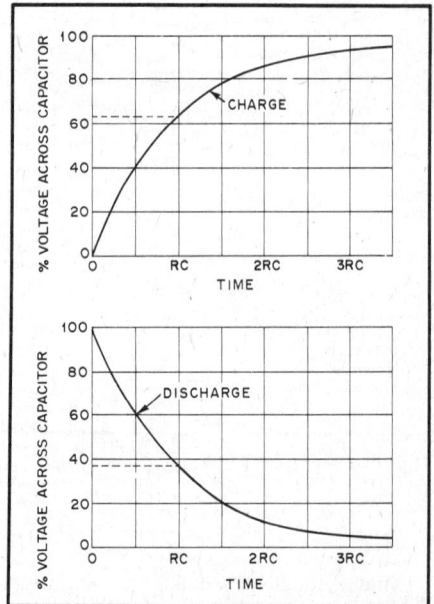

Fig. 29 — How the voltage across a capacitor rises, with time, when charged through a resistor. The lower curve shows the way in which the voltage decreases across the capacitor terminals on discharging through the same resistor.

Eddy-current and hysteresis losses in iron increase rapidly as the frequency of the alternating current is increased. For this reason, ordinary iron cores can be used only at power and audio frequencies — up to, say, 15,000 hertz. Even so, a very good grade of iron or steel is necessary if the core is to perform well at the higher audio frequencies. Iron cores of this type are completely useless at radio frequencies.

Inductances in Series and Parallel

When two or more inductors are connected in series (Fig. 26) the total inductance is equal to the sum of the individual inductances, *provided the coils are sufficiently separated so that no coil is in the magnetic field of another.* That is,

$$L_{total} = L1 + L2 + L3 + L4 + \ldots$$

If inductors are connected in parallel (Fig. 26) — and the coils are separated sufficiently, the total inductance is given by

$$L_{total} = \cfrac{1}{\cfrac{1}{L1} + \cfrac{1}{L2} + \cfrac{1}{L3} + \cfrac{1}{L4} + \ldots}$$

and for two inductances in parallel,

$$L = \frac{L1 \times L2}{L1 + L2}$$

Thus the rules for combining inductances in series and parallel are the same for resistances, if the coils are far enough apart so that each is unaffected by another's magnetic field. When this is not so the formulas given above cannot be used.

Mutual Inductance

If two coils are arranged with their axes on the same line, as shown in Fig. 27, a current sent through coil 1 will cause a magnetic field which "cuts" coil 2. Consequently, an emf will be induced in coil 2 whenever the field strength is changing. This induced emf is similar to the emf of self-induction, but since it appears in the *second* coil because of current flowing in the *first,* it is a "mutual" effect and results from the *mutual inductance* between the two coils.

If all the flux set up by one coil cuts all the turns of the other coil, the mutual inductance has its maximum possible value. If only a small part of the flux set up by one coil cuts, the turns of the other the mutual inductance is relatively small. Two coils having mutual inductance are said to be *coupled.*

The ratio of actual mutual inductance to the maximum possible value that could theoretically be obtained with two given coils is called the *coefficient of coupling* between the coils. It is frequently expressed as a percentage. Coils that have nearly the maximum possible (coefficient = 1 or 100 percent) mutual inductance are said to be *closely,* or *tightly,* coupled, but if the mutual inductance is relatively small the coils are said to be *loosely* coupled. The degree of coupling depends upon the physical spacing between the coils and how they are placed with respect to each other. Maximum coupling exists when they have a common axis and are as close together as possible (one wound over the other). The coupling is least when the coils are far apart or are placed so their axes are at right angles.

The maximum possible coefficient of coupling is closely approached only when the two coils are wound on a closed iron core. The coefficient with air-core coils may run as high as 0.6 or 0.7 if one coil is wound over the other, but will be much less if the two coils are separated.

Although unity coupling is suggested by Fig. 27, such coupling is possible only when the coils are wound on a closed magnetic core.

Time Constant: Capacitance and Resistance

Connecting a source of emf to a capacitor causes the capacitor to become charged to the full emf practically instantaneously, if there is no resistance in the circuit. However, if the circuit contains resistance, as in Fig. 28A, the resistance limits the current flow and an appreciable length of time is required for the emf between the capacitor plates to build up to the same value as the emf of the source. During this "building-up" period, the current gradually decreases from its initial value, because the increasing emf stored on the capacitor offers increasing opposition to the steady emf of the source.

Theoretically, the charging process is never really finished, but eventually the charging current drops to a value that is smaller than anything that can be measured. The *time constant* of such a circuit is the length of time, in seconds, required for the voltage across the capacitor to reach 63 percent of the applied emf (this figure is chosen for mathematical reasons). The voltage across the capacitor rises with time as shown by Fig. 29.

The formula for time constant is

$$T = RC$$

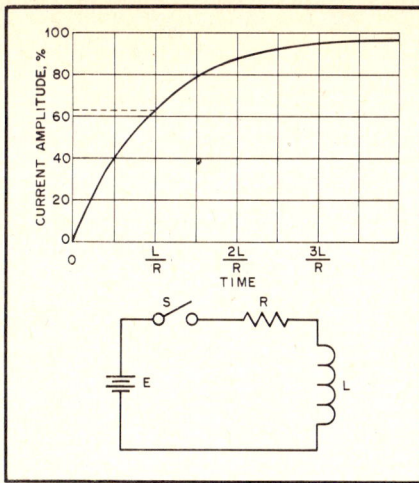

Fig. 30 — Time constant of an LR circuit.

where

T = time constant in seconds
C = capacitance in farads
R = resistance in ohms

Example: The time constant of a 2-μF capacitor and a 250,000-ohm (0.25 M) resistor is

$$T = RC = 0.25 \times 2 = 0.5 \text{ second}$$

If the applied emf is 1000 volts, the voltage between the capacitor plates will be 630 volts at the end of 1/2 second.

If C is in microfarads and R in megohms, the time constant also is in seconds. These units usually are more convenient.

If a charged capacitor is *discharged* through a resistor, as indicated in Fig. 28B, the same time constant applies. If there were no resistance, the capacitor would discharge instantly when S was closed. However, since R limits the current flow the capacitor voltage cannot instantly go to zero, but it will decrease just as rapidly as the capacitor can rid itself of its charge through R. When the capacitor is discharging through a resistance, the time constant (calculated in the same way as above) is the time, in seconds, that it takes for the capacitor to lose 63 percent of its voltage; that is, for the voltage to drop to 37 percent of its initial value.

Example: If the capacitor of the example above is charged to 1000 volts, it will discharge to 370 volts in 1/2 second through the 250 kΩ resistor.

Inductance and Resistance

A comparable situation exists when resistance and inductance are in series. In Fig. 30, first consider L to have no resistance and also assume that R is zero. Then closing S would tend to send a current through the circuit. However, the instantaneous transition from no current to a finite value, however small, represents a very rapid *change* in current, and a *back emf* is developed by the self-inductance of L that is practically equal and opposite to the applied emf. The result is that the initial current is very small.

The back emf depends upon the change in current and would cease to offer opposition if the current did not continue to increase. With no resistance in the circuit (which would lead to an infinitely large current, by Ohm's Law) the current would increase forever, always growing just fast enough to keep the emf of self-induction equal to the applied emf.

When resistance is in series, Ohm's Law sets a limit to the value that the current can reach. The back emf generated in L has only to equal the difference between E and the drop across R, because that difference is the voltage actually applied to L. This difference becomes smaller as the current approaches the final Ohm's Law value. Theoretically, the back emf never quite disappears and so the current never quite reaches the Ohm's Law value, but practically the differences become unmeasurable after a time. The time constant of an inductive circuit is the time in seconds required for the current to reach 63 percent of its final value. The formula is

$$T = \frac{L}{R}$$

where

T = time constant in seconds
L = inductance in henrys
R = resistance in ohms.

The resistance of the wire in a coil acts as if it were in series with the inductance.

Example: A coil having an inductance of 20 henrys and a resistance of 100 ohms has a time constant of

$$T = \frac{L}{R} = \frac{20}{100} = 0.2 \text{ second}$$

if there is no other resistance in the circuit. If a dc emf of 10 volts is applied to such a coil, the final current, by Ohm's Law, is

$$I = \frac{E}{R} = \frac{10}{100} = 0.1 \text{ A or } 100 \text{ mA}$$

The current would rise from 0 to 63 milliamperes in 0.2 second after closing the switch.

An inductor cannot be "discharged" in the same way as a capacitor, because the magnetic field disappears as soon as current flow ceases. Opening S does not leave the inductor "charged." The energy stored in the magnetic field instantly returns to the circuit when S is opened. The rapid disappearance of the field causes a very large voltage to be induced in the coil — ordinarily many times larger than the voltage applied, because the induced voltage is proportional to the

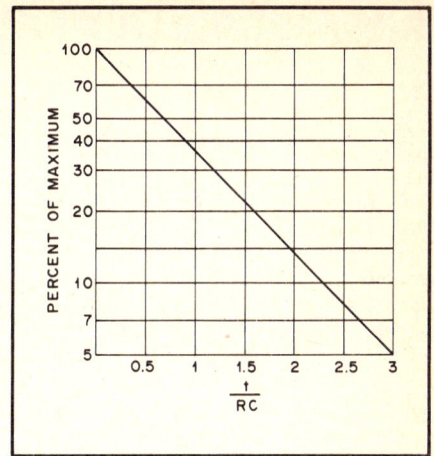

Fig. 31 — Voltage across capacitor terminals in a discharging RC circuit, in terms of the initial charged voltage. To obtain time in seconds, multiply the factor t/RC by the time constant of the circuit.

speed with which the field changes. The common result of opening the switch in a circuit such as the one shown is that a spark or arc forms at the switch contacts at the instant of opening. If the inductance is large and the current in the circuit is high, a great deal of energy is released in a very short period of time. It is not at all unusual for the switch contacts to burn or melt under such circumstances. The spark or arc at the opened switch can be reduced or suppressed by connecting a suitable capacitor and resistor in series across the contacts.

Time constants play an important part in numerous devices, such as electronic keys, timing and control circuits, and shaping of keying characteristics of vacuum tubes. The time constants of circuits are also important in such applications as automatic gain control and noise limiters. In nearly all such applications a resistance-capacitance (RC) time constant is involved, and it is usually necessary to know the voltage across the capacitor at some time interval larger or smaller than the actual time constant of the circuit as given by the formula above. Fig. 31 can be used for the solution of such problems, since the curve gives the voltage across the capacitor, in terms of percentage of the initial charge, for percentages between 5 and 100, at any time after discharge begins.

Example: A 0.01-μF capacitor is charged to 150 volts and then allowed to discharge through a 0.1-megohm resistor. How long will it take the voltage to fall to 10 volts? In percentage, 10/150 = 6.7 percent. From the chart, the factor corresponding to 6.7 percent is 2.7. The time constant of the circuit is equal to RC = 0.1 × 0.01 = 0.001. The time is therefore 2.7 × 0.001 = 0.0027 second, or 2.7 milliseconds.

Alternating Currents

In picturing current flow it is natural to think of a single, constant force causing the electrons to move. When this is so, the electrons always move in the same direction through a path or *circuit* made up of conductors connected together in a continuous chain. Such a current is called a *direct current*, abbreviated *dc*. It is the type of current furnished by batteries and by certain types of generators.

It is also possible to have an emf that periodically reverses. With this kind of emf the current flows first in one direction through the circuit and then in the other. Such an emf is called an alternating emf, and the current is called an *alternating current* (abbreviated *ac*). The reversals (alternations) may occur at any rate from a few per second up to several billion per second. Two reversals make a *cycle*; in one cycle the force acts first in one direction, then in the other, and then returns to the first direction to begin the next cycle. The number of cycles in one second is called the *frequency* of the alternating current. The inverse of frequency, or the time duration of one cycle is the *period* of the current.

The difference between direct current and alternating current is shown in Fig. 32. In these graphs the horizontal axis measures time, increasing toward the right away from the vertical axis. The vertical axis represents the amplitude or strength of the current, increasing in either the up or down direction away from the horizontal axis. If the graph is *above* the horizontal axis the current is flowing in one direction through the circuit (indicated by the + sign) and if it is *below* the horizontal axis the current is flowing in the reverse direction through the circuit (indicated by the − sign). Fig. 32A shows that, if we close the circuit − that is, make the path for the current complete − at the time indicated by *X, the current instantly takes the amplitude indicated by the height A.* After that, the current continues at the same amplitude as time goes on. This is an ordinary *direct* current.

If Fig. 32B, the current starts flowing with the amplitude *A* at time *X*, continues at that amplitude until time *Y* and then instantly ceases. After an interval *YZ* the current again begins to flow and the same sort of start-and-stop performance is repeated. This is an *intermittent* direct current. We could get it by alternately closing and opening a switch in the circuit. It is a *direct* current because the *direction* of current flow does not change; the graph is always on the + side of the horizontal axis. The intermittent direct current illustrated has an ac component, however, which can be isolated by an electrical circuit called a *filter*. Filtering is discussed in greater detail in later sections.

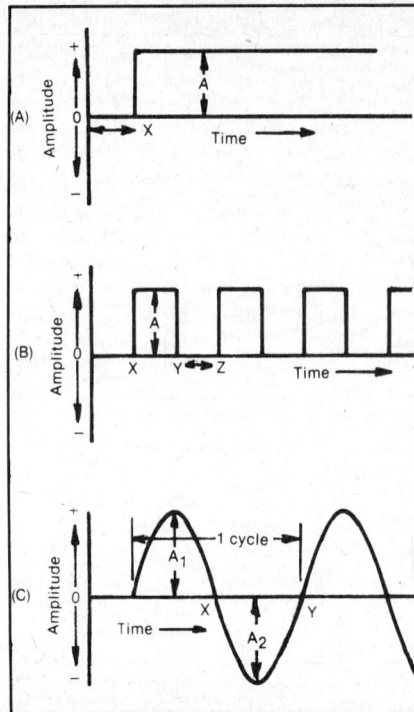

Fig. 32 — Three types of current flow. A—direct current; B—intermittent direct current; C—alternating current.

In Fig. 32 the current starts at zero, increases in amplitude as time goes on until it reaches the amplitude *A1* while flowing in the + direction, then decreases until it drops to zero amplitude once more. At that time *(X)* the *direction* of the current flow reverses; this is indicated by the fact that the next part of the graph is below the axis. As time goes on the amplitude increases, with the current now flowing in the − direction, until it reaches amplitude *A2*. Then the amplitude decreases until finally it drops to zero *(Y)* and the direction reverses once more. This is an *alternating current*.

Waveforms

The type of alternating current shown in Fig. 32C is known as a *sine wave*. An electrodynamic machine called an alternator generates this waveshape because the current induced in the stator winding is proportional to the sine of the angle the winding makes with the magnetic flux lines produced by the rotating field. It is also possible to generate a sine wave electronically. The variations in many ac waves are not so smooth, nor is one half-cycle necessarily just like the preceding one in shape. However, these *complex* waves can be shown to be the sum of two or more sine waves of frequencies that are exact integral (whole-number) multiples of some lower frequency. The lowest frequency is called the *fundamental*, and the

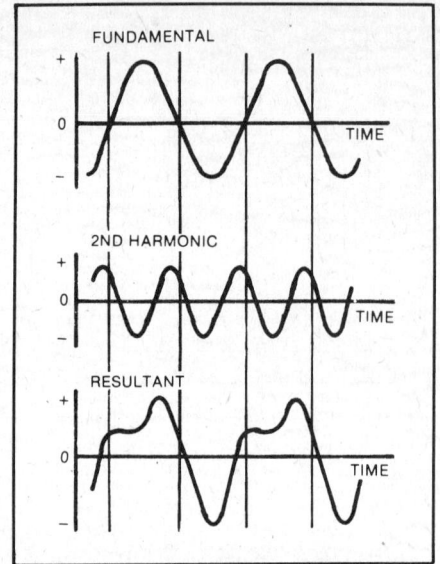

Fig. 33 — A complex waveform. A fundamental (top) and second harmonic (center) added together, point by point at each instant, result in the waveform shown at the bottom. When the two components have the same polarity at a selected instant, the resultant is the simple sum of the two. When they have opposite polarities, the resultant is the *difference;* if the negative-polarity component is larger, the resultant is negative at that instant.

higher frequencies are called *harmonics*.

Fig. 33 shows how a fundamental and a second harmonic (twice the fundamental) might add to form a complex wave. Simply by changing the relative amplitudes of the two waves, as well as the times at which they pass through zero amplitude, as infinite number of waveshapes can be constructed from just a fundamental and second harmonic. More complex waveforms can be constructed if more harmonics are used.

When two or more sinusoidal or complex signals that are not necessarily harmonically related are applied to a common load resistor, the resultant waveform is the sum of the instantaneous voltages. If the two signals have significantly different frequencies and amplitudes, they are easily distinguishable as components of a *composite* wave. The illustration in Fig. 34 is an example of this phenomenon. Two signals having equal amplitudes and nearly equal frequencies combine to produce a composite wave that is not so simply analyzed. Shown in Fig. 35 are two signals having a frequency relationship of 1.5:1. When the positive peaks coincide, the resultant amplitude is twice that of either tone. Similarly, when the maximum negative excursion of one signal corresponds with the maximum positive excursion of the other, the resultant amplitude is the algebraic sum, or zero. The negative peaks never coincide;

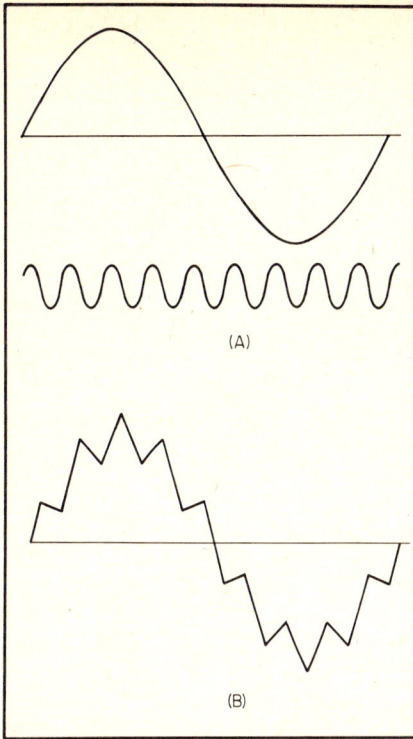

(A)

(B)

Fig. 34 — Two tones of dissimilar frequency and amplitude (A) are easily recognizable in the composite waveform (B).

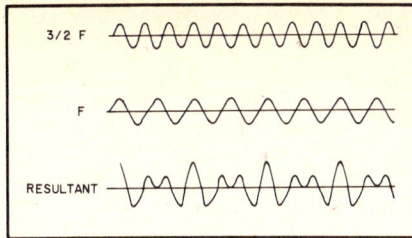

3/2 F

F

RESULTANT

Fig. 35 — The graphic addition of equal amplitude signals fairly close in frequency illustrates the phenomenon of beats. The beat note

$$\left(\frac{3}{2}F - F = \frac{1}{2}F\right)$$

is visible in the resultant waveform.

Frequency	Classification	Abbrev.
10 to 30 kHz	Very-low frequencies	vlf
30 to 300 kHz	Low frequencies	lf
300 to 3000 kHz	Medium frequencies	mf
3 to 30 MHz	High frequencies	hf
30 to 300 MHz	Very-high frequencies	vhf
300 to 3000 MHz	Ultrahigh frequencies	uhf
3 to 30 GHz	Superhigh frequencies	shf
30 to 300 GHz	Extremely-high freq.	ehf

Wavelength

Radio waves travel at the same speed as light — 300,000,000 meters or about 186,000 miles per second in space. They can be set up by a radio-frequency current flowing in a circuit, because the rapidly changing current sets up a magnetic field that changes in the same way, and the varying magnetic field in turn sets up a varying electric field. And whenever this happens, the two fields radiate at the speed of light.

Suppose an rf current has a frequency of 3,000,000 cycles per second. The field will go through complete reversals (one cycle) in 1/3,000,000 second. In that same period of time the fields — that is, the wave — will move 300,000,000/3,000,000 meters, or 100 meters. By the time the wave has moved that distance the next cycle has begun and a new wave has started out. The first wave, in other words, covers a distance of 100 meters before the beginning of the next, and so on. This distance is the *wavelength*.

The longer the time of one cycle — that it, the lower the frequency — the greater the distance occupied by each wave and hence the longer the wavelength. The relationship between wavelength and frequency is shown by the formula

$$\lambda = \frac{300,000}{f}$$

where

  λ = wavelength in meters
  f = frequency in kilohertz

or

$$\lambda = \frac{300}{f}$$

where

  λ = wavelength in meters
  f = frequency in megahertz

Example: The wavelength corresponding to a frequency of 3650 kilohertz is

$$\lambda = \frac{300,000}{3650} = 82.2 \text{ meters}$$

Phase

The term phase essentially means "time," or the *time interval* between the instant when one thing occurs and the instant when a second related thing takes place. The later event is said to *lag* the earlier, while the one that occurs first is said to *lead*. In ac circuits the current amplitude changes continuously, so the concept of phase or time becomes important. Phase can be measured in the ordinary time units, such as the second, but there is a more convenient method: Since each ac cycle occupies exactly the same amount of time as every other cycle of the same frequency, we can use the cycle itself as the time unit. Using the cycle as the time unit makes the specification or measurement of phase independent of the frequency of the current, so long as only one frequency is under consideration at a time. When two or more frequencies are to be considered, as in the case where harmonics are present, the phase measurements are made with respect to the lowest, or fundamental, frequency.

The time interval or "phase difference" under consideration usually will be less than one cycle. Phase difference could be measured in decimal parts of a cycle, but it is more convenient to divide the cycle into 360 parts or *degrees*. A phase degree is therefore 1/360 of a cycle. The reason for this choice is that with sine-wave alternating current the value of the current at any instant is proportional to the sine of the angle that corresponds to the number of degrees — that is, length of time — from the instant the cycle began. There is no actual "angle" associated with an alternating current. Fig. 36 should help make this method of measurement clear.

Measuring Phase

The phase difference between two currents of the same frequency is the time or

therefore this composite waveform is not symmetrical about the zero axis. Notice the periodic variation in the amplitude or *envelope* of the composite waveform. This variation has a frequency equal to the difference or *beat* between the two tones.

FREQUENCY AND WAVELENGTH

Frequencies ranging from about 15 to 15,000 cycles per second (cps, hertz or Hz) are called *audio* frequencies, because the vibrations of air particles that our ears recognize as sounds occur at a similar rate. Audio frequencies (abbreviated *af*) are used to actuate loudspeakers and thus create sound waves.

Frequencies above about 15,000 cps are called *radio* frequencies (*rf*) because they are useful in radio transmission. Frequencies all the way up to and beyond 100,000,000,000 cps have been used for radio purposes. At radio frequencies it becomes convenient to use a unit larger than the cycle. Three such units are the *kilohertz*, which is equal to 1000 cycles (or Hz) and is abbreviated *kHz*, the *megahertz*, which is equal to 1,000,000 hertz or 1000 kilohertz, and is abbreviated *MHz*, and the *gigahertz*, which is equal to 1,000,000,000 hertz or 1000 MHz and is abbreviated *GHz*.

Various radio frequencies are divided into classifications. These classifications, listed below, constitute the *frequency spectrum* as far as it extends for radio purposes at the present time.

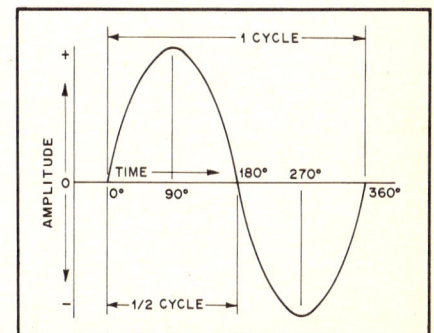

Fig. 36 — An ac cycle is divided off into 360 degrees that are used as a measure of time or phase.

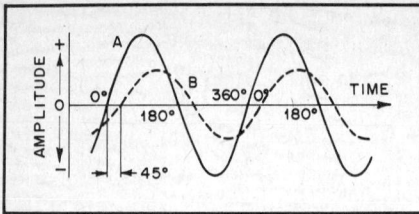

Fig. 37 — When two waves of the same frequency start their cycles at slightly different times, the time difference or phase difference is measured in degrees. In this drawing wave B starts 45 degrees (one-eighth cycle) later than wave A, and so lags 45 degrees behind A.

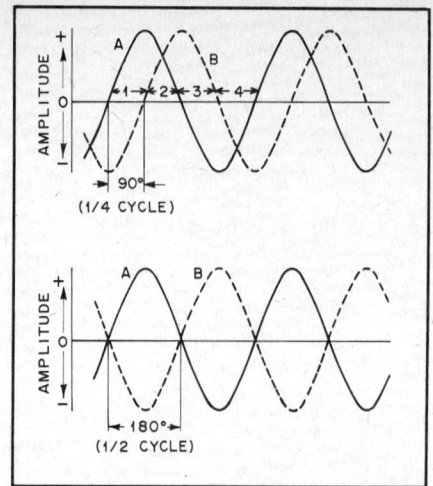

Fig. 38 — Two important special cases of phase difference. In the upper drawing, the phase difference between A and B is 90 degrees; in the lower drawing the phase difference is 180 degrees.

angle difference between corresponding parts of cycles of the two currents. This is shown in Fig. 37. The current labeled A leads the one marked B by 45 degrees, since A's cycles begin 45 degrees earlier in time. It is equally correct to say that B lags A by 45 degrees.

Two important special cases are shown in Fig. 38. In the upper drawing B lags 90 degrees behind A; that is, its cycle begins just one-quarter cycle later than that of A. When one wave is passing through zero, the other is just at its maximum point.

In the lower drawing A and B are 180 degrees out of phase. In this case it does not matter which one is considered to lead or lag. B is always positive while A is negative, and vice versa. The two waves are thus *completely* out of phase.

The waves shown in Figs. 38 and 39 could represent current, voltage, or both. A and B might be two currents in separate circuits, or A might represent voltage and B current in the same circuit. If A and B represent two currents in the *same* circuit (or two voltages in the same circuit) the

total or *resultant* current (or voltage) also is a sine wave, because adding any number of sine waves of the same frequency always gives a sine wave also of the same frequency.

Phase in Resistive Circuits

When an alternating voltage is applied to a resistance, the current flows exactly in step with the voltage. In other words, the voltage and current are *in phase*. This is true at any frequency if the resistance is "pure" — that is, is free from the reactive effects discussed in the next section. Practically, it is often difficult to obtain a purely resistive circuit at radio frequencies, because the reactive effects become more pronounced as the frequency is increased.

In a purely resistive circuit, or for purely resistive parts of circuits, Ohm's Law is just as valid for ac of any frequency as it is for dc.

Reactance: Alternating Current in Capacitance

In Fig. 39 a sine-wave ac voltage having a maximum value of 100 is applied to a capacitor. In the period OA, the applied voltage increases from 0 to 38; at the end of this period the capacitor is charged to that voltage. In interval AB the voltage increases to 71; that is, 33 volts additional. In this interval a *smaller* quantity of charge has been added than in OA, because the voltage rise during interval AB is smaller. Consequently the average current during AB is smaller than during OA. In the third interval, BC, the voltage rises from 71 to 92, an increase of 21 volts. This is less than the voltage increase during AB, so the quantity of electricity added is less; in other words, the average current during interval BC is still smaller.

In the fourth interval, CB, the voltage increases only 8 volts; the charge added is smaller than in any preceding interval and therefore the current also is smaller.

By dividing the first quarter cycle into a very large number of intervals, it could be shown that the current charging the capacitor has the shape of a sine wave, just as the applied voltage does. The current is largest at the beginning of the cycle and becomes zero at the maximum value of the voltage, so there is a phase difference of 90 degrees between the voltage and current. During the first quarter cycle the current is flowing in the normal direction through the circuit, since the capacitor is being charged. Hence the

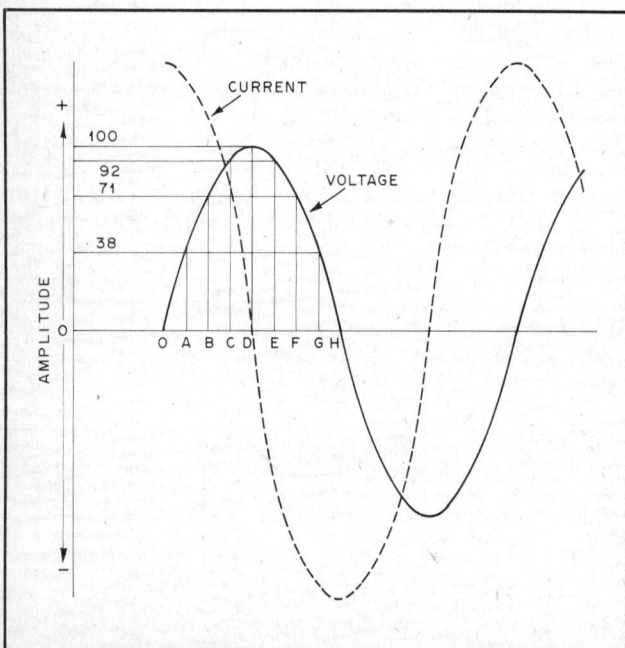

Fig. 39 — Voltage and current phase relationships when an alternating voltage is applied to a capacitor.

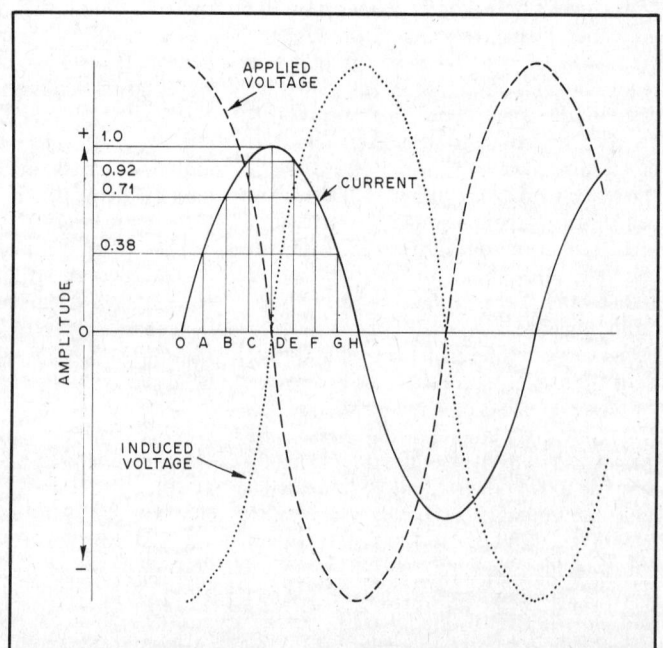

Fig. 40 — Phase relationships between voltage and current when an alternating voltage is applied to an inductance.

current is positive, as indicated by the dashed line in Fig. 39.

In the second quarter cycle — that is, in the time from D to H, the voltage applied to the capacitor decreases. During this time the capacitor *loses* its charge. Applying the same reasoning, it is plain that the current is small in interval DE and continues to increase during each succeeding interval. However, the current is flowing *against* the applied voltage because the capacitor is discharging into the circuit. The current flows in the *negative* direction during this quarter cycle.

The third and fourth quarter cycles repeat the events of the first and second, respectively, with this difference — the polarity of the applied voltage has reversed, and the current changes to correspond. In other words, an alternating current flows in the circuit because of the alternate charging and discharging of the capacitance. As shown in Fig. 39, the current starts its cycle 90 degrees before the voltage, so the current in a capacitor leads the applied voltage by 90 degrees.

Capacitive Reactance

The quantity of electric charge that can be placed on a capacitor is proportional to the applied emf and the capacitance. This amount of charge moves back and forth in the circuit once each cycle, and so the *rate* of movement of charge — that is, the current — is proportional to voltage, capacitance and frequency. If the effects of capacitance and frequency are lumped together, they form a quantity that plays a part similar to that of resistance in Ohm's Law. This quantity is called *reactance*, and the unit for it is the ohm, just as in the case of resistance. The formula for it is

$$X_C = \frac{1}{2\pi fC}$$

where
  X_C = capacitive reactance in ohms
  f = frequency in hertz
  C = capacitance in farads
  π = 3.14

Although the unit of reactance is the ohm, there is no power dissipation in reactance. The energy stored in the capacitor in one quarter of the cycle is simply returned to the circuit in the next.

The fundamental units (cycles per second, farads) are too cumbersome for practical use in radio circuits. However, if the capacitance is in microfarads (μF) and the frequency is in megahertz (MHz), the reactance will come out in ohms in the formula.

Example: The reactance of a capacitor of 470 pF (0.00047 μF) at a frequency of 7150 kHz (7.15 MHz) is

$$X = \frac{1}{2\pi fC} = \frac{1}{6.28 \times 7.15 \times 0.00047}$$
$$= 47.4 \text{ ohms}$$

Inductive Reactance

When an alternating voltage is applied to a *pure* inductance (one with no resistance — all *practical* inductors have resistance) the current is again 90 degrees out of phase with the applied voltage. However, in this case the current *lags* 90 degrees behind the voltage — the opposite of the capacitor current-voltage relationship.

The primary cause for this is the *back emf* generated in the inductance, and since the amplitude of the back emf is proportional to the rate at which the current changes, and this in turn is proportional to the frequency, the amplitude of the current is inversely proportional to the applied frequency. Also, since the back emf is proportional to inductance for a given rate of current change, the current flow is inversely proportional to inductance for a given applied voltage and frequency. (Another way of saying this is that just enough current flows to generate an induced emf that equals and opposes the applied voltage.)

The combined effect of inductance and frequency is called *inductive reactance*, also expressed in ohms, and the formula for it is

$$X_L = 2\pi fL$$

where
  X_L = inductive reactance in ohms
  f = frequency in hertz
  L = inductance in henrys
  π = 3.14

Example: The reactance of a coil having an inductance of 8 henrys, at a frequency of 120 hertz, is

$$X_L = 2\pi fL = 6.28 \times 120 \times 8$$
$$= 6029 \text{ ohms}$$

In radio-frequency circuits the inductance values usually are small and the frequencies are large. If the inductance is expressed in millihenrys and the frequency in kilohertz, the conversion factors for the two units cancel, and the formula for reactance may be used without first converting to fundamental units. Similarly, no conversion is necessary if the inductance is in microhenrys and the frequency is in megahertz.

Example: The reactance of a 15-microhenry coil at a frequency of 14 MHz is

$$X_L = 2\pi fL = 6.28 \times 14 \times 15$$
$$= 1319 \text{ ohms}$$

The resistance of the wire of which the coil is wound has no effect on the reactance, but simply acts as though it were a separate resistor connected in series with the coil.

Ohm's Law for Reactance

Ohm's Law for an ac circuit containing *only* reactance is

$$I = \frac{E}{X} \qquad E = IX \qquad X = \frac{E}{I}$$

where
  E = emf in volts
  I = current in amperes
  X = reactance in ohms

The reactance in the circuit may, of course, be either inductive or capacitive.

Example: If a current of 2 amperes is flowing through the capacitor of the earlier example (reactance = 47.4 ohms) at 7150 kHz, the voltage drop across the capacitor is

$$E = IX = 2 \times 47.4 = 94.8 \text{ volts}$$

If 420 volts at 120 hertz is applied to the 8-henry inductor of the earlier example, the current through the coil will be

$$I = \frac{E}{X} = \frac{420}{6029} = .0697 \text{ A}$$
$$= 69.7 \text{ mA}$$

Reactance Chart

The accompanying chart, Fig. 41, shows the reactance of capacitances from 1 pF to 100 μF, and the reactance of inductances from 0.1 μH to 10 henrys, for frequencies between 100 hertz and 100 megahertz. The approximate value of reactance can be read from the chart or, where more exact values are needed, the chart will serve as a check on the order of magnitude of reactances calculated from the formulas given above, and thus avoid "decimal-point errors."

Reactances in Series and Parallel

When reactances of the same kind are connected in series or parallel the resultant reactance is that of the resultant inductance or capacitance. This leads to the same rules that are used when determining the resultant resistance when resistors are combined. That is, for series reactances of the same kind the resultant reactance is

$$X = X1 + X2 + X3 + X4$$

and for reactances of the same kind in parallel the resultant is

$$X = \frac{1}{\frac{1}{X1} + \frac{1}{X2} + \frac{1}{X3} + \frac{1}{X4} + \cdots}$$

or for two in parallel,

$$X = \frac{X1 \times X2}{X1 + X2}$$

The situation is different when reactances of opposite kinds are combined.

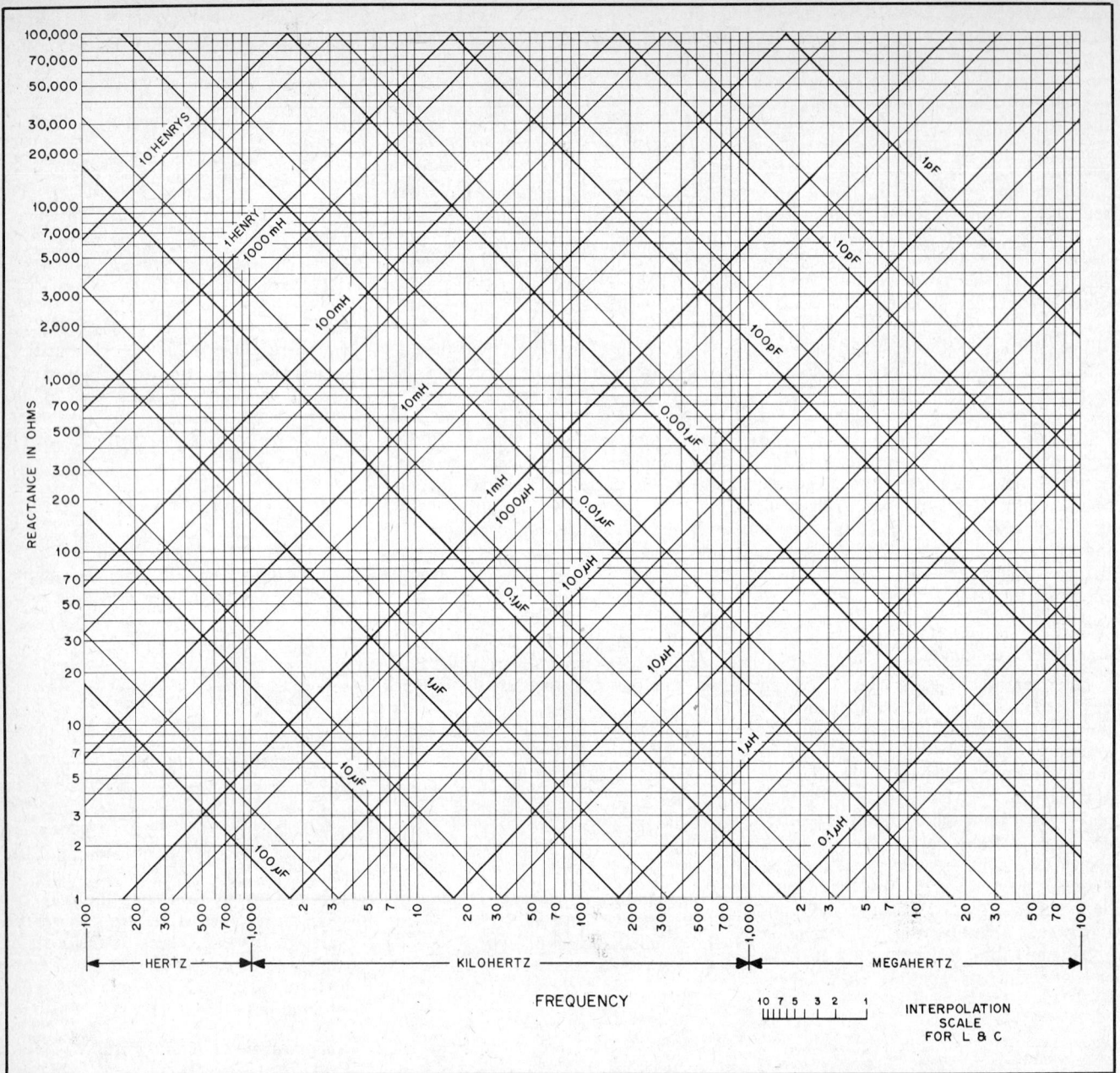

Fig. 41—Inductive and capacitive reactance vs. frequency. Heavy lines represent multiples of 10, intermediate light lines multiples of five: e.g., the light line between 10 μH and 100 μH represents 50 μH; the light line between 0.1 μF and 1 μF represents 0.5 μF, etc. Intermediate values can be estimated with the help of the interpolation scale. Reactances outside the range of the chart may be found by applying appropriate factors to values within the chart range. For example, the reactance of 10 henrys at 60 Hz can be found by taking the reactance to 10 henrys at 600 Hz and dividing by 10 for the 10-times decrease in frequency.

Since the current in a capacitance leads the applied voltage by 90 degrees and the current in an inductance lags the applied voltage by 90 degrees, the voltages at the terminals of opposite types of reactance are 180 degrees out of phase in a series circuit (in which the current has to be the same through all elements), and the currents in reactances of opposite types are 180 degrees out of phase in a parallel circuit (in which the same voltage is applied to all elements). The 180-degree phase relationship means that the currents or voltages are of opposite polarity, so in the series circuit of Fig. 42A the voltage E_L across the inductive reactance X_L is of opposite polarity to the voltage E_C across the capacitive reactance X_C. Thus if we call X_L "positive" and X_C "negative" (a common convention) the applied voltage E_{AC} is $E_L - E_C$. In the parallel circuit at B the total current, I, is equal to $I_L - I_C$, since the currents are 180 degrees out of phase.

In the series case, therefore, the resultant reactance of X_L and X_C is

$$X = X_L - X_C$$

and in the parallel case (Fig. 42B),

$$X = \frac{-X_L X_C}{X_L - X_C}$$

Note that in the series circuit the total reactance is negative if X_C is larger than X_L; this indicates that the total reactance is capacitive in such a case. The resultant reactance in a series circuit is always smaller than the larger of the two individual reactances.

In the parallel circuit, the resultant reactance is negative (i.e., capacitive) if

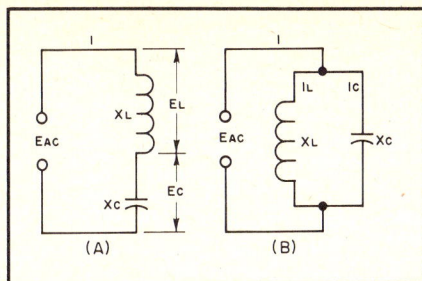

Fig. 42 — Series and parallel circuits containing opposite kinds of reactance.

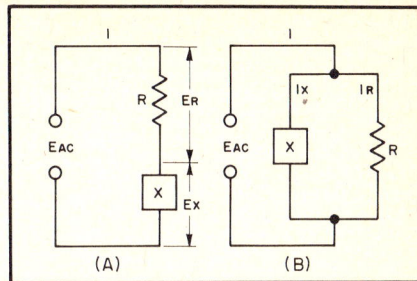

Fig. 43 — Series and parallel circuits containing resistance and reactance.

X_L is larger than X_C, and positive (inductive) if X_L is smaller than X_C, but in every case is always larger than the smaller of the two individual reactances.

In the special case where $X_L = X_C$, the total reactance is zero in the series circuit and infinitely large in the parallel circuit.

Reactive Power

In Fig. 42A the voltage drop across the inductor is larger than the voltage applied to the circuit. This might seem to be an impossible condition, but it is not; the explanation is that while energy is being stored in the inductor's magnetic field, energy is being returned to the circuit from the capacitor's electric field, and vice versa. This stored energy is responsible for the fact that the voltages across reactances in series can be larger than the voltage applied to them.

In a resistance the flow of current causes heating and a power loss equal to I^2R. The power in a reactance is equal to I^2X, but is not a "loss"; it is simply power that is transferred back and forth between the field and the circuit but not used up in heating anything. To distinguish this "nondissipated" power from the power which is actually consumed, the unit of reactive power is called the *volt-ampere-reactive*, or *var*, instead of the watt. Reactive power is sometimes called "wattless" power.

Impedance

When a circuit contains both resistance and reactance the combined effect of the two is called *impedance,* symbolized by the letter Z. (Impedance is thus a more general term than either resistance or reactance, and is frequently used even for circuits that have only resistance or reactance although usually with a qualification — such as "resistive impedance" to indicate that the circuit has only resistance, for example.)

The reactance and resistance comprising an impedance may be connected either in series or parallel, as shown in Fig. 43. In these circuits the reactance is shown as a box to indicate that it may be either inductive or capacitive. In the series circuit the current is the same in both elements, with (generally) different voltages appearing across the resistance and reactance. In the parallel circuit the same voltage is applied to both elements, but different currents flow in the two branches.

Since in a resistance the current is in phase with the applied voltage while in a reactance it is 90 degrees out of phase with the voltage, the phase relationship between current and voltage in the circuit as a whole may be anything between zero and 90 degrees, depending on the relative amounts of resistance and reactance.

Series Circuits

When resistance and reactance are in series, the impedance of the circuit is

$$Z = \sqrt{R^2 + X^2}$$

where

Z = impedance in ohms
R = resistance in ohms
X = reactance in ohms

The reactance may be either capacitive or inductive. If there are two or more reactances in the circuit they may be combined into a resultant by the rules previously given, before substitution into the formula above; similarly for resistances.

The "square root of the sum of the squares" rule for finding impedance in a series circuit arises from the fact that the voltage drops across the resistance and reactance are 90 degrees out of phase, and so combine by the same rule that applies in finding the hypotenuse of a right-angled triangle when the base and altitude are known.

Parallel Circuits

With resistance and reactance in parallel, as in Fig. 43B, the impedance is

$$Z = \frac{RX}{\sqrt{R^2 + X^2}}$$

where the symbols have the same meaning as for series circuits.

Just as in the case of series circuits, a number of reactances in parallel should be combined to find the resultant reactance before substitution into the formula above; similarly for a number of resistances in parallel.

Equivalent Series and Parallel Circuits

The two circuits shown in Fig. 43 are equivalent if the same current flows when a given voltage of the same frequency is applied, and if the phase angle between voltage and current is the same in both cases. It is in fact possible to "transform" any given series circuit into an equivalent parallel circuit, and vice versa.

Transformations of this type often lead to simplification in the solution of complicated circuits. However, from the standpoint of practical work the usefulness of such transformations lies in the fact that the impedance of a circuit may be modified by the addition of *either* series or parallel elements, depending on which happens to be most convenient in the particular case. Typical applications are considered later in connection with tuned circuits and transmission lines.

A series RX circuit can be converted into its parallel equivalent by means of the formula

$$R_p = \frac{R_s^2 + X_s^2}{R_s} \text{ and}$$

$$X_p = \frac{R_s^2 + X_s^2}{X_s}$$

where the subscripts *p* and *s* represent the parallel- and series-equivalent values, respectively. If the parallel values are known, the equivalent series circuit can be found from

$$R_s = \frac{R_p}{1 + \left(\frac{R_p}{X_p}\right)^2} \text{ and}$$

$$X_s = \frac{R_s R_p}{X_p} \ .$$

Circuits containing reactance and resistance in any series or parallel combination are called *complex* circuits. The term "complex" means that the numerical resistance and reactance values can't be combined arithmetically because the voltages (in series circuits) and currents (in parallel circuits) are not in phase. Complex notation for a series RX circuit has the form $R \pm jX$, where $j = \sqrt{-1}$. The reactive portion of the impedance is called the *imaginary* component, because the square root of a negative number can be represented only by a mathematical operator. This is so because squaring a positive or negative number always produces a positive result.

If the reactance is inductive, the sign of the j operator is positive; a negative sign indicates a capacitive reactance. The resistive part of the impedance is called the *real* component.

The *magnitude* of the impedance (in series-equivalent form) is represented by $Z = \sqrt{R^2 + X^2}$. Magnitude is simply a numerical quantity expressing the ratio of voltage to current at the terminals of the complex circuit — it provides no information about the type or amount of reactance present.

If parallel circuits are to be expressed in $R \pm jX$ form, the R and X components must first be transformed into their series-equivalent values. A useful complex notation for parallel circuits expresses the components in terms of conductance and susceptance, the resultant being admittance. These concepts are treated in detail in the section on radio frequency circuits. For a thorough explanation of complex circuits with practical examples, see Hall, "A Simple Approach to Complex Circuits," July 1977 *QST*.

Another way to represent a complex impedance is to indicate the magnitude and the phase angle in the *polar* form $Z \angle \theta$. Given any two of the three quantities R, X and Z, the phase angle can be determined by

$$\theta = \arctan \frac{X}{R} \text{ or}$$

$$\theta = \arcsin \frac{X}{Z} \text{ or}$$

$$\theta = \arccos \frac{R}{Z}$$

Inductive reactances are assigned positive X values which lead to positive values of θ. Conversely, capacitive reactance causes θ to be negative. Since the current in an inductor lags the applied voltage (voltage leads the current), defining θ as the angle by which the voltage at the terminals of the complex impedance *leads* the current makes for consistency — a negative value of θ indicates the angle by which the voltage *lags* the current. Note that the arc cosine formula will not produce a sign for the phase angle unless the sign of the reactance is known.

Ohm's Law for Impedance

Ohm's Law can be applied to circuits containing impedance just as readily as to circuits having resistance or reactance only. The formulas are

$$I = \frac{E}{Z}$$

$$E = IZ$$

$$Z = \frac{E}{I}$$

where E = emf in volts
I = current in amperes
Z = impedance in ohms

Fig. 44 shows a simple circuit consisting of a resistance of 75 ohms and a reactance of 100 ohms in series. From the formula previously given, the impedance is

$$Z = \sqrt{R^2 + X_L^2}$$

$$= \sqrt{(75)^2 + (100)^2} = 125$$

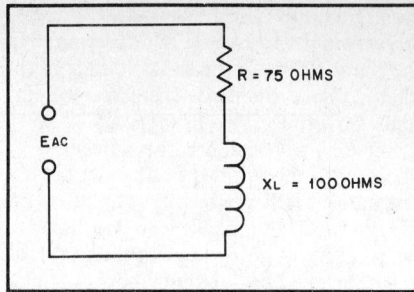

Fig. 44 — Circuit used as an example for impedance calculations.

If the applied voltage is 250, then

$$I = \frac{E}{Z} = \frac{250}{125} = 2 \text{ amperes}$$

This current flows through both the resistance and reactance, so the voltage drops are

$$E_R = IR = 2 \times 75 = 150 \text{ volts}$$

$$E_{XL} = IX_L = 2 \times 100 = 200 \text{ volts}$$

The simple arithmetical sum of these two drops, 350 volts, is greater than the applied voltage because the two voltages are 90 degrees out of phase. Their actual resultant, when phase is taken into account, is

$$\sqrt{(150)^2 + (200)^2} = 250 \text{ volts}$$

Power Factor

In the circuit of Fig. 44 an applied emf of 250 volts results in a current of 2 amperes, giving an apparent power of $250 \times 2 = 500$ watts. However, only the resistance actually consumes power. The power in the resistance is

$$P = I^2R = (2)^2 \times 75 = 300 \text{ watts}$$

The ratio of the power consumed to the apparent power is called the *power factor* of the circuit, and in this example the power factor would be 300/500 = 0.6. Power factor is frequently expressed as a percentage; in this case, it would be 60 percent.

"Real" or dissipated power is measured in watts; apparent power, to distinguish it from real power, is measured in volt-amperes. It is simply the product of volts and amperes and has no direct relationship to the power actually used up or dissipated unless the power factor of the circuit is known. The power factor of a purely resistive circuit is 100 percent or 1, while the power factor of a pure reactance is zero. In this illustration, the reactive power is

$$VAR = I^2X = (2)^2 \times 100$$
$$= 400 \text{ volt-amperes}$$

An equivalent definition of power factor is

$$\frac{R}{Z}$$

or $\cos \theta$. Since power factor is always rendered as a positive number, the value must be followed by the words "leading" or "lagging" to identify the phase of the voltage with respect to the current. Specifying the numerical power factor is not always sufficient. For example, many dc-to-ac power inverters can safely operate loads having a large net reactance of one sign but only a small reactance of the opposite sign.

Reactance and Complex Waves

It was pointed out earlier in this chapter that a complex wave (a "nonsinusoidal" wave) can be resolved into a fundamental frequency and a series of harmonic frequencies. When such a complex voltage wave is applied to a circuit containing reactance, the current through the circuit will not have the same wave shape as the applied voltage. This is because the reactance of an inductor and capacitor depend upon the applied frequency. For the second-harmonic component of a complex wave, the reactance of the inductor is twice and the reactance of the capacitor one-half their respective values at the fundamental frequency; for the third harmonic the inductor reactance is three times and the capacitor reactance one-third, and so on. Thus the circuit impedance is different for each harmonic component.

Just what happens to the current wave shape depends upon the values of resistance and reactance involved and how the circuit is arranged. In a simple circuit with resistance and inductive reactance in series, the amplitudes of the harmonic currents will be reduced because the inductive reactance increases in proportion to frequency. When capacitance and resistance are in series, the harmonic current is likely to be accentuated because the capacitive reactance becomes lower as the frequency is raised. When both inductive and capacitive reactance are present the shape of the current wave can be altered in a variety of ways, depending upon the circuit and the "constants," or the relative values of L, C and R, selected.

This property of nonuniform behavior with respect to fundamental and harmonics is an extremely useful one. It is the basis of "filtering," or the suppression of undesired frequencies in favor of a single desired frequency or group of such frequencies.

AC Waveform Measurements

The time dependence of alternating current raises questions about defining and measuring values of voltage, current and power. Because these parameters change from one instant to the next, one might wonder, for example, which point on the cycle characterizes the voltage or current for the entire cycle. Viewing a single-tone (that is, a pure sine wave) signal on an oscilloscope, the easiest dimension to measure is the total vertical displacement, or *peak-to-peak* voltage. This value, abbreviated pk-pk, is important in evaluating the signal-handling capability of a linear processing device such as an electronic amplifier or ferromagnetic transformer. If the steepest part of the waveform has a potential of zero, the signal has equal positive and negative excursions and no dc bias. The oscilloscope measurement of the maximum positive or negative excursion, or maximum instantaneous potential, is called the *peak* (pk) voltage, and in a symmetrical waveform it has half the value of the peak-to-peak amplitude. Insulators, air gaps and capacitor dielectrics must withstand the peak value of an ac voltage. In a well-designed ac-to-dc power supply the rectified dc output voltage will be nearly equal to the peak ac voltage.

When an ac voltage is applied to a resistor, the resistor will dissipate energy in the form of heat, just as if the voltage were dc. The dc voltage that would cause identical heating in the ac-excited resistor is the *root-mean-square* (rms) value of the ac voltage. The rms voltage of any waveform can be determined with the use of integral calculus, but for a pure sine wave the following relationships hold:

$$V_{pk} = V_{rms} \times \sqrt{2} \approx V_{rms} \times 1.414$$

$$\text{and} \quad V_{rms} = \frac{V_{pk}}{\sqrt{2}} \approx V_{pk} \times 0.707.$$

Unless otherwise specified or obvious from context, ac voltage is rendered as an rms value. For example, the household 117-Vac outlet provides 117 V_{rms}, 165.5 V_{pk} and 331 V_{pk-pk}.

An electrodynamic instrument such as a meter movement responds to the *average* value of an ac waveform. Again, integral calculus is required for computation of the average value of the general (complex) wave, but for a sinusoidal signal the peak, rms and average signals are related by the

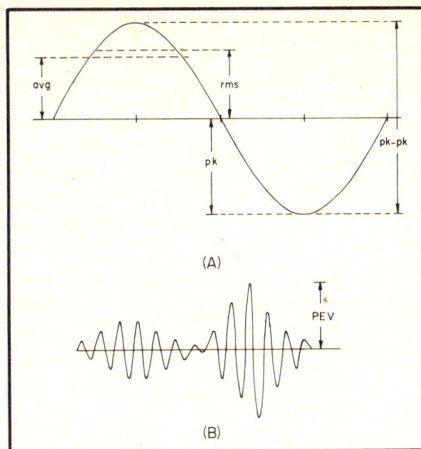

Fig. 45 — Ac voltage and current measurements. The sine-wave parameters are illustrated at A, while B shows the peak envelope voltage for a composite waveform.

formulas:

$$V_{avg} \approx V_{pk} \times 0.636 = V_{rms} \times 0.899$$

and

$$V_{pk} \approx \frac{V_{avg}}{0.636} = V_{avg} \times 1.572.$$

Thus, our 117 Vac outlet provides an average voltage of 105.2.

Part A of Fig. 45 illustrates the four voltage parameters of a sine wave. The most accurate way to determine the rms voltage of a complex wave is to measure the heat produced by applying the complex voltage to a known resistance and measure the dc voltage required to produce the same heat. However, some modern electronic voltmeters provide accurate rms readings by performing mathematical operations on the waveform. The ratio of peak voltage to rms voltage of an ac signal is called the *crest factor*. From the relationships presented earlier, the crest factor of a sine wave is $\sqrt{2}$.

The significant dimension of a multi-tone signal is the *peak envelope voltage*, shown in part B of Fig. 45. PEV is impor-

tant in calculating the power in a modulated signal, such as that from an amateur ssb voice transmitter.

All that has been said about voltage measurements applies also to current (provided the load is resistive) because the waveshapes are identical. However, the terms rms, average and peak have different meanings when they refer to ac *power*. The reason is that while voltage and current are sinusoidal functions of time, power is the *product* of voltage and current, and this product is a sine *squared* function. The mathematical operations that define rms, average and so on will naturally yield different results when applied to this new function. The relationships between ac voltage, current and power follow:

Rms voltage × rms current = average power ≠ rms power

The average power used to heat a resistor is equal to the dc power required to produce the same heat. Rms power is a mathematical curiosity only and has no physical significance. Many audio amplifiers have power ratings in rms watts, but this is a misnomer; the figures specified are really *average* watts.

Peak voltage × peak current = peak power = 2 × average power. Unfortunately, the definition given above for peak power conflicts with the meaning of the term when it is used in radio work. Peak power output of a radio transmitter is the power averaged over the rf cycle having the greatest amplitude. Modulated signals are not purely sinusoidal because they are composites of two or more tones. However, the cycle-to-cycle variation is small enough that sine-wave measurement techniques produce accurate results. In the context of radio signals then, peak power means "maximum average" power. *Peak envelope power* (PEP) is the parameter most often used to express the maximum signal-handling capability of a linear amplifier. To compute the PEP of a waveform such as that sketched in Fig. 45B, multiply the PEV by 0.707 to obtain the rms value, square the result and divide by the load resistance.

Transformers

Two coils having mutual inductance constitute a *transformer*. The coil connected to the source of energy is called the *primary* coil, and the other is called the *secondary* coil.

The usefulness of the transformer lies in the fact that electrical energy can be transferred from one circuit to another

without direct connection, and in the process can be readily changed from one voltage level to another. Thus, if a device to be operated requires, for example, 7 volts ac and only a 440-volt source is available, a transformer can be used to change the source voltage to that required. A transformer can be used only with ac,

since no voltage will be induced in the secondary if the magnetic field is not changing. If dc is applied to the primary of a transformer, a voltage will be induced in the secondary only at the instant of closing or opening the primary circuit, since it is only at these times that the field is changing.

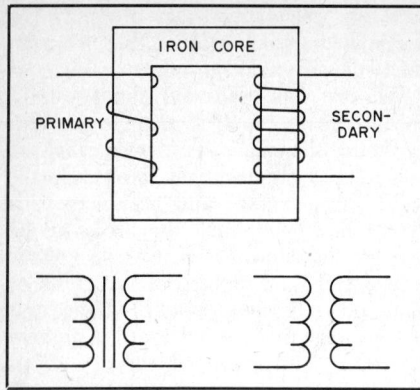

Fig. 46 — The transformer. Power is transferred from the primary coil to the secondary by means of the magnetic field. The lower symbol at left indicates an iron-core transformer, the right one an air-core transformer.

The Iron-Core Transformer

As shown in Fig. 46, the primary and secondary coils of a transformer may be wound on a core of magnetic material. This increases the inductance of the coils so that a relatively small number of turns may be used to induce a given value of voltage with a small current. *A closed core* (one having a continuous magnetic path) such as that shown in Fig. 41 also tends to ensure that practically all of the field set up by the current in the primary coil will cut the turns of the secondary coil. However, the core introduces a power loss because of hysteresis and eddy currents, so this type of construction is normally practicable only at power and audio frequencies. The discussion in this section is confined to transformers operating at such frequencies.

Voltage and Turns Ratio

For a given varying magnetic field, the voltage induced in a coil in the field will be proportional to the number of turns in the coil. If the two coils of a transformer are in the same field (which is the case when both are wound on the same closed core) it follows that the induced voltages will be proportional to the number of turns in each coil. In the primary the induced voltage is practically equal to, and opposes, the applied voltage, as described earlier. Hence,

$$E_s = \left(\frac{n_s}{n_p} \right)$$

where E_s = secondary voltage
E_p = primary applied voltage
n_s = number of turns on secondary
n_p = number of turns on primary

The ratio, n_s/n_p is called the secondary-to-primary *turns ratio* of the transformer.

Example: A transformer has a primary of 400 turns and a secondary of 2800 turns, and an emf of 117 volts is applied to the primary.

$$E_s = \left(\frac{n_s}{n_p} \right) E_p = \frac{2800}{400} \times 117 = 7 \times 117$$
$$= 819 \text{ volts}$$

Also, if an emf of 819 volts is applied to the 2800-turn winding (which then becomes the primary) the output voltage from the 400-turn winding will be 117 volts.

Either winding of a transformer can be used as the primary, provided the winding has enough turns (enough inductance) to induce a voltage equal to the applied voltage without requiring an excessive current flow.

Effect of Secondary Current

The current that flows in the primary when no current is taken from the secondary is called the *magnetizing current* of the transformer. In any properly designed transformer the primary inductance will be so large that the magnetizing current will be quite small. The power consumed by the transformer when the secondary is "open" — that is, not delivering power — is only the amount necessary to supply the losses in the iron core and in the resistance of the wire with which the primary is wound.

When power is taken from the secondary winding, the secondary current sets up a magnetic field that opposes the field set up by the primary current. But if the induced voltage in the primary is to equal the applied voltage, the original field must be maintained. Consequently, the primary must draw enough additional current to set up a field exactly equal and opposite to the field set up by the secondary current.

In practical calculations on transformers it may be assumed that the entire primary current is caused by the secondary "load." This is justifiable because the magnetizing current should be very small in comparison with the primary "load" current at rated power output.

If the magnetic fields set up by the primary and secondary currents are to be equal, the primary current multiplied by the primary turns must equal the secondary current multiplied by the secondary turns. From this it follows that

$$I_p = \left(\frac{n_s}{n_p} \right) I_s$$

where I_p = primary current
I_s = secondary current
n_p = number of turns on primary
n_s = number of turns on secondary

Example: Suppose that the secondary of the transformer in the previous example is delivering a current of 0.2 ampere to a load. Then the primary current will be

$$I_p = \left(\frac{n_s}{n_p} \right) I_s = \frac{2800}{400} \times 0.2 = 7 \times 0.2$$
$$= 1.4 \text{ ampere}$$

Although the secondary voltage is higher than the primary voltage, the secondary current is lower than the primary current, and by the same ratio.

Power Relationships; Efficiency

A transformer cannot create power; it can only transfer it and change the emf. Hence, the power taken from the secondary cannot exceed that taken by the primary from the source of applied emf. There is always some power loss in the resistance of the coils and in the iron core, so in all practical cases the power taken from the source will exceed that taken from the secondary. Thus,

$$P_o = nP_i$$

where P_o = power output from secondary
P_i = power input to primary
n = efficiency factor

The efficiency, n, always is less than 1. It is usually expressed as a percentage; if n is 0.65, for instance, the efficiency is 65 percent.

Example: A transformer has an efficiency of 85 percent as its full-load output of 150 watts. The power input to the primary at full secondary load will be

$$P_i = \frac{P_o}{n} = \frac{150}{0.85} = 176.5 \text{ watts}$$

A transformer is usually designed to have the highest efficiency at the power output for which it is rated. The efficiency decreases with either lower or higher outputs. On the other hand, the *losses* in the transformer are relatively small at low output but increase as more power is taken. The amount of power that the transformer can handle is determined by its own losses, because these heat the wire and core. There is a limit to the temperature rise that can be tolerated, because a too-high temperature either will melt the wire or cause the insulation to break down. A transformer can be operated at reduced output, even though the efficiency is low, because the actual loss will be low under such conditions.

The full-load efficiency of small power transformers such as are used in radio receivers and transmitters usually lies between about 60 and 90 percent, depending upon the size and design.

Leakage Reactance

In a practical transformer not all of the magnetic flux is common to both windings, although in well-designed transformers the amount of flux that "cuts" one coil and not the other is only a small percentage of the total flux. This *leakage flux* causes an emf of self-induction; consequently, there are small amounts of *leakage inductance* associated with both windings of the transformer. Leakage inductance acts in exactly the same way as an equivalent amount of ordinary inductance in-

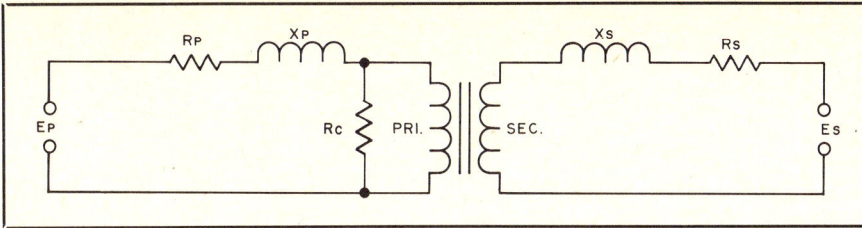

Fig. 47 — The equivalent circuit of a transformer includes the effects of leakage inductance and resistance of both primary and secondary windings. The resistance R_c is an equivalent resistance representing the core losses. Since these are comparatively small, their effect may be neglected in many approximate calculations.

serted in series with the circuit. It has, therefore, a certain reactance, depending upon the amount of leakage inductance and the frequency. This reactance is called *leakage reactance*.

Current flowing through the leakage reactance causes a voltage drop. This voltage drop increases with increasing current, hence it increases as more power is taken from the secondary. Thus, the greater the secondary current, the smaller the secondary terminal voltage becomes. The resistances of the transformer windings also cause voltage drops when current is flowing; although these voltage drops are not in phase with those caused by leakage reactance, together they result in a lower secondary voltage under load than is indicated by the turns ratio of the transformer.

At power frequencies (60 Hz) the voltage at the secondary, with a reasonably well-designed transformer, should not drop more than about 10 percent from open-circuit conditions to full load. The drop in voltage may be considerably more than this in a transformer operating at audio frequencies because the leakage reactance increases directly with the frequency. The various transformer losses are modeled in Fig. 47.

Impedance Ratio

In an ideal transformer — one without losses or leakage reactance — the following relationship is true:

$$Z_p = Z_s \left[\frac{N_p}{N_s} \right]^2$$

where
Z_p = impedance looking into primary terminals from source of power
Z_s = impedance of load connected to secondary
N_p/N_s = turns ratio, primary to secondary

That is, a load of any given impedance connected to the secondary of the transformer will be transformed to a different value "looking into" the primary from the source of power. The impedance transformation is proportional to the square of the primary-to-secondary turns ratio.

Example: A transformer has a primary-to-secondary turns ratio of 0.6 (primary has 6/10 as many turns as the secondary) and a load of 3000 ohms is connected to the secondary. The impedance looking into the primary then will be

$$Z_p = Z_s \left[\frac{N_p}{N_s} \right]^2 = 3000 \times (0.6)^2$$

$$= 3000 \times 0.36 = 1080 \text{ ohms}$$

By choosing the proper turns ratio, the impedance of a fixed load can be transformed to any desired value, within practical limits. If transformer losses can be neglected, the transformed or "reflected" impedance has the same phase angle as the actual load impedance; thus, if the load is a pure resistance, the load presented by the primary to the source of power also will be a pure resistance.

The above relationship may be used in practical work even though it is based on an "ideal" transformer. Aside from the normal design requirements of reasonably low internal losses and low leakage reactance, the only requirement is that the primary have enough inductance to operate with low magnetizing current at the voltage applied to the primary.

The primary impedance of a transformer — *as it appears to the source of power* — is determined wholly by the load connected to the secondary and by the turns ratio. If the characteristics of the transformer have an appreciable effect on the impedance presented to the power source, the transformer is either poorly designed or is not suited to the voltage and frequency at which it is being used. Most transformers will operate quite well at voltages from slightly above to well below the design figure.

Impedance Matching

Many devices require a specific value of load resistance (or impedance) for optimum operation. The impedance of the actual load that is to dissipate the power may differ widely from this value, so a transformer is used to change the actual load into an impedance of the desired value. This is called *impedance matching*. From the preceding,

$$\frac{N_p}{N_s} = \sqrt{\frac{Z_p}{Z_s}}$$

where
N_p/N_s = required turns ratio, primary to secondary
Z_p = primary impedance required
Z_s = impedance of load connected to secondary

Example: A vacuum-tube af amplifier requires a load of 5000 ohms for optimum performance, and is to be connected to a loudspeaker having an impedance of 10 ohms. The turns ratio, primary to secondary, required in the coupling transformer is

$$\frac{N_p}{N_s} = \sqrt{\frac{Z_p}{Z_s}} = \sqrt{\frac{5000}{10}} = \sqrt{500} = 22.4$$

The primary therefore must have 22.4 times as many turns as the secondary.

Impedance matching means, in general, adjusting the load impedance — by means of a transformer or otherwise — to a desired value. However, there is also another meaning. It is possible to show that any source of power will deliver its maximum possible output when the impedance of the load is equal to the internal impedance of the source. The impedance of the source is said to be "matched" under this condition. The efficiency is only 50 percent in such a case; just as much power is used up in the source as is delivered to the load. Because of the poor efficiency, this type of impedance matching is limited to cases where only a small amount of power is available and heating from power loss in the source is not important.

Transformer Construction

Transformers usually are designed so that the magnetic path around the core is as short as possible. A short magnetic path means that the transformer will operate with fewer turns, for a given applied voltage, than if the path were long. A short path also helps to reduce flux leakage and therefore minimizes leakage reactance.

Two core shapes are in common use, as shown in Fig. 48. In the shell type both windings are placed on the inner leg, while in the core type the primary and secondary windings may be placed on separate legs, if desired. This is sometimes done when it is necessary to minimize capacitive effects between the primary and secondary, or when one of the windings must operate at very high voltage.

Core material for small transformers is usually silicon steel, called "transformer iron." The core is built up of laminations, insulated from each other (by a thin coating of shellac, for example) to prevent the flow of eddy currents. The laminations are interleaved at the ends to make the

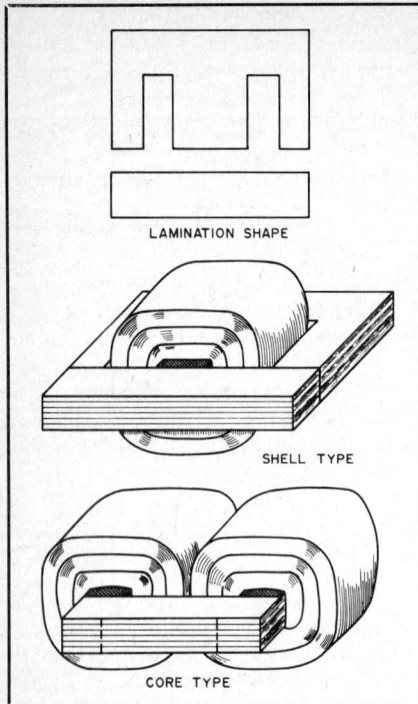

Fig. 48 — Two common types of transformer construction. Core pieces are interleaved to provide a continuous magnetic path.

LAMINATION SHAPE

SHELL TYPE

CORE TYPE

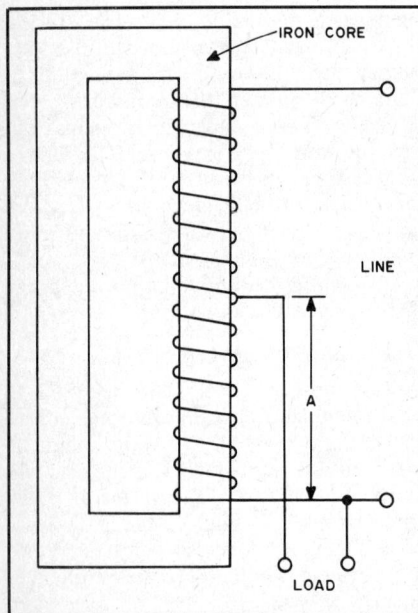

IRON CORE

LINE

A

LOAD

Fig. 49 — The autotransformer is based on the transformer principle, but uses only one winding. The line and load currents in the common winding (A) flow in opposite directions, so that the resultant current is the difference between them. The voltage across A is proportional to the turns ratio.

magnetic path as continuous as possible and thus reduce flux leakage.

The number of turns required in the primary for a given applied emf is determined by the size, shape and type of core material used, and the frequency. The number of turns required is inversely proportional to the cross-sectional area of the

core. As a rough indication, windings of small power transformers frequently have about six to eight turns per volt on a core of 1-square-inch (645 sq. mm) cross section and have a magnetic path 10 or 12 inches (254 or 305 mm) in length. A longer path or smaller cross section requires more turns per volt, and vice versa.

In most transformers the coils are wound in layers, with a thin sheet of treated-paper insulation between each layer. Thicker insulation is used between coils and between coils and core.

Autotransformers

The transformer principle can be utilized with only one winding instead of two, as shown in Fig. 49; the principles just discussed apply equally well. A one-winding transformer is called an *autotransformer*. The current in the common section (A) of the winding is the difference between the line (primary) and the load (secondary) currents, since these currents are out of phase. Hence, if the line and load currents are nearly equal, the common section of the winding may be wound with comparatively small wire. This will be the case only when the primary (line) and secondary (load) voltages are not very different. The autotransformer is used chiefly for boosting or reducing the power-line voltage by relatively small amounts. Continuously variable autotransformers are commercially available under a variety of trade names; "Variac" and "Powerstat" are typical examples.

Ferromagnetic Transformers and Inductors

The design concepts and general theory of transformers which is presented earlier in this chapter apply also to transformers which are wound on ferromagnetic core materials (ferrite and powdered iron). As is the case with stacked cores made of laminations in the classic I and E shapes, the core material has a specific permeability factor which determines the inductance of the windings versus the number of wire turns used. Both ferrite and powdered-iron materials are manufactured with a wide range of μ_i (initial permeability) characteristics. The value chosen by the designer will depend upon the intended operating frequency and the desired bandwidth of a given broadband transformer.

Core-Types in Common Use

For use in radio-frequency circuits especially, a suitable core type must be chosen to provide the Q required by the designer. The wrong core material destroys the Q of an rf type of inductor.

Toroid cores are useful from a few hundred hertz well into the uhf spectrum. Tape-wound steel cores are employed in some types of power supplies — notably dc-to-dc converters. The toroid core is doughnut shaped, hence the name *toroid* (Fig. 50). The principal advantage to this

Fig. 50 — An assortment of toroid cores. A ferrite rod is placed at the top of the picture for comparison. The two light-colored, plastic encased toroids at the upper left are tapewound types (Hypersil steel) are suitable for audio and dc-to-dc converter transformers. The wound toroid at the right center contains two toroid cores which have been stacked atop one another to increase the power capability.

Fig. 51 — Breakaway view of a pot-core assembly (left) and an assembled pot core (right).

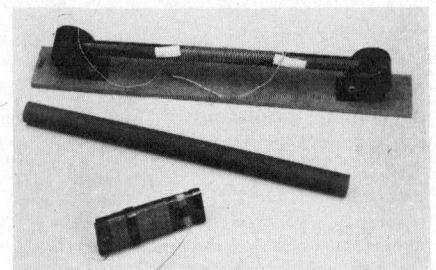

Fig. 52 — A bc-band ferrite rod loop antenna is at the top of the picture (J.W. Miller Co.). A blank ferrite rod is seen at the center and a flat bc-band ferrite loop antenna is in the lower foreground.

type of core is the self-shielding characteristic. Another feature is the compactness of a transformer or inductor, which is possible when using a toroidal format. Therefore, toroids are excellent not only in dc-to-dc converters, but at audio and radio frequencies up to at least 1000 MHz, assuming the proper core material is selected for the range of frequencies over which the device must operate. Toroid cores are available from micro-miniature sizes well up to several inches in diameter. The latter can be used, as one example, to build a 20-kW balun for use in antenna systems.

Another form taken in ferromagnetic transformers and inductors is the "pot-core" or "cup-core" device. Unlike the toroid, which has the winding over the outer surface of the core material, the pot-core winding is inside the ferromagnetic material (Fig. 51). There are two cup-shaped halves to the assembly, both made of ferrite or powdered iron, which are connected tightly together by means of a screw which is passed through a center hole. The wire for the assembly is wound on an insulating bobbin which fits inside the two halves of the pot-core unit. The advantage to this type of construction is that the core permeability can be chosen to ensure a minimum number of wire turns for a given value of inductance. This reduces the wire resistance and increases the Q as opposed to an equivalent inductance which is wound on a core that has relatively low permeability. By virtue of the winding being contained inside the ferrite or powdered-iron pot core, shielding is excellent.

Still another kind of ferromagnetic-core inductor is found in today's technology — the solenoidal type (Fig. 52). Transformers and inductors fabricated in this manner consist of a cylindrical, oval or rectangular rod of material over which the wire winding is placed. This variety of device does not have a self-shielding trait. Therefore it must be treated in the same manner as any solenoidal-wound inductor (using external shield devices). An example of a ferrite-rod inductor is the built-in loop antennas found in portable radios and direction finders.

Core Size

The cross-sectional area of ferromagnetic core is chosen to prevent saturation from the load seen by the transformer. This means that the proper thickness and diameter are essential parameters to consider. For a specific core the maximum operational ac excitation can be determined by

$$B_{max (ac)} = \frac{E_{rms} \times 10^8}{4.44 \, f N_p A_e} \text{ (gauss)}$$

where

A_e = equivalent area of the magnetic path in cm^2
E_{rms} = applied voltage
N_p = number of core turns
f = operating frequency in Hz
B_{op} = flux density in gauss

The foregoing equation is applicable to inductors which do not have dc flowing in the winding along with ac. When both ac and dc flows

$$B_{max(total)} = \frac{E_{rms} \times 10^8}{4.444 \, f N_p A_e} + \frac{N_p I_{dc} A_L}{10 A_e}$$

where

I_{dc} = the dc current through the winding
A_L = the manufacturer's index for the core being used

The latter can be obtained for the core in use by consulting the manufacturer's data sheet.

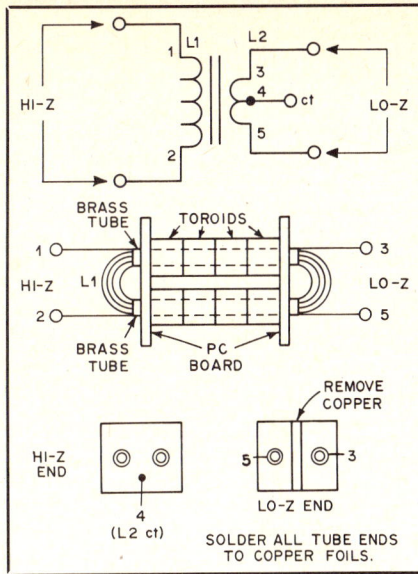

Fig. 53 — Schematic and pictorial representations of one type of "conventional" transformer. This style is used frequently at the input and output ports of rf power amplifiers which use transistors. The magnetic material consists of two rows of 950-mu toroid cores for use from 1.8 to 30 MHz. The primary and secondary windings are passed through the center holes of the toroid-stack rows as shown.

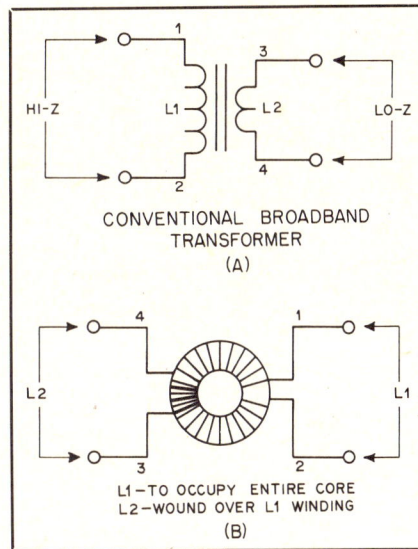

Fig. 54 — Another conventional transformer. Primary and secondary windings are wound over the outer surface of a toroid core.

Types of Transformers

The most common ferromagnetic transformers used in amateur radio work are the narrow-band, broadband, conventional and transmission-line varieties. *Narrow-band* transformers are used when selectivity is desired in a tuned circuit, such as an audio peaking or notching circuit, a resonator in an rf filter, or a tuned circuit associated with an rf amplifier. *Broadband* transformers are employed in circuits which must have uniform response over a substantial spread of frequency, as in a 2- to 30-MHz

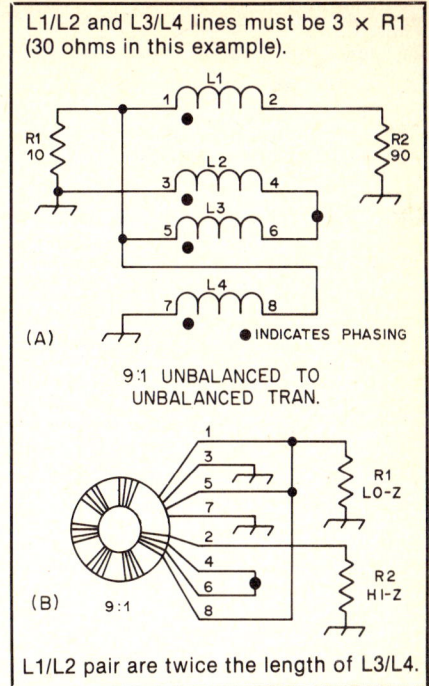

Fig. 55 — Schematic and pictorial presentations of a transmission-line transformer in which the windings need to be configured for a specific impedance.

broadband amplifier. In such an example the reactance of the windings should be at least four times the impedance the winding is designed to look into. Therefore, a transformer which has a 300-ohm primary and a 50-ohm secondary load should have winding reactances (X_L) of at least 1200 ohms and 200 ohms, respectively. The windings, for all practical purposes, can be regarded as rf chokes, and the same rules apply. The permeability of the core material plays a vital role in designing a good broadband transformer. The performance of the transformer at the low-frequency end of the operating range depends on the permeability. That is, the μ_e (effective permeability) must be high enough in value to provide ample winding reactance at the low end of the operating range. As the operating frequency is increased, the effects of the core tend to disappear progressively until there are scarcely any core effects at the upper limit of the operating range. For this reason it is common to find a very low frequency core material utilized in a transformer that is contained in a broadband circuit which reaches well into the upper hf region, or even into the vhf spectrum. By way of simple explanation, at high frequency the low-frequency core material becomes inefficient and tends to vanish electrically. This desirable trait makes possible the use of ferromagnetics in broadband applications.

Conventional transformers are those that are wound in the same manner as a power transformer. That is, each winding is made from a separate length of wire, with one winding being placed over the

Table 6
Powdered-Iron Toroidal Cores — A_L Values (μH/100 turns)

Core Size	41-Mix Green $\mu = 75$	3-Mix Grey $\mu = 35$ 0.05-0.5 MHz	15-Mix Rd & Wh $\mu = 25$ 0.1-2 MHz	1-Mix Blue $\mu = 20$ 0.5-5 MHz	2-Mix Red $\mu = 10$ 1-30 MHz	6-Mix Yellow $\mu = 8$ 10-90 MHz	10-Mix Black $\mu = 6$ 60-150 MHz	12-Mix Gn & Wh $\mu = 3$ 100-200 MHz	0-Mix Tan $\mu = 1$ 150-300 MHz
T-200	755	360	NA	250*	120	100*	NA	NA	NA
T-184	1640	720	NA	500*	240	195	NA	NA	NA
T-157	970	420	360*	320*	140	115	NA	NA	NA
T-130	785	330	250*	200	110	96	NA	NA	15.0
T-106	900	405	345*	325*	135	116	NA	NA	19.0*
T- 94	590	248	200*	160	84	70	58	32	10.6
T- 80	450	180	170	115	55	45	32*	22	8.5
T- 68	420	195	180	115	57	47	32	21	7.5
T- 50	320	175	135	100	49*	40	31	18	6.4
T- 44	229	180	160	105	52*	42	33	NA	6.5
T- 37	308	120*	90	80	40*	30	25	15	4.9
T- 30	375	140*	93	85	43	36	25	16	6.0
T- 25	225	100	85	70	34	27	19	13	4.5
T- 20	175	90	65	52	27	22	16	10	3.5
T- 16	130	61	NA	44	22	19	13	8	3.0
T- 12	112	60	50*	48	20*	17*	12	7.5	3.0

NA — Not available in that size.

$\text{Turns} = 100\sqrt{L_{\mu H} \div A_L \text{ Value (above.)}}$

All frequency figures optimum.

*Updated values (1979) from Micrometals, Inc.

Number of Turns vs. Wire Size and Core Size
Approximate maximum of turns — single layer wound enameled wire

Wire Size	T-200	T-130	T-106	T-94	T-80	T-68	T-50	T-37	T-25	T-12
10	33	20	12	12	10	6	4	1		
12	43	25	16	16	14	9	6	3		
14	54	32	21	21	18	13	8	5	1	
16	69	41	28	28	24	17	13	7	2	
18	88	53	37	37	32	23	18	10	4	1
20	111	67	47	47	41	29	23	14	6	1
22	140	86	60	60	53	38	30	19	9	2
24	177	109	77	77	67	49	39	25	13	4
26	223	137	97	97	85	63	50	33	17	7
28	281	173	123	123	108	80	64	42	23	9
30	355	217	154	154	136	101	81	54	29	13
32	439	272	194	194	171	127	103	68	38	17
34	557	346	247	247	218	162	132	88	49	23
36	683	424	304	304	268	199	162	108	62	30
38	875	544	389	389	344	256	209	140	80	39
40	1103	687	492	492	434	324	264	178	102	51

Physical Dimensions

Core Size	Outer Dia. (in.)	Inner Dia. (in.)	Height (in.)	Cross Sect. Area cm²	Mean Length cm	Core Size	Outer Dia. (in.)	Inner Dia. (in.)	Height (in.)	Cross Sect. Area cm²	Mean Length cm
T-200	2.000	1.250	0.550	1.330	12.97	T- 50	0.500	0.303	0.190	0.121	3.20
T-184	1.840	0.950	0.710	2.040	11.12	T- 44	0.440	0.229	0.159	0.107	2.67
T-157	1.570	0.950	0.570	1.140	10.05	T- 37	0.375	0.205	0.128	0.070	2.32
T-130	1.300	0.780	0.437	0.733	8.29	T- 30	0.307	0.151	0.128	0.065	1.83
T-106	1.060	0.560	0.437	0.706	6.47	T- 25	0.255	0.120	0.096	0.042	1.50
T- 94	0.942	0.560	0.312	0.385	6.00	T- 20	0.200	0.088	0.067	0.034	1.15
T- 80	0.795	0.495	0.250	0.242	5.15	T- 16	0.160	0.078	0.060	0.016	0.75
T- 68	0.690	0.370	0.190	0.196	4.24	T- 12	0.125	0.062	0.050	0.010	0.74

Inches × 25.4 = mm.   Courtesy of Amidon Assoc., N. Hollywood, CA 91607 and Micrometals, Inc.

previous one with suitable insulation in between (Figs. 53 and 54). A *transmission-line transformer* is, conversely, one that uses windings which are configured to simulate a piece of transmission line of a specific impedance. This can be achieved by twisting the wires together a given number of times per inch, or by laying the wires on the core (adjacent to one another) at a distance apart which provides a two-wire line impedance of a particular value. In some applications these windings are called *bifilar*. A three-wire winding is known as *trifilar* one, and so forth (Fig. 55). It can be argued that a transmission-line transformer is more efficient than a conventional one, but in practice it is difficult to observe a significant difference in the performance characteristics. An interesting technical paper on the subject of toroidal broadband transformers was published by Sevick, W2FMI.[1] The classic reference work on the subject is by Ruthroff.[2]

Ferrite Beads

Another form of toroidal inductor is the *ferrite bead*. This component is available in various μ_i values and sizes, but most beads are less than 0.25-inch (6.3-mm) diameter. Ferrite beads are used principally as vhf/uhf parasitic sup-

[1]Sevick, "Simple Broadband Matching Networks," *QST*, January 1976.

[2]Ruthroff, "Some Broadband Transformers," *Proc. IRE,* Vol. 47, August 1959, p. 137.

pressors at the input and output terminals of amplifiers. Another practical application for them is in decoupling networks which are used to prevent unwanted migration of rf energy from one section of a circuit to another. They are used also in suppressing RFI and TVI in hi-fi and television sets. In some circuits it is necessary only to place one or more beads over a short length of wire to obtain ample inductive reactance for creating an rf choke. A few turns of small-diameter enameled wire can be looped through the larger beads to increase the effective inductance. Ferrite beads are suitable as low-Q base impedances in solid-state vhf and uhf amplifiers. The low-Q characteristics prevents self-oscillation that might occur if a high Q solenoidal rf choke were used in place of one made from beads. Miniature broadband transformers are sometimes fashioned from ferrite beads. For the most part, ferrite beads can be regarded as small toroid cores.

Number of Turns

The number of wire turns used on a toroid core can be calculated by knowing the A_L of the core and the desired inductance. The A_L is simply the *inductance index* for the core size and permeability being used. Table 6 provides information of interest concerning a popular assortment of powdered-iron toroid cores. The complete number for a given core is composed of the core-size designator in the upper left column, plus the corresponding mix number. For example, a half-inch diameter core with a no. 2 mix would be designated at a T-50-2 unit. The A_L would be 49 and the suggested operating frequency would be from 1 to 30 MHz. The μ_i for that core is 10.

The required number of wire turns for a specified inductance on a given type of core can be determined by

$$\text{Turns} = 100 \sqrt{\text{desired L } (\mu H) \div A_L}$$

where A_L is obtained from Table 6. The table also indicates how many turns of a particular wire gauge can be close wound to fill a specified core. For example, a T-68 core will contain 49 turns of no. 24 enameled wire, 101 turns of no. 30 enameled wire, and so on. Generally speaking, the larger the wire gauge the higher the unloaded Q of the toroidal inductor. The inductance values are based on the winding covering the entire circumference of the core. When there is space between the turns of wire, some control over the net inductance can be effected by compressing the turns or spreading them. The inductance will increase if compression is used and will decrease when the turns are spread farther apart.

Table 7 contains data for ferrite cores. The number of turns for a specified inductance in mH versus the A_L can be determined by

$$\text{Turns} = 1000 \sqrt{\text{desired L } (mH) \div A_L}$$

where the A_L for a specific core can be taken from Table 7. Thus, if one required a 1-mH inductor and chose a no. FT-82-43 toroid core, the number of turns would be

$$\text{Turns} = 1000 \sqrt{1 \div 557}$$
$$= 1000 \sqrt{0.001795}$$
$$= 1000 \times 0.0424 = 42.4 \text{ turns}$$

For an FT-82 size core no. 22 enameled wire would be suitable as indicated in Table 6 (using the T-80 core size as the nearest one to an FT-82). If the toroid core has rough edges (untumbled), it is suggested that insulating tape (3M glass epoxy tape or Mylar tape) be wrapped through the core before the wire is added.

This will prevent the rough edges of the core from abrading the enameled wire.

The inductance of a toroidal coil with known A_L is

$$L = A_L \left(\frac{N}{100}\right)^2$$

L and A_L must be in the same units. Tables 8 and 9 cross-reference the ferrite toroidal cores offered by several sources.

Checking RF Toroidal Devices

The equations given previously will provide the number of wire turns needed for a particular inductance, plus or minus 10 percent. However, slight variations in core permeability may exist from one production run to another. Therefore, for circuits which require exact values of inductance it is necessary to check the toroid winding by means of an RCL

Table 7
Ferrite Toroids A_L — Chart (mH per 1000 turns) Enameled Wire

Core Size	63-Mix μ=40	61-Mix μ=125	43-Mix μ=950	72-Mix μ=2000	75-Mix μ=5000
FT- 23	7.9	24.8	189.0	396.0	990.0
FT- 37	17.7	55.3	420.0	884.0	2210.0
FT- 50	22.0	68.0	523.0	1100.0	2750.0
FT- 82	23.4	73.3	557.0	1172.0	2930.0
FT-114	25.4	79.3	603.0	1268.0	3170.0

Number turns = $1000 \sqrt{\text{desired L (mH)} \div A_L \text{ value (above)}}$

Ferrite Magnetic Properties

Property	Unit	63-Mix	61-Mix	43-Mix	72-Mix	75-Mix
Initial Perm. (μi)		40	125	950	2000	5000
Maximum Perm.		125	450	3000	3500	8000
Saturation Flux Density @ 13 oer	Gauss	1850	2350	2750	3500	3900
Residual Flux Density	Gauss	750	1200	1200	1500	1250
Curie Temp.	°C	500	300	130	150	160
Vol. Resistivity	ohm/cm	1×10^8	1×10^8	1×10^5	1×10^2	5×10^2
Opt. Freq. Range	MHz	15-25	.2-10	.01-1	.001-1	.001-1
Specific Gravity		4.7	4.7	4.5	4.8	4.8
Loss Factor $\frac{1}{u \, O}$		9.0×10^{-5} @ 25 MHz.	2.2×10^{-5} @ 2.5 MHz.	2.5×10^{-5} @ .2 MHz.	9.0×10^{-6} @ .1 MHz.	5.0×10^{-6} @ .1 MHz.
Coercive Force	Oer.	2.40	1.60	0.30	0.18	0.18
Temp. Co-eff of initial Perm.	%/°C 20-70°C	0.10	0.10	0.20	0.60	

Ferrite Toroids — Physical Properties

Core Size	OD	ID	Height	A_e	I_e	V_e	A_S	A_W
FT- 23	0.230	0.120	0.060	0.00330	0.529	0.00174	0.1264	0.01121
FT- 37	0.375	0.187	0.125	0.01175	0.846	0.00994	0.3860	0.02750
FT- 50	0.500	0.281	0.188	0.02060	1.190	0.02450	0.7300	0.06200
FT- 82	0.825	0.520	0.250	0.03810	2.070	0.07890	1.7000	0.21200
FT-114	1.142	0.748	0.295	0.05810	2.920	0.16950	2.9200	0.43900

OD - Outer diameter (inches) A_e - Effective magnetic cross-sectional area (in)²
ID - Inner diameter (inches) I_e - Effective magnetic path length (inches)
Hgt - Height (inches) V_e - Effective magnetic volume (in)³
A_W - Total window area (in)² A_S - Surface area exposed for cooling (in)²

Inches × 25.4 = mm.   Courtesy of Amidon Assoc., N. Hollywood, CA 91607

Table 8

Ferrite Toroid Cores — Size Cross-Reference

(inches) OD	ID	Thickness	Amidon	Fair-Rite	Indiana General	Ferroxcube	Magnetics, Inc.
0.100	0.050	0.050	——	——		——	40200TC
0.100	0.070	0.030	——	701	F426-1	——	——
0.155	0.088	——	——	801	F2062-1	——	40502
0.190	0.090	0.050	——	——		213T050	——
0.230	0.120	0.060	FT-23	101	F303-1	1041T060	40601
0.230	0.120	0.120	——	901			
0.300	0.125	0.188	——	——	F867-1	——	40705
0.375	0.187	0.125	FT-37	201	F625-9	266T125	41003
0.500	0.281	0.188	FT-50	301	——	768T188	——
0.500	0.312	0.250	——	1101	F627-8	——	41306
0.500	0.312	0.500	——	1901			
0.825	0.520	0.250	FT-82	601	——	——	——
0.825	0.520	0.468	——	501	——	——	——
0.870	0.500	0.250	——	401	——	——	——
0.870	0.540	0.250	——	1801	F624-19	846T250	42206
1.000	0.500	0.250	——	1501	F2070-1	——	42507
1.000	0.610	0.250	——	1301	——	——	——
1.142	0.748	0.295	FT-114	1001	——	K300502	42908
1.225	0.750	0.312	——	1601	——	——	——
1.250	0.750	0.375	——	1701	F626-12	——	——
1.417	0.905	0.591	——	——	——	K300501	——
1.417	0.905	0.394	——	——	——	K300500	——
1.500	0.750	0.500	——	——	——	528T500	43813
2.000	1.250	0.750	——	——	——	400T750	——
2.900	1.530	0.500	——	——	——	144T500	——
3.375	1.925	0.500	——	——	F1707-15	——	——
3.500	2.000	0.500	——	——	F1707-1	——	——
5.835	2.50	0.625	——	——	F1824-1	——	——

Table 9

Ferrite Toroid Cores — Permeability Cross-Reference

μ_o	Amidon	Fair-Rite	Indiana General	Ferroxcube	Magnetics, Inc.
16	——	——	Q3		
20	——	68		——	——
40	FT — 63	63, 67	Q2		
100	——	65			
125	FT — 61	61	Q1	4C4	
175	——	62			
250	FT — 64	64	——	——	——
300	——	83			
375	——	31			
400	——	——	G	——	——
750	——	——		3D3	A
800	——	33		——	——
850	——	43	H	——	——
950	FT — 43	——	TC-3	——	——
1400	——	——		——	C
1200	——	34		——	——
1500	——	——	TC-7		
1800	FT — 77	77		3B9	——
2000	FT — 72	72	TC-9		S, V, D
2200	——	——	05		——
2300	——	——		3B7	G
2500	FT — 73	73	TC-12		
2700	——	——	——	3E (3C8)	——
3000	——	——	05P	3C5	F
4700	——	——	06		
5000	FT — 75	75		3E2A	J
10,000	——	——	——		W
12,500	——	——	——	3E3	

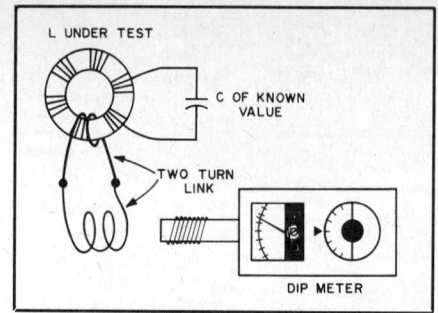

Fig. 56 — Method for checking the inductance of a toroid winding by means of a dip meter, known capacitance value and a calibrated receiver. The self-shielding properties of a toroidal inductor prevent dip meter readings when the instrument is coupled directly to the toroid. Sampling is done by means of a coupling link as illustrated.

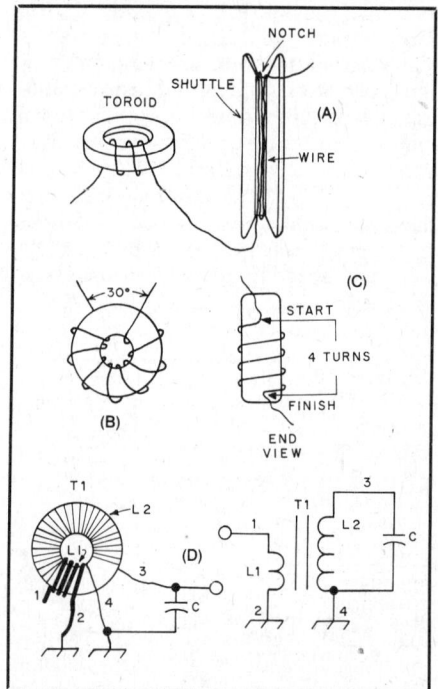

Fig. 57 — (A) Illustation of a homemade winding shuttle for toroids. The wire is stored on the shuttle and the shuttle is passed through the center hole of the toroid, again and again, until the required number of turns is in place. (B) It is best to leave a 30° gap between the ends of the toroid winding. This will reduce the distributed capacitance considerably. (C) Edgewise view of a toroid core, illustrating the method for counting the turns accurately. (D) The low-impedance winding of a toroidal transformer is usually wound over the large winding, as shown. For narrow-band applications the link should be wound over the cold end of the main winding (see text).

bridge or an RX meter. If these instruments are not available, close approximations can be had by using a dip meter, standard capacitor (known value, stable type, such as a silver mica) and a calibrated receiver against which to check the dipper frequency. Fig. 56 shows how to couple a dip meter to a completed toroid for testing. The coupling link in the illustration is necessary because the toroid has a self-shielding characteristic. The latter makes it difficult, and often impossible, to secure a dip in the meter reading when coupling the instrument directly to the toroidal inductor or transformer. The inductance can be determined by X_L since $X_L = X_c$ at resonance. Therefore,

$$X_c = \frac{1}{2\pi f C} \quad \text{and} \quad L_{(\mu H)} = \frac{X_L}{2\pi f}$$

where X_c is the reactance of the known capacitor value, f is in MHz and C is in μF. Using an example, where f is 3.5 MHz (as noted on a dip meter) and C is 100 pF, L is determined by

$$X_c = \frac{1}{6.28 \times 3.5 \times 0.0001} = 455 \text{ ohms}$$

$$L = X_c \text{ at resonance,}$$

$$L_{(\mu H)} = \frac{455}{6.28 \times 3.5} = 20.7 \ \mu H$$

It is assumed, for the purpose of accuracy, that the dip-meter signal is checked for precise frequency by means of a calibrated receiver.

Practical Considerations

Amateurs who work with toroidal inductors and transformers are sometimes confused by the winding instructions given in construction articles. For the most part, winding a toroid core with wire is less complicated than it is when winding a cylindrical single-layer coil.

When many turns of wire are required, a homemade winding shuttle can be used to simplify the task. Fig. 57A illustrates how this method may be employed. The shuttle can be fashioned from a piece of circuit-board material. The wire is wound on the shuttle after determining how many inches are required to provide the desired number of toroid turns. (A sample turn around the toroid core wil reveal the wire length per turn.) Once the shuttle is loaded, it is passed through the toroid center again and again until the winding is completed. The edges of the shuttle should be kept smooth to prevent abrasion of the wire insulation.

How to Wind Toroids

The *effective inductance* of a toroid coil or a transformer winding is dependent in part upon the distributed capacitance between the coil turns and between the ends of the winding. When a large number of turns are used (e.g., 500 or 1000), the distributed capacitance can be as great as 100 pF. Ideally, there would be no distributed or ''parasitic'' capacitance, but this is not possible. Therefore, the unwanted capacitance must be kept as low as possible in order to take proper advantage

of the A_L factors discussed earlier in this section. The greater the distributed capacitance the more restrictive the transformer or inductor becomes when applied in a broadband circuit. In the case of a narrow-band application, the Q can be affected by the distributed capacitance. The pictorial illustration at Fig. 57B shows the inductor turns distributed uniformly around the toroid core, but a gap of approximately 30° is maintained between the ends of the winding. This method is recommended to reduce the distributed capacitance of the winding. The closer the ends of the winding are to one another, the greater the unwanted capacitance. Also, in order to closely approximate the desired toroid inductance when using the A_L formula, the winding should be spread over the core as shown. When the turns of the winding are not close wound, they can be spread apart to *decrease* the effective inductance (this lowers the distributed C). Conversely, as the turns are pushed closer together, the effective inductance is *increased* by virtue of the greater distributed capacitance. This phenomenon can be used to advantage during final adjustment of narrow-band circuits in which toroids are used.

The proper method for counting the turns on a toroidal inductor is shown in Fig. 57C. The core is shown as it would appear when stood on its edge with the narrow dimension toward the viewer. In this example a four-turn winding has been placed on the core.

Some manufacturers of toroids recommend that the windings on toroidal transformers be spread around all of the core in the manner shown in Fig. 57B. That is, the primary and secondary windings should each be spread around most of the core. This is a proper method when winding conventional broadband transformers. However, it is not recommended when narrow band transformers are being

built. It is better to place the low-impedance winding (L1 of Fig. 57D at the ''cold'' or grounded end of L2 on the core. This is shown in pictorial and schematic form at Fig. 57D. The windings are placed on the core in the same rotational sense, and L1 is wound over L2 at the grounded end of L2. The purpose of this winding method is to discourage unwanted *capacitive* coupling between the windings — an aid to the reduction of spurious energy (harmonics, etc.) which might be present in the circuit where T1 is employed.

In circuits which have a substantial amount of ac and/or dc voltage present in the transformer windings, it is prudent to use a layer of insulating material between the toroid core and the first winding. Alternatively, the wire can have high-dielectric insulation, such as Teflon. This procedure will prevent arcing between the winding and the core. Similarly, a layer of insulating tape (3-M glass tape, mylar or Teflon) can be placed between the primary and secondary windings of the toroidal transformer (Fig. 57D). Normally, these precautions are not necessary at impedance levels under a few hundred ohms at rf power levels below 100 watts.

Once the inductor or transformer is wound and tested for proper performance, a coating or two of high-dielectric cement should be applied to the winding(s) of the toroid. This will protect the wire insulation from abrasion, hold the turns in place and seal the assembly against moisture and dirt. Polystyrene Q Dope is excellent for the purpose.

The general guidelines given for toroidal components can be applied to pot cores and rods when they are used as foundations for inductors or transformers. The important thing to remember is that all of the powdered-iron and ferrite core materials are brittle. They break easily under stress.

The Decibel

It is useful to appraise signal strength in terms of relative loudness as registered by the ear. For example, if a person estimates that a signal is ''twice as loud'' when the transmitter power is increased from 10 watts to 100 watts, he or she will also estimate that a 1000-watt signal is twice as loud as a 100-watt signal: The human ear has a *logarithmic* response.

This fact is the basis for the use of the relative-power unit called the decibel (dB). A decibel is one-tenth of a bel, the unit of sound named for Alexander Graham Bell. A change of one decibel in power level is just detectable as a change in loudness under ideal conditions. The number of decibels corresponding to a given power ratio is given by:

$$dB = 10 \ \log_{10} \frac{P_2}{P_1}$$

Voltage and Current Ratios

Note that the decibel is based on *power* ratios. Voltage or current ratios can be used, but only when the impedance is the same for both values of voltage, or current. The gain of an amplifier cannot be expressed correctly in dB if it is based on the ratio of the output voltage to the input voltage unless both voltages are measured across the same value of impedance. When the impedance at both points of measurement is the same, the following formula may be used for voltage or current ratios:

$$dB = 20 \ \log \frac{V_2}{V_1} \text{ or } 20 \ \log \frac{I_2}{I_1}$$

where V = voltage
   I = current

If the voltage formula above is applied to an amplifier, where V2 is output voltage and V1 is the input voltage, a positive decibel value indicates amplifier gain. On the other hand, applying the formula to a resistor network would result in a negative decibel value, signifying a loss.

When the decibel value is known, the numerical ratio can be calculated from:

$$\frac{P_2}{P_1} = \text{antilog} \ \frac{dB}{10}$$

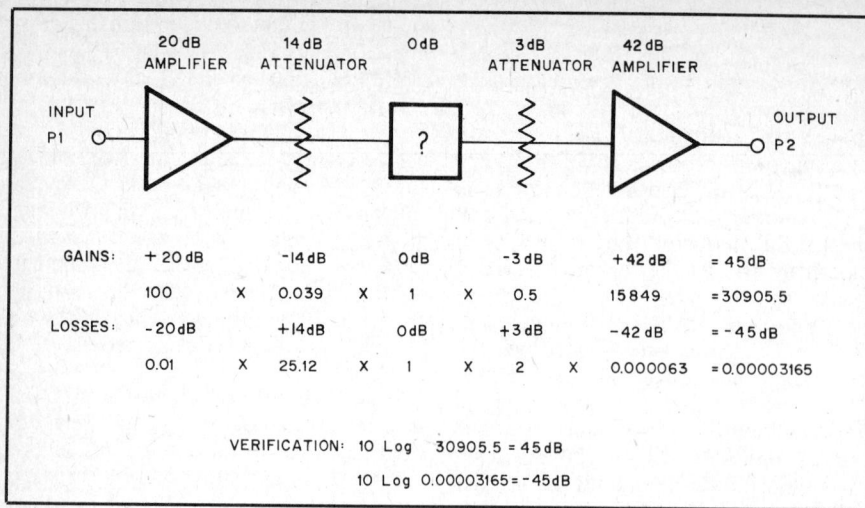

Fig. 58 — How to compute the composite gain of a system. Either numerical ratios or decibels may be used, but all units must be consistent in gain or loss. The method is explained in the text.

or $\dfrac{V_2}{V_1} = $ antilog $\dfrac{dB}{20}$.

Many mathematics textbooks contain tables of logarithms, but these numbers can be produced very quickly with a slide rule or inexpensive calculator. In any case, it is convenient to memorize the decibel values for a few of the common power and voltage ratios. For power changes, a numerical ratio of 2 is 3 dB, 4 is 6 dB, 10 is 10 dB, 100 is 20 dB, 1000 is 30 dB, and so on. When voltage changes are considered, doubling the voltage causes a 6-dB increase, a numerical ratio of 10 is worth 20 dB, 100 is 40 dB and so on. One can interpolate between known ratios to estimate a gain or loss within one decibel. Inverting a numerical ratio simply reverses the algebraic sign of the decibel value. For example, a voltage gain of 10 corresponds to 20 dB, while a gain of 1/10 (which is a *loss* of 10) corresponds to − 20 dB.

In a system of cascaded gain and loss blocks where numerical ratios are specified for each block, the overall system gain or loss can be calculated this way: Convert the ratios to all gains or all losses (gains become fractional losses or losses become fractional gains). The overall numerical gain or loss will be the product of the individual figures, and the decibel value can be derived as before. If the individual gains and losses are given in decibels, the procedure is again to convert to all losses or gains. Only the algebraic signs need to be changed; that is, losses become negative gains or gains become negative losses. The overall decibel gain or loss is the algebraic sum of the individual figures, and this can be converted to a numerical ratio if desired. Fig. 58 illustrates both methods.

The decibel is a relative unit. When using decibels to specify an absolute voltage, current or power level, the decibel value must be qualified by a reference level. For example, when discussing sound intensity, a reference lvel of 0 dB corresponds to an acoustical field strength of 10^{-16} W/cm^2, the normal human hearing threshold at 600 Hz. A lion's roar at 20 feet might have a sound intensity of 90 dB, and the threshold of pain occurs at 130 dB. Thus, the human ear/brain has a dynamic range of 130 dB, or a ratio of ten trillion to one.

In radio work, power is often rendered in dBW (0 dBW = 1 watt) or dBm (0 dBm = 1 milliwatt). With this system, 2 kilowatts equals + 63 dBm or + 33 dBW, and 5 microwatts equals − 23 dBm or − 53 dBW. Voltages are sometimes given as decibel values with respect to one volt or one microvolt; 2 millivolts equals + 66 dBμV or − 54 dBV. Antenna gain is specified with respect to some standard reference element such as an isotropic radiator or a dipole. The measurement units are the dBi and dBd.

Radio Frequency Circuits

The designer of amateur equipment needs to be familiar with radio-frequency circuits and the various related equations. This section provides the basic data for most amateur circuit development.

Resonance in Series Circuits

Fig. 59 shows a resistor, capacitor and inductor connected in series with a source of alternating current, the frequency of which can be varied over a wide range. At some *low* frequency the capacitive reactance will be much larger than the resistance of R, and the inductive reactance will be small compared with either the reactance of C or the resistance of R. (R is assumed to be the same at all frequencies.) On the other hand, at some very *high* frequency the reactance of C will be very small and the reactance of L will be very large. In either case the current will be small, because the net reactance is large.

At some intermediate frequency, the reactances of C and L will be equal and the voltage drops across the coil and capacitor will be equal and 180 degrees

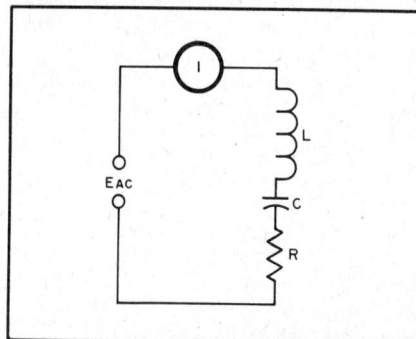

Fig. 59 — A series circuit containing L, C and R is "resonant" at the applied frequency when the reactance of C is equal to the reactance of L.

out of phase. Therefore, they cancel each other completely and the current flow is determined wholly by the resistance, R. At that frequency the current has its largest possible value, assuming the source voltage to be constant regardless of frequency. A series circuit in which the inductive and capacitive reactances are equal is said to be *resonant*.

The principle of *resonance* finds its most extensive application in radio-frequency circuits. The reactive effects associated with even small inductances and capacitances would place drastic limitations on rf circuit operation if it were not possible to "cancel them out" by supplying the right amount of reactance of the opposite kind—in other words, "tuning the circuit to resonance."

Resonant Frequency

The frequency at which a series circuit is resonant is that of which $X_L = XC$. Substituting the formulas for inductive and capacitive reactance gives

$$f = \frac{1}{2\pi \sqrt{LC}}$$

where
f = frequency in hertz
L = inductance in henrys
C = capacitance in farads
π = 3.14

These units are inconveniently large for radio-frequency circuits. A formula using

more appropriate units is

$$f = \frac{10^6}{2\pi \sqrt{LC}}$$

where

  f = frequency in kilohertz (kHz)
  L = inductance in microhenrys (μH)
  C = capacitance in picofarads (pF)
  π = 3.14

Example: The resonant frequency of a series circuit containing a 5-μH inductor and a 35-pF capacitor is

$$f = \frac{10^6}{2\pi \sqrt{LC}} = \frac{10^6}{6.28 \times \sqrt{5 \times 35}}$$

$$= \frac{10^6}{6.28 \times 13.23} = \frac{10^6}{83.08} = 12{,}036 \text{ kHz}$$

The formula for resonant frequency is not affected by resistance in the circuit.

Resonance Curves

If a plot is drawn on the current flowing in the circuit of Fig. 59 as the frequency is varied (the applied voltage being constant) it would look like one of the curves in Fig. 60. The shape of the *resonance curve* at frequencies near resonance is determined by the ratio of reactance to resistance.

If the reactance of either the coil or capacitor is of the same order of magnitude as the resistance, the current decreases rather slowly as the frequency is moved in either direction away from resonance. Such a curve is said to be *broad*. On the other hand, if the reactance is considerably larger than the resistance the current decreases rapidly as the frequency moves away from resonance and the circuit is said to be *sharp*. A sharp circuit will respond a great deal more readily to the resonant frequency than to frequencies quite close to resonance; a broad circuit will respond almost equally well to a group or band of frequencies centering around the resonant frequency.

Both types of resonance curves are useful. A sharp circuit gives good *selectivity* — the ability to respond strongly (in terms of current amplitude) at one desired frequency and discriminate against others. A broad circuit is used when the apparatus must give about the same response over a band of frequencies rather than to a single frequency alone.

Q

Most diagrams of resonant circuits show only inductance and capacitance; no resistance is indicated. Nevertheless, resistance is always present. At frequencies up to perhaps 30 MHz this resistance is mostly in the wire of the coil. Above this frequency energy loss in the capacitor (principally in the solid dielectric which must be used to form an insulating support for the capacitor plates) also becomes a factor. This energy loss is equivalent to resis-

tance. When maximum sharpness or selectivity is needed the object of design is to reduce the inherent resistance to the lowest possible value.

The value of the reactance of either the inductor or capacitor at the resonant frequency of a series-resonant circuit, divided by the *series* resistance in the circuit, is called the Q (quality factor) of the circuit, or

$$Q = \frac{X}{r}$$

where

  Q = quality factor
  X = reactance of either coil or capacitor in ohms
  r = series resistance in ohms

Example: The inductor and capacitor in a series circuit each have a reactance of 350 ohms at the resonant frequency. The resistance is 5 ohms. Then the Q is

$$Q = \frac{X}{r} = \frac{350}{5} = 70$$

The effect of Q on the sharpness of resonance of a circuit is shown by the curves of Fig. 61. In these curves the frequency change is shown in percentage above and below the resonant frequency. Qs of 10, 20, 50 and 100 are shown; these values cover much of the range commonly used in radio work. The *unloaded Q* of a circuit is determined by the inherent resistances associated with the components.

Voltage Rise at Resonance

When a voltage of the resonant frequency is inserted in series in a resonant circuit, the voltage that appears across either the inductor or capacitor is considerably higher than the applied voltage. The current in the circuit is limited only by the resistance and may have a relatively high value; however, the same current flows through the high reactances of the inductor and capacitor and causes large voltage drops. The ratio of the reactive voltage to the applied voltage is equal to the ratio of reactance to resistance. This ratio is also the Q of the circuit. Therefore, the voltage across either the inductor or capacitor is equal to QE, where E is the voltage inserted in series. This fact accounts for the high voltages developed across the components of series-tuned antenna couplers.

Resonance in Parallel Circuits

When a variable-frequency source of constant voltage is applied to a parallel circuit of the type shown in Fig. 62 there is a resonance effect similar to that in a series circuit. However, in this case the "line" current (measured at the point indicated) is *smallest* at the frequency for which the inductive and capacitive reactances are equal. At that frequency the current through L is exactly canceled by

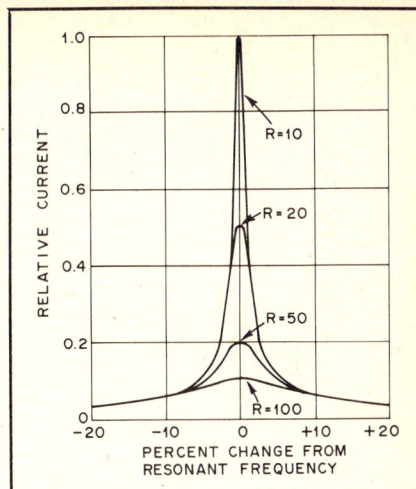

Fig. 60 — Current in a series-resonant circuit with various values of series resistance. The values are arbitrary and would not apply at all circuits, but represent a typical case. It is assumed that the reactances (at the resonant frequency) are 1000 ohms. Note that at frequencies more than plus or minus 10 percent away from the resonant frequency, the current is substantially unaffected by the resistance in the circuit.

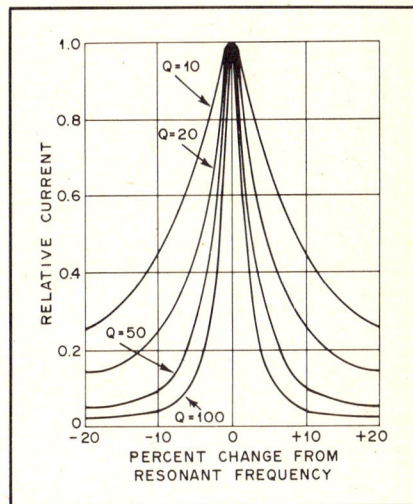

Fig. 61 — Current in series-resonant circuits having different Qs. In this graph the current at resonance is assumed to be the same in all cases. The lower the Q, the more slowly the current decreases as the applied frequency is moved away from resonance.

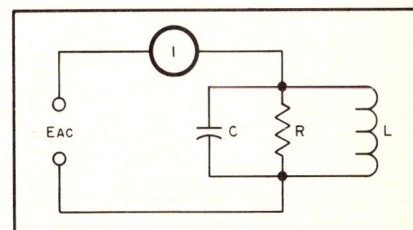

Fig. 62 — Circuit illustrating parallel resonance.

the out-of-phase current through C, so that only the current taken by R flows in the line. At frequencies *below* resonance the current through L is larger than that through C, because the reactance of L is

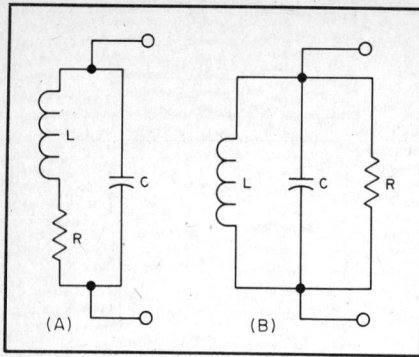

Fig. 63 — Series and parallel equivalents when the two circuits are resonant. The series resistance, r, in A is replaced in B by the equivalent parallel resistance ($R = X^2c/r = X^2L/r$) and vice versa.

smaller and that of C higher at low frequencies; there is only partial cancellation of the two reactive currents and the line current therefore is larger than the current taken by R alone. At frequencies *above* resonance the situation is reversed and more current flows through C than through L, so the line current again increases. The current at resonance, being determined wholly by R, will be small if R is large and large if R is small.

The resistance R shown in Fig. 62 is not necessarily an actual resistor. In many cases it will be the series resistance of the coil "transformed" to an equivalent parallel resistance (see later). It may be antenna or other load resistance coupled into the tuned circuit. In all cases it represents the total effective resistance in the circuit.

Parallel and series resonant circuits are quite alike in some respects. For instance, the circuits given at A and B in Fig. 63 will behave identically, when an external voltage is applied, if (1) L and C are the same in both cases, and (2) R multiplied by r, equals the square of the reactance (at resonance) of either L or C. When these conditions are met the two circuits will have the same Q. (These statements are approximate, but are quite accurate if the Q is 10 or more). The circuit at A is a *series* circuit if it is viewed from the "inside" — that is, going around the loop formed by L, C and r — so its Q can be found from the ratio of X to r.

Thus, a circuit like that of Fig. 63A has an equivalent *parallel impedance* (at resonance) of

$$R = \frac{X^2}{r}$$

where X is the reactance of either the inductor or the capacitor. Although R is not an actual resistor, to the source of voltage the parallel-resonant circuit "looks like" a pure resistance of that value. It is "pure" resistance because the inductive and capacitive currents are 180 degrees out of phase and are equal; thus there is no reactive current in the line. In a practical

circuit with a high-Q capacitor, at the resonant frequency the parallel impedance is

$$Z_R = QX$$

where Z_R = resistive impedance at resonance
Q = quality factor of inductor
X = reactance (in ohms) of either the inductor or capacitor

Example: The parallel impedance of a circuit with a coil Q of 50 and having inductive and capacitive reactance of 300 ohms will be

$$Z_R = QX = 50 \times 300 = 15,000 \text{ ohms}$$

At frequencies off resonance the impedance is no longer purely resistive because the inductive and capacitive currents are not equal. The off-resonant impedance therefore is complex, and is lower than the resonant impedance for the reasons previously outlined.

The higher the circuit Q, the higher the parallel impedance. Curves showing the variation of impedance (with frequency) of a parallel circuit have just the same shape as the curves showing the variation of current with frequency in a series circuit. Fig. 64 is a set of such curves. A set of curves showing the relative response as a function of the departure from the resonant frequency would be similar to Fig. 61. The −3 dB bandwidth (bandwidth at 0.707 relative response) is given by

$$\text{Bandwidth } -3 \text{ dB} = f_o/Q$$

where f_o is the resonant frequency and Q the circuit Q. It is also called the "half-power" bandwidth, for ease of recollection.

Parallel Resonance in Low-Q Circuits

The preceding discussion is accurate for Qs of 10 or more. When the Q is below 10, resonance in a parallel circuit having resistance in series with the coil, as in Fig. 63A, is not so easily defined. There is a set of values for L and C that will make the parallel impedance a pure resistance, but with these values the impedance does not have its maximum possible value. Another set of values for L and C will make the parallel impedance a maximum, but this maximum value is not a pure resistance. Either condition could be called "resonance," so with low-Q circuits it is necessary to distinguish between *maximum impedance* and *resistive impedance* parallel resonance. The difference between these L and C values and the equal reactances of a series-resonant circuit is appreciable when the Q is in the vicinity of 5, and becomes more marked with still lower Q values.

Q of Loaded Circuits

In many applications of resonant circuits the only power lost is that dissipated in the resistance of the circuit

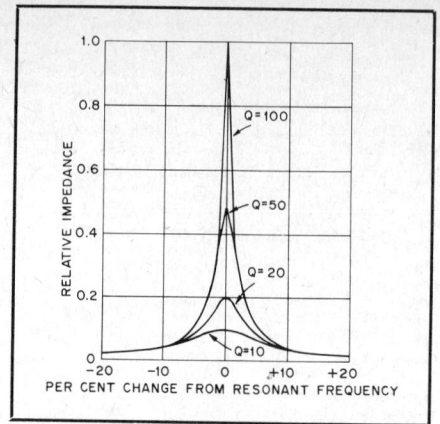

Fig. 64 — Relative impedance of parallel-resonant circuits with different Qs. These curves are similar to those in Fig. 61 for current in a series-resonant circuit. The effect of Q on impedance is most marked near the resonant frequency.

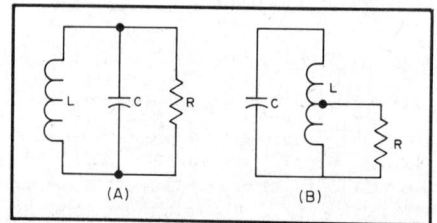

Fig. 65 — The equivalent circuit of a resonant circuit delivering power to a load. The resistor R represents the load resistance. At B the load is tapped across part of L, by which transformer action is equivalent to using a higher load resistance across the whole circuit.

itself. At frequencies below 30 MHz most of this resistance is in the coil. Within limits, increasing the number of turns in the coil increases the reactance faster than it raises the resistance, so coils for circuits in which the Q must be high are made with relatively large inductance for the frequency.

However, when the circuit delivers energy to a load (as in the case of the resonant circuits used in transmitters) the energy consumed in the circuit itself is usually negligible compared with that consumed by the load. The equivalent of such a circuit is shown in Fig. 65A, where the parallel resistor represents the load to which power is delivered. If the power dissipated in the load is at least ten times as great as the power lost in the inductor and capacitor, the parallel impedance of the resonant circuit itself will be so high compared with the resistance of the load that for all practical purposes the impedance of the combined circuit is equal to the load resistance. Under these conditions the Q of a parallel resonant circuit loaded by a resistive impedance is

$$Q = \frac{R}{X}$$

where
R = parallel load resistance (ohms)
X = reactance (ohms)

Example: A resistive load of 3000 ohms is connected across a resonant circuit in which the inductive and capacitive reactances are each 250 ohms. The circuit Q is then

$$Q = \frac{R}{X} = \frac{3000}{250} = 12$$

The "effective" Q of a circuit loaded by a parallel resistance becomes higher when the reactances are decreased. A circuit loaded with a relatively low resistance (a few thousand ohms) must have low-reactance elements (large capacitance and small inductance) to have reasonably high Q.

Impedance Transformation

An important application of the parallel-resonant circuit is as an impedance-matching device in the output circuit of a vacuum-tube rf power amplifier. There is an optimum value of load resistance for each type of tube or transistor and set of operating conditions. However, the resistance of the load to which the active device is to deliver power usually is considerably lower than the value required for proper device operation. To transform the actual load resistance to the desired value the load may be tapped across part of the coil, as shown in Fig. 65B. This is equivalent to connecting a higher value of load resistance across the whole circuit, and is similar in principle to impedance transformation with an iron-core transformer. In high-frequency resonant circuits the impedance ratio does not vary exactly as the square of the turns ratio, because all the magnetic flux lines do not cut every turn of the coil. A desired reflected impedance usually must be obtained by experimental adjustment.

When the load resistance has a very low value (say below 100 ohms) it may be connected in series in the resonant circuit (as in Fig. 63A, for example), in which case it is transformed to an equivalent parallel impedance as previously described. If the Q is at least 10, the equivalent parallel impedance is

$$Z_R = \frac{X^2}{r}$$

where Z_R = resistive parallel impedance at resonance
X = reactance (in ohms) of either the coil or capacitor
r = load resistance inserted in series

If the Q is lower than 10 the reactance will have to be adjusted somewhat, for the reasons given in the discussion of low-Q circuits, to obtain a resistive impedance of the desired value.

While the circuit shown in Fig. 65B will usually provide an impedance step-up as with an iron-core transformer, the network has some serious disadvantages for some applications. For instance, the common connection provides no dc isolation and the common ground is sometimes troublesome in regards to ground-loop currents. Consequently, a network in which only mutual magnetic coupling is employed is usually preferable. However, no impedance step-up will result unless the two coils are coupled tightly enough. The equivalent resistance seen at the input of the network will always be lower regardless of the turns ratio employed. However, such networks are still useful in impedance-transformation applications if the appropriate capacitive elements are used. A more detailed treatment of matching networks and similar devices will be taken up in the next section.

Unfortunately, networks involving reactive elements are usually narrowband in nature and it would be desirable if such elements could be eliminated in order to increase the bandwidth. With the advent of ferrites, this has become possible and it is now relatively easy to construct actual impedance transformers that are both broadband and permit operation well up into the vhf portion of the spectrum. This is also accomplished in part by tightly coupling the two (or more) coils that make up the transformer either by twisting the conductors together or winding them in a parallel fashion. The latter configuration is sometimes called a *bifilar* winding, as discussed in the section on ferromagnetic transformers.

Coupled Circuits and Filters

Two circuits are said to be coupled when a voltage or current in one network produces a voltage or current in the other one. The network where the energy originates is often called the *primary* circuit and the network that receives the energy is called the *secondary* circuit. Such coupling is often of a desirable nature since in the process, unwanted frequency components or noise may be rejected or isolated and power transferred from a source to a load with greatest efficiency. On the other hand, two or more circuits may be coupled inadvertently and undesirable effects produced. While a great number of coupling-circuit configurations are possible, one very important class covers so many practical applications that analysis of it will be covered in detail.

Ladder Networks

Any *two* circuits that are coupled can be drawn schematically as shown in Fig. 66A. A voltage source represented by E_{ac} with a source resistance R_p and a source reactance X_p is connected to the input of the coupling network, thus forming the primary circuit. At the output, a load reactance X_s and a load resistance R_s are connected as shown to form the secondary circuit. The circuit in the box could consist of an infinite variety of resistors,

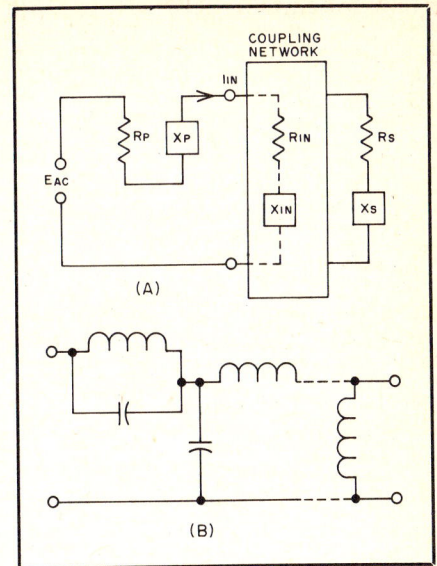

Fig. 66 — A representative coupling circuit (A) and ladder network (B).

capacitors, inductors, and even transmission lines. However, it will be assumed that the network can be reduced to a combination of series and shunt elements consisting only of inductors and capacitors as indicated by the circuit shown in Fig. 66B. For obvious reasons, the circuit is often called a *ladder network*. In addition, if there are no resistive elements present, or if such elements can be neglected, the network is said to be *dissipationless*.

If a network is dissipationless, all the power delivered to the input of the network will be dissipated in the load resistance R_s. This effect leads to important simplifications in computations involved in coupled networks. The assumption of a dissipationless network is usually valid with transmitting circuits since even a small network loss (0.5 dB) will result in considerable heating at the higher power levels used in amateur applications. On the other hand, coupled circuits used in some receiving stages may have considerable loss. This is because the network may have some advantage and its high loss can be compensated by additional amplification in another stage. However, such devices form a relatively small minority of coupled networks commonly encountered and only the dissipationless case will be considered in this section.

Effective Attenuation and Insertion Loss

The most important consideration in any coupled network is the amount of power delivered to the load resistance, R_s, from the source, E_{ac}, with the network present. Rather than specify the source voltage each time, a comparison is made with the maximum available power from any source with a given primary resistance, R_p. The value of R_p might be considered as the impedance level associated

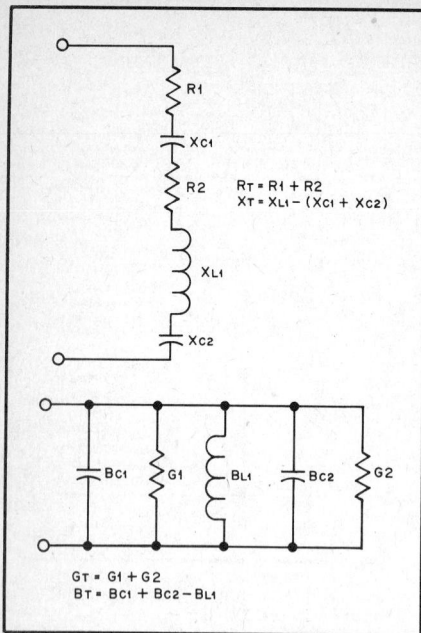

Fig. 67 — Resistances and reactances add in series circuits while conductances and susceptances add in parallel circuits. (Formulas shown are for numerical values of X and B.)

$$R_T = R_1 + R_2$$
$$X_T = X_{L1} - (X_{C1} + X_{C2})$$

$$G_T = G_1 + G_2$$
$$B_T = B_{C1} + B_{C2} - B_{L1}$$

with a complex combination of sources, transmission lines, coupled networks, and even antennas. Typical values of R_p are 52, 75, 300 and 600 ohms. The maximum available power is given by

$$P_{max} = \frac{E_{ac}^2}{4R_p}$$

If the network is also dissipationless, the power delivered to the load resistance, R_s, is just the power "dissipated" in R_{in}. This power is related to the input current by

$$P_o = I_{in}^2 R_{in}$$

and the current in terms of the other variables is

$$I_{in} = \frac{E_{ac}}{\sqrt{(R_p + R_{in})^2 + (X_p + X_{in})^2}}$$

Combining the foregoing expressions gives a very useful formula for the ratio of power delivered to a load in terms of the maximum available power. This ratio expressed in decibels is given by

$$Attn = -10 \log\left(\frac{P_o}{P_{in}}\right) =$$
$$-10 \log\left[\frac{4R_{in}R_p}{(R_p + R_{in})^2 + (X_p + X_{in})^2}\right]$$

and is sometimes called the *effective attenuation*.

In the special case where X_p and X_s are either zero or can be combined into a coupling network, and where R_p is equal to R_s, the effective attenuation is also equal to the *insertion loss* of the network. The insertion loss is the ratio of the power delivered to the load with the coupling network in the circuit to the power delivered to the load with the network absent. Unlike the effective attenuation which is always positive when defined by the previous formula, the insertion loss can take on negative values if R_p is not equal to R_s or if X_p and X_s are not zero. In effect, the insertion loss would represent a *power gain* under these conditions. The interpretation of this effect is that maximum available power does not occur with the coupling network out of the circuit because of the unequal source and load resistances and the non-zero reactances. With the network in the circuit, the resistances are now "matched" and the reactances are said to be "tuned out." The action of the coupling network in this instance is very similar to that of a transformer (which was discussed in a previous section) and networks consisting of "pure" inductors and capacitors are often used for this purpose. Such circuits are often referred to as matching networks. On the other hand, it is often desired to deliver the greatest amount of power to a load at some frequencies while rejecting energy at other frequencies. A device that accomplishes this action is called a *filter*. In the case of unequal source and load resistance, it is often possible to combine the processes of filtering and matching into one network.

Solving Ladder-Network Problems

From the last section it is evident that if the values of R_{in} and X_{in} of Fig. 66A can be determined, the effective attenuation and possibly the insertion loss are also easily found. Being able to solve this problem has wide applications in rf circuits. For instance, design formulas for filters often include a simplifying assumption that the load resistance is constant with frequency. In the case of many circuits, this assumption is not true. However, if the value of R_s and X_s at any particular frequency is known, the attenuation of the filter can be determined even though it is improperly terminated.

Unfortunately, while the solution to any ladder problem is possible from a theoretical standpoint, practical difficulties are encountered as the network complexity increases. Many computations to a high degree of accuracy may be required, making the process a tedious one. Consequently, the availability of a calculator or similar computing device is recommended. The approach used here is adapted readily to any calculating method including the use of an inexpensive pocket calculator.

Susceptance and Admittance

The respective reactances of an inductor and a capacitor are given by

$$X_L = 2\pi fL \qquad X_C = \frac{-1}{2\pi fC}$$

In a simple series circuit, the total resistance is just the sum of the individual

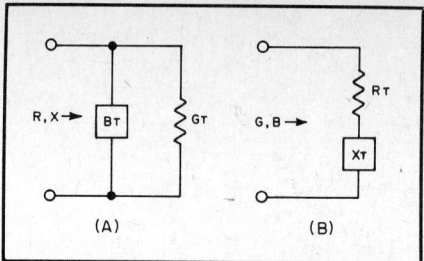

Fig. 68 — Application of conversion formulas can be used to transform a shunt conductance and susceptance to a series-equivalent circuit A. The converse is illustrated at B.

resistances in the network and the total reactance is the sum of the reactances. However, it is important to note the sign of the reactance. Since capacitive reactance is negative and inductive reactance is positive, it is possible that the sum of the reactances might be zero even though the individual reactances are not zero. In a series circuit, it will be recalled that the network is said to be resonant at the frequency where the reactances cancel.

A complementary condition exists in a parallel combination of circuit elements and it is convenient to introduce the concepts of *admittance*, *conductance* and *susceptance*. In the case of a simple resistance, the conductance is just the reciprocal. That is, the conductance of a 50-ohm resistance is $1/50$ or 2×10^{-2}. The reciprocal unit of the ohm is the mho. For simple inductances and capacitances, the formulas for the respective reciprocal entities are

$$B_L = \frac{-1}{2\pi fL} \qquad B_C = 2\pi fC$$

and are defined as susceptances. In a parallel combination of conductances and susceptances, the total conductance is the sum of the individual conductances, and the total susceptances is the sum of the individual susceptances, taking the respective signs of the latter into account. A comparison between the way resistance and reactance add and the manner in which conductance and susceptance add is shown in the example of Fig. 67. An entity called admittance can be defined in terms of the total conductance and total susceptance by the formula

$$Y = \sqrt{G_T^2 + B_T^2}$$

and is often denoted by the symbol Y. If the impedance of a circuit is known, the admittance is just the reciprocal. Likewise, if the admittance of a circuit is known, the impedance is the reciprocal of the admittance. However, conductance, reactance, resistance and susceptance are not so simply related. If the total resistance and total reactance of a series circuit are known, the conductance and susceptance of the circuit are related to the latter by the formulas

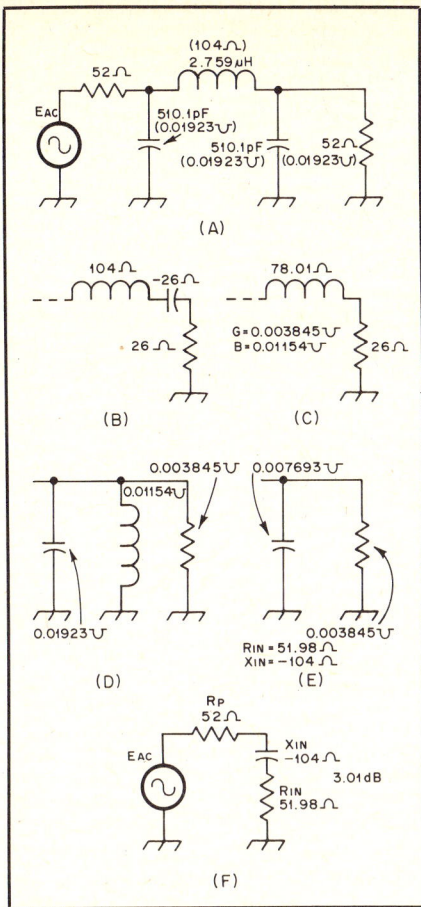

Fig. 69 — Problem illustrating network reduction to find insertion loss.

$$G = \frac{R_T}{R_T{}^2 + X_T{}^2} \qquad B = \frac{-X_T}{R_T{}^2 + X_T{}^2}$$

On the other hand, if the total conductance and total susceptance of a parallel combination are known, the equivalent resistance and reactance can be found from the formulas

$$R = \frac{G_T}{G_T{}^2 + B_T{}^2} \qquad X = \frac{-B_T}{G_T{}^2 + B_T{}^2}$$

The relations are illustrated in Fig. 68A and Fig. 68B respectively. While the derivation of the mathematical expressions will not be given, *the importance of the sign change cannot be stressed too highly.* Solving network problems with a calculator is merely a matter of bookkeeping, and failure to take the sign change associated with the transformed reactance and susceptance is the most common source of error.

A Sample Problem

The following example illustrates the manner in which the foregoing theory can be applied to a practical problem. A filter with the schematic diagram shown in Fig. 69A is *supposed* to have an insertion loss at 6 MHz of 3 dB when connected between a 52-ohm load and a source with a 52-ohm primary resistance (both X_p and X_s are zero). Since this is a case where the effective attenuation is equal to the insertion loss, the previous formula for effective attenuation applies. Therefore, it is required to find R_{in} and X_{in}.

Starting at the output, the values for the conductance and susceptance of the parallel RC circuit must be determined first. The conductance is just the reciprocal of 52 ohms and the previous formula for capacitive susceptance gives the value shown in parentheses in Fig. 69A. (The upside-down Ω is the symbol for mho.) The next step is to apply the formulas for resistance and reactance in terms of the conductance and susceptance and the results give a 26-ohm resistance in series with a -26-ohm capacitive reactance as indicated in Fig. 69B. The reactance of the inductor can now be added to give a total reactance of 78.01 ohms. The conductance and susceptance formulas can now be applied and the results of both of these operations is shown in Fig. 69C. Finally, adding the susceptance of the 510.1-pF capacitor (Fig. 69D) gives the circuit at Fig. 69A and applying the formulas once more gives the value of R_{in} and X_{in} (Fig. 69F). If the latter values are substituted into the effective attenuation formula, the insertion loss and effective attenuation are 3.01 dB, which is very close to the value specified. The reader might verify that the insertion loss is 0.167, 0.37 and 5.5 dB at 3.5, 4.0 and 5.5 MHz respectively. If a plot of insertion loss versus frequency was constructed it would give the frequency response of the filter.

Frequency Scaling and Normalized Impedance

Quite often, it is desirable to be able to change a coupling network at one frequency and impedance level to another one. For example, suppose it was desired to move the 3-dB point of the filter in the preceding illustration from 6 to 7 MHz. An examination of the reactance and susceptance formulas reveals that multiplying the frequency by some constant k and dividing both the inductance and capacitance by the same value of k leaves the equations unchanged. Thus, if the capacitances and inductance in Fig. 63A are multiplied by 6/7, all the reactances and susceptances in the new circuit will now have the same value at 7 MHz that the old one had at 6 MHz.

It is common practice with many filter tables especially, to present all the circuit components for a number of designs at some convenient frequency. Translating the design to some desired frequency is simply accomplished by multiplying all the components by some constant factor. The most common frequency used is the value of f such that $2\pi f$ is equal to 1.0. This is sometimes called a radian frequency of 1.0 and corresponds to 0.1592 Hz. To change a "one-radian" filter to a new frequency f_o (in Hz), all that is necessary is to multiply the inductances and capacitances by $0.1592/f_o$.

In a similar manner, if one resistance (or conductance) is multiplied by some factor n, all the other resistances (or conductances) and reactances (or susceptances) must be multiplied by the same factor in order to preserve the network characteristics. For instance, if the secondary resistance, R_s is multiplied by n, all circuit inductances must be multiplied by n and the circuit capacitances *divided* by n (since capacitive reactance varies as the inverse of C). If, in addition to converting the filter of Fig. 69A to 7 MHz from 6 MHz, it was also desired to change the impedance level from 52 to 600 ohms, the inductance would have to be multiplied by (6/7)(600/52) and the capacitances by (6/7)(52/600).

Using Filter Tables

In a previous example, it was indicated that the frequency response of a filter could be derived by solving for the insertion loss of the ladder network for a number of frequencies. The question might be asked if the converse is possible. That is, given a desired frequency response, could a network be found that would have this response? The answer is a qualified yes and the technical nomenclature for this sort of process is network synthesis. Frequency responses can be "cataloged" and, if a suitable one can be found, the corresponding network elements can be determined from an associated table. Filters derived by network synthesis and similar methods (such as optimized computer designs) are often referred to as "modern filters" even though the theory has been in existence for years. The term is useful in distinguishing such designs from those of an older approximate method called *image-parameter* theory.

Butterworth Filters

Filters can be grouped into four general categories as illustrated in Fig. 70A. Low-pass filters have zero insertion loss up to some critical frequency (f_c) or cutoff frequency and then provide high rejection above this frequency. (The latter condition is indicated by the shaded lines in Fig. 70.) Band-pass filters have zero insertion loss between two cutoff frequencies with high rejection outside of the prescribed "bandwidth." (Band-stop filters reject a band of frequencies while passing all others.) And high-pass filters reject all frequencies below some cutoff frequency.

The attenuation shapes shown in Fig. 70A are ideal and can only be approached or approximated in practice. For instance, if the filter in the preceding problem was used for low-pass purposes in an 80-meter transmitter to reject harmonics on 40 meters, its performance would leave a lot to be desired. While insertion loss at 3.5

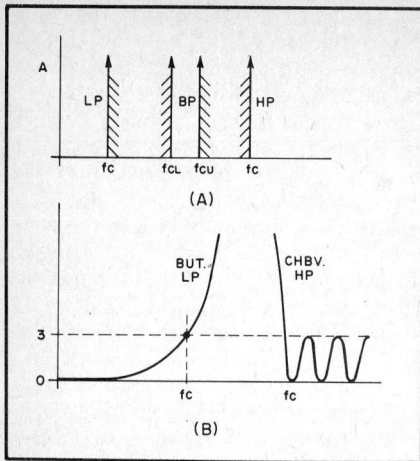

Fig. 70 — Ideal filter response curves are shown at A and characteristics of practical filters are shown at B.

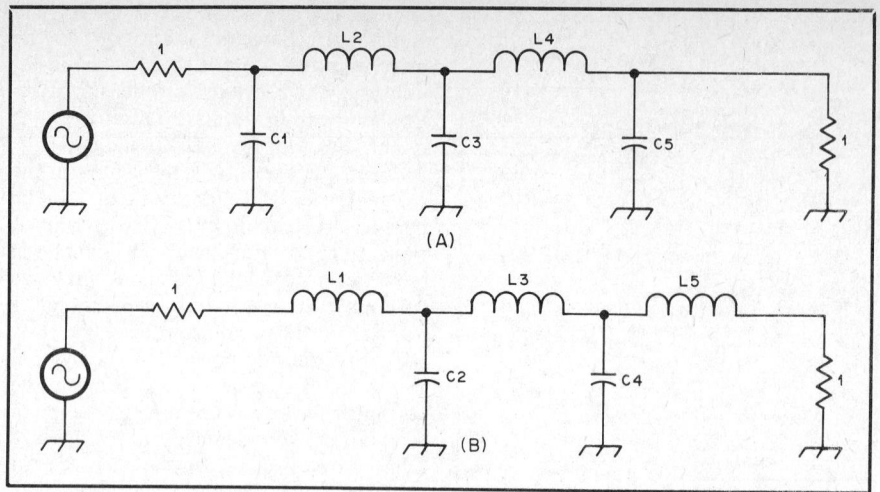

Fig. 71 — Schematic diagram of a Butterworth low-pass filter. (See Table 10 for element values.)

MHz was acceptable, it would likely be too high at 4.0 MHz and rejection would probably be inadequate at 7.0 MHz.

Fortunately, design formulas exist for this type of network and form a class called *Butterworth* filters. The name is derived from the shape of the curve for insertion-loss vs. frequency and is sometimes called a *maximally flat response*. A formula for the frequency response curve is given by

$$A = 10 \log_{10} \left[1 + \left(\frac{f}{f_c} \right)^{2k} \right]$$

where

f_c = the frequency for an insertion loss of 3.01 dB
k = the number of circuit elements

The shape of a Butterworth low-pass filter is shown in the left-hand portion of Fig. 70B. (Another type that is similar in nature, only one that allows some "ripple" in the passband, is also shown in Fig. 70B. Here, a high-pass characteristic illustrates a *Chebyshev* response.)

As can be seen from the formula, increasing the number of elements will result in a filter that approaches the "ideal" low-pass shape. For instance, a 20-element filter designed for a 3.01-dB cutoff frequency of 4.3 MHz, would have an insertion loss at 4 MHz of 0.23 dB and 84.7 dB at 7 MHz. However, practical difficulties would make such a filter very hard to construct. Therefore, some compromises are always required between a theoretically perfect frequency response and ease of construction.

Element Values

Once the number of elements, k, is determined, the next step is to find the network configuration corresponding to k. (Filter tables sometimes have sets of curves that enable the user to select the desired frequency response curve rather than use a formula. Once the curve with the fewest number of elements for the specified passband and stop-band insertion loss is found, the filter is then fabricated around the corresponding value of k.) Table 10 gives normalized element values for values of k from 1 to 10. This table is for 1-ohm source and load resistance (reactance zero) and a 3.01-dB cutoff frequency of 1 radian/second (0.1592 Hz). There are two possible circuit configurations and these are shown in Fig. 71. Here, a five-element filter is given as an example with either a shunt element next to the load (Fig. 71A) or a series element next to the load (Fig. 71B) Either filter will have the same response.

After the values for the 1-ohm, 1-radian/second "prototype" filter are found, the corresponding values for the actual frequency/impedance level can be determined (see the section on frequency and impedance scaling). The prototype inductance and capacitance values are multiplied by the ratio $(0.1592/f_c)$ where f_c is the actual 3.01-dB cutoff frequency. Next, this number is multiplied by the load resistance in the case of an inductor and divided by the load resistance if the element is a capacitance. For instance, the filter in the preceding example is for a three-element design (k equal to 3) and the reader might verify the values for the components for an f_c of 6 MHz and load resistance of 52 ohms.

High-Pass Butterworth Filters

The formulas for change of impedance and frequency from the 1-ohm, 1-radian/second prototype to some desired level can also be conveniently written as

$$L = \frac{R}{2\pi f_c} L_{\text{prototype}} \qquad C = \frac{1}{2\pi f_c R} C_{\text{prototype}}$$

Table 10
Prototype Butterworth Low-Pass Filters

Fig. 71A	C1	L2	C3	L4	C5	L6	C7	L8	C9	L10
Fig. 71B	L1	C2	L3	C4	L5	C6	L7	C8	L9	C10
k										
1	2.0000									
2	1.4142	1.4142								
3	1.0000	2.0000	1.0000							
4	0.7654	1.8478	1.8478	0.7654						
5	0.6180	1.6180	2.0000	1.6180	0.6180					
6	0.5176	1.4142	1.9319	1.9319	1.4142	0.5176				
7	0.4450	1.2470	1.8019	2.0000	1.8019	1.2470	0.4450			
8	0.3902	1.1111	1.6629	1.9616	1.9616	1.6629	1.1111	0.3902		
9	0.3473	1.0000	1.5321	1.8794	2.0000	1.8794	1.5321	1.0000	0.3473	
10	0.3129	0.9080	1.4142	1.7820	1.9754	1.9754	1.7820	1.4142	0.9080	0.3129

where

    R = the load resistance in ohms
    f_c = the desired 3.01-dB frequency in Hz

Then L and C give the actual circuit-element values in henrys and farads in terms of the prototype element values from Table 10.

However, the usefulness of the low-pass prototype does not end here. If the following set of equations is applied to the prototype values, circuit elements for a high-pass filter can be obtained. The filters are shown in Fig. 72A and Fig. 72B which correspond to Fig. 71A and Fig. 71B in Table 10. The equations for the actual high-pass circuit values in terms of the low-pass prototype are given by

$$C = \frac{1}{R2\pi f_c C_{prot.}} \qquad L = \frac{R}{2\pi f_c L_{prot.}}$$

and the frequency response curve can be obtained from

$$A = 10\log\left[1 + \left(\frac{f_c}{f}\right)^{2k}\right]$$

For instance, a high-pass filter with three elements, a 3.01-dB f_c of 6 MHz and 52 ohms, has a C1 and C3 of 510 pF and an L2 of 0.6897 μH. The insertion loss at 3.5 and 7 MHz would be 14.21 and 1.45 dB respectively.

Butterworth Band-Pass Filters

Band-pass filters can also be designed through the use of Table 10. Unfortunately the process is not as straightforward as it is for low- and high-pass filters if a practical design is to be obtained. In essence, a low-pass filter is resonated to some "center frequency" with the 3.01-dB cutoff frequency being replaced by the filter bandwidth. The ratio of the bandwidth to center frequency must be relatively large, otherwise component values tend to become unmanageable.

While there are many variations of specifying such filters, a most useful approach is to determine an upper and lower frequency for a given attenuation. The center frequency and bandwidth are then given by

$$f_o = \sqrt{f_1 f_2} \qquad BW = f_2 - f_1$$

If the bandwidth specified *is not* the 3.01-dB bandwidth (BW_c), the latter can be determined from

$$BW_c = \frac{BW}{\left[10^{.1A} - 1\right]^{\frac{1}{2K}}}$$

in the case of a Butterworth response or from tables of curves. A is the required attenuation at the cutoff frequencies. The upper and lower cutoff frequencies (f_{cu}

and f_{cl}) are then given by

$$f_{cl} = \frac{-BW_c + \sqrt{(BW_c)^2 + 4f_o^2}}{2}$$

$$f_{cu} = f_{cl} + BW_c$$

A somewhat more convenient method is to pick a 3.01-dB bandwidth (the wider the better) around some center frequency and compute the attenuation at other frequencies of interest by using the transformation:

$$\frac{f}{f_c} = \left| \left(\frac{f}{f_o} - \frac{f_o}{f}\right) \frac{f_o}{BW_c} \right|$$

which can be substituted into the insertion-loss formula or table of curves.

As an example, suppose it is desired to build a band-pass filter for the 15-meter Novice band in order to eliminate the possibility of radiation on the 14- and 28-MHz bands. For a starting choice, 16 and 25 MHz will be picked as the 3.01-dB points giving a 3-dB bandwidth of 9 MHz. For these two points, f_o will be 20 MHz. It is common practice to equate the number of branch elements or filter resonators to certain mathematical entities called "poles" and the number of poles is just the value of k for purposes of discussion here. For a three-pole filter (k of 3), the insertion loss will be 12.79 and 11.3 dB at 14 and 28 MHz, respectively.

C1, C3 and L2 are then calculated for a 9-MHz low-pass filter and the elements for this filter are resonated to 20 MHz as shown in Fig. 73A. The response shape is plotted in Fig. 73B and it appears to be unsymmetrical about f_o. In spite of this fact, such filters are called symmetrical band-pass filters and f_o is the "center frequency."

If the response is plotted against a logarithmic frequency scale, the symmetry

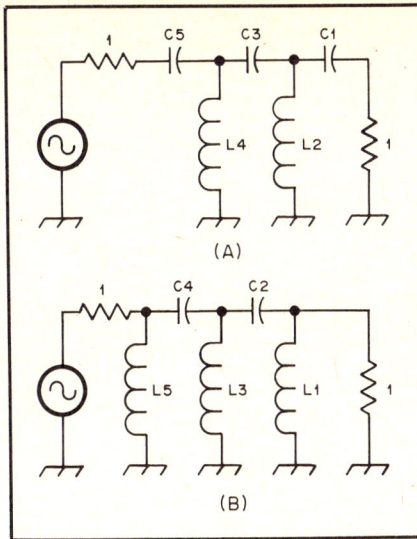

Fig. 72 — Network configuration of a Butterworth high-pass filter. The low-pass prototype can be transformed as described in the text.

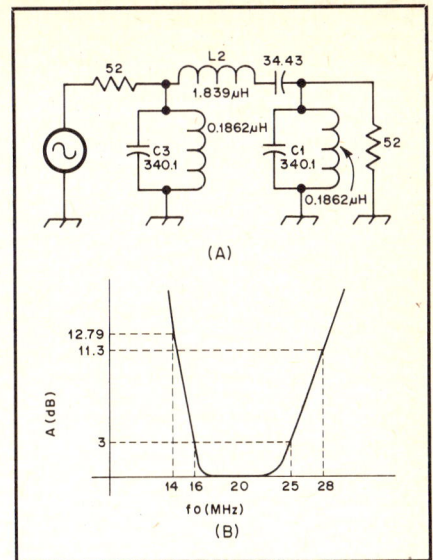

Fig. 73 — A Butterworth band-pass filter. (Capacitance values are in picofarads.)

will become apparent. Consequently, using a logarithmic plot is helpful in designing filters of this type.

Examination of the component values reveals that while the filter is practical, it is a bit untidy from a construction standpoint. Rather than using a single 340.1-pF capacitor, paralleling a number of smaller valued units would be advisable. Encountering difficulty of this sort is typical of most filter designs, consequently, some tradeoffs between performance, complexity and ease of construction are usually required.

Coupled Resonators

A problem frequently encountered in rf circuits is that of a coupled resonator. Applications include simple filters, oscillator tuned circuits, and even antennas. The circuit shown in Fig. 74A is illustrative of the basic principles involved. A series RLC circuit and the external terminals ab are "coupled" through a common capacitance, Cm. Applying the formulas for conductance and susceptance in terms of series reactance and resistance gives

$$G_{ab} = \frac{R_r}{R_r^2 + X^2}$$

$$B_{ab} = B_{cm} - \frac{X}{R_r^2 + X^2}$$

The significance of these equations can be seen with the aid of Fig. 74B. At some point, the series inductive reactance will cancel the series capacitive reactance (at a point slightly below f_o where the conductance curve reaches a peak). Depending upon the value of the coupling susceptance, B_m, it is possible that a point can be found where the total input susceptance is zero. The input conductance at

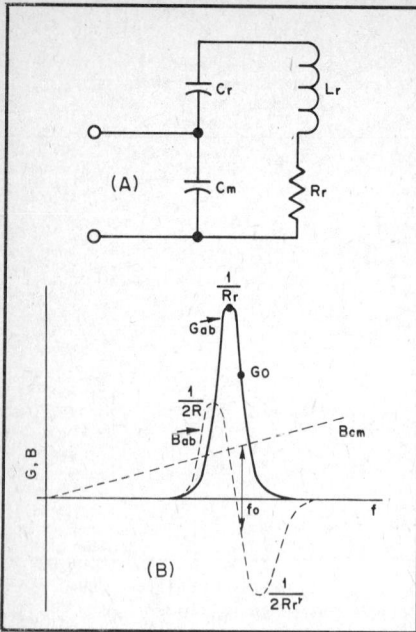

Fig. 74 — A capacitively coupled resonator is shown at A. See text for explanation of figure shown at B.

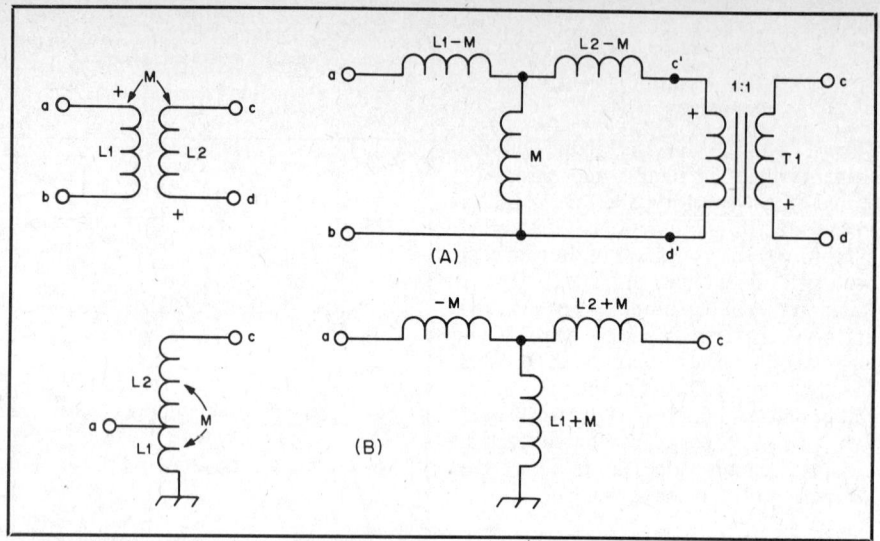

Fig. 76 — Two types of magnetically coupled circuits. At A, only mutual magnetic coupling exists while the circuit at B contains a common inductance also. Equivalents of both circuits are shown at the right which permit the application of the ladder-network analysis discussed in this section. (If the sign of voltage is unimportant, T1 can be eliminated.)

this frequency, f_o, is then G_o.

Since G_o is less than the conductance at the peak of the curve, $1/G_o$ or R_o is going to be greater than R_r. This effect can be applied when it is desired to match a low-value load resistance (such as found in a mobile whip antenna) to a more practical value. Suppose R_r and C_r in Fig. 74A are 10 ohms and 21 pF, respectively, and represent the equivalent circuit of a mobile antenna. Find the value of L_r and C_m which will match this antenna to a 52-ohm feed line at a frequency of 3900 kHz. Substituting the foregoing values into the formulas for input conductance gives

$$\frac{1}{52} = \frac{10}{10^2 + X^2}$$

Solving for X (which is the *total* series reactance) gives a value of 20.49 ohms. The reactance of a 21-pF capacitor at 3900 kHz is 1943.3 ohms so the inductive reactance must be 1963.7 ohms. While

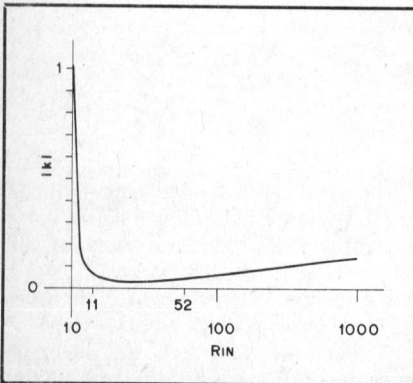

Fig. 75 — Variation of k with input resistance for circuit for Fig. 74.

either a positive or negative reactance will satisfy the equation for G_{ab}, a positive value is required to tune out B_{cm}. If the coupling element was a shunt inductor, the total reactance would have to be capacitive or negative in value.) Thus, the required inductance value for L_r will be 80.1 μH. In order to obtain a perfect match, the input susceptance must be zero and the value of B_{cm} can be found from

$$0 = B_{cm} - \frac{20.49}{10^2 + (20.49)^2}$$

giving a susceptance value of 0.04 mhos, which corresponds to a capacitance of 1608 pF.

Coefficient of Coupling

If the solution to the mobile whip-antenna problem is examined, it can be seen that for a given frequency R_r, L_r and C_r, only one value of C_m results in an input load that appears as a pure resistance. While such a condition might be defined as resonance, the resistance value obtained is not necessarily the one required for maximum transfer of power.

A definition that is helpful in determining how to vary the circuit elements in order to obtain the desired input resistance is called the *coefficient of coupling*. The coefficient of coupling is defined as the ratio of the common or mutual reactance and the square root of the product of two specially defined reactances. If the mutual reactance is capacitive, one of the special reactances is the sum of the series capacitive reactances of the primary mesh (with the resonator disconnected) and the other one is the sum of the series capacitive reactances of the resonator (with the primary disconnected). Applying this definition to the circuit of Fig. 74A, the coefficient of coupling, k, is given by

$$k = \sqrt{\frac{C_r}{C_r + C_m}}$$

How meaningful the coefficient of coupling will be depends upon the particular circuit configuration under consideration and which elements are being varied. For example, suppose the value of L_r in the mobile-whip antenna problem was fixed at 100 μH and C_m and C_r were allowed to vary. (It will be recalled that C_r is 21 pF and represents the antenna capacitance. However, the total resonator capacitance could be changed by adding a series capacitor between C_m and the antenna. Thus, C_r could be varied from 21 pF to some lower value but not a higher one.)

A calculated plot of k versus input resistance, R_{in}, is shown in Fig. 75. Note the unusually high change in k when going from resistance values near 10 ohms to slightly higher ones.

Similar networks can be designed to work with any ratio of input resistance and load resistance but it is evident small ratios are going to pose difficulties. For larger ratios, component tolerances are more relaxed. For instance, C_m might consist of switchable fixed capacitors with C_r being variable. With a given load resistance, C_m essentially sets the value of the reactance and thus the input resistance while C_r and L_r provide the required reactance for the conductance formula. However, if L_r is varied, k varies also. Generally speaking, higher values of L_r (and consequently circuit Q) require lower values of k.

At this point, the question arises as to the significance and even the merit of such definitions as coefficient of coupling and Q. If the circuit element values are known, and if the configuration can be resolved into a ladder network, important properties such as input impedance and attenuation can be computed directly for any frequency. On the other hand, circuit information might be obscured or even lost by attempting to attach too much importance to an arbitrary definition. For example, the plot in Fig. 75 merely indicates C_m and C_r are changing with respect to one another. But it doesn't illustrate how they are changing. Such information is important in practical applications and even a simple table of C_m and C_r vs. R_{in} for a particular R_r would be much more valuable than a plot of k.

Similar precautions have to be taken with the interpretation of circuit Q. Selectivity and Q are simply related for single resonators and circuit components, but the situation rapidly deteriorates with complex configurations. For instance, adding loss or resistance to circuit elements would seem to contradict the idea that low-loss or high-Q circuits provide the best selectivity. However, this is actually done in some filter designs to improve frequency response. In fact, the filter with the added loss has identical characteristics to one with "pure" elements. The method is called predistortion and is very useful in designing filters where practical considerations require the use of circuit elements with parasitic or undesired resistance.

As the frequency of operation is increased, discrete components become smaller until a point is reached where other forms of networks have to be used. Here, entities such as k and Q are sometimes the only means of describing such networks. Another definition of Q that is quite useful in this instance is that it is equal to the ratio of 2π (energy stored per rf cycle)/(energy lost per rf cycle.)

Mutually Coupled Inductors

A number of very useful rf networks involve coupled inductors. In a previous section, there was some discussion on iron-core transformers which represent a special case of the coupled-inductance problem. The formulas presented apply to instances where the coefficient of coupling is very close to 1.0. While it is possible to approach this condition at frequencies in the rf range, many practical circuits work at values of k that are considerably less than 1.0. A general solution is rather complex but many practical applications can often be simplified and solved through use of the ladder-network method. In particular, the sign of the mutual inductance must be taken into account if there are a number of coupled circuits or if the phase of the voltage between two coupled circuits is important.

The latter consideration can be illustrated with the aid of Fig. 76A. An exact circuit for the two mutually coupled coils on the left is shown on the right. T1 is an "ideal" transformer that provides the "isolation" between terminals ab and

cd. If the polarity of the voltages between these terminals can be neglected, the transformer can be eliminated and just the circuit before terminals cd substituted. A second circuit is shown in Fig. 76B. Here, it is assumed that the winding sense doesn't change between L1 and L2. If so, then the circuit on the right of Fig. 76B can be substituted for the tapped coil shown at the left.

Coefficients of coupling for the circuits in Fig. 76A and 76B are given by

$$k = \frac{M}{\sqrt{L_1 L_2}}$$

$$k = \frac{L_1 + M}{\sqrt{L_1(L_1 + L_2 + 2M)}}$$

If L1 and L2 do not have the same value, an interesting phenomenon takes place as the coupling is increased. A point is reached where the mutual inductance exceeds the inductance of the smaller coil. The interpretation of this effect can be illustrated with the aid of Fig. 77. While all the flux lines (as indicated by the dashed lines) associated with L1 also encircle turns of L2, there are additional ones that encircle extra turns of L2, also. Thus, there are more flux lines for M than there are for L1. Consequently, M becomes larger than L1. Normally, this condition is difficult to obtain with air-wound coils but the addition of ferrite material greatly increases the coupling. As k increases so that M is larger than L1 (Fig. 77), the network begins to behave more like a transformer and for a k of 1, the equivalent circuit of Fig. 77A yields the transformer equations of a previous section. On the other hand, for small values of k, the network becomes merely three coils arranged in a "T" fashion. One advantage of the circuit of Fig. 76A is that there is no direct connection between the two coils. This property is important from an isolation standpoint and can be used to suppress unwanted currents that are often responsible for RFI difficulties.

Piezoelectric Crystals

A somewhat different form of resonator consists of a quartz crystal between two conducting plates. If a voltage is applied to the plates, the resultant electric field causes a mechanical stress in the crystal. Depending upon the size and "cut" of the crystal, a frequency will exist at which the crystal begins to vibrate. The effect of this mechanical vibration is to simulate a series RLC circuit as in Fig. 74A. There is a capacitance associated with the crystal plates which appears across the terminals (C_m in Fig. 74A). Consequently, this circuit can also be analyzed with the aid of Fig. 74B. At some frequency (f1 in Fig. 78), the series reactance is zero and G_{ab} in the preceding formula will just be $1/Rr$. Typical values for Rr range from 10 k-ohm and higher. However, the

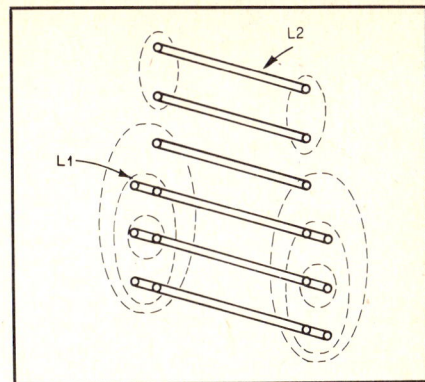

Fig. 77 — Diagram illustrating how M can be larger than one of the self inductances. This represents the transition from lightly coupled circuits to conventional transformers since an impedance step-up is possible without the addition of capacitive elements.

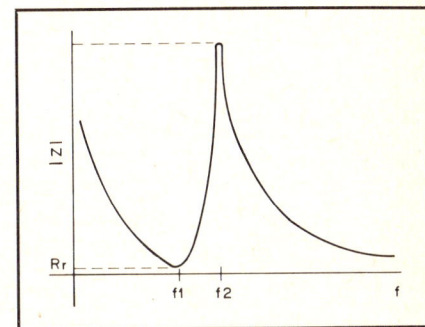

Fig. 78 — Frequency response of a quartz-crystal resonator. The minimum value is only approximate since holder capacitance is neglected.

equivalent inductance of the mechanical circuit is normally extremely high (over 10,000 henrys in the case of some low-frequency units) which results in a very high circuit Q (30,000). Above f_1, the reactance is "inductive" and at f_2, the susceptance of the series resonator is just equal to the susceptance of the crystal holder, B_{cm}. Here the total susceptance is zero. Since B_{cm} is usually very small, the equivalent series susceptance is also small. This means the value for X in the susceptance formula will be very large and consequently G_{ab} will be small, which corresponds to a high input resistance. A plot of the magnitude of the impedance is shown in Fig. 78. The dip at F_1 is called the series-resonant mode and the peak at f_2 is referred to as the parallel-resonant or "anti-resonant" mode. When specifying crystals for oscillator applications, the type of mode must be given along with external capacitance across the holder or type oscillator circuit to be used. Otherwise, considerable difference in actual oscillator frequency will be observed. The effect can be used to advantage and the frequency of a crystal oscillator can be "pulled" with an external reactive element or even frequency modulated with a device that converts voltage or current fluctuations into changes in reactance.

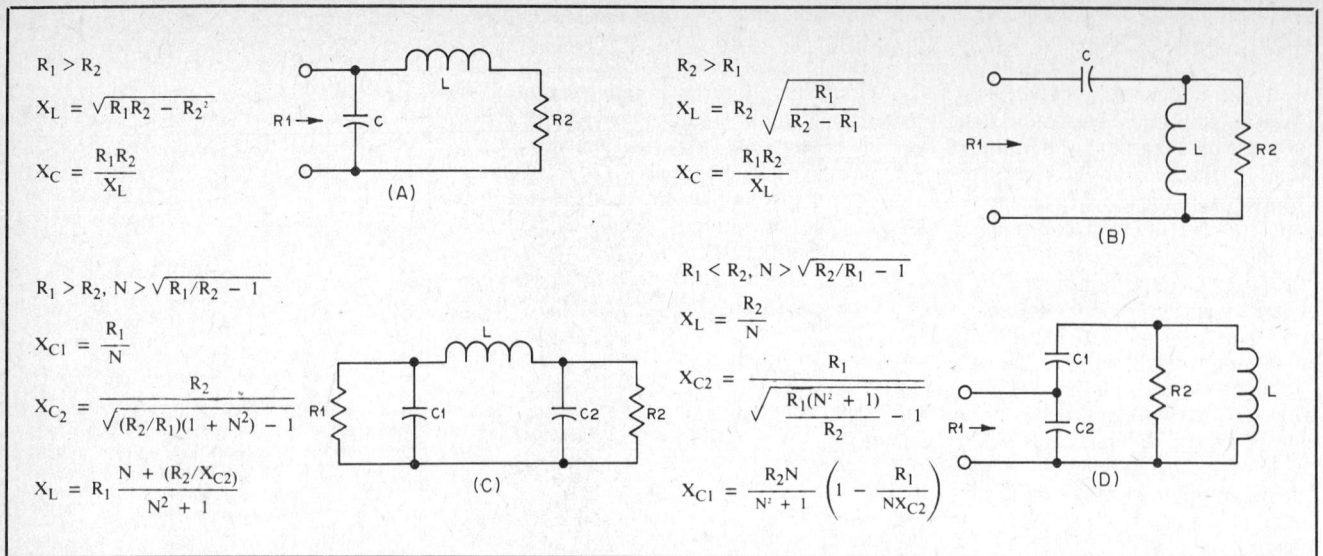

$R_1 > R_2$

$X_L = \sqrt{R_1 R_2 - R_2^2}$

$X_C = \frac{R_1 R_2}{X_L}$

(A)

$R_2 > R_1$

$X_L = R_2 \sqrt{\frac{R_1}{R_2 - R_1}}$

$X_C = \frac{R_1 R_2}{X_L}$

(B)

$R_1 > R_2, N > \sqrt{R_1/R_2 - 1}$

$X_{C1} = \frac{R_1}{N}$

$X_{C2} = \frac{R_2}{\sqrt{(R_2/R_1)(1 + N^2) - 1}}$

$X_L = R_1 \frac{N + (R_2/X_{C2})}{N^2 + 1}$

(C)

$R_1 < R_2, N > \sqrt{R_2/R_1 - 1}$

$X_L = \frac{R_2}{N}$

$X_{C2} = \frac{R_1}{\sqrt{\frac{R_1(N^2 + 1)}{R_2} - 1}}$

$X_{C1} = \frac{R_2 N}{N^2 + 1}\left(1 - \frac{R_1}{N X_{C2}}\right)$

(D)

Fig. 79 — Four matching networks that can be used to couple a source and load with different resistance values. (Although networks are drawn with R1 appearing as the source resistance, all can be applied with R2 at the source end. Also, all formulas with capacitive reactance are for the numerical or absolute value.)

Matching Networks

In addition to filters, ladder networks are frequently used to match one impedance value to another one. While there are many such circuits, a few of them offer particular advantages such as simplicity of design formulas or minimum number of elements. Some of the more popular ones are shown in Fig. 79. Shown at Fig. 79A and 79B, are two variations of an "L" network. These networks are relatively simple to design.

The situation is somewhat more complicated for the circuits shown at 79C and 79D. For a given value of input and output resistance, there are many networks that satisfy the conditions for a perfect match. The difficulty can be resolved by introducing the "dummy variable" labeled N.

From a practical standpoint, N should be selected in order to optimize circuit component values. Either values of N that are too low or too high result in networks that are hard to construct.

The reason for this complication is as follows. Only two reactive elements are required to match any two resistances. Consequently, adding a third element introduces a redundancy. This means one element can be assigned a value arbitrarily and the other two components can then be found. For instance, suppose C2 in Fig. 79C is set to some particular value. The parallel combination of C2 and R2 can then be transformed to a series equivalent (see Fig. 80). Then, L could be found by breaking it down into two components, L' and L''. One component (L'') would tune out the remaining capacitive reactance of the output series equivalent circuit. The network is then reduced to the one shown in Fig. 79A and the other component (L') of L along with the value for C1 could be determined from formulas (Fig. 79A). Adding the two inductive components would give the actual inductive reactance required for match in the circuit of Fig. 79C.

As mentioned before, it is evident an infinite number of networks of the form shown in Fig. 79C exist since C2 can be assigned any value. Either a set of tables or a family of curves for C1 and L in terms of C2 could then be determined from the foregoing method and as illustrated in Fig. 80. However, similar data along with other information can be obtained by approaching the problem somewhat differently. Instead of setting one of the element values arbitrarily and finding the other two, a third variable is contrived and in the case of Fig. 79C and Fig. 79D is labeled N. All three reactances are then expressed in terms of the variable N.

The manner in which the reactances change with variation in N for two representative circuits of the type shown in Fig. 79C is shown in Fig. 81. The solid curve is for an R1 of 3000 ohms and R2 equal to 52 ohms. The dashed curve is for the same R2 (52 ohms) but with R1 equal to 75 ohms. For values of N very close to the minimum specified by the inequality (Fig. 79C), X_{C2} becomes infinite which means C_2 approaches zero. As might be expected, the values of X_L and X_{C1} at this point are approximately those of an L network (Fig. 79A) and could be determined by means of the formulas in Fig. 79A for the corresponding values of R1 and R2.

The plots shown in Fig. 81 should give a general idea of the optimum range of component values. The region close to the left-hand portion should be avoided since there is little advantage to be gained over an L network, while an extra component is required. For very high values of N, the capacitance values become large without producing any particular advantage either. A good design choice is an N a few percent above the minimum specified by the inequality.

Quite often, one of the elements is fixed with either one or the other or both of the remaining two elements variable. In many

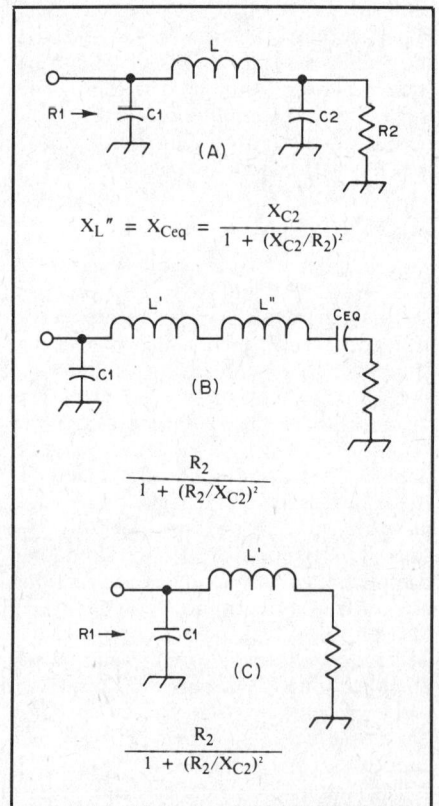

$X_L'' = X_{Ceq} = \frac{X_{C2}}{1 + (X_{C2}/R_2)^2}$

(A)

(B)

$\frac{R_2}{1 + (R_2/X_{C2})^2}$

(C)

$\frac{R_2}{1 + (R_2/X_{C2})^2}$

Fig. 80 — Illustration of the manner in which the network of Fig. 79C can be reduced to the one of Fig. 79A assuming C2 is assigned some arbitrary value. (The formulas shown are for numerical reactance values.)

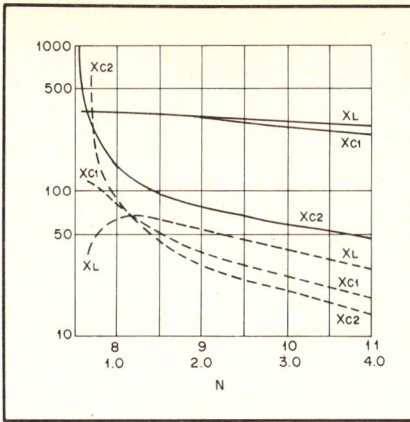

Fig. 81 — Network reactance variation as a function of dummy variable N. Solid curves and values of N from 8 to 11 are for an input resistance of 3000 ohms and an output resistance of 52 ohms. The dashed curves are for a similar network with an input and output resistance of 75 and 52 ohms, respectively. Values of N from 1 to 4 are for the latter curves.

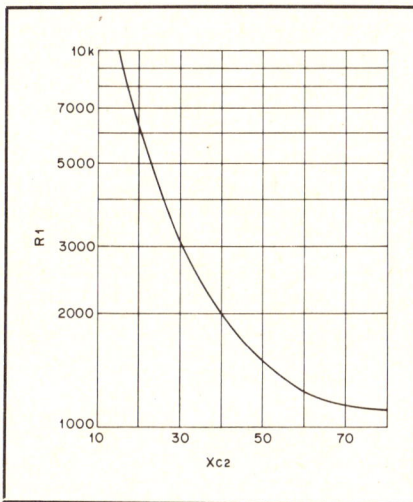

Fig. 82 — Input resistance vs. output reactance for an output resistance of 52 ohms. The curve is for a fixed inductor of 219 ohms (Fig. 79C). X_{C1} varies from 196 to 206 ohms.

amateur transmitters, it is the inductor that remains fixed (at least for a given band) while C1 and C2 (Fig. 79C) are made variable. While this system limits the bandwidth and matching capability somewhat, it is still a very useful approach. For instance, the plot shown in Fig. 82 indicates the range of input resistance values that can be matched for an R2 of 52 ohms. The graph is for an inductive reactance of 219 ohms. X_{C1} varies from 196 to 206 ohms over the entire range of R1 (or approximately 20 percent). However, X_{C2} varies from 15 to almost 100 ohms as can be seen from the graph.

Since C2 more or less sets the transformed resistance, it is often referred to as the "loading" control on transmit-

ters using the network of Fig. 79C, with C1 usually labeled "Tune." While the meaning of the latter term should be clear, the idea of loading in a matching application perhaps needs some explanation. For small values of X_{C2} (very large C2), the transformed resistance is very high. Consequently, a source that was designed for a much lower resistance would deliver relatively little power. However, as the resistance is lowered, increasing amounts of current will flow resulting in more power output. Then, the source is said to be "loaded" more heavily.

Similar considerations such as those discussed for the network of Fig. 79C also exist for the circuit of Fig. 79D. Only the limiting L network for the latter is the one shown in Fig. 79B. The circuit of Fig. 79C is usually called a pi network and as pointed out, it is used extensively in the output stage of transmitters. The circuit of Fig. 79D has never been given any special name, but it is quite popular in both antenna and transistor-matching applications.

The plot shown in Fig. 81 is for fixed input and output resistances with the reactances variable. Similar figures can be plotted for other combinations of fixed and variable elements. An interesting case is for X_L and R1 fixed with R2, X_{C1}, and X_{C2} variable. A lower limit for N also exists for this plot only instead of an L network, the limiting circuit is a network of three equal reactances. A feature of this circuit is that the output resistance is the ratio of the square of the reactance and the input resistance. An analogous situation exists with a quarter-wavelength transmission-line transformer. The output resistance is the ratio of the square of the characteristic impedance of the line and the input resistance. Consequently, the special case where all the reactances are equal in the circuit of Fig. 79C is the lumped-constant analog of the quarter-wavelength transformer. It has identical phase shift (90 degrees) along with the same impedance-transforming properties.

Frequency Response

In many instances, a matching network performs a dual role in transforming a resistance value while providing frequency rejection. Usually, matching versatility, component values, and number of elements are the most important considerations. But a matching network might also be able to provide sufficient selectivity for some application, thus eliminating the need for a separate circuit such as a filter.

It will be recalled that Q and selectivity are closely related for simple RLC series and parallel circuits. Bandwidth and the parameter N of Fig. 79 are approximately related in this manner. For values of N much greater than the minimum specified by the inequality N and Q can be considered to mean the same thing for all

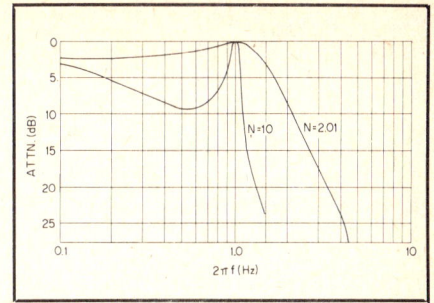

Fig. 83 — Frequency response of the network of Fig. 79C for two values of N.

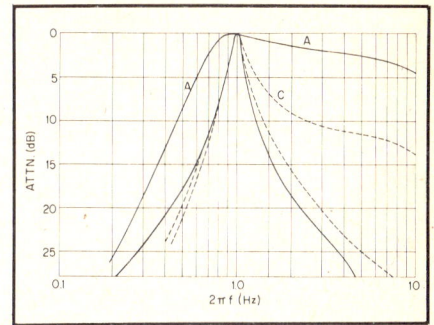

Fig. 84 — Frequency response of the circuit of Fig. 79D (see text).

practical purposes. However, the frequency response of networks that are more complex than simple RLC types is usually more complicated also. Consequently, some care is required in the interpretation of N or Q in regard to frequency rejection. For instance, a simple circuit has a frequency response that results in increasing attenuation for increasing excursions from resonance. That is not true for the pi network as can be seen from Fig. 83. For slight frequency changes below resonance, the attenuation increases as in the case of a simple RLC network. At lower frequencies, the attenuation decreases and approaches 2.55 dB. This plot is for a resistance ratio of 5:1, and the low-frequency loss is just caused by the mismatch in source and load resistance. Thus, while increasing N improves the selectivity near resonance, it has little effect on response for frequencies much farther away.

A somewhat different situation exists for the circuit of Fig. 79D. At frequencies far from resonance, either a series capacitance provides decoupling at the lower fequencies or a shunt capacitance causes additional mismatch at the higher ones. This circuit, then, has a response resembling those of simple circuits unlike the pi network. Curves a and b of Fig. 84 are for a resistance ratio of 5:1 with N equal to 2.01 for curve a. Curve b is for an N of 10. Curves c and d are for a resistance ratio of 50:1 with N equal to 7.04 and 10 respectively.

Chapter 3

Radio Design Technique and Language

Many amateurs desire to construct their own radio equipment and some knowledge of design procedures becomes important. Even when some commercially manufactured equipment is used, these techniques may still be required in setting up peripheral equipment. Also, an applicant for an Amateur Radio license might be tested on material in this area.

"Pure" vs. "Impure" Components

In the chapter on electrical laws and circuits, it is assumed that the components in an electrical circuit consist solely of elements that can be reduced to a resistance, capacitance or inductance. However, such elements do not exist in nature. An inductor always has some resistance associated with its windings and a carbon-composition resistor becomes a complicated circuit as the frequency of operation is increased. Even conductor resistance must be taken into account if long runs of cable are required.

In many instances, the effects of these "parasitic" components can be neglected and the actual device can be approxi-

mated by a "pure" element such as a resistor, capacitor, inductor or a short circuit in the case of an interconnecting conductor. In other cases, the unwanted component must be taken into account. However, it may be possible to break the element down into a simple circuit consisting of single elements alone. Then, the actual circuit may be analyzed by means of the basic laws discussed in the previous chapter. It may be also possible to make a selection such that the effects of the residual element are negligible.

However, there are other parasitic elements that may not only be difficult to remove but will affect circuit operation adversely as well. In fact, such considerations often set a limit on how stringent a design criterion can be tolerated. For instance, it is a common practice to connect small-value capacitors in various parts of a complex circuit, such as a transmitter or receiver, for bypassing purposes. A bypass capacitor permits energy below some specified frequency to pass a given point while providing rejection to energy at higher frequencies. In essence, the capacitor is used in a crude form of filter. In more complicated filter designs, capacitors may be required for complex functions (such as matching) in addition to providing a low reactance to ground.

An equivalent circuit of a capacitor is shown in Fig. 1A. Normally, the series resistance, R_s, can be neglected. On the other hand, the upper frequency limit of the capacitor is limited by the series inductance, L_s. In fact, above the point where L_s and C_p form a resonant circuit, the capacitor actually appears as an inductor at the external terminals. As a result, it becomes useless for bypassing purposes. This is why it is common practice to use two capacitors in parallel for bypassing, as shown in Fig. 2. At first inspection, this might appear as superfluous duplication. But the "self-resonant"

frequency of a capacitor is lower for high-capacitance units than it is for smaller-value ones. Thus, C1 in Fig. 2 provides a low reactance for low frequencies such as those in the audio range while C2 acts as a bypass for frequencies above the self-resonant frequency of C1.

RF Leakage

Although the capacitor combination shown in Fig. 2 provides a low-impedance path to ground, it may not be very effective in preventing rf energy that travels along the conductor from point 1 from reaching point 2. At dc and low-frequency applications, a circuit must always form a closed path in order for a current to flow. Consequently, two conductors are required if power is to be delivered from a source to a load. In many instances, one of the conductors may be common to several other circuits and constitutes a local ground.

However, as the frequency of operation is increased, a second type of coupling mechanism is possible. Power may be transmitted *along a single conductor*. (Although the same effect is possible at low frequencies, unless circuit dimensions are extremely large, such transmission effects can be neglected.) The conductor acts as a waveguide in much the same manner as a large conducting surface, such as the earth, will permit propagation of a radio wave close to its boundary with the air. This latter type of propagation is often called a ground wave and is important up to and slightly above the standard a-m broadcast band. At higher frequencies, the conductivity of the earth is such to attenuate ground-wave propagation.

A mode similar to ground-wave propagation that can travel along the boundary of a single conductor is illustrated by the dashed lines in Fig. 2. As with the wave traveling close to the earth, a poor conducting boundary will cause attenuation.

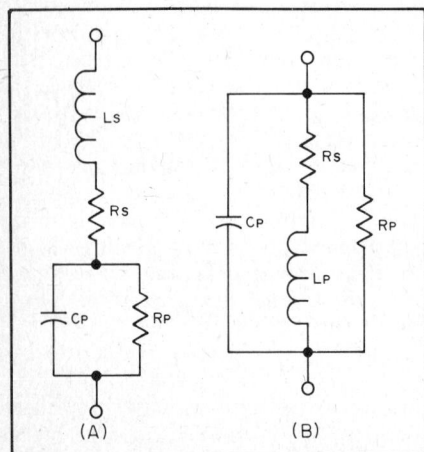

Fig. 1 — Equivalent circuit of a capacitor is shown at A, and for an inductor in B.

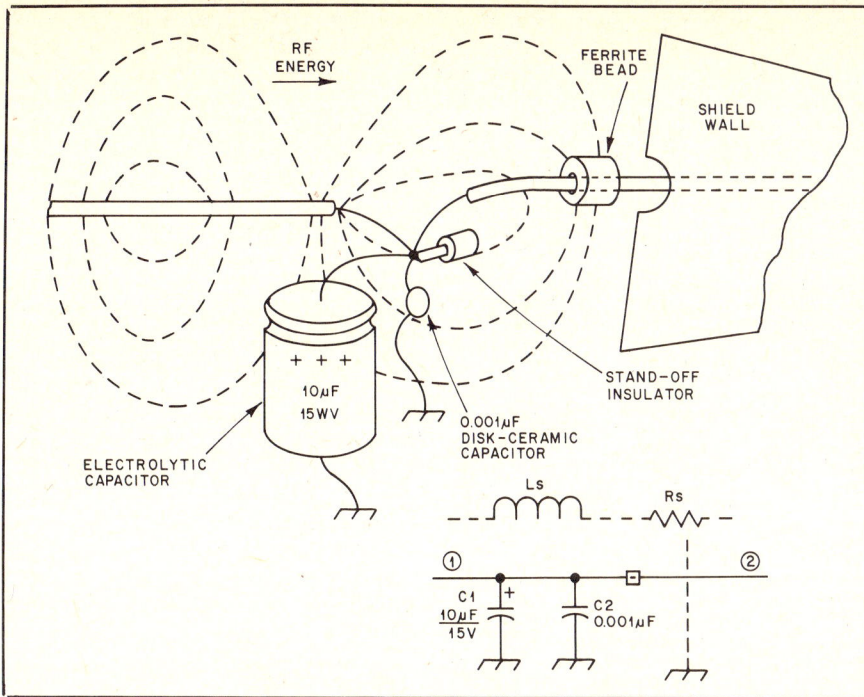

Fig. 2 — A bypassing arrangement that affords some measure of isolation (with the equivalent circuit shown in the inset). Dashed lines indicate a mode of wave travel that permits rf energy to leak past the bypass circuitry and should be taken into account when more stringent suppression requirements are necessary. (L_s and R_s in the inset represent the equivalent circuit of the ferrite bead.)

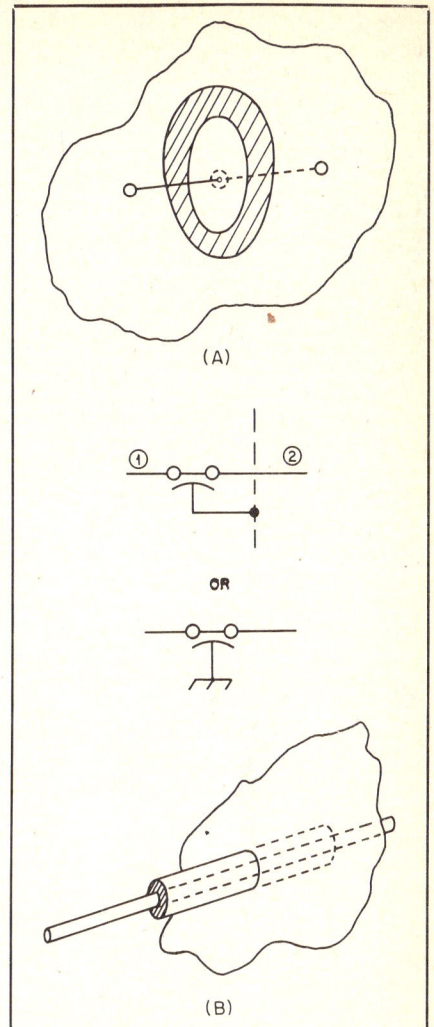

Fig. 3 — A superior type of bypassing arrangement to that shown in Fig. 2. Concentric conductors provide a low-inductance path to ground and better rejection of unwanted single-wire wave modes.

This is why a ferrite bead is often inserted over the exit point of a conductor from an area where rf energy is to be contained or excluded. In addition to loss (particularly in the vhf range), the high permeability of the ferrite introduces a series-inductive reactance as well. Finally, the shield wall provides further isolation.

While the techniques shown in Fig. 2 get around some of the deficiencies of capacitors that are used for bypass purposes, the resulting suppression is inadequate for a number of applications. Examples would be protection of a VFO to surrounding rf energy, a low-frequency receiver with a digital display, and suppression of radiated harmonic energy from a transmitter. In each of these cases, a very high degree of isolation is required. For instance, a VFO is sensitive to voltages that appear on dc power supply lines and a transmitter output with a note that sounds "fuzzy" or rough may result. Digital displays usually generate copious rf energy in the low-frequency spectrum. Consequently, a receiver designed for this range presents a situation where a strong source of emission is in close proximity to very sensitive receiving circuits. A similar case exists with transmitters operating on a frequency that is a submultiple of a fringe-area TV station. In the latter two instances, the problem is not so severe if the desired signal is strong enough to "override" the unwanted energy. Unfortunately, this is not the case normally and stringent measures are required to isolate the sensitive circuits from the strong source.

A different type of bypass-capacitor configuration is often used with associated shielding for such applications, as shown in Fig. 3. In order to reduce the series inductance of the capacitor, and to provide better isolation between points 1 and 2, either a disk-type (Fig. 3A) or a coaxial configuration (Fig. 3B) is employed. The circuit diagram for either configuration is shown in the inset. While such "feedthrough" capacitors are always connected to ground through the shield, this connection is often omitted on drawings. Only a connection to ground is shown, as in the inset in Fig. 3.

Dielectric Loss

Even though capacitors are usually high-Q devices, the effect of internal loss can be more severe than the case of a coil. This is because good insulators of electricity are usually good insulators of heat also. Therefore, heat generated in a capacitor must be conducted to the outside via the conducting plates to the capacitor leads. In addition, most capacitors are covered with an insulating coating that further impedes heat conduction. The problem is less severe with capacitors using air as a dielectric for two reasons. The first advantage of air over other dielectrics is that the loss in the presence of an alternating electric field is extremely small. Secondly, any heat generated by currents on the surface of the conducting plates is either conducted away by air currents or through the mass of the metal.

The dielectric loss in a capacitor can be represented by R_p as shown in Fig. 1A. However, if a dc ohmmeter was placed across the terminals of the capacitor, the reading would be infinite. This is because dielectric loss is an ac effect. Whenever an alternating electric field is applied to an insulator, there is a local motion of the electrons in the individual atoms that make up the material. Even though the electrons are not displaced as they would be in a conductor, this local motion requires the expenditure of energy and results in a power loss.

Consequently, some care is required in the application of capacitors in moderate to high-power circuits. The applied voltage should be such that rf-current ratings are not exceeded for the particular frequency of operation. This is illustrated in Fig. 4. A parallel-resonant circuit consisting of L_p, C_p and R_L is connected to a voltage source, V_s, through a coupling capacitor, C_c. It is also assumed the R_L is much greater than either the inductive or capacitive reactance taken alone. This condition would be typical of that found in most rf-power amplifier circuits employing vacuum tubes.

Since the inductive and capacitive reactance of L_p and C_p cancel at resonance, the load presented to the source would be just R_L. This would mean the current through C_c would be much less than the current through either C_p or L_p. The effect of such "current rise" is similar to the voltage rise at resonance discussed in the previous chapter. Even though the current at the input of the parallel-resonant circuit is small, the currents that flow in the elements that make up the circuit can be quite large.

The requirements for C_c then, would be rather easy to satisfy in regards to current rating and power dissipation. On the other hand, C_p would ordinarily be restricted to air-variable types although some experiments have been successful using Teflon as a dielectric.[1] Generally speaking, the coupling capacitor should have a low reactance (at the lowest frequency of operation) in comparison to the load presented by the tuned circuit. The effect of the coupling-capacitor reactance could then be compensated by slightly retuning the parallel-resonant circuit.

Inductors

Similar considerations to those discussed in the previous sections exist with inductors also, as shown in Fig. 1B. Since an inductor usually consists of a coil of wire, there will be a resistance associated with the wire material and this component is represented by R_s (Fig. 1B). In addition, there is always a capacitance associated with conductors in proximity as illustrated in Fig. 5. While such capacitance is distributed throughout the coil, it is a convenient approximation to consider an equivalent capacitance, C_p, exists between the terminals (Fig. 1B). Finally, inductors are often wound on materials that have high permeability in order to increase the inductance. Thus, it is possible to build an inductor with fewer turns and smaller in size than an equivalent coil with an "air core."

Unfortunately, high-permeability materials presently available have considerable loss in the presence of an rf field. It will be recalled a similar condition existed with the dielectric in a capacitor. Consequently, in addition to the wire resistance, a loss resistance is associated with the core and represented by R_p. (See Fig. 1B.) Since this loss is more or less independent of the current through the coil but dependent upon the applied voltage, it is represented by a parallel resistance.

RF Transformers

Although the term transformer might be applied to any network that "trans-

[1]DeMaw and Dorbuck, "Transmitting Variables," *QST* February 1975.

formed" a voltage or an impedance from one level to another one, the term is usually reserved for circuits incorporating mutual magnetic coupling. Examples would be i-f transformers, baluns, broad-band transformers, and certain antenna matching networks (see chapter 2). Of course, many devices used at audio and power frequencies are also transformers in the sense used here and have been covered in a previous chapter.

Networks that use mutual magnetic coupling exclusively have attractive advantages over other types in many common applications. A principal advantage is that there is no direct connection between the input and output terminals. Consequently, dc and ac components of current are separated easily thus eliminating the need for coupling capacitors. Perhaps even more importantly, it is also possible to isolate rf currents because of the lack of a common conductor. Quite often, an hf receiver in an area where strong local broadcast stations are present will suffer from "broadcast harmonics" and possibly even rectified audio signals getting into sensitive af circuits. In such cases, complicated filters sometimes prove ineffective while a simple tuned rf transformer clears up the problem completely. This is because the unwanted bc components are prevented from flowing on the receiver chassis along with being rejected by the tuned-transformer filter characteristic.

A second advantage of coupled circuits using mutual magnetic coupling exclusively is that analysis is relatively simple compared to other forms of coupling although exact synthesis is somewhat complicated. That is, finding a network with some desired frequency response would be quite difficult in the general case.

However, circuits using mutual-magnetic coupling usually have very good out-of-band rejection characteristics when compared to networks incorporating other forms. (A term sometimes applied to transformer or mutual-magnetic coupling is *indirect coupling*. Circuits with a single resistive or reactive element for the common impedance are called direct-coupled networks. Two or more elements in the common impedance are said to comprise complex coupling.) For instance, relatively simple band-pass filters are possible with mutual-magnetic coupling and are highly recommended for vhf-transmitter multiplier chains. For receiving, such filters are often the main source of selectivity. Standard a-m and fm broadcast receivers would be examples where intermediate-frequency (i-f) transformers derive their band-pass characteristics from mutually coupled inductors.

A third advantage of mutually coupled networks is that practical circuits with great flexibility particularly in regard to matching capabilities are possible. For

this reason, variable-coupling matching networks or those using "link coupling" have been popular for many years. In addition to matching flexibility, these circuits are good band-pass filters and can also provide isolation between antenna circuits and those of the transmitter.

Design Formulas

A basic two-mesh circuit with mutual magnetic coupling is shown in Fig. 6. The reactance, X, is arbitrary and could be either inductive or capacitive. However, it is convenient to combine it with the secondary reactance (X_{LS}) since this makes the equations somewhat more compact. Hence, the total secondary reactance is defined by

$$X_s = 2\pi fL_s + X$$

The primary reactance and mutual reactance are also defined respectively as

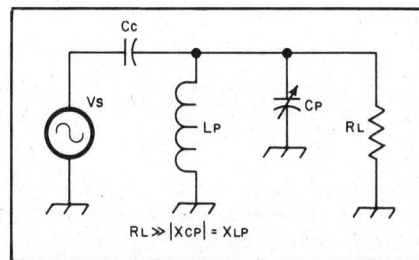

Fig. 4 — Consideration of capacitor voltage and current ratings should be kept in mind in moderate-power applications.

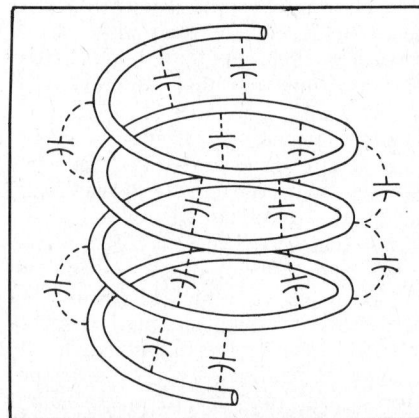

Fig. 5 — Distributed capacitance (indicated by dashed lines) affects the operation of a coil at high frequencies.

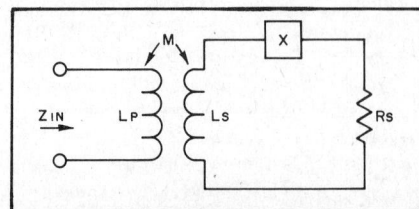

Fig. 6 — Basic magnetically coupled circuit.

$$X_p = 2\pi f L_p, \quad X_m = 2\pi f M$$

A set of equations for the input resistance and reactance is given by

$$R_{in} = \frac{R_s X_m^2}{R_s^2 + X_s^2}$$

$$X_{in} = X_p - \frac{X_m^2 X_s}{R_s^2 + X_s^2}$$

This permits reducing the two-mesh circuit of Fig. 6 to the single-mesh circuit of Fig. 7.

Double-Tuned Circuits

A special case occurs if the value of X_s is zero. This could be accomplished easily by tuning out the inductive reactance of the secondary with an appropriate capacitor or by varying the frequency until a fixed capacitor and the secondary inductance resonated. Under these conditions, the input resistance and reactance would be

$$R_{in} = \frac{X_m^2}{R_s}, \quad X_{in} = X_p$$

Then, in order to make the input impedance purely resistive, a second series capacitor could be used to cancel the reactance of X_p. The completed network is shown in Fig. 8 with C1 and C2 being the primary and secondary series capacitors.

If X_m could be varied, it is evident that the secondary resistance could be transformed to almost any value of input resistance. Usually, the desired resistance, would be made equal to the generator resistance, R_g, for maximum power transfer. It might also be selected to satisfy some design goal, not necessarily related to maximum power transfer. This brings up a minor point but one that can cause considerable confusion. Normally, in transmitting circuits, the "unloaded Q" of the reactive components would be very high and the series parasitic resistances (discussed in a previous section) could be neglected. However, if it is not desired to do so, how should these resistances be taken into account? If maximum power transfer is the goal, the series resistance of the primary coil would be added to the generator resistance, R_g, and the transformed secondary resistance would be made equal to this sum.

On the other hand, a more common case requires the *total* input resistance to be equal to some desired value. For instance, an amplifier might provide optimum efficiency or harmonic suppression when terminated in a particular load resistance. Transmission lines also require a given load resistance in order to be "matched." In such cases, the series resistance of the primary coil would be

subtracted from the actual resistance desired and the transformed resistance made equal to this difference. As an example, suppose an amplifier required a load resistance of 3000 ohms, and the primary-coil resistance was 100 ohms. Then, the transformed resistance must be equal to 2900 ohms. (In either case, the secondary coil resistance is merely added to the secondary load resistance and the sum substituted for R_s.)

Coefficient of Coupling

Although the equations for the input impedance can be solved in terms of the mutual reactance, the transforming mechanism involved becomes somewhat clearer if the coefficient of coupling is used instead. The coefficient of coupling, k, in terms of the corresponding reactances of inductances is

$$k = \frac{X_m}{\sqrt{X_p X_s}} = \frac{L_m}{\sqrt{L_p L_s}}$$

Then, the input resistance becomes

$$R_{in} = \frac{k^2 X_p X_s}{R_s}$$

The primary and secondary Qs are defined as

$$Q_p = \frac{X_p}{R_g}, \quad Q_s = \frac{X_s}{R_s}$$

where a "loaded" Q is assumed. This would mean R_s included any secondary-coil loss. For maximum-power transfer, R_g would be the total primary resistance which consists of the generator and coil resistance.

The coefficient of coupling under these conditions reduces to a rather simple formula

$$k_c = \frac{1}{\sqrt{Q_p Q_s}}$$

However, if it is desired to make the input resistance some particular value (as in the case of the previous example), the coefficient of coupling is then

$$k_c = \sqrt{\frac{R_{in} - R_p}{X_p Q_s}}$$

If the primary "loss" resistance is zero, both formulas are identical.

At values of k less than k_c, the input resistance is lower than either the prescribed value or for conditions of maximum power transfer. Higher values of k result in a higher input resistance. For this reason, k_c is called the critical coefficient of coupling. If k is less than k_c, the circuit is said to be undercoupled and for k greater than k_c, an overcoupled condition results. A plot of attenuation vs. frequency for the three cases is shown in Fig.

9. Critical coupling gives the flattest response although greater bandwidth can be obtained by increasing k to approximately 1.5 k_c. At higher values, a pronounced dip occurs at the center or resonant frequency.

In the undercoupled case, a peak occurs at the resonant frequency of the primary and secondary circuit but the transformed resistance is too low and results in a mismatch. As the coupling is decreased still further, very little power is transferred to the secondary circuit and most of it is dissipated in the primary-loss and generator-source resistances. On the other hand, an interesting phenomenon occurs with the overcoupled case. It will be recalled that the transformed resistance is too high at resonance because the coefficient of coupling is greater than the critical value. However, a special case occurs if the primary and secondary circuits are identical which also means the transformed resistance, R_{in} must equal R_s.

The behavior of the circuit under these conditions can be analyzed with the aid of Fig. 7. Assuming the Q of both circuits is high enough, the reactance, X_s, increases very rapidly on either side of resonance. If this variation is much greater than the variation of X_m with frequency, a frequency exists on each side of resonance where the ratio of X_m^2 and $R_s^2 + X_s^2$ is 1.0. Consequently, R_{in} is equal to R_s and the transformed reactance is $-X_s$. Since the primary and secondary resonators are identical, the reactances cancel because of the minus sign. The frequency plot for a k of 0.2 (k_c is 0.1) is shown in Fig. 9. If the primary and secondary circuits are not identical, a double-hump response still

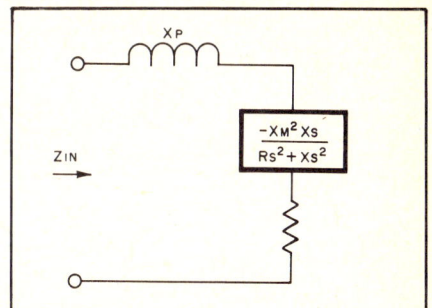

Fig. 7 — Equivalent single-mesh network of the two-mesh circuit of Fig. 6.

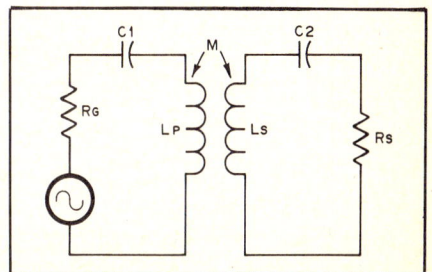

Fig. 8 — Double-tuned series circuits with magnetic coupling.

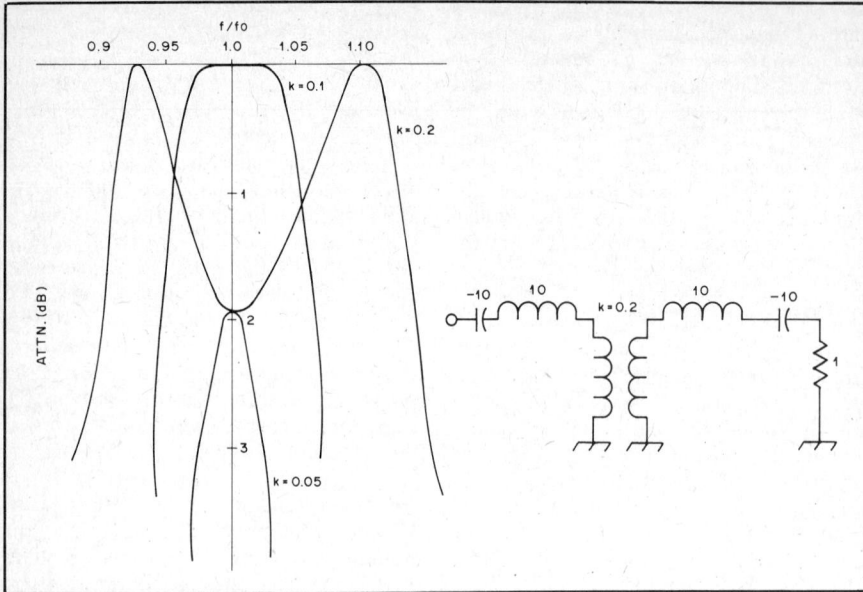

Fig. 9 — Response curves for various degrees of coupling coefficient k. The critical coefficient of coupling for the network shown in the inset is 0.1. Lower values give a single response peak (but less than maximum power transfer) while "tighter" coupling results in a double-peak response.

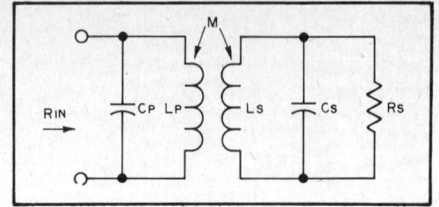

Fig. 10 — Coupled network with parallel-tuned circuits or "i-f" transformer.

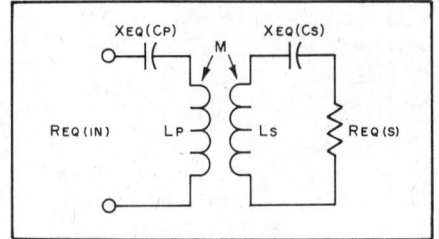

Fig. 11 — Equivalent series circuit of the parallel network shown in Fig. 10. This transformation is only valid at single frequencies and must be revalued if the frequency is changed.

occurs but the points where the transformed resistance is equal to the desired value, the reactances are not the same numerically. Consequently, there is attenuation at peaks unlike the curve of Fig. 9.

Other Circuit Forms

While the coupled network shown in Fig. 8 is the easiest to analyze, it is not commonly encountered in actual circuits. As the resistance levels are increased, the corresponding reactances become very large also. In transmitting circuits, extremely high voltages are then developed across the coils and capacitors. For high-impedance circuits, the circuit shown in Fig. 10 is often used. Although the frequency response is somewhat different than the circuit of Fig. 8 (in fact, the out-of-band rejection is greater), a matching network can be designed based upon the previous analysis for the series circuit.

This is accomplished by changing the parallel primary and secondary circuits to series equivalents. (It should be emphasized that this transformation is good at one frequency only.) The equivalent circuit of the one shown in Fig. 10 is illustrated in Fig. 11 where the new resistance and reactance of the secondary are given by

$$R_{eq}(S) = \frac{R_s}{1 + \gamma^2}$$

$$X_{eq}(C_s) = \frac{-R_s\gamma}{1 + \gamma^2}$$

$$\gamma = R_s/X_c$$

A similar set of transformations exists for the primary circuit also. In most instances, where one high-impedance load is matched to another one, R_s in Fig. 10 is much greater than the reactance of C_s and C_p. This simplifies the transformations and approximate relations are given by

$$R_{eq}(S) \cong \frac{X_c^2}{R_s}$$

$$X_{eq}(C_s) \cong X_c$$

As an example, suppose it was desired to match a 3000-ohm load to a 5000-ohm source using a coupled inductor with a 250-ohm (reactance) primary and secondary coil. Assume the coupling can be varied. Determine the circuit configuration and the critical coefficient of coupling.

Since the load and source resistance have a much higher numerical value than the reactance of the inductors, a parallel-tuned configuration must be used. In order to tune out the inductive reactance, the equivalent series capacitive reactance must be −250 ohms. Since both R_s and R_p are known, the exact formulas could be solved for γ and R_{eq}. However, because the respective resistances are much greater than the reactance, the simplified approximate formulas can be used. This means the primary and secondary equivalent capacitive reactances are −250 ohms. The equivalent secondary resistance is $(250)^2/3000$ or 20.83 ohms, resulting in a secondary Q of 250/20.83 or 12. (A formula could be derived directly for the Q from the approximate equations.) The equivalent primary resistance and Q are 12.5 ohms and 20, respectively. Substituting the values for Q into the formula for the critical coefficient of coupling gives $1/\sqrt{(20)(12)}$ or 0.065.

Double-tuned coupled circuits of the type shown in Fig. 10 are widely used in radio circuits. Perhaps the most common example is the i-f transformer found in a-m and fm bc sets. Many communications receivers have similar transformers although the trend has been toward somewhat different circuits. Instead of achieving selectivity by means of i-f transformers (which may require a number of stages), a single filter with quartz-crystal resonators is used instead. (The subject of receivers is treated in a later chapter.)

Single-Tuned Circuits

In the case of double-tuned circuits, separate capacitors are used to tune out the inductive components of the primary and secondary windings. However, examination of the equivalent circuit of the coupled coil shown in Fig. 7 suggests an alternative. Instead of a separate capacitor, why not "detune" a resonant circuit slightly and "reflect" a reactance of the proper sign into the primary in order to tune out the primary inductance. Since the transformation function (shown in the box in Fig. 7) reverses the sign of the secondary reactance, it is evident X_s must be *inductive* in order to tune out the primary inductance.

This might seem to be a strange result but it can be explained with the following reasoning. From a mathematical point of view, the choice of the algebraic sign of the transformed reactance is perfectly arbitrary. That is, a set of solutions to the equations governing the coupled circuit is possible assuming either a positive or negative sign for the transformed reactance. However, if the positive sign is chosen, the transformed resistance *would be negative*. But from a physical point of

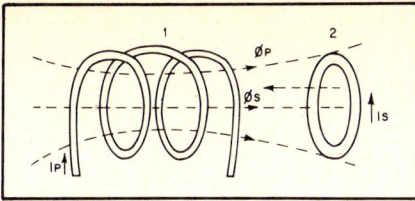

Fig. 12 — A coil coupled magnetically to a "shorted" turn provides insight to coils near solid shield walls.

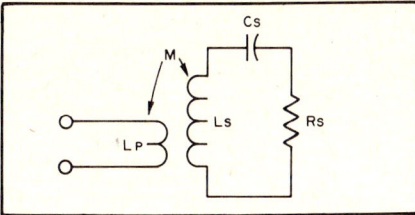

Fig. 13 — "Link" coupling can be used to analyze a number of important circuits.

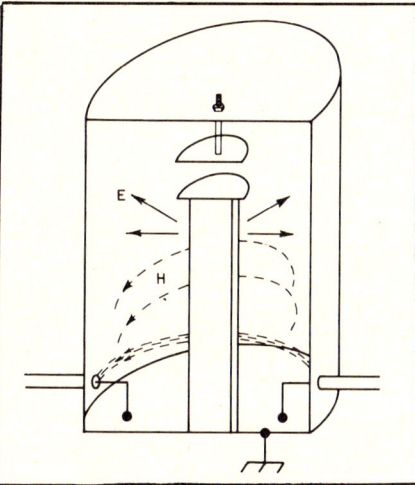

Fig. 14 — A vhf/uhf circuit which can be approximated by a link-coupled network using "conventional" components.

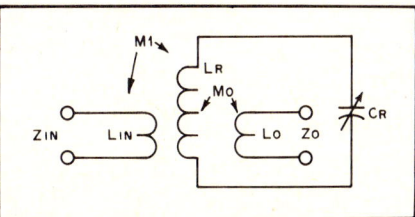

Fig. 15 — Equivalent low-frequency analog of the circuit shown in Fig. 14.

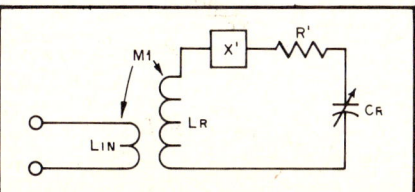

Fig. 16 — The network of Fig. 15 can be reduced with the transformation shown in Fig. 7.

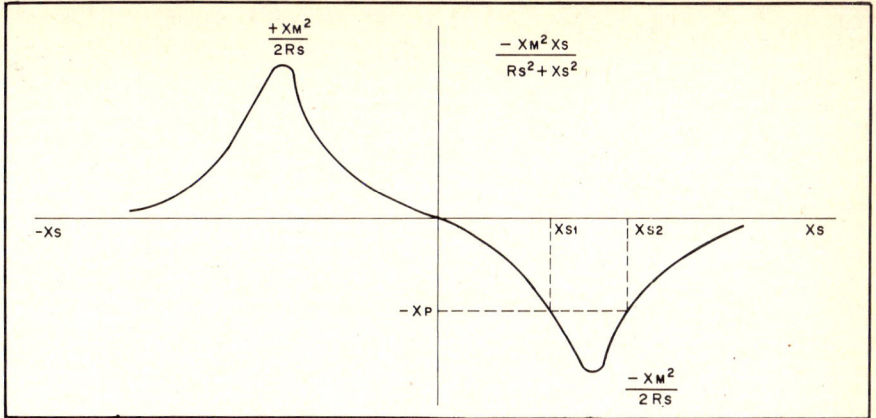

Fig. 17 — "Reflected" reactance into the primary of a single-tuned circuit places restraints on resistances that can be matched. This gives rise to a general rule that high-Q secondary circuits require a lower coefficient of coupling than low-Q ones.

view, this is a violation of the conservation of energy since it would imply the secondary resistance acts as a source of energy rather than an energy "sink." Consequently, the solution with the negative resistance does not result in a physically realizable network.

The foregoing phenomenon has implications for circuits one might not normally expect to be related to coupled networks. For instance, consider coil 1 (Fig. 12) in proximity to the one-turn "shorted" coil 2. A time-varying current in coil 1 will induce a current in coil 2. In turn, the induced current will set up a magnetic field of its own. The question is will the induced field aid or oppose the primary field. Since the energy in a magnetic field is proportional to the square of the flux, the induced field must oppose the primary field, otherwise the principle of the conservation of energy would be violated as it was with the "negative" resistance. Consequently, the induced current must always be in a direction such that the induced field opposes changes in the generating field. This result is often referred to as Lenz's Law.

If, instead of a one-turn loop, a solid shield wall was substituted, a similar phenomenon would occur. Since the total flux (for a given current) would be less with the shield present than it would be in the absence of the shield, the equivalent coil inductance is decreased. That is why it is important to use a shield around a coil that is big enough to reduce the effect of such coupling. Also, a shield made from a metal with a high conductivity such as copper or aluminum is advisable, otherwise a loss resistance will be coupled into the coil as well.

Link Coupling

An example of a very important class of single-tuned circuits is shown in Fig. 13. The primary inductor consists of a small coil either in close proximity or wound over one end of a larger coil. Two resonators can be coupled in this manner although there may be considerable separation (and no mutual coupling between the larger coils) hence the term "link" coupling. While this particular method is seldom used nowadays, the term is still applied to the basic configuration shown in Fig. 13. Applications would be antenna-matching networks, output stages for amplifiers and, especially important, many circuits used at vhf that have no direct hf equivalent.

The cavity resonators used in repeater duplexers are one form of vhf circuit that uses link coupling. A cross-sectional view of a representative type is shown in Fig. 14. Instead of ordinary coils and capacitors, a section of coaxial transmission line comprises the resonant circuit. The frequency of the resonator may be varied by adjusting the tuning screw which changes the value of the capacitor. Energy is coupled into and out of the resonator by means of two small, one-turn loops. Current in the input loop causes a magnetic field (shown by dashed lines). If the frequency of the generating field is near one of the resonant "modes" of the configuration, an electric field will also be generated (shown by solid lines). Finally, energy may then be coupled out of the resonator by means of a second loop.

A low-frequency equivalent circuit of the resonator is shown in Fig. 15. However, the circuit can only be used to give an approximate idea of the actual frequency response of the cavity. At frequencies not close to the resonant frequency, the mathematical laws governing resonant circuits are different from those of "discrete" components used at hf. Over a limited frequency range, the resonator can be approximated by the series LC circuit shown in Fig. 15.

Applying the formulas for coupled networks shown in Fig. 7 to the two-link circuit of Fig. 15, the output link and load can be transformed to an equivalent series resistance and reactance as shown in Fig. 16. In most instances, the reactance, X_s, in

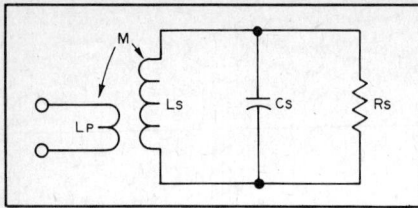

Fig. 18 — Single-tuned circuit with a parallel RC secondary.

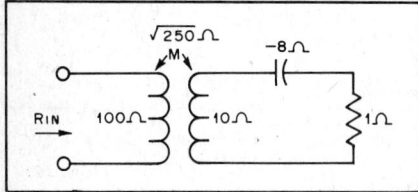

Fig. 19 — Text example of a single-tuned circuit.

the formula is just the reactance of the output link. Since the two-link network has been reduced to a single coupled circuit, the formulas can be applied again to find the input resistance and reactance.

Analysis of Single-Tuned Circuits

Single-tuned circuits are very easy to construct and adjust experimentally. If desired, the tuned circuit consisting of L_s, C_s, and perhaps the load, R_s, can be constructed first and tuned to the "natural" resonant frequency

$$f_o = \frac{1}{2\pi \sqrt{L_cC_s}}$$

Then, the primary inductor, which may be a link or a larger coil, is brought into proximity of the resonant circuit. The resonant frequency will usually shift upward. For instance, a coil and capacitor combination was tuned to resonance by means of a grid-dip oscillator (see the chapter on measurements) at a frequency of 1.8 MHz. When a two-turn link was wound over the coil, and coupled to the grid-dip oscillator the resonant frequency had increased to 1.9 MHz. A three-turn link caused a change to 2 MHz.

Quite often an actual load may be an unknown quantity, such as an antenna, and some insight into the effects of the various elements is helpful in predicting single-tuned circuit operation. Usually, as in the case of most matching networks, R_s (Fig. 7) and the input resistance are specified with the reactive components being the variables. Unfortunately, the variables in the case of mutually coupled networks are not independent of each other which complicates matters somewhat.

Examination of the equivalent circuit shown in Fig. 7 would indicate the first condition is that the reactance reflected from the secondary into the primary be sufficient to tune out the primary reactance. Otherwise, even though the proper resistance transformation is obtainable, a reactive component would always be

present. A plot of the reflected reactance as a function of X_s is shown in Fig. 17. From mathematical considerations (which will not be discussed) it can be shown that the maximum and minimum of the curve have a value equal to $X_m^2/2R_s$. Consequently, this value must be greater than or equal to X_p in order that a value of X_s exists such that the reflected reactance will cancel X_p. In the usual case where $X_m^2/2R_s$ is greater than X_p, it is interesting to note that *two values* of X_s exist where X_p and the reflected reactance cancel. This means there are two cases where the input impedance is purely resistive and R_s could be matched to either one of two source resistances if so desired. The value of X_s at these points is designated as X_{s1} and X_{s2}.

On the other hand, a high value of R_s requires X_m to be large also. This could be accomplished by increasing the coefficient of coupling or by increasing the turns on the secondary coil. Increasing the turns on the primary also will cause X_m to be higher but X_p will increase also. This is somewhat self-defeating since X_m^2 is proportional to X_p.

An alternate approach is to use the parallel configuration of Fig. 18. The approximate equivalent series resistance of the parallel combination is then $X(C_s)^2/R_s$ and the reactance is approximately $X(C_s)$. (See diagram and text for Fig. 11.) This approach is often used in multiband antenna systems. On some frequencies, the impedance at the input of the feed line is high so the circuit of Fig. 18 is employed. This is referred to as parallel tuning. If the impedance is very low, the circuit of Fig. 13 is used and is called series tuning.

As an example, suppose a single-tuned circuit is to be used to match a 1-ohm load to a 50-ohm source as shown in Fig. 19. It might be pointed out at this juncture that coupling networks using mutual magnetic coupling can be scaled in the same manner that filter networks are scaled (as discussed in chapter 2). For instance, the circuit of Fig. 19 could be scaled in order to match a 50-ohm load to a 2500-ohm source merely by multiplying all the reactances by a factor of 50.

The input resistance and reactance of the circuit of Fig. 19 are plotted in Figs. 20 and 21, respectively. As pointed out earlier, there are two possible points where the reactance is zero and this circuit could be used to match the 1-ohm load to either a 50-ohm or 155-ohm source. Assuming a 50-ohm source was being used, the attenuation plot as a function of frequency would be given by the solid curve in Fig. 22.

With slight modification to include the effect of the source, the transformation of Fig. 7 can be applied to the primary side of the coupled circuit shown in Fig. 19. This is illustrated in Fig. 23. The complete circuit is shown at Fig. 23A and the network with the transformed primary

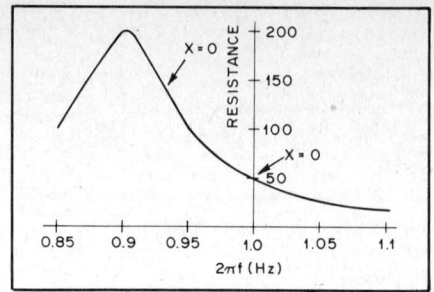

Fig. 20 — Input resistance of the Fig. 19 circuit as a function of frequency.

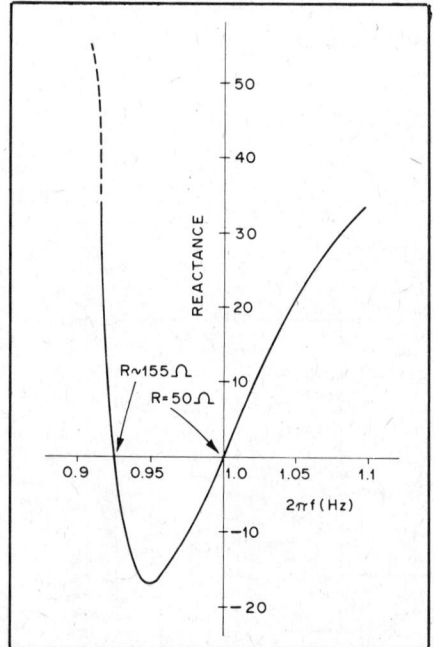

Fig. 21 — Input reactance of the Fig. 19 network. Note two "resonant" frequencies (where reactance is zero).

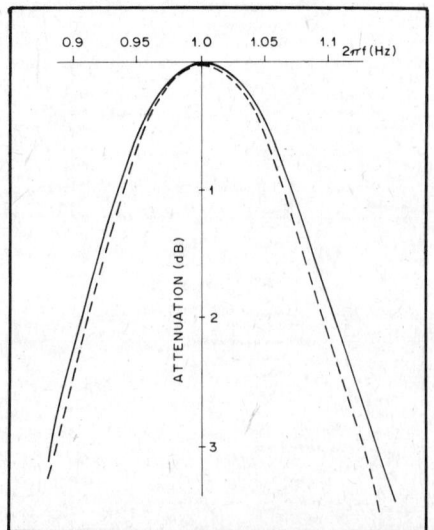

Fig. 22 — Response of the circuit shown in Fig. 19.

resistance and reactance is shown in Fig. 23B.

In a lossless transformer, the maximum available power at the secondary must be the same as that of the original source on the primary side, neglecting the effects of reactance. That is, the power delivered to

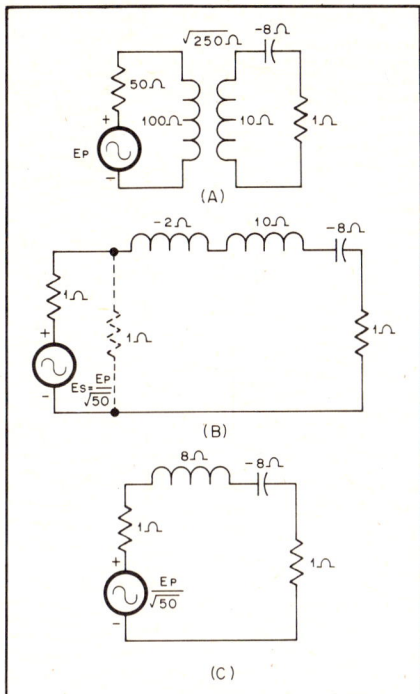

Fig. 23 — The transformation of Fig. 7 applied to the primary side of the circuit of Fig. 19.

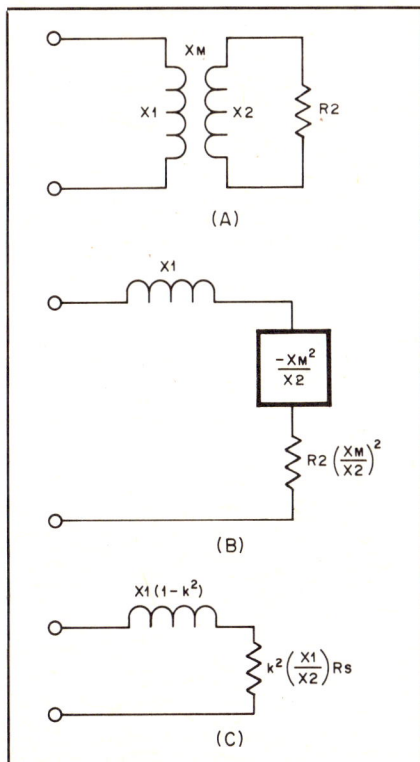

Fig. 24 — Equivalent-circuit approximation of two coupled coils.

a 1-ohm resistance (shown as a dashed line in Fig. 23B) must be the same as that delivered to a 50-ohm load in Fig. 23A. This assumes that the rest of the circuit has been disconnected in either case. In order to fulfill this requirement, the original source voltage must be multiplied by the square root of the ratio of the new and old source resistance.

The single-mesh transformed network is shown in Fig. 23C and it is interesting to compare the response of an RLC series circuit that actually possessed these element values at resonance with the circuit of Fig. 19. For comparison, the response of such a circuit is shown in Fig. 22 as a dashed curve and it can be seen that it differs only slightly from the coupled-circuit curve. The reason for the similarity is that even though the transformation of the primary resistance and reactance also changes with frequency, the effect is not that great in the present case.

Broadband RF Transformers

The "sensitivity" of the frequency characteristic of the transformation shown in Fig. 7 depends mostly on the ratio of X_s to R_s. However, if X_s is much greater than R_s, the transformed reactance can be approximated by

$$\frac{-X_m^2 X_s}{R_s^2 + X_s^2} \cong \frac{-X_m^2}{X_s}$$

and the resistance becomes

$$\frac{R_s X_m^2}{R_s^2 + X_s^2} \cong R_s \frac{X_m^2}{X_s^2}$$

Applying this approximation to the general coupled circuit shown in Fig. 24A results in the transformed network of Fig. 24B. The coefficient of coupling for the

circuit of Fig. 24A is

$$k = \frac{X_m}{\sqrt{X_1 X_2}}$$

and the network shown in Fig. 24B in terms of the coefficient of coupling is illustrated in Fig. 24C. For k equal to 1.0, the input reactance is zero and the input resistance is given by

$$R_{in} = \left(\frac{X_1}{X_2}\right) R_2 = \left(\frac{L_1}{L_2}\right) R_2 \cong \left(\frac{N_1}{N_2}\right) R_2$$

where N_1 and N_2 are the number of turns on coil 1 and 2, respectively. From maximum-power transfer considerations, such as those discussed for the circuit of Fig. 23, the voltage transfer ratio becomes

$$e_2 = \left(\frac{N_2}{N_1}\right) e_1$$

It will be recalled that the foregoing equations occurred in the discussion of the "ideal transformer" approximation in Chapter 2. It was assumed then that the leakage reactance and magnetizing current were negligible. The effects on circuit operation of these variables are shown in Fig. 25. The curves were computed for various load resistances (R_s) using the exact equations shown in Fig. 7.

X_1 and X_2 are assumed to be 100 and 10 ohms, respectively, with the solid curves for a k of 1.0 and the dashed reactance curve for k equal to 0.99 (the resistance curve for the latter value is the same as the one for k equal to 1.0). The ideal-transformer representation can be modified slightly to approximate the curve of Fig. 25 as shown in Fig. 26. The shunt reactance, X_{mag} is called the magnetizing

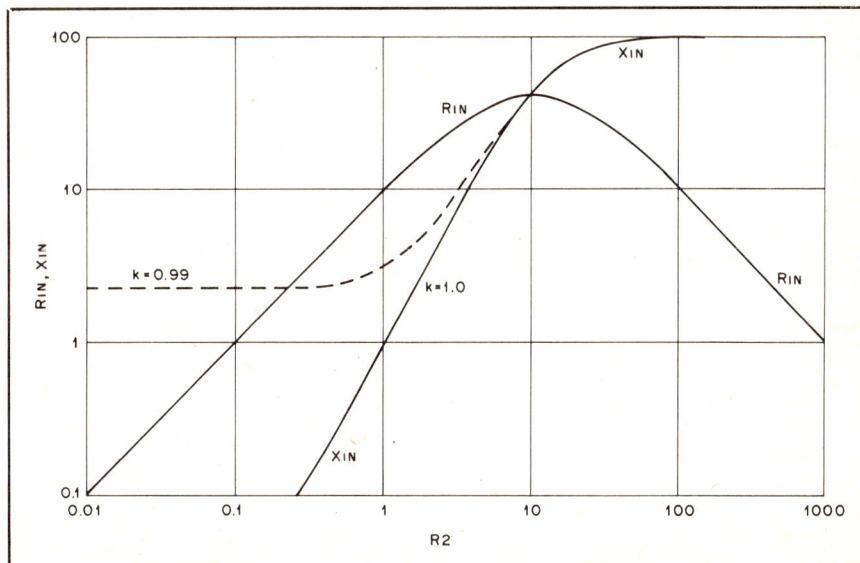

Fig. 25 — Input resistance and reactance as a function of output load resistance for X_1 and X_2 equal to 100 ohms and 10 ohms respectively (Fig. 24).

reactance and X_L is referred to as the leakage reactance.

Unfortunately, the two reactances are not independent of each other. That is, attempts to change one reactance so that its effect is suppressed causes difficulties in eliminating the effects of the other reactance. For instance, increasing X_1, X_m, and X_2 will increase X_{mag} which is desirable. However, examination of Fig. 24C reveals that the coefficient of coupling, k, will have to be made closer to 1.0. Otherwise, the leakage reactance increases since it is proportional to X_1.

High-Permeability Cores

As a consequence of the interaction between the leakage reactance and the magnetizing reactance, transformers that approach ideal conditions are extremely difficult (if not impossible) to build using techniques common in air-wound or low-permeability construction. In order to build a network that will match one resistance level to another one over a wide range of frequencies, ideal-transformer conditions have to be approached quite closely. Otherwise, considerable inductive reactance will exist along with the resistive component as shown in Fig. 25.

One approach is to use a core with a higher permeability than air. Familiar examples would be power transformers and similar types common to the af range. However, when an inductor configuration contains materials of more than one permeability, the analysis relating to Fig. 24C has to be modified somewhat. The manner in which the core affects the circuit is a bit complicated although even a qualitative idea of how such transformers work is very useful.

First, consider the coupled coils shown in Fig. 27. For a given current, I_1, a number of "flux lines" are generated that link both coil 1 and coil 2. Note that in coil 1, not all of the flux lines are enclosed by all the turns. The inductance of a coil is equal to the ratio of the sum of flux lines linking each turn and the generating current or

$$L_1 = \frac{\Lambda_{TOTAL}}{I_1}$$

where for the example shown in Fig. 27, Λ_{TOTAL} is given by

$$\Lambda_{TOTAL} = \Lambda_1 + \Lambda_2 + \Lambda_3 + \Lambda_4 + \Lambda_5$$

Counting up the number of flux linkages in coil 1 gives

$$\Lambda_{TOTAL} = 5 + 5 + 7 + 7 + 5 = 29$$

If all the flux lines linked all the turns, Λ_{TOTAL} would be 35 so L_1 is 29/35 or 83 percent of its maximum possible value. Likewise, if all the flux (7 lines) generated in coil one linked all the turns of coil 2, the maximum number of flux linkages would be the number of turns on coil 2 times 7 or

28. Since only three lines link coil 2, the mutual inductance is 3 × 4/28 or 43 percent of maximum.

Assuming both coils are "perfect," if a current I_1 produced 7 flux lines in a five-turn coil, then the same current in a four-turn coil would produce (4/5)(7) flux lines, since the flux is proportional to the magnetizing current times the number of turns. Consequently, the maximum flux linkages in coil 1 from a current of the same value as I_1 but in coil 2 instead would be (4/5)(7)(5) or 28. Therefore, it can be seen that the mutual inductance is independent of the choice of coil used for the primary or secondary. That is, a voltage produced in one coil by a current in the other one would be the same if the coils were merely interchanged. (This result has been used implicitly on a number of previous occasions without proof.) In addition, the maximum flux linkages in coil 2 produced by a current, I_1, would be (4/5)(7)(4). As an exercise, substitute the maximum inductance values into the formula for the coefficient of coupling and show that k is 1.0.

The next step is to consider the effect of winding coils on a form with a magnetic permeability much higher than that of air. An example is illustrated in Fig. 28 and the configuration shown is called a toroidal transformer. Since the flux is proportional to the product of the permeability and the magnetizing current, the flux in the core shown in Fig. 28 will be much greater than the coil configuration of Fig. 27. However, not all of the flux is confined to the core. As can be seen in Fig. 28, some of the flux lines never penetrate the core (see lines marked a in Fig. 28) while others enclose all the windings of coil 1 but not coil 2 (see line marked b). The significance of these effects is as follows. The total flux linkage produced by the current, I_1, is

$$\Lambda_{TOTAL} = \Lambda_{air} + \Lambda_{core}$$

and dividing both sides of the equation by I_1 gives

$$L_T = L_{air} + L_{core}$$

Consequently, the circuit of Fig. 24 can be represented as shown in Fig. 29A. For X_2 much greater than the load resistance, the approximate network of Fig. 29B can replace the one of Fig. 29A.

At first sight, it might seem as though little advantage has been gained by introducing the core since the formulas are much the same as those of Fig. 24C. However, the reactances associated with the core can be made very high by using a material with a high permeability. Also, even though there may be some "leakage" from the core as indicated by line b in Fig. 28, it is ordinarily low and the coefficient of coupling in the core can be considered 1.0 for all practical purposes. This is

especially true at af and power frequencies with transformers using iron cores where the permeability is extremely high. This means the magnetizing reactance can be made very high without increasing the leakage reactance accordingly as is the case with the circuit in Fig. 24C. Therefore, ideal transformer conditions are considered to exist in the core and the final circuit can be approximated by the one shown in Fig. 29C.

Bifilar and Twisted-Pair Windings

Although the core helps alleviate some of the problems with leakage and mag-

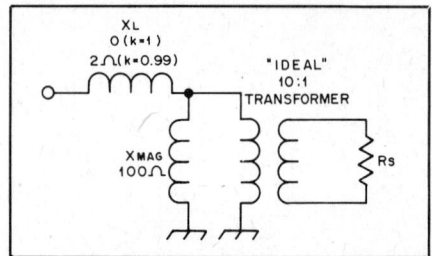

Fig. 26 — Approximate network for the curves of Fig. 25.

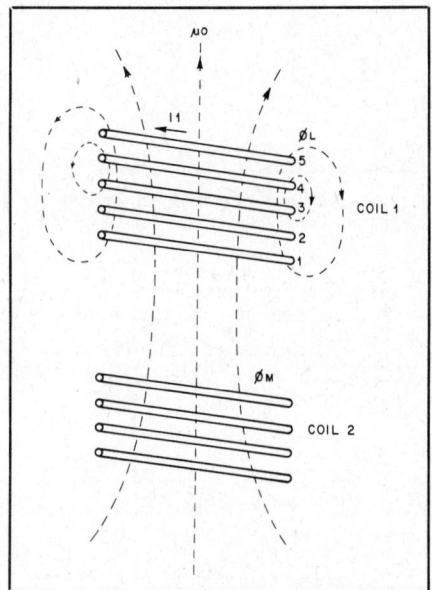

Fig. 27 — Coupled coils showing magnetic flux lines.

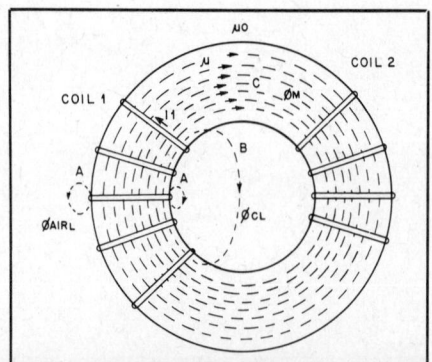

Fig. 28 — Toroidal transformer.

netizing reactance, the residual parasitic elements must still be made as low as possible. This is especially important in matching applications as the following example illustrates. A transformer has a primary and secondary leakage reactance of 1 ohm and 0.1 ohm, respectively, with a coefficient of coupling of 1.0 in the core. X_1 and X_2 are 1000 ohms and 100 ohms.

A plot similar to the one of Fig. 25 is shown in Fig. 30 along with a curve for

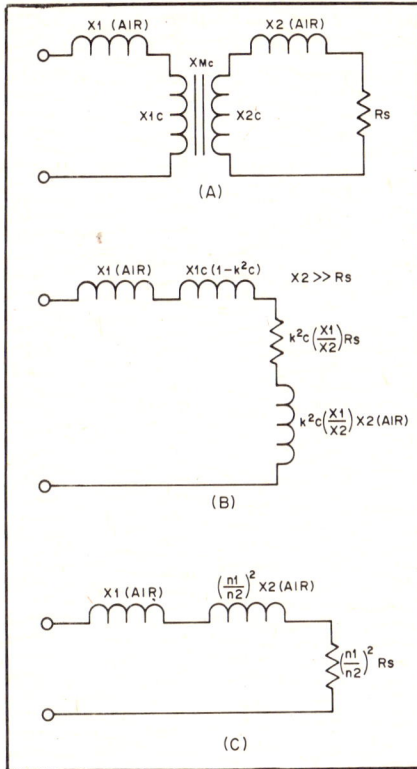

Fig. 29 — Effect of a high-permeability core on transformer equivalent circuit.

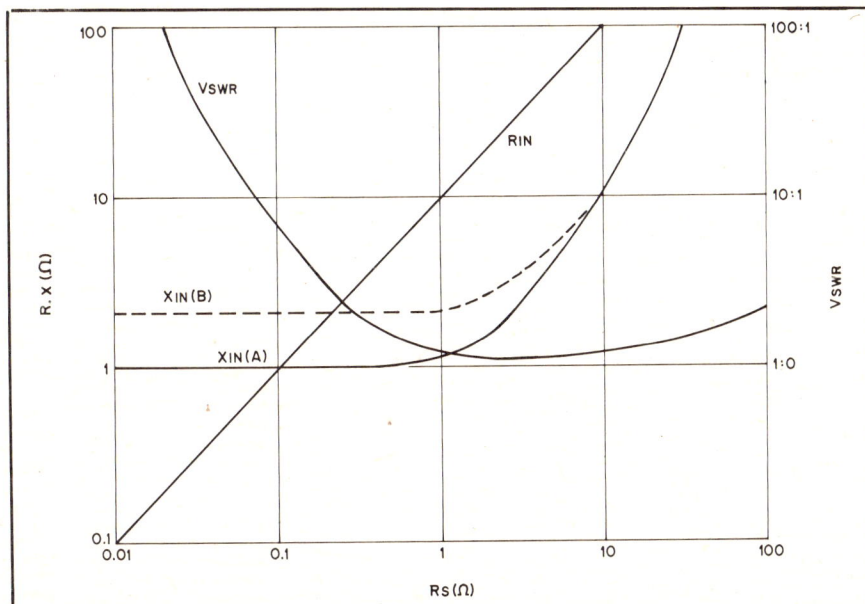

Fig. 30 — Curve for transformer problem discussed in the text.

voltage-standing-wave ratio (VSWR). These results are based on the exact equations and it can be seen that the approximate relations shown in Fig. 29C are valid up to 1 ohm or so. Curve A (Fig. 30) only includes the effect of the secondary reactance and illustrates the manner in which the reactance is transformed. Curve B is the total input reactance which merely requires the addition of 1 ohm. The VSWR curve includes the effect of the latter. Useful range of the transformer is between 1 and 10 ohms with rapid deterioration in VSWR outside of these values. (The VSWR curve is for a characteristic impedance equal to 10 times the secondary resistance. For instance, the transformer would be useful in matching a 5-ohm load to a 50-ohm line.)

As mentioned previously, these difficulties are less pronounced at audio frequencies since the permeabilities normally encountered in iron-core transformers are so high, the actual inductance of the winding itself is small in comparison to the component represented by the core. That is, a small number of turns of wire wound on a core may actually be the equivalent of a very large coil. However, materials suitable for rf applications have much lower permeabilities and a narrower range of matching values is likely to be the result (such as in the example of Fig. 30). Therefore, other means are required in keeping the parasitic elements as low as possible. Either that, or less conventional transformer designs are used.

One approach is shown in Fig. 31. Instead of separating the windings on the core as shown in Fig. 28, they are wound in parallel fashion. This is called a bifilar winding although a more common approach to achieve the same purpose is to

twist the wires together. Either way, there are a number of advantages (and some disadvantages) to be gained. Referring to Fig. 27, the fact that not all the flux lines linked all of the turns of a particular coil meant the self inductance was lower than if all the turns were linked. Since the separation between turns of a particular coil is quite large in the configuration of Fig. 31, the flux linkage between turns is quite low. This means the corresponding leakage inductance is reduced accordingly. However, the coupling between both coils is increased because of the bifilar winding (flux line A) in Fig. 31 which also tends to reduce the leakage inductance of either coil.

On the other hand, the capacitance between windings is increased considerably as indicated by B in Fig. 31. As a result, the coupling between windings is both electrical and magnetic in nature. Generally speaking, analysis of the problem is quite complicated. However, a phenomenon usually associated with such coupling is that it tends to be directional. That is, energy transferred from one winding to another one propagates in a preferred direction rather than splitting equally.

Directional Coupling

Two conductors are oriented side by side over a conducting plane as shown in Fig. 32. A current I in conductor 1 will induce a current I_m in conductor 2 because of magnetic coupling. The actual value of the current will depend upon the external circuitry attached to the conductors but it will be assumed that the two of them extend to infinity in both directions.

Since capacitive coupling exists also, a second set of current components denoted by I_c will also flow. The result is that a wave traveling toward the *right* in conductor 1 will produce a wave traveling toward the *left* in conductor 2. Such coupling is called contradirectional coupling since the induced wave travels in the opposite direction to the generating wave.

This is the principle behind many practical devices and ones that are quite common in amateur applications. In adjusting a load such as an antenna, it is desirable to insure that energy is not reflected back to the transmitter. Otherwise, the impedance presented to the transmitter output may not be within range of permissible values. A directional coupler is useful in determining how much power is reflected as indicated in Fig. 33. Energy originating from the transmitter and flowing to the right causes a voltage to be produced across the resistor at the left. On the other hand, a wave traveling from the right to the left produces a voltage across the right-hand resistor. If both of these voltages are sampled, some idea of the amount of power reflected can be determined. (The subject of reflected-power is taken up in more detail in the

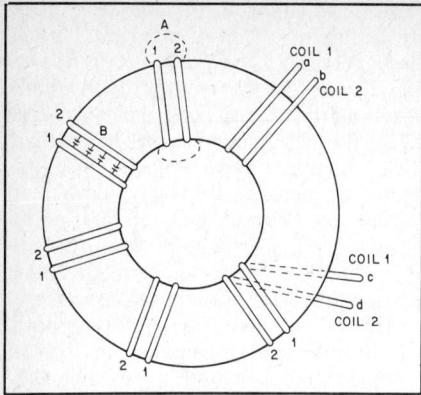

Fig. 31 — Bifilar-wound transformer on toroidal core.

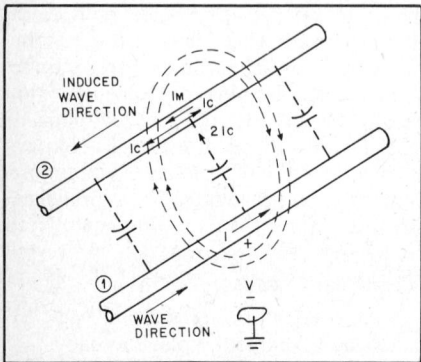

Fig. 32 — Effect of distributed capacitance on transformer action.

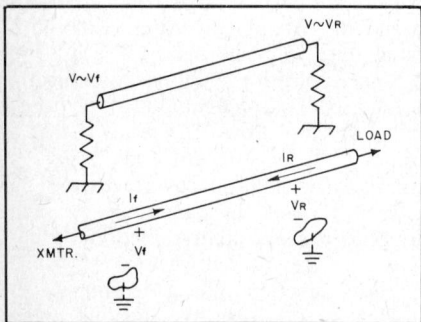

Fig. 33 — Basic configuration for a directional-coupler type VSWR detector.

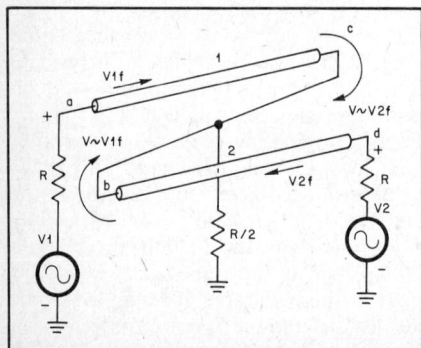

Fig. 34 — Directional-coupler hybrid combiner.

chapter on transmission lines.)

In some situations, the coupling described can be very undesirable. For instance, the lines shown in Fig. 33 might be conductors on a circuit board in a piece of equipment. As a result, the coupling between lines can cause "feedback" and because of its directional nature, it can be very difficult to suppress with conventional methods. Therefore, it is good design practice to use "double-sided" board (board with conductive foil on both sides) so that a ground plane of metal is in close proximity to the conductors. This tends to confine the fields to the region in the immediate vicinity of wires.

Transmission-Line Transformers

In effect, sections of transmission line in close proximity act as transformers with the unique feature that the coupling is directional. For instance, if only magnetic coupling was present in the configuration of Fig. 33, power would be divided equally between the resistors at either end of the "secondary" section of transmission line. As another example of directional effects, the network shown in Fig. 34 can be used to couple two sources to a common load without "cross-coupling" of power from one source to the other one. (This assumes the sources have the same frequency and phase. Otherwise, a resistance of value 2R must be connected from points a to d.) Such a configuration is called a hybrid combiner and is often used to combine the outputs of two solidstate amplifiers in order to increase the powerhandling capability. This permits the use of less expensive low-power devices rather than very expensive high-power ones. Even though more devices are required, it is still simpler since the difficulties in producing a high-power transistor increase in a greater proportion as the power level is raised.

The manner in which the circuit shown in Fig. 34 operates is as follows. A wave from the generator on the left end of line 1 travels toward the right and induces a wave in line 2 that travels toward the left and on into the load. No wave is induced in line 2 that travels toward the right except for a small fraction of power.

A similar situation exists with the second generator connected at the right end of line 2. A wave is induced in line 1 that travels toward the right. Since the load is also connected to the right end of line 1, power in the induced wave will be dissipated here with little energy reaching the generator at the left end of line 1. In order to "simulate" a single load (since there are two generators involved), the value of the load resistance must be half of the generator resistance. Assuming that two separate resistors of value R were connected to the ends of the line, it would be possible to connect them together without affecting circuit operation. This is because the voltage across both resistors is

of the same phase and amplitude. Consequently, no additional current would flow if the two resistors were paralleled or combined into a single resistor of R/2.

Extending the Low-Frequency Range

As might be expected, the coupling mechanism illustrated in Figs. 32 through 34 is highly dependent on dimensions such as conductor spacing and line length. For instance, maximum coupling of power from the primary wave to the induced wave occurs when the "secondary" line is a quarter-wavelength long[2] or some odd multiple of a quarter-wavelength. This would normally make such couplers impractical for frequencies in the hf range. However, by running the leads through a ferrite core as shown in Fig. 35, lower-frequency operation is possible. Although the transformer of Fig. 35A is seldom used, it illustrates the manner in which the conductors are employed electrically in the more complicated configurations of Fig. 31 and Fig. 35B. Also, the relationship between the parallel-line coupler in Fig. 34 and the "loaded" version of Fig. 35A is easier to visualize.

Recalling an earlier problem discussion (Fig. 28), a set of coupled coils wound on a high-permeability core can be broken down into combinations of two series inductances. One inductance represents the path in air while the other one includes the effects of the flux in the core. As

[2]Oliver, "Directional Electromagnetic Couplers," *Proceedings of the I.R.E.*, Vol. 42, p. 1686-1692; November, 1954.

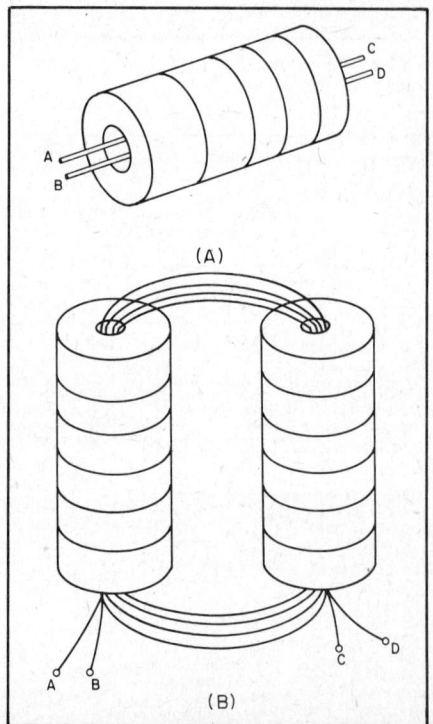

(A)

(B)

Fig. 35 — Transmission-line transformers with ferrite cores.

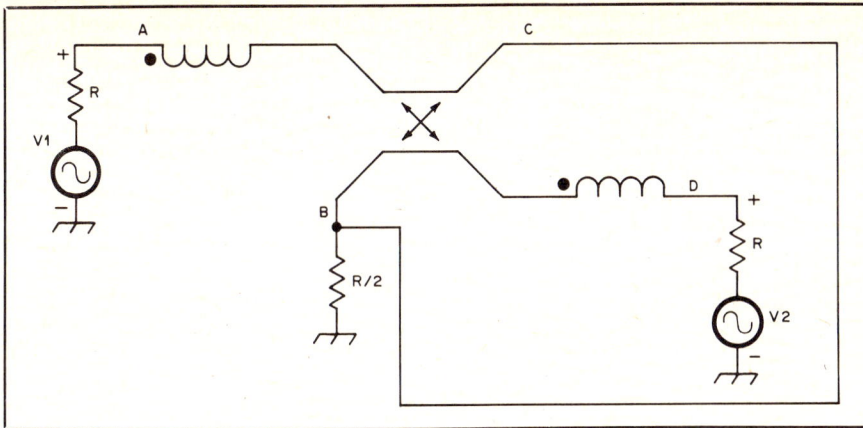

Fig. 36 — Equivalent circuit of transmission-line transformer in the presence of the core. Dots indicate winding sense of coils. A positive current into a dotted end of one coil will produce a voltage in the other coil because of mutual coupling. The polarity of this voltage will be such that dotted end of the "secondary" coil will be positive. (See text for crossed-arrow symbol in the middle of the parallel lines.)

before, it is assumed that the coefficient of coupling in the core is 1.0.

If the hybrid combiner of Fig. 34 is wound on a core (such as those of Fig. 31 or Fig. 35), the low-frequency range of the entire system is increased considerably. The equivalent circuit showing the effect of the core on the air-wound coupler is illustrated in Fig. 36. (The symbol in the middle of the parallel lines is the standard one for a directional coupler.) At the higher frequencies, most core materials decrease in permeability so the operation approaches that of the original air-wound coupler and the inductance produced by the core can be neglected. At the low end of the frequency range, the line lengths are

usually too short to provide much coupling or isolation. Therefore, the circuit can be represented by the set of coupled coils shown in Figs. 37 and 38.

For a current I_{12} flowing from a source 1 over to the mesh that includes source 2, the mutual-reactance components *add* to the self inductance of each coil. Consequently, a large reactance appears in series between the two sources which effectively isolates them. On the other hand, currents from both sources that flow through the load resistor R/2 produce fluxes that cancel and the voltages produced by the self- and mutual-reactance terms *subtract*. If both sources have the same amplitude and

phase, currents I_1 and I_2 must be identical because of the symmetry involved. However, if the coefficient of coupling is 1.0, the self and mutual-reactance must be equal. Therefore, the voltage across either coil is zero since the terms subtract and a low-impedance path exists between both sources and the load.

Other Transformer Types

The hybrid combiner is only one application of a combination transmission-line or directional-coupler transformer and conventional coupled-coil arrangement. With other variations, the low-frequency isolation is accomplished in the same manner. Mutual-reactance terms add to the self reactance to provide isolations for some purpose with cancellation of reactive components in the path for the desired coupling. Very good bandwidth is possible with a range from bc frequencies to uhf in the more esoteric designs. Models that cover all the amateur hf bands can be constructed easily.[3]

Unfortunately, there is also a tendency to expect too much from such devices on occasion. Misapplication or poor design often results in inferior performance. For instance, as indicated in an earlier example (Fig. 30), actual impedance levels were important along with the desired transforming ratio. Using a transformer for an impedance level that it was not intended for resulted in undesirable reactive components and improper transforming ratio. However, when applied properly, the transformers discussed in the previous sections can provide bandwidth characteristics that are obtainable in no other way.

Another transformer type is shown in Fig. 39A. The windings of the coils are such that the voltages across the inductors caused by the desired current are zero. This is because the induced voltages produced by the current in the mutual-reactance terms just cancel the voltage drop caused by the current flowing in the self reactances of either coil. (Assuming that the coefficient of coupling is 1.0.) However, an impedance connected to ground at point c would be in series with the self reactance (X_L) of the coil connected between points a and c. But there would be no induced voltage to counter the voltage drop across this coil. Therefore, if X_L is large, very little current would flow in the impedance Z and it would effectively be isolated from the source.

In fact, terminal c could be grounded as shown in Fig. 39B. The voltage drop across the coil from a to c would then be equal to V_1. However, the induced voltage in the coil connected between points b and d would also be V_1 assuming unity coupling (k equal to 1.0). Although the

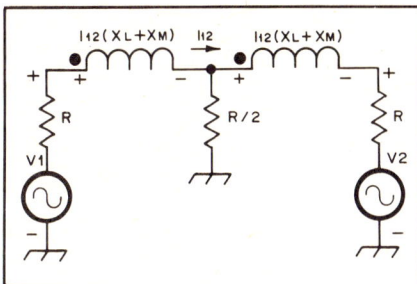

Fig. 37 — Low-frequency equivalent circuit of hybrid combiner showing isolation of sources.

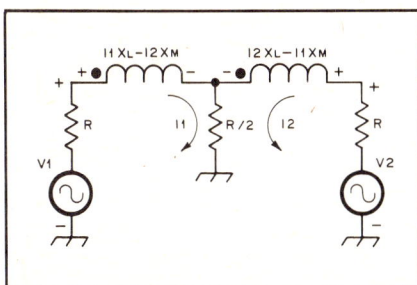

Fig. 38 — Desired coupling mode of hybrid combiner.

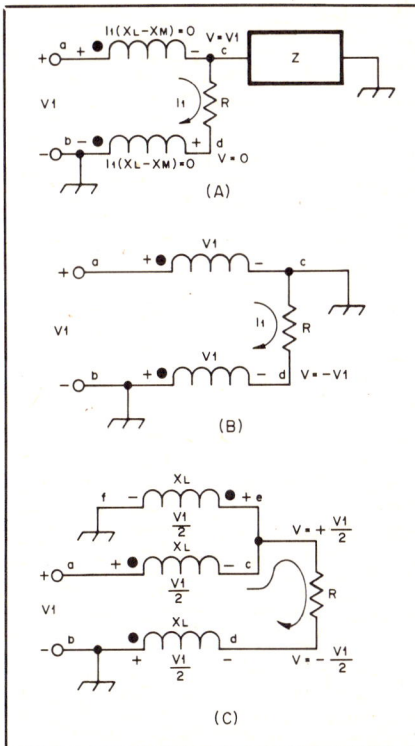

Fig. 39 — Other applications of transmission-line transformers.

[3]Ruthroff, "Some Broad-Band Transformers," *Proceedings of the I.R.E.,* Vol. 97, pp. 1337-1342; August, 1959.

voltage drop produced by the inductors around the mesh through which I_1 flows is still zero, point d is now at potential $-V_1$ and a phase reversal has taken place. For this reason, the configuration shown in Fig. 39B is called a phase-reversal transformer.

Baluns

The circuit shown in Fig. 39A is useful in isolating a load from a grounded source. This is often required in many applications and the device that accomplishes this goal is called a balun (*bal*anced to *un*balanced) transformer. Baluns may also be used in impedance transforming applications along with the function of isolation and a "1:1 balun" such as the one shown in Fig. 39A means the impedance at the input terminals ab will be the same as the load connected across terminals cd. Other transforming ratios are possible such as 4:1 with the appropriate circuit connections.

One disadvantage of the network of Fig. 39A is that although the load is isolated from the source, the voltages at the output are not balanced. This is important in some applications such as diode-ring mixers where a "push-pull" input is required and so the circuit of Fig. 39C is used. A third coil connected between points e and f is wound on the same core as the orignal transformer (Fig. 39A). This coil is connected so that a voltage across it produces a flux that adds to that produced by the coil between a and c. Assuming that both coils are identical, the voltage drop across either one must be the same or half the applied voltage. However, since the coil between b and d is also coupled to this combination (and is an identical coil), the induced voltage must also be $V_1/2$. Consequently, the end of the load connected to points c and e is at a potential of $+V_1/2$ with respect to ground while point d is $-V_1/2$ with respect to ground when the input voltage has the

Table 1

Impedance (Ohms)

Wire Size	Twists per Inch				
	2-1/2	5	7-1/2	10	12-1/2
no. 20	43	39	35		
no. 22	46	41	39	37	32
no. 24	60	45	44	43	41
no. 26	65	57	54	48	47
no. 28	74	53	51	49	47
no. 30			49	46	47

Measured at 14.0 MHz

This chart illustrates the impedance of various two-conductor lines as a function of the wire size and number of twists per inch.

Table 2

Attenuation (dB) per Foot

Wire Size	Twists per Inch				
	2-1/2	5	7-1/2	10	12-1/2
no. 20	0.11	0.11	0.12		
no. 22	0.11	0.12	0.12	0.12	0.12
no. 24	0.11	0.12	0.12	0.13	0.13
no. 26	0.11	0.13	0.13	0.13	0.13
no. 28	0.11	0.13	0.13	0.16	0.16
no. 30			0.25	0.27	0.27

Measured at 14.0 MHz

Attenuation in dB per foot for the same lines as shown in Table 1.

polarity shown. Therefore, this circuit not only isolates the load from the source but provides a balanced voltage also.

Either the circuit of Fig. 39A or Fig. 39C can be used if only isolation is desired. However, the network shown in Fig. 39C is more difficult to design and construct since the reactance of the coils between points a and f must be very high throughout the frequency range of the transformer. With both transformers, the coefficient of coupling must also be very close to 1.0 in order to prevent undesirable reactance in

series with the load. This problem can be offset somewhat by reducing X_L slightly (by using fewer turns) but this is counter to the requirement of large X_L in the circuit of Fig. 39C. Isolation is reduced in both cases although no detrimental effect on input impedance results in the transformer of Fig. 39A by reducing X_L.

Twisted Pairs — Impedance and Attenuation

Twisted pairs of wire are often used in the construction of broadband rf transformers. The question often arises as to what size conductors and what number of twists per inch should be used. To help answer these questions the information contained in Tables 1 and 2 was developed. Table 1 illustrates the approximate impedance for various sized conductors with different numbers of twists per inch. These values are based on laboratory measurements and should be accurate to within an ohm or two. Enameled copper wire was used for each pair. The information shown in Table 2 is the attenuation per foot for the same twisted pairs of wire. Information is not included for twists per inch greater than 7-1/2 for the no. 20 wire since this results in an unusable tight pair. Likewise, the information for twists per inch less than 7-1/2 for no. 30 wire is omitted since these pairs are extremely loose.

As a general rule the wire size can be selected based on the size core to be used and the number of turns that are required. The number of twists per inch can be selected according to the impedance level of transmission line that is needed. For applications where moderate levels of power are to be handled (such as in the low- and medium-level stages of a solid-state transmitter), smaller wire sizes should be avoided. For receiver applications, very small wire can be used. It is not uncommon to find transformers wound with pairs of no. 32 wire and smaller.

Nonlinear and Active Networks

Almost all the theory in previous sections has dealt with so-called passive components. Passive networks and components can be represented solely by combinations of resistors, capacitors and inductors. As a consequence, the power output at one set of terminals in a passive network cannot exceed the total power input from sources connected to other terminals in the circuit. This assumes all the sources are at one frequency. Similar considerations hold true for any network, however, it is possible for energy to be converted from one frequency (including dc) to other ones. While the total power input must still equal the total power

output, it is convenient to consider certain elements as controllable sources of power. Such devices are called amplifiers and are part of a more general class of circuits called active networks. An active network generally possesses characteristics that are different than those of simple RLC circuits although the goal in many instances is to attempt to represent them in terms of passive elements and generators.

Nonlinearity

Two other important attributes of passive RLC elements are that they are linear and bilateral. A two-terminal

element such as a resistor is said to be bilateral since it doesn't matter which way it is connected in a circuit. Semiconductor and vacuum-tube devices such as triodes, diodes, transistors and integrated circuits (ICs) are all examples where the concept of a bilateral element breaks down. (For readers with limited backgrounds in the basic operation of vacuum tubes, recommended study would be *The Radio Amateur's License Manual* and *Understanding Amateur Radio*. Both publications contain fundamental treatments of vacuum-tube principles and are available from The American Radio Relay League.) The manner in which the device is connected in a circuit and the polarity of the voltages involved are very important.

An implication of the failure to satisfy the bilateral requirements is that such devices are nonlinear in the strictest sense. Linearity means that the amplitude of a voltage or current is related to other voltages and currents in a circuit by a single proportionality constant. For instance, if all the voltages and currents in a circuit were doubled, a single remaining voltage or current would be doubled also. That is, it couldn't change by a factor of one half or three no matter how complex the network might be. Likewise, if all the polarities of the currents and voltages in a circuit are reversed, the polarity of a remaining voltage or current must be reversed also. Finally, if all the generators or sources in a linear network are sine waves at a single frequency, any voltage or current produced by these sources must also be a sine wave at the same frequency too.

Consequently, if a device is sensitive to the polarity of the voltage applied to its terminals, it doesn't meet the requirements of a bilateral element or a linear one either. However, because of the extreme simplicity of the mathematics of linear circuits as compared to the general nonlinear case, there is tremendous motivation in being able to represent a nonlinear circuit by a linear approximation. Many devices exhibit linear properties over part of their operating range or may satisfy some but not all of the requirements of linear circuits. Such devices in these categories are sometimes termed piece-wise linear. Either that, or they are just referred to simply as linear. For instance, a linear mixer doesn't satisfy the rule that a voltage or current must be at the same frequency as the generating source(s). However, since the desired output voltage (or current) varies in direct proportion to the input voltage (or current), the term linear is applied to distinguish the mixer from types without this "quasi-linear" property.

Harmonic-Frequency Generation

In a circuit with only linear components, the only frequencies present are those generated by the sources themselves. However, this is not true with nonlinear elements. One of the properties of nonlinear networks mentioned earlier is that energy at one frequency (including dc) may become converted to other frequencies. In effect, this is how devices such as transistors and vacuum tubes are able to amplify radio signals. Energy from the dc power supply is converted to energy at the desired signal frequency. Therefore, a greater amount of signal power is available at the output of the network of an active device than at the input.

On the other hand, such frequency generation may be undesirable. For instance, the output of a transmitter may

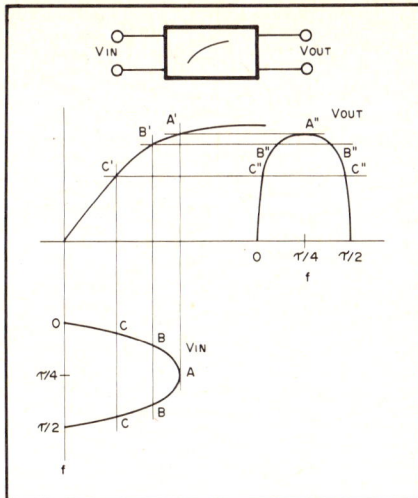

Fig. 40 — Nonlinear transfer characteristic (see text discussion).

have energy at frequencies that could cause interference to nearby receiving equipment. Filters and similar devices must be used to suppress this energy as much as possible.

The manner in which this energy is produced is shown in Fig. 40. A sine-wave at the input of a nonlinear network (V_{in}) is "transformed" into the output voltage waveform (V_{out}) illustrated. If the actual device characteristic is known, the waveform could be constructed graphically. It could also be tabulated if the output voltage as a function of input voltage was available in either tabular or equation form. (Only one-half of the period of a sine-wave is shown in Fig. 40 for clarity.)

Although the new waveform retains many of the characteristics of the original sine-wave, some transformations have taken place. It has zero value when t is either 0 or T/2 and attains a maximum at T/4. However, the fact that the curve is flattened somewhat means energy at the original sine-wave frequency has been converted to other frequencies. It will be recalled that the sum of a number of sine waves at one frequency result in another sine-wave at the same frequency. Therefore, it must be concluded that the waveform of Fig. 40 has more than one frequency component present since it is no longer a sine-wave.

One possible "model" for the new waveform is shown in Fig. 41A. Instead of one sine-wave at a single frequency, there are two generators in series with one generator at three times the "fundamental" frequency ω where ω is 2π f(Hz). If the two sine waves are plotted point by point, the dashed curve of Fig. 41B results. While this curve doesn't resemble the one of Fig. 40 very closely, the general symmetry is the same. It would take an infinite number of generators to represent the desired curve exactly, but it is evident all the frequencies must be odd multiples of the

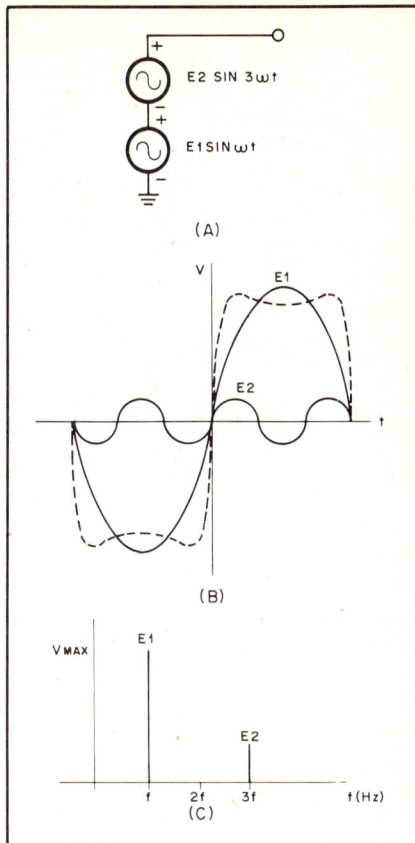

Fig. 41 — Harmonic analysis and spectrum.

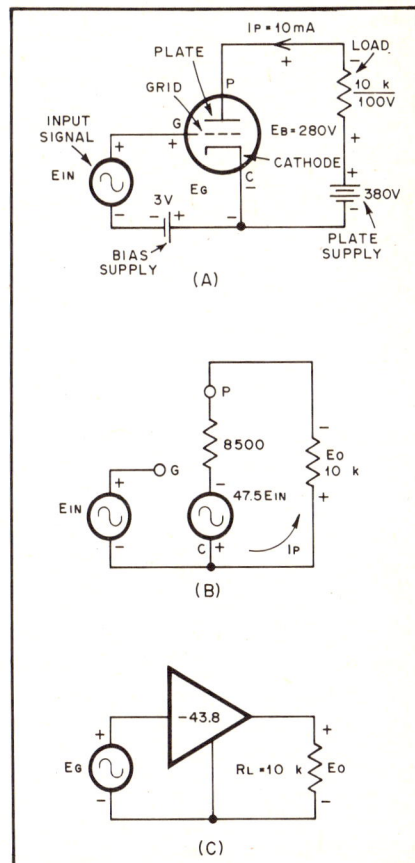

Fig. 42 — Basic triode amplifier and equivalent circuit.

fundamental. Even multiples would produce a lopsided curve which might be useful for representing other types of waveforms.

In either case, the multiples have a specific name and are called harmonics. There is no "first" harmonic (by definition) with the second, third and fourth multiples designated as the second, third and fourth harmonics. Thus the dashed curve of Fig. 41 is the sum of the fundamental and third harmonic.

Analyzing waveforms such as those of Fig. 40 is a very important subject. A plot of harmonic amplitude such as that shown in Fig. 41C is called the spectrum of the waveform and can be displayed on an instrument called a spectrum analyzer. If the mathematical equation or other data for the curve is known, the harmonics can also be determined by means of a process called Fourier Analysis.

Linear Approximations of Nonlinear Devices

Nonlinear circuits may have to be analyzed graphically as in the previous example. There are many other instances where only a graphical method may be practical such as in power-amplifier problems. However, a wide variety of applications permit a different approach. A model is derived from the nonlinear characteristics using linear elements to approximate the more difficult nonlinear problem. This model is then used in more complicated networks instead of the nonlinear characteristics which simplifies analysis considerably.

The following example illustrates how this is accomplished and although a vacuum-tube application is considered, a similar process is employed in solving semiconductor problems as well. However, there are some additional factors involved in semiconductor design that do not apply to vacuum tubes. Device characteristics of early transistors were less uniform than those of tubes although this is much less of a problem than it was formerly. In fact, much of the analysis required with vacuum tubes is unnecessary with modern solid-state components since many of the problems have already been "solved" before the device leaves the counter at the radio store. That is, amplifiers such as those in integrated circuits have the peripheral elements built in and there is no need to determine the gain or other parameters such as the values of bias resistors.

The Triode Amplifier

A simple network using a triode vacuum type is shown in Fig. 42A and a typical set of characteristic curves is illustrated in Fig. 43A. The first chore in finding a suitable linear approximation for the triode is to determine an optimum operating point. Generally speaking, a point in the center of the set of curves is desirable and is indicated by point Q in

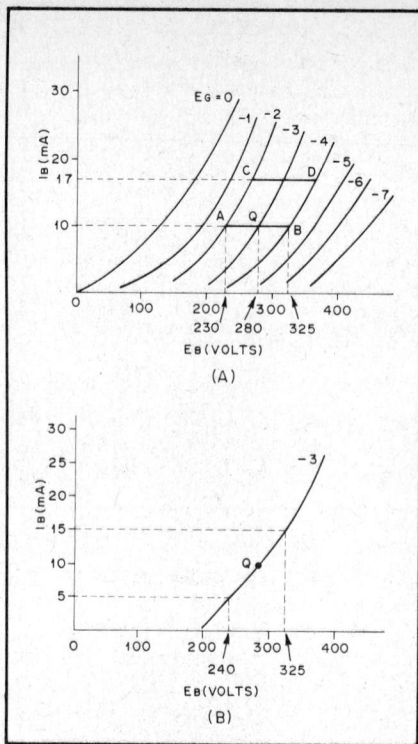

Fig. 43 — Triode characteristics and derivation of small-signal parameters.

Fig. 43A. (Other areas are often picked for power-amplifier operation but the goal here is to find a point where the maximum voltage swing is possible without entering regions where the non-linearities affect the linear approximation.)

In the particular operating point chosen, the cathode-to-grid voltage is -3, the cathode-to-plate voltage is 280, and the plate current is 10 mA. It is assumed that the input-signal source in Fig. 42A is a "short circuit" at dc and a 3-V battery connected as shown results in a dc voltage of -3 being applied to the grid at all times. Such a battery is called a bias battery or bias supply.

The next step is to determine how the plate voltage varies with grid voltage (e_g) for a constant plate current. Assuming that the characteristic curves were completely linear, this would permit evaluation of an equivalent ac voltage generator as shown in Fig. 42B. For a constant plate current of 10 mA, the plate voltage changes from 325 (point b) to 230 (point a) when the grid voltage is changed from -4 to -2 (Fig. 43A).

These numbers can be used to compute the amplification factor (μ) of the triode which is

$$\mu = \frac{325 - 230}{(-4) - (-2)} = -47.5$$

Quite often, a set of characteristics will not be published for a triode and only the amplification factor will be given along with a typical operating point. However,

note that the amplification factor is *negative*. This means that for an increase in the signal voltage (e_{in}), the controlled generator decreases in voltage. Consequently, there is a 180 degree phase shift between the input voltage and the controlled source. (Note the polarity of the generator shown in Fig. 42B.)

In order to complete the equivalent generator circuit, the source "impedance" must be computed. This is accomplished by determining how the plate voltage varies with plate current at constant grid voltage as shown in Fig. 43B. The plate resistance is then

$$r_p = \frac{325 - 240}{(15 - 5) \times 10^{-3}} = 8500 \text{ ohms}$$

which must be considered to be in series with the controlled source of Fig. 42B.

It should be pointed out at this juncture that the reasoning *why* the foregoing procedure is valid has not been presented. That is, why was the amplification factor defined as the ratio of a change in plate voltage to change in grid voltage at constant current? Unfortunately, the mathematics involved although not difficult is somewhat sophisticated. Some knowledge of the subject of partial differential equations is required for the theoretical derivation of these parameters. However, an intuitive idea can be obtained from the following.

If the characteristics were completely linear, instead of being nonlinear as shown, the equivalent generator would be unaffected by *changes* in plate current but only by changes in grid voltage. For instance, if the plate current was increased from 10 to 17 mA (Fig. 43A), the amplification factor would be the equivalent of the change in voltage represented by the line cd divided by -2. However, since the length of cd is almost the same as that of ab (the difference in plate voltage for a -2-V change at 10 mA), it can be concluded μ doesn't change very much. Not at least in the center region of the characteristics.

Similar considerations hold for the plate resistance, r_p. It wouldn't matter if the curve for -4 or -2 V was picked (Fig. 43B), since the change in plate voltage vs. plate current would be approximately the same. Entities such as μ and r_p are often called *incremental* or *small-signal* parameters. This means they are valid for small ac voltages or currents around some operating point but less so for large variations in signal or for regions removed from the specified operating point. Also, such parameters are not closely related to *dc voltage* characteristics. For instance, a "static" plate resistance could be defined as the ratio of plate voltage to plate current. For the -3-V operating point chosen, the static plate resistance would be 280 divided by 10^{-3} or 2.8 MΩ. This is considerably different from the small-signal plate resistance determined previously which was 8500 ohms.

Amplifier Gain

The ratio of the variation in voltage across the load resistance to change in input voltage is defined as the gain of the amplifier. For the equivalent circuit shown in Fig. 42B, this ratio would be

$$A = \frac{e_o}{e_{in}}$$

In order to solve for the gain, the first step is to determine the incremental plate current. This is just the source voltage divided by the total resistance of the circuit mesh or

$$i_p = \frac{47.5 e_{in}}{10 + 8.5} \text{ mA}$$

The output voltage is then

$$e_o = i_p 10$$

and combining the two foregoing equations gives

$$A = \frac{e_o}{e_{in}} = \frac{(47.5)(10)}{10 + 8.5} = 25.67$$

It is somewhat inconvenient to have the input and output voltages defined with opposite polarities as shown in Fig. 42B. Therefore, the gain becomes negative as illustrated in the triangle in Fig. 42C. A triangle is the standard way of representing an amplifier stage in "block-diagram" form. The amplifier gain depends of course on the load resistance, R_L, and a general formula for the gain of the circuit of Fig. 42B is

$$A = \frac{-\mu R_L}{r_p + R_L}$$

Feedback

Being able to eliminate the equivalent circuit and use only one parameter such as the gain permits analysis of more complicated networks. A very important application occurs when part of the

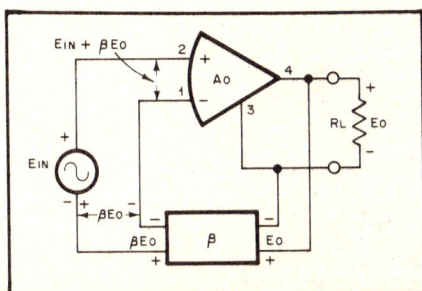

Fig. 44 — Network illustrating voltage feedback.

output energy of an amplifier is returned to the input circuit and gets amplified again. Since energy is being "fed back" into the input, the general phenomenon is called feedback. The manner in which feedback problems are analyzed is illustrated in Fig. 44. The output voltage is "sampled" by a network in the box marked beta and multiplied by this term. This transformed voltage then appears in series with the input voltage, e_{in} which is applied to the input terminals of the amplifier (triangle with A_o). A_o is defined as the open-loop gain. It is the ratio of the voltage that appears between terminals 3 and 4 when a voltage is applied to terminals 1 and 2. The circuit of Fig. 44 is an example of voltage feedback and a similar analysis holds for networks incorporating current feedback.

The closed-loop gain, A_c, can then be found by inspection of Fig. 44. From the diagram, the output voltage must be

$$e_o = A_o (e_{in} + \beta e_o)$$

rearranging terms gives

$$e_o (1 - \beta A_o) = A_o e_{in}$$

and the closed-loop gain is defined by

$$A_c = \frac{A_o}{1 - \beta A_o} = \frac{e_o}{e_{in}}$$

Cathode Bias

As an application of the feedback concept, consider the amplifier circuit shown in Fig. 45. It will be recalled that a bias battery was required in the previous example and a method of eliminating this extra source is to insert a small-valued resistor in series with the cathode lead to ground (Fig. 45A). In terms of the amplifier block diagram, the circuit of Fig. 45B results. The next task is to evaluate the open-loop gain and the value of β.

With the exception of the cathode resistor, the circuit of Fig. 45 is the same as that of Fig. 42. Consequently, the ac plate current must be

$$i_p = \frac{-\mu e_{12}}{r_p + R_L + R_c}$$

The open-loop gain can then be determined and is

$$\frac{e_o}{e_{12}} = A_o = \frac{-\mu R_L}{r_p + R_L + R_c}$$

Next, β is determined from the expression for output voltage

$$e_o = i_p R_L$$

and the feedback voltage which is

$$e_f = i_p R_c$$

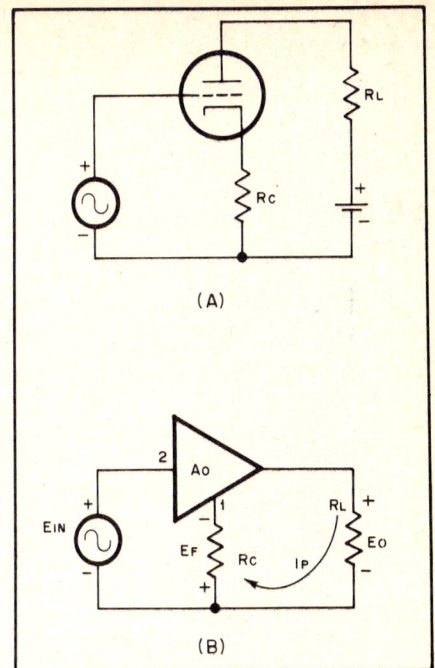

Fig. 45 — Feedback example of an amplifier with cathode bias.

β is then

$$\beta = \frac{e_f}{e_o} = \frac{i_p R_c}{i_p R_L} = \frac{R_c}{R_L}$$

Note that β is positive since if the path 1 to 2 is considered, the feedback voltage is *added* to the input signal. Substituting the values of β and A_o into the feedback equation gives

$$A_c = \frac{A_o}{1 - \frac{R_c}{R_L} A_o}$$

which after some manipulation becomes

$$A_c = \frac{-\mu R_L}{r_p + R_L + (1 + \mu) R_c}$$

Comparison of this equation with the one for the previous circuit with no cathode resistor reveals that the gain has *decreased* because of the term $(1 + \mu) R_c$ in the denominator. Such an effect is called *negative* or *degenerative* feedback.

On the other hand, if the feedback was such that the gain *increased, regenerative* or *positive* feedback would result. Positive feedback can be either beneficial or detrimental in nature and the study of feedback is an important one in electronics. For instance, frequency generation is possible in a circuit called an oscillator. But on the other hand, unwanted oscillation or *instability* in an amplifier is very undesirable.

Oscillators

A special case of feedback occurs if the term

$$1 - \beta A_o$$

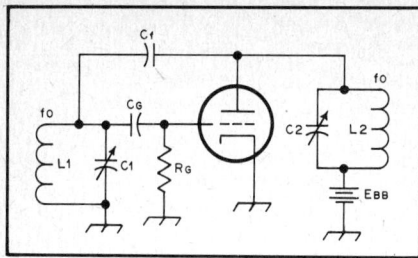

Fig. 46 — Tuned-plate tuned-grid oscillator.

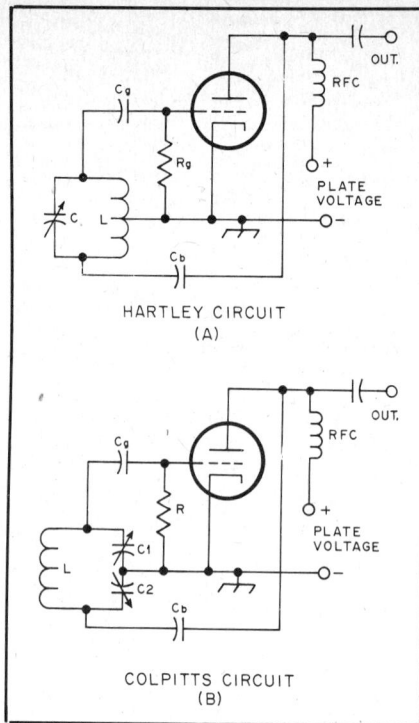

HARTLEY CIRCUIT
(A)

COLPITTS CIRCUIT
(B)

Fig. 47 — Hartley and Colpitts oscillators.

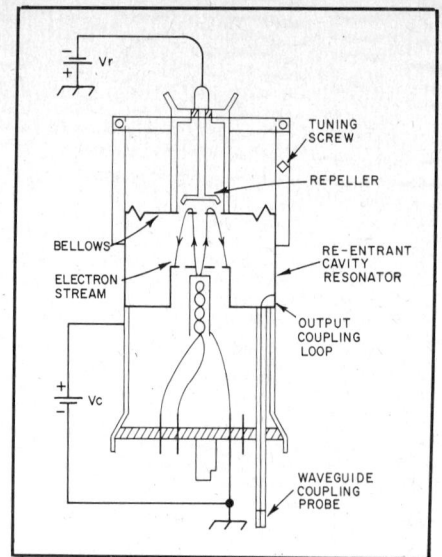

Fig. 48 — Cross-sectional view of a typical reflex klystron oscillator. Such types as the 732 may still be available on occasion in surplus sales.

becomes zero. This would mean the closed-loop gain would become infinite. An implication of this effect is that a very small input signal would be amplified and fed back and amplified again until the output voltage became infinite. Either that, or amplifier output would exist with *no* signal input. Random noise could "trigger" the input into producing output.

Of course, an infinite output voltage is a physical impossibility and circuit limitations such as the nonlinearities of the active device would alter the feedback equation. For instance, at high output voltage swings, the amplifier would either "saturate" (be unable to supply more current) or "limit" (be cutoff because the grid was too negative) and A_o would decrease.

Tuned-Plate Tuned-Grid Oscillator

It should be stressed that it is the *product* of βA_o that must be 1.0 for oscillations to occur. In the general case, both β and A_o may be complex numbers unlike those of the cathode-bias problem just discussed. That is, there is a phase shift associated with A_o and β with the phase shift of the product being equal to the sum of the individual phase shifts associated with each entity.

Therefore, if the total phase shift is 180 degrees and if the amplitude of the product is 1.0, oscillations will occur. At low frequencies, these conditions normally are the result of the effects of reactive components. A typical example is shown in Fig. 46 and the configuration is called a tuned-plate tuned-grid oscillator. If the input circuit consisting of L1 and C1 is tuned to a frequency f_o, with the output circuit (L2, C2) tuned to the same frequency, a high impedance to ground will exist at the input and output of the amplifier. Consequently, a small capacitance value represented by C_f is capable of supplying sufficient voltage feedback from the plate to the grid.

At other frequencies, or if either circuit is detuned, oscillations may not occur. For instance, off-resonant conditions in the output tank will reduce the output voltage and in effect, reduce the open-loop gain to the point where oscillations will cease. On the other hand, if the input circuit is detuned far from f_o, it will present a low impedance in series with the relatively high reactance of C_f. The

voltage divider thus formed will result in a small-valued β and the conditions for oscillations will not be fulfilled. However, for conditions near f_o, both the amplitude and phase of the βA_o product will be correct for oscillations to occur.

Under some conditions, the voltage across the tank circuit may be sufficient to cause the grid to be driven positive with respect to the cathode and grid current will flow through C_g. During the rest of the rf cycle, C_g will discharge through R_g causing a negative bias voltage to be applied to the grid. This bias voltage sets the operating point of the oscillator and prevents excessive current flow.

Miscellaneous Oscillator Circuits

Two other common type of oscillators are shown in Fig 47. In Fig. 47A, feedback voltage is applied across a tapped inductor while in Fig 47B, the voltage is applied across a capacitor instead. Quite often, a tuned plate circuit is not employed and an rf choke coil provides a high impedance load instead.

So-called "conventional" components such as tubes, transistors, ICs, resistors, inductors and capacitors are suitable up to and including the uhf range. However, at higher frequencies and for higher power levels in the uhf range, physical restrictions on the size of such components make them impractical. Consequently, a different approach is required. All the components necessary for a particular application may be included in the active device itself. This is true in the klystron oscillator shown in Fig. 48. Here the feedback action takes place inside of the

tube and in the electron stream. Electrons emitted from the cathode are accelerated and "modulated" on the first pass through the cavity resonator (which replaces the conventional tuned circuit used at lower frequencies). The electrons are then turned around by the repeller electrode and pass through the cavity again. On entering the cavity, the phase of the ac field there is such that the stream is retarded. However, this means that energy must be given up to the cavity and on out to the external circuit. As a result, the oscillations in the cavity are sustained.

Similar effects are employed in other microwave oscillators and amplifiers. Motional energy in the electron stream is transferred to a desired ac field. In doing so, dc energy in the power supply is converted to useful ac energy at the microwave frequency.

Glossary of Radio Terms

It is not unusual for an inexperienced radio amateur to be confused by some of the terms which are taken for granted by those who have been involved with electronics for many years. Many of the "strange" words which are found in this *Handbook* and other amateur publications are listed here for the convenience of the layman in the radio field. Although this glossary is far from being all-inclusive, it does contain the most-used words which may cause confusion. A complete dictionary of modern electronics words and expressions is available to those wishing to complement the radio library — *IEEE Standard Dictionary of Electrical and Electronics Terms* by John Wiley & Sons, Inc., New York, NY 10016.

The Terms

active — As used in *active filter* or *active device:* A device or circuit which requires an operating voltage. (See *passive*.)

analog — A term used in computer work, meaning a system which operates with numbers represented by directly measurable quantities (analog readout-mechanical dial system. See *digital*).

attenuator — A passive network that reduces the power level of a signal without introducing appreciable distortion.

balun — Balanced to unbalanced-line transformer.

bank wound — Pertaining to a coil (inductor) which has two or more layers of wire, each being wound over the top of the preceeding one. (See *solenoid*.)

bandpass — A circuit or component characteristic which permits the passage of a single band of frequencies while attenuating those frequencies which lie above and below that frequency band.

band-reject — A circuit or device which rejects a specified frequency band while passing those frequencies which lie above and below the rejected band (opposite of *band-pass*). Sometimes called "band-rejection," as applied to a filter.

bandwidth — The frequency width of circuit or component, such as a bandpass filter or tuned circuit. Usually measured at the half-power points of the response curve (-3 dB points).

base loading — Applies to vertical antennas for mobile and fixed-location use; an inductance placed near the ground end of a vertical radiator to change the electrical length. With variations the inductor aids in impedance matching.

bifilar — Two conducting elements used in parallel; two parallel wires wound on a coil form, as one example.

bilateral — Having two symmetrical sides or terminals; a filter (as one example) which has a 50-ohm characteristic at each port, with either port suitable as the input or output one.

bias — To influence current to flow in a specified direction by means of dc voltage; forward bias on a transistor stage, or grid bias on a tube type of amplifier.

binary — Relating to two logical elements; a system of numbers having two as its base.

bit — An abbreviation of a binary digit; a unit of storage capacity. Relates primarily to computers.

blanker — A circuit or device which momentarily removes a pulse or signal so that it is not passed to the next part of a circuit; a noise blanker. Not to be confused with a *clipper*, which clips part of a pulse or waveform.

bridge — An electrical instrument used for measuring or comparing inductance, impedance, capacitance or resistance by comparing the ratio of two opposing voltages to a known ratio; to place one component in parallel with another; to join two conductors or components by electrical means.

broadband — A device or circuit that is broadband has the capability of being operated over a broad range of frequencies. A broadband antenna is one example.

byte — A sequence of adjacent binary digits operated upon as a unit — usually shorter than a word.

cascade — One device or circuit which directly follows another; two or more similar devices or circuits in which the output of one is fed to the input of the succeeding one (tandem).

cascode — Two-stage amplifier having a grid-driven (or common-emitter or source) input circuit and a grounded-grid (or common base or gate) output circuit.

chip — Slang term for an integrated circuit, meaning a chip of semiconductor material upon which an IC is formed.

clamp — A circuit which maintains a predetermined characteristic of a wave at each occurrence so that the voltage or current is "clamped" or held at a specified value.

clipper — A device or circuit which limits the instantaneous value of a wave form or pulse to a predetermined value (see *blanker*).

closed loop — A signal path which includes a forward route, a feedback path, and a summing point which provides a closed circuit. In broad terms, an amplifying circuit which is providing voltage or power gain while being terminated correctly at the input and output ports, inclusive of feedback.

cold end — The circuit end of a component which is connected to ground or is bypassed for ac or rf voltage (the grounded end of a coil or capacitor).

common-mode signal — The instantaneous algebraic average of two signals applied to a balanced circuit, both signals referred to a common reference.

composite — Made up of a collection of distinct components; a complete ("composite") circuit rather than a discrete part of an overall circuit.

conversion loss/gain — Relating to a mixer circuit from which less output energy is taken than is supplied at the input-signal port (loss); when a mixer delivers greater signal output than is supplied to the input-signal port (gain).

converter — A circuit used to convert one frequency to another frequency. In a receiver the converter stage *converts* the incoming signals to the imtermediate frequency.

core — An element made of magnetic material, serving as part of a path for magnetic flux.

damping — A progressive reduction in the amplitude of a wave with respect to time (usually referenced to microseconds or milliseconds); a device or network added to a circuit to "damp" unwanted oscillations.

decay time — The period of time during which the stored energy or information "decays" to a specified value less than its initial value, such as the discharge time of a timing network.

decibel (dB) — One tenth of a bel. The number of decibels denotes the ratio of two amounts of *power* being 10 times the logarithm to the base 10 of this ratio. Also, the number of decibels denoting the ratio of two amounts of *voltage* being 20 times the logarithm to the base 10 of this ratio.

decoder — A device used for decoding an encoded message. One such circuit would be a decoder used for decoding the output signal of a Touch-Tone pad.

differential amplifier — An amplifier that has an output signal which is proportional to the algebraic difference between two input signals (sometimes called a "difference amplifier").

digital — Relating to data which is rendered in the form of digits; digital readout or display (see *analog*).

diplexing — The simultaneous transmission or reception of two signals while using a common antenna, made possible by using a "diplexer." Used in TV broadcasting to transmit visual and aural carriers by means of a single antenna.

discrete — A single device or circuit (a transistor as opposed to an IC) (see *composite*).

dish — An antenna reflector for use at vhf and higher which has a concave shape. For example, a part of a sphere or paraboloid.

Doppler — The phenomenon evidenced by the change in the observed frequency of a wave in a transmission system caused by a time rate of change in the effective length of the path of travel between the source and the point of observation.

drift — A change in componentor circuit parameters over a period of time.

drive — Rf energy applied at the input of an rf amplifier (rf driving power or voltage).

dummy load — A dissipative but essentially nonradiating device having impedance characteristics simulating those of the substituted device.

duplex — Simultaneous two-way independent transmission and reception in both directions.

duplexer — A device which permits simultaneous transmission and reception of related signal energy while using a common antenna (see *diplexing*)

dynamic range — Difference in dB or dBm between the overload level and minimum discernible signal level (MDS) in a system, such as a receiver. Parameters include desensitization point and distortion products as referenced to the receiver noise floor.

EME — Earth-moon-earth. Communications carried on by bouncing signals off the lunar surface. Commonly referred to as moonbounce.

empirical — Not based on mathematical

design procedures; experimental endeavor during design or modification of a circuit. Founded on case-history experience or intuition.

enabling — The preparation of a circuit for a subsequent function (enabling pulse or signal).

encoder — A device for enabling a circuit; to express a character or message by means of a code while using an encoder. Using a tone or tones to activate a repeater, as one example, in which case a Touch-Tone pad could be the encoder.

excitation — Signal energy used to drive a transmitter stage (see *drive*). Voltage applied to a component to actuate it, such as the field coil of a relay.

Faraday rotation — Rotation of the plane of polarization of an electromagnetic wave when traveling through a magnetic field. In space communications this effect occurs when signals traverse the ionosphere.

feedback — A portion of the output voltage being fed back to the input of an amplifier. Description includes ac and dc voltage which can be used separately or together, depending on the particular circuit.

feedthrough — Energy passing through a circuit or component, but not usually desired. A type of capacitor which can be mounted on a chassis or panel wall to permit *feeding through* a dc voltage while bypassing it to ground at ac or rf. Sometimes called a "coaxial capacitor."

ferromagnetic — Material which has a relative permeability greater than unity and requires a magnetizing force. (Ferrite and powdered-iron rods and toroids).

finite — Having a definable quantity; a finite value of resistance or other electrical measure.

flip-flop — An active circuit or device which can assume either of two stable states at a given time, as dictated by the nature of the input signal.

floating — A circuit or conductor which is above ac or dc ground for a particular reason. Example: A floating ground bus which is not common to the circuit chassis.

gate — A circuit or device, depending upon the nature of the input signal, which can permit the passage or blockage of a signal or dc voltage.

GDO — Abbreviation for a grid-dip oscillator (test instrument). Correct for only a tube-type of dip meter.

ground loop — A circuit-element condition (pc-board conductor, metal chassis or metal cabinet wall) which permits the unwanted flow of ac from one circuit point to another.

half-power point — The two points on a response curve which are 3 dB lower in level than the peak power. Sometimes called the "3 dB bandwidth."

Hall effect — The change of the electric conduction caused by the component of the magnetic field vector normal to the current density vector, which instead of being parallel to the electric field, forms an angle with it.

high end — Refers generally to the "hot" (rf or dc) end of a component or circuit; the end opposite the grounded or bypassed end (see *cold end*).

high level — The part of a circuit which is relatively high in power output and consumption as compared to the small-signal end of a circuit. Example: A transmitter PA stage is the high-level amplifier, as might be the driver also.

high-pass — Related mainly to filters or networks which are designed to pass energy above a specified frequency, but attenuate or block the passage of energy below that frequency.

high-Z — The high-impedance part of a circuit; a high-impedance microphone; a high-impedance transformer winding.

hot end — see *high end*.

hybrid — A combination of two generally unlike things; a circuit which contains transistors and tubes, for example.

ideal — A theoretically perfect circuit or component; a lossless transformer or device that functions without any faults.

insertion loss — That portion of a signal, current or voltage which is lost as it passes through a circuit or device. The loss of power through a filter or other passive network.

interpolate — To estimate a value between two known values.

leakage — The flow of signal energy beyond a point at which it should not be present. Example: Signal leakage across a filter because of poor layout (stray coupling) or inadequate shielding.

linear amplification — The process by which a signal is amplified without altering the characteristic of the input waveform. Class A, AB and B amplifiers are generally used for linear amplification.

load — A circuit or component that receives power; the power which is delivered to such a circuit or component. Example: A properly matched antenna is a load for a transmitter.

loaded — A circuit is said to be *loaded* when the desired power is being delivered to a load.

logic — Decision-making circuitry of the type found in computers.

long wire — A horizontal wire antenna which is one wavelength or greater in size. A long piece of wire (less than one wavelength) does not qualify as a long wire.

low end — See *cold end*.

low level — Low-power stage or stages of a circuit as referenced to the higher-power stages (see *high level*).

low pass — A circuit property which permits the passage of frequencies below a specified frequency, but atte-

nuates or blocks those frequencies above that frequency (see *high pass*).

low-Z — Low impedance (see *high-Z*).

mean — A value between two specified values; an intermediate value.

master oscillator — The primary oscillator for controlling a transmitter or receiver frequency. Can be a VFO (variable-frequency oscillator), VXO (variable crystal oscillator), PTO (permeability-tuned oscillator), PLL (phase locked loop), LMO (linear master oscillator) or frequency synthesizer.

modulation index — The ratio of the frequency deviation of the modulated wave to the frequency of the modulating signal.

narrowband — A device or circuit that can be operated only over a narrow range of frequencies. Low-percentage bandwidth.

network — A group of components connected together to form a circuit which will conduct power, and in most examples effect an impedance match. Example: An LC matching network between stages of a transistorized transmitter.

noise figure — Of a two-port transducer the ratio of the total noise power to the input noise power, when the input termination is at the standard temperature of 290 K.

nominal — A theoretical or designated quantity which may not represent the actual value. Sometimes referred to as the "ball-park value."

op amp — Operational amplifier. A high-gain, feedback-controlled amplifier. Performance is controlled by external circuit elements. Most op amps are used as dc amplifiers.

open loop — A signal path which does not contain feedback (see *closed loop*).

parameter — The characteristic behavior of a device or circuit, such as the operating characteristics of a 2N5109 transistor.

parametric amplifier — Synonym for "reactance amplifier." An inverting parametric device for amplifying a signal without frequency translation from input to output.

parasitic — Unwanted condition or quantity, such as *parasitic oscillations* or *parasitic capacitance*; additional to the desired characteristic.

passive — Operating without an operating voltage. Example: An LC filter which contains no amplifiers, or a diode mixer.

PEP — Peak envelope power; maximum amplitude that can be achieved with any combination of signals.

permeability — A term used to express relationships between magnetic induction and magnetic force.

pill — Slang expression for a transistor or an IC.

PL — Private line, such as a repeater which is accessed by means of a specified tone.

PLL — Phase-locked loop type of oscillator.

port — The input or output terminal of a circuit or device.

prototype — A first full-scale working version of a circuit design.

Q_L — Loaded Q of a circuit.

Q_u — Unloaded Q of a circuit.

quagi — An antenna consisting of both full-wavelength loops (quad) and Yagi elements.

resonator — A general term for a high-Q resonant circuit, such as an element of a filter.

return — That portion of a circuit which permits the completion of current flow, usually to ground — a "ground return."

ringing — The generation of an audible or visual signal by means of oscillation or pulsating current; the annoying sound developed in some audio filters when the Q is extremely high.

ripple — Pulsating current. Also, the gain depressions which exist in the flat portion of a bandpass response curve (above the -3 dB points on the curve). Example: Passband ripple in the nose of an i-f filter response curve.

rise time — The time required for a pulse or waveform to reach the peak value from some smaller specified value.

rms — Root mean square. The square root of the mean of the square of the voltage or current during a complete cycle.

rotor — A moving rotary component within a rotation-control device. Not to be confused with a *rotator*, which is the total assembly.

saturation — A condition which exists when a further change in input produces no additional output (a saturated amplifier).

selectivity — A measure of circuit capability to separate the desired signal from those at other frequencies.

shunt — A device placed in parallel with or across part of another device. Examples: Meter shunts, shunt-fed ver-

tical antennas and a capacitor placed (shunted) across another capacitor.

solenoidal — A single-layer coil of wire configured to form a long cylinder.

spectral purity — An emission which contains essentially the desired signal component, with all mixing products and harmonics attenuated greatly.

standing-wave ratio — The ratio of the amplitude of a standing wave at an antinode to the amplitude at a node. The SWR of a uniform feed line is expressed as

$$\frac{1 + \rho}{1 - \rho}$$

where ρ is the reflection coefficient.

strip — General term for two or more stages of a circuit which in combination perform a particular function. Examples: A local-oscillator strip, an audio strip or an i-f strip.

subharmonic — A frequency that is an integral *submultiple* of a frequency to which it is referred. A misleading term which implies that subharmonic energy can be created along with harmonic energy (not true). More aptly, a 3.5-MHz VFO driving a 40-meter transmitter, with 3.5-MHz leakage at the output, qualifying as a subharmonic.

tank — A circuit consisting of inductance and capacitance, capable of storing electrical energy over a band of frequencies continuously distributed about a single frequency at which the circuit is said to be resonant, or tuned.

toroidal — Doughnut-shaped physical format, such as a toroid core.

transducer — A device which is used to transport energy from one system (electrical, mechanical or acoustical) to another. Example: A loudspeaker or phonograph pickup.

transceiver — A combination transmitter and receiver which uses some parts of the circuit for both functions.

transverter — A converter which permits

transmitting and receiving at a specified frequency apart from the capability of the transceiver to which it is connected as a basic signal source. Example: A 2-meter transverter used in combination with an hf-band transceiver.

Transmatch — An LC network used to effect an impedance match between a transmitter and a feed line to an antenna. Not an "antenna tuner" or "antenna coupler."

trap — A device consisting of L and C components which permits the blockage of a specified frequency while allowing the passage of other frequencies. Example: A wave trap or an antenna trap.

trifilar — Same as *bifilar*, but with three parallel conductors.

trigger — To initiate action in a circuit by introducing an energy stimulus from an external source, such as a scope trigger.

U — Symbol for unrepairable assembly, such as an integrated circuit. (U1, U2, etc.)

unloaded — The opposite condition of *loaded*.

varactor — A two-terminal semiconductor device (diode) which exhibits a voltage-dependent capacitance. Used primarily as a tuning device or frequency multiplier at vhf and uhf.

VCO — Voltage-controlled oscillator. Uses tuning diodes which have variable dc applied to change their junction capacitances.

VSWR — Voltage standing-wave ratio. (See *SWR*.)

VU — Volume Unit.

VXO — Variable crystal oscillator.

Zener diode — Named after the inventor. A diode used to regulate voltage or function as a *clamp* or *clipper*.

Z — Symbol for a device or circuit which contains two or more components. Example: A parasitic suppressor which contains a resistor and an inductor in parallel (Z1, Z2, etc.). Z is also the symbol for impedance.

Chapter 4

Solid-State Fundamentals

The electrical characteristics of solid-state devices such as diodes and transistors are dependent upon phenomena that take place at the atomic level. While semiconductors can be employed without a complete knowledge of these effects, some understanding is helpful in various applications. Electrons, which are the principal charge carriers in both vacuum tubes and semiconductors, behave much differently in either of the two circumstances. In free space, an electron can be considered as a small charged solid particle. On the other hand, the presence of matter affects this picture greatly. For instance, an electron attached to an atom has many properties similar to those of rf energy in tuned circuits. It has a frequency and wavelength that depend upon atomic parameters just as the frequency associated with electrical energy in a tuned circuit depends upon the values of inductance and capacitance.

A relation between the energy of an electron in an atomic "orbit" and its associated frequency is given by

$$f(Hz) = \frac{E \text{ (joules)}}{6.625 \times 10^{-27}}$$

where the constant in the denominator is called Planck's constant. This equation is

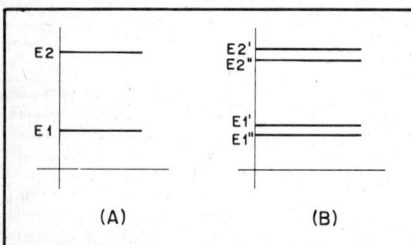

Fig. 1 — Energy-level diagram of a single atom is shown at A. At B, the levels split when two atoms are in close proximity.

quite important when an electron is either raised or falls between two different energy "states." For instance, when an electron drops from one level to a lower one, energy is emitted in the form of electromagnetic radiation. This is the effect that gives the characteristic glow to neon tubes, mercury-vapor rectifiers, and even light-emitting diodes. The frequency of the emitted radiation is given by the foregoing formula where E is the difference in energy. However, if an electron receives enough energy such that it is torn from an atom, a process called ionization is said to occur (although the term is also loosely applied to transitions between any two levels). If the energy is divided by the charge of the electron (-1.6×10^{19} coulombs), the equivalent in voltage is obtained.

A common way of illustrating these energy transitions is by means of the energy-level diagram shown in Fig. 1A. It should be noted that unlike ordinary graphical data, there is no significance to the horizontal axis. In the case of a single atom, the permitted energy can only exist at discrete levels (this would be characteristic of a gas at low pressure where the atoms are far apart). However, if a single atom is brought within close proximity of another one of similar type, the single energy levels split into pairs of two that are very close together (Fig. 1B). The analogy between tuned circuits and electron energy levels can be carried even further in this case.

Consider the two identical circuits that are coupled magnetically as shown in Fig. 2A. Normally, energy initially stored in C1 would oscillate back and forth between L1 and C1 at a single frequency after the switch was closed. However, the presence of the second circuit consisting of L2 and C2 (assume L1 equals L2 and C1 equals C2) results in the waveform shown in Fig. 2B. Energy also oscillates back and forth

between the two circuits and the current then consists of components at two slightly different frequencies. The effect is similar to the splitting of electron energy levels when two atoms are close enough to interact.

Conductors, Insulators and Semiconductors

Solids are examples of large numbers of atoms in close proximity. As might be expected, the splitting of energy levels continues until a band structure is reached. Depending upon the type of atom, and the physical arrangement of the component atoms in the solid, three basic conditions can exist. In Fig. 3A, the two discrete energy levels have split into two bands. All the states in the lower band are "occupied" by electrons while the ones in the higher energy band are only partially filled.

In order to impart motion to an electron, the expenditure of energy is required. This means an electron must then be raised from one energy state to a higher one. Since there are many permitted states in upper level of Fig. 3A that are both unoccupied and close together, electrons in this level are relatively free to move about. Consequently, the material is a conductor. In Fig. 3B, all the states in the lower level are occupied, there is a big gap between this level and the next higher one, and the upper level is empty. This means if motion is to be imparted to an electron, it must be raised from the lower level to the upper one. Since this requires considerable energy, the material is an insulator. (The energy-level representation gives an insight into the phenomena of breakdown. If the force on an electron in an insulator becomes high enough because of an applied field, it can acquire enough energy to be raised to the upper level. When this happens, the material goes into a conducting state.)

A third condition is shown in Fig. 4. In

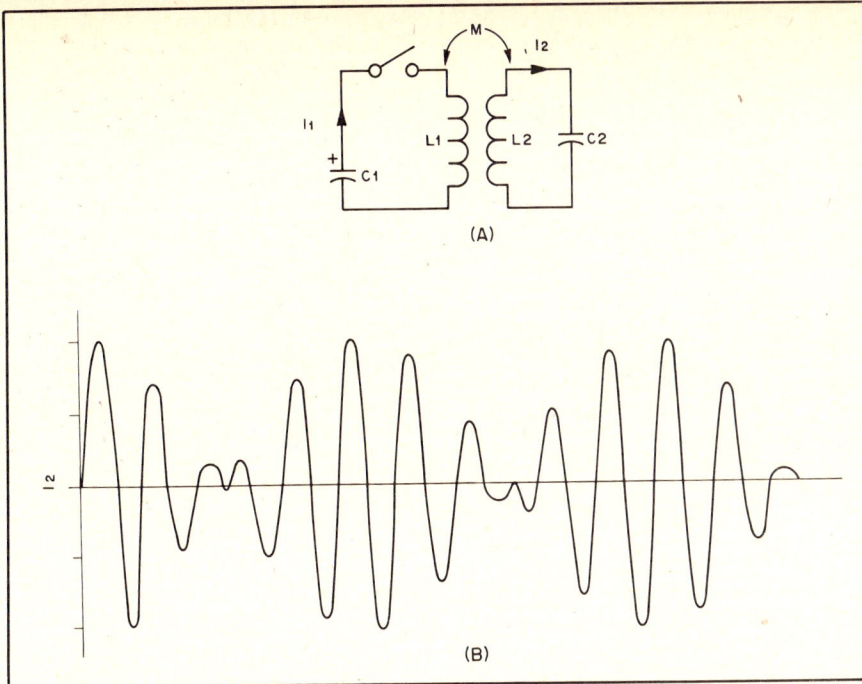

Fig. 2 — Electrical-circuit analog of coupled atoms.

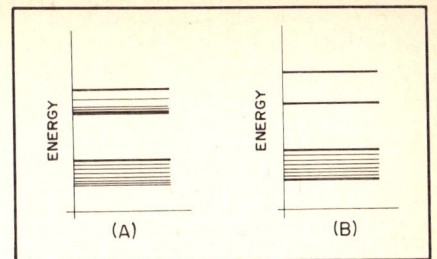

Fig. 3 — The energy level of a conductor is illustrated at A. A similar level for an insulator is depicted at B.

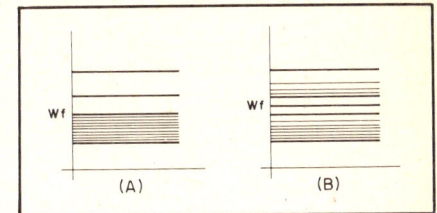

Fig. 4 — Semiconductor energy-level representation.

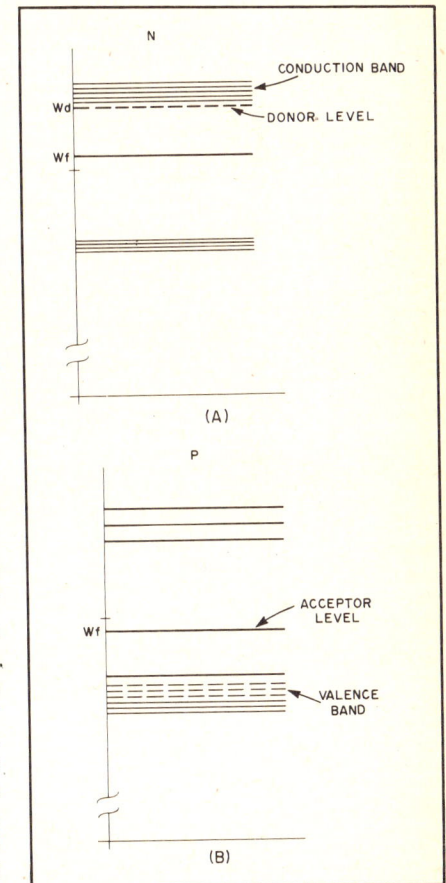

Fig. 5 — The effects on the energy level if impurity atoms are introduced.

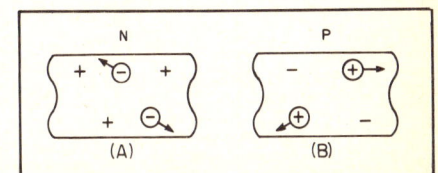

Fig. 6 — N- and p-type semiconductors.

the material associated with this diagram, the upper level is unoccupied but is very close to the occupied one. Hence, under conditions where the random electron motion is low (low temperature), the material acts as an insulator (Fig. 4A). However, as the random or thermal motion increases, some electrons acquire enough energy to move up to states in the upper level. Consequently, both levels are partially occupied as shown in Fig. 4B. The line marked W_f represents a statistical entity related to the "average" energy of electrons in the material and is called the Fermi Level. At absolute zero (no thermal motion), W_f is just at the top of the lower energy level. As electrons attain enough energy to move to the upper level, W_f is approximately halfway between the two levels.

The PN Junction

The material for the diagram shown in Fig. 4 is called an intrinsic semiconductor and examples are the elements germanium and silicon. As such, the materials do not have any rectifying properties by themselves. However, if certain elements are mixed into the intrinsic semiconductor in trace amounts, a mechanism for rectification exists. This is shown in Fig. 5A. If an element with an occupied energy level such as arsenic is introduced into germanium, a transformation in conductivity takes place. Electrons in the new occupied level are very close to the upper partially filled band of the intrinsic germanium. Consequently, there are many extra charge carriers available when thermal energy is sufficient to raise some of the electrons in the new level to the partially filled one. Germanium with an

excess of mobile electrons is called an n-type semiconductor.

By introducing an element with an empty or unoccupied energy level near the lower partially filled level (such as boron), a somewhat different transformation in conductivity occurs. This is shown in Fig. 5B. Electrons from the lower level can move into the new unoccupied level if the thermal energy is sufficient. This means there is an excess of *unoccupied* states in the germanium lower energy level. Germanium treated this way is called a p-type semiconductor.

A physical picture of both effects is shown in Fig. 6. The trace elements or impurities are spread throughout the intrinsic crystal. Since the distance of separation is much greater for atoms of the trace elements than it is for ones of the intrinsic crystal, there is little interaction between the former. Because of this lack of "coupling," the distribution of energy states is a single level rather than a band. In Fig. 6A, atoms of the trace element are represented by the + signs since they have lost an electron to the higher energy level. Consequently, such elements are called donors. In Fig. 6B, the impurity atoms that have "trapped" an electron in the new state are indicated by the − signs. Atoms of this type are called acceptor impurities.

While it is easy to picture the extra free electrons by the circled "minus" charges in Fig. 6A, a conceptual difficulty exists with the freed "positive" charges shown in Fig. 6B. In either case, it is the motion of electrons that is actually taking place and the factor that is responsible for any current. However, it is convenient to consider that a positive charge carrier exists called a hole. It would seem as though a dislocation in the crystal-lattice structure was moving about and contributing to the total current.

If a section of n-type material is joined to another section made from p-type, a one-way current flow results. This is shown in Fig. 7. A positive potential applied to the p-type electrode attracts any electrons that diffuse in from the n-type end. Likewise, holes migrating from the p-type end into the n-type electrode are attracted to the negative

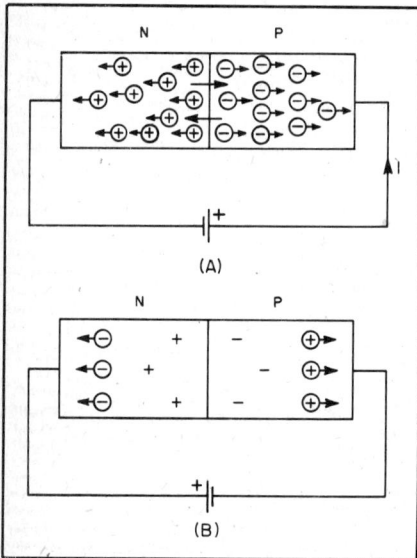

Fig. 7 — Elementary illustration of current flow in a semiconductor diode.

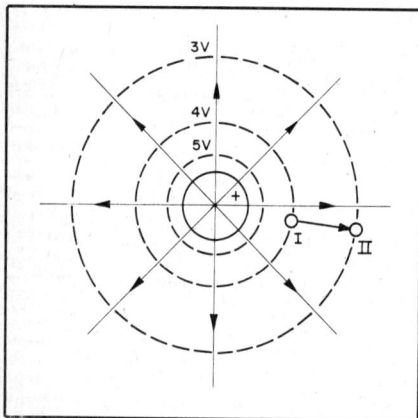

Fig. 8 — Potential diagram of an electron in atomic orbit.

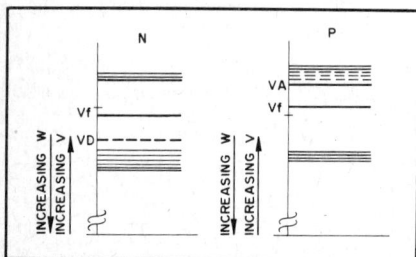

Fig. 9 — Energy-level diagram in terms of potential.

terminal. Note that the diagram indicates not all the carriers reach the terminals. This is because some carriers combine with ones of the opposite sign while enroute. In the case of a diode, this effect doesn't present much of a problem since the total current remains the same. Other carriers take the place of those originally injected from the opposite regions. However, such recombination degrades the performance of transistors considerably and will be discussed shortly.

If a voltage of the opposite polarity to that of a Fig. 7A is applied to the terminals, the condition in Fig. 7B results. The mobile charge carriers migrate to each end as shown leaving only the fixed charges in the center near the junction. Consequently, little current flows and the pn junction is "back biased." It can be seen that the pn junction constitutes a diode since current can flow readily only in one direction. While this simple picture suffices for introductory purposes, proper treatment of many important effects in semiconductors requires a more advanced analysis than the elementary model affords. Returning to Figs. 3, 4 and 5, it would be convenient if the diagrams were in terms of voltage rather than energy. As pointed out earlier, the relation between energy and voltage associated with an electron is given by

$$W = eV = (-1.6 \times 10^{-19})V$$

Because the electron has been assigned a minus charge, a somewhat upside-down world results. However, if it is kept in mind that it requires the *expenditure* of energy to move an electron from a point of *higher* potential to one at a *lower* value, this confusion can be avoided. As an illustration, suppose an electron is moved from an atomic orbit indicated by I in Fig. 8 to orbit II. This would mean the electron would have had to been moved against the force of attraction caused by the positive nucleus resulting in an *increase* in potential energy. (In other words, orbit II is at a higher energy level than orbit I.) However, note that the electrostatic potential around the nucleus decreases with distance and that orbit II is at a lower potential than orbit I.

Consequently, the energy-level diagram in terms of voltage becomes inverted as shown in Fig. 9. It is now possible to approach the problem of the pn junction diode in terms of the energy-level diagrams presented previously. If a section of n-type and p-type material is considered separately, the respective energy (or voltage) levels would be the same. However, if the two sections were joined together and connected by an external conductor as shown in Fig. 10, a current would flow initially. This is because the voltage corresponding to the statistical entity referred to previously (Fermi Level)

is not the same for p- and n-type materials at the same temperature. At the Fermi Level, the probability that a particular energy state is occupied is one half. For n-type material, the Fermi Level is shifted upward toward the "conduction band" (Fig. 5A). In a p-type material, it is shifted downward toward the "valence band." Although the theory behind the Fermi Level and definitions concerning the conduction and valence bands won't be dealt with here, it is sufficient to know that the band structure shifts so that the Fermi Levels are the same in both parts of the joined sections (Fig. 10).

The reasoning behind this effect is as follows. Consider conditions for hole flow only for the moment. Since there is an excess of holes in the p region (Fig. 10), there is a tendency for them to move over into the adjacent n region because of diffusion. The process of diffusion is demonstrated easily. If a small amount of dye is dropped into some water, it is concentrated in a small area at first. However, after a period of time has passed, it spreads out completely through the entire volume.

Once the holes diffuse into the n region, they recombine with the electrons present and produce a current in the external terminals denoted by I_D (Fig. 10). But a paradox results because of this current. If S_1 is opened so that I_D flows through R, where does the energy that is transferred (irreversibly) to this resistance come from? In effect, it represents a perpetual-motion dilemma or else the semiconductor will cool down since the diffusion process is the result of a form of thermal motion. Both conclusions are against the laws of physics, so a third alternative is necessary. It is then assumed that the Fermi Levels align so that the potential across the terminals becomes zero, and no current will flow in the external circuit.

However, if the Fermi Levels are the same, the conduction and valence bands in either section will no longer align. As a consequence, a difference in potential between the two levels exists and is indicated by V_B in Fig. 10. *The formation of this junction or barrier voltage is of prime importance in the operation of pn-junction devices.* Note that holes in the p region must overcome the barrier voltage which impedes the flow of the diffusion current. It will also be recalled that both holes and electrons were generated in the intrinsic semiconductor because of thermal effects (Fig. 4B). The addition of either donor or acceptor atoms modifies this effect somewhat. If donor atoms are present (n-type material), fewer holes are generated. On the other hand, if acceptor atoms represent the impurities, fewer electrons are generated in comparison to conditions in an intrinsic semiconductor. In the case of p-type material, holes predominate and are termed the majority carriers. Since there are fewer electrons in p-type material, they are termed the minority carriers.

Referring to Fig. 10A, there are some holes in the n region (indicated by the + signs) because of the foregoing thermal effects. Those near the junction will experience a force caused by the electric field associated with the barrier voltage. This field will produce a flow of holes into the p region and the current is denoted by I_T. Such a current is called a drift current as compared to the diffusion current I_D. Under equilibrium conditions, the two currents are equal and just cancel each other. This is consistent with the assumption that no current flows in the external circuit because of the fact that the Fermi Levels are the same and no voltage is produced.

So far, only conditions for the holes in the upper (or conduction) band have been considered, but identical effects take place with the motion of electrons in the lower energy band (valence band). Since the flow of charge carriers is in opposition, but because holes and electrons have opposite signs, the currents *add*.

The Forward-Biased Diode

If an external emf is applied to the diode terminals as shown in Fig. 10B, the equilibrium conditions no longer exist and the Fermi Level voltage in the right-hand region is shifted upward. This means the barrier voltage is decreased and considerable numbers of carriers may now *diffuse* across the junction. Consequently, I_D becomes very large while I_T decreases in value because of the decrease in barrier voltage. The total current under "forward-bias" conditions then becomes

$$I = I_s \left(e^{\frac{qV_x}{kt}} - 1 \right)$$

where

q = 1.6×10^{-19} coulombs (the fundamental charge of an electron),
k = 1.38×10^{-23} joules/Kelvin (Boltzmann's constant),
t = junction temperature in Kelvins,
e = 2.718 (natural logarithmic base)
V_x = applied emf, and I_s = reverse-bias saturation current.

This equation is discussed in greater detail in the section dealing with common silicon diodes.

The Reverse-Biased Diode

If the source, V_x, is reversed as shown in Fig. 10C, the barrier voltage is

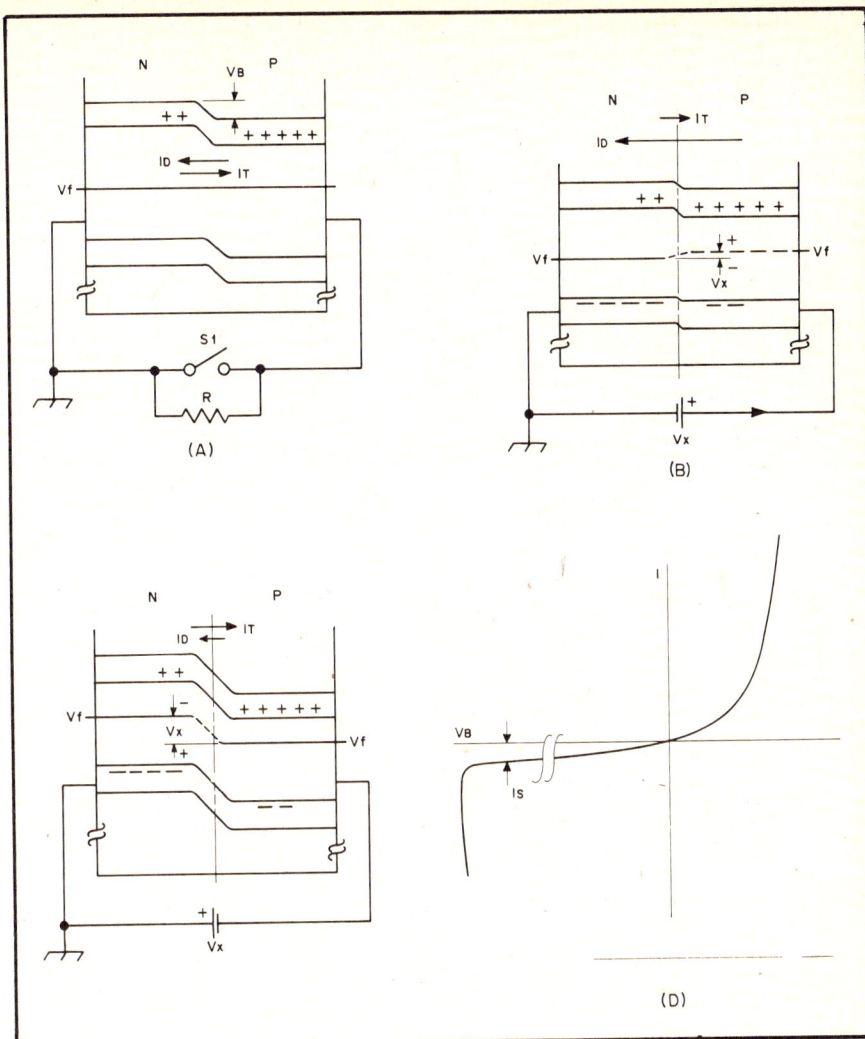

Fig. 10 — Energy-level diagrams for unbiased (A), forward-biased (B), and reversed-biased diode (C). Illustration D shows the resultant characteristics of the diode.

increased. Consequently, charge carriers must overcome a large "potential hill" and the diffusion current becomes very small. However, the drift current caused by the thermally generated carriers returns to the value it had under equilibrium conditions. For large values of V_x, the current approaches I_s, defined as the reverse saturation current, I_s is the sum of I_T and its counterpart in the lower or "valence" band. Finally, the characteristic curves of the forward- and reversed-bias diode can be constructed and are shown in Fig. 10D.

It is obvious that I_s should be as small

as possible in a practical diode since it would only degrade rectifier action. Also, since it is the result of the generation of thermal carriers, it is quite temperature sensitive which is important when the diode is part of a transistor. If the reverse voltage is increased further, an effect called avalanche breakdown occurs as indicated by the sudden increase in current at V_b. In such an instance, the diode might be damaged by excessive current. However, the effect is also useful for regulator purposes and devices used for this purpose are called Zener diodes.

Semiconductor Devices and Practical Circuits

The radio amateur may be more interested in the practical aspects of semiconductors than he or she might be in the theoretical considerations that are covered in the previous section. This portion of the chapter provides a practical treatment of how diodes, transistors and ICs perform in actual circuits. Greater coverage of the chemistry and in-depth theory of solid-state devices can be obtained by reading some of the texts referenced throughout the chapter. *Solid State Design for the Radio Amateur* (an ARRL publication) is recommended as a primer on basic semiconductor theory. The book also contains many pages of proven construction projects which use semiconductor devices.

Semiconductor Diodes

The vacuum-tube diode has been replaced in modern equipment designs. Semiconductor diodes are more efficient because they do not consume filament power. They are much smaller than tube diodes. In low-level applications they operate cooler than tubes do. Solid-state diodes are superior to tube types with respect to operating frequency. The former are suitable into the microwave region: Most vacuum-tube diodes are not practical at frequencies above 50 MHz.

Semiconductor diodes fall into two main categories, structurally. Although they can be made from silicon or germanium crystals, they are usually classified as p-n *junction* diodes or *point-contact* diodes. These formats are illustrated in Fig. 12. Junction diodes are used from dc to the microwave region, but point-contact diodes are intended primarily for rf applications: the internal capacitance of a point-contact diode is considerably less than that of a junction diode designed for the same circuit application. As the operating frequency is increased the unwanted internal and external capacitance of a diode becomes more troublesome to the designer. Where a given junction type of diode may exhibit a capacitance of several picofarads, the point-contact device will have an internal capacitance of 1 pF or less.

Selenium Diodes

Power rectifiers made from selenium were in common use in ac power supplies before 1965. Today they are manufactured for replacement purposes only. Selenium diodes are characterized by high forward voltage drop (which increases with age) and high reverse leakage current. The voltage drop causes the device to dissipate power, and a typical rectifier stack has large cooling fins. An additional shortcoming of selenium rectifiers is that they sometimes emit toxic fumes when they burn out. When replacing selenium diodes with silicon units, be certain that the filter capacitors (and the entire equipment) can withstand the higher output voltage. Some early solar cells were made of selenium, but silicon devices have taken over this area, too.

Germanium Diodes

The germanium diode is characterized by a relatively large current flow when small amounts of voltage are applied in the "forward" direction (Fig. 11). Small currents will flow in the reverse (back) direction for much larger applied voltages. A representative curve is shown in Fig. 13. The dynamic resistance in either the forward or back direction is determined by the change in current that occurs, at any given point on the curve, when the applied voltage is changed by a small amount. The forward resistance will

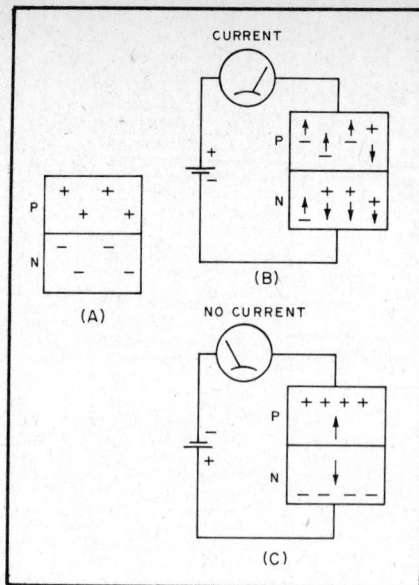

Fig. 11 — A p-n junction (A) and its behavior when conducting (B) and nonconducting (C).

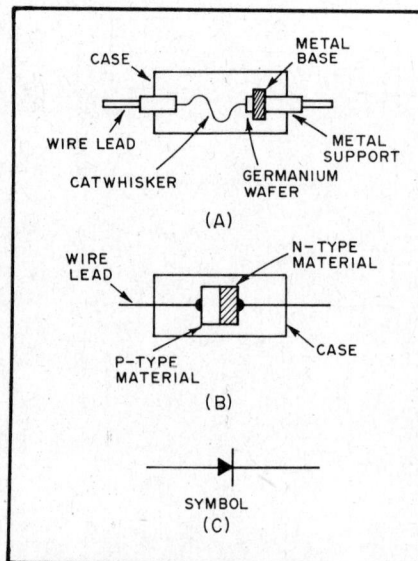

Fig. 12 — A point-contact type of diode is seen at A. A junction diode is depicted at B and the diode symbol is at C.

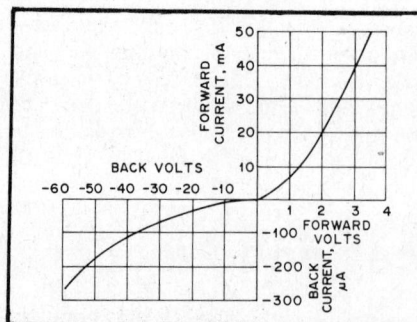

Fig. 13 — Typical point-contact diode (germanium) characteristic curve. Because the back current is much smaller than the forward current, a different scale is used for back voltage and current.

vary somewhat in the region of very small applied voltages. However, the curve is mostly straight, indicating a relatively constant dynamic resistance. For small applied voltages the resistance is on the order of 200 ohms or less. The back resistance exhibits considerable variation and is dependent upon the specific applied voltage during the test. It may vary from a few thousand ohms to well over a megohm. The back resistance of a germanium diode is considerably lower than that of a silicon diode. The latter is greater than a megohm in most instances, but the germanium diode is normally less than a megohm.

Common Silicon Diodes

Apart from the fact that silicon p and n materials are used in the formation of a silicon junction diode, the characteristics of these devices are similar to those of germanium diodes. The voltage/current curves of Fig. 13 are representative.

The junction barrier voltage for silicon diodes is somewhat higher (approximately 0.7 volt) than that of a germanium diode. The latter is on the order of 0.3 volt. The majority of the diodes in use today fall into the silicon class. They are rugged and reliable from rf small-signal applications to dc power use.

Silicon diodes are available in ratings of 1000 volts (PRV) or greater. Many of these diodes can accommodate dc in excess of 100 amperes. The primary rule in preventing damage to any diode is to operate the device within the maximum ratings specified by the manufacturer. The device temperature is one of the important parameters. Heat sinks are used with diodes that must handle large amounts of power, thereby holding the diode junction temperature at a safe level.

The behavior of junction diodes under varying temperatures is of interest to designers of circuits that must perform over some temperature range. The relationship between forward bias current, forward bias voltage and temperature is defined by the classic *diode equation:*

$$I_f = I_s \left(e^{\frac{qV}{kt}} - 1 \right)$$

where q is the fundamental electronic charge (1.6×10^{-19} coulombs), V is the bias potential, k is Boltzmann's constant (1.38×10^{-23} joules/Kelvin), (Kelvin = °Celsius + 273), t is the junction temperature in Kelvins, I_s is the reverse-bias saturation current, I_f is the forward-bias current, and e is the natural logarithmic base (2.718). The ratio q/k is approximately 11,600, so the diode equation can be written:

$$I_f = I_s \left(e^{\frac{11,600V}{t}} - 1 \right) \qquad \text{(Eq. 1)}$$

It is useful to have an expression for the voltage developed across the junction when the forward current is held constant. To obtain such an expression we must solve the diode equation for V. Expanding the right side of Eq. 1 yields:

$$I_f = I_s e^{\frac{11,600V}{t}} - I_s \qquad \text{(Eq. 2)}$$

Adding I_s to both sides gives:

$$I_f + I_s = I_s e^{\frac{11,600V}{t}} \qquad \text{(Eq. 3)}$$

Dividing through by I_s produces:

$$\frac{I_f}{I_s} + 1 = e^{\frac{11,600V}{t}} \qquad \text{(Eq. 4)}$$

which implies

$$\frac{11,600V}{t} = \ln\left(\frac{I_f}{I_s} + 1\right) \qquad \text{(Eq. 5)}$$

Multiplying each term by $\frac{t}{11,600}$ leaves:

$$V = \frac{t}{11,600} \ln\left(\frac{I_f}{I_s} + 1\right) \qquad \text{(Eq. 6)}$$

The undetermined quantity in Eq. 6 is I_s, the reverse saturation current. In ordinary silicon signal diodes this current approximately doubles with each 4.5 Kelvin-temperature increase. A mathematical expression for this behavior as a function of temperature is:

$$I_{s(t)} = 2I_{s(t-4.5)} \qquad \text{(Eq. 7)}$$

At room temperature (300 Kelvins), the reverse saturation current is on the order of 10^{-13} amperes. Eq. 7 describes a phenomenon similar to radioactive decay, where the 4.5-Kelvin current-doubling interval is analogous to the half-life of a radioactive substance. This equation with the given initial condition sets up an initial-value problem, the solution of which is:

$$I_{s(t)} = 10^{-13} e^{\frac{(t-300)\ln 2}{4.5}} \qquad \text{(Eq. 8)}$$

Substituting this expression for I_s into Eq. 6 produces the diode voltage drop as a function of temperature for a constant current:

$$V_{(t)} = \frac{t}{11,600} \times$$

$$\ln\left[\frac{I_f}{10^{-13} e^{\frac{(t-300)\ln 2}{4.5}}} + 1\right] \qquad \text{(Eq. 9)}$$

The temperature coefficient of the junction potential can be obtained from the partial derivative of V with respect to t,

but it's a simple matter (with the aid of a pocket calculator) to extract the information directly from Eq. 9. If the forward current is fixed at 1 milliampere, the diode drop at room temperature is 0.5955 volts. This potential decreases at an initial rate of 2 millivolts per Kelvin. The temperature coefficient gradually increases to 3 millivolts per Kelvin at 340 Kelvins. While the temperature curve isn't linear, it is gradual enough to be considered linear over small intervals. When the bias current is increased to 100 milliamperes, the room temperature junction potential increases to 0.7146 volts as might be expected, but the temperature coefficient stays well-behaved. The initial potential decrease is 1.6 millivolts per Kelvin, and this value increases to 2.5 millivolts per Kelvin at 340 Kelvins.

The significance of the very minor dependence of temperature coefficient on bias current is that it isn't necessary to use an elaborate current regulator to bias diodes used in temperature compensation applications. The equations defining the behavior of junction diodes are approximations. Some of the voltages were expressed to five significant figures so the reader can verify his calculations, but this much precision exceeds the accuracy of the approximations.

Diodes as Switches

Solid-state switching is accomplished easily by using diodes or transistors in place of mechanical switches or relays. The technique is not a complicated one at dc and audio frequencies when large amounts of power are being turned on and off, or transferred from one circuit point to another.

Examples of shunt and series diode switching are given in Fig. 14. The illustration at A shows a 1N914 rf-switching type of diode as a shunt on-off element between C1 and ground. When +12 volts are applied to D1 through R1, the diode saturates and effectively adds C1 to the oscillator tank circuit. R1 should be no less than 2200 ohms in value to prevent excessive current flow through the diode junction.

Series diode switching is seen in Fig. 14B. In this example the diode, D1, is inserted in the audio signal path. When S1 is in the ON position the diode current path is to ground through R2, and the diode saturates to become a closed switch. When S1 is in the OFF state R1 is grounded and +12 volts are applied to the diode cathode. In this mode D1 is back biased (cut off) to prevent audio voltage from reaching the transistor amplifier. This technique is useful when several stages in a circuit are controlled by a single mechanical switch or relay. Rf circuits can also be controlled by means of series diode switching.

A significant advantage to the use of

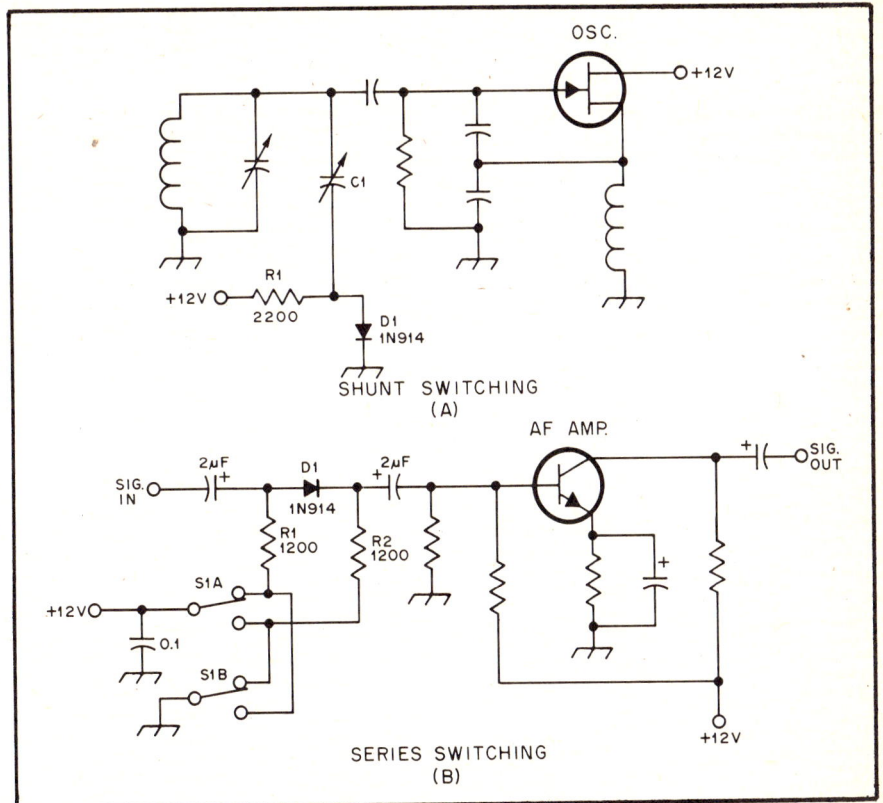

Fig. 14 — A silicon-switching diode, D1, is used at A to place C1 in the circuit. At B is seen a series switch with D1 in the signal path.

diode switching is that long signal leads are eliminated. The diode switch can be placed directly at the circuit point of interest. The dc voltage which operates it can be at some convenient remote point. The diode recovery time (switching speed) must be chosen for the frequency of operation. In other words, the higher the operating frequency the faster the switching speed required. For dc and audio applications one can use ordinary silicon power-supply rectifier diodes.

Diodes as Gates

Diodes can be placed in series with dc leads to function as gates. Specifically, they can be used to allow current to flow in one direction only. An example of this technique is given in Fig. 15A.

A protective circuit for the solid-state transmitter is effected by the addition of D1 in the 13.6-volt dc line to the equipment. The diode allows the flow of positive current, but there will be a drop of approximately 0.7 volt across the diode, requiring a supply voltage of 14.3. Should the operator mistakenly connect the supply leads in reverse, current will not flow through D1 to the transmitter. In this application the diode acts as a gate. D1 must be capable of passing the current taken by the transmitter, without over-heating.

A power type of diode can be used in shunt with the supply line to the transmitter for protective purposes. This method is illustrated in Fig. 15B. If the supply polarity is crossed accidentally, D1 will draw high current and cause F1 to open. This is sometimes referred to as a "crowbar" protection circuit. The primary advantage of circuit B over circuit A is that there is no voltage drop between the supply and the transmitter.

Diodes as Voltage References

Zener diodes are discussed later in this chapter. They are used as voltage references or regulators. Conventional junction diodes can be used for the same purposes by taking advantage of their barrier-voltage characteristics. The greater the voltage needed, the higher the number of diodes used in series. Some examples of this technique are given in Fig. 16. At A the diode (D1) establishes a fixed value of forward bias (0.7 V) for the transistor, thereby functioning as a regulator. R1 is chosen to permit a safe amount of current to flow through the diode junction while it is conducting at the barrier voltage.

The circuit of Fig. 16B shows two diodes inserted in the emitter return of a relay-driver transistor. D1 and D2 set up a cutoff voltage of approximately 1.4. This reduces the static current of the transistor when forward bias is not provided at the transistor base. If too much static current flows the relay may not drop out when the forward bias decays across the timing network. The more sensitive the relay the

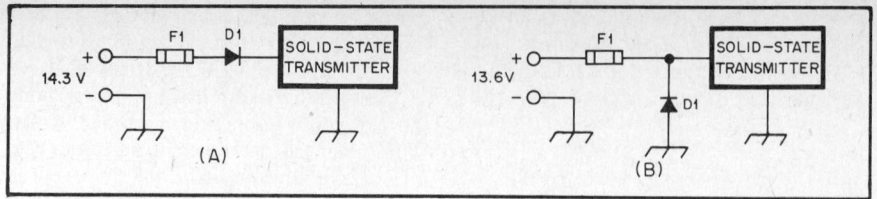

Fig. 15 — D1 at A protects the equipment if the supply leads are cross-polarized in error. At B the fuse will blow if the power supply is connected for the wrong polarity.

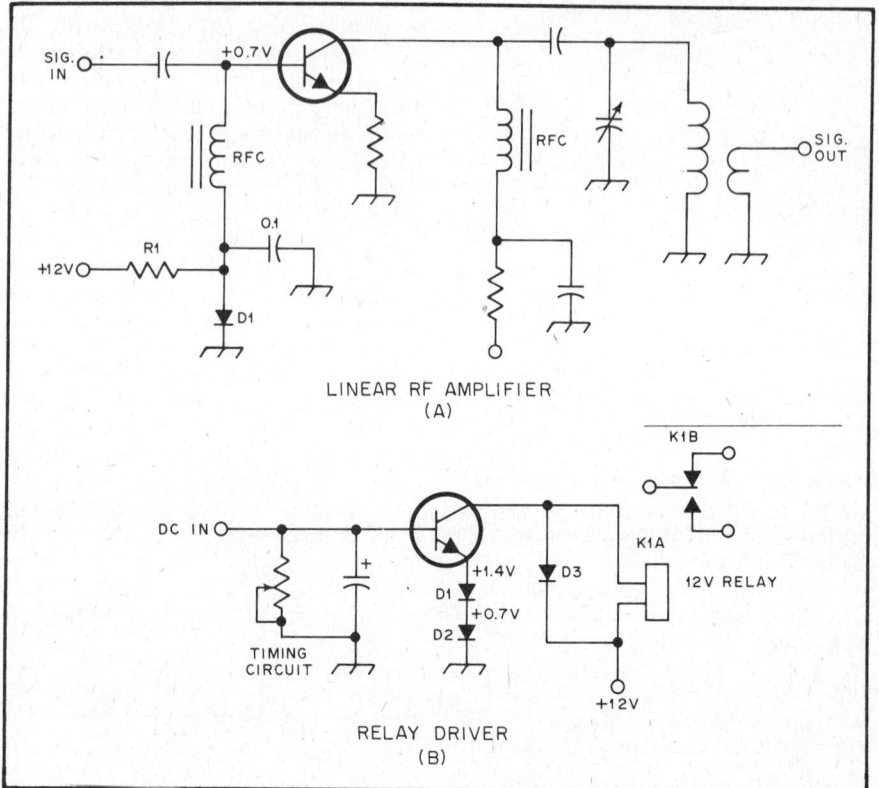

LINEAR RF AMPLIFIER
(A)

RELAY DRIVER
(B)

Fig. 16 — D1 establishes a 0.7-volt bias reference at A. Approximately 1.4 volts of emitter bias are established by connecting D1 and D2 in series at illustration B.

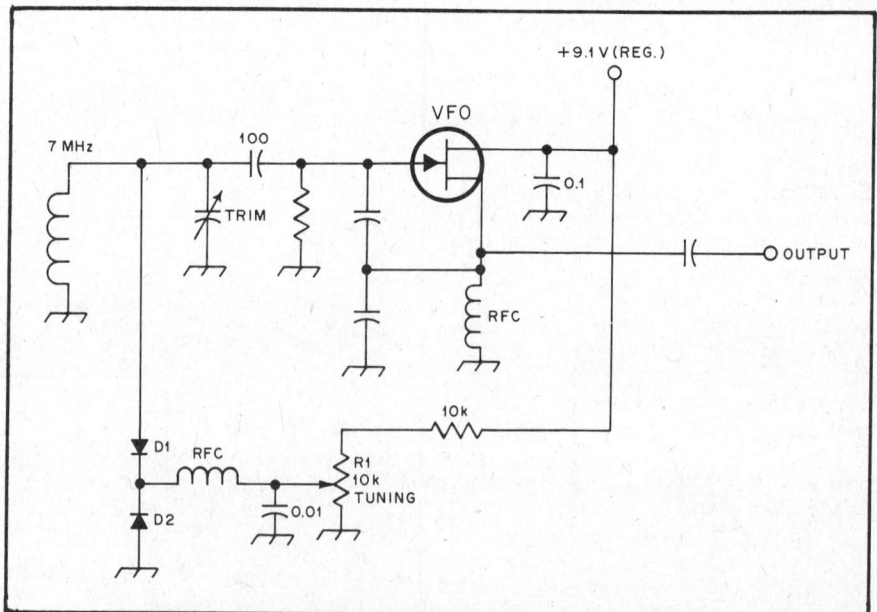

Fig. 17 — High-speed switching diodes of the 1N914 variety can be connected back to back and used as tuning diodes. As the reverse voltage is varied by means of R1, the internal capacitance of the diodes will change.

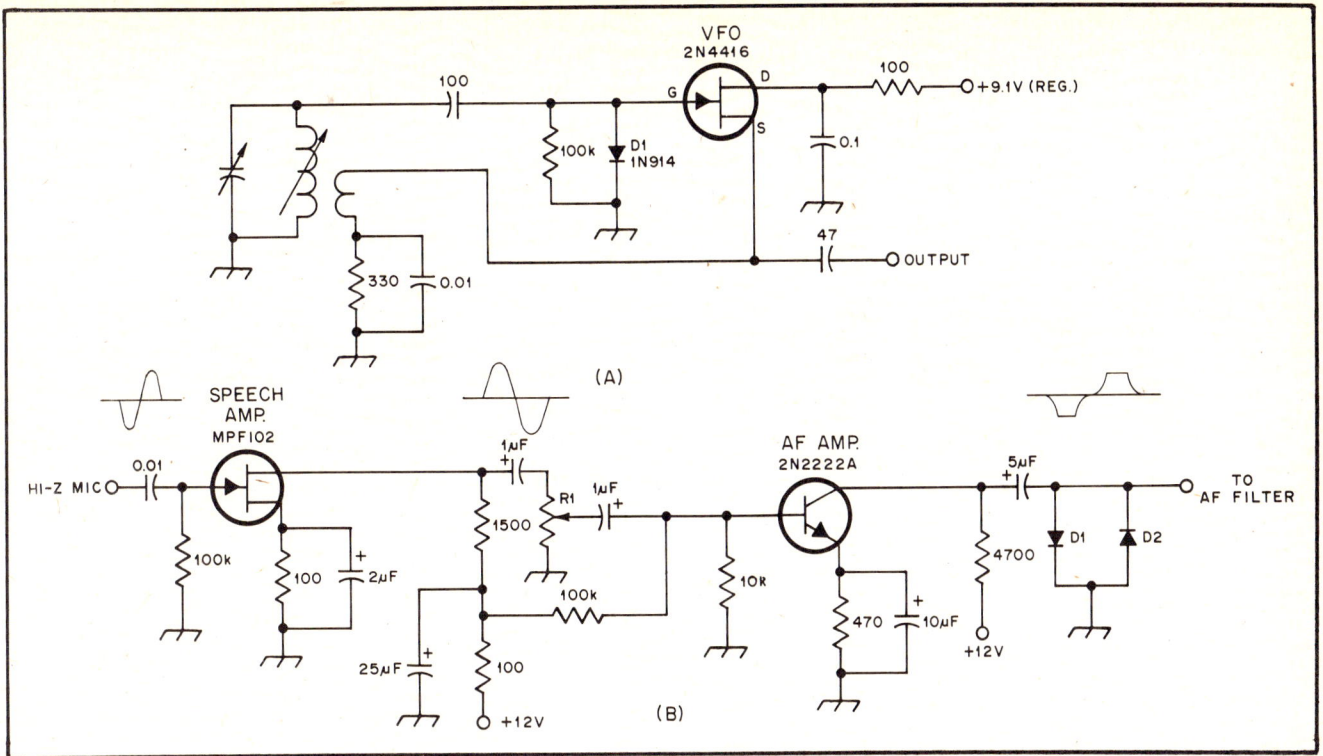

Fig. 18 — D1 serves as a bias stabilization device at A (see text). At B, D1 and D2 are employed as clippers to flatten the positive and negative af peaks. Clipping will occur at roughly 0.7 volt if silicon diodes are used. Audio filtering is required after the clipper to remove the harmonic currents caused by the diode action.

greater the chance for such a problem. D1 and D2 prevent relay dropout problems of this variety. D3 is used as a transient suppressor. A spike will occur when the relay coil field collapses. If the amplitude of the spike is great enough, the transient, while following the dc bus in a piece of equipment, can destroy transistors and diodes elsewhere in the circuit. In this application the diode (D3) can be regarded as a *clamp,* since it clamps the spike at approximately 0.7 volt.

Using Diodes as Capacitors

Later in this chapter there is a discussion about VVC (voltage-variable capacitor) diodes. They are known also as tuning diodes and Varicap diodes. It is possible, however, to use ordinary silicon diodes as voltage-variable capacitors. This is accomplished by taking advantage of the inherent changes in diode junction capacitance as the reverse bias applied to them is changed. The primary limitation in using high-speed switching diodes of the 1N914 variety is a relatively low maximum capacitance. At a sacrifice to low minimum capacitance, diodes can be used in parallel to step up the maximum available capacitance. An example of two 1N914 silicon diodes in a diode tuning circuit is given in Fig. 17. As R1 is adjusted to change the back bias on D1 and D2, there will be a variation in the junction capacitance. That change will alter the VFO operating frequency. The junction capacitance increases as the back bias is

lowered. In the circuit shown here the capacitance will vary from roughly 5 pF to 15 pF as R1 is adjusted. The diodes used in circuits of this kind should have a high Q and excellent high-frequency characteristics. Generally, tuning diodes are less stable than mechanical variable capacitors are. This is because the diode junction capacitance will change as the ambient temperature varies. This circuit is not well suited to mobile applications because of the foregoing trait.

Diode Clippers and Clamps

The previous mention of diode clamping action (D3 in Fig. 16) suggests that advantage can be taken of the characteristic barrier voltage of diodes to clip or limit the amplitude of a sine-wave. Although there are numerous applications in this general category, diode clippers are more familiar to the amateur in noise limiter, audio limiter and audio compressor circuits. Fig. 18 illustrates some typical circuits which employ small-signal diodes as clamps and clippers. D1 in Fig. 18A functions as a bias clamp at the gate of the FET. It limits the positive sine-wave swing at approximately 0.7 V. Not only does the diode tend to regulate the bias voltage, it limits the transconductance of the FET during the positive half of the cycle. This action restricts changes in transistor junction capacitance. As a result, frequency stability of the oscillator is enhanced and the generation of harmonic currents is greatly minimized.

The circuit of Fig. 18B shows how a pair of diodes can be connected in back-to-back fashion for the purpose of clipping the negative and positive sine-wave peaks in an audio amplifier. If germanium diodes are used at D1 and D2 (1N34As or similar) the audio will limit at roughly 0.3 V. With silicon diodes (1N914 or rectifier types) the voltage will not exceed 0.7 V. R1 serves as the clipping-level control. An audio gain control is normally used after the clipper filter, along with some additional gain stages. The output of the clipper must be filtered to restore the sine-wave if distortion is to be avoided. Diode clippers generate considerable harmonic currents, thereby requiring an RC or LC type of audio filter.

Diode Frequency Multipliers

Designers of rf circuits use small-signal diodes as frequency multipliers when they want to minimize the number of active devices (tubes or transistors) in a circuit. The primary disadvantage of diode multipliers is a loss in gain compared to that which is available from an active multiplier. Fig. 19 contains examples of diode frequency multipliers. The circuit at A is useful for obtaining odd or even multiples of the driving voltage. The efficiency of this circuit is not high, requiring that an amplifier be used after the diode multiplier in most applications. Resonator L1/C1 must be tuned to the desired output frequency.

A diode frequency doubler is seen at B in Fig. 19. It functions like a full-wave power-supply rectifier, where 60-Hz energy is transformed to 120-Hz by virtue of the diode action. This circuit will cause a loss of approximately 8 dB. Therefore, it is shown with a succeeding amplifier stage. If reasonable circuit balance is maintained, the 7-MHz energy will be down some 40 dB at the output of D1 and D2 — prior to the addition of L1 and C1. Additional suppression of the driving energy is realized by the addition of resonator L1/C1. T1 is a trifilar-wound toroidal transformer. At this frequency (7 MHz) a 0.5-inch diameter ferrite core (permeability of 125) will suffice if the trifilar winding contains approximately 10 turns. Additional information on this subject is given in *Solid State Design for the Radio Amateur*.

Diode Detectors and Mixers

Diodes are effective as detectors and mixers when circuit simplicity and strong-signal handling capability are desired. Impedance matching is an important design objective when diodes are used as detectors and mixers. The circuits are lossy, just as is the case with diode frequency multipliers. A diode detector or mixer will exhibit a conversion loss of 7 dB or more in a typical example. Therefore, the gain before and after the detector or mixer must be chosen to provide an acceptable noise figure for the overall circuit in which the diode stage is used. This is a particularly critical factor when diode mixers are used at the front end of a receiver. A significant advantage in the use of diode mixers and detectors is that they are broadband in nature, and they provide a wide dynamic range. Hot-carrier diodes are preferred by some designers for these circuits, but the 1N914 class of switching diodes provide good performance if they are matched for a similar resistance before being placed in the circuit.

Fig. 20 illustrates some examples of diode detectors. A basic a-m detector is seen at A. The circuit at B is that of a two-diode product detector. R1 and the two bypass capacitors serve as an rf filter to keep signal and BFO energy out of the following af amplifier stage. A four-diode product detector is illustrated at C. T1 is a trifilar-wound broadband transformer. The characteristic input impedance of T1 is 50 ohms. An rf filter follows this detector also. BFO injection voltage for the detectors at B and C should be between 8 and 10 volts pk-pk for best detector performance.

Circuits for typical diode mixers are given in Fig. 21. Product detectors are also mixers except for the frequencies involved. The output energy is at audio frequencies rather than at some rf intermediate frequency. The examples at A and B can be compared to those at C

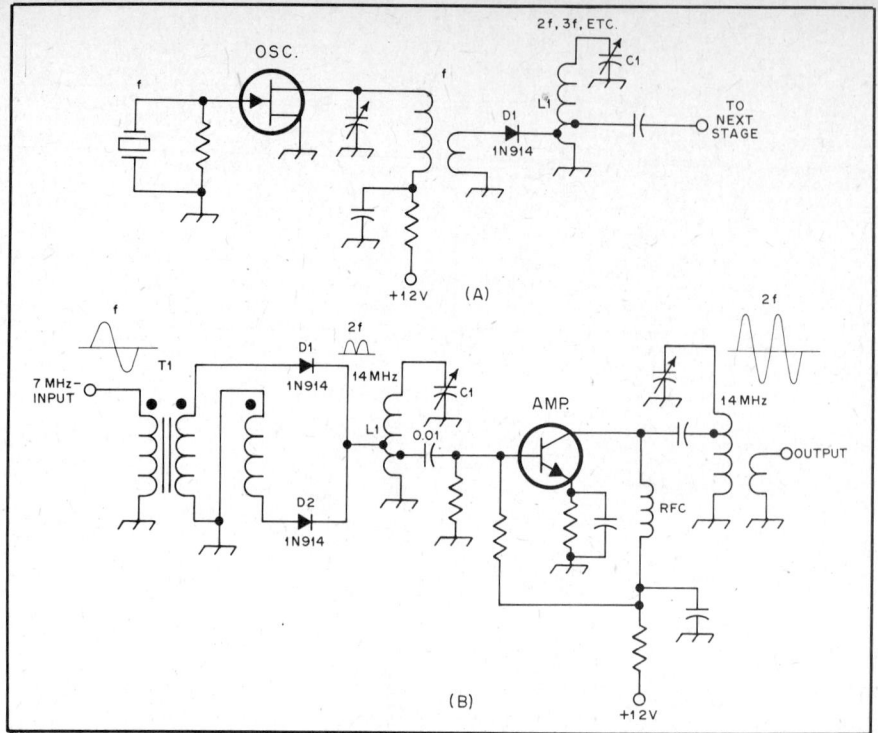

Fig. 19 — A simple diode frequency multiplier is shown at A. A balanced diode frequency doubler is seen at B. T1 is a trifilar-wound broadband toroid transformer.

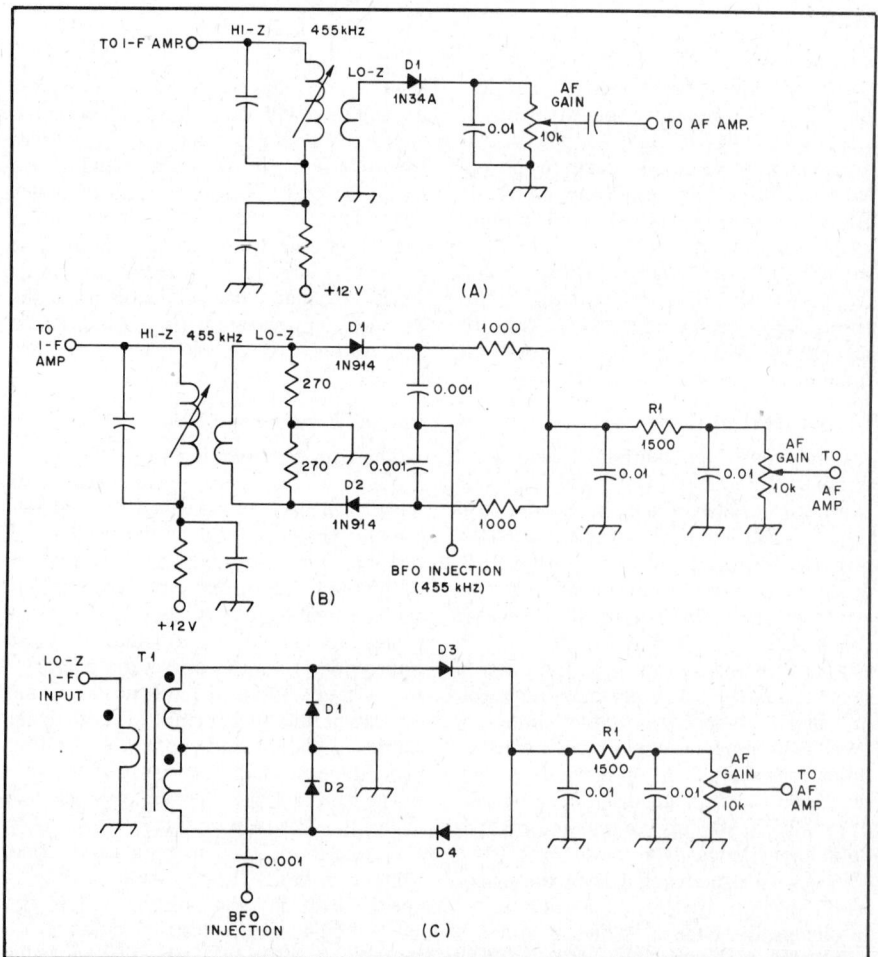

Fig. 20 — D1 at A is used as a simple a-m detector. Two versions of diode product detectors are illustrated at B and C. BFO injection for B and C should be approximately 10 volts pk-pk for best detector performance.

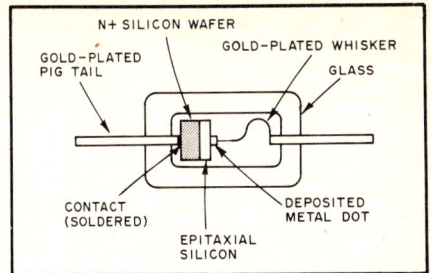

Fig. 22 — Cross-sectional representation of a hot-carrier diode (HCD).

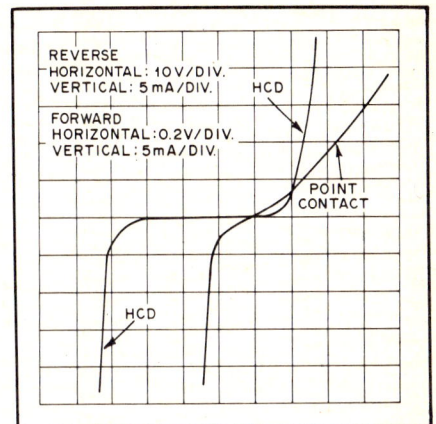

Fig 23 — Forward and reverse characteristics of a hot-carrier diode as compared to a p-n junction diode.

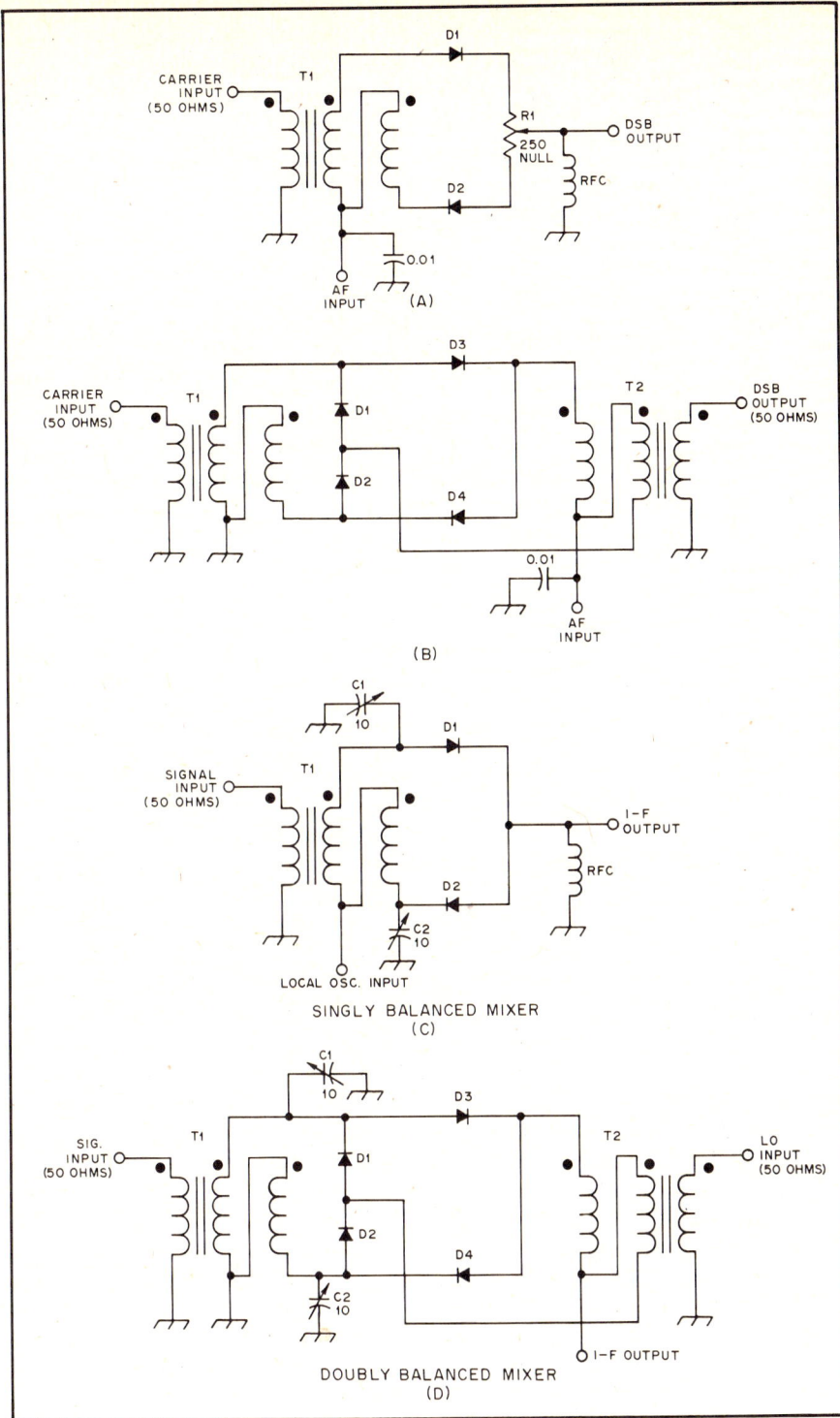

Fig. 21 — The examples at A and B are for use in balanced modulators. The similarity between these and balanced mixers is shown at C and D.

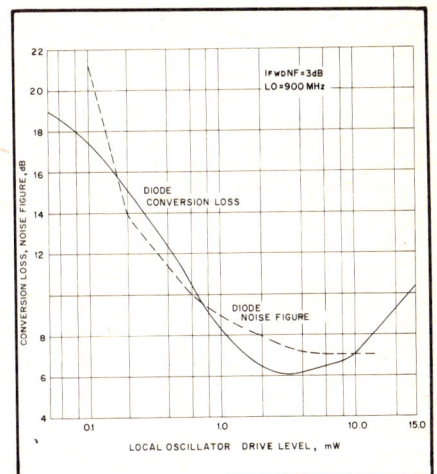

Fig. 24 — Noise figure and the conversion loss of a typical HCD that has no bias applied.

and D for the purpose of illustrating the similarity between balanced modulators and mixers. It is evident that product detectors, balanced modulators and mixers are of the same family. The diodes in all examples can be hot-carrier types or matched silicon switching diodes of the 1N914 class.

C1 and C2 in Fig. 21C and D are used for balancing purposes. They can be employed in the same manner with the circuits at A and B. The transformers in each illustration are trifilar-wound

toroidal types. They provide a broadband circuit characteristic.

Hot-Carrier Diodes

One of the more recent developments in the semiconductor field is the *hot-carrier diode,* or "HCD." It is a metal-to-semiconductor, majority-carrier conducting device with a single rectifying junction. The carriers are typically high-mobility electrons in an n type of semiconductor material. The HCD is particularly useful in mixers and detectors

at vhf and higher. Notable among the good features of this type of diode are its higher operating frequency and lower conduction voltage compared to a p-n junction diode such as the 1N914.

When compared to a point-contact diode, the HCD is mechanically and electrically superior. It has lower noise, greater conversion efficiency, larger square-law capability, higher breakdown voltage, and lower reverse current. The internal capacitance of the HCD is markedly lower than that of a p-n junction diode

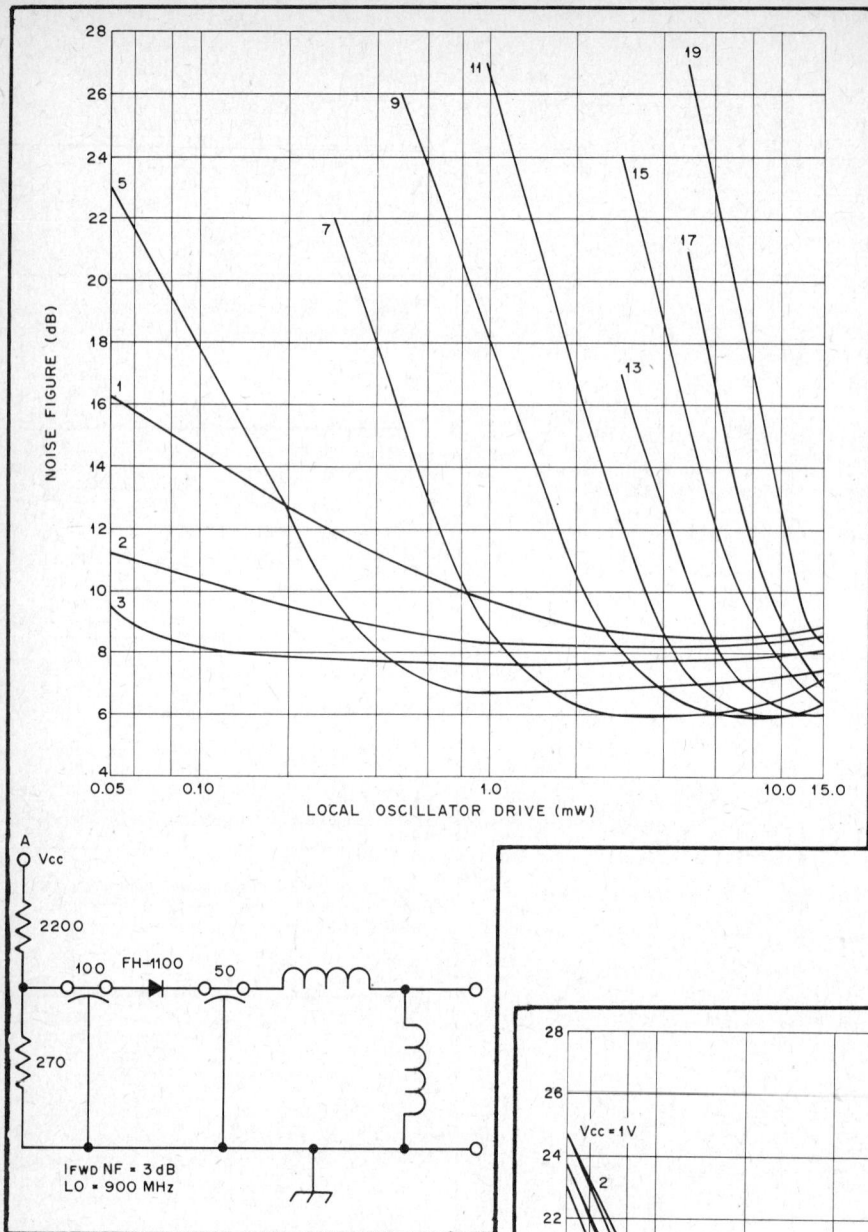

Fig. 25 — Curves for hot-carrier diode noise figure versus local-oscillator drive power. The bias currents are in mA as measured at point A in the representative test circuit.

versus LO drive for an HCD mixer are given in Fig. 26. The test circuit used for the curves of Fig. 25 applies. The curve numbers indicate mA measured at point A. Further information on hot-carrier diodes can be found in the Fairchild *Application Note APP-177* and in the Hewlett Packard *Application Note 907*.

Varactor Diodes

Mention was made earlier in this chapter of diodes being used as voltage-variable capacitors, wherein the diode junction capacitance can be changed by varying the reverse bias applied to the diode. Manufacturers have designed certain diodes for this application. They are called Varicaps (variable capacitor diodes) or varactor diodes (variable reactance diodes). These diodes depend upon the change in capacitance which occurs across their *depletion* layers. They are not used as rectifiers.

Varactors are designed to provide various capacitance ranges from a few picofarads to more than 100 pF. Each one has a specific minimum and maximum capacitance, and the higher the maximum amount the greater the minimum value. Therefore, the amateur finds it necessary to tailor his circuits for the midrange of the capacitance curve. Ideally, he will choose the most linear portion of the curve. Fig. 27A shows typical capacitance-voltage curves for three varactor diodes.

A representative circuit of a varactor diode is presented in Fig. 28. In this

and it is less subject to temperature variations.

Fig. 22 shows how the diode is structured internally. A typical set of curves for an HCD and a p-n junction diode are given in Fig. 23. The curves show the forward and reverse characteristics of both diode types.

Fig. 24 illustrates the noise figure and conversion loss of an HCD with no bias applied. When forward bias is applied to the diode, the noise figure will change from that which is seen in Fig. 24. Curves for various bias amounts are seen in Fig. 25. The numbers at the ends of the curves signify the amount of current (in mA) flowing into the test circuit at point A.

A set of curves showing conversion loss

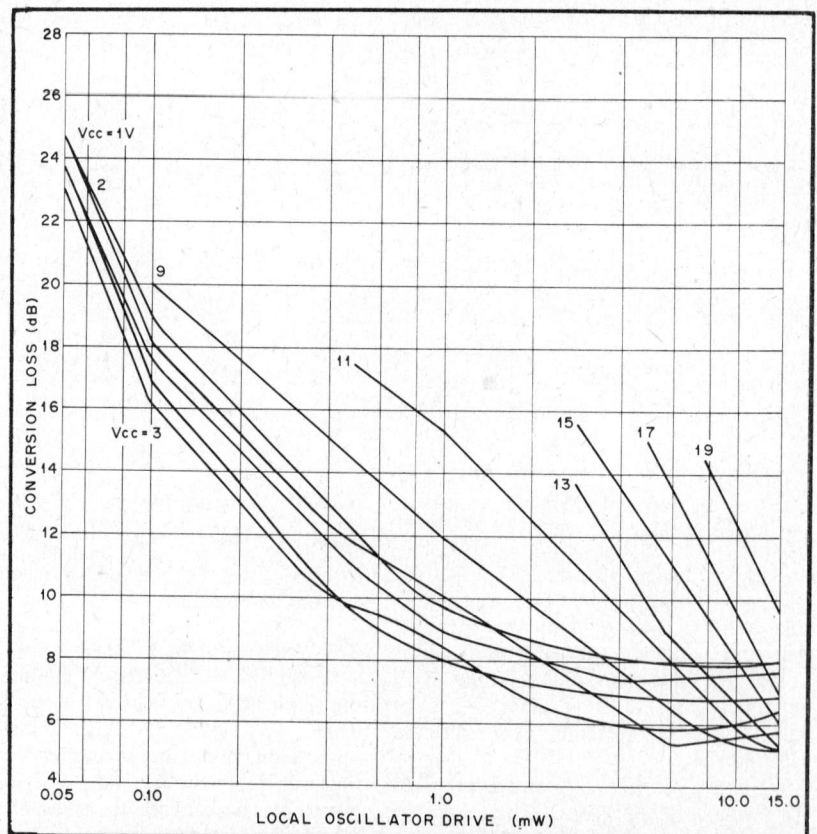

Fig. 26 — Local-oscillator drive power versus conversion loss for a specified bias amount. Bias currents are in mA as measured at point A of the circuit in Fig. 25.

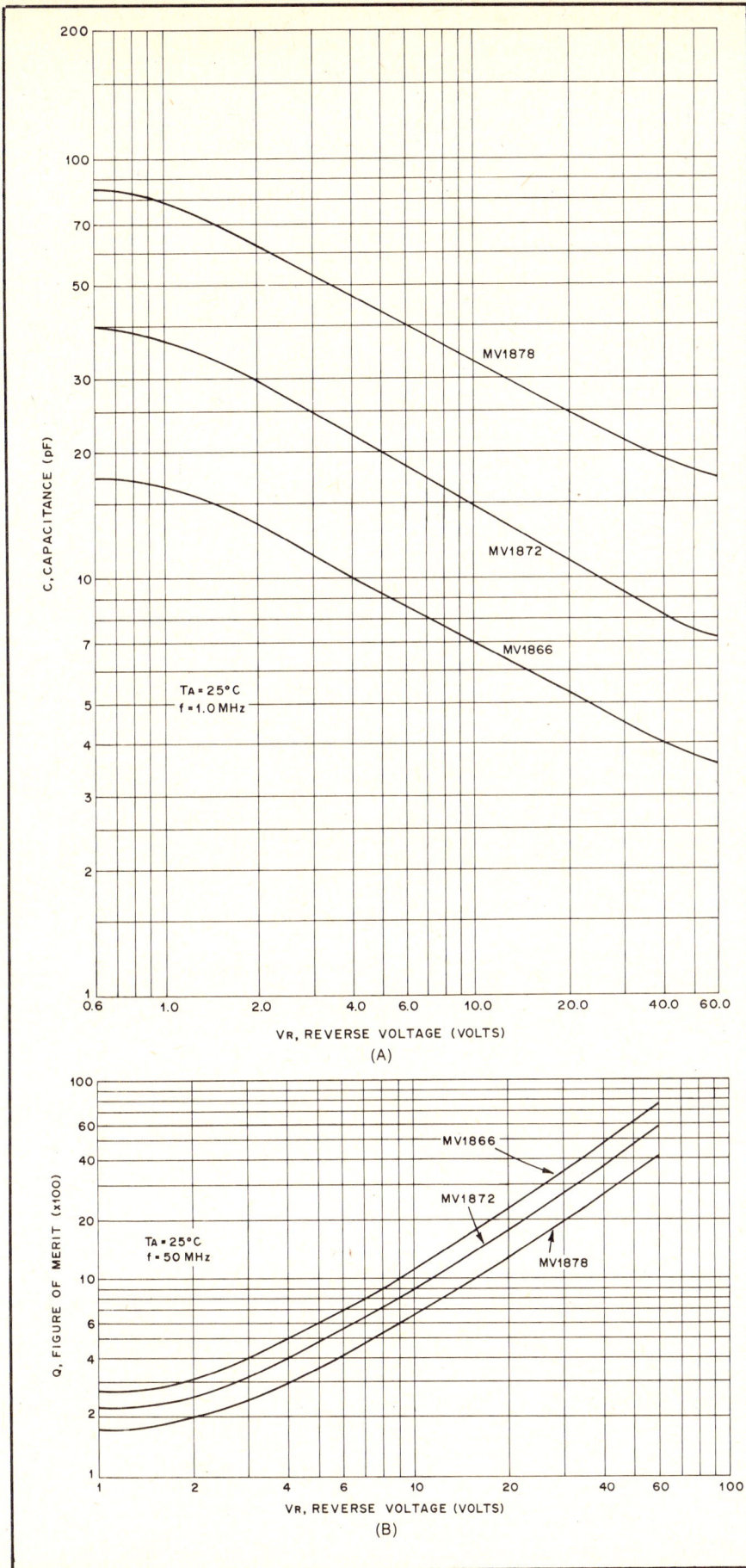

Fig. 28 — Representative circuit of a varactor diode showing case resistance, junction resistance and junction capacitance.

equivalent circuit the diode junction consists of C_j (junction capacitance) and R_j junction resistance). The bulk resistance is shown as R_s. For the most part R_j can be neglected. The performance of the diode junction at a particular frequency is determined mainly by C_j and R_s. As the operating frequency is increased, the diode performance degrades, owing to the transit time established by C_j and R_s.

An important characteristic of the varactor diode is the Q, or figure of merit. The Q of a varactor diode is determined by the ratio of its capacitive reactance (X_j) and its bulk resistance, R_s, just as is true of other circuit elements, such as coils and capacitors, where $Q = X/R_s$ at a specified frequency. Fig. 27B characterizes the Q of three Motorola varactor diodes (versus reverse bias) at 50 MHz.

Present-day varactor diodes operate into the microwave part of the spectrum. They are quite efficient as frequency multipliers at power levels as great as 25 watts. The efficiency of a correctly designed varactor multiplier exceeds 50 percent in most instances. Fig. 29 illustrates the basic circuit of a frequency multiplier which contains a varactor diode. D1 is a single-junction device which serves as a frequency tripler in this example. FL1 is required in order to assure reasonable purity of the output energy. It is a high-Q strip-line resonator. Without FL1 in the circuit there would be considerable output energy at 144, 288 and 864 MHz. Similar circuits are used as doublers, quadruplers and higher.

A Motorola MV104 tuning diode is used in the circuit of Fig. 30. It contains two varactor diodes in a back-to-back arrangement. The advantage in using two diodes is reduced signal distortion, as compared to a one-diode version of the same circuit. Reverse bias is applied equally to the two diodes in the three-terminal device. R1 functions as an rf isolator for the tuned circuit. The reverse bias is varied by means of R2 to shift the operating frequency. Regulated voltage is as important to the varactor as it is to the FET oscillator if reasonable frequency stability is to be assured. Varactor diodes are often used to tune two or more circuits at the same time (receiver rf amplifier, mixer and oscillator), using a single potentiometer to control the capacitance of the diodes. It is worth mentioning that some Zener diodes and selected silicon

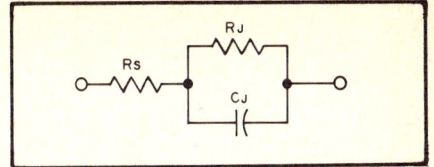

Fig. 27 — Reverse voltage respective to diode capacitance of three Motorola varactor diodes (A). Reverse voltage versus diode Q for the varactors at A are shown at B.

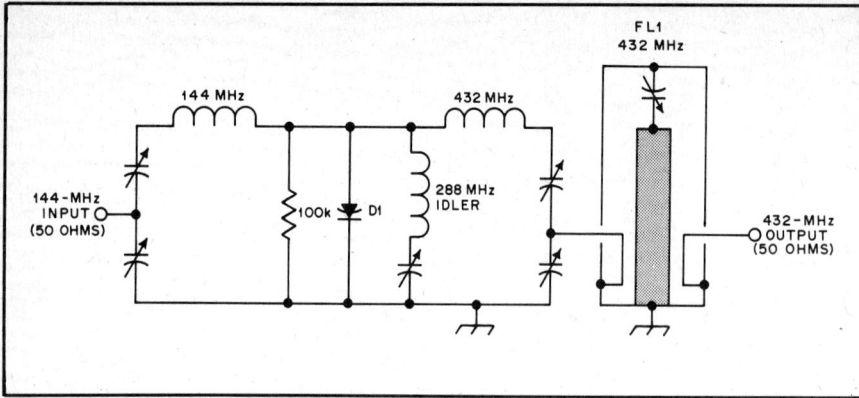

Fig. 29 — Typical circuit for a varactor-diode frequency tripler.

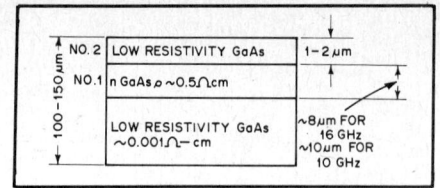

Fig. 30 — Example of a varactor-tuned VFO. D1 contains two varactors, back to back (see text).

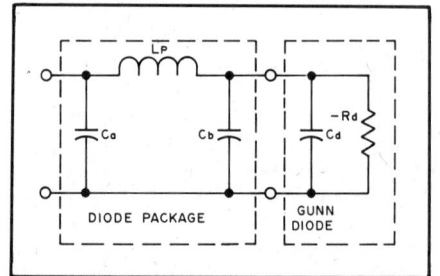

Fig. 31 — Cross-sectional illustration of Gunn diode.

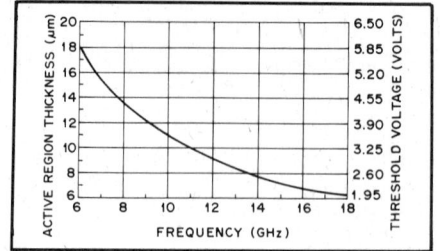

Fig. 32 — Active region thickness versus frequency of a Gunn diode.

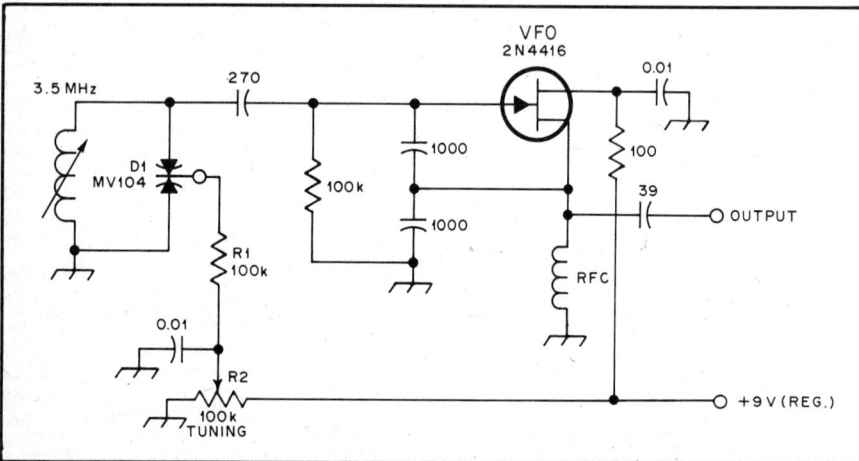

Fig. 33 — Equivalent circuit of a Gunn diode. The parasitic reactances of the diode package are included.

power-supply rectifier diodes will work effectively as varactors at frequencies as high as 144 MHz. If a Zener diode is used in this manner it must be operated below its reverse breakdown voltage point. The stud-mount variety of power supply diodes (with glass headers) are reported to be the best candidates as varactors, but not all diodes of this type will work effectively: Experimentation is necessary.

Gunn Diodes

Gunn diodes are named after the developer, J. B. Gunn, who was studying carrier behavior at IBM Corp. in 1963. During that period he discovered what is known today as the "Gunn effect."

Recently, semiconductor devices of the "bulk effect" variety have become so practical that in areas where advanced technology is practiced they are commonplace. Among these newer devices are the TDO (tunnel-diode oscillator), the ADO (avalanche-diode oscillator) and the TEO (transferred-electron oscillator). With some of the present-day techniques and solid-state devices it is possible to generate useful power up beyond 100 GHz.

Amateurs have been using Gunn diodes

at 10 GHz, but it is important to realize that these diodes are available for other microwave frequencies. An in-depth treatment of the technology and chemistry of Gunn diodes is provided in the *Gunn Diode Handbook* by Microwave Associates, Incorporated.

Fig. 31 shows a cross-sectional representation of a slice of the material from which Gunn diodes are made. Layer no. 1 is the active region of the device. The thickness of this layer depends on the chosen frequency of operation. For the 10-GHz band it is approximately 10 μm (10^{-6} meters) thick. The threshold voltage is roughly 3.3 volts. At 16 GHz the layer would be formed to a thickness of 8 μm, and the threshold voltage would be about 2.6 volts.

Layer no. 2 is grown epitaxially and is doped to provide low resistivity. This layer is grown on the active region of the semiconductor, but it is not essential to the primary operation of the diode. It is used to ensure good ohmic contact and to prevent metalization from damaging the n-layer of the diode.

The composite wafer of Fig. 31 is metalized on both sides to permit bonding

into the diode package. This process of metalization also ensures a low electrical and thermal resistance. The completed chip is bonded to a gold-plated copper pedestal, with layer no. 2 next to the heat sink. A metal ribbon is connected to the back side of the diode to provide for electrical contact.

The curve in Fig. 32 shows the relationship of the diode active-region thickness to the frequency of operation. The curve illustrates an approximation because the actual thickness of the active region depends on the applied bias voltage and the particular circuit used. The input power to the diode must be 20 to 50 times the desired output power. Thus the efficiency from dc to rf is on the order of two to five percent.

The resonant frequency of the diode assembly must be higher than the operating frequency to allow for parasitic C and L components which exist. Fig. 33 shows the equivalent circuit of a packaged Gunn diode. Assuming a diode natural resonant frequency of 17 GHz, the following approximate values result: L_p = 0.25 nH, C_a = 0.15 pF and C_b = 0.15 pF. Additional components exist within the

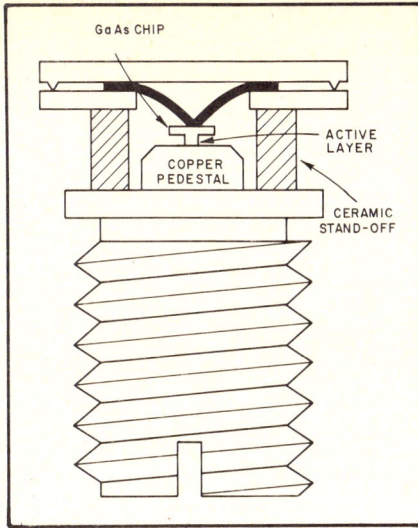

Fig. 34 — Illustration of a packaged Gunn diode as seen in literature from Microwave Associates.

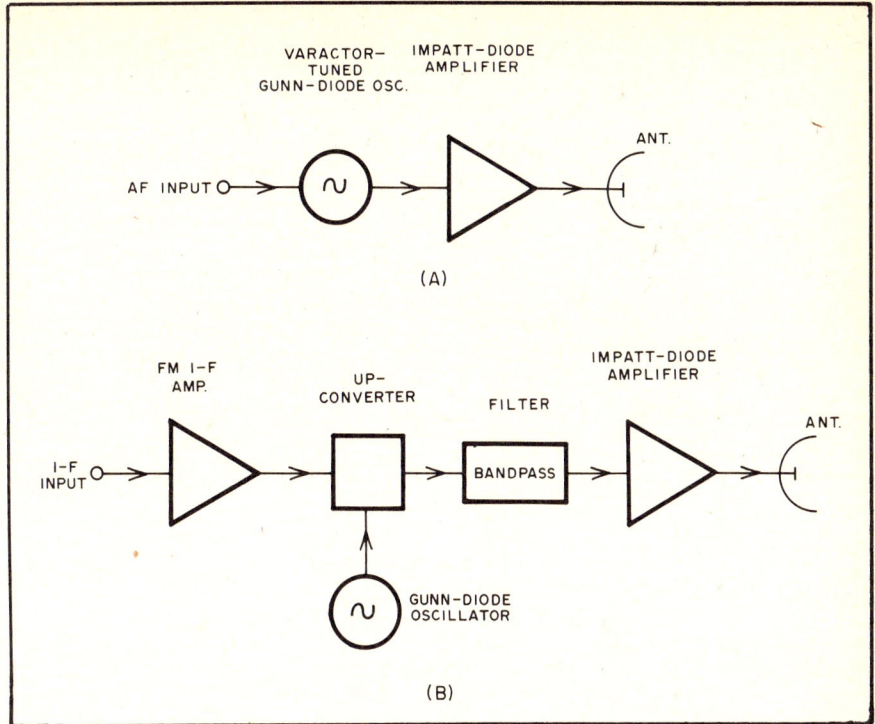

Fig. 35 — Block diagram of a simple Gunn-diode transmitter for fm (A) and an up-converter Gunn-diode transmitter (B).

Additional components exist within the diode chip. They are represented by C_d (capacitance) and $-R_d$ (negative resistance). These quantities, plus the stray resonances in the diode holder and bias leads in the microwave cavity, have a direct bearing on the electrical behavior of the Gunn oscillator. A cross-sectional representation of a packaged Gunn diode is shown at Fig. 34.

Presently, Gunn diodes are useful for generating powers between 0.1 and 1 watt. As the technology advances these power limits will increase. IMPATT (impact-avalanche transit time) diodes are useful as microwave amplifiers after a Gunn diode signal source. IMPATT diodes are also capable of providing power output in the 0.1 to 1-watt class. Fig. 35 shows block diagrams of two Gunn-diode systems. In each example an IMPATT diode is used as an amplifier. Fig. 35A shows a direct fm transmitter which employs a varactor-deviated Gunn-diode oscillator as a signal source. Fm is provided by applying audio to the bias lead of the varactor diode. The latter is coupled to the Gunn-diode cavity. Fig. 35B illustrates a microwave relay system in which a Gunn diode is used as an LO source. Essentially, the equipment is set up as a heterodyne up-converter transmitter. The upper sideband from the mixer is amplified at microwave frequency by means of an IMPATT diode.

PIN Diodes

A PIN diode is formed by diffusing heavily doped p + and n + regions into an almost intrinsically pure silicon layer, as illustrated in Fig. 36. In practice it is impossible to obtain intrinsically pure material and the I layer can be considered to be a lightly doped n region. Characteristics of the PIN diode are primarily determined by the thickness, area and semiconductor nature of the chip, especially that of the I region. Manufacturers design for controlled thickness I regions having long carrier lifetime and high resistivity. Carrier lifetime is basically a measure of the delay before an average electron and hole recombine. In a pure silicon crystal the theoretical delay is on the order of several milliseconds, although impurity doping can reduce the effective carrier lifetime to microseconds or nanoseconds.

When forward bias is applied to a PIN diode, holes and electrons are injected from the p + and n + regions into the I region. These charges do not immediately recombine. Rather, a finite quantity of charge always remains stored and results in a lowering of the I-region resistivity. The amount of stored charge depends on the recombination time (carrier lifetime) and the level of the forward-bias current. The resistance of the I region under forward-bias conditions is inversely proportional to the charge and depends on the I-region width and mobility of the holes and electrons of the particular semiconductor material. Representative graphs of resistance vs. forward bias level are shown in Fig. 37A and B for low-level receiving and high-power transmitting PIN diodes.

When a PIN diode is at zero or reverse bias, there is essentially no charge, and the intrinsic region can be considered as a low-loss dielectric. As with an ordinary pn junction there is a reverse breakdown or Zener region where the diode current increases rapidly as the reverse voltage increases. For the intrinsic region to remain in a low-loss state, the maximum instantaneous reverse or negative voltage must not exceed the breakdown voltage. Also, the positive voltage excursion must not cause thermal losses to exceed the diode dissipation rating.

At high radio frequencies when a PIN diode is at zero or reverse bias, the diode appears as a parallel plate capacitor, essentially independent of reverse voltage. It is the value of this capacitance that limits the effective isolation that the diode can provide. PIN diodes intended for high isolation and not power-handling capability are designed with as small a geometry as possible to minimize the capacitance.

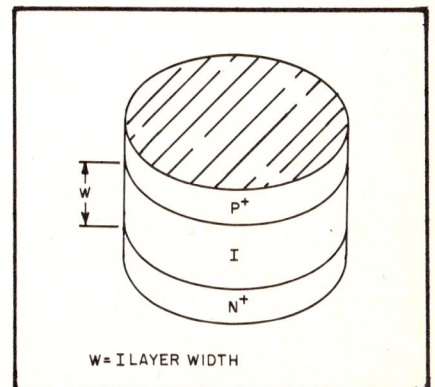

Fig. 36 — The PIN diode is constructed by diffusing p + and n + regions into an almost intrinsically pure silicon layer. Thus the name PIN diode.

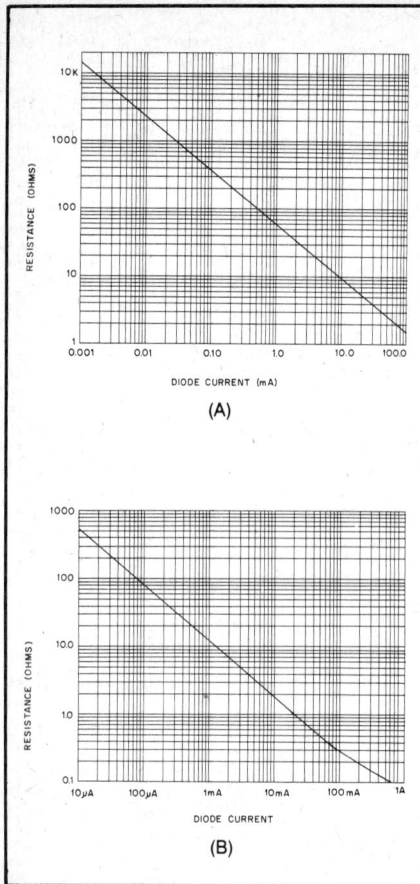

Fig. 37 — At A is a graph comparing diode resistance to forward-bias current for a PIN diode intended for low-level receiver applications. At B is a similar graph for a diode capable of handling over 100 watts of rf.

Manufacturers of PIN diodes supply data sheets with all necessary design data and performance specifications. Key parameters are diode resistance (when forward biased), diode capacitance, carrier lifetime, harmonic distortion, reverse voltage breakdown and reverse leakage.

PIN diodes are used in many applications, such as rf switches, attenuators and various types of phase-shifting devices. Our discussion will be confined to switch and attenuator applications since these are the most likely to be encountered by the amateur. The simplest type of switch that can be created with a PIN diode is the series spst type. The circuit is shown in Fig. 38A. C1 functions as a dc blocking capacitor and C2 is a bypass capacitor. In order to have the signal from the generator flow to the load, a forward bias must be applied to the bias terminal. The amount of insertion loss caused by the diode is determined primarily by the diode bias current. Fig. 38B illustrates an spdt type of switch arrangement which uses essentially two spst switches with a common connection. For a generator current to flow into the load resistor at the left, a bias voltage is applied to bias terminal 1. For signal to flow into the load at the right a bias must be applied to terminal 2. In

practice it is usually difficult to achieve more than 40 dB isolation with a single diode switch at uhf and microwaves. Better performance, in excess of 100 dB, is achievable using compound switches. Compound switches are made up of two or more diodes in a series/shunt arrangement. Since not all diodes are biased for the same state, some increase in bias-circuit complexity results.

One general class of switches used in connection with transceive applications requires that a common antenna be connected to either the receiver or transmitter during the appropriate receive or transmit states. When PIN diodes are used as switching elements in these applications, higher reliability, better mechanical ruggedness and faster switching speeds are achieved relative to the electromechanical relay. A basic approach is shown in Fig. 39A and B where a PIN diode is used in series with the transmit line and another diode in shunt with the receive line. A single bias supply is used to turn on the series diode during transmit while also turning on the shunt diode to protect the receiver. The quarter-wave line between the two diodes is necessary to isolate the low resistance of the receiver diode from the antenna connection. During receive periods both diodes are effectively open circuited, allowing signal energy to be applied to the receiver. At B is the same basic circuit, although the quarter-wave line has been replaced with a lumped element section.

Two of the more common types of attenuators using PIN diodes are shown in Fig. 40. The type at A is referred to as a Bridged Tee, while the circuit at B is the common pi type. Both are useful as very broadband devices. It is interesting to note that the useful upper frequency of these attenuators is often dependent on the bias circuit isolation rather than the PIN diode characteristics.

Light-Emitting Diodes (LEDs)

The primary component in optoelectronics is the LED. This diode contains a p-n junction of crystal material which produces luminescense around the junction when forward bias current is applied. LED junctions are made from gallium arsenide (GaAs), gallium phosphide (GaP), or a combination of both materials (GaAsP). The latter is dependent upon the color and light intensity desired. Today, the available LED colors are red, green and yellow.

Some LEDs are housed in plastic which is affixed to the base header of a transistor package. Other LEDs are contained entirely in plastic packages which have a dome-shaped head at the light-emitting end. Two wires protrude from the opposite end (positive and negative leads) for applying forward bias to the device.

There are countless advantages to the use of LEDs. Notable among them are the

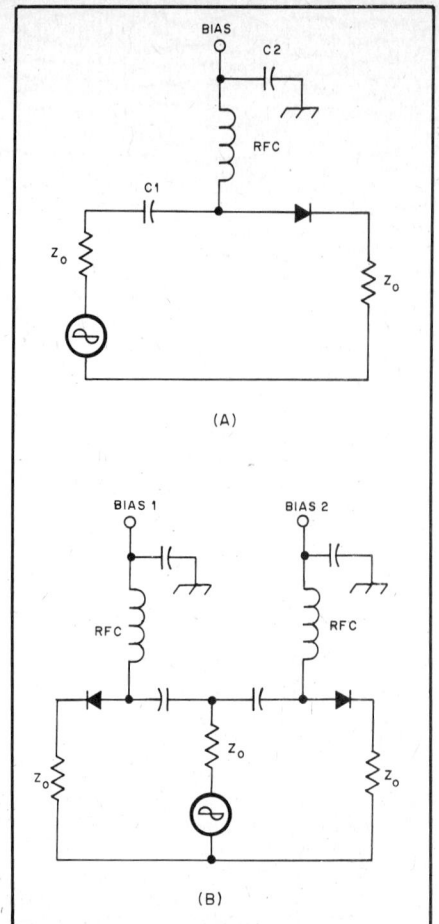

Fig. 38 — At A a PIN diode is used as an spst switch. At B, two diodes form an spdt switching arrangement.

low current drain, long life (sometimes 50 years, as predicted), and small size. They are useful as visual indicators in place of incandescent panel lamps. One of their greatest applications is in digital display units, where arrays of tiny LEDs are arranged to provide illuminated segments in numeric-display assemblies.

The forward bias current for a typical LED ranges between 10 and 20 mA for maximum brilliance. An applied voltage of 1-1/2 to 2 is also typical. A 1000-ohm resistor in series with a 12 volt source will permit the LED to operate with a forward current of approximately 10 mA (IR drop = 10 V). A maximum current of 10 mA is suggested in the interest of longevity for the device.

LEDs are also useful as reference diodes, however unique the applications may seem. They will regulate dc at approximately 1.5 V.

The following are definitions and terms used in optics to characterize the properties of an LED.

Incident flux density is defined as the amount of radiation per unit area (expressed as lumens/cm² in photometry; watts/cm² in radiometry). This is a measure of the amount of flux received by a detector measuring the LED output.

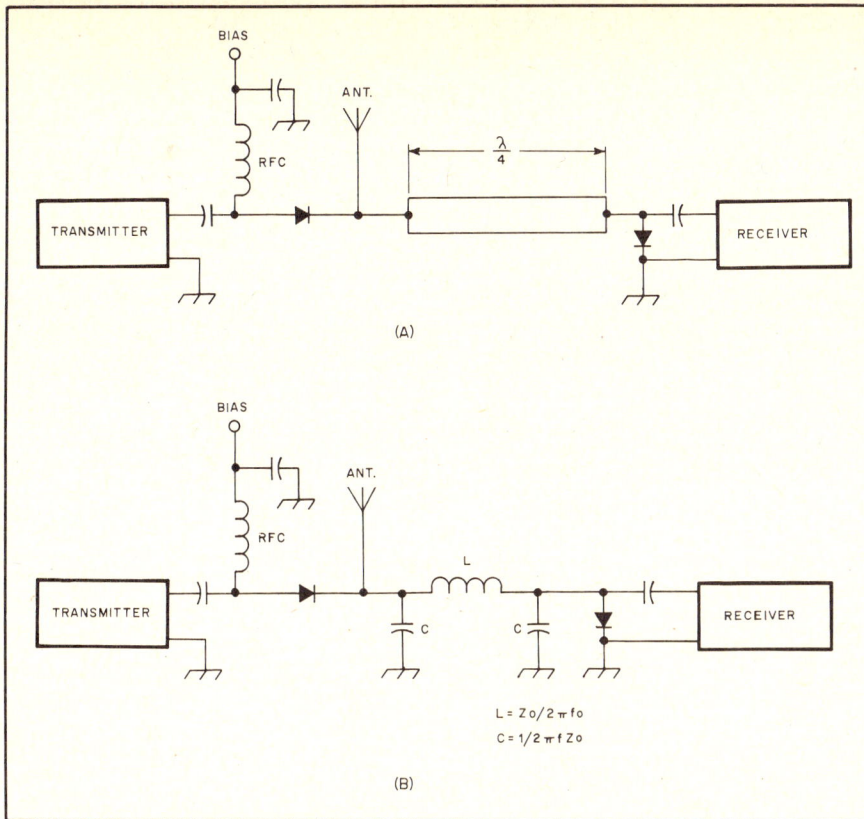

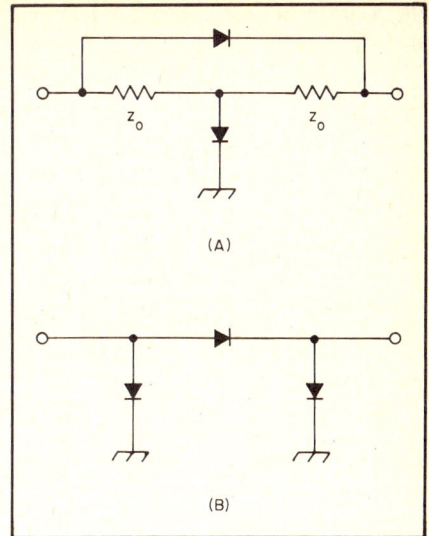

Fig. 40 — Two types of PIN diode attenuator circuits. The circuit at A is called a Bridged Tee and the circuit at B is a pi type. Both exhibit very broadband characteristics.

Fig. 39 — PIN diodes are used to transfer a common antenna to either a transmitter or receiver. A voltage applied to the bias terminal will switch the system to the transmit mode connecting the output of the transmitter to the antenna. At the same time the diode across the receiver input is forward biased to a low-resistance state to protect the input stage of the receiver. The quarter-wave line isolates the low resistance of the receiver diode from the common antenna connection. At B the quarter-wave line is replaced with a lumped-element equivalent.

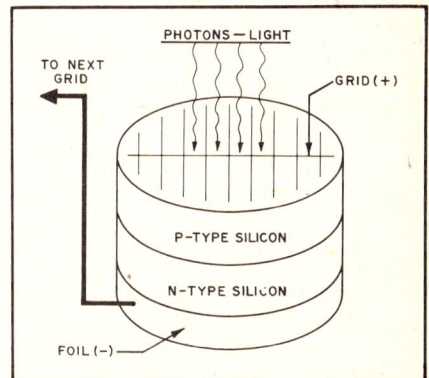

Fig. 41 — A solar-energy diode cell. Electrons flow when light strikes the upper surface. The bottom of the cell is coated with foil to collect current for the load, or for the succeeding cell in series-connected arrays of cells.

Emitted flux density is also defined as radiation per unit area and is used to describe light reflected from a surface. This measure of reflectance determines the total radiant or luminous emittance.

Source intensity defines the flux density which will appear at a distant surface and is expressed as lumens/steradian (photometry) or watts/steradian (radiometry).

Luminance is a measure of photometric brightness and is obtained by dividing the luminous intensity at a given point by the projected area of the source at the same point. Luminance is a very important rating in the evaluation of visible LEDs.

While luminance is equated with photometric brightness, it is inaccurate to equate luminance as a figure of merit for brightness. The only case where this rating is acceptable is when comparing *physically* identical LEDs. Different LEDs are subject to more stringent examination. Manufacturers do not use a set of consistent ratings for LEDs (such as optical flux, brightness and intensity). This is because of the dramatic differences in optical measurements between point- and area-source diodes. Point-source diodes are packaged in a clear epoxy or set within a transparent glass lens. Area-source diodes must employ a diffusing lens to spread the flux over a wider viewing area and hence have much less point intensity (luminance) than the point-source diodes.

Solar-Electric Diodes

Sunlight can be converted directly into electricity by a process known as *photovoltaic conversion*. For this purpose a solar cell is used. It relies on the photoelectric properties of a semiconductor. Practically, the solar cell is a large-area p-n junction diode. The greater the area of the cell, the higher the output current will be. A dc voltage output of approximately 0.5 is obtained from a single cell. Numerous cells can be connected in series to provide 6, 9, 12, 24 or whatever low voltage is required. In a like manner, cells can be connected in parallel to provide higher output current, overall.

The solar diode cell is built so that light can penetrate into the region of the p-n junction, Fig. 41. Most modern solar cells use silicon material. Impurities (doping) are introduced into the silicon material to establish excess positive or negative charges which carry electric currents. Phosphorous is used to produce n-type silicon. Boron is used as the dopant to produce p-type material.

Light is absorbed into the silicon to generate excess holes and electrons (one hole/electron pair for each photon absorbed). When this occurs near the p-n junction, the electric fields in that region will separate the holes from the electrons. This causes the holes to increase in the p-type material. At the same time the electrons will build up in the n-type material. By making direct connection to the p and n regions by means of wires, these excess charges generated by light (and separated by the junction) will flow into an external load to provide power.

Approximately 0.16 A can be secured from each square inch of solar-cell material exposed to bright sunlight. A 3-1/2 inch diameter cell can provide 1.5 A of output current. The efficiency of a solar cell (maximum power delivered to a load versus total solar energy incident on the cell) is typically 11 to 12 percent.

Arrays of solar cells are manufactured for all manner of practical applications. A storage battery is used as a buffer between the solar panel and the load. A p-n junction diode should be used between the solar-array output and the storage battery to prevent the battery from discharging back into the panel during dark periods. An article on the subject of solar cells and their amateur applications was written by DeMaw ("Solar Power for the Radio Amateur," August 1977 *QST*) and should be of interest to those who wish to utilize solar power. *Solar Electric Generator Systems,* an application pamphlet by Solar Power Corp. of N. Billerica, MA 01862, contains valuable information on this subject.

Fig. 42 shows the voltage/current curves to, a Model E12-01369-1.5 solar array manufactured by Solar Power Corp. It can be seen that temperature has an effect on the array performance.

Tunnel Diodes

One type of semiconductor diode having no rectifying properties is called a tunnel diode. The bidirectional conduction of the device is a result of heavily doped p and n regions with a very narrow junction. The Fermi level lies within the conduction band for the n side and within the valence band for the p side. A typical current-vs.-voltage curve for a tunnel diode is sketched in Fig. 43. When the forward bias potential exceeds about 30 mV, increasing the voltage causes the current to *decrease*, resulting in a negative resistance characteristic. This effect makes the tunnel diode capable of amplification and oscillation. At one time tunnel diodes were expected to dominate in microwave applications, but other devices soon surpassed tunnel diodes in performance. The two-terminal oscillator concept had great fad appeal, and some amateurs built low-power transmitters based on tunnel diodes. In the 1960s the Heath Company marketed a dip meter that used a tunnel diode oscillator. Tunnel diodes are not widely used in new designs; this material is included only for completeness.

Zener Diodes

Zener diodes have, for the most part, replaced the gaseous regulator tube. They have been proved more reliable than tube types of voltage regulators, are less expensive and far smaller in size.

These diodes fall into two primary classifications: Voltage regulators and voltage-reference diodes. When they are used in power supplies as regulators, they provide a nearly constant dc output voltage even though there may be large changes in load resistance or input voltage. As a reference element the Zener diode utilizes the voltage drop across its junction when a specified current passes through it in the reverse-breakdown direction (sometimes called the *Zener direction*). This "Zener voltage" is the value established as a reference. There-

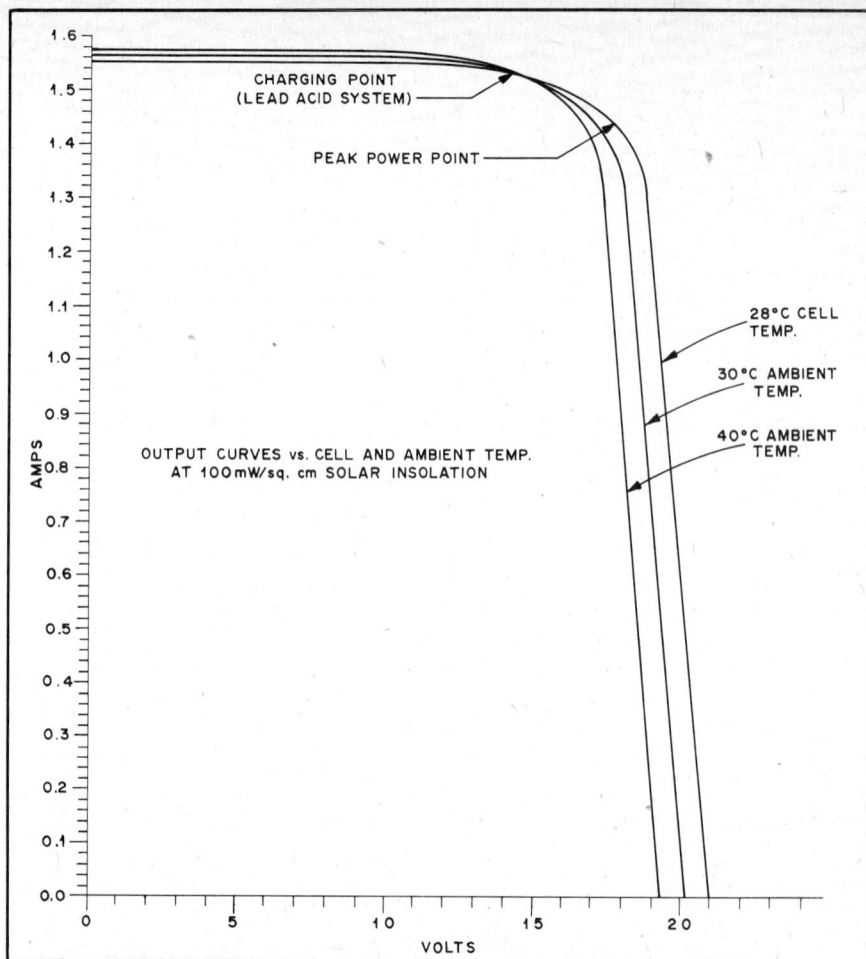

Fig. 42 — Voltage/current/temperature curve for a Solar Power Corp. array which contains 36 solar-electric cells in series. The curves are for a model E12-01369-1.5 solar panel.

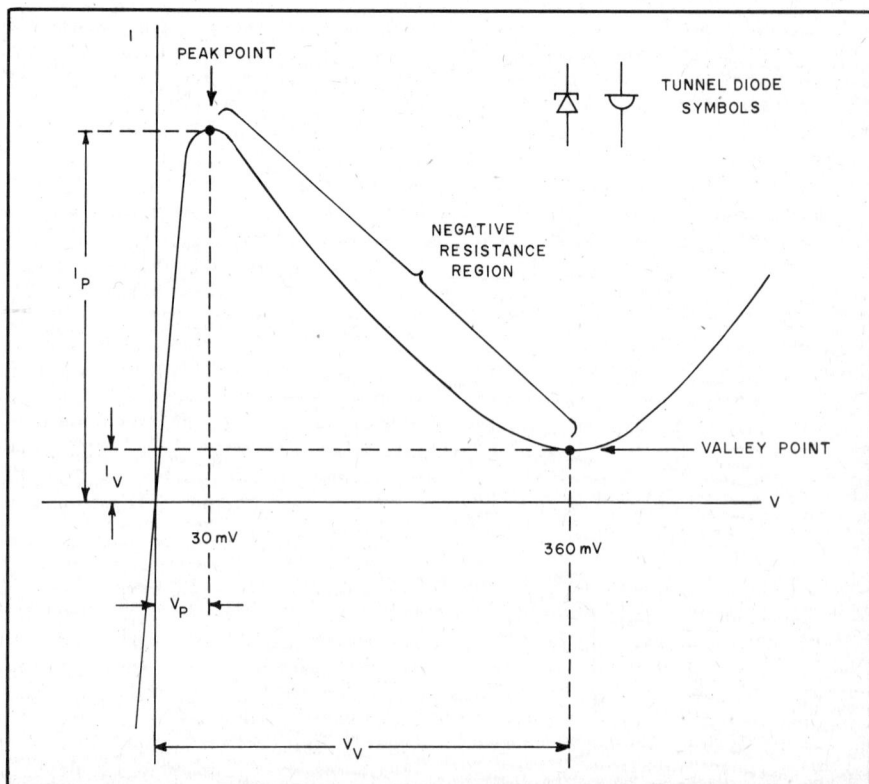

Fig. 43 — Schematic symbol and current-vs.-voltage characteristic for a tunnel diode.

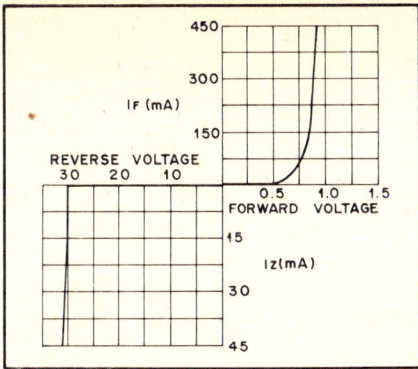

Fig. 44 — Typical characteristics of a Zener diode (30 V).

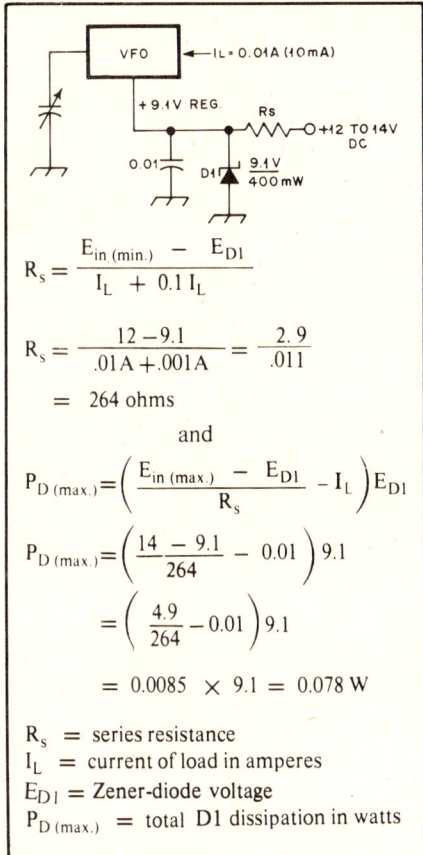

$$R_s = \frac{E_{in\,(min.)} - E_{D1}}{I_L + 0.1\,I_L}$$

$$R_s = \frac{12 - 9.1}{.01A + .001A} = \frac{2.9}{.011}$$

$$= 264 \text{ ohms}$$

and

$$P_{D\,(max.)} = \left(\frac{E_{in\,(max.)} - E_{D1}}{R_s} - I_L \right) E_{D1}$$

$$P_{D\,(max.)} = \left(\frac{14 - 9.1}{264} - 0.01 \right) 9.1$$

$$= \left(\frac{4.9}{264} - 0.01 \right) 9.1$$

$$= 0.0085 \times 9.1 = 0.078 \text{ W}$$

R_s = series resistance
I_L = current of load in amperes
E_{D1} = Zener-diode voltage
$P_{D\,(max.)}$ = total D1 dissipation in watts

Fig. 45 — Example of how a shunt type of Zener diode regulator is used. The equations show how to calculate the value of the series resistor and the diode power dissipation. In this example a 400-mW Zener diode will suffice (D1).

fore, if a 6.8-volt Zener diode was set up in the foregoing manner, the resultant reference voltage would be 6.8.

At the present time it is possible to purchase Zener diodes which are rated for various voltages between 2.4 and 200. The power ratings range from 1/4 to 50 watts. Fig. 44 shows the characteristics of a Zener diode designed for 30-volt operation.

Fig. 45 shows how to calculate the series resistance needed in a simple shunt

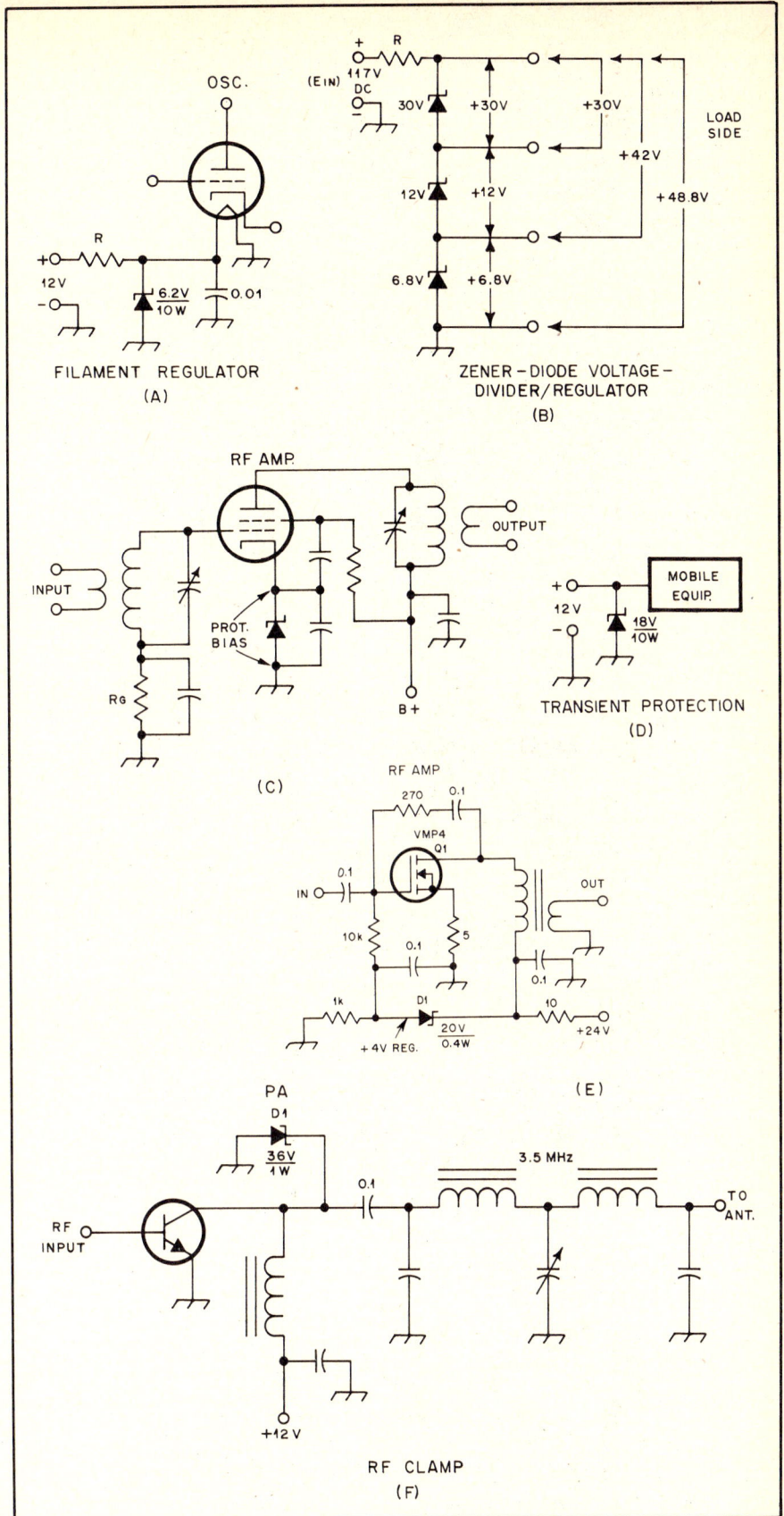

Fig. 46 — Practical examples of Zener diode applications. The circuit at A is useful for stabilizing the filament voltage of oscillators. Zener diodes can be used in series to obtain various levels of regulated voltage (B). Fixed-value bias for transmitter stages can be obtained by inserting a Zener diode in the cathode return (C). At D an 18-volt Zener diode prevents voltage spikes from harming a mobile transceiver. A Zener-diode series regulator (20-V drop) is shown at E and an rf clamp is seen at F. D1 in the latter circuit will clamp at 36 volts to protect the PA transistor from dc voltage spikes and extreme sine-wave excursions at rf. This circuit is useful in protecting output stages during no-load or short-circuit conditions.

regulator which employs a Zener diode. An equation is included for determining the wattage rating of the series resistor. Additional data on this subject is given in chapter 7 of *Solid State Design for the Radio Amateur.*

Some practical applications for Zener diodes are illustrated in Fig. 46. In addition to the shunt applications given in the diagram, Zener diodes can be used as series elements when it is desired to provide a gate that conducts at a given voltage. These diodes can be used in ac as well as dc circuits. When they are used in an ac type of application they will conduct at the peak voltage value or below, depending upon the voltage swing and the voltage rating of the Zener diode. For this reason they are useful as audio and rf clippers. In rf work the reactance of the diode may be the controlling factor above approximately 10 MHz with respect to the performance of the rf circuit and the diode.

Most Zener diodes which are rated higher than 1 watt in dissipation are contained in stud-mount packages. They should be affixed to a suitable heat sink to prevent damage from excessive junction temperatures. The mounting techniques are the same as for power rectifiers and high-wattage transistors.

Reference Diodes

While ordinary Zener diodes are useful as voltage regulators, they don't exhibit the thermal stability required in precision reference applications. A reverse-biased semiconductor junction has a positive temperature coefficient of barrier potential, and a forward-biased junction has a negative coefficient. The way to temperature-compensate a Zener diode is to connect one or more common silicon diodes in series with it. When this is done as part of the manufacturing process, the resulting component is termed a *reference diode.* A 1N3499 6.2-volt reference diode will maintain a temperature coefficient of 0.0005 percent per degree over the range of 0 to 75° C. Reference diodes work best when operated at a few milliamperes of current from a high-impedance or constant-current bias source. As the name implies, these diodes aren't suited for circuits where power is taken directly from the device. Reference diodes can't be tested with an ohmmeter because two junctions are back-to-back — the instrument can't supply enough voltage to overcome the Zener barrier potential.

Step-Recovery Diodes

One device characterized by extremely low capacitance and short storage time is the *step-recovery diode* (SRD), sometimes called a "snap" diode. These diodes are used as frequency multipliers well into the microwave spectrum. Switching the device in and out of forward conduction is the multiplication mechanism, and the power

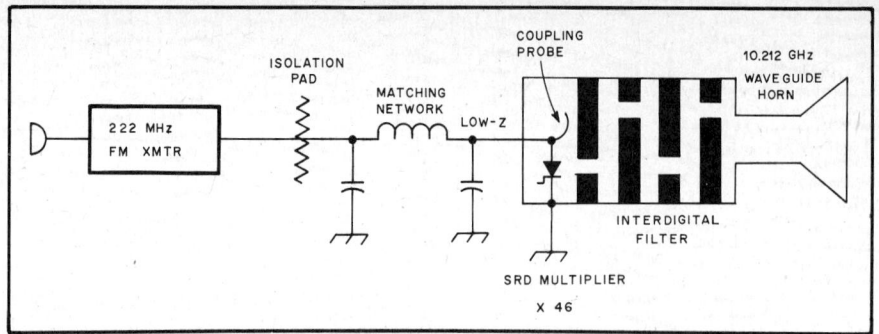

Fig. 47 — Step-recovery diode frequency multiplier for 10 GHz. The matching network elements are represented as lumped components but would take the form of Microstrip in an actual design.

efficiency is inversely proportional to the frequency multiple. Very high orders of multiplication are possible with step-recovery diodes, and one use for this feature is in a *comb generator* — an instrument used to calibrate the frequency axis of a spectrum analyzer.

A single harmonic of the excitation frequency can be selected by an interdigital filter or cavity resonator. A 1-watt, 220-MHz fm transmitter could drive a snap diode multiplier (× 46) and filter combination to an output of about 10 milliwatts in the 10-GHz band — a typical and effective power level at that frequency. A representative system of this variety is suggested in Fig. 47. The exciter should be well isolated from the SRD and its matching network to prevent parasitic oscillations.

Current-Regulator Diodes

A JFET with its gate shorted to its source or connected below a source resistor will draw a certain current whose value is almost entirely independent of the applied potential. The current (I_{DSS} in FET terminology) is also quite stable with temperature. Semiconductor manufacturers take advantage of these properties and package the JFET circuit of Fig. 48A in a two-terminal package and call it a constant-current diode. A special symbol, given in Fig. 48B, is assigned to this type of diode. The 1N5305 diode approaches an ideal current generator, in that it draws two milliamperes over the range of 1.8 to 100 volts. Constant-current diodes find application in ohmmeters, ramp generators and precision voltage references.

BIPOLAR TRANSISTORS

The word "transistor" was chosen to describe the function of a three-terminal p-n junction device which is able to amplify signal energy (current). The inherent characteristic is one of "transferring current across a resistor." The transistor was invented by Shockley, Bardeen and Brattain at Bell Labs in 1947 and has become the standard amplifying device in electronic equipment. In rf and

audio applications it is practical to obtain output power in excess of 1000 watts by using several amplifier blocks and hybrid power combiners. The primary limitation at the higher power levels is essentially a practical or economic one: Low voltage, high-current power supplies are required, and the cost can exceed that of a high-voltage, moderate-current supply of the variety which would be employed with a vacuum-tube amplifier of comparable power. The primary advantages obtained through the use of solid-state power amplifiers are compactness and reliability.

In small-signal applications the transistor outweighs the vacuum-tube in performance. The former is more efficient, operates cooler, has much longer life, is considerably smaller in size, and is less expensive. A naive designer might insist that "tubes are better," but the transistor ranks no. 1 in the industry at this time.

There remains in isolated instances a belief that transistors are hard to tame, noisier than tubes and that they are subject to damage at the flick of a switch. None of this is true. A transistorized circuit which is designed and operated correctly is almost always capable of exceeding an equivalent vacuum-tube circuit in all respects. An understanding of

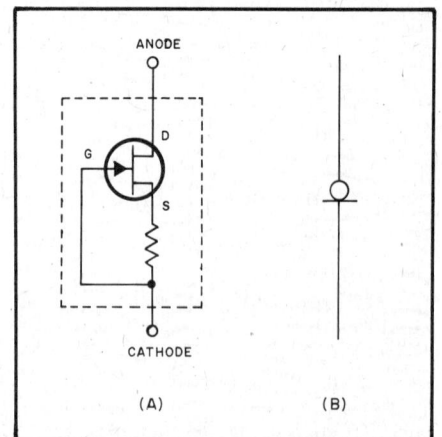

Fig. 48 — At A, an n-channel JFET connected as a constant-current source. At B, the schematic symbol for the circuit in A when it is packaged as a two-terminal device.

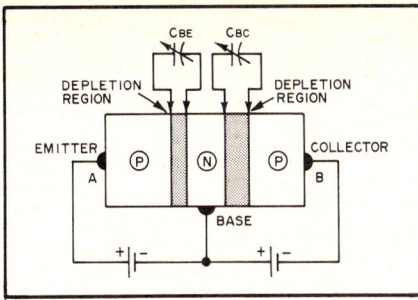

Fig. 49 — Illustration of a junction pnp transistor. Capacitances C_{be} and C_{bc} vary with changes in operating and signal voltage (see text).

how transistors function will help to prevent poor circuit performance: The fundamentals outlined in this chapter are provided for the amateur designer so that the common pitfalls can be avoided.

Fig. 49 shows a "sandwich" made from two layers of p-type semiconductor material with a thin layer of n-type between. There are in effect two pn junction diodes back-to-back. If a positive bias is applied to the p-type material at the left, current will flow though the left-hand junction, the holes moving to the right and the electrons from the n-type material moving to the left. Some of the holes moving into the n-type material will combine with the electrons there and be neutralized, but some of them also will travel to the region of the right-hand junction.

If the pn combination at the right is biased negatively, as shown, there would normally be no current flow in this circuit. However, there are now additional holes available at the junction to travel to point B and electrons can travel toward point A, so a current can flow even though this section of the sandwich is biased to prevent conduction. Most of the current is between A and B and does not flow out through the common connection to the n-type material in the sandwich.

A semiconductor combination of this type is called a *transistor,* and the three sections are known as the *emitter, base* and *collector,* respectively. The amplitude of the collector current depends principally upon the amplitude of the emitter current; that is, the collector current is controlled by the emitter current.

Between each p-n junction exists an area known as the *depletion,* or *transition region.* It is similar in characteristics to a dielectric layer, and its width varies in accordance with the operating voltage. The semiconductor materials either side of the depletion region constitute the plates of a capacitor. The capacitance from base to emitter is shown as C_{be} (Fig. 49), and the collector-base capacitance is represented as C_{bc}. Changes in signal and operating voltages cause a nonlinear change in these junction capacitances, which must be taken into account when designing some circuits.

A base-emitter resistance, rb', also exists. The junction capacitance, in combination with rb', determines the useful upper frequency limit (f_T or fa) of a transistor by establishing an RC time constant.

Power Amplification

Because the collector is biased in the back direction the collector-to-base resistance is high. On the other hand, the emitter and collector currents are substantially equal, so the power in the collector circuit is larger than the power in the emitter circuit (P = I²R, so the powers are proportional to the respective resistances, if the currents are the same). In practical transistors emitter resistance is of the order of a few hundred ohms while the collector resistance is hundreds or thousands of times higher, so power gains of 20 to 40 dB or even more are possible.

Types

The transistor may be one of the types shown in Fig. 43. The assembly of p- and n-type materials may be reversed, so that pnp and npn transistors are both possible.

The first two letters of the npn and pnp designations indicate the respective polarities of the voltages applied to the emitter and collector in normal operation. In a pnp transistor, for example, the emitter is made positive with respect to both the collector and the base, and the collector is made negative with respect to both the emitter and the base.

Manufacturers are constantly working to improve the performance of their transistors — greater reliability, higher power and frequency ratings, and improved uniformity of characteristics for any given type number. One such development provided the *overlay transistor,* whose emitter structure is made up of several emitters which are joined together at a common case terminal. This process lowers the base-emitter resistance, rb', and improves the transistor input time constant. The latter is determined by rb' and the junction capacitance of the device. The overlay transistor is extremely useful in vhf and uhf applications. It is capable of high-power operation well above 1000 MHz. These transistors are useful as frequency doublers and triplers, and are able to provide an actual power gain in the process.

Another multi-emitter transistor has been developed for use from hf through uhf. It should be of interest to the radio amateur. It is called a *balanced-emitter transistor* (BET), or "ballasted" transistor. The transistor chip contains several triode semiconductors whose bases and collectors are connected in parallel. The various emitters, however, have built-in emitter resistors (typically about 1 ohm) which provide a current-limiting safety factor during overload periods, or under conditions of significant mismatch. Since the emitters are brought out to a single

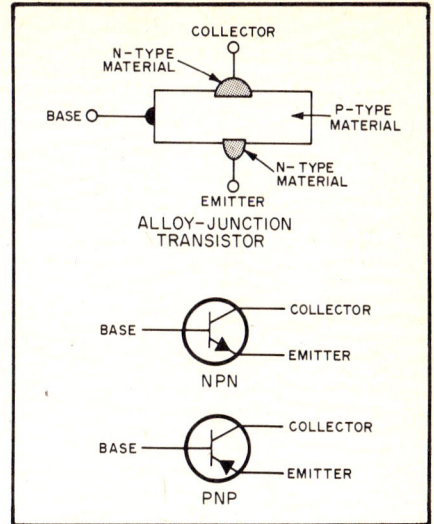

Fig. 50 — Pictorial and schematic representations of junction transistors. By way of analogy the base, collector and emitter can be compared to the grid, plate and cathode of a triode tube, respectively.

case terminal the resistances are effectively in parallel, thus reducing the combined emitter resistances to a fraction of an ohm. (If a significant amount of resistance were allowed to exist it would cause degeneration in the stage and would lower the gain of the circuit.)

Most modern transistors are of the junction variety. Various names have been given to the several types, some of which are junction alloy, mesa, and planar. Though their characteristics may differ slightly, they are basically of the same family and simply represent different physical properties and manufacturing techniques.

Transistor Characteristics

An important characteristic of a transistor is its *beta* (β), or *current-amplification factor,* which is sometimes expressed as h_{FE} (static forward-current transfer ratio) or h_{fe} (small-signal forward-current transfer ratio). Both symbols relate to the grounded-emitter configuration. Beta is the ratio of the collector current to the base current

$$\beta = \frac{I_c}{I_b}$$

Thus, if a base current of 1 mA causes the collector current to rise to 100 mA the beta is 100. Typical betas for junction transistors range from as low as 10 to as high as several hundred.

A transistor's *alpha* (α) is the ratio of the collector to the emitter current. Symbols h_{FB} (static forward-current transfer ratio) and h_{fb} (small-signal forward-current transfer ratio), common-base hookup, are frequently used in connection with gain. The smaller the base current, the closer the collector current comes to being equal to that of the emitter, and the closer alpha comes to being 1. Alpha for a junction transistor is usually between 0.92 and 0.98.

Transistors have frequency characteristics which are of importance to circuit designers. Symbol f_T is the *gain bandwidth product* (common-emitter) of the transistor. This is the frequency at which the gain becomes unity, or 1. The expression "alpha cutoff" is frequently used to express the useful upper-frequency limit of a transistor, and this relates to the common-base hookup. Alpha cutoff is the point at which the gain is 0.707 its value at 1000 Hz.

Another factor which limits the upper frequency capability of a transistor is its *transit time*. This is the period of time required for the current to flow from emitter to collector, through the semiconductor base material. The thicker the base material, the greater the transit time. Hence, the thicker the base material the more likelihood there will be of phase shift of the signal passing through it. At frequencies near and above f_T or alpha cutoff, partial or complete phase shift can occur. This will give rise to *positive feedback* because the internal capacitance, C_{be}, feeds part of the in-phase collector signal back to the base. The positive feedback can cause instability and oscillation, and in most cases will interlock the input and output tuned circuits of an rf amplifier so that it is amost impossible to tune them properly. This form of feedback can be corrected by using what is termed "unilateralization." Conventional positive feedback can be nullified by using neutralization, as is done with vacuum-tube amplifiers.

Characteristic Curves

The operating principles of transistors can be shown by a series of characteristic curves. One such set of curves is shown in Fig. 51. It shows the collector current vs. collector voltage for a number of fixed values of emitter current. Practically, the collector current depends almost entirely on the emitter current and is independent of the collector voltage. The separation between curves representing equal steps of emitter current is quite uniform, indicating that almost distortionless output can be obtained over the useful operating range of the transistor.

Another type of curve is shown in Fig. 52, together with the circuit used for obtaining it. This also shows collector current vs. collector voltage, but for a number of different values of base current. In this case the emitter element is used as the common point in the circuit. The collector current is not independent of collector voltage with this type of connection, indicating that the output resistance of the device is fairly low. The base current also is quite low, which means that the resistance of the base-emitter circuit is moderately high with this

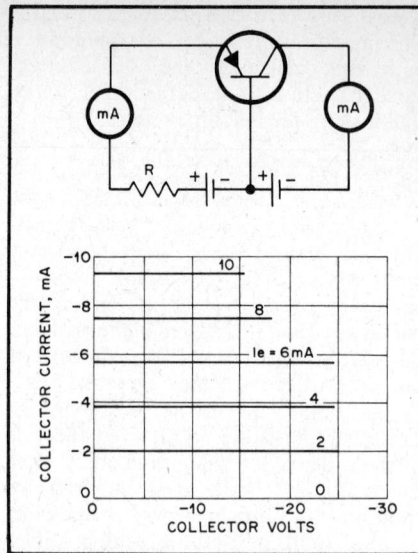

Fig. 51 — Typical collector-current versus collector-voltage characteristics of a junction transistor for various emitter-current values. Because the emitter resistance is low, a current-limiting resistor (R) is placed in series with the source current. The emitter current can be set at a desired value by adjustment of this resistance.

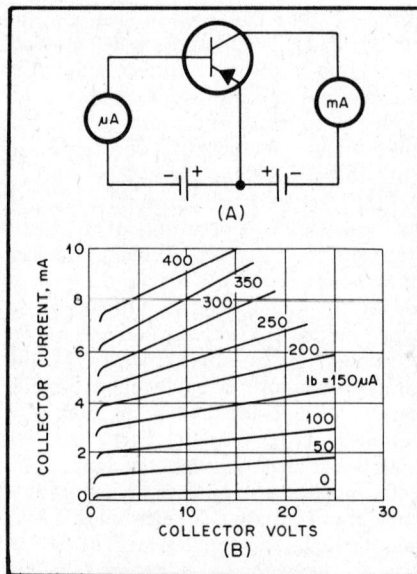

Fig. 52 — Collector current versus collector voltage for various values of base current in a junction transistor. The illustration at A shows how the measurements are made. At B is a family of curves.

method of connection. They may be constrasted with the high values of emitter current shown in Fig. 51. An actual oscillograph of a characteristic family of curves for a small-signal transistor is shown in Fig. 53. It was obtained by means of a curve tracer.

Transistor Amplifiers

Amplifier circuits used with transistors fall into one of three types, known as the *common-base*, *common-emitter*, and *common-collector* circuits. These are

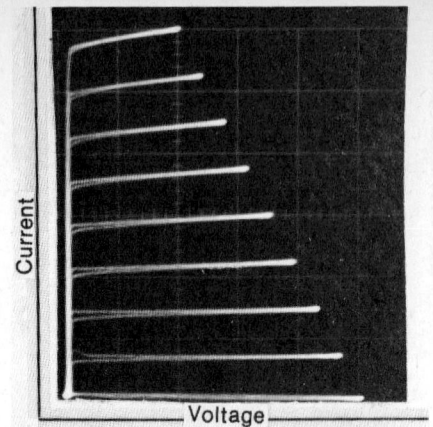

Fig. 53 — Curve-tracer display of a small-signal transistor characteristics.

shown in Fig. 54 in elementary form. The three circuits correspond approximately to the grounded-grid, grounded-cathode and cathode-follower circuits, respectively, used with vacuum tubes.

The important transistor *parameters* in these circuits are the *short-circuit current transfer ratio,* the *cut-off frequency,* and the *input* and *output impedances.* The short-circuit current transfer ratio is the ratio of a small change in output current to the change in input current that causes it, the output circuit being short-circuited. The cutoff frequency was discussed earlier in this chapter. The input and output impedances are, respectively, the impedance which a signal source working into the transistor would see, and the internal output impedance of the transistor (corresponding to the plate resistance of a vacuum tube, for example).

Common-Base Circuit

The input circuit of a common-base amplifier must be designed for low impedance, since the emitter-to-base resistance is of the order of $26/I_e$ ohms, where I_e is the emitter current in milliamperes. The optimum output load impedance, R_L, may range from a few thousand ohms to 100,000, depending upon the requirements.

In this circuit the phase of the output (collector) current is the same as that of the input (emitter) current. The parts of these currents that flow through the base resistance are likewise in phase, so the circuit tends to be regenerative and will oscillate if the current amplification factor is greater than one.

Common-Emitter Circuit

The common-emitter circuit shown in Fig. 54 corresponds to the ordinary grounded-cathode vacuum-tube amplifier. As indicated by the curves of Fig. 52, the base current is small and the input impedance is therefore fairly high — several thousand ohms in the average case. The

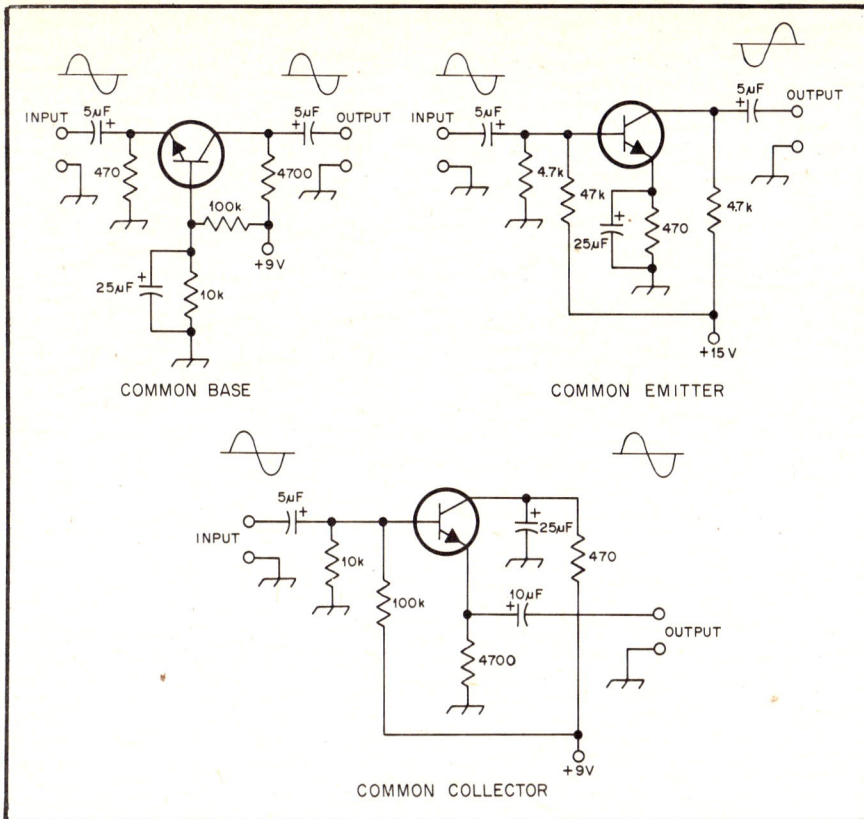

Fig. 54 — Basic transistor amplifiers. Observe the input and output phase relationships for the various configurations.

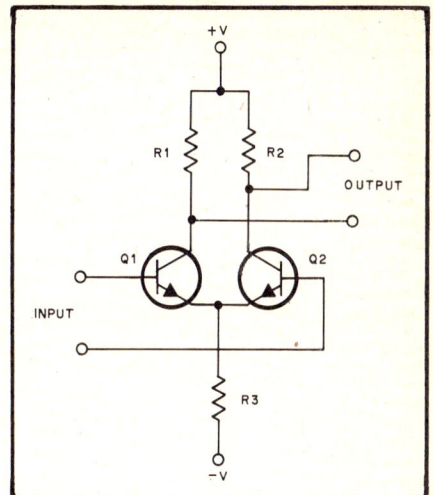

Fig. 54A — Differential amplifier. This arrangement can be analyzed as a composite of the common-collector and common-base circuits.

collector resistance is some tens of thousands of ohms, depending on the signal source impedance. The common-emitter circuit has a lower cutoff frequency than does the common-base circuit, but it gives the highest power gain of the three configurations.

In this circuit the phase of the output (collector) current is opposite to that of the input (base) current so such feedback as occurs through the small emitter resistance is negative and the amplifier is stable.

Because the common-emitter amplifier circuit illustrated in Fig. 54 is one of the most often seen applications for a bipolar transistor, a brief analysis and discussion of the design procedure is appropriate. The divider network biasing the base supplies an open-circuit potential of 1.36 V. Connecting the base has little loading effect if the transistor beta is high (a fair assumption). A beta of 100, for example, causes the 470-ohm emitter resistor to present a dc base resistance of 47 kΩ, which is a negligible shunt. The base-emitter junction drops 0.6 V, so the potential across the emitter resistor is $1.36 - 0.6 = 0.76$ V. This voltage causes 1.6 mA to flow in the emitter and the 470-ohm resistor. Since the transistor alpha is nearly unity, the collector current is also 1.6 mA, which drops 7.52 volts across the collector load resistor. The quiescent (no signal) collector voltage is therefore $15 - 7.52 = 7.48$ V. This value allows the maximum undis-

torted output voltage swing between cutoff and saturation.

Assuming the emitter bypass capacitor has negligible reactance at the operating frequency, the emitter is at ac ground. Because of the high alpha mentioned earlier, any emitter current variation caused by an input signal will also appear in the collector circuit. Since the current variation is the same, the voltage gain is the ratio of the collector load resistance to the (internal) emitter resistance. For small signals, this emitter resistance can be approximated by $R_e = 26/I_e$, where R_e is the emitter-to-base junction resistance, and I_e is the emitter current in milliamperes. In our example, $I_e = 1.6$, so $R_e = 16.25\ \Omega$. The voltage gain, then, is 289, which is 49 dB. The ac base impedance is given by βR_e. Using the previous values for beta and emitter resistance results in a base impedance of 1625 Ω. The circuit input impedance is found by shunting the base impedance with the bias resistors. The result in this case is about 1189 Ω.

If the emitter bypass capacitor is omitted, the external resistor dominates the gain equation, which becomes: $A_v = R_L/R_E$, where A_v is the voltage gain, R_L is the collector load resistance as before, and R_E is the unbypassed emitter resistance. Without the bypass capacitor, the common-emitter circuit in Fig. 54 exhibits a numerical voltage gain of 10 (20 dB). The base impedance becomes βR_E, or 47 kΩ.

This value is swamped by the bias network for a circuit input impedance of 3.91 kΩ. The emitter resistor has introduced 29 dB of degenerative feedback to the circuit, stabilizing the gain and impedance values over a wide frequency range. A dc beta of 100 was assumed in this example, and for convenience this value was also assigned at the operating frequency. In reality, however, beta decreases with increasing frequency, as noted in the section on transistor characteristics. Degenerative feedback overcomes this effect to a large extent.

Common-Collector Circuit (Emitter-Follower)

Like the vacuum-tube cathode follower, the common-collector transistor amplifier has high input impedance and low output impedance. The latter is approximately equal to the impedance of the signal input source multiplied by $(1 - \alpha)$. The input resistance depends on the load resistance, being approximately equal to the load resistance divided by $(1 - \alpha)$. The fact that input resistance is directly related to the load resistance is a disadvantage of this type of amplifier if the load is one whose resistance or impedance varies with frequency.

The current transfer ratio with this circuit is

$$\frac{1}{1 - \alpha}$$

and the cutoff frequency is the same as in the grounded-emitter circuit. The output and input currents are in phase.

Differential Amplifier Circuit

An important variation of the fundamental amplifier types is the *differential amplifier*, drawn in Fig. 54A. The output voltage is proportional to the difference (with respect to ground) between the voltages applied to the input terminals. With the proper choice of operating conditions, several differential amplifier stages of the type shown can be cascaded directly. Fig. 54A shows the circuit in its classic balanced form, but many circuits use differential amplifiers in a single-ended configuration. When only a single input and output terminal is required, R1 could be a short circuit and the Q2 base could be grounded. Under these circumstances the differential amplifier can be understood as an emitter-follower driving a common-base stage. The output is taken between the Q2 collector and ground. R3 establishes the current in Q1 and Q2, which should be equal under static conditions.

Differential amplifiers work best when R3 is replaced by some type of constant-current source. One type of current regulator has been discussed in the diode section, and current sources made from bipolar transistors are covered later.

With a current source biasing Q1 and Q2 the input signal cannot modulate the total collector current; only the *ratio* of the currents varies. One beneficial result of the constant-current bias is that a higher impedance is presented to the driving signal.

Bipolar Transistor Dissipation

Apart from the characteristics mentioned earlier, it is necessary to consider the matters of collector dissipation, collector voltage and current and emitter current. Variations in these specifications are denoted by specific parameter symbols which appear later in the chapter. The maximum dissipation ratings of transistors, as provided on the manufacturer's data sheets, tend to confuse some amateurs. An acceptable rule of thumb is to select a transistor which has a maximum dissipation rating of approximately twice the dc input power of the circuit stage. That is, if a 5-watt dc input is contemplated, choose a transistor with a 10-watt or greater rating. When power levels in excess of a few hundred mW are necessary there is a need for heat sinking. A sink is a metal device which helps to keep the transistor cool by virtue of heat transfer from the transistor case to the sink. At power levels below 5 watts it is common practice to employ clip-on heat sinks of the crown variety. For powers greater than 5 watts it is necessary to use large-area heat sinks which are fashioned from extruded aluminum. These sinks have cooling fins on one or more of their surfaces to hasten the cooling process. Some high-power, solid-state amplifiers employ cooling fans from which the air stream is directed on the metallic heat sink. Regardless of the power level or type of heat sink used, silicone heat-transfer compound should always be used between the mating surfaces of the transistor and the heat sink. Another rule of thumb is

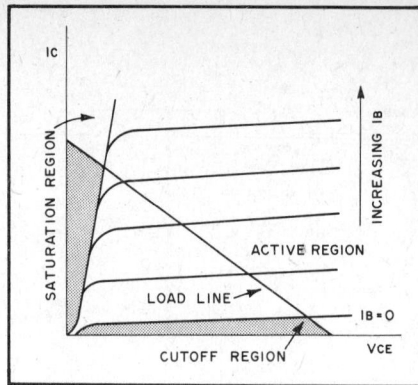

Fig. 55 — Typical characteristic for the collector of an npn transistor which shows the three primary regions involved during switching.

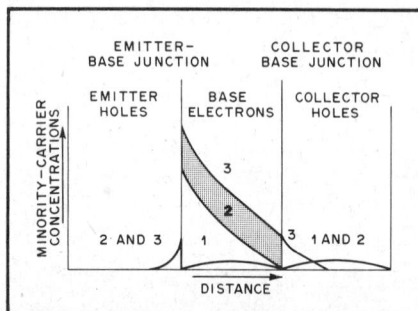

Fig. 56 — Illustration of the minority-carrier concentrations in an npn transistor. No. 1 shows the cutoff region. No. 2 is the active region at the threshold of the saturation region. No. 3 is in the saturation region.

offered: If the heat-sink-equipped transistor is too warm to touch with comfort, the heat sink is not large enough in area.

Excessive junction heat will destroy a transistor. Prior to destruction the device may go into *thermal runaway*. During this condition the transistor becomes hotter and its internal resistance lowers. This causes an increase in emitter/collector and emitter/base current. This increased current elevates the dissipation and further lowers the internal resistance. These effects are cumulative: Eventually the transistor will be destroyed. A heat sink of proper size will prevent this type of problem. Excessive junction temperature will eventually cause the transistor to become open. Checks with an ohmmeter will indicate this condition after a failure.

Excess collector voltage will also cause immediate device failure. The indication of this type of failure, as noted by means of an ohmmeter, is a shorted junction.

Bipolar Transistor Applications

Silicon transistors are the most common types in use today, although a few germanium varieties are built for specific applications. Collector voltages as great as 1500 can be accommodated by some of the high-power silicon transistors available now. Most small-signal transistors will safely handle collector voltages of 25 or greater. Generally speaking, transistors

in the small-signal class carry dissipation ratings of 500 mW or less. Power transistors are normally classed as 500-mW and higher devices. The practical applications for all of these semiconductors range from dc to the microwave spectrum.

Bipolar Transistor Switches

Our present-day technology includes the use of solid-state switches as practical alternatives to mechanical switches. When a bipolar transistor is used in a switching application it is either in an *on* or *off* state. In the on state a forward bias is applied to the transistor, sufficient in level to saturate the device. The common-emitter format is used for nearly all transistor switches. Switching action is characterized by large-signal nonlinear operation of the device. Fig. 55 shows typical output characteristics for an npn switching transistor in the common-emitter mode. There are three regions of operation — *cutoff, active* and *saturation*. In the cutoff region the emitter-base and collector-base junctions are reverse biased. At this period the collector current is quite small and is comparable to the leakage current, I_{ceo}, I_{cev} or I_{cbo}.

Fig. 56 illustrates the minority-carrier concentration relative to an npn transistor. During cutoff the concentration is zero at both junctions because they are reverse biased (curve no. 1).

The emitter-base junction is forward biased in the active region. At this time the collector-base junction is reverse biased. Fig. 55 shows a load line along which switching from the cutoff to the active region is done. The transit time (speed) through the active region is dependent upon the transistor frequency-response characteristics. Thus, the higher the frequency rating of the device, the faster the switching time. Curve no. 2 in Fig. 56 depicts the minority-carrier concentration of the active region.

In the saturation region the emitter-base and collector-base junctions are forward biased. During this period the forward voltage drop across the emitter-base junction $V_{BE(sat)}$ is larger than it is across the collector-base junction. This results in a collector-emitter voltage termed $V_{CE(sat)}$. Series resistances present in the emitter and collector legs of the circuit contribute to the determination of $V_{CE(sat)}$. Since the collector in this state is forward biased, additional carriers are injected into the base. Some also reach the collector. Curve no. 3 of Fig. 56 shows this minority-carrier concentration. Fig. 57 contains the circuit for a basic saturated-transistor switch.

It is extremely important to make certain that none of the transistor voltage ratings is exceeded during the "off" period: The minimum emitter-base breakdown voltage, $V_{(BR)EBO}$, must not exceed $V_{BE(off)}$. Also, the minimum collector-

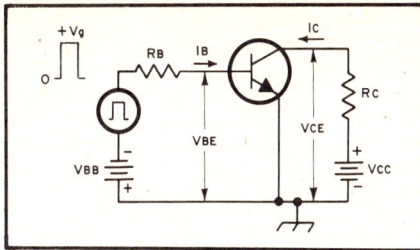

Fig. 57 — Circuit for a transistor switching circuit (saturated).

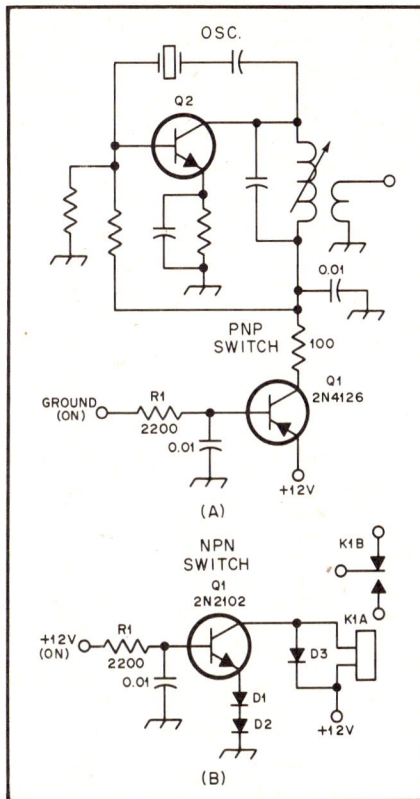

Fig. 58 — Examples of practical switching circuits. A pnp switch is used to key an oscillator at A. When R1 is grounded the switching transistor is forward biased to saturation, thereby permitting current to flow from the dc supply line to Q2. The circuit at B shows Q1 as a relay-driver npn switch. When +12 volts is applied to the base of Q1 it is forward biased to saturation, permitting current to flow through the field coil of K1A. D1 and D2 are included to reduce the static collector current of Q1, which in some instances could cause K1A to remain closed after forward bias was removed from Q1. D3 serves as a spike suppressor when the field of K1A collapses.

base breakdown voltage, $V_{(BR)CBO}$, should be no greater than $V_{cc} + V_{BE(off)}$. Finally, the minimum collector-to-emitter breakdown voltage, $V_{(BR)CERL}$, must be greater than V_{cc}. As is true in any transistor application, the junction temperature must be maintained at a safe value by whatever means necessary.

A transistor switch can be turned on by means of a pulse (Fig. 57) or by application of a dc forward bias. Typical circuits for the latter are given in Fig. 58. The circuit at A illustrates how a pnp tran-

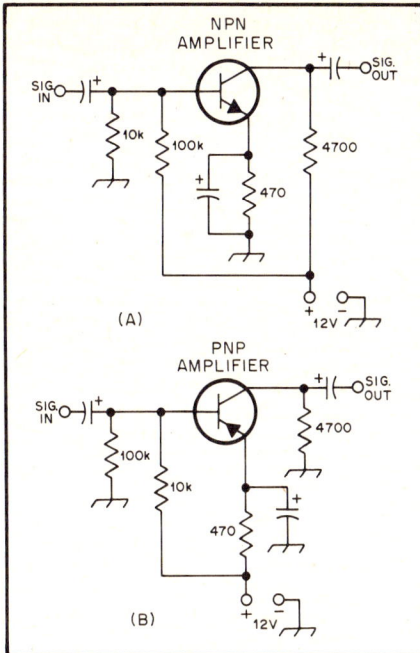

Fig. 59 — Examples of npn and pnp amplifiers operating from a power supply with a negative ground.

sistor can be used as a low-power switch to turn oscillator Q2 on and off. In the "on" state R1 is grounded. This places the bipolar switch, Q1, in a saturated mode, thereby permitting current to flow to Q2. A transistor switch of the type shown at A of Fig. 58 can be used to control more than one circuit stage simultaneously. The primary criterion is that the switching transistor be capable of passing the combined currents of the various stages under control. The method seen at A is often used in keying a transmitter.

An npn transistor switch is shown in Fig. 58B. If desired, it can be "slaved" to the circuit of Fig. 58A by attaching R1 of circuit B to the collector of Q1 in circuit A. Because an npn device is used at B, a positive forward bias must be applied to the base via R1 to make the transistor saturate. When in that state, current flows through the relay (K1A) field coil to actuate the contacts at K1B. D3 is connected across the relay coil to damp inductive

spikes which occur when the relay-coil field collapses. D1 and D2 may not be necessary. This will depend on the sensitivity of the relay and the leakage current of Q1 in the off state. If there is considerable leakage, K1 may not release when forward bias is removed from Q1. D1 and D2 will elevate the emitter to approximately 1.4 volts, thereby providing sufficient reverse bias to cut off Q1 in the off state. It can be seen from the illustrations in Fig. 58 that either npn or pnp transistors can be used as electronic switches.

Transistor Audio Amplifiers

Bipolar transistors are suitable for numerous audio-amplifier applications from low-level to high power. It is common practice to use all npn or all pnp devices, regardless of the polarity of the power supply. In other circuits a mixture of the two types may be found, especially when direct-coupled or complimentary-symmetry stages are included. Fig. 59 shows how pnp or npn stages can be used with power supplies which have positive or negative grounds. The essential difference in the circuits concerns returning various elements to the negative or positive sides of the power supply. The illustrations show that all one needs to do to use either type of device with the same power supply is to interchange the resistor connections. The same principle applies when using npn or pnp transistors with a power supply which has a positive ground. Knowledge of how this is done enables the designer to mix npn and pnp devices in a single circuit. This basic technique is applicable to any type of transistor circuit — rf, audio or dc.

Some basic low-level audio amplifiers are shown in Fig. 60. These stages operate in the Class A mode. The input impedance of these circuits is low — typically between 500 and 1500 ohms. For the most part the output impedance is established by the value to the collector load resistor. A matching transformer can be used at the input of these stages (Fig. 60B) when it is necessary to use, for example, a high-impedance microphone with one of them. T1 serves as a step-down transformer.

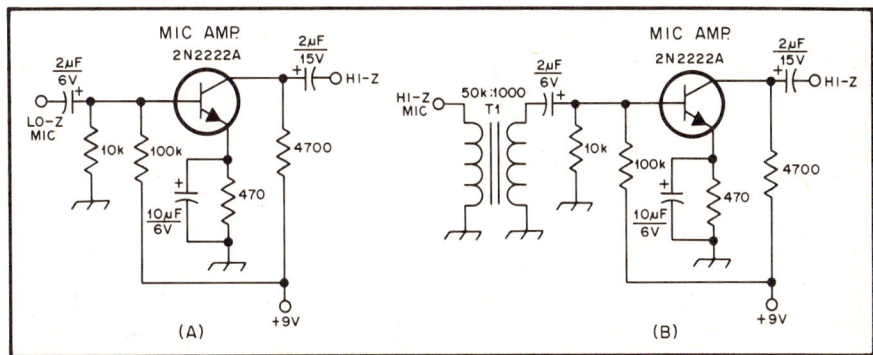

Fig. 60 — RC and transformer-coupled audio amplifiers suitable for high- and low-impedance microphones.

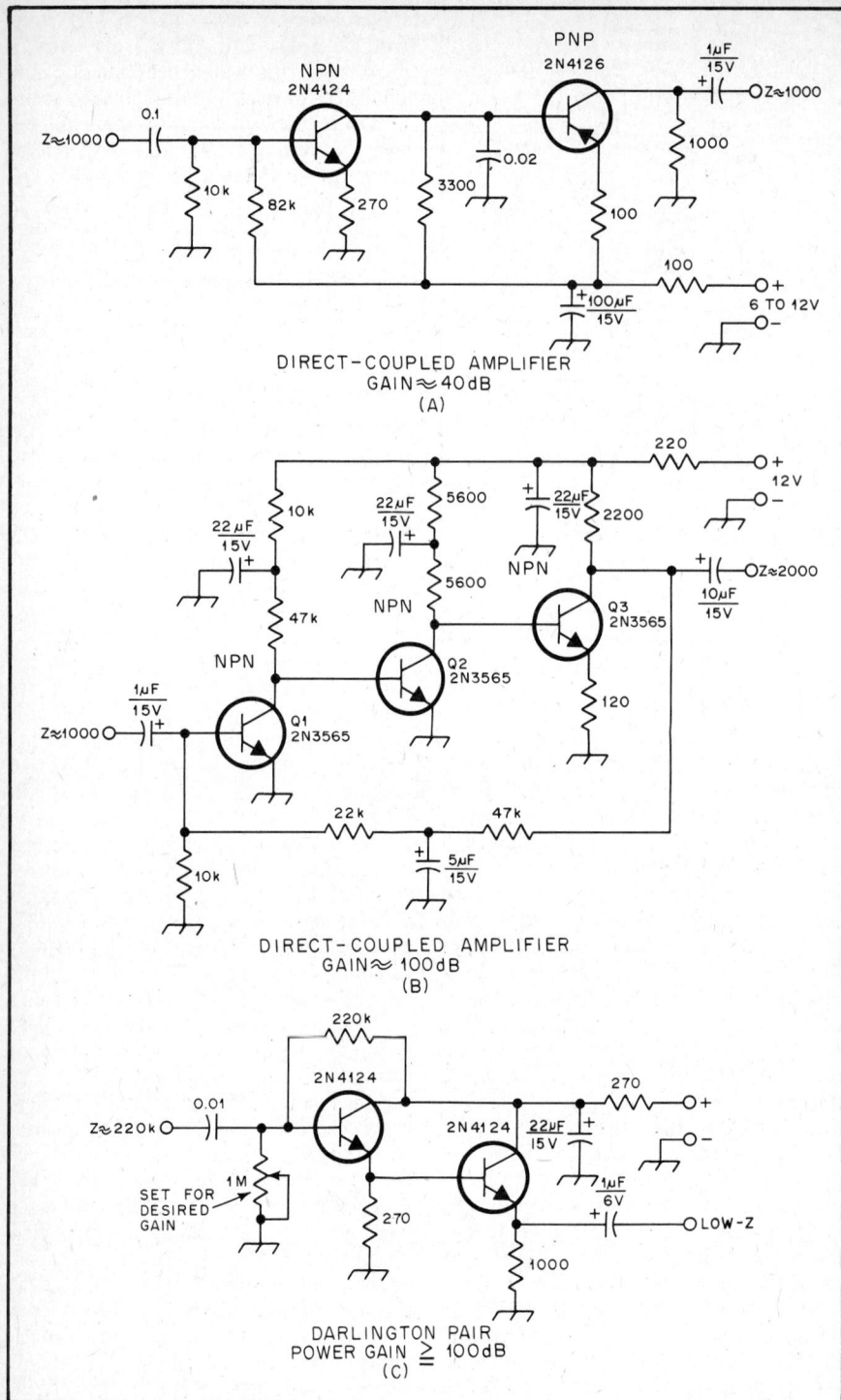

DIRECT-COUPLED AMPLIFIER
GAIN ≈ 40 dB
(A)

DIRECT-COUPLED AMPLIFIER
GAIN ≈ 100 dB
(B)

DARLINGTON PAIR
POWER GAIN ≥ 100 dB
(C)

Fig. 61 — Practical examples of direct-coupled audio amplifiers.

amplifiers. Other considerations are noise figure, purity of the amplified signal and dynamic range.

Although bipolar transistors can be used as rf amplifiers for receiver front ends, they are not found there in most of the high-performance receivers: Field-effect transistors are more often the designer's choice because of their high input impedance and good dynamic-range traits. A correctly designed bipolar rf input stage can exhibit good dynamic range, however. It is necessary to operate a fairly husky low-noise transistor in Class A, using a relatively high standing collector current — 50 to 100 mA, typically.

Some rf and i-f amplifier circuits which employ bipolar transistors are shown in the examples of Fig. 62. When used with the appropriate L and C networks they are suitable for either application. At A in Fig. 62 the transistor base is tapped near the cold end of the input tuned circuit to provide an impedance match. The collector is tapped down on the output tuned circuit to provide a proper match. If it is desired, the base and collector taps can be moved even farther down on the tuned circuits. This will result in a deliberate mismatch. The technique is sometimes used to aid stability and/or lower the stage gain. The circuit at B is operated in the common-base mode. Taps are shown on the input and output coils for impedance-matching purposes.

Broadband amplifiers with heavy negative feedback are useful as small-signal rf and i-f amplifiers. An example is shown at C in Fig. 62. Not only is negative feedback applied (collector to base), but degenerative feedback is obtained by virtue of the unbypassed 10-ohm emitter resistor. The use of feedback ensures an unconditionally stable stage. As the operating frequency is decreased the negative feedback increases because the feedback-network reactance becomes lower. This is important if reasonably constant gain is desired over a wide range of frequencies, say, from 1.8 to 30 MHz. This form of gain compensation is necessary because as the operating frequency of a given transistor is decreased, the gain increases. Typically, the gain will increase on the order of 6 dB per octave. Therefore, the probability of instability (self-oscillations) becomes a major consideration at low frequencies in an uncompensated rf amplifier. The circuit of Fig. 62C operates stably and has a characteristic input and output impedance of approximately 50 ohms. The broadband 4:1 transformer in the collector circuit is required to step down the collector impedance to 50 ohms. Design information on this type of circuit is provided in the ARRL book, *Solid State Design for the Radio Amateur*. A bandpass type of filter is needed at the amplifier input. Another can be used at the output of the

Some direct-coupled audio amplifiers are shown in Fig. 61. The circuit at A combines pnp and npn devices to provide a compatible interface between them. Three npn stages are in cascade at B to provide high gain. This circuit is excellent for use in direct-conversion receivers, owing to the need for very high gain after the detector. At C is a Darlington pair — so named after the person who developed the configuration. The principle advantages of this circuit are high gain, high in-

put impedance, and low output impedance.

Transistor RF Amplifiers

In most respects small-signal rf amplifiers are similar in performance to those used in audio applications. However, to effect maximum stable amplification some important design measures are necessary. Furthermore, the matter of proper impedance matching becomes more important than it is in simple audio

4:1 transformer if desired. The transistor used in any of the amplifiers of Fig. 62 should have an f_T which is five to ten times greater than the highest operating frequency of the stage. The 2N5179 has an f_T in excess of 1000 MHz, making it a good device up to 148 MHz for this application.

Transistor RF Power Amplifiers

Rf power amplifiers which use bipolar transistors fall into two general categories — Class C and linear. The latter is used for a-m and ssb signal amplification and the class of operation is A or AB. These amplifiers are designed for narrow or wideband applications, depending on the purpose for which the stage or stages will be used. Class C bipolar-transistor amplifiers are used for fm and cw work.

Most wideband amplifiers contain ferrite-loaded broadband transformers at the input and output ports. The output transformer is followed by a multipole low-pass filter for each band of operation. This is necessary to attenuate harmonic currents so that they will not be radiated by the antenna system. Although this type of filtering is not always needed with a narrowband amplifier (the networks provide reasonable selectivity), filters should be used in the interest of spectral purity. Two-section filters of the half-wave or low-pass T variety are entirely suitable for harmonic reduction at the 50-ohm output ports of amplifiers.

One of the principal difficulties encountered by amateurs who design and build their own high-power, solid-state amplifiers is instability at some point in the power range. That is, an amplifier driven to its maximum rated output may be stable when terminated properly, but when the drive level is reduced it is apt to break into self-oscillation at the operating frequency, at vhf, or perhaps at very low frequencies. Part of the problem is caused by an increase in beta as the collector current is decreased. This elevates the amplifier gain to encourage instability. Also, solid-state amplifiers are designed for a specific network impedance at a specified power-output level. When the drive is reduced the collector and base impedances increase. This causes a mismatch. An increase in the loaded Q of the networks may also result — a situation that encourages instability. Therefore, it is best to design for a specified power output and adhere closely to that level during operation.

Solid-state power amplifiers should be operated just below their saturation points for best efficiency and stability. That is the point which occurs when no additional rf output can be obtained with increased driving power. Some designers recommend that, for example, a 28-volt transistor be used for 12-volt operation: Saturation will take place at a level where the transistors are relatively safe from

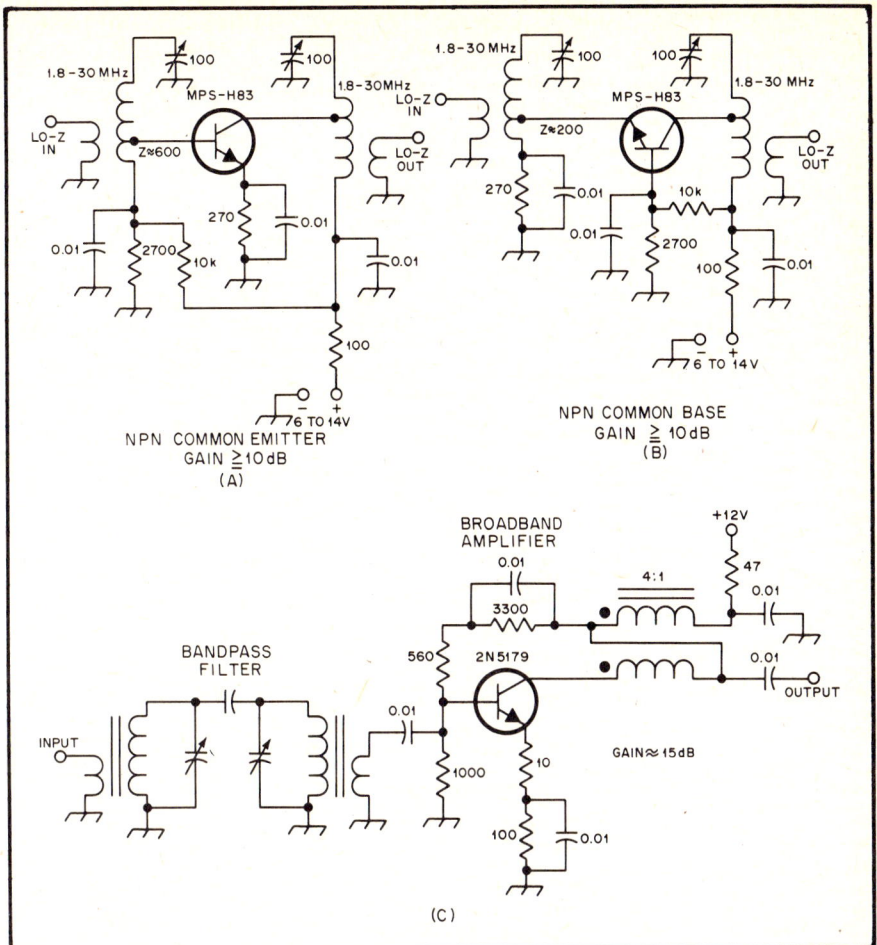

Fig. 62 — Illustrations of common-emitter and common-base rf amplifiers for narrow-band use. The circuit at C is that of a fed-back broadband amplifier which has a bandpass filter at the input.

damage if a significant output mismatch is present. Stability is usually better under these same conditions, although the gain of the transistors will be considerably lower than would be the case if equivalent types of 12-volt devices were used.

Fig. 63 shows two single-ended amplifiers of typical design. At A there is a broadband input transformer which steps down the 50-ohm source to the low base impedance of Q1. Most power stages have a base impedance of 5 ohms or less. Although there are a number of suitable tuned networks that can be used to effect the desired impedance match, the use of T1 eliminates components and the sometimes complex calculations required for the design of a proper network. When the actual base impedance of Q1 is unknown (it varies with respect to drive level and operating frequency), empirical adjustment of the T1 turns ratio will permit close matching. An SWR indicator can be used between T1 and the signal source to indicate a matched condition. This test should be made with the maximum intended drive applied.

To continue the discussion relating to Fig. 63A, a 10-ohm resistor (R1) is bridged across the secondary of T1 to aid stability. This measure is not always necessary. It will depend on the gain of the transistor, the layout and the loaded Q of T1. Other values of resistance can be used. A good rule of thumb is to employ only that value of resistance which cures instability. It must be remembered that R1 is in parallel with the transistor input impedance. This will have an effect on the turns ratio of T1. When excessive driving power is available, a deliberate mismatch can be introduced at the input to Q1 by reducing the number of secondary turns. If that is done, R1 can often be eliminated. The shortcoming which results from this technique is that the driving source will not be looking into a 50-ohm termination. T1 is normally a ferrite-loaded transformer — toroidal or solenoidal. The core material for operation from 1.8 to 30 MHz is typically the 950-μ_i (Initial permeability) type. The primary winding of T1 (and other broadband transformers should be approximately four times the terminal impedance with respect to reactance. Therefore, for a

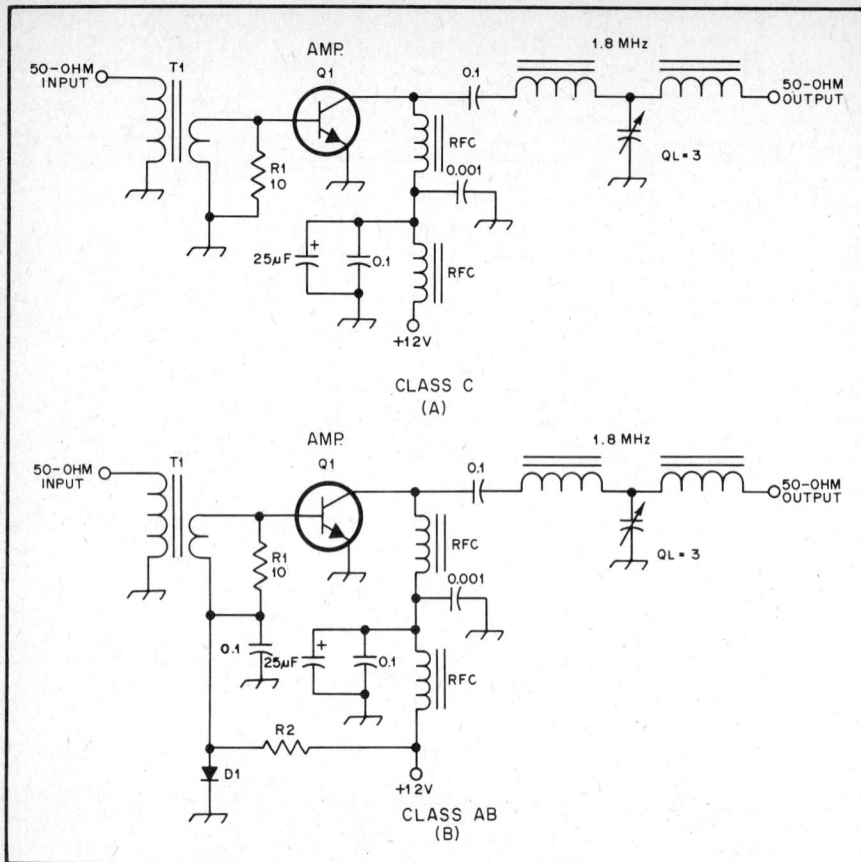

Fig. 63 — Circuits for rf power amplifiers. At A is a Class C type. The circuit at B is biased for linear amplification.

of matching network should be kept below 4 in the interest of amplifier stability. Information on this and other types of tuned matching networks is given in the *ARRL Electronics Data Book*. Data are also given in that volume concerning broadband transformer design.

Fig. 63B shows the same general amplifier. The difference is in the biasing. The circuit at A is set up for Class C. It is driven into the cutoff region during operation. At B there is a small amount of forward bias applied to Q1 (approximately 0.7 volt) by means of the barrier voltage set by D1, a silicon power diode. D1 also functions as a simple bias regulator. R2 should be selected to provide fairly substantial diode current. The forward bias establishes linearity for the amplifier so that ssb and a-m driving energy can be amplified by Q1 with minimum distortion.

Although some transistors are designed especially for linear amplification, any power transistor can be used for the purpose. Once the proper biasing point is found for linear amplification with a Class C type of transistor, an investigation of linearity versus output power should be undertaken. Some Class C transistors are incapable of delivering as much power (undistorted) in Class AB as they can under Class C conditions. Most power transistors intended for linear amplification have built-in, degenerative-feedback resistors at the emitter sites. This technique aids linearity. Depending on the package style of a Class C type of transistor, an emitter-feedback resistor can be added externally. Such resistors are usually on the order of 1 ohm.

A broadband Class C amplifier is shown in Fig. 64A. T1 and T2 are 4:1 broadband transformers connected in series to provide an impedance step down of 16:1. For most applications this arrangement will provide an acceptable

50-ohm load characteristic the primary-winding reactance of T1 should be roughly 200 ohms.

Two rf chokes are shown in Fig. 63A. These are necessary to assure ample dc-lead decoupling along with the related bypass capacitors. The upper rf choke serves also as a collector load impedance. The reactance should be four or five times the collector impedance. Three values of

bypass capacitors are used to ensure effective decoupling at vhf, hf and lf. If the decoupling is inadequate, rf from the amplifier can flow along the 12-volt bus to other parts of the transmitter, thereby causing instability of one or more of the stages. A simple low-pass T network is used for matching the collector to the 50-ohm load. It also suppresses harmonic energy. The loaded Q of this general type

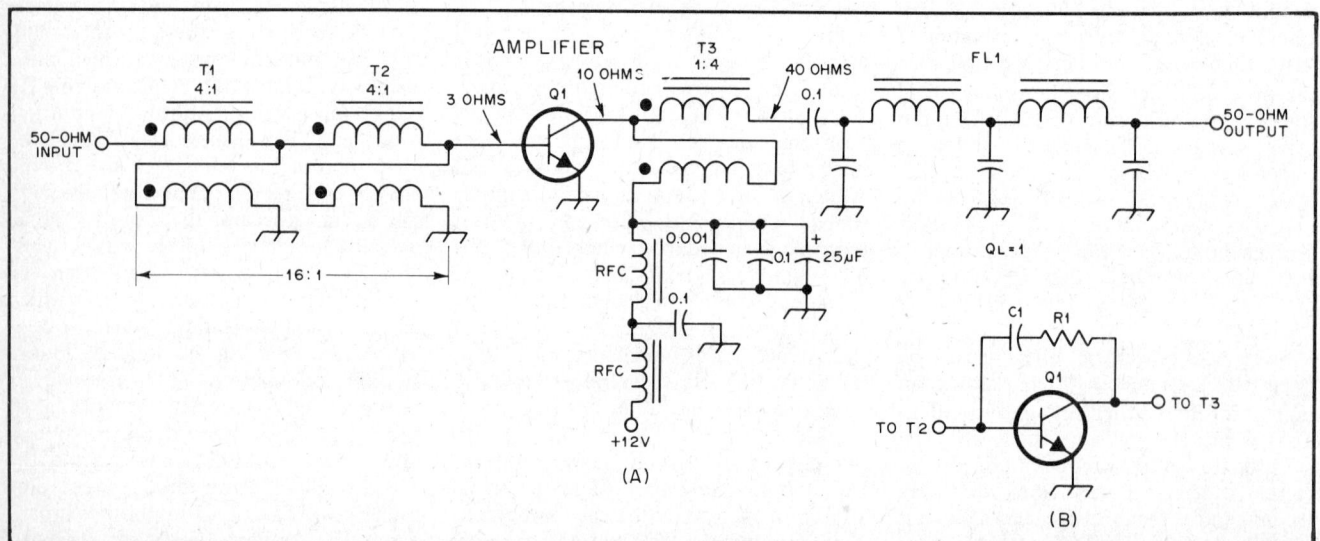

Fig. 64 — Broadband transformers are employed at A for impedance matching. FL1 suppresses harmonic currents at the amplifier output. In the examples at B are feedback components C1 and R1 (see text).

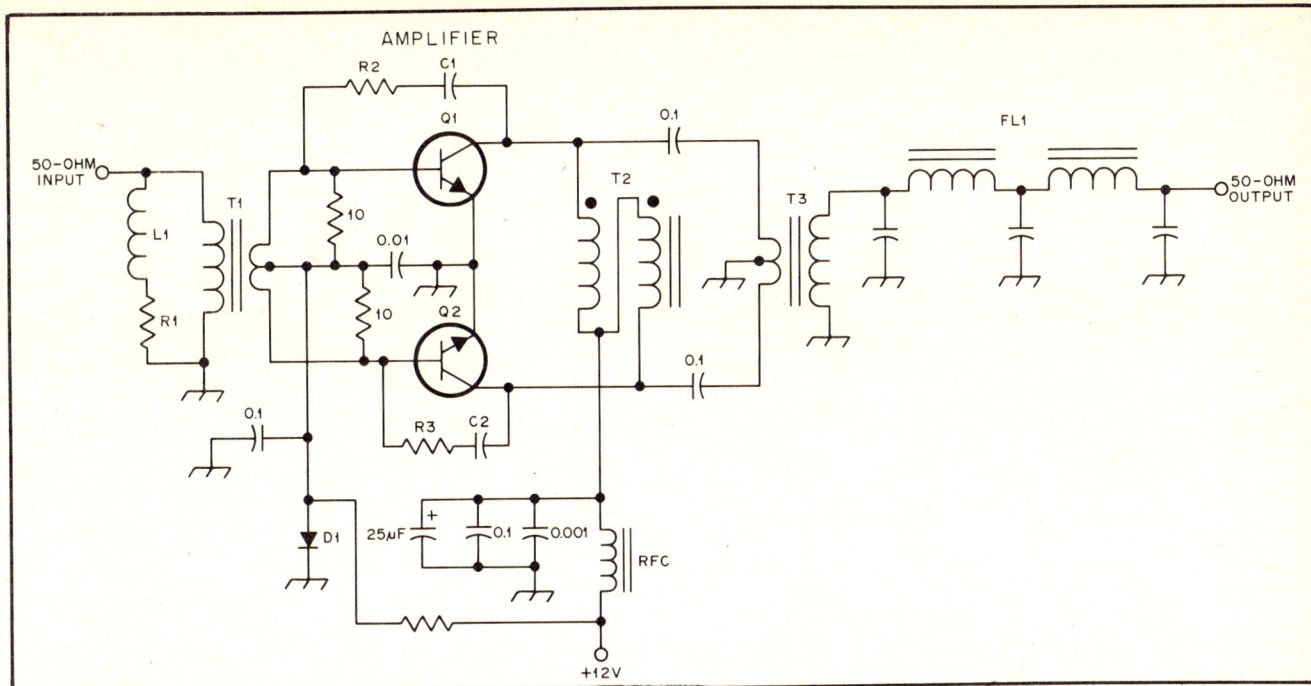

Fig. 65 — Example of a fed-back, push-pull, rf power amplifier set up for broadband service from 1.8 to 30 MHz. The circuit is biased for linear amplification.

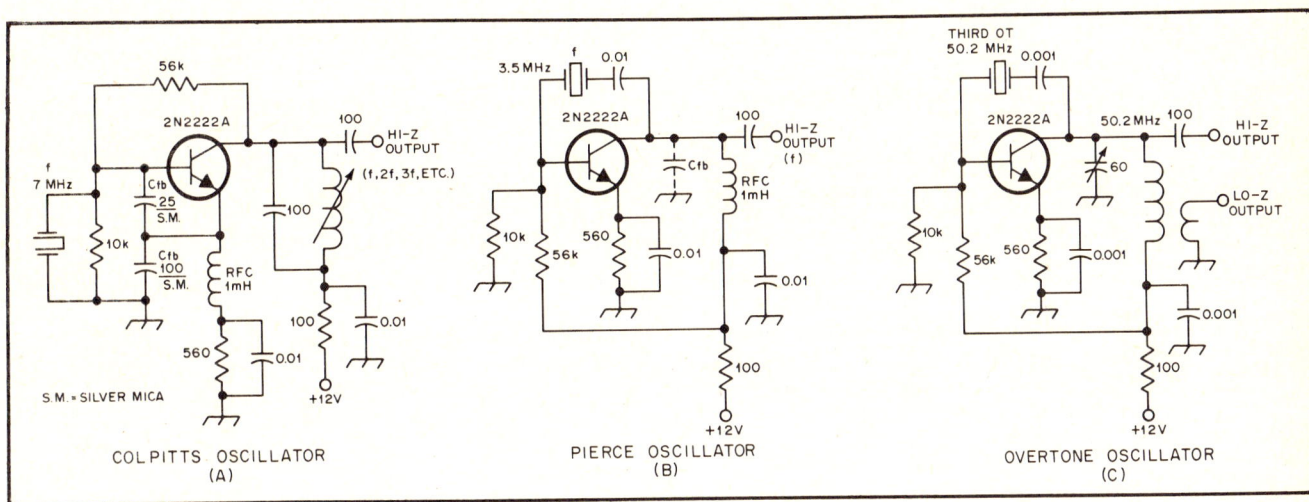

Fig. 66 — Examples of transistor oscillators which use crystal control.

match between 50 ohms and the base impedance of Q1. In the example we have assumed a base impedance of approximately 3 ohms.

T3 serves as a collector load and a step-up transformer. It is useful to use a step-up transformer when the collector impedance is low (25 ohms or less). This enables the designer to work with filter-component values (FL1) that are more practical than would be the case if an attempt was made to match 10 ohms to 50 ohms with the filter network. FL1 in this example is a double pi-section low-pass type (half-wave filter). It is designed to match 40 ohms to 50 ohms and has a loaded Q of 1.

Feedback can be applied to stabilize the amplifier. This is seen in Fig. 64B. C1 and

R1 are chosen to reduce the amplifier gain by whatever amount is necessary to provide stability and the broadband characteristics desired. C1 serves as a dc blocking capacitor.

A push-pull broadband linear amplifier is illustrated at Fig. 65. When additional frequency compensation is desired (beyond that available from a negative-feedback network) L1 and R1 can be added across the amplifier input. They are selected to roll off the driving power toward the low end of the amplifier operating range. As the frequency is reduced, L1 represents a lower reactance, thereby permitting some of the drive power to be dissipated in R1.

T1 is a conventional broadband transformer (not a transmission-line type) with

a turns ratio set for matching 50 ohms to the base load presented by Q1 and Q2. T2, another broadband transformer, is used to provide balanced dc feed to the collectors. T3 is another broadband transformer which is wound for lowering the collector-to-collector impedance to 50 ohms. FL1 is designed for a bilateral impedance of 50 ohms in this example.

Bipolar-Transistor Oscillators

Transistors function well as crystal-controlled or LC oscillators. RC oscillators are also practical when a bipolar transistor is used as the active element. The same circuits used for tube-type oscillators apply when using transistors. The essential difference is that transistor oscillators have lower input and output

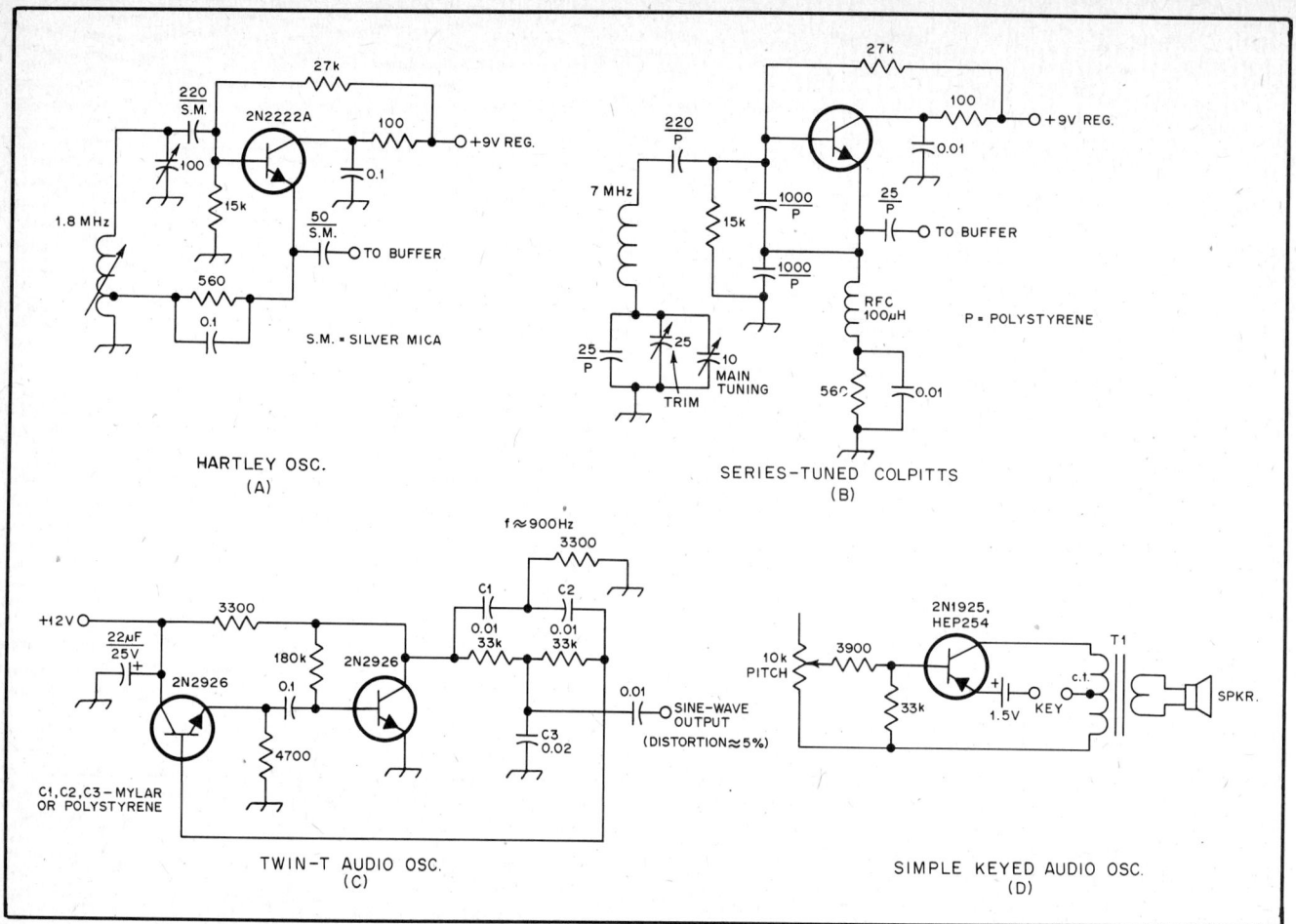

Fig. 67 — The circuits at A and B are VFOs for use in transmitters or receivers. Audio oscillators are shown at C and D.

impedances, operate at low voltages, and deliver low output power — usually in the mW range. The greater the oscillator power, the greater the heating of the transistor junction and other circuit elements. Therefore, in the interest of oscillator stability it is wise to keep the dc input power as low as practical. The power level can always be increased by means of subsequent amplifier stages at minor cost.

Some representative examples of crystal-controlled oscillators are provided in Fig. 66. At A is an oscillator that can be used to obtain output at f (the crystal frequency), or at multiples of f. The circuit at B illustrates a Pierce type of oscillator for fundamental output at 3.5 MHz. C_{fb} may be necessary with some crystals to provide ample feedback to cause oscillation. The value of C_{fb} will depend on the operating frequency and the gain of the transistor. Typically for 1.8 to 20-MHz crystals (fundamental mode) the capacitance value ranges from 25 to 100 pF. The higher values are typical at the lower end of the frequency range. In Fig. 66C is an overtone oscillator. The collector tuned circuit must be able to resonate slightly above the crystal overtone frequency in order to ensure oscillation. Low-impedance output can be had by

means of the link shown. Alternatively, a capacitive divider can be placed across the inductor to provide a low-Z tap-off point. The trimmer should be retained in parallel with the inductor to permit resonating the circuit.

Some typical rf and audio oscillators are seen in Fig. 67. The circuit at A obtains feedback by means of the emitter tap on the tuned circuit. Approximately 25 percent of the oscillator rf power is used as feedback. The tap point on this type of oscillator is between 10 and 25 percent of the total coil turns. The designer should use the smallest amount of feedback that will provide reliable oscillator performance with the load connected.

Fig. 67B illustrates a series-tuned Colpitts oscillator, although this general circuit is often referred to as a "series-tuned Clapp" oscillator. It is very stable when polystyrene capacitors are used in the feedback and tuned circuits. Silver-mica capacitors can be used as substitutes at a slight sacrifice in drift stability (long term).

A twin-T audio oscillator is shown at C in Fig. 67. It is a very stable type of circuit which delivers a clean sine-wave output. Mylar or polystyrene capacitors should be used for best stability.

A simple feedback circuit is effected by

means of T1 in Fig. 67D. T1 is a small transistor output transformer with a center-tapped primary and an 8-ohm secondary. This circuit is excellent for use as a code-practice or side-tone oscillator.

All of the rf oscillators described in these examples should be followed by one or more buffer stages to prevent frequency changes resulting from load variations occurring after the oscillator chain.

Transistor Mixers

Much of the modern equipment used by amateurs contains mixers which utilize FETs or diode rings. Good dynamic range is offered by those two circuits. However, there is no reason why a bipolar mixer can't be used to obtain satisfactory results if care is taken with the operating parameters and the gain distribution in the receiver or transmitter where they are used. The bipolar transistors used in receiver mixers should be selected according to noise figure (low) and dynamic range (high). The signal applied to it should be kept as low as possible, consistent with low-noise operation. Most semiconductor manufacturers specify certain transistors for mixer service. Although this does not mean that other types of bipolar transistors can't be used for mixing, it is wise to select a device that is designed for that class of service.

Fig. 68 contains examples of three basic types of transistor mixers. At A is seen the most common one. It is found in simple circuits such as transistor a-m broadcast-band receivers. As an aid to dynamic range, the mixers of Fig. 68 can be used without rf amplifier stages ahead of them for frequencies up to and including 7 MHz: The noise in that range (ambient from the antenna) will exceed that of the mixer.

The primary limitation in the performance of the mixer of Fig. 68A is that the local-oscillator voltage is injected at the base. This does not afford good LO/input-signal isolation. The unfavorable result can be oscillator "pulling" with input load changes, and/or radiation of the LO energy via the antenna if the front-end selectivity is marginal or poor. The advantage of the circuit is that it requires less injection voltage than the one at B, where emitter injection is used.

At Fig. 68B is the same basic mixer, but with LO voltage applied to the emitter. This technique requires slightly higher levels of LO energy, but affords greater LO isolation from the mixer input port.

A singly balanced bipolar-transistor mixer is illustrated in Fig. 68 at C. R1 is adjusted to effect balance. This circuit could be modified for emitter injection by changing R1 to 1000 ohms, replacing the 220-ohm resistors with 1-mH rf chokes, and injecting the LO output at the junction of the two 0.01-μF capacitors. The center tap of the input transformer (base winding) would then be bypassed by means of a 0.01-μF capacitor.

Other Uses for Bipolar Transistors

It is possible to take advantage of the junction characteristics of small-signal transistors for applications which usually employ diodes. One useful technique is that of employing transistors as voltage-variable capacitors (varactors). This method is seen in Fig. 69. The collector-base junction of Q1 and Q2 serve as diodes for tuning the VFO. In this example the emitters are left floating. A single transistor could be used, but by connecting the pair in a back-to-back arrangement they never conduct during any part of the rf cycle. This minimizes loading of the oscillator. The junction capacitance is varied by adjusting the tuning control, R1. In this circuit the tuning range is approximately 70 kHz.

A bipolar-transistor junction can be used as a Zener diode in the manner shown in Fig. 70. Advantage is taken of the reverse-breakdown characteristic of Q1 to establish a fixed reference level. Most transistors provide Zener-diode action between 6 and 9 volts. The exact value can be determined experimentally.

When the base and collector of a bipolar transistor are connected and a for-

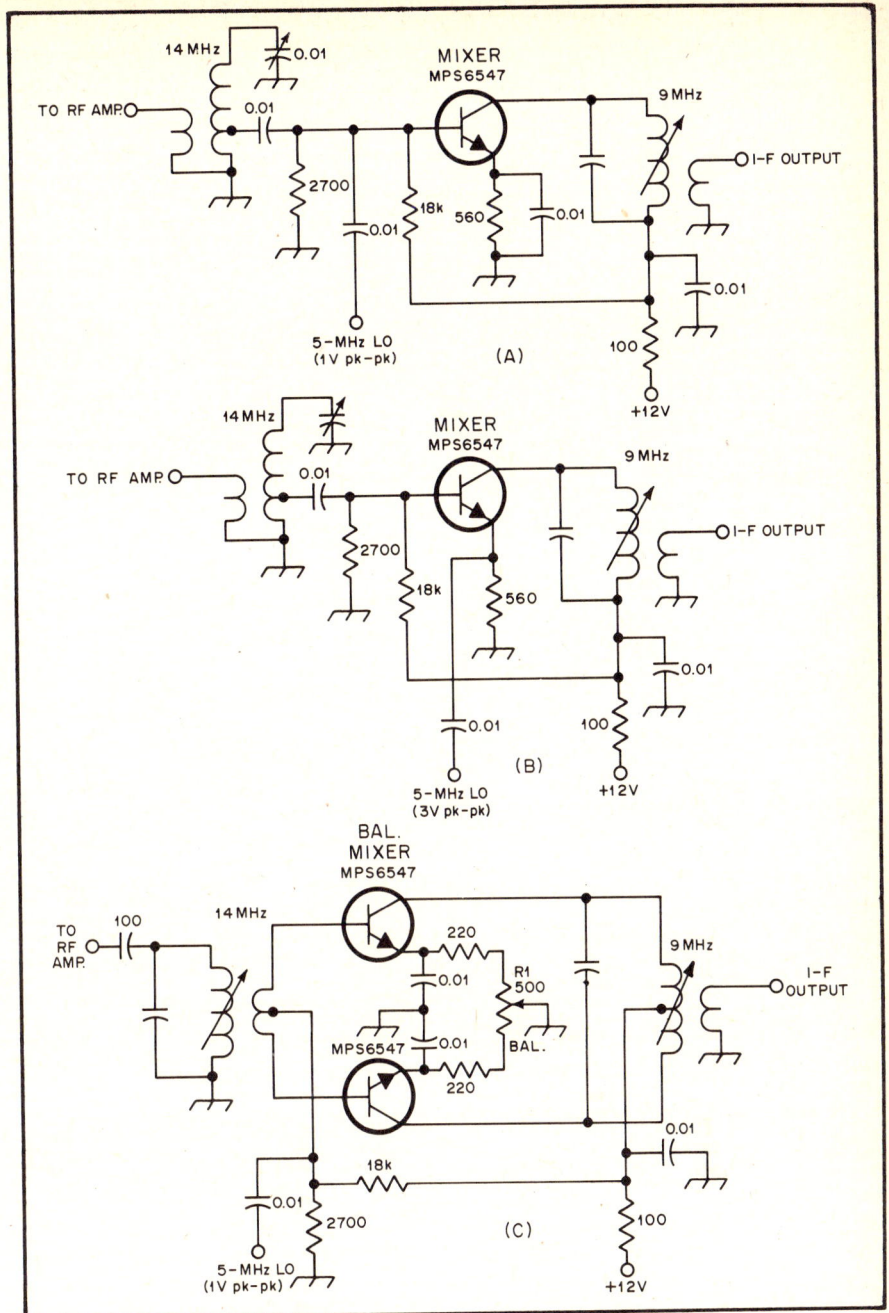

Fig. 68 — Some typical bipolar-transistor mixers. Their characteristics are discussed in the text.

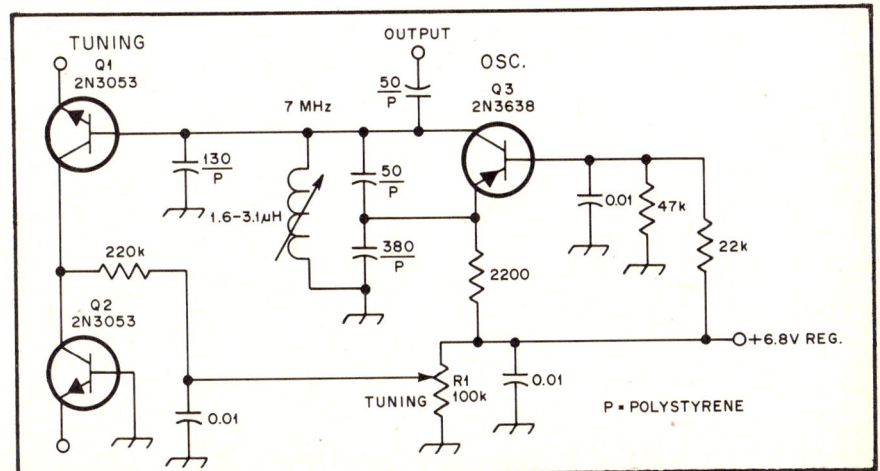

Fig. 69 — Bipolar transistors serve as varactor tuning diodes in this circuit (Q1 and Q2).

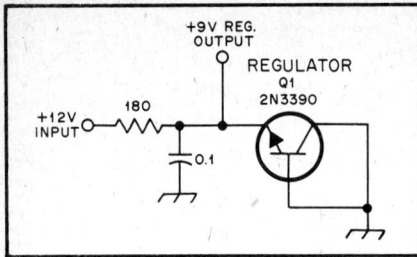

Fig. 70 — A bipolar transistor will function as a Zener diode when connected as shown here.

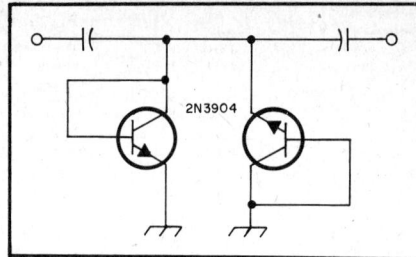

Fig. 71 — A peak clipper circuit using bipolar transistors connected in the superdiode configuration.

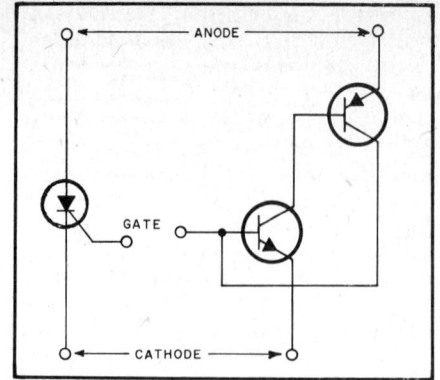

Fig. 73 — An SCR and its discrete functional near-equivalent.

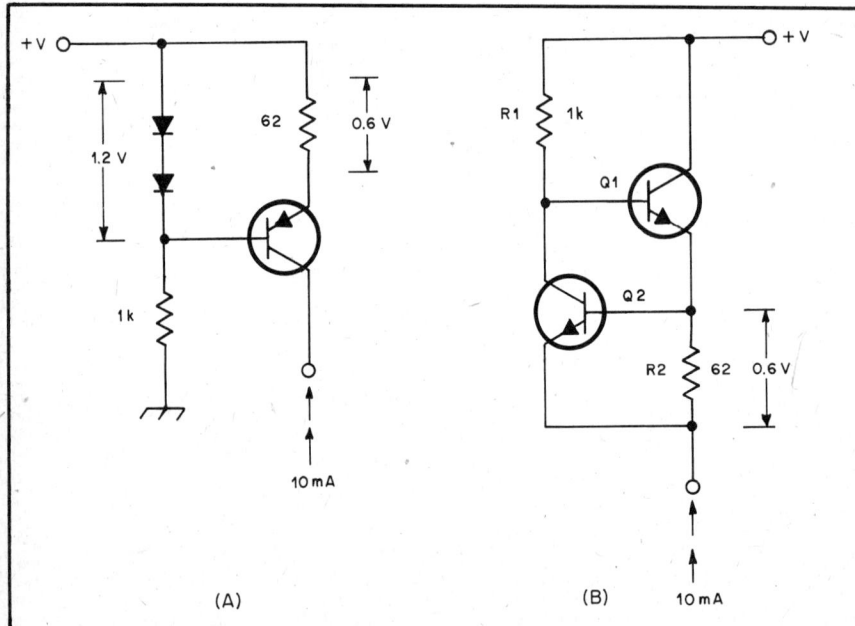

Fig. 72 — Constant-current generators made with bipolar transistors. In A, the reference voltage established by the diodes is converted to a current by the emitter resistor. A two-transistor feedback arrangement is employed at B. The functions of both circuits are explained in the text.

ward bias applied to the base-emitter junction, a *superdiode* results. If the collector were left open, the base-emitter junction would behave like an ordinary diode. With the collector tied to the base, the diode current rises much more rapidly with applied voltage because of the amplification provided by the transistor action. Two cross-connected super-diodes form the basis for a highly effective peak clipper or hard limiter. Fig. 71 illustrates the application. Npn transistors are shown, but pnp units will yield identical performance.

Constant-Current Generators

The curves in Figs. 51 to 53 show that the collector current of a bipolar transistor is essentially independent of the collector-to-emitter potential when the device is biased in its active region. Fig. 72A illustrates a constant-current source (or sink, if actual electron flow is considered) using a pnp transistor. A fairly constant 1.2-volt potential drop is maintained across the diode string. The base-emitter junction introduces a diode drop,

so the emf applied to the 62-Ω resistor is a constant 0.6 volt. A constant voltage across a resistor forces a constant current. This current flows in the emitter, and the high alpha causes the collector current to be nearly the same.

The circuit of Fig. 72B works in a similar manner. R1 biases Q1 into conduction. When the emf developed by R2 reaches 0.6 volts, Q2 begins to conduct, shunting base drive away from Q1 and limiting its collector current.

A device that passes an arbitrary current independent of the applied voltage presents an infinite dynamic impedance to the driving signal. This feature makes the constant-current generator valuable in several applications. One use for the circuits of Fig. 72 is in the bias control circuit of a differential amplifier. Either configuration can be used to establish the proper amplifier current while providing the tightest possible coupling between the emitters of the differential pair. Another way to employ a constant-current circuit is to use it as an active load for the collector of a transistor amplifier stage. The in-

finite dynamic impedance of the current source causes the amplifier to exhibit very high voltage gain. When the amplifier is an npn transistor, the current source must be a pnp device, and vice-versa.

Thyristors

Two complementary bipolar transistors connected as in Fig. 73 form the solid-state analog of the latching relay — a trigger pulse applied to the base of Q2 will initiate current flow in both devices. This current is limited only by the external circuit resistance and continues independent of the trigger signal until the main source is interrupted. Four-layer semiconductors (pnpn or npnp) having this property are known as *thyristors* or *silicon controlled rectifiers* (SCRs). SCRs find use in power supply overvoltage protection circuits (crowbars), electronic ignition systems, alarms, solid-state commutating systems for dc motors and a host of other applications. Two complementary SCRs fabricated in parallel, with a common gate terminal, form a *triac*. These are used to switch alternating currents. The most common application of the triac is in incandescent light dimmers. Triacs have sensitive gates, and prolonging the trigger signal or injecting excessive gate current can cause excessive heating. In circuits operating on 117-volt ac, a *diac* is used to trigger a triac. A diac is a bidirectional current-limiting diode. Structurally, it can be compared to a triac without a gate. A motor speed control illustrating the use of triacs and diacs is drawn in Fig. 74.

Unijunction Transistors

An unusual three-terminal semiconductor device is the unijunction transistor (UJT), sometimes called a *double-base diode*. The elements of a UJT are base 1, base 2 and emitter. The single rectifying junction is between the emitter and the silicon substrate. The base terminals are ohmic contacts, meaning that the current is a linear function of the applied voltage. Current flowing between the bases sets up a voltage gradient along the substrate. In operation, the direction of flow causes the

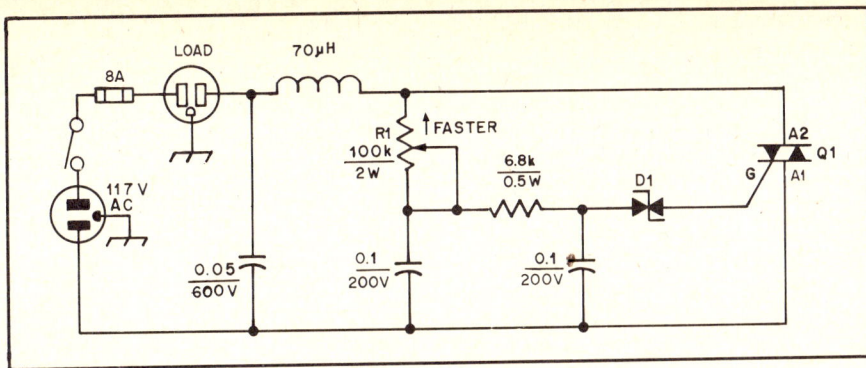

Fig. 74 — Schematic diagram of motor-speed control.
Q1 — Triac (silicon bidirectional thyristor), 8-A, 200-V (Motorola MAC2-4 or HEP340 or equiv.).
D1 — Diac (silicon bilateral trigger), 2-A, 300-mW.

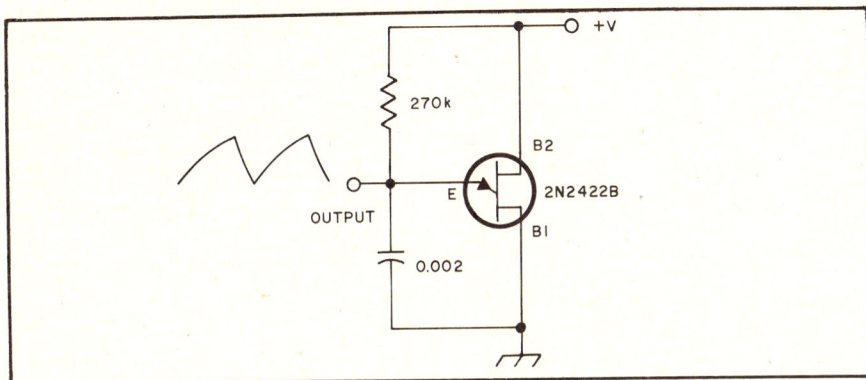

Fig. 76 — Profile and symbol for an n-channel junction field-effect transistor. In a p-channel device, all polarities are reversed and the gate arrow points *away* from the substrate.

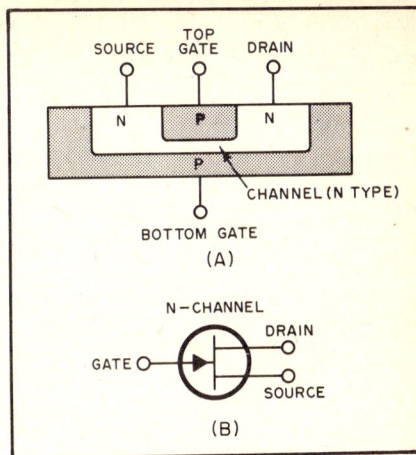

Fig. 75 — A relaxation oscillator based on a unijunction transistor. The frequency of oscillation is approximately 1500 Hz.

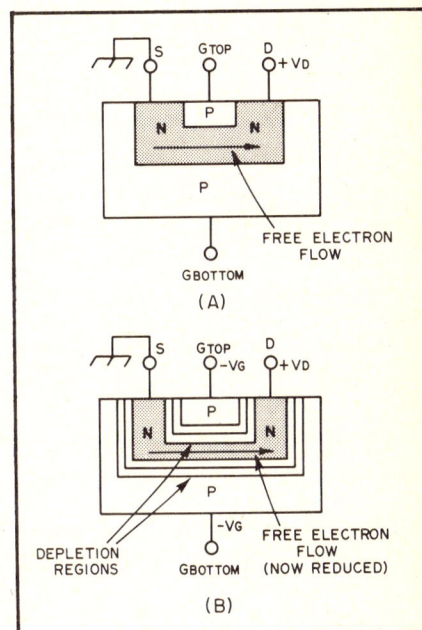

Fig. 77 — Operation of a JFET under applied bias. A depletion region (light shading) is formed, compressing the channel and increasing the resistance to current flow.

emitter junction to be reverse biased. The relaxation oscillator circuit (the most common UJT application) of Fig. 75 illustrates the function of the UJT. When the circuit is energized, the capacitor charges through the resistor until the emitter voltage overcomes the reverse bias. As soon as current flows in the emitter, the resistance of the base 1 region decreases dramatically, discharging the capacitor. The decreased base 1 resistance alters the voltage distribution along the substrate, establishing a new bias point for the emitter junction. As more and more emitter current flows, the majority carrier injection builds a space charge in the base 1 region, which causes the emitter current to cease. Current is again available to charge the capacitor and the cycle repeats. If the resistor were replaced by a constant-current source, the output waveform would be a linear ramp instead of a sawtooth. The UJT schematic symbol resembles that of an n-channel JFET — the angled emitter distinguishes the unijunction transistor.

Field-Effect Transistors

Field-effect transistors are assigned that name because the current flow in them is controlled by varying *electric field* which is brought about through the application of a *voltage* that controls the electrode known as the *gate*. The analogy for a

bipolar transistor is that in the latter the current flow is controlled by the *current* applied to the *base* electrode.

There are two essential types of field-effect transistors (FETs) in use today. They are the *junction FET* and the *MOSFET*. The former is most commonly called a JFET. It has no insulation between its elements, just as is the case with bipolar transistors. The MOSFET has a thin layer of oxide between the gate or gates and the drain-source junction. The term MOSFET is derived from *metal-oxide silicon* field-effect transistor. The basic characteristic of the two types are similar — high input impedance and good dynamic range. These characteristics apply to small-signal FETs. Power FETs, which will be treated later, have different characteristics. Although some MOS-FETs have but one gate, others have two gates. Single-gate FETs can be equated practically to a triode vacuum tube. The gate represents the grid, the anode is similar to the drain, and the cathode is like the source. The input impedance of FETs is a megohm or greater. The noise figure of an FET is quite low, making them ideal as preamplifiers for audio and rf well into the uhf region. Nearly all of the MOSFETs manufactured today have built-in gate-protective Zener diodes. Without this provision the gate insulation can be perforated easily by small static

charges on the user's hands or by the application of excessive voltages. The protective diodes are connected between the gate (or gates) and the source of the FET.

The Junction FET

As was stated earlier, field-effect transistors are divided into two main groups: Junction FETs and MOSFETs. The basic JFET is shown in Fig. 76.

The reason for the terminal names will become clear later. A dc operating condition is set up by starting a current flow between source and drain. This current flow is made up of free electrons since the semiconductor is n-type in the channel, so a positive voltage is applied at the drain. This positive voltage attracts the negative-

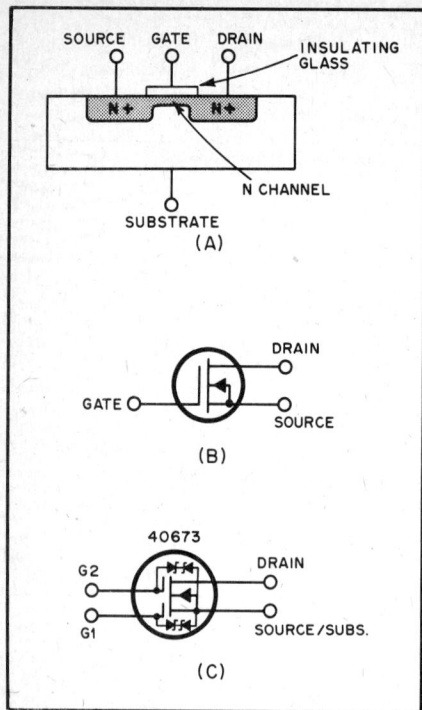

SOURCE GATE DRAIN   INSULATING
                        GLASS

N+   N+

N CHANNEL

SUBSTRATE
(A)

DRAIN

GATE O

SOURCE

(B)

40673

G2   DRAIN

G1   SOURCE/SUBS.

(C)

Fig. 78 — Profile and symbol for a MOSFET.

ly charged free electrons and the current flows (Fig. 77). The next step is to apply a gate voltage of the polarity shown in Fig. 77. Note that this reverse-biases the gates with respect to the source, channel, and drain. This reverse-bias gate voltage causes a depletion layer to be formed which takes up part of the channel, and since the electrons now have less volume in which to move the resistance is greater and the current between source and drain is reduced. If a large gate voltage is applied the depletion regions meet, causing *pinch off*, and consequently the source-drain current is reduced nearly to zero. Since the large source-drain current changes with a relatively small gate voltage, the device acts as an amplifier. In the operation of that JFET, the gate terminal is never forward biased, because if it were the source-drain current would all be diverted through the forward-biased gate junction diode.

The resistance between the gate terminal and the rest of the device is very high, since the gate terminal is always reverse biased, so the JFET has a very high input resistance. The source terminal is the *source* of current carriers, and they are *drained* out of the circuit at the drain. The gate *opens* and *closes* the amount of channel current which flows in the pinch-off region. Thus the operation of an FET closely resembles the operation of the vacuum tube with its high grid-input impedance.

MOSFETs (Metal-Oxide Semiconductors)

The other large family which makes up

field-effect transistors is the insulated-gate FET, or MOSFET, which is pictured schematically in Fig. 78. In order to set up a dc operating condition, a positive polarity is applied to the drain terminal. The substrate is connected to the source, and both are at ground potential, so the channel electrons are attracted to the positive drain. In order to regulate this source-drain current, voltage is applied to the gate contact. The gate is insulated from the rest of the device by a layer of very thin dielectric material, so this is not a p-n junction between the gate and the device — thus the name insulated gate. When a negative gate polarity is applied, positive-charged holes from the p-type substrate are attracted toward the gate and the conducting channel is made more narrow; thus the source-drain current is reduced. When a positive gate voltage is connected, the holes in the substrate are repelled, the conducting channel is made larger, and the source-drain current is increased. The MOSFET is more flexible since either a positive or negative voltage can be applied to the gate. The resistance between the gate and the rest of the device is extremely high because they are separated by a layer of thin dielectric. Thus the MOSFET has an extremely high input impedance. In fact, since the leakage through the insulating material is generally much smaller than through the reverse-biased p-n gate junction in the JFET, the MOSFET has a much higher input impedance. Typical values of R_{in} for the MOSFET are over a million megohms. There are both single-gate and dual-gate MOSFETs available. The latter has a signal gate, gate 1, and a control gate, gate 2. The gates are effectively in series making it an easy matter to control the dynamic range of the device by varying the bias on gate 2. Dual-gate MOSFETs are widely used as agc-controlled rf and i-f amplifiers, as mixers and product detectors, and as variable attenuators. The isolation between the gates is relatively high in mixer service. This reduces oscillator "pulling" and reduces oscillator radiation. The forward *transadmittance* (transconductance, or g_m) of dual-gate MOSFETs is as high as 40,000 micromhos.

FET Characteristics

The characteristic curves for the FETs described above are shown in Figs. 79 and 80. The drain-source current is plotted against drain-source voltage for given gate voltages.

The dynamic characteristics of an FET are most heavily influenced by dynamic mutual conductance or *transconductance.* This parameter is defined as the ratio of drain current change to the small gate-to-source voltage change that caused it. Mathematically, the relationship is expressed: $G_m = \Delta I_D / \Delta E_{GS}$, where Δ represents a small change or increment.

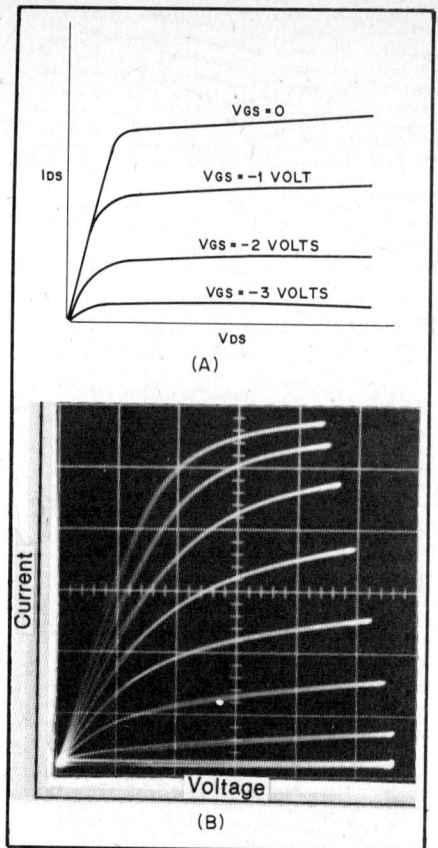

VGS = 0

IDS   VGS = -1 VOLT

VGS = -2 VOLTS

VGS = -3 VOLTS

VDS
(A)

Current

Voltage
(B)

Fig. 79 — At A are typical JFET characteristic curves. The picture at B shows an actual oscillograph of the family of curves produced by a curve tracer.

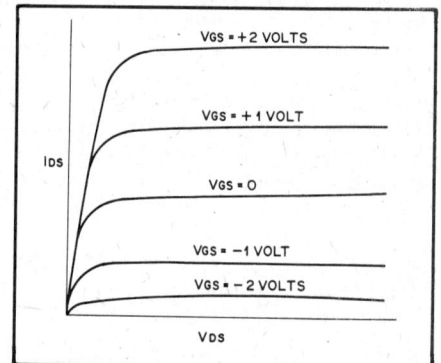

VGS = +2 VOLTS

VGS = +1 VOLT

IDS   VGS = 0

VGS = -1 VOLT

VGS = -2 VOLTS

VDS

Fig. 80 — Typical characteristic curves for a MOSFET.

Typical general-purpose JFETs for small-signal rf and audio work have G_m values in the neighborhood of 5000 micromhos, while some units designed for CATV service feature transconductance over 13,000 μU. As reported above, 40,000 μU (40 mU) is the transconductance figure for the "hottest" dual-gate MOSFETs. Some JFETs intended for analog switching also have 40 mU or more of transconductance to achieve low ON resistance. The newer power FETs boast transconductance figures on the order of one mho.

Transconductance is of great importance in calculating the gain and impedance values of FET circuits. In common-source and common-gate amplifiers with no degeneration, the numerical voltage gain is given by $A_v = G_m R_L$, where A_v is the gain, G_m is the transconductance in mhos (or siemens) and R_L is the drain load resistance in ohms. Also, the source impedance of a common-gate or common-drain (source follower) amplifier is approximately $1/G_m$.

Classifications

Field-effect transistors are classed into two main groupings for application in circuits, *enhancement mode* and *depletion mode*. The enhancement-mode devices are those specifically constructed so that they have *no* channel. They become useful only when a gate voltage is applied that causes a channel to be formed. IGFETs (insulated gate FET) can be used as enhancement-mode devices since both polarities can be applied to the gate without the gate becoming forward biased and conducting.

A depletion-mode unit corresponds to Figs. 76 and 77, shown earlier, where a channel exists with no gate voltage applied. For the JFET we can apply a gate voltage and deplete the channel, causing the current to decrease. With the MOSFET we can apply a gate voltage of either polarity so the device can be depleted (current decreased) or enhanced (current increased).

To sum up, a depletion-mode FET is one which has a channel constructed; thus it has a current flow for zero gate voltage. Enhancement-mode FETs are those which have no channel, so no current flows with zero gate voltage.

Power FETs

FETs capable of handling substantial amounts of power are available for use from dc through the vhf spectrum. They are known under more than one name — *vertical FETs, MOSPOWER FETs* and *VMOS FETs*. The power FET (MOS-POWER™ FET) was introduced in 1976 by Siliconix, Inc. The device enabled designers to switch a current of 1 ampere in less than four nanoseconds. The transfer characteristic of the power FET is a linear one. It can be employed as a linear power amplifier or a low-noise, small-signal amplifier with high dynamic range. With this kind of FET there is no thermal runaway, as is the case with power types of bipolar transistors. Furthermore, there is no secondary breakdown or minority-carrier storage time. The latter makes them excellent for use in switching amplifiers (Class D service). Of particular interest to amateurs is the immunity of power FETs to damage from a high SWR (open or short condition). These devices can be operated in Class A, AB, B or C. Zero bias results in Class C operation.

Fig. 81 illustrates the manner in which a MOSPOWER FET is formed. These de-

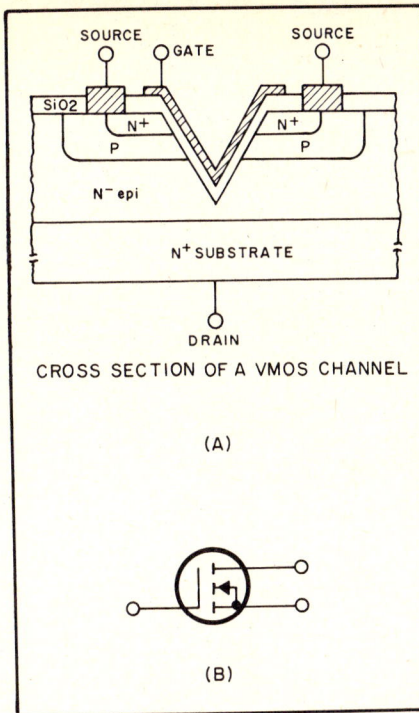

CROSS SECTION OF A VMOS CHANNEL

(A)

(B)

Fig. 81 — Profile and symbol for a power FET (VMOS enhancement-mode type).

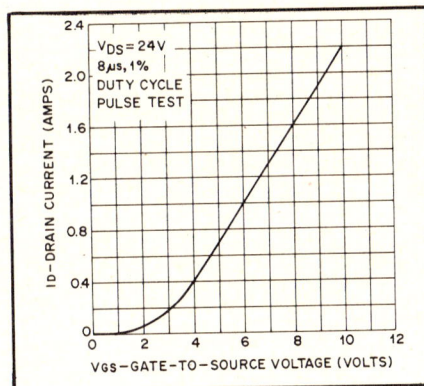

Fig. 82 — Curve showing relationship between gate-source voltage and drain current of a power FET.

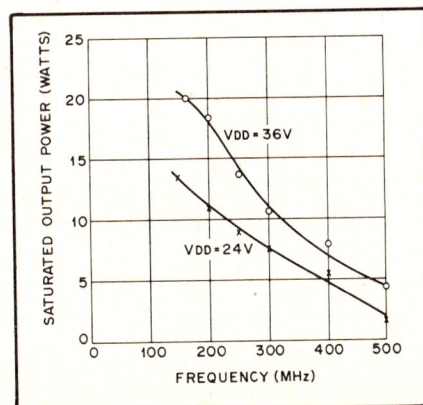

Fig. 83 — Curves for 24- and 36-volt operation of a power FET.

vices operate in the enhancement mode. The current travels vertically. The source is on top of the chip but the drain is on the backside of the chip. In this vertical structure there are four layers of material (N+, P, N− and N+). This device offers high current density, high source/drain breakdown capability and low gate/drain feedback capacitance. These features make the transistor ideal for hf and vhf use.

Fig. 82 depicts the drain current as being linearly proportional to the gate-to-source voltage. The more conventional JFET exhibits a square-law response (drain current being proportional to the square of the gate-to-source voltage).

As an example, the Siliconix VMP-4 power FET can provide a power just short of 20 watts (saturated) at 160 MHz. Fig. 83 shows curves for this device respective to saturated output power versus frequency. In this case both the input and output impedances of the transistor are matched conjugately. An advantage to this device over the power bipolar transistor is that these impedances are barely affected by the drive levels applied. In wideband amplifier service the MOSPOWER FET can be operated with complete stability. In-depth data on these devices is given in the Siliconix application note, TA-76-1.

GaAsFETs

For low-noise amplification at uhf and microwaves, the state of the art is defined by field-effect transistors fabricated from gallium arsenide. Also used in LEDs and microwave diodes, gallium arsenide is semiconductor *compound*, as opposed to silicon and germanium, which are *elements*. This compound exhibits greater carrier mobility (the electrons can move more freely) than silicon or germanium, hence the transit time is reduced and high-frequency performance improved in GaAs FETs. GaAs FETs are classified as depletion-mode junction devices. The gate is made of gold or aluminum, the latter type being susceptible to damage from static charges.

GaAs FETs are available for both small-signal and power applications. The power devices have noise figures almost as low as those specified for the small-signal types, and naturally exhibit greater dynamic range and ruggedness. Several semiconductor manufacturers throughout the free world offer gallium-arsenide field-effect transistors in various noise figure, frequency and power ratings. In the U.S., Hewlett-Packard and Microwave Semiconductor Corp. feature units usable up through Ku-band. Representative type numbers are HFET-2201 and MSC H001, respectively. In Great Britain, the Plessey GAT5 and GAT6 devices feature low-noise performance up to 14 GHz. The Nippon Electric Company of Japan is also competing strongly for leadership in the GaAs FET market. A

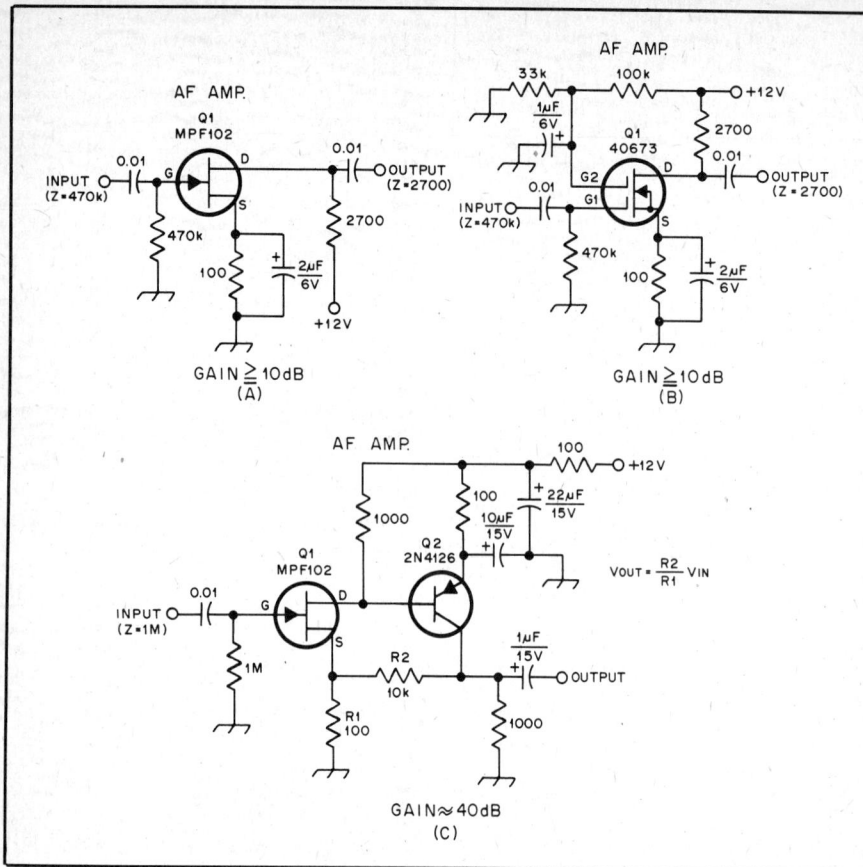

Fig. 84 — Some typical audio amplifiers which employ FETs.

practical GaAsFET preamplifier is featured in the vhf/uhf receiving chapter. For more background information on GaAs FETs, see Wade, "Introduction to GaAs Field-Effect Transistors," *Ham Radio*, January 1978, and Wade and Katz, "Low-Noise GaAs FET UHF Pre-amplifiers," *QST*, June 1978.

Practical FET Circuits

Small-signal FETs can be used in the same general types of circuits given earlier for bipolar transistors. The primary obstacle in some types of amplifier circuits is instability. Certain precautions are necessary to prevent unwanted self-oscillations, but they do not differ markedly from those techniques applied when working with triode tubes.

In Fig. 84 are examples of FET audio amplifiers. The circuit at A shows a simple RC coupled stage with a gain of 10 dB or greater. The input and output impedances are set by the gate and drain resistors. The circuit at B in Fig. 84 is similar to that at A, except that a dual-gate MOSFET is used as the active device (Q1). A positive bias is supplied to gate 2 by means of a resistive divider. In the circuit of Fig. 84C, a pnp transistor is combined with a JFET to provide a direct-coupled pair. This configuration provides high gain. The amount of gain is set by the ratio of R1 and R2. Again, the input and output im-

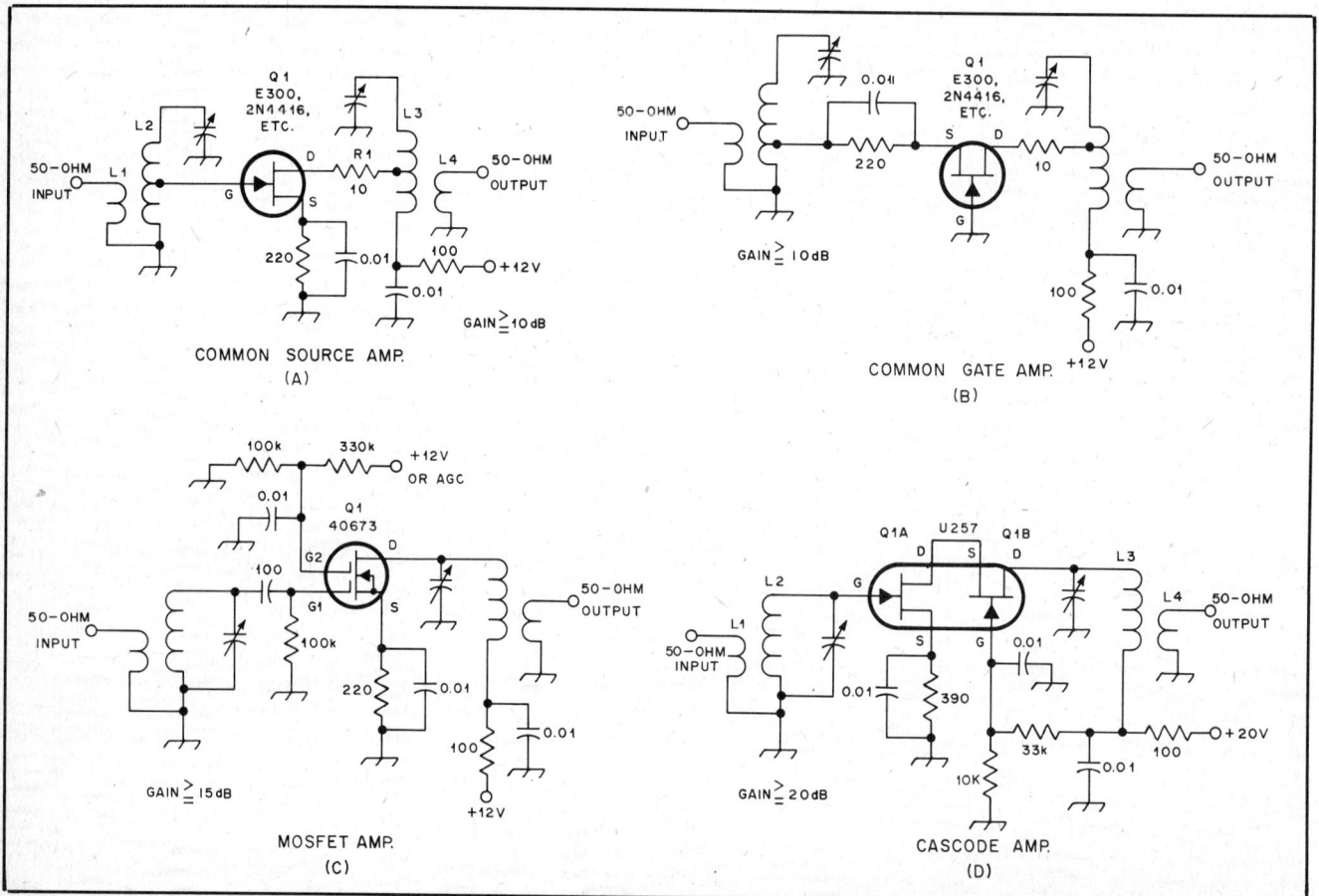

Fig. 85 — Examples of FET amplifiers suitable for rf or i-f applications.

pedances are determined for the most part by the values of the input and output resistors, 1 megohm and 1000 ohms, respectively.

RF and I-F Amplifiers

Small-signal rf and i-f amplifiers which use FETs are capable of good dynamic range and will exhibit a low noise figure. It is for these reasons that many designers prefer them to bipolar transistors. Fig. 85 contains examples of FET rf or i-f amplifiers. In the example at A the gate and drain elements of Q1 are tapped down on L2 and L3 to provide stability. This represents an intentional mismatch, which causes a slight sacrifice in stage gain. The 10-ohm drain resistor (R1) is used only if vhf parasitic oscillations occur.

At B in Fig. 85 is seen a common-gate FET amplifier. The source is tapped well

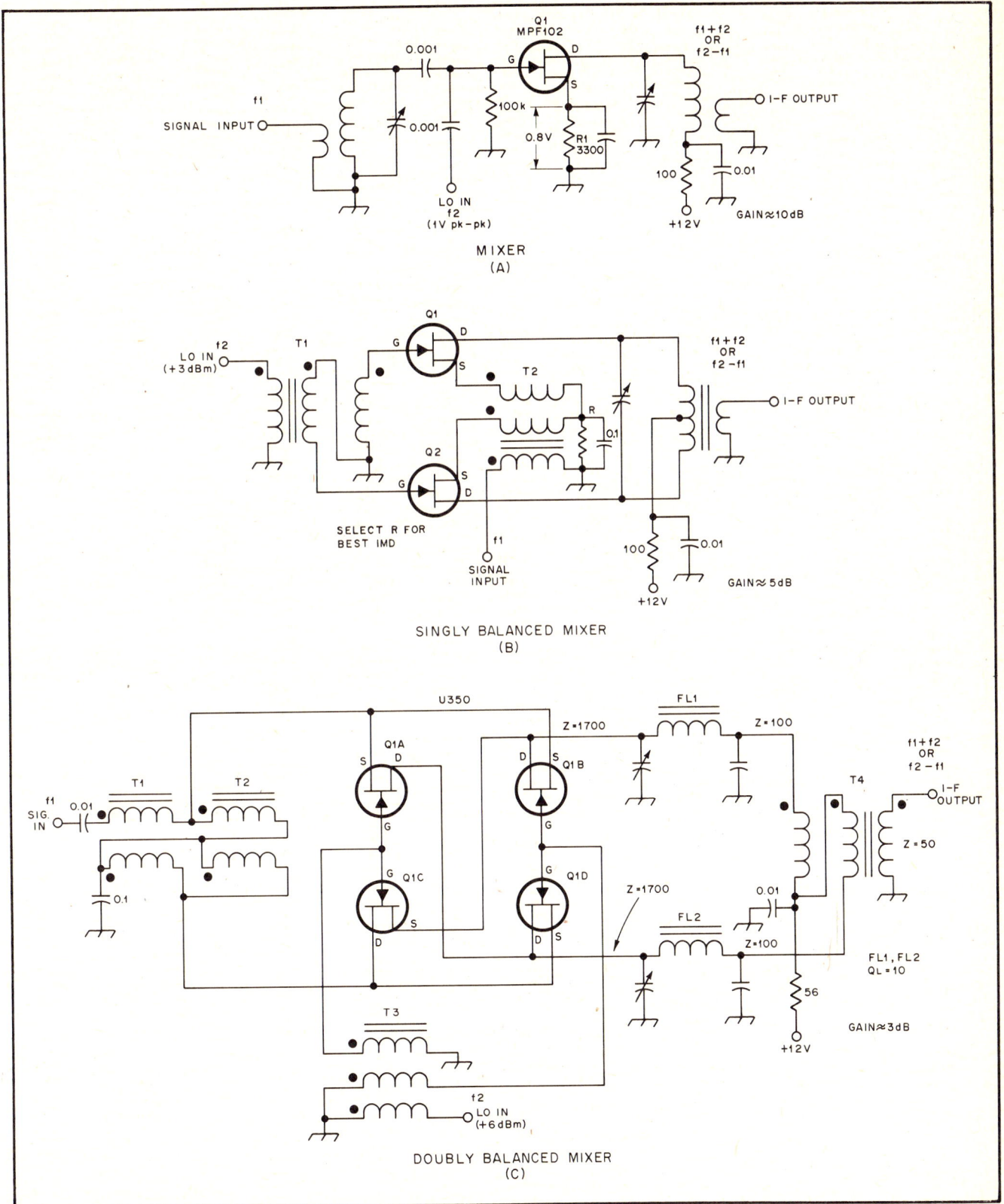

MIXER
(A)

SINGLY BALANCED MIXER
(B)

DOUBLY BALANCED MIXER
(C)

Fig. 86 — Various JFET mixers. See text for data.

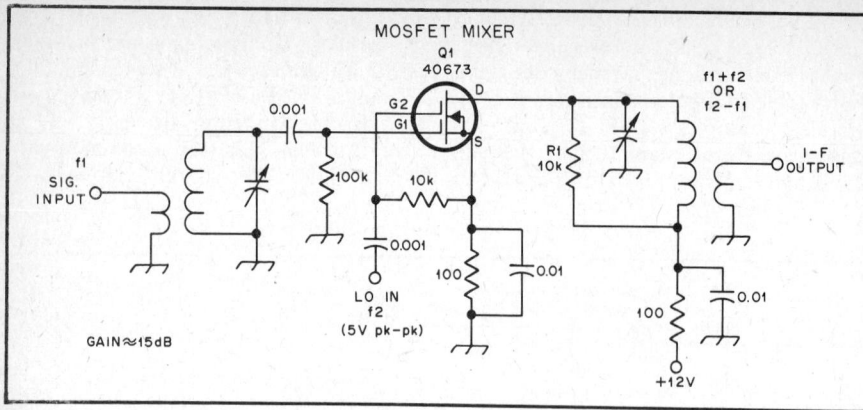

Fig. 87 — A dual-gate MOSFET single-ended mixer.

down on the input tuned circuit to effect an impedance match. This circuit is characterized by its excellent stability, provided the gate lead is returned to ground by the shortest path possible. This type of circuit will have slightly less gain capability than the common-source example at A.

In Fig. 85 at C is an illustration of a dual-gate MOSFET amplifier. Provided the input and output tuned circuits are well isolated from one another there is less chance for self-oscillation than with a JFET. A positive bias is applied to gate 2, but agc voltage can be used in place of a fixed-value voltage if desired. This circuit can provide up to 25 dB of gain.

Fig. 85D shows the configuration of a cascode rf amplifier in which a dual JFET (siliconix U257) is specified. The advantage in using the dual FET is that both transistors have nearly identical characteristics, owing to the fact that they are fabricated on a common substrate. Two separate JFETs can be used in this circuit if the one nearest to V_{DD} has an I_{DSS} higher than its mate. This ensures proper dc bias for cascode operation. Unmatched FETs require special forward-biasing techniques and ac-coupling measures that aren't seen in this circuit.

Cascode amplifiers are noted for their high gain, good stability, and low noise figure. With the circuit shown the noise figure at 28 MHz is approximately 1.5 dB. Short leads are necessary, and shielding between the tuned circuits is recommended in the interest of stability. Careful layout will permit the use of toroidal inductors at L2 and L3. These components should be spaced apart and mounted at right angles to one another in order to reduce unwanted infringing magnetic fields. Agc can be applied to this amplifier by routing the control voltage to the gate of Q1B.

FET Mixers

There are three types of FET mixers in common use today — single-ended, singly

balanced and doubly balanced. In all cases there is an advantage to using active devices in place of passive ones (diodes). This assures a conversion gain which helps minimize the number of gain stages required in a given circuit.

A single-ended JFET mixer requires 0 dbm of LO injection power. It can provide several decades of bandwidth and has a good IMD characteristic. The latter is far superior to most bipolar single-ended mixers. The major shortcoming is very poor isolation between the three mixer ports (rf, LO and i-f). A typical single-ended mixer using a JFET is seen in Fig. 86A. Optimum tradeoff between conversion gain and IMD occurs near the point where the self-bias is 0.8 V. LO injection voltage will be on the order of 1 (pk-pk) to provide good mixer performance. Conversion gain with this mixer will be approximately 10 dB.

Fig. 86B illustrates a singly balanced JFET mixer. A broadband transformer (T1) provides a low-impedance source for the LO and supplies injection voltage in push-pull to the gates of Q1 and Q2. The latter should be matched FETs or a dual FET such as the U430 by Siliconix. This mixer provides between 10 and 20 dB of isolation between the mixer ports. The signal is applied in parallel across the sources of Q1 and Q2 by means of broadband transformer T2. Output at the i-f is taken from a balanced tuned circuit.

A doubly balanced FET mixer is shown at C in Fig. 86. Broadband transformers are used throughout, with FL1 and FL2 providing low-pass selectivity at the mixer output. The filters also provide an impedance stepdown between the drains of Q1 and T4 (1700 ohms to 100 ohms). LO injection is supplied to the gates and signal input is to the sources. Port-to-port isolation with this mixer is on the order of 30 dB or greater. Bandwidth is one octave. In-depth information on this type of circuit is given in the Siliconix application note AN-73-4.

Fig. 87 contains the circuit of a typical

dual-gate MOSFET single-ended mixer. Its performance characteristics are similar to those of the mixer at Fig. 86A. The primary exception is that the port-to-port isolation is somewhat better by virtue of the gate no. 2 isolation from the remainder of the electrodes. This mixer and all other active FET mixers require a fairly low drain-load impedance in the interest of good IMD. If the drain tuned circuit is made high in terms of impedance (in an effort to improve conversion gain) the drain-source peak signal swing will be high. This will lead to a change in junction capacitance (varactor effect) and the generation of harmonic currents. The result is distortion. Of primary significance is the condition called "drain-load distortion." This malady occurs when excessive signal levels overload the drain circuit. The result is degraded IMD and cross-modulation effects. R1 in Fig. 87 is used to decrease the drain-load impedance by means of swamping. A value to 10,000 ohms is suitable for a 40673 MOSFET mixer. Some JFETs require a lower drain load for optimum performance. Values as low as 5000 ohms are not unusual. This form of overloading is more pronounced at low dc drain-voltage levels, such as 6 or 8.

FET Crystal Oscillators

A group of crystal-controlled FET oscillators is presented in Figs. 88 and 89. At Fig. 88A is an overtone type. The tuned circuit in the drain is resonated slightly higher than the crystal frequency to assure reliable oscillation. The circuit at B is a variation of the one at A, but performs the same function. A Pierce type of triode oscillator is shown in Fig. 89 at A. It is suitable for use with fundamental types of crystals. A Colpitts oscillator appears at B in Fig. 89. C_{fb} in these circuits are feedback capacitors. C_{fb} in the circuit at C is chosen experimentally. Typically, it will be from 100 to 500 pF, depending on the transistor characteristics and the crystal activity.

FET VFOs

The principle of operation for FET VFOs is similar to that which was discussed in the section on bipolar transistor oscillators. The notable difference is the impedance level at the device input. The circuits of Fig. 90 all have high-impedance gate terminals. Furthermore, fewer parts are needed than is true of bipolar transistor equivalent circuits: There is no resistive divider for applying forward bias.

The circuits of Fig. 90A and B are identical except for the biasing of gate 2 at B. Both circuits illustrate oscillators. The source tap on L1 should be selected to provide approximately 25 percent of the oscillator power as feedback. D1 in each example is used to stabilize the gate bias. It acts as a diode clamp on positive-going

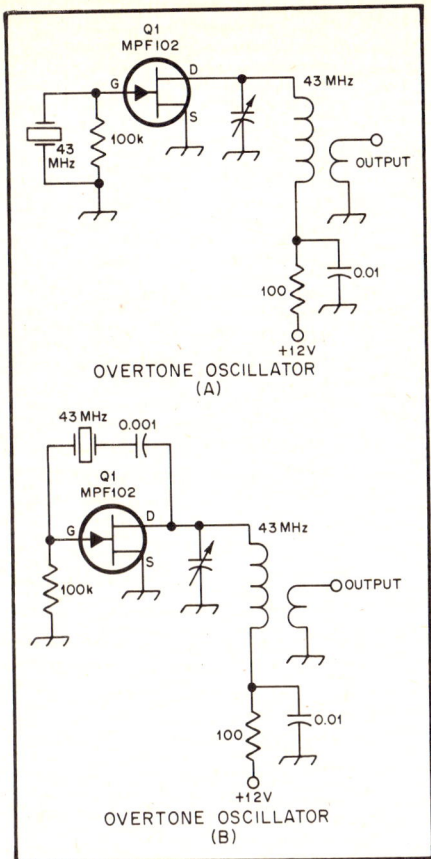

Fig. 88 — Overtone crystal oscillators using FETs.

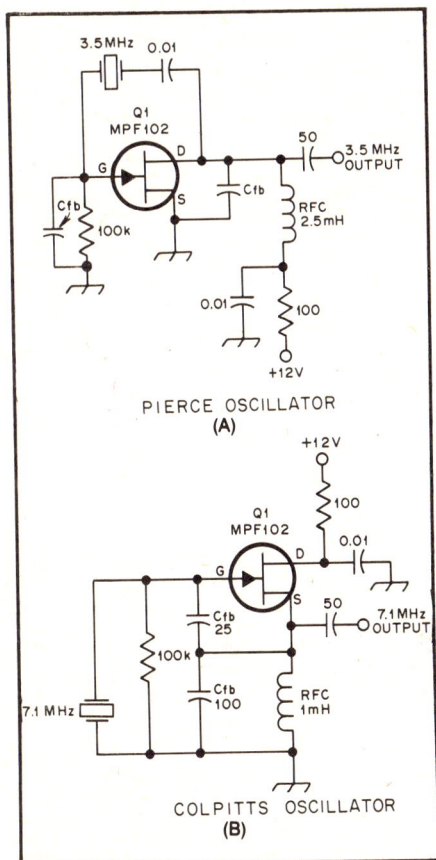

Fig. 89 — Fundamental-mode FET crystal oscillators.

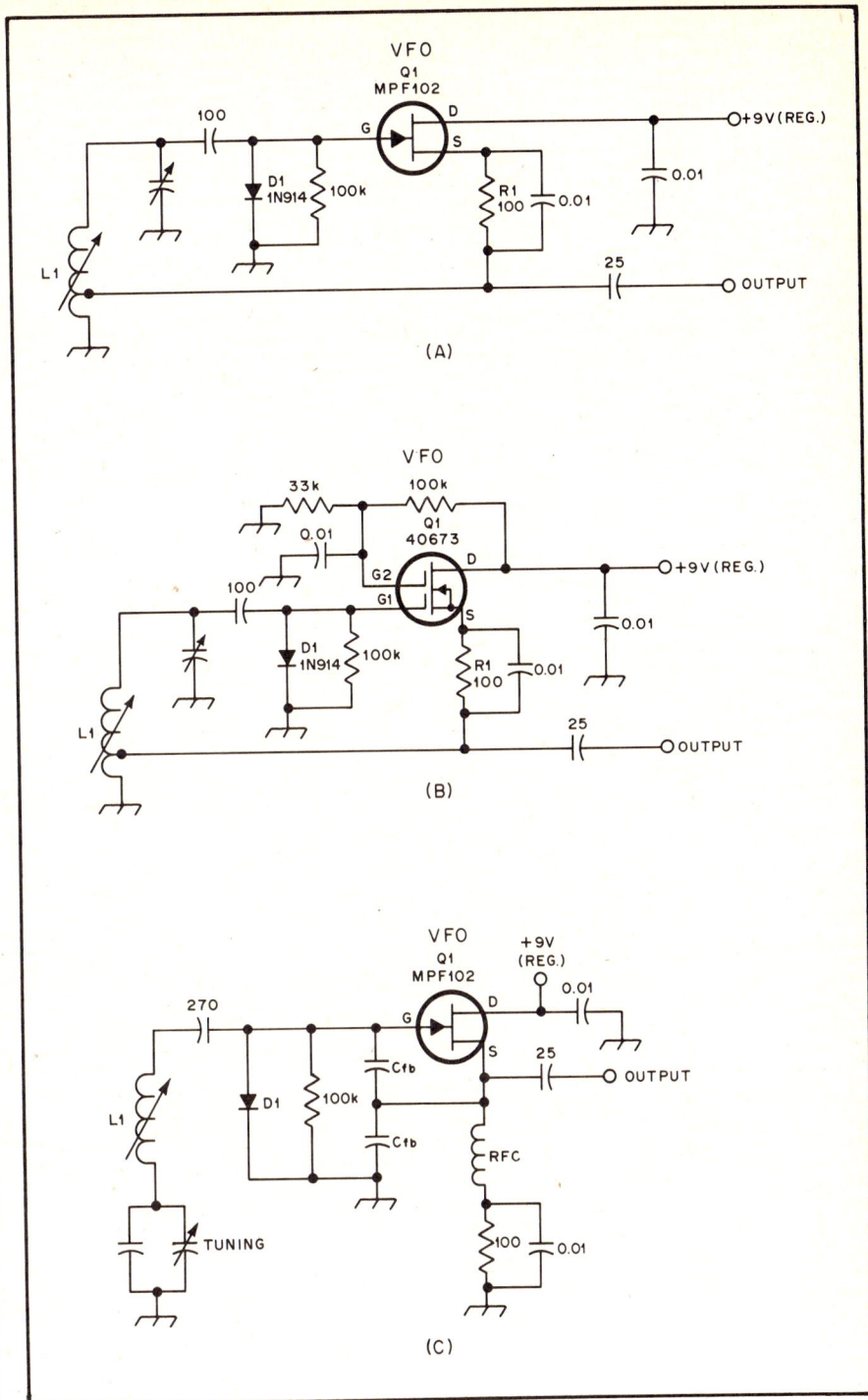

Fig. 90 — Three examples of VFOs in which FETs are used.

excursions of the signal. This aids oscillator stability and reduces the harmonic output of the stage. The latter is reduced as a result of the positive swing of the sine-wave being limited by D1, which in turn limits the device transconductance on peaks. This action reduces changes in junction capacitance, thereby greatly restricting the varactor action which generates harmonic currents. D1 is most effective when source-bias resistors are included in the circuit (R1).

Shown in Fig. 90 at C is a series-tuned Colpitts VFO which uses a JFET. This is an exceptionally stable VFO if careful design and component choice is applied. All of the fixed-value capacitors in the rf parts of the circuit should be temperature-stable. Polystyrene capacitors are recommended, but dipped silver-mica capacitors will serve adequately as a second choice. Preferably, L1 should be a rigid air-wound inductor. A slug-tuned inductor can be used if the coil Q is high. In such cases the slug should occupy the least amount of coil space possible: Tempera-

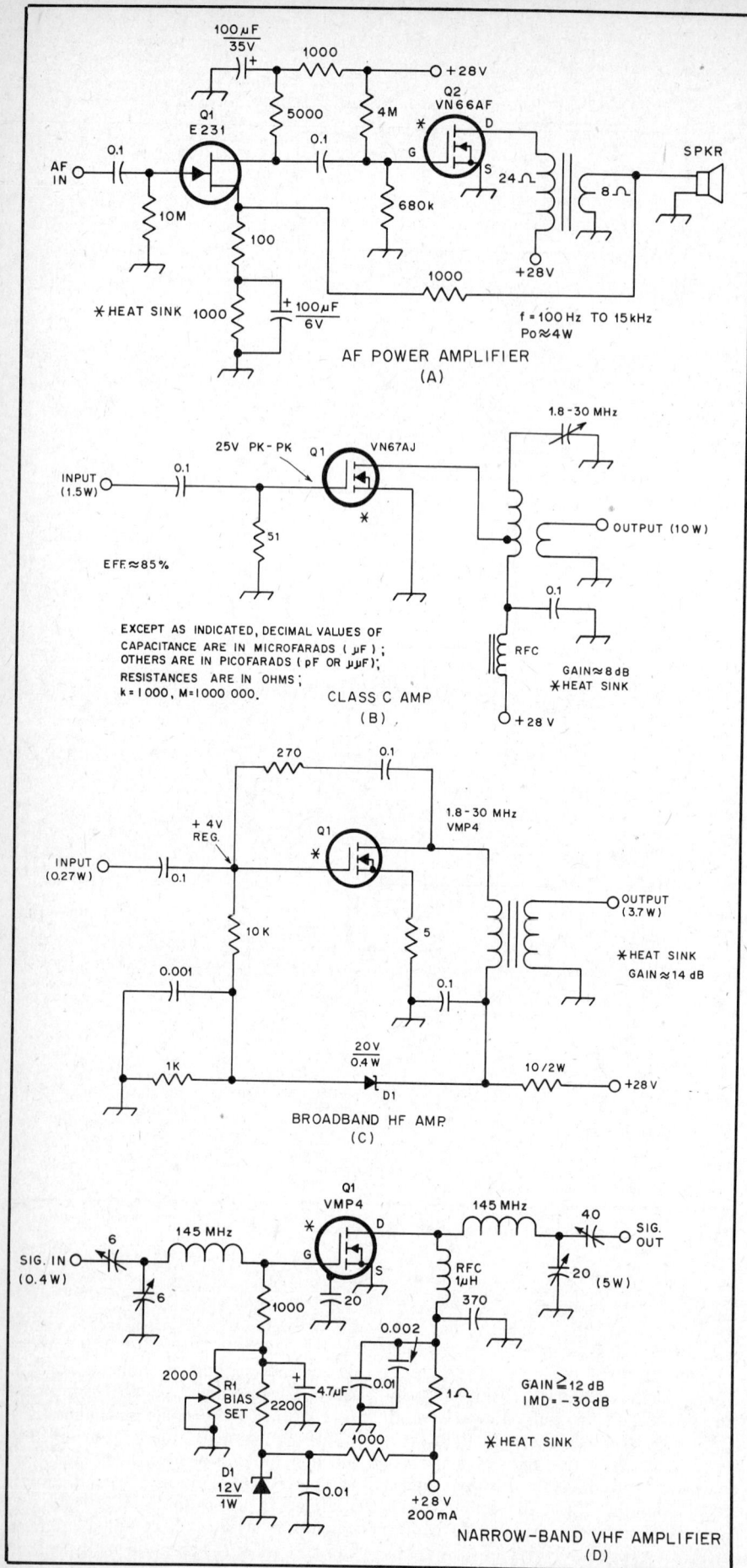

Fig. 91 — Examples of power FETs in three amplifier circuits.

AF POWER AMPLIFIER
(A)

EXCEPT AS INDICATED, DECIMAL VALUES OF CAPACITANCE ARE IN MICROFARADS (μF); OTHERS ARE IN PICOFARADS (pF OR $\mu\mu$F); RESISTANCES ARE IN OHMS; k = 1000, M = 1000 000.

CLASS C AMP
(B)

BROADBAND HF AMP.
(C)

NARROW-BAND VHF AMPLIFIER
(D)

ture changes have a marked effect on ferrite or powdered-iron slugs, which can change the coil inductance markedly. C_{fb} of Fig. 90C are on the order of 1000 pF each for 3.5-MHz operation. They are proportionally lower in capacitance as the operating frequency is increased, such as 680 pF at 7 MHz, and so on.

Power FET Examples

Fig. 91 contains examples of three amplifiers which employ power FETs. The circuit at A is an audio amplifier which can deliver 4 watts of output. At 3 watts of output the distortion is approximately two percent. Feedback is employed to aid the reduction of distortion.

A Class C amplifier is seen at B in Fig. 91. The VN67AJ is capable of a saturated output near 15 watts at 30 MHz. In this circuit the power output is considerably less. A medium output power of 7 to 10 watts is suggested. The gain is approximately 8 dB over the frequency range specified (with the appropriate drain tank). If proper layout techniques are used this amplifier is unconditionally stable.

A broadband hf linear amplifier is shown at C.

A narrow-band linear vhf power amplifier is shown at D in Fig. 91. Power output is 5 watts PEP. IMD is −30 dB. It is interesting to realize that this same amplifier is suitable as a high-dynamic-range preamplifier for a vhf receiver. In this application the noise figure is on the order of 2.5 dB and the gain is 11 dB.

Other FET Uses

Fig. 92 contains illustrations of additional practical uses for JFETs. The circuit at A shows a Schmitt trigger. It is emitter-coupled and provides a comparator function. Q1 places very light loading on the measured input voltage. Q2 has high beta to enable the circuit to have a fast transition action and a distinct hysteresis loop. Additional applications of this type are found in *Linear Applications* by National Semiconductor Corp.

A simple FET dc voltmeter with high input impedance is seen in Fig. 92B. Multiplier resistances are given for a full-scale range of 2 or 20 volts. Meter accuracy is quite good, with a linear reading provided by M1.

A push-push frequency doubler is shown at C in Fig. 92. The input frequency (f) is applied to the gates of Q1 and Q2 in push-pull. Output from the doubler is taken with the connected drains in parallel. R1 is adjusted for best waveform purity at 2f. The efficiency of this Class C doubler is on par with that of a straight-through Class C amplifier. Careful adjustment will nearly eliminate frequency component f at the doubler output.

Linear Integrated Circuits

There are two general types of ICs

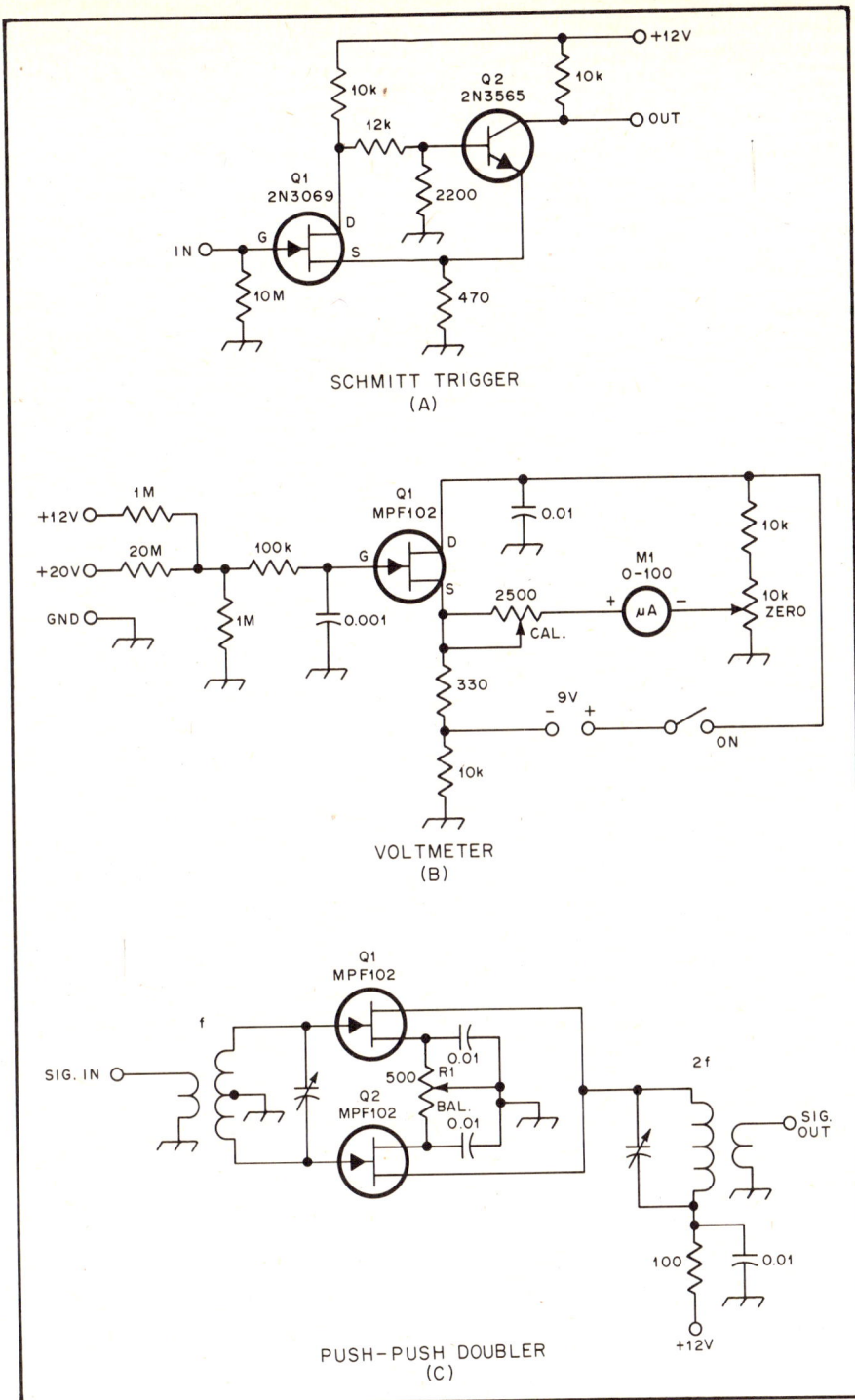

SCHMITT TRIGGER
(A)

VOLTMETER
(B)

PUSH-PUSH DOUBLER
(C)

Fig. 92 — JFETs are useful in additional kinds of circuits. Here are three such examples.

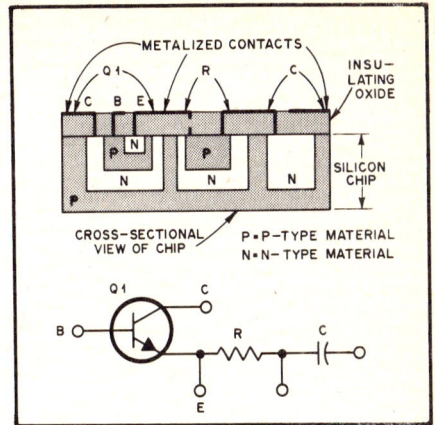

Fig. 93 — Pictorial and schematic representation of a simple IC.

mental conditions. This provides an inherent balance in their performance traits — a condition which is nearly impossible to realize with closely matched discrete transistors. Therefore, when changes in IC temperature take place, the parameters of the transistors on the chip change in unison — a distinct advantage when the IC is used in, say, a balanced modulator, mixer or push-push doubler.

Most of the theory given earlier for bipolar transistors applies to ICs, so it will not be repeated here. Rather, the text will provide data on practical applications of ICs. Linear ICs are so-called because in most applications where they are used the performance mode is a linear one. This does not mean, however, that they can't be used in a nonlinear mode, such as Class C. The biasing will determine the operating mode, Class A through and including Class C.

IC Structures

The basic IC is formed on a uniform chip of n-type or p-type silicon. Impurities are introduced into the chip, their depth into it being determined by the diffusion temperature and time. The geometry of the plane surface of the chip is determined by masking off certain areas, applying photochemical techniques, and providing a coating of insulating oxide. Certain areas of the oxide coating are then opened up to allow the formation of interconnecting leads between sections of the IC. When capacitors are formed on the chip, the oxide serves as the dielectric material. Fig. 93 shows a representative three-component IC in both pictorial and schematic form. Most integrated circuits are housed in TO-5 type cases, or in flat-pack epoxy blocks. ICs may have as many as 12 or more leads which connect to the various elements on the chip.

Some of the present-day ICs are called *LSI* chips. The term LSI means *large-scale integration.* Such devices may contain the equivalent of several conventional ICs,

(integrated circuits). The first variety, which we are addressing at the moment, are called *linear ICs.* The other group are known generally as *logic ICs.* These devices will be discussed later in the chapter.

ICs are characterized by the term "microcircuit." In essence they are composed of numerous — sometimes hundreds — of bipolar and /or field-effect transistors on a single silicon chip (substrate). Along with the individual transistors formed on the substrate are diodes, capacitances and resistances. Some ICs contain only diodes. Others may contain only resistors. The principal advantages of ICs are their compactness over an equivalent number of discrete transistors, and the fact that all of the devices on the substrate are evenly matched in characteristics. That is the result of the manufacturing process, whereby all of the IC transistors are formed from a single slice of semiconductor material under the same environ-

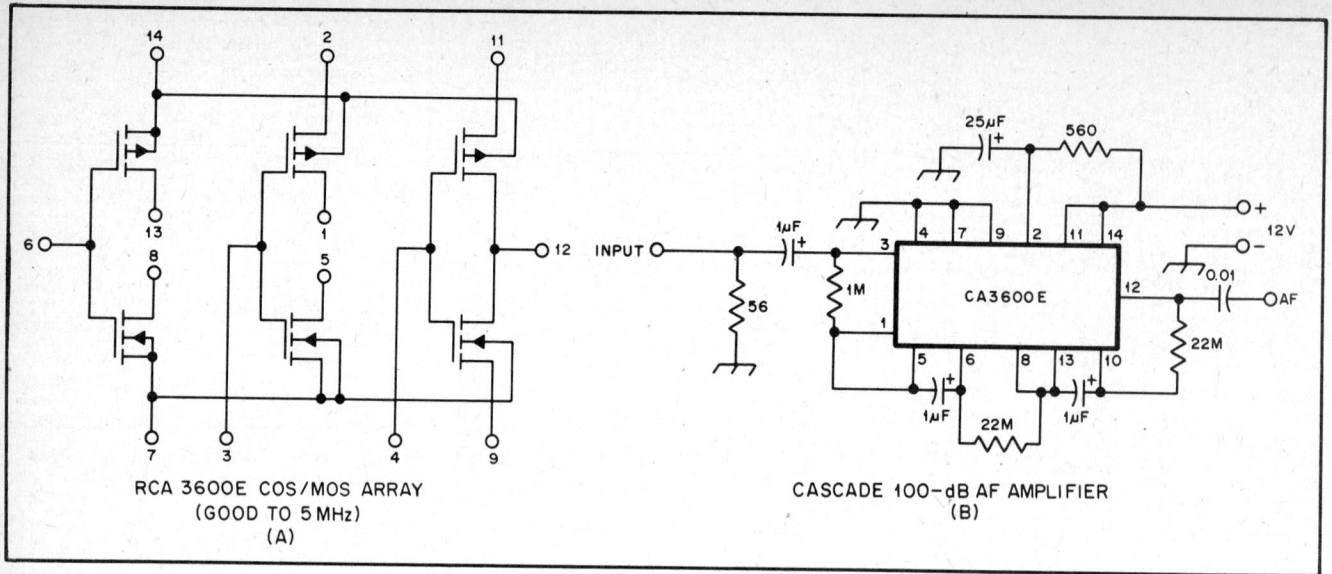

RCA 3600E COS/MOS ARRAY
(GOOD TO 5 MHz)
(A)

CASCADE 100-dB AF AMPLIFIER
(B)

Fig. 94 — The diagram at A shows the internal workings of a CMOS IC. A 100-dB audio amplifier which employs the CA3600E is shown at B.

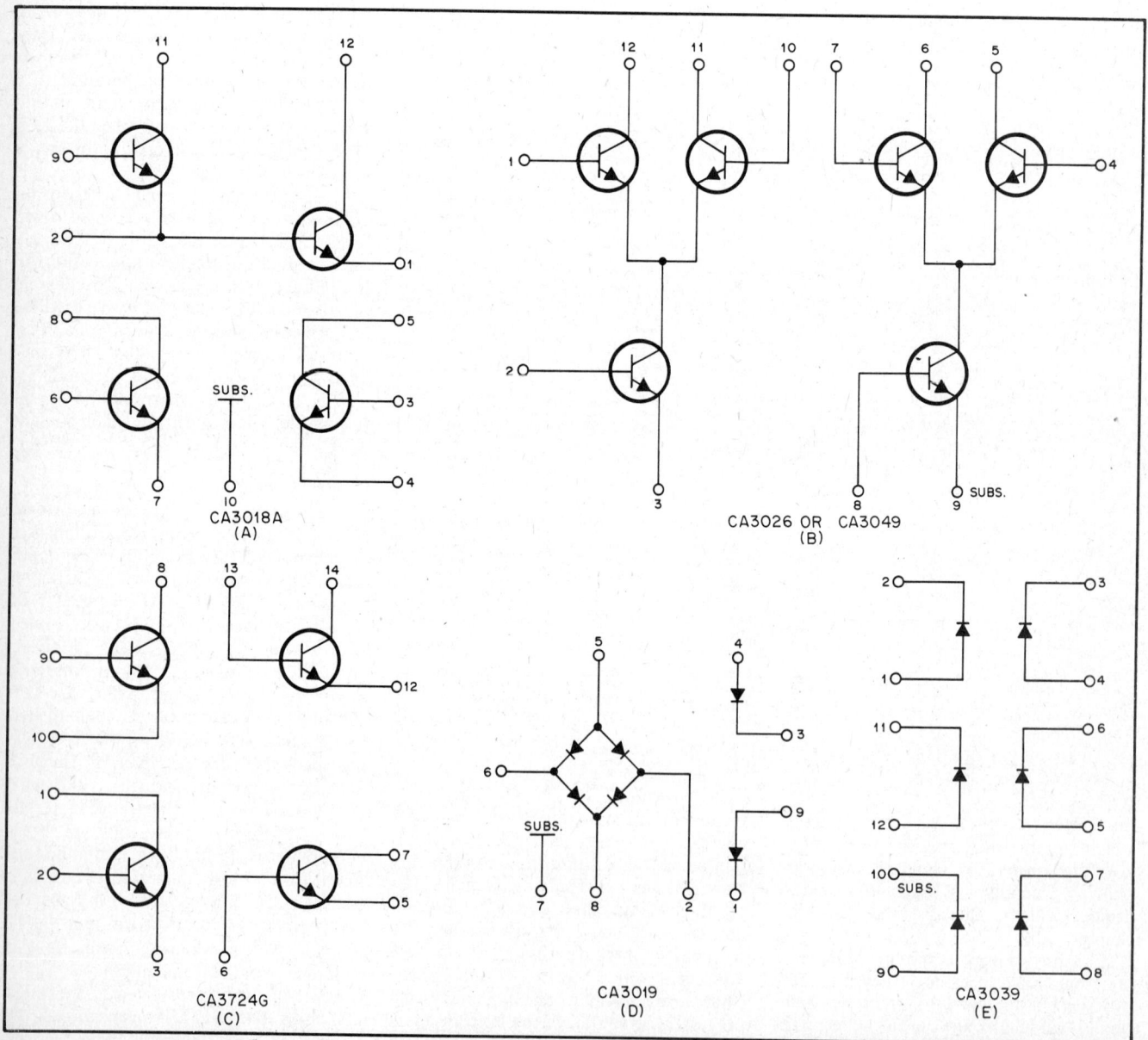

CA3018A
(A)

CA3026 OR CA3049
(B)

CA3724G
(C)

CA3019
(D)

CA3039
(E)

Fig. 95 — Various transistor and diode-array ICs. The configurations suggest a variety of amateur applications.

and can have dozens of dual-in-line package (DIP) connector pins. LSI ICs are used in electronic organs, digital clocks, electronic calculators, and so on. Essentially, they are just super-size ICs.

Some Practical Considerations

In circuits where slight extra lead lengths can be tolerated, it is prudent to install the ICs in sockets rather than solder them into the pc board directly. In amateur work there is an occasional need to replace an IC during circuit development for a typical one-shot design. This is particularly pertinent when bargain-house ICs are purchased: Many have defects, and the task of removing an IC that is not in a socket is a task that no builder finds delightful.

When using ICs for rf work it is best to install them in a low-profile type of IC socket (minimum lead length type). The thicker sockets are suitable for dc and audio applications, where lead length is not likely to be a critical factor. Excessive lead length can cause instability. This is brought on by having numerous high-gain devices packaged physically close to one another on the common substrate: High gain and stray lead coupling set the stage for self-oscillation!

CMOS ICs

The term CMOS means that the IC is a *complimentary metal-oxide silicon* type of integrated circuit. Essentially, the internal workings of the device are not unlike those of MOSFETs, the latter of which were treated earlier in this chapter: MOSFETS are formed on the CMOS IC substrates.

CMOS devices consume very low power — an advantage in battery-operated equipment, especially. The transit time (propagation delay) through the FET gates of a CMOS IC is very short — ideal in logic circuits. It ranges from 25 to 50 ns in most devices. This does not imply that CMOS ICs aren't useful in linear applications: Some are designed primarily for the linear amplification of audio and rf energy (CA3600E, for one). Another salient feature of CMOS chips is low noise. Because FETs are used in these ICs the input impedance is high, making them more suitable than bipolar ICs for interfacing with comparable impedance levels outside the IC package. Fig. 94 shows the diagram of a CA3600E CMOS IC along with a block-symbol circuit for its use as a high-gain audio amplifier.

Array ICs

One branch of the linear-IC family is known as the *IC array* group. A short course on these and other linear ICs was given by DeMaw in *QST* for January through March 1977. Basically, the IC array is a substrate which contains a number of individual diodes or npn bipolar transistors. They differ from conventional ICs

by virtue of having each of the transistors independent from one another. Each transistor base, emitter and collector is brought out of the IC package by means of its own single pin. This enables the designer to treat each transistor as a discrete device, with the advantage that each transistor has nearly identical electrical characteristics (f_T, beta, dissipation rating, etc.). Some array ICs have f_T ratings as high as 1200 MHz, with maximum collector dissipation ratings as high as 1 watt. Schematic illustrations of some popular RCA array ICs are seen in Fig. 95.

Subsystem ICs

A branch of the linear-IC family tree is the *subsystem IC*. It is a conventional-package integrated circuit, but contains nearly as much circuitry as an LSI chip. Some of these devices represent the entire active-device circuitry for an fm or a-m radio receiver. The designer needs only to add essential outboard components (front-end tuned circuits, i-f transformers, tuning meter, and audio power amplifier) to realize a composite piece of equipment. Other subsystem ICs may contain only the i-f amplifiers, product detectors, agc loops, and audio pre-amplifiers. This style of IC is sold by such manufacturers as RCA, National Semiconductor and Plessey. Fig. 96 illustrates an example of this kind of device — the RCA CA3089E which is designed expressly for use in wide- or narrow-band fm receivers. It features a quadrature detector, and contains amplifiers, limiters, squelch circuit, metering circuit and an af amplifier. Those interested in

compact portable amateur receivers should find these devices especially interesting.

Practical Examples

The main disadvantage in the use of IC symbols in circuit diagrams is that the internal workings are not shown. This makes the designer work with a collection of "magic boxes." Fortunately, IC manufacturers publish data books which show the block symbols and pin arrangements versus the schematic diagrams of the active devices on the chips. This

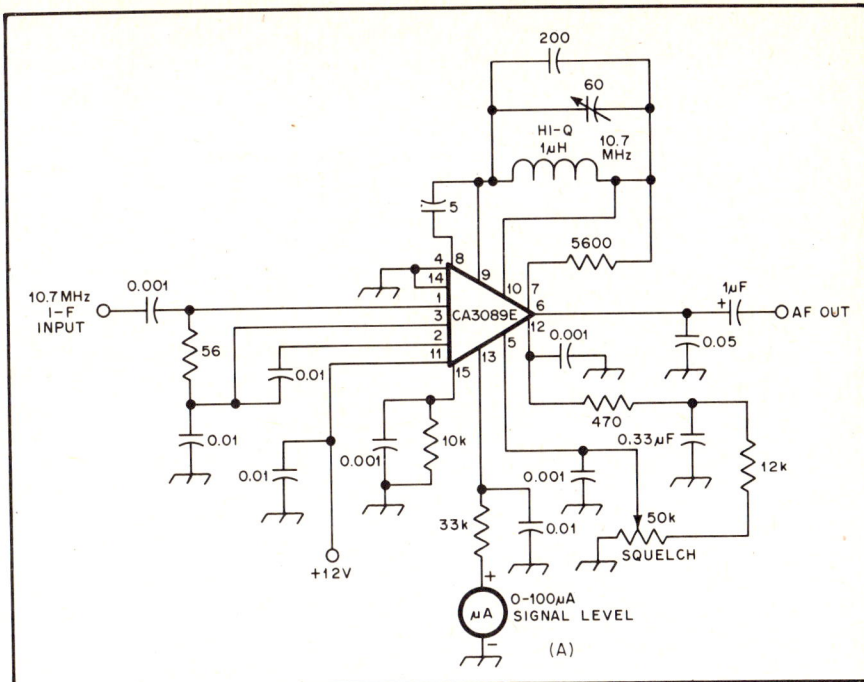

Fig. 96 — Example of a subsystem IC used as the heart of a narrow-band fm receiver.

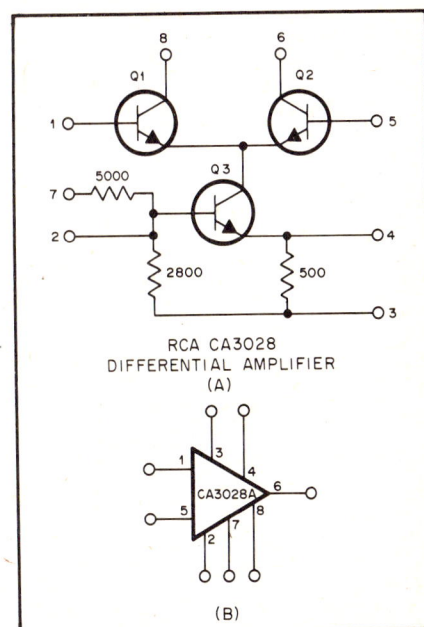

Fig. 97 — Schematic and block examples of an RCA CA3028A IC.

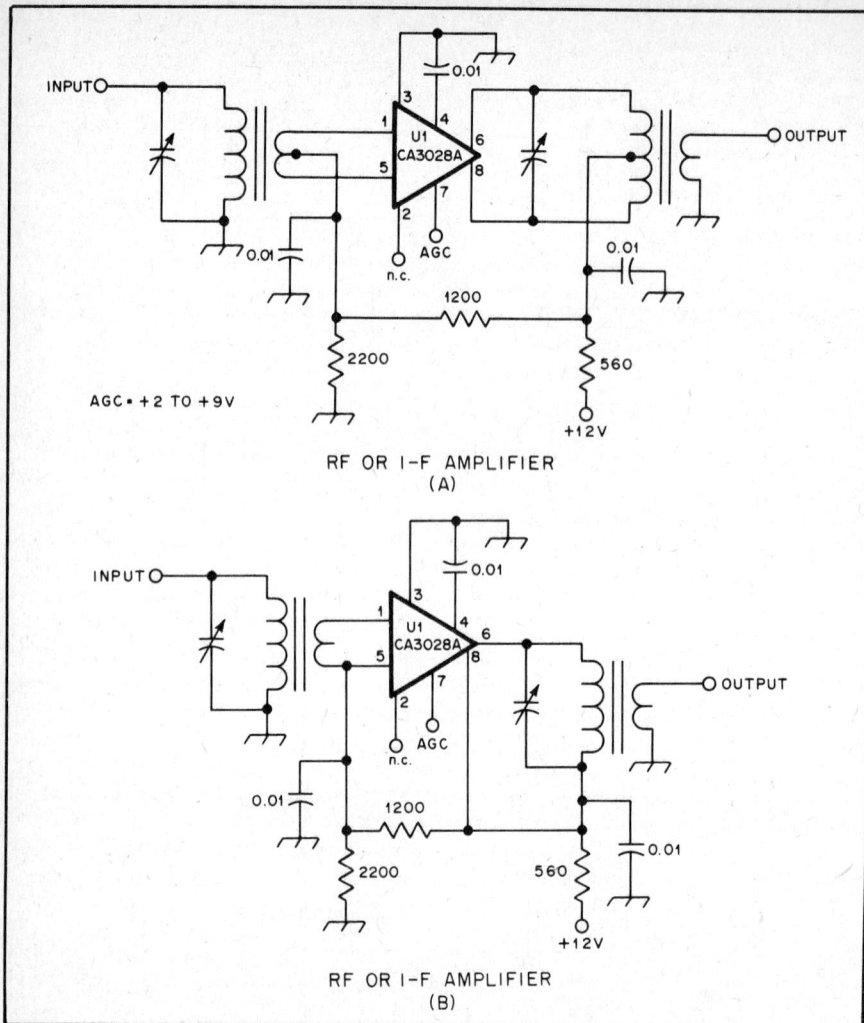

Fig. 98 — Rf and i-f amplifiers using the CA3028A IC. The example at A is balanced for ac and dc, whereas the circuit at B is balanced only for dc conditions.

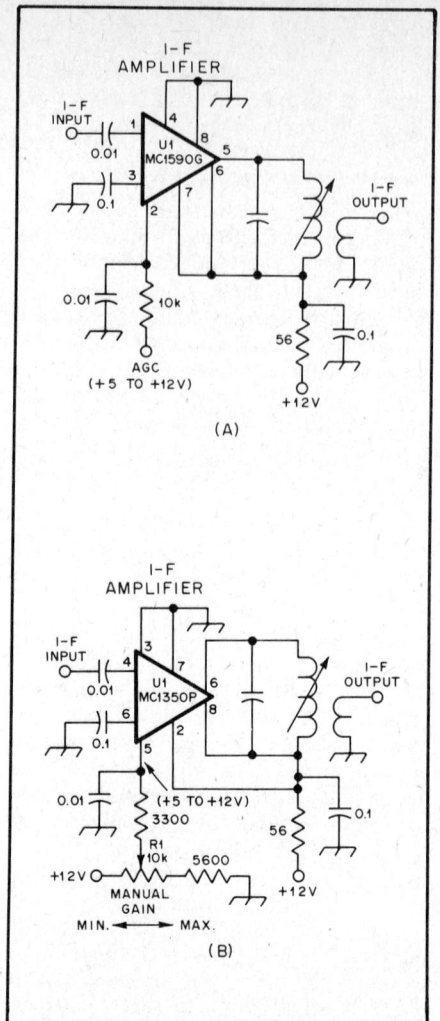

Fig. 99 — Circuit examples for Motorola IC i-f amplifiers.

permits the amateur to understand what the circuit configuration is before the design work is started. It is beyond practicality to include the schematic diagrams of the ICs used in this book, but we will show the circuit of the RCA CA3028A, because it is used frequently in the following section. Fig. 97 contains the block and schematic representation of this IC.

RF and I-F Amplifiers

Nearly every manufacturer of ICs produces chips that are suitable for use as rf/i-f amplifiers, mixers, detectors, oscillators and audio amplifiers. The circuits of Fig. 98 are examples of CA3028A rf or i-f amplifiers to which agc is applied. Maximum gain occurs when the agc voltage (IC forward bias) is at its highest potential. The IC is nearly cut off when the agc level drops below 2 volts. The circuit of Fig. 98A functions as a differential amplifier, as does the one at B. The basic difference is that dc and ac balance are featured at A, whereas only dc balance is effected at B. The gain of either stage is

approximately 40 dB. Pin 2 of U1 is left floating, but is used for LO injection when the CA3028A is employed as a mixer or product detector. A Motorola MC1550G is similar to the RC shown in Fig. 98. A MC1590G is a more suitable IC for i-f amplification when greater amounts of stage gain or agc control are desired.

An example of an MC1590G amplifier is given in Fig. 99A. It is shown with agc applied to pin 2. The lower the agc voltage the higher the stage gain. This is the opposite condition from that of the CA3028A of Fig. 98, where the gain increases with elevated agc voltage. The MC1350P of Fig. 99B is the low-cost version of the MC1590G. It is shown with manual control of the gain (R1), but agc voltage can be applied instead.

IC Mixers

Examples of IC active mixers are given in Fig. 100. At A is seen a singly balanced mixer formed by the differential transistor pair in a CA3028A. A doubly balanced

mixer is illustrated at B in Fig. 100. The MC1496G contains two differential transistor pairs to permit the doubly balanced configuration. This circuit does not exactly follow the suggested one by Motorola. It has been optimized for use as a transmitting mixer by W7ZOI and KL7IAK (*Solid State Design for the Radio Amateur*, 1st edition, page 204). There are numerous other ICs which can be used as mixers. Examples of many practical circuits are given in the ARRL book just referenced.

The circuit arrangements for product detectors and balanced modulators are similar to those shown in Fig. 100. They will not be described in this text, because the primary difference between them and a regular mixer lies in the frequencies of the signals mixed (af versus rf) and the frequency of the resulting output energy.

IC Audio Amplifiers

Practically every IC manufacturer offers a line of audio ICs. Some are for use as low-noise preamplifiers and others are capable of delivering up to a few watts of output to a loudspeaker. Most of the audio-power ICs are designed for looking directly into an 8- or 16-ohm load without the need for a matching transformer. Because these circuits are relatively mundane in nature they shall not be offered here as illustrative examples. Practical applications for audio ICs can be found in the construction projects elsewhere in this volume. Manufacturer's data sheets also provide definitive information on the use of these devices.

Operational Amplifiers

An operational amplifier (op amp) is a high-gain, direct-coupled differential amplifier whose characteristics are chiefly determined by components external to the amplifier unit. Op amps can be assembled from discrete transistors, but better thermal stability results from fabricating the circuit on a single silicon chip. Integrated circuit op amps are manufacturered with bipolar, JFET and MOSFET devices, either exclusively or in combination.

A design based on discrete components is shown in Fig. 101. Circuits of this variety were in common use before the advent of inexpensive IC-fabrication technology. The input stage consists of a differential pair biased by a constant current source. The terminal marked "−" is the inverting input and the one marked "+" is the noninverting input. The next stage, the pnp transistor, provides most of the voltage gain. High gain is realized through the use of a constant current source for the pnp collector load. The frequency response is determined by the collector-to-base capacitance of the pnp stage. This capacitance is fixed internally in some IC op amps and connected externally to others. A pair of emitter followers in a complementary symmetry arrangement forms the output buffer. A more comprehensive discussion of operational amplifiers is given by Woodward in "A Beginner's Look at Op Amps," April and June 1980 *QST*.

The most common application for op amps is in negative feedback circuits operating from dc to perhaps a few hundred kHz. Provided the device has sufficient open-loop gain, the amplifier transfer function is determined almost solely by the external feedback network. The differential inputs allow for both inverting and noninverting circuits. Fig. 102 shows these configurations and gives their transfer equations. R_L does not appear in the equations, implying that the output impedance is zero. This condition results from the application of heavy negative feedback. Most IC op amps have built-in current limiting. This feature protects the IC from damage caused by short circuits, but also limits the values of load resistance for which the output impedance is zero. Most op amps work best with load resistances of at least 2 kΩ.

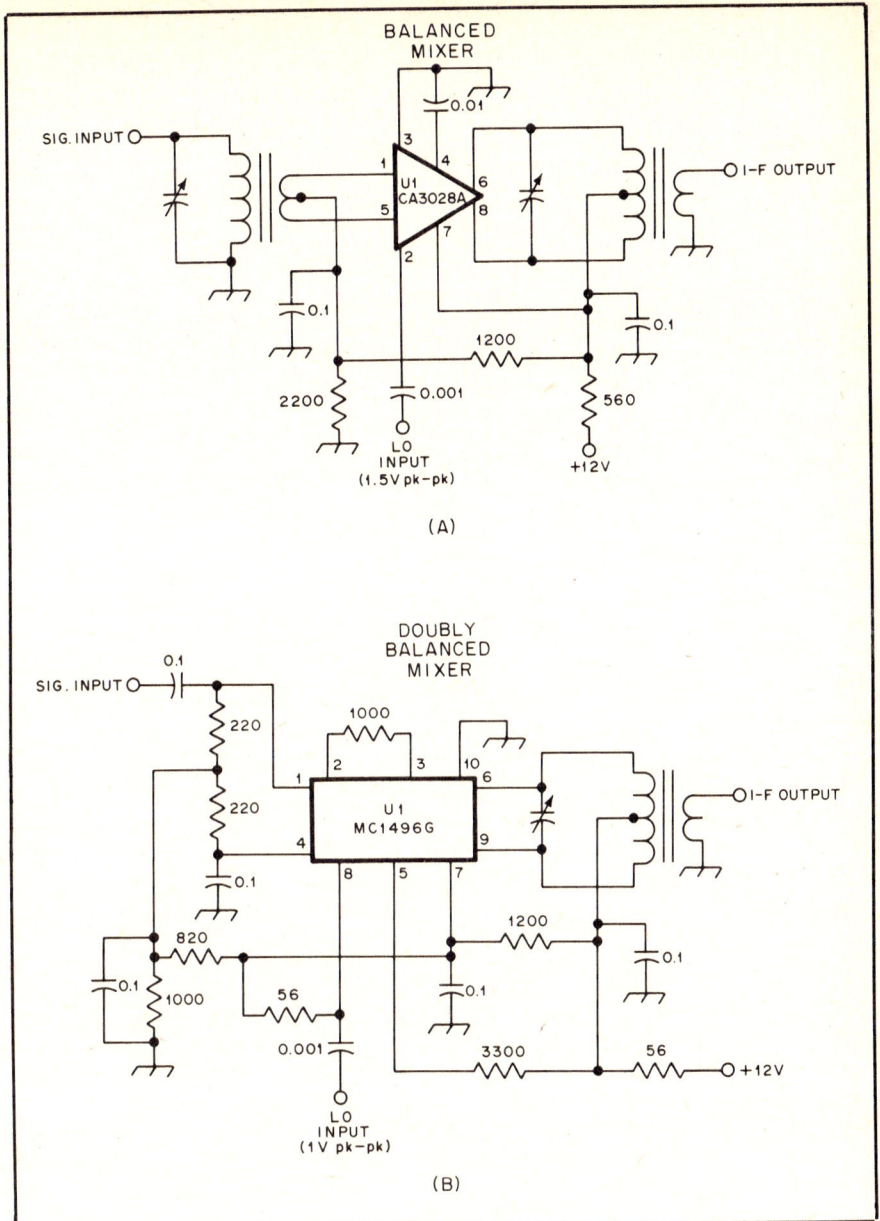

Fig. 100 — Two types of ICs are shown as mixers. The one at A is a singly balanced mixer.

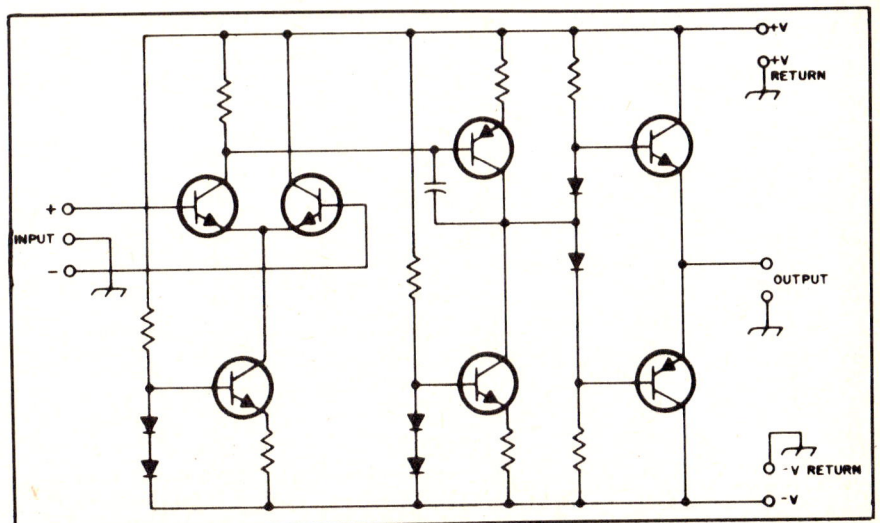

Fig. 101 — An operational amplifier assembled from discrete components. IC op amps contain more transistors (for current limiting and other peripheral functions), but the circuit topology is similar to that shown here.

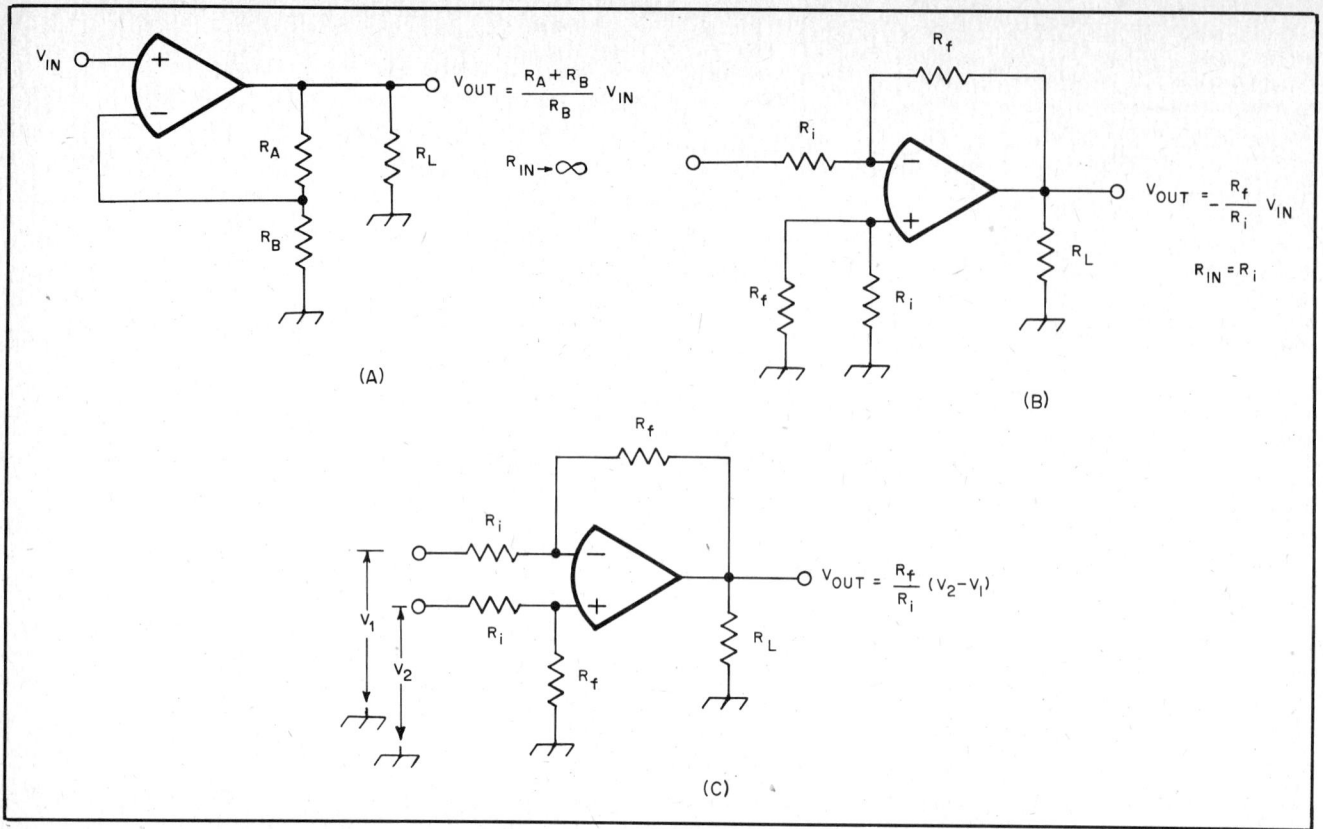

Fig. 102 — The standard negative-feedback op-amp circuits with their transfer equations. At A is a noninverting amplifier, at B, an inverting amplifier, and a differential amplifier is shown at C.

Since the op amp magnifies the difference between the voltages applied to its inputs, applying negative feedback has the effect of equalizing the input voltages. In the inverting amplifier configuration the feedback action combined with Kirchhoff's current law establishes a zero impedance, or virtual ground at the junction of R_f and R_i. The circuit input impedance is just R_i. Negative feedback applied to the noninverting configuration causes the input impedance to approach infinity.

The virtual ground at the inverting input terminal of an inverting operational amplifier circuit allows several currents to be summed without interaction. This principle can be used to advantage by the amateur wishing to simplify his or her sta-

tion control system. An example of a summing amplifier is given in Fig. 103. The circuit shown allows the operator to monitor the outputs of several receivers with one loudspeaker. The 3.9-Ω resistors simulate the loudspeaker in each receiver. An inverter follows the summing amplifier to restore the antivox signal to the proper phase. Fig. 104 shows another ap-

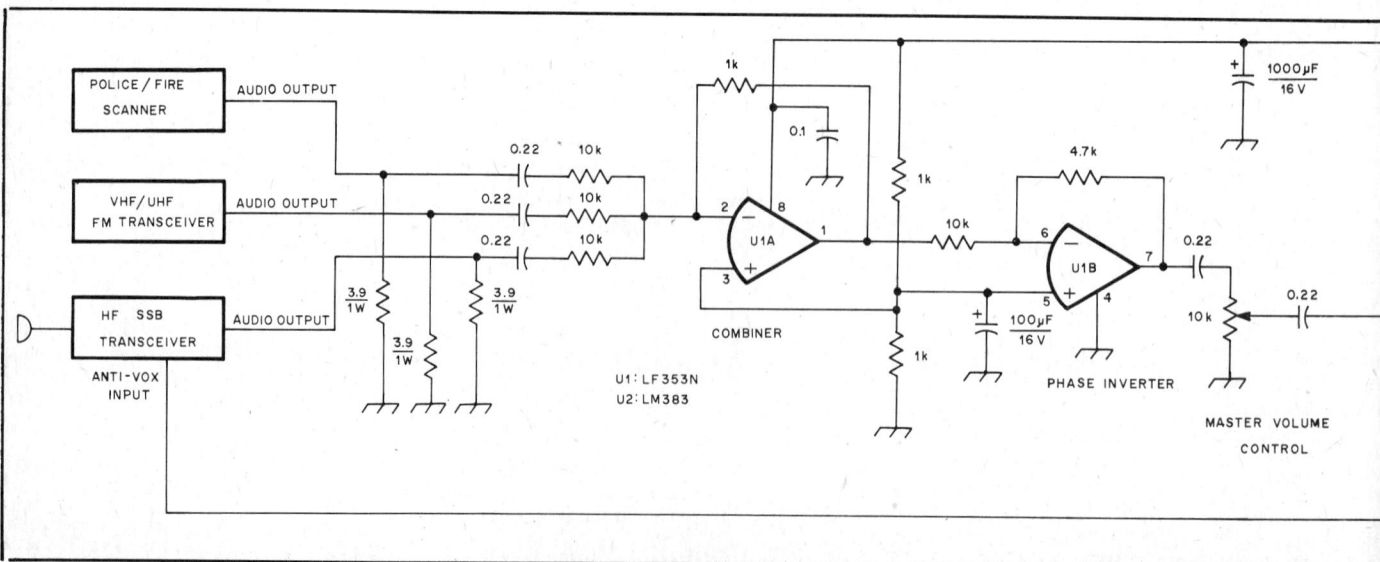

Fig. 103 — An amateur application for a summing amplifier — an audio combiner.

plication for a summing amplifier, a D/A converter. An FET-input operational amplifier can operate with the high-value resistors required by CMOS digital ICs while maintaining low offset and drift errors.

ICs intended for op-amp service can also be used in open-loop or positive feedback applications. Connecting one input to a fixed reference voltage as in Fig. 105A forms a comparator. The open-loop gain of the IC is so high that it acts more like a switch than an amplifier. When the voltage applied to the free input terminal is less than the reference voltage, the IC output stays near one of the power rails. If the input voltage exceeds the reference, the output swings to the opposite rail. A comparator with positive feedback, or *hysteresis*, is called a Schmitt trigger. A Schmitt trigger is illustrated in Fig. 105B. The potential on the noninverting input terminal depends on the output state as well as the reference voltage.

The Norton Amplifier

An unusual type of op amp is the *Norton*, named for the network theorem on which its operation is based. Fig. 106 shows a simplified diagram of the input stage of a Norton amplifier. The noninverting input makes use of D1 and Q1 in a *current mirror* configuration. When input current is applied to Q1, it steals base drive from Q2, the inverting input. This amplifier must have input current to operate, hence it is *not* a high-impedance device. In the inverting-amplifier configuration the numerical voltage gain is R_f/R_i, but the noninverting input terminal must be returned to the *positive supply* through a resistance of $2R_f$ to equalize the input currents. Any attempt to use this type of IC as a voltage follower is doomed to failure — the input stage will be destroyed by excessive current. The chief

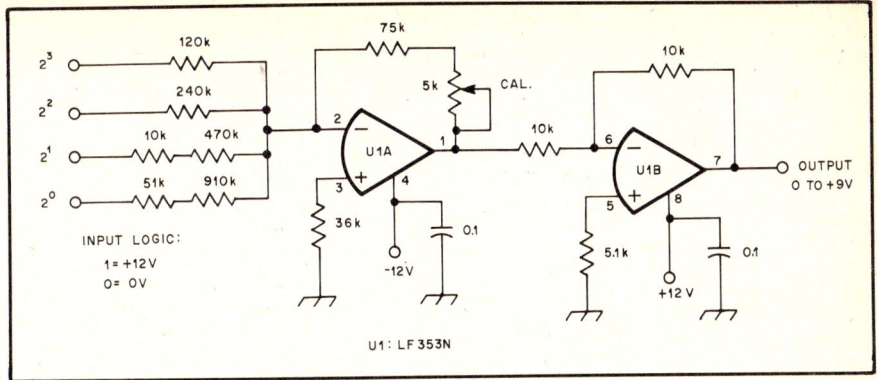

Fig. 104 — BCD D/A converter suitable for connection to a B-series CMOS driving source.

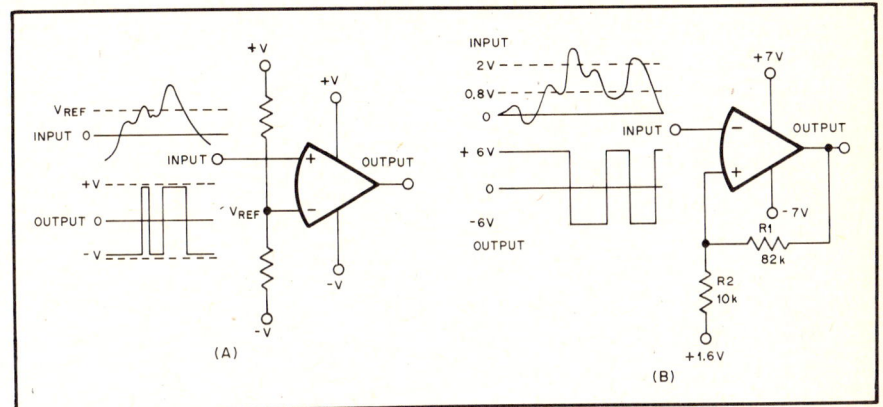

Fig. 105 — (A) Differential voltage comparator. Either inverting or noninverting circuits may be used. (B) Schmitt trigger. The constants shown here are suitable for connecting +5-V TTL to ±7-V CMOS logic.

usefulness of Norton amplifiers is in single-supply applications where the dc level of the signal is very near ground. The ssb chapter of this *Handbook* features a VOX circuit using the LM3900 Norton op amp.

Op Amps as Audio Filters

One of the more common uses to which op amps are put can be seen in the RC active audio-filter field. Op amps have the distinct advantage of providing gain and variable parameters when used as audio filters. Passive filters which contain L and C elements are generally committed to some fixed-value frequency, and they exhibit an insertion loss. Finally, op amps contribute to the attainment of miniaturization which is seldom possible while employing bulky inductors in a passive type of audio filter.

Although RC active filters can be built with bipolar transistors, the modern approach is to utilize operational amplifiers (op amps). The use of an op-amp IC, such as a type 741, results in a compact filter pole which will provide stable operation. Only five connections are made to the IC, and the gain of the filter section, plus the frequency characteristic, is determined by the choice of components external to the IC.

Although there are numerous applications for RC active filters, the

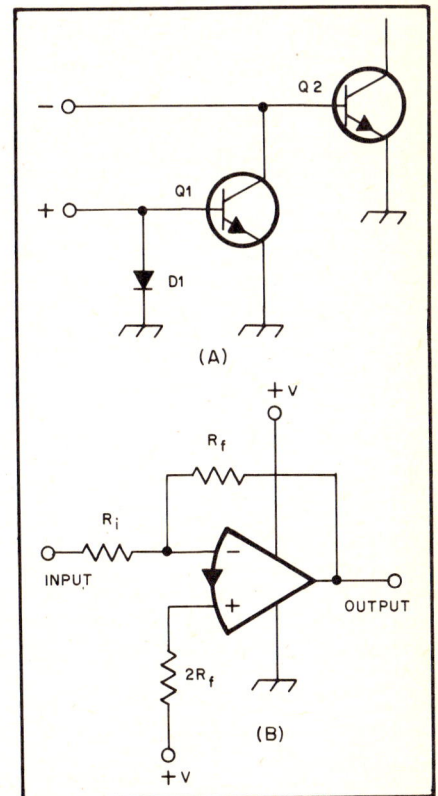

Fig. 106 — (A) Input circuit of a Norton operational amplifier. (B) Norton op amp connected as an inverting amplifier. Note the special symbol used to denote a Norton IC.

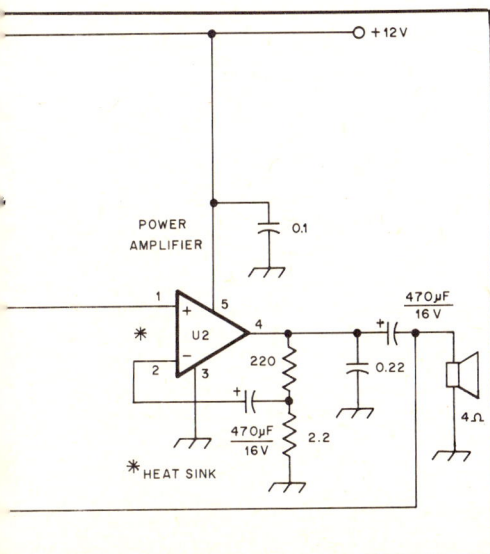

RC ACTIVE BANDPASS FILTER

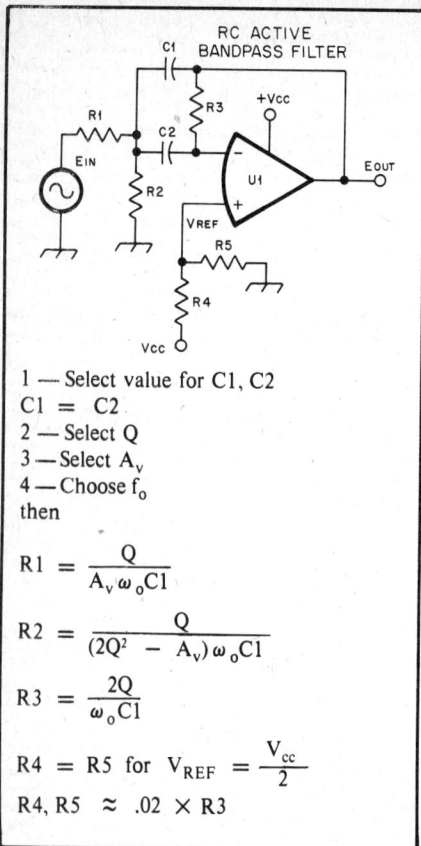

1 — Select value for C1, C2
C1 = C2
2 — Select Q
3 — Select A_v
4 — Choose f_o
then

$$R1 = \frac{Q}{A_v \omega_o C1}$$

$$R2 = \frac{Q}{(2Q^2 - A_v) \omega_o C1}$$

$$R3 = \frac{2Q}{\omega_o C1}$$

R4 = R5 for $V_{REF} = \dfrac{V_{cc}}{2}$

R4, R5 ≈ .02 × R3

Fig. 107 — Basic circuit for an RC active band-pass filter. One pole is shown along with the fundamental equations for finding the resistance values needed.

$f_o = 900\,Hz$ $A_v = 1$ $\omega_o = 2\pi f_{o(Hz)}$

Q = 5   C1, C2 = 0.00068 μF

$$R1 = \frac{5}{(1)\,(6.28 \times 900)\,(0.00068 \times 10^{-6})}$$
$$= 1{,}300{,}948 \text{ ohms}$$

$$R2 = \frac{5}{[2(25) - 1]\,[6.28 \times 900\,(0.00068 \times 10^{-6})]} = 26{,}550 \text{ ohms}$$

$$R3 = \frac{10}{(6.28 \times 900)\,(0.00068 \times 10^{-6})} = 2{,}601{,}896 \text{ ohms}$$

R4, R5 ≈ 2,601,896 × 0.02 = 52,018 ohms

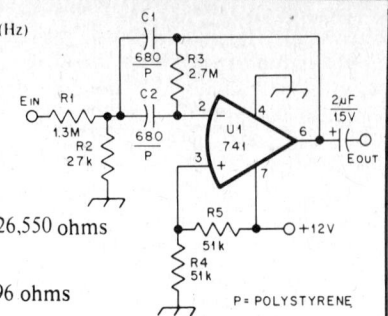

Fig. 108 — A design example based on the circuit of Fig. 107.

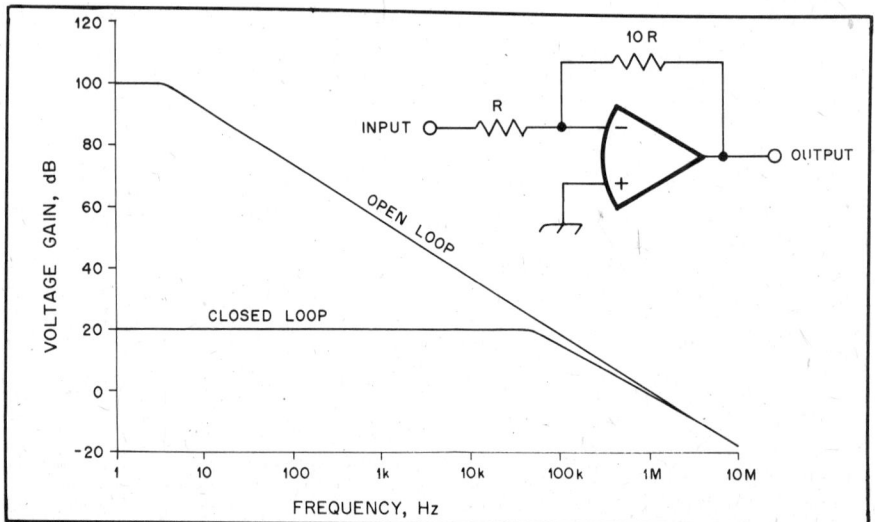

Fig. 109 — Open-loop gain and closed-loop gain as a function of frequency. The vertical distance between the curves is the feedback or gain margin.

principal use in amateur work is that of establishing selectivity at audio frequencies. One or two poles may be used as a band-pass or low-pass section for improving the passband characteristics during ssb or a-m reception. Up to four filter poles are frequently employed to acquire selectivity for cw or RTTY reception. The greater the number of poles, up to a practical limit, the sharper the skirt response of the filter. Not only does a well-designed RC filter help to reduce QRM, it improves the signal-to-noise ratio in some receiving systems.

Considerable design data is found in the National Semiconductor Corp. application note AN72-15 in which a thorough treatment of Norton amplifiers is given, centering on the LM3900 current-differencing type of op-amp. Design information is given for high-, low-, and band-pass types of RC active audio filters. The simplified design data presented here is based on the technique used in AN72-15.

Fig. 107 shows a single band-pass-filter pole and gives the equations for obtaining the desired values for the resistors once the gain, Q, f_o and C1-C2 capacitor values are chosen. C1 and C2 are equal in value and should be high-Q, temperature-stable components. Polystyrene capacitors are excellent for use in this part of the circuit.

Disc-ceramic capacitors *are not recommended*. R4 and R5 are equal in value and are used to establish the op-amp reference voltage. This is $V_{cc}/2$.

C1 and C2 should be standard values of capacitance. The filter design is less complicated when C1 and C2 serve as the starting point for the equations. Otherwise some awkward values for C1 and C2 might result. The resistance values can be "fudged" to the nearest standard value after the equations have been worked. The important consideration is that matched values must be used when more than one filter pole is employed. For most amateur work it will be satisfactory to use five percent, 1/2-watt composition resistors. If the resistor and capacitor values are not held reasonably tight in tolerance for a multipole filter, the f_o for each pole may be different, however slight. The result is a wide nose for the response, or even some objectionable passband ripple.

Fig. 108 illustrates the design of a single-pole band-pass filter. An arbitrary f_o of 900 Hz has been specified, but for cw

reception the operator may prefer something much lower — 200 to 700 Hz. An A_v (gain) of 1 (unity) and a Q of 5 are stated. Both the gain and Q can be increased for a single-section filter if desired, but for a multisection RC active filter it is best to restrict the gain to 1 or 2 and use a maximum Q of 5. This will help prevent unwanted filter "ringing" and audio instability.

C1 and C2 are 680-pF polystyrene capacitors. Other standard values can be used from, say, 500 to 2000 pF. The limiting factor will be the resultant resistor values. For certain design parameters and C1-C2 values, unwieldy resistance values may result from the equations. If this happens, select a new value for C1 and C2.

The resistance values assigned to R1 through R5, inclusive, are the nearest standard values to those obtained from the equations. The principal effect from this is a slight alteration of f_o and A_v.

In a practical application the RC active filter should be inserted in the low-level audio stages. This will prevent overloading the filter during the reception of strong signals. The receiver af gain control should be used between the audio preamplifier and the input of the RC active filter for best results. If audio-derived agc is used in the receiver, the RC active filter will give best performance when it is contained within the agc loop. Information on other types of active filters is given by Bloom in July 1980 *QST*.

Important Op Amp Specifications

Construction projects in the amateur literature call for the 709 and 741 more than any other type of operational amplifier, not because these devices are ideal for every application, but because until recently, they were the only ones commonly available to the electronics hobbyist. Questions of drift, offset, bandwidth, slew rate and noise were academic; the only practical alternative to the 709 or 741 was circuitry made from discrete transistors. A much wider selection of op amps is available today, and the amateur designer can choose the components best suited to the application. Also, the performance of some existing circuits can be upgraded by replacing 709s and 741s with improved devices. To this end, a brief survey of op amp specifications is in order.

Offset voltage is the potential between the amplifier input terminals in the closed-loop condition. Ideally, this voltage would be zero. Offset results from imbalance between the differential input transistors. Values range from millivolts in ordinary consumer-grade devices to only nanovolts in premium Mil-spec units. The temperature coefficient of offset voltage with respect to time is *drift*. A few microvolts per degree Celsius (at the input) is a typical drift specification.

There are two types of noise associated with operational amplifiers. Burst, or *popcorn* noise is a low-frequency pulsing, usually below 10 Hz. The amplitude of this noise is approximately an inverse function of temperature. The other noise is sometimes called *flicker*, and is a wideband signal whose amplitude varies inversely with frequency. For some analytical purposes, drift is considered as a very low frequency noise component. Op amps that have been optimized for offset, drift and noise are called instrumentation amplifiers. The latest instrumentation amplifier is the National Semiconductor LM10, designed by Robert Widlar, the acknowledged "father of the IC op amp." The architecture of the LM10 is different from any other device, but the practical applications are the same.

The small-signal bandwidth of an op amp is the frequency range over which the open-loop voltage gain is at least unity. This specification depends mostly on the frequency compensation scheme (for example, the capacitor in Fig. 101). Fig. 109 shows how the maximum closed-loop gain varies with frequency. The power bandwidth of an operational amplifier is a function of *slew rate*, and is always less than the small-signal value. Slew rate is a measurement of output voltage swing per unit time. Values from 0.8 to 13 volts per microsecond are typical of modern devices.

The hobbyist should maintain a supply of inexpensive 741 and 301 op amps for breadboarding, but should also be prepared to use improved devices in the final design. In an active filter for example, a 741 will demonstrate whether or not the circuit is working, but a low-noise, wide-bandwidth device will give higher performance, especially in receiving service. An abbreviated table of operational amplifier specifications is given in chapter 23. Most of the devices listed are available from hobby electronics stores or the mail-order firms listed in chapter 17.

Digital-Logic Integrated Circuits

Digital logic is the term used to describe an overall design procedure in which "on" and "off" are the important words, not "amplification," "detection," and other terms commonly applied to most amateur equipment. It is "digital" because it deals with discrete events that can be characterized by digits or integers, in contrast with linear or analog systems in which an infinite number of levels may be encountered. It is "logic" because it follows mathematical laws in which "effect" predictably follows "cause."

Logic systems can be implemented by mechanical, hydraulic, pneumatic or electrical means. The linkage that causes automotive windshield wipers to complete their cycle after they have been turned off is an example of a mechanical logic system. It is entirely possible to understand complex digital subjects such as computer architecture without any electrical knowledge. For this reason, digital logic theory is usually treated separately from electrical theory. Digital designers are often highly specialized. It is fairly common to hear of an "electrical engineer" who can design an incredibly complex logic system but cannot get the power supply working! Digital designers sometimes tend to be somewhat chauvinistic, holding the view that anything worth doing is worth doing digitally. While it is true that almost any function can be implemented digitally, in some cases an analog approach may be simpler or more cost-effective. The radio amateur, as a well-rounded communications expert, is in a position to choose the technology best suited to his needs, without prejudice. In recent years, Amateur Radio equipment has made increasing use of digital electronics, and this trend can be expected to continue. Today's repeater control systems, keyers, Morse code/RTTY readers, frequency counters and frequency synthesizers depend heavily on digital techniques. In the future, digital electronics will make further inroads in Amateur Radio communications, particularly in the area of signal processing.

The fundamental principle of digital electronics is that a device can have only two logical states: "on" and "off." This system is perfectly suited to binary (base or radix 2) arithmetic, which uses only two numerals: 0 and 1. The simplest digital devices are switches and relays. Some pre-1950 computers were built almost entirely of relays. Low speed and rapid wear were the objections to mechanical devices, so the next generation of digital instruments used electron tubes as the switching elements. Physical size and power consumption were the factors that limited the complexity of digital circuits using tubes. Modern semiconductor technology allows digital systems of tremendous complexity to be built at a small fraction of the cost of previous methods.

Combinational Logic

The three logical operations are "and," "or" and "not." An AND gate may be assembled with two relays as shown in Fig. 110A. In order to have voltage at the output (C), we must energize A *and* B. If we connect the contacts in parallel rather than in series, an OR gate results (Fig. 110B). The "not," "complement" or "inverse" function may be implemented with a normally-closed relay contact as illustrated in Fig. 110C. If we apply voltage to A we will have no voltage at C, and vice versa. With the proper system of AND and NOT gates or OR and NOT gates, any logical or arithmetic function may be synthesized. AND or OR gates are often combined with inverters in IC packages and called NAND and NOR gates.

A special combination of gates called an "exclusive OR" has an output only if the two inputs are complementary. This combination is used frequently enough to be packaged specially and assigned a fundamental symbol.

Logic systems have polarity. If the highest voltage level represents a binary

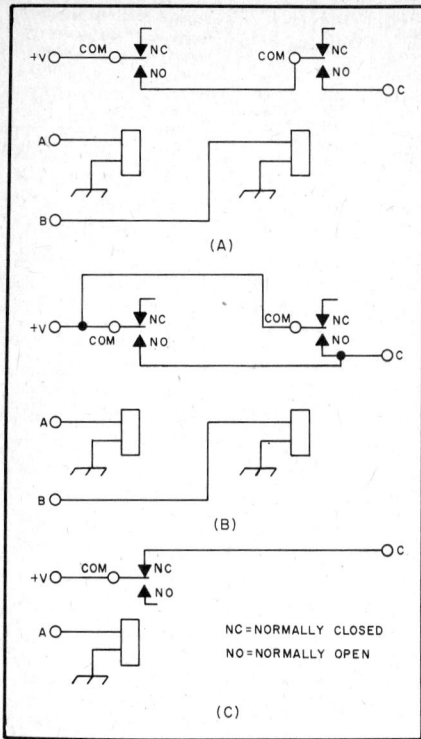

Fig. 110 — Relay models of the three logical operations. A: AND gate B: OR gate C: inverter.

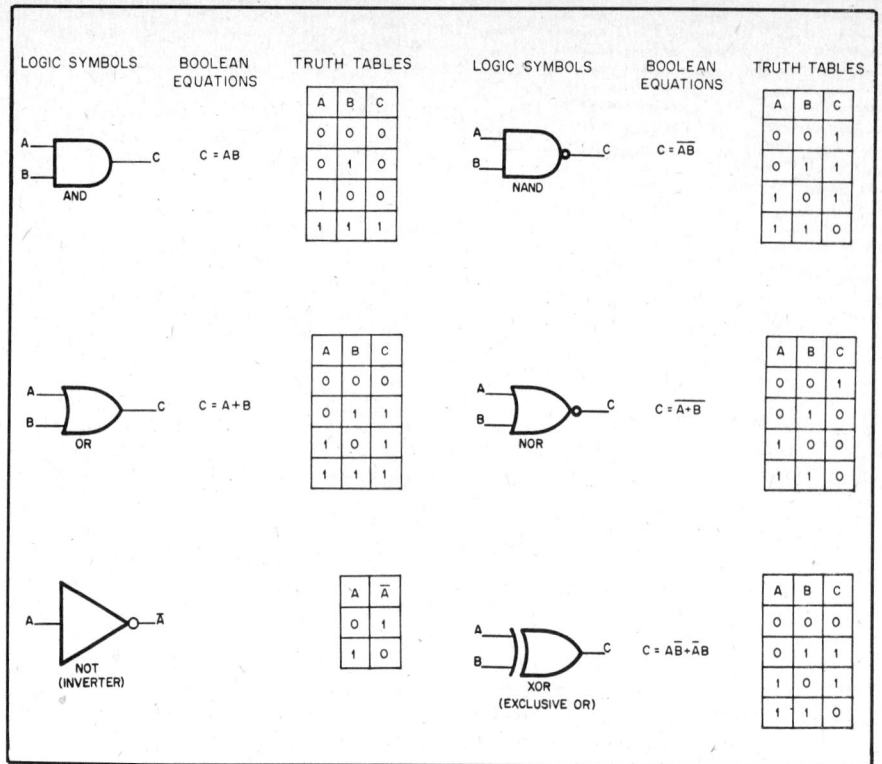

Fig. 111 — The standard logic symbols, with corresponding Boolean equations and truth tables.

AND $C = AB$

A	B	C
0	0	0
0	1	0
1	0	0
1	1	1

NAND $C = \overline{AB}$

A	B	C
0	0	1
0	1	1
1	0	1
1	1	0

OR $C = A + B$

A	B	C
0	0	0
0	1	1
1	0	1
1	1	1

NOR $C = \overline{A + B}$

A	B	C
0	0	1
0	1	0
1	0	0
1	1	0

NOT (INVERTER)

A	$\overline{A}$
0	1
1	0

XOR (EXCLUSIVE OR) $C = A\overline{B} + \overline{A}B$

A	B	C
0	0	0
0	1	1
1	0	1
1	1	0

one and the lowest level represents a zero, the logic is said to be positive. If the opposite representation is used, the logic is negative.

Since each input or output of a digital network can have only two possible states, it is possible to list all of the input combinations and their corresponding outputs, thus completely characterizing the operation. Such a list is called a truth table.

Each type of gate is assigned a distinctive schematic symbol. The AND gate symbol has a straight edge on the input side and a blunt convex edge on the output side. The OR gate is characterized by a concave edge on the input side and a sharp cusp on the output side. A small circle at the output of a gate signifies that an inversion has taken place.

Digital systems may be designed with *Boolean algebra*. Circuit functions may be defined by algebraic equations. The symbology and laws of Boolean algebra are somewhat different from those of ordinary algebra. The " + " symbol is used to indicate the "or" function. "And" is represented by "·" or juxtaposition of the variables. A bar with a variable indicates that it has been inverted. Fig. 111 shows the symbols for the common logic functions with their associated Boolean equations and truth tables. Positive logic is assumed. With the exception of the "exclusive OR," all of the gates may be expanded to any number of inputs. There is no universally-accepted definition for an exclusive OR gate with more than two inputs.

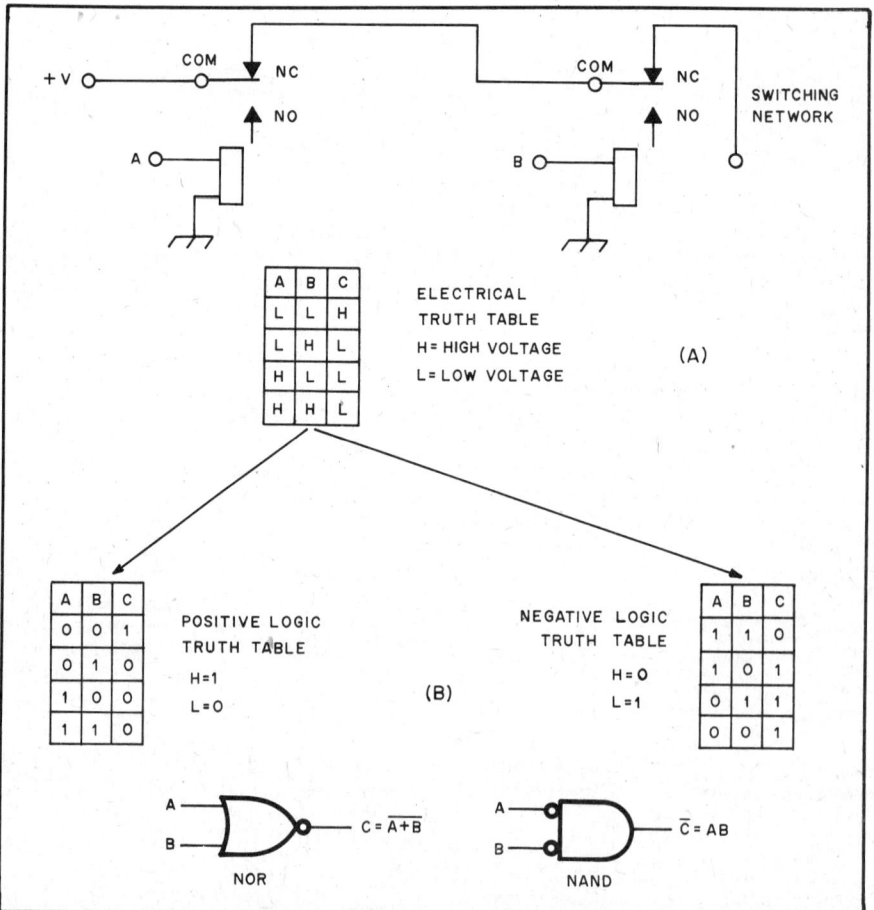

ELECTRICAL TRUTH TABLE H = HIGH VOLTAGE L = LOW VOLTAGE (A)

A	B	C
L	L	H
L	H	L
H	L	L
H	H	L

POSITIVE LOGIC TRUTH TABLE H = 1 L = 0 (B)

A	B	C
0	0	1
0	1	0
1	0	0
1	1	0

NEGATIVE LOGIC TRUTH TABLE H = 0 L = 1

A	B	C
1	1	0
1	0	1
0	1	1
0	0	1

NOR $C = \overline{A + B}$

NAND $\overline{C} = AB$

Fig. 112 — At A, combinational logic implemented with relays, shown with the electrical truth table. Assigning 1 and 0 to the electrical states as in B leads to two schematic symbols, one for positive logic and one for negative logic. The two symbols are electrically equivalent; depending on the application, one may represent the logical operation being performed better than the other.

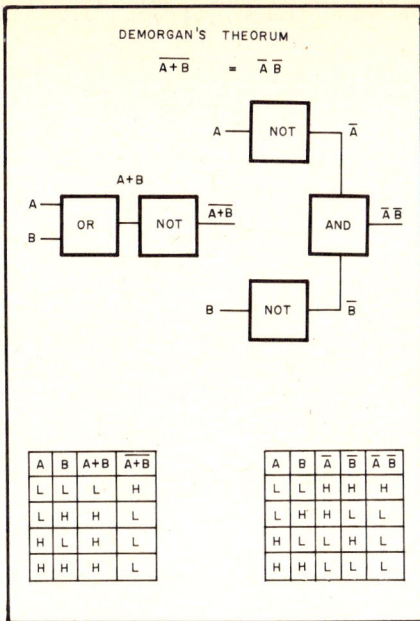

Fig. 113 — One of the fundamental theorums of combinational logic. The block diagrams and their corresponding expanded truth tables verify the theorum and show the relationship between the schematic symbols in Fig. 112. DeMorgan's Theorum can also be stated as $\overline{AB} = \overline{A} + \overline{B}$.

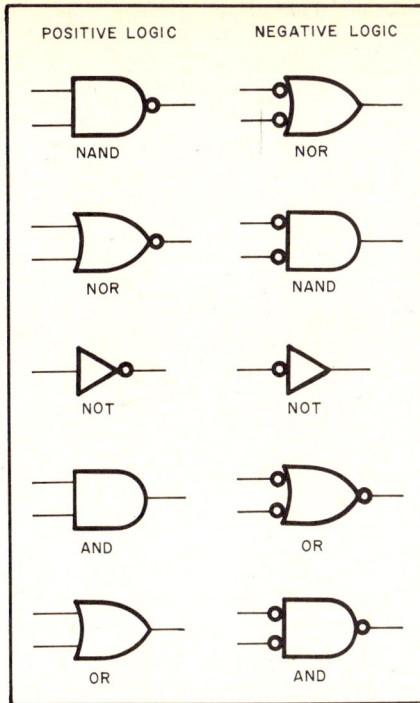

Fig. 114 — Positive- and negative-logic symbols for the common functions. The horizontally opposite gates are electrically identical.

The Boolean algebra associated with logic networks can sometimes be simplified through the use of negative logic. Consider a circuit having two inputs and one output, and suppose a HIGH level output is desired only when both inputs are LOW. A search through the truth tables of Fig. 111 shows the NOR gate to have the proper characteristics for our application. However, the phrasing of the problem (the words *only* and *both*) strongly suggests the AND (or NAND) function. A negative-logic NAND is functionally equivalent to a positive-logic NOR gate, and the NAND symbol better expresses the circuit function in the application just described. Small circles (called state indicators) on the input side of a gate signify negative logic. Fig. 112 traces the evolution of the electromechanical switching circuit into a NOR or NAND gate, depending on the logic convention chosen.

DeMorgan's Theorem, one of the most important results from Boolean algebra, justifies the conversion from one logic convention to the other. An application of this theorum appears in Fig. 113. When viewed as a continuation of Fig. 112, the block representation of the Boolean equations clarifies the negative logic symbology.

Finally, Fig. 113 gives detailed electrical truth tables showing identical output states for any combination of inputs. A complete chart of equivalent symbolic representations is given in Fig. 114.

A circuit made of the fundamental gates and configured in such a way that the output is a function of the present static input levels only is called a *combinational* logic circuit. Pulses and transitions are not considered in the design of a combinational circuit.

Sequential Logic

A circuit in which the output state is a function not only of the input levels but also of past output states is a *sequential* logic circuit. Conventional truth tables are not generally applicable to sequential circuits because a certain input condition may not have a unique output state. The simpler sequential circuits are sometimes defined by a modified truth table showing input transitions and output state progressions. State tables, flow diagrams and timing charts are the tools used to design complex sequential machines.

The dependence on previous output states implies a requirement for *memory*. The simplest memory element is a special type of bistable multivibrator (flip-flop) called a latch. A D (for data) flip-flop is often used as a latch. A flip-flop can store one bit (binary digit) of information. A typical D flip-flop with its truth table is shown in Fig. 115A. The logic level at D is transferred to Q on the positive transition of the clock pulse. The Q output will retain this logic level regardless of any changes at D until the next positive clock transition. The D throughput is said to be *synchronous* because it is actuated by the clock signal. The flip-flop shown also has

set and reset (S and R) inputs. These inputs are *asynchronous* because they are independent of, and in fact overide, the clock and data inputs.

Fig. 115B shows a common application for a D flip-flop, a modulus-two frequency divider. The sequence of events is illustrated by the timing diagram. Several of these flip-flops may be cascaded in a single IC package and called a counter. The states of the Q outputs can be read as a binary code, indicating the number of clock pulses received in an interval.

An RS flip-flop is shown in Fig. 116. Two inverting gates connected in this fashion form a *regenerative* switching circuit. The accuracy of the accompanying truth table depends on the input states occuring in the order given. The output corresponding to an input of 11 could easily be the complement of that shown if it followed a 00 input state. An important rule in the design of sequential logic circuits is that the simultaneity of events cannot be depended upon.

The RS flip-flop is the simplest type. Its outputs change directly as a result of changes at its inputs. The type T flip-flop "toggles," "flips," or changes its state during the occurrence of a T pulse, called a clock pulse. The T flip-flop can be considered as a special case of the J-K flip-flop. Although there is some disagreement in the nomenclature, a J-K flip-flop is generally considered to be a toggled or clocked R-S flip-flop. It may also be used as a storage element. The J input is frequently called the "set" input; the K is called the "clear" input (not to be confused with the clock input). The clock input is called c. A clear-direct or C_D input which overrides all other inputs to clear the flip-flop to 0 is provided in most J-K flip-flop packages.

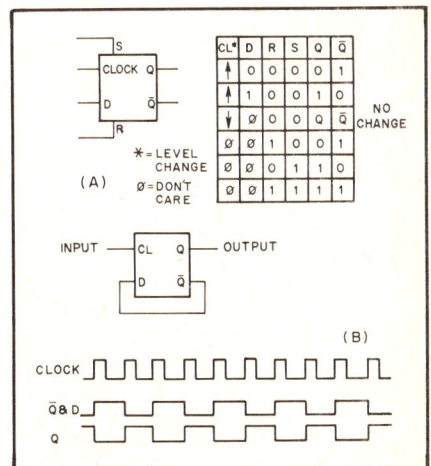

Fig. 115 — A D flip-flop. In A, set and reset ("jam") inputs are provided. Note that the functional truth table shows Q and $\overline{Q}$ both in the high state for one combination of R and S. While this appears contradictory, it is the standard way of defining the operation of this type of flip-flop.

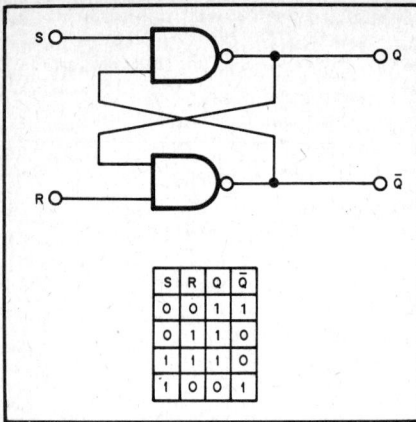

S	R	Q	Q̄
0	0	1	1
0	1	1	0
1	1	1	0
1	0	0	1

Fig. 116 — A regenerative switching circuit called an RS flip-flop. The circuit could be implemented with NOR gates, in which case the first defined input state would be 11.

There are essentially two types of flip-flop inputs, the dc or level-sensitive type, and the "ac" or transition-sensitive type. It should not be concluded that an ac input is capacitively coupled. This was true for the discrete-component flip-flops but capacitors of relatively large value just do not fit into microcircuit dimensions. The construction of an ac input uses the "master-slave" principle, where the actions of a master flip-flop driving a slave flip-flop are combined to produce a shift in the output level during a transition of the input.

Semiconductor Memories

While simple systems of flip-flops can be used to store a small number of bits, efficient filing of large amounts of information calls for special-purpose devices. Semiconductor memories are classified by their operating characteristics, organization and size.

When specifying memory size or organization the symbol "k" refers to 1024 bits, bytes or words. Thus a 64-k bit memory contains 65,536 bits of storage (sometimes called a 65-k memory in error!). A *byte* is a fundamental fraction of a word and most often refers to a collection of eight bits. A *word* may be any number of bits, depending on the application and system. Common word lengths are 8, 16, 32, 36, 60 and 64 bits.

A memory IC of a given capacity may be organized in a number of ways. A 4-k memory may be organized 4k × 1, 1k × 4, 512 × 8 and so on, with the second number designating the number of bits that can be accessed simultaneously.

Several sets of operating characteristics are used to classify memories. If the locations in an IC can be accessed in any order it is said to be a *random access* device. Almost all semiconductor memories are random access devices.

The other generic access mode is *serial* access. Examples of serial access memories are shift registers, CCD (charge-coupled device) memories, and most mechanical storage devices such as magnetic tape. Serial devices introduce a variable access delay, called latency, which depends on the internal state of the device when an address is presented to it.

Unless specified otherwise, it is assumed that a memory device can be written into as well as read. Semiconductor memories which can be written in are usually "volatile," meaning they lose their contents if the power is removed. A special class of memory, the *Read Only Memory* (ROM), is not volatile. Some are mask-programmed during manufacture — this type of device is what is usually called a ROM. Another device of read-only memory is programmed in the field and is called a *Programmable Read-Only Memory* (PROM). A PROM is manufactured with all bits in one state, and the user creates bits of the opposite state by an irreversible process, such as blowing fuses or destroying transistors in the IC. The manufacturer's specified programming technique must be followed exactly if a reliable result is to be obtained. *Erasable* PROMs (EPROMs) can be returned to their unprogrammed state by exposing the IC to ultraviolet light through a window in the package. Another type of PROM is the *Electrically Alterable* PROM (EAPROM). These devices are programmed in a nondestructive, reversible manner, usually in the normal operating circuit. They can retain stored data for up to 10 years even when power is removed (storage time is reduced at high temperatures). Individual words, and sometimes blocks of words, can be erased and rewritten. This device is sometimes called a "read-mostly memory."

Semiconductor RAMs may be volatile even with power applied — these are called *dynamic* RAMs and must have the stored data "refreshed" at regular intervals (100 ms or less). The refresh interval and technique varies significantly from one device type to another. RAMs that do not need to be refreshed are called *static* RAMs. Dynamic RAMs tend to have larger storage capacity and slower access than static RAMs. Some static RAMs have "dynamic read-out" circuits which limit the length of time data remains valid on the output pins and imposes minimum times between successive readings. Many RAMs may have their supply voltages reduced without loss of data while they are not being accessed, thus reducing power consumption.

Large memory arrays are often used for the generation and conversion of information codes. One IC can be programmed to convert the five-level RTTY code to the eight-level ASCII code popular in computer devices. National Semiconductor manufactures a single IC which generates the entire 56-character eight-level code. Several ICs are now available for character generation where letters and numerals are produced for display on an oscillograph screen.

Microprocessors

An important new (from the 1970s) class of integrated circuit is the *microprocessor*. A microprocessor combined with a few other ICs and input/output devices forms a *microcomputer*. Today, practically every IC manufacturer produces microprocessors, either of its own design or as a second source. Some recent pieces of Amateur Radio equipment incorporate microprocessors for channel sequencing and other functions. The microprocessor portion of such equipment is part of a *special purpose computer*, meaning the input and/or output isn't available for general use or programming.

More and more amateurs are using *general-purpose computers* for radio-related activities. A variety of architectures is possible, but the most basic configuration is illustrated in Fig. 117. In a typical amateur set-up, the program and data are input through a keyboard or cassette recorder, and a CRT display or printer serves as an output device. The capabilities of most computer systems can be enhanced by supplementing the internal storage unit with additional memory.

A computer is a machine and is incapable of independent thought or action. The machine can only do what it is instructed or *programmed* to do. There are programs called text editors, which translate invalid instructions into the nearest valid ones, but even here the machine isn't thinking, because the text editor program was written by a human.

Amateur computing is a hobby quite distinct from Amateur Radio, but there are some worthwhile computer applications in radio. Some of these are Morse code and RTTY encoding and decoding, SSTV character generation, aural readout (any format) of digital displays and satellite commanding. Of course, a microcomputer system can be used for routine filing and record keeping. Some hams keep their station logs by computer, and can instantly retrieve information about previous contacts. A computer can relieve the tedium of sorting tasks (for example, the index of this *Handbook* was organized with the aid of a microcomputer system), but the most exciting applications are to things that weren't possible before the personal computer era. Some innovations that need to be developed by radio/computer enthusiasts are video bandwidth compression (MSTV), high-speed data communication and weak-signal enhancement. One method for transmitting moving images in a narrow bandwidth is to send only those picture elements that differ from the previous frame. This technique, known as *digital refreshment*, is a sophisticated job for a microcomputer system. The high-speed data communication made possible by computers may allow a complete vhf/uhf contact in a single meteor burst. Existing practice with this propagation mode sometimes requires hours to exchange call signs and signal reports.

The current state of the art in amateur EME work requires many kilowatts of erp and ultra-low-noise receivers to obtain barely perceptible lunar echos. By statistical analysis of the receiver output, a microcomputer could possibly pull a lunar echo out of the noise, thereby easing the station gain requirements.

Amateurs are becoming more interested in *computer-aided design*. This technique is especially useful for designs requiring many iterative calculations, such as interstage matching networks in solid-state transmitters. The tables of filters appearing elsewhere in this *Handbook* were generated by a computer.

QST publishes articles on nontrivial applications of microcomputers to Amateur Radio. A bibliography of *QST* articles from 1975 to the present on microprocessors and microcomputers is available for an s.a.s.e. from ARRL. A three-part series entitled *Meet the Microprocessor*, by Thomas and Belter, appeared in August, September and October 1976 *QST*.

Digital System Design Considerations

Digital engineers use a variety of graphical and analytical tools to design logic systems. Once a circuit having the desired performance is found, the engineer works to minimize the number of components through the application of switching theory. As noted in the introduction to this digital logic section, switching theory is a subject unto itself and cannot be treated justly in this *Handbook*.

Switching theory is based on ideal switches, and real electrical devices don't always emulate ideal switches well enough to synthesize a logic design with switching theory alone. Therefore, a digital designer must consider the electrical characteristics of the logic elements he's using. Propagation delay and transmission line reflections become significant factors as the speed of the logic system increases. The ultimate application of the system also influences the design. For example, an attempt to design a frequency synthesizer strictly as a "number cruncher" is doomed to failure. The reason is that any solid-state digital device is also an *analog* device having a finite transfer function. This transfer function can cause a device to act as an amplifier, multiplier or mixer as well as a switch. A frequency synthesizer designed only as a logical machine will be rife with spurious outputs and noise that can't be predicted from switching theory.

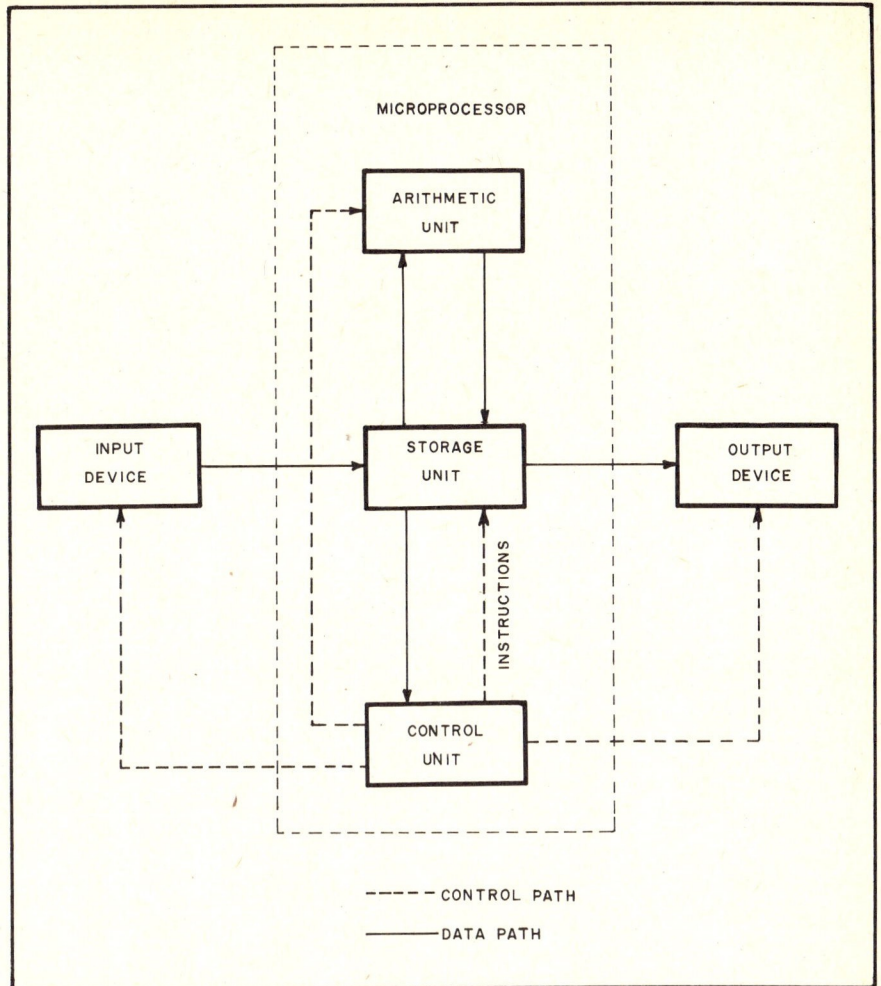

Fig. 117 — The basic parts of a digital computer.

Stray coupling between sections, small transients on the power supply, and junction noise can pollute the final output spectrum without affecting the logical function.

Particular attention should be paid to power supply decoupling. In general, a few 0.01 μF disc ceramic bypass capacitors will prevent the switching transients of one IC from changing the state of another. If the logic system is to be used with radio equipment, more extensive measures may be necessary. For a logic system to have *electromagnetic compatibility* (EMC), it must not radiate energy into a radio receiver or be adversely affected by energy radiated from a nearby transmitter. The EMC problem is receiving increased attention as more radio amateurs acquire personal computers for their stations.

Logic Device Fabrication Technology

Nearly all modern logic systems use integrated circuits. The ICs are classified according to the complexity of the circuit on the chip. ICs having 15 or more active devices fall into the MSI or medium-scale integration category. The simple gate and flip-flop circuits use MSI technology. Dual-inline packages with 14 or 16 pins are common for these circuits. Specialized circuits using 100 or more active devices are classified as LSI, for large-scale integration. Some serial-to-parallel converters, arithmetic logic units and data converters use LSI technology. Many LSI packages occupy four times the pc-board area of standard MSI packages. Up to 64 pins are used on some LSI devices. The very latest fabrication technology has enabled the creation of VLSI, or very large-scale integration systems. These circuits have over 1000 active devices on a single chip. Third-generation microcomputers, frequency counters and a wide variety of "smart" instruments will make use of this technology.

Digital-Logic IC Families

There are several families or types of ICs that are seeing widespread use. Each family has its own inherent advantages and disadvantages. Each is geared to its

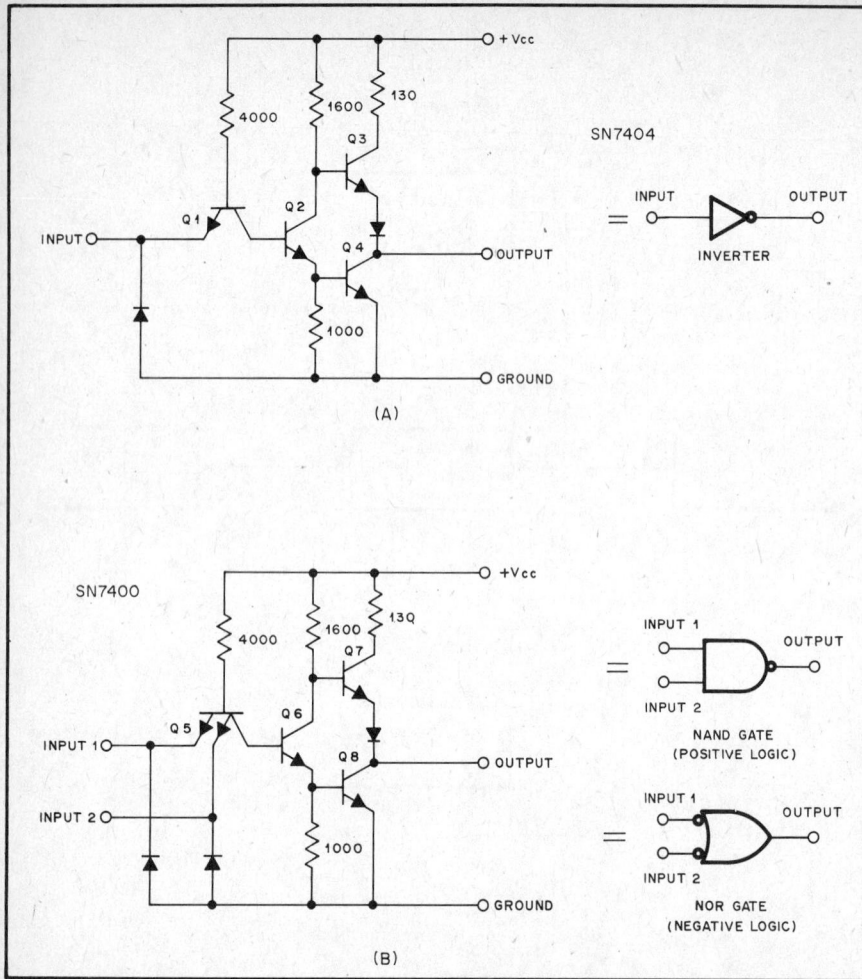

Fig. 118 — TTL circuits and their equivalent logic symbols (see text). Indicated resistor values are typical. Identification of transistors is for text reference only; these are *not* discrete components.

own particular market, meeting a specific set of needs. RTL (resistor-transistor logic) and DTL (diode-transistor logic) are obsolete and are no longer used in new designs. They are manufactured for exact replacement purposes only.

Transistor-Transistor Logic — TTL

TTL is one of the bipolar logic families. Also known as T²L (T squared L), this family has a variety of circuit configurations. Some devices have "open collector" outputs, and these may be "wire ORed." Open collector outputs are useful for interfacing with other logic families or discrete components. Although + 5 volts is the recommended power supply for TTL, open collectors can be connected to a different voltage through the external load resistor, within the limits specified for the device. Most TTL devices have "totem pole," or "active pull-up" output stages, and these cannot be wired ORed. Typically, the outputs are capable of sinking more current than they can source. This situation is of importance only when interfacing devices outside a particular TTL subfamily. TTL devices have a fanout (number of inputs that can be driven by a single output) of 10 within a subfamily. If TTL inputs are left open, they assume a "high" logic state, but greater noise immunity will be realized if pull-up resistors are used. When operated with a + 5-volt supply, any input voltage level between 2.0 and 5.5 is defined to be high. A voltage less than 0.8 is an input low. TTL ICs output a minimum high level of 2.4 volts, and a maximum low level of 0.4 volts. The switching transients generated by TTL devices appear on the supply line and can cause false triggering of other devices. For this reason, the power bus should have several bypass capacitors per pc board.

"Plain" TTL ICs are identified by 5400- or 7400-series numbers and operate at speeds up to 35 MHz. Two commonly used TTL devices are represented schematically in Fig. 118. High-speed ICs (50 MHz) are identified by 54H00- or 74H00-series numbers. These ICs consume more power than their ordinary counterparts. The 54L00- and 74L00-series of devices are designed for lower power consumption that the standard types. These ICs typically dissipate one milliwatt per gate, or about one-tenth of that dissipated by standard TTL. Operating speed is the tradeoff for the lower power, and the maximum speed for this subfamily is 3 MHz.

The subfamilies discussed so far operate as saturated switches. The 54S00- and 74S00-series have Schottky diode clamps that keep the transistors out of saturation. Some ICs of this series are useful up to 125 MHz. The power dissipation is about twice that of standard TTL. A commonly used subfamily combining low power dissipation with fairly high speed is the 54LS00- and 74LS00-series. The dissipation and speed for this series are 2 mW and 45 MHz, respectively.

One section of a type 7404 hex inverter is represented schematically in Fig. 118A. A LOW level applied to the input will cause Q1 to conduct current. This will cause Q2 to be near cutoff, in turn biasing the "totem pole" arrangement of Q3 into saturation and Q4 near cutoff. As a result the output level will be HIGH, about 1 volt below V_{cc}. If the signal at the input is HIGH the conduction state of each transistor reverses and the output drops nearly to ground potential (LOW). The input diode protects the circuitry by clamping any negative potential to approximately − 0.7 volt, limiting the current in Q1 to a safe value. Note that this protection is effective only against transients. The output diode is required to ensure that Q3 is cut off when the output is LOW.

The circuit in Fig. 118B, one section of a type 7400 gate is very similar to that of A. The difference is that Q5 is a multiple-emitter transistor with one input to each emitter. A LOW level at either input will turn on Q5, causing the output to go HIGH.

Emitter-Coupled Logic — ECL

ECL has the highest speed of any of the logic forms. Some ECL devices can operate at frequencies higher than 1.2 GHz. This family is different from the other forms of bipolar logic in that the transistors operate in a non-saturating mode that is analogous to that of some linear devices. The typical logic swing is only 800 mV. ECL devices are characterized for use with a − 5.2-volt power supply, but operation from other supplies is possible. If the Vcc terminal is connected to + 2.0 volts and the Vee terminal connected to − 3.2 volts, the device can drive a 50-ohm load directly with respect to ground. The power output obtained this way is about 0 dBm. ECL ICs dissipate a great deal of power, and heat sinking is sometimes necessary. The ECL family finds use in uhf frequency synthesizers and counters, as well as in computers. Some highly specialized ICs have capacitively coupled inputs, and therefore have minimum as well as maximum toggle rates.

There are several ECL subfamilies being produced. Speed, power dissipation, and the ability to drive transmission lines cannot be optimized simultaneously, so different versions are offered to allow the designer to choose the tradeoffs best suited to his application. ECL subfamilies are compatible, but only over a limited temperature range. The differences between subfamilies are mostly in resistance values, and the presence or absence of input and output pull-down resistors.

A significant feature of ECL gates is that complementary output functions are available from each circuit. The circuit of Fig. 119, for example, is a NOR/OR gate. Q1 or Q2, together with Q3 forms a differential amplifier. When the Q2 collector goes HIGH, the Q3 collector goes LOW, and these levels appear at the emitters of the output buffers, Q5 and Q6. The circuitry associated with Q4, D1 and D2 is a bias generator. The reference voltage established at the base of Q3 determines the input switching threshold.

Metal-Oxide Semiconductors — MOS

The logic families using all n-channel or all p-channel field-effect transistors are used extensively in microprocessors, digital watches and calculators. Where entire functions can be synthesized on a single chip, this technology is quite useful. Ordinary NMOS and PMOS gate packages are not very popular. Most general-purpose logic networks are now made with complementary metal-oxide semiconductor (CMOS) ICs. This family has p-channel and n-channel transistors on the same chip. Only one of each complementary pair is turned on at any time, so the power dissipation is negligible except during logic transitions. A notable feature of CMOS devices is that the logic levels swing to within a few millivolts of the power supply voltages. The input switching threshold is approximately one half the power supply voltage ($V_{DD} - V_{SS}$). This characteristic contributes to high immunity to noise on the input signal or power supply. CMOS input current drive requirements are miniscule, so the fanout capability is tremendous, at least in low-speed systems. For high speed systems, the input capacitance increases the dynamic power dissipation and limits the fanout.

Four subfamilies of CMOS logic ICs are being produced at present. The 4000A series is the original commercial line and operates with power supplies from 3 to 12 volts. A subfamily having some improved characteristics is the 4000B (for buffered) series. The B series can be powered from supplies up to 18 volts. This feature makes the devices especially attractive for automotive applications. The output impedance of buffered ICs is independent of the input state. An unbuffered series, designated 4000UB meets all the B series specifications except that the logic outputs are not buffered and the input logic levels must be within 20 percent of the power

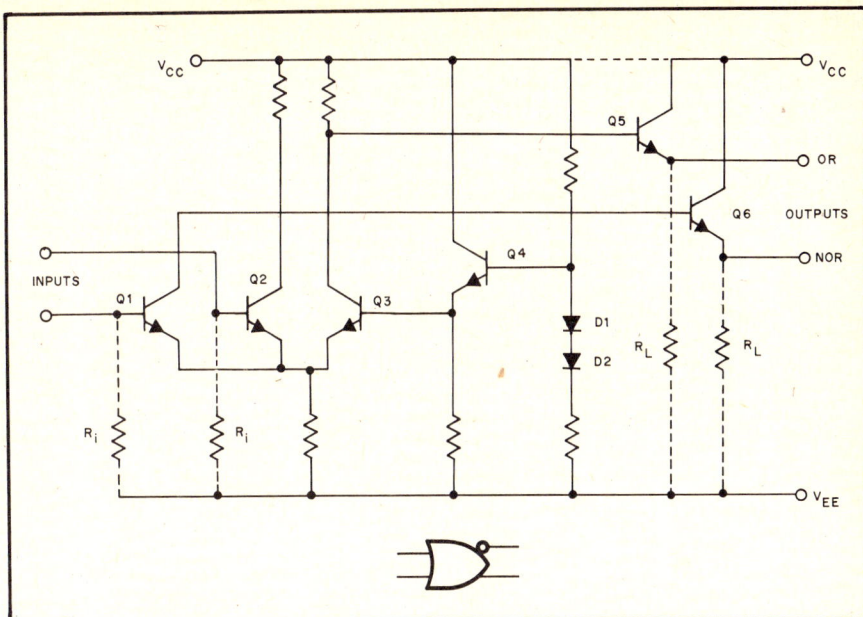

Fig. 119 — Circuit topology of the ECL family. The functions of the various components are explained in the text. Complementary outputs call for the modified logic symbol.

supply terminal voltages. Several tradeoffs must be considered when choosing between buffered and unbuffered ICs. The buffered devices have greater noise immunity and drive capability, but the speed is low compared to the unbuffered types. Some special-purpose 4000 series ICs have tri-state output circuits. The third state is neither HIGH or LOW, but is a high-impedance condition which allows several outputs to be paralleled for wire ORing or multiplexing. The 74C00 series is designed to be a plug-in replacement for low-power TTL devices in some applications. Some CMOS devices can function at speeds greater than 15 MHz.

A simplified diagram for a CMOS logic inverter is given in Fig. 120. Some of the diodes in the input and output protection circuits are inherently part of the manufacturing process. Even with the protection circuits, CMOS ICs are susceptible to damage from static charges.

Certain precautions have become

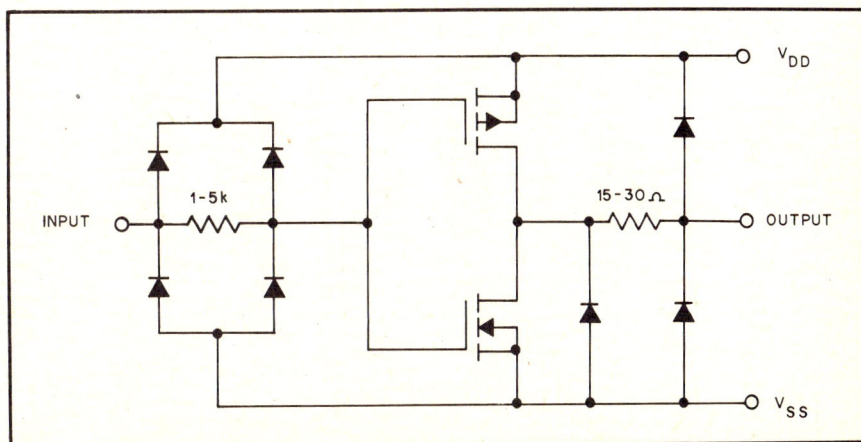

Fig. 120 — Internal structure of a CMOS gate.

accepted for handling these devices. The pins should not be inserted in styrofoam as is commonly done with bipolar integrated circuits unless the styrofoam is wrapped in aluminum foil. The 3M Company manufactures a spongy conductive material for this purpose under the trade name Velostat. Before removing a CMOS IC from its protective material, make certain that your body is grounded. A conductive bracelet connected to the ground terminal of a 3-wire ac outlet through a 10-MΩ resistor is adequate for this purpose. In industry, extreme protective measures are sometimes taken, such as blowing ionized air over an assembly area. The amateur experimenter needn't go to that extent; common sense will enable him or her to use CMOS logic without destroying many devices.

Special Digital ICs

In addition to the logic families mentioned above, other families are being developed (sometimes on a speculative basis) and are seeing comparatively limited use. One such family is integrated injection logic, or I^{2}L. Other families use various techniques in the production of metal-oxide semiconductor (MOS) devices. H-MOS (for high performance) is a scaled-down silicon gate MOS process; V-MOS is an anisotropically etched double-diffused MOS process, and D-MOS is a planar double-diffused process. SOS is a complementary silicon-on-sapphire process. The proponents of each of these families claim theirs is the best way of achieving higher speeds and densities at lower power dissipation on smaller chips which cost less.

Interconnecting Logic Families

Each semiconductor logic family has its own advantages in particular applications.

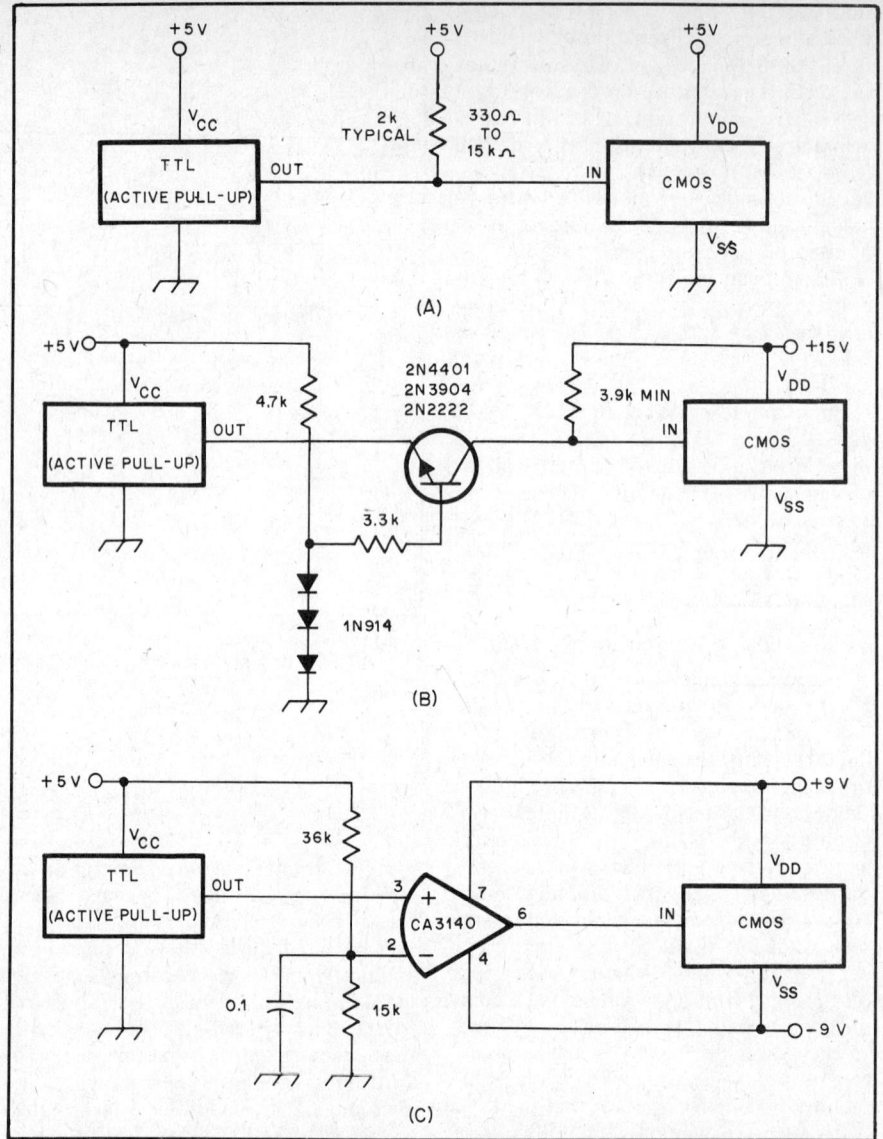

Fig. 121 — Methods for driving CMOS loads from TTL sources. The circuit complexity depends on the power supply voltages. The operation of these circuits is discussed in the text.

Table 1

Electrical Characteristics of the Common Logic Families

	Standard TTL (active pull-up) $V_{CC} = +5.0$ V	Schottky TTL (74S) $V_{CC} = +5.0$ V	High-speed TTL (74H) $V_{CC} = +5.0$ V	Low-power TTL (74L) $V_{CC} = +5.0$ V	Schottky low-power TTL (74LS) $V_{CC} = +5.0$ V	ECL III (1600 series) $V_{CC} = 0$ V, $V_{EE} = -5.2$ V	ECL 10 k
minimum HIGH input voltage	2.0 V	2.0 V	2.0 V	2.0 V	2.0 V	−1.095 V	−1.105 V
maximum HIGH input current	40 μA	50 μA	50 μA	20 μA	20 μA	—	—
maximum LOW input voltage	0.8 V	0.8 V	0.8 V	0.8 V	0.8 V	−1.485 V	−1.475 V
maximum LOW input current	1.6 mA	2.0 mA	2.0 mA	400 μA	400 μA	—	—
minimum HIGH output voltage	2.4 V	2.5 V	2.4 V	2.4 V	2.5 V	−0.9 V (30 mA)	−0.825 V (30 mA)
maximum HIGH output current	800 μA	1.0 mA	1.0 mA	400 μA	400 μA	40 mA	50 mA
maximum LOW output voltage	0.4 V	0.5 V	0.4 V	0.3 V	0.5 V	−1.75 V	−1.725 V
maximum LOW output current	16 mA	20 mA	20 mA	4.0 mA	8.0 mA	Open emitter — pull-down resistor required	

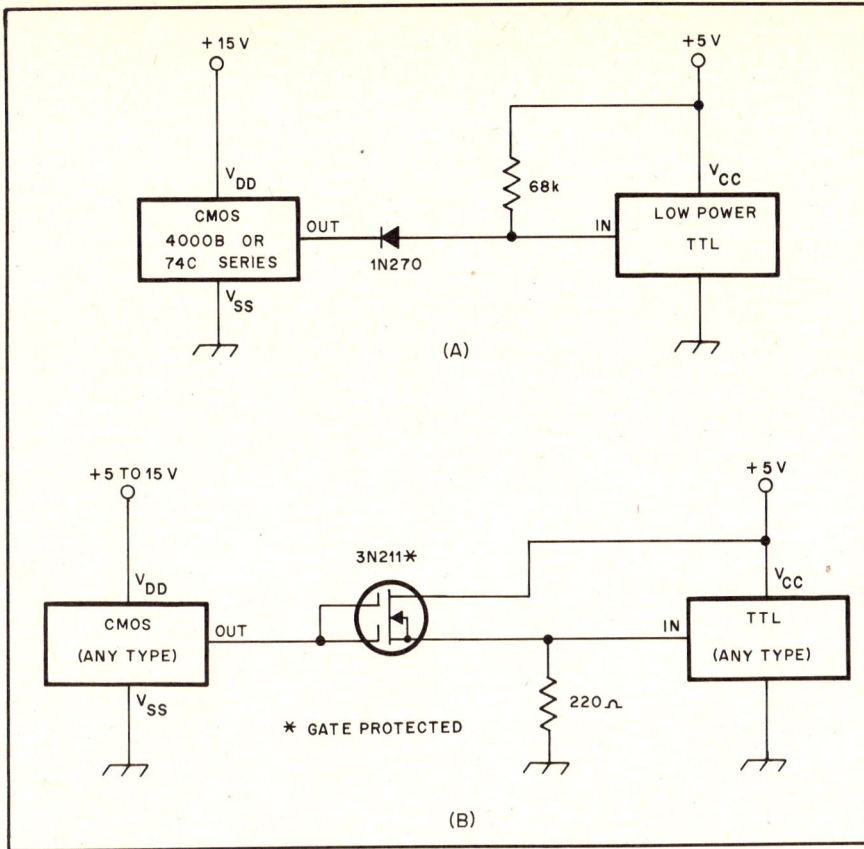

Fig. 122 — CMOS-to-TTL interface circuits. When both devices operate from a +5-volt supply, the diode in A can be eliminated. The circuit in B exhibits maximum flexibility with respect to supply voltages and logic subfamilies.

satisfactory results, even if they have a common power supply. The buzzword *interface* is often used to describe the integration of two types of logic into a compatible system. There are a number of ICs intended especially for mating different logic families. The CD4049UB and CD4050B hex buffers are designed to drive TTL gates from CMOS input signals. TTL-to-ECL and ECL-to-TTL conversion can be implemented with the N1017 and N1068 integrated circuits manufactured by Signetics and others. Unfortunately, these components aren't always conveniently available to the small investor, so logic interface must sometimes be accomplished by other means.

A knowledge of the circuit topologies and input/output characteristics will allow the designer to concoct reliable digital interstage networks. Typical internal structures have been illustrated for each common logic family. The input/output characteristics of the common logic families are listed in Table 1. This information was compiled from various industrial publications and is intended only as a guide. Certain ICs may have characteristics that vary from the values given. The following section discusses some specific logic conversions. Often more than one conversion scheme is possible, depending on whether the designer wishes to optimize power consumption or speed. Usually one must be traded off for the other. Where an electrical connection between two logic systems isn't possible, an optical isolator can always be used.

TTL-Driving CMOS

A CMOS gate is easily driven by a TTL device when both are powered by a +5-volt source. The totem-pole output structure of most TTL ICs prevents a HIGH output level of sufficient potential to properly activate the CMOS input. A pull-up resistor connected from the interface point to the power bus will remedy this problem. The maximum usable value for this component is 15 kΩ, but the circuit capacitance will reduce the maximum possible speed of the CMOS gate. Lower values will generate a more favorable RC product at the expense of increased power dissipation. A standard TTL gate can drive a pull-up resistor of 330 Ω, but a low-power version is limited to 1.2 kΩ for a minimum. The resistor pull-up technique is illustrated in Fig. 121A.

When the CMOS device is operating on a power supply other than +5 volts, the TTL interface is more complex. The common-base level shifter of Fig. 121B will translate a TTL output signal to a +15-volt CMOS signal while preserving the full noise immunity of both gates. An operational amplifier configured as a comparator, as in Fig. 121C, makes an excellent converter from TTL to CMOS using dual power supplies. An FET op amp is shown because the output voltage

For example, the highest frequency stages in a uhf counter or frequency synthesizer would use ECL. After the frequency has been divided down to less than 25 MHz, the speed of ECL is unnecessary, and the expense and power dissipation is unjustified. TTL is the obvious choice for the signal-processing operations in this frequency range. The programming functions have practically no speed constraints, so considerable power can be saved by using CMOS in that part of the system.

Each of these logic families has its own input voltage and current requirements, so they can't be randomly intermixed with

CMOS (4000A) Vcc		CMOS (4000B) Vcc			CMOS (74C00) Vcc		
+5.0 V	+10.0 V	+5.0 V	+10.0 V	+15.0 V	+5.0 V	+10.0 V	+15.0 V
3.5 V	7.0 V	4.0 V	8.0 V	12.5 V	3.5 V	8.0 V	—
—	—	—	—	1.0 µA	—	—	1.0 µA
1.5 V	3.0 V	1.0 V	2.0 V	2.5 V	1.5 V	2.0 V	—
—	—	—	—	1.0 µA	—	—	1.0 µA
4.95 V	9.95 V	4.95 V	9.95 V	14.95 V	2.4 V (360 µA)	9.0 V (10 µA)	—
(-------------------------------------- no load --------------------------------------)							—
300 µA (2.5 V)	250 µA (9.5 V)	1.6 mA (2.5 V)	1.3 mA (9.5 V)	3.4 mA (13.5 V)	1.75 mA (0 V)	8.0 mA (0 V)	—
0.05 V	0.05 V	0.05 V	0.05 V	0.05 V	0.4 V (360 µA)	1.0 V (360 µA)	—
(-------------------------------------- no load --------------------------------------)							—
300 µA (0.4 V)	600 µA (0.5 V)	500 µA (0.4 V)	1.3 mA (0.5 V)	3.4 mA (1.5 V)	1.75 mA (5 V)	8.0 mA (10 V)	—

of this type can usually swing closer to the power rails than a bipolar unit. Where the pulse rate is below 10 kHz or so, a 741 type of amplifier may be used. The Schmitt trigger configuration of Fig. 105B will further enhance the noise immunity at the interface point, but will invert the logic. An additional inverter (either linear or digital) placed on either side of the interface will resurrect the logic to its "true" form.

CMOS Driving TTL

The 4049UB and 4050B devices already mentioned can drive two standard TTL loads when a common +5-volt supply is used. Most A-series CMOS ICs can't sink enough current to drive TTL gates to a reliable LOW input state. Gates from the more-modern B-series can drive one low-power TTL load directly. The 74C00 family is capable of direct connection to low-power TTL with a fanout of two. The drive capability of CMOS gates can be increased by connecting identical gates in parallel, but this practice is not recommended unless all the gates are contained in a single IC package.

Fig. 122A shows a simple method for driving a TTL load from a CMOS source operating with a higher voltage power supply. The diode blocks the high voltage from the CMOS gate when it is in the HIGH output state. A germanium diode is used because its lower forward voltage drop provides higher noise immunity for the TTL device in the LOW state. The 68-kΩ resistor pulls the input HIGH when the diode is back biased. While TTL inputs assume the HIGH state when left open, floating the input is a poor practice because the gate is sensitive to noise in this condition.

Standard TTL inputs draw 1.6 mA in the LOW state. A pull-down resistor for this purpose can be no larger than 220 Ω. To pull this resistor up to an acceptable HIGH level requires 10 mA, which is beyond the capabilities of most CMOS devices. When a pull-down resistor is used, a dual-gate MOSFET having high transconductance makes a good buffer between CMOS and TTL systems. This scheme is diagrammed in Fig. 122B. The CMOS power supply voltage isn't critical when this system is used, because the output impedance of the CMOS device is high compared to the pull-down resistance, and the protective diodes in the FET can handle more current than the CMOS IC can provide. In fact, this circuit can also be used with split supplies, provided the positive CMOS output excursion is at least 5 volts.

TTL Driving ECL

When a common power supply is used, the resistor network of Fig. 123A will allow a standard TTL gate to drive an ECL input at the maximum TTL speed. Although shown with V_{CC} connected to

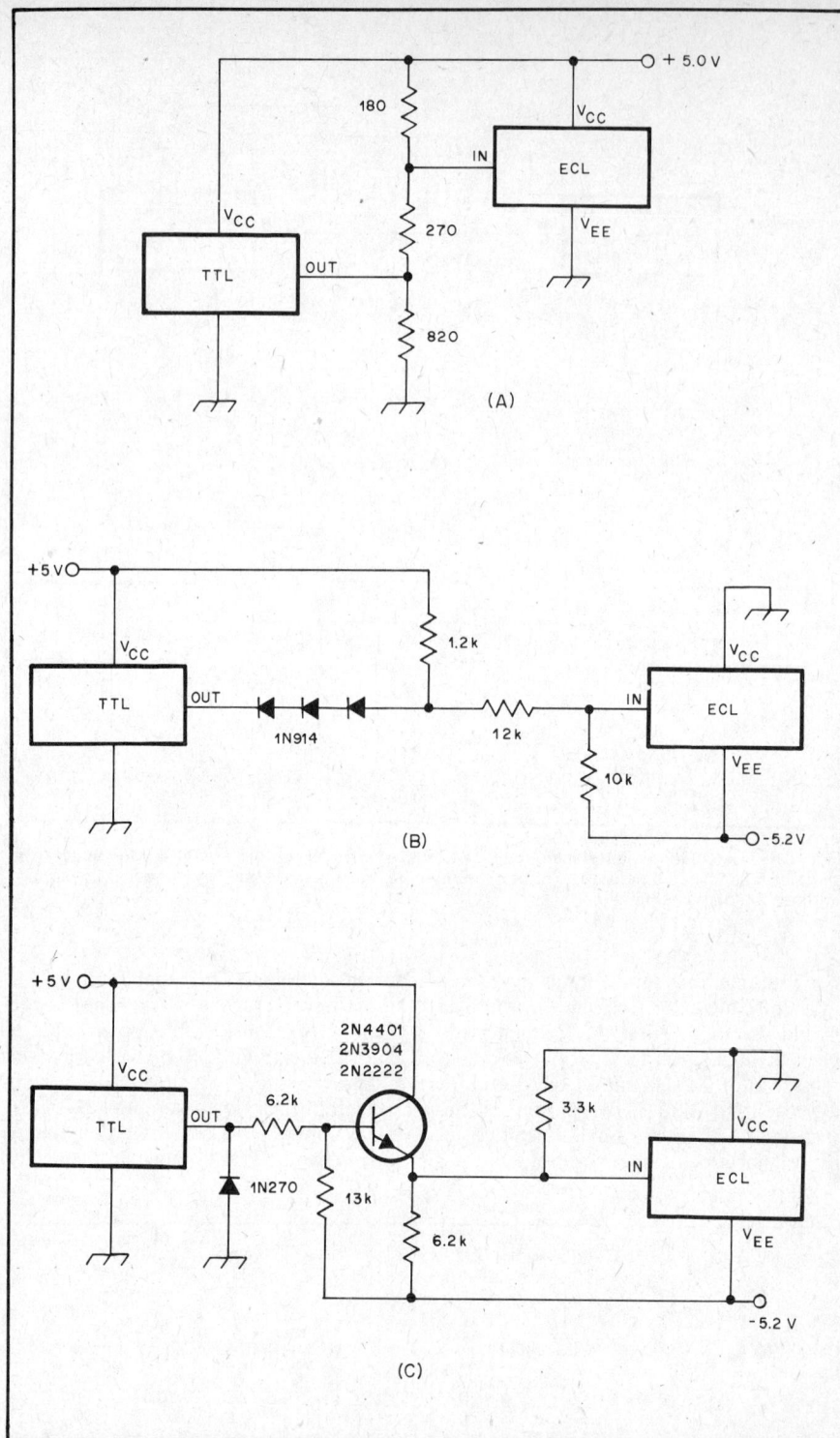

Fig. 123 — These circuits will allow TTL gates to drive ECL systems using various power supply arrangements. Each is discussed in the text.

+5V and V_{EE} grounded, the same circuit will work with V_{CC} connected to ground and V_{EE} (and the "ground" terminal of the TTL device) connected to −5V. This arrangement provides full noise immunity for the ECL system. Where speed is not a consideration, a TTL output can be connected directly to an ECL input if a pull-up resistor is used.

Independent TTL and ECL systems can

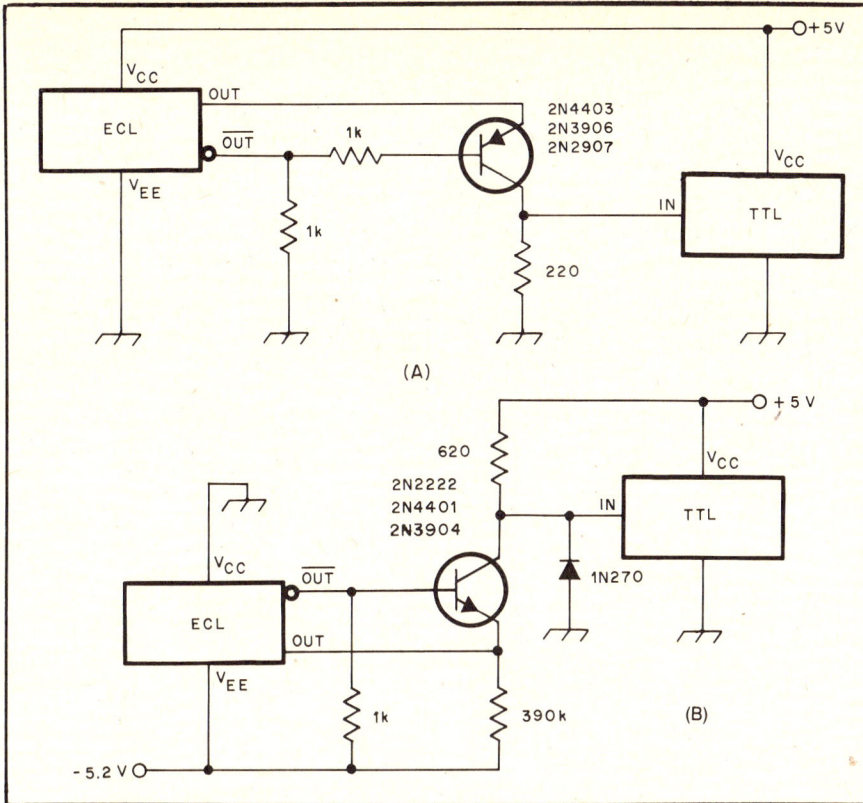

Fig. 124 — Bipolar transistors are used in these ECL-to-TTL translators.

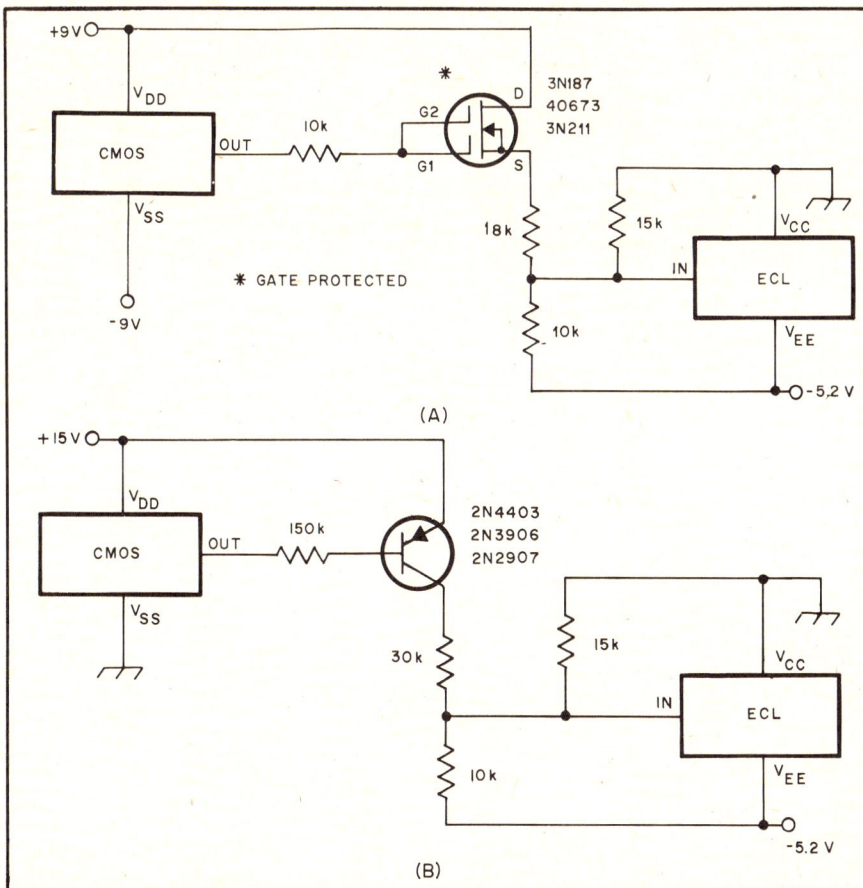

Fig. 125 — Split-supply CMOS logic can drive ECL through a MOSFET, as in A. When the CMOS system is powered from a single high-voltage supply, the bipolar transistor circuit of B can be used. Note that the logic is inverted when this method is used.

be coupled by the circuits drawn in Fig. 123B and C. In B, the TTL gate is divorced from the voltage divider network when the output is HIGH. In this state the junction of the 1.2-kΩ and 12-kΩ resistors assumes a potential of nearly +5 volts. When the TTL output goes low, the anode end of the diode string is pulled down to about +2.5V. This 2.5-volt logic swing is attenuated and shifted to the proper non-saturating ECL levels by the resistor network.

An emitter follower stage is used in C. The −1.8-volt potential at the ECL input established by the resistor network prevents the transistor from turning on when the TTL output is LOW. A germanium diode provides a stiff voltage reference in the LOW state and prevents excessive conduction in the upper transistor of the IC output structure. The voltage translation process is similar to that in part B of the figure. Returning the collector to +5V rather than ground keeps the transistor well out of saturation.

ECL Driving TTL

The complementary output of ECL gates can be used to advantage in converting to TTL levels. Modern ECL ICs have emitter-follower outputs that are ideal for switching the base-emitter junctions of bipolar transistors. For coupling logic systems having a common 5- or 5.2-volt power supply, the pnp transistor and pull-up resistor combination of Fig. 124A may be used. A positive supply is shown, but the system will also work with negative supplies. The circuit in Fig. 124B will condition −5.2-volt ECL signals to drive +5-volt TTL gates. Transposing the out and $\overline{out}$ connections of the ECL device will effect a logic inversion with the translation. This technique can also be applied to flip-flops, which have Q and $\overline{Q}$ outputs.

CMOS Driving ECL

Speed is rarely a consideration when mating a relatively slow logic family to one that is very fast — the system cannot be faster than the slowest logic element used. The speed of ECL ICs comes from keeping the transistors out of saturation, and it is for this reason that the defined input logic swing is only about 400 mV. However, the input levels can be anywhere within the range of the power supply without damaging the device. Negligible input current is required for either logic state, so when a common 5- or 5.2-volt power supply is used, CMOS can drive ECL directly.

A variety of circuits can be used between CMOS and ECL systems having different power supplies. The scheme illustrated in Fig. 125A is useful when a split power supply is used for the CMOS logic. The advantage of using a MOSFET converter is that the fanout (to other CMOS devices) is not compromised. Fig.

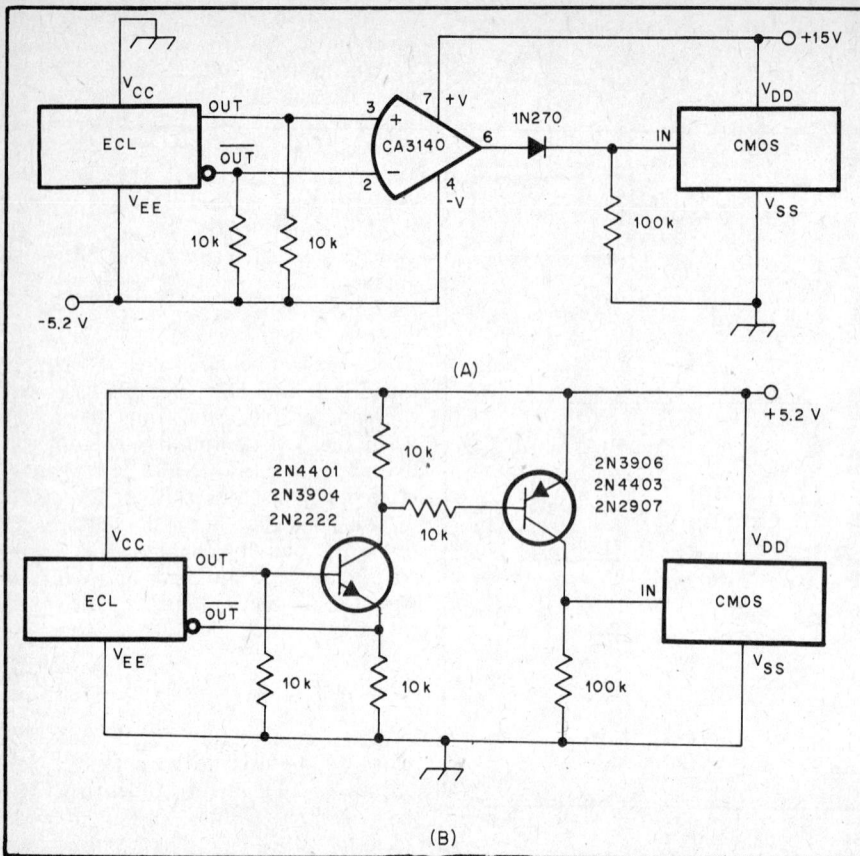

Fig. 126 — Interface circuits for ECL-driving CMOS. A method useful when different power supplies are used is illustrated in A. The diode prevents the −5.2-volt LOW level from damaging the CMOS device. When a common supply is available, the two-transistor amplifier/translator of B may be used.

125B shows a +15-volt CMOS system driving a −5.2-volt ECL gate through a pnp transistor. Altering some of the resistance values will make this circuit work with split-supply CMOS as well. This conversion method results in a logic inversion, but that problem can usually be remedied at the ECL output.

ECL-Driving CMOS

Some voltage amplification is required if an ECL gate is to drive CMOS. When the ECL supply is negative and the CMOS supply is positive, the circuit of Fig. 124B, illustrated for ECL-to-TTL conversion may be used. All of the resistors can be made much larger with CMOS for reduced power consumption.

The differential comparator arrangement in Fig. 126A is another good translation method. If the CMOS system has split power supplies, the −V terminal of the op amp should be returned to V_{SS}. With split CMOS supplies the op amp can be connected directly to the CMOS input; the resistor and diode are unnecessary. If complementary ECL outputs aren't available, one of the comparator inputs should be biased to some potential between the two ECL logic levels.

Fig. 126B shows a way to obtain the required CMOS logic swing when both families are powered from the same source. This npn-pnp saturated amplifier will also work when a common negative supply is used.

Abbreviated Semiconductor Symbol List

Field-Effect Transistor Symbols

A	— Voltage amplification
C_c	— Intrinsic channel capacitance
C_{ds}	— Drain-to-source capacitance (includes approximately 1-pF drain-to-case and interlead capacitance)
C_{gd}	— Gate-to-drain capacitance (includes 0.1-pF interlead capacitance)
C_{gs}	— Gate-to-source interlead and case capacitance
C_{iss}	— Small-signal input capacitance, short circuit
C_{rss}	— Small-signal reverse transfer capacitance, short circuit
g_{fs}	— Forward transconductance
g_{is}	— Input conductance
g_{os}	— Output conductance
I_D	— Dc drain current
$I_{DS(OFF)}$	— Drain-to-source OFF current
I_{GSS}	— Gate leakage current
r_c	— Effective gate series resistance
$r_{DS(ON)}$	— Drain-to-source ON resistance
r_{gd}	— Gate-to-drain leakage resistance
r_{gs}	— Gate-to-source leakage resistance
V_{DB}	— Drain-to-substrate voltage
V_{DS}	— Drain-to-source voltage
V_{GB}	— Dc gate-to-substrate voltage
V_{GB}	— Peak gate-to-substrate voltage
V_{GS}	— Dc gate-to-source voltage
V_{GS}	— Peak gate-to-source voltage
$V_{GS(OFF)}$	— Gate-to-source cutoff voltage
Y_{fs}	— Forward transadmittance ≈ g_{fs}
Y_{os}	— Output admittance
Y_L	— Load admittance

Bipolar Transistor Symbols

C_{ibo}	— Input capacitance, open circuit (common base)
C_{ieo}	— Input capacitance, open circuit (common emitter)
C_{obo}	— Output capacitance, open circuit (common base)
C_{oeo}	— Output capacitance, open circuit (common emitter)
f_c	— Cutoff frequency
f_T	— Gain-bandwidth product (frequency at which small-signal forward current-transfer ratio, common emitter, is unity or 1)
g_{me}	— Small-signal transconductance (common emitter)
h_{FB}	— Static forward-current transfer ratio (common base)
h_{fb}	— Small-signal forward-current transfer ratio, short circuit (common base)
h_{FE}	— Static forward-current transfer ratio (common emitter)
h_{fe}	— Small-signal forward-current transfer ratio, short circuit (common emitter)
h_{IE}	— Static input resistance (common emitter)
h_{ie}	— Small-signal input impedance, short circuit (common emitter)
I_b	— Base current
I_c	— Collector current
I_{CBO}	— Collector-cutoff current, emitter open

Bipolar Transistor Symbols, (continued)

I_{CEO}	— Collector-cutoff current, base open
I_E	— Emitter current
MAG	— Maximum available amplifier gain
P_{CE}	— Total dc or average power input to collector (common emitter)
P_{OE}	— Large-signal output power (common emitter)
R_L	— Load resistance
R_s	— Source resistance
V_{BB}	— Base-supply voltage
V_{BC}	— Base-to-collector voltage
V_{BE}	— Base-to-emitter voltage
V_{CB}	— Collector-to-base voltage
V_{CBO}	— Collector-to-base (emitter open)
V_{CC}	— Collector-supply voltage
V_{CE}	— Collector-to-emitter voltage
V_{CEO}	— Collector-to-emitter voltage (base open)
$V_{CE(sat)}$	— Collector-to-emitter saturation voltage
V_{EB}	— Emitter-to-base voltage
V_{EBO}	— Emitter-to-base voltage (collector open)
V_{EE}	— Emitter-supply voltage
Y_{fe}	— Forward transconductance
Y_{ie}	— Input admittance
Y_{oe}	— Output admittance

OPTICAL ELECTRONIC DEVICES

The conductivity of any semiconductor junction is enhanced when it is illuminated. The discussion of solar cells introduced the photovoltaic phenomenon. Photoelectric semiconductor devices used for signaling rather than power generation fall into the optoelectronic category. Before photodiodes and phototransistors were commonly available, amateur experimenters improvised by removing portions of the device envelopes and installing lenses to focus light on the junctions. An experimental optical communications system is sketched in Fig. 127.

A large variety of photoelectric semiconductors exists today, including bipolar phototransistors, photoFETs, photodiodes (pn junction, PIN and varactor), light-activated silicon-controlled rectifiers (LASCRs) and optocouplers. An *optocoupler*, or *optoisolator*, is an LED and a phototransistor in a common IC package. These devices often represent the cleanest way to mate solid-state circuits operating at widely differing voltages. One such application might be a low-voltage dc power supply regulated in the ac primary circuit. Units having several kilovolts of isolation are available. The output circuits of some optocouplers are designed to drive digital logic circuitry with a minimum of additional components.

The figure of merit for an optocoupler is the ratio of the LED current to the phototransistor collector current. A Darlington output transistor is used in some devices to establish a more favorable transfer characteristic. The phototransistor base lead is brought out of some packages for controlling the transistor when the LED is not energized.

A circuit illustrating the use of an optocoupler appears in Fig. 128. The keying circuit of a vacuum-tube type of transmitter is another place where an optocoupler could be used to advantage. An optocoupler can also be used to key an afsk generator from a teleprinter loop.

One class of optocoupler leaves the LED and photodector exposed. These devices are used extensively in punched-card readers for electronic data processing systems. Another use is in automotive ignition systems as a replacement for mechanical breaker points. An *optical shaft-encoder* is an array of open optocouplers chopped by a rotating wheel. When a shaft encoder is used to address a frequency synthesizer, the operator can adjust the frequency in a manner that has the "feel" of an ordinary VFO.

An optoisolator packaged with a triac forms a *solid-state relay* that can replace electromechanical units is most applications. The advantages of this scheme include freedom from contact bounce, arcing, mechanical wear and noise. Solid-

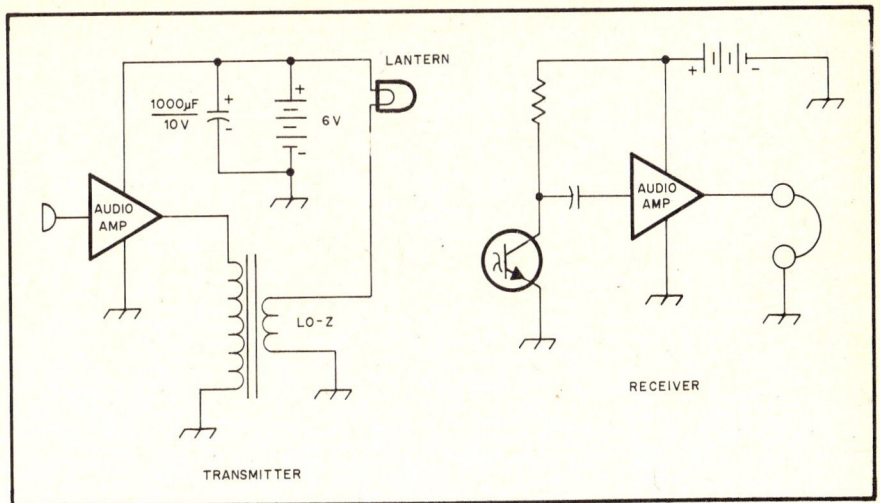

Fig. 127 — A visible light (or infrared) voice communications system. A phototransistor in the receiver recovers the amplitude-modulated signal from the transmitter.

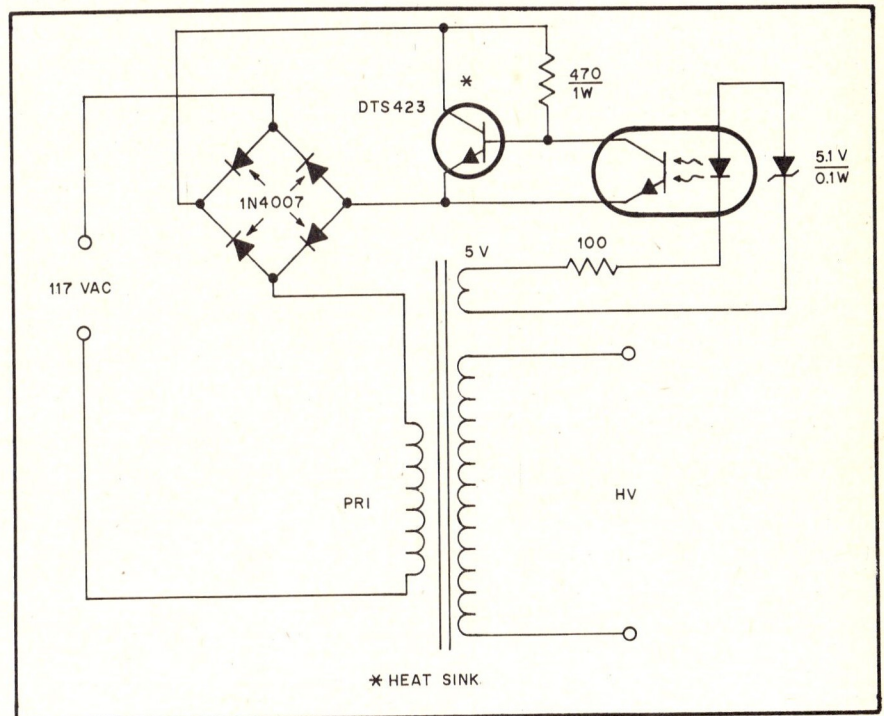

Fig. 128 — Line voltage regulation is a representative linear application for an optoisolator. The circuit shown here can protect high-voltage rectifiers and filter capacitors from ac surges.

state relays capable of switching 10 A at 117 V from CMOS control signals are manufactured by International Rectifier Company and others.

Solid-State Displays

Digital readout devices comprise an important branch of optoelectronics. The advantage of digital readout is that there is no interpretation error, as in an analog readout such as a clock, speedometer or ammeter. The digital readout can be no more accurate than the circuitry driving it,

but the elimination of visual uncertainty allows greater precision. A digital display is an array of light sources that can be energized in various combinations to form symbols. Some of the light source arrangements are illustrated in Fig. 129. The dot-matrix system is the most versatile, but the seven-segment format is the one most used by amateurs. It displays the decimal digits with high readability. Each segment is identified by a letter, and Fig. 130 shows the standard layout. In addition to the segments, some displays

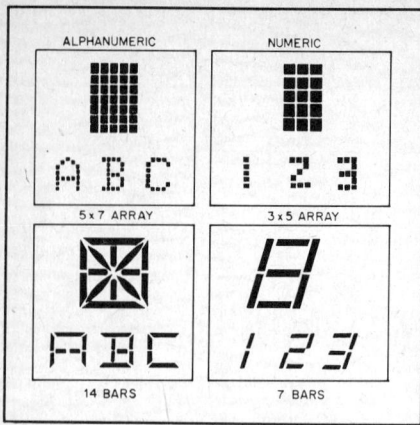

Fig. 129 — Various formats for symbolic displays.

ALPHANUMERIC
NUMERIC
5 x 7 ARRAY
3 x 5 ARRAY
14 BARS
7 BARS

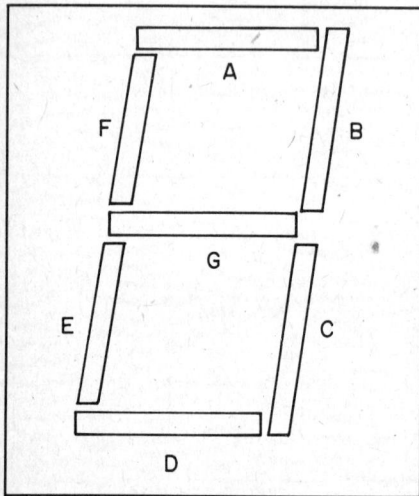

Fig. 130 — Segment identification and layout of a seven-segment readout device.

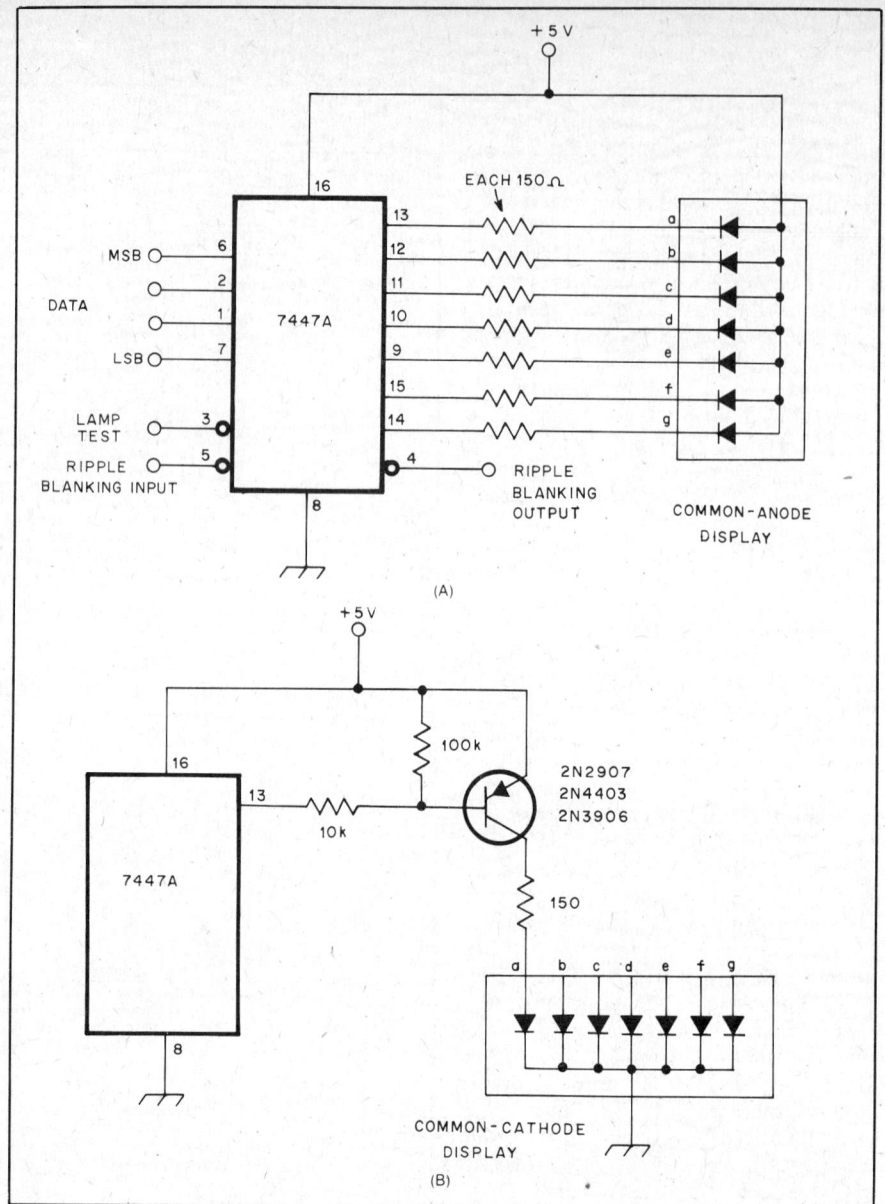

Fig. 131 — At A, a 7447A decoder/driver connected to a common-anode LED display. The current drain is 20 mA per segment. At B is a method for using the same decoder with a common-cathode device.

contain a decimal point on one side of the character array. Another type of display is the *bar graph*. This device can be thought of as a fast-responding incremental analog meter. Some Amateur Radio applications for the bar graph device might be receiver S-meters or PEP-output indicators.

LED Readouts

Displays made from light-emitting diodes are the ones most commonly used because of their good readability and relative ease of circuit integration. Character heights of 0.3 and 0.6 inches are standard, although larger units are available. LED readouts are manufactured with one element of each segment common. The character to be displayed is usually encoded in BCD form, so it is necessary to employ some combinational logic to illuminate the proper segments. The circuitry to accomplish this is called a *decoder*. Various seven-segment decoders are manufactured to drive common-cathode and common-anode devices. Some of these devices contain advanced features. The decoder IC most available to hobbyists is the 7447A, manufactured

by Texas Instruments and others. This is an open-collector TTL device designed to pull down common-anode displays through external current-limiting resistors. A 7447A will also drive common-cathode displays if external transistors are used. Fig. 131 shows the connections for both types of display.

The dc illumination method shown is the easiest to implement, but higher light output with lower energy consumption can be had by pulsing the display. A flash rate of 100 Hz will be imperceptible because of the persistence of human vision. As more digits are added to a display, using a decoder/driver for each digit becomes unfeasible from an economic and pc-board real estate point of view. A technique called *multiplexing* allows a single decoder/driver IC to be time-shared

among several digits. The multiplexer logic, which is usually contained on a single IC, scans the input data lines and sequentially routes the data for each digit into the decoder. The displays are wired in parallel; that is, all of the "a" segments are connected together, and so on. The common element of each digit is enabled at the proper time by the multiplexing circuit that selects the input data. With this system, only one digit is energized at any instant, a factor that greatly eases the power-supply requirements. In order to maintain the brightness of each digit, the current to each LED segment must be increased. A typical display should be operated at a peak current of 100 mA for each segment, with a pulse duration of 50 to 250 microseconds at a 20-percent duty cycle.

Chapter 5

Ac-Operated Power Supplies

Power-line voltages have been "standardized" throughout the U.S. at 117 and 234 volts in residential areas where a single phase voltage is supplied. These figures represent nominal voltages. "Normal" line voltage in a particular area may be between approximately 110 and 125 volts, but generally will be above 115 volts. In many states the service is governed by a PUC (public utilities commission). The voltage average across the country is approximately 117.

The ac-current capability of the service is a factor of line length from the dwelling to the nearest pole transformer, plus the conductor size of the line. Many older homes are supplied with a 60-ampere service while most new homes have 100 amperes. Houses equipped with electric heat will have services ranging from 150 to 200 amperes.

The electrical power required to operate Amateur Radio equipment is usually taken from the ac lines when the equipment is operated where power is available. For mobile operation the source of power is almost always the car storage battery.

Dc·voltages used in transmitters, receivers and other related equipment are derived from the commercial ac lines by using a transformer-rectifier-filter system. The transformer changes the ac voltage to a suitable value and the rectifier converts the ac to pulsating dc. A filter is used to smooth out these pulsations to an acceptably low level. Essentially pure direct current is required to prevent 60- or 120-Hz hum in most pieces of amateur equipment. Transmitters must be operated from a pure dc supply as dictated by federal regulations. If a constant voltage is required under conditions of changing load or ac-line voltage, a regulator is used following the filter.

When the prime power source is dc (a battery), the dc is used directly or is first changed to ac and is then followed by the transformer-rectifier-filter combination. The latter system has lost considerable popularity with the advent of low-voltage semiconductor devices.

Transformerless power supplies are used in some applications (notably ac-dc radios and some television receivers). Supplies of this sort operate directly from the power line, making it necessary to connect the chassis or common-return point of the circuit directly to one side of the ac line. This type of power supply represents a shock hazard when the equipment is connected to other units in the amateur station or when the chassis is exposed. For safety reasons, an isolation transformer should be used with such equipment.

Power-Line Considerations: Connections

In most residential systems, three wires are brought in from the outside to the distribution board, while in a few older systems there are only two wires. In the three-wire system, the third wire is the *neutral,* which is grounded. The voltage between the two wires normally is 234, while half of this voltage appears between each of these wires and neutral, as indicated in Fig. 1A. In systems of this type the 117-volt household load is divided as evenly as possible between the two sides of the circuit, half of the load being connected between one wire and the neutral, while the other half of the load is connected between the other wire and neutral. Heavy appliances, such as electric stoves and heaters are designed for 234-volt operation and therefore are connected across the two ungrounded wires. While both ungrounded wires should be fused, a fuse should never be used in the neutral wire, nor should a switch be used in this side of the line. The reason for this is that opening the neutral wire does not disconnect the equipment. It simply leaves the equipment on one side of the 234-volt circuit in series with whatever load may be across the other side of the circuit, as shown in Fig. 1B. Furthermore, with the neutral open, the voltage will then be divided between the two sides in inverse proportion to the load resistance, the voltage on one side dropping below normal, while it soars on the other side, unless the loads happen to be equal.

The usual line running to baseboard outlets is rated at 15 amperes. Considering the power consumed by filaments, lamps, transmitter, receiver and other auxiliary

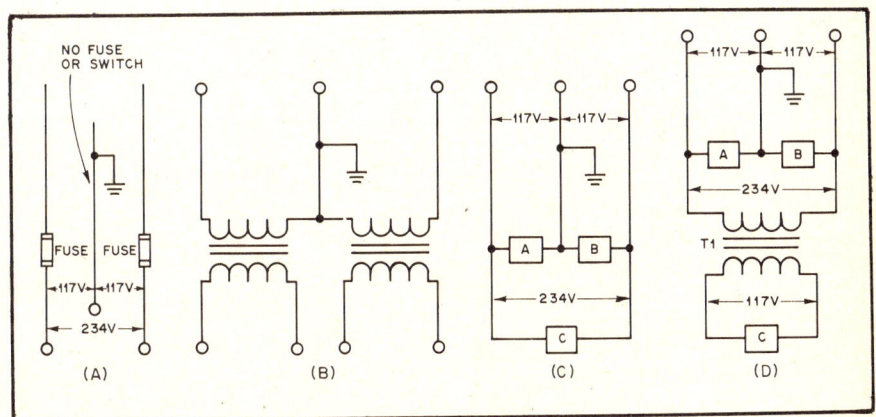

Fig. 1 — Three-wire power-line circuits. At A — Normal three-wire-line termination. No fuse should be used in the grounded (neutral) line. B — A switch in the neutral does not remove voltage from either side of the line. C — Connections for both 117- and 234-volt transformers. D — Operating a 117-volt plate transformer from the 234-volt line to avoid light blinking. T1 is a 2:1 step-down transformer.

equipment, it is not unusual to find this 15-A rating exceeded by the requirements of a station of only moderate power. It must also be kept in mind that the same branch may be in use for other household purposes through another outlet. For this reason, and to minimize light blinking when keying or modulating the transmitter, a separate heavier line should be run from the distribution board to the station whenever possible. A 3-volt drop in line voltage will cause noticeable blinking of lights.

If the system is of the three-wire, 234-V type, the three wires should be brought into the station so that the load can be distributed to keep the line balanced. The voltage across a fixed load on one side of the circuit will increase as the load current on the other side is increased. The rate of increase will depend upon the resistance introduced by the neutral wire. If the resistance of the neutral is low, the increase will be correspondingly small. When the currents in the two circuits are balanced, no current flows in the neutral wire and the system is operating at maximum efficiency.

Light blinking can be minimized by using transformers with 234-volt primaries in the power supplies for the keyed or intermittent part of the load, connecting them across the two ungrounded wires with no connection to the neutral, as shown in Fig. 1C. The same can be accomplished by the insertion of a step-down transformer with its primary operating at 234 volts and secondary delivering 117 volts. Conventional 117-volt transformers may be operated from the secondary of the step-down transformer (see Fig. 1D).

When a special heavy-duty line is to be installed, the local power company should be consulted as to local requirements. In some localities it is necessary to have such a job done by a licensed electrician, and there may be special requirements to be met. Some amateurs terminate the special line to the station at a switch box, while others may use electric-stove receptacles as the termination. The power is then distributed around the station by means of conventional outlets at convenient points. All circuits should be properly fused.

Three-Wire 117-V Power Cords

To meet the requirements of state and national codes, electrical tools, appliances and many items of electronic equipment now being manufactured to operate from the 117-volt line must be equipped with a three-conductor power cord. Two of the conductors carry power to the device in the usual fashion, while the third conductor is connected to the case or frame.

When plugged into a properly wired mating receptacle, the three-contact polarized plug connects this third conductor

to an earth ground, thereby grounding the chassis or frame of the appliance and preventing the possibility of electrical shock to the user. All commercially manufactured items of electronic test equipment and most ac-operated amateur equipment are being supplied with these three-wire cords. Adapters are available for use where older electrical installations do not have mating receptacles. For proper grounding, the lug of the green wire protruding from the adapter must be attached underneath the screw securing the cover plate of the outlet box where connection is made, and the outlet box itself must be grounded.

Fusing

All transformer primary circuits should be properly fused. To determine the approximate current rating of the fuse or circuit breaker to be used, multiply each current being drawn from the supply in amperes by the voltage at which the current is being drawn. Include the current taken by bleeder resistances and voltage dividers. In the case of series resistors, use the source voltage, not the voltage at the equipment end of the resistor. Include filament power if the transformer is supplying filaments. After multiplying the various voltages and currents, add the individual products. Then divide by the line voltage and add 10 or 20 percent. Use a fuse or circuit breaker with the nearest larger current rating.

Line-Voltage Adjustment

In certain communities trouble is sometimes experienced from fluctuations in line voltage. Usually these fluctuations are caused by a variation in the load on the line. Since most of the variation comes at certain fixed times of the day or night, such as the times when lights are turned on at evening, they may be taken care of by the use of a manually operated compensating device. A simple arrangement is shown in Fig. 2A. A tapped transformer is used to boost or buck the line voltage as required. The transformer should have a secondary varying between 6 and 20 volts in steps of 2 or 3 volts and its secondary should be capable of carrying the full load current.

The secondary is connected in series

with the line voltage and, if the phasing of the windings is correct, the voltage applied to the primaries of the transmitter transformers can be brought up to the rated 117 volts by setting the transformer tap switch on the right tap. If the phasing of the two windings of the transformer happens to be reversed, the voltage will be reduced instead of increased. This connection may be used in cases where the line voltage may be above 117 volts. This method is preferable to using a resistor in the primary of a power transformer since it does not affect the voltage regulation as seriously. The circuit of 2B illustrates the use of a variable autotransformer (Variac) for adjusting line voltage.

Constant-Voltage Transformers

Although comparatively expensive, special transformers called *constant-voltage transformers* are available for use in cases where it is necessary to hold line voltage and/or filament voltage constant with fluctuating supply-line voltage. These are static-magnetic voltage regulating transformers operating on principles of ferro-resonance. They have no tubes or moving parts, and require no manual adjustments. These transformers are rated over a range of less than 1 volt-ampere (VA) at 5 volts output up to several thousand VA at 117 or 234 volts. On the average they will hold their output voltages within one percent under an input voltage variation of ±15 percent.

Safety Precautions

All power supplies in an installation should be fed through a single main power-line switch so that all power may be cut off quickly, either before working on the equipment, or in case of an accident. Spring-operated switches or relays are not sufficiently reliable for this important service. Foolproof devices for cutting off all power to the transmitter and other equipment are shown in Fig. 3. The arrangements shown in Figs. 3A and B are similar circuits for two-wire (117-volt) and three-wire (234-volt) systems. S is an enclosed double-throw switch of the sort usually used as the entrance switch in house installations. J is a standard ac outlet and P a shorted plug to fit the outlet. The switch should be located

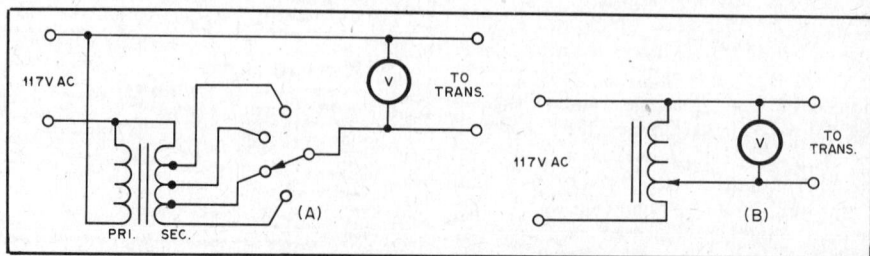

Fig. 2 — Two methods of transformer primary control. At A is a tapped transformer which may be connected so as to boost or buck the line voltage as required. At B is indicated a variable transformer or autotransformer (Variac) which feeds the transformer primaries.

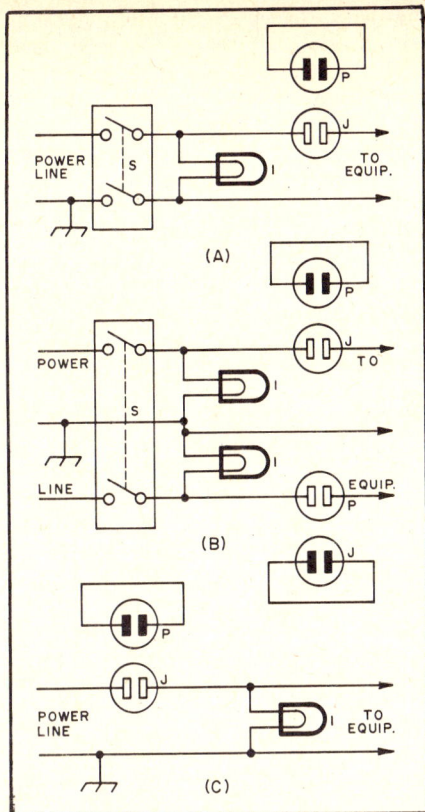

Fig. 3 — Reliable arrangements for cutting off all power to the transmitter. S is an enclosed double-pole power switch, J a standard ac outlet, P a shorted plug to fit the outlet and I a red lamp.

A is for a two-wire 117-volt line, B for a three-wire 234-volt system, and C a simplified arrangement for low-power stations.

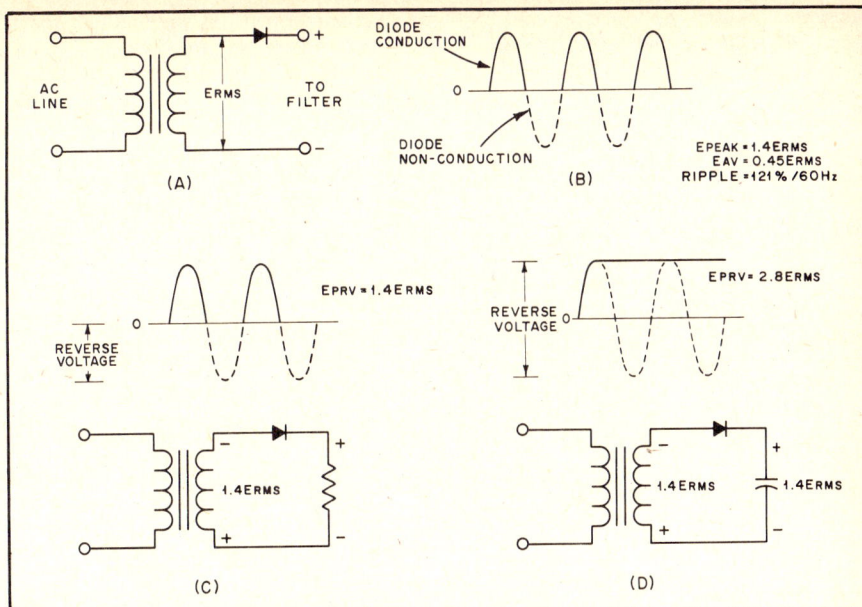

Fig. 4 — Half-wave rectifier circuit. A illustrates the basic circuit and B displays the diode conduction and nonconduction periods. The peak-reverse voltage impressed across the diode is shown at C and D with a simple resistor load at C and a capacitor load at D. E_{prv} for the resistor load is 1.4 E_{rms} and 2.8 E_{rms} for the capacitor load.

prominently in plain sight, and members of the household should be instructed in its location and use. I is a red lamp located alongside the switch. Its purpose is not so much to serve as a warning that the power is on as it is to help in identifying and quickly locating the switch should it become necessary for someone else to cut the power off in an emergency.

The outlet J should be placed in some corner out of sight where it will not be a temptation for children or others to play with. The shorting plug can be removed to open the power circuit if there are others around who might inadvertently throw the switch while the operator is working on the rig. If the operator takes the plug with him, it will prevent someone from turning on the power in his absence and either hurting himself or the equipment or perhaps starting a fire. Of utmost importance is the fact that the outlet J *must* be placed in the *ungrounded* side of the line.

Those who are operating low power and feel that the expense or complication of the switch isn't warranted can use the shorted-plug idea as the main power switch. In this case, the outlet should be located prominently and identified by a signal light, as shown in Fig. 3C.

The test bench should be fed through

the main power switch, or a similar arrangement at the bench, if the bench is located remotely from the transmitter.

A bleeder resistor with a power rating which gives a considerable margin of safety should be used across the output of all transmitter power supplies, so that the filter capacitors will be discharged when the high-voltage is turned off.

Rectifier Circuits: Half-Wave

Fig. 4 shows a simple half-wave rectifier circuit. As pointed out in the semiconductor chapter a rectifier (in this case a semiconductor diode) will conduct current in one direction but not the other. During one half of the ac cycle the rectifier will conduct and current will flow through the rectifier to the load (indicated by the solid line in Fig. 4B). During the other half cycle the rectifier is reverse biased and no current will flow (indicated by the dotted line in Fig. 4B) to the load. As shown, the output is in the form of pulsed dc and current always flows in the same direction. A filter can be used to smooth out these variations and provide a higher average dc voltage from the circuit. This idea will be covered in the next section on filters.

The average output voltage — the voltage read by a dc voltmeter — with this circuit (no filter connected) is 0.45 times the rms value of the ac voltage delivered by the transformer secondary. Because the frequency of the pulses is rather low (one pulsation per cycle), considerable filtering is required to provide adequately smooth dc output. For this reason the circuit is usually limited to applications where the current required is small, as in a transmitter bias supply.

The peak reverse voltage (PRV), the voltage that the rectifier must withstand when it isn't conducting, varies with the load. With a resistive load it is the peak ac voltage (1.4 E_{rms}) but with a capacitor filter and a load drawing little or no current it can rise to 2.8 E_{rms}. The reason for this is shown in Figs. 4C and 4D. With a resistive load as shown at C the amount of reverse voltage applied to the diode is that voltage on the lower side of the Zero-axis line or 1.4 E_{rms}. A capacitor connected to the circuit (shown at D) will store the peak positive voltage when the diode conducts on the positive pulse. If the circuit is not supplying any current the voltage across the capacitor will remain at that same level. The peak reverse voltage impressed across the diode is now the sum of the voltage stored in the capacitor plus the peak negative swing of voltage from the transformer secondary. In this case the PRV is 2.8 E_{rms}.

Full-Wave Center-Tap Rectifier

A commonly used rectifier circuit is shown in Fig. 5. Essentially an arrangement in which the outputs of two half-wave rectifiers are combined, it makes use of both halves of the ac cycle. A transformer with a center-tapped secondary is required with the circuit.

The average output voltage is 0.9 times the rms voltage of half the transformer secondary; this is the maximum that can be obtained with a suitable choke-input filter. The peak output voltage is 1.4 times the rms voltage of half the transformer secondary; this is the maximum voltage that can be obtained from a capacitor-input filter.

As can be seen in Fig. 5C the PRV

impressed on each diode is independent of the type load at the output. This is because the peak reverse voltage condition occurs when diode A conducts and diode B does not conduct. The positive and negative voltage peaks occur at precisely the same time, a different condition than exists in the half-wave circuit. As diodes A and B cathodes reach a positive peak (1.4 E_{rms}), the anode of diode B is at a negative peak, also 1.4 E_{rms}, but in the opposite direction. The total peak reverse voltage is therefore 2.8 E_{rms}.

Fig. 5B shows that the frequency of the output pulses is twice that of the half-wave rectifier. Comparatively less filtering is required. Since the rectifiers work alternately, each handles half of the load current: The current rating of each rectifier need be only half the total current drawn from the supply.

Two separate transformers, with their primaries connected in parallel and secondaries connected in series (with the proper polarities), may be used in this circuit. However, if this substitution is made, the primary volt-ampere rating must be reduced to about 40 percent less than twice the rating of one transformer.

Full-Wave Bridge Rectifier

Another commonly used rectifier circuit is illustrated in Fig. 6. In this arrangement, two rectifiers operate in series on each half of the cycle, one rectifier being in the lead to the load, the other being in the return lead. As shown in Figs. 6A and B, when the top lead of the transformer secondary is positive with respect to the bottom lead diodes A and C will conduct while diodes B and D are reverse biased. On the next half cycle when the top lead of the transformer is negative with respect to the bottom diodes B and D will conduct while diodes A and C are reverse biased.

The output wave shape is the same as that from the simple center-tap rectifier circuit. The maximum output voltage into a resistive load or choke-input filter is 0.9 times the rms voltage delivered by the transformer secondary; with a capacitor filter and a light load the output voltage is 1.4 times the secondary rms voltage.

Fig. 6C shows the peak reverse voltage to be 2.8 E_{rms} for each pair of diodes. Since the diodes are connected in series each diode has 1.4 E_{rms} as the reverse voltage impressed across it. Each pair of diodes works alternately so each handles half of the load current. The rectifier in this circuit should have a minimum current rating of one half the total load current to be drawn from the supply.

Filtering

The pulsating dc waves from the rectifiers are not sufficiently constant in amplitude to prevent hum corresponding to the pulsations. Filters are required be-

tween the rectifier and the load to smooth out the pulsations into an essentially constant dc voltage. Also, the design of the filter depends to a large extent on the dc voltage output, the voltage regulation of the power supply, and the maximum load current that can be drawn from the supply without exceeding the peak-current rating of the rectifier. Power supply filters are low-pass devices using series inductors and shunt capacitors.

Load Resistance

In discussing the performance of power-supply filters, it is sometimes convenient to express the load connected to the output terminals of the supply in terms of resistance. The load resistance is equal to the output voltage divided by the total current drawn, including the current drawn by the bleeder resistor.

Voltage Regulation

The output voltage of a power supply always decreases as more current is

drawn, not only because of increased voltage drops on the transformer, filter chokes and the rectifier (if high-vacuum rectifiers are used) but also because the output voltage at light loads tends to soar to the peak value of the transformer voltage as a result of charging the first capacitor. By proper filter design the latter effect can be eliminated. The change in output voltage with load is called *voltage regulation* and is expressed as a percentage.

$$\text{Percent regulation} = \frac{100 \, (E_1 - E_2)}{E_2}$$

where

E_1 = the no-load voltage
E_2 = the full-load voltage

A steady load, such as that represented by a receiver, speech amplifier or unkeyed stages of a transmitter, does not require good (low) regulation as long as the proper voltage is obtained under load conditions. However, the filter capacitors

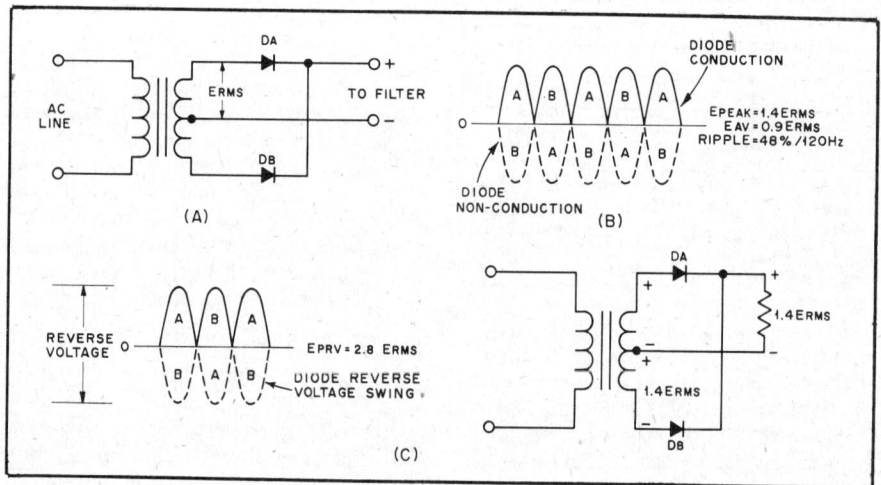

Fig. 5—Full-wave center-tap rectifier circuit. A illustrates the basic circuit. Diode conduction is shown at B with diodes A and B alternately conducting. The peak-reverse voltage for each diode is 2.8 E_{rms} as depicted at C.

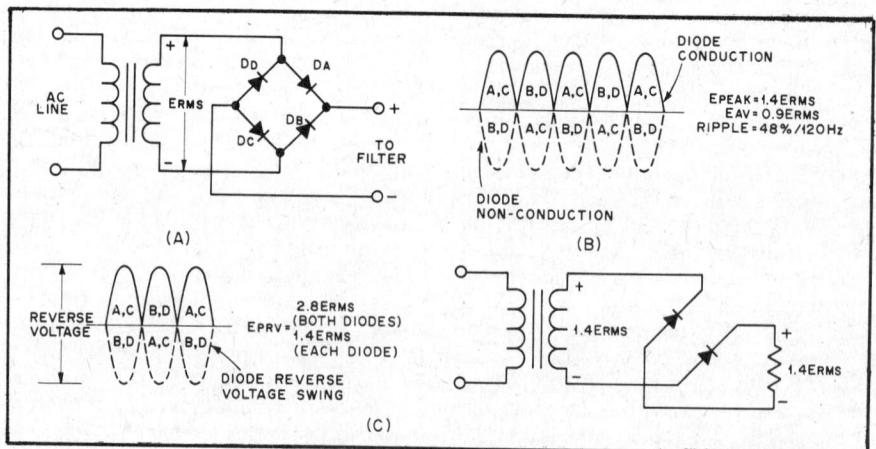

Fig. 6 — Full-wave bridge rectifier circuit. The basic circuit is illustrated at A. Diode conduction and nonconduction times are shown at B. Diodes A and C conduct on one half of the input cycle while diodes B and D conduct on the other. C displays the peak-reverse voltage for one-half cycle. Since this circuit uses two diodes essentially in series, the 2.8 E_{rms} is divided between two diodes, or, 1.4 E_{rms} PRV for each diode.

must have a voltage rating safe for the highest value to which the voltage will soar when the external load is removed.

A power supply will show more (higher) regulation with long-term changes in load resistance than with short temporary changes. The regulation with long-term changes is often called the *static regulation,* to distinguish it from the *dynamic regulation* (short temporary load changes). A load that varies at a syllabic or keyed rate, as represented by some audio and rf amplifiers, usually requires good dynamic regulation (15 percent or less) if distortion products are to be held to a low level. The dynamic regulation of a power supply is improved by increasing the value of the output capacitor.

When essentially constant voltage regardless of current variation is required (for stabilizing an oscillator, for example), special voltage-regulating circuits described later in this chapter are used.

Bleeder

A bleeder resistor is a resistance connected across the output terminals of the power supply. Its functions are to discharge the filter capacitors as a safety measure when the power is turned off and to improve voltage regulation by providing a minimum load resistance. When voltage regulation is not of importance, the resistance may be as high as 100 ohms per volt. The resistance value to be used for voltage-regulating purposes is discussed in later sections. From the consideration of safety, the power rating of the resistor should be as conservative as possible, since a burned-out bleeder resistor is more dangerous than none at all!

Ripple Frequency and Voltage

Pulsations at the output of the rectifier can be considered to be the resultant of an alternating current superimposed on a steady direct current. From this viewpoint, the filter may be considered to consist of shunt capacitors which short-circuit the ac component while not interfering with the flow of the dc component. Series chokes will readily pass dc but will impede the flow of the ac component.

The alternating component is called *ripple.* The effectiveness of the filter can be expressed in terms of percent ripple, which is the ratio of the rms value of the ripple to the dc value in terms of percentage.

$$\text{Percent ripple (rms)} = \frac{100\,E_1}{E_2}$$

where
  E_1 = the rms value of ripple voltage
  E_2 = the steady dc voltage

Any multiplier or amplifier supply in a code transmitter should have less than five percent ripple. A linear amplifier can tolerate about three percent ripple on the plate voltage. Bias supplies for linear amplifiers, and modulator and modulated-amplifier plate supplies, should have less than one percent ripple. VFOs, speech amplifiers and receivers may require a ripple reduction to 0.01 percent.

Ripple frequency is the frequency of the pulsations in the rectifier output wave — the number of pulsations per second. The frequency of the ripple with half-wave rectifiers is the same as the frequency of the line supply — 60 Hz with 60-Hz supply. Since the output pulses are doubled with a full-wave rectifier, the ripple frequency is doubled — to 120 Hz with a 60-Hz supply.

The amount of filtering (values of inductance and capacitance) required to give adequate smoothing depends upon the ripple frequency, with more filtering being required as the ripple frequency is lowered.

Type of Filter

Power-supply filters fall into two classifications, capacitor input and choke input. Capacitor-input filters are characterized by relatively high output voltage in respect to the transformer voltage. Advantage of this can be taken when silicon rectifiers are used or with any rectifier when the load resistance is high. Silicon rectifiers have a higher allowable peak-to-dc ratio than do thermionic rectifiers. This permits the use of capacitor-input filters at ratios of input capacitor to load resistance that would seriously shorten the life of a thermionic rectifier system. When the series resistance through a rectifier and filter system is appreciable, as when high-vacuum rectifiers are used, the voltage regulation of a capacitor-input power supply is poor.

The output voltage of a properly designed choke-input power supply is less than would be obtained with a capacitor-input filter from the same transformer. Generally speaking, a choke-input filter will permit a higher load current to be drawn from a thermionic rectifier without exceeding the peak rating of the rectifier.

Capacitive-Input Filters

Capacitive-input filter systems are shown in Fig. 7. Disregarding voltage drops in the chokes, all have the same characteristics except in respect to ripple. Better ripple reduction will be obtained when LC sections are added as shown in Figs. 7B and C.

Output Voltage

To determine the approximate dc voltage output when a capacitive-input filter is used, the graphs shown in Fig. 8 will be helpful. An example of how to use the graph is given below.
  Example:
Full-wave rectifier (use graph at B)
Transformer rms voltage = 350
Load resistance = 2000 ohms
Series resistance = 200 ohms
Input capacitance = 20 μF

$$\frac{R_s}{R} = \frac{200}{2000} = 0.1 \quad \frac{RC}{1000} = \frac{2000 \times 20}{1000} = 40$$

From curve 0.1 and RC = 40, the dc voltage is (350 × 1.06) = 370.

In many cases it is desirable to know the amount of capacitance required for a power supply given certain performance criteria. This is especially true when designing a power supply for an application such as powering a solid-state transceiver. The following example should give the builder a good handle on how to arrive at circuit values for a power supply using a single capacitor filter.

Fig. 9 is the circuit diagram of the power supply to be used.
  Requirements:
Output voltage = 12.6
Output current = 1 ampere
Maximum ripple = 2 percent
Load regulation = 5 percent
The rms secondary voltage of T1 must be the desired output voltage plus the voltage drops across D2 and D4 divided by 1.41.

$$E_{SEC} = \frac{12.6 + 1.4}{1.41} = 9.93$$

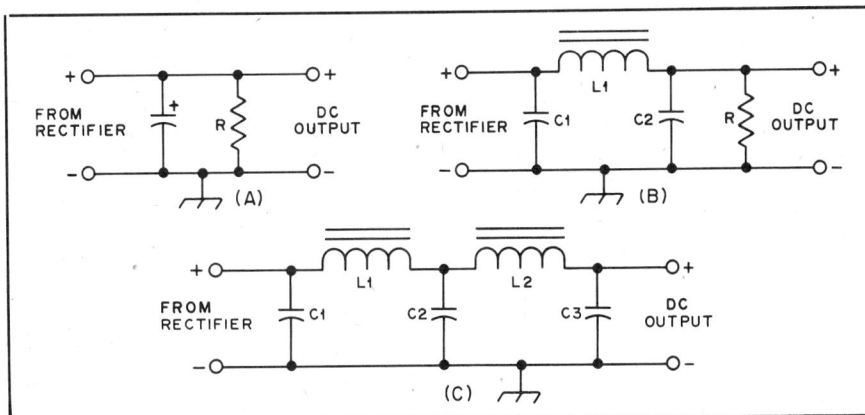

Fig. 7 — Capacitive-input filter circuits. At A is a simple capacitor filter. B and C are single- and double-section filters, respectively.

In practice the nearest standard transformer (10 V) would work fine. Alternatively, the builder could wind his own transformer, or remove secondary turns from a 12-volt transformer to obtain the desired rms secondary voltage.

A two percent ripple referenced to 12.6 volts is 0.25 V rms. The peak-to-peak value is therefore $0.25 \times 2.8 = 0.7$ V. This value is required to calculate the required capacitance for C1.

Also needed for determining the value of C1 is the time interval (t) between the full-wave rectifier pulses which is calculated as follows:

$$t = \frac{1}{f_{(Hz)}} = \frac{1}{120} = 8.3 \times 10^{-3}$$

where t is the time between pulses and f is the frequency in Hz. Since the circuit makes use of a full-wave rectifier a pulse occurs twice during each cycle. With half-wave rectification a pulse would occur only once a cycle. Thus 120 Hz is used as the frequency for this calculation.

C1 is calculated from the following equation:

$$C_{(\mu F)} = \left[\frac{I_L t}{E_{rip(pk-pk)}} \right] 10^6$$

$$= \left[\frac{1A \times 8.3 \times 10^{-3}}{0.7} \right] 10^6$$

$$= 11,857 \ \mu F$$

where I_L is the current taken by the load. The nearest standard capacitor value is 12,000 μF. It will be an acceptable one to use, but since the tolerance of electrolytic capacitors is rather loose, the builder may elect to use the next larger standard value.

Diodes D1-D4, inclusive, should have a PRV rating of at least two times the transformer secondary peak voltage. Assuming a transformer secondary rms value of 10 volts, the PRV should be at least 28 volts. Four 50-volt diodes will provide a margin of safety. The forward current of the diodes should be at least twice the load current. For a 1-A load, the diodes should be rated for at least 2 A.

The load resistance, R_L, is determined by E_o/I_L, which in this example is $12.6/1 = 12.6$ ohms. This factor must be known in order to find the necessary series resistance for five-percent regulation. Calculate as follows:

$$R_{S(max)} = \text{Load regulation} \left(\frac{R_L}{10} \right)$$

$$= 0.05 \left(\frac{12.6}{10} \right) = 0.063 \ \text{ohm}$$

Therefore, the transformer secondary dc resistance should be no greater than 0.063

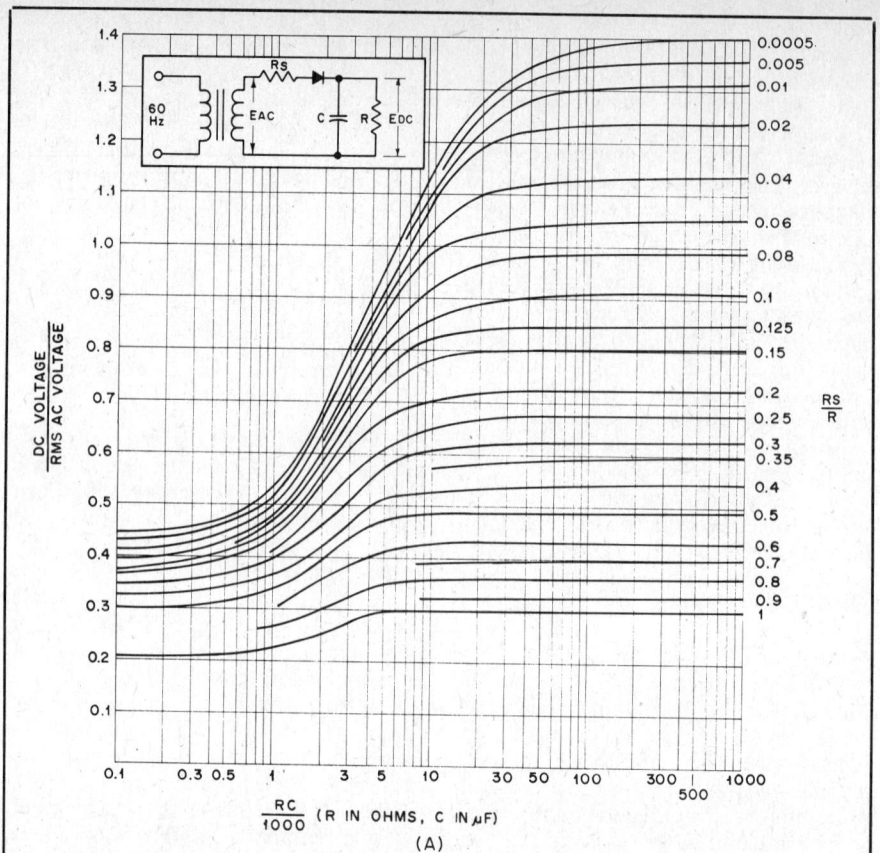

These curves are adapted from those published by Otto H. Schade in "Analysis of Rectifier Operation," *Proceedings of the I.R.E.*, July 1943.

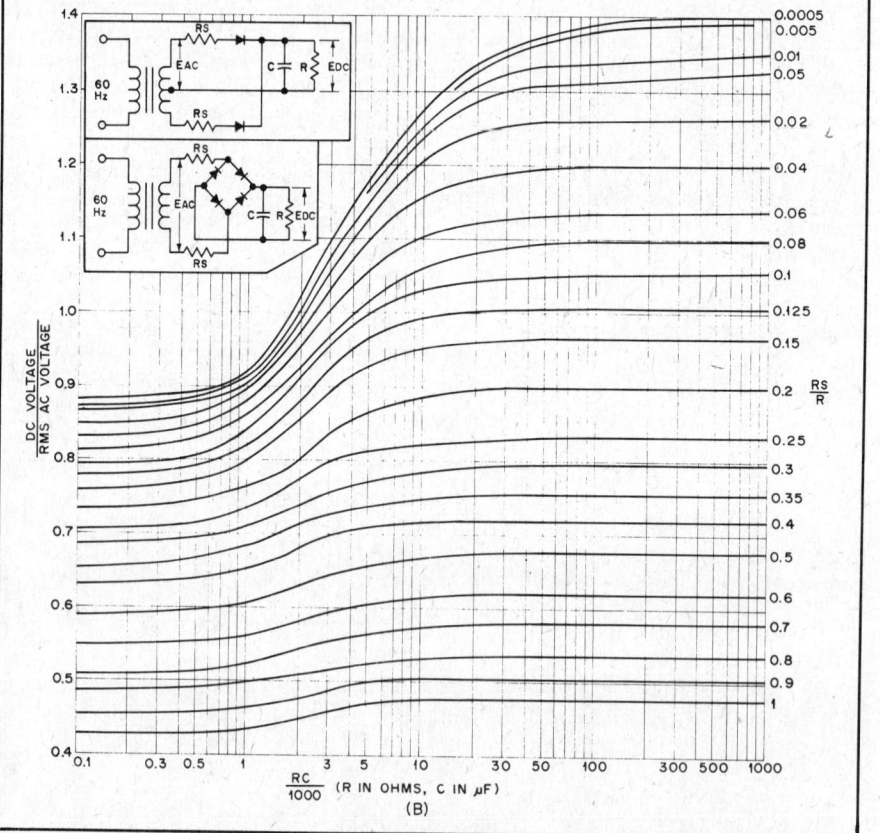

Fig. 8 — Dc output voltages from a half- and full-wave rectifier circuit as a function of the filter capacitance and load resistance (half-wave shown at A and full-wave shown at B). R_s includes transformer winding resistance and rectifier forward resistance. For the ratio R_s/R, both resistances are in ohms; for the RC product, R is in ohms and C is in μF.

ohm. The secondary current rating should be equal to or greater than the $I_L = 1$ ampere.

C1 should have a minimum working voltage of 1.4 times the output voltage. In the case of this power supply the capacitor should be rated for at least 18 volts.

Choke-Input Filters

With thermionic rectifiers better voltage regulation results when a choke-input filter, as shown in Fig. 10, is used. Choke input permits better utilization of the thermionic rectifier, since a higher load current can be drawn without exceeding the peak current rating of the rectifier.

Minimum Choke Inductance

A choke-input filter will tend to act as a capacitive-input filter unless the input choke has at least a certain minimum value of inductance called the *critical* value. This critical value is given by

$$L_{crit} \text{ (henrys)} = \frac{E \text{ (volts)}}{I \text{ (mA)}}$$

where E = the supply output voltage
I = the current being drawn through the filter.

If the choke has at least the critical value, the output voltage will be limited to the average value of the rectified wave at the input to the choke when the current drawn from the supply is small. This is in contrast to the capacitive-input filter in which the output voltage tends to soar toward the peak value of the rectified wave at light loads.

Minimum-Load — Bleeder Resistance

From the formula above for critical inductance, it is obvious that if no current is drawn from the supply, the critical inductance will be infinite. So that a practical value of inductance may be used, some current must be drawn from the supply at all times the supply is in use. From the formula we find that this minimum value of current is

$$I \text{ (mA)} = \frac{E \text{ (volts)}}{L_{crit}}$$

In the majority of cases it will be most convenient to adjust the bleeder resistance so that the bleeder will draw the required minimum current. From the formula, it may be seen that the value of critical inductance becomes smaller as the load current increases.

Swinging Chokes

Less costly chokes are available that will maintain at least the critical value of inductance over the range of current likely to be drawn from practical supplies. These chokes are called *swinging chokes*. As an example, a swinging choke may have an inductance rating of 5/25 H and a current rating of 200 mA. If the supply delivers 1000 volts, the minimum load current should be 1000/25 = 40 mA. When the full

load current of 200 mA is drawn from the supply, the inductance will drop to 5 H. The critical inductance for 200 mA at 1000 volts is 1000/200 = 5 H. Therefore the 5/25 H choke maintains the critical inductance at the full current rating of 200 mA. At all load currents between 40 mA and 200 mA, the choke will adjust its inductance to the approximate critical value.

Output Voltage

Provided the input-choke inductance is at least the critical value, the output voltage may be calculated quite closely by:

$$E_o = 0.9E_t - (I_B + I_L) \times (R1 + R2) - E_r$$

where

E_o = output voltage
E_t = rms voltage applied to the rectifier (rms voltage between center-tap and one end of the secondary in the case of the center-tap rectifier)
I_B = bleeder current (A)
I_L = load current (A)
R_1 = first filter choke resistance
R_2 = second filter choke resistance
E_r = voltage drop across the rectifier.

The various voltage drops are shown in Fig. 10. At no load I_L is zero; hence the no-load voltage may be calculated on the basis of bleeder current only. The voltage regulation may be determined from the no-load and full-load voltages using the formulas previously given.

Output Capacitor

Whether the supply has a choke- or capacitor-input filter, if it is intended for use with a Class A af amplifier, the reactance of the output capacitor should be low for the lowest audio frequency; 16

μF or more is usually adequate. When the supply is used with a Class B amplifier (for modulation or for ssb amplification) or a cw transmitter, increasing the output capacitance will result in improved dynamic regulation of the supply. However, a region of diminishing returns can be reached, and 20 to 30 μF will usually suffice for any supply subjected to large changes at a syllabic (or keying) rate.

Resonance

Resonance effects in the series circuit across the output of the rectifier, formed by the first choke and first filter capacitor, must be avoided, since the ripple voltage would build up to large values. This not only is the opposite action to that for which the filter is intended, but may also cause excessive rectifier peak currents and abnormally high peak-reverse voltages. For full-wave rectification the ripple frequency will be 120 Hz for a 60-Hz supply, and resonance will occur when the product of choke inductance in henrys times capacitor capacitance in microfarads is equal to 1.77. At least twice this product of inductance and capacitance should be used to ensure against resonance effects. With a swinging choke, the minimum rated inductance of the choke should be used. If too high an LC filter product is used, the resonance may occur at the radio-telegraph keying or voice syllabic rate, and large voltage excursions (filter bounce) may be experienced at that rate.

Ratings of Filter Components

In a power supply using a choke-input filter and properly designed choke and bleeder resistor, the no-load voltage across the filter capacitors will be about

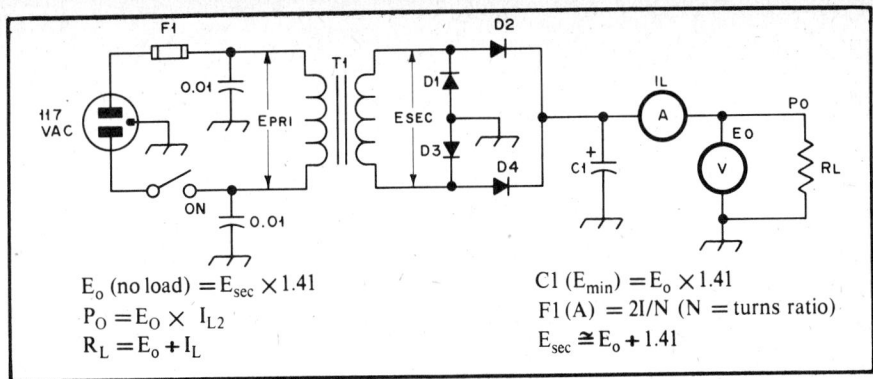

Fig. 9 — This figure illustrates how to design a simple unregulated power supply. See text for a thorough discussion.

E_o (no load) = $E_{sec} \times 1.41$
$P_O = E_O \times I_{L2}$
$R_L = E_o \div I_L$

C1 (E_{min}) = $E_o \times 1.41$
F1(A) = 2I/N (N = turns ratio)
$E_{sec} \cong E_o + 1.41$

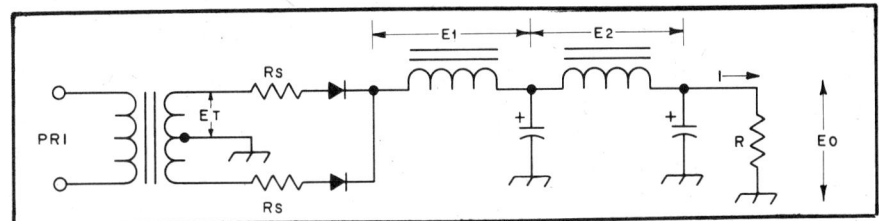

Fig. 10 — Diagram showing various voltage drops that must be taken into consideration in determining the required transformer voltage to deliver the desired output voltage.

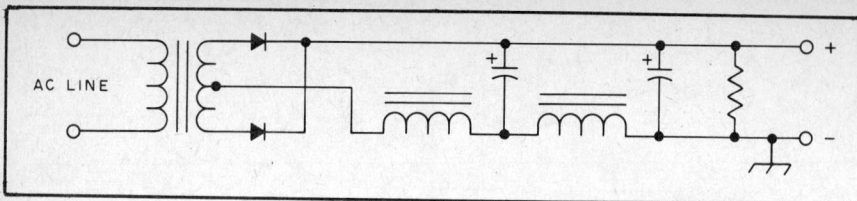

Fig. 11 — In most applications, the filter chokes may be placed in the negative instead of the positive side of the circuit. This reduces the danger of a voltage breakdown between the choke winding and core.

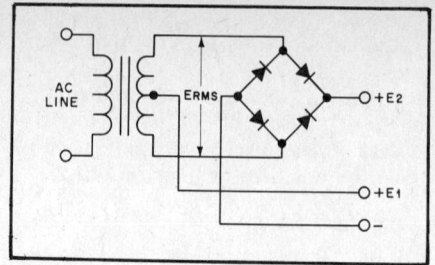

Fig. 12 — The "economy" power supply circuit is a combination of the full-wave and bridge-rectifier circuits.

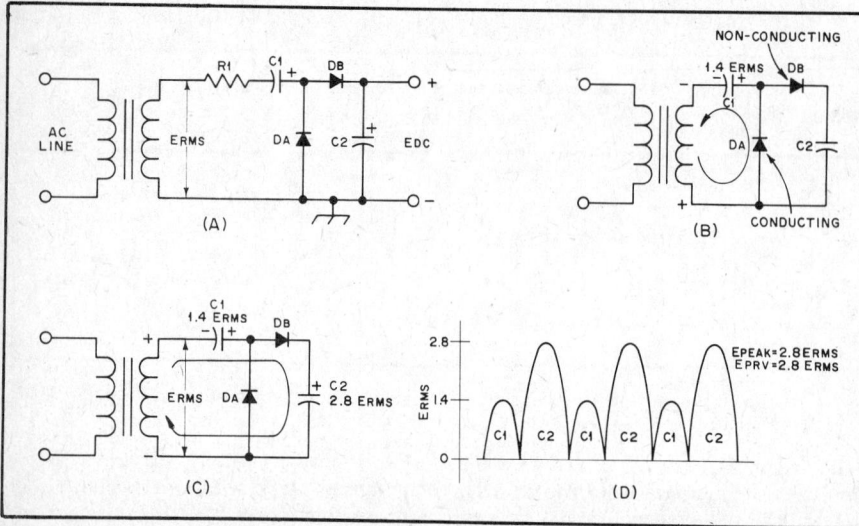

Fig. 13 — Illustrated at A is a half-wave voltage-doubler circuit. B displays how the first half cycle of input voltage charges C1. During the next half cycle (shown at C) capacitor C2 is charged with the transformer secondary voltage plus that voltage stored in C1 from the previous half cycle. D illustrates the levels to which each capacitor is charged throughout the cycle.

nine-tenths of the ac rms voltage. Nevertheless, it is advisable to use capacitors rated for the *peak* transformer voltage. This large safety factor is suggested because the voltage across the capacitors can reach this peak value if the bleeder should burn out and there is no load on the supply.

In a capactive-input filter, the capacitors should have a working-voltage rating at least as high, and preferably somewhat higher, than the peak voltage from the transformer. Thus, in the case of a center-tap rectifier having a transformer delivering 550 volts each side of the center tap, the minimum safe capacitor voltage rating will be 550 × 1.41 or 775 volts. An 800-volt capacitor should be used, or preferably a 1000-volt unit.

Filter Capacitors in Series

Filter capacitors are made in several different types. Electrolytic capacitors, which are available for peak voltages up to about 800, combine high capacitance with small size, since the dielectric is an extremely thin film of oxide on aluminum foil. Capacitors of this type may be connected in series for higher voltages, although the filtering capacitance will be reduced to the resultant of the two capacitances in series. If this arrangement

is used, it is important that *each* of the capacitors be shunted with a resistor of about 100 ohms per volt of supply voltage applied to the individual capacitors, with an adequate power rating. These resistors may serve as all or part of the bleeder resistance. Capacitors with higher voltage ratings usually are made with a dielectric of thin paper impregnated with oil. The working voltage of a capacitor is the voltage that it will withstand continuously.

Filter Chokes

Filter chokes or inductances are wound on iron cores, with a small gap in the core to prevent magnetic saturation of the iron at high currents. When the iron becomes saturated its permeability decreases, and consequently the inductance also decreases. Despite the air gap, the inductance of a choke usually varies to some extent with the direct current flowing in the winding; hence it is necessary to specify the inductance at the current which the choke is intended to carry. Its inductance with little or no direct current flowing in the winding will usually be considerably higher than the value when full load current is flowing.

Negative-Lead Filtering

For many years it has been almost

universal practice to place filter chokes in the positive leads of plate power supplies. This means that the insulation between the choke winding and its core (which should be grounded to chassis as a safety measure) must be adequate to withstand the output voltage of the supply. This voltage requirement is removed if the chokes are placed in the negative lead as shown in Fig. 11. With this connection, the capacitance of the transformer secondary to ground appears in parallel with the filter chokes tending to bypass the chokes. However, this effect will be negligible in practical application except in cases where the output ripple must be reduced to a very low figure. Such applications are usually limited to low-voltage devices such as receivers, speech amplifiers and VFOs where insulation is no problem and the chokes may be placed in the positive side in the conventional manner. In higher-voltage applications, there is no reason why the filter chokes should not be placed in the negative lead to reduce insulation requirements. Choke terminals, negative capacitor terminals and the transformer center-tap terminal should be well protected against accidental contact, since these will assume full supply voltage to chassis should a choke burn out or the chassis connection fail.

The "Economy" Power Supply

In many transmitters of the 100-watt class, an excellent method for obtaining plate and screen voltages without wasting power in resistors is by the use of the "economy" power-supply circuit. Shown in Fig. 12, it is a combination of the full-wave and bridge-rectifier circuits. The voltage at E1 is the normal voltage obtained with the full-wave circuit, and the voltage at E2 is that obtained with the bridge circuit. The *total* dc power obtained from the transformer is, of course, the same as when the transformer is used in its normal manner. In cw and ssb applications, additional power can usually be drawn without excessive heating, especially if the transformer has a rectifier filament winding that isn't being used.

Half-Wave Voltage Doubler

Fig. 13 shows the circuit of half-wave voltage doubler. Figs. 13B, C and D illustrate the circuit operation. For clarity, assume the transformer voltage polarity at the moment the circuit is activated is that shown at B. During the first negative half cycle D_A conducts (D_B is in a nonconductive state), charging C1 to the peak rectified voltage (1.4 E_{rms}). C1 is charged with the polarity shown at B. During the positive half cycle of the secondary voltage, D_A is cut off and diode D_B conducts charging capacitor C2. The amount of voltage delivered to C2 is the sum of peak secondary voltage of the transformer plus the voltage stored in C1 (1.4 E_{rms}). On the next negative half cycle, D_B is nonconducting and C2 will discharge into the load. If no load is connected across C2 the capacitors will remain charged — C1 to 1.4 E_{rms} and C2 to 2.8 E_{rms}. When a load is connected to the output of the doubler, the voltage across C2 drops during the negative half cycle and is recharged up to 2.8 E_{rms} during the positive half cycle.

The output waveform across C2 resembles that of a half-wave rectifier circuit in that C2 is pulsed once every cycle. The drawing at Fig. 13D illustrates the levels to which the two capacitors are charged throughout the cycle. In actual operation the capacitors will not discharge all the way to zero as shown.

Full-Wave Voltage Doubler

Shown in Fig. 14 is the circuit of a full-wave voltage doubler. The circuit operation can best be understood by following Figs. 14B, C and D. During the positive half cycle of transformer secondary voltage, as shown at B, D_A conducts charging capacitor C1 to 1.4 E_{rms}. D_B is not conducting at this time.

During the negative half cycle, as shown at C, D_B conducts charging capacitor C2 to 1.4 E_{rms} while D_A is nonconducting. The output voltage is the sum of the two capacitor voltages which will be 2.8 E_{rms} under no-load conditions. Fig. 14D illustrates that each capacitor alternately receives a charge once per cycle. The effective filter capacitance is that of C1 and C2 in series, which is less than the capacitance of either C1 or C2 alone.

Resistors R in Fig. 14A are used to limit the surge current through the rectifiers. Their values are based on the transformer voltage and the rectifier surge-current rating, since at the instant the power supply is turned on the filter capacitors look like a short-circuited load. Provided the limiting resistors can withstand the surge current, their current-handling capacity is based on the maximum load current from the supply. Output voltages approaching twice the peak voltage of the transformer can be obtained with the voltage doubling circuit shown in Fig. 14. Fig. 15 shows how the voltage depends upon the ratio of the series resistance to the load resistance, and the load resistance times the filter capacitance. The peak reverse voltage across each diode is 2.8 E_{rms}.

Voltage Tripling and Quadrupling

A voltage-tripling circuit is shown in Fig. 16A. On one half of the ac cycle C1 and C3 are charged to the source voltage through D1, D2 and D3. On the opposite half of the cycle D2 conducts and C2 is charged to twice the source voltage, because it sees the transformer plus the charge in C1 as its source. (D1 is cut off during this half cycle.) At the same time, D3 conducts, and with the transformer and the charge in C2 as the source, C3 is charged to three times the transformer voltage.

The voltage-quadrupling circuit of

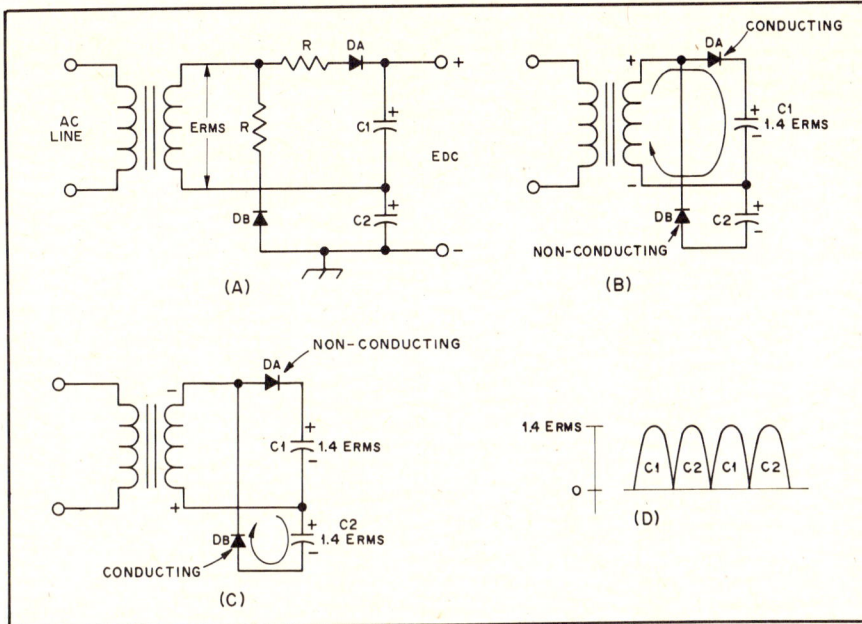

Fig. 14 — A full-wave voltage doubler is displayed at A. One half cycle is shown at B and the next half cycle at C. Each capacitor receives a charge during every cycle of input voltage. D illustrates how each capacitor is alternately charged.

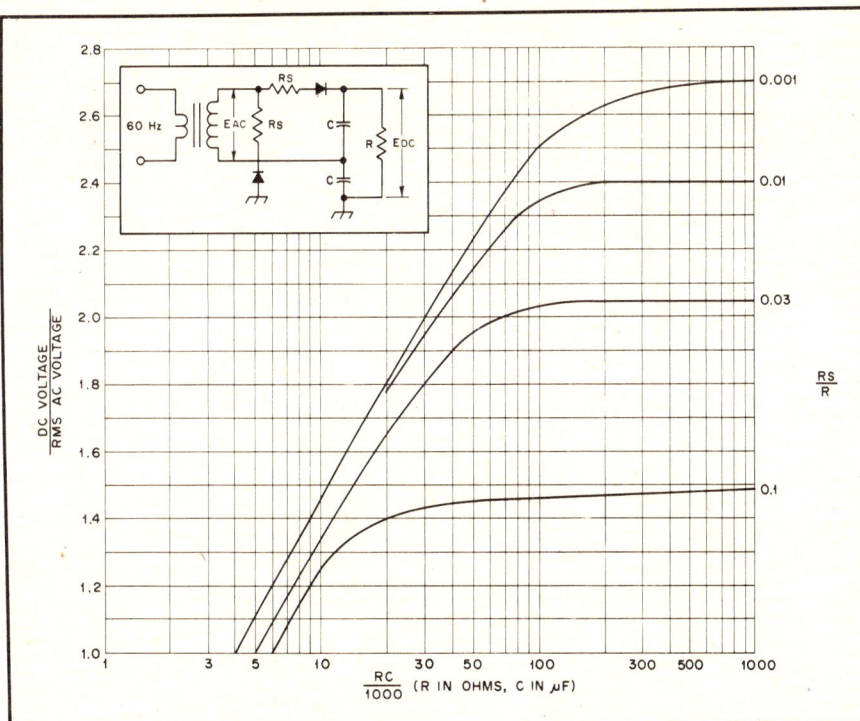

Fig. 15 — Dc output voltages from a full-wave voltage-doubling circuit as a function of the filter capacitances and load resistance. For the ratio R_s/R and for the RC product, resistances are in ohms and capacitance is in microfarads. Equal resistance values for R_s and equal capacitance values for C are assumed. These curves are adapted from those published by Otto H. Schade in "Analysis of Rectifier Operation," *Proceedings of the I.R.E.*, July 1943.

Fig. 16B works in substantially similar fashion. In either of the circuits of Fig. 16, the output voltage will approach an exact multiple of the peak ac voltage when the output current drain is low and the capacitance values are high.

In the circuits shown, the negative leg of the supply is common to one side of the transformer. The positive leg can be made common to one side of the transformer by reversing the diodes and capacitors.

Plate and Filament Transformers: Volt-Ampere Rating

The number of volt-amperes delivered by a transformer depends upon the type of filter (capacitor or choke input) used, and upon the type of rectifier used (full-wave center tap, or full-wave bridge). With a capacitive-input filter the heating effect in the secondary is higher because of the high ratio of peak-to-average current. The volt-amperes handled by the transformer may be several times the watts delivered to the load. With a choke-input filter, provided the input choke has at least the critical inductance, the secondary volt-amperes can be calculated quite closely by the equation:

$$\text{(Full-wave ct) Sec VA} = \frac{0.707 \, E \, I}{1000}$$

$$\text{(Full-wave bridge) Sec VA} = \frac{E \, I}{1000}$$

where

E = *total* rms voltage of the secondary (between the outside ends in the case of a center-tapped winding)

I = dc output current in milliamperes (load current plus bleeder current)

The primary volt-amperes will be somewhat higher because of transformer losses.

Broadcast and Television Replacement Transformers

Small power transformers of the type sold for replacement in broadcast and television receivers are usually designed for service in terms of use for several hours continuously with capacitor-input filters. In the usual type of amateur transmitter service, where most of the power is drawn intermittently for periods of several minutes with equivalent intervals in between, the published ratings can be exceeded without excessive transformer heating.

With a capacitor-input filter, it should be safe to draw 20 to 30 percent more current than the rated value. With a choke-input filter, an increase in current of about 50 percent is permissible. If a bridge rectifier is used, the output voltage will be approximately doubled. In this case, it should be possible in amateur transmitter service to draw the rated current, thus obtaining about twice the rated output power from the transformer.

This does not apply, of course, to amateur transmitter plate transformers, which usually are rated for intermittent service.

Rewinding Power Transformers

Although the home winding of power transformers is a task that few amateurs undertake, the rewinding of a transformer secondary to give some desired voltage for powering filaments or a solid-state device is not difficult. It involves a matter of only a small number of turns and the wire is large enough to be handled easily. Often a receiver power transformer with a burned-out high-voltage winding or the power transformer from a discarded TV set can be converted into an entirely satisfactory transformer without great effort and with little expense. The average TV power transformer for a 17-inch or larger set is capable of delivering from 350 to 450 watts, continuous duty. If an amateur transmitter is being powered, the service is not continuous, so the ratings can be increased by a factor of 40 or 50 percent without danger of overloading the transformer.

The primary volt-ampere rating of the transformer to be rewound, if known, can be used to determine its power-handling capability. The secondary volt-ampere rating will be 10 to 20 percent less than the primary rating. The power rating may also be determined approximately from the cross-sectional area of the core which is *inside* the windings. Fig. 17 shows the method of determining the area, and Fig. 18 may be used to convert this information into a power rating.

Before disconnecting the winding leads from their terminals, each should be marked for identification. In removing the core laminations, care should be taken to note the manner in which the core is assembled, so that the reassembling will be done in the same manner. Most transformers have secondaries wound over the primary, while in some the order is reversed. In case the secondaries are on the inside, the turns can be pulled out from the center after slitting and removing the fiber core.

The turns removed from one of the original filament windings of known voltage should be carefully counted as the winding is removed. This will give the number of turns per volt and the same figure should be used in determining the number of turns for the new secondary. For instance, if the old filament winding was rated at 5 volts and had 15 turns, this is 15/5 = 3 turns per volt. If the new secondary is to deliver 18 volts, the required number of turns on the new winding will be 18 × 3 = 54 turns.

In winding a transformer, the size of wire is an important factor in the heat developed in operation. A cross-sectional area of 1000 circular mils per ampere is conservative. A value commonly used in

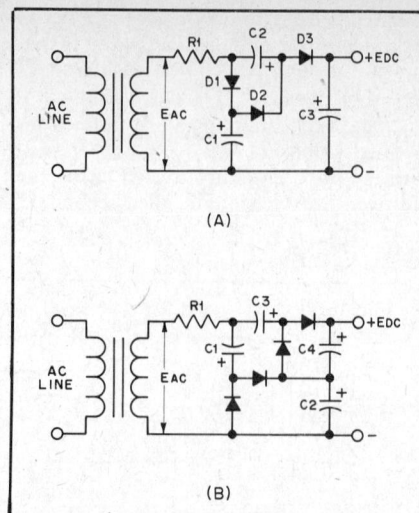

Fig. 16 — Voltage-multiplying circuits with one side of transformer secondary common. (A) Voltage tripler; (B) voltage quadrupler.

Capacitances are typically 20 to 50 μF depending upon output current demand. Dc ratings of capacitors are related to E_{peak} (1.4 E_{ac}).
C1 — Greater than E_{peak}
C2 — Greater than 2E_{peak}
C3 — Greater than 3E_{peak}
C4 — Greater than 2E_{peak}

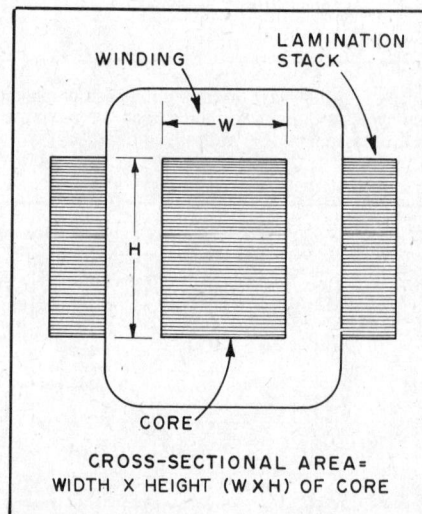

Fig. 17 — Cross-sectional drawing of a typical power transformer. Multiplying the height (or thickness of the laminations) by the width of the central core area in inches gives the value to be applied to Fig. 18.

amateur-service transformers is 700 cmil/A. The larger the cmil/A figure, the cooler the transformer will run. The current rating in amperes of various wire sizes is shown in the copper-wire table in another chapter. If the transformer being rewound is a filament transformer, it may be necessary to choose the wire size carefully to fit the small available space. On the other hand, if the transformer is a power unit with the high-voltage winding removed, there should be plenty of room for a size of wire that will conservatively handle the required current.

After the first layer of turns is put on during rewinding, secure the ends with cellulose tape. Each layer should be insulated from the next; ordinary household waxed paper can be used for the purpose, a single layer being adequate. Sheets cut to size beforehand may be secured over each layer with tape. Be sure to bring all leads out the same side of the core so the covers will go in place when the unit is completed. When the last layer of the winding is put on, use two sheets of waxed paper, and then cover those with vinyl electrical tape, keeping the tape as taut as possible. This will add mechanical strength to the assembly.

The laminations and housing are assembled in just the opposite sequence to that followed in disassembly. Use a light coating of shellac between each lamination. During reassembly, the lamination stack may be compressed by clamping in a vise. If the last few lamination strips cannot be replaced, it is better to omit them than to force the unit together.

Rectifier Ratings: Semiconductors

Silicon rectifiers are being used almost exclusively in power supplies for amateur equipment. Types are available to replace high-vacuum and mercury-vapor rectifiers. The semiconductors have the advantages of compactness, low internal voltage drop, low operating temperature and high current-handling capability. Also, no filament transformers are required.

Silicon rectifiers are available in a wide range of voltage and current ratings. In peak reverse voltage ratings of 600 or less, silicon rectifiers carry current ratings as high as 400 amperes, and at 1000 PRV the current ratings may be several amperes or so. The extreme compactness of silicon types makes feasible the stacking of several units in series for higher voltages. Standard stacks are available that will handle up to 10,000 PRV at a dc load current of 500 mA, although the amateur can do much better, economically, by stacking the rectifiers himself.

Protection of Silicon Power Diodes

The important specifications of a silicon diode are

1) PRV (or PIV), the peak reverse (or peak inverse) voltage.

2) I_o, the average dc current rating.

3) I_{REP}, the peak repetitive forward current.

4) I_{SURGE}, the peak one-cycle surge current. The first two specifications appear in most catalogs. The last two often do not, but they are very important.

Since the rectifier never allows current to flow more than half the time, when it does conduct it has to pass at least twice the average direct current. With a capacitor-input filter, the rectifier conducts much less than half the time, so that when it does conduct, it may pass as much

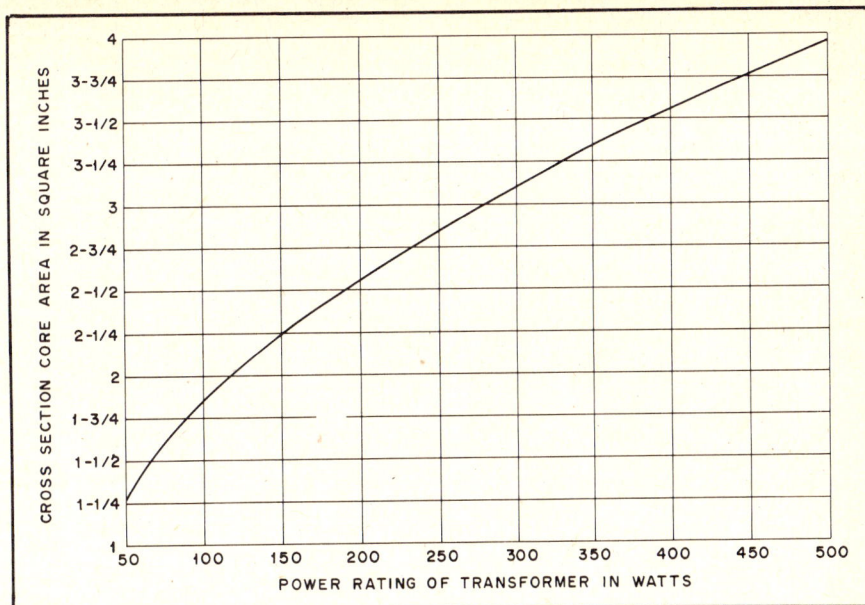

Fig. 18 — Power-handling capability of a transformer versus cross-sectional area of core.

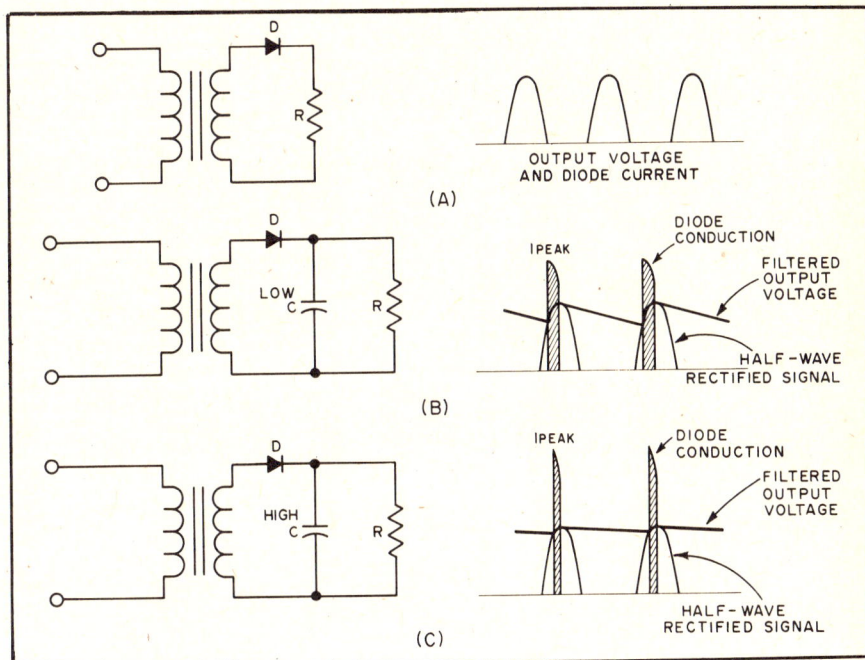

Fig. 19 — The circuit shown at A is a simple half-wave rectifier with a resistive load. The waveform shown to the right is that of output voltage and diode current. B illustrates how the diode current is modified by the addition of a capacitor filter. The diode conducts only when the rectified voltage is greater than stored capacitor voltage. Since this time period is usually only a short portion of a cycle, the peak current will be quite high. C shows an even higher peak current. This is due to the larger capacitor which effectively shortens the conduction period of the diode.

as 10 to 20 times the average dc current, under certain conditions. This is shown in Fig. 19. At A is a simple half-wave rectifier with a resistive load. The waveform to the right of the drawing shows the output voltage along with the diode current. At B and C there are two periods of operation to consider. After the capacitor is charged to the peak-rectified voltage a period of diode nonconduction elapses while the output voltage discharges through the load.

As the voltage begins to rise on the next positive pulse a point is reached where the rectified voltage equals the stored voltage in the capacitor. As the voltage rises beyond that point the diode begins to supply current. The diode will continue to conduct until the waveform reaches the crest, as shown. Since the diode must pass a current equal to that of the load over a short period of a cycle the current will be high. The larger the capacitor for a given load,

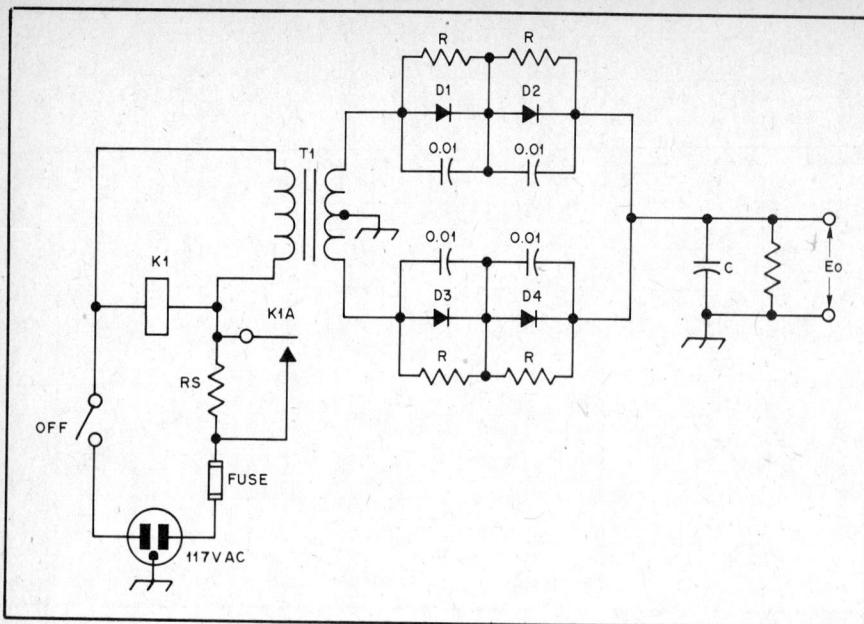

Fig. 20 — The primary circuit of T1 shows how a 117-volt ac relay and a series dropping resistor, R_s, can provide surge protection while C charges. When silicon rectifiers are connected in series for high-voltage operation, the inverse voltage does not divide equally. The reverse voltage drops can be equalized by using equalizing resistors, as shown in the secondary circuit. To protect against voltage "spikes" that may damage an individual rectifier, each rectifier should be bypassed by a 0.01-μF capacitor. Connected as shown two 400-PRV silicon rectifiers can be used as an 800-PRV rectifier, although it is preferable to include a safety factor and call it a "750-PRV" rectifier. The rectifiers, D1 through D4, should be the same type (same type number and ratings).

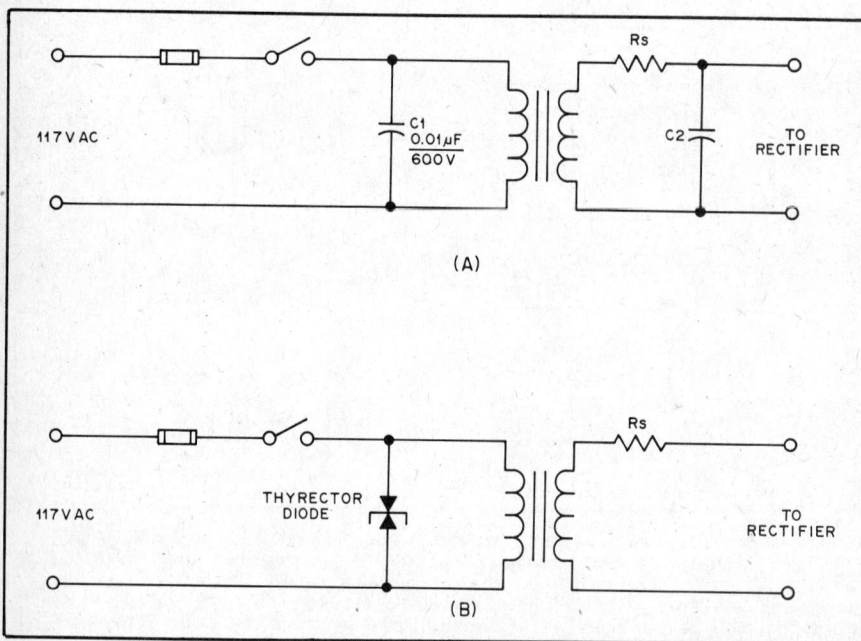

Fig. 21 — Methods of suppressing line transients. See text.

the shorter the diode conduction time and the higher the peak repetitive current (I_{REP}).

When the supply is first turned on, the discharged input capacitor looks like a dead short, and the rectifier passes a very heavy current. This is I_{SURGE}. The maximum I_{SURGE} rating is usually for a duration of one cycle (at 60 Hz), or about 16.7 milliseconds.

If a manufacturer's data sheet is not available, an educated guess about a diode's capability can be made by using these rules of thumb for silicon diodes of the type commonly used in amateur power supplies:

Rule 1) The maximum I_{REP} rating can be assumed to be approximately four times the maximum I_o rating.

Rule 2) The maximum I_{SURGE} rating can be assumed to be approximately 12 times the maximum I_o rating. (This should provide a reasonable safety factor.)

Silicon rectifiers with 750-mA dc ratings, as an example, seldom have 1-cycle surge ratings of less than 15 amperes; some are rated up to 35 amperes or more.) From this then, it can be seen that the rectifier should be selected on the basis of I_{SURGE} and not on I_o ratings.

Thermal Protection

The junction of a diode is quite small, hence it must operate at a high current density. The heat-handling capability is, therefore, quite small. Normally, this is not a prime consideration in high-voltage, low-current supplies. When using high-current rectifiers at or near their maximum ratings (usually 2-ampere or larger stud-mount rectifiers), some form of heat sinking is necessary. Frequently, mounting the rectifier on the main chassis — directly, or by means of thin mica insulating washers — will suffice. If insulated from the chassis, a thin layer of silicone grease should be used between the diode and the insulator, and between the insulator and the chassis to assure good heat conduction. Large high-current rectifiers often require special heat sinks to maintain a safe operating temperature. Forced-air cooling is sometimes used as a further aid. Safe case temperatures are usually given in the manufacturer's data sheets and should be observed if the maximum capabilities of the diode are to be realized.

Surge Protection

Each time the power supply is activated, assuming the input filter capacitor has been discharged, the rectifiers must look into what represents a dead short. Some form of surge protection is usually necessary to protect the diodes until the input capacitor becomes nearly charged. Although the dc resistance of the transformer secondary can be relied upon in some instances to provide ample surge-current limiting, it is seldom enough on high-voltage power supplies to be suitable. Series resistors can be installed between the secondary and the rectifier strings, but are a deterrent to good voltage regulation. By installing a surge-limiting device in the primary circuit of the plate transformer, the need for series resistors in the secondary circuit can be avoided. A practical method for primary-circuit surge control is shown in Fig. 20. The resistor, R_s introduces a voltage drop in the primary feed to T1 until C is nearly charged. Then, after C becomes partially charged, the voltage drop across R_s lessens and allows K1 to pull in, thus applying full primary power to T1 as K1A shorts out R_s. R_s is usually a 25-watt resistor whose resistance is somewhere between 15 and 50 ohms, depending upon the power supply characteristics.

Transient Problems

A common cause of trouble is transient voltages on the ac power line. These are short spikes, mostly, that can temporarily increase the voltage seen by the rectifier to values much higher than the normal transformer voltage. They come from distant lightning strokes, electric motors turning on and off, and so on. Transients cause unexpected, and often unexplained, loss of silicon rectifiers.

It's always wise to suppress line transients, and it can be easily done. Fig. 21 A shows one way. C1 looks like 280,000 ohms at 60 Hz, but to a sharp transient (which has only high-frequency components), it is an effective bypass. C2 provides additional protection on the secondary side of the transformer. It should be 0.01 µF for transformer voltages of 100 or less, and 0.001 µF for high-voltage transformers.

Fig. 21B shows another transient-suppression method using selenium suppressor diodes. The diodes do not conduct unless the peak voltage becomes abnormally high. Then they clip the transient peaks. General Electric sells protective diodes under the trade name, "Thyrector." Sarkes-Tarzian uses the descriptive name, "Klipvolt."

Transient voltages can go as high as twice the normal line voltage before the suppressor diodes clip the peaks. Capacitors cannot give perfect suppression either. Thus, it is a good idea to use power-supply rectifiers rated at about twice the expected PRV.

Diodes in Series

Where the PRV rating of a single diode is not sufficient for the application, similar diodes may be used in series. (Two 500-PRV diodes in series will withstand 1000 PRV, and so on.) When this is done, a resistor and a capacitor should be placed across each diode in the string to equalize the PRV drops and to guard against transient voltage spikes, as shown in Fig. 22A. Even though the diodes are of the same type and have the same PRV rating, they may have widely different back resistances when they are cut off. The reverse voltage divides according to Ohm's Law, and the diode with the higher back resistance will have the higher voltage developed across it. The diode may break down.

If we put a swamping resistor across each diode, R as shown in Fig. 22A, the resultant resistance across each diode will be almost the same, and the back voltage will divide almost equally. A good rule of thumb for resistor size is this: Multiply the PRV rating of the diode by 500 ohms. For example, a 500-PRV diode should be shunted by 500 × 500, or 250,000 ohms.

The shift from forward conduction to high back resistance does not take place instantly in a silicon diode. Some diodes take longer than others to develop high back resistance. To protect the "fast" diodes in a series string until all the diodes are properly cut off, a 0.01-µF capacitor

should be placed across each diode. Fig. 22A shows the complete series-diode circuit. The capacitors should be non-inductive, ceramic disk, for example, and should be well matched. Use 10-percent-tolerance capacitors if possible.

Diodes in Parallel

Diodes can be placed in parallel to increase current-handling capability. Equalizing resistors should be added as shown in Fig. 22B. Without the resistors, one diode may take most of the current. The resistors should be selected to have about a 1-volt drop at the expected peak current.

Voltage Dropping Resistor

Certain plates and screens of the various tubes in a transmitter or receiver often require a variety of operating voltages differing from the output voltage of an available power supply. In most cases, it is not economically feasible to provide a separate power supply for each of the required voltages. If the current drawn by an electrode (or combination of electrodes operating at the same voltage) is reasonably constant under normal operating conditions, the required voltage may be obtained from a supply of higher voltage by means of a voltage-dropping resistor in series, as shown in Fig. 23A. The value of the series, resistor, R1, may be obtained from Ohm's Law,

$$R = \frac{E_d}{I}$$

where

E_d = voltage drop required from the supply voltage to the desired voltage.
I = total rated current of the load

Example: The plate of the tube in one stage and the screens of the tubes in two other stages require an operating voltage of 250. The nearest available supply voltage is 400 and the total of the rated plate and screen currents is 75 mA. The required resistance is

$$R = \frac{400 - 250}{.075} = \frac{150}{.075} = 2000 \text{ ohms}$$

The power rating of the resistor is obtained from P (watts) = I^2R = $(0.075)^2$ × (2000) = 11.2 watts. A 20-watt resistor is the nearest safe rating to be used.

Voltage Dividers

The regulation of the voltage obtained in this manner obviously is poor, since any change in current through the resistor will cause a directly proportional change in the voltage drop across the resistor. The regulation can be improved somewhat by connecting a second resistor from the low-voltage end of the first to the negative power-supply terminal, as shown in Fig. 23B. Such an arrangement constitutes a

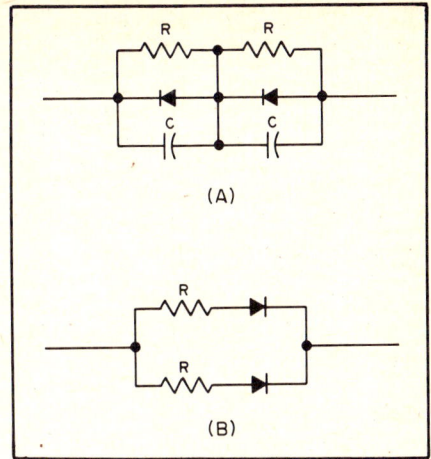

Fig. 22 — A — Diodes connected in series should be shunted with equalizing resistors and spike-suppressing capacitors. B — Diodes connected in parallel should be series current equalizing resistors.

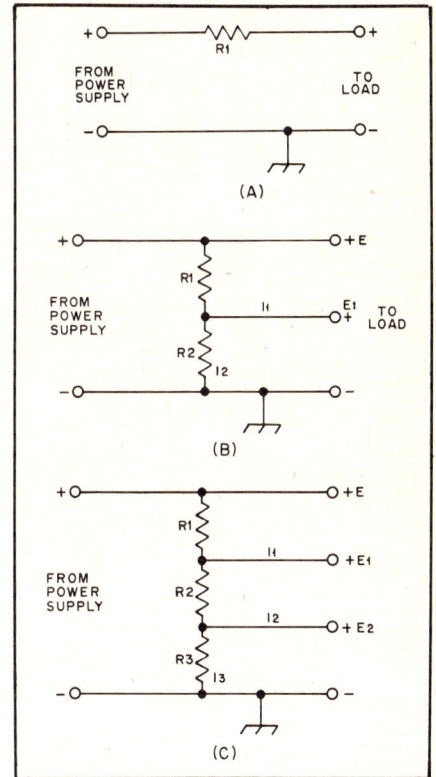

Fig. 23 — A — Series voltage-dropping resistor. B — Simple voltage divider.

$$R2 = \frac{E1}{I2} \qquad R1 = \frac{E - E1}{I1 + I2}$$

I2 must be assumed.

C — Multiple divider circuit.

$$R3 = \frac{E2}{I3} \qquad R2 = \frac{E1 - E2}{I1 + I3}$$

$$R1 = \frac{E - E1}{I1 + I2 + I3}$$

I3 must be assumed.

voltage divider. The second resistor, R2, acts as a constant load for the first, R1, so that any variation in current from the tap becomes a smaller percentage of the total current through R1. The heavier the current drawn by the resistors when they alone are connected across the supply, the better will be the voltage regulation at the tap.

Such a voltage divider may have more than a single tap for the purpose of obtaining more than one value of voltage. A typical arrangement is shown in Fig. 23C. The terminal voltage is E, and two taps are provided to give lower voltages, E1 and E2, at currents I1 and I2 respectively. The smaller the resistance between taps in proportion to the total resistance, the lower is the voltage between the taps. The voltage divider in the figure is made up of separate resistances, R1, R2 and R3. R3 carries only the bleeder current, I3; R2 carries I2 in addition to I3; R1 carries I1, I2 and I3. To calculate the resistances required, a bleeder current, I3, must be assumed; generally it is low compared with the total load current (10 percent or so). Then the required values can be calculated as shown in the caption of Fig. 23, I being in decimal parts of an ampere.

The method may be extended to any desired number of taps, each resistance section being calculated by Ohm's Law using the needed voltage drop across it and the total current through it. The power dissipated by each section may be calculated by multiplying I and E or I² and R.

Voltage Stabilization: Gaseous Regulator Tubes

There is frequent need for maintaining the voltage applied to a low-voltage low-current circuit at a practically constant value, regardless of the voltage regulation of the power supply or variations in load current. In such applications, gaseous regulator tubes (0B2/VR105, 0A2/VR150, etc.) can be used to good advantage. The voltage drop across such tubes is constant over a moderately wide current range. Tubes are available for regulated voltages near 150, 105, 90 and 75 volts.

The fundamental circuit for a gaseous regulator is shown in Fig. 24. The tube is connected in series with a *limiting resistor*, R1, across a source of voltage that must be higher than the starting voltage. The starting voltage is about 30 to 40 percent higher than the operating voltage. The load is connected in parallel with the tube. For stable operation, a minimum tube current of 5 to 10 mA is required. The maximum permissible current with most types is 40 mA; consequently, the load current cannot exceed 30 to 35 mA if the voltage is to be stabilized over a range from zero to maximum load. A single VR tube may also be used to regulate the

voltage to a load current of almost any value as long as the *variation* in the current does not exceed 30 to 35 mA. If, for example, the average load current is 100 mA, a VR tube may be used to hold the voltage constant provided the current does not fall below 85 mA or rise above 115 mA.

The value of the limiting resistor must lie between that which just permits minimum tube current to flow and that which just passes the maximum permissible tube current when there is no load current. The latter value is generally used. It is given by the equation:

$$R = \frac{(E_s - E_r)}{I}$$

where

R = limiting resistance in ohms
E_s = voltage of the source across which the tube and resistor are connected.
E_r = rated voltage drop across the regulator tube.
I = maximum tube current in amperes (usually 40 mA, or 0.04 A)

Two tubes may be used in series to give a higher regulated voltage than is obtainable with one, and also to give two values of regulated voltage. Regulation of the order of one percent can be obtained with these regulator tubes when they are operated within their proper current range. The capacitance in shunt with a VR tube should be limited to 0.1 μF or less. Larger values may cause the tube drop to oscillate between the operating and starting voltages.

Zener Diode Regulation

A Zener diode (named after Dr. Carl Zener) can be used to stabilize a voltage source in much the same way as when the gaseous regulator tube is used. The typical circuit is shown in Fig. 25A. Note that the cathode side of the diode is connected to the positive side of the supply. The electrical characteristics of a Zener diode under conditions of forward and reverse voltage are given in chapter 4.

Zener diodes are available in a wide variety of voltages and power ratings. The voltages range from less than two to a few hundred, while the power ratings (power the diode can dissipate) run from less than 0.25 watt to 50 watts. The ability of the Zener diode to stabilize a voltage is dependent upon the conducting impedance of the diode, which can be as low as one ohm or less in a low-voltage, high-power diode to as high as a thousand ohms in a low-power, high-voltage diode.

Diode Power Dissipation

Unlike gaseous regulator tubes, Zener diodes of a particular voltage rating have varied maximum current capabilities, depending upon the power ratings of each of the diodes. The power dissipated in a diode is the product of the voltage across

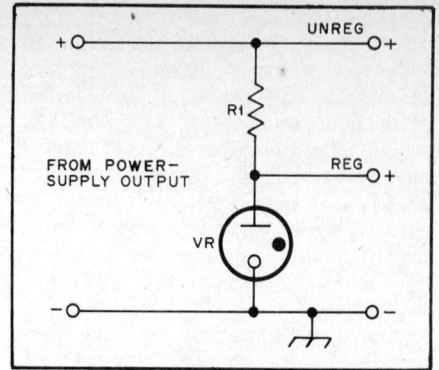

Fig. 24 — Voltage stabilization circuit using a VR tube. A negative-supply output may be regulated by reversing the polarity of the power-supply connections and the VR-tube connections from those shown here.

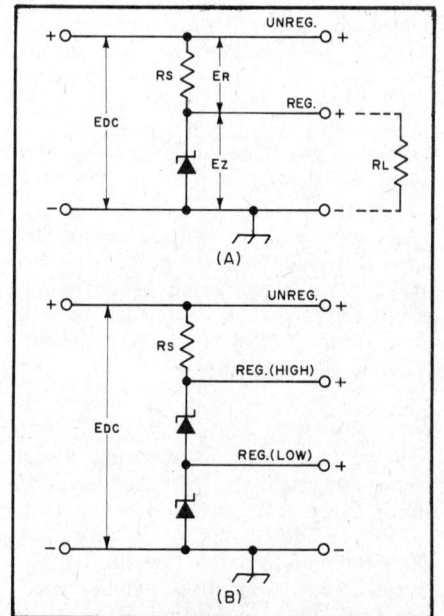

Fig. 25 — Zener-diode voltage regulation. The voltage from a negative supply may be regulated by reversing the power-supply connections and the diode polarities.

it and the current through it. Conversely, the maximum current a particular diode may safely conduct equals its power rating divided by its voltage rating. Thus, a 10-V, 50-W Zener diode, if operated at its maximum dissipation rating, would conduct 5 amperes of current. A 10-V 1-W diode, on the other hand, could safely conduct no more than 0.1A, or 100 mA. The conducting impedance of a diode is its voltage rating divided by the current flowing through it, and in the above examples would be 2 ohms for the 50-W diode, and 100 ohms for the 1-W diode. Disregarding small voltage changes which may occur, the conducting impedance of a given diode is a function of the current flowing through it, varying in inverse proportion.

The power-handling capability of most Zener diodes is rated at 25°C, or approximately room temperature. If the diode is

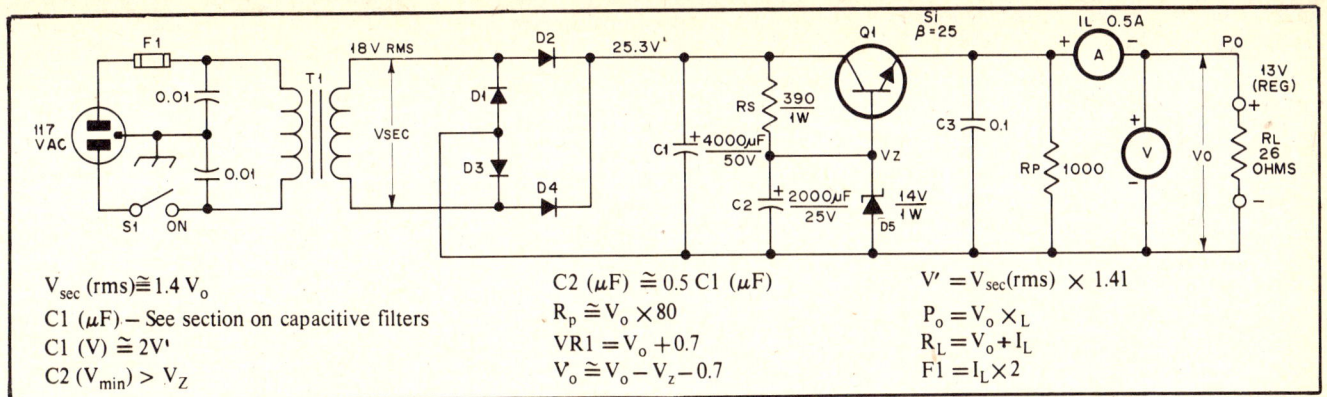

Fig. 26 — Illustration of a power supply with regulation. A pass transistor, Q1, is used to extend the range of the Zener-diode regulator.

$V_{sec}(rms) \cong 1.4 V_o$
C1 (μF) — See section on capacitive filters
$C1 (V) \cong 2V'$
$C2 (V_{min}) > V_Z$

$C2 (\mu F) \cong 0.5\, C1 (\mu F)$
$R_p \cong V_o \times 80$
$VR1 = V_o + 0.7$
$V_o \cong V_o - V_Z - 0.7$

$V' = V_{sec}(rms) \times 1.41$
$P_o = V_o \times L$
$R_L = V_o + I_L$
$F1 = I_L \times 2$

operated in a higher ambient temperature, its power capability must be derated. A typical 1-watt diode can safely dissipate only 1/2 watt at 100°C.

Limiting Resistance

The value of R_S in Fig. 25 is determined by the load requirements. If R_S is too large the diode will be unable to regulate at large values of I_L, the current through R_L. If R_S is too small, the diode dissipation rating may be exceeded at low values of I_L. The optimum value for R_S can be calculated by:

$$R_S = \frac{E_{DC\,(min)} - E_Z}{1.1\ I_{L\,(max)}}$$

When R_S is known, the maximum dissipation of the diode, P_D, may be determined by

$$P_D = \left[\frac{E_{DC(max)} - E_Z}{R_S} - I_{L\,(min)} \right] E_Z$$

In the first equation, conditions are set up for the Zener diode to draw 1/10 the maximum load current. This assures diode regulation under maximum load.

Example: A 12-volt source is to supply a circuit requiring 9 volts. The load current varies between 200 and 350 mA.

$E_Z = 9.1$ V (nearest available value)

$$R_S = \frac{12 - 9.1}{1.1 \times 0.35} = \frac{2.9}{0.385} = 7.5 \text{ ohms}$$

$$P_D = \left[\frac{12 - 9.1}{7.5} - 0.2 \right] 9.1$$

$$= 0.185 \times 9.1 = 1.7 \text{ W}$$

The nearest available dissipation rating above 1.7 W is 5; therefore, a 9.1-V 5-W Zener diode should be used. Such a rating, it may be noted, will cause the diode to be in the safe dissipation range even though the load is completely disconnected $[I_L\,(min) = 0]$.

Obtaining Other Voltages

Fig. 25B shows how two Zener diodes may be used in series to obtain regulated voltages not normally obtainable from a single Zener diode, and also to give two values of regulated voltage. The diodes need not have equal breakdown voltages, because the arrangement is self equalizing. However, the current-handling capability of each diode should be taken into account. The limiting resistor may be calculated as above, taking the sum of the diode voltages as E_Z, and the sum of the load currents as I_L.

Electronic Voltage Regulation

Several circuits have been developed for regulating the voltage output of a power supply electronically. While more complicated than the VR-tube and Zener-diode circuits, they will handle higher voltage and current variations, and the output voltage may be varied continuously over a wide range.

Voltage regulators fall into two basic types. In the type most commonly used by amateurs, the dc supply delivers a voltage higher than that which is available at the output of the regulator, and the regulated voltage is obtained by dropping the voltage down to a lower value through a dropping "resistor." Regulation is accomplished by varying either the current through a fixed dropping resistance as changes in input voltage or load currents occur (as in the VR-tube and Zener-diode regulator circuits), or by varying the equivalent resistive value of the dropping element with such changes. This latter technique is used in electronic regulators where the voltage-dropping element is a vacuum tube or a transistor, rather than an actual resistor. By varying the dc voltage at the grid or current at the base of these elements, the conductivity of the device may be varied as necessary to hold the output voltage constant. In solid-state regulators the series-dropping element is called a pass transistor. Power transistors are available which will handle several amperes of current at several hundred volts, but solid-state regulators of this type are usually operated at potentials below 100 volts.

The second type of regulator is a switching type, where the voltage from the dc source is rapidly switched on and off (electronically). The average dc voltage available from the regulator is proportional to the duty cycle of the switching wave form, or the ratio of the on time to the total period of the switching cycle. Switching frequencies of several kilohertz are normally used to avoid the need for extensive filtering to smooth the switching frequency from the dc output.

The above information pertains essentially to voltage regulators. A circuit can also be constructed to provide current regulation. Such regulation is usually obtained in the form of current limitation — to a maximum value which is either preset or adjustable, depending on the circuit. Relatively simple circuits, such as described later, can be used to provide current limiting only. Current limiting circuitry may also be used in conjunction with voltage regulators.

Discrete Component Regulators

The previous section outlines some of the limitations when using Zener diodes as regulators. Greater current amounts can be accommodated if the Zener diode is used as a reference at low current, permitting the bulk of the load current to flow through a series pass transistor (Q1 of Fig. 26). An added benefit in using a pass transistor is that of reduced ripple on the output waveform. This technique is commonly referred to as "electronic filtering."

Q1 of Fig. 26 can be thought of as a simple emitter-follower dc amplifier. It increases the load resistance seen by the Zener diode by a factor of beta (β). In this circuit arrangement D5 is required to supply only the base current for Q1. The net result is that the load regulation and ripple characteristics are improved by a factor of beta. Addition of C2 reduces the ripple even more, although many simple supplies such as this do not make use of a capacitor in that part of the circuit.

The primary limitation of this circuit is that Q1 can be destroyed almost immediately if a severe overload occurs at R_L. The fuse cannot blow fast enough to protect Q1. In order to protect Q1 in case of an accidental short at the output, a current limiting circuit is required. An example of a suitable circuit is shown in Fig. 27.

It should be mentioned that the greater the value of transformer secondary voltage, the higher the power dissipation in Q1. This not only reduces the overall efficiency of the power supply, but requires stringent heat sinking at Q1.

Design Example

Example: Design a regulated, well-filtered, 13-volt dc supply capable of delivering 0.5 A, using the circuit of Fig. 26. Calculate the ratings for all components. A standard 18-volt secondary transformer is to be used.

Information on calculating the transformer, diode and input capacitor ratings were given earlier in this chapter and will not be repeated here. In order to calculate the value or R_S in Fig. 26 the base current of Q1 must be known. The base current is approximately equal to the emitter current of Q1 in amperes divided by beta. The transistor beta can be found in the manufacturer's data sheet, or measured with simple test equipment (beta = I_c/I_b). Since the beta spread for a particular type of transistor — 2N3055 for example, where it is specified as 25 to 70 — is a fairly unknown quantity, more precise calculations for Fig. 26 will result if the transistor beta is tested before the calculations are done. A conservative approach is to design for beta minimum of the transistor used. Calculating I_b:

$$I_b = \frac{0.5}{25} = 0.02A = 20\,mA$$

As pointed out earlier, in order for D5 to regulate properly it is necessary that a fair portion of the current flowing through R_S should be drawn by D5. The resistor will have 0.02 A flowing through it as calculated above (base current of Q1). A conservative amount of 10 mA will be used for the Zener diode current bringing the total current through R_S to 0.03 A or 30 mA. From this, the value of R_S can be calculated as follows:

$$R_S = \frac{(V' - V_Z)}{I_{R_S}} = \frac{(25.3 - 14)}{0.03} = 376\ ohms$$

The nearest standard ohmic value for R_S is 390. The wattage ratings for R_S and D5 can be obtained with the aid of the formulas given earlier for Zener-diode regulators.

The power rating for Q1 will be calculated next. The power dissipation of Q1 is equal to the emitter current times the collector-to-emitter voltage. Calculate as follows:

$$P_{Q1} = I_E \times V_{CE}$$

where
  V_{CE} = the desired $V' - (V_Z - V_{BE})$, and V_{BE} is approximately 0.7 V for a silicon transistor.

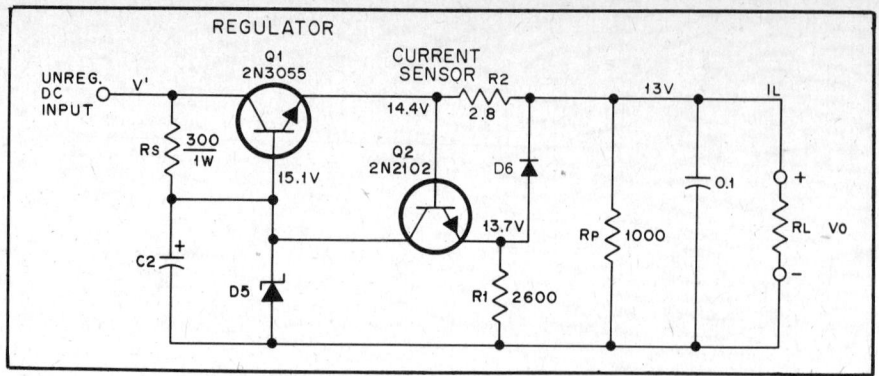

Fig. 27 — Overload protection for a regulated supply can be effected by addition of a current-overload protective circuit.

Therefore:
  $P_{Q1} = 0.5A \times 12V = 6$ watts

It is a good idea to choose a transistor for Q1 that has at least twice the rating calculated. In this example a transistor with a power dissipation rating 12 watts or more would be used.

The 0.01-μF capacitors at the primary of T1 serve two functions. They act as transient suppressors and help prevent rf energy from entering the power-supply regulator.

Current Limiting for Discrete-Component Regulators

Damage to Q1 of Fig. 26 can occur when the load current exceeds the safe amount. Fig. 27 illustrates a simple current-limiter circuit that will protect Q1. All of the load current is routed through R2. A voltage difference will exist across R2, the amount being dependent upon the exact load current at a given time. When the load current exceeds a predetermined safe value, the voltage drop across R2 will forward bias Q2 and cause it to conduct. Since D6 is a silicon diode, and because Q2 is a silicon transistor, the combined voltage drops through them (roughly 0.7 V each) will be 1.4 V. Therefore the voltage drop across R2 must exceed 1.4 V before Q2 can turn on. This being the case, R2 is chosen for a value that provides a drop of 1.4 V when the maximum safe load current is drawn. In this instance 1.4 volts will be seen when I_L reaches 0.5A.

When Q2 turns on, some of the current through R_S flows through Q2, thereby depriving Q1 of some of its base current. This action, depending upon the amount of Q1 base current at a precise moment, cuts off Q1 conduction to some degree, thus limiting the flow of current through it.

High-Current-Output Regulators

When a single pass transistor is not available to handle the current which may be required from a regulator, the current-handling capability may be increased by connecting two or more pass transistors in parallel. The circuits at B and C of Fig. 28 show the method of connection. The

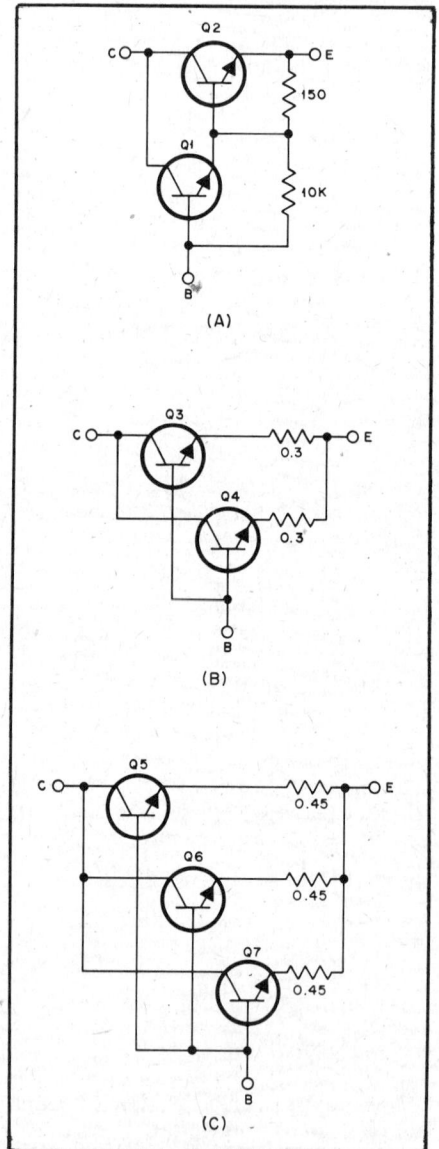

Fig. 28 — At A, a Darlington-connected pair for use as the pass element in a series-regulating circuit. At B and C, the method of connecting two or more transistors in parallel for high current output. Resistances are in ohms. The circuit at A may be used for load currents from 100 mA to 5 A, at B for currents from 6 to 10 A, and at C for currents from 9 to 15 A.
Q1 — Motorola MJE 340 or equivalent.
Q2-Q7, incl. — Power transistor such as 2N3055 or 2N3772.

resistances in the emitter leads of each transistor are necessary to equalize the currents.

Fixed-Voltage IC Regulators

The modern trend in regulators is toward the use of three-terminal devices commonly referred to as three-terminal regulators. Inside each regulator is a reference, a high-gain error amplifier, sensing resistors and transistors, and a pass element. Some of the more sophisticated units have thermal shutdown, over-voltage protection and current foldback. Many of the regulators currently on the market are virtually destruction-proof. Several supplies using these ICs are featured in the construction section of this chapter.

Three-terminal regulators (a connection for unregulated dc input, regulated dc output and ground) are available in a wide range of voltage and current ratings. Fairchild, National and Motorola are perhaps the three largest suppliers of these regulators at present. It is easy to see why regulators of this sort are so popular when one considers the low price and the number of individual components they can replace. The regulators are available in several different package styles — TO-3, TO-39, TO-66, TO-220 and dual in-line (DIP), to name just a few.

Three-terminal regulators are available as positive or negative types. In most cases, a positive regulator is used to regulate a positive voltage and a negative regulator a negative voltage. However, depending on the systems ground requirements, each regulator type may be used to regulate the "opposite" voltage.

Figs. 29A and B illustrate how the regulators are used in the conventional mode. Several regulators can be used with a common-input supply to deliver several voltages with a common ground. Negative regulators may be used in the same manner. If no other common supplies operate off the input supply to the regulator, the circuits of Figs. 29C and D may be used to regulate positive voltages with a negative regulator and vice versa. In these configurations the input supply is floated; neither side of the input is tied to the system ground.

When choosing a three-terminal regulator for a given application the important specifications to look for are maximum output current, maximum output voltage, minimum and maximum input voltage, line regulation, load regulation and power dissipation.

In use, these regulators require an adequate heat sink since they may be called on to dissipate a fair amount of power. Also, since the chip contains a high-gain error amplifier, bypassing of the input and output leads is essential to stable operation (See Fig. 30). Most manufacturers recommend bypassing the input and output directly at the leads

where they protrude through the heat sink. Tantalum capacitors are usually recommended because of their excellent bypass capabilities up into the vhf range.

Adjustable-Voltage IC Regulators

Relatively new on the electronic scene are high-current, adjustable voltage regulators. These ICs require little more than an external potentiometer for an adjustable voltage range from 5 to 24 volts at up to 5 amperes. The unit price on these items is currently around $6 making them ideal for a test bench power supply. An adjustable-voltage power supply using the Fairchild 78HG series of regulator is described in the construction section of this chapter. The same precautions should be taken with these types of regulators as with the fixed-voltage units. Proper heat sinking and lead bypassing is essential for proper circuit operation.

A 12-Volt 3-Ampere Power Supply

Shown in Fig. 31 is a no-frills 12-volt supply capable of continuous operation at the 3-ampere level. Many low-power hf transceivers and most vhf-fm transceivers require voltages and currents on this order. Power supplies of this type purchased from the manufacturers can be quite costly. Described here is a very simple to build and relatively inexpensive (around $20 using all new components) alternative.

The schematic diagram for the power supply is shown in Fig. 32. As can be seen, the circuit is simplicity itself. A transformer, two diodes, three capacitors and a regulator form the heart of the supply. Binding posts, a pilot light, fuse and on-off switch complete the design.

Ac from the mains is supplied to the transformer-primary winding through the fuse in one leg, and the on-off switch in the other. The secondary circuit feeds a full-wave rectifier circuit which is filtered by C1. This unregulated voltage is routed to the input terminal of the regulator IC which is bypassed directly at the case with a 2-μF tantalum capacitor. The case of the IC is connected to ground. A 2-μF tantalum capacitor is also used at the output terminal of the regulator to prevent unwanted oscillation of the error amplifier inside the IC. A pilot light attached to the regulated output indicates when the supply is in use.

The regulator has built-in thermal shut down and over-current protection. Short circuiting the output of the supply will cause no damage. A wide margin of conservative component rating was used in the design of this supply. It should be possible to run the supply for hours on end at its maximum rating.

Construction

Rather than using an expensive cabinet, the power supply is housed on an aluminum chassis measuring 5 × 9-1/2 × 3

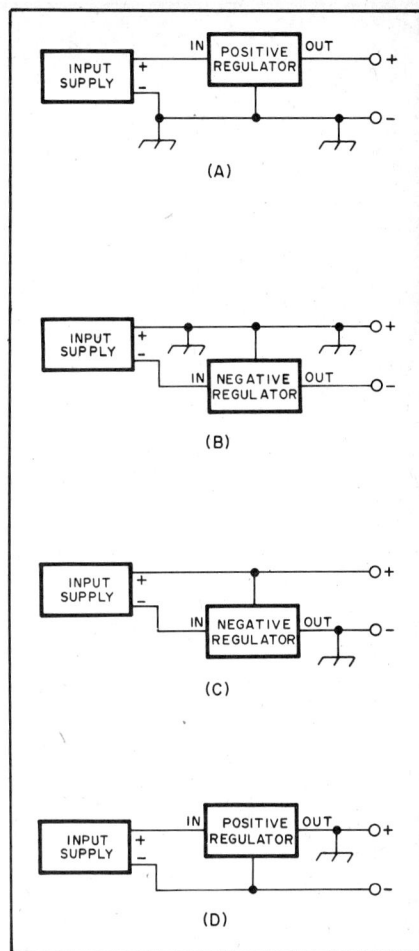

Fig. 29 — A and B illustrate the conventional manner in which three-terminal regulators are used. C and D show how one polarity regulator can be used to regulate the opposite polarity voltage.

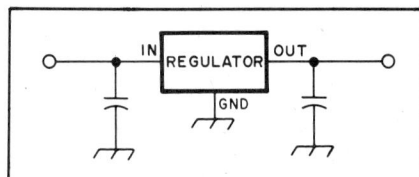

Fig. 30 — Three-terminal regulators require careful bypassing directly at the case. Here, both the input and output leads are bypassed.

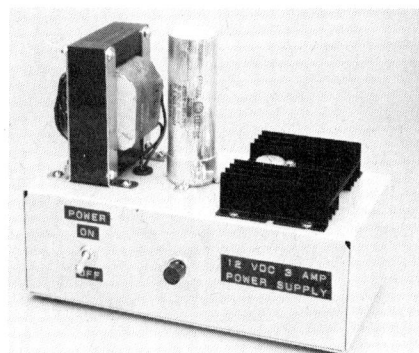

Fig. 31 — Exterior view of the 12-volt, 3-ampere, no-frills power supply.

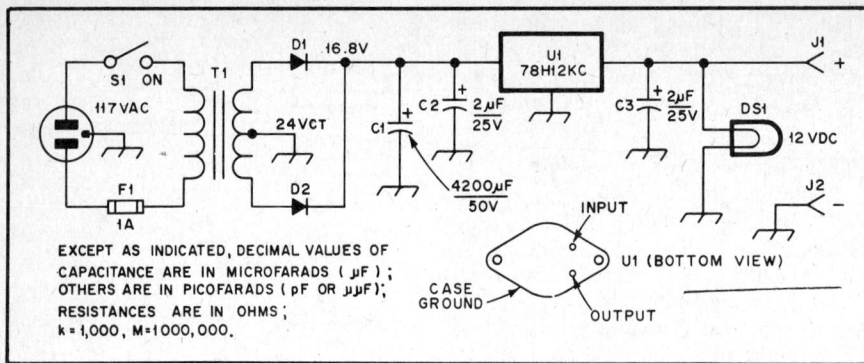

Fig. 32 — Schematic diagram of the 12-volt, 3- ampere power supply.

C1 — Electrolytic capacitor, 4200 μF, 50 V, General Electric 86F166M or equiv.
C2, C3 — Tantalum capacitor, 2 μF, 50 V.
D1, D2 — Silicon diode, 50 V, 6 A, HEP RO100 or equiv.
DS1 — Pilot light assembly, 12 V.
F1 — Fuse, 1 A.

J1, J2 — Binding post.
S1 — Spst toggle.
T1 — Power transformer; primary 117 V, secondary 24 V ct, Stancor P-8663 or equiv.
U1 — Voltage regulator, Fairchild 78H12KC or equiv.

inches (127 × 241 × 76 mm). Mounted atop the chassis is the power transformer, filter capacitor and regulator. The regulator is attached to a heat sink that measures 3 × 4-1/2 × 1 inches (76 × 114 × 25 mm). Two tantalum capacitors, not visible in the pictures, are mounted at the IC terminals on the underside of the sink. Since good ground connections are required to prevent IC oscillations, remove the anodizing from the heat sink where it will contact the chassis.

The layout of the underside of the chassis can be seen in Fig. 33. Two binding posts (one red and one black) and the fuse holder are mounted on the rear apron. The on-off switch and pilot light occupy a portion of the front panel. Dymo tape labels complete the front panel.

A 300- to 400-Watt 12-Volt Supply

Most modern hf transceivers in the 100-watt output class have solid-state output stages that require a nominal 13.6-volt power source. This voltage is available directly from vehicular electrical systems, so manufacturers don't generally built ac power supplies into these transceivers. This approach leads to lightweight compact units for mobile service, but the

fixed-station operator must obtain an external ac supply. These "accessories" cost upwards of $200 and sometimes offer only marginal performance. The heat sinks on most transceivers are adequate for low-duty-cycle ssb operation. A fan is required for SSTV, RTTY, slow-speed cw, or even speech-processed ssb. Matching accessory power supplies are subject to the same limitations and may fall down under these conditions, even if externally cooled. The combined shortcomings of the transceiver and power supply discourage many operators from trying modes other than ssb. A sturdier power supply can go a long way toward increasing the flexibility of a solid-state amateur station.

The power supply described in this section is a heavy-duty unit capable of 300 watts continuous duty or 400 watts PEP. The cost will depend on the availability of components, but should be significantly less than those sold to match a product line. Complete output metering, over-voltage shutdown and current limiting are features of the design.

Design Information

The generous power rating is made possible by the heavy-duty transformers. In the unit pictured, the transformers were each specified for 40 V at 15 A, and had a cross-sectional area of nearly 4 in² (2500 mm²). The secondaries were wound with no. 11 (2 mm) wire. Turns were removed from each secondary winding until each produced an open-circuit output of 19.5 V rms (at the minimum expected line voltage). With a 20-A dc load, the ac output potential drops to 18.8 V. This information is provided because the exact transformers used are no longer available. The transformer output voltage is very important in high-power applications. If the voltage applied to the regulatory circuitry is higher than necessary, excessive heating of the series

pass transistors will result. On the other hand, the bottom of the filter ripple voltage must be high enough to maintain the necessary voltage differential across the regulator. Since the power capability of this supply is limited by the dissipation of the pass transistors rather than the transformers, every effort was made to optimize the input to the regulator. An "odd couple" of computer-grade aluminum filter capacitors, totaling more than 0.1 F, supplies 22.3 V dc (under a 20-A load) to the regulator at less than 1 V pk-pk ripple. The price for low preregulator ripple is extremely high peak rectifier current. A bridge package rated for 25 amperes continuous duty proved unequal to the task; individual 35 A diodes perform adequately.

Voltage regulation is handled by an IC regulator that is supplemented by an array of eight 2N3055 power transistors. These transistors came from various sources and showed unequal current distribution, despite the 0.1-Ω spreading resistor in each emitter. Although higher value resistors would correct this condition, their power rating would have to be increased, as would the regulator input voltage. Transistors from a single manufacturing lot should be sufficiently uniform for this application. An additional 2N3055, Q4, drives the array of eight from the regulator in a Darlington configuration. Q2 monitors the voltage developed across the current-sense resistor, reducing the regulator output drive when the load current exceeds 30 A. If the unit has been supplying 20 A for more than 10 minutes, the current-limiting point folds back to about 24 A as a result of heating in the sense resistor. While the power supply as a whole is current limited by Q2, the regulator IC is not. If the collector of Q4 were to open, for example, the regulator would attempt to drive the pass transistors alone, through the Q4 base-emitter junction. The IC maximum rated current (500 mA) could be exceeded under these conditions without activating Q2. Q1 senses base drive to the Darlington array and

Fig. 33 — Interior view of the power supply.

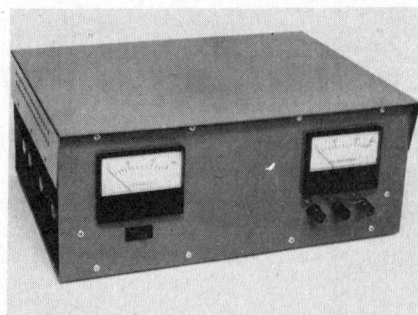

Fig. 34 — 3/4 view of the 300- to 400-watt power supply showing four of the series pass transistors. The various sheet metal panels are fastened together with angle brackets.

limits the maximum regulator output current to about 300 mA. This additional protection is not shown in the IC manufacturer's applications literature. R3, R4 and R9 set the output voltage over an adjustment range of 12 to 14 V. The power supply includes a "crowbar" over-voltage protection citcuit consisting of Q5, D1 and the associated resistor network. R10 forms a divider with the cathode-to-gate resistor internal to most SCRs. If your device does not include such a resistor, one can be installed externally. Should the output voltage exceed the threshold of D1, plus some level determined by R10, Q5 will fire, short-circuiting the output terminals. Q5 has a continuous current rating of 25 A. While this may seem underspecified, the component can withstand a surge of many times the continuous figure, and won't be called upon to pass the full current for more than a few milliseconds. As soon as Q5 fires, it pulls the output voltage below the conduction threshold of Q3, which allows the input voltage to shut down the regulator. Without this feature, both the series pass transistors and the SCR would be destroyed in a short time from excessive dissipation.

This supply has one other feature — remote sensing. When heavy loads are connected through long cables, the cable voltage drop degrades the load regulation. This problem can be circumvented by including the cable within the regulator feedback loop. To accomplish this, remove the jumper between the output and sense terminals, and connect the sense terminal directly to the load (the sense lead wire size isn't critical). This connection can't compensate for resistance in the return (ground) lead, but an extra chassis-to-chassis connection can usually be made to reduce the return resistance to an insignificant value. The pk-pk output ripple at full load is 5 mV. A complete schematic diagram is given in Fig. 35.

Mechanical Details

One assembly method can be seen in Figs. 34 and 36, although this is by no means the only workable solution. The unit shown uses a 16-1/2 × 12 × 1/8-in. (419 × 305 × 3 mm) aluminum plate for a foundation. This is expensive material if purchased new — a framework fashioned from angle stock may be an expedient alternative. Transformers suitable for this application are quite heavy and require a firm supporting structure. The pass transistor heat sinks form the sides of the enclosure. These sinks, like most of the components, are surplus items measuring 4 × 12 inches (102 × 305 mm) with fins protruding 1/2 in. from each flat surface. Asymmetrically spaced transistor mounting holes explain the surplus status of the sinks. The dimensions quoted represent a minimum, rather than maximum radiation area, especially considering that the fins would be more efficient if mounted vertically. A more conservative approach might be to mount an additional heat sink in the rear panel area and use 9 or 12 pass transistors. The emitter spreading resistors are mounted on the transistor heat sinks. Q4 should be separately heat sinked, although the sink shown is larger than necessary. A "top hat" and chassis bracket similar to that of Q5 would be sufficient.

The rectifier diode anodes are common to the mounting studs. For this reason, two of the studs must be sleeved and the cases and nuts must be insulated from the base plate with mica washers. If a suitably rated bridge assembly can be obtained at a reasonable price, it will simplify the mounting while providing superior thermal conduction. In any case, be sure to use mica washers with all semiconductors whose cases are above ground and apply heat sink compound to all mounting surfaces.

Liberal use of terminal strips simplifies the wiring and troubleshooting, should it be necessary. Use no. 10 or no. 12 wire for the high-current circuits. The electronic components that don't generate appreciable heat are contained on a printed circuit card. The etching pattern and parts placement guide are presented in Fig. 37. Most of the components for this power supply were purchased from mail-order firms that advertise in the Amateur Radio press.

A 1.2- to 15-Volt, 5-Ampere Supply

The power supply shown in Figs. 38 to 42 is intended for general purpose, test-bench applications. The output is adjustable from 1.2 to 17 volts at currents up to 6 amperes. Metering is provided for voltage levels up to 15 volts and current levels up to 5 amperes. Most of the components used in this supply are of the junk-box variety with the possible exception of U2, the three-terminal voltage regulator. The circuit will tolerate fairly wide component substitutions and still offer good performance. The majority of the circuit components are mounted on a 2-3/4 × 4-1/2 (70 × 114 mm) circuit board. All controls, including the mains fuse are located on the front panel for easy access.

The Circuit

Two power transformers are used in parallel to feed U1, the full-wave bridge rectifier assembly. The transformers specified are rated at 2 amperes each. The prospective builder might question the wisdom of using only 4 amperes worth of transformer in a 5-ampere supply. This is a valid question. With a 5-ampere load connected to the output of the supply, the transformers deliver more than their rated secondary voltage and do not become unreasonably warm to the touch even after continuous-duty operation. If transformers of different manufacture are used it might be wise to select units having a higher current rating — just to be sure.

S2 is included in the design so that either half or all of the secondary voltage may be applied to U1. This feature is included so that the dissipation of the pass transistor may be reduced when using the supply with low-voltage, high-current loads. The graph displayed in Fig. 41 can be used as a guide in selecting the HI or LO mode of operation.

The regulator consists of a pass transistor "wrapped around" an adjustable voltage regulator. Circuit operation can be understood by noting the values of R3 and R2. The majority of the three-terminal regulator current will flow through R3 and D1. The offset voltage in D1 is approximately equal to the emitter-base potential of Q1. Because of this, the voltage drop across R3 will be the same as that across R2. Since the ohmic value of R2 is 0.33 of R3, three times as much current will flow in Q1 as in U2. The net result is that the current capability of the overall circuit is increased by a factor of four. Also, the current-limiting characteristics of the three-terminal regulator are transferred directly to the composite circuit.

M1 and its associated shunt resistor are placed at the input to the regulator circuit so that the voltage drop across the resistor will not adversely alter the supply voltage regulation. The relatively small current drawn by the regulator circuitry does not seriously affect the meter accuracy. M2 measures supply output voltage.

Construction

The power supply is housed in a homemade enclosure that was fabricated from sheet aluminum. Dimensions of the enclosure are 5-1/2 × 6 × 8 inches (140 × 152 × 203 mm), although any cabinet that will house the components may be used. Circuit board pattern and layout information is given in Fig. 42. The completed circuit board is mounted vertically to the chassis using spade lugs and no. 6 hardware. A small heat sink for the LM317K regulator was made from a scrap piece of aluminum. A Radio Shack 276-1364 sink designed for the TO-3 package will work fine. The pass transistor is mounted to a larger heat sink which is bolted to the rear panel of the power supply. Here, a Motorola MS-10 was used. A suitable substitute would be the Radio Shack 276-1361. Bear in mind that the transistor must be insulated from the sink. Use a small amount of heat-sink compound between the transistor and the sink for a good thermal bond.

Since the power supply can deliver up to 6 amperes, fairly heavy wire should be used for those runs carrying the bulk of the current. No. 18 plastic wire was used in this unit and it appeared to be adequate.

The completed power supply may be "crowbarred" without worry of regulator or pass transistor destruction. Perhaps the only precaution that should be mentioned is that of the exposed collector of the pass transistor. Although no damage will occur if the case is shorted to ground it will cause the loss of output voltage. This could occur if the power supply is mounted on a test bench with a number of leads dangling behind the unit. A simple fix for this would be to mount a plastic TO-3 transistor cover over the case.

A Deluxe 5- to 25-Volt, 5-Ampere Supply

The power supply illustrated in Figs. 43 and 45 and schematically at Fig. 44 might be termed a rich man's power supply. The unit shown can supply voltages from 5 to 25 at currents up to 5 amperes. With ther-

mal and short-circuit protection it is virtually destruction proof. A digital panel meter is used to monitor voltage and current, selectable by a front-panel switch. Although we termed this a "rich man's supply", it will cost far less to construct

Fig. 36 — Interior of the heavy-duty supply. The current-sense resistors are mounted on a simple pc board that is elevated on ceramic standoffs in the front center. The control pc board is fastened to the base plate by spade bolts. The large internal heat sink is for Q4.

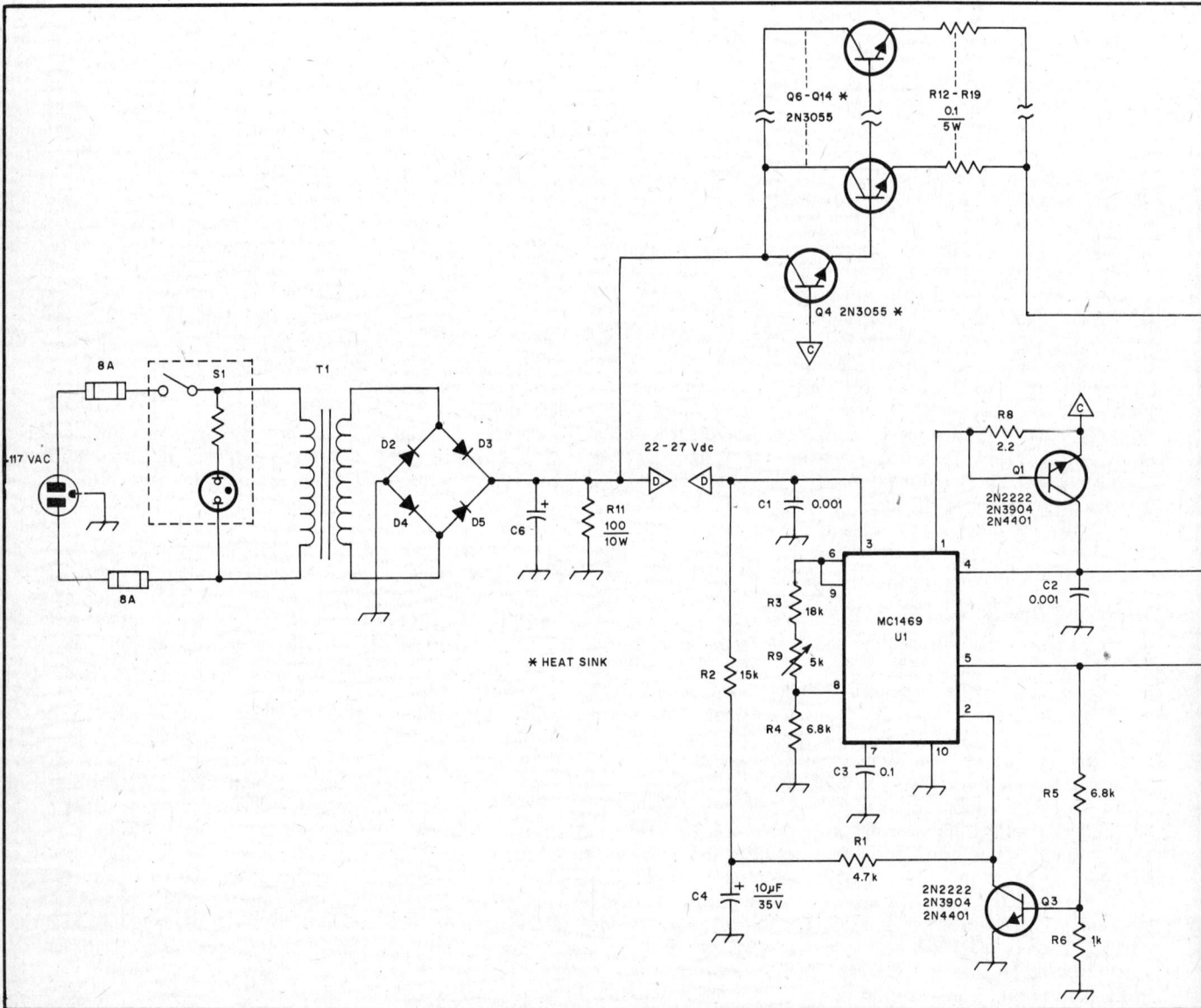

Fig. 35 — Schematic diagram of the 300- to 400-watt power supply.

C6 — Filter capacitor or capacitors totaling 0.1 F or more at 30 V or greater.
D2-D5 — Silicon rectifier diodes, anode stud type, 35 A, 100 PRV. 1N1184^R or equiv.
Q5 — Silicon control rectifier, 25 A, 50 V.
R20 — 5 0.1-Ω, 5-W resistors in parallel.
S1 — Spst, 10 A, 125 V ac, built-in pilot light

optional.
T1 — Combination of transformers capable of supplying 18.8 V ac at a 20-A dc load.

this unit as compared to a ready-made supply with the same features. Cost, using all new components, will be on the order of $75. The most expensive single item in the supply is the digital panel meter, which sells in single lot quantities for around $40 at present. As more companies start manufacturing these items the prices should drop significantly.

The digital readout, however, is not much more expensive than two high-quality meters. The prospective builder should consider this when choosing between the digital panel meters and two analog panel meters. Voltage measurements are read directly off the panel meter in volts. Current is measured in amperes with a reading of 0.05 equal to 50 mA.

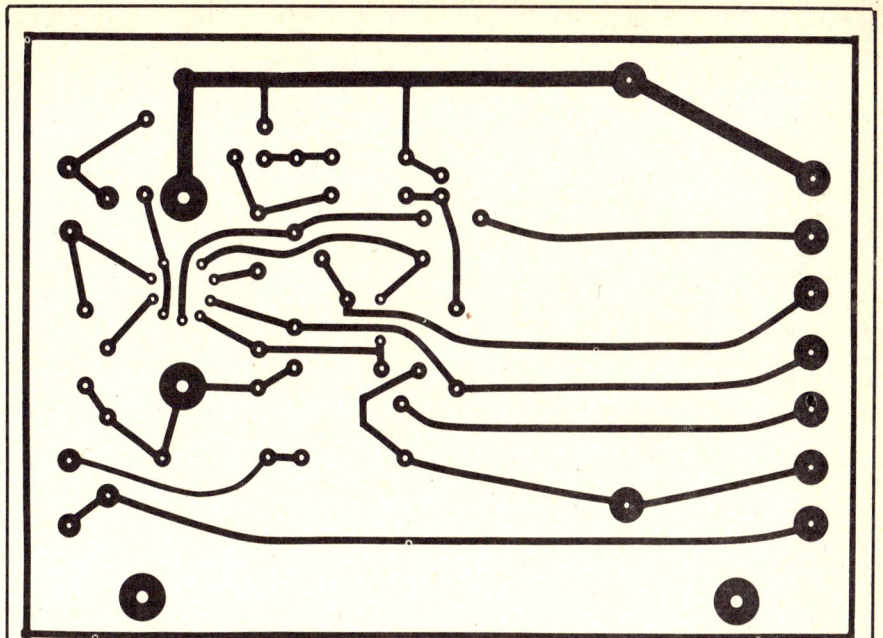

(A)

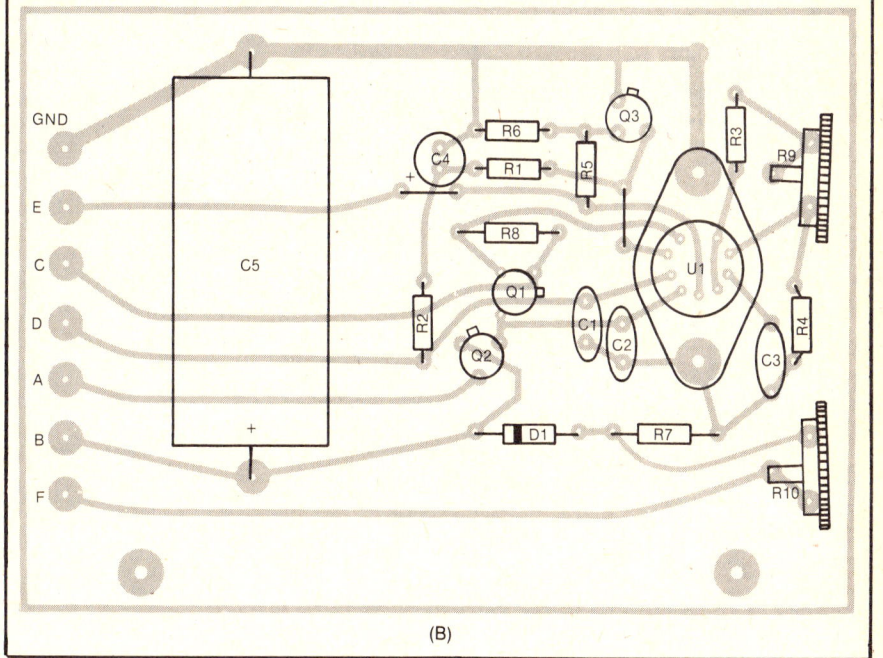

(B)

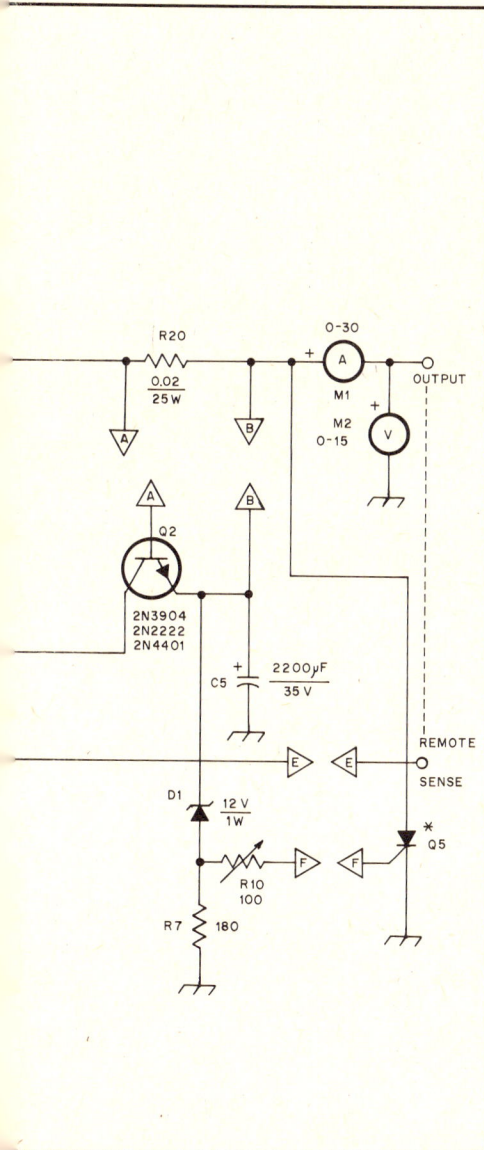

Fig. 37 — At A, the etching pattern for the control pc card. B is the parts-placement guide. The black lines are an x-ray view of the foil side.

Fig. 38 — A simple 1.2- to 15-volt, 5-ampere power supply. All controls are mounted on the front panel for easy access. The milliammeter reading is multiplied by 100 to obtain the true output current.

Circuit Details

The circuit diagram of the power supply is shown in Fig. 44. T1 is a 36-volt, center-tapped transformer rated at 6 amperes. D1 and D2 are used in a full-wave rectifier providing dc output to the filter capacitor, C3, a 34,000-μF, 50-volt electrolytic of the computer-grade variety. The unregulated voltage is fed to U1, a Fairchild 78HGKC regulator, the heart of the supply. This chip is rated for 5-A continuous duty when used with an adequate heat sink. R1 and R2 form a voltage divider which sets the output voltage of the supply. R1 is a ten-turn potentiometer. U1 is bypassed with 2.2-μF tantalum capacitors directly at the input and output pins.

Z1, as outlined earlier, is a digital panel meter. Connections to the meter are made through a special edge connector supplied with the readout. U2 is used to supply a regulated 5 volts for powering the digital panel meter. The input and ground leads of this regulator are attached to the input (non-regulated) side of U1.

R4, R5 and R6 form a divider circuit to supply the digital meter with an output voltage reading. R5 is made adjustable so that the meter can be calibrated. R3 is a current-sensing resistor which is placed in the negative lead of the supply. This resistor is used on the input side of the regulator (U1) so as not to affect the voltage regulation of the power supply at high load currents. Any voltage dropped across the resistor will be made up by the regulator, so the output voltage will remain unchanged. Notice that U2 is placed to the left or at the input side of the regulator. This is so the current drawn by the readout will not affect current readings taken at the load. Sections A and B of S2 are used to switch the meter between the voltage and current sensors. S3C is used to switch the decimal point in the digital panel meter to read correctly for both voltage and current.

As shown in the schematic, a single-point ground is used for the supply. Used in many commercial supplies, this technique provides better voltage regulation and stabilization than the "ground it anywhere" attitude. In this supply, the single-ground point is at the front panel binding post labelled MINUS. All leads that are to be connected to ground should go only to that point.

Construction

The deluxe power supply is housed in a homemade enclosure that measures 9 × 11 × 5-1/4 inches (229 × 279 × 133 mm). U1 is mounted to a large heat sink (3 × 5 × 2 inches; 76 × 127 × 51 mm) which is attached to the rear apron of the supply. The front panel sports the digital-panel meter, power switch, binding posts, fuse holder, voltage-adjust potentiometer and meter-selector switch. Although a circuit board is shown in the photograph as supporting R4, R5, R6, R7, D1, D2, C1 and C2, these items could just as well be mounted on terminal strips. For this reason a board pattern is not supplied.

The front and rear panels are spray painted white and the cover is blue. Dymo labels are used on the front panel to identify each of the controls. Cable lacing of the various leads adds to the clean appearance of the supply.

SWITCHING REGULATORS

The electronic voltage regulator circuits discussed up to this point have been of the *linear* variety; that is, a series or shunt control element has varied its effective resistance in direct proportion to load or input changes. A second class of regulator circuit becoming increasingly important in electronics is the *switching* regulator. In a switching regulator the pass element oscillates between fully off and fully on, and the transitions between these two states are made as short as possible. Operating a control device (usually a transistor) in the switching mode significantly reduces the heating, as voltage and current are not applied simultaneously. The chief application of switching supplies has been to low-voltage, high-current loads (such as computers), where high efficiency is difficult to realize in a linear circuit. However, the switching technique can also be applied to high-voltage situations. Al Helfrick, K2BLA, demonstrated a 2000-volt, 1-ampere switching supply no larger than a shoe box to the ARRL technical staff.

Conducted and radiated RFI are the major shortcomings attributed to switching power supplies, and careful engineering is required to reduce the out-

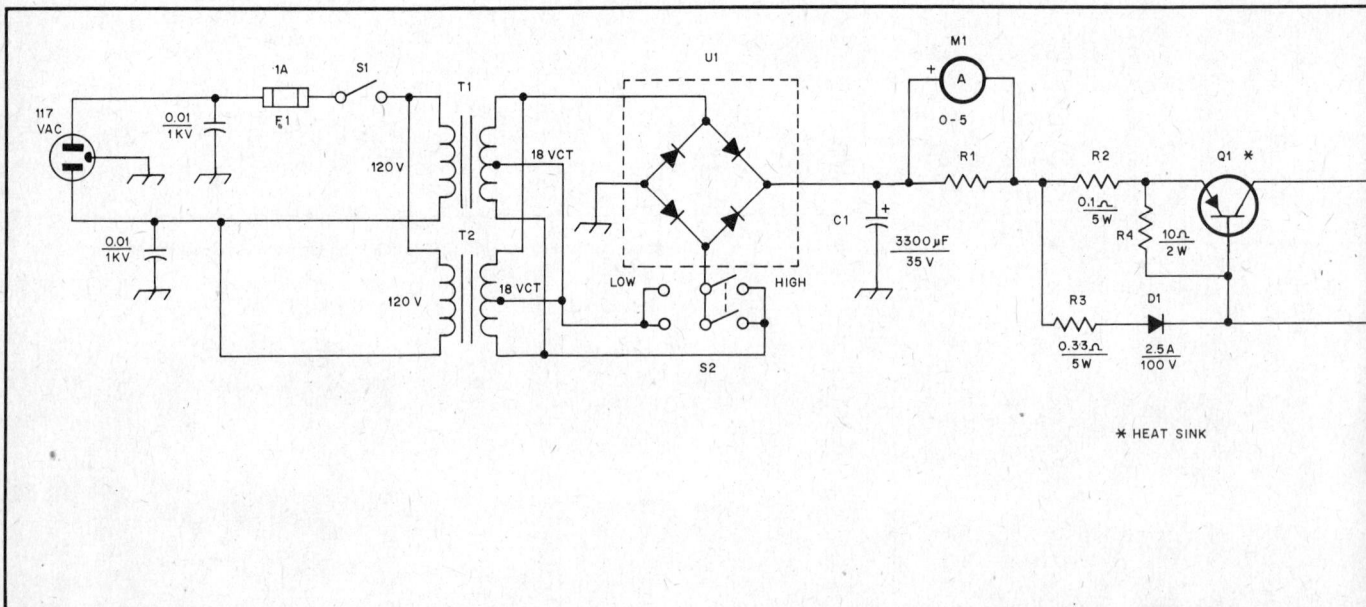

Fig. 39 — Schematic diagram of the 5-ampere power supply. Component designations on the schematic diagram but not shown in the parts list are for text or placement-guide reference only.

C1 — 3300 μF, 35 volt, axial leads.
J1, J2 — Binding posts.
M1 — 0-50 mA, Calectro DI-914 or equiv.
M2 — 0-15 volt, Calectro DI-920 or equiv.

Q1 — Silicon PNP power, Radio Shack 276-2043 or equiv.
R1 — Meter shunt, 13 inches no. 22 enameled wire wound on a high-value, 1-watt resistor.

(Resistor used only as a form for the wire).
R6 — 2500 ohms, 2 watts, panel mount.
S1 — Spst, toggle.
S2 — Dpdt, toggle. Both sections connected in parallel.

Fig. 40 — Inside view of the power supply. Component placement is not at all critical, however the layout shown here provides a neat appearance.

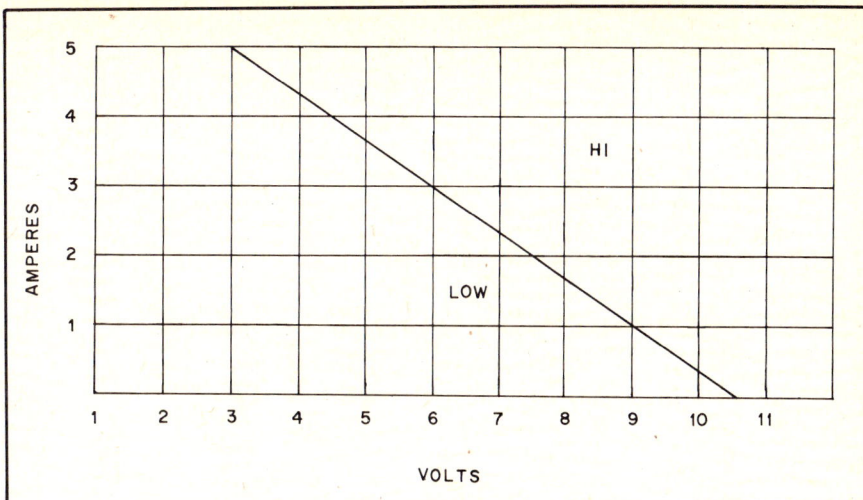

Fig. 41 — For voltage and current requirements that fall to the left of the diagonal line, the power supply may be operated in the LO mode. Pass-transistor dissipation will be reduced when the supply is operated in this manner.

put noise to an acceptable level. What constitutes an acceptable level depends on the electronic environment and the sensitivity of the load. In general, a digital computer would be far more tolerant of a noisy power supply than would be a frequency synthesizer or a small-signal audio amplifier. Also, switching supplies are not as forgiving as linear supplies in terms of component substitutions and design shortcuts. One can assemble a linear power supply from a collection of available components without a rigorous design procedure and often achieve a workable (if not optimum) result. An acceptable method is to power up the unit, check ripple, noise, line and load regulation, and, finally, the heat sink temperature. Constructing and testing a switching supply is not as straightforward

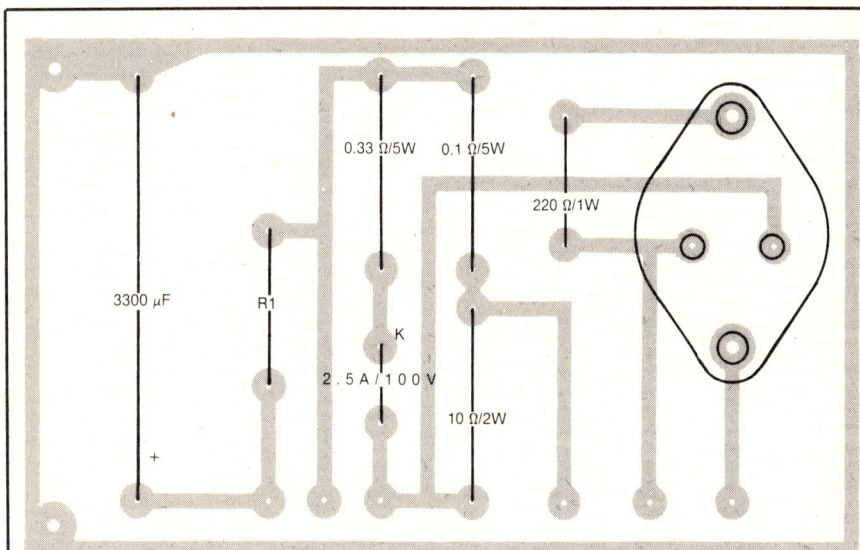

(A)

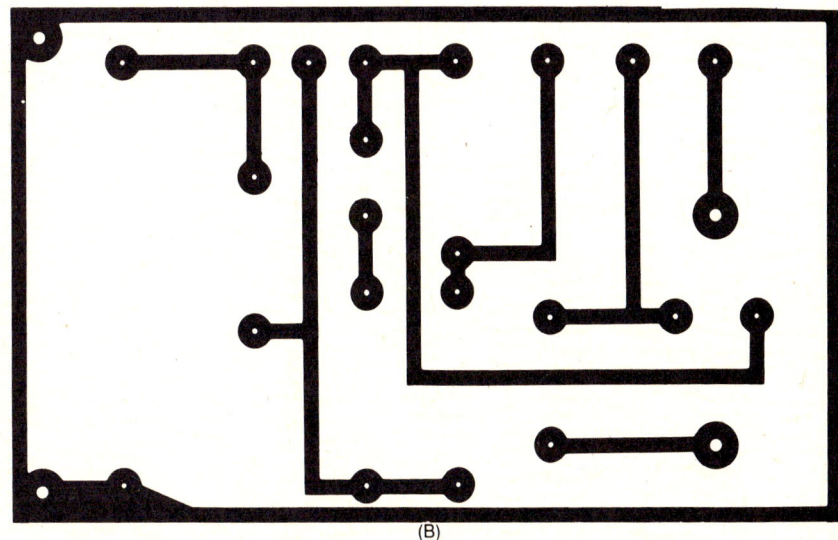

(B)

Fig. 42 — At A, circuit board layout pattern as viewed from the component side of the board. B is the full-scale etching pattern for the power supply circuit board. Black areas represent unetched copper as seen from the foil side of the board.

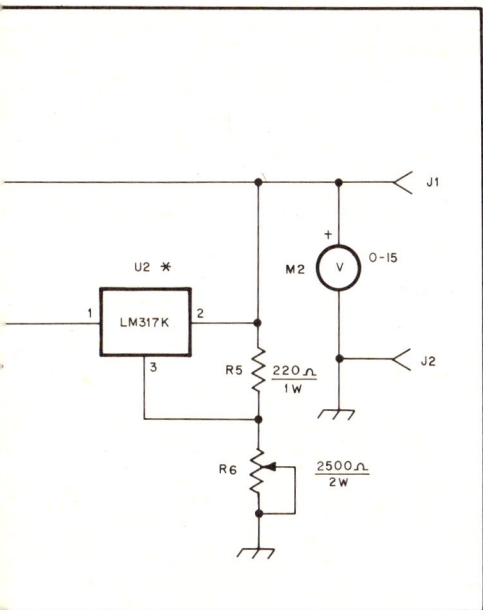

T1, T2 — 117-V primary, 18-V ct secondary. Radio Shack 273-1515 or equiv.
U1 — Bridge-rectifier assembly, 50 V, 25 A.
U2 — Regulator, LM-317K

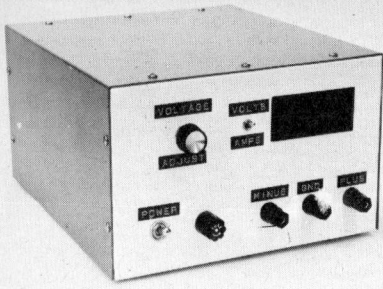

Fig. 43 — Front view of the deluxe 5- to 25-volt, 5-ampere power supply.

Fig. 45 — Interior view of the deluxe power supply.

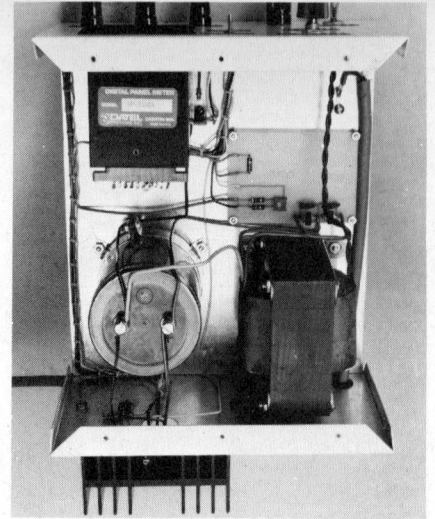

because the operation is complex and the failure modes can be subtle.

Fig. 46 illustrates the switching regulator concept. While the switch is closed (as in A of Fig. 46) the inductor current builds up through the load resistor, storing energy. Note the polarity of the voltages. The diode is reverse-biased when the switch is closed. After the current has built up to the maximum value drawn by the load, the switch may be opened, as shown in Fig. 46B. The inductor tends to resist any change in current flow. When the switch is opened, the interruption of battery current causes the magnetic field in the inductor to collapse, inducing a large emf. The polarity of the emf is such to maintain current in the original direction, so the armature of the open switch becomes negative with respect to the load end of the inductor. Without

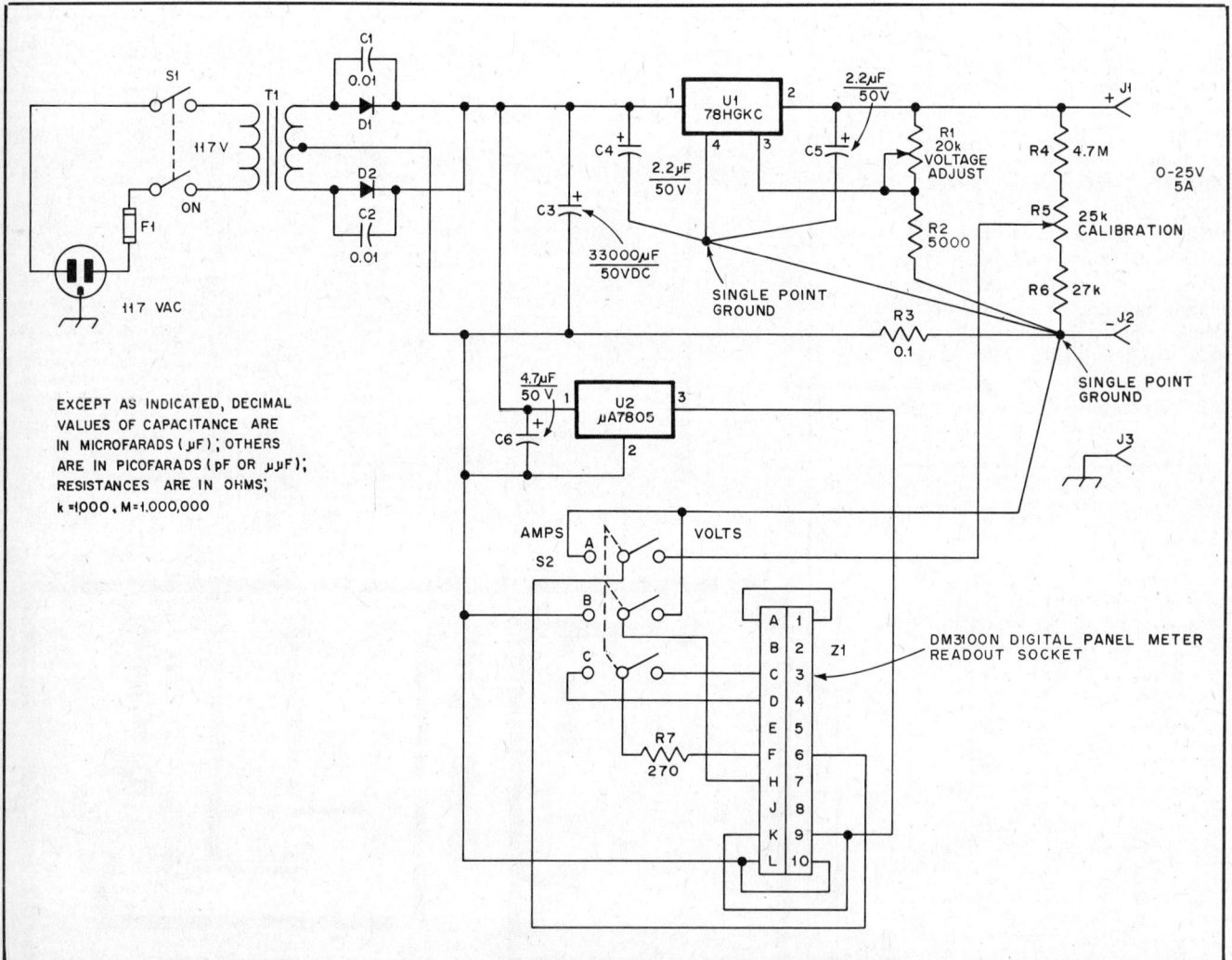

Fig. 44 — Schematic diagram of the deluxe power supply. All resistors are half-watt carbon types unless noted otherwise. Capacitors are disc ceramic unless noted otherwise. Numbered components not appearing in the parts list are for text reference only.

C3 — Electrolytic capacitor, 34,000 μF, 50 V. Sprague 36D343G050DF2A or equiv.
C4, C5 — Tantalum capacitor, 2.2 μF, 50 V.
C6 — Tantalum capacitor, 4.7 μF, 50 V.
D1, D2 — Silicon rectifier, 100 V, 12 A.
F1 — Fuse, 2 A.

J1-J3, incl. — Binding post.
R1 — Potentiometer, 20-kΩ, linear, 10 turn. Clarostat type 731A or equiv.
R2 — Resistor, 0.1Ω, circuit board mount.
S1 — Toggle switch, dpst.
S2 — Toggle switch, 3pdt.
T1 — Power transformer; primary 117 V,

secondary 36 V ct, 6 A. Stancor P-8674 or equiv.
U1 — Regulator, Fairchild 78HGKC or equiv.
U2 — Regulator, μ A7805 or equiv.
Z1 — Digital panel meter. Datel DM3100N or equiv.

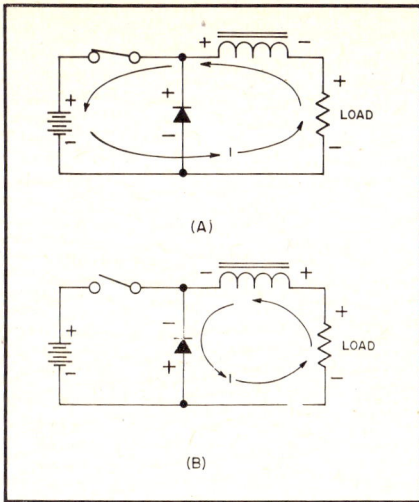

Fig. 46 — Inductive flyback switching circuit used in voltage regulators. At A, the switch is closed, charging the inductor through the load resistor. Opening the switch as in B causes the inductor voltage to reverse polarity as it attempts to maintain current through the load. The discharge path is through the diode. Actual electron flow is depicted by the arrows.

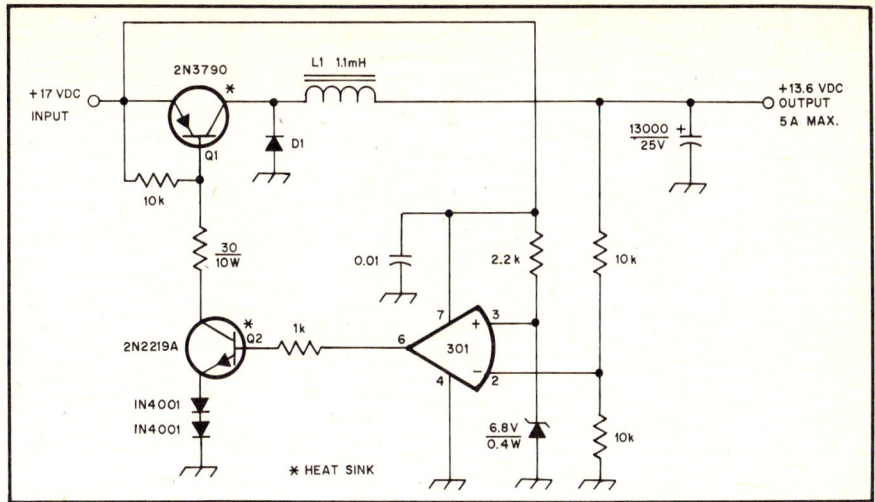

Fig. 47 — A switching regulator suitable for use with medium-power amateur transceivers. At full load the switching rate is approximately 50 kHz. Resistance values are in ohms and capacitance values are in microfarads.

D1 — 6A, 600 PIV, anode common to case, ECG 5863 or equiv.
L1 — 33 turns no. 14 enam. on Amidon FT240-43 toroidal core.

Q1 — Heat sink — 4 × 3 inches (102 × 76 mm) with 1/2-inch (13-mm) fins extruded from flat surfaces. Radio Shack 276-1361 or equiv.
Q2 — Heat sink — TO-5 clip-on unit.

the diode in the circuit, the emf would increase until it jumped across the switch in an effort to maintain the current. However, the diode (called a "catch diode") becomes forward biased by the inductor emf polarity reversal, clamping the switched end of the inductor essentially to ground. The negative-going transient at the switched end is transformed into a positive-going spike at the load. If all components were ideal, no transients would occur; the inductor would simply maintain a constant current through (and therefore a constant voltage across) the load. In practice, a large capacitor is connected across the load to bypass the switching transients.

To achieve regulation, a switching type of power supply must have a voltage reference and feedback loop similar to those employed in a linear regulator. A transistor usually serves as the control element (switch), but some of the more ad-

vanced designs use thyristors in the ac primary circuit, completely obviating the usual (heavy and lossy) transformer-rectifier circuit.

The design of a switching regulator is not trivial. Many factors enter into the calculations, including the magnetic properties of the inductor and the dynamic switching characteristics of the semiconductors. The switching performance is important to avoid "hot spots." This phenomenon results from too slow a device being used to switch heavy current. Sometimes the current is concentrated in a small portion of the junction rather than distributed evenly. This effect is manifested by the sudden failure of the circuit after a period of perfect operation. The failed transistor is usually cool to the touch.

Drawn in Fig. 47 is a practical 13.6-volt, 5-ampere switching power supply suitable for use with transceivers in the 25-watt

output class. Circuit simplicity and easy parts acquisition were the major design goals. The supply was breadboarded and tested in the ARRL laboratory. A 10-watt output 220-MHz fm transceiver was used as a load with no apparent degradation of the rf output spectral purity. The line and load regulation are acceptable, and the unit is slightly more efficient than an equivalent linear supply. With the heat sink specified, the unit can withstand 3-minute fm transmissions interspersed with 1-minute listening periods. Because of the simple circuit configuration, conventional pre-regulation components are used. Therefore, this power supply offers no particular economic advantage over one using linear regulator. This project was intended to demonstrate the switching principle as applied to voltage regulation, and it serves that purpose well. However, it is not in any way represented to be a state-of-the-art design.

Chapter 6

HF Transmitting

Even though some modern transmitters and transceivers contain only solid-state devices, it is still practical to use a hybrid circuit that contains a mixture of tubes and active semiconductor stages. Typically, the unit has transistors, diodes and ICs up to the driver stage of the transmitter. At that point one will find a tube driver which is used to supply rf power to a tube type of amplifier. The latter might consist of a pair of 6146Bs or two sweep tubes.

The principal advantage of tube amplifiers is that they are somewhat less subject to damage from excessive drive levels and mismatched loads. However, a properly designed solid-state driver and PA section should be immune to output mismatch damage, provided an SWR-protection circuit has been included in the transmitter. A solid-state amplifier is slightly more difficult to design and have work correctly than is a tube amplifier of equivalent power. This is because purity of emissions is harder to achieve when transistor power stages are employed. Transistors generate considerably more harmonic energy than tubes do, and the former are prone to self-oscillation at lf, vlf and audio frequencies unless some careful design work is done. This is not generally true of tube amplifiers.

If one is to ignore the foregoing problems and concentrate mainly on cost and convenience, transistors may have the edge over tubes. A 13.6-volt design can be operated directly from an automotive or solar-electric supply, whereas a tube amplifier requires a high-voltage power supply for mobile, portable and fixed-station use. When an ac power source is required, the cost of a high-voltage, medium-current supply for tubes versus a low-voltage, high-current power source for transistors is similar, provided new components are used in both. At power-output levels in excess of approximately 150 watts the transistor-amplifier power

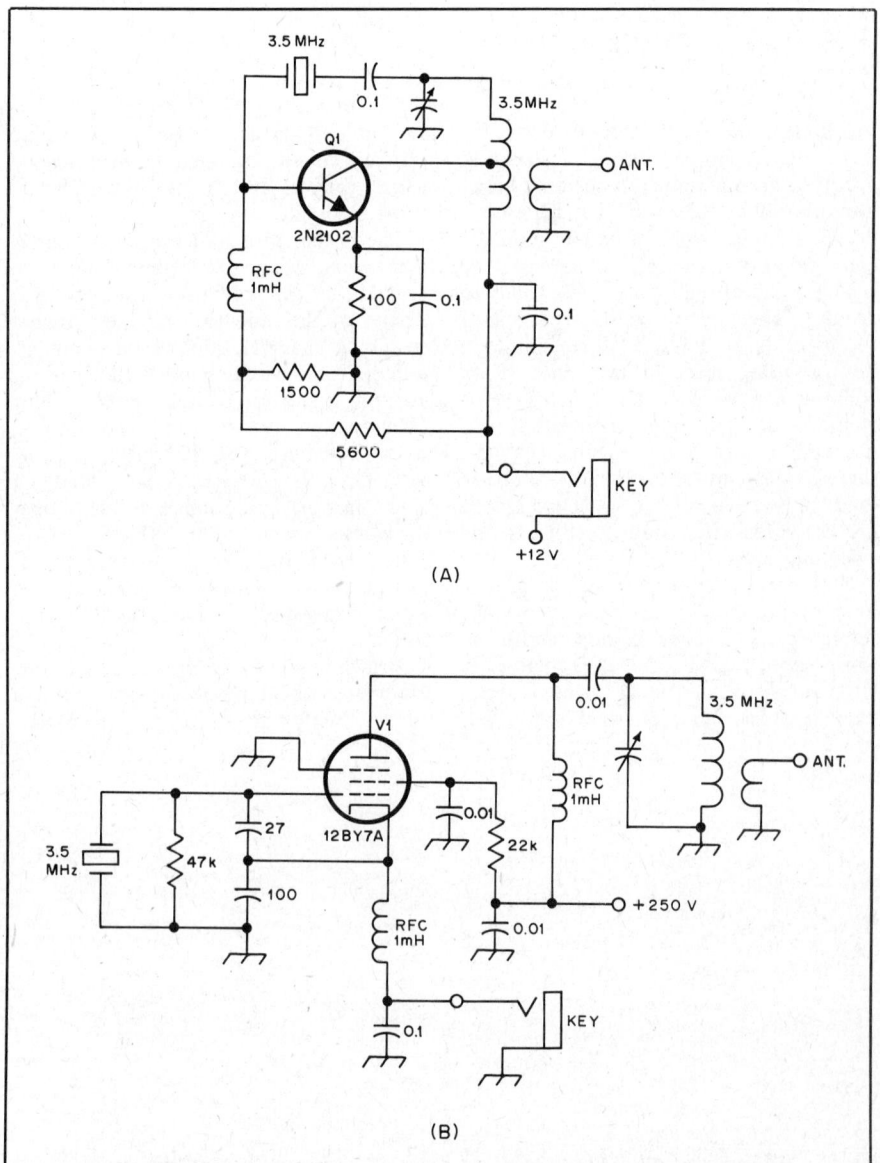

Fig. 1 — A transistor oscillator is shown at A. The example at B illustrates a tube type of crystal oscillator.

supply becomes rather expensive because of complex regulator-circuit requirements. For this reason it is the choice of most amateurs to utilize vacuum tubes in high-power hf and vhf amplifiers. The number of power transistors required (plus combiners) to generate a 1-kW signal may run considerably higher in cost than a tube or tubes for an amplifier of equivalent power. The price of large heat sinks versus a cooling fan may place the solid-state amplifier in a prohibitive class also.

The decision to buy or build a transmitter is founded on some basic considerations: Cost compared to features; professional equipment appearance contrasted to that of homemade apparatus; the knowledge and satisfaction gained from building equipment, as weighed against buying store-bought gear and simply becoming an operator. The judgment must fit the amateur's objectives and affluence. Home-built transmitters are usually easier to service than commercial ones because the builder knows the circuit layout and how each stage functions. Furthermore, the cost of maintenance is markedly lower for homemade equipment than for most factory-built gear. But the greatest significance to home-built circuits is the *knowledge* gained from constructing a project and the pride that goes with using it on the air!

Frequency Generation

The most basic type of transmitter is one which contains a single stage, is crystal controlled, and is designed for cw operation. A circuit example is given in Fig. 1. This kind of transmitter is not especially suitable for use on the air because it is somewhat inefficient and is prone to generating a chirpy cw signal unless loaded rather lightly. But, the same circuit is entirely acceptable when followed by an isolating stage (buffer/amplifier) as shown in Fig. 2. The second stage not only builds up the power level, but it gives the oscillator a relatively constant load to look into. The latter helps to prevent oscillator pulling and attendant chirping of the cw note.

Fig. 3 illustrates the basic types of transmitters for cw and RTTY work. The drawing at A represents the general circuit given in Fig. 2. Illustration B is an expansion of that circuit and includes a frequency multiplier. A heterodyne type of generator, which is currently popular for multiband transmitters and transceivers, is shown as the exciter section of a transmitter in drawing C. A frequency synthesizer is shown as the rf generator at D.

For operation on a-m, any of the lineups given in Fig. 3 are suitable, provided a modulator is added. It is used to modulate the operating voltage to the PA stage, or in some designs the operating voltage to the PA and the stage immediately before it.

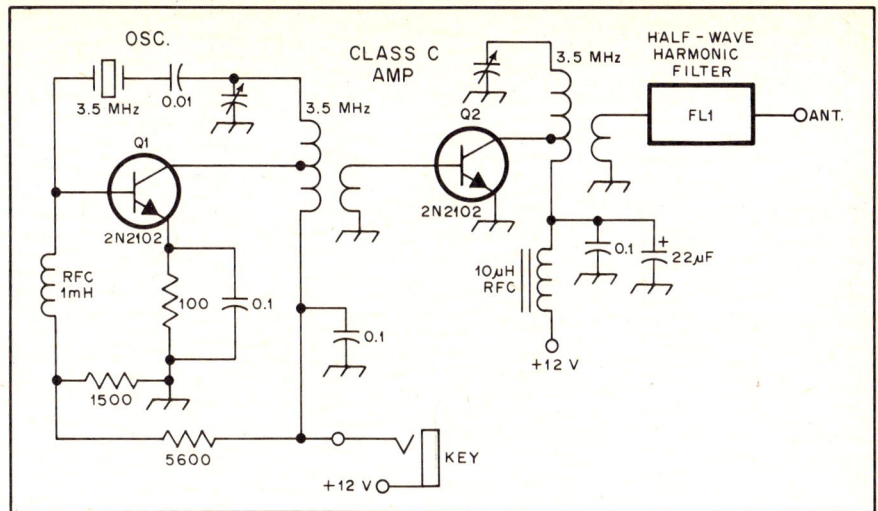

Fig. 2 — Circuit example of a simple, solid-state cw transmitter.

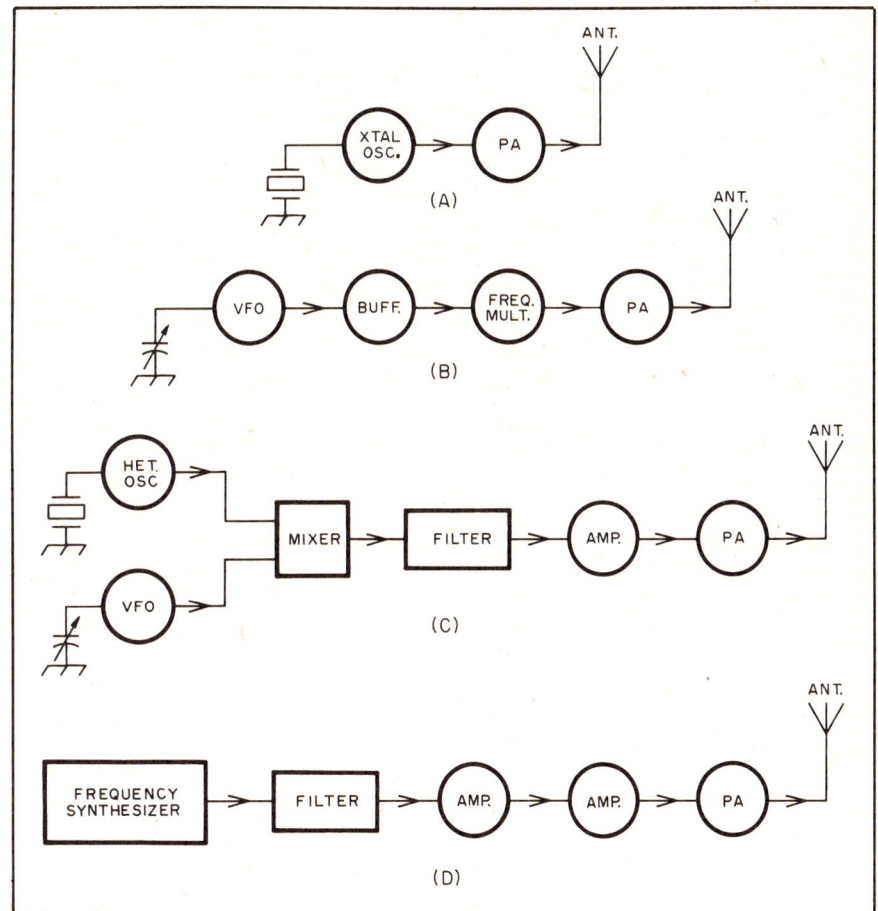

Fig. 3 — Block diagrams of various transmitters which use specific frequency-generation methods.

The block diagram of Fig. 4 outlines the functional stages of a ssb transmitter. Z1 can be a simple VFO, a heterodyne generator (Fig. 3C) or a frequency synthesizer. The essential difference between this type of transmitter and one that would be used for cw/RTTY is that the rf amplifiers must operate in the Class A, AB or B mode (linear) rather than the Class C mode which is suitable for cw work. However, linear amplifiers are entirely satisfactory for any transmission mode at a sacrifice in efficiency. Once the ssb signal is generated it can not be passed through a frequency multiplier. All post-filter stages must operate straight through. Class C amplifiers are generally used in fm transmitters as well as in cw and RTTY transmitters. Fm operators who are heard to say, "I'm running my transceiver into a

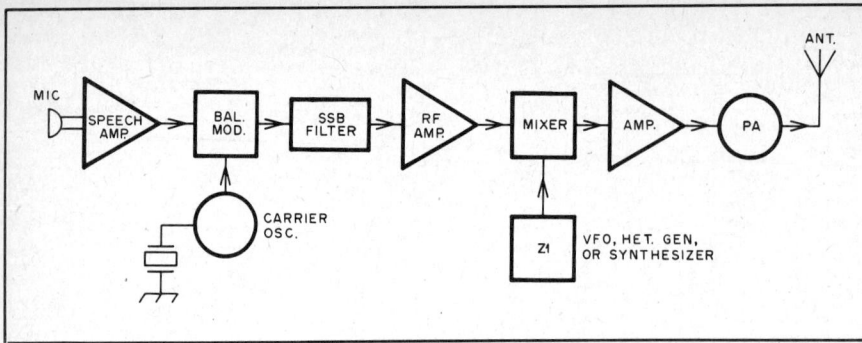

Fig. 4 — Block diagram of a heterodyne type of ssb transmitter.

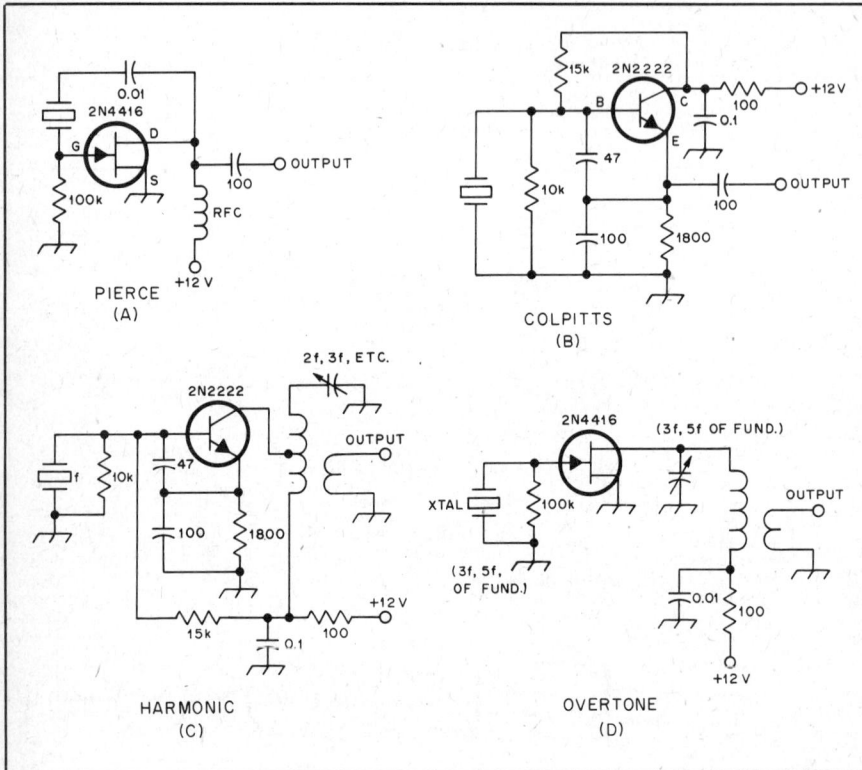

Fig. 5 — Four types of popular solid-state crystal oscillators.

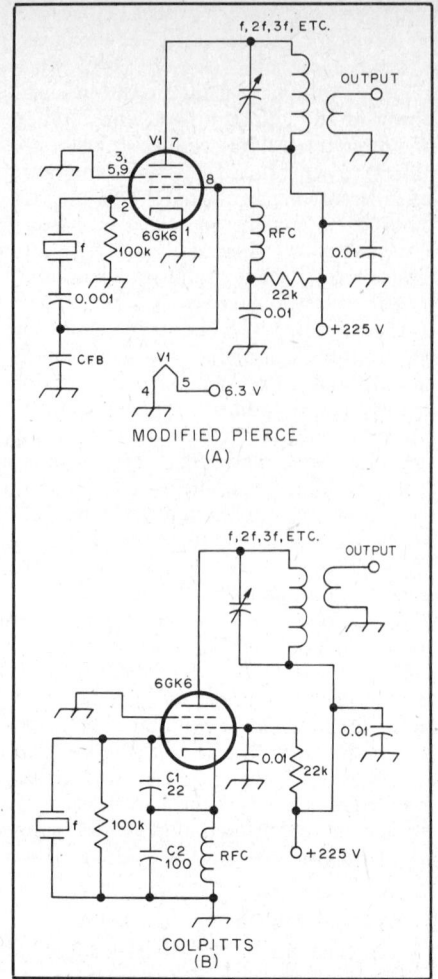

Fig. 6 — Two common tube types of crystal oscillators.

linear," are missing the technical facts: The amplifier is probably a Class C one, which is very *non-linear* in operation.

Crystal Oscillators

A crystal-controlled oscillator uses a piece of quartz which has been ground to a particular thickness, length and width. For the most part, the thickness determines the frequency at which the crystal oscillates, irrespective of the stray capacitance in the immediate circuit of the crystal. The stray capacitance does have some effect on the operating frequency, but overall it can be considered minor. The power available from such an oscillator is restricted by the heat (caused by circulating rf current) the crystal can withstand before fracturing. The circulating current is determined by the amount of feedback required to ensure excitation. Excessive heating of the crystal

causes frequency drift. The extent of the drift is related to the manner in which the quartz crystal is cut and the actual heat at a given point in time. It is for these reasons that the amount of feedback used should be held to only that level which provides quick oscillator starting and reliable operation under load. The power necessary to excite a successive stage properly can be built up inexpensively by means of low-level amplifiers.

The active element in an oscillator can be a tube, transistor or IC. Some common examples of solid-state oscillators are shown in Fig. 5. A triode Pierce oscillator which employs a JFET is illustrated at A. A bipolar transistor is used at B to form a Colpitts oscillator. The example in Fig. 5C shows a means by which to extract the harmonic of a crystal by tuning the collector circuit to the desired harmonic. Unless a bandpass filter is used after the

tuned circuit, various harmonics of the crystal frequency will appear in the output. Therefore, if good spectral purity is desired it is necessary to use a double-tuned collector tank to obtain a bandpass characteristic, or to employ the tank circuit shown and follow it with a harmonic filter.

An overtone oscillator is depicted in Fig. 5 at D. The crystal oscillates at an odd multiple of the fundamental cut — usually the third or fifth harmonic. In this example the drain tank is tuned approximately to the desired overtone. Oscillation will begin when the tank is tuned slightly above the overtone frequency. A high-Q tuned circuit is necessary.

Vacuum-tube crystal oscillators are presented in Fig. 6. A modified Pierce oscillator is shown at A. In this case the screen grid of V1 functions as the plate of a triode tube. Feedback is between the screen and control grids. C_{fb} may be required to ensure the desired feedback voltage. In a typical oscillator the value of C_{fb} will range from 10 to 100 pF for oscillators operating from 1.8 to 20 MHz. At lower frequencies the feedback capacitor may require a higher value.

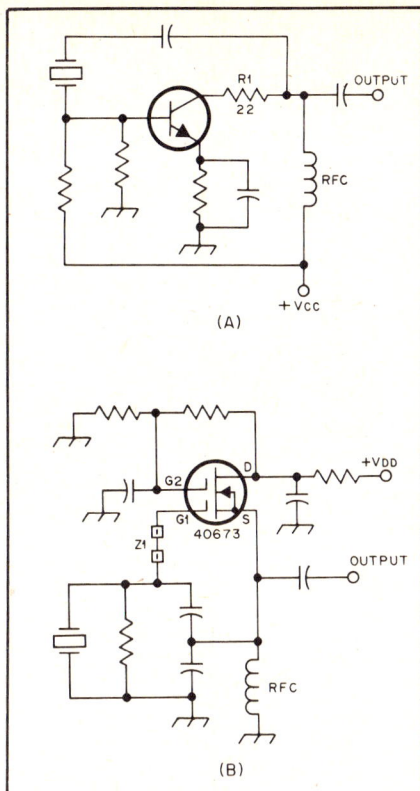

Fig. 7 — Two methods for suppressing vhf and uhf parasitic oscillations. R1 at A damps the parasitics and Z1 at B (ferrite beads) serves that purpose.

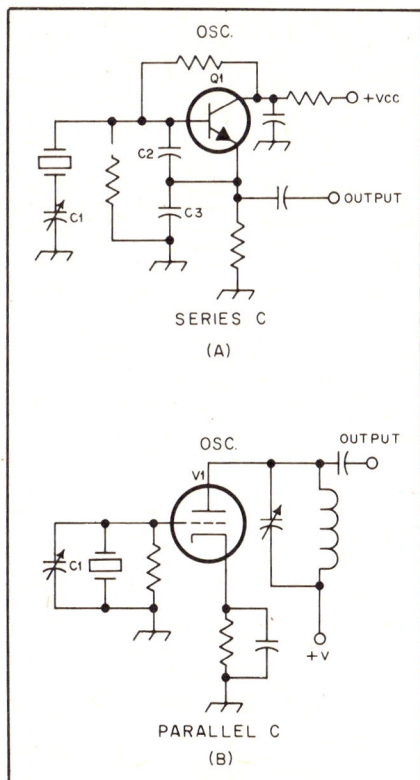

Fig. 8 — The crystal-oscillator operating frequency can be shifted slightly by means of trimmer capacitors as shown at A and B. A series hookup (A) is used with transistors to help compensate for the relatively high input capacitance of the transistor.

A Colpitts style of tube oscillator is illustrated in Fig. 6 at B. The feedback is between the grid and cathode by means of a capacitive divider (C1 and C2). The plate tank can be tuned to the crystal frequency or harmonics thereof. In the interest of good oscillator stability it is suggested that the supply voltage to the circuits of Figs. 5 and 6 be regulated. This is especially significant in the case of harmonic or overtone oscillators where small amounts of drift are multiplied by the chosen harmonic factor.

The usual cause of erratic oscillation, or no oscillation at all, is excessive loading on the oscillator output (succeeding stage of circuit), insufficient feedback, or a sluggish crystal. Concerning the latter, a crystal which is not ground to a uniform thickness and feathered carefully around the edges may be difficult to make oscillate. Attempts by inexperienced amateurs to grind their own crystals may lead to this condition.

Some crystal oscillators develop unwanted vhf self-oscillations (parasitics) even though the circuits may be functioning normally otherwise. The result will be a vhf waveform superimposed on the desired output waveform when the rf voltage is viewed by means of an oscilloscope. Parasitics can cause TVI and specific problems elsewhere in the circuit with which the oscillator is used. Two simple methods for preventing vhf parasitics are shown in Fig. 7. The technique at A calls for the insertion of a low-value resistance (R1) in the collector lead as close to the transistor body as possible. Typical resistance values are 10 to 27 ohms. The damping action of the resistor inhibits vhf oscillation. An alternative to the use of resistance for swamping vhf oscillation is illustrated at B in Fig. 7. One or two high-mu miniature ferrite beads ($\mu_i = 950$) are placed near the transistor body in the lead to gate 1. The beads can be used in the drain lead when a tuned circuit or rf choke are used in that part of the circuit. However, when the drain is at ac ground, as shown at B, it does not constitute part of the feedback circuit. Ferrite beads can be used in the base or collector lead of the circuit of Fig. 7A rather than employing R1. Similarly, R1 can be used at gate 1 of the oscillator.

It is necessary in some applications of crystal oscillators to ensure spot accuracy of the operating frequency. Various reactances are present in most oscillator circuits, causing the operating frequency to differ somewhat from that for which the crystal was manufactured. Addition of a trimmer capacitor will permit "rubbering" the crystal to a specified frequency within its range. This procedure is sometimes referred to as "netting" a crystal.

Fig. 8 shows two circuits in which a trimmer capacitor might be used to compensate for differences in the operating frequency of the oscillator. At A the

series capacitor (C1) is connected between the low side of the crystal and ground. The series hookup is used to help offset the high input capacitance of the oscillator. The input capacitance consists of the series value of feedback capacitors C2 and C3 plus the input capacitance (C_{in}) of Q1. Conversely, the input capacitance of the circuit at B in Fig. 8 is quite low because a triode tube is employed. In this kind of circuit the trimmer capacitor is used in a parallel manner as shown. The choice between series and parallel trimming will depend on the active device used and the amount of input capacitance present. This rule applies to tube oscillators as well as those which use transistors.

Crystal Switching

Although several crystals for a single oscillator can be selected by mechanical means, a switch must be contained in the rf path. This can impose severe restrictions on the layout of a piece of equipment. Furthermore, mechanical switches normally require that they be operated from the front panel of the transmitter or receiver. That type of format complicates the remote operation of such a unit. Also, the switch leads can introduce unwanted reactances in the crystal circuit. A better technique is illustrated in Fig. 9, where D1 and D2 — high-speed silicon switching diodes — are used to select one of two or more crystals from some remote point. As operating voltage is applied to one of the diodes by means of S1, it is forward biased into "hard" conduction, thereby completing the circuit between the crystal trimmer and ground. Some schemes actually call for reverse-biasing the unused diode or diodes when they are not activated. This ensures almost complete cutoff, which may not be easy to achieve in the circuit shown because of the existing rf voltage on the anodes of D1 and D2.

Variable-Frequency Oscillators

The theory and general application of variable-frequency oscillators is treated in chapter 8. The circuit principles are the same regardless of the VFO application.

Some additional considerations pertain to the use of VFOs in transmitters as compared to a VFO contained in a receiver. Generally, heating of the interior of a transmitter cabinet is greater than in a receiver. This is because considerably more power is being dissipated in the former. Therefore, greater care must be given to oscillator long-term stability. Temperature-compensating capacitors are often needed in the frequency-determining portion of the oscillator to level off the long-term stability factor. Some oscillators are designed for use with a temperature-control oven for the purpose of maintaining a relatively constant ambient temperature in the oscillator compartment — even while the equipment

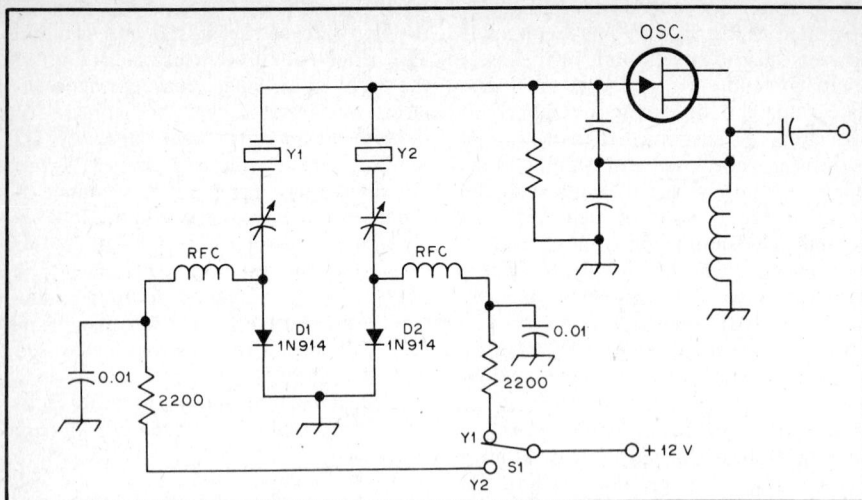

Fig. 9 — A method for selecting one of two (or several) crystals by means of diode switching. D1 and D2 are the switches.

is otherwise turned off.

Another design matter related to a transmitter-contained VFO is rf shielding of the oscillator and the attendant low-level buffer/isolation stages that follow it. Fairly high levels of stray rf can be present in a transmitter and some of that energy may migrate to the oscillator section by means of stray radiation or conduction along wiring leads or circuit-board elements. Thus, it is important to provide as much physical and electrical isolation as possible. The VFO should be housed in a rigid metal box. All dc leads entering the enclosure require rf-decoupling networks that are effective at all frequencies involved in the transmitter design. The VFO box needs to be fastened securely to the metal chassis on which it rests to ensure good electrical contact. Excessive stray rf entering the VFO circuitry can cause severe instability and erratic oscillator operation.

Fundamental Stability Considerations

Apart from the recommendations given in the foregoing text for VFO stability, there are some specific measures which must be taken when designing an oscillator of this type. The form upon which the VFO coil is wound is of special significance with regard to stability. Ideally, the use of magnetic core material should be avoided. Therefore, powdered iron, brass, copper and ferrite slugs, or toroid cores for that matter, are not recommended when high stability is required. The reason is that the properties of such core materials are affected by changes in temperature and can cause a dramatic shift in the value of inductance which might not occur if an air-core coil was employed. Furthermore, some styles of slug-tuned inductors are subject to mechanical instability in the presence of vibration. This can cause severe frequency jumping and a frequent need for recalibration of the VFO readout versus operating frequency.

Regardless of the format selected for the VFO coil, the finished product should be coated with two or three applications of polystyrene cement (Q dope) or similar low-loss dopant. This will keep the coil turns secured in a permanent position — an aid to mechanical stability.

The VFO coil should be mounted well away from nearby conducting objects (cabinet walls, shield cans, and so on) to prevent frequency shifts which are likely to occur if the chassis or cabinet are stressed during routine handling or mobile operation. Movement of the chassis, cabinet walls and other nearby conductive objects can (if the coil is close by) change the coil inductance. Furthermore, the proximity effects of the conductive objects present an undefined value of capacitance between the coil and these objects. Changes in spacing will alter that capacitance, causing frequency shifts of an abrupt nature.

It follows that all forms of mechanical stability are of paramount importance if the VFO is to be of "solid" design. Thus, the trimmer or padder capacitors that are used in the circuit should be capable of remaining at their preset values despite temperature changes and vibration. For this reason it is not wise to utilize ceramic or mica trimmers. Air-dielectric variable capacitors of the pc-board-mount subminiature type are recommended.

The main tuning element (capacitor or permeability tuner) needs to have substantial rigidity. It should be mounted in place in a secure manner. Variable capacitors used as main-tuning elements should be of the double-bearing variety. They should rotate easily (minimum torque) in order to minimize mechanical stress of the VFO assembly when they are being adjusted from the front panel of the equipment. Variable capacitors with plated brass plates are preferred over those which have aluminum plates. The aluminum is more subject to physical changes in the presence of temperature variations

than is the case with brass. The VFO tuning-capacitor rotor must be grounded at both ends as a preventive measure against instability. Some designers have found that a 1/8- to 1/4-inch (3.2- to 6.4-mm) thick piece of aluminum or steel plate serves as an excellent base for the VFO assembly. It greatly reduces instability which can be caused by stress on the main chassis of the equipment. The VFO module can be installed on shock mounts to enhance stability during mobile operation.

Concerning Electrical Stability

Apart from the mechical considerations just discussed, the relative quality of the components used in a VFO circuit is of great importance. Fig. 10 contains three illustrations of basic solid-state tunable oscillators which are suggested for amateur applications. The numbered components have a direct bearing on the short- and long-term stability of the VFO. That is, the type of component used at each specified circuit point must be selected with stability foremost in mind. The fixed-value capacitors, except for the drain bypass, should be temperature-stable types. NPO ceramic capacitors are recommended for frequencies up to approximately 10 MHz. A second choice is the silver-mica capacitor (dipped or plain versions). Silver micas tend to have some unusual drift characteristics when subjected to changes in ambient temperature. Some increase in value while others decrease. Still others are relatively stable. It is often necessary to experiment with several units of a given capacitance value in an effort to select a group of capacitors that are suitably temperature stable. The same is not true of polystyrene capacitors. When used with the commonly available slug-tuned coils, the temperature characteristics of the polystyrene capacitors and those of the inductor tend to cancel each other. This results in excellent frequency stability. If toroidal cores are used they should be made of SF-type powdered iron material, (Amidon mix 6). This material has a low temperature coefficient and when used with NPO type ceramic capacitors produces very low drift oscillators. Ordinary disc-ceramic capacitors are unsuitable for use in stable VFOs. Those with specified temperature characteristics (N150 and similar) are useful, however, in compensating for drift.

The circuit of Fig. 10A is capable of very stable operation if polystyrene capacitors are used at C3 through C8, inclusive. A test model for 1.8 to 2.0 MHz exhibited only 1 hertz of drift from a cold start to a period some two hours later. Ambient temperature was 25°C. Q1 can be any high-g_m JFET for use at vhf or uhf. Capacitors C1 through C4 are used in parallel as a means to distribute the rf current among them. A single fixed-value capacitor in that part of the circuit would tend to change value versus time because of the rf heating within it. Therefore, a distinct advantage exists when several capacitors can be used in parallel at such points in a VFO circuit. The same concept is generally true of C5, C6 and C7. In the interest of stability, C5 should be the smallest value that will permit reliable oscillation. Feedback capacitors C6 and C7 are typically the same value and have an X_c of roughly 60. Therefore, a suitable value for a 1.9-MHz VFO would be 1500 pF.

C8 of Fig. 10A should be the smallest capacitance value practical with respect to ample oscillator drive to the succeeding stage — generally a buffer or amplifier. The smaller the value of C8, the less the chance for oscillator pulling during load changes. D1 is a gate-clamping diode for controlling the bias of the FET. The function of this stabilizing diode is treated in chapter four. Basically, it limits the positive swing of the sine wave. This action restricts the change in Q1 junction capacitance to minimize harmonic generation and changes in the amount of C associated with L1.

The reactance of RFC can be on the order of 10-kΩ. The drain bypass, C9, should have a maximum X_c of 10 ohms to ensure effective bypassing at the operating frequency. Ideally, an X_c of 1 ohm would be used (0.1 μF at 1.5 MHz). D2 is used to provide 9.1 volts, regulated, at the drain of Q1. Lower operating voltages aid stability through reduced rf-current heating, but at the expense of reduced oscillator output.

A Hartley oscillator is shown in Fig. 10B. This circuit offers good stability also, and is one of the better circuits to use when the tank is parallel tuned. The tap on L1 is usually between 10 and 25 percent of the total coil turns, tapped above the grounded end. This ensures adequate feedback for reliable oscillation. The higher the FET g_m the lower the feedback needed. Only that amount of feedback which is necessary to provide oscillation should be used: Excessive feedback will cause instability and prohibitive rf heating of the components. Most of the rules for the circuit of Fig. 10A apply to the oscillator in Fig. 10B.

Parallel tuning of the kind used in Fig. 10B and C are suitable for use below, say, 6 MHz, although the circuit at B can be used successfully into the vhf region. However, the Colpitts oscillators of A and C in Fig. 10 have large amounts of shunt capacitance caused by C6 and C7 of A, and C5 and C6 of C. The smaller the coupling capacitor between L1 and the gate, the less pronounced this effect is. The net result is a relatively small value of inductance at L1, especially with respect to Fig. 10C, which lowers the tank impedance and may prevent oscillation (high C-to-L ratio). The series-tuned circuit of Fig. 10A solves the shunt-C problem nicely by requiring considerably

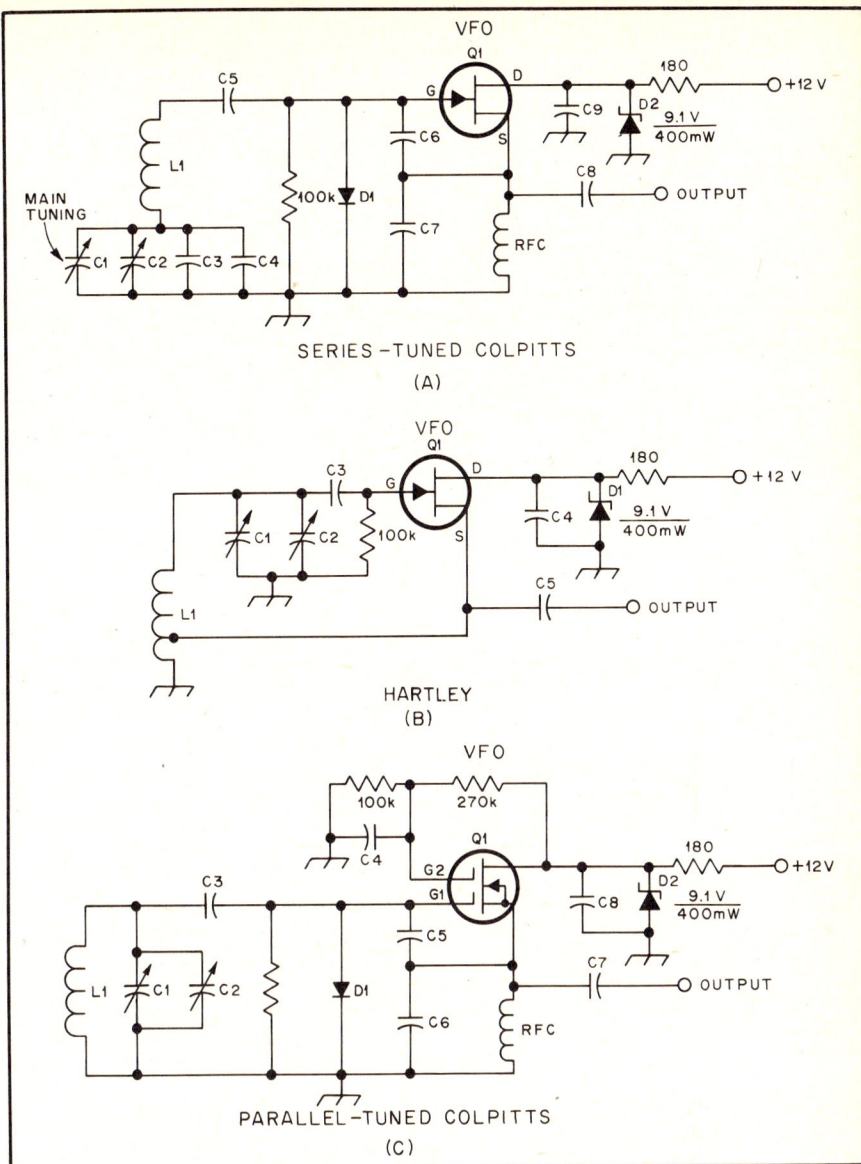

Fig. 10 — Three common types of VFOs for use in receivers and transmitters.

greater inductance at L1 than would be acceptable in the circuit of Fig. 10C. The circuit at A resembles the popular "Clapp" circuit of the early 1950s.

A suitable transistor for Q1 of Fig. 10C is an RCA 40673. The Texas Instruments 3N211 is also ideal, as it has an extremely high g_m — approximately 30,000. Dual-gate MOSFETs are suitable for the circuits of Fig. 10A and B if biased as shown at C. Also, they can be used as single-gate FETs by simply connecting gates 1 and 2 together. No external bias is required if this is done. Gate 2 of Q1 (Fig. 10C) should be bypassed with a low-reactance capacitor (C4), as is the rule for the drain bypassing of all three examples given in Fig. 10.

Bipolar transistors are satisfactory for use in the three VFOs just discussed. The principal disadvantage attendant to the use of bipolars in these circuits is the low base impedance and higher device input

capacitance. Most FETs exhibit an input C of approximately 5 pF, but many bipolar transistors have a substantially higher capacitance, which tends to complicate a VFO design for the higher operating frequencies. The uhf small-signal transistors, such as the 2N5179, are best suited to the circuits under discussion.

Load Isolation for VFOs

Load changes after the oscillator have a pronounced effect on the operating frequency. Therefore, it is imperative to provide some form of buffering (isolating stage or stages) between the oscillator and the circuit to which it will be interfaced. The net effect of load changes, however minor, is a change in reactance which causes phase shifts. The latter affects the operating frequency to a considerable degree. Therefore, the more isolating stages which follow the oscillator (up to a

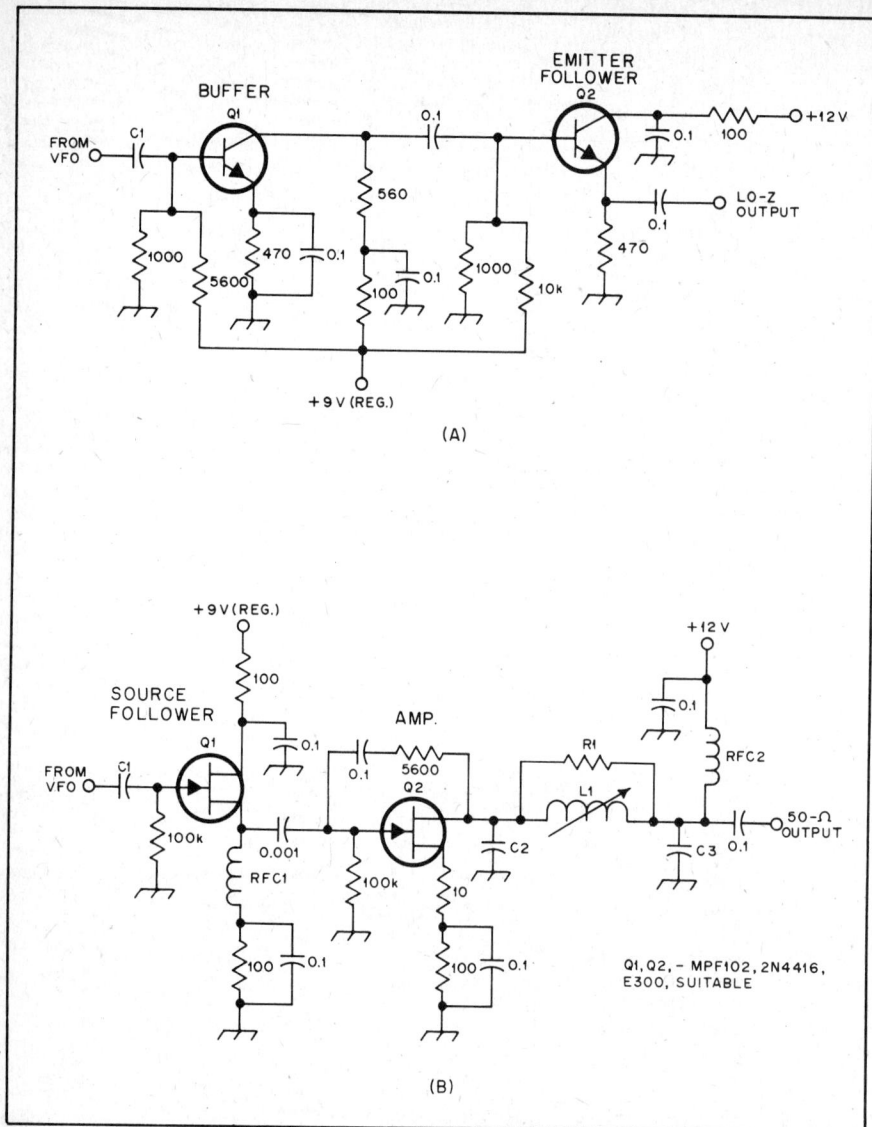

Fig. 11 — VFO buffer and buffer/amplifier sections which provide isolation between the oscillator and the VFO-chain load. The circuit at B is recommended for most applications.

practical number, of course), the less likelihood of load shifts being reflected back to the oscillator.

Buffer stages can perform double duty by affording a measure of rf amplification, as needed. But, care must be taken to avoid introducing narrow-band networks in the buffer/amplifier chain if considerable frequency range is planned, e.g., 5.0 to 5.5 MHz. If suitable broadband characteristics are not inherent in the design, the oscillator-chain output will not be constant across the desired tuning range. This could seriously affect the conversion gain and dynamic range of a receiver mixer, or lower the output of a transmitter in some parts of a given band.

Fig. 11A illustrates a typical RC coupled VFO buffer with broadband response. C1 is selected for minimum coupling to the oscillator, consistent with adequate drive to Q1. Q1 and Q2 should have high f_T and medium beta to ensure a slight rf-voltage gain. Devices such as the 2N2222A and 2N5179 are suggested.

Q2 of Fig. 11A operates as an emitter-follower. The rf-voltage output will be approximately 0.9 of that which is supplied to the base. In a typical VFO chain, using an oscillator such as the one in Fig. 10A, this buffer strip will deliver approximately 1 volt pk-pk across the 470-ohm emitter resistor of Q2.

A somewhat better circuit is offered in Fig. 11B. Q1 is a JFET which has a high input impedance (1-MΩ or greater) by virtue of the FET-device characteristic. This minimizes loading of the oscillator. RFC1 is chosen to resonate broadly with the stray circuit capacitance (roughly 10 pF) at the midrange frequency of the LO chain. Although this does not introduce significant selectivity, it does provide a rising characteristic in the rf-voltage level at the source of Q1.

Q2 functions as a fed-back amplifier with shunt feedback and source degeneration. The feedback stabilizes the amplifier and makes it broadband. The drain tank is designed as a pi network with a loaded

Q of 3. The transformation ratio is on the order of 20:1 (1000-ohm drain to 50-ohm load). R1 is placed across L1 to further broaden the network response. The 50-ohm output level is recommended in the interest of immunity to load changes: The higher the output impedance of a buffer chain the greater the chance for oscillator pulling with load changes. Pk-pk output across C3 should be on the order of 3 volts when using the oscillator of Fig. 10A.

Other VFO Criteria

Apart from the stability considerations just treated, purity of emissions from VFOs is vital to most designers. It is prudent to minimize the harmonic output of a VFO chain and to ensure that vhf parasitic energy is not being generated within the LO system.

The pi-network output circuit of Fig. 11B helps reduce harmonics because it is a low-pass network. Additional filtering can be added at the VFO-chain output by inserting a half-wave filter with a loaded Q of 1 (X_L and $X_C = 50$ for a 50-ohm line).

Vhf parasitics are not uncommon in the oscillator or its buffer stages, especially when high f_T transistors are employed. The best preventive measures are keeping the signal leads as short as possible and adding parasitic suppressors as required. The parasitic energy can be seen as a superimposed sine wave riding on the VFO output waveform when a high-frequency scope is used.

A low-value resistor (10 to 22 ohms) can be placed directly at the gate or base of the oscillator transistor to stop parasitic oscillations. Alternatively, one or two ferrite beads ($950\mu_i$) can be slipped over the gate or base lead to resolve the problem. If vhf oscillations occur in the buffer stages, the same preventive measures can be taken.

VFO noise should be minimized as much as possible. A high-Q oscillator tank will normally limit the noise bandwidth adequately. Resistances placed in the signal path will often cause circuit noise. Therefore, it is best to avoid the temptation to control the rf excitation to a given LO stage by inserting a series resistor. The better method is to use small-value coupling capacitors.

A Practical VFO

The circuit of Fig. 12 is for a high stability 1.8- to 1.9-MHz VFO. Although the VFO frequency is for 160-meter operation, other tuning ranges up to 10 MHz are possible with this circuit. The design guidelines offered in the previous section will be useful in altering the circuit to other frequencies. A close approximation can be had by simply taking the X_L and X_C values for the components specified in Fig. 12 and determining from those reactances the new values for the frequency of interest. The tuning range will be determined by the capacitance

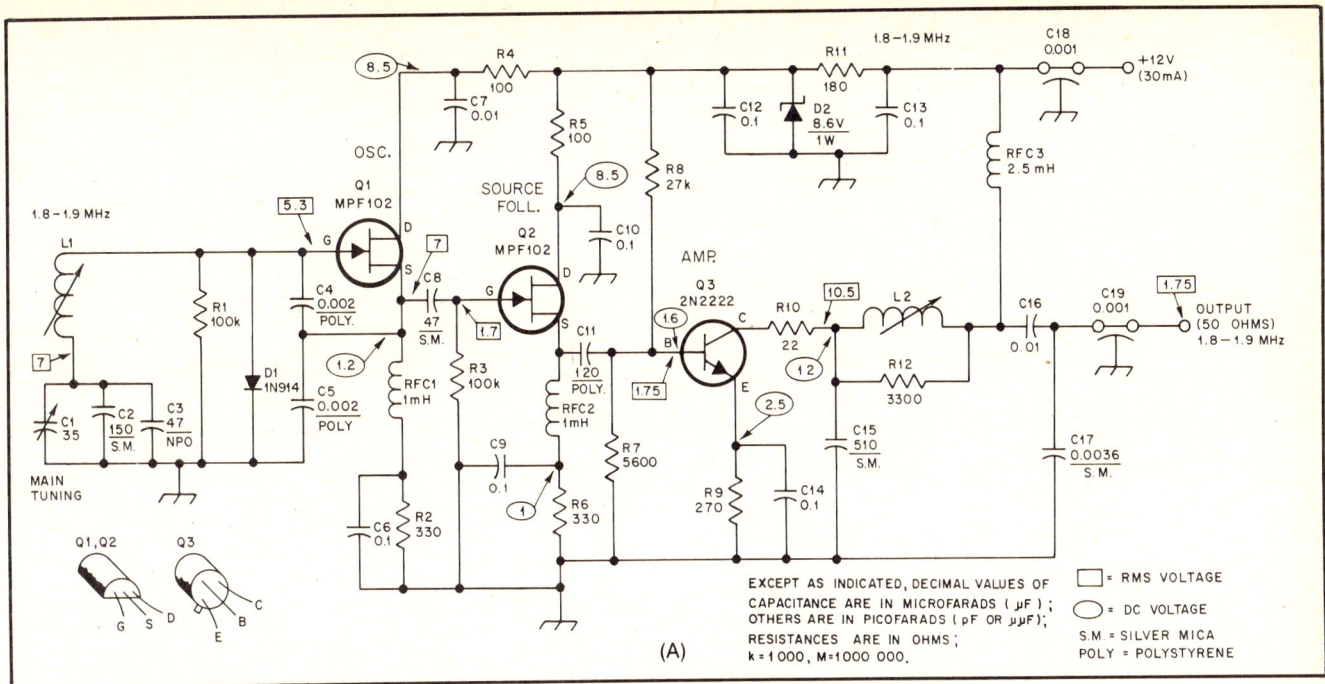

Fig. 12 — Circuit of a proven VFO chain which has exceptional stability under fairly constant room temperature. Although it is shown for operation from 1.8 to 1.9 MHz, it can be modified and used as high as 10 MHz. L1 is a 25- to 58-µH inductor (Miller 43A47CB1); L2 is a 10- to 18.7-µH inductor (Miller 23A 155RPC).

value of C1. Data on precise component values for other frequency ranges are not available from the ARRL. Further information on VFO design and the general circuit of Fig. 12 was provided by DeMaw in June 1976 *Ham Radio*.

Premixing

It is difficult to build a variable-frequency oscillator for operation above 10 MHz with drift of only a few Hz. A scheme called premixing shown in Fig. 13A, may be used to obtain VFO output in the 10- to 50-MHz range. The output of a highly stable VFO is mixed with energy from a crystal-controlled oscillator. The frequencies of the two oscillators are chosen so that spurious outputs generated during the mixing process do not fall within the desired output range. A band-pass filter at the mixer output attenuates out-of-band spurious energy. The charts given in chapter 8 can be used to choose oscillator combinations which will have a minimum of spurious outputs.

PLL

Receivers and transmitters of advanced design are now using phase-locked loops (PLLs) to generate highly stable local oscillator energy as high as the microwave region. The PLL has the advantage that no mixing stage is used in conjunction with the output oscillator, so the output energy is quite "clean." The Kenwood TS-820, the Collins 651S-1, and the National HRO-600 currently use PLL high-frequency oscillator systems.

The basic diagram of a PLL is shown in

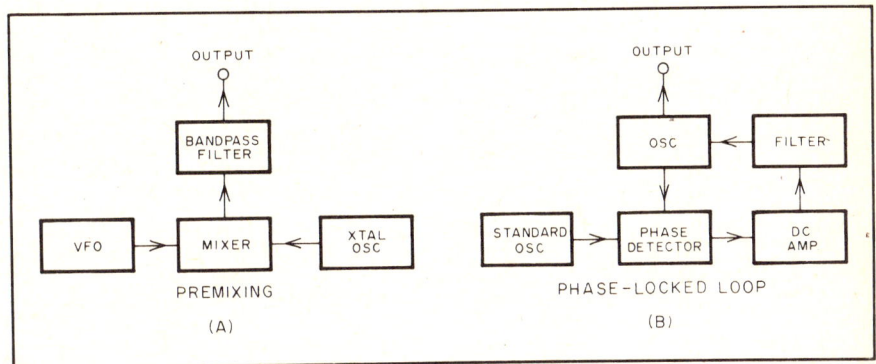

Fig. 13 — Block diagrams of the (A) premixing and (B) phase-locked loop methods.

Fig. 13B. Output from a voltage-controlled oscillator (VCO) and a frequency standard are fed to a phase detector which produces an output voltage equal to the difference in frequency between the two signals. The *error* voltage is amplified, filtered, and applied to the VCO. The error voltage changes the frequency of the VCO until it is locked to the standard. The bandwidth of the error-voltage filter determines the frequency range over which the system will remain in phase lock.

Three types of phase-locked loops are now in use. The simplest type uses harmonics of a crystal standard to phase-lock an HFO, providing the injection for the first mixer in a double-conversion receiver. A typical circuit is given in Fig. 14. Complete construction details on this PLL were given in January 1972 *QST*. A second type of phase-locked loop uses a

stable mf VFO as the standard which stabilizes the frequency of an hf or vhf VCO. This approach is used in the receiver described by Fischer in March 1970 *QST*.

The other PLL system also uses a crystal-controlled standard, but with programmable frequency dividers included so that the VCO output is always locked to a crystal reference. The frequency is changed by modifying the instructions to the dividers; steps of 100 Hz are usually employed for hf receivers while 10-kHz increments are popular in vhf gear.

VFO Dials

One of the tasks facing an amateur builder is the difficulty of finding a suitable dial and drive assembly for a VFO. A dial should provide a sufficiently slow rate of tuning — 10 to 25 kHz per knob revolution is considered optimum —

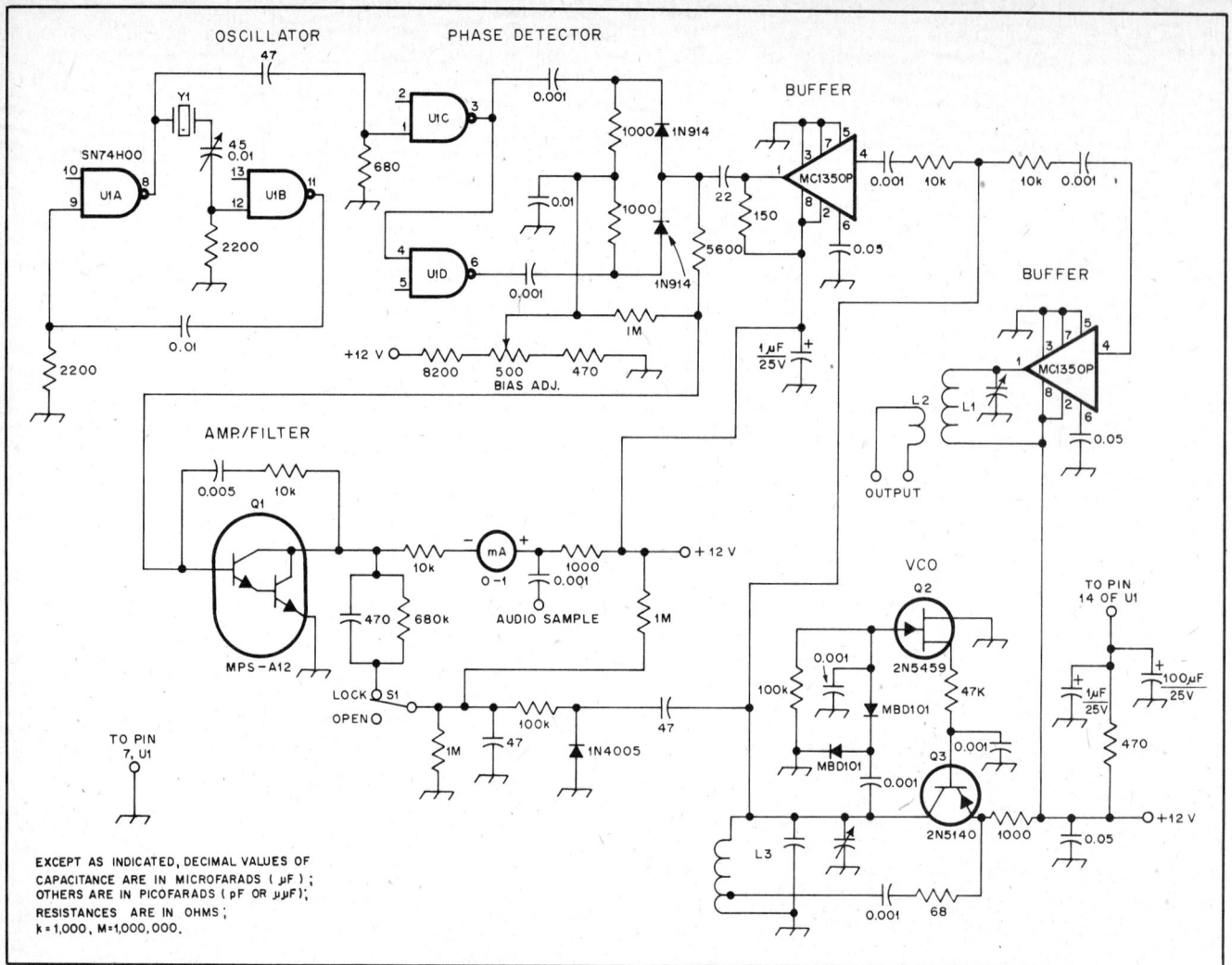

Fig. 14 — A practical PLL for a crystal-controlled HFO. Y1 is chosen so the harmonic content is ample at the desired output frequency. A 200-kHz crystal is fine to 40 MHz, a 500-kHz one is suitable to 60 MHz and a 1-MHz crystal is good for use to 80 MHz. L1 and L3 are resonant at the output frequency.

without backlash. Planetary drives are popular because of their low cost; however, they often develop objectional backlash after a short period of use. Several types of two-speed drives are available. They are well suited to home-made amateur equipment. The Eddystone 898 precision dial has long been a favorite with amateurs, although the need to elevate the VFO far above the chassis introduces some mechanical-stability problems. If a permeability-tuned oscillator (PTO) is used, one of the many types of turn counters made for vacuum variable capacitors or rotary inductors may be employed.

Linear Readout

If linear-frequency readout is desired on the dial, the variable capacitor must be only a small portion of the total capacitance in the oscillator tank. Capacitors tend to be very nonlinear near the ends of rotation. A gear drive providing a 1.5:1 reduction should be employed so that only the center of the capacitor range is used. Then, as a final adjustment, the plates of the capacitor must be filed until

linear readout is achieved. In a PTO, the pitch of the oscillator coil winding may be varied so that linear frequency change results from the travel of the tuning slug. Such a VFO was described in July 1964 *QST*. A different approach was employed by Lee (November 1970 *QST*), using a variable-capacitance diode (Varicap) as the VFO tuning element. A meter which reads the voltage applied to the Varicap was calibrated to indicate the VFO frequency.

Electronic Dials

An electronic dial consists of a simplified frequency counter which reads either the VFO or operating frequency of a transmitter or receiver. The advantage of an electronic dial is the excellent accuracy (to 1 hertz, if desired) and the fact that VFO tuning does not have to be linear. The readout section of the dial may use neon-glow tubes called *Nixies* (a tradename of the Burroughs Corp.), or a seven-segment display using incandescent lamps, filament wires in a vacuum tube, or LCDs (liquid crystal display), or LEDs (light-

emitting diodes). The use of MSI and LSI circuits, some containing as many as 200 transistors on a single chip, reduces the size required for an electronic dial to a few square inches of circuit-board space.

A typical counter circuit is given in Fig. 15. The accuracy of the counter is determined by a crystal standard which is often referred to as a *clock*. The output from a 100-kHz calibration oscillator, the type often used in receivers and transceivers, may be employed if accuracy of 100 Hz is sufficient. For readout down to 1 Hz, a 1- to 10-MHz AT-cut crystal should be chosen, because the type of high-accuracy crystal exhibits the best temperature stability. The clock output energy is divided in decade-counter ICs to provide the pulse which opens the input gate of the counter for a preset time. The number of rf cycles which pass through the gate while it is open are counted and stored. Storage is used so that the readout does not blink. At the end of each counting cycle the information that has been stored activates the display LEDs, which present the numbers counted until another count

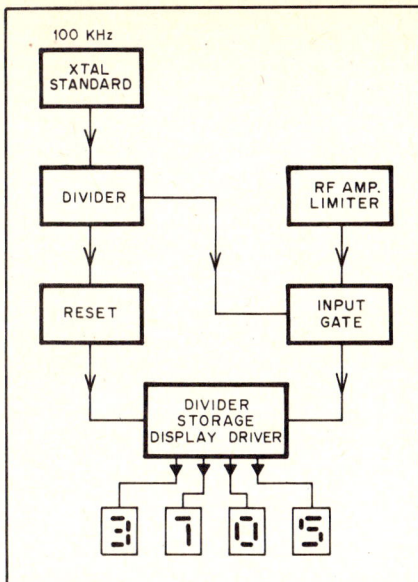

Fig. 15 — Frequency counter block diagram.

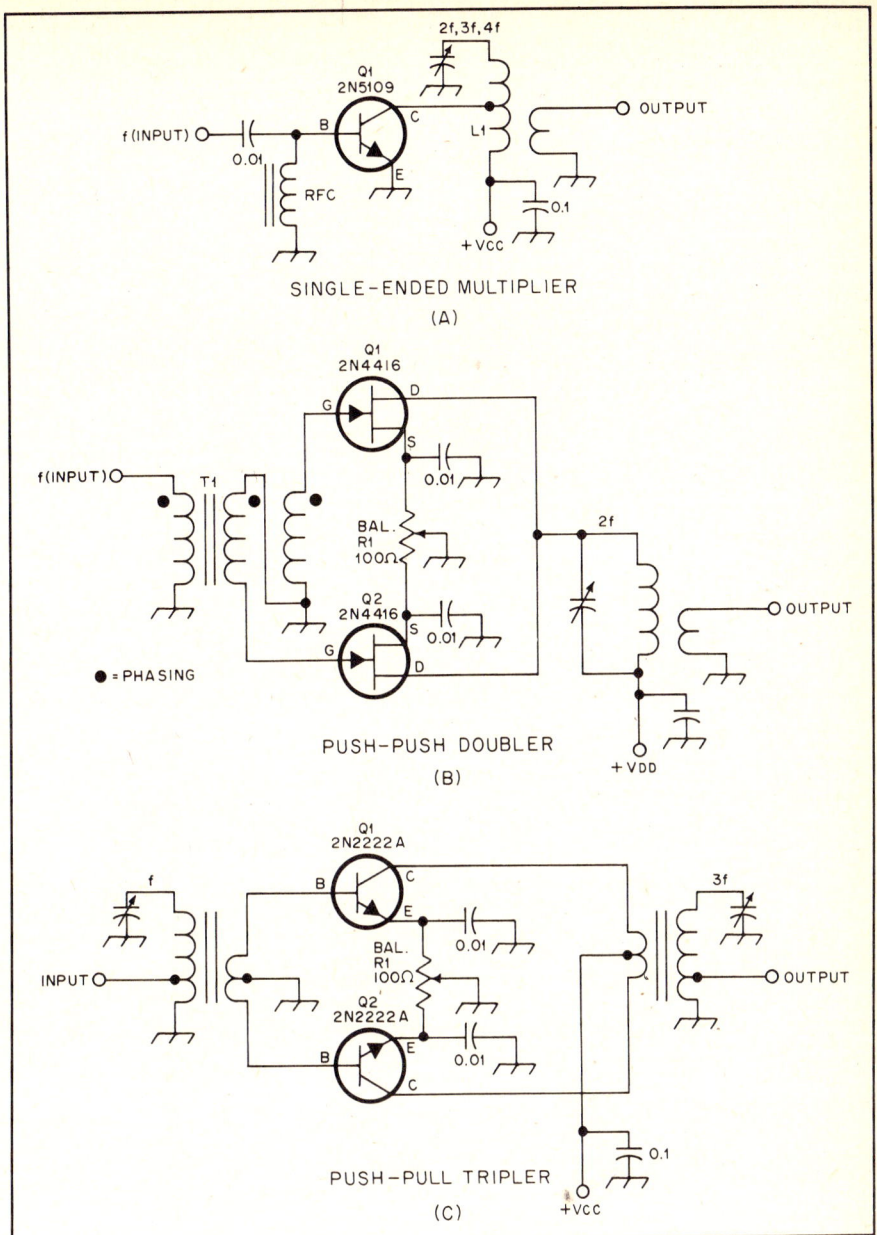

Fig. 16 — Single-ended multiplier (A), push-push doubler (B) and push-pull tripler (C).

cycle is complete. A complete electronic dial arranged to be combined with an existing transmitter or receiver was described in October 1970 *QST*. Also, Macleish et al. reported an adapter which allows a commercially made frequency counter to be mated with ham gear so that the counter performs as an electronic dial (May 1971 *QST*).

Frequency Multiplication

It may be necessary to use frequency multipliers at some point after the VFO or other frequency generator in a transmitter. When this need is present, the circuits of **Fig. 16** can be applied. Of course, vacuum-tube multipliers are entirely suitable if the design is not one which uses semiconductors. The fundamental principles for frequency multiplication are applicable to tubes and transistors alike. The requisite is that of operating the devices in *Class C*. Although a transistor circuit may be seen with forward bias applied to a frequency multiplier, the stage must be driven hard enough to override the bias and operate Class C. Forward bias is sometimes used in a multiplier stage (solid state) to lower the excitation requirements. Negative voltage (reverse bias) is often used on the grid of a vacuum-tube multiplier, but forward bias is not.

The circuit of Fig. 16A is probably the least suitable for frequency multiplication. Typically, the efficiency of a doubler of this type is 50 percent, a tripler is 33 percent, and a quadrupler is 25 percent. Additionally, harmonics other than the one to which the output tank is tuned will appear in the output unless effective band-pass filtering is applied. The collector tap on L1 of Fig. 16A is placed at a point which offers a reasonable compromise between power output and spectral purity: The lower the tap with respect

to V_{cc}, the lighter the collector loading on L1 and the greater the filtering action of the tuned circuit. The tradeoff is, however, a reduction in output power as the mismatch of the collector to the load increases.

A push-push doubler is shown at Fig. 16B. Because of the conduction angle of this type of circuit the efficiency is similar to that of a straight amplifier which operates in Class C. Also, the driving frequency (f) will be well attenuated at the doubler output if electrical balance and component symmetry are ensured. A 12AU7A tube will work nicely in this type of circuit well into the vhf region. T1 in this example is a trifilar-wound, broadband toroidal transformer. It drives the gates of Q1 and Q2 in push pull (opposite phase). The drains are in parallel and are

tuned to 2f. R1 is used to establish electrical balance between Q1 and Q2. R1 is set while the doubler is being fully driven. Diode doublers can be used in a similar circuit, but the subject will not be treated here (see chapter four).

A push-pull tripler is illustrated in Fig. 16C. Once again the matter of electrical balance and symmetry is important to good operation. The circuit discriminates against even harmonics, thereby aiding spectral purity. The efficiency is somewhat better than a tripler using the circuit of Fig. 16A. If vacuum tubes are used in the circuits of Fig. 16, the input ports should employ high-impedance tuned circuits for best performance.

Output Filtering

Output purity from oscillators, multi-

Table 1
Chebyshev High-Pass and Low-Pass Filters-Attenuation (dB)

No. poles, ripple	VSWR	2 f_c	3 f_c	4 f_c	5 f_c	6 f_c	7 f_c
3 pole, 1 dB	2.66	22.46	34.05	41.88	47.85	52.68	56.74
3 pole, 0.1 dB	1.36	12.24	23.60	31.42	37.39	42.22	46.29
3 pole, 0.01 dB	1.10	4.08	13.73	21.41	27.35	32.18	36.24
3 pole, 0.001 dB	1.03	0.63	5.13	11.68	17.42	22.20	26.25
5 pole, 1 dB	2.66	45.31	64.67	77.73	87.67	95.72	102.50
5 pole, 0.1 dB	1.36	34.85	54.21	67.27	77.21	85.26	92.04
5 pole, 0.01 dB	1.10	24.82	44.16	57.22	67.17	75.22	82.00
5 pole, 0.001 dB	1.03	14.94	34.16	47.22	57.16	65.22	71.99
7 pole, 1 dB	2.66	68.18	95.29	113.57	127.49	138.77	148.26
7 pole, 0.1 dB	1.36	57.72	84.83	103.11	117.03	128.31	137.80
7 pole, 0.01 dB	1.10	47.68	74.78	93.07	106.99	118.27	127.75
7 pole, 0.001 dB	1.03	37.68	64.78	83.06	96.98	108.26	117.75
9 pole, 1 dB	2.66	91.06	125.91	149.42	167.32	181.82	194.01
9 pole, 0.1 dB	1.36	80.60	115.45	138.96	156.86	171.36	183.55
9 pole, 0.01 dB	1.10	70.56	105.41	128.91	146.81	161.31	173.51
9pole, 0.001 dB	1.03	60.55	95.40	118.91	136.91	151.31	163.50

Note: For high-pass filter configuration $2f_c$ becomes $f_c/2$, etc.

Table 2
Chebyshev Low-Pass Filter — T Configuration

No. poles, ripple	L1	L2	L3	L4	L5	C1	C2	C3	C4
3 pole, 1 dB	16.10	16.10				3164.3			
3 pole, 0.1 dB	8.209	8.209				3652.3			
3 pole, 0.01 dB	5.007	5.007				3088.5			
3 pole, 0.001 dB	3.253	3.253				2312.6			
5 pole, 1 dB	16.99	23.88	16.99			3473.1	3473.1		
5 pole, 0.1 dB	9.126	15.72	9.126			4364.7	4364.7		
5 pole, 0.01 dB	6.019	12.55	6.019			4153.7	4153.7		
5 pole, 0.001 dB	4.318	10.43	4.318			3571.1	3571.1		
7 pole, 1 dB	17.24	24.62	24.62	17.24		3538.0	3735.4	3538.0	
7 pole, 0.1 dB	9.400	16.68	16.68	9.400		4528.9	5008.3	4528.9	
7 pole, 0.01 dB	6.342	13.91	13.91	6.342		4432.2	5198.4	4432.2	
7 pole, 0.001 dB	4.690	12.19	12.19	4.690		3951.5	4924.1	3951.5	
9 pole, 1 dB	17.35	24.84	25.26	24.84	17.35	3562.5	3786.9	3786.9	3562.5
9 pole, 0.1 dB	9.515	16.99	17.55	16.99	9.515	4591.9	5146.2	5146.2	4591.9
9 pole, 0.01 dB	6.481	14.36	15.17	14.36	6.481	4542.5	5451.2	5451.2	4542.5
9 pole, 0.001 dB	4.854	12.81	13.88	12.81	4.854	4108.2	5299.0	5299.0	4108.2

Component values normalized to 1 MHz and 50 ohms. L in µH; and C in pF.

Table 3
Chebyshev Low-Pass Filter — Pi Configuration

No. poles, ripple	C1	C2	C3	C4	C5	L1	L2	L3	L4
3 pole, 1 dB	6441.3	6441.3				7.911			
3 pole, 0.1 dB	3283.6	3283.6				9.131			
3 pole, 0.01 dB	2002.7	2002.7				7.721			
3 pole, 0.001 dB	1301.2	1301.2				5.781			
5 pole, 1 dB	6795.5	9552.2	6795.5			8.683	8.683		
5 pole, 0.1 dB	3650.4	6286.6	3650.4			10.91	10.91		
5 pole, 0.01 dB	2407.5	5020.7	2407.5			10.38	10.38		
5 pole, 0.001 dB	1727.3	4170.5	1727.3			8.928	8.928		
7 pole, 1 dB	6896.4	9847.4	9847.4	6896.4		8.85	9.34	8.85	
7 pole, 0.1 dB	3759.8	6673.9	6673.9	3759.8		11.32	12.52	11.32	
7 pole, 0.01 dB	2536.8	5564.5	5564.5	2536.8		11.08	13.00	11.08	
7 pole, 0.001 dB	1875.7	4875.9	4875.9	1875.7		9.879	12.31	9.879	
9 pole, 1 dB	6938.3	9935.8	10,105.	9935.8	6938.3	8.906	9.467	9.467	8.906
9 pole, 0.1 dB	3805.9	6794.5	7019.9	6794.5	3805.9	11.48	12.87	12.87	11.48
9 pole, 0.01 dB	2592.5	5743.5	6066.3	5743.5	2592.5	11.36	13.63	13.63	11.36
9 pole, 0.001 dB	1941.7	5124.6	5553.2	5124.6	1941.7	10.27	13.25	13.25	10.27

Component values normalized to 1 MHz and 50 ohms. L in µH; C in pF.

bandwidth must be adequate for the tuning range of the VFO in order to prevent attenuation of the output energy within the desired band. For this reason, a low-pass type of filter is used in preference to a bandpass one by some designers.

The information contained in Figs. 17-19 and in Tables 1 to 5 will allow the builder to select an appropriate Chebyshev filter design to fulfill a particular need. Information is included for both high-pass and low-pass filters with 1, 0.1, 0.01 and 0.001 db passband ripple. These figures correspond to VSWRs of 2.66, 1.36, 1.10 and 1.03 respectively. Additionally, information is provided for both "T" and "pi" types of filter configurations.

The filters are "normalized" to a frequency of 1 MHz and an input and output impedance of 50 ohms. In order to translate the designs to other frequencies, all that is necessary is to divide the component values by the new frequency in MHz. (The 1-MHz value represents a "cutoff" frequency. That is, the attenuation increases rapidly above this frequency for the low-pass filter or below this frequency for the high-pass filter. This effect should not be confused with the variations in attenuation in the passband.) For instance, if it is desired to reduce harmonics from a VFO at frequencies above 5 MHz (the new cutoff frequency), the inductance and capacitance values would be divided by 5.0.

Other impedance levels can also be used by multiplying the inductor values by the ratio $Z_o/50$ and the capacitor values by $50/Z_o$, where Z_o is the new impedance. This factor should be applied in addition to the ones for frequency translation.

In order to select a suitable filter design the builder must determine the amount of attenuation required at the harmonic frequencies (for the low-pass case) or "subharmonic" frequencies (for the high-pass application). Additionally, the builder must determine the maximum permissible amount of passband ripple and therefore the VSWR of the filter. With this information the builder can refer to Table 1 to select an appropriate filter design. The attenuation values given here are theoretical and assume perfect components, no coupling between filter sections and no signal leakage around the filter. A "real life" filter should follow these values fairly close down to the 60- or 70-dB attenuation level. At this point the theoretical response will likely be degraded somewhat by the factors just mentioned. Once the filter design has been selected the builder can refer to Tables 2-5 to obtain the normalized component values.

In many cases the calculated capacitor values will be sufficiently close to a standard value so that the standard-value item may be used. Alternatively, a combination of fixed-value silver-mica capacitors

pliers and amplifiers is of paramount importance to the performance of numerous circuits. In the interest of compliance with current FCC regulations, wherein all spurious emissions from a transmitter must be 40 dB or greater below the peak power of the desired signal, filtering is important. The type of filter used — band-pass, notch, low-pass or high-pass — will depend on the application. Band-pass filters afford

protection against spurious responses above and below the amateur band for which they have been designed. Low-pass filters attenuate energy above the desired output frequency, while high-pass filters reduce energy below the band of interest. It is common practice to include a harmonic filter at the output of a VFO chain to ensure purity of the driving voltage to a mixer or amplifier stage. The filter

Table 4
Chebyshev High-Pass Filter — T Configuration

No. poles, ripple	C1	C2	C3	C4	C5	L1	L2	L3	L4
3 pole, 1 dB	1573.0	1573.0				8.005			
3 pole, 0.1 dB	3085.7	3085.7				6.935			
3 pole, 0.01 dB	5059.1	5059.1				8.201			
3 pole, 0.001 dB	7786.9	7786.9				10.95			
5 pole, 1 dB	1491.0	1060.7	1491.0			7.293	7.293		
5 pole, 0.1 dB	2775.6	1611.7	2775.6			5.803	5.803		
5 pole, 0.01 dB	4208.6	2018.1	4208.6			6.098	6.098		
5 pole, 0.001 dB	5865.7	2429.5	5865.7			7.093	7.093		
7 pole, 1 dB	1469.2	1028.9	1028.9	1469.2		7.160	6.781	7.160	
7 pole, 0.1 dB	2694.9	1518.2	1518.2	2694.9		5.593	5.058	5.593	
7 pole, 0.01 dB	3994.1	1820.9	1820.9	3994.1		5.715	4.873	5.715	
7 pole, 0.001 dB	5401.7	2078.0	2078.0	5401.7		6.410	5.144	6.410	
9 pole, 1 dB	1460.3	1019.8	1002.7	1019.8	1460.3	7.110	6.689	6.689	7.110
9 pole, 0.1 dB	2662.4	1491.2	1443.3	1491.2	2662.4	5.516	4.922	4.922	5.516
9 pole, 0.01 dB	3908.2	1764.1	1670.2	1764.1	3908.2	5.576	4.647	4.647	5.576
9 pole, 0.001 dB	5218.3	1977.1	1824.6	1977.1	5218.3	6.657	4.780	4.780	6.657

Component values normalized to 1 MHz and 50 ohms. L in µH; and C in pF.

Table 5
Chebyshev High-Pass Filter — Pi Configuration

No. poles, ripple	L1	L2	L3	L4	L5	C1	C2	C3	C4
3 pole, 1 dB	3.932	3.932				3201.7			
3 pole, 0.1 dB	7.714	7.714				2774.2			
3 pole, 0.01 dB	12.65	12.65				3280.5			
3 pole, 0.001 dB	19.47	19.47				4381.4			
5 pole, 1 dB	3.727	2.652	3.727			2917.3	2917.3		
5 pole, 0.1 dB	6.939	4.029	6.939			2321.4	2321.4		
5 pole, 0.01 dB	10.52	5.045	10.52			2439.3	2439.3		
5 pole, 0.001 dB	1.466	6.074	1.466			2837.3	2837.3		
7 pole, 1 dB	7.159	5.014	5.014	7.159		1469.2	1391.6	1469.2	
7 pole, 0.1 dB	6.737	3.795	3.795	6.737		2237.2	2023.1	2237.2	
7 pole, 0.01 dB	9.985	4.552	4.552	9.985		2286.0	1949.1	2286.0	
7 pole, 0.001 dB	13.50	5.195	5.195	13.50		2564.1	2057.7	2564.1	
9 pole, 1 dB	3.651	2.549	2.507	2.549	3.651	2844.1	2675.6	2675.6	2844.1
9 pole, 0.1 dB	6.656	3.728	3.608	3.728	6.656	2206.5	1968.9	1968.9	2206.5
9 pole, 0.01 dB	9.772	4.410	4.176	4.410	9.772	2230.5	1858.7	1858.7	2230.5
9 pole, 0.001 dB	13.05	4.943	4.561	4.943	13.05	2466.3	1911.8	1911.8	2466.3

Component values normalized to 1 MHz and 50 ohms. L in µH; and C in pF.

Fig. 18 — Here is a photograph of a 7-pole low-pass filter designed with the information contained in Table 3. The filter is housed in a small aluminum Minibox.

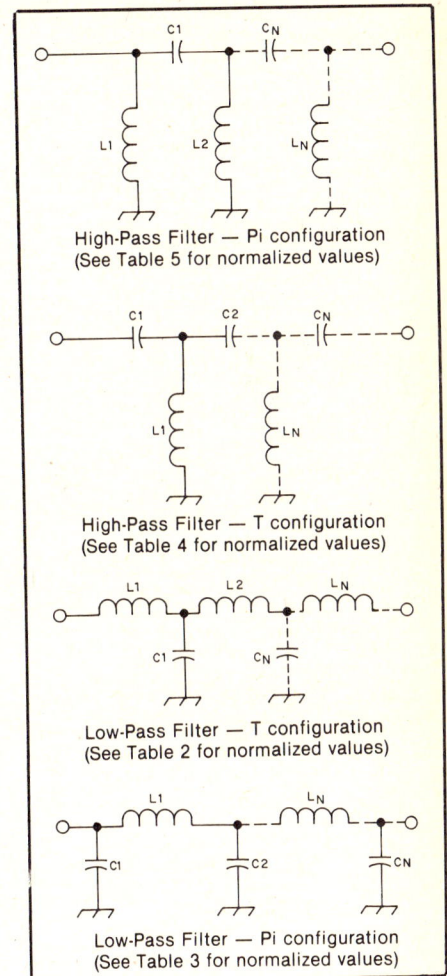

High-Pass Filter — Pi configuration
(See Table 5 for normalized values)

High-Pass Filter — T configuration
(See Table 4 for normalized values)

Low-Pass Filter — T configuration
(See Table 2 for normalized values)

Low-Pass Filter — Pi configuration
(See Table 3 for normalized values)

Fig. 19 — Shown here are the four filter types discussed in the text and Tables 1-5.

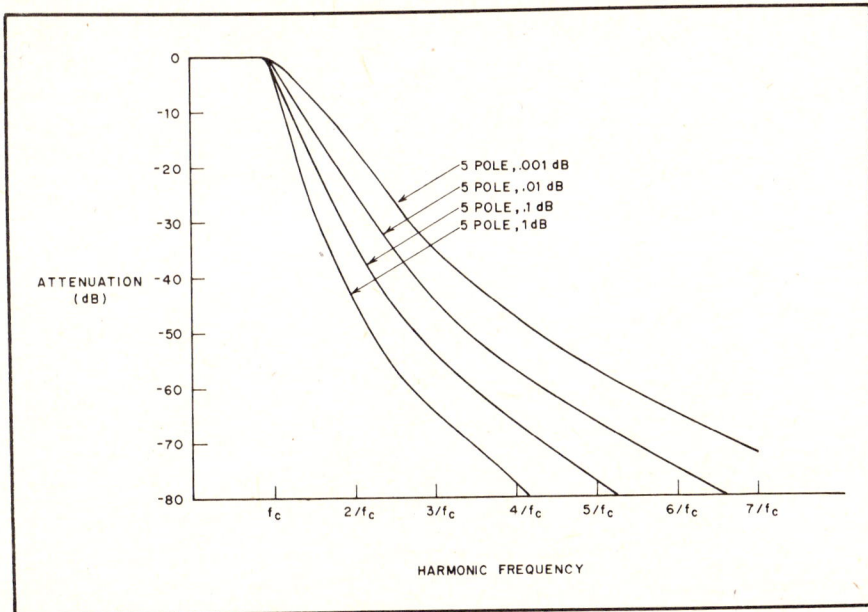

Fig. 17 — A representative drawing of the attenuation levels that could be expected from a 5-pole, low-pass filter designed from the information contained in Tables 2 or 3. The exact amount of attenuation (theoretical) can be obtained from Table 1. This drawing shows how passband ripple and roll-off slope are interrelated.

and mica compression trimmers can be used in parallel to obtain the chart values. Toroidal inductors, because of their self-shielding properties, are ideal for use in these filters. Miniductor stock can also be used. However, it is much bulkier and will not offer the same degree of shielding between filter sections. Disc-ceramic or paper capacitors are not suitable for use in rf filters. Standard mica or silver-mica types are recommended.

Fig. 18 shows a filter that was designed with the information contained in Table

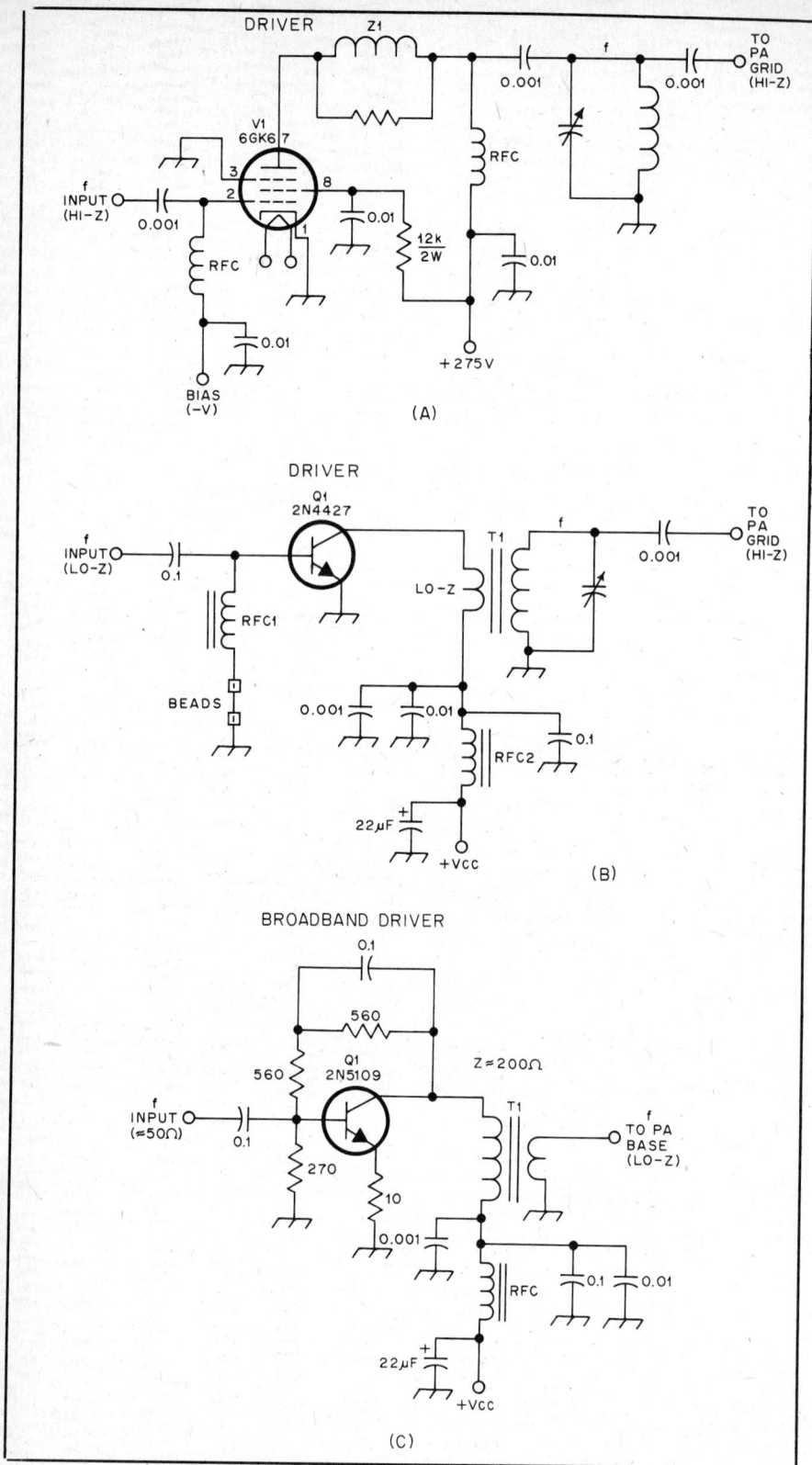

Fig. 20 — Circuit examples of transistor and tube driver stages for use in transmitters.

tain a mixture (hybrid) of tubes and semiconductors, while other circuits have no vacuum tubes at all. If tubes are used in a hybrid circuit, they are generally restricted to the driver and PA sections of the transmitter. There is no particular reason why tubes should be used in preference to power transistors for output powers up to, say, 150 watts, despite the prevailing myth that tubes are more rugged, operate more stably, and produce less spurious output. It is true that transistors are less tolerant than tubes to SWR levels in excess of 2:1, but a correctly designed transistor amplifier can be operated safely if SWR-protection circuitry is included. Furthermore, spectral purity can be just as good from a solid-state amplifier as it is from a tube type of amplifier. A harmonic filter normally follows a solid-state power stage, whereas this measure may not be required when tubes are used in the amplifier. Amplifier IMD (third- and fifth-order products) in solid-state power stages which operate linearly is fully as acceptable as that which is observed in most tube types of linear amplifiers. Typically, if a design is correct, the IMD will be on the order of -33 dB from the reference power value.

The major area of concern when designing a solid-state driver or PA section is to prevent low-frequency self-oscillations. Such parasitics tend to modulate the carrier and appear as spurious responses within the amplifier passband. The low-frequency parasitics occur as a result of the extremely high gain exhibited by hf and vhf transistors at the low-frequency end of the spectrum. The theoretical gain increase for a given transistor is 6 dB per octave as the operating frequency is lowered. The same is not true of vacuum tubes. Therefore, it is necessary to employ quality decoupling and bypassing in the circuit. It is similarly important to use low-Q, low-inductance rf chokes and matching networks to discourage low-frequency tuned-base, tuned-collector oscillations. The suppression concepts just discussed are illustrated in Fig. 20 at B and C. In the circuit at B there are two 950-mu ferrite beads added over the pigtail of RFC1 to swamp the Q of the choke. Three bypass capacitors (0.001, 0.01 and 0.1μF) are used with RFC2 of Fig. 20B to provide effective rf decoupling from vhf to mf. A 22-μF capacitor is used near RFC2 to bypass the $+V_{cc}$ line at low frequency and audio. This method is recommended for each high-gain solid-state stage in a transmitter.

Driver Circuits

The circuits of Fig. 20A and 20B are typical of those which would be employed to excite a tube type of PA stage. The 6GK6 tube driver at A can be biased for Class C or Class AB operation, making it suitable for cw or ssb service. Of course,

3. It is a 7-element, low-pass type of pi configuration. The unit is housed in a small aluminum Minibox and makes use of BNC connectors for the input and output connections. Some practical low-pass filter values are given later in this chapter.

Driver Stages

The choice between tubes and transistors in low-level amplifier and driver stages will depend upon the nature of the composite transmitter. Some designs con-

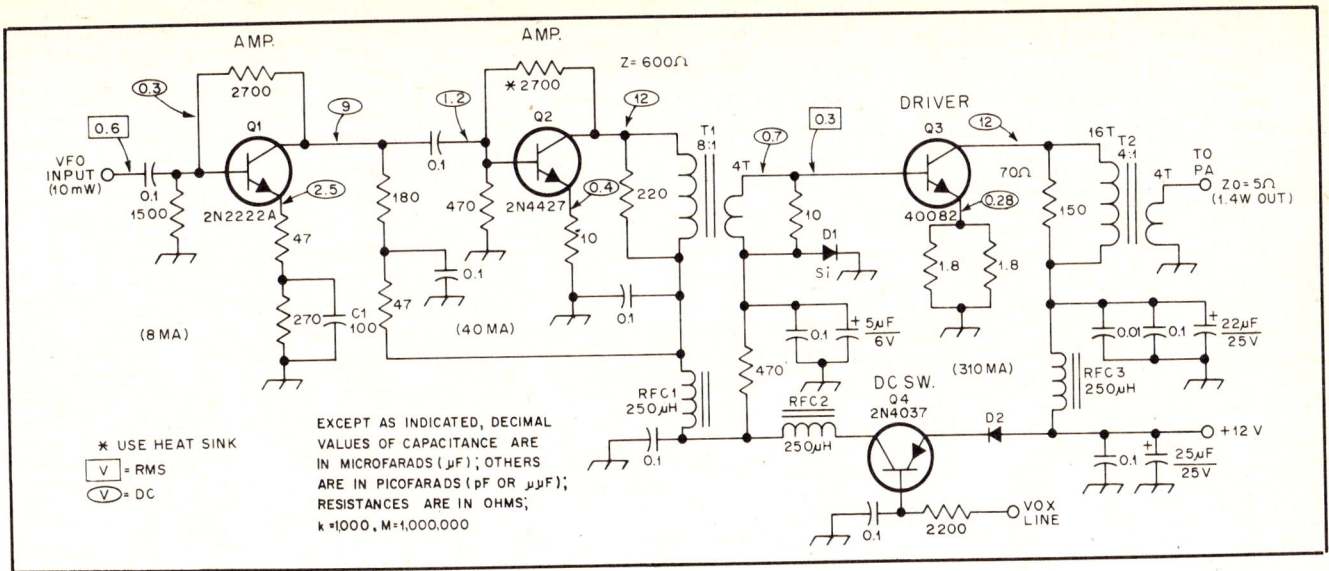

Fig. 21 — Practical circuit for a three-stage broadband amplifier/driver. See text.

the AB mode would be suitable for cw and ssb, and would require considerably less excitation power than would the same stage operating in Class C. Other tubes that perform well in this circuit are the 6CL6, 12BY7A and 5763. The output tank is designed for high impedance in order to interface properly with the high-impedance grid of the PA. It may be necessary to include a neutralization circuit with this type of amplifier, especially if careful layout is not used. The high transconductance of the 6GK6 series encourages self-oscillation near the operating frequency. Z1 is a parasitic choke which should be included as a matter of course to prevent vhf parasitics.

A transistor amplifier which is suitable for driving a Class C tube PA is presented in Fig. 20B. Q1 operates Class C, so it is not satisfactory for amplifying ssb energy. However, forward bias (approximately 0.7 volt) can be added to move the operating curve into the Class AB (linear) region, thereby making the stage suitable for ssb signal amplification. A 1.5-ohm resistor can be added between the emitter and ground to help prevent thermal runaway and to introduce degeneration (feedback) for enhancing stability. No bypass capacitor would be used from emitter to ground if this were done. T1 is a narrow-band toroidal rf transformer that has a turns ratio suitable for transforming the collector impedance to the grid impedance (determined by the value of the grid resistor of the PA) of the final amplifier. The secondary winding of T1 is tuned to resonance at the operating frequency. Approximately 1 watt of power output can be taken from Q1 in the hf region when a 12-volt V_{cc} is used. This is ample power for driving a pair of 6146B tubes in Class AB1.

A broadband type of solid-state driver is shown in Fig. 20C. The tradeoff for

broadband operation (1.8 to 30 MHz in this example) is a reduction in maximum available gain (MAG). Therefore, the output power from Q1 of Fig. 20C will be less than 1 watt. The stage operates Class A, making it linear. The emitter is unbypassed to provide emitter degeneration. Shunt feedback is used between the base and collector to enhance stability and contribute to the broadband characteristic of the circuit. T1 is a broadband conventional transformer wound on a toroid core. The turns ratio is adjusted to match the approximate 200-ohm collector impedance to the base impedance of the transistor PA stage. The latter is typically less than 5 ohms. Heat sinks are required for the transistors of Fig. 20B and C. The primary of T1 should have a reactance of roughly four times the collector impedance. This is related to the lowest proposed operating frequency. Therefore, for 1.8 MHz the primary winding would be 70 μH (X_L = 800 ohms). This can be achieved easily by using an FT50-43 Amidon core. The primary advantage to a broadband driver is that it need not be band-switched or peaked by means of a front-panel control. The transistor selected for broadband service should have a very high f_T rating. It needs to have high beta as well. Transistors designed for uhf service are excellent as hf-band amplifiers when broadbanding is contemplated. Neutralization is not necessary when using bipolar-transistor amplifiers.

A practical three-stage broadband amplifier strip is shown schematically in Fig. 21. With an input level of 10 mW it is possible to obtain 1.4 watts of output from 3.5 to 29 MHz. A keying transistor (Q4) is included for turning the amplifier off by means of a VOX, or for keying it during cw operation.

Rms and dc voltages are noted on the diagram of Fig. 21 to aid in

troubleshooting. Overall gain for the strip at 7 MHz is 31 dB, with slight gain variations elsewhere in the passband. T1 consists of 30 turns of no. 28 enameled wire (primary) on an FT50-43 toroid. The secondary has four turns on no. 28 wire. T2 uses 16 turns of no. 28 enameled wire (primary) looped through an Amidon BLN-43-302 ferrite balun core. The secondary contains four turns of no. 28 wire. RFC1, RFC2 and RFC3 are 250-μH units. They are made by winding 20 turns of no. 28 enameled wire on FT37-43 toroid cores. D1 and D2 are 1-A, 50 PRV rectifier diodes. This driver was designed to excite a Motorola MRF449A PA stage to a power-output range from 15 to 30 watts. C1 at the emitter of Q1 can be selected to provide the overall gain needed in this strip. The value given at C1 proved suitable for the ARRL version of this amplifier. The final value will depend on the gain of the individual transistors acquired for this circuit.

Coupling Between Transmitter Stages

Correct impedance matching between a stage and its load provides maximum transfer of power. The load can be an antenna or a succeeding stage in a transmitter. Thus, the output impedance of a stage must be matched to the input of the following stage. Various forms of coupling networks are popular for use in tube or transistor circuits. The choice will depend on a number of considerations — available driving power versus tolerable mismatch, selectivity required and the impedance levels being matched. When working with transistors, the collector impedance can be approximated by

$$Z = \frac{V_{cc}^2}{2P_o}$$

where Z is in ohms and P_o is the power

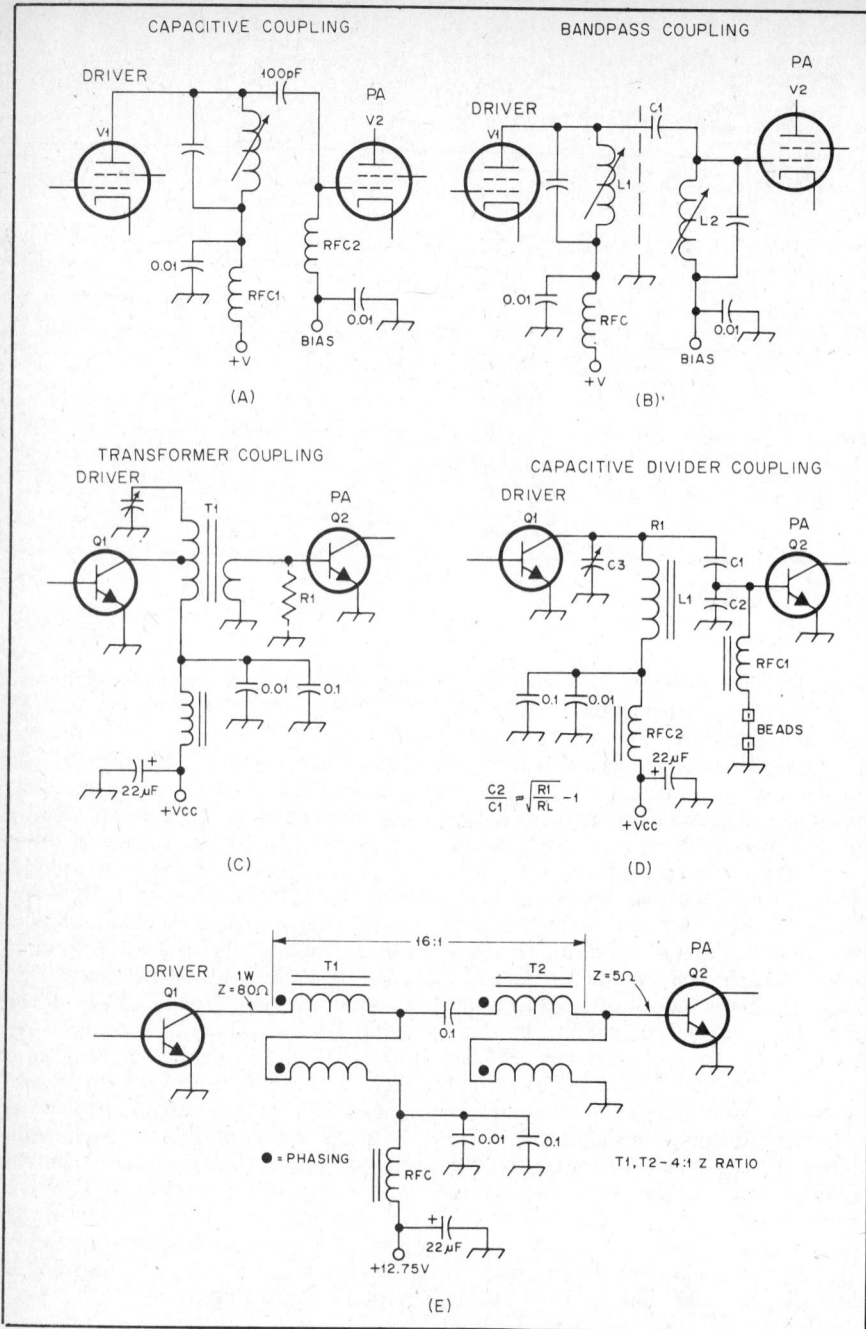

Fig. 22 — Typical coupling methods for use between amplifier stages. See text.

CAPACITIVE COUPLING

(A)

BANDPASS COUPLING

(B)

TRANSFORMER COUPLING

(C)

CAPACITIVE DIVIDER COUPLING

$$\frac{C2}{C1} = \sqrt{\frac{R1}{RL}} - 1$$

(D)

T1, T2 — 4:1 Z RATIO

● = PHASING

(E)

penalty for using a low-Q resonant network is poor selectivity: There is little attenuation of harmonic or other spurious energy. Conversely, tube stages operate at relatively high impedance levels (plate and grid) and can be neutralized easily (not true of transistors). This permits the employment of high-Q networks between stages, which in turn provide good selectivity. Most solid-state amplifiers use matching networks with loaded Qs of 5 or less. Tube stages more commonly contain networks with loaded Qs of 10 to 15. The higher the Q, up to a practical limit, the greater the attenuation of frequencies other than the desired ones. In all cases, the input and output capacitances of tubes and transistors must be included in the network constants, or to use the engineering vernacular, "absorbed" into the network. The best source of information on the input and output capacitances of power transistors is the manufacturer's data sheet. The capacitance values are dependent upon the operating frequency and power level of the transistor — a very complex set of curves. Most tube data sheets list specific values of input and output capacitance, which do not vary with the operating frequency or power level.

The interstage coupling method shown in Fig. 22A is a common one when vacuum tubes are employed. The driver plate has a tuned circuit which is resonant at the operating frequency. A low-value coupling capacitor (100 pF in this example) routes the drive from the plate of V1 to the grid of V2 across a high-impedance element, RFC2. The other choke, RFC1, is used as part of the decoupling network for the supply voltage to V1.

Band-pass coupling between tube stages is demonstrated at Fig. 22B. C1 has a very small capacitance value and is chosen to provide a single-hump response when the two resonators (L1 and L2) are peaked to the operating frequency. The principal advantage to this circuit over that of Fig. 22A is greater purity of the driving energy to V2 by virtue of increased selectivity. As an alternative to capacitive coupling (C1), link coupling can be used between the cold ends of L1 and L2. Similar band-pass networks are applicable to transistor stages. The collector and base of the two stages would be tapped down on L1 and L2 to minimize loading. This helps preserve the loaded Q of the tuned circuits, thereby aiding selectivity.

A common form of transformer coupling is seen at Fig. 22C. T1 is usually a toroidal inductor for use up to approximately 30 MHz. At higher frequencies it is often difficult to provide a secondary winding of the correct impedance ratio respective to the primary winding. Depending on the total number of transformer turns used, the secondary might call for less than one turn, which is impractical. However, for most of the spectrum up to

output from the stage. However, determining the input impedance of the base of the following stage is difficult to do without expensive laboratory equipment. Generally, when the PA delivers in excess of 2 watts of output power, the base impedance of that stage will be less than 10 ohms — frequently just 1 or 2 ohms. For this reason some kinds of LC matching networks do not lend themselves to the application. Furthermore, without being able to predict the precise input impedance of a transistor power amplifier, it becomes desirable to use what is sometimes referred to as a "sloppy" matching network. This is an LC network

in which both the inductance and capacitance elements are variable to allow latitude of adjustment while securing a matched condition. On the other hand, some designers purposely introduce a mismatch between stages to control the power distribution and aid stability. When this technique is used it is necessary to have more driving power than would be needed under a matched condition. An intentional mismatch results in a tradeoff between gain and the desired end effect of introducing a mismatch.

In the interest of stability it is common practice to use low-Q networks between stages in a solid-state transmitter. The

30 MHz this technique is entirely satisfactory. The primary tap on T1 is chosen to transform the collector impedance of Q1 to the base impedance of Q2 by means of the turns ratio between the tapped section and the secondary winding of the transformer. R1 may be added in shunt with the secondary to stabilize Q2 if there is a tendency toward self-oscillation. The value used will be in the 5- to 27-ohm range for most circuits. The rule of thumb is to use just enough resistance to tame the instability.

A method for coupling between stages by means of a capacitive divider is illustrated in Fig. 22D. The net value of C1 and C2 in series must be added to the capacitance of C3 when determining the inductance required for resonance with L1. The basic equation for calculating the capacitance ratio of C1 and C2 is included in the diagram. RFC1 serves as a dc return for the base of Q2. The Q of the rf choke is degraded intentionally by the addition of two 950-mu ferrite beads. This aids stability, as discussed earlier in this chapter. An advantage to using this type of circuit is that vhf and uhf parasitics are discouraged and harmonic currents are attenuated when C2 is fairly high in capacitance. This is not true of the circuit in Fig. 22C.

When the impedance levels to be matched are of the proper value to permit employing specific-ratio broadband transformers, the circuit of Fig. 22E is useful. In this example two 4:1 transformers are used in cascade to provide a 16:1 transformation ratio. This satisfies the match between the 80-ohm collector of Q1 and the 5-ohm base of Q2. The shortcoming of this technique is the lack of selectivity between stages, but the advantage is in the broadband characteristic of the coupling system. The phasing dots on the diagram near T1 and T2 indicate the correct electrical relationship of the transformer windings.

Network Equations

The three networks shown in Figs. 24, 25 and 26 will provide practical solutions to many of the impedance-matching problems encountered by amateurs. In each of the figures it is assumed that the output impedance being matched is lower than the input impedance of the following stage. If this is not the case the network (the circuit elements between the points marked A and B) can be turned around to provide the correct transformation.

Normally, the output impedance of a transistor is given as a resistance in parallel with a capacitance, C_{out}. To use the design equations for these three networks, the output impedance must first be converted from the parallel form (R_{out} and C_{out}) to the equivalent series form (R_s and C_s). These equivalent circuits and the equations for conversion are given in Fig. 23. Often the output capacitance is small

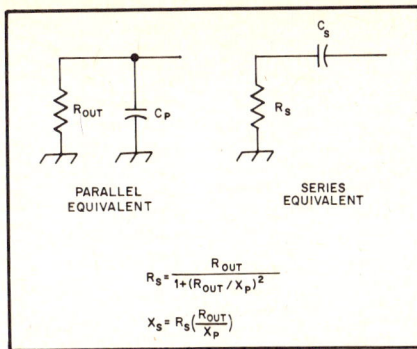

Fig. 23 — Parallel and series equivalent circuits and the formulas used for conversion.

$$R_s = \frac{R_{OUT}}{1 + (R_{OUT}/X_P)^2}$$

$$X_s = R_s\left(\frac{R_{OUT}}{X_P}\right)$$

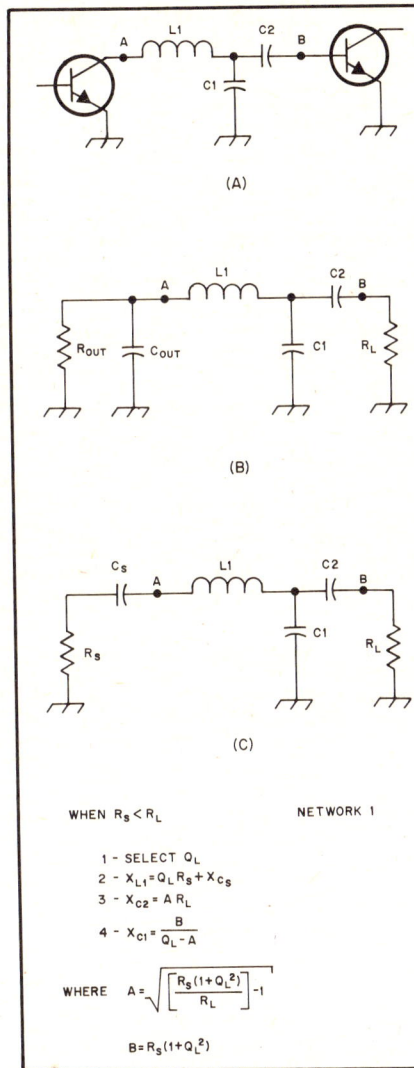

Fig. 24 — Circuit and mathematical solution for matching network no. 1. Actual circuit (A), parallel equivalent (B) and series equivalent (C).

WHEN $R_s < R_L$ NETWORK 1

1 - SELECT Q_L
2 - $X_{L1} = Q_L R_s + X_{Cs}$
3 - $X_{C2} = A R_L$
4 - $X_{C1} = \frac{B}{Q_L - A}$

WHERE $A = \sqrt{\left[\frac{R_s(1+Q_L^2)}{R_L}\right]-1}$

$B = R_s(1+Q_L^2)$

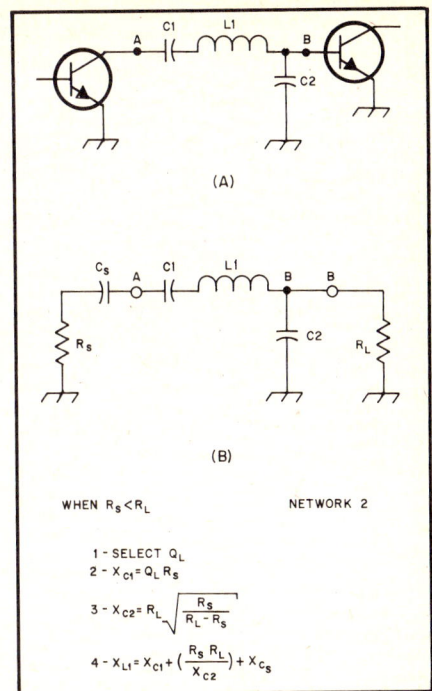

Fig. 25 — Circuit and equations for network no. 2. Actual circuit (A) and series equivalent (B).

WHEN $R_s < R_L$ NETWORK 2

1 - SELECT Q_L
2 - $X_{C1} = Q_L R_s$
3 - $X_{C2} = R_L \sqrt{\frac{R_s}{R_L - R_s}}$
4 - $X_{L1} = X_{C1} + \left(\frac{R_s R_L}{X_{C2}}\right) + X_{Cs}$

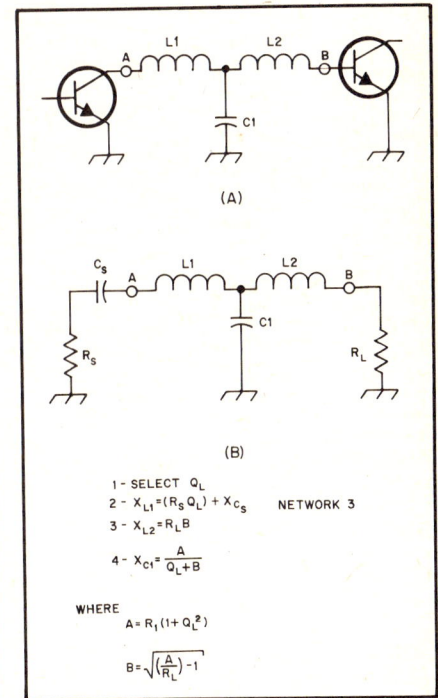

Fig. 26 — Network-solution equations and circuit for network no. 3. Actual circuit (A) and series equivalent (B).

1 - SELECT Q_L
2 - $X_{L1} = (R_s Q_L) + X_{Cs}$ NETWORK 3
3 - $X_{L2} = R_L B$
4 - $X_{C1} = \frac{A}{Q_L + B}$

WHERE
$A = R_1(1 + Q_L^2)$

$B = \sqrt{\left(\frac{A}{R_L}\right) - 1}$

enough that it may be neglected; the resulting error is compensated for by using variable components in the network.

The low-pass T network (Fig. 26) has the advantage of matching a wide range of impedances with practical component values. Some designers feel that of the various networks used in solid-state work, the T network is best in terms of collector efficiency. The harmonic suppression afforded by the T network varies with the transformation ratio and the Q_L of the network. For stages feeding an antenna

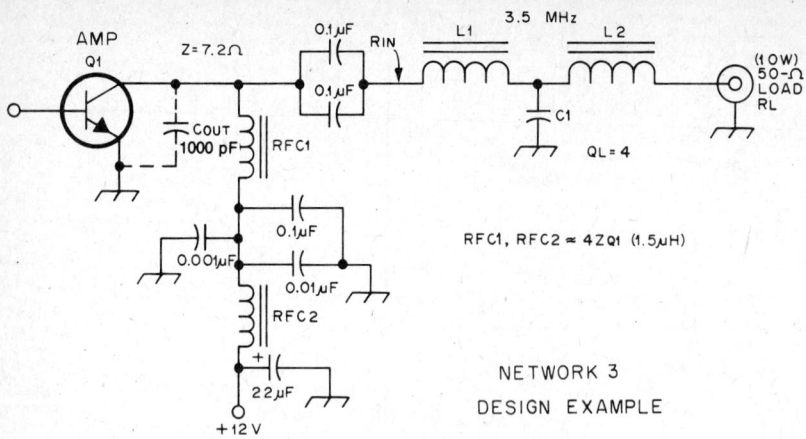

$$R_{out} \approx \frac{V_{cc}^2}{2P_o} = \frac{144}{20} = 7.2 \ \Omega$$

$$X_{C_{out}} = \frac{1}{(2\pi \times 3.5 \times 10^6 \times 1000 \times 10^{-12})} = 45.5 \ \Omega$$

$$R_s = \frac{R_p}{1 + (R_p/X_p)^2} = \frac{7.2}{1 + (7.2/45.5)^2} = 7.0 \ \Omega$$

$$X_{c_s} = R_s \left(\frac{R_p}{X_p}\right) = 7.0 \left(\frac{7.2}{45.5}\right) = 1.1 \ \Omega$$

$$R_L = 50 \ \Omega$$

$$Q_L = 4$$

$$X_{L1} = (R_s Q_L) + X_{c_s} = (7.0 \times 4) + 1.1 = 29.1 \ \Omega$$

$$\therefore L1(\mu H) = \frac{X_{L1}}{2\pi \ f(MHz)} = \frac{29.1}{2\pi \times 3.5} = 1.3 \ \mu H$$

$$A = R_s(1 + Q_L^2) = 7.0 \ (1 + 4^2) = 119$$

$$B = \sqrt{\left(\frac{A}{R_L}\right) - 1} = \sqrt{\left(\frac{119}{50}\right) - 1} = 1.17$$

$$X_{L2} = R_L B = 50 \times 1.17 = 58.7 \ \Omega$$

$$\therefore L2(\mu H) = \frac{X_{L2}}{2\pi \ f(MHz)} = \frac{58.7}{2\pi \times 3.5} = 2.7 \ \mu H$$

$$X_{C1} = \frac{A}{(Q_L + B)} = \frac{119}{(4 + 1.17)} = 23.0 \ \Omega$$

$$\therefore C1 \ \mu F = \frac{1}{2\pi \ f \ X_{C1}} = \frac{1}{2\pi \times 3.5 \times 23.0} = 0.002 \ \mu F$$

Fig. 27 — A practical example of network no. 3 and the solution to the network design.

additional harmonic suppression will normally be needed. This is also true for networks 1 and 2. These three networks are covered in detail in the Motorola *Application Note AN-267*. Another excellent paper on the subject was written by Becciolini — *Motorola Application Note AN-271*.

The equations for networks 1, 2 and 3 were taken from *AN-267*. That paper contains computer solutions to these networks and others, with tabular information for various Qs and source impedances. A fixed load value of 50 ohms is the base for the tabular data.

A design example for network 3 is given in Fig. 27. The solutions for the other two networks follow the same general trend,

so examples for networks 1 and 2 will not be given. In Fig. 27 the component "C$_{out}$" is taken from the manufacturer's data sheet. If it is not available it can be ignored at the expense of a slight mathematical error in the network determination. By making C1 variable the network can be made to approximate the correct transformation ratio. At the lower frequencies C1 will be fairly large in value. This may require a fixed-value silver-mica capacitor being used in parallel with a mica compression trimmer to obtain the exact value of capacitance needed. The equations will seldom yield standard values of capacitance.

L1 and L2 of Fig. 27 can be wound on

powdered-iron toroid cores of suitable cross-sectional area for the power involved. This is explained in an earlier chapter of this book. L1 and L2 should be separated from one another by mounting them apart and at right angles. Alternatively, a shield can be used between the inductors. This will prevent unwanted capacitive and inductive coupling effects between the input and output terminals of the network. Despite the self-shielding nature of toroidal inductors, some coupling is possible when they are in close proximity.

Stable Operating Conditions

Purity of emissions and longevity of the active devices in a tube or transistor circuit depend heavily upon stability during operation. The subject of power-lead decoupling has already been treated, wherein bypassing for vhf, hf and lf is essential in the dc leads to each transistor amplifier stage. The bypass capacitors are used in combination with low-inductance rf chokes in most instances. Although the same concept can be applied to tube types of amplifiers, the possibility of self-oscillations at frequencies lower than vhf are not as pronounced. For the most part, tube amplifiers will operate stably if input-output shielding is provided for high-gain stages (grid circuitry shielded from plate circuitry). Depending upon the inter-electrode capacitances of tubes, a neutralization circuit may be necessary. This will cancel *positive feedback* and prevent regeneration. It involves sampling a small amount of the output energy (opposite phase of the input energy) and feeding it back to the amplifier input, thereby cancelling the unwanted in-phase (positive) feedback. A typical circuit is given in Fig. 28. L2 provides a 180-phase reversal because it is center tapped. C1 is used between the plate and the lower half of the grid tank to permit cancellation of the unwanted feedback voltage. C1 is set for the approximate value of the grid-plate capacitance of the tube, the value of which can usually be found on the tube data sheet. C1 is adjusted in one of two ways: It is set at a value which results in no change in tube grid current as the plate tank is tuned through its range. Alternatively, operating voltages are applied to the tube, but no drive is used. A scope or sensitive rf meter is connected to the plate tank and C1 is adjusted for zero output signal — indicating that self-oscillation is not taking place. *Extreme care must be exercised when the tube operating voltages are present!* Always keep the probe of the measuring instrument connected to a circuit point which does not contain dc voltage. Sampling at L4 is recommended for the circuit of Fig. 28. C1 needs to have ample plate spacing to prevent voltage breakdown when the amplifier is operating.

All leads which conduct rf energy

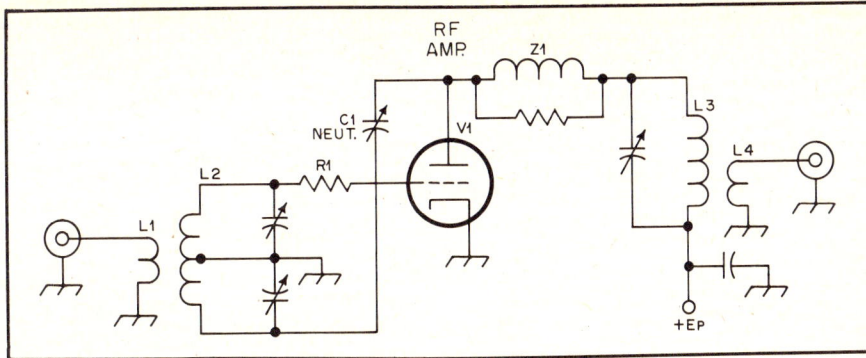

Fig. 28 — Example of neutralization of a single-ended rf amplifier.

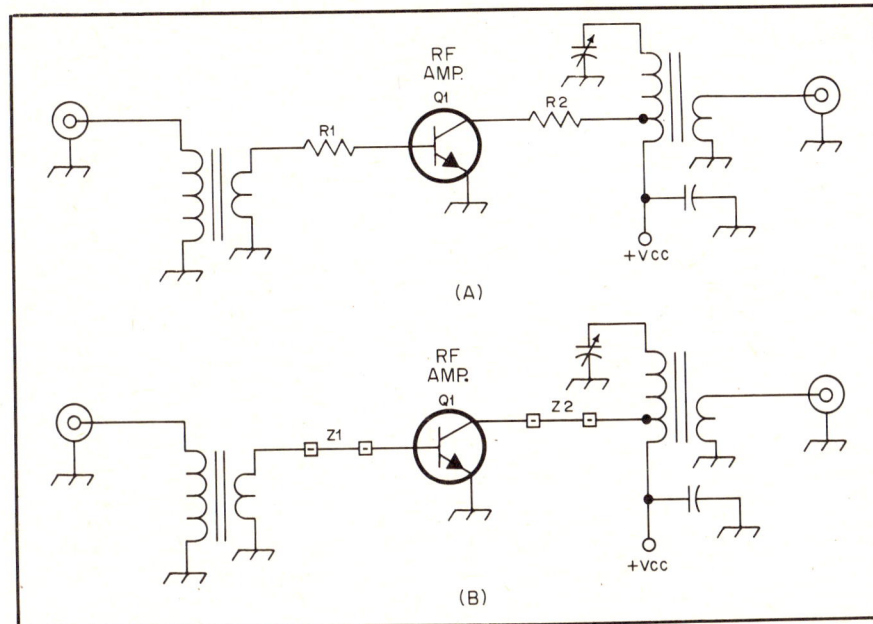

Fig. 29 — Suppression methods for vhf and uhf parasitics in solid-state amplifiers.

should be kept as short as possible in an amplifier circuit, and likewise with the pigtails of bypass capacitors. This applies to tube or transistor amplifiers.

Z1 of Fig. 28 is a vhf parasitic choke. Such a network will damp self-oscillations at vhf and uhf by acting as a series impedance which breaks up the usual unwanted vhf/uhf circuit path. Z1 consists of a non-inductive resistor between 51 and 100 ohms. A coil is wound around the resistor body to provide a broadband rf choke which presents a high impedance at vhf and higher, but looks like a low reactance in the hf region and lower. A typical parasitic suppressor for a power level up to 150 watts contains 6 to 8 turns of no. 20 wire wrapped around a 56-ohm, 1-watt composition resistor. The coil ends are soldered to the resistor pigtails near the body of the resistor. Z1 is then placed as close to the tube plate pin or cap as possible. For higher rf powers it is practical to use a high-wattage Globar resistor, or a 25-watt noninductive (NIT) power resistor around which a suitable number of wire turns (no. 14 or 12 wire) have been wound. Additional parasitic

suppression can be had by inserting a low-ohmage resistor (10 to 51 ohms) in series with the tube input, near the tube socket. This is illustrated by R1 of Fig. 28. Vhf or uhf parasitics can be detected by means of a high-frequency scope, or by probing the plate tank with a sensitive wavemeter which tunes from 30 MHz and higher.

Parasitic oscillations can be prevented in solid-state amplifiers by using a small amount of resistance in the base or collector lead of low-power amplifiers (Fig. 29A). The value of R1 or R2 is typically between 10 and 22 ohms. Resistors are seldom necessary at both points in a circuit, but can be used effectively at either point. R1 or R2 should be located as close to the transistor as practical.

At power levels in excess of approximately 0.5 Watt, the method of parasitic suppression shown in Fig. 29B is suggested. The voltage drop across a resistor would be prohibitive at the higher power levels, so one or more ferrite beads can be substituted (Z1 and Z2). A permeability of 125 will suffice. The beads need not be used at

both circuit locations. Generally, the lowest power port is best suited for the suppression devices discussed here. This means that the resistor or ferrite beads should be located at the base terminal of the transistor.

Because of the rising gain characteristic of bipolar transistors as the frequency is lowered, shunt and degenerative feedback are often used to prevent instability. The net effect is that in the regions where low-frequency self-oscillations are most likely to occur, the feedback increases by nature of the feedback network. The heavier the feedback, the lower the amplifier gain. In the circuit of Fig. 30 C1 and R3 provide negative feedback which increases progressively as the frequency is lowered. The network has some effect at the desired operating frequency — a gain/stability tradeoff — but has a pronounced effect at the lower frequencies. The values for C1 and R3 are chosen experimentally in most instances, the precise values being dependent upon the operating frequency of the amplifier and the amount of feedback voltage available from the tap-off point. C1 will usually be between 220 pF and 0.0015 μF for hf-band amplifiers. R3 may be a value from 51 to 5600 ohms. A rule of thumb that may prove helpful is to use a network that reduces the stage gain by approximately 1.5 dB at the lowest operating frequency.

R2 of Fig. 30 provides emitter degeneration at low frequencies when the bypass capacitor, C2, is chosen for adequate rf bypassing at the intended operating frequency. Below the desired frequency, C2 becomes progressively less effective as the frequency is lowered, thereby increasing the degenerative feedback caused by R2. This lowers the amplifier gain. R2 in a power stage is seldom greater than 10 ohms in value, and may be as low as 1 ohm. It is important to consider that under some operating and layout conditions R2 can cause instability. This form of feedback should be used only in those circuits where unconditional stability can be achieved.

Solid-state amplifiers that are built on pc boards can be made stable (in addition to the foregoing measures) by utilizing double-clad pc board material. The copper on the component side of the board is used as a ground-plane surface by removing the copper around each hole where a component is to be mounted. This ground plane is made electrically common to the ground elements on the etched side of the board. Such a technique helps prevent unwanted ac ground loops which can cause feedback and instability. Furthermore, the nonground etched elements form low-capacitance bypass capacitors against the ground-plane surface of the board. This aids in reducing the chance for vhf and uhf parasitic oscillations.

R1 of Fig. 30 is useful in swamping the input of an amplifier. This reduces the chance for low-frequency self-oscillations, but has a minor effect on the amplifier performance in the desired operating range. Values from 3 to 27 ohms are typical. When they are in shunt with the normal (low) base impedance of a power amplifier, they contribute only slightly to the lowering of the device input impedance. The rule of thumb is to use the largest value of resistance that will ensure stability. R1 should be located as close to the transistor base terminal as possible. The pigtails must be kept short to prevent stray inductances from forming. It is helpful to use two resistors in parallel to limit the amount of inductive reactance introduced by a single resistor.

C3 of Fig. 30 can be added to some power amplifiers to damp vhf/uhf parasitic oscillations. The capacitor should be low in reactance at vhf and uhf, but must present a high reactance at the operating frequency. The exact value selected will depend upon the collector impedance. A reasonable rule of thumb is to use an X_c of 10 times the collector impedance at the operating frequency. Silver-mica or ceramic-chip capacitors are suggested for this application. For example, a 3.5-MHz amplifier with a 10-ohm collector impedance would use a capacitor with an X_c of 100 ohms. C1 would be 454 pF under this rule. At 150 MHz the same capacitor would have an X_c of only 2.3 ohms, making it an effective vhf and higher bypass element. An additional advantage is seen in the bypassing action for vhf and uhf harmonic energy in the collector circuit. C3 should be placed as close to the collector terminal as possible, using short leads. The effects of C3 in a broadband amplifier are relatively insignificant at the operating frequency. However, when a narrow-band collector network is employed, the added

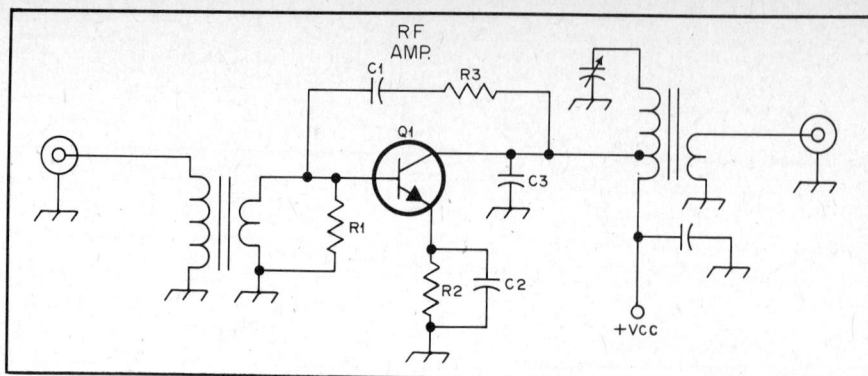

Fig. 30 — Illustration of shunt feedback in a transistor amplifier. Components C1 and R3 comprise the feedback network.

capacitance of C3 must be absorbed into the network design in the same manner as the C_{out} of the transistor.

Broadband Transformers

The usefulness of toroidal broadband transformers is practically beyond description in this text. But, some of the more popular transformer configurations are presented here for those who wish to employ them in matching networks associated with solid-state devices and tubes. It is important to realize that broadband transformers are best suited to low-impedance applications, say, up to a few hundred ohms. They should be thought of as devices which can transform one impedance to another, in terms of the transformation ratio they make possible. They should not be regarded as devices which are built for some specified pair of impedances, such as 200 ohms to 50 ohms in the case of a 4:1 transformer. The term "balun," despite its misuse, pertains only to a broadband transformer which converts a *bal*anced condition to one which is *un*balanced, or vice versa. The often-heard

expressions "bal-oon" and "bal-um" are not correct.

The broadband transformers illustrated in Figs. 31, 32 and 33 are suitable for use in solid-state circuits, as matching devices between circuit modules and in antenna-matching networks. For low power levels the choice of core material is often ferrite. Powdered-iron is more often the designers preference when working with fairly high levels of power. The primary objection of some, respective to the use of ferrite at high power, is damage to the core material during saturation and overheating. This can alter the permeability factor of the core material permanently. Powered-iron is more tolerant in this regard.

Fig. 31 shows two types of 4:1 transformers, plus a method for connecting two of them in series to effect a 16:1 transformation. The circuit at E is often used between a 50-ohm source and the base of an rf power transistor.

Two styles of 9:1 transformer are seen in Fig. 32 at A and C. They are also found at the input to transistor amplifiers and between the collector and the load. The

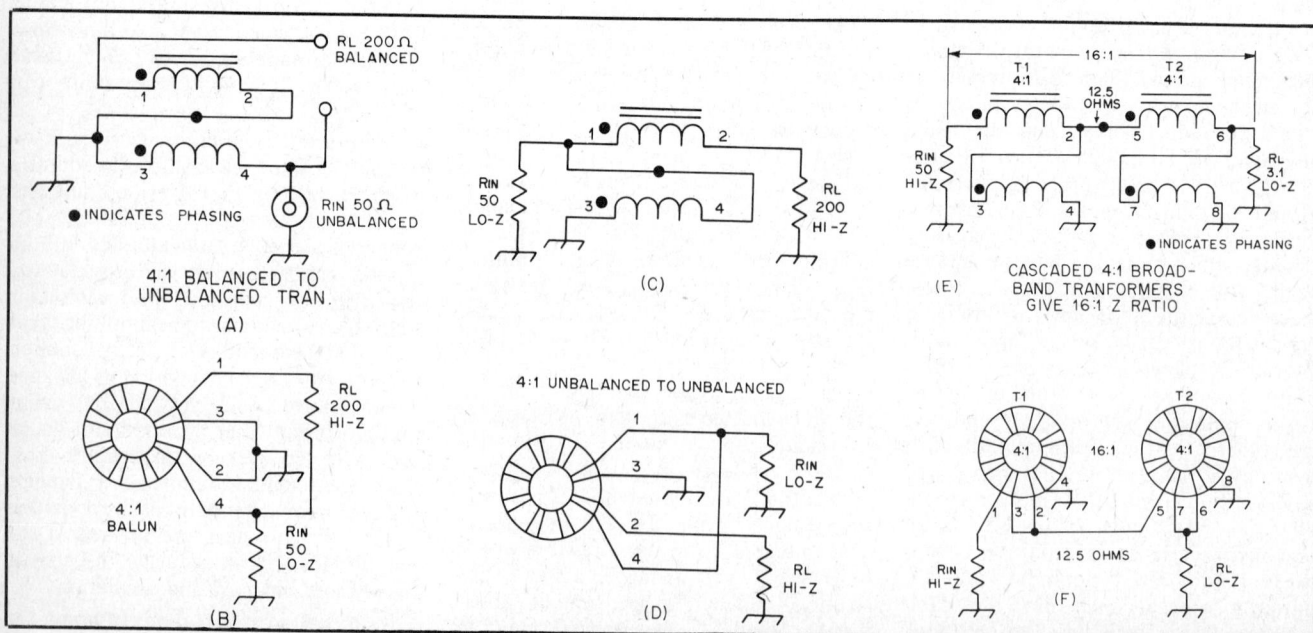

Fig. 31 — Circuit illustrations of 4:1 broadband transformers.

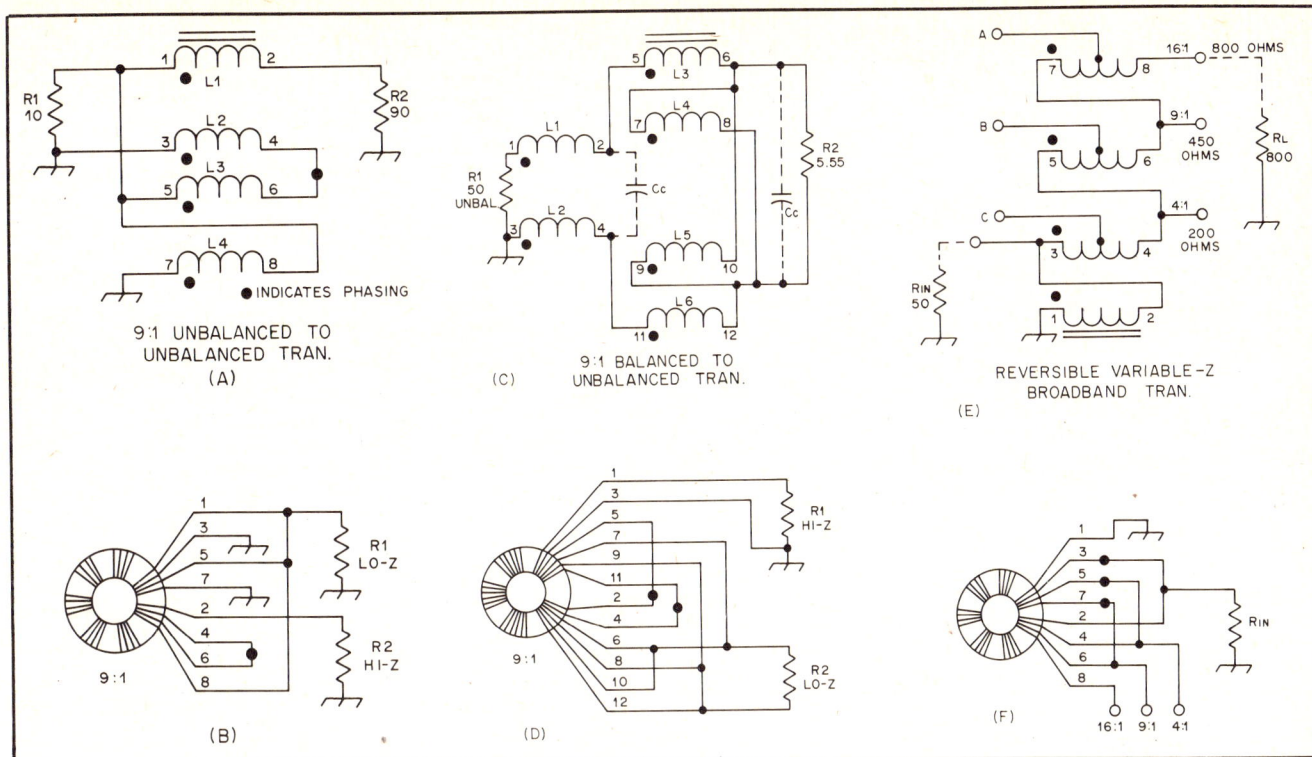

Fig. 32 — Circuit examples of 9:1 broadband transformers (A and C) and a variable-impedance transformer (E).

variable-ratio transformer of Fig. 32 C is excellent for obtaining a host of impedance transformations. This transformer was developed by W2FMI for use in matching ground-mounted vertical antennas.

Phase-reversal, 1:1 balun and hybrid-combiner transformers are shown in Fig. 33. The circuit at E of Fig. 33 is useful when it is necessary to feed two signals to a single load. When the input signals are on different frequencies the power is split evenly between R3 and R4. When the input voltages are on the same frequency (as with two transistor amplifiers feeding a single load), with the amplitudes and phase identical, all of the power is delivered to R4.

RF Power Amplifier Circuitry

In addition to proper tank and output-coupling circuits, an rf amplifier must be provided with suitable operating voltages and an rf driving or excitation voltage. All rf amplifier tubes require a voltage to operate the filament or heater (ac is usually permissible), and a positive dc voltage between the plate and filament or cathode (plate voltage). Most tubes also require a negative dc voltage (biasing voltage) between control grid (grid no. 1) and filament or cathode. Screen-grid tubes require in addition a positive voltage (screen voltage or grid no. 2 voltage) between screen and filament or cathode.

Biasing and plate voltages may be fed to the tube either in series or parallel with the associated rf tank circuit as discussed in the chapter on electrical laws and circuits.

It is important to remember that true plate, screen or biasing voltage is the voltage between the particular electrode and filament or cathode. Only when the cathode is directly grounded to the chassis may the electrode-to-chassis voltage be taken as the true voltage. The required rf driving voltage is applied between grid and cathode.

Plate power input is the dc power input to the plate circuit (dc plate voltage × dc plate current). Screen power input likewise is the dc screen voltage × the dc screen current.

Plate dissipation is the difference between the rf power delivered by the tube to its loaded plate tank circuit and the dc plate power input. The screen, on the other hand, does not deliver any output power, and therefore its dissipation is the same as the screen power input.

Transmitting-Tube Ratings

Tube manufacturers specify the maximum values that should be applied to the tubes they produce. They also publish sets of typical operating values that should result in good efficiency and normal tube life.

The same transmitting tube may have different ratings, depending upon the manner in which the tube is to be operated and the service in which it is to be used. These different ratings are based primarily upon the heat that the tube can safely dissipate. Some types of operation, such as with grid or screen modulation, are less efficient than others, meaning that the tube must dissipate more heat. Other types of operation, such as cw or single-sideband phone are intermittent in nature, resulting in less average heating than in other modes where there is a continuous power input to the tube during transmissions. There are also different ratings for tubes used in transmitters that are in almost constant use (CCS — Continuous Commercial Service), and for tubes that are to be used in transmitters that average only a few hours of daily operation (ICAS — Intermittent Commercial and Amateur Service). The latter are the ratings used by amateurs who wish to obtain maximum output with reasonable tube life.

Maximum Tube Ratings

Maximum ratings, where they differ from the values given under typical operating values, are not normally of significance to the amateur except in special applications. No single maximum value should be used unless all other ratings can simultaneously be held within the maximum values. As an example, a tube may have a maximum plate-voltage rating of 2000, a maximum plate-current rating of 300 mA, and a maximum plate-power-input rating of 400 watts. Therefore, if the maximum plate voltage of 2000 is used, the plate current should be limited to 200 mA (instead of 300 mA) to stay within the maximum power-input rating of 400 watts.

Maximum Transistor Ratings

Transistor data sheets specify a maxi-

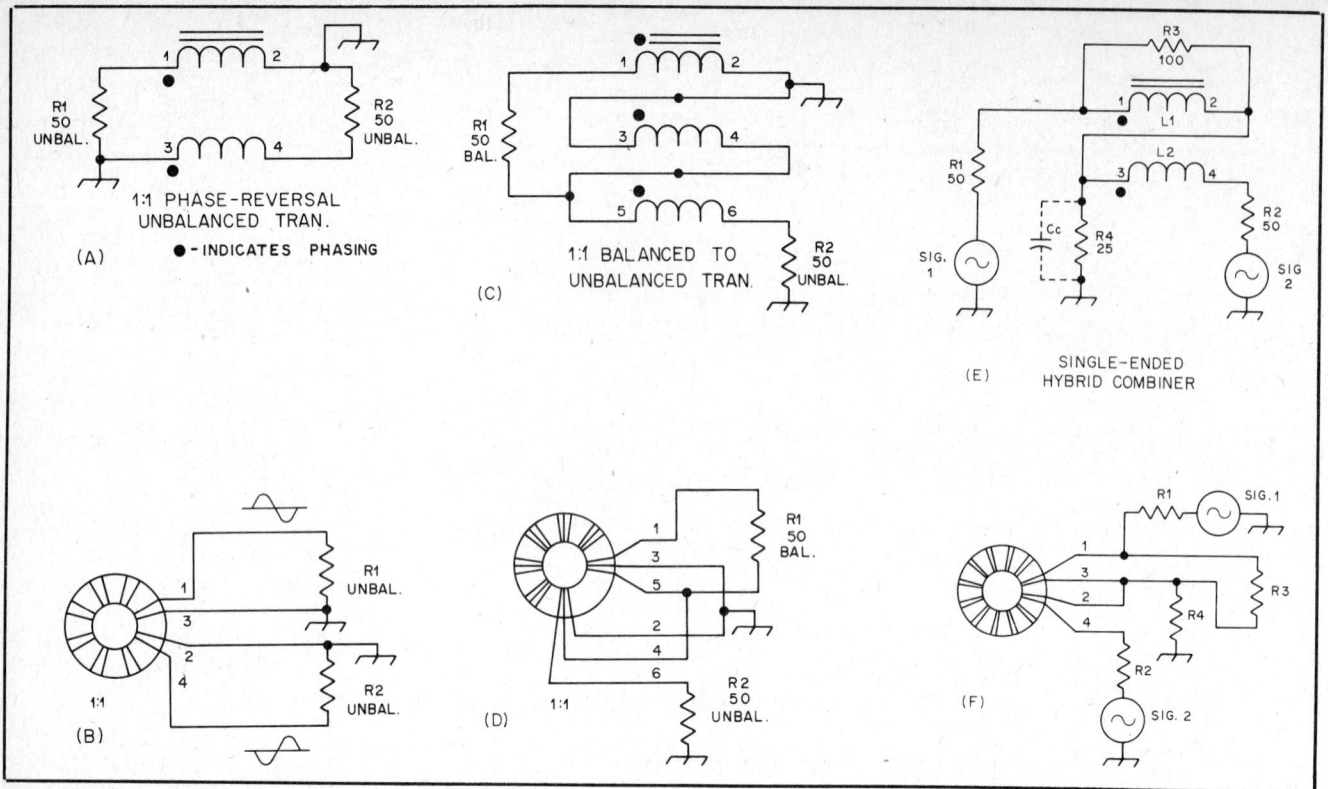

Fig. 33 — Assorted broadband transformers.

mum operating voltage for several conditions. Of special interest to amateurs is the V_{ceo} specification (collector-to-emitter voltage, with the base open). When a transistor is called upon to handle an ac signal, the collector-to-emitter voltage can rise to twice the dc supply. Thus, if a 12-volt supply is used, the transistor should have a V_{ceo} of 24 or greater to prevent damage. If that same transistor is amplitude-modulated (as in the PA of an a-m transmitter), a collector-emitter voltage swing (theoretical) as great as four times the supply voltage can occur. A transistor chosen for this application should have a V_{ceo} of 48 or greater.

The f_T rating of a common-emitter transistor amplifier is based on the point at which the transistor gain is unity (1) with respect to operating frequency. In the interest of predictable performance and amplifier stability it is best to select a transistor that was designed for a particular frequency range. When this is not practical, the f_T should be roughly 5 to 10 times the operating frequency. Therefore, a suitable transistor for use at 3.5 MHz would have an f_T between 17.5 and 35 MHz. If a much higher f_T were selected, say, 250 MHz, the published rf performance curves for the device would be quite inaccurate at 3.5 MHz, and the transistor gain would be extremely high compared to the rated gain at the intended operating frequency of the device.

Power transistor gain is normally specified as "typical" dB. This information applies to some specified operating frequency and is by no means all-inclusive from, for example, medium frequency up to the vhf spectrum. The frequency at which a particular gain figure applies is stated on the transistor data sheet. Generally, the gain will be higher below that frequency and it will decrease above that frequency. Gain information is useful in predicting how much output power can be obtained for a given input power; i.e., a 13-dB gain transistor delivering an output of 10 watts would require a driving power of 0.5 W (Gain$_{(dB)}$ = 10 log [P2/P1])

Power dissipation for a transistor is expressed symbolically as P_D. This maximum rating is based on a case temperature of 25°C. For example, a total device dissipation of 30 watts might be specified at a case temperature of 25°C. If greater temperatures were expected, the transistor would have to be derated in mW per degree C. A Motorola MRF215 would be derated 177 mW per additional degree C. The effectiveness of the transistor heat sink plays an important role in maximum power utilization of a given device. It is not unusual to see a cooling fan used in combination with a large heat sink to aid in lowering the transistor case temperature: Heat is one of the worst enemies of power transistors.

A rule of thumb for selecting a P_D rating which is suitable for a given rf power output amount is to choose a transistor which has a maximum dissipation of twice the desired output power. Hence, a 20-watt transistor would be picked for use in a 10-watt-output amplifier. Some manufacturers push the power margin a bit harder, utilizing a transistor which delivers a power output which is as great as 3/4 the P_D rating. So close a safety margin is somewhat risky for inexperienced builders.

Sources of Tube Electrode Voltages: Filament or Heater Voltage

The heater voltage for the indirectly heated cathode-type tubes found in low-power classifications may vary 10 percent above or below rating without seriously reducing the life of the tube. But the voltage of the higher-power, filament-type tubes should be held closely between the rated voltage as a minimum and five percent above rating as a maximum. Make sure that the plate power drawn from the power line does not cause a drop in filament voltage below the proper value when plate power is applied.

Thoriated-type filaments lose emission when the tube is overloaded appreciably. If the overload has not been too prolonged, emission sometimes may be restored by operating the filament at rated voltage with all other voltages removed for a period of 10 minutes, or at 20 percent above rated voltage for a few minutes.

Plate Voltage

Dc plate voltage for the operation of rf amplifiers is most often obtained from a transformer-rectifier-filter system (see power-supply chapter) designed to deliver the required plate voltage at the required current. However, batteries or other dc-generating devices are sometimes used in certain types of operation (see portable-mobile chapter).

Bias and Tube Protection

Several methods of obtaining bias are shown in Fig. 34. At A, bias is obtained by the voltage drop across a resistor (R1) in the grid dc return circuit when rectified grid current flows. The proper value of resistance may be determined by dividing the required biasing voltage by the dc grid current at which the tube will be operated. Then, so long as the rf driving voltage is adjusted so that the dc grid current is the recommended value, the biasing voltage will be the proper value. The tube is biased only when excitation is applied, since the voltage drop across the resistor depends upon grid-current flow. When excitation is removed, the bias falls to zero. At zero bias most tubes draw power far in excess of the plate-dissipation rating. It is advisable to make provision for protecting the tube when excitation fails by accident, or by intent as it does when a precending stage in a cw transmitter is keyed.

If the maximum cw ratings shown in the tube tables are to be used, the input should be cut to zero when the key is open. Aside from this, it is not necessary that plate current be cut off completely but only to the point where the rated dissipation is not exceeded. In this case plate-modulated phone ratings should be used for cw operation, however.

With most tubes this protection, plus the required operating bias, can be supplied by obtaining all bias from a source of fixed voltage, as shown in Fig. 34B.

Fixed bias may be obtained from dry batteries or from a power pack (see power-supply chapter). If dry batteries are used, they should be checked periodically, since even though they may show normal voltage, they eventually develop a high internal resistance.

In Fig. 34C and D, bias is obtained from the voltage drop across a Zener diode in the cathode (or filament center-tap) lead. Operating bias is obtained by the voltage drop across VR1 as a result of plate (and screen) current flow. The Zener-diode wattage rating is twice the product of the maximum cathode current times the developed bias. Therefore, a tube requiring 15 volts of bias during a maximum cathode-current flow of 100 mA would dissipate 1.5 W in the Zener diode. The diode rating, to allow a suitable safety factor, would be 3 W or greater. The circuit of Fig. 34D illustrates how D1 would be used with a cathode-driven (grounded-grid) amplifier as opposed to the grid-driven example at C.

Transistor Biasing

Solid-state power amplifiers generally operate in Class C or Class AB. When some bias is desired during Class C operation (Fig. 35A), a resistance of the appropriate value can be placed in the emitter return as shown. Most transistors will operate in Class C without adding bias externally, but in some instances the amplifier efficiency can be improved by means of emitter bias. Reverse bias supplied to the *base* of the Class C transistor should be avoided because it will lead to internal breakdown of the device during peak drive periods. The destruction is frequently a cumulative phenomenon, leading to gradual destruction of the transistor junction.

A simple method for Class AB biasing is seen in Fig. 35B. D1 is a silicon diode which acts as a bias clamp at approximately 0.7 V. The forward bias establishes linear-amplification conditions. That value of bias is not always optimum for a specified transistor in terms of IMD. Variable bias of the type illustrated in Fig. 35C permits the designer sufficient variance to locate the best operating point respective to linearity.

Screen Voltage for Tubes

For cw and fm operation, and under certain conditions of phone operation (amplitude modulation) the screen may be operated from a power supply of the same type used for plate supply, except that voltage and current ratings should be appropriate for screen requirements. The screen may also be operated through a series resistor or voltage-divider from a source of higher voltage, such as the plate-voltage supply, thus making a separate supply for the screen unnecessary. Certain precautions are necessary, depending upon the method used.

It should be kept in mind that screen current varies widely with both excitation and loading. If the screen is operated from a fixed-voltage source, the tube should never be operated without plate voltage and load, otherwise the screen may be damaged within a short time. Supplying the screen through a series dropping resistor from a higher-voltage source, such as the plate supply, affords a measure of protection, since the resistor causes the screen voltage to drop as the current increases, thereby limiting the power drawn by the screen. However, with a resistor, the screen voltage may vary considerably with excitation, making it necessary to check the voltage at the screen terminal under actual operating conditions to make sure that the screen voltage is normal. Reducing excitation will cause the screen current to drop, increasing the voltage; increasing excitation will have the opposite effect. These changes are in addition to those caused by changes in bias and plate loading, so if a screen-grid tube is operated from a series resistor or a voltage divider, its voltage should be checked as one of the final adjustments after excitation and loading have been set.

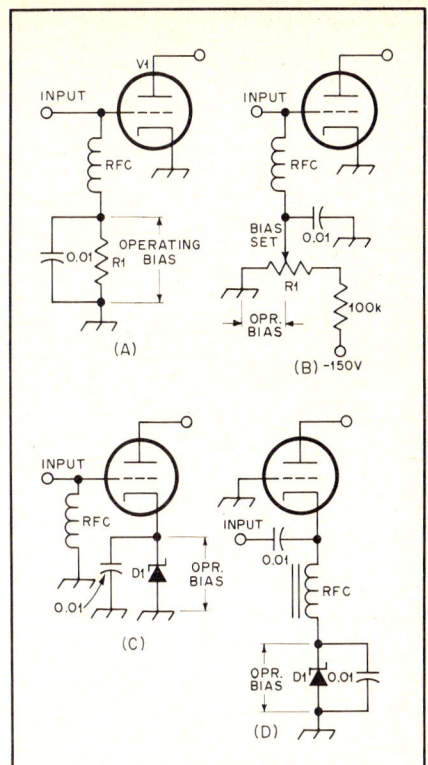

Fig. 34 — Various techniques for providing operating bias with tube amplifiers.

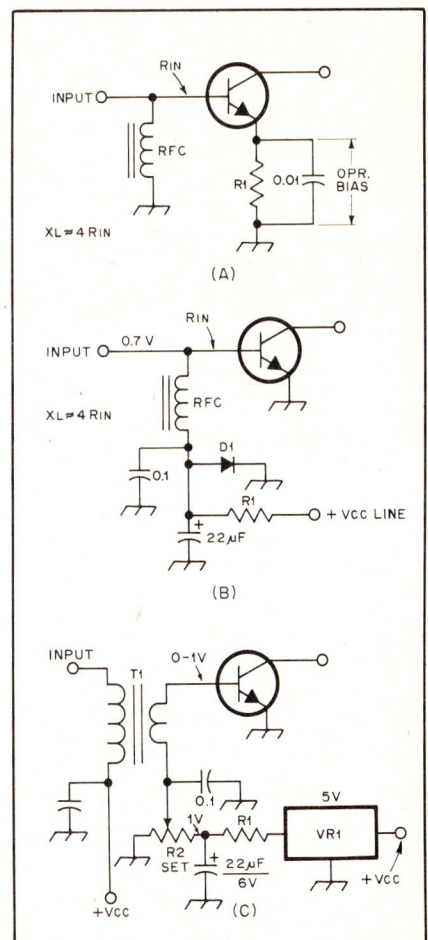

Fig. 35 — Biasing methods for use with transistor amplifiers.

An approximate value for the screen-voltage dropping resistor may be obtained by dividing the voltage *drop* required from the supply voltage (difference between the supply voltage and rated screen voltage) by the rated screen current in decimal parts of an ampere. Some further adjustment may be necessary, as mentioned above, so an adjustable resistor with a total resistance above that calculated should be provided.

Protecting Screen-Grid Tubes

Considerably less grid bias is required to cut off an amplifier that has a fixed-voltage screen supply than one that derives the screen voltage through a high value of dropping resistor. When a "stiff" screen voltage supply is used, the necessary grid cutoff voltage may be determined from an inspection of the tube curves or by experiment.

Feeding Excitation to the Grid

The required rf driving voltage is supplied by an oscillator generating a voltage at the desired frequency, either directly or through intermediate amplifiers, mixers, or frequency multipliers.

The grid of an amplifier operating under Class C conditions must have an exciting voltage whose peak value exceeds the negative biasing voltage over a portion of the excitation cycle. During this portion of the cycle, current will flow in the grid-cathode circuit as it does in a diode circuit when the plate of the diode is positive in respect to the cathode. This requires that the rf driver supply power. The power required to develop the required peak driving voltage across the grid-cathode impedance of the amplifier is the rf driving power.

The tube tables give approximate figures for the grid driving power required for each tube under various operating conditions. These figures, however, do not include circuit losses. In general, the driver stage for any Class C amplifier should be capable of supplying at least three times the driving power shown for typical operating conditions at frequencies up to 30 MHz and from three to 10 times at higher frequencies.

Since the dc grid current relative to the biasing voltage is related to the peak driving voltage, the dc grid current is commonly used as a convenient indicator of driving conditions. A driver adjustment that results in rated dc grid current when the dc bias is at its rated value, indicates proper excitation to the amplifier when it is fully loaded.

In coupling the grid input circuit of an amplifier to the output circuit of a driving stage the objective is to load the driver plate circuit so that the desired amplifier grid excitation is obtained without exceeding the plate-input ratings of the driver tube.

Driving Impedance

The grid-current flow that results when the grid is driven positive in respect to the cathode over a portion of the excitation cycle represents an average resistance across which the exciting voltage must be developed by the driver. In other words, this is the load resistance into which the driver plate circuit must be coupled. The approximate grid input resistance is given by

Input impedance (ohms)

$$= \frac{\text{driving power (watts)}}{\text{dc grid current (mA)}^2} \times 620,000$$

For normal operation, the driving power and grid current may be taken from the tube tables. Since the grid input resistance is a matter of a few thousand ohms, an impedance step-up is necessary if the grid is to be fed from a low-impedance transmission line.

Cooling: Tubes

Vacuum tubes must be operated within the temperature range specified by the manufacturer if long tube life is to be achieved. Tubes with glass envelopes rated at up to 25 watts of plate dissipation may be run without forced-air cooling, if a moderate amount of cooling by convection can be arranged. If a perforated-metal enclosure is used, and a ring of 1/4-inch diameter holes are placed around the tube socket, normal air flow can be relied upon to remove excess heat at room temperatures.

For tubes with greater plate dissipation, or those operated with plate currents in excess of the manufacturer's ratings (often the case with TV sweep tubes) forced air cooling with a fan or blower is needed. Fans, especially those designed for cooling hi-fi cabinets, are preferred because they operate quietly. However, all fans lose their ability to move air when excessive back pressure exists. For applications where a stream of air must be directed through a tube socket, a blower is usually required.

One method for directing a flow of air around a tube envelope or through tube cooling fins involves the use of a pressurized chassis. This system is shown in Fig. 36. A blower is attached to the chassis and forces air up through the tube socket and around the tube. A chimney (not shown in this drawing) is used to guide the air around the tube as it leaves the socket. A chimney will prevent the air from being dispersed as it hits the envelope or cooling fins, concentrating the flow for maximum cooling.

Most manufacturers rate tube cooling requirements for continuous-duty operation. The manufacturer's literature will indicate the required cubic feet per minute (CFM) of air flow at some particular back pressure. Back pressure is the pressure that is built up inside the airtight chassis when the blower is operational. Forced air entering the chassis from the blower can escape only through the tube socket/tube/chimney assembly. Since this assembly represents a certain amount of resistance to the flow of air, an amount of pressure is built up inside the chassis. The exact amount of pressure will depend on the blower and the tube socket/tube/chimney characteristics. Blowers vary in their ability to work against back pressure so the matter of blower selection should not be taken lightly.

Values of CFM and back pressure for some of the more popular tubes, sockets and chimneys are given in Table 6. Back pressure is specified in inches of water and can be easily measured as indicated in Figs. 36 and 37, by means of a manometer. A manometer is nothing

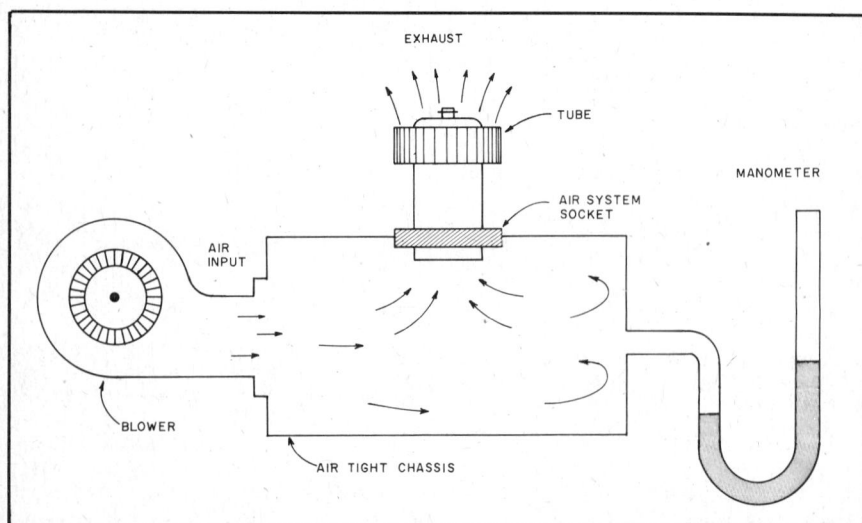

Fig. 36 — Air is forced into the chassis by the blower and exits through the tube socket. The manometer is used to measure system back pressure, which is an important factor in determining the proper size blower.

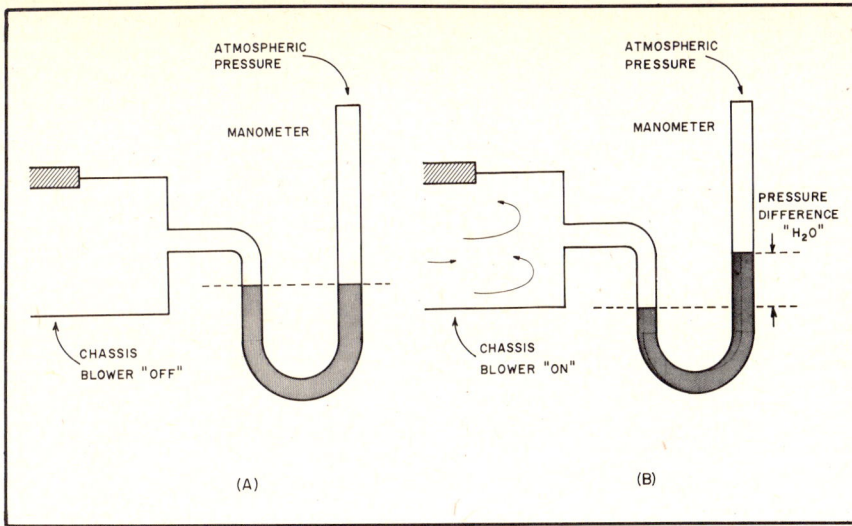

Fig. 37 — At A the blower is "off" and the water will seek its own level in the manometer. At B the blower is "on" and the amount of back pressure in terms of inches of water can be measured as indicated.

Table 6
Specifications of Some Popular Tubes, Sockets and Chimneys

Tube	CFM	Back Pressure (inches)	Socket	Chimney
3-400Z/8163	13	0.13	SK-400, SK-410	SK-416
3-500Z	13	0.082	SK-400, SK-410	SK-406
3-1000Z/8164	25	0.38	SK-500, SK-510	SK-516
3CX1500/8877	35	0.41	SK-2200, SK-2210	SK-2216
4-250A/5D22	2	0.1	SK-400, SK-410	SK-406
4-400A/8438	14	0.25	SK-400, SK-410	SK-406
4-1000A/8166	20	0.6	SK-500, SK-510	SK-506
4CX250R/7850W	6.4	0.59	SK-600, SK-600A, SK-602A, SK-610, SK-610A, SK-611, SK-612, SK-620, SK-620A SK-621, SK-630	SK-606 SK-626
4CX300A/8167	7.2	0.58	SK-700, SK-710, SK-711A, SK-712A, SK-740, SK-760 SK-761, SK-770	SK-606
4CX350A/8321	7.8	1.2	Same as 4CX250R	
4CX1000A/8168 4CX1500/8660	25	0.2	SK-800B, SK-810B, SK-890B	SK-806
8874	8.6	0.37		

These values are for sea-level elevation. For locations well above sea-level (Denver, Colorado, for example), add an additional 20% to the figure listed.

Table 7
Blower Performance Specifications

Wheel Dia.	Wheel Width	RPM	Free Air	Back Pressure (inches) 0.1	0.2	0.3	0.4	0.5	Cutoff	Stock No.
2″	1″	3160	15	13	4	—	—	—	0.22	2C782
3″	1-15/32″	3340	54	48	43	36	25	17	0.67	4C012
3″	1-7/8″	3030	60	57	54	49	39	23	0.60	4C440
3″	1-7/8″	2880	76	70	63	56	45	8	0.55	4C004
3-13/16″	1-7/8″	2870	100	98	95	90	85	80	0.80	4C443
3-13/16″	2-1/2″	3160	148	141	135	129	121	114	1.04	4C005

more than a piece of clear tubing, open at both ends and fashioned in the shape of a "U." The manometer is temporarily connected to the chassis and is removed after the measurements are completed. As shown in the diagrams, a small amount of water is placed in the tube. At Fig. 37 the blower is "off" and the water will seek its own level. At B, the blower is "on" (socket, tube and chimney in place) and the pressure difference, in terms of inches of water, is measured. For most applica-

tions a standard rules can be used for the measurement and the results will be sufficiently accurate.

Table 7 illustrates the performance specifications for one particular brand of blowers. These are Dayton blowers which are available through W. W. Grainger outlets throughout the U.S. Blowers with similar wheel dimensions of different manufacture likely have similar characteristics. If in doubt about specifications contact the manufacturer of

the unit. As an example, assume that an amplifier is to be built using a 3-1000Z tube. A blower capable of supplying 25 CFM at a back pressure of 0.38 inches of water is required. Referring to Table 7 it appears that the second blower listed would be suitable, although it may be marginal since it can only supply 25 CFM into a back pressure of 0.4 inches of water. The next larger size would provide a margin of safety.

When a pair of tubes is used, the CFM rating is doubled, but the back pressure remains the same as that for one tube. A pair of 3-1000Z tubes, for example, would require 50 CFM at a back pressure of 0.38 inches of water. In this case the fifth blower listed in the Table would be suitable since it can supply 85 CFM at a back pressure of 0.4 inches of water. Always choose a blower that can supply at least the required amount of air. Smaller blowers will almost certainly lead to shortened tube life.

Table 6 also contains the part numbers for air-system sockets and chimneys to be used with the tubes that are listed. The builder should investigate which of the sockets listed for the 4CX250R, 4CX300A, 4CX1000A and 4CX1500A best fits the circuit needs. Some of the sockets have certain tube elements grounded internally through the socket. Others have elements bypassed to ground through capacitors that are integral parts of the sockets.

An efficient blower is required when using the external-anode tubes, such as the 4CX250R. Such tubes represent a trade-off which allows high-power operation with a physically small device at the expense of increased complexity in the cooling system. Other types of external-anode tubes are now being produced for conductive cooling. Electrical insulators which are also excellent thermal conductors, such as AlSiMg (aluminum-silicon-magnesium compound) and Be (beryllium), couple the tube to a heat sink. Requirements for the heat dissipator are calculated in the same way as for power transistors, as outlined below. Similar tubes are made with special anode structures for water or vapor cooling, allowing high-power operation without producing an objectionable noise level from the cooling system.

Transistor Cooling

Some bipolar power transistors have the collector connected directly to the case of the device, as the collector must dissipate most of the heat generated when the transistor is in operation. Others have the emitter connected to the case. However, even the larger case designs cannot conduct heat away fast enough to keep the operating temperature of the device functioning within the *safe area*, the maximum temperature that a device can stand without damage. Safe area is usually

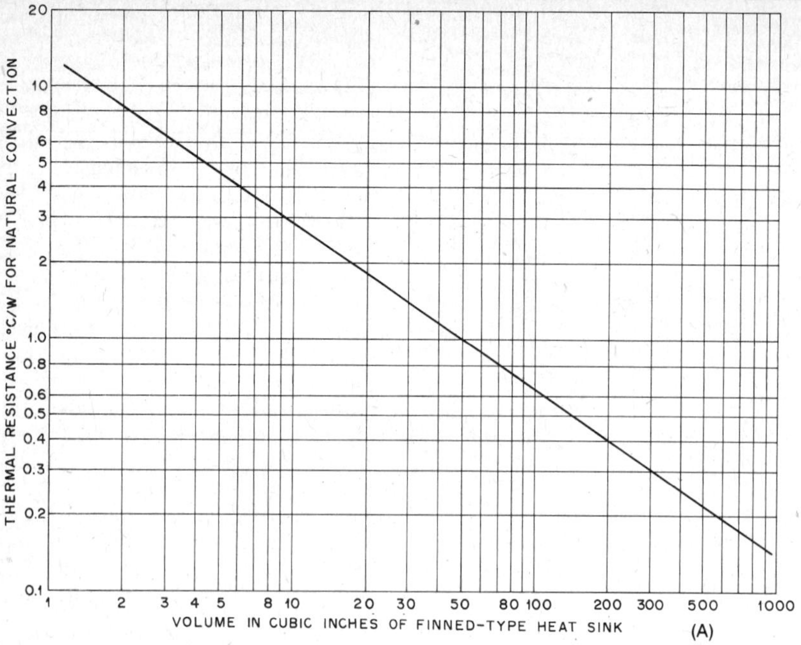

Device Case	5 W	10 W	25 W	50 W	100 W
TO-5	17.2	7.2	1.2	0.71	0.35
TO-44	1.2	9.2	0.44	n/a	n/a

(B)

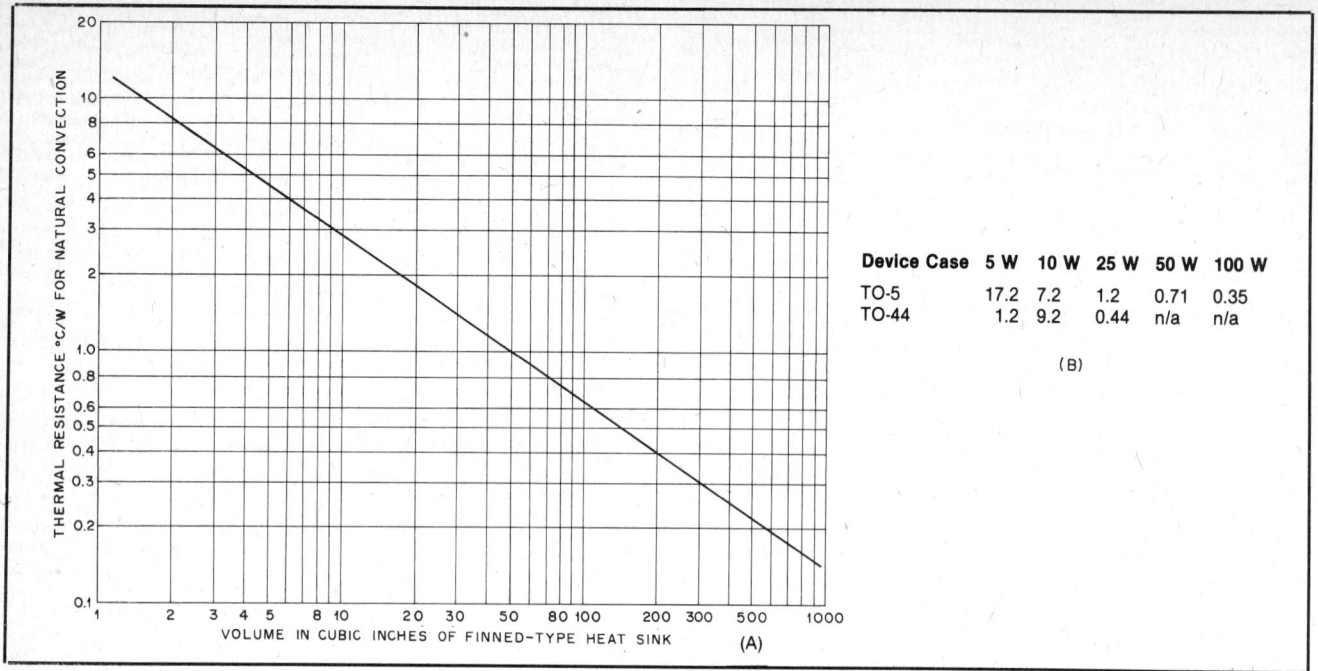

Fig. 38 — Heat-sink thermal resistance versus size. The sink volume can be determined by multiplying the height and cross-sectional area. At B are numbers which show the approximate thermal resistance needed for operating at various power levels with adequate cooling. TO-5 and TO-44 case styles are listed.

specified in a device data sheet, often in graphical form. Germanium power transistors may be operated at up to 100° C while the silicon types may be run at up to 200° C. Leakage currents in germanium devices can be very high at elevated temperatures; thus, for power applications silicon transistors are preferred.

A thermal sink, properly chosen, will remove heat at a rate which keeps the transistor junction temperature in the safe area. For low-power applications a simple clip-on heat sink will suffice, while for 100-watts input power a massive cast-aluminum finned radiator will be necessary. In general, the case temperature of a power transistor must be kept below the point at which it will cause discomfort when touched. Silicone grease should be used between the transistor body and the heat sink to aid heat transfer.

Heat-Sink Design

Simple heat sinks, made as described in the Construction Practices chapter, can be made more effective (by 25 percent or more) by applying a coat of flat-black paint. Finned radiators are most effective when placed where maximum air flow can be achieved — outside a case with the fins placed vertically. The size of a finned heat sink required to give a desired *thermal resistance*, a measure of the ability to dissipate heat, is shown in Fig. 38A. Fig. 38B is a simplified chart of the thermal resistance needed in a heat sink for transistors in TO-5 and TO-44 cases.

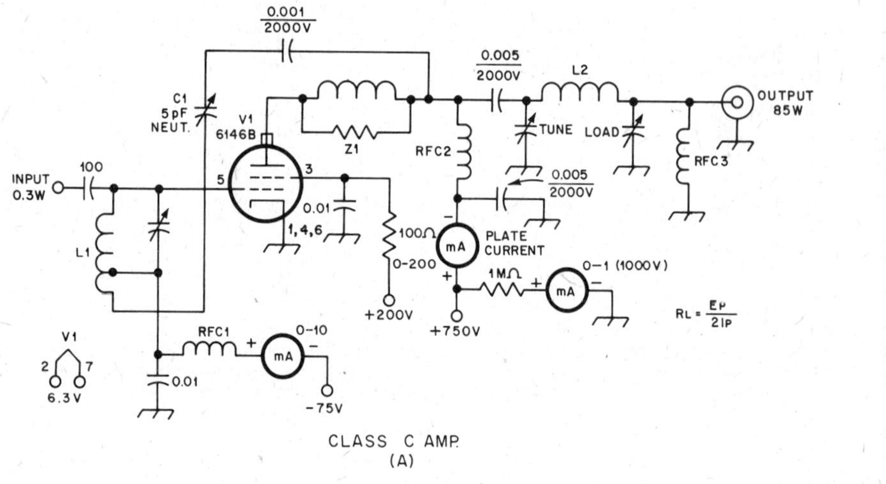

Fig. 39 — Circuit examples of typical single-ended and parallel tube amplifiers.

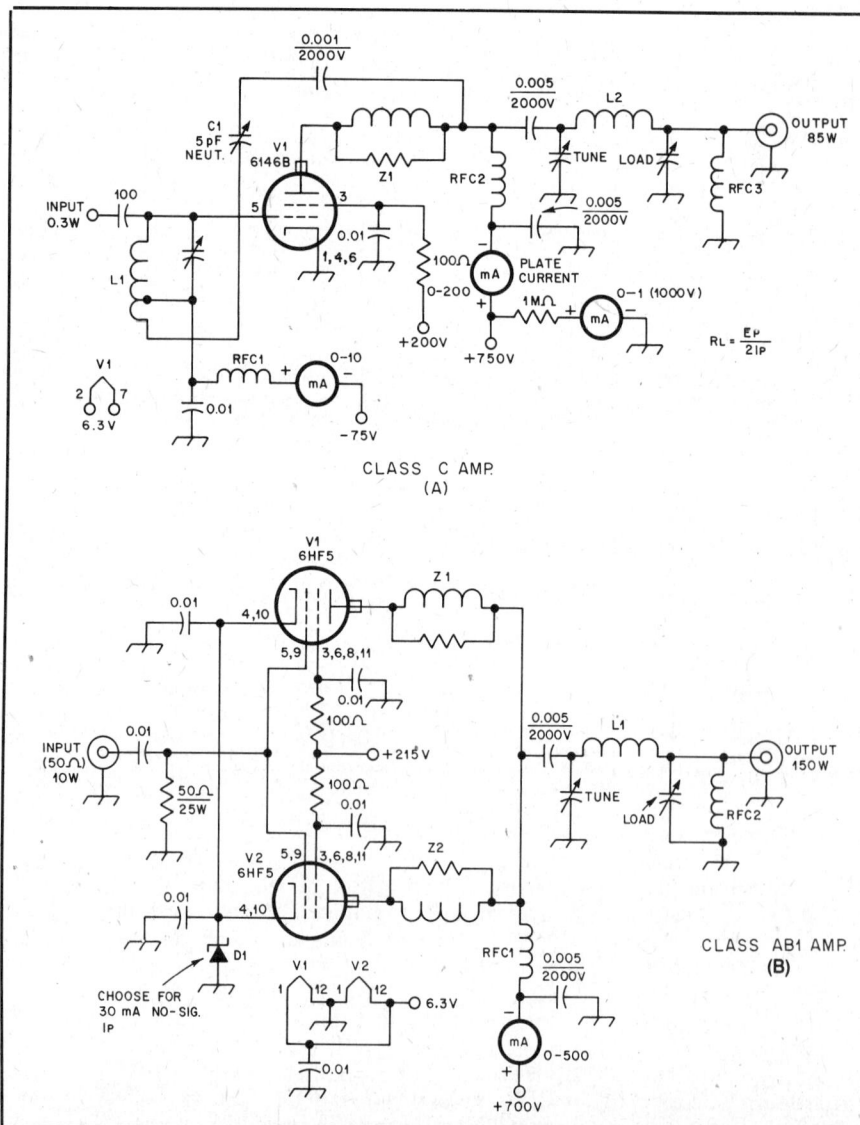

These figures are based on several assumptions, so they can be considered a *worst-case* situation. Smaller heat sinks may be usuable.

The thermal design of solid-state circuits has been covered in April 1972 *QST*. The surface contact between the transistor case and the heat sink is extremely important. To keep the sink from being "hot" with dc, a mica insulator is usually employed between the transistor case and the heat dissipator. Newer types of transistors have a case mounting bolt insulated from the collector so that it may be connected directly to the heat sink.

Output Power from Transmitters: CW or FM

In a cw or fm transmitter, any class of amplifier can be used as an output or intermediate amplifier. (For reasonable efficiency, a frequency multiplier *must* be operated Class C.) Class C operation of the amplifier gives the highest efficiency (65 to 75 percent), but it is likely to be accompanied by appreciable harmonics and consequent TVI possibilities. If the excitation is keyed in a cw transmitter, Class C operation of subsequent amplifiers will, under certain conditions, introduce key clicks not present on the keyed excitation (see chapter on Code Transmission). The *peak envelope power* (PEP) input or output of any cw (or fm) transmitter is the "key-down" input or output.

A-M

In an amplitude-modulated phone transmitter, plate modulation of a Class C output amplifier results in the highest output for a given input to the output stage. The efficiency is the same as for cw or fm with the same amplifier, from 65 to 75 percent. (In most cases the manufacturer rates the *maximum allowable input* on plate-modulated phone at about 2/3 that of cw or fm.) A plate-modulated stage running 100 watts input will deliver a carrier output from 65 to 75 watts, depending upon the tube, frequency and circuit factor. The PEP output of any a-m signal is four times the carrier output power, or 260 to 300 watts for the 100-watt input example.

Grid- (control or screen) modulated output amplifiers in a-m operation run at a carrier efficiency of 30 to 35 percent, and a grid-modulated stage with 100 watts input has a carrier output of 30 to 35 watts. (The PEP output, four times the carrier output, is 120 to 140 watts.)

Running the legal input limit in the United States, a plate-modulated output stage can deliver a carrier output of 650 to 750 watts, while a screen- or control-grid-modulated output amplifier can deliver only a carrier of 300 to 350 watts.

SSB

Only *linear* amplifiers can be used to amplify ssb signals without prohibitive distortion, and this limits the choice of output amplifier operation to Classes A, AB_1, AB_2 and B. The efficiency of operation of these amplifiers runs from about 20 to 65 percent. In all but Class A operation the indicated (by plate-current meter) input will vary with the signal, and it is not possible to talk about relative inputs and outputs as readily as it is with other modes. Therefore linear amplifiers are rated by PEP (input or output) at a given distortion level, which indicates not only how much ssb signal they will deliver but also how effective they will be in amplifying an a-m signal.

Linear Amplifiers for A-M

In considering the practicality of adding a linear output amplifier to an existing a-m transmitter, it is necessary to know the carrier output of the a-m transmitter and the PEP output rating of the linear amplifier. Since the PEP output of an a-m signal is four times the carrier output, it is obvious that a linear with a PEP output rating of only four times the carrier output of the a-m transmitter is no amplifier at all. If the linear amplifier has a PEP output rating of eight times the a-m transmitter carrier output, the output power will be doubled and a 3-dB improvement will be obtained. In most cases a 3-dB change is *just discernible* by the receiving operator.

By comparison, a linear amplifier with a PEP output rating of four times an existing ssb, cw or fm transmitter will *quadruple* the output, a 6-dB improvement, it should be noted that the linear amplifier must be rated for the mode (ssb, cw or fm) with which it is to be used.

Grounded-Grid Amplifiers

The preceding discussion applies to vacuum-tube amplifiers connected in a grounded-cathode or grounded-grid circuit. However, there are a few points that apply only to grounded-grid amplifiers.

A tube operated in a given class(AB_1, B, C) will require more driving power as a grounded-grid amplifier than as a grounded-cathode amplifier. This is not because the grid losses run higher in the grounded-grid configuration but because some of

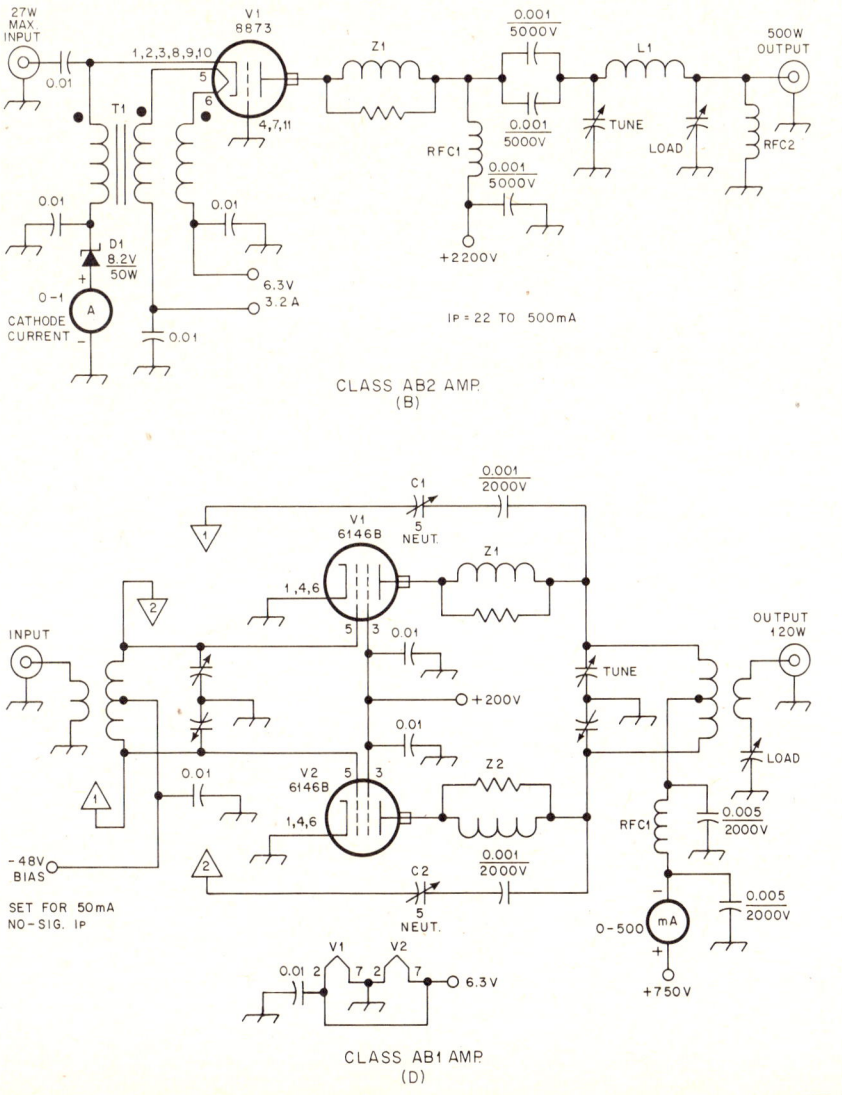

CLASS AB2 AMP.
(B)

CLASS AB1 AMP.
(D)

the driving power is coupled directly through the tube and appears in the plate load circuit. Provided enough driving power is available, this increased requirement is of no concern in cw or linear operation. In a-m operation, however, the fed-through power prevents the grounded-grid amplifier from being fully modulated (100 percent).

Amplifier Circuits: Parallel and Single-Ended Amplifiers

The circuits for parallel-tube amplifiers are the same as for a single tube, similar terminals of tubes being connected together. The grid impedance of two tubes in parallel is half that of a single tube. This means that twice the grid tank capacitance should be used for the same Q.

The plate load resistance is halved so that the plate-tank capacitance for a single tube also should be doubled. The total grid current will be doubled, so to maintain the same grid bias, the grid-leak resistance should be half that used for a single tube. The required driving power is doubled. The capacitance of a neutralizing capacitor should be doubled and the value of the screen dropping resistor should be cut in half.

In treating parasitic oscillation, it is often necessary to use a choke in each plate lead, rather than one in the common lead. This avoids building in a push-pull type of vhf resonance, which may cause inefficient operation at higher frequencies. See Fig. 39B.

Two or more transistors are often operated in parallel to achieve high output power, because several medium-power devices often cost less than a single high-power type. When parallel operation is used, precautions must be taken to insure that equal drive is applied to each transistor. Otherwise, one transistor may "hog" most of the drive and exceed its safe ratings.

In practice, it is not wise or necessary to use transistors in parallel. A push-pull circuit, such as that of Fig. 43 is preferable and it tends to cancel even harmonics — a benefit. Alternatively, single-ended amplifiers can be joined to deliver power to a single load by means of hybrid combiners. This technique was illustrated by Granberg in April and May 1976 QST. He used combiners to parallel the outputs of four 300-watt, push-pull, solid-state amplifiers. Fig. 33E shows the circuit of a hybrid-combiner transformer.

A typical single-ended tube amplifier which employs a 6146B in Class C is shown in Fig. 39A. Neutralization is provided by means of C1. L1 has a tap near the ac-ground end to provide a small amount of feedback voltage of the correct phase for neutralization. Meters are placed in the appropriate circuit points for monitoring the important voltages and currents.

Fig. 39B shows the circuit of a single-ended amplifier which operates Class AB$_1$ in grounded-grid fashion. T1 is a broadband, trifilar transformer which keeps the cathode and filaments above ac ground to provide a method for driving the cathode of V1. Operating bias is developed by inserting D1 in the cathode return. Z1 is a vhf parasitic suppressor. RFC2 functions as a safety device in the event the plate blocking capacitors short and dc flows into the load (antenna or Transmatch). The rf choke permits high dc current to flow to ground, blowing the power supply fuses and destroying the choke. It should have an XL which is at least 10 times the load resistance.

TV sweep tubes used in parallel are seen at Fig. 39B. Each plate lead contains a parasitic choke (Z1 and Z2). D1 is chosen to provide the necessary idling current for the class of operation desired. The unique feature of this circuit is that the control grids of V1 and V2 are tied together and driven across a 50-ohm resistor. This method eliminates the need for a tuned circuit or matching transformer at the amplifier input. Additionally, by strapping the grids to a low impedance (50 ohms), amplifier stability can be realized without the need for neutralization. The

100-ohm resistors in the screen-grid leads are used to discourage vhf oscillations and to help equalize the screen currents of the tubes. Several sweep tubes can be parallel-connected as shown to obtain a 1-kW-dc input linear amplifier. Attention must be paid to selecting a set of tubes with nearly matched dynamic characteristics. If this is not done, one or more of the tubes may draw the major part of the current during the driven period. This would cause them to operate in excess of their safe plate-dissipation ratings and be destroyed.

The circuit concepts shown in Fig. 39 are applicable to all types of transmitting tubes and power levels. Specific types of tubes are shown merely to provide practical examples during this treatment.

Grounded-Grid Amplifiers

Fig. 40A shows the input circuit of a grounded-grid triode amplifier. In configuration it is similar to the conventional grounded-cathode circuit except that the grid, instead of the cathode, is at ground potential. An amplifier of this type is characterized by a comparatively low input impedance and a relatively high driver power requirement. The additional driver

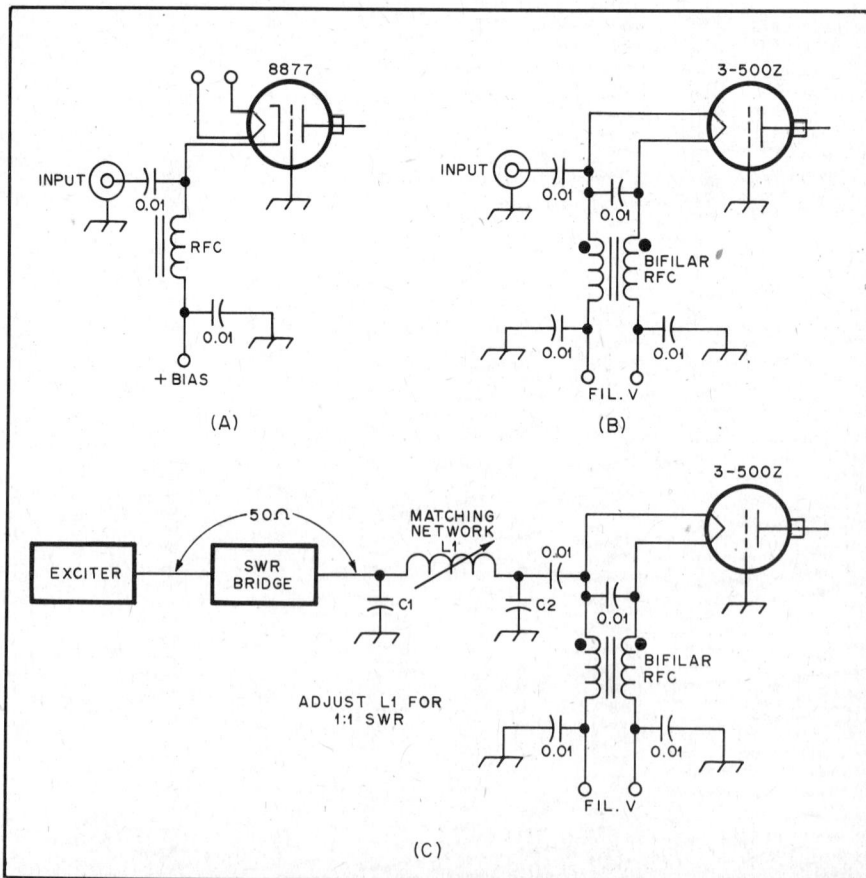

Fig. 40 — Methods for driving grounded-grid amplifiers.

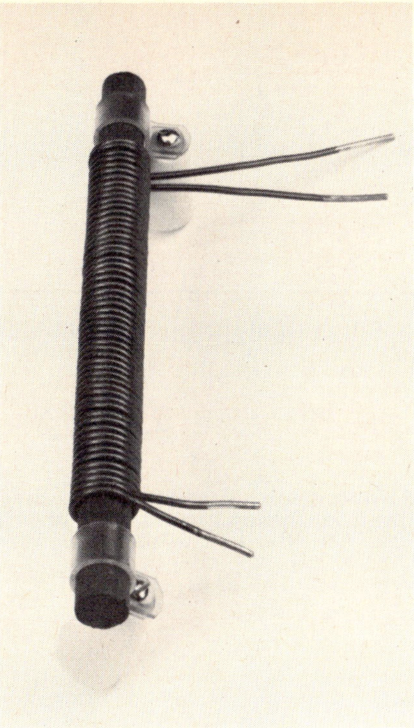

Fig. 41 — A 30-A filament choke for use with grounded-grid amplifiers. It contains 28 turns of no. 10 enameled wire, closewound in bifilar fashion on a 7-inch (178-mm) by 1/2-inch (13-mm) ferrite rod. The core permeability can be 950 or 125.

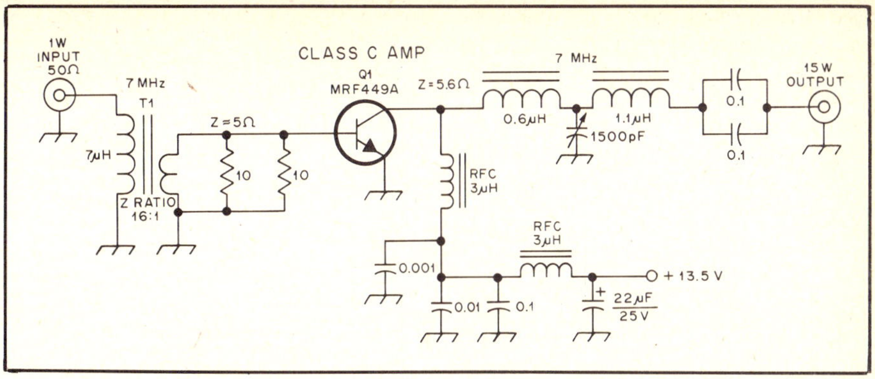

Fig. 42 — Typical circuit for a single-ended, Class C transistor power amplifier.

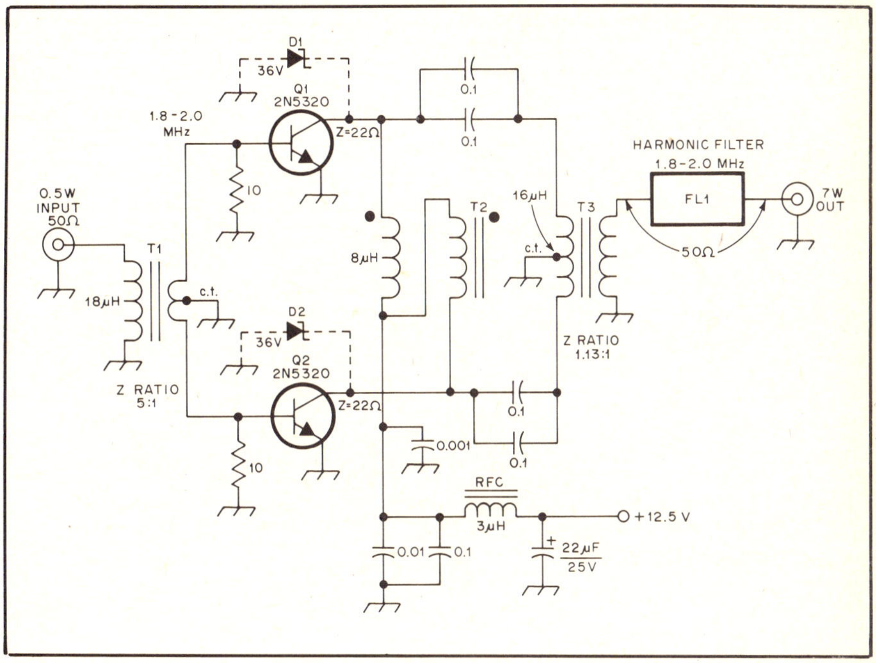

Fig. 43 — Circuit example of a broadband, push-pull, solid-state power amplifier.

power is not consumed in the amplifier but is "fed through" to the plate circuit where it combines with the normal plate output power. The total rf power output is the sum of the driver and amplifier output powers less the power normally required to drive the tube in a grounded-cathode circuit.

Positive feedback is from plate to cathode through the plate-cathode capacitance of the tube. Since the grounded-grid is interposed between the plate and cathode, this capacitance is small, and neutralization usually is not necessary.

In the grounded-grid circuit the cathode must be isolated for rf from ground. This presents a practical difficulty especially in the case of a filament-type tube whose filament current is large. In plate-modulated phone operation the driver power fed through to the output is not modulated.

The chief application for grounded-grid amplifiers in amateur work below 30 MHz is in the case where the available driving power far exceeds the power that can be used in driving a conventional grounded-cathode amplifier.

Screen-grid tubes are also used sometimes in grounded-grid amplifiers. In some cases, the screen is simply connected in parallel with the grid and the tube operates as a high-μ triode. In other cases, the screen is bypassed to ground and operated at the usual dc potential. Since the screen is still in parallel with the

grid for rf, operation is very much like that of a triode except that the positive voltage on the screen reduces driver-power requirements.

In indirectly-heated cathode tubes, the low heater-to-cathode capacitance will often provide enough isolation to keep rf out of the heater transformer and the ac lines. If not, the heater voltage must be applied through rf chokes.

In a directly-heated cathode tube, the filament must be maintained above rf ground. This can be done by using a bifilar-wound filament choke (Fig. 40B and C). With this method, a double solenoid (often wound on a ferrite core) is generally used, although separate chokes can be used, or a toroid core of large cross-sectional area can be used. A typical filament choke is shown in Fig. 41.

The input impedance of a grounded-grid power stage is usually between 30 and 150 ohms. A high-C, low-Q pi-section network can be used to obtain an SWR of 1:1 between the exciter and the amplifier. This is shown in Fig. 40C. The input net-

work provides benefit other than impedance matching — a reduction in the IM distortion produced by the stage when amplifying an ssb signal.

Transistor Amplifiers

Fig. 42 contains the circuit of a typical single-ended transistor amplifier. It is shown for Class C operation. T1 is a conventional toroidal broadband transformer which matches the 50-ohm driver load to the 5-ohm base of Q1. The primary of T1 requires sufficient reactance to look like four times the 50-ohm source impedance. A 7-μH winding satisfies the need at 7 MHz. The collector circuit employs a T network to transform the 5.6-ohm collector to a 50-ohm load. The collector rf choke is followed by a second one, which with the associated bypass capacitors decouples the amplifier from the 13.5-volt power supply.

A push-pull, broadband, solid-state amplifier circuit is seen in Fig. 43. As shown, it is biased for Class C operation. However, if linear amplification were desired, the center tap of T1 could be lifted from ground (but bypassed with a capacitor) and forward bias applied at that point. If that were done, the 10-ohm, base-swamping resistors would be returned to the transformer center tap instead of being grounded as shown.

T2 of Fig. 43 is a phase-reversal choke which places the collectors of Q1 and Q2 in the correct phase (180 degrees apart). T3 is a conventional transformer which matches the 44-ohm collector-to-collector impedance to a 50-ohm harmonic filter, FL1. The collector coupling capacitors are pairs of 0.1-µF capacitors in parallel. This method will pass more current with less capacitor heating than would be the case if only one capacitor were used at each point in the circuit. Ceramic chip capacitors are recommended. D1 and D2 may be added as protection against dc voltage spikes on the 13.5-V line. Also, if the amplifier should break into self-oscillation, the Zener diodes will limit the collector swing and prevent damage to the transistors. The diodes are helpful also when the amplifier is not terminated in a proper load. ARRL lab tests show that Zener diodes used in the manner indicated have no significant effect on amplifier performance, and they do not enhance the generation of harmonic currents. The reason is that the diodes are not conducting under normal conditions. They have been proven effective as high as 30 MHz, and may function satisfactorily into the vhf region. Matching networks and their solutions can be found earlier in this chapter.

RF Power-Amplifier Tanks and Coupling for Tubes

Tank Q

Rf power-amplifiers used in amateur transmitters are operated under Class C or AB conditions (see chapter on tube fundamentals). The main objective, of course, is to deliver as much fundamental power as possible into a load, R, without exceeding the tube ratings. The load resistance, R, may be in the form of a transmission line to an antenna, or the input circuit of another amplifier. A further objective is to minimize the harmonic energy (always generated by an amplifier) fed into the load circuit. In attaining these objectives, the Q of the tank circuit is of importance. When a load is coupled inductively, the Q of the tank circuit will have an effect on the coefficient of coupling necessary for proper loading of the amplifier. In respect to all of these factors, a tank Q of 10 to 20 is usually considered optimum. A much lower Q will result in less efficient operation of the amplifier tube, greater harmonic output, and greater difficulty in coupling inductively to a load. A much higher Q will result in higher tank current with increased loss in the tank coil.

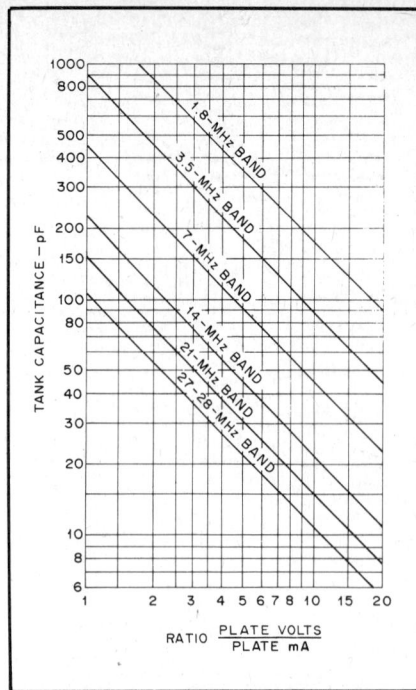

Fig. 44 — Chart showing plate tank capacitance required for a Q of 10. Divide the tube plate voltage by the plate current in milliamperes. Select the vertical line corresponding to the answer obtained. Follow this vertical line to the diagonal line for the band in question, and thence horizontally to the left to read the capacitance. For a given ratio of plate voltage/plate current, doubling the capacitance shown doubles the Q. When a split-stator capacitor is used in a balanced circuit, the capacitance of each section may be one half the value given by the chart.

Efficiency of a tank circuit is determined by the ratio of loaded Q to unloaded Q by the relationship:

$$\text{Eff.} = 100 \left(1 - \frac{Q_L}{Q_U}\right)$$

where Q_L is the loaded Q and Q_U is the unloaded Q.

The Q is determined (see chapter on electrical laws and circuits) by the L/C ratio and the load resistance at which the tube is operated. The tube load resistance is related, in approximation, to the ratio of the dc plate voltage to dc plate current at which the tube is operated and can be computed from

Class-A Tube:

$$R_L = \frac{\text{Plate Volts}}{1.3 \times \text{Plate Current}}$$

Class-B Tube:

$$R_L = \frac{\text{Plate Volts}}{1.57 \times \text{Plate Current}}$$

Class-C Tube:

$$R_L = \frac{\text{Plate Volts}}{2 \times \text{Plate Current}}$$

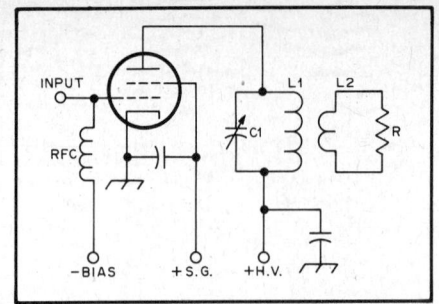

Fig. 45 — Inductive-link output coupling circuits.
C1 — Plate tank capacitor — see text and Fig. 44 for capacitance.
L1 — To resonate at operating frequency with C1. See LC chart and inductance formula in electrical-laws chapter, or use ARRL Lightning Calculator.
L2 — Reactance equal to line impedance. See reactance chart and inductance formula in electrical-laws chapter, or use ARRL Lightning Calculator.
R — Representing load.

Transistor:

$$R_L = \frac{(\text{Collector Volts})^2}{2 \times \text{Power Output (Watts)}}$$

Parallel-Resonant Tank

The amount of C that will give a Q of 10 for various ratios is shown in Fig. 44. For a given plate-voltage/plate-current ratio, the Q will vary directly as the tank capacitance, twice the capacitance doubles the Q, and so on. For the same Q, the capacitance of *each section* of a split-stator capacitor in a balanced circuit should be half the value shown.

These values of capacitance include the output capacitance of the amplifier tube, the input capacitance of a following amplifier tube if it is coupled capacitively, and all other stray capacitances. At the higher plate-voltage/plate-current ratios, the chart may show values of capacitance, for the higher frequencies, smaller than those attainable in practice. In such a case, a tank Q higher than 10 is unavoidable.

Inductive-Link Coupling: Coupling to Flat Coaxial Lines

When the load R in Fig. 45 is located for convenience at some distance from the amplifier, or when maximum harmonic reduction is desired, it is advisable to feed the power to the load through a low-impedance coaxial cable. The shielded construction of the cable prevents radiation and makes it possible to install the line in any convenient manner without danger of unwanted coupling to other circuits.

If the line is more than a small fraction of a wavelength long, the load resistance at its output end should be adjusted, by a matching circuit if necessary, to match the impedance of the cable. This reduces losses in the cable and makes the coupling

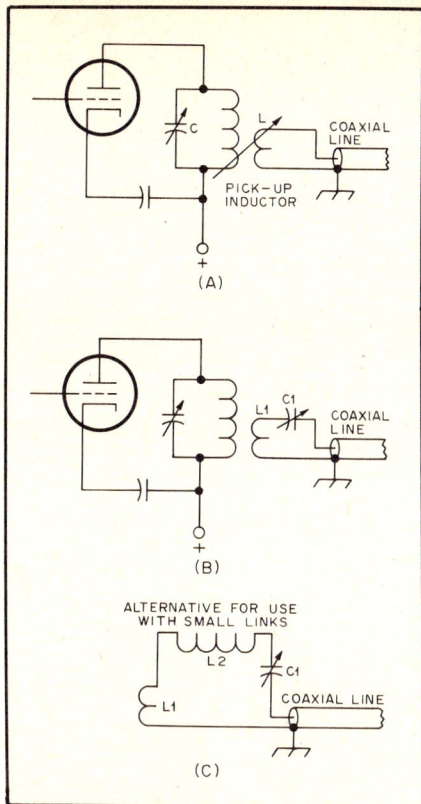

Fig. 46 — With flat transmission lines, power transfer is obtained with looser coupling if the line input is tuned to resonance. C1 and L1 should resonate at the operating frequency. See table for maximum usable value of C1. If the circuit does not resonate with maximum C1 or less, inductance of L1 must be increased or added in series at L2.

Table 8

Capacitance in pF for Coupling to Flat Coaxial Lines with Tuned Coupling Circuit

Frequency Band (MHz)	Characteristic Impedance of Line	
	52 (ohms)	75 (ohms)
3.5	450	300
7	230	150
14	115	75
21	80	50
28	60	40

'Capacitance values are maximum usable.
Note: Inductance in circuit must be adjusted to resonate at operating frequency.

adjustments at the transmitter independent of the cable length.

Assuming that the cable is properly terminated, proper loading of the amplifier will be assured, using the circuit of Fig. 46A, if

1) The plate tank circuit has reasonably higher value of Q. A value of 10 is usually sufficient.

2) The inductance of the pickup or link coil is close to the optimum value for the frequency and type of line used. The opti-

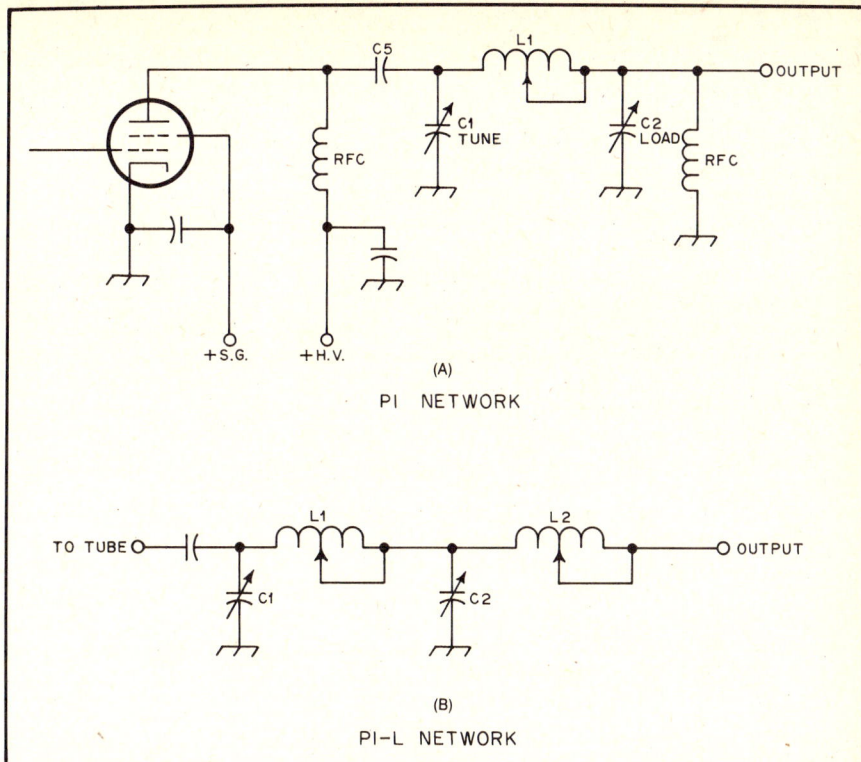

Fig. 47 — Examples of pi (A) and pi-L (B) networks.

mum coil is one whose self-inductance is such that its reactance at the operating frequency is equal to the characteristic impedance, Z_o, of the line.

3) It is possible to make the coupling between the tank and pickup coils very tight.

The second in this list is often hard to meet. Few manufactured link coils have adequate inductance even for coupling to a 50-ohm line at low frequencies.

If the line is operating with a low SWR, the system shown in Fig. 46A will require tight coupling between the two coils. Since the secondary (pickup coil) circuit is not resonant, the leakage reactance of the pickup coil will cause some detuning of the amplifier tank circuit. This detuning effect increases with increased coupling, but is usually not serious. However, the amplifier tuning must be adjusted to resonance, as indicated by the plate-current dip, each time the coupling is changed.

Tuned Coupling

The design difficulties of using "untuned" pickup coils, mentioned above, can be avoided by using a coupling circuit tuned to the operating frequency. This contributes additional selectivity as well, and hence aids in the suppression of spurious radiations.

If the line is flat the input impedance will be essentially resistive and equal to the Z_o of the line. With coaxial cable, a circuit of reasonable Q can be obtained with

practicable values of inductance and capacitance connected in series with the line's input terminals. Suitable circuits are given in Fig. 46 at B and C. The Q of the coupling circuit often may be as low as 2, without running into difficulty in getting adequate coupling to a tank circuit of proper design. Larger values of Q can be used and will result in increased ease of coupling, but as the Q is increased the frequency range over which the circuit will operate without readjustment becomes smaller. It is usually good practice, therefore, to use a coupling-circuit Q just low enough to permit operation, over as much of a band as is normally used for a particular type of communication, without requiring retuning.

Capacitance values for a Q of 2 and line impedances of 52 and 75 ohms are given in the accompanying table. These are the *maximum* values that should be used. The inductance in the circuit should be adjusted to give resonance at the operating frequency. If the link coil used for a particular band does not have enough inductance to resonate, the additional inductance may be connected in series as shown in Fig. 46C.

Characteristics

In practice, the amount of inductance in the circuit should be chosen so that, with somewhat loose coupling between L1 and the amplifier tank coil, the amplifier plate current will increase when the variable capacitor, C1, is tuned through the value of capacitance given by the table. The coupling between the two coils should then be increased until the amplifier loads normally, without changing the setting of C1. If the transmission line is flat over the entire frequency band under consideration, it should not be necessary to readjust C1 when changing frequency, if the values given in the table are used. However, it is unlikely that the line actually will be flat over such a range, so some readjustment of C1 may be needed to compensate for changes in the input impedance of the line. If the input impedance variations are not large, C1 may be used as a loading control, no changes in the coupling between L1 and the tank coil being necessary.

The degree of coupling between L1 and the amplifier tank coil will depend on the coupling circuit Q. With a Q of 2, the coupling should be tight — comparable with the coupling that is typical of "fixed-link" manufactured coils. With a swinging link it may be necessary to increase the Q of the coupling circuit in order to get sufficient power transfer. This can be done by increasing the L/C ratio.

Pi and Pi-L Output Tanks

A pi-section and pi-L tank circuit may also be used in coupling to an antenna or transmission line, as shown in Fig. 47. The optimum values of capacitance and inductance are dependent upon values of amplifier power input and output load resistance.

Values for L and C may be taken directly from Tables 9 and 10 if the output load resistance is the usual 52 ohms. It should be borne in mind that these values apply only where the output load is resistive, i.e., where the antenna and line have been matched.

Output-Capacitor Ratings

The voltage rating of the output capacitor will depend upon the SWR. If the load is resistive, receiving-type air capacitors should be adequate for amplifier input powers up to 2-kW PEP when feeding 52-75-ohm loads. In obtaining the larger capacitances required for the lower frequencies, it is common practice to switch one or more fixed capacitors in parallel with the variable air capacitor. While the voltage rating of a mica or ceramic capacitor may not be exceeded in a particular case, capacitors of these types are limited in current-carrying capacity. Postage-stamp silver-mica capacitors should be adequate for amplifier inputs over the range from about 70 watts at 28 MHz to 400 watts at 14 MHz and lower. The larger mica capacitors (CM-45 case) having voltage ratings of 1200 and 2500 volts are usually satisfactory for inputs varying from about 350 watts at 28 MHz to 1 kW at 14 MHz and lower. Because of these current limitations, particularly at the higher frequencies, it is advisable to use as large an air capacitor as practicable, using the micas only at the lower frequencies.

Table 9
Pi-Network Values for Various Plate Impedances with a Loaded Q of 12

	MHz	1500(12)	2000(12)	2500(12)	3000(12)	3500(12)	4000(12)	5000(13)	6000(14)	8000(16)
C1	3.5	364	273	218	182	156	136	118	106	91
	7	182	136	109	91	78	68	59	53	46
	14	91	68	55	46	39	34	30	27	23
	21	61	46	36	30	26	23	20	18	15
	28	46	34	27	23	20	17	15	13	11
C2	3.5	1755	1455	1242	1076	940	823	766	736	716
	7	877	728	621	538	470	411	383	368	358
	14	439	364	310	269	235	206	192	184	179
	21	293	243	207	179	157	137	128	123	119
	28	279	182	155	135	117	103	96	92	90
L1	3.5	6.59	8.57	10.52	12.45	14.35	16.23	18.56	20.55	23.81
	7	3.29	4.29	5.26	6.22	7.18	8.12	9.28	10.28	11.90
	14	1.64	2.14	2.63	3.11	3.59	4.06	4.64	5.14	5.95
	21	1.10	1.43	1.75	2.07	2.39	2.71	3.09	3.43	3.97
	28	0.82	1.07	1.32	1.56	1.79	2.03	2.32	2.57	2.98

These component values are for use with the circuit of Fig. 45A and were provided by W6FFC.

Table 10
Pi-L-network values for various plate impedance and frequencies. These values are based on a loaded Q of 12.

Zin (Ohms)	Freq. (MHz)	C1 (pF)	L1 (µH)	C2 (pF)	L2 (µH)
1500	3.50	403.	7.117	1348.	4.518
1500	4.00	318.	7.117	991.	4.518
1500	7.00	188.	3.900	596.	2.476
1500	7.30	174.	3.900	543.	2.476
1500	14.00	93.	1.984	292.	1.259
1500	14.35	89.	1.984	276.	1.259
1500	21.00	62.	1.327	191.	0.843
1500	21.45	59.	1.327	185.	0.843
1500	28.00	48.	0.959	152.	0.609
1500	29.70	43.	0.959	134.	0.609
2000	3.50	304.	9.086	1211.	4.518
2000	4.00	239.	9.086	894.	4.518
2000	7.00	142.	4.978	534.	2.476
2000	7.30	131.	4.978	490.	2.476
2000	14.00	70.	2.533	264.	1.259
2000	14.35	67.	2.533	249.	1.259
2000	21.00	47.	1.694	173.	0.843
2000	21.45	45.	1.694	167.	0.843
2000	28.00	36.	1.224	135.	0.609
2000	29.70	32.	1.224	120.	0.609
2500	3.50	244.	11.010	1115.	4.518
2500	4.00	191.	11.010	827.	4.518
2500	7.00	114.	6.033	493.	2.476
2500	7.30	105.	6.033	453.	2.476
2500	14.00	56.	3.069	240.	1.259
2500	14.35	53.	3.069	230.	1.259
2500	21.00	38.	2.053	158.	0.843
2500	21.45	36.	2.053	154.	0.843
2500	28.00	29.	1.483	127.	0.609
2500	29.70	26.	1.483	111.	0.609
3000	3.50	204.	12.903	1042.	4.518
3000	4.00	159.	12.903	777.	4.518
3000	7.00	94.	7.070	468.	2.476
3000	7.30	87.	7.070	426.	2.476
3000	14.00	47.	3.597	222.	1.259
3000	14.35	44.	3.597	217.	1.259
3000	21.00	32.	2.406	146.	0.843
3000	21.45	30.	2.406	145.	0.843
3000	28.00	24.	1.738	115.	0.609
3000	29.70	21.	1.738	105.	0.609
3500	3.50	174.	14.772	997.	4.518
3500	4.00	136.	14.772	738.	4.518
3500	7.00	81.	8.094	444.	2.476
3500	7.30	75.	8.094	404.	2.476
3500	14.00	40.	4.118	215.	1.259

Broadcast-receiver replacement-type capacitors can be obtained reasonably. Their voltage insulation should be adequate for inputs of 1000 watts or more.

More About Stabilizing Amplifiers

A straight amplifier operates with its input and output circuits tuned to the same frequency. Therefore, unless the coupling between these two circuits is brought to the necessary minimum, the amplifier will oscillate as a tuned-plate, tuned-grid circuit. Care should be used in arranging components and wiring of the two circuits so that there will be negligible opportunity for coupling external to the tube or transistor itself. Complete shielding between input and output circuits usually is required. All rf leads should be kept as short as possible, and particular attention should be paid to the rf return paths from input and output tank circuits

to emitter or cathode. In general, the best arrangement using a tube is one in which the cathode connection to ground, and the plate tank circuit are on the same side of the chassis or other shielding. The "hot" lead from the input tank (or driver plate tank) should be brought to the socket through a hole in the shielding. Then when the grid tank capacitor or bypass is grounded, a return path through the hole to cathode will be encouraged, since transmission-line characteristics are simulated.

Screen-Grid Tube Neutralizing Circuits

The plate-grid capacitance of screen-grid tubes is reduced to a fraction of a picofarad by the interposed grounded screen. Nevertheless, the power sensitivity of these tubes is so great that only a very small amount of feedback is necessary to start oscillation. To assure a stable

amplifier, it is usually necessary to load the grid circuit, or to use a neutralizing circuit.

The capacitive neutralizing system for screen-grid tubes is shown in Fig. 48A. C1 is the neutralizing capacitor. The capacitance should be chosen so that at some adjustment of C1,

$$\frac{C1}{C3} = \frac{\text{Tube grid-plate capacitance (or } C_{gp})}{\text{Tube input capacitance (or } C_{IN})}$$

The grid-cathode capacitance must include all strays directly across the tube capacitance, including the capacitance of the tuning-capacitor stator to ground. This may amount to 5 to 20 pF. In the case of capacitance coupling, the output capacitance of the driver tube must be added to the grid-cathode capacitance of the amplifier in arriving at the value of C1.

Neutralizing a Screen-Grid Amplifier Stage

There are two general procedures available for indicating neutralization in a screen-grid amplifier stage. If the screen-grid tube is operated with or without grid current, a sensitive output indicator can be used. If the screen-grid tube is operated with grid current, the grid-current reading can be used as an indication of neutralization. When the output indicator is used, both screen and plate voltages must be removed from the tubes, but the dc circuits from the plate and screen to cathode must be completed. If the grid-current reading is used, the plate voltage may remain on but the screen voltage must be zero, with the dc circuit completed between screen and cathode.

The immediate objective of the neutralizing process is reducing to a minimum the rf-driver voltage fed from the input of the amplifier to its output circuit through the grid-plate capacitance of the tube. This is done by adjusting carefully, bit by bit, the neutralizing capacitor or link coils until an rf indicator in the output circuit reads minimum, or the reaction of the unloaded plate-circuit tuning on the grid-current value is minimized.

The wavemeter shown in the Measurements chapter makes a sensitive neutralizing indicator. The wavemeter coil should be coupled to the output tank coil at the low-potential or "ground" point. Care should be taken to make sure that the coupling is loose enough at all times to prevent burning out the meter or the rectifier. The plate tank capacitor should be readjusted for maximum reading after each change in neutralizing.

When the grid-current meter is used as a neutralizing indicator, the screen should be grounded for rf and dc, as mentioned above. There will be a change in grid current as the unloaded plate tank circuit is tuned through resonance. The neutrali-

Z_{in} (Ohms)	Freq. (MHz)	C1 (pF)	L1 (μH)	C2 (pF)	L2 (μH)
3500	14.35	38.	4.118	206.	1.259
3500	21.00	27.	2.755	136.	0.843
3500	21.45	25.	2.755	138.	0.843
3500	28.00	21.	1.989	106.	0.609
3500	29.70	18.	1.989	99.	0.609
4000	3.50	153.	16.621	947.	4.518
4000	4.00	119.	16.621	706.	4.518
4000	7.00	71.	9.107	418.	2.476
4000	7.30	65.	9.107	387.	2.476
4000	14.00	35.	4.633	204.	1.259
4000	14.35	33.	4.633	197.	1.259
4000	21.00	23.	3.099	137.	0.843
4000	21.45	22.	3.099	132.	0.843
4000	28.00	18.	2.238	107.	0.609
4000	29.70	16.	2.238	95.	0.609
5000	3.50	123.	20.272	872.	4.518
5000	4.00	95.	20.272	658.	4.518
5000	7.00	57.	11.108	387.	2.476
5000	7.30	52.	11.108	360.	2.476
5000	14.00	29.	5.651	186.	1.259
5000	14.35	27.	5.651	183.	1.259
5000	21.00	19.	3.780	125.	0.843
5000	21.45	18.	3.780	123.	0.843
5000	28.00	15.	2.730	95.	0.609
5000	29.70	13.	2.730	89.	0.609
6000	3.50	103.	23.873	829.	4.518
6000	4.00	80.	23.873	621.	4.518
6000	7.00	48.	13.081	368.	2.476
6000	7.30	44.	13.081	340.	2.476
6000	14.00	24.	6.655	172.	1.259
6000	14.35	22.	6.655	173.	1.259
6000	21.00	16.	4.452	117.	0.843
6000	21.45	15.	4.452	116.	0.843
6000	28.00	13.	3.215	87.	0.609
6000	29.70	11.	3.215	84.	0.609
8000	3.50	78.	30.967	747.	4.518
8000	4.00	60.	30.967	569.	4.518
8000	7.00	36.	16.968	337.	2.476
8000	7.30	33.	16.968	312.	2.476
8000	14.00	18.	8.632	165.	1.259
8000	14.35	17.	8.632	159.	1.259
8000	21.00	12.	5.775	104.	0.843
8000	21.45	11.	5.775	106.	0.843
8000	28.00	9.	4.171	86.	0.609
8000	29.70	8.	4.171	77.	0.609

Operating Q - 12. Output load - 52 ohms.   Computer data provided by Bill Imamura, JA6GW.

zing capacitor (or inductor) should be adjusted until this deflection is brought to a minimum. As a final adjustment, screen voltage should be returned and the neutralizing adjustment continued to the point where minimum plate current, maximum grid current and maximum screen current occur simultaneously. An increase in grid current when the plate tank circuit is tuned slightly on the high-frequency side of resonance indicates that the neutralizing capacitance is too small. If the increase is on the low-frequency side, the neutralizing capacitance is too large. When neutralization is complete, there should be a slight decrease in grid current on either side of resonance.

Grid Loading

The use of a neutralizing circuit may often be avoided by loading the grid circuit if the driving stage has some power capability to spare. Loading by tapping the grid down on the grid tank coil (or the plate tank coil of the driver in the case of capacitive coupling), or by a resistor from grid to cathode is effective in stabilizing an amplifier.

Low-Frequency Parasitic Oscillation

The screening of most transmitting screen-grid tubes is sufficient to prevent low-frequency parasitic oscillation caused by resonant circuits set up by rf chokes in grid and plate circuits. When rf chokes are used in both grid and plate circuits of a triode amplifier, the split-stator tank capacitors combine with the rf chokes to form a low-frequency parasitic circuit, unless the amplifier circuit is arranged to prevent it. Often, a resistor is substituted for the grid rf choke, which will produce the desired result. This resistance should be a least 100 ohms. If any grid-leak resistance is used for biasing, it should be substituted for the 100-ohm resistor.

Component Ratings: Output Tank Capacitor Voltage

In selecting a tank capacitor with a spacing between plates sufficient to prevent voltage breakdown, the peak rf voltage across a tank circuit under load, but without modulation, may be taken conservatively as equal to the dc plate or collector voltage. If the dc supply voltage also appears across the tank capacitor, this must be added to the peak rf voltage, making the total peak voltage twice the dc supply voltage. If the amplifier is to be plate-modulated, this last value must be doubled to make it four times the dc plate voltage, because both dc and rf voltages double with 100-percent amplitude modulation. At the higher voltages, it is desirable to choose a tank circuit in which the dc and modulation voltages do not appear across the tank capacitor, to permit the use of a smaller capacitor with less plate spacing.

Capacitor manufacturers usually rate their products in terms of the peak voltage

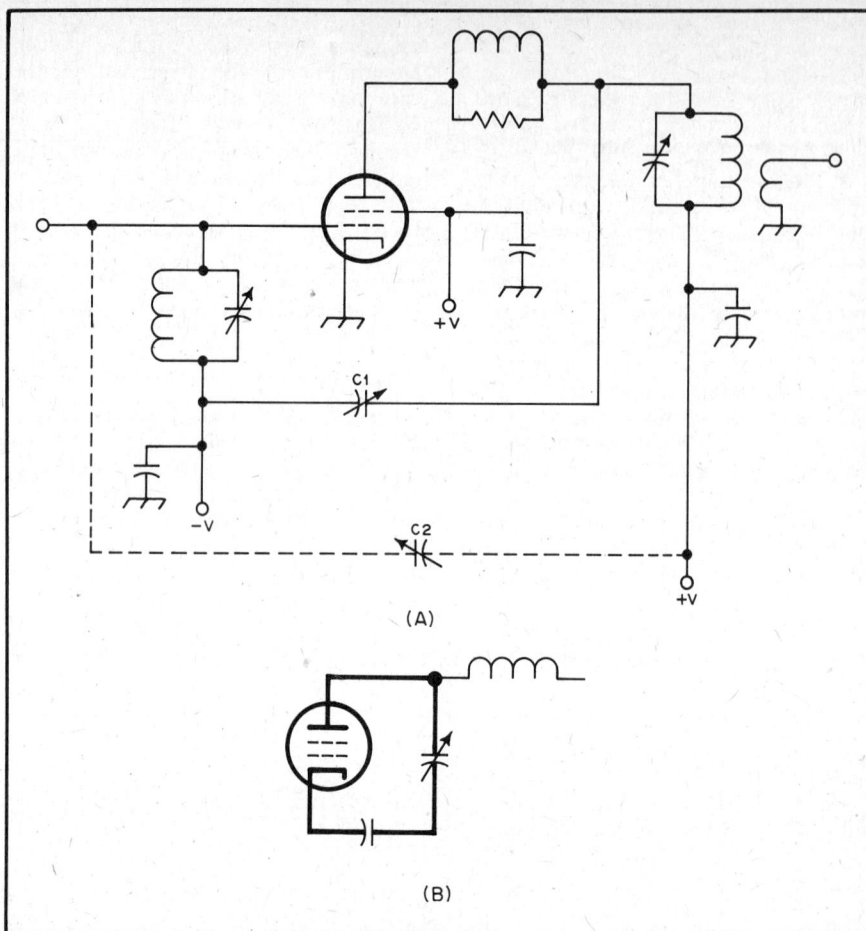

Fig. 48 — A neutralization circuit may use either C1 or C2 to cancel the effect of the tube grid-plate capacitance (A). The circuit at B shows the usual vhf-parasitic circuit in bold lines.

Table 11

Typical Tank-Capacitor Plate Spacings

Spacing Inches (mm)	Peak Voltage	Spacing Inches (mm)	Peak Voltage	Spacing Inches (mm)	Peak Voltage
0.015 (0.4)	1000	0.07 (1.8)	3000	0.175 (4.4)	7000
0.02 (0.5)	1200	0.08 (2.8)	3500	0.25 (6.3)	9000
0.03 (0.8)	1500	0.125 (3.0)	4500	0.35 (8.9)	11000
0.05 (1.3)	2000	0.15 (3.8)	6000	0.5 (12.7)	13000

between plates. Typical plate spacings are shown in Table 11.

Output tank capacitors should be mounted as close to the tube as temperature considerations will permit, to make possible the shortest capacitive path from plate to cathode. Especially at the higher frequencies, where minimum circuit capacitance becomes important, the capacitor should be mounted with its stator plates well spaced from the chassis or other shielding. In circuits where the rotor must be insulated from ground, the capacitor should be mounted on ceramic insulators of size commensurate with the plate voltage involved and — most important of all, from the viewpoint of safety to the operator — a well-insulated coupling should be used between the capacitor shaft and the dial. *The section of the shaft attached to the dial should be well grounded.* This can be done conveniently through the use of panel shaft-bearing units.

Tank Coils

Tank coils should be mounted at least their diameter away from shielding to prevent a marked loss in Q. Except perhaps at 28 MHz it is not important that the coil be mounted quite close to the tank capacitor. Leads up to 6 or 8 inches are permissible. It is more important to keep the tank capacitor as well as other components out of the immediate field of the coil. For this reason, it is preferable to mount the coil so that its axis is parallel to the capacitor shaft, either alongside the capacitor or above it.

Table 12

Wire Sizes for Transmitting Coils for Tube Transmitters

Power Input (Watts)	Band (MHz)	Wire Size
1000	28-2	6
	14-7	8
	3.5-1.8	10
500	28-21	8
	14-7	12
	3.5-1.8	14
150	28-21	12
	14-7	14
	3.5-1.8	18
75	28-21	14
	14-7	18
	3.5-1.8	22
25 or less*	28-21	18
	14-7	24
	3.5-1.8	28

*Wire size limited principally by consideration of Q.

There are many factors that must be taken into consideration in determining the size of wire (see table 12) which should be used in winding a tank coil. The considerations of form factor and wire size which will produce a coil of minimum loss are often of less importance in practice than the coil size that will fit into available space or that will handle the required power without excessive heating. This is particularly true in the case of screen-grid tubes where the relatively small driving power required can be easily obtained even if the losses in the driver are quite high. It may be considered preferable to take the power loss if the physical size of the exciter can be kept down by making the coils small.

Transistor output circuits operate at relatively low impedances because the current is quite high. Coils should be made of heavy wire or strap, with connections made for the lowest possible resistance. At vhf, stripline techniques are often employed, as the small inductance values required for a lumped inductance become difficult to fabricate.

RF Chokes

The characteristics of any rf choke will vary with frequency, from characteristics resembling those of a parallel-resonant circuit of high impedance, to those of a series-resonant circuit, where the impedance is lowest. In between these extremes, the choke will show varying amounts of inductive or capacitive reactance.

In series-feed circuits, these characteristics are of relatively small importance because the rf voltage across the choke is negligible. In a parallel-feed circuit, however, the choke is shunted across the tank circuit, and is subject to the full tank rf voltage. If the choke does not present a sufficiently high impedance, enough power will be absorbed by the choke to cause it to burn out.

To avoid this, the choke must have a sufficiently high reactance to be effective at the lowest frequency, and yet have no series resonances near the higher-frequency bands.

A VXO-CONTROLLED TRANSMITTER FOR 3.5 TO 21 MHz

The variable-crystal oscillator (VXO) represents a convenient method for generating a highly stable, adjustable-frequency signal. With the circuit shown in Fig. 50 the frequency spans shown in Table 13 can be realized. This circuit makes use of fundamental-type crystals only and is therefore limited to frequencies below roughly 25 MHz. Most crystals produced for frequencies above 25 MHz are overtone types and will not work in this circuit. Since the operator is apt to favor one portion of a band over another, a few crystals are all that is necessary for coverage of a part of the band. The transmitter, as shown, is rated for 6 watts of output while running at the 10-watt dc input level.

Circuit Description

The schematic diagram of the transmitter is displayed at Fig. 50. Q1 and associated components comprise a Colpitts variable-frequency crystal oscillator. C1 is used to adjust the frequency of the oscillator and C2 is used to limit the span of the oscillator. If no limit is provided the oscillator can operate "on its own" and no longer be under the control of the crystal — an undesirable circumstance. Supply voltage is fed to the oscillator only during transmit and spot periods. This prevents the oscillator from interfering with received stations operating on the same frequency.

Output energy from the oscillator is routed to Q2, a grounded-base amplifier. This stage provides some gain, but more important, a high degree of isolation between the oscillator and the driver stage. Pulling and chirp are virtually nonexistent.

The driver stage uses a broadband amplifier that operates Class A. Keying of this stage is accomplished by ungrounding the base and emitter resistors. C10 is used to shape the waveform. Although the keying is rather hard there is no evidence of clicks.

Two MRF472 transistors are used in parallel for the power amplifier. These transistors were designed for the Citizens Band service and work nicely at 21 MHz. Each transistor is rated for 4-watts output which gives a margin of safety when operated at the 6-watt output level. L2 is used as a dc ground for the bases making the transistors operate Class C.

The low output impedance at the collectors is stepped up by a broadband transformer to the 50-ohm impedance level. A five-pole Chebyshev filter is used to assure a clean output signal. This transmitter exceeds current FCC specifications regarding spectral purity. D2 is used to clamp the collector voltage waveform should the transmitter be operated into an

Fig. 49 — Photographs of the completed 6-watt, VXO-controlled transmitter. Miniature coaxial cable (RG-174/U) is used for connections between the circuit board, connectors and switches for all runs carrying rf energy.

open circuit or high SWR antenna system. The transmitter is designed to operate into a load that is close to 50 ohms resistive.

S1 is used as the transmit/receive switch. One section transfers the antenna to an accompanying receiver or the output of the transmitter strip. Another section is used to activate the VXO during transmit and the third section is provided for receiver muting purposes.

Construction

The majority of the circuit components are mounted on a double-sided, printed-circuit boards. One side of the board is etched with the circuit pattern and the other side is left as a ground plane. A small amount of copper is removed from around each hole on the ground-plane side of the board to prevent the leads from shorting to ground. A scale etching pattern and parts layout guide is shown in Fig. 52.

Affixed to the front panel are the transmit/receive switch, spot switch, and the tuning capacitor. The rear apron supports the antenna and mute jacks, key jack and binding posts.

A homemade cabinet measuring 3 × 6 × 8-1/4 inches (76 × 150 × 210 mm) was used in the construction of this transmitter. The builder may elect to build his or her own cabinet from sheet aluminum or circuit-board material. The layout is not critical except that the lead from the circuit board to C1 should be kept as short as possible—an inch or two (25 to 51 mm) is fine.

The final transistors are heat sinked to the ground plane of the circuit board using mica washers and silicone compound. With the normal transmit duty cycle this heat sink is sufficient. If for some reason the prospective builder plans long key-down periods, an additional heat sink connected to the tops of the transistors would be helpful.

The only adjustment needed is that of setting the VXO limit capacitor. This can be done with the aid of a receiver. Using a 21-MHz fundamental-type crystal, adjust the capacitor for a maximum frequency spread of approximately 10-15 kHz. If too much frequency spread is available, increase the amount of capacitance. Make a final check with the receiver by listening to the keyed signal from the transmitter. It should be steady and chirp free. The spectral display of the transmitter is shown in Fig. 51.

LOW-POWER VMOS TRANSMITTER FOR 3.5 TO 28 MHz

Power FETs have a distinct advantage over power bipolar transistors: They are virtually destruct-proof when handled correctly. They do not go into thermal runaway, are not subject to secondary breakdown, and are immune to the potentially damaging effects of operating into incorrect loads. Therefore, anything from

a dead short to a wide-open load will cause no damage to a VMOS power FET. Damage can occur as a result of excessive operating voltages on the gate or drain, just as when using too much voltage on the grids or plate of a vacuum tube. The same rules apply to drain current (excessive dissipation) versus plate current. Finally, excessive heat will ruin a transistor as it will when tubes are permitted to run too hot.

It is convenient to regard a VMOS device in the same fashion as we might perceive a triode vacuum tube. The major difference is that VMOS transistors use lower operating voltages than tubes do, and the input and output impedances are low, comparatively speaking. One can even dip the drain current during tuneup, and monitor the current by means of a drain meter.

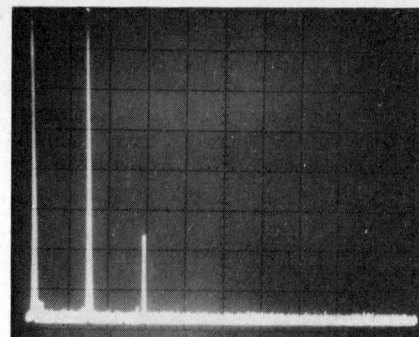

Fig. 51 — Spectral display of the VXO-controlled transmitter. Here the transmitter is operated in the 20-meter band. The second harmonic is down 56 dB relative to the fundamental output. Similar presentations were obtained on each of the other bands. This transmitter complies with the current FCC specifications regarding spectral purity.

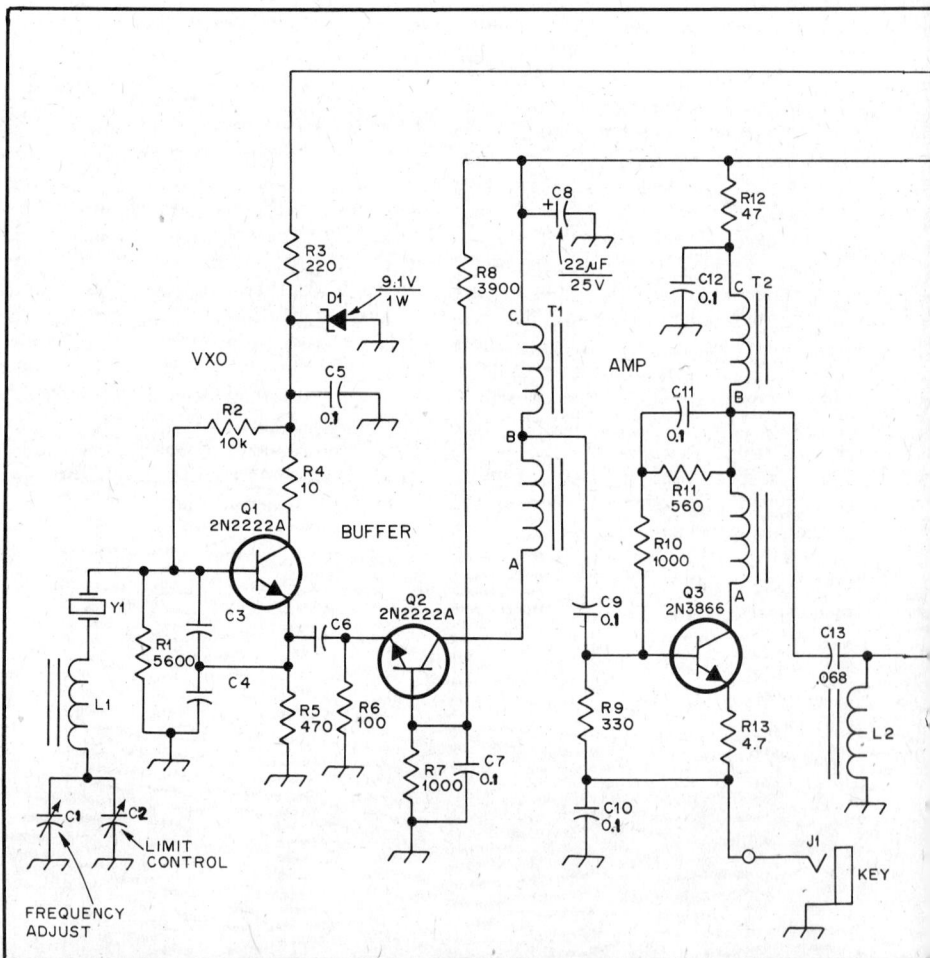

Fig. 50 — Schematic diagram of the VXO-controlled transmitter. All resistors are 1/4-watt carbon types unless noted otherwise. All resistors are mylar or disc ceramic unless otherwise noted. Polarized capacitors are electrolytic or tantalum. Q4 and Q5 must be heat sinked to the circuit board.

C1, C2, C3, C4, C6, C17 and C18 — See Table 13.
D1 — Zener diode, 9.1 V, 1 W.
D2 — Zener diode, 36 V, 1 W.
J1, J2 — Binding post.
J3 — Key jack.
L1, L3, L4, L5 — See Table 13.

L2 — 8 turns no. 26 enameled wire on FB73-801 ferrite bead.
M1 — 0-1 mA meter, Calector DI-91Z or equiv.
Q1, Q2, Q3 — Transistor, 2N2222A or equiv.
Q3 — Transistor, 2N3866 or equiv.
Q4, Q5 — Transistor, MRF-472 or equiv.

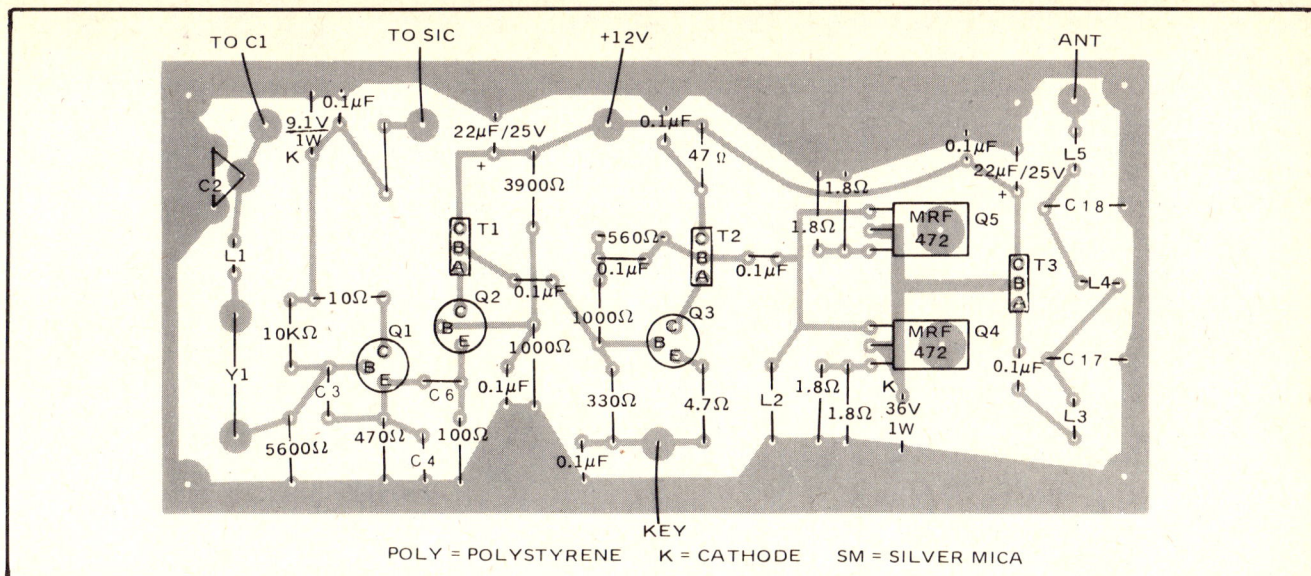

Fig. 52 — Scale etching pattern and parts layout guide for the 6-watt transmitter. Gray areas represent unetched copper. One side of the board is a complete ground plane. This view is from the circuit foil side of the board.

POLY = POLYSTYRENE   K = CATHODE   SM = SILVER MICA

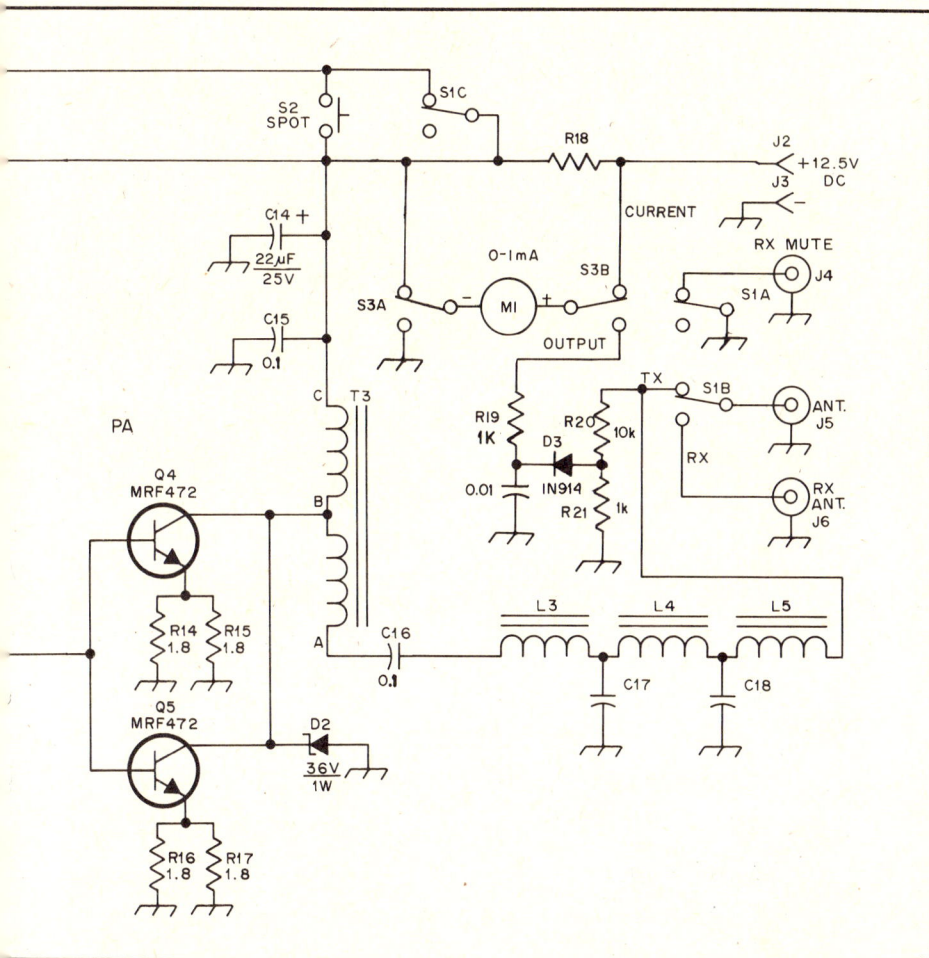

Transmitter Circuit

A block diagram of the transmitter is provided in Fig. 54. It shows the switching method used to change from the transmit to receive modes, plus the keying circuit. Q5 and the related components are built as a separate module. In this model a small piece of circuit board is used. It has a group of 1/4-inch (6.3 mm) squares cut into the copper-clad side by means of a hacksaw. The pads serve as solder terminals. The two resistors and three capacitors in the base lead of Q5 are as key-shaping components.

Fig. 55 shows the interior of the transmitter. The keying transistor is at the upper right, the PA module is just below it and the oscillator/doubler pc board is at the left of the first two. Although Fig. 54 indicates that S4B is used as a receiver-muting switch, it has not been wired into the unit shown, and no muting jack has been included on the back panel. The U-shaped main chassis measures 5 × 7 × 2 inches (127 × 178 × 51 mm), the width being the larger dimension. The chassis and perforated cover are homemade from aluminum stock which is 1/16 inch (1.6 mm) thick. The crystal switch (S1) is

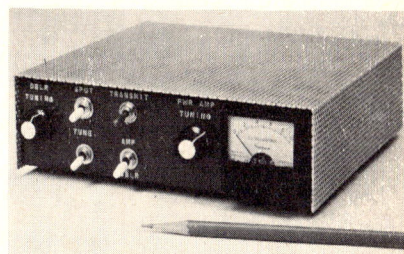

R14 - R17, incl. — Fixed resistor, 1.8 ohms, 1/2 W.
R18 — meter shunt, 13-1/2 inches no. 26 enamel wire wound on a high-value 1-watt resistor.
S1 — Toggle switch, 3pdt.
S2 — Push-button switch, spst, normally open.

S3 — Toggle switch, dpdt.
T1, T2 — Broadband transformer, 10 turns no. 24 enamel wire, bifilar wound on an FT-37-43 core.
T3 — Broadband transformer, 10 bifilar turns no. 24 enameled wire on an FT-50-43 core.
Y1 — See Table 13.

Fig. 53 — Exterior of the W1FB VMOS transmitter. The basics of this unit were treated in May 1979 *QST*.

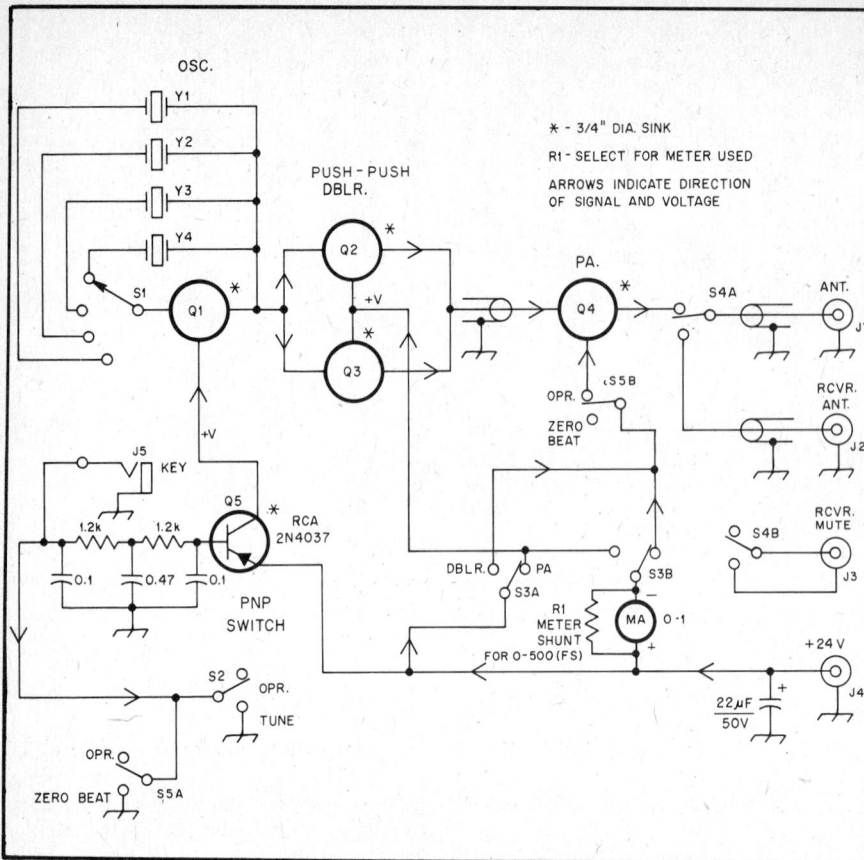

Fig. 54 — Block diagram of the VMOS transmitter. The arrows indicate the direction of dc and the signal. Q5 is the keying switch. It may be eliminated by keying the B + line to Q1 directly. J1-J4, incl. are single-hole mount phono jacks. J5 is a two-circuit phone jack. M1 is a 500-mA instrument. A 0-1 mA meter can be used by adding a suitable shunt at R1. A Simpson no. 2121 (1-1/2 inch — 38 mm) diameter meter is seen in Fig. 53. S2-S5, incl. are miniature toggle switches.

Fig. 55 — Interior view of the transmitter. Details are given in the text.

mounted on the rear lip of the chassis.

Four enhancement-mode FETs are used in the transmitter. Q1, Q2 and Q3 (Fig. 56) are Siliconix VN66AK devices in TO-39 cases. Supertex VN0106N-2 VMOS FETs are suitable as direct substitutes.[1] Crown heat sinks are required on all three transistors (Thermalloy 2215B or equiv.).

To ensure operation as an oscillator, Q1 requires a forward voltage on the gate. A Pierce oscillator is used at Q1, with one winding of broadband transformer T1 being the drain impedance. C1, C2 and C3 are feedback capacitors. Source capacitor Cs controls the amount of degenerative feedback in the oscillator. It may or may not be required. This depends upon the type of crystals used at Y1-Y4 (crystal activity). If low oscillator output is noted, or if a chirpy cw note is obtained, Cs should be included. Use only that amount of capacitance which will ensure proper oscillation. Typical values range from 10 pF to 100 pF.

T1 supplies drive to push-push doubler Q2/Q3 at a phase difference of 180°. The drains of the doubler are in parallel and tapped toward the B + end of L1 to effect an impedance match between the doubler and the gate of the PA transistor. Fig. 57 shows the board pattern and parts layout to scale.

Circuit details for the PA stage are given in Fig. 58. A Siliconix VN67AJ or Supertex VN1206N-1 (TO-3 case) is employed at Q4. It is insulated from its homemade U-shaped heat sink (1 × 2 × 1-1/4 inch — 25 × 51 × 32 mm) by means of a standard TO-3 size mica wafer. The Q4 mounting screws must pass through holes large enough to prevent contact between the transistor case (common to the drain), the heat sink and circuit ground. Final checks should be made with an ohmmeter before applying power. Silicone grease is spread on both sides of the mica washer before assembly, plus on the bottom surface of the heat sink.

A half-wave low-pass filter is used at the PA input to suppress harmonic currents in the drive to Q4. It is designed for a bilateral impedance of 150 ohms and a Q_L of 1. The 35 pF input capacitance (C_{iss}) of Q4 is absorbed into the network at C7.

RFC1-RFC4 of Fig. 56 and RFC6 of Fig. 58 are 950-mu miniature ferrite beads. They are necessary to damp vhf parasitic oscillations. Drive is applied to the PA gate across a 150-ohm resistor if Class C operation is desired. For Class AB use, R1 is added and the gate-to-ground resistor is changed to 160 Ω.

[1]Siliconix Incorporated, 2201 Laurelwood Rd., Santa Clara, CA 95054. Tel. 408-988-8000.
Supertex, Inc., 1225 Bordeaux Dr., Sunnyvale, CA 94086. Tel. 408-744-0100. Order VMOS devices from Sue Short. A $2 handling fee is required for orders less than $100.

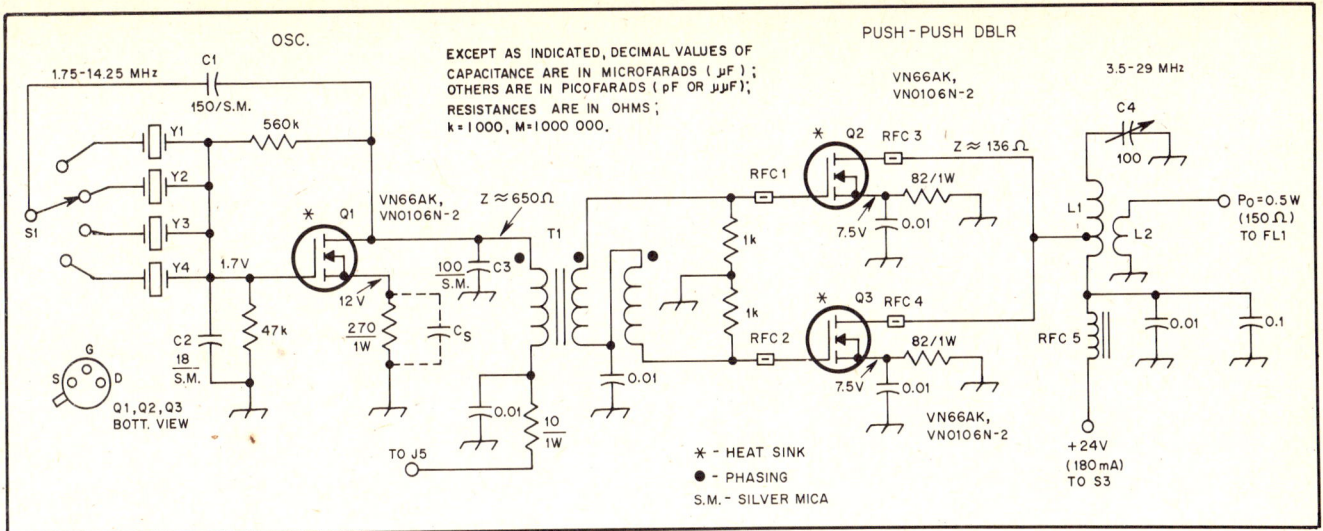

Fig. 56 — Schematic diagram of the VMOS exciter strip. Fixed-value capacitors are disc ceramic unless otherwise noted. Resistors are 1/2-W composition unless indicated otherwise.

C1-C3, incl. — Silver-mica feedback capacitor.
Cs — See text.
C4 — Miniature air variable, 100 pF (Hammarlund MAPC-100-B or equiv.) Arco 424 mica trimmer can be used. If trimmer is used, mount it on the pc board.
L1, L2 — See Table 14.

RFC1-RFC4, incl. — Miniature 950-mu ferrite bead by Amidon Assoc.
RFC5 — 10 turns no. 20 enam. wire on Amidon T50-43 (950 mu, 0.5-in. diam.) ferrite toroid.
S1 — Single-pole, 4-position, single-wafer phenolic switch.
T1 — See Table 14.

Y1-Y4, incl — Fundamental crystal at one half the desired operating frequency. Sockets are F-605 pc mount. These and the crystals are type GP, 30 pF load capacitance in HG-6/U type cases.

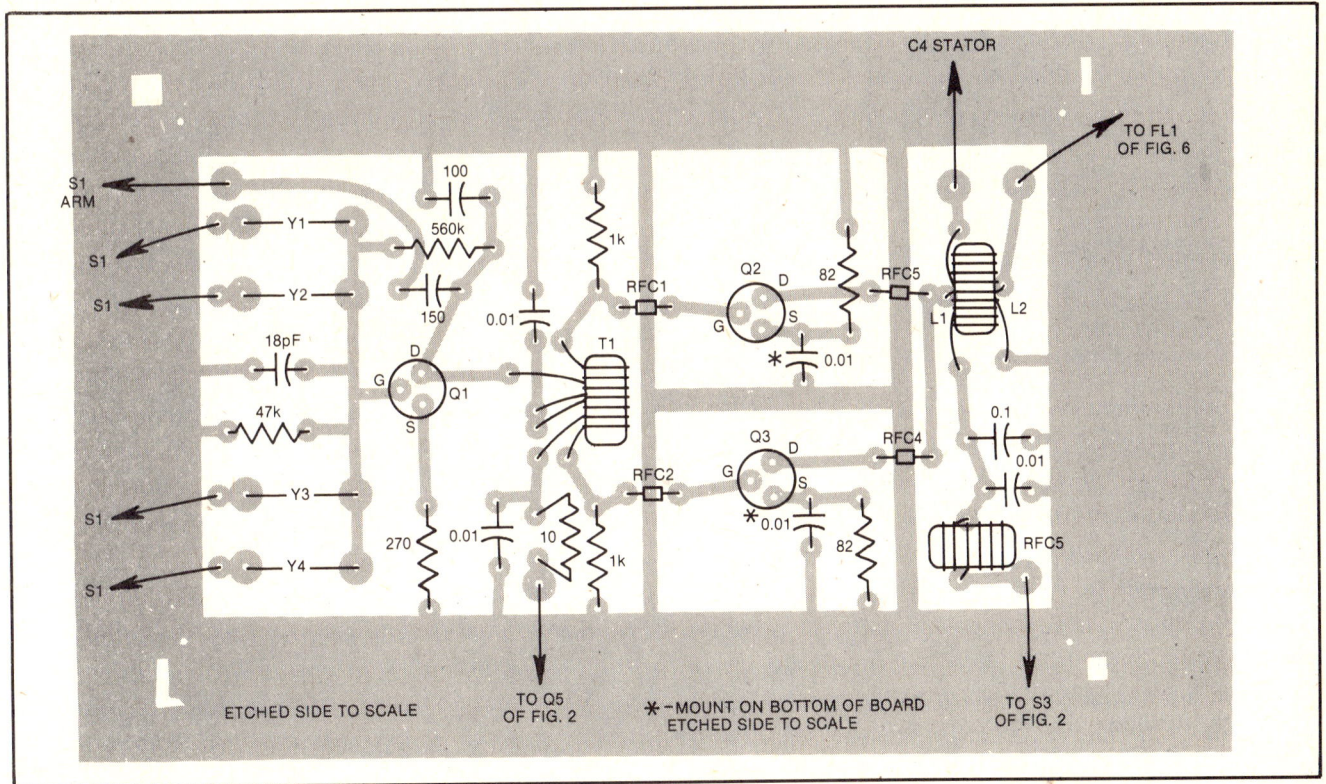

Fig. 57 — Scale pattern and layout of the exciter board. Parts marked with an asterisk (*) are mounted on the etched side of the board.

The PA drain is tapped near the B + end of L5 to provide an impedance match to the load. FL2 removes excessive harmonic currents from the transmitter output, making the spectral purity considerably better than that required by present FCC regulations. A spectral oscillograph is shown at Fig. 59. It was taken while using an HP analyzer. Q4 was operating in Class AB, which provides somewhat better spectral purity than is obtained under Class C conditions.

L1, L2, L5, L6, FL1 and FL2 must be built for the band of operation. Details for the various component values are provided in Table 14.

C12 of Fig. 58 must be added between the rf modules if Class AB operation is planned. It prevents the gate bias from being short-circuited to ground.

If a VN67AJ is not readily available for

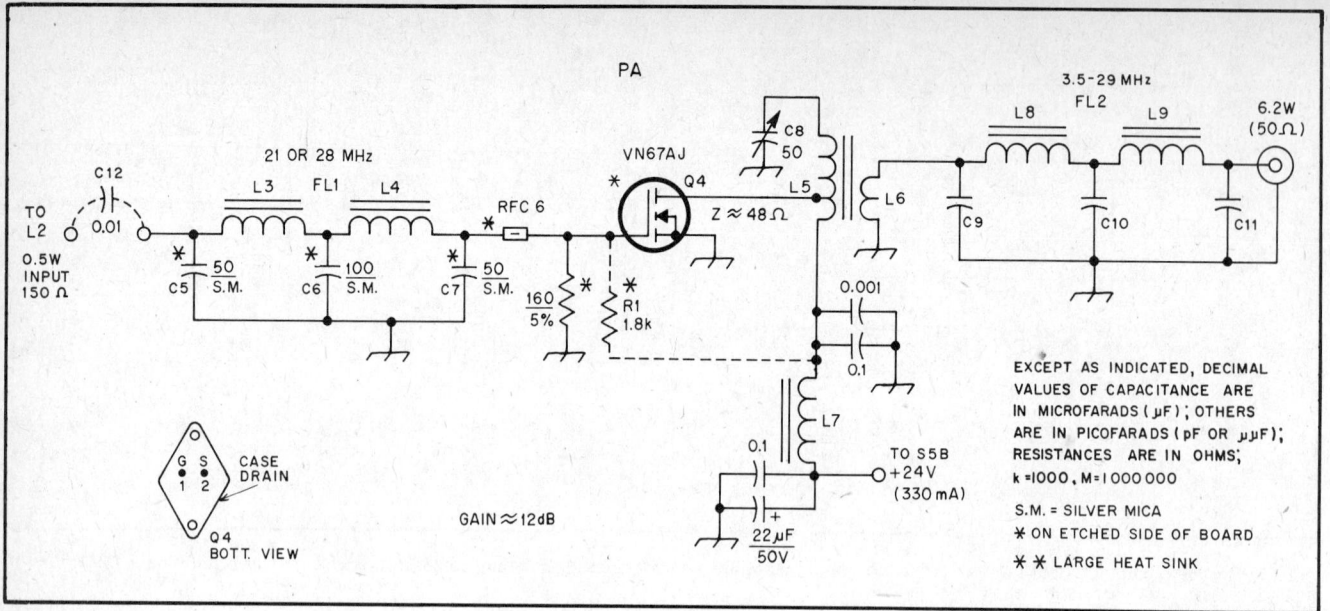

Fig. 58 — Schematic diagram of the VMOS power amplifier. Fixed-value capacitors are disc ceramic unless otherwise noted. Resistors are 1-watt composition.

C8 — 50-pF miniature air variable (Hammarlund MAPC-50-B or equiv.). Arco 424 mica trimmer suitable. If used, mount on pc board. See Table 10.

C5, C6, C7, C9, C10, C11 — See Table 10.
L3, L4, L5, L6, L8, L9 — See Table 10.

L7 — 10 turns no. 20 enam. wire on Amidon FT-50-43 ferrite toroid (0.5-in. dia, 950 mu).
RFC6 — Miniature 950-mu ferrite bead.

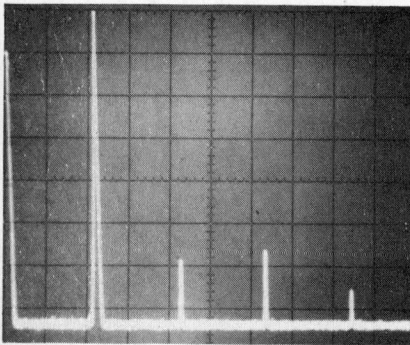

Fig. 59 — Spectral display of the transmitter output while delivering 6.5 watts to a 50-Ω dummy load. Horizontal scale is 10 MHz/div. Vertical scale is 10 dB/div. The white line at the left is the zero reference of the analyzer. The three responses to the right of the full-scale carrier blip are the second, third and fourth harmonics, respectively. All spurs are −56 dB or greater below peak carrier power.

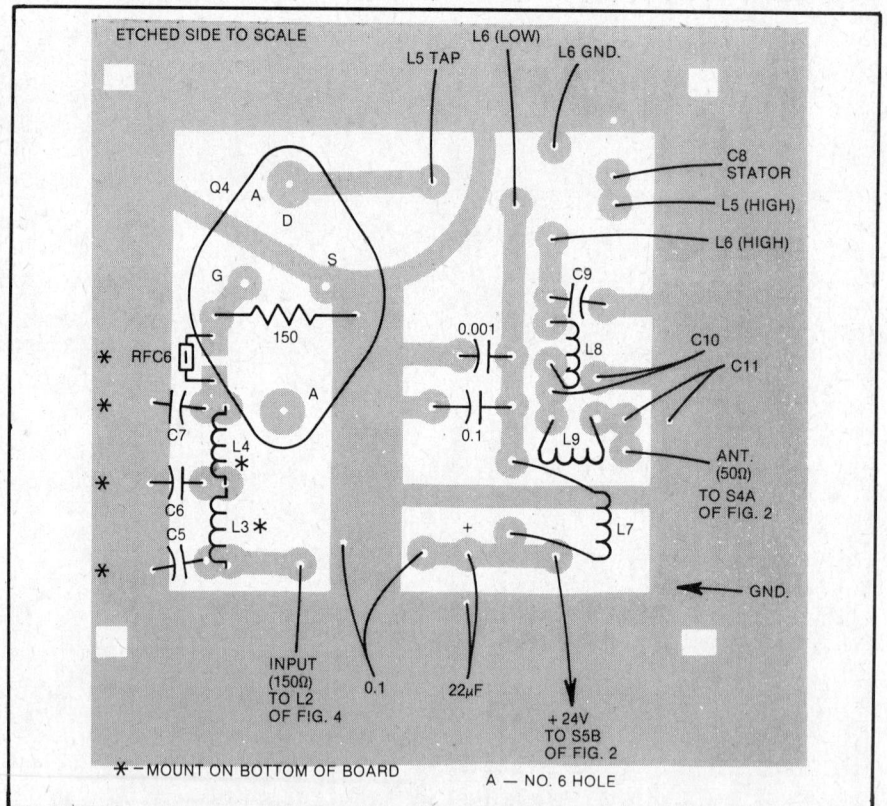

Fig. 60 — Scale pattern and parts layout for the VMOS PA module. Parts marked by an asterisk (*) are mounted on the etched side of the board.

use at Q4, a Supertex VN0106N can be used as a direct substitute. VMOS devices with a built-in gate-source Zener diode should be avoided in this transmitter. The Zener diode increases the C_{iss}, limits the upper-frequency characteristics of the transistor and clamps the rf drive voltage at +15. A Siliconix VN89AA was tested at Q4. Power output at 21 MHz was approximately 1.5 watts lower than when using the VN67AJ. This was caused by the built-in Zener diode. Performance was otherwise satisfactory.

The oscillator (Q1) operates at half frequency in order to restrict chirp on the cw

note. The tank circuits of Q2/Q3 and Q4 suppress the oscillator frequency by some 80 dB when they are tuned *exactly* to resonance.

A scale pattern and parts layout for the PA circuit board is given in Fig. 60. Both rf modules use double-clad pc board. The foil on the component side

Table 13

Component Values for the VXO-Controlled, 6-Watt Transmitter

Band	C1	C2	C3, C4	C6	C17, C18	L1	L3, L5	L4	VXO Range
80 M	365	—	220	100	820	52 Turns T-37-2	29 Turns T-502	35 Turns T-502	3-5 kHz
40 M	365	—	100	—	470	40 Turns T-37-2	21 Turns T-50-2	25 Turns T-50-2	6-8 kHz
30 M (10 MHz)	150	—	68	50	330	30 Turns T-37-2	18 Turns T-50-2	22 Turns T-50-2	8-10 kHz
20 M	50	10	50	50	240	35 Turns T-37-6	16 Turns T-50-6	19 Turns T-50-6	10-12 kHz
15 M	50	10	33	33	150	27 Turns T-37-6	15 Turns T-50-6	18 Turns T-50-6	12-14 kHz

serves as a ground plane to reduce unwanted rf ground loops — an aid to stability.[2]

Assembly Notes

The pc boards are supported above the chassis on metal standoff posts. This provides clearance between the chassis and the etched sides of the boards. It also ensures a secure grounding method for the ground foils on the boards.

All rf leads between the boards and their related panel controls are shielded. Miniature RG-174/U 50-Ω coax cable was used in the version described here. The shield braids are grounded at each end of each cable.

[2]Boards, negatives and parts kits are available from Circuit Board Specialists, P. O. Box 969, Pueblo, CO 81002. Tel. 213-722-2000.

The front and rear panels of the transmitter are sprayed with Aerosol-can enamel paint of the builder's choice. The aluminum surfaces are first abraded with sandpaper to aid the paint in adhering to the metal surfaces. Soap and hot water should be applied liberally to the metal surfaces prior to painting. Moisture and finger marks should not exist on the metal to be painted. Dymo tape labels are used to identify the controls. The tape should be of a color that matches the painted surfaces, thereby imparting a professional decor.

The top cover can be made of solid aluminum if desired. Metal L brackets can be mounted on the right and left edges of the chassis bottom to permit affixing the cover at two points. No. 6 sheet-metal screws are suitable for the purpose.

Operation

Checkout should include visual inspection for unwanted solder bridges on the pc boards. Check also to make certain that no poor solder joints exist. Finally, test the B+ lines by means of an ohmmeter to ensure that no short circuits are present.

A 24-volt regulated power supply is re-

Table 14

Coil and Capacitor Data for the 5-Band VMOS Power FET Transmitter

Band	T1	L1	L2	L3, L4,	L5	L6	L8, L9	C5 (pF)	C6 (pF)	C7 (pF)	C8 (pF)	C9, C11 (pF)	C10 (pF)
80	15 trifilar turns no. 28 enam. wire (10 twists per inch) on an FT50-43 ferrite toroid core.	24µH. 60 turns no. 28 enam. wire on T68-2 toroid core. Tap 37 turns above RFC5 end.	15 turns no. 28 ins. wire over RFC5 end of L1.	6µH. 37 turns of no. 28 enam. wire on T37-2 toroid core.	21µH. 36 turns of no. 24 enam. wire on T68-2 toroid core. Tap at 8 turns.	10 turns of no. 22 ins. wire over L7 end of L5.	2µH. 21 turns no. 26 enam. wire on T37-2 toroid core.	270	560	220	140	820	1500
40		7µH. 35 turns of no. 24 enam. wire on T68-2 toroid core. Tap at 20 turns above RFC5 end.	8 turns of no. 24 ins. wire over RFC5 end of L1.	3.22µH. 27 turns of no. 26 enam. wire on T37-2 toroid core.	6µH. 32 turns of no. 24 enam. wire on T68-2 toroid core. Tap 6 turns above L7.	9 turns of no. 22 ins. wire over L7 end of L5.	1µH. 15 turns of no. 22 enam. wire on T37-2 toroid core.	150	300	120	140	470	910
20	15 trifilar turns (10 twists per inch) of no. 28 enam. wire on an FT50-61 ferrite toroid core.	2µH. 20 turns of no. 22 enam. wire on T68-6 toroid core. Tap 11 turns above RFC5 end.	5 turns of no. 24 ins wire over RFC5 end of L1.	1.65µH. 20 turns of no. 26 enam. wire on T37-2 toroid core.	1.8µH. 19 turns of no. 22 enam. wire on T68-6 toroid core. Tap 4 turns above L7.	5 turns of no. 22 ins. wire over L7 end of L5.	0.55µH. 11 turns of no. 22 enam. wire on T37-2 toroid core.	75	150	39	100	220	470
15		0.7µH. 12 turns of no. 22 enam. wire on T68-6 toroid core. Tap 7 turns above RFC5 end.	3 turns of no. 22 ins. wire over RFC5 end of L1.	1.1µH. 17 turns of no. enam. wire on T37-2 toroid core.	1µH. 14 turns of no. 20 enam. wire on T68-6 toroid core. Tap 3 turns above L7.	4 turns of no. 22 ins. wire over L7 end of L5.	0.37µH. 9 turns of no. 22 enam. wire on T37-2 toroid core.	50	100	18	50	150	300
10		0.5µH. 10 turns of no. 22 enam. wire on T68-6 toroid core. Tap 6 turns above RFC5.	2 turns of no. 22 ins. wire over RFC5 end of L1.	0.8µH. 14 turns of no. 22 enam. wire on T37-2 toroid core.	0.8µH. 13 turns of no. 20 enam. wire on T68-6 toroid core. Tap 3 turns above L7.	3 turns of no. 22 ins. wire over L7 end of L5.	0.28µH. 8 turns of no. 22 enam. wire on T37-2 toroid core.	39	75	5	50	100	200

All cores other than for T1 are powdered iron. Cores and ferrite beads for this circuit were furnished via courtesy of Amidon Assoc., N. Hollywood, CA 12033.

quired for this circuit. It should be capable of delivering at least 600 mA of current continuously. It needs to be well filtered so that a T9 cw note will result.

Positive keying is used if a solid-state keyer is employed. A bug or straight key is suitable if a keyer is not desired.

Tuneup entails adjusting the doubler tank for a dip in drain current. Alternatively, C4 can adjust for maximum drain current (optimum drive) at Q4. Next, C8 is adjusted for a dip in drain current at Q4.

This transmitter is designed to work into a 50-Ω load. If the antenna SWR is much above 1.5:1, a Transmatch is recommended to assure that FL2 looks into the required 50 Ω. A proper match will result in maximum power transfer to the antenna. Output from this transmitter will vary from 6 to 7 watts in Class AB (5 to 6 watts for Class C). Amplifier efficiency for Class C operation is roughly 85 percent, whereas an efficiency of 79 percent was observed during Class AB conditions. The standing drain current of Q4 is approximately 40 mA for Class AB service. for Class C use it is approximately zero.

A 140-W SOLID-STATE LINEAR AMPLIFIER

Solid-state technology has progressed to the point where devices capable of producing up to 100-watts output in the hf range are available at reasonable cost. This unit, designed by K7ES, applies the aforementioned technology to a broad-band linear-amplifier design. Power output is 140 watts from 1.6 to 30 MHz. The amplifier described here originally appeared in *QST* for June and July 1977.

The Circuit

The MRF454 transistor is specified for a continuous power output of 70 watts. The maximum continuous current allowable is 15 A. Maximum dissipation is limited to 250 watts. The minimum recommended collector idling current is 100 mA per device.

Fig. 62 shows the basic circuit. The bias supply uses active components rather than the common diode clamping scheme. Some advantages of this system are line-voltage regulation capability, low standby current, and a wide range of voltage adjustment. With the component values shown the bias is adjustable from 0.5 to 0.9 volt. The 0.5-ohm resistor between pins 1 and 10 of the MC1723 sets the current-limiting point of the chip at 1.35A. Temperature-compensating diode D1 is added for bias tracking as the power devices heat up. This circuit prevents thermal runaway. The base-emitter junction of a 2N5190 is used as this diode. Physical dimensions of the 2N5190 allow it to be used as the center standoff post of the circuit board, ensuring adequate thermal bonding of the diode junction to the heat sink.

Fig. 61 — The completed 140-watt amplifier board and heat sink. D1 is mounted on the underside of the circuit board sandwiched between the board and the heat sink.

Device input impedances vary across the frequency range this amplifier is designed for. An input frequency-correction network is employed to give a gain flatness response of better than 1 dB across the band. R1, R2 and the associated 5600-pF shunt capacitors comprise the input-correction network. With the negative feedback applied via R3 and R4 through L5, the equivalent of an attenuator is formed with frequency-sensitive characteristics. At 30 MHz the input power loss is 1 to 2 dB, increasing to 10 to 12 dB at 1.6 MHz. The gain variation of the rf transistors is the reciprocal of this, resulting in a gain flatness of approximately 1 dB. Input VSWR is better than 1.75 across the band.

The input transformer, T1, and the output transformer, T3, are of the same type. The low-impedance windings consist of two pieces of metal tubing which are shorted electrically at one end — the opposite ends being the connections of the windings (Fig. 63A). The multiturn, high-impedance windings are wound through the tubing so that the low- and high-impedance winding connections are at opposite ends of the transformer. Alternatively, copper braid can be substituted for the metal tubing. The braid can be taken from the coaxial cable of the proper diameter. This is exemplified in Fig. 63B.

The primary and secondary coupling coefficient is determined by the length-to-diameter ratio of the metal tubing or braid, plus the gauge and insulation thickness of the wire used for the high-impedance winding. A high coupling coefficient is necessary for good high-frequency response. Both transformers are loaded with ferrite material to provide sufficient low-frequency inductance.

The collector choke (T2) provides an artificial center-tap for T3. This produces even-harmonic cancellation. Additionally, T2 is a low-impedance point that supplies

negative feedback voltage through a separate one-turn winding. The characteristic impedance of ac and bd (windings on T2) should be approximately equal to the collector-to-collector impedance of Q1 and Q2, but it is not critical. For physical convenience a bifilar winding is used. The center-tap of T2 is really bc, but for stabilization purposes, b and c are separated by rf chokes which are bypassed individually.

Construction

A scale pc-board template is shown in Fig. 64. Double-sided board is used, and patterns are shown for each side of the board.

All parts are mounted on the circuit board. The MRF454s are soldered to the board, but the flanges are bolted tightly to the heat sink. Apply a thin layer of silicone grease to Q1, Q2 and D1 where they will contact the heat sink. This aids in efficient heat transfer. The board is attached to the heat sink by means of five screws. They are tapped into the heat sink: Four are at the board corners, and the fifth goes through the board, through D1, and into the heat sink.

A large heat sink is required for proper dissipation of heat. A nine-inch (229-mm) length of Thermalloy 6153 or a seven-inch (178-mm) length of Aavid Engineering 60140 extrusion is suitable for 100-percent duty cycle operation (such as RTTY or SSTV operation).

Tune Up and Operation

Since this is a broadband amplifier, no tuning adjustments are necessary. One need only set the bias control so that the amplifier draws 200 mA of quiescent current. A regulated power supply capable of delivering 13.6 volts dc at 25 A is required for amplifier operation.

If direct operation into an antenna is expected, filtering of the output is necessary to meet FCC regulations for spectral purity. The filters shown in Fig. 65 will provide more than sufficient harmonic attenuation with less than 1 dB of loss at the operating frequency.

Collector efficiency is in the vicinity of 50 percent. IMD products are typically 32 to 34 dB below PEP. Power gain is approximately 15 dB, which means that approximately 5 watts will drive the amplifier to the rated output of 140 watts. The spectral displays for harmonics and IMD are shown in Fig. 66.

THE "1/4-GALLON" AMPLIFIER — A WARC UPDATE

Some Novice class licensees own transmitters that are not capable of running the maximum legal-power input for that class license — 250 watts. This amplifier was designed primarily with this in mind. The 250-watt input level also makes this amplifier compatible with the proposed power limit for the new 30-meter WARC band. The necessary information on winding the tank coil for

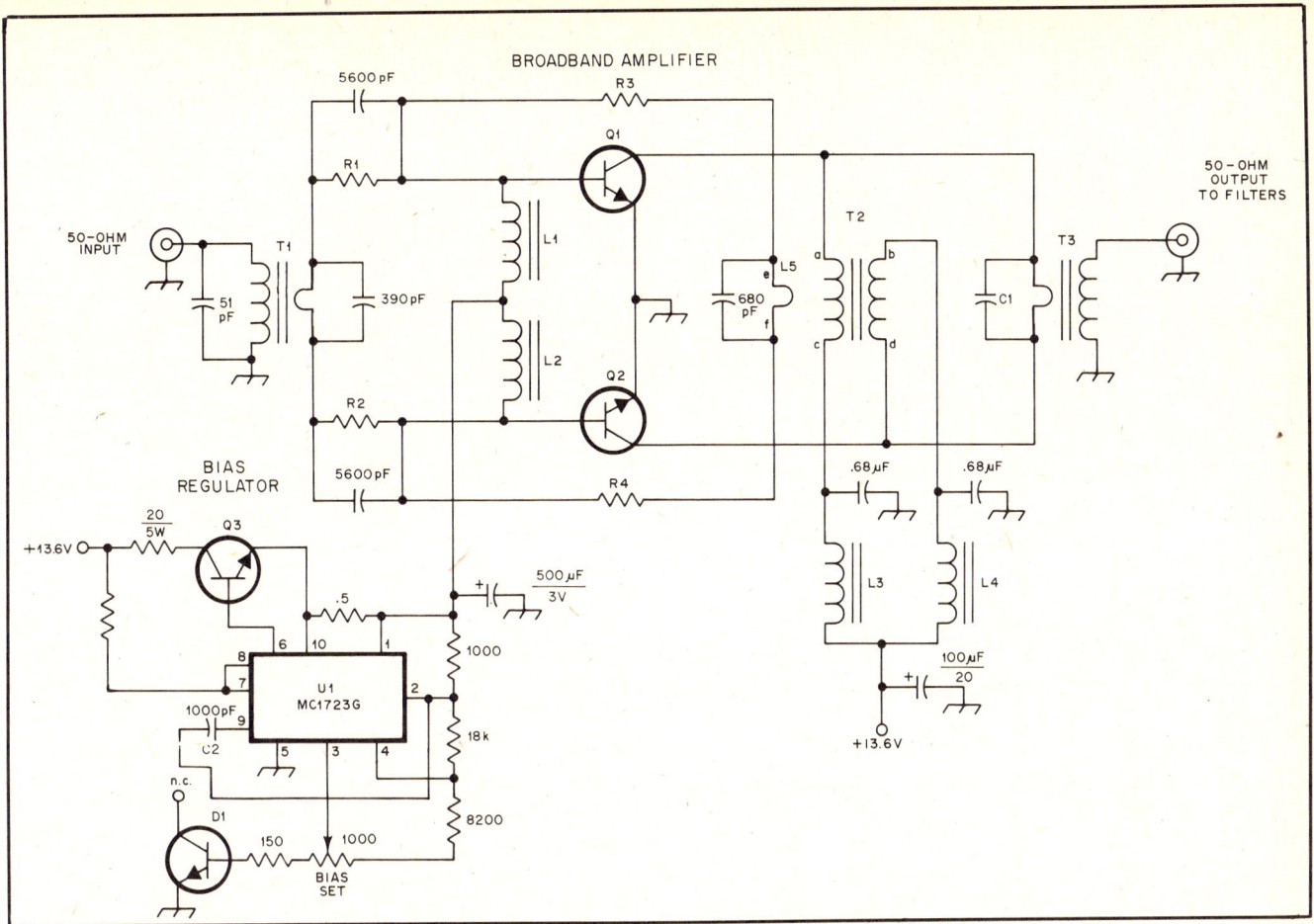

BROADBAND AMPLIFIER

BIAS REGULATOR

Fig. 62 — Circuit diagram of the 140-watt amplifier. All capacitors except part of C1, C2 and the electrolytic types are ceramic chips. Capacitors with values higher than 82 pF are Union Carbide type 1225 or Varadyne size 14. Others are type 1813 or size 18, respectively.

C1 — 1760 pF (two 470-pF chip capacitors in parallel with an 820-pF silver mica).
C2 — 1000-pF disc ceramic.
D1 — 2N5190 or equiv.
L1, L2 — Ferrite choke, Ferroxcube VK200 19/4B.
L3, L4 — Two Fair-Rite Products ferrite beads (2673021801 or equiv.) on no. 16 wire.
L5 — 1 turn through toroid of T2.
Q1, Q2 — MRF454.

Q3 — 2N5989 or equiv.
R1, R2 — Two 3.6 ohm, 1/2 W, in parallel.
R3, R4 — Two 5.6 ohm, 1/2 W, in parallel.
T1 — Primary: 3 turns no. 22 insulated wire. Secondary: Braid or tubing loop. Core material: Stackpole 57-1845-24B, Fair-Rite Products 2873000201 or two Fair-Rite Products 0.375-inch OD × 0.2-inch ID × 0.4 inch (9.5 × 5.1 × 10.2 mm), Material-77 beads for type A (Fig. 63A) transformer.

T2 — 6 turns no. 18 enameled wire, bifilar wound. Ferrite core: Stackpole 57-9322, Indiana General F627-8 Q1 or equiv.
T3 — Primary: Braid or tubing loop. Secondary: 4 turns no. 18 insulated wire. Core material: Two Stackpole 57-3238 ferrite sleeves (7D material) or a number of toroids with similar magnetic characteristics and 0.175-inch sq. (113-mm sq.) total cross-sectional area.
U1 — Motorola MC1723G or equiv.

10-MHz operation is given in Fig. 68. Approximately 25 watts of power are required to drive the amplifier to 165 watts output on 80 meters. If a 75-watt transmitter is used to drive the amplifier, the transmitter output should be reduced to a level just sufficient to drive the amplifier to its full-power input. This can usually be accomplished by lowering the drive to the transmitter output stage and reloading.

The Circuit

The main ingredient in this amplifier is a pair of 6KD6 television-sweep tubes (see Fig. 68. Although the tubes are rated at 33 watts of plate dissipation, they can handle temporary overloads of at least 100 watts without sustaining permanent damage. These tubes were chosen over 811s or 572Bs because they can often be "liberated" from old television sets or can be purchased new from most TV service shops.

D11, an 11-volt Zener diode, provides cathode bias for the tubes and establishes the operation as Class B. That class of amplifier requires less driving power than does a Class C stage for the same power output. It is easier to reduce the output from a transmitter that has more than enough power to drive the amplifier than it is to boost the output from a transmitter that provides insufficient drive. Class B

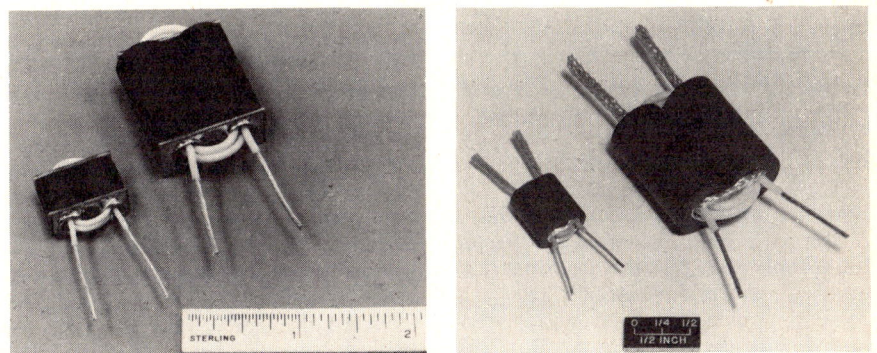

Fig. 63 — Shown are the two methods of constructing the transformers as outlined in the text. At the left, the one-turn loop is made from brass tubing; at the right, a piece of coaxial cable braid is used for the loop.

TOP (layout labels):
L4 L3
OUT GND
T3
C1 C1 C1
C a d C
E Q1 E b T2 c E Q2 E
B L5 L5 B
R3 R4
R3 R4
R1 R2
R1 R2
110μF/20V
500μF/3V
L1 L2 Q3
T1
+ +
150Ω
20Ω/5W
1000Ω 0.5Ω
1000pF C2
1000 18K 1000Ω
8200
U1
SSB AMPLIFIER 1.6 – 30 MHZ
INP POS NEG

BOTTOM (layout labels):
OUTP GND
0.68μF 0.68μF
680pF
D1
B C E
5600pF 5600pF
390pF
51pF

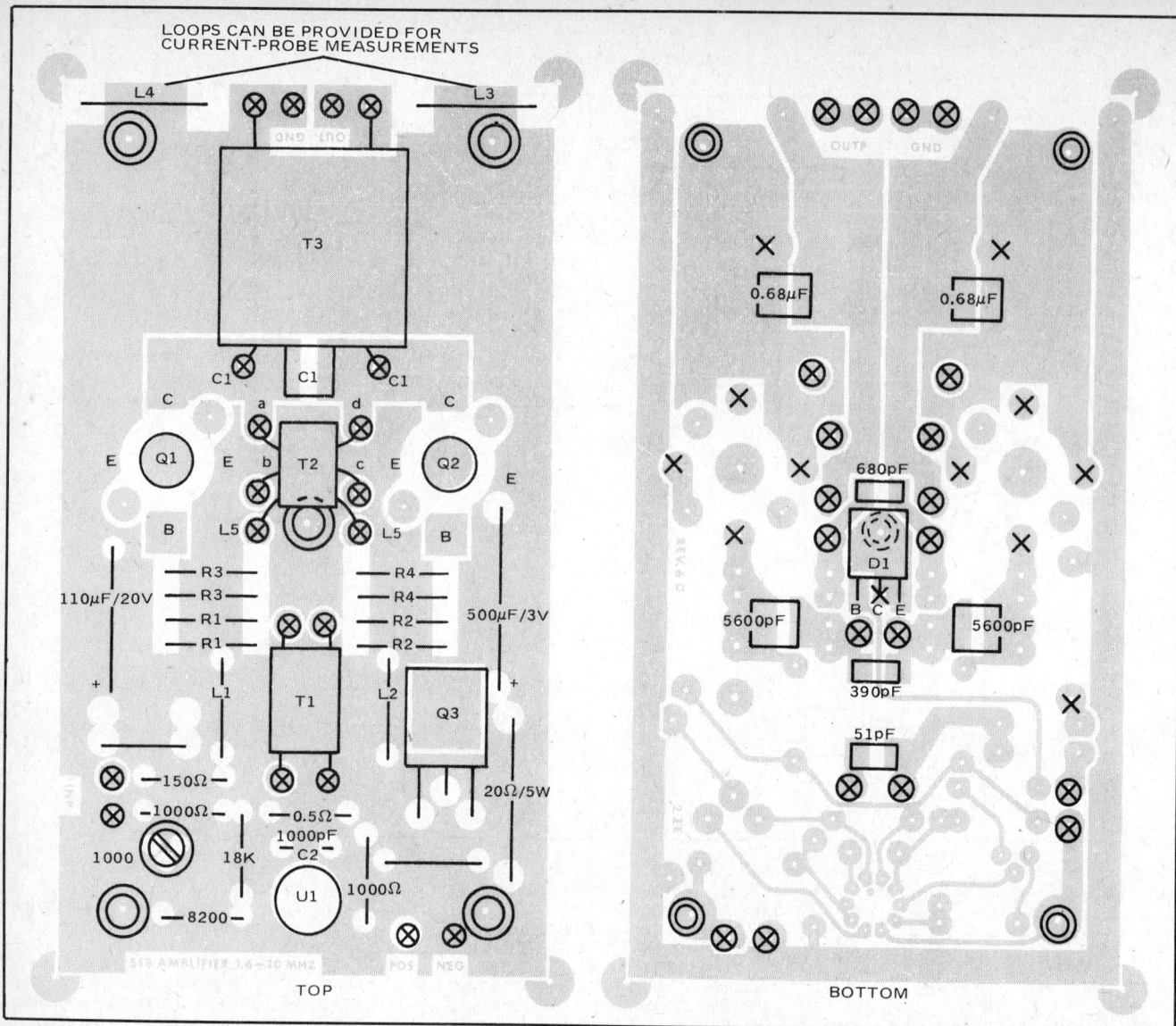

Fig. 64 — Actual-size layouts for the amplifier. Gray areas represent unetched copper. The pattern at the left is for the top side of the board and the one at the right for the bottom of the board.

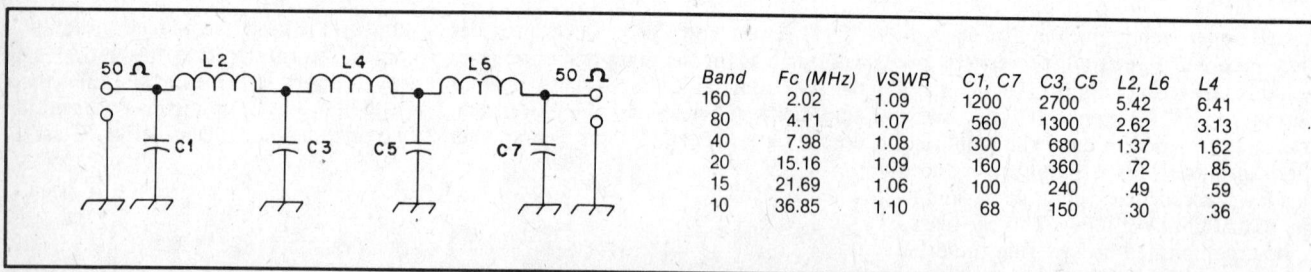

Band	Fc (MHz)	VSWR	C1, C7	C3, C5	L2, L6	L4
160	2.02	1.09	1200	2700	5.42	6.41
80	4.11	1.07	560	1300	2.62	3.13
40	7.98	1.08	300	680	1.37	1.62
20	15.16	1.09	160	360	.72	.85
15	21.69	1.06	100	240	.49	.59
10	36.85	1.10	68	150	.30	.36

(Filter schematic labels: 50 Ω, L2, L4, L6, 50 Ω, C1, C3, C5, C7)

Fig. 65 — This table shows the values for seven-pole Chebyshev low-pass filters suitable for use with the amplifier. These filters have been designed around standard capacitance values for each of the capacitors. Capacitance values are in pF and inductance values are in μH. This information was prepared by Ed Wetherhold, W3NQN.

operation was chosen for that reason.

The power supply uses an old TV-set transformer that has three secondaries: 600, 6.3 and 5.0 volts. The 600-volt winding supplies the full-wave bridge rectifier with ac energy. Dc output from the rectifier assembly is filtered by means of three 330-μF capacitors. The 47-kΩ resistors across each of the capacitors equalize the voltages across the capacitors and drain the charge when the power supply is turned off. Equalizing resistors are used across each of the diodes to ensure that the reverse voltage will divide equally between the two diodes in each leg. The capacitors across each diode offer spike protection. Each 6KD6 draws 2.85 amperes of filament current. Both filaments are connected in parallel across the 6.3-volt transformer winding. The 5-volt winding is connected in series with the 6.3-volt winding; the total (11.3 volts) is rectified, filtered, and used to power the relay.

A 0-50 μA meter is used to measure the plate voltage and current. The meter reads 0-1000 in the plate-volts position and

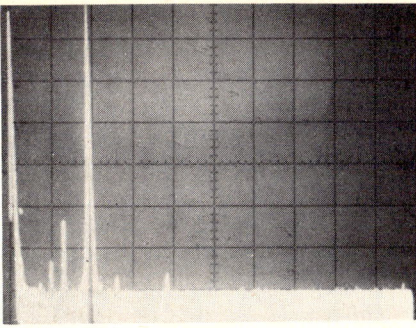

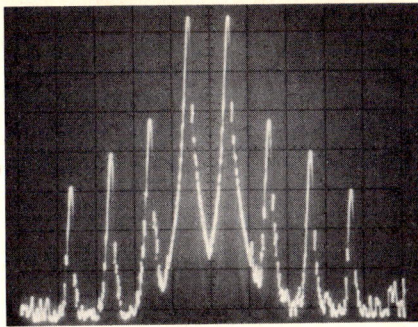

Fig. 66 — Spectral photographs of the output from the 140-watt amplifier. The display at the left was taken with the amplifier operating on the 80-meter band. At the right is a close look at the IMD products generated by the amplifier. Third-order products are down some 32 to 34 dB below PEP.

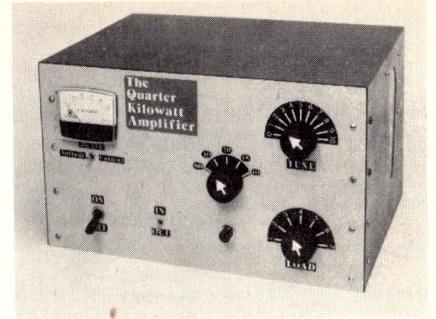

Fig. 67 — The completed quarter-kilowatt amplifier.

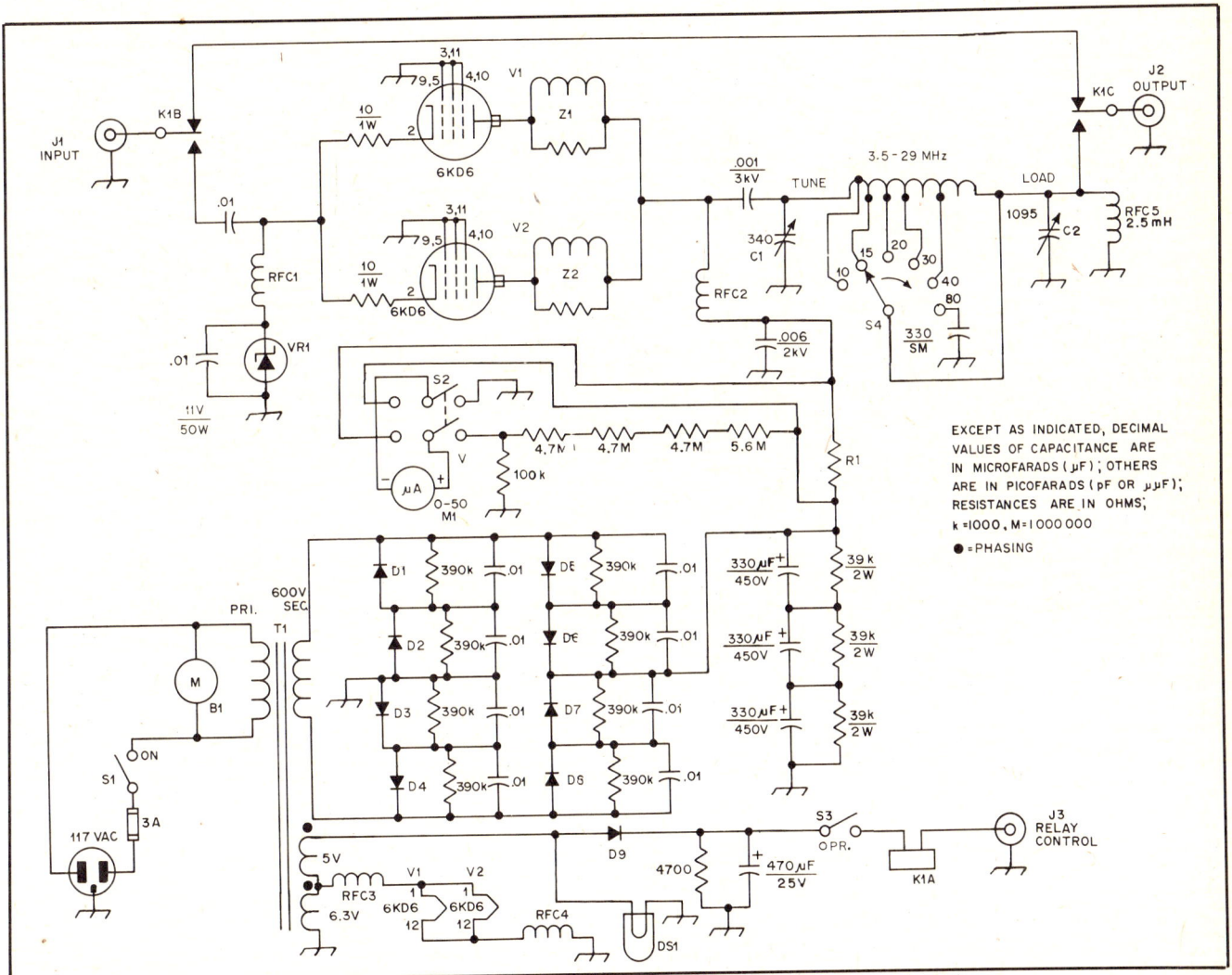

Fig. 68 — Schematic diagram of the 1/4-kilowatt amplifier. Resistors are the 1/2-watt composition type unless otherwise specified. Fixed-value capacitors are disc ceramic unless otherwise noted. Polarized capacitors are electrolytic.

B1 — 117-V ac blower.
C1 — Variable capacitor, 340 pF maximum, (Millen 19335 or equiv.)
C2 — Variable capacitor, 1095 pF maximum (surplus 3-section 365-pF variable).
D1 - D8, incl. — Silicon diodes, 1000 Volt, 2 A.
D9 — Silicon diode, 50 V, 1 A.
D10 — Pilot lamp assembly, 12 V.
D11 — Zener diode, 11 V, 50 W.
J1, J2 — Coaxial connector, SO-239.
J3 — Connector, phono type.
K1 — Dpdt relay, 12-V field, 2-A contacts.

L1 — 16 turns of no. 14 wire spaced 8 turns per inch. The entire coil is used for 80-meter operation and is tapped for the other bands as follows: 7 turns for 40 meters; 10 turns for 30 meters; 13 turns for 20 meters; 14 turns for 15 meters; 15 turns for 10 meters. Turns are measured from C2 end of coil.
M1 — Panel meter 0-50 μA, Calectro D1-910 or equiv.
R1 — Meter shunt, 10 feet (3.048 m) no. 24 enam. wire wound on a large ohmic value 2-watt composition resistor.

RFC1 — 20 turns no. 24 enam. wire on an Amidon FT-82-72 core.
RFC2, RFC5 — Rf choke, 2.5 mH, 500 mA.
RFC3, RFC4 — 15 turns no. 14 enam. wire on an Amidon FT-82-72 core.
S1 — Spst toggle switch, 4 A.
S2 — Dpdt toggle switch, low current type.
S3 — Spst toggle switch, low current type.
S4 — Single-pole, 5-position ceramic rotary switch, single-wafer type.
T1 — Television transformer (see text).
Z1, Z2 — 5 turns no. 18 enam. wire on a 47-ohm, 1-watt resistor.

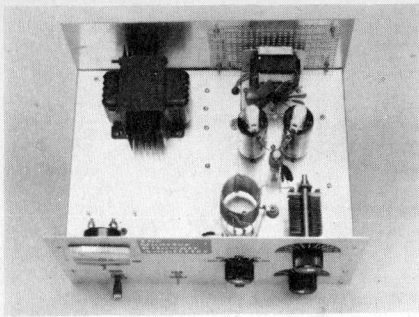

Fig. 69 — An inside view of the amplifier. C1 is located at the lower right. C2 is mounted below the chassis and is connected to L1 using a feedthrough insulator.

0-500 mA in the plate-current position. A 0.26-ohm shunt is placed in the high-voltage lead to facilitate metering of the plate current. One should be careful if working near the meter with the power on, as full plate potential will be present between each of the meter leads and ground. *Caution: Turn off and unplug the amplifier before making any changes or adjustments.*

Construction

Perhaps the best way to classify this amplifier would be to call it a "junker type" amplifier. Every attempt was made to keep the amplifier as simple and easy to build as possible. The only critical values are those for the amplifier plate-tank circuitry. Reasonable parts substitutions elsewhere in the circuit should have little or no effect on the performance of the amplifier. For example, if 330-μF filter capacitors are not available, 250- or 300-μF units could be substituted. They should be rated at 450 volts or greater to provide a margin of safety. If a 0.001-μF plate-blocking capacitor is not on hand and a 0.005-μF unit is, use it. Builders often attempt to match parts exactly to the type specified in a schematic or parts list. A few projects are this critical in nature, but the majority, including this one, are not.

The transformer used in this amplifier was garnered from an old TV set. Any hefty transformer with a high-voltage secondary between 550 and 700 volts should be adequate. Most of these transformers will have multiple low-voltage secondaries suitable for the tube filaments and relay requirements.

The chassis used to house the amplifier happened to be on hand and measured 3 × 10 × 14 inches (76 × 254 × 356 mm). No doubt the amplifier could be constructed on a smaller chassis. The beginner is cautioned not to attempt to squeeze too much in too small a space.

The front, rear, side and top panels are constructed from sheet aluminum and help to keep the amplifier "rf tight." Any

air-flow openings are "screened" with perforated aluminum stock. The front-panel meter opening is shielded by means of an aluminum enclosure (a small Minibox would serve quite nicely). The on-off power switch, pilot light, meter switch, band switch, tuning and loading controls, and amplifier in-out switch are all located on the front panel. On the rear panel are the amplifier input and output connections, relay control jack and the fuse holder. As can be seen from the photograph, a fan is located near the tube envelopes to keep them cool during operation.

Setup and Operation

Attach the transmitter output to the amplifier input connection. Then, join the output of the amplifier to a 50-ohm noninductive dummy load. Connect the relay control line to the transmitter or external antenna-relay contacts. Then plug in the line cord and turn the power switch to the ON position. With the meter switch in the PLATE VOLTS position, the reading on the meter should be approximately 425, which corresponds to 850 volts. If the power transformer used has a high-voltage secondary other than 600 volts, the reading will vary accordingly. If no plate voltage is indicated by the meter, check your wiring for possible errors or defective components. Next, place meter in the PLATE CURRENT position, the band switch to the 80-meter band, and apply a small amount of drive to the amplifier — enough to make the meter read 50 mA (5 on the meter scale). With the plate-tank loading control fully meshed, quickly adjust the plate tuning capacitor for a dip in plate current. Apply more drive (enough to make the meter read 100 mA), advance the loading control approximately one-eighth turn and readjust the plate tuning control for a dip in the plate current. Continue this procedure until the plate-current maximum dip is approximately 300 mA. The final value of plate current at which the amplifier should be run depends on what the plate voltage is under load. In our case this value was 800 volts. Therefore, the amount of current corresponding to 250 watts input is approximately 310 mA. (I = P/E, I = 250/800, I = 312.5 mA.) The same tune-up procedure should be followed for each of the other bands. The amplifier efficiency on 80 through 20 meters is approximately 65 percent, dropping to 60 percent on 15 meters. On 10 meters, efficiency is slightly less than 50 percent. Poor efficiency on the higher bands is caused primarily by the high-output capacitance characteristics of sweep tubes.

AN ECONOMY 2-kW AMPLIFIER

The amplifier described here was built with mostly junk-box components. A pair of 4-400A tubes are run in grounded-grid fashion and can develop 2-kW PEP input

Fig. 70 — Underside of the amplifier. Component layout is not particularly critical.

when driven from a 100-watt exciter. Approximately 40 watts of drive is required for 1-kW operation and 100 watts for 2-kW operation. The amplifier makes use of Eimac SK-410 air-system sockets and SK-406 chimneys. A blower, which is mounted external to the amplifier chassis, forces air through a length of automobile defroster hose, approximately 2 inches (51 mm) in diameter and is flexible, through the pressurized chassis and out the air system sockets and chimneys. By mounting the blower away from the amplifier, and therefore the operating position, this source of noise can be greatly reduced, if not eliminated.

The Circuit

The schematic diagram of the amplifier is shown in Fig. 72. Relay K1 is used to switch "around" the amplifier during receive periods or when it is desired to bypass the amplifier. S4 allows the operator to switch around the amplifier while leaving it in a standby condition, ready for operation. Filament voltage is fed through a heavy duty bifilar-wound choke which uses a 950 mu ferrite rod 1/2 inch (13 mm) in diameter and 7 inches (178 mm) long of material. Power is fed directly into the filament without the aid of an input matching network. The input

Fig. 71 — This is a photograph of the front panel of the 4-400A amplifier. The amplifier is relatively compact, measuring 10 × 14 × 10 inches (254 × 356 × 254 mm). The PEP input for this amplifier is 2 kW.

Fig. 72 — Schematic diagram of the amplifier. A pair of 4-400A tetrodes is used.

C1, C4 — Transmitting ceramic, 1000 pF.
C2 — Variable, 150 pF, 4500 V. Johnson 154-15 or equiv.
C3 — Variable, 1095 pF, receiving type, 3 section, 365 pF per section used here.
J1, J2 — Coaxial connector, builder's choice.
J3 — Phono connector.
K1 — Relay, dpdt, 12-volt field.
K2 — Relay, spdt, 12-volt field.
L1 — Copper tubing, 3/16-inch (4.7 mm) diameter. 6-1/2 turns at 2-inch (51 mm) diameter tapering into 6 turns at 3-inch (76 mm) diameter. Tap at 5-1/4 turns for 10 meters, 6-1/4 turns for 15 meters, 9-1/4 turns for 20 meters. (10 meters, 2.3 µH; 15 meters 3.2 µH; 20 meters, 4.9 µH. Note: stray inductances must be subtracted from these figures.)
L2 — 14-1/2 turns no. 12 wire, 6 turns per inch. B&W 3033 or equivalent. Tap at 4-3/4 turns from L1 end for 40 meters. Use entire coil for 80 meters. (40 meters, 9.6 µH; 80 meters, 17.5 µH. Note: stray inductances must be subtracted from these figures.)
M1 — 0-150 mA panel meter. Simpson 06400 or equiv.
R1 — Grid meter shunt, 0.33 ohms. Wind 2 feet (610 mm) no. 32 wire on a large value 1- or 2-watt resistor.
R2 — Plate-meter shunt, 0.0733 ohms. Wind 1 foot, 1 inch (330 mm) no. 32 wire on a large value 1- or 2-watt resistor.
RFC1 — Filament choke, 20 bifilar turns on no. 12 enameled or plastic covered wire on a 950 mu, 7 × 1/2 inch (178 × 13 mm) ferrite rod.
RFC2 — Plate choke, B&W 800 or equiv.
RFC3 — Choke, 2.5 mH, 300 mA.
S1 — Band switch, 5 position. Millen 51001 or equiv.
S2 — Toggle switch, spdt.
S3, S4 — Toggle switch, spst.
T1 — Filament transformer, 5 volts, 30 amperes, 117-volt primary. Stancor P6492 or equiv.
T2 — Transformer, 12.6 volts, 300 mA.
Z1, Z2 — Parasitic suppressor, 2 turns 5/16 inch (8 mm) wide copper strap wound around three 100-ohm, 2-watt resistors in parallel.

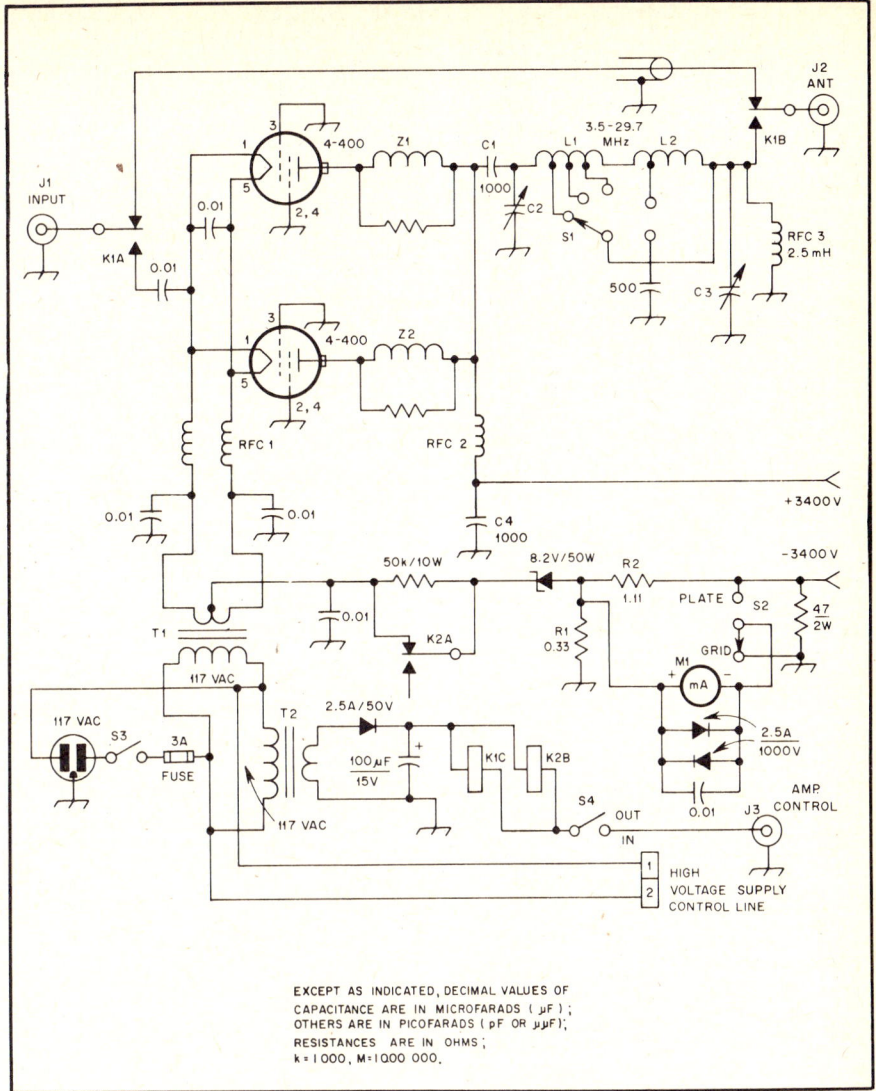

EXCEPT AS INDICATED, DECIMAL VALUES OF CAPACITANCE ARE IN MICROFARADS (µF); OTHERS ARE IN PICOFARADS (pF OR µµF); RESISTANCES ARE IN OHMS; k = 1 000, M = 1 000 000.

VSWR is such that most transmitters should have no difficulty in driving the amplifier.

The 8.2-volt Zener diode develops bias and allows the tubes to run in Class AB.

The 50-kΩ/10 watt resistor in conjunction with K2 cut the amplifier off during receive or standby periods. A single 0-150 mA panel meter is used to monitor either plate or grid current — selectable with a front-panel switch, S2. For plate-current measurements the meter reads 0-1.5 amperes (utilizing meter shunt R2) and reads 0-450 mA for grid-current measurements (using meter shunt R1).

Fig. 73 — A top view of the amplifer. The transmit/receive relay can be seen between the filament transformer and one of the 4-400A tubes.

Fig. 74 — Photograph of the underside of the chassis. Component placement is not critical. The builder should plan a layout that suits his or her components.

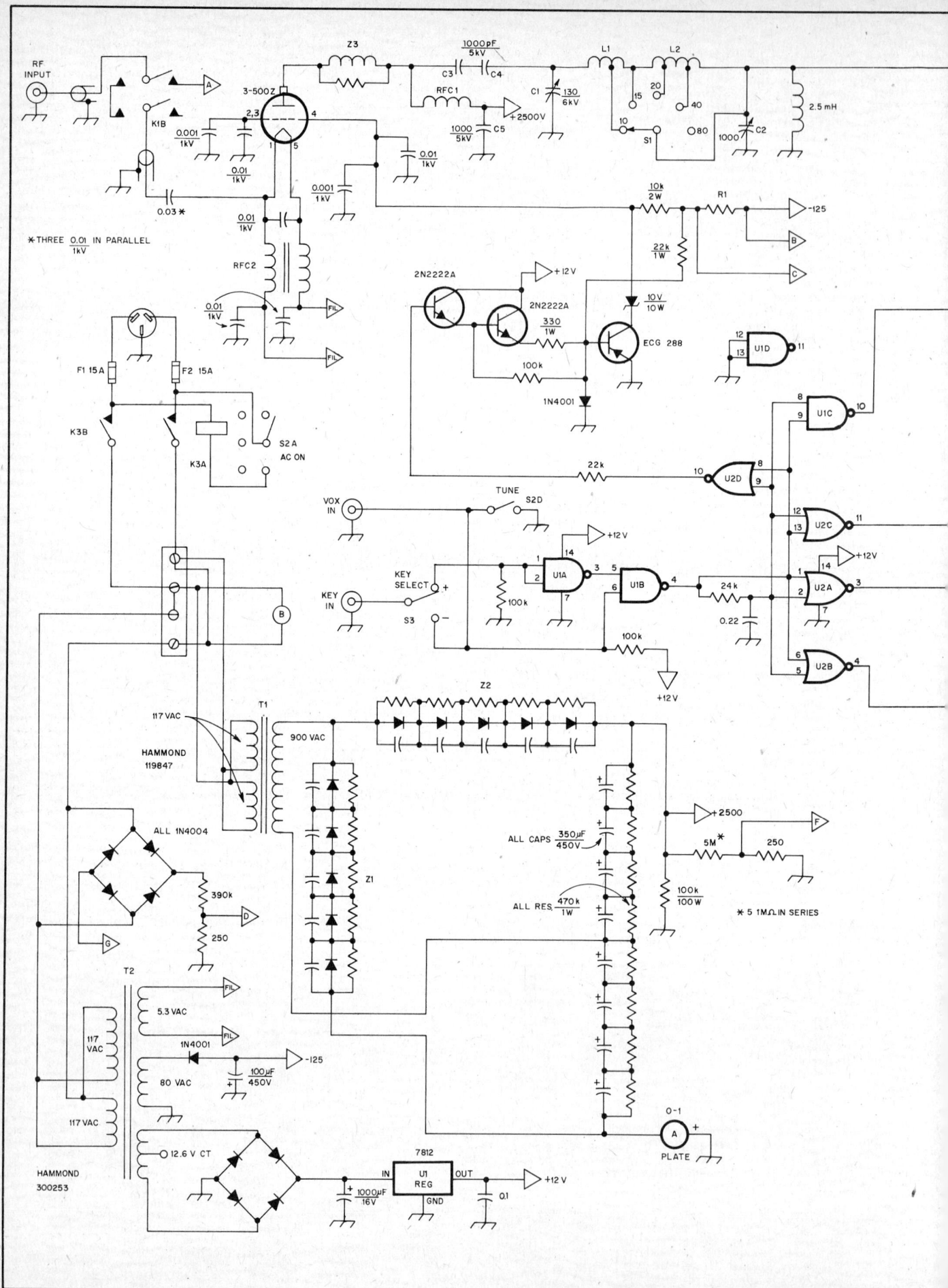

RF INPUT

K1B

A

0.001
1kV

0.03 *

* THREE 0.01 IN PARALLEL
1kV

3-500Z

2,3   4

1   5

0.01
1kV

0.01
1kV

0.001
1kV

Z3

RFC1

1000
5kV   C5

C3   C4

1000pF
5kV

+2500V

C1 130
6kV

L1   L2

15
20

10
40

S1   80

1000
C2

2.5 mH

RFC2

0.01
1kV

FIL

FIL

F1 15A   F2 15A

K3B

K3A

S2A
AC ON

B

10k
2W   R1

22k
1W

-125

B

C

2N2222A

+12V

2N2222A

330
1W

100k

1N4001

10V
10W   ECG 288

12
13   U1D   11

8
9   U1C   10

22k

10   U2D   8
9

VOX IN

TUNE   S2D

KEY SELECT

KEY IN

S3   +

-

100k

1
2   U1A   3
7   14

5
6   U1B   4

24k

0.22

100k

+12V

+12V

12
13   U2C   11

1
2   U2A   3
7   14   +12V

6
5   U2B   4

T1

117 VAC

HAMMOND
119847

900 VAC

Z2

Z1

ALL 1N4004

390k

250

G

D

T2

5.3 VAC

1N4001

80 VAC

100µF
450V   -125

117 VAC

117 VAC

12.6 V CT

HAMMOND
300253

FIL

FIL

ALL CAPS 350µF
450V

ALL RES. 470k
1W

+2500   F

5M *   250

100k
100W

* 5 1MΩ IN SERIES

0-1

A
PLATE   +

7812

IN   U1
REG   OUT   +12V

GND

1000µF
16V   0.1

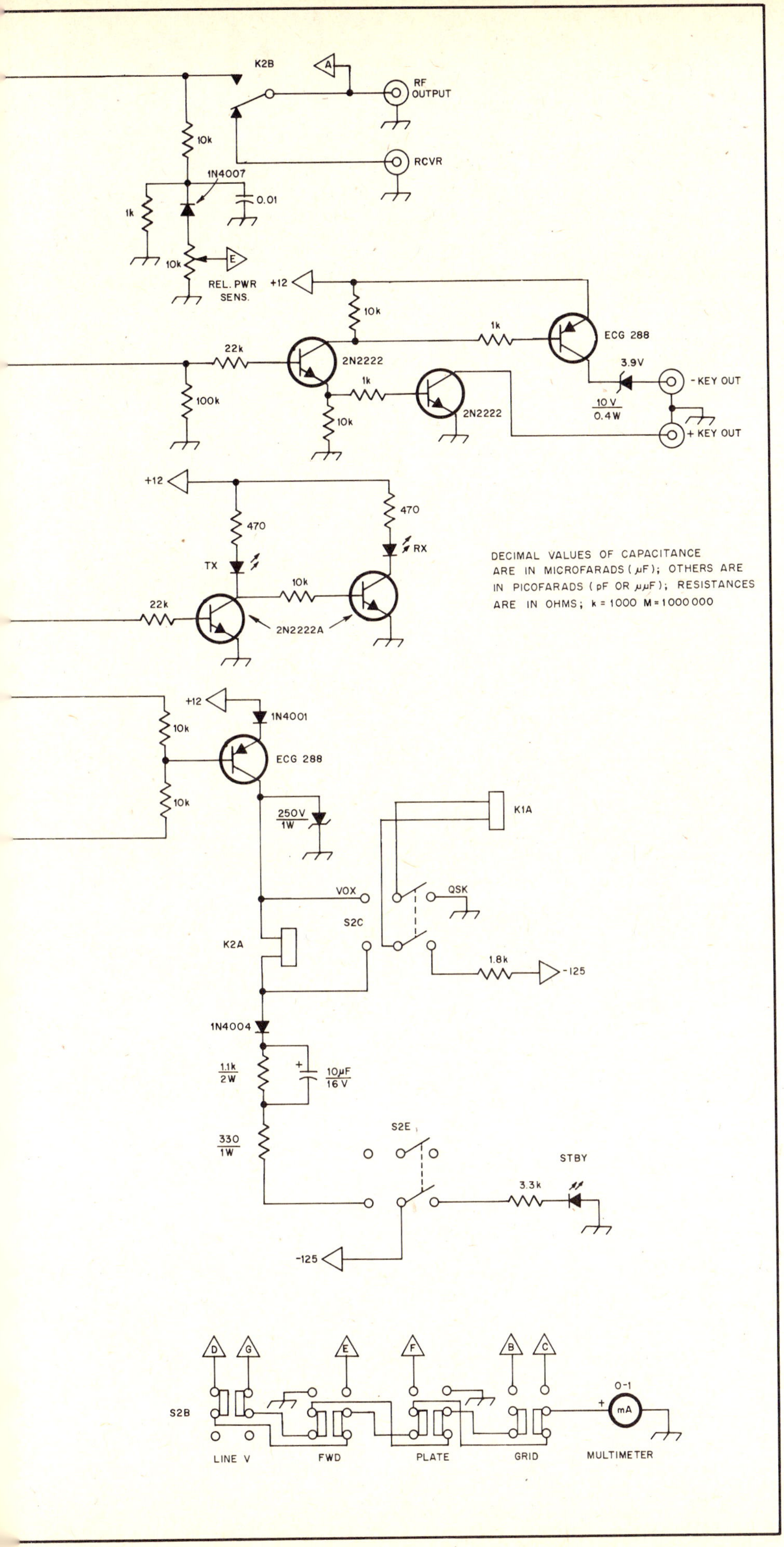

Fig. 75 — Schematic diagram of the full-break-in kilowatt amplifier.

C1 — 130-pF, 6-kV air variable (Millen 16510 or equiv.).
C2 — 1000-pF air variable (Millen 2152 or equiv.).
C3-4 — 1000-pF, 5-kV transmitting ceramic (Centralab 800 series).
K1 — Dpdt relay, 10-A contacts, 24-V coil.
K2 — Spdt vacuum relay, 18-A, 2.5-kV contacts, 24-V coil. Kilovac type HC1/S43.
K3 — Dpst relay, 15-A contacts, 115-V coil.
L1 — 4 turns of 1/8-inch copper tubing, 1-1/4 inches long and 1-3/4 inches in diameter, tapped 1-1/4 turns from L2 end.
L2 — 25 turns of no. 12 copper wire, 2-1/2 inches in diameter, 6 tpi (B and W 3029 Mini-ductor stock.) Tapped (from C2 end) at 10-3/4 turns for 40 meters and 19-1/2 turns for 20 meters.
R1 — 12-1/2 inches of no. 26 enamel wound on 1-megohm, 2-watt resistor (0.043 ohms).
RFC1 — 180-μF, 500-mA plate choke (B and W Model 800 or equiv.).
RFC2 — Bifilar filament choke (B and W FC-25A or equiv.).
S1 — Single-pole, 5-position rf switch, 20-A contacts, 13 kV (Millen 51001 or equiv.).
S2 — 10 section push-lock switch.
S3 — Spdt toggle switch.
Z1, Z2 — 5 series-connected 1N4007 diodes, each shunted by a 0.01-μF disc ceramic and a 470-kΩ, 1/2-watt resistor.
Z3 — 3 turns of 1/8-inch copper-tubing around three paralleled 100-Ω, 1-watt resistors.

A small power supply capable of operating K1 and K2 is included in the amplifier. An amplifier control jack located on the rear apron of the amplifier is grounded for amplifier operation.

The pi network consists of two coils — one wound from 3/16-inch (4.7 mm) copper tubing and one wound from large Miniductor stock. The coil made from copper tubing is wound in two different diameters. It starts out from the 4-400A plates at the smaller diameter (for 10 and 15 meters) and progresses to the larger diameter (for 20 meters) all with one continuous piece of tubing. A portion of the copper tubing is used on 40 meters along with a length of Miniductor stock. The builder may wish to end the copper tubing coil at the 20-meter tap position and continue on from that point with the Miniductor stock for 40 and 80 meters. The inductance values required for each band are given in Fig. 72 so that the builder may adapt junk-box coils for the tank circuitry.

K1 is a plug-in type of relay with contacts that are rated for 10 amperes of current. This relay plugs into an octal socket that is located between the filament transformer and one of the 4-400A tubes. K2 is a small spdt relay that was garnered from the junk box and is of unknown origin. Any similar relay should work fine.

Construction

The amplifier was built on a 10 × 14 × 3-inch (254 × 356 × 76-mm) aluminum chassis. A heavy duty front panel was constructed from a discarded rack panel and measures 10 × 14 inches (254 × 356 mm) making the overall dimensions of the

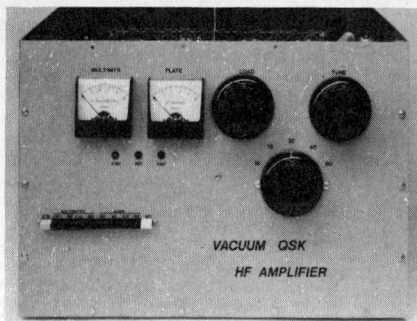

Fig. 76 — Front panel of the 3-500Z QSK amplifier.

Fig. 77 — Top view of the 3-500Z amplifier. The top section of the amplifier is divided into two compartments. One contains the power supply and blower and the other holds the rf components.

Fig. 78 — Bottom view of the 3-500Z amplifier.

amplifier 14 × 10 × 10 inches (356 × 254 × 254 mm).

Layout of the components is not especially critical, however the builder should strive to keep the tank-circuitry leads as short as possible to avoid stray capacitances and inductances. The general layout of this amplifier can be seen in the accompanying photographs.

A flange made from Plexiglass tubing

and flat 1/4-inch Plexiglass stock is used to connect to the defroster hose. Part of this flange is visible in the photograph of the underneath of the chassis. An alternative to building your own flange would be to purchase one from a supplier of central vacuum cleaning systems. Of course, there is no reason the builder couldn't mount the blower directly on the rear panel of the amplifier.

A perforated aluminum cover (not shown in these photographs) is used to enclose the top of the amplifier. Good contact between this cover and the mating surfaces is important for effective shielding. Make sure to remove any paint to ensure a good metal-to-metal contact.

Operation

This amplifier is designed to operate at the 1-kW level for cw operation and at the 2-kW PEP level for ssb. This amplifier was optimized for the 1500-watt level, thereby providing reasonable efficiency at both the 1 and 2 kW levels without the need for switching the power-supply voltage. With 3400 volts on the plate, the 1 kW plate current should be 294 mA and the grid current should be roughly 100 mA. At 2 kW the plate current will be 588 mA and the grid current approximately 200 mA. Idling plate current will run approximately 120 mA. Efficiency on all bands should be roughly 60 percent.

A 1-kW AMPLIFIER WITH FULL BREAK-IN

Break-in operation (QSK) has been a popular mode among cw operators for many years. Previous circuit designs for QSK power amplifiers have used vacuum tubes as T-R switches. That system works well, but can be difficult to adjust and is prone to generating TVI. Two other methods for providing QSK operation are available to the amateur today. One is the use of PIN diodes for rf switching, and the other uses a high-speed vacuum relay. The PIN-diode method becomes quite complex and expensive at the 1-kW level. On the other hand, vacuum relays can be purchased at prices attractive to the amateur. Sequencing circuits for controlling the relay can be made from simple logic gates.

The Circuit

The schematic diagram of the amplifier is shown in Fig. 75. A single 3-500Z develops 1 kW input on cw and 1 kW PEP on ssb. The output network is a conventional pi design for the hf amateur bands 3-30 MHz. No tuned input circuit is used.

Bias and input/output switching for the amplifier are controlled by six CMOS gates. When in the QSK mode, keying pulses from a keyer or other keying device are directed to either U1 pin 1, or U2 pin 6. The resultant output pulses from U1 pin 4 are presented to one of the inputs of gates U1C, U2A, B, C and D. The other inputs of these gates are connected to U1 pin 4 via

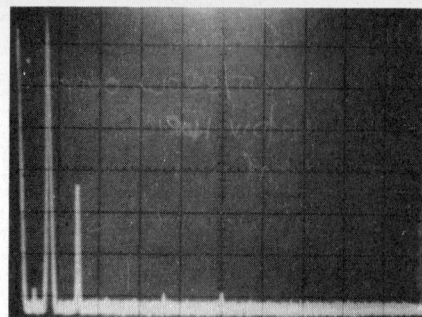

(A)

(B)

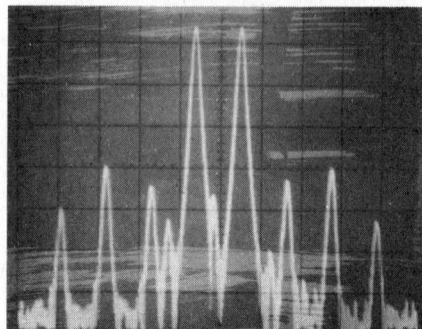

Fig. 79 — Spectral displays of the output of the 3-500Z amplifier. At A is shown the worst-case harmonic spectrum. The second harmonic is 44 dB below the fundamental (1-kW input on 40 meters). At B is the two-tone IMD spectrum. Third-order products are 42 dB below the PEP output and the fifth-order products are down 40 dB (1-kW PEP input on 20 meters).

an RC network. The output states of U1C and U2 will change with a delay equal to the time constant of the RC network. Where a NAND gate is shown, its output state will not change until one RC time constant after an input state change. A NOR adds a "tail" of one RC time constant to the output pulse. The time constant of R1C1 is 5 ms, which places a 5-ms delay between contact closure of the vacuum relay and the application of rf. This sequencing is necessary to prevent "hot-switching" the relay. For VOX operation, the VOX contacts on the exiter relay are connected to the relay control circuitry, and an additional open-frame relay is added to the circuit to bypass the amplifier during receive operation.

Metering and Switching

Amplifier plate current is read directly on a 1-A movement. Other functions are monitored by a multimeter using a 1-mA movement. The functions monitored by the multimeter are grid current (0-200 mA), plate potential (0-5000 V), relative power and line voltage. The line volts position measures the line voltage on a relative scale. This function is useful for checking line voltage drop at full power input. All control and meter functions are selected by a 10-position push-lock switch. The metering functions have a different interlock from the control functions. The switch

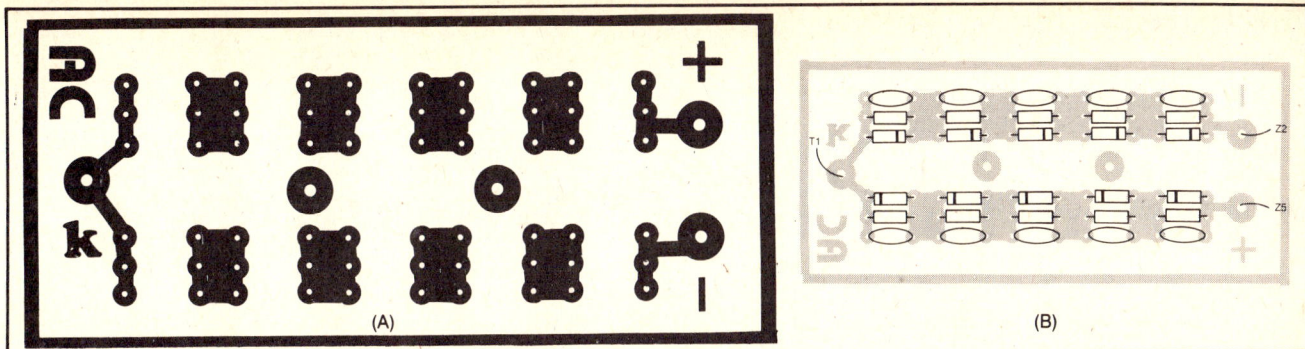

Fig. 80 — Full-scale etching pattern for the power supply circuit board. Black areas represent unetched copper as seen from the foil side of the board (A). At B is the parts-placements guide. Gray areas represent an x-ray view of the foil as seen from the component side of the board.

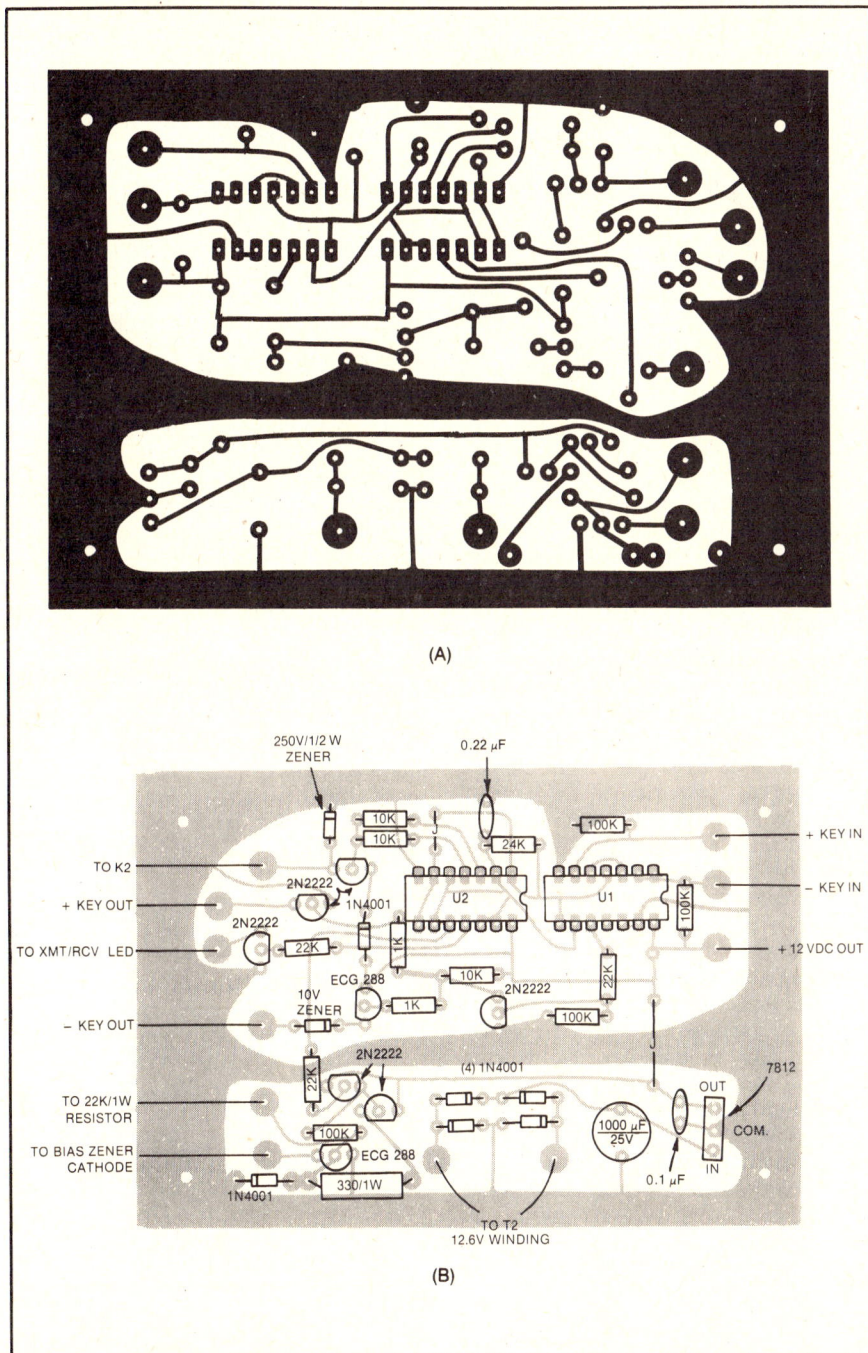

(A)

(B)

Fig. 81 — Full-scale etching pattern for the logic board. Black areas represent unetched copper as seen from the foil side of the board (A). At B is a parts-placement guide for the logic board seen from the component side of the board. The gray area is an x-ray view of the foil.

used in the amplifier was purchased as a surplus item from Mendelson Electronics in Dayton, Ohio. Other switches of the same type are available as new units.

Construction

The amplifier is built around a 17 × 17 × 3-inch (417 × 417 × 74-mm) aluminum chassis. The cabinet top, bottom, sides, back and front are constructed from 1/16-inch (1.5-mm) aluminum sheet. All circuits carrying rf are completely shielded to prevent TVI or instability. An aluminum sheet separates the plate compartment from the power supply section. The blower was purchased at a flea market for a bargain price, and provides more than enough air to cool the tube envelope. Air flow is from the back of the chassis over the transformers, under the chassis and through the Eimac air system socket and chimney. Component layout can be seen in the photographs.

Connections and Switch Functions

The amplifier is designed to work with transceivers or transmitter/receiver combinations. Rf from the exiter is connected to the RF INPUT jack. If the exiter is equipped for QSK operation, the receiving antenna input is connected to the RECEIVE jack on the amplifier.

The normally open VOX contacts from the exiter connect to the VOX IN on the amplifier for keying the amplifier during ssb and semi-break-in cw operation. For QSK operation, the keying device is connected to the KEY IN jack. The KEY OUT jack connects to the key jack on the exiter.

Operation

The amplifier should be tuned for maximum output power using the relative power mode of the multimeter. Watch the plate and grid currents to ensure that the manufacturer's ratings for the tube are not exceeded. The amplifier is designed for continuous-duty service at full power; after 10 minutes of key-down operation no excessive heating was noted. QSK operation with the amplifier is smooth and trouble-free at speeds in excess of 50 wpm.

Chapter 7

VHF and UHF Transmitting

The frequencies above 50 MHz were once a world apart from the rest of amateur radio, in equipment required, in modes of operation and in results obtained. Today these worlds blend increasingly. Thus, if the reader does not find what he needs in these pages to solve a transmitter problem, it will be covered in the hf transmitting chapter. This chapter deals mainly with aspects of transmitter design and operation that call for different techniques in equipment for 50 MHz and up.

SSB/CW vs. FM

Whenever vhf operators gather, the subject of fm vs. ssb and cw is bound to come up. Because of their mode differences, the two types of operation are segregated on the lower four amateur vhf bands. Actually, both forms of communications have their advantages and disadvantages. They are better discussed subjectively among operators. Here we are only interested in the different requirements each mode places upon the transmitting equipment used. In general, equipment used for fm is of the oscillator-multiplier type. Because ssb cannot be passed satisfactorily through a frequency-multiplication stage, generation of vhf ssb signals requires the use of one or more mixer stages. Vhf cw may be generated by either method. Recently, manufacturers of synthesized as well as crystal-controlled amateur fm transceivers have been using a combination of both approaches. The multimode vhf transceiver, which offers the operator a choice of cw, ssb, fm and often a-m, is a reality. Here again we find both approaches to signal generation.

The Oscillator-Multiplier Approach

This type of transmitter, which may be used for fm or cw, generally starts with a crystal oscillator operating in the hf range, followed by one or more frequency-multiplier stages and at least one amplifier

stage. While relatively simple to construct, such transmitters can be a cause of much grief unless the builder takes precautions to prevent undesired multiples of the oscillator and the multiplier stages from being radiated. For frequencies below 450 MHz the transmitting mixer is not difficult to construct and is recommended for most applications. Spurious-signal radiation is much easier to prevent with the latter, although it does not lend itself to compact fm equipment design. For operation on the higher amateur uhf bands, the oscillator-multiplier approach offers definite advantages and is recommended at present. Fig. 1 shows how the harmonics of a 144-MHz signal may be multiplied to permit operation on amateur microwave bands. Stability at 144 MHz is easy to achieve with the current technology, making stable microwave signals simple to generate. Varactor diodes are used as frequency-multiplying devices. They are installed in resonant cavities constructed from double-sided pc-board material. Operation will probably be crystal controlled, as even the best transceiver/transverter combination used to generate the 144-MHz signal may create problems when the output is multiplied in frequency 40 times! A frequency synthesizer with a stable reference oscillator may be used to generate the 144-MHz signal, but its output should be well filtered to eliminate noise.

Although spurious outputs of the various multiplier stages may not cause harmful interference, that is no excuse for not removing them. In most cases, the Q of successive cavities will suffice. A band-pass filter may be used to filter the final multiplier stage. Construction details of a 432-to-1296 MHz frequency multiplier using switching diodes are presented later in this chapter.

Transmitting Mixers

With the possible exception of the

power levels involved, there is no reason to consider transmitting mixers differently than their receiving counterparts. One thing to keep in mind is that many deficiencies in the transmit mixer will show up on the air. Receiver-mixer troubles are *your* problem. Transmit-mixer troubles become everyone's problem!

A trio of popular types of transmitting mixers is shown in Fig. 2. The doubly balanced diode mixer at A may be built using either discrete components, or the phase relationship between ports may be established using etched-circuit strip lines. Miniature DBMs are available at low cost from several manufacturers. They offer an almost-foolproof method of generating vhf ssb. Another popular mixer uses a pair of FETs in a singly balanced configuration. If care is taken in construction and adjustment, local-oscillator rejection will be adequate with this circuit. To be safe, a series-tuned trap, designed to attenuate the LO leakage even further, should follow this stage. A typical FET balanced mixer is shown in Fig. 2B.

Finally, we see a typical vacuum-tube mixer (Fig. 2C). Because it can handle more power, the tube mixer has endured at vhf. Its higher output, when compared to most solid-state mixers, reduces the number of subsequent amplifier stages

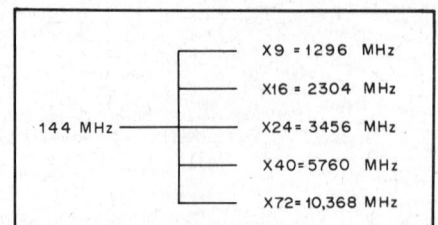

Fig. 1 — The harmonic relationships of most microwave bands to the 2-meter band are diagrammed here. The 15-mm (24 GHz) band bears no easily utilized integral relationship to 144 MHz.

needed to reach a specific power level. Apart from feeling more comfortable with tubes, this is the only advantage available from using them as mixers, at least on the lower vhf bands.

High-Level Transmitting Mixers

When designing a transmitting converter for vhf, the tradeoffs between the advantages of mixing at a low power level, such as in a diode-ring mixer, and using several stages of linear amplification must be weighed against the cost of amplifying devices. Linear uhf transistors are still relatively expensive. On 432 MHz and above, it may be desirable to mix the i-f and local-oscillator signals at a fairly high level. This method makes it unnecessary to use costly linear devices to reach the same power level. High-level mixing results in a slightly more distorted signal than it is usually possible to obtain with conventional methods, so it should be used only when essential. Fig. 3 gives the schematic diagram of a typical 432-MHz high-level mixer. V1 is the final amplifier tube of a retired commercial 450-MHz fm transmitter. The oscillator and multiplier stages now produce local-oscillator injection voltage, which is applied to the grid as before. The major change is in the cathode circuit. Instead of being directly at ground, a parallel LC circuit is inserted and tuned to the i-f. In this case a 10-meter i-f was chosen. With the exception of 144 MHz any amateur band could serve as the i-f. Two meters is unsuitable because the third harmonic of the i-f would appear at the output, where it would combine with the desired signal. In fact, some additional output filtering is needed with this circuit. A simple strip-line filter, such as appears in the ARRL *Radio Amateur's VHF Manual,* will do the job. The original crystal in the transmitter is replaced with one yielding an output at the desired local-oscillator frequency, then the intermediate stages are retuned.

One disadvantage of the high-level mixer is the relatively large amount of local-oscillator injection required. In most cases it is simpler to mix at a lower level and use linear amplifiers than to construct the local-oscillator chain. On the higher bands, it may be feasible to generate local-oscillator energy at a lower frequency and use a passive varactor mixer to reach the injection frequency. Here again, the previous caveats pertaining to diode-multiplier spurious outputs pertain. If the local-oscillator injection is impure, the mixer output will be also. Considerable theory concerning mixers is found in chapters 4 and 8 of this publication and in *Solid State Design for the Radio Amateur,* an ARRL publication.

Designing for SSB and CW

The almost universal use of ssb for voice work in the hf range has had a major impact on equipment design for the vhf

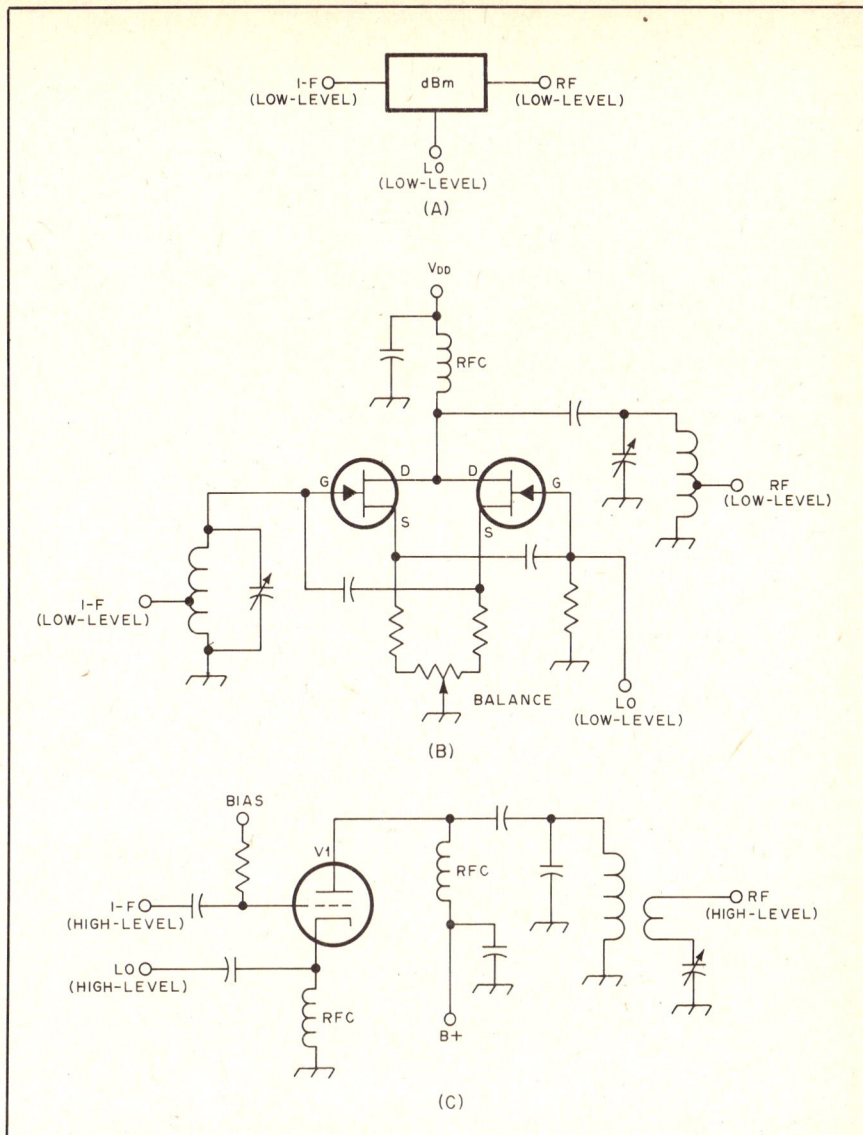

Fig. 2 — A trio of commonly used vhf transmitting mixers. At A, perhaps the simplest, a commercially available diode doubly balanced mixer. Rf output is low, requiring the use of several stages of amplification to reach a useful level. At B, a singly balanced mixer using FETs. Adjustment of this circuit is somewhat critical to prevent the local-oscillator signal from leaking through. A mixer of this type can supply slightly more output than a diode mixer. (3 dBm, as opposed to 1 dBm for the mixer shown at A). At C, a high-level mixer using a vacuum-tube triode is shown. V1 in this case might be a 2C39 or 7289. With the correct circuit constants this mixer could provide an ssb output of 15 watts on 1296 MHz. Power input would be about 100 watts! In addition, 10 watts of LO and 5 watts of i-f drive would be needed. Despite these requirements, such a circuit provides a relatively low-cost means of generating high-level microwave ssb. Spurious outputs at the LO and image frequencies will be quite strong. To attenuate them a strip-line or cavity filter should be used at the mixer output.

and even uhf bands. Many amateurs have a considerable investment in hf sideband gear. This equipment provides accurate frequency calibration and good mechanical and electrical stability. It is effective in cw as well as ssb communication: These qualities being attractive to the vhf operator, it is natural for him to look for ways to use his hf gear on frequencies above 50 MHz.

Increasing use is currently being made of vhf accessory devices, both ready made and homebuilt. This started years ago with the vhf converter, for receiving. Rather similar conversion equipment for transmitting has been widely used since

ssb began taking over the hf bands. Today the hf trend is to one-package stations, transceivers. The obvious move for many vhf operators is a companion box to perform both transmitting and receiving conversion functions. Known as *transverters,* these are offered by several manufacturers. They are also relatively simple to build, and are thus attractive projects for the homebuilder of vhf gear.

Transverter vs. Separate Units

It does not necessarily follow that what is popular in hf work is ideal for vhf use. Our bands are wide, and piling-up in a

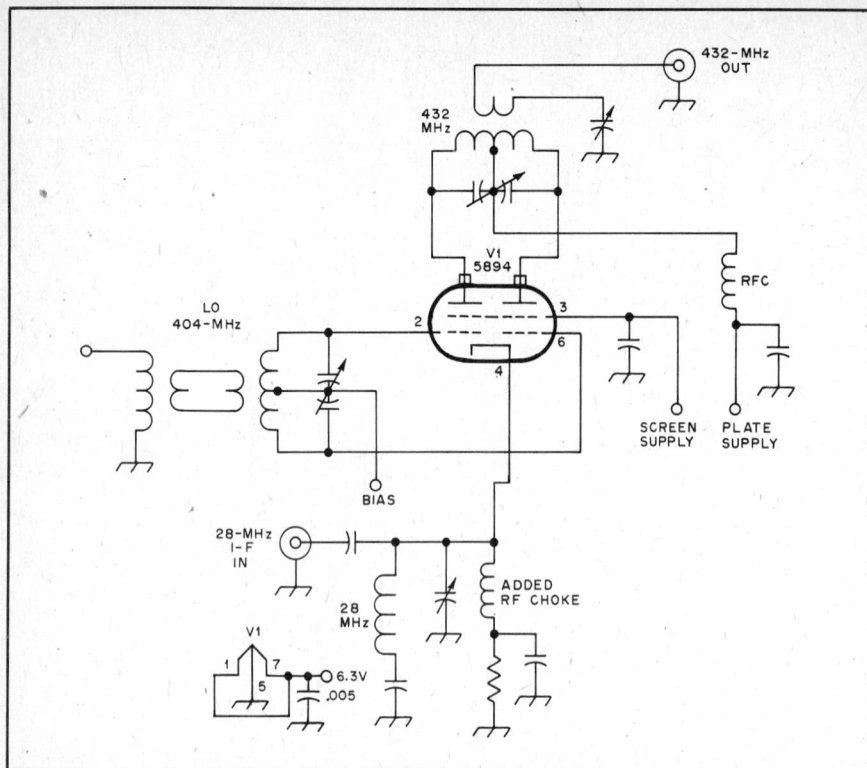

Fig. 3 — Partial schematic diagram of a 70-cm (432-MHz) mixer, built from a converted fm transmitter. The original oscillator-multiplier-driver stages of the unit now provide LO injection. A strip-line filter should be used at the output of the mixer to prevent radiation of spurious products.

Amplifier Design and Operation

All amplifiers in vhf transmitters once ran Class C, or as near thereto as available drive levels would permit. This was mainly for high-efficiency cw and quality high-level amplitude modulation. Class C is now used mostly for cw or fm, and in either of these modes the drive level is completely uncritical, except as it affects the operating efficiency. The influence of ssb techniques is seen clearly in current amplifier trends. Today Class AB1 is popular and most amplifiers are set up for linear amplification, for ssb and — to a lesser extent — a-m. The latter is often used in connection with small amplitude-modulated vhf transmitters, having their own built-in audio equipment. Where a-m output is available from the ssb exciter, it is also useful with the Class AB1 linear amplifier, for only a watt or two of driver output is required.

There is no essential circuit difference between the AB1 linear amplifier and the Class C amplifier; only the operating conditions are changed for different classes of service. Though the plate efficiency of the AB1 linear amplifier is low in a-m service, this type of operation makes switching modes a very simple matter. Moving toward the high efficiency of Class C from AB1 for cw or fm service is accomplished by merely raising the drive from the low AB1 level. In AB1 service the efficiency is typically 30 to 35 percent. No grid current is ever drawn. As the grid drive is increased, and grid current starts to flow, the efficiency rises rapidly. In a well-designed amplifier it may reach 60 percent, with only a small amount of grid current flowing. Unless the drive is run well into the Class C region, the operating conditions in the amplifier can be left unchanged, other than the small increasing of the drive, to improve the efficiency available for cw or fm. No switching or major adjustments of any kind are required for near-optimum operation on ssb, a-m, fm or cw, if the amplifier is designed primarily for AB1 service. If high-level a-m were to be used, there would have to be major operating-conditions changes, and very much higher available driving power.

Tank-Circuit Design

Except in compact low-powered transmitters, conventional coil-and-capacitor circuitry is seldom used in transmitter amplifiers for 144 MHz and higher frequencies. U-shaped loops of sheet metal or copper tubing, or even copper-laminated circuit board, generally give higher Q and circuit efficiency at 144 and 220 MHz. At 420 MHz and higher, coaxial tank circuits are effective. Resonant cavities are used in some applications above 1000 MHz. Examples of all types of circuits are seen later in this chapter. Coil and capacitor circuits are common in 50-MHz amplifiers, and in low-powered, mobile and portable equipment for 144 and even 220 MHz.

Stabilization

Most vhf amplifiers, other than the grounded-grid variety, require neutralization if they are to be satisfactorily stable. This is particularly true of AB1 amplifiers, which are characterized by very high power sensitivity. Conventional neutralization is discussed in chapter 6. An example is shown in Fig. 4A.

A tetrode tube has some frequency where it is inherently neutralized. This is likely to be in the lower part of the vhf region for tubes designed for hf service. Neutralization of the opposite sense may be required in such amplifiers, as in the example shown in Fig. 4B.

Conventional screen bypassing methods may be ineffective in the vhf range. Series-tuning the screen to ground, as in 4C, may be useful in this situation. A critical combination of fixed capacitance and lead length may accomplish the same result. Neutralization of transistorized amplifiers is not generally practical, at least where bipolar transistors are used.

Parasitic oscillation can occur in vhf amplifiers, and, as with hf circuits, the oscillation is usually at a frequency considerably higher than the operating frequency, and it cannot be neutralized out. Usually it is damped out by methods illustrated in Fig. 5. Circuits A and B are commonly used in 6-meter transmitters. Circuit A may absorb sufficient fundamental energy to burn up in all but low-power transmitters. A better approach is to use the selective circuit illustrated at B. The circuit is coupled to the plate tank circuit and tuned to the parasitic frequency. Since a minimum amount of the fundamental energy will be absorbed by the trap, heating should no longer be a problem.

At 144 MHz and higher, it is difficult to construct a parasitic choke that will not be resonant at or near the operating frequency. Should uhf parasitics occur, an effective cure can often be realized by shunting a 56-ohm, 2-watt resistor across a small section of the plate end of the tuned circuit as shown in Fig. 5C. The resistor should be attached as near the plate connector as practical. Such a trap can often be constructed by bridging the resistor across a portion of the flexible strap-connector that is used in some transmitters to join the anode fitting to the plate-tank inductor.

Instability in solid-state vhf and uhf amplifiers can often be traced to oscillations in the lf and hf regions. Because the gain of the transistors is very high at the

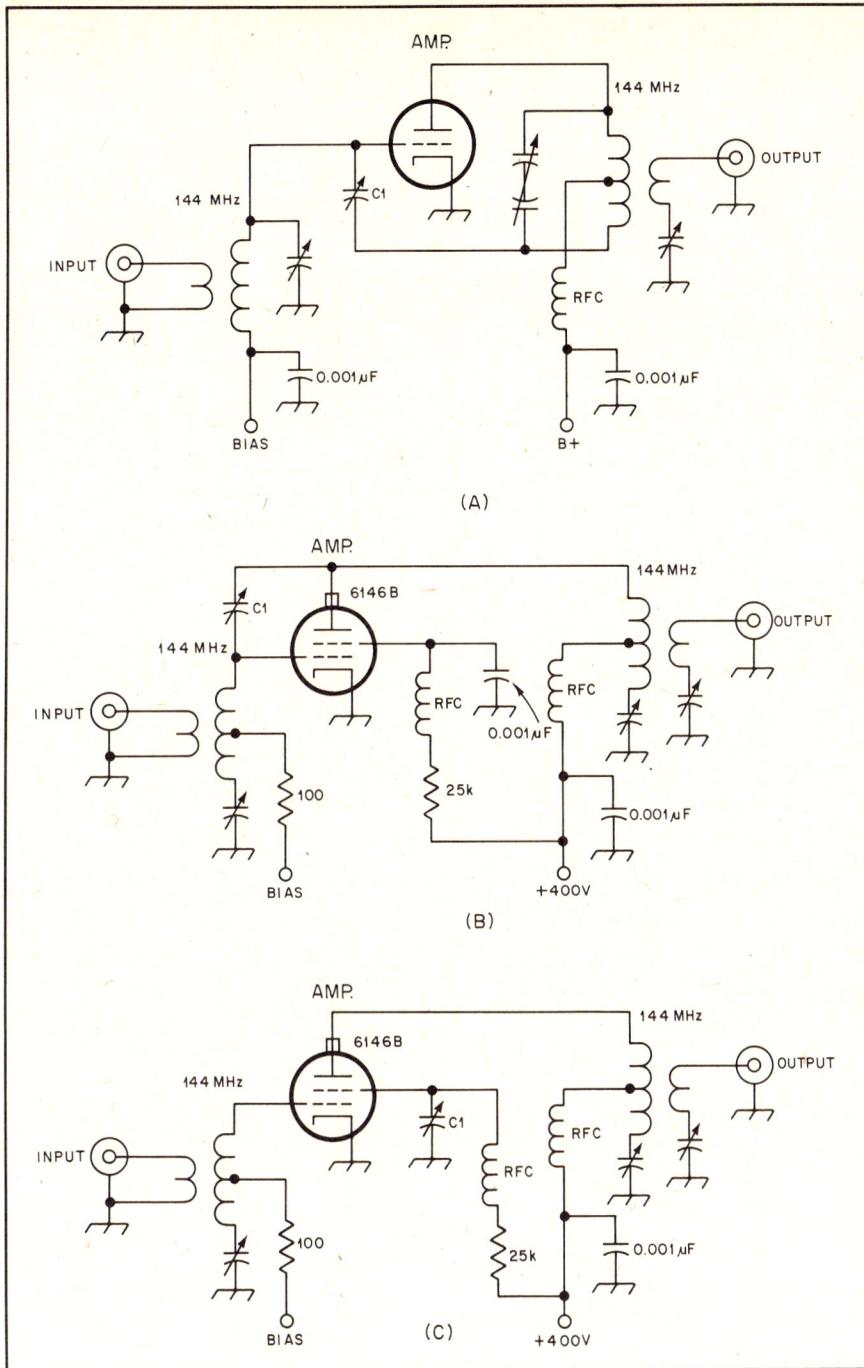

Fig. 4 — Representative circuits for neutralizing vhf single-ended amplifiers. The same techniques are applicable to stages that operate in push-pull. At A, C1 is connected in the manner that is common to most vhf or uhf amplifiers. The circuits at B and C are required when the tube is operated above its natural self-neutralizing frequency. At B, C1 is connected between the grid and plate of the amplifier. Ordinarily, a short length of stiff wire can be soldered to the grid pin of the tube socket, then routed through the chassis and placed adjacent to the tube envelope, and parallel to the anode element. Neutralization is effected by varying the placement of the wire with respect to the anode of the tube, thus providing variable capacitance at C1. The circuit at C is a variation of the one shown at B. It too is useful when a tube is operated above its self-neutralizing frequency. In this instance, C1 provides a low-Z screen-to-ground path at the operating frequency. RFC in all circuits shown are vhf types and should be selected for the operating frequency of the amplifier.

lower frequencies, instability is almost certain to occur unless proper bypassing and decoupling of stages is carried out. Low-frequency oscillation can usually be cured by selecting a bypass-capacitor value that is effective at the frequency of oscillation and connecting it in parallel with the vhf bypass capacitor in the same part of the circuit. It is not unusual, for example, to employ a 0.1-μF disk ceramic in parallel with a 0.001-μF disk capacitor in such circuits as the emitter, base, or collector return. The actual values used will depend upon the frequencies involved. An additional stabilization method for solid-state amplifiers is shown in Fig. 5E.

Capacitive reactance of C is chosen to be very low at the parasitic frequency. R then appears as a swamping resistor, damping the oscillation.

VHF TVI Causes and Cures

The principal causes of TVI from vhf transmitters are:

1) Adjacent-channel interference in channels 2 and 3 from 50 MHz.

2) Fourth harmonic of 50 MHz in channels 11, 12 or 13, depending on the operating frequency.

3) Radiation of unused harmonics of the oscillator or multiplier stages. Examples are 9th harmonic of 6 MHz, and 7th harmonic of 8 MHz in channel 2; 10th harmonic of 8 MHz in channel 6; 7th harmonic of 25-MHz stages in channel 7; 4th harmonic of 48-MHz stages in channel 9 or 10; and many other combinations. This may include i-f pickup, as in the cases of 24-MHz interference in receivers having 21-MHz i-f systems, and 48-MHz trouble in 45-MHz i-fs.

4) Fundamental blocking effects, including modulation bars, usually found only in the lower channels, from 50-MHz equipment.

5) Image interference in channel 2 from 144 MHz, in receivers having a 45-MHz i-f.

6) Sound interference (picture clear in some cases) resulting from rf pickup by the audio circuits of the TV receiver.

There are other possibilities, but nearly all can be corrected completely, and the rest can be substantially reduced.

Items 1, 4 and 5 are receiver faults, and nothing can be done at the transmitter to reduce them, except to lower the power or increase separation between the transmitting and TV antenna systems. Item 6 is also a receiver fault, but it can be alleviated at the transmitter by using fm or cw instead of a-m phone.

Treatment of the various harmonic troubles, Items 2 and 3, follows the standard methods detailed elsewhere in this *Handbook*. The prospective builder of new vhf equipment should become familiar with TVI prevention techniques and incorporate them in new construction projects.

Use as high a starting frequency as possible, to reduce the number of harmonics that might cause trouble. Select crystal frequencies which do not have harmonics in local TV channels. Example: The 10th harmonic of 8-MHz crystals used for operation in the low part of the 50-MHz band falls in channel 6, but 6-MHz crystals for the same band have no harmonic in that channel.

If TVI is a serious problem, use the lowest transmitter power that will do the job at hand. Keep the power in the multiplier and driver stages at the lowest practical level, and use link coupling in preference to capacitive coupling. Plan for complete shielding and filtering of the rf

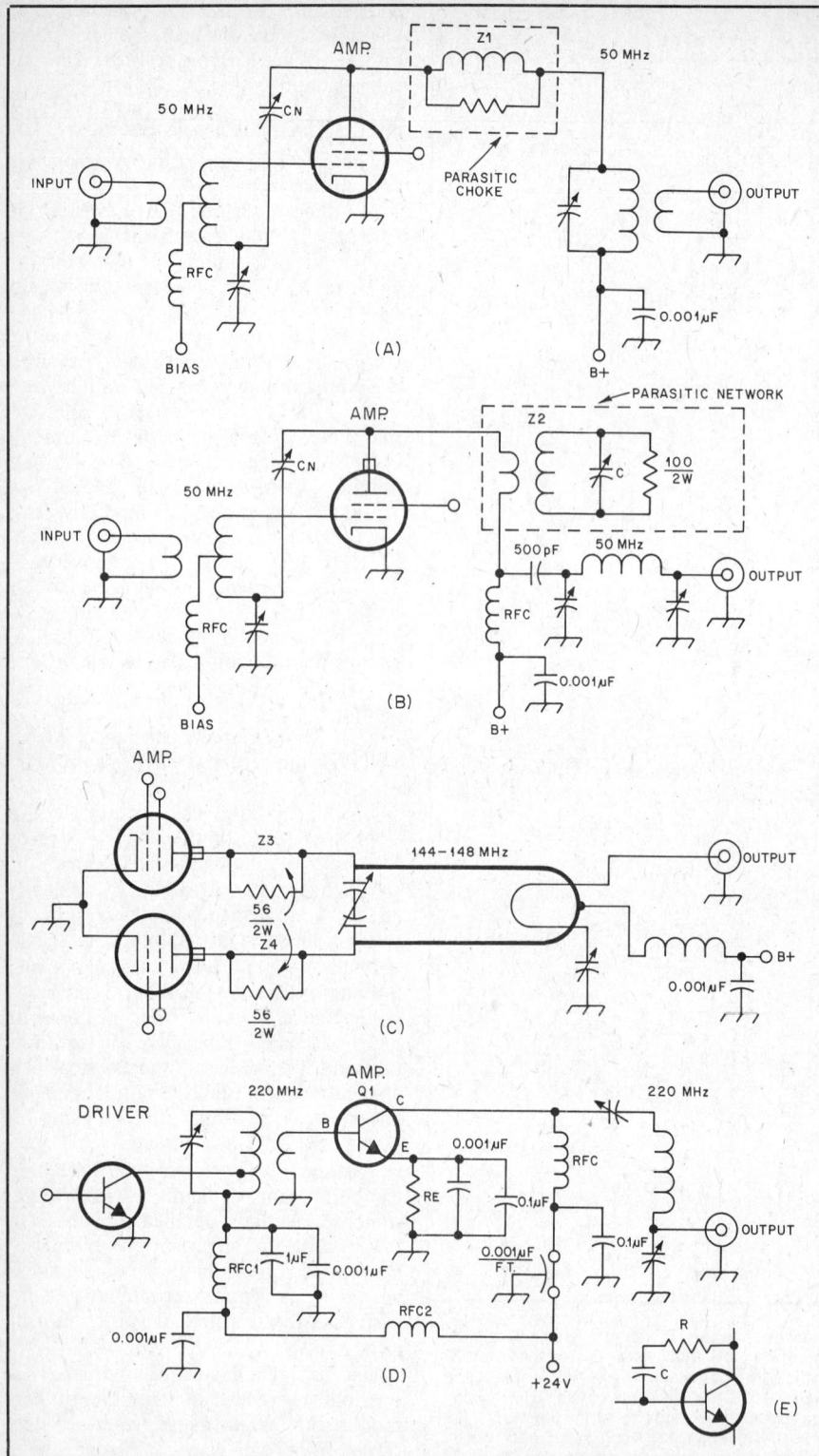

Fig. 5 — Representative circuits for vhf parasitic suppression are shown at A, B and C. At A, Z1 (for 6-meter operation) would typically consist of 3 or 4 turns of no. 13 wire wound on a 100-ohm 2-watt non-inductive resistor. Z1 overheats in all but very low power circuits. The circuit at B, also for 6-meter use, is more practical where heating is concerned. Z2 is tuned to resonance at the parasitic frequency by C. Each winding of Z2 consists of two or more turns of no. 14 wire — determined experimentally — wound over the body of a 100-ohm 2-watt (or larger) noninductive resistor. At C, an illustration of uhf parasitic suppression as applied to a 2-meter amplifier. Noninductive 56-ohm 2-watt resistors are bridged across a short length of the connecting lead between the tube anode and the main element of the tank inductor, thus forming Z3 and Z4.

The circuit at D illustrates how bypassing for both the operating frequency and lower frequencies is accomplished. Low-frequency oscillation is discouraged by the addition of the 0.1μF disc ceramic capacitors. RFC1 and RFC2 are part of the decoupling network used to isolate the two stages. This technique is not required in vacuum-tube circuits. At E, a capacitor with low reactance at the parasitic frequency is connected in series with a 1/2-watt carbon resistor. At 144 MHz, C is typically 0.001μF. R may be between 470 and 2200 ohms.

sections of the transmitter, should these steps become necessary.

Use coaxial line to feed the antenna system, and locate the radiating portion of the antenna as far as possible from TV receivers and their antenna systems.

A complete discussion of the problems and cures for interference is in the ARRL publication, *Radio Frequency Interference*.

THREE-WATT TRANSMITTING CONVERTER FOR 6 METERS

The linear transmitting converter shown in Figs. 6 through 10 is a simple, low-cost way to extend the coverage of any 28-MHz transmitter to the 6-meter band. Output power is 3 watts PEP and the 28-MHz drive requirement is 1 mW. This drive level is compatible with the transverter outputs found on many current hf transceivers. By selecting the appropriate resistor values, the input attenuator can easily be adjusted for drive levels as high as 100 mW.

Particular care was taken during the design stage to ensure that spurious emissions are at a minimum. The IMD performance is such that this unit can be used to drive a high-power linear amplifier without generating excessive adjacent-channel interference.

Circuit Description

LO injection is supplied to the mixer by a crystal oscillator operating at 22 MHz. The output of the oscillator is filtered to reduce harmonics to approximately −40 dBc and is then applied to a power splitter, T2. One port of the splitter feeds the mixer and the other output is connected to J1. This output can be used to supply LO to a receive converter for transceive operation. A 3-dB attenuator between the splitter and the mixer provides a wideband termination for the mixer LO port. The power delivered to the mixer is +7 dBm.

A commercial diode-ring module is used for the mixer. The excellent balance of this type of mixer reduces the bandpass filter requirements following the mixer. The IMD performance is also very good; the third-order products are 40-dB below each tone of the two-tone output signal when the input signal is at the recommended level of −10 dBm (each tone). The attenuator at the i-f port prevents overdriving the mixer. The 20-dB pad shown in Fig. 7 allows a driving signal of 30 to 40 mW (PEP) to be used. This attenuator should be adjusted if other drive levels are desired.

The mixer is followed by a broadband amplifier, 6-dB pad and a 3-pole bandpass filter. The amplifier has a gain of nearly 20 dB at 50 MHz and provides a good termination for the mixer. Signal level at the filter output is about −10 dBm. Two more stages of class-A amplification follow the band-pass filter, bringing the signal level up to the +16 dBm necessary to drive the power-amplifier stage.

The power amplifier uses a Motorola MRF476 transistor operating class AB. The MRF476, a low-cost device in a plastic TO-220 package, will deliver a minimum of 3 watts PEP with good linearity at 50 MHz. A cw output of up to 4 watts can be obtained from this converter, but the key-down time should be limited to 60 seconds. This is because of the relatively small heat sink used in this unit. The power-amplifier input and output networks are both designed to match the transistor to 50 ohms. Their design was based on data supplied in the *Motorola RF Data Manual* (2nd edition). The amplifier output is filtered by a 7-pole low-pass filter designed for a cutoff frequency of 56 MHz and a ripple factor of 0.17 dB.

Construction

The converter is constructed in two sections: the low-level section consisting of the LO, mixer, band-pass filter, and class-A amplifiers and the power-amplifier/low-pass filter section. No printed circuit board is needed for the low-level section; rather, the components are mounted directly on a piece of unetched copper-clad board. The component leads going to ground support the components above the board. Among the many advantages to this style of construction is the ease with which the circuits can be modified. Also, if a printed circuit board is used, the components used to lay out the board, such as variable capacitors, are likely to be the only types that can be used. This makes it difficult for the builder to use his junk box.

Note that the LO and its filter are completely enclosed by a shield made of double-sided circuit board material. This was found to be necessary to ensure that 22-MHz LO energy does not find its way into the amplifier stages. It was expected that a 1-inch-high shield around the LO would provide sufficient isolation — this is not the case; a complete enclosure is required. It is recommended that a shield enclosure be constructed around the remaining portion of the low-level section. This will prevent feedback from the power-amplifier board to the class-A stages. It was not necessary to do this in the prototype shown in the photographs, but is good practice when the two boards are mounted in the same enclosure (such as an aluminum chassis).

The power-amplifier board is constructed on a printed board (Figs. 9 and 10). It is not necessary that the board be etched. The layout shown can be used as a guide and the copper clad cut away with a sharp Xacto knife. After cutting through the copper with the knife, applying heat from a soldering iron allows the foil to be lifted off easily. The heat sink for the

Fig. 6 — Six-meter transmitting converter and 3-W PEP amplifier.

MRF476 is made from a 3-1/2 × 1-inch (89 × 25-mm) strip of 1/16-inch (1.6-mm) aluminum folded into a U shape. Use a *small* amount of heat-sink compound between the transistor body and the heat sink. The MRF476, unlike many rf transistors, has the collector connected to the mounting tab; thus the heat sink is hot for both rf and dc. The center pin of the transistor is also connected to the collector, but it is not used as connection to the collector is made through the mounting tab. The unused pin should be cut off at the point where it becomes wider (about 1/8 inch from the transistor body). The other pins are bent down to make connection to the circuit board at this same point. Connection between the two boards can be made with a short length of RG-174/U coaxial cable.

Tune-Up and Operation

The low-level stages are aligned before connecting the power-amplifier stage. A 50-ohm resistor is connected through the 0.01-μF blocking capacitor to the output

of the last class-A amplifier and 12 volts is applied to the oscillator and class-A amplifiers. The oscillator is adjusted by setting C2 at mid-range and tuning C3 for maximum output as measured with a VTVM and rf probe connected to pin 8 of the mixer. C2 can then be adjusted to bring the frequency to exactly 22 MHz. If a frequency counter with a high-impedance input is used, it should be connected to pin 8 of the mixer. If the counter has a 50-ohm input it can be connected to J1. The adjustment of C2 and C3 will interact so the procedure should be repeated. J1 should always be connected to a load close to 50 ohms. If a receive converter is not being used, a 50-ohm resistor, mounted in a BNC connector, should be connected to J1.

The band-pass filter is aligned by applying a 28.7-MHz cw signal to the input of the converter and adjusting C5, C6 and C7 for maximum output at the output of the last amplifier. Check the frequency response of the filter by varying the frequency of the input signal from 28 to 29

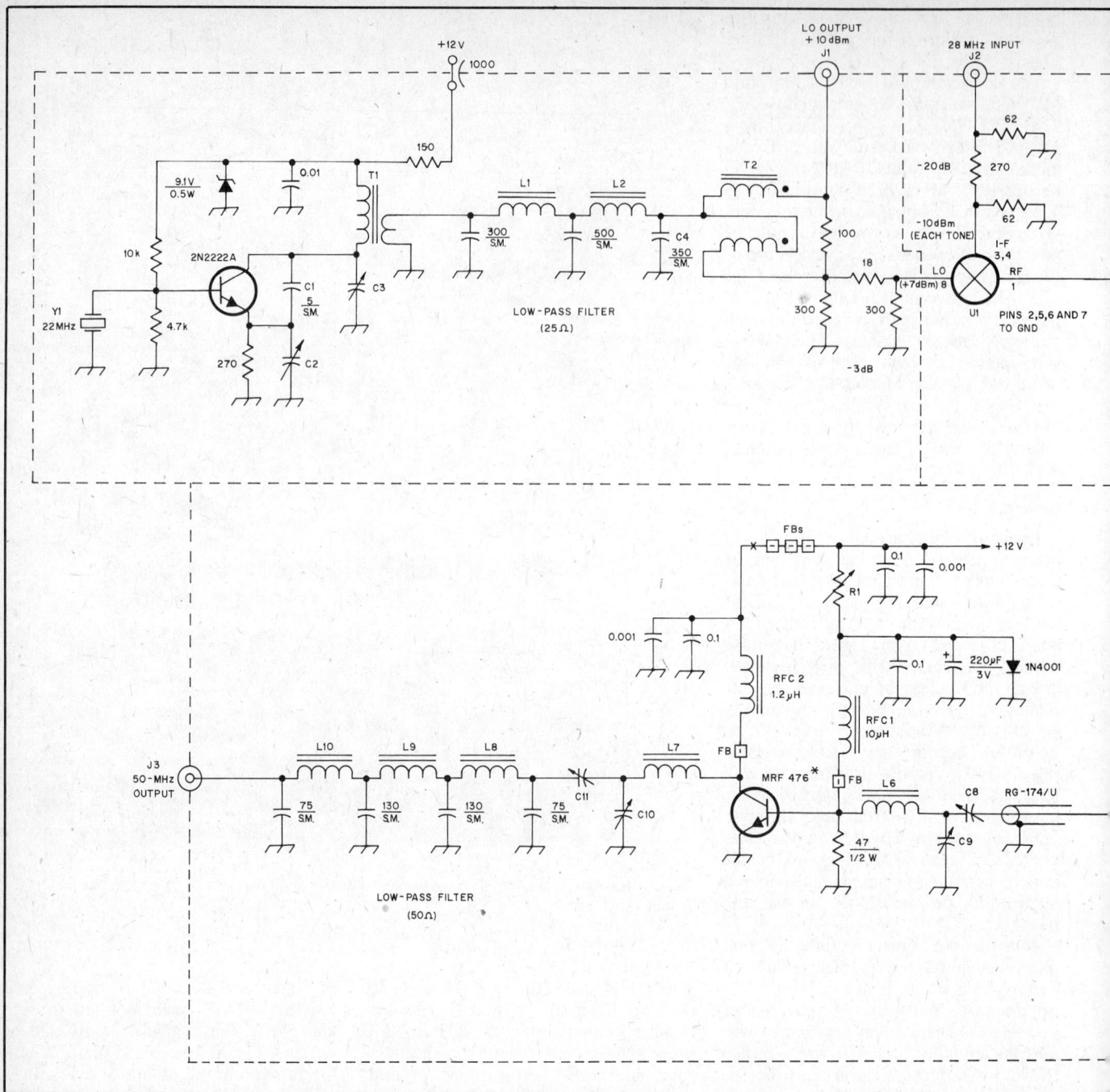

Fig. 7 — Schematic diagram of the 6-meter transmitting converter.

C2-C7, incl. — 2-25 pF air or foil dielectric trimmer.
C8, C9, C11 — 15-115 pF mica trimmer (Arco 406 or equiv.).
C10 — 1.5-20 pF mica trimmer (Arco 402 or equiv.).
FB — Ferrite bead, Amidon FB101-43.
L1, L2 — 7 turns, no. 28 enamel on T25-6 core.
L3-L5, incl. — 12 turns, no. 28 enamel on T37-6 core.
L6 — 6-1/2 turns, no. 26 enamel on T37-12 core.
L7 — 9-1/2 turns, no. 22 enamel on T50-12 core.
L8, L10 — 8 turns, no. 22 enamel on T50-12 core.
L9 — 8 turns, no. 22 enamel on T50-12 core.

MHz. The output power should not vary more than 1 dB within the passband of the filter. The 50-ohm resistor at the output of the last class-A amplifier should now be removed and the power amplifier connected through a short length of coaxial cable.

To align the input and output networks of the final, connect the amplifier output to a wattmeter and 50-ohm load. Attach the amplifier to a 12-volt, 1-A power supply (maximum current drain is approximately 750 mA for the complete converter) and place a milliammeter in the collector supply line at the point marked X in Fig. 7. With no drive applied, turn on the dc power and adjust R1 for a collector current of 35 to 40 mA. Apply just enough drive, at 29 MHz, to the input of the converter to get an indication on the wattmeter and peak C8 through C11 for maximum output. Now increase the drive slowly, keeping the networks peaked, until the output power is 4 watts. Do not

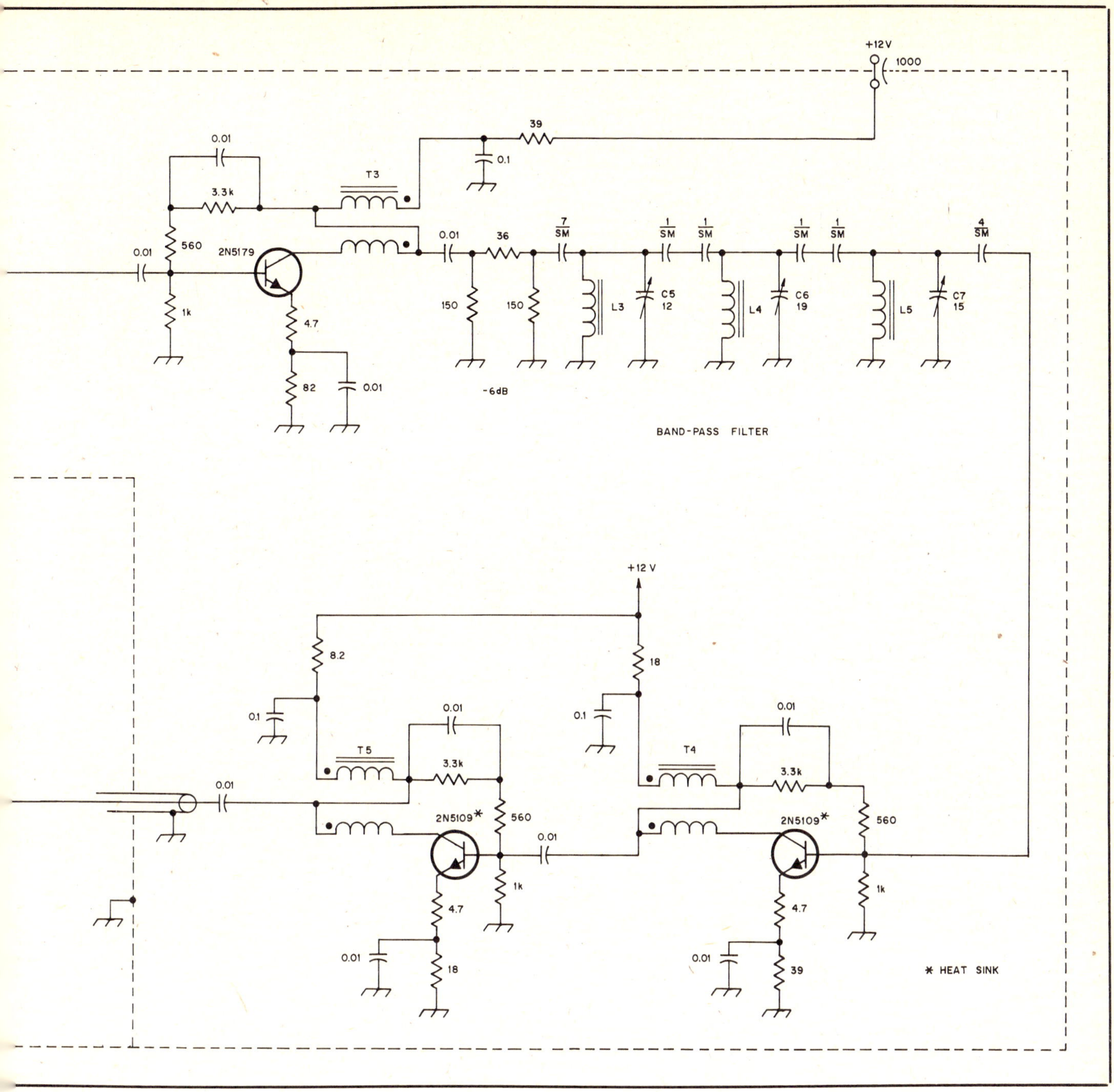

BAND-PASS FILTER

0.01
3.3k
T3
560
2N5179
0.01
1k
4.7
82
0.01
39
0.1
+12V
1000
0.01
36
150
150
7 / SM
L3
C5 12
1 / SM
1 / SM
L4
C6 19
1 / SM
1 / SM
L5
C7 15
4 / SM
-6dB

+12V
8.2
0.1
T5
0.01
3.3k
2N5109*
560
0.01
1k
4.7
18
0.01
0.01
18
0.1
T4
0.01
3.3k
2N5109*
560
1k
4.7
39
0.01
* HEAT SINK
0.01

Compress winding to cover half of core.
R1 — 220-ohm, 1-W resistor in series with a 500-ohm, 1/2-W trimmer.
RFC1 — 10-μH solenoid type rf choke (Miller 74F105AP or equiv.).

RFC2 — 1.2-μH solenoid type rf choke (Miller 74F126AP or equiv.).
T1 — 27 turns primary, 5 turns secondary, no. 28 enamel on T37-6 core.
T2 — 5 bifilar turns, no. 26 enamel on

FT37-61 core.
T3 — 7 bifilar turns, no. 28 enamel on FT23-43 core.
T4, T5 — 5 bifilar turns, no. 26 enamel on FT37-61 core.

maintain the output at 4 watts any longer than is necessary. It is important that the output network be adjusted while the output is at the 3.5- to 4-watt level. If it is adjusted at a lower level the IMD performance will be degraded. When the output is adjusted as described above the third-order IMD products should be at least 32 dB below the 3-watt PEP level (Fig. 8C).

A 2-KW PEP AMPLIFIER FOR 50 TO 54 MHZ

A number of manufacturers sell low-power-output (10 watts) transceivers and transverters for 6-meter operation. This particular power-output level is not in line with that required to drive the popular grounded-grid amplifiers. Even the high-mu triodes, such as those of the 8874 family, require at least 25 watts of drive for 1 kilowatt of input. Over 50 watts of drive may be required for 2-kilowatt service. This leaves two options for owners of 10-watt-output rigs. One solution is to

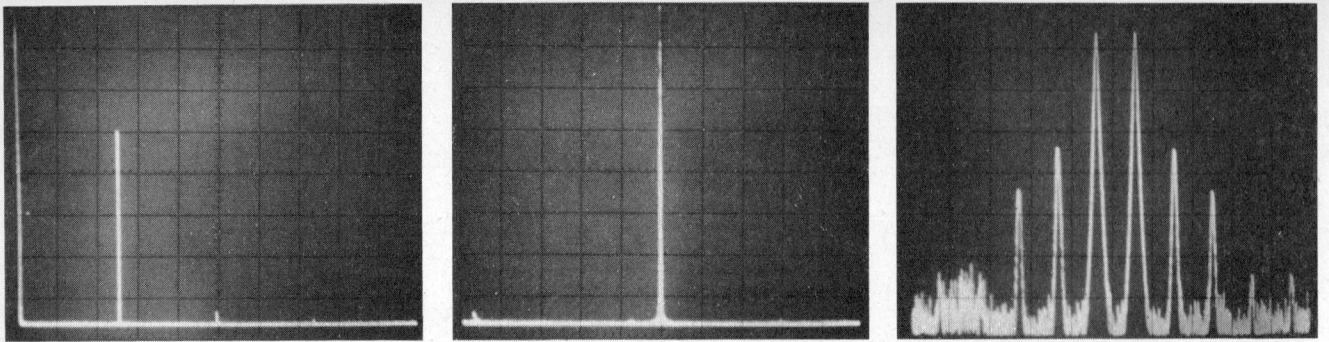

Fig. 8 — Spectral analysis of 6-meter transmitting converter. Display at A shows the fundamental and second harmonic. The harmonic is 73 dB below the 3-watt carrier. The fundamental has been notched 30 dB to prevent overload of the spectrum analyzer. Each horizontal division is 20 MHz. A "close-in display" is shown in B: Each horizontal division is 2 MHz and the center frequency is 50 MHz. The 2 × 28-MHz spurious product (which falls in channel 2) is attenuated more than 75 dB relative to the carrier power. The two-tone IMD spectrum is shown in C: The third-order products are 34 dB below PEP. Each horizontal division is 2 kHz. In all three displays each vertical division is 10 dB.

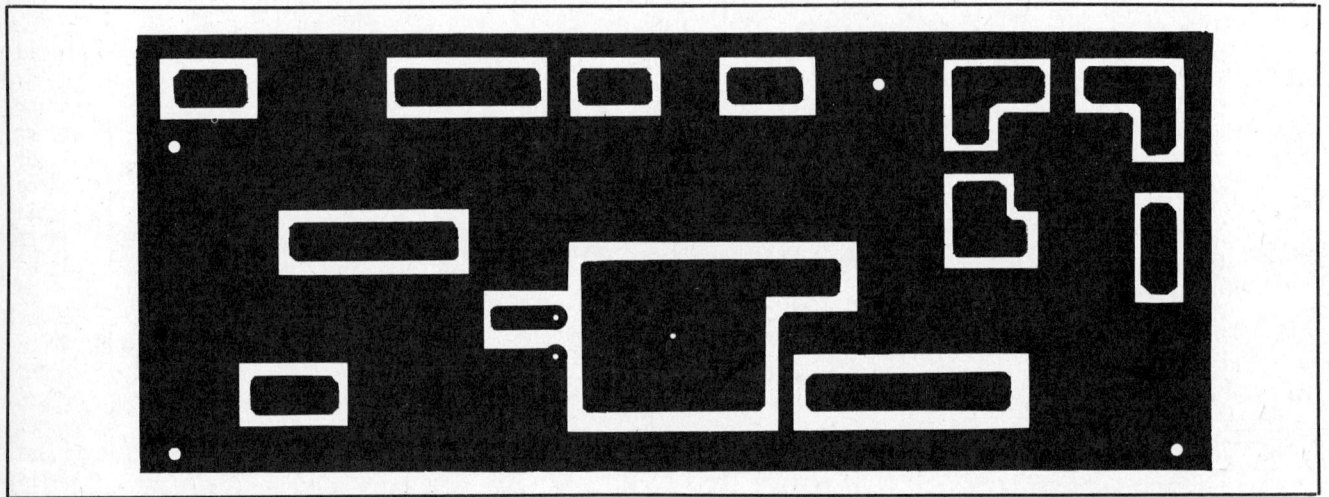

Fig. 9 — Full-scale etching pattern for the power amplifier section of the 6-meter transmitting converter. Black areas represent unetched copper. This view is from the foil side of the board.

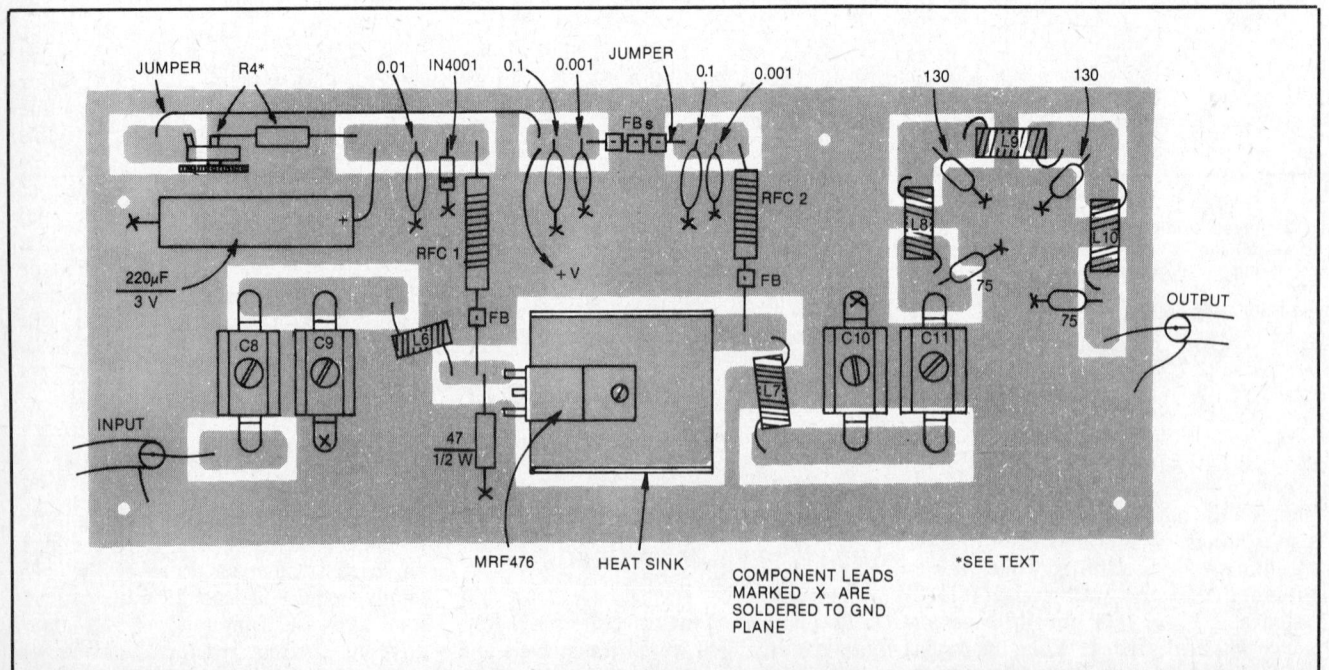

Fig. 10 — Parts-placement guide for Fig. 9. All components mount on the foil side of the board.

use a solid-state "brick" amplifier to drive the grounded-grid amplifier. This results in complex relay-switching systems, not to mention the cost of the amplifier and a high-current supply to power it. The other approach is to use a tetrode such as the 4CX1000A in a grounded-cathode amplifier. In the circuit described, 10 watts of drive will provide 2 kilowatts of input power, thereby obviating an intermediate amplifier stage.

Circuit Description

The schematic diagram of the amplifier is shown in Fig. 12. As the 4CX1000A is a high-mu tube, care must be taken in the design and construction of the amplifier to prevent instability. In the circuit shown, this was accomplished without the need for neutralization — one of the drawbacks commonly associated with the use of these tubes.

The input network is series tuned and a link couples power from the exciter to the tube. Series tuning was chosen primarily because of the large input capacitance of the tube (roughly 100 pF). The tube socket and strays add an additional 20 pF. This tuning method places the tube input capacitance in series with the tuning capacitor, effectively reducing the circuit capacitance. A number of other networks were tried, including one somewhat exotic current-feed method, but none proved as simple and effective as the one shown. Bias voltage is shunt fed to the grid through R1. This resistor also heavily swamps the grid circuit and helps to provide amplifier stability without the need for neutralization. Should it be desired to increase the gain of the amplifier for use with lower power rigs, the value of R1 can be increased and the network components changed as appropriate. The cathode is grounded through short, heavy leads at the base of the SK-800B tube socket. A screen bypass capacitor is an integral part of this socket. The capacitor is 1500 pF, is rated for 400 volts and is of the mylar-film variety. Capacitors are connected between ground and each of the three filament-connection points.

The plate circuit used in this amplifier is of the common pi-network variety, with two exceptions. First, the internal tube plate capacitance, which is on the order of 10-pF, comprises what would be considered the "tuning" capacitor of the network. Since the tube capacitance is fixed, the inductor is used as an adjustable element. A compressible/expandable coil is used here. This arrangement allows the amplifier to cover the entire 6-meter band, whereas other designs using shorted-turn and slug-tuned techniques will cover little more than 1-MHz of tuning range. Additionally, the compressing and expanding of the coil does not have an adverse effect on the inductor Q, as typically occurs with the other systems. The tube capacitance dictates the circuit Q, which is roughly 14.

Fig. 11 — Front-panel view of the 2-kW, 6-meter amplifier.

A single, multisecondary transformer provides power for the filament, bias and screen supplies. Although the filament winding is rated for 6.3 V at 8.8 A, the transformer delivers 6.0 V at 9.0 A — precisely that required by the 4CX1000A. The transformer was run for a period of 10 hours under these conditions and no appreciable heating of the wires or core was noted. Transformers of other manufacture or different voltage/current ratings should be carefully checked.

The high-voltage secondary of the transformer provides energy for the bias and screen supplies. Zener diodes limit the dissipation of the bias potentiometer. R6 cuts the amplifier off during receive periods. For transmit, a short is placed at J5.

The same transformer secondary also feeds the screen supply. Here, a full-wave rectifier provides the dc energy. Q1, D3, R1 and R2 form a regulator that limits the current that can be drawn through the transistor to 40 mA. This extra circuitry is desirable in the event of high-voltage supply failure. No more than 40 mA of current can be drawn through the regulator, ensuring that the screen is protected. Without such a safety device, a failure of the high-voltage supply would mean that the screen dissipation might be pushed well above the specified 12-watt maximum rating. D4-D9, inclusive, make up the screen-voltage regulator.

M1, located in the - HV lead, continuously monitors cathode current. Meter shunts are provided in the bias and screen supplies. M2 can be switched between grid and screen by means of S2. This meter reads 0 to 10 mA for grid and 0 to 100 mA for screen current.

Construction

Many of the construction details can be seen in the accompanying photographs. The general layout is not especially critical, although good isolation between the grid and plate circuitry is a must. To this end, the Eimac SK-800B socket is used. The built-in screen bypass capacitor is an important factor in amplifier stability. Other sockets without this capacitor should be avoided.

The amplifier is built on a 10 × 12 × 3 inch (254 × 305 × 76 mm) aluminum

chassis that is attached to an 8-3/4 inch (222 mm) high, 19 inch (483 mm) wide rack panel. Sheet-aluminum panels are made so that the overall height of the amplifier enclosure is 7-1/2 inches (191 mm).

The grid compartment is constructed from a second aluminum chassis that measures 7 × 7 × 2 inches (178 × 178 × 51 mm). It is easiest to cut the hole for the tube socket with the smaller chassis bolted in place. In this manner exact alignment can be guaranteed. A heavy-duty hand nibbler tool was used to make the large hole. All leads entering the grid compartment pass through 1000-pF feed-through capacitors to ensure an rf-tight enclosure. The amplifier input connection is through a UG-625B/U (female) chassis-mount BNC connector, directly into the grid compartment. The general placement of the input-network components can be seen in the photograph. A perforated aluminum cover is placed over the smaller grid-compartment chassis. This piece must allow adequate air flow for proper tube cooling. It is a good idea to check air flow with and without this piece in place in order to determine whether the material is suitable.

The remainder of the bottom chassis houses the screen and bias power-supply components and the control circuitry. Many of the parts are contained on a single-sided circuit board. The board etching pattern and parts-layout diagram are given in Figs. 18 and 19. Zener diodes are mounted on an aluminum plate and the circuit board. While the aluminum provides some heat-sink action, cooling for the diodes is supplied primarily by the system blower, as the diode stack is mounted in the air flow path. Cooling for the tube is by means of a blower that is mounted external to the chassis. A plastic flange (Newtone 366, used with central vacuum cleaning systems) is suitable for use with 2-inch (51-mm) automobile heater/defroster hose. A length of this hose connects the pressurized chassis to the blower, which can be mounted at some convenient location. For ultra-quiet operation the blower can be mounted in a closet or in an adjoining room. Copper or brass screening is used to cover the flange opening, thus maintaining an rf-tight chassis. Information on how to select a suitable blower is given in the Hf Transmitting chapter of this *Handbook*. Connection to the meters is through 1000-pF feedthrough capacitors.

The plate compartment is somewhat less "busy" than the underside of the amplifier. The only component that needs explanation is the inductor. Details of the construction are shown in Fig. 15, and a photograph is shown in Fig. 16. Two pieces of 3/8 inch (10 mm) Delrin plate are used as the stationary end pieces. A 1/4 inch (6.4 mm) control bushing is used at the center of each piece. A third piece of

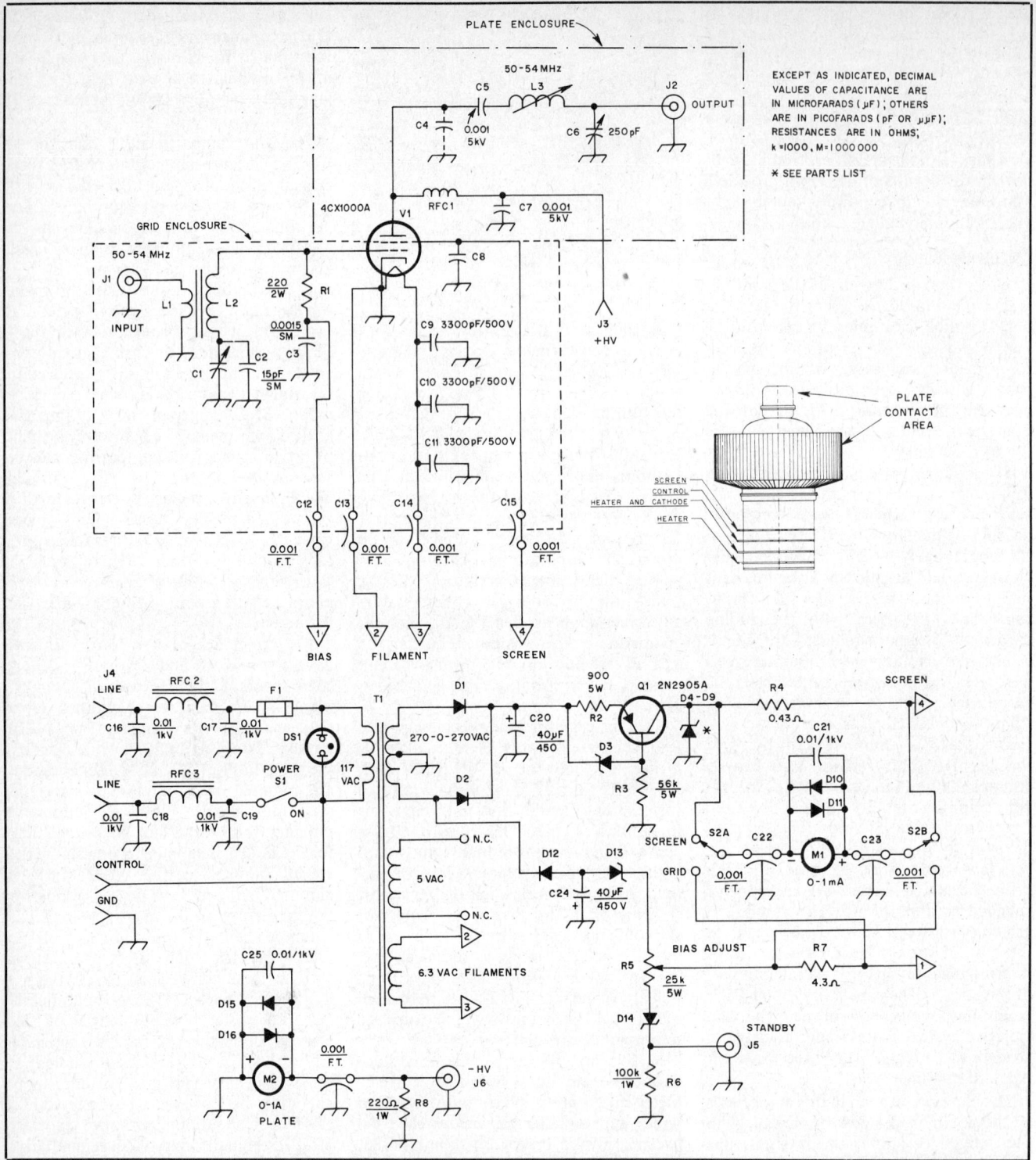

Fig. 12 — Schematic diagram of the 6-meter amplifier. Parts designations shown on the schematic but not called out in the parts list are for text reference only.

C1 — Miniature variable, panel mount, 32 pF maximum.
C2 — Silver mica, 15 pF, mounted directly across C1.
C4 — Tube internal plate capacitance.
C5, C7 — Transmitting capacitor, 0.001 μF, 5000 volt. Centralab 858S-1000 or equiv.
C6 — Transmitting variable, 250 pF maximum.
C8 — Screen bypass built into SK-800B socket.
C20, C24 — Electrolytic, 40 μF, 450 volt.
D1, D2, D10-D12, incl., D15, D16 — Silicon, 2.5 ampere, 1000 volt.
D3 — Zener diode, 45 volt, 1 watt.
D4-D9, incl. — Zener diode, stud mount, 56 volt, 10 watt.
D13 — Zener diode, 200 volt, 5 watt.

D14 — Zener diode, 30 volt, 1 watt.
DS1 — Neon indicator built into S1.
F1 — Fuse, 1 ampere.
J1 — Coaxial connector, BNC chassis mount, UG-625B/U or equiv.
J2 — Coaxial connector, type N chassis mount, UG-58/U or equiv.
J3 — High-voltage connector, Millen 37001 or equiv.
J4 — 4 conductor.
J5, J6 — Phono connector.
L1 — 2 turns no. 24 enam. wire wound over L2.
L2 — 10 turns no. 24 enam. wire on a T-50-12 core.
L3 — See text and drawing.

M1 — 0-1 mA, Simpson 15070 or equiv.
M2 — 0-1 A, Simpson 15101 or equiv.
Q1 — 2N2905A or equiv.
R4 — Meter shunt, 1′ 7-1/2″ (41.4 mm) no. 34 enam. wire wound on a high-value, 1/2-watt resistor.
R5 — Potentiometer, 25 kΩ, 5 watt.
R7 — Meter shunt, 4.3 Ω, 1/4 watt.
RFC1 — 36 turns no. 24 enam. wire on a 1″ (25.4 mm) diameter Teflon rod.
RFC2, RFC3 — 6 turns no. 22 enam. wire on an FT-50-43 core.
S1 — Spst, rocker type with built-in neon indicator.
S2 — Rotary, 2 pole, 2 position.
T1 — Stancor P8356 or equiv. See text.

Fig. 13 — Photograph of the underside of the amplifier. The grid-compartment chassis is normally enclosed by a perforated aluminum cover which was removed for this photograph.

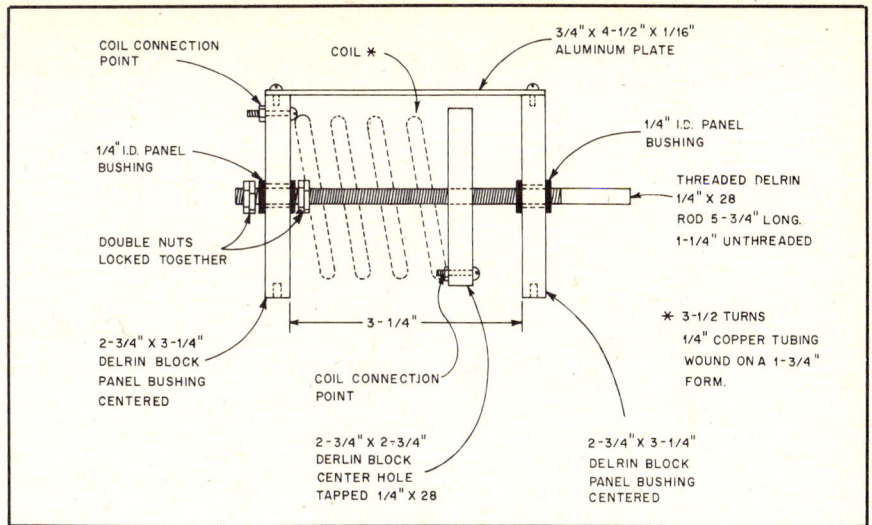

Fig. 15 — Dimensional drawing of the inductor assembly. Delrin can be obtained from most plastic supply houses. Check the Yellow Pages of your local telephone directory for dealers.

Fig. 14 — This is the plate compartment of the amplifier. Details of the inductor are given in the text and in additional photographs and drawings.

Fig. 16 — Photograph of the assembled adjustable inductor. Make certain to use only plated brass hardware for pieces that come into contact with the coil.

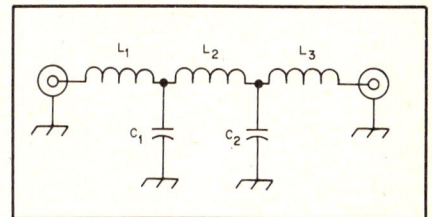

Fig. 17 — Schematic diagram of the low-pass filter that is used after the amplifier. Good quality capacitors such as silver-mica types should be used.
C1, C2 — 82 pF silver mica, 1000 V.
L1, L3 — 4 turns no. 14 enam. wire close wound on a 5/16 in. (8 mm) form.
L2 — 5 turns no. 14 enam. wire close wound on a 7/16 in. (11 mm) form

Delrin stock is used for the movable plate. This piece is tapped to accommodate the 1/4-28 thread of the shaft that is used to move the plate. The shaft is "double nutted" on each side of the rear support to prevent the shaft from moving in relation to either the front or rear supports. As the front-panel knob is turned, only the movable plate changes position to compress or expand the coil. The exact dimension of the coil is fairly critical, so it should be made as close to specification as possible. Once the coil has been wound it can be silver plated to prevent oxidation.

As a point of interest, the amplifier front panel was painted orange. Black press-on lettering was used to label the front panel controls and a light coat of clear lacquer was applied to protect the lettering.

Operation

It is a good idea initially to check out the various tube-element voltages without the tube in the socket. The screen voltage may turn out to be different than the nominal 336 because of poor Zener-diode voltage tolerances. As long as the voltage is under 375 no problems should be encountered. Operation of the protective circuitry can be verified by loading the output of the screen supply with appropriate high-wattage resistors. Ensure

that the bias control will allow adjustment between the values shown on the schematic. Set the voltage to roughly — 150 so that the tube will be cut off when it is first turned on.

Eimac recommends that the 4CX1000A heater voltage be applied for a period of not less than three minutes before other operating voltages are applied. Whether this procedure is really necessary or not is a good question. Several well-known and respected amplifier manufacturers that use this tube do not follow this philosophy.

Connect a transmitter capable of supplying 10 watts of power to J1 through an SWR indicator. Also, connect a dummy load and power meter to J2. After the amplifier has warmed up for several minutes, short J5 and adjust the bias control for an idling current of 50 mA. Apply a small amount of drive power and adjust L3 and C6 for maximum power output. At this time adjust C1 for minimum reflected power as indicated by the input

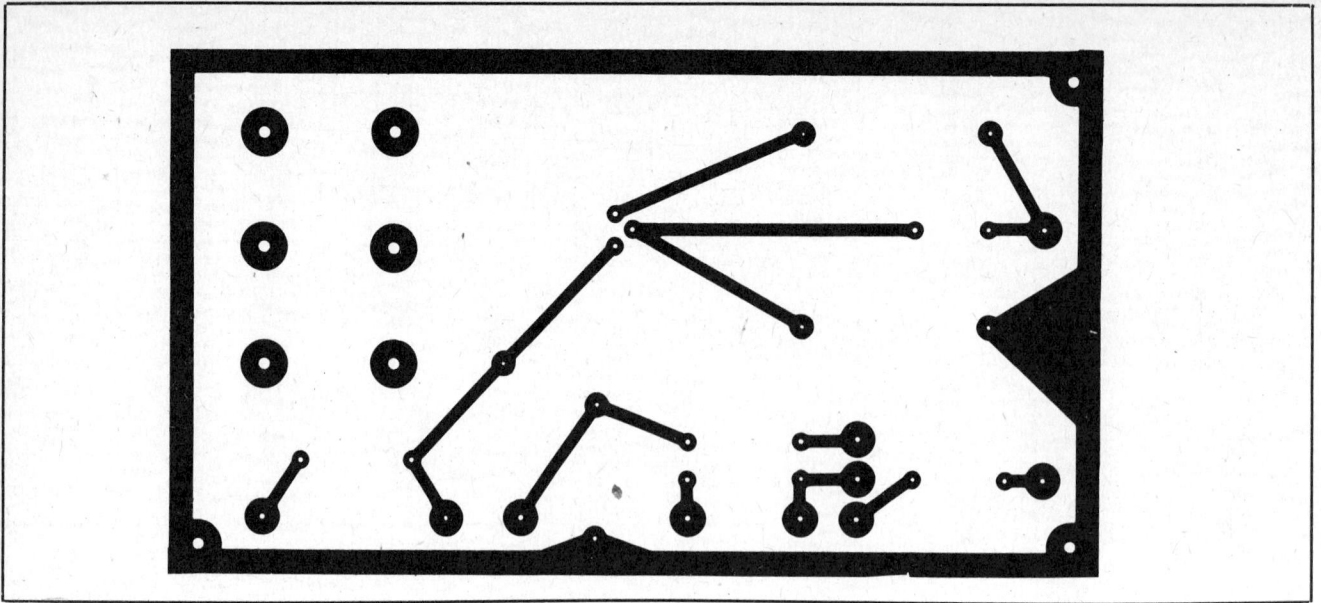

Fig. 18 — Full scale etching pattern for the printed-circuit board. Black areas represent unetched copper.

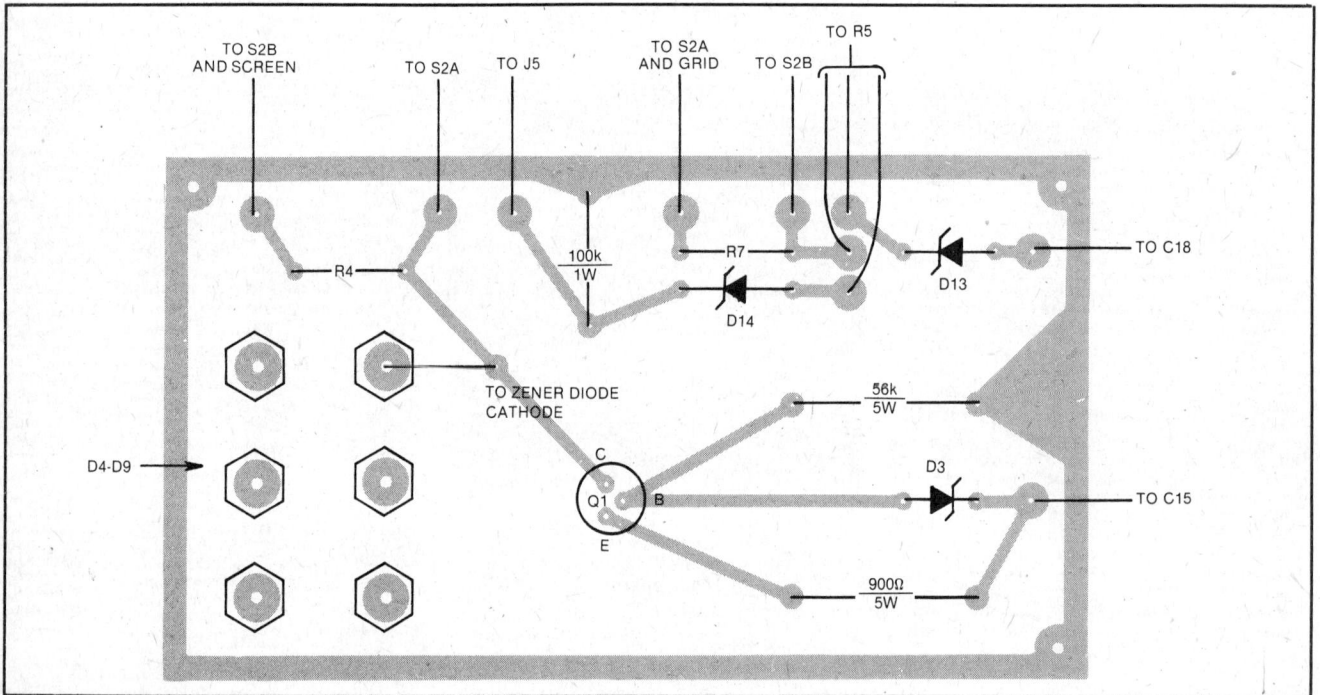

Fig. 19 — Parts-placement guide for the printed-circuit board as shown from the component side.

SWR indicator. Apply additional drive power and continue adjusting L3 and C6 for maximum power output. Use the operating parameters given in Table 1 as a guide. For the amplifier described here an adjustable high-voltage supply was used. The voltage and current levels for 1-kW and 2-kW operation were chosen to equalize the plate impedance, thus requiring only a minimal change in the settings of L3 and C6 for cw and ssb operation. If the fixed-voltage supply is used to

power this amplifier, similar efficiencies should be attained.

As with many amplifier designs for vhf and uhf operation, additional low-pass filtering is required to meet the FCC spectral-purity requirements (all spurious − 60 dB or greater below peak power). A suitable filter circuit is shown in Fig. 17. The unit was constructed in a small aluminum Minibox and is mounted directly at the output connector by means of a double-male adapter. With this filter the

amplifier easily exceeds the FCC requirements. A spectral photograph is shown in Fig. 20.

A LOW-DRIVE 2-METER PA

This amplifier will provide a 200-watt output with as little as 2 watts of drive in linear service. With more drive, more power output can be had (up to 350 watts), but at higher power levels the amplifier components get intolerably warm.

Table 1
Operating Parameters

	1 kW	2 kW
Plate voltage	2100 Vdc	3000 Vdc
Plate current (single tone)	480 mA	667 mA
Plate current (idling)	50 mA	50 mA
Power input	1000 W	2000 W
Power output	620 W	1250 W
Efficiency	62 %	62.5 %
Drive power	4 W	9 W

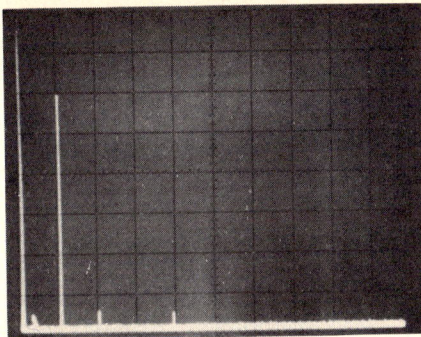

Fig. 20 — Spectral photograph of the amplifier and filter adjusted for 1-kilowatt operation. Each horizontal division represents 50 MHz and each vertical division is 10 dB. This amplifier complies with current FCC requirements for spectral purity.

The Circuit

A 4CX250 tetrode is used in the grounded-cathode circuit. The 4CX250 is a high-mu tube, so high gain is inherent to the design. The cathode is grounded directly through tabs in the Eimac SK-630 socket, reducing feedback possibilities. The screen is grounded through a low-inductance bypass capacitor built into the tube socket. The socket also has a built-in screen-ring shield. All these measures help eliminate the greatest problem with amplifiers of this type: feedback and subsequent self oscillation.

The grid circuit is a simple tuned line made from no. 14 bus wire. A link couples power to this line. At the high-impedance, tube end of the line, bias voltage is shunt fed to the grid through a 2700-ohm resistor. This resistor also swamps the input heavily, assuring amplifier stability without neutralization.

The plate circuit is series tuned. Series tuning places the tube output capacitance in series with the tuning capacitor, effectively reducing tank circuit capacitance. This allows for a larger tank coil which might otherwise become unworkably small if parallel tuning were used. High voltage is fed to the tube at the low-impedance point of the tank coil through an rf choke. Power output is coupled through a variable link, reducing harmonic content. Series tuned traps at the second and third harmonics ensure clean spectral response, far surpassing FCC requirements.

The power supply uses a full-wave bridge rectifier circuit in the high voltage circuit. Series-dropping resistors lower the high voltage to the correct value for the tube screen. Screen voltage is regulated by a string of Zener diodes. Series LEDs in the bias and screen-voltage lines provide warning of excess current flow. Bias is switched between cutoff (− 120 volts) and − 50 volts regulated during transmit operation.

Construction

Despite the compact design, few precautions are necessary. Caution with respect to high-voltage leads is mandatory, of course. The plate circuitry is entirely enclosed by a shielded box, which also prevents air leakage. The grid circuit should be installed so that no other circuit is in close proximity to the tuned line, link or tuning capacitor. Cooling air is blown into the plate compartment through a screened hole. Several screw, nut and flat-washer combinations guarantee good electrical contact to the screen. A standard Eimac ceramic tube chimney is mounted on a wooden standoff. The chimney fits over the tube anode in an inverted style. The hole in the top cover is screened in the same way as is the fan inlet. Cool air enters through the fan hole and passes through the tube anode cooler. Tube-warmed air exits through the chimney and out the screened hole in the top. A small portion of the cooling air passes through the tube socket, cooling the tube base before flowing through the underside of the chassis.

The output traps are built into a box fabricated from copper-clad circuit board. The easiest approach to mounting it is by means of a double-male UHF-type connector to the jack on the rear of the plate compartment.

An input attenuator may be required to reduce transceiver output power down to the 2-watt level necessary to drive the amplifier. Standard composition resistors can be used in attenuator construction. The attenuator is built in a separate box from the amplifier.

Tune-Up and Operation

Upon completion, all wiring should be thoroughly checked for mistakes. *The high voltage present can be lethal if not treated carefully.* Be sure no wires are touching anything they shouldn't.

Primary voltage should initially be applied through a variable line transformer. This allows the voltage to be brought up slowly so that if something is awry it can be noticed before any damage can occur. Once the primary has been safely brought up to its normal voltage, voltage checks on the tube should be made. Screen voltage should be about 325 and bias should be about − 130 V. AC filament voltage should be about 6.

Shorting J2 places the amplifier in the transmit mode. With no drive applied, adjust bias for an idling plate current of 50 mA. This establishes class of operation AB2. Adjustment of the warning-LED potentiometers requires the use of temporary load resistors. Remove power from the amplifier. Temporarily connect a 150-kΩ resistor from the grid terminal to ground. Turn on the power. Do not short J2. Adjust R13 until the grid-sensing LED just comes on. Turn the power off again. Connect a string of 10 1000-ohm, 1-watt resistors from the screen terminal of the tube socket to ground. Turn the power on again. Adjust R26 until the screen sensing LED just comes on. Turn the power off and disconnect the temporary load resistors.

Connect a source of drive to J1 through an SWR indicator. A 50-ohm dummy load should be connected to J3. Applying a small amount of drive, adjust grid capacitor C for a dip in SWR on the indicator. The SWR may not be close to 1:1. If not, readjust the position of L1 with respect to L2. Recheck the SWR. Continue the process until the input SWR is close to 1:1. Since no voltage is applied to the amplifier, adjustment is very simple. In the amplifier shown, the L1 adjustment was optimum when placed approximately 1/8-inch (3-mm) from and parallel to L2.

Reconnect the system so that a power-indicating device is inserted in the line between the dummy load and J3. Apply primary power. Short J2 and apply approximately 2 watts of drive. Adjust the grid capacitor for maximum plate current. Do not key the rig for longer than about 20 seconds at a time. Peak the tuning and the loading controls alternately for maximum output. Since the amplifier is not neutralized, the plate-current dip will not coincide with maximum output.

If a calibrated wattmeter is available, output should be measured at about 200 watts with a plate current of 200 mA.

Fig. 21 — A 500-watt amplifier for the 2-meter band, complete with power supply, is housed in this cabinet. The hole plug covers a former location of a control.

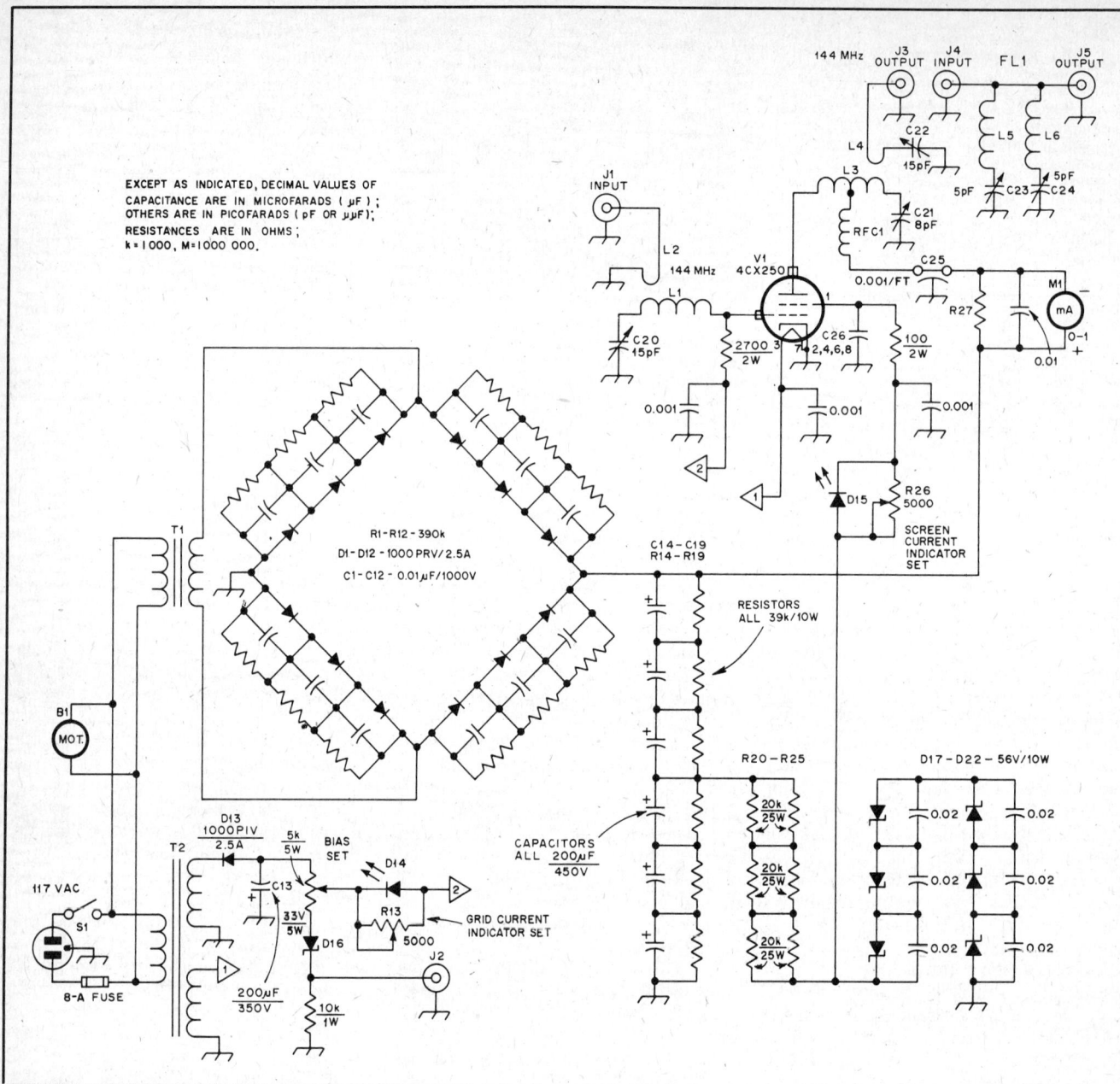

Fig. 22 — Schematic diagram and parts list of the 144-MHz amplifier. Inches × 25.4 = mm.

B1 — Blower, 15 ft³/min.
C1-C12, incl. — 0.01 µF disc ceramic, 100-Vdc.
C13 — 100 µF, 350 Vdc.
C14-C19, incl. — 200 µF electrolytic, 450-Vdc.
C20, C22 — 15-pF air variable, Hammarlund HF-15-X or equiv.
C21 — 5-pF air variable, Hammarlund HFA-25-B with 2 middle rotor plates and two end stator plates removed, or equiv.
C23, C24 — 5-pF air variable, E. F. Johnson 160-0104-001 or equiv.
C25 — 0.001 µF, 4-kV feedthrough capacitor, Erle 2498 or equiv.
C26 — Screen bypass capacitor built into Eimac SK-630A tube socket.
D1-D13, incl. — 1000 PIV, 2.5-A silicon diodes.

D14, D15 — Light emitting diode (V_F = 1.6 V, I_F = 60 mA).
D16 — 33-V, 5-W Zener diode.
D17-D22, incl. — 56-V, 10-W Zener diode.
F1 — 10-A fuse.
J1, J3, J4, J5 — Type SO-239.
J2 — Phono jack, panel mount.
L1, L2 — See text and Fig. 24.
L5, L6 — 3 turns no. 18 tinned, 1/4-inch ID, 3/8 inch long.
L3 — 3-1/2 turns no. 10, 1-1/4 inch ID, 2 inch long.
L4 — 1 turn no. 14 enameled, 1-inch ID.
M1 — 0-1 mA meter.
R1, R12, incl. — 390 kΩ, 1/2-W.

R13, R26 — 5 kΩ, 2-W potentiometer.
R14-R19, incl. — 39 kΩ, 10-W.
R20-R25, incl. — 20 kΩ, 25-W.
R27 — Meter shunt, 0.05555 ohms, 3.375 feet (1.03 m) no. 22 enam. wire wound on any large-value, 2-watt resistor.
RFC1 — 20 turns no. 24 enam. wound on 100 k, 1-W resistor.
S1 — Spst, 10A.
T1 — Primary 117 V ac, secondary 1250 V ac, 500 mA, Hammond 720 or equiv.
T2 — Primary 117 V ac, secondary 125 V ac, 50 mA; 6.3 V ac, 2.0 A, Stancor PA-8421 or equiv.
V1 — Eimac 4CX250B.

Efficiency is about 58 percent. These are the normal operating parameters. If either of the indicator LEDs turns on, either the amplifier is not tuned properly, or there is too much drive, or some equipment failure has taken place.

Adjustment of the traps requires use of a wavemeter or a dip meter in the wavemeter function. Coupling the wavemeter to the dummy load should indicate some harmonic energy at twice the fundamental and three times the fundamental. Adjust one trap capacitor for minimum harmonic energy at the second harmonic and the other for minimum third-

Fig. 23 — Top view of the 2-meter amplifier. Voltage-dropping resistors and Zener diodes may be seen on the left-hand side of the chassis. The small transformer next to the plate compartment is T2. Details of the wood spacer may be seen at right.

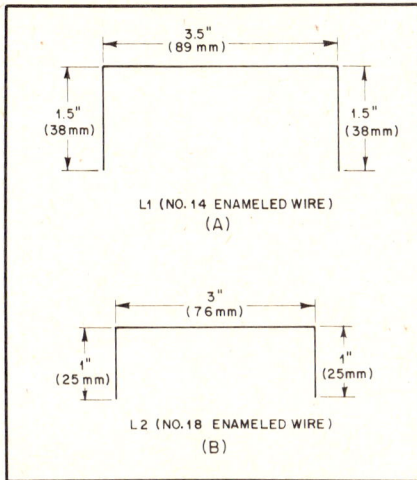

Fig. 24 — Formation details for L1 and L2.

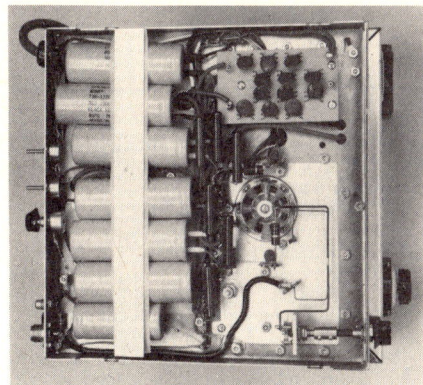

Fig. 25 — Bottom view of the amplifier. A thin sheet of Teflon has been placed between the electrolytic capacitors and their aluminum mounting strap. At upper right is the power supply rectifier board. Details of the grid circuit are also visible.

harmonic energy.

A 2-KW PEP AMPLIFIER FOR 144 MHZ

Large external-anode triodes in a cathode-driven configuration offer outstanding reliability, stability and ease in obtaining high power at 144 MHz. The selection is somewhat limited and they are not inexpensive. Data on the recently introduced 3CX1500A7/8877, a high-mu, external-anode power triode, appeared very promising. A reasonable heater requirement (5 V at 10 A) and an inexpensive socket and chimney combination made the tube even more attractive.

The techniques employed in the design and construction of the cathode-driven 3CX1500A7/8877 amplifier described here (Figs. 26 to 29) have removed many of the mechanical impositions of other designs. Those interested in obtaining complete constructional details should refer to the two-part article appearing in December 1973 and January 1974 QST.

Input Circuit

The plate tank operates with a loaded Q on the order of 40 at 2-kW PEP and 80 at 1 kW. Typical loaded Q values of 10 to 15 are used in hf amplifiers. In comparison, we are dealing with a relatively high loaded Q, so losses in the strip-line tank-circuit components must be kept very low. To this end, small-diameter Teflon rods are used as mechanical drive for the tuning capacitor and for physical support as well as mechanical drive for the output-coupling capacitor. The tuning vane or flapper capacitor is solidly grounded through a wide flexible strap of negligible inductance, directly to the chassis in close proximity to the grid-return point. A flexible-strap arrangement, similar to that of the tuning capacitor, is used to connect the output coupling capacitor to the center pin of a type-N coaxial connector mounted in the chassis base. Ceramic (or Teflon) pillars, used to support the air strip line, are located under the middle set of plate-line dc isolation bushings. This places these pillars well out of the intense rf field associated with the tube, or high-impedance end of the line. In operation, plate tuning and loading is quite smooth and stable, so a high-loaded Q is apparently not bothersome in this respect.

In this amplifier, output coupling is accomplished by the capacitive probe method. As pointed out by Knadle[1]: "Major advantages of capacitive probe coupling are loading linearity and elimination of moving contact surfaces."

Capacitive-probe coupling is a form of "reactive transformation matching" whereby the feed-line (load) impedance is transformed to the tube resonant-load impedance (R_o) of 1800 ohms (at the 2-kW level) by means of a series reactance (a capacitor in this case). At the 1-kW level, R_o is approximately twice that at the 2-kW PEP level. Therefore, the series coupling capacitor should be variable and of sufficient range to cover both power levels.

[1]Knadle, "A Strip-line Kilowatt Amplifier for 432 MHz," QST, April and May 1972.

Fig. 26 — Front-panel layout of the 2-meter kilowatt amplifier.

Fig. 27 — The placement of input-circuit components and supporting bracket may be seen in this bottom view. When the bottom cover is in place, the screened air inlet allows the blower to pull air in, pressurizing the entire under-chassis area. The Minibox on the rear apron is a housing for the input reflectometer circuit.

Fig. 28 — The tube and plate line is in place, with the top and side of the compartment removed for clarity. The plate-tuning vane is at bottom center. A bracket is attached to the side panel to support the rear of the Teflon rod supporting the tuning vane. The coil at the opposite end of the plate line is RFC1, connected between the high-voltage-bypass plate and the top section of the plate-line sandwich. Items outside the tube enclosure include the filament transformer, blower motor, relays, and a power supply to operate a VOX-controlled relay system.

Formulas to calculate the transformation values have been presented in *QST*.[2]

The electromechanical method of probe coupling used in this amplifier is easy to assemble and provides good electrical performance. Also, it has no moving-contact surfaces and enables placement of the output coupling, or loading, control on the front panel of the amplifier for ease in adjustment.

Support Electronics

The grid- and cathode-metering circuits employed are conventional for cathode-driven amplifiers. The multimeter, a basic 0-1 mA movement, is switched to appropriate monitoring points.

An rf-output monitor is a virtual necessity in vhf amplifiers to assure maximum power transfer to the load while tuning. Most capacitive-probe output coupling schemes presented to date do not lend themselves to built-in relative-output monitoring circuits. In this amplifier, one of these built-in circuits is achieved quite handily. The circuit consists of a 10:1 resistive voltage divider, diode rectifier, filter and adjustable indicating instrument. Two 7500-ohm, 2-watt carbon resistors are located in the plate compartment connected between the type-N rf-output connector and a BNC connector. A small wire was soldered to the center pin of the BNC connector, inside a Minibox, with the 1500-ohm, 1-watt composition resistor and the rectifier diode joined at this point. Relative output voltage is fed, via feedthrough capacitors, to the level-setting potentiometer and multimeter switch.

A calibrated string of 2-watt composition resistors, totaling 5 MΩ, was installed to facilitate "on-the-spot" determination of power input, and to attest to the presence or absence of high voltage in the plate tank circuit. A full-scale range of 5000 volts is obtained with the 0-1 mA meter. If desired, the builder may use 10 500-kΩ, 2-watt, 1-percent resistors for the string, and reasonable accuracy will be obtained. Of course this monitor feature may be eliminated if other means are used to measure and monitor plate voltage.

Testing and Operation

The amplifier is unconditionally stable, with no parasitics. To verify this, a zero-bias check for stability was made. This involved shorting out the Zener diode in the cathode return lead, reducing bias to essentially zero volts. Plate voltage was applied, allowing the tube to dissipate about 885 watts. The input and output circuits were then tuned through their ranges with no loads attached. There was no sign of output on the relative output

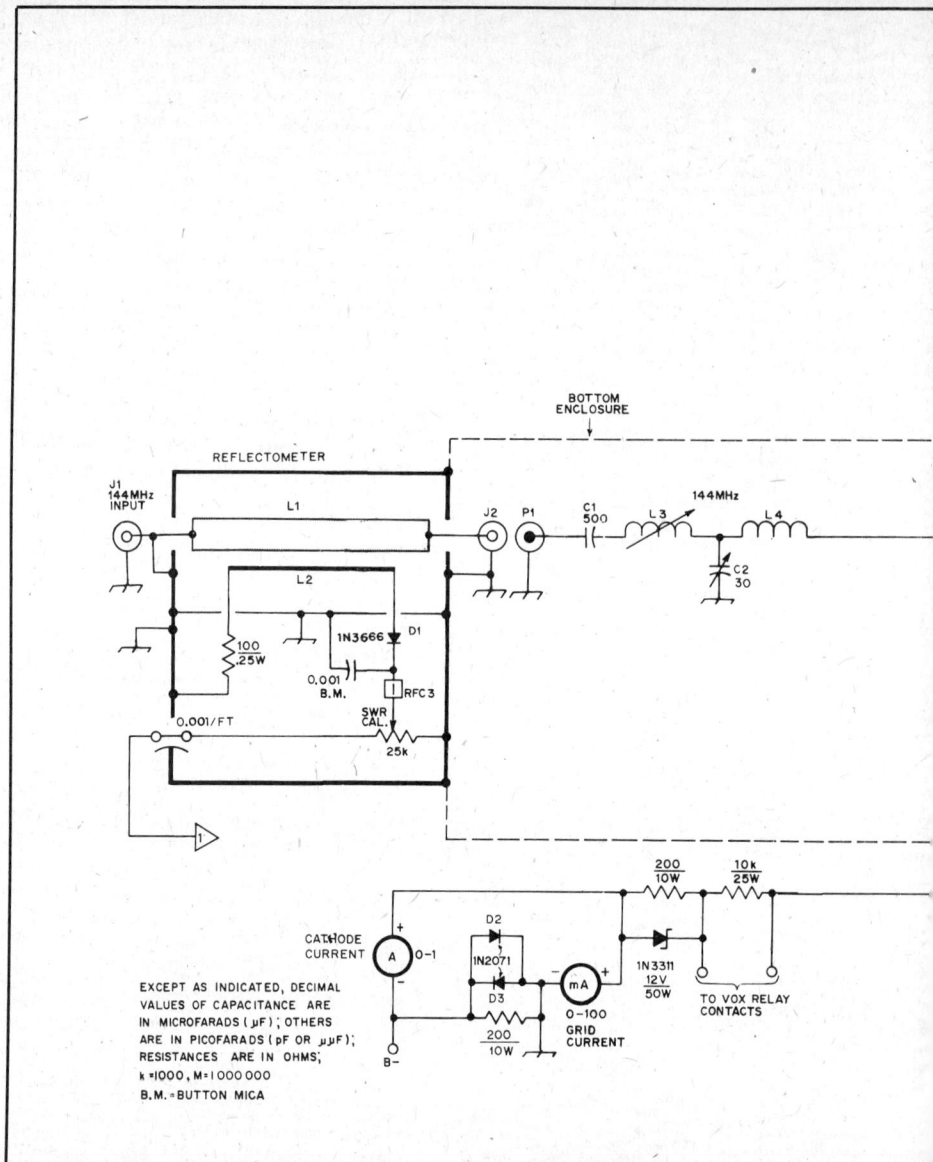

Fig. 29 — Schematic diagram of the 144-MHz amplifier. Included is information for the input reflectometer used as an aid to tuning the cathode circuit for low SWR. C7, C8 and C9 are fabricated as described in the text and Fig. 26. Inches × 25.4 = mm.

B1 — Blower, Fasco 59752-IN or Dayton 2C610. Wheel diameter is 3-13/16 inches.
C2 — 5- to 30-pF air variable, Hammarlund HF-30-X or equiv.
C3, C4, C5, C6 — 0.1 μF, 600-V, 20-A feedthrough capacitor. Sprague 8OP3 or equiv.
J1, J2, J6 — Type BNC.
J3 — Type N.
J4 — Coaxial panel jack, UG-22B/U, Amphenol

82-62 or equiv.
J5 — HV connector, James Millen 37001 or equiv.
L1 — Double-sided pc board, 1-1/4 × 4-7/16 inches.
L2 — 4-1/4 inches of no. 18 wire. L1 and L2 are part of the input reflectometer circuit.
L3 — 6 turns no. 18 enam., 5/8-in. long on 3/8-in. dia form (white slug).
L4 — 3 turns no. 14 enam., 5/8-in. long ×

meter and no change in the plate and grid currents. As with most cathode-driven amplifiers, there is a slight interaction between grid and plate currents during normal tune-up under rf-applied conditions. This should not be misconstrued as amplifier instability.

Tolerances of the Zener diode used in the cathode return line will result in values of bias voltage and idling plate currents other than those listed in Table 2. The 1N3311, a 20-percent tolerance unit, is rated at 12 volts nominal but actually

operates at 10 volts in this amplifier (within the 20-percent tolerance).

All testing and actual operation of this amplifier was conducted with a Raytrack high-voltage power supply used in conjunction with the author's 6-meter amplifier. The power supply control and output cable harness was moved from one amplifier to the other, depending on the desired frequency of operation.

Drive requirements were measured for plate power-input levels of 1000 and 1600 watts with a Bird Model 43 Thruline

[2]Belcher, "Rf Matching Techniques, Design and Example," *QST*, October 1972.

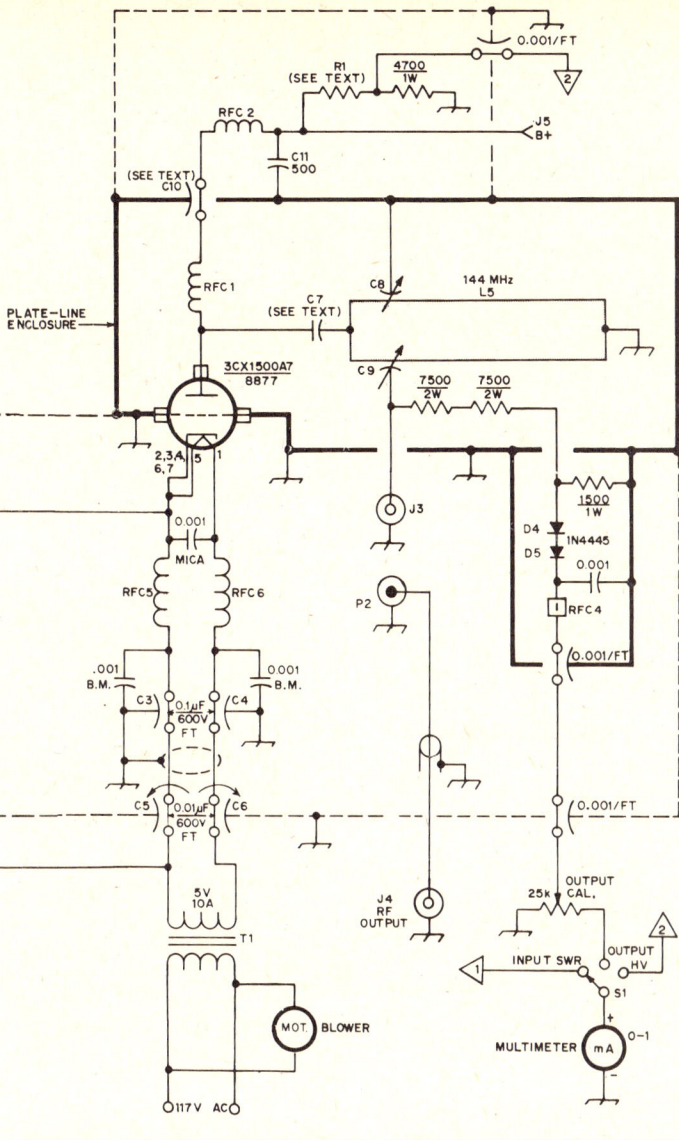

Table 2
Performance Data

	1 kW	1.6 kW
Power input, watts	1000	1600
Plate voltage	2600	2450
Plate current (single tone)	385 mA	660 mA
Plate current (idling)	50 mA	50 mA
Grid bias	−10 V	−10 V
Grid current (single tone)	35 mA	54 mA
Drive power, watts	18	41
Efficiency (apparent)	59.5%	61.8%
Power gain (apparent)	15.2 dB	13.9 dB
Power output, watts	595	1000

To commence routine operation, the variable capacitor in the input circuit should be set at the point where lowest input VSWR was obtained during the "cold-tube" initial tune-up. The ability of the plate tank to resonate at 144-145 MHz with the top cover in place should be verified with a grid-dip meter, via a one-turn link attached to the rf output connector. Top and bottom covers are then secured. *As with all cathode driven amplifiers, excitation should never be applied when the tube heater is activated and plate voltage is removed.* Next, turn on the tube heater and blower simultaneously, allowing 90 seconds for warm-up. A plate potential between 2400 and 3000 volts then may be applied and its presence verified on the multimeter. The power supply should be able to deliver 800 mA or so. With the VOX relay actuated, resting current should be indicated on the cathode meter. A small amount of drive is applied and the plate tank circuit tuned for an indication of maximum relative power output. The cathode circuit can now be resonated, tuning for minimum reflected power on the reflectometer, and not for maximum drive power transfer. Tuning and loading of the plate-tank circuit follows the standard sequence for any cathode driven amplifier. Resonance is accompanied by a moderate dip in plate/cathode current, a rise in grid current and a considerable increase in relative power output. Plate-current dip is not absolutely coincident with maximum power output, but it is very close. Tuning

9/16-in. ID. Lead length to L3 is 5/8-in. Lead
length to cathode bus is 3/4-in.
L5 — Air-dielectric stripline. See text.
P1 — Type BNC.
P2 — Type N.
R1 — Meter range multiplier. Ten 500-kΩ, 2-W
composition resistors in series.
RFC1 — 7 turns no. 16 tinned, 1/2-in. ID ×
1-in. long.
RFC2 — 18 turns no. 18 enam., close wound

on 1-MΩ, 2-watt compostiton resistor.
RFC3, RFC4 — Each 2 ferrite beads on com-
ponent leads.
RFC5, RFC6 — 10 turns no. 12 enam., bifilar
wound, 5/8-in. dia.
S1 — Single-pole, three position rotary switch,
non-shorting contacts.
T1 — 5-V, 10-A secondary, center tap not used,
Stancor P-6135 or equiv.

wattmeter and a slug of known accuracy. Output power was measured simultaneously with drive requirements at the 1000 and 1600 watt plate power input levels. A second Bird model 43 with a 1000-watt slug was used to measure amplifier output into a Bird 1000-watt Termaline load. A 2500-watt slug would be necessary to determine output power at the 2-kW input level, so I stopped at the 1000-watt output point and worked backwards to calculate apparent stage gain and efficiency.

Efficiency measurements also were made employing the "tube air-stream heat-differential" method. Several runs were made at 885 watts static dc and normal rf input. Apparent efficiencies of 62 to 67 percent were noted. These values were about five percent higher than the actual power output values given in Table 1. Both efficiency measurement schemes serve to confirm that the amplifier is operating at the upper limit of the theoretical 50 to 60 percent efficiency range for typical Class AB2 amplifiers.

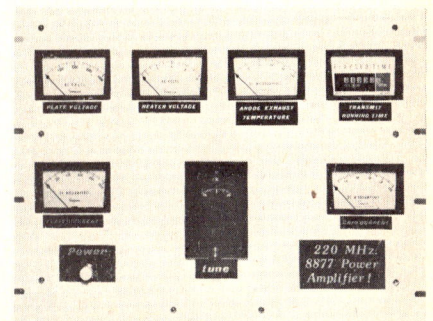

Fig. 30 — Front-panel layout of the 220-MHz kilowatt.

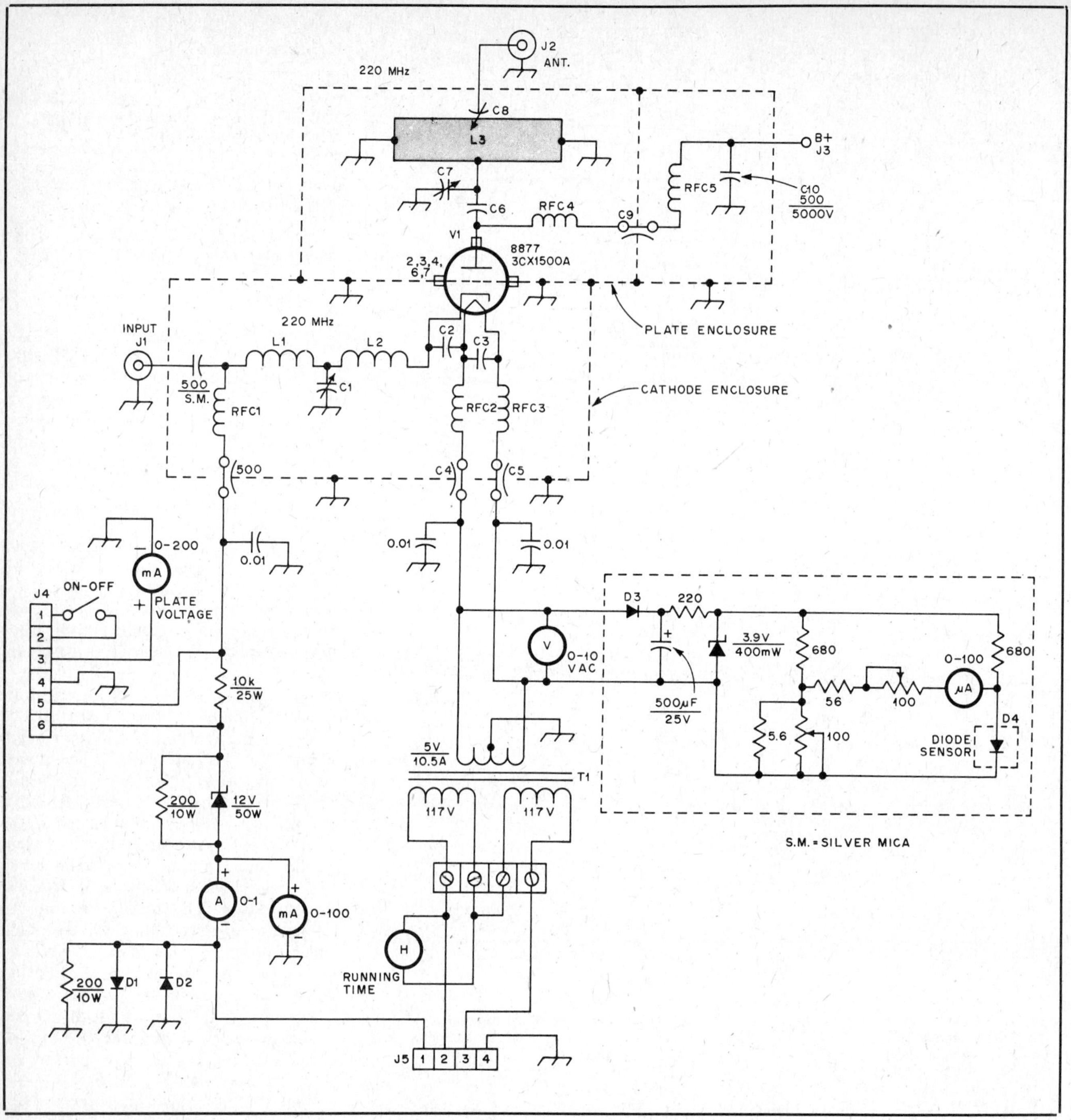

Fig. 31 — Schematic diagram of the 220-MHz amplifier. Unless otherwise specified, all capacitors are disc ceramic and resistors are 1/2-watt carbon composition. Inches × 25.4 = mm.

C1 — Air variable, 15 pF.
C2, C3 — Button mica, 500-pF, 500-V rating.
C4-C9, inclusive — Teflon capacitor (use 10-mil Teflon sheet).
C10 — Doorknob capacitor, 500 pF, 5-kV rating.
D1-D4 — 1000 PRV, 3A.
J1, J2 — Coaxial receptacle, type N.

J3 — High-voltage connector (Millen).
L1 — 3 turns no. 14, 1/4-inch ID, 3/4-inch long.
L2 — 1/4-inch wide, 2-3/8 inch long copper flashing strap.
L3 — Plate inductor (see Fig. 32).
RFC1 — 8 turns no. 18 enam. 1/2-inch dia.,

3/4-inch long.
RFC2, RFC3 — 10 turns no. 18 enam. bifilar wound on 3/4-inch Teflon rod close wound.
RFC4 — 5 turns no. 16 enam. wound on 1-MΩ, 2-watt composition resistor.
T1 — Filament transformer, 5.0 V at 10.5 A.

and output-loading adjustments should be for maximum efficiency and output as indicated on the output meter. Final adjustment for lowest VSWR at amplifier input should be done when the desired plate input-power level has been reached.

A 220-MHZ HIGH-POWER AMPLIFIER

Circuits for 220-MHz power amplifiers have long been designed around the external-anode tetrode. While these tubes offer high gain, instability problems have caused many builders considerable consternation over the years. Multiple-tube amplifiers are often necessary to obtain the high power levels many moon-bouncers and weak-signal specialists require. Push-pull amplifiers have been tried with moderate success, and recently parallel-tube designs have found favor.[3]

Modern computer-aided tube designs have brought forth high-μ triodes such as the 3CX1500A7/8877, a 1500-watt dissipation external-anode triode with maxi-

[3]Knadle, "A Strip-line Kilowatt Amplifier for 432 MHz," *QST*, April and May 1972.

mum ratings good through 250 MHz. The ceramic insulation allows a heavy flow of rf current through the tube, with no loss of stability in a properly designed circuit. Low heater requirements (5 V at 10.5 A) add to the appeal of the 8877. This amplifier employs the 3CX1500A7/8877 in a cathode-driven circuit. The grid is grounded directly to the chassis, adding to the stability. The amplifier (Figs. 30 to 34) is unconditionally stable — more so than some amplifiers built for the rf region.

Circuit Details

The input circuit consists of a T network. Medium values of Q were chosen to provide high efficiency. Both the cathode and the heater are operated at the same rf potential; the heater is held above rf ground by the impedance of the filament choke. The plate tank is a pair of quarter-wavelength striplines placed symmetrically about the tube.[4] This arrangement permits a more uniform flow of current through the anode, preventing "hot spots" on the anode conducting surface. Additionally, tube output capacitance is effectively halved, as one-half the tube capacitance (13 pF) is used to load each stripline. Striplines act as low-pass circuit elements even with the high unloaded-Q conditions found at 220 MHz. Linear inductors also offer control of odd-mode harmonics. No spurious responses could be found in this amplifier up through the 900-MHz region.

A strip-line impedance can be varied by changing its width and relation to its ground planes. Physical dimensions of the tube limit the position of the stripline above one ground plane. In order to utilize commercially available chassis, the stripline was placed 1-1/4 inch (32-mm) above one side of an inverted 4-inch (102-mm) high chassis. This means that approximately 75 percent of the rf current flows through the chassis, but only 25 percent flows through the top shield cover. The small percentage flowing through the top reduces the effect of any mechanical anomalies associated with a removable cover.

For quarter-wavelength lines, the ratio of line impedance to reactance should be between 1.5 and 2.0 for the best bandwidth. Taking stray capacitance into account, expected tuning capacitance and tube output capacitance gives a value of 55 ohms for X_C. Values of line impedance versus line length for resonance at 222 MHz were computed on a programmable calculator for impedances between 30 and 100 ohms. These were plotted on a graph. Final dimensions were determined using this system, choosing dimensions that fell into the middle of the graph, thus allowing for any unpredicted effects.

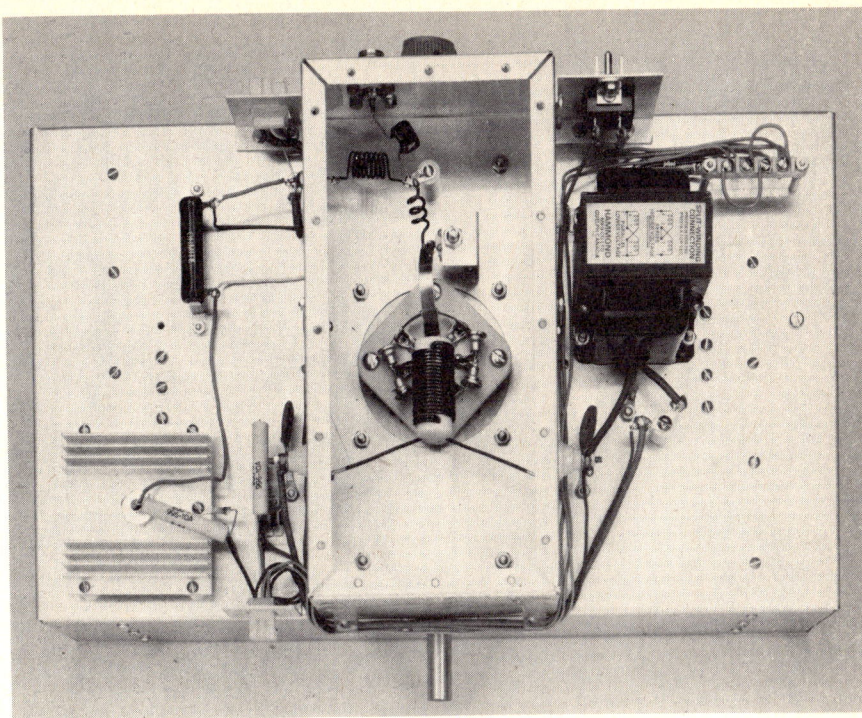

Fig. 32 — Bottom view of the amplifier. RFC2 and RFC3 can be seen above tube socket (bifilar winding). Copper strap is L2 shown connected to C1. Small coil is L1 and larger coil is RFC1. The grid of the tube should be grounded to the chassis with finger stock similar to that used in the plate line. Component mounted on the heat sink at left is the Zener diode used for biasing purposes.

The plate blocking capacitor consists of a sandwich of brass plate and the stripline, with Teflon sheet as the dielectric. This forms a very low-loss, high-voltage capacitor. The plate bypass capacitor is built along the same principles. A piece of circuit board was once again sandwiched with Teflon sheet to the side wall of the chassis. This technique is used effectively throughout as an inexpensive bypass or feedthrough capacitor at vhf.

Amplifier output is coupled through a capacitive probe. Transformation of the load impedance to the tube resonant-load impedance is achieved by means of a series reactance (the loading capacitor). The tuning capacitor is solidly grounded by means of a flexible strap of negligible inductance. Mechanical details were described by Sutherland.[5]

A rather elaborate metering system is employed. Although all of the meters provide useful data, only the plate and grid meters are necessary for proper amplifier use. At a repeater site where key-down service is the rule rather than the exception, measurement of heater usage and voltage provide data requisite to tube replacement. The anode exhaust-temperature metering circuit takes advantage of a thermal property of semiconductors. As the temperature changes the forward resistance of a diode changes in a nearly linear manner. The diode sensor is made a part of a bridge circuit, allowing calibrated operation. Calibration may be determined by packing the diode in ice for the low point (0° C) and immersing it in boiling water for the high point (100° C). The amount of heat dissipated by the tube is inversely proportional to the efficiency for a given power input. Low heat dissipation yields longer tube life.

High-power amplifiers require considerable attention to cooling. The plate compartment is pressurized by air from an external blower, and holes in the chassis allow a portion of this air to pass through the grid and cathode structure. Most of the air flows through the anode, a handmade Teflon chimney, then out the top cover. Aluminum screening is tightly bonded around these two openings. No radiated rf could be detected around the chassis except within one inch of the anode exhaust hole.

To commence operation, the input should be adjusted for minimum VSWR with no voltages applied. The covers should be in place whenever voltage is present. Drive should *never* be applied without plate voltage and a load connected if the filament is energized. Cooling air must always be supplied whenever the filament is turned on.

After a 60-second warmup small amounts of drive may be applied. The plate circuit is then tuned for maximum output indication. The drive level is then increased. Tuning and loading follow the

[4]Barber, Rinaudo, Orr and Sutherland, "Modern Circuit Design for VHF Transmitters," *CQ*, November and December 1965.

[5]Sutherland, "High-Performance 144 MHz Power Amplifier," *Ham Radio*, August 1971.

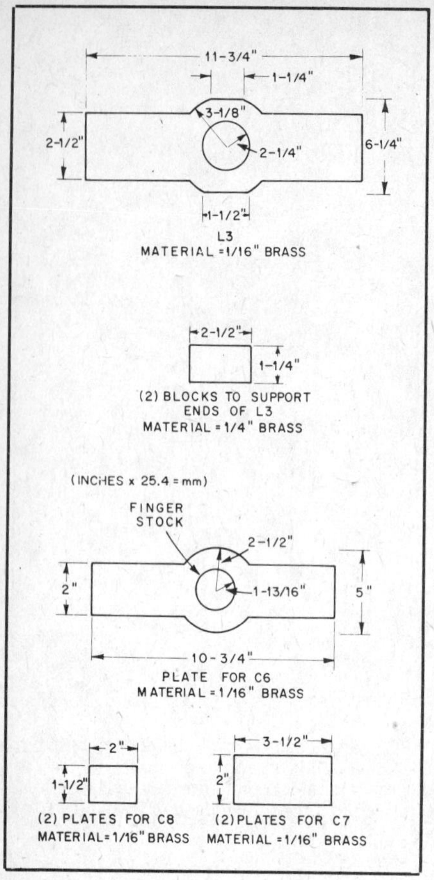
11-3/4"
1-1/4"
3-1/8"
2-1/2"
2-1/4"
6-1/4"
1-1/2"
L3
MATERIAL =1/16" BRASS
2-1/2"
1-1/4"
(2) BLOCKS TO SUPPORT ENDS OF L3
MATERIAL = 1/4" BRASS
(INCHES x 25.4 = mm)
FINGER STOCK
2-1/2"
2"
1-13/16"
5"
10-3/4"
PLATE FOR C6
MATERIAL = 1/16" BRASS
2"
1-1/2"
3-1/2"
2"
(2) PLATES FOR C8
MATERIAL=1/16" BRASS
(2) PLATES FOR C7
MATERIAL =1/16" BRASS

Fig. 33 — Construction details of plate line and associated components. Inches × 25.4 = mm.

Fig. 34 — Interior view of the amplifier. Contact of the tube with "hot side" of C6 is accomplished with suitable finger stock (available from the tube manufacturer). This conductor, in conjunction with a similar one separated by the Teflon insulator, forms the L3/C6 combination. The entire assembly is sandwiched together by means of four insulated bushings (approximately 3/4-inch or 19-mm diameter). Placement of bushings is not critical. RFC4 can be seen at the right connected to C9. C8 is seen at the center of the photo and has a nominal spacing of 1 inch (25 mm) to similar plate soldered to L3. Tuning capacitor, C7, can be seen at the right connected to C9. C8 is seen at the bottom center of the photo and has a nominal spacing of 1 inch (25 mm) to similar plate soldered to L3. Tuning capacitor, C7, can be seen in upper center of photo (see Fig. 33 for dimensions). Drive mechanism can be of builder's choice or see reference 5.

FILAMENT
OPERATE
STANDBY
METER
LOAD
PLATE
TUNE
PLATE MILLIAMPERES

Fig. 35 — The high-power uhf amplifier. The toggle switches control filament power and standby/operate functions respectively. Multimeter function is selected with the switch located between the meters, while the plate tuning and loading controls are at the right. Modern knobs and homemade meter faces give the amplifier a commercial appearance.

normal procedure for any cathode-driven amplifier: Adjustments are made for maximum output and efficiency. When the desired plate output power has been achieved, the input circuit should be adjusted for minimum input VSWR.

A GROUNDED-GRID KILOWATT AMPLIFIER FOR 432 MHZ

An Eimac 8874 high-mu triode was selected for use in this amplifier. Triodes offer a simpler design approach than multigrid tubes, such as those in the popular 4CX250 family. No screen or bias supplies are required and stability is all but guaranteed. The only price that must be paid for these conveniences is the added drive-power requirement. Approximately 25 watts of energy is required to drive this amplifier to the 1-kW input level. The amplifier will deliver over 500 watts of output when adjusted for operation at 1 kW input. This amplifier was originally described in the October 1979 issue of *QST* by Stephen Powlishen, K1FO.

Circuit Description

A schematic diagram of the 432-MHz kilowatt is given in Fig. 36. W1 is a half-wavelength stripline which is tuned and loaded by C6 and C7 respectively. Plate choke RFC4 is connected at the approximate electrical center of the plate line. C8 functions as the plate-bypass capacitor. The half-wavelength cathode line is comprised of W2, L2 and C2. L1 and C1 serve to match the tube input impedance to the amplifier 50-ohm input. As the grid is grounded for dc as well as rf, D1 is used to develop operating bias at the cathode. R3 is switched in to supply near-cutoff bias during standby periods. M1 is used solely to monitor plate current in the high-voltage supply negative-return lead. M2 is switched to read grid current, high voltage and relative output. The latter function is by means of an external line sampler.[6] With the exception of the multimeter functions the metering and bias circuits are similar to those in the 220-MHz amplifier.[7]

Separate coaxial relays attached to the input and output terminals allow the amplifier to be switched in and out of the line in a manner popular with rf amplifiers. Time-delay relay K1 prevents the amplifier from being switched into service for 90 seconds after the tube heater is energized, allowing the element to reach operating temperature. A normally closed contact of K2 applies full voltage to the heater during standby periods. The voltage is reduced during operation as recommended by the manufacturer.

Construction

Place and cathode-compartment construction is from 0.032-inch (0.8-mm) thick aluminum sheet attached to 1/2-inch (13-mm) aluminum angle stock. Some angle stock may be anodized, giving the surface a dull appearance. This material must be lightly sanded to remove the anodized metal, which is a poor conductor. Holes are drilled in the angle stock to allow attachment of the covers; these are tapped for no. 4-40 screws. Details of the 10.5 × 4 × 3-inch (267 × 102 × 76-mm) plate compartment may be seen in the top view photo. Construction of the cathode compartment is similar, and may be seen in the photo of the underside. It measures 4 × 4 × 1-3/4 inches (102 × 102 × 44 mm). The aluminum brackets holding the rf enclosures to the front

[6]McMullen, "The Line Sampler," *QST*, April 1972.
[7]Sutherland, "High Power Linear Amplifier for 220 MHz," *Ham Radio*, December 1971.

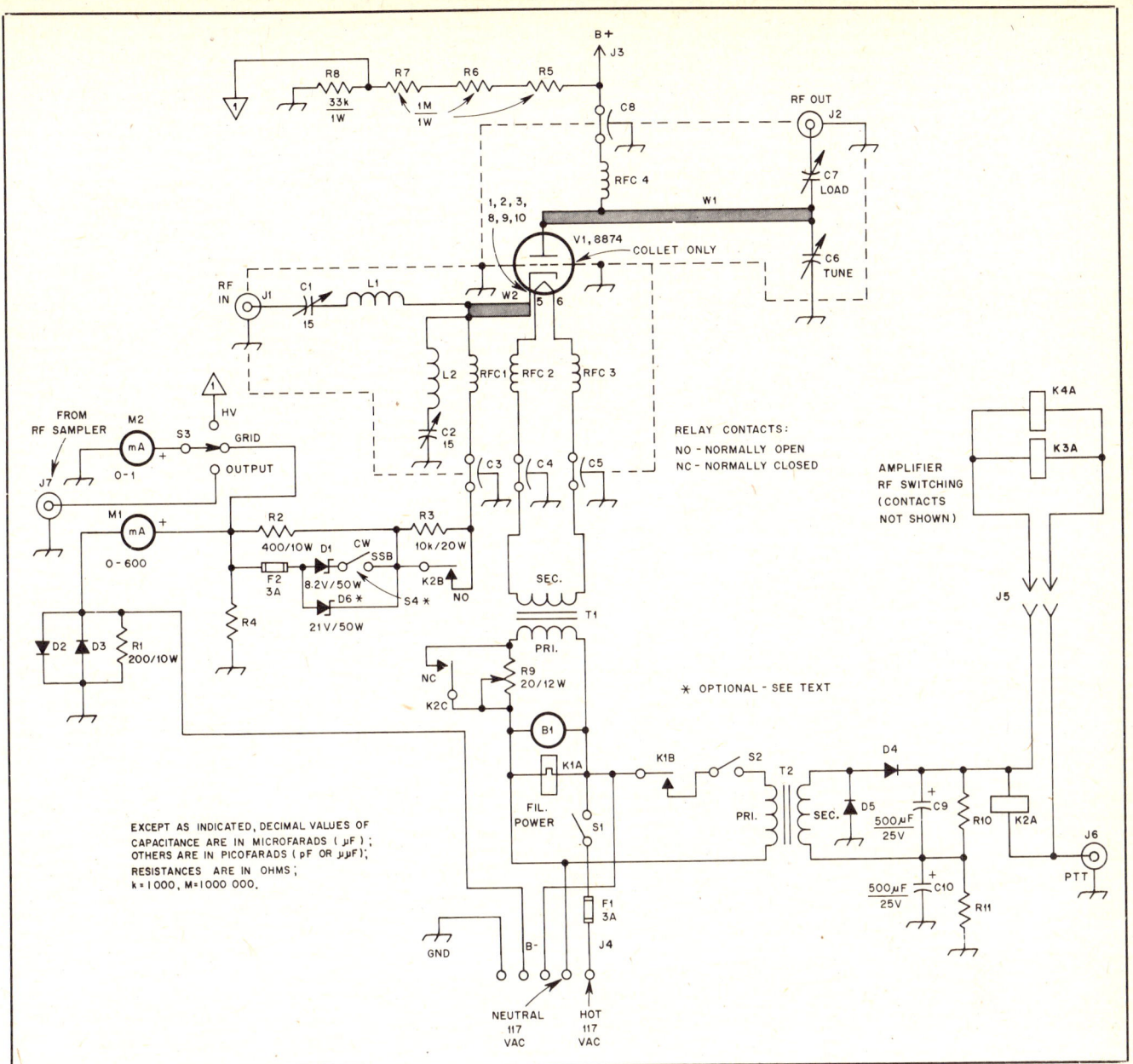

Fig. 36 — Schematic diagram of the amplifier.

B1 — 50-ft^3 (1.4-m^3)/min blower, Ripley Sk2754-2A or equiv.

C1, C2 — Air-variable capacitor, 15 pF, E. F. Johnson 189-0565-001, 160-0107-001 or equiv.

C3-C5, incl. — Feedthrough capacitor, 500 pF, 300 V.

C6-C8, incl. — Homemade "flapper" capacitor. Details of construction in text and Fig. 41.

C9, C10 — Electrolytic capacitor, 500 μF, 25 V.

D1 — 50-watt, 8.2-volt Zener diode, IR Z-3307-C or equiv.

D2-D5, incl. — 1-A 1000-PIV diode, 1N4007 or equiv.

D6 — 50-watt, 21-volt Zener diode (optional — see text).

F1, F2 — 3AG fuses.

J1 — Chassis mount BNC female connector, UG-1094/U.

J2 — Chassis mount N female connector, UG-58A/U.

J3 — High-voltage connector, Millen 37001.

J4, J5 — Power connectors, as available.

J6, J7 — RCA phono jacks.

K1 — Time-delay relay, 90 second, normally open contact, Amperite 115N090T.

K2 — Control relay, 28-volt coil, 1-A 4pdt contacts.

K3, K4 — Coaxial relays equipped with suitable connectors. K4 should have N connectors, K3 may be BNC or N.

L1 — 3-1/2 turns no. 16 enam. wire, 3/4-inch (19 mm) long, 1/4-inch (6 mm) diameter.

L2 — 1-1/2 turns no. 16 enam. wire, 5/8-inch (16 mm) long, 1/4-inch (6 mm) diameter.

M1 — 1-mA meter movement with shunt to provide 600-mA full-scale deflection.

M2 — 1-mA meter movement with shunts to provide 90-mA (grid current) and 3-kV (plate voltage) full-scale deflection.

R4 — Grid-current shunt.

RFC1 — 10 turns no. 18 enam. wire, close wound, 1/4 inch (6 mm) diameter.

RFC2, RFC3 — 10 turns no. 16 enam. wire, close wound, 1/4 inch (6-mm) diameter.

RFC4 — 5 turns no. 16 wire, one inch (25 mm) long, 1/4 inch (6-mm) diameter.

S1 — Toggle switch, spst.

S2 — Toggle switch, spst.

S3 — Rotary switch, single pole, three position.

S4 — Toggle switch, spst (optional, see text).

T1 — Filament transformer, 6.3-volt, 3-A, Stancor P-6466 or equiv.

T2 — Transformer, 12.6 volts, 1 A.

panel also serve as end covers for the compartments. Compartment spacing from the panel is four inches (102 mm). A 5-1/4 × 19-inch (133 × 483-mm) rack panel is used.

The plate line was fabricated from a piece of 1/16-inch (1.6-mm) thick brass. Fig. 39 gives detailed information for making the line. In addition to brass, lines were made from copper, both unplated and silver plated, with no discernible difference in efficiency. Double-sided G-10 printed circuit board would probably work as well. Best thermal stability was obtained with the unplated solid-copper line. The line is supported by 1.5-inch (38-mm) long ceramic insulators, although standoffs made of Teflon will

Fig. 37 — Top view of the amplifier, with the plate compartment cover removed. The tube, plate line (W2) and RFC4 may be seen at the top of the photo. Note the large number of holes drilled in the plate compartment to receive the cover hold-down screws. A tight seal is required to prevent rf and air leaks.

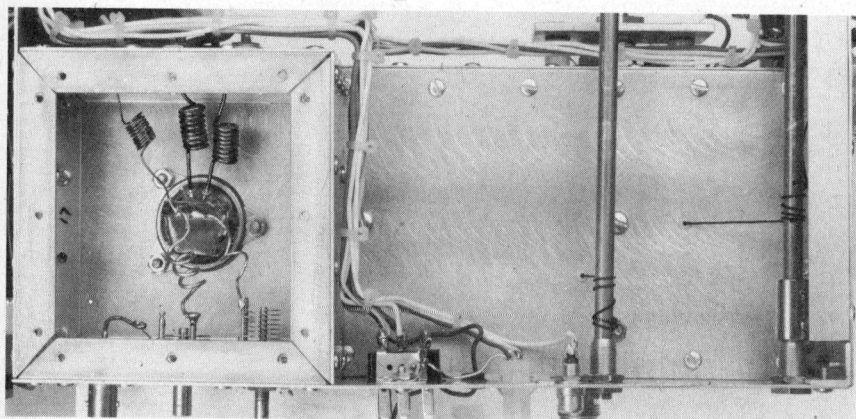

Fig. 38 — This bottom view shows the cathode compartment and the shafts for C6 and C7. A cover is placed over the cathode compartment during tune-up and operation.

also serve. C6 and C7 are made from beryllium-copper sheet. Details of their construction appear in Fig. 41. These "flappers" are moved with fishing line which is tied to 1/4-inch (6.4-mm) fiber shafts. These shafts may be seen in the underside view.

The anode collet (Eimac no. 008294) is secured to the bottom of W1 with standard 60/40 solder. Use no. 4-40 screws and nuts to hold the collet in place during the soldering operation. The grid collet (Eimac no. 882931) is attached to the chassis with eight no. 4-40 machine screws and nuts. A poor ground connection for the grid will greatly increase the amplifier drive requirements or make the unit totally inoperative.

C8, the plate-bypass capacitor, is made from two brass plates, one mounted on either side of the plate compartment. A 0.005-inch (0.13-mm) thick piece of Teflon sheet is used for the dielectric material. While this Teflon thickness may seem inadequate, it is rated at 1000 volts per mil (0.03 mm) thickness. It is necessary to coat the dielectric with Dow Corning type DC-4 silicone grease to fill in any imperfections in the surface that might allow a leakage path and subsequent capacitor breakdown. This silicone grease has dielectric properties similar to Teflon. A no. 8 (4-mm) brass screw is used to hold the plates in place, and also acts as the high-voltage feedthrough terminal. A 3/8-inch (10-mm) diameter

washer was sliced from a Teflon rod and used to center the screw in the hole. Fig. 41 gives details of the remaining metalwork.

An enclosure attached to the rear wall houses the meter dropping resistors and provides a protective hood over the high-voltage terminal. I made this cover 3 × 4 × 1.5 inches (76 × 102 × 38 mm) in size, but the dimensions are not critical. As a final note on construction, it is necessary to isolate the shaft of C1 from ground, if the rotor is connected to the shaft of the capacitor. Rf potential at this point is low, allowing the capacitor to be mounted on a small piece of plastic if an insulated unit is not available.

Cooling the Amplifier

This amplifier is thermally stable; that is, heat-induced warping of tuned-circuit components and the resulting decrease in power output is minimal. A major reason is no doubt the effective cooling system used. The cathode compartment is pressurized with a medium-sized blower. Any convenient unit capable of supplying 50 ft³/min. (1.4 m³/min.) may be used. A piece of copper window screen is attached to the side cover with aluminum solder, to shield the air inlet. Air flows from the cathode compartment through the socket and into the plate compartment, providing some cooling of the grid area of the tube as well. A chimney is made of 0.01-inch (0.25-mm) Teflon sheet, 1.5 × 12 inches (38 × 305 mm) in size. A piece of 1-5/8-inch (41-mm) OD copper pipe was used as a form to make the chimney. The Teflon is held together with RTV (room-temperature vulcanizing) adhesive. Air in the plate compartment must now flow through the anode cooling fins to escape. The air outlet is built on a 2-1/4-inch (57-mm) square copper plate. A 1-5/8-inch (41-mm) diameter hole is made in the plate and a piece of copper window screening is soldered over it. On the side opposite the screening is soldered a 3/8-inch (9.5-mm) long piece of 1-1/2-inch (38-mm) copper pipe. This pipe has an outside diameter of 1-5/8 inches (41 mm) and should fit snugly into the hole. The Teflon chimney will be held firmly in place and no air should leak from the box without passing through the anode cooler.

Operation

Adjust R9 to place maximum resistance in series with the tube heater. Apply heater power and allow two minutes for the element to reach operating temperature. Now energize K2 and adjust R9 to place 5.7 volts at the socket pins. Apply plate voltage (about 2000 volts). Idling plate current should be approximately 30 mA. Apply drive and adjust its level to bring the plate current up to 150 mA. Adjust C6 (plate tuning) for maximum output. Input capacitors C1 and C2 may then

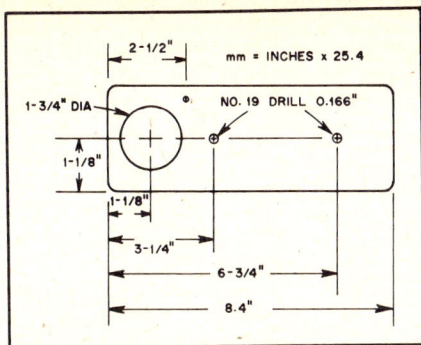

Fig. 39 — Dimensions of the plate line are given here. The line may be constructed from 1/16-inch (1.6-mm) thick copper or brass. Corners of the line should be filed to give a 3/16-inch (5-mm) radius.

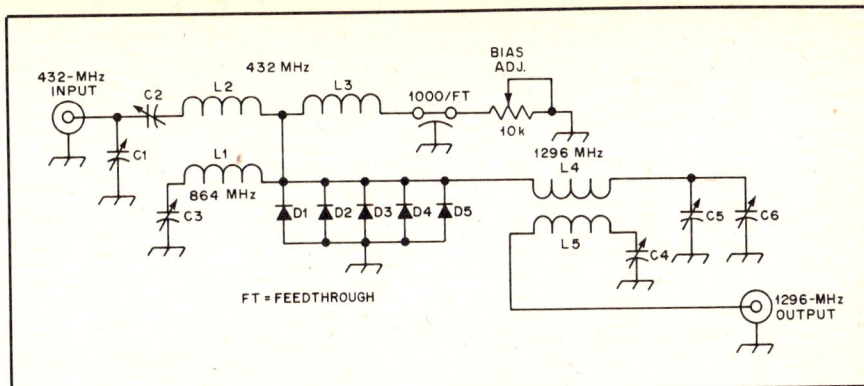

Fig. 42 — Schematic diagram and parts list for the 432- to 1296-MHz tripler.
C1-C3 — Air-variable capacitors, 3-11 pF.
C4-C6 — Piston-trimmer capacitors, 1-5 pF.
D1-D5 — 1N914 diodes or equiv.
R1 — 10-kΩ potentiometer, linear taper.

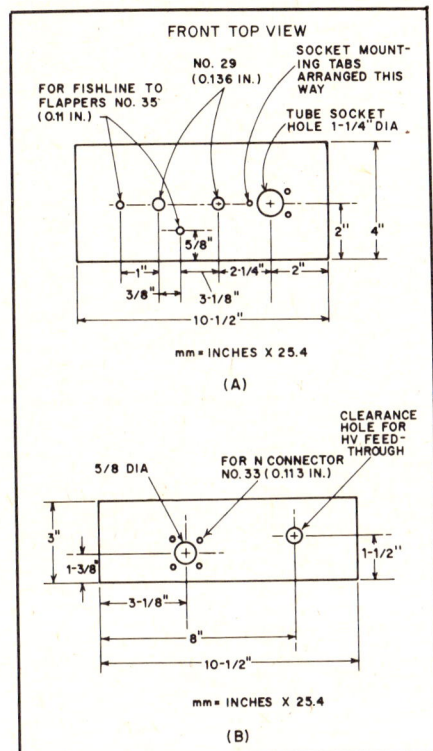

Fig. 40 — At A, dimensions of the plate compartment bottom cover. At B, dimensions of the rear panel of the amplifier.

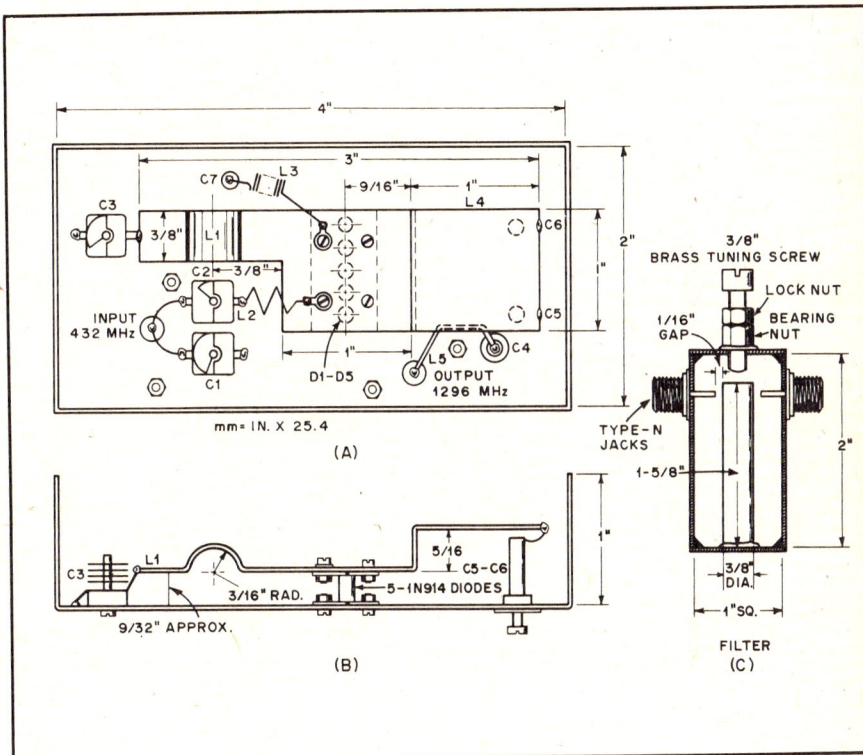

Fig. 43 — Construction information for the tripler and filter. At A, a top view of the tripler, showing the position of the components. At B, a side view of the tripler, showing the installation of D1-D5. At C, a cavity filter designed to remove undesired harmonics generated in the tripler.
L2 — 3 turns no. 24 enameled wire space wound on a 3/32-inch form (remove form after winding).
L3 — 12 turns no. 24 enameled wire close wound on a 3/32-inch form (remove form after winding).

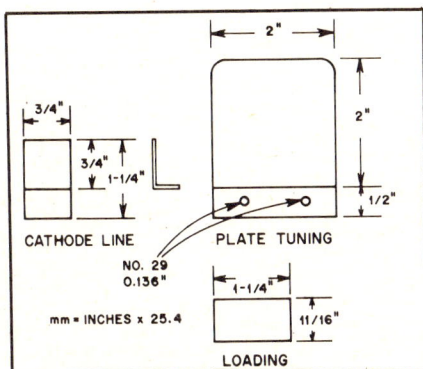

Fig. 41 — Dimensions of the cathode line and the flappers used to tune and load the plate circuit are given here. Additional information is contained in the text.

be coarsely adjusted for maximum plate current. Simultaneously increase drive and adjust plate tuning and loading for maximum output until input power reaches one kilowatt or the desired level. The input circuit may be adjusted for minimum reflected power when the proper drive level is established.

Conclusions

From a cold start, the amplifier reaches full output five seconds after drive is applied. After the first transmission, full output is obtained in one second, with no further drift noticed. This amplifier has been operated for several months without need for retuning.

AN INEXPENSIVE DIODE MULTIPLIER FOR 23 CM

Instead of expensive and hard-to-find varactors this tripler, designed by G8AZM, uses computer switching diodes. It has been described in several RSGB publications.

A schematic diagram of the tripler is shown in Fig. 42. The unit is fabricated from sheet brass or copper. Construction details are shown in Fig. 43A and B. To increase the dissipation ability of the diodes a heat sink is fabricated, allowing up to 6 watts drive at 432 MHz and 2 watts output at 1296 MHz. The five diodes are connected between two 1 × 5/8-inch (25

× 16-mm) plates to form a stack, one side of which is bolted to the chassis. The other side is fastened to L1. Great care must be taken when soldering the diodes. Lead length should be as short as possible. To reduce the risk of damaging the diodes, the leads and the holes in the plates should be pretinned and a *hot* soldering iron used for the minimum time necessary to make each joint. The top and bottom surfaces of the stack should then be made flat by filing, and then with emery paper. It will be helful to place the emery paper grit-side up on a hard surface, such as glass, and then draw the plate across its surface. Mating surfaces of the chassis and L1 should also be carefully flattened to ensure good thermal contact. Output loop L5 is made of no. 18 wire, 1/2-inch (13-mm) long, placed as near to L1 as possible.

The output filter, shown in Fig. 43C, is a simple cavity resonator. It may be built from thin brass or copper sheet. Before drive is applied to the tripler, the filter should be aligned. This may be done by connecting it to a receiver and peaking on a weak signal. Insertion loss should be less than 1 dB. The lock nut should be tightened and the filter connected between the tripler and a load.

Adjust R1 to its centerpoint and connect a high-impedance dc voltmeter across it. When 5-watts drive at 432 MHz is applied to the tripler, C1, C2 and C3 should be adjusted for maximum meter deflection (about 20 volts). R1 and all tuning capacitors should then be adjusted for maximum rf output.

Chapter 8

Receiving Systems

How good should receiver performance be? A suitable answer might be, "As good as is possible, consistent with the state of the amateur art and the money available to the purchaser." That opens up a wide area for debate, but the statement is not meant to imply that a receiver has to be costly or complex to provide good performance: Some very basic, inexpensive homemade receivers offer outstanding performance.

For many years the evolution of commercial amateur receivers seemed to stagnate except for the window dressing and frills added to the front panels. Emphasis was placed on "sensitivity" (whatever was *really* meant by that term) in the advertising. Some amateurs concluded, as a result of the strong push for *sensitive* receivers, that the mark of a good unit was seen when atmospheric noise on the hf bands could push the S-meter needle up to an S2 or S3. Very little thought, if any, was given to the important parameters of a receiver — high dynamic range, fine readout resolution and frequency stability. Instead, countless receivers were placed on the market with 5- or 10-kHz dial increments and excessive amounts of front-end gain. The latter caused the mixer (or mixers) to collapse in the presence of moderate and strong signals. Double-conversion superheterodyne receivers were for a long time the choice of manufacturers and amateurs. The second i-f was often 100 or 50 kHz, thereby enabling the designer to get fairly reasonable orders of selectivity by means of high-Q i-f transformers. That concept predated the availability of crystal-lattice and mechanical filters. The low-frequency second i-f dictated the use of a double-conversion circuit in order to minimize image responses.

Single-conversion receivers offer much cleaner performance in terms of spurious responses and dynamic range. They are being offered by some manufacturers, and although performance is acceptable in many instances, there is considerable room for improvement. At least there is only *one* mixer to cause intermodulation distortion (IMD) and overloading problems in a single-conversion superheterodyne receiver. A strong doubly balanced mixer (DBM) and careful gain distribution in such a receiver can yield superb performance if a proper design effort is put forth. Of course, the local oscillator should be stable and low in noise components to further enhance performance. Thus far, not many commercially built amateur receivers meet the foregoing criteria. In terms of dynamic range, some manufactured receivers exhibit an MDS (minimum discernible signal) of −145 dBm (referenced to the noise floor), blocking of the desired signal does not occur (1 dB of compression) until the adjacent test signal is some 116-dB above the noise floor, and the two-tone IMD dynamic range is on the order of 85 dB. Greater detail concerning this measurement technique is given in Chapter 16. A receiver with the approximate figures just given is considered to be an acceptable one for use where fairly strong signals prevail. However, it is possible to improve those numbers considerably; it has been done by amateurs who designed and built their own receivers. Examples are W7ZOI's "Competition Grade Receiver" (March and April 1974 *QST*) and the W1CER receiver described in June and July 1976 *QST*.

The foregoing suggests strongly that amateurs should consider designing and building their own receivers. Certainly, such an endeavor is within the capability of many experimenters. The satisfaction derived from such an effort can't be measured. The following sections of this chapter are written for those who wish to acquire a better understanding of how a practical receiver operates. Design data and related philosophy are included for those who are inspired toward developing a homemade receiver.

Sensitivity

One of the least understood terms among amateurs is *sensitivity*. In a casual definition the word refers to the ability of a receiver to respond to incoming signals. It is proper to conclude from this that the better the sensitivity, the more responsive the receiver will be to weak signals. The popular misconception is that the greater the receiver front-end gain, the higher the sensitivity. An amateur who subscribes to this concept can ruin the performance of a good receiver by installing a high-gain preamplifier ahead of it. Although this will cause the S meter to read much higher on all signals, it can actually *degrade* the receiver sensitivity if the preamplifier is of inferior design (noisy).

A true measure of receiver sensitivity is obtained when the input signal is referenced to the noise generated within the receiver. Since the significant noise generated inside a receiver of good design originates in the rf and mixer stages (sometimes in the post-mixer amplifier), a low-noise front end is vital to high sensitivity. The necessary receiver gain can be developed after the mixer — usually in the i-f amplifier section. The internal noise is generated by the thermal agitation of electrons inside the tubes, transistors or ICs. It is evident from the foregoing discussion that a receiver of high sensitivity could be one with relatively low front-end gain. This thought should be kept in mind as we enter the discussion of dynamic range and noise figure.

Noise Figure

The lower the receiver *noise figure* (NF), the more sensitive it is. Receiver noise figures are established primarily in the rf amplifier and/or mixer stages. Low-noise active devices (tubes or transistors) should be used in the receiver front end to help obtain a low noise figure. The unwanted noise, in effect, masks the weaker signals and makes them difficult to copy. Noise generated in the receiver front end is amplified in the succeeding stages along with the signal energy. Therefore, it is in the interest of sensitivity that internal noise should be kept as low as possible.

Some amateurs confuse external noise (man-made and atmospheric, which comes in on the antenna) with receiver noise during discussions of noise figure. Although the ratio of the external noise to the incoming signal level has a lot to do with reception, external noise does not relate to this general discussion. It is because external noise levels are quite high on 160, 80, 40 and 20 meters that emphasis is seldom placed on a low receiver noise figure for those bands. However, as the operating frequency is increased from 15 meters up through the microwave spectrum, the matter of receiver noise becomes a primary consideration. At these higher frequencies the receiver noise almost always exceeds that from external sources, especially at 2 meters and above.

Noise-Figure Measurements

Amateurs can use a thermal noise source for determining receiver noise figure. The resistance of the noise-generator output must match that of the receiver input, 50 ohms to 50 ohms, for example. Fig. 1 shows a setup for making these measurements. The first reading is taken with the noise generator turned off. The receiver audio gain is adjusted for a convenient noise reading in dB, as observed on the audio power meter. The noise generator is turned on next, and its output is increased until a convenient power ratio, expressed by N_2/N_1, is observed. The ratio N_2/N_1 is referred to as the "Y-factor" and this noise figure measurement technique is commonly called the Y-factor method. From the Y-factor and output power of the noise generator the noise figure can be calculated.

$$NF = ENR - 10 \log_{10} (Y - 1)$$

where:

NF = noise figure in dB
ENR = excess noise ratio of the noise generator in dB
Y = output noise ratio, N_2/N_1.

The excess noise ratio of the generator is

$$ENR = 10 \log_{10} \left(\frac{P_2}{P_1} - 1 \right)$$

where:

P_2 = noise power of the generator
P_1 = noise power from a resistor at 290 Kelvins.

If a thermal diode such as a 5722 tube is used as the noise source, and if the circuit is operated in the "temperature-limited

mode" (portion of the tube curve where saturation occurs, dependent upon cathode temperature and plate voltage), the (excess) dB can be calculated by

$$(excess) dB = 10 \log (20R_d I_d)$$

where

R_d = the noise source output resistance
I_d = the diode current in amperes

Most manufacturers of amateur communications receivers rate the noise characteristics with respect to signal input. A common expression is S + Noise/Noise, or the signal-to-noise ratio. Usually the sensitivity is given as the number of μVs required for an S + Noise/Noise ratio of 10 dB. Sensitivity can also be expressed in terms of the minimum discernable signal (MDS) or noise floor of the receiver. (See chapter 16).

Typically, the noise figure for a good receiver operating below 30 MHz runs about 5 to 10 dB. Lower noise figures can be obtained, but they are of no real value because of the external noise arriving from the antenna. It is important to remember also that optimum noise figure in an rf amplifier does not always coincide with maximum stage gain, especially at vhf and higher. It is for this reason that actual noise measurements are important to peak performance.

Selectivity

Many amateurs regard the expression "selectivity" as equating to the ability of a receiver to separate signals. This is a fundamental truth, particularly with respect to i-f selectivity which has been established by means of high-Q filters (LC, crystal, monolithic or mechanical). But in a broader sense, *selectivity* can be employed to reject unwanted signal energy in any part of a receiver — the front end, i-f section, audio circuit or local-oscillator chain. Selectivity is a relative term, since the degree of bandwidth can vary from a few hertz to more than a megahertz, depending on the design objectives. Therefore, it is not uncommon to hear terms like "broadband filter" or "narrow-band filter."

The degree of selectivity is determined by the *bandwidth* of a filter network. The bandwidth is normally specified for the minus 3-dB points on the filter response curve; the frequencies where the filter output power is half the peak output power elsewhere in the passband. The difference in frequency between a minus

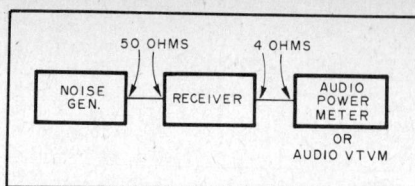

Fig. 1 — Block diagram of a noise measurement setup.

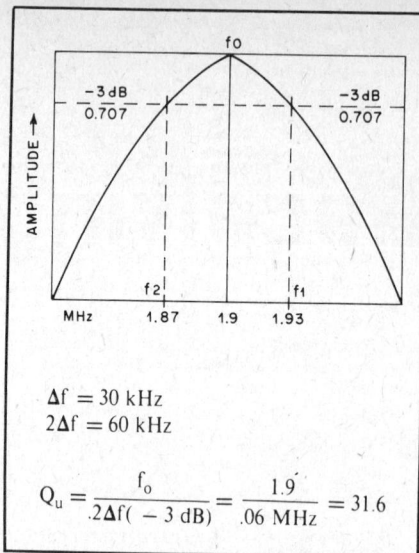

$$\Delta f = 30 \text{ kHz}$$
$$2\Delta f = 60 \text{ kHz}$$

$$Q_u = \frac{f_o}{.2\Delta f(-3 \text{ dB})} = \frac{1.9}{.06 \text{ MHz}} = 31.6$$

Fig. 2 — A curve and equation for determining the unloaded Q of a tuned circuit.

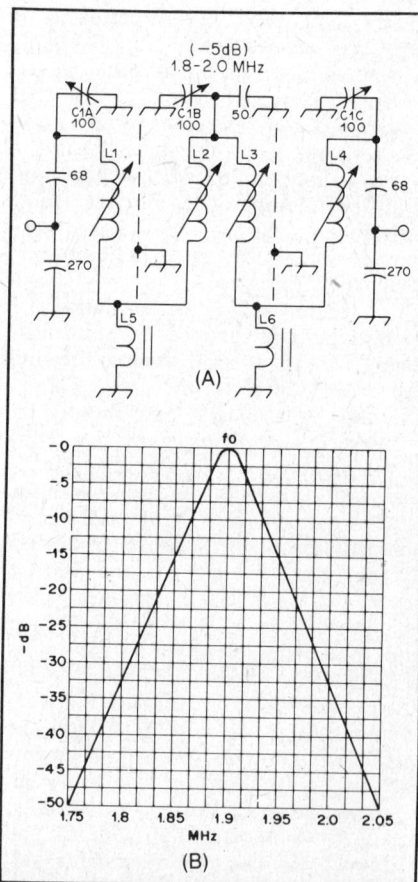

Fig. 3 — A tunable Cohn type of filter is shown at A. L5 and L6 are the bottom-coupling inductors (1.45 μH). L1 and L4 are 70 μH and L2, L3 are 140 μH. A response curve for the tunable filter is given at B.

3-dB point and the filter center frequency is known as Δf. The bandwidth of the filter then becomes $2 \Delta f$. Fig. 2 illustrates this principle and shows how the unloaded Q of a tuned circuit or resonator relates to the bandwidth characteristic.

If a tuned circuit is used as a filter, the

higher its loaded Q, the greater the selectivity. To make the skirts of the response curve steeper, several high-Q resonators can be used in cascade. This aids the selectivity by providing greater rejection of signals close in frequency to the desired one. The desirable effect of cascaded filter sections can be seen in Fig. 3. The circuit is that of a tunable Cohn type of three-pole filter for use in the front end of a 160-meter receiver. The response curve is included to illustrate the selectivity obtained.

An ideal receiver with selectivity applied to various significant parts of the circuit might be structured something like this:

a) Selective front end for rejecting out-of-band signals to prevent overloading and spurious responses.

b) Selective i-f circuit (two i-f filters: one for 2.4-kHz ssb bandwidth and one for 400-Hz cw bandwidth).

c) RC active or passive audio filter for audio selectivity to reduce wideband noise and provide audio selectivity in the range from 400 to 2500 Hz (ssb), or a very narrow bandwidth, such as 650 to 750 Hz, for cw.

d) Selective circuits or filters in the local oscillator chain to reject all mixer injection energy other than the desired frequency.

This illustrates clearly that selectivity does not simply mean the ability of a receiver to separate one amateur signal from another that is nearby in frequency, or to "separate the stations." More specifically, it means that selectivity can be used to select one frequency or band of frequencies while rejecting others. Practical applications of selective circuit elements will be found later in the chapter.

Dynamic Range

Here is another term which seems to confuse some amateurs and even some receiver manufacturers. The confusion concerns true dynamic range (as treated briefly at the start of this chapter) and the agc control range in a receiver. That is, if a receiver agc circuit has the capability of controlling the overall receiver gain by some 100 dB from a no-signal to a large-signal condition, a misinformed individual might claim that the dynamic range of the receiver is 100 dB. A receiver with a true dynamic range of 100 dB would be a very fine piece of equipment, indeed!

Dynamic range relates specifically to the amplitude levels of multiple signals that can be accommodated during reception.[1] This is expressed as a numeric ratio, generally in decibels. The present state of the receiver art provides *optimum dynamic ranges* of up to 100 dB. This is the maximum dynamic range attainable when the distortion products are at the sen-

[1] Hayward, "Defining and Measuring Receiver Dynamic Range," *QST*, July, 1975.

sitivity limit of the receiver. Simply stated, dynamic range is the dB difference (or ratio) between the largest tolerable receiver input signal (without causing audible distortion products) and the minimum discernible signal (sensitivity).

Poor dynamic range can cause a host of receiving problems when strong signals appear within the front-end passband. Notable among the maladies is cross modulation of the desired signal. Another effect is desensitization of the receiver from a strong unwanted signal. Spurious signals may appear in the receiver tuning range when a strong signal is elsewhere in the band. This is caused by IMD products from the mixer. Clearly, strong signals cause undesired interference and distortion of the desired signal when a receiver's dynamic range is poor. Design features of importance to high dynamic range receivers will be appearing in the theory sections of this chapter. The four terms which have been defined in this section were thus treated to enable the reader to better understand the material that follows.

Detection and Detectors

Detection (demodulation) is the process of extracting the signal information from a modulated carrier wave. When dealing with an a-m signal, detection involves only the rectification of the rf signal. During fm reception, the incoming signal must be converted to an a-m signal for detection.

Detector sensitivity is the ratio of desired detector output to the input. Detector linearity is a measure of the ability of the detector to reproduce the exact form of the modulation on the incoming signal. The resistance or impedance of the detector is the resistance or impedance it presents to the circuits it is connected to. The input resistance is important in receiver design, since if it is relatively low it means that the detector will consume power, and this power must be furnished by the preceding stage. The signal-handling capability means the ability to accept signals of a specified amplitude without overloading or distortion.

Diode Detectors

The simplest detector for a-m is the diode. A germanium or silicon crystal is an imperfect form of diode (a small current can usually pass in the reverse direction), but the principle of detection in a semiconductor diode is similar to that in a vacuum-tube diode.

Circuits for both half-wave and full-wave diodes are given in Fig. 4. The simplified half-wave circuit at Fig. 4A includes the rf tuned circuit, L2C1, a coupling coil, L1, from which the rf energy is fed to L2C1, and the diode, D1, with its load resistance, R1, and bypass capacitor, C2.

The progress of the signal through the

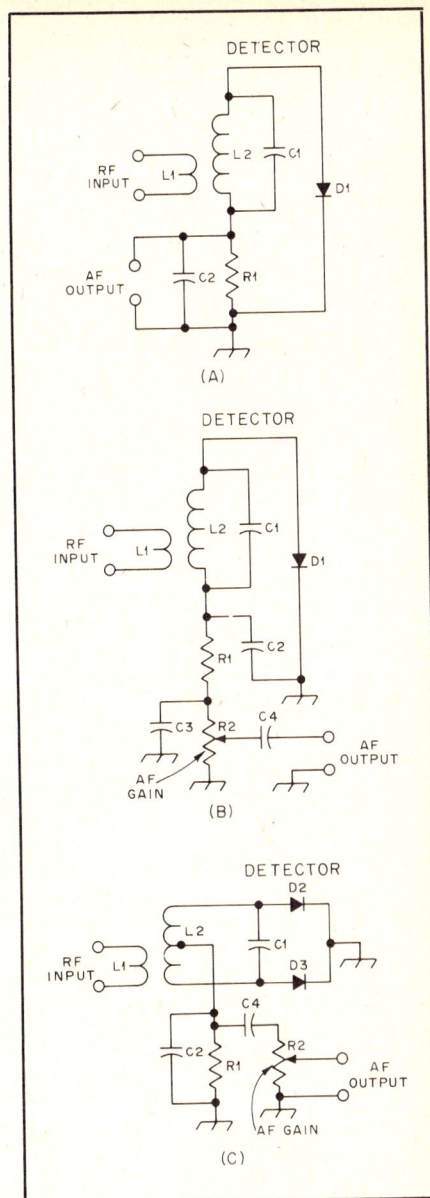

Fig. 4 — Simplified and practical diode detector circuits. A, the elementary half-wave diode detector; B, a practical circuit with rf filtering and audio output coupling; C, full-wave diode detector, with output coupling indicated. The circuit, L2C1, is tuned to the signal frequency; typical values for C2 and R1 in A and C are 250 pF and 250 kΩ, respectively; in B, C2 and C3 are 100 pF each; R1, 50 kΩ; and R2 250 kΩ. C4 is 0.1 µF.

detector or rectifier is shown in Fig. 5. A typical modulated signal as it exists in the tuned circuit is shown at A. When this signal is applied to the rectifier, current will flow only during the part of the rf cycle when the anode is positive with respect to cathode, so that the output of the rectifier consists of half-cycles of rf. These current pulses flow in the load circuit comprised of R1 and C2, the resistance of R1 and the capacitance of C2 being so proportioned that C2 charges to the peak value of the rectified voltage on each pulse and retains enough charge between pulses so that the voltage across R1 is smoothed out, as shown in C. C2

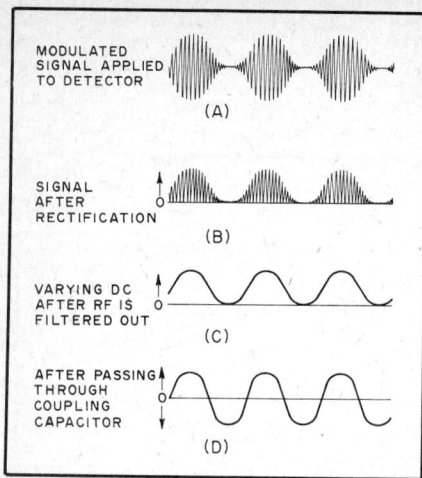

Fig. 5 — Illustrations of the detection process.

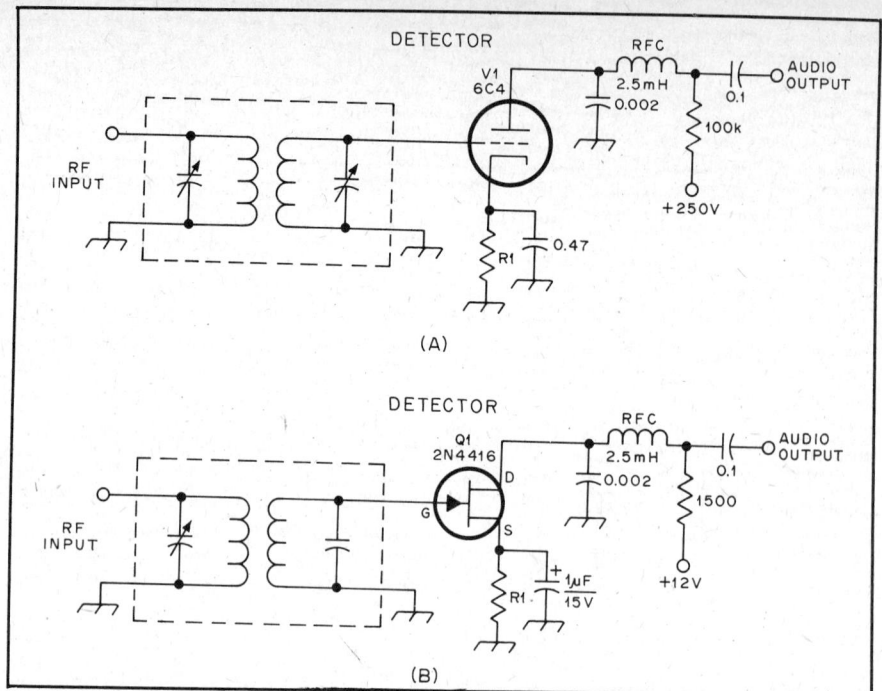

Fig. 6 — Plate-detection circuits. In each example the input circuit is tuned to the signal frequency. Typical R1 values for the tube circuit at A are 1000 to 5600 ohms. For the FET circuit at B, R1 is on the order of 100 to 3900 ohms.

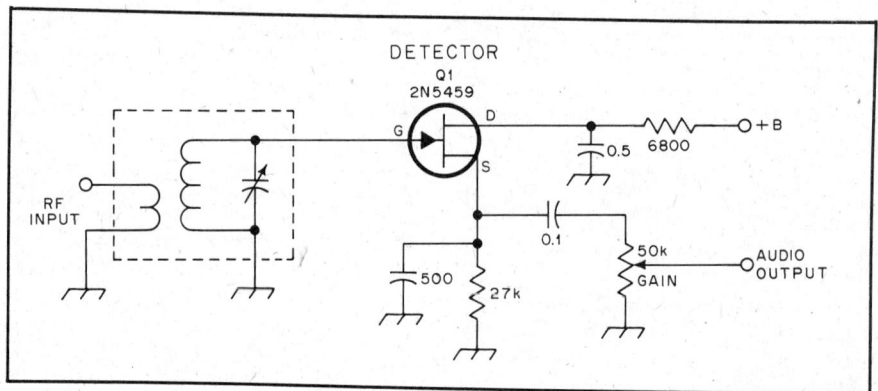

Fig. 7 — An infinite-impedance detector.

thus acts as a filter for the radio-frequency component of the output of the rectifier, leaving a dc component that varies in the same way as the modulation on the original signal. When this varying dc voltage is applied to a following amplifier through a coupling capacitor (C4 in Fig. 4), only the *variations* in voltage are transferred, so that the final output signal is ac, as shown in D.

In the circuit at 4B, R1 and C2 have been divided for the purpose of providing a more effective filter for rf. It is important to prevent the appearance of any rf voltage in the output of the detector, because it may cause overloading of a succeeding amplifier stage. The audio-frequency variations can be transferred to another circuit through a coupling capacitor, C4. R2 is usually a "potentiometer" so that the audio volume can be adjusted to a desired level.

Coupling from the potentiometer (volume control) through a capacitor also avoids any flow of dc through the moving contact of the control. The flow of dc through a high-resistance volume control often tends to make the control noisy (scratchy) after a short while.

The full-wave diode circuit at Fig. 4C differs in operation from the half-wave circuit only in that both halves of the rf cycle are utilized. The full-wave circuit has the advantage that rf filtering is easier than in the half-wave circuit. As a result, less attenuation of the higher audio frequencies will be obtained for any given degree of rf filtering.

The reactance of C2 must be small compared to the resistance of R1 at the radio frequency being rectified, but at audio frequencies must be relatively large compared to R1. If the capacitance of C2 is too large, response at the higher audio frequencies will be lowered.

Compared with most other detectors, the gain of the diode is low, normally running around 0.8 in audio work. Since the diode consumes power, the Q of the

tuned circuit is reduced, bringing about a reduction in selectivity. The loading effect of the diode is close to one half the load resistance. The detector linearity is good, and the signal-handling capability is high.

Plate Detectors

The plate detector is arranged so that rectification of the rf signal takes place in the plate circuit of the tube or the drain of an FET. Sufficient negative bias is applied to the grid to bring the plate current nearly to the cutoff point, so that application of a signal to the grid circuit causes an increase in average plate current. The average plate current follows the changes in the signal in a fashion similar to the rectified current in a diode detector.

In general, transformer coupling from the plate circuit of a plate detector is not satisfactory, because the plate impedance of any tube is very high when the bias is

near the plate-current cutoff point. The same is true of a JFET or MOSFET. Impedance coupling may be used in place of the resistance coupling shown in Fig.6. Usually 100 henrys or more of inductance is required.

The plate detector is more sensitive than the diode because there is some amplifying action in the tube or transistor. It will handle large signals, but is not so tolerant in this respect as the diode. Linearity, with the self-biased circuits shown, is good. Up to the overload point the detector takes no power from the tuned circuit, and so does not affect its Q and selectivity.

Infinite-Impedance Detector

The circuit of Fig. 7 combines the high signal-handling capabilities of the diode detector with the low distortion and, like the plate detector, does not load the tuned circuit it connects to. The circuit resembles that of the plate detector, except that the load resistance, 27-kΩ, is connected between source and ground and thus is common on both gate and drain circuits, giving negative feedback for the audio frequencies. The source resistor is bypassed for rf but not for audio, while the drain circuit is bypassed to ground for both audio and radio frequencies. An rf filter can be connected between the source and the output coupling capacitor to eliminate any rf that might otherwise appear in the output.

The drain current is very low with no signal, increasing with signal as in the case of the plate detector. The voltage drop across the source resistor consequently increases with signal. Because of this and the large initial drop across this resistor, the gate usually cannot be driven positive, with respect to the source, by the signal.

Product Detectors

A *product detector* is similar in function to a balanced or *product modulator*. It is also similar to a mixer. In fact, the latter is sometimes called a "first detector" in a receiver circuit. Product detectors are used principally for ssb and cw signal detection. Essentially, it is a detector whose output is approximately equal to the product of the beat-frequency oscillator (BFO) and the rf signals applied to it. Output from the product detector is at audio frequency. Some rf filtering is necessary at the detector output to prevent unwanted i-f or BFO voltage from reaching the audio amplifier which follows the detector. LC or RC rf decoupling networks are satisfactory, and they need not be elaborate. Fig. 9 illustrates this type of filtering.

Diode Product Detectors

The product detectors shown in Fig. 8 are called "passive." The term means that the devices used do not require an operating voltage. Active devices (transistors, ICs and tubes) do require an operating voltage. Passive mixers and detectors exhibit a *conversion loss,* whereas active detectors can provide *conversion gain.* Passive detectors usually require a substantially greater level of BFO injection voltage than is the case with active detectors. Therefore, the primary drawbacks to the use of diodes in these circuits are the loss in gain and the high injection level required. A typical conversion loss for a two-diode detector (Fig. 8A) is 5 dB. The four-diode detectors have a loss of approximately 8 dB. The BFO injection level for each of the diode detectors shown in Fig. 8 is +13 dBm, or 20 mW. Since the terminal impedance of the detector is roughly 50 ohms, an rms BFO voltage of 1, or a pk-pk voltage of 2.8 is required.

The advantages to the use of diodes in a product detector are circuit simplicity, low cost, broadband characteristics, low

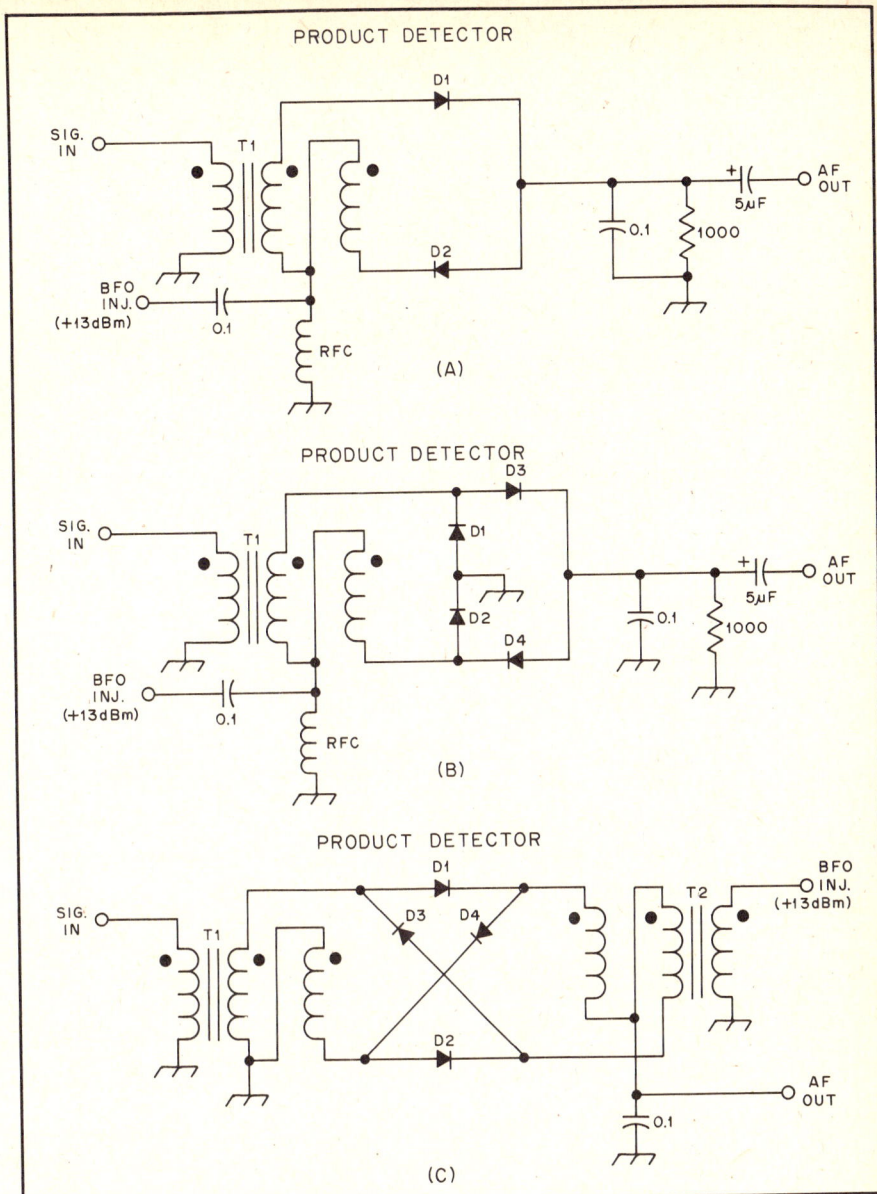

Fig. 8 — Examples of diode product detectors. Singly balanced types are shown at A and B. A doubly balanced version is illustrated at C.

noise figure and good port-to-port signal isolation. This type of detector is excellent at the input of a direct-conversion receiver (to be treated later in the chapter).

The transformers shown in the circuits of Fig. 8 are broadband, toroidal-wound types. The black dots near the windings of T1 and T2 indicate the phasing required. The core material is ferrite and the windings are trifilar. Core permeability can be 950 for most applications, although some designers use cores with less initial permeability. An Amidon FT-50-43 is entirely suitable as a transformer core for the circuits shown. Fifteen trifilar turns are ample for each transformer.

High-speed silicon switching diodes are satisfactory for use in the circuits of Fig. 8. They should be as closely matched as possible for forward and back resistance. Closely matched diodes can be had by

using a diode-array IC, such as the RCA CA3019 or 3039. Hot-carrier diodes are excellent for the circuits shown. Matched 1N914s are the choice of many amateur designers.

A singly balanced detector is seen at A in Fig. 8. An improved singly balanced detector is shown at B. Two diodes have been added to improve the circuit balance while presenting a more symmetrical load to the BFO. The result is better isolation between the BFO and i-f input ports.

Two broadband transformers are used to provide the doubly balanced detector of Fig. 8C. The advantage with this configuration is that all three ports are isolated from one another effectively.

Simple Active Product Detectors

Fig. 9 contains two examples of single-ended active detectors which em-

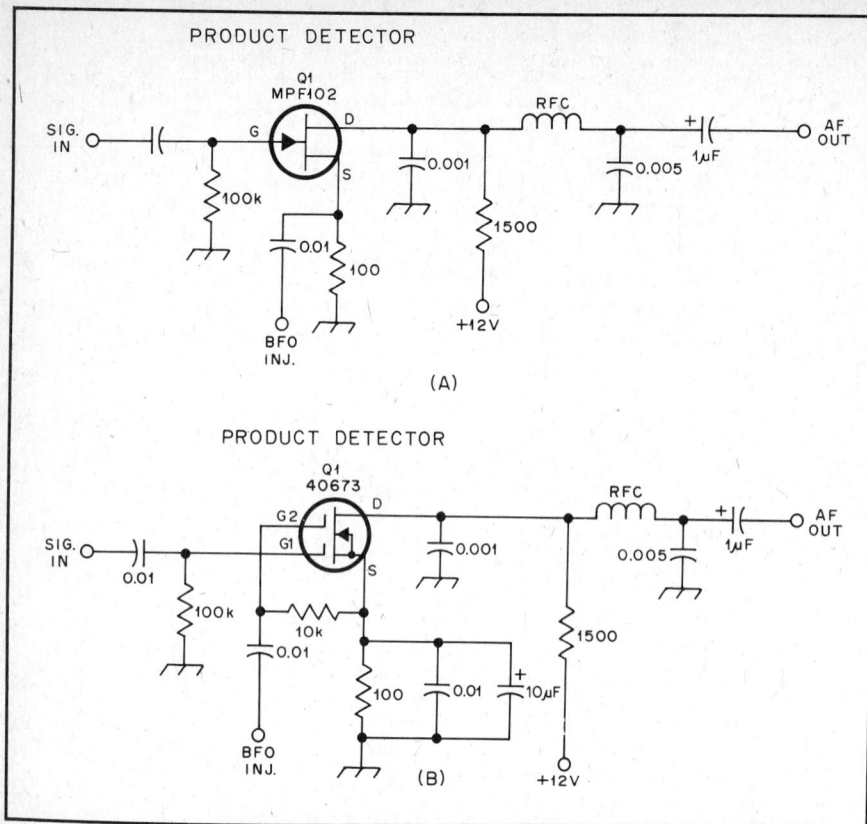

Fig. 9 — Active product detectors. A JFET example is provided at A and a dual-gate MOSFET type is at B.

ploy FETs. They are quite acceptable for use in simple receivers which do not require high performance characteristics. The circuit at A uses a JFET which has BFO injection voltage supplied across the source resistor. Because the source is not bypassed, instability can occur if the circuit is used as a mixer which has an i-f that is close to the signal frequency. This problem is not apt to become manifest when the output is at audio frequency. Slightly more injection power is needed for circuit A than is necessary for the detector at B. An rms voltage of roughly 0.8 is typical (6.5 mW).

The detector of Fig. 9B operates in a similar fashion to that of A, but the BFO is injected on control gate no. 2. Approximately 1 volt rms is needed (0.1 mW). FETs with proper injection levels and moderate signal-input amounts have excellent IMD characteristics. Generally, they are preferred to single-ended, bipolar-transistor detectors. The circuits at A and B contain rf chokes and bypass capacitors in the drain leads to minimize the transfer of BFO energy to the succeeding audio stage. The bypass capacitors are useful also for rolling off the unwanted high-frequency audio components.

Active Balanced Product Detectors

Examples of active IC product de-tectors are given in Fig. 10. A singly balanced version is shown at A. It uses an RCA differential-pair IC. Except for the conversion gain it provides, it performs similarly to the singly balanced diode detector of Fig. 8B. Doubly balanced active detectors are seen at B and C of Fig. 10. These ICs contain two sets of differential amplifiers each. The "diff amps" are cross-connected in the examples shown to obtain doubly balanced circuits. The virtues of these detectors are similar to the equivalent four-diode types, but they exhibit several dB of conversion gain. The MC1496G is made by Motorola and the CA3102E is an RCA device.

FM Detectors

The first type of fm detector to gain popularity was the frequency discriminator. The characteristic of such a detector is shown in Fig. 11. When the fm signal has no modulation, and the carrier is at point zero, the detector has no output. When audio input to the fm transmitter swings the signal higher in frequency, the rectified output increases in the positive direction. Over a range where the discrimination is linear (shown as the straight portion of the line), the conversion of fm to a-m which is taking place will be linear.

A practical discriminator circuit is shown in Fig. 12A. The fm signal is converted to a-m by means of transformer T1. The voltage induced in the T1 secondary is 90 degrees out of phase with the current in the primary. The primary signal is introduced through a center tap on the secondary, coupled through a capacitor. The secondary voltages combine on each side of the center tap so that the voltage on one side leads the primary signal while the other side lags by the same amount. When rectified, these two voltages are equal and of opposite polarity, resulting in zero-voltage output. A shift in input frequency causes a shift in the phase of the voltage components that result in an increase of output amplitude on one side of the secondary, and a corresponding decrease on the other side. The differences in the two changing voltages, after rectification, constitute the audio output.

RCA developed a circuit that has now become standard in entertainment radios and which eliminated the need for a preceding limiter stage. Known as the ratio detector, this circuit is based on the idea of dividing a dc voltage into a ratio which is equal to the ratio of the amplitudes from either side of a dis-criminator transformer secondary. With a detector that responds only to ratios, the input signal may vary in strength over a wide range without causing a change in the level of output voltage — fm can be detected, but not a-m. In an actual ratio detector, Fig. 12B, the dc voltage required is developed across two load resistors, shunted by an electrolytic capacitor. Other differences include the two diodes, which are wired in series-aiding rather than series-opposing, as in the standard discriminator circuit. The recovered audio is taken from a tertiary winding which is tightly coupled to the primary of the transformer. Diode-load resistor values are selected to be lower (5000 ohms or less) than for the discriminator.

The sensitivity of the ratio detector is one half that of the discriminator. In general, however, the transformer design values for Q, primary-secondary coupling, and load will vary greatly, so the actual performance differences between these two types of fm detectors are usually not significant. Either circuit can provide excellent results.

A crystal discriminator is shown in Fig. 12C. This provides an adjustment-free discriminator by virtue of the quartz resonator. The components without as-signed values are selected to give the desired bandwidth.

TRF Receivers

Tuned-radio-frequency receivers have little value in Amateur Radio today, but in the early days they were suitable for the reception of spark and a-m signals. They consisted mainly of a couple of stages of selective rf amplification, an a-m type of

Fig. 11 — Characteristic of an fm discriminator.

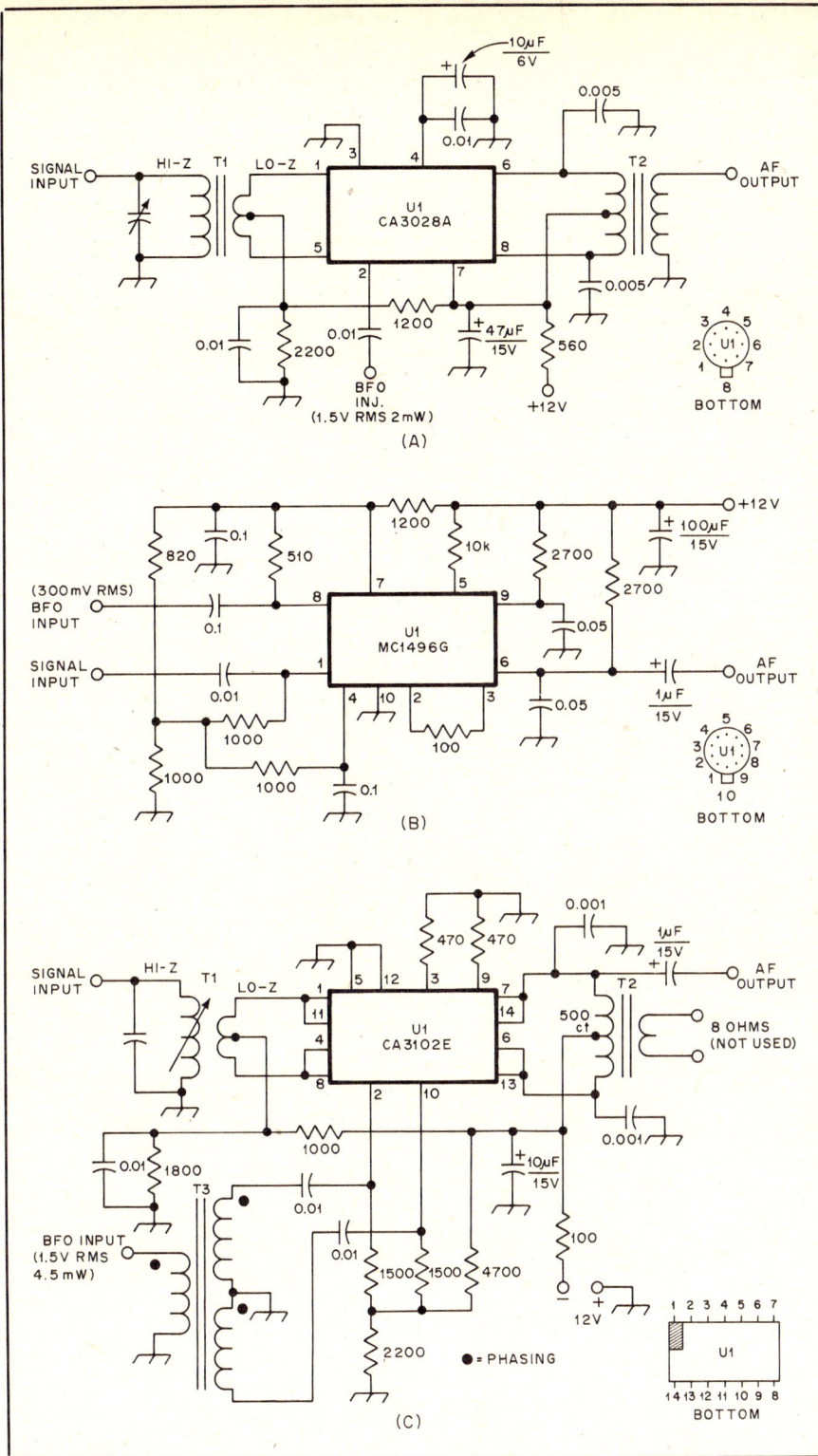

Fig. 10 — Examples of IC product detectors. At A is a singly balanced version, while those at B and C are doubly balanced.

uses for the TRF receiver are restricted mainly to reception of a-m broadcast signals, for hi-fi reception and for field-strength indicators of cw or a-m signals.

Superregenerative receivers were quite popular among vhf and uhf amateurs in the '30s, '40s and early '50s. The principle of operation was an oscillating detector which had its oscillation interrupted (quenched) by a low-frequency voltage slightly above the audible range (20 to 50 kHz being typical). Some superregenerative detectors employed a so-called self-quenching trait, brought about by means of an RC network of the appropriate time constant. The more esoteric "supergenny" or "rushbox" detectors used an outboard quench oscillator. This type of circuit was more sensitive than the straight regenerative detector, but was best suited for reception of a-m and wide-band fm signals. Because of the quenching action and frequency, the detector response was extremely broad, making it unsuitable for narrow-band signals versus audio recovery. High-Q input tuned circuits helped make them more selective, but a typical superregenerative receiver which used a tuned cavity at the detector input could accommodate only 10 1000-μV, 30-percent modulated a-m signals in a range from 144 to 148 MHz without signal overlap. These tests were performed in the ARRL laboratory with the 10 signals separated from one another by equal amounts.

A major problem associated with the use of regenerative and superregenerative receivers was oscillator (detector) radiation. The isolation between the detector and the antenna was extremely poor, even when an rf amplifier was employed ahead of the detector. In many instances the radiated energy could be heard for several miles, causing intense interference to other amateurs in the community.

Direct-Conversion Receivers

A more satisfactory type of simple receiver is called the *direct-conversion* or

detector and an audio amplifier. Variations were developed as *regenerative* and *superregenerative* receivers. The straight regenerative detector was simply a self-oscillating detector which provided increased sensitivity (similar in function to a product detector) and a beat note for cw reception. Amplitude-modulated signals could be copied, if they were loud, when the regeneration control was set for a non-oscillating condition. For weak-signal a-m reception the regeneration control was advanced to increase the detector sensitivity and the signal was tuned in at zero beat, thereby eliminating the heterodyne from the carrier. Present-day

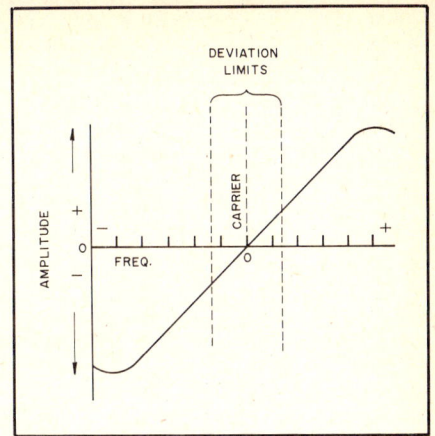

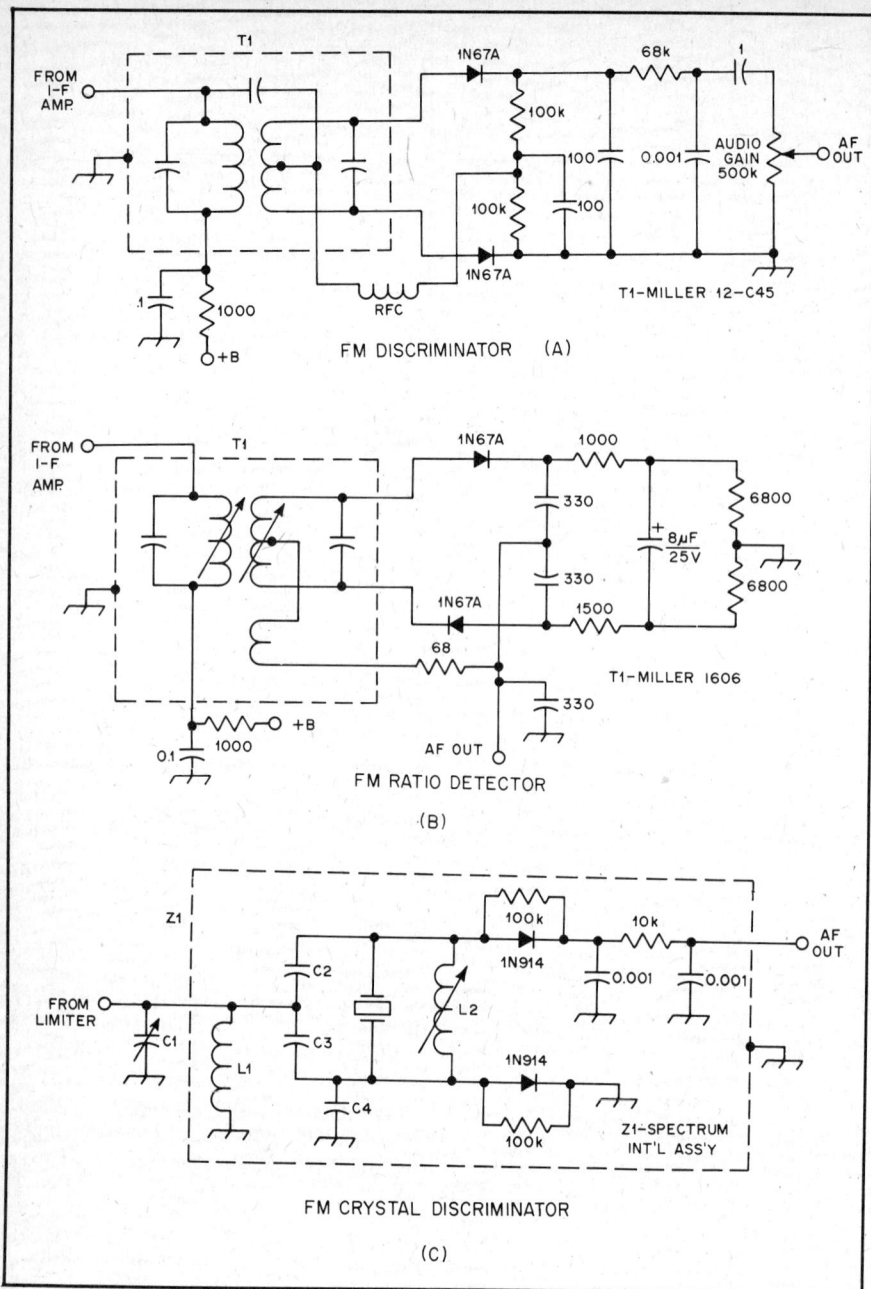

Fig. 12 — Popular types of fm detectors.

selectivity is provided by a tuned circuit. T1 is a broadband, trifilar-wound toroid transformer. It is tapped on the input tuned circuit at the approximate 50-ohm point. An rf filter is used after D1 and D2 to prevent LO energy from being passed on to the audio amplifier.

Fig. 13B illustrates an active singly balanced IC detector. The input impedance across pins 1 and 5 is roughly 1000 ohms. However, the secondary winding of the input tuned circuit can be made lower than 1000 ohms to reduce the signal amount to the detector. This will ensure improved dynamic range through a deliberate mismatch. Such a practice is useful when an rf amplifier precedes the detector. For maximum sensitivity when no rf amplifier is included, it is more practical to use a 1000-ohm transformation from the 50-ohm antenna (larger link at the detector input). An audio transformer is used at the detector output. The primary winding should have low dc resistance to provide dc balance between the collectors of the differential-amplifier pair in the IC. Alternatively, a center-tapped primary can be used. If this is done, pin 8 should be connected to one end of the winding and the B+ fed to the center tap. The impedance between pins 6 and 8 is approximately 8000 ohms.

In order to obtain ample headphone volume during reception of weak signals it is necessary to use an audio amplifier which has between 80 and 100 dB of gain. The first af amplifier should be a low-noise type, such as a JFET. The audio-gain control should follow the first audio amplifier. Selectivity for ssb and cw reception can be had by including a passive or RC active audio filter after the gain control. Fig. 13C contains a circuit which shows a typical direct-conversion receiver in its entirety. As was stated earlier, the detector is operating as a product detector rather than a mixer, and the VFO is serving as a *BFO*. The difference frequency between the incoming 7-MHz signal and the 7-MHz BFO injection voltage is at audio frequency (zero i-f). This is amplified by means of Q1, filtered through a passive LC audio network, then amplified by two 40-dB op-amp stages. It is possible to copy a-m signals with this type of receiver by tuning the signal in at zero beat.

Direct-conversion receivers of the type illustrated in Fig. 13 provide double-signal reception. That is, a cw beat note will appear either side of zero beat. This is useful during sideband reception, wherein the upper sideband is received on one side of zero beat and the lower sideband will appear on the opposite side of zero beat. QRM will be greater, of course, with this kind of receiver because there is no rejection of the unwanted sideband. Some designers have contrived elaborate circuits which, by means of phasing networks, provide single-signal reception. Unfor-

synchrodyne type. Although there is a distinct possibility of signal radiation, it is considerably lower in level than with regenerative receivers. This results from better isolation between the antenna and the source of the oscillation. A modern direct-conversion receiver uses a separate oscillator and a balanced or doubly balanced detector. Both features help to reduce unwanted radiation.

The detection stage of this receiver is actually a product detector that operates at the desired signal frequency. The product-detector circuits described earlier in the chapter are suitable in this kind of receiver. A tuned rf amplifier is useful ahead of the detector at 14 MHz and higher, but it is seldom necessary from 160 through 40 meters. This is because the atmospheric and man-made noise from the resonant antenna usually exceeds that of the detector below 14 MHz. When no rf stage is used, it is desirable to include a tuned network ahead of the detector.

Fig. 13 shows typical front ends for direct-conversion receivers. One circuit (A) employs a passive detector. The other (B) contains an active detector. The latter is desirable in the interest of increased gain.

The circuit of Fig. 13A shows a singly balanced passive detector. Front-end

tunately, the circuit becomes nearly as complex as that of superheterodyne. The benefits obtained are probably not worth the effort.

Direct-conversion receivers are not especially suitable above 14 MHz because it is difficult to secure adequate BFO stability at so high a frequency. A practical solution to the problem is the employment of a heterodyne BFO chain in which a 5-MHz VFO is heterodyned with crystal-controlled oscillators. Direct-conversion receivers are ideal for use in simple transceivers because the BFO can be used also as the frequency source for the transmitter, provided the appropriate frequency offset is included between transmit and receive to permit copy of ssb and cw signals without readjusting the BFO.

Characteristic Faults

A major difficulty connected with direct-conversion receivers is *microphonics*. The effect is noted when the operating receiver is bumped or moved. An annoying ringing sound is heard in the receiver output until the mechanical vibration ceases. The simple act of peaking the front end or adjusting the volume control can set off a microphonic response. This trait is caused by the extremely high gain needed in the audio amplifier. Slight electrical noises in the receiver front end, caused by small vibrations, are amplified many times by the audio channel. They are quite loud by the time they reach the speaker or phones. The best precautionary measure to reduce microphonics is to make all of the detector and BFO circuit leads and components as rigid as possible. Addition of an rf amplifier stage ahead of the detector will also help by virtue of increasing the front-end gain. This reduces the amount of audio gain needed to copy a signal, thereby diminishing the loudness of the microphonics.

The other common problem inherent in direct-conversion receivers is hum (Fig. 14). The fault is most pronounced when an ac type of power supply is used. The hum becomes progressively worse as the operating frequency is increased. For the most part, this is caused by ac ground loops in the system. The ac modulates the BFO voltage, and the hum-modulated energy is introduced in the detector directly, as well as being radiated and picked up by the antenna. The most practical steps toward a cure are to affix an effective earth ground to the receiver chassis and power supply, use a battery power supply, and feed the antenna with coaxial cable. End-fed wire antennas increase the possibility of hum if they are voltage fed (high impedance at the receiver end). Decoupling of the ac power supply leads (dc leads to the receiver) is also an effective preventive measure for hum. This was explained by Hayward,

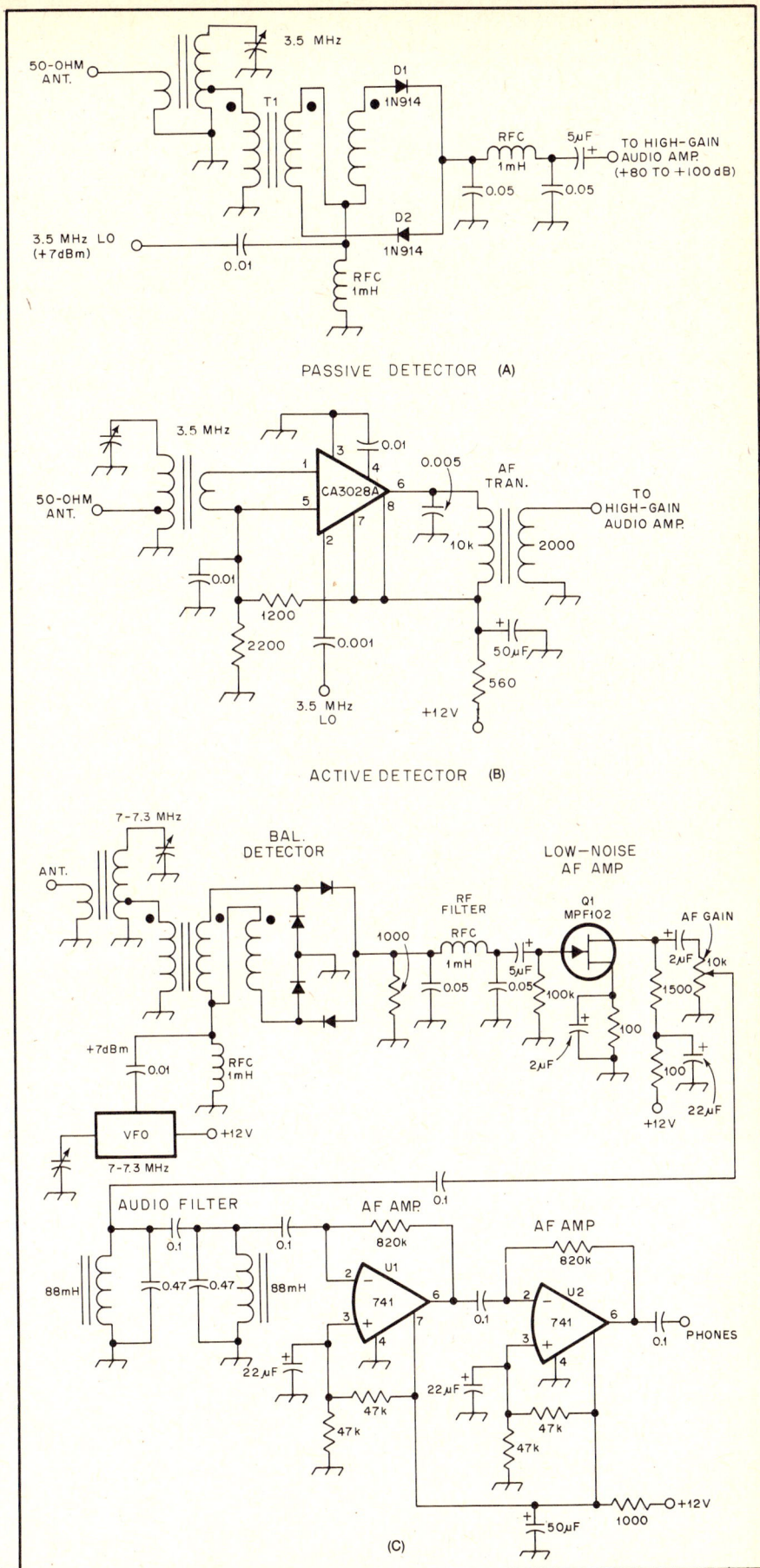

Fig. 13 — Typical detectors which can be used in the front ends of direct-conversion receivers. A passive diode detector is shown in A. The active detector (B) provides considerable conversion gain. An example of a practical direct-conversion receiver for 40 meters is shown in C.

W7ZOI, in July 1977 *QST,* page 51. The cure is to add a toroidal decoupling choke, bifilar wound, in the plus and minus dc leads from the power supply. This will prevent high-impedance rf paths between the power supply and receiver. The effect is to prevent BFO energy from entering the power supply, being modulated by the rectifier diodes and being reradiated by the ac line. This form of buzz is called "common-mode hum."

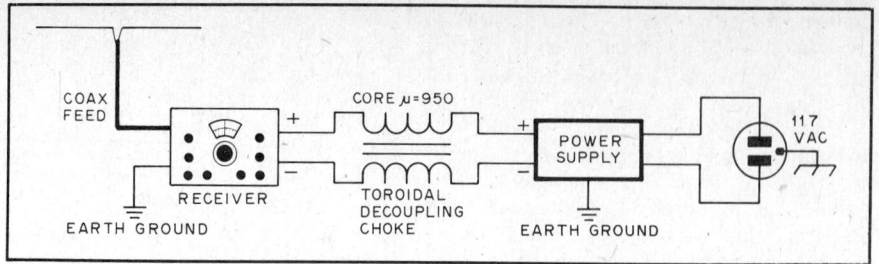

Fig. 14 — A method for eliminating common-mode hum in a direct-conversion receiver, as described by W7ZOI.

Superheterodyne Receivers

Nearly all of the present-day communications receivers are structured as *superheterodyne* types. Fig. 15 shows a simple block diagram of a single-conversion superheterodyne circuit. This basic design has been popular since the 1930s, and only a few general circuit enhancements have been introduced in recent years. Sophisticated versions of this type of receiver use various alternatives to the circuits indicated in the block diagram. The local oscillator, for example, might utilize a phase-locked loop or synthesizer type of LO chain rather than a straight VFO. Digital readout is used in some models in place of the more traditional analog readout method. Rf types of noise blankers (often very complex) are chosen by some designers in preference to simple shunt audio noise limiters. An assortment of techniques is being used to improve the overall selectivity of these receivers — elaborate i-f filtering, RC active or LC passive audio filters. However, the basic circuit concept remains unchanged. The advancement of greatest significance in recent years is the changeover from vacuum

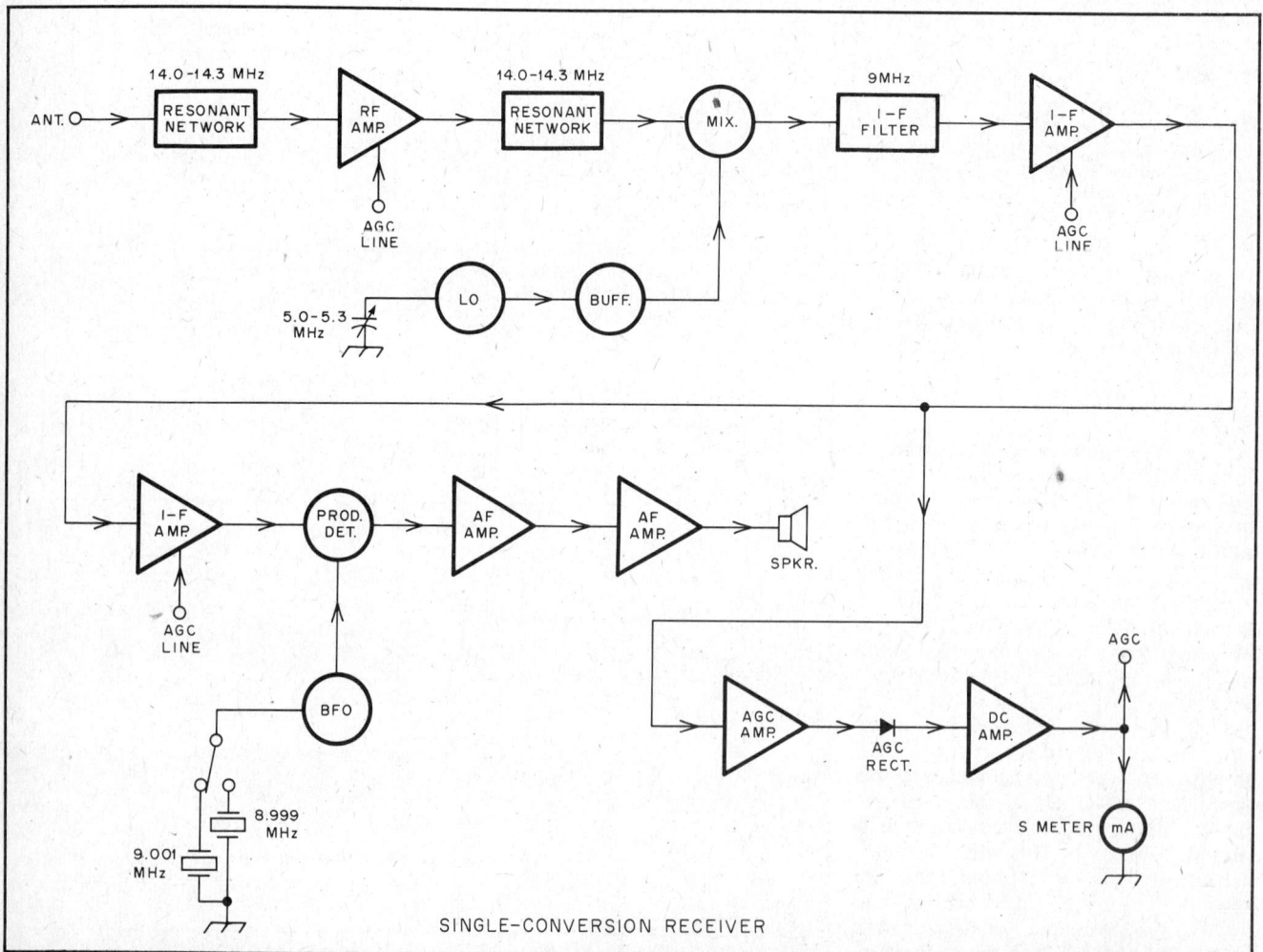

Fig. 15 — Block diagram of a single-conversion superheterodyne receiver for 20 meters. The arrows indicate the direction of signal and voltage components.

tubes to semiconductors. This has increased the life span of the of the equipment, improved overall efficiency, aided stability (reduced heating), and contributed to greater ruggedness and miniaturization.

Some manufacturers still produce double- or multiconversion superheterodyne receivers, but the circuits are similar to that of Fig. 15. Multiconversion receivers have a second mixer and LO chain for the purpose of making the second i-f lower than the first. This helps to increase the overall selectivity in some designs, but it often degrades the receiver dynamic range through the addition of a second mixer. Multiconversion receivers are more prone to spurious responses than is the case with single-conversion designs, owing to the additional oscillator and mixing frequencies involved. The "cleanest" performance is obtained from properly designed single-conversion receivers.

Circuit Function

In the example of Fig. 15 it is assumed that the receiver is adjusted to receive the 20-meter band. Front-end selectivity is provided by the resonant networks before and after the rf amplifier stage. This part of the receiver is often called the *preselector*, meaning that it affords a specific degree of front-end selectivity at the operating frequency. The rf amplifier increases the level of the signal from the antenna before it reaches the mixer. The amount of amplification is set by the designer, consistent with the overall circuit requirements (gain distribution). Generally, the gain will be from a few dB to as much as 25 dB.

When the incoming signal reaches the mixer it is heterodyned with the local-oscillator frequency to establish an i-f (intermediate frequency). The i-f can be the *sum* or the *difference* of the two frequencies. In the example given, the i-f is the difference frequency, or 9 MHz.

An i-f filter (crystal lattice or ceramic monolithic) is used after the mixer. At low intermediate frequencies (455 kHz and similar), mechanical filters are often used. The i-f filter sets the *overall receiver selectivity*. For ssb reception it is usually 2.1 kHz wide at the 3-dB points of the filter response curve. For cw reception it is between 200 and 500 Hz in bandwidth, depending upon the design objective. Wider filters are available for a-m reception.

Output from the i-f filter is increased by one or more amplifier stages. The overall gain of most i-f strips varies from 50 to 100 dB. The amount of signal gain is determined by the design objective, the type of amplifier devices used, and the number of gain stages.

The amplified i-f energy is routed to a product detector where it is mixed with the beat-frequency oscillator output. This produces an audio-frequency voltage

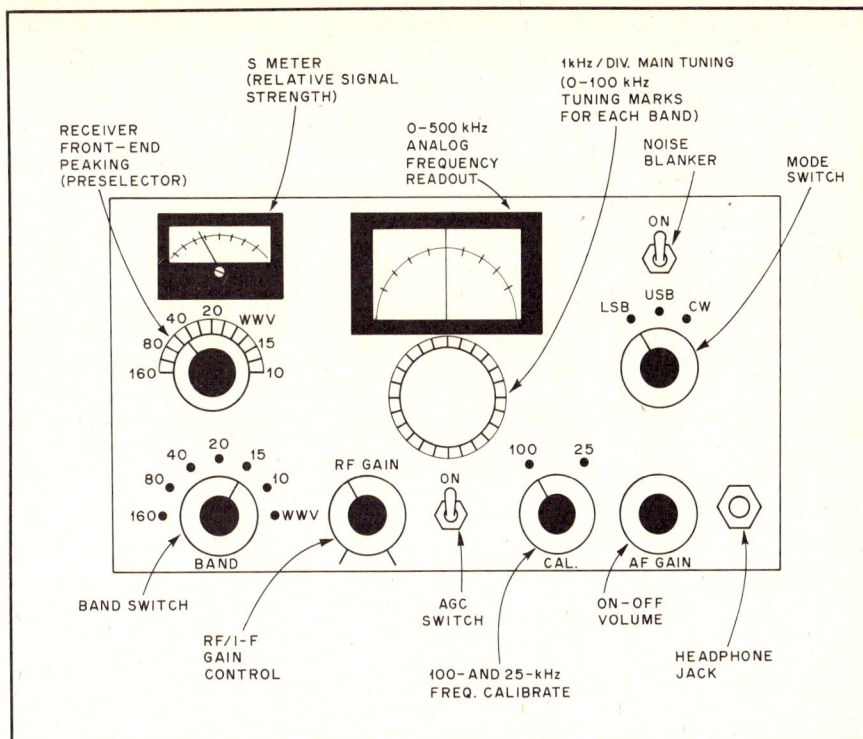

Fig. 16 — Layout of a typical modern amateur receiver.

which is amplified and fed to a speaker or headphones. The BFO is adjusted for reception of the upper or lower sideband, depending on which is appropriate at the time. In either case the BFO frequency is offset slightly from the center frequency of the i-f filter. For ssb reception it is usually offset approximately 1.5 kHz, in which case it falls on the slope of the i-f response curve. For cw reception the BFO is offset approximately 700 Hz from the i-f filter center frequency to produce a 700-Hz peak audio tone in the speaker. Other values of cw offset are common, but 700 Hz is preferred by many cw operators.

The overall gain of the receiver can be adjusted manually (by means of a panel-mounted control) or automatically. The latter is accomplished by means of an *agc* circuit. Energy can be sampled from the i-f amplifier output or the audio amplifier. Depending on the method used, the resultant agc is called *i-f derived* or *audio derived*. There are many arguments pro and con about which method is best. They shall not be considered here. In Fig. 15 the agc voltage is sampled from the i-f strip, amplified by the agc amplifier and then rectified to provide a dc control voltage. A dc amplifier is used to drive the agc terminals of the rf and i-f amplifiers. It can be used also to operate an S meter for observing relative signal-strength levels. When the incoming signal is weak the gain-controlled stages operate fully. As the incoming signal becomes stronger the agc circuit starts lowering the gain of the rf and i-f stages, thereby leveling the audio output at the speaker. A well-designed agc system will provide a uniform level of audio output (at a given af-gain control

setting) over an incoming signal-level variation of 100 dB. The net effect is to prevent overloading of some of the receiver stages and to protect the operator from the startling effect of tuning from a weak signal to an extremely loud one. Fig. 16 shows the front panel and controls for a typical amateur-band superheterodyne receiver.

Local Oscillators

A good communications receiver contains oscillators that operate in a stable and spectrally pure manner. Poor oscillator performance can spoil the best of receivers even though all other parts of the circuit are functioning in elegant fashion. Not only should the oscillator be stable with regard to short- and long-term drift, it should have minimum noise in the output (at least 80 dB below the peak value of the fundamental energy) and be reasonably free of spurious responses. Concerning the latter, it is not difficult to design an oscillator which has all harmonics attenuated by 60 or 70 dB. Another important characteristic of an oscillator is quick starting when operating voltage is applied.

Oscillator instability can result from a host of poor design practices. To improve the stability characteristics it is useful to observe the following:

1) Use regulated operating voltages (well filtered).

2) Avoid whenever possible the use of magnetic core material in the oscillator tank coil. Air-wound* or ceramic-form coils are best if they are rigid.

3) Use temperature-stable, fixed-value capacitors in the frequency-determining

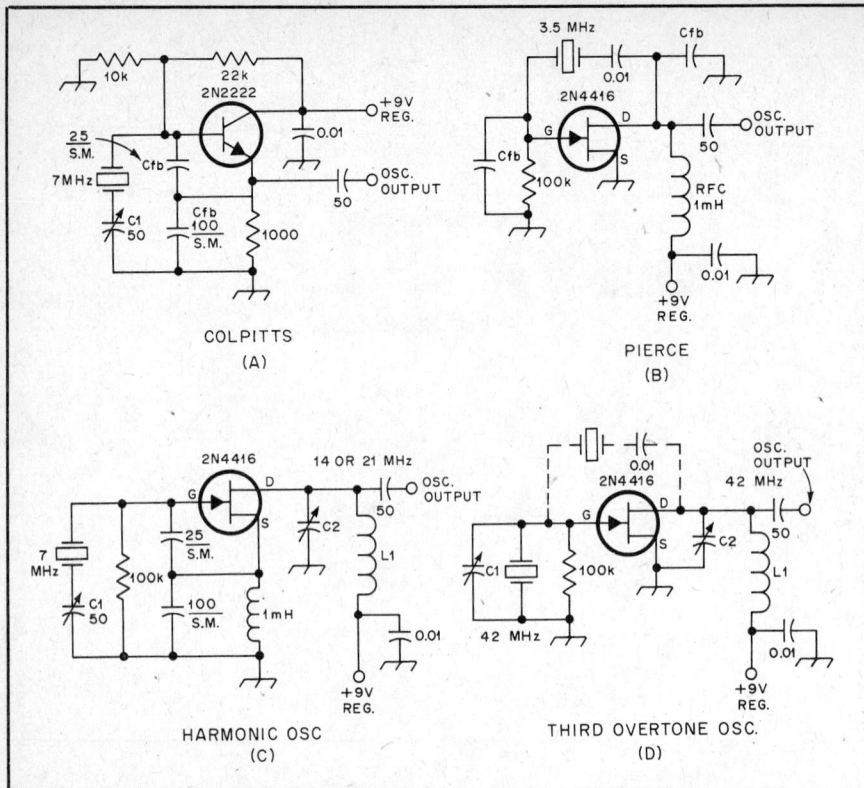

Fig. 17 — Practical examples of crystal-controlled oscillators.

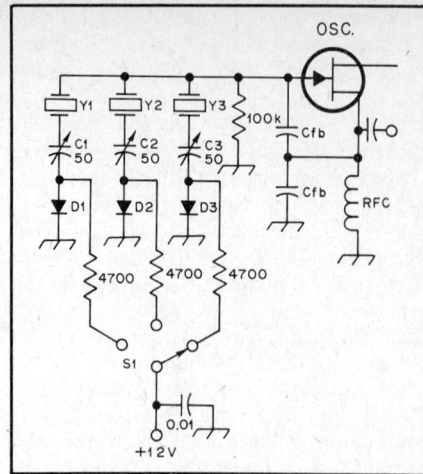

Fig. 18 — Method for changing crystals by means of diode switching.

part of the circuit. Polystyrene and silver-mica capacitors are recommended.

4) Ensure that all mechanical and electrical components are secured rigidly in their part of the circuit. This will lessen the chance for mechanical instability.

5) Build the oscillator on a firm, flex-free chassis.

6) When practical, enclose the oscillator in its own shield compartment and use rf filtering in the dc supply leads. Needless to say, the more constant the ambient temperature surrounding the oscillator, the greater will be the frequency stability.

Precautions should be taken to ensure that the oscillator in a receiver looks into a constant load impedance. Even minute load changes will cause phase shifts which can affect the oscillator frequency. The effect is more pronounced with VFOs than it is with crystal-controlled oscillators. Because of these conditions it is good design practice to couple very lightly to the oscillator stage. The power level can be increased by adding one or more buffer/amplifiers before the oscillator signal is supplied to the mixer or detector.

Changes in operating voltage will result in frequency shifts. It is for this reason that regulated voltage is recommended for oscillators. Zener diodes are adequate for the purpose.

Magnetic cores, such as those in slug-tuned coils, change their properties with variations in ambient temperature, thereby causing inductance changes which can severely affect the oscillator fre-quency. Furthermore, mechanical instability can result if the slugs are not affixed securely in the coil forms. Toroidal inductors are similarly unsuitable for use in stable VFOs.

Oscillator noise can be held to an acceptable level by employing high-Q tuned circuits. The higher the tank Q, the narrower the bandwidth, and hence, the lower the noise output voltage. Excessive LO noise will have a serious effect on mixer performance.

High amounts of harmonic current in the LO-chain output can cause unwanted mixer injection. If the receiver front-end selectivity is not of high magnitude, spurious signals from outside the band of interest will be heard along with the desired ones. Harmonic energy can degrade the performance of some kinds of mixers, making it worthwhile to use suitable filtering at the LO-chain output.

Crystal Oscillators

Although there is a wide variety of crystal-controlled oscillator circuits that provide acceptable performance in amateur equipment, only a few of the popular ones will be highlighted here. In the circuits offered as illustrations, the feedback must be ample to assure quick starting of the oscillator. Some circuits function quite well without the addition of external feedback components (internal capacitance within the transistor or tube being adequate). Other circuits need external feedback capacitors. Poor-quality (sluggish) crystals generally require larger amounts of feedback to provide operation which is comparable with that of lively crystals. Some surplus crystals are sluggish, as can be the case with those which have been reground or etched for a different operating frequency. Therefore, some experimentation with feedback voltage may be necessary when optimizing a given circuit. As a rule of thumb it is necessary to use one-fourth of the oscillator output power as feedback power, to ensure oscillation.

Fig. 17 shows four common types of oscillator. The same circuits can be used with tubes by applying the appropriate operating voltages. C1 is included for adjusting the crystal to the frequency for which it has been ground. In circuits where considerable shunt capacitance is present (Fig. 17A and C) the trimmer is usually connected in series with the crystal. When there is minimal parallel capacitance (approximately 6 pF in the circuits at B and D, Fig. 17) the netting trimmer can be placed in parallel with the crystal. Whether a series or parallel trimmer is used will depend also on the type of crystal used (load capacitance and other factors).

Feedback capacitance (C_{fb}) for the circuit at B in Fig. 17 must be found experimentally. Generally, a value of 100 pF will suffice for operation from 3.5 to 20 MHz. As the operating frequency is lowered it may require additional capacitance. The drain rf choke should be self-resonant below operating frequency.

A third-overtone crystal is illustrated at Fig. 17D. Satisfactory operation can be had by inserting the crystal as shown by the dashed lines. This method is especially useful when low-activity crystals are used in the overtone circuit. However, C1 will have little effect if the crystal is connected from gate to drain, as shown. C2-L1 is adjusted slightly above the desired overtone frequency to ensure fast starting of the oscillator. The circuits shown in Fig. 17 can be used with dual-gate MOSFETs

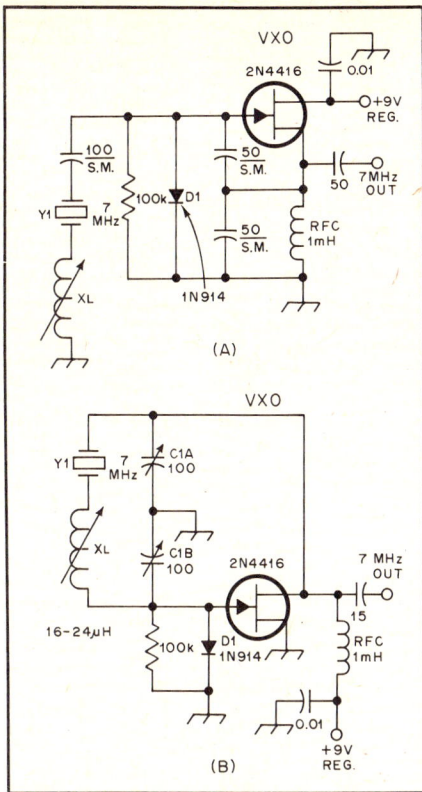

Fig. 19 — Circuits for two types of VXOs.

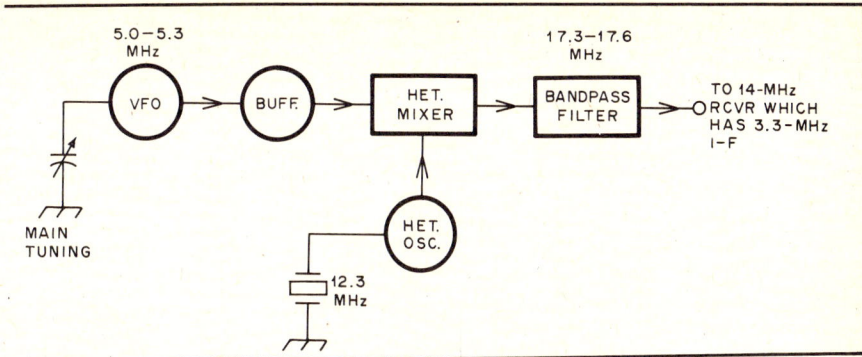

Fig. 20 — Technique for heterodyne frequency generation in a receiver.

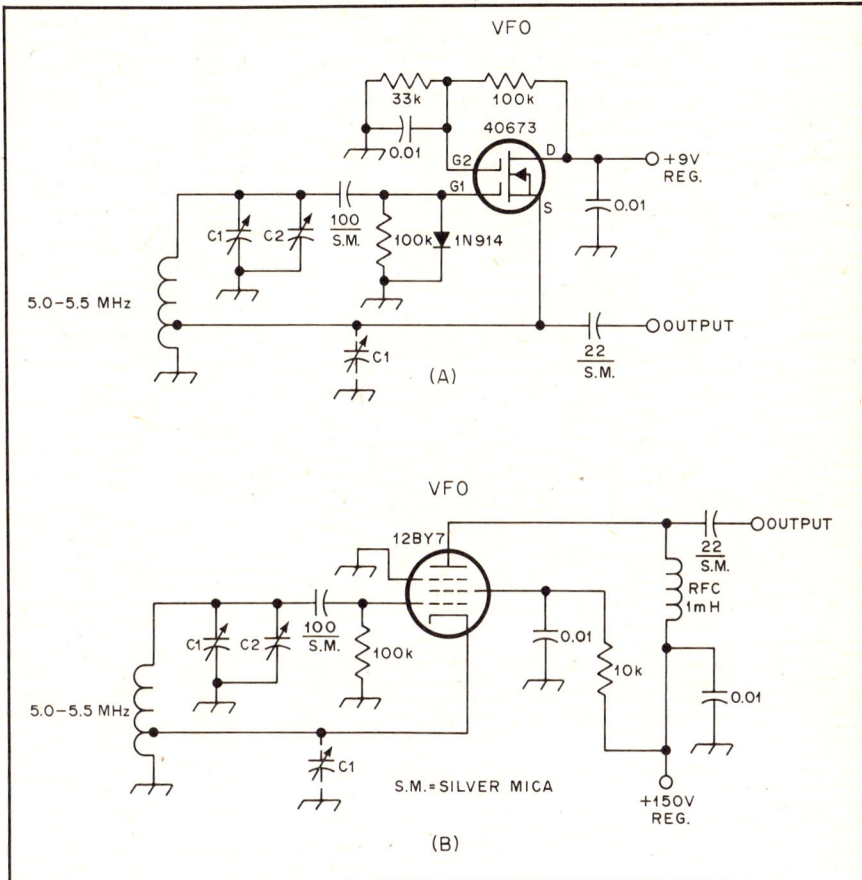

Fig. 21 — Examples of Hartley VFOs.

also, assuming that gate 2 is biased with a positive 3 to 4 volts.

A large number of crystals can be switched by means of silicon diodes in the manner shown at Fig. 18. The advantage of this technique is that the switching is done at dc, thereby permitting the control point to be a considerable distance from the oscillator circuit. D1 through D3, inclusive, are 1N914 diodes or equivalent.

VXO Circuits

Variable-frequency crystal oscillators (VXOs) are useful in place of conventional crystal oscillators when it is necessary to "rubber" the crystal frequency a few kHz. AT-cut crystals in HC-6/U type holders seem to provide the greatest frequency change when used in a VXO. To obtain maximum frequency shift it is vital to reduce stray circuit capacitance to the smallest possible amount. This calls for low-capacitance switches, low minimum-capacitance variable capacitors, and the avoidance of crystal sockets. The crystals should be spaced well away from nearby metal surfaces and circuit components to further reduce capacitance effects. The higher the crystal fundamental frequency, the greater the available frequency swing. For example, a 3.5-MHz crystal might be moved a total of 3 kHz, whereas a 7-MHz crystal could be shifted 10 kHz. Although some amateurs claim shifts as great as 50 kHz at 7 MHz, the circuit under those conditions is no longer operating as a true VXO.

Rather, it has gone into the VFO mode. In a situation of that kind the high-stability traits of a VXO are lost.

Fig. 19 contains a simple VXO circuit at A. By adjusting X_L the operator can shift the crystal frequency. The range will start at the frequency for which the crystal is cut and move lower. D1 in both circuits is included to stabilize the FET bias and reduce the transistor junction capacitance during the peak of the positive rf-voltage swing. It acts as a clamp, thereby limiting the transistor g_m at peak-voltage periods. This lowers the junction capacitance and provides greater VXO swing. D1 also reduces harmonic output from the VXO by restricting the nonlinear change in

transistor junction capacitance — a contributing factor to the generation of harmonic currents. Clamp diodes are used for the same purpose in conventional FET VFOs. The circuit of Fig. 19A will provide a swing of approximately 5 kHz at 7 MHz.

An improved type of VXO is presented at B in Fig. 19. Depending upon the exact characteristics of the crystal used at Y1, swings as great as 15 kHz are possible. X_L is set initially for a reactance value that will provide the maximum possible frequency shift when C1 is tuned through its range. The frequency shift should be only that which corresponds to true VXO control, even though greater range can be had after the circuit ceases to be a highly stable one. X_L is not adjusted again. A buffer stage should be used after either of the VXO circuits to prevent frequency pulling during load changes.

VXOs of this general type are useful in portable transmitters and receivers when full band coverage is sacrificed in exchange for stability and simplicity. Output from VXOs can be multiplied several times to provide LO energy for vhf and uhf receivers and transmitters. When that is done it is possible to realize 100 kHz or more of frequency change at 144 MHz.

VFO Circuits

Variable-frequency oscillators are similar in performance to the VXOs which were described in the foregoing text. The essential difference is that greater frequency coverage is possible, and no crystals are used. The practical upper frequency limits for good stability range between 7 and 10 MHz. For operation at higher frequencies it is better to employ a heterodyne type of VFO. This calls for a VFO operating at , say, 5 MHz. The VFO output is heterodyned in a mixer with energy from a crystal-controlled oscillator to provide a resultant sum or difference frequency at the desired LO-chain output frequency. A block diagram is given at Fig. 20 to illustrate the concept. Most modern receivers employ this style of local-oscillator circuit. The heterodyne oscillator has a crystal for each amateur band accommodated by the receiver. The crystals and appropriate bandpass filters are switched by means of a panel-mounted control. The band-pass filter (Fig. 20) is desirable in the interest of preventing 5- and 12.3-MHz energy from reaching the receiver mixer. A doubly balanced mixer is recommended if minimum unwanted energy is desired at the mixer output.

Some typical VFOs are shown in Figs. 21 and 22. A vacuum-tube Hartley is compared to a similar one which utilizes a dual-gate MOSFET in Fig. 21. The capacitor shown in dashed lines (C1) can be used in that part of the circuit rather than at the low end of the tank coil if greater bandspread is desired. C1 is the main-tuning capacitor and C2 is the padder for calibrating the oscillator. The coil tap is approximately 25 percent of the total number of turns for proper feedback.

A comparison is drawn in Fig. 22 between a bipolar transistor and JFET version of a series-tuned Colpitts oscillator (sometimes called a "series-tuned Clapp"). This type of oscillator can be made very stable by using polystyrene capacitors in the frequency-determining part of the circuit. Silver-mica capacitors are satisfactory if they have been graded out for temperature stability, but most run-of-the-mill, silver-mica units have unpredictable characteristics in this regard. Some are very stable, while others from the same lot may exhibit positive or negative drift characteristics. NPO ceramic capacitors are considerably better in this respect. These VFO circuits can be scaled to other operating frequencies by using the values shown to determine the reactances of the capacitors. This information will enable the designer to select approximate values in pF for other frequencies.

Fig. 23 illustrates a composite VFO which has a buffer stage that is followed by an amplifier. D1 can be included to provide the necessary frequency offset when switching the receiver from upper to lower ssb. This is necessary in order to eliminate the need to readjust the receiver calibration dial when changing sidebands. C1 is adjusted for the desired offset

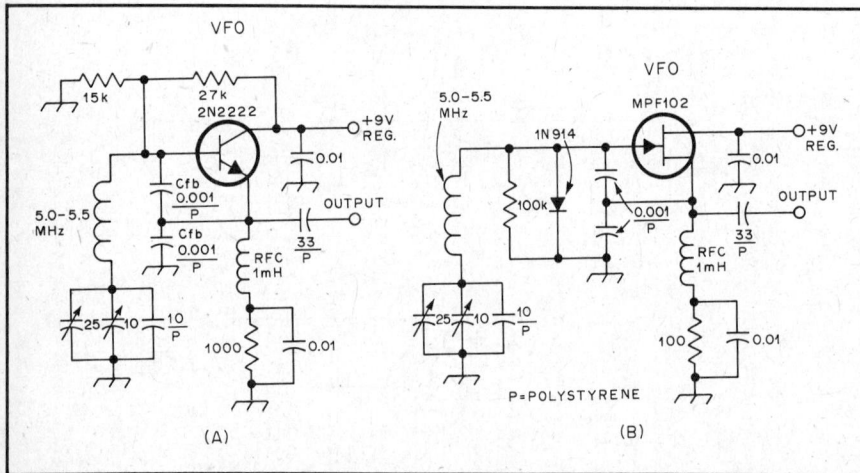

Fig. 22 — Colpitts VFOs. A bipolar transistor type is seen at A, while an FET version is given at B.

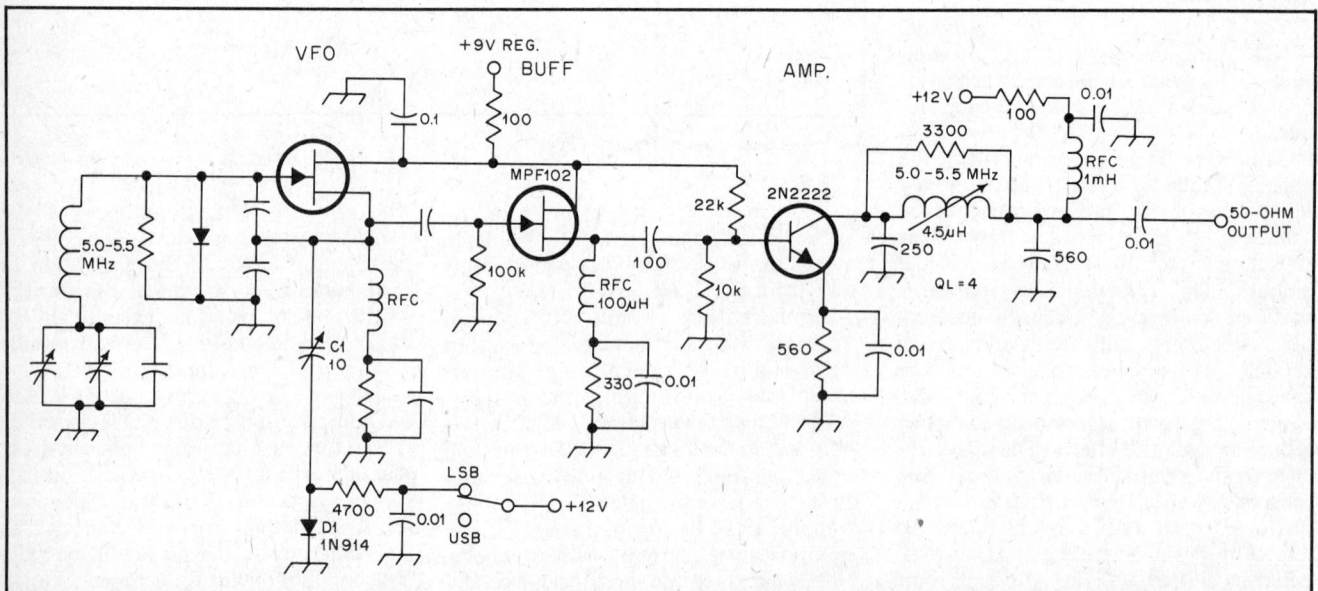

Fig. 23 — Suggested circuit for a stable series-tuned Colpitts VFO. Buffering follows the oscillator to increase the output level and provide load isolation.

amount. The pi-network output from the amplifier stage is designed to transform 500 ohms to 50 ohms. The low-impedance output is desirable in the interest of minimum frequency pulling from load changes. A 3300-ohm swamping resistor is used across the pi-network inductor to broadband the tuned circuit and to prevent any tendency toward instability when a high-impedance load is attached to the circuit. Long-term drift measurements with this type of circuit at the frequency specified indicated a maximum shift of 60 Hz over a three-hour period. Output was measured at approximately 1 volt rms across 50 ohms.

Receiver Front Ends

The designer has a number of options available when planning the input section of a receiver. The band-pass characteristics of the input tuned circuits are of considerable significance if strong out-of-band signals are to be rejected — an ideal design criterion. Most of the commercial receivers available to the amateur use tuned circuits which can be adjusted from the front panel of the equipment. The greater the network Q, the sharper the frequency response, and hence, the better the adjacent-frequency rejection. For a given network design the bandwidth doubles for each octave higher. That is, an 80-meter front-end network may have a 3-dB bandwidth of 100 kHz for a given Q and load factor. At 40 meters the same type of network would be 200-kHz wide at the 3-dB points of the response curve. It is for this reason that most receivers have a tunable front-end section (preselector). If fixed-tuned filters were used, at least two such filters would be necessary to cover from 3.5 to 4 MHz or 1.8 to 2.0 MHz, this would complicate the design and cost of the equipment.

Fig. 24 shows the two concepts just discussed. The circuit at A covers all of the 80-meter band, and if selective enough offers some in-band rejection. A pair of Butterworth band-pass filters might be used at FL1 and FL2 of Fig. 24 to cover all of the 80-meter band. A lot of additional components would be required, and the in-band rejection of unwanted signals would be less than in the case of circuit A. The principal advantage to the circuit at B is that front-panel peaking adjustments would not be necessary once the trimmers in the filters were set for the desired response. A similar tuned circuit for either example in Fig. 24 would be used between the rf amplifier and the mixer.

Regardless of the type of LC input network used, a built-in step attenuator is worth considering. It can be used for measuring changes in signal level, or to reduce overloading effects when strong signals appear in the receiver passband. Fig. 25 shows how this can be done. The example at A is suitable for simple

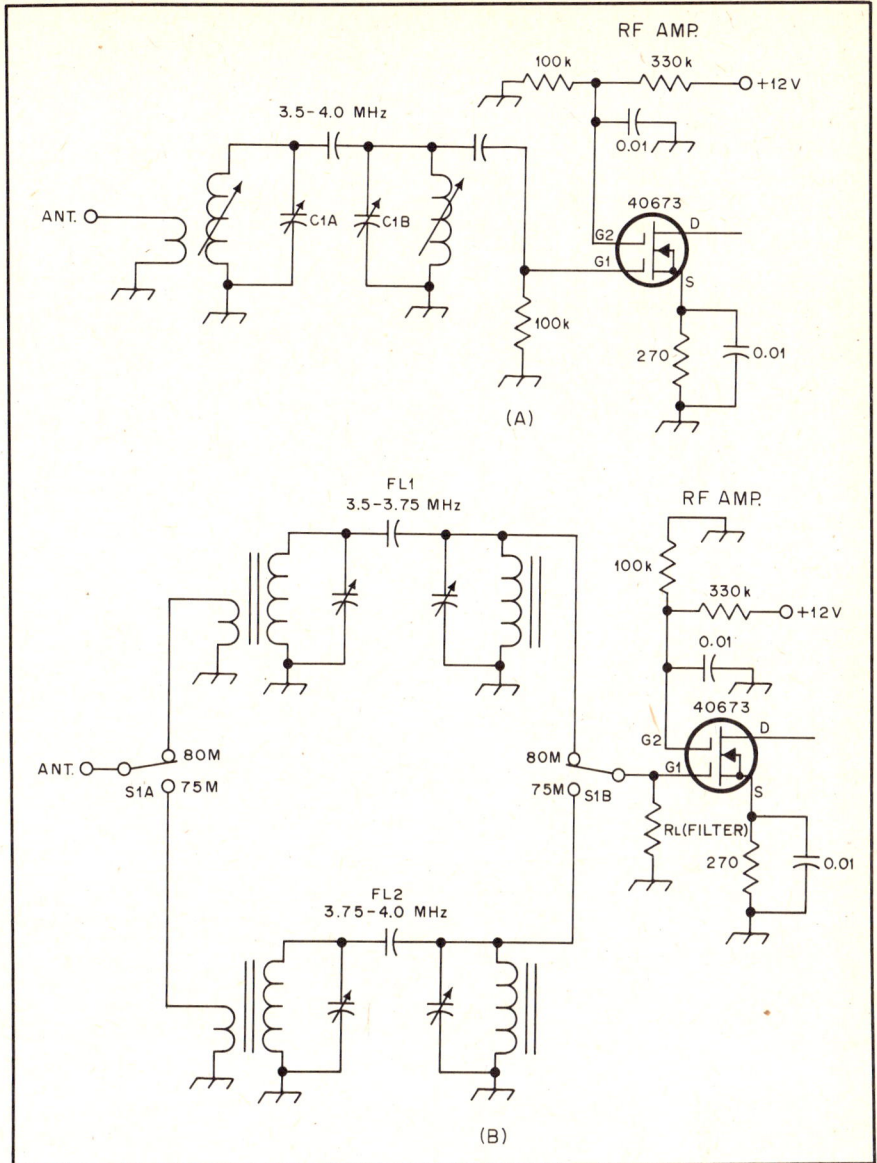

Fig. 24 — Method for selecting band-pass filters for 75 and 80 meters at the input to an rf amplifier.

receivers when calibration in dB is not a requisite, and when maintaining an impedance match between the tuned circuit and the antenna is not vital. The circuit at B is preferred because the pads are of 50 ohm impedance. In the circuit shown there are three steps available: 6, 12 and 18 dB, depending on how the switches are thrown. The resistance values specified are the closest standard ones to the actual values needed to provide precisely 6 or 12 dB of attenuation. For amateur work the accuracy is adequate. Front-end attenuators are useful when vhf converters are used ahead of the station receiver. If the converters have a significant amount of overall gain they can degrade the dynamic range of the main receiver when strong signals are present. The attenuators can be set to simulate a condition of unity gain through the converter, thereby aiding receiver dynamic range. It is worth remembering, however, that an attenuator

used at the *input* of a receiver when no converter is attached will degrade receiver sensitivity and noise figure. A receiver used frequently for antenna and received-signal dB measurements might have several 3-dB pads included, thereby providing greater resolution during measurements.

RF Amplifiers

It was implied earlier in the chapter that rf amplifiers are useful primarily to improve the receiver noise figure. When atmospheric and man-made noise levels exceed that of the mixer it should be possible to realize better dynamic range by not having an rf amplifier. The gain of the rf stage, when one is used, should be set for whatever level is needed to override the mixer noise. Sometimes that is only a few dB. A good low-noise active device should be employed as the rf amplifier in such instances. For hf-band work 40673s,

MPF102s and 2N4416s are good. Most modern receiving tubes specified for use as rf amplifiers are suitable also. For vhf work the Siliconix E300 and U310 FETs are excellent low-noise devices.

A well-designed receiver should not have agc applied to the rf amplifier. The best noise figure and rf-stage dynamic range will result when agc is not applied. This is because the agc voltage changes the operating characteristics of the rf amplifier from Class A to a less linear mode.

Fig. 26 shows some typical rf amplifiers for use in amateur receivers. Tube-type circuits have not been included because they are not much in vogue, nor do they offer any particular advantages over solid-state amplifiers.

The circuit at A in Fig. 26 is likely to be the least subject to self-oscillation of the four examples given. The common-gate hookup helps to ensure stability if the gate lead is kept as short as is physically possible. The gain from a common-gate amplifier of this type is lower than that of a common-source amplifier. However, gains up to 15 dB are entirely typical. The drain of the FET need not be tapped down on the drain coil, but if it is there

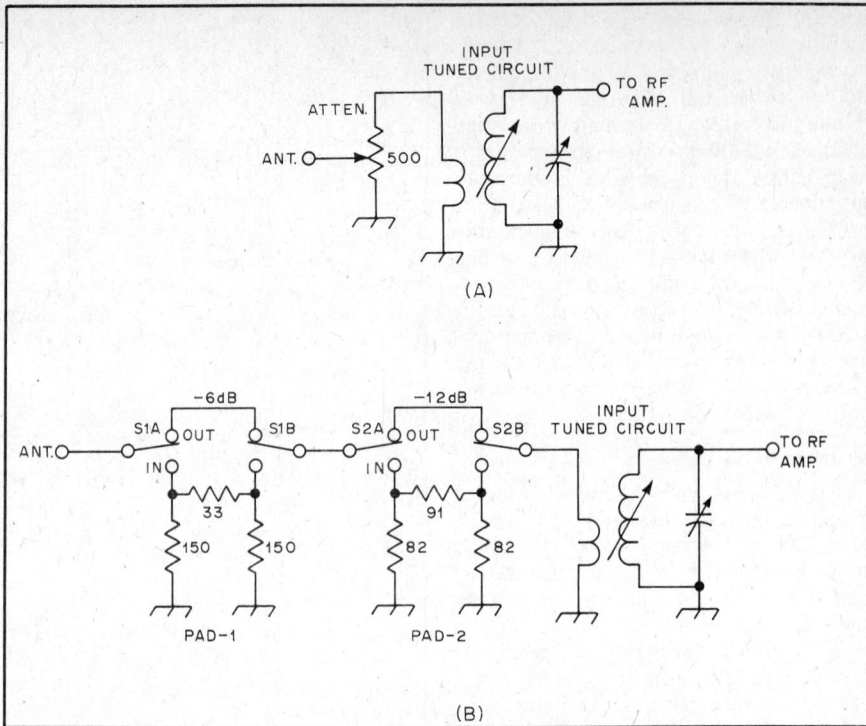

Fig. 25 — Front-end attenuators. A simple type is given at A and a step-attenuator version is seen at B.

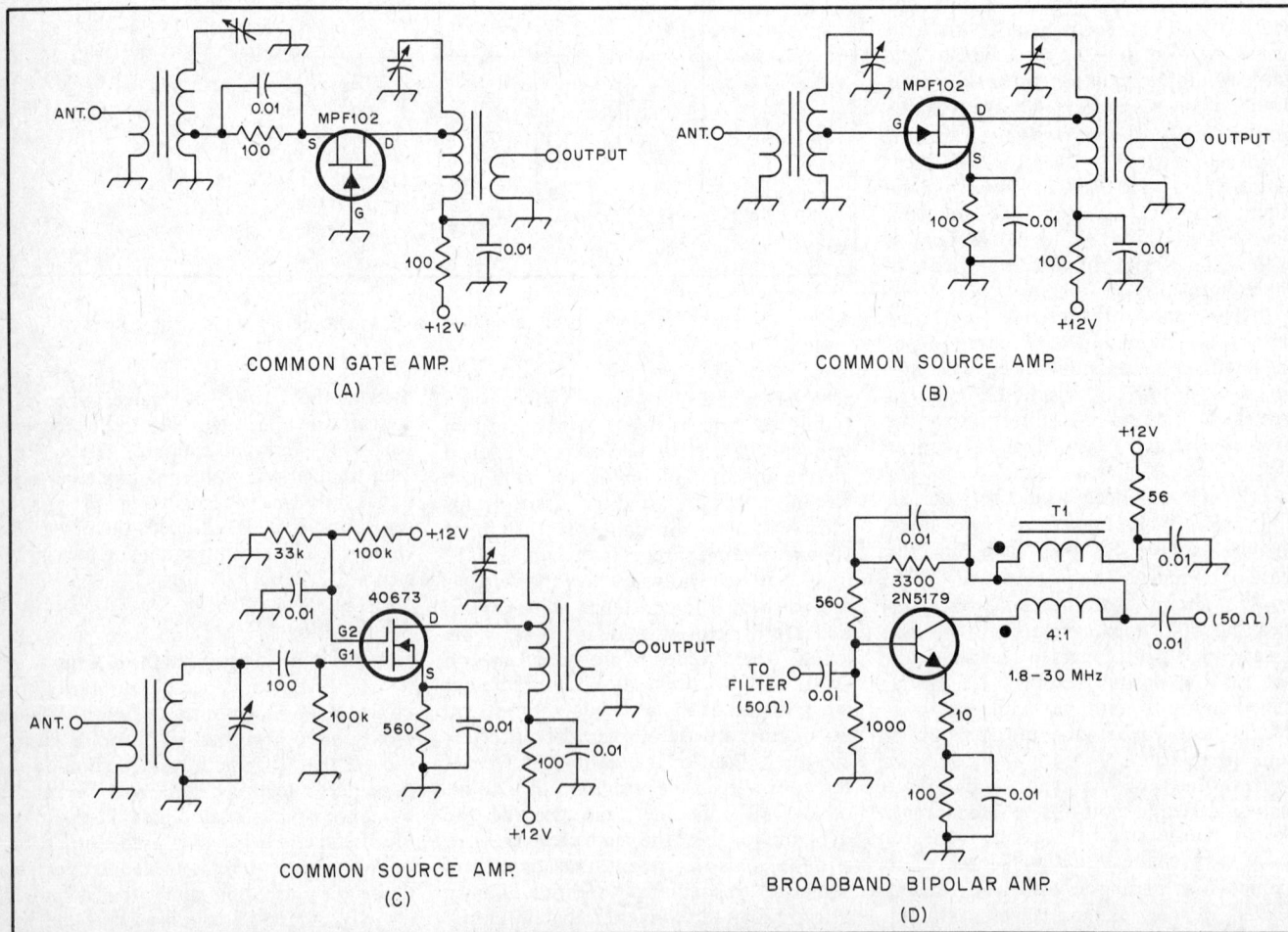

Fig. 26 — Narrowband rf amplifiers are shown from A to C. A fed-back broadband rf amplifier is seen at D.

will be less loading on the tuned circuit, thereby permitting somewhat greater tuned-circuit selectivity: The lower the drain tap, the less the stage gain.

All of the FET amplifiers in Fig. 26 are capable of providing low-noise operation and good dynamic range. The common-source circuits at illustrations B and C can provide up to 25 dB of gain. However, they are more prone to instability than is the circuit at A. Therefore, the gates are shown tapped down on the gate tank: Placing the input at a low impedance point on the tuned circuit will discourage self-oscillation. The same is true of the drain tap. JFETs will hold up under considerable rf input voltage before being damaged. Laboratory tests of the MPF102 showed that 80 volts pk-pk (gate to source) were required to destroy the device. However, in the interest of good operating practices the pk-pk voltage should be kept below 10. Tapping the gate down on the input tuned circuit will result in lower levels of pk-pk input voltage, in addition to aiding stability.

A broadband bipolar-transistor rf amplifier is shown in Fig. 26 at D. This type of amplifier will yield approximately 16 dB of gain up to 148 MHz, and it will be unconditionally stable because of the degenerative feedback in the emitter and the negative feedback in the base circuit. A broadband 4:1 transformer is used in the collector to step the impedance down to approximately 50 ohms at the amplifier output. A 50-ohm characteristic exists at the input to the 2N5179 also. A band-pass filter should be used at the input and output of the amplifier to provide selectivity. The 4:1 transformer helps to assure a collector load of 200 ohms, which is preferred in an amplifier of this type. This style of amplifier is used in CATV applications where the transformation from collector to load is 300 to 75 ohms.

Receiver Mixers

One of the most important parts of a high-performance receiver is the mixer. It is at this point where the greater consideration for dynamic range exists. For best receiver performance the mixer should receive only enough preamplifier signal to overcome the mixer noise. When excessive amounts of signal energy are permitted to reach the mixer there will be desensitization, cross-modulation and IMD products in the mixer. When these effects are severe enough the receiver can be rendered useless. Therefore, it is advantageous to utilize what is often called a "strong mixer." That is one which can handle high signal levels without being adversely affected.

Generally speaking, diode-ring passive mixers fare the best in this regard. However, they are fairly noisy and require considerably more LO injection than is the case with active mixers. For the less sophisticated types of receivers it is

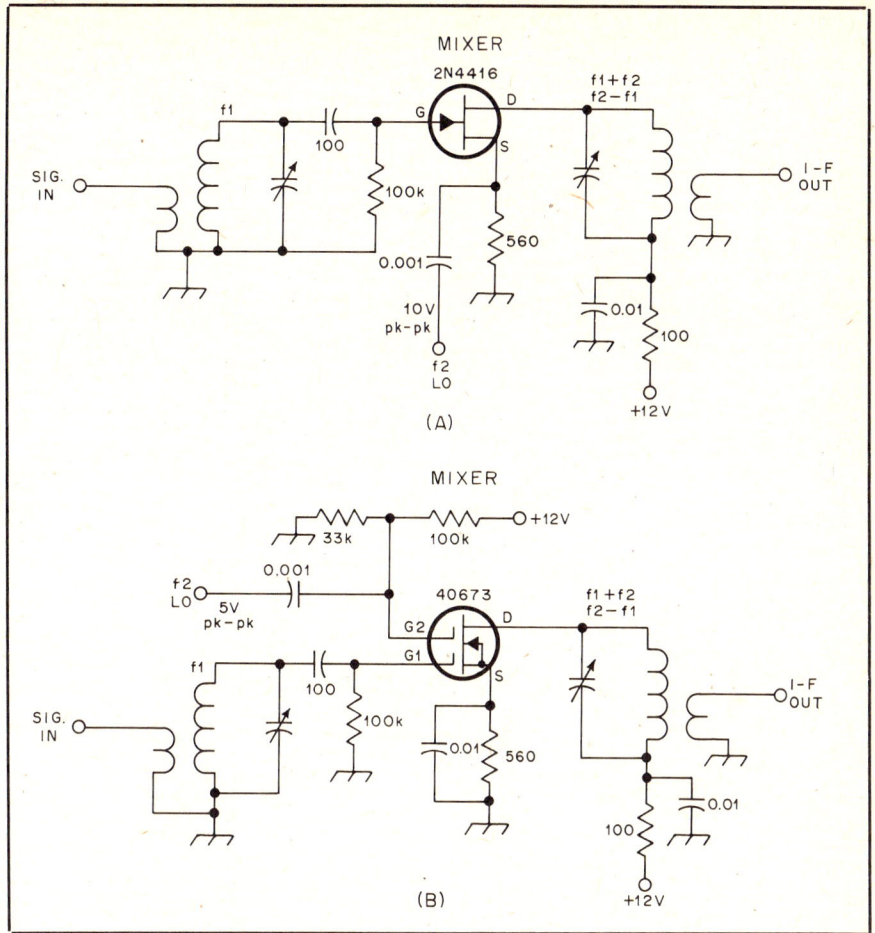

Fig. 27 — Two styles of active mixers using FETs.

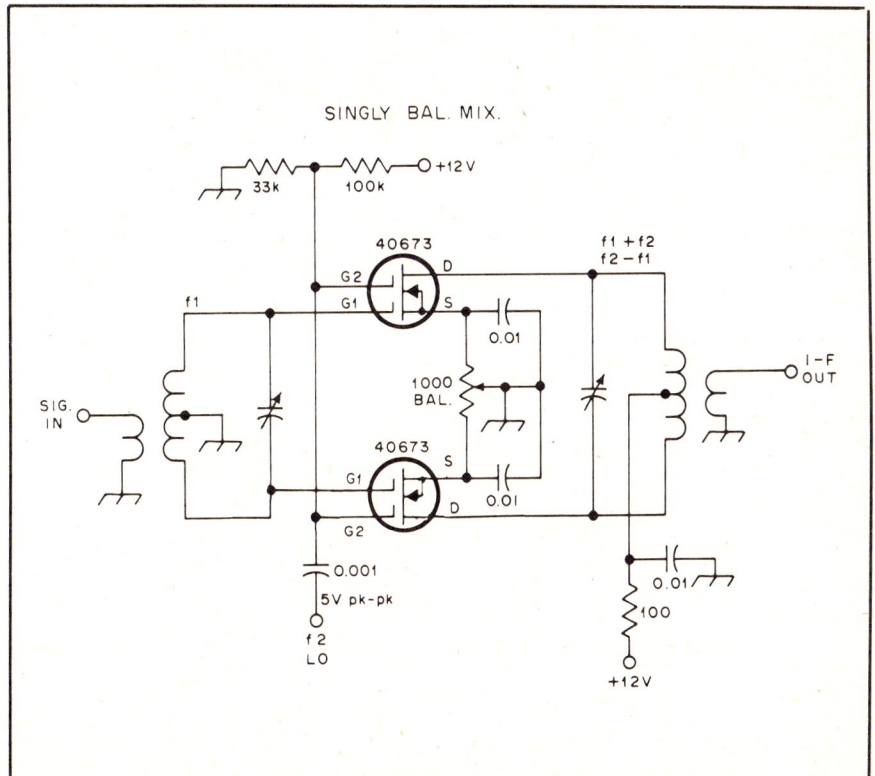

Fig. 28 — An active singly balanced FET mixer.

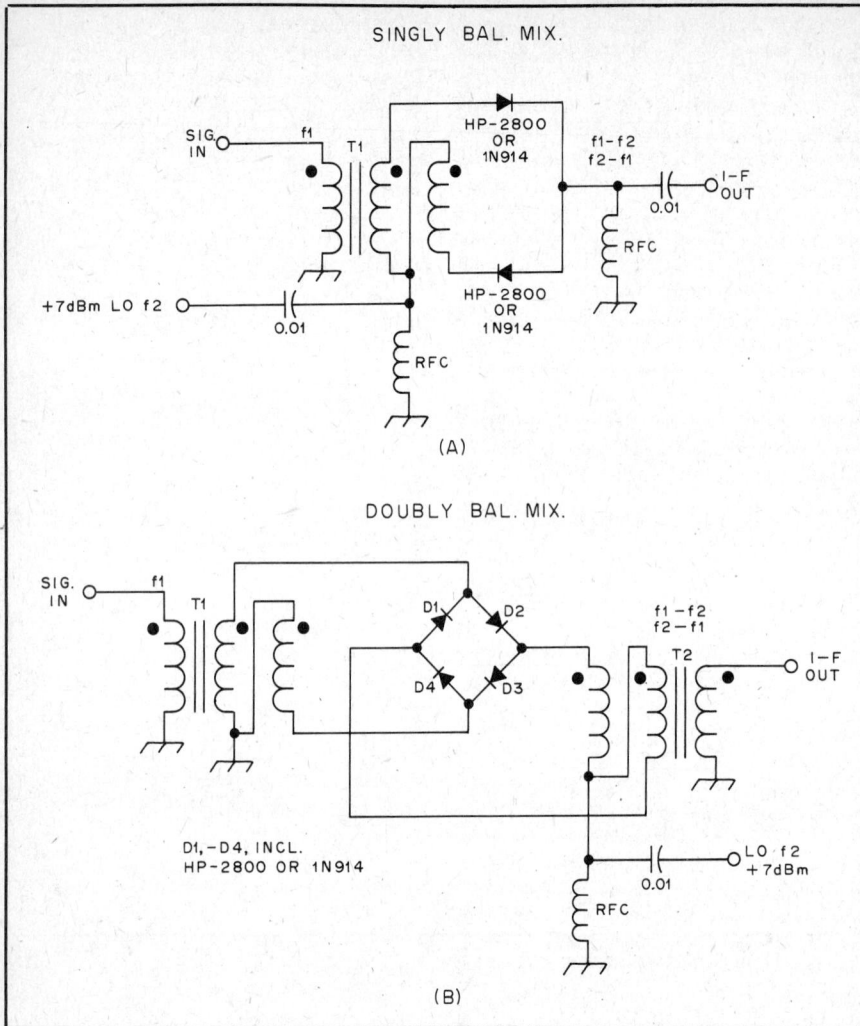

SINGLY BAL. MIX.

(A)

DOUBLY BAL. MIX.

D1,–D4, INCL.
HP-2800 OR 1N914

(B)

Fig. 29 — Singly and doubly balanced diode mixers.

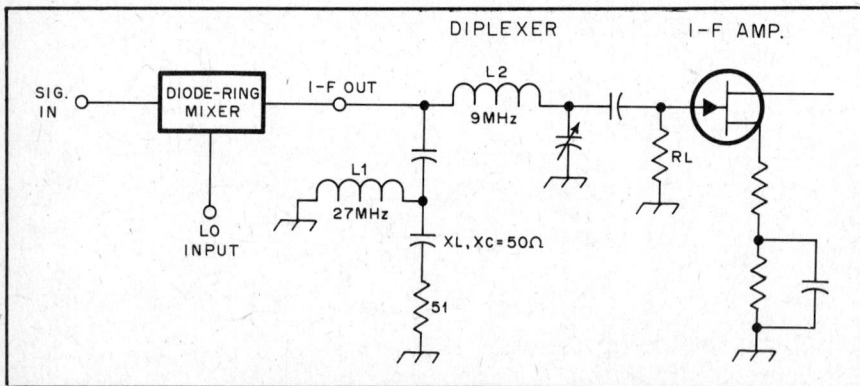

Fig. 30 — Method for diplexing the mixer output to improve the IMD characteristics.

An LC, crystal-lattice, or mechanical type of band-pass filter is almost always used after the mixer or the post-mixer amplifier. This helps to establish the overall selectivity of the receiver. It also rejects unwanted mixer products that fall outside the passband of the filter.

In the interest of optimum mixer performance, the LO energy supplied to it should be reasonably clean with respect to frequencies other than the desired LO one. Many designers, for this reason, use a band-pass filter between the LO output and the mixer input. Excessive LO noise will seriously degrade receiver performance. LO noise should be 80 dB or more below the peak level of the desired LO frequency. Excessive noise will appear as noise sidebands in the receiver output.

Typical Mixer Circuits

Fig. 27 shows two single-ended active mixers which offer good performance. The example at A employs a JFET with LO injection supplied to the source across a 560-ohm resistor. This injection mode requires somewhat more LO power than would be used if injection was done at the gate. However, there is less occasion for LO pulling when source injection is used, and there is better isolation between the LO and antenna than would be the case with gate injection.

The circuit at B in Fig. 27 is similar to that of A with regard to general performance. The major difference is that a dual-gate MOSFET is used to permit injection of the LO energy at gate 2. Since there is considerable signal isolation between gates 1 and 2, LO pulling is minimized and antenna-LO isolation is good.

A singly balanced active mixer is illustrated at Fig. 28. Two 40673 dual-gate MOSFETS are connected in push-pull, but with the LO frequency injected in parallel at gate 2 of each device. A potentiometer is used in the sources of the transistors to permit circuit balance. This mixer offers performance superior to those shown in Fig. 27.

One of the least complicated or expensive mixers is the two-diode version (singly balanced) seen in Fig. 29A. A trifilar-wound broadband toroidal transformer is used at the mixer input. The shortcoming of this mixer over the one seen at B is that signal isolation between all three mixer ports is not possible. A better version is that at B in Fig. 29. In this case all three mixer ports are well isolated from one another. This greatly reduces the probability of spurious responses in the receiver. Conversion loss with these mixers is approximately 8 dB. The impedance of the mixer ports is approximately 50 ohms.

Improved IMD characteristics can be had from a diode-ring mixer, by placing a diplexer after the mixer as seen in Fig. 30. The diplexer consists of a high-pass network (L1) and a low-pass

entirely adequate to use single-ended active mixers, provided the gain distribution between the antenna and mixer is proper for the mixer device used. Field-effect transistors are preferred by most designers; bipolar-transistor mixers are seldom used.

The primary advantage of an active mixer is that it has conversion gain rather than loss. This means that the stages following the mixer need not have as much gain as when diode mixers are used. A typical doubly balanced diode mixer will have a conversion loss of some 8 dB, whereas an FET active mixer may exhibit a conversion gain as great as 15 dB. The cost of gain stages in this era is relatively small. This easily justifies the use of strong passive mixers in the interest of high dynamic range.

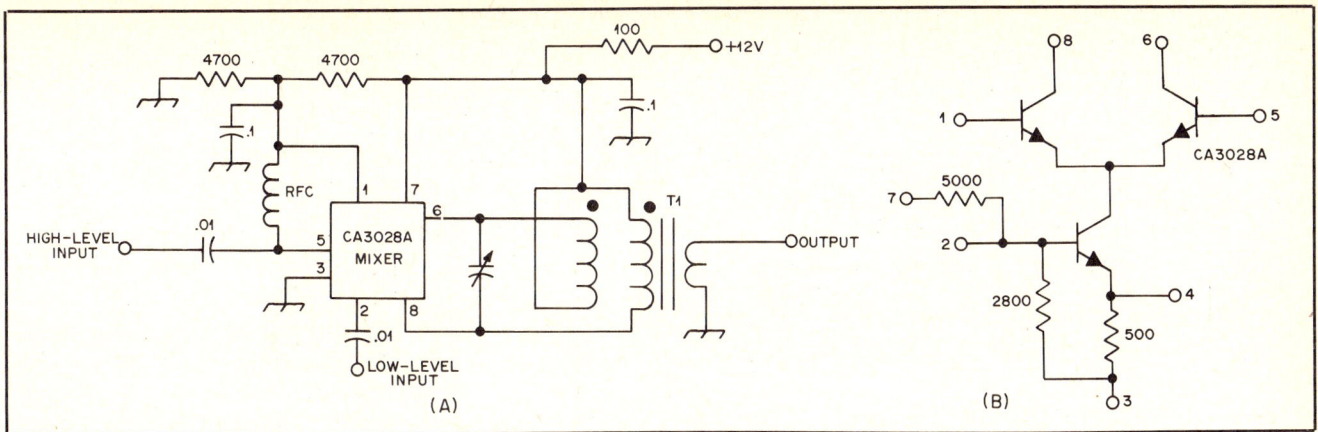

Fig. 31 — CA3028A singly balanced mixer. The circuit for the IC is given at B.

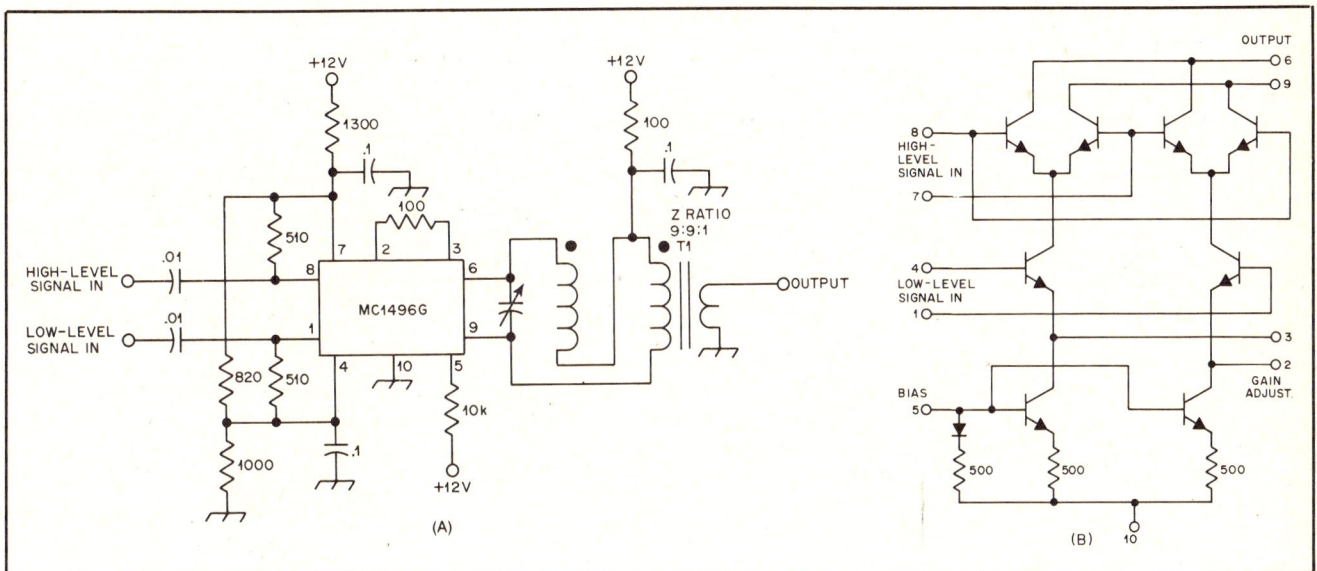

Fig. 32 — MC1496G doubly balanced mixer and circuit of the IC.

one (L2). L2 is tuned to the i-f and serves as a matching network between 50 ohms and R_L, the FET gate resistor. L1 and the associated series capacitors are tuned to three times the i-f and terminated in 50 ohms. This gives the mixer a proper resistive termination without degrading the 9-MHz i-f. The high-pass network has a loaded Q of 1.

IC Mixers

Although there are numerous ICs available for use as mixers, only three are shown here. Fig. 31 shows a CA3028A singly balanced active mixer. The diagram at B shows the inner workings of the IC. The LO is injected at pin 2 of the IC. Conversion gain is on the order of 15 dB.

Fig. 32A illustrates a doubly balanced IC active mixer which employs an MC1496G. A schematic diagram of the IC is shown at B. The performance of this mixer is excellent, but it is not as strong a mixer as that of Fig. 29. However, it has good conversion gain and a fairly low noise figure.

A relatively new addition to the IC mixer market is the Plessey SL6400C programmable high-level doubly-balanced mixer. Fig. 33 shows the SL6400C in a broadband circuit. This configuration produced a conversion gain of 8 dB and a third-order output intercept of approximately +22 dBm[2]. The single-sideband noise figure was found to be 11 dB. If the mixer input and output ports are terminated in 50 ohms, rather than 200 ohms as shown in Fig. 33, the typical conversion gain will be −1 dB.

I-F Amplifiers

The amount of i-f amplification used in a receiver will depend upon how much signal level is available at the input to the i-f strip. Sufficient gain is needed to ensure ample audio output consistent with driving headphones or a speaker. Another consideration is the amount of agc-initiated i-f gain range. The more i-f stages used (a maximum of two is typical) the

[2]Referenced to one tone of a two-tone test.

greater the gain change caused by agc action. The range is on the order of 80 dB when two CA3028A ICs are used in the i-f strip. A pair of MC1590G ICs will provide up to 120 dB of gain variation with agc applied.

Nearly all modern receiver circuits utilize ICs as i-f amplifiers. Numerous types of ICs are available to provide linear rf and i-f amplification at low cost. The CA3028A and MC1590G ICs are the most popular ones for amateur work because they are easy to obtain and are relatively low in cost. With careful layout techniques either device will operate in a stable manner. Bypassing should be done as near to the IC pins as possible. Input and output circuit elements must be separated to prevent mutual coupling which can cause unstable operation. If IC sockets are used they should be the low-profile variety with short socket conductors.

Fig. 34 contains examples of bipolar transistor and FET i-f amplifiers. Typical component values are given. A CA2038A

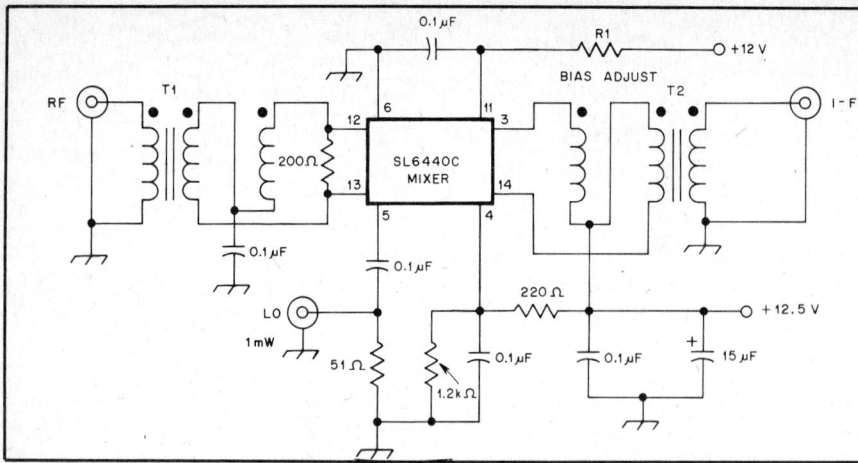

Fig. 33 — Plessey SL6400C doubly balanced mixer. R1 is selected for a bias current of about 12 mA. T1 and T2 are broadband transformers wound on ferrite toroid cores.

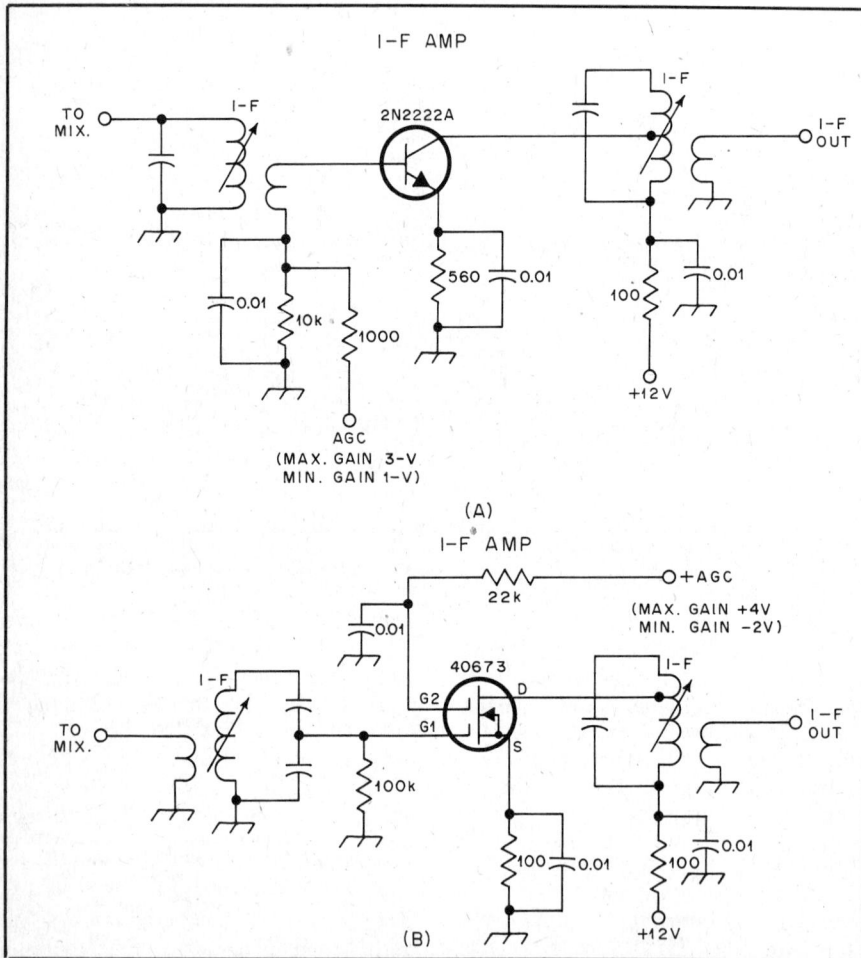

Fig. 34 — Methods for applying agc to a bipolar i-f amplifier (A) and a dual-gate MOSFET i-f stage (B).

ended operation, as shown. The values are doubled when either device is operated in push-pull with respect to input and output tuned circuits.

Choice of Frequency

The selection of an intermediate frequency is a compromise between conflicting factors. The lower the i-f, the higher the selectivity and gain, but a low i-f brings the image nearer the desired signal and hence decreases the image ratio. A low i-f also increases pulling of the oscillator frequency. On the other hand, a high i-f is beneficial to both image ratio and pulling, but the gain is lowered and selectivity is harder to obtain by simple means.

An i-f of the order of 455 kHz gives good selectivity and is satisfactory from the standpoint of image ratio and oscillator pulling at frequencies up to 7 MHz. The image ratio is poor at 14 MHz when the mixer is connected to the antenna, but adequate when there is a tuned rf amplifier between antenna and mixer. At 28 MHz and on the very high frequencies, the image ratio is very poor unless several rf stages are used. Above 14 MHz, pulling is likely to be bad without very loose coupling between mixer and oscillator. Tuned-circuit shielding also helps.

With an i-f of about 1600 kHz, satisfactory image ratios can be secured on 14, 21 and 28 MHz with one rf stage of good design. For frequencies of 28 MHz and higher, a common solution is to use double conversion, choosing one high i-f for image reduction (9 MHz is frequently used) and a lower one for gain and selectivity. A popular i-f at present is 3.3 to 3.4 MHz, which is used by some commerical designers as the last i-f in double-conversion receivers.

In choosing an i-f it is wise to avoid frequencies on which there is considerable activity by the various radio services, since such signals may be picked up directly by the i-f wiring. Shifting the i-f or better shielding are the solutions to this interference problem.

Fidelity: Sideband Cutting

Amplitude modulation of a carrier generates sideband frequencies numerically equal to the carrier frequency plus and minus the modulation frequencies present. If the receiver is to give a faithful reproduction of modulation that contains, for instance, audio frequencies up to 5000 Hz, it must at least be capable of amplifying equally all frequencies contained in a band extending from 5000 Hz above or below the carrier frequency. In a superheterodyne, where all carrier frequencies are changed to the fixed intermediate frequency, the i-f amplification must be uniform over a band 5-kHz wide, when the carrier is set at one edge. If the carrier is set in the *center*, a 10-kHz band is required. The signal-frequency circuits

IC, connected for differential-amplifier operation, is shown in Fig. 35 as an i-f amplifier. Up to 40 dB of gain is possible with this circuit. The IC is useful up to 120 MHz and has a low noise figure.

A Motorola MC1590G IC will provide up to 50 dB of stage gain when used as an i-f amplifier. An example of the circuit is given in Fig. 36. Agc operates in the reverse of that which is applied to a CA3028A. With the latter the gain will be maximum with maximum agc voltage. An MC1590G delivers maximum gain at the low agc voltage level.

With both amplifiers (Figs. 35 and 36) the input impedance is on the order of 1000 ohms. The output impedance is close to 4000 ohms. These values are for single-

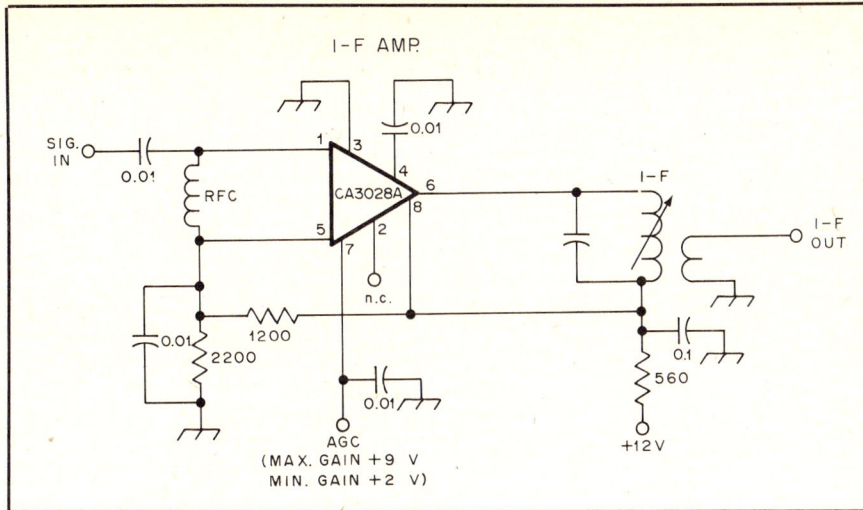

Fig. 35 — An IC type of i-f amplifier with agc applied.

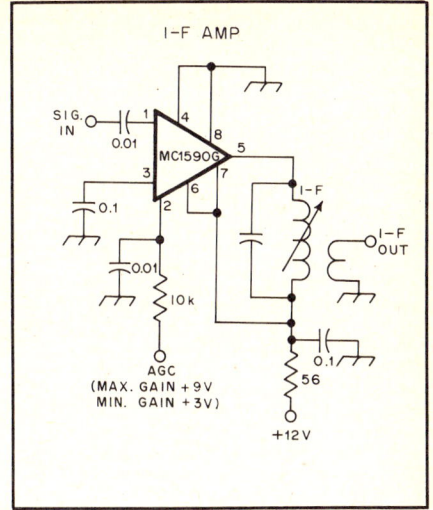

Fig. 36 — Agc is applied to an MC1590G IC.

usually do not have enough overall selectivity to affect materially the "adjacent-channel" selectivity, so that only the i-f amplifier selectivity need be considered.

If the selectivity is too great to permit uniform amplification over the band of frequencies occupied by the modulated signal, some of the sidebands are "cut." While sideband cutting reduces fidelity, it is frequently preferable to sacrifice reproduction naturalness in favor of communications effectiveness.

The selectivity of an i-f amplifier, and hence, the tendency to cut sidebands, increases with the number of tuned circuits and also is greater with lower intermediate frequencies. From the communication standpoint, sideband cutting is never serious with two-stage amplifiers at frequencies as low as 455 kHz. A two-stage i-f amplifier at 85 or 100 kHz will be sharp enough to cut some of the higher-frequency sidebands if good transformers are used. However, the cutting is not at all serious, and the gain in selectivity is worthwhile in crowded amateur bands as an aid to QRM reduction.

I-F Selectivity

The most significant selectivity in a receiver is that which separates signals and reduces QRM, assuming that selectivity in other parts of a receiver is ignored but correct. Narrow-response filters are used after the last mixer or post-mixer amplifier to establish the overall selectivity of a receiver. Most receivers which use a second i-f of 455 kHz contain mechanical filters. Table 1 lists various mechanical filters which are manufactured by Collins Radio Co.

At intermediate frequencies above 500 kHz it is common practice to use crystal filters. These can be designed with just one crystal (Fig. 37A), or with two or more crystals. Fig. 37B illustrates a two-crystal, half-lattice filter and a cascaded

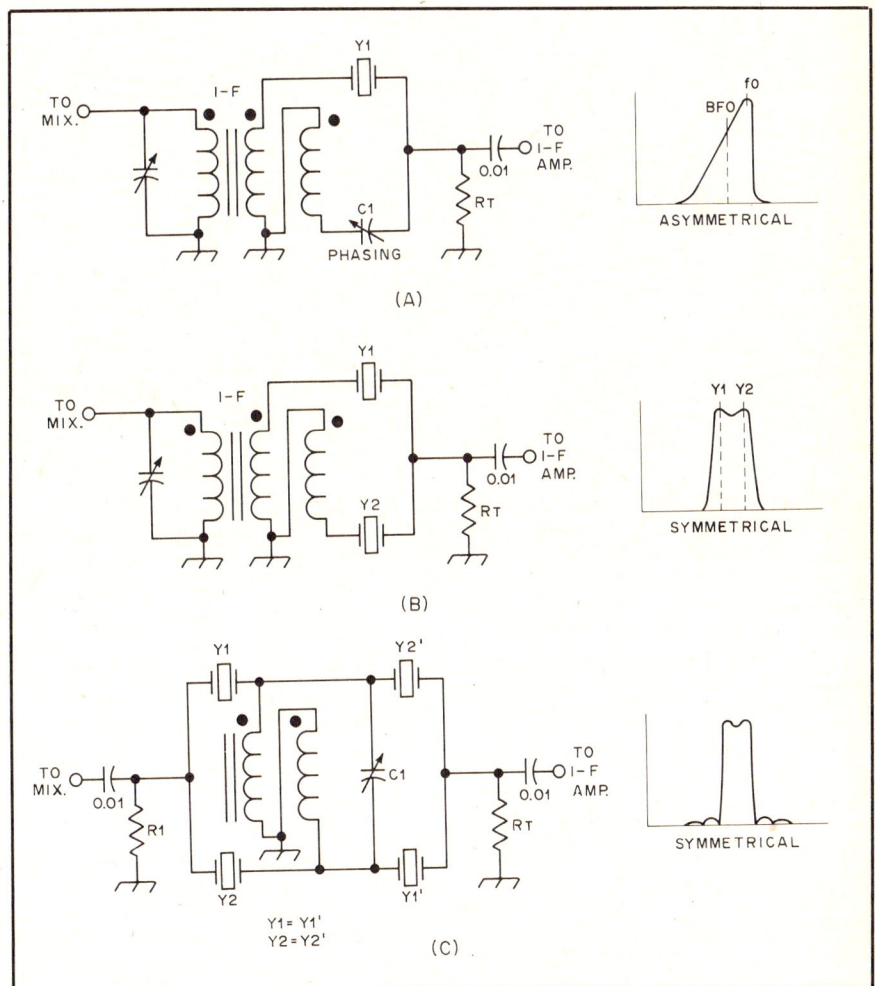

Fig. 37 — A comparison between crystal i-f filters. The selectivity is increased as crystals are added.

half-lattice filter is shown at C of Fig. 37.

The single-crystal example shown at A of Fig. 37 is best suited for simple receivers intended mainly for cw use. C1 is adjusted to provide the bandpass characteristic shown adjacent to the circuit. When the BFO frequency is placed on the part of the low-frequency slope (left) which gives the desired beat note respective to f_o (approximately 700 Hz), single-signal reception will result. To the right of f_o in Fig. 37A the response drops sharply to reduce output on the unwanted side of zero beat, thereby making single-signal

Courtesy of Collins Radio Co.

Table 1

Part & Type Numbers	Min. 3 dB BW @ 25°C (kHz)	Min. 4 dB BW OTR (kHz)	Max. 60 dB BW @ 25°C (kHz)	Max. 60 dB BW OTR (kHz)	Max. RV @ 25°C (dB)	Max. RV OTR (dB)	Max. IL @ 25° (dB)	Max. IL COTR (dB)	Min. 60 dB SBR (kHz)	S&L −5% ohms	Res. Cap. +5% (pF)
526-9689-010	0.375	0.375	3.5	4.0	3.0	4.0	10.0	12.0	445-F60L	2000	350
F455FD-04									F60H-465		350
526-9690-010	1.2	1.2	8.7	9.5	3.0	4.0	10.0	12.0	445-F60L	2000	350
F455FD-12									F60H-465		350
F526-9691-010	1.9	1.9	5.4	5.9	3.0	4.0	10.0	12.0	445-F60L	2000	330
F455FD-19									F60H-465		330
526-9692-010	2.5	2.5	6.5	7.0	3.0	4.0	10.0	12.0	445-F60L	2000	510
F455FD-25									F60H-465		510
526-9693-010	2.9	2.9	7.0	8.0	3.0	4.0	10.0	12.0	445-F60L	2000	510
F455FD-29									F60H-465		510
526-9694-010	3.8	3.8	9.0	10.0	3.0	4.0	10.0	12.0	445-F60L	2000	1000
F455FD-38									F60H-465		1000
526-9695-010	5.8	5.8	14.0	15.0	3.0	4.0	10.0	12.0	445-F60L	2000	1100
F455FD-58									F60H-465		1100

OTR = Operating Temperature Range, RV = Ripple Voltage, IL = Insertion Loss, SBR = Stop Band Range, S & L = Source and Load

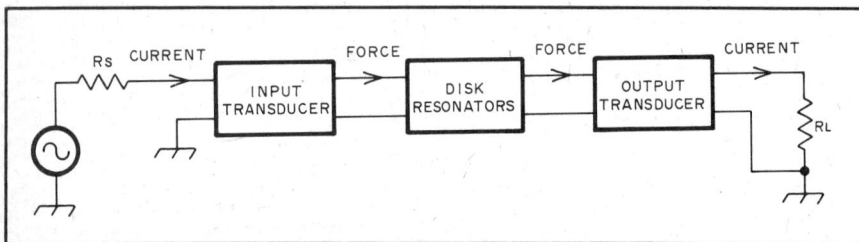

Fig. 38 — Block diagram of a mechanical filter (Collins Radio).

reception possible. If no i-f filter was used, or if the BFO frequency fell at f_o, nearly equal response would exist either side of zero beat (double-signal response) as is the case with direct-conversion receivers. QRM on the unwanted-response side of the i-f passband would interfere with reception. The single-crystal filter shown is capable of at least 30 dB of rejection on the high-frequency side of zero beat. The filter termination R_T, has a marked effect on the response curve. It is necessary to experiment with the resistance value until the desired response is obtained. Values can range from 1500 to 10,000 ohms.

A half-lattice filter is shown at B in Fig. 37. The response curve is symmetrical and there is a slight dip at center frequency. The dip is minimized by proper selection of R_T. Y1 and Y2 are separated in frequency by the amount needed to obtain cw or ssb selectivity. The bandwidth at the 3-dB points will be approximately 1.5 times the crystal-frequency spacing. For upper or lower sideband reception Y1 and Y2 would be 1.5 kHz apart, yielding a 3-dB bandwidth of approximately 2.25 kHz. For cw work a crystal spacing of 0.4 kHz would result in a bandwidth of roughly 600 Hz. The skirts of the curve are fairly wide with a single half-lattice filter, which uses crystals in the hf region.

The skirts can be steepened by placing two half-lattice filters in cascade, as shown in Fig. 37C. R1 and R_T must be selected to provide minimum ripple at the center of the passband. The same rule for frequency spacing between the crystals applies. C1 is adjusted for a symmetrical response.

The circuits of Fig. 37 can be built easily and inexpensively by amateurs. The transformers shown are tuned to center frequency. They are wound bifilar or trifilar on ferrite or powdered-iron cores of appropriate frequency characteristics.

An illustration of how a mechanical filter operates is provided in Fig. 38. Perhaps the most significant feature of a mechanical filter is the high Q of the resonant metallic disks it contains. A Q figure of 10,000 is the nominal value obtained with this kind of resonator. If L and C constants were employed to acquire a bandwidth equivalent to that possible with a mechanical filter, the i-f would have to be below 50 kHz.

Mechanical filters have excellent frequency-stability characteristics. This makes it possible to fabricate them for fractional bandwidths of a few hundred hertz. Bandwidths down to 0.1 percent can be obtained with these filters. This means that a filter having a center frequency of 455 kHz could have a bandwidth as small as 45.5 Hz. By inserting a wire through the centers of several resonator disks, thereby coupling them, the fractional bandwidth can be made as great as 10 percent of the center frequency. The upper limit is governed primarily by occurrence of

unwanted spurious filter responses adjacent to the desired passband.

Mechanical filters can be built for center frequencies from 60 to 600 kHz. The main limiting factor is disk size. At the low end of the range the disks become prohibitively large, and at the high limit of the range the disks become too small to be practical.

The principle of operation is seen in Fig. 38. As the incoming i-f signal passes through the input transducer it is converted to mechanical energy. This energy is passed through the disk resonators to filter out the undesired frequencies, then through the output transducer where the mechanical energy is converted back to the original electrical form.

The transducers serve a second function: They reflect the source and load impedances into the mechanical portion of the circuit, thereby providing a termination for the filter.

Mechanical filters require external resonating capacitors which are used across the transducers. If the filters are not resonated there will be an increase in insertion loss, plus a degradation of the passband characteristics. Concerning the latter, there will be various unwanted dips in the nose response (ripple), which can lead to undesirable effects. The exact amount of shunt capacitance will depend on the filter model used. The manufacturer's data sheet specifies the proper capacitor values.

Most bipolar transistor i-f amplifiers have an input impedance of 1000 ohms or less. There are situations where the output impedance of the stage preceding the filter is similarly low. In circuits of this variety it is best to use series-resonating capacitors in preference to parallel ones. Stray circuit capacitance, including the input and output capacitances of the stages before and after the filter, should be subtracted from the value specified by the manufacturer.

Collins mechanical filters are available with center frequencies from 64 to 500 kHz and in a variety of bandwidths. Insertion loss ranges from 2 dB to as much as 12 dB, depending on the style of filter used. Of greatest interest to amateurs are the 455-kHz mechanical filters specified as F455. They are available in bandwidths of 375 Hz, 1.2 kHz, 1.9 kHz, 2.5 kHz, 2.9 kHz, 3.8 kHz and 5.8 kHz. Maximum insertion loss is 10 dB, and the characteristic impedance is 2000 ohms. Different values of resonating capacitance are required for the various models, spreading from 350 to 1100 pF. Although some mechanical filters are terminated internally, this series requires external source and load termination of 2000 ohms. The F455 filters are the least expensive of the Collins line.

Most modern receivers have selectable i-f filters to provide suitable bandwidths for ssb and cw. Most of the commercial receivers use a 500- or 600-Hz bandwidth filter for cw and a 2.1- or 2.4-kHz bandwidth for ssb. The input and output ends of a filter should be well isolated from one another if the filter characteristics are to be realized. Leakage across a filter will negate the otherwise good performance of the unit. The problem becomes worse as the filter frequency is increased. Mechanical switches are not recommended above 455 kHz for filter selection because of leakage across the switch wafers and sections. Diode switching is preferred by most designers. The switching diodes for the filter that is out of the circuit are usually back-biased to ensure minimum leakthrough.

In the interest of reducing wideband noise from the i-f amplifier strip it is **worthwhile to use a second filter which has exactly the same center frequency as the first. The second filter is placed at the end of the i-f strip, ahead of the product detector. This is shown in Fig. 39.** The technique was described by W7ZOI in March and April 1974 *QST*. The second filter, FL2, has somewhat wider skirts than the first, FL1. An RC active audio filter after the product detector has a similar effect, but the results are not quite as spectacular as when two i-f filters are used. The overall signal-to-noise ratio is greatly enhanced by this method.

Automatic Gain Control

Automatic regulation of the gain of the receiver in inverse proportion to the signal strength is an operating convenience in reception, since it tends to keep the output level of the receiver constant regardless of input-signal strength. The average rectified dc voltage, developed by the received signal across a resistance in a detector circuit, is used to vary the bias on the rf and i-f amplifier stages. Since this voltage is proportional to the average amplitude of the signal, the gain is reduced as the signal strength becomes greater. The control will be more complete and the output more constant as the

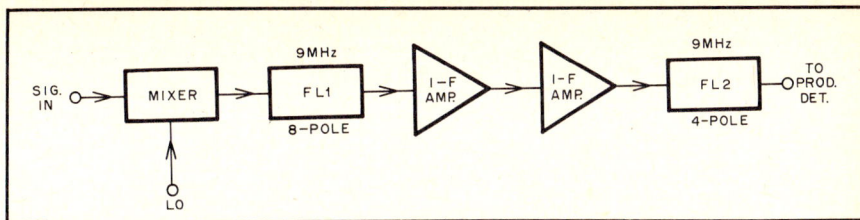

Fig. 39 — Crystal i-f filters can be used at both ends of the i-f strip. FL2 greatly reduces wideband i-f noise.

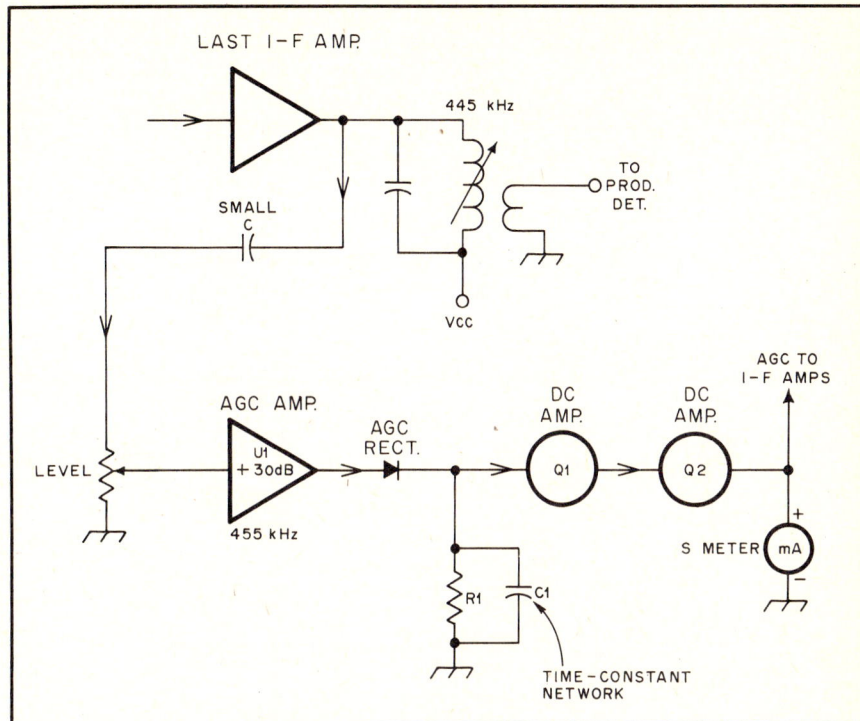

Fig. 40 — A system for developing receiver agc voltage.

number of stages to which the agc bias is applied is increased. Control of at least two stages is advisable.

Various schemes from simple to extravagant have been conceived to develop agc voltage in receivers. Some perform poorly because **the attack time of the circuits is wrong for cw work, resulting in** "clicky" or "pumping" agc. The first significant advance toward curing the problem was presented by Goodman W1DX, "Better AVC for SSB and Code Reception," January 1957 *QST*. He coined the term, "hang" avc, and the technique has been adopted by many amateurs who have built their own receiving equipment. The objective is to make the agc take hold as quickly as possible to avoid the ailments mentioned in the foregoing text.

For best receiver performance the i-f filters should be contained within the agc loop, which strongly suggests the use of rf-derived agc. Most commercial receivers follow this rule. However, good results can be obtained with audio-derived agc,

despite the tendency toward a clicky response. If *RC* active audio filters are used to obtain receiver selectivity, they should be contained within the audio-agc loop if possible.

Fig. 40 illustrates the general concept of an agc circuit. Rf energy is sampled from the output of the last i-f by means of light coupling. This minimizes loading on the tuned circuit of the i-f amplifier. The i-f energy is amplified by the agc amplifier, then converted to dc by means of an agc rectifier. R1 and C1 are selected to provide a suitable decay time constant (about 1 second for ssb and cw). Q1 and Q2 function as dc amplifiers to develop the dc voltage needed for agc control of the i-f (and sometimes, rf) amplifier stages. The developed agc voltage can be used to drive an S meter. A level control can be placed at the input of the agc amplifier to establish the signal input level (receiver front end) which turns on the agc system. Most designers prefer to have this happen when the received signal level is between 0.25 and 1 μV. The exact

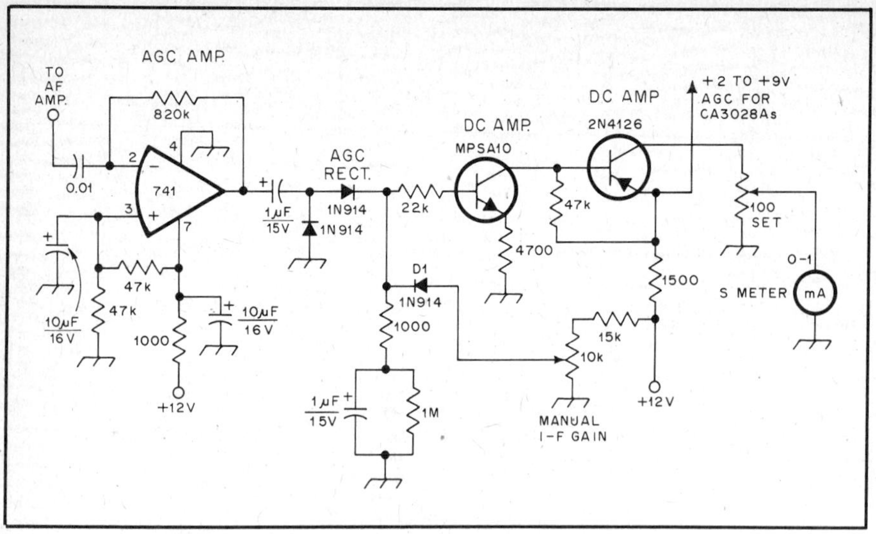

Fig. 41 — A practical circuit for developing agc voltage for a CA3028A i-f amplifier.

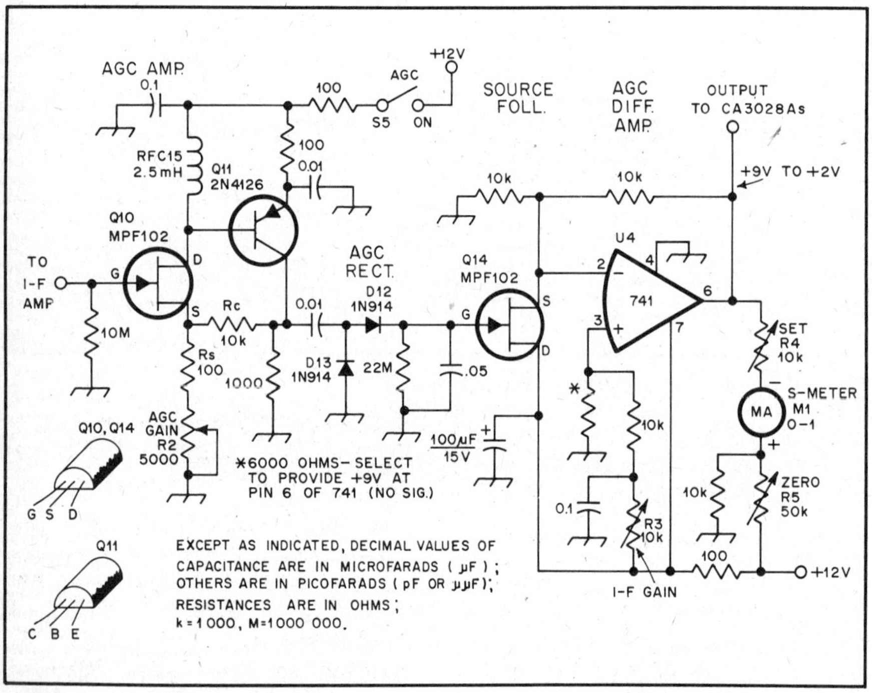

Fig. 42 — An agc system for CA3028A i-f amplifiers. An op-amp is used as difference amplifier to provide agc voltage while operating an S meter.

parameters are based somewhat on subjectivity.

An af-derived agc loop is shown in Fig. 41. It is suitable for use with CA3028A i-f amplifier ICs. Provision is made for manual i-f gain control. D1 functions as a gating diode to prevent the manual-control circuitry from affecting the normal agc action. This circuit was first used in a receiver described by DeMaw in June and July 1976 *QST*.

An rf-derived agc system is seen in Fig. 42. It operates on a similar principle as that of Fig. 41, except that an op-amp is used in place of the discrete bipolar dc amplifiers of Fig. 41. Current changes are sampled across the 10-kΩ FET source resistor by means of the op-amp difference amplifier. With the values of resistance given, the output dc swing of the op-amp is the desired +2 to +9 volts for controlling CA3028A i-f amplifiers. This system was also used in the DeMaw receiver.

It certainly is not essential to have agc in a receiver. If the operator is willing to adjust the gain manually, good performance is certain to result. Agc is mainly an operator convenience: It prevents loud signals from blasting out of the speaker or headphones when the operator tunes the band at a given af-gain setting.

Beat-Frequency Oscillators

The circuits given for crystal-controlled oscillators earlier in this chapter are suitable for use in BFO circuits. A beat oscillator generates energy which is supplied to a product detector for reception of cw and ssb signals. The BFO frequency is offset by the appropriate amount with respect to the center frequency of the i-f filter. For example, a BFO used during cw reception is usually some 700 Hz above or below the i-f center frequency. During ssb reception the offset is slightly more — approximately 1.5 kHz above or below the i-f center frequency, depending upon the need for upper or lower sideband operation. Typically, the BFO is placed roughly 20 dB down on the slope of the i-f passband curve for ssb reception or transmission.

Table 2

Application	SSB Tran.	SSB Rec.	CW or Digital Data	A-M	A-M	CW	FM
Filter type	XF-9A	XF-9B	XF-9NB	XF-9C	XF-9D	XF-9M	XF-9E
No. of crystals	5	8	8	8	8	4	8
6-dB bandwidth	2.5 kHz	2.4 kHz	0.5 kHz	3.75 kHz	5.0 kHz	0.5 kHz	12 kHz
Passband ripple	<1 dB	<2 dB	<0.5 dB	<2 dB	<2 dB	<1 dB	<2 dB
Insertion loss	<3 dB	<3.5 dB	<6.5 dB	<3.5 dB	<3.5 dB	<5 dB	<3 dB
Term. impedance	500Ω	500Ω	500Ω	500Ω	500Ω	500Ω	1200Ω
Ripple capacitors	30 pF	30 pF	30 pF	30 pF	30 pF	30 pF	30 pF
Shape factor	6:50 dB	6:60 dB	6:60 dB	6:60 dB	6:60 dB	6:60 dB	6:60 dB
	1.7	1.8	2.2	1.8	1.8	4.4	1.8
Stop-band atten.	>45 dB	>100 dB	>90 dB	>100 dB	>100 dB	>90 dB	>90 dB

Courtesy of Spectrum International

A BFO need not be crystal controlled. It can use a VFO type of circuit, or it can be tuned by means of a Varactor diode (see chapter 4 for a discussion of semiconductor theory). Elimination of the crystals represents a cost savings to the builder, but frequency stability may not be as good as when crystal control is employed.

When the BFO is operated at frequencies above 3 MHz it is helpful to use a buffer stage after the oscillator to minimize the effects of pulling. Furthermore, if a passive product detector is used in the receiver, a substantial amount of BFO output power will be required — approximately +7 dBm. The buffer/amplifier helps to boost the oscillator output to satisfy the requirement.

S Meters

Signal-strength meters are useful when there is a need to make comparative readings. Such might be the case when another operator asks for a comparison between two antennas he is testing. Because S meters are relative-reading instruments, signal reporting based on the amount of needle deflection is generally without meaning. No two receivers render the same reading for a given signal, unless by coincidence. This is because the gain distribution within an amateur receiver varies from band to band. Since most S meters are activated from the agc line in a receiver, what might be S9 on one ham band could easily become S6 or 10 dB over S9 on another band. A receiver that rendered accurate readings on each band it covered would be extremely esoteric and complex.

An attempt was made by at least one receiver manufacturer in the early 1940s to establish some significant numbers for S meters. S9 was to be equivalent to 50 μV, and each S unit would have been equal to 6 dB. The scale readings above S9 were given in dB. The system never took hold in the manufacturing world, probably for the reasons given earlier in this section.

In addition to the example shown in Fig. 42, some typical S-meter circuits are offered in Fig. 43. The example at C can be used with rf- or audio-derived agc.

Noise Reduction

In addition to active-device and circuit noise, much of the noise interference experienced in reception of high-frequency signals is caused by domestic or industrial electrical equipment and by automobile ignition systems. The interference is of two types in its effects. The first is the "hiss" type, consisting of overlapping pulses similar in nature to the receiver noise. It is largely reduced by high selectivity in the receiver, especially for code reception. The second is the "pistol-shot" or "machine-gun" type, consisting of separated impulses of high amplitude.

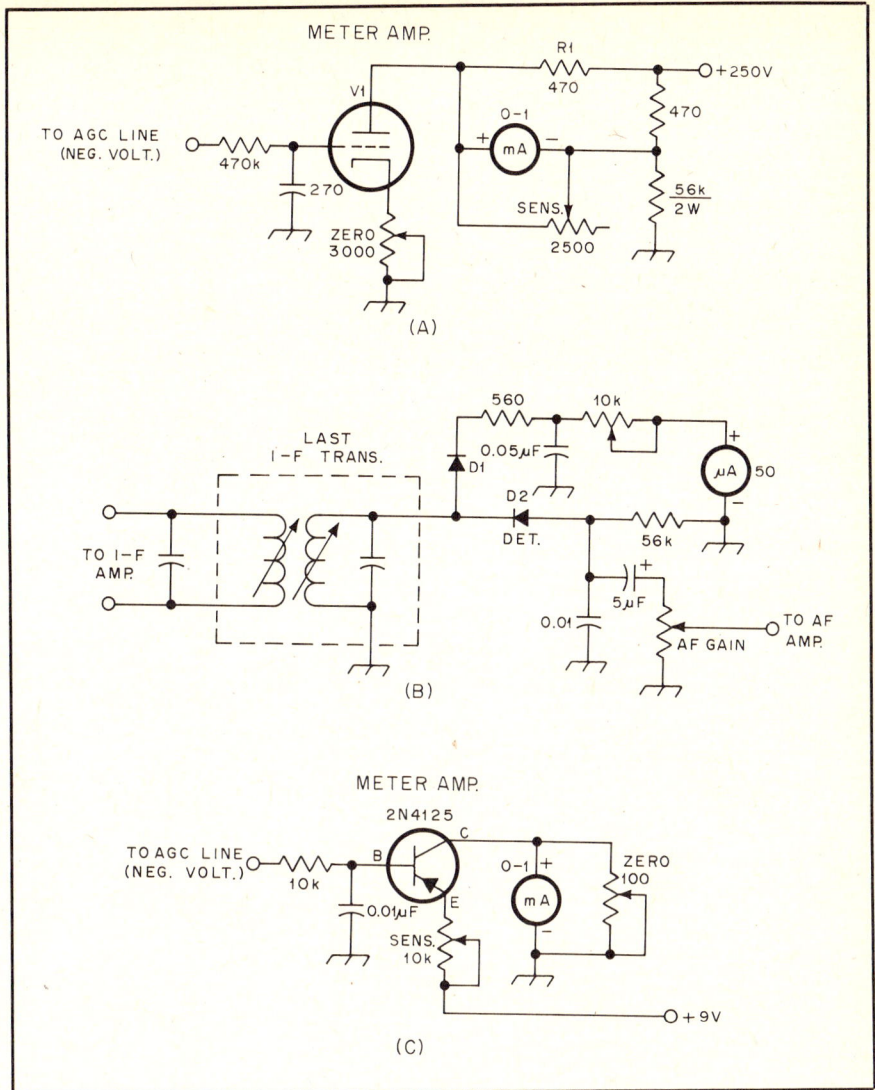

Fig. 43 — Various methods for using an S meter. At A, V1 is a meter amplifier. As the agc voltage increases the plate current decreases to lower the voltage drop across R1. An up-scale meter reading results as the current through the meter increases. At B, the i-f energy is rectified by means of D1 to deflect the meter. A 10-kΩ control sets the meter sensitivity. At C, the negative agc voltage forward biases the transistor to cause an increase in collector current, thereby deflecting the meter upwards with signal increases.

The hiss type of interference usually is caused by commutator sparking in dc and series-wound ac motors, while the shot type results from separated spark discharges (ac power leaks, switch and key clicks, ignition sparks, and the like).

The only known approach to reducing tube, transistor and circuit noise is through the choice of low-noise, front-end, active components and through more overall selectivity.

Impulse Noise

Impulse noise, because of the short duration of the pulses compared with the time between them, must have high amplitude to contain much average energy. Hence, noise of this type strong enough to cause much interference generally has an instantaneous amplitude much higher than that of the signal being received. The general principle of devices intended to reduce such noise is to allow the desired signal to pass through the receiver unaffected, but to make the receiver inoperative for amplitudes greater than that of the signal. The greater the amplitude of the pulse compared with its time of duration, the more successful the noise reduction.

Another approach is to "silence" (render inoperative) the receiver during the short duration time of any individual pulse. The listener will not hear the "hole" because of its short duration, and very effective noise reduction is obtained. Such devices are called "blankers" rather than "limiters."

In passing through selective receiver circuits, the time duration of the impulses is increased, because of the bandwidth of the circuits. Thus, the more selectivity ahead of the noise-reducing device, the more difficult it becomes to secure good

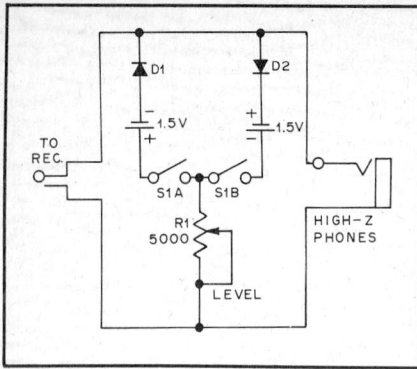

Fig. 44 — A simple audio limiter/clipper. R1 sets the bias on the diodes for the desired limiting level.

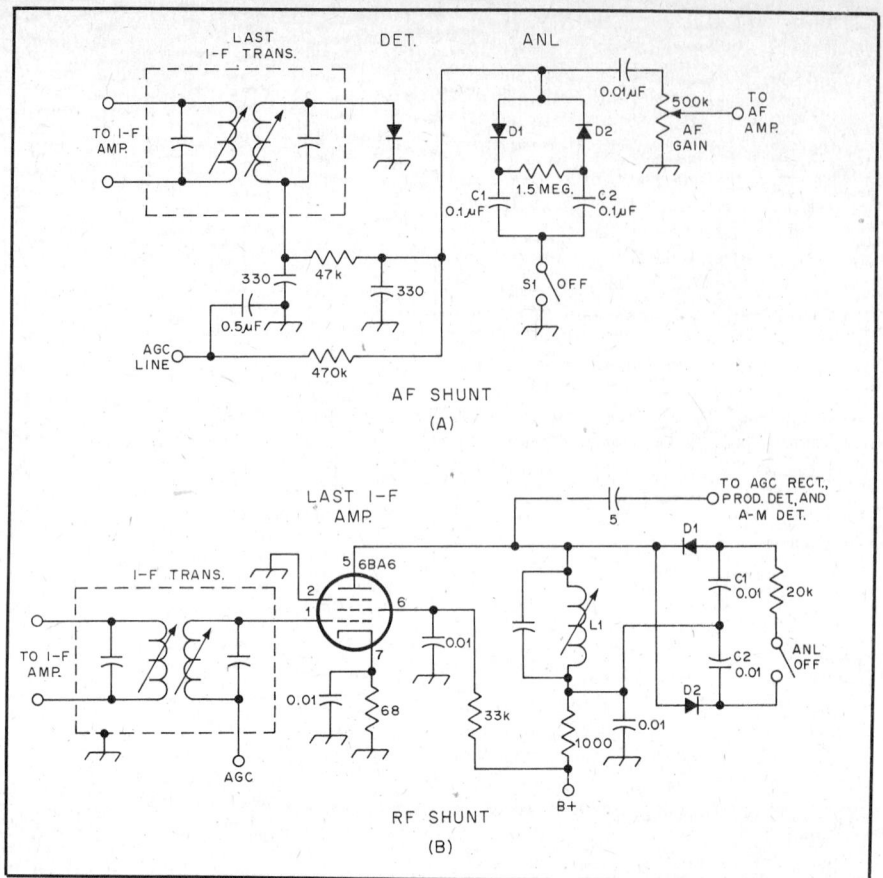

Fig. 45 — Examples of rf and audio anl circuits. Positive and negative clipping takes place in both circuits. The circuit at A is self-adjusting.

pulse-type noise suppression.

Audio Limiting

A considerable degree of noise reduction in code reception can be accomplished by amplitude-limiting arrangements applied to the audio-output circuit of a receiver. Such limiters also maintain the signal output nearly constant during fading. These output-limiter systems are simple, and they are readily adaptable to most receivers without any modification of the receiver itself. However, they cannot prevent noise peaks from overloading previous stages.

Noise-Limiter Circuits

Pulse-type noise can be eliminated to an extent which makes the reception of even the weakest of signals possible. The noise pulses can be clipped, or limited in amplitude, at either an rf or af point in the receiver circuit. Both methods are used by receiver manufacturers; both are effective.

A simple audio noise limiter is shown at Fig. 44. It can be plugged into the headphone jack of the receiver and a pair of headphones connected to the output of the limiter. D1 and D2 are wired to clip both the positive and negative peaks of the audio signal, thus removing the high spikes of pulse noise. The diodes are back-biased by 1.5-volt batteries permitting R1 to serve as a clipping-level control. This circuit also limits the amount of audio reaching the headphones. When tuning across the band, strong signals will not be ear-shattering and will appear to be the same strength as the weaker ones. S1 is open when the circuit is not in use to prevent battery drain. D1 and D2 can be germanium or silicon diodes, but 1N34As or 1N914s are generally used. This circuit is usable only with high-impedance headphones.

The usual practice in communications receivers is to use low-level limiting, Fig. 45. The limiting can be carried out at rf or af points in the receiver, as shown. Limiting at rf does not cause poor audio quality as is sometimes experienced when using series or shunt af limiters. The latter limits the normal af signal peaks as well as the noise pulses, giving an unpleasant audio quality to strong signals.

In a series-limiting circuit, a normally conducting element (or elements) is connected in the circuit in series and operated in such a manner that it becomes nonconductive above a given signal level. In a shunt limiting circuit, a non-conducting element is connected in shunt across the circuit and operated so that it becomes conductive above a given signal level, thus short-circuiting the signal and preventing its being transmitted to the remainder of the amplifier. The usual conducting element will be a forward-biased diode, and the usual nonconducting element will be a back-biased diode. In many applications the value of bias is set manually by the operator; usually the clipping level will be set at about 1 to 10 volts.

The af shunt limiter at A, and the rf shunt limiter a B operate in the same manner. A pair of self-biased diodes are connected across the af line at A, and across an rf inductor at B. When a steady cw signal is present the diodes barely conduct, but when a noise pulse rides in on the incoming signal, it is heavily clipped because capacitors C1 and C2 tend to hold the diode bias constant for the duration of the noise pulse. For this reason the diodes conduct heavily in the presence of noise and maintain a fairly constant signal output level. Considerable clipping of cw signal peaks occurs with this type of limiter, but no apparent deterioration of the signal quality results. L1 at C is tuned to the i-f of the receiver. An i-f transformer with a conventional secondary winding could be used in place of L1, the clipper circuit being connected to the secondary winding; the plate of the 6BA6 would connect to the primary winding in the usual fashion.

I-F Noise Silencer

The i-f noise silencer circuit shown in Fig. 46 is designed to be used ahead of the high-selectivity section of the receiver. Noise pulses are amplified and rectified, and the resulting negative-going dc pulses are used to cut off an amplifier stage during the pulse. A manual "threshold" control is set by the operator to a level that only permits rectification of the noise pulses that rise above the peak amplitude of the desired signal. The clamp transistor, Q3, short circuits the positive-going pulse "overshoot." Running the 40673

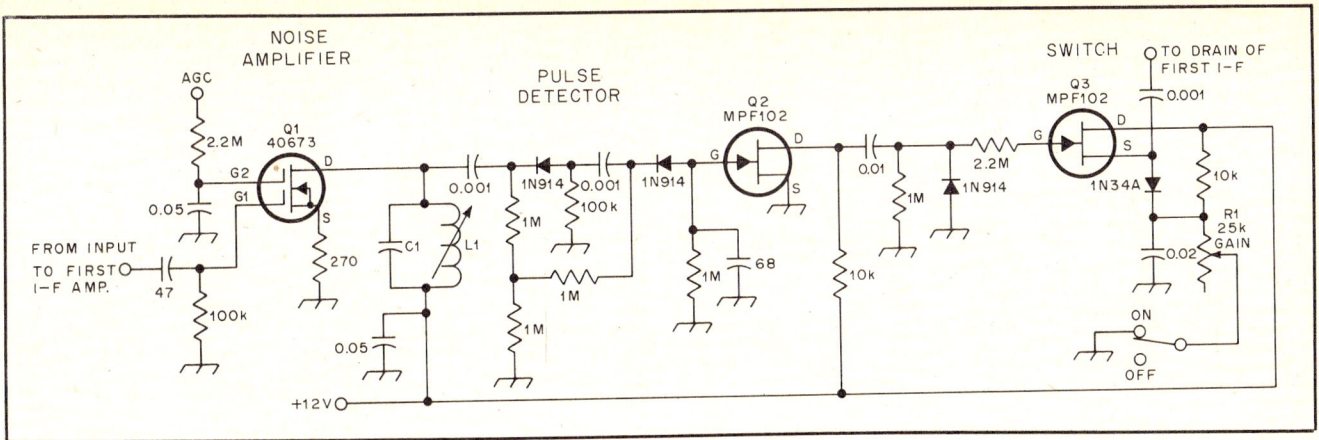

Fig. 46 — Diagram of a noise blanker. C1 and L1 are tuned to the receiver i-f.

controlled i-f amplifier at zero gate 2 voltage allows the direct application of agc voltage. See July 1971 *QST* for additional details.

Passive CW Audio Filters

Even though a receiver may have narrow-band i-f filtering for cw reception (200 to 600 Hz, typically), the addition of an audio filter can be of value to the operator. If a post i-f amplifier crystal or mechanical filter is included in the receiver design, a subsequent audio filter won't do much toward enhancing performance. But, when only one i-f filter is used (ahead of the i-f strip), audio filters of narrow bandwidth can greatly improve the receiver noise figure. This is because the inherent wide-band noise from the i-f amplifiers has the potential of degrading the signal-to-noise ratio of the receiver. Similarly, many receiver audio amplifiers will pass high and low frequencies that are not needed in communications work. A well-designed passive or active audio filter will greatly reduce wide-band noise and will restrict the audio bandwidth of the receiver — a significant aid to reception during weak-signal work and when the QRM level is high.

Fig. 47 contains the circuit of a 10-element, high-performance Chebyshev band-pass filter. It was designed by Ed Wetherhold, W3NQN, for use in this edition of the *Handbook*. The filter consists of two stacks of the commonly available 88-mH telephone-type toroidal inductors and one modified 88-mH inductor which is externally mounted on the end of one of the stacks. Table 3 provides component values and pertinent data for seven different center frequencies between 540 and 940 Hz. The center frequency selected will depend on the operator's peak-frequency preference.

Although the inductance of these surplus toroids is frequently given as 88 mH, this is only a nominal value. Those inductors having their two 22-mH windings on separate halves of the toroidal

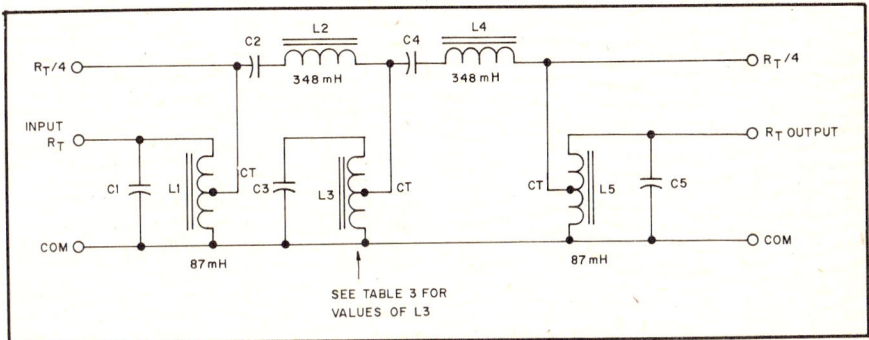

Fig. 47 — Schematic diagram of the W3NQN high-performance passive cw filter. All inductors (except L3) are the standard surplus 88 mH toroidal inductors with their two 22 mH windings connected in series aiding. L2 and L4 consist of four series-connected 87 mH inductors (4 × 87 mH = 348 mH). L1, 3 and 5 are center tapped.

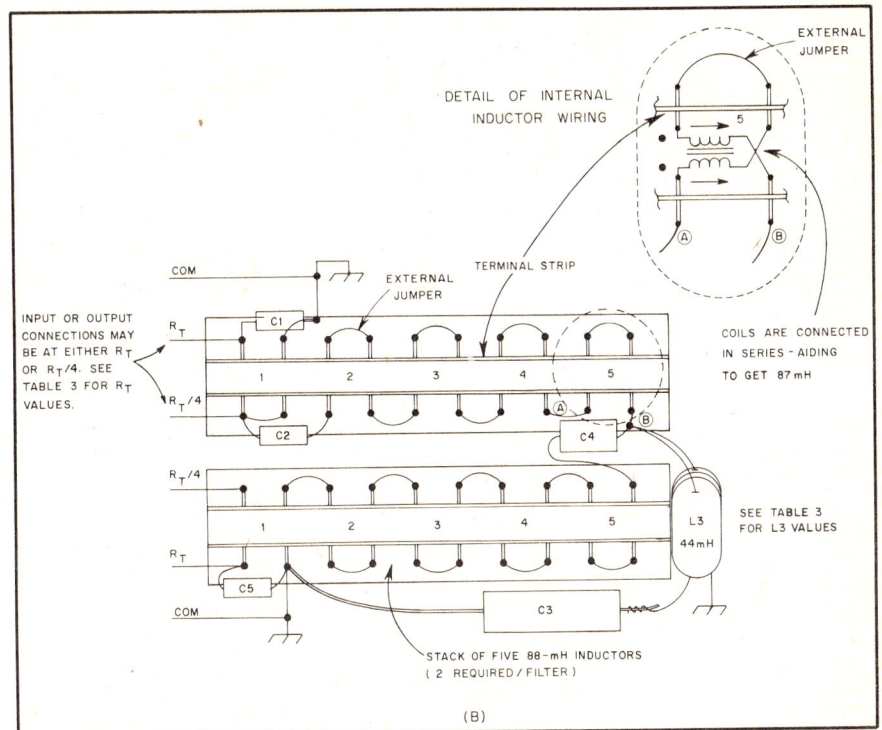

Fig. 48 — Pictorial details of how two standard 5-inductor assemblies can be wired to provide the filter in Fig. 47. Inductor L3 is a standard 88 mH surplus inductor that is modified by removing turns to give the L3 inductance specified in Table 3. See Fig. 49 to find how many turns must be removed to give the desired L3 value.

Table 3

CW Band-pass Filter Designs Using 88-mH Surplus Inductor Assemblies.
Component Values and Design Parameters of 5-Resonator Band-pass Filters[1]

F-CNTR[2] (Hz)	C1,5 (µF)	C3 (µF)	C2,4[3] (µF)	L3[4] (mH)	R-TERM[5] (OHMS)	BW-3[6] (Hz)	FL-3 (Hz)	FH-3 (Hz)	FL-30 (Hz)	FH-30[7] (Hz)
939.3	0.33	0.56	0.0825	51.3	1915	337	786	1123	709	1245
864.0	0.39	0.68	0.0975	49.9	1701	305	725	1030	652	1144
787.1	0.47	0.82	0.1175	49.9	1549	278	660	938	594	1042
750.4	0.52	1.00	0.129	45.0	1327	251	635	886	569	989
721.1	0.56	1.00	0.1400	48.7	1381	252	606	858	545	954
654.3	0.68	1.22	0.1700	48.5	1246	228	550	778	495	866
595.9	0.82	1.50	0.2050	47.6	1112	205	502	707	451	788
539.6	1.00	1.50	0.250	58.1	1367	206	446	652	406	717
539.6	1.00	2.00	0.2500	43.5	927	177	458	635	410	710

Notes:

[1] All tabulated designs are based on L1 = L5 = 87 mH, and L2 = L4 = 348 mH.

[2] Use the tabulated capacitor and L3 values to obtain the desired passband center frequency, F-CNTR.

[3] C2,4 is always one quarter of C1.

[4] To get inductor L3, remove an equal number of turns from each winding of a standard 88-mH inductor. See Fig. 49 to find the number of turns to remove.

[5] For satisfactory operation, the filter input and output terminals must be terminated within ±20% of the listed R-TERM value.

[6] The BW-3 frequency is the 3-dB bandwidth of the filter.

[7] The FL-3, FH-3, FL-30 and FH-30 frequencies correspond to the lower and upper 3 and 30-dB attenuation levels of the filter response.

core produce somewhat less than the expected 88 mH inductance when connected in series-aiding fashion. This is the result of less than perfect coupling between the two windings. The actual inductance varies between 85.7 and 87.4 mH depending on the closeness of coupling. For this filter design a value of 87.0 mH was used, as it closely approximates (within 2%) any inductance value that might be encountered.

The required inductance value for L3 is obtained by removing turns from a standard 88-mH inductor. Fig. 49A shows the number of turns to remove for inductance values between 54 and 40 mH. Earlier filter designs required a 44-mH inductor for L3[3,4]. This restriction often resulted in C3 being a non-standard value. The 44-mH inductors, while a standard surplus value, have become difficult to obtain, consequently it is now more convenient to modify an 88-mH inductor. With L3 no longer restricted in value, it is possible to use only standard value capacitors for C3. This makes construction much more convenient.

The filters constructed by Wetherhold are comprised of two inductor stacks having sheet metal covers. The metal covers allow the stacks to be soldered together and the mounting straps and capacitor leads are soldered directly to the covers. The external inductor (L3) is fastened to the end of the stack with GE RTV 108 Silicone Rubber Adhesive. An improved method of mounting L3, recommended by NØARQ, uses Scotch® Mounting Tape, Cat. #114. A 1-1/2-inch piece of the double-sided tape is placed

[3] Wetherhold, "Modern Design of a CW Filter Using 88- and 44-mH Surplus Inductors," *QST*, Dec., 1980.

[4] Wetherhold, "High-Performance CW Filter," *Ham Radio*, April, 1981.

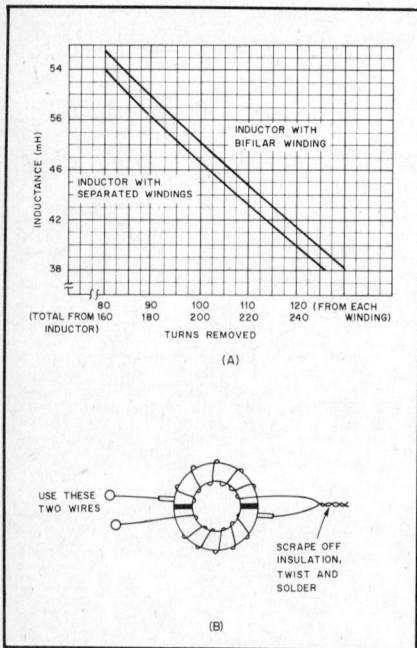

Fig. 49 — Inductance versus turns removed for typical 88-mH surplus inductor with windings in series-aiding connection (A). Series-aiding connection method to obtain 88-mH (B).

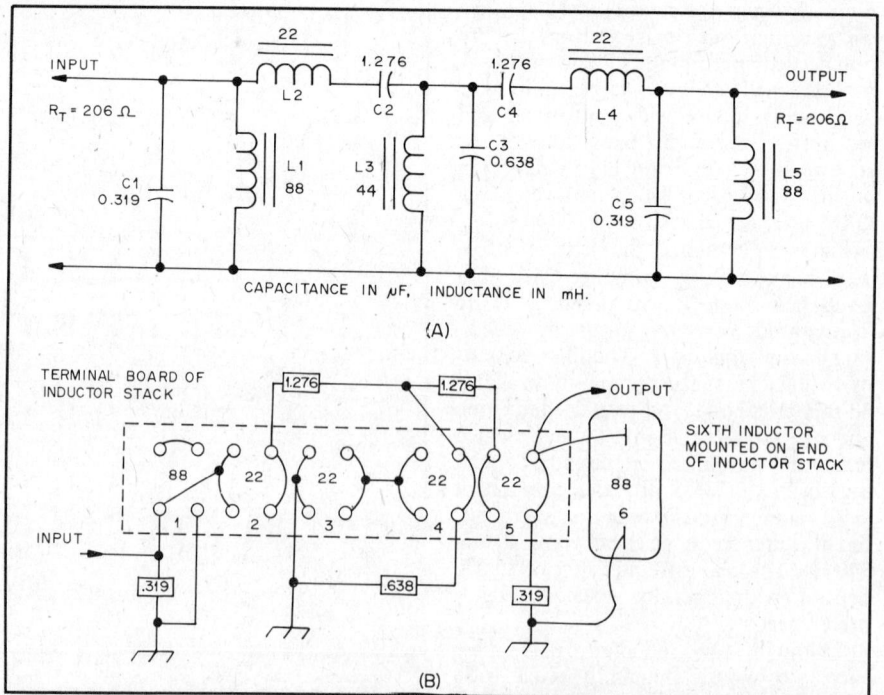

Fig. 50 — Schematic diagram of the ssb bandpass filter (A). Shown in B is a pictorial wiring diagram of the terminal board on the inductor stack.

over the end of the stack and pressed firmly in place. The protective liner is removed and the inductor is pressed into the sticky tape, the ends of the tape are pushed into the sides of the inductor to secure it to the stack.

The filter is conveniently installed between the receiver audio output and the headphones, but attention must be given to the termination of the filter. If the receiver output and headset are high impedance (600 to 1000 ohms), fixed-value resistors can be inserted to provide the exact termination required. If the receiver and headset are low impedance (4 to 8 ohms), then two 8- to 1000-ohm matching transformers, such as Radio Shack #273-1380, will be needed. The insertion loss of the filter is less than 3 dB, and the slight drop in signal level is easily compensated for by increasing the audio output level.

Surplus inductors are available at no charge (except for handling and shipping expenses) to those who wish to build this filter. These inductors are being made available to the amateur fraternity through the courtesy and cooperation of the Chesapeake and Potomac Telephone Company of Maryland. Write to E.E. Wetherhold, 102 Archwood Ave., Annapolis, MD 21401 for additional information. *Be sure to include a stamped self-addressed envelope.*

A PASSIVE AUDIO FILTER FOR SSB

While audio filters are most often used during cw reception, the ssb operator can also benefit from their use. Shown in Figs. 50 and 51 is a passive band-pass filter designed by Ed Wetherhold, W3NQN, for 'phone operation. This filter appeared in Dec. 1979 *QST*.

All of the inductors are the surplus 88-mH toroidal type with their windings wired either in series or parallel to get the required 88 or 22 mH of inductance. The 0.319-μF capacitors were selected from several 0.33-μF capacitors which were about 3 percent on the low side. The 0.638-μF value was obtained with a single 0.68-μF capacitor which was about 6 percent on the low side. The 1.276-μF values were obtained by paralleling selected 1-μF and 0.33-μF capacitors.

Fig. 51 shows the measured and calculated attenuation reponses of the filter. The difference between the measured and calculated responses at the low frequency side of the passband is probably caused by the much lower Q of the inductors at these frequencies.

The necessary termination resistance of this filter is 206 ohms. While this is not a standard value, it should not be too difficult for most amateurs to accommodate. If low-impedance headphones are used, a matching transformer can be used to provide the correct termination.

RC Active Audio Filters

The active type of audio filter is more

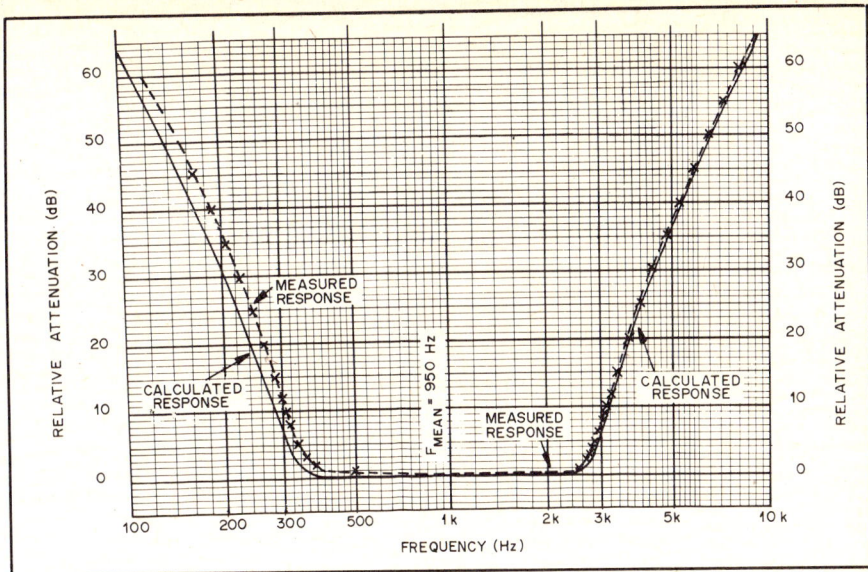

Fig. 51 — Response curves of the ssb band-pass filter.

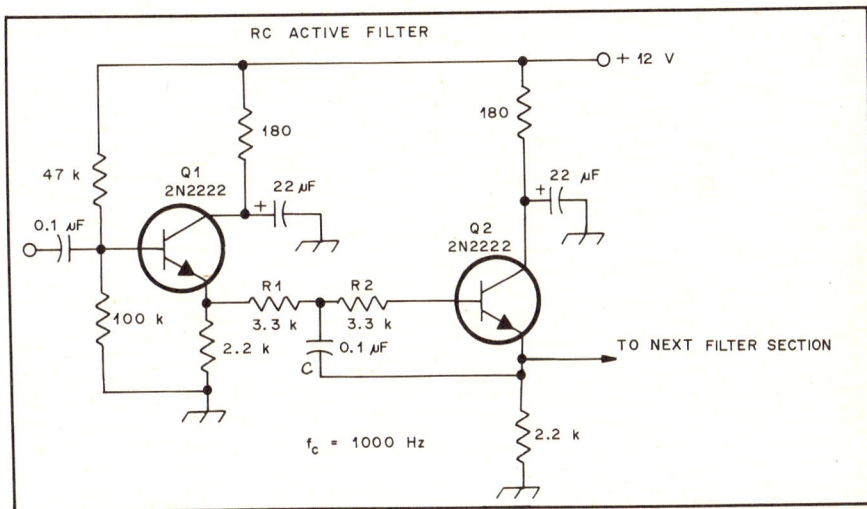

Fig. 52 — Circuit example of one pole section of an RC active audio filter which uses discrete active devices, Q1 and Q2.

popular than the passive type shown in Figs. 47 and 50. The primary advantages of active filters are (1) unity gain or greater (passive filters have some insertion loss), and (2) they can be built more compactly than LC filters can. Another advantage of RC active filters is that they can be made with variable Q and variable center or cutoff frequencies. These two features can be controlled at the panel of the receiver by means of potentiometers.

Most RC active filters are designed for a gain of 1 to 5. A recommended gain amount is 2 for most amateur applications. The more filter sections placed in cascade, the better the skirt selectivity. The maximum number of usable RC filter sections is typically 4. The minimum acceptable number is 2 for cw work, but a single section RC active filter is often suitable for ssb reception in simple receivers. As the Q and number of filter sections increases there is a strong tendency toward "ringing." This becomes manifest in the speaker or earphones as a

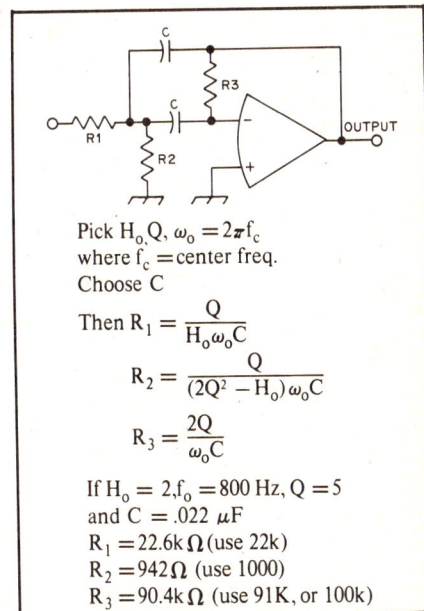

Pick $H_o, Q, \omega_o = 2\pi f_c$
where $f_c =$ center freq.
Choose C

Then $R_1 = \dfrac{Q}{H_o \omega_o C}$

$R_2 = \dfrac{Q}{(2Q^2 - H_o)\omega_o C}$

$R_3 = \dfrac{2Q}{\omega_o C}$

If $H_o = 2, f_o = 800$ Hz, $Q = 5$
and $C = .022\ \mu F$
$R_1 = 22.6$kΩ (use 22k)
$R_2 = 942\Omega$ (use 1000)
$R_3 = 90.4$kΩ (use 91K, or 100k)

Fig. 53 — Equations for designing an RC active audio filter.

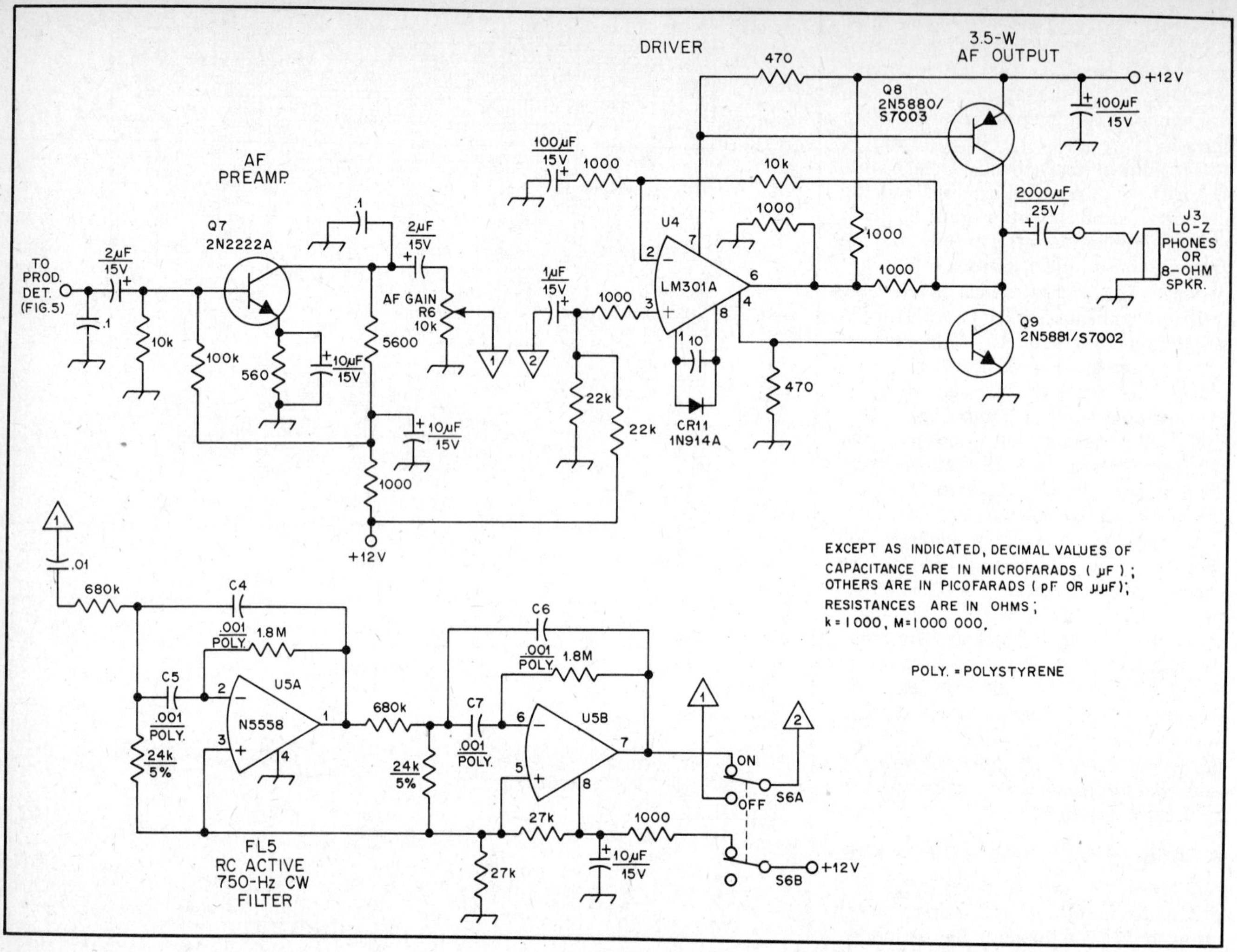

Fig. 54 — Practical circuit for a two-pole cw RC active filter, showing how it can be switched into and out of the audio channel of a receiver.

howling sound which can be most unpleasant to hear. The same is true of passive audio filters which have extremely high loaded-Q values.

Op amp ICs are used as the active devices in most RC active filters. The 741, LM301 and 747 types of ICs are suggested for that application. However, discrete devices can be used with equal success if the builder so desires. Fig. 52 shows one section of an active filter which uses transsistors. Q1 serves as a source-follower at the input and Q2 is one section of the filter. Cascaded additional filter sections would consist of the circuit which is common to Q2. The values of R1 and R2 would be changed to modify the f_c of the

filter. The lower the resistance value the higher the f_c. A dual potentiometer could be used in place of R1 and R2 to provide frequency variations.

Design data for RC active filters which use ICs is given in Fig. 53. One pole is shown. The term H_o is the desired voltage gain of the filter. Gains between unity (1) and 2 are the most common.

High-Q, stable capacitors are imperative to proper filter performance. Polystyrene capacitors are recommended for use at C of Figs. 52 and 53. The frequency-determining resistors and capacitors should be as close to the design

values as possible. Variations greater than 5 percent in resistance and capacitance in a multipole filter will widen the 3-dB bandwidth and cause dips in the nose of the response curve. In other words, f_c should be exactly the same for all filter sections in an ideal example.

A practical example of a two-pole RC active filter which uses a dual op-amp IC is given in Fig. 54. It is switched in and out of the audio amplifier by means of S1. As shown, the filter represents the minimum acceptable design for most cw work. A three- or four-section filter of this type would be more desirable for cw work under adverse band conditions (QRM or weak signals).

Simple Receivers for Beginners

AN 80- OR 40-METER DIRECT CONVERSION RECEIVER

Good results are obtainable with the basic direct-conversion receiver presented here. The circuit is without frills, making it easy to construct and operate. Information is given for operation on 80 or 40 meters, with only four transistors, one IC and three diodes. No attempt has been made to provide cw selectivity, but reception of cw and ssb signals is entirely adequate for this first effort at receiver building.

The circuit of Fig. 56 is designed for headphone output. The overall receiver sensitivity is suitable for comfortable reception of even the weaker signals when a resonant antenna is used. The front-end filter, L1/L2/C1/C2/C3, is designed to work into a 50-ohm antenna. This network is fixed tuned, so it does not have to be adjusted across the tuning range once it has been preset for midrange in the coverage of the main-tuning dial. If a 50-ohm antenna is not available, the user can install a small antenna-matching circuit between the receiver and the antenna system (a small Transmatch) to provide the proper termination for FL1. A rough approximation of the proper settings for the tuner can be found by adjusting for maximum sensitivity while listening to a weak signal. A better method is to adjust the matching network by means of a transmitter and SWR indicator. Then the receiver can be connected in place of the transmitter and FL1 adjusted as outlined later.

Q1 of Fig. 56 operates as a fed-back broadband amplifier. Output from Q1 is routed to the product detector (Q2) through T1, driving the source of Q2. VFO injection is on the detector gate. Q3 is the audio preamplifier. It is direct

Fig. 55 Front view of the beginner's receiver. The panels are made from double-sided pc board material.

coupled to the drain of the product detector. An additional 40 dB of audio gain is provided by U1, an operational-amplifier IC. A 330-ohm resistor is in series with the headphone line to permit the use of 8- and 16-ohm phones. Without the resistor the IC tends to self-oscillate at audio frequencies. The resistor is not required when the headphones have an impedance of 600 ohms or greater.

A protection diode, D2, is included to prevent accidental damage to the receiver if the power supply is attached with the wrong polarity.

The VFO uses only one transistor. Voltage to the drain of the FET, Q4, is regulated at 9.1 volts by means of a Zener diode, D2. D1 helps to stabilize the oscillator by limiting the positive swing of the sine wave during oscillation. Table 4 contains the correct values for the VFO parts, respective of the operating band, which do not have specified values on the schematic diagram of Fig. 56.

Construction

The receiver is built from pieces of double-sided pc board, except for the

Table 4
Component Data for the Direct-Conversion Receiver

Band	L1, L2 (μH)	C1 (pF)	C2 (pF)	C3 (pF)	C4 (pF)	C5 (pF)	C6 (pF)	C7, C8 (pF)
40 M	2 20 ts. no. 20 enam. on T68-6 toroid core. Use 2- turn link on L1.	5	240 nom. Arco 427	198 nom. Arco 427	42 s.m.	10 max.	200 poly. or s.m.	560 poly. or s.m.
80 M	8.7 40 ts. no. 24 enam. on T68-6 toroid core. Use 2- turn link on L1.	7	184 nom. Arco 427	116 nom. Arco 427	68 s.m.	15 max.	300 poly. or s.m.	950 poly. or s.m.

S.M. = silver mica. Poly = polystyrene. L3 nom. inductance for 40 meters = 4.5 μH (J. W. Miller 42A476CBI or equivalent). L3 for 80 meters has nom. inductance of 9.6 μH (J. W. Miller 42A105CBI or equivalent). T1 consists of 15 bifilar turns of no. 24 enam. wire on an Amidon FT 50-43 toroid core. Twist wires 6 times per inch before winding. C2 and C3 are miniature mica compression trimmers.

etched-circuit board which is *single* sided. The overall layout is larger than need be, in order to make the project less difficult for inexperienced builders to assemble. The VFO is enclosed in a separate compartment made from three pieces of double-sided pc board. The side pieces measure 2-1/2 × 3 inches (64 × 76 mm). The rear wall of the VFO box is 2 × 2-1/2 inches (51 × 64 mm). The rear corners of the box are joined by flowing solder along the inner seams. A 1-1/2 × 2-inch (38 × 51 mm) plate is soldered inside the front of the compartment, 1/2 inch (13 mm) from the front panel. This plate is also made from double-sided pc board. It is the mounting plate for the main-tuning capacitor. Three no. 6 spade bolts hold the VFO box to the main circuit board.

The front and rear panels of the receiver are 4-3/4 inches (127 mm) high and 6 inches (625 mm) wide. The etched-circuit board is soldered to the panels 3/4 inch (19 mm) up from the lower edges of the panels. After the panels are in place (they should be drilled for the attached parts first — see photographs), the leading edges of the VFO box can be tack-soldered to the front panel at two points to ensure rigidity. The rear panel can be reinforced by connecting a small home-made bracket between the inner wall of the panel and the etched-circuit board at some convenient point.

A 1-inch (25-mm) diameter hole is made in the front panel, centered on the shaft of the tuning capacitor. This will permit ample leeway when mounting the vernier-drive mechanism.

A blank area is provided near the rear of the etched-circuit board. This was done to allow room for accessories to be added, such as an RC active audio filter or a small crystal-controlled converter to permit reception of the 20- or 15-meter bands. Alternatively, an audio amplifier could be placed in that spot to provide speaker volume. Designs for such accessory items are left to the skills of the builder.

All solder joints must be done with care to prevent cold-solder joints and unwanted bridges (shorts) from one circuit foil to another. Final inspection with a magnifying glass is recommended. If a radio club adopts this receiver as a group project it is suggested that the members trade receivers for the purpose of inspection before applying the operating voltage.

Fig. 58 contains a scale template for the circuit board, plus a parts-placement guide. The general layout scheme of the receiver is apparent when viewing the photographs. D3 is connected between J3 and the pc board 12-volt bus.

Checkout and Operation

The VFO tuning range should be checked first. With operating voltage from two 6-volt lantern batteries (series

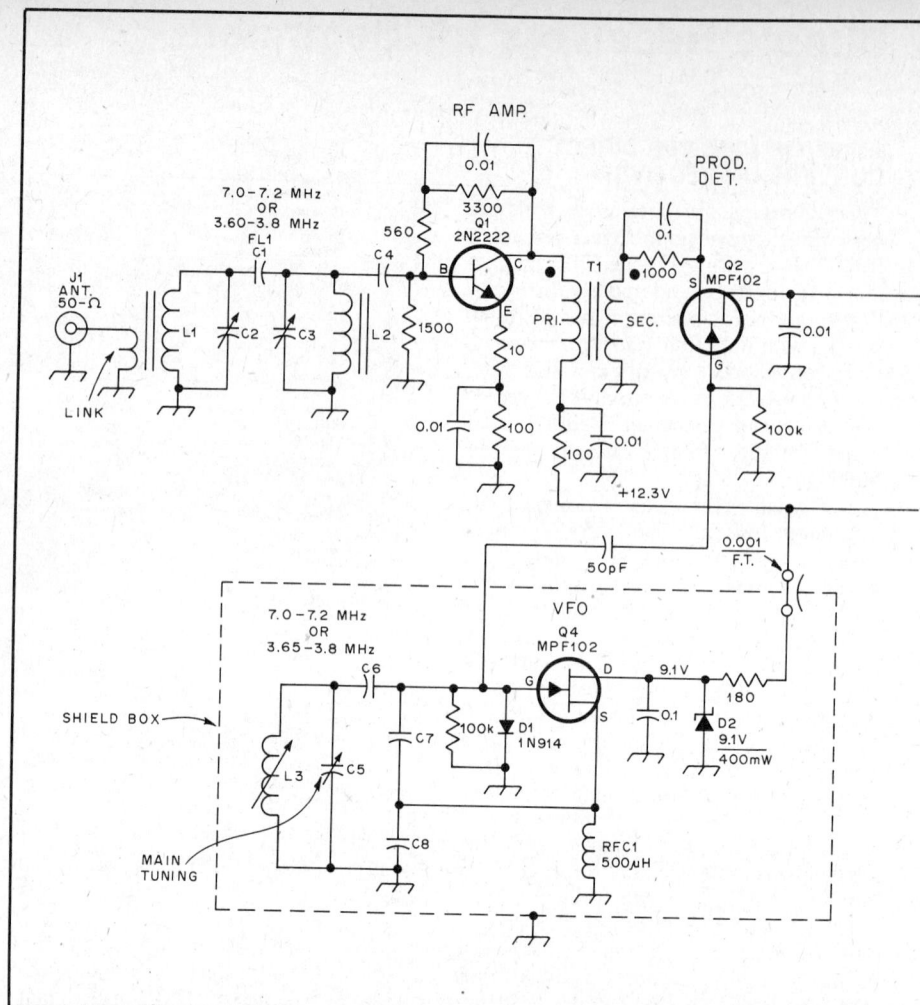

Fig. 56 — Schematic diagram of the direct-conversion receiver. Capacitors are disc ceramic unless noted otherwise. Polarized capacitors are electrolytic. Resistors are 1/4- or 1/2-watt composition. Numbered components which do not have assigned values are so designated for layout purposes. Others appear in Table 3. FT indicates feedthrough type.

C2, C3 — Mica compression trimmer, Arco 427 or equiv.
C5 — Miniature air variable, Hammarlund HF-15 or similar for 80-meter version. HF-10 or similar for 40-meter version.
D1 — Silicon switching diode, 1N914 or equiv.
D2 — Zener diode, 9.1-V, 400 mW or 1 watt.
D3 — Silicon rectifier diode, 50 PRV, 1 A

connected), 10 size D flashlight cells in series, or a 12-volt regulated dc power supply, place a short length of wire near the VFO circuit. The opposite end of the wire is connected to the antenna post of a calibrated amateur receiver. Look for the VFO signal in the 80- or 40-meter bands (depending upon which model is built). Once it is located, indicating that the oscillator is working, set C5 for maximum capacitance. Tune the receiver being used for calibration purposes to the low-frequency end of the VFO range (3.6 or 7.0 MHz) and adjust the slug of L3 in the homemade receiver until a signal is heard. When C5 is set for minimum capacitance the signal should appear at approximately 3.8 or 7.2 MHz. As an alternative to the

foregoing alignment method, a frequency counter can be coupled to the gate lead of Q2 for direct VFO readout.

All that remains in the alignment procedure is to set C2 and C3 of FL1 for peak signal response in the center of the VFO tuning range. If a 50-ohm signal generator is available it should be used for the purpose. If not, a 56-ohm resistor can be connected across J1 and a random length of antenna wire attached to the antenna post of the receiver. The trimmers of FL1 should then be set for maximum response of a weak signal near the center of the VFO tuning range. The 56-ohm resistor will provide a suitable termination for FL1 if the antenna wire is not a quarter wavelength long or some multiple thereof.

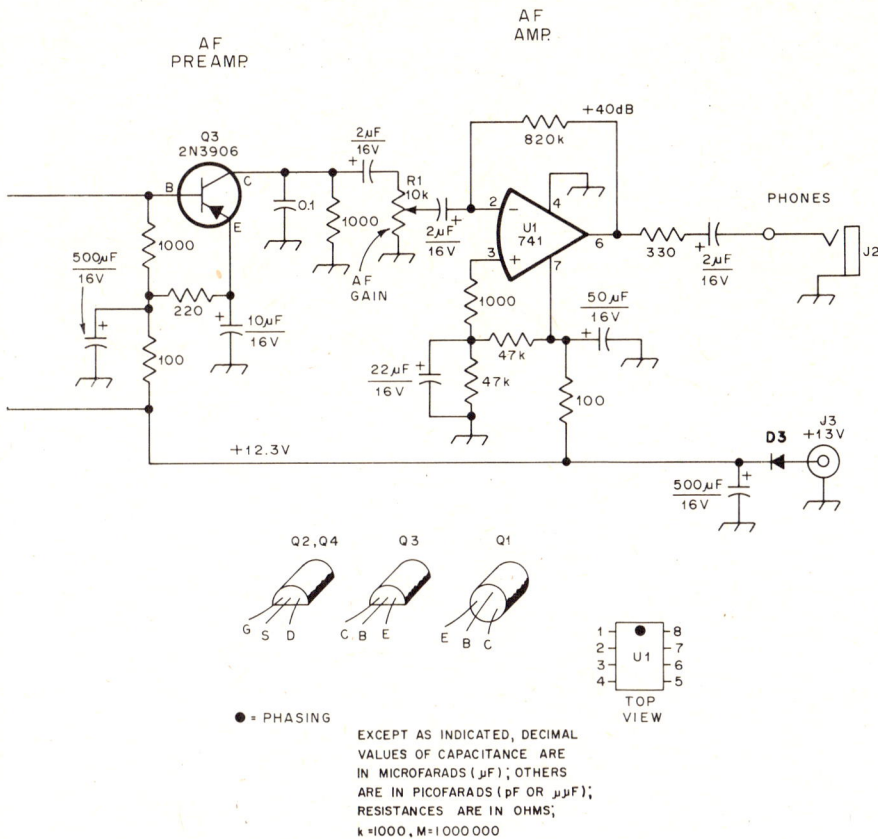

AF
PREAMP.

AF
AMP.

Fig. 57 — Top view of the receiver showing the input filter at the upper right, the VFO and its enclosure at the lower center and the audio-amplifier IC at the lower left. A Radio Shack vernier drive is used to turn the VFO tuning capacitor. A panel-reinforcing bracket is visible at the upper right.

• = PHASING

EXCEPT AS INDICATED, DECIMAL VALUES OF CAPACITANCE ARE IN MICROFARADS (μF) ; OTHERS ARE IN PICOFARADS (pF OR μμF); RESISTANCES ARE IN OHMS; k =1000, M =1 000 000

suitable.
J1 — SO-239.
J2 — Two-circuit phone jack.
J3 — Single-hole mount phono jack.
Q2, Q4 — Motorola MPF102 JFET.

R1 — 10-kΩ carbon control, audio taper preferred.
U1 — 741 op amp, 8-pin dual in line. Mount directly on pc board or use 8-pin IC socket.

Circuit boards and negatives for this receiver are available from Circuit Board Specialists, Box 969, Pueblo, CO 81002. Complete parts kits are also available from the same supplier.

A Simple Superhet for 75-Meter SSB

Circuit elaboration is not always essential to good receiver performance. This is particularly true when the builder desires compact equipment for portable operation. The simplicity concept is enhanced further by the low current drain which can be realized when only the bare essentials are designed into the circuit.

This superheterodyne receiver represents the most basic approach that will provide acceptable selectivity, immunity to front-end overloading and sensitivity. Five transistors, four diodes and one IC comprise the semiconductor count in the design. A supply voltage of 11 to 14 is suitable. The current drain is on the order of 50 mA maximum.

Fig. 60 shows the schematic diagram of this receiver. A fixed-tuned Butterworth front-end filter permits coverage from 3.75 to 4.0 MHz without the need for retuning the filter. A high-transconductance dual-gate MOSFET, Q1, serves as the mixer. Conversion gain with this device (3N211) is very high, owing to the g_m of 30,000 micromhos. A Collins Radio CB-type mechanical filter is shown as the i-f selectivity element, FL2. This part was chosen because of the low cost. However,

any of the Collins 455-kHz mechanical filters designed for ssb bandwidth can be used, provided the insertion loss is low. The only circuit changes necessary would be modification of the terminating resistances, the filter resonating capacitors, and the BFO crystals (Y1 and Y2). This model calls for 2700 ohms at each end of the filter, plus two 360 pF resonating capacitors. The filter bandwidth is 2.2 kHz at the 3-dB points. The 60-dB bandwidth is 5.5 kHz. A lower-cost 455-kHz filter alternative is offered in the modification diagram of Fig. 61. A J. W. Miller 8814 transformer/filter is specified. It contains a monolithic filter that provides a 4-kHz bandwidth at the 3-dB points of the response curve. It is suggested that the Miller 8814 be used in place of a mechanical filter to ensure maximum overall receiver gain and reduced cost. An extra stage of i-f amplification may be required with some mechanical filters in order to have sufficient headphone output on weak signals.

Output from FL2 is routed to a single-stage i-f amplifier, Q2, another 3N211 FET. The gain of this amplifier, plus that of the audio-amplifier IC U1 is controlled manually by means of a dual control, R1A/R1B. The bias on gate 2 of Q2 is varied at R1A to set the i-f gain level. In order to obtain a wide range of control it is necessary to place gate 2 at a volt or two minus with respect to gate 1. This is achieved by "bootstrapping" the stage with D1, an LED which conducts at

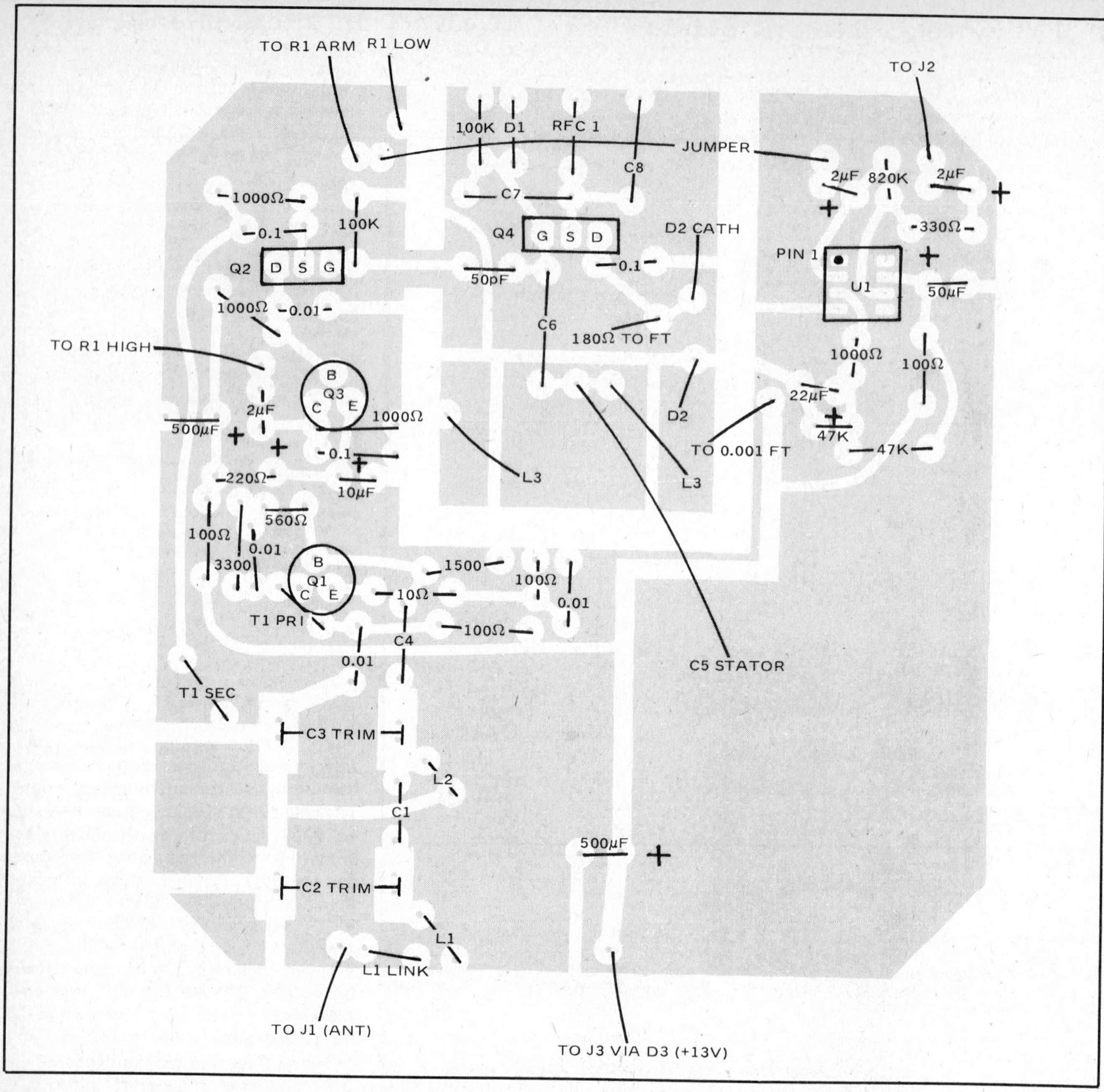

Fig. 58 — Scale pattern of the pc board for the direct-conversion receiver showing parts placement. Shown from the component side of the board.

roughly 1.5 volts.[5] Therefore, when R1A has its arm at ground, gate 2 of Q2 is effectively at −1.5 volts (minimum gain). D1 serves purely as a reference diode in this instance. Alternatively, two 1N914 diodes can be used in series from source to ground to provide a reference of roughly 1.4 volts.

A 3N211 FET serves as the VFO. Gates 1 and 2 are connected together to simulate a single-gate transistor. Injection for the mixer is taken from the gate of Q3 in order to realize a 3-volt pk-pk level at gate

2 of Q1. A pure sine wave is available at that take-off point. Some oscillator pulling (slight) will be observed when FL1 is aligned. However, once the front-end filter is tuned the effect will not be noticed. A VFO buffer/amplifier would resolve this condition, but is not necessary in practice if circuit simplicity is to be keynoted. D2 is used as a switching diode to offset the VFO frequency when changing from upper to lower sideband. This eliminates the need to readjust the main-tuning dial of the receiver. This part of the circuit need not be included if dial calibration is not essential when changing

Fig. 59 — Outside view of the simple superhet receiver. The front and rear panels are made from double-sided circuit board.

[5]The LED should be installed so that it illuminates when R1 is set for maximum receiver gain.

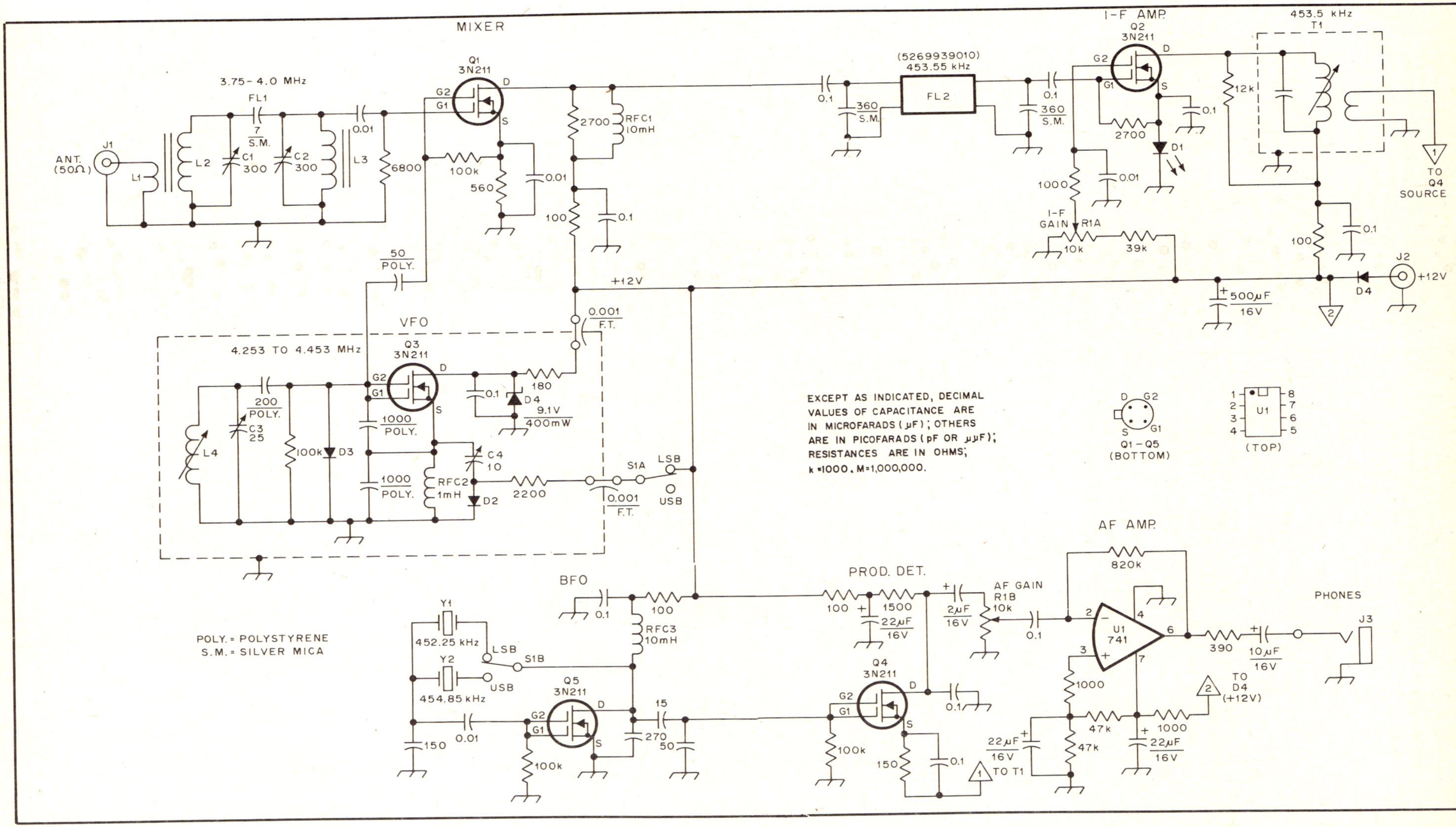

Fig. 60 — Schematic diagram of the ssb receiver. Fixed-value capacitors are disc ceramic unless noted otherwise. Polarized capacitors are electrolytic. Fixed-value resistors are 1/4- or 1/2-watt composition.

C1, C2 — Mica compression trimmer. 300 pF max. Arco 427 or equiv.

C3 — Miniature 25-pF air variable. Hammarlund HF-25 or similar.

C4 — Circuit-board mount subminiature air variable or glass piston trimmer, 10 pF mix. NP0; miniature ceramic trimmer suitable as second choice.

D1 — LED, any color or size. Used only as 1.5-V reference diode.

D2, D3 — Silicon switching diode, 1N914 or equiv.

D4 — Polarity-guarding diode. Silicon rectifier, 50 PRV, 1A.

D5 — Zener diode, 9.1 volts, 400 mW or 1 watt.

FL1 — Band-pass filter (see text)

FL2 — Collins Radio CB-type mechanical filter, Rockwell International no. 5269939010, 453.33 kHz center freq.

J1 — SO-239.

J2 — Single-hole-mount phono jack.

J3 — Two-circuit phone jack.

L1 — Two turns no. 24 insulated wire over ground end of L2.

L2, L3 — 40 turns no. 24 enam. wire on T68-6 toroid core.

L4 — Slug-tuned inductor, 3.6 to 8.5 range, J. W. Miller 42A686CBI or equivalent suitable. Substitutes should have Q of 100 or greater at 4 MHz and be mechanically rigid.

Q1-Q5, incl. — Texas Instruments 3N211 FET.

R1 — Dual control, 10-kΩ per section, linear taper. Allen Bradley type JD1N200P or similar. Separate controls can be used by providing extra hole in front panel.

RFC1, RFC3 — 10-mH miniature rf choke,

J. W. Miller 70F102AI or equiv.

RFC2 — 1-mH miniature rf choke, J. W. Miller 70F103AI or equiv.

S1 — Two-pole, two-position phenolic or ceramic wafer switch.

T1 — 455-kHz miniature i-f transformer (see text). J. W. Miller no. 2067.

U1 — 8-pin dual-in-line 741 op amp.

Y1, Y2 — International Crystal Co. type GP, 30-pF load capacitance, HC-6/U style of holder. Lsb 452.25 kHz, and usb 454.85 kHz.

EXCEPT AS INDICATED, DECIMAL VALUES OF CAPACITANCE ARE IN MICROFARADS (μF); OTHERS ARE IN PICOFARADS (pF or μμF); RESISTANCES ARE IN OHMS; k=1000, M=1,000,000.

POLY = POLYSTYRENE
S.M. = SILVER MICA

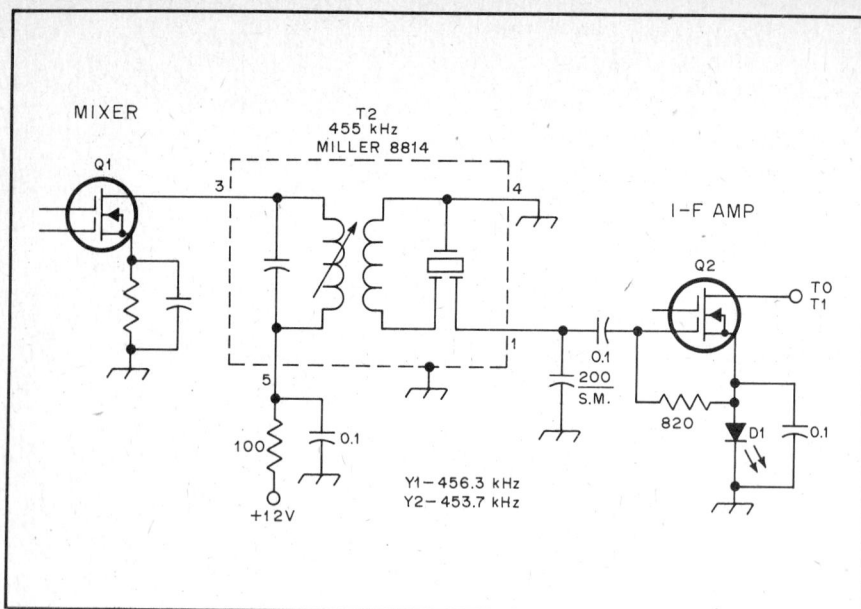

Fig. 61 — Circuit modifications for inclusion between the mixer and i-f amplifier of the circuit in Fig. 60. T2 is a J. W. Miller no. 8814 i-f filter transformer (see text).

Fig. 62 — Top view of the simple superhet receiver showing the locations of the various components. The VFO and its enclosure are at the center of the etched-circuit board adjacent to the front panel. The layout seen here differs in some areas from the artwork in Fig. 63 due to late circuit changes. The speaker on the rear panel is not used.

sidebands.

A 3N211 is employed at Q4 (gates tied together) as the product detector. Output from the i-f amplifier is fed to the source of Q4. BFO energy is supplied to Q4 from another 3N211 (Q5). Y1 and Y2 are selectable to permit upper and lower sideband reception. Since lower-sideband transmissions are the general rule on 75 meters, the builder may elect to use only the lsb crystal and eliminate S1 and the VFO offset circuitry.

Audio gain is provided by U1, a 40-dB op amp. The receiver output is adequate for weak signals (1 V or greater) with the arrangement shown in Fig. 54. Addition of a 0.5 or 1-watt audio IC would enable the builder to employ a speaker rather than phones, but the current drain of the receiver would be considerably higher.

Construction Notes

The etched-circuit board is the single-sided variety (copper on only one side). Double-sided pc board material is used for the front panel, rear panel and the VFO enclosure. Constructional details and dimensions follow closely those given for the direct-conversion beginner's receiver treated earlier in this chapter. Therefore, that information will not be repeated here. Pc boards, negatives and parts kits for this receiver are available from Circuit Board Specialists, listed in the text for the foregoing beginner's receiver. An adapter pc board is available from that supplier to permit using the modification shown in Fig. 61.

Fig. 63 shows the scale pc pattern and parts placement for this receiver. The panels are soldered to the ground foil which forms the perimeter of the etched-circuit board.

Adjustment and Use

FL1 is designed to be terminated in a 50-ohm load. The antenna or signal generator used during adjustment of C1 and C3 should provide a 50-ohm termination. Tune in a weak signal near 3875 kHz and adjust C1 and C2 for maximum signal output in the headphones. There may be some interaction, so repeat this step two or three times. This assumes that the VFO has been calibrated by means of the slug in L4 to provide mixer injection at 4453 kHz when C3 is set for minimum capacitance (plates unmeshed).

T1 is peaked last for maximum signal output from the receiver, again using a weak-signal source. This transformer is adjusted for resonance at the center of the FL2 passband — 453.55 kHz.

Offset trimmer C4 of Fig. 60 should be set as follows. Tune in a weak signal for zero beat, with S1 in the upper-sideband position. Switch S2 to lower sideband and adjust C4 for zero beat as heard in the headphones. The main-tuning dial should not be moved during this part of the alignment.

The reason FL2 does not have a center frequency of 455 kHz is because it was designed by Collins for use as a lower-sideband filter. An upper-sideband filter is used with it in CB radios. That permits the use of a single BFO crystal at 455 kHz for either sideband. The passband of this filter is symmetrical, just as is true of the regular 455-kHz Collins mechanical filters in amateur receivers. FL2 was chosen for this project because the price is roughly one-third that of the mechanical filters used by most amateur builders.

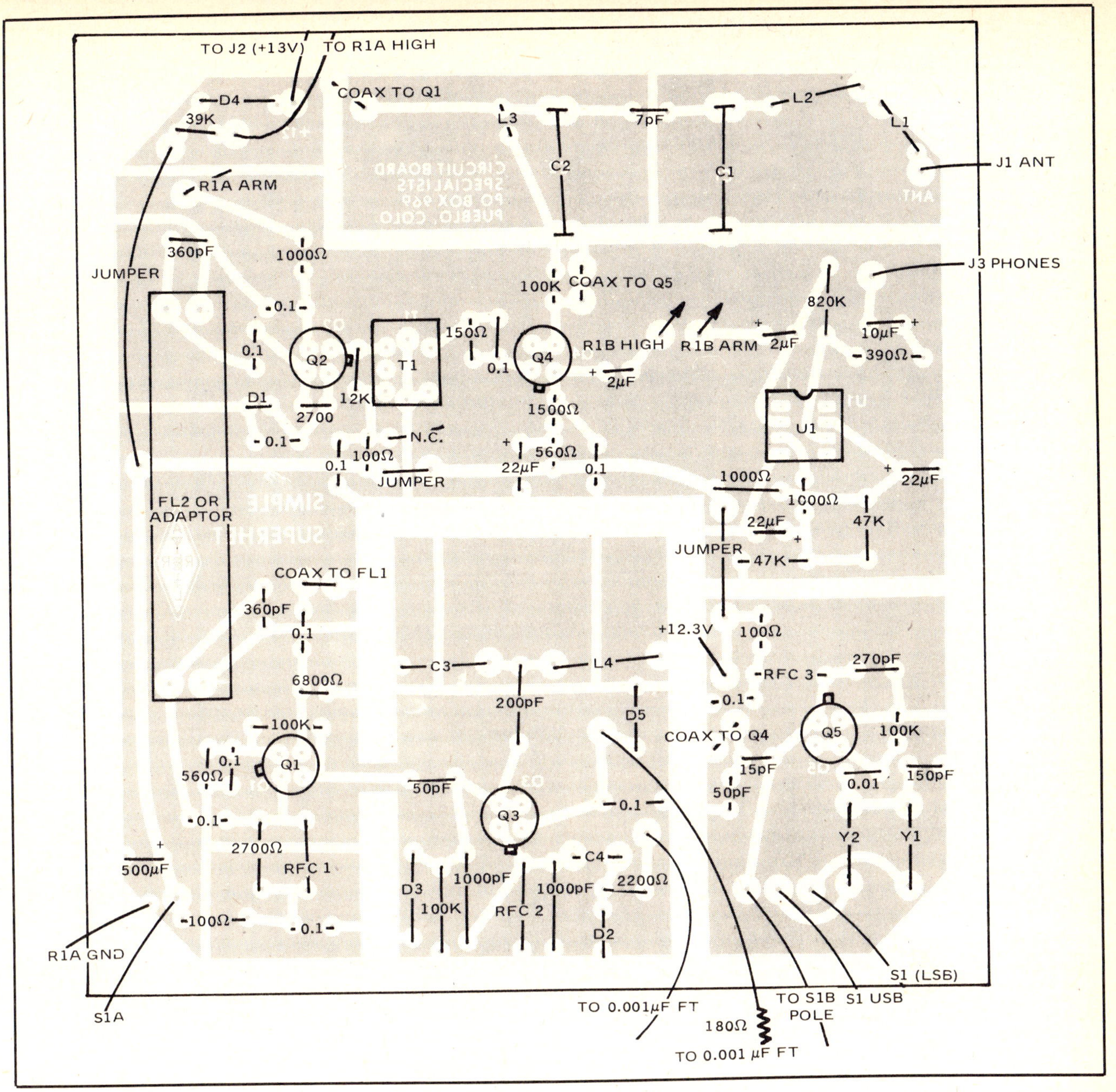

TO J2 (+13V) TO R1A HIGH

COAX TO Q1

D4
39K

L3   7pF   L2   L1

C2   C1   J1 ANT

R1A ARM

360pF   1000Ω

JUMPER

0.1   100K   COAX TO Q5   J3 PHONES

820K

0.1   150Ω   10µF

Q2   T1   Q4   R1B HIGH   R1B ARM   2µF   390Ω

D1   12K   0.1   2µF

2700   1500Ω   U1

0.1   560Ω

0.1   100Ω   22µF   0.1   1000Ω   22µF

FL2 OR ADAPTOR   N.C.   JUMPER   1000Ω   47K

22µF   JUMPER   47K

COAX TO FL1   +12.3V   100Ω

360pF   270pF

0.1   C3   L4   RFC 3

6800Ω   200pF   D5   0.1

COAX TO Q4   Q5   100K

100K   15pF   0.01   150pF

0.1   560Ω   Q1   50pF   Y2   Y1

0.1   50pF   Q3   0.1

500µF   2700Ω   RFC 1   C4   2200Ω

D3   1000pF   1000pF

100Ω   100K   RFC 2   D2

R1A GND   S1 (LSB)

S1A   TO 0.001µF FT   TO S1B S1 USB   POLE

180Ω   TO 0.001µF FT

Fig. 63 — Scale template of the circuit board and the parts-placement guide for the superhet receiver.

A High-Performance Communications Receiver

The receiver described here is the work of W7ZOI and K5IRK and was presented in Nov. 1981 *QST*. A progressive system was used in the *QST* version, starting with a simple (but very useable) direct conversion receiver and concluding with a multiband superheterodyne. This approach is highly recommended for those who may lack construction experience. Shown in Figs. 64 through 73 is the receiver in its final form. The overall layout of the receiver is shown in the block diagram, Fig. 65. On 80-meters the receiver functions as a single conversion superheterodyne, with reception of the higher bands provided by high-performance crystal-controlled converters. Note that all of the converters use the same mixer module, switching only the converter filter and crystal oscillator modules. As shown, the 40-meter converter does not use an rf amplifier — it is unnecessary and improved dynamic range results from its omission. The lower noise figure obtained with the rf stage is required on the bands above 7 MHz, however.

Many criteria were used in the design of this receiving system, but first and foremost were simplicity and ease of duplication. To this end, readily available components were used throughout. Alternative components are suggested where appropriate and the circuits are insensitive

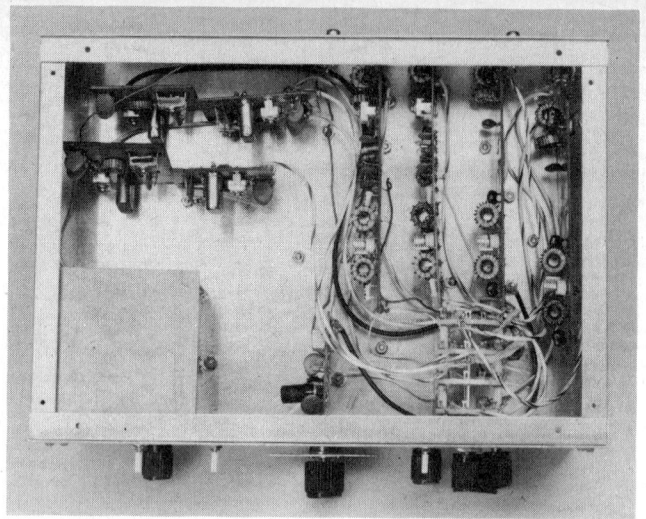

Fig. 64 — A 5-band, ssb version of the high-performance communications receiver constructed by K5IRK. In the top view the VFO is located in the center with the input filter, mixer and i-f boards to the left. The board at the far right contains the product detector and audio stages. The two boards to the right of the VFO are active audio filters (A). The bottom view shows the converter and oscillator boards. The BFO is contained in the shielded box at the left and the mixer board is at the lower center (B).

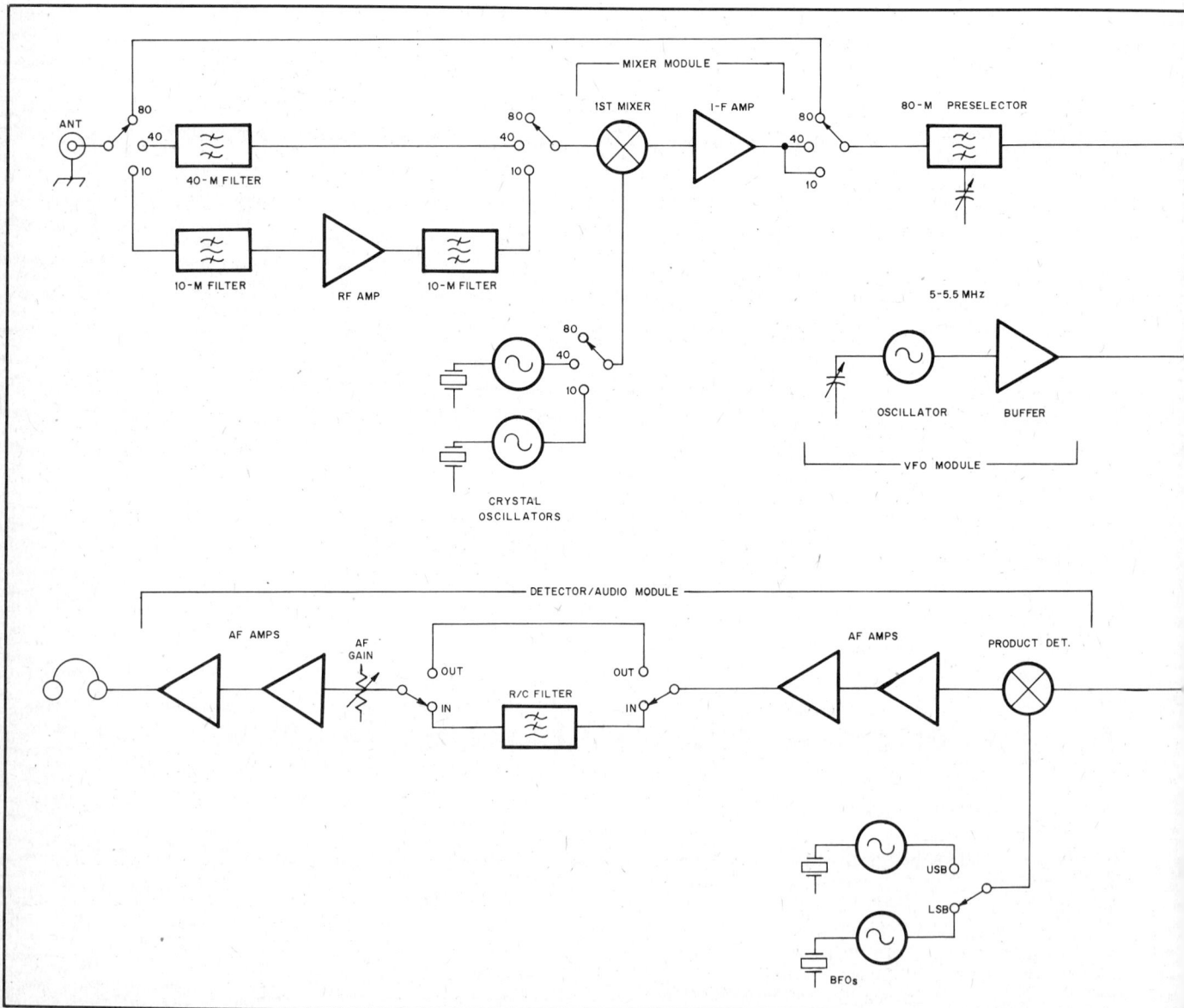

Fig. 65 — Block diagram of the W7ZOI/K5IRK communications receiver.

to transistor type, allowing freedom in substitution. This does not in any way imply that the performance has been compromised; indeed, this receiver can equal the strong-signal performance of many of the high-priced receivers on the market today.

Circuit Description: The Converter, Filter and RF Amplifier

Preselection for the individual converter sections is provided by the circuit shown in Fig. 66. The optional rf amplifier is shown in A and the version without the amplifier is shown in B. The same circuit board layout can be used for both versions.

Each filter module uses two types of filters. The first is a 5-pole low-pass, necessary to prevent spurious responses from vhf TV and fm broadcast signals. The second filter provides the majority of the front-end selectivity, it is a double-tuned circuit comprised of L7 and L8 and their related capacitors. A variable capacitor is used at C15 because the small, non-standard values required here are difficult to obtain; a 1- to 5-pF variable is readily available and can be preset to the value given in Table 5 or adjusted during alignment.

The rf amplifier uses a dual gate MOSFET and modified input low-pass filter. The first section is a simple low-pass filter while the second section is a pi network that transforms from 50 to 2000 ohms with a Q of 10. This provides a near optimum driving impedance for the amplifier. The output uses a broadband transformer to provide a 50-ohm output impedance, ensuring proper termination for the following double-tuned circuit.

The filters may be aligned with a signal generator or crystal calibrator. If a calibrator is used, the input of the receiver should be terminated with a 50-ohm resistor. Initially, C15 is set near minimum capacitance and the receiver is tuned to the center of the band. C14 and C17 are then adjusted for maximum response. C15 is then increased and C14 and C17 are again peaked for maximum response. The filter bandwidth is estimated by observing the response as the receiver is tuned toward the band edges. This procedure is repeated until the desired bandwidth is realized. The input pi network used with the rf amplifier is adjusted by setting C22 for maximum response at the center of the band.

Mixer Module

The two mixer modules used in the composite receiver are identical, each being comprised of a doubly balanced diode ring mixer, U2 of Fig. 67, followed by a 9-MHz i-f amplifier. The i-f amplifier, Q9, is one of the more critical stages in the receiver. It must have a reasonable noise figure, low IMD, and the input and output impedances must be 50 ohms. A bipolar transistor with negative feedback is used to establish the gain and impedances and the 6 dB pad at the output preserves the input and output impedances of the stage. The moderately high bias current used ensures low distortion.

The transistor type used for Q9 *is critical*. It should have an f_T of at least 500 MHz. The 2N5109, 2N3866 and 2SC1252 are all suitable.

Amplifier gain, including the loss of the pad, is about 16 dB. The mixer has a loss of about 6 dB, leaving a net module gain of 10 dB. The amplifier output intercept is about +30 dBm. Careful measurements have shown that a diplexer is not required between the mixer and this amplifier.

Crystal Oscillator Module

Shown in Fig. 68 is the circuit used for all of the crystal oscillators in the receiver. One or two of the modules are used for the BFO and one module is used with each converter. The circuit is a Hartley oscillator with the crystal in each series with the feedback tap from the coil. A trimmer capacitor in series with the crystal adjusts the operating frequency of the oscillator. If a single BFO is used, the 12-volt operating bias is applied through the output link as shown in Fig. 68B. When more than one module is used, as with the converters, operating voltage is applied through the bandswitch, Fig. 68C. Only the oscillator in use has power applied to it. This circuit will deliver an output power of about +10 dBm, which is more than enough to drive the diode mixers. Adjustment of the oscillators is best done with the mixer attached. C11 is tuned for maximum output and proper starting of the oscillator. The series capacitor is then adjusted for the correct operating frequency. This capacitor may be eliminated in those modules used with

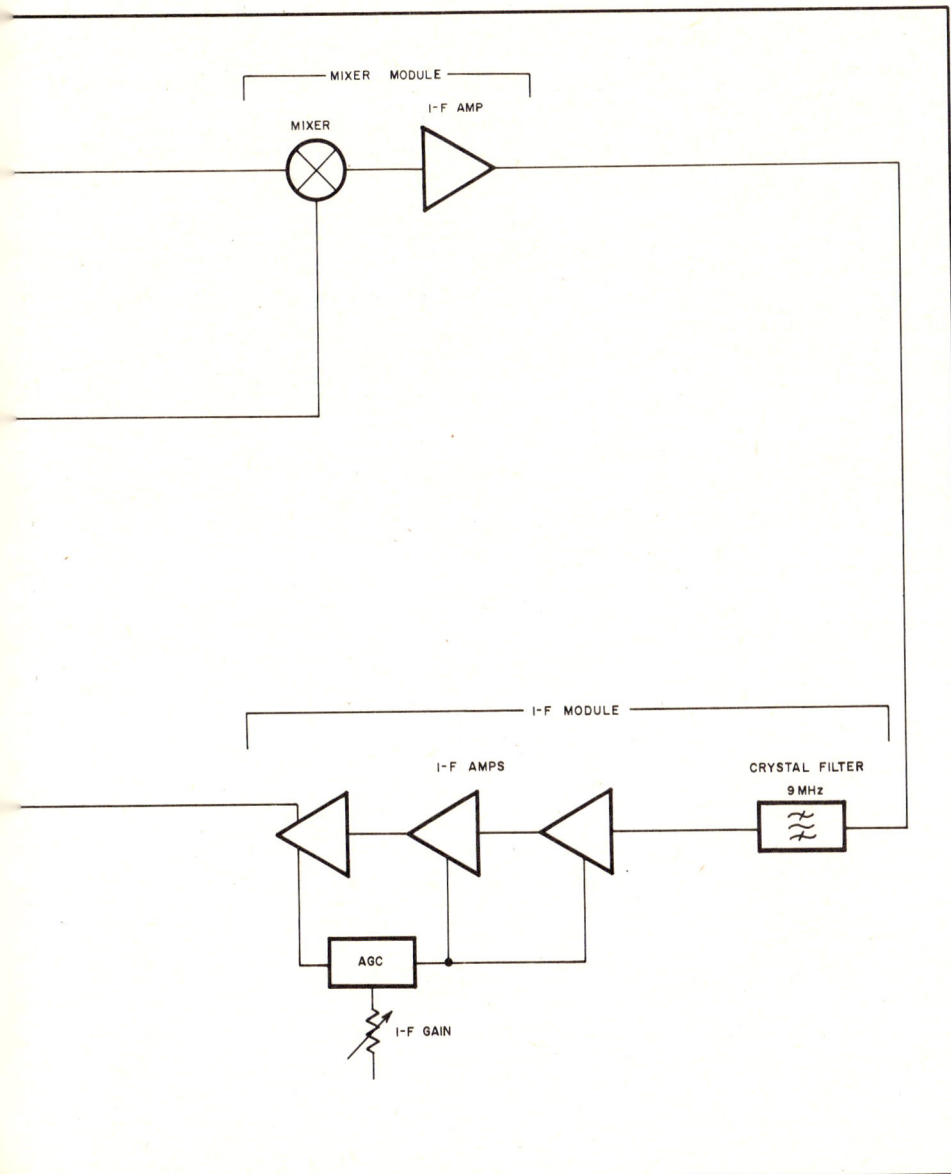

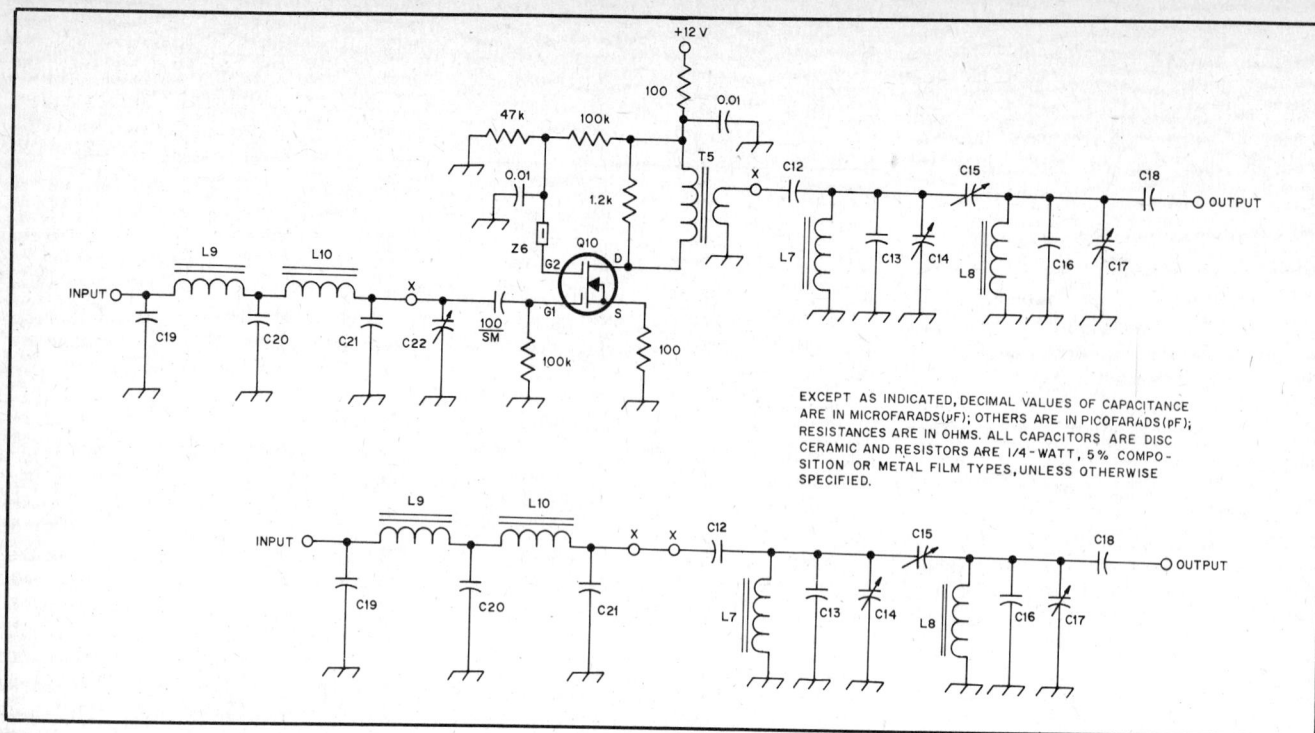

Fig. 66 — Circuit diagram of the converter filter and optional rf amplifier used in the W7ZOI/K5IRK receiver. The circuit using the rf amplifier is shown in A, while the configuration without the amplifier is shown in B.

C12, C13, C16, C18-21 — Silver mica or
  ceramic, see Table 5 for values.
C14, C17, C22 — Mica compression trimmer or
  similar variable, see Table 5 for values.

C15 — Air variable, 1 to 5 pF.
L7-10 — No. 22 enamel wire wound on Amidon
  T50-6 core, see Table 5 for number of turns.
Q10 — Dual gate MOSFET, 40673, 3N211,
  3SK40 or similar.

T5 — Ferrite transformer, 20 turns primary,
  4 turns secondary, on Amidon FT37-43 core.
Z6 — Ferrite bead on lead of Q10. Amidon
  type FB43-101.

the converters, the crystal is then connected directly to the tap on T3.

80-Meter Preselector Filter

The 80-meter preselector filter is shown in Fig. 69. It consists of two cascaded filters: The first is a 7-pole high-pass (3-MHz cutoff) composed of the components between the two 650-pF capacitors. This filter suppresses spurious responses from a-m broadcast signals. The second part of the filter, while basically a low-pass type, was *designed* for a very pronounced peak, resulting in a sharp, band-pass-like response. C6 is a 365-pF broadcast replacement type capacitor mounted on the front panel.

VFO Module

The variable frequency oscillator, Fig. 70, uses a JFET in a Hartley circuit followed by a dual-gate MOSFET buffer. For best temperature stability, type SF material (Amidon -6 code) is used for the inductor core, as this material has a lower temperature coefficient than the usual slug-tuned inductor. All of the capacitors in the tuned circuit should be NPO ceramics, as they have the lowest temperature coefficient of any readily available type. Silver mica and polystyrene types should not be used in this circuit.

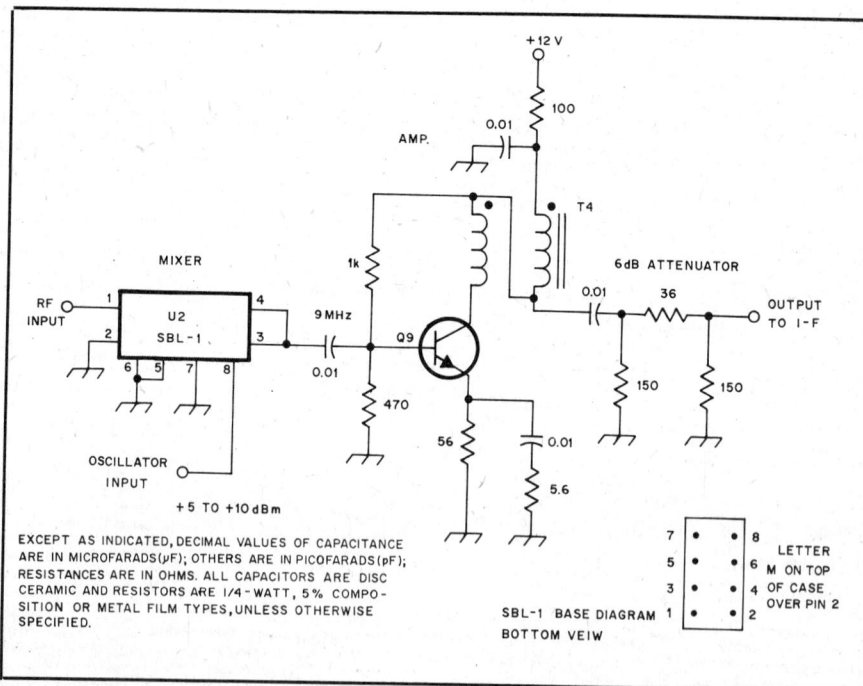

Fig. 67 — Diagram of the mixer module. Two of these modules are used in the completed receiver. One is for 80-meter input, with its output at 9 MHz. The second is used with the converters, its output is at 80 meters.

Q9 — TO-39 CATV type bipolar transistor,
  f_t = 1 GHz or greater. 2N5109,
  2SC1252 2SC1365 or 2N3866 suitable. A
  small heat sink is used on this transistor.
T4 — Broadband ferrite transformer, 10 bifilar

turns, no. 28 enamel on Amidon FT37-43 core
U2 — Mini Circuit Lab double balanced mixer,
  type SBL-1. Type SRA-1 is also suitable, as
  are similar units from other manufacturers.

CRYSTAL OSCILLATOR

Q8 C10 C11 T3
270 Y1 2.2k OUTPUT
60
0.01 100
0.01 0.01
0.01
4.7k
1k

EXCEPT AS INDICATED, DECIMAL VALUES OF CAPACITANCE ARE IN MICROFARADS (µF); OTHERS ARE IN PICOFARADS(pF); RESISTANCES ARE IN OHMS. ALL CAPACITORS ARE DISC CERAMIC AND RESISTORS ARE 1/4 WATT, 5% COMPOSITION OR METAL FILM TYPES, UNLESS OTHERWISE SPECIFIED.

(A)

+12 V
39
220µF / 20V 0.01
RFC 2
FROM OSCILLATOR 0.01 TO MIXER

(B)

+12 V
39
220µF / 20V 0.01
RFC 3
FROM INDIVIDUAL OSCILLATORS S1 0.01 TO MIXER

(C)

HIGH-PASS FILTER   PEAKED LOW-PASS FILTER

650/SM 420/SM 420/SM 650/SM L5 L6
INPUT OUTPUT TO MIXER
L2 L3 L4 5000/SM C5 5000/SM
C6 365 PRESELECT

EXCEPT AS INDICATED, DECIMAL VALUES OF CAPACITANCE ARE IN MICROFARADS(µF); OTHERS ARE IN PICOFARADS(pF); RESISTANCES ARE IN OHMS. ALL CAPACITORS ARE DISC CERAMIC AND RESISTORS ARE 1/4-WATT, 5% COMPOSITION OR METAL FILM TYPES, UNLESS OTHERWISE SPECIFIED.

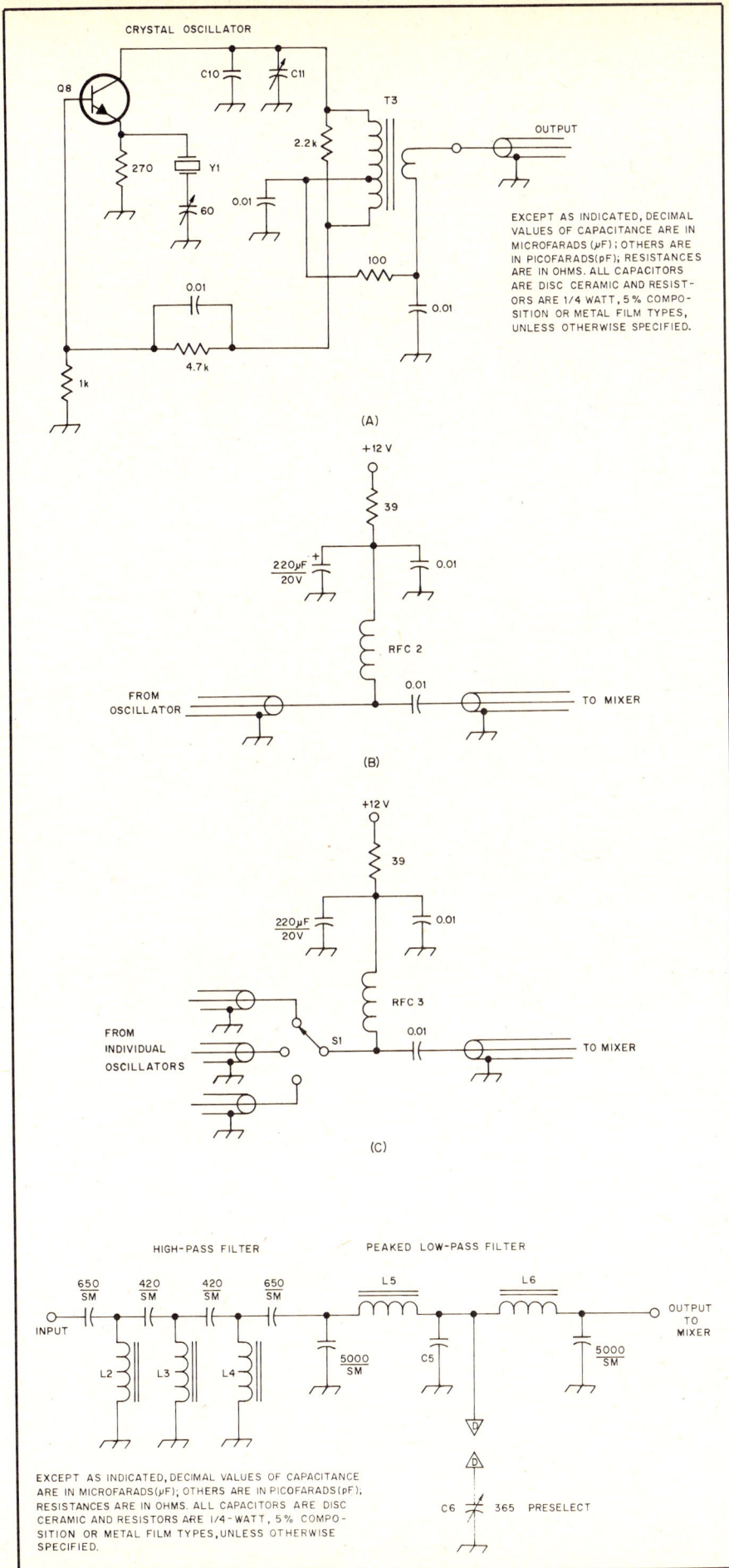

Fig. 68 — Crystal oscillator module used for BFO and converter oscillators.
C10 — Silver mica or ceramic. See Table 6 for values.
C11 — Mica compression or similar trimmer. See Table 6 for values.
Q8 — General purpose NPN, 2N3904, 2N2222A or similar.
RFC2, RFC3 — 20 turns no. 28 enamel on Amidon FT37-43 core.
S1 — Part of bandswitch or sideband selection switch. See text.
T3 — No. 22 wire on Amidon core, see Table 6 for number of turns and type of core.
Y1 — Series-resonant crystal. See Table 6 for frequency. For 9-MHz BFO applications, a KVG type XF-903 can be used for either usb or lsb.

The resonator (tuned circuit) should be lightly loaded; to this end, the coupling capacitor to the gate of the JFET is kept as small as possible. If the specified 2.7-pF NPO ceramic cannot be obtained, a small air variable of similar value can be substituted. Following these precautions will ensure excellent stability. Typical warm-up drift is under 200 Hz over a period of about 10 minutes. After warm-up, drift is no more than 10 or 20 Hz in a 5-minute period.

The buffer stage, Q7, is conventional with the exception of the broadband output transformer, T2. The buffer provides good isolation for the oscillator and an output power of between +5 and +8 dBm.

Intermediate Frequency Amplifier

The heart of the receiver is its i-f section. This design uses an i-f of 9 MHz, with selectivity provided by a crystal filter of the builder's choice. The circuitry shown (Fig. 71) is designed for a filter requiring 500-ohm input and output terminations.

The input is a pi-network which transforms the 50-ohm source impedance (from the mixer module) to the 500 ohms required by the filter. The filter output is terminated in a 560-ohm resistor.

The majority of the i-f gain is provided by two dual-gate MOSFETs, Q11 and Q12. The bias on these stages is shifted upward by a pair of silicon diodes. This extends the gain control range as the gate 2 bias is altered. The last i-f stage is a

Fig. 69 — 80-meter preselector filter. The input section is a high-pass filter and the output section forms a peaked low-pass filter. The variable capacitor is mounted on the front panel.
C5 — 560-pF, silver mica.
C6 — 365-pF or larger broadcast replacement type air variable.
L2, L4 — 21 turn, no 22 enamel wire on Amidon type T50-2 core.
L3 — 20 turns, no. 22 enamel wire on Amidon T50-2 core.
L5, L6 — 30 turns, no. 22 enamel wire on Amidon T68-2 core.

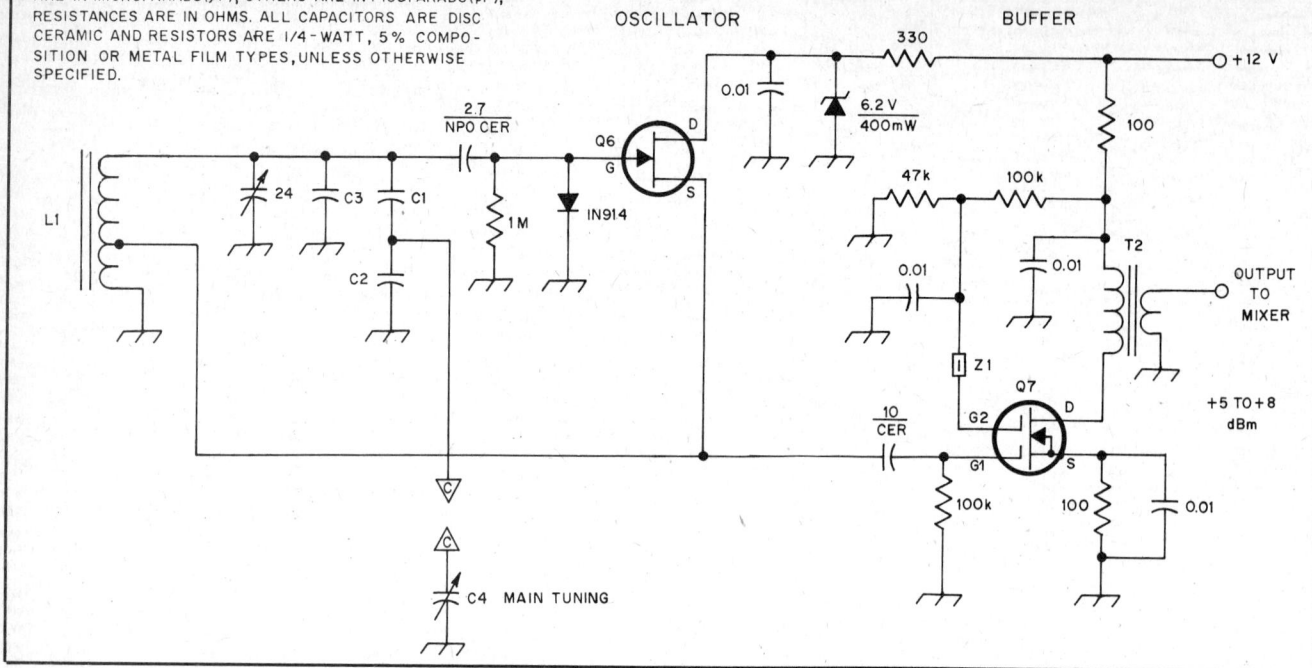

EXCEPT AS INDICATED, DECIMAL VALUES OF CAPACITANCE ARE IN MICROFARADS (μF); OTHERS ARE IN PICOFARADS (pF); RESISTANCES ARE IN OHMS. ALL CAPACITORS ARE DISC CERAMIC AND RESISTORS ARE 1/4-WATT, 5% COMPOSITION OR METAL FILM TYPES, UNLESS OTHERWISE SPECIFIED.

Fig. 70 — Schematic diagram of the VFO used in the W7ZOI/K5IRK receiver. This circuit will function well from 2.5 to 10 MHz. The tuning range using the components listed is 5.0 to 5.5 MHz.

C1 — 100-pF NPO ceramic.
C2 — 82-pF NPO ceramic.
C3 — 126-pF NPO ceramic.
C4 — Air variable, 365-pF broadcast replacement type.
L1 — 35 turns no. 28 enamel wire on Amidon

T50-6 core. Tap 8 turns from ground. Approximately 4.9-μH total inductance.
Q6 — General purpose JFET, MPF-102, 2N4416 TIS-88, 2SK19GR or similar.
Q7 — Dual gate MOSFET, 40673, 3N140 3N211,

3SK40 or similar.
T2 — Ferrite transformer, 18 turn primary, 5 turn secondary, no. 28 enamel wire on Amidon FT37-43 core.
Z1 — Ferrite bead on lead of Q7. Amidon FB43-101 or similar.

differential pair of PNP transistors, Q13 and Q14. Outputs are available from each of the collectors. The one from Q14 is routed through coaxial cable to the product detector.

The other i-f output drives the detector, D1. When large signals are present, the detected voltage from D1 appears at the base of Q16, discharging the timing capacitor, C24. The voltage change at C24 is coupled to the line through a diode, and reduces the gain of Q11 and Q12. R10, the "agc set", is adjusted for a dc potential of 0.4- to 0.5-volts at the base of Q16. This adjustment is made with the agc on, but no signals present. When measured with a high impedance voltmeter, the agc line should show about 6 volts at maximum gain.

Two transistor switches are contained in the amplifier. Q17 is used to defeat the agc. It is activated by a positive voltage applied to the "agc off" line. The other switch, Q15, is attached directly to the agc line. A positive voltage applied to its input shorts the agc line to ground, muting the receiver. The extra diodes allow muting to occur quickly while not discharging C24. The i-f returns rapidly to full gain after muting periods.

The agc response is more than adequate and overshoot is minimal. Recovery time is relatively independent of signal level. The recovery time may be shortened by

Table 5
Component Data for W7ZOI/K5IRK Receiver Converter Filters

Parts for filter without amplifier:

F,MHz	C19 C21	C20	L9 L10	C12 C18	C13 C16	C14 C17	C15	L7 L8
7.1	430	860	17	42	50	180	4.6	25
10.6	300	600	13	32	50	180	4.1	17
14.2	220	430	12	20	—	180	2.3	17
18.2	180	360	10	22	50	180	3.9	10
21.2	150	300	10	18	—	180	3.0	10
24.2	130	270	9	14	—	180	2.1	10
28.5	110	220	8	12	—	180	1.6	10

Parts for filter with RF amplifier:

F,MHz	C19	C20	C21	C22	L9	L10
10.6	300	680	33	50	13	29
14.2	220	500	22	50	12	25
18.2	180	390	—	50	10	22
21.2	150	330	—	50	10	20
24.2	130	300	—	50	9	19
28.5	110	250	—	50	8	17

Note: Other filter parts are identical to that without the amplifier. Values listed for capacitors are capacitance in pF. Values listed for inductors are the number of turns of wire required.

Table 6
Component Data for W7ZOI/K5IRK Receiver Crystal Oscillator Module

Y1	Band	C10	C11	T3			
				Core type	primary turns	tap turns	secondary turns
3.3 MHz	40	100 pF	90 pF	T68-2	65	13	10
9	BFO	56	60	T50-6	35	7	6
10.5	20	56	60	T50-6	30	7	6
11	20/40	22	60	T50-6	30	7	6
17.5	15	33	60	T50-6	23	5	4
24.5	10/15	—	60	T50-6	20	4	4
32	10	—	60	T50-6	15	3	3
6.5	30	100	60	T50-6	35	7	6
14.5	17	33	60	T50-6	23	5	4
20.5	12	—	60	T50-6	20	4	4

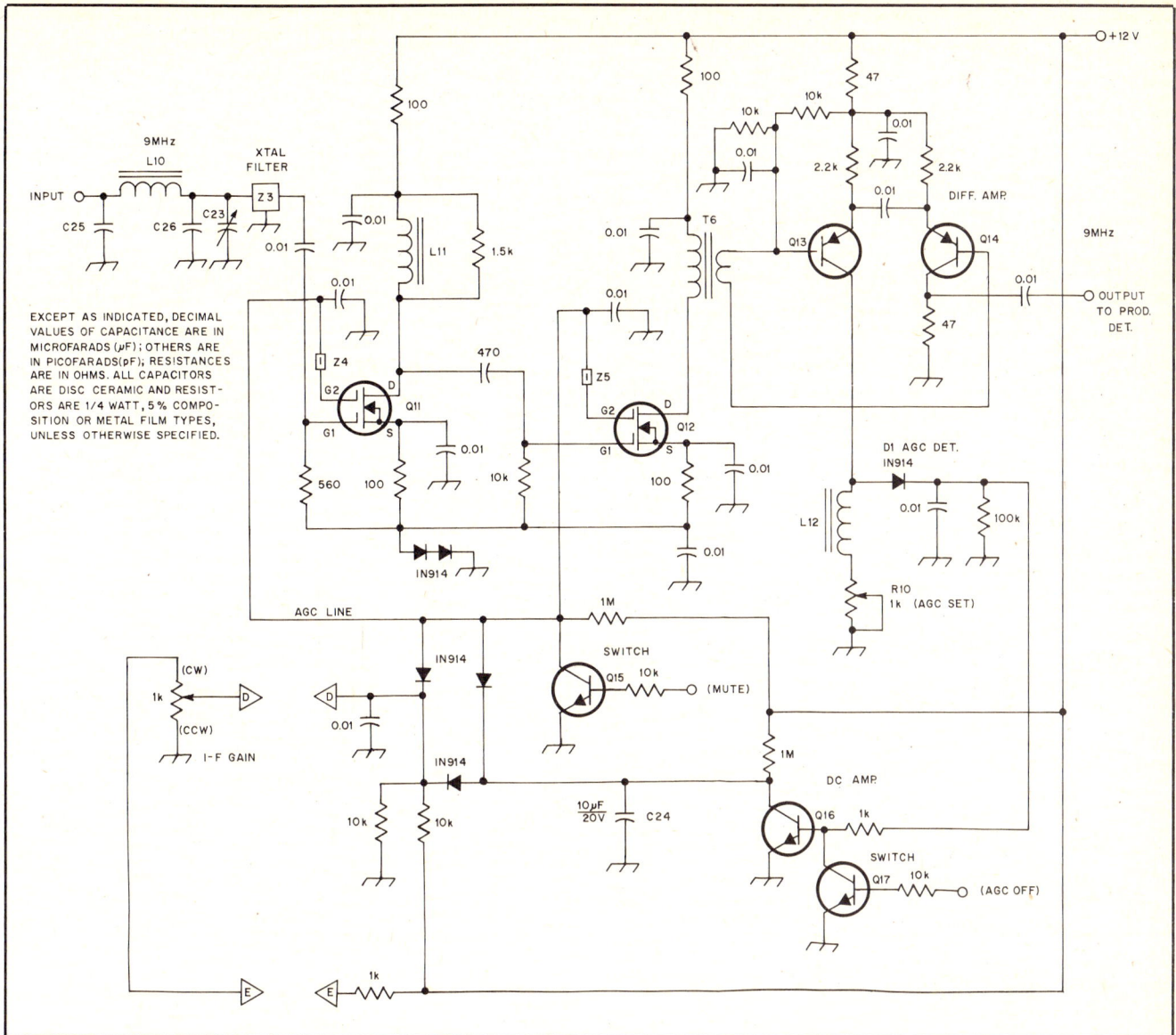

Fig. 71 — Schematic diagram of the 9-MHz i-f amplifier.

C23 — 90-pF mica compression trimmer or similar variable.
C25 — 470-pF silver mica or ceramic.
C26 — 130-pF silver mica or ceramic.
L10 — 23 turns, no. 22 enamel wire on Amidon T50-6 core.
L11, L12 — 15 turns, no. 22 enamel wire on

Amidon FT37-43 core.
R10 — 1-kΩ PC-board mount variable resistor linear taper.
T6 — Ferrite transformer, 15 turn primary, 3 turn secondary, no. 28 enamel wire on Amidon FT37-43 core.

Z3 — 9-MHz crystal filter, 500 ohm impedance level: For ssb, KVG type XF-9B, Yaesu XF-92A or Fox-Tango Corp. eq. For cw, KVG type XF-9M or XF-9NB, Yaesu XF-90C or Fox-Tango Corp. eq.
Z4, Z5 — Ferrite bead on lead of MOSFET, Amidon FB43-101 or similar.

decreasing the value of C24 or the associated 1-megohm resistor.

Detector Audio Module

The i-f section of the receiver is followed by the detector and audio amplifiers, shown in Fig. 72. The detector used here is a doubly balanced diode ring, a Mini-Circuits Labs SBL-1. Mixers from other manufacturers or homemade equivalents will work as well. The excellent balance provided by this type of mixer helps eliminate problems with the agc system caused by BFO leakage.

The detector output is applied to a diplexer network formed by RFC1 and the related components. This network ensures

that the detector is properly terminated at all frequencies from audio to vhf.

The first audio amplifier is somewhat unusual in that it uses the common-base configuration, and when biased for an emitter current of about 0.5 mA, provides the 50-ohm input impedance necessary to properly terminate the detector. The second audio stage, Q2, is a direct-coupled PNP amplifier. The receiver may be muted by shorting the collector of Q2 to ground. This is done by applying a positive voltage to the muting input to saturate Q5. The output of Q2 drives the audio gain control which is mounted on the front panel. If the optional RC active filter is used, it is connected between the

output of Q2 and the gain control.

Q3 functions as a common-emitter amplifier while Q4 is an emitter follower. Q4, biased for an emitter current of about 30 mA, provides sufficient audio output to drive low-impedance (4 to 16 ohms) headphones. If high-impedance headphones are to be used, a step-up transformer should be used to increase the system voltage gain. This is shown in Fig. 72B. An auxiliary input to the audio amplifier is provided for injection of a side tone signal for cw monitoring.

The optional RC active filter shown in Fig. 73 can be used to improve selectivity during cw reception, especially when the i-f filter is of a bandwidth suitable for ssb

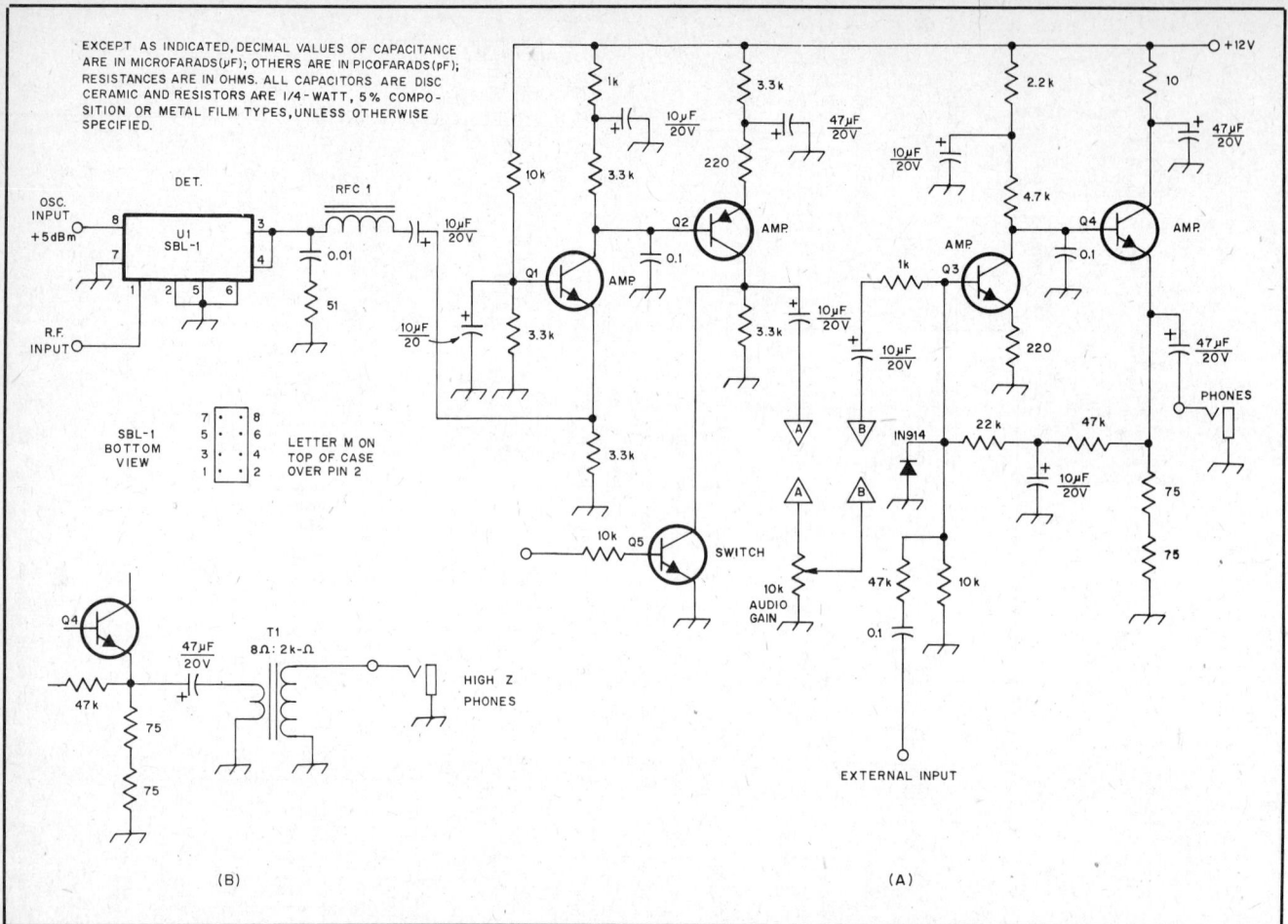

EXCEPT AS INDICATED, DECIMAL VALUES OF CAPACITANCE ARE IN MICROFARADS (μF); OTHERS ARE IN PICOFARADS (pF); RESISTANCES ARE IN OHMS. ALL CAPACITORS ARE DISC CERAMIC AND RESISTORS ARE 1/4-WATT, 5% COMPOSITION OR METAL FILM TYPES, UNLESS OTHERWISE SPECIFIED.

Fig. 72 — Schematic diagram of the product detector and audio amplifier. If high impedance headphones are to be used, the output circuit shown in B is recommended.

Q1, Q3 — Low noise NPN, 2N3565 or similar.
Q2 — General purpose PNP, 2N3906 or simi-
lar.
Q4 — TO-5 to TO-39 NPN, 2N3053 or similar
with small heat sink.
R3 — 10-kΩ audio taper.

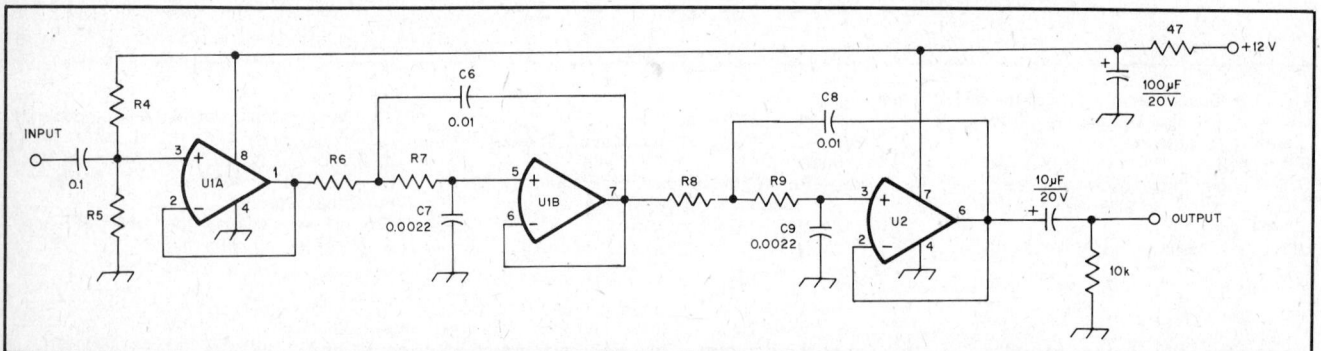

Fig. 73 — Schematic diagram of the optional R-C active audio filter. A 4-pole low-pass filter is shown here. Additional sections may be added for improved performance.

C6, C8 — 0.01-μF, 10% or better tolerance, ceramic or polystyrene.
C7, C9 — 0.0022-μF, 10% or better tolerance, ceramic or polystyrene.
R4-9 — 33-kΩ for cw or 15-kΩ for ssb filter.
U1 — 1458, or similar, dual op-amp.
U2 — 741, or similar, op-amp.

operation. As shown (Fig. 73) the filter has a single pole of high-pass filtering and four poles of low-pass response. The cutoff frequency is about 1-kHz for cw or 2-kHz for ssb. The filter bandwidth is determined through proper choice of the resistor values. Values for both bandwidths are given in Fig. 73. The filter may be expanded to many more sections for improved skirt selectivity.

Construction

Construction details, such as the type of cabinet and dial drive, are left to the discretion of the builder. Band switching is not critical, as all of the switched points occur at low impedance levels. A multi-wafer rotary switch will serve the purpose

well. Small coaxial cable, such as RG-174/U, should be used for all signal lines.

In the version built by K5IRK (shown in the photographs), the bulk of the 80-meter part of the receiver is mounted above the chassis. The VFO is contained in an aluminum box, providing both shielding and mechanical strength. The BFOs are below the chassis in a box constructed of scrap circuit-board material. The converters are mounted below the chassis.

Homemade, etched circuit boards were used in the K5IRK model, while "ugly-but-quick" breadboards were utilized by W7ZOI in the construction of his receiver.[6],[7] "Ugly boards" are easily built with scraps of unetched circuit-board material serving as a ground foil. The circuitry is supported by those components that are normally grounded. Additional support is provided by suitable tie points. Large-value resistors serve well for this purpose, especially in rf circuits where the impedance level is low. All of the circuitry for the receiver was initially breadboarded using this method. While not as "professional" in appearance, the performance was virtually identical to that of later circuit board versions. In the few cases where a performance difference could be detected, the ugly breadboards were superior, usually a result of improved grounding.

Performance

It should be emphasized that, although relatively simple, this receiver is not a toy. It features excellent stability, selectivity consistent with the filter used by the builder, adequate sensitivity and a dynamic range that rivals or exceeds that of many of the commercially available equivalents. The only major compromise is the utilization of dual conversion on the higher bands. The penalty is small, because the gain distribution has been carefully planned.

System measurements were made on the receiver at various stages of development. This data is summarized in Table 7. Both measured data and calculated results are presented to give the prospective builder some feel for the performance to be expected. The table shows the system noise figure, input intercept, minimum detectable signal and the two-tone dynamic range. Both cw and ssb bandwidths are considered in both single- and dual-conversion designs, with and without an rf amplifier. Measurements and calculations generally agree within 1 dB.

Table 7 reveals no surprises. The nature of the tradeoff between single and dual conversion is well illustrated, as is the effect of adding an rf amplifier. The system showing the largest dynamic range is the single-conversion design without an rf amplifier. It should be noted that this data pertains only to the modules described. Changes in gain, noise figure or intercept of any stage will change the results. The table should not be used for general comparisons.

The dual-conversion systems are about

[6]Circuit board templates are available from ARRL for $1.00 and a large s.a.s.e.
[7]Circuit boards, negatives and many parts for this receiver will be available from Circuit Board Specialists, P.O. Box 969, Pueblo, CO 81002.

Table 7
Measured and Calculated Receiver Performance Characteristics

Circuit	RF Amp.	Bandwidth, Hz	NF, dB	IP$_{in}$, dBm	MDS, dBm	DR, dB
Single Conv.*	no	500	16	+18	−131	99
Single Conv.	no	2500	16	+18	−124	94
Single Conv.	yes	500	5	+2	−142	96
Single Conv.	yes	2500	5	+2	−135	92
Dual Conv.*	no	500	18	+12	−129	94
Dual Conv.*	no	2500	18	+12	−122	89
Dual Conv.	yes	500	6	−2	−141	92
Dual Conv.*	yes	2500	6	−2	−141	88

RF amplifier assumed to have a 3 dB noise figure, a 15 dB gain and a +22 dBm output intercept. Circuits marked with (*) are measured cases. All measurements done at 14 MHz.

5 dB "weaker" than the single conversions one; this is a fairly typical situation. It should be understood that this observation applies only to dual-conversion systems with a wide bandwidth first i-f section. Modern systems using a crystal filter at the first i-f will display performance much the same as a single-conversion receiver, even if they utilize a multiplicity of conversions. No gain compression was measureable in the single-conversion model with no rf amplifier with an input signal of -10 dBm. VFO phase noise was measured to be -152 dBc/Hz at a spacing of 10 kHz.

Operationally, the receiver is pleasing to use, offering a clean, crisp sound that is not always found in the commercial equivalents. Of greatest significance is that the receiver should be easily duplicated at a cost well under that of similar "appliance."

HIGH-PERFORMANCE RECEIVER DESIGN CONCEPTS

The Amateur Radio design technology is changing so rapidly today that it is impossible to publish a high-performance receiver circuit which remains timely at the time the work is committed to print. As new components and active devices are introduced to the market, better designs become possible. These advances make obsolete many of the circuits found in the current amateur literature. Therefore, this section of chapter 8 is devoted to design objectives, circuit techniques and some practical examples. This will serve as the basis for individual designs which can be carried out by the more experienced amateur.

The interest in building homemade receivers of the more complex variety has waned in a tragic fashion during the past decade. This has been brought on by an increased interest in operating and through the availability of sophisticated receivers and transceivers found on the commercial market. For this reason it seems prudent to devote this portion of the *Handbook* to design approaches. The information given here is based on circuit and performance investigations in the ARRL laboratory. It is slanted toward the practical side of design and application in

order to be of greater use to amateurs who have no formal background in electronics.

Performance Objectives

What should an amateur look for in terms of high performance when building a receiver? A subjective outlook would call for a lot of "bells and whistles" with which to play, but a discerning operator is interested in *performance* under all of the adverse conditions one might encounter in the course of operating an amateur station. The following are representative of the major considerations in receiver performance:

1) *High dynamic range.* This is the ability of the receiver to perform well in the presence of strong signals within and outside the amateur band of interest. Poor dynamic range results in cross-modulation effects, receiver desensitization and spurious responses from the mixer which appear in the tuning range as additional signals (mixer IMD).

2) *Good selectivity.* This feature includes the receiver front end (rf amplifier and mixer) along with the i-f and audio selectivity. The objective (ideal) is to have the receiver pass only those frequencies to which it is tuned, while rejecting all others. This Utopian goal can not be realized, but it can be approached closely enough to ensure good performance.

3) *Low noise figure.* The noise figure should be such that it is somewhat below the level of the receiver antenna noise under typical "quiet" band conditions. This means that the noise generated within the receiver — notably the early stages — should be kept to an absolute minimum so that it does not mask (degrade) weak incoming signals.

4) *High order of stability.* All of the receiver oscillators, crystal-controlled or LC types, need to be drift free in an ideal circuit. Since this is practically impossible, maximum drift (long term) should not exceed 50 to 100 Hz in a good design. The greater the i-f selectivity, the more important the oscillator stability. Self-oscillations should not be allowed to take place in any part of a receiver.

5) *Wide-range agc.* The agc circuit should engage at low signal levels and hold the receiver output at a constant

plateau over a wide range of input-signal levels. For example, the audio output should remain constant in amplitude over a range of input signal from less than a microvolt to better than 10,000 μV, depending on the external noise level which reaches the receiver front end. The agc attack time should be set so that "pumping" and "clicking" is not noted when strong signals are received.

6) *Local oscillator.* Not only must the local oscillator be stable, it needs to have low noise and good spectral purity at the output. Ideally, the LO noise floor should be 80 to 100 dB below the peak output voltage. Spurs and harmonics in the output should fall at least 50 dB below peak output. LO output energy must be confined to the mixer by means of appropriate shielding and filtering.

7) *I-f amplifiers.* An i-f amplifier strip needs to have sufficient gain to drive the detector and provide ample excitation to the audio channel. The design should include active devices which can ensure a collective 80 dB or greater agc swing over the input-signal range mentioned in item 5.

Wide-band noise is generated within most i-f amplifier chains. An improvement in receiver "noise bandwidth" can be realized by adopting the W7ZOI filter "tailending" scheme which calls for use of a second i-f filter immediately after the last i-f amplifier. The second filter can have slightly greater bandwidth than the filter used ahead of the first i-f amplifier. The tailend filter will reduce the wide-band noise components. This technique is discussed in greater detail in *Solid State Design for the Radio Amateur,* by The ARRL.

8) *I-f filters.* In order for a filter to function as such, there must be *some* insertion loss (IL) if a passive network is being used. The IL is typically highest when a mechanical filter is used. This factor must be taken into account when planning the receiver gain distribution. Most mechanical filters have an IL of 8 dB or greater, whereas a well designed crystal-lattice filter has a characteristic IL of less than 5 dB.

In some designs the i-f filter is the limiting factor in achieving high performance in mixer IMD. This is because of the movement of the mechano-electrical contacts within the filter, which generate IM products which are independent of those in the mixer. Laboratory investigations indicated that mechanical filters were somewhat worse than crystal filters in this respect, limiting the receiver IMD profile to roughly 95 dB. The crystal holders in lattice filters must be able to provide positive electrical contact with the quartz element and the circuit to minimize the generation of IM products.

Careful attention must be given to correct filter termination and input/output resonance to ensure minimum passband ripple (unwanted dips in the nose of the response curve): Most filters require a specified external terminal capacitance to resonate the input and output transformers within the filter module. Similarly, each filter has a characteristic input and output impedance which must be matched to the source and load.

9) *Detector and audio channel.* An otherwise excellent receiver can be spoiled by an inferior product detector or audio-amplifier strip. The detector must be able to handle the highest output signal from the i-f strip without saturating. Although active product detectors are sometimes used, they are the most prone to the foregoing malady. The preference of most designers is a passive diode detector of the singly or doubly balanced variety. This type of detector can handle high signal levels with large amounts of BFO injection. Since the detector is the lowest-level part of the audio channel, hum and noise should be minimal at that point in the circuit. Passive detectors do not need operating voltages; hence one primary source of hum is avoided.

A low-noise audio preamplifier should follow the detector in a quality design. It should be able to withstand the maximum output from the detector at peak receiver input signal without operating in a non-linear manner. The audio gain control can be used to the best advantage when it is located *after* the af preamplifier. If it is used ahead of the preamplifier, the noise figure of the audio channel may be degraded at low settings of the gain control.

Audio shaping is normally applied to the af channel to provide a low-frequency rolloff at some frequency well above 60 or 120 Hz. This greatly reduces the chance of power-supply ripple appearing in the audio output. Also, there is little need for low-frequency response below, say, 300 Hz in a communications receiver. Similarly, the high-frequency response should be restricted so that rolloff starts around 2000 Hz. A satisfactory tailoring of the audio passband can often be done by proper selection of the R and C components in that part of the circuit.

All of the audio stages must operate as linearly as possible up to peak signal levels. This will minimize distortion and aid weak-signal reception greatly. It is prudent, therefore, to use an output stage which is capable of delivering greater undistorted power output than will ever be needed. Cross-over distortion is to be avoided also. The effects of this are most apparent under weak-signal conditions. The signal has a "fuzzy" sound when this type of distortion is present. Some of the audio-power ICs have significant cross-over distortion which can not be corrected. This is because the biasing is done within the IC, and it can't be changed. For this reason it is helpful to use discrete devices in the audio channel. This enables the designer to bias the amplifiers for minimum distortion.

Tuned audio amplifiers can be used to provide steep skirts outside the desired passband. An example of a simple application of this, using a single pot-core inductor with variable Q, was described by K1TX in April 1979 *QST.* Various types of passive LC filters can be used to obtain cw or ssb selectivity at audio.

RC active audio filters with variable Q and adjustable peak frequency offer an excellent means for limiting the audio bandwidth, minimizing wide-band noise and reducing QRM. Ideally, these filters should be contained in the low-level part of the audio channel rather than at the receiver output. This will prevent overloading of the filter, which can impair the performance and introduce intolerable amounts of distortion.

10) *Structural considerations:* There can be considerable latitude in the mechanical approach one takes when laying out a high-performance receiver. Aesthetics have no place in this discussion. We will address the matter of structure versus performance and leave the beauty of the front panel to the builder.

The major points of concern are rigidity of the overall assembly and shielding against incidental pickup and radiation. The chassis and panels should be strong enough to prevent undue stress on the pc boards during flexing or vibration. In a like manner the local oscillator should be relatively immune to any mechanical stress which is imposed on the receiver.

An excellent assembly technique is one that uses a modular approach for the various key circuit assemblies in the receiver. Each module is contained in its own shield box. All signal leads entering and leaving the various modules are made from RG-174/U or similar coaxial cable. **The shield braid is grounded at each end of each cable. Leads which carry dc are decoupled where they leave the module shield.** LC or RC decoupling networks are suitable in most instances. Feedthrough capacitors can be mounted on the box walls of each module to serve as terminal connections for the dc voltages, while functioning also as parts of the decoupling networks. Since 50-Ω miniature cable is suggested for interconnecting leads, the points to which they connect in the circuit should be designed for a like impedance. This form of modular construction and shielding greatly reduces the chances for "birdies" by keeping rf energy where it belongs. It also prevents unwanted external signals from being picked up by low-level parts of the circuit. Miniboxes or die-cast aluminum boxes are excellent for use in modular work. But, homemade enclosures can be fashioned inexpensively from pieces of double-sided pc board. Modular construction permits the amateur to try new circuits within the receiver without disrupting the remainder of the circuitry.

11) *"Bells and whistles"*: This discussion does not include such themes as synthesizers, i-f passband tuning, noise blankers, computer-programmed functions and digital frequency readout. These are primarily matters of whim and subjectivity, however useful they might be.

For reasons of practicality the builder must decide whether he will use analog or digital readout of the receiver frequency. There are two disadvantages attendant to analog systems: (1) Quality dial mechanisms are scarce and highly expensive. (2) Readout resolution is usually poor if more than 200 kHz of any band is covered. The major advantages of analog frequency readout are reduced circuit complexity, lower cost (sometimes) and less current drain from the receiver power supply. Heating is diminished also — a definite benefit to stability.

A frequency counter and a digital display, on the other hand, permit 500-kHz frequency spreads with good resolution. A shaft encoder is needed for synthesized LO systems to avoid thumbwheel frequency selection. But, it is easy to use parts of the synthesizer circuit for the frequency counter, thereby making the two circuits compatible. In this type of system, or in one which has a conventional LO and a counter, a 10:1 vernier drive without detectable backlash is almost mandatory to keep the tuning rate within practical limits.

It should be stressed here that counters can create noise and spurious responses if they aren't designed and used correctly. Careful shielding and filtering must be applied to prevent the counter from affecting other parts of the receiver circuit.

The same general considerations apply to synthesizers. The design must be carried out with care to minimize phase noise, which can degrade the mixer noise figure and the ultimate i-f selectivity.

Three *QST* articles are offered as references on high-performance receivers. They contain information which will be of value to the amateur designer.[1]

RF Amplifiers

When it is deemed necessary to use an rf amplifier ahead of the receiver mixer, thought must be given to gain, linearity, signal-handling ability and noise figure. The choice between bipolar transistors and FETs is another consideration. An rf amplifier should not be necessary in a properly designed receiver, even if a passive mixer is used, provided the input networks are not highly lossy or poorly matched to the source and load. This rationale applies to frequencies up to approximately 14 MHz. At 20 meters and

[1]Wes Hayward, W7ZOI, "A Competition-Grade Receiver," *QST*, March and April 1974. Doug DeMaw, W1FB, "His Eminence the Receiver," *QST*, June and July 1976. Jay Rusgrove, W1VD "Human Engineering the Station Receiver," *QST*, January 1979.

higher an rf amplifier may be needed to ensure an acceptable receiver noise figure.

As a general rule the designer should use no more gain in the rf stage than is necessary to obtain an acceptable noise figure. The higher the stage gain, the greater the *sensitivity*. But, more gain than is needed will degrade the receiver dynamic range markedly, by virtue of the mixer being fed larger amounts of input signal than if no rf amplifier was used. So, even at the very early part of a receiver it is vital to pay attention to *gain distribution*. This fundamental rule applies from stage to stage throughout the receiver.

There should be sufficient selectivity ahead of the rf amplifier (and in most instances between it and the mixer) to restrict passage of signals outside the amateur band of interest. This will greatly reduce the probability of unwanted images in the tuning range. Furthermore, it will help prevent very strong out-of-band commercial signals from entering the receiver front end and impairing performance. This form of selectivity is called "preselection." It can take the form of LC circuits which are very narrow in bandwidth, and tracked manually from the front panel. Alternatively, fixed-tuned LC filters can be used to provide selective circuits. A bandpass type of filter or tuned circuit is the choice of most designers because rejection is offered *above* and *below* the frequency band of interest.

The choice between small-signal FETs and bipolar transistors in an rf preamplifier is more than arbitrary. FETs exhibit low noise figures at hf and they consume less dc power than bipolars for an equivalent output intercept. Generally speaking, FETs are less subject to blocking in the presence of strong input signals.

Bipolar transistors, on the other hand, have rather well defined input and output impedances and can be used more easily with negative feedback than is true of FETs. These features make them ideal for ensuring a proper and constant filter termination (filters must be terminated correctly in order to perform in a proper manner). A common-source FET which operates in the hf spectrum can not meet the foregoing condition. The use of feedback (negative and degenerative) in a bipolar-transistor preamplifier makes possible a low noise figure and a good input and output match.

Bipolar transistors which are designed for CATV and uhf oscillator work, such as the 2N5179 (biased for about 20 mA) and the 2N5109 (biased for roughly 50 mA) are excellent for use as rf amplifiers ahead of a mixer.

Fig. 74 shows a high performance rf amplifier which uses a VMOS vhf power FET. (Oxner, May 1979 *QST*, p. 23). The circuit is designed with feedback and is structured for a source and load impedance of 50Ω. The stage gain is determined by the designer's needs. Once this

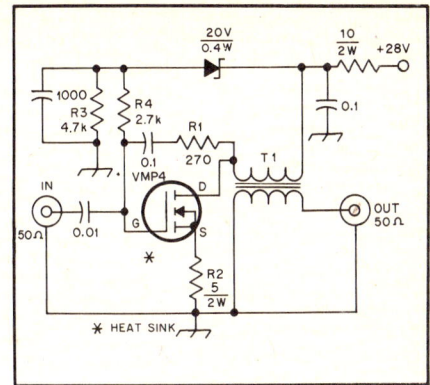

Fig. 74 — Diagram of the Class A large-signal rf amplifier which uses a VMOS power FET. T1 has 9-1/2 turns of no. 30 enam. wire on a Stackpole no. 57-9130 ferrite balun core.

parameter is chosen the values for R1 and R2 can be obtained from

$$R1 = \frac{\sqrt{R_S R_L}}{2} \cdot$$
$$\left[\sqrt{G} + \sqrt{G + 4\left(1 + \sqrt{G} \cdot \frac{R_S + R_L}{\sqrt{2} \, R_S R_L}\right)} \right]$$

(Eq. 1)

$$R2 = \left(\frac{R_S R_L}{R1}\right) - \frac{1}{G_M}$$

(Eq. 2)

where

G is the desired stage gain for the amplifier, and

G_M is the forward transconductance value of the transistor expressed in mhos. (Y_{21} real)

The equations don't yield standard resistance values in most instances. In an amateur application the nearest standard value will often suffice. R2 of Fig. 74 consists of six 30-Ω resistors in parallel to obtain 5Ω. Three are soldered from one source tab to ground, and the other three go from the remaining source tab to ground. This helps reduce stray inductance in that part of the circuit. A power gain of approximately 13 dB results with the component values shown. Noise figure is 4 dB at 30 MHz. A 1-dB saturation power output of 3.7 watts was observed, indicating the suitability of this type of circuit for high signal-handling applications.

The VMP4 is a fairly expensive Siliconix transistor. It is likely that one of the lower-priced pieces, such as the VN66AK, would provide good service at hf in the circuit of Fig. 74.

High-Performance Mixers

Doubly balanced diode-ring mixers (DBMs) of the type discussed early in this chapter are often used to obtain high

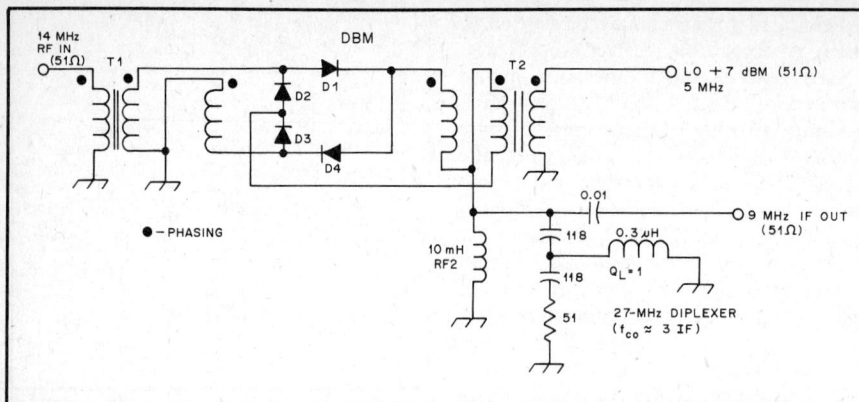

Fig. 75 — Practical circuit for a doubly balanced diode-ring mixer. The components are discussed in the text.

dynamic range. Among the advantages are low noise (diode mixers generate very little noise) and broadband characteristics. The mixer noise figure is approximately the conversion loss of the diode ring — typically 7 to 8 dB. The balanced mixer circuit provides port-to-port isolation which is not possible with single-ended or singly balanced mixers. This feature can aid the mixer IMD and help to minimize spurious responses resulting from the LO energy entering other parts of the receiver circuit.

The main shortcomings of diode mixers are the high level of LO injection needed (approximately + 7 dBm for most) and the necessity of proper mixer termination, especially at the i-f output port. This type of mixer is subject to harmonic mixing — another trait which the designer must deal with.

Some high-level diode-ring mixers are available commercially. They require a high amount of injection power (+ 17 dBm for acceptable performance). Laboratory analysis suggests that high-level mixers misbehave as a result of diode imbalance at specific current levels. The effect is one of the IMD not dropping 3 dB when the input tones are lowered 1 dB in level. This phenomenon could be caused in part by saturation of the broadband input and output transformers at specific lower levels. (A thorough discussion concerning diode mixers and their behavior is presented in the League's book, *Solid State Design for the Radio Amateur*, chapter 6.) Fig. 75 shows a practical circuit for a DBM. It includes a diplexer at the i-f port to establish a 51-Ω termination for the mixer. This offers an improvement to the IMD level by a few dB over a similar mixer with no diplexer.

The diodes can be HP2800 hot-carrier types. Carefully matched 1N914s are sometimes used as substitutes. T1 and T2 are broadband toroidal transformers (baluns). For wideband use in the hf spectrum the cores should have a high permeability. A 0.37-inch (9.4-mm) diameter ferrite core (Amidon FT37-43)

with a mu of 950 will work nicely. Ten trifilar turns of no. 30 enamel wire can be used for the windings. Output intercept for this circuit is typically + 13 dBm with the LO injection at + 7 dBm. This provides an input intercept of 20 dBm (output intercept plus the 7 dB conversion loss = 20 dBm). Calculations for a high-level diode mixer, assuming a + 17 dBm LO level (recommended), the output intercept will be + 23 dBm. Again, assuming a 7-dB conversion loss the input intercept becomes quite desirable — + 30 dBm. This is based on the respective performances of the commercially available SRA-1 and SRA-1H DBMs. It can be seen from the foregoing that better mixer performance can be realized at the higher LO-injection levels. The actual LO power applied will depend upon the ability of the diodes to handle the current.

Singly Balanced VMOS Mixer

The circuit of an experimental active mixer with high-level capability is shown in Fig. 76. Two VMOS power FETs are employed in a singly balanced format. Laboratory measurements of the circuit between T1 and T2 (50-Ω terminations) at 14 MHz, with a LO frequency of 5 MHz and an i-f of 9 MHz, yielded a mixer conversion gain of 6 dB. Output intercept is + 23 dBm and the input intercept checks out at + 17 dBm. Indications are that with further experimentation with VMOS devices, mixer biasing and LO injection power the input intercept could be improved to at least + 25 dBm in an optimized case. The circuit of Fig. 76 was biased for a total mixer current of 50 mA with LO power (+ 15 dBm) applied. The use of Siliconix VMP-4 transistors should lead to even better mixer performance. A photograph of the assembled experimental mixer with its post-mixer amplifier and filter is shown in Fig. 77.

Fig. 76 shows that a broadband linear post-mixer amplifier is used. It is followed by a 6-dB pad and diode clamps. The pad provides a constant load for the amplifier, Q3, and stabilizes the filter termination

(FL3). The diodes prevent damage to FL3 in the presence of very strong receiver input signals. FL1 filters the LO output to ensure a clean injection voltage to the mixer. The VMOS balanced mixer exhibits a noise figure of roughly 8 dB at 14 MHz.

I-F Amplifiers

The criteria for i-f amplifiers are pretty well defined in the philosophy section of this part of the chapter. The choice of active devices for i-f strips usually leads to ICs. The Motorola MC1590G or MC1350P are commonly found in high performance receivers. These and the RCA CA3028A IC offer good gain and agc range with low noise figures. So, the choice will depend in part on availability and whim. Normally, just two IC stages are used in an i-f strip.

Dual-gate MOSFETs are used as i-f amplifiers by some amateurs, but at least four stages are needed to approach the gain of two ICs designed for the application. Furthermore, FETs do not provide the agc range of ICs unless the control gates can be made to swing slightly negative. The usual gate no. 2 agc range is from − 2 volts to + 4 volts for full control.

Fig. 78 contains the circuit of an i-f strip which uses two Collins mechanical filters in the "tail-ending" scheme discussed earlier. The ssb filter is at the front end of the strip and the narrower (400 Hz) filter is diode-switched into the circuit for low-noise cw reception. For phone-only reception both filters can be of a 2.1-kHz bandwidth, and both would remain in the circuit at all times. At considerably greater expense one could have a cw and a ssb filter at the output end of the i-f strip. They would be selected by means of diode switching to permit a tail-end filter to be in the circuit for either mode. The photograph in Fig. 79 was taken before some circuit changes were made, so it does not conform exactly to the diagram in Fig. 78. It does, however, illustrate the recommended layout for good filter isolation.

During ssb operation FL2 is shorted across by means of D1 and D2. Q2 is used to equalize the overall gain of the receiver when the modes are changed. It compensates for the 10 dB of insertion loss caused by FL2. During ssb operation the gain of Q2 is reduced by virtue of S1C and R1.

Agc amplifier Q1 is used to prevent loading across T2. The i-f energy is sampled at the drain of Q2 so that the agc will be relatively constant for both cw and ssb operation. Sampling ahead of Q2 would cause a 10-dB differential in the agc action between the cw and ssb modes. FL1 should have an input termination of 2000Ω. The characteristic input impedance of U1 provides a suitable termination for FL1 — roughly 2000Ω. T2 should have a 20:1 impedance step-down ratio for going into a diode type of pro-

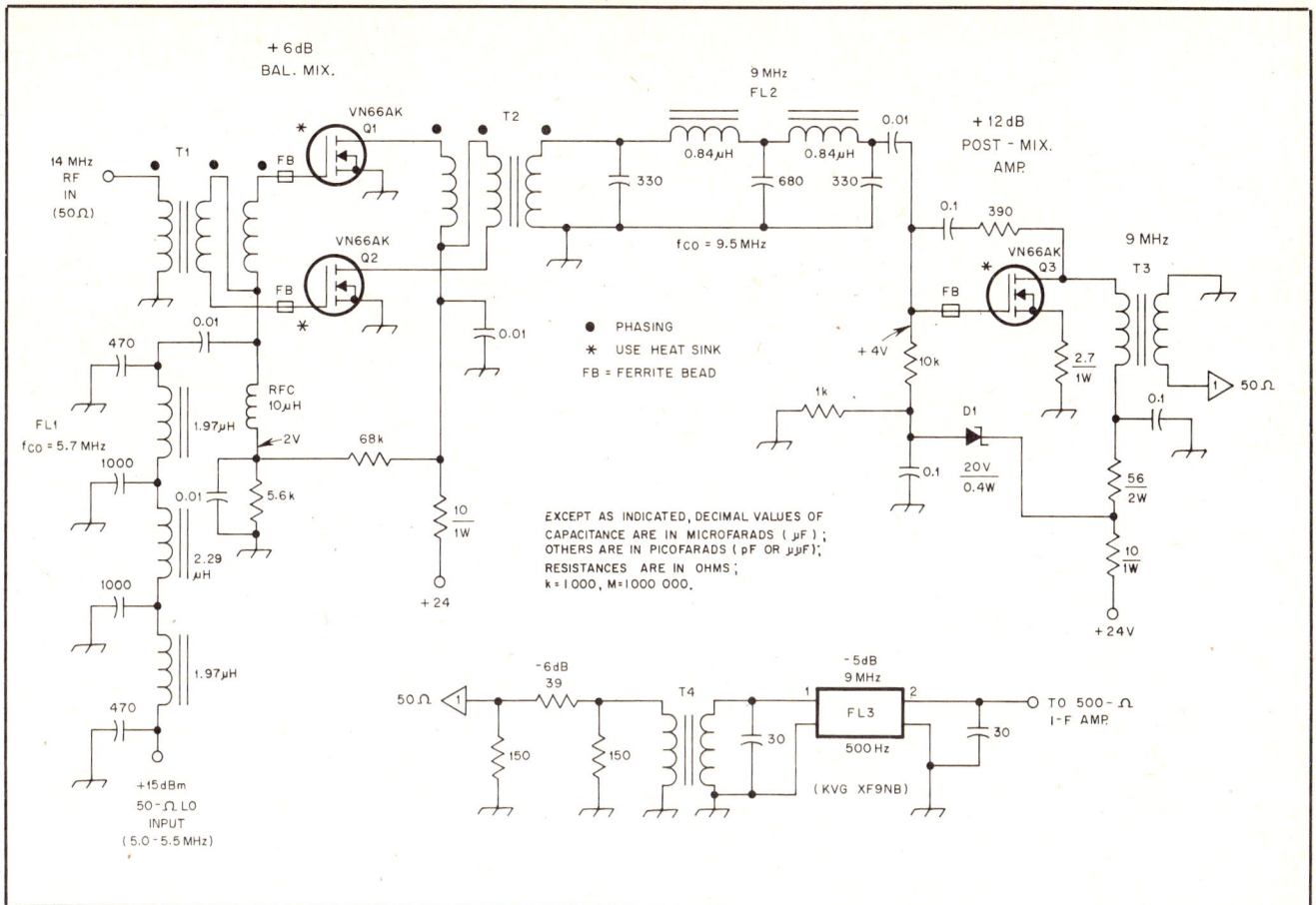

Fig. 76 — Circuit details are for a VMOS power FET singly balanced mixer. L1 and L2 have 13 turns of no. 24 enam. on a T50-6 toroid core. L4 contains 21 turns of no. 26 enam. wire on a T50-2 core. T1 and T2 have 12 trifilar turns of no. 26 enam. wire on an FT50-61 toroid core. T4 contains 7 primary turns of no. 26 enam. and 21 secondary turns of no. 26 enam. on an FT50-61 toroid core. FL3 is a Spectrum International 8-pole crystal lattice filter. Bandwidth is 500 Hz.

duct detector. The value is not critical.

Universal BFO

Fig. 80 provides a diagram of a BFO which can be tailored for use at any of the popular intermediate frequencies. The constants shown are for use at 455 kHz. For higher operating frequencies it will be necessary only to modify the feedback capacitors of the oscillators (C1-C5, inclusive). The higher the operating frequency the lower the capacitance value. The division ratio established by C4 and C5 should be maintained at all frequencies on which this circuit is used. This will prevent the 50-Ω input impedance of Q4 from loading the oscillators and preventing them from functioning.

This circuit was designed especially for use with a diode-ring product detector, which requires a substantial amount of injection power. A 50-Ω pad can be placed in the output of T1 if lower injection levels are needed. T1 should be terminated in approximately 50Ω for the best performance.

Y1, Y2 and Y3 provide the proper BFO

Fig. 77 — Photograph of the experimental VMOS high-level mixer. *Circuit boards for this and other modules shown photographically in this section of chapter 8 are available from Circuit Board Specialists, Box 969, Pueblo, CO 81002.*

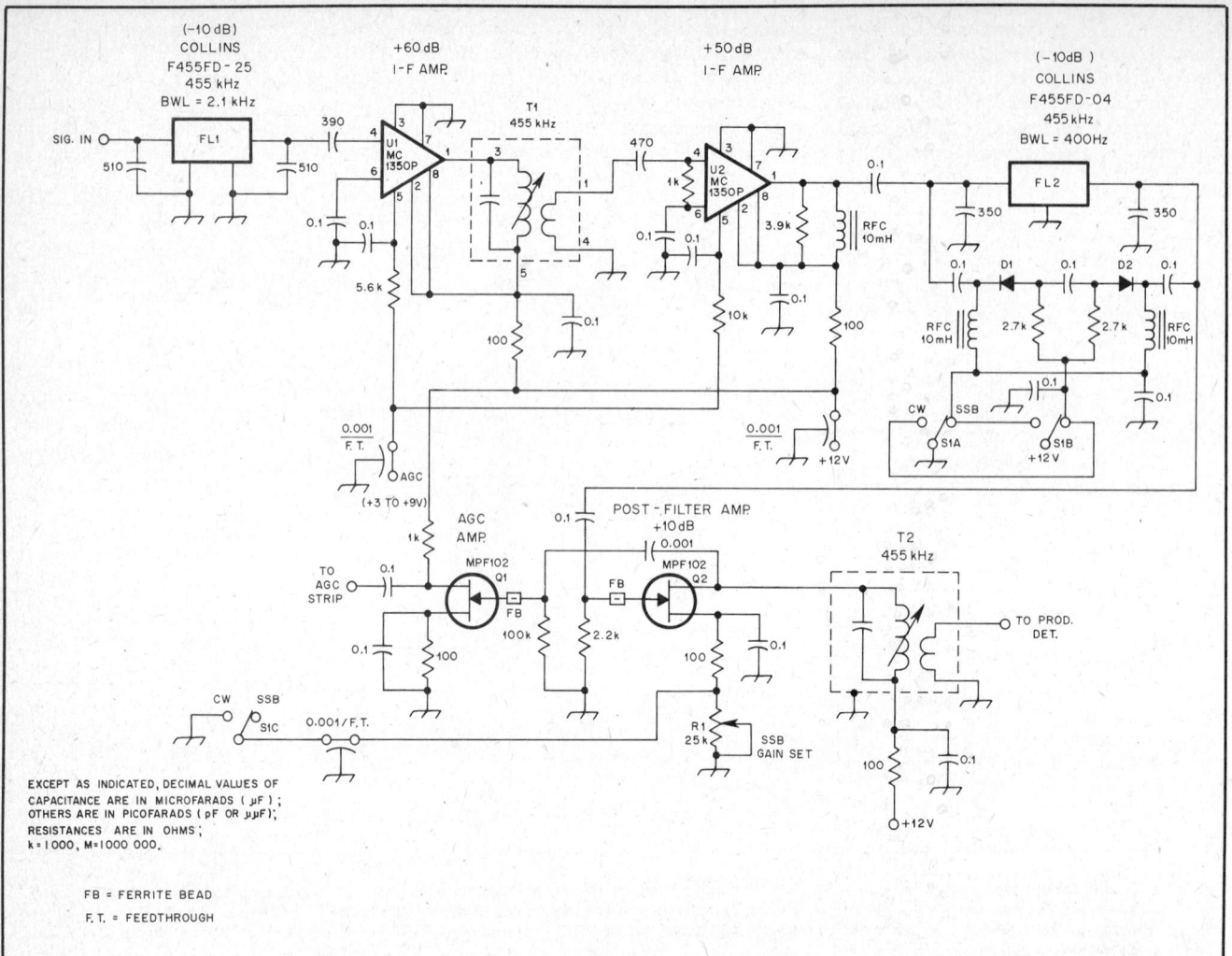

Fig. 78 — Schematic diagram of a 455-kHz i-f strip which uses filter tail-ending. See the text.

frequencies for upper sideband, lower sideband and cw. JFETs can be substituted for the 40673s at Q1, Q2 and Q3.

The assembled BFO is seen in Fig. 81. The shield compartment is made from sections of double-sided pc board. A feedthrough capacitor serves as a terminal for the 12-volt line to the BFO and helps to prevent rf energy from leaving the BFO module via the 12-volt lead.

Local-Oscillator Structure

The local oscillator system shown in Fig. 82 is an assembled version of the practical VFO shown in chapter 6. The compartment is made from pc-board sections. A U-shaped press-fit aluminum cover is used to enclose the top of the box. This kind of shielding is important for the prevention of stray radiation from the VFO into other parts of the receiver circuit. The enclosed module tends to prevent rapid internal changes in ambient temperature — an aid to frequency stability. Double-sided etched-circuit boards are not recommended for LC oscillators. This is because the etched foils and the ground

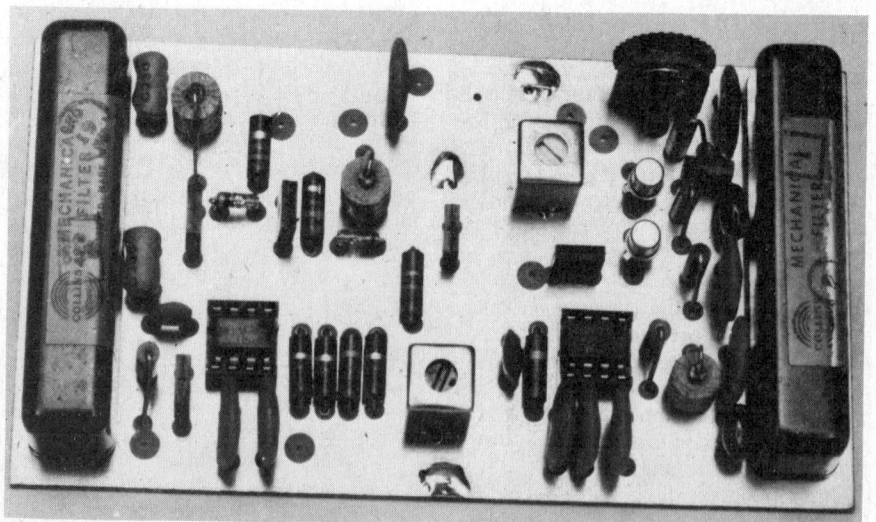

Fig. 79 — Photograph of the assembled i-f strip from Fig. 78. Double-sided pc-board material is used.

plane form numerous low-value capacitors, none of which is stable: The pc-board insulating material represents a poor dielectric material for stable

capacitors.

A High-Performance AGC Circuit

Fig. 83 contains the circuit of an i-f strip

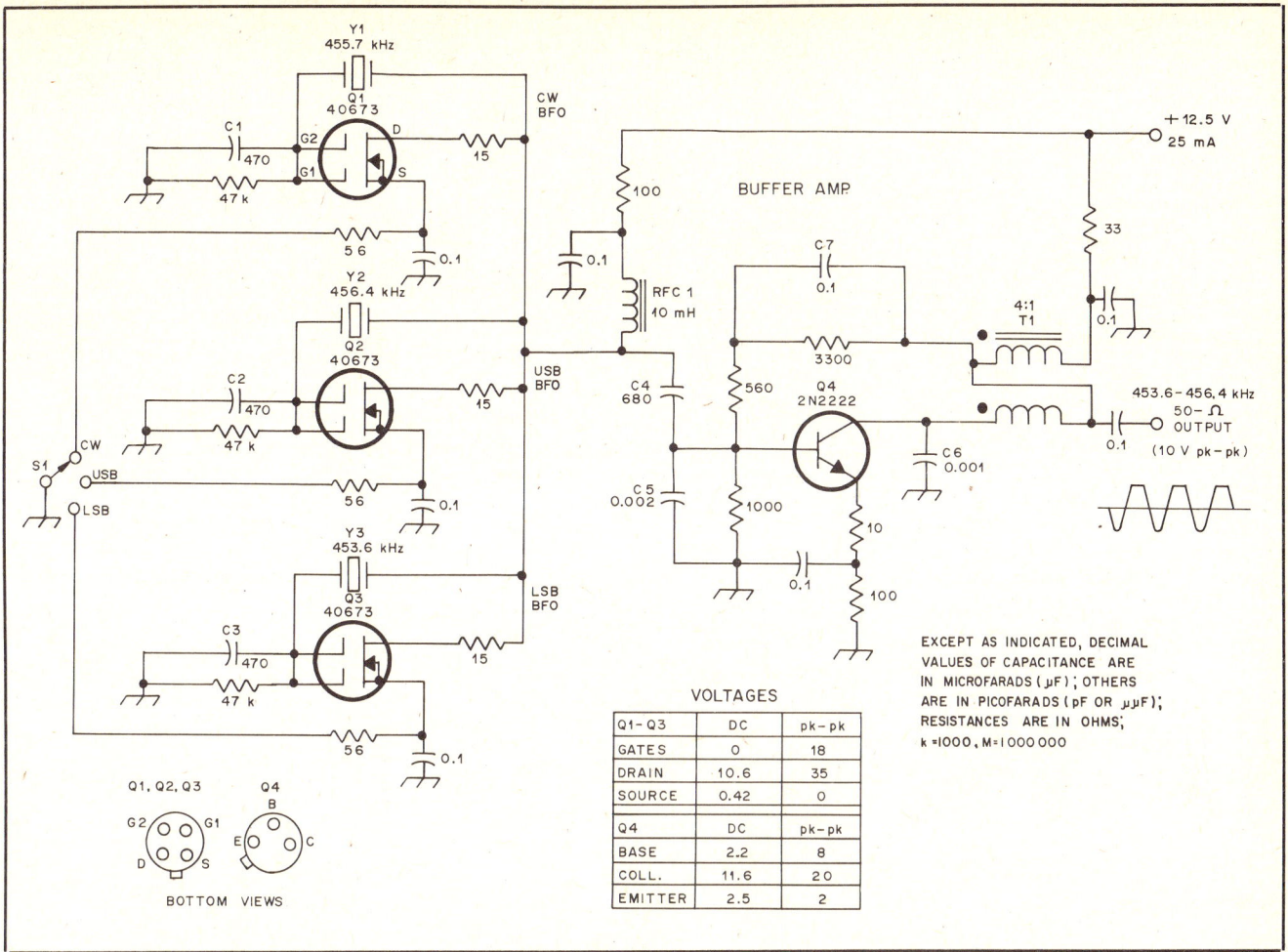

Fig. 80 — A 3-channel BFO for universal use in receivers. T1 contains 25 bifilar turns of no. 30 enam. wire on an FT50-43 toroid core (950 mu).

VOLTAGES		
Q1– Q3	DC	pk–pk
GATES	0	18
DRAIN	10.6	35
SOURCE	0.42	0
Q4	DC	pk–pk
BASE	2.2	8
COLL.	11.6	20
EMITTER	2.5	2

EXCEPT AS INDICATED, DECIMAL
VALUES OF CAPACITANCE ARE
IN MICROFARADS (μF) ; OTHERS
ARE IN PICOFARADS (pF OR رسF) ;
RESISTANCES ARE IN OHMS;
k =1000, M=1000 000

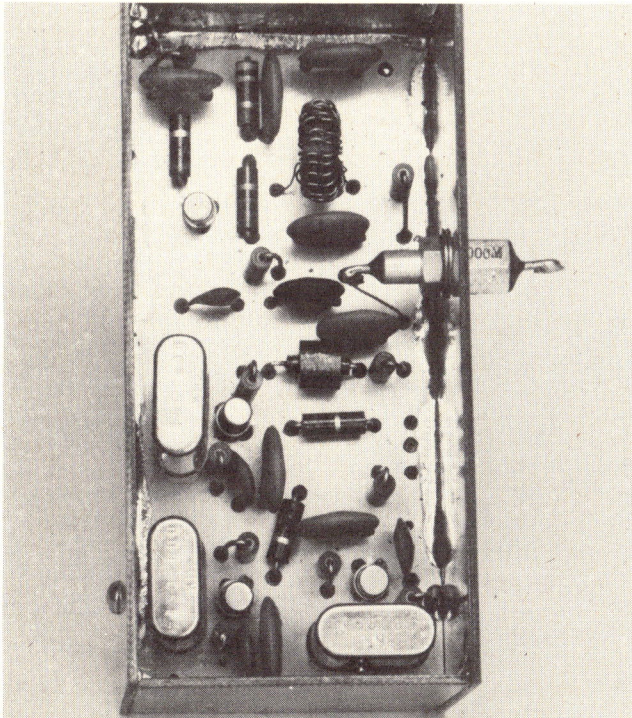

Fig. 81 — Interior view of the BFO module. Double-sided pc-board sections serve as the box walls.

Fig. 82 — Photograph of an assembled local oscillator showing how it should be packaged for good stability. Circuit details are discussed in the text and early in chapter 8.

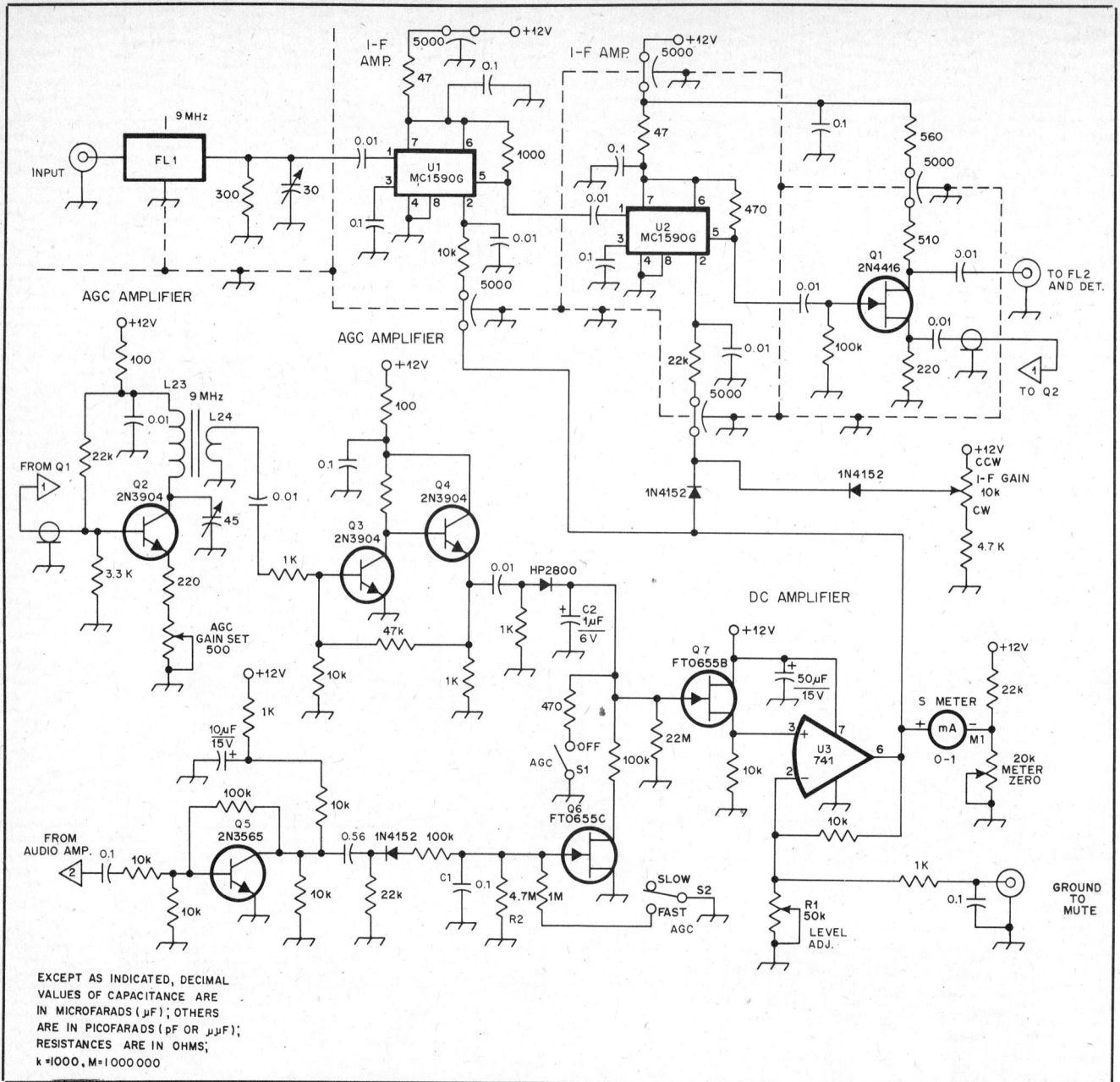

Fig. 83 — Circuit details of the W7ZOI i-f amplifier and high-performance agc system. This circuit provides full-hang agc characteristics.

and agc chain which offers excellent performance. This circuit was designed by W7ZOI for use in his Competition Grade CW Receiver. The complete receiver circuit is found in *Solid State Design for the Radio Amateur*, chapter 9.

This agc circuit is suitable for use with the i-f strip of Fig. 78. It employs a full "hang" action. The agc is defeated by means of S1. The time constant is select-

able at S2. R1 at U3 should be set for +5 volts at pin 6 of U3 with the agc off. With Q5 and Q6 as part of the circuit, the receiver is practically silent after a strong signal disappears from the passband. But, after a timing period associated with network C1-R2, the receiver will return to full gain in roughly 50 milliseconds. This is very advantageous when loud pulses of

noise enter the receiver. The effect is similar to that of a noise blanker. A detailed description of this type of agc circuit can be found in chapter 5 of *Solid State Design for the Radio Amateur*, by the ARRL.

A less complex agc circuit for use with RCA CA3028A i-f amplifiers is provided in Fig. 42 of this chapter. It does not incorporate the hang feature used in Fig. 83.

Receiver Accessories

A number of receiver accessories have become popular either through necessity or through the operating convenience that they provide. Notable among these items are noise blankers, audio filters, preamplifiers, frequency converters and interference filters. Although only relatively few amateurs choose to build their own receivers from the ground up, many more are inclined to try their hand at a somewhat less complex station accessory — especially if it is something that is needed for successful station operation. A broadcast-band filter is a good example. The circuit and construction information for such a filter appear later in this section.

A Stable 10-Meter Preamp

When receivers run out of gas in terms of front-end gain and noise figure, it usually happens on 15 and 10 meters. A typical symptom is the need to carry the audio gain wide open to copy a weak signal. This problem can often be resolved by adding a preamplifier between the receiver input and the antenna.

It is not beneficial to add gain ahead of a receiver that has sufficient sensitivity. The end product may be excessive front-end gain, leading to the demise of dynamic range. If the latter is degraded significantly, cross modulation, IMD and desensitization will become manifest when strong signals are present in and near the band of interest. In other words, don't use a preamp if your receiver performs satisfactorily now. All that will be achieved is a higher S meter reading and a higher ambient noise level when no signals are present.

Circuit Description

A major problem experienced by some builders who use common-source FETs in rf amplifiers is instability. Despite careful layout and input-output isolation, instability seems to occur. A quick solution might be to change the circuit to a common-gate (grounded gate) type. Stability is relatively easy to obtain with the latter, but a tradeoff in gain will accompany modification. A common-gate JFET amplifier usually provides between 10 and 14 dB of gain, whereas a common-source version will yield up to 25 dB of gain in some circuits.

An alternative to using a common-gate configuration was highlighted in *Solid State Design for the Radio Amateur*, where Hayward and DeMaw gave design data for fed-back bipolar-transistor rf amplifiers. The same principles apply to FET amplifiers, where gain is traded for bandwidth and stability.

A design objective with any amplifier

Fig. 84 — Exterior view of the enclosed preamplifier. A U-shaped, press-fit aluminum cover is used.

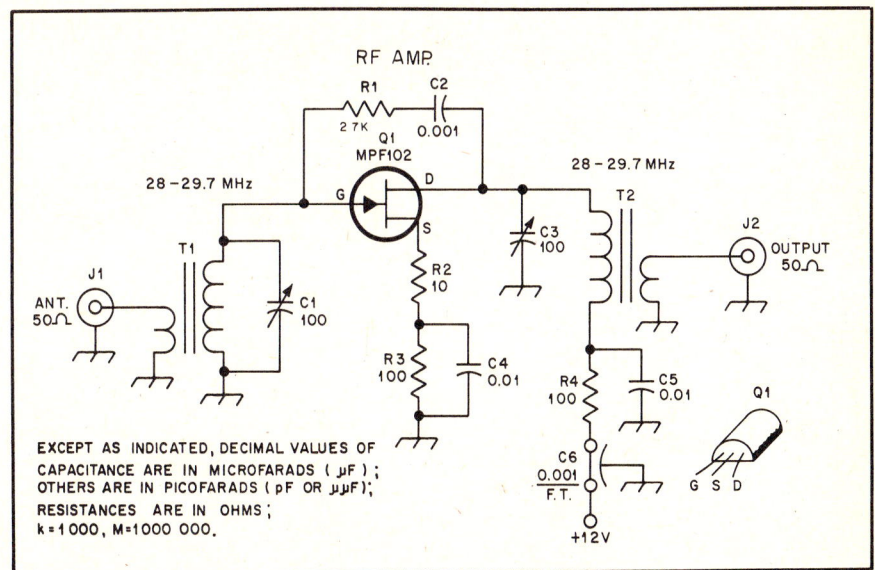

Fig. 85 — Schematic diagram of the 10-meter preamplifier. Data for 15-meter operation can be found in the text. Fixed-value capacitors are disc ceramic. Resistors are 1/4-watt composition types. Numbered components not appearing in the parts list are so identified for pc layout purposes only.

C1, C3 — Mica compression trimmer, 100 pF max. Elmenco 423 suitable.
J1, J2 — Coaxial connector of builder's choice.
Q1 — Motorola JFET, MPF102 or equiv. (see text).
T1, T2 — 0.6 μH with 1-turn link. Wind 12 turns of no. 24 enameled wire on Amidon or Palomar T50-6 powdered-iron toroid core. Spread turns evenly around core and cement in place.

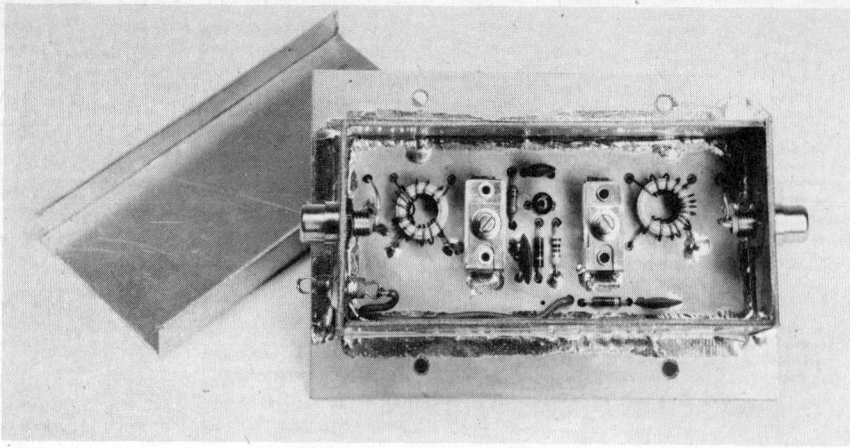

Fig. 86 — Inside view of the preamplifier.

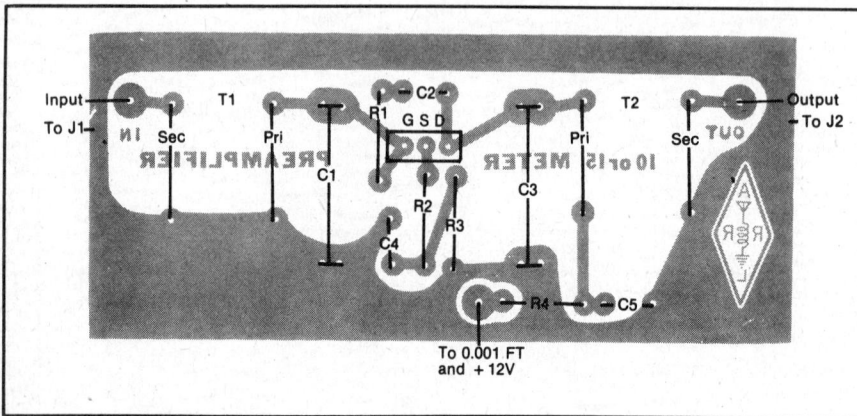

Fig. 87 — Scale pattern and layout for the pc board.

should be unconditional stability — no self-oscillations at any frequency, regardless of the load connected to the amplifier. Properly applied feedback ensures unconditional stability. Circuits of this kind are ideal for amateur builders who have limited practical experience in the workshop.

Fig. 85 shows the preamplifier circuit. T1 is a toroidal input transformer which is tuned to the operating frequency by means of C1. R1 and C2 form a shunt feedback-network which aids stability by lowering the gain. R2 provides additional stability by introducing degenerative feedback. R2 is not bypassed for rf.

The drain circuit contains a second tuned transformer, T2. R4 and C5 form a decoupling network in the 12-volt supply line. This helps prevent unwanted signal energy from entering the preamplifier via the power-supply leads.

Construction Data

A printed-circuit layout is given to scale in Fig. 87. Pc-board material (double sided) is used as a shield enclosure for the preamplifier. The etched-circuit board is also double sided, with the copper on the component side serving as a ground plane. This aids stability. The outer dimensions of the box are (HWD) 1-1/2 × 1-3/4 × 3-15/16 inches (38 × 44 × 100 mm). A 3 × 4-1/2 inch (76 × 115 mm), pc-board base is used as the bottom cover, and a U-shaped, press-fit aluminum cover serves as a top shield for the assembly.

The pc board is soldered in place inside the box (Fig. 86) after the circuit has been built and tested. A single-hole mount phono jack is used at each end of the box to provide input and output connections for the 50-ohm coaxial cables of the system. A 0.001-µF feedthrough capacitor is mounted at one end of the box to allow routing of the 12-volt supply to the preamplifier.

Performance

A 3-dB bandwidth of 1 MHz is characteristic of this preamplifier. The measured gain is 15 dB. Stability is excellent under all conditions, including an open-loop situation (no termination at either end of the circuit). The noise figure is under 2 dB at 30 MHz. A slightly better noise figure and increased dynamic range might be possible with a Siliconix U310

JFET. However, at 10 and 15 meters either the MPF102 or 2N5484 are entirely suitable. A 2N4416 FET would be an acceptable substitute.

Operation on 15 meters can be had by adding two turns of wire to the main windings of T1 and T2. No other changes are necessary. The circuit constants given in Fig. 77 are for 10-meter operation. A parts kit, pc board or pc negative are available from Circuit Board Specialists, P.O. Box 969, Pueblo, CO 81002. This circuit is suitable for use on 20, 40 and 80 meters by merely changing the T1/C1 and T2/C3 circuits for the desired frequency.

A HIGH-DYNAMIC-RANGE NOISE BLANKER

The two most violent sources of pulse interference are discharges during a lightning storm and noise generated by jamming stations and pulse radar stations. A nuisance called the "woodpecker" is a several-megawatt over-the-horizon pulse radar system that apparently has its origin in the U.S.S.R. This system produces pulses up to several hundred microvolts at the receiver input and interferes with communications. Naturally occurring noise discharges such as lightning add to the man-made noise sources to make a noise blanker a necessity in modern communications receivers.

In general, the rise and decay times of man-made and naturally occurring noise pulses are substantially faster than the rise and decay times of desired signals. This phenomenon can be used to differentiate between the two types of interference. It is therefore desirable to build a pulse receiver that can become part of the existing receiver system without degrading the overall receiver performance.

The noise blanker example of Fig. 88 is based on a publication by M. Martin, Hahn-Meitner Institute, Berlin, West Germany.[9,10] Martin's circuit is very involved and expensive. The simpler version published here can be added to any hf receiver with a first i-f between 9 and 70 MHz. It is assumed that the receiver has no rf preamplifiers and that the amplifier following the mixer has a low enough noise figure to make such a preamplifier unnecessary. This circuit originally appeared in an article by Rohde in the May 1980 issue of QST.

The noise blanker uses a Siemens TCA440 IC that incorporates all the elements of a single-conversion receiver. The i-f chosen is about 2 MHz and the

[9]Martin, "Moderner Stocraustaster mit hoher luter-modulationsfestigkeit," cq-DL magazine, July 1978, p. 300.
[10]Martin, "Grossignalfester Stocraustaster fuer Kurzwellen-und UKW-Empfaenger mit grossem Dynamikbereich," UKW-Berishte, Feb. 1979, p. 74.

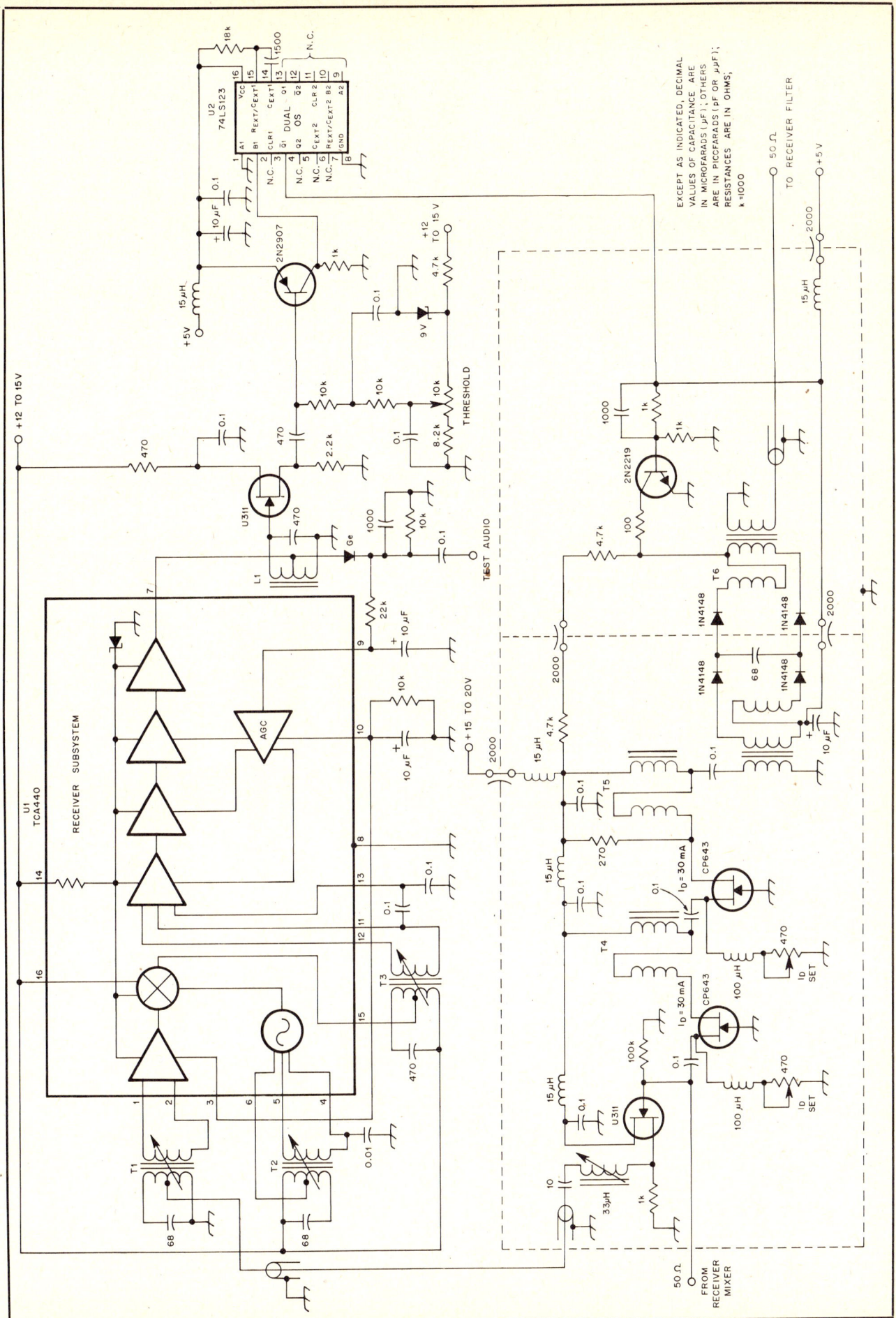

Fig. 88 — A high-performance noise blanker.

Fig. 89 — Photograph of the space-perception cw filter. Press-on lettering was used to label the control and jack functions.

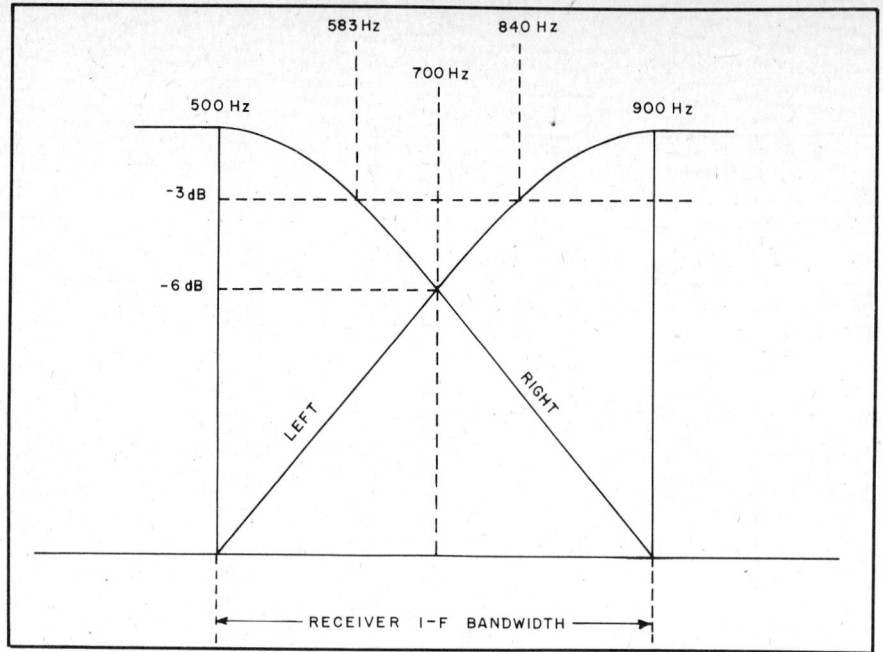

Fig. 90 — Relative frequency response of the two filter sections. The left ear receives information from the low-pass filter and the right ear receives signals from the high-pass filter.

values of the input coils are selected for an input frequency of 9 MHz. [Editor's Note: This circuit is provided for tutorial purposes. Further design information on the TCA440 is contained in "Designed Examples of Semiconductor Circuits," Siemans Corp., Issue 1975/76. Copies of the material relevant to the TCA440 are available from Siemens Corp., IC Component Group, 186 Wood Ave. South, Iselin, NJ 08830.]

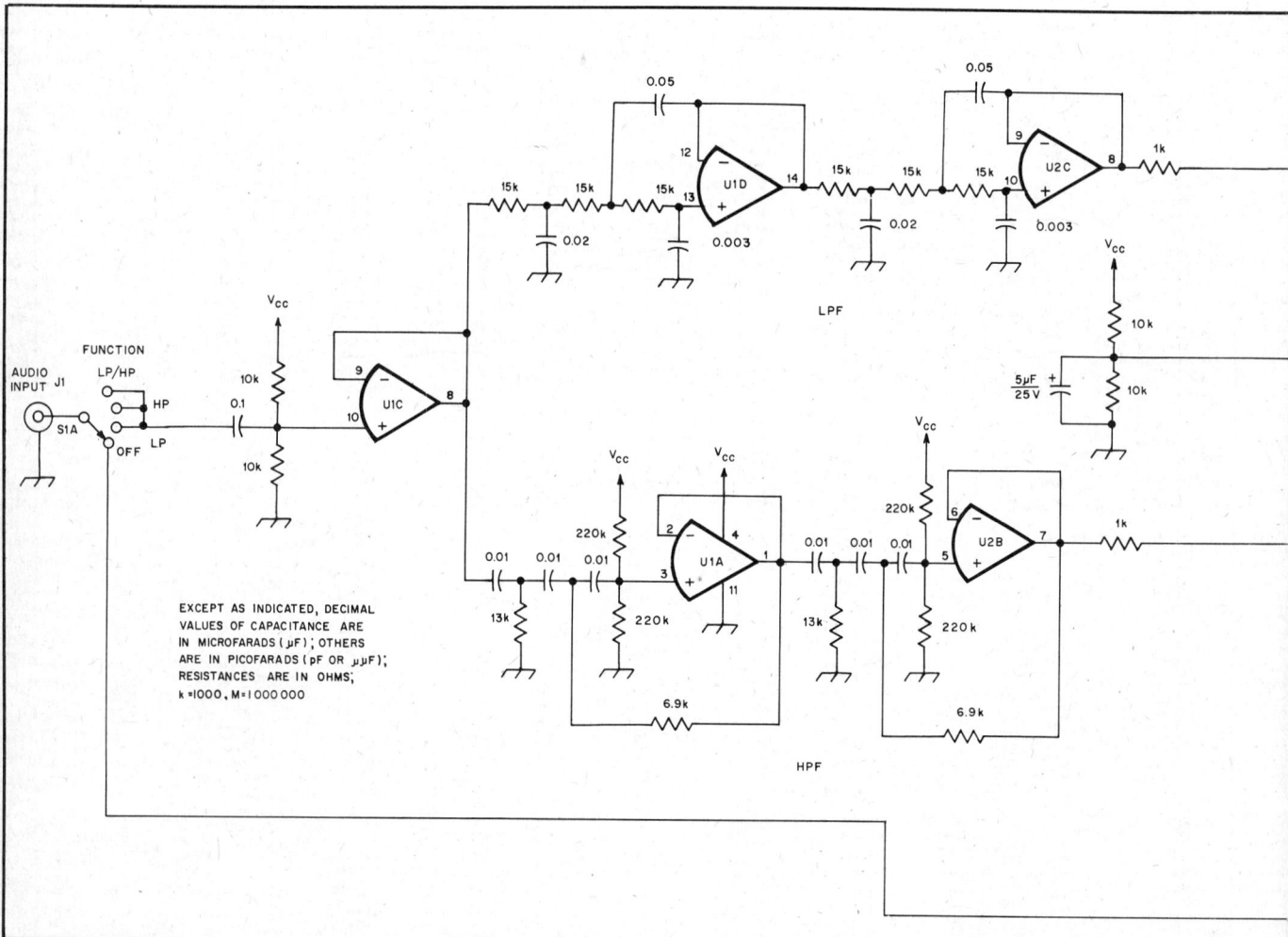

Fig. 92 — Schematic diagram of the cw filter. All resistors are 1/4-watt, 5-percent types. All polarized capacitors are electrolytic. Others are miniature ceramic types.

J1 — Phono connector.
J2 — Stereo headphone connector.
J3 — Miniature power-supply connector.
S1 — Rotary, 4 pole, 4 position.
T1, T2 — Primary 1000 ohm, secondary 8 ohm.
U1, U2 — TL084C BiFET quad op-amp, Radio Shack 276-1714 or equiv.

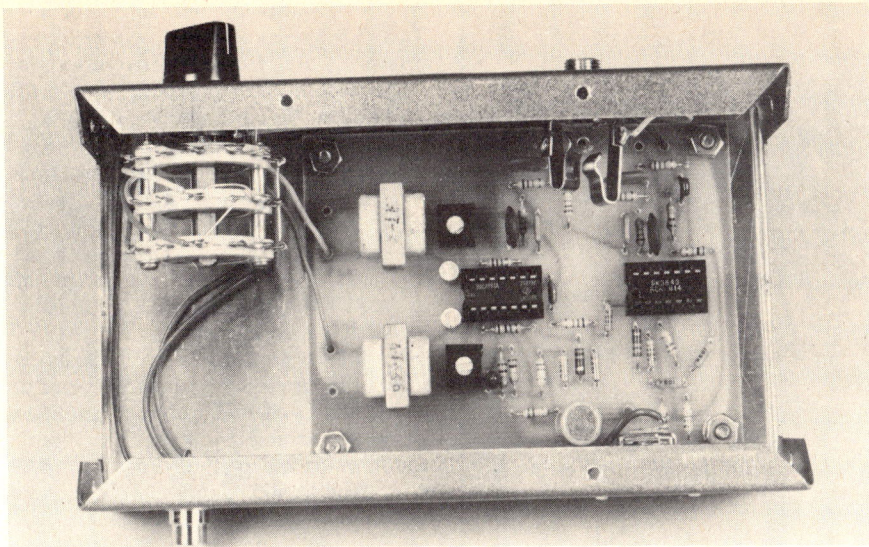

Fig. 91 — Interior view of the completed filter.

The 9-MHz signal is taken from the mixer of the receiver and fed to a CP643 amplifier and a BF246C source follower. The source follower drives a series tuned circuit. The signal is then applied to the TCA440 single-conversion receiver and converted to the 2-MHz i-f. An external germanium diode provides fixed agc

voltage to pin 9 of the TCA440. An audio test output is available to monitor the agc action of this receiver section. The 2-MHz i-f output is taken from a BF246 source follower and drives a BC177 with an adjustable-trigger threshold. The 74LS173 IC has the proper rise and decay times to drive the four-diode switching gate via a 2N2219 driver.

It was determined that the intercept point of this arrangement is about 26 dBm and the switching gate has a depth of about 80 dB. This is sufficient to suppress interference since the rf signal-to-noise ratio very rarely exceeds 60 dB. Practical experiments have shown this noise blanker to be superior to other configurations. The famous "woodpecker" completely disappears when it is switched in. The circuit layout should not be too critical. However, some care is required to build the switching gate without leakage. Good balance is required. Slightly better performance can be expected using HP 3081 PIN diodes, but this is achieved at considerably higher expense.

A SPACE-PERCEPTION CW FILTER

The circuit described here was originally conceived as an experiment to test the ability of the brain/ear subsystem to differentiate between tones of different pitch. An example of where this would occur is in a "pileup" situation where many stations are transmitting at the same time on or about the same frequency. When these signals are of the same relative amplitude, it is difficult to separate them. It was thought that if the brain/ear system was given more information than just frequency difference between tones, that separation of the individual signals would be enhanced. Initial tests with several users of this filter indicate that this is likely the case.

The filter detailed here is actually comprised of two filters — a high-pass and a low-pass. The output from the low-pass filter is fed to the left ear and the information from the high-pass filter is applied to the right ear. Stereo headphones are required. The frequency response is tailored as shown in Fig. 90. As can be seen, the roll-off is such that at 700 Hz (the center of the passband) each tone is down 6 dB. It is assumed that the receiver i-f bandwidth is limited to approximately 400 Hz for the filter described.

With this filter in the receive line the user will notice that signals move from one ear to the other when tuning across the signals. Assume that the receiver produces a low-pitched note, increasing in audio frequency as the receiver is tuned through the signal. As the signal is first heard (at a 500-Hz tone) it will appear predominantly in the left earpiece. As the signal is tuned and the pitch rises, more of the signal will be heard in the right ear. When the 700-Hz note is approached, the tones will be equal in each ear. If the

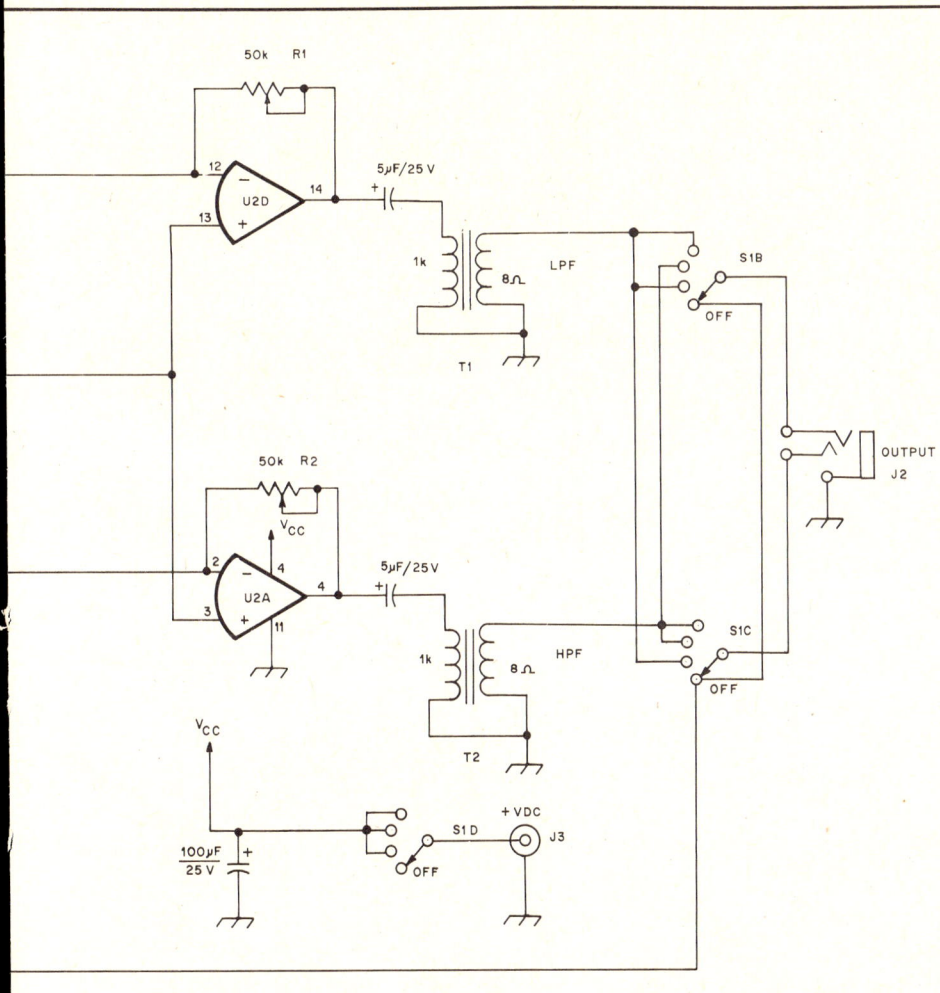

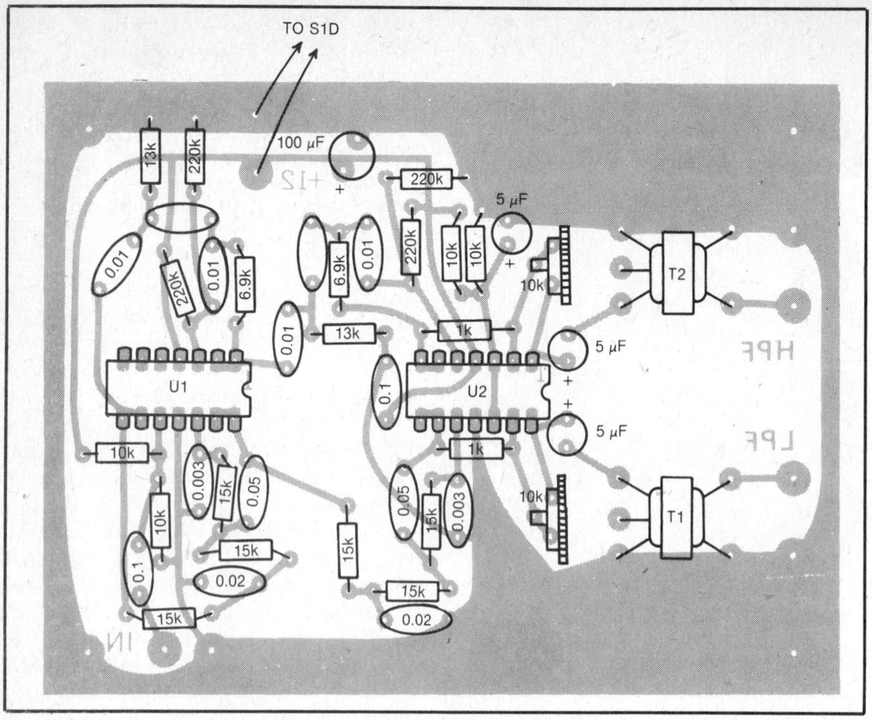

Fig. 93 — Parts-placement guide for the pc board as viewed from the component side of the board.

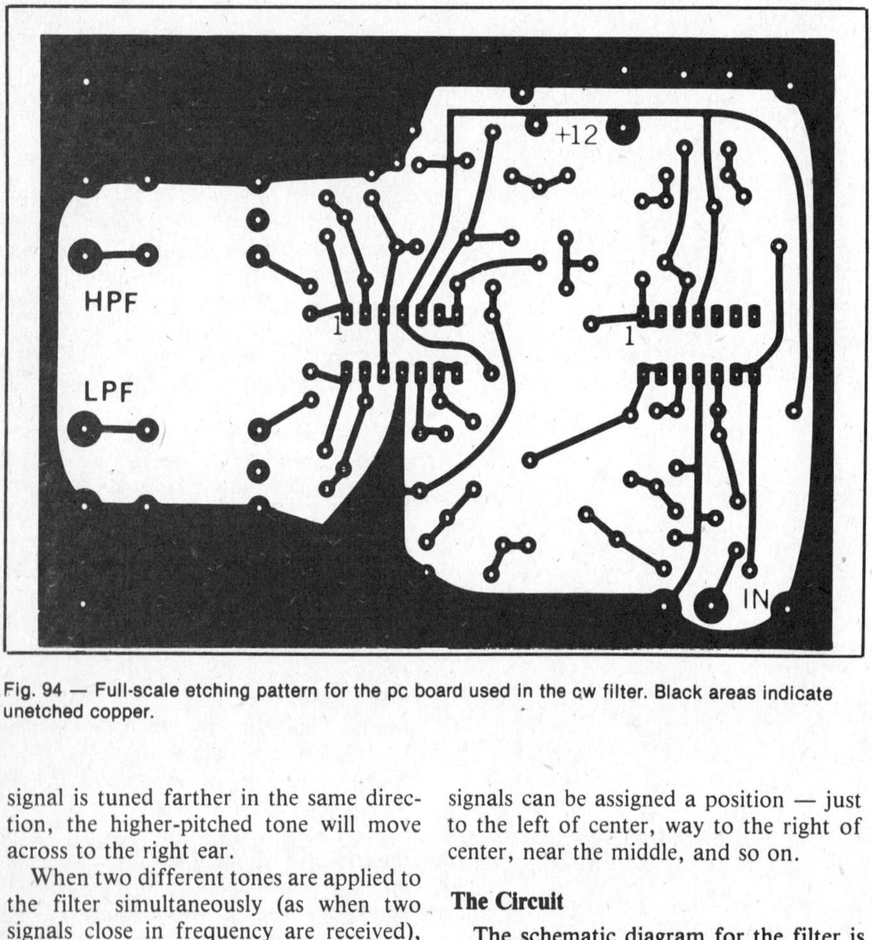

Fig. 94 — Full-scale etching pattern for the pc board used in the cw filter. Black areas indicate unetched copper.

signal is tuned farther in the same direction, the higher-pitched tone will move across to the right ear.

When two different tones are applied to the filter simultaneously (as when two signals close in frequency are received), the filter will split the signals, placing the lower-pitched tone more in the left ear and the higher-pitched tone in the right ear. By doing this, the signals are separated not only in tone, but also in relative position. With a little practice,

signals can be assigned a position — just to the left of center, way to the right of center, near the middle, and so on.

The Circuit

The schematic diagram for the filter is shown in Fig. 92. Two BiFET quad op-amps are used for the filter sections and headphone drivers. U1C is a follower used to feed the high- and low-pass branches. U1D and U2C form the low-pass section. Each filter section is a 3-pole Butterworth

type with 18-dB attenuation per octave. Two filters in cascade produce a 36-dB per octave roll-off. The cutoff frequency was selected as 583 Hz to produce the -6 dB response at 700 Hz. Section U2D is an adjustable-gain headphone driver.

U1A and U2B form the high-pass filter section with characteristics similar to those of the low-pass circuit. The roll-off frequency of this section is 840 Hz as this produces the -6 dB response at the desired 700-Hz frequency. U2A is the low-pass driver section.

No pains were taken to impedance-match the system for various positions of the function switch. This results in a slight change in volume from the filter "out" condition to any of the filter "in" positions. The small changes in volume were not found to be objectionable. A power supply was not included in the unit as the station 12-volt regulated supply could be used. Alternatively, a power supply similar to those used with small tape recorders and the like may be used. These units have the transformer and other power-supply components in a molded assembly that plugs into a wall outlet.

Construction and Adjustment

Construction of the filter is simplied through the use of a printed-circuit board. The parts-placement guide and etching pattern for this board are shown in Figs. 93 and 94. The filter is assembled in a homemade aluminum enclosure that measures 6-3/4 × 2-1/4 × 3-1/2 inches (171 × 57 × 89 mm). Construction is not critical and any cabinet that will house the components may be used. No doubt the unit could be made smaller than the one shown in the photographs. The rotary switch is mounted to the left of the board, and hookup wire is used to make connections throughout the unit. Other types of connectors may be used for the input, output and power connections. The ones shown here were chosen for use in the writer's station.

Adjustment of the filter involves setting the level controls for each of the two channels. This is accomplished by first inserting a pair of stereo headphones at J2 and making a connection between the output of the station receiver and J1. Apply a regulated 10 to 15 volts dc at J3. Adjust the receiver so that a 700-Hz note is applied to the filter. With the filter in the 1p/hp (low-pass/high pass) mode, adjust controls R1 and R2 so that the tone is of equal volume in each ear. It should be noted that the gains of the driver amplifiers should not be adjusted too high, as there might be some hum pickup in the low-pass section. Experimentation with the settings of R1 and R2 should provide an optimum gain/noise setting.

A BC- BAND ENERGY-REJECTION FILTER

Inadequate front-end selectivity, of

bipolar-transistor rf amplifier and mixer stages which perform poorly, can result in unwanted cross-talk and overloading from adjacent commercial or amateur stations. A simple cure for this problem is to install between the antenna and receiver a filter that will sufficiently attenuate the out-of-band signals but pass those signals of interest with little or no attenuation. If the receiver is designed for reception of frequencies below *and* above the broadcast band, a 550- to 1600 kHz band-stop filter will be required. However, if reception is desired only below *or* above the broadcast band, then a less complex low *or* high-pass filter will suffice. Because a majority of ham receivers are used for reception above 1600 kHz, a high-pass filter will generally be preferable to the band-reject filter. For the same number of components, the high-pass filter performance is superior to that of the band-reject type.

Since the power level of broadcast stations can be quite high, the stop-band attenuation of the high-pass filter should also be high, preferably in excess of 60 dB. The cutoff frequency should be selected so less than 1 dB of attenuation occurs above 1800 kHz, the start of the 160-meter band. Receivers are generally designed to present a 50-ohm load to the antenna, and the filter should also be designed for the same impedance level. The rate of attenuation rise, VSWR, pass-band ripple, and number of filter components are all interrelated and many design choices are possible. In the high-pass design to be discussed, the maximum VSWR of the filter was selected to be 1.353. To obtain adequate stop-band attenuation and a reasonable rate of attenuation rise, a filter of 10 elements was considered necessary. Finally, to simplify construction, only those designs permitting the use of standard-value capacitors were considered.

Building the Filter

The filter layout, schematic diagram and response curve, the component values used, the toroidal-inductor winding specifications are all shown in Fig. 96. The standard-value capacitors used are listed under the filter schematic diagram. Note that all standard values are within 2.8 percent of the design values. Since the maximum deviation between the actual capacitance used and the design value will be only 5.3 percent, there should be little or no difficulty in obtaining the desired response. If the attenuation peaks (f2, f4 and f6) at 0.677, 1.293 and 1.111 MHz are not obtained, a slight squeezing or separating of the toroidal-inductor winding should be all that is required to tune the series-resonant circuits. Note that series circuit C6-L6 should resonate at f6 = 1.111 MHz, but from the response curve it actually resonated at about 1.130 MHz. This frequency error of about 2 percent is small

enough to ignore. The A_S value was selected to be 58.3 dB, and examination of the response curve shows the measured filter response to be in good agreement. The measured values of cutoff frequency (at the attenuation level of 0.0988 dB) and the measured value of fA_S (the frequency where A_S is first reached) are also in good agreement with the calculated values. The measured pass-band loss was less than 0.8 dB from 1.8 to 10 MHz. Between 10 and 100 MHz, the insertion loss of the filter gradually increased to 2 dB. The measured input impedance versus frequency was in good agreement with the calculated input impedance between 1.7 and 4.2 MHz. (The frequency range above 4.2 MHz was not tested.) Over the range tested, the input impedance of the filter remained within the 37 to 67.7 ohms input-impedance window (equivalent to a maximum VSWR of 1.353).

Construction of the filter is relatively simple, as shown in the photograph, and no difficulty should be experienced if the Mallory SXM polystyrene capacitors are used. These capacitors have a standard tolerance of 2.5 percent and are available

Fig. 95 — The filter is built on perfboard in a 2 × 2 × 5-inch Minibox. The filter can be made smaller if desired, and phono connectors can be used in place of the BNC fittings shown here.

through all Mallory distributors. The Micro-metals iron powder T50-2 toroidal cores are available through either Amidon or Palomar Engineers. This material originally appeared in a *QST* article by Wetherhold, W3NQN, in February 1978.

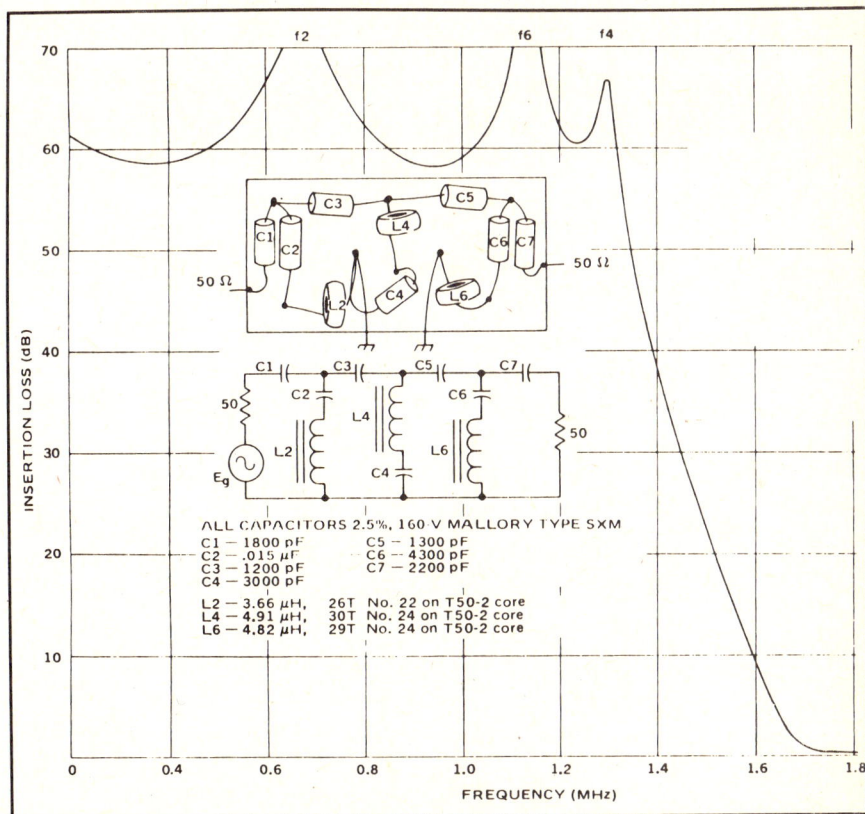

ALL CAPACITORS 2.5%, 160-V MALLORY TYPE SXM
C1 — 1800 pF   C5 — 1300 pF
C2 — .015 µF   C6 — 4300 pF
C3 — 1200 pF   C7 — 2200 pF
C4 — 3000 pF

L2 — 3.66 µH,   26T   No. 22 on T50-2 core
L4 — 4.91 µH,   30T   No. 24 on T50-2 core
L6 — 4.82 µH,   29T   No. 24 on T50-2 core

Fig. 96 — Filter-response curve, insertion loss, layout and schematic diagram. Terminal impedance is 50 ohms for this 1.7-MHz, high-pass filter.

Chapter 9

VHF and UHF Receiving Techniques

Adequate receiving capability is essential in vhf and uhf communications, whether the station is a transceiver or a combination of separate transmitting and receiving units, and regardless of the modulation system used. Transceivers and fm receivers are treated separately in this *Handbook,* but their performance involves basic principles that apply to all receivers for frequencies above 30 MHz. Important attributes are good signal-to-noise ratio (low noise figure), adequate gain, stability, and freedom from overloading and other spurious responses.

Except where a transceiver is used, the vhf station often has an hf-band communications receiver for lower bands, with a crystal-controlled converter for the vhf band in question ahead of it. The receiver serves as a tunable i-f system, complete with detector, noise limiter, BFO and audio amplifier. Unless one enjoys work with communications receivers, there may be little point in building this part of the station. Thus our concern here will be mainly with converter design and construction.

Choice of a suitable communications receiver for use with converters should not be made lightly, however. Several degrees of selectivity are desirable: 500 Hz or less for cw, 2 to 3 kHz for ssb, 4 to 8 kHz for a-m phone and 12 to 36 kHz for fm phone are useful. The special requirements of fm phone are discussed in chapter 13. Good mechanical design and frequency stability are important. Image rejection should be high in the range tuned for the converter output. This may rule out 28 MHz with receivers of the single-conversion type having 455-kHz i-f systems.

Broadband receiving gear of the surplus variety is a poor investment at any price, unless one is interested only in local work. The superregenerative receiver,

though simple to build and economical to use, is inherently lacking in selectivity. With this general information in mind, this section will cover vhf and uhf receiver "front end," stage by stage.

RF AMPLIFIERS

Signal-to-Noise Ratio (S/N): The limiting factor in the reception of signals is noise. Noise can be classified into two broad forms, random and nonrandom. Nonrandom noise such as interfering signals are reduced or eliminated through techniques aimed at directly filtering or otherwise suppressing detection of the unwanted signals. This is only possible because the nonrandom noises are discrete in nature and are relatively predictable.

Random noise is generated by sources both internal and external to the receiver. The external noise problem varies considerably with frequency of reception. Below about 25 MHz, man-made, atmospheric, and galactic noise picked up by the antenna is usually far greater than the noise generated within the receiver. In a majority of cases, noise output from a receiver tuned to the hf range drops dramatically when the antenna is disconnected from the receiver. When this is observed, it clearly demonstrates that reception is limited by external noise.

Since the ionosphere is less active at 50 MHz, atmospheric noise is of less concern than at hf. Even in a quiet location, however, external noise usually overrides receiver noise in a well-designed system. Above 100 MHz, external noise other than man-made is rarely a problem in weak-signal work. The noise characteristics of a receiving system become important as they are the primary limitation in weak-signal work. Unfortunately, circuit design and component choice be-

come increasingly critical with respect to signal-to-noise ratio as the operating frequency rises.

Noise Temperature, Noise Factor and Noise Figure

Noise, in the context of this discussion, is produced by the movement of electrons in any substance (such as resistors, transistors and FETs) that has a temperature above absolute zero ($-273°$ C or 0 K). Electrons move in a random fashion colliding with relatively immobile ions that make up the bulk of the material. The final effect is that, in most substances there is no net current in any particular direction on a long-term average, but rather a series of random pulses. These pulses produce what is called thermal agitation noise, thermal noise or Johnson noise.

As the currents caused by electron movement increase with temperature, so does the noise power. Also, as the pulses are random, they spread out over a broad frequency spectrum. As it turns out, if we examine the power contained in a given passband, the value of that power is independent of the center frequency of the passband. This is expressed as

$$p = kTB$$

where p is the thermal noise power, k is Boltzmann's constant (1.374×10^{-23} joule per K), T is absolute temperature in K and B is the bandwidth in hertz. Notice that the power is directly proportional to temperature, and at 0 K the noise power is zero.

Active devices normally exhibit noise temperatures different from their ambient temperatures. The thermal noise produced by a semiconductor device will limit its ability to respond to input signals below the level of the internally generated noise. Noise temperature, noise factor and noise figure are all measures of this device noise. The results are expressed in terms of temperature, ratios and decibels, respectively.

Consider a 50-ohm termination connected to the input of a device with the termination cooled to absolute zero. There would be no noise produced by this source, and the noise output from the device would be that of the internally generated noise. If the termination were now heated to a temperature that would raise the output noise of the device by 3 dB (thermal agitation noise equal to the internally generated noise of the device) and the temperature of the termination measured, the *effective input noise temperature* (T_E) of the device would be this value. The noise temperature specification is independent of bandwidth and is directly proportional to noise power. For example, if we were to halve the noise temperature we would double the signal-to-noise ratio.

In order to convert a noise temperature measurement to noise figure an intermediate calculation is required — *noise factor* (f). Noise factor is by definition the ratio of the total output noise power to the input noise power when the termination is at the standard temperature of 290 K (17° C). The noise power caused only by the input noise of the termination is simply the noise power of the source multiplied by the gain of the device. Mathematically

$$N_{power\ input} = GkBT_o$$

where G is the gain of the device and T_o is 290 K. The *total* noise caused by the input noise of the termination and the internally generated noise is simply the sum of the two noise sources multiplied by the gain of the device, or

$$N_{power\ total} = GkB(T_o + T_E)$$

where T_E is the effective input noise temperature and T_o is 290 K. The noise factor (f) is calculated as

$$f = \frac{N_{power\ total}}{N_{power\ input}}$$

$$= \frac{GkBT_o + GkBT_E}{GkBT_o} = 1 + \frac{T_E}{T_o}$$

where T_E is the effective input noise temperature and T_o is 290 K. Noise figure can then be calculated as follows:

$$NF = 10 \log_{10} f = 10 \log_{10} (1 + \frac{T_E}{T_o})$$

where noise figure is expressed in decibels. Should the noise figure of the device be known and it is desired to find the noise

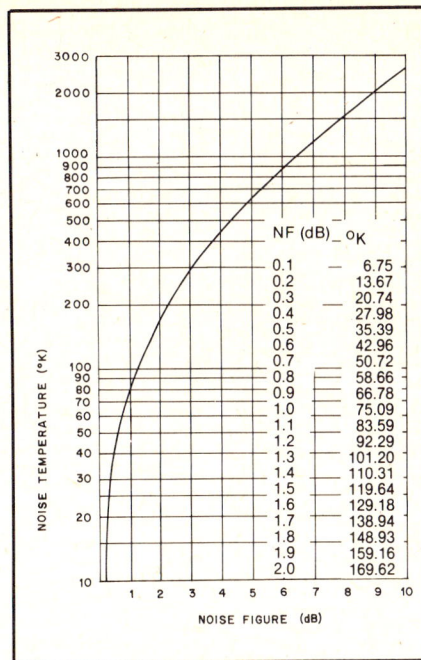

Fig. 1 — Relationship between noise figure and noise temperature.

temperature, the equation can be rearranged as follows:

$$T_E = 290\ [antilog(NF/10) - 1]K$$

where noise figure is expressed in dB. A graph illustrating the relationship between noise figure and noise temperature is given in Fig. 1.

Noise factor can also be represented in terms of signal-to-noise ratios as

$$f = \frac{S/N\ at\ input}{S/N\ at\ output}$$

and noise figure can be found from

$$NF = 10 \log_{10} f$$

$$= 10 \log_{10} \frac{S/N\ at\ input}{S/N\ at\ output}$$

A vhf receiving system consists of an interconnection of individual stages, some noisier than others. Each stage's noise contribution to the reduction of signal-to-noise ratio can be expressed as a noise figure. How much the noise figure of a particular stage affects system noise figure depends on the gain of the stages between that stage and the antenna. That is, if a stage's gain is sufficiently large, its noise figure will tend to override or "mask" the noise contribution of the stage following it. Mathematically, the noise factor of a receiving system can be expressed as

$$F = f_1 + \frac{f_2 - 1}{G_1} + \frac{f_3 - 1}{G_1 G_2} + \cdots + \frac{f_n - 1}{G_n \ldots G_2 G_1}$$

where

f_n = noise factor of the nth stage
G_n = gain of the nth stage

Brief analysis of this equation shows that the first stage of a receiving system is the most important with regard to noise figure. If the gain of this and succeeding stages is greater than unity, the denominator of each successive term becomes greater. The numerical value of terms beyond the second or third approaches zero and can be ignored.

It might seem that the more gain an rf amplifier has, the better the signal-to-noise ratio and therefore the better the reception. This is not necessarily true. The primary function of an rf amplifier is to establish the noise figure of the system. One good rf stage is usually adequate unless the mixer is a passive type with loss instead of gain. Two rf stages are the usual maximum requirement.

Once the system noise figure is established, any further gain necessary to bring a signal to audible levels may be obtained from intermediate-frequency stages or in the audio channel. Use of the minimum gain necessary to set the overall receiver noise figure is desirable in order to avoid overloading and spurious signals in subsequent stages.

Further examination of the equation points out the desirability of mounting the first stage of the receiver system at the antenna. The transmission line from the antenna to the receiver can be considered as a stage in the receiving system. The first stage of a receiving system makes the major contribution of noise figure to the system, so it is highly desirable that the first stage be a low-noise amplifier with gain. A transmission line is a "lossy" amplifier, and if placed as the first stage of a receiving system, automatically limits the system noise figure to that of the transmission line, *at best*. If the first rf amplifier is placed before the lossy transmission line stage, at the antenna, the amplifier gain will tend to mask the noise added by the transmission line.

Stability

Excessive gain or undesired feedback may cause amplifier instability. Oscillation may occur in unstable amplifiers under certain conditions. Damage to the active device from overdissipation is only the most obvious effect of oscillation. Deterioration of noise figure, spurious signals generated by the oscillation, and reradiation of the oscillation through the antenna, causing interference to other services (i.e., RFI), can also occur from amplifier instability.

Neutralization or other forms of feedback may be required in rf amplifiers to reach stability. Amplifier neutralization is achieved by feeding energy from the amplifier output circuit back to the input in such an amount and phase as to cancel out the effects of device internal capacitance and other unwanted input-output coupling. Care in termination of both the input and output can produce stable results from an otherwise unstable amplifier. Attention to proper grounding and proper isolation of the input from the output by means of shielding also can yield stable operating conditions.

Overloading and Spurious Signals

Normally, the rf amplifier is not a significant contributor to overloading problems in vhf receiving systems. The rf amplifiers in the first or second stage of a receiving system operate in a linear service and if properly designed require a substantial signal input to cause deviation from linearity. Overloading usually occurs in the naturally nonlinear mixer stages. Images and other responses to out-of-band signals can be reduced or eliminated by proper filtering at the amplifier input.

In general, unwanted spurious signals and overloading increase as the signal levels rise at the input to the offending stage. Consequently, minimum gain prior to the stage minimizes overloading. Since noise figure may suffer at reduced gain, a compromise between optimum noise figure and minimum overloading must often be made. Especially in areas of high amateur activity, sacrificing noise figure somewhat may result in increased weak-signal reception effectiveness if the lower noise-figure system is easily overloaded.

Typical Circuits

Common circuits for rf amplifiers are illustrated in Figs. 2 through 5. Examples of amplifier construction are given later in this chapter. The termination impedances of both the input and output of these examples are low (50 ohms), lending them well to preamplifier service. Preamplifiers are useful for improving the noise figure of existing equipment.

The choice of active device has a profound effect on the weak-signal performance of an rf amplifier. Although tubes can be used on the vhf and uhf bands, their use is seldom seen, as solid-state devices provide far better performance at lower cost. Bipolar transistors can provide excellent noise figures up through 4 GHz if chosen and used properly. The JFET is usable through the 432-MHz band, although the most commonly available ones drop off in performance quickly beyond that frequency. Dual-gate MOSFETs also are usable through 432 MHz. The GaAs FET, though somewhat costly, provides superior noise figures past the 1296 MHz band.

Most rf amplifiers for use below the 432-MHz band use FETs rather than bipolars. Unless bipolar transistors are run at relatively high standing currents they are prone to overloading from strong signals. Additionally, their lower terminating impedances can present somewhat awkward design considerations to the builder. The FET minimizes these problems while presenting acceptable

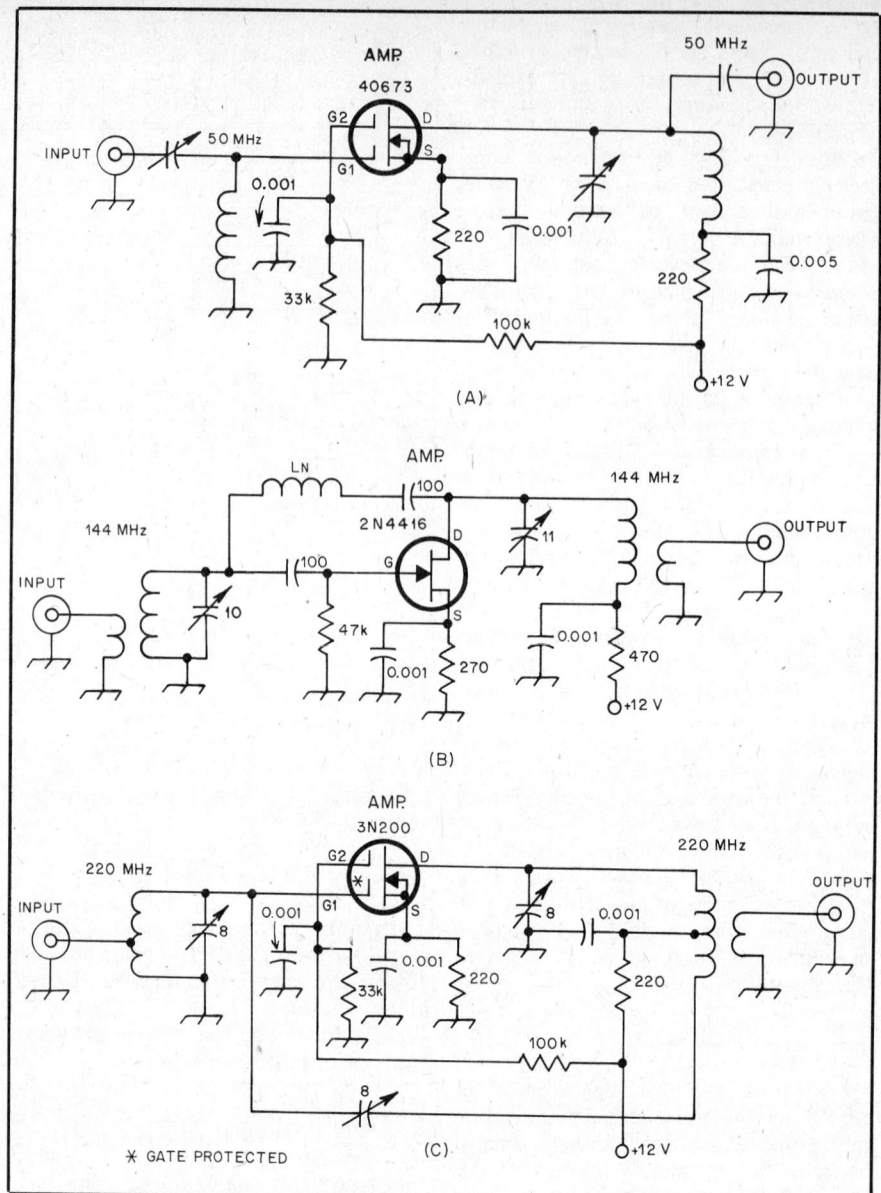

Fig. 2 — Typical grounded-source rf amplifiers. The dual-gate MOSFET, A, is useful below 500 MHz. The junction FET, B, and neutralized MOSFET, C, work well on all vhf bands. Except where given, component values depend on frequency.

noise figure.

At 432 MHz and above, inexpensive FETs cannot provide the low noise figure attainable from bipolars. The wavelength at these higher frequencies also allows the convenient use of tuned lines rather than conventional coils, easing the possible design difficulties of the lower terminating impedances of bipolars.

The input network of an rf amplifier should be as low in loss as possible, if a low noise figure is desired. Since any loss before the first stage is effectively added to the noise figure, it is well to keep these losses to a minimum. High-selectivity circuits often have significant losses and should be avoided at the front end. L networks usually provide the least loss while assuring proper impedance matching. High-quality components should

also be employed in the input circuit to further reduce the losses.

It should be pointed out that the terminating impedance of transistors for optimum noise figure is usually not the same as that for optimum power transfer (gain). This complicates the designing and tuning procedures somewhat, but careful measurements and adjustment can compensate for these shortcomings. The dual-gate MOSFET has different internal geometry, so optimum noise match is virtually identical to optimum gain match. This means that adjusting a dual-gate MOSFET amplifier for maximum gain usually provides best noise figure.

Some examples of common-source amplifiers are shown in Fig. 2. Many properly designed dual-gate MOSFET amplifiers do not require neutralization to

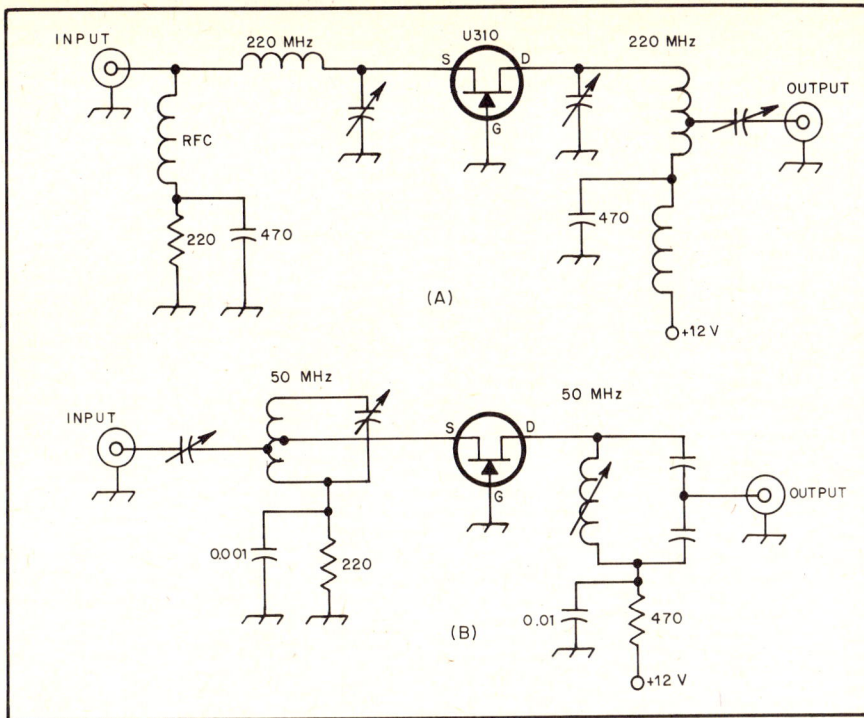

Fig. 3 — Grounded-gate FET preamplifier tends to have lower gain and broader frequency response than other amplifiers described.

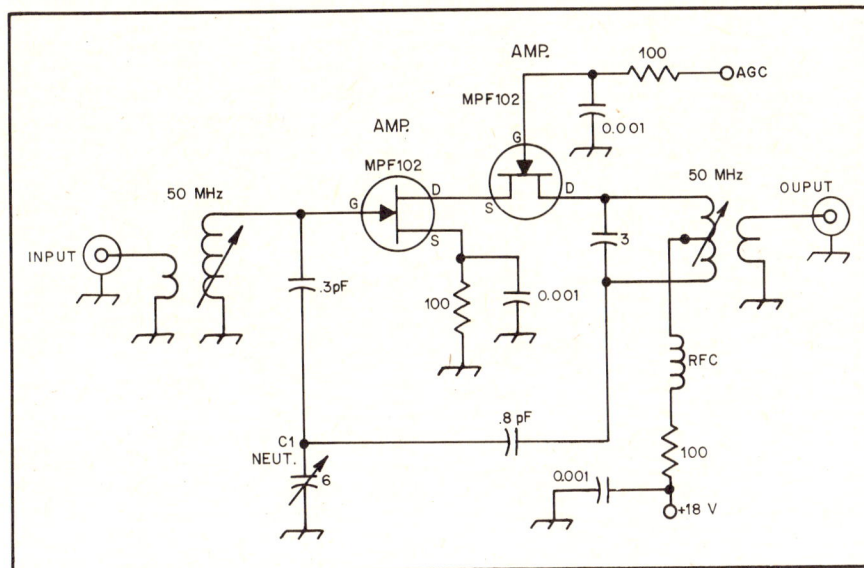

Fig. 4 — Cascode amplifier circuit combines grounded-source and grounded-gate stages, for high gain and low noise figure. Though JFETs are shown, the cascode principle is useful with MOSFETs as well.

achieve stability and best noise figure. An example of this approach is shown in Fig. 2A. Neutralization may be required; Fig. 2C shows capacitive neutralization applied to dual-gate MOSFET amplifier. Common-source JFET amplifiers usually require neutralization to attain satisfactory operation. Inductive neutralization as shown in Fig. 2B is commonly used.

Using the gate as the common stage element introduces negative feedback and eliminates the need for neutralization in a common-gate amplifier, shown in Fig. 3. The feedback reduces the stage gain and lowers the input impedance, increasing the bandwidth of the stage. An additional benefit of common-gate amplifiers is reduced susceptibility to overload as compared to common-source amplifiers.

The cascode circuit of Fig. 4 combines the common-source and the common-gate amplifiers, securing some of the advantages of each. Increased gain over a single stage is its greatest asset.

Fig. 5 shows typical bipolar amplifiers for the uhf range. Fig. 5A illustrates a common-emitter amplifier, analogous to the common-source FET amplifier. The common-base amplifier of Fig. 5B can similarly be compared to a common-gate FET amplifier.

Front-End Protection

The first amplifier of a receiver is susceptible to damage or complete burnout through application of excessive voltage to its input element by way of the antenna. This can be the result of lightning discharges (not necessarily in the immediate vicinity), rf leakage from the station transmitter through a faulty send-receive relay or switch, or rf power from a nearby transmitter and antenna system. Bipolar transistors often used in low-noise uhf amplifiers are particularly sensitive to this trouble. The degradation may be gradual, going unnoticed until the receiving sensitivity has become very poor.

No equipment is likely to survive a direct hit from lightning, but casual damage can be prevented by connecting diodes back-to-back across the input circuit. Either germanium or silicon vhf diodes can be used. Both have thresholds of conduction well above any normal signal level, about 0.2 volt for germanium and 0.6 volt for silicon. The diodes used should have fast switching times. Computer diodes such as the 1N914 and hot-carrier types are suitable. A check on weak-signal reception should be made before and after connection of the diodes.

RF Selectivity

Ever-increasing occupancy of the radio spectrum brings with it a parade of receiver overload and spurious responses. Overloading problems can be minimized by the use of high dynamic range receiving techniques, but spurious responses such as the receive image must be filtered out before mixing occurs. Conventional tuned circuits cannot provide the selectivity necessary to eliminate the plethora of signals found in most ham neighborhoods. Other filtering techniques must be used.

Although some amateurs use quarter-wavelength coaxial cavities on 50, 144 and 220 MHz, the helical resonators shown in Fig. 6 are usually a better choice as they are smaller and easier to build. In the frequency range from 30 to 100 MHz, where it's difficult to build high-Q inductors, and because coaxial cavities are very large, the helical resonator is an excellent choice. At 50 MHz, for example, a capacitance-tuned, quarter-wavelength coaxial cavity with an unloaded Q of 3000 would be about 4 inches (100 mm) in diameter and nearly 5 feet (1.5 m) long. On the other hand, a helical resonator with the same unloaded Q is about 8.5 inches (216 mm) in diameter and 11.3 inches (287 mm) long. Even at 432 MHz, where coaxial cavities are common, the use of a

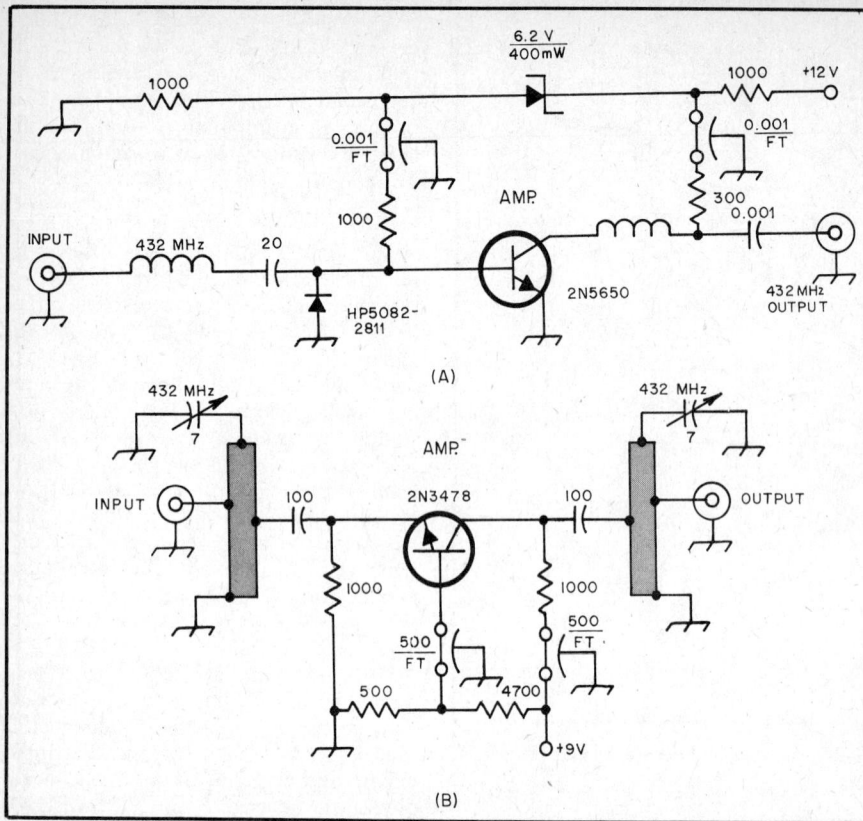

(A)

(B)

Fig. 5 — Examples of uhf amplifiers using bipolar transistors.

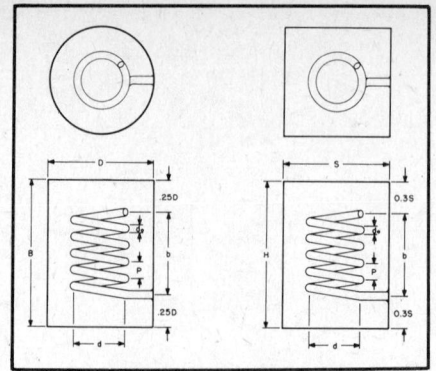

Fig. 6 — Round and square helical resonators, showing principal dimensions. Diameter, D (for side, S) is determined by the desired unloaded Q. Other dimensions are expressed in terms of D (or S) as described in the text.

helical resonator will result in substantial size reductions. The following design information on helical resonators originally appeared in a June 1976 *QST* article by W1HR.

The helical resonator has often been described simply as a coil surrounded by a shield, but it is actually a shielded, resonant section of helically wound transmission line with relatively high characteristic impedance and low axial propagation velocity. The electrical length is about 94 percent of an axial quarter wavelength, or 84.6 electrical degrees. One lead of the helical winding is connected directly to the shield and the other end is open circuited as shown in Fig. 6. Although the shield may be any shape, only round and square shields will be considered in this section.

Design

The unloaded Q of a helical resonator is determined primarily by the size of the shield. For a round resonator with a copper coil on a low-loss form, mounted in a copper shield, the unloaded Q is given by

$$Q_u = 50D \sqrt{f_c}$$

where D = inside diameter of the shield in inches ($\times$ 2.54 = mm) and f_o = frequency in MHz.

If the shield can is square, assume D to be 1.2 times the width of one side. This formula, which includes the effects of

losses and imperfections in practical materials, yields values of unloaded Q which are easily attained in practice. Silver plating of the shield and coil will increase the unloaded Q by about three percent over that predicted by the equation. At vhf and uhf, however, it is more practical to increase slightly the shield size (i.e., increase the selected Q_u by about three percent before making the calculation). The fringing capacitance at the open-circuited end of the helix is about 0.15D pF (i.e., approximately 0.3 pF for a shield two inches, or 51 mm, in diameter).

Once the required shield size has been determined, the total number of turns, N, winding pitch, P, and characteristic impedance, Z_o, for round and square helical resonators with air dielectric between the helix and shield, are given by

$$N = \frac{1908}{f_o D} \quad N = \frac{1590}{f_o S} \quad P = \frac{f_o D^2}{2312} f_o$$

$$Z_o = \frac{99,000}{f_o D} \quad Z_o = \frac{82,500}{f_o S} \quad P = \frac{f_o S^2}{1606}$$

In these equations dimensions D and S are in inches, and f_o is in MHz. The design nomograph for round helical resonators in Fig. 7, which can be used with slide-rule accuracy, is based on these formulas.

Although there are many variables to consider when designing helical resonators, certain ratios of shield size and length, and coil diameter and length, will provide optimum results. For helix diameter, d = 0.55D, or d = 0.66S. To

determine helix length, b = 0.825D or (b = 0.99S). For shield length, B = 1.325D and H = 1.60S.

Calculation of these dimensions is simplified by the design chart of Fig. 8. Note that these ratios result in a helix with a length 1.5 times its diameter, the condition for maximum Q. The shield is about 60 percent longer than the helix — although it can be made longer — to completely contain the electric field at the top of the helix and the magnetic field at the bottom.

It should be mentioned that the winding pitch, P, is used primarily to determine the required conductor size. During actual construction the length of the coil is adjusted to that given by the equations for helix length. Conductor size ranges from 0.4P to 0.6P for both round and square resonators and is plotted graphically in Fig. 9.

Obviously, an area exists (in terms of frequency and unloaded Q) where the designer must make a choice between a conventional cavity (or lumped LC circuit) and a helical resonator. At the higher frequencies, where cavities might be considered, the choice is affected by shape factor; a coaxial resonator is long and relatively small in diameter, while the length of a helical resonator is not much greater than its diameter. A second consideration is that point where the winding pitch, P, is less than the radius of the helix (otherwise the structure tends to be nonhelical). This condition occurs when the helix has fewer than three turns ("upper limit" on the design nomograph of Fig. 7).

Construction

To obtain as high an unloaded Q as possible, the shield should not have any seams parallel to the axis of the helix. This is usually not a problem with round resonators because large-diameter copper tubing is used for the shield, but square resonators require at least one seam and usually more. However, the effect on unloaded Q is minimal if the seam is silver

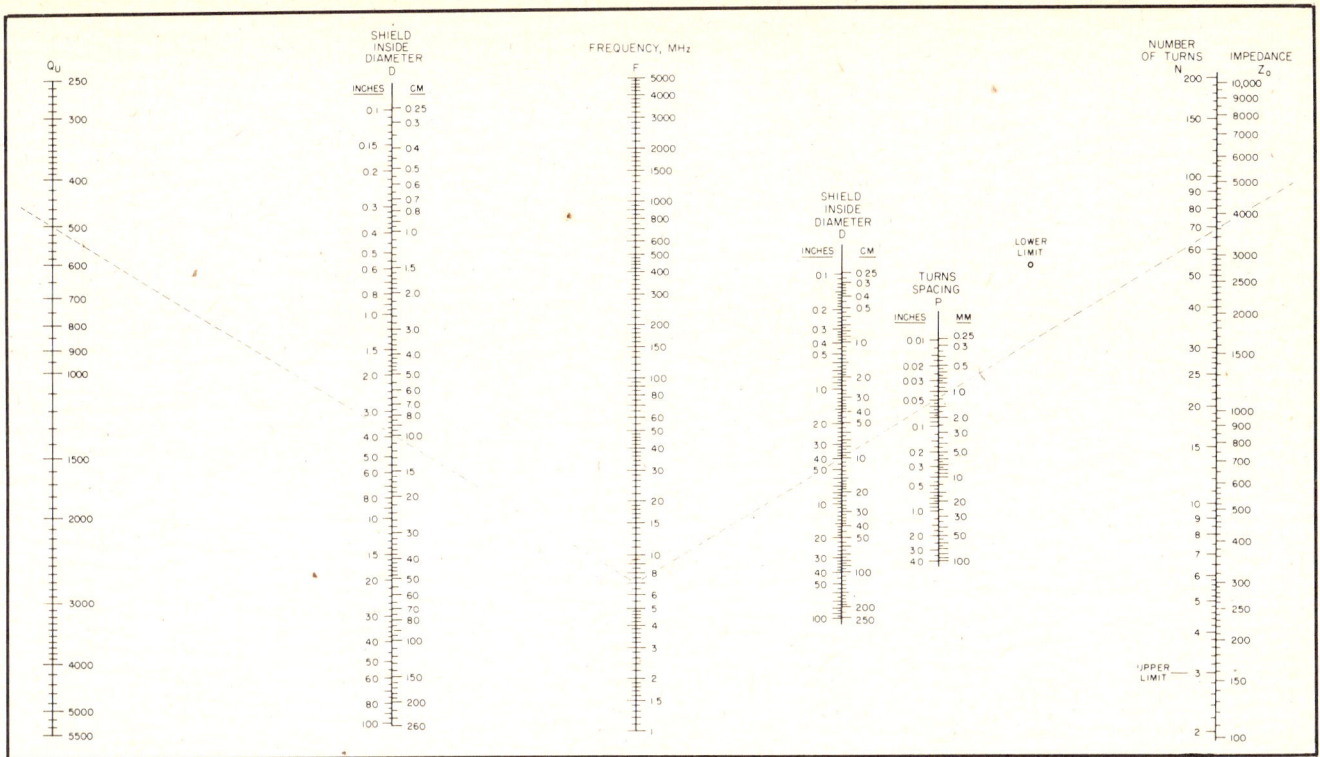

Fig. 7 — Design nomograph for round helical resonators. After selecting unloaded Q_u, required shield diameter is indicated by index line from Q_u scale to frequency scale (dashed index line shown here indicates a shield of about 3.8 inches (97 mm) for an unloaded Q of 500 at 7 MHz). Number of turns, N, winding pitch, P, and characteristic impedance, Z_o, are determined by index line from the frequency scale through previously determined shield diameter on right-hand side of the chart (index line indicates P = 0.047 inch, 1 mm, N = 70 turns, and Z_o = 3600 ohms).

soldered carefully from one end to the other.

Best results are obtained when little or no dielectric is used inside the shield. This is usually no problem at vhf and uhf because the conductors are large enough that a supporting coil form is not required. The lower end of the helix should be soldered to the inside of the shield at a point directly opposite from the bottom of the coil.

Although the external field is minimized by the use of top and bottom covers, the top and bottom of the shield may be left open with negligible effect on frequency or unloaded Q. If covers are provided, however, they should make good electrical contact with the shield. In those resonators where the helix is connected to the bottom cover, that cover must be soldered solidly to the shield to minimize losses.

Tuning

A helical resonator designed from the nomograph of Fig. 7, if carefully built, will resonate very close to the design frequency. Resonance can be adjusted over a *small* range by slightly compressing or expanding the helix. If the helix is made slightly longer than that called for in Fig. 8, the resonator can be tuned by pruning the open end of the coil. However, neither of these methods is recommended for wide frequency excursions because any major deviation in helix length will degrate the unloaded Q of the resonator.

Most helical resonators are tuned by means of a brass tuning screw or high quality air-variable capacitor across the open end of the helix. Piston capacitors also work well, but the Q of the tuning capacitor should ideally be several times the unloaded Q of the resonator. Varactor diodes have sometimes been used where remote tuning is required, but varactors can generate unwanted harmonics and other spurious signals if they are excited by strong, nearby signals.

When a helical resonator is to be tuned by a variable capacitor, the shield size is based on the chosen unloaded Q at the operating frequency. Then the number of turns, N and the winding pitch, P, are based on resonance at $1.5f_o$. Tune the resonator to the desired operating frequency, f_o.

Insertion Loss

The insertion loss (dissipation loss), IL, in dB, of all tuned resonant circuits is given by

$$IL = 20 \log \left(\frac{1}{1 - Q_d/Q_u} \right) dB$$

where Q_d = loaded Q
Q_u = unloaded Q.

This is plotted in Fig. 10. For the most practical cases ($Q_d > 5$) this can be closely approximated by $IL \cong 9.0 (Q_d/Q_u)$ dB. The selection of a *loaded* Q for a tuned circuit is dictated primarily by the required

selectivity of the circuit. However, to keep dissipation loss to 0.5 dB or less (as is the case for low-noise vhf receivers), the *unloaded* Q must be at least 18 times the loaded Q. Although this may be difficult to achieve in practice, it points up the necessity of considering both selectivity *and* insertion loss before choosing the unloaded Q of any resonant tuned circuit.

Coupling

Signals may be coupled into and out of helical resonators with inductive loops at the bottom of the helix, capacitive probes at the top of the helix, direct taps on the coil, or any combination of these. Although the correct tap point can be calculated easily, coupling by loops and probes must be determined experimentally. When only one resonator is used, the input and output coupling is often provided by probes. For maximum isolation the probes are positioned on opposite sides of the resonator.

When coupling loops are used, the plane of the loop should be perpendicular to the axis of the helix and separated a small distance from the bottom of the coil. For resonators with only a few turns, the plane of the loop can be tilted slightly so it is parallel with the slope of the adjacent conductor. Helical resonators with inductive coupling (loops) will exhibit more attenuation to signals *above* the resonant frequency (as compared to attenuation below resonance) whereas resonators with capacitive coupling (probes) exhibit more

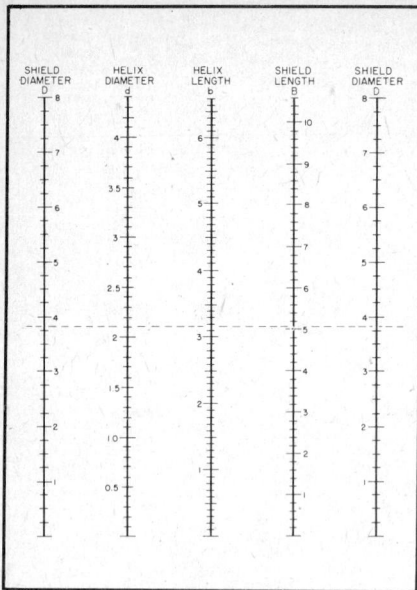

Fig. 8 — Helical-resonator design chart. After the shield diameter has been determined, helix diameter, d, helix length, b, and shield length, B, can be determined with this graph. Index line indicates that a shield diameter of 3.8 inches (97 mm) requires helix mean diameter of 2.1 inches (53 mm), helix length of 3.1 inches (79 mm), and shield length of 5 inches (127 mm).

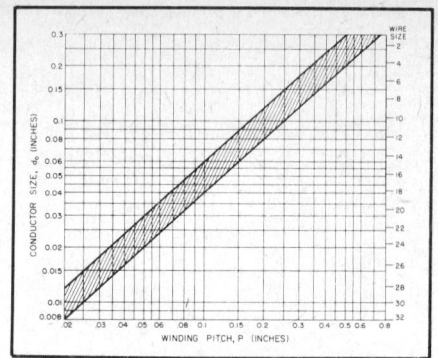

Fig. 9 — Helix conductor size vs. winding pitch, P. A winding pitch of 0.047 inch (1 mm), for example, dictates a conductor diameter between 0.019 and 0.028 inch (number 22 or 24 AWG).

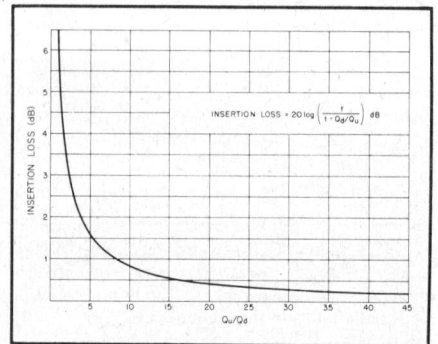

Fig. 10 — Insertion loss of all tuned resonant circuits is determined by the ratio of loaded to unloaded Q as shown here.

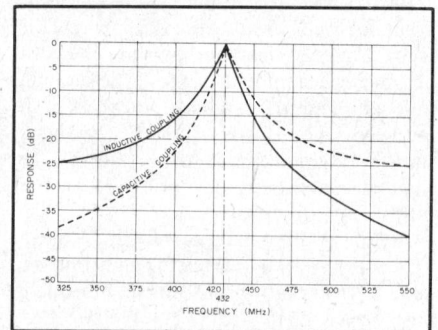

Fig. 11 — Response curve for a single-resonator 432-MHz filter showing the effects of capacitive and inductive input/output coupling. Response curve can be made symmetrical on each side of resonance by combining the two methods (inductive input and capacitive output or vice versa).

attenuation *below* the passband as shown for a typical 432-MHz resonator in Fig. 11. This characteristic may be a consideration when choosing a coupling method. The passband can be made more symmetrical by using a combination of coupling methods (inductive input and capacitive output, for example).

If more than one helical resonator is required to obtain a desired bandpass characteristic, adjacent resonators may be coupled through apertures in the shield wall between the two resonators. Unfortunately, the size and location of the aperture must be found empirically, so this method of coupling is not very practical unless you're building a large number of identical units.

Since the loaded Q of a resonator is determined by the external loading, this must be considered when selecting a tap (or position of a loop or probe). The ratio of this external loading, R_b, to the characteristic impedance, Z_o, for a quarter-wavelength resonator is calculated from

$$K = \frac{R_b}{Z_o} = 0.785 \left(\frac{1}{Q_d} - \frac{1}{Q_u} \right)$$

Even when filters are properly designed and built, they may be rendered totally ineffective if not installed properly. Leakage around a filter can be quite high at vhf and uhf where wavelengths are short. Proper attention to shielding and good grounding is mandatory for minimum leakage. Poor coaxial cable shield connection into and out of the filter is one of the greatest offenders with regard to filter leakage. Proper dc lead bypassing throughout the receiving system is good practice, especially at vhf and above. Ferrite beads placed over the dc leads may help to reduce leakage even further.

Proper termination of a filter is a necessity if minimum loss is desired from the filter. Most vhf rf amplifiers optimized for noise figure do *not* have a 50-ohm terminating input impedance. As a result, any filter attached to the input of an rf amplifier optimized for noise figure will not be properly terminated, and the filter's loss may rise substantially. As this loss is directly added to the rf amplifier's noise figure, prudent consideration should be made of filter choice and placement in the receiver.

MIXERS

Conversion of the received energy to a lower frequency, so that it can be amplified more efficiently than would be possible at the signal frequency, is a basic principle of the superheterodyne receiver. The stage in which this is done may be called a "converter," or "frequency converter," but we will use the more common term, *mixer*, to avoid confusion with *converter*, as applied to a complete vhf receiving accessory. Mixers perform similar functions in both transmitting and receiving circuits, and mixer theory and practice are treated in considerable detail elsewhere in this *Handbook*.

A receiver for 50 MHz or higher usually has at least two such stages; one in the vhf or uhf converter, and usually two or more in the communications receiver that follows it. We are concerned here with the first mixer.

The ideal mixer would convert any signal input to it to another chosen frequency with no distortion, and would have a noise figure of 0 dB. Unfortunately a mixer such as that only exists in a dream world. The mixer that has a 0 dB noise figure (or equivalent loss) has yet to be conceived. This means that the proper use of rf amplification and perhaps post-mixer amplification is necessary for maximum receiver performance with regard to sensitivity. Improving sensitivity is the less difficult of the mixer failings to mend.

Because the mixer operates in a nonlinear mode, reduction of distortion becomes a major design problem. As the mixer input level is increased, a point is reached where the output no longer increases linearly with input. A phenomenon known as *compression* occurs. When the compression point is reached, the sensitivity of the mixer is reduced *for every signal in the passband*. This is manifested as desensing. Different types of mixers characteristically reach their compression points at different input levels, so proper mixer choice can minimize this type of distortion. Any amplifier before the mixer will increase the input levels to the mixer, lowering the point where the input level to the *receiving system* will cause compression. It behooves the builder not to use more gain than is necessary to establish system noise figure prior to the mixer.

If more than one signal is present in the passband going into the mixer, they may mix with each other to produce spurious responses known as intermodulation distortion (IMD) products. As the input levels further increase, higher-order IMD products may appear, seemingly filling the

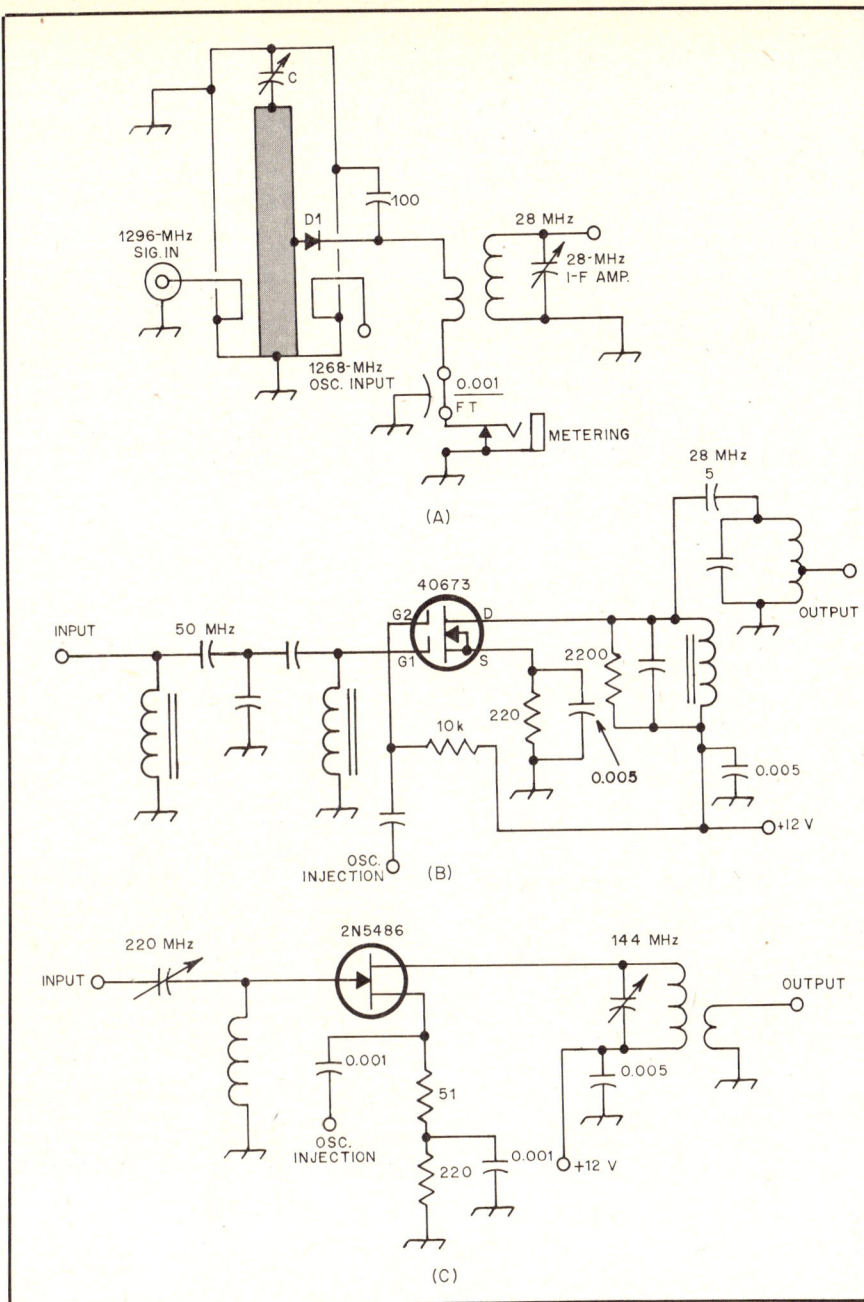

Fig. 12 — Examples of single-ended mixers. The diode mixer, A, is usable through the microwave region. FET mixers, B and C, offer conversion gain and low noise figure.

The signal and the heterodyning frequency are fed into the mixer and the mixer output includes both the sum and difference frequencies of the two. In the case of the circuit shown in Fig. 12A the difference frequency is retained, so the 1296-MHz input signal is converted down to 28 MHz. The sum frequency is filtered out by the 28-MHz tuned circuits.

A quality diode (such as the hot-carrier type) has a fairly low noise figure up through the microwave region. Since most active mixers fall off in performance above 500 MHz, the diode mixer is the one most commonly found in amateur microwave service. Unfortunately, all diode mixers have conversion loss. The loss must be added to the noise figure of the stage following the mixer to determine the system noise figure. A low-noise stage following the mixer is necessary for good weak-signal reception. The noise figure of most communications receivers is far higher than what is needed for a low noise figure system, if no rf amplification is used.

Bipolar transistors are not good square-law type devices, and thus are not favored for single-ended applications. Their major use is in switching-type mixers of the balanced variety.

Field-effect transistors have good square-law response and are very popular vhf mixers. The dual-gate MOSFET is probably the most common mixer found in vhf amateur equipment. The MOSFET can provide considerable conversion gain, while at the same time maintaining a reasonable noise figure. MOSFET overload characteristics are suitable for the vast majority of applications. Local-oscillator energy can be applied at one of the MOSFET gates, effectively isolating the local oscillator from the other signals. The gate impedance is high, so relatively little injection is needed for maximum conversion gain. A typical example is shown in Fig. 12B.

JFETs are close to the MOSFET in mixer performance but are more difficult to apply in practical hardware. As with the MOSFET, input impedance to a JFET mixer is high, and substantial conversion gain is available. JFET bias for mixer service is critical and must be adjusted for best results. The output impedance of a JFET is lower than a dual-gate MOSFET; typically around 10 kΩ. Although other possibilities exist, local-oscillator injection should be made at the JFET source for best results. The source is a low-impedance point, so considerably more local-oscillator power is required than if a dual-gate MOSFET were used as mixer. Noise figures as low as 4 dB are possible with circuits like that shown in Fig. 12C.

The injection level of the local oscillator affects mixer performance. Raising LO level increases conversion gain in an FET mixer. The local-oscillator signal should be as large as possible without pushing the

passband. Proper mixer operating conditions will alleviate IMD problems, and also reduce gain-compression problems.

A third type of distortion is *cross modulation*. This is most readily observed on a-m signals. When the carrier is on, cross modulation is evidenced by modulation characteristics of another signal being superimposed on the received carrier. Techniques to improve IMD characteristics also improve cross modulation performance.

A problem inherent to all mixing systems is *image* generation. Whenever two signals are mixed, components are produced at the sum and difference of the two signal frequencies, and at multiples of these frequencies. For receiving applications, amateurs typically want to detect only one of the mixing products, usually the first order mixing product. Filtering must be applied to separate the desired signal from the rest. Post-mixer filtering is not adequate, as input images can be mixed to the same intermediate frequency as the desired signal. Input filtering discriminates against these images and prevents unwanted out-of-band signals from possibly overloading the mixer.

Single-ended Mixers

Most mixers are single-ended. The simplest type of mixer is the diode mixer.

FET into its pinchoff region. The gate junction of the FET should never conduct in mixer applications. Increased IMD products result from either of the afore-mentioned conditions and should be carefully avoided. The local-oscillator energy should be as pure as possible. Distorted injection energy not only increases IMD production but also increases stage noise figure.

Proper termination of the output of an FET mixer optimizes overload performance. If the impedance seen at the drain of an FET mixer is too high at any of the mixer product frequencies, large voltage excursions can occur on the FET drain. If the voltage excursion on the drain is large enough, output distortion will be evident. Often these high-voltage excursions occur at frequencies outside the desired passband, causing distortion from signals not even detectable by the receiver. A resistor within the output matching network is often used to limit the broadband impedance to a suitably low level.

Balanced Mixers

Use of more than one device in either a singly or doubly balanced mixer offers many advantages over a single-ended mixer. The balance prevents energy injected into a mixer port from appearing at another port. The implications of this are significant when minimum mixer distortion is sought. The port-to-port isolation inhibits any signals other than the mixing products from reaching any other stages further along in the system where they might be mixed, causing undesirable signals. The usually large local-oscillator signal is kept away from the rf amplifier stages where it might cause gain compression because of its magnitude. Any amplitude-modulated noise found on the local oscillator signal is suppressed from the mixer output, where it might be later detected. In a singly balanced mixer only one port, usually the local-oscillator input, is isolated from the other two. A doubly balanced mixer isolates all three ports from each other.

The most common balanced mixer uses diodes. The disadvantages presented earlier with respect to single-ended diode mixers apply here also. A singly balanced diode mixer is shown in Fig. 13A. Hot-carrier diodes are normally used for D1 and D2, as they can handle high currents, have a low noise figure, and are available for use up through several gigahertz.

The doubly balanced mixer is more common today. Commercial modules, very reasonably priced, are often used instead of homemade circuits. Large-scale manufacturing can usually offer electrical balance not so easily attained with the homemade units. Isolation of 35 to 40 dB is typical at vhf, with only 6 to 7 dB of conversion loss. High local-oscillator

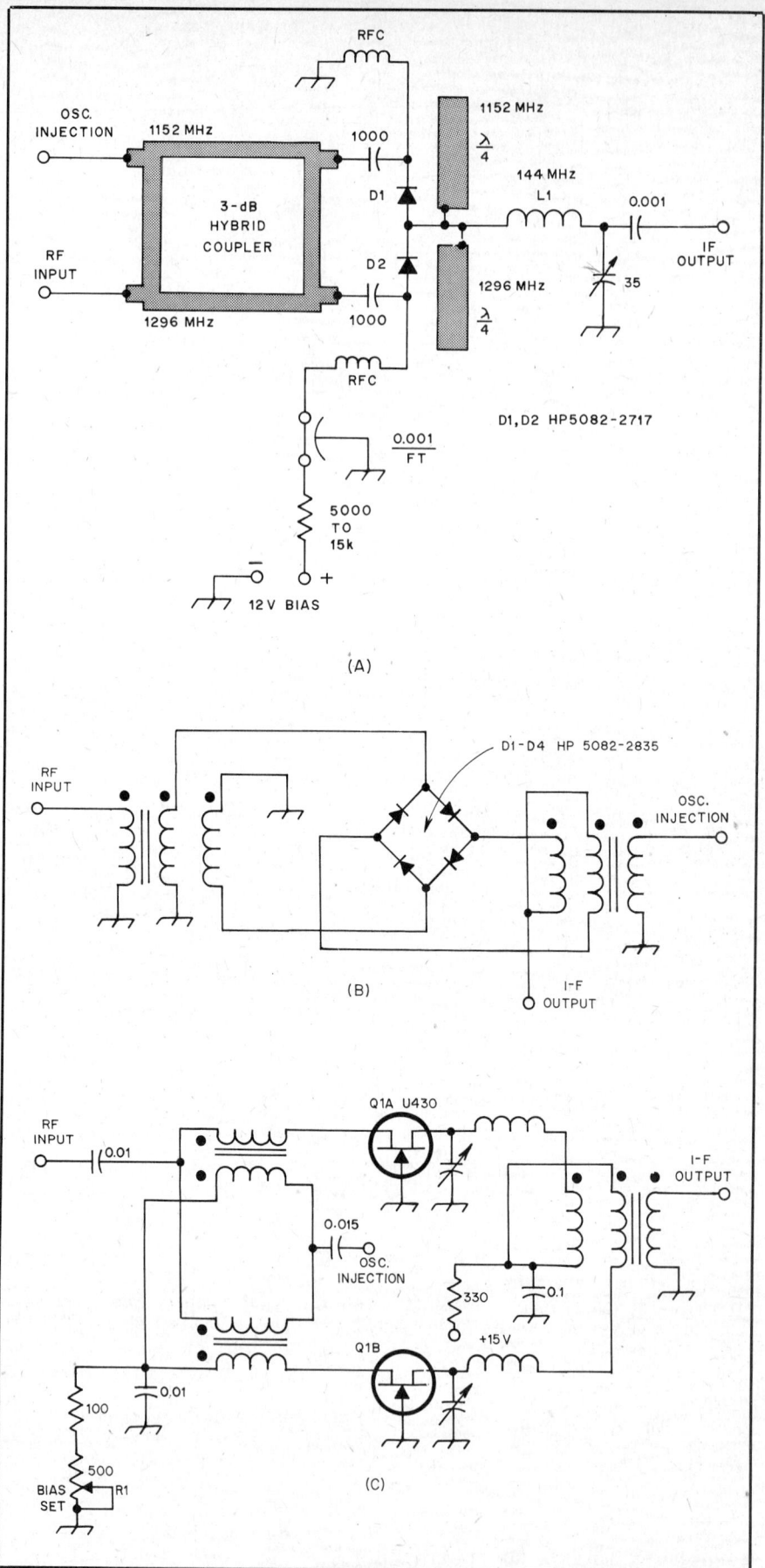

Fig. 13 — Balanced mixers for vhf and uhf. The singly balanced mixer, A, provides isolation of the local oscillator from the output. The double balanced diode mixer, B, has all ports isolated from each other, and is broadband throughout vhf. A special dual JFET is used in C to give high dynamic range with low noise figure.

injection is needed to reach optimum performance with these mixers. Proper broadband termination of all the mixer ports is necessary to prevent unwanted signals from being reflected into the mixer "rat race," only to emerge at another port. The i-f port (shown in Fig. 13B) is the most critical with respect to termination and should be terminated at 50 + j0 ohms. Transmission line transformers provide the necessary phase shift, as half the bridge is fed 180 degrees out of phase with respect to the other half. These can be wound on ferrite toroid forms to effect a broadband response. Careful winding of the transformers improves balance in the circuit, which in turn improves port-to-port isolation.

Active devices can be used very effectively in balanced mixers. Both FETs and bipolars can be used successfully. Active balanced mixers offer all the benefits of balanced diode mixers plus the added advantage of conversion gain rather than loss. Less rf amplification is needed to establish low system noise figure because of this conversion gain than would be needed with a diode mixer. Low gain prior to the mixer keeps mixer input levels low, maximizing mixer overload resistance. High dissipation active devices can be used, yielding better mixer performance than is available from diode balanced mixers. Fig. 13C shows a dual FET which has been specially designed for mixer applications. R1 allows for electrical balance adjustment in the circuit. A sharp null in local oscillator output at the mixer output can be observed when R1 is set to the optimum point, showing electrical balance has been achieved.

Injection Stages

Oscillator and multiplier stages supplying heterodyning energy to the mixer should be as stable and free of unwanted frequencies as possible. Proper application of crystal control gives stability pursuant to needs. Two major influencing factors control oscillator stability: temperature and operating voltage. As the temperature of a component changes, its internal geometry changes somewhat as the constituent materials expand or contract. When the geometry changes, the internal capacitance often changes, affecting the resonant frequency of the tuned circuits controlling oscillator frequency. Use of quality components which have good temperature characteristics helps in this regard. Minimum power should be extracted from the oscillator as excessive heat dissipation within either the crystal or the transistor will cause internal capacitance changes in those units, moving the resonant frequency. Voltage to the transistor should be regulated for best stability. Simple Zener diode regulation is sufficient or a three-terminal regulator IC can be used.

Any unwanted injection frequencies will mix with signals present in the mixer,

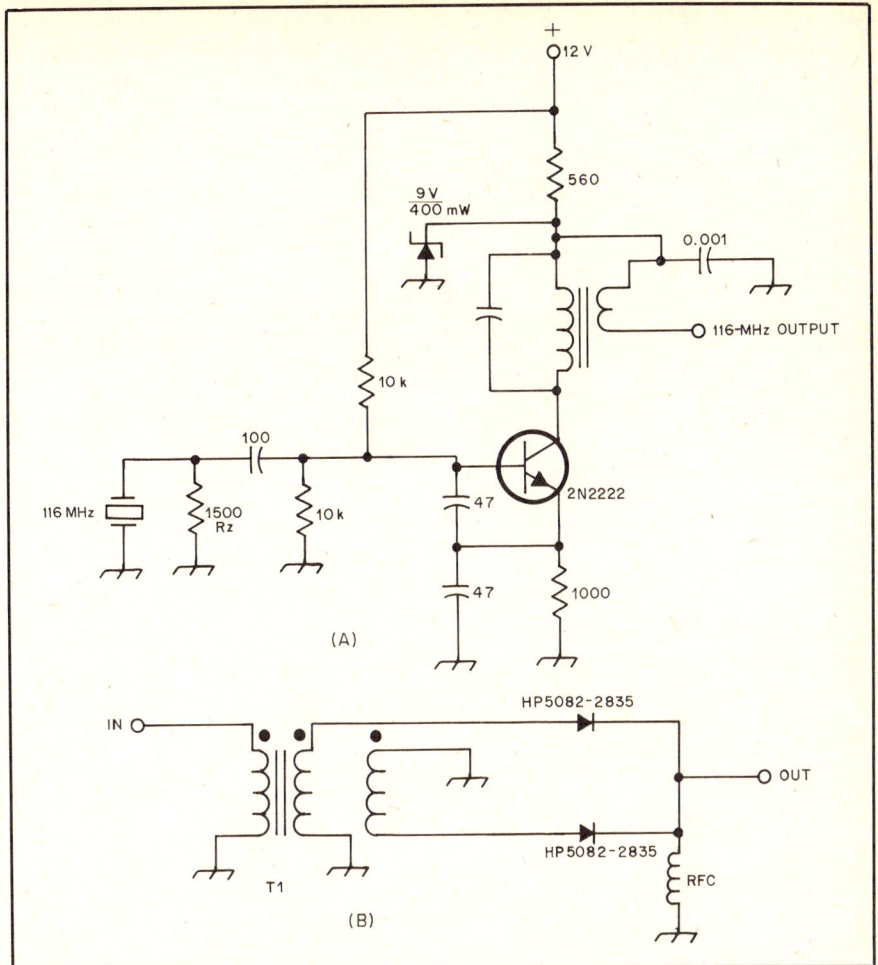

Fig. 14 — Typical crystal oscillator for vhf use, A. The diode frequency doubler, B, provides good rejection of the fundamental signal.

creating spurious outputs at the mixer output. A clean local oscillator will prevent these unwanted outputs. The oscillator chain output can be heavily filtered to cut down the harmonic content of the oscillator, but good planning and design will minimize the unwanted energy, making the filtering job less demanding. A high-frequency crystal in the oscillator minimizes the number of times the fundamental oscillator frequency has to be multiplied to reach the converter injection frequency. Proper use of doublers rather than triplers can eliminate any odd oscillator frequency multiples, so a low-pass filter at the output only has to filter the fourth harmonic and beyond. A band-pass filter would be needed at the output of a tripler to eliminate the second harmonic and the higher ones. Finally, good shielding and power-line filtering should be used throughout to prevent any stray radiation from reaching the mixer or causing RFI problems elsewhere.

Fig. 14A shows a typical circuit useful for providing the 116-MHz injection energy necessary to convert a 144-MHz signal down to 28 MHz. R_z dampens the crystal action somewhat, assuring that the proper overtone is the actual oscillation frequency. The collector tank network is parallel tuned and can be wound on a toroid core to reduce radiation. The output is link coupled from the tank, minimizing harmonic coupling. This oscillator would be followed by a buffer to bring the signal up to that level needed and to purify the oscillator signal further.

A similar oscillator could be used in a 220-MHz converter. Since crystals are not available at 192 MHz, the frequency required for conversion to 28-MHz converter output, the most logical approach is to use a 96-MHz oscillator and double its output. Fig. 14B shows a diode frequency doubler suitable for the application. The phase-shifting transformer can be made from a trifilar winding on a ferrite core. Hot-carrier diodes allow the use of a doubler like this up through at least 500 MHz. There is a loss of about 8 dB through the doubler, so amplification is needed to raise the injection signal to the appropriate level. Fundamental energy is down by as much as 40 dB from the second harmonic with a balanced diode doubler such as this. All of the odd harmonics are well down in amplitude also, all without tuned circuits. A low-pass filter can be used to eliminate the

Fig. 15 — Two versions of the preamplifier. The one in the box is for 2-meter use. Toroids are used in the six-meter version (right) and in the ten-meter preamplifier (not shown). Input is at the right on both units. The extra rf choke and feedthrough capacitor on the right end of the Minibox are for decoupling a crystal-current metering circuit that is part of a 2304-MHz mixer.

undesired harmonics from the output.

MOSFET PREAMPLIFIERS FOR 10, 6 AND 2 METERS

If an hf or vhf receiver lacks gain, or has a poor noise figure, an external preamplifier can improve its ability to detect weak signals. This preamplifier uses an RCA 40673 dual-gate MOSFET. Designs for using this device as a mixer or as a preamplifier abound and many of them are excellent.

When it comes to simplicity, small size, good performance, low cost, and flexibility, a design by Gerald C. Jenkins, W4CAH, certainly qualifies.

The preamplifier really shines in pepping up the performance of some of the older 10-meter receivers that many have pressed into service. A 6-meter version is also very useful for any of the modes of communication available on that band.

In Fig. 16 the voltage dropping resistor, R4 and the Zener diode, D1, may be of the value necessary to obtain 9 to 12 V dc for operation of the unit. By increasing the resistance and dissipation rating of R4 and D1, the preamplifier may be operated from the 150- to 200-V supply found in many tube-type receivers.

The layout of the board is so simple that it is hardly worth the effort of making a negative for the photo-etch process. A Kepro resist-marking pen was used with success on several boards. Another approach — and one that is highly recommended — is to cover the copper with masking tape, transfer the pattern with carbon paper, then cut away the tape to expose the part to be etched. On small, simple boards the masking-tape method is hard to beat.

The pc board may be mounted in almost any small enclosure. Construction is not tricky or difficult. It should take only a few minutes to complete the unit after the board is prepared. The board is fastened in the enclosure by means of one metal standoff post and a no. 4 screw and nut. Input and output connectors are not critical; phono-type jacks may be used in the interest of low cost.

Adjustment is so easy that it almost needs no description. After connecting the amplifier to a receiver, simply tune the input (C1) and the output (C4) for maximum indication on a weak signal. One possible area of concern might be that the toroids used in the 10- and 6-meter versions are not always uniform in permeability, as purchased from various suppliers. However, it is an easy matter to add capacitance or remove a turn as required to make the circuits resonate at the correct frequency.

LOW-NOISE 220-MHZ JFET PREAMPLIFIER

At 220 MHz cosmic noise is below 1.4 dB. This preamplifier uses a 2N5245 or 2N5486 JFET to achieve a noise figure as low as 1.3 dB. The JFET operates grounded-source with inductive neutralization. The preamplifier was first described by WB6NMT in March 1972 QST.

The preamplifier is built on double-clad circuit board, mounted on spacers inside an LMB T-F770 aluminum box, 2-3/4 × 2-1/8 × 1-5/8 (70 × 54 × 42 mm) inches in size. A shield of circuit board stock divides the amplifier as indicated by the broken line in the circuit diagram. SMA or N-type fittings are recommended for the input and output connectors. E. F. Johnson manufactures an inexpensive line of SMA connectors which are highly desirable for low-noise connections at vhf and uhf. Top grade glass trimmers or miniature air variables should be used for C1, C2 and C4 for best results.

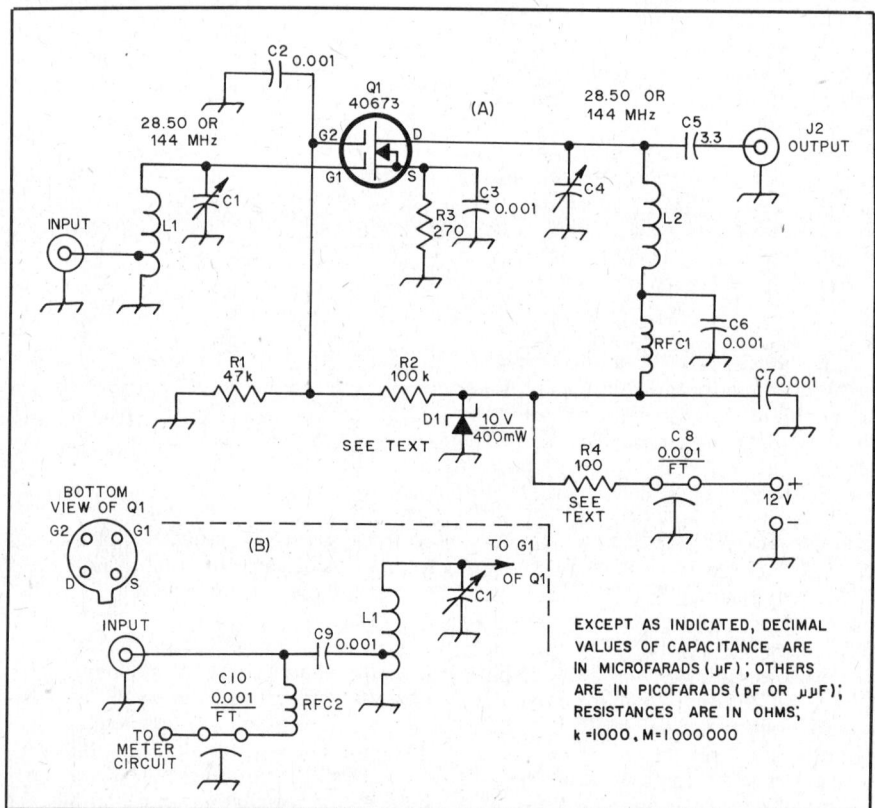

Fig. 16 — Schematic diagram for the preamplifier. Part designations not listed below are for pc board placement purposes. Alternative input circuit for use with microwave diode mixer is shown at B.

C1, C4 — See Table 1.
C2, C3, C5, C6, C7, C9 — Disc ceramic.
C8 — 0.001 feedthrough capacitor.
J1, J2 — Coaxial connectors. Phono-type, BNC or SO-239 acceptable.

L1, L2 — See Table 1.
R4 — 3 turns no. 28 enam. on ferrite bead. A 220-ohm, 1/2-watt resistor may be substituted.
RFC2 — 33 µH, iron-core inductor. Millen J300-33 or J. W. Miller 7OF335A1.

Table 1

	28 MHz	50 MHz	144 MHz
L1	17 turns no. 28 enam. on Amidon T-50-6 core. Tap at 6 turns from ground end.	12 turns no. 26 enam. on Amidon T-37-10 core. Tap at 5 turns from ground end.	5 turns no. 20 tinned 1/2-inch ID × 1/2-inch long. Tap at 2 turns from ground end.
L2	Same as L1, without tap.	Same as L1, without tap.	4 turns no. 20 tinned like L1, without tap.
C1, C4	15 to 60-pF ceramic trimmer. Erie 538-002F.	1.8- to 16.7-pF air variable. E. F. Johnson 189-506-005.	1.5- to 11.6-pF air variable. E. F. Johnson 189-504-005.

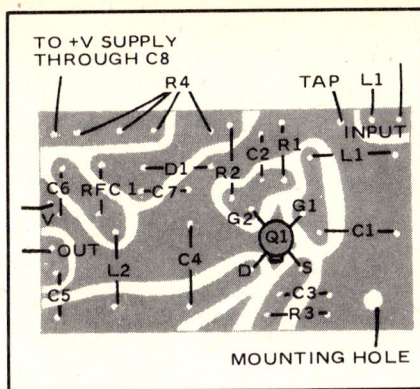

Fig. 17 — Full-scale layout and parts placement guide for the pc board. Foil side shown.

The dc voltage for the preamplifier is fed through one arm of a coaxial T fitting at the receiver input. This assumes use of some sort of blocking capacitor in the receiver input, to prevent grounding the dc through a coupling loop or tap on a grounded tuned circuit. The rf choke in the preamplifier circuit, RFC1, and the one used at the receiver input (to isolate the dc from the rf) are not critical. Any reasonably good vhf choke should do. If you're still willing to take the losses involved in the line, and you want to use the preamp at the receiver input, leave RFC1 out of the circuit, and connect the dc as shown in Fig. 18.

Adjustment

First set R1 for about 5 mA current

drain, at 12 to 15 volts dc. Touch the neutralizing coil, L2. If there is any change in current, the stage is oscillating. Keeping contact with L2 (to prevent oscillation), readjust R1 for 5 mA. Using a strong 220-MHz test signal, adjust C4 for maximum signal indication. Set C1 to minimum capacitance, and peak C2. Increase C1 slowly until signal no longer rises, then back off one turn and readjust C2 and C4 for maximum signal.

Now reverse the preamplifier, connecting J1 to the receiver input, and feeding the signal to J2. With the dc still applied, tune L2 to minimum signal feed-through. If L2 has an ungrounded brass slug, the amplifier attenuation should be about 50 dB. Drain current should remain at 5 mA.

Connect the amplifier normally, and repeat the process outlined above, until the tuning of C4 remains nearly constant. Finally, adjust C1 for best signal-to-noise ratio (lowest noise figure) and readjust C2. This should yield a noise figure of 1.5 to 2 dB, and gain of 12 to 18 dB, depending on the transistor used. Often the lower-gain condition will also give the best noise figure.

LOW-NOISE GaAs FET PREAMPLIFIERS FOR 432 AND 1296 MHZ

Gallium-arsenide field-effect transistors (GaAs FETs) have recently come into use as low-noise microwave amplifiers. Amateur experimentation has shown that they can provide excellent performance on the uhf and lower microwave amateur bands.

(A)

(B)

Fig. 19 — At A, 432-MHz GaAs FET preamplifier built by K2UYH. The transistor is mounted at the central shield by soldering the source lead directly to the copper foil. The drain lead of the transistor passes through a hole in the shield. At B, a 1296-MHz GaAs FET preamplifier built by WA2ZZF. In this model, the transistor is connected to striplines etched on glass-epoxy board. SMA-type coaxial connectors are shown although type N or BNC connectors may be used.

These devices are rather expensive, particularly the ones characterized as C-band and X-band (4-12 GHz) microwave low-noise amplifiers. However, other GaAs FETs, characterized as *power amplifiers* for low and medium-power (up to 1/4 watt) microwave applications will provide almost the same noise figure at uhf and are being made available to amateurs. The power devices also have wide dynamic range, providing less intermodulation distortion and lower susceptibility to burnout. The receiver preamplifiers to be described are relatively simple to construct and have sufficient tuning range for almost any GaAs FET available. They were first described by K2UYH and WA2ZZF in June 1978 *QST*.

Construction

These preamps for 432 MHz (Fig. 19A) and 1296 MHz (Fig. 19B) use power GaAs FETs made by Microwave Semiconductor Corp.; however, devices made by NEC (Nippon Electric Co.) perform at least as well, and many similar devices will also certainly work. Construction details are shown in the photographs and schematic diagrams. The 432-MHz preamp is built in a 2-1/4 × 1-1/2 × 1-inch (57 × 38 × 25-mm) box made of double-sided printed-circuit board. A cover plate is recommended but does not significantly affect tuning. The GaAs FET source is soldered to the central shield board with

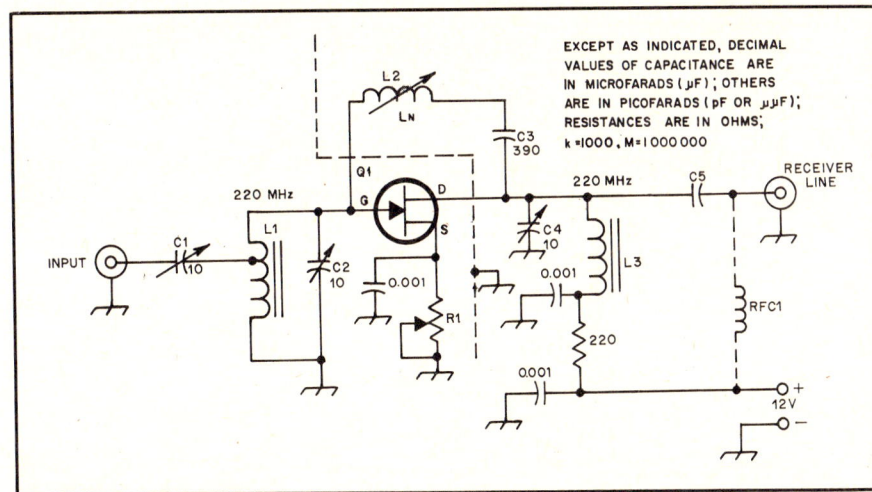

Fig. 18 — Circuit and parts information for the WB6NMT 220-MHz preamplifier.

C1 — 0.8 or 1 to 10 pF glass trimmer, Johanson 2950 of JFD VAM or MVM series.
C2 — Like C1, or Corning Direct Traverse CGW. 0.8 to 10 pF.
C3 — 390-pF silver mica.
C4 — Like C1, C2 or less-expensive type with 1 to 10-pF range.
C5 — Experiment with values 1 to 5 pF, for maximum gain in system as it will be used.
J1, J2 — SMA or N-type connector.
L1 — 4 turns no. 22 enam. or Micrometals T-37-0 toroidal core (Amidon Associates). Tap

1 turn from top, subject to adjustment for lowest nf. Air-wound coils also usable, but toroids preferred.
L2 — 9 turns no. 28 enam. on 1/4-in. (6.3 mm) slug-tuned form (Miller 4500, brass slug). Do not ground the slug.
L3 — Like L1, but no tap.
Q1 — 2N5245, 2N5486, MPF-107, TIS-88.
R1 — 200- or 250-ohm control.
RFC1 — Vhf rf choke, 0.8 to 3 µH. Use only when preamp is antenna mounted (see text).

the drain lead projecting through a hole. Several other versions have been constructed; in one of these, the wire inductors are replaced by straps placed parallel to the bottom plate, and spaced approximately 1/8 inch (3 mm) above it; a typical strap dimension would be 3 inches (76.2 mm) long by 1/2 inch (13 mm) wide. The 1296-MHz preamp is constructed in a 2-3/4 × 2-1/8 × 1-5/8-inch (70 × 54 × 41-mm) Minibox (BUD CU-3000A or equivalent). The GaAs FET is bolted between two pieces of 1/16-inch (1.6-mm) printed-circuit board, using 0-80 screws (available at many hobby shops). The lead height is just right to sit on top of the 50-ohm lines printed on these boards. The ground connection for the tuning capacitors is provided by mounting screws and by copper foil soldered around one edge of each board. The ground plane sides of the board are smoothly tinned to reduce copper-to-aluminum corrosion.

Handling Precautions

The MSC GaAs FETs have static-resistant gold gates, and are only susceptible to damage from overvoltage or excess heating. Some other types, particularly those of Japanese manufacture, have aluminum gates which are very sensitive to static burnout, and should be handled in the same manner as unprotected MOS devices. In any case, work quickly when soldering the devices and use a grounded or cordless soldering iron. After assembly, the Zener diodes shown should protect the device in normal operation. Of course, it should be realized that these devices are physically small and require reasonably careful handling.

Adjustment and Performance

Normal operating voltages are $V_{DS} = 1.5$ to 3V, $V_{GS} = -0.5$ to -2V; gate current is negligible and may be supplied from a battery. Peak the tuning capacitors on a strong signal, then trim them and adjust the drain and gate voltages with the aid of a noise-figure meter or weak-signal source. Minimum noise figure occurs near the tuning for maximum gain. Output tuning should have little effect, but the noise figure is sensitive to the input tuning and gate voltage; varying the drain voltage should give a broad peaking of S/N. Drain current is controlled by gate voltage. After peaking up the preamp, drain current will probably be between 20 and 100 mA.

It should be emphasized that these devices have extremely high gain at uhf and will readily oscillate unless adequate precautions are taken. Stability is obtained by the use of the resistor connected *directly* from the drain to ground, at the expense of some gain reduction. The values shown should provide adequate stability if good bypassing is used; gain will be around 20 dB at 432 MHz and 15 dB at 1296 MHz. Any increase in the

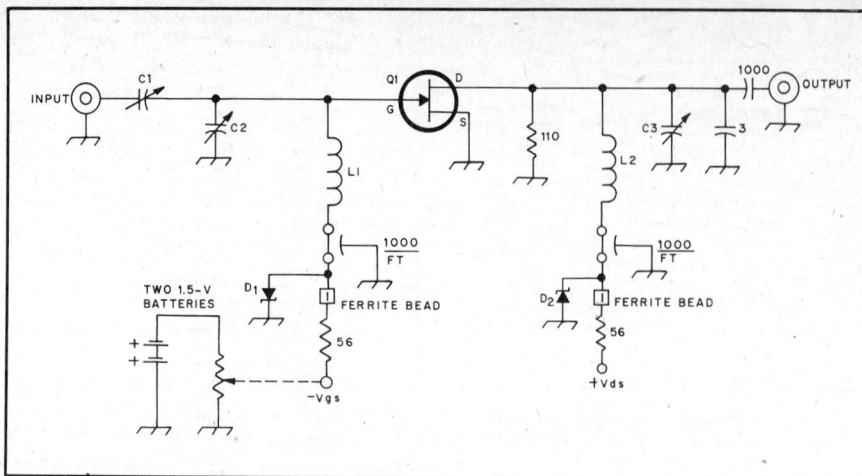

Fig. 20 — Schematic diagram of the 432-MHz preamplifier.
C1 — 0.03- to 3.5-pF piston trimmer (Johanson or JFD).
C2, C3 — 0.8- to 10-pF piston trimmer (Johanson or JFD).
D1, D2 — Zener diode, 5.6 volts (4.7 to 6.2 volts usable).
L1 — 1 turn no. 18 wire (see photo) or stripline (see text).
L2 — no 18 wire, 0.9 in. (23 mm) long.
Q1 — GaAs FET (see text).

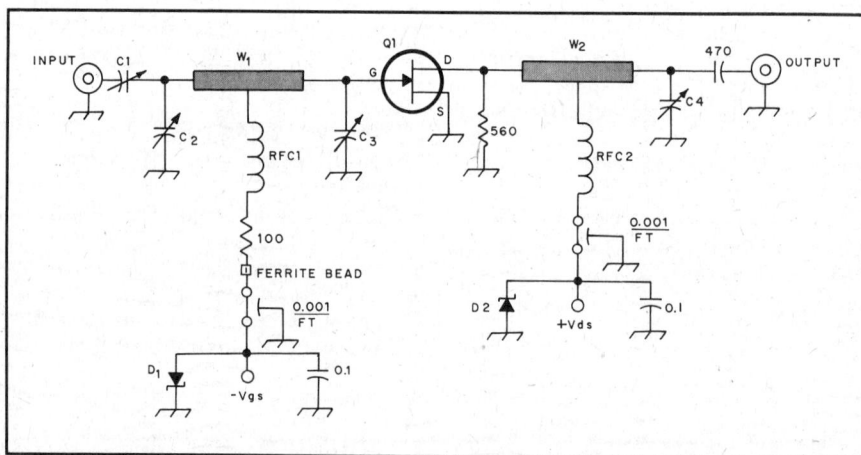

Fig. 21 — Schematic diagram of the 1296-MHz preamplifier.
C1, C2, C4 — 0.8- to 10-pF piston trimmer (Johanson or JFD). Note: C1 may be replaced by a fixed low-inductance capacitor of 10-pF or more.
C3 — 0.3- to 3.5-pF piston trimmer (Johanson or JFD).
D1, D2 — Zener diode, 5.6 V (4.7 to 6.2 V usable).
Q1 — GaAs FET (see text).
RFC1 — 3 turns, 1/16-in (1.6-mm) ID, in lead of resistor, spaced wire diameter.
RFC2 — 5 turns no. 32 wire, 1/16-in (1.6-mm) ID, spaced two wire diameters.
W1 — 50-ohm microstripline, 0.105 in. (2.7 mm) wide by 0.9 in (23 mm) long on 1/16-in. (1.6-mm) thick double-sided G-10 printed-circuit board.
W2 — 50-ohm microstripline, 0.105 in. (2.7 mm) wide by 1.1-in (28 mm) long on 1/16-in. (1.6-mm) thick double-sided G-10 printed-circuit board.

value of these stabilizing resistors is at your own risk!

Typical noise figures to be expected with these preamps are on the order of 1 dB at 432 MHz and 3 dB or less at 1296 MHz. The devices are capable of even better performance than this; significant improvements are obtainable at 1296 MHz with attention to good uhf construction techniques and low-loss circuitry. However, the circuits shown are easily reproduced and still provide excellent performance.

DOUBLY BALANCED MIXERS

Advances in technology have, in recent years, provided the amateur builder with many new choices of hardware to use in the building of receivers, converters, or preamplifiers. The broadband doubly balanced mixer package is a fine example of this type of progress, and as amateurs gain an understanding of the capabilities of this device, they are incorporating this type of mixer in many pieces of equipment, especially receiving mixers. The combined mixer/amplifier described here was presented originally in March 1975 *QST* by K1AGB.

Mixer Comparisons

Is a DBM really better than other

Table 2

Manufacturer	Relcom	Anzac	MCL	MCL	MCL	MCL
Model	M6F	MD-108	SRA-1	SRA-1H	RAY-1	MA-1
Frequency Range (MHz)						
LO	2-500	5-500	5-500	5-500	5-500	1-2500
rf	2-500	5-500	5l-500	5-500	5-500	1-2500
i-f	DC-500	DC-500	DC-500	DC-500	DC-500	1-1000
Conversion loss	9 dB max.	7.5 dB max.	6.5 dB typ.	6.5 dB typ.	7.5 dB typ.	8.0 dB typ.
Mid-range						
Isolation, LO-RF	34-40 dB min.	40 dB min.	45 dB typ.	45 dB typ.	40 dB typ.	40 dB typ.
Mid-range LO i-f	25-35 dB min.	35 dB min.	40 dB typ.	40 dB typ.	40 dB typ.	40 dB typ.
Total input power	50 mW	400 mW	500 mW	500 mW	1 W	50 mW
LO power requirement:	+ 7 dBm (5 mW)	+ 7 dBm (5 mW)	+ 7 dBm (5 mW)	+ 17 dBm (50 mW)	+ 23 dBm (200 mW)	+ 10 dBm (10 mW)
Signal 2-dB compression level	Not spec.	Not spec.	+ 1 dBm	+ 10 dBm	+ 15 dBm	+ 7 dBm
Impedance, all ports	50 ohms	50 ohms	50 ohms	50 ohms	50 ohms	50 ohms

All specifications apply only at stated LO power level.

Relcom, Division of Watkins-Johnson, 3333 Hillview Ave., Palo Alto, CA 94304.
Anzac Electronics, 39 Green Street, Waltham, MA 02154.
MCL — Mini-Circuits Laboratory, 2625 East 14th St., Brooklyn, NY 11235.

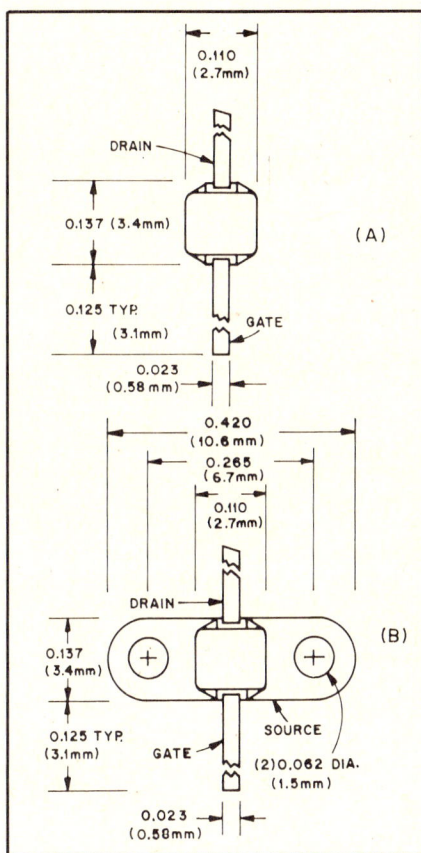

Fig. 22 — Dimensional information for the GaAs FET packages supplied by MSC. At A, case style 98, top view, and at B, top view of case style 97. Drain and source leads are spaced 0.065 in. (1.65 mm) above the bottom of the case. MSC designation for these case styles is Flipac.

types? What does it offer, and what are its disadvantages? To answer these questions, a look at more conventional "active" (voltages applied) mixing techniques and some of their problems is in order. The reader is referred to an article in *QST* [1] dealing with mixers. Briefly reiterated, common single-device active mixers with gain at vhf and uhf are beset with problems of noise, desensitization and small local-oscillator (LO) isolation from

[1] DeMaw and McCoy; 'Learning to Work With Semi-conductors," Part IV, *QST*, July 1974.

the r-f and i-f "ports." As mixers, most devices have noise figures in excess of those published for them as rf amplifiers and will not provide sufficient sensitivity for weak-signal work. To minimize noise, mixer-device current is generally maintained at a low level. This can reduce dynamic range, increasing overload potential, as defined in the terminology appendix. Gain contributions of rf amplifiers (used to establish a low system noise figure) further complicate the overload problem. LO-noise leakage to the rf and i-f ports adversely affects system performance. Mixer dynamic range can be limited by conversion of this noise to i-f, placing a lower limit on mixer system sensitivity. Generally 20 dB of mixer midband interport isolation is required, and most passive DBMs can offer greater than 40 dB.

A commercially manufactured doubly balanced diode mixer offers performance predictability, circuit simplicity and flexibility. Closely matched Schottky-barrier hot-carrier diodes, commonly used in most inexpensive mixers of this type, provide outstanding strong-signal mixer performance (up to about 0 dBm at the rf input port) and add little (0.5 dB or so) to the mixer noise figure. Essentially, diode conversion loss from rf to i-f, listed in Table 2 represents most of the mixer contribution to system noise figure. Midband isolation between the LO port and the rf and i-f ports of a DBM is typically > 35 dB — far greater than that achievable with conventional single device active-mixing schemes. This isolation is particularly advantageous in dealing with low-level local-oscillator harmonic and noise content. Of course, selection of LO devices with low audio noise figures, and proper rf filtering in the LO putput, will reduce problems from this source.

Often-listed disadvantages of a diode DBM are (a) conversion loss, (b) LO power requirements, and (c) i-f-interface problems. The first two points are closely interrelated. Conversion loss necessitates some low-noise r-f amplification to establish a useful weak-signal system noise figure. Additional LO power is fairly easy

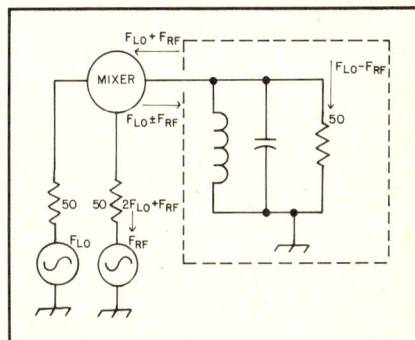

Fig. 23 — The i-f port of a doubly balanced mixer is matched at fLO − $f$rf and reactive at fLO ± $f$rf. In this configuration conversion loss, rf compression and desensitization levels can vary ± 3dB while harmonic modulation and third-order IMD products can vary ± 20 dB.

to generate, filter, and measure. If we accept the fact that more LO power is necessary for the DBM than is used in conventional single-device active mixing circuits, we leave only two real obstacles to be overcome in the DBM, those of conversion loss and i-f output interfacing.

To minimize conversion loss in a DBM, the diodes are driven by the LO beyond their square-law region, producing an output spectrum which in general includes the terms [2]:

1) Fundamental frequencies fLO and $f$rf
2) All of their harmonics
3) The desired i-f output, fLO ± $f$rf
4) All higher order products of nfLO ± mfrf, where n and m are integers.

The DBM, by virtue of its symmetry and internal transformer balance, suppresses a large number of the harmonic modulation products. In the system described here, fLO is on the low side of $f$rf, therefore, numerically, the desired i-f output is $f$rf − fLO. Nonetheless, the term fLO + $f$rf appears at the i-f-output port equal in amplitude to the desired i-f signal, and this unused energy must be effectively terminated to obtain no more than the specified mixer-conversion loss.

[2] See appendix.

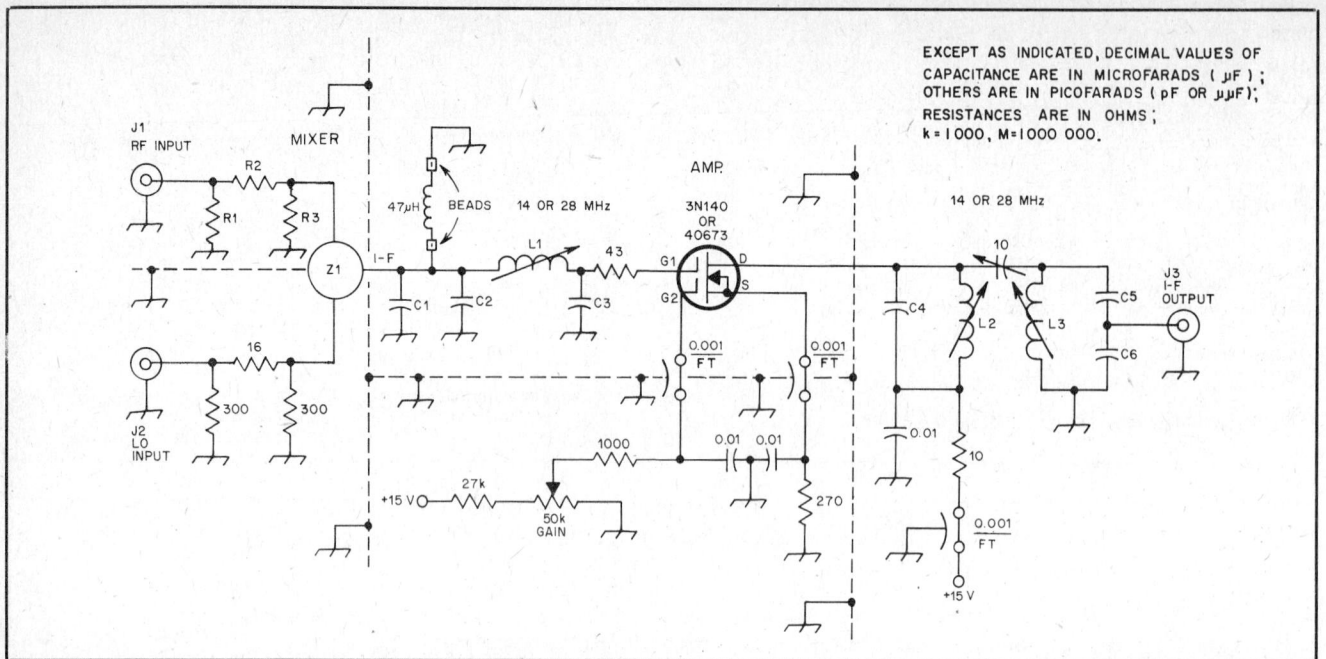

Fig. 24 — A schematic diagram for the doubly-balanced mixer and i-f post amplifier. The i-f can be either 14 or 28 MHz. Parts values are given in Table 3.

This is not the image frequency, $fLO - fi$-f, which will be discussed later.

In any mixer design, all rf port signal components must be bypassed effectively for best conversion efficiency (minimum loss). Energy not "converted" by mixing action will reduce conversion gain in active systems, and increase conversion loss in passive systems such as the diode DBM. Rf bypassing also prevents spurious resonances and other undesired phenomena from affecting mixer performance. In this system, rf bypassing at the i-f-output port will be provided by the input capacitance of the i-f interface. The DBM is not a panacea for mixing ills, and its effectiveness can be reduced drastically if all ports are not properly terminated.

DBM Port Terminations

Most DBM-performance inconsistencies occur because system source and load impedances presented to the mixer are not matched at all frequencies encountered in normal operation. The terminations (attenuator pads) used in conjunction with test equipment by manufacturers to measure published performance characteristics are indeed "broadband" matched. Reactive mixer terminations can cause system problems, and multiple reactive terminations can usually compound these problems to the point where performance is very difficult to predict. Let's see how we can deal with reactive terminations.

The I-F Port

The i-f port is very sensitive to mismatch conditions. Reflections from the mixer/i-f amplifier interface (the pi network in Fig. 24) can cause the conversion loss to vary as much as 6 dB.

Also greatly affected are third-order inter-modulation-product ratio and the suppression of spurious signals, both of which may vary ±10 dB or more. It is ironic that the i-f port is the most sensitive to a reactive termination, as this is a receiving system point where sharp-skirted filters are often desired.

Briefly, here is what happens with a reactive i-f port termination. Fig. 23 shows a DBM with "high side" LO injection and an i-f termination matched at $fLO - frf$ but reactive to $fLO + frf$. The latter term re-enters the mixer, again combines with the LO and produces terms that exit at the rf port, namely $2fLO + frf$, a dc term, and $fLO + frf - fLO$ (the original rf-port input frequency). This condition affects conversion loss, as mentioned earlier, in addition to rf-port VSWR, depending on the phase of the reflected signal. The term $2fLO + frf$ also affects the harmonic spectrum resulting in spurious responses.

One solution to the i-f-interface problem is the use of a broadband 50-ohm resistive termination, like a pad, to minimize reflections. In deference to increased post-conversion system noise figure, it seemed impractical to place such a termination at the mixer i-f output port. While a complimentary filter or diplexer (high-pass/low-pass filters appropriately terminated) can be used to terminate both $frf + fLO$ and $frf - fLO$[3], a simpler method can be used if $frf + fLO$ is less than 1 GHz and $frf + (fLO)/(frf - fL0) \geqslant$ 10. Place a short-circuit termination to $frf + fLO$, like a simple lumped capacitance,

[3]Presentation and calculation format of these terms is based on "low-side" LO injection. See the appendix for explanation.

directly at the mixer i-f terminal. This approach is easiest for the amateur to implement and duplicate, so a form of it was tried — with success. In our circuit, C1 serves a dual purpose. Its reactance at $frf + fLO$ is small enough to provide a low-impedance "short-circuit" condition to this term for proper mixer operation. Additionally, it is part of the input reactance of the mixer i-f-amplifier interface. Fortunately the network impedance-transformation ratio is large enough, and in the proper direction, to permit a fairly large amount of capacitance (low reactance) at the mixer i-f-output port. The capacitor, in its dual role, must be of good quality at vhf/uhf (specifically $frf + fLO$), with short leads, to be effective. The mixer condition $(frf + fLO)/frf - fLO) \geqslant 10$ is met at 432 and 220 MHz with a 404/192-MHz LO (28-MHz i-f) and on 144 MHz with a 130-MHz LO (144-MHz i-f). At 50 MHz, with a 36 MHz LO, we are slightly shy of the requirement, but no problems were encountered in an operating unit. The pi-type interface circuit assures a decreasing impedance as i-f operation departs from midband, thereby lessening IMD problems.

The LO Port

The primary effect of a reactive LO source is an increase in harmonic modulation and third-order IMD products. If the drive level is adequate, no effect is noted on conversion loss, rf compression and desensitization levels. A reactive LO source can be mitigated by simply padding the LO port with a 3- or 6-dB pad and increasing the LO drive a like amount. If excess LO power is not available, matching the LO source to the mixer will improve performance. This method is acceptable for single-frequency LO applications, when appropriate test equipment is available to evaluate matching results. For simplicity, a 3-db pad was incorporated at the LO-input port as an interface in both versions of the mixer. Thus the LO port is presented with a reasonably broadband termination, and is relatively insensitive to applied frequency, as long as it is below about 500 MHz. This implies that frequencies other than amateur assignments may be covered — and such is indeed the case when appropriate LO frequencies and rf amplifiers are used. Remotely located LOs, when adjusted for a 50-ohm load, can be connected to the mixer without severe SWR and reflective-loss problems in the transmission line.

Broadband mixers exhibit different characteristics at different frequencies, owing to circuit resonances and changes in diode impedances resulting from LO power-level changes. Input impedances of the various ports are load dependent, even though they are isolated from each other physically, and by at least 35 dB electrically. At higher frequencies, this effect is more noticeable, since isolation tends to drop as frequency increases. For this reason, it is important to maintain the LO power at its appropriate level, once other ports are matched.

The RF Port

A reactive rf source is not too detrimental to system performance. This is good, since the output impedance of most amateur preamplifiers is seldom 50 ohms resistive. A 3-dB pad is used at the rf port in the 50- and 144-MHz mixer to 14 MHz, and a 2-dB pad is used in the 220/432-MHz mixer to 28 MHz, although they add directly to mixer noise figure. Rf inputs between about 80 and 200 MHz are practical in the 14-MHz i-f-output model, while the 28-MHz-output unit is most useful from 175 to 500 MHz. Mixer contribution to system noise figure will be almost completely overcome by a low-noise rf amplifier with sufficient gain and adequate image rejection.

Image Response

Any broadband mixing scheme will have a potential image-response problem. In most amateur vhf/uhf receiver systems (as in these units) single-conversion techniques are employed, with the LO placed below the desired rf channel for non-inverting down-conversion to i-f. Conversion is related to both i-f and LO frequencies and, because of the broadband nature of the DBM, input signals at the rf image frequency (numerically $f_{LO} - f_{i-f}$ in our case) will legitimately appear inverted at the i-f-output port, unless proper filtering is used to reduce them at the mixer rf-input port. For example, a 144-MHz converter with a 28-MHz i-f output (116-MHz LO) will have rf

Table 3

DBM I-F Amplifier Parts List

	14 MHz i-f output	28 MHz i-f output
C1	470 pF JFD 471J or equal.	300 pF JFD 301J or equal.
C2	390 pF SM	not used.
C3	180 pF SM	51 pF SM
C4	39 pF SM	18 pF SM
C5	56 pF SM	27 pF SM
C6	300 pF SM	150 pF SM
L1	9 turns no. 18 enam., close wound on a 3/8-inch (9.5 mm) diameter red-slug coil form.	9 turns no. 24 enam., close wound on a 1/4-inch (6.3 mm) diameter green-slug coil form.
L2	18 turns no. 26 enam., close wound on a 3/8-inch diameter red-slug coil form.	12 turns no. 26 enam., close wound on a 1/4-inch diameter green-slug coil form.
	Tap down 7 turns from top for 3N140 drain connection. See text.	No tap used.
L3	Same as L2 but no tap. spaced 1 1/8-inch (29 mm) center-to-center with L2.	Same as L2, spaced 1 inch (25 mm) center-to-center with L2.
R1, R3	300 ohm 1/4 W, carbon.	430 ohm, 1/4 W, carbon.
R2	16 ohm, 1/4 W, carbon.	11 ohm, 1/4 W, carbon.

Ferrite beads can be replaced by a 10-ohm, 1/4 W carbon resistor at one end of the choke, if desired.

SM = Silver Mica.

image-response potential in the 84 to 88-MHz range. TV channel 6 wideband-fm audio will indeed appear at the i-f-output port near 28 MHz unless appropriate rf-input filtering is used to eliminate it. While octave-bandwidth vhf/uhf "imageless mixer" techniques can improve system noise performance by about 3 dB (image noise reduction), and image signal rejection by 20 dB — and much greater with the use of a simple gating scheme — such a system is a bit esoteric for our application. Double or multiple-conversion techniques can be used to advantage, but they further complicate an otherwise simple system. Image noise and signal rejection will depend on the effectiveness of the filtering provided in the rf-amplifier chain.

Mixer Selection

The mixer used in this system is a Relcom M6F, with specifications given in Table 2. Suitable substitute units are also presented. The M6F is designed for printed-circuit applications (as are the recommended substitutes), and the lead pins are rather short. While mixers are available with connectors attached, they are more expensive. The simple package is suggested as, aside from less expense, improved interface between mixer and i-f amplifier is possible because of the short leads. The combining of mixer and i-f amplifier in one converter package was done for that reason. Along these lines,

the modular-construction approach permits good signal isolation and enables the mixer-amplifier/i-f system to be used at a variety of rf and LO-input frequencies, as mentioned earlier.

Most commonly available, inexpensive DBMs are not constructed to take advantage of LO powers much above +10 dBm (10 mW). To do so requires additional circuitry which could degrade other mixer characteristics, specifically conversion loss and interport isolation. The advantage of higher LO power is primarily one of improved strong-signal-handling performance. At least one manufacturer advertises a moderately priced "high-level" receiving DBM that can use up to +23 dBm (200 mW) LO power, and still retain the excellent conversion loss and isolation characteristics, shown in Table 2. The usefulness of mixers with LO power requirements above the commonly available +7 dBm (5 mW) level in amateur receiving applications may be a bit moot, as succeeding stages in most amateur receivers will likely overload before the DBM. Excessive over-design is not necessary.

In general, mixer selection is based on the lowest practical LO level requirement that will meet the application, as it is more economical and results in the least LO leakage within the system. As a first-order approximation, LO power should be 10 dB greater than the highest anticipated input-signal level at the rf port. Mixers with LO requirements of +7 dBm are quite adequate for amateur receiving applications.

Application Design Guidelines

While the material just presented only scratches the surface in terms of DBM theory and utilization in amateur vhf/uhf receiving systems, some practical solutions to the non-ideal mixer-port-termination problem have been offered. To achieve best performance from most commercially manufactured broadband DBM in amateur receiver service, the following guidelines are suggested:

1) Choose i-f and LO frequencies which will provide maximum freedom from interference problems. Don't "guesstimate"; go through the numbers!

2) Provide a proper i-f-output termination (most critical).

3) Increase the LO-input power to rf-input power ratio to a value that will provide the required suppression of any in-band interfering products. The specified LO power (+7 dBm) will generally accomplish this.

4) Provide as good an LO match as possible.

5) Include adequate pre-mixer rf-image filtering at the rf port.

When the mixer ports are terminated properly, performance usually in excess of published specifications will be achieved — and this is more than adequate for most amateur vhf/uhf receiver mixing applications.

THE COMBINED DBM/I-F AMPLIFIER

A low-noise i-f amplifier (2 dB or less) following the DBM helps ensure an acceptable system noise figure when the mixer is preceeded by a low-noise rf amplifier. A pi-network matching system used between the mixer i-f-output port and gate 1 of the 3N140 transforms the nominal 50-ohm mixer-output impedance to a 1500-ohm gate-input impedance (at 28 MHz) specifically for best noise performance. The network forms a narrowband mixer/i-f-output circuit which serves two other important functions: It helps achieve the necessary isolation between rf-and i-f signal components, and serves as a 3-pole filter, resulting in a monotonic decrease in match impedances as the operating i-f departs from mid-band. This action aids in suppression of harmonic-distortion products.

The combined DBM/i-f amplifier is shown schematically in Fig. 24 and pictorially in the photographs. In the 14-MHz model, the 3N140 drain is tapped down on its associated inductance to provide a lower impedance for better strong-signal-handling ability. The 3N140 produces about 19 dB gain across a 700-kHz passband, flat within 1 dB between 13.8 and 14.5 MHz. A 2-MHz passband is used for the 28-MHz model, and the device drain is connected directly to the high-impedance end of its associated inductance. Both amplifiers were tuned independently of their respective mixers, and checked for noise figure as well as gain. With each i-f amplifier pretuned and connected to its mixer, signals were applied to the LO and rf-input ports. The pi-network inductance in the i-f interface was adjusted carefully to see if performance had been altered. No change was noted. I-f gain is controlled by the externally accessable potentiometer. Passband tuning adjustments in the drain circuit are best made with a sweep generator, but single-signal tuning techniques will be adequate. While there should be no difficulty with the non-gate-protected 3N140, a 40673 may be substituted directly if desired.

DBM/I-F Amplifier IMD Evaluation

Classical laboratory IMD measurements were made on the DBM/i-f amplifier using two −10 dBm signals, closely spaced in the 144-MHz range. The LO power used was +7 dBm. Conversion loss was 5 dB and the calculated third-order output intercept point was +15 dBm. In operation, as simulated by these test conditions, equivalent output signal levels at J3 would be strong enough to severely overload most amateur receivers. Perhaps the early Collins 75A series, R390A and those systems described by

Sabin[4] and Hayward[5] would still be functioning well.

A high-performance, small-signal, vhf/uhf receiving amplifier optimized for IMD reduction and useful noise figure is only as good as any succeeding receiving-system stage, in terms of overload. The DBM/i-f-amplifier combination presented significantly reduces common first-mixer overload problems, leaving the station receiver as the potentially weak link in the system. When properly understood and employed, the broadband DBM followed by a selective low-noise i-f amplifier can be a useful tool for the amateur vhf/uhf receiver experimenter.

APPENDIX

Mixer Terminology

$f$rf — rf input frequency
fLO — local-oscillator input frequency
fi-f — i-f output frequency

By convention, mixing signals and their products are referred to the LO frequency for calculations. In the mixer system presented, $f$rf is always above fLO, so we will refer our signals to $f$rf, with the exception of Fig. 23 which uses the fLO reference.

Overload

A generic term covering most undesired operating phenomena associated with device non-linearity.

Harmonic Modulation Products

Output responses caused by harmonics of fLO and $f$rf and their mixing products.

RF Compression Level

The absolute single-signal rf input-power level that causes conversion loss to increase by 1 dB.

RF Desensitization Level

The rf input power of an interfering signal that causes the small-signal conversion loss to increase by 1 dB, i.e., reducing a weak received signal by 1 dB.

Intermodulation Products

Distortion products caused by multiple rf signals and their harmonics mixing with each other and the LO, producing new output frequencies.

Mixer Intermodulation Intercept Point

Because mixers are nonlinear devices, all signals applied will generate others. When two signals (or tones), F1 and F2, are applied simultaneously to the rf-input port, additional signals are generated and appear in the output as fLO ± (nF1 + mF2). These signals are most troublesome when $n ± m$ is a low odd number, as the resulting product will lie close to the desired output. For $n − 1$ (or 2) and $m − 2$ (or 1), the result is three (3), and is called the two-tone/third-order intermodulation prod-

uts. When F1 and F2 are separated by 1 MHz, the third-order products will lie 1 MHz above and below the desired outputs. Intermodulation is generally specified under anticipated operating conditions since performance varies over the broad mixer-frequency ranges. Intermodulation products may be specified at levels required (i.e., 50 dB below the desired outputs for two 0-dBm input signals) or by the intercept point.

The intercept point is a fictitious point determined by the fact that an increase of level of two input tones by 10 dB will cause the desired output to increase by 10 dB, but the third-order output will increase by 30 dB. If the mixer exhibited no compression, there would be a point at which the level of the desired output would be equal to that of the third-order product. This is called the third-order intercept point.

Noise Figure

Noise figure is a relative measurement based on excess noise power available from a termination (input resistor) at a particular temperature (290 K). When measuring the NF of a doubly balanced mixer with an automatic system, such as the HP-342A, a correction may be necessary to make the meter reading consistent with the accepted definition of receiver noise figure.

In a broadband DBM, the actual noise bandwidth consists of two i-f passbands, one on each side of the local-oscillator frequency (fLO + fi-f and fLO − fi-f). This double sideband (dsb) i-f response includes the rf channel and its image. In general, only the rf channel is desired for further amplification. The image contributes nothing but receiver and background noise.

When making an automatic noise-figure measurement using a wideband noise source, the excess noise is applied through both sidebands in a broadband DBM. Thus the instrument meter indicates NF as based on both sidebands. This means that the noise in the rf and image sidebands is combined in the mixer i-f-output port to give a double contribution (3 dB greater than under ssb conditions). For equal rf-sideband responses, which is a reasonable assumption, and in the absence of preselectors, filters, or other image rejection elements, the automatic NF meter readings are 3 dB lower than the actual NF for DBM measurements.

The noise figure for receivers (and most DBM) is generally specified with only one sideband for the useful signal. As mentioned in the text, most DBM diodes add no more than 0.5 dB (in the form of NF) to conversion loss, which is generally measured under single-signal rf-input (ssb) conditons. Assuming DBM conversion efficiency (or loss) to be within specifications, there is an excellent proba-

[4]Sabin, "The Solid-State Receiver," *QST*, July 1970.
[5]Hayward, "A Competition-Grade CW Receiver," *QST*, March and April 1974.

bility that the ssb NF is also satisfactory. Noise figure calculations in the text were made using a graphical solution of the well known noise-figure formula:

$$f_T = f_1 + \frac{f_2 - 1}{g_1}$$

converted to dB.

Improved Wide Band I-F Responses

The following information was developed in achieving broad-band performance in the mixer-to-amplifier circuitry. In cases where only a small portion of a band is of interest the original circuit values are adequate. For those who need to receive over a considerable portion of a band, say one to two MHz, a change of some component will provide improved performance over a broad range while maintaining an acceptable noise figure.

The term "nominal 50-ohm impedance" applied to diode DBM ports is truly a misnomer, as their reflective impedance is rarely 50 ohms + j0 and a VSWR of 1 is almost never achieved. Mixer performance specified by the manufacturer is measured in a 50-ohm broadband system, and it is up to the designer to provide an equivalent termination to ensure that the unit will meet specifications. Appropriate matching techniques at the rf and LO ports will reduce conversion loss and LO-power requirements. Complex filter synthesis can improve the i-f output match. However, if one does not have the necessary equipment to evaluate his efforts, they may be wasted. Simple, effective, easily reproduced reduced circuitry was desired as long as the trade-offs were acceptable, and measurements indicate this to be the case.

The most critical circuit in the combined unit is the interface between mixer and i-f amplifier. It must be low-pass in nature to satisfy vhf signal component bypassing requirements at the mixer i-f port. For best mixer IMD characteristics and low conversion loss, it must present to the i-f port a nominal 50-ohm impedance at the desired frequency, and this impedance value must not be allowed to increase as i-f operation departs from midband. The impedance at the i-f amplifier end of the interface network must be in the optimum region for minimum cross-modulation and low noise. A dual-gate device offers two important advantages over most bipolars. Very little, if any, power gain is sacrificed in achieving best noise figure, and both parameters (gain and NF) are relatively independent of source resistance in the optimum region. As a result, the designer has a great deal of flexibility in choosing a source impedance. In general, a 3:1 change in source resistance results in only a 1-dB change in NF. With minimum cross-modulation as a prime system consideration, this 3:1 change (reduction) in source resistance implies a 3:1 improvement in cross-modulation and total harmonic distortion.

Tests on the 3N201 dual-gate MOSFET have shown device noise performance to be excellent for source impedances in the 1000- to 2000-ohm region. For optimum noise and good cross-modulation performance, the nominal 50-ohm mixer i-f output impedance is stepped up to about 1500 ohms for i-f amplifier gate 1, using the familiar low-pass pi network. This is a mismatched condition for gate 1, as the device input impedance for best gain in the hf region is on the order of 10 kΩ. Network loaded-Q values in the article are a bit higher than necessary, and a design for lower Q_L is preferred. Suggested modified component values are listed in Fig. 25. High-frequency attenuation is reduced somewhat, but satisfactory noise and b a n d w i d t h performance is more easily obtained. Coil-form size is the same, so no layout changes are required for the modification. Components in the interface must be of high Q and few in number to limit their noise contribution through losses. The 28-MHz values provide satisfactory interface network performance over a 2-MHz bandwidth. A higher Q_L in the 28-MHz interface can be useful if one narrows the output network and covers only a few hundred kilohertz bandwidth, as is commonly done in 432-MHz weak-signal work.

Device biasing and gain control methods were chosen for simplicity and adequate performance. Some sort of gain adjustment is desirable for drain-circuit overload protection. It is also a handy way to "set" the receiver S meter. A good method for gain adjustment is reduction of the gate-2 bias voltage from its initial optimum-gain bias point (greater than +4 V dc), producing a remote-cutoff characteristic (a gradual reduction in drain current with decreasing gate bias). The initial gain-reduction rate is higher with a slight forward bias on gate 1, than for $V_{g1s} = 0$. Input and output circuit detuning resulting from gain reduction (Miller effect) is inconsequential as the gate-1 and drain susceptances change very little over a wide range of V_{g2s} and I_D at both choices of i-f. Best intermodulation figure for the 3N201 was obtained with a small forward bias on

gate 1, and the bias-circuit modification shown in Fig. 25 may be tried, if desired.

CONVERTERS FOR 50, 144 AND 220 MHZ

The converters here were originally designed by the Rochester VHF Group and appeared in August 1973 *QST*. The design was the basis for a club project so the same board is used for all three converters, with only slight modifications for each band. Design features include

1) Low noise figure.

2) State-of-the-art freedom from cross modulation.

3) Sufficient gain to override the front-end-noise of most receivers.

4) Double-tuned bandpass interstage and output circuits to achieve a flat response over a 2-MHz portion of either band.

5) Filtering of the local oscillator chain in the 144- and 220-MHz models to reduce spurious responses.

6) Small size and low power consumption.

7) Freedom from accidental mistuning during the life of the converter.

Other points considered were such things as freedom from the necessity of neutralization and the use of moderately priced transistors.

Circuit Design

Schematic diagrams for the three versions are shown in Figs. 26, 27 and 28. The configuration of the rf and mixer portions of the circuit are virtually identical for all three with the values of the frequency-determining components being scaled appropriately. The major difference between the converters is a change in the local oscillator chain. A minor change in the method of interstage coupling was necessary to prevent stray-capacitance effects from making the alignment critical on the 50-MHz converter.

All inductors in the 50-MHz model and the 28-MHz output circuit are wound on Amidon toroid cores. The tuned circuits are aligned by spreading or compressing the turns around the toroid core. After alignment the coils are glued in place with

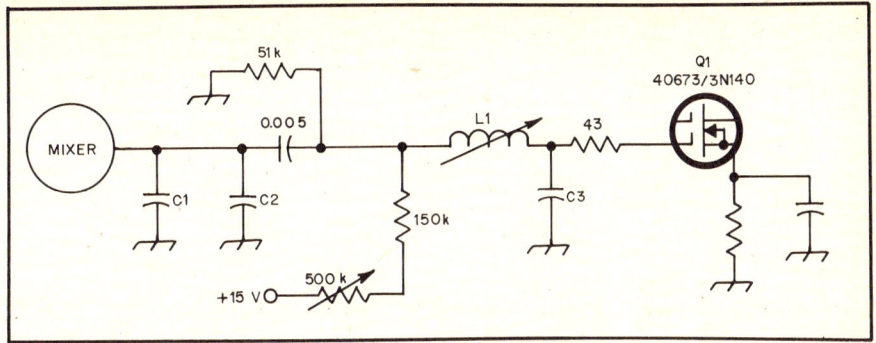

Fig. 25 — Suggested changes in the mixer-to 3N140 pi-network interface circuit, producing lower Q_L and better performance.

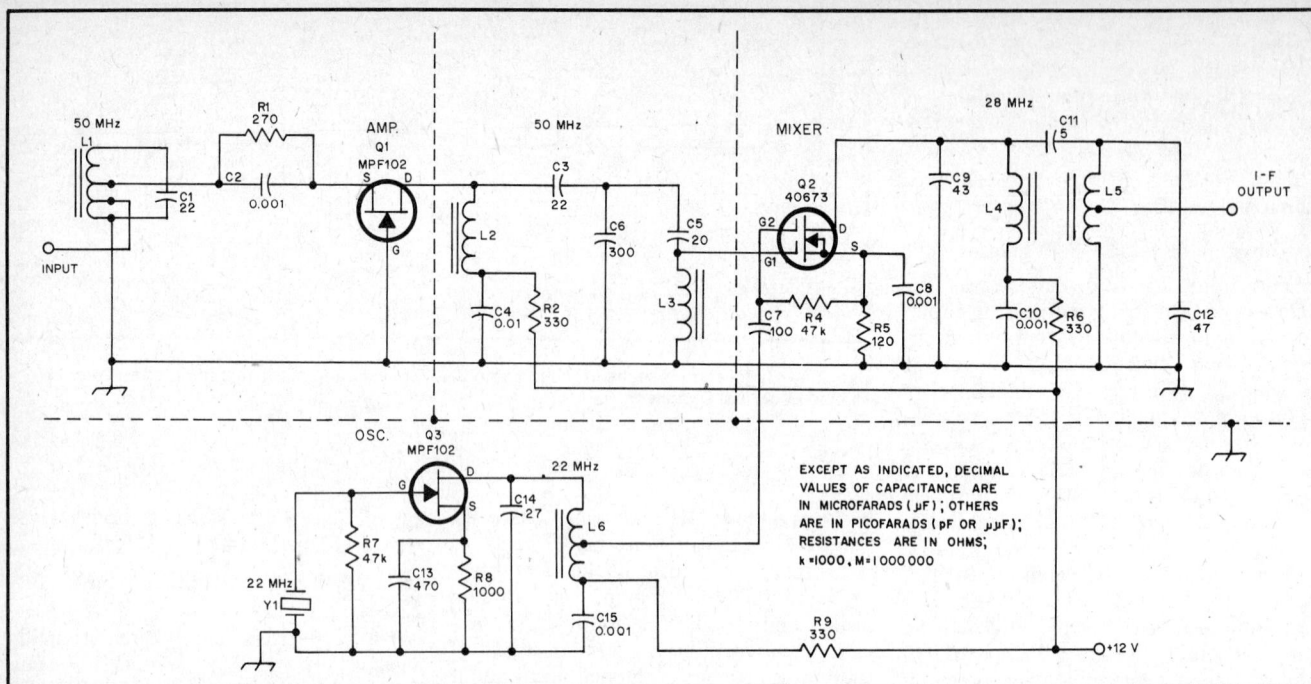

Fig. 26 — Schematic diagram of the 50-MHz converter. All resistors are 1/4-watt composition. C2, C8, C10 and C15 are 0.001 µF disc ceramic. C4 is 0.01-µF disc ceramic. All other capacitors are dipped mica.

L1-L6, incl. — All no. 28 enam. wire wound on Amidon T-25-6 cores as follows: L1, 14 turns tapped at 4 turns and 6 turns; L2, 13 turns; L3, 12 turns; L4, 18 turns; L5, 18 turns tapped at 4 turns from cold end; L6, 26 turns tapped at 6 turns from hot end.

Y1 — 22-MHz crystal. International Crystal Mfg. Co. type EX.

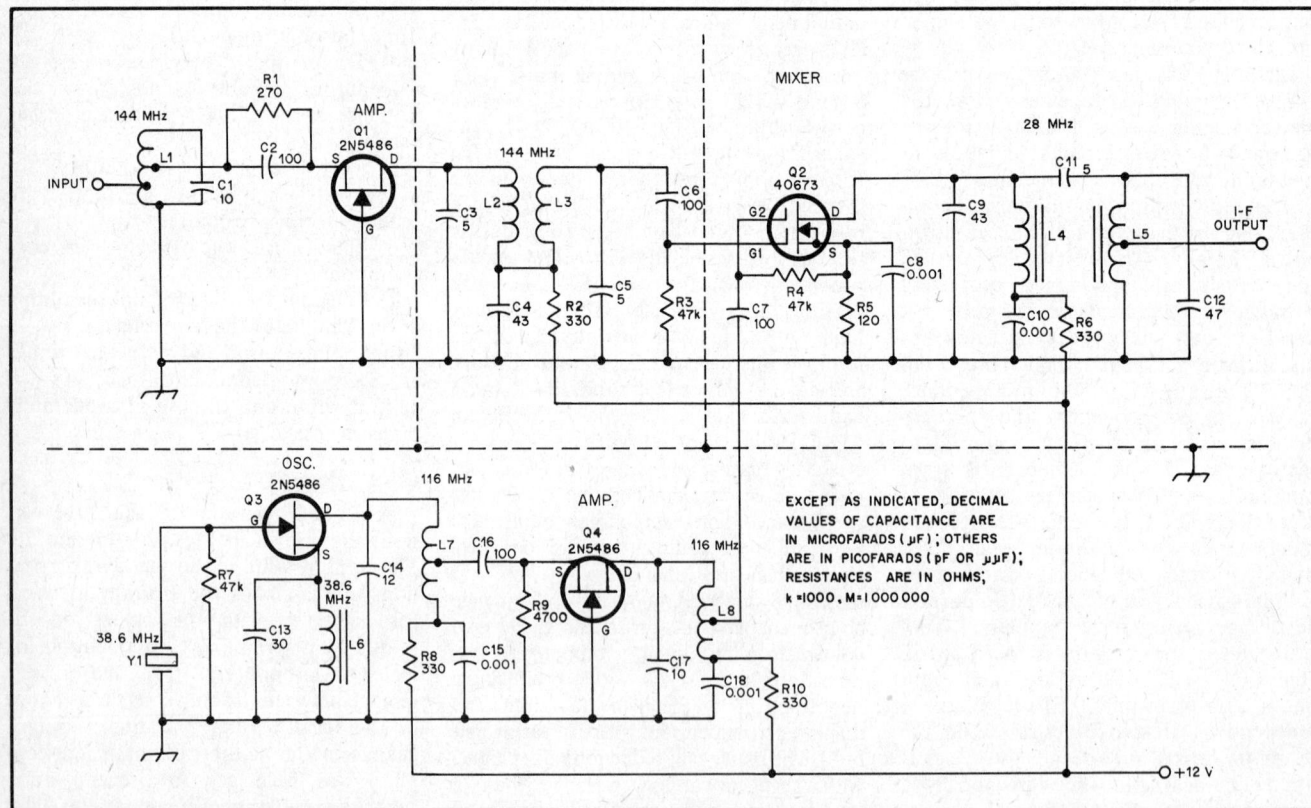

Fig. 27 — Schematic diagram of the 144-MHz converter. All resistors are 1/4-watt composition. C8, C10, C15 and C18 are 0.001-µF disc ceramic. All other capacitors are dipped mica units.

L1, L2, L3, L7, L8 — All no. 20 enam. wire formed by using the threads of a 1/4-20 bolt as a guide. L1, 5 turns tapped at 1-3/4 turns and 3/4 turn from cold end; L2, 5 turns; L3, 4 turns; L7 and L8, 5 turns tapped at 2 turns from hot end.

L4 — 18 turns no. 28 enam. wound on Amidon T-25-6 core.

L5 — 18 turns like L4, tapped at 4 turns from cold end.

L6 — 16 turns no. 28 enam. wound on Amidon T-25-10 core.

Y1 — 38.666-MHz crystal. International Crystal Mfg. Co. type EX.

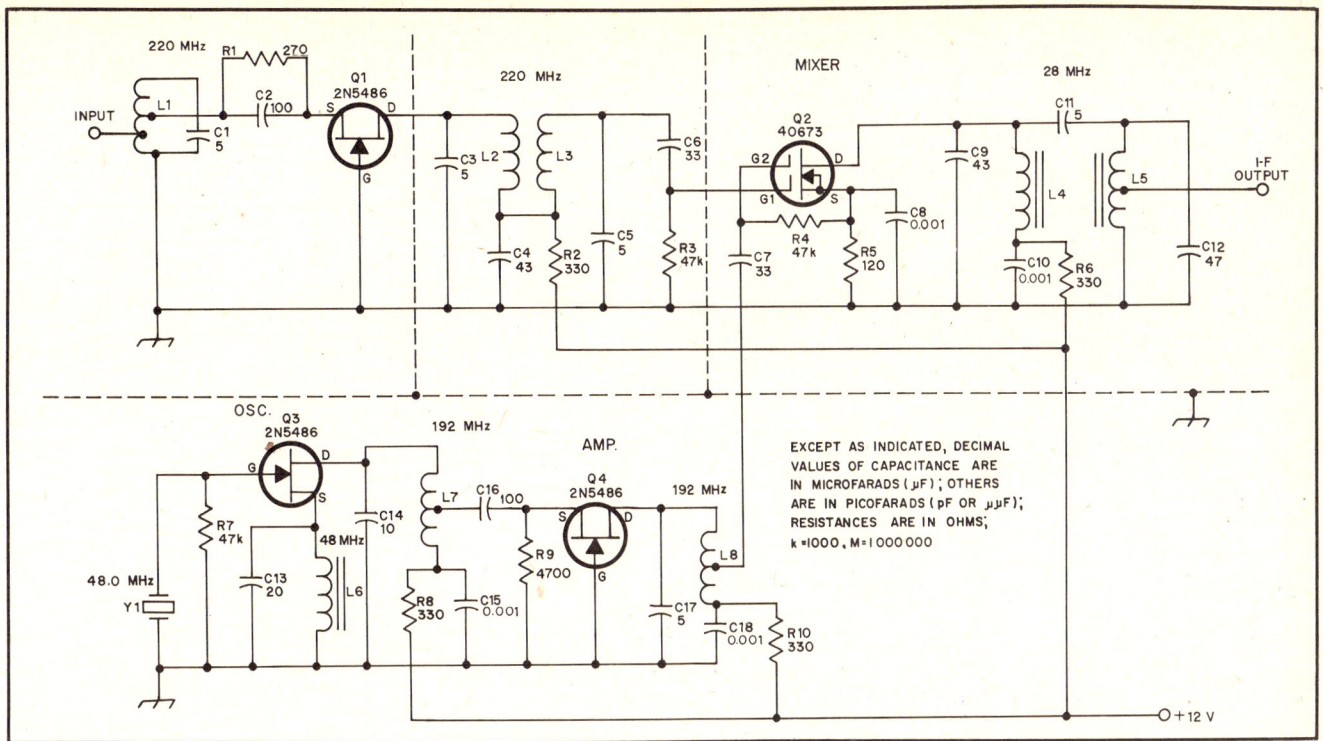

Fig. 28 — Schematic diagram of the 220-MHz converter. All resistors are 1/4-watt composition. C8, C10, C15 and C18 are 0.001-μF disc ceramic. All other capacitors are dipped mica units.

L1, L2, L3, L7, L8 — All no. 20 enam. wire formed by using the threads of a 1/4-20 bolt as a guide. L1, 4 turns tapped at 1-1/2 turns and 3/4 turn from cold end; L2, 4 turns; L3, 2 turns; L7, 3 turns tapped at 1-1/2 turns from cold end, and L8, 4 turns tapped at 2-1/2 turns from hot end.

L4 — 18 turns no. 28 enam. wound on Amidon T-25-6 core.

L5 — 18 turns like L4, tapped at 4 turns from cold end.

L6 — 18 turns no. 28 enam. wound on Amidon T-25-10 core.

Y1 — 48.000 MHz crystal. International Crystal Mfg. Co. type EX.

Silastic compound (sold as bathtub caulk).

The rf amplifier, Q1, is used in a grounded-gate configuration. The input circuit is tapped to provide a proper match between the antenna and source of the FET while maintaining a reasonable Q. The 50-MHz interstage coupling network consists of C3, C5, L2 and L3. Band-pass coupling is controlled by the capacitive T network of C3 and C5 in ratio with C6. A 40673 dual-gate MOSFET is used in the mixer circuit (Q2). Gate 1 receives the signal, while gate 2 has the local-oscillator injection voltage applied to it through C7. A slight amount of positive bias is applied to gate 2 through R2. A top-coupled configuration, using toroid inductors, serves as the 28-MHz output circuit of both converters.

The oscillator circuit in the 50-MHz model is straighforward, relying on the drain-to-gate capacitance of the FET for feedback. A tap at four turns from the hot end of the toroid winding provides the injection to the mixer through capacitor C7. In the 144- and 220-MHz converter the rf stage is identical to the 50-MHz version except for the tuning networks. L1, L2, and L3 are air wound, self-supporting, and are formed initially by winding wire around the threads of a 1/4-20 bolt. The turns of L1 are spread to permit adding taps prior to mounting on the board. The degree of interstage coupling in the two-meter model is

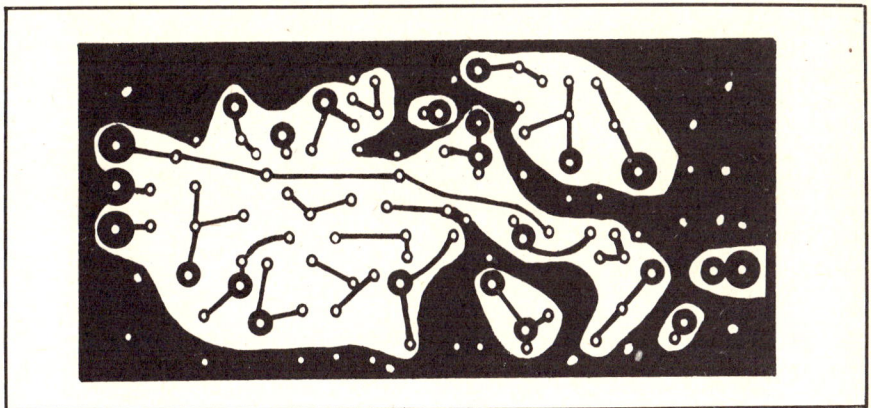

Fig. 29 — Scale-size layout for the pc board. The same pattern is used for each band. Foil side shown here.

controlled by the positions of L2 and L3. Since they are mounted at right angles, the coupling is very light. By changing the angle between these two coils, the passband may be optimized.

In the 144- and 220-MHz converter oscillator stages, Q3 is changed to an oscillator/tripler by replacing the source bias resistor with L6. Replace bypass capacitor, C13, with a suitable value to resonate L6 near the crystal frequency. Source-to-gate capacitance provides the feedback in this case. The drain tank is modified to provide output at the third harmonic, thus eliminating the need for a separate tripler stage. Q4 is used as an

isolation amplifier running at very low current level (as controlled by R9) to provide attenuation of the adjacent harmonics. This stage is not needed for amplification of the oscillator signal but without the additional filtering, severe "birdies" may result from nearby fm or TV stations. In all three versions, a number of printed-circuit pads will be left over when construction is completed. These are the result of providing both bands on a common pc layout. For example, the isolation amplifier following the oscillator is not used on 50 MHz. Therefore, this stage is bypassed by a jumper wire from L6 to C7. Five

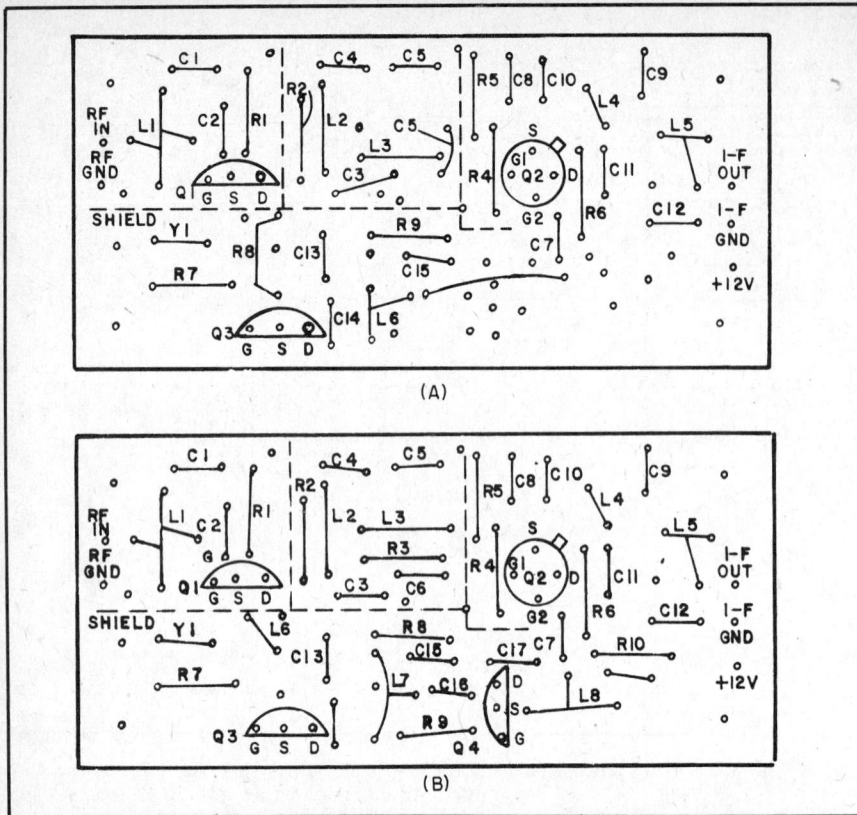

Fig. 30 — Parts-placement guide for the 50-MHz converter, A, and the 144- and 220-MHz converters, B. View is from the foil side of the board. Dashed lines show the location of shields that are soldered to short pieces of wire which project through holes in the pc board. The shields may be fabricated from sheet brass or copper, or scraps of copper-clad board material.

additional holes are located in the ground area along the centerline of the board and between rf and mixer stages. Component lead clippings are soldered into these holes to provide a mounting for the shield partitions, which are soldered to the wires where they extend through the board. Fig. 30 shows the parts layout for the three converters. Notice that one lead of C3 must reach past the ground hole and connect to the foil. R3 is not used on the 50-MHz converter.

Alignment and Test

Perhaps the most difficult task in the project was the test and tune-up of the finished converter. A single test setup using a sweep generator, diode probe, and oscilloscope is a necessity to assure the flat response over the tuning range. Commercial attenuators can be used to calibrate each converter by the substitution method.

Tuning of the air-wound rf circuit for 144 and 220 MHz is accomplished by spreading or compressing the turns of the coils. After alignment, the windings are secured by a bead of Silastic compound along the coil to hold the turns in place.

The transistors used in the rf stage are also subject to some variation in noise figure. When this occurs, an rf FET should be carefully traded with an oscillator FET, since performance of the

FET as an oscillator usually is satisfactory.

Small ceramic trimmers can be used in place of the fixed-value mica capacitors in the tuned circuits of these converters. The midrange of the trimmer should be approximately the value of the mica capacitors replaced. This procedure may simplify the tuning process of the converters where a sweep generator setup is not available. A little careful tweaking should give a reasonably flat response.

If trimmers are used, the rf input circuit should be tuned to the center of the desired response, 50.5 MHz as an example. This circuit tunes broadly and is not too critical. The rf interstage circuits should be stagger tuned, one at 50.0 MHz and the other at 51.0 MHz. As an example, the output i-f circuits can be tuned in a manner similar to the interstage circuits.

A HIGH-PERFORMANCE 2-METER CONVERTER

Top performance is a requisite for vhf receivers during contests and when the operator is engaged in weak-signal work. The criteria are low noise, frequency stability and freedom from unwanted spurious responses in the overall receiver. The latter can be assured by using an LO chain which has proper spectral purity, and by using narrow-band filtering at the receiver front end. Additional benefits are

obtained when utilizing filtering at the i-f output of a vhf converter. This circuit, developed by W1FB, complies with the foregoing design and performance objectives. It is intended for use on ssb, cw and a-m, notably below 146 MHz. The dynamic range of this converter is excellent, owing in part to the use of a doubly balanced diode-ring mixer that is followed by a diplexer. The diplexer assures that the mixer is terminated in its characteristic impedance — 50 ohms.

An overall gain of 15 dB was measured for the composite converter. All spurious energy at the LO-chain output is −72 dB or greater below the desired output (116 MHz). A two-pole Butterworth filter is used at the i-f output of the converter. This 28-MHz filter prevents unwanted energy from the converter from reaching the tunable i-f main receiver — a further aid to the elimination of birdies. The converter has a 3-dB bandwidth of 200 kHz. This is established by the highly selective four-resonator input filter which is similar to a helical resonator. The converter can be adjusted for any 200-kHz segment of the 2-meter band. The example given here was designed primarily for use from 144.0 to 144.2 MHz. In this case the tunable i-f is 28.0 to 28.2 MHz.

RF Amplifiers and Filter

Fig. 31 shows the front-end section of the converter. C1 was not needed in the prototype model, but can be included in the interest of obtaining the lowest possible noise figure. The tap point to which C1 is connected is approximately 50 ohms without C1 in the circuit. L1 and C2 comprise the input resonator, FL1. Q1 operates in the common-gate mode. The source is tapped on L1 to provide an impedance match. A Siliconix E300 JFET is used at Q1 and Q2 in the interest of low noise and high dynamic range. It is rated by the manufacturer as having a typical noise figure of 1.3 dB at 100 MHz. The dynamic range is specified as 100 dB or greater. Power gain is listed as 17 to 20 dB in the common-gate mode. In this circuit the gain is set at approximately 10 dB per stage by means of the tap points on the resonators and the value of source-bias resistance. The 10-ohm resistors used in the drain leads of each transistor reduce the gain slightly. They are included for suppression of parasitic oscillations.

FL2 consists of two aperture-coupled resonators. The aperture is set for an insertion loss of roughly 5 dB in order to obtain the desired 200-kHz overall front-end selectivity. FL3 is the fourth resonator. Output is taken at 50 ohms by means of a 560-pF coupling capacitor. This circuit is unconditionally stable. The gate lead must be made as short as possible to ensure stability. It is returned to ground on the *inner conductor* of the

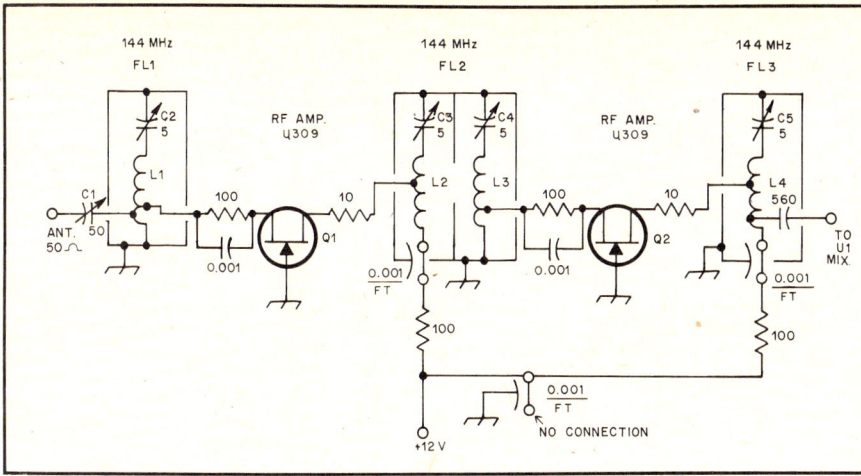

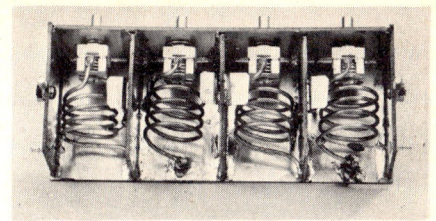

Fig. 32 — Close-up view of the interior of the rf front-end amplifier and band-pass filters. Spring-brass tabs are soldered to the divider walls to assure good electrical contact with the aluminum side plate (see text). A third 0.001-µF feedthrough capacitor was added to the lower edge of the third compartment from the left after this photograph was taken. It serves as a tie point for the 100-ohm decoupling resistors (see Fig. 31).

Fig. 31 — Schematic diagram of the converter front end. Fixed-value capacitors are disc ceramic. Feedthrough types are used where indicated. Resistors are 1/2-watt composition.

C1 — 50-pF ceramic trimmer (see text).
C2-C5, incl. — 5-pF capacitor. Subminiature air variable of glass piston trimmer of high-Q type. E.F. Johnson 160-102 suitable. Johnson 193-4-1 (13 pF max.) used in this model.
L1-L4, incl. — 4 turns no. 16 bare or silver-plated copper wire, 5/8-inch ID × 1-1/4 inches (16 × 32 mm) long. Q1, Q2 sources tapped 1/2 turn above ground. Input and output taps are 1/4 turn from ground. Drain taps are 1/2 turn from high end of coils.
Q1, Q2 — Siliconix E300 JFET. 2N4416 suitable at slight reduction in performance.

pc-board double-clad module wall. Three 0.001-µF feedthrough capacitors and two 100-ohm resistors serve as the 12-volt decoupling elements. Q1 and Q2 are mounted on the outside wall of the rf-amplifier module.

Mixer and Post-Mixer Amplifier

U1 and Q3 of Fig. 33 are contained in the second module. U1 is a four-diode doubly balanced mixer assembly. It has a conversion loss of approximately 8 dB. This requires the inclusion of a post-mixer i-f amplifier, Q3. The latter is set for a gain of roughly 8 dB. LO injection is supplied to U1 at 116 MHz. The LO power is +7 dBM (approximately 0.5 volt rms across 50 ohms).

L5, the 51-ohm resistor and two 39-pF capacitors form the diplexer in combination with the L network (L6/C6). L5 and the 39-pF capacitors comprise a high-pass network with a loaded Q of 1 (X_L and X_C = 50). The cutoff frequency of the network is three times the i-f (84 MHz). The 51-ohm resistor serves as a termination for the mixer. L6 and C6, by virtue of the low-pass characteristic, help prevent 116- and 144-MHz energy from reaching Q3. This network is designed to match 50 ohms to 2200 ohms.

Q3 is another E300 FET, chosen for the low-noise characteristic. An unbypassed 10-ohm resistor is used in the source to cause degeneration in the interest of stability. A two-pole Butterworth band-pass filter is used at the output of Q3. It has a 3-dB bandwidth of 500 kHz and is tuned for a center frequency of 28.250 MHz. The filter input characteristic is 12,000 ohms. The 10-pF output coupling capacitor provides a 50-ohm terminal impedance for looking into a 50-ohm tunable i-f receiver. The circuit of Fig. 33 is contained in a module fashioned from single-sided pc board. The copper surfaces are inside the box.

Local-Oscillator Section

A simple overtone oscillator is used at

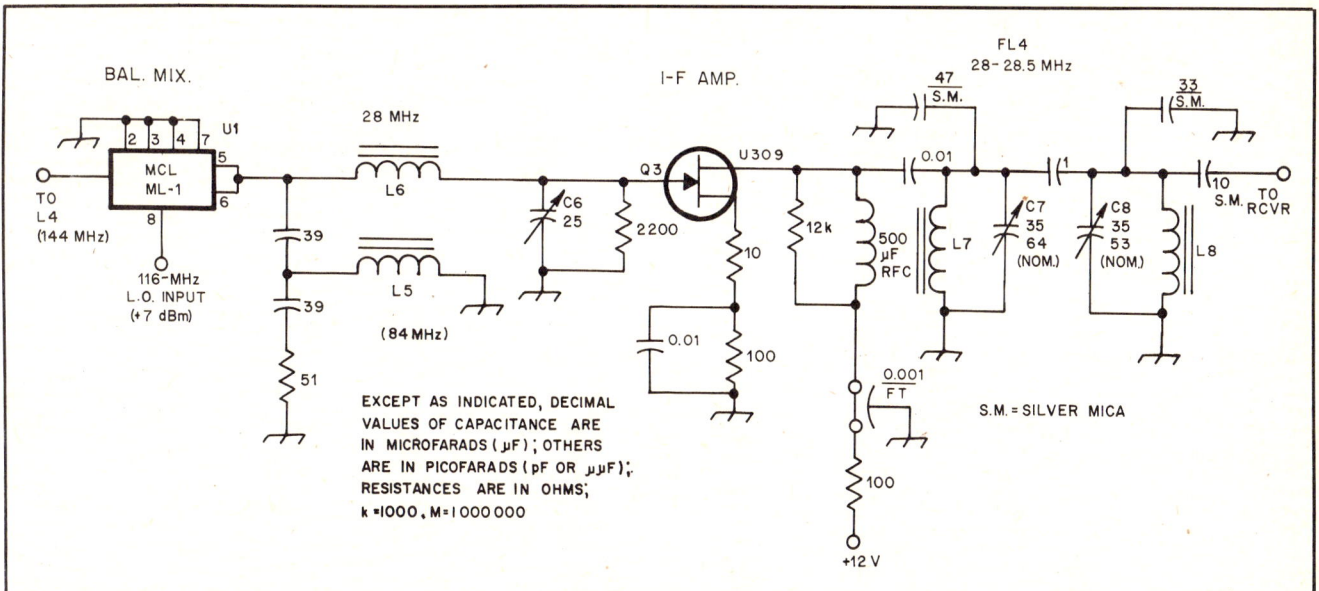

Fig. 33 — Schematic diagram of the passive mixer, diplexer and post-mixer amplifier. Fixed-value capacitors are disc ceramic except those shown as feedthrough types. Resistors are 1/2-watt composition.

C6-C8, incl. — 35-pF ceramic or Mylar trimmer. Solder 47- and 33-pF silver mica capacitors across pc foils for C7 and C8, respectively.
L5 — 5 turns no. 24 enam. wire on T-50-6 toroid core (0.09 µH).
L6 — 21 turns no. 24 enam. wire on T-50-6 toroid core (1.8 µH).
Q3 — Siliconix E300 JFET or 2N4416.
U1 — Diode-quad doubly balanced mixer. MCL SRA-1 or SBL1, or Cimarron CM-1 suitable.

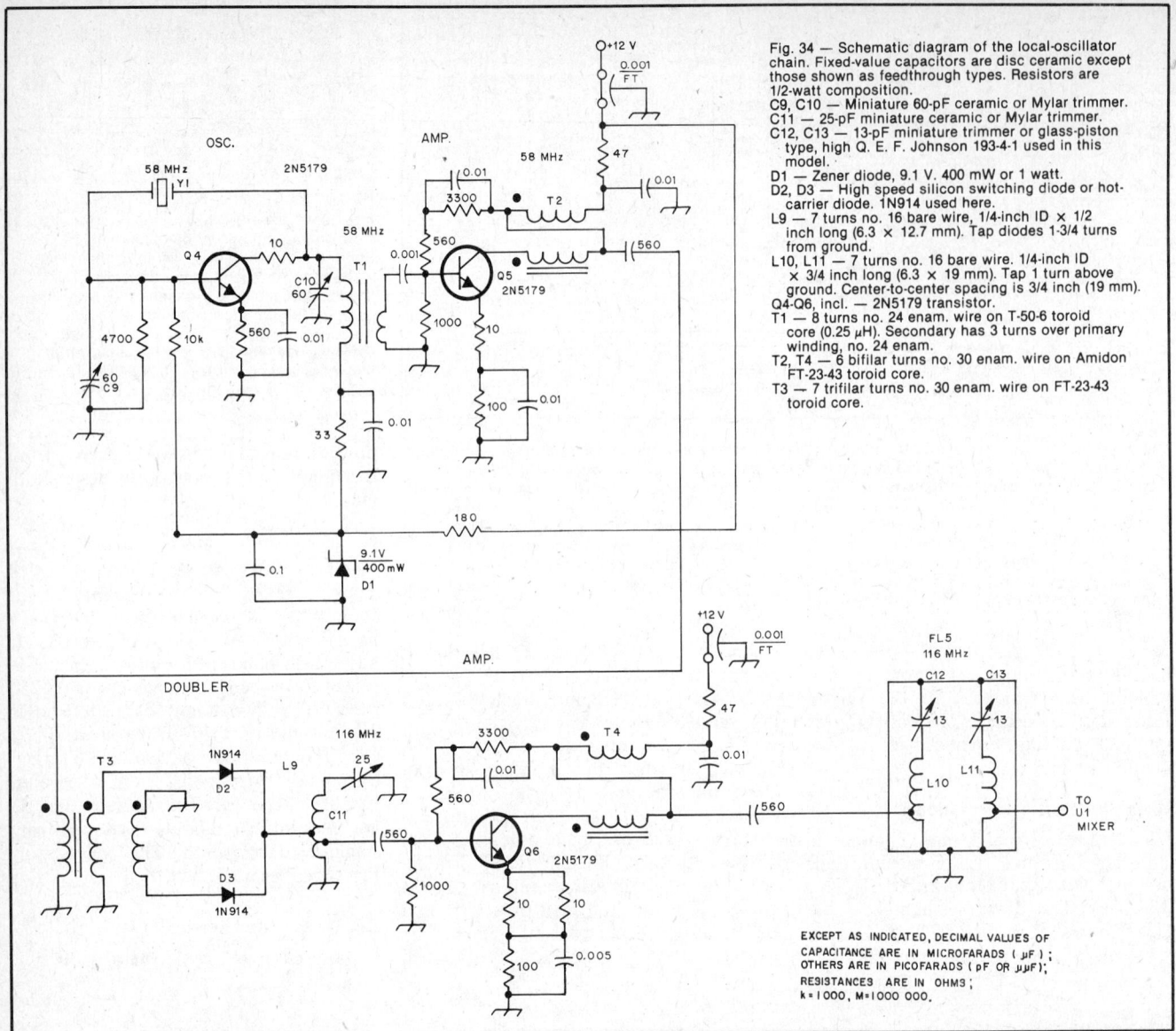

Fig. 34 — Schematic diagram of the local-oscillator chain. Fixed-value capacitors are disc ceramic except those shown as feedthrough types. Resistors are 1/2-watt composition.
C9, C10 — Miniature 60-pF ceramic or Mylar trimmer.
C11 — 25-pF miniature ceramic or Mylar trimmer.
C12, C13 — 13-pF miniature trimmer or glass-piston type, high Q. E. F. Johnson 193-4-1 used in this model.
D1 — Zener diode, 9.1 V. 400 mW or 1 watt.
D2, D3 — High speed silicon switching diode or hot-carrier diode. 1N914 used here.
L9 — 7 turns no. 16 bare wire, 1/4-inch ID × 1/2 inch long (6.3 × 12.7 mm). Tap diodes 1-3/4 turns from ground.
L10, L11 — 7 turns no. 16 bare wire. 1/4-inch ID × 3/4 inch long (6.3 × 19 mm). Tap 1 turn above ground. Center-to-center spacing is 3/4 inch (19 mm).
Q4-Q6, incl. — 2N5179 transistor.
T1 — 8 turns no. 24 enam. wire on T-50-6 toroid core (0.25 μH). Secondary has 3 turns over primary winding, no. 24 enam.
T2, T4 — 6 bifilar turns no. 30 enam. wire on Amidon FT-23-43 toroid core.
T3 — 7 trifilar turns no. 30 enam. wire on FT-23-43 toroid core.

EXCEPT AS INDICATED, DECIMAL VALUES OF CAPACITANCE ARE IN MICROFARADS (μF); OTHERS ARE IN PICOFARADS (pF OR μμF); RESISTANCES ARE IN OHMS; k = 1000, M = 1000 000.

the beginning of the LO chain (Fig. 34). C9 can be adjusted to shift the oscillator frequency by a small amount. Greater range for netting the crystal can be obtained by inserting a small amount of inductive reactance between C6 and Y1.

A tuned toroidal transformer, T1, is adjusted for resonance at approximately 58 MHz. C10 is set for reliable starting of the oscillator (consistent with high output) when the operating voltage is turned on. Zener-diode regulation (D1) is used to aid oscillator stability.

Output from Q4 is routed to a fed-back, broadband, Class A amplifier, Q5. The feedback provides a 50-ohm input characteristic and contributes to excellent stability of the 58-MHz amplifier. Negative feedback and emitter degeneration are used at Q5 and Q6 for this purpose. T2 is a broadband 4:1 toroidal transformer. It transforms the collector impedance of Q5 to 50 ohms. Output is taken at this point and fed to a separate module which contains T3 through FL5.

The output from Q5 is used to drive a two-diode balanced doubler, D2 and D3. A trifilar-wound broadband toroidal transformer (T3) couples the energy to the doubler. L9 and C11 comprise a 116-MHz resonator which serves as an impedance transformer between the diodes and Q6. This resonator also suppresses energy at other than 116 MHz. The diodes and the base of Q6 are tapped close to ground on L9 to effect an impedance match. The stage gain from L9 to the output of FL5 was measured as 16 dB. The gain is needed to ensure a +7-dBm injection level at U1, the mixer.

Another broadband step-down transformer, T4, is used to effect an impedance match. It transforms the collector impedance of Q6 to the 50-ohm input impedance of band-pass filter FL5. The latter greatly suppresses the 58-MHz energy passing through Q6. It also rejects the harmonics of the 116-MHz LO chain. Fig. 35 shows the spectral output of the LO chain as viewed on an HP analyzer. It can

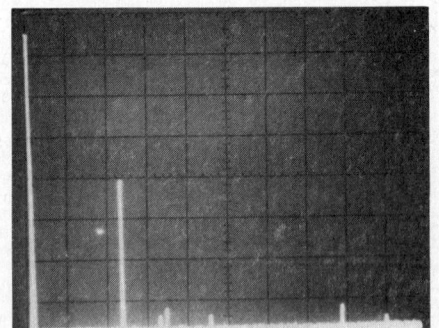

Fig. 35 — Spectral output of the 116-MHz LO chain showing all spurious responses at −72 dB or greater. The carrier has been suppressed by means of a 116-MHz trap to prevent front-end overload of the analyzer. In effect, it is at full scale. The vertical line at the far left is a zero-reference response from within the analyzer. Vertical scale is 10 dB/div. and the horizontal scale is 50 MHz/div.

be seen that all spurious responses are 72 dB or greater below the desired 116-MHz energy level. Imagine that the carrier amplitude is full scale when comparing the levels of the spurs. The 116-MHz carrier has been suppressed by means of a trap to prevent front-end overloading of the analyzer. This has no effect on the accuracy of the spur-level readings. The full-scale line at the far left in Fig. 35 is the zero-reference blip from the analyzer. It should be ignored.

Q4, Q5 and the related circuitry are contained in a module made from single-sided pc board. The copper foil is on the inside of the box. D2, D3, Q6 and related components are in a separate container. Double-sided pc board is used.

Construction Data

Modular construction is used in this design so that various portions of the circuit can be isolated from one another in an effective manner. Another benefit to this style of construction is that the builder can experiment with other circuits (substitutes) and install them without disrupting the complete converter. Most of the circuit boards are much larger than necessary. This was done with a view toward possible revisions to the circuits of the first model. Those who are skilled at layout work may want to compress the circuits somewhat.

Small Teflon press-fit feedthrough bushings are used as input and output terminals for the various modules. They were purchased at a flea market, so the original source is not known. Any low-loss miniature bushing should be suitable as a substitute. Alternatively, the RG-174 miniature coaxial cables can be routed directly into and out of the modules for connection to the circuits. For attachment to the +12-volt supply, each module has 0.001-μF feedthrough capacitors. These components are mounted on the box walls. Small Teflon feedthrough bushings are used at Q1 and Q2 to permit circuit connections for the FET leads inside the module. The FETs are installed on the outer wall of the front-end module.

All of the modules are affixed to the main chassis 7 × 11 × 2 inches (180 × 280 × 50 mm) by means of no. 6 spade bolts. The latter are bolted to the side walls of the modules. The rf amplifier assembly uses the chassis surface as part of the box shield. An aluminum plate is attached to the side wall of the module by means of spade bolts. This type of construction was used to ease assembly of the front end filter and amplifiers. Spring-brass fingers are soldered to the compartment dividers to assure solid contact to the aluminum side plate. The aperture size between L2 and L3 of FL2 is 1 × 5/8 inch (25 × 16 mm). Each resonator compartment measures 1-1/4 × 1-1/4 × 2 inches (32 × 32 × 50 mm). The coils are centered in the compartment.

Dimensions for the mixer/post-mixer amplifier assembly are 5-3/4 × 1-7/8

inches (155 × 50 mm), length and width. The box height is 1-1/2 inches (38 mm). This container and the one for the low-level stages of the LO chain do not have copper on the outside. Therefore it is necessary to provide a grounding contact for the press-fit U-shaped aluminum covers. Shim-brass or flashing-copper strips are located opposite one another on the upper lips of the side walls. The strips are approximately 1-1/2 inches long (38 mm) and 3/4 inch (19 mm) wide. They are soldered to the inner walls of the box, then bent over the edges and down the outside of the box walls. This provides a ground contact for the box covers.

Single-sided glass-epoxy pc-board material is used for all of the etched circuits. A ground-bus copper strip is retained around the entire perimeter of each board. This permits the builder to solder the boards into the shield boxes.

The low-level section of the LO chain is housed in a box which is 5-1/4 inches (135 mm) long, 2-1/4 inches (58 mm) wide and 2 inches (51 mm) high. The last half of the LO circuit is contained in an enclosure which is 4-1/8 inches (105 mm) long, 1-1/2 inches (38 mm) wide and 1-1/4 inches (32 mm) high. The end compartment houses the band-pass filter. It is 1-1/2 × 1-1/2 inches (38 × 38 mm) square. The depth of the compartment is 1 inch (25 mm). The lower surface of the filter compartment is part of the main pc board, the end of which has not been stripped of copper.

The inner and outer copper surfaces of the modules which use double-clad pc board sides should be connected together to assure electrical contact. This can be done by running short lengths of bare wire through the box walls at four points per wall, then soldering the wires in place on each surface (inner and outer). Alternatively, angle stock can be made of flashing copper and soldered across each corner of the box. The U-shaped lids will complete the electrical contacts when installed.

Alignment

Choose a 200-kHz segment of the 2-meter band that suits your purposes. Place a 2-meter signal at the midpoint of that tuning range (144.1 MHz for coverage from 144.0 to 144.2 MHz). Connect the converter to a receiver tuned and peaked at 28.1 MHz. Apply power to the converter and locate the 2-meter signal in the receiver tuning range. It should fall close to 28.1 MHz. Observe the receiver S meter and adjust all of the converter tuned circuits for maximum meter reading. This procedure should be repeated two or three times, using the weakest 2-meter signal that will provide needle deflection on the S meter.

This circuit should provide a noise figure of less than 4 dB with careful adjustment. C1 of Fig. 31 may be required

to achieve this result. It was not used in the ARRL model because accurate noise-figure measuring equipment was not available. However, without the input capacitor, a 0.1-μV signal into the converter provided a loud cw response in the tunable i-f receiver, roughly equivalent to an RST of 559. The test receiver was a Kenwood TS-820 with a 500-Hz i-f filter. A similar front end was built earlier, using 2N4416 FETs. When used with a 2-meter fm receiver it provided 20 dB of quieting with a 0.18 μV input signal.

Power Supply

A well-filtered 12-volt regulated dc supply is recommended for use with this converter. The maximum current required is less than 100 mA. A supply using a three-terminal regulator IC would be ideal.

INTERDIGITAL CONVERTER FOR 1296 OR 2304 MHZ

In a world where rf spectrum pollution is becoming more serious, even into the microwave region, it is almost as important to keep unwanted signals out of a receiver as it is to prevent radiation of spurious energy. An interdigital filter was described some years ago, featuring low insertion loss, simplicity of construction, and reasonable rejection to out-of-band signals. It could be used in either transmitters or receivers.

This twice-useful principle has now been put to work again — as a mixer. Again, the ease of construction and adaptation leads many to wonder that it had not been thought of before. It was first described by W2CQH in January 1974 QST.

A Filter and Mixer

A layout of the microwave portions of both converters is shown in Fig. 36. The structure consists of five interdigitated round rods, made of 3/8-inch (9.5 mm) OD brass or copper tubing. They are soldered to two sidewalls and centrally located between two ground-planes made of 1/16-inch (1.6 mm) sheet brass or copper-clad epoxy fiberglass. One ground plane is made larger than the microwave assembly and thus provides a convenient mounting plate for the remainder of the converter components.

The sidewalls are bent from 0.032-inch thick sheet brass or they can be made from 1/4 × 3/4-inch (6 × 19 mm) brass rod. One edge of each sidewall is soldered to the larger ground plane. The other edge is fastened to the smaller ground plane by 4-40 machine or self-tapping screws, each located over the centerline of a rod. The sidewall edges should be sanded flat, before the ground plane is attached, to assure continuous electrical contact. Note that no end walls are required since there are no electric fields in these regions.

Electrically, rods A, B, and C comprise a one-stage, high-loaded-Q ($Q_L = 100$),

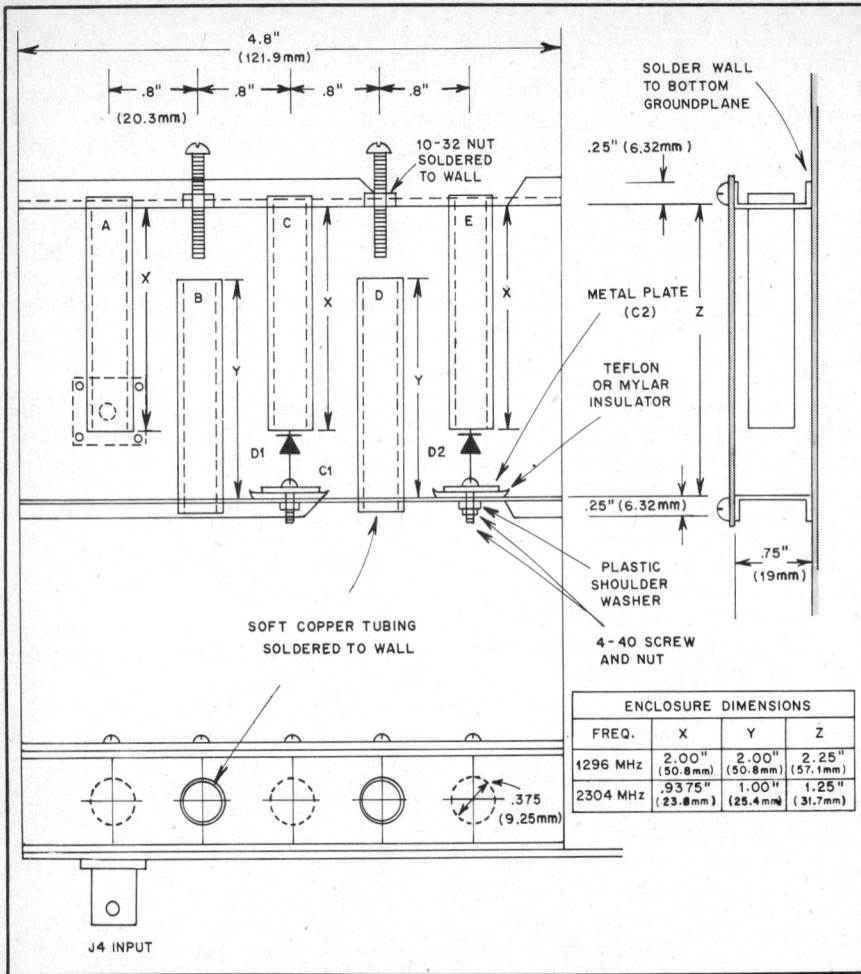

Fig. 36 — Dimensions and layout for the filter and mixer portions of the interdigital converters. The signal input is to the left rod, labeled "A." Local-oscillator injection is through the diode to rod "E." D1 is the mixer diode, connected to the center rod in the assembly.

ENCLOSURE DIMENSIONS

FREQ.	X	Y	Z
1296 MHz	2.00" (50.8mm)	2.00" (50.8mm)	2.25" (57.1mm)
2304 MHz	.9375" (23.8mm)	1.00" (25.4mm)	1.25" (31.7mm)

Table 4

Converter Specifications

	1296 MHz	2304 MHz
Noise figure	5.5 dB	6.5 dB
Conversion gain	20 dB	14 dB
3-dB bandwidth	2 MHz	7 MHz
Image rejection	18 dB	30 dB
I-f output	28 MHz	144 MHz

interdigital filter[6] which is tuned to the incoming signal frequency near 1296 or 2304 MHz. The ungrounded end of rod A is connected to a BNC coaxial connector filter input. Rod B is the high-Q resonator and is tuned by a 10-32 machine screw. Rod C provides the filter output-coupling section to the mixer diode, D1.

The original mixer diode was a Hewlett-Packard 5082-2577 Schottkey-barrier type which is no longer available. The 5082-2817 and MA-4853 (Microwave Associates) are recommended substitutes. The cheaper 5082-2853 can be used instead, but this substitution will increase the 2304-MHz mixer noise figure by approximately 3 dB.

One pigtail lead of the mixer diode is tack-soldered to a copper disc on the ungrounded end of rod C. Care should be taken to keep the pigtail lead as short as possible. If rod C is machined from solid brass stock, then it is feasible to clamp one of the mixer-diode leads to the rod end with a small setscrew. This alternative method facilitates diode substitution and was used in the mixer models shown in the photographs.

Fig. 36 also shows that the other end of D1 is connected to a homemade 30-pF bypass capacitor, C1, which consists of a 1/2-inch-square copper or brass plate clamped to the sidewall with a 4-40 machine screw. The dielectric material is a small sheet of 0.004-inch (0.1 mm) thick Teflon or mylar. A 4-40 screw passes through an oversize hole and is insulated from the other side of the wall by a small plastic shoulder washer.

In the first converter models constructed by the author and shown in the photographs, C1 was a 30-pF button mica unit soldered to the flange of a 3/8-inch diameter threaded panel bearing (H. H. Smith No. 119). The bearing was then screwed into a threaded hole in the sidewall. This provision made it convenient to measure the insertion loss and bandwidth of the interdigital filters since the capacitor assembly could be removed and replaced with a BNC connector.

Rods C, D, and E comprise another high loaded-Q (Q_L = 100) interdigital filter tuned to the local oscillator (LO)

Fig. 37 — The converter for 1296 MHz. This unit was built by R.E. Fisher, W2CQH. While the mixer assembly (top center) in this model has solid brass walls, it can be made from lighter material, as explained in the text and shown in Fig. 36. The i-f amplifier is near the center, just above the mixer-current-monitoring jack, J1. A BNC connector at the lower left is for 28-MHz output. The local oscillator and multiplier circuits are to the lower right. Note that L6 is very close to the chassis, just above the crystal. The variable capacitor near the crystal is an optional trimmer to adjust the oscillator to the correct frequency.

[6]Fisher, "Interdigital Bandpass Filters for Amateur VHF/UHF Applications," *QST*, March 1968.

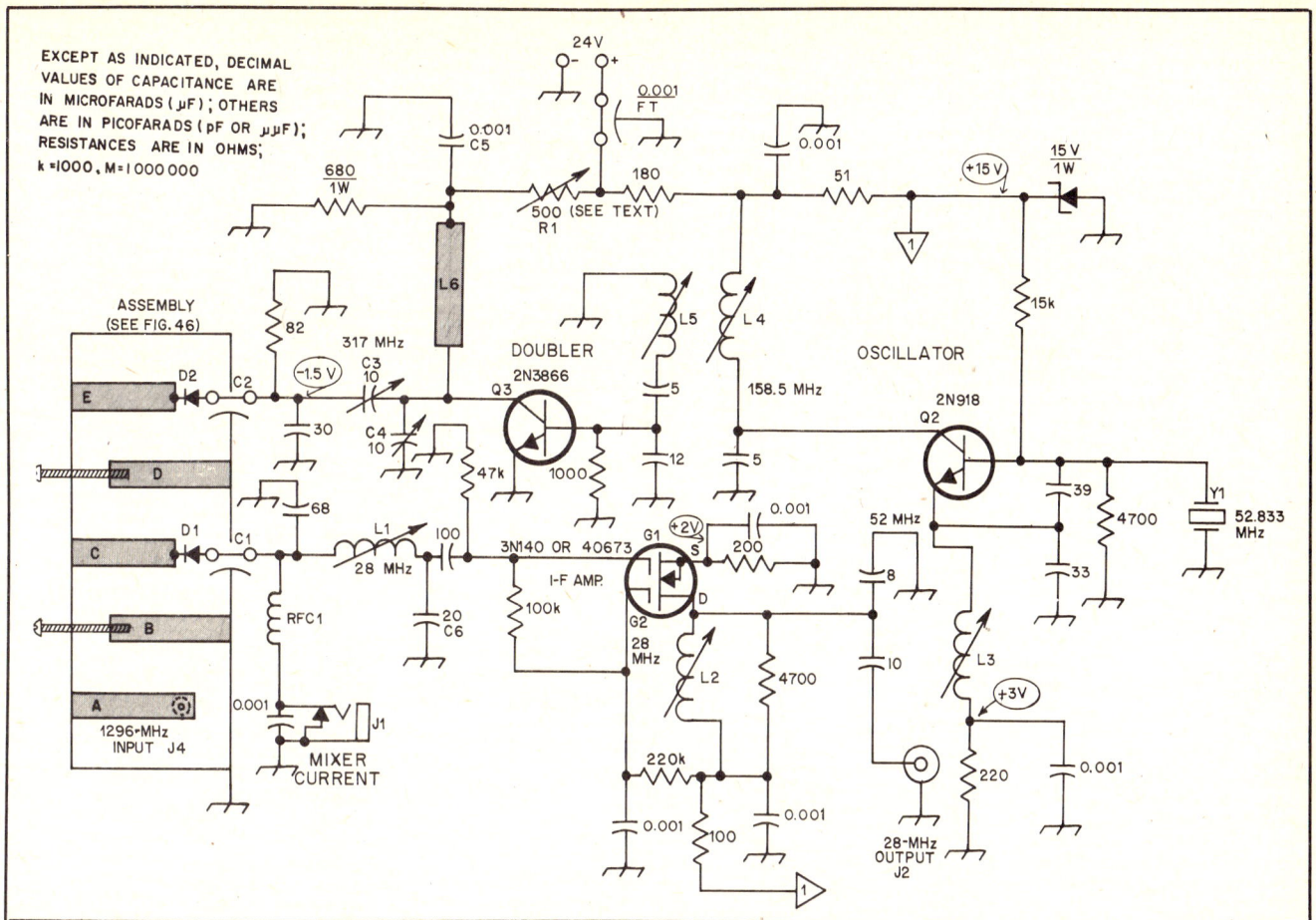

Fig. 38 — Schematic diagram of the 1296-MHz converter with oscillator and multplier sections included. Dimensions for the filter and mixer assembly are given in Fig. 36.

C1, C2 — 30-pF homemade capacitor. See text and Fig. 36.
C3, C4 — 0.8- to 10-pF glass trimmer, Johanson 2945 or equiv.
C5 — 0.001-μF button mica.
C6 — 2- to 20-pF air variable, E.F. Johnson 189-507-004 or equiv.

D1 — Hewlett Packard 5082-2817 or 5082-2835.
D2 — Hewlett Packard 5082-2811 or 5082-2835.
J1 — Closed-circuit jack.
J2 — Coaxial connector, type BNC acceptable.
L1, L2 — 18 turns no. 24 enam. on 1/4-inch (6.3-mm) OD slug-tuned form (1.5 μH nominal).

L3 — 10 turns like L1 (0.5 μH).
L4, L5 — 6 turns like L1 (0.2 μH).
L6 — Copper strip, 1/2-inch wide × 2-1/2 inches (12.7 × 63.5 mm) long. See text and photographs.
RFC1 — 33 μH, J.W. Miller 74F33SA1 or equiv.

frequency. This filter passes only the fourth harmonic (1268 or 2160 MHz) from the multiplier diode, D2. The two filters have a common output-coupling section (rod C) and their loaded Qs are high enough to prevent much unwanted coupling of signal power from the antenna to the multiplier diode and LO power back out to the antenna.

The multiplier diode is connected to the driver circuitry through C2, a 30-pF bypass capacitor identical to C1. D2 is a Hewlett-Packard 5082-2811 although the 5082-2835 works nearly as well. Fifty milliwatts drive at one-quarter of the LO frequency is sufficient to produce 2 mA of mixer diode current, which represents about 1 milliwatt of the local-oscillator injection. A Schottky-barrier was chosen over the more familiar varactor diode for the multiplier because it is cheaper, more stable, and requires no idler circuit.

Fig. 38 shows the schematic diagram of the 1296 to 28 MHz converter. All components are mounted on a 7 × 9-inch

(178 × 229 mm) sheet of brass or copper-clad epoxy-fiberglass board. As mentioned earlier, this mounting plate also serves as one ground plane for the microwave mixer. When completed, the mounting plate is fastened to an inverted aluminum chassis which provides a shielded housing.

Oscillator and Multipliers

The nonmicrowave portion of the converter is rather conventional. Q1, a dual-gate MOSFET, was chosen as the 28-MHz i-f amplifier since it can provide 25 dB of gain with a 1.5 dB noise figure. The mixer diode is coupled to the first gate of Q1 by a pi-network matching section. It is most important that the proper impedance match be achieve between the mixer and i-f amplifier if a low noise figure is to be obtained. In this case, the approximately 30-ohm output impedance of the mixer must be stepped up to about 1500 if Q1 is to yield its rated noise figure of 1.5 dB. It is for this reason

that a remote i-f amplifier was not employed, as is the case with many contemporary uhf converters.

Q2 functions in an oscillator-tripler circuit which delivers about 10 milliwatts of 158.5-MHz drive to the base of Q3. The emitter coil, L3, serves mainly as a choke to prevent the crystal from oscillating at its fundamental frequency. Coils L4 and L5, which are identical, should be spaced closely such that their windings almost touch.

Q3 doubles the frequency to 317 MHz, providing about 50 milliwatts drive to the multiplier diode. It is important that the emitter lead of Q3 be kept extremely short; 1/4-inch (6.3 mm) is probably too long. L6, the strip-line inductor in the collector circuit of Q3, consists of a 1/2 × 2-1/2-inch (12.7 × 63.5-mm) piece of flashing copper spaced 1/8-inch (3-mm) above the ground plane. The cold end of L6 is bypassed to ground by C5, a 0.001-μF button mica capacitor.

The multiplier circuits are tuned to

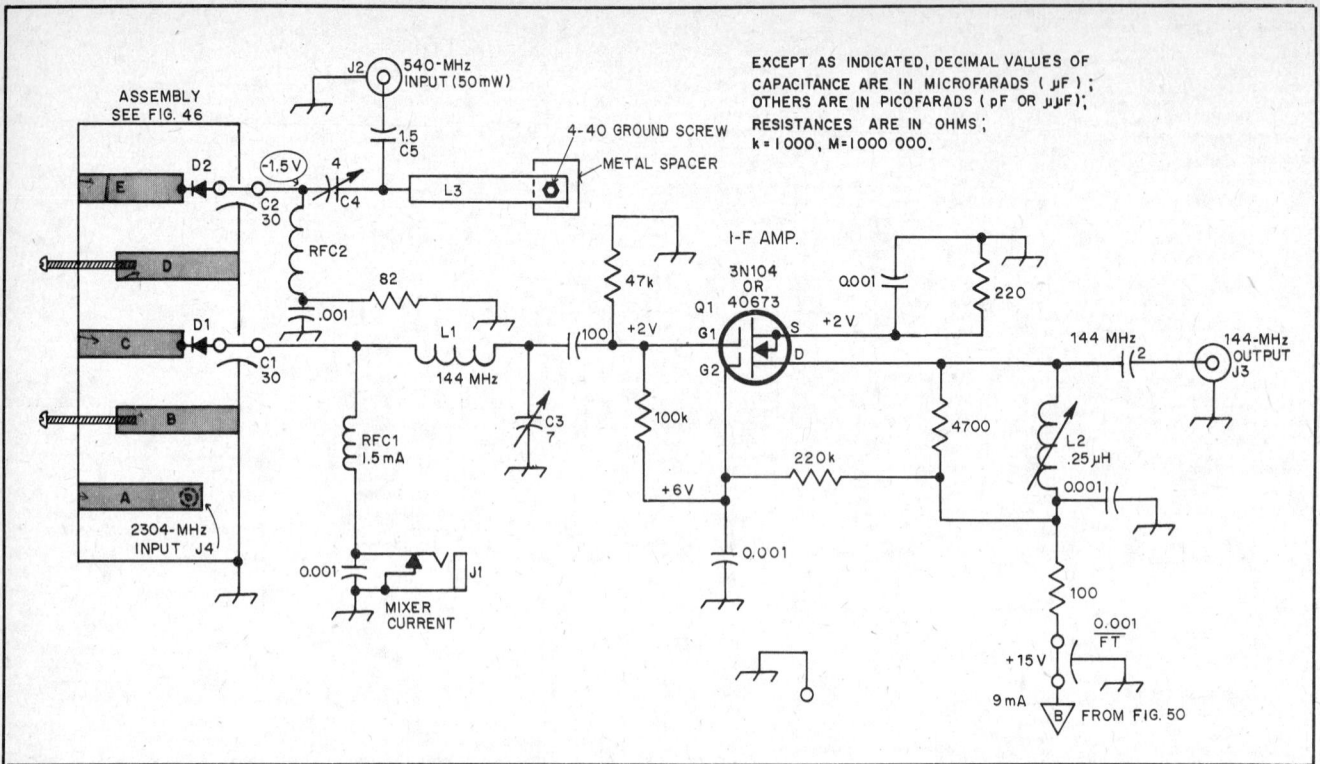

Fig. 39 — Schematic diagram of the 2304-MHz version of the converter, with the i-f amplifier. The oscillator and multiplier circuits are constructed separately.

C1, C23 — 30-pF homemade capacitor (see text).
C3, C4, C5 — 0.8- to 10-pF glass trimmer, Johanson 2945 or equiv.
D1 — Hewlett Packard 5082-2817 or 5082-2835.

D2 — Hewlett Packard 5082-2811 or 5082-2835.
J1 — Closed-circuit jack.
J2, J3, J4 — Coaxial connector, type BNC.
L1 — 5 turns no. 20 enam. 1/4-inch ID × 1/2-inch long. (6.3 × 12.7 mm).
L2 — 6 turns no. 24 enam. on 1/4-inch OD

slug-tuned form (0.25 μH).
L3 — Copper strip 1/2 × 2-11/16 inches (13 × 69 mm). See text and photographs.
RFC1 — Ohmite Z-144 or equiv.
RFC2 — Ohmite Z-460 or equiv.

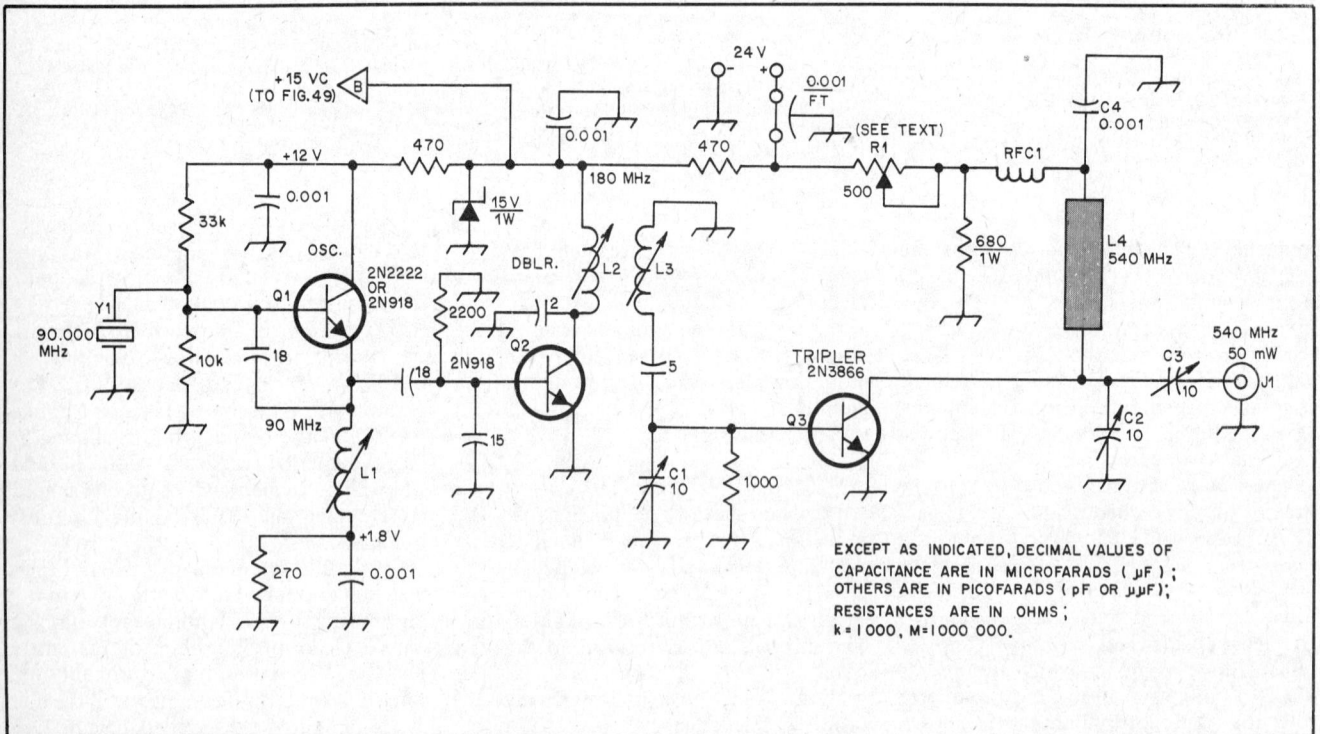

Fig. 40 — Schematic diagram of the oscillator and multiplier for the 2304 MHz converter. As explained in the text, a fixed-value resistor may be substituted for R1 after the value that provides proper performance has been found.

C1, C2, C3 — 0.8- to 10-pF glass trimmer, Johanson 2945 or equiv.
C4 — 0.001-μF button mica.
J1 — Coaxial connector, type BNC or equiv.

L1 — 10 turns no. 24 enam. on 1/4-in. OD slug-tuned form.
L2, L3 — 3 turns like L1.

L4 — Copper strip 1/2 × 1-1/2 in. (13 × 38 mm). Space 1/8 inch (3.2 mm) from chassis.
RFC1 — 10 turns no. 24 enam. 1/8-in. ID, close wound.

resonance in the usual manner by holding a wavemeter near each inductor being tuned. Resonance in the Q3 collector circuit is found by touching a VTVM probe (a resistor must be in the probe) to C2 and adjusting the Johanson capacitors until about −1.5 volts of bias is obtained. The 317- to 1268-MHz multiplier cavity is then resonated by adjusting the 10-32 machine screw until maximum mixer current is measured at J1. When resonance is found, R1 should be adjusted so that about 2 mA of mixer current is obtained. As an alternative to mounting a potentiometer in the converter, once a value of resistance has been found that provides correct performance it can be measured and the nearest standard fixed-value resistor substituted. Some means of adjusting the collector voltage on the multiplier stage must be provided initially to allow for the nonuniformity of transistors.

A 2304-MHz Version

Fig. 39 and 40 show the schematic diagrams of the 2304-MHz converter and multiplier. The mixer and i-f preamplifier was built on a separate chassis since, at the time of their construction, a multiplier chain from another project was available. An i-f of 144 MHz was chosen, although 50 MHz would work as well. An i-f output of 28 MHz, or lower, should not be used since this would result in undesirable interaction between the mixer and multiplier interdigital filters.

The 2304-MHz mixer and i-f amplifier section, shown in Fig. 39, is very similar to its 1296-MHz counterpart. Q1, the dual-gate MOSFET, operates at 144-MHz and thus has a noise figure about 1-dB higher than that obtainable at 28 MHz.

The multiplier chain, Fig. 40, has a separate oscillator for improved drive to the 2N3866 output stage. Otherwise the circuitry is similar to the 1296-MHz version.

References

Fisk, "Double-Balanced Mixers," *Ham Radio,* March 1968.
Ress, "Broadband Double-Balanced Modulator," *Ham Radio,* March 1970.

Chapter 10

Mobile, Portable and Emergency Equipment

A major justification for the existence of Amateur Radio in the USA is to provide a pool of experienced operators in time of national or community need. When the call for emergency communications is voiced by cities, towns, counties, states or the federal government, mobile and portable radio equipment is pressed into service where needed. Aside from the occasional need for disaster and emergency communications provided by amateurs, a great deal of pleasure and challenge can reward the amateur when operating portable or mobile under normal conditions. In this regard, most mobile operation is carried out today by means of narrow-band fm and repeaters. The major repeater frequencies are 146,

Fig. 1 — This may represent the ultimate in portable vhf operation. The aggregate antenna gain represented by this setup could be very effective in time of emergency. This installation was built and operated by N6NB/K6YNB, who with a similar system, set a single-operator national record during a vhf contest, earning more than 60,000 points. The antenna shown is effective also for EME work.

220 and 440 MHz. It is expected that this reliable service mode will soon include widespread occupancy of the 1215-MHz band and higher.

Mobile hf-band operation still appeals to numerous amateurs because it eliminates the constrictions imposed by vhf and uhf repeaters, their operators and their normal coverage contours. When operating mobile on ssb or cw with hf transceivers, worldwide contacts are possible for those who enjoy that style of communication.

High-power mobile operation has become entirely practical from the automotive 13.6-volt dc battery system. This results from the use of transistorized equipment. With only two transistors, a linear power amplifier can deliver 150 watts or more of rf output. For the most part, dynamotors, vibrator packs and dc-to-dc converters are things of the past, making mobile and portable operation much less expensive and much more efficient.

QRP Operation

Low-power operation has taken a significant jump forward in recent years and the ardent core of the movement almost qualifies as a cult. The basic concept is to do things the "hard way," proving that power levels of less than 10 watts are entirely effective when reasonably good antennas are used. QRP is a relative term. To the station running 1 watt, 10 watts is QRO (high power). It is the equivalent of a 100-watt station using a kilowatt amplifier for a 10-dB gain. For operating awards and contests, the ARRL definition of QRP is 10 watts input (or 5 watts measured output). The expression "QRPp" has been adopted by some low-power enthusiasts to mean "very low power." It is not recognized by ARRL.

It should be said that QRP in its proper use as an International Q Code symbol means "Decrease power" or "Shall I

decrease power?" Thus, going from 800 watts to 500 watts is an act of QRP. Conversely, QRO means "increase power."

Power levels as low as a few milliwatts are often suitable for emergency work when the cw mode is employed in the hf bands. Similarly, ssb and fm transmissions on vhf and uhf at the mW level are effective over line-of-sight paths. The advantage of this type of QRP equipment is long-term operation from batteries. Numerous examples of equipment designed for QRP work are provided in the League's book, *Solid State Design for the Radio Amateur.*

Electrical-Noise Elimination

One of the most significant deterrents to effective signal reception during mobile or portable operation is electrical impulse noise from the automotive ignition system. The problem also arises during the

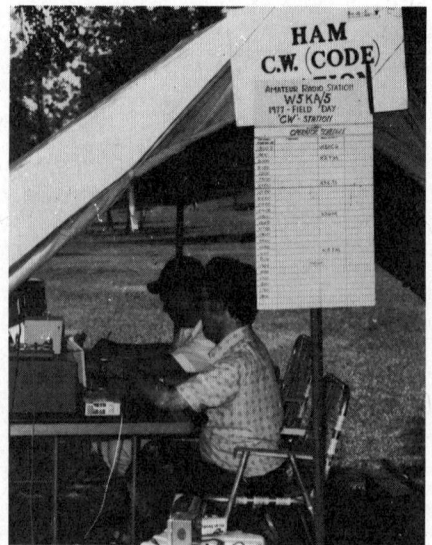

Fig. 2 — WB5MKX and K5GM demonstrate the practicality of portable operation from a remote site as they operate W5KA/5 during an ARRL Field Day contest.

use of gasoline-powered portable ac generators. This form of interference can completely mask a weak signal, thus rendering the station ineffective. Most electrical noise can be eliminated by taking logical steps toward suppressing it. The first step is to clean up the *noise source* itself, then utilize the receiver's built-in noise-reducing circuit as a last measure to knock down any noise pulses from passing cars, or from other man-made sources.

Spark-plug Noise

Most vehicles manufactured prior to 1975 were equipped with simple Kettering inductive-discharge ignition systems. A variety of noise-suppression methods were devised for these systems, including resistor spark plugs, clip-on suppressors, resistive high-voltage cable and even complete shielding. Resistive high-voltage cable and resistor plugs provide the greatest noise reduction for the least expense and effort. While almost all vehicles produced after 1960 had resistance cable as standard equipment, such cable develops microscopic cracks in the insulation and segmentation of the conductor after a few years of service. These defects can defeat the suppression ability of the cable before the engine performance degrades noticeably. Two years is a reasonable replacement interval for spark-plug cable. Older editions of this Handbook described complete shielding methods for inductive-discharge ignitions.

Late-model automobiles employ sophisticated high-energy electronic ignition systems in an attempt to reduce exhaust pollution and increase fuel mileage. With increased sophistication comes greater sensitivity to modification — solutions to RFI caused by older Kettering systems cannot be uniformly applied to the modern electronic ignitions.

Such fixes may be ineffective at best, and at worst may impair the engine performance. One should thoroughly understand an ignition system before attempting to modify it. One of the significant features of capacitive discharge systems, for example, is extremely rapid voltage rise, which combats misfire caused by fouled spark plugs. Rapid voltage rise depends on a low RC time constant presented to the output transformer. For this reason, high-voltage suppression cable designed for capacitive-discharge systems is wound with monel wire. It exhibits a distributed resistance of only about 600 ohms per foot, as contrasted with 10 kilohms per foot for carbon-center cable used with inductive-discharge systems. Increasing the RC product by shielding or installing improper spark-plug cable could seriously compromise the capacitive-discharge circuit operation.

Ferrite beads represent a possible means for RFI reduction in newer vehicles. Both primary and secondary ignition leads are candidates for beads. Install them liberally, then load test the engine for adequate spark energy.

The plane sheet metal surfaces and cylindrical members such as exhaust pipes often exhibit resonance in one of the amateur bands. Such resonances encourage reradiation of spark impulse energy. Bonding these structural members with heavy braid can reduce the level of spark noise inside the vehicle and out. Other types of noise to be described later can also be helped by bonding. Here are the main areas to bond. 1) Engine to frame; 2) air cleaner to engine block; 3) exhaust lines to frame; 4) battery ground terminal to frame; 5) hood to fire wall; 6) steering column to frame; 7) bumpers to frame; and 8) trunk lid to frame.

In the United States, automobile manufacturers voluntarily comply with an RFI standard devised by the Society of Automotive Engineers, SAE J551, most recently revised in May 1980. Basically, the standard requires that the strength of the field radiated by a random vehicle be no greater than 14 dB above 1 microvolt per meter per kilohertz from 20 to 80 MHz, sloping up to 25 dB in the 400- to 1000-MHz range. The receiving antenna is located 33 feet (10 m) from the vehicle and is oriented for measurements in both the vertical and horizontal planes. SAE J551 has been adopted as law in Canada. Automotive RFI standards are also under active development in Europe. The group studying the situation there is CISPR (Comite International Special des Perturbations Radioelectrique). The SAE is cooperating with CISPR to promote international standardization. FCC docket 20654 is a Notice of Inquiry concerning the automotive RFI issue. The manufacturers and users of radios in the land mobile service (just above the 2-meter band), represented by the Electronics Industries Association, responded to the NOI by stating that even vehicles complying with J551 can seriously degrade mobile vhf communications. SAE J551 was never intended to protect vehicular receivers — only fixed receivers located near roadways.

On the whole, modern automobiles are cleaner from an RFI standpoint than those of 20 years ago. The interference problem, however, at least at vhf and uhf, persists because present-day receivers are about 10 dB more sensitive than their predecessors.

Useful tips for solving ignition noise problems can be found in *Giving Two-way Radio Its Voice,* published by the Champion Spark Plug Company. This well-prepared publication covers noise-suppression fundamentals, preliminary procedures, methods for pinpointing interference, and techniques for solving noise problems in automotive and marine environments. A copy of this booklet can be obtained for $1 from Champion Spark Plug Co., Box 910, Toledo, OH 43661. Some automotive parts distributors also stock this publication.

Charging-System Noise

Noise from the vehicular battery-charging circuit can interfere with both transmission and reception of radio signals. The charging system of a modern automobile consists of a belt-driven three-phase alternator and a solid-state voltage regulator. Solid-state regulators are a great improvement over the electromechanical vibrating contact units of earlier times. Interference from the charging system can affect a receiver in two ways: Rf radiation can be picked up by the antenna, and noise can be conducted directly into the circuits via the power cable. "Alternator whine" is a common form of conducted interference. It has the greatest effect on vhf fm communications, because synthesized carrier generators and local oscillators are easily frequency modulated by power supply voltage fluctuations. The alternator ripple is most noticeable when transmitting, because the machine is more heavily loaded in that condition. If the ripple amplitude is great enough, alternator whine will be imparted to all incoming signals by the LO.

Conducted noise can be minimized by connecting the radio power leads directly to the battery, as this is the point in the electrical system having the lowest impedance. If the regulator is adjustable, set the voltage no higher than is necessary to ensure complete battery charging. Radio equipment manufacturers combat voltage variations by internally regulating critical circuits wherever possible.

Both conducted and radiated noise can be suppressed by filtering the alternator leads. Coaxial capacitors (about 0.5 μF) are suitable, but *don't connect a capacitor to the field.* The field lead can be shielded or loaded with ferrite beads if necessary. A parallel-tuned LC trap in this lead may be effective against radiated noise. Such a trap in the output lead must be made of no. 10 wire or larger, as some alternators conduct up to 100 amperes. The alternator slip rings should be kept clean to prevent excess arcing. An increase in "hash" noise may indicate that the brushes need to be replaced.

Instrument Noise

Some automotive instruments are capable of creating noise. Among these gauges and senders are the heat- and fuel-level indicators. Ordinarily, the addition of a 0.5 μF coaxial capacitor at the sender element will cure the problem.

Other noise-generating accessories are turn signals, window-opener motors, heating-fan motors and electric windshield-wiper motors. The installation of a 0.25-μF capacitor will usually eliminate their interference noise.

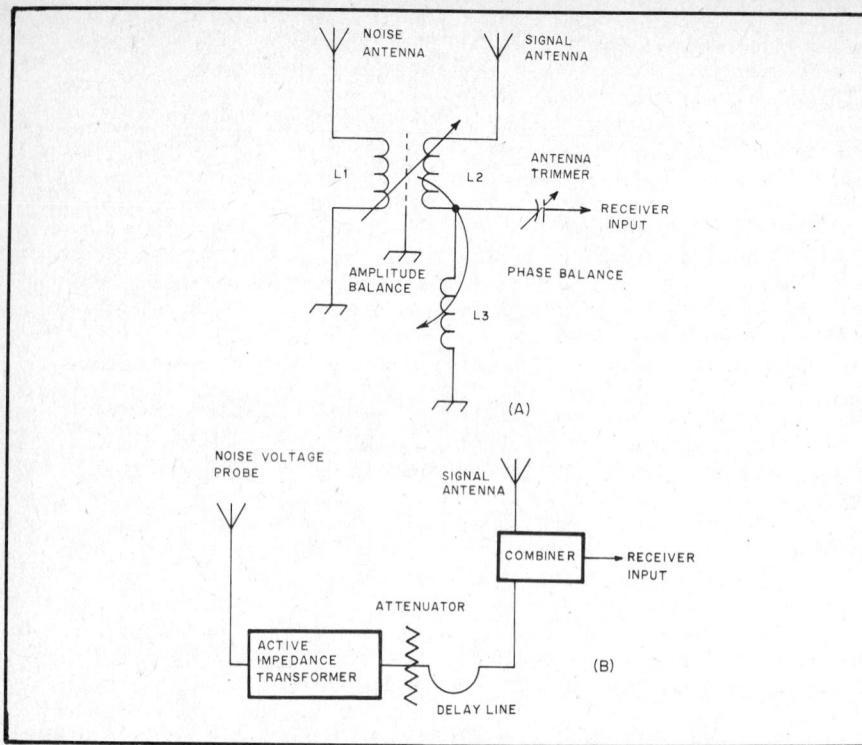

Fig. 3 — Automotive noise-cancelling systems. At A, the circuit used in the BC-342 hf receiver. At B, a suggested broadband noise-cancelling scheme.

Faraday shield between L1 and L2 ensures that the coupling is purely magnetic. The coupling between L2 and L3 is purely electrostatic. Adjusting the coil coupling causes the noise to null. The block diagram of Fig. 3B illustrates a more modern broadband approach to noise cancellation. A short wire near the ignition coil couples impulse energy into the active impedance transformer, which is simply an FET source follower stage. The amplitude and phase of the noise are controlled by the attenuator and delay line, respectively. The signal combiner can be a hybrid ferrite transformer at hf or a transmission line multicoupler at vhf.

BATTERY POWER

(The following material was assembled by Dave Geiser, WA2ANU.) The availability of solid-state equipment makes practical the use of battery power under portable or emergency conditions. Hand-held transceivers and instruments are obvious applications, but even fairly powerful transceivers (100 W or so output) may be practical users of battery power. Solid-state kilowatt mobile amplifiers exist, but these are intended for operating from an auxiliary battery that is constantly charged. The lower-power equipment can be powered from either of two types of batteries, the "primary" battery intended for one-time use, and the storage (or "secondary") battery that may be recharged many times.

A battery is a group of chemical cells, usually series connected to give some desired multiple of the cell voltage. Each assortment of chemicals used in the cell gives a particular nominal voltage, and this must be taken into account to make up a particular battery voltage.

Primary Batteries

The most common primary cell is the carbon-zinc flashlight type, in which chemical oxidation converts the zinc into salts and electricity. When there is no current flow, the oxidation stops until the next time current is required. Some chemical action does continue, so eventually stored batteries will degrade or dry out to the point where the battery will no longer supply the desired current. If this has happened without battery use, the time taken for the degradation is called shelf life.

The carbon-zinc battery has a nominal voltage of 1.5 volts, as does its "heavy duty" or "industrial" brother. These latter types are capable (for a given size) of producing more milliampere hours and less voltage drop than a carbon-zinc battery of the same size, and also have longer shelf life. Alkaline primary batteries have even better characteristics and will retain more capacity at low temperatures. Nominal voltage is 1.5 volts.

Lithium primary batteries have a nominal voltage of about 3 volts per cell

Corona-Discharge Noise

Some mobile antennas are prone to corona build-up and discharge. Whip antennas which come to a sharp point will sometimes create this kind of noise. This is why most mobile whips have steel or plastic balls at their tips. But, regardless of the structure of the mobile antenna, corona buildup will frequently occur during or just before a severe electrical storm. The symptoms are a high-pitched "screaming" noise in the mobile receiver, which comes in cycles of one or two minutes duration, then changes pitch and dies down as it discharges through the front end of the receiver. The condition will repeat itself as soon as the antenna system charges up again. There is no cure for this condition, but it is described here to show that it is not of origin within the electrical system of the automobile.

Electronic Noise Reduction

When all electrical noise generated within a vehicle has been eliminated, the mobile operator can be annoyed by RFI from passing vehicles. Some measures can be taken in the receiver to reduce or reject impulse noise. (Noise limiters and noise blankers are discussed in the hf receiving chapter.) The placement of a noise blanker in the receiver is important. The blanking circuit must be placed ahead of the sharp selectivity, otherwise the i-f filter will stretch the noise pulses, and they cannot be blanked without destroying a major portion of the received intelligence.

The Soviet "woodpecker" over-the-horizon radar has inspired some serious development work on noise blankers that don't degrade receiver dynamic range. Receivers for vhf fm service are generally designed for optimum noise figure at the expense of resistance to overload. Recent advances in rf amplifier design have proven that low noise figure and high dynamic range are not mutually exclusive. A high-performance noise blanker is useless if the front end of the receiver overloads on the noise pulses. A helical resonator at the receiver input affords some protection against noise overload because it restricts the total noise energy delivered to the front end.

Some fm receivers suffer from impulse noise because of inadequate a-m rejection. The cure for this ailment is to ensure hard limiting in the i-f stages and to use a detector that is inherently insensitive to amplitude variations.

Particularly troublesome vehicular impulse noise can sometimes be cancelled at the receiver input. The technique involves sampling the noise voltage from a separate "noise antenna" and adjusting its phase and amplitude to cancel the noise delivered by the "signal antenna." For this system to be effective, the signal antenna must be positioned to provide the best possible signal-to-noise ratio, and the noise antenna located close to the noise source and effectively shielded from the desired signal. Fig. 3A shows the noise cancellation circuit used in some models of the BC-342, a WW II receiver. The

and by far the best capacity, discharge, shelf-life and temperature characteristics. Their disadvantages are high cost and the fact that they cannot be readily replaced by other types in an emergency.

Silver oxide (1.5 V) and mercury (1.4 V) batteries are very good where nearly constant voltage is desired at low currents for long periods. Their main use (in subminiature versions) is in hearing aids, though they may be found in other quantity uses such as household smoke alarms.

Rechargeable or Storage Batteries

Many of the chemical reactions in primary batteries are theoretically reversible if current is passed through the battery in the reverse direction. For instance, zinc may be plated back onto the negative electrode of a zinc-carbon battery. Recharging of primary batteries should not be done for two reasons: It may be dangerous because of heat generated within sealed cells, and where there may be some success, both the charge and life are limited. In the zinc-carbon example, the zinc may not replate in the locations that had been oxidized. Pinholes in the case result, with consequent fluid leakage that will damage the using equipment.

One type of alkaline battery is rechargeable, and is so marked. If the recommended charging rate is not marked on such a battery, the manufacturer's advice should be asked. In a number of cases the manufacturer markets chargers and recommends that only those should be used.

The most-common small-storage battery is the nickel-cadmium type, with a nominal voltage of 1.2 V per cell. Carefully used, these are capable of 500 or more charge and discharge cycles, compared to 50 or so for alkaline types. The nickel-cadmium battery must not be fully discharged for best life. Where there is more than one cell in the battery, the most-discharged cell may suffer polarity reversal. All storage batteries have discharge limits, and nickel-cadmium types should not be discharged more than 1.2 V below nominal battery voltage.

The most widely used storage battery is the lead-acid type. In automotive service the battery is usually expected to discharge partially at a very high rate, and then to be recharged promptly while the alternator is also carrying the electrical load. If the conventional auto battery is allowed to discharge fully from its nominal 2 V per cell to 1.75 V per cell, only about 50 cycles of charge and discharge may be expected, with reduced storage capacity.

The most attractive battery for extended high-power electronic application is the so-called "deep-cycle" battery (intended for such use as powering electrical fishing motors and the accessories in recreational vehicles). The size 24 and 27 batteries furnish a nominal 12 volts and are about the size of small and medium automotive batteries. These batteries may furnish between 1000 and 1200 watt-hours per charge at room temperature, and when properly cared for may be expected to last more than 200 cycles. They often have lifting handles, screw terminals as well as the conventional truncated-cone automotive terminals, and may be fitted with accessories such as plastic carrying cases, with or without built-in chargers.

Lead-acid batteries are also available with jelled electrolyte. These types may be mounted in any position if sealed, but some are position-sensitive if vented.

Lead-acid batteries with liquid electrolyte usually fall into one of three classes — conventional with filling holes and vents, permitting the addition of distilled water lost from evaporation or during high-rate charge or discharge, "maintenance-free" from which gas may escape but water cannot be added, and sealed. Generally, the deep-cycle batteries have filling holes and vents.

Battery Capacity

The common rating of battery capacity is ampere hours, a product of current drain and time. The symbol "C" is commonly used; C/10, for example, would be the current available for 10 hours continuously. The value of C changes with the discharge rate and might be 110 at 2 amperes but only 80 at 20 amperes. Fig. 4 gives capacity-to-discharge rate for two standard-size lead-acid batteries. Capacity may vary from 35 mA hours for some of the small hearing-aid batteries to over 100 ampere hours for a size 27 deep-cycle storage battery.

The primary cells, being sealed, usually benefit from intermittent (rather than continuous) use. The resting period allows completion of chemical reactions needed to dispose of by-products of the discharge.

All batteries will fall in output voltage as discharge proceeds. "Discharged" condition for a 12-V lead-acid battery, for instance, should not be less than 10.5 volts. (It is also good to keep a running record of hydrometer readings, but the conventional readings of 1.265 charged and 1.100 discharged apply only to a long, low-rate discharge. Heavy loads may discharge the battery with little reduction in the hydrometer reading.)

Batteries that become cold have less of their charge available, and some attempt to keep a battery warm before use is worthwhile. The battery may lose 70% or more of its capacity at cold extremes but recover with warmth.

All batteries have some tendency to freeze, but those with full charges are less susceptible. A fully-charged lead-acid battery is safe to -30° F (-26° C) or colder.

Storage batteries may be warmed somewhat by charging. Blowtorches or other flame should never be used to heat any type of battery.

A practical limit of discharge occurs when the load will no longer operate satisfactorily on the lower output voltage near the "discharged" point. Much gear intended for "mobile" use may be designed for an average of 13.6 V and a peak of perhaps of 15 V, and not operate well below 12 V. For full use of battery charge the gear should operate well (if not at full power) on as little as 10.5 V with a nominal 12 to 13.6-V rating.

Somewhat the same condition may be seen in the replacement of carbon-zinc cells by nickel-cadmium storage cells. Eight of the former will give 12 V, while ten of the same-size nickel-cadmium units are required. If a 10-cell battery-holder is used, the equipment should be designed for 15 V in case the carbon-zinc units are plugged in.

Deep-cycle and nickel-cadmium storage batteries should be run to the end of their useful charge before recharging, and then charged fully. Both types will tolerate a light continual charge (trickle) and the sealed nickel-cadmium types tolerate a near-full charging rate continuously.

Discharge Planning

Transceivers usually drain a battery at two or three rates: one for receiving, one perhaps for transmit standby, and one for key-down or average voice transmit. Considering just the first and last of these (assuming the transmit standby equal to receive), average 2-way cw communication would require the low rate 3/4 of the time and the high rate 1/4 of the time. The ratio may vary somewhat with voice. The user may calculate the percentage of battery charge used in an hour by the combination (sum) of rates. If, for example, 20% of the battery capacity is used, the battery will provide 5 hours of communications per charge. In most actual

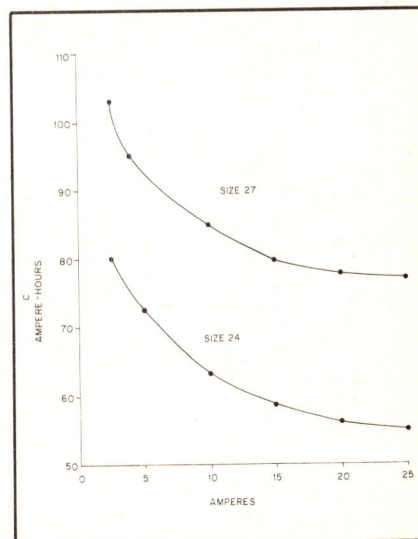

Fig. 4 — Output capacity as a function of discharge rate for two sizes of lead-acid batteries.

traffic and DX-chasing situations the time spent listening should be much greater than that spent transmitting.

Caring for Storage Batteries

In addition to the precautions given above, the following are recommended. (Your manufacturer's advice will probably be more applicable.)

Gas escaping from storage batteries *may be explosive*. Keep flame away.

Dry-charged storage batteries should be given electrolyte and allowed to soak for at least half an hour. They then should be charged at perhaps a 15 A rate for 15 minutes or so. The capacity of the battery will build up slightly for the first few cycles of charge and discharge, and then have fairly constant capacity for many cycles. Slow capacity decrease may then be noticed.

No battery should be subjected to unnecessary heat, vibration or physical shock. The battery should be kept clean. Frequent inspection for leaks is a good idea. Leaking or spraying electrolyte should be cleaned from the battery and surroundings. The electrolyte is chemically active and electrically conductive, and may ruin electrical equipment. Acid may be neutralized with sodium bicarbonate (baking soda), and alkalies may be neutralized with a weak acid such as vinegar. Both neutralizers will dissolve in water, and themselves be quickly washed off. Do not let any of the neutralizer enter the battery.

Keep a record of the battery usage, and include the last output voltage and (for lead-acid storage batteries) the hydrometer reading. This allows prediction of useful charge remaining, and the recharging or procuring of extra batteries, thus minimizing failure of battery power during an excursion or emergency.

Charging Storage Batteries

The rated full charge of a storage battery, C, is expressed in ampere-hours. Since no battery is perfect, more charge than this must be offered the battery for full-charge. If, for instance, the charge rate is 0.1 C (the "10-hour" rate), 12 or more hours may be needed for the charge.

Four common classes of charge rate are standby (or trickle), slow (or overnight), quick (or "rapid") and fast. The standby charge may be on the order of 30 to 100 mA for a C of 100 Ah, with the slow charge 10 A for the same C, the quick charge 30 A and the fast charge 100 A. Note that one battery is not designed for all of these charge rates. Deep cycle lead-acid and sealed nickel-cadmium cells are best charged at a slow rate, while automotive and some nickel-cadmium types may safely be given quick charges. (This depends on the amount of heat generated within each cell, and cell venting to prevent DX pressure build-up.) Some batteries have built-in temperature sens-

Fig. 5 — The WA7ARK mobile NiCad charger. The power cable has an in-line fuse and is terminated by a cigarette-lighter accessory plug.

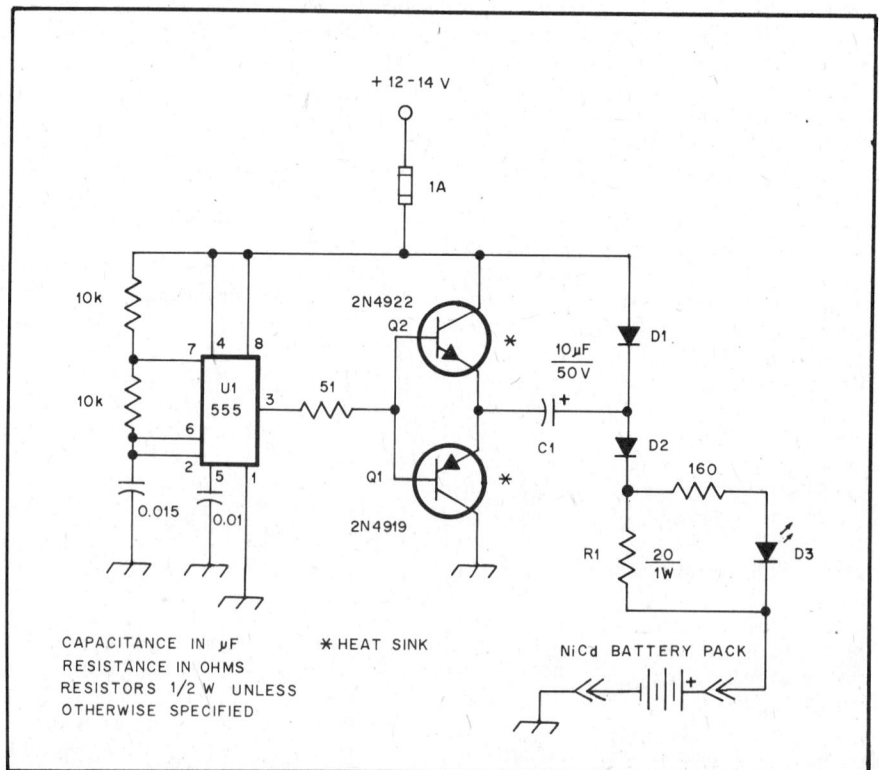

Fig. 6 — Schematic diagram for the mobile NiCad charger.
D1, D2 — Silicon rectifiers, 1 A, 50 V (1N4001 or equiv.)
D3 — General-purpose LED.
Q1 — Pnp audio transistor, 25 V, 2 A (2N4919,

TIP32, or equiv.)
Q2 — Npn audio transistor, 25 V, 2 A (2N4922, RS 276-2020, or equiv.)
U1 — 555 timer.

ing, used to stop or reduce charging before the heat rise becomes a danger. Quick and fast charges do not usually allow gas recombination, so some of the battery water will escape in the form of gas. If the water level falls below a certain point, acid hydrometer readings are no longer reliable. If the water level falls to plate level, permanent battery damage may result.

Overcharging in moderation causes little loss of battery life, and some nickel-cadmium batteries may be left on continual charge in storage. A timer on chargers of lead-acid batteries prevents excessive overcharge if set to make up for your recorded discharge plus perhaps 20%. Some chargers will switch over automatically to an acceptable standby charge.

A VEHICULAR NiCAD BATTERY CHARGER

Charging a 12-volt NiCad battery pack from an automotive electrical system is difficult because the vehicle voltage is only a few tenths of a volt higher than the NiCad voltage. The problem is to control the charging current under varying engine speeds. The device pictured in Fig. 5 solves the problem electronically, making charging a NiCad pack in your car easy and convenient.

The circuit, shown in Fig. 6, designed by Mike Mladejovsky, WA7ARK, first appeared in *The Microvolt*. The basis of this charger is the familiar capacitive voltage doubler circuit used in conventional dc power supplies. The voltage doubler is driven by a chopper consisting of a 555 timer followed by a complementary pnp-npn emitter-follower buffer. The 555 is programmed to oscillate at about 3.3 kHz; its output is high for about 200 μs and low for about 100 μs.

While the 555 output is low, Q1 conducts, precharging C1 to the supply voltage minus the conduction thresholds of D1 and Q1. When the timer output goes high, Q2 conducts, causing the charge in C1 to be added to the supply voltage. C1 discharges through Q2, D2 and R1 into the NiCad battery.

R1 was chosen to provide an average charging current of about 150 mA at 12.6 V input. The LED indicator shows that the battery is being charged. The charging current depends on the supply voltage. This varies from about 12.5 V in a parked vehicle to about 14.8 V when the engine is running, causing the charging current to vary from about 140 mA to about 185 mA. This will charge a 500 mAh battery pack in five or four hours, respectively. If your battery requires a lower charging current, you can scale R1 (doubling R1 halves the charging rate, and so on.)

Ed Kalin, K1RT, designed the pc artwork and built the unit pictured in Fig. 5.

Figs. 7 and 8 are the etching pattern and parts-placement guide for the charger. The Plexiglas assembly shown is only one of several possible packaging schemes. For maximum convenience and utility, build a compartment and connector to fit your battery pack.

SOLAR-ELECTRIC POWER

Although solar-electric arrays are quite expensive when purchased new, surplus individual cells and groups of cells (arrays) can be bought inexpensively on occasion. Photons from the sun strike the p-n junctions of the cells to generate 0.5 volt per cell (see chapter 4). The current rating of an individual cell is dependent upon the diameter of the cell. Typical production units deliver 100 mA, 600 mA, 1 A or 1.5 A. Cells with higher current ratings are manufactured, but are quite costly. Table 1 lists some solar battery suppliers.

A solar-electric panel generally contains 36 cells wired in series. This provides approximately 18 volts dc (no-load conditions) at peak sunlight. The current capability of the panel is determined by the diameter of the cells. Greater amounts of current output can be had by paralleling like panels. That is two 1.5-A panels can be operated in parallel to deliver 3 amperes of current, and so on.

The usual operating system has the array output routed through a regulator to a storage battery. The regulator prevents overcharging of the battery. The station equipment takes its power from the battery. Most automotive 12-volt batteries are suitable for use with solar-electric panels. NiCad batteries are satisfactory also. Fig. 9 shows a solar array in a frame. The cells are wired in series.

Fig. 10 shows a solar-electric system suitable for low- or high-power operation. If the current drain is less than the capacity of the solar bank (1.5 A in this case), the load can be powered from the solar cells through the regulator circuit. For heavier loads, the current is taken from the storage batteries, which are charged by the solar array. The circuit of Fig. 10 was designed by John Akiyama, W6PQZ, and was described by John Halliday, W5PIZ in August 1980 *QST*. In the same issue, Doug Blakeslee, N1RM, described an electronic switch to automatically disconnect storage batteries from a solar system when full charge (13.5 V) is reached. The circuit is shown in Fig. 11. U1, D4 and D5 establish a 6.2-V reference for comparator U2. A voltage divider composed of R1, R3 and R7 scales the battery voltage down to the reference value, while R4 provides hysteresis to prevent oscillation. When the battery potential exceeds the comparator threshold, U2 goes high, turning off Q1 and Q2. The LED, D6, indicates that the battery is being charged.

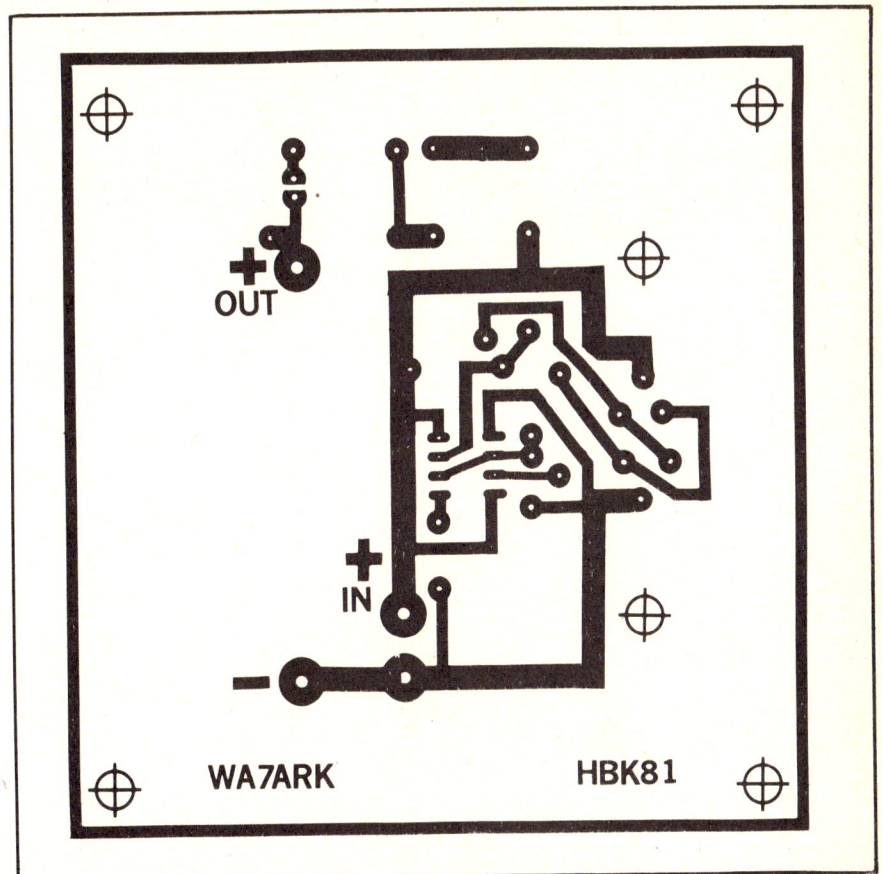

Fig. 7 — Etching pattern for the mobile NiCad charger pc board. The foil side is shown; black represents copper.

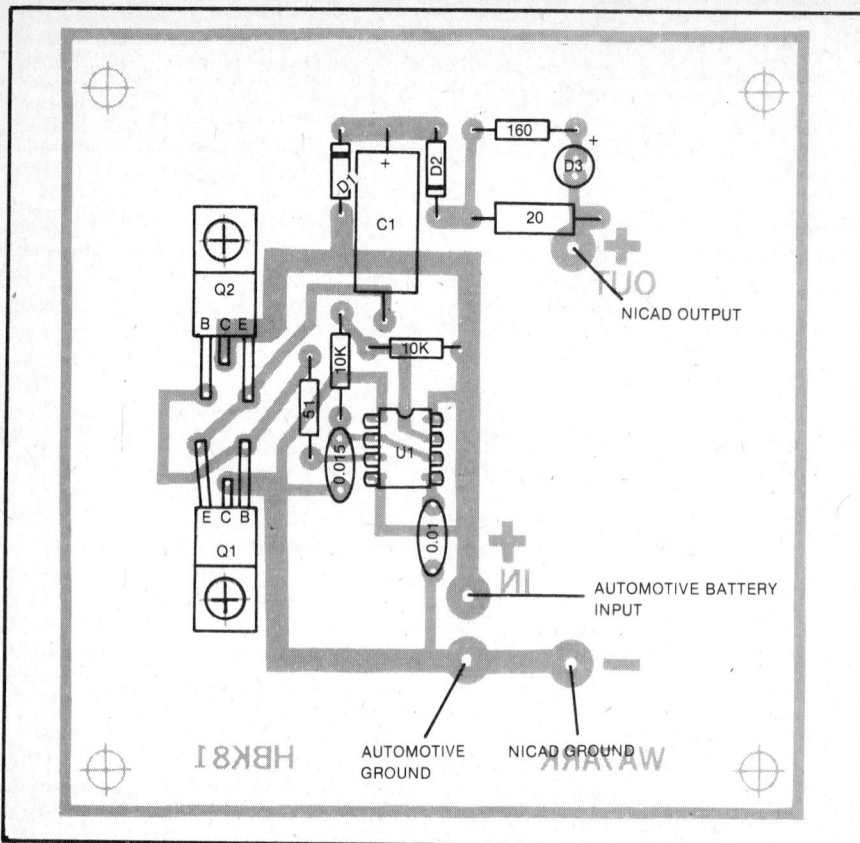

Fig. 8 — Parts-placement guide for the mobile NiCad charger pc board. The component side is shown with an x-ray view of the foil.

Table 1

Some Solar Battery Manufacturers and Distributors

Solar Power Corporation
c/o Lindberg Company
4163 Montgomery, NE
Albuquerque, NM 87109
Tel. 505-881-1006

Solarex Corporation
1335 Piccard Dr.
Rockville, MD 20850
Tel. 301-948-0202

Applied Solar Energy
15251 E. Don Julian Rd.
City of Industry, CA 91746
Tel. 213-968-6581

Solec International
12533 Chadron Ave.
Hawthorne, CA 90250
Tel. 213-325-6215

Fig. 9 — Solar-electric arrays are excellent for short- or long-term field and emergency use to power amateur stations. A 14-volt, 1.5-A solar panel and two automobile batteries in parallel can provide many after-dark hours of operation with typical 100-watt hf-band transceivers of the solid-state variety.

PORTABLE AC POWER SOURCES

There are two popular sources of ac power for use afield. The first is what is referred to as a dc-to-ac converter, or more commonly, an *inverter*.

The ac output voltage is a square wave. Therefore, some types of equipment can not be operated satisfactorily from the inverter. Certain types of motors are among those items which require a sine-wave output. Fig. 12 shows a picture of one style of commercial inverter. Heat sinks are used to cool the switching transistors. The unit shown is available from Heath Company in kit form. It delivers 117 volts of ac at 175 watts continuous power rating. The primary voltage is 6 or 12 dc.

When sine-wave output is required from a portable ac power supply, gasoline-engine alternators are used. They are available with ratings of several kilowatts, or as little as 500 watts. One of the larger units is shown in Fig. 13 where WB9QPI has just completed a maintenance run for the WØOHU/Ø Field Day group.

Alternators powered by internal-combustion engines have been used for years to supply 117/235 ac independently of the commercial mains. Such combinations range from tiny units powered by two-cycle or four-cycle gasoline engines in the low-wattage class to giant multicylinder diesels capable of supplying

megawatts of power. Perhaps the most practical power range for most purposes would be in the neighborhood of 2 kW. Larger units tend to become too heavy for one person to lift and handle easily while smaller generators lack sufficient power output for many applications. A 2-kW alternator is quite heavy but is capable of supplying power for just about any large power tool. It is roughly the equivalent of having a single 15-A outlet in an ordinary electric service. Of course, it will handle moderate-power amateur equipment with ease.

Maintenance Checklist

Although more complicated maintenance chores should be performed by qualified service personnel, many simple measures which will prolong the life of the alternator can be done at home or afield. Perhaps the best plan is to log the dates of when the unit was used and the operating time in hours. Also included in the log would be dates of maintenance and type of service performed. Oil changes, when gasoline was purchased for emergency purposes, and similar data would fall under this category.

Important points that are common to all types of generators are indicated for a typical one in Fig. 14 (Consult the manufacturer's manual for additional instructions that might apply to a particular model.) The following checklist relates to

the numbers on the drawing.

1) Use the proper grade of fuel. Newer models will burn either "no-lead" or regular leaded gasoline. Do not use premium or so-called "hi-test" grades unless the owner's manual recommends it. Such fuels have a high lead content for proper burning in high-compression automobile engines and are generally unsuitable for small, low-compression engines found in most alternator combinations. Check the owner's manual to determine whether oil must be mixed with the gasoline. While *two-cycle* models require an oil-gas mixture, most generators have a *four-cycle* engine that burns ordinary gasoline with no extra additives. Gasoline for emergency purposes should only be stored in small amounts and rotated on a regular basis. Older stock can be burned in a car (that uses the same grade of gas as the generator) since storing gasoline for any length of time is inadvisable. The more volatile components evaporate, leaving excess amounts of a varnish-like substance that will clog carburetor passages. Also, be sure gasoline containers are of an approved type with a

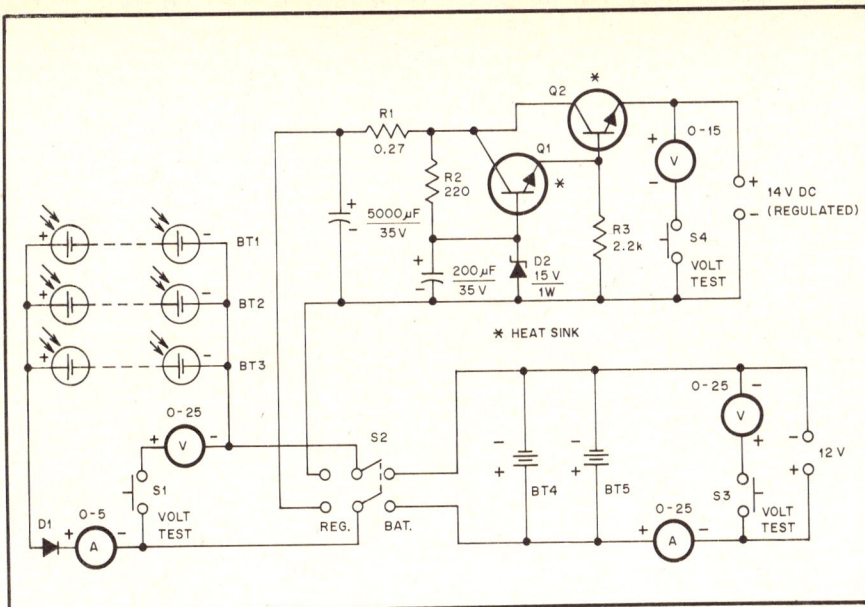

Fig. 10 — Schematic diagram of solar power supply. Note that battery charging circuit does *not* employ a regulator or switch to shut off charging current once the storage battery reaches full charge state. Because the output of the solar panels is, at most, 1-1/2 ampere and the storage batteries are full-size automobile batteries, the danger of damage from overcharging is not great. Anyone contemplating higher current solar batteries or smaller storage batteries should give serious consideration to a regulator and/or an automatic cutoff switch for the charging circuit. (See Fig. 11)

BT1, BT2, BT3 — 20-V, 1/2-ampere solar panels by Spectrolab.
BT4, BT5 — 12-V, lead-acid automobile batteries.
D1 — Motorola MR 752/7414 or any diode with at least 2-ampere capacity and with at least 50 PIV.
Q1 — Npn silicon 90-W transistor, power switching, TIP31, Radio Shack 276-2020 or

equiv.
Q2 — Npn silicon 115 W transistor, power switching, 2N3055, Radio Shack 276-2041 or equiv.
R1 — 0.27 Ω,1 watt.
R2 — 220 Ω, 1 watt, carbon composition.
R3 — 2.2 k-Ω, 1 watt, carbon composition.
S1, S2, S3 — Spst, momentary contact switch.
S2 — Dpdt knife switch.

Fig. 12 — Photograph of a commercial dc-to-ac inverter that operates from 6 to 12 volts dc and delivers 117 volts ac (square wave) at 175 watts.

clean interior, free of rust or other foreign matter. Similar considerations apply to the gas tank on the engine itself.

The majority of difficulties with small engines are related to fuel problems in some way. Dirty fuel or water in the gasoline is one source, with carburetor trouble because of the use of old gas being another common cause. Except for minor adjustments recommended in the instruction manual, it is seldom necessary to touch the carburetor controls. Avoid the temptation to make such adjustments in the case of faulty operation. *Follow the recommendations* in this guide so that

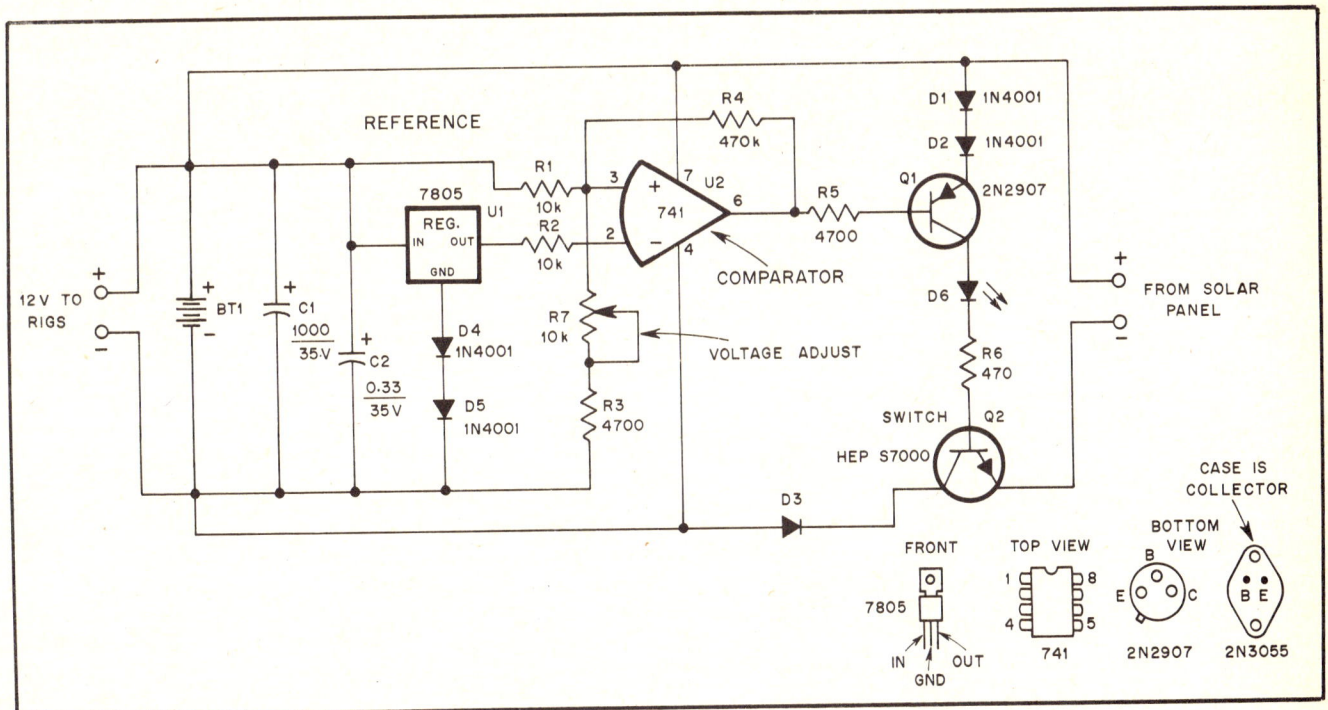

Fig. 11 — Schematic diagram of the electronic switch. Resistances are in ohms; k = 1000; capacitance values are in microfarads (μF).

BT1 — Automotive storage battery, lead-acid type.
C1 — 1000-μF, 35-V electrolytic.
C2 — 0.33-μF, 35-V.
D3 — Silicon diode, PIV of 50 or more, cur-

rent rating sufficient to pass full output of the solar panel.
D6 — Light-emitting diode, any type.
Q2 — Low-frequency power transistor; 2N3055, HEP S7000, or equivalent. Use heat sink of

9 in.² (5800 mm²) or more.
R7 — 10 kΩ, 1/2 watt, carbon control, linear taper, pc mount.
U1 — 3 terminal, 5-volt regulator.
U2 — Op amp, any of the 741 family usable.

Fig. 13 — Large gasoline generators of the kW and higher class are excellent for powering several amateur stations from a complex field site. Maintenance, as discussed in the text, is a vital matter to ensure reliable operation. Here, WB9QPI has just finished a maintenance check of the group's Field Day power plant.

Some manufacturers recommend a high-detergent oil that comes in various service grades such as MS, SD, and similar types. Examine the top or side of the cans in which the oil is sold and see if the letters correspond to those recommended by the engine manufacturer.

3) The carburetor mixes gasoline with air, which is then burned in the engine. Before entering the carburetor, the air must be filtered so that it is free of dust and other foreign matter that might otherwise be drawn into cylinder(s). Particles that do get by the air filter are picked up by the oil. That should be changed more often if the alternator is operated in a dusty location. Also, it is important to clean the air filter frequently. It contains a foam-like substance which can be cleaned in kerosene and then soaked in fresh motor oil. Squeeze excess oil from the filter before replacing. Also, consult the instruction manual for further recommendations.

4) Once the gas/air mixture enters the cylinder, it is compressed by the piston into a very small volume and ignited by the spark plug. During the rapid burning that then occurs, the expansion caused by the resulting heat forces the piston down and delivers the mechanical power to the alternator.

As might be expected, proper operation of the ignition system is an important factor in engine performance. Power for the spark is supplied by a device called a magneto that is normally installed on the front of the engine. The magneto seldom requires servicing and such work should only be done by those qualified to do so. (This is one reason why the magneto is often located under a flywheel that is difficult to remove by the inexperienced.)

On the other hand, faulty spark plugs are the usual cause of ignition problems. Special equipment is required to test a spark plug properly, but an easier solution is to have a new one handy. In fact, keep *two* spare plugs on hand. Spark plug life can be notoriously short on occasion. However, repeated plug failure is also abnormal and other causes such as a poor gas/air mixture might be the culprit.

Replace the spark plug with a type similar to the one that came with the alternator or a substitute recommended by the manufacturer. Some models have resistor-type plugs which are desirable for ignition-noise suppression. Resistor plugs are usually indicated by an R prefix. For instance the resistor version of a Champion CJ-8 would be an RC-J8.

5) Little maintenance is required in regard to the exhaust system. In some forested areas, a spark-arrester type of muffler is required, so be sure that your unit is so equipped *before* contemplating operation in such a location. "Quiet hours" may also be imposed in some places during the nighttime hours if generator exhaust noise is too loud.

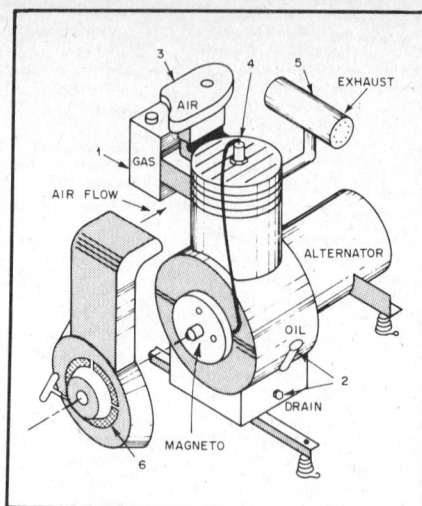

Fig. 14 — The numbers indicate the primary maintenance points of a large power generator (see text for details).

Two very important safety precautions should be observed with regard to the exhaust system. *Never* operate an alternator in closed surroundings such as a building. Dangerous gases are emitted from the exhaust which are highly toxic. Secondly, *never* refuel an engine while it is running or if the exhaust system is still very hot. Unfortunately, this last precaution is disregarded by many, which is extremely foolish. (Experienced service station operators will refuse to refuel an automobile with the motor running, which is often prohibited by law.) Don't become an unnecessary statistic.

6) Most alternators are air-cooled as opposed to the water-cooled radiator system of the automobile. A fan on the front of the engine forces air over the cylinder and an unobstructed entrance for this air flow is necessary. Avoid operating the alternator in areas where obstruction to this flow might result (such as in tall grass). Alternators should be operated such that a sufficient amount of air circulation is present for cooling, caburetion and exhaust.

Storage

Proper maintenance of an alternator when it is not being used is just as important as during the time it is in operation. The usual procedure is to run the engine dry of gasoline, drain the crankcase and fill it with fresh oil, and remove the spark plug. Then pour a few tablespoons of oil into the cylinder and turn the engine over a few times with the starter and replace the plug. But *never* crank the engine with the plug removed and the *ignition* or *start* switch in the *on* or *run* position. The resulting no-load high voltage might cause damage to the magneto. It is also a good idea to ground the spark-plug wire to the engine frame with a clip lead in case the switch is accidentally activated.

more complicated maintenance procedures (such as carburetor overhaul) are not required.

2) Another important factor often neglected in maintenance of alternator engines is oil. While lubrication is one job oil has to perform, there are other considerations as well. The engine oil in the crankcase also collects a large amount of solid combustion products, bits of metal worn away by the moving parts, and any dust or other foreign matter that enters the carburetor intake. For instance, it is especially important to observe the manufacturer's recommendations concerning the length of time the engine may be operated before an oil change is required during the break-in period. If you ever have the opportunity to examine the oil from a new engine, you will note a metallic sheen to it. This is from the excessive amount of metal that is worn away. After the break-in period, much less metal is abraded and the oil doesn't have to be replaced as often.

The oil level should be checked frequently during engine operation. Each time fuel has to be added the oil should be checked also. When storing an alternator, it is also wise to drain the oil and replace it with fresh stock. This is because one of the combustion products is sulfur which forms sulfuric acid with water dispersed in the oil. The acid then attacks the special metal in the bearing surfaces causing pitting and premature replacement.

Also note the grade and weight of oil recommended by the manufacturer. Unlike their larger counterparts in the automobile, most small engines do not have oil filters, which is another reason why required changes are more frequent.

Moisture is the greatest enemy of an iron product — such as a generator, in storage. The coating of oil helps retard rust formation here which might actually weld the two surfaces when the engine is restarted, resulting in premature wear. Consequently it is important to store the alternator in an area of low humidity.

Although the maintenance procedures outlined may seem like a chore, the long-term benefits include low repair costs and like-new performance. Engines for alternator combinations must be able to handle a variety of loads while maintaining a constant speed in order to keep the output frequency constant. A mechanical governor performs this latter function by metering the fuel supplied to the engine under different load conditions. However, the system cannot function properly with an engine in poor mechanical condition because of lack of proper maintenance.

Grounds

Newer generators are supplied with a three-wire outlet and the ground connection should go to the plug as shown in Fig. 15. On older types, the ground would have to be connected separately to the generator frame and then to the common terminal in the junction box. A pipe or rod can then be driven into the ground and a wire connection made to either a clamp supplied with the rod or by means of a C-clamp for larger sizes of pipe. From an ignition-noise-suppression standpoint, the ground is desirable along with safety considerations when power tools are being used.

The ground connection goes to the green wire in commercially made three-wire conduit. Conduit purchased from an electrical store comes with a color-coded insulation and the colored wires should be connected as shown in Fig. 15. Consult the owner's manual for the generator for further details on power hookup that might apply to your particular model.

HF MOBILE ANTENNAS

The antenna is perhaps the most important item in the successful operation of a mobile installation. Mobile antennas, whether designed for single or multiband use, should be securely mounted to the automobile, as far from the engine compartment as possible (for reducing noise pickup), and should be carefully matched to the coaxial feed line connecting them to the transmitter and receiver. All antenna connections should be tight and weatherproof. Mobile loading coils should be protected from dirt, rain and snow if they are to maintain their Q and resonant frequency. The greater the Q of the loading coil, the better the efficiency, but the narrower will be the bandwidth of the antenna system.

Though bumper-mounted mobile

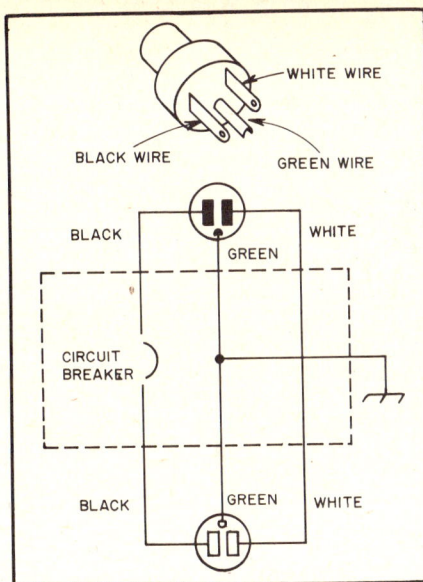

Fig. 15 — A simple accessory that provides overload protection for generators that do not have such provisions built in.

antennas are favored by some, it is better to place the antenna mount on the rear deck of the vehicle, near the rear window. This locates the antenna high and in the clear, assuring less detuning of the system when the antenna moves to and from the car body. *Never use a base-loaded antenna on a bumper mount.* Many operators avoid cutting holes in the car body for fear of devaluation when selling the automobile. Such holes are easily filled, and few car dealers lower the trade-in price because of the holes.

The choice of base or center loading a mobile antenna has been a matter of controversy for many years. In theory, the center-loaded whip presents a slightly higher base impedance than does the base-loaded antenna. However, with proper impedance-matching techniques employed there is no discernible difference in per-formance between the two methods. A base-loading coil requires fewer turns of wire than one for center loading, and this is an electrical advantage because of reduced coil losses. A base-loaded antenna is more stable during wind loading and sway. If a homemade antenna system is contemplated, either system will provide good results, but the base-loaded antenna may be preferred for its mechanical advantages.

Loading Coils

There are many commercially built antenna systems available for mobile operation, and some manufacturers sell the coils as separate units. Air-wound coils of large wire diameter are excellent for use as loading inductors. Large Miniductor coils can be installed on a solid phenolic rod and used as loading coils. Miniductors, because of their turns spacing, are easy to adjust when resonating the mobile antenna, and provide excellent Q. Phenolic-impregnated paper or fabric tubing of large diameter is suitable for making homemade loading coils. It should be coated with liquid fiberglass, inside and out, to make it weatherproof. Brass insert plugs can be installed in each end, their centers drilled and tapped for a standard 3/8 × 24 (9.5 mm × 24) thread to accommodate the mobile antenna sections. After the coil winding is pruned to resonance it should be coated with a high-quality, low-loss compound to hold the turns securely in place, and to protect the coil from the weather. Liquid polystyrene is excellent for this. Hobby stores commonly stock this material for use as a protective film for wall plaques and other artwork. Details for making a home-built loading coil are given in Fig. 16.

Impedance Matching

Fig. 17 illustrates the shunt-feed

Table 2
Approximate Values for 8-foot Mobile Whip

Base Loading

f(kHz)	Loading L(μH)	RC(Q50) Ohms	RC(Q300) Ohms	RR Ohms	Feed R* Ohms	Matching L(μH)
1800	345	77	13	0.1	23	3
3800	77	37	6.1	0.35	16	1.2
7200	20	18	3	1.35	15	0.6
14,200	4.5	7.7	1.3	5.7	12	0.28
21,250	1.25	3.4	0.5	14.8	16	0.28
29,000	—	—	—	—	36	0.23

Center Loading

f(kHz)	Loading L(μH)	RC(Q50) Ohms	RC(Q300) Ohms	RR Ohms	Feed R* Ohms	Matching L(μH)
1800	700	158	23	0.2	34	3.7
3800	150	72	12	0.8	22	1.4
7200	40	36	6	3	19	0.7
14,200	8.6	15	2.5	11	19	0.35
21,250	2.5	6.6	1.1	27	29	0.29

RC = Loading-coil resistance; RR = Radiation resistance.
*Assuming loading coil Q = 300, and including estimated ground-loss resistance.
Suggested coil dimensions for the required loading inductance are shown in a following table.

Table 3
Suggested Loading-Coil Dimensions

Req'd L(μH)	No. Turns	Wire Size	Dia. In.	Length In.
700	190	22	3	10
345	135	18	3	10
150	100	16	2-1/2	10
77	75	14	2-1/2	10
77	29	12	5	4-1/4
40	28	16	2-1/2	2
40	34	12	2-1/2	4-1/4
20	17	16	2-1/2	1-1/4
20	22	12	2-1/2	2-3/4
8.6	16	14	2	2
8.6	15	12	2-1/2	3
4.5	10	14	2	1-1/4
4.5	12	12	2-1/2	4
2.5	8	12	2	2
2.5	8	6	2-3/8	4-1/2
1.25	6	12	1-3/4	2
1.25	6	6	2-3/8	4-1/2

To obtain dimensions in millimeters, multiply inches by 25.4.

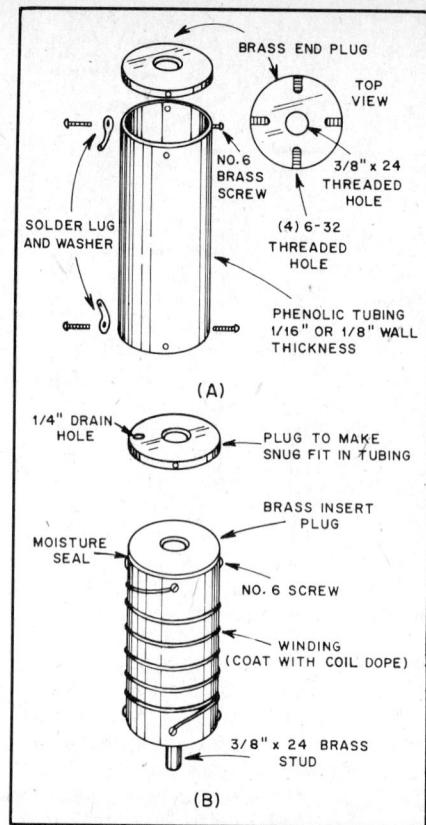

Fig. 16 — Details for making a home-built mobile loading coil. A breakdown view of the assembly is given at A. Brass end plugs are snug-fit into the ends of the phenolic tubing, and each is held in place by four 6-32 brass screws. Center holes in the plugs are drilled and tapped for 3/8-24 thread. The tubing can be any diameter from one to four inches (25 to 100 mm). The larger diameters are recommended. Illustration B shows the completed coil. Resonance can be obtained by installing the coil, applying transmitter power, then pruning the turns until the lowest SWR is obtained. Pruning the coil for maximum field-strength-meter indication will also serve as a resonance indication.

$$L_M = \left[\frac{\sqrt{R_A (R_O - R_A)}}{2\pi f \, (kHz)} \right] \times 10^3 \; \mu H$$

where

R_A = the antenna feed-point impedance

R_O = the characteristic impedance of the transmission line.

As an example, if the radiation resistance is 20 ohms and the line is 50-ohm coaxial cable, then at 4000 kHz,

$$C_M = \left[\frac{\sqrt{20 \, (50 - 20)}}{(6.28)(4000)(20)(50)} \right] \times 10^9$$

$$= \left[\frac{\sqrt{600}}{(6.28)(4)(2)(5)} \right] \times 10^4$$

$$= \frac{24.5}{251.2} \times 10^4 = 975 \; pF$$

$$L_M = \left[\frac{\sqrt{20 \, (50 - 20)}}{(6.28)(4000)} \right] \times 10^3$$

$$= \frac{\sqrt{600}}{25.12} = \frac{24.5}{25.12} = 0.97 \; \mu H$$

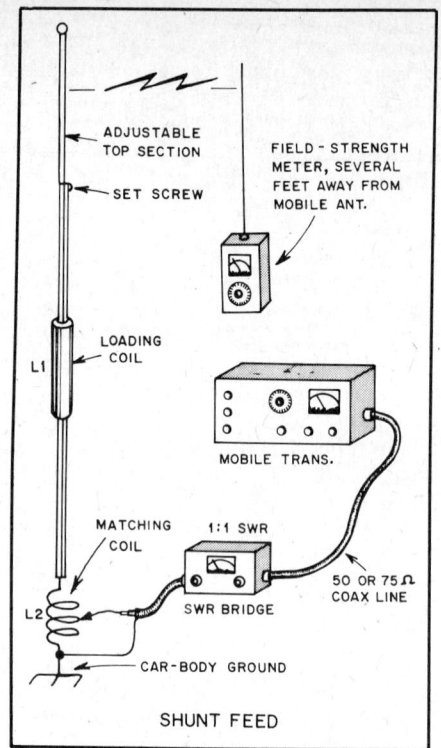

Fig. 17 — A mobile antenna using shunt-feed matching. Overall antenna resonance is determined by the combination of L1 and L2. Antenna resonance is set by pruning the turns of L1, or adjusting the top section of the whip, while observing the field-strength meter or SWR indicator. Then, adjust the tap on L2 for lowest SWR.

The chart of Fig. 19 shows the capacitive reactance of C_M and the inductive reactance of L_M necessary to match various antenna impedances to 50-ohm coaxial cable. The chart assumes the antenna element has been resonated.

In practice, L_M need not be a separate inductor. Its effect can be duplicated by adding an equivalent amount of inductance to the loading coil, regardless of whether the loading coil is at the base or at the center of the antenna.

Adjustment

In adjusting this system, at least part of C_M should be variable, the balance being made up of combinations of fixed mica capacitors in parallel as needed. A small, one-turn loop should be connected between C_M and the chassis of the car, and the loading coil should then be adjusted for resonance at the desired frequency as indicated by a dip meter coupled to the loop at the base. Then the transmission line should be connected, and a check made with an SWR indicator connected at the transmitter end of the line.

With the line disconnected from the antenna again, C_M should be readjusted and the antenna returned to resonance by readjustment of the loading coil. The line should be connected again, and another check made with the SWR bridge. If the SWR is less than it was on the first trial,

method of obtaining a match between the antenna and the coaxial feed line. For operation on 75 meters with a center-loaded whip, L2 will have approximately 18 turns of no. 14 wire, spaced one wire thickness between turns, and wound on a 1-inch (25-mm) diameter form. Initially, the tap will be approximately 5 turns above the ground end of L2. Coil L2 can be inside the car body, at the base of the antenna, or it can be located at the base of the whip, outside the car body. The latter method is preferred. Since L2 helps determine the resonance of the overall antenna, L1 should be tuned to resonance in the desired part of the band with L2 in the circuit. The adjustable top section of the whip can be telescoped until a maximum reading is noted on the field-strength meter. The tap is then adjusted on L2 for the lowest reflected-power reading on the SWR bridge. Repeat these two adjustments until no further increase in field strength can be obtained; this point should coincide with the lowest SWR. The number of turns needed for L2 will have to be determined experimentally for 40- and 20-meter operation. There will be proportionately fewer turns required.

Matching with an L Network

Any resonant mobile antenna that has a feed-point impedance less than the characteristic impedance of the transmission line can be matched to the line by means of a simple L network, as shown in Fig. 18. The network is composed of C_M and L_M. The required values of C_M and L_M may be determined from the following:

$$C_M = \left[\frac{\sqrt{R_A (R_O - R_A)}}{2\pi f \, (kHz) \, R_A R_O} \right] \times 10^9 \; pF$$

and

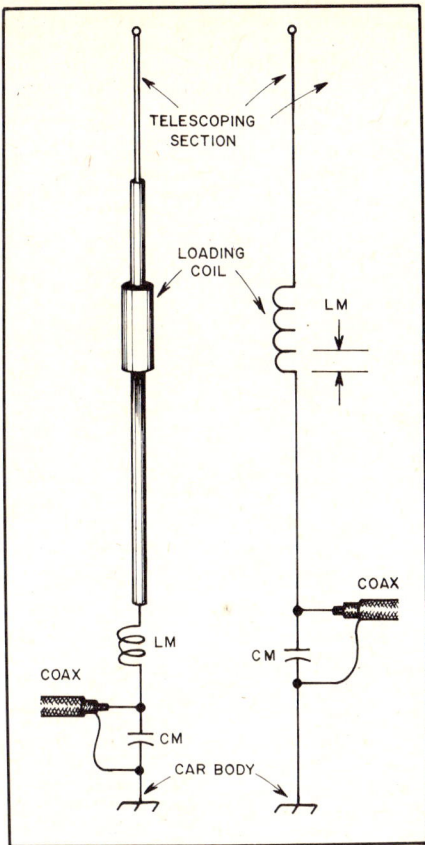

Fig. 18 — A whip antenna may also be matched to coax line by means of an L network. The inductive reactance of the L network can be combined in the loading coil, as indicated at the right.

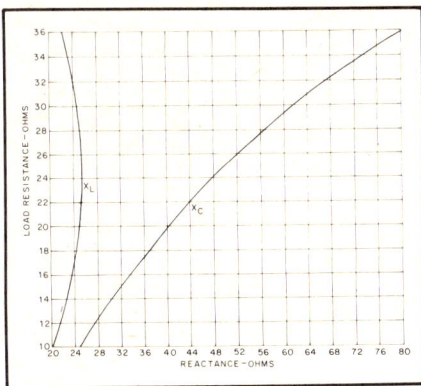

Fig. 19 — Curves showing inductive and capacitive reactances required to match a 50-ohm coax line to a variety of antenna resistances.

C_M should be readjusted in the same direction until the point of minimum SWR is found. Then the coupling between the line and the transmitter can be adjusted for proper loading. It will be noticed from Fig. 19 that the inductive reactance varies only slightly over the range of antenna resistances likely to be encountered in mobile work. Therefore, most of the necessary adjustment is in the capacitor. The one-turn loop at the base should be removed at the conclusion of the adjustment and slight compensation made at the loading coil to maintain resonance.

Top-Loading Capacitance

Because the coil resistance varies with the inductance of the loading coil, the resistance can be reduced, beneficially, by reducing the number of turns on the coil. This can be done by adding capacitance to that portion of the mobile antenna that is *above* the loading coil (Fig. 20). To achieve resonance, the inductance of the coil is reduced proportionally. "Capacity hats," as they are often called, can consist of a single stiff wire, two wires or more, or a disc made up from several wires like the spokes of a wheel. A solid metal disc can also be used. The larger the capacity hat, in terms of volume, the greater the capacitance. The greater the capacitance, the smaller the amount of inductance needed in the loading coil for a given resonant frequency.

There are two schools of thought concerning the attributes of center-loading and base-loading. It has not been established that one system is superior to the other, especially in the lower part of the hf spectrum. For this reason both the base and center-loading schemes are popular. Capacity-hat loading is applicable to either system. Since more inductance is required for center-loaded whips to make them resonant at a given frequency, capacity hats should be particularly useful in improving their efficiency.

VHF QUARTER-WAVELENGTH VERTICAL

Ideally, the vhf vertical antenna should be installed over a perfectly flat plane reflector to assure uniform omnidirectional radiation. This suggests that the center of the automobile roof is the best place to mount it. Alternatively, the flat portion of the auto rear-trunk deck can be used, but will result in a directional pattern because of car-body obstruction. Fig. 21 illustrates at A and B how a Millen high-voltage connector can be used as a roof mount for a 144-MHz whip. The hole in the roof can be made over the dome light, thus providing accessibility through the upholstery. RG-59/U and matching section L, Fig. 21C, can be routed between the car roof and the ceiling upholstery and brought into the trunk compartment, or down to the dashboard of the car. Some operators install an SO-239-type coax connector on the roof for mounting the whip. The method is similar to that of Fig. 21.

It has been established that quarter-wavelength vertical antennas for mobile work through repeaters are not as effective as 5/8-wavelength verticals are. The 1/4-wavelength types cause considerably more "picket fencing" (rapid flutter) of

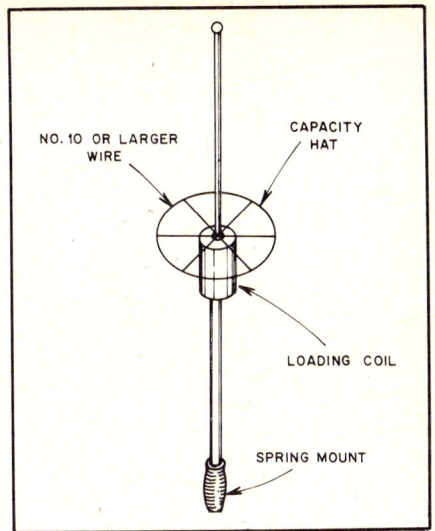

Fig. 20 — A capacitance "hat" can be used to improve the performance of base- or center-loaded whips. A solid metal disc can be used in place of the skeleton disc shown here.

the signal than is the case with the 5/8-wavelength type of antenna. The flutter that takes place when vertical polarization is used is caused by vertical conductive objects being between the mobile antenna (near field) and the station being worked (or the repeater). As the vehicle moves past these objects there is a momentary blockage or partial blockage of the signal path.

2-METER 5/8-WAVELENGTH VERTICAL

Perhaps the most popular vertical antenna for fm mobile and fixed-station use is the 5/8-wavelength vertical. As compared to a 1/4-wavelength vertical, it has some gain over a dipole. Additionally, the so-called "picket-fencing" type of flutter which results when the vehicle is in motion is greatly reduced when a 5/8-wavelength radiator is employed, as discussed earlier.

This style of antenna is suitable for mobile or fixed-station use because it is small, omnidirectional and can be used with radials or a solid-plane ground (such as is afforded by the car body). If radials are used, they need be only 1/4 wavelength or slightly longer. A 5 percent increase in length over 1/4 wavelength is suggested for the radial wires or rods.

Construction

The antenna shown here is made from low-cost materials. Fig. 22A and B shows the base coil and aluminum mounting plate. The coil form is a piece of low-loss solid rod, such as Plexiglas or phenolic. The dimensions for this and other parts of the antenna are given in Fig. 23. A length of brazing rod is used as the whip section.

The whip should be 47 inches (1.2 m) long. However, brazing rod comes in standard 36-inch (0.9-m) lengths, so it is

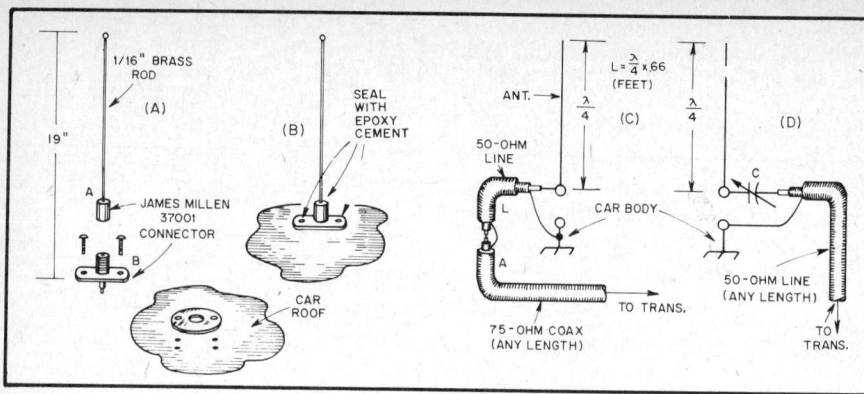

Fig. 21 — At A and B, an illustration of how a quarter-wavelength vertical antenna can be mounted on a car roof. The whip section should be soldered into the cap portion of the Millen connector, then screwed to the base socket. This handy arrangement permits removing the antenna when desired. Epoxy cement should be used at the two mounting screws to prevent moisture from entering the car. Diagrams C and D are discussed in the text.

necessary to solder an 11-inch (279-mm) extension to the top of the whip. A piece of no. 10 copper wire will suffice. Alternatively, a stainless-steel rod can be purchased to make a 47-inch whip. Shops that sell CB antennas should have such rods for replacement purposes on base-loaded antennas. The limitation one can expect with brazing rod is the relative fragility of the material, especially when the threads are cut for screwing the rod into the base-coil form. Excessive stress can cause the rod to break where it enters the coil form. The problem is complicated somewhat in this design by the fact that a spring is not used at the antenna mounting point. Innovators can find all manner of solutions to the problems just outlined by changing the physical design and using different materials when constructing the overall antenna. The main purpose of this description is to provide dimensions and tune-up data.

The aluminum mounting bracket must be shaped to fit the car with which it will be used. The bracket can be used to effect a "no-holes" mount with respect to the exterior portion of the car body. The inner lip of the vehicle trunk (or hood for front mounting) can be the point where the bracket attaches by means of no. 6 or no. 8 sheet-metal screws. The remainder of the bracket is bent so that when the trunk lid or car hood is raised and lowered, there is no contact between the bracket and the moving part. Details of the mounting unit are seen in Fig. 23 at B. A 14-gauge metal thickness (or greater) is recommended for best rigidity.

There are 10-1/2 turns of no. 10 or no. 12 copper wire wound on the 3/4-inch (19-mm) diameter coil form. The tap on L1 is placed approximately four turns below the whip end. A secure solder joint is imperative.

Tune-up

After the antenna has been affixed to the vehicle, insert an SWR indicator in the 50-Ω transmission line. Turn on the

2-meter transmitter and experiment with the coil-tap placement. If the whip section is 47 inches long, an SWR of 1:1 can be obtained when the tap is at the right location. As an alternative to the foregoing method of adjustment, place the tap at four turns from the top of L1, make the whip 50 inches long and trim the whip length until an SWR of 1:1 is secured. Keep the antenna free of conductive objects or human bodies during tune-up, as conductive objects will detune the antenna and spoil the match. This antenna was described more completely in June 1979 *QST,* page 15.

A 5/8-WAVELENGTH 220-MHz MOBILE ANTENNA

This antenna, Figs. 24 and 25, was developed to fill the gap between a homemade 1/4-wavelength mobile antenna and a commercially made 5/8-wavelength model. There have been other antennas made using modified CB models. This still presents the problem of cost in acquiring the original antenna. The major cost in this setup is the whip portion. This can be any tempered rod that will spring easily.

Construction

The base insulator portion is constructed of 1/2-inch (13-mm) Plexiglas rod. A few minutes work on a lathe was sufficient to shape and drill the rod. The bottom 1/2-inch (13-mm) of the rod is turned down to a diameter of 3/8 inch (9.5-mm). This portion will now fit into a PL-259 uhf connector. A hole, 1/8-inch (3-mm) diameter, is drilled through the center of the rod. This hole will contain the wires that make the connections between the center conductor of the connector and the coil tap. The connection between the whip and the top of the coil is also run through this opening. A stud is force-fitted into the top of the Plexiglas rod. This allows the whip to be detached from the insulator portion.

The coil should be initially wound on a form slightly smaller than the base in-

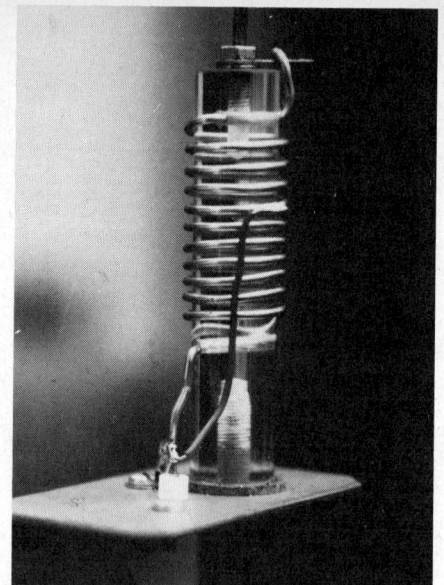

(A)

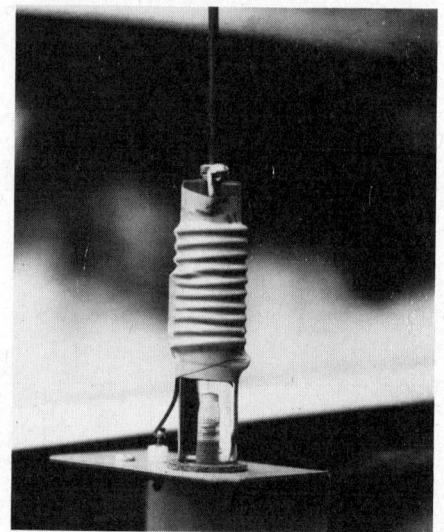

(B)

Fig. 22 — (A) A photograph of the 5/8-wavelength vertical base section. The matching coil is affixed to an aluminum bracket which screws onto the inner lip of the car trunk. (B) The completed assembly. The coil has been wrapped with vinyl electrical tape to prevent dirt and moisture from degrading performance.

sulator. When the coil is transferred to the Plexiglas rod it will keep its shape and will not readily move. After the tap point has been determined, a longitudinal hole is drilled into the center of the rod. A no. 22 wire can then be inserted through the center of the insulator into the connector. This method is also used to attach the whip to the top of the coil. After the whip has been fully assembled a coating of epoxy cement is applied. It seals the entire assembly and provides some additional strength. During a full winter's use there was not any sign of cracking or mechanical failure. The adjustment procedure is the same as for the 2-meter antenna just described.

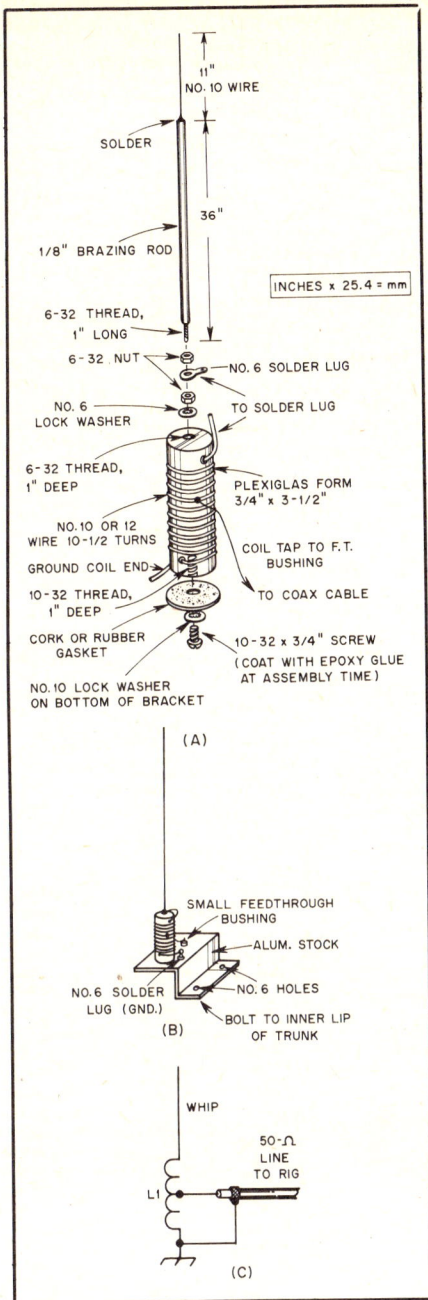

Fig. 23 — Structural details for the 2-meter antenna are provided at A. The mounting bracket is shown at B and the equivalent circuit is given at C.

SIMPLE ANTENNAS FOR HF PORTABLE OPERATION

The typical portable hf antenna is a random-length wire flung over a tree and end-fed through a Transmatch. QRP Transmatches can be made quite compact, but each additional piece of equipment necessary makes portable operation less attractive. The station can be simplified by using resonant impedance-matched antennas for the bands of interest. Perhaps the simplest antenna of this type is the half-wave dipole, center-fed with 50- or 75-ohm coax. Unfortunately, RG-58, RG-59 or RG-8 cable is

quite heavy and bulky for backpacking, and the miniature cables such as RG-174 are too lossy. A practical solution to the coax problem is to use folded dipoles made from lightweight TV twin lead. The characteristic impedance of this type of dipole is near 300 ohms, but it can be transformed to a 50-ohm source or load by means of a simple matching stub.

Fig. 26 illustrates the construction method and important dimensions for the twin-lead dipole. A silver-mica capacitor is shown for the reactive element, but an open-end stub of twin lead can serve as well, provided it is dressed at right angles to the transmission line for some distance. The stub method has the advantage of easy adjustment of the system resonant frequency.

To preserve the balance of the feeder, a 1:1 balun must be used at the end of the feed line. In most applications the balance is not important, and the twin lead can be connected directly to a coaxial output jack, one lead to the center contact, and one lead to the shell. Because of its higher impedance, a folded dipole exhibits a wider bandwidth than a single-conductor type. The antennas described here are not as broad as a standard folded dipole because the impedance transformation mechanism is frequency selective. However, the bandwidth should be adequate. An antenna cut for 14.175 MHz, for example, will present a VSWR of less than 2:1 over the entire 20-meter band.

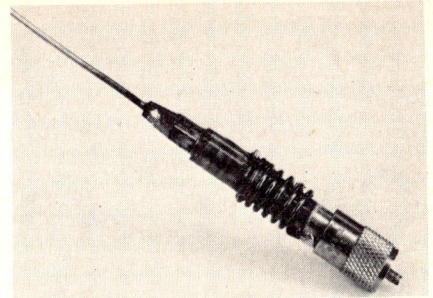

Fig. 24 — Photograph of the 220-MHz 5/8-wavelength mobile antenna. The bottom end of the coil is soldered to the coaxial connector.

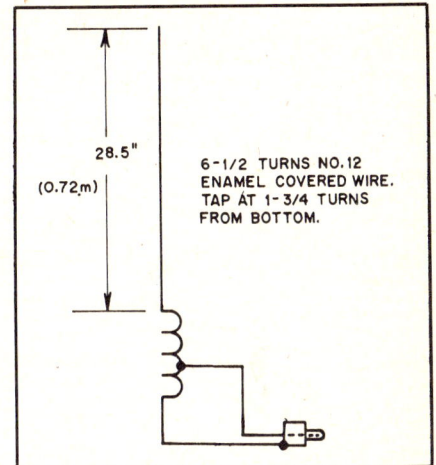

Fig. 25 — Diagram of the 220-MHz mobile antenna.

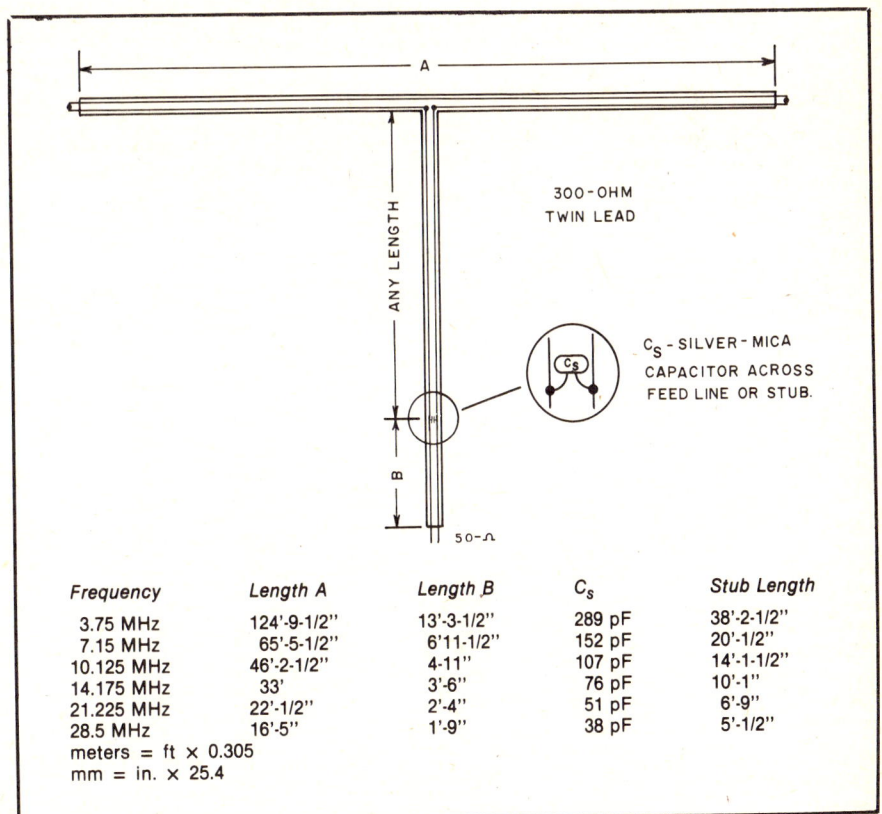

Frequency	Length A	Length B	C_S	Stub Length
3.75 MHz	124'-9-1/2"	13'-3-1/2"	289 pF	38'-2-1/2"
7.15 MHz	65'-5-1/2"	6'11-1/2"	152 pF	20'-1/2"
10.125 MHz	46'-2-1/2"	4'-11"	107 pF	14'-1-1/2"
14.175 MHz	33'	3'-6"	76 pF	10'-1"
21.225 MHz	22'-1/2"	2'-4"	51 pF	6'-9"
28.5 MHz	16'-5"	1'-9"	38 pF	5'-1/2"

meters = ft × 0.305
mm = in. × 25.4

Fig. 26 — A twin-lead folded dipole makes an excellent portable antenna that is easily matched to 50-ohm stations.

Chapter 11

Code Transmission

Radiotelegraphy is a popular medium of amateur communications because it is highly effective and relatively uncomplicated. The process by which a radio signal is interrupted to generate a coded message of dits and dahs is called on-off keying or make-break keying. On-off keying may also be used in Radioteletype transmission, although this method is seldom used today. In the early years of Amateur Radio, keying a transmitter consisted of inserting a telegraph key in any convenient power lead — even the ac input! Such a haphazard approach today is an invitation to trouble, from the standpoints of signal quality and safety. Our cw subbands may shrink in the future. Successful operation under crowded conditions and adherence to FCC regulations concerning stability and purity of emissions requires clean signals. In this chapter we define good keying, analyze keying defects and show how radiotelegraphy can be implemented with modern technology.

On-off keying is a form of amplitude modulation and, as such, generates sidebands whose spacing from the carrier is proportional to the keying speed. A keying speed of 12 wpm corresponds to an information rate of 5 Hz, which theoretically requires 10 Hz of bandwidth. An untreated keying waveform, however, approaches *square-wave* modulation, consisting of the keying frequency plus *all of the odd harmonics*. These harmonics create sidebands extending many kilohertz either side of the carrier. They are called *key clicks*. Limiting the rise and decay times of the keying waveform to not less than 5 ms restricts the bandwidth of the transmitted signal to 100 Hz, theoretically allowing keying speeds up to 120 wpm. In practice, the keying at this speed would sound too "soft," but a 5 ms time constant is adequate for speeds up to 40 to 50 wpm.

To illustrate the key-click problem we used a square-wave model. In real life, the situation may be worse because the keying envelope often has a sharp spike on the leading edge and decays exponentially throughout the duration of the pulse. This type of wave is even richer in harmonics than a square wave. The severity of the problem is a function of power supply load regulation (see chapter 5). Note that in a class C amplifier the change in power output related to plate or collector voltage regulation is proportional to the *square* of the voltage change (see chapter 6). The power supply for a solid-state transmitter may be regulated easily and inexpensively

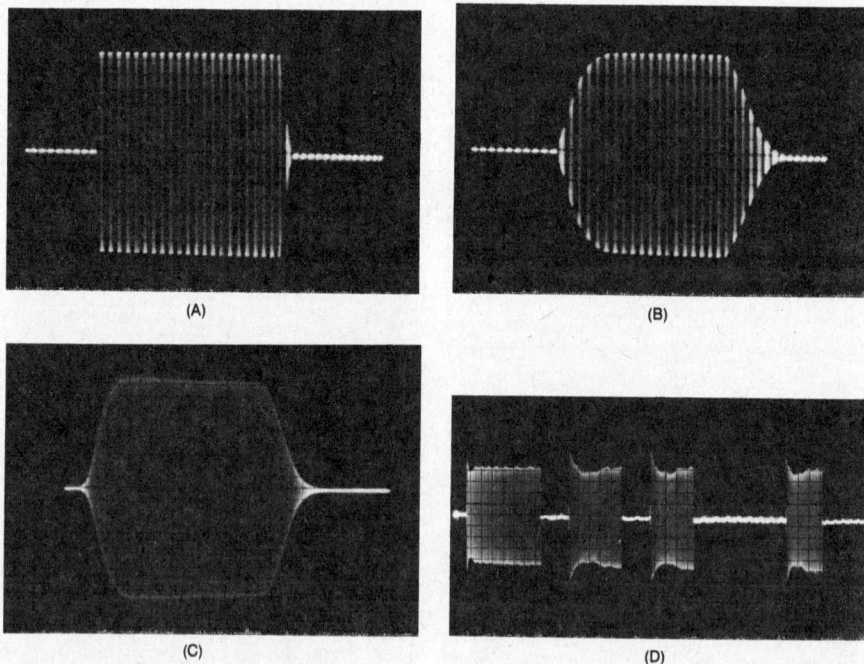

Fig. 1 — These photos show cw signals as observed on an oscilloscope. At A is a dot generated at a 46-baud rate with no intentional shaping, while at B the shaping circuits have been adjusted for approximately 5-ms rise and decay times. Vertical lines are from a 1-kHz signal applied to the Z or intensity axis for timing. Shown at C is a shaped signal with the intensity modulation of the pattern removed. For each of these photos, sampled rf from the transmitter was fed directly to the deflection plates of the oscilloscope.

A received signal having essentially no shaping is shown at D. The spike at the leading edge is typical of poor power-supply regulation, as is also the immediately following dip and rise in amplitude. The clicks were quite pronounced. This pattern is typical of many observed signals, although not by any means a worst case. The signal was taken from the receiver's i-f amplifier (before detection) using a hand-operated sweep circuit to reduce the sweep time to the order of one second. (Photos from October and November 1966 *QST*.)

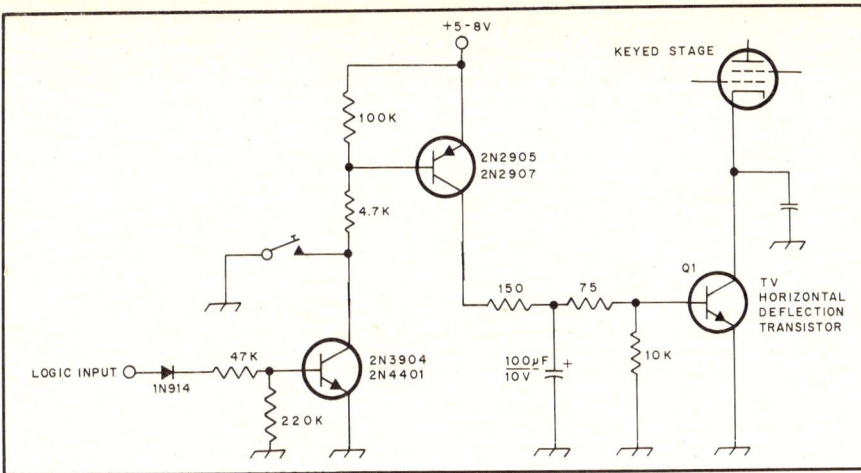

Fig. 2 — Cathode keying. Envelope shaping is accomplished by means of the RC network. Q1 must be able to withstand the plate voltage of the keyed stage. Some suitable types are: DTS-423, 2N6457 (400V), SDT 13305 (500V), DTS-801 (800V), MJ12010 (950V), 2SC1308K, ECG 238 (1500V). These are high-energy devices and are capable of switching any value of plate current the tube is likely to draw. For plate voltages below about 350, the 2N3439 is adequate (and much less expensive).

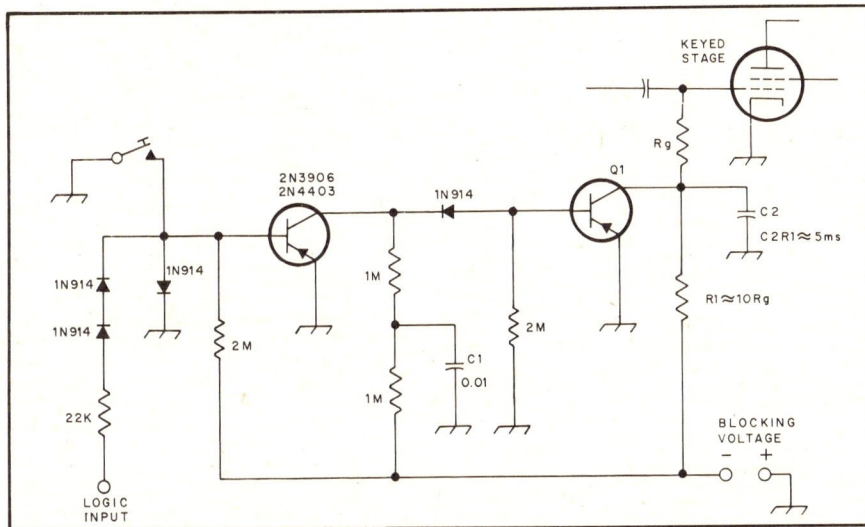

Fig. 3 — Blocked-grid keying. The rise time of the keying pulse is determined by C1 and its associated network. The decay time is governed by the R1C2 product. R_g is the existing grid leak. Typical values for R1 and C2 are 220 kΩ and 0.022 µF. Some transistors suitable for Q1 are: 2N5415 (200V), MM4003 (250V), MJE350 (300V), 2N5416, RCS882 (350V), 2N6213 (375V), 2N6214 (425V).

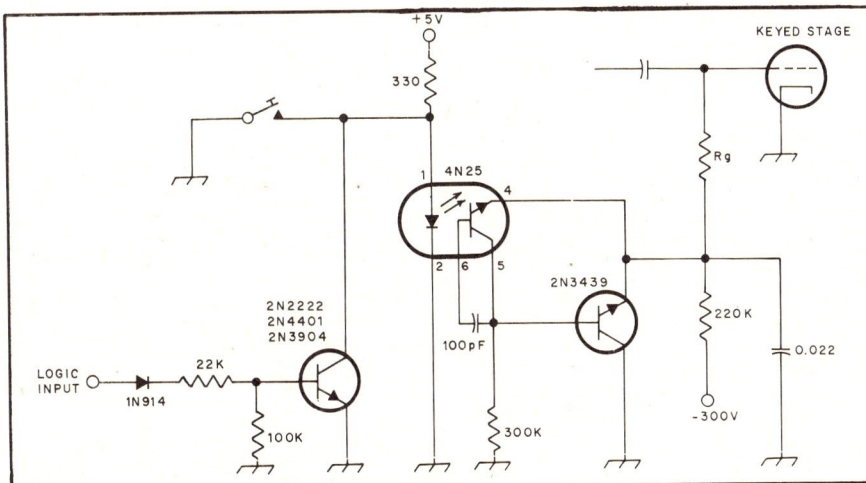

Fig. 4 — If a suitable high-voltage pnp transistor cannot be obtained, an npn unit can be used with an optical isolator. The rise time of the keying envelope is controlled by the "integrating" capacitor connected to the base of the phototransistor.

by electronic means, but such a scheme applied to a high-power tube type of transmitter is costly and dissipates considerable power. Fortunately, the plate voltage waveform can be corrected with a passive circuit (see Dome, May 1977 *QST*). Note also that the power source must be nearly pure dc to ensure that the transmitter output signal is not broadened by hum modulation.

Once the power supply voltage has been brought under control, it is a simple matter to shape the keying envelope with an RC network. The figures in this section illustrate the application of time-constant circuits to various keying methods.

When a circuit carrying current is opened or closed mechanically, a spark is generated. This spark causes the circuit to radiate energy throughout the electromagnetic spectrum. When a transmitter is keyed manually or through a relay, the spark at the contacts can cause local BCI, *but this spark has no effect on the rf output signal*. A simple filter (0.01-µF capacitor in series with 10 ohms) across the key or relay contacts will usually reduce the local clicks to a tolerable level. Solid-state switching methods significantly reduce the current and voltage that must be switched mechanically, thereby reducing local clicks and enhancing operator safety. Modern transistorized transmitters incorporate this type of keying. With proper device selection, solid-state keying may be implemented in older tube types of designs as well.

Amplifier tubes may be keyed in the cathode (filament transformer center-tap for directly heated types), grid-bias supply or screen. Transistors should be keyed in one leg of the collector supply. The low impedance of rf power transistor circuits usually requires the emitter to be grounded as directly as possible; therefore, no solid-state analog of cathode keying exists. Similarly, blocked-grid keying has no transistor equivalent, because a reverse bias sufficient to cut the stage off in the presence of heavy excitation would cause breakdown of the base-emitter junction.

Mechanical contacts frequently bounce several times before stabilizing in the closed state. The beginnings of keying pulses formed by bouncing contacts are poorly defined. This defect can degrade the readability of a code signal under adverse conditions. Relays and semiautomatic keys are especially prone to this malady. The circuit of Fig. 9 will help clean up the pulses generated by mechanical contacts.

A satisfactory code signal can be amplified by means of a *linear* amplifier without affecting the keying characteristics. If, however, the signal is amplified by one or more nonlinear stages (e.g., a class C multiplier or amplifier), the signal envelope will be modified, possibly introducing significant key clicks. It is possible to compensate for this effect by using longer-than normal rise and decay times in the exciter and letting the amplifier modify the signal to an acceptable one. Any clicks generated by a linear amplifier are likely to be the result of low-frequency parasitic oscillations.

A change in frequency at the beginning of a keying pulse is called a *chirp*. If the oscillator isn't keyed, chirp is the result of changing dc operating potentials or changing rf load conditions on the oscillator. The voltage to the oscillator can be regulated easily, as most transmitters use fairly low power oscillators. If the oscillator frequency is pulled by the loading effect of subsequent keyed stages, better load isolation is indicated. Chapter 6 gives a thorough treatment of buffering techniques.

If break-in operation is desired (see below), it may be necessary to key the transmitter's oscillator. Oscillators may be keyed by the same methods used for amplifiers, but greater care is required to obtain good results. In general, the goals of clickless and chirpless oscillator keying are mutually exclusive. This is because a key-click filter will cause the operating voltage to be applied *slowly*, thereby creating a chirp. Crystal oscillators may be keyed satisfactorily if active crystals are used. A keyed oscillator may exhibit a continuous frequency change during a keying pulse. This defect is called a *yoop*. It is caused by rf heating in the tank circuit or crystal. Yoop is usually an indication of a faulty component or excessive oscillator power.

Break-in operation with a VFO-controlled transmitter usually dictates some form of *differential* keying. In this system, the oscillator is turned on as quickly as possible and the amplifier is keyed after the oscillator has stabilized. When the key is released, the oscillator operates until the amplifier output has decayed to zero. Shaping of the keying envelope is accomplished in the amplifier keying circuit. In the past, break-in operation was implemented with VR tubes and relays. The complexity of such systems frightened many hams away from this convenient mode. With modern circuitry, break-in can be simple, quiet and safe. One method of break-in keying is shown in Fig. 10.

A few notes concerning oscillator keying are in order. Do not attempt to key an oscillator unless it is stable while running free, at the same time the other stages are keyed. If the oscillator frequency is multiplied, chirps and yoops will be multiplied also. The transmitter signal should be checked at the highest operating frequency. It may be found that the signal quality is satisfactory on 160 meters but leaves something to be desired on 10. Modern transmitters and transceivers provide multiband operation by the heterodyne method (see chapter 6) and should be stable on all bands.

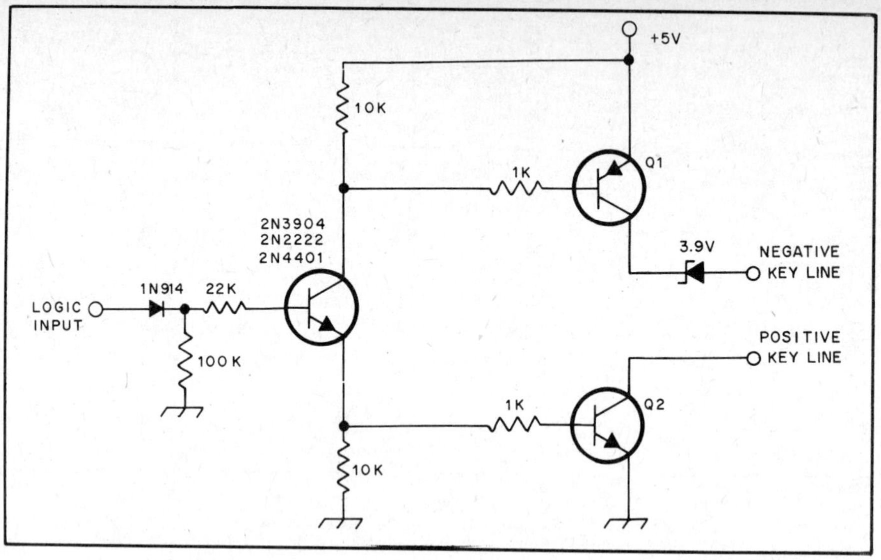

Fig. 5 — Circuit to interface digital logic with positive or negative key lines. Q1 and Q2 must be able to withstand the expected negative and positive keying voltages and currents.

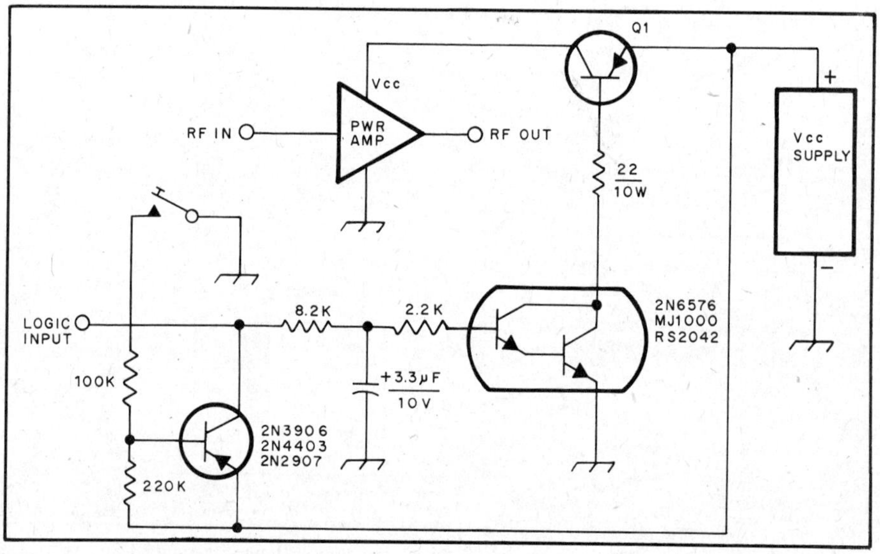

Fig. 6 — Keying circuit for solid-state Class C amplifiers in the 100-watt class, such as those sold for vhf fm service. Q1 must be able to pass the amplifier collector current without dropping too much voltage. Types 2N6246, SK3173 and RS2043 are good for currents up to 15 amperes.

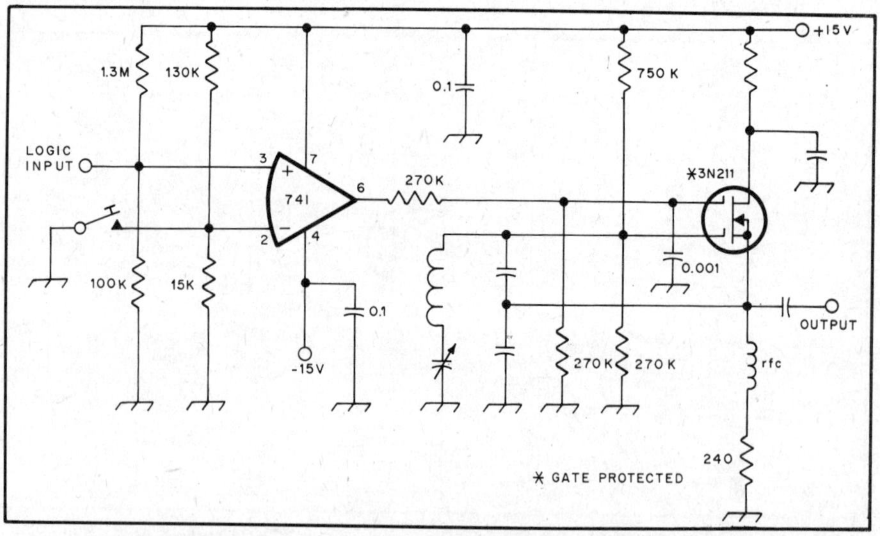

Fig. 7 — Keying a dual-gate MOSFET oscillator. The 741 op-amp is used as a comparator. With the input resistors shown, the circuit can be triggered by any +5-volt logic device.

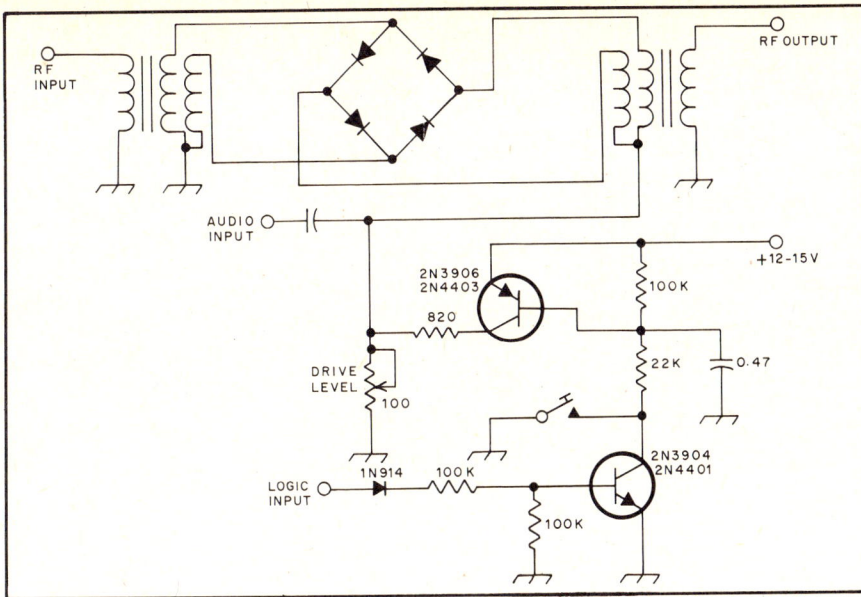

Fig. 8 — Keying a doubly balanced modulator in a cw-ssb transmitter.

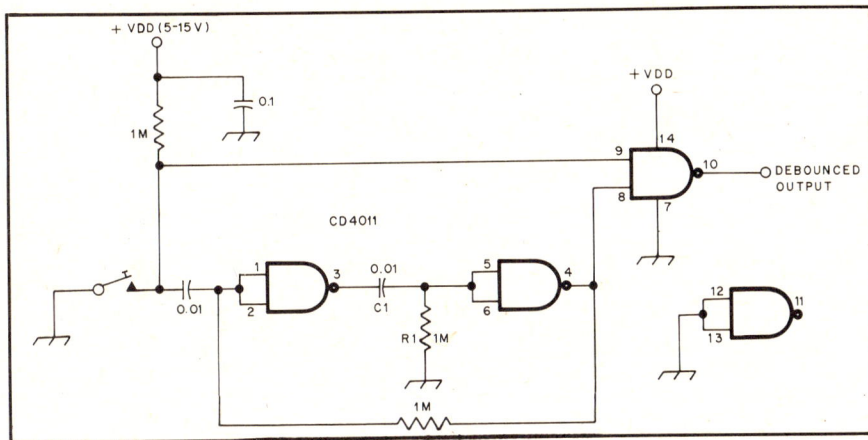

Fig. 9 — Debouncing circuit for hand keys and relay contacts. The minimum dit length is determined by the R1C1 product.

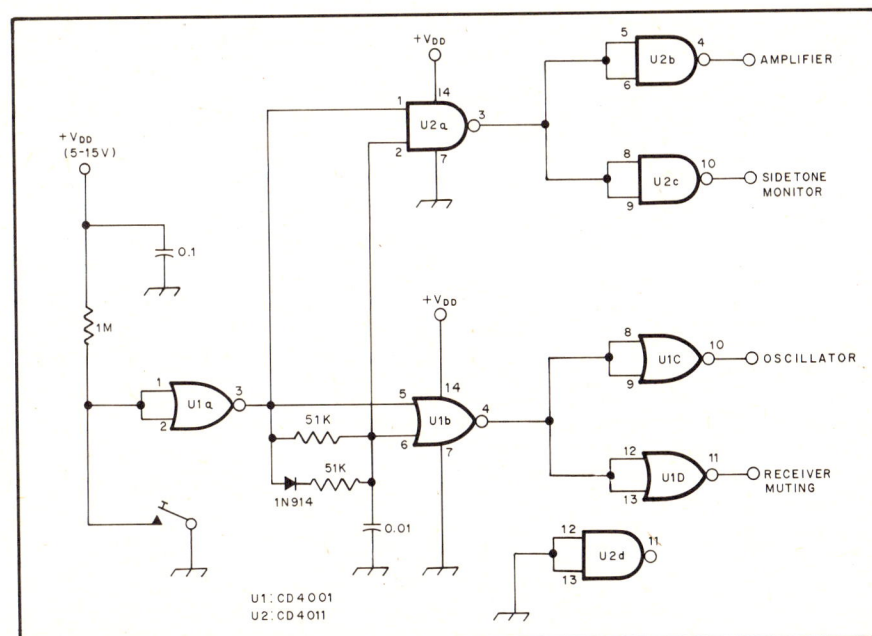

Fig. 10 — Differential (sequential) keying system for fast break-in with oscillator-multiplier transmitters.

If the transmitter oscillators run continuously, they may be audible as a *backwave* between keying pulses. A strong backwave may indicate the need for neutralizing one or more transmitter stages. In general, if the backwave conforms to the -40 dB spurious signal rule, it won't be objectionable.

The figures in this section illustrate methods by which various solid-state and thermionic devices may be keyed. In these circuits, the armature of the hand key is at ground potential and the voltages across the key are imperceptible. The current through the key is generally less than one milliampere. A neon bulb with a proper series resistor across the key will alert the operator to junction breakdown of the high-voltage transistors. As long as safety is given due consideration, the key-at-ground convention need not be followed, but this standardization is useful for equipment interconnections. Digital control is shown in all of the examples. This feature simplifies the simultaneous keying of transmitter stages, T-R switches, sidetone oscillators and muting systems. The ICs used to perform the control functions are very plentiful and inexpensive. These systems use a logic "one" to indicate a key-down condition.

Break-In

Break-in (QSK) is a system of radiotelegraph transmission in which the station receiver is sensitive to other signals between the transmitted keying pulses. This capability is very important to traffic handlers, but can be used to great advantage in ragchewing as well. Break-in gives cw communication the dimension of more natural conversation.

Most commercially manufactured transceivers feature a "semi break-in" mode in which the first key closure actuates the VOX relay. The VOX controls are usually adjusted to hold the relay closed between letters. With proper VOX adjustment, it is possible for the other operator to break your transmission between words, but this system is a poor substitute for true break-in.

Separate Antennas

The simplest way to implement break-in is to use a separate antenna for receiving. If the transmitter power is low (below 50 watts or so) and the isolation between transmitting and receiving antennas is good, this method can be satisfactory. Best isolation is obtained by mounting the antennas as far apart as possible and at right angles. Smooth break-in involves protecting the receiver from permanent damage by the transmitter power and assuring that the receiver will "recover" fast enough to be sensitive between keying pulses. If the receiver recovers fast enough but the transmitter clicks are bothersome (they may be caused by receiver overload and so exist only in the receiver) their effect on the operator can be minimized through the use of an output limiter. The separate antenna method is most useful on the 160-, 80- and 40-meter bands, where the directional effects of the antennas aren't pronounced.

Switching a Common Antenna

When powers above about 50 watts are used, where two antennas are not available, or when it is desired to use the same antenna for transmitting and receiving (a "must" when directional antennas are used), special treatment is required for quiet break-in operation on the transmitter frequency.

Vacuum relays or reed switches may be used to switch the antenna between the transmitter and receiver in step with the keying. This method is satisfactory for power levels up to the legal limit, but the relays are expensive and the system timing is critical.

Perhaps the most modern and elegant approach is the use of PIN diodes to switch the antenna. These devices are available in power ratings up to about 100 watts, but are quite expensive at present. There are no keying-speed constraints when PIN diodes are used, and if the proper devices are selected, the spectral purity of the output signal won't be affected. The important electrical parameter in this regard is *carrier lifetime*.

An easy and economical way to implement break-in with a single antenna is to use an electronic *T-R switch*. With such a device the antenna is connected to the transmitter at all times. In the most common type of electronic T-R switch, a tube is used to couple the antenna to the receiver. When the transmitter is keyed, the rf output causes the tube to draw grid current through a high-resistance grid leak. The high negative bias thus developed cuts off the plate current, limiting the signal delivered to the receiver.

Unfortunately, when the grid circuit is driven into rectification, harmonics are generated. A commercially manufactured low-pass filter after the T-R switch can help to eliminate TVI caused by the harmonics, but the lower-order harmonics may cause interference to other communications. Another common shortcoming of T-R switches is that the transmitter output circuit may "suck out" the received signal. In a transmitter having a high-impedance tuned output tank circuit, both of these problems can be circumvented by connecting the T-R switch to the *input* side of the tank circuit. With this configuration, the grid rectification harmonics are suppressed and the received signal peaked by means of the tuned circuit. Fig. 11 shows a MOSFET T-R switch that works on this principle.

A T-R switch for use external to any transmitter is shown in Fig. 12. The tube is grid-block keyed, and the fixed-value bias

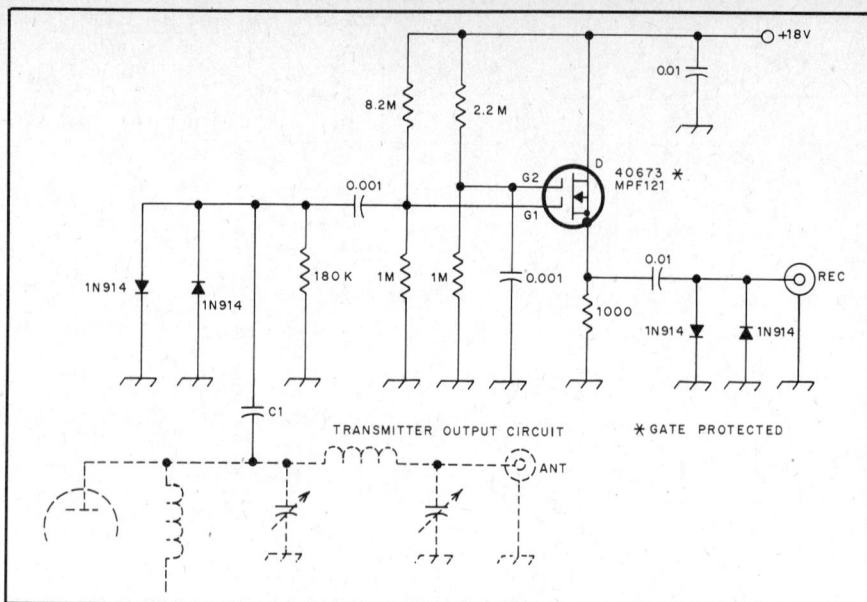

Fig. 11 — A T-R switch can be connected to the input side of the transmitter pi network. For powers up to about 100 W, C1 can be a 5-10 pF, 1000-V mica unit. For high-power operation a smaller "gimmick" capacitor made from a short length of coaxial cable should be used.

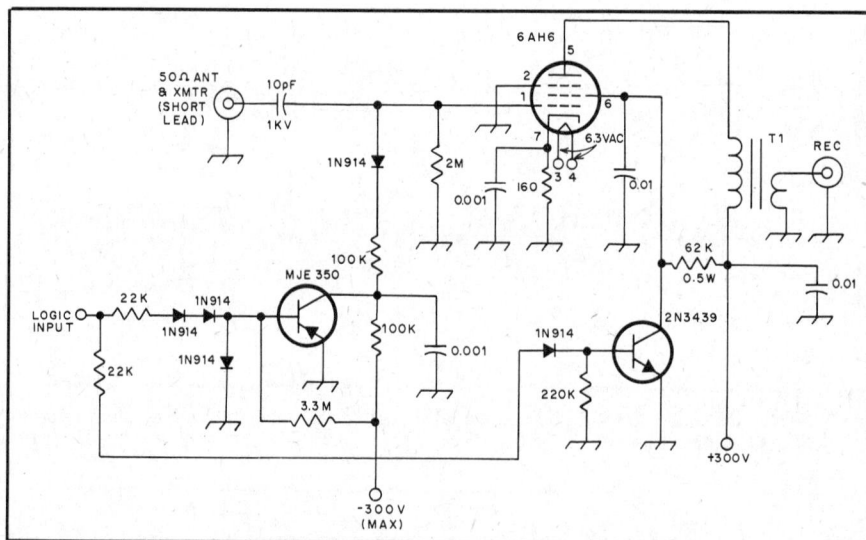

Fig. 12 — An external T-R switch. The primary of T1 is 50 turns of no. 30 enameled wire on an FT37-43 toroid core. The secondary is 15 turns of no. 30 wound over the primary.

prevents any grid current flow at power levels up to 800 watts. This power figure assumes a 50-ohm system with a unity VSWR. The circuit can withstand peak rf voltages up to 300. The power capability must be derated if the impedance at the point of connection is higher than 50 ohms. Although the signal path has a diode, it is effectively "linearized" by the high-value series resistor, and should not significantly degrade the spectral purity of the transmitter output.

External T-R switches should be well shielded and the power leads carefully filtered. In general, the coaxial cable to the transmitter should be as short as possible, but some experimenting may be necessary to eliminate "suckout" of the

received signal. It is commonly stated that electronic T-R switches are usable only with transmitters having Class C output stages because the "diode noise" generated by the resting current of a linear amplifier will mask weak signals. Actually, the class-of-service designation is not related to key-up conditions, so there is no reason that a linear amplifier can't be biased off during key-up periods.

Reduction of Receiver Gain During Transmission

For absolutely smooth break-in operation with no clicks or thumps, means must be provided for momentarily reducing the gain through the receiver. A muting function completely disables the receiver

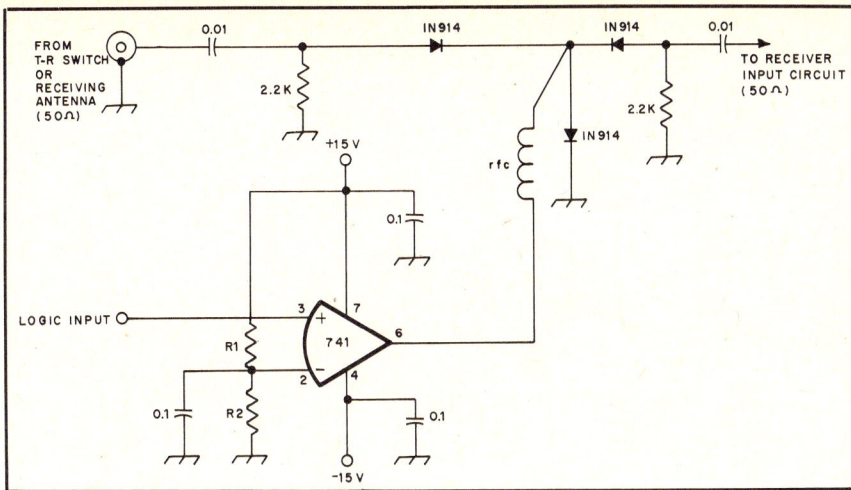

Fig. 13 — A diode attenuator for receiver gain reduction during keying. The logic threshold is determined by $\dfrac{R1 + R2}{R2}$ (+VCC). For +15-volt CMOS logic, R1 = R2 = 1M. For +5-volt TTL operation, R1 = 130 kΩ, R2 = 15k.

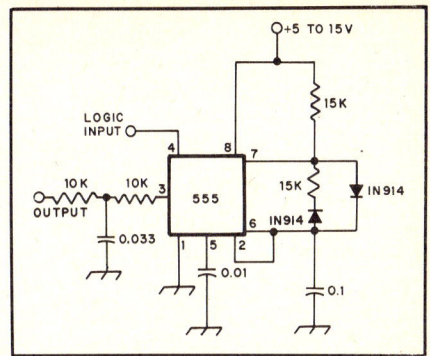

Fig. 15 — A 555 universal timer used as a sidetone generator. Pin 4 is taken to ground to interrupt the tone. The frequency of oscillation is about 500 Hz with the constants shown.

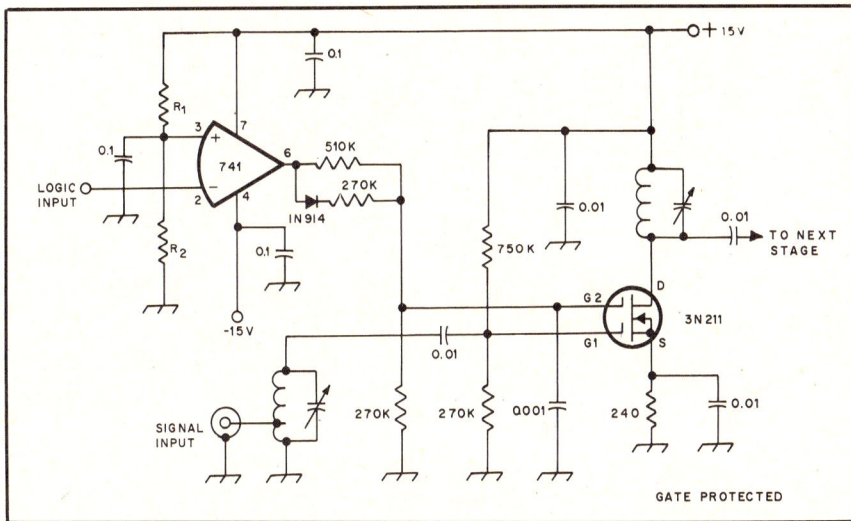

Fig. 14 — Gain-reduction circuit for receivers using a fixed-bias dual-gate MOSFET in the first stage. As much as 40 dB of attenuation is possible with this method. The logic threshold is calculated as in Fig. 13.

audio during key-down periods. Assuming the transmitter signal at the receiver is held below the damage level, muting the audio output can be an effective means of achieving smooth break-in, *provided no agc is used.* Agc systems suitable for cw operation are characterized by long "hang" times. Unless the transmitter signal in the receiver is at a level similar to that of the other station, the agc system will seriously desense the receiver, rendering the break-in system useless. A diode attenuator suitable for use with T-R switches or separate antennas is shown in Fig. 13. If the receiver uses a dual-gate MOSFET with no agc in the first stage, the method of Fig. 14 may be used.

Monitoring

If the receiver output is muted, an audio sidetone oscillator must be used to monitor one's sending. A 555 timer connected as an astable multivibrator is commonly used for this purpose. This device delivers rectangular output pulses, and the resulting signal often sounds quite raucous. A variation of the standard 555 circuit appears in Fig. 15. The diodes maintain the symmetry of the waveform independently of the pitch and the RC filter removes many of the objectional harmonics. A keying monitor can be powered by the rf output of the transmitter. Such a circuit is shown in Fig. 16. Keying monitors often have built-in loudspeakers, but it is less expensive and more convenient to inject the monitor signal into the audio output stage of the receiver. With this system one always hears his sidetone from the same source (speaker or headphones) as the other station's signal.

If the audio output isn't muted, the receiver can be used to monitor one's keying, provided both stations are on the same frequency. Some DX operators transmit and listen on separate frequencies. When using your receiver as a monitor, you should be careful about drawing any conclusions concerning the quality of your signal. The signal reaching the receiver must be free of any line voltage effects induced by the transmitter. To be certain of your signal quality you should listen to your station from a distance. Trading stations with a nearby amateur is a good way to make signal checks.

Keying Speeds

In radiotelegraphy the basic code element is the dot, or unit pulse. The time duration of a dot and a space together is that of two unit pulses. A dash is three unit pulses long. The space between letters is three unit pulses; the space between words or groups is seven unit pulses. A speed of one *baud* is one pulse per second.

Assuming that a speed key is adjusted to give the proper dot, space and dash values mentioned above, the code speed can be found from

$$\text{Speed (wpm)} = \frac{\text{dots/min}}{2.5} =$$

$2.4 \times \text{dots/sec.}$

For example, a properly adjusted electronic key gives a string of dots which are counted as 10 dots per second. Speed = $2.4 \times 10 = 24$ wpm. Many modern electronic keyers use a clock or pulse-generator circuit which feeds a flip-flop dot generator. For these keyers the code speed may be determined directly from the clock frequency

$$\text{Speed (wpm)} = 1.2 \times \text{clock frequency (Hz)}$$

For a quick and simple means of determining the code speed, send a continous string of dashes and count the number of dashes which occur in a five-second period. This number, to a close approximation, is the code speed in words per minute. A method for checking the speed of a Morse keyboard is to send a continuous string of zeros, with proper interletter spacing. Most keyboards will automatically insert the correct space if the key is released and reactivated before the end of the character. If zeros are sent for one minute, the speed is approximately

$$\text{Speed (wpm)} = (\text{zeros/min}) \times 0.44.$$

A Single IC Keyer

Fig. 17 shows a circuit for an electronic keyer built around a single 8044 CMOS integrated circuit. Features designed into this IC include contact debouncing, rf immunity and self-completing character generation. A single- or dual-lever paddle may be used. If operated with a dual-lever paddle, the 8044 is capable of *iambic* operation (an alternating series of dots and dashes is produced when both contacts are closed). A dot memory is included to prevent loss of dots caused by the operator leading the keyer. The circuit shown incorporates a weight control to alter the dot-space ratio. This feature allows partial compensation for delays and slow rise times in the transmitter.

The 8044 has a built-in sidetone generator with adjustable pitch. At the keyer output, a high logic level represents the key-down state. This output can source enough current to turn on most npn transistors. With the timing components shown, the speed range is about 8-50 wpm. The quiescent current drain of this keyer is on the order of 50 microamperes, so no on-off switch is required. This keyer is ideal for mounting inside a transceiver for station compactness. The M-suffix version of the 8044 has an 18-pin package and contains circuitry for driving an analog speed readout. The price class is $15 for the 8044 and $20 for the 8044M. These devices are marketed by Curtis Electro Devices, Inc., Box 4090, Mountain View, CA 94040.

An Inexpensive Buffered Morse Keyboard

Until recently, the price of commercially manufactured keyboards and the cost of many home-built units have made the keyboard keyer an expensive station component. But, with the explosive growth of home computers, large supplies of surplus keyboards and digital-logic material have been made available at reasonable prices. The Morse keyboard described in earlier Handbooks was a cost-effective keyer that worked well but lacked some of the deluxe features found on commercial units. This revised version is an extension of the previous design, with a 32-character buffer memory added. An investment of about $50 (1980 prices) can secure the parts for an alert shopper. Al Helfrick, K2BLA, did the electrical design work. The pc-board development, mechanical design and construction were performed in the ARRL laboratory.

Circuit Description

A 64-kHz scan oscillator drives a binary counter. The Q outputs of the counter are used to program the multiplexers. Each combination of scanning lines is selected in sequence until a key-switch closure is detected. When the feedback loop is closed the scan oscillator is inhibited. At this point the binary number represented by the counter-output states is loaded into the first-in/first-out (FIFO) registers. This parallel data ripples down the first set of FIFOs directly into the second set. If all four FIFOs are full they inhibit further data entry. The parallel data from the FIFOs is then converted to serial form by

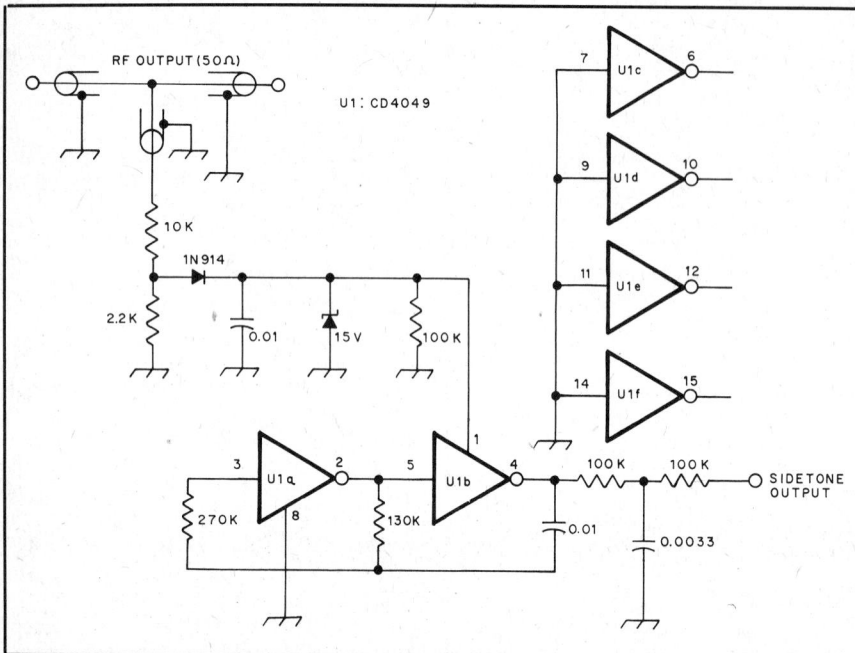

Fig. 16 — An rf-powered keying monitor suitable for power outputs from about 20 to 100 W. This circuit should be installed inside the transmitter or a shielded enclosure to minimize RFI.

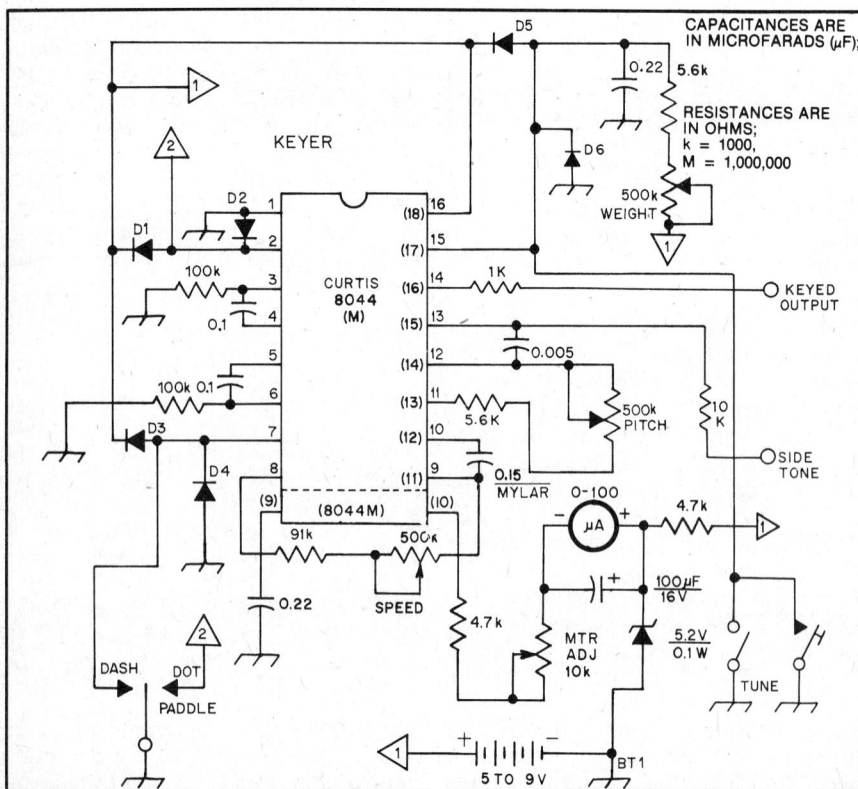

Fig 17 — Circuit diagram for the keyer. D1-D6, incl., 1N270 or equiv. All potentiometers are linear taper and resistors are 1/2- or 1/4-watt. The speed meter must have a moving coil type of movement. Pin numbers in parentheses are for the 8044M IC.

the two shift registers.

The seven-bit (six data plus one carry out) data format has 128 possible combinations. By connecting the key switches to the proper scanning lines (given in Table 1), the combinations corresponding to the valid Morse characters are selected. In the keyer, binary zero represents dot, and binary one represents dash. An extra zero is added as an "end bit," and the characters are sent from right to left. All of the characters must be represented by seven-bit binary numbers where all of the unused bits to the left of the end bit are ones. An optional circuit "examines" the contents of the FIFOs and displays the buffer status on three LEDs. The complete circuit is given in Figs. 18, 21 and 22. This circuit description is necessarily brief. Helfrick's original *QST* article (January 1978) contains a more comprehensive treatment of the keyboard circuitry.

Hardware Assembly

The keyer circuits are assembled on two 6 × 4-1/2-inch (152 × 114-mm) double-sided pc boards. One board holds the key-switch encoder and buffer, while the other holds the shift registers and output decoding circuitry. Figs. 23 through 26 contain the etching patterns and component-placement guides for these boards. The two boards are designed to be stacked one over the other. A small third circuit board holds the optional sidetone oscillator and buffer indicator. This is a single-sided board and its etching pattern and parts layout appear in Fig. 27 and 28. The buffer indicator draws 100 times the current of the main circuit; therefore a separate power switch should be used to disable the indicator. All of the ICs are CMOS units. The oscillator circuits using gates require B-series devices, but any series can be used elsewhere. The keyboard will work on any dc voltage from 5 to 15; a regulated supply isn't necessary. Penlight cells will work fine, but if the optional circuitry is included, the large C or D cells should be used.

Figs. 19 and 20 reveal some of the construction details. An Apollo cabinet houses the assembly, but a homemade enclosure will work as well. The dimensions and mounting arrangements will be dictated by the keyboard the builder obtains.

Operating the Keyboard

Sending Morse code with this keyer is very much like typing. Character spacing is automatic (provided one types in step with or ahead of the output), and word spaces are made with the space bar. With the buffer the operator may type up to 32 characters ahead of the output. The unit has a "dump" switch, so if the operator must abort a transmission, pushing the button will instantly terminate the code output.

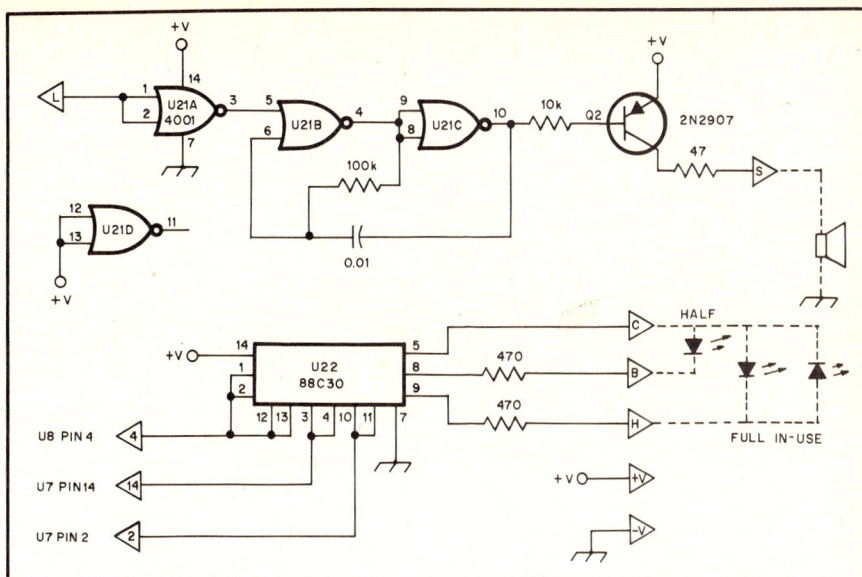

Fig. 18 — Optional sidetone generator and buffer status indicator.
U21 — CD4001B quad 2-input NOR gate. U22 — 88C30 dual differential line driver.

Table 1
Keyswitch Connections

Character	From	Connect To
A	B7	A1Ø
B	B6	A1
C	B6	A5
D	B7	A1
E	B7	A12
F	B7	A4
G	B7	A3
H	B6	AØ
I	B7	A8
J	B6	A14
K	B7	A5
L	B6	A2
M	B7	A11
N	B7	A9
O	B7	A7
P	B6	A6
Q	B6	A11
R	B7	A2
S	B7	AØ
T	B7	A13
U	B7	A4
V	B6	A8
W	B7	A6
X	B6	A9
Y	B6	A13
Z	B6	A3
1	B5	A14
2	B5	A12
3	B5	A8
4	B5	AØ
5	B4	AØ
6	B3	A1
7	B4	A3
8	B4	A7
9	B4	A15
SK	B2	A8
AR	B4	A1Ø
KN	B4	A13
AS	B4	A2
BT	B5	A1
AA	B6	A1Ø
Comma	B3	A3
Period	B2	A1Ø
Question Mark	BØ	A12

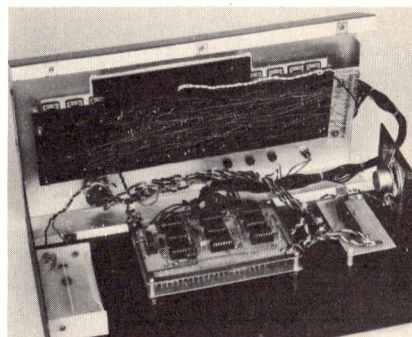

Fig. 19 — Interior view of the keyboard. Heavy bus wire is used for the pc-board interconnections. A more recent layout (in the same cabinet) places the boards side-by-side for easier access.

Fig. 20 — The completed Morse keyboard measures 14-1/2 × 8-1/4 × 3-1/4 inches (368 × 210 × 83 mm).

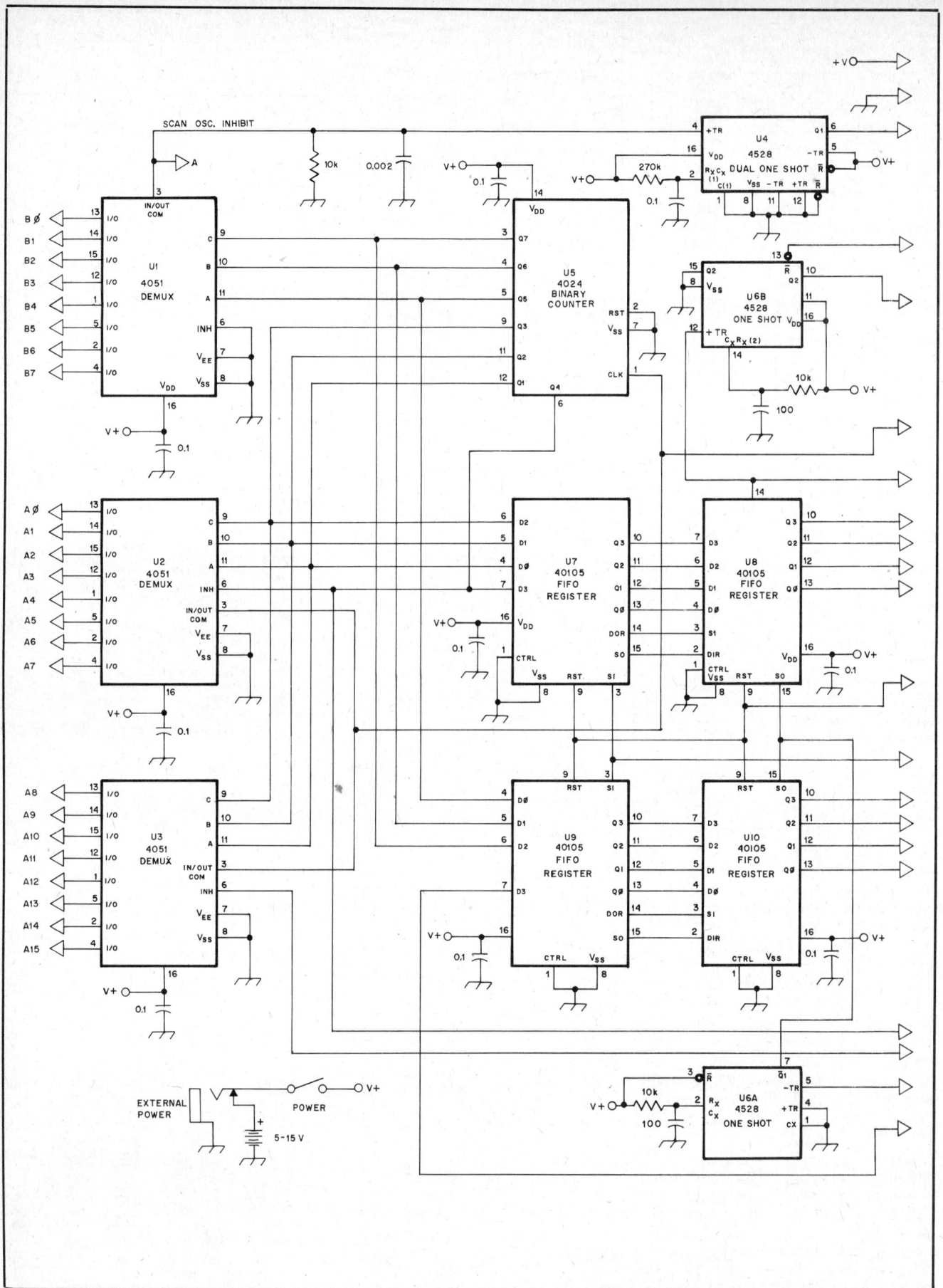

Fig. 21 — Keyswitch encoder and buffer circuits for the K2BLA keyer.
U1, U2, U3 — CD4051B decoder/multiplexer.
U4, U6 — MC14528, MC14538 or CD4098 dual monostable multi-vibrator.

U5 — CD4024B binary counter.
U7, U8, U9, U10 — CD40105 F1FO register.

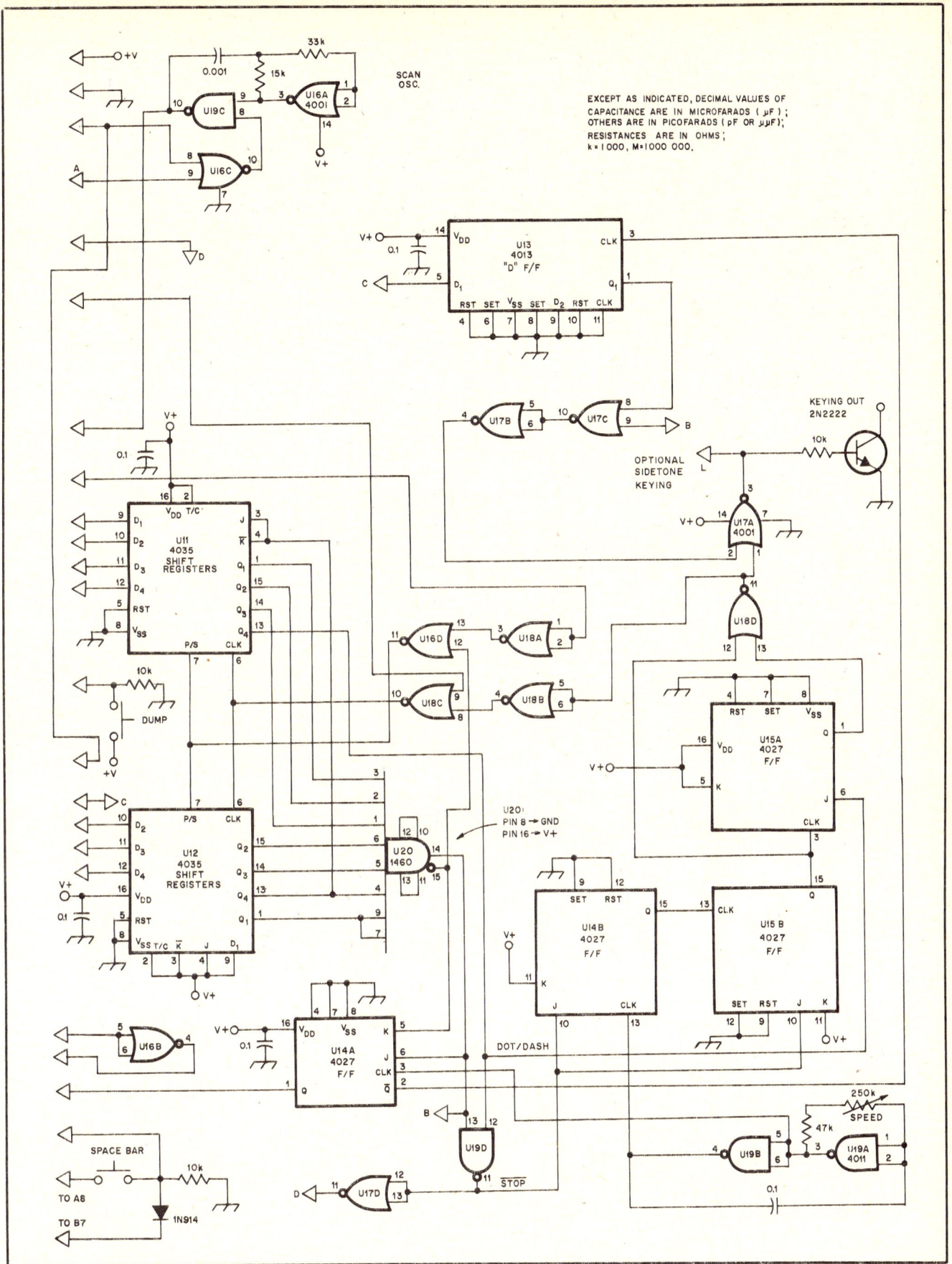

Fig. 22 — Decoder and output circuits for the keyboard. The open-collector output is suitable for transmitters having positive key lines up to 40 V. Arrangements for other key lines are given earlier in the chapter.

U11, U12 — CD4035B parallel shift register
U13 — CD4013B dual D flip-flop
U14, U15 — CD4027B dual JK flip-flop
U16, U17, U18 — CD4001B quad 2-input NOR gate
U19 — CD4011B quad 2-input NAND gate
U20 — MC14501 8-input NAND/AND gate

Code Transmission 11-10

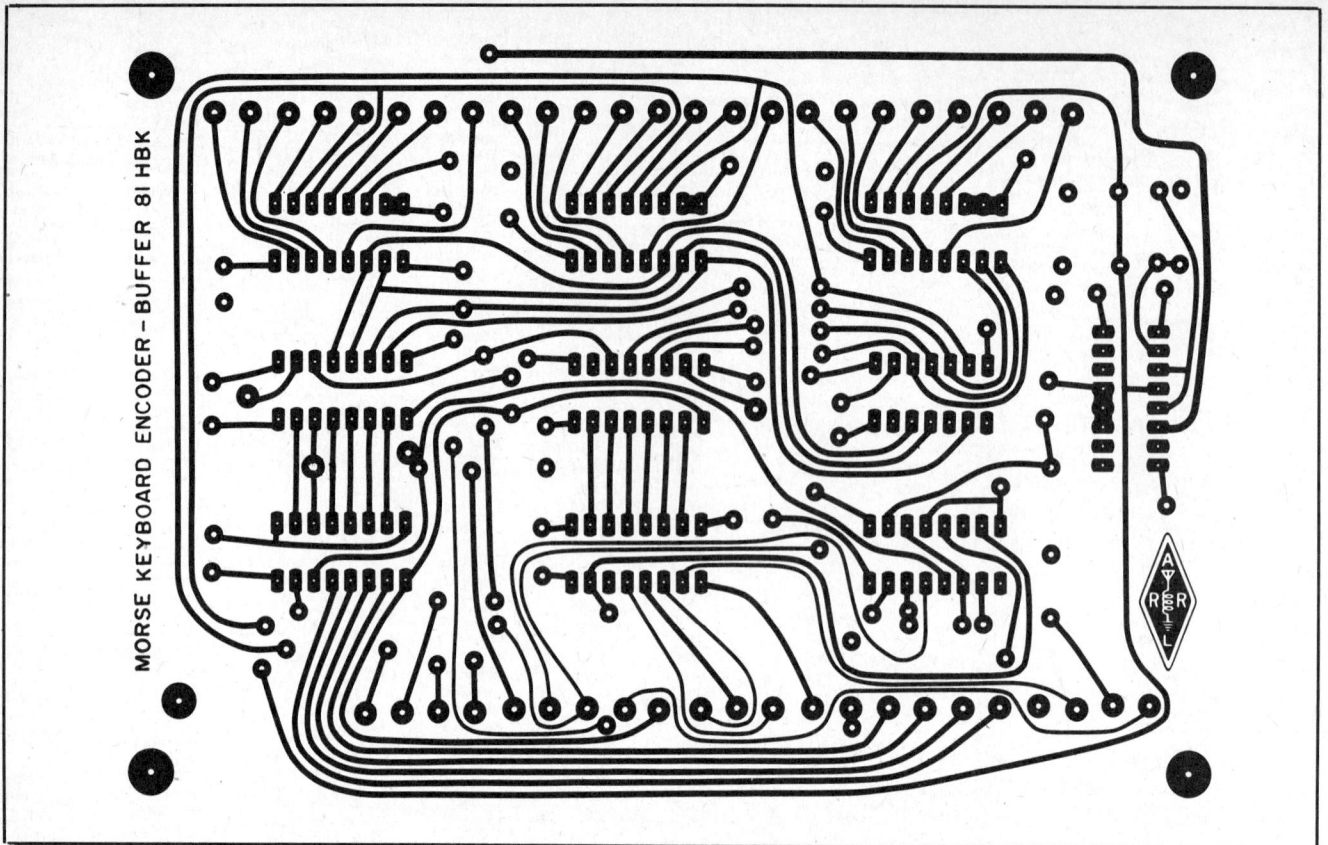

Fig. 23 — Bottom-side etching pattern for the encoder/buffer circuit board. Black represents unetched copper.

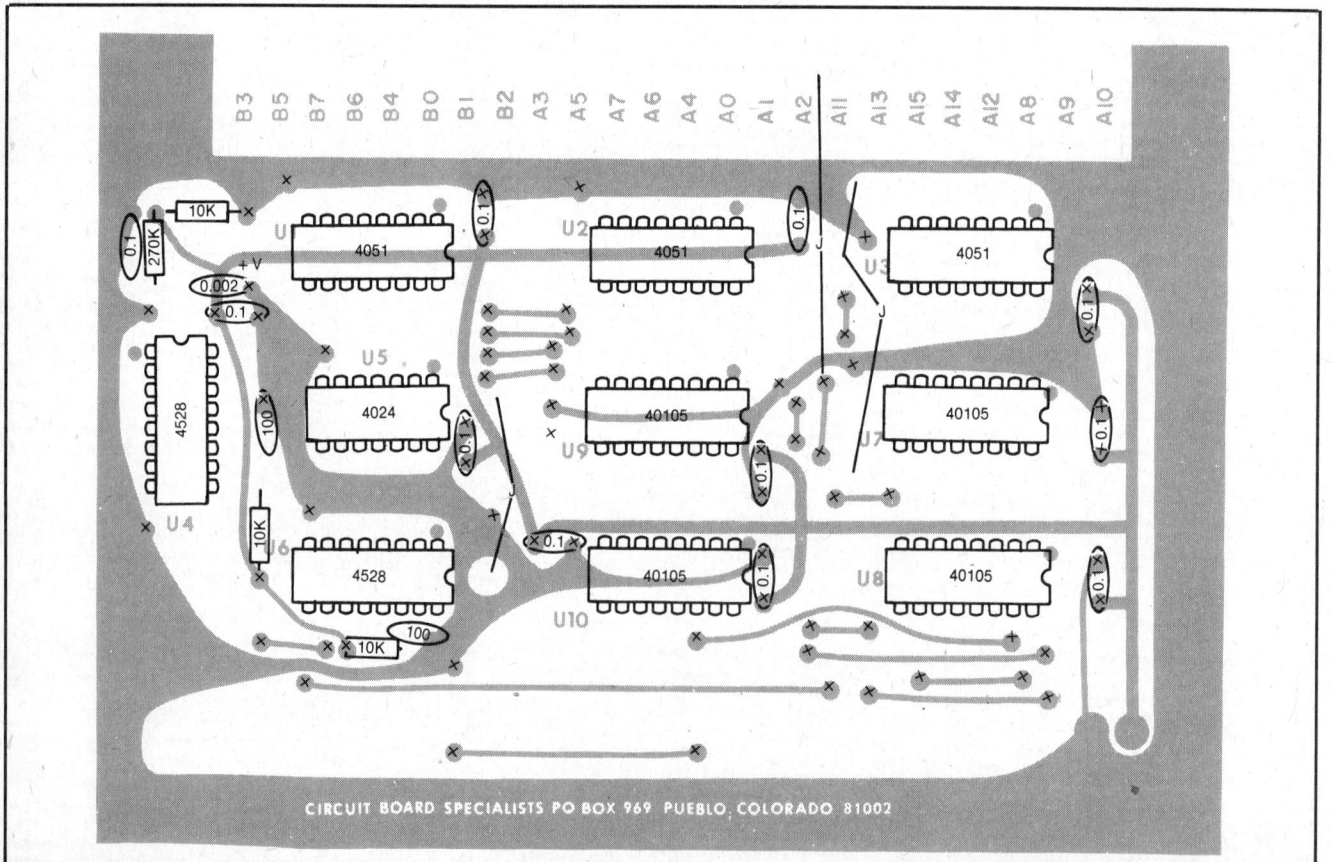

Fig. 24 — Component-side etching pattern and component-placement guide for the encoder/buffer board. X = feedthrough connection. J = wire jumper.

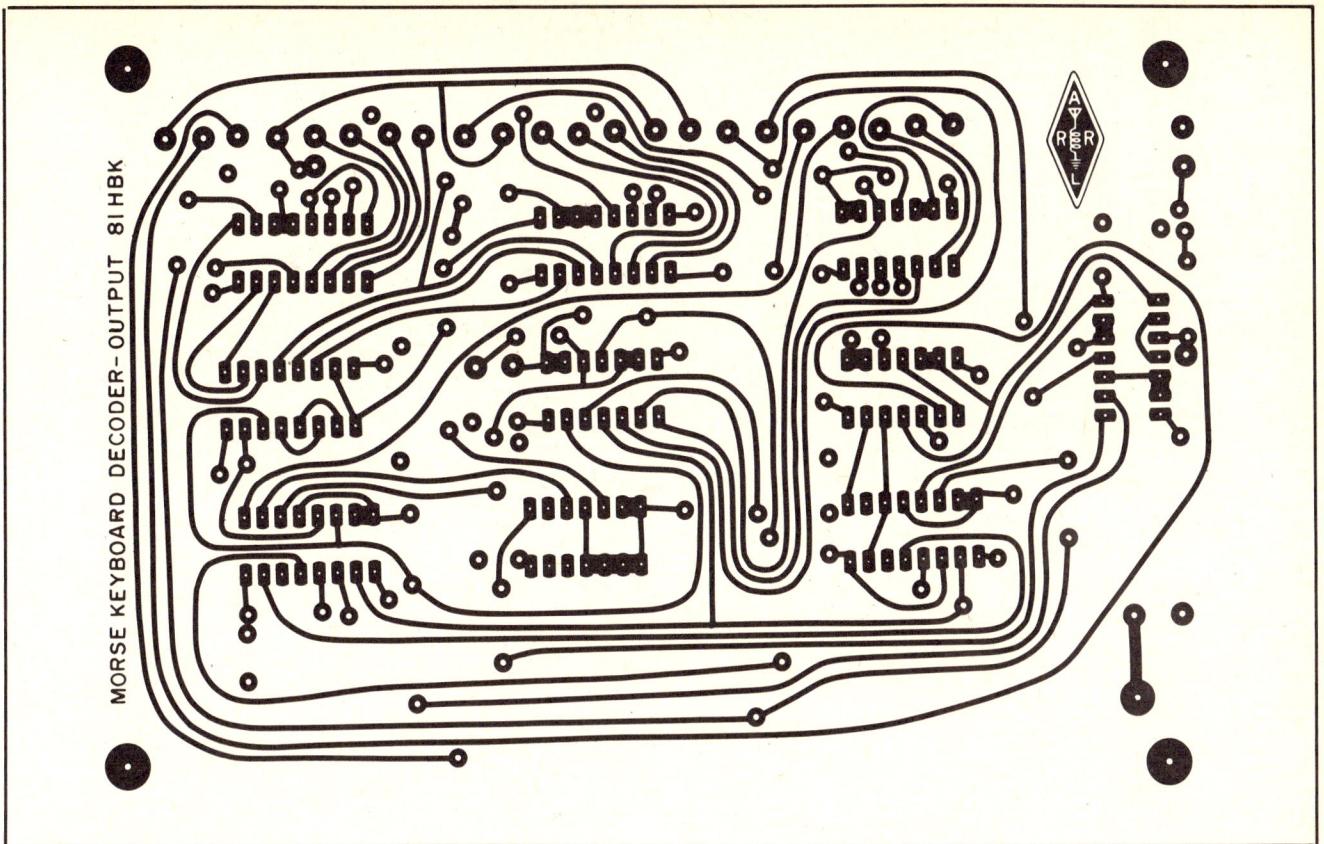

Fig. 25 — Bottom-side etching pattern for the decoder/output board.

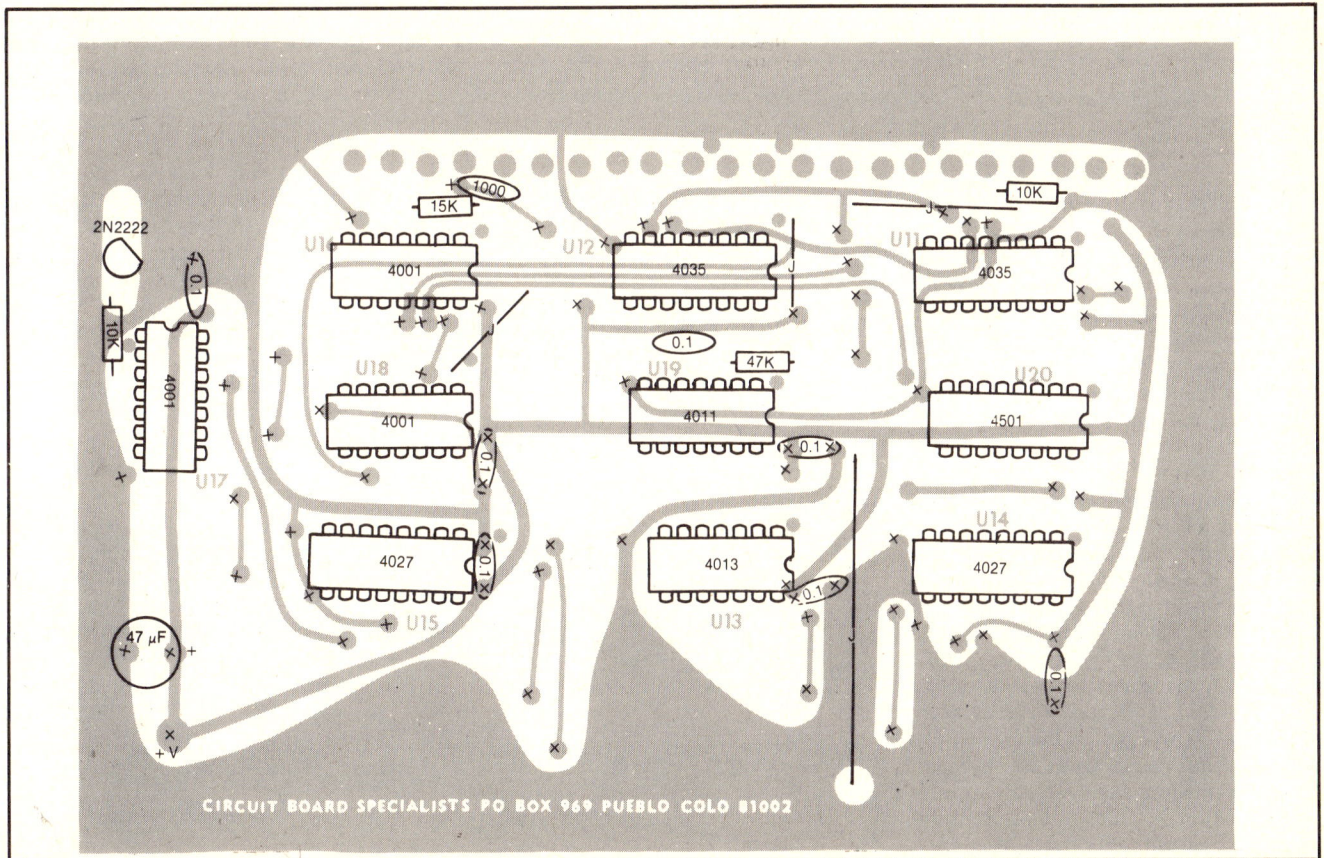

Fig. 26 — Component-side etching pattern and component-placement guide for the decoder/output board. X = feedthrough connection. J = wire jumper.

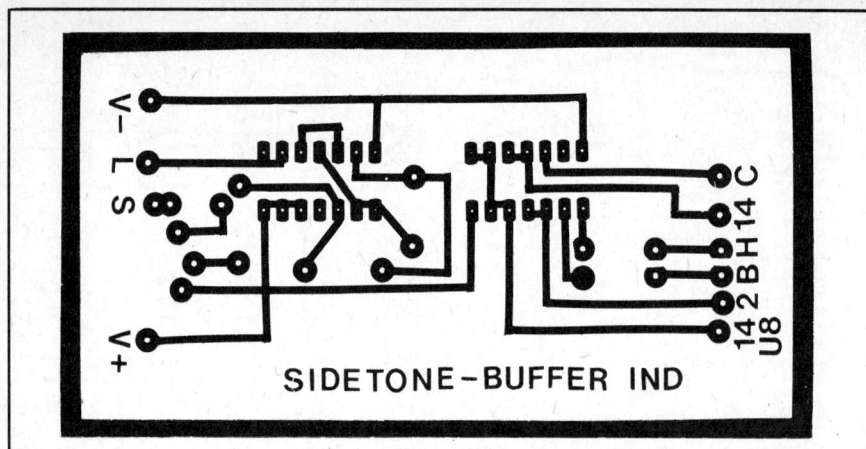

Fig. 27 — Etching pattern for the buffer indicator/sidetone generator board, shown from the foil side.

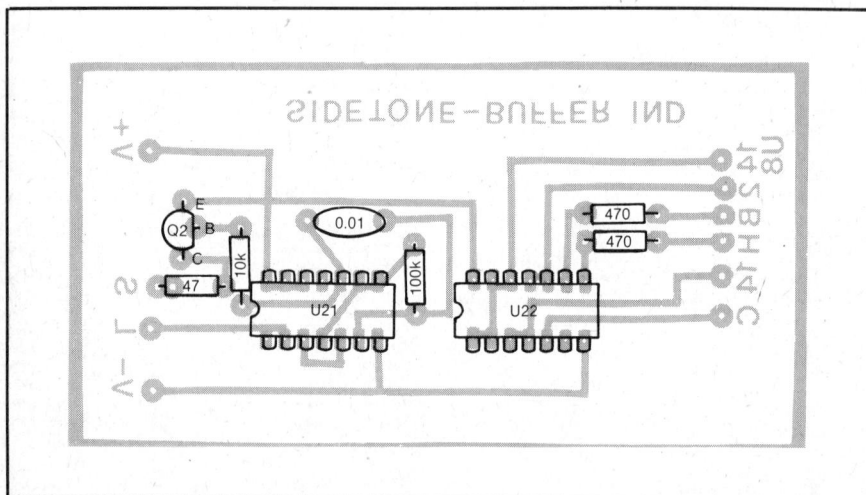

Fig. 28 — Component-placement guide for the buffer indicator/sidetone generator board, with an x-ray view of the foil.

A PIN Diode T-R Switch

The T-R switch system described here is usable with most any 100-W (output) power level transmitter/receiver or transceiver/receiver combination. This system is designed for flexible operation and interconnection to various pieces of commercial or home-made equipment. For the simplest setup, no modifications to the transmitter or receiver are necessary. All that's required is to plug the station equipment into the system. The only limitation associated with this simple setup is the recovery speed of the receiver. If the receiver agc time constant is fairly fast, it should be possible to hear signals between characters at keying speeds of up to 25 wpm. If the receiver agc is turned off, or set for very fast recovery, signals can be heard between characters at speeds of up to 50 wpm. If you prefer, the receiver can be muted during characters and two outputs, the + mute and − mute, are provided for this purpose. Several transmitters were tried with this system and it was possible to use them without modification, so long as the final amplifier was

biased off under key-up conditions. No background hash was noticable.

The PIN diode T-R system is not plagued by problems commonly associated with some other systems. First, "suckout" (receiver desensing) has been eliminated, as has the problem of critical interconnecting line lengths. Also, since the saturated diode technique has been abandoned there should be no chance for

TVI. No high-priced vacuum relays are used. No amplifiers are placed ahead of the receiver that could affect receiver performance. In short, the system described here provides excellent performance and suffers none of the ills of earlier systems.

The Circuit

A schematic diagram of the circuit is shown in Fig. 30. The diagram is divided into two sections, as is the actual circuitry. That portion of the circuit to the left of the dotted line is intended to be mounted at the operating position for easy access. Circuitry to the right of the line can be mounted remotely, perhaps behind the station equipment.

The transmitter is connected to the antenna through a quarter-wavelength, lumped-constant circuit. S1 selects the appropriate circuit for the frequency in use. Quarter-wave circuits are required to prevent "suckout" of the received signal. "Suckout" occurs with tube-type transmitters when the high-impedance end of the transmitter pi network becomes unloaded; during receive periods, for example. As the pi network is one type of impedance-inverting network, the high resistance presented by the non-operational tube causes the low-impedance end of the network to approach 0 ohms. The quarter-wavelength lumped-constant sections provided in the T-R system serve to step the nearly zero impedance level of the transmitter output up to an almost infinitely high impedance that will not reduce the received signal level. As shown in the schematic diagram, the antenna is connected directly to J3 which feeds the PIN diode switch section of the T-R circuit.

The components located between J4 and J5 comprise the switch that protects the receiver from the transmitted signal. A "T" configuration is used, with D1 and D2 connected in series and D3 in shunt. Combination switches provide superior attenuation performance to either the series or shunt elements alone. Approximately 50 dB of isolation from the antenna to the receiver connection is provided throughout the rf range. These results

Fig. 29 — Exterior view of the PIN Diode T-R Switch. The box at the right is mounted at the operating position. The box at the left can be mounted remotely.

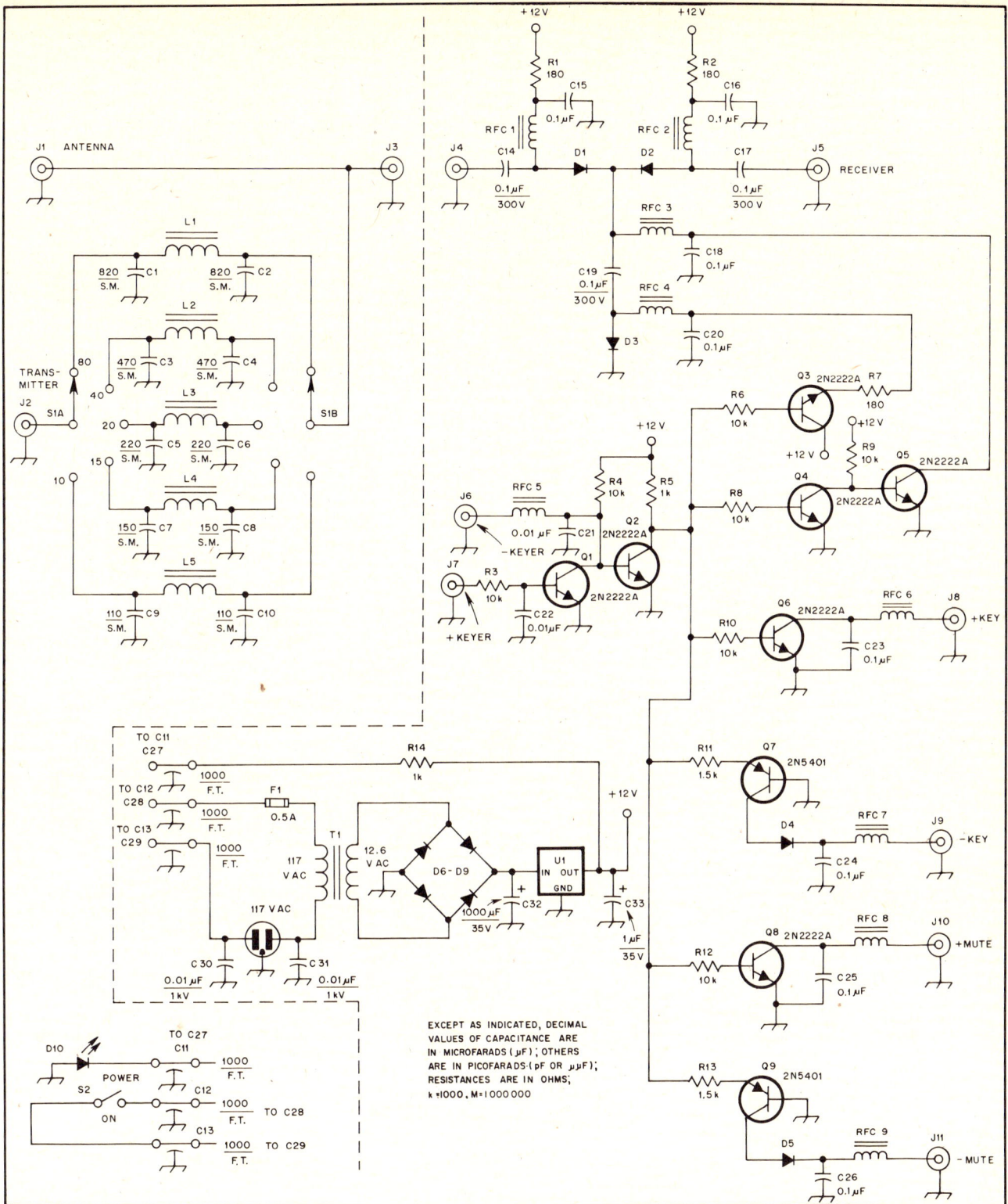

Fig. 30 — Schematic diagram of the T-R switch system. All resistors are 1/4-watt composition types. All capacitors are miniature ceramic, 50-volt types unless polarization is indicated. Polarized types are aluminum electrolytic or tantalum. Component designations listed in the schematic, but not called out in the parts list, are for text or layout reference only.

C1, C2 — Mica, 820 pF, 500 V.
C3, C4 — Mica, 470 pF, 500 V.
C5, C6 — Mica, 220 pF, 500 V.
C7, C8 — Mica, 150 pF, 500 V.
C9, C10 — Mica, 110 pF, 500 V.
C32 — Electrolytic, 1000 μF, 35 V.
C33 — Tantalum, 1 μF, 35 V.
D1-D3, incl. — PIN diode, Unitrode 1N5763 or equiv.
D4-D9, incl. — Power, 100 PRV, 1 A.
D10 — Light-emitting diode.
F1 — Fuse, 1/2 A.
J1-J5, incl. — Rf connector, female (builder's

choice).
J6, J7 — Phone, 1/4 inch or builder's choice.
J8-J11, incl. — Phono or builder's choice.
L1 — Toroid, 20 turns no. 18 enam. wire on a T-80-2 core.
L2 — Toroid, 15 turns no. 18 enam. wire on a T-80-2 core.
L3 — Toroid, 11 turns no. 18 enam. wire on a T-80-6 core.
L4 — Toroid, 9 turns no. 18 enam. wire on a T-80-6 core.
L5 — Toroid, 8 turns no. 18 enam. wire on a

T-80-6 core.
RFC1, RFC2 — Toroid choke, 20 turns no. 26 enam. wire on a FT-37-75 core.
RFC3-RFC9, incl. — Toroid choke, 26 turns no. 30 enam. wire on an FT-23-75 core.
S1 — Rotary wafer, 2 sections, 5 positions, ceramic.
S2 — Toggle, spst.
T1 — Miniature power, primary 117 V, secondary 12 V at 300 mA. Radio Shack 273-1385 or equiv.
U1 — Three-terminal regulator, 12-V output. Radio Shack RS-7812 or equiv.

Fig. 31 — Interior view of the control head. The quarter-wave sections are mounted to the single-sided pc board.

Fig. 32 — Inside view of the remotely mounted portion of the system. Short lengths of wire are used to attach the connectors to the appropriate circuit board foils. All power supply components are mounted on the circuit board.

should be reproducible if the same type of PIN diodes are used and the board layout shown is followed closely.

The station keyer (or straight key, bug or keyboard) is connected to either J6 or J7 depending on the output keying potential. Q1 and Q2 provide a suitable signal for driving Q3 through Q9. Q3 controls D3 and turns the diode on during transmission and off during receive. Q4 and Q5 control D1 and D2, biasing the diodes on for receive and off for transmit. At first glance it might appear that some simplification of the switching diode and control circuitry might be possible. Since it was desired to power the system from a 12-V dc source (for portable operation), and high negative voltages could not be used to reverse bias the diodes during transmit, an unusual arrangement was devised. Hence, the more complicated circuit. J8 and J9 are provided for keying the transmitter; one of the two outputs should be suitable for almost any transmitter. J10 and J11 are for muting the station receiver during transmission. Again, dual-polarity outputs are provided. Choose the one applicable to the equipment in use.

The power supply is depicted at the bottom center of the schematic diagram. Power to the system is routed through an on/off switch that is mounted at the control head. An LED indicator is included as a reminder that power is switched on. Connection from the control head to the remote unit is made via feedthrough capacitors at each box. These capacitors ensure that each cabinet remains rf-tight. The power supply is of the usual variety in all respects. A step-down transformer, full-wave rectifier, filter and three-terminal regulator provide the necessary voltage. A fuse is included in one leg of the ac line.

Construction

As mentioned earlier, the T-R switch system is constructed in two enclosures. The circuitry intended for mounting at the operating position is built into a Bud Minibox that measures 3 × 4 × 5 inches (76 × 102 × 127 mm). The part number of this box is CU-3005A. This enclosure houses the circuit-board mounted quarter-wave-length sections, rotary switch, on/off power switch, LED indicator, coaxial connectors and feedthrough capacitors. The circuit board etching and parts-placement patterns are shown in Figs. 33 and 34. An interior view of this unit is shown in Fig. 31. The rotary-switch wafers are positioned to line up closely with the appropriate circuit board connection points. An extra ceramic spacer is inserted in each of the switch section support rods to provide the needed separation. Short lengths of no. 18 tinned wire are used for the connections from the board to the switch contacts. RG-58A/U cable is used to make the connection from the antenna coaxial connector to the front wafer. The cable braid is soldered to ground lugs at each end.

The second enclosure used for the T-R switch system is constructed from sheet aluminum and measures 2-1/8 × 4-1/8 × 7 inches (54 × 105 × 178 mm). An ideal commercial enclosure would be the Bud CU-247 die-cast aluminum box. These boxes are extremely rugged and rf tight. The power supply, PIN-diode switch and control circuitry are mounted on a circuit board. Circuit board etching and parts-layout information are shown in Figs. 33 and 34. Double-sided circuit-board material was used, with the top side of the board left substantially unetched to act as a ground plane. Copper must be removed from around circuit-hole locations for components that are not connected to ground. This can be accomplished in the etching process with the aid of the top-side etching pattern. Alternatively, the copper can be removed from around holes with a large drill. Do not remove copper from around holes where component leads are grounded. Many of the component ground connections are not made on the pattern (bottom) side of the board. These components must be soldered on the top side to complete the gound connection.

Garden-variety components are used, with the exception of the PIN diodes. The diodes are Unitrode 1N5763 types, which can be obtained from many supply houses. All of the rf chokes are hand wound on small ferrite cores. Since encapsulated chokes are relatively expensive and cores are not, the time spent winding the chokes can result in reduced cost.

Circuit Checkout

Interconnection of the two modules requires four lengths of hookup wire, each long enough to reach between the two units when installed in their operating positions. The wires are twisted, cable-tied or laced together. Wires of different colors will help distinguish the connections and prevent possible surprises the first time power is applied! Connections are as follows: C11 to C27, C12 to C28 and C13 to C29. The fourth wire, ground, connects the two boxes. A coaxial cable is used to connect J3 and J4.

Connection to the station equipment is a simple matter. The keyer is plugged into either J6 or J7. If the keyer provides a

RCVR (J5) RECEIVE LINE (J4) – MUTE (J11) + MUTE (J10) – KEY (J9) + KEY (J8) – KEYER (J6) + KEYER (J7)

0.01µF/300V

0.1µF D3 RFC 1 0.1µF RFC 9 RFC 8 RFC 7 RFC 6 RFC 5 10K 0.01µF

RFC 2 D2 D1 RFC 3 0.1µF 0.1µF 0.1µF 0.1µF B

RFC 4 D5 C D4 C B Q 2 E

E B Q 5 C Q9 B Q8 C Q7 E Q6 B C Q 1 E

B C E C 1.5kΩ 10kΩ 1.5kΩ 10kΩ 1kΩ 10kΩ C

180Ω 10kΩ 180Ω 1µF/35V/TANT.

Q 4 B Q 3 C 10kΩ 180Ω 0.01µF 12-VOLT REGULATOR

E IN

0.01µF/300V 10kΩ + OUT

SEC. D6-D9

F1 1000µF 35V ELECTROLYTIC

0.01µF 1kV 0.01µF 1kV FUSE T1 XFMR

PRI. 1kΩ

TO AC PLUG TO FEEDTHROUGH CAPACITORS C27-C29

TO S1A TO S1A TO S1A TO S1A TO S1A

110pF S.M. 150pF S.M. 220pF S.M. 470pF S.M. 820pF S.M.

L 5 L 4 L 3 L 2 L 1

110pF S.M. 150pF S.M. 220pF S.M. 470pF S.M. 820pF S.M.

TO S1B TO S1B TO S1B TO S1B TO S1B

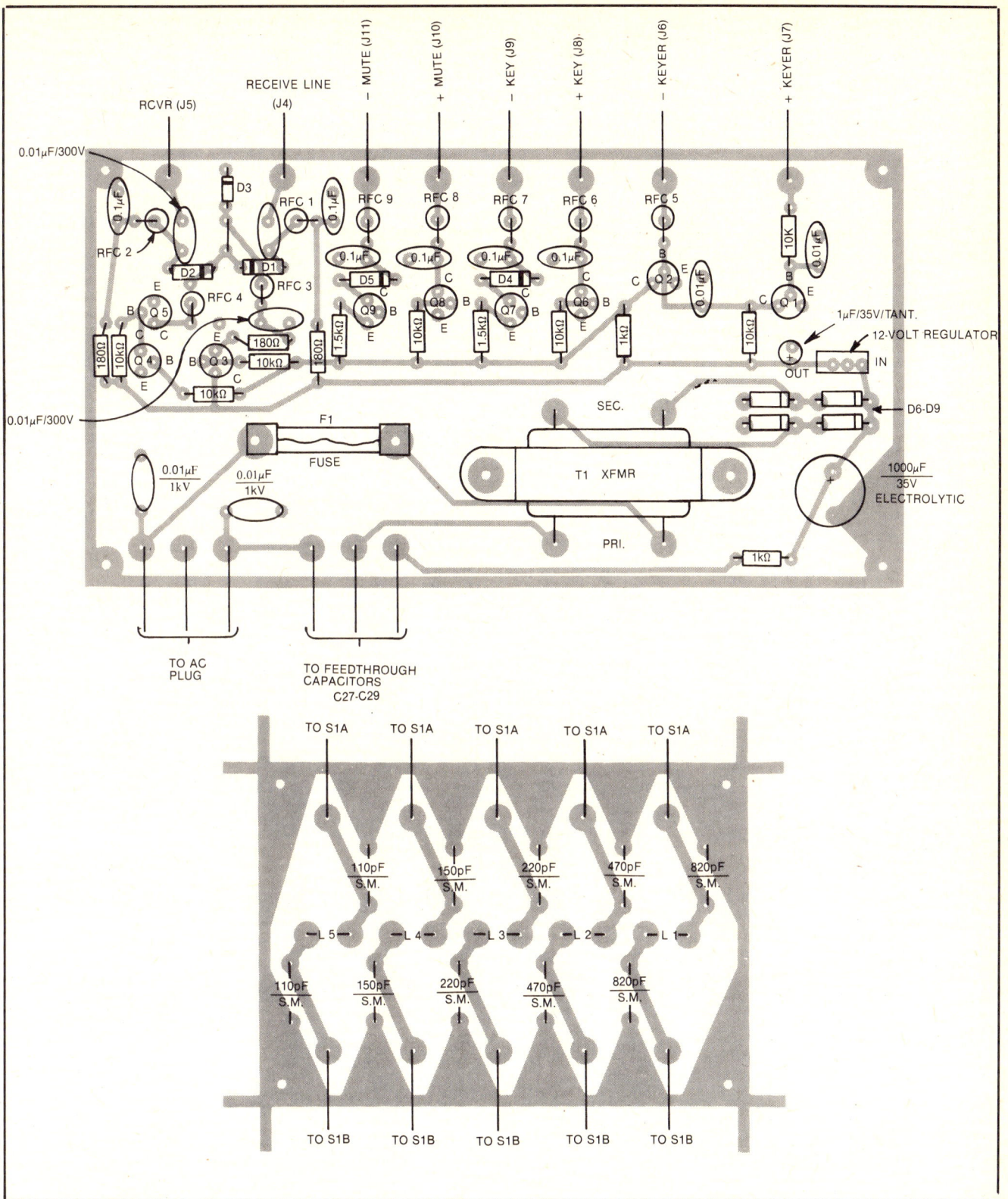

Fig. 33 — Parts-layout patterns of the two printed-circuit boards. Each board is shown from the component side with an X-ray view of the foil.

positive voltage when keyed, use J7. If the keyer provides a ground, use J6. Connect the antenna to J1 and the station receiver to J5. Do not connect the transmitter at this time. A check of the system operation can now be made. If all is in order at this point, signals should be heard in the receiver. Actuating the keyer should cause the signals to become inaudible. The exact amount of attenuation can be measured using a calibrated signal generator or a step attenuator and received signals. Attenuation should be on the order of 50 dB. If no measurement equipment is available, a received signal and the receiver S-meter may be used. A strong signal should become almost completely buried in the receiver noise when the keyer is activated. Connect the transmitter output to J2 and install a cable between the transmitter key jack and J8 or J9. If muting of the receiver is desired, make the appropriate connection at J10 or J11.

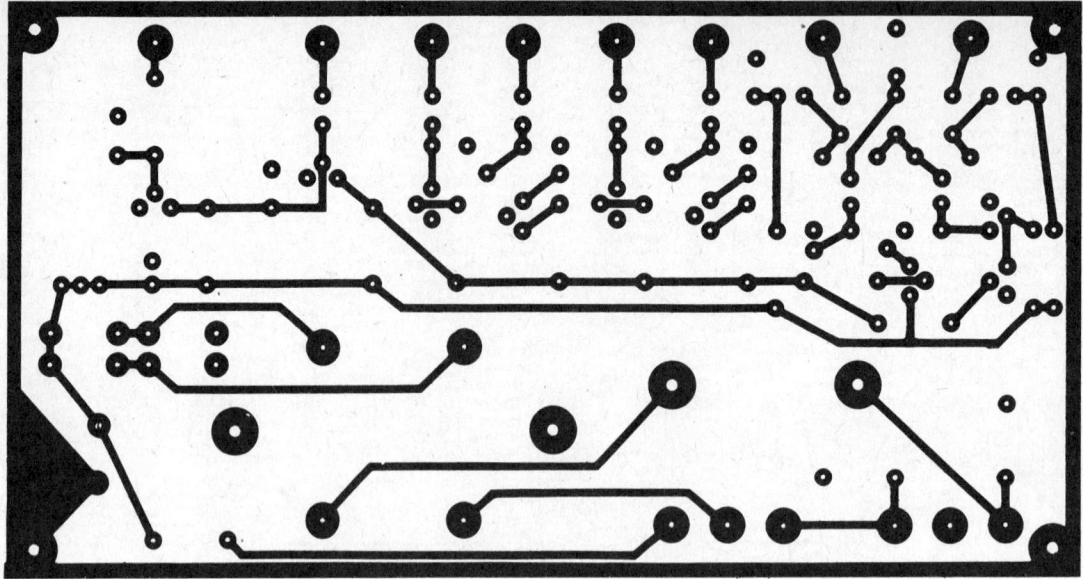

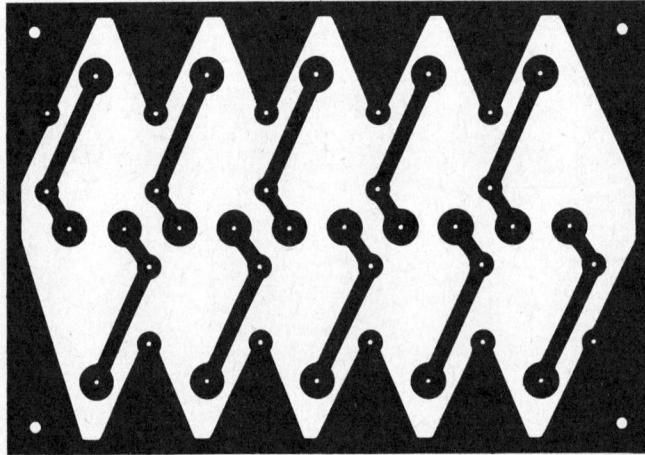

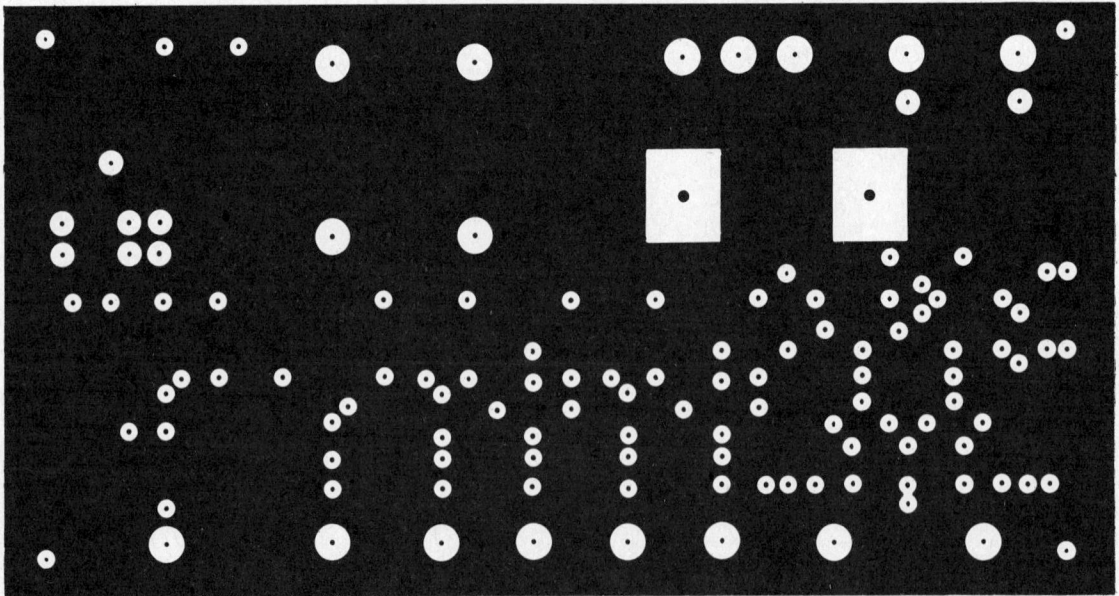

Fig. 34 — Etching patterns for the two printed-circuit boards. The smaller board is single-sided while the large one is double-sided. Patterns are provided for both sides of the board. Patterns are to scale. Black areas represent unetched copper.

Chapter 12

Single-Sideband Transmission

On the high-frequency amateur bands, single-sideband is the most widely used radiotelephony mode. Since ssb is a sophisticated (or simplified, depending on one's point of view) form of *amplitude modulation,* it is worthwhile to take a brief look at some a-m fundamentals. Modulation is a mixing process. When rf and af signals are combined in a standard a-m transmitter (such as one used for commercial broadcasting) four output signals are generated: the original rf signal, called the *carrier,* the original af signal, and two *sidebands,* whose frequencies are the sum and difference of the original rf and af signals, and whose amplitudes are proportional to that of the original af signal. The sum component is called the *upper sideband.* It is *erect,* in that increasing the frequency of the modulating audio signal causes a corresponding increase in the frequency of the rf output signal. The difference component is called the *lower sideband,* and is *inverted,* meaning an increase in the modulating frequency results in a decrease in the output frequency. The amplitude and frequency of the carrier are unchanged by the modulation process, and the original af signal is rejected by the rf output network. The rf *envelope* as viewed on an oscilloscope has the shape of the modulating waveform.

Fig. 1B shows the envelope of an rf signal that is modulated 20 percent by an af sine wave. The envelope varies in amplitude because it is the vector sum of the carrier and the sidebands. A spectrum analyzer or selective receiver will show the carrier to be constant. The spectral photograph also shows that the bandwidth of an a-m signal is twice the highest frequency component of the modulating wave.

An amplitude-modulated signal cannot be frequency multiplied without special processing because the phase/frequency relationship of the components of the modulating waveform would be severely distorted. For this reason, once an a-m signal has been generated at a fixed frequency, it can be moved in frequency only by heterodyning.

All of the intelligence is contained in the sidebands, but two-thirds of the rf power is in the carrier. The carrier serves only to demodulate the signal in the receiver. If this carrier is suppressed in the transmitter and reinserted in the proper phase in the receiver, several significant communications advantages accrue. If the reinserted carrier is strong compared to the incoming double-sideband signal, *exalted carrier* reception is achieved in which distortion caused by frequency-selective fading is greatly reduced. A refinement of this technique, called *synchronous detection* uses a phase-locked loop to enhance the rejection of interference. Also, the lack of a transmitted carrier eliminates the heterodyne interference common to adjacent a-m signals. Perhaps the most important advantage of eliminating the carrier is that the overall efficiency of the transmitter is increased. The power consumed by the carrier can be put to better use in the sidebands. The power in the carrier is continuous and an a-m transmitter requires a heavy-duty power supply. A *dsb* (double sideband) transmitter having the same power output as an a-m transmitter can use a much lighter power supply because the duty cycle of voice operation is low.

Balanced Modulators

The carrier can be suppressed or nearly eliminated by using a balanced modulator or an extremely sharp filter. In ssb transmitters it is common practice to use both

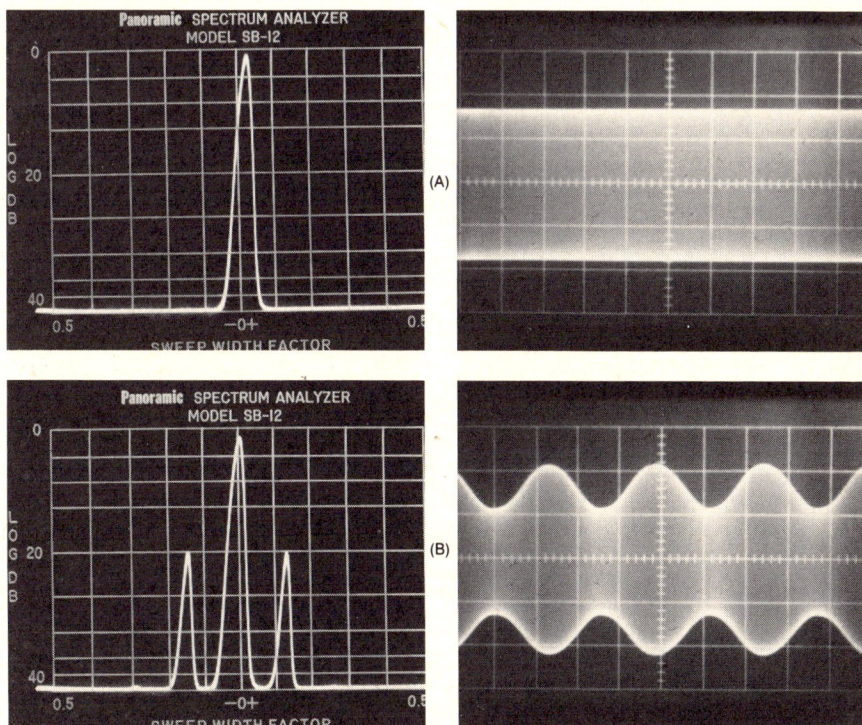

Fig. 1 — Electronic displays of a-m signals in the frequency and time domains. (A) Unmodulated carrier or single-tone ssb signal. (B) Full-carrier a-m signal with single-tone sinusoidal modulation.

Table 1

A-M Emission Types

Type of transmission	Supplementary characteristics	Symbol
With no modulation		A∅
Telegraphy without the use of a modulating audio frequency (by on-off keying).		A1
Telegraphy by the on-off keying of an amplitude modulating audio frequency or audio frequencies or by the on-off keying of the modulated emission (special case: an unkeyed emission amplitude modulated).		A2
Telephony	Double sideband	A3
	Single sideband, reduced carrier	A3A
	Single sideband, suppressed carrier	A3J
	Two independent sidebands	A3B
Facsimile (with modulation of main carrier either directly or by a frequency modulated subcarrier).		A4
Facsimile	Single sideband, reduced carrier	A4A
Television	Vestigial sideband	A5C
Multichannel voice-frequency telegraphy	Single sideband, reduced carrier	A7A
Cases not covered by the above, e.g., a combination of telephony and telegraphy.	Two independent sidebands	A9B

(See Table 2.)

Table 2

Emission Types Possible with an SSB or ISB Transmitter

Symbol	Audio Input
A∅	Single steady tone
A1	On-off keying of a single tone.
A3	Speech.
A3A	Speech.
A3B	Speech (two channels).
A3J	Speech.
A9B	SSTV and speech, RTTY and speech.
F1	Two alternating constant-amplitude tones (RTTY).
F5	Frequency-varying, constant-amplitude tone (SSTV).

devices. The basic principle of any balanced modulator is to introduce the carrier in such a way that it does not appear in the output, but so that the sidebands will. The type of balanced-modulator circuit chosen by the builder will depend upon the constructional considerations, cost, and the active devices to be employed.

In any balanced-modulator circuit there will be no output with no audio signal. When audio is applied, the balance is upset, and one branch will conduct more than the other. Since any modulation process is the same as "mixing" in receivers, sum and difference frequencies (sidebands) will be generated. The modulator is not balanced for the sidebands, and they will appear in the output.

In the rectifier-type balanced modulators shown in Fig. 2, at A and B, the diode rectifiers are connected in such a manner that, if they have equal forward resistances, no rf can pass from the carrier source to the output circuit via either of the two possible paths. The net effect is that no rf energy appears in the output. When audio is applied, it unbalances the circuit by biasing the diode (or diodes) in one path, depending upon the instantaneous polarity of the audio, and hence some rf will appear in the output. The rf in the output will appear as a double-sideband suppressed-carrier signal.

In any diode modulator, the rf voltage should be at least six to eight times the peak audio voltage for minimum distortion. The usual operation involves a fraction of a volt of audio and several volts of rf. Desirable diode characteristics for balanced modulator and mixer service include: low noise, low forward resistance, high reverse resistance, good temperature stability, and fast switching time (for high-frequency operation). Fig. 3 lists the different classes of diodes, giving the ratio

of forward-to-reverse resistance of each. This ratio is an important criterion in the selection of diodes. Also, the individual diodes used should have closely matched forward and reverse resistances; an ohmmeter can be used to select matched pairs or quads.

One of the simplest diode balanced modulators in use is that of Fig. 2A. Its use is usually limited to low-cost portable equipment in which a high degree of carrier suppression is not vital. A ring balanced modulator, shown in Fig. 2B, offers good carrier suppression at low cost. Diodes D1 through D4 should be well matched and can be 1N270s or similar. C1 is adjusted for best rf phase balance as evidenced by maximum carrier null. R1 is also obtainable. It may be necessary to adjust each control several times to secure optimum suppression.

Varactor diodes are part of the unusual circuit shown in Fig. 2C. This arrangement allows single-ended input of near-equal levels of audio and carrier oscillator. Excellent carrier suppression, 50 dB or more, and a simple method of unbalancing the modulator for cw operation are features of this design. D1 and D2 should be rated at 20 pF for a bias of − 4 V. R1 can be adjusted to cancel any mismatch in the diode characteristics, so it isn't necessary that the varactors be well matched. T1 is wound on a small-diameter toroid core. The tap on the primary winding of this transformer is at the center of the winding.

A bipolar-transistor balanced modulator is shown in 2D. This circuit is similar to one used by Galaxy Electronics and uses closely matched transistors at Q1 and Q2. A phase splitter (inverter), Q3, is used to feed audio to the balanced modulator in push-pull. The carrier is supplied to the circuit in parallel and the output is taken

in push-pull. D1 is a Zener diode and is used to stabilize the dc voltage. Controls R1 and R2 are adjusted for best carrier suppression.

The circuit at E offers superior carrier suppression and uses a 7360 beam-deflection tube as a balanced modulator. This tube is capable of providing as much as 60 dB of carrier suppression. When used with mechanical or crystal-lattice filters the total carrier suppression can be as great as 80 dB. Most well-designed balanced modulators can provide between 30 and 50 dB of carrier suppression; hence the 7360 circuit is highly desirable for optimum results. The primary of transformer T1 should be bifilar wound for best results.

Vacuum-tube balanced modulators can be operated at high power levels and the double-sideband output can be used directly into the antenna. Past issues of *QST* have given construction details on such transmitters (see, for example, Rush, "180-Watt D.S.B. Transmitter," *QST*, July 1966).

IC Balanced Modulators

Integrated circuits (ICs) are presently available for use in balanced-modulator and mixer circuits. A diode array such as the RCA CA3039 is ideally suited for use in circuits such as that of Fig. 4A. Since all diodes are formed on a common silicon chip, their characteristics are extremely well matched. This fact makes the IC ideal in a circuit where good balance is required. The hot-carrier diode also has closely matched characteristics and excellent temperature stability. Using broadband toroidal-wound transformers, it is possible to construct a circuit similar to that of Fig. 5 which will have 40 dB of carrier suppression without the need for balance controls. T1 and T2 consist of trifilar windings, 12 turns of no. 32 enam. wire wound on a 1/2-inch (13-mm) toroid core. Another device with good inherent balance is the special IC made for modulator/mixer service, such as the Motorola MC1496G or Signetics S5596. A sample circuit using the MC1496 can be seen in Fig. 4B. R1 is adjusted for best carrier balance. The amount of energy delivered

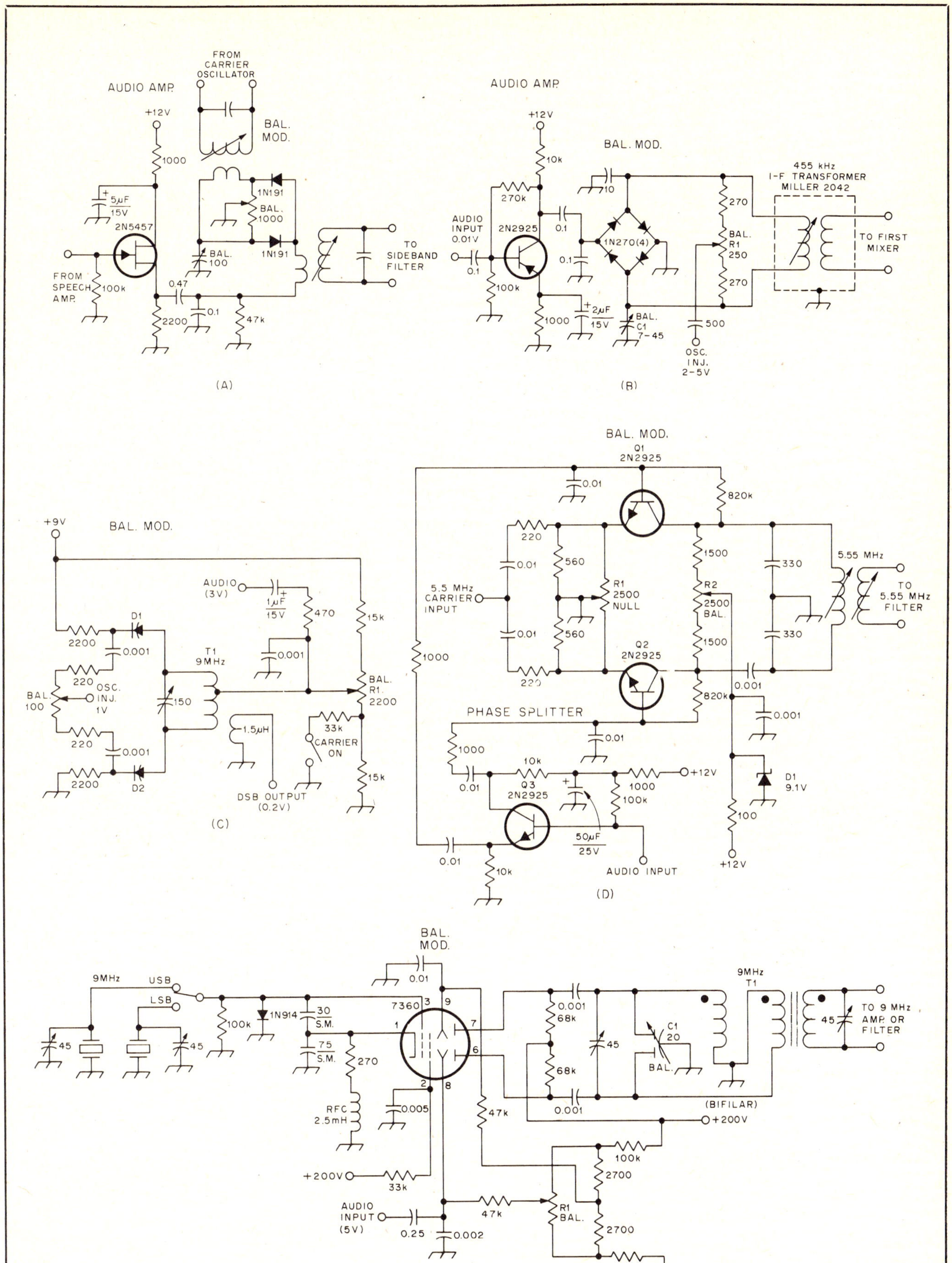

Fig. 2 — Typical circuits of balanced modulators. Representative parts values are given and should serve as a basis for designing one's own equipment.

Diode Type	Ratio M = 1,000,000
Point-contact germanium (1N98)	500
Small-junction germanium (1N270)	0.1 M
Low-conductance silicon (1N457)	48 M
High-conductance silicon (1N645)	480 M
Hot-carrier (HPA-2800)	2000 M

Fig. 3 — Table showing the forward-to-reverse resistance ratio for the different classes of solid-state diodes.

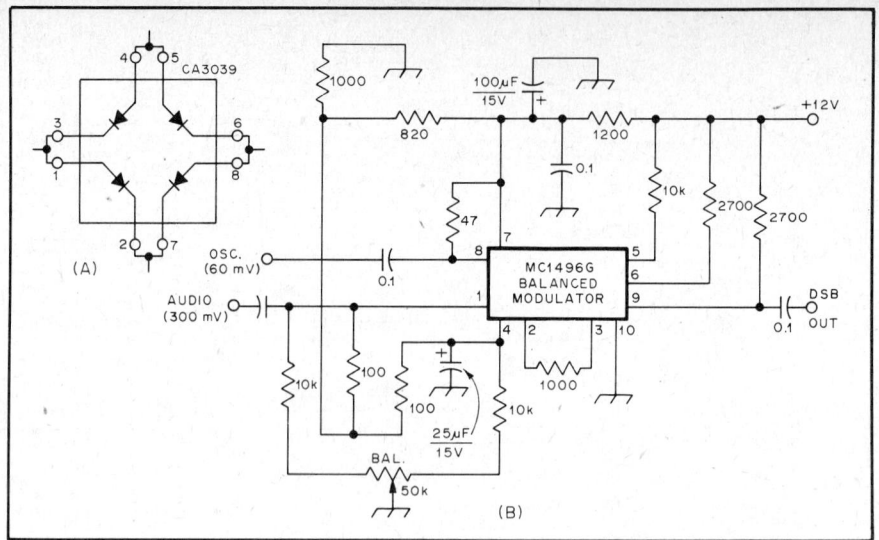

Fig. 4 — Additional balanced-modulator circuits in which integrated circuits are used.

from the carrier generator affects the level of carrier suppression; 100 mV of injection is about optimum, producing up to 55 dB of carrier suppression. Additional information on balanced modulators and other ssb-generator circuits is given in the texts referenced at the end of this chapter.

SINGLE-SIDEBAND EMISSION

A further improvement in communications effectiveness can be obtained by transmitting only one of the sidebands. When the proper receiver bandwidth is used, a single-sideband signal will show an effective gain of up to 9 dB over an a-m signal of the same peak power. Because the redundant information is eliminated, the required bandwidth of an ssb signal is half that of a comparable a-m or dsb emission. Unlike dsb, the phase of the local carrier generated in the receiver is unimportant.

Generating the SSB Signal: Filter Method

If the dsb signal from the balanced modulator is applied to a bandpass filter, one of the sidebands can be greatly attenuated. Because a filter cannot have infinitely steep skirts, to obtain adequate suppression of the unwanted sideband the response of the filter must begin to roll off within about 300 Hz of the phantom carrier. This effect limits the ability to transmit bass frequencies, but as will be shown in the section on speech processing, these frequencies have little communications value. The filter rolloff can be used to obtain an additional 20 dB of carrier suppression. The bandwidth of an ssb filter is selected for the specific application. For voice communications, typical values are 1800 to 4000 Hz.

Fig. 6 illustrates two variations of the filter method of ssb generation. The heterodyne oscillator is represented as a simple VFO, but may be a premixing system or synthesizer. The scheme at B is perhaps less expensive than that of A, but the heterodyne oscillator frequency must be shifted when changing sidebands if the dial calibration is to be maintained. The ultimate sense (erect or inverted) of the final output signal is influenced as much by the relationship of the heterodyne-oscillator frequency to the fixed ssb frequency as by the filter or carrier frequency selection. The heterodyne-oscillator frequency must be chosen to allow the best

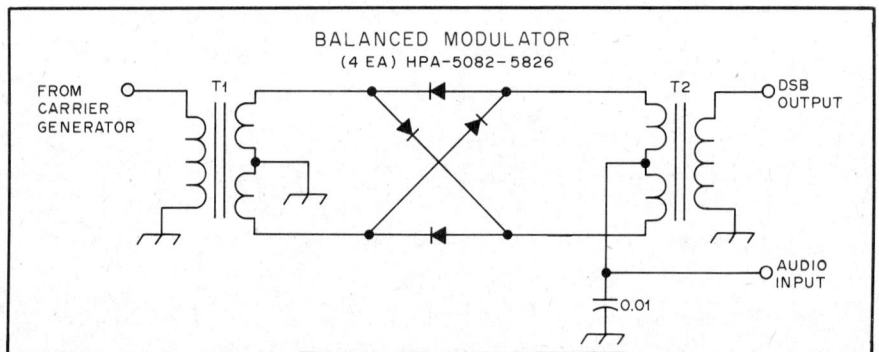

Fig. 5 — Balanced-modulator design using hot-carrier diodes.

Table 3

Guidelines for Amateur SSB Signal Quality

Parameter	Suggested Standard
Carrier suppression	At least 40-dB below PEP.
Opposite-sideband suppression	At least 40-dB below PEP.
Hum and noise	At least 40-dB below PEP.
Third-order intermodulation distortion	At least 30-dB below PEP.
Higher-order intermodulation distortion.	At least 35-dB below PEP.
Long-term frequency stability	At most 100-Hz drift per hour.
Short-term frequency stability	At most 10-Hz pk-pk deviation in a 2-kHz bandwidth.

image rejection. This consideration requires the heterodyne-oscillator frequency to be above the fixed ssb frequency on some bands and below it on others. To reduce circuit complexity, early amateur filter types of ssb transmitters did not include a sideband selection switch. The result was that the output was lsb on 160, 75 and 40 meters, and usb on the higher bands. This convention persists despite the flexibility of most modern amateur ssb equipment.

Filter Types

For carrier frequencies in the 50- to 100-kHz region, a satisfactory filter can be made up of lumped-constant LC sections. High quality components and careful adjustment are required for good results with this type of filter. An alternate possibility is a "synthesized" filter comprised of high-performance operational amplifiers used as gyrators or "active inductors." A further drawback of ssb generation in this frequency range is that multiple conversion is necessary to reach the desired output frequency with adequate suppression of spurious mixing products.

Mechanical filters are an excellent choice for ssb generation in the 400- to 500-kHz region. These filters are described in some detail in the receiving chapter. For wide dynamic range receiving

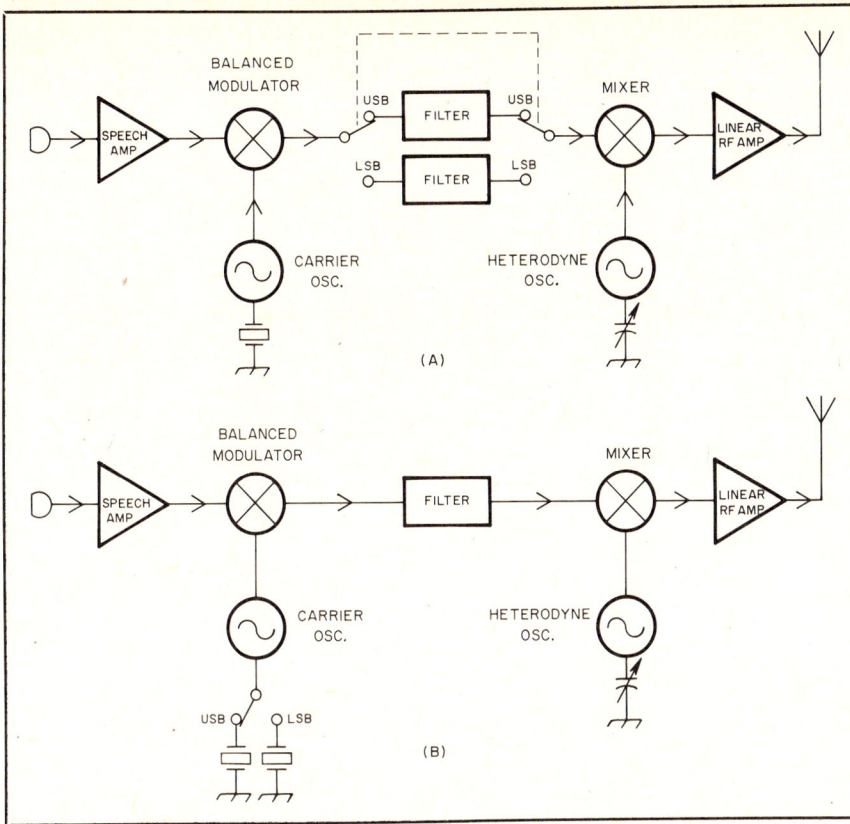

Fig. 6 — The filter method of ssb generation. Two sideband selection schemes are commonly used.

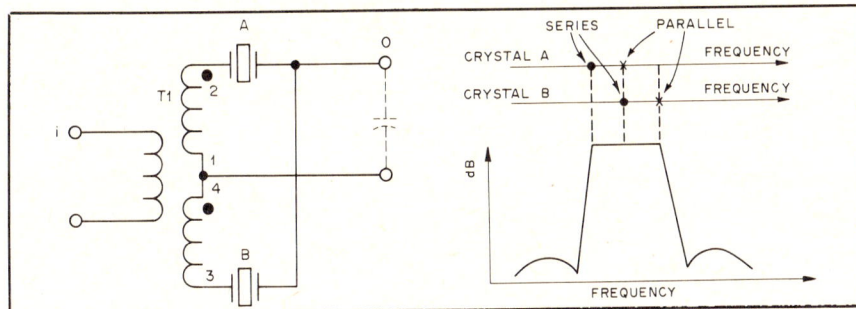

Fig. 7 — The half-lattice crystal filter. Crystals A and B should be chosen so that the parallel-resonant frequency of one is the same as the series-resonant frequency of the other. Very tight coupling between the two halves of the secondary of T1, is required for optimum results. The theoretical attenuation-vs.-frequency curve of a half-lattice filter shows a flat passband between the lower series-resonant frequency and higher parallel-resonant frequency of the pair of crystals.

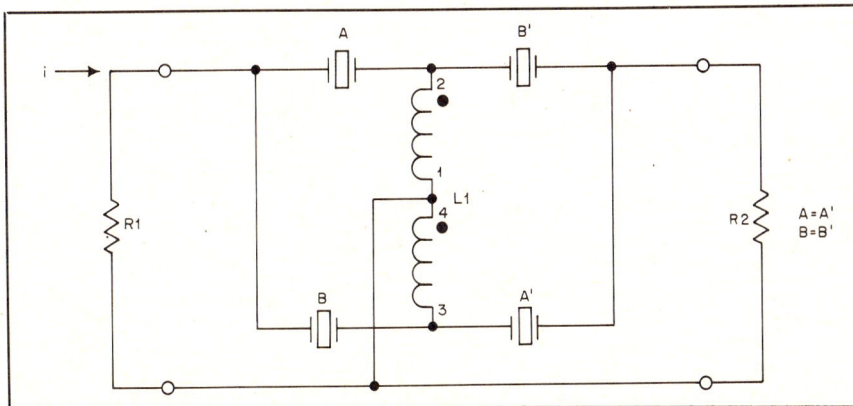

Fig. 8 — Half-lattice filters cascaded in a back-to-back arrangement. The theoretical curve of such a filter has increased skirt selectivity and fewer spurious responses, as compared with a simple half lattice, but the same passband as the simple circuit.

applicatons, the more modern types using piezoelectric transducers are preferred for lowest intermodulation distortion. In transmitters, where the signal levels can be closely predicted, the types using magnetostrictive transducers are entirely suitable.

Quartz crystal filters are commonly used in systems in which the ssb signal is generated in the high-frequency range. Some successful amateur designs have also employed crystals at 455 kHz. Generally, four or more crystal elements are required to obtain adequate selectivity for ssb transmission. Crystal-filter design is a sophisticated subject, and the more esoteric aspects are beyond the scope of this *Handbook*. The discussion of piezoelectric crystal theory in the Electrical Laws and Circuits chapter is sufficient background material for the general understanding of the concepts outlined in this section.

A fundamental crystal filter section is the half-lattice, shown in Fig. 7. The passband of this type of filter is slightly wider than the frequency spacing between the crystals. The antiresonant (parallel resonant) frequency or *pole* of the low-frequency crystal must be equal to the series-resonant frequency or *zero* of the high-frequency crystal. Such a filter is useful for casual receiving purposes, but the ultimate stopband attenuation is poor, and numerous spurious responses will exist just outside the passband. Cascading two of these sections back-to-back, as in Fig. 8, will greatly suppress these parasitic resonances and steepen the skirts without materially affecting the passband. An important factor in the design of this type of filter is the coefficient of coupling between the two halves of the transformer. The coupling must approach unity for proper operation. A twisted-pair or bifilar winding on a high-permeability ferrite core most nearly approximates this ideal. Some crystal filters have tuned input and output transformers. The flatness of the passband is heavily dependent on the terminating resistances. Lattice filters exhibit fairly symmetrical response curves and can be used for lsb or usb selection by means of placing the carrier frequency on the upper or lower skirt.

An asymmetrical filter is shown in Fig. 9. Good unwanted sideband suppression can be obtained with only two crystals using this approach. The crystals are ground for the same frequency. The potential bandwidth here is only half that obtained with a half-lattice design. The maximum bandwidth of almost any crystal filter can be increased by using plated crystals intended for overtone operation.

The home construction of crystal filters can be very time-consuming, if not expensive. The reason for this is that one must experiment with a large number of crystals to produce a filter with satisfactory per-

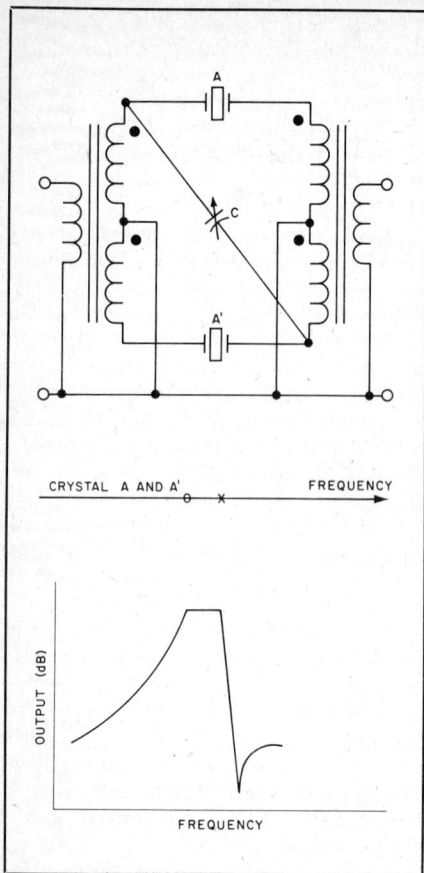

Fig. 9 — An asymmetrical filter and theoretical attenuation curve.

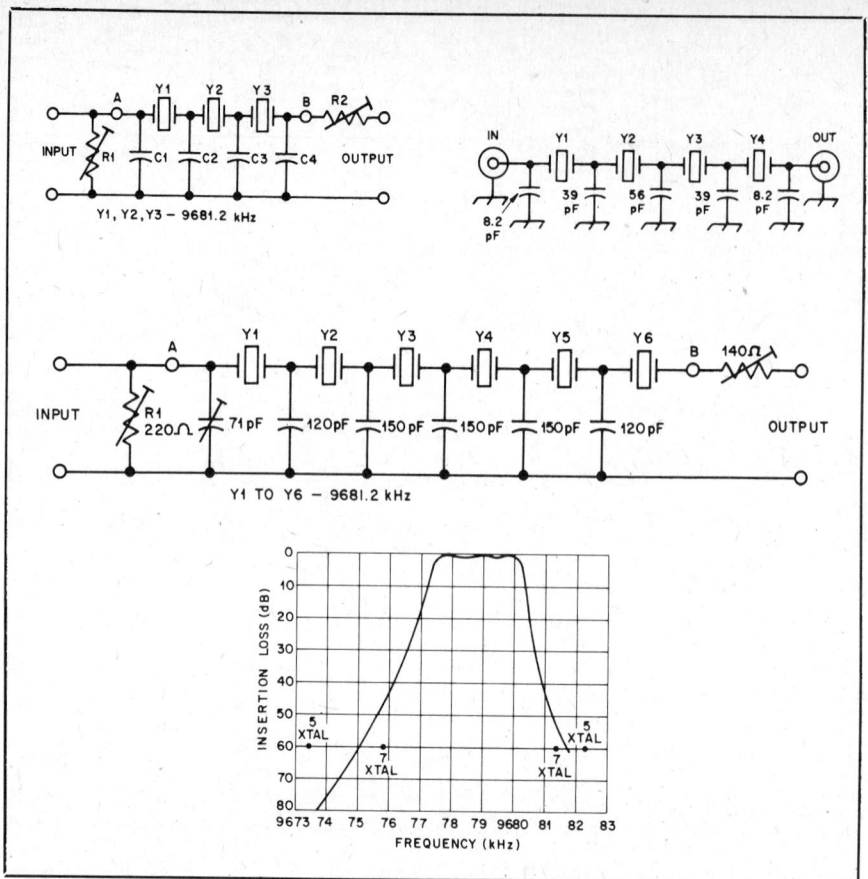

Fig. 10 — Some ladder filters based on CB crystals, with the response that can be expected from the 6-pole unit.

formance. Crystal grinding and etching can be a fascinating and highly educational activity, but most home builders would prefer to spend their time on other aspects of equipment design. High-quality filters are available from several manufacturers in the $50 to $100 price class. Most amateurs who build their own ssb equipment adopt a "systems engineering" approach and design their circuits around filters of known performance. Some filter suppliers are listed in the Construction and Data Tables chapter. It is still worthwhile to have an appreciation for the basic design ideas, however, for many of the less expensive filters can be improved markedly by the addition of a couple of crystals external to the package. The technique is to steepen the skirts by grouping sharp notches on either side of the passband.

An important exception to the above commentary is the *ladder filter*. Although this type of filter is treated in textbooks, it has received attention in the amateur literature only recently. The significant feature of ladder filters is that all of the crystals are ground for the same frequency. Low-cost CB crystals are ideally suited to this application. Representative designs by F6BQP and G3JIR are given in Fig. 10. Filter sections of this type can be cascaded for improved shape factor with

very little effect on the 3-dB bandwidth. Ladder filters having six or more elements are suitable for ssb transmitting and receiving service. In general, the bandwidth is inversely proportional to the values of the shunt capacitors and directly proportional to the terminal impedances. Table 4 lists the frequencies of the CB channels. Overtone crystals for CB service have fundamental resonance at approximately one third of the listed frequency.

Filter Applications

The important considerations in circuits using bandpass filters are impedance matching and input/output isolation. The requirements for the latter parameter are less severe in transmitting applications than they are for receiving, but with proper layout and grounding, the opposite sideband suppression should be determined by the shape factor rather than signal leakage. The filter must be terminated with the proper impedances to ensure a smooth bandpass response.

Fig. 11A shows a typical ssb generator using a KVG (see *QST* ads) crystal filter. The grounded-gate JFET presents a broadband 50-ohm termination to the balanced modulator and transforms the impedance to the 500 ohms required by the filter. The dc return for the source of the JFET is through the output

Table 4

CB Frequencies

Channel	Frequency (MHz)	Channel	Frequency (MHz)
1	26.965	21	27.215
2	26.975	22	27.225
3	26.985	23	27.255
4	27.005	24	27.235
5	27.015	25	27.245
6	27.025	26	27.265
7	27.035	27	27.275
8	27.055	28	27.285
9	27.065	29	27.295
10	27.075	30	27.305
11	27.085	31	27.315
12	27.105	32	27.325
13	27.115	33	27.335
14	27.125	34	27.345
15	27.135	35	27.355
16	27.155	36	27.365
17	27.165	37	27.375
18	27.175	38	27.385
19	27.185	39	27.395
20	27.205	40	27.405

transformer of the modulator. The tank circuit is broadly resonant at 9 MHz and rejects any spurious signals generated in the modulator that might be propagated through the filter. Crystal filters should be isolated from any dc voltages present in the circuit.

A circuit using a Collins mechanical

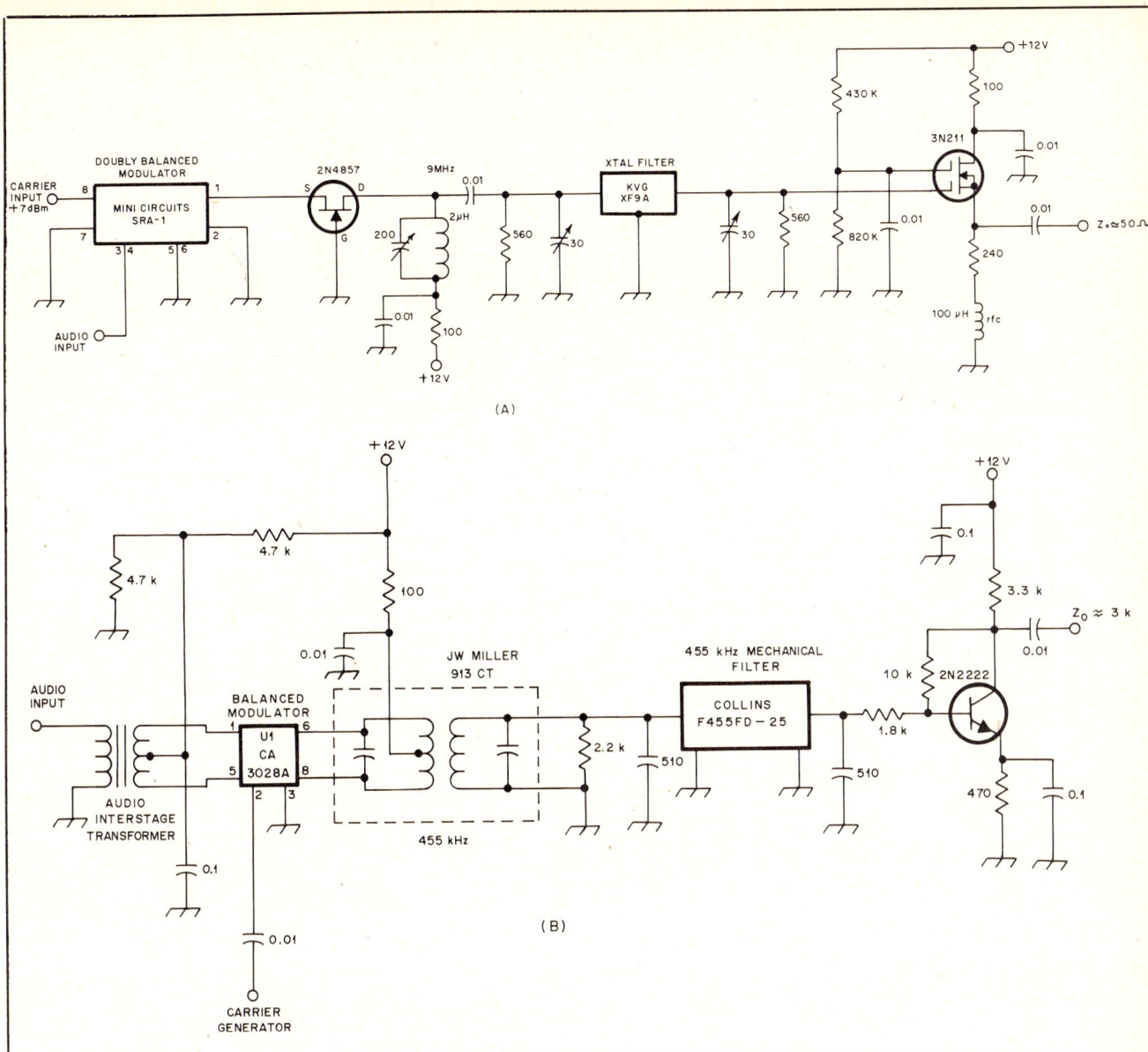

Fig. 11 — Connecting a packaged filter into an ssb generator. (A) 9-MHz crystal filter. (B) 455-kHz mechanical filter.

filter is illustrated in Fig. 11B. The i-f transformer prevents spurious responses and removes dc bias. The output terminating network does double duty as the bias network for the transistor amplifier stage. The filter output transformer is the dc return for the base circuit. This technique is legitimate so long as the current is limited to 2 mA.

SSB Generation: The Phasing Method

Fig. 12 shows another method for obtaining a single-sideband signal. The audio and carrier signals are each split into components separated 90° in phase and applied to balanced modulators. When the dsb outputs of the modulators are combined, one sideband is reinforced and the other is cancelled. The figure shows sideband selection by means of transposing the audio leads, but the same result can be had by means of switching the carrier leads.

The phasing method was used in many pre-1960 amateur ssb exciters, but became less popular after the introduction of relatively inexpensive high-performance bandpass filters. The phase shift and amplitude balance of the two channels must be very accurate if the unwanted sideband is to be adequately attenuated. Table 5 shows the required phase accuracy of one channel (af or rf) for various levels of opposite sideband suppression. The numbers given assume perfect amplitude balance and phase accuracy in the other channel. It can be seen from the table that a phase accuracy of ±1° must be maintained if the signal quality is to satisfy the criteria tabulated at the beginning of this chapter. It is difficult to achieve this level of overall accuracy over the entire speech band. Note, however, that speech has a complex spectrum with a large gap in the octave from 700 to 1400 Hz. The phase-accuracy tolerance can be loosened to

Table 5

Unwanted Sideband Suppression as a Function of Phase Error

Phase Error (degrees)	Suppression (dB)
0.125	59.25
0.25	53.24
0.5	47.16
1.0	41.11
2.0	35.01
3.0	31.42
4.0	28.85
5.0	26.85
10.0	20.50
15.0	16.69
20.0	13.93
30.0	9.98
45.0	6.0

±2° if the peak deviations can be made to occur within the spectral gap.

The major advantage of the phasing system is that the ssb signal can be generated at the operating frequency

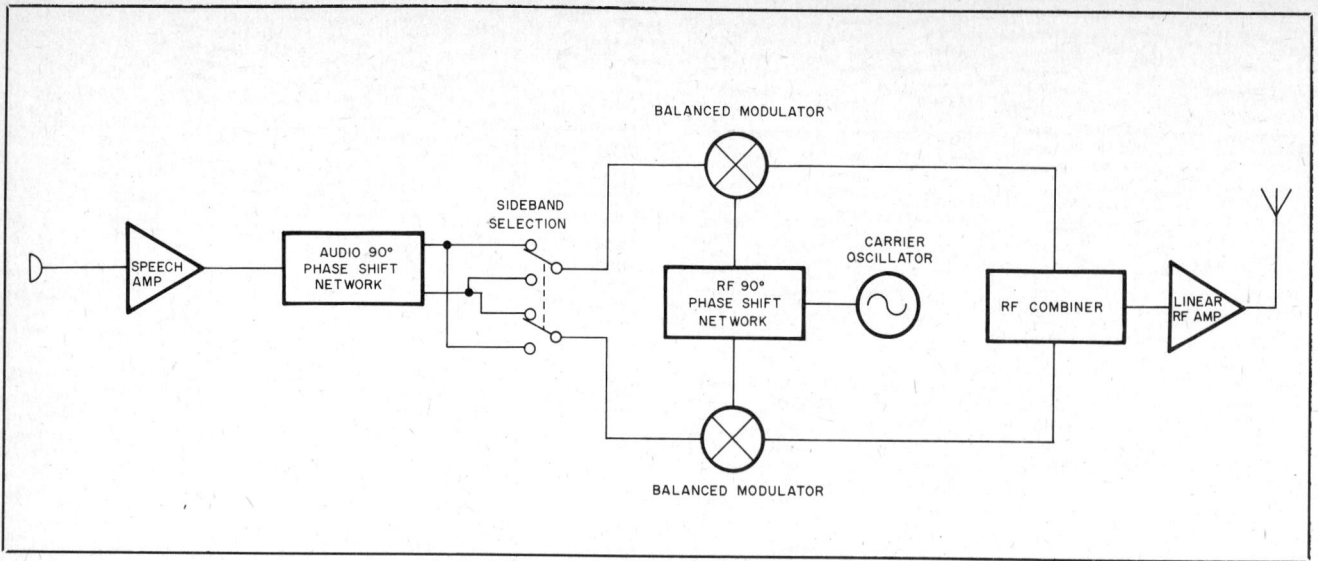

Fig. 12 — The phasing system of ssb generation.

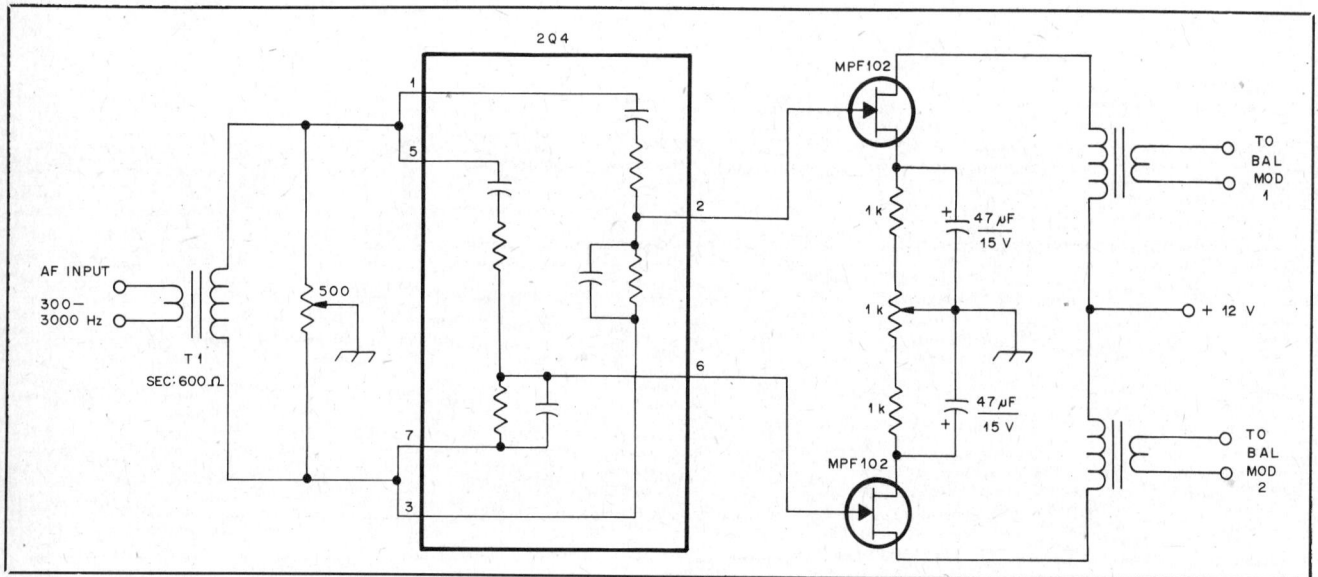

Fig. 13 — A circuit using the B&W 2Q4 audio phase shift network.

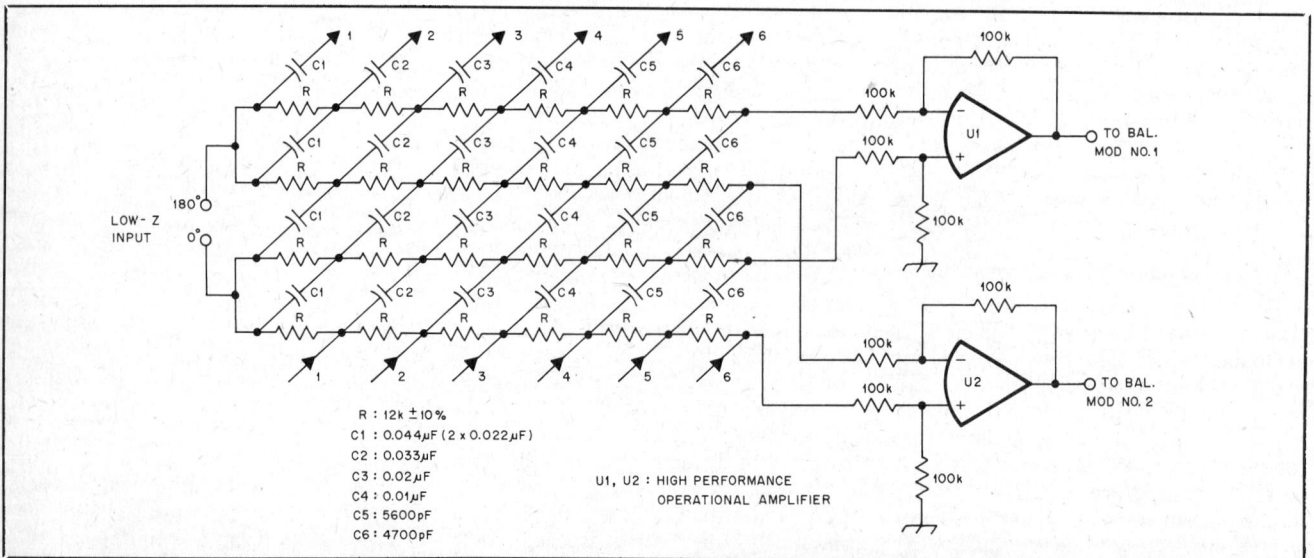

R : 12k ±10%
C1 : 0.044µF (2 x 0.022µF)
C2 : 0.033µF
C3 : 0.02µF
C4 : 0.01µF
C5 : 5600pF
C6 : 4700pF

U1, U2 : HIGH PERFORMANCE
OPERATIONAL AMPLIFIER

Fig. 14 — A high-performance audio phase shifter made from ordinary loose-tolerance components.

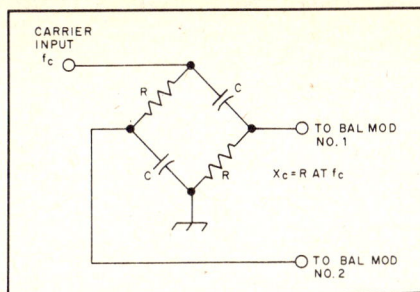

Fig. 15 — A simple rf phase shifter. One of the capacitors can be variable for precise alignment.

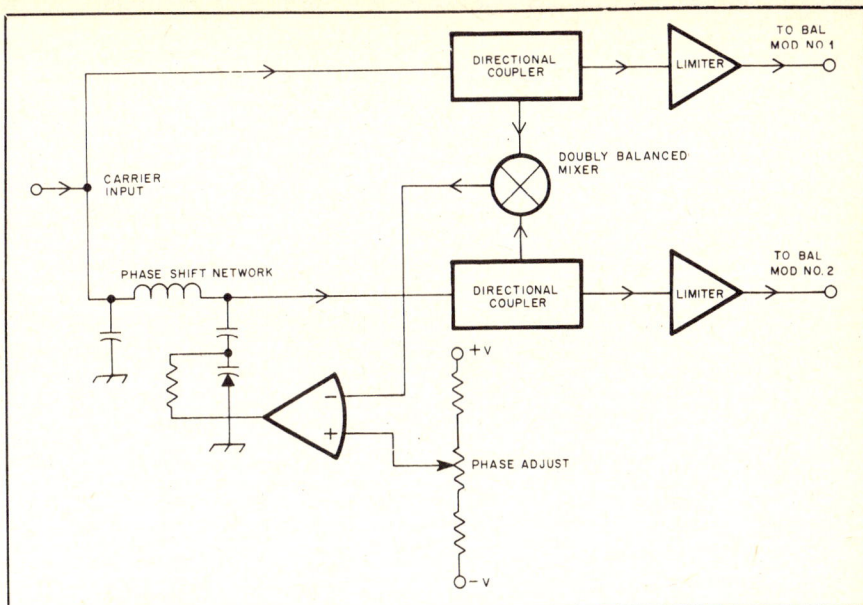

Fig. 16 — Block diagram of a phase-locked-loop phase-shifting system capable of maintaining quadrature over a wide bandwidth. The doubly balanced mixer is used as a phase detector.

without the need for heterodyning. Phasing can be used to good advantage even in fixed-frequency systems. A loose-tolerance (± 4°) phasing exciter followed by a simple two-pole crystal filter can generate a high-quality signal at very low cost.

Audio Phasing Networks

It would be difficult to design a two-port network having a quadrature (90°) phase relationship between input and output with constant-amplitude response over a decade of bandwidth. A practical approach, pioneered by Robert Dome, W2WAM, is to use two networks having a *differential* phase shift of 90°. This differential can be closely maintained in a simple circuit if precision components are used. The 350/2Q4 audio phase shift network manufactured by Barker and Williamson is such a circuit. The price class is $20. The 2Q4 is a 1950 vintage component but it is still useful. A modern design using this device is given in Fig. 13. The insertion loss of the 2Q4 is 30 dB and the phase shift accuracy is ± 1.5° over the 300-3000 Hz speech band.

The tolerances of the components can be relaxed considerably if several phase-shift sections are cascaded. A sixth-order network designed by HA5WH is shown in Fig. 14. Using common ± 10-percent tolerance components, this phase shifter provides approximately 60 dB of opposite sideband attenuation over the range of 300 to 3000 Hz.

Numerous circuits have been developed to synthesize the required 90° phase shift electronically. Active-filter techniques are used in most of these systems, but precision components are needed for good results. An interesting phasing system described in *Electronics* for April 13, 1978, makes use of a tapped analog delay line. These "bucket brigade" devices are becoming available at reasonable prices on the surplus market.

RF Phasing Networks

If the ssb signal is to be generated at a fixed frequency, the rf phasing problem is trivial; any method that produces the proper phase shift can be used. If the signal is produced at the operating fre-

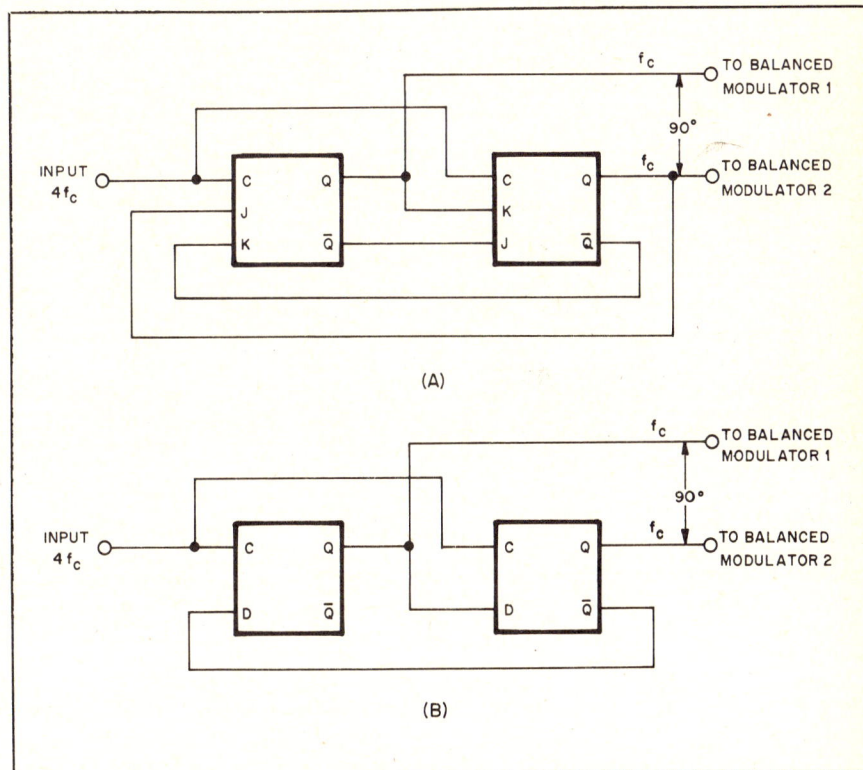

Fig 17 — Digital rf phase shift networks. The circuit at A uses JK flip flops, and the circuit at B uses D flip flops. In each case the desired carrier frequency must be quadrupled before it is processed by the phase shift network.

quency, problems similar to those in the audio networks must be overcome.

A differential rf phase shifter is shown in Fig. 15. The amplitudes of the quadrature signals won't be equal over an entire phone band, but this is of little consequence as long as the signals are strong enough to saturate the modulators.

Where percentage bandwidths are

small, such as in the 144.1- to 145-MHz range, the rf phase shift can be obtained conveniently by means of transmission-line methods. If one balanced modulator feed line is made an electrical quarter wavelength longer than the other, the two signals will be 90° out of phase. It is important that the cables be properly terminated.

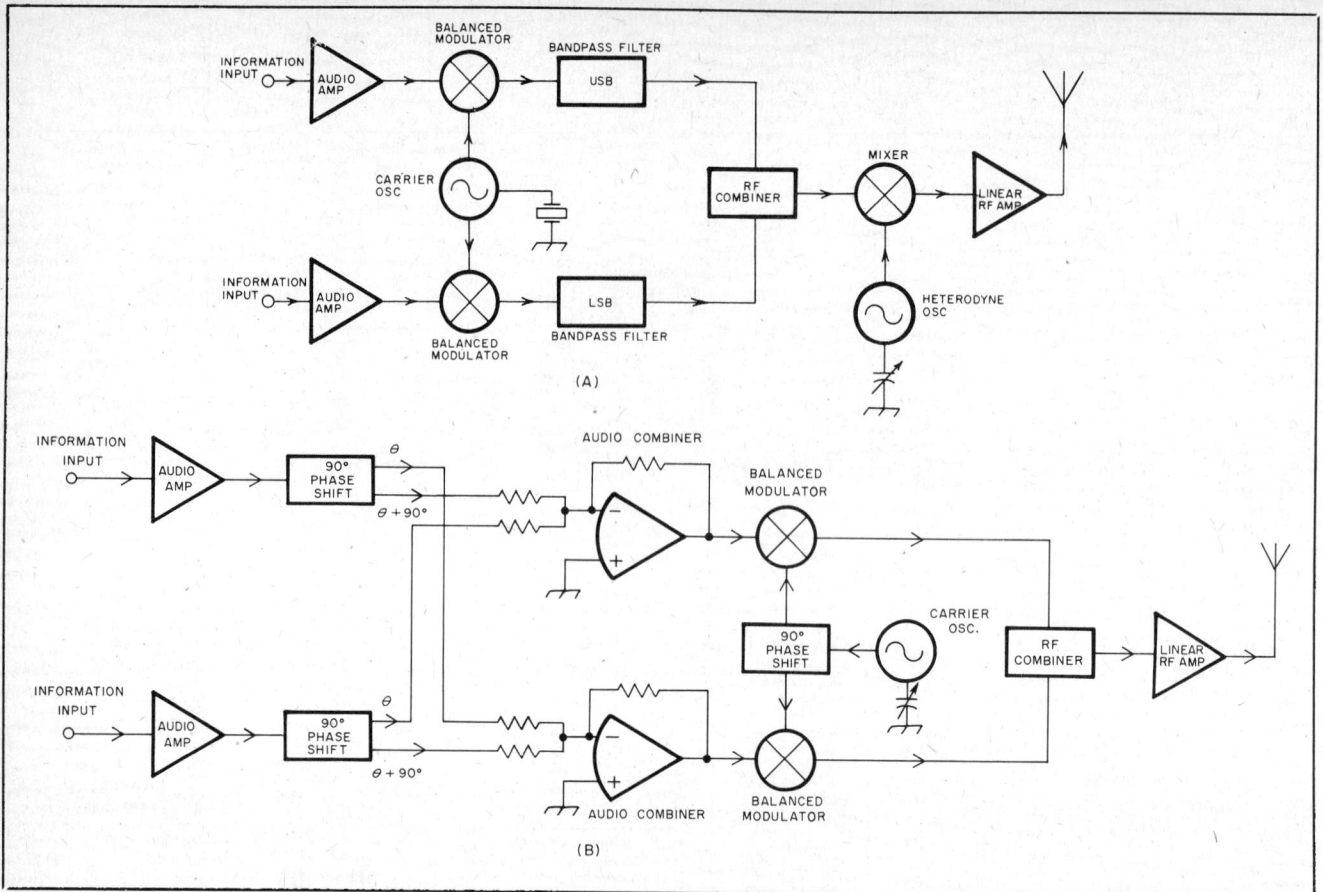

Fig. 18 -- Independent-sideband generators. (A) Filter system. (B) Phasing system. The block marked "rf combiner" can be a hybrid combiner or a summing amplifier.

One method for obtaining a 90° phase shift over a wide bandwidth is to generate the quadrature signals at a fixed frequency and heterodyne them individually to any desired operating frequency. Quadrature hybrids having multioctave bandwidths are manufactured commercially, but they cost hundreds of dollars. Another practical approach is to use two VFOs in a master-slave phase-locked loop system. Many phase detectors lock the two signals in phase quadrature. A doubly balanced mixer has this property. One usually thinks of a phase-locked loop as having a VCO locked to a reference signal, but a phase differential can be controlled independently of the oscillator. The circuit in Fig. 16 illustrates this principle. Two digital phase shifters are sketched in Fig. 17. If ECL ICs are used, this system can work over the entire hf spectrum.

Other SSB Modes

An ssb transmitter is simply a frequency translator. Any frequency- or amplitude-varying signal (within the bandwidth capabilities of the transmitter) applied to the input will be translated intact (although frequency inversion takes place in lsb) to the chosen radio frequency. If amplitude-limited tones corresponding to the video information of a slow-scan television picture are fed into the microphone input, F5 emission will result. Two alternating tones from an afsk RTTY keyer will cause the transmitter to produce an F1 signal. A keyed audio tone will be translated into an A1 signal. This technique is a perfectly legitimate way to operate cw with an ssb transceiver, and is simpler than the more traditional method of upsetting the balanced modulator for carrier leakage. One can vary the transmitting frequency independently of the receiving frequency by means of changing the audio tone. The strength of the tone determines the transmitter power output. Good engineering practice requires that the tone be frequency-stable and that the total harmonic distortion be less than one percent. Also, the carrier and opposite sideband must be suppressed at least 40 dB. Of course the rise and decay times of the audio envelope must be controlled to avoid key clicks. This subject is treated in detail in Chapter 11.

Independent Sideband Emission

If two ssb exciters, one usb and the other lsb, share a common carrier oscillator, two channels of information can be transmitted from one antenna. Methods for isb generation in filter and phasing transmitters are shown in Fig. 18. May 1977 *QST* carried an article on converting the popular Drake TR4-C to isb.

Many commercially manufactured ssb transceivers have provisions for controlling the transmit or receive frequency with an external VFO or receiver. With slight modification it should be possible to slave two transceivers to a single VFO for isb operation. The oscillators in the transceivers must be aligned precisely.

The most obvious amateur application for independent sideband is the transmission of slow-scan television with simultaneous audio commentary. On the vhf bands, other combinations are possible, such as voice and code or SSTV and RTTY.

The Speech Amplifier

The purpose of a speech amplifier is to raise the level of audio output from a microphone to that required by the modulator of a transmitter. In ssb and fm transmitters the modulation process takes place at low levels, so only a few volts of audio are necessary. One or two simple voltage-amplifier stages will suffice. A-m transmitters often employ high-level plate modulation requiring considerable audio power, compared to ssb and fm. The microphone-input and audio voltage-amplifier circuits are similar in all three types of phone transmitters, however.

When designing speech equipment it is necessary to know (1) the amount of

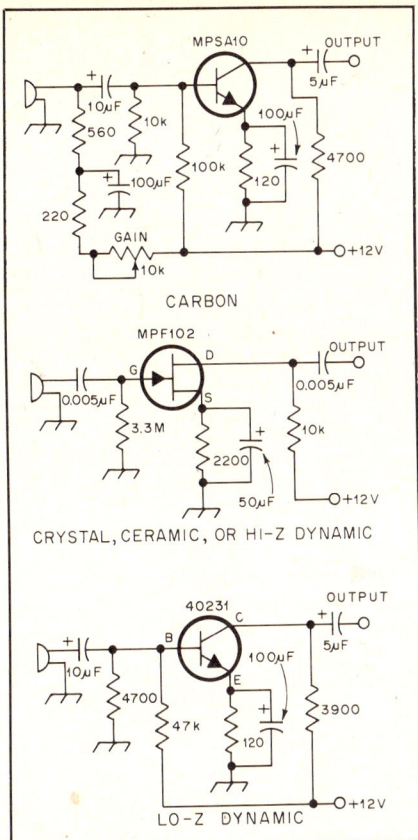

Fig. 19 — Speech circuits for use with standard-type microphones. Typical parts values are given.

audio power the modulation system must furnish, and (2) the output voltage developed by the microphone when it is spoken into from normal distance (a few inches) with ordinary loudness. It then becomes possible to choose the number and type of amplifier stages needed to generate the required audio power without overloading or undue distortion anywhere in the system.

Microphones

The level of a microphone is its electrical output for a given sound intensity. The level varies somewhat with the type. It depends to a large extent on the distance from the sound source and the intensity of the speaker's voice. Most commercial transmitters are designed for the median level. If a high-level mic is used, care should be taken not to overload the input amplifier stage. Conversely, a microphone of too low a level must be boosted by a preamplifier.

The frequency response (fidelity) of a microphone is its ability to convert sound uniformly into alternating current. For high articulation it is desirable to reproduce a frequency range of 200-3500 Hz. When all frequencies are reproduced equally, the microphone is considered "flat." Flat response is highly desirable as

peaks (sharp rises in the reproduction curve) limit the swing or modulation to the maximum drive voltage, whereas the usable energy is contained in the flat part of the curve. A microphone must be terminated in its specified load impedance if the designed frequency response is to be realized.

Microphones are generally omnidirectional, and respond to sound from all directions, or unidirectional, picking up sound from one direction. If a microphone is to be used close to the operator's mouth, an omnidirectional microphone is ideal. If, however, speech is generated a foot (0.31 m) or more from the microphone, a unidirectional microphone will reduce reverberation by a factor of 1.7:1. Some types of unidirectional microphones have a proximity effect in that low frequencies are accentuated when the microphone is too close to the mouth.

Carbon Microphones

The carbon microphone consists of a metal diaphragm placed against a cup of loosely packed carbon granules. As the diaphragm is actuated by the sound pressure, it alternately compresses and decompresses the granules. When current is flowing through the button, a variable dc will correspond to the movement of the diaphragm. This fluctuating dc can be used to provide voltage corresponding to the sound pressure. The output of a carbon microphone is extremely high, but nonlinear distortion and instability has reduced its use.

Piezoelectric Microphones

Piezoelectric microphones make use of the phenomenon by which certain materials produce a voltage by mechanical stress or distortion of the material. A diaphragm is coupled to a small bar of material such as Rochelle salt or ceramic made of barium titanate or lead zirconium titanate. The diaphragm motion is thus translated into electrical energy. Rochelle-salt crystals are susceptible to high temperatures, excessive moisture, or extreme dryness. Although the output level is higher, their use is declining because of their fragility.

Ceramic microphones are impervious to temperature and humidity. The output level is adequate for most modern amplifiers. They are capacitive devices and the output impedance is high. The load impedance will affect the low frequencies. To provide attenuation, it is desirable to reduce the load to 0.25 megohm or even lower, to maximize performance when operating ssb, thus eliminating much of the unwanted low-frequency response.

Dynamic Microphones

The dynamic microphone somewhat resembles a dynamic loudspeaker. A lightweight coil, usually made of aluminum wire, is attached to a diaphragm. This coil is suspended in a magnetic circuit. When sound impinges on the diaphragm, it moves the coil through the magnetic field, generating an alternating voltage.

Electret Microphones

The electret microphone has recently appeared as a feasible alternative to the carbon, piezoelectric or dynamic microphone. An electret is an insulator which has a quasi-permanent static electric charge trapped in or upon it. The electret operates in a condenser fashion which uses a set of biased plates whose motion, caused by air pressure variations, creates a changing capacitance and an accompanying change in voltage. The electret acts as the plates would, and being charged, it requires no bias voltage. A low voltage provided by a battery used for an FET impedance converter is the only power required to produce an audio signal.

Electrets traditionally have been susceptible to damage from high temperatures and high humidity. New materials and different charging techniques have lowered the chances of damage, however. Only in extreme conditions (such as 120°F or 49°C at 90 percent humidity) are problems present. The output level of a typical electret is higher than that of a standard dynamic microphone.

Microphone Amplifiers

The circuit immediately following the audio input establishes the signal-to-noise ratio of the transmitter. General-purpose ICs such as the 709 and 741 op amps are widely used in speech amplifiers, but they are fairly noisy, so it is best to precede them with a lower-noise discrete device (FET or bipolar transistor). The circuits in Fig. 19 fulfill this requirement.

Voltage Amplifiers

The important characteristics of a voltage amplifier are its *voltage gain,* maximum undistorted *output voltage,* and its *frequency response.* The voltage gain is the voltage-amplification ratio of the stage. The output voltage is the maximum af voltage that can be secured from the stage without distortion. The amplifier frequency response should be adequate for voice reproduction; this requirement is easily satisfied.

The voltage gain and maximum undistorted output voltage depend on the operating conditions of the amplifier. The output voltage is in terms of *peak* voltage rather than rms; this makes the rating independent of the waveform. Exceeding the peak value causes the amplifier to distort, so it is more useful to consider only peak values in working with amplifiers.

A circuit suitable for use as a microphone preamplifier or the major gain block of a speech system is shown in Fig. 20. The response rolls off below 200 Hz to reduce hum pickup. Ordinary 741 op amps can be used in stages following the preamp, provided the voltage gain is held to about 20 (26 dB).

Gain Control

A means for varying the overall gain of the amplifier is necessary for keeping the final output at the proper level for modulating the transmitter. The common method of gain control is to adjust the value of ac voltage applied to the input of one of the amplifiers by means of a voltage divider or potentiometer.

The gain-control potentiometer should be near the input end of the amplifier, at a point where the signal voltage level is so low there is no danger that the stages ahead of the gain control will be overloaded by the full microphone output. In a high-gain amplifier it is best to operate the first stage at maximum gain, since this gives the best signal-to-hum ratio. The control is usually placed in the input circuit of the second stage.

Remote gain control can be accomplished with an electronic attenuator IC, such as the Motorola MFC6040. A dc voltage varies the gain of the IC from +6 to −85 dB, eliminating the need for shielded leads to a remotely located volume control. A typical circuit is shown in Fig. 21.

Phase Inversion

Some balanced modulators and phase shifters require push-pull audio input. The obvious way to obtain push-pull output from a single-ended stage is to use a transformer with a center-tapped secondary. *Phase inverter* or *phase splitter* circuits can accomplish the same task electronically. A differential amplifier can be used to convert a single-ended input to a push-pull output. Two additional phase splitter circuits are shown in Fig. 23.

Speech-Amplifier Construction

Once a suitable circuit has been selected for a speech amplifier, the construction problem resolves itself into avoiding two difficulties — excessive hum, and unwanted feedback. For reasonably humless operation, the hum voltage should not exceed about one percent of the maximum audio output voltage — that is, the hum and noise should be at least 40 dB below the output level.

Unwanted feedback, if negative, will reduce the gain below the calculated value; if positive, is likely to cause self-oscillation or "howls." Feedback can be minimized by isolating each stage with decoupling resistors and capacitors, by avoiding layouts that bring the first and last stages near each other, and by shielding of "hot" points in the circuit, such as high-impedance leads in low-level stages.

If circuit-board construction is used, high-impedance leads should be kept as short as possible. All ground returns should be made to a common point. A

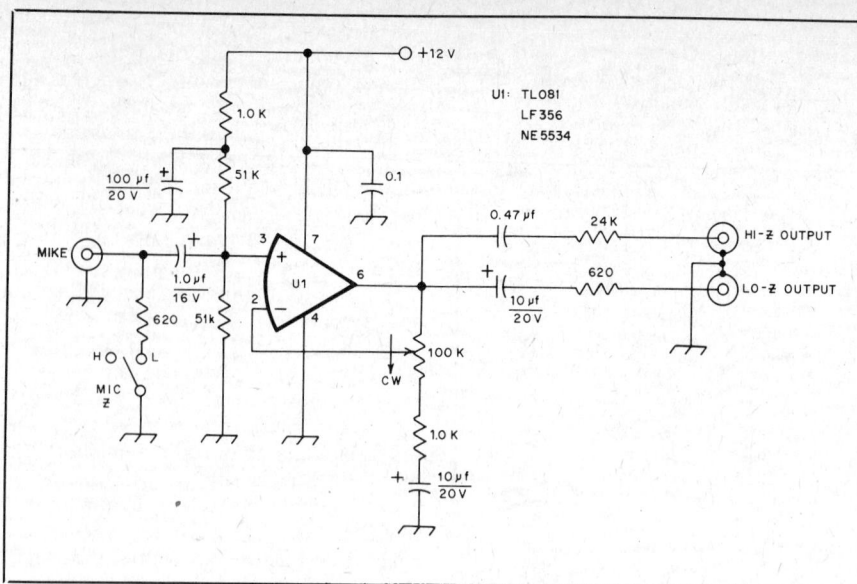

Fig. 20 — A speech amplifier suitable for microphone or interstage use. The input and output impedances can be tailored to match a wide range of loads. Maximum gain of this circuit is 40 dB.

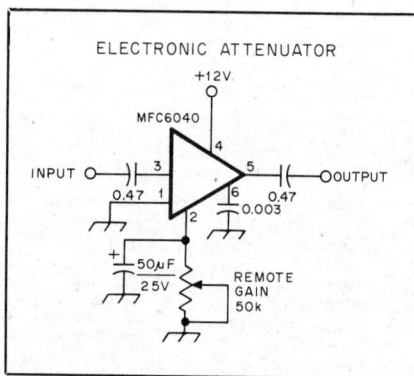

Fig. 21 — A dc voltage controls the gain of this IC, eliminating the need for shielded leads to the gain control.

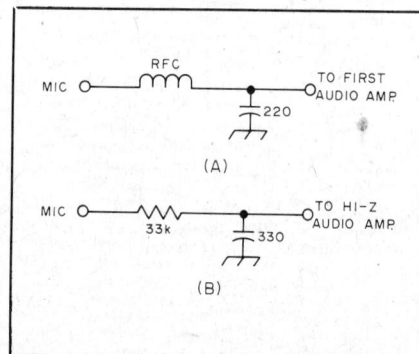

Fig. 22 — Rf filters using LC (A) and RC (B) components, which are used to prevent feedback caused by rf pickup on the microphone lead.

good ground between the circuit board and the metal chassis is necessary. Complete shielding from rf energy is always required for low-level solid-state audio circuits. The microphone input should be decoupled for rf with a filter, as shown in Fig. 22. At A, an rf choke with a high impedance over the frequency range of the transmitter is employed. For high-impedance inputs, a resistor may be used in place of the choke.

When using paper capacitors as bypasses, be sure that the terminal marked "outside foil," often indicated with a black band, is connected to ground. This utilizes the outside foil of the capacitor as a shield around the "hot" foil. When paper or mylar capacitors are used for coupling between stages, always connect the outside foil terminal to the side of the circuit having the lower impedance to ground.

Driver and Output Stages

The most-commonly-used balanced modulators and transmitting mixers have power outputs too low for consistently effective communications. Most modern grounded-grid linear amplifiers require 30 to 100 watts of exciter output power to drive them to their rated power input. An exciter output amplifier serves to boost the output power to a useful level while providing additional selectivity to reject spurious mixing products.

Two stages are usually required to obtain the necessary power. The stage preceding the output amplifier is called the driver. Some tubes that work well as drivers are the 6CL6, 12BY7, 6EH7 and 6GK6. Since all of these tubes are capable of high gain, instability is sometimes encountered in their use. Parasitic suppression should be included as a matter of course. Some form of neutralization is recommended. Driver stages should be operated in Class A or AB1 to minimize distortion. The higher quiescent dissipation can be easily handled at these power levels. The new VMOS power FETs are well suited to ssb driver circuits.

The exciter output amplifier can be a

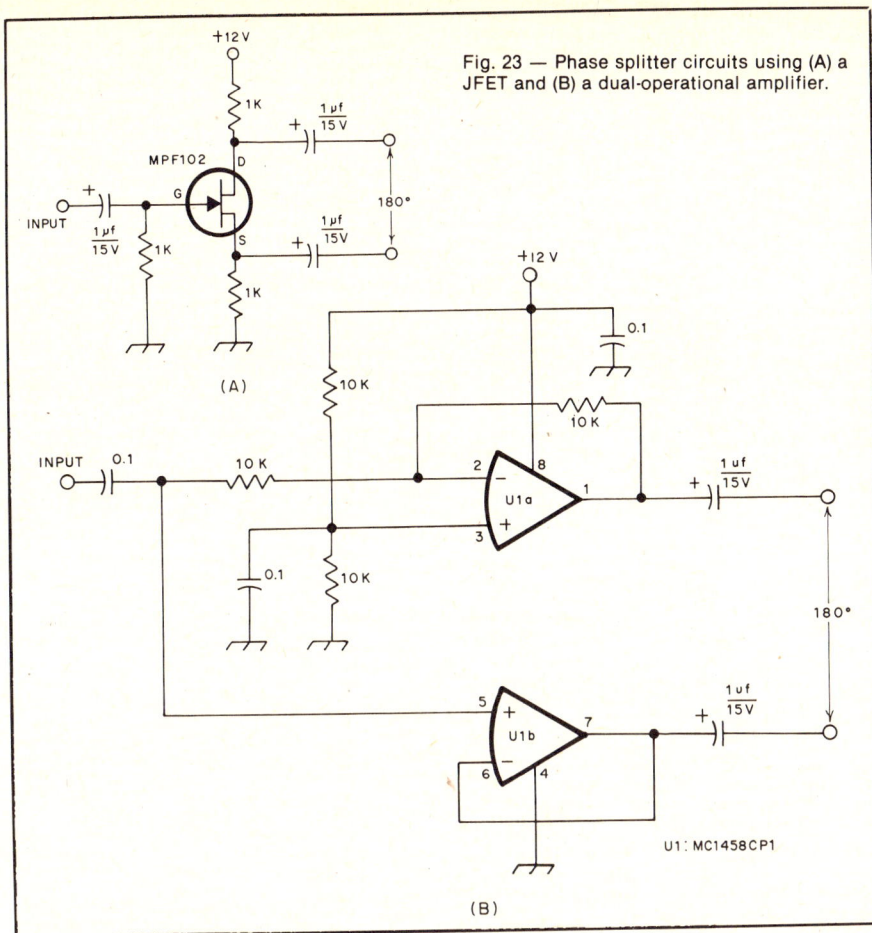

Fig. 23 — Phase splitter circuits using (A) a JFET and (B) a dual-operational amplifier.

U1: MC1458CP1

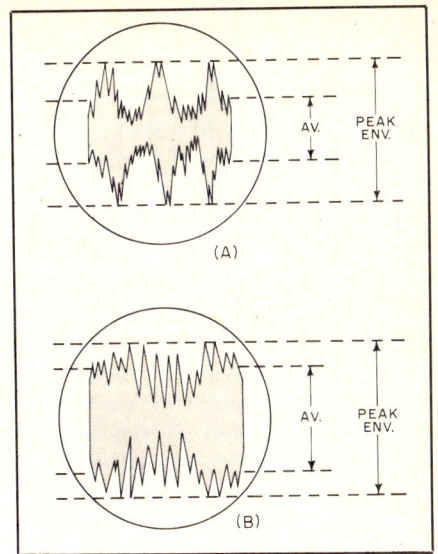

Fig. 24 — (A) Typical ssb voice-modulated signal might have an envelope of the general nature shown, where the rf amplitude (current or voltage) is plotted as a function of time, which increases to the right horizontally. (B) Envelope pattern after speech processing to increase the average level of power output.

TV horizontal sweep tube. Some sweep tubes are capable of lower distortion than others, but if not overdriven most of them are satisfactory for amateur use, yielding IMD levels between −26 and −30 dB, typically. Some types suitable for AB1 service are 6DQ5, 6GB5, 6GE5, 6HF5, 6JE6, 6JS6, 6KD6, 6LF6, 6KG6 and 6LQ6. A genuine transmitting tube such as a 6146B can be operated in the higher efficiency Class-AB2 or B modes for the same distortion produced by sweep tubes in AB1. Transmitting tubes have the additional advantages of uniformity and ruggedness. Linear amplifiers, including those using solid-state devices, are treated in detail in chapter 6.

POWER RATINGS OF SSB TRANSMITTERS

Fig. 24A is more or less typical of a few voice-frequency cycles of the modulation envelope of a single-sideband signal. Two amplitude values associated with it are of particular interest. One is the *maximum peak amplitude*, the greatest amplitude reached by the envelope at any time. The other is the *average amplitude,* which is the average of all the amplitude values contained in the envelope over some significant period of time, such as the time of one syllable of speech.

The power contained in the signal at the maximum peak amplitude is the basic transmitter rating. It is called the *peak-envelope power,* abbreviated PEP. The peak-envelope power of a given transmitter is intimately related to the distortion considered tolerable. The lower the signal-to-distortion ratio the lower the attainable peak-envelope power, as a general rule. For splatter reduction, an S/D ratio of 25 dB is considered a border-line minimum, and better figures are desirable.

The signal power, S, in the standard definition of S/D ratio is the power in *one* tone of a two-tone test signal. This is 6 dB below the peak-envelope power in the same signal. Manufacturers of amateur ssb equipment usually base their published S/D ratios on PEP, thereby getting an S/D ratio that looks 6 dB better than one based on the standard definition. In comparing distortion-product ratings of different transmitters or amplifiers, first make sure that the ratios have the same base.

Peak vs. Average Power

Envelope peaks occur only sporadically during voice transmission, and have no direct relationship to meter readings. The meters respond to the amplitude (current or voltage) of the signal averaged over several cycles of the modulation envelope. (This is true in practically all cases, even though the transmitter rf output meter may be *calibrated* in watts. Unfortunately, such a calibration means little in voice transmission since the meter can be calibrated in watts only by using a sine-wave signal — which a voice-modulated signal definitely is not.)

The ratio of peak-to-average amplitude varies widely with voice of different characteristics. In the case shown in Fig. 24A the average amplitude, found graphically, is such that the peak-to-average ratio of amplitudes is almost 3:1. The ratio of peak *power* to average *power* is something else again. There is no simple relationship between the meter reading and actual average power, for the reason mentioned earlier.

DC Input

FCC regulations require that the transmitter power be rated in terms of the dc input to the final stage. Most ssb final amplifiers are operated Class AB1 or AB2, so that the plate current during modulation varies upward from a "resting" or no-signal value that is generally chosen to minimize distortion. There will be a peak-envelope value of plate current that, when multiplied by the dc plate voltage, represents the instantaneous tube power input required to produce the peak-envelope output. This is the "peak-envelope dc input" or "PEP input." It does not register on any meter in the transmitter. Meters cannot move fast enough to show it — and even if they did, the eye couldn't follow. What the plate meter *does* read is the plate current averaged over several modulation-envelope cycles. This multiplied by the dc plate voltage is the number of watts input required to produce the *average* power output described earlier.

In voice transmission the power input and power output are both continually varying. The power input peak-to-average ratio, as the power-output peak-to-average ratio, depends on the voice characteristics. Determination of the input ratio is further complicated by the fact that there is a resting value of dc plate input even when there is no rf output. *No exact figures are possible.* However, experience has shown that for many types of voices and for ordinary tube operating conditions where a moderate value of resting current is used, the ratio of PEP input to average input (during a modulation peak) will be in the neighborhood of 2:1. That is why many amplifiers are rated for a PEP input of 2 kilowatts even though the maximum legal input is 1 kilowatt.

PEP Input

The 2-kW PEP input rating can be interpreted in this way: The amplifier can handle dc peak-envelope inputs of 2 kW, presumably with satisfactory linearity. But it should be run up to such peaks if — and *only* if — in doing so the dc plate current (the current that shows on the plate meter) multiplied by the dc plate voltage does not at any time exceed 1 kW. On the other hand, if your voice has characteristics such that the dc peak-to-average ratio is, for example, 3:1, you should not run a greater dc input during peaks than 2000/3, or 660 watts. Higher dc input would drive the amplifier into nonlinearity and generate splatter.

If your voice happens to have a peak-to-average ratio of less than 2:1 with this particular amplifier, you cannot run more than 1 kW dc input even though the envelope peaks do not reach 2 kW.

It should be apparent that the dc input rating (based on the *maximum* value of dc input developed during modulation, of course) leaves much to be desired. Its principal virtues are that it can be measured with ordinary instruments, and that it is consistent with the method used for rating the power of other types of emission used by amateurs. The meter readings offer no assurance that the transmitter is being operated within linearity limits, unless backed up by oscilloscope checks using *your* voice.

It should be observed, also, that in the case of a grounded-grid final amplifier, the 1-kW input permitted by FCC regulations must include the input to the driver stage as well as the input to the final amplifier itself. Both inputs are measured as described above.

Speech Processing

Four basic systems, or a combination thereof, can be used to reduce the peak-to-average ratio, and thus, to raise the average power level of an ssb signal. They are compression or clipping of the af wave before it reaches the balanced modulator,

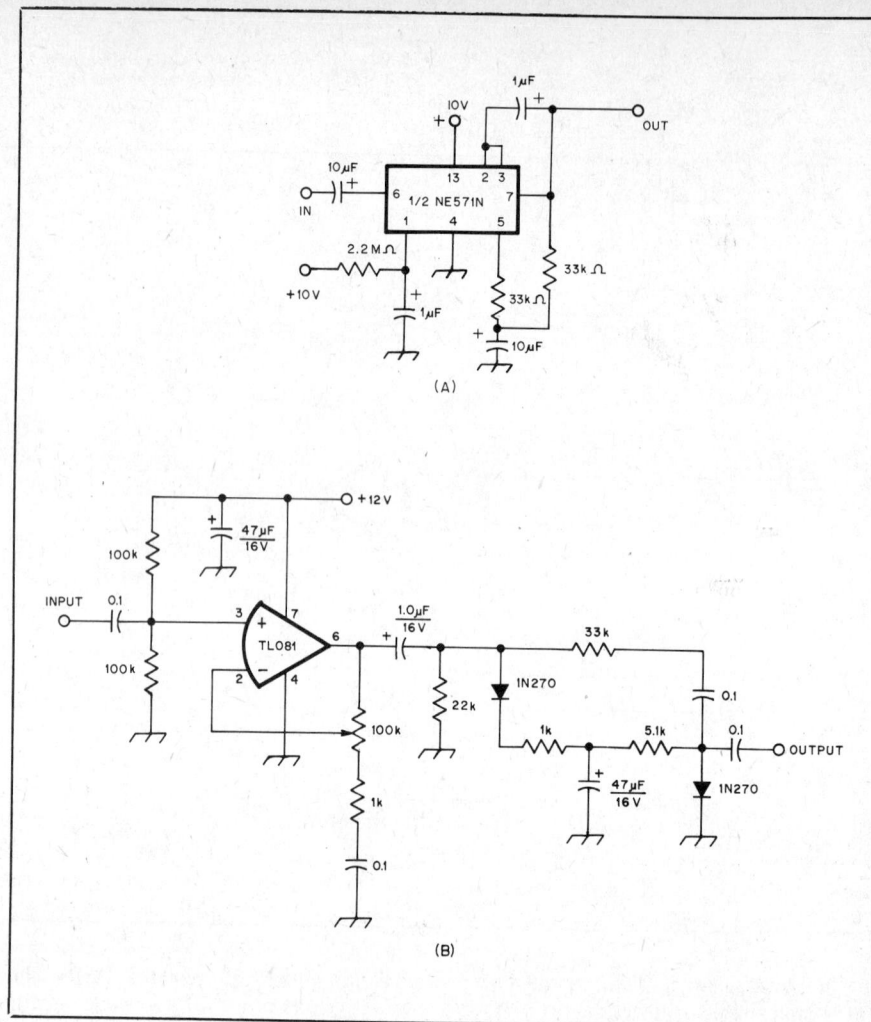

Fig. 25 — Typical solid-state compressor circuits.

and compression or clipping of the rf waveform after the ssb signal has been generated. One form of rf compression, commonly called *alc* (automatic level control) is almost universally used in amateur ssb transmitters. Audio processing is also used to increase the level of audio power contained in the sidebands of an a-m transmitter and to maintain constant deviation in an fm transmitter. Both compression and clipping are used in a-m systems, while most fm transmitters employ only clipping.

Volume Compression

Although it is obviously desirable to keep the voice level as high as possible, it is difficult to maintain constant voice intensity when speaking into the microphone. To overcome this variable output level, it is possible to use automatic gain control that follows the *average* (not instantaneous) variations in speech amplitude. This can be done by rectifying and filtering some of the audio output and applying the rectified and filtered dc to a control electrode in an early stage in the amplifier.

The circuit of Fig. 25A works on this agc principle. One section of a Signetics NE571N IC is used. The other section can be connected as an expander to restore the dynamic range of received signals that have been compressed in transmission. This system is used in the VBC 3000 nbvm baseband transceiver. Operational transconductance amplifier ICs such as the CA3080 are also well suited for speech compression service.

When an audio agc circuit derives control voltage from the output signal, the system is closed loop. If short attack time is necessary, the rectifier/filter bandwidth must be opened up to allow syllabic modulation of the control voltage. This allows some of the voice-frequency signal to enter the control terminal, causing distortion and instability. Because the syllabic frequency and speech-tone frequencies have relatively small separation, the simpler feedback agc systems compromise fidelity for fast response.

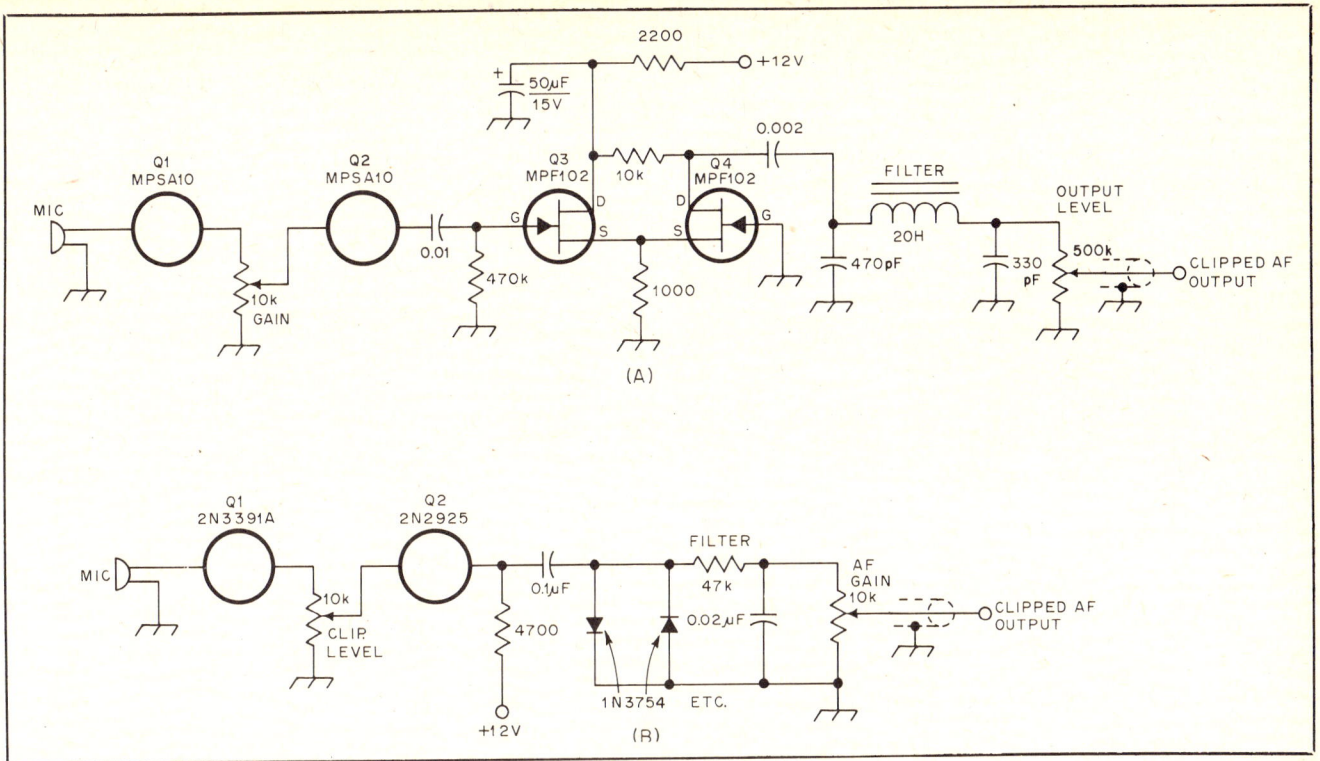

Fig. 26 — This drawing illustrates use of JFETs or silicon diodes to clip positive and negative voice peaks.

Loop dynamics problems in audio agc systems can be sidestepped by eliminating the loop and using a forward-acting system. The control voltage is derived from the *input* of the amplifier rather than the output. Eliminating the feedback loop allows unconditional stability, but the trade-off between response time and fidelity remains. Care must be taken to avoid excessive gain between the signal input and control voltage output. Otherwise, the transfer characteristic can reverse; that is, an increase in input level can cause a *decrease* in output. A simple forward-acting compressor is shown in part B of Fig. 25.

To test or adjust a speech compressor, one would apply a single steady tone to the input and measure the input and output signal levels over the dynamic range of the instrument. The single-tone test may indicate gross distortion, but one cannot judge the speech performance by this result. Provided the release time is longer than the time between syllables, the compressor will generate distortion only during brief signal peaks.

Speech Clipping and Filtering

In speech waveforms the average power content is considerably less than in a sine wave of the same peak amplitude. If the high amplitude peaks are clipped off, the remaining waveform will have a considerably higher ratio of average power to peak amplitude. Although clipping distorts the waveform and the result therefore does not sound exactly like the original, it is possible to secure a worth-

while increase in audio power without sacrificing intelligibilty. Once the system is properly adjusted *it will be impossible to overdrive* the modulator stage of the transmitter because the maximum output amplitude is fixed.

By itself, clipping generates high-order harmonics and therefore will cause splatter. To prevent this, the audio frequencies above those needed for intelligible speech must be filtered out *after* clipping and *before* modulation. The filter required for this purpose should have relatively little attenuation below about 2500 Hz, but high attenuation for all frequencies above 3000 Hz.

There is a loss in naturalness with "deep" clipping, even though the voice is highly intelligible. With moderate clipping levels (6 to 12 dB) there is almost no change in "quality" but the voice power is increased considerably.

Before drastic clipping can be used, the speech signal must be amplified several times more than is necessary for normal modulation. Also, the hum and noise must be much lower than the tolerable level in ordinary amplification, because the noise in the output of the amplifier increases in proportion to the gain.

In the circuit of Fig. 26B a simple diode clipper is shown following a two-transistor preamplifier section. The 1N3754s conduct at approximately 0.7 volt of audio and provide positive- and negative-peak clipping of the speech waveform. A 47-kΩ resistor and a 0.02-μF capacitor follow the clipper to form a simple R-C filter for attenuating the high-

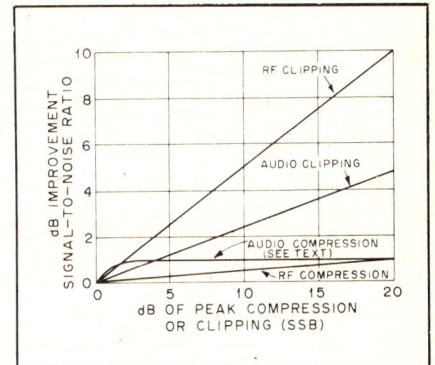

Fig. 27 — The improvement in received signal-to-noise ratio achieved by the simple forms of signal processing.

frequency components generated by the clipping action, as discussed earlier. Any top-hat or similar silicon diodes can be used in place of the 1N3754s. Germanium diodes (1N34A type) can also be used, but will clip at a slightly lower peak audio level.

SSB Speech Processing

Compression and clipping are related, as both have fast attack times, and when the compressor release is made quite short, the effect on the waveform approaches that of clipping. Speech processing is most effective when accomplished at radio frequencies, although a combination of af clipping and compression can produce worthwhile results. The advantage of an outboard audio speech pro-

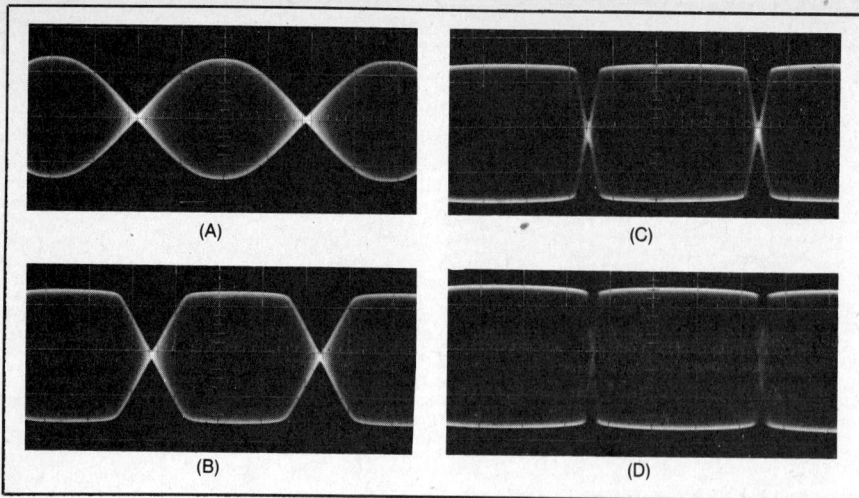

Fig. 28 — Two-tone envelope patterns with various degrees of rf clipping. All envelope patterns are formed using tones of 600 and 1000 Hz. (A) At clipping threshold; (B) 5 dB of clipping; (C) 10 dB of clipping; (D) 15 dB of clipping.

cessor is that no internal modifications are necessary to the ssb transmitter with which it will be used.

To understand the effect of ssb speech processing, review the basic rf waveforms shown in Fig. 24A. Without processing, they have high peaks but low average power. After processing, Fig. 24B, the amount of average power has been raised considerably. Fig. 27 shows an advantage of several dB for rf clipping (for 20 dB of processing) over its nearest competitor.

Investigations by W6JES reported in January 1969 *QST* show that, observing a transmitted signal using 15 dB of audio clipping from a remote receiver, the intelligibility threshold was improved nearly 4 dB over a signal with no clipping. Increasing the af clipping level to 25 dB gave an additional 1.5 dB improvement in intelligibility. Audio compression was found to be valuable for maintaining relatively constant average-volume

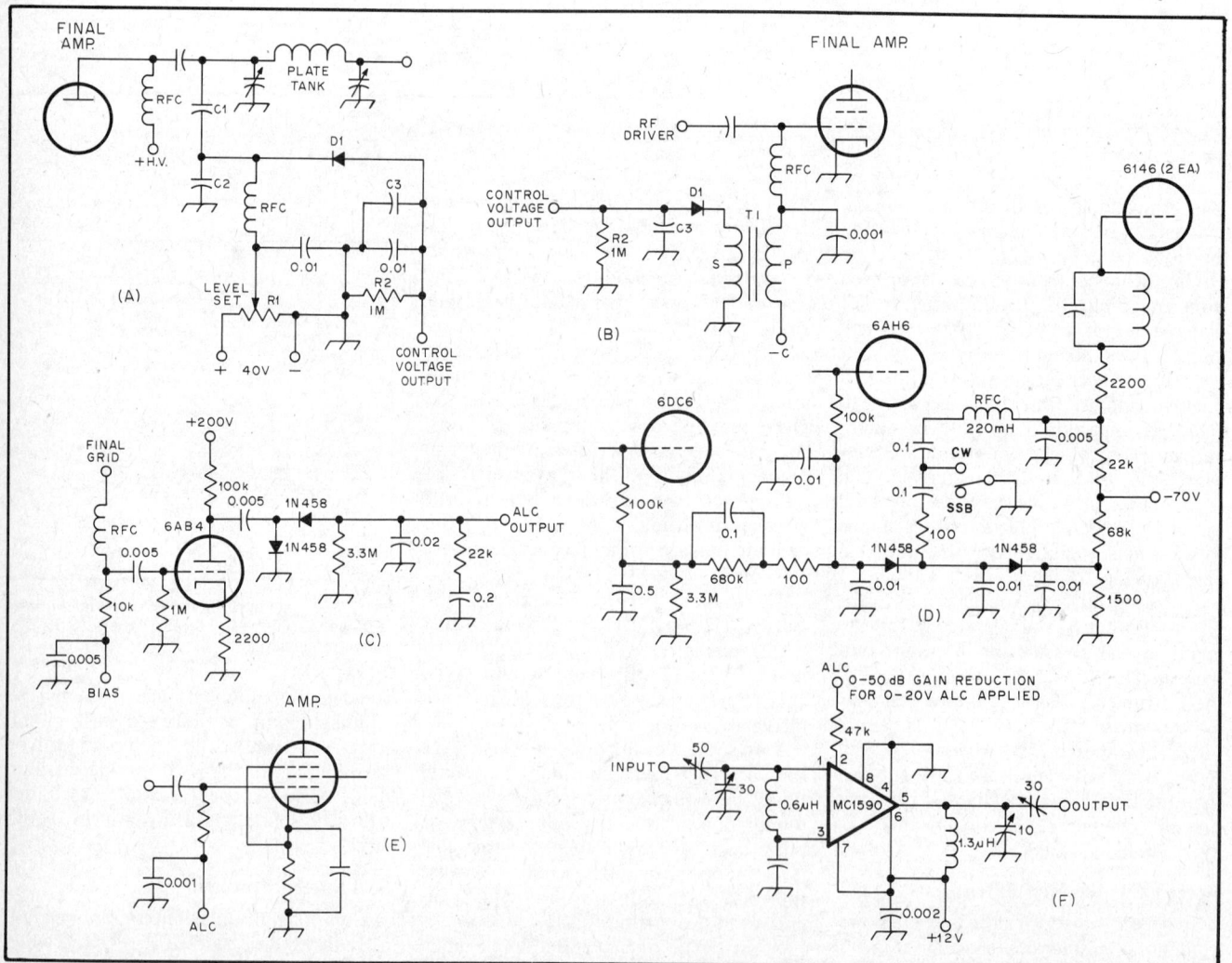

Fig. 29 — (A) Control voltage obtained by sampling the rf output voltage of the final amplifier. The diode back bias, 40 volts or so maximum, may be taken from any convenient positive voltage source in the transmitter. R may be a linear control having a maximum resistance of the order of 50 kΩ. D1 may be a 1N34A or similar germanium diode.
(B) Control voltage obtained from grid circuit of a Class AB1 tetrode amplifier. T1 is an interstage audio transformer having a turns ratio, secondary to primary, of 2 or 3 to 1. An inexpensive transformer may be used, since the primary and secondary currents are negligible. D1 may be a 1N34A or similar; time constant R2C3 is discussed in the text.
(C) Control voltage is obtained from the grid of a Class AB1 tetrode amplifier and amplified by a triode audio stage.
(D) Alc system used in the Collins 32S-3 transmitter.
(E) Applying control voltage to the tube or (F) linear IC-controlled amplifier.

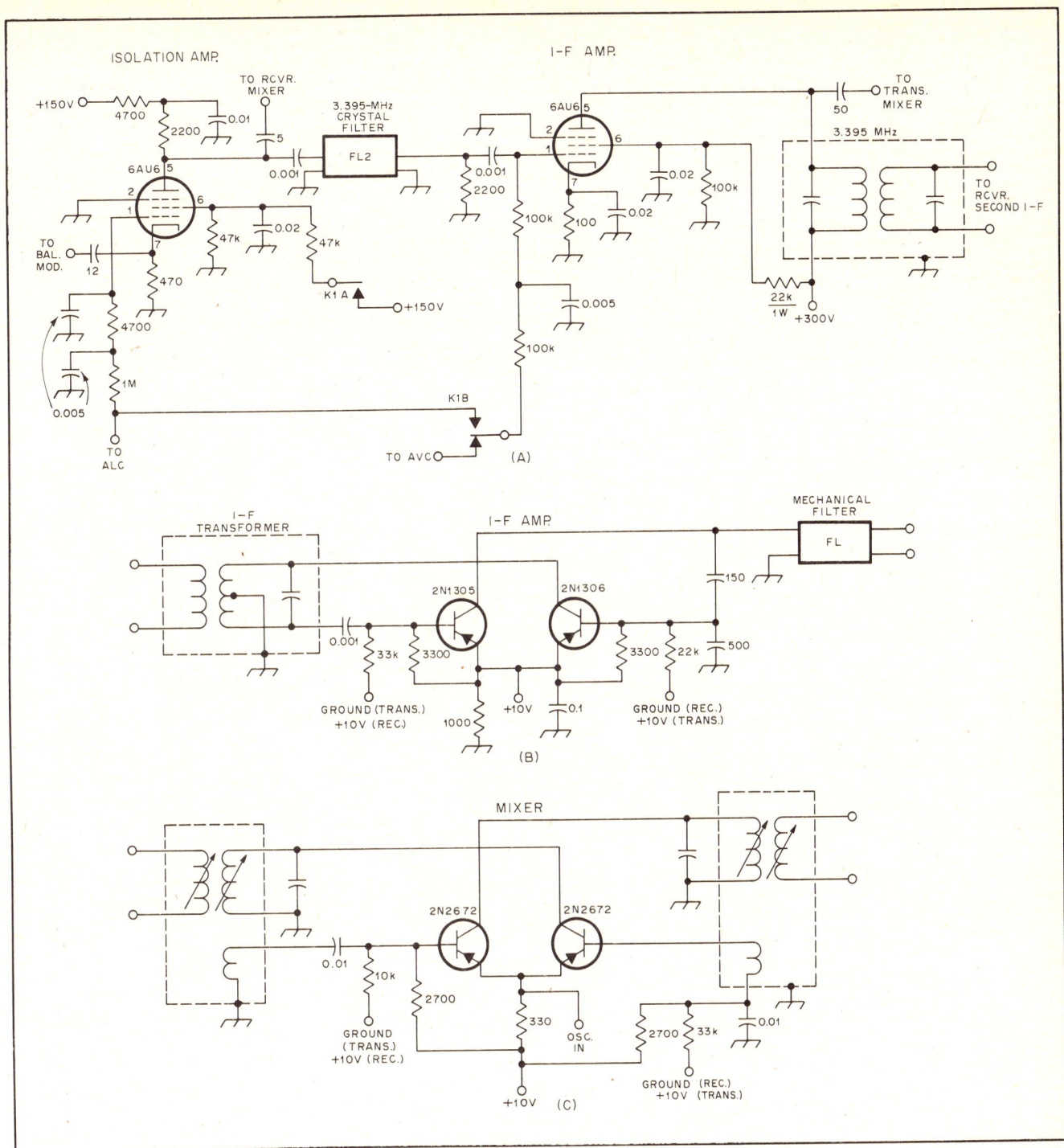

Fig. 30 — Transceiver circuits where a section is made to operate on both transmit and receive. See text for details.

speech, but such a compressor added little to the intelligibilty threshold at the receiver, only about 1-2 dB.

Evaluation of rf clipping from the receive side with constant-level speech, and filtering to restore the original bandwidth, resulted in an improved intelligibility threshold of 4.5 dB with 10 dB of clipping. Raising the clipping level to 18 dB gave an additional 4-dB improvement at the receiver, or 8.5-dB total increase. The improvement of the intelligibility of a weak ssb signal at a distant receiver can thus be substantially improved by rf clip-

ping. The effect of such clipping on a two-tone test pattern is shown in Fig. 28.

Automatic level control, although a form of rf speech processing, has found its primary application in maintaining the peak rf output of an ssb transmitter at a relatively constant level, hopefully below the point at which the final amplifier is overdriven when the audio input varies over a considerable range. These typical alc systems, shown in Fig. 29, by the nature of their design time constants offer a limited increase in transmitted average-to-PEP ratio. A value in the region of 2-5

dB is typical. An alc circuit with shorter time constants will function as an rf syllabic compressor, producing up to 6 db improvement in the intelligibility threshold at a distant receiver. The Collins Radio Company uses an alc system with dual time constants (Fig. 29D) in their S/Line transmitters, and this has proven to be quite effective.

Heat is an extremely important consideration in the use of any speech processor which increases the average-to-peak power ratio. Many transmitters, in particular those using television sweep tubes, simply are not built to stand the effects of increased average input, either in the final-amplifier tube or tubes or in the power supply. If heating in the final tube is the limiting factor, adding a cooling fan may be a satisfactory answer.

SINGLE-SIDEBAND TRANSCEIVERS

A *transceiver* combines the functions of transmitter and receiver in a single package. In contrast to a packaged "transmitter-receiver," it utilizes many of the active and passive elements for both transmitting and receiving. Ssb transceiver operation enjoys widespread popularity for several justifiable reasons. In most designs the transmissions are on the same (suppressed-carrier) frequency as the receiver is tuned to. The only practical way to carry on a rapid multiple-station "roundtable" or net operation is for all stations to transmit on the same frequency. Transceivers are ideal for this, since once the receiver is properly set the transmitter is also. Transceivers are by nature more compact than separate transmitter and receiver setups and thus lend themselves well to mobile and portable use.

Although the many designs available on the market differ in detail, there are of necessity many points of similarity. All of them use the filter type of sideband generation, and the filter unit furnishes the receiver i-f selectivity as well. The carrier oscillator doubles as the receiver (fixed) BFO. One or more mixer or i-f stage or stages will be used for both transmitting and receiving. The receiver S meter may become the transmitter plate-current or output voltage indicator. The VFO that sets the receiver frequency also determines the transmitter frequency. The same signal-frequency tuned circuits may be used for both transmission and reception, including the transmitter pi-network output circuit.

Usually the circuits are switched by a multiple-contact relay, which transfers the antenna if necessary and also shifts the biases on several stages. Most commercial designs offer *VOX* (voice-controlled operation) and *MOX* (manual operation). Which is preferable is a controversial subject; some operators like VOX and others prefer MOX.

Circuits

The use of a filter-amplifier combination common to both the transmitter and receiver is shown in Fig. 30A. This circuit is used by the Heath Company in several of their transceiver kits. When receiving, the output of the hf mixer is coupled to the crystal filter, which, in turn, feeds the first i-f amplifier. The output of this stage is transformer-coupled to the second i-f amplifier. During transmit, K1 is closed, turning on the isolation amplifier that links the balanced modulator to the bandpass filter. The single-sideband output

from the filter is amplified and capacitance-coupled to the transmitter mixer. The relay contacts also apply alc voltage to the first i-f stage and remove the screen voltage from the second i-f amplifier, when transmitting.

Bilateral amplifier and mixer stages, first used by Sideband Engineers in their SBE-33, also have found application in other transceiver designs. The circuits shown in Fig. 30B and C are made to work in either direction by grounding the bias divider of the input transistor, completing the bias network. The application of these designs to an amateur transceiver for the 80- through 10-meter bands is given in the Fifth Edition of *Single Sideband for the Radio Amateur*.

The complexity of a multiband ssb transceiver is such that most amateurs buy them fully built and tested. There are, however, some excellent designs available in the kit field, and any amateur able to handle a soldering iron and follow instructions can save himself considerable money by assembling an ssb transceiver kit.

Some transceivers include a feature that permits the receiver to be tuned a few kHz either side of the transmitter frequency. This consists of a voltage-sensitive capacitor, which is tuned by varying the applied dc voltage. This can be a useful device when one or more of the stations in a net drift slightly. The control for this function is usually labeled RIT for receiver incremental tuning. Other transceivers include provision for a crystal-controlled transmitter frequency plus full use of the receiver tuning. This is useful for "DXpeditions" where net operation (on the same frequency) may not be desirable.

Testing a Sideband Transmitter

There are three commonly used methods for testing an ssb transmitter. These include the wattmeter, oscilloscope, and spectrum-analyzer techniques. In each case, a two-tone test signal is fed into the mic input to simulate a speech signal. From the measurements, information concerning such quantities as PEP and intermodulation-distortion-product (IMD) levels can be obtained. Depending upon the technique used, other aspects of transmitter operation (such as hum problems and carrier balance) can also be checked.

As might be expected, each technique has both advantages and disadvantages and the suitability of a particular method will depend upon the desired application. The wattmeter method is perhaps the simplest one but it also provides the least amount of information. Rf wattmeters suitable for single-tone or cw operation may not be accurate with a two-tone test signal. A suitable wattmeter for the latter case must have a reading that is proportional to the actual power consumed by the load. The reading must be indepen-

dent of signal waveform. A thermocouple ammeter connected in series with the load would be a typical example of such a system. The output power would be equal to I^2R, where I is the current in the ammeter and R is the load resistance (usually 50 ohms). In order to find the PEP output with the latter method (using a two-tone test input signal), the power output is multiplied by two.

A spectrum analyzer is capable of giving the most information (of the three methods considered here), but it is also the most costly method and the one with the greatest chance of misinterpretation. Basically, a spectrum analyzer is a receiver with a readout which provides a plot of signal amplitude vs. frequency. The readout could be in the form of a paper chart but usually it is presented as a trace on a CRT. For a spectrum analyzer to provide accurate information about a signal, that signal must be well within the linear dynamic range of the analyzer. For a thorough explanation of the function and application of this instrument see Rusgrove, "Spectrum Analysis — One Picture's Worth a . . .," August 1979 *QST*.

Two-Tone Tests and Scope Patterns

A very practical method for amateur applications is to use a two-tone test signal (usually audio) and sample the transmitter output. The waveform of the latter is then applied directly to the vertical-deflection plates in an oscilloscope. An alternative method is to use an rf probe and detector to sample the waveform and apply the resulting audio signal to the vertical-deflection amplifier input.

If there are no appreciable nonlinearities in the amplifier, the resulting envelope will approach a perfect sine-wave pattern (see Fig. 31A). As a comparison, a spectrum-analyzer display for the same transmitter and under the same conditions is shown in Fig. 31B. In this case, spurious products can be seen which are approximately 30 dB below the amplitude of each of the tones.

As the distortion increases, so does the level of the spurious products and the resulting waveform departs from a true sine-wave function. This can be seen in Fig. 31C. One of the disadvantages of the scope and two-tone test method is that a relatively high level of IMD-product voltage is required before the waveform seems distorted to the eye. For instance, the waveform in Fig. 31C doesn't seem too much different from the one in Fig. 31A but the IMD level is only 17 dB below the level of the desired signal (see analyzer display in Fig. 31D). A 17- to 20-dB level corresponds to approximately 10-percent distortion in the voltage waveform. Consequently, a "good" waveform means the IMD products are at least 30 db below the desired tones. Any *noticeable* departure from the waveform in Fig. 31A should be

(A)

Panoramic SPECTRUM ANALYZER MODEL SB-12

(B)

(C)

Panoramic SPECTRUM ANALYZER MODEL SB-12

(D)

(E)

Panoramic SPECTRUM ANALYZER MODEL SB-12

(F)

Fig. 32 — Waveform of an amplifier with a single-tone input showing flattopping and crossover distortion.

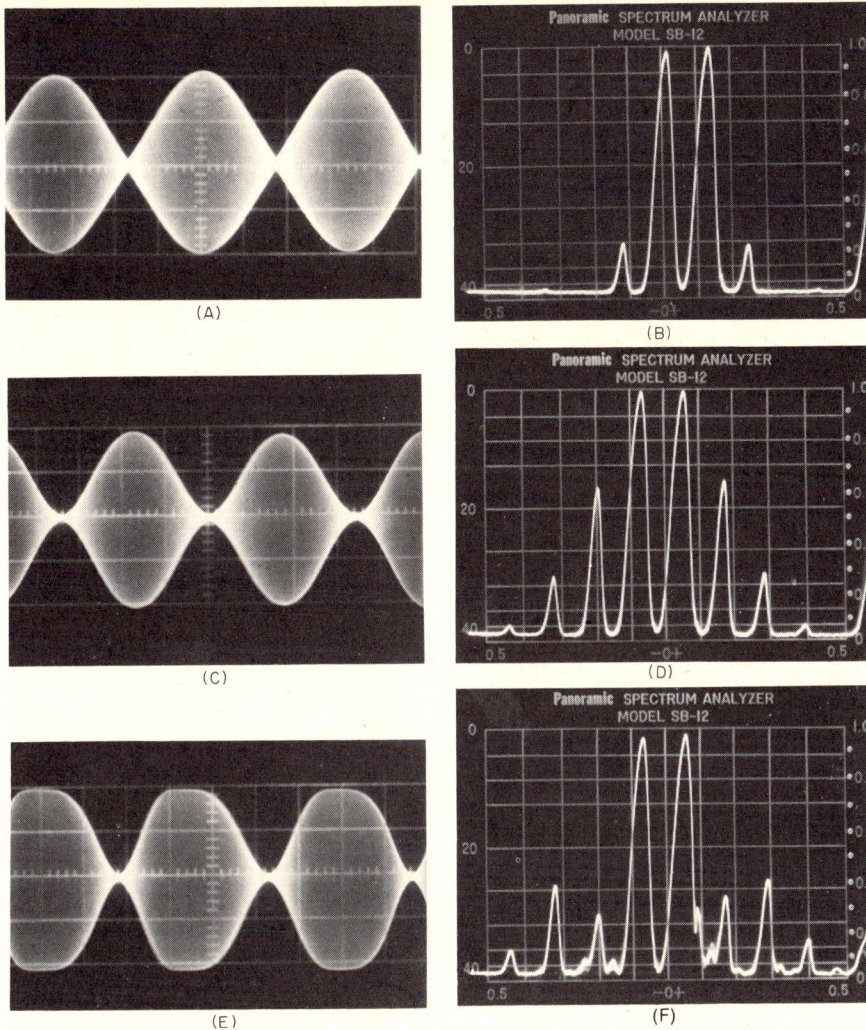

Fig. 31 — Scope patterns for a two-tone test signal and corresponding spectrum-analyzer displays. The pattern in A is for a properly adjusted transmitter and consequently the IMD products are relatively low as can be seen on the analyzer display. At C, the PA bias was set to zero idling current and considerable distortion can be observed. Note how the pattern has changed on the scope and the increase in IMD level. At E, the drive level was increased until the flattopping region was approached. This is the most serious distortion of all since the width of the IMD spectrum increases considerably causing splatter (F).

suspect and the transmitter operation should be checked.

The relation between the level at which distortion begins for the two-tone test signal and an actual voice signal is a rather simple one. The maximum deflection on the scope is noted (for an acceptable two-tone test waveform) and the transmitter is then operated such that voice peaks are kept below this level. If the voice peaks go above this level, a type of distortion called "flattopping" will occur and results are shown for a two-tone test signal in Fig. 31E. IMD-product levels raise very rapidly when flattopping occurs. For instance, third-order product levels will increase 30 db for every 10-dB increase in desired output as the flattopping region is approached, and fifth-order terms will increase by 50 dB (per 10 dB).

Interpreting Distortion Measurements

Unfortunately, considerable confusion has grown concerning the interpretation and importance of distortion in ssb gear. Distortion is a very serious problem when high spurious-product levels exist at frequencies removed from the passband of the desired channel but is less serious if such products fall within the bandwidth of operation. In this former case, such distortion may cause needless interference to other channels ("splatter") and should be avoided. This can be seen quite dramatically in Fig. 31F when the flattopping region is approached and the fifth and higher order terms increase drastically.

On the other hand, attempting to suppress in-band products more than necessary is not only difficult to achieve but may not result in any noticeable increase in signal quality. In addition, measures required to suppress in-band IMD often cause problems at the expense of other qualities such as efficiency. This can lead to serious difficulties such as shortened

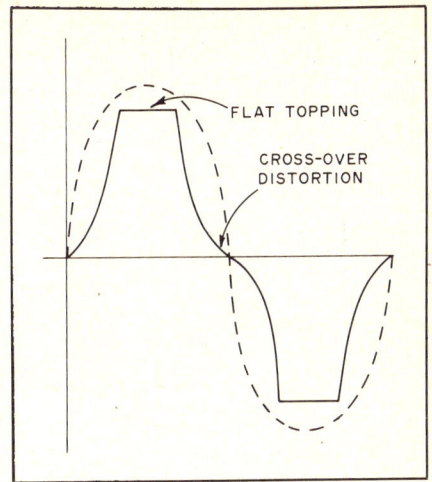

tube life or transistor heat-dissipation problems.

The two primary causes of distortion can be seen in Fig. 32. While the waveform is for a single-tone input signal, similar effects occur for the two-tone case. As the drive signal is increased, a point is reached where the output current (or voltage) cannot follow the input and the amplifier saturates. This condition is often referred to as flattopping (as mentioned previously). It can be prevented by ensuring that excessive drive doesn't occur and the usual means of accomplishing this is by alc action. The alc provides a signal that is used to lower the gain of earlier stages in the transmitter.

The second type of distortion is called "crossover" distortion and occurs at low signal levels. (See Fig. 32.) Increasing the idling plate or collector current is one way of reducing the effect of crossover distortion in regards to producing undesirable components near the operating frequency. Instead, the components occur at frequencies considerably removed from the operating frequency and can be eliminated by filtering.

As implied in the foregoing, the effect of distortion, frequencywise, is to generate components which add or subtract in order to make up the complex waveform. A more familiar example would be the harmonic generation caused by the nonlinearities often encountered in amplifiers. However, a common misconception which should be avoided is that IMD is caused by fundamental-signal components beating with harmonics. Generally speaking, no such simple relation exists. For instance, single-ended stages have relatively poor second-harmonic suppression but with proper biasing to increase the idling current, such stages can have very good IMD-suppression qualities.

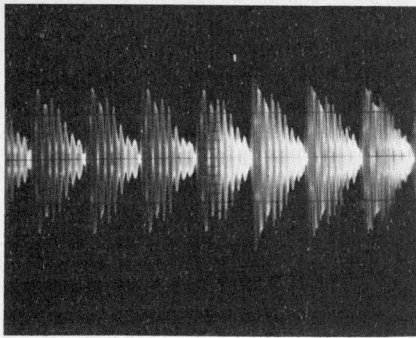

Fig. 33 — Speech pattern of the word "X" in a properly adjusted ssb transmitter.

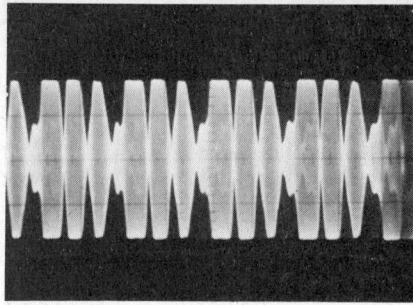

Fig. 34 — Severe clipping (same transmitter as Fig. 33 but with high drive and alc disabled).

Fig. 35 — Exterior view of the outboard speech processor. The lack of external adjustments makes the unit easy to use and difficult to abuse.

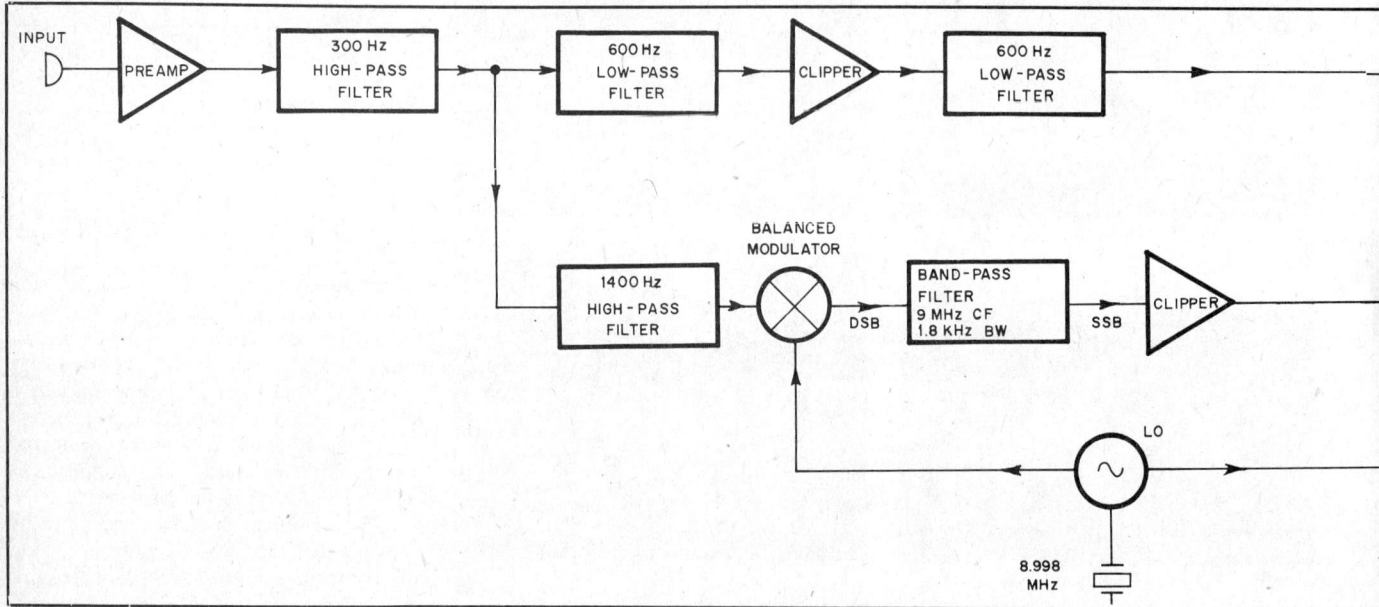

Fig. 36 — Block diagram of the processor. The separation of the formant channels illustrates the "hybrid" concept.

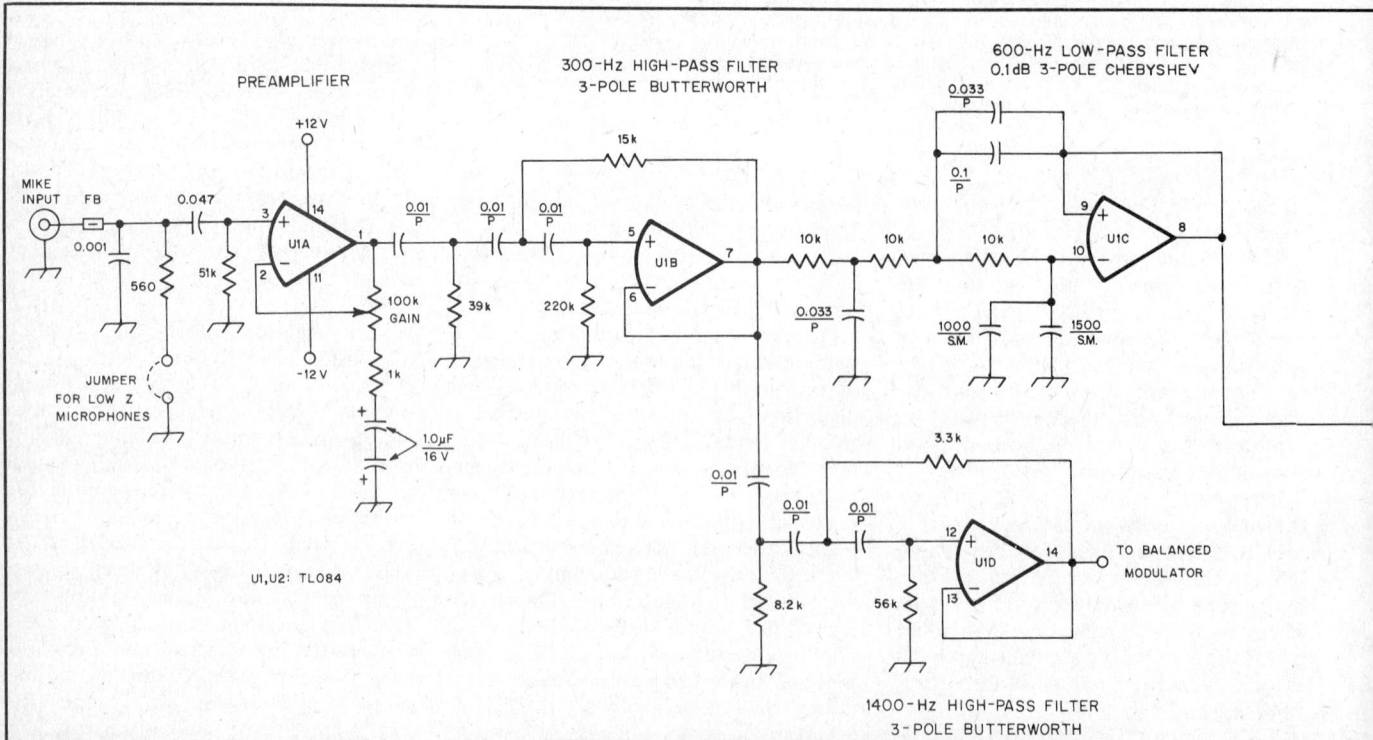

Fig. 37 — Schematic diagram of the audio board. All potentiometers are linear taper pc-mount types. See text for complete component information.

However, a definite mathematical relation does exist between the desired components in an ssb signal and the "distortion signals." Whenever nonlinearities exist, products between the individual components which make up the desired signal will occur. The mathematical result of such multiplication is to generate other signals of the form (2f1 − f2), (3f1), (5f2 − f1) and so on. Hence the term intermodulation-distortion products. The "order" of such products is equal to the sum of the multipliers in front of each frequency component. For instance, a term such as (3f1 − 2f2) would be called a

fifth-order term since 3 + 2 is equal to 5. In general, the third, fifth, seventh, and similar "odd-order" terms are the most important ones since some of these fall near the desired transmitter output frequency and can't be eliminated by filtering. As pointed out previously, such terms do not normally result from fundamental components beating with harmonics. An exception would be when the fundamental signal along with its harmonics is applied to another nonlinear stage such as a mixer. Components at identical frequencies as the IMD products will result.

When two equal tones are applied to an

amplifier and the result is displayed on a spectrum analyzer, the IMD products appear as "pips" off to the side of the main signal components (Fig. 31). The amplitudes associated with each tone and the IMD products are merely the dB difference between the particular product and one tone. However, each desired tone is 3 dB down from the *average* power output and 6 dB down from the *PEP* output.

Since the PEP represents the most important quantity as far as IMD is concerned, relating IMD-product levels to PEP is one logical way of specifying the "quality" of a transmitter or amplifier in regard to low distortion. For instance, IMD levels are referenced to PEP in "Product Review" specifications of commercially made gear in *QST*. PEP output can be found by multiplying the PEP input by the efficiency of the amplifier. The input PEP for a two-tone test signal is given by

$$PEP = E_p I_p \left(1.57 - 0.57 \frac{I_o}{I_p} \right)$$

where

E_p = the plate voltage
I_p = the average plate current
I_o = the idling current

Generally speaking, most actual voice patterns will look alike (in the presence of distortion) except in the case where severe flattopping occurs. This condition is not too common since most rigs have an alc system which prevents overdriving the amplifiers. However, the voice pattern in a properly adjusted transmitter usually has a "Christmas tree" shape when

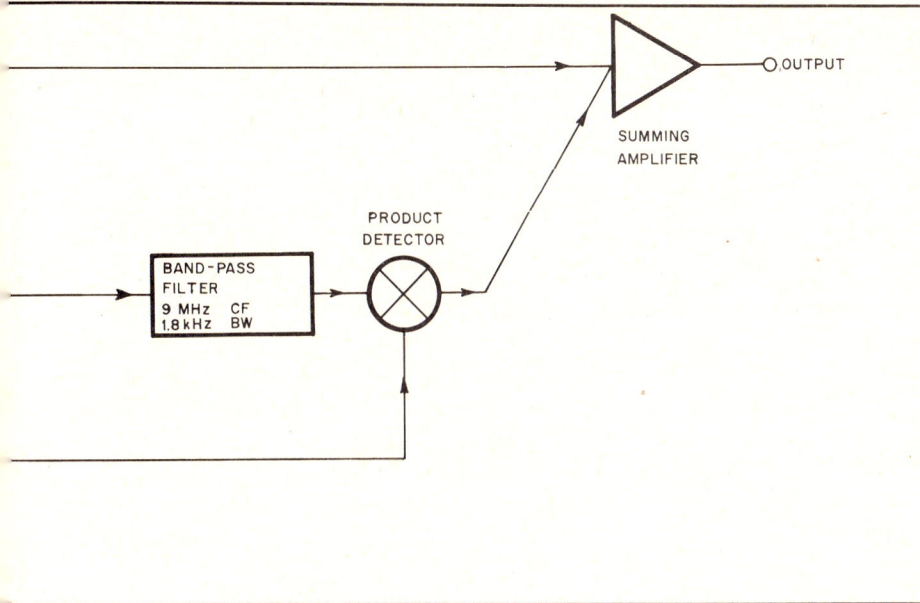

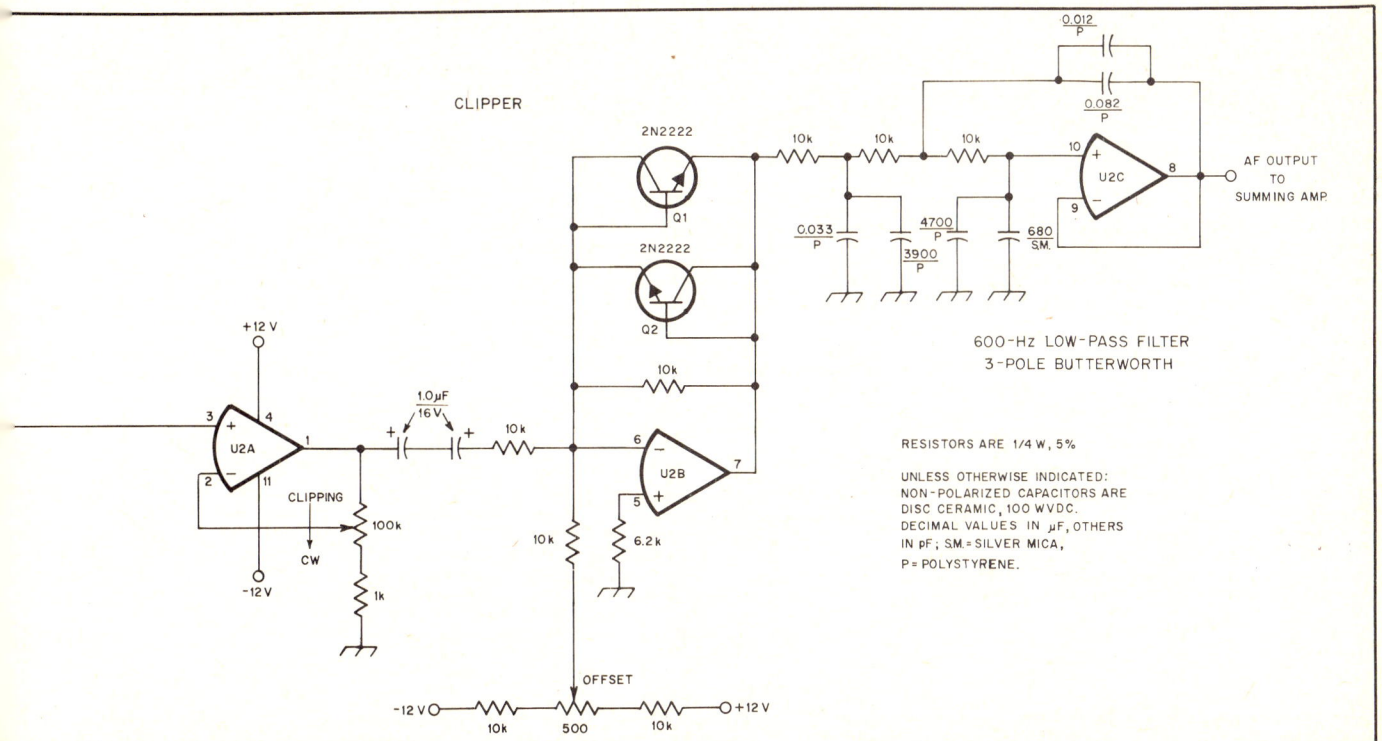

observed on a scope; an example is shown in Fig. 33.

AN OUTBOARD SPEECH PROCESSOR FOR SSB

The unit pictured in Fig. 35 features a combination of processing techniques borrowed from various successful designs of the past. Fig. 36 is a functional block diagram of the system. The most significant speech formants (300 to 600 Hz and 1400 to 2500 Hz) are separated by means of active filters and processed independently. Independent formant pro-cessing permits a higher degree of processing for a given amount of distortion. Additionally, the design is highly flexible, allowing precise adjustment to an individual's voice characteristics for maximum effectiveness.

Since the low-frequency formant occupies only one octave of spectrum width, it can be clipped at audio (baseband) without excessive distortion. Clipping a sinusoidal waveform generates only odd harmonics. The third harmonic of the lowest voice frequency falls outside the filter passband, so the amount of audio clipping employed is limited by intermodulation distortion rather than harmonic distortion.

Second-formant energy containing most of the speech articulation is translated to 9 MHz for rf clipping. The narrow-bandwidth crystal filters used at this frequency have much steeper skirts than would be obtainable by ordinary means in the baseband. After rf processing, the second-formant signal is heterodyned back to the baseband for recombination with the first-formant information.

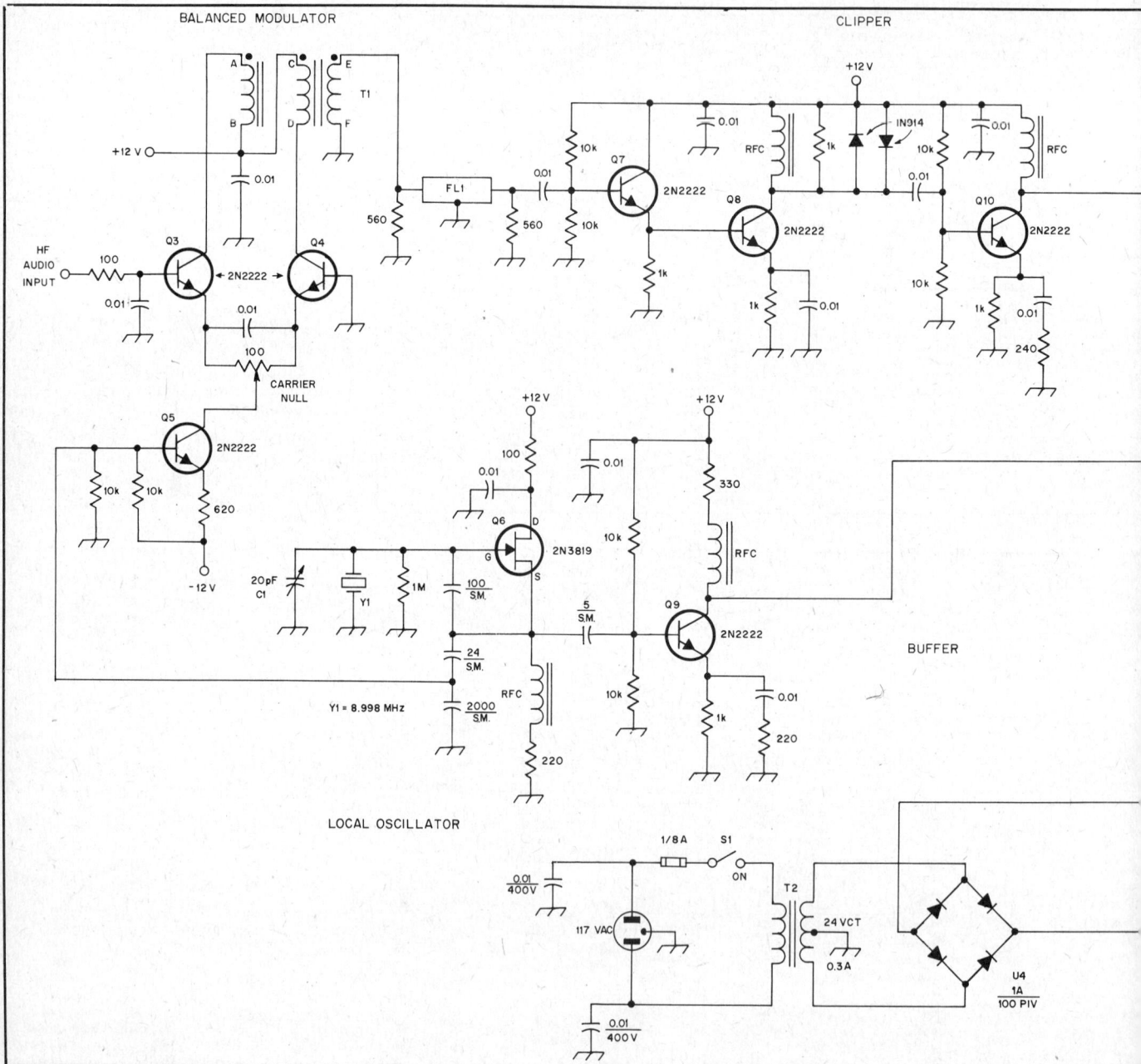

Fig. 38 — Schematic diagram of the rf board and power supply.

C1 — Pc-mount trimmer capacitor, E. F. Johnson 189-506-105 or equiv.

FL1, FL2 — Crystal filter, 9.0 MHz center frequency, 1.9 kHz bandwidth, 500 ohm terminations. Fox-Tango TF-90H1.8.

RFC — 10 turns no. 26 enameled wire on FT37-72 toroidal core. (approx. 50 μH)

T1 — 10 trifilar turns no. 26 enameled wire on FT37-72 core.

Y1 — Quartz crystal, 8.998 MHz, parallel mode, 32 pF. Spectrum International XF-901 or equiv.

Figs. 37 and 38 are the schematic diagrams for the af and rf sections. Substituting wider 2.4- or 2.7 kHz-bandwidth crystal filters for the units specified will allow a somewhat simpler circuit and construction in that only the rf processing section need be built. The economic advantage of such a scheme is small, because the crystal filters are the most expensive components in the processor. Filters suitable for this application are in the $55 price class, regardless of bandwidth. In any event, the balanced modulator must be supplied audio from a zero-impedance source that swings symmetrically about ground.

Construction

The processor is housed in a 6-3/4-inch (171-mm)-square × 2-1/2 -inch (63-mm)-high box fashioned from pc-board material. The copper sides face inward, and the seams are soldered full length for shielding against rf from the transmitter. Some construction details are revealed in Fig. 39. Flanges soldered to the top cover provide anchors for sheet-metal screws through the sides, ensuring mechanical and electrical integrity. Miniature coaxial cable is used for audio connections to the front and rear panels and between boards. A shield wall is visible in Fig. 39. This is to reduce LO leakage around the balanced modulator. The rf board is copper clad on both sides, and the top side is used for a ground plane. Be sure to solder all ground connections top and bottom where applicable.

There are no unusual components used in the processor. The two FETs are stocked by Radio Shack, as are the operational amplifier ICs and bipolar

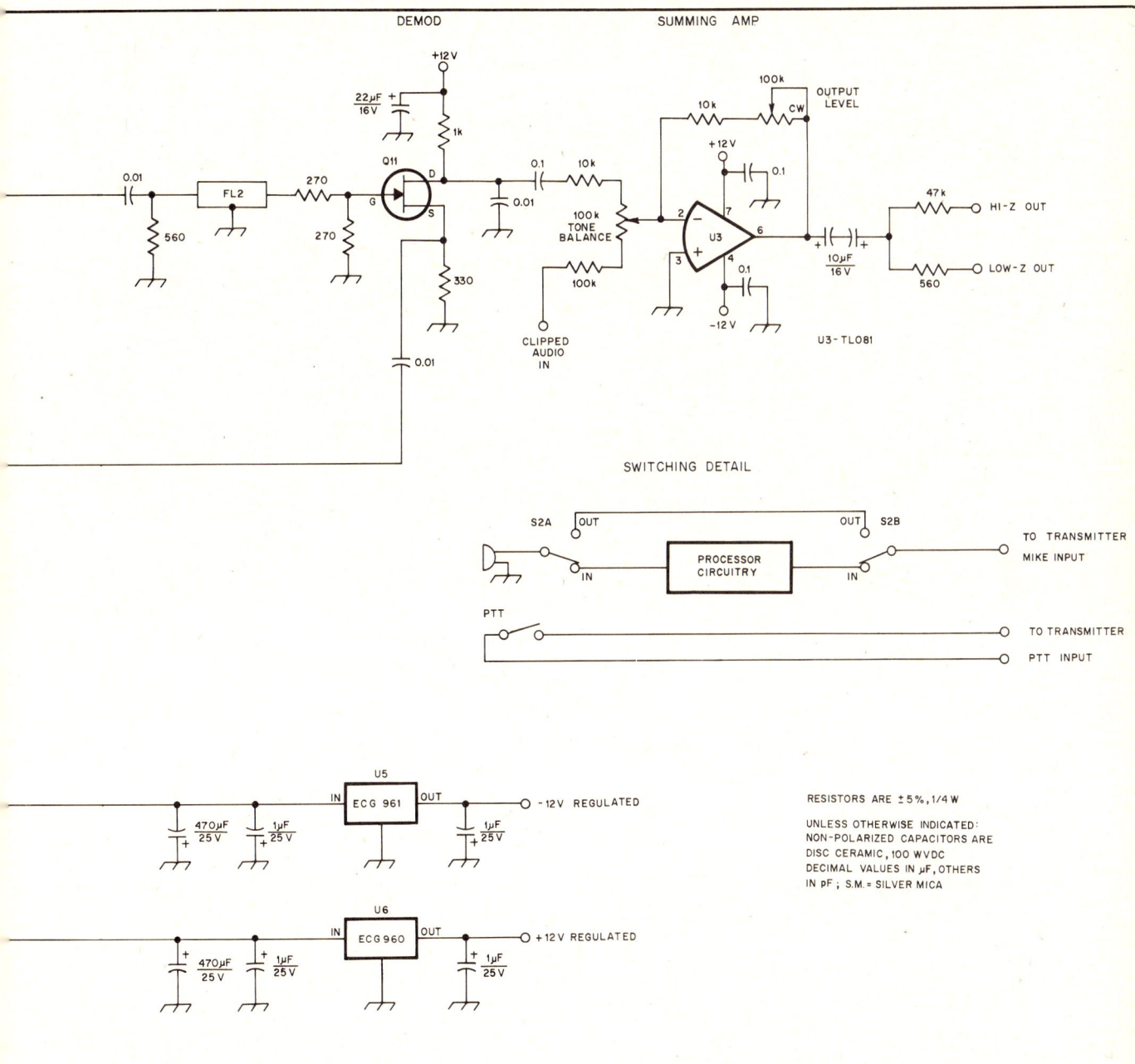

Fig. 39 — Interior view of the speech processor, showing the rf board. The audio board nests beneath the rf board. The power supply components and microphone connectors are mounted on the rear panel.

transistors. Perhaps the most difficult items to procure are the polystyrene capacitors. Mylar units can be substituted if necessary, and the values specified can be synthesized from series and parallel combinations. Try to hold the capacitance values within 10 percent of nominal for predictable filter response.

Adjustment and Operation

A calibrated audio sine-wave generator is required for initial setup. An indicating device is also needed — an oscilloscope is the most useful instrument for this purpose, but an audio voltmeter can be used if the operator can interpret it properly.

The first step is to make sure the power supplies are functioning. Most of the adjustments must be made with the boards separated. Set the LO frequency to 8.99795 MHz. The demodulator source resistor is the best place to connect a counter. With the audio input short-circuited, adjust the CARRIER NULL potentiometer for minimum signal at the input of FL1. Now apply a 2000-Hz signal having amplitude equal to the station microphone output with a normal whistle. Adjust the preamplifier GAIN control to produce the desired amount of clipping, measured at the demodulator drain. The amount of clipping is the input signal reduction necessary to produce a slight reduction in output. A reasonable value is 10 dB.

Move the signal generator to 500 Hz and set the CLIPPING control for slight clipping, measured at the output of the clipper stage. Adjust the OFFSET potentiometer for symmetrical clipping. After symmetry has been achieved, the audio clipping can be set to the desired level. There is room for considerable experimentation with this level, but a good

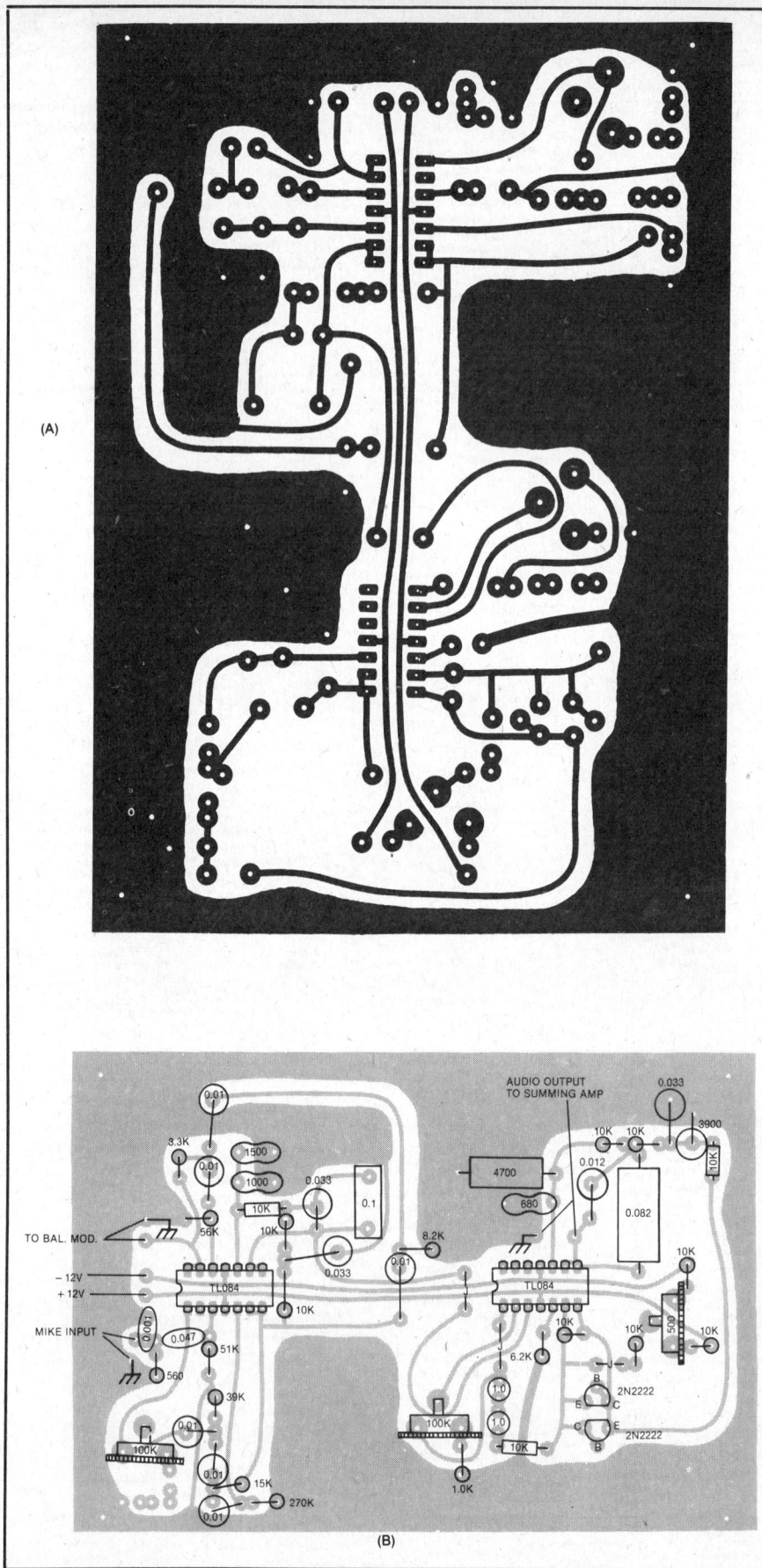

Fig. 40 — (A) Full-scale etching pattern for the audio processing board, shown from the foil side. Black represents unetched copper. This pc board is clad on one side only. (B) Parts-placement guide for the audio board. The component side is shown with an x-ray view of the foil. J = wire jumper.

starting point is equal to the rf clipping. Once the rf and af clipping have been set, they can be varied simultaneously by changing the preamplifier GAIN adjustment. The OUTPUT LEVEL potentiometer is normally adjusted to produce an output equal to the microphone input at the clipping threshold.

Because considerable gain is needed to effect clipping, this processor (like all others) will amplify background noise from clocks and blowers. To realize the communications advantage possible with this processor, it must be used in a quiet operating room. This unit is purposely devoid of external knobs to encourage proper adjustment with instruments and to combat the all-too-prevalent amateur philosophy "if a little is good, more is better, and too much is just right."

A Sideband-Generator Module

A practical circuit for a simple ssb generator is given in Fig. 42. Output is at 9 MHz. This circuit can be followed by appropriate stages of the designer's choice, thereby making it possible to heterodyne the 9-MHz energy to a desired amateur band. Circuit design information for the additional stage necessary to build a complete exciter is found elsewhere in this book and in *Solid State Design for the Radio Amateur*.

A 741 op amp functions as the audio amplifier in Fig. 42. Output from U1 is supplied to the gates of Q1. Q1 and Q2 are used as a balanced modulator. MOSFETs are used to prevent changes in gate-source capacitance when the audio level is increased by means of R1. JFETs will not work properly at Q1 and Q2 because the junction capacitance changes with increased audio drive, thereby unbalancing the modulator.

T1 is a broadband trifilar-wound toroidal transformer. It provides the necessary 180° phase difference for the drains of Q1 and Q2 while coupling the balanced modulator output to the i-f preamplifier, Q3.

R2 compensates for differences in operating level between Q1 and Q2. This balancing control makes it unnecessary to use matched transistors in the balanced modulator. R2 is adjusted for maximum carrier suppression, which will be on the order of 45 to 50 dB, according to lab tests with a spectrum analyzer. The S meter of a general-coverage receiver will suffice for adjustment of the carrier null.

Q4 is a variable dc attenuator. For cw operation, S2 is placed in the cw position and R3 is advanced until the desired carrier level is obtained. As the collector current of Q4 increases with elevated forward bias, the source voltage of Q2 is shifted to permit carrier insertion during cw use.

Q3 functions as an i-f preamplifier and helps compensate for the 5-dB insertion loss of FL1. It operates in Class A. The filter must be terminated in a load

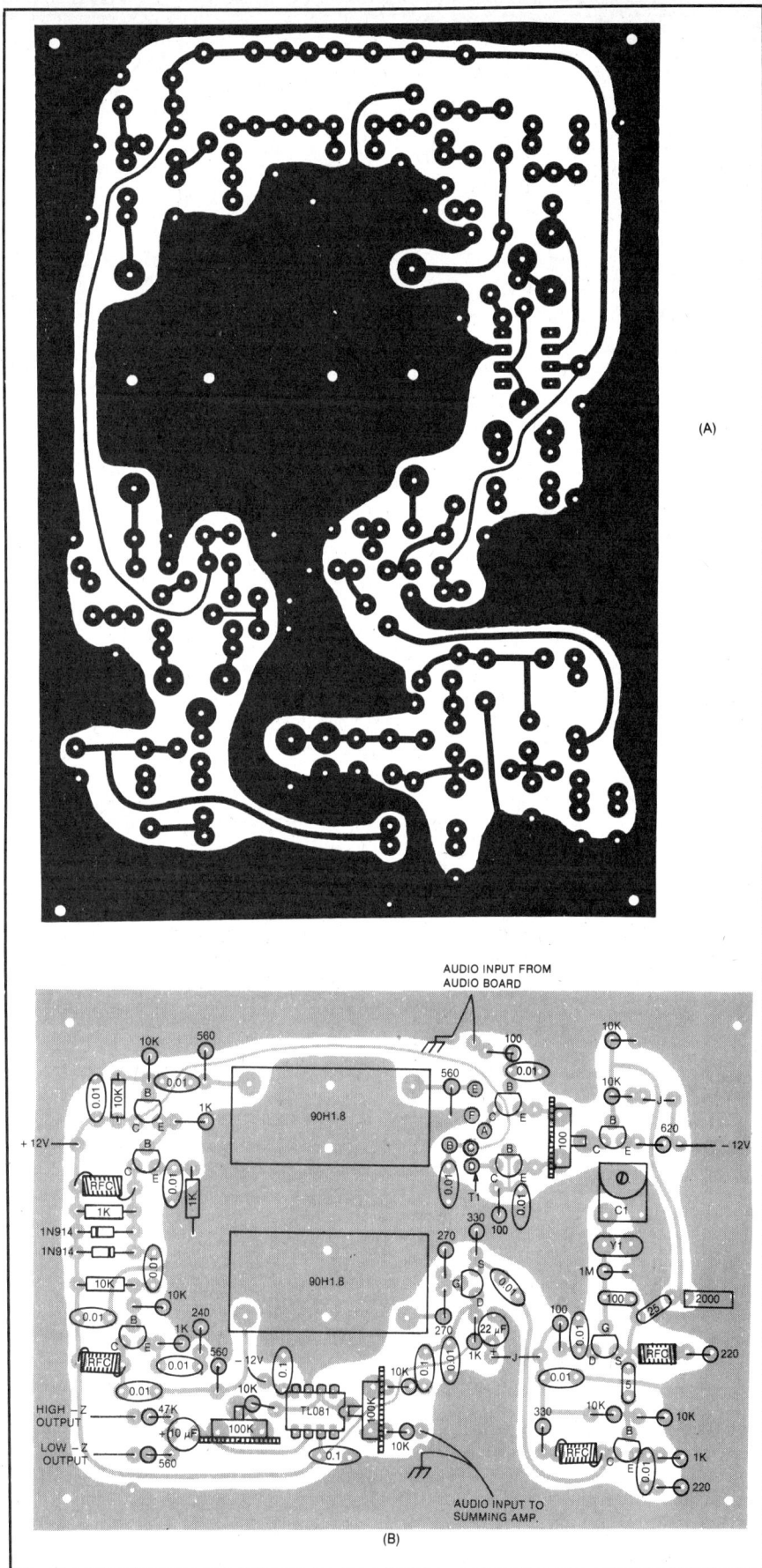

(A)

(B)

Fig. 41 — (A) Full-scale etching pattern for the rf board, shown from the foil side. Black represents unetched copper. This pc board is clad on both sides. (B) Parts-placement guide for the rf board. The component side is shown with an x-ray view of the foil. J = wire jumper. The circled letters correspond to the terminals of T1. (See Fig. 38)

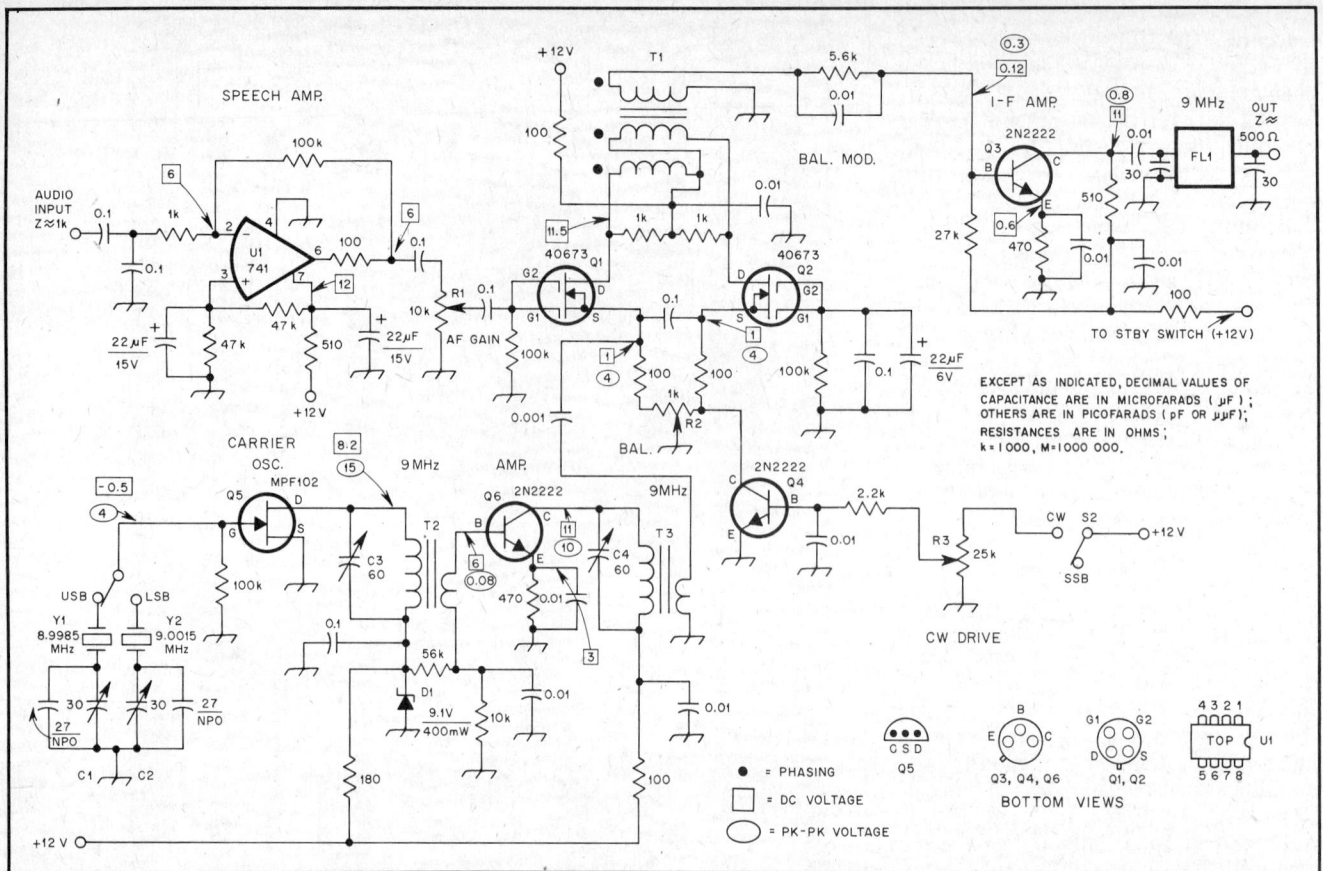

Fig. 42 — Schematic diagram of a practical 9-MHz ssb generator. Fixed-value capacitors are disc ceramic unless otherwise noted. Polarized capacitors are aluminum or tantalum. Fixed-value resistors are 1/2 watt composition.

C1, C2 — Miniature 30-pF trimmer. NP0 ceramic preferred.
C3, C4 — Miniature 60-pF trimmer. Mica compression type suitable.
D1 — 9.1-V, 400-mA zener diode.
FL1 — Spectrum International 9-MHz crystal-lattice filter. Type XF-9A. (see *QST* ads).
R1 — 10-kΩ audio-taper control, panel mounted.

R2 — 1000-Ω pc-board-mount control.
R3 — 25-kΩ linear-taper control, panel mounted.
S1, S2 — Single-pole, double-throw miniature switch, panel mounted.
T1 — 15 trifilar turns of no. 26 enam. wire (twist 10 times per inch) on an FT-50-61 toroid core (μ_e = 125, dia = 0.5 inch/13 mm).
T2 — 10-μH primary. 44 turns no. 26 enam.

wire on a T50-2 iron core. μ_e = 10, dia = 0.5 inch. Link has 10 turns of no. 30 insulated wire over D1 end of primary.
T3 — 10-μH. 44 turns of no. 26 enam. wire on a T50-2 iron core. Link has 22 turns of no. 30 insulated wire over cold end of primary.
Y1, Y2 — Crystals to match FL1. Obtain from filter manufacturer or International Crystal Mfg. Co.

resistance of 500 ohms for proper pass-band response.

The carrier generator consists of Q5 and Q6. Two crystals are used at Q5 to permit operation on upper or lower sideband. C1 and C2 are adjusted so that the carrier is placed at the correct point on the filter (FL1) curve. This is approximately 20 dB down from the peak response. The trimmers can be set while listening to the ssb signal on a communications receiver. They are adjusted for best "naturalness" of the operator's voice, consistent with adequate rejection of the unwanted sideband.

Q6 amplifies the 9-MHz output of Q5 to provide 4 volts pk-pk (1.4 volts rms) of injection on the sources of Q1 and Q2.

Circuit boards and negatives for this circuit can be obtained from Circuit Board Specialists, Box 969, Pueblo, CO 81002. Figs. 43 and 44 are etching pattern and parts-placement guide, respectively.

A Modern Solid-State VOX

Voice-operated T-R control is a great convenience in ssb operation. The unit described here is compact and uses inexpensive components. It is ideal for inclusion in a home-built transceiver or exciter, or retrofitting to commercial gear that does not have VOX. The performance of this unit is improved over previously published versions because of a modification suggested by W7KGZ.

Circuit Description

The schematic diagram of the VOX device is shown in Fig. 45. Three of the LM3900 sections have been configured as high-gain audio amplifiers. U1A and U1B amplify the signal from the microphone. U1C functions as an amplifier for audio sampled at the station speaker. Coupling capacitors in the audio stages have been chosen to reduce response below 300 Hz. This will minimize hum problems.

Outputs from the microphone and speaker amplifiers are capacitively coupled to rectifier stages which convert the audio signals to varying dc voltages. Germanium diodes, because of their lower

threshold voltage, have been used as audio rectifiers instead of silicon units. The outputs of the two rectifier stages are summed resistively by means of R6 and R7, and applied to the inverting input of a voltage comparator, U1D. The output of U1D remains high (approximately 0.5 volt less than the supply voltage) so long as the voltage at the noninverting input is less than the 0.2-volt reference applied to the inverting input. Whenever the input exceeds the reference, the output of the comparator goes low — to near the ground or common potential. Voltage output from the microphone-signal rectifier is positive and, thus, will cause the comparator to switch as soon as the reference is exceeded. Because the speaker-signal rectifier produces negative voltage, it will not trigger the comparator. If the outputs of the two rectifiers are equal, as will happen when the microphone is picking up audio from the speaker, the resulting voltage from the summing network will be zero and the comparator will not trigger. The ability to reject speaker audio is usually called the *antivox* function.

The positive-to-ground transition of the comparator output starts the timing cycle of the 555. The length of the time cycle is determined by the values used for R9 and C1. The time delay produced is identical each time the microphone signal stops. Q1 allows the 555 to be retriggered continuously. One of the major difficulties of earlier VOX circuits was that the capacitor discharge circuits were used where the capacitor would not always be fully charged, so the time delay produced would vary.

The 555 has a current-switching capability of 200 mA, sufficient to directly drive either a relay or a solid-state switching arrangement. D5 is included to protect the IC from transients generated when switching an inductive load such as a relay coil.

Components and Construction

The VOX unit is constructed on a 2-3/8 × 2-3/4-inch (60 × 70-mm) etched circuit board. The photo indicates that one-third of the board real estate is unused, so a smaller version is possible. The type of controls and relay employed will be determined by the builder's individual requirements. This unit uses pc-mount controls which are aligned on the board so that they may be accessed through small holes in the rear panel of the transceiver. If panel-mount controls are desired, Mallory MLC units may be used for R4, R5 and R8.

The VOX device is small enough so it can be mounted inside most rigs. If a separate VOX unit is needed, a small utility box or Minibox will make an appropriate housing. Rf interference can cause trouble, so the unit should be shielded in any application where rf fields may be present. The bypass capacitors for the audio inputs are located on the circuit board. If the leads from the audio connectors are more than a few inches long, the bypass capacitors and their associated ferrite-bead chokes should be mounted at the connectors.

No provision has been made for mounting the relay on the circuit board, as the type of relay will depend on how the VOX device will be used. Any 12-volt relay which requires less than 200 mA of current can be employed. When the VOX relay must drive a second relay, such as the antenna relay in a transceiver, the fast operating time of a reed relay is needed to prevent clipping of the first syllable spoken. The total close time of all relays connected in tandem should be 10 milliseconds or less. If the VOX relay will perform all switching functions directly, a miniature control relay such as the Potter & Brumfield R10 series is appropriate. These relays are available in 2-, 4- and 6-pole versions, part numbers R10-E1-Y2-185, R10-E1-Y4-V185 and R10-E1-Y6-V90, respectively.

Fig. 43 — Full-scale etching pattern for the ssb generator pc board, shown from the foil side. Black represents unetched copper.

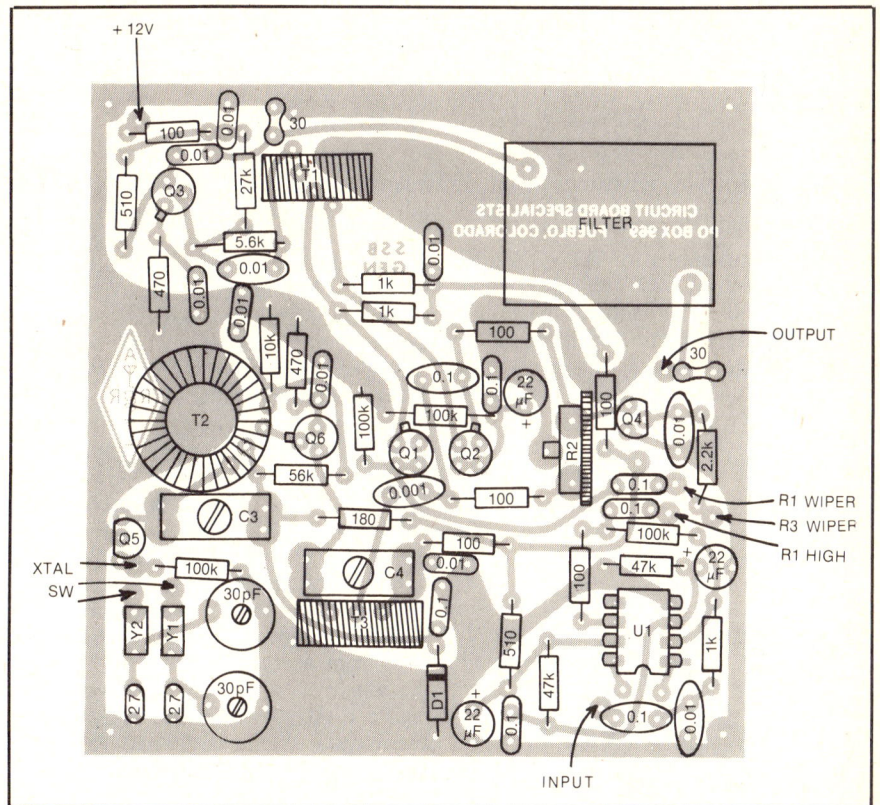

Fig. 44 — Parts-placement guide for the ssb generator from the component side with an x-ray view of the foil.

The circuit board is designed for 1/4-watt resistors which are mounted flat. If 1/2-watt units are used, they must be positioned vertically. Care must be employed when mounting and soldering the germanium diodes. If the leads are bent too close to the body of the diode, breakage can result. If excessive heat is applied to the diode, it can be damaged, so use a heat sink (such as a small alligator clip) when soldering. Assure that proper polarity is observed when installing the

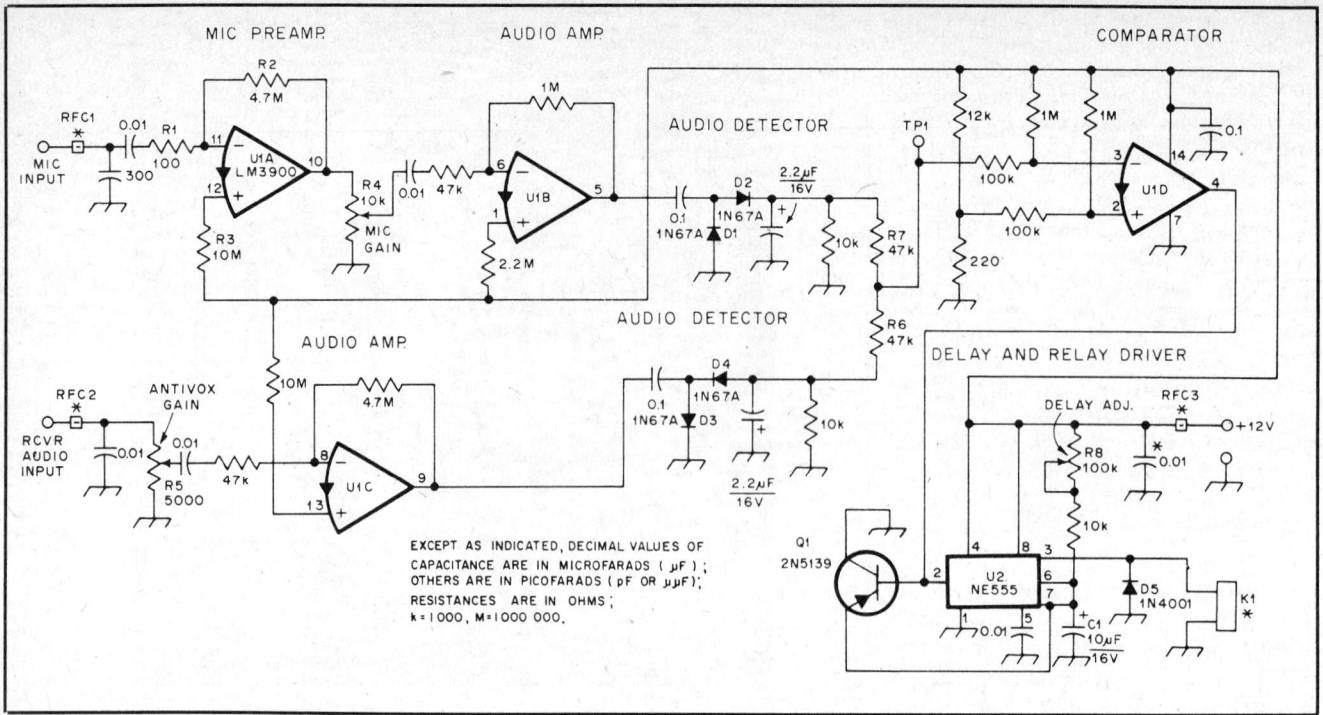

Fig. 45 — Schematic diagram of the VOX unit. Unless otherwise specified, resistors are 1/4-watt composition. Capacitors with polarity marked are plastic-encapsulated tantalum; others are disc ceramic.

C1 — For text reference.
D1-D4, incl. — Germanium diode, 1N34A, 1N67 or equivalent.
D5 — Silicon diode, 50 PRV or more, 1N4001 or

similar.
K1 — Miniature type, 12-volt coil (see text).
Q1 — 2N5139 silicon pnp.
R1-R3, incl., R6, R7 — For text reference.

R4, R5, R8 — Miniature control (see text).
RFC1-RFC3, incl — Ferrite bead.
U1 — National Semiconductor LM3900.
U2 — Type 555 timer.

diodes and tantalum capacitors.

Installation and Operation

Typical connections for the VOX unit are shown in Fig. 47. Shielded cable should be used for all audio connections. Audio for the antivox function can be sampled at the station speaker or at the phone-patch output (which is a feature of many commercial transceivers). If VOX operation of a cw rig is desired, connect the output of a sidetone monitor to the

microphone input of the VOX unit. The mic gain control should be set so that the VOX relay closes each time a word is spoken. The delay control should be adjusted to fit individual speech patterns and operating habits. The delay time must be long enough that the VOX relay will drop out only during a pause in speech. There are two methods of setting the antivox gain control. The first way is simply to advance the control until audio from the speaker does not trip the VOX unit. A more scientific approach is to connect a voltmeter to TP1. With no audio input, the meter should read only the comparator reference voltage, approximately 0.1 volt. Tune the receiver to provide a steady tone signal, such as the heterodyne note from a crystal calibrator. Advance the antivox control until the voltmeter registers only the reference voltage. The antivox gain should be set with the audio from the speaker slightly louder than is necessary during normal operation.

SSB Selected Bibliography

Single Sideband for the Radio Amateur, American Radio Relay League, Fifth Edition, 1970.
Hennebury, *Single Sideband Handbook,* Technical Material Corporation, 1964.
Pappenfus, Bruene and Schoenike, *Single Sideband Principles and Circuits,* McGraw-Hill, 1964.
Amateur Single Sideband, by Collins Radio Company, 1962.
Newland, "A Safe Method for Etching Crystals" *QST,* January 1958.
Kosowsky, "High Frequency Crystal Filter Design Techniques and Applications," *Proceedings of the IRE,* February 1958.
Weaver and Brown, "Crystal Lattice Filters," *QST,* June 1951.
Good, "A Crystal Filter for Phone Reception," *QST,* October 1951.
Burns, "Sideband Filters Using Crystals," *QST,* November 1954.
Morrison, "Cascaded Half-Lattice Crystal Filters for Phone and C.W. Reception," *QST,* May 1954.
Vester, "Surplus-Crystal High Frequency Filters," *QST,* January 1959.
Vester, "Mobile S.S.B. Transceiver," *QST,* June 1959.

Fig. 46 — the VOX unit shown here was designed and built by N1RM. It originally appeared in March 1976 QST.

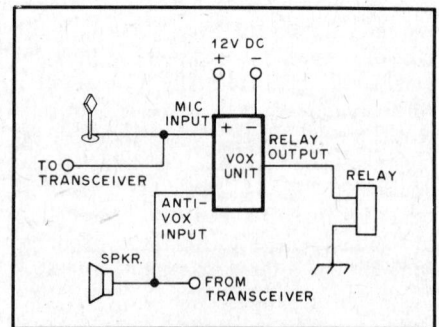

Fig. 47 — Typical connections to the VOX unit.

Chapter 13

Frequency Modulation and Repeaters

Methods of radiotelephone communication by frequency modulation were developed in the 1930s by Major Edwin Armstrong in an attempt to reduce the problems of static and noise associated with receiving a-m broadcast transmissions. The primary advantage of fm, the ability to produce a high signal-to-noise ratio when receiving a signal of only moderate strength, has made fm the mode chosen for mobile communications services and quality broadcasting. The disadvantages, the wide bandwidth required and the poor results obtained when an fm signal is propagated via the ionosphere (because of phase distortion), has limited the use of frequency modulation to the 10-meter band and the vhf/uhf section of the spectrum.

Fm has some impressive advantages for vhf operation, especially when compared to a-m. With fm the modulation process takes place in a low-level stage and remains the same, regardless of transmitter power. The signal may be frequency multiplied after modulation, and the PA stage can be operated Class C for best efficiency, as the "final" need not be linear.

In recent years there has been increasing use of fm by amateurs operating around 29.6 MHz in the 10-meter band. The vhf spectrum now in popular use includes 52 to 54 MHz, 146 to 148 MHz, 222 to 225 MHz, and 440 to 450 MHz. The subject of fm and repeaters is covered in great depth in the ARRL publication, *FM and Repeaters for the Radio Amateur*.

Frequency and Phase Modulation

It is possible to convey intelligence by modulating any property of a carrier, including its frequency and phase. When the frequency of the carrier is varied in accordance with the variations in a modulating signal, the result is *frequency modulation* (fm). Similarly, varying the phase of the carrier current is called *phase modulation* (pm).

Frequency and phase modulation are not independent, since the frequency cannot be varied without also varying the phase, and vice versa.

The effectiveness of fm and pm for communication purposes depends almost entirely on the receiving methods. If the receiver will respond to frequency and phase changes but is insensitive to amplitude changes, it will discriminate against most forms of noise, particularly impulse noise such as that set up by ignition systems and other sparking devices. Special methods of detection are required to accomplish this result.

Modulation methods for fm and pm are simple and require practically no audio power. Also, since there is no amplitude variation in the signal, interference to broadcast reception resulting from rectification of the transmitted signal in the audio circuits of the bc receiver is substantially eliminated.

Frequency Modulation

Fig. 2 is a representation of frequency modulation. When a modulating signal is applied, the carrier frequency is increased during one half cycle of the modulating signal and decreased during the half cycle of opposite polarity. This is indicated in the drawing by the fact that the rf cycles occupy less time (higher frequency) when the modulating signal is positive, and more time (lower frequency) when the modulating signal is negative. The change in the carrier frequency (*frequency deviation*) is proportional to the instantaneous amplitude of the modulating signal. Thus, the deviation is small when the instantaneous amplitude of the modulating signal is small, and is greatest when the modulating signal reaches its peak, either positive or negative.

As shown in the drawing, the amplitude of the signal does not change during modulation.

Phase Modulation

If the phase of the current in a circuit is changed there is an instantaneous frequency change during the time the phase is being shifted. The amount of frequency

Fig. 1 — The most effective repeaters are situated well above average terrain and obstacles which could be in the signal path.

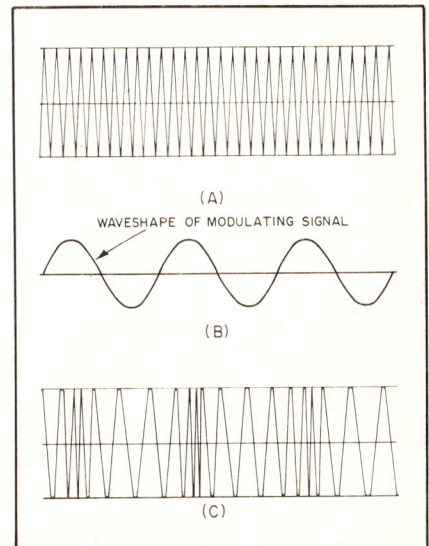

Fig. 2 — Graphical representation of frequency modulation. In the unmodulated carrier at A, each rf cycle occupies the same amount of time. When the modulating signal, B, is applied, the radio frequency is increased and decreased according to the amplitude and polarity of the modulating signal.

change, or deviation, depends on how rapidly the phase shift is accomplished. It is also dependent upon the total amount of the phase shift. In a properly operating pm system the amount of phase shift is proportional to the instantaneous amplitude of the modulating signal. The rapidity of the phase shift is directly proportional to the frequency of the modulating signal. Consequently, the frequency deviation in pm is proportional to both the amplitude and frequency of the modulating signal. The latter represents the outstanding difference between fm and pm, since in fm the frequency deviation is proportional only to the amplitude of the modulating signal.

FM and PM Sidebands

The sidebands set up by fm and pm differ from those resulting from a-m in that they occur at integral multiples of the modulating frequency on either side of the carrier rather than, as in a-m, consisting of a single set of side frequencies for each modulating frequency. An fm or pm signal therefore inherently occupies a wider channel than a-m.

The number of "extra" sidebands which occur in fm and pm depends on the relationship between the modulating frequency and the frequency deviation. The ratio between the frequency deviation, in hertz, and the modulating frequency, also in hertz, is called the *modulating index*. That is

$$\text{modulation index} = \frac{\text{carrier frequency deviation}}{\text{modulating frequency}}$$

Example: The maximum frequency deviation in an fm transmitter is 3000 Hz either side of the carrier frequency. The modulation index when the modulation frequency is 1000 Hz is

$$\text{Modulation index} = \frac{3000}{1000} = 3$$

At the same deviation with 3000 Hz modulation the index would be 1; at 100 Hz it would be 30, and so on.

In pm the modulation index is constant regardless of the modulating frequency; in fm it varies with the modulating frequency, as shown in the above example. In an fm system the ratio of the *maximum* carrier-frequency deviation to the *highest* modulating frequency used is called the *deviation ratio*.

Fig. 3 shows how the amplitudes of the carrier and the various sidebands vary with the modulation index. This is for single-tone modulation; the first sideband (actually a pair, one above and one below the carrier) is displaced from the carrier by an amount equal to the modulating frequency, the second is twice the modulating frequency away from the carrier, and so on. For example, if the modulating frequency is 2000 Hz and the carrier frequency is 29,500 kHz, the first sideband pair is at 29,498 kHz and 29,502 kHz, the second pair is at 29,496 kHz and 29,504 kHz, the third at 29,494 kHz and 29,506 kHz, and so on. The amplitudes of these sidebands depend on the modulation index, not on the frequency deviation.

Note that as shown in Fig. 3, the carrier strength varies with the modulation index. (In amplitude modulation the carrier strength is constant; only the sideband amplitude varies.) At a modulation index of approximately 2.4, the carrier disappears entirely. It then becomes "negative" at a higher index, meaning that its phase is reversed compared to the phase without modulation. In fm and pm the energy that goes into the sidebands is taken from the carrier, the *total* power remaining the same regardless of the modulation index.

Since there is no change in amplitude with modulation, an fm or pm signal can be amplified without distortion by an ordinary Class C amplifier. The modulation can take place in a very low-level stage and the signal can then be amplified by either frequency multipliers or straight-through amplifiers.

If the modulated signal is passed through one or more frequency multipliers, the modulation index is multiplied by the same factor that the carrier frequency is multiplied. For example, if modulation is applied on 3.5 MHz and the final output is on 28 MHz, the total frequency multiplication is eight times, so if the frequency deviation is 500 Hz at 3.5 MHz it will be 4000 Hz at 28 MHz. Frequency multiplication offers a means for obtaining practically any desired amount of frequency deviation, whether or not the modulator itself is capable of giving that much deviation without distortion.

Bandwidth

FCC amateur regulations (97.61) limit the bandwidth of F3 (frequency and phase modulation) to that of an a-m transmission having the same audio characteristics below 29.0 MHz and in the 50.1- to 52.5-MHz frequency segment. Greater bandwidths are allowed from 29.0 to 29.7 MHz and above 52.5 MHz.

If the modulation index (with single-tone modulation) does not exceed 0.6 or 0.7, the most important extra sideband, the second, will be at least 20 dB below the unmodulated carrier level. This should represent an effective channel width about equivalent to that of an a-m signal. In the case of speech, a somewhat higher modulation index can be used. This is because the energy distribution in a complex wave is such that the modulation index for any one frequency component is reduced as compared to the index with a sine wave having the same peak amplitude as the voice wave.

The chief advantage of fm or pm for

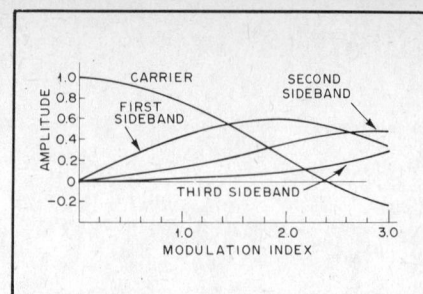

Fig. 3 — How the amplitude of the pair of sidebands varies with the modulation index in an fm or pm signal. If the curves were extended for greater values of modulation index the carrier amplitude would go through zero at several points. The same statement also applies to the sidebands.

frequencies below 30 MHz is that it eliminates or reduces certain types of interference to broadcast reception. Also, the modulating equipment is relatively simple and inexpensive. However, assuming the same unmodulated carrier power in all cases, narrow-band fm or pm is not as effective as a-m *with the methods of reception used by many amateurs*. To obtain the benefits of the fm mode, a good fm receiver is required. As shown in Fig. 3, at an index of 0.6 the amplitude of the first sideband is about 25 percent of the unmodulated-carrier amplitude; this compares with a sideband amplitude of 50 percent in the case of a 100 percent modulated a-m transmitter. When copied on an a-m receiver, a narrow-band fm or pm transmitter is about equivalent to a 100-percent modulated a-m transmitter operating at one-fourth the carrier power. On a suitable (fm) receiver, fm is as good or better than a-m, watt for watt.

The deviation standard now is ±5 kHz, popularly known as *narrow band*. For a while after WW II, 2.5- to 3-kHz deviation ("sliver band") was used on 10 meters and the vhf bands. During the '60s and early '70s 15 kHz was extensively used since many amateur rigs were commercial surplus. Narrow-band deviation developed as a middle ground between audio quality and spectrum conservation. The rule-of-thumb for determination of bandwidth requirements for an fm system is

$$2(\Delta F) + F_{A\,max}$$

where

ΔF = 1/2 total frequency deviation

$F_{A\,max}$ = maximum audio frequency (3 kHz for communication purposes)

Thus, for narrow-band fm, the bandwidth equals (2)5 + 3 or 13 kHz. Wide-band systems need a 33-kHz receiver bandwidth.

Comparison of FM and PM

Frequency modulation cannot be applied to an amplifier stage, but phase modulation can; pm is therefore readily adaptable to transmitters employing oscillators of high stability such as the

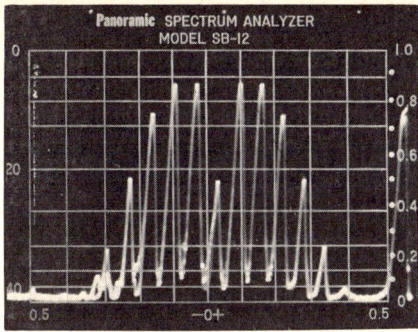

Fig. 4 — Output frequency spectrum of a narrow-band fm transmitter modulated by a 1-kHz tone.

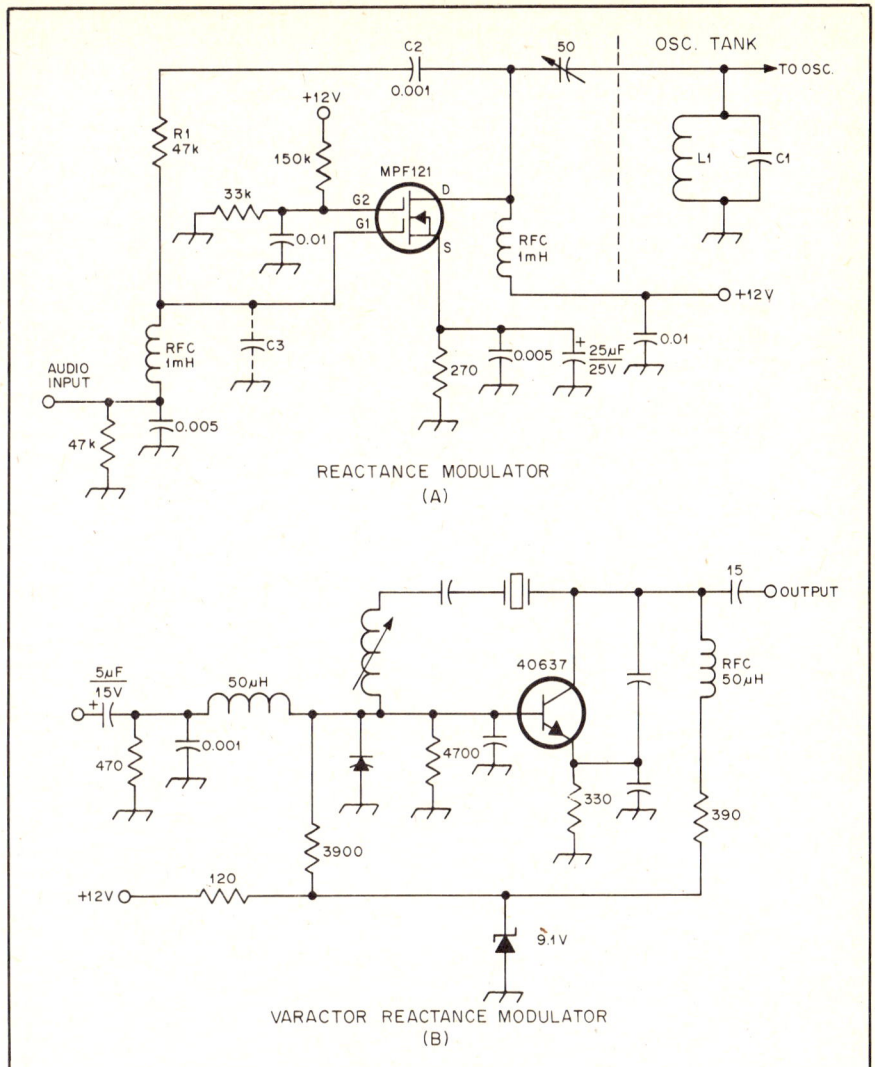

Fig. 5 — Reactance modulators using (A) a high-transconductance MOSFET and (B) a varactor diode.

crystal-controlled type. The amount of phase shift that can be obtained with good linearity is such that the maximum practicable modulation index is about 0.5. Because the phase shift is proportional to the modulating frequency, this index can be used only at the highest frequency present in the modulating signal, assuming that all frequencies will at one time or another have equal amplitudes. Taking 3000 Hz as a suitable upper limit for voice work, and setting the modulation index at 0.5 for 3000 Hz, the frequency response of the speech-amplifier system above 3000 Hz must be sharply attenuated, to prevent excess splatter. (See Fig. 4.) Also, if the "tinny" quality of pm as received on an fm receiver is to be avoided, the pm must be changed to fm, in which the modulation index decreases in inverse proportion to the modulating frequency. This requires shaping the speech-amplifier frequency-response curve in such a way that the output voltage is inversely proportional to frequency over most of the voice range. When this is done the maximum modulation index can only be used to some relatively low audio frequency, perhaps 300 to 400 Hz in voice transmission, and must decrease in proportion to the increase in frequency. The result is that the maximum linear frequency deviation is only one or two hundred Hz, when pm is changed to fm. To increase the deviation for narrow band requires a frequency multiplication of eight or more.

It is relatively easy to secure a fairly large frequency deviation when a self-controlled oscillator is frequency-modulated directly. (True frequency modulation of a crystal-controlled oscillator results in only very small deviations and so requires a great deal of frequency multiplication.) The chief problem is to maintain a satisfactory degree of carrier stability, since the greater the inherent stability of the oscillator the more difficult it is to secure a wide frequency swing with linearity.

Frequency Modulation Methods: Direct FM

A simple, satisfactory device for producing fm in the amateur transmitter is

the reactance modulator. This is a vacuum tube or transistor connected to the rf tank circuit of an oscillator in such a way as to act as a variable inductance or capacitance.

Fig. 5A is a representative circuit. Gate 1 of the modulator MOSFET is connected across the oscillator tank circuit, C1/L1, through resistor R1 and blocking capacitor C2. C3 represents the input capacitance of the modulator transistor. The resistance of R1 is made large compared to the reactance of C3, so the rf current through R1/C3 will be practically in phase with the rf voltage appearing at the terminals of the tank circuit. However, the voltage across C3 will lag the current by 90 degrees. The rf current in the drain circuit of the modulator will be in phase with the grid voltage, and consequently is 90 degrees behind the current through C3, or 90 degrees behind the rf tank voltage. This lagging current is drawn through the oscillator tank, giving the same effect as though an inductance were connected across the tank. The frequency increases in proportion to the amplitude of the lagging plate current of the modulator. The audio voltage, introduced through a

radio-frequency choke, varies the transconductance of the transistor and thereby varies the rf drain current.

The modulated oscillator usually is operated on a relatively low frequency, so that a high order of carrier stability can be secured. Frequency multipliers are used to raise the frequency to the final frequency desired.

A reactance modulator can be connected to a crystal oscillator as well as to the self-controlled type as shown in Fig. 5B. However, the resulting signal can be more phase-modulated than it is frequency-modulated, for the reason that the frequency deviation that can be secured by varying the frequency of a crystal oscillator is quite small.

The sensitivity of the modulator (frequency change per unit change in grid voltage) depends on the transconductance of the modulator transistor. It increases when R1 is made smaller in comparison with C3. It also increases with an increase in L/C ratio in the oscillator tank circuit. However, for highest carrier stability it is desirable to use the largest tank capacitance that will permit the desired devia-

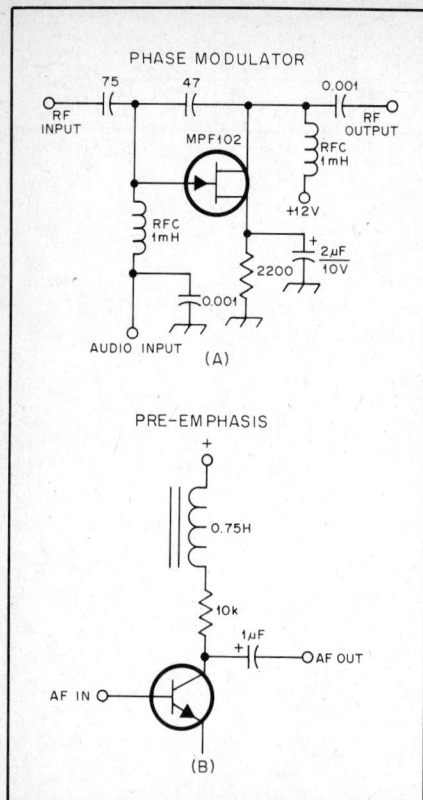

PHASE MODULATOR

(A)

PRE-EMPHASIS

(B)

Fig. 6 — (A) The phase-shifter type of phase modulator. (B) preemphasis and (C) deemphasis circuits.

tion to be secured while keeping within the limits of linear operation.

A change in *any* of the voltages on the modulator transistor will cause a change in rf drain current, and consequently a frequency change. Therefore it is advisable to use a regulated power supply for both modulator and oscillator.

Indirect FM

The same type of reactance-tube circuit that is used to vary the tuning of the oscillator tank in fm can be used to vary the tuning of an amplifier tank and thus vary the phase of the tank current for pm. Hence the modulator circuit of Fig. 5A or 6A can be used for pm if the reactance transistor or tube works on an amplifier tank instead of directly on a self-controlled oscillator. If audio shaping is used in the speech amplifier, as described above, fm instead of pm will be generated by the phase modulator.

The phase shift that occurs when a circuit is detuned from resonance depends on the amount of detuning and the Q of the circuit. The higher the Q, the smaller the amount of detuning needed to secure a given number of degrees of phase shift. If the Q is at least 10, the relationship between phase shift and detuning (in kHz either side of the resonant frequency) will be substantially linear over a phase-shift range of about 25 degrees. From the standpoint of modulator sensitivity, the tuned circuit Q on which the modulator operates should be as high as possible. On the other hand, the effective Q of the circuit will not be very high if the amplifier is delivering power to a load since the load resistance reduces the Q. There must therefore be a compromise between modulator sensitivity and rf power output from the modulated amplifier. An optimum Q figure appears to be about 20; this allows reasonable loading of the modulated amplifier and the necessary tuning variation can be secured from a reactance modulator without difficulty. It is advisable to modulate at a low power level.

Reactance modulation of an amplifier stage usually results in simultaneous amplitude modulation because the modulated stage is detuned from resonance as the phase is shifted. This must be eliminated by feeding the modulated signal through an amplitude limiter or one or more "saturating" stages — that is, amplifiers that are operated Class C and driven hard enough so that variations in the amplitude of the input excitation produce no appreciable variations in the output amplitude.

For the same type of reactance modulator, the speech-amplifier gain required is the same for pm as for fm. However, as pointed out earlier, the fact that the actual frequency deviation increases with the modulating audio frequency in pm makes it necessary to cut off the frequencies above about 3000 Hz before modulation takes place. If this is not done, unnecessary sidebands will be generated at frequencies considerably away from the carrier.

Speech Processing for FM

The speech amplifier preceding the modulator follows ordinary design, except that no power is taken from it and the af voltage required by the modulator grid usually is small — not more than 10 or 15 volts, even with large modulator tubes, and only a volt or two for transistors. Because of these modest requirements, only a few speech stages are needed; a two-stage amplifier consisting of two bipolar transistors, both resistance-coupled, will more than suffice for crystal ceramic or Hi-Z dynamic microphones.

Several forms of speech processing produce worthwhile improvements in fm system performance. It is desirable to limit the peak amplitude of the audio signal applied to an fm or pm modulator, so that the deviation of the fm transmitter will not exceed a preset value. This peak limiting is usually accomplished with a simple audio clipper placed between the speech amplifier and modulator. The clipping process produces high-order harmonics which, if allowed to pass through to the modulator stage, would create unwanted sidebands. Therefore, an audio low-pass filter with a cut-off frequency between 2.5 and 3 kHz is needed at the output of the clipper. Excess clipping can cause severe distortion of the voice signal. An audio processor consisting of a compressor and a clipper, such as described in chapter 12, has been found to produce audio with a better sound (i.e., less distortion) than a clipper alone.

To reduce the amount of noise in some fm communications systems, an audio shaping network called *preemphasis* is

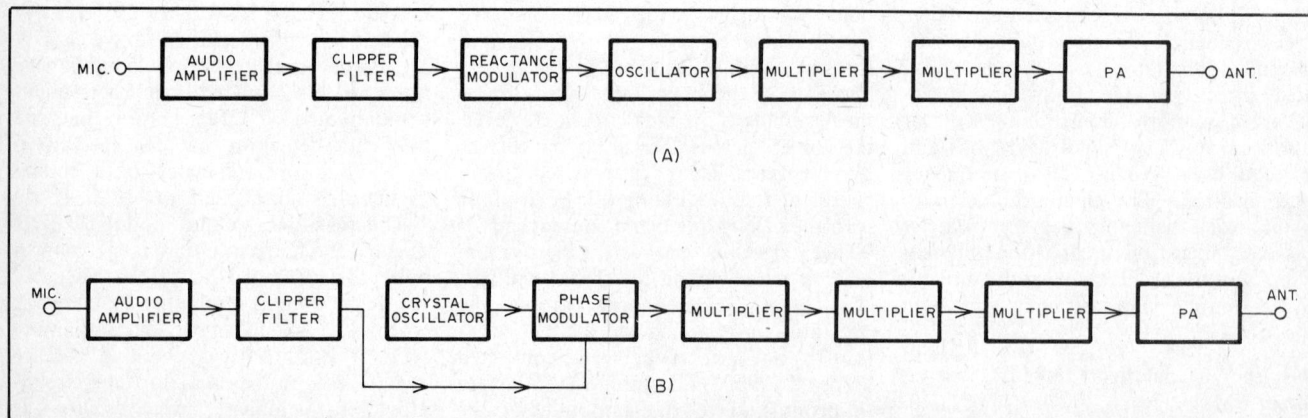

(A)

(B)

Fig. 7 — Block diagrams of typical fm exciters.

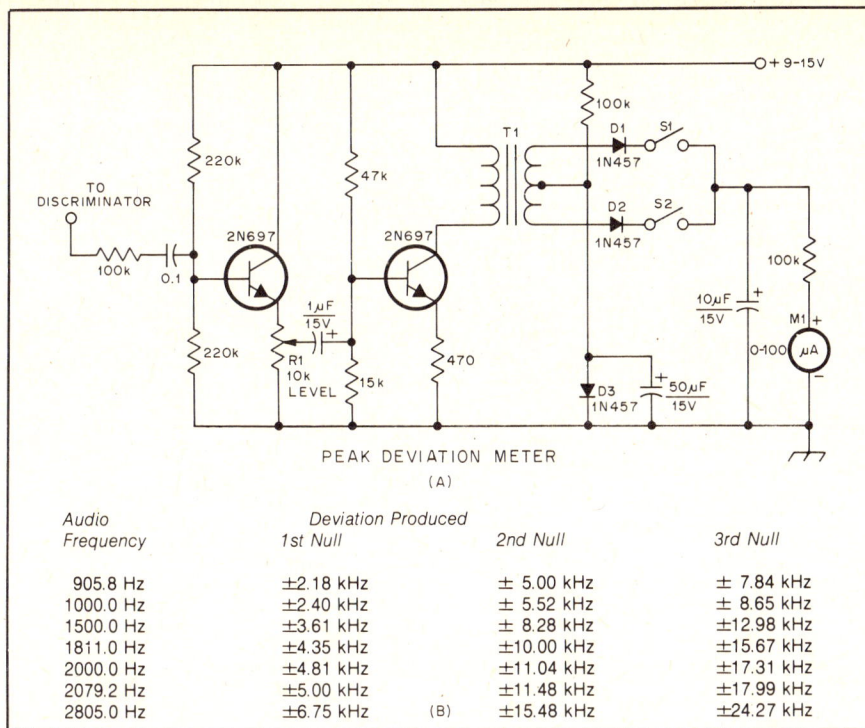

PEAK DEVIATION METER

(A)

Audio Frequency	Deviation Produced		
	1st Null	2nd Null	3rd Null
905.8 Hz	±2.18 kHz	± 5.00 kHz	± 7.84 kHz
1000.0 Hz	±2.40 kHz	± 5.52 kHz	± 8.65 kHz
1500.0 Hz	±3.61 kHz	± 8.28 kHz	±12.98 kHz
1811.0 Hz	±4.35 kHz	±10.00 kHz	±15.67 kHz
2000.0 Hz	±4.81 kHz	±11.04 kHz	±17.31 kHz
2079.2 Hz	±5.00 kHz	±11.48 kHz	±17.99 kHz
2805.0 Hz	±6.75 kHz (B)	±15.48 kHz	±24.27 kHz

Fig. 8 — (A) Schematic diagram of the deviation meter. Resistors are 1/2-watt composition and capacitors are ceramic, except those with polarity marked, which are electrolytic. D1-D3, incl., are high-speed silicon switching diodes. R1 is a linear-taper composition control, and S1, S2 are spst toggle switches. T1 is a miniature audio transformer with 10-kΩ primary and 20-kΩ center-tapped secondary (Triad A31X). (B) Chart of audio frequencies which will produce a carrier null when the deviation of an fm transmitter is set for the values given.

added at the transmitter to proportionally attenuate the lower audio frequencies, giving an even spread to the energy in the audio band. This results in an fm signal of nearly constant energy distribution. The reverse process, called *deemphasis*, is accomplished at the receiver to restore the audio to its original relative proportions. Sample circuits are shown in Fig. 6.

FM Exciters

FM exciters and transmitters take two general forms. One, shown at Fig. 7A, consists of a reactance modulator which shifts the frequency of an oscillator to generate an fm signal directly. Successive multiplier stages provide output on the desired frequency, which is amplified by a PA stage. This system has a disadvantage in that, if the oscillator is free running, it is difficult to achieve sufficient stability for vhf use. If a crystal-controlled oscillator is employed, because the amount that the crystal frequency is changed is kept small, it is difficult to achieve equal amounts of frequency swing.

The indirect method of generating fm shown in Fig. 7B is currently popular. Shaped audio is applied to a phase modulator to generate fm. Since the amount of deviation produced is very small, a large number of multiplier stages is needed to achieve wide-band deviation at the operating frequency. In general, the

system shown at A will require a less complex circuit than that at B, but the indirect method (B) often produces superior results.

Testing an FM Transmitter

Accurate checking of the operation of an fm or pm transmitter requires different methods than the corresponding checks on an a-m or ssb set. This is because the common forms of measuring devices either indicate amplitude variations only (a milliammeter, for example), or because their indications are most easily interpreted in terms of amplitude.

The quantities to be checked in an fm transmitter are the linearity and frequency deviation and the output frequency, if the unit uses crystal control. The methods of checking differ in detail.

Frequency Checking

Crystal-controlled, channelized fm operation requires that a transmitter be held within a few hundred hertz of the desired channel even in the wide-band system. Having the transmitter on the proper frequency is particularly important when operating through a repeater. The rigors of mobile and portable operation make a frequency check of a channelized transceiver a good idea at three-month intervals.

Frequency meters generally fall into two

categories, the heterodyne type and the digital counter. Today the digital counter is used almost universally; units counting to over 500 MHz are available at relatively low cost in kit form. Even less expensive low-frequency counters can be employed by using a *prescaler*, a device which divides an input frequency by a preset ratio, usually 10 or 100. Many prescalers may be used at 148 MHz or higher, using a counter with a 2-MHz (or more) upper frequency limit. If the counting system does not have a sufficient upper frequency limit to measure the output of an fm transmitter directly, one of the frequency-multiplier stages can be sampled to provide a signal in the range of the measurement device. Alternatively, a crystal-controlled converter feeding an hf receiver which has accurate frequency readout can be employed, if a secondary standard is available to calibrate the receiving system.

Deviation and Deviation Linearity

A simple deviation meter can be assembled following the diagram of Fig. 8A. This circuit was designed by K6VKZ. The output of a wide-band receiver discriminator (before any deemphasis) is fed to two amplifier transistors. The output of the amplifier section is transformer coupled to a pair of rectifier diodes to develop a dc voltage for the meter, M1. There will be an indication on the meter with no signal input because of detected noise, so the accuracy of the instrument will be poor on weak signals.

To calibrate the unit, signals of known deviation will be required. If the meter is to be set to read 0-15 kHz, then a 7.5-kHz deviation test signal should be employed. R1 is then adjusted until M1 reads half scale, 50 μA. To check the peak deviation of an incoming signal, close both S1 and S2. Then, read the meter. Opening first one switch and then the other will indicate the amount of positive and negative deviation of the signal, a check of deviation linearity.

Measurement of Deviation Using Bessel Functions

Using a mathematical relationship known as the *Bessel Function* it is possible to predict the points at which, with certain audio-input frequencies and predetermined deviation settings, the carrier output of an fm transmitter will disappear completely. Thus, by monitoring the carrier frequency with a receiver, it will be possible to identify by ear the deviation at which the carrier is nulled. A heterodyne signal at either the input or receiver i-f is required so that the carrier will produce a beat note which can easily be identified. Other tones will be produced in the modulation process, so some concentration is required by the operator when making the test. With an audio tone selected from the chart (Fig. 8B), advance the deviation control slowly until the first

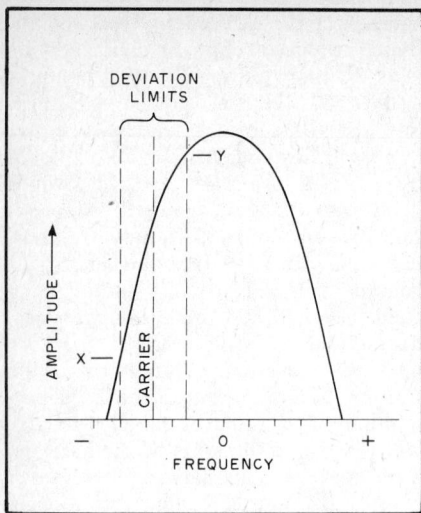

Fig. 9 — Fm detector characteristics. Slope detection, using the sloping side of the receivers selectivity curve to convert fm to a-m for subsequent detection.

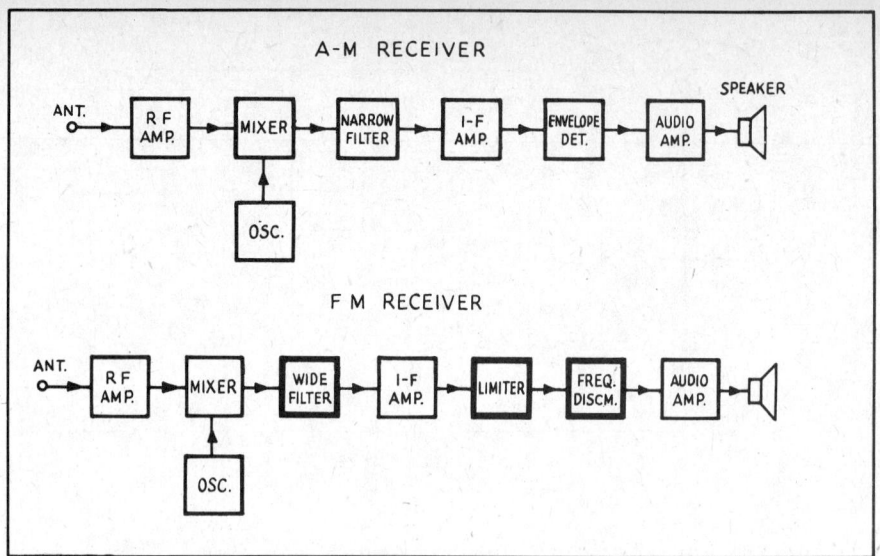

Fig. 10 — Block diagrams of (A) an a-m, (B) an fm receiver. Dark borders outline the sections that are different in the fm set.

FM Filters

Manufacturer	Model	Center Frequency	Nonimal Bandwidth	Ultimate Rejection	Impedance (r) In	Out	Insertion Loss	Crystal Discriminator
KVG (1)	XF-9E	9.0 MHz	12 kHz	90 dB	1200	1200	3 dB	XD9-02
KVG (1)	XF-107A	10.7 MHz	12 kHz	90 dB	820	820	3.5 dB	XD107-01
KVG (1)	XF-107B	10.7 MHz	15 kHz	90 dB	910	910	3.5 dB	XD107-01
KVG (1)	XF-107C	10.7 MHz	30 kHz	90 dB	2000	2000	4.5 dB	XD107-01
Heath Dynamics (2)	-	21.5 MHz	15 kHz	90 dB	550	550	3 dB	-
Heath Dynamics (2)	-	21.5 MHz	30 kHz	90 dB	1100	1100	2 dB	-
Clevite (3)	TCF4-12D3CA	445 kHz	12 kHz	60 dB	40k	2200	6 dB	-
Clevite (3)	TCF4-18G45A	455 kHz	18 kHz	50 dB	40k	2200	6 dB	-
Clevite (3)	TCF6-30D55A	455 kHz	30 kHz	60 dB	20k	1000	5 dB	-

Fig. 11 — A list of fm-bandwidth filters that are available to amateurs. Manufacturer's addresses are as follows: (1) Spectrum International, P. O. Box 1084, Concord, MA 01742; (2) Heath Dynamics, Inc., 6050 N. 52nd Ave., Glendale, AZ 85301, tel. 602-934-5234; (3) Semiconductor Specialists, Inc., P. O. Box 66125, O'Hare International Airport, Chicago, IL 60666.

null is heard. If a higher-order null is desired, continue advancing the control further until the second, and then the third, null is heard. Using a carrier null beyond the third is generally not practical.

For example, if a 905.8-Hz tone is used, the transmitter will be set for 5-kHz deviation when the second null is reached. The second null achieved with a 2805-Hz audio input will set the transmitter deviation at 15.48 kHz. The Bessel-function approach can be used to calibrate a deviation meter, such as the unit shown in Fig. 8A.

Reception of FM Signals

Receivers for fm signals differ from others principally in two features — there is no need for linearity preceding detection (it is, in fact, advantageous if amplitude variations in signal and background noise can be "washed out") and the detector must be capable of converting frequency variations of the incoming signal into amplitude variations.

Frequency-modulated signals can be received after a fashion on any ordinary receiver. The receiver is tuned to put the carrier frequency partway down on one side of the selectivity curve. When the frequency of the signal varies with modulation it swings as indicated in Fig. 9, resulting in an a-m output varying between X and Y. This is then rectified as an a-m signal.

With receivers having steep-sided selectivity curves, the method is not very satisfactory because the distortion is quite severe unless the frequency deviation is small, since the frequency deviation and output amplitude is linear over only a small part of the selectivity curve.

The FM Receiver

Block diagrams of an a-m/ssb and an fm receiver are shown in Fig. 10. Fundamentally, to achieve a sensitivity of less than 1 μV, an fm receiver requires a gain of several million — too much total gain to be accomplished with stability on a single frequency. Thus, the use of the superheterodyne circuit has become standard practice. Three major differences will be apparent from a comparison of the two block diagrams. The fm receiver employs a wider-bandwidth filter, a different detector, and has a limiter stage added between the i-f amplifier and the detector.

Otherwise the functions, and often the circuits, of the rf, oscillator, mixer and audio stages will be the same in either receiver.

In operation, the noticeable difference between the two receivers is the effect of noise and interference on an incoming

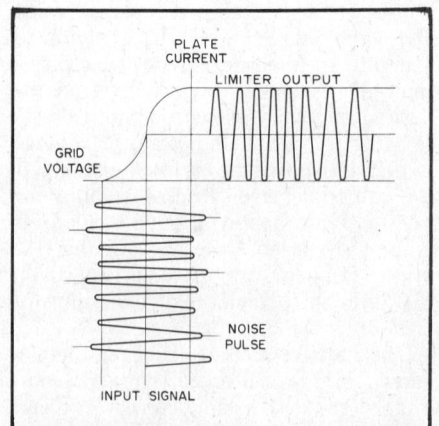

Fig. 12 — Representation of limiter action. Amplitude variations on the signal are removed by the diode action of the grid- and plate-current saturation.

Fig. 13 — (A) Input wave form to a limiter stage shows a-m and noise. (B) The same signal, after passing through two limiter stages, is devoid of a-m components.

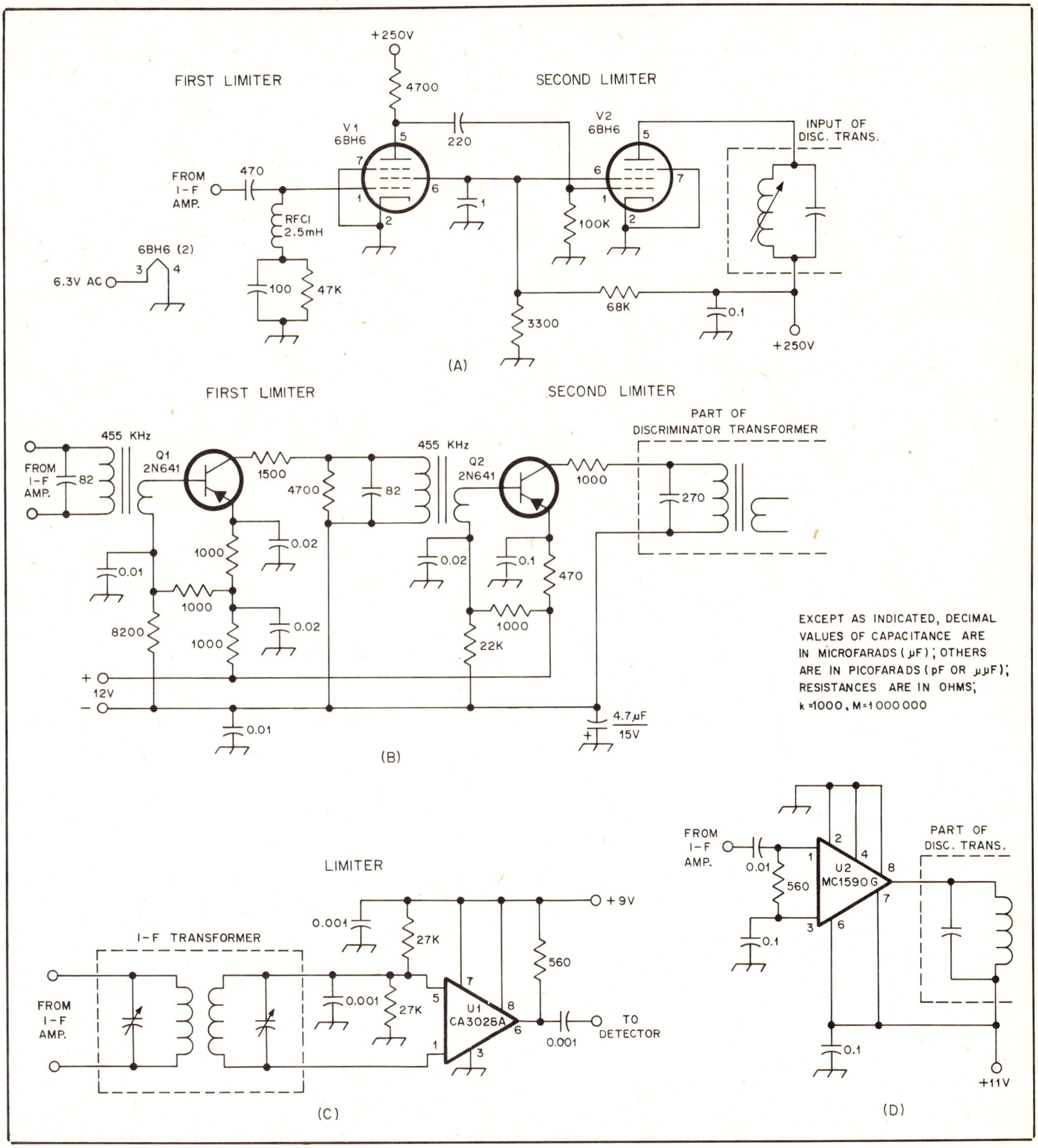

signal. From the time of the first spark transmitters, "rotten QRN" has been a major problem for amateurs. The limiter and discriminator stages in an fm set can eliminate a good deal of impulse noise, except noise which manages to acquire a frequency-modulation characteristic. Accurate alignment of the receiver i-f system and phase tuning of the detector are required to achieve good noise suppression. Fm receivers perform in an unusual manner when QRM is present,

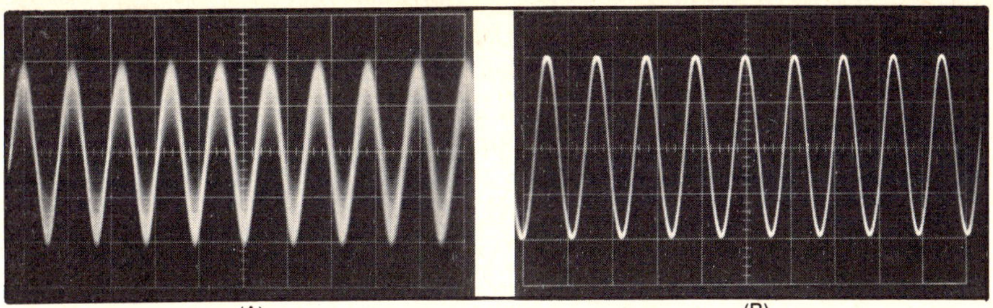

Fig. 14 — Typical limiter circuits using (A) tubes, (B) transistors, (C) a differential IC, (D) a high-gain linear IC.

exhibiting a characteristic known as the *capture effect*. The loudest signal received, even if it is only two or three times stronger than other stations on the same frequency, will be the only transmission demondulated. By comparison, an S9 a-m or cw signal can suffer noticeable interference from an S2 carrier.

Bandwidth

Most fm sets that use tubes achieve i-f selectivity by using a number of over-coupled transformers. The wide bandwidth and phase-response characteristic needed in the i-f system dictate careful design and alignment of all interstage transformers.

For the average ham, the use of a high-selectivity filter in a homemade receiver offers some simplification of the alignment task. Following the techniques used in ssb receivers, a crystal or ceramic filter should be placed in the circuit as close as possible to the antenna connector — at the output of the first mixer, in most cases. Fig. 11 lists a number of suitable filters that are available to amateurs. Prices for these filters are in the $50 range. Experimenters who wish to "roll their own" can use surplus hf crystals, as outlined in the ssb chapter, or ceramic resonators.

One item of concern to every amateur fm user is the choice of i-f bandwidth for his receiver. Deviation of 5 kHz is now standard on the amateur bands. A wide-band receiver can receive narrowband signals, suffering only some loss of audio in the detection process. Naturally, it also will be subject to adjacent-channel interference, especially in congested areas.

Limiters

When fm was first introduced, the main selling point used for the new mode was the noise-free reception possibilities. The circuit in the fm receiver that has the task of chopping off noise and amplitude modulation from an incoming signal is the *limiter*. Most types of fm detectors respond to both frequency and amplitude

variations of the signal. Thus, the limiter stages preceding the detector are included to "cleanse" the signal so that only the desired frequency modulation will be demodulated. This action can be seen in Fig. 13.

Limiter stages can be designed using tubes, transistors, or ICs. For a tube to act as a limiter, the applied B voltages are chosen so that the stage will overload easily, even with a small amount of signal input. A sharp-cutoff pentode such as the 6BH6 is usually employed with little or no bias applied. As shown in Fig. 12, the input signal limits when it is of sufficient amplitude so that diode action of the grid and plate-current saturation clip both sides of the input signal, producing a constant-amplitude output voltage.

Obviously, a signal of considerable strength is required at the input of the limiter to assure full clipping, typically several volts for tubes, 1 volt for transistors, and several hundred microvolts for ICs. Limiting action should start with an rf input of 0.2 μV or less, so a large amount of gain is required between the antenna terminal and the limiter stages. For example, the Motorola 80D has eight tubes before the limiter, and the solid-state receivers use nine transistor stages to get sufficient gain before the first limiter. The new ICs offer some simplification of the i-f system, as they pack a lot of gain into a single package.

When sufficient signal arrives at the receiver to start limiting action, the set *quiets* — that is, the background noise disappears. The sensitivity of an fm receiver is rated in terms of the amount of input signal required to produce a given amount of quieting, usually 20 dB. Use of solid-state devices allow receivers to achieve 20 dB quieting with 0.15 to 0.5 μV of input signal.

A single tube or transistor stage will not provide good limiting over a wide range of input signals. Two stages, with different input time constants, are a minimum requirement. The first stage is set to handle impulse noise satisfactorily while

the second is designed to limit the range of signals passed on by the first. At frequencies below 1 MHz it is useful to employ untuned RC-coupled limiters which provide sufficient gain without a tendency toward oscillation.

Fig. 14A shows a two-stage limiter using sharp-cutoff tubes, while 14B has transistors in two stages biased for limiter service. The base bias on either transistor may be varied to provide limiting at a desired level. The input-signal voltage required to start limiting action is called the *limiting knee*, referring to the point at which collector (or plate) current ceases to rise with increased input signal. Modern ICs have limiting knees of 100 mV for the circuit shown in Fig. 14C, using the RCA CA3028A or Motorola MC1550G, or 200 mV for the MC1590G of Fig. 14D. Because the high-gain ICs such as the CA3076 and MC1590G contain as many as six or eight active stages which will saturate with sufficient input, one of these devices provides superior limiter performance compared to a pair of tubes or transistors.

Detectors

The first type of fm detector to gain popularity was the frequency discriminator. The characteristic of such a detector is shown in Fig. 15. When the fm signal has no modulation, and the carrier is at point 0, the detector has no output. When audio input to the fm transmitter swings the signal higher in frequency, the rectified output increases in the positive direction. When the frequency swings lower the output amplitude increases in the negative direction. Over a range where the discriminator is linear (shown as the straight portion of the line), the conversion of fm to a-m that is taking place will be linear.

A practical discriminator circuit is shown in Fig. 16. The fm signal is converted to a-m by transformer T1. The voltage induced in the T1 secondary is 90 degrees out of phase with the current in the primary. The primary signal is

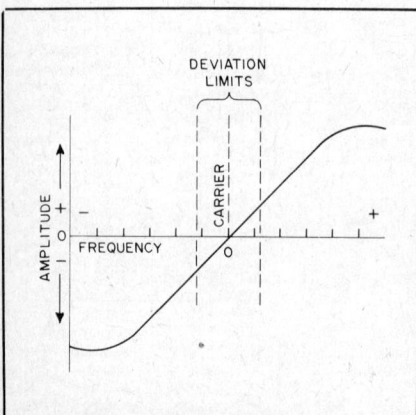

Fig. 15 — The characteristic of an fm discriminator.

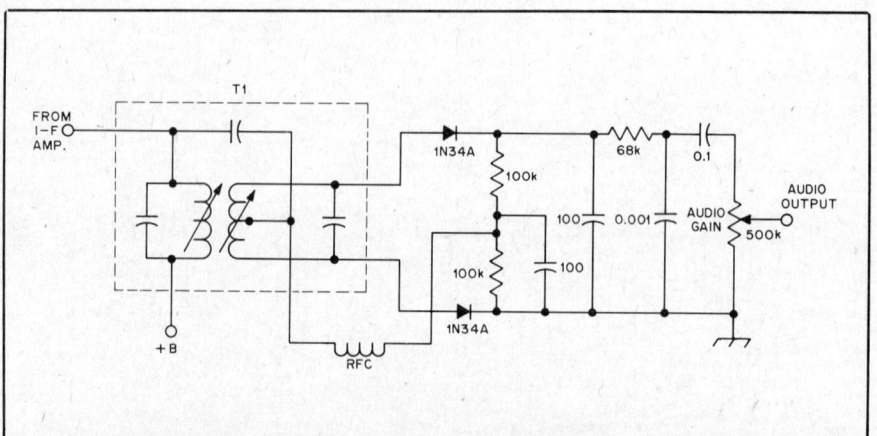

Fig. 16 — Typical frequency-discriminator circuit used for fm detection. T1 is a Miller 12-C45.

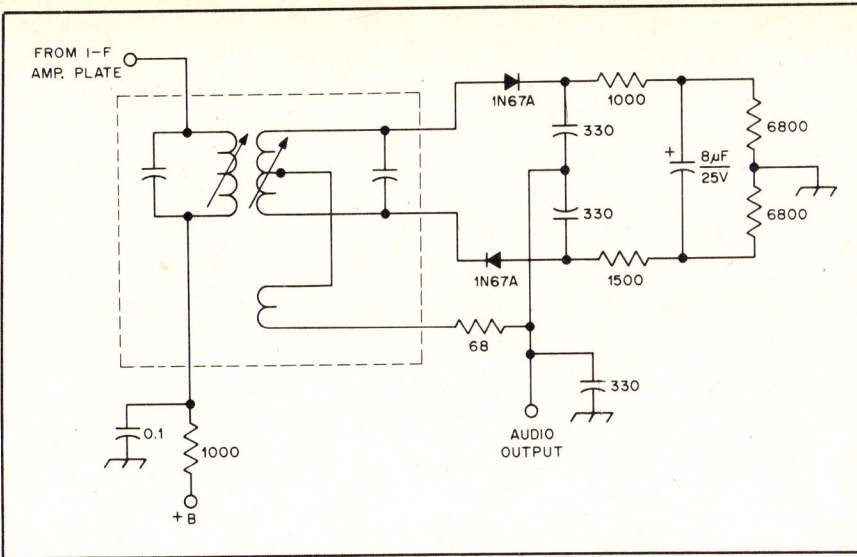

Fig. 17 — A ratio detector of the type often used in entertainment radio and TV sets. T1 is a ratio-detector transformer such as the Miller 1606.

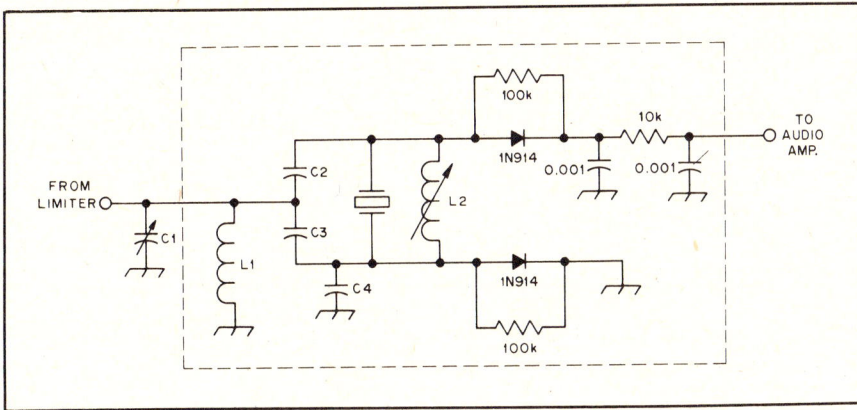

Fig. 18 — Crystal discriminator, C1 and L1 are resonant at the intermediate frequency. C2 is equal in value to C3. C4 corrects any circuit imbalance so that equal amounts of signal are fed to the detector diodes.

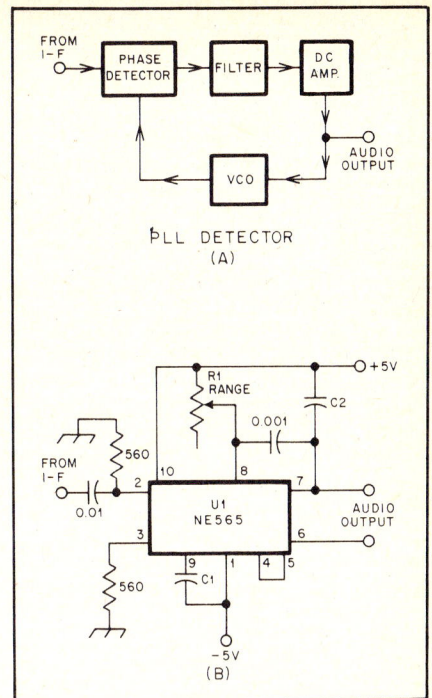

Fig. 19 — (A) Block diagram of a PLL demodulator. (B) Complete PLL circuit.

introduced through a center tap on the secondary, coupled through a capacitor. The secondary voltages combine on each side of the center tap so that the voltage on one side leads the primary signal while the other side lags by the same amount. When rectified, these two voltages are equal and of opposite polarity, resulting in zero-voltage output. A shift in input frequency causes a shift in the phase of the voltage components that results in an increase of output amplitude on one side of the secondary, and a corresponding decrease on the other side. The differences in the two changing voltages, after rectification, constitute the audio output.

In the search for a simplified fm detector, RCA developed a circuit that has now become standard in entertainment radios which eliminated the need for a preceding limiter stage. Known as the *ratio detector*, this circuit is based on the idea of dividing a dc voltage into a ratio which is equal to the ratio of the amplitudes from either side of a dis-

criminator-transformer secondary. With a detector that responds only to ratios, the input signal may vary in strength over a wide range without causing a change in the level of output voltage — fm can be detected, but not a-m. In an actual ratio detector, Fig. 17, the dc voltage required is developed across two load resistors, shunted by an electrolytic capacitor. Other differences include the two diodes, which are wired in series aiding rather than series opposing, as in the standard discriminator circuit. The recovered audio is taken from a tertiary winding which is tightly coupled to the primary of the transformer. Diode-load resistor values are selected to be lower (5000 ohms or less) than for the discriminator.

The sensitivity of the ratio detector is one-half that of the discriminator. In general, however, the transformer design values for Q, primary-secondary coupling, and load will vary greatly, so the actual performance differences between these two types of fm detectors are usually not

significant. Either circuit can provide excellent results. In operation, the ratio detector will not provide sufficient limiting for communications service, so this detector also is usually preceded by at least a single limiting stage.

Other Detector Designs

The difficulties often encountered in building and aligning LC discriminators have inspired research that has resulted in a number of adjustment-free fm detector designs. The *crystal discriminator* utilizes a quartz resonator, shunted by an inductor, in place of the tuned-circuit secondary used in a discriminator transformer. A typical circuit is shown in Fig. 18. Some commercially made crystal discriminators have the input-circuit inductor, L1, built in (C1 must be added) while in other types both L1 and C2 must be supplied by the builder. Fig. 18 shows typical component values; unmarked parts are chosen to give the desired bandwidth. Sources for crystal discriminators are listed in Fig. 11.

The PLL

Since the *phase-locked loop* (PLL) was reduced to a single IC package, this circuit is revolutionizing some facets of receiver design. Introduction by Signetics of a PLL in a single flat-pack IC, followed by Motorola and Fairchild (who are making the PLL in separate building-block ICs), allows a builder to get to work with a minimum of bother.

A basic phase-locked loop consists of a phase detector, a filter, a dc amplifier, and a voltage-controlled oscillator (VCO). The VCO runs at a frequency close to that of an incoming signal. The phase detector produces an error voltage if any frequency difference exists between the VCO and the i-f signal. This error voltage is applied to the VCO. Any changes in the frequency of the incoming signal are sensed at the detector and the error voltage readjusts the VCO frequency so that it remains locked to the intermediate frequency. The bandwidth of the system is determined by a filter on the error-voltage line.

Because the error voltage is a copy of the audio variations originally used to shift the frequency of the transmitter, the PLL functions directly as an fm detector. The sensitivity achieved with the Signetics NE565 PLL is good — about 1 mV for a typical circuit. No transformers or tuned circuits are required. The PLL bandwidth is usually two to ten percent of the i-f for fm detection. Components R1/C1 set the VCO to near the desired frequency. C2 is the loop-filter capacitor which determines the capture range — that range of frequencies over which the loop will acquire lock with an input signal, initially starting out of lock. The NE565 has an upper frequency limit of 500 kHz; for higher frequencies, the NE561, which is usable up to 30 MHz, can be employed.

Preamplifiers for Increased Sensitivity

Some surplus, homemade and commercial new equipment for the vhf and uhf fm bands need additional receiver gain and noise-figure improvement for weak-signal work. Too much gain can seriously degrade the receiver dynamic range, so care must be exercised when adding a preamplifier ahead of an existing receiver front end. The temptation of some inexperienced amateurs is to use a preamp which has a gain of 25 dB or greater. As a result, strong local signals can overload the receiver and cause severe mixer IMD. The two preamplifiers described here are tailored for useful but not excessive gain amounts. They should enhance the performance of receivers or converters that are marginal in terms of overall gain and noise figure. They should not have a serious effect on the receiver dynamic range.

The 2-meter version shown in Fig. 20 utilizes a single Siliconix U310 JFET in a common-gate circuit. This helps to ensure stability and provide a gain of over 10 dB. The U310 is well known for its low noise up to 450 MHz (about 3 dB at 450 MHz and 1.5 dB at 144 MHz). This transistor also has excellent dynamic-range characteristics (in excess of 100 dB). A less costly substitute is the Siliconix E300, which comes in a plastic case. The performance traits are approximately the same, but stability may be harder to realize because the E300 has no metal case which can be grounded automatically when the gate is grounded.

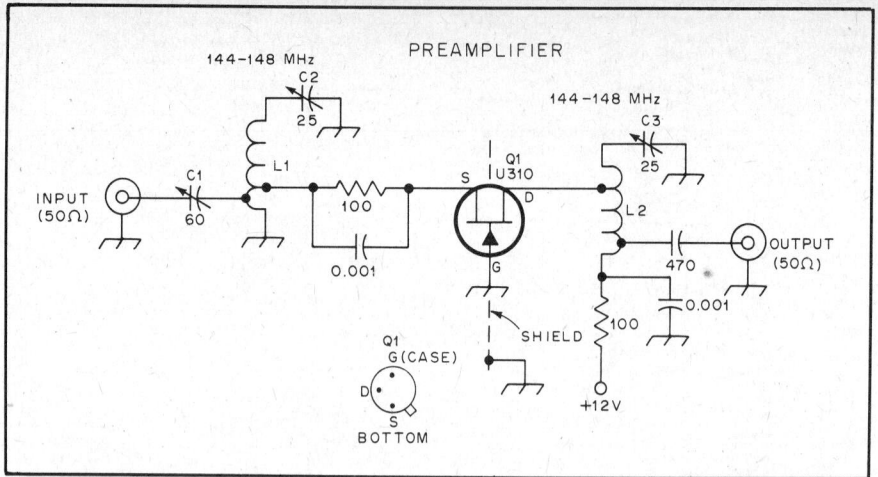

Fig. 20 — Schematic diagram of the low-noise 2-meter preamp. Fixed-value capacitors are disc ceramic. Resistors are 1/4- or 1/2-watt composition. See text for data on the trimmers. L1 has 5 turns of no. 20 wire, 3/4-inch (19-mm) long with an ID of 1/4 inch (6.3 mm). C1 tap approx. 1/2 turn from ground. Q1 tap approx. 1 turn from ground. L2 has same dimensions except for Q1 tap, which is approx. 1 turn from C3 end. See text for Q1 data.

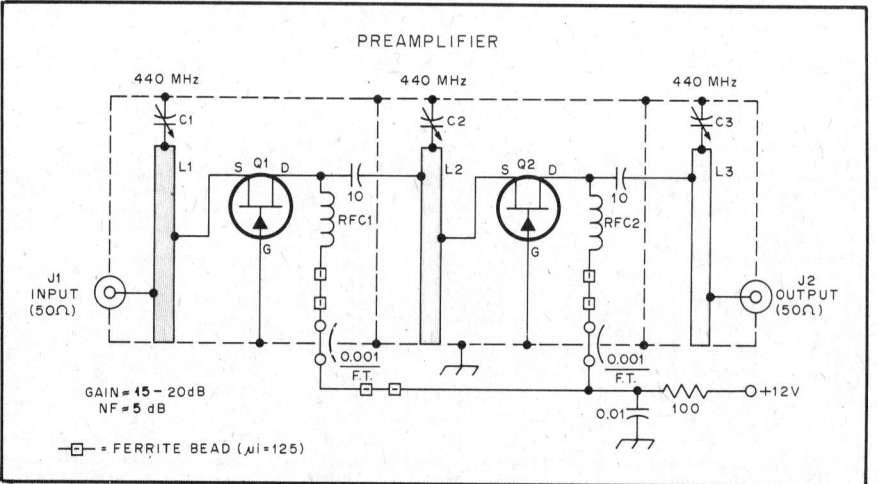

Fig. 21 — Schematic diagram of the 440-MHz preamp. The 10-pF capacitors are silver mica. FT indicates feedthrough capacitor.

C1-C3, incl. — 1.4 to 9.2-pF miniature air variable, Johnson 189-0563-001 or equiv.
J1, J2 — BNC-type connector soldered to case outer wall.
L1-L3, incl. — 2-5/8 × 1/4-inch (67 × 6.3-mm) brass strip. Input and output taps on L1 and L3 are approx. 1/2 inch (13 mm) up from ground (see text). Attach Q1/Q2 drain taps approx. 1/4 inch (6.3 mm) below C2 and C3 ends of line. Q1, Q2 source taps on L1 and L2 are approx. 3/4 inch (19 mm) up from ground.
Q1, Q2 — Siliconix JFET.
RFC1, RFC2 — 420-MHz choke J. W. Miller 4584 or equiv. Ferrite beads assoc. with these chokes are Amidon miniature type.

C1 of Fig. 20 is adjusted for lowest noise figure. The C1 coil tap can be adjusted also if further improvement is needed. C2 and C3 should be high-Q trimmers for best performance. Miniature ceramic trimmers should be suitable for use at C1, C2 and C3. Ideally, Teflon trimmers or small air variables would be used at those circuit points.

If L1 and L2 are at right angles to one another and spaced well apart, it may not be necessary to use a shield divider across Q1 as shown in Fig. 20. However, a small piece of copper, brass or double-sided pc board should be easy to add to the circuit board.

A strip-line preamplifier for use at 440 MHz is shown in Figs. 21 and 22. The maximum attainable gain is roughly 20 dB with the circuit shown. Noise figure should be better than 5 dB when the taps onto L1 are optimized. Some experimenting will be necessary. The loaded Q of the three resonators can be increased by moving the Q1/Q2 taps closer to the ground ends of each line. The tradeoff is in preamplifier gain. The increased Q may be important when gain requirements aren't too great (as in a repeater installation), but when rejection of out-of-band commercial signals are vital to good performance. L1, L2 and L3 are silver plated

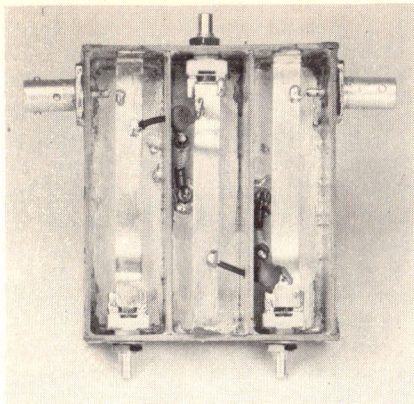

Fig. 22 — Interior view of the 440-MHz preamplifier. The center strip line is reversed from the end ones to prevent excessive lead lengths from Q1 and Q2. The box dimensions are 3 × 3-1/2 × 1 inch (76 × 89 × 25 mm). The internal shields are 3 × 15/16 inches (76 × 24 mm). All mating surfaces of the box walls are soldered inside and outside. Both sides of the dividers are soldered to the inner surface of the box.

in this model. The double-sided pc-board housing for the amplifier is also silver plated, as are the two internal divider walls which isolate the tuned circuits from one another.

Amplifier stability is excellent when the gate leads of Q1 and Q2 are made as short as possible. Additional insurance against instability is provided by the decoupling circuits in the drain leads of Q1 and Q2. A press-fit aluminum or brass cover (U-shaped) is used to enclose the open side of the preamplifier case. Craftsmen may elect to make the housing and divider walls from sections of 1/16-inch (1.6-mm) brass, using silver solder to join the mating surfaces. This unit is suitable for use anywhere in the 420- to 450-MHz band.

A SELECTIVE 2-METER PREAMPLIFIER

A preamplifier used ahead of a surplus receiver as part of a repeater must be based on something more than casual design if good performance is to be realized. Special attention must be paid to selectivity, and the noise figure should be low enough to assure the kind of sensitivity desired by most repeater operators — 0.2 µV or less for 20 dB of quieting. Transient and rf-burnout protection are the other criteria for successful use at the repeater site.

While some groups have had success with solid-state preamplifiers, others have decried the reliability of transistorized preamps, mainly because of overloading, IMD, and susceptibility to device damage from static discharges and line transients. Certainly tube-type equipment is less subject to catastrophic failure from the foregoing causes, but a properly designed solid-state preamplifier can hold its own when competing against a vacuum-tube equivalent. The unit treated here (designed by W1FB) meets the design specifications of most amateur repeaters. Precautions have been taken to prevent the usual problems inherent in homemade preamps.

Design

JFETs were chosen for use in the preamp over gate-protected MOSFETs because the former can sustain up to 80 volts pk-pk, gate to source, before being damaged. Protected MOSFETs are rated at 20 volts maximum. Furthermore, the employment of JFETs eliminated four resistors and two capacitors, all of which would have been required in the gate-2 biasing circuits of the MOSFETs.

In the interest of eliminating the need for those sometimes-tricky neutralization circuits, the common-gate configuration was chosen. Common-gate amplifiers provide somewhat less gain than do the common-source types — approximately 10 dB less gain per stage, but by using two stages in common-gate the gain of the preamplifier is more than adequate for most applications. The circuit of Fig. 23 should exhibit a gain of between 15 and 20 dB, depending upon the transconductances of the two FETs picked from the supplier's shelf.

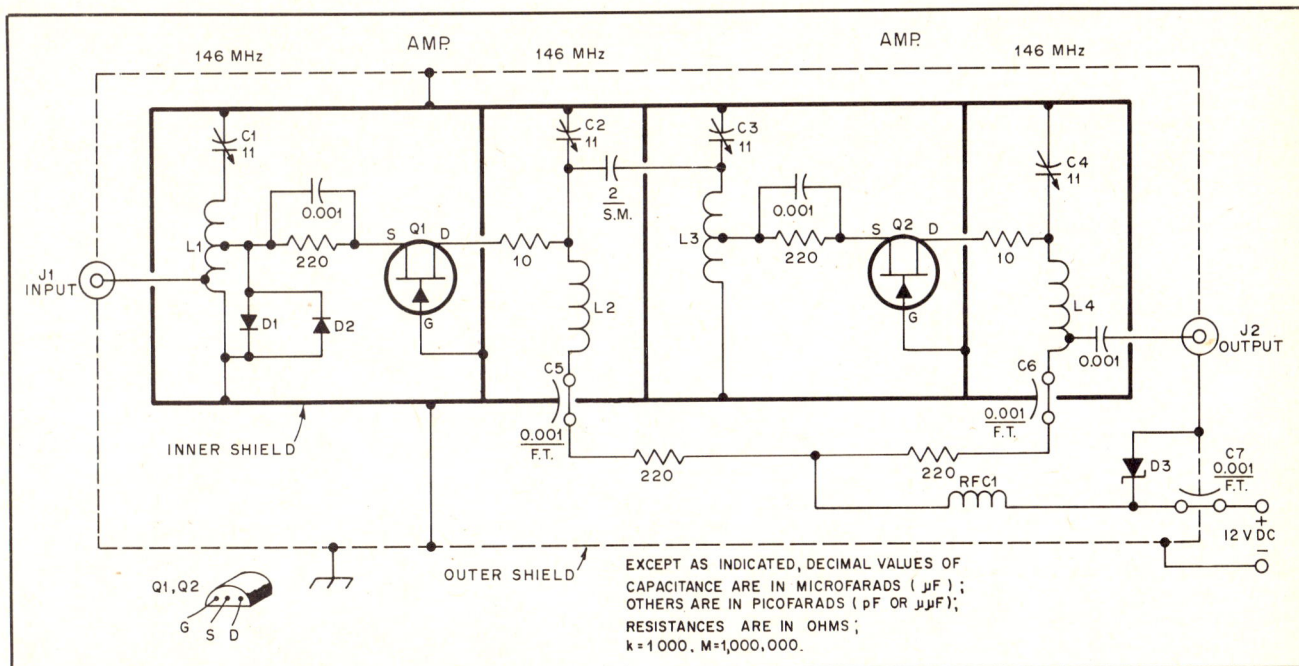

Fig. 23 — Circuit diagram of the preamplifier. Heavy lines indicate the pc-board shield box and dividers. The outer shield box is shown in dashed lines. Fixed-value capacitors are disc ceramic unless otherwise noted. S.M. indicates silver mica. Resistors are 1/2 watt carbon.

C1-C4, incl. — 11-pF subminiature air variable, E. F. Johnson 189-564. Piston trimmers or Johnson 160-0104-001 suitable also.
C5-C7, incl. — Feedthrough capacitor.
D1, D2 — High-speed silicon switching diode, 1N914 or equivalent.
D3 — 15-volt, 1-watt Zener diode.
J1, J2 — Coaxial connector of builder's choice. (Type BNC used in this model.)

L1, L3 — 3-1/2 turns no. 14 tinned bus wire, 1/2-inch ID × 3/4-inch (13 × 19 mm) long. Tap source at 1-3/4 turns from trimmer end. Tap L1 also at 1/2 turn from ground.
L2, L4 — 3-1/3 turns no. 14 tinned bus wire, 1/2-inch ID × 3/4-inch (13 × 19 mm) long. Tap L4 1/2 turn above C6.

Q1, Q2 — Vhf or uhf JFET (see text). Keep gate lead as short as possible, 1/8 inch (3 mm) or less.
RFC1 — 144-MHz rf choke, approximately 2.7 µH. James Millen 34300-2.7 or equivalent. Alternatively, wind 20 inch (508 mm) of no. 30 enam. wire on the body of a 2700-ohm 1-watt carbon resistor. Use pigtails as anchor points for ends of winding.

Motorola 2N5484s (MPF106) are used at Q1 and Q2. Alternatively, U310, E300, 2N4416 or 2N4417 devices can be used if one is willing to pay a bit more money. MPF102s would probably do a good job in the circuit of Fig. 23. Since the 2N5484s are designed for use into the 400-MHz region, they seemed likewise choices for low-noise operation. To assure good selectivity and thereby offer reasonable immunity to nearby out-of-band commercial signals, high-Q tuned circuits were employed. The section between Q1 and Q2 is a bandpass type, lightly coupled by means of a 2-pF silver-mica capacitor. Lighter coupling will provide greater selectivity, but with an attendant loss in gain. Aperture coupling can be used in place of the method shown. If so, the aperture size must be adjusted to establish the gain and selectivity desired by the user.

Source bias is used in each stage to prevent the amplifiers from saturating in the presence of strong signals. The sources are tapped down on their respective tuned circuits to provide impedance matching.

In the prototype unit the drains were tapped down on the tuned circuits — a method used to achieve stability. As an aid to stability each stage has a 10-ohm resistor between its drain and the related tuned circuit. However, the gain of the preamplifier was somewhat less when using the 10-ohm resistors and connecting them to the stators of C2 and C4.

Decoupling networks are used between the stages (220-ohm resistors and 0.001-μF bypass capacitors) to prevent interstage coupling along the 12-volt supply line. Filtering at rf is provided by using RFC1 and another 0.001-μF feedthrough capacitor, thus helping to prevent unwanted rf from entering the preamplifier on the dc supply line.

Rf burnout protection is offered by two 1N914 diodes connected from the source tap on L1 to ground. The diodes are located at an impedance point which is higher than that of the 50-ohm antenna tap. This means the diodes will conduct sooner at the source tap because the rf-voltage level from static discharge or abnormally strong signals will always be greater at the tap point than at the 50-ohm terminal. No change in amplifier performance could be noted after adding the diodes.

Protection from any abrupt increase in supply voltage brought about by ac-line transients is afforded by the use of a 15-volt, 1-watt Zener diode (D3) which is connected between the 12-volt supply line and chassis ground.

Construction

To provide for adequate shielding against RFI, two boxes are used in the construction of the preamp. The inner box is made from double-sided copper-clad pc board. It measures 4-1/2 × 1-7/8 × 1-1/4 inches (114 × 48 × 32 mm). The box walls

and the partitions are soldered in place by using a 100-watt soldering iron with a small-diameter tip. The metal surfaces of the pc-board sections are silver plated, though the plating is not necessary as far as circuit performance is concerned. The plating does, however, retard tarnishing and make soldering somewhat easier.

The input and output rf connections, and those between the compartments, are made by means of small Teflon push-in feedthrough terminals which were obtained as surplus. The source-bias resistors and bypass capacitors are attached to Teflon standoff posts. Satisfactory substitutes for the feedthrough bushings can be fashioned from short lengths of RG-59/U coax with the vinyl jacket and shield braid removed. Epoxy cement can be used to hold the homemade bushings in place.

Once the circuit is assembled in its pc-board enclosure the subassembly can be installed in a Minibox which measures 5-1/4 × 3 × 2-1/8 inches (133 × 76 × 54 mm).

Adjustment

Connect the preamplifier ahead of the fm receiver with which it is to be used. Apply 12 volts dc to the preamp, then supply a low-level signal to Q1 via J1. Peak each tuned circuit for maximum response by observing the 1st-limiter current reading of the fm receiver. The unit should then be ready to use.

This preamplifier was used ahead of a Motorola five-pipe Sensicon receiver during all tests. The "barefoot" fm receiver provided 20 dB of quieting with a 0.4-μV input signal. With the preamp installed it was possible to obtain 20 dB of

quieting with somewhat less than 0.1 μV of input signal.

A SYNTHESIZED 2-METER TRANSMITTER

Today, because of the large number of 2-meter repeaters across the country, the synthesized transmitter has become "standard" on the 2-meter band. The 12-watt transmitter shown in Figs. 24 through 31 covers the entire 2-meter band in 5-kHz steps, has no significant spurious outputs and has no audible reference whine or microphonics. The transmitter, designed and built by Albert Helfrick, K2BLA, was described in September 1980 QST.

Circuit Description

Several features of this transmitter are unique. First, the VCO in the synthesizer operates at the same frequency as the transmitter output stage. That implies that there are no mixers or multipliers to produce spurious outputs, which greatly simplifies the tuning of the power amplifiers.

A dual-modulus (or "pulse-swallowing") divider is a type of programmable divider which allows very high frequencies to be divided by any integer with the use of only one high-frequency integrated circuit. The heart of the circuit, shown in Fig. 25, is U6, a dual-modulus, divide-by-10/11 ECL integrated circuit. The output of this chip drives two presettable TTL down-counters, called the main counter and the swallow counter. Both counters are preset at the same time, and each is clocked down simultaneously with the output of the prescaler. The swallow counter necessarily has a smaller number preset than the main counter, and thus

Fig. 24 — Front panel on the synthesized 2-meter transmitter showing the ON-OFF switch and the frequency-selector switches. The microphone connector and accessory jack are located on the rear panel.

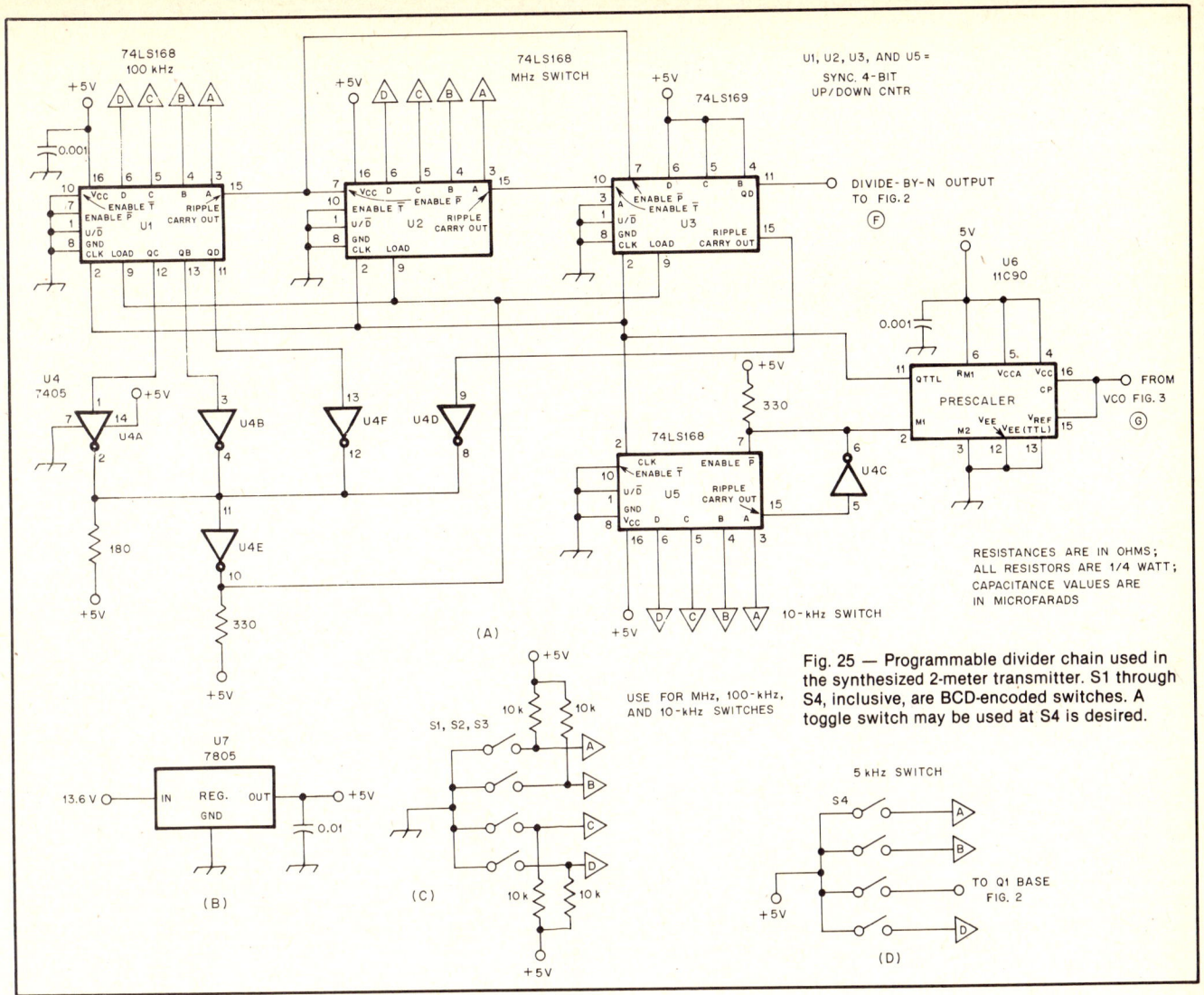

Fig. 25 — Programmable divider chain used in the synthesized 2-meter transmitter. S1 through S4, inclusive, are BCD-encoded switches. A toggle switch may be used at S4 is desired.

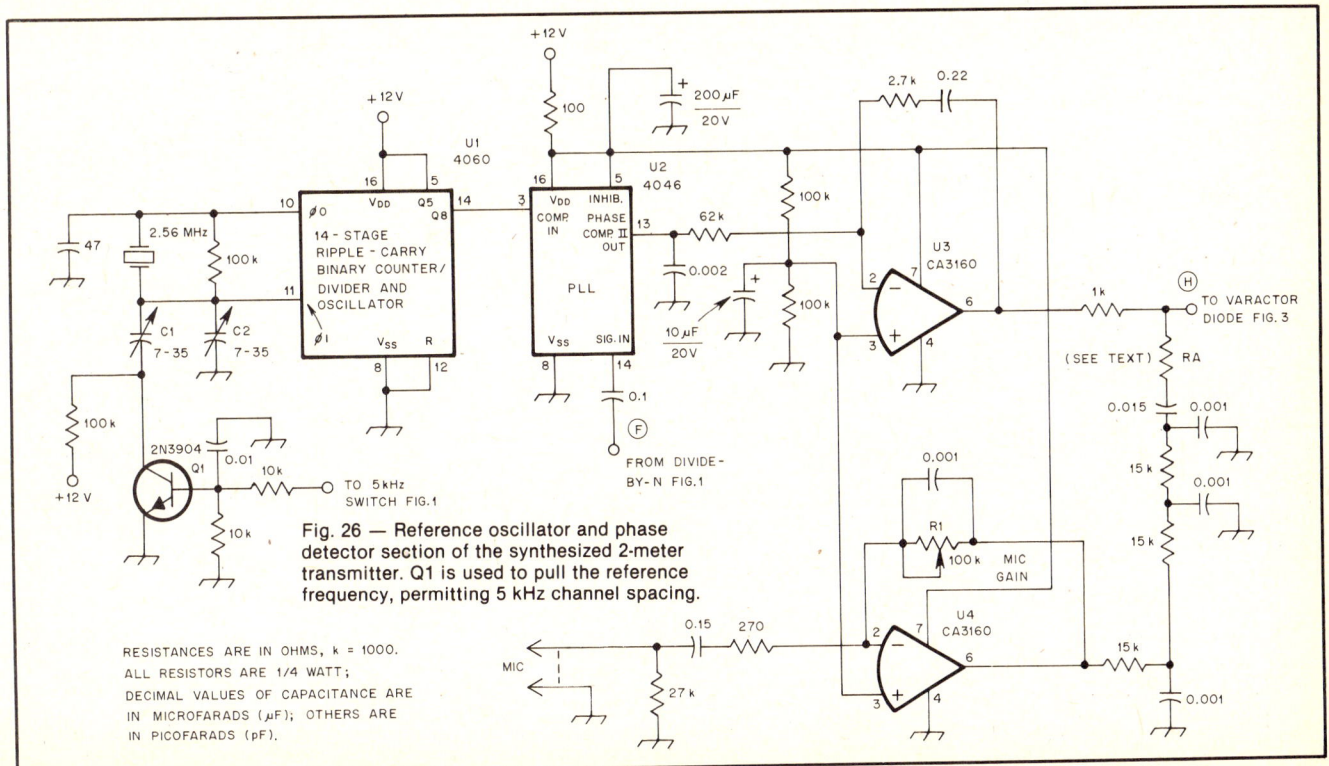

RESISTANCES ARE IN OHMS; ALL RESISTORS ARE 1/4 WATT; CAPACITANCE VALUES ARE IN MICROFARADS

Fig. 26 — Reference oscillator and phase detector section of the synthesized 2-meter transmitter. Q1 is used to pull the reference frequency, permitting 5 kHz channel spacing.

RESISTANCES ARE IN OHMS, k = 1000. ALL RESISTORS ARE 1/4 WATT; DECIMAL VALUES OF CAPACITANCE ARE IN MICROFARADS (μF); OTHERS ARE IN PICOFARADS (pF).

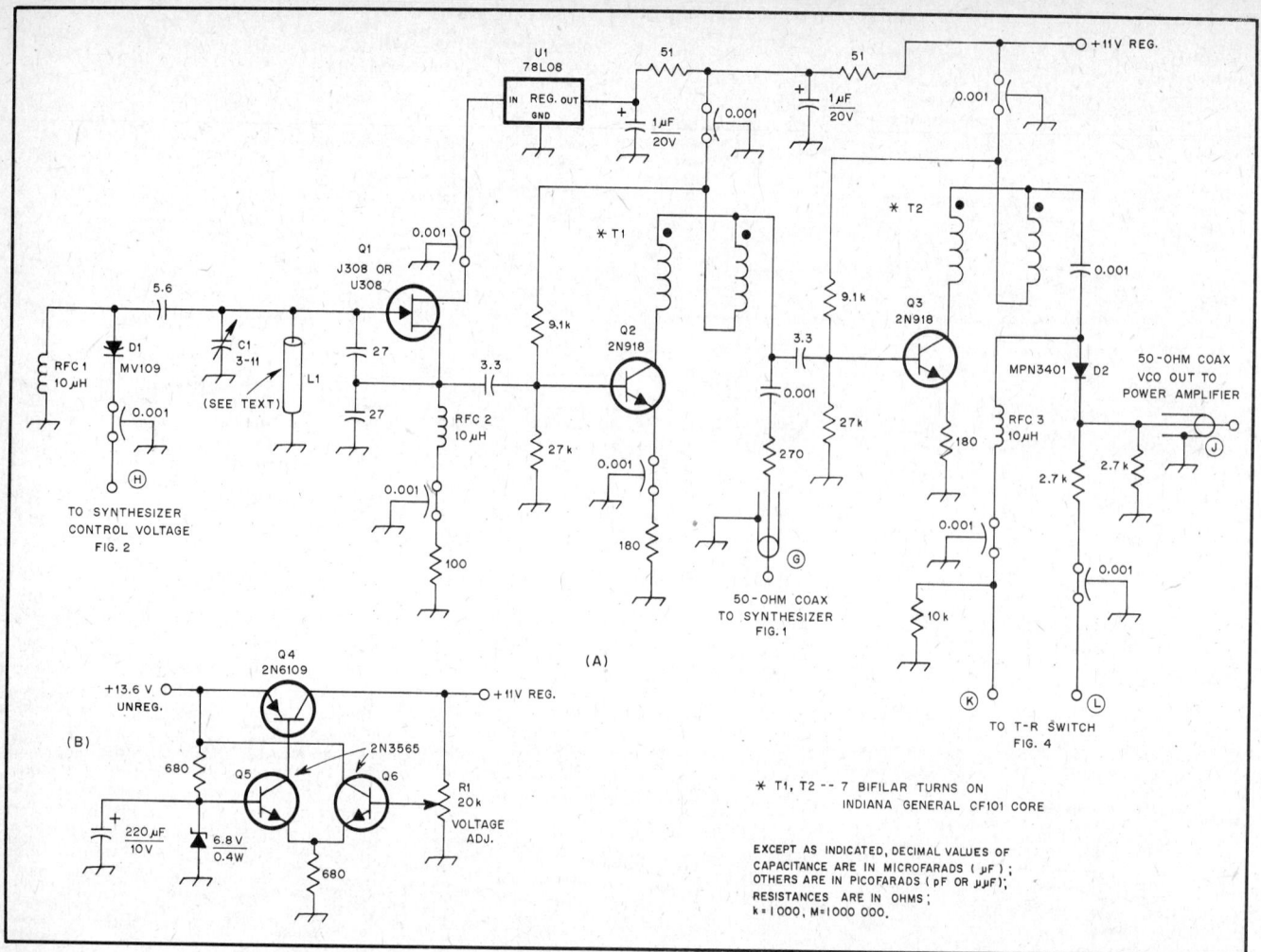

Fig. 27 — The VCO and buffer of the synthesized transmitter are shown at A, while the 11-volt regulator circuit may be seen at B; the regulator was included to reduce the possibility of "alternator whine" modulation.

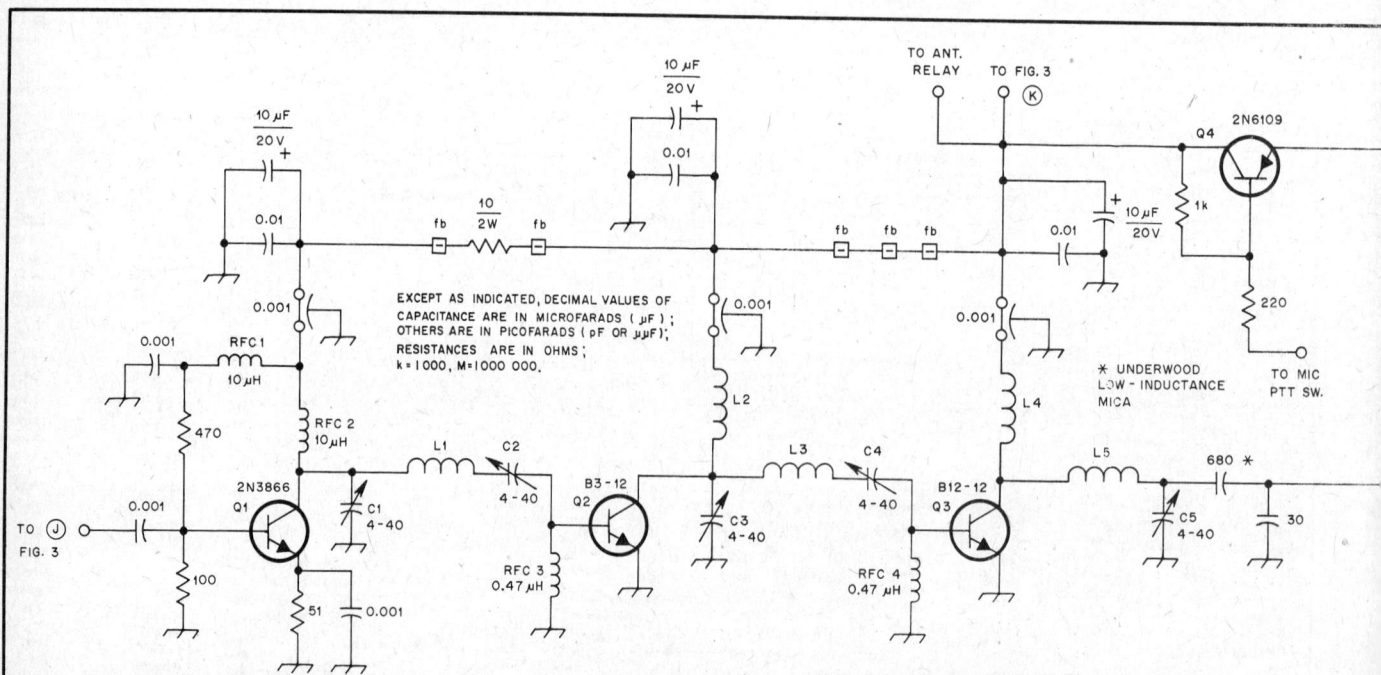

Fig. 28 — Diagram of the synthesized transmitter's amplifier stages. The ferrite beads may be any commonly available type.
C1-C5, incl. — Mica compression trimmers.
L1 — 2 turns no. 20, 1/4 in. (6 mm) ID.
L2 — 5 turns no. 20, 1/4 in. (6 mm) ID.
L3 — 1 turn no. 18, 1/4 in. (6 mm) ID.
L4 — 2-1/2 turns no. 18, 3/8 in. (9.5 mm) ID.
L5 — 1 turn no. 18, 3/8 in. (9.5 mm) ID.

reaches zero before the main counter does. Upon reaching zero, the swallow counter switches the mode of the ECL prescaler from divide-by-11 to divide-by-10 operation, and the main counter continues toward zero. When the main counter reaches zero, both counters are preset to their respective numbers and the process begins again.

If M is the number preset into the main counter and N is the number preset into the swallow counter, the total number of input cycles to the ECL prescaler is

$$P = 11N + (M - N)10$$
$$= 10M + N$$

The output frequency of a properly locked phase-locked loop is the reference frequency times the programmable divider ratio, or in this case

$$F_{out} = (10 \text{ kHz}) (10M + N)$$

The values of both M and N are selectable; that is , N goes from zero to nine and is selected by the tens-of-kHz switch, while M goes from 1440 to 1479 and is selected from the MHz and hundreds-of-kHz switches. The two most significant digits of the number M (i.e., 14) are permanently wired into the counter and are not selectable.

The Loop Reference Frequency

A second unique feature is the use of a 10-kHz loop reference frequency even though the channel spacing is 5 kHz. There are several ways of keeping all traces of the reference frequency out of

the transmitter output, one of these being the use of a carefully designed and implemented loop filter. A filter of this sort must not degrade the lockup time, capture range, lock range or loop stability. If the high-frequency capability of the loop is reduced too much, excessive noise and microphonics can become a problem. The use of a higher loop-reference frequency allows a wider loop bandwidth with the same level of reference sidebands. Quite simply, the higher the reference frequency, the better.

Normally, if a 10-kHz reference frequency were used with a phase-locked loop, only frequencies that were exact multiples of 10 kHz could be generated, which would not be acceptable for 2-meter fm operation. The method of obtaining 5-kHz resolution while using a 10-kHz reference is called "reference pulling." See Fig. 26. This "pulling" involves shifting the 10-kHz reference frequency very slightly so that the 144-MHz output frequency moves up by about 5 kHz. This technique does not produce an exact 5-kHz shift on all channels, with the greatest error being at the band edges. If the frequency shift were set at exactly 5 kHz at the center of the band, however, the error at the band edges would be 40 Hz, hardly enough to cause problems with even the fussiest repeaters.

The VCO

The last and most important feature of this transmitter is the low-microphonic VCO, Fig. 27. Designing a phase-locked loop that is to be modulated for fm use is a very difficult task. Changing the frequency of the VCO to provide modulation is exactly what the phase-locked loop is supposed to prevent — i.e., any change in frequency caused by noise, microphonics or other disturbances. It is necessary to restrict the capabilities of the loop so that it will not cancel any attempt to modulate the VCO. This implies that any modulation introduced from other sources, such as noise within the VCO circuit, from external sources or microphonics, will not be appreciably removed by the loop, either. Therefore, the VCO must be constructed to have a low level of microphonics and phase noise, and be free of modulation from noise on the power-supply lines.

To accomplish these objectives, the VCO is constructed rigidly and mounted in a cast-aluminum box. The oscillator has a small amount of inductance and a large amount of capacitance, which reduces the effect of stray capacitance from surrounding objects. The inductor itself is a small piece of coaxial cable (details appear later). The cable is totally shielded from its environment and is not affected by nearby components, as would be its wire-wound counterpart. Last, sufficient power-line filtering and a separate voltage regulator are provided for the oscillator.

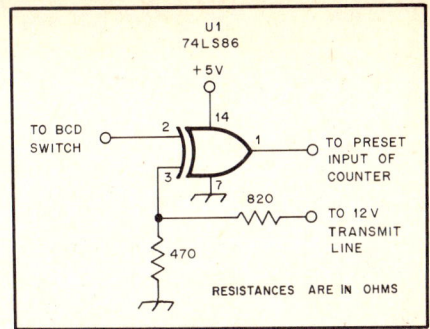

Fig. 29 — Optional transmit/receive frequency-offset circuit for the synthesized transmitter.

Construction

Fig. 28 is the diagram for the amplifier stages. The entire transmitter is mounted inside a 7-1/2 × 6 × 3-in. (190 × 150 × 76-mm) aluminum box. Only an ON-OFF switch, pilot lamp and the frequency-selector switches are mounted on the front panel. Connections to the microphone and external equipment (such as a receiver and antenna changeover relay) are made at the rear of the enclosure.

The VCO, two buffer amplifiers, PIN diode switch, and the voltage regulator are all mounted in a 1-1/2 × 2- × 4-in. (38- × 50- × 100-mm) cast-aluminum box. Since the synthesizer operates continuously, a PIN diode switch is employed to prevent energy from leaking through the power amplifier and being radiated by the antenna. There is also a potential for interference with the companion receiver when operating simplex. During repeater operation, the receiver frequency is different from the transmitter frequency, so that the low-level leakage will not be a problem. An optional offset circuit may be added (as shown in Fig. 29) to eliminate any potential interference during simplex operation. This circuit causes the synthesizer to be shifted up or down 5 kHz between transmit and receive modes. If the operating frequency ends with the digit 5, the frequency is shifted down during periods of receive, and shifted up 5 kHz if the operating frequency ends with a zero. If the receiver filter is broad, the 5-kHz shift may not be sufficient. In this case, the same circuit may be applied to the one bit of the tens-of-kHz switch instead. This will cause a 10-kHz shift to occur.

The VCO Inductor

The coaxial cable used for the VCO inductor is a piece of miniature, semi-rigid, 50-ohm cable with a tetrafluoroetheleyne (TFE) dielectric. This cable is extremely expensive and difficult to locate. The cost is not hard to bear since so little of the cable is used. If semi-rigid cable of any sort is not available, a piece of miniature, flexible cable may be substituted with a slight increase in microphonics; types

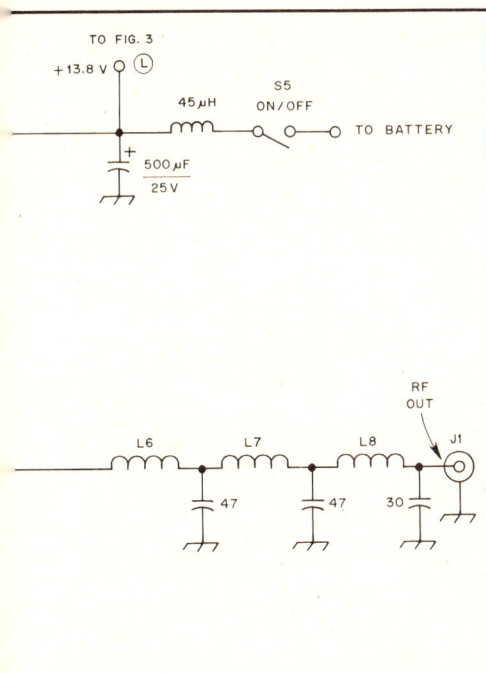

L6, L8 — 5-1/2 turns no. 18, 3/16 in. (4.8 mm) ID.
L7 — 4-1/2 turns no. 18, 3/16 in. (4.8 mm) ID.

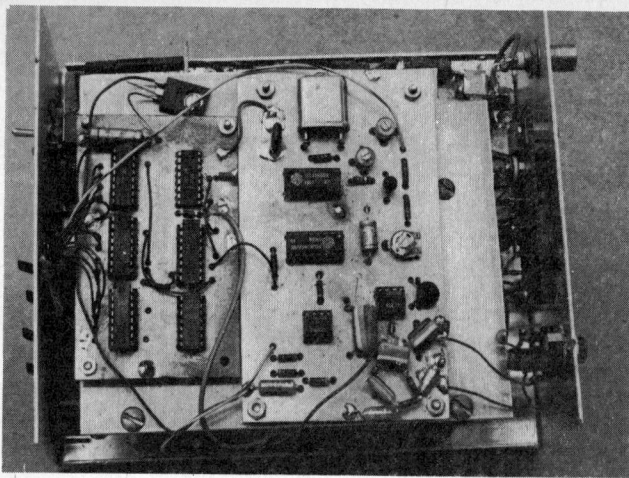

Fig. 30 — Top view of the 2-meter synthesized transmitter. The programmable divider and the reference oscillator are visible on the top shelf. The VCO and power amplifier are mounted on the bottom of the enclosure.

Fig. 31 — Side view of the transmitter showing the power amplifier. The transistors are mounted on the aluminum box for heat sinking.

which may be employed are RG-188/U or RG-143/U. A larger diameter cable may be used with excellent results. Ensure that any cable used has TFE insulation so that it may be soldered to the pc board without melting or distorting the dielectric. Both semi-rigid and flexible cables must be *firmly* soldered to the pc board to be effective.

The synthesizer components other than the VCO and the speech amplifier are mounted on two double-sided pc boards, one containing the digital circuits and the other the reference oscillator/divider and the analog circuitry.

The speech amplifier is designed for use with a low-impedance dynamic microphone. The normal clipping characteristics of the CA3160 op amp are utilized for modulation limiting. The MOSFET output of the amplifier provides a predictable clipping level for the speech amplifier. The unwanted harmonics of the clipped waveform are removed with an R-C lowpass filter. A similar op amp is used for the loop amplifier/filter.

An 11-volt regulator is included for use with the speech amplifier and the loop amplifier. It was desired to use the regulator to reduce the susceptibility of the transmitter to be modulated by input voltage noise, producing so-called "alternator whine." It is important to note that a 3-terminal regulator is not suitable at this point because of an input-output voltage differential of only 1 volt at worst. Most 3-pin regulators have a drop-out voltage of 2 volts or more.

The Crystal Oscillator and Divider

The 2.56-MHz crystal oscillator drives a CMOS ripple counter that divides the crystal frequency down to 10 kHz (see Fig. 2). A variable capacitor is switched across the crystal with a transistor (Q1). The output frequency is raised about 5

kHz by switching this capacitor out of the circuit with the 5-kHz switch. The transistor switch is connected to the "4" bit of the switch which raises the output frequency 5 kHz for numbers of 4 or greater. The switch may be mechanically prevented from going beyond the digit 5 if desired.

The Power Amplifier

A three-stage, broadly tuned power amplifier provides about 12 watts of output. The first stage operates Class A and boosts the 50-mW signal from the VCO to about 500 mW. The second stage, operating Class C, develops 3 watts of drive for the Class C final amplifier. A low-pass filter is used to remove any harmonic energy appearing at the output. All bypassing and tuning components throughout the power amplifier are specially selected for low inductance. Failure to use these special components will result in instabilities and possible component destruction. No pc-board pattern was produced. Instead, small islands were cut into double-sided pc-board material with a sharp knife and the copper peeled away.

Switching Circuitry

A solid-state switching circuit is used for switching the PIN diode and supplying the Vcc to the power amplifier stage. In addition, a terminal is provided at the back of the transmitter cabinet for use with an external changeover relay and receiver muting.

Check Out and Tune-Up

After the unit has been wired and checked to ensure correct assembly, power may be applied. Before depressing the PTT switch, check for overheating components or other unwanted symptoms. Once it has been determined that no

catastrophic errors exist, adjust capacitor C1 while watching the VCO control voltage at point H (see Fig. 3). If the loop is locked, the voltage at H should change with the capacitor setting. With the frequency-selector switches set to 146 MHz, set C1 for a VCO control voltage of 5.5 volts. Check to see if the synthesizer is on the proper frequency by means of a frequency counter or by listening for the synthesizer signal with a 2-meter receiver. The frequency counter should be coupled to the VCO by removing the coaxial lead to the power amplifier and keying the mike PTT switch. While monitoring the synthesizer frequency and the VCO control voltage at point H, set the frequency-selector switches to 144 Mhz and 147.99 MHz, and determine that the loop remains locked. Set the switches to 146.005 MHz and adjust C26 for a counter reading of exactly 146.005 MHz. Move the switches to the 146.000 MHz position and adjust C25 for that frequency-counter reading.

Disconnect the counter and reconnect the power amplifier. With the frequency selector at 146.000 MHz, adjust C20 through C24, inclusive, for maximum power output. A wattmeter may be used as an indicator of power output. In the early stages of tuning up, the power output will be quite low, so careful attention must be paid to the wattmeter readings. The final power output should be on the order of 12 watts with an input current of approximately 2.5 A.

Microphone Gain and Modulation Levels

Set the mike gain and modulation levels with the aid of a deviation meter if possible. Without the aid of such a device, a cut-and-try approach may be used by asking for reports on the air or by using a receiver and comparing the modulation level of the transmitter to that of other stations heard. The microphone-gain potentiometer may be set so that clipping occurs only on loud voice peaks and so that normal speech produces peaks of about 10 volts peak-to-peak. The actual deviation is set by the value of R_A which can be anywhere between 0 and 15 kΩ.

Future Considerations

The pulse-swallowing synthesizer is especially suited to receiver applications because of the ease with which the i-f off-set can be programmed. A matching receiver could be constructed using a similar synthesizer and a VCO in the 133.3- to 137.3-MHz range for low-side injection. The receiver could be a simple, single-conversion affair with a 10.7-MHz i-f and a monolithic crystal filter.

Other dual-modulus schemes can be used such as 20/21, 40/41 and 63/64. The major advantage of the higher divisors is that higher frequencies (even into the uhf region) can be brought down to the frequency range of standard TTL devices. It is entirely possible to build a synthesized 1-GHz fm transceiver with only a few more components than are used in this 2-meter synthesized unit.

2-METER SOLID-STATE FM POWER AMPLIFIER

The majority of the commercially made 2-meter fm transceivers available today have rf power-output levels of 1 to 15 watts. There are many occasions when an fm operator would like to have a little more power to be able to work over greater distances. Described here is a 50-watt output amplifier for the 2-meter band. This amplifier makes use of a single transistor and operates directly from a 13.6-volt vehicular electrical system.

Circuit Description

The amplifier circuit shown in Fig. 33 utilizes a single 2N6084 transistor operated in a Class C, zero-bias configuration. This mode of operation has the advantages of high collector efficiency at full

Fig. 32 — An end view of the breadboard version of the 50-watt 2-meter amplifier. The input circuit is at the lower right, and the output network is at the upper left.

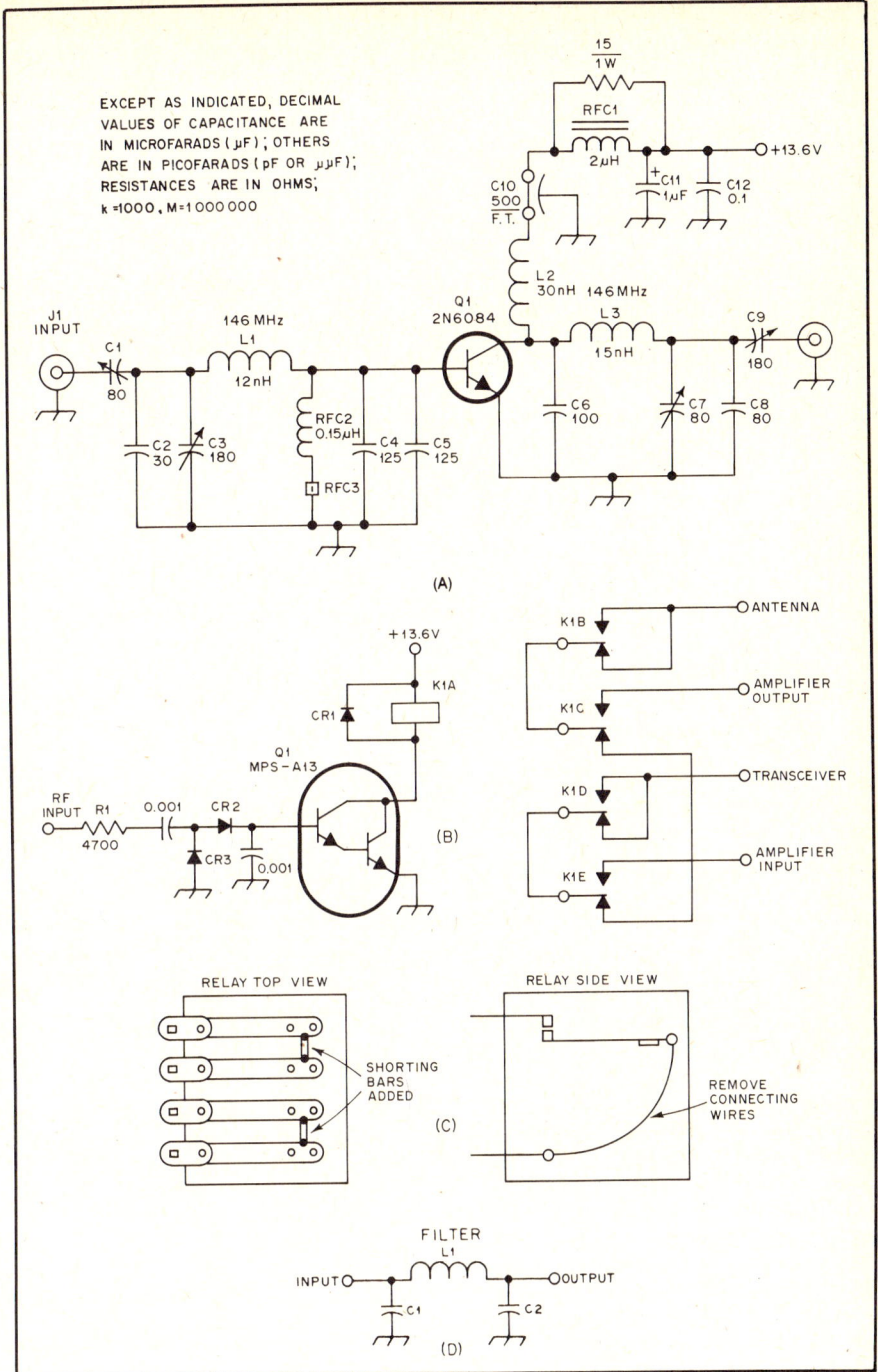

Fig. 33 — (A) Diagram of the amplifier which provides 40 to 50 watts output. Capacitors are mica unless otherwise noted. The heat sink is a Thermalloy 6169B, Allied Electronics no. 957-2890. (B) COR circuit. Capacitors are disc ceramic. (C) The COR relay is modified by removing the connecting wires from all four wiper arms and adding two shorting bars, as shown. Only the stationary-contact connections are used. (D) Pi-section output filter, C1 and C2 are 39-pF mica capacitors, Elemenco 6ED3900J03 or equiv. and L1 consists of 2 turns of no. 18 tinned wire, 1/4 inch ID, 0.2 inch (61 × 5 mm) long (approximately 44 nH).

C1, C7 — 5- to 80-pF compression trimmer, Arco 462 or equiv.
C2, C4, C5, C6, C8 — Mica button, Underwood J-101.
C3, C9 — 9- to 180-pF compression trimmer, Arco 463 or equiv.
C10 — Feedthrough type.
C11 — Tantalum.
C12 — Ceramic disc.
D1 — 100-PRV or more, 500-mA or more silicon diode (Motorola 1N4001 or equiv.).
D2, D3 — High-speed, low-capacitance 100-PRV silicon diode (Motorola MSD7000 dual package used here).
J1, J2 — Coaxial connector, panel mount.
K1 — 4pdt open-frame relay, 12-V contacts

(Comar CRD-1603-4S35 or equiv., Sigma 67R4-12D also suitable), modified as described above.
L1 — 12 nH, no. 10 tinned wire, 1-1/4-inch (32-mm) long straight conductor.
L2 — 30 nH, 1-3/4 turns, no. 10 tinned wire, 3/8-inch ID, 3/4 inch (10 × 19 mm) long.
L3 — 15 nH, no. 14 tunned wire, 3/4-inch (19-mm) long straight conductor.
L4 — 2 turns of no. 18 tinned wire 1/4-inch ID, 0.2-inch (6 × 5 mm) long (approximately 44 nH).
Q1 — Motorola silicon power transistor.
Q2 — Npn silicon Darlington transistor, H_{FE} of 5000 or more, Motorola MPS-A13 or equiv.

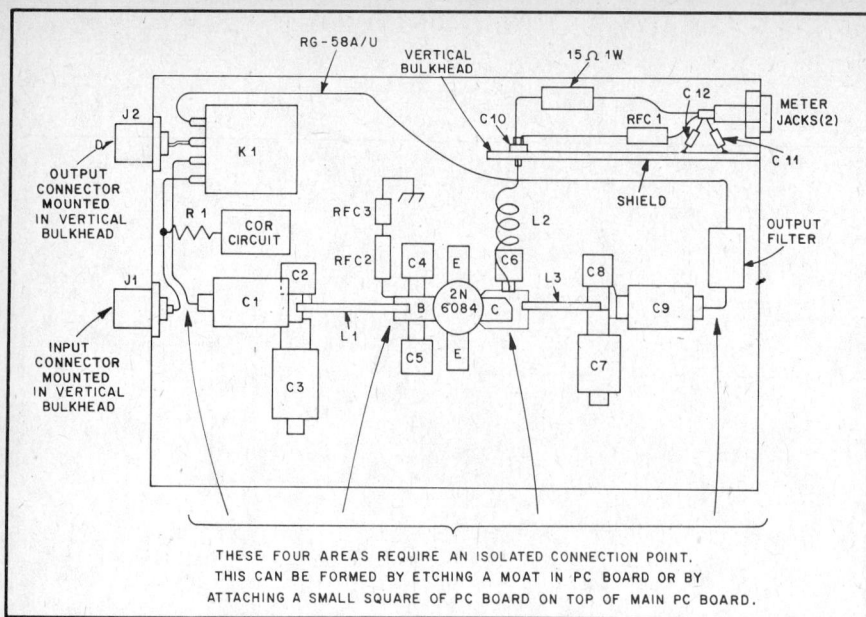

Fig. 34 — Parts-layout diagram for the 50-watt amplifier (not to scale). A 4 × 6-inch (102 × 152-mm) pc board is used as the base.

output and zero dc current drain when no rf driving signal is applied. The reader should note that zero-bias operation yields an amplifier that is not "linear." It operates Class C and is designed for fm or cw operation only; it would produce objectional distortion and splatter if used to amplify a-m or ssb signals.

The amplifier operates directly from an automobile electrical system, so no additional power supply is required for mobile operation. The input and output tuned circuits are designed to match the impedances of the transistor to a 50-ohm driving source and to a 50-ohm antenna system, respectively. Since both the input and output impedances of the transistor are extremely low (in the 1- to 5-ohm region), the matching networks employed are somewhat different than those used with tubes. The networks chosen for the amplifier are optimized for low-impedance matching.

The elaborate decoupling network used in the collector dc feed is for the purpose of assuring amplifier stability with a wide variety of loads and tuning conditions. The 2N6084 transistor is conservatively rated at 40 watts output (approximately 60 watts dc input). The amplifier can be driven to power-output levels considerably higher than 40 watts, but it is recommended that it be kept below 50 watts output. If the transmitter or transceiver has more than 10 watts of output, an attenuator should be used at the amplifier input to keep the power output below 50 watts.

Construction Details

The usual precautions for building a solid-state amplifier are followed. These include proper mechanical mounting of the transistor, emitter grounding, heat sinking, and decoupling of the supply voltage leads. The fixed-value mica capacitors, Underwood[1] type J-101, are special mica units designed for high-frequency applications. The core for RFC1 and the rf bead used for RFC3 are Ferroxcube products.[2]

The amplifier is constructed on a pc board that is bolted to a heat sink. A few islands can be etched on the board for tie points. A complex foil pattern is not required. In the amplifier shown in the photograph and pictorial layout (Figs. 32 and 34) islands were etched only for input and output tie points. Circuit-board islands may also be etched for the transistor base and collector leads. However, an interesting alternative method was used in the author's breadboard amplifier. The base and collector islands were formed by attaching small pieces of pc board to the top of the main board. This procedure added a few tenths of a pF of capacitance at the connection points, so if you choose to etch islands directly on the main board you may want to increase the value of C6 slightly. (The values of C4 and C5 are not critical.)

A word about the care of a stud-mount rf power transistor: Two of the most important mounting precautions are (1) to assure that there is no upward pressure (in the direction of the ceramic cap) applied to the leads, and (2) that the nut on the mounting stud is not overtightened. The way to accomplish item 1 is to install the nuts *first* and solder the leads to the circuit later. For item 2, the recommended stud torque is 6 inch-pounds. For those who don't have a torque wrench in the shack, remember that it is better to undertighten than to overtighten the mounting nut.

The transistor stud is mounted through a hole drilled in the heat sink. A thermal compound, such as Dow Corning 340 heat-sink grease, should be used to decrease the thermal resistance from transistor case to heat sink. See the excellent article by White in April 1971 *QST* for details of heat-sink design.

Series impedance in the emitter circuit can drastically reduce the gain of the amplifier. Both transistor emitter leads should be grounded as close to the transistor body as is practical.

The wiring for the dc voltage feeder to the collector should have extremely low dc resistance. Even a drop of 1 volt can significantly reduce the power output of the amplifier. A good goal is less than 0.5-volt drop from the car battery to the transistor collector. With operating currents of several amperes, the total dc resistance should be only a fraction of an ohm. A standard commercially made heat sink is used for the 50-watt amplifier, and it is adequate for amateur communications. Forced-air cooling across the heat sink should be used for any application requiring longterm key-down operation at 40 watts or more of output.

Tune-Up Procedure

Generally, the best way to tune a transistor power amplifier is for maximum rf power output. If this approach results in exceeding the power ratings of the transistor, then the power output should be reduced by reducing the drive level, not detuning the final. In the case of an outboard PA stage, such as decribed here, both the input and output networks can be tuned for maximum rf output, if the driving source has an output impedance of approximately 50 ohms. However, a better procedure consists of tuning the output tank circuit for maximum rf output and tuning the input circuit for minimum SWR as measured between the exciter and the final amplifier. This tune-up procedure has the added advantage of assuring that the amplifier presents a 50-ohm load to the exciter. A dc ammeter to check collector current is a useful tune-up aid. Since tuning is for peak output, a Monimatch-type SWR bridge is adequate for the job. The best tuning procedure is to monitor simultaneously both output power (absolute or relative) and the SWR between the exciter and amplifier.

First, apply dc voltage with no rf drive. No collector current should flow. Then apply a low level of rf drive — perhaps 25 percent or less of the rated 10 watts maximum drive — and tune the input network for maximum indicated collector current. The networks may not tune to resonance at this low drive level, but you should at least get an indication of proper operation by smooth tuning and lack of any erratic behavior in the collector-current reading.

Fig. 35 — Photograph of the completed 430- to 450-MHz amplifier.

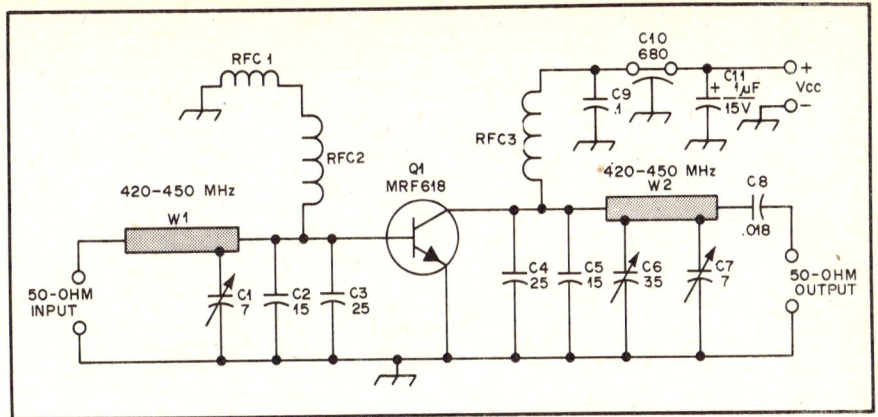

Fig. 36 — Schematic diagram of the 15-watt, 430- to 450-MHz amplifier.

C1, C7 — 0.9- to 7-pF mica compression trimmer, Arco 400.
C2, C5 — Unelco 15-pF mica.
C3, C4 — Unelco 25-pF mica.
C6 — 3- to 35-pF mica compression trimmer, Arco 403.
C8 — 0.018 μF chip capacitor. ATC or equiv. (a 250-pF Unelco mica or a 0.001-μF Erie Redcap may work as a substitute).
C9 — 0.1-μF disc ceramic.

C10 — 680-pF feedthrough capacitor.
C11 — 1-μF, 15-V tantalum.
L1, L4 — 50-ohm microstrip line, 2.3 inches long, 0.110-inch wide.
RFC1 — Ferrite bead on cold lead of L2.
RFC2 — 8 turns no. 22 enam., 1/8-inch ID, close wound.
RFC3 — 4 turns no. 22 enam., 1/4-inch ID, close wound.

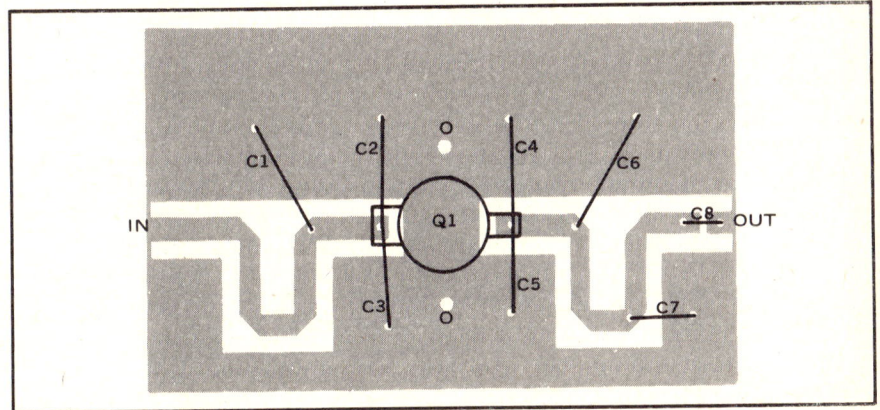

Fig. 37 — A parts-placement guide for the 430- to 450-MHz amplifier board. Be sure to provide rf-connecting paths between the top and bottom ground surfaces, as explained in the text.

Gradually increase the drive until full rated output is reached.

A SOLID-STATE PA FOR 440 MHz

Whether the application is by a person using a hand-held transceiver as a mobile rig or by an experimenter building a repeater, a medium-power amplifier would come in quite handy for increasing the effective range of his station. This article describes a compact, inexpensive, 10-dB gain power amplifier which is simple enough for nearly any experimenter to build. This circuit originally appeared in February 1977 QST in an article by Olsen, WA7CNP.

Circuit Analysis

The circuit employed, Fig. 36, is essentially a basic narrow-band amplifier capable of being tuned over a broad range of frequencies — 430 to 450 MHz. Input-match and collector-load transformations are accomplished by using multiple L sections comprised of 50-ohm microstrip-line and mica-compression variable capacitors. The active device is the Motorola MRF618 — an internally matched, 12.5-volt, controlled-Q transistor designed for application from 420 to 512 MHz.

Construction

The amplifier is built on double-sided G10 glass-epoxy board. Fig. 37 is a 1:1 drawing of the board layout. Care should be taken in etching to maintain the line width of the microstrip at 0.110 inch for a Z_o of 50 ohms.

After the board has been etched, the first step is to cut the hole in the board for the transistor heat sink (flange). The transistor flange can be used as a stencil for laying out the hole. When the hole has been formed, the next thing is to ensure that a good rf path is continuous from the ground plane on one side to that on the other. Connections can be made with copper or brass eyelets crimped and soldered to both sides of the board. If no eyelets are available, these connections may be made by drilling a no. 50 hole through the board, inserting a piece of no. 18 wire

through the hole, and soldering and trimming both sides flush with the board. Be sure that there is one such connection made under each Unelco capacitor and alongside the microstrip line and dc feed point.

Next, the Unelco capacitors are mounted as closely to the transistor package as possible, and at the same time they double as mounting surfaces for the transistor emitter leads. Connections to the input and output lines may be made with 2- to 5-mil copper strap or foil.

The transistor can be mounted at this time. The holes in the transistor flange are made to clear no. 4-40 screws. Drill and tap two no. 42 holes in the heat sink, using the flange as a drill guide. Next, clean the heat sink and bottom of the transistor flange so that foreign matter will not prevent the transistor from seating properly on the heat sink. Apply a very small amount of thermal compound to the

flange and bolt it firmly to the heat sink. The transistor leads may then be soldered to the circuit.

The base-return choke and dc collector-feed circuit may now be put into place. C10 may easily be installed by drilling a 0.192-inch hole in a small copper strap (0.02-inch thick) and then putting a right-angle bend in the strap so that the capacitor can be mounted upright above the board. Be sure to make the base return and dc-feed connections as close to the transistor package as possible.

Now comes the remaining tuning elements and output dc-isolation capacitor. Fig. 37 shows the mounting position. Care should be taken to solder the ground taps of the trimmer to the ground plane to reduce the amount of lead inductance inherent in these capacitors at uhf. C8 should be mounted after cutting a small break in the output line near the end. This capacitor need not be a 0.018-μF chip, but

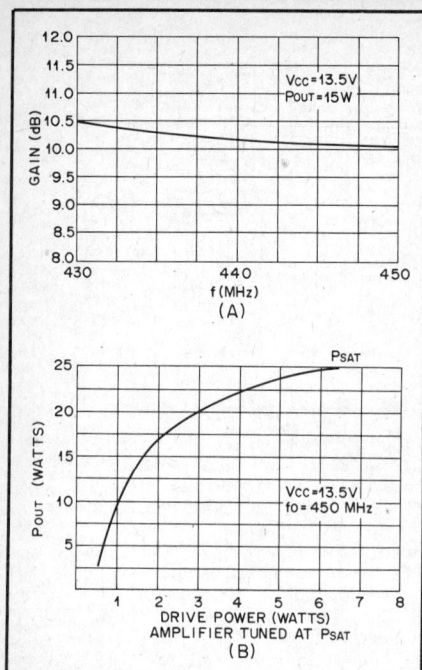

Fig. 38 — These gain and output-power graphs show what performance can be expected from an amplifier utilizing the MRF618.

Fig. 39 — This typical 144-MHz amateur repeater uses GE Progress-Line transmitter and receiver decks. Power supplies and metering circuits have been added. The receiver located on the middle deck is a 440-MHz control receiver, also a surplus GE unit. A preamplifier, similar to that shown in Fig. 30, has been added to the 2-meter receiver to improve the sensitivity so that a 0.2-μV input signal will produce 20 dB of quieting.

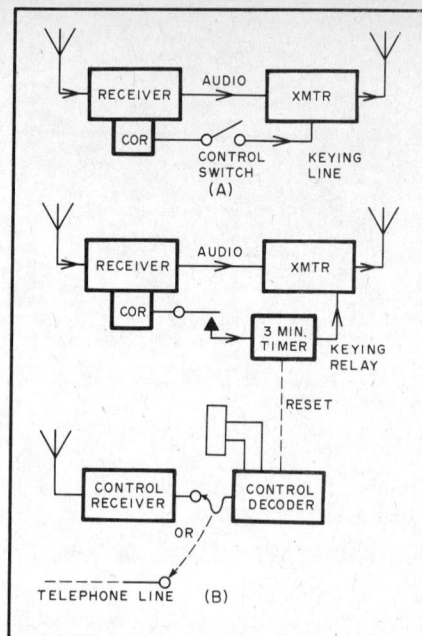

Fig. 40 — Simple repeaters. The system at A is for local control; remote control is shown at B.

care should be taken in choosing a substitute that will not grossly affect the output load characteristic of the circuit (see parts list). Depending upon your application, any 50-ohm outside-world connection may be used, ranging from a piece of coax to RCA phono plugs.

Tune-Up

Tuning is simple: Apply low power (about 3/4 watt) to the input and tune the input capacitor until a small amount of collector current begins to flow. Then tune the output capacitors for peak output. Switch back and forth between input and output, and tune until the desired operating conditions are achieved. Fig. 38 shows examples of typical data taken in the lab. You will find that operating frequencies greater than 1 MHz away from the tune-up frequency can be used without the necessity for further adjustment.

REPEATERS

A repeater is a device which retransmits received signals in order to provide improved communications range and coverage. This communications enhancement is possible because the repeater can be located at an elevated site which has coverage that is superior to that obtained by most stations. A major improvement is usually found when a repeater is used between vhf mobile stations, which normally are severely limited by their low antenna heights and resulting short communications range. This is especially true where rough terrain exists.

The simplest repeater consists of a re-

ceiver with its audio output directly connected to the audio input of an associated transmitter tuned to a second frequency. But, certain additional features are required to produce a workable repeater. These are shown in Fig. 40A. The "COR" or carrier-operated relay is a device connected to the receiver squelch circuit which provides a relay contact closure to key the transmitter when an input signal of adequate strength is present. As all amateur transmissions require a licensed operator to control the emissions, a "control" switch is provided in the keying path so that the operator can exercise his duties. This repeater, as shown, is suitable for installation where an operator is present, such as the home of a local amateur with a superior location, and would require no special licensing under existing rules.

In the case of a repeater located where no licensed operator is available, provisions must be made to control the equipment over a telephone line or a radio circuit on 220 MHz or higher. Fig. 40B shows the simplest system of this type. The control decoder may be variously designed to respond to simple audio tones, dial pulsed tones, or even "Touch-Tone" signals. If a leased telephone line with dc continuity is used, control voltages may be sent directly, requiring no decoder. A three-minute timer to disable the repeater transmitter is provided for fail-safe operation. This timer resets during pauses between transmissions and does not interfere with normal communications. The system just outlined is

suitable where all operation is to be through the repeater and where the frequencies to be used have no other activity.

Remote Base Stations

The remote base, like the repeater, utilizes a superior location for transmission and reception, but is basically a simplex device. That is, it transmits and receives on a single frequency in order to communicate with other stations also operating on that frequency. The operator of the remote base listens to his hilltop receiver and keys his hilltop transmitter over his 220-MHz or higher control channels (or telephone line). Fig. 41A shows such a system. Control and keying features have been omitted for clarity. In some areas of high activity, repeaters have all but disappeared in favor of remote bases because of the interference to simplex activity caused by repeaters unable to monitor their output frequency from the transmitter location.

A Complete System

Fig. 41B shows a repeater that combines the best features of the simple repeater and the remote base. Again, necessary control and keying features have not been shown in order to simplify the drawing, and make it easier to follow. This repeater is compatible with simplex operation on the output frequency because the operator in control monitors the output frequency from a receiver at the repeater site between transmissions. The control operator may also operate the system as a remote base. This type of

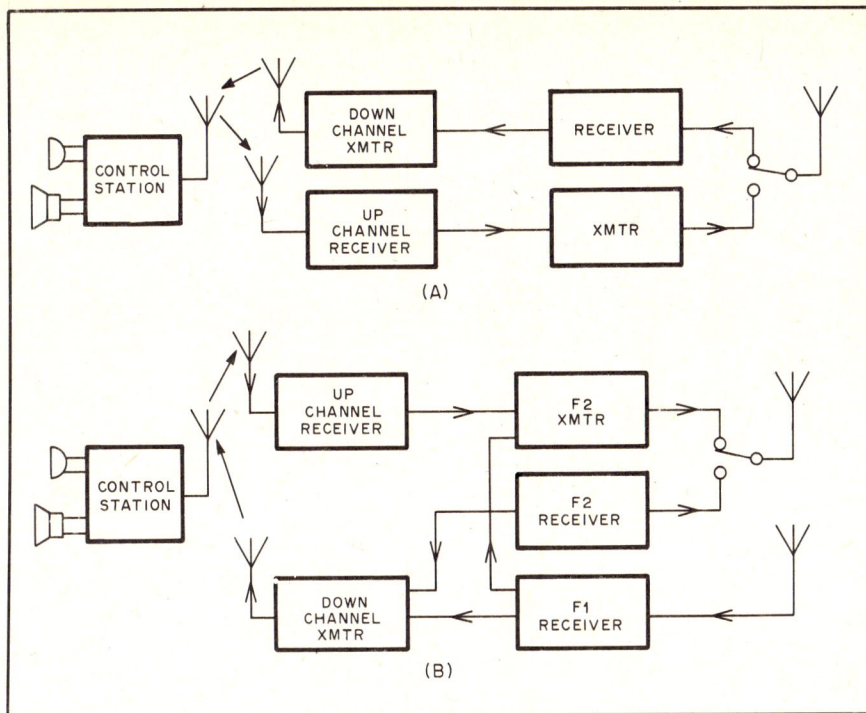

Fig. 41 — A remote base is shown at A. A repeater with remote-base operating capability is shown at B. Control and keying circuits are not shown. Telephone-line control may be substituted for the radio-control channels shown.

Table 1
EIA Standard Subaudible Tone Frequencies

Reed	Freq. (Hz)	Reed	Freq. (Hz)
L1	67.0	2A	114.8
WZ	69.3	2B	118.8
L2	71.9	3	123.0
WA	74.4	3B	131.8
L3	77.0	4	136.5
WB	79.7	4A	141.3
L4	82.5	4B	146.2
YA	85.4	5	151.4
L4A	88.5	5A	156.7
ZZ	91.5	5B	162.2
L5	94.8	6	167.9
1	100.0	6A	173.8
1A	103.5	6B	179.9
1B	107.2	7	186.2
2	110.9	7A	192.8

has shaped the voice audio. Table 1 lists the EIA-standard frequencies.

Practical Repeater Circuits

Because of their proven reliability, commercially made transmitter and receiver decks are generally used in repeater installations. Units designed for repeater or duplex service are preferred because they have the extra shielding and filtering necessary to hold mutual interference to a minimum when both the receiver and transmitter are operated simultaneously.

Wide-band noise produced by the transmitter is a major factor in the design of any repeater. The use of high-Q tuned circuits between each stage of the transmitter, plus shielding and filtering throughout the repeater installation, will hold the wideband noise to approximately 80 dB below the output carrier. However, this is not sufficient to prevent *desensitization* — the reduction in sensitivity of the receiver caused by noise or rf overload from the nearby transmitter — if the antennas for the two units are placed physically close together.

Desensitization can easily be checked by monitoring the limiter current of the receiver with the transmitter switched off, then on. If the limiter current increases when the transmitter is turned on, then the problem is present. Only physical isolation of the antennas or the use of high-Q tuned cavities in the transmitter and receiver antenna feedline will improve the situation.

Antenna Considerations

The ultimate answer to the problem of receiver desensing is to locate the repeater transmitter a mile or more away from the receiver. The two can be interconnected by telephone line or uhf link. Another effective approach is to use a single antenna with a *duplexer,* a device that provides up to 120 dB of isolation between the transmitter and receiver. High-Q cavities in the duplexer prevent transmitted signal energy and wideband noise from degrading the sensitivity of the

system is almost mandatory for operation on one of the national calling frequencies, such as 146.52 MHz, because it minimizes interference to simplex operation and permits simplex communications through the system with passing mobiles who may not have facilities for the repeater-input frequency.

The audio interface between the repeater receivers and transmitters can, with some equipment, consist of a direct connection bridging the transmitter microphone inputs across the receiver speaker outputs. This is not recommended, however, because of the degradation of the audio quality in the receiver-output stages. A cathode follower connected to each receiver's first squelch-controlled audio amplifier stage provides the best results. A repeater should maintain a flat response across its audio passband to maintain the repeater intelligibility at the same level as direct transmissions. There should be no noticeable difference between repeated and direct transmissions. The intelligibility of some repeaters suffers because of improper level settings which cause excessive clipping distortion. The clipper in the repeater transmitter should be set for the maximum system deviation, 5 kHz, usually. Then the receiver level driving the transmitter should be set by applying an input signal of known deviation below the maximum, and adjusting the receiver audio gain to produce the same deviation at the repeater output. Signals will then be repeated linearly up to the maximum desired deviation. The only incoming signal that

should be clipped in a properly adjusted repeater is an overdeviated signal.

The choice of repeater input and output frequencies must be carefully made. In general, check with the appropriate volunteer frequency coordinator, who is listed in the *ARRL Repeater Directory,* since about 4000 repeaters are operating across the U.S. and Canada. Some 10-meter repeaters are operational, and most use 100-kHz separation. A popular arrangement on 6 meters uses 52.525 MHz as either the input or output, with several choices for the other half. Many stations, however, are moving toward either a 600-kHz or 1-MHz offset. On 2 meters the standard is 600 kHz. The 220-MHz band uses 1.6-MHz separation. On the 450-MHz band it is 5 MHz. The choice and usage is a matter for local agreement.

In some cases where there is overlapping geographical coverage of repeaters using the same frequencies, special methods for selecting the desired repeater have been employed. One technique requires the user to transmit automatically a 0.5-second burst of a specific audio tone at the start of each transmission. Different tones are used to select different repeaters. Standard tone frequencies are 1800, 1950, 2100, 2250 and 2400 Hz.

Because of growing congestion among same- and adjacent-channel repeaters, an increasingly popular access method is a continuous subaudible tone. Popularly known as PL (Motorola trademark for Private Line), it must be applied to a transmitter after the clipper/filter stage

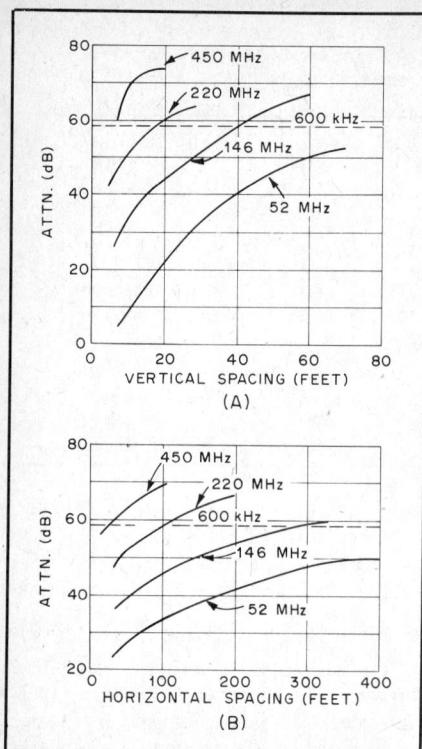

Fig. 42 — Charts to calculate the amount of isolation achieved by (A) vertical and (B) horizontal spacing of repeater antennas. If 600-kHz separation between the transmitted and received frequencies is used, approximately 58-dB attenuation (indicated by the dotted line) will be needed. (Feet × 0.3048 = meters.)

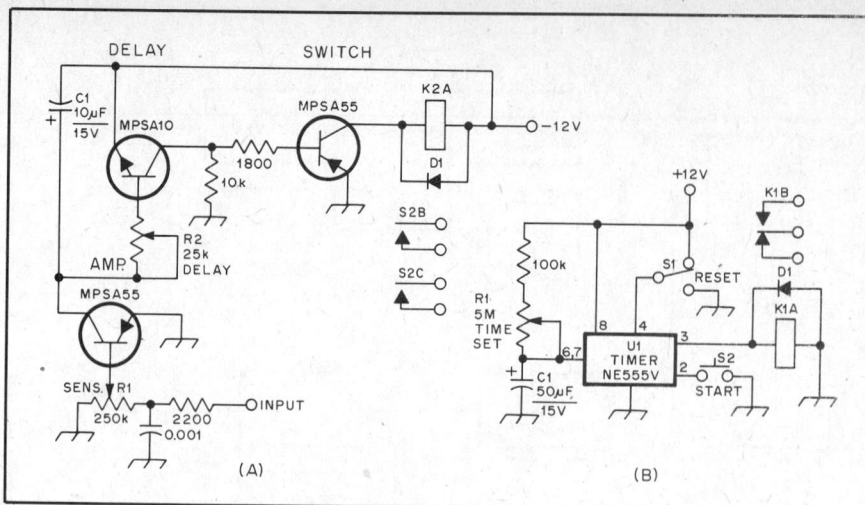

Fig. 43 — (A) COR circuit for repeater use. R2 sets the length of time that K1 will stay closed after the input voltage disappears. K1 may be any relay with a 12-volt coil, although the long-life reed type is preferred. D1 is a silicon diode. (B) Timer circuit using a Signetics NE555. R1, C1 set the timer range. C1 should be a low-leakage type capacitor. S1, S2 could have their contacts paralleled by the receiver COR for automatic *START* and *RESET* controlled by an incoming signal.

receiver, even though the transmitter and receiver are operating on a single antenna simultaneously. A commercially made duplexer is very expensive, and constructing a unit requires extensive metalworking equipment and test facilities.

If two antennas are used at a single site, a minimum spacing of the two antennas is required to prevent desensing. Fig. 42 indicates the spacing necessary for repeaters operating in the 50-, 144-, 220- and 420-MHz bands. An examination of Fig. 42 will show that vertical spacing is far more effective than is horizontal separation for vertically polarized antennas. The chart assumes unity-gain antennas will be used. If some type of gain antenna is employed the pattern of the antennas will be a modifying factor.

Control

Two connections are needed between the repeater receiver and transmitter, audio and transmitter control. The audio should be fed through an impedance-matching network to ensure that the receiver output circuit has a constant load while the transmitter receives the proper input impedance. Filters limiting the audio response to the 300- to 3000-Hz band are desirable, and with some gear an audio-compensation network may be required. A typical COR (carrier-operated relay) circuit is shown in Fig. 43A. This

unit may be operated by the grid current of a tube limiter or the dc output of the noise detector in a solid-state receiver.

Normally, a repeater is given a "tail"; a timer holds the repeater transmitter on for a few seconds after the input signal disappears. This delay prevents the repeater from being keyed on and off by a rapidly fading signal. Other timers keep each transmission to less than three minutes duration (an FCC requirement), turn on identification, and control logging functions. A simple timer circuit is shown in Fig. 43B.

Touch-Tone Control

From the inception of automatic dialing, signaling from telephone instruments was accomplished using dc pulses. This signaling method required direct wired connections, as a dc path was needed. For transmission via a radio circuit, the dc pulses had to be converted to a keyed audio tone. In the early 1960s the Bell Telephone Companies introduced a new, faster tone-coded dialing system which was given the registered trade name *Touch-Tone*. Because the tone signals of the Touch-Tone system could be transmitted over any audio carrier or radio circuit, many amateurs have adopted the telephone-company system for control of fm remote-base stations and repeaters.

Because two tones are used for each function in the Touch-Tone system, reliability is excellent even when used on radio circuits that are noisy or fading. Another factor that has made Touch-Tone popular with repeater groups is that many use autopatch connections to the public telephone network. By ordering a Touch-Tone line for the repeater autopatch, the same encoders and decoders

Table 2
Touch-Tone Audio Frequencies

Low Tone (Hz)	High Tone			
	1209 Hz	1336 Hz	1477 Hz	1633 Hz
697	1	2	3	F_o
770	4	5	6	F
852	7	8	9	I
941	*	0	#	P

Fig. 44 — This Western Electric Touch-Tone encoder has been mounted in a 4 × 4 × 2-inch (102 × 102 × 51-mm) utility box (Bud AU-1083). Encoders are sold by most telephone supply houses, including Telephone Equipment Co., P. O. Box 596, Leesburg, FL 32748. Tel. 904-728-2730.

can be used for the phone patch and repeater control.

Encoders

Touch-Tone information is coded in tone pairs, using two of eight possible tones for digits zero through nine and six special functions. The audio frequencies used are given in Table 2. The tones are divided into the low group, 697, 770, 852 and 941 Hz; and the high group, 1209, 1336, 1477 and 1633 Hz. One tone from each group is used for each function. For residential and business telephones, a 12-button encoder pad consisting of digits zero through nine and symbols pound # and star * are employed. A typical encoder is shown in Fig. 44, and the connections for pads manufactured by Western Electric and Automatic Electric are shown in Fig. 45. The telephone pads will work with as little as nine volts or as much as 24 volts dc applied. Either high- or low-impedance output may be employed, as shown in Figs. 45C and D.

A circuit diagram of a typical telephone-company pad is given in Fig. 46. Individual models will vary slightly, but the basic circuit used in all models is the same. A single transistor produces two tones. Two LC circuits are used, one for the high tone group and one for the low tones. Some people are bothered by the use of a single transistor to generate two audio frequencies, so the lower tone can be considered the frequency of oscillation while the high tone is called a parasitic oscillation, for purposes of explanation.

A HOMEMADE TOUCH-TONE ENCODER

To be compatible with all repeaters and telephone systems, a Touch-Tone signal must be accurate and stable in frequency, and have a nearly sinusoidal waveform. Simpler encoders than the one described here can be built, but they will not provide its high performance.

This encoder features internal voltage regulation, allowing power to be taken from the rig it is used with; there is no need to depend on separate batteries for power. When a tone pair is selected by pressing the keyboard switch, the transmitter is automatically keyed. When the key switch is released, a delay timer keeps the transmitter on long enough for the next tone pair to be selected. It's no longer necessary to hang onto the push-to-talk switch while fumbling with the Touch-Tone pad, and there are no squelch tails between digits. It has a low-impedance audio output which is electronically disconnected from the transmitter audio system when no keyboard switches are pressed. The encoder may be connected to the mic input of transceivers having either high- or low-impedance mike inputs — with negligible loading of the transmitter audio circuitry. The audio frequencies are crystal-controlled, meaning there is no drift. This circuit was originally described by Hejhall, K7QWR, in the February 1979 issue of *QST*.

Theory of Operation

Fig. 47 is a schematic diagram of the encoder. Tone generation is performed by U1, a CMOS IC. High-frequency tones from pin 15 are mixed with their low-

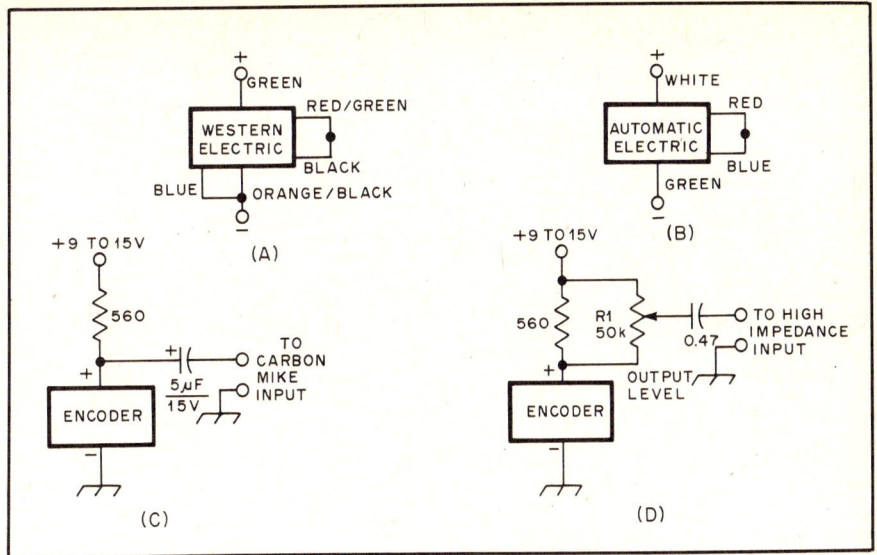

Fig. 45 — Typical connections for the encoders manufactured by Western Electric (A) and Automatic Electric (B). If low-impedance output is needed to drive a carbon-microphone input, the circuit at C can be employed for either encoder. Likewise, the circuit at D will provide a high-impedance output. R1 can be any miniature composition control; the types made for mounting on circuit boards are ideal.

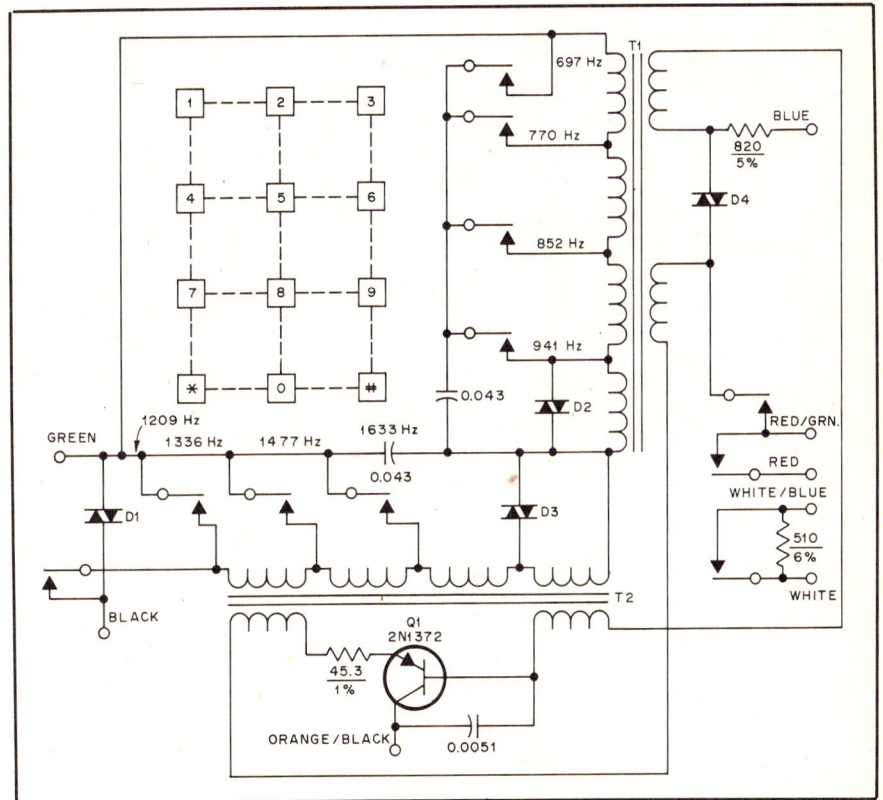

Fig. 46 — Diagram of a typical Western Electric Touch-Tone generator. T1 and T2 are special multi-winding transformers manufactured by Sangamo Electric and others. D1-D4, incl. are silicon varistors.

frequency counterparts from pin 2, and passed through the level control, R1, before reaching emitter follower Q1. Q1 performs an impedance transformation, providing the low-impedance output mentioned previously. Q2, Q4 and Q5 are used as switches. Q2 forces the audio-output impedance high when no keyboard switches are depressed, preventing the en-

coder from loading the transmitter mike input. Q4 and Q5 are operated as a Darlington pair, keying the transmitter push-to-talk (PTT) line when a keyboard switch is pressed. A single-package Darlington pair was originally used in this application, but its saturated collector voltage was high enough to prevent transmitter keying in some transceivers. Substituting

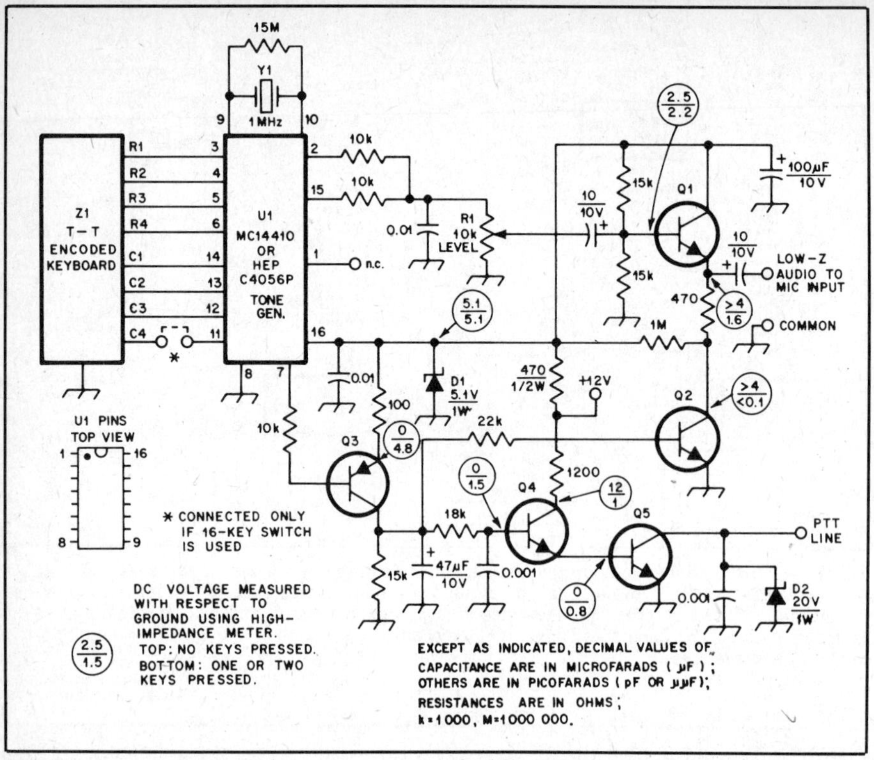

Fig. 47 — Schematic diagram of the K7QWR Touch-Tone encoder. Any properly encoded keyboard may be used with this circuit, but the units specified will plug directly into a row of Molex pins soldered to the circuit board. If the encoder is constructed on a printed-circuit board there should be no difficulties. Should you experience problems, voltage levels at various points in the circuit are included on the schematic diagram.

D1 — 5.1-volt, 400-mW Zener diode, 1N4733, HEP Z0406 or equivalent.
D2 — 20-volt, 1-watt Zener diode, 1N4747, HEP Z0421 or equivalent.
Q1, Q2, Q4 — Silicon npn transistor, 2N4123, HEP S0036 or equivalent.
Q3 — Silicon pnp transistor, 2N4125, HEP S0037 or equivalent.
Q5 — Silicon npn transistor, 2N4401, HEP S0015 or equivalent.
R1 — Circuit-board-mounted trimmer potentiometer, 10 kΩ, linear taper.
U1 — Integrated-circuit Touch-Tone encoder, Motorola MC14410 or HEP C4056P.
Y1 — 1-MHz crystal in HC-18/U holder. Frequency tolerance is 0.1 percent; series re-

sistance and load capacitance are typically 540 Ω and 7 pF, respectively. Available from Data Signal, Inc., 2403 Commerce Ln., Albany, GA 31707. Price is approximately $6, plus postage.
Z1 — Touch-Tone encoding keyboard. The circuit-board layout will accommodate Digitran keyboards KL0054 (12-key) or KL0049 (16-key). They are available from distributors in single lot quantities. For the name of the nearest distributor, contact Bob Privell at Digitran, 855 South Arroyo Pkwy., Pasadena, CA 91105, or call him at 213-449-3110. At the time of this writing, the keyboards cost approximately $6 and $7.50, respectively.

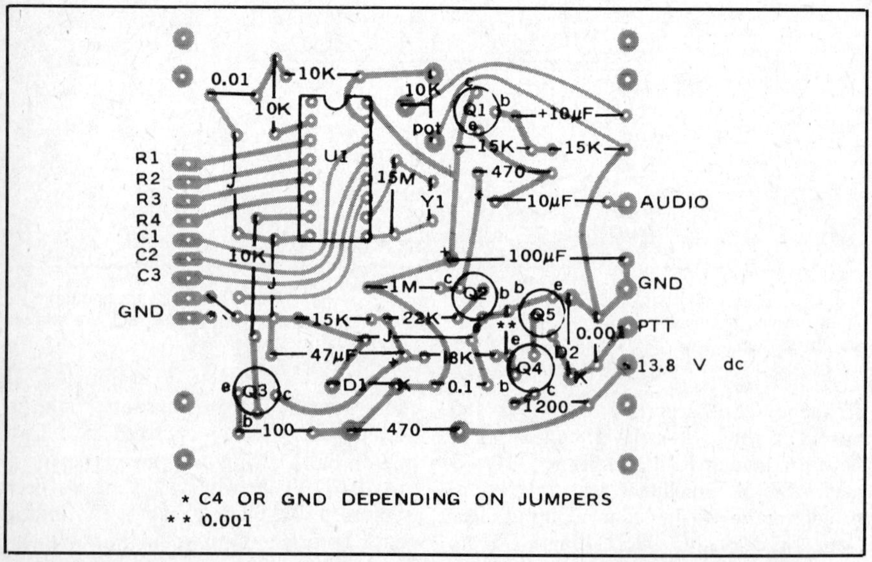

discrete transistors solved the problem.

Q2 and Q4 are driven by Q3, which is turned on by pulses from pin 7 of U1 when a keyboard switch is depressed.

U1 requires a 5-volt supply for proper operation. This is provided by the 470-ohm resistor and 1N4733 Zener diode, D1. With the exception of Q4, the remainder of the encoder circuit was also designed to operate from a 5-volt supply.

The length of time the transmitter remains keyed after a keyboard switch is released is determined by the value of C1, connected to the collector of Q3. On the prototype unit, a value of 25 μF provided a delay of just under one second. If you prefer a longer drop-out time, increase the value of this capacitor. Lowering its value will decrease drop-out time.

Construction and Testing

The prototype was built on a piece of perforated board, but a pc board is preferable. A commercially available circuit board is shown in Figs. 48 and 49. U1 should be installed in a socket. The 0.001-μF disc capacitors connected to the base of Q4 and collector of Q5 should be installed as near the transistors as possible. Their function is to bypass rf from the transmitter, which can cause Q4 and Q5 to latch up in the keyed position.

A few simple checks will tell whether the circuit is functioning properly. The following tests may be performed before connecting the encoder to the radio, using only a 12-volt power supply, a high-impedance dc voltmeter, and a scope if one is available.

First ensure that D1 is regulating the encoder supply voltage at +5.1 V dc ±10 percent. U1 may be damaged if more than 6 volts is applied to pin 16.

The two operating states for the encoder are (1) no keyboard buttons depressed and (2) one or more buttons depressed. Connect the 12-V dc supply and measure the voltage at the test points shown on the schematic diagram. Voltages measured should be in accordance with those shown.

If any voltages are incorrect, look for wiring errors. If the collector voltage of Q2 is not at least 4 volts with no buttons pressed, the problem may be a leaky transistor at Q1 or Q2. If a scope is available it may be used to inspect the audio output. Pressing any one button should produce a signal, while depressing any two buttons simultaneously should produce a single tone.

Installation

The electrical portion of the installation simply involves running four wires from the encoder to the transceiver: +12 V, ground, push-to-talk (PTT) and audio output. Shielded audio cable is recommended for the audio output, which is connected to the transceiver mike input.

Fig. 48 — If the circuit board is used, this parts overlay will guide you when installing components. Circuit boards are available from Lea Engineering, 1230 E. Layola Dr., Tempe, AZ 85282, for $5.50 each.

The PTT lead is connected to the hot side of the mike PTT switch. The +12-V and ground leads are self-explanatory. The encoder PTT circuit is designed for rigs with an antenna relay coil which is connected to the +12-V bus and the PTT switch. The latter grounds the cold side of the relay coil during transmit. Assure that your rig has this type of PTT circuit and that the relay coil draws less than 300 mA. Most of the popular vhf and uhf fm ham rigs have this type of PTT circuit. The mechanical details of the installation are left to the discretion of the reader.

Since the encoder will not load the audio system, it should not be necessary to change the setting of any transmitter mike-level controls. Adjust only R1 in the encoder for proper tone deviation. The prototype unit has provided excellent performance on both a Tempo VHF/One 2-meter rig and a Kenwood TR-8300 uhf rig.

A UNIVERSAL TOUCH-TONE DECODER

The control unit described here is extremely reliable, flexible and immune to false signals. Any number of control functions can be built into this modular unit. Starting with a simple, single-digit, on/off control, it may be expanded to provide up to 45 different control functions, including a three-digit on/off command. The application of the decoder system described here is not limited to repeater use. With a little ingenuity one might adapt the simpler systems to turn on house lights or open garage doors.[3] Also included is a voltage-to-frequency converter.

The heart of the system is the NE567 tone decoder. Note the unique method of interconnection as shown in Fig. 50.[4] In other systems, seven ICs are used to provide all the decoding functions. These decoders may respond to false signals and are critical of input tone levels, however. In this unit 24 ICs are used, two for each digit (0 to 9) and two each for the asterisk (*) and pound (#) signs. This may at first seem to be a waste of ICs, but the selectivity of the decoders is greatly enhanced and this arrangement allows the use of other capabilities of the IC. This circuit was originally described in March 1980 QST by WA0UZO.

Circuit Description

Refer to Fig. 50. U1 is used to decode the higher frequency (f1) of the Touch-Tone pair (see Fig. 51A). When U1 receives the correct tone, the output (pin 8) will supply a low to U2, pin 7, enabling it to decode the lower frequency of the pair (f2). Upon reception of the frequency pair, the output of U2 will go low.

This low is used to "latch" the digit into the system. D1, D2, R4 and R5 are used for this purpose. If the latch feature is not desired, omit these components. To

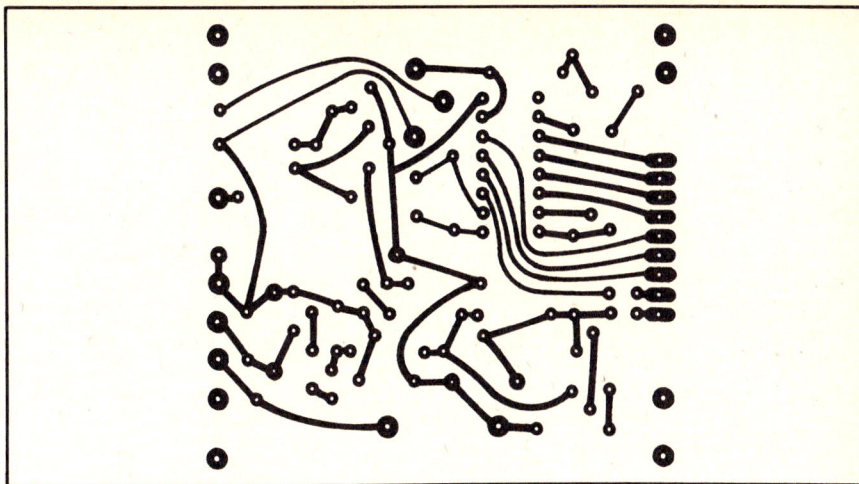

Fig. 49 — Circuit board etching pattern for the Touch-Tone encoder. The board is single sided, shown at actual size from the foil size, with black representing copper.

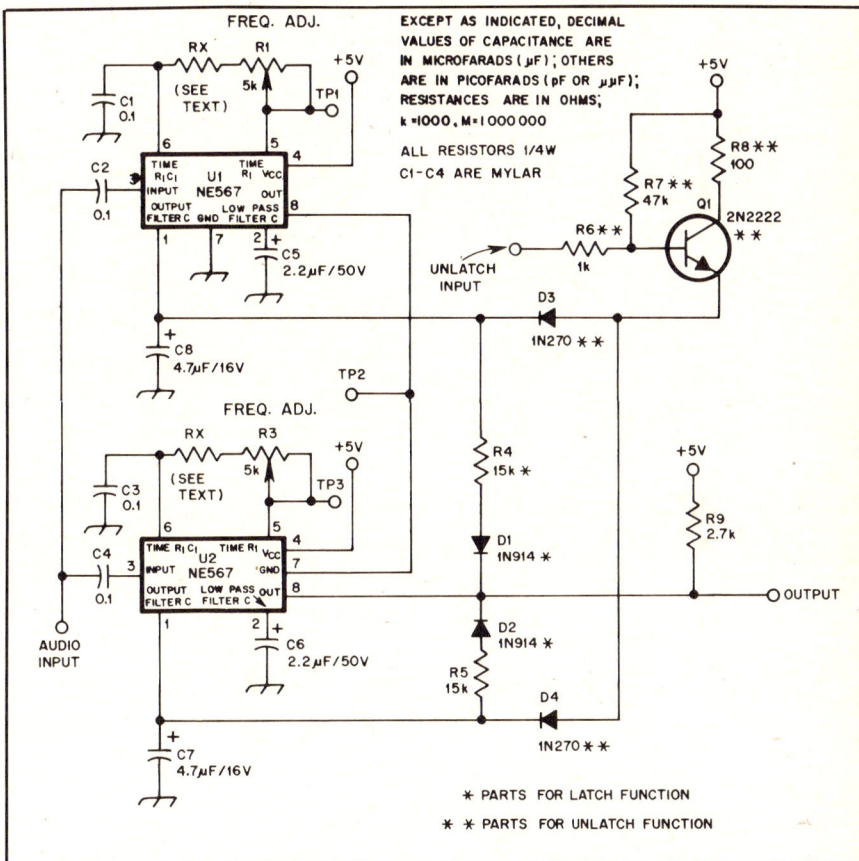

Fig. 50 — The circuit diagram of the Universal Touch-Tone Decoder. A pair of ICs is used to provide better reliability and immunity to "falsing." If desired, the capacitor at pin 1 of each IC may be increased to 100 μF to provide a two-second decoding delay.

"unlatch" the unit, Q1 is used. When the base of Q1 is low (grounded), the latch is enabled. If the base is high (ungrounded) the system will unlatch.

Note that the decoders may be built one

Fig. 51 — The layout of the 12-key pad and the frequencies associated with each line and row are shown at A. Fig. 51B gives the values required for RX for the various frequencies to which the decoders of Fig. 50 are tuned.

DIGIT FREQUENCIES				VALUE OF RX	
F1	1209	1336	1477	F2	FREQ.	VALUE
1	2	3	697	1209	6.8k
4	5	6	770	1336	5.6k
7	8	9	852	1477	5.1k
*	0	#	941	697	13k
				770	12k
				852	10k
				941	9.1k
(A)				(B)	

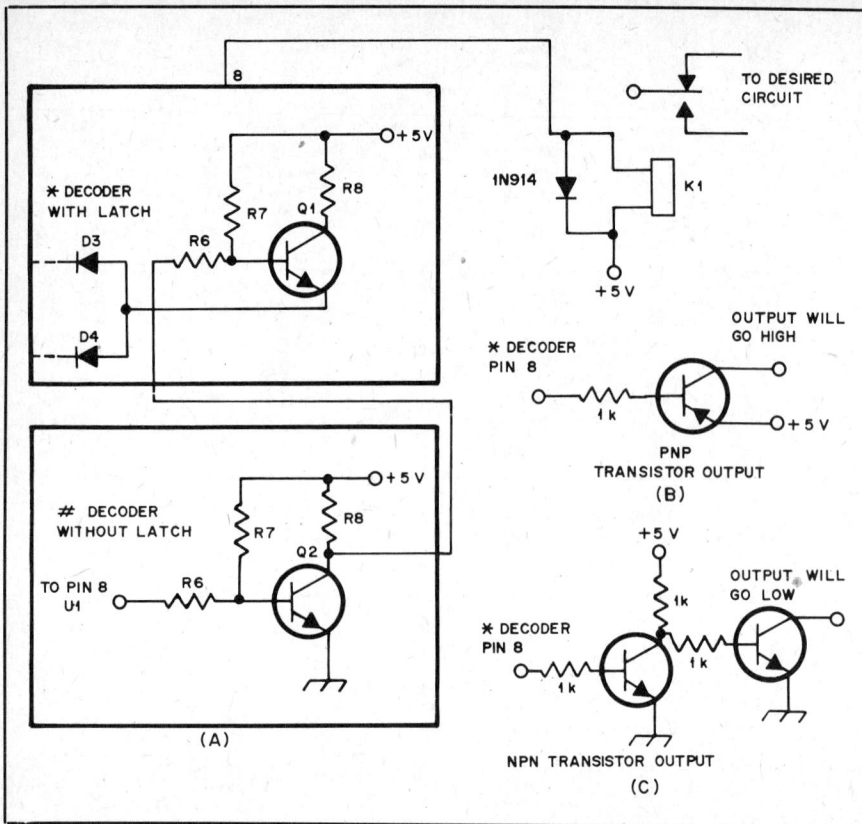

Fig. 52 — A simple two-button, on/off decoder. A relay is shown at A, but transistor switches may be substituted as at B and C.

Construction

To construct a single-digit decoder, select a frequency pair from Fig. 51A and the appropriate resistor value for RX from Fig. 51B. Mount the components on the board with the exception of those re-

Fig. 53 — With a little ingenuity, this simple version of the decoder can find many uses.

quired for the latch/unlatch circuitry. Install short wires at TP1, TP2 and TP3 for attaching test leads. Apply power to the circuit and connect a frequency counter to TP1. Use a low-value capacitor (approximately 300 pF) between the counter and TP1 to prevent the counter from loading the IC. Adjust R1 to provide the correct chosen frequency.

To adjust U2, a signal source at f1 is required. A Touch-Tone pad may be connected to the audio-input point of Fig. 50. The pad will generate a single-frequency tone (f1) when two buttons in a vertical row are pressed simultaneously. Any two coincidentally pressed buttons in a horizontal line will generate f2. Feed f1 into the decoder and adjust the amplitude of the tone so that TP2 goes low. With the counter at TP3, adjust R3 for f2 with f1 still applied. Now, when the digit corresponding to the frequency pair (f1/f2) is pressed with the output of the pad applied to the decoder, the output of U2 (pin 8) will go low. When the tones are removed, pin 8 will return to a high.

Install the components associated with the latch function. Now, when the frequency pair is recognized by the decoder, the output of U2 will go low and remain low after the tones are removed. Mount the unlatch function components and ground the base of Q1. You should note that pin 8 will go low when the tones are received and remain low until the base of

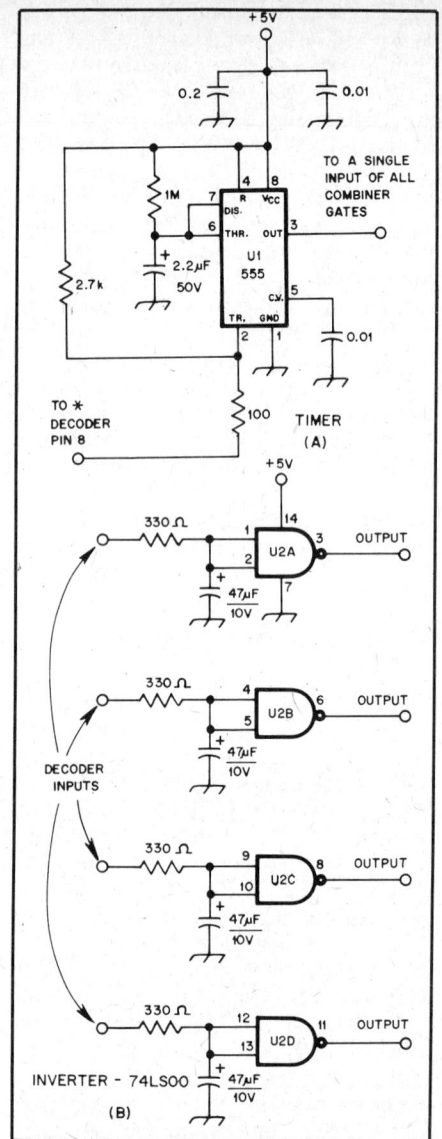

Fig. 54 — The diagram of the TIMER-INVERTER board. The timer is used to provide a "window" through which the control data must be passed. Only one inverter IC is shown, but there are actually three on the board. The 47-μF capacitors at the input to the gates slow down the action of the inverters and prevent system "falsing" because of voltage "spikes."

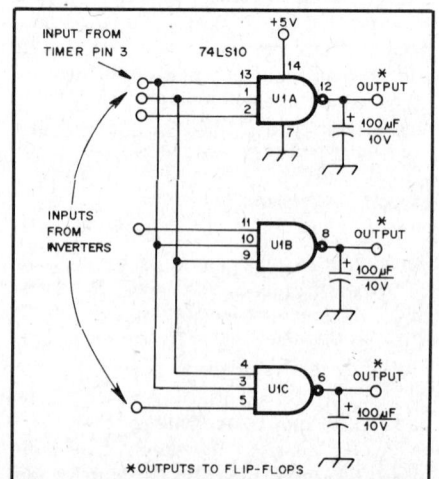

Fig. 55 — There are three 74LS10 ICs on each COMBINER board although only one is shown.

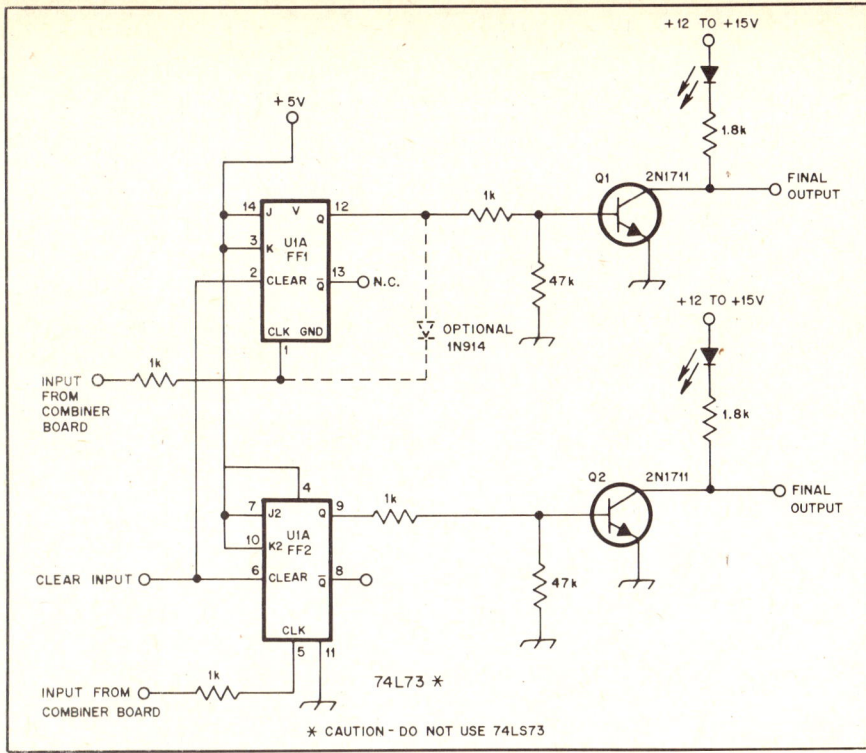

Fig. 56 — A diagram of one section of the FINAL board. FF1 and FF2 are both part of the same IC, a dual J-K flip-flop. Five of these dual flip-flops and 10 of their accompanying output transistors are mounted on each board. See Fig. 57 and the text concerning the installation of the optional diode shown in dashed lines.

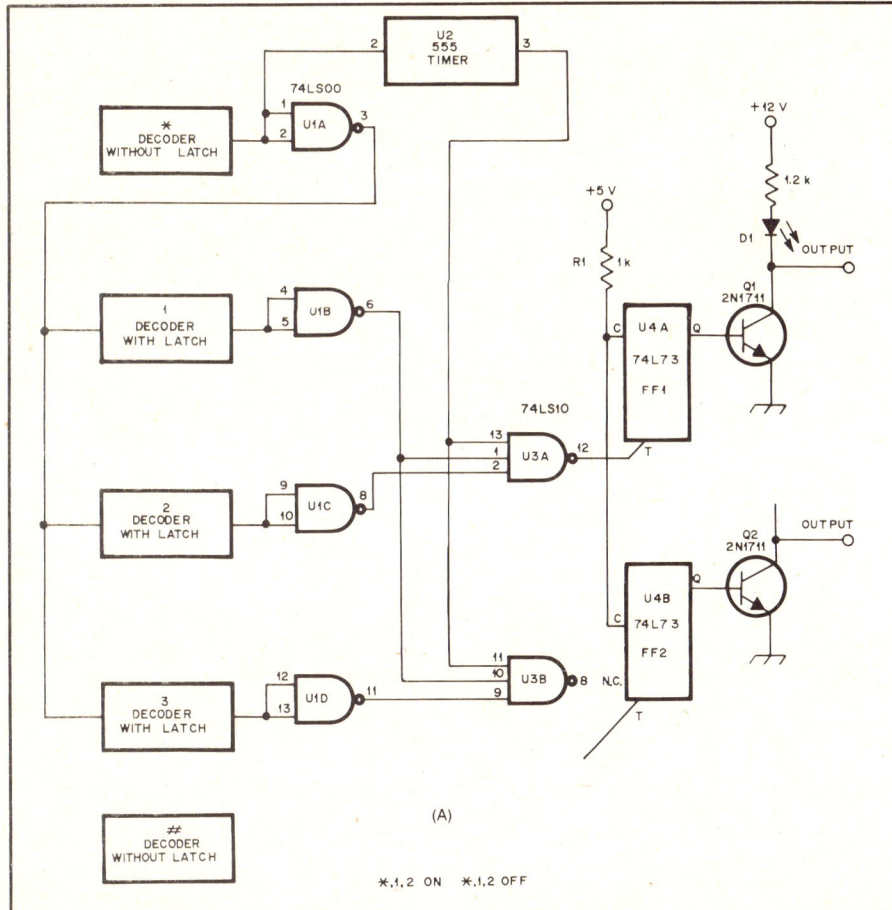

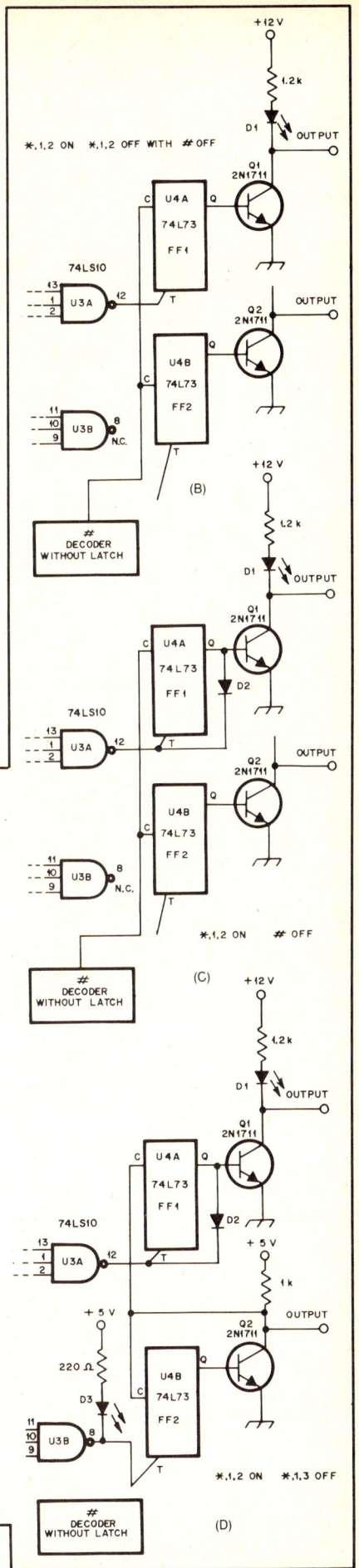

Fig. 57 — The four circuits diagrammed here are discussed in the text. The most simple circuit (A) does have a weak point, while that at D is the most reliable.

Q1 is ungrounded or taken high.

By combining two single-digit decoders, a simple ON/OFF function can be constructed. The decoder for the ON digit is built with the latched output while the OFF tone decoder is wired without the latch function. The necessary connections between the two decoders are shown in Fig. 52. A relay is shown at the output although either a relay or transistor may be employed. When the ON digit is received the relay will close and remain closed (latched) until the OFF digit is received.

An Expanded System

A complete system offering up to 45 different control functions and using a three-digit entry code can be constructed by combining the desired number of tone decoder circuits with some additional logic. Use of either the * or # sign as the first entry for a control function is recommended, especially if the repeater is equipped with an autopatch. In this manner, numbers alone cannot initiate a control function. Personal preference is to use the * symbol to initiate the command and the # sign as an "all clear"; this also permits system reset.

To carry the logic required for the larger system, other circuit boards are needed. All are of the 28-pin plug-in variety. A timer is also needed; an NE555 serves nicely. The timer, activated by the * decoder, (constructed without a latch) will open a "window" through which the other two digits must be passed. This "window" will remain open for only two seconds. The * is also used to reset all other decoders. This was done so that if any of the decoders have been accidently activated, no command function will be carried out.

It is necessary to invert all the lows supplied by the decoders. This operation is performed by SN74LS00 quad, two-input NAND gates. The input pins of the gates are tied together, thereby creating inverters. The timer and inverter diagram is shown in Fig. 54. Only one TIMER-INVERTER board is required for any system of up to 45 functions.

To combine the digits and generate one output, a COMBINER board is used. Each board contains three SN74LS10 ICs. Fig. 55 shows the make-up of a single 74LS10 which incorporates three triple-input NAND gates. Each board furnishes nine functions; five boards are used in a 45-function decoding system.

The last board required is (appropriately) the FINAL board, which contains the latches for the desired functions and the transistor drivers. Five SN74LS73 dual J-K flip-flops and 10 2N1711 transistors are mounted on each FINAL board. Five of these boards are used in a 45-function system with 5 functions left open for possible use later. The circuit for a single SN74LS73 is shown in Fig. 56. A complete system will require an INVERTER board and at least one COMBINER board and one FINAL board.

To contain the system, a card cage can be constructed out of double-sided printed-circuit-board material. The function-control outputs are brought out to card sockets on the rear of the case. This allows everything to be disconnected easily for servicing. LEDs are mounted on the front panel of the enclosure to provide an indication of the status of all the functions. A Touch-Tone pad can also be installed on the front panel. This pad can be switched in for local system checks.

Fig. 57A shows how the system is configured for a *, 1, 2 ON/OFF function. Note that the output of the * decoder is used to start the timer and reset all the decoders. The output of U2 is applied to one input of the three-input gate, U3A. The second input of the gate is satisfied by the output of the "1" decoder and the third input by the "2" decoder. This forces the output of U3A low. This low toggles the J-K flip-flop U4A (note that the clear or C input of U4A is held high through R1). The Q output of U4A will go high and remain high. This causes transistor Q1 (the output transistor) to conduct. The corresponding LED will glow, indicating that the function has been carried out. If desired, a relay could be used at the output of Q1. With this simple system, one must use the same codes (*,1,2) to turn off the function. This is not a sound idea since one cannot tell (from a remote point) whether the function was being keyed on or off. A better method is shown in Fig. 57B. The clear (C) input of U4 is connected to the # decoder. The basic action of the decoder will be the same as before, but now, use of the # key will ensure the function is in its off state. Extra insurance may be obtained through the addition of D2 as shown in Fig. 57C. This will prevent the same code (*,1,2) from turning the function off; now the *only* way this may be done is with the # key.

In Fig. 57D, a system is shown which uses an ON code of *,1,2 and OFF code of *,1,3.

It is best to use a number of decoder chips and separate inputs to each chip for a couple of reasons. Some of the decoder audio inputs can be connected to the repeater receiver so that a number of users can key them up, as in autopatch use. Other decoders can be used with a 450-MHz control receiver and still others

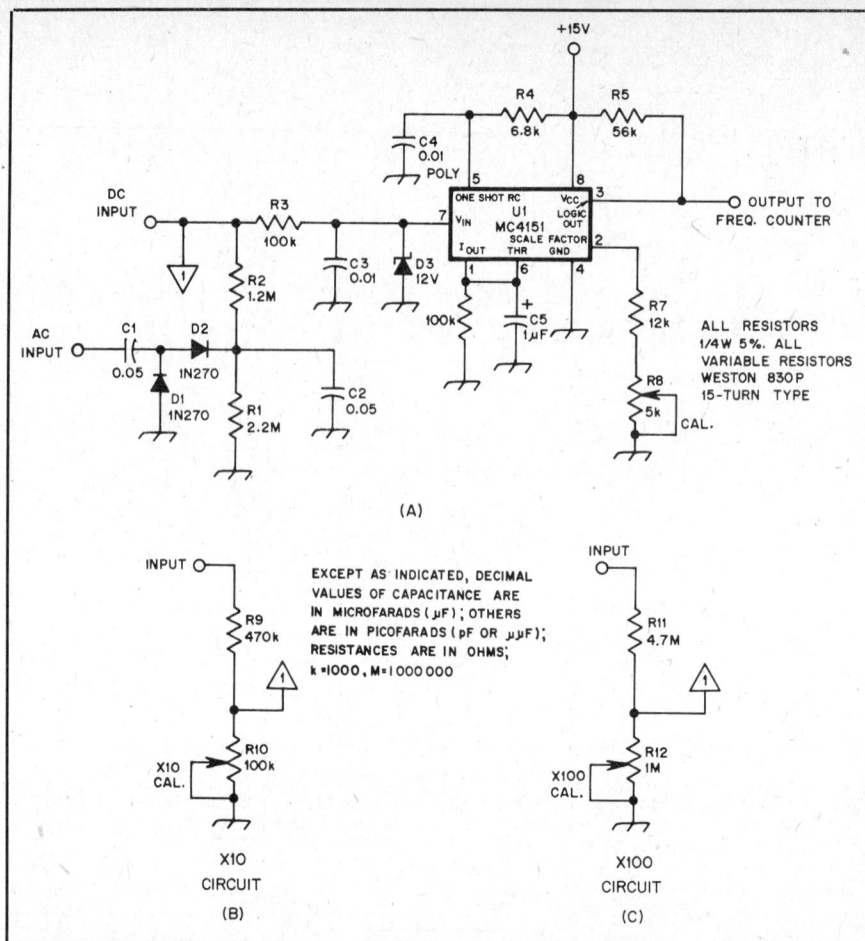

Fig. 58 — The voltage-to-frequency converter is shown at A. Both dc and ac voltage amplitudes may be read out on a frequency counter at the output of the IC. The circuits at B and C may be used as ×10 and ×100 multipliers at the input to the converter.

properly coupled to the telephone line for other uses.

A highly accurate voltage-to-frequency converter circuit is presented in Fig. 58A. Calibration is straightforward. Couple a frequency counter to the output of the converter and connect a +12V source to the input. Adjust the calibrate potentiometer for a reading of 12 kHz, as read on the frequency counter. A +1.5-V source should provide a reading of 1.5 kHz, and so on. Provisions have been made on the printed circuit board for inclusion of a ×10 and ×100 multiplier circuit. These additions are shown in Fig. 58B and C, respectively. The ac-to-dc converter permits measurement of ac voltage and will be read as an rms voltage on the frequency counter. This novel voltage-to-frequency converter circuit can be utilized in many ways such as providing a digital readout of signal levels. Or, if your repeater is equipped for telemetry, you could use this circuit to provide readout of a variety of data inputs.

GROUND-PLANE ANTENNAS FOR 144, 220 AND 420 MHz

For the fm operator living in the primary coverage area of a repeater the ease of construction and low cost of a quarter-wavelength ground-plane antenna make it an ideal choice. Three different types of construction are detailed in

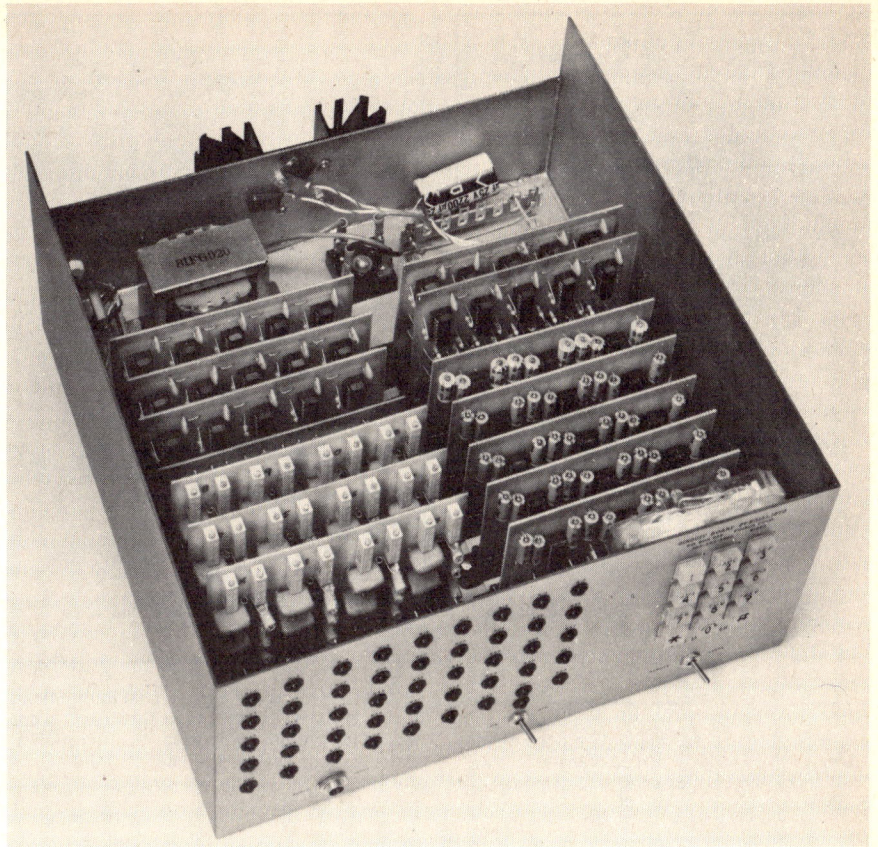

Fig. 60 — The completed 45-function Touch-Tone decoder. Double-sided printed-circuit board soldered along the seams makes a sturdy enclosure.

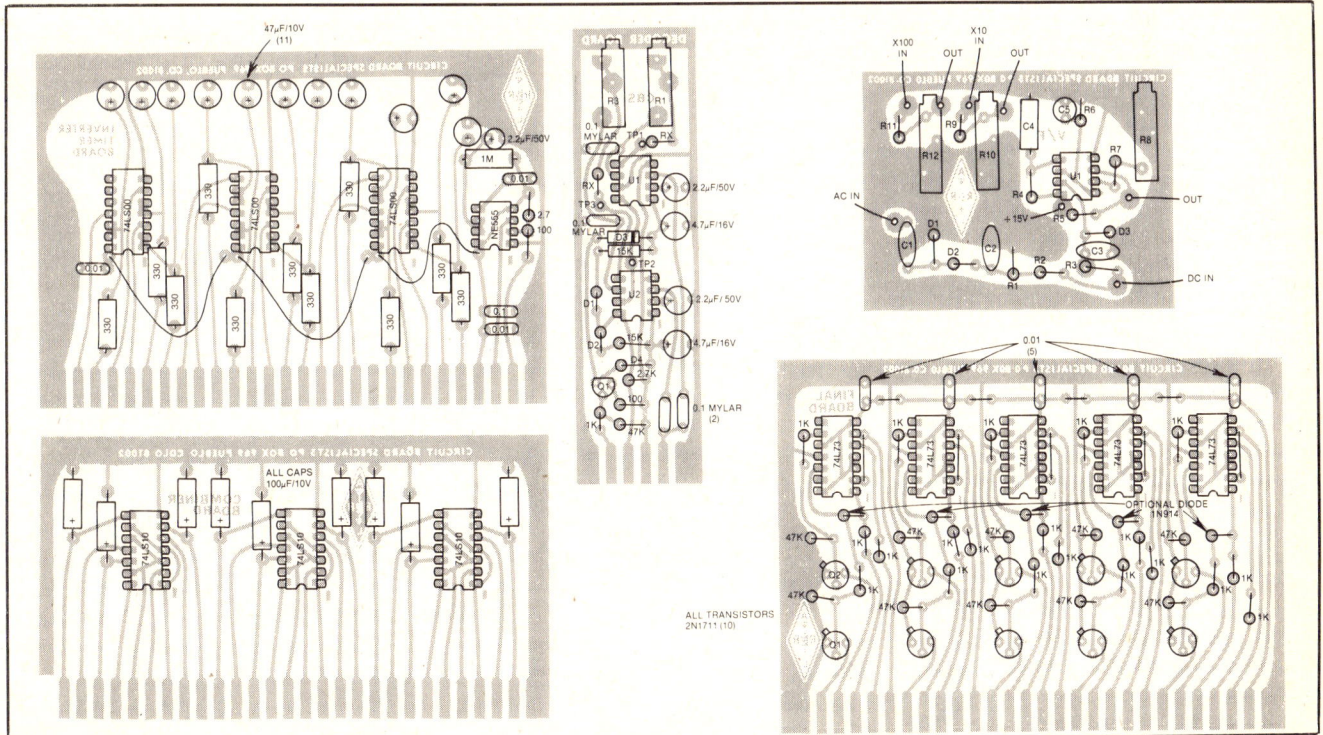

Fig. 59 — Half-size component placement guides for the pc boards. Clockwise from upper left: INVERTER-TIMER, DECODER, V/F CONVERTER, FINAL, COMBINER. See note 5 of the appendix.

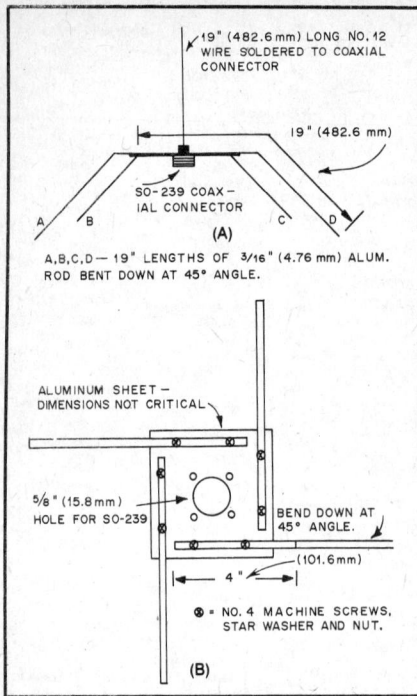

Fig. 61 — These drawings illustrate the dimensions for the 144 MHz ground-plane antenna. The radials are bent down at a 45-degree angle.

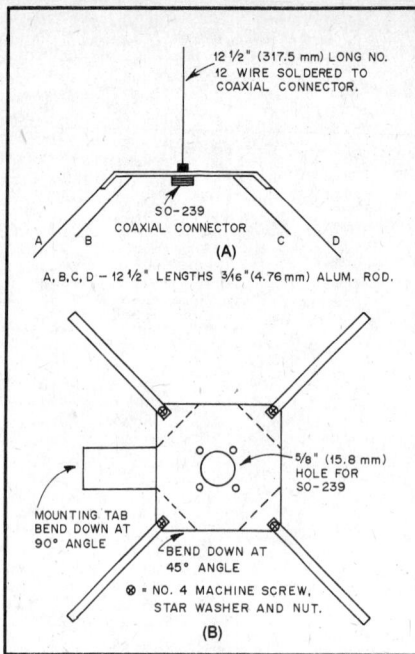

Fig. 62 — Here is the dimensional information for the 220 MHz ground-plane antenna. The corners of the aluminum plate are bent down at a 45-degree angle rather than bending the aluminum rod as in the 144-MHz model. Either method is suitable for these antennas.

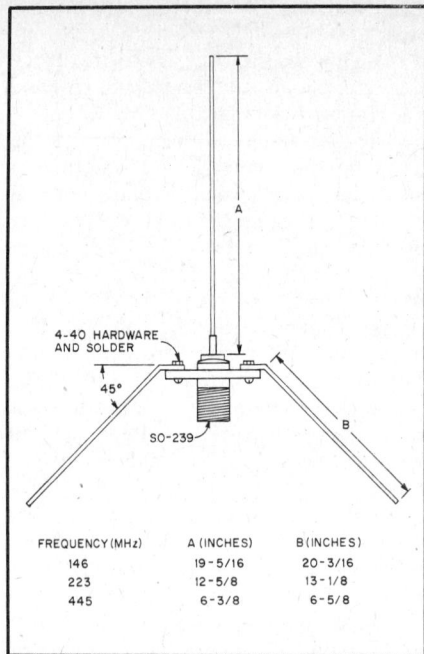

FREQUENCY (MHz)	A (INCHES)	B (INCHES)
146	19-5/16	20-3/16
223	12-5/8	13-1/8
445	6-3/8	6-5/8

Fig. 63 — Simple ground-plane antenna for the 146-, 220-, and 440-MHz bands. The vertical element and radials are 3/32- or 1/16-inch brass welding rod. Although 3/32-inch rod is preferred for the 2-meter antenna, 10 or 12 ga. copper wire can also be used.

Figs. 61 through 64; the choice of construction method will depend upon the materials at hand and the desired style of antenna mounting.

The 2-meter model shown in Fig. 61 uses a flat piece of sheet aluminum, to which the radials are connected with machine screws. A 45-degree bend is made in each of the radials. This bend can be made with the aid of an ordinary bench vise. An SO-239 chassis connector is mounted at the center of the aluminum plate with the threaded part of the connector facing down. The vertical portion of the antenna is made of no. 12 copper wire soldered directly to the center pin of the SO-239 connector.

The 220-MHz version, Fig. 62, uses a slightly different technique for mounting and sloping the radials. In this case the corners of the aluminum plate are bent down at a 45-degree angle with respect to the remainder of the plate. The four radials are held to the plate with matching machine screws, lockwashers and nuts. A mounting tab has been included in the design of this antenna as part of the aluminum base. A compression-type hose clamp could be used to secure the antenna to a mast. As with the 2-meter version the vertical portion of the antenna is soldered directly to the SO-239 connector.

A very simple method of construction, shown in Figs. 63 and 64, requires nothing more than an SO-239 connector and some 4-40 hardware. A small loop, formed at one end of the radial, is used to attach the radial directly to the mounting holes of the coaxial connector. After the radial is fastened to the SO-239 with 4-40 hardware, a large soldering iron or propane torch is used to solder the radial and the mounting hardware to the coaxial connector. The radials are bent to a 45-degree angle and the vertical portion is soldered to the center pin to complete the antenna. The antenna can be mounted by passing the feed line through a mast of 3/4 inch (19 mm) ID plastic or aluminum tubing. A compression hose clamp can be used to secure the PL-259 connector, attached to the feed line, in the end of the mast. Dimensions for the 144-, 220- and 440-MHz bands are given in Fig. 63.

If these antennas are to be mounted outside it is wise to apply a small amount of RTV sealant or similar material around the area of the center pin of the connector to prevent water from entering the connector and coax line.

Appendix

[1] Underwood mica capacitors are available from Alpha Electronic Laboratories, 2302 Oakland Gravel Rd., Columbia, MO 65201.
[2] Ferroxcube components can be purchased from Elna Ferrite Laboratories, Inc., 9 Pine Grove St., Woodstock, NY 12498.
[3] If such operation is intended, one should pay particular attention to sections 19.34(b), 19.35(c), 97.61, 97.89 and 97.99 of the Communications Act of 1934. See The Radio Amateur's License Manual ARRL, $4.
[4] FM and Repeaters for the Radio Amateur, first edition, p. 119.
[5] Circuit boards and parts kits are available from Circuit Board Specialists, P. O. Box 969, Pueblo, CO 81002. Printed circuit board etching patterns are available from the ARRL for 50 cents and an s.a.s.e.

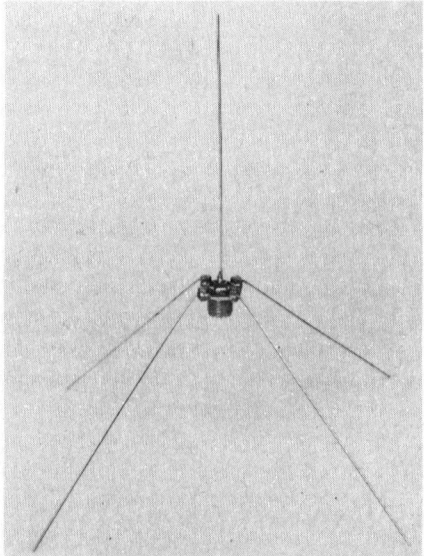

Fig. 64 — A 440-MHz ground-plane constructed using only an SO-239 connector, 4-40 hardware and 1/16-inch brass welding rod.

Chapter 14

Specialized Communications Systems

Among the qualities unique to Amateur Radio is its diversity. Virtually all hams start with either cw, ssb or fm. Beyond these modes, however, the electromagnetic spectrum provides a medium for communication methods which are limited only by one's imagination. Deeply ingrained in the basis and purpose of this hobby is the *amateur* concept — a person who pursues experimentation solely for the advancement of the art. His tools are intuition, trial and error, and the experience of others.

As radio began, the specialized technique of the day was simply hearing a spark transmission. Then came cw, followed by a-m, fm and ssb. The specialized techniques of today include space communications via OSCAR (Orbiting Satellite Carrying Amateur Radio) and EME (earth-moon-earth); visual transmission by SSTV (slow-scan television), ATV (fast-scan TV), RTTY (radioteletype) and FAX (facsimile); voice techniques of NBVM (narrow-band voice modulation) and pulse; and interface to nonradio devices such as microcomputers and telephone lines.

Use your imagination; see what *you* can do. In addition to the surveys presented here, some good resources are the *ARRL Operating Manual, QST* and AMSAT's *Orbit* Magazine.

SATELLITE COMMUNICATIONS

Large antenna arrays and comparatively sophisticated equipment may rule out some forms of space communications for many amateurs, but the amateur satellite program puts the excitement of vhf DX within the reach of every amateur: The possibilities grow with every new launch!

It began in 1961 with the successful launch of OSCAR 1, the world's first nongovernmental satellite. OSCAR rode "piggyback" on a regular launch, with the tiny ham satellite serving as ballast. The small breadbox-size satellite, built from out-of-pocket expenses of $64, transmitted the Morse code letters "HI"

Fig. 1 — After several years of work, the first high-altitude Phase II spacecraft was prepared for launch on May 23, 1980. As thousands of amateurs listened to the launch net, the Ariane launch fell into the Atlantic soon after liftoff. Work began almost immediately on a Phase III B satellite, scheduled for launch during 1982.

on 2 meters at speeds corresponding to the internal spacecraft temperature. This continued for three weeks before its on-board batteries were finally exhausted. The nearly identical OSCAR 2 transmitted signals for 18 days after its launch in 1962.

The first active communications satellite in the OSCAR series was OSCAR 3, launched in March, 1965. More than 100 stations in 16 countries helped make satellite history with OSCAR 3, the first free-access satellite. OSCAR 4 followed later that year and achieved another communications first despite a bad orbit — the first U.S.-to-U.S.S.R. satellite contact. OSCAR 5, constructed by a team at Melbourne University, was launched in 1970, the first launch coordinated by a newly formed group of amateurs in the Washington, DC area: AMSAT. The spacecraft internal batteries powered its 2-meter and 10-meter beacons for more than six weeks.

AMSAT-OSCAR 7 joined the series late in 1974, providing another major step forward for the amateur satellite program: it is scheduled to remain on continuously, alternating between two different modes of operation. Though it has recently suffered from periods of "falsing," occasional, unpredictable mode switching, OSCAR 7 is still providing long range vhf communications in its seventh year of service. A cooperative international effort, OSCAR 7 was constructed in module form by amateurs in West Germany, Canada and Australia as well as in the U.S.

The latest in the Phase II (low orbit, long life) OSCAR series of amateur satellites, AMSAT-OSCAR 8, was launched in March of 1978, joining its predecessors, though at a slightly lower altitude. A joint effort by radio amateurs in Canada, Japan, the United States and West Germany, this satellite is intended to continue support for the OSCAR Education Program while providing another spacecraft for experimentation and routine communication. Several weeks after launch, the ARRL assumed operations responsibility to free AMSAT personnel for intensive work on the AMSAT Phase III-A project. The OSCAR 8 transponders and other on-board systems appear to be functioning flawlessly; continued good health and useful operation is anticipated for years to come.

Late in 1978 satellite users throughout the world were pleased to welcome two additional low circular orbit satellite entries by the Soviet Union: Radio 1 and Radio 2. Intended for educational and communication use with very low power ground stations, RS 1 and RS 2 are no longer routinely available. Those who have communicated through them are grateful for the work of our fellow radio amateurs in the Soviet Union and hope these satellites mark the beginning of continued active involvement.

More than a challenging means of communication for hams, the amateur satellites have been involved in a host of unique experiments. Using the Phase II AMSAT-OSCAR series, amateurs have demonstrated the effectiveness of satellites in pinpointing emergency locator transmitters similar to those that are carried aboard downed aircraft, and in relaying complex medical data such as electrocardiograms from coast to coast, simulating in-transit work from a disaster site. Finally, when OSCARs 6 and 7 and 8 have been in close proximity, satellite to satellite linkups have been achieved: another milestone for free access satellites. In the future, with the greatly extended access time and insignificant Doppler shift near AMSAT Phase III's apogee, and with the flixibility that results from its sophisticated on-board computer, many more experiments are planned in areas of remote-store-and-forward data transmission, computer software exchange and emergency communication.

In recognition of the experimental and educational potential of these free-access satellites, NASA has been most generous in providing "secondary payload" launch opportunities to AMSAT to get its OSCARs in orbit. NASA has a particularly strong interest in the OSCAR Education Program, designed to bring satellite and space technology into classrooms throughout the world. Using readily available commercial equipment, students are experiencing firsthand the unique aspects of space communications: Doppler shift, orbital mechanics, Faraday rotation, telemetry decoding and much more. (If you are interested in participating in this program locally, contact OSCAR Education at ARRL headquarters.)

Altitude, Time and Range

The determining factor in the maximum theoretical range of satellite communications is the height of the satellite. Fig. 2 can be used to determine the range for the low orbital characteristics of early satellites in which the altitude is assumed to be constant throughout each orbit. To determine when you can hear the satellite, draw a circle with a radius equal to the map range from Fig. 2. For OSCAR 7 this is about 2450 miles (4000 km) for its 910-mile (1490-km) high orbit and for OSCAR 8, orbiting at an altitude of 560 miles (910 km), the map range is about 2000 miles (3265 km). When a given satellite passes through your range circle, it is within range of your location and you should be able to hear it.

With the Phase III elliptical-orbit satellites, communication range is, of course, still determined by altitude, but the altitude continuously varies between a perigee of 915 miles (1500 km) and apogee of about 22,300 miles (35,900 km). Near apogee, the satellite illuminates, fully, that half of the earth's surface directly below. This enables reliable, very long

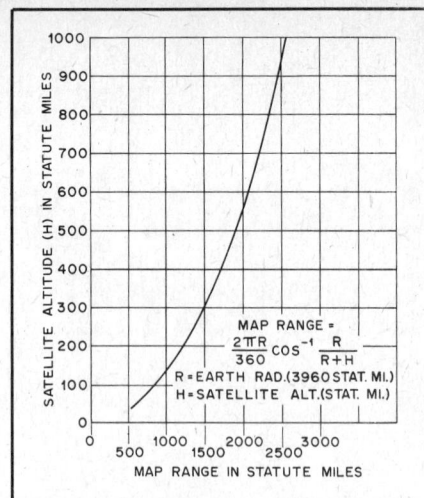

Fig. 2 — Satellite altitude above earth versus ground station map range (statute miles).

$$\text{MAP RANGE} = \frac{2\pi R}{360} \cos^{-1} \frac{R}{R+H}$$
$$R = \text{EARTH RAD.}(3960 \text{ STAT. MI.})$$
$$H = \text{SATELLITE ALT.}(\text{STAT. MI.})$$

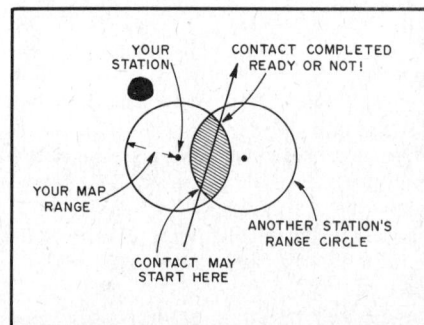

Fig. 3 — Satellite passes through the range of two stations, enabling contact.

distance vhf communication for long periods of time.

The maximum range of two-way communication through the satellite is illustrated in Fig. 3. The greater the range circle overlap for the two stations, the longer the time that these stations can remain in contact. With low-orbit satellites, communication at maximum range may last less than one minute between stations whose range circles overlap only a small amount; the effective range, therefore, is slightly less than twice the radius of your range circle. The key here is in mutual access of the satellite with the other stations. As you can see, at times with Phase III satellites, you'll be able to communicate with stations halfway around the world.

The length of time the satellite is within range of your station is determined, as is the range, by the height of the orbit. As many satellite users can tell you, the higher the orbit, the slower the satellite moves, and, the longer the satellite will be in range. With the Phase II low, nearly circular orbit satellites, the altitude is assumed to be constant. With the Phase III high elliptical orbit, however, as the altitude varies, so does the speed of the

Table 1
Amateur Radio Spacecraft Orbital Parameters

Satellite	Inclination (Deg.)	Apogee (km)	Perigee (km)	Period (min)	Increment (Deg. W)
OSCAR 7	101.4467	1471.11	1448.11	114.9422	28.7374
OSCAR 8	98.8668	940.30	905.91	103.2026	25.8023

These parameters are based on Project OSCAR predictions dated August 1981.

Satellite	Orbit	Equator Crossing	Long W. (Deg.)
OSCAR 7	30,773	0118	100.1
OSCAR 8	17,444	0057	77.8

Equator-crossing times are ascending (south to north) and in UTC.

Table 2
Spacecraft Frequencies

OSCAR 7	Uplink	Downlink	Beacon
Mode A	145.850 - 145.950 MHz	29.400 - 29.500 MHz	29.502 MHz
Mode B	432.125 - 432.175 MHz	145.975 - 145.925 MHz	145.972 MHz

OSCAR 8	Uplink	Downlink	Beacon
Mode A	145.850 - 145.950 MHz	29.400 - 29.500 MHz	29.402 MHz
Mode J	145.900 - 146.000 MHz	435.100 - 435.200 MHz	435.095 MHz

Formulas for calculating downlink frequencies. x = downlink frequency.

OSCAR 7	
Mode A	x = uplink frequency − 116.450 MHz ± Doppler shift
Mode B	x = uplink frequency − 578.100 MHz ± Doppler shift

OSCAR 8	
Mode A	x = uplink frequency − 116.458 MHz ± Doppler shift
Mode J	x = uplink frequency − 581.106 MHz ± Doppler shift

Note: A minus sign in front of the downlink frequency indicates that the passband of the satellite is inverted in that mode. This means that signals transmitted up to the satellite at the low end of the uplink passband will appear at the high end of the downlink passband.

Additionally, upper-sideband signals transmitted on the uplink will appear as lower-sideband signals on the downlink.

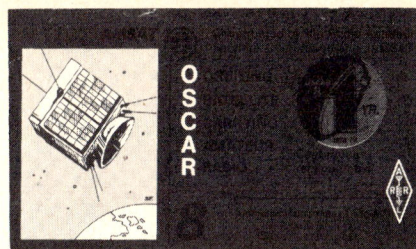

Fig. 4 — The OSCAR 8 QSL card you'll receive for submitting OSCAR 8 telemetry reports to ARRL headquarters.

makes copying the code easy for even the beginner, many of whom tape record the code at 7-1/2 ips and play it back at half speed (3-3/4 ips). With the use of decoding charts, one can compute any of the parameters being monitored and get a feel for the condition of the satellite at that time. Keeping track of the telemetry over time, you'll be able to discern patterns as the satellite goes from darkness to sunlight or spends a different percentage of time in sunlight as the seasons change. The Radio Amateur Satellite Corporation (AMSAT), which oversees the construction of present amateur satellites, is always looking for telemetry reports from OSCAR 7 as is the ARRL for reports from OSCAR 8. The first such report received from each station is rewarded with a handsome QSL as shown in Fig. 4. One final note: AMSAT Phase III, thanks to its on-board computer, will be able to transmit any of over 60 possible satellite parameters in many different transmission modes. Thus, the format on the general beacon will not be quite so obvious. We can expect, however, that telemetry will frequently be in Morse code, and with the proper decoding charts, a great deal of pertinent information will be readily accessible.

Using the Satellites

The first step in using a satellite is, of course, knowing where it is. The simplest way to determine this is with an ARRL OSCARLOCATOR. Briefly, you need one reference point each day, which is usually the first time in a given UTC day that the satellite passes over the equator in a northerly direction. These data are reference orbit EQX (equator crossing) time and longitude. They can be found in a variety of places: W1AW bulletins, AMSAT's *Orbit* magazine and *QST*. Armed with one reference point and the locator, you can determine the approximate location of and bearing to the satellite anywhere in the Northern Hemisphere.

A similar way to get a rough estimate of when to listen if you haven't access to an OSCARLOCATOR, is to find your "EQX window." Given your station latitude and longitude and the previously described range circle, you can estimate the range of EQX longitudes that will

satellite. AMSAT Phase III will travel slowest at apogee where its coverage is greatest, meaning that it spends a greater amount of time in those portions of its orbit where communications range is greatest! The time it takes the satellite to make a complete revolution around the earth is called its period.

The Spacecraft

Present communications satellites are functionally integrated systems. Rechargeable batteries, solar cell arrays, voltage regulators, command decoders, antenna-deployment mechanisms, stabilization systems, sensors, telemetry encoders and even on-board computers and kick motors each serve a unique and indispensable purpose. But to the radio amateur interested in communication through OSCAR, of primary importance is the transponder. These transponders receive signals over a given segment of one amateur band and retransmit each signal over another segment in another band. For example, OSCAR 8 receives signals between 145.850 and 145.950 MHz on its

Mode A transponder and retransmits the signals between 29.4 and 29.5 MHz. Other modes and other transponders may have different passbands, though their operation is the same. The use of a transponder rather than a channelized repeater allows more stations to use the satellite at one time. In fact, the number of different stations using OSCAR at any one time is limited only by mutual interference, and the fact that the output power of the satellite (a couple of watts on the low orbit satellites and about 50 watts on Phase III) is divided among the users.

Each satellite transponder is equipped with a telemetry beacon, which continuously transmits status reports on a variety of satellite parameters, such as internal temperature, current drain, power generation from the solar cells, and more. (See Table 2 for beacon frequencies.) The telemetry information is used to monitor the "health" of the satellite and diagnose any operating difficulty.

The telemetry from OSCARs 7 and 8 is sent in Morse encoded numbers at approximately 20 wpm. Its repetitive format

bring the satellite in range of your location. In general, for amateurs in the United States and the lower latitudes of Canada, those orbits crossing the equator between your longitude and 15 degrees or so east, will pass close to your QTH. Now, using the reference orbit data, knowing that the satellite is travelling from south to north on its reference orbit, and knowing that each successive orbit will cross the equator at a point x degrees (the increment, also published in *QST*) further west and y minutes (the period) later than the previous orbit, you can approximate when to begin listening. With this method, allow 10 to 15 minutes for error.

Tracking high elliptical orbit satellites is another matter entirely — much more complex or simpler depending on your approach. Tracking information and announcements of the availability of Phase-III type tracking devices will appear in *QST* and *Orbit*. Keep in mind that Phase-III satellites will be in range for very long periods. Once you hear the 2-meter downlink, periodically peaking the signal with steerable antennas as the satellite moves slowly across the Northern Hemisphere should be a simple task.

Receiving

Receiving is an excellent way to get started in satellite work. As in any endeavor, you have several alternatives based on your present station, your QTH restrictions and your budget. To receive the Mode A down-link, for example, you'll need an antenna (10 meter) and an hf receiver with adequate sensitivity on 10 meters. Most hams already have the needed equipment and can listen to OSCAR with very little effort. What is adequate sensitivity? Compare the noise level produced in your receiver with the antenna connected, and then with a dummy load or other 50-ohm nonreactive load. If the noise level is higher on the antenna, you have the needed sensitivity. If not, a low-noise preamplifier for 10 meters should make a world of difference.

What is the ideal antenna for 10-meter satellite reception? If you already have a beam, a quarter-wavelength vertical or dipole, try them. You may find that the beam works best when the satellite is low on the horizon, and performs adequately when it's overhead; you will have to aim the beam, however. A vertical antenna may be adequate for low satellite elevations, but may exhibit a null when OSCAR is overhead. A dipole may perform just the opposite. Two antennas that are especially well suited to OSCAR work are the turnstile and the full-wave loop. Both exhibit broad patterns and work well from horizon to horizon. Examples of both these antennas appear later in this chapter.

Mode B and Mode J reception (2 meters and 70 cm respectively) present even more

alternatives. The key points toward successful reception at vhf and uhf are minimizing the losses between your antenna and receiver, getting the antennas out in the open, and having adequate sensitivity. There are many 2-meter receivers and several 2-meter and 70-cm receive converters on the market to choose from. Keeping in mind that OSCAR operation is predominantly cw and ssb, and given your present station and plans for the future, the choice is yours. Many of the commercially available receivers in the vhf/uhf range have relatively poor noise figures, in the range of 6 dB or more. With today's modern devices, however, it is not unreasonable to shoot for a noise figure of 2 dB or so. If your receiver performs less than adequately and you've done everything to minimize losses (short runs of high quality coaxial cable, properly soldered coaxial connectors — preferably N-type at 435 MHz, and proper sealing against corrosion), consider building or buying a low-noise preamplifier. Alternatively, you could borrow one from a friend to ensure that the improvement in performance warrants the expense.

The shorter wavelength at vhf and uhf means that multi-element gain antennas are more compact, easier to build and more easily aimed than hf arrays. The excellent performance of the OSCAR 7 Mode B transponder showed, however, that in some instances little more than a quarter wavelength ground-plane antenna may be necessary. Generally, a beam with modest gain will pay large dividends when the going gets rough.

Transmitting

After listening to the OSCARs for a short period of time you'll probably want to try making contacts through the satellite. Once you know what the satellite "wants to see," you're faced with several choices. Currently, AMSAT requests that you use a maximum of 100 watts erp (effective radiated power) on OSCARs 7 and 8 (10 watts erp on Monday UTC QRP

days) and approximately 1000 watts erp for the upcoming Phase III at apogee. Mode A and J up-link frequencies are in the 2-meter band, and Mode B is in the 70-cm band.

Effective radiated power can be calculated as the power leaving your antenna and is equivalent to the power output of your transmitter minus losses in your feed line, all times the gain of your antenna. For example, 10 watts of output from your transmitter at 145.950 MHz, through 100 feet of RG-8/U (approximately 3 dB loss results in 5 watts reaching the antenna) into a 13-dB gain beam would yield 100 watts erp. The same erp would result from an arrangement with 10 watts output into a 10-dB gain amplifier (100 watts), 100 feet of RG-8/U and a 3-dB gain antenna. Alternatively, a 10-watt transmitter fed into a 13-dB gain amplifier, 100 feet of RG-8/U, and a ground plane, dipole or turnstile antenna will also produce the same erp. Where is the difference? Given your present station and future budget, one of the alternatives will probably be best. Vhf/uhf amplifiers and their associated power supplies are a good deal more expensive than gain antennas which you can easily build yourself. The trade off lies in the fact that the higher the gain of an antenna, the narrower its beamwidth, and the more accuracy you'll need in tracking. A typical 10-dB gain, 2-meter antenna has a beamwidth of roughly 26°: an arc through which OSCARs 7 and 8 will pass quickly. At the very least you'll have to track the satellite in azimuth (keeping the antenna at a fixed 30-degree elevation above the horizon) or more likely resort to tracking in both azimuth and elevation.

For the AMSAT Phase III satellite you'll need close to a kilowatt erp at 435 MHz to span the 35,900-km distance at apogee. Practicality limits the choice here, however, and a reasonable approach would suggest a 100-watt signal through as short a feed line as possible into a 13-dB gain antenna. Several 70-cm solid-state

Table 3

AMSAT Nets and Bulletins Schedule

The following AMSAT Nets meet regularly to disseminate information to newcomers and to keep regular satellite users in communication with one another.

USA-East Coast Net	Wednesdays	0100Z	3850 kHz LSB	Net Control WA3NAN
USA-Mid States Net	Wednesdays	0200Z	3850 kHz LSB	Net Control WØCY
USA-West Coast Net	Wednesdays	0300Z	3850 kHz LSB	Net Control W6CG/W6DOW
International Net	Sundays	1800Z	14,280 kHz USB	Net Control WA3NAN
International Net	Sundays	1900Z	21,280 kHz USB	Net Control WA3NAN

Bulletins of general interest to those interested in amateur satellites are transmitted regularly on OSCAR-7 and OSCAR-8 reference orbits, at approximately 10 minutes after ascending node. These bulletins are transmitted on a downlink frequency of approximately 29,490 kHz, 145.960 MHz and 435.160 MHz and can be received over most of eastern North America.

Glossary of Satellite Terminology

AMSAT — The Radio Amateur Satellite Corporation, a nonprofit organization located in Washington, DC; has overseen the OSCAR program since the launch of OSCAR 5. (AMSAT, P. O. Box 27, Washington, DC 20044.)

AOS — Acquisition of signal — The time you can first hear satellites, usually just after it rises above the horizon.

Apogee — That point in a satellite orbit where it is farthest above the earth.

Area Coordinators — Volunteers in the AMSAT organization who coordinate satellite activity in their regions. Most states have at least one; many countries are also represented.

Ascending Node — The point where the satellite crosses the equator travelling from the south to the north.

ASSC — Amateur Satellite Service Council — A coordinating body comprising equal representation from Project OSCAR, AMSAT and ARRL.

Az-el mount — Antenna mount that allows antenna positioning in both the azimuth and elevation planes.

Azimuth — Direction (side-to-side in the horizontal plane) from a given point on earth; usually specified in degrees (N = 0°, E = 90°, S = 180°, W = 270°).

Circular Polarization — A special case in which the electric field component of a transmitted radio wave is equal in both the vertical and horizontal planes and effectively rotates. The sense of polarization, whether right-hand circular or left-hand circular, is determined from behind the antenna, looking out along its axis of propagation.

Codestore — A special system that allows digital information (Morse Code, e.g.) to be placed in on-board electronic memory storage by ground stations for later retransmission.

COSMAC 1802 — A CMOS, 8 bit microprocessor made by RCA. Its low power consumption, high noise immunity, wide temperature tolerance and flexible (to the programmer) architecture, make it well suited to use in orbital hardware. The 1802 was the heart of the on-board Phase III-A computer.

Descending Node — The point where the satellite crosses the equator traveling from north to the south.

Desense — A problem characteristic of Mode J operation in which the strong 2-m uplink signal overloads a low-noise 70-cm preamplifier or converter.

Doppler Effect — An apparent shift in frequency caused by satellite movement toward or away from your location.

Down-link — The frequency of signals transmitted from the satellite to earth.

Earth sensor — A device to be used on the AMSAT Phase III satellite that will enable the computer to determine the spacecraft orientation to the earth.

Eccentricity — That orbital parameter used to describe how much an elliptical orbit deviates from a circle; eccentricity values vary between 0 and 1: e = 0 for a circle.

Elliptical Orbit — Those orbits in which the satellite path traces an ellipse with the earth at one focus.

Elevation — Direction (up-and-down in the vertical plane) from a given point on earth usually specified in degrees (0° = plane of the earth's surface at your location; and 90° = straight up, perpendicular to the plane of the earth, overhead).

EQX — Equator crossing, usually specified in time (UTC) of crossing, and in degrees west longitude (0-360°).

erp — Effective radiated power — system power output after transmission-line losses and antenna gain are considered.

ESA — European Space Agency — Agency responsible for the AMSAT Phase III-A launch.

Experiment Day — Routinely scheduled days during which the satellite is closed to casual use and reserved for scientific and educational experimentation; a secondary purpose is to allow for battery recharge.

Falsing — Unscheduled, unintentional and undesirable mode switching initiated by conditions in the spacecraft.

Geostationary orbit — An orbit at such an altitude (22,300 miles) and in such a direction (W to E) over the equator that the satellite appears to be fixed above a given point.

Groundtrack — The imaginary line traced on the surface of the earth by a satellite subsatellite point.

IHU — Integrated housekeeping unit — Phase III's on-board computer that will manage many of the routine in-flight tasks automatically.

Inclination — The angle at which the satellite crosses the equator at its ascending node; also the highest latitude reached in an orbit. An orbit crossing directly over the North Pole would have an inclination of 90°, east of the pole less than 90° and west of the pole greater than 90°.

Increment — The number of degrees longitude that the satellite appears to move westward at the equator with each orbit, caused by the earth's rotation under the satellite during each orbit. (The earth rotates 360° in a 24-hour period.)

Kick Motor — A one-shot motor on-board the Phase III satellite that will be fired at perigee a few weeks after launch to adjust

---

amplifiers in the 100-watt class are now on the market or you can build a suitable amplifier yourself. The narrow beamwidth of high-gain antennas will be much less a problem with the Phase III satellite elliptical orbit as the spacecraft moves across the sky at a comparatively slow speed (± 3 hours at apogee).

Satellite Hints and Kinks

Satellite communications confronts the newcomer with many unfamiliar phenomena that will challenge his or her operating techniques. Here are a few explanations and "tricks of the trade" to help make your satellite work as effective and enjoyable as possible.

Circular Polarization

When considering a travelling wavefront of rf energy, we can describe it in terms of polarization. Radio waves are made up of both electric and magnetic fields which are perpendicular to each other and to the direction of propagation. The classification of polarization is determined by the plane in which the electric component lies: If the plane of the electric component is vertical, the polarization is vertical; if horizontal, the polarization is horizontal. In the case where components exist in both the vertical and horizontal planes, polarization elliptical, and circular polarization is the special case where the vertical and horizontal components are equal; the field effectively rotates.

The sense of polarization, whether right-hand circularly polarized (clockwise) or left-hand circularly polarized (counter-clockwise), is determined, by convention, as though viewed from behind the antenna, looking out along the axis of propagation. The important factor for our purposes, however, is that one achieves maximum down-link signal strength when the circularly polarized antenna has the same sense as the incoming wave. Cross polarization, on the other hand, can result in 30 dB or more attenuation of the signal.

How does this affect OSCAR operation? The 2-meter antennas on OSCARs 7, 8 and Phase III satellites are circularly polarized. Theoretically, to maximize station performance, you'll want antennas that are circularly polarized in the same sense as those on the spacecraft. A better choice would be an antenna that is switchable between right-hand and left-hand circular polarization. This is because as the downlink signal from the satellite passes through the ionosphere the polarization sense may switch. This effect is commonly referred to as Faraday rotation. An example of a switchable right-hand and left-hand polarized antenna is given later in this chapter. This does not mean that circularly polarized antennas are a requirement — in fact it would be a better idea to begin with simple antennas and modify them to suit your needs after you've had a little experience.

The 10-meter antennas on OSCARs 7 and 8 are dipoles that were deployed after orbital insertion. With a linearly polarized 10-meter, ground-station antenna, you'll notice periodic fading as the satellite spins on its axis: maximum signal strength occurs when the antennas are parallel, minimum when perpendicular. With a full-wave horizontal loop or a circularly-polarized turnstile the effect of this fading will be minimized.

Full Duplex

In satellite communications you can continuously monitor your transmitted signal through the satellite, as the downlink signal from the satellite is on a different band from the up-link. You can thus evaluate both the strength and quality of your own signal throughout contacts. It is through this full-duplex capability that one is able to adjust for Doppler shift (see Doppler Shift below).

To locate yourself in the satellite passband, select an appropriate up-link frequency, calculate the approximate downlink frequency, transmit a string of dits and tune a few kHz either side of the

the orbit to the desired final perigee and inclination.

LHCP — Left-hand circular polarization — counterclockwise.

LOS — Loss of signal — The time when the satellite passes out of range.

Mode A — Transponders with 2-meter uplink and 10-meter downlink.

Mode A/J — Simultaneous operation of the Mode A and Mode J transponders on AMSAT-OSCAR 8.

Mode B — Transponders with 70-cm uplink and 2-meter downlink.

Mode C — Equivalent to Mode B with less power output; of no discernable difference than Mode B to the user.

Mode D — Battery-recharge mode; transponders off.

Mode J — Transponders with 2-meter uplink and 70-cm downlink.

NASA — National Aeronautics and Space Administration — U.S. Government agency that has provided "piggyback" launch opportunities for AMSAT OSCARs 5, 6, 7 and 8 in recognition of the OSCAR program's contributions.

OSCAR — Orbiting Satellite Carrying Amateur Radio; there have been eight Amateur Radio satellites named OSCAR at the end of 1980 and two Soviet Amateur Radio satellites designated RS1 and RS2.

OSCAR Education Program — A special program that brings live demonstrations of the OSCAR satellites to classrooms, helping teach students physics, space science, astronomy and related subjects. Teachers use ARRL curriculum materials to structure their courses around the OSCAR satellites.

OSCARLOCATOR — A satellite tracking device consisting of a ranging oval and ground-tracks superimposed on a polar projection map.

Pass — An orbit of the satellite.

Passband — The range of frequencies handled by a satellite transponder.

Perigee — That point in a satellite orbit where it passes closest to earth.

Period — The time it takes for a complete orbit, usually measured from one EQX to the next. The higher the altitude, the longer the period.

Phase I — The term given to the earliest, short-lived OSCAR satellites that were not equipped with solar cells. When their batteries were depleted, they ceased operating.

Phase II — The term given to low altitude, long-lived satellites. Equipped with solar panels that powered the spacecraft systems and recharged its batteries, these satellites have been shown to be capable of lasting up to five years. (OSCARs 6, 7 and 8).

Phase III — Extended-range, high-orbit satellites, typically in either elliptical orbit, as AMSAT Phase III-A, or in geostationary orbit.

Power budget — A determination of how much power is actually available to operate the on-board satellite systems, taking into account such things as solar cell surface area, solar cell efficiency and angle toward the sun. A positive power budget means that ample power will be available to power the desired systems; a negative power budget means that periods of shutdown and recharge must be periodically scheduled.

Precession — An effect that will be characteristic of the AMSAT Phase III orbit; the satellite apogee after the firing of the perigee kick motor will occur at about 37° N latitude, but will gradually rise higher to 57° N latitude within a few years. Then, gradually, it will move lower in latitude until after five years or

so, the apogee will occur near the equator.

Project OSCAR — California-based group, among the first to recognize the potential of space for Amateur Radio; responsible for OSCARs 1 through 4.

QRP test — Special orbits set aside for operating through the satellites while using a maximum of 10 watts erp; output powers of less than 1 watt have proven effective in some cases.

Reference orbit — The orbit beginning with the first ascending node during a given day UTC.

RHCP — Right hand circular polarization — clockwise.

RS1 and RS2 — The first two Soviet Amateur Radio satellites.

Secondary payload — Usually smaller packages that in essence share the launch with the primary payload which is the main purpose for the launch. OSCAR 8, for example, was a secondary payload "hitchhiker" to the LANDSAT C Earth Resources Technology Satellite primary payload; secondary payloads, in effect, substitute for ballast weight.

Spin modulation — Periodic amplitude-fade and-peak resulting from Phase III 60-rpm spin; the effect is a 3-Hz "modulation" of the passband.

SSC — Special service channels — Frequencies in the downlink passband of AMSAT Phase III that are designated for authorized, scheduled use in such areas as education, data exchange, scientific experimentation, bulletins and official traffic.

SSB — Subsatellite point — That point directly beneath a satellite on the surface of the earth at a given instant; usually defined in terms of latitude and longitude.

Sun sensor — A device to be used on AMSAT Phase III to determine the spacecraft orientation to the sun.

---

calculated down-link frequency. Once you've located your signal in the passband, you'll know approximately how much to change the transmit frequency to "hit" a given receive frequency. Simply change your transmit frequency by the proper amount — with the transmitter off! Swishing the passband with the key down is inconsiderate of the other users.

Doppler Shift

Doppler shift is caused by the relative motion between you and the satellite. As the satellite is moving toward you, the frequency of down-link signals will increase by a small amount. As the satellite passes overhead and begins to move away from you, there will be a sudden drop in frequency of a few kilohertz, much the same way as the tone of a car horn or train whistle drops as the vehicle moves past you. This Doppler effect will be different for stations located at different distances from the satellite. The result is that signals passing through the satellite move slowly around the calculated down-link frequency. Locating your own signal is a little more difficult than simply computing the relation between input and output frequency, as the rather hard-to-predict effects of Doppler must be taken into account.

Mode-A Down Link

When first listening to the OSCAR satellites, one is likely to tune to the Mode-A transponder down-link between 29.4 and 29.5 MHz. This may prove disappointing in that the signal may have a warbling quality, reminiscent of weak hf DX. If this is the case, check the remainder of the 10-meter band. At this point in the sunspot cycle with high solar activity, you may find the band "wide open." This means that the ionosphere is bending terrestrially originated 10-meter signals back down to earth and long distance communication at 28 MHz is successful. The OSCAR satellites orbit hundreds of miles above the ionosphere and so 29-MHz down-link signals are also bent by the ionosphere — but away from the earth! You'll usually find that the down-link signal degradation is worse at low elevations where the angle of incidence with the ionosphere is shallow; with the satellites at high elevations, the angles are much steeper and the probability of a signal penetrating the ionosphere is much greater.

Mode-J Desensing

As most members of the Mode J Club (details in Fig. 6) will tell you, the strong 2-meter up-link signal from your station

will overload or "desense" most 435-MHz low-noise preamplifiers and converters. If your transmit and receive antennas are located close to each other your 2-meter up-link signal may obliterate the 435-MHz down-link while you are transmitting!

The problem is common and the solution is, fortunately, simple. For about $5 you can build an effective 70-cm cavity filter from materials readily available at your neighborhood hardware or plumbing

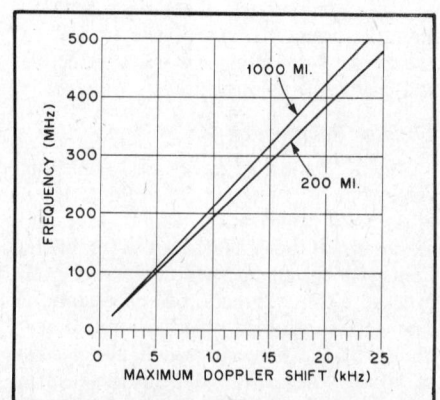

Fig. 5 — Satellite transmitter frequency versus Doppler shift for satellite in 200- or 1000-statute-mile orbits. For a translator, use the difference between uplink and downlink frequencies as the "frequency."

Fig. 6 — To become a member of the Mode J club, first complete eight Mode J contacts. QSL cards are not required. Just list the call sign of each station worked, date, orbit number and station equipment used for the contacts. Send this information along with $3 in U.S. funds, a one-time charge to cover certificate and newsletter costs, to: Mode J Club, c/o Larry Roberts, W9MXC, AMSAT Area Coordinator, 3300 Fernwood, Alton, IL 62002. To receive the Mode J Club Newsletter, send an s.a.s.e. to W9MXC for each issue. Details on the Mode J Club were outlined in the January 1979 issue of QST.

retailer. This filter, placed in the receive line, has a very-narrow passband and only 0.4 to 0.5 dB of insertion loss. An example of such a filter is shown later in this chapter.

Spin Modulation

Spin modulation is a phenomenon that has emerged with the planning of the AMSAT Phase III type of satellite. As the satellite orbits overhead, the on-board computer pulses an electromagnet which works against the earth's magnetic field. This spins the spacecraft at approximately 1 revolution per second, thereby stabilizing it. A side effect, however, is the relatively rapid (3 Hz) periodic fade-enhancement of the transmitted signal amplitude — called spin modulation. It is important to note that the passband is not electronically modulated in the sense to which amateurs are accustomed; rather, the apparent modulation is a residual effect of physical rotation.

Operationally, using linear antennas will deepen the fades to a point where it may become annoying. Circularly polarized antennas of the proper sense will minimize the effect and allow normal communications to continue with little disturbance.

On the Horizon

The future holds much promise for those interested in Amateur Radio satellite communication. A second Phase III is already being prepared to replace the one lost because of launch failure of the Ariane LO2 mission on May 23, 1980. The University of Surrey in the U.K. is now building an experimental satellite that will have beacons on 7, 14, 21, 28, 145 and 435 MHz, SSTV cloud-cover camera,

codestore, synthesized-speech telemetry, magnetometer, particle/radiation detectors, and two experimental beacons on 1.3 and 10.4 GHz. This launch is scheduled for September 1981. Amateurs of the Soviet Union are also planning the launch of their RS 3 and RS 4 satellites. Several groups in Japan, Canada and the United States are also experimenting with terrestrial transponders, some of which may someday achieve geostationary orbit.

A Full-Wave Horizontal Loop

Described here is an antenna that is ideally suited for the reception of 10-meter downlink signals from OSCAR 7 and 8. It consists of a horizontal loop of wire, one wavelength in circumference, that is mounted one-eighth wavelength or higher above ground. This arrangement provides for excellent reception of signals when the satellite is roughly 30 degrees above the horizon. Maximum response favors the satellite when it is nearly or directly overhead.

The impedance of the loop is in the vicinity of 100 ohms and a quarter-wave matching section of RG-59/U or RG-11/U is used to lower the impedance to somewhere near 50 ohms. Make sure to include the velocity factor for the coaxial cable when calculating the length of this transformer.

The loop can also be arranged in a triangle or some other polygonal configuration with little deterioration in the performance. Use whatever supports are available and try to open the loop up as much as possible. Details of this antenna are shown in Fig. 7.

A 146-MHz Turnstile Antenna

Here is a simple and effective 146-MHz antenna suitable for use with Modes A, B and J. The antenna, called a turnstile-reflector array, can be built very inexpensively and put into operation without the need for test equipment. The information contained here was extracted from a QST article by Davidoff, which appeared in the September 1974 issue.

Experience with several OSCAR satellites has shown that rapid fading can be a severe problem for satellite communicators. Fortunately, the ground station has control over two important parameters affecting fading — cross polarization between the ground-station antenna and OSCAR antenna, and nulls in the ground-station antenna pattern. Fading because of cross polarization can be reduced by using a circularly polarized ground station antenna. Fading because of radiation pattern nulls can be overcome by (1) using a rotatable, tiltable array, and continuously tracking the satellite or (2) using an antenna with a broad null-free pattern. The turnstile reflector array solves this problem since it is circularly polarized and produces a balloon-like pattern.

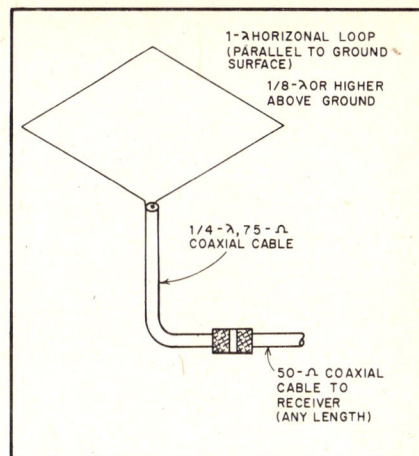

Fig. 7 — This is a drawing of a horizontal loop antenna that is ideally suited for the reception of Mode A downlink signals.

Construction

The mast used to support the two dipoles was constructed from wood and is 2 inches (50 mm) square and 8 feet (2.4 m) long. Dipoles are formed from no. 12 copper wire, aluminum rod, or tubing. The reflecting screen is 20-gauge hexagonal chicken wire, 1-inch (25-mm) mesh, stapled to a four-foot-square frame made from furring strips. Hardware cloth can be used in place of the chicken wire. Corner bracing of the reflector screen will help provide mechanical stability. Spar varnish applied to the wooden members will help extend the life expectancy.

Dimensions for the two dipole antennas and the phasing network are shown in Fig. 10. Spacing between the dipole antennas and the reflecting screen affect the antenna pattern. This is illustrated in Fig. 9, choose the pattern that best suits your needs and construct the antenna accordingly. Alternative methods of constructing turnstile arrays can be found in current editions of the ARRL *Antenna Book*.

Circularly Polarized Antennas for 2-Meter and 70-Cm Satellite Work

The antenna described here provides switchable right-hand or left-hand circular polarization and positioning for both azimuth and elevation. This system makes use of commercially available antennas (KLM 144-148-9 and KLM 420-450-14), rotators (Alliance U-100 and CDE TR-44) and relays (Inline Instruments type 101) which are combined in a system that offers total flexibility. This setup is suited for operation through OSCAR 7 and 8 and also the Phase III satellite. As shown in the accompanying photographs, the whole assembly is built on a heavy-duty, TV type of tripod so that is may be roof-mounted. The idea for this system came from Clarke Greene, K1JX.

System Outline

The antennas displayed in Fig. 11 are

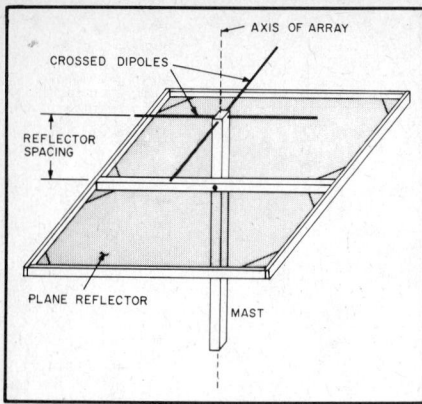

Fig. 8 — The turnstile-reflector (TR) array consists of crossed dipoles above a screen reflector.

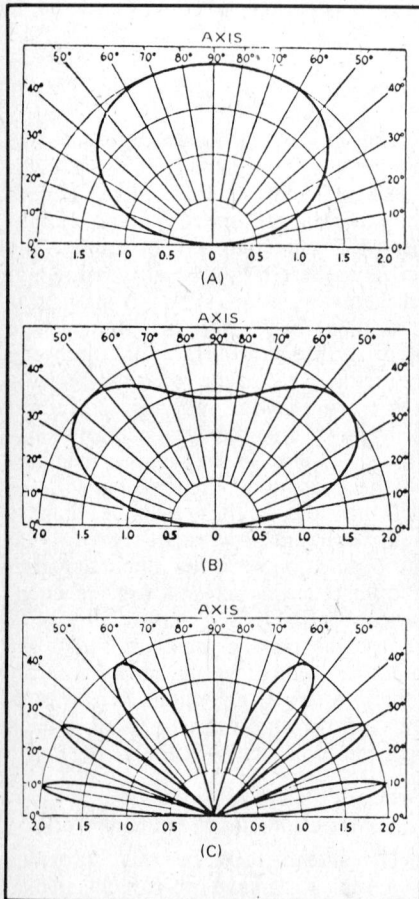

Fig. 9 — Elevation patterns for dipoles mounted over a ground plane. Pattern A is for spacing of 0.25 wavelength, B is for 0.37, and C is for 1.5-wavelength spacing.

actually two totally separate systems sharing the same azimuth and elevation positioning systems. Each system is identical in the way it performs — one system for 2 meters and one for 70 cm. This arrangement is quite handy for Mode B and J work since both antenna systems are tracked together automatically. Individual control lines allow independent control of the polarization sense for each system. This is mandatory, as often a different polarization sense is required for the

uplink and downlink. Also, througout any given "pass" of the satellite the sense is apt to switch at least several times.

Mechanical Details

The TR-44 rotator is mounted inside the tripod by means of a rotator plate of the type commonly used with a top section of Rohn 25 tower. U-bolts around the tripod legs secure the plate to the tripod. A length of 1-inch (25-mm) galvanized water pipe (used as the mast) extends from the top of the rotator out through a homemade aluminum bearing at the peak of the tripod. Since a relatively small diameter mast is used, several pieces of shim material are required between it and the body of the rotator to assure that it will be aligned in the bearing through 360 degrees of rotation. This is covered in detail in any TR-44, CD-45, Ham-M or Ham-IV rotator instruction sheet.

The Alliance U-100 elevation rotator is mounted to the 1-inch (25-mm) water-pipe mast by means of an 1/8-inch (3.2-mm) aluminum plate. TV U-bolt hardware provides a perfect fit for this mast material. The cross-arm that supports the two 2-meter and 70-cm antennas is a piece of 1-1/4-inch (31.8-mm) thick fiberglass rod, 6 feet (1.82 m) in length. Other materials can be used. However, most cannot match the strength of fiberglass. This should be a consideration if you live in an area that is frequented by ice storms. Although it is relatively expensive (about $3 per foot), one piece should last a lifetime.

Electrical Details

Since the antenna systems are identical, this description will apply to either. A simple way of obtaining a circularly polarized pattern is to use two Yagi antennas with the elements mounted at right angles to each other and feed the antennas 90 degrees out of phase. In many cases this is accomplished by mounting the horizontal and vertical elements on the same boom. It is also possible to use two separate antennas mounted apart from each other as shown in the photograph. One advantage of this system is that the weight distribution on each side of the elevation rotator is equal. As long as the separation between antennas is small, performance should be as good as when having both sets of elements on a single boom.

In order to obtain circular polarization one antenna must be fed 90 degrees out of phase with respect to the other. For switchable right-hand and left-hand polarization some means must be included to shift a 90-degree phasing line in series with either antenna. Such a scheme is shown in Fig. 13. Since two antennas are essentially connected in parallel, the feed impedance will be half that of either antenna alone. The antennas used in this system have a 50-ohm feed impedance. For this reason both antennas make use of

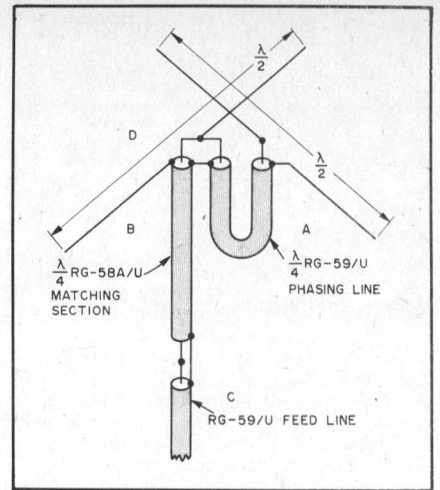

Fig. 10 — Dimensions and connections for the turnstile antenna. The phasing line is 13.3 inches (338 mm) of RG-59/U coax. A similar length of RG-58/U cable is used as a matching section between the turnstile and the feed line.

Fig. 11 — A circularly-polarized antenna system for satellite communications on 144 and 432-MHz. The array is assembled from KLM log-periodic yagis.

Fig. 12 — The polarization sense of the antenna is controlled by the coaxial relays and phasing lines. The 144 and 432-MHz systems are controlled independently.

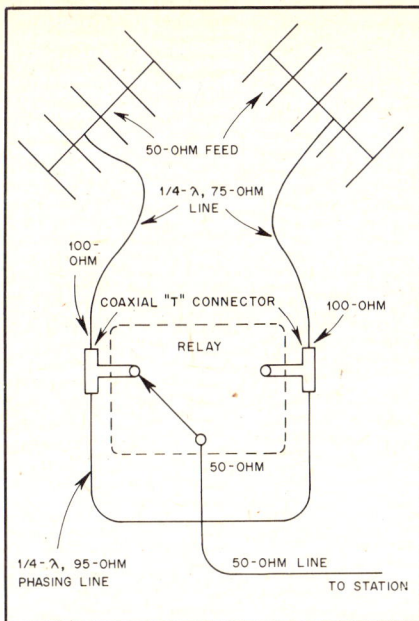

Fig. 13 — A drawing of the switchable polarization antenna system complete with cable specifications. When calculating the length of individual cable be sure to include the velocity factor of the cables.

a quarter-wavelength transformer between the antenna and the relay. This quarter wavelength of 75-ohm line steps up the 50-ohm impedance of each antenna to roughly 100 ohms. As shown in the drawing, each fixed contact of the relay is also connected to the quarter-wavelength

(90-degree) section of cable that acts as the phasing line. The phasing line was constructed from RG-133/U cable, which has a characteristic impedance of 95 ohms. This provides a very close match to the 100-ohm impedance of the system. If RG-133/U proves difficult to locate, RG-63/U (125-ohm impedance) may be used with a slightly higher mismatch. As can be seen in the drawing the phasing line is always in series with the system feed point and one of the antennas. As shown, the antenna on the left receives energy 90 degrees ahead of the one on the right. If the relay were switched, just the opposite would be true.

In reality it is not necessary to use single quarter wavelengths of line. For example, the 75-ohm impedance transforming lines between each antenna and the relay could be any odd multiple of one quarter wavelength, such as 3/4, 5/4, 7/4 wavelength, etc. The same is true for the 95- or 125-ohm phasing line. One must keep track while using different lengths for the phasing line. This is especially true when figuring out which position of the relay will yield right- or left-hand polarization. The builder is apt to find that it will be necessary to use one of the odd multiples of a quarter wavelength

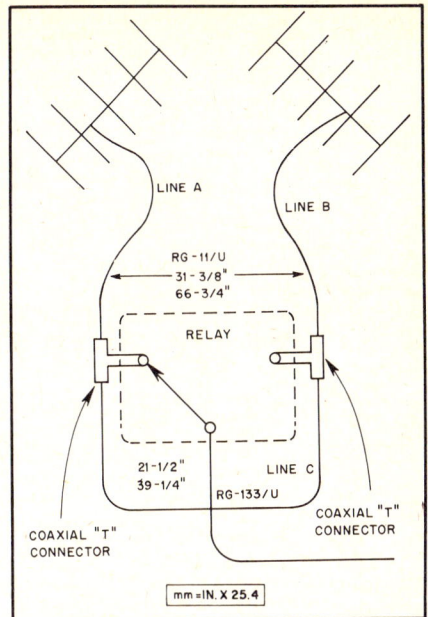

Fig. 14 — This is a drawing of the basic antenna system for switchable right- or left-hand circular polarization. The quarter-wavelength lines between the antennas and the relay step up the antenna 50-ohm impedance to 100 ohms. The phasing line is made from 95-ohm coaxial cable so as to provide a good match to the 100-ohm system. See text for a detailed description of the system. The shorter lengths are for 435.15 MHz and the longer lengths are for 145.925 MHz.

since a single quarter wavelength of line, when the velocity factor is taken into consideration, will be extremely short. The lengths used in this particular system are shown in Fig. 14. The builder should try to use the shortest lengths practical, since the higher the multiple of quarter wavelengths of line the narrower the SWR bandwidth will become.

A Cavity for 435 MHz

If your 435-MHz receiving system is quite sensitive, chances are that you might suffer desensing problems related to the 145-MHz uplink signal. This "Mode J filter" should eliminate the problem. The filter is inserted between the 435-MHz antenna and the 435-MHz preamplifier or converter. The insertion loss is roughly 0.4 to 0.5 dB and the "nose" of the filter is approximately 7 MHz wide. One important feature of this filter is the excellent roll-off characteristic lower in frequency than the passband. This allows a 50-dB attenuation (or more) of the 145-MHz uplink signal frequency.

Most small plumbing businesses can supply you with the materials needed for construction. A local plumber cut several pieces of the 3/4- and 3-inch copper pipe and charged the writer only a few dollars. Circuit-board material can be used for the top and bottom plates. Silver plating all of the parts is a good idea to protect against poor contacts. However, this is not mandatory. The dimensions for the filter are given in Fig. 15.

Parts List

Piece No.

1	Pipe, Copper 3" dia. 5" long	Cut ends square. Drill or punch for connectors 3-3/4" from bottom.
2	Pipe, Copper 3/4" dia. 4" long	Solder to center of 10
3	Disc, Copper 3/4" dia. 1/16" — 1/8" thick	Drill thru center. Solder solid hook up wire between disc and connector to space disc 3/16" from pc 2.
4	Disc, Copper 3/4" dia. 1/16" — 1/8" thick	Drill thru center. Solder solid hook up wire between disc and connector to space disc 3/16" from pc 2.
5	Connector, Coax	BNC, SMA or N type. Solder to prevent turning. For large connector use chassis punch
6	Connector, Coax	BNC, SMA or N type. Solder to prevent turning. For large connector use chassis punch.
7	Nut, Brass 1/4"—20 hex	
8	Nut, Brass 1/4"—20 hex	
9	P/C Board, double sided. Top 4" × 4".	Drill hole in center to clear 1/4-20 bolt Solder 7 and 8 each side of hole. (Use bolt 11 to hold nuts in place when soldering.)
10	P/C Board, double sided. Bottom 4" × 4".	Solder 2 in center.
11	Bolt, Brass 1/4-20 × 3"	Insert thru 12 then thru 7 and 8.
12	Locking Nut, Brass 1/4-20 Hex	To hold piece 11 after resonance adjustment.

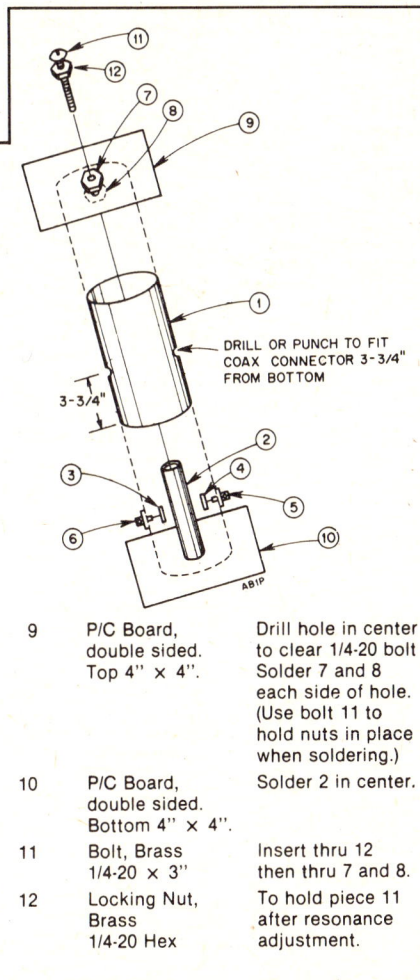

Fig. 15 — Details of the "Mode J desense filter."

EARTH-MOON-EARTH

Popularly known as moonbounce, EME is the second-most popular method of space communication after OSCAR. The concept is straightforward: Stations which can simultaneously see the moon communicate by reflecting vhf and uhf signals off the lunar surface. Unlike OSCAR, though, the two stations have a relatively stable target and may be separated by virtually 180 degrees of arc on the earth's surface, which translates to more than 11,000 miles (17,700 km).

There is a trade-off, though; since the moon's mean distance from earth is 239,000 miles (385,000 km), path losses are huge when compared to "local" vhf work. Thus, each station on an EME circuit demands the most out of the transmitter, antenna, receiver and operator skills. Even with all those factors in an optimum state, the signal in the headphones may be barely perceptible above the noise. Nevertheless, for any type of amateur communication over a distance of 500 miles (800 km) or more at 432 MHz, for example, moonbounce comes out the winner over terrestrial methods when various factors are figured on a balance sheet.

EME thus presents amateurs with the ultimate challenge in strengthening radio systems. Before amateur involvement the only other known moon relay circuit was operated by the U.S. Navy between Washington, DC and Hawaii. Their 400 megawatts of effective radiated power carried four multiplexed RTTY channels. The first two-way amateur link took place between the Eimac Radio Club, W6HB, and the Rhododendron Swamp VHF Society, W1BU, on 1296 MHz in July 1960. Only a few amateurs heard anything more than their own echoes during the next few years. Hams at government and private institutions began conducting tests with other hams by using very large arrays such as the 150-foot (45.7-m) steerable dish at WA6LET (Stanford University) or the 1000-foot (305-m) parabolic surface at KP4BPZ (Arecibo). Amateur-to-amateur contacts did not become established until the early 70s, a notable effort being between VE7BBG and WA6HXW. Activity spread to all continents — except South America. In July 1976, the Mt. Airy VHF Club of Philadelphia (Packrats) staged an expedition to Barraquilla, Colombia, which allowed K2UYH to become the first amateur to work all continents on 432 MHz.

Through the efforts of these early pioneers and others, the state of the art has progressed such that most of the components for an EME station on 144 or 432 MHz are now commercially available. Whether a prospective EME-er chooses that route or builds all the gear, some design considerations must be taken because it is weak-signal work.

1) Transmissions must be made on cw or ssb with as close to the maximum legal input as possible.

2) The antenna should have at least 20 dB of gain over a dipole.

3) As with an OSCAR antenna system, rotators are needed for both azimuth and elevation. Since the half-power beamwidth of a high-gain antenna is quite sharp, the rotators must have an appropriate accuracy.

4) Transmission-line losses should be held to a minimum.

5) The receiving system should have a very low noise figure and sharp filters.

Don't let these requirements scare you! Most EME-ers started out as listeners, and in the first, second and third ARRL International EME Competition operators with nothing more than a single Yagi, preamplifier and multimode transceiver were hearing the stronger stations. For those who are seriously interested in assembling a complete station, the Eimac Division of Varian has assembled a comprehensive package on the technical details. Write to Eimac, 301 Industrial Way, San Carlos, CA 94070.

A short section about operating techniques is offered as a guide to the beginner. It should be noted, however, that the details differ from one band to another to some degree. Such differences are slight, and should cause no great concern. Perhaps as the ranks grow an accepted universal operating procedure will evolve.

EME Scheduling

The best days to schedule are *usually* when the moon is at *perigee* (closest to the earth) since the path loss is typically 2 dB less than when the moon is at *apogee* (farthest from the earth). The moon's perigee and apogee dates may be determined from Tables 4 and 5 by inspecting the column headed SD (semi-diameter of the moon in minutes of arc). An SD of 1653 equates to an approximate earth-to-moon distance of 225,000 miles (362,000 km), typical perigee, and an SD of 14.7 to an approximate distance of 252,500 miles (406,400 km), typical apogee. If the semi-diameters are located on Fig. 16, the EME path losses in decibels may be determined for the most popular amateur frequencies.

The moon follows a sine-wave orbit pattern. Hence, the day-to-day path changes at apogee and perigee are minor. The greatest changes take place at the time when the moon is traversing between apogee and perigee. However, several other factors must be considered for optimum scheduling aside from the path losses.

If perigee occurs near the time of a new moon, one to two days will be unusable since proximity of the moon to the sun's orbit will cause increased sun-noise pickup. Therefore, schedules should be avoided when the moon is within 10° of the sun (and farther if your antenna has a

wide beam or strong side lobes).

Asterisks beside the SD values in Tables 4 and 5 indicate the dates of a new moon, times when sun noise will be highest. The moon's orbit follows a cycle of 18 to 19 years, so the relationships between perigee and new moon will not be the same from one year to the next.

Low moon declinations and low aiming elevation generally produce poor results and should be avoided if possible. Conversely, high moon declinations and high elevation angles should yield best results. Good results are usually obtained when both stations are using similar elevation angles, since then both stations are looking through comparable electron densities. Generally, low elevation angles increase antenna-noise pickup and increase tropospheric absorption, especially above 420 MHz, where the galactic noise is very low. This situation cannot be avoided when one station is unable to elevate the antenna above the horizon or when there is a great terrestrial distance between stations. Ground gain (gain obtained when the antenna is aimed at the horizon) has been used very effectively at 144 MHz, but has been more elusive above 420 MHz. It is hoped that current tests on 144 and 432 MHz, using this mode of propagation, will yield more predictable results.

Usually, signals are stronger in the fall and winter months and weaker in the summer. Also, signals are generally better at night than during the day. This may be attributable to decreased ionization or less Faraday rotation.

Whenever the moon crosses the galactic plane (twice a month for three to five days each occurrence), the sky temperature will be higher. Hence, some degradation (1 to 2 dB) may be observed, especially above 420 MHz where the normal background sky temperature is lower. Areas of the sky to avoid are the constellations Orion and Gemini at northern declinations and Scorpius and Sagittarius at southern declinations. Positions of the moon with respect to these constellations can be checked with *Sky and Telescope* magazine or the *Nautical Almanac*. The galactic plane is biased toward southern declinations, which will cause southerly declinations to be less desirable (with respect to noise) than are northern declinations.

Finally, the time of the day and the day of the week must be considered since most of us have to work for a living and cannot always be available for schedules. Naturally, weekends and evenings are preferred, especially when perigee occurs on a weekend.

General Considerations

It helps to know your own EME window as accurately as possible. This can be determined with the help of information contained in a later section of this chapter. Most EME operators determine their local window and translate it into

Table 4

Greenwich hour angle (GA) and declination (decl.) of the moon at 0000 UTC for each day of the month for January through June 1982. The 2nd and 4th columns indicate hourly increments for GHA and decl., respectively. The 5th column indicates the semi-diameter (SD) of the moon in minutes of arc. An asterisk beside the SD value indicates the date of a new moon. See text for full explanation.

JANUARY

DAY	GHA	HR. INCR.	DECL.	HR. INCR.	SD
1	114.11	14.526	-10.35	0.190	15.5
2	102.73	14.523	-5.80	0.205	15.7
3	91.27	14.510	-0.87	0.212	15.9
4	79.53	14.489	4.23	0.209	16.1
5	67.27	14.459	9.24	0.193	16.3
6	54.29	14.425	13.86	0.161	16.5
7	40.49	14.392	17.74	0.115	16.6
8	25.90	14.371	20.50	0.056	16.6
9	10.81	14.369	21.84	-0.009	16.6
10	355.66	14.387	21.62	-0.072	16.4
11	340.96	14.421	19.90	-0.124	16.3
12	327.06	14.461	16.93	-0.161	16.0
13	314.13	14.499	13.07	-0.184	15.8
14	302.10	14.529	8.64	-0.195	15.5
15	290.80	14.551	3.96	-0.196	15.3
16	280.03	14.563	-0.76	-0.190	15.1
17	269.54	14.566	-5.32	-0.177	14.9
18	259.13	14.562	-9.57	-0.159	14.8
19	248.60	14.551	-13.40	-0.136	14.8
20	237.82	14.536	-16.67	-0.108	14.7
21	226.69	14.520	-19.26	-0.074	14.8
22	215.18	14.506	-21.03	-0.035	14.8
23	203.33	14.497	-21.88	0.007	14.9
24	191.26	14.495	-21.72	0.050	15.0
25	179.13	14.498	-20.51	0.091	15.1*
26	167.09	14.506	-18.32	0.130	15.2
27	155.24	14.516	-15.21	0.162	15.3
28	143.61	14.523	-11.33	0.186	15.5
29	132.16	14.525	-6.86	0.202	15.6
30	120.75	14.519	-2.00	0.210	15.8
31	109.21	14.506	3.04	0.207	15.9

FEBRUARY

DAY	GHA	HR. INCR.	DECL.	HR. INCR.	SD
1	97.36	14.484	8.00	0.193	16.0
2	84.98	14.456	12.63	0.167	16.2
3	71.91	14.425	16.64	0.127	16.3
4	58.12	14.400	19.70	0.076	16.3
5	43.71	14.386	21.52	0.016	16.4
6	28.98	14.390	21.91	-0.045	16.3
7	14.34	14.411	20.82	-0.102	16.3
8	0.20	14.442	18.38	-0.147	16.1
9	346.81	14.477	14.84	-0.179	15.9
10	334.25	14.508	10.56	-0.196	15.7
11	322.43	14.532	5.85	-0.202	15.5
12	311.21	14.549	1.00	-0.198	15.3
13	300.37	14.556	-3.75	-0.186	15.1
14	289.73	14.557	-8.22	-0.169	15.0
15	279.09	14.550	-12.28	-0.146	14.8
16	268.30	14.539	-15.78	-0.118	14.8
17	257.24	14.525	-18.60	-0.085	14.8
18	245.85	14.512	-20.64	-0.048	14.8
19	234.13	14.501	-21.78	-0.007	14.9
20	222.15	14.495	-21.94	0.037	15.0
21	210.02	14.494	-21.06	0.080	15.1
22	197.88	14.498	-19.15	0.120	15.3
23	185.84	14.505	-16.25	0.156	15.4*
24	173.97	14.512	-12.51	0.184	15.6
25	162.25	14.515	-8.08	0.204	15.7
26	150.60	14.513	-3.19	0.213	15.8
27	138.90	14.503	1.94	0.212	15.9
28	126.98	14.487	7.03	0.199	16.0

MARCH

DAY	GHA	HR. INCR.	DECL.	HR. INCR.	SD
1	114.67	14.465	11.79	0.173	16.1
2	101.83	14.441	15.95	0.136	16.1
3	88.41	14.419	19.21	0.088	16.1
4	74.46	14.406	21.32	0.032	16.1
5	60.21	14.406	22.08	-0.027	16.1
6	45.96	14.420	21.43	-0.083	16.1
7	32.04	14.444	19.44	-0.131	16.0
8	18.71	14.473	16.30	-0.167	15.9
9	6.06	14.501	12.28	-0.191	15.7
10	354.07	14.523	7.70	-0.203	15.6
11	342.64	14.540	2.80	-0.203	15.4
12	331.59	14.549	-2.05	-0.195	15.2
13	320.76	14.551	-6.74	-0.179	15.1
14	309.98	14.547	-11.05	-0.157	14.9
15	299.11	14.539	-14.82	-0.130	14.8
16	288.05	14.528	-17.93	-0.097	14.8
17	276.72	14.516	-20.26	-0.060	14.8
18	265.11	14.507	-21.71	-0.020	14.8
19	253.27	14.501	-22.20	0.022	14.9
20	241.27	14.498	-21.67	0.065	15.0
21	229.22	14.500	-20.10	0.107	15.2
22	217.21	14.503	-17.53	0.145	15.4
23	205.29	14.509	-14.05	0.178	15.6
24	193.45	14.508	-9.77	0.203	15.8
25	181.64	14.504	-4.90	0.218	15.9*
26	169.73	14.494	0.33	0.221	16.1
27	157.59	14.478	5.63	0.211	16.2
28	145.07	14.457	10.69	0.187	16.2
29	132.04	14.435	15.18	0.150	16.2
30	118.48	14.416	18.78	0.101	16.2
31	104.47	14.406	21.20	0.045	16.2

APRIL

DAY	GHA	HR. INCR.	DECL.	HR. INCR.	SD
1	90.22	14.409	22.27	-0.014	16.1
2	76.05	14.425	21.93	-0.070	16.0
3	62.25	14.450	20.24	-0.119	15.9
4	49.05	14.478	17.39	-0.157	15.8
5	36.52	14.505	13.63	-0.183	15.6
6	24.63	14.526	9.23	-0.199	15.5
7	13.27	14.542	4.45	-0.204	15.4
8	2.26	14.550	-0.45	-0.200	15.2
9	351.46	14.551	-5.25	-0.188	15.1
10	340.69	14.547	-9.75	-0.168	15.0
11	329.82	14.539	-13.78	-0.142	14.9
12	318.76	14.529	-17.18	-0.110	14.8
13	307.46	14.519	-19.82	-0.074	14.8
14	295.90	14.510	-21.59	-0.034	14.8
15	284.15	14.505	-22.41	0.008	14.8
16	272.27	14.504	-22.22	0.050	14.9
17	260.37	14.506	-21.02	0.092	15.0
18	248.52	14.510	-18.81	0.131	15.2
19	236.77	14.513	-15.67	0.165	15.4
20	225.09	14.513	-11.70	0.194	15.7
21	213.40	14.508	-7.04	0.215	15.9
22	201.58	14.495	-1.88	0.225	16.1
23	189.47	14.476	3.53	0.223	16.3*
24	176.90	14.451	8.87	0.205	16.4
25	163.73	14.425	13.79	0.171	16.5
26	149.92	14.401	17.89	0.123	16.5
27	135.55	14.389	20.84	0.064	16.4
28	120.88	14.392	22.37	0.001	16.3
29	106.28	14.411	22.39	-0.059	16.1
30	92.14	14.440	20.99	-0.110	16.0

MAY

DAY	GHA	HR. INCR.	DECL.	HR. INCR.	SD
1	78.70	14.473	18.35	-0.150	15.8
2	66.06	14.504	14.75	-0.178	15.6
3	54.17	14.529	10.48	-0.195	15.5
4	42.87	14.546	5.80	-0.203	15.3
5	31.99	14.556	0.94	-0.201	15.2
6	21.32	14.557	-3.89	-0.192	15.1
7	10.70	14.553	-8.50	-0.175	15.0
8	359.97	14.544	-12.70	-0.152	14.9
9	349.02	14.532	-16.33	-0.122	14.8
10	337.79	14.521	-19.26	-0.087	14.8
11	326.28	14.512	-21.34	-0.047	14.7
12	314.56	14.507	-22.47	-0.006	14.7
13	302.73	14.507	-22.61	0.037	14.8
14	290.90	14.512	-21.73	0.078	14.9
15	279.18	14.518	-19.86	0.117	15.0
16	267.61	14.524	-17.06	0.151	15.2
17	256.18	14.526	-13.43	0.181	15.4
18	244.80	14.523	-9.08	0.205	15.7
19	233.34	14.511	-4.16	0.220	15.9
20	221.61	14.492	1.12	0.225	16.2
21	209.41	14.464	6.53	0.216	16.4
22	196.54	14.430	11.72	0.192	16.5
23	182.87	14.397	16.32	0.149	16.6*
24	168.40	14.373	19.90	0.092	16.6
25	153.35	14.367	22.10	0.026	16.6
26	138.15	14.381	22.72	-0.040	16.4
27	123.29	14.412	21.75	-0.099	16.2
28	109.17	14.450	19.38	-0.144	16.0
29	95.98	14.489	15.93	-0.175	15.8
30	83.70	14.520	11.72	-0.194	15.6
31	72.20	14.544	7.06	-0.203	15.4

JUNE

DAY	GHA	HR. INCR.	DECL.	HR. INCR.	SD
1	61.24	14.557	2.20	-0.202	15.2
2	50.61	14.562	-2.66	-0.194	15.1
3	40.10	14.559	-7.32	-0.180	14.9
4	29.52	14.550	-11.63	-0.158	14.9
5	18.73	14.538	-15.44	-0.131	14.7
6	7.64	14.525	-18.58	-0.098	14.7
7	356.24	14.514	-20.92	-0.059	14.7
8	344.56	14.507	-22.34	-0.018	14.7
9	332.73	14.506	-22.77	0.025	14.7
10	320.88	14.511	-22.18	0.066	14.8
11	309.16	14.520	-20.58	0.105	14.9
12	297.64	14.529	-18.06	0.140	15.0
13	286.34	14.536	-14.70	0.169	15.2
14	275.21	14.538	-10.64	0.193	15.4
15	264.11	14.531	-6.00	0.210	15.6
16	252.87	14.516	-0.95	0.219	15.9
17	241.25	14.491	4.30	0.217	16.1
18	229.03	14.457	9.51	0.202	16.3
19	215.99	14.418	14.35	0.170	16.5
20	202.01	14.381	18.42	0.121	16.7
21	187.16	14.358	21.32	0.057	16.7*
22	171.76	14.357	22.69	-0.012	16.7
23	156.32	14.378	22.39	-0.079	16.5
24	141.39	14.416	20.51	-0.133	16.4
25	127.37	14.458	17.32	-0.172	16.1
26	114.36	14.497	13.20	-0.195	15.9
27	102.28	14.527	8.52	-0.206	15.6
28	90.94	14.548	3.58	-0.206	15.4
29	80.09	14.558	-1.37	-0.199	15.2
30	69.49	14.560	-6.14	-0.185	15.0

GHA (Greenwich hour angle) and declination. This information is a constant, so once it is determined it is usable by other stations just as one would use UTC. Likewise, it helps to know the window of the station to be scheduled. Most EME stations are limited in some way by local obstructions, antenna-mounting constraints, geographical considerations, and the like. Therefore, the accuracy of each station's EME window is very important for locating common windows and setting schedule times.

A boresight of some type is practically mandatory in order to align your antenna accurately with the moon. Most antenna systems exhibit some pattern skewing which must be accounted for. A simple calibration method is to peak your antenna on received sun noise and then align the boresight tube on the sun. The boresight of the antenna is now calibrated and can be used to aim the antenna at the moon. **Readers are cautioned against using a telescope or other device employing lenses as a boresight device!** Even the best of optical filters will not eliminate the

Table 5

GHA and decl. of the moon at 0000 UTC for each day of the month for July through December 1982. Hourly increments for GHA, decl., the moon's SD and new moon (*) are also indicated.

JULY

DAY	GHA	HR. INCR.	DECL.	HR. INCR.	SD
1	58.92	14.554	-10.57	-0.164	14.9
2	48.23	14.543	-14.52	-0.138	14.8
3	37.26	14.530	-17.83	-0.107	14.7
4	25.98	14.517	-20.39	-0.070	14.7
5	14.39	14.508	-22.06	-0.029	14.7
6	2.57	14.505	-22.76	0.014	14.7
7	350.68	14.508	-22.43	0.056	14.8
8	338.87	14.516	-21.08	0.096	14.9
9	327.26	14.527	-18.78	0.132	14.9
10	315.92	14.538	-15.62	0.162	15.1
11	304.82	14.544	-11.73	0.186	15.2
12	293.87	14.544	-7.28	0.203	15.4
13	282.92	14.535	-2.42	0.212	15.6
14	271.75	14.516	2.67	0.213	15.8
15	260.14	14.488	7.78	0.202	16.0
16	247.85	14.451	12.63	0.179	16.2
17	234.69	14.412	16.92	0.139	16.4
18	220.57	14.378	20.26	0.085	16.6
19	205.63	14.359	22.29	0.019	16.6
20	190.24	14.363	22.73	-0.050	16.6*
21	174.96	14.388	21.52	-0.113	16.5
22	160.29	14.426	18.81	-0.161	16.4
23	146.52	14.467	14.94	-0.193	16.2
24	133.72	14.502	10.31	-0.209	15.9
25	121.77	14.528	5.29	-0.213	15.7
26	110.45	14.545	0.18	-0.207	15.4
27	99.53	14.552	-4.78	-0.193	15.2
28	88.79	14.551	-9.40	-0.172	15.0
29	78.02	14.544	-13.54	-0.146	14.9
30	67.08	14.533	-17.05	-0.115	14.8
31	55.87	14.520	-19.82	-0.080	14.7

AUGUST

DAY	GHA	HR. INCR.	DECL.	HR. INCR.	SD
1	44.35	14.510	-21.73	-0.040	14.7
2	32.59	14.504	-22.69	0.003	14.8
3	20.68	14.504	-22.62	0.046	14.8
4	8.79	14.511	-21.52	0.087	14.9
5	357.05	14.521	-19.42	0.125	15.0
6	345.54	14.532	-16.42	0.157	15.1
7	334.30	14.540	-12.66	0.182	15.2
8	323.27	14.544	-8.29	0.200	15.3
9	312.33	14.541	-3.49	0.210	15.5
10	301.30	14.529	1.55	0.211	15.6
11	289.99	14.508	6.61	0.202	15.8
12	278.17	14.478	11.46	0.182	16.0
13	265.64	14.443	15.82	0.148	16.1
14	252.27	14.409	19.38	0.101	16.3
15	238.07	14.383	21.80	0.042	16.4
16	223.27	14.374	22.81	-0.024	16.5
17	208.25	14.385	22.24	-0.088	16.5
18	193.50	14.412	20.13	-0.143	16.4
19	179.39	14.447	16.70	-0.184	16.3*
20	166.11	14.480	12.29	-0.208	16.1
21	153.64	14.509	7.30	-0.218	15.9
22	141.85	14.528	2.06	-0.216	15.7
23	130.53	14.540	-3.12	-0.203	15.4
24	119.48	14.543	-8.00	-0.184	15.2
25	108.51	14.539	-12.41	-0.158	15.0
26	97.45	14.531	-16.19	-0.126	14.9
27	86.20	14.521	-19.22	-0.091	14.8
28	74.71	14.512	-21.40	-0.051	14.8
29	62.98	14.505	-22.63	-0.009	14.8
30	51.11	14.504	-22.85	0.034	14.8
31	39.19	14.507	-22.02	0.077	14.9

SEPTEMBER

DAY	GHA	HR. INCR.	DECL.	HR. INCR.	SD
1	27.37	14.515	-20.18	0.116	15.0
2	15.73	14.524	-17.38	0.151	15.1
3	4.32	14.533	-13.75	0.180	15.2
4	353.10	14.537	-9.44	0.200	15.4
5	341.99	14.536	-4.64	0.212	15.5
6	330.85	14.527	0.45	0.215	15.6
7	319.51	14.511	5.62	0.207	15.8
8	307.77	14.487	10.58	0.187	15.9
9	295.46	14.458	15.08	0.156	16.0
10	282.45	14.429	18.81	0.112	16.1
11	268.73	14.405	21.49	0.057	16.2
12	254.45	14.394	22.86	-0.005	16.2
13	239.90	14.399	22.75	-0.067	16.3
14	225.47	14.418	21.14	-0.123	16.2
15	211.50	14.445	18.18	-0.169	16.2
16	198.19	14.475	14.12	-0.200	16.1
17	185.58	14.500	9.32	-0.217	15.9*
18	173.58	14.519	4.10	-0.221	15.8
19	162.04	14.531	-1.20	-0.213	15.6
20	150.77	14.535	-6.32	-0.196	15.4
21	139.60	14.533	-11.03	-0.171	15.2
22	128.40	14.527	-15.14	-0.140	15.0
23	117.05	14.519	-18.51	-0.105	14.9
24	105.50	14.511	-21.02	-0.065	14.8
25	93.77	14.506	-22.58	-0.022	14.8
26	81.90	14.505	-23.12	0.021	14.8
27	70.01	14.508	-22.61	0.064	14.8
28	58.20	14.514	-21.07	0.105	14.9
29	46.54	14.522	-18.56	0.142	15.1
30	35.07	14.529	-15.16	0.173	15.2

OCTOBER

DAY	GHA	HR. INCR.	DECL.	HR. INCR.	SD
1	23.76	14.532	-11.00	0.198	15.4
2	12.54	14.530	-6.24	0.215	15.6
3	1.26	14.522	-1.09	0.221	15.7
4	349.78	14.505	4.23	0.217	15.9
5	337.91	14.483	9.43	0.199	16.0
6	325.49	14.456	14.21	0.169	16.1
7	312.43	14.429	18.25	0.125	16.1
8	298.72	14.408	21.25	0.070	16.1
9	284.51	14.400	22.93	0.009	16.1
10	270.10	14.406	23.15	-0.052	16.1
11	255.85	14.426	21.90	-0.109	16.1
12	242.06	14.452	19.29	-0.155	16.0
13	228.92	14.480	15.57	-0.189	15.9
14	216.45	14.505	11.02	-0.211	15.8
15	204.56	14.522	5.96	-0.220	15.7
16	193.09	14.532	0.68	-0.218	15.6
17	181.86	14.535	-4.55	-0.205	15.4*
18	170.71	14.532	-9.47	-0.184	15.3
19	159.49	14.526	-13.89	-0.155	15.1
20	148.10	14.517	-17.61	-0.120	15.0
21	136.50	14.509	-20.49	-0.080	14.9
22	124.72	14.504	-22.41	-0.037	14.8
23	112.83	14.504	-23.31	0.006	14.8
24	100.93	14.509	-23.16	0.050	14.8
25	89.14	14.516	-21.97	0.091	14.9
26	77.53	14.525	-19.79	0.129	15.0
27	66.14	14.533	-16.70	0.162	15.1
28	54.93	14.536	-12.82	0.190	15.3
29	43.80	14.534	-8.26	0.211	15.5
30	32.62	14.524	-3.21	0.223	15.7
31	21.19	14.506	2.15	0.225	15.9

NOVEMBER

DAY	GHA	HR. INCR.	DECL.	HR. INCR.	SD
1	9.34	14.481	7.55	0.214	16.1
2	356.88	14.450	12.68	0.187	16.2
3	343.68	14.419	17.18	0.146	16.3
4	329.72	14.395	20.67	0.090	16.3
5	315.19	14.385	22.84	0.026	16.3
6	300.42	14.393	23.47	-0.038	16.2
7	285.85	14.416	22.55	-0.098	16.1
8	271.84	14.448	20.21	-0.146	16.0
9	258.60	14.481	16.70	-0.182	15.9
10	246.14	14.509	12.34	-0.205	15.8
11	234.35	14.529	7.43	-0.216	15.6
12	223.05	14.540	2.24	-0.217	15.5
13	212.02	14.544	-2.97	-0.209	15.4
14	201.07	14.540	-7.97	-0.191	15.2
15	190.03	14.531	-12.57	-0.166	15.1*
16	178.78	14.520	-16.55	-0.134	14.9
17	167.26	14.510	-19.75	-0.095	14.9
18	155.50	14.503	-22.04	-0.053	14.8
19	143.57	14.502	-23.31	-0.009	14.8
20	131.63	14.507	-23.51	0.035	14.7
21	119.80	14.517	-22.66	0.077	14.8
22	108.20	14.529	-20.81	0.115	14.8
23	96.89	14.539	-18.05	0.149	14.9
24	85.83	14.546	-14.48	0.177	15.1
25	74.94	14.547	-10.23	0.200	15.3
26	64.07	14.539	-5.43	0.216	15.5
27	53.01	14.522	-0.24	0.224	15.8
28	41.55	14.496	5.14	0.221	16.0
29	29.45	14.461	10.44	0.204	16.3
30	16.51	14.422	15.33	0.170	16.4

DECEMBER

DAY	GHA	HR. INCR.	DECL.	HR. INCR.	SD
1	2.63	14.386	19.40	0.119	16.6
2	347.89	14.364	22.25	0.054	16.6
3	332.63	14.364	23.54	-0.016	16.5
4	317.36	14.386	23.15	-0.083	16.4
5	302.62	14.422	21.16	-0.138	16.3
6	288.73	14.462	17.85	-0.178	16.1
7	275.81	14.498	13.59	-0.203	15.9
8	263.76	14.525	8.72	-0.215	15.7
9	252.36	14.542	3.55	-0.217	15.5
10	241.37	14.550	-1.66	-0.210	15.3
11	230.56	14.548	-6.70	-0.195	15.2
12	219.72	14.541	-11.38	-0.172	15.1
13	208.70	14.529	-15.51	-0.143	15.0
14	197.39	14.516	-18.94	-0.107	14.9
15	185.77	14.506	-21.50	-0.066	14.8*
16	173.91	14.501	-23.09	-0.022	14.7
17	161.94	14.504	-23.62	0.022	14.7
18	150.04	14.514	-23.08	0.065	14.7
19	138.37	14.527	-21.52	0.104	14.7
20	127.02	14.541	-19.02	0.138	14.8
21	116.01	14.553	-15.72	0.166	14.9
22	105.27	14.559	-11.73	0.189	15.1
23	94.68	14.557	-7.19	0.206	15.2
24	84.05	14.546	-2.25	0.216	15.5
25	73.14	14.524	2.94	0.218	15.7
26	61.72	14.491	8.16	0.208	16.0
27	49.51	14.450	13.16	0.185	16.3
28	36.32	14.406	17.60	0.144	16.5
29	22.05	14.368	21.07	0.087	16.6
30	6.87	14.348	23.15	0.017	16.7
31	351.22	14.354	23.54	-0.056	16.7

hazard from solar radiation when viewed directly. A simple piece of tubing of small diameter and two or three feet long can serve the purpose in this instance. A symmetrical spot of light cast upon a piece of paper near the back end of the tube will indicate alignment.

A remote readout (such as a syncro or selsyn) is a highly recommended convenience. Accuracies of ± 2° are usually necessary and can be attained with syncros. A remote readout is particularly important for scheduling when the moon is within 45° of the sun or when the sky is overcast. Very few of us are not bothered by occasional fog, rain, snow or overcast. Aiming the antenna blindly seldom pays off.

Locating the Moon

The moon orbits the earth once in approximately 28 days, a lunar month. Because the plane of the moon's orbit is tilted from the earth's equatorial plane, the moon swings in a sine-wave pattern both north and south of the equator. The angle of departure of the moon's position at a given time from the equatorial plane is termed *declination*. Declination angles of the moon, which are continually changing (a few degrees a day), indicate the latitude on the earth's surface where the moon will be at zenith. For this presentation, positive declination angles are used when the moon is north of the equator, and negative angles when south.

The longitude on the earth's surface where the moon will be at zenith is related to the moon's Greenwich Hour Angle, abbreviated G.H.A. or GHA. "Hour angle" is defined as the angle in degrees to the west of the meridian. If the GHA of the moon were zero degrees, it would be directly over the Greenwich meridian. If the moon's GHA were 15 degrees, the moon would be directly over the meridian, which is designated as 15° W. longitude on a globe. As one can readily understand, the GHA of the moon is continually changing, too, because of the orbital velocity of the moon and because of the earth's rotation inside the moon's orbit. The moon's GHA changes at the rate of approximately 347° per day.

GHA and declination are terms that may be applied to any celestial body. *The Nautical Almanac*[1] and other publications list the GHA and decl. of the sun and moon (as well as for other celestial bodies that may be used for navigation) for every hour of the year. Tables 4 through 7 are based on information contained in *The Nautical Almanac* for the same year as this edition of the *Handbook*. This information may be used to point an antenna with precision, rather than merely looking up in the sky and pointing one's antenna by eye. Tables 6 and 7 are for the sun. That information may be useful for boresighting an antenna array, as explained earlier in this chapter. These tables indicate the position of the sun and moon at 0000 UTC for each day of the year (GHA and DECL). Also shown are hourly increments (HR. INCR.) for both GHA and decl., which may be used to interpolate the position of the sun or moon for any time on a given date. Tables 4 and 5 further indicate the semi-diameter (SD) of the moon, as previously discussed.

Using the Tables

The hourly increment (HR. INCR.) information from Tables 4 through 7 may be used to make linear interpolations for the positions of the sun or moon for any time on a given date. Because the orbital velocity of the moon is not constant, linear interpolations will yield some small error, but such error will likely be negligible for Amateur Radio purposes. Worst-case conditions exist when apogee or

[1]*The Nautical Almanac for the Year****, where ****indicates the calendar year for the data. This annual publication is printed by the U.S. Government Printing Office, Washington, DC. It is available from the Superintendent of Documents and from many dealers of marine products.

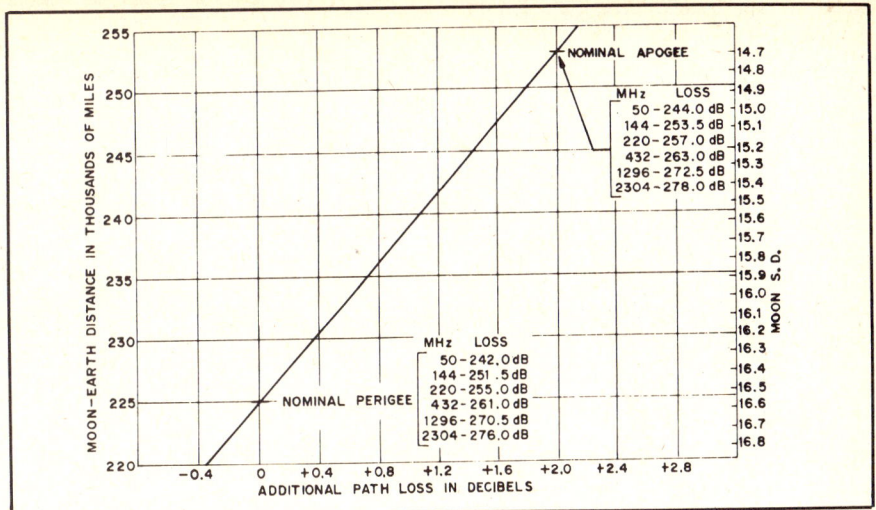

Fig. 16 — Variations in EME path loss can be determined from this graph. SD refers to the semi-diameter of the moon, which is indicated for each day of the year in Tables 4 and 5.

perigee occurs near mid-day on the date in question. Under such conditions the total angular error in the position of the moon as determined by this procedure may be as much as a sixth of a degree. Because it takes a full year for the earth to orbit the sun, the similar error for determining the position of the sun will be no more than a few hundredths of a degree.

Assume that we wish to know the GHA and decl. of the moon at 1115 UTC on June 7, 1982. From Table 4 we see that the GHA on that date is 356.24. In the column to the right we see that the moon's GHA is changing at an average rate of 14.514 degrees per hour on that date. At 1115 UTC on June 7, 11.25 hours will have elapsed since the clock time was 0000 UTC. Multiplying the hourly increment by 11.25 tells us that the GHA of the moon will have increased by 163.28° in that time. Adding this 163.28 to the GHA at 0000 UTC yields 519.52°; we subtract 360° to bring the value into range, and the result indicates that the GHA of the moon at 1115 UTC on June 7 is 159.52°. From the same line in the table we see that the decl. of the moon at 0000 UTC is −20.92°, and the hourly increment is −0.059. Using the same procedure, we multiply −0.059 times 11.25, yielding −0.66. This tells us that the declination at 1115 UTC is 0.66° more negative than it was at 0000 UTC; −20.92 + (−0.66) = −21.58° declination at 1115 UTC on June 7, 1981. The moon's semi-diameter on June 7 is 14.7 minutes of arc. The procedure for using Tables 6 and 7 for determining the sun's position is identical. The result will permit accurate calibration of remote indicators during boresighting.

In using the tables, remember that negative declination angles denote that the sun or moon is south of the equator. As a matter of information, negative hourly increment figures indicate the sun or moon

is moving south, while positive figures indicate northward movement. In interpolating for positions at times other than 0000 UTC, however, it is necessary only to observe algebraic signs listed in the tables and to add the values algebraically.

If a polar mount (a system having one axis parallel to the earth's axis) is used, information from Tables 4 through 7 may be used directly to point the antenna array. The local hour angle (LHA) is simply the GHA plus or minus the observer's longitude (plus if east long., minus if west). The LHA is the angle west of the observer's meridian at which the celestial body is located. The LHA and declination information may be translated to your EME window by taking local obstructions and any other constraints into account.

Azimuth and Elevation

An antenna system which is positioned in azimuth (compass direction) and elevation (angle above the horizon) is called an az-el system. For such a system, some additional work will be necessary to convert the data from Tables 4 through 7 into useful information. The GHA and decl. information may be converted into azimuth and elevation angles with the mathematical equations that follow. An electronic calculator that treats trigonometric functions may be used.

Determining az-el data from equations follows a procedure similar to calculating great-circle bearings and distances for two points on the earth's surface. There is one additional factor, however. Visualize two observers on opposite sides of the earth who are pointing their antennas at the moon. Imaginary lines representing the boresights of the two antennas will converge at the moon at an angle of approximately 2°. Now assume these observers aim their antennas at some distant star.

Table 6

GHA and decl. of the sun at 0000 UTC for each day of the month for January through June 1982. Hourly increments for GHA and decl. are also indicated. Partial eclipses of the sun occur on January 25 and June 21.

JANUARY

DAY	GHA	HR. INCR.	DECL.	HR. INCR.
1	179.18	14.995	-23.05	0.003
2	179.06	14.995	-22.96	0.004
3	178.94	14.995	-22.87	0.004
4	178.83	14.995	-22.78	0.004
5	178.71	14.995	-22.67	0.005
6	178.60	14.995	-22.56	0.005
7	178.49	14.996	-22.44	0.005
8	178.38	14.996	-22.31	0.005
9	178.28	14.996	-22.18	0.006
10	178.18	14.996	-22.04	0.006
11	178.07	14.996	-21.89	0.007
12	177.98	14.996	-21.73	0.007
13	177.88	14.996	-21.57	0.007
14	177.79	14.996	-21.40	0.007
15	177.70	14.996	-21.22	0.008
16	177.61	14.996	-21.04	0.008
17	177.52	14.997	-20.85	0.008
18	177.44	14.997	-20.65	0.008
19	177.36	14.997	-20.45	0.009
20	177.28	14.997	-20.24	0.009
21	177.21	14.997	-20.02	0.009
22	177.14	14.997	-19.80	0.010
23	177.07	14.997	-19.57	0.010
24	177.01	14.997	-19.33	0.010
25	176.95	14.998	-19.09	0.010
26	176.89	14.998	-18.85	0.010
27	176.83	14.998	-18.60	0.011
28	176.78	14.998	-18.34	0.011
29	176.74	14.998	-18.07	0.011
30	176.69	14.998	-17.81	0.011
31	176.65	14.998	-17.53	0.012

FEBRUARY

DAY	GHA	HR. INCR.	DECL.	HR. INCR.
1	176.61	14.999	-17.25	0.012
2	176.58	14.999	-16.97	0.012
3	176.55	14.999	-16.68	0.012
4	176.52	14.999	-16.38	0.012
5	176.50	14.999	-16.08	0.013
6	176.48	14.999	-15.78	0.013
7	176.46	14.999	-15.47	0.013
8	176.45	15.000	-15.16	0.013
9	176.44	15.000	-14.85	0.013
10	176.44	15.000	-14.53	0.014
11	176.43	15.000	-14.20	0.014
12	176.43	15.000	-13.87	0.014
13	176.44	15.000	-13.54	0.014
14	176.44	15.000	-13.20	0.014
15	176.45	15.000	-12.86	0.014
16	176.46	15.001	-12.52	0.014
17	176.48	15.001	-12.17	0.015
18	176.50	15.001	-11.82	0.015
19	176.52	15.001	-11.47	0.015
20	176.54	15.001	-11.11	0.015
21	176.57	15.001	-10.75	0.015
22	176.60	15.001	-10.39	0.015
23	176.63	15.001	-10.03	0.015
24	176.66	15.002	-9.66	0.015
25	176.70	15.002	-9.29	0.015
26	176.74	15.002	-8.92	0.016
27	176.78	15.002	-8.55	0.016
28	176.82	15.002	-8.17	0.016

MARCH

DAY	GHA	HR. INCR.	DECL.	HR. INCR.
1	176.87	15.002	-7.79	0.016
2	176.92	15.002	-7.41	0.016
3	176.97	15.002	-7.03	0.016
4	177.02	15.002	-6.65	0.016
5	177.08	15.002	-6.26	0.016
6	177.13	15.003	-5.88	0.016
7	177.19	15.003	-5.49	0.016
8	177.25	15.003	-5.10	0.016
9	177.31	15.003	-4.71	0.016
10	177.38	15.003	-4.32	0.016
11	177.44	15.003	-3.93	0.016
12	177.51	15.003	-3.54	0.016
13	177.57	15.003	-3.14	0.016
14	177.64	15.003	-2.75	0.016
15	177.71	15.003	-2.35	0.016
16	177.78	15.003	-1.96	0.016
17	177.85	15.003	-1.56	0.016
18	177.93	15.003	-1.17	0.016
19	178.00	15.003	-0.77	0.016
20	178.07	15.003	-0.38	0.016
21	178.15	15.003	0.02	0.017
22	178.22	15.003	0.41	0.016
23	178.29	15.003	0.81	0.016
24	178.37	15.003	1.20	0.016
25	178.45	15.003	1.60	0.016
26	178.52	15.003	1.99	0.016
27	178.60	15.003	2.38	0.016
28	178.67	15.003	2.77	0.016
29	178.75	15.003	3.16	0.016
30	178.82	15.003	3.55	0.016
31	178.90	15.003	3.94	0.016

APRIL

DAY	GHA	HR. INCR.	DECL.	HR. INCR.
1	178.97	15.003	4.33	0.016
2	179.05	15.003	4.72	0.016
3	179.12	15.003	5.10	0.016
4	179.20	15.003	5.48	0.016
5	179.27	15.003	5.86	0.016
6	179.34	15.003	6.24	0.016
7	179.41	15.003	6.62	0.016
8	179.48	15.003	7.00	0.016
9	179.55	15.003	7.37	0.015
10	179.62	15.003	7.74	0.015
11	179.69	15.003	8.11	0.015
12	179.76	15.003	8.48	0.015
13	179.82	15.003	8.85	0.015
14	179.89	15.003	9.21	0.015
15	179.95	15.003	9.57	0.015
16	180.01	15.002	9.93	0.015
17	180.07	15.002	10.28	0.015
18	180.13	15.002	10.63	0.015
19	180.18	15.002	10.98	0.014
20	180.24	15.002	11.33	0.014
21	180.29	15.002	11.67	0.014
22	180.34	15.002	12.01	0.014
23	180.39	15.002	12.35	0.014
24	180.44	15.002	12.68	0.014
25	180.48	15.002	13.01	0.014
26	180.53	15.002	13.34	0.013
27	180.57	15.002	13.66	0.013
28	180.61	15.002	13.98	0.013
29	180.64	15.001	14.29	0.013
30	180.68	15.001	14.60	0.013

MAY

DAY	GHA	HR. INCR.	DECL.	HR. INCR.
1	180.71	15.001	14.91	0.013
2	180.74	15.001	15.21	0.012
3	180.77	15.001	15.51	0.012
4	180.80	15.001	15.81	0.012
5	180.82	15.001	16.10	0.012
6	180.84	15.001	16.38	0.012
7	180.86	15.001	16.66	0.012
8	180.88	15.001	16.94	0.011
9	180.89	15.001	17.21	0.011
10	180.91	15.000	17.48	0.011
11	180.92	15.000	17.74	0.011
12	180.92	15.000	18.00	0.010
13	180.93	15.000	18.25	0.010
14	180.93	15.000	18.49	0.010
15	180.93	15.000	18.74	0.010
16	180.93	15.000	18.97	0.010
17	180.92	15.000	19.20	0.009
18	180.92	15.000	19.43	0.009
19	180.91	15.000	19.65	0.009
20	180.90	14.999	19.86	0.009
21	180.88	14.999	20.07	0.008
22	180.87	14.999	20.28	0.008
23	180.85	14.999	20.47	0.008
24	180.83	14.999	20.66	0.008
25	180.80	14.999	20.85	0.007
26	180.78	14.999	21.03	0.007
27	180.75	14.999	21.20	0.007
28	180.72	14.999	21.37	0.007
29	180.69	14.999	21.53	0.006
30	180.66	14.999	21.69	0.006
31	180.62	14.998	21.84	0.006

JUNE

DAY	GHA	HR. INCR.	DECL.	HR. INCR.
1	180.59	14.998	21.98	0.006
2	180.55	14.998	22.12	0.005
3	180.51	14.998	22.25	0.005
4	180.47	14.998	22.37	0.005
5	180.43	14.998	22.49	0.005
6	180.38	14.998	22.60	0.004
7	180.34	14.998	22.70	0.004
8	180.29	14.998	22.80	0.004
9	180.25	14.998	22.89	0.003
10	180.20	14.998	22.97	0.003
11	180.15	14.998	23.05	0.003
12	180.10	14.998	23.12	0.003
13	180.05	14.998	23.18	0.002
14	180.00	14.998	23.24	0.002
15	179.94	14.998	23.29	0.002
16	179.89	14.998	23.33	0.001
17	179.84	14.998	23.36	0.001
18	179.78	14.998	23.39	0.001
19	179.73	14.998	23.42	0.001
20	179.67	14.998	23.43	0.000
21	179.62	14.998	23.44	0.000
22	179.56	14.998	23.44	0.000
23	179.51	14.998	23.44	-0.000
24	179.45	14.998	23.42	-0.001
25	179.40	14.998	23.40	-0.001
26	179.35	14.998	23.38	-0.001
27	179.29	14.998	23.35	-0.002
28	179.24	14.998	23.31	-0.002
29	179.19	14.998	23.26	-0.002
30	179.14	14.998	23.21	-0.003

The boresight lines now may be considered to be parallel, each observer having raised his antenna in elevation by approximately 1°. The reason for the necessary change in elevation is that the earth's diameter in comparison to its distance from the moon is significant. The same is not true for distant stars, or for the sun.

Equations for az-el calculations are:

$$\sin E = \sin L \sin D + \cos L \cos D \cos LHA \quad \text{(Eq. 1)}$$

$$\tan F = \frac{\sin E - K}{\cos E} \quad \text{(Eq. 2)}$$

$$\cos C = \frac{\sin D - \sin E \sin L}{\cos E \cos L} \quad \text{(Eq. 3)}$$

where

E = elevation angle for the sun
L = your latitude (negative if south)
D = declination of the celestial body
LHA = local hour angle = GHA plus or minus your longitude (plus if east long., minus if west long.)
F = elevation angle for the moon

Table 7

GHA and decl. of the sun at 0000 UTC for each day of the month for July through December 1982. Hourly increments for GHA and decl. are also indicated. Partial eclipses of the sun occur on July 20 and December 15.

JULY

DAY	GHA	HR. INCR.	DECL.	HR. INCR.
1	179.09	14.998	23.15	-0.003
2	179.04	14.998	23.08	-0.003
3	178.99	14.998	23.01	-0.003
4	178.95	14.998	22.93	-0.004
5	178.90	14.998	22.84	-0.004
6	178.86	14.998	22.75	-0.004
7	178.82	14.998	22.65	-0.004
8	178.78	14.998	22.54	-0.005
9	178.74	14.998	22.43	-0.005
10	178.70	14.999	22.31	-0.005
11	178.67	14.999	22.18	-0.006
12	178.64	14.999	22.05	-0.006
13	178.60	14.999	21.91	-0.006
14	178.57	14.999	21.76	-0.006
15	178.55	14.999	21.61	-0.007
16	178.52	14.999	21.46	-0.007
17	178.50	14.999	21.29	-0.007
18	178.47	14.999	21.12	-0.007
19	178.45	14.999	20.95	-0.008
20	178.44	14.999	20.77	-0.008
21	178.42	14.999	20.58	-0.008
22	178.41	15.000	20.39	-0.008
23	178.40	15.000	20.19	-0.009
24	178.39	15.000	19.98	-0.009
25	178.39	15.000	19.77	-0.009
26	178.38	15.000	19.56	-0.009
27	178.38	15.000	19.34	-0.009
28	178.39	15.000	19.11	-0.010
29	178.39	15.000	18.88	-0.010
30	178.40	15.000	18.65	-0.010
31	178.41	15.001	18.40	-0.010

AUGUST

DAY	GHA	HR. INCR.	DECL.	HR. INCR.
1	178.42	15.001	18.16	-0.010
2	178.44	15.001	17.91	-0.011
3	178.45	15.001	17.65	-0.011
4	178.47	15.001	17.39	-0.011
5	178.50	15.001	17.12	-0.011
6	178.52	15.001	16.85	-0.011
7	178.55	15.001	16.58	-0.012
8	178.58	15.001	16.30	-0.012
9	178.61	15.001	16.02	-0.012
10	178.65	15.002	15.73	-0.012
11	178.68	15.002	15.44	-0.012
12	178.72	15.002	15.14	-0.013
13	178.77	15.002	14.84	-0.013
14	178.81	15.002	14.53	-0.013
15	178.86	15.002	14.23	-0.013
16	178.91	15.002	13.91	-0.013
17	178.96	15.002	13.60	-0.013
18	179.01	15.002	13.28	-0.013
19	179.06	15.002	12.96	-0.014
20	179.12	15.002	12.63	-0.014
21	179.18	15.003	12.30	-0.014
22	179.24	15.003	11.97	-0.014
23	179.30	15.003	11.63	-0.014
24	179.37	15.003	11.29	-0.014
25	179.43	15.003	10.95	-0.014
26	179.50	15.003	10.61	-0.015
27	179.57	15.003	10.26	-0.015
28	179.65	15.003	9.91	-0.015
29	179.72	15.003	9.56	-0.015
30	179.79	15.003	9.20	-0.015
31	179.87	15.003	8.84	-0.015

SEPTEMBER

DAY	GHA	HR. INCR.	DECL.	HR. INCR.
1	179.95	15.003	8.48	-0.015
2	180.03	15.003	8.12	-0.015
3	180.11	15.003	7.76	-0.015
4	180.19	15.003	7.39	-0.015
5	180.27	15.003	7.02	-0.015
6	180.36	15.004	6.65	-0.015
7	180.44	15.004	6.28	-0.016
8	180.53	15.004	5.90	-0.016
9	180.61	15.004	5.53	-0.016
10	180.70	15.004	5.15	-0.016
11	180.79	15.004	4.77	-0.016
12	180.87	15.004	4.39	-0.016
13	180.96	15.004	4.01	-0.016
14	181.05	15.004	3.63	-0.016
15	181.14	15.004	3.24	-0.016
16	181.23	15.004	2.86	-0.016
17	181.31	15.004	2.47	-0.016
18	181.40	15.004	2.09	-0.016
19	181.49	15.004	1.70	-0.016
20	181.58	15.004	1.31	-0.016
21	181.67	15.004	0.92	-0.016
22	181.76	15.004	0.53	-0.016
23	181.84	15.004	0.14	-0.016
24	181.93	15.004	-0.25	-0.016
25	182.02	15.004	-0.64	-0.016
26	182.11	15.004	-1.03	-0.016
27	182.19	15.004	-1.42	-0.016
28	182.28	15.004	-1.81	-0.016
29	182.36	15.003	-2.19	-0.016
30	182.44	15.003	-2.58	-0.016

OCTOBER

DAY	GHA	HR. INCR.	DECL.	HR. INCR.
1	182.53	15.003	-2.97	-0.016
2	182.61	15.003	-3.36	-0.016
3	182.69	15.003	-3.75	-0.016
4	182.76	15.003	-4.13	-0.016
5	182.84	15.003	-4.52	-0.016
6	182.92	15.003	-4.90	-0.016
7	182.99	15.003	-5.29	-0.016
8	183.06	15.003	-5.67	-0.016
9	183.13	15.003	-6.05	-0.016
10	183.20	15.003	-6.43	-0.016
11	183.27	15.003	-6.81	-0.016
12	183.33	15.003	-7.19	-0.016
13	183.39	15.003	-7.57	-0.016
14	183.45	15.002	-7.94	-0.015
15	183.51	15.002	-8.31	-0.015
16	183.57	15.002	-8.68	-0.015
17	183.62	15.002	-9.05	-0.015
18	183.67	15.002	-9.42	-0.015
19	183.72	15.002	-9.78	-0.015
20	183.77	15.002	-10.14	-0.015
21	183.81	15.002	-10.50	-0.015
22	183.85	15.002	-10.86	-0.015
23	183.89	15.001	-11.21	-0.015
24	183.92	15.001	-11.56	-0.015
25	183.95	15.001	-11.91	-0.014
26	183.98	15.001	-12.25	-0.014
27	184.01	15.001	-12.60	-0.014
28	184.03	15.001	-12.93	-0.014
29	184.05	15.001	-13.27	-0.014
30	184.07	15.001	-13.60	-0.014
31	184.08	15.000	-13.93	-0.014

NOVEMBER

DAY	GHA	HR. INCR.	DECL.	HR. INCR.
1	184.09	15.000	-14.25	-0.013
2	184.10	15.000	-14.57	-0.013
3	184.10	15.000	-14.89	-0.013
4	184.10	15.000	-15.20	-0.013
5	184.10	15.000	-15.51	-0.013
6	184.10	15.000	-15.82	-0.013
7	184.09	14.999	-16.12	-0.012
8	184.07	14.999	-16.41	-0.012
9	184.05	14.999	-16.70	-0.012
10	184.03	14.999	-16.99	-0.012
11	184.01	14.999	-17.27	-0.012
12	183.98	14.999	-17.55	-0.011
13	183.95	14.999	-17.82	-0.011
14	183.91	14.998	-18.08	-0.011
15	183.88	14.998	-18.34	-0.011
16	183.83	14.998	-18.60	-0.010
17	183.79	14.998	-18.85	-0.010
18	183.74	14.998	-19.09	-0.010
19	183.69	14.998	-19.33	-0.010
20	183.63	14.998	-19.57	-0.009
21	183.57	14.997	-19.79	-0.009
22	183.51	14.997	-20.01	-0.009
23	183.44	14.997	-20.23	-0.009
24	183.37	14.997	-20.44	-0.008
25	183.30	14.997	-20.64	-0.008
26	183.23	14.997	-20.84	-0.008
27	183.15	14.997	-21.03	-0.008
28	183.07	14.996	-21.21	-0.007
29	182.98	14.996	-21.39	-0.007
30	182.89	14.996	-21.56	-0.007

DECEMBER

DAY	GHA	HR. INCR.	DECL.	HR. INCR.
1	182.80	14.996	-21.72	-0.007
2	182.71	14.996	-21.87	-0.006
3	182.62	14.996	-22.02	-0.006
4	182.52	14.996	-22.16	-0.006
5	182.42	14.996	-22.30	-0.005
6	182.31	14.996	-22.43	-0.005
7	182.21	14.996	-22.55	-0.005
8	182.10	14.995	-22.66	-0.004
9	181.99	14.995	-22.76	-0.004
10	181.88	14.995	-22.86	-0.004
11	181.77	14.995	-22.95	-0.003
12	181.65	14.995	-23.04	-0.003
13	181.53	14.995	-23.11	-0.003
14	181.42	14.995	-23.18	-0.002
15	181.30	14.995	-23.24	-0.002
16	181.18	14.995	-23.29	-0.002
17	181.05	14.995	-23.34	-0.002
18	180.93	14.995	-23.37	-0.001
19	180.81	14.995	-23.40	-0.001
20	180.68	14.995	-23.42	-0.001
21	180.56	14.995	-23.44	-0.000
22	180.44	14.995	-23.44	0.000
23	180.31	14.995	-23.44	0.000
24	180.19	14.995	-23.43	0.001
25	180.06	14.995	-23.41	0.001
26	179.94	14.995	-23.39	0.001
27	179.81	14.995	-23.35	0.002
28	179.69	14.995	-23.31	0.002
29	179.57	14.995	-23.26	0.002
30	179.45	14.995	-23.20	0.003
31	179.33	14.995	-23.14	0.003

K = 0.01657, a constant (see text which follows)

C = true azimuth from north if sin LHA is negative; if sin LHA is positive, then the azimuth = 360 − C

Assume our location is 50° N. lat., 100° W. long. Further assume that the GHA of the moon is 140° and its declination is 10°. To determine the az-el information we first find the LHA, which is 140 − 100 or 40°. Then we solve Eq. 1:

sin E
= sin 50 sin 10 + cos 50 cos 10 cos 40
sin E = 0.61795 and E = 38.2°

Solving Eq. 2 for F, we proceed. (The value for sin E has already been determined in Eq. 1.)

$$\tan F = \frac{0.61795 - 0.01657}{\cos 38.2} = 0.76489$$

From this, F, the moon's elevation angle, is 37.4°

We continue by solving Eq. 3 for C. (The value for sin E has already been determined.)

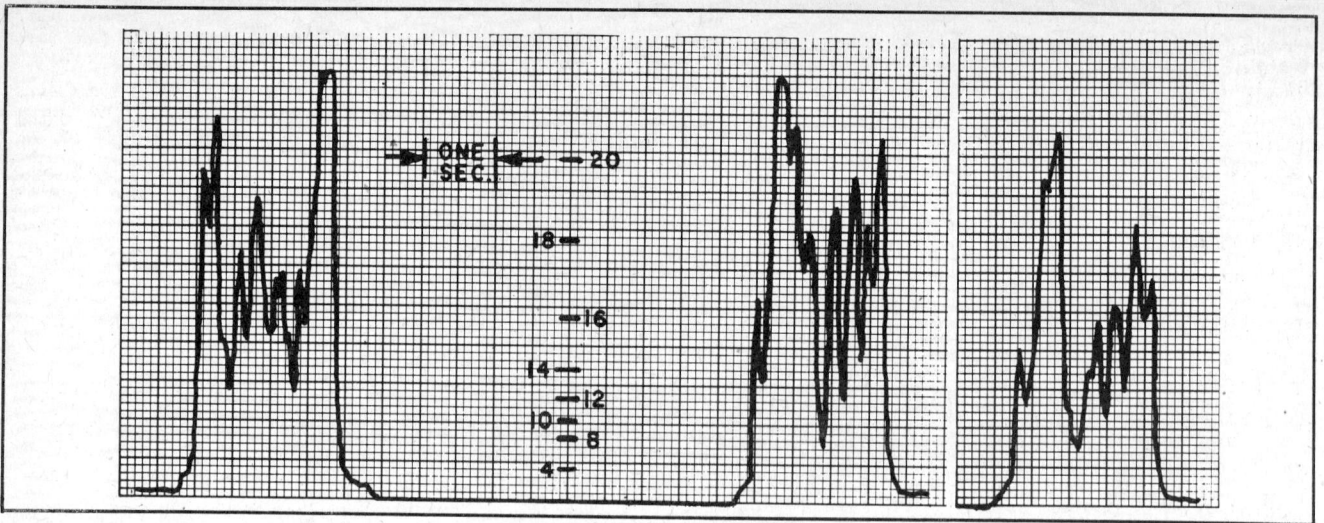

Fig. 17 — Chart recording of moon echoes received at W2NFA on July 26, 1973, at 1630 UTC. Antenna gain 44 dBi; transmitting power 400 watts and system temperature 400 K.

$$\cos C = \frac{\sin 10 - 0.61795 \sin 50}{\cos 38.2 \cos 50}$$
$$= -0.59308$$

C therefore equals 126.4°. To determine if C is the actual azimuth, we find the polarity for sin LHA, which is sin 40° and has a positive value. The actual azimuth then is 360 − C = 233.6°.

If az-el data is being determined for the sun, use of Eq. 2 may be omitted; Eq. 2 takes into account the nearness of the sun. The solar elevation angle may be determined from Eq. 1 alone. In the above example, this angle is 38.2°.

The mathematical procedure is the same for any location on the earth's surface. Remember to use negative values for southerly latitudes. If solving Eq. 1 or 2 yields a negative value for E or F, this indicates the celestial body is below the horizon.

The above equations may also be used to determine az-el data for man-made satellites, but a different value for the constant, K, must be used. K is defined as the ratio of the earth's radius to the distance from the earth's center to the satellite.

The value for K as given above, 0.01657, is based on an average earth-moon distance of 239,000 miles (384,630 km). The actual earth-moon distance varies from approximately 225,000 mi (362,100 km) to 253,000 mi (407,200 km). This change in distance, if taken into account, yields a change in elevation angle of approximately 0.1° when the moon is near the horizon. For greater precision in determining the correct elevation angle for the moon, the moon's distance from the earth may be taken as:

$$D = -15074.5 \times SD + 474,332$$
where

D = moon's distance in miles
SD = moon's semi-diameter, from

Table 4 or 5

Eqs. 1, 2 and 3 are readily adaptable for use with programmable calculators or with computers. Listings of such programs are available from ARRL Hq. for the Hewlett-Packard 25 and similar calculators,[2] and in BASIC language for the Radio Shack TRS-80 Level II computer.[3] Be sure to include a stamped envelope with your request (or an IRC for addresses outside the U.S.), and don't forget to mention which program listing you desire.

Libration Fading of EME Signals

One of the most troublesome aspects of receiving a moonbounce signal besides the enormous path loss and Faraday rotation fading, is *libration fading*. This section will deal with libration (pronounced *lie-bray-shun*) fading, its cause and effects, and possible measures to minimize it.

Libration fading of an EME signal is characterized in general as a fluttery, rapid, irregular fading not unlike that observed in tropospheric scatter propagation. Fading can be very deep, 20 dB or more, and the maximum fading rate will depend on the operating frequency. At 1296 MHz the maximum fading rate is about 10 Hz, and scales directly with frequency.

On a weak cw EME signal, libration fading gives the impression of a randomly keyed signal. In fact on very slow cw telegraphy the effect is as though the keying is being done at a much faster speed.

[2]For the HP-25 program, send a return business-size envelope with postage and 50 cents for handling (or IRCs for foreign addresses) to ARRL Hq., Dept. TD-EME, 225 Main St., Newington, CT 06111.

[3]For the TRS-80 program send a return business-size envelope with postage and 50 cents for handling (or IRCs for foreign addresses) to ARRL Hq., Dept. TD-MOONTRAK, 225 Main St., Newington, CT 06111.

On very weak signals only the peaks of libration fading are heard in the form of occasional short bursts or "pings."

Fig. 17 shows samples of a typical EME echo signal at 1296 MHz. These recordings, made at W2NFA, show the wild fading characteristics with sufficient S/N ratio to record the deep fades. Circular polarization was used to eliminate Faraday fading; thus these recordings are of libration fading only. The recording bandwidth was limited to about 40 Hz to minimize the higher sideband-frequency components of libration fading which persist but are much smaller in amplitude. For those who would like a better statistical description, libration fading is Raleigh distributed. In the recordings shown by Fig. 16, the average signal-return level computed from path loss and mean reflection coefficient of the moon is at about the + 15 dB S/N level.

It is clear that enhancement of echoes far in excess of this average level is observed. This point should be kept clearly in mind when attempting to obtain echoes or receive EME signals with marginal equipment. The probability of hearing an occasional peak is quite good since random enhancement as much as 10 dB is possible. Under these conditions however, the amount of useful information which can be copied will be near zero. The enthusiastic newcomer to EME communications will be stymied by this effect since he knows that he can hear the signal strong enough on peaks to copy but can't make any sense out of what he tries to copy.

What causes libration fading? Very simply, multipath scattering of the radio waves from the very large (2000-mile diameter) and rough moon surface combined with the relative motion between earth and moon called librations.

To understand these effects, assume

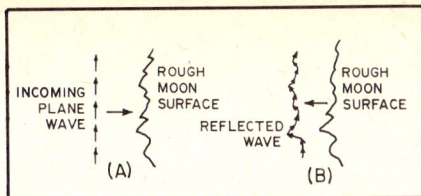

Fig. 18 — How the rough surface of the moon reflects a plane wave as one having many field vectors.

as shown in Fig. 19. Then over a period of time we will observe that this marker wanders around within a small area. All this means is that the surface of the moon as seen from the earth is not quite fixed but changes slightly as different areas of the periphery are exposed because of this rocking motion. Moon libration is very slow (on the order of 10^{-7} radians per second) and can be determined with some difficulty from published moon ephemeris tables.

Although the libration motions are very small and slow, the large surface area of the moon, having nearly an infinite number of scattering points (small area), means that even these slight geometric movements can alter the total summation of the returned multipath echo by a significant amount. Since the librations of the earth and moon are calculable, it is only logical to ask if there ever occurs a time when the total libration is zero or near zero? The answer is yes, and it has been observed and experimentally verified on radar echoes that minimum *fading rate* (not depth of fade) is coincident with minimum total libration. Calculation of minimum total libration is at best tedious and can only be done successfully by means of a digital computer. It is a problem in extrapolation of rates of change in coordinate motion and in small differences of large numbers.

EME Operating Techniques

Most EME signals tend to be near the threshold of readability, a condition caused by a combination of path loss, Faraday rotation and libration fading. This weakness and unpredictability of the signal has led to the development of techniques for exchange of EME information that differs from those used for normal vhf work — the usual RST reporting would be jumbled and meaningless for many EME contacts. Dashes are often chopped into pieces, a string of dots would be incomplete, and complicated words would make no sense at all.

Unfortunately, there is no universal agreement as to procedures for all the bands, although there is similarity. Two-meter operators generally use the "T M O R" system, while those on 432 MHz use a similar system but applied at somewhat different levels of readability. The meanings, and typical use, of each part of the sequence are given in Tables 8 through 11.

At the moment, there is no widespread system in use for bands other than 144 and 432 MHz. There are so few participants on 50 and 220 MHz that presumably they will have no difficulty in arranging techniques by correspondence prior to schedules for EME tests. The amount of operation on 1296 MHz is low, but on the increase. Perhaps an operating technique can evolve that will be acceptable to those on any band. The important

first that the earth and moon are stationary (no libration) and that a plane wave front arrives at the moon from your earth-bound station as shown in Fig. 18A.

The reflected wave shown in Fig. 17B consists of many scattered contributions from the rough moon surface. It is perhaps easier to visualize the process as if the scattering were from many small individual flat mirrors on the moon which reflect small portions (amplitudes) of the incident wave energy in different directions (paths) and with different path lengths (phase). Those paths directed toward the moon arrive at your antenna and appear as a collection of small wave fronts (field vectors) of various amplitudes and phases. The vector summation of all these coherent (same frequency) returned waves (and there is a near-infinite array of them) takes place at the feed point of your antenna (the collecting point in the antenna system). The level of the final summation as measured by a receiver can, of course, have any value from zero to some maximum. Remember now that we assumed the earth and moon were stationary, which means that the final summation of these multipath signal returns from the moon will be one *fixed* value. The condition of relative motion between earth and moon being zero is a rare event which will be discussed later in this section.

Consider now that the earth and moon are moving relative to each other (as they are in nature), so that the incident radio wave "sees" a slightly different surface of the moon from moment to moment. Since the lunar surface is very irregular, the reflected wave will be equally irregular, changing in amplitude and phase from moment to moment. The resultant continuous summation of the varying multipath signals at your antenna feed point produces the effect called libration fading of the moon-reflected signal.

The term libration is used to describe small perturbations in the movement of celestial bodies. Earth libration consists mainly of its diurnal rotation; moon libration consists mainly of its 28-day rotation which appears as a very slight rocking motion with respect to an observer on earth. This rocking motion can be visualized as follows: Place a marker on the surface of the moon at the center of the moon disc, which is the point closest to the observer,

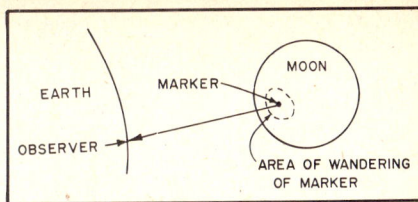

Fig. 19 — The moon appears to "wander" in its orbit about the earth. Thus, a fixed marker on the moon's surface will appear to move about in a circular area.

Table 8
Signal Reports Used on 144-MHz EME

T — Signal just detectable
M — Portions of calls copied
O — Complete call set has been received
R — Both "O" report and call sets have been received
SK — End of contact

Table 9
Signal Reports Used on 432-MHz EME

T — Portions of calls copied
M — Complete calls copied
O — Good signal — solid copy (possibly enough for ssb work)
R — Calls and reports copied
SK — End of contact

Table 10
144-MHz Procedure—2-Minute Sequence

Period	1-1/2 minutes	30 seconds
1	Calls (W6XXX de W1XXX)	
2	W1XXX de W6XXX	T T T T
3	W6XXX de W1XXX	O O O O
4	RO RO RO RO	de W1XXX k
5	R R R R R	de W6XXX k
6	QRZ? EME	de W1XXX k

Table 11
432-MHz Procedure—2-1/2-Min. Sequence

Period	2 minutes	30 seconds
1	VE7BBG de K2UYH	
2	K2UYH de VE7BBG	
3	VE7BBG de K2UYH	T T T
4	K2UYH de VE7BBG	M M M
5	RM RM RM RM	de K2UYH k
6	R R R R R	de VE7BBG SK

consideration is that an exchange of information takes place. This information should include three basic parts: calls of both stations, signal reports, and confirmation that previous information was received.

In the schedule sequence for both 144 and 432 MHz, the initial period starts on the hour, but because of the difference in sequence lengths for the two bands, schedules starting on the half hour will not be the same. On two meters, there are 15 sequence periods to the half hour, which would make the period 0030 to 0032 an "even" sequence. This could make a difference, depending on which operator was assigned an "odd" or "even" sequence. Note that *odd* or *even* refers to the sequence number, not the minutes designated within that sequence.

On 432 MHz, there are 12 sequence periods to the half hour. The eastern-most station calls first, and since two 2-1/2-minute periods fill a 5-minute space, it works out conveniently that the eastern (or first) station will call starting with every five minute mark, and start listening 2-1/2 minutes later. Thus a schedule starting at 0030 would be an "odd" period, although operators on 432 MHz seldom label them as such. It is convenient for the operators to simply start with the eastern-most station calling on the hour or half hour, unless arranged otherwise.

Of course there is much room for change in these arrangements, but they do serve as vital guidelines for schedules. As signals become stronger, the rules can be relaxed to a degree, and after many contacts, stations can often ignore them completely, if the signals are strong enough.

Calls are often extremely difficult to hear in their entirety. A vital dit or dah can be missing, which can render a complex call unreadable. To copy both calls completely requires much patience and a good ear. Both calls must be copied, because even though most work is by timed schedules, there can be last-minute substitutions because of equipment trouble at one station, unexpected travel, plan changes or the like, which make it impossible for the scheduled station to appear. Thus, rather than have one station spend the entire period listening, only to find that no one was there, a system of standby stations is becoming more popular. This is good, because nothing will demoralize a newcomer faster than several one-sided schedules.

An exchange of signal reports is a useful and required bit of information: useful because it helps in evaluating your station performance and the prevailing conditions at the time, required because it is a "non-prearranged" exchange, thereby requiring that you copy what was sent as part of the contact. Obviously there are other things that could be included in an "exchange of unknown information," and when conditions permit stronger signals, many operators do include names, elaborate on the signal reports, arrange next schedule times, and so on. Unfortunately, such exchanges are rare.

Confirmation is essential for completion of the exchange. There is no way that you can be sure that the other operator copied what you sent until you hear him say so. That final R or "roger" means that he has copied your information, and your two-way contact is complete.

Sending speed is usually in the 5- to 10-wpm range, although it can be adjusted according to conditions and operator skill. Characters sent too slowly tend to become chopped up and confusing. High-speed cw is hard to copy at marginal signal levels for most amateurs, and the fading that is typical of an EME path can make it well nigh impossible to decipher the content.

Other Modes

Only a few stations have the capability of sending (and receiving) signals of a strength sufficient to allow experimentation with other than cw. There have been some ssb contacts and echoes of RTTY and fm signals have been heard, but no two-way communication by these latter modes has been accomplished to date. In general, only the stations with large parabolic reflectors try these more difficult means of EME work. Such installations are often "borrowed" from some research program for the amateur endeavors.

Frequencies

Most amateur EME work is conducted within a few kHz of some convenient spot frequency. On 144 MHz there is some room to move about, but most operation is very near the low edge, consistent with the ability of the station to stay within the band.

Operation on 432 MHz is generally within 1 or 2 kilohertz of that frequency, with a few stations going as far afield as 431.997 or 432.003 for general schedules. It is not unknown for a pair of stations to move up 10 or 15 kHz for a contact, while others are on the nominal ".000" spot. There has been a movement afoot recently to reach a gentleman's agreement to avoid any short-range, local, or non-DX operation between 432.0 and 432.025 MHz — a concept that most EME enthusiasts heartily approve of.

For operation on 1296 MHz, most stations are within a very few kilohertz of that spot frequency. Many devices, tubes and transistors, would work much better at the low end of the 1215-MHz band, but the 1296.0 spot became popular because it was convenient to triple from an existing 432-MHz exciter.

Of course, it is obvious that as the number of stations on EME increases, the frequency spread must become greater. Since the moon is in convenient locations only a few days out of the month, and only a certain number of stations can be scheduled for EME during a given evening, the answer will be in the use of simultaneous schedules, spaced a few kilohertz apart. The time may not be too far away — QRM has already been experienced on each of our three most active EME frequencies.

EME Net Information

An EME net meets on weekends at 1600 and 1700 UTC for the purpose of arranging schedules and exchanging pertinent in-

Fig. 20 — The EME antenna system used at K1ZZ — four Cushcraft 2-meter Boomers with associated stacking and wiring harness. This system is mounted atop a 70-foot Rohn 25 tower.

formation. The frequency of operation is 14.345 MHz.

Antenna Requirements

The tremendous path loss incurred over an EME circuit places stringent requirements on the station performance. Low-noise receiving equipment, maximum legal power and large antenna arrays are required for successful EME operation. Although it may be possible to copy some of the better-equipped stations while using a single high-gain Yagi antenna, it is doubtful whether such an antenna could provide reliable two-way communication. Antenna gain of at least 20 dB is required for reasonable success. Generally speaking, more antenna gain will yield the most noticeable improvement in station performance, as the increased gain will aid both the received and transmitted signals.

Several types of antennas have become popular among EME enthusiasts. Perhaps the most popular antenna for 144-MHz work is an array of either four or eight long-boom (14- to 15-dB gain) Yagis. The four-Yagi array would provide approximately 20-dB gain, and the eight-antenna system would show an approximate 3 dB increase over the four-antenna array. At 432 MHz, eight or 16 long-boom Yagis are used. Yagi antennas are available commercially or can be constructed from readily available materials. Information on maximum-gain Yagi antennas is presented in the VHF and UHF Antennas chapter of this volume. The dimensions presented are based on figures developed by the National Bureau of Standards for Yagi design. At least one manufacturer has used the NBS design information for their latest series of high-performance antennas.

A moderately sized Yagi array has the advantage that it is relatively easy to construct and can be positioned in azimuth and elevation with commercially available equipment. Matching and phasing lines present no particular problems. The main disadvantage of a Yagi array is that the polarization plane of the antenna cannot be conveniently changed. One way around this would be to use cross-polarized Yagis and a relay switching system to select the desired polarization. This represents a considerable increase in system cost and complexity. Polarization shift at 144 MHz is fairly slow and the added complexity of the cross-polarized antenna system may not be worth the effort. At 432 MHz, where the shift is at a somewhat faster rate, an adjustable polarization system offers a definite advantage over a fixed one.

A photograph of the Yagi antenna system used at K1ZZ is shown in Fig. 20. The system consists of four, 2-meter Cushcraft Boomer antennas, mounted on a 70-foot, Rohn 25 tower. A CDE Ham-III rotator is used for positioning the antenna in azimuth and a TET KR-500 rotator is used for elevation control. The gain of this array is approximately 20 dB, taking into account phasing line losses.

Quagi antennas (made from both quad and Yagi elements) are also popular for EME work. Slightly more gain per unit boom length is possible as compared to the conventional Yagi. Additional information on the quagi is presented in the VHF and UHF Antennas chapter of this book.

The collinear is another popular type of antenna for EME work. A 40-element collinear array has approximately the same frontal area as an array of four Yagis. The collinear array would produce approximately 1 to 2 dB less gain. Of course the depth dimension of the collinear array is considerably less than for the long-boom Yagis. An 80-element collinear would be marginal for EME communications, providing approximately 19-dB gain. Many operators choosing this type of antenna use 160-element or larger systems. As with Yagi and quagi antennas, the collinear cannot be easily adjusted for polarity changes. From a constructional standpoint there may be little difference in complexity and material costs between the collinear and Yagi arrays.

The parabolic dish is another antenna that is used extensively for EME work. Unlike the other antennas described, the major problems associated with dish antennas are mechanical ones. Dishes 20 feet in diameter are required for successful EME operation on 432 MHz. Structures of this size and wind/ice loading place a severe strain on the mounting/positioning systems. Extremely rugged mounts are required for large dish antennas, especially when used in windy locations. Several aspects of the parabolic dish antennas make the extra mechanical problems worth the effort. For example, the dish antenna is inherently broadband and may be used on several different bands by

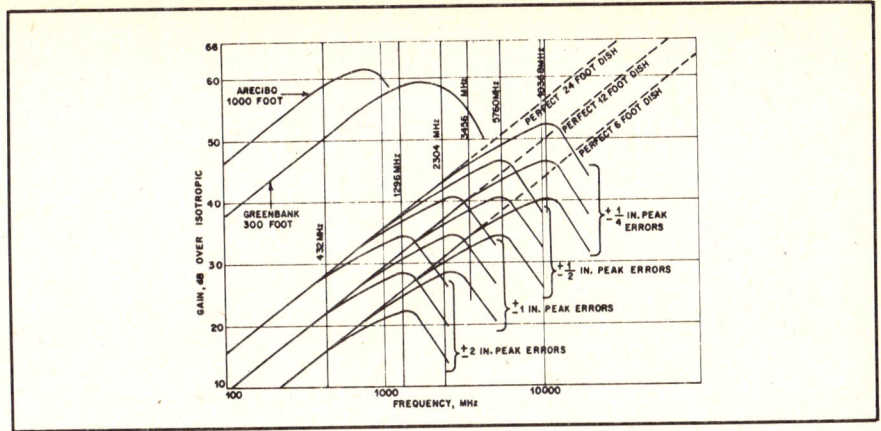

Fig. 21 — Parabolic-antenna gain versus size, frequency and surface errors. All curves assume 60-percent aperture efficiency and 10-dB power taper. Reference: J. Ruze, British *IEE*.

simply changing the feed. The graph at **Fig. 21 relates antenna gain, frequency and size**. As can be seen, an antenna that is suitable for 432 MHz work is also usable for each of the higher amateur bands. Additional gain is available as the frequency of operation is increased. Another advantage of this antenna is in the feed system. The polarization of the feed, and therefore the polarization of the antenna, can be adjusted with little difficulty. It should be a relatively easy matter to devise a system whereby the feed could be rotated remotely from the shack. Changes in polarization of the signal could be compensated for at the operating position! As polarization changes can account for as much as 30 dB of signal attenuation, the rotatable feed could make the difference between working a station and not. A photograph of the parabolic dish antenna used at K2UYH is shown in **Fig. 22. More information on parabolic dish antennas is available in the ARRL** *Antenna Book*.

Antennas suitable for EME work are by no means limited to the types described thus far. Rhombics, quad arrays, helixes and others have been used. These types have not gained the popularity of the Yagi, quagi, collinear and parabolic dish, however.

Receiver Requirements

A low-noise receiving setup is essential for successful EME work. Since many of the signals to be copied on EME are barely, and not always, out of the noise, a low-noise-figure receiver is a must. The mark to shoot for at 144 MHz is something under 2 dB, as the cosmic noise will then be the limiting factor in the system. Noise figures of this level are relatively easy to achieve, even with inexpensive devices that are available.

As low a noise figure as can be attained will be usable at 432 MHz. Noise figures on the order of 0.5 dB are possible with

Fig. 22 — A newcomer to EME stands in awe of the K2UYH 28-foot dish.

GaAs FETs. As most GaAs FETs are currently still quite expensive and somewhat fragile, many builders choose the more rugged bipolar, which offers a noise figure just under 1 dB.

Since the loss in the transmission line that connects the antenna to the preamplifier adds directly to the system noise figure, most serious EME operators mount a preamplifier at the top of the tower or directly at the antenna. If an exceptionally good grade of transmission line is available, it is possible to obtain almost as good results with the preamplifier located in the shack. Two relay/preamplifier switching systems are sketched in Fig. 23. The system at A

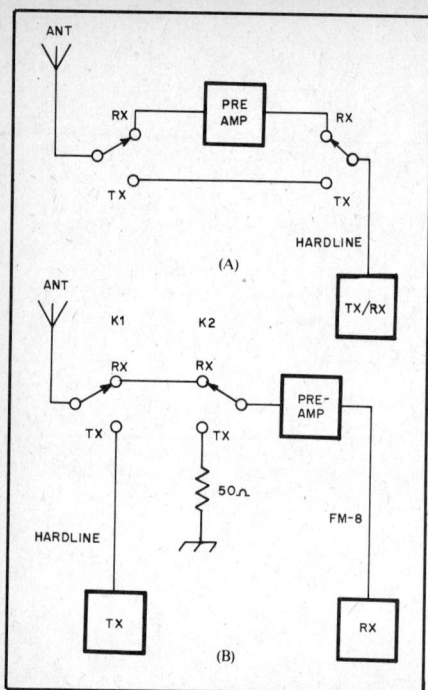

Fig. 23 — Two systems for switching a preamplifier in and out of the receive line. At A, a single length of cable is used for both the transmit and receive line. At B is a slightly more sophisticated system that uses two separated transmission lines. See text for details.

makes use of two relays and a single transmission line for both transmit and receive. The preamplifier is simply switched "in" for receive and "out" for transmit.

The system outlined at Fig. 23B also uses two relays, but the circuit is somewhat more sophisticated. Two transmission lines are used, one for the receive line and one for the transmit line. Also, a 50-ohm termination is provided. Since relays with high isolation in the vhf/uhf frequency range are difficult and expensive to obtain, two relays with a lower-isolation factor may be used. When the relays are switched for the transmit mode, K1 connects the antenna to the transmit line. K2 switches the preamplifier into the 50-ohm termination. Hence, two relays provide the isolation between the transmitter connection and the preamplifier. If independent control of K2 is provided for, the preamplifier can be switched between the 50-ohm termination and the antenna during receive. This feature is especially useful when making sun-noise measurements to check system performance. For this measurement the antenna is directed toward the sun and the preamplifier is alternately switched between the 50-ohm load and the antenna. The dB difference can be recorded and used as a reference when checking system improvements. The complete circuit for this relay system is presented later in this chapter.

As the preamplifier is mounted ahead of the transmission line to the receiver, a cable of mediocre performance can be used. The loss of the cable, as long as it is within reason, will not add appreciably to the system noise figure. Information contained in the VHF/UHF Receiving chapter of this book explains how to calculate system noise figures. Foam-type RG-8 cable is acceptable for runs up to 100 feet at 144 MHz.

It is important to get as much transmitter power as possible to the antenna. For this reason Hardline, Heliax or similar low-loss cable is specified for the transmit line.

Transmitter Requirements

In many EME installations the antenna gain is not much above the minimum required for communications. It is highly likely that the maximum legal limit of power will be required for successful EME work on up through 432 MHz. Since many contacts may require long, slow sending, the transmitter/amplifier should have adequate cooling. Also, an amplifier with some power to spare rather than an amplifier running "flat out" is desirable. This is especially important should ssb communication be attempted. An amplifier run all out on ssb will likely produce large amounts of odd-order IMD products that fall within the band. While the splatter produced will not affect your communications, it will certainly affect that of others close in frequency!

Remote Preamplifier Switching System

The preamplifier-switching system described here is intended primarily for EME applications. Serious vhf and uhf operators may wish to consider similar systems for terrestrial work, as a tower-mounted preamplifier usually means a noticeable reduction in system noise figure. The thoughts behind this system are outlined in the previous section, entitled Receiver Requirements.

The Circuit

The relay switching system is separated into two parts. One section is mounted at the tower, and the other, the control circuitry and power supply, is mounted at the station. A length of four-conductor, TV-rotator cable can be used to connect the two units. The package that is mounted at the tower consists of two rf coaxial relays and a preamplifier. The schematic diagram is shown in Fig. 25. As can be seen, K1 is used to switch the antenna between the transmit line and a line which connects with K2. K2 switches the preamplifier to either the antenna or to a 50-ohm termination. Two relays provide more than adequate isolation between the preamplifier and the transmit line. Additionally, K2 can be switched between the antenna and the termination independently of K1. This allows for sun-noise measurements when the system is in the receive mode.

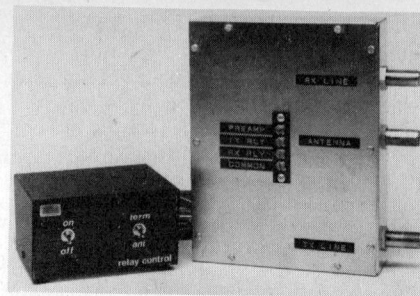

Fig. 24 — The preamplifier relay switching system suitable for EME work. The relay box is mounted at the antenna and the control box is located in the station. A length of four-conductor wire connects the two units.

The portion of the system mounted at the station is essentially a power supply and control circuitry. A line isolation transformer is used to power the 117-volt relay coils. S2 controls the action of K2, which is either connected in parallel with K1 in the ANTENNA position, or activated separately for the TERMINATION position. A pilot light, fuse and ON/OFF switch are provided in this design.

Construction

The items to be mounted at the tower are enclosed in an ordinary chassis and bottom plate assembly. A photograph, shown in Fig. 26, indicates the general layout. Short lead lengths are used throughout. Bulkhead uhf feedthrough connectors are used to ensure an rf- and watertight enclosure. The item shown at the bottom center of the enclosure is a commercial preamplifier. Power for the preamplifier (12 V dc) is fed through the fourth wire of the four-wire cable that connects the two modules (tower and station) together. When all components are properly mounted, the chassis is sealed with silicone rubber (RTV). A terminal block provides for connection to the four-conductor cable. Two 0.01-μF capacitors are mounted across the relay coils at terminal block TB3.

The station circuitry is mounted in a small aluminum cabinet. A neon indicator, on/off power and termination switch are mounted to the front panel. The fuse and interconnection terminal blocks TB1 and TB2 are mounted on the rear apron. Component layout is not at all critical.

Additional Thoughts

Although the circuit described here performed flawlessly for many months, one change might be considered by the prospective builder. The change would involve rewiring the relays so that the preamplifier would be automatically switched into the termination when the system is de-energized. This would protect the preamplifier from static or nearby lightning strikes. As the circuit is presently shown, the preamplifier will remain con-

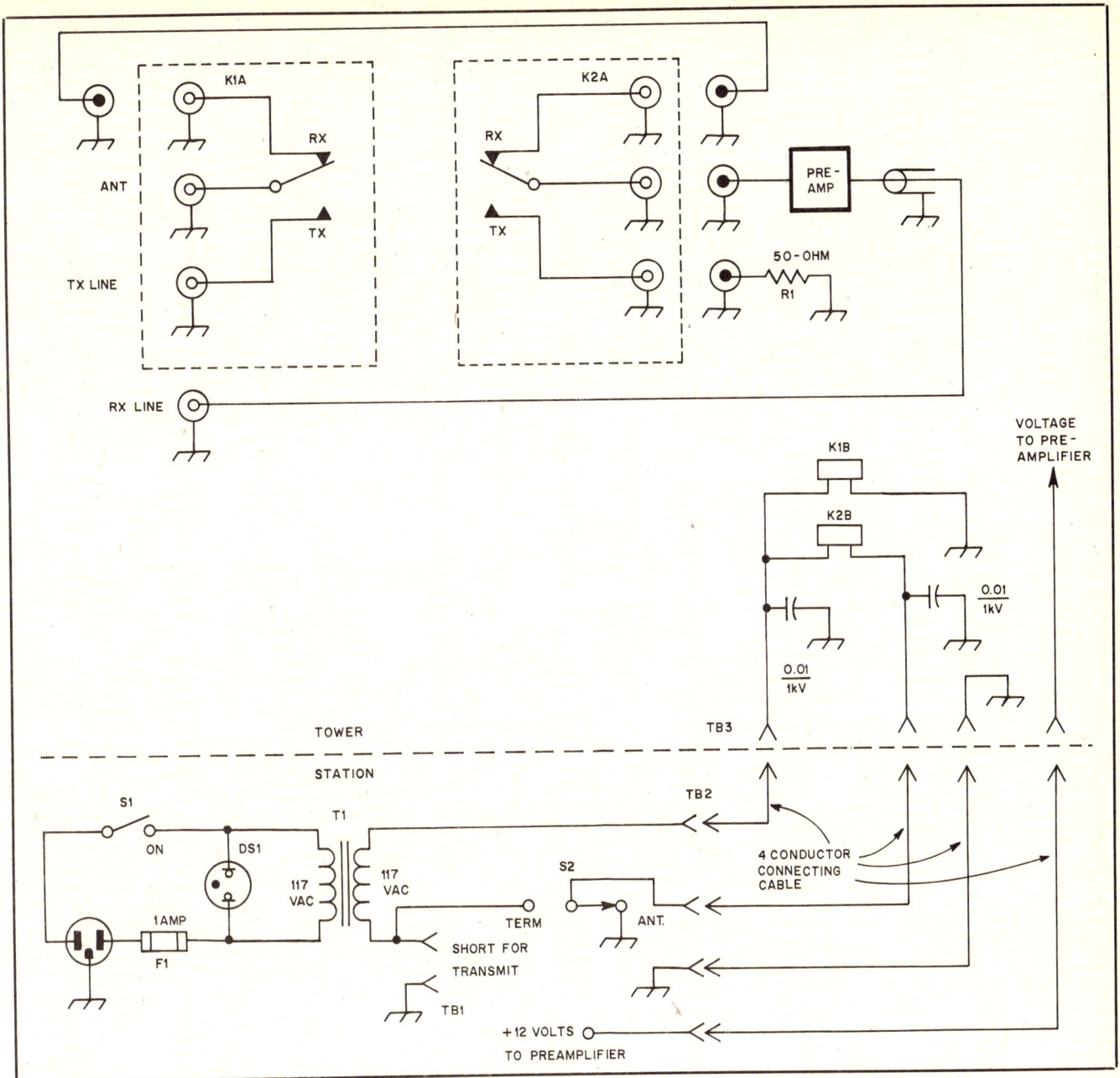

Fig. 25 — Schematic diagram of the preamplifier switching system. The diagram is divided into two parts; the top portion is for the circuitry at the antenna and the bottom is that for use in the station.

DS1 — Neon indicator light with built-in dropping resistor.

K1, K2 — Rf-coaxial relays suitable for the frequency range to be used.

R1 — Termination, 50-ohms, noninductive.

Resistor built into a PL-259 connector.
S1 — Toggle, spst.
S2 — Toggle, spdt.
T1 — Isolation transformer, 115-Vac primary,

115-Vac secondary, 15 VA. Stancor P-6411 or equiv.
TB1-TB3, incl. — Terminal block, screw connection, four terminals.

Fig. 26 — Interior view of the package that is mounted at the antenna. The object at the bottom center of the chassis is a commercial preamplifier.

nected to the antenna when the power is switched off. Although no damage has occurred to the preamplifier used by the author, some GaAS FET amplifiers may not be able to tolerate the voltage levels produced by nearby lightning storms.

10-GHz GUNNPLEXER COMMUNICATIONS

Communications on the amateur 10-GHz band has been simplified greatly with the advent of the Microwave Associates Gunnplexer™ transceivers. It is in-teresting to note that a similar communications system as little as 10 years ago would have required, literally, a rack full of equipment. The Gunnplexer transceiver will fit conveniently in your hand and is operated from a single 12-volt

Fig. 27 — View of the Microwave Associates Gunnplexer. A protective resistor and diode are connected from the detector output to ground. A 50-μF electrolytic capacitor is connected from the Gum diode terminal to ground. The Gunnplexer is designed to mate a UG-39/U waveguide flange when the horn antenna is removed.

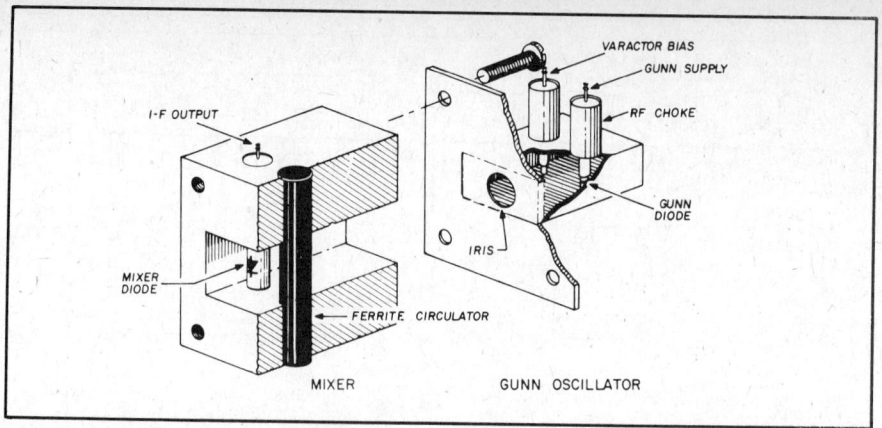

Fig. 29 — Drawing of the Gunnplexer assembly. The Gunn oscillator assembly consists of a resonant cavity in which is mounted the Gunn diode and varactor diode. The cylinders contain the quarter-wave choke sections, which connect to the diodes. The Gunn oscillator assembly bolts to the mixer assembly, which contains the detector diode and ferrite circulator. A horn antenna bolts to the front of the mixer assembly. (Reproduced with permission from *Ham Radio* Magazine, January 1979.)

power supply, either ac line operated or batteries! This makes the Gunnplexer ideal for fixed or portable operation.

The Gunnplexer is most often used in wide-band fm systems and is ideal for audio, full-color video and data transfer. With suitable peripheral equipment, systems for full-duplex audio, color video and up to several megabit data transfer are possible. Such high data rates should allow direct computer-to-computer memory transfer that would not be practical (or legal) on

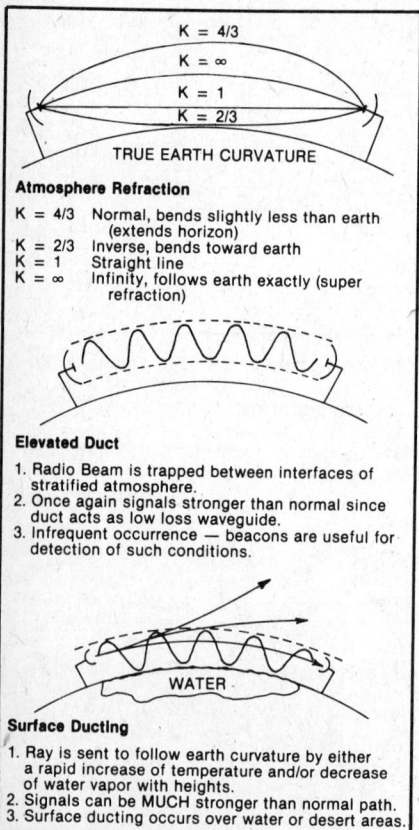

Fig. 28 — Signal propagation on 10 GHz can take any of the several forms outlined here. Straight line-of-sight propagation (K = 1) will be the form most often encountered.

other lower-frequency amateur bands.

Communications at 10 GHz is normally limited to line-of-sight paths although several other forms of propagation exist. Several of these are shown in Fig. 28. A combination of propagation modes has allowed the Italians to set the world distance record at 476 miles (767 km). It is anticipated that most communications will occur over line-of-sight paths and information is included in a following section on how to calculate expected signal levels taking path loss into account.

The Gunnplexer, or microwave hardware portion of the communication system, can be tower mounted provided it is enclosed in a weather-tight enclosure. Shielded cable, such as RG-58/U, can be used between the Gunnplexer and the "electronics," which may be mounted inside at some convenient location. In many cases the tower mounting will be mandatory in order to provide a reasonable line-of-sight path. Trees (with leaves) make excellent microwave absorbers!

The Gunnplexer

The heart of the Gunnplexer is a Gunn diode oscillator, named after its inventor, John Gunn of IBM. A more detailed discussion of the Gunn diode is presented in chapter 4. Refer to the cut-away drawing of the Gunnplexer, Fig. 29, for the following discussion. The Gunn diode is mounted with a varactor diode in a resonant cavity. When a regulated voltage is applied to the Gunn diode it oscillates and the frequency of oscillation is determined by the capacitance of the varactor diode and two mechanical tuning screws. The mechanical tuning screws can be likened to coarse tuning controls and are factory set for the appropriate tuning range. The voltage applied to the varactor diode (1 to 20 V dc) tunes the frequency electronically a minimum of 60 MHz. Power is coupled out of the cavity through a small iris which has been designed as somewhat of a

compromise between maximum power output and isolation from changes in diode impedance and load.

The Gunn oscillator is also used to provide the local oscillator signal for the detector diode. A ferrite circulator couples an appropriate amount of energy into the low-noise Schottky mixer diode and isolates the transmitter and receiver. As the Gunn oscillator functions as both the transmitter and the receiver local oscillator, the i-f at each end must be at the same frequency. Also, the frequencies of the Gunnplexers must be separated by the i-f. This is illustrated in Fig. 30. Intermediate frequencies of 30 MHz are more or less standard for audio work in the U.S. Both 45 and 70 MHz are used for video and high-speed data work.

As can be seen from Fig. 30, the Gunnplexer communications system is full duplex. In other words, both parties can talk and listen at the same time, without throwing any switches. This is something that may take a while to get used to as most amateurs are programmed for VOX or PTT operation. In short, it is the ultimate break-in system!

One detail of the Gunnplexer that does require some specific attention is frequency control. The Gunnplexer has a frequency stability specification of −350 kHz frequency change per degree Celsius temperature change. This does not pose much of a problem with wide-bandwidth applications such as video or data transfer. However, for relatively narrowband audio work (200 kHz and less) some form of afc, phase lock or other frequency control scheme is required. In most cases simple afc circuitry is sufficient and quite easy to implement. The electrical characteristics of the Gunnplexer are given in Table 12.

Communications Range

The effective communciations range of a Gunnplexer system depends on a number of factors, including transmitter

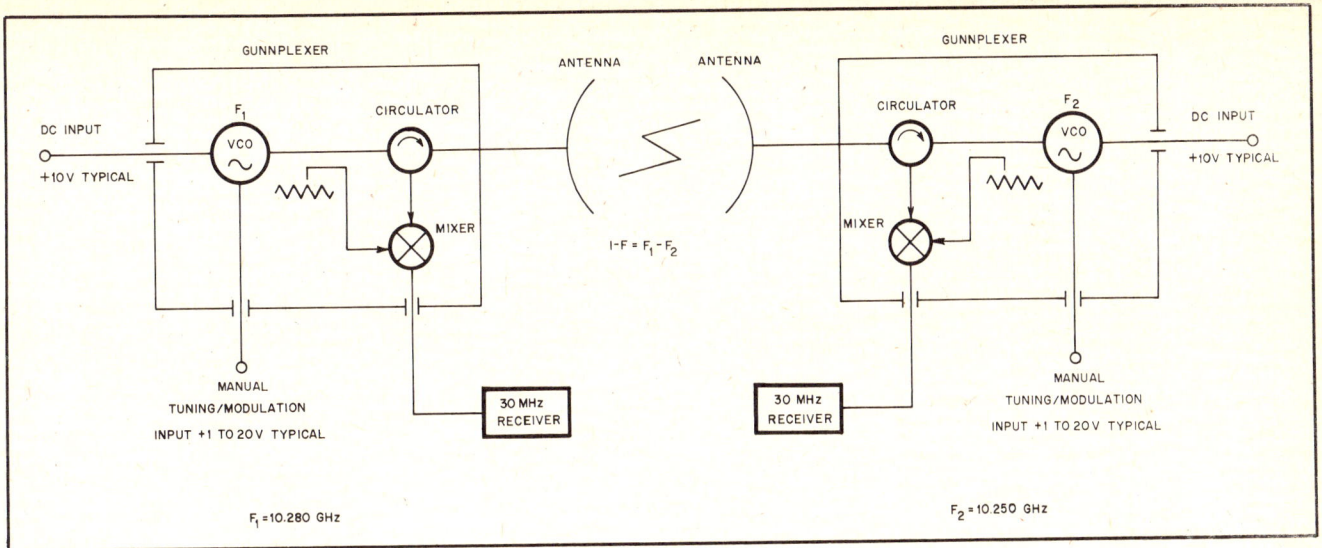

Fig. 30 — This drawing depicts a Gunnplexer communications system running full duplex. The VCOs are offset by the desired i-f, which in this case is 30 MHz.

power, path loss, receiver noise figure, receiver bandwidth, antenna gain and desired carrier-to-noise ratio.

Power output from the Gunnplexer can be measured by using a waveguide-to-coax transition and a power meter. If this equipment is not available the power rating of the Gunnplexer can be used. In most cases this will represent a conservative estimate as most Gunnplexers generate several decibels of power in excess of their rating. For range calculation the power is expressed in dBm and is calculated as follows:

$$dBm = 10 \log P(mW)$$

where P is the power output.

Since most communications will occur over a line-of-sight path, the expected attenuation of signal strength can be readily calculated as follows:

$$dB = 92.5 \text{ dB} + 20 \log f(GHz) + 20 \log D \text{ (kilometers)}$$
$$dB = 96.6 \text{ dB} + 20 \log f(GHz) + 20 \log D \text{ (miles)}$$

where f is the operating frequency and D is the distance between sites. From these equations it can be seen that the first kilometer of station separation will yield 112.7 dB. Note that each time the distance is doubled the path loss increases by only 6 dB. For example, two-mile separation would be 122.8 dB loss and four-mile separation would be 128.8 dB path loss.

Unless sophisticated equipment is available to measure the receiver noise figure, the manufacturer's specification must be used. In most cases the 12 dB figure turns out to be a conservative number as long as a good low-noise i-f receiver is used. The receiver bandwidth factor can be calculated as follows:

$$dB = 10 \log BW_{kHz}$$

where BW is the receiver bandwidth. A chart of various receiver bandwidth factors is given in Table 13.

Antenna gain is a known quantity. The horn antennas supplied with the Gunnplexers are rated for 17 dBi. Optional 2-foot (0.6-m) parabolic antennas have 32

Table 12
Gunnplexer Specifications @ T_A = 25° C

Electrical Characteristics

RF Center Frequency	10.250 GHz[1]
Tuning	
Mechanical	± 50 MHz
Electronic	60 MHz min.
Linearity	1 to 40%
Frequency Stability	− 350 kHz/°C max.
RF Power vs Temperature and Tuning Voltage	6 dB max.
Frequency Pushing	15 MHz/V max.
Input Requirements DC GUNN Voltage Range	+ 8.0 to + 10.0 VDC[2]
Maximum Operating Current	500 mA
Tuning Voltage	+ 1 to + 20 volts
Noise Figure[3]	<12 dB
RF Output Power[1]	

Model	P out (Mw)
MA87141-1	10 min. 15 typ.
MA87141-2	20 min. 25 typ.
MA87141-3	35 min. 40 typ.

NOTES:
[1] Tuning voltage set at 4.0 volts.
[2] Operating voltage specified within this range on each unit.
[3] 1.5 dB i-f NF at 30 MHz.

Table 13
Bandwidth Factors for Microwave Range Calculations (see text)

Bandwidth	Attenuation (dB)
30 MHz	44.8
15 MHz	41.8
10 MHz	40
5 MHz	37
1 MHz	30
500 kHz	27
200 kHz	23
100 kHz	20
50 kHz	17
25 kHz	14
10 kHz	10
5 kHz	7
2 kHz	3

dBi and 4-foot (1.2-m) dishes have 38 dBi. Remember to consider antenna gain at each end of the link.

Two final factors are required for communication range calculation. There are the thermal noise floor and the carrier-to-noise ratio. The thermal noise floor is − 144 dBm and is set by the laws of nature and in fact determines the ultimate sensitivity of receivers. A discussion of this factor is presented in chapter 9. When calculating the effective communications range, the carrier-to-noise ratio can be chosen and the resulting distance between the stations computed. Alternatively, the distance separation can be specified and the resulting carrier-to-noise ratio can be computed. The latter technique will be illustrated.

To make the communication range calculation it is necessary to take into account the power, system gains and system losses. For a hypothetical link assume the following: transmitter power 15 mW, transmitter and receiver antenna gain 17 dBi, 10 mile path, 12 dB receiver noise figure and 200 kHz receiver bandwidth. Calculation is as follows:

	Individual Item	System Total
Transmitter power 15 mW	+ 11.8 dBm	+ 11.8 dBm
Transmitter antenna gain	+ 17 dB	+ 28.8 dBm
Path loss 5 miles	− 130.8 dB	− 102 dBm
Receiver antenna gain	+ 17 dB	− 85 dBm
Receiver noise figure	− 12 dB	− 97 dBm
Receiver bandwidth factor	− 23 dB	− 120 dBm
Thermal noise floor		− 144 dBm
Carrier-to-noise ratio		+ 24 dB

A similar calculation could be made with a specified carrier-to-noise ratio, obtaining distance as the result.

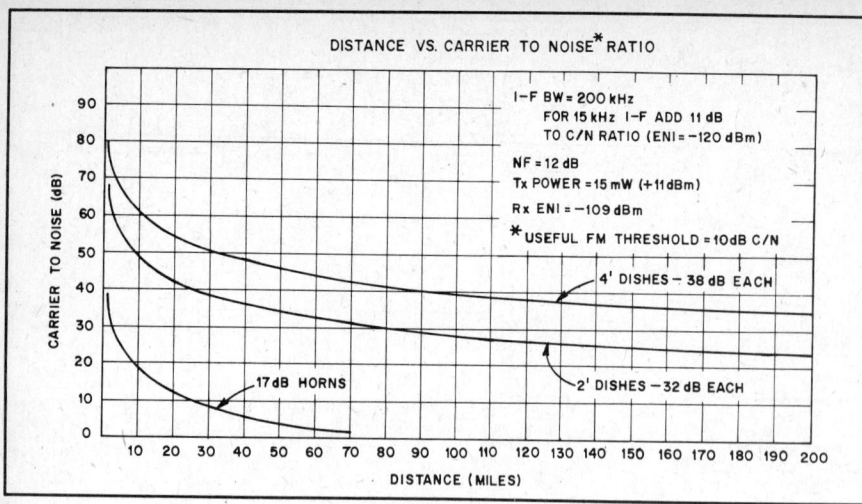

Fig. 31 — This graph shows distance vs. carrier-to-noise ratio.

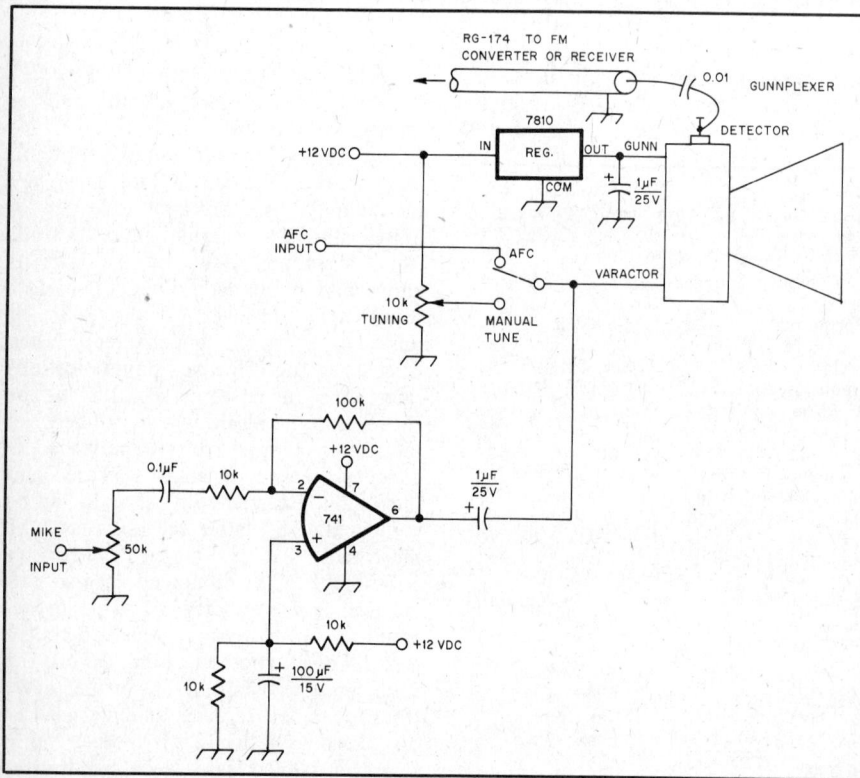

Fig. 32 — Here is a complete "bare minimum" communications system using the Gunnplexer transceivers. An inexpensive automobile fm converter or receiver is used for the receiver i-f. A 7810 voltage regulator is used for the Gunn diode supply and a simple 741 op amp serves as the microphone stage. The frequency of operation is set by the potentiometer that biases the varactor diode. A 10-turn potentiometer provides a comfortable tuning rate.

The 35-mW version represents a 5.44 dB increase over the 10-mW unit. Perhaps the easiest way to obtain a substantial amount of system gain is with a larger antenna. As mentioned earlier, a 2-foot parabolic antenna has 32 dBi gain as compared to the 17 dBi horn supplied with the Gunnplexer. That is a 15-dB increase. If 2-foot dishes are used at each end of the link a total gain of 30 dB will be realized over the 17 dB horns! Fig. 31 illustrates how the additional gain of these antennas increases system performance. The use of 2-foot dish antennas places a stringent requirement on antenna alignment as these antennas have half-power beamwidths of 4 degrees.

Bare Minimum Audio Communications System

The simplest of communications systems using the Gunnplexer transceivers can be formed with two 88- to 108-MHz fm receivers, two Gunnplexers, two microphones and associated amplifiers, and two sources of 12 V dc. A diagram of such a system is shown in Fig. 32. Some of the low-cost fm converters and receivers for automotive use make good i-f strips for this communications system. Also, the afc signal developed in the converter or receiver can be routed to the Gunnplexer varactor diode to lock the two units together. The microphone amplifier can be a single 741 operational amplifier as shown in the diagram.

There are two shortcomings with this system. The first involves the use of the fm broadcast band as the i-f. If mountain-top DXing is planned, it is likely that strong fm broadcast stations will be received no matter how short the lead between the detector and the fm converter or receiver is made. It will be necessary to select that part of the band where there are no strong signals present. The second item involves afc. Since it is possible to tune the Gunnplexers over quite a range with the varactor diode tuning, it is possible to tune them on either side of each other. This means that the single polarity afc system in the receiver or converter will work for only one combination. If the units are operated so that the afc polarity is incorrect, the afc will push the received signal out of the receiver passband. If this happens, simply tune the two Gunnplexers to produce the "other" i-f signal. Afc lock should then be obtained.

A High-Performance Audio Communications System

The high-performance Gunnplexer audio system described here was developed by Advanced Receiver Research,[4] Burlington, Connecticut, and is sold as an amateur/commercial product. The information is included here for those who wish to construct their own system.

[4]Advanced Receiver Research, Box 1242 Burlington, CT 06013, tel. 203-584-0776.

There are several methods for increasing the distance of effective communications using Gunnplexer transceivers. In general these are: higher power transmitter, higher gain antennas, lower noise receiver and narrower bandwidth receiver.

The receiver noise figure is limited principally by the noise figure obtainable from the detector diode (assuming the i-f does not add appreciable noise to the system). Significant reduction in noise figure is not likely with currently manufactured devices. Narrowing the receiver bandwidth to less than 200 kHz for audio work will require crystal or phase locking techniques to keep the two units "locked up." Although these systems are not particularly difficult to build and make operational, they are somewhat more involved than simple afc techniques.

The largest increases in distances can be achieved with higher power and larger antennas. Gunnplexers are manufactured with power options of 10, 20 and 35 mW.

This Gunnplexer support system has been specifically designed for use with the Microwave Associates Gunnplexers. The board contains a complete 30-MHz fm receiver, diode-switched i-f filters, dual-polarity afc system, Gunn-diode regulator and modulators for phone and cw. The system is suitable for fixed, portable or mobile operation. Power-supply requirements are a nominal 13 volts at 250 mA (this includes the current drawn by the Gunnplexer assembly). The circuit is specified for operation over the temperature range −25 to +65° C.

Theory of Operation

The circuit diagram is shown in Fig. 34, and an interconnection wiring guide in Fig. 33. Signal energy arriving at the i-f input (pins 24 and 25) is routed to Q1, a low-noise rf amplifier. A band-pass filter, with a 3-dB bandwidth of 2 MHz, is located between the rf amplifier and mixer, Q2. LO injection for the mixer is provided by a crystal controlled 40.7 MHz oscillator stage. Output from the mixer, at 10.7 MHz, is applied to either of two diode-switched ceramic filters. Supply voltage applied to pin 23 selects FL1 (supplied) and supply voltage applied to pin 22 selects FL2 (optional). Output from the filter is fed to i-f amplifier state Q3. Amplified 10.7 MHz energy is routed to the fm subsystem chip, U1, a CA3189E. Detected audio passes through an af gain control (pins 16, 17 and 18) to U2, the audio output amplifier. This stage has sufficient power to drive headphones and/or a speaker. Squelch control is available by U5A. U5B inverts the amplified afc signal so that either polarity afc may be selected by S3. Overall afc gain is controlled by R_c. U5C sums the afc information along with the manual tuning voltage and this composite signal is applied to the varactor terminal of the Gunnplexer. U5D functions as a meter driver for center tune (discriminator zero) and manual tuning voltage indication. Input to the driver is selected by S1.

U3 is a microphone amplifier boosting the output from a microphone to a level suitable for modulating the Gunnplexer. U4 generates an approximate 500 Hz tone for mcw operation. Voice and mcw information is applied to the varactor of the Gunnplexer along with the AFC/tuning voltage. U6 is a three-terminal positive voltage regulator that provides an accurate 10-volt supply for the Gunn diode. A 1N4001 diode is included to protect against inadvertent application of reverse-polarity voltage.

It is advisable to mount the board in a well-shielded enclosure since the receiver is quite sensitive. The Gunnplexer unit can be mounted (atop a tower) and connected to the board through three lengths of coaxial cable. Although the Gunn diode and varactor line carry only dc and low-level audio signals, shielded cable is a must as even small voltage pickups can produce relatively large modulating voltages (which are undesirable). In cases of severe pickup, small rf chokes may be required directly at the Gunnplexer Gunn diode and varactor connections of the Gunnplexer. Lengths of cable of up to several hundred feet between the Gunnplexer and the board have been used with no problems. For long i-f connection runs it is desirable to install a 30-MHz preamplifier at the Gunnplexer to prevent the long cable run from adversely affecting the system noise figure.

The Gunnplexer system is capable of complete duplex communications, so good separation between the microphone and speaker are mandatory. The use of headphones for this type of communications is generally advisable.

Alignment

1) Connect a signal generator capable of delivering a 30-MHz signal to the i-f input connection.

2) Adjust the output level of the generator to a point where the signal strength meter just begins to move up scale.

3) Adjust C_a, C_b and C_c for maximum indication on the meter.

4) Temporarily disconnect the signal

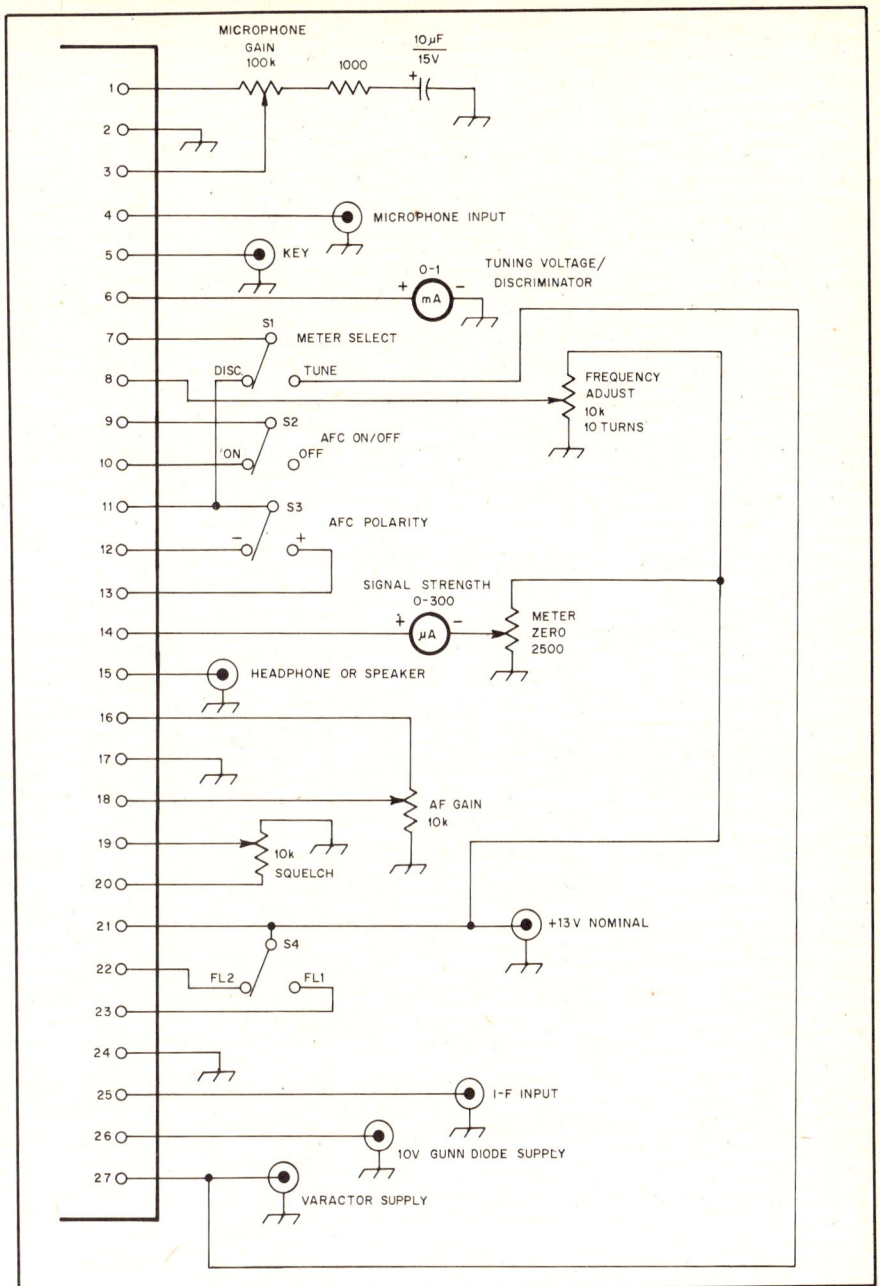

Fig. 33 — Interconnection diagram for the high-performance audio Gunnplexer communications system.

generator from the input.

5) Place S1 in the discriminator position.

6) While switching S3 between the + and − afc positions, adjust R_b so that the same meter reading is obtained in both the switch positions. This reading should be mid scale on the discriminator meter.

7) Reconnect the signal generator to the input and center the signal in the receiver passband.

8) While switching S1 between the afc on and off positions, adjust C_d so that the same meter reading is obtained for both positions. This reading should be mid scale on the discriminator meter.

9) Adjust R_c for the desired afc gain/locking range by shifting the signal generator frequency away from the passband center.

10) R_c is adjusted while in communication with another station. Advance the

level of this control to the point where distortion occurs. Back off the setting of the control to the point where distortion is no longer present.

AMATEUR TELEVISION

Many amateurs rationalize their failure to participate in television by indicating a lack of equipment. In prior years such reasoning could be considered acceptable. Recent development in solid-

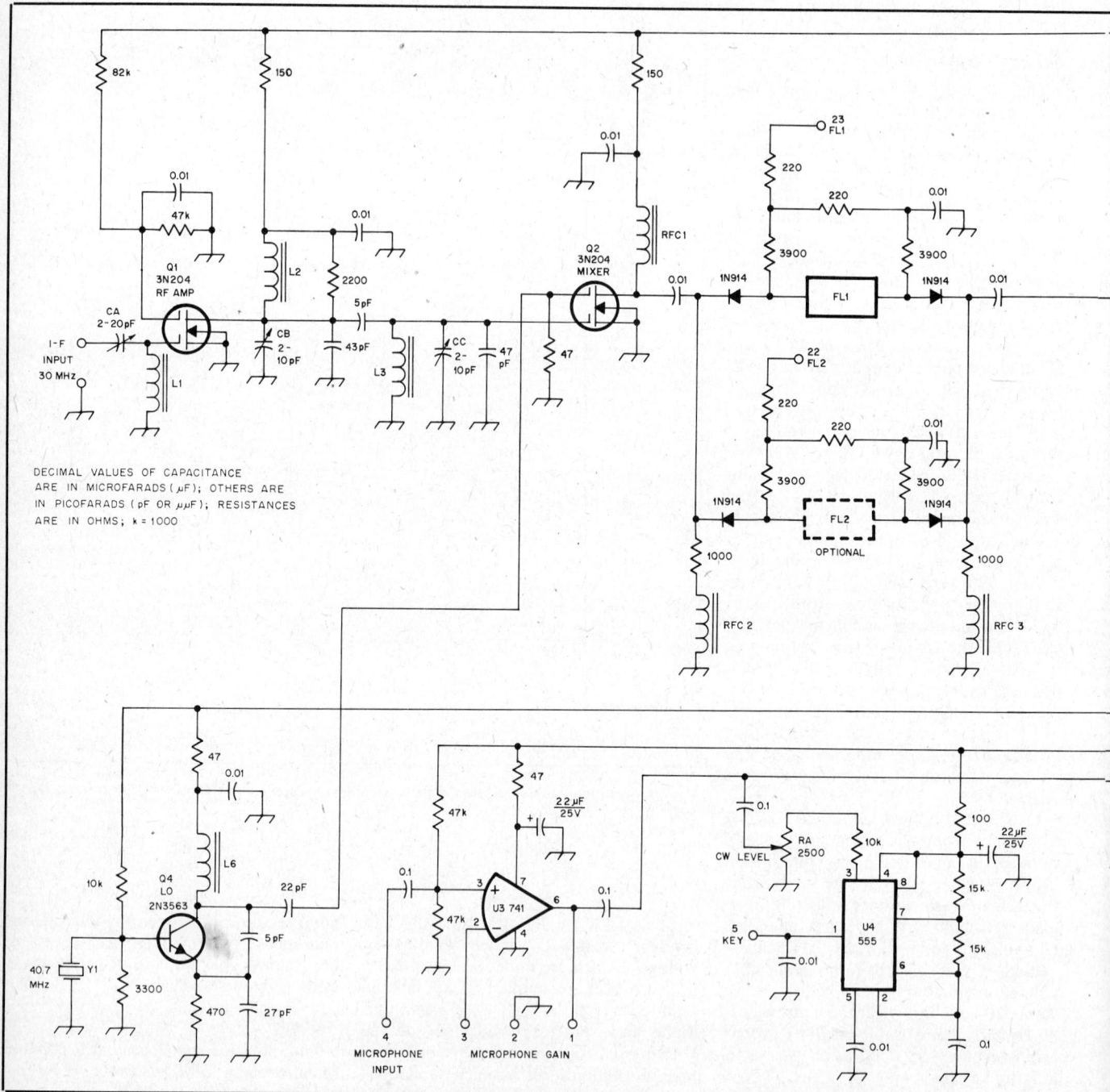

Fig. 34 — Schematic diagram of the high-performance audio communications system. All capacitors specified in picofarads are silver mica DM5 variety. All resistors are quarter-watt film types.

FL1 — Murata SFJ10. 7 mA, 10.7 MHz.
L1 — 25 turns no. 30 enam. wire on a T25-10 core.
L2, L3 — 13 turns no. 28 enam. wire on a T25-6 core.

L4, L5 — 25 turns no. 30 enam. wire on a T25-20 core.
L6 — 16 turns no. 28 enam. wire on a T25-10 core.
RFC1 — 10 turns no. 28 enam. wire on an FT23-43 core.

RFC2, RFC3 — 10 turns no. 28 enam. wire on an FT-23-72 core.
RFC4 — 11 turns no. 28 enam. on an FT23-43 core.
Y1 — 40.7 MHz, third overtone.

state products, however, make such rationalization today as weak as jail-house coffee. Now, in this semiconductor age, there is little difficulty in obtaining essential components, nor does one have to worry about putting the arm on a rich uncle to pick up the tab. An amateur television station can even be assembled in an evening! The material presented here is from an article by Ruh, which appeared in April 1978 *QST*.

When widespread interest in television sprouted after World War II, bandwidth technicalities required a split of TV operations in the amateur bands. Slow-scan television (SSTV) is used in the hf bands where interference must be minimized. Experiments are currently being made with the transmission of color pictures by SSTV. Various techniques are being used, but in essence the process involves the sending of three separate frames of the

same picture with a red, a green, and a blue filter successively placed in front of the camera lens for each of the three frames. At the receiving end of the circuit, corresponding filters are used and each frame is photographed on color film. After a tricolor exposure is made, the photograph is developed and printed in the normal manner. The use of "instant" camera equipment with color film is popular in this work because it affords

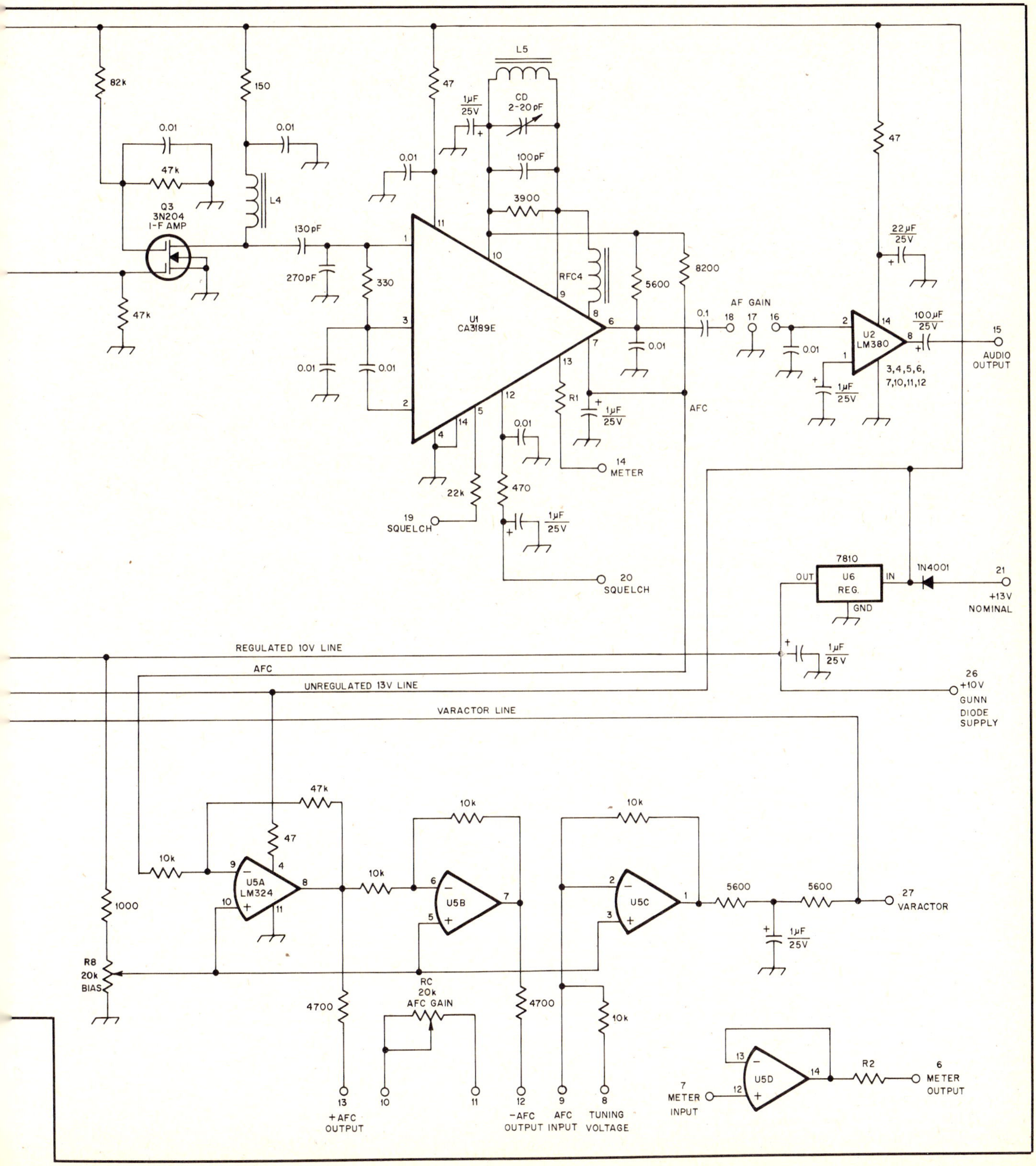

82k 150 47
0.01 0.01 1µF 25V
47k L5 CD 2-20 pF
Q3 3N204 I-F AMP L4 100 pF
130 pF 3900
47k 270 pF 330
RFC4
U1 CA3189E 5600 8200
0.01 AF GAIN
0.01 0.01 0.1 18 17 16 U2 LM380 100µF 25V 15 AUDIO OUTPUT
4 14 5 13 3,4,5,6,7,10,11,12
22k 470 0.01 1µF 25V 22µF 25V
19 SQUELCH 1µF 25V 14 METER 47
20 SQUELCH AFC 7810 U6 REG. OUT IN GND 1N4001 21 +13V NOMINAL
REGULATED 10V LINE 1µF 25V
AFC 26 +10V GUNN DIODE SUPPLY
UNREGULATED 13V LINE
VARACTOR LINE
47k 10k 10k
47 10k U5B 6 7 U5C 2 1 5600 5600 27 VARACTOR
10k U5A LM324 9 8 5 3 1µF 25V
1000 10 11 RC 20k AFC GAIN 4700 10k
R8 20k BIAS 4700
13 +AFC OUTPUT 10 11 -AFC OUTPUT 9 AFC INPUT 8 TUNING VOLTAGE
U5D 13 14 R2 6 METER OUTPUT
7 METER INPUT 12

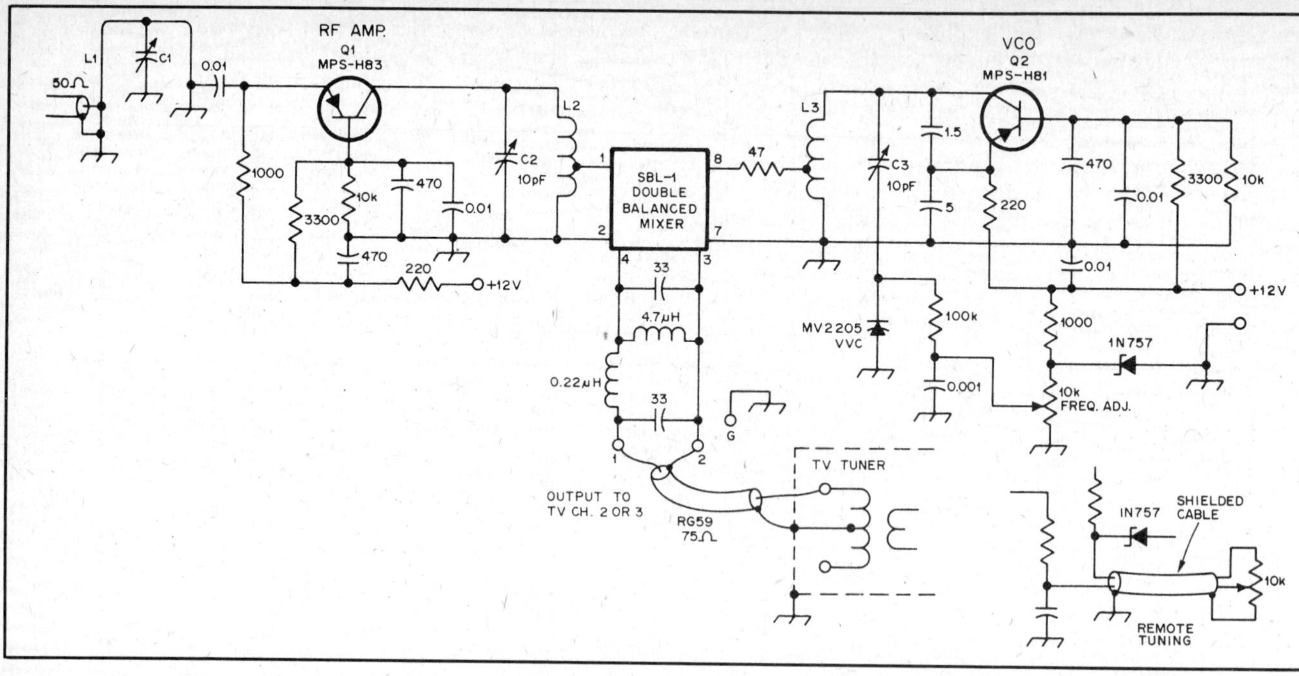

Fig. 35 — The circuit of the TVC-1 converter consists of a low-noise, high-gain, rf amplifier, a doubly balanced mixer and a Varicap-tuned VCO. Remote tuning of the converter permits it to be mounted on the antenna if desired. For remote operation, the remote tuning circuit replaces the 10-kΩ potentiometer shown below the oscillator transistor, Q2. C2 and C3 are Arco 400, 1-10 pF, capacitors. L2 consists of 1-1/2 turns no. 22 wire, 1/4-inch diameter, tapped 3/8 inch from the lower end. L3 is a hairpin loop, 1/2 inch across the bottom and 5/8 inch high. It is made with no. 22 bus wire.

on-the-spot processing. Color reproduction by this technique can be quite good. More advanced experiments in color SSTV reception are being contemplated by storing frame data in digital memory and by applying picture information to a standard color TV set, along with encoded red-green-blue data, via an rf modulator.

Fast-scan TV (ATV) is permitted on the bands above 420 MHz where wider bandwidths needed for this mode can be accommodated. This article is concerned with the latter method.

A Converter for Reception

Understandably a station that transmits well but is deficient in receiving for want of a good receiver leaves much to be desired. Efforts by some amateurs to modify the home-TV uhf tuner have proved rather disappointing. Others have geared up homemade devices that failed to live up to expectations. What is the alternative?

P. C. Electronics, which produces the units one needs for an ATV station, has resolved the matter of providing a means for obtaining excellent reception. The TVC-1 converter, available in ready-made form, is tailor-made for the 439-MHz enthusiast. It fits nicely inside a TV set and presents no problem if one wishes to mount it on the antenna.

The TVC-1, sensitive and selective, performs exceptionally well, yet it is an example of simplicity. Performance is enhanced by a commercially manufactured, doubly balanced mixer and a voltage-controlled oscillator that is tuned by a

10-kΩ potentiometer. The latter feature enables the converter to be tuned remotely by a similar potentiometer. Such remote-control operation would be useful if the converter is to be mounted on the antenna or some location apart from the operating position.

Physically, the TVC-1 components are mounted on a very small pc board. There is a choice of a 50- or 75-ohm input. A

coaxial-cable connector is provided for the output on channel 3. The output signal may be fed to an ordinary television set for viewing the picture and hearing the sound. No modification of the TV set is required. Therefore the TV receiver may be used in the normal manner for home entertainment without any inconveniences.

As manufactured, the converter module

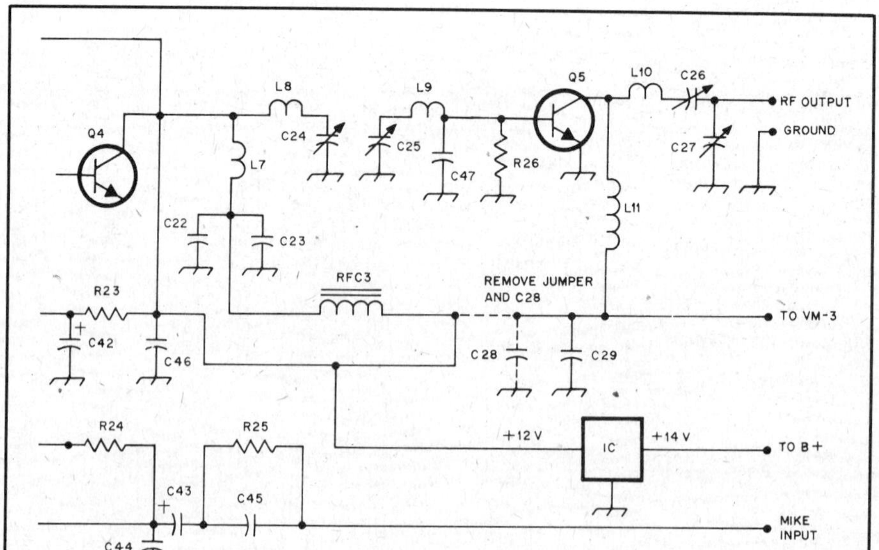

Fig. 36 — Two minor changes are all that need be made to modify the TX-432 transmitter for ATV. C28 and the jumper connecting it with RFC3 are removed. The line input at the lower right of the diagram is not used.

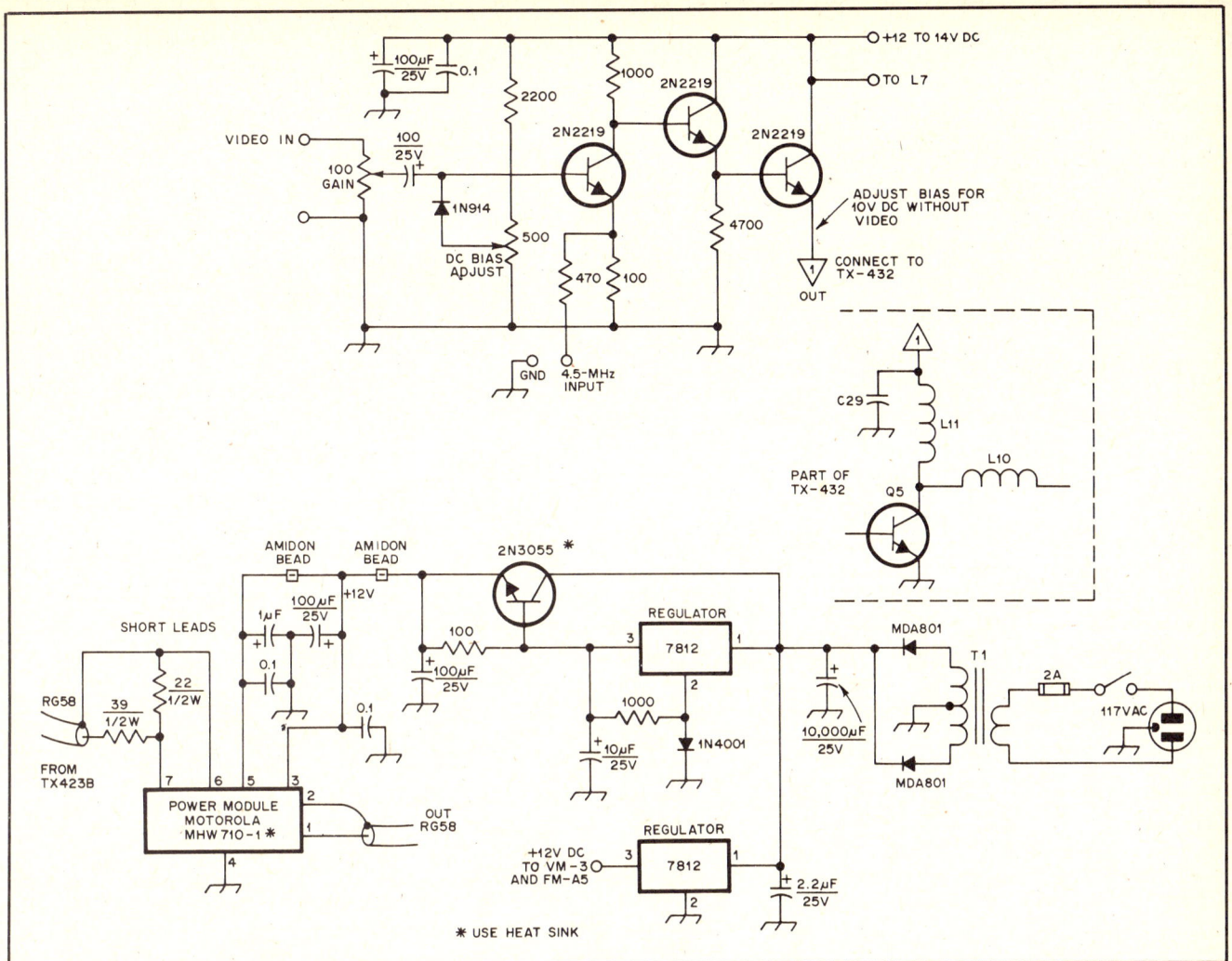

Fig. 37 — The VM-3 solid-state video modulator. The wide-bandwidth capability of the modulator enables it to resolve 64 character-per-line signals from TV typewriters and microcomputers. The 3-dB-down point is typically 8 MHz, more than enough for color and sound plus greater resolution than broadcast. The upper portion of the drawing illustrates the modulator circuit. The lower portion covers the power source and the mHW-7-10-1 amplifier. T1 consists of two 25-V ct filament transformers with primaries in series and secondaries paralleled. Triad no. F-41X or Radio Shack no. 273-1512 may be used.

is supplied pretuned and ready to connect to the ATV system. The TVC-1 should be installed in a shielded chassis or enclosure. Power for the module is to be supplied by a regulated 12-V source. The power supply must be turned off when transmitting.

A length of 50- or 75-ohm cable should be used for connecting the converter to the TV tuner. Remove or unclip the twin-lead at the tuner. The coaxial-cable shield is then connected to one of the antenna terminals. The other terminal is left open.

To adjust the converter, tune the television set to either channel 2 or 3 (whichever is not used locally). Fine-tune the selected channel to minimize any signals from a commercial TV station. Then connect the ATV antenna and swing the converter across the band to locate a nearby ATV station. After the station has been found by tuning with the 10-kΩ potentiometer, fine adjustments are made with C1. When the latter is properly set, tuning will be good for ± 10 MHz. For installations

where the converter is mounted on the antenna, the 10-kΩ frequency control on the VCO may be replaced by the remote tuning circuit shown in Fig. 35. Use of this arrangement allows the tuning adjustments to be made from the operating position. Shielded cable is recommended for the connection between the VCO and the remote 10-kΩ potentiometer.

The pnp transistors shown in the diagram are no longer available. However, the scheme can be implemented with npn devices simply by using a power supply having a positive ground and reversing the Zener and varactor diodes. Some inexpensive transistors worth investigating for this application are the 2N2857, 2N3478, 2N5109 and 2N5179.

Transmitting Equipment

Basic units for transmission of ATV include a video modulator, a sound-subcarrier device, a transmitter and a power amplifier. The VHF Engineering TX-432

and Motorola MHW-710 rf-module power amplifier serve as the rf strip. Both have been adequately described in other articles which appeared in *73* for August, 1976 and in *A-5 Magazine* for March 1977. The circuits, in articles by Bruce Brown, WB4YTU, are presented here for convenience of the readers.

Video modification of the VHF Engineering TX-432 strip is simple. One capacitor, C28, and the 12-V dc bus jumper to the pad connecting C28, C29 and L11, are disconnected. The strip is supplied with + 12 V for all stages except the final. L11 is the final-stage rf choke which is connected to the B + line and is the feed point for the video/audio signals. C28, an electrolytic capacitor on the TX-432 pc board, is not used. C29 is a small disc capacitor that is retained for rf decoupling. Fig. 36 shows the TX-432 modifications.

When tuning the TX-432 for operation below 444 MHz, one may find substitu-

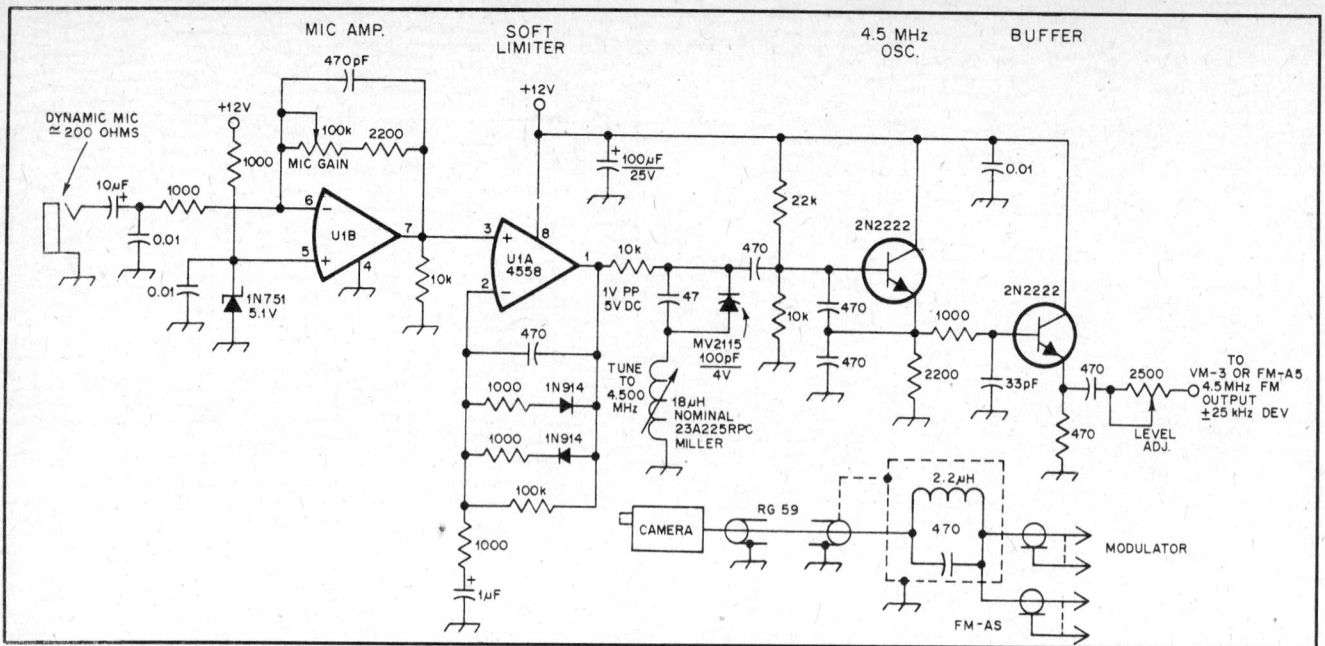

Fig. 38 — This ATV sound-subcarrier generator module permits both voice and video to be transmitted. The trap shown in the diagram is necessary to isolate the capacitance of the long coaxial cable from the FM-A5 output. The capacitance would act as a bypass to the 4.5-MHz signals without the trap. U1 is either a Motorola MC1458CP1 or Raytheon RC4558DN operational amplifier.

tion of Arco no. 402 compression capacitors for the Arco no. 400s desirable. Improved heat dissipation may be achieved by replacing the thin metal heat sinks furnished with the TX-432. Solid-aluminum TO-5 heat sinks are better suited for the purpose.

Fast-Scan Modulator

The VM-3 fast-scan modulator, another P.C. Electronics unit, was developed to be used mainly for supplying video to the TX-432B exciter. Because of the wide bandwidth of the VM-3, it has the capability of resolving 64-character TV typewriter and microcomputer graphics. The 3-dB point is typically 8 MHz which is more than enough for color and sound plus greater resolution than found in broadcast TV. Another feature is a separate input for 4.5-MHz subcarrier for sound transmission. An ATV operator will appreciate this advantage. Fig. 37 illustrates the connections to be made for adding the modulator to the installation. The VM-3 is sold for about $20.

Subcarrier System for Sound

One of the better subcarrier systems available is also produced by P. C. Electronics. The FM-A5 utilizes a stable 4.5-MHz oscillator that is modulated by a Varicap diode driver from an IC audio amplifier. The unit has sufficient gain to fully modulate the transmitter to 25-kHz deviation even with an inexpensive microphone placed at a distance of 25 feet or more. Provision is made for microphone sensitivity and subcarrier rf level control to customize the operation

according to the operator's liking. Moreover, the FM-A5 incorporates a soft limiter to prevent overdeviation. Distortion is extremely low, providing broadcast-quality sound to accompany the pictures.

This subcarrier module is designed for feeding the VM-3 video modulator directly or it may be connected to any 75-ohm video coax line with the addition of a 4.5-MHz trap. The trap, as shown in Fig. 38 is necessary for isolating the capacity of the long coaxial line from the FM-A5 output. That capacity would act as a bypass to the 4.5-MHz signal without the trap.

Output from the FM-A5 is adjustable to match the camera video level. Nominal subcarrier level is 0.5 to 1.0 V peak-to-peak. In some cases there may be other band-pass attenuation in the transmitter and modulator that could require more adjustment.

Under operating conditions, the oscillator should be adjusted to within 10 kHz. The receiving station should be tuned to the high-frequency side of a signal for best sound with picture. The transmitter should also be peaked to the high side.

Frequency response of the FM-A5 is rolled off just short of 300 Hz and just above 3000 Hz for best voice communication. Deviation is fixed at the 25-kHz broadcast standard. P. C. Electronics has priced this subcarrier generator in the $25 class.

The MHW-710 Module

To give the transceiver an energy boost, the rf amplifier uses a Motorola

MHW-710 rf module. As mentioned earlier, only a few parts (as indicated in Figs. 37 and 39) are needed to complete the amplifier stage. The module may not be a stock item at some Motorola parts dealers, but it can be ordered from the source indicated in Table 14. It is sold for about $54. Fig. 39 is an interconnection diagram for assembling an ATV transceiver from the modules described. A packaging scheme is shown in Fig. 40.

Table 14
Where to Buy Components

Component	Source
Audio subcarrier unit FM A5	P.C. Electronics, 2522 S. Paxson, Arcadia, CA 91006
Video I-D generator TVID-1 Video Modulator VM-3	
MHW-710 power amplifier	Motorola parts dealers or Regency Electronics, 7701 Records, Indianapolis, IN 46226
Power supply	Godbout Electronics, P. O. Box 2355, Oakland Airport, CA 94614
Regulators	Poly Paks, Box 942R, Lynnfield, MA 01940
TX-432 transmitter	VHF Engineering 320 Water St., Binghamton, NY 13902

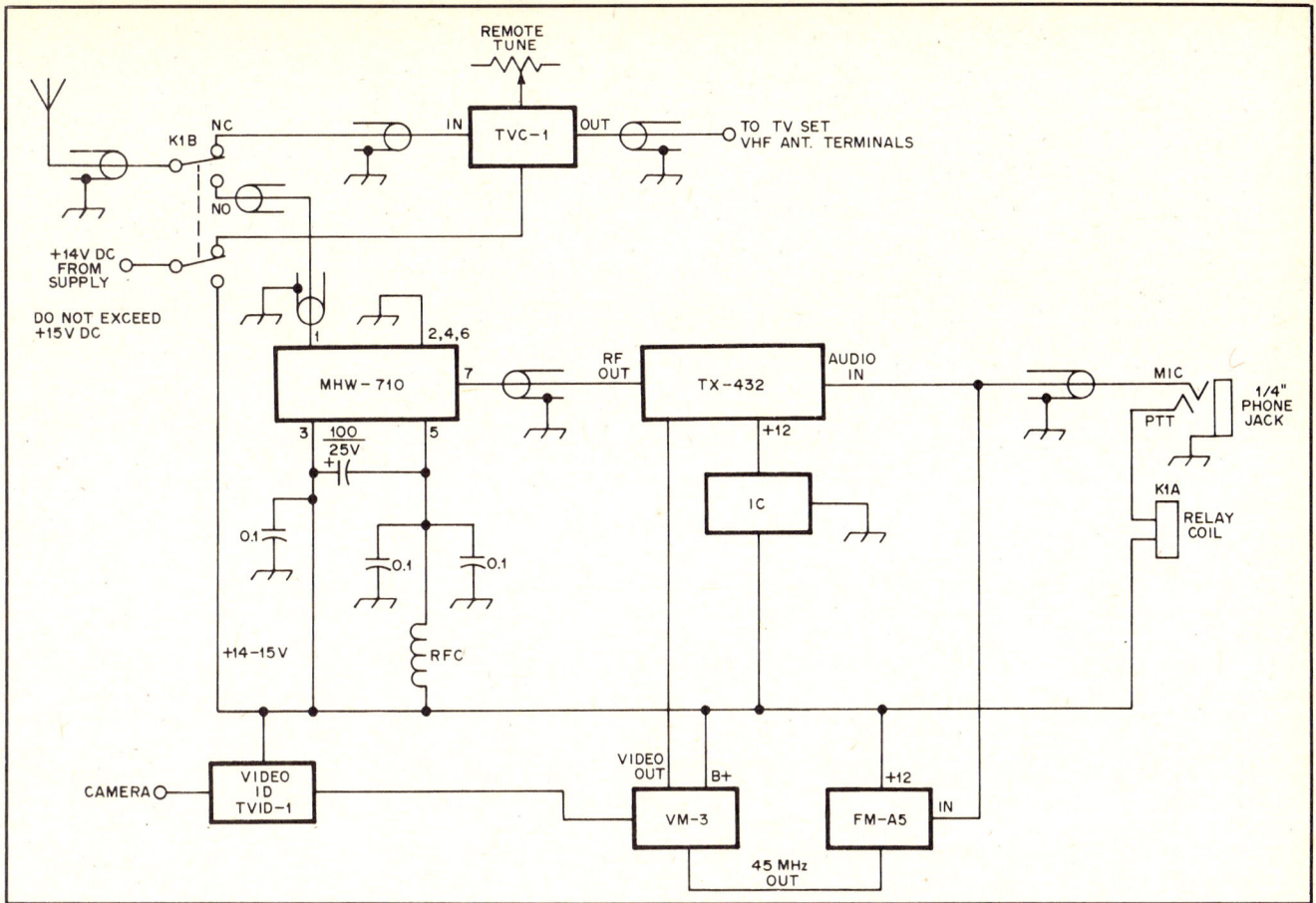

Fig. 39 — Block diagram of the ATV transceiver. Principal units of the circuit are the VHF Engineering TX-432 transmitter. Motorola MHW-710 amplifier, P. C. Electronics TVC-1 converter, VM-3 video modulator, FM-A5 audio subcarrier unit and video i-d generator. Output of the converter is connected to a standard TV set. The IC is a type 7812 12-V dc regulator.

SLOW-SCAN TELEVISION

Fast-scan TV signals take up more than 5 MHz of bandwidth. Since this is more kHz than in all the amateur bands below 6 meters, it is obvious that if we want to work TV-DX on the hf bands we will have to modify the TV signal a bit.

Slow-scan TV (SSTV) is, just as its name implies, a TV signal with a very slow scan rate. While a regular fast-scan TV signal produces 30 frames per second, it takes eight seconds to send one SSTV frame. Thus, motion pictures are impossible. If ATV is analogous to watching home movies by radio, then SSTV resembles a photographic slide show on the air. In addition, SSTV picture definition is four times coarser than fast-scan TV. Table 15 summarizes the video SSTV format used by amateurs.

But these disadvantages are more than balanced by the fact that SSTV can be used in any amateur phone band above 3.5 MHz. Anyone you can work with a good signal on ssb can be worked via slow scan. Many DX stations are now equipped for picture transmission, and more than one amateur has worked over 100 countries over SSTV!

The signal that comes out of a SSTV camera is a variable frequency audio tone — high tones for bright areas and low tones for dark. To send SSTV over the air, you just feed this tone into the microphone jack of any ssb transmitter. (SSTV on double sideband a-m or fm is illegal on the hf bands.) To receive, you tune in the signal on an ssb receiver and feed the audio into the SSTV monitor.

All you need to get started is an ssb station, a monitor (the slow-scan "TV set") and a camera. You don't even need the camera if you already have a tape recorder.

Recent advances have led to the development of fast-scan converters. On receive, such a device converts the incoming audio to a signal that is usable by a conventional fast-scan video monitor. Similarly, on transmit the converter

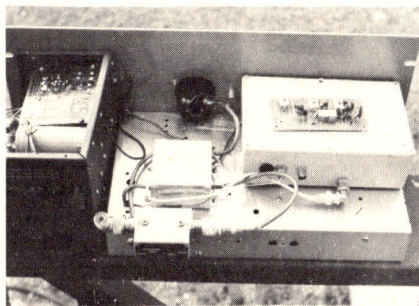

Fig. 40 — Top view of the ATV transceiver. The power supply is at the left. A coaxial relay and Bird power sensor are near the center of the back apron. The TVC-1 tunable converter has been placed atop a Janel converter box for size comparison. A 12-V regulator chip has been provided for the converter because the main supply furnishes 15 volts to enable the power amplifier to reach full output.

Table 15
Amateur Slow-Scan Standards

	60-Hz Areas	50-Hz Areas
Sweep Rates:		
Horizontal	15 Hz	16 2/3 Hz
	(60 Hz/4)	(50 Hz/3)
Vertical	8 sec.	7.2 sec.
No. of Scanning		
Lines	120	120
Aspect Ratio	1:1	1:1
Direction of Scan:		
Horizontal	Lt to Rt	Lt to Rt
Vertical	Top to Bot.	Top to Bot.
Sync Pulse Duration:		
Horizontal	5 millisec.	5 millisec.
Vertical	30 millisec.	30 millisec.
Subcarrier Freq.:		
Sync	1200 Hz	1200 Hz
Black	1500 Hz	1500 Hz
White	2300 Hz	2300 Hz
Req. Trans.		
Bandwidth	1.0-2.5 kHz	1.0-2.5 kHz

Fig. 40A — A typical slow-scan TV picture.

changes the output of a fast-scan camera to a standard slow-scan signal.

SSTV is legal anywhere in the Advanced and Extra Class voice segments of the 75, 40, 20 and 15 meter bands and all voice bands above 28 MHz. The standard calling frequencies are 3845, 7171, 14,230, 21,340 and 28,680 kHz with 75 and 20 meters being the most popular bands.

SSTV signals must be tuned in properly so the picture will come out with the proper brightness and the 1200-Hz synchronization pulses will be detected. If the signal is not "in sync," the picture will appear wildly skewed. The easiest way to tune SSTV is to wait for the transmitting operator to say something on voice and then tune him in while he is talking. With experience you may find you are able to zero in on an SSTV signal by listening to the sync pulses and by watching for proper synchronization on the screen. Many SSTV monitors are equipped with tuning aids of various kinds.

If you want to record slow-scan pictures off the air, there are two ways of doing it. One is to tape record the audio signal for playback later. The other is to take a picture of the image right from the SSTV screen. Polaroid cameras equipped with a closeup lens enable you to see the results shortly after the picture is taken. If you want to do this without darkening the room lights, you'll have to fabricate a light-tight hood to fit between the camera and the monitor screen.

On any amateur transmission, the legal identification must be made by voice or cw. Sending "This is WA0XYZ" on the screen is *not* sufficient. Most stations intersperse the pictures with comments anyway, so voice i-d is not much of a problem. Otherwise SSTV operating procedures are quite similar to those used on ssb.

As with RTTY, the station transmitter must be tuned for 100-percent duty cycle, since the SSTV emission is a constant tone. Only the frequency is changing.

For more information about amateur TV, see *Amateur Television Magazine*, c/o Mike Stone, WB0QCD. A *QST* television bibliography is available for an s.a.s.e. from ARRL.

NARROW-BAND VOICE MODULATION

The December 1977 issue of *QST*[5] heralded a new and unique system for conserving communications bandwidth. This technique, now implemented and in the production stage, works at baseband (audio) rather than at intermediate or radio frequencies (i-f or rf). Thus it is applicable to virtually all types of analog and digital transmission systems. The system includes the newly developed frequency compandor[6] and the well-known, but not extensively used, amplitude compandor.[7] Use of both devices within the same baseband system offers significant improvements in adjacent-channel rejection and signal-to-noise ratio (SNR).

The transceive baseband system operates on the audio waveform just after the microphone but before the speaker. The frequency compandor filters the essential parts of speech and down converts this information electronically on transmission, thus providing a significant reduction in transmitted bandwidth. A narrower bandwidth signal causes less interference to others operating in the same band. It also allows the use of a sharper and narrower receive filter, which greatly reduces adjacent-channel interference.

What Is Speech?

One can better understand how the frequency compandor works by considering the composition of speech. Acoustically, human speech consists predominantly of two types of sounds — voiced and unvoiced.

[5]Harris, R. W. and Gorski, J. C., "A New Era in Voice Communications," *QST*, December 1977.
[6]A frequency compandor compresses signal bandwidth on transmission and expands signal bandwidth on reception.
[7]An amplitude compandor compresses signal amplitude on transmission and expands signal amplitude on reception.

Voiced sounds originate by passing air from the speaker's lungs through the larynx (voice box), a passage in the human throat with the opening obstructed by vocal cords.[8] As air is passed by these cords, they vibrate, causing puffs of air to escape into the aural cavity, which consists of the throat, nasal cavity and mouth. Studies indicate that the acoustic waveform produced by the vocal cords has many harmonics of the fundamental vibration. Because of the irregular shape of the aural cavity, the spectral-amplitude distribution of the harmonics tends to show peaks at distinct points. As speech is produced, changes occur in the aural cavity shape, thus changing the spectral location of these peaks.

Fig. 41 shows a spectrogram, or voice print, of the utterance "digital communication." The vertical axis represents frequency (80-8000 Hz), and the horizontal axis represents time (0-1.5 s). Darkness of the bands indicates amplitude or voice strength. The fine structure of amplitude peaks that are very close together in the horizontal dimension is a measurement of vocal-cord vibration (fundamental frequency).

Notice the rather strong amplitude concentrations below 4000 Hz. These are the spectral peaks referred to above and are called formants. The first three formants are shown in Fig. 41 at the beginning of the utterance. Proper processing of these three formants is a major concern of bandwidth conservation in speech.

Unvoiced sounds occur when there is no vocal-tract excitation. Sounds such as clicks, hisses and popping are caused by the speaker using his tongue, lips and teeth. These sounds, or evidence of their occurrence by formant extensions into or from a voiced sound, are very important to the intelligibility of speech. Spectral-amplitude distributions of unvoiced

[8]Flanagan, J. L., *Speech Analysis Synthesis and Perception*, 2nd Ed., Springer-Verlag, 1972.

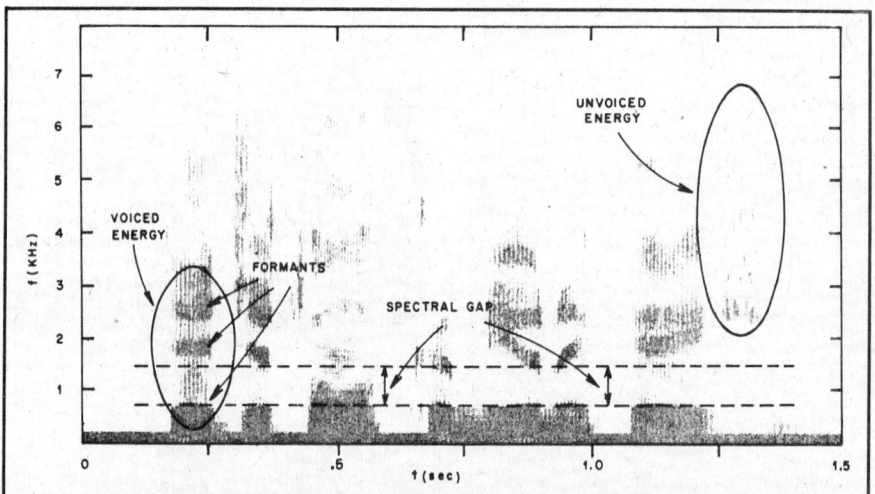

Fig. 41 — A spectrogram or voice print of the utterance "digital communication." The vertical axis represents frequency (80-8000 Hz), and the horizontal axis represents time (0-1.5 s).

sounds are generally above 1500 Hz and are "noise-like," in that very little periodic structure is present (see Fig. 41).

Briefly, speech is the continuous production of voiced and unvoiced sounds, with appropriate pauses to add clarity and distinctness. Measurements performed on voices from different speakers indicate that the first three formants lie predominantly below 2500 Hz. Speech consisting of these three formants is of good quality, both from an intelligibility and "listenability" standpoint. Sufficient information as to the existence of some unvoiced sounds appears to lie in this range. For example, to produce an "s" sound, the frequency range must extend to approximately 4000 Hz, but this is not usually required for intelligibility since contextural clues provide sufficient evidence for the listener to "hear" an "s."

Evidence from theory and that gained through practice (amateur communications) indicate a bandwidth of 300 to 2500 Hz is adequate for good quality speech.

Audio Bandwidth Reduction and the Frequency Compandor

To explain the approach taken consider the spectrogram in Fig. 41. Notice that there are natural gaps between the first and second, and second and third formants. There is little energy present in these gaps.

After extensive listening tests and consideration of various filtering and mixing combinations it was found that the first formant is not as essential to intelligibility as the second and third. Furthemore, the gap between the first and second formants is wider then between the second and third and it is more constant with time. As a result the system shown in Fig. 42 was developed.

To understand how the system works note that two bands of speech are preserved, the first from dc to 600 Hz (most communications transceivers limit the low end to 350 Hz) and second from f_1 to 2500 Hz; f_1 corresponds to the low end of the second formant and is variable depending on the transmission and reception low-pass filter (LPF) cutoff frequency f_T. In equation form

$$f_1 = 3100 - f_T$$

For example, the two filter options provided by the first commercial system will be $f_T = 1600$ Hz and $f_T = 2100$ Hz. Both the transmission and reception filters have a 1.3 shape factor. Thus the narrow system with a transmission bandwidth of 1600 Hz is designed to preserve speech from 350-600 Hz, which is the first formant approximation, and from 1500-2500 Hz, which is the band of contiguous second and third formants. The wider system with a 2100-Hz transmission bandwidth preserves speech from 350-600 Hz system, but also preserves the region from

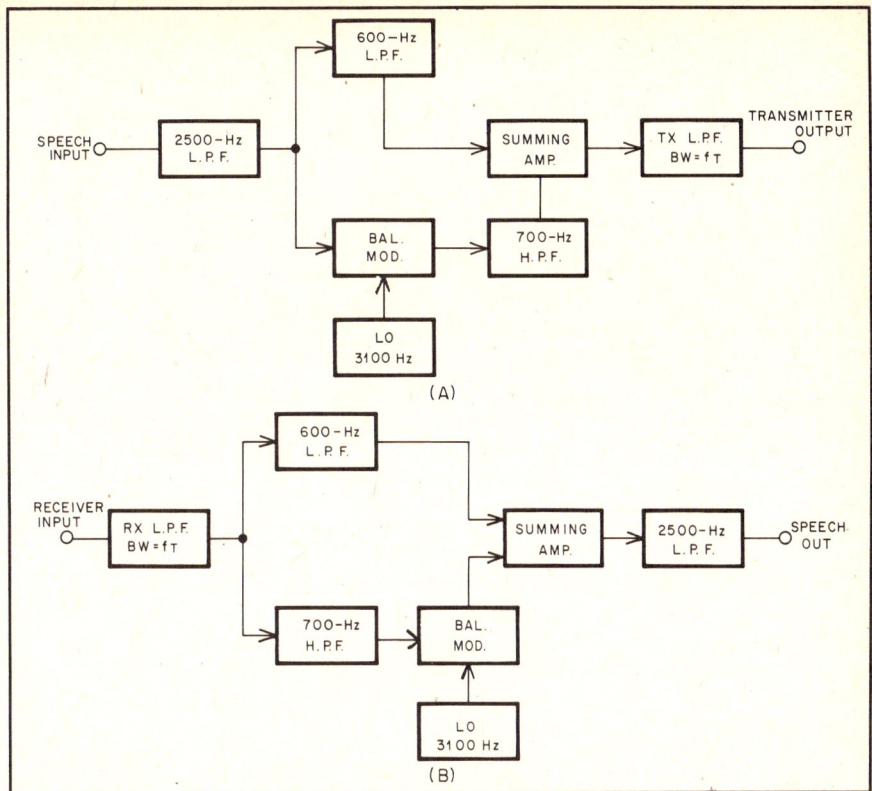

Fig. 42 — Block diagram of the basic frequency compandor scheme. The circuit at A is used to compress the audio bandwidth on transmit. The circuit at B expands this signal for receive.

1000-2500 Hz which includes more of the lower end of the second formant.

Operationally, the first formant, 350-600 Hz, passes essentially straight through the system. The second and third formants are inverted and down converted for transmission, then reinverted and up converted on reception. Use of the 700-Hz high-pass filter (HPF) aids in eliminating potential distortion products caused by high frequencies mixed low on transmit and low frequencies mixed high on receive.

SPREAD SPECTRUM COMMUNICATIONS

Intentionally broadbanding a radio signal by some means other than increased modulation by the baseband information is called spread spectrum (SS) transmission. Scattering the rf energy over a broad frequency spectrum and "descattering" or refocusing the signal at the receiver can result in several communications advantages, including signal-to-noise ratio enhancement and interference rejection.

Spread spectrum development began in the late 1940s, the objective then being secure, jam-proof military communications. John P. Costas, W2CRR was the first person to recognize non-military applications for SS. In 1956 he submitted an article, "SSB Better than A-M?" for publication in *QST*. This manuscript challenged the conventional wisdom that

narrowband communications systems are inherently superior to wideband ones. *QST* declined to publish the article, probably not because the ideas were heretical, but more likely because ultra wideband schemes were nearly impossible to implement with the hardware then available to most amateurs. Somewhat ironically, the first published paper suggesting the use of SS on the amateur bands appeared in the professional literature. "Poisson, Shannon, and the Radio Amateur," by Costas, was presented in *Proceedings of the IRE* for December 1959. (Poisson and Shannon developed mathematical models for communications systems — all analytical studies in communication theory and information theory are based on their results.) The following statements are excerpted from Costas' paper to induce the reader to study the entire work. "Congested band operation as found in the amateur service presents an interesting problem in analysis which can only be solved by statistical methods. Consideration is given to the relative merits of two currently popular modulation techniques, ssb and dsb. It is found that in spite of the bandwidth economy of ssb this system can claim no over-all advantage with respect to dsb for this service. It is further shown that there are definite advantages to the use of very broadband techniques in the amateur service."

In the two decades since Costas' exposi-

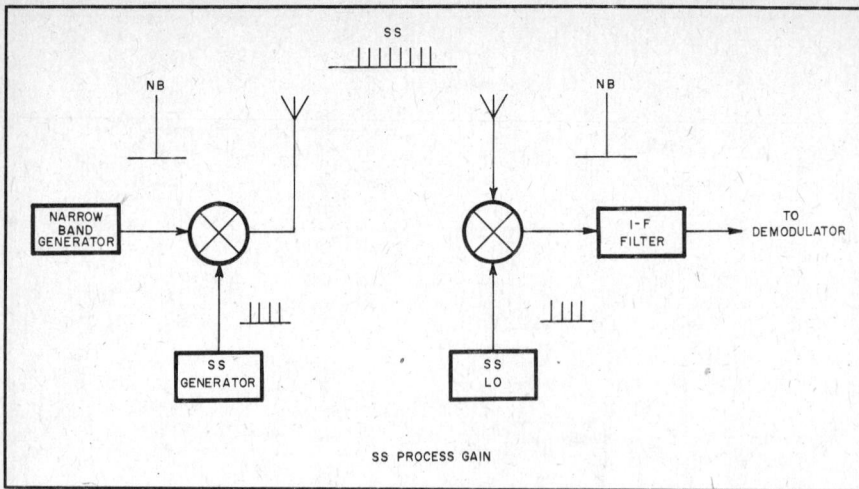

Fig. 43 — This sketch shows a simple method for spreading a baseband signal over a wide frequency spectrum. Carrying this technique to its logical extreme (infinitely many spectral lines having infinitesimal spacing) would generate an apparent white noise spectrum having nearly zero energy at any discrete frequency.

tion, congestion in the amateur bands has increased dramatically — possibly even beyond Costas' imagination. This, coupled with technological advances which bring very broadband communications systems within practical reach, has caused renewed amateur interest in SS. November 1980 *QST* featured an article by Paul Rinaldo, W4RI, entitled "Spread Spectrum and the Radio Amateur." Most of the material that follows is taken from that article.

Spread-Spectrum Fundamentals

SS systems employ radio-frequency bandwidths that greatly exceed the bandwidth necessary to convey the intelligence. Bandwidths for SS systems generally run from 10 to 100 times the information rate. By spreading the power over a wide band, the amount of energy in any particular hertz or kilohertz is very much smaller than for conventional narrow-band modulation techniques. Depending upon the transmitter power level and the distance from the transmitter to the receiver, the

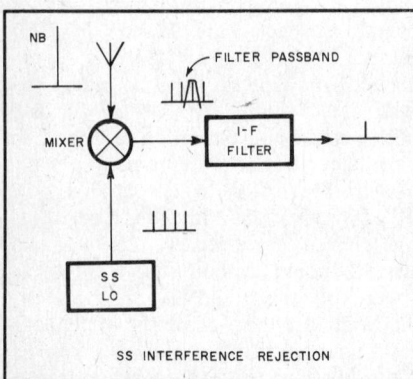

Fig. 44 — Unwanted (or improperly coded) signals and noise are spread by the SS LO, causing only a small portion of the unwanted energy to be presented to the demodulator.

SS signal may be below the noise level.

SS systems also use coding sequences to modulate and demodulate the transmission. Receivers with the wrong code will not demodulate the encoded SS signal and will be highly immune to interference from it. On the other hand, receivers with the right code are able to add all the spread energy in a constructive way to reproduce the intended modulation. In fact, the use of coherent correlation can yield some process gain. Changing the code to another sequence effectively creates a new "channel" on which a private conversation can take place. Many good code combinations could be made available on a single chip and selected by means of thumbwheel switches on the SS transceiver.

Process gain is the signal-to-noise ratio advantage achieved by SS techniques. Fig. 43 illustrates the process-gain mechanism by way of the simplest possible example. An ordinary narrowband signal is heterodyned with (modulated by) a wideband rf signal. A comb generator is shown in this example, and was also used by Costas for his analysis of amateur spectrum utilization. The codes used in actual SS communications systems are much more complex. The mixer output consists of a set of discrete identical narrowband signals, two corresponding to each spectral line from the comb generator. The energy from the original narrowband generator is distributed among the comb lines. An identical comb generator serves as the receiver LO. Each incoming line of the SS signal is offset from the LO comb generator by the i-f, so the receiving mixer reconstructs the original narrowband signal for conventional i-f processing. Any noise or unwanted discrete narrowband signal present at the receiver input is "chopped up and spread out" by the SS LO, and only a small fraction of the un-

wanted energy falls within the i-f passband. This effect is shown in Fig. 44.

Types of Spread Spectrum

There are four basic types of spread spectrum; direct sequence, frequency hopping, pulse-fm and time hopping. In addition, there are hybrids consisting of combinations of two or more of the above basic types.

Direct Sequence (DS): Direct sequence SS is produced by modulation of a carrier with a digitized code stream. This type of modulation is also known by the terms pseudo-noise (PN), phase hopping (PH), direct spread, or direct code. Phase-shift keying (psk) is usually used to produce the marks and spaces, but frequency-shift keying (fsk) could also be used. The wide rf bandwidth arises from the use of a high-speed code. Of course, if the transmitter were allowed to rest on the mark frequency, there would be a steady carrier in one place whenever there is no modulation. This would produce interference to a narrow-band user on that frequency. It would also pose problems for other SS users of the same band, particularly if they did the same thing. So it is conventional for SS systems to include techniques to continue a pseudo-random code sequence even during intervals when intelligence is not being transmitted.

The power spectrum for a DS signal (as might be seen on a spectrum analyzer) is not uniform across the band, but has a main lobe and sets of sidelobes as illustrated in Fig. 45. The bandwidth of the main lobe as measured from null to null is two times the clock rate of the code sequence. The bandwidth of the side lobes is equal to the clock rate. To receive a DS signal, the receiver must collapse or "despread" it to the original bandwidth of the information. This is done by using a replica of the code sequence used by the transmitter.

Frequency Hopping (FH): As the name implies, frequency hopping is simply jumping to a number of different frequencies in an agreed sequence. The code sequence is usually at a slower rate than for direct sequence and is normally slower than the information rate. The hopping rate may also be determined by practical

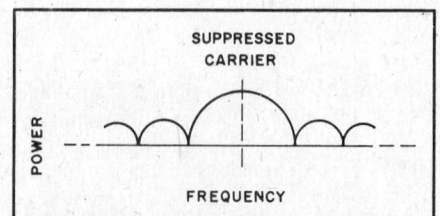

Fig. 45 — Power vs. frequency for a direct-sequence-modulated spread-spectrum signal. The envelope assumes the shape of a $\frac{\sin x^2}{x}$ curve. With proper modulating techniques, the carrier is suppressed.

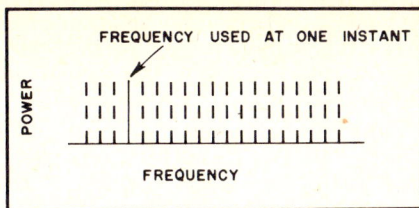

Fig. 46 — Power vs. frequency for frequency-hopping spread-spectrum signals. Emissions jump around in pseudo-random fashion to discrete frequencies.

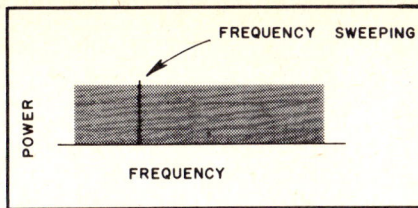

Fig. 47 — Power vs. frequency for chirp spread-spectrum signals. The carrier is repeatedly swept, continuously, from one end to the other in a given band.

considerations, such as how long it takes for a particular frequency synthesizer to settle down on a new frequency.

Actual modulation of the frequencies uses normal narrow-band techniques such as frequency modulation. At any instant, an FH transmitter is emitting all of its power on a specific frequency slot and potentially could interfere with someone else using a narrow-band system on that frequency. However, the FH dwell time on that particular frequency is so short that most narrow-band users would not be bothered. Mutual interference between two or more FH users sharing the same band could be extremely low, depending upon the design of the code sequences. Fig. 46 illustrates the power spectrum for an FH signal.

Pulse-FM (Chirp): A chirp spread-spectrum system sweeps its carrier frequency over a wide band at a known rate. Again, conventional narrow-band modulation of the sweeping carrier is used to convey the intelligence. The receiver uses a matched, dispersive filter to compress the signal to a narrow band. Chirp systems typically do not use a code sequence to control the sweep generator. Sweep time can be largely independent of the information rate. Normally a linear-sweep pulse is used, similar to that produced by a sweep generator. The power spectrum for a chirp system is illustrated in Fig. 47.

Time-Hopping (TH): Time hopping is a form of pulse modulation using a code sequence to control the pulse. As in other pulse techniques, the transmitter is not on full time and can have a duty cycle of 50% or less. Several systems can share the same channel and function as a time-division multiple-access (TDMA) system. TH is more vulnerable to interference on its center frequency than other SS systems. Seldom seen in its pure form, TH is typically used in hybrid systems using frequency hopping as well.

Hybrids: In addition to the TH/FH hybrid system just mentioned, there are also DS/FH and DS/TH combinations. Hybrid systems are typically designed to accommodate a large number of users and to provide a higher immunity to interference. They also produce better results at practical code sequence rates governed, for example, by how fast a frequency synthesizer can be switched. Also, hybrids can produce greater spreads than those which are practical for pure SS systems.

Some Considerations

Synchronization: In the design of a spread-spectrum system, usually the toughest problem is synchronization of the code sequence at the receiver with that of the incoming signal. If sync is not attained, even just one bit off, nothing but noise can be heard. The problem becomes worse when more than two stations are trying to communicate in a net. This is because of the different propagation delays between stations; i.e., it takes a different time for a signal to travel over paths A-B, A-C, or B-C if the stations are not equidistant. These differences may be only slight but just enough to degrade the signal-to-noise ratio of the received signal. In addition to the time uncertainty related to propagation, there is also a frequency uncertainty in trying to keep oscillators at two or more stations from drifting.

Because the stations cannot be expected to synchronize on their own with no reference, it is normal for at least one station to transmit an initial reference for sync purposes. Upon reception, the receiving stations can generate the code sequence at a rate different from the code sequence used at the transmitter. Eventually, the two code streams will slide into phase with one another and may then be locked up. After initial synchronization, maintaining sync presents another problem which can be solved in different ways. One is to use a code sequence preamble at the beginning of each transmission. Another is to use ultra-stable clocks at all stations to ensure that the code-sequence clock frequency does not change. Numerous other schemes have been devised and implemented with varying degrees of difficulty. The exception is that chirp systems do not have this problem because the matched filter used in demodulation inherently achieves sync on each pulse transmitted.

Transmitter and Receiver Design: One difference between SS and conventional rf equipment is that SS requires transmitters and receivers that have 10 to 100 times the bandwidth of narrow-band systems. That may pose some problems at lower frequencies, but in the 420-MHz band the

amateur television (ATV) experimenters already have equipment that can handle wideband signals. The transmitter design, which should be well within amateur capability, amounts to taking care in broadbanding the rf stages after modulation to maintain amplitude linearity, and in keeping the antenna system VSWR very low. Receivers must not only have wideband front ends but must also have good dynamic range and linearity to handle both the desired signal and any interference. Where an i-f is used, the frequency chosen must be higher than for conventional transceivers. In practice, 70 MHz is a common SS i-f. Components (such as filters) are available for this frequency to build SS i-f modems (modulator/demodulators).

Amateur SS Experimentation

In March 1981, W4RI of Amateur Radio Research and Development Corporation (AMRAD) and 28 other stations received Special Temporary Authorization from FCC to experiment with SS techniques. The group is conducting frequency-hopping operation on the hf bands and direct-sequence SS on uhf. The uhf experiments will include an evaluation of SS as a means to combat libration fading on EME circuits. As this material is being prepared (July 1981), FCC has just adopted a Notice of Inquiry and Notice of Proposed Rulemaking (docket 81-414) which would permit Advanced and Extra Class licensees to experiment with SS techniques on the vhf bands without special authorization.

Selected Bibliography

Reading material on spread spectrum may be difficult to obtain for the average amateur. Below are references that can be mail ordered. Spread spectrum papers have also been published in IEEE Transactions on Communications, on Aerospace and Electronic Systems and on Vehicular Technology.

Dixon, *Spread Spectrum Systems,* 1976, Wiley-Interscience, 605 Third Ave., New York, NY 10016, $29.50

Dixon, *Spread Spectrum Techniques,* IEEE Service Center, 445 Hoes La., Piscataway, NJ 08854, IEEE member prices $19.45 clothbound, $12.95 paperbound; nonmembers $29.95 clothbound.

Brumbaugh, et al., *Spread Spectrum Technology,* a series of papers presented at the 1980 Armed Forces Communications Electronics Association show printed in the August 1980 issue of *Signal,* available from AFCEA, Skyline Center, 5205 Leesburg Pike, Falls Church, VA 22041.

Current published searches on spread spectrum are available from the U.S. Department of Commerce, National Technical Information Service, Springfield, VA 22161 for $30 each:

Spread Spectrum Communications (99), May 79 NTIS/PS-79/0494/9.

Spread Spectrum Communications (188), May 79 (EI) NTIS/PS-79/0495/6.

COHERENT CW

While spectrum management has received much attention in the recent Amateur Radio literature, the problems and possibilities of "more QSOs per kilohertz" were first recognized more than half a century ago. Frederick Emmons Terman, 6FT, presented his vision of narrow-band communications in

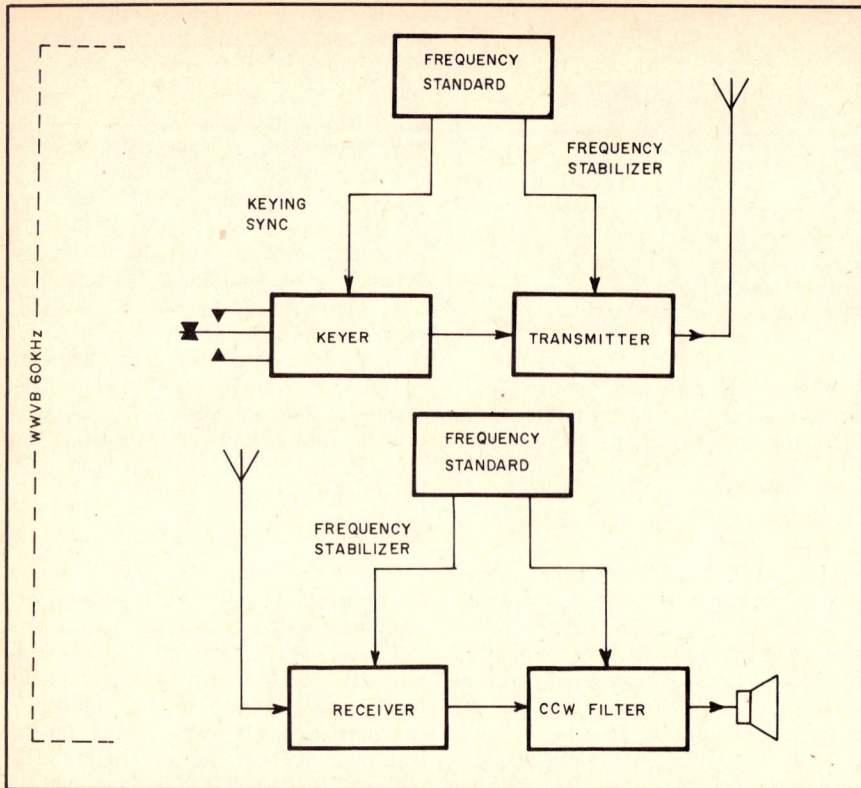

Fig. 48 — A ccw communications link.

[Figure labels: FREQUENCY STANDARD, FREQUENCY STABILIZER, KEYING SYNC, KEYER, TRANSMITTER, WWVB 60kHz, FREQUENCY STANDARD, FREQUENCY STABILIZER, RECEIVER, CCW FILTER]

''Some Possibilities of Intelligence Transmission When Using a Limited Band of Frequencies,'' published in *Proceedings of the Institute of Radio Engineers*, January 1930.

An important part of Terman's paper deals with the theoretical bandwidth required for radiotelegraphy. As early as 1927, the Bell Telephone Company had reported successful experiments with 200-wpm Baudot TTY communication in a 50-Hz bandwidth over undersea cables. The bandwidth reduction resulted from synchronization of the transmitter and receiver. In proposing the application of synchronized telegraphy to radio, Terman addressed the frequency stability problem attendant to any reduced-bandwidth system by suggesting that transmitter and receiver oscillators be locked to one of the standard frequencies broadcast by NBS. He noted that given a reference frequency, any other frequency can be derived by means of multiplication, division or heterodyning. Only the word ''synthesizer'' was missing!

Technology made giant leaps in the next 45 years. In September 1975 *QST*, Raymond Petit, W7GHM, described the experiments of some radio amateurs with a mode he called ''coherent cw.'' Petit did not acknowledge Terman's paper, so we must conclude that he rediscovered the wheel. In any case, ccw is an idea whose time has come. Adrian Weiss, WØRSP, disclosed some of the technical details of the ccw system in June and July 1977 *CQ*.

The presentation contains some errors, but the astute reader will be able to recognize the significant principles.

As discussed in Chapter 11 of this *Handbook*, the bandwidth required for transmitting a radiotelegraph signal is directly proportional to the keying rate. For a speed of 12 wpm the unit pulse length is 0.1 second. Since a dot and a space each require 0.1 second, a string of dots at 12 wpm is a square wave having a fundamental frequency of 5 Hz. To preserve the square-wave characteristic of the emission, an ssb transmission bandwidth of at least 15 Hz is required. A baseband (or dc wire telegraph) receiver needs a similar bandwidth for conventional information recovery. Terman reported that with synchronization techniques, the receiver bandwidth could be reduced to 1.5 to 2.0 times the keying rate. In conventional (Morse) radiotelegraphy, the intelligence is ultimately received as an audio tone. Even a 15-Hz bandwidth filter centered on, say, 500 Hz, would require a Q of 33, causing intolerable ringing.

The ringing problem can be overcome with time-domain processing at both ends of the communications path. The transmitter is stabilized to within 1 Hz of the proper frequency by phase-locking to a reference standard. Precisely timed keying pulses are derived from the same reference standard. A similar reference standard stabilizes the receiver frequency and synchronizes the audio output filter. The receiver output is sampled at twice the

ACTIVITY CODE

GOVERNMENT EXCLUSIVE

GOVERNMENT AND NON-GOVERNMENT SHARED

NON-GOVERNMENT ONLY

BAND ALLOTMENT CODE

FIXED
MOBILE
AERONAUTICAL FIXED
AERONAUTICAL MOBILE
FIXED SATELLITE
MARITIME MOBILE
RADIOLOCATION
LAND MOBILE
RADIO ASTRONOMY
MOBILE SATELLITE
SPACE OPERATIONS
AERONAUTICAL MOBILE SATELLITE
SPACE RESEARCH
MARITIME MOBILE SATELLITE
RADIONAVIGATION
BROADCASTING
AERONAUTICAL RADIONAVIGATION
BROADCASTING SATELLITE
MARITIME RADIONAVIGATION
AMATEUR
RADIONAVIGATION SATELLITE
AMATEUR SATELLITE
AERONAUTICAL RADIONAVIGATION SATELLITE
CITIZENS
MARITIME RADIONAVIGATION SATELLITE
INTER SATELLITE
STANDARD FREQUENCY
STANDARD FREQUENCY SATELLITE
METEOROLOGICAL AIDS
METEOROLOGICAL SATELLITE
EARTH EXPLORATION SATELLITE

the paddle-actuated clock in the keyer must be replaced by a continuous pulse train from the frequency standard. Coordinating one's paddle movements with the "metronome" requires a different keying technique. A buffered keyboard (controlled by the standard) is the ideal ccw sending instrument.

When more stations have ccw capability, the mode may prove highly useful for emergency communications. Another possibility for ccw is in EME work. The signal-to-noise ratio advantage should reduce the station gain requirements. Charles Woodson, W6NEY, played a stereophonic demonstration tape at ARRL headquarters of a 20-meter ccw contact with JR1ZZR using a power of 100 mW. While this type of contact isn't too unusual in itself when conditions are good and interference is absent, the channel recorded without the ccw filter was unreadable, while the ccw channel was perfect copy. Woodson publishes a ccw newsletter whenever there is new information to report. In addition, Petit Logic Systems, P.O. Box 51, Oak Harbor, WA 98277 markets kits and accessories for ccw operation.

RADIOTELETYPE

Radioteletype (abbreviated RTTY) is a form of telegraphic communication employing typewriterlike machines for (1) generating a coded set of electrical impulses when a typewriter key corresponding to the desired letter or symbol is pressed, and (2) converting a received set of such impulses into the corresponding printed character. The message to be sent is typed out in much the same way that it would be written on a typewriter, but the printing is done at the distant receiving point. The teletypewriter at the sending point may also print the same material.

The teleprinter machines used for RTTY are far too complex mechanically for home construction, and if purchased new would be highly expensive. However, used teletypewriters in good mechanical condition are available at quite reasonable prices. These are machines retired from commercial service but capable of entirely satisfactory operation in amateur work. They may be obtained from several sources on condition that they will be used purely for amateur purposes and will not be resold for commercial use.

Some dealers and amateurs around the country make it known by advertising that they handle parts or may be a source for machines and accessory equipment. QST's Ham Ads and other publications often show good buys in equipment as amateurs move about, obtain newer equipment, or change interests.

Periodic publications are available which are devoted exclusively to amateur RTTY. They carry timely technical articles and operating information, as well as classified ads. Over the years QST has

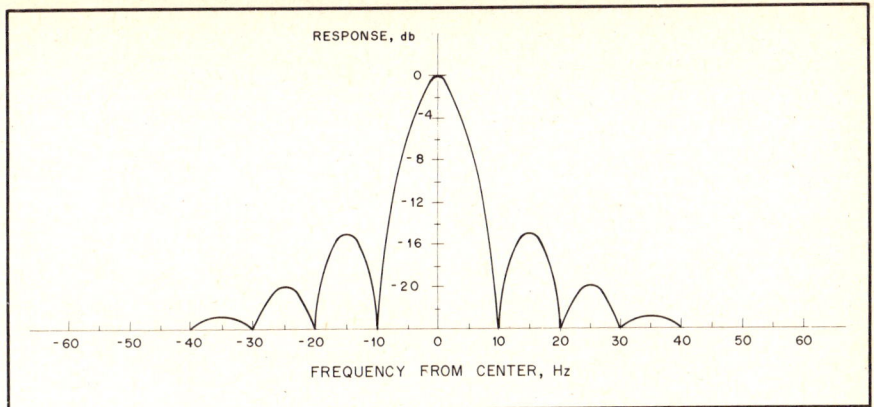

Fig. 50 — Amplitude vs. frequency response of the receiving filter.

carried a number of articles on all aspects of RTTY. For a list of surplus equipment dealers, information on publishers of RTTY periodicals, and a bibliography of all articles on RTTY which have appeared in QST, write to ARRL, 225 Main Street, Newington, CT 06111. U.S. residents should enclose a stamped business-size envelope bearing a return address with their request.

Types of Machines

There are two general types of machines, the *page printer* and the *tape printer*. The former prints on a paper roll about the same width as a business letterhead. The latter prints on paper tape, usually gummed on the reverse side so it may be cut to letter-size width and pasted on a sheet of paper in a series of lines. The page printer is the more common type in the equipment available to amateurs.

The operating speed of most machines is such that characters are sent at the rate of either 60, 67, 75 or 100 wpm depending on the gearing ratio of a particular machine. Current FCC regulations allow amateurs the use of any of these four speeds. Interchangeable gears permit most machines to operate at these speeds. Ordinary teletypewriters are of the *start-stop* variety, in which the pulse-forming mechanism (motor driven) is at rest until a typewriter key is depressed. At this time it begins operating, forms the proper pulse sequence, and then comes to rest again before the next key is depressed to form the succeeding character. The receiving mechanism operates in similar fashion, being set into operation by the first pulse of the sequence from the transmitter. Thus, although the actual transmission speed cannot exceed about 60 wpm (or whatever maximum speed the machine is geared for), it can be considerably slower, depending on the typing speed of the operator.

It is also possible to transmit by using perforated tape. This has the advantage that the complete message may be typed

out in advance of actual transmission, at any convenient speed; when transmitted, however, it is sent at the machine's normal maximum speed. A special tape reader, called a *transmitter-distributor*, and tape perforator are required for this process. A *reperforator* is a device that may be connected to the conventional teletypewriter for punching tape when the machine is operated in the regular way. It may thus be used either for an original message or for "taping" an incoming message for later retransmission.

Start, Data, Stop Pulses

In a teleprinter machine, the normal "rest" condition of the selector-magnet solenoids is with loop current on. Interruption of the loop current releases the selector magnet, allowing rotation of a cam in the machine. Transmission of a TTY character begins with a space pulse (current off), called the *start* pulse. The start pulse signals to the machine that reception of a character has begun. Immediately after the start pulse, a series of *data* pulses is transmitted with mark or space condition as indicated by the encoding for the desired character. The number of data pulses used to represent the letters, numbers and symbols varies

Fig. 51 — This is the first complete amateur station to be built for coherent cw operation. Assembled by Andy McCaskey, WA7ZVC, it consists of a modified Ten-Tec PM-2 transceiver and homemade modules which provide for the control and processing of signals as required for coherent-cw operation.

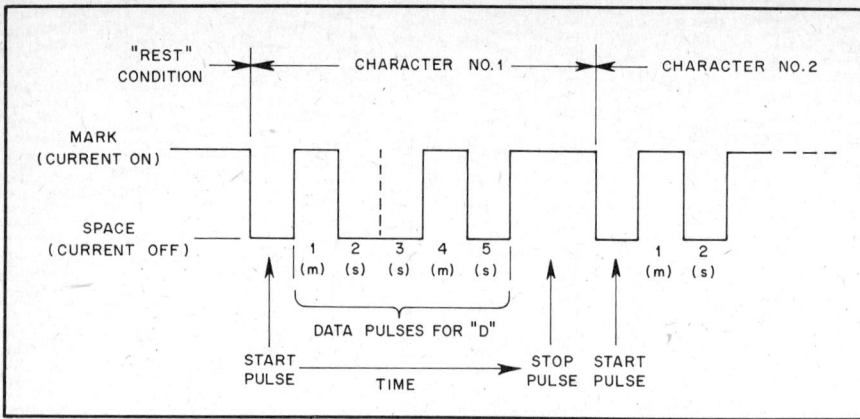

Fig. 52 — Time sequence of typical Baudot character, the letter D.

with the TTY code being used; Baudot code uses five data pulses, ASCII uses eight. Immediately after the last data pulse, a *stop* pulse is included which is always a mark pulse. The stop pulse, therefore, always occurs in a fixed time after the start pulse (after five data pulses in Baudot and eight in ASCII). The stop pulse gives the machine a "rest time" to prepare for the beginning of the next character, maintaining receive machine synchronization with the transmitted signal. The time length of the start and each data pulse are the same and are often called the unit-pulse or select-pulse time. The stop-pulse length varies from code to code and even with speeds within a code, as will be explained later. In general, the minimum stop-pulse length can be one or two times as long as the unit-pulse time; stop pulses may be as long as desired since the machine is "at rest" until the next start pulse is received. This type of TTY code that uses start, data and stop pulses in the construction of each character is called an asynchronous or start-stop serial code. Other codes also in commercial use include synchronous serial codes, in which start and stop pulses are not attached to the data pulses for each character, and parallel data codes, in which each data pulse is assigned a separate wire to and from the terminal device. Such codes are found in common use with computer and line-printer devices. FCC regulations currently authorize amateurs to use either the Baudot or the ASCII serial asynchronous TTY codes.

The Baudot TTY Code

One of the first data codes used with mechanical printing machines uses a total of five data pulses to represent the alphabet, numerals and symbols. This code is commonly called the Baudot or Murray telegraph code, after the work done by these two pioneers. Although commonly called the Baudot code in the United States, a similar code is usually called the Murray code in other parts of the world and is formally defined as the International Telegraphic Alphabet No. 2 Baudot Code in part 97.69 of the FCC Rules and Regulations. This standard defines the codes for letters, numerals and the slant or fraction bar but allows variations in the choice of code combinations for punctuation. U.S. amateurs have generally adopted a version of the so-called "Military Standard" code arrangement for punctuation, largely because of the ready availability of military surplus machines in the post-1945 years. Amateurs in other countries (particularly in Europe) have standardized on the International Consultative Committee for Telephone and Telegraph (CCITT) No. 2 code arrangement, which is similar to the U.S. standard but has minor symbol and code-arrangement differences.

Since each of the five data pulses can be in either a mark or space condition (two possible states per pulse), a total of $2 \times 2 \times 2 \times 2 \times 2 = 2^5 = 32$ different code combinations are possible. Since it is necessary to provide transmission of all 26 letters, 10 numerals and punctuation, the 32 code combinations are not sufficient. This problem is solved by using the codes twice; once in the *letters* (LTRS) case and again in the *figures* (FIGS) case. Two special characters, LTRS and FIGS, are used to indicate to the printer whether the following characters will be of the letters or figures case. The printer has a latching mechanism that "remembers" or stores the last received LTRS or FIGS character so that it remains in the last received case until changed. Control operations such as LTRS, FIGS, carriage return (CR), line feed (LF), space bar (SP) and blank (BLNK = no print or carriage movement) are assigned to both the LTRS and FIGS case so that they can be sent in either case. The remaining 26 code combinations have different letter or numeral/symbol meanings, depending upon whether preceded by a LTRS or FIGS character.

Keyboards on Baudot machines such as the Teletype Corp. models 15 and 28 differ from standard typewriter keyboards, having only three rows of keys with the related letter and number/symbol on each keytop — Q and 1, K and (, and so on). The typist soon discovers this difference! Newer electronic terminals such as the HAL DS2000 and DS3100 have standard keyboard arrangements and automatically insert LTRS or FIGS characters as they are needed. The Baudot code itself is restricted to upper-case letters only since insufficient codes are available to represent lower-case letters.

The Baudot code has seen extensive commercial use throughout the world and is still actively utilized for international wire, press and weather communications. Because of the ready availability of Baudot mechanical equipment, this code will continue to be quite popular among radio amateurs. Nevertheless, the lack of code space for control, extended punctuation or lower-case letters is a severe limitation of the five-unit Baudot code. These limitations are particularly inconvenient in computer-terminal applications, even though various serial and parallel data-coding schemes have been used with computers. Fig. 52 shows a time diagram of typical Baudot characters, and Table 16 shows the Baudot data code for both the U.S. and CCIT No. 2 alphabet. Notice that the waveform drawing of Fig. 52 shows the *current* waveform, with mark represented by the upper deflection. Also, the bits in Fig. 52 are arranged in a left-to-right order, as would be observed on an oscilloscope. The bits in Table 16, however, are arranged in *descending* order (b5 to b1), conforming to the standard binary representation. Thus the letter D shown in Fig. 52 would be written as the binary character 01001.

ASCII

In 1968, the American National Standards Institute (ANSI) adopted the American National Standard Code for Information Interchange (ASCII), ANSI Standard X3.4-1968. This code uses seven data pulses to specify the letter, number, symbol or control operation desired. An eighth data pulse, called the *parity* bit, is provided for optional error checking. As with the Baudot code, the ASCII standard as approved for U.S. amateur use is asynchronous and serial with both start and stop pulses.

Whereas the five-unit Baudot code was arranged by Murray so that the most frequently used letters are represented by the least number of mark holes punched in paper tape, ASCII has been arranged to optimize computer applications. The code has been particularly designed for rapid collation of alphanumeric lists, one data bit difference between upper- and lower-case letters, and isolation of all control operations from printing operations. A time diagram of a typical ASCII character is shown in Fig. 53. Table 17 shows the ASCII data code. As noted for the Baudot-waveform drawing, Fig. 53 shows

Table 16
The Baudot Data Code

Bit Number 54321	*Letters*	*U.S. Figures*	*CCITT No. 2 Figures*
00000	BLANK	BLANK	BLANK
00001	E	3	3
00010	LF	LF	LF
00011	A	—	—
00100	SPACE	SPACE	SPACE
00101	S	BELL	'
00110	I	8	8
00111	U	7	7
01000	CR	CR	CR
01001	D	$	WRU
01010	R	4	4
01011	J	'	BELL
01100	N	,	,
01101	F	!	!
01110	C	:	:
01111	K	(	(
10000	T	5	5
10001	Z	"	+
10010	L	)	)
10011	W	2	2
10100	H	#	£
10101	Y	6	6
10110	P	Ø	Ø
10111	Q	1	1
11000	O	9	9
11001	B	?	?
11010	G	&	&
11011	FIGS	FIGS	FIGS
11100	M	.	.
11101	X	/	/
11110	V	;	=
11111	LTRS	LTRS	LTRS

Note: FIGS-H (10100) may also be used for MOTOR STOP function. "1" = mark = hole in punched tape

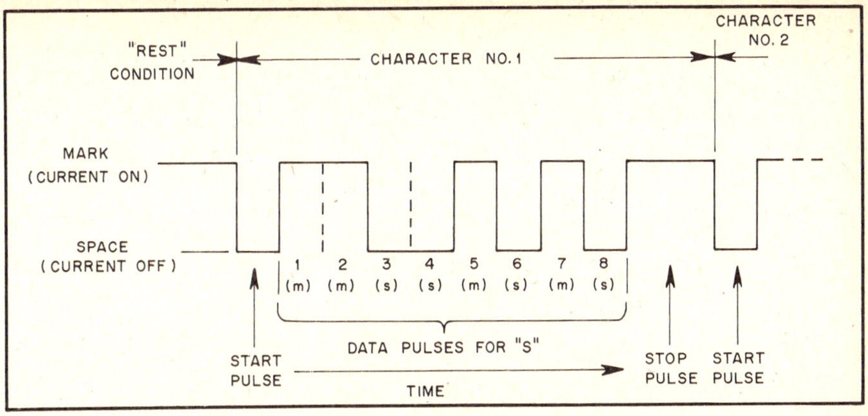

Fig. 53 — Time sequence of typical ASCII character, the letter s. The eighth or parity bit may be set for any of four conditions: (1) always mark, (2) always space, (3) odd parity or (4) even parity. All four choices are in common usage.

Table 17
The ASCII Data Code

7		0	0	0	0	1	1	1	1
6		0	0	1	1	0	0	1	1
5		0	1	0	1	0	1	0	1
4321									
0000		NUL	DLE	SPC	Ø	@	P	\	p
0001		SOH	DC1	!	1	A	Q	a	q
0010		STX	DC2	"	2	B	R	b	r
00011		ETX	DC3	#	3	C	S	c	s
0100		EOT	DC4	$	4	D	T	d	t
0101		ENQ	NAK	%	5	E	U	e	u
0110		ACK	SYN	&	6	F	V	f	v
0111		BEL	ETB	'	7	G	W	g	w
1000		BS	CAN	(	8	H	X	h	x
1001		HT	EM	)	9	I	Y	i	y
1010		LF	SUB	*	:	J	Z	j	z
1011		VT	ESC	+	;	K	[	k	{
1100		FF	FS	,	<	L	\	l	/
1101		CR	GS	–	=	M	]	m	}
1110		SO	RS	.	>	N	^	n	~
1111		SI	US	/	?	O	_	o	DEL

ACK = acknowledge
BEL = signal bell
BS = backspace (←)
CAN = cancel
CR = carriage return
DC1 = device control 1
DC2 = device control 2
DC3 = device control 3
DC4 = device control 4
DEL = (delete)
DLE = data link escape
ENQ = enquiry (WRU)
EM = end of medium
EOT = end of trans.
ESC = escape
ETB = end of block
ETX = end of text
FF = form feed (home)
FS = file separator
GS = group separator
HT = horizontal tab (→)
LF = line feed (↓)
NAK = not acknowledge
NUL = null
RS = record separator
SI = shift in
SO = shift out
SOH = start of heading
SPC = space
STX = start of text
SUB = substitute
SYN = synchronous idle
US = unit separator
VT = vertical tab (↑)

Note: "1" = mark = hole in punched tape

the loop current with mark represented by the upward deflection. Also the bits in Table 17 are arranged in binary number order (b7 to b1). Thus, the letter s in Fig. 53 would be written as the binary number 0101 0011, with the eighth (parity) bit set to space (0), as indicated.

As can be seen from the code table, many more punctuation symbols are included in ASCII than in Baudot. ASCII also includes a large number of control characters designed for print control of the terminal itself, formatting of data to the computer, and control of other hardware devices by the terminal. Although these control functions are defined by the ANSI definition, variations in the use of the control characters abound in the differing commercial applications.

The keyboards of both mechanical and electronic ASCII terminals are arranged similar to the "standard" typewriter keyboard, thus minimizing any retraining required when an operator moves from a typewriter to a terminal. The "extra" ASCII keys are arranged around the periphery of the standard keyset if they are provided at the terminal keyboard.

A common abbreviation of the full 128-character ASCII code restricts the alphabetic letters to upper-case only, often called CAPS-LOCK or CAPLK. In general, these terminals transmit the upper-case ASCII code for a letter whether the SHIFT key is used or not; they may or may not be capable of transmitting all of the control codes. These terminals usually print (or display) the upper-case letter when either the upper- or lower-case letter ASCII code is received. The Teletype model 33 is an example of a popular upper-case-only ASCII terminal. Other terminals, such as the Teletype model 43 or the HAL DS3100 ASR, have user-selectable upper/lower case or upper-case-only (CAPSLK) transmit/receive features.

The optional eighth data bit may be set to four conditions: (1) always mark, (2) always space, (3) odd parity, or (4) even parity. All four choices are in common usage. Simple non error-detecting terminals usually set the eighth bit to be always a mark or space (usually space). Parity is sometimes used with computer and data interconnections where error-detection is desired. When used, the parity bit is controlled so as to set the total number of mark data bits in the ASCII

Table 18
Baudot Data Rates and Speeds

Baud Rate	Data Pulse (ms)	Stop Pulse (ms)	WPM	Common Name
45.45	22.0	22.0	65.00	Western Union
	22.0	31.0	61.33	"60 speed"
	22.0	33.0	60.61	45 baud
50.00	20.0	30.0	66.67	European; 50 baud
56.92	17.57	25.00	76.68	"75 speed"
	17.57	26.36	75.89	57 baud
74.20	13.47	19.18	100.00	"100 speed"
	13.47	20.21	98.98	74 baud
100.0	10.00	15.00	133.33	100 baud

Table 19
ASCII Data Rates

Baud Rate	Data Pulse (ms)	Stop Pulse (ms)	CPS	WPM
110	9.091	9.091	11.0	110
	9.091	18.182	10.0	100
150	6.667	6.667	15.0	150
300	3.333	3.333	30.0	300
600	1.667	1.667	60.0	600
1200	0.8333	0.8333	120	1200
1800	0.5556	0.5556	180	1800
2400	0.4167	0.4167	240	2400
4800	0.2083	0.2083	480	4800
9600	0.1041	0.1041	960	9600
19200	0.0520	0.0520	1920	19200

$$\text{CPS} = \text{characters per second} = \frac{1}{\text{START} + 8\,(\text{DATA}) + \text{STOP}}$$

$$\text{WPM} = \text{words per minute} = \frac{\text{CPS}}{6} \times 60$$

= number of 5-letter-plus-space groups per minute.

character to be always even or odd (even or odd parity). For example, if odd parity is used with the ASCII character C (first seven bits = 100 0011), the eighth parity bit will be set to space to give an odd number (3) of data bits (0100 0011). Conversely, the odd-parity eight-bit code for the letter B would be 1100 0010. (Logic convention has it that lowest order bits are placed to the right; thus the bit order in the binary representation is 8765 4321.) Upon reception, the receiving terminal simply counts the number of mark pulses in each ASCII 8-bit character. If an odd number is counted, it is *assumed* that no errors occurred. Notice, however, that even if a bit error is detected, there is insufficient data to determine which bit was wrong, and therefore no error *correction* is provided by the parity check itself. Also, if there are two bit errors in the same ASCII character, the parity count will still be odd, and no error indication is given even though two errors occurred. Thus, parity checking will not give complete error detection and does not provide for error correction. Some applications require more sophisticated error detection and correction schemes. Even parity works in a similar manner, except that the eighth bit is chosen to make the total number of mark pulses even rather than odd. The U.S. amateur regulations do not specify a requirement for use of the eighth data bit; it may be set to mark, space, odd or even parity, depending upon the preference of the operator and the capability of his equipment. Relatively simple terminals do not provide parity options; more sophisticated equipment such as the DS3100 ASR do.

Speeds and Baud Rates

The transmission rate of Baudot TTY signals is usually specified in words per minute, much like that used for telegraph codes. Actually, the speed is given in the *approximate* number of five-letter-plus-space combinations transmitted in a *continuous sequence* of start-stop characters in a one-minute interval. Convenient choices of gear ratios and motor-shaft speeds have resulted in the use of noninteger wpm rates. Common usage, however, has rounded the exact speeds to easily remembered numbers. Thus, "60 speed" Baudot is actually sent at 61.33 wpm and "75 speed" is really 76.67 wpm.

A major problem occurs with the use of words per minute as a TTY speed specification because of the varying length of stop pulses in use. For example, "60 speed" Baudot TTY has 22-ms-long start and data pulses and a 31-ms stop pulse; the Western Union "65 speed" also has 22-ms start and data pulses, but the stop pulse is also 22-ms long; electronic terminals commonly use 22-ms start and data pulses and 33-ms stop pulses (1.5 times the data-pulse width). All of these three codes are compatible and may be received on the same printer or terminal since the stop-pulse length is a *minimum* time. The common factor between these codes is the 22-ms length of the data, or unit pulse. Therefore, a new data-rate specification has been adopted, the *baud* rate, which is the reciprocal of the data- or unit- or select-pulse width:

Baud rate = 1/t, where t = length of unit pulse.

Using this definition, all three of the above codes have a data rate of 45.45 baud, commonly abbreviated to "45 baud."

As noted above, the length of the stop pulse varies between codes, being from 1.0 to 2.0 times as long as the unit (or data) pulse; multipliers of 1.0, 1.42, and 1.5 are commonly used with the Baudot codes. Standard Baudot data rates and speeds are shown in Table 18.

U.S. amateurs are authorized to use all of the Baudot data rates shown in Table 18 with the exception of 100 baud. This rate has seen limited commercial use in Europe. The 45-baud data rate is by far the most popular worldwide amateur data rate. A limited amount of amateur use of 74 baud ("100 speed") has been noted on the high-frequency bands. Most commercial RTTY transmissions on high frequencies use 50, 57 and 74 baud, with little 45-baud activity.

ASCII data rates are commonly specified as a baud rate, although a character-per-second (cps) or words-per-minute (wpm) rate may also be given. The lowest standard ASCII data rate in common usage is 110 baud. ASCII characters sent at 110 baud are usually sent with a 2-unit-wide stop pulse, although the 1-unit stop pulse may also be found in some applications. Above 110 baud, it is common to make the stop pulse one unit pulse in length. The standard ASCII data rates commonly used with asynchronous serial transmission are shown in Table 19.

The ASCII data rates up to 300 baud are authorized for U.S. amateur use on frequencies between 3.500 and 21.250 MHz. Data rates up to 1200 baud are permitted between 28 and 225 MHz; up to 19,600 baud may be used above 420 MHz. The 110-baud rate is by far the most practical for 3.5 to 21.5 MHz use, again because of the ready availability of equipment as well as the increased susceptibility of the higher data rates to noise, static, interference and so forth. Vhf fm amateur activity finds 110 and 300 baud useful for terminal-to-terminal communications, and 300 and 1200 baud for computer-related activities such as exchanging programs and the like.

Loop Circuits

As discussed earlier, the printing mechanisms use solenoids or selector magnets to sense the presence (mark) or absence (space) of the loop current. The letters typed on the sending keyboard are encoded with proper mark and space pulses by mechanically driven keyboard contacts. Since the keyboards and selector magnets of both machines are series connected, text typed on one keyboard is reproduced on both printers. Connection of the keyboard directly to its associated printer is called a local loop and results in what is called half duplex (HDX), giving local copy of transmitted text, termed local echo.

Selector magnets have been designed for mark loop currents of 60 or 20 mA dc, with 60 mA being by far the most common for older machines such as the Teletype Corp. model 15 or 28. Newer Baudot machines and most ASCII machines and terminals use electronic interface circuits that accept a wide range of loop currents (10 to 120 mA for the HAL DS3100, for example); a 20-mA loop current is quite commonly used with ASCII terminals.

Since the dc resistance of the machine selector magnets is rather low (100 to 300 ohms, typically), it would at first seem that a low-voltage loop supply could be used. However, the inductance of the magnet is usually quite high (on the order of 4 henrys for a model 15), causing a delay in the current rise time. This, in turn, delays the selector magnet response to a mark pulse, distorting the signal. This distortion can be severe enough to cause misprinting of received text, particularly if other forms of distortion are present (such as caused by variations in the radio signal). The effect of this inductive distortion is reduced considerably if the L/R ratio (L is solenoid inductance and R is total loop resistance) is reduced by increasing R. Increasing R requires that the dc voltage be increased to maintain the required 60-mA loop current. In general, the higher the loop voltage and loop resistance used, the lower the distortion. In practice, loop power-supply voltages between 100 and 300 Vdc are common; 130- and 260-volt supplies were often used with model 15, 19 and 28 Teletype machines. Modern TTY systems use a 150- to 200-volt loop power supply and a 2000- to 3000-ohm loop resistance to set the 60-mA loop current. Because of the related keying circuitry, the demodulator unit of a good RTTY system usually includes the loop power supply and current-limiting resistor.

On the other hand, the newer ASCII machines (such as the Teletype Corp. models 33, 35 and 43) are available with a wide variety of input/output (I/O) interfaces. These devices usually include a high-current, low-voltage selector-magnet assembly (500 mA, 10 to 30 volts is typical), an internal magnet-driver transistor and power supply, and an electronic interface to the data connections. These machines may be supplied with a 20-mA

loop-interface circuit or other interface standards.

RS-232 Data Interface

The EIA has defined a new standard for interconnection of terminals, modems and computers in *EIA Standard RS-232-C* (August 1969). This interface standard specifies *voltage* levels for mark and space, rather than the current levels used in a loop circuit. The basic RS-232-C voltage ranges are shown in Fig. 54.

Note that mark, normally considered to be a logic 1, is represented by a negative voltage and the logic 0 by a positive voltage. Thus, the RS-232 interface standard can be thought of as a polar voltage standard. Also, note that RS-232 voltage levels really are *not* transistor-transistor logic (TTL) compatible (in spite of some claims you may read to the contrary)! Users should carefully examine "pseudo RS-232-compatible" equipment to make sure no damage will be caused to their RS-232-interfaced equipment. The RS-232-C standard also includes definition of control signals to pass between originating and receiving devices, and even defines the connector-pin assignment. The RS-232 connector signal and pin assignments are shown in Table 20.

The control signals such as DATA TERMINAL READY (DTR), REQUEST TO SEND (RTS), and so forth, are often called *handshaking* signals; they provide status indicators between data devices (terminal to and from a modulator/demodulator or modem, for example). All control-signal voltage and impedance levels also conform to the RS-232-C standard shown in Fig. 54. Control signals are considered to be active or on when the voltage is positive; a negative-voltage control signal is off or inactive. The terms "mark" and "space" are usually not used when describing RS-232 control signals. Thus a positive voltage on pin 5 (CB = CLEAR TO SEND) is the terminal's signal to the data circuit that the terminal is ready to receive data. Also, an open-circuit or no-voltage condition on an RS-232 signal is interpreted as a "mark" or "active" condition.

Other Data-Connection Standards

A number of other data-connection standards may be found in use in commercial equipment. Chief among these is an interface that is compatible with the popular TTL integrated circuits, a TTL-compatible interface. TTL interconnections are particularly useful when directly interfacing computers or other digital devices. Although TTL and RS-232 interface standards are not directly compatible, a number of line-receiver and line-driver integrated circuits are available.

A common data standard used in U.S. military applications is the MIL-188 Data Standard. MIL-188 is very similar to the RS-232 standard, with the exception that logic voltages are inverted — mark is

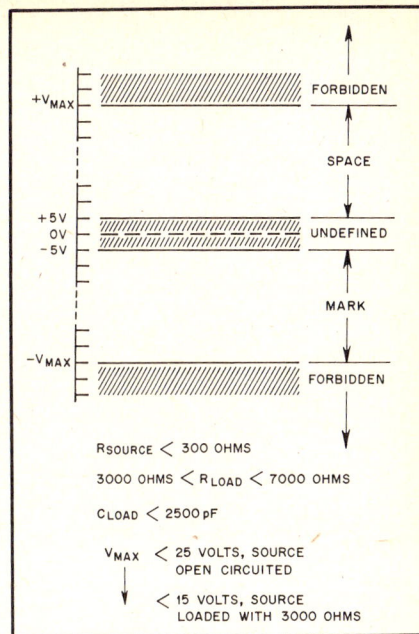

Fig. 54 — RS-232-C voltage standards. Note that these voltages are *not* compatible with TTL integrated circuits.

represented by positive voltages and space by negative voltages.

Other data interfaces are to be found for CMOS logic integrated-circuit connections. Interfaces for both + 5-volt and + 12-volt CMOS operation are in use.

Additional System Requirements

To be used in radio communication, the pulses (dc) generated by the teletypewriter must be utilized in some way to key a radio transmitter so they may be sent in proper sequence and usable form to a distant point. At the receiving end the incoming signal must be converted into dc pulses suitable for operating the printer. These functions, shown in block form in Fig. 55, are performed by electronic units known respectively as the *frequency-shift keyer* or RTTY modulator and *receiving converter* or RTTY *demodulator*.

The radio transmitter and receiver are quite conventional in design. Practically all the special features needed can be incorporated in the keyer and converter, so that most ordinary amateur equipment is suitable for RTTY with little or no modification.

Transmission Methods

It is quite possible to transmit teleprinter signals by ordinary "on-off" or "make-break" keying such as is used in regular hand-keyed cw transmissions. In practice, however, *frequency-shift keying* is preferred because it gives definite pulses on both mark and space, which is an advantage in printer operation. Also, since fsk can be received by methods similar to those used for fm reception, there is considerable discrimination against noise,

Table 20

RS-232-C Interface Connector

Pin Number	Circuit	Description	Abbreviation
1	AA	Protective Ground	PG
2	BA	Transmitted Data	TXD
3	BB	Received Data	RXD
4	CA	Request to Send	RTS
5	CB	Clear to Send	CTS
6	CC	Data Set Ready	DSR
7	AB	Signal Ground (Common Return)	SG
8	CF	Received Line Signal Detector	CD
9	—	Reserved for data set testing	
10	—	Reserved for data set testing	
11	—	Unassigned	
12	SCF	Secondary Received Line Sig. Detc.	
13	SCB	Secondary Clear to Send	
14	SBA	Secondary Transmit Data	
15	DB	Transmit Signal Clock	TXC
16	SBB	Secondary Received Data	
17	DD	Receive Signal Clock	RXC
18	—	Unassigned	
19	SCA	Secondary Request to Send	
20	CD	Data Terminal Ready	DTR
21	CG	Signal Quality Indicator	
22	CE	Ring Indicator	RI
23	CH/CI	Data Rate Selector (DTE/DCE Source)	
24	DA	Transmit Signal Clock to Modem	
25	—	Unassigned	

Use of a 25-pin connector is assumed but not defined in the RS-232-C standard. Commercial usage has seen the adoption of the TRW-Cinch D-Subminiature connector series (DB25P and DB25S), also manufactured by many other firms (Amphenol 17-10250 and 17-20250, for example). In general, the male pin connector (DB25P or 17-20250) is installed on the data terminal, but other arrangements may be found on some equipment.

Fig. 56 — Many RTTY operators now use all-electronic systems.

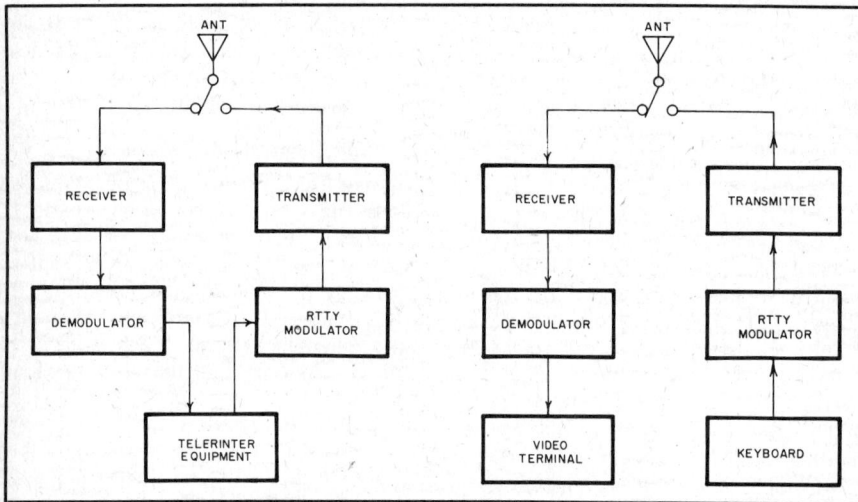

Fig. 55 — At left is a block diagram of an RTTY system using surplus teleprinter equipment. At right, a modern all-electronic RTTY station setup.

both natural and manmade, distributed uniformly across the receiver's passband, when the received signal is not too weak. Both factors make for increased reliability in printer operation.

Frequency-Shift Keying

On the vhf bands where A2 and F2 transmission is permitted, *audio frequency-shift keying* (afsk) is generally used. In this case the rf carrier is transmitted continuously, the pulses being transmitted by frequency-shifted tone modulation. The audio frequencies used have been more-or-less standardized at 2125 and 2295 Hz, the shift being 170 Hz. With afsk, the lower audio frequency is customarily used for mark and the higher for space.

Below 50 MHz, F1 or fsk emission must be used. The carrier is on continuously, but its frequency is shifted to represent marks and spaces. General practice with fsk is to use a frequency shift of 170 Hz, although FCC regulations permit the use of any value of frequency shift up to 900 Hz. The smaller values of shift have been shown to have a signal-to-noise-ratio advantage. The nominal transmitter frequency is the mark condition and the frequency is shifted 170 Hz (or whatever shift may have been chosen) lower for the space signal.

RTTY with SSB Transmitters

Amateurs operating RTTY in the hf bands are using audio tones fed into the microphone input of an ssb transmitter. With properly designed and constructed equipment which is correctly adjusted, this provides a satisfactory method of obtaining F1 emission. The user should make certain, however, that audio distortion, carrier, and unwanted sidebands are not present to the degree of causing interference in receiving equipment of good engineering design. *The user should also make certain that the equipment is capable of withstanding the higher-than-normal average power involved.* The RTTY signal is transmitted with a 100-percent duty cycle, i.e., the average-to-peak power ratio is 1, while ordinary speech waveforms generally have duty cycles in the order of 25 percent or less. Many ssb transmitters such as those using sweep-tube final amplifiers, are designed only for low-duty-cycle use. Power-supply components, such as the plate-voltage transformer, may also be rated for light-duty use only. As a general rule when using ssb equipment for RTTY operation, the dc input power to the final PA stage should be no more than twice the plate dissipation rating of the PA tube or tubes.

Bibliography

A bibliography of RTTY articles from *QST* is available from ARRL Hq. for a large s.a.s.e.

American National Standards Institute, Inc., 1430 Broadway, New York, NY 10018

ANSI X3.4-1977 American National Standard Code for Information Interchange

ANSI X3.15-1976 American National Standard for Bit Sequencing of the American National Standard Code for Information Interchange in Serial-By-Serial Data Transmissions

ANSI X3.16-1976 American National Standard Character Structure and Character Parity Sense for Serial-By-Data Communication in the American National Standard Code for Information Interchange

ANSI X3.28-1976 American National Standard Procedures for the Use of Control Characters of American National Standard Code for Information Interchange in Specified Data Communications Links

Electronic Industries Association, EIA Engineering Department, Standards Orders, 2001 Eye St., Washington, DC 20006

EIA Standard RS-232-C Interface Between Data Terminal Equipment and Data Communication Equipment Employing Serial Binary Data Interchange

EIA Standard RS-269-B Synchronous Signaling Rates for Data Transmission

EIA Standard RS-363 Standard for Specifying Signal Quality for Transmitting and Receiving Data Processing Terminal Equipments Using Serial Data Transmission at the Interface with Non-Synchronous Data Communication Equipment

EIA Standard RS-404 Standard for Start-Stop Signal Quality Between Data Terminal Equipment and Non-Synchronous Data Communication Equipment

EIA Industrial Electronics Bulletin No. 9: Application Notes for EIA Standard RS-232-C

Goacher and Denny, *The Teleprinter Handbook*, Chapters 1, 2 and 3. Radio Society of Great Britain, 35 Doughty St., London WC1N 2AE, 1973.

Guentzler, "RTTY for the Beginner," pp. 3-24, *RTTY Journal*, P. O. Box RY, Cardiff by the Sea, California 92007, 1975

Kretzman, *The New RTTY Handbook*, Chapters 1 and 2, Cowan Publishing Corp., 14 Vanderventer Ave., Port Washington, NY, 1962

Schwartz, "An RTTY Primer," *CQ Magazine*, issues of August 1977, November 1977, February 1978, May 1978, July 1978 and August 1978

Tucker, *RTTY from A to Z*, Chapters 1, 2, 3 and 11, Cowan Publishing Corp., 14 Vanderventer Ave., Port Washington, NY, 1970

Reference Data for Radio Engineers, Fifth Edition, Chapter 30, Howard W. Sams & Co., Inc., a subsidiary of International Telephone and Telegraph Corporation, Indianapolis, Kansas City and New York, 1968

PACKET RADIO

Computers process information very rapidly. Most of a computer's time is wasted waiting for a human to enter data or read output. Therefore many users, each having an input/output device (terminal) may sequentially access (time share) a computer. Because of the relative slowness of the human input/output compared to the processing speed, each user has the illusion of exclusive access to the computer. The multiplexing, or time-sharing system is "transparent" to the user.

Each user's data is partitioned into individual bursts called *packets*. A packet is a group of ASCII characters (information) surrounded by control signals and error-detection features. The control signals help recognize the presence of a packet and tell any intervening switching equipment where the packet should be sent. A typical packet is diagrammed in Fig. 57. A packet is similar to a message format. Beside carrying smaller payload, the header and trailer components are designed to be read by computer, not by human operators. The computer, in this case, can be either a home computer programmed to perform this function or a packet controller — a single-purpose microcomputer board dedicated to this task. There are advantages to the packet-controller board approach, such as (a) taking advantage of packet controller chips on the market, (b) keeping the hardware costs low by not tying up the personal computer and (c) avoiding the necessity of generating new software for every type of computer as changes are made.

Following this philosophy, a typical vhf Amateur Radio packet station would look like that in Fig. 58. The terminal in this case could be either a cathode-ray tube (CRT) or printer and could operate in either ASCII or Baudot code. The Terminal Node Controller (TNC) of the type designed by Doug Lockhart, VE7APU, can be programmed by means of programmable read-only memories (PROMs) to handle serial or parallel communication with a wide variety of terminals, including computers. The other side of the TNC manages the line — sending and receiving packets in High-Level Data Link Control (HDLC) format.

Example Packet Transmissions

Assume that the source station wishes to send a two-page message to a destination station using packets. The transmission might be broken up into 48 packets, each containing the address of the source and destination, an information (data) field containing a part of the total message and a frame check sequence (FCS) for error detection. The source station would enter his message into a computer terminal attached to a Terminal Node Controller. The TNC would accept the message as input, break it up into packets, send the packets over the transmission medium (radio, in this case) and receive an acknowledgment of correct reception from the destination station for each packet sent. The destination station would also employ a TNC to receive the packets, acknowledge correctly received packets (ASCII ACK) or request retransmission of any bad packets (ASCII NAK or negative acknowledgment). Bad packets are detected using the FCS. An FCS is appended to each packet by the transmitting station. The receiving station computes what the FCS should be and compares that with the FCS supplied with the packet. If the two agree, the chances are very great that the packet is error free. If the two answers disagree, the destination station knows that the packet is bad and requests retransmission of that packet only.

If two stations happen to send packets simultaneously, the result is a "collision." There are numerous so-called contention schemes to avoid collisions, but they happen even in the best packet networks. In this case, the TNC performs a carrier-sense check (to see if anyone is using the channel). To reduce the possibility of two TNC boards hearing nothing and bursting packets at exactly the same time, a variable time delay is built in. Because the time delay at each TNC (user) is changing, repeated collisions between the same pair of users should not occur.

Local Repeaters

In packet terminology, Local Area Network (LAN) is used to designate a number of terminals within a small geographical area which are able to talk to one another through a common channel. That may be coaxial cable or radio. This type of repeater may be quite different from the usual ham 2-meter fm variety. Because local area network packet repeaters are still highly experimental, those now operating or in the construction stage in the U.S. and Canada represent different approaches. As examples of two implementations of local area networks using the same Vancouver TNC boards and HDLC protocol, let us look briefly at

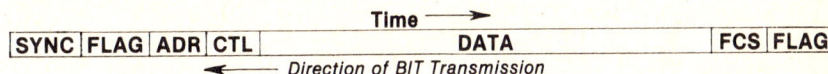

SYNC — First packet in a group of packets contains 16 bits of alternating zeros and ones.
FLAG — 8 bits, always 01111110 (7E hex).
ADR — Address of the sending station, either assigned dynamically by the Station Node at sign-on or hard-coded into the Terminal Interface Program (8 bits).
CTL — 8 bits containing control information for handling the packet.
DATA — From 0 (supervisory packet) to 255 bytes of data in ASCII.
FCS — Frame Check Sequence — 16 bits, computed by the sending station and checked by the receiving station.
FLAG — 8 bits, always 01111110 (7E hex).

Fig. 57 — Format for a typical packet.

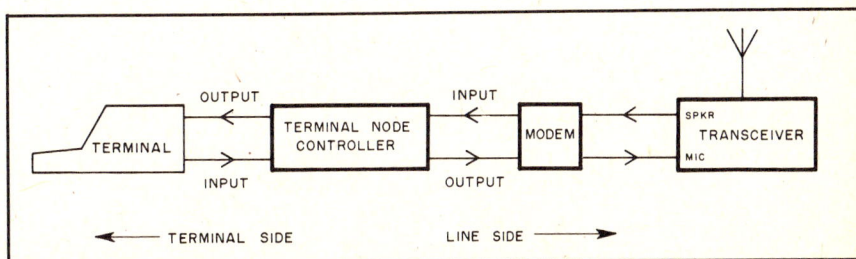

Fig. 58 — Block diagram showing a typical vhf packet radio station. The arrows indicate the direction of data flow.

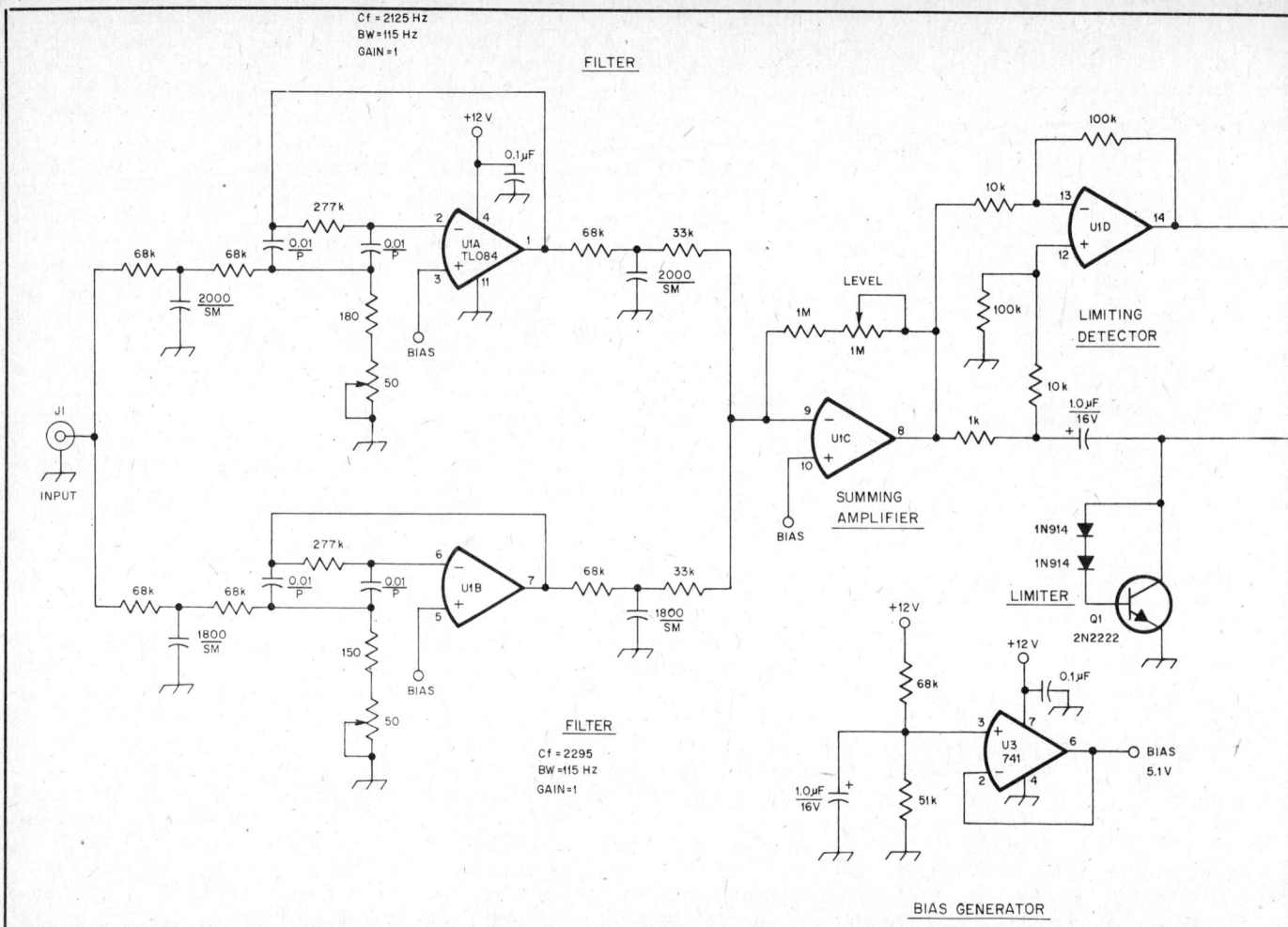

Fig. 59 — Schematic diagram of the general-purpose RTTY terminal unit. Component designations are for text reference only. K1 is a dpdt relay with a 12-V coil and contacts rated for 5 A at 117 V. Unless specified otherwise, all fixed-value resistors are 1/4-watt, 5% tolerance composition type. Except as indicated, decimal values of capacitance are in microfarads (μF); others are in picofarads (pF); resistances are in ohms; k = 1000, M = 1,000,000.

San Francisco and Washington, DC.

the KA6M/R local area network packet repeater was activated on December 10, 1980. It is a single-frequency repeater that accepts packets, performs an error check on them and retransmits them when the packets contain no errors. The repeater itself uses a Z-80 microprocessor driving a custom-built board containing a Western Digital 1933 HDLC chip. Bell 202 1200-baud modems are used both at the repeater and by members of the net. Individual stations are using Vancouver TNC boards.

The WD4IWG/R repeater is a 2-meter fm voice repeater that is also used by local AMRAD members for data communications. So, what comes in is repeated on the output frequency at the same time. This approach has the advantages of (1) using an existing repeater and (2) being able to sense the repeater output for presence of carrier before transmitting, avoiding collisions when the other station cannot be heard directly. Like those local networks in Vancouver, San Francisco and

Hamilton, Ontario, the Washington, DC group is using the Vancouver TNC boards.

The Wider Network

The eventual goal is to tie these and many other local areas together to form a larger packet network. The focus, at the moment, is on interconnecting the various groups in Canada and the U.S. Computer enthusiasts are using the acronym AMNET to designate this wider network, which may go much beyond North America.

More Information

The following publications are recommended for radio amateurs interested in packet communications:
I. Hodgson, "An Introduction to Packet Radio," *Ham Radio*, June 1979.
I. Hodgson and R. Rouleau, *Packet Radio*. Blue Ridge Summit, PA: Tab Books, Inc., 1981.
M. Schwartz, *Computer-Communication Network Design and Analysis*. Englewood Cliffs, NJ: Prentice-Hall, Inc., 1977.
Additionally, these organizations regularly publish newsletters with substantial packet information:
Amateur Radio Research and Development Corp. (AMRAD), monthly *AMRAD Newsletter* ($12).

Gerald Adkins, N4GA, 1206 Livingston St. North, Arlington, VA 22205.
Vancouver Amateur Digital Communications Group (V.A.D.G.C.), *The Packet* ($10). Don Oliver, VE7AOG, 818 Rondeau St., Coquitlam, BC V3J 5Z3 Canada.
Hamilton Area Packet Network (H.A.P.N.), *I-Frame de VE3PKT* ($10). Stu Beal, VE3MWM, 2391 Arnold Cres., Burlington, ON L7P 4J2 Canada.

A GENERAL-PURPOSE RTTY TERMINAL UNIT

Experiments with some different circuits and techniques for RTTY demodulation led to the development of the terminal unit (TU) pictured in Figs. 60 and 61. The goal was self-contained equipment capable of interconnection between the audio lines of any radio equipment and the logic lines of any RTTY input/output device. That goal was met in the device described here, although as in any electronic system, room for improvement exists. The photographs reveal the developmental and evolutionary nature of the project. The TU is designed for signaling rates up to 110 baud, but is restricted to 170-Hz shift. That one limitation notwithstanding, the circuit descrip-

FILTER   2125 Hz MARK   SHMITT TRIGGER   FREQUENCY DOUBLER

+12 V   0.1 µF   100 k

680   LIMITING

68 k   68 k   277 k   0.01 P   0.01 P   6 4   U2B   7   33 k   2000   12 + U2D 14   13 −   100 k   100 pF SM   1 U4A 4070 3   2

2000 SM   180   5 11   BIAS   50

Q2 2N2222   1N914   1N914

560 pF SM   130 k   0.01 P   0.01 P   9 + U2C 8   10   750 pF SM   2 − U2A TL084 1   3 +   100 k   100 pF SM   5 U4B 6 4

10 k   BIAS   +12 V   100 k   S1   2295 ONLY   DIVERSITY   2125 ONLY   100 k

100 k   10 k   BIAS

150   50   2295 Hz SPACE   BIAS

FILTER   SHMITT TRIGGER   FREQUENCY DOUBLER

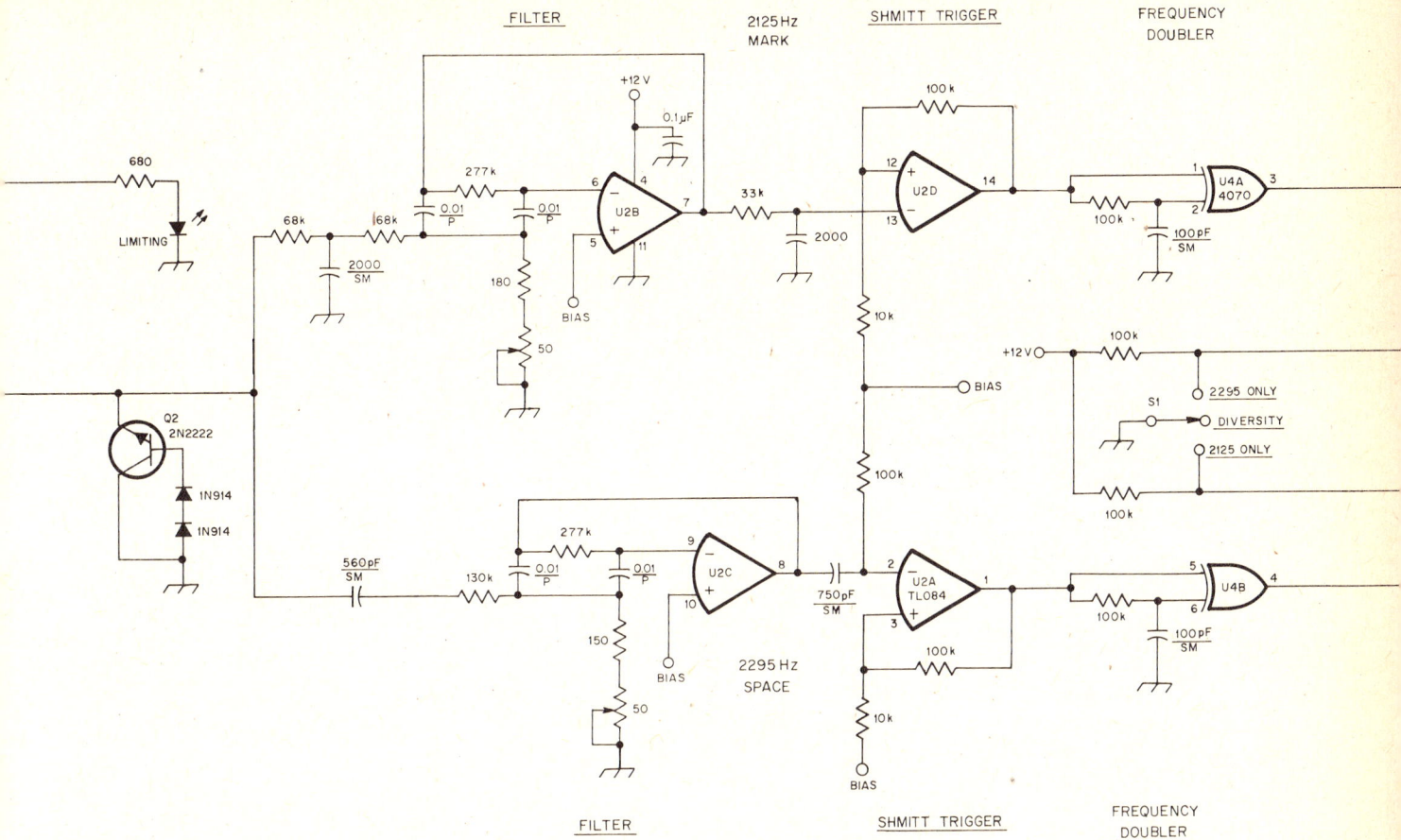

tion that follows will show the TU to be truly versatile — an RTTY enthusiast should be able to use it with any combination of radio, teleprinter and/or computer equipment. If one plans to install the unit in a permanent, dedicated system, however, some of the versatility can be sacrificed for reduced cost and complexity. Some features of the general-purpose TU are true mark/space diversity, selectable loop current, RS-232, TTL and CMOS compatibility, auto-start, selectable data sense and afsk outputs.

The Electronics

Fig. 59 in its various parts is the schematic diagram. Audio from the low-impedance output of the station receiver is applied in parallel to two active bandpass filters, tuned to the mark and space tones. The filter outputs are summed and applied to a common limiter stage. The limiting detector is a differential amplifier that senses current in the limiter diodes. After limiting, the audio signal is again separated into mark and space channels for conver-

sion to CMOS logic levels. A pair of Schmitt triggers performs the sinusoidal-to-rectangular waveform conversion. The bias level and hysteresis are set to allow triggering about 6 dB below limiting. Demodulators from the past boasted dynamic range much greater than this, but modern communications receivers have tight agc, rendering wide demodulator signal range redundant.

Tone detection in each channel is accomplished by a retriggerable monostable multivibrator. The multivibrator period is set slightly longer than the triggering waveform period, so that the multivibrator output remains in one state while the input signal is present. U5 performs the detection function in Fig. 59. The monostable period must not be set any longer than necessary to provide a steady output level, otherwise the detected output will be excessively long with respect to the input tone burst, causing distortion of the RTTY characters. This requirement is eased somewhat by doubling the tone frequencies prior to detection. The shorter monostable period re-

quired for double-frequency operation reduces the pulse stretching by half. Exclusive-OR gates U4A and B, with their RC delay networks double the tone frequencies. The envelope detectors also drive the mark and space indicators, which serve as tuning aids. Both channels are normally used. However, either tone detector can be disabled by means of S1.

This demodulator achieves automatic diversity in an unusual manner. Except for the common limiter stage, the mark and space channels are independent. Normally, the lack of a mark tone can be interpreted as a space and vice versa. The filter responses and Schmitt-trigger sensitivities are tailored to recognize one tone at a time. The envelope detectors are connected to produce identical outputs, that is, the mark detector goes LOW when the 2125-Hz tone is present, and the space detector goes HIGH when the 2295-Hz tone is present. A simple OR gate would combine the detector outputs and provide copy if one tone were lost, but if one tone were jammed, the other tone would be locked out by such a

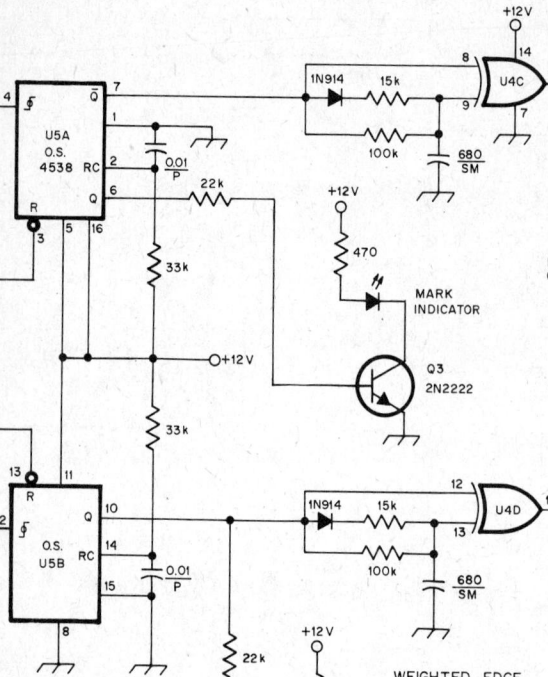
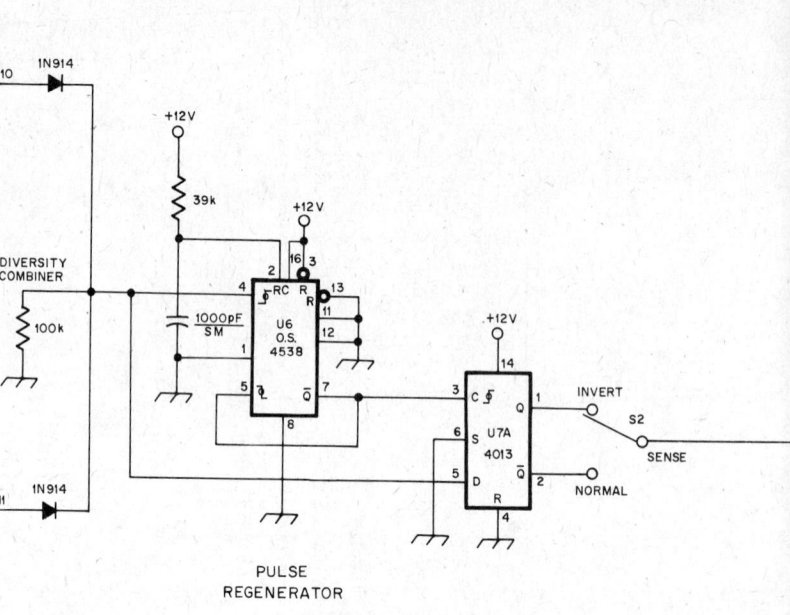
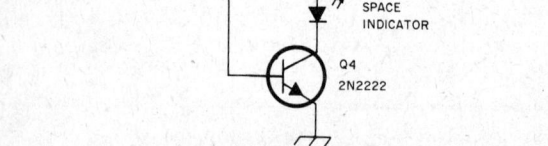
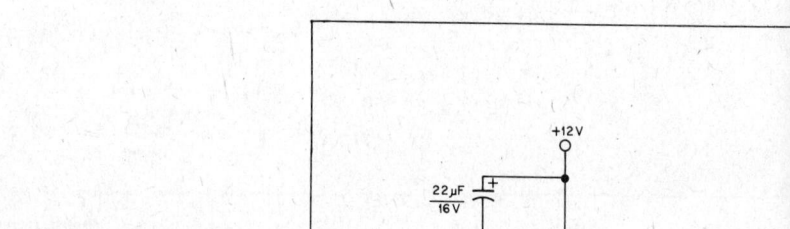
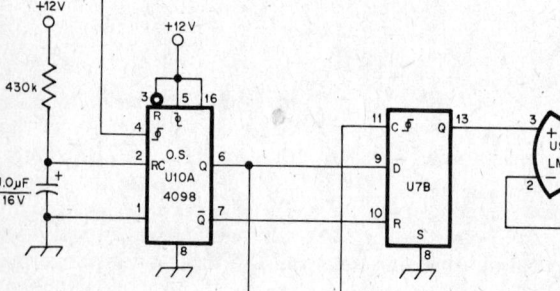
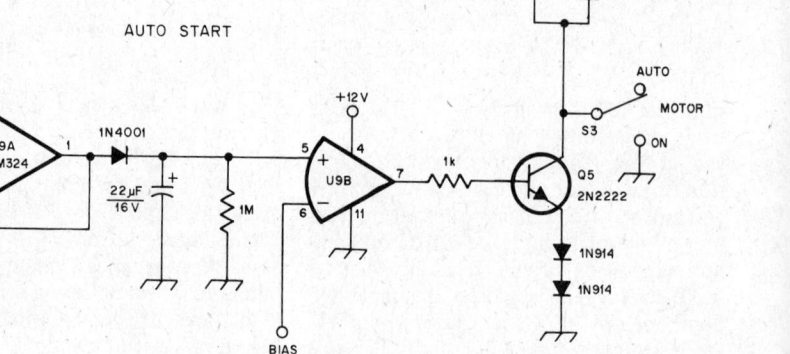
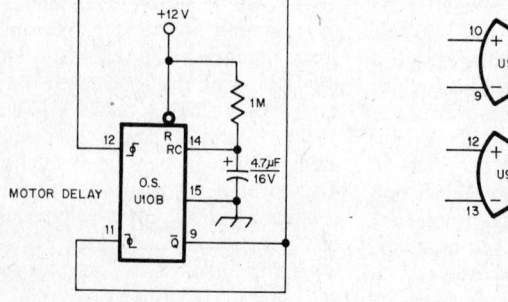
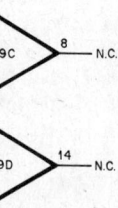

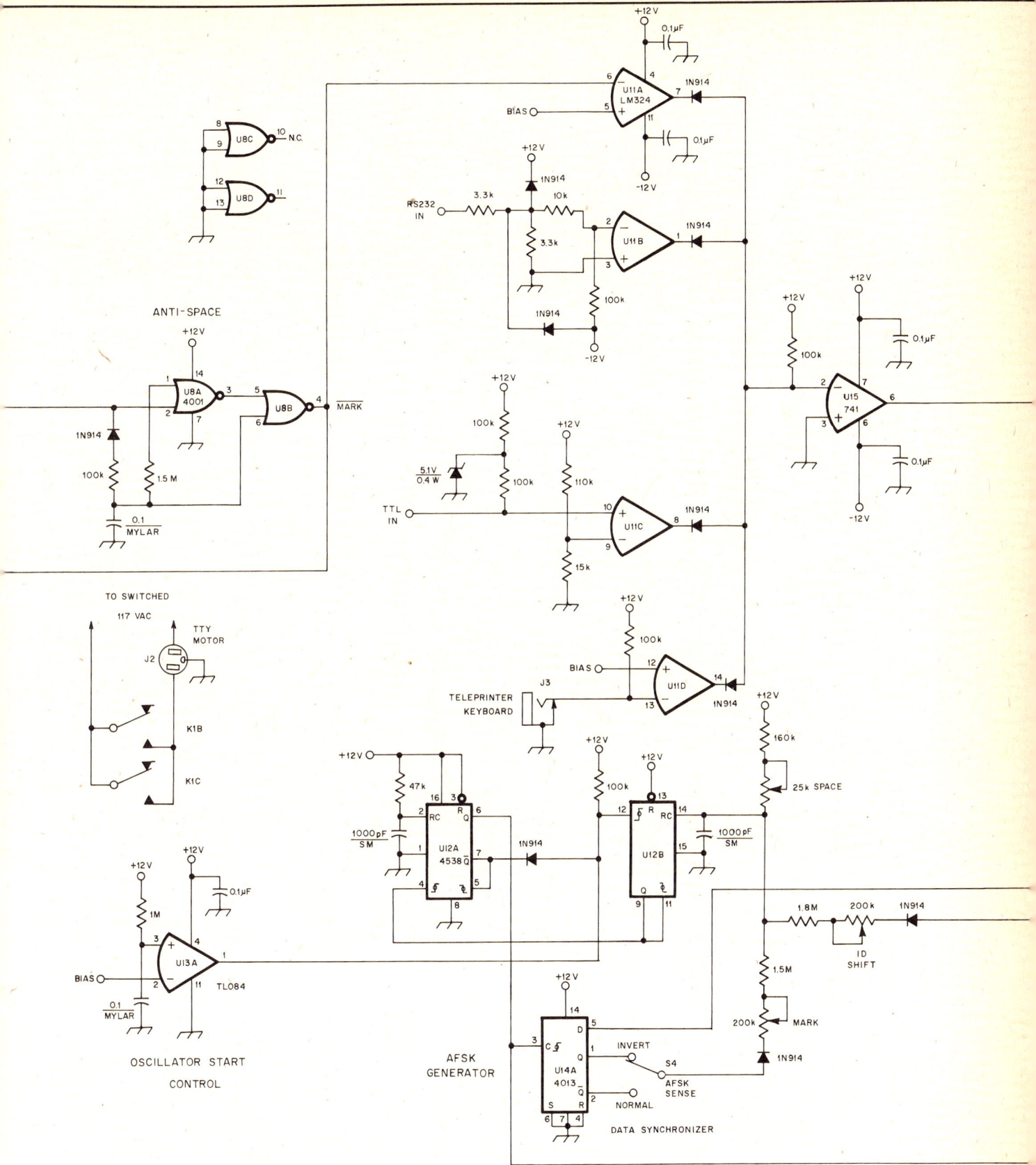

+12 V
0.1μF
6
4
U11A
LM324
7
1N914
BIAS
5
11
0.1μF
−12 V

8
10 N.C.
9 U8C
12
13 U8D 11

+12 V
1N914
RS232 IN
3.3k
10k
2
U11B
1
1N914
3.3k
3
100k
1N914
−12 V

+12 V
0.1μF
+12 V
100k
7
U15
741
6
3
0.1μF
−12 V

ANTI-SPACE

+12 V
1
14
U8A
4001
3
5
U8B
4
MARK
2
7
6
1N914
100k
1.5 M
0.1 MYLAR

+12 V
100k
5.1V 0.4 W
100k
+12 V
110k
TTL IN
10
U11C
8
1N914
9
15k

TO SWITCHED
117 VAC

TTY MOTOR
J2
K1B
K1C

+12 V
100k
BIAS
12
U11D
14
1N914
13

TELEPRINTER KEYBOARD J3

+12 V
160k
25k SPACE

+12 V
1M
0.1μF
+12 V
3
4
U13A
1
BIAS
2
11
TL084
0.1 MYLAR

+12 V
47k
16
3
2
RC
R
Q
6
1
U12A
4538
Q
7
1N914
4
Q
5
8

+12 V
100k
12
R
13
RC
14
U12B
15
Q
9
Q
11
1000pF SM

1000pF SM

1.8M
200k
1N914
ID SHIFT
1.5M
200k
MARK
1N914

OSCILLATOR START CONTROL

AFSK GENERATOR

+12 V
14
5
D
Q
1
INVERT
3
C
U14A
4013
Q
2
S4
AFSK SENSE
S
R
NORMAL
6
7
4

DATA SYNCHRONIZER

160 V
LOOP SUPPLY

+12V

0.1 µF

220 k

27 k
1W

J5
TELEPRINTER
MAGNETS
J6

Q8
2N3439

Q9
WEP744 OR
MJE340
* HEATSINK

100 k
1W

40 µF
350 V

RS232 OUT
MARK

U16
TIL111

5.1V
0.4 W
60 mA

15V
0.4 W
20 mA

220
1W

S6

PRINT

STBY S6

+12V

0.1µF

75k

Q6
2N2222

TTL OUT
MARK

51k

100k

1N914

FSK +

U17
TIL111

FSK −

MJE340
E C B

TOP
VIEW

WEP744

E
B
C

BASE VIEW

360

10k

Q7
2N2222

1N914

1N914

KEYED CMOS OUTPUT
MARK

100 k

7812

+12V

OUT   U18   IN

470

0.1

GND

220µF
35V

POWER

LM320T-12

−12V

OUT   U19   IN

0.1

GND

220µF
35V

LOW − PASS FILTER
3 − POLE CHEBYSHEV
0.1 − dB RIPPLE, 2400 Hz

+12 V

ID
KEYER

+12V

1M

7
6
U13B
5

J4
KEY

BIAS

10k

39k

1N914   1N914

ATTENUATOR

20k

100k

910 pF
SM

100k

100k

62pF
SM

3300 pF
P

10
U13C
8
9

100k

BIAS

100k

÷ 2

11
C
13
Q

8
S

U14B
10
R

9
D
12
Q̄

CW

2000pF
SM

50k

1.0µF
16V

TONE LEVEL

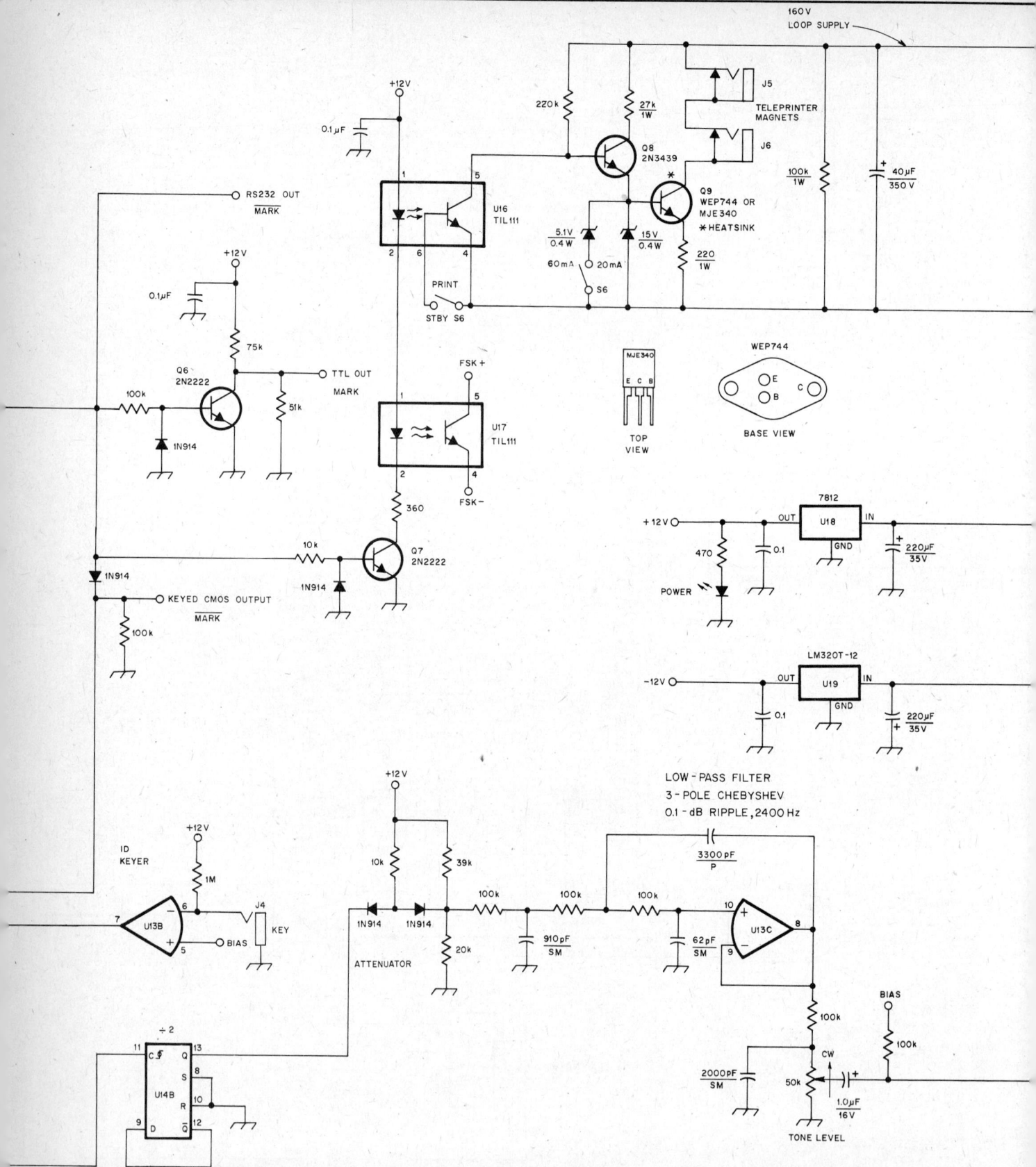

simple scheme. Diversity reception in the presence of jamming requires an AND gate, but this configuration precludes copying one tone if the other is lost. No simple array of combinational logic can provide automatic immunity to tone loss or tone jamming. This demodulator achieves that objective by examining the transitions between mark and space, and reconstructing the keying waveform from the transitions. U4C and D generate short pulses corresponding to each logic transition. A diode-gated RC delay network at the input of each exclusive OR gate generates a 10-µs pulse corresponding to each positive-going transition and a 70-µs pulse corresponding to each negative-going transition. If the tone input to an envelope detector is lost,

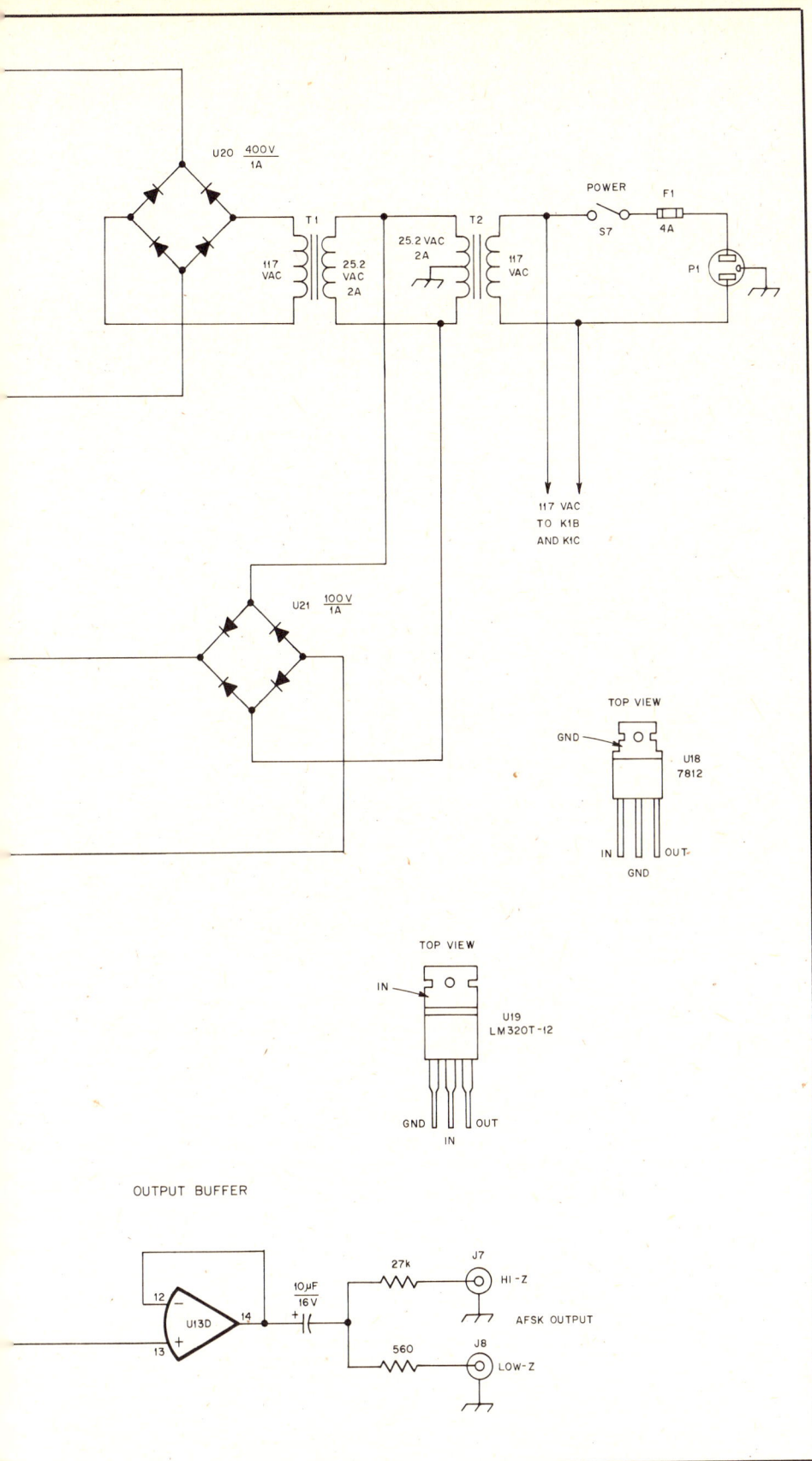

U20 400V 1A

117 VAC

T1

25.2 VAC 2A

25.2 VAC 2A

T2

POWER

S7

F1

4A

117 VAC

P1

117 VAC TO K1B AND K1C

U21 100V 1A

TOP VIEW

GND

U18 7812

IN   OUT

GND

TOP VIEW

IN

U19 LM320T-12

GND   OUT

IN

OUTPUT BUFFER

12

−

13

+

U13D

14

10 μF 16 V

+

27k

J7

HI-Z

AFSK OUTPUT

560

J8

LOW-Z

---

the detector output will be a steady LOW level. If the tone input is jammed, the output will be a steady HIGH level. Neither of these conditions will trigger the edge generator — it responds only to transitions. The edge generator responds to both fault conditions by outputting a LOW level, so the two channels can be combined in a diode OR gate. The keying waveform is regenerated by a monostable multivibrator (U6) working with a D flip-flop (U7A) as a pulse-width discriminator. A 40-μs time constant is used in the monostable. This period is between those output by the edge generators. Pulses that are 70 μs long are clocked through U7A when U6 times out (40 μs after the rising edge of the trigger pulse), setting Q of U7A HIGH. Pulses that are only 10 μs long disappear from D before U6 times out, so Q of U7A is clocked LOW 40 μs after the rising edge. S2 selects one of the complimentary outputs from U7A to establish the polarity of the data.

U8 and the associated circuitry prevent the teleprinter from running "open" if the transmitted signal hangs up in the space condition. Data is passed intact through the anti-space circuit as long as the space condition doesn't last more than 150 ms. After 150 ms the RC circuit charges to the HIGH state, which forces U8B LOW. The diode and 100-kΩ resistor quickly rearm the circuit during periods of mark.

A combination of analog and digital circuitry is used in the auto-start section. U10A is a retriggerable monostable multivibrator having a 4-second period. If the trigger input receives a space pulse at least every half second, Q of U10A (and therefore D of U7B) will remain HIGH. Five seconds after the first data pulse, U10B times out, clocking the D signal through to Q of U7B. If the data rate was high enough (2 CPS), Q goes HIGH, charging a 22-second RC circuit through buffer U9A. U9B is a comparator that switches the relay driver Q5. This relay (K1) energizes the teleprinter motor through the ac outlet (J2) on the rear panel. The relay opens approximately 22 seconds after the input data has ceased.

The heart of the afsk generator is two monostable multivibrators connected to form an astable circuit. The oscillator runs at twice the standard mark and space frequencies. U13A and its time-constant circuit ensure that the oscillator will start whenever the power is applied. Frequency-shift keying is accomplished by U14A, which alters the RC period of U12B when keyed from the logic interface, but only at the oscillator waveform zero crossings. Synchronizing the keying in this way prevents "RTTY clicks," which can cause adjacent-channel interference. U14B divides the oscillator frequency by two and forces a 50-percent duty cycle. At this point the afsk signal swings between ground and +12 V. A diode and resistor network attenuates this signal and shifts the dc level so that the square wave applied to the output filter swings from +4 V to +8 V. This circuit is necessary to give the filter some dynamic headroom. The low-pass filter (U13C) attenuates the harmonics of the square wave to a level that will not cause harmful interference when used with a transmitter of good design. The TONE LEVEL control allows a maximum signal of about 1 V pk-pk from U13D, the output buffer. This is sufficient to modulate any transmitter. A Morse keyer connected to J4 will shift the oscillator frequency by means of U13B for station identification.

Fig. 60 — This RTTY terminal unit can be used with any combination of equipment. It was built by Ed Kalin, K1RT, in the ARRL laboratory. Circuit design is by W1RN.

Fig. 61 — Underside view of the TU chassis. The transformers are mounted directly to the rack panel. The auto-start relay and socket are mounted next to the teleprinter motor outlet on the rear apron. This photo was taken before final assembly to show the construction more clearly.

The versatility claimed for this TU is largely because of the logic interface section. A quad op amp (U11) used for comparators diode ORed to another comparator (U15) allows RTTY to be copied or generated by any device having RS-232, TTL, CMOS or current-loop capability. A radio transceiver and three types of terminals may be connected to the TU, and each can communicate with the others without sacrificing local copy. If the builder desires to incorporate this universal interface capability, he or she must make one concession: The electromechanical teleprinter keyboard must be separated from the selector magnets. One can easily fabricate an adapter cable for reversion to the "standard" series connection. If the keyboard is left in the current loop, a positive feedback situation will be generated in which the system will latch up in the space condition.

The teleprinter selector magnets are keyed through optoisolator U16 and Q8 and Q9. Optical isolation and a floating loop supply allow any teleprinter to be used, regardless of ground polarity or lack thereof. Q9, the keying transistor, is connected as a constant-current source. S6 selects the loop current by changing the reference voltage at the base of Q9. S5 inhibits loop keying in the "standby" position by short-circuiting the base-emitter junction of the phototransistor in U16. Direct fsk of a transmitter is available at U17, within the 30-V, 150-mW limits of the phototransistor.

Construction

This TU was built using wire-wrap techniques. Fig. 61 shows two perf boards, but this can be a single-board assembly if desired. The passive components are soldered to DIP headers, which are plugged into IC sockets. Be sure to isolate J5 and J6 from the chassis. Plexiglas was used for that purpose in the unit pictured. Most of the electronics are housed in a 7- × 12- × 3-inch (178 × 305 × 76-mm) aluminum chassis attached to a 5-1/4-inch (133-mm) rack panel. Although the chassis could serve adequately as a heat sink for the keying transistor, a separate outboard assembly was chosen to keep the heat away from the sensitive timing and tuning components. Keep the CMOS ICs in their pro-

tective material until some preliminary tests have been performed and all other components installed. All of the potentiometers are pc-mount types; there are no external adjustments.

Making it Work

As with most electronic projects, it's best to test the power supplies before installing any ICs. The logic interface can be tested with a VOM. When testing this circuit, ground pin 6 of U11. The teleprinter should provide local copy with the keyboard plugged into J3. One of the teleprinter current loop jacks can be used to monitor the loop current. It's a simple matter to check the other logic inputs and outputs. A calibrated signal generator or a known afsk generator should be used to set the active filters to the proper tone frequencies. Set the LEVEL control to light the limiting indicator at a comfortable receiver volume. With the filters properly tuned, the Schmitt triggers should begin to generate rectangular pulses before limiting occurs.

A scope is needed to test the remaining demodulator circuits. With steady tones in both channels, the envelope detectors should show a steady dc output and the mark and space indicators should glow. If the envelope detectors show any pulses at the outputs, longer RC time constants are called for.

Connect an afsk generator to the demodulator input to test the diversity circuits. With the afsk generator toggling uniformly between mark and space, monitor the output of each weighted edge generator. Pins 10 and 11 of U4 should produce alternating pulses of 10 and 70 μs duration. The square wave modulation applied to the afsk generator should be reproduced at pin 2 of U7. Check the period of U6 at pin 7 — it should be LOW for 40 μs at a time. To check the anti-space function, set the audio generator to produce a steady mark tone and observe that pin 4 of U8 is LOW. Now switch the generator to the space tone and note that pin 4 goes HIGH for an instant and returns to the LOW state. Toggling the generator between the two tones at a reasonable rate should cause the modulation waveform to be reproduced at the anti-space output.

The simplest way to test the afsk generator is to disconnect the wiper of S4 and connect the anode of the diode to ground. Adjust the 25-kΩ potentiometer to produce a 2125-Hz tone at the afsk output. Now connect the diode to +12 V and adjust the 200-kΩ potentiometer to shift the tone up to 2295 Hz. Reconnect S4. Afsk signals applied to the demodulator input should be regenerated at the afsk output jacks. Check the output signal with an oscilloscope to be sure it is a clean sine wave, and that the mark and space tones have equal amplitude. Adjust the potentiometer in the ID keyer circuit to produce the desired shift when identifying.

Chapter 15

Interference with Other Services

Radio Frequency Interference (RFI) has probably been with us since the first amateur stations came on the air some 70 years ago. Fed by the technology that developed during and following WW II, the problem has become an increasing source of irritation between radio operators and their neighbors. Home-entertainment electronics devices now abound, with most families owning at least one television receiver, an a-m or fm radio, and any one of several audio devices (such as a phonograph, an intercom, an electronic guitar, or an electronic organ). Given the innate perversity of these objects to intercept radio signals, it should surprise no one to learn that RFI is one of the most difficult problems amateurs face in their day-to-day operations.

How Serious is the RFI Problem?

In one year alone, the FCC received 150,000 RFI complaints, up more than 200 percent from the number of complaints received in 1970. Of these, the great majority involved interference to home-entertainment equipment. Most important, nearly all of these would never have come to the Commission's attention if the manufacturers had corrected design deficiencies in their home-entertainment products at the time of manufacture. It is of interest to note that more than 60 percent of the interference cases reported were related to television interference (TVI).

In the case of television interference, FCC experience shows that 90 percent of the problems experienced can only be cured at the television receiver. Further, when it comes to audio equipment, the *only* cure for RFI is to treat the audio device experiencing the interference. There is nothing an amateur can do to his transmitter which will stop a neighbor's phonograph from acting as a short-wave receiver. It should be emphasized that phonographs and hi-fi units are not designed to be receivers, but simply audio devices.

It is clear, therefore, that almost all RFI problems experienced with home-entertainment devices result from basic design deficiencies in this equipment. The few small components or filters which would prevent RFI are often left out of otherwise well-designed products as manufacturers attempt to reduce costs, and hence, to reduce the prices of their products.

The Solution — Consumer Protection

Given the present unacceptable situation, what can we as amateurs do to help the consumer resolve the RFI problem? One step which should certainly be taken is to advise our friends and neighbors to inquire, before they make a purchase of an electronic device, whether the product has been certified for operation in the presence of a radio transmitter. Manufacturers must be made to recognize that RFI protection of their home-entertainment equipment has become essential. Further, where interference is being experienced, the consumer should be encouraged to contact the manufacturer of his equipment and to request that the *manufacturer* furnish the components or services necessary to eliminate RFI.

What Are Manufacturers Doing Today?

Many responsible manufacturers have a policy of supplying filters for eliminating television interference when such cases are brought to their attention. A list of those manufacturers, and a more thorough treatment of the RFI problem, can be obtained by writing the ARRL. If a given manufacturer is not listed, it is still possible that he can be persuaded to supply a filter; this can be determined by writing either directly to him or to the Electronic Industries Association (EIA).[1]

With respect to audio devices, some manufacturers will supply modified sche-

matic diagrams showing the recommended placement of bypass capacitors and other components to reduce rf susceptibility. One large American manufacturer of hi-fi equipment has in some cases supplied the necessary components free of charge, although no consistent policy has been evident and the consumer must still pay to have a serviceman install the components.

While these are encouraging developments, it appears likely that meaningful and widespread corrective action by equipment designers will require both pressure from consumers and establishment of suitable government standards.

Voluntary after-the-fact measures on the part of manufacturers simply are not enough. It is a foregone conclusion that as long as the inclusion of additional components for susceptibility reduction increases a manufacturer's cost, however slightly, there will be reluctance to take steps to improve equipment designs by the manufacturers themselves. What appears to be necessary, therefore, is federal legislation giving the FCC the authority to regulate the manufacture of home-entertainment devices and thus protect the consumer.

It's Up to Us

If requests to manufacturers of home-entertainment equipment for those components and installation services necessary to relieve RFI problems are to be successful, each of us, when faced with an RFI problem, must make known our position to the manufacturers involved. While a respectful request for assistance will bring more cooperation than a blunt demand, do not hesitate to let the manufacturers know that they have a responsibility to the consumer for correcting the design deficiencies that are causing the problem. Before casting the first stone, however, make sure you're not sitting in a glass house. Certainly, if your own television receiver experiences no interference while you are on the air, it is

[1]Electronic Industries Association, 2001 Eye Street, N.W., Washington, DC 20006. Attention: Director of Consumer Affairs.

most likely that interference to a more distant television receiver is not the fault of your transmitter.

All of the above is not to say, however, that we should not continue to assist in resolving RFI problems. Radio amateurs have typically sought to assist their neighbors in correcting RFI problems, even where those problems were in no way attributable to the performance of the transmitter. Ultimately, of course, it is the manufacturers' responsibility to correct those deficiencies which lead to the interception of radio signals. But in the interest of good neighborhood relations, we must continue to provide this assistance wherever older equipment designs are in use.

THE DEVELOPING RFI PICTURE

Several factors are combining to compound and deteriorate the RFI situation. On the consumer electronics side of the ledger is the explosive increase in home computers and video games that operate in conjunction with a TV receiver on a locally "unused" channel. Additionally, the proliferation of cable television (CATV — Community Antenna Television) networks not only increases the total public viewing time, but also aggravates the total RFI susceptibility by increasing the number of channels. The land mobile vhf bands were once the exclusive domain of serious communications systems using professional equipment. This service did not pose a serious RFI problem for amateurs. However, many homes today are blessed with "scanners." These radios are entertainment devices, and therefore are not designed for severe electromagnetic environments. While the yet-to-be-allocated amateur bands acquired at WARC-79 will certainly enhance the versatility and effectiveness of the service, they will also multiply the number of unfortunate relationships between Amateur Radio and consumer frequencies.

Previous editions of this *Handbook* contained a simple chart showing the frequency relationships between TV channels and amateur harmonics. That chart and its attendant commentary are obsolete and have been replaced by the comprehensive (and potentially demoralizing) overview presented in Fig. 1.

VHF Television: Frequency Effects

Fig. 2 shows the placement of the picture and sound carriers in the standard TV channel. In channel 2, for example, the picture carrier frequency is 54 + 1.25 = 55.25 MHz and the sound carrier frequency is 60 − 0.25 = 59.75 MHz. The second harmonic of 28,010 kHz (56,020 kHz or 56.02 MHz) falls 56.02 − 54 = 2.02 MHz above the low edge of the channel and is in the region marked "severe" in Fig. 2. On the other hand, the second harmonic of 29,500 kHz (59,000 kHz or 59 MHz) is 59 − 54 = 5 MHz from the low

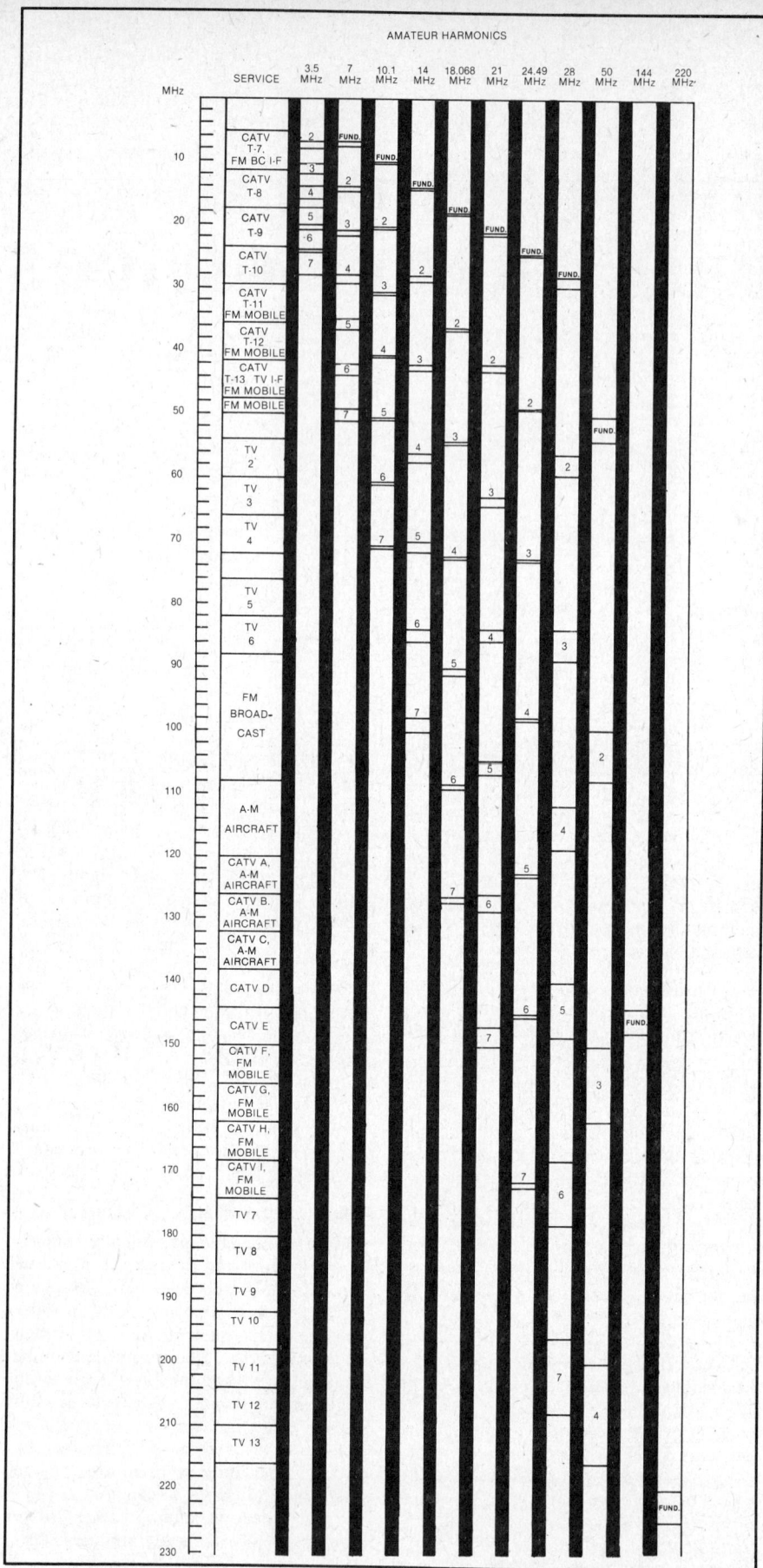

Fig. 1 — Relationship of hf and vhf amateur bands to frequencies used in consumer electronic devices. The CATV channels are used in "closed" systems, but experience has shown these systems to have poor isolation from outside signals.

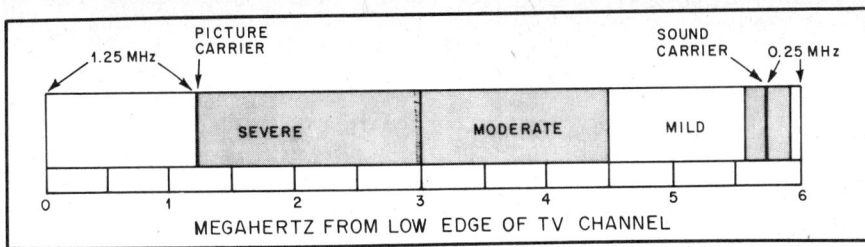

Fig. 2 — Location of picture and sound carriers in a monochrome television channel, showing the relative intensity of interference as the location of the interfering signal within the channel is varied without changing its strength. The three regions are not actually sharply defined as shown in this drawing, but merge into one another gradually.

Fig. 3 — "Cross-hatching," caused by the beat between the picture carrier and an interfering signal inside the TV channel.

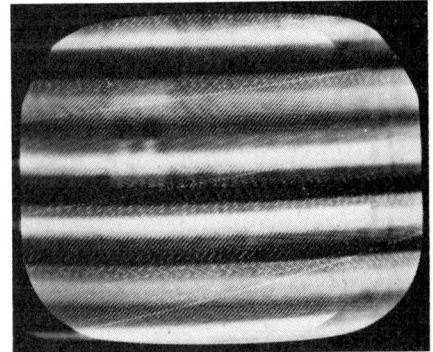

Fig. 4 — "Sound bars" or "modulation bars" accompanying amplitude modulation of an interfering signal. In this case the interfering carrier is strong enough to destroy the picture, but in mild cases the picture is visible through the horizontal bars. Sound bars may accompany modulation even though the unmodulated carrier gives no visible cross-hatching.

edge of the channel and falls in the region marked "mild." Interference at this frequency has to be about 100 times as strong as the 56,020 kHz to cause effects of equal intensity. Thus, an operating frequency that puts a harmonic near the picture carrier requires about 40 dB more harmonic suppression in order to avoid interference, as compared with an operating frequency that puts the harmonic near the upper edge of the channel.

For a region of 100 kHz or so either side of the sound carrier there is another "severe" region where a spurious radiation will interfere with reception of the sound program and this region also should be avoided. In general, a signal of intensity equal to that of the picture carrier will not cause noticeable interference if its frequency is in the "mild" region shown in Fig. 2, but the same intensity in the "severe" region will utterly destroy the picture.

Interference Patterns

The visible effects of interference vary with the type and intensity of the interference. Complete "blackout," where the picture and sound disappear completely, leaving the screen dark, occurs only when the transmitter and receiver are quite close together. Strong interference ordinarily causes the picture to be broken up, leaving a jumble of light and dark lines, or turns the picture "negative" — the normally white parts of the picture turn black and the normally black parts turn white. "Cross-hatching" — diagonal bars or lines in the picture — accompanies the latter, usually, and also represents the most common type of less severe interference. The bars are the result of the beat between the harmonic frequency and the picture carrier frequency. They are broad and relatively few in number if the beat frequency is comparatively low — near the picture carrier — and are numerous and very fine if the beat frequency is very high — toward the upper end of the channel. Typical cross-hatching is shown in Fig. 3. If the frequency falls in the mild region in Fig. 2, the cross-hatching may be so fine as to be visible only on close inspection of the picture, in which case it may simply cause the apparent brightness of the screen to change when the transmitter carrier is thrown on and off.

Whether or not cross-hatching is visible, an amplitude-modulated transmitter may cause "sound bars" in the picture. These look about as shown in Fig. 4. They result from the variations in the intensity of the interfering signal when modulated. Under most circumstances modulation bars will not occur if the amateur transmitter is frequency- or phase-modulated. With these types of modulation the cross-hatching will "wiggle" from side to side with the modulation.

Except in the more severe cases, there is seldom any effect on the sound reception when interference shows in the picture, unless the frequency is quite close to the sound carrier. In the latter event the sound may be interfered with even though the picture is clean.

Reference to Fig. 1 will show whether or not harmonics of the frequency in use will fall in any television channels that can be received in the locality. It should be kept in mind that not only harmonics of the final frequency may interfere, but also harmonics of any frequencies that may be present in buffer or frequency-multiplier stages. In the case of 144-MHz transmitters, frequency-multiplying combinations that require a doubler or tripler stage to operate on a frequency actually in a low-band vhf channel in use in the locality should be avoided.

Harmonic Suppression

Effective harmonic suppression has three separate phases:

1) Reducing the amplitude of harmonics generated in the transmitter. This is a matter of circuit design and operating conditions.

2) Preventing stray radiation from the transmitter and associated wiring. This requires adequate shielding and filtering of all circuits and leads from which radiation can take place.

3) Preventing harmonics from being fed into the antenna.

It is impossible to build a transmitter that will not generate *some* harmonics, but it is obviously advantageous to reduce their strength, through circuit design and choice of operating conditions, by as large a factor as possible before attempting to prevent them from being radiated. Harmonic radiation from the transmitter itself or from its associated wiring obviously

will cause interference just as readily as radiation from the antenna, so measures taken to prevent harmonics from reaching the antenna will not reduce TVI if the transmitter itself is radiating harmonics. But once it has been found that the transmitter itself is free from harmonic radiation, devices for preventing harmonics from reaching the antenna can be expected to produce results.

Reducing Harmonic Generation

Since reasonably efficient operation of rf power amplifiers always is accompanied by harmonic generation, good judgment calls for operating all frequency-multiplier stages at a very low power level. When the final output frequency is reached, it is desirable to use as few stages as possible in building up to the final output power level and to use tubes that require a minimum of driving power.

Circuit Design and Layout

Harmonic currents of considerable amplitude flow in both the grid and plate circuits of rf power amplifiers, but they will do relatively little harm if they can be effectively bypassed to the tube cathode. Fig. 5 shows the paths followed by harmonic currents in an amplifier circuit;

because of the high reactance of the tank coil there is little harmonic current in it, so the harmonic currents simply flow through the tank capacitor, the plate (or grid) blocking capacitor, and the tube capacitances. The lengths of the leads forming these paths is of great importance, since the inductance in this circuit will resonate with the tube capacitance at some frequency in the vhf range (the tank and blocking capacitances usually are so large compared with the tube capacitance that they have little effect on the resonant frequency). If such a resonance happens to occur at or near the same frequency as one of the transmitter harmonics, the effect is just the same as though a harmonic tank circuit had been delib-

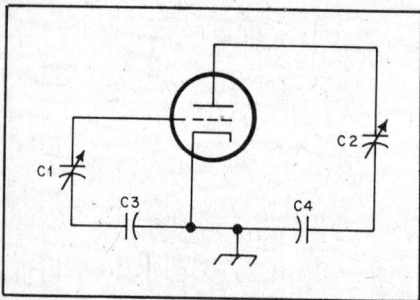

Fig. 5 — A vhf resonant circuit is formed by the tube capacitance and the lead inductances through the tank and blocking capacitors. Regular tank coils are not shown, since they have little effect on such resonances. C1 is the grid tuning capacitor and C2 is the plate tuning capacitor. C3 and C4 are the grid and plate blocking or bypass capacitors, respectively.

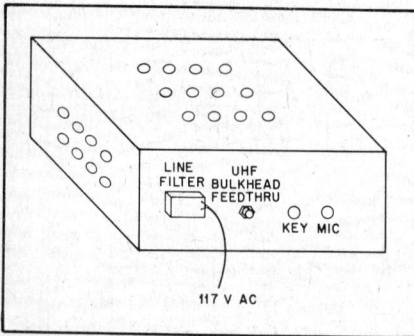

Fig. 6 — Construction details of the rf enclosure. For the model shown, thin aluminum sheet metal was used to form a box 12 × 28 × 20 inches (HWD) (304 × 711 × 508 mm). Small holes were drilled for ventilation and a fan might be advisable if temperature rise is considered a problem. Feed-through connectors can be of the builder's choice but ac conduits and control leads should be filtered. For key and mic leads, bypass with 0.001-μF disk-ceramic capacitors and install a small ferrite bead (if available). A commercially manufactured line filter was used, although a homemade one of the "brute force" type would also be suitable.

Although dimensions and material are not critical, the cabinet should be deep enough to form as much of an overlap as possible between the front of the equipment panel and the front of the cabinet. It is important that all leads be run through the rear of the cabinet.

erately introduced; the harmonic at that frequency will be tremendously increased in amplitude.

Such resonances are unavoidable, but by keeping the path from plate to cathode and from grid to cathode as short as is physically possible, the resonant frequency usually can be raised above 100 MHz in amplifiers of medium power. This puts it between the two groups of television channels.

It is easier to place grid-circuit vhf resonances where they will do no harm when the amplifier is link-coupled to the driver stage, since this generally permits shorter leads and more favorable conditions for bypassing the harmonics than is the case with capacitive coupling. Link coupling also reduces the coupling between the driver and amplifier at harmonic frequencies, thus preventing driver harmonics from being amplified.

The inductance of leads from the tube to the tank capacitor can be reduced not only by shortening but by using flat strip instead of wire conductors. It is also better to use the chassis as the return from the blocking capacitor or tuned circuit to cathode, since a chassis path will have less inductance than almost any other form of connection.

The vhf resonance points in amplifier tank circuits can be found by coupling a grid-dip meter covering the 50-250 MHz range to the grid and plate leads. If a resonance is found in or near a TV channel, methods such as those described above should be used to move it well out of the TV range. The grid-dip meter also should be used to check for vhf resonances in the tank coils, because coils made for 14 MHz and below usually will show such resonances. In making the check, disconnect the coil entirely from the transmitter and move the grid-dip meter coil along it while exploring for a dip in the 54- to 88-MHz band. If a resonance falls in a TV channel that is in use in the locality, changing the number of turns will move it to a less-troublesome frequency.

Operating Conditions

Grid bias and grid current have an important effect on the harmonic content of the rf currents in both the grid and plate circuits. In general, harmonic output increases as the grid bias and grid current are increased, but this is not necessarily true of a *particular* harmonic. The third and higher harmonics, especially, will go through fluctuations in amplitude as the grid current is increased, and sometimes a rather high value of grid current will minimize one harmonic as compared with a low value. This characteristic can be used to advantage where a particular harmonic is causing interference, remembering that the operating conditions that minimize one harmonic may greatly increase another. For equal operating

conditions, there is little or no difference between single-ended and push-pull amplifiers in respect to harmonic generation. Push-pull amplifiers are frequently trouble-makers on even-order harmonics because with such amplifiers the even-harmonic voltages are in phase at the ends of the tank circuit and hence appear with equal amplitude across the whole tank coil if the center of the coil is not grounded. Under such circumstances the even harmonics can be coupled to the output circuit through stray capacitance between the tank and coupling coils. This does not occur in a single-ended amplifier having an inductively coupled tank if the coupling coil is placed at the cold end or with a pi-network tank.

Some TVI Tests

One of the difficulties in solving TVI problems, particularly in fringe areas, is the number of possible causes and their elusive nature. A "cure" seems to be found only to have the problem return with renewed severity the next day. Consequently, some tests were performed by the ARRL in order to isolate the causes, if possible. Although the results weren't encouraging in regard to certain aspects, one bright spot was some "fall-out" in the way of additional suppression measures previously neglected.

Test Procedures

A ham experiencing some TVI in a fringe area (on his own set) generously agreed to be the "patient" in the tests. A large screened enclosure was transported to his location. It could contain the TV set along with a smaller version of the screen enclosure (Fig. 6) for the transmitter. Other equipment included a gasoline-powered generator that could power either the TV set or transmitter. There is always the possibility that feedback of rf energy through the power line (or "conducted interference") is a factor in a TVI problem. By running the equipment on separate power systems, some idea of the importance of this type of coupling would be ascertained.

Various low-pass filters, high-pass filters, and power harnesses made up the rest of the equipment list. Checks on various TV channels indicated the most serious problem resulted from third-harmonic energy on channel 3 during 15-meter operation. Tests were performed with the rig inside and out of its shielded enclosure, the TV set inside and out of the larger enclosure, and with either the TV set or transmitter on independent power.

Test Results

Previous checks in the lab revealed that almost all currently manufactured amateur transmitters and transceivers emit harmonics in the form of "chassis radiation" to varying degrees. It should be pointed out that no outstanding "saints" were found in this area but mostly "sinners." Once this energy escapes from the transmitter cabinet, it can be conducted to the antenna or back through the power line via a single-conductor waveguide type of mode. This mode is very similar to the propagation of rf energy over a two-dimensional conducting surface in the form of a ground wave. But the important thing to keep in mind is that devices such as filters, traps and grounds are ineffective since the rf energy flows *around* the suppression network. The only effective measure is adequate shielding.

As might be expected, the field tests verified the importance of this aspect. *The*

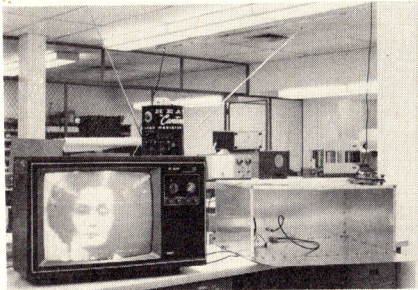

Fig. 7 — Lab simulation of TVI tests discussed in the text. A dummy load placed next to the "rabbit ears" served as the transmitting antenna.

Fig. 8 — Severe interference occurred with the setup shown above with the transmitter out of the enclosure. Interference was about the same with the leads running from the front of the cabinet instead of through the rear connectors.

Fig. 9 — The interference was either reduced considerably (as shown in this photo) or eliminated completely, depending upon TV signal strength, with all the leads exiting from the rear of the cabinet.

only test that indicated appreciable reduction of TVI was the one with the transmitter placed inside of its shielded enclosure. In all other tests, there was no improvement or the change was so slight as to be inconclusive. Some residual interference still remained with the rig enclosed in the shield. This was likely caused by rectification in the external environment. One unexpected result was that no noticeable difference was observed with the door of the enclosure opened or closed. In fact, with the door partially closed and touching the shield at only a few points the TVI became worse!

Conclusions

The transmitter power for these tests was approximately 180 watts. Considerable work has yet to be performed to determine the important factors at higher levels. However, there is hope for the ham experiencing TVI because of chassis radiation. With the door open, it is believed that the enclosure acts as a waveguide below cutoff and still offers some measure of suppression while permitting access to the controls. In discussing the tests with other amateurs experiencing TVI, reports from the field were favorable when similar measures were tried.

Some further experimentation along these lines is in order. For instance, former shielding theory advocated the use of *high-conductivity* materials. Newer methods often rely on the dissipation of unwanted rf energy in lower conductivity materials such as steel. Although rf energy can penetrate deeper into low-conductivity metals, and greater thicknesses are required to provide the same isolation (as that of copper, for example), other problems are simplified. Unwanted rf energy must be dissipated *somewhere* and when a good conductor is used for a shielding enclosure there is a greater tendency for this energy to leak out through doors, conduits and other points of entry. By dissipating energy internally on the shield walls, there is less chance for it to leak out. (However, if the unexpected attenuation with the door open was actually caused by a waveguide-below-cutoff effect as speculated, high-conductivity material near the door opening would be advisable.)

Other Results

With the TV set in its shielded enclosure and with power fed through a commercially manufactured line filter, there was no difference in TVI with the rig *or* TV set on independent power. Rf signal energy from the TV antenna was fed into the shield enclosure through a commercially manufactured high-pass filter. Little change was noted when these measures were eliminated and the set operated on the same power service as the transmitter, without a high-pass filter and outside of the shield.

It should not be concluded that such

measures will be equally ineffective under all circumstances. However, the claims of some manufacturers are open to question. Items such as power supplies that eliminate TVI, and similar nostrums have come to our attention. Consequently, common sense is in order in judging whether or not a particular device will prove effective in eliminating interference or how it should be employed.

Suppression Practices

As the test results reported in the previous section reveal, complete elimination of TVI is often not a simple process. It seldom happens that a single measure such as installing a high-pass filter at the TV set will cure the problem. Rather, a number of methods must be applied simultaneously. The principal factor in any TVI situation is the ratio of TV signal strength to interference level. This includes interference of all types such as ignition noise, random or thermal noise (which isn't really interference but sets the minimum signal that permits "snow-free" reception), and unwanted signals that fall within the TV channel. A signal-to-interference ratio greater than approximately 35 to 40 dB is required for good picture quality.

In this regard, an area frequently overlooked in TVI difficulties is the TV-set antenna. A poor antenna with little gain in the direction of the TV station, old and corroded wire and connections (which can *cause* the harmonic generation by rectification of a "clean" signal generated in a nearby amateur transmitter), may result in a TVI situation that is impossible to solve. For instance, the "simulated" lab tests illustrated in the photographs were performed with a dummy load next to a set of "rabbit ears," which comprised the TV-set antenna. With a good outdoor system, the TVI would not have been present even though there was a leakage from the transmitter cabinet. Generally speaking, if the picture quality on the TV set experiencing the interference is poor to begin with, even sophisticated suppression measures are likely to prove futile.

Grounds

Grounding of equipment has long been considered to be a first step in eliminating interference. While the method is very effective in the mf range and below, for all practical purposes it is useless in suppressing vhf energy. This is because even short lengths of wire have considerable reactance at vhf. For instance, suppose a length of wire by itself has an inductance of $1\,\mu H$. At 550 kHz, the reactance would be about 3.46 ohms. On the other hand, the same wire would have a reactance of over 300 ohms at 56 MHz, which is the frequency range of TV channel 2. (Actually, the impedance of a wire becomes a more complicated entity to define at vhf.

The delay effects along the wire are similar to those on the surface of an antenna. Consequently, the wire might even appear as an open circuit rather than as a ground as the electrical length approaches a quarter wavelength.)

From a shock-hazard point of view, grounding is important. *However, never connect a ground for any reason to the chassis of a TV set.* This is because many TV sets derive their operating voltages directly from the ac-service line. Although a schematic diagram of a TV set may indicate a "power transformer" is being used, caution should be exercised to be sure it is actually being employed for this purpose. Quite often, the only voltage the transformer is supplying is for the TV picture tube filament.

Shielding

Effective shielding is perhaps the single most important measure in preventing or curing any RFI problem. However, as pointed out in previous sections, unwanted rf energy must be dissipated. The task becomes harder to perform when the spacing between the source of energy and the boundaries of the shield diminish. Consequently, the use of a double shield (as used in the tests) is one way of reducing residual radiation from the primary shielding surface.

In order to obtain maximum effectiveness of a particular shielding measure, no breaks or points of entry should be permitted. Small holes for ventilation purposes usually do not degrade shielding effectiveness. But even here, a honeycomb type of duct is often employed when maximum isolation is required. (A parallel bundle of small tubing has very high attenuation since each tube by itself acts as a waveguide below cutoff.)

The isolation of a coaxial cable can be degraded considerably unless the ends of the shield are terminated properly. A braid should be soldered so that it completely encloses the inner conductor(s) at the connector junction. For instance, the practice of twisting the braid and point soldering it to the base of a connector may result in a 20-dB degradation in isolation. Normally, this effect is not serious if the cable is run through an area where sensitive circuits don't exist. However, the isolation afforded by a filter can be reduced considerably in circuits where such cable breaks occur.

One instance where a shield break causes a serious problem is in the connection between the antenna terminals on a TV set and the tuner. Newer sets have a 75-ohm coaxial input along with a balun for 300-ohm line. However, because many TV sets have direct connections to the ac line, a decoupling network is used. The shielded lead to the tuner is broken and a capacitor is connected in series with the braid. This provides a low-impedance path for rf energy while presenting a high

impedance at 60 Hz. Consequently, because of the cable break, high-pass filters at the antenna input terminals are not as effective as those built into the tuner itself.

Coax Shield Chokes

As mentioned previously, vhf currents flowing on the *outside* of coaxial cables are frequently the cause of RFI. Figs. 10, 11 and 12 show techniques for reducing or eliminating conducted chassis radiation from coaxial cables. The cardboard tube stuffed with steel wool in Fig. 10 works on the absorption principle. The steel wool is very lossy and dissipates the rf energy on the shield. The tube pictured is 18 inches (0.45 m) long. A longer tube would be more effective. Fig. 11 shows a choke wound on a ferrite toroidal core. RG-8X cable is ideal for hf transmitters because of its high flexibility and power-handling capability. It can be spliced into an RG-8 transmission line with connectors and adapters. Another coax radiation-suppression device is illustrated in Fig. 12. If the plate is at least a half wavelength (at the harmonic frequency to be suppressed) on its smallest dimension, it will provide a very effective barrier. Large pieces of sheet metal are expensive, so the baffle can be made from a sheet of cardboard or Masonite covered with aluminum foil. The ideal placement of any of these chokes will vary with the standing wave pattern on the coax shield, but in general they should be close to the transmitter. Like all RFI remedies, the effectiveness of these devices varies widely with each interference situation. Therefore, one should not expect miracles.

Capacitors at RF

Capacitors are common elements found in almost any piece of electronics gear. However, some precautions are necessary when they are employed in RFI-preventive purposes such as in filters and bypassing applications. In particular, lead inductance may be sufficient to resonate with the capacitor proper and cause the entire combination to have a high inductive reactance rather than the desired capacitive reactance.

This effect is illustrated in the accompanying photographs. The response curve shown in Fig. 13A is for a 10-MHz low-pass filter arranged in a "pi" configuration. However, this particular circuit realization required some large-valued capacitors. Using ordinary capacitor types resulted in an unwanted resonance as evidenced by the sharp dip in the response curve at approximately 15 MHz. However, by going to the equivalent "T" configuration (see the section on filters in the chapter on electrical laws and circuits), a circuit realization for the desired response required much smaller capacitance values. The curve shown in Fig. 13B approximated this response quite closely and no effects

Fig. 10 — This coax shield decoupler is made with steel wool stuffed into an IBM copier tube. This size was selected for photographic purposes, but to be truly effective, the device should be about twice as long.

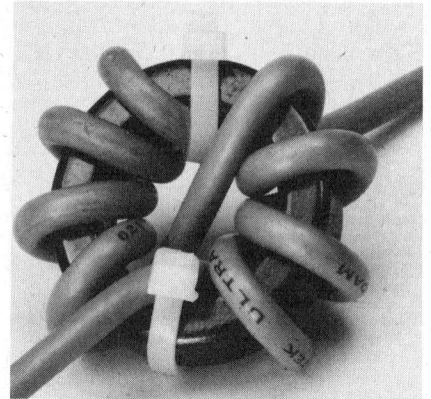

Fig. 11 — Winding the cable on a ferrite toroid is a highly effective shield current suppressor in some cases. Reversing the winding as pictured allows more turns with less shunt capacitance. RG-8X is used in this model, but RG-58 will suffice for moderate power applications. The most important property of the cable is complete shielding — avoid "bargain" cable having less than 95-percent braid coverage.

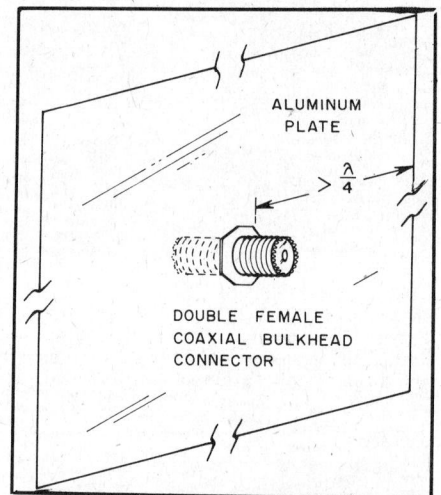

Fig. 12 — A large metallic baffle effectively inhibits waves propagating on the outside of a coaxial cable. For vhf TV channel 2, the smallest effective baffle is a 9-foot-diameter circle, but the required size decreases linearly with frequency.

of parasitic inductance were noticeable. When designing filters, it is advisable to compute the component values for as many configurations as possible in order to determine which one results in the most practical elements. If large capacitance values are unavoidable, either special low-inductance types should be used or a number of ordinary smaller-valued

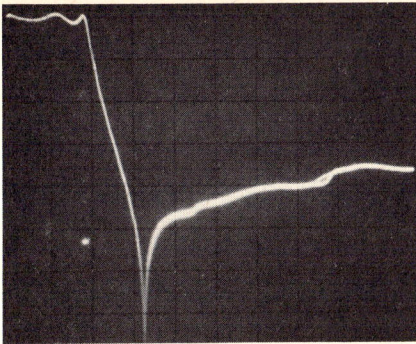

(A)

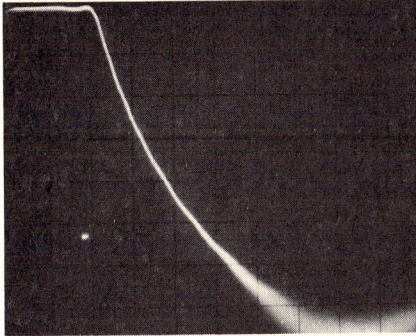

(B)

Fig. 13 — Stray lead inductance of a capacitor can degrade filter performance.

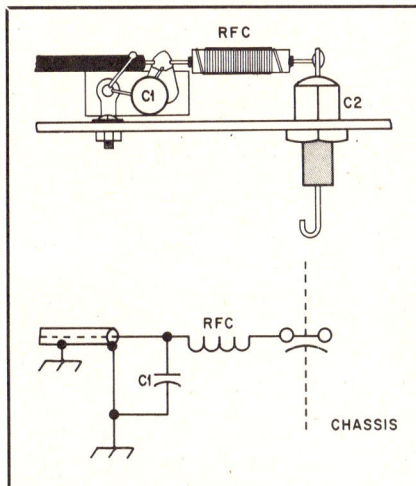

Fig. 14 — Additional lead filtering for harmonics or other spurious frequencies in the high vhf TV band (174-216 MHz).
C1 — 0.001-μF disc ceramic
C2 — 500- or 1000-pF feed-through bypass (Centralab FT-1000. Above 500 volts, substitute Centralab 858S-500.)
RFC — 14 inches (356 mm) no. 26 enamel close-wound on 3/16-inch (4.76-mm) dia form or composition resistor body.

capacitors can be paralleled to reduce the effect of lead inductance.

A very desirable capacitor (C2) from an RFI point of view is shown in Fig. 14. Instead of having two or more plates arranged in a parallel fashion, the conductors are coaxial and are separated by the dielectric. Such feedthrough capa-

citors are highly recommended for conducting leads in and out of circuits where the radiation of harmonic energy is possible. In addition, the rfc illustrated in Fig. 14 could either consist of a small coil wound over a composition resistor as shown or it could be a ferrite bead on a straight piece of wire.

Decoupling from the AC Line

Direct feedback of rf energy into the ac power service is usually not a problem with modern transmitting equipment. However, currents induced on the antenna feed line may flow on the transmitter chassis and back into the ac line. A rig "hot" with rf or even the presence of "broadcast harmonics" while receiving may mean a problem of this sort. In the case where an antenna is being used that requires a ground (such as an end-fed wire), never use any part of the ac conduits, water systems, or other conductors in a building. It is always advisable to have a separate ground system for the antenna itself.

It is also good practice to use an antenna-matching network with no direct connection between the transmitter and antenna feed line. Any matching network that uses mutual-magnetic coupling exclusively will fulfill this requirement. Antenna pattern is another factor to consider and if possible, a type should be used that directs the minimum possible signal into other dwellings. For instance, ground-mounted vertical antennas have considerable low-angle radiation while a dipole directs energy at angles *below* the horizontal plane. A vertical ground plane or beam mounted on as high a tower as

practical will generally be better from an RFI and TVI standpoint than antennas closer to the ground.

FCC Rules Concerning RFI

Part 97.73 of the U.S. amateur regulations specifies the amateur's responsibility for signal purity:

(a) Except for a transmitter or transceiver built before April 15, 1977 or first marketed before January 1, 1978, the mean power of any spurious emission or radiation from an amateur transmitter, transceiver, or external radio frequency power amplifier being operated with a carrier frequency below 30 MHz shall be at least 40 decibels below the mean power of the fundamental without exceeding the power of 50 milliwatts. For equipment of mean power less than five watts, the attenuation shall be at least 30 decibels.

(b) Except for a transmitter or transceiver built before April 15, 1977 or first marketed before January 1, 1978, the mean power of any spurious emission or radiation from an amateur transmitter, transceiver, or external radio frequency power amplifier being operated with a carrier frequency above 30 MHz but below 235 MHz shall be at least 60 decibels below the mean power of the fundamental. For a transmitter having a mean power of 25 watts or less, the mean power of any spurious radiation supplied to the antenna transmission line shall be at least 40 decibels below the mean power of the fundamental without exceeding the power of 25 microwatts, but need not be reduced below the power of 10 microwatts.

(c) Paragraphs (a) and (b) of this section notwithstanding, all spurious emissions or radiation from an amateur transmitter, transceiver, or external radio frequency power amplifier shall be reduced or eliminated in accordance with good engineering practice.

(d) If any spurious radiation, including

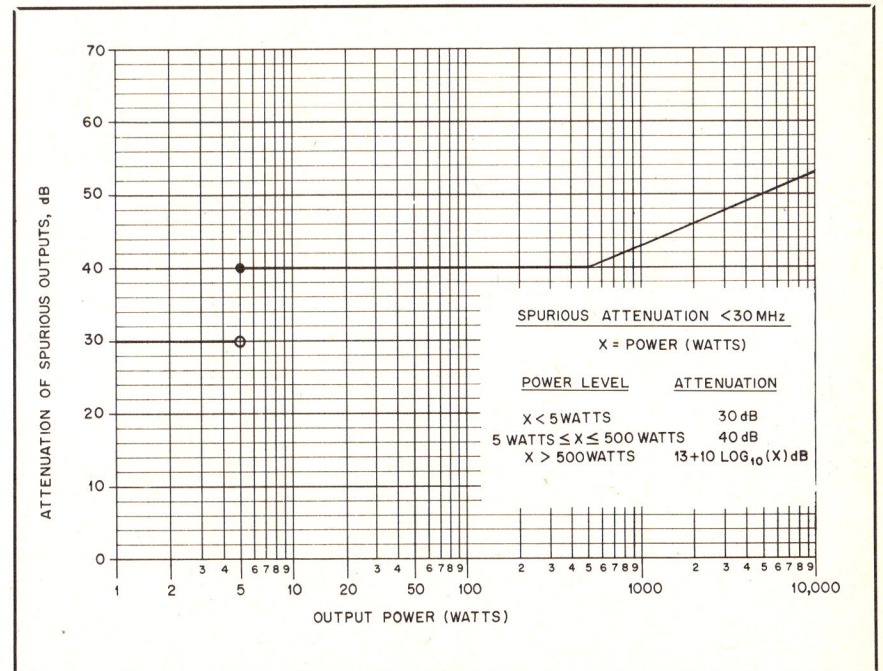

Fig. 15 — The FCC specifies that spurious signals generated by transmitting equipment must be reduced well below the level of the fundamental. This graph illustrates exactly how far the spurious components must be reduced. This applies to amateur transmitters operating below 30 MHz.

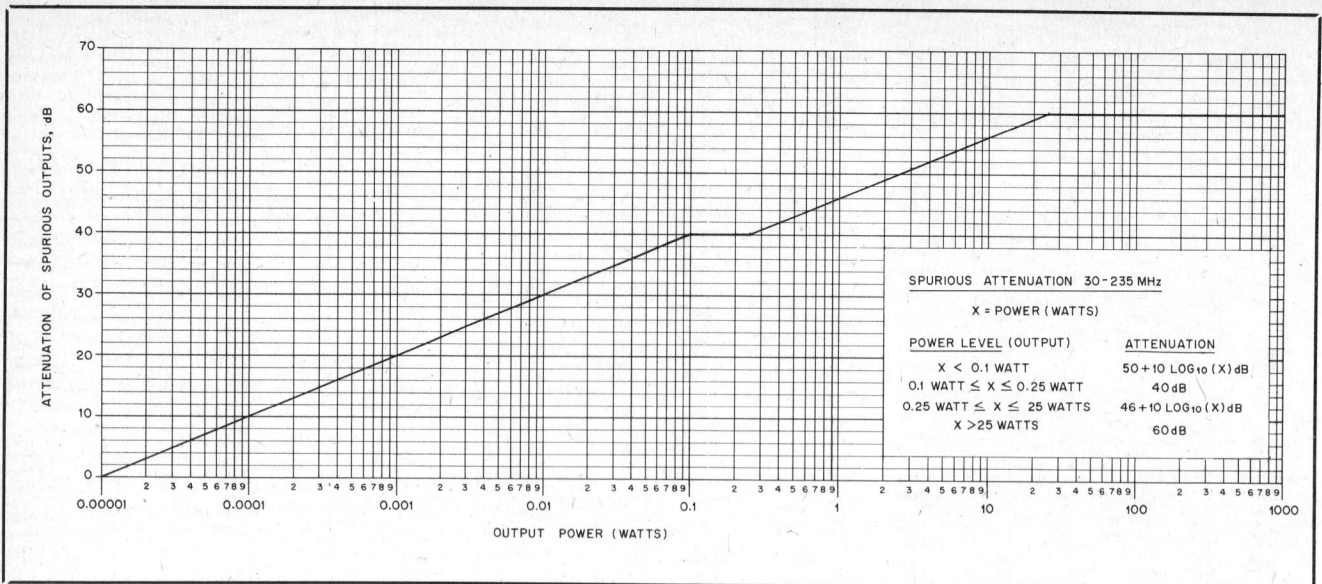

Fig. 16 — This graph illustrates to what level spurious-output energy must be reduced for equipment designed to operate in the 30- to 235-MHz range.

chassis or power line radiation, causes harmful interference to the reception of another radio station, the licensee may be required to take steps to eliminate the interference in accordance with good engineering practice.

NOTE: For the purposes of this section, a spurious emission or radiation means any emission or radiation from a transmitter, transceiver, or external radio frequency power amplifier which is outside of the authorized Amateur Radio Service frequency band being used.

The numerical limits cited in 97.73 are interpreted graphically in Figs. 15 and 16. Note, however, that paragraphs (c) and (d) go beyond absolute limits in defining the amateur's obligation.

Filters and Interference

The judicious use of *filters*, along with other suppression measures such as shielding, has provided solutions to interference problems in widely varying applications. As a consequence, considerable attention has been given to the subject over the years that has resulted in some very esoteric designs. Perhaps the most modern approach is the optimization and/or realization for a particular application of a filter by means of a digital computer. However, there are a number of other types with component values cataloged in tabular form. Of these, the most important ones are the so-called Chebyshev and elliptic-function filters. (Butterworth filters are often considered a special case of Chebyshev types with a ripple factor of zero.)

Elliptic-function filters might be considered optimum in the sense that they provide the sharpest rolloff between the passband and stopband. Computed values for a low-pass filter with a 0.1-dB ripple in the passband and a cutoff frequency of 30.6 MHz are shown in Fig. 17. The filter is supposed to provide an attenuation of 35 dB above 40 MHz. An experimental

model was built and the response is shown in Fig. 18. As can be seen, the filter came quite close to the design goals. Unfortunately, as with most of the designs in this section, alignment of the more complicated filters requires some sort of sweep-generator setup. This is the only practical way of "tweaking" a filter to the desired response. While building a sweep setup is not beyond the talents of an advanced experimenter, the lack of one is an obstacle in the home construction of filters.

An Absorptive Filter

The filter shown in Fig. 19 not only provides rejection by means of a low-pass section, it also includes circuitry that absorbs harmonic energy. A high-pass section consists of L1, L2, C1 and C2 is terminated in a 50-ohm "idler load" and this combination performs the latter function. The advantages of this technique are that degradation of filter rejection caused by antenna mismatch at the harmonic frequency are not as severe (with a filter of this type) and the transmitter is terminated in a resistive load at the harmonic.

Construction and Test Techniques

If good performance above 100 MHz is not a necessity, this filter can be built using conventional fixed capacitors. Teflon-dielectric pc board can be purchased from Alaska Microwave Labs, listed in Chapter 17. Regular fiberglass-insulated board is satisfactory for low power. One such filter has been used with an SB-100 transceiver running 100 watts. Although the Q of the fiberglass capacitors will be lower than that of Teflon-dielectric capacitors, this should not greatly affect the type of filter described here.

Test equipment needed to build this filter at home includes a reasonably ac-

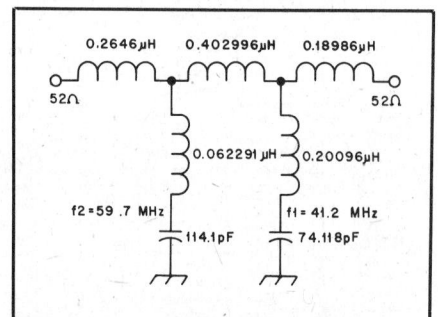

Fig. 17 — Schematic diagram showing component values of an experimental elliptic function filter.

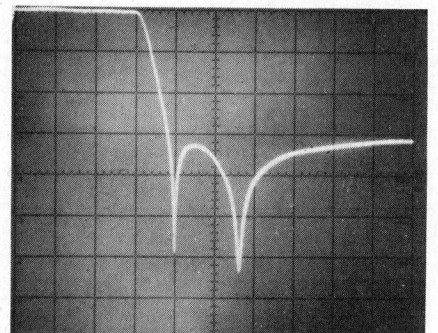

Fig. 18 — Response curve of the filter shown in Fig. 17. Vertical scale represents 10 dB/div. and horizontal scale is 10 MHz/div.

curate grid-dip oscillator, an SWR bridge, a reactance chart or the ARRL Lightning Calculator (for L, C and f), a 50-ohm dummy load, and a transmitter.

Once the value of a given capacitor has been calculated, the next step is to determine the capacitance per square inch of the double-clad circuit board you have. This is done by connecting one end of a coil of known inductance to one side of the circuit board, and the other coil lead

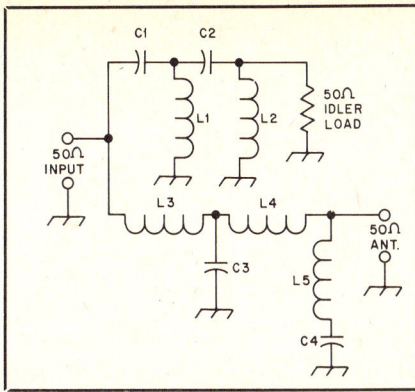

Fig. 19 — Schematic diagram of the absorptive filter. The pc-board used is MIL-P-13949D, FL-GT-.062 in, C-2/2-11017, Class 1, Grade A. Polychem Bud Division. Capacitance between copper surfaces is 10-pF per square inch. Values are as follows for a design cutoff frequency of 40 MHz and rejection peak in channel 2:

C1 — 52 pF   C4 — 15 pF   L3 — 0.3 µH
C2 — 73 pF   L1 — 0.125 µH   L4 — 0.212 µH
C3 — 126 pF   L2 — 0.52 µH   L5 — 0.55 µH

to the other side of the circuit board. Use the grid-dip oscillator, coupled lightly to the coil, to determine the resonant frequency of the coil and the circuit-board capacitor. When the frequency is known, the total capacitance can be determined by working the Lightning Calculator or by looking the capacitance up on a reactance chart. The total capacitance divided by the number of square inches on one side of the circuit board gives the capacitance per square inch. Once this figure is determined, capacitors of almost any value can be laid out with a ruler!

High voltages can be developed across capacitors in a series-tuned circuit, so the copper material should be trimmed back at least 1/8 inch (3 mm) from all edges of a board, except those that will be soldered to ground, to prevent arcing. This should not be accomplished by filing, since the copper filings would become imbedded in the board material and just compound the problem. The capacitor surfaces should be kept smooth and sharp corners should be avoided.

If the filter box is made of double-clad fiberglass board, both sides should be bonded together with copper stripped from another piece of board. Stripped copper foil may be cleaned with a razor blade before soldering. To remove copper foil from a board, use a straight edge and a sharp scribe to score the thin copper foil. When the copper foil has been cut, use a razor blade to lift a corner. This technique of bonding two pieces of board can also be used to interconnect two capacitors when construction in one plane would require too much area. Stray inductance must be minimized and sufficient clearance must be maintained for arc-over protection.

Capacitors with Teflon dielectric have

been used in filters passing up to 2 kW PEP. One further word of caution: No low-pass filter will be fully effective until the transmitter with which it is used is properly shielded and all leads filtered.

The terminating loads for the high-pass section of the filter can be made from 2-watt, 10-percent tolerance composition resistors. Almost any dissipation rating can be obtained by suitable series-parallel combinations. For example, a 16-watt, 50-ohm load could be built as shown in Fig. 19. This load should handle the harmonic energy of a signal with peak fundamental power of 2 kilowatts. With this load, the harmonic energy will see an SWR under 2:1 up to 400 MHz. For low power (under 300 watts PEP), a pair of 2-watt, 100-ohm resistors is adequate.

This filter was originally described by Weinrich and Carroll in November 1968 *QST*. The component values given here were calculated by Keith Wilkinson, ZL2BJR.

How Much Harmonic Suppression is Needed?

While it's a fairly simple matter to install a low-pass filter and hope for good results, approaching the problem scientifically can reduce unnecessary expense and aggravation. Rasmussen and Gerue presented an orderly method for determining the required harmonic attenuation in "Harmonic TVI — A New Look at an Old Problem," *QST*, September 1975.

The following factors affect the tolerable harmonic levels:

T_P — Transmitter power output in dBm.
T_A — Transmitter harmonic attenuation.
G_T — Gain of the transmitting antenna at the harmonic frequency,
G_R — Gain of the TV receiving antenna at the interfering frequency,
S_A — Attenuation (path loss) from the transmitting antenna to the receiving antenna,
S_S — TV signal strength at the TV receiver input,
S_R — Signal-to-interference ratio at the TV receiver needed to preclude visual or audio interference.

The relationship between these factors, when expressed in dB, is

$$H_R = (T_P + T_A + G_T + G_R + S_A) - S_S + S_R$$

where H_R is the further transmitter harmonic reduction required (in dB).

The first step in determining H_R is to find the harmonic relationship between the bands you use and the frequencies of the weaker TV signals in your area. The harmonics of the hf amateur bands that fall in the lower vhf TV channels are given in Fig. 1. Then, for each combination of amateur band and TV channel concerned, determine H_R, using the following information:

T_P — Peak transmitter power output in dB above 1 mW. 25 watts is 44 dBm.

Table 1

TV Receiving Antenna Gain, dB, vs Angle to Transmitter

	315-45°	45-135°	135-225°	225-315°
Fringe-Area Antenna	+10	<−10	<0	<−10
Lower-Grade Antenna	+3	<0	<0	<0

Table 2

Path Loss, dB, Transmitting Antenna to Receiving Antenna

Distance in Meters	10	20	30	40	50	100
44-MHz I-F	25	31	34	37	39	45
TV 2	27	33	37	39	41	47
TV 3	28	34	38	40	42	48
TV 4	29	35	39	41	43	49
TV 5	30	36	40	42	44	50
TV 6	31	37	41	43	45	51

Table 3

TV Received-Signal Power

TV Channel	Class B	Class A
2	−55 dBm	−34 dBm
3	−56 dBm	−35 dBm
4	−57 dBm	−35 dBm
5	−58 dBm	−37 dBm
6	−59 dBm	−38 dBm

Add 3 dB for each doubling of the power (1000 watts is 60 dBm).

T_A — Transmitter harmonic attenuation, can be estimated from specifications given for commercial transmitters.

G_T — Conservative estimate 0 dB.

G_R — Estimate relative bearing from the TV antenna to the amateur antenna, and use Table 1.

S_A — Estimate distance between the antennas, and use Table 2 for path loss for the TV channel concerned.

S_S — TV signal level from Table 3 for TV channel and distance to TV station. For under 50 miles, use Class A; for over 50 miles, use Class B.

S_R — Signal/interference ratio required can be as high as 40 dB, with an operating frequency whose harmonic is within 1 MHz of the picture carrier frequency, or as low as 20 dB, if the harmonic is more than 2 MHz from both picture and sound carriers. 35 dB is a suggested reasonable value.

Let us assume a station operating on 15 meters with 1-kW PEP output. The third harmonic (in channel 3) is down 40 dB. The antenna is a triband beam pointed at the TV antenna, with a separation of 40 meters (130 feet). Fringe-area reception applies. The numbers are:

$T_P = +60$
$T_A = -50$
$G_T = 0$
$G_R = -10$

$$S_A = -40$$
$$S_S = -56$$
$$S_R = +35$$

$$H_R = (+60) + (-50) + (0) + (+10)$$
$$+ (-40) + (-56) + (+35) = +71$$

The indicated need for 71 dB of additional attenuation confirms our hypothetical situation as a particularly bad one: fringe-area reception with a signal very close to the TV receiver noise level.

Assuming good engineering design for minimum harmonic generation and adequate shielding to prevent radiation of harmonic power that is always present in any transmitter, the way to reduce harmonic radiation further is to use a low-pass filter. A good filter will offer at least 50 dB of additional attenuation, when used with a well-designed transmitter working into a proper load. This should be enough to solve most harmonic-TVI problems. With a situation like the above worst-case conditions, relocation of the amateur antenna or, in some instances, raising it well above the plane of residential TV antennas, may help. Operation at reduced transmitter power levels should be tried, though this does not always result in lowered harmonic-power radiation. The reverse may be true, depending on transmitter operating conditions.

Filters For VHF Transmitters

High rejection of unwanted frequencies is possible with the tuned-line filters of Fig. 20. Examples are shown for each band from 50 through 450 MHz. Construction is relatively simple, and the cost is low. Standard boxes are used for ease of duplication.

The filter of Fig. 21 is selective enough to pass 50-MHz energy and attenuate the seventh harmonic of an 8-MHz oscillator that falls in TV channel 2. With an insertion loss at 50 MHz of about 1 dB, it can provide up to 40 dB of attenuation to energy at 57 MHz in the same line.

The filter uses a folded line in order to keep it within the confines of a standard chassis. The case is a 6 × 17 × 3-inch (152 × 432 × 76-mm) chassis (Bud AC-433) with a cover plate that fastens in place with self-tapping screws. An aluminum partition down the middle of the assembly is 14 inches (356-mm) long, and the full height of the chassis, 3 inches (76-mm).

The inner conductor of the line is 32 inches (813-mm) long and 13/16-inch (21-mm) wide, of 1/16-inch (2-mm) brass, copper or aluminum. This was made from two pieces of aluminum spliced together to provide the 32-inch (813-mm) length. Splicing seemed to have no ill effect on the circuit Q. The sides of the "U" are 2-7/8 inches (73-mm) apart, with the partition at the center. The line is supported on ceramic standoffs. These were shimmed

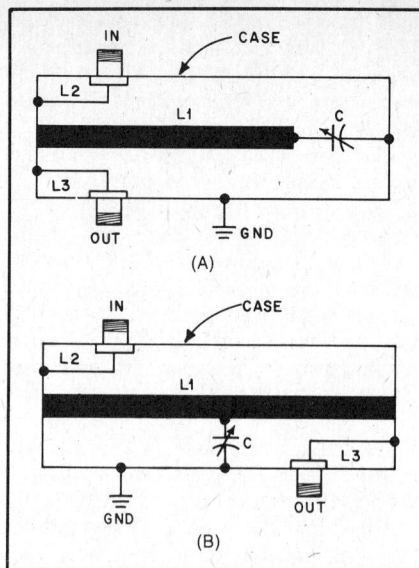

Fig. 20 — Equivalent circuits for the strip-line filters. At A, the circuit for the 6- and 2-meter filters are shown. L2 and L3 are the input and output links. These filters are bilateral, permitting interchanging of the input and output terminals. At B, the representative circuit for the 220- and 432-MHz filters. These filters are also bilateral.

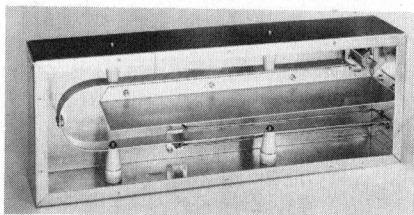

Fig. 21 — Interior of the 50-MHz strip-line filter. Inner conductor of aluminum strip is bent into U shape, to fit inside a standard 17-inch (432-mm) chassis.

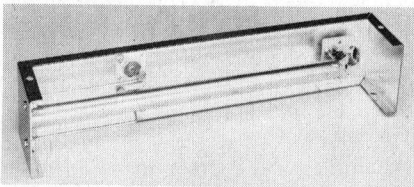

Fig. 22 — The 144-MHz filter has an inner conductor of 1/2-inch (13-mm) copper tubing 10 inches (254-mm) long, grounded to the left end of the case and supported at the right end by the tuning capacitor.

Fig. 23 — A half-wave strip line is used in the 220-MHz filter. It is grounded at both ends and tuned at the center.

up with sections of hard wood or bakelite rod, to give the required 1-1/2-inch (38-mm) height.

The tuning capacitor is a double-spaced variable (Hammarlund HF-30-X) mounted 1-1/2 inches (38-mm) from the right end of the chassis. Input and output coupling loops are of no. 10 or 12 wire, 10 inches (254-mm) long. Spacing away from the line is adjusted to about 1/4 inch (6-mm).

The 144-MHz model is housed in a 2-1/4 × 2-1/2 × 12-inch (57 × 64 × 305-mm) Minibox (Bud CU-2114-A).

One end of the tubing is slotted 1/4-inch (6-mm) deep with a hacksaw. This slot takes a brass angle bracket 1-1/2-inches (38-mm) wide, 1/4-inch (6-mm) high, with a 1/2-inch (13-mm) mounting lip. This 1/4-inch (6-mm) lip is soldered into the tubing slot, and the bracket is then bolted to the end of the box, so as to be centered on the end plate.

The tuning capacitor (Hammarlund HF-15-X) is mounted 1-1/4 inches (32-mm) from the other end of the box, in such a position that the inner conductor can be soldered to the two stator bars.

The two coaxial fittings (SO-239) are 11/16 inch (17-mm) in from each side of the box, 3-1/2 inches (89-mm) from the left end. The coupling loops are no. 12 wire, bent so that each is parallel to the center line of the inner conductor, and about 1/8 inch (3-mm) from its surface. Their cold ends are soldered to the brass mounting bracket.

The 220-MHz filter uses the same size box as the 144-MHz model. The inner conductor is 1/16-inch (2-mm) brass or copper, 5/8-inch (16-mm) wide, just long enough to fold over at each end for bolting to the box. It is positioned so that there will be 1/8 inch (3-mm) clearance between it and the rotor plates of the tuning capacitor. The latter is a Hammarlund HF-15-X, mounted slightly off-center in the box, so that its stator plates connect to the exact mid-point of the line. The 5/16-inch (8-mm) mounting hole in the case is 5-1/2 inches (140-mm) from one end. The SO-239 coaxial fittings are 1 inch (25-mm) in from opposite sides of the box, 2 inches (51-mm) from the ends. Their coupling links are no. 14 wire, 1/8 inch (3-mm) from the inner conductor of the line.

The 420-MHz filter is similar in design, using a 1-5/8 × 2 × 10-inch (41 × 51 × 254-mm) Minibox (Bud CU-2113-A). A half-wave line is used, with disc tuning at the center. The discs are 1/16-inch (2-mm) brass, 1-1/4-inch (32-mm) diameter. The fixed one is centered on the inner conductor, the other mounted on a no. 6 brass lead-screw. This passes through a threaded bushing, which can be taken from the end of a discarded slug-tuned form. An advantage of these is that usually a tension device is included. If there is none, use a lock nut.

Type N coaxial connectors were used on the 420-MHz model. They are 5/8 inch (16-mm) in from each side of the box, and

1-3/8 inches (35-mm) in from the ends. Their coupling links of no. 14 wire are 1/16 inch (2-mm) from the inner conductor.

Adjustment and Use

If you want the filter to work on both transmitting and receiving, connect the filter between antenna line and SWR indicator. With this arrangement you need merely adjust the filter for minimum reflected power reading on the SWR bridge. This should be zero, or close to it, if the antenna is well-matched. The bridge should be used, as there is no way to adjust the filter properly without it. If you insist on trying, adjust for best reception of signals on frequencies close to the ones you expect to transmit on. This works only if the antenna is well matched.

When the filter is properly adjusted (with the SWR bridge) you may find that reception can be improved by retuning the filter. Don't do it if you want the filter to work best on the job it was intended to do: The rejection of unwanted energy, transmitting or receiving. If you want to improve reception with the filter in the circuit, work on the receiver input circuit. To get maximum power out of the transmitter and into the line, adjust the transmitter output coupling, not the filter. If the effect of the filter on reception bothers you, connect it in the line from the antenna relay to the transmitter only.

Summary

The methods of harmonic elimination outlined here have been proved beyond doubt to be effective even under highly unfavorable conditions. It must be emphasized once more, however, that the problem must be solved one step at a time, and the procedure must be in logical order. It cannot be done properly without two items of simple equipment: A grid-dip meter and wavemeter covering the TV bands, and a dummy antenna.

To summarize:

1) Take a critical look at the transmitter on the basis of the design considerations outlined under "Reducing Harmonic Generation."

2) Check all circuits, particularly those connected with the final amplifier, with the grid-dip meter to determine whether there are any resonances in the TV bands. If so, rearrange the circuits so the resonances are moved out of the critical frequency region.

3) Connect the transmitter to the dummy antenna and check with the wavemeter for the presence of harmonics on leads and around the transmitter enclosure. Seal off the weak spots in the shielding and filter the leads until the wavemeter shows no indication at any harmonic frequency.

4) At this stage, check for interference with a TV receiver. If there is interference, determine the cause by the methods

described previously and apply the recommended remedies until the interference disappears.

5) When the transmitter is completely clean on the dummy antenna, connect it to the regular antenna and check for interference on the TV receiver. If the interference is not bad, a Transmatch or matching circuit installed as previously described should clear it up. Alternatively, a low-pass filter may be used. If neither the Transmatch nor filter makes any difference in the interference, the evidence is strong that the interference, at least in part, is being caused by receiver overloading because of the strong fundamental-frequency field about the TV antenna and receiver. A Transmatch and/or filter, installed as described above, will invariably make a difference in the intensity of the interference if the interference is caused by transmitter harmonics alone.

6) If there is still interference after installing the Transmatch and/or filter, and the evidence shows that it is probably caused by a harmonic, more attenuation is needed. A more elaborate filter may be necessary. However, it is well at this stage to assume that part of the interference may be caused by receiver overloading. Take steps to alleviate such a condition before trying highly elaborate filters and traps on the transmitter.

Harmonics by Rectification

Even though the transmitter is completely free from harmonic output it is still possible for interference to occur because of harmonics generated outside the transmitter. These result from rectification of fundamental-frequency currents induced in conductors in the vicinity of the transmitting antenna. Rectification can take place at any point where two conductors are in poor electrical contact, a condition that frequently exists in plumbing, downspouting, BX cables crossing each other, and numerous other places in the ordinary residence. It also can occur at any exposed vacuum tubes in the station, in power supplies, speech equipment, or other items which may not

be enclosed in the shielding about the rf circuits. Poor joints anywhere in the antenna system are especially bad, and rectification also may take place in the contacts of antenna changeover relays. Another common cause is overloading the front end of the communications receiver when it is used with a separate antenna (which will radiate the harmonics generated in the first tube) for break-in.

Rectification of this sort will not only cause harmonic interference but also is frequently responsible for cross-modulation effects. It can be detected in greater or less degree in most locations, but fortunately the harmonics thus generated are not usually of high amplitude. However, they can cause considerable interference in the immediate vicinity in fringe areas, especially when operation is in the 28-MHz band. The amplitude decreases rapidly with the order of the harmonic, the second and third being the worst. It is ordinarily found that even in cases where destructive interference results from 28-MHz operation the interference is comparatively mild from 14 MHz, and is negligible at still lower frequencies.

Nothing can be done at either the transmitter or receiver when rectification occurs. The remedy is to find the source and eliminate the poor contact either by separating the conductors or bonding them together. A crystal wavemeter (tuned to the fundamental frequency) is useful for hunting the source by showing which conductors are carrying rf and, comparatively, how much.

Interference of this kind is frequently intermittent since the rectification efficiency will vary with vibration, weather and so on. The possibility of corroded contacts in the TV receiving antenna should not be overlooked, especially if it has been up a year or more.

TV Receiver Deficiencies

When a television receiver is quite close to the transmitter, the intense rf signal from the transmitter's fundamental may overload one or more of the receiver circuits to produce spurious responses

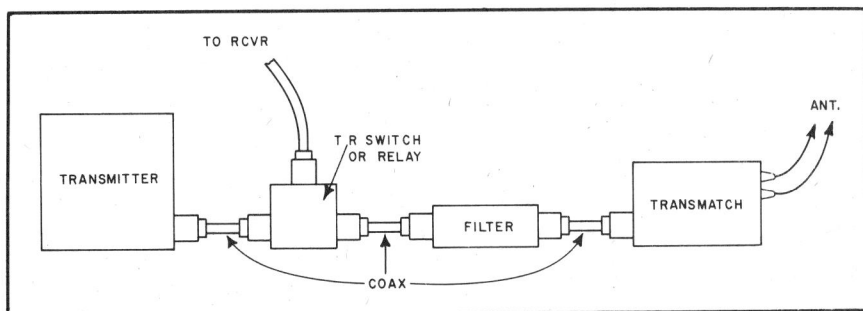

Fig. 24 — The proper method of installing a low-pass filter between the transmitter and a Transmatch. If the antenna is fed through coax, the Transmatch can be eliminated, but the transmitter and filter must be completely shielded. If a TR switch is used, it should be installed between the transmitter and low-pass filter. TR switches can generate harmonics themselves, so the low-pass filter should follow the TR switch.

which cause interference.

If the overload is moderate, the interference is of the same nature as harmonic interference; it is caused by harmonics generated in the early stages of the receiver and, since it occurs only on channels harmonically related to the transmitting frequency, it is difficult to distinguish from harmonics actually radiated by the transmitter. In such cases additional harmonic suppression at the transmitter will do no good, but any means taken at the receiver to reduce the strength of the amateur signal reaching the first tube will effect an improvement. With very severe overloading, interference also will occur on channels *not* harmonically related to the transmitting frequency, so such cases are easily identified.

Cross-Modulation

Under some circumstances overloading will result in cross-modulation or mixing of the amateur signal with that from a local fm or TV station. For example, a 14-MHz signal can mix with a 92-MHz fm station to produce a beat at 78 MHz and cause interference in channel 5, or with a TV station on channel 5 to cause interference in channel 3. Neither of the channels interfered with is in harmonic relationship to 14 MHz. Both signals have to be on the air for the interference to occur, and eliminating either at the TV receiver will eliminate the interference.

There are many combinations of this type, depending on the band in use and the local frequency assignments to fm and TV stations. The interfering frequency is equal to the amateur fundamental frequency either added to or subtracted from the frequency of some local station, and when interference occurs in a TV channel that is not harmonically related to the amateur transmitting frequency, the possibilities in such frequency combinations should be investigated.

I-f Interference

Some TV receivers do not have sufficient selectivity to prevent strong signals in the intermediate-frequency range from forcing their way through the front end and getting into the i-f amplifier. The once-standard intermediate frequency of, roughly, 21 to 27 MHz, is subject to interference from the fundamental-frequency output of transmitters operating in the 21-MHz band. Transmitters on 28 MHz sometimes will cause this type of interference as well.

A form of i-f interference peculiar to 50-MHz operation near the low edge of the band occurs with some receivers having the standard "41-MHz" i-f, which has the sound carrier at 41.25 MHz and the picture carrier at 45.75 MHz. A 50-MHz signal that forces its way into the i-f system of the receiver will beat with the i-f picture carrier to give a spurious signal

on or near the i-f sound carrier, even though the interfering signal is not actually in the nominal passband of the i-f amplifier.

There is a type of i-f interference unique to the 144-MHz band in localities where certain uhf TV channels are in operation. It affects only those TV receivers in which double-conversion type plug-in uhf tuning strips are used. The design of these strips involves a first intermediate frequency that varies with the TV channel to be received and, depending on the particular strip design, this first i-f may be in or close to the 144-MHz amateur band. Since there is comparatively little selectivity in the TV signal-frequency circuits ahead of the first i-f, a signal from a 144-MHz transmitter will "ride into" the i-f, even when the receiver is at a considerable distance from the transmitter. The channels that can be affected by this type of i-f interference are

Receivers with 21-MHz second i-f	Receivers with 41-MHz second i-f
Channels 14-18, incl.	Channels 20-25, incl.
Channels 41-48, incl.	Channels 51-58, incl.
Channels 69-77, incl.	Channels 82 and 83.

If the receiver is not close to the transmitter, a trap of the type shown in Fig. 27 will be effective. However, if the separation is small the 144-MHz signal will be picked up directly on the receiver circuits and the best solution is to readjust the strip oscillator so that the first i-f is moved to a frequency not in the vicinity of the 144-MHz band. This has to be done by a competent technician.

I-f interference is easily identified since it occurs on all channels — although sometimes the intensity varies from channel to channel — and the cross-hatch pattern it causes will rotate when the receiver's fine-tuning control is varied. When the interference is caused by a harmonic, overloading or cross modulation, the structure of the interference pattern does not change (its intensity may change) as the fine-tuning control is varied.

High-Pass Filters

In all of the above cases the interference can be eliminated if the fundamental signal strength can be reduced to a level that the receiver can handle. To accomplish this with signals on bands below 30 MHz, the most satisfactory device is a high-pass filter having a cutoff frequency between 30 and 54 MHz installed at the tuner input terminals of the receiver. Circuits that have proved effective are shown in Figs. 25 and 26. Fig. 26 has one more section than the filters of Fig. 25 and thus has somewhat better cutoff characteristics. All the circuits given are designed to have little or no effect on the TV signals, but will attenuate all signals lower in frequency than about 40 MHz. These filters preferably should be constructed in some sort of shielding container, although shielding is not always necessary. The dashed lines in Fig. 26 show how individual filter coils can be shielded from each other. The capacitors can be tubular ceramic units centered in holes in the partitions that separate the coils.

Simple high-pass filters cannot always

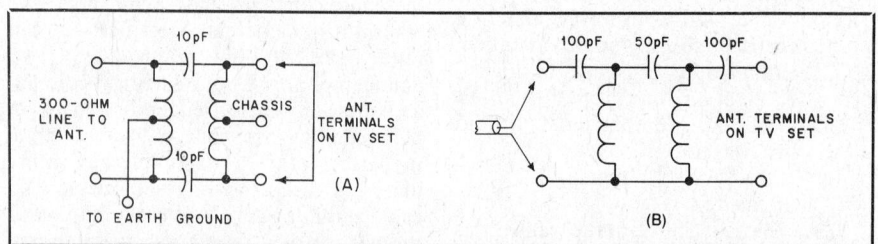

Fig. 25 — High-pass filters for installation at the TV receiver antenna terminals. Each coil 8 turns no. 14, 3/4-in. dia. 1-in. long (19 × 25 mm) tapped at center. B — For 75-ohm coaxial line. Each coil 3 turns no. 14, 3/4-in. (19-mm) dia. 8 turns per inch. *Important:* Do not use a direct ground on the chassis of a transformerless receiver. Ground through a 0.001-μF mica capacitor.

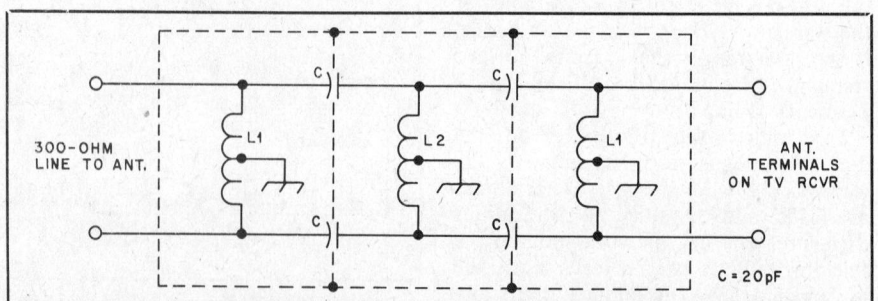

Fig. 26 — Another type of high-pass filter for 300-ohm line. The coils may be wound on 1/8-inch (3-mm) diameter plastic knitting needles. L1 = 40 turns no. 30 enam. close wound, 1/8-in. (3-mm) dia. L2 = 22 turns no. 30 enam. close-wound, 1/8-inch (3-mm) dia. *Important:* Do not use a direct ground on the chassis of a transformerless receiver. Ground through a 0.001-μF mica capacitor.

be applied successfully in the case of 50-MHz transmissions, because they do not have sufficiently sharp cutoff characteristics to give both good attenuation at 50-54 MHz and no attenuation above 54 MHz. A more elaborate design capable of giving the required sharp cutoff has been described (Ladd, "50-MHz TVI — Its Causes and Cures," *QST*, June and July, 1954). This article also contains other information useful in coping with the TVI problems peculiar to 50-MHz operation. As an alternative to such a filter, a high-Q wave trap tuned to the transmitting frequency may be used, suffering only the disadvantage that it is quite selective and therefore will protect a receiver from overloading over only a small range of transmitting frequencies in the 50-MHz band. A trap of this type is shown in Fig. 27. These "suck-out" traps, while absorbing energy at the frequency to which they are tuned, do not affect the receiver operation otherwise. The assembly should be mounted near the input terminals of the TV tuner and its case should be grounded to the TV set chassis. The traps should be tuned for minimum TVI at the transmitter operating frequency. An insulated tuning tool should be used for adjustment of the trimmer capacitors, since they are at a "hot" point and will show considerable body-capacitance effect.

High-pass filters are available commercially at moderate prices. In this connection, it should be understood by all parties concerned that while an amateur is responsible for *harmonic* radiation from his transmitter, it is no part of his responsibility to pay for or install filters, wave traps or other devices that may be required at the receiver to prevent interference caused by his *fundamental* frequency. Proper installation usually requires that the filter be installed right at the input terminals of the rf tuner of the TV set and not merely at the external antenna terminals, which may be at a considerable distance from the tuner. The question of cost is one to be settled between the set owner and the organization with which he deals. Don't overlook the possibility that the manufacturer of the TV receiver may supply a high-pass filter free of charge.

If the fundamental signal is getting into the receiver by way of the line cord a line filter such as those shown in Fig. 28 may help. To be most effective it should be installed inside the receiver chassis at the point where the cord enters, making the ground connections directly to the chassis at this point. It may not be so helpful if placed between the line plug and the wall socket unless the rf is actually picked up on the house wiring rather than on the line cord itself.

Antenna Installation

Usually, the transmission line between the TV receiver and the actual TV antenna will pick up a great deal more energy from

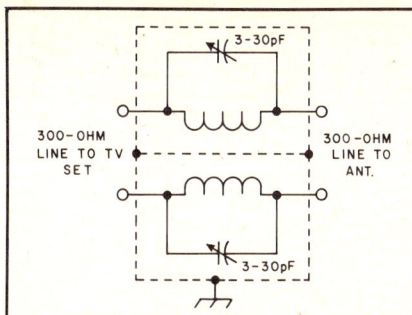

Fig. 27 — Parallel-tuned traps for installation in the 300-ohm line to the TV set. The traps should be mounted in an aluminum Minibox with a shield partition between them, as shown. For 50 MHz, the coils should have nine turns of no. 16 enamel wire, close wound to a diameter of 1/2 inch (13-mm). The 144-MHz traps should contain coils with a total of six turns of the same type wire, close-wound to a diameter of 1/4 inch (6 mm). Traps of this type can be used to combat fundamental-overload TVI on the lower-frequency bands as well.

a nearby transmitter than the television receiving antenna itself. The currents induced on the TV transmission line in this case are of the "parallel" type, where the phase of the current is the same in both conductors. The line simply acts like two wires connected together to operate as one. If the receiver's antenna input circuit were perfectly balanced it would reject these "parallel" or "unbalanced" signals and respond only to the true transmission-line ("push-pull") currents; that is, only signals picked up on the actual antenna would cause a receiver response. However, no receiver is perfect in this respect, and many TV receivers will respond strongly to such parallel currents. The result is that the signals from a nearby amateur transmitter are much more intense at the first stage in the TV receiver than they would be if the receiver response were confined entirely to energy picked up on the TV antenna alone. This situation can be improved by using shielded transmission line — coax or, in the balanced form, "twinax" — for the receiving installation. For best results the line should terminate in a coax fitting on

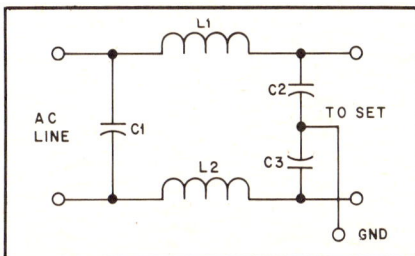

Fig. 28 — "Brute-force" ac line filter for receivers. The values of C1, C2 and C3 are not generally critical; capacitances from 0.001 to 0.01 µF can be used. L1 and L2 can be a 2-inch (51-mm) winding of no. 18 enameled wire on a 1/2-inch (13-mm) diameter form. In making up such a unit for use external to the receiver, make sure that there are no exposed conductors to offer a shock hazard.

the receiver chassis, but if this is not possible the shield should be grounded to the chassis right at the antenna terminals.

The use of shielded transmission line for the receiver also will be helpful in reducing response to harmonics actually being radiated from the transmitter or transmitting antenna. In most receiving installations the transmission line is very much longer than the antenna itself, and is consequently far more exposed to the harmonic fields from the transmitter. Much of the harmonic pickup, therefore, is on the receiving transmission line when the transmitter and receiver are quite close together. Shielded line, plus relocation of either the transmitting or receiving antenna to take advantage of directive effects, often will reduce overloading, as well as harmonic pickup, to a level that does not interfere with reception.

UHF Television

Harmonic TVI in the uhf TV band is far less troublesome than in the vhf band. Harmonics from transmitters operating below 30 MHz are of such high order that they would normally be expected to be quite weak; in addition, the components, circuit conditions and construction of low-frequency transmitters are such as to tend to prevent very strong harmonics from being generated in this region. However, this is not true of amateur vhf transmitters, particularly those working in the 144-MHz and higher bands. Here the problem is quite similar to that of the low vhf TV band with respect to transmitters operating below 30 MHz.

There is one highly favorable factor in uhf TV that does not exist in the most of the vhf TV band: If harmonics are radiated, it is possible to move the transmitter frequency sufficiently (within the amateur band being used) to avoid interfering with a channel that may be in use in the locality. By restricting operation to a portion of the amateur band that will not result in harmonic interference, it is possible to avoid the necessity of taking extraordinary precautions to prevent harmonic radiation.

The frequency assignment for uhf television consists of 70 channels (14 to 83, inclusive) in 6-MHz steps, beginning at 470 MHz and ending at 890 MHz. The harmonics from amateur bands above 50 MHz span the uhf channels as shown in Table 4. Since the assignment plan calls for a minimum separation of six channels between any two stations in one locality, there is ample opportunity to choose a fundamental frequency that will move a harmonic out of range of a local TV frequency.

Color Television

The color TV signal includes a subcarrier spaced 3.579545 MHz from the regular picture carrier (or 4.83 MHz from the low edge of the channel) for transmitting the color information. Harmonics which fall in the color subcarrier region can be expected to cause break-up of color in the received picture. This modifies the chart of Fig. 2 to introduce another "severe" region centering around 4.8 MHz measured from the low-frequency edge of the channel. Hence with color television reception there is less opportunity to avoid harmonic interference by choice of operating frequency. In other respects the problem of eliminating interference is the same as with black-and-white television.

Hi-Fi Interference

Since the introduction of stereo and high-fidelity receivers, interference to this type of home-entertainment device has become a severe problem for amateurs. Aside from placing the amateur antenna as far as possible from any hi-fi installation, there is little else that can be done at the amateur's ham shack. Most of the hi-fi gear now being sold has little or no filtering to prevent rf interference. In other words, corrective measures must be done at the hi-fi installation.

Hi-Fi Gear

Hi-fi gear can consist of a simple amplifier, with record or tape inputs and speakers. The more elaborate installations may have a tape deck, record player, fm and a-m tuners, an amplifier and two or more speakers. These units are usually connected together by means of shielded leads, and in most cases the speakers are positioned some distance from the amplifier via long leads. When such a setup is operated near an amateur station, say within a few hundred feet, there are two can reach the hi-fi installation to cause important paths through which rf energy interference.

Step number one is to try determining how the interference is getting into the hi-fi unit. If the volume control has no effect on the level of interference or very slight effect, the audio rectification of the amateur signal is taking place past the volume control, or on the output end of the amplifier. This is by far the most common type. It usually means that the amateur signal is being picked up on the speaker leads, or possibly on the ac line, and is then being fed back into the amplifier.

Experience has shown that most of the rf gets into the audio system via the speaker leads or the ac line, mostly the speaker leads. The amateur may find that on testing, the interference will only show up on one or two bands, or all of them. In hi-fi installations speakers are sometimes set up quite some distance from the amplifier. If the speaker leads happen to be resonant near an amateur band in use, there is likely to be an interference problem. The speaker lead will act as a resonant antenna and pick up the rf. One

Table 4

Harmonic Relationship — Amateur VHF Bands and the UHF TV Channels

Amateur Band	Harmonic	Fundamental Freq. Range	Channel Affected
144 MHz	4th	144.0-144.5	31
		144.5-146.0	32
		146.0-147.5	33
		147.5-148.0	34
	5th	144.0-144.4	55
		144.4-145.6	56
		145.6-146.8	57
		146.8-148.0	58
	6th	144.0-144.33	79
		144.33-145.33	80
		145.33-147.33	81
		147.33-148.0	82
220 MHz	3rd	220-220.67	45
		220.67-222.67	46
		222.67-224.67	47
		224.67-225	48
	4th	220-221	82
		221-222.5	83
420 MHz	2nd	420-421	75
		421-424	76
		424-427	77
		427-430	78
		430-433	79
		433-436	80

easy cure is to bypass the speaker terminals at the amplifier chassis. Use 0.01 to 0.03-μF disk capacitors from the speaker terminals directly to chassis ground; see Fig. 29. Try 0.01μF and see if that does the job. In some amplifiers 0.03 μF is required to eliminate the rf. Be sure to install bypasses on *all* the speaker terminals. In some instances, it may appear that one of each of the individual speaker terminals is grounded to the chassis. However, some amplifiers have the speaker leads above ground on the low side, for feedback purposes. If you have a circuit diagram of the amplifier you can check, but in the absence of a diagram, bypass all the terminals. If you can get into the amplifier, you can use the system shown in Fig. 30A.

In this system, two rf chokes are installed in series with the speaker leads from the output transformers, or amplifier output, to the speakers. These chokes are simple to make and help keep rf out of the amplifier. In particularly stubborn cases, use shielded wire for the speaker leads, grounding the shields at the amplifier chassis and still using the bypasses on the terminals. When grounding, all chassis used in the hi-fi installation should be bonded together and connected to a good earth ground (such as a water pipe) if at all possible. It has been found that grounding sometimes eliminates the interference. On the other hand, don't be discouraged if grounding doesn't appear to help. Even with the bypassing and filtering grounding may make the difference.

Fig. 30B shows the method for filtering the ac line at the input of the amplifier chassis. The choke dimensions are the

same as those given in Fig. 30A. Be sure that the bypasses are rated for ac because the dc types have been known to short out.

Antenna Pickup

If the hi-fi setup includes an fm installation, and many of them do, there is the possibility of rf getting into the audio equipment by way of the fm antenna. Chances for this method of entry are very good and precautions should be taken here to prevent the rf from getting to the equipment. A TV-type high-pass filter can prove effective in some cases.

Turntables and Tape Decks

In the more elaborate hi-fi setups, there may be several assemblies connected together by means of patch cords. It is a good idea when checking for RFI to disconnect the units, one at a time, observing any changes in the interference. Not only disconnect the patch cords connecting the pieces together, but also unplug the ac line cord for each item as you make the test. This will help you determine which section is the culprit.

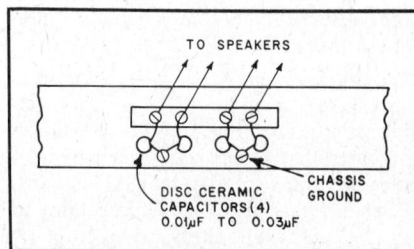

Fig. 29 — The disc capacitors should be mounted directly between the speaker terminals and chassis ground, keeping the leads as short as possible.

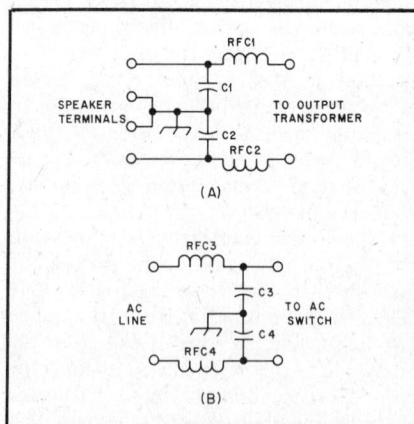

Fig. 30 — At A, the method for additional speaker filter, and at B, filtering and ac-line input. In both cases, these installations should be made directly inside the amplifier chassis, keeping the leads as short as possible.
C1, C2 — 0.01- to 0.03-μF disc ceramic.
C3, C4 — 0.01 disc ceramic, ac type.
RFC1 through RFC4 — 24 turns no. 18 enamel-covered wire, close-spaced and wound on a 1/4-inch (6-mm) diameter form (such as a pencil).

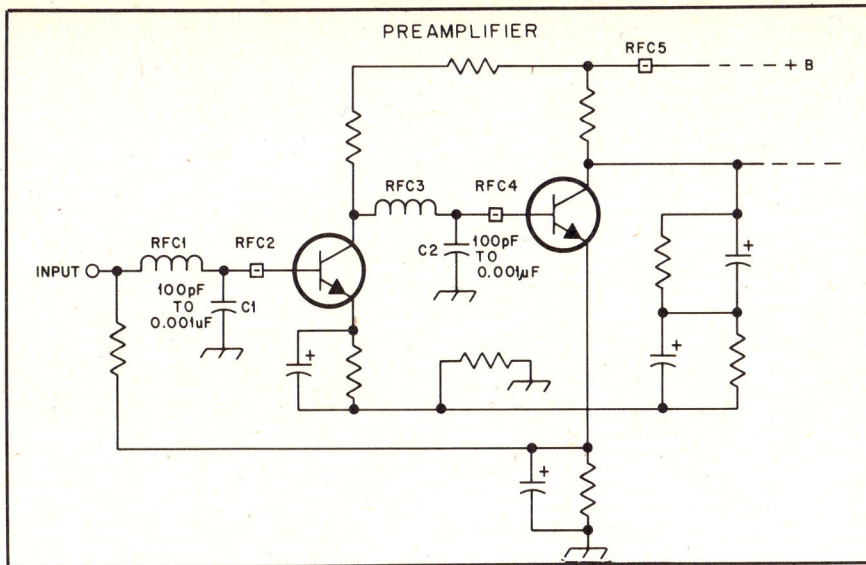

Fig. 31 — Typical circuit of a solid-state preamplifier.

Patch cords are usually, *but not always*, made of shielded cable. The lines *should* be shielded, which brings up another point. Many commercially available patch cords have poor shields. Some have wire spirally wrapped around the insulation, covering the main lead, rather than braid. This method provides poor shielding and could be the reason for RFI problems.

Record-player tone-arm connections to the cartridge are usually made with small clips. The existence of a loose clip, particularly if oxidation is present, offers an excellent invitation to RFI. Also, the leads from the cartridge and those to the amplifier are sometimes resonant at vhf, providing an excellent receiving antenna for rf. One cure for unwanted rf pickup is to install ferrite beads, one on each cartridge lead. Check all patch-cord connections for looseness or poor solder joints. Inferior connections can cause rectification and subsequent RFI.

Tape decks should be treated the same as turntables. Loose connections and bad solder joints all can cause trouble. Ferrite beads can be slipped over the leads to the recording and play-back pickup heads. Bypassing of the tone-arm or pickup-head leads is also effective, but sometimes it is difficult to install capacitors in the small area available. Disc capacitors (0.001 μF) should be used as close to the cartridge or pickup head as possible. Keep the capacitor leads as short as possible.

Preamplifiers

There are usually one or more preamplifiers used in a hi-fi amplifier. The inputs to these stages can be very susceptible to RFI. Fig. 31 illustrates a typical preamplifier circuit. In this case the leads to the bases of the transistors are treated for RFI with ferrite beads by the addition of RFC2 and RFC4. This is a very effective method for stopping RFI when vhf energy is the source of the trouble.

Within the circuit of a solid-state audio system, a common offender can be the emitter-base junction of a transistor. This junction operates as a forward-biased diode, with the bias set so that a change of base current with signal will produce a linear but amplified change in collector current. Should rf energy reach the junction, the bias could increase, causing nonlinear amplification and distortion as the result. If the rf level is high it can completely block (saturate) a transistor, causing a complete loss of gain. Therefore, it may be necessary to reduce the transmitter power output in order to pinpoint the particular transistor stage that is affected.

In addition to adding ferrite beads it may be necessary to bypass the base of the transistor to chassis ground, C1 and C2, Fig. 31. A suitable value is 100 pF, and keep the leads short! As a general rule, the capacitor value should be as large as possible without degrading the high-frequency response of the amplifier. Values up to 0.001 μF can be used. In severe cases, a series inductor (RFC1 and RFC3) may be required, such as the Ohmite Z-50 or Z-144, or their equivalents (7 and 1.8 μH, respectively). Fig. 31 shows the correct placement for an inductor, bypass capacitor, and ferrite bead. Also, it might help to use a ferrite bead in the plus-B lead to the preamplifier stages (RFC5 in Fig. 31). Keep in mind that Fig. 31 represents only one preamplifier of a stereo set. *Both* channels may require treatment.

FM Tuners

There is often an fm tuner used in a hi-fi installation. Much of the interference to tuners is caused by fundamental overloading of the first stage (or stages) of the tuner, effected by the amateur's signal. The cure is the installation of a high-pass filter, the same type used for TVI. The filter should be installed as close as possible to the antenna input of the tuner. The high-pass filter will attenuate the amateur *fundamental* signal, thus preventing overloading of the front end.

Shielding

Lack of shielding on the various components in a hi-fi installation can permit rf to get into the equipment. Many units have no bottom plates, or are installed in plastic cases. One easy method of providing shielding is to use aluminum foil. Make sure the foil doesn't short circuit the components, and connect it to chassis ground.

Interference with Standard Broadcasting

Transmitter Defects

Out-of-band radiation is something that must be cured at the transmitter. Parasitic oscillations are a frequently unsuspected source of such radiations, and no transmitter can be considered satisfactory until it has been thoroughly checked for both low- and high-frequency parasitics. Very often parasitics show up only as transients, causing key clicks in cw transmitters and "splashes" or "burps" on modulation peaks in a-m transmitters. Methods for detecting and eliminating parasitics are discussed in the transmitter chapter.

In cw transmitters the sharp make and break that occurs with unfiltered keying causes transients that, in theory, contain frequency components through the entire radio spectrum. Practically, they are often strong enough in the immediate vicinity of the transmitter to cause serious interference to broadcast reception. Key clicks can be eliminated by the methods detailed in the chapter on keying.

BCI is frequently made worse by radiation from the power wiring or the rf transmission line. This is because the signal causing the interference, in such cases, is radiated from wiring that is nearer the broadcast receiver than the antenna itself. Much depends on the method used to couple the transmitter to the antenna, a subject that is discussed in the chapters on transmission lines and antennas. If it is at all possible the antenna itself should be placed so that it is not in close proximity to house wiring, telephone and power lines, and similar conductors.

The BC Set

Most present day receivers use solid-state active components, rather than tubes. A large number of the receivers in use are battery powered. This is to the amateur's advantage because much of the bc interference an amateur encounters is due to ac line pickup. In the case where the bc receiver is powered from the ac line, whether using tube or solid-state components, the amount of rf pickup must be reduced or eliminated. A line filter such as is shown in Fig. 28 often will help accomplish this. The values used for the coils and capacitors are in general not critical. The effectiveness of the filter may depend considerably on the ground connection used, and it is advisable to use a short ground lead to a cold-water pipe if at all possible. The line cord from the set should be bunched up to minimize the possibility of pick-up on the cord. It may be necessary to install the filter inside the receiver, so that the filter is connected between the line cord and the set wiring, in order to get satisfactory operation.

Cross-Modulation

With phone transmitters, there are occasionally cases where the voice is heard whenever the broadcast receiver is tuned to a bc station, but there is no interference when tuning between stations. This is cross-modulation, a result of rectification in one of the early stages of the receiver. Receivers that are susceptible to this trouble usually also get a similar type of interference from regular broadcasting if there is a strong local bc station and the receiver is tuned to some *other* station.

The remedy for cross modulation in the receiver is the same as for images and oscillator-harmonic response — reduce the strength of the amateur signal at the receiver by means of a line filter.

The trouble is not always in the receiver, since cross modulation can occur in any nearby rectifying circuit — such as a poor contact in water or steam piping, gutter pipes, and other conductors in the strong field of the transmitting antenna — external to both receiver and transmitter. Locating the cause may be difficult, and is best attempted with a battery-operated portable broadcast receiver used as a "probe" to find the spot where the interference is most intense. When such a spot is located, inspection of the metal structures in the vicinity should indicate the cause. The remedy is to make a good electrical bond between the two conductors having the poor contact.

Handling BCI Cases

Tune the receiver through the broadcast band to see whether the interference tunes like a regular bc station. If so, image or oscillator-harmonic response is the cause. If there is interference only when a bc station is tuned in, but not between stations, the cause is cross-modulation. If the interference is heard at all settings of the tuning dial, the trouble is pickup in the audio circuits. In the latter case, the receiver's volume control may or may not affect the strength of the interference, depending on the means by which your signal is being rectified.

Organs

The electronic organ is an RFI problem

area. All of the techniques outlined for hi-fi gear hold true in getting rid of RFI in an organ. Two points should be checked — the speaker leads and the ac line. Many organ manufacturers have special servicemen's guides for taking care of RFI. However, to get this information you or the organ owner must contact the manufacturer, not the dealer or distributor. Don't accept the statement from a dealer or serviceman that there is nothing that can be done about the interference.

P-A Systems

The cure for RFI in p-a systems is almost the same as that for hi-fi gear. The one thing to watch for is rf on the leads that connect the various stations in a p-a system. These leads should be treated the same as speaker *leads* and bypassing and filtering should be done at *both* ends of the lines. Also, watch for ac-line pickup of rf.

Telephone Interference

Telephone interference may be cured by connecting a bypass capacitor (about $0.001\ \mu F$) across the microphone unit in the telephone handset. The telephone companies have capacitors for this purpose. When such a case occurs, get in touch with the repair department of the phone company, giving the particulars. Section 500-150-100 of the Bell System Practices *Plant Series* gives detailed instructions. This section discusses causes and cures of telephone interference from radio signals. It points out that interference can come from corroded connections, unterminated loops, and other sources. It correctly points out that the rf can be picked up on the drop wire coming into the house, and also on the wiring within the house, but the rf detection usually occurs inside the phone. The detection usually takes place at the varistors in the compensation networks, and/ or at the receiver noise suppressor and the carbon microphone. But interference suppression should be handled two ways: Prevent the rf from getting to the phone, and prevent it from being rectified.

The telephone companies (Bell System) have two devices for this purpose. The first is a 40BA capacitor, which is installed at the service entrance protector, and the second is the 1542A inductor, which is installed at the connector block. According to the practices manual, the 40BA bypasses rf picked up on the drop wire coming into the house from the phone, and the 1542A suppresses rf picked up on the inside wiring. These are mentioned because in very stubborn cases they may be necessary. But the telephone should be modified first.

Since there are several different series of phones, they will be discussed separately: 500 series — These are the desk and

wall phones most commonly in use. They come in several different configurations, but all use 425-series compensation network. The letter designation can be A, B, C, D, E, F, G or K, and all these networks contain varistors. The network should be replaced with a 425J, in which the varistors are replaced by resistors. Also, $0.01\text{-}\mu F$ disc-ceramic capacitors should be placed across the receiver suppressor. The suppressor is a diode across the receiver terminals. The carbon microphone in the handset should be bypassed with a $0.01\text{-}\mu F$ ceramic capacitor.

Series 1500, 1600, 1700 — These are the "Touch-Tone" phones, and the cure is similar to that for the 500 series, except that the network is a 4010B or -D, and should be replaced with a 4010E.

Trimline series — These are the "Princess" series phones. The practice manual says that these should be modified by installing bypass capacitors across *all* components in the set which may act as demodulators. This statement is rather vague, but evidently the telephone company is aware of a solution.

At the end of section 500-150-100 is an ordering guide for special components and sets, as follows:

Ordering Guide
Capacitor, 40BA
Inductor, 1542A
    -49 Gray, -50 Ivory
    Set, Telephone, -rf Modified
    Set, Telephone Hand, 220A, -rf Modified
    Set, Telephone Hand, 2220B, -rf Modified
    Set, Hand G, -rf Modified
    Dial — (Touch-Tone dial only), -rf Modified.
The type "G" handset is the one used with the 500 and Touch-Tone series phones. Also, Mountain Bell has put out an "Addendum 500-150-100MS, Issue A, January 1971" to the practices manual, which states that items for rf modified phones should be ordered on nonstock form 3218, as follows:
(Telephone Set type)
Modified for BSP 500-150-100
for Radio Signal Suppression

Additional Information

In response to the many hundreds of thousands of RFI-related complaints it has received in recent years, the FCC has produced a booklet designed to show how to solve common RFI problems before they become serious. Entitled *How to Identify and Resolve Radio-TV Interference Problems,* it is available for $1.50 from Consumer Information Center, Dept. 051F, Pueblo, CO 81009. Make check payable to Superintendent of Documents. The ARRL publication *Radio Frequency Interference* which sells for $3, covers all aspects of RFI and includes the complete FCC booklet.

Chapter 16

Test Equipment and Measurements

Measurement and testing seemingly go hand in hand, but it is useful to make a distinction between "measuring" and "test" equipment. The former is commonly considered to be capable of giving a meaningful quantitative result. For the latter a simple indication of "satisfactory" or "unsatisfactory" may suffice; in any event, the accurate calibration associated with real measuring equipment is seldom necessary, for simple test apparatus.

Certain items of measuring equipment that are useful to amateurs are readily available in kit form, at prices that represent a genuine saving over the cost of identical parts. Included are volt-ohm-milliammeter combinations, vacuum-tube and transistor voltmeters, oscilloscopes and the like. The coordination of electrical and mechanical design, components, and appearance make it far preferable to purchase such equipment than to attempt to build one's own.

However, some test gear is either not available or can easily be built. This chapter considers the principles of the more useful types of measuring equipment and concludes with the descriptions of several pieces that not only can be built satisfactorily at home but which will facilitate operation of the amateur station.

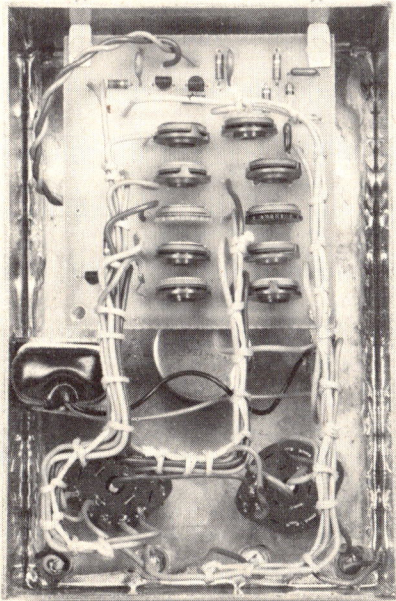

The Direct-Current Instrument

In measuring instruments and test equipment suitable for amateur purposes the ultimate "readout" is generally based on a measurement of direct current. A meter for measuring dc uses electromagnetic means to deflect a pointer over a calibrated scale in proportion to the current flowing through the instrument.

In the *D'Arsonval* type a coil of wire, to which the pointer is attached, is pivoted between the poles of a permanent magnet. When current flows through the coil it sets up a magnetic field that interacts with the field of the magnet to cause the coil to turn. The design of the instrument is usually such as to make the pointer deflection directly proportional to the current.

A less expensive type of instrument is the *moving-vane* type, in which a pivoted soft-iron vane is pulled into a coil of wire by the magnetic field set up when current flows through the coil. The farther the vane extends into the coil the greater the magnetic pull on it, for a given change in current. This type of instrument thus does not have "linear" deflection — the intervals of equal current are crowded together at the low-current end and spread out at the high-current end of the scale.

Current Ranges

The *sensitivity* of an instrument is usually expressed in terms of the current required for full-scale deflection of the pointer. Although a very wide variety of ranges is available, the meters of interest in amateur work have basic "movements" which will give maximum deflection with currents measured in microamperes or milliamperes. They are called *microammeters* and *milliammeters*, respectively.

Thanks to the relationships between current, voltage and resistance expressed by Ohm's Law, it becomes possible to use a single low-range instrument — e.g., 1 milliampere or less full-scale pointer deflection — for a variety of direct-current measurements. Through its ability to measure current, the instrument can also be used indirectly to measure voltage. Likewise, a measurement of *both* current

You can build this FET Volt-ohmmeter with the information that appears in this chapter. See page 16-5.

and voltage will obviously yield a value of resistance. These measurement functions are often combined in a single instrument — the *volt-ohm-milliammeter* or "VOM," a multirange meter that is one of the most useful pieces of measuring and test equipment an amateur can possess.

Accuracy

The accuracy of a dc meter of the D'Arsonval type is specified by the manufacturer. A common specification is "2 percent of full scale," meaning that a 0-100 microammeter, for example, will be correct to within 2 microamperes at any part of the scale. There are very few cases in amateur work where accuracy greater than this is needed. However, when the instrument is part of a more complex measuring circuit, the design and components of which all can cause error, the overall accuracy of the complete device is always less.

Extending the Current Range

Because of the way current divides between two resistances in parallel, it is possible to increase the range (more specifically, to decrease the sensitivity) of a dc micro- or milliammeter to any desired extent. The meter itself has an inherent resistance — its *internal resistance*— which determines the full-scale current through it when its rated voltage is applied. (This rated voltage is of the order of a few millivolts.) By connecting an external resistance in parallel with the internal resistance, as in Fig. 1, the current will divide between the two, with the meter responding only to that part of the current which flows through the internal resistance of its movement. Thus it reads only part of the total current; the effect is to make more total current necessary for a full-scale meter reading. The added resistance is called a shunt.

It is necessary to know the meter's internal resistance before the required value for a shunt can be calculated. It may vary from a few ohms to a few hundred, with the higher resistance values associated with higher sensitivity. When known, it can be used in the formula below to determine the required shunt for a given current multiplication:

$$R = \frac{R_m}{n - 1}$$

where:
    R = the shunt
    R_m = internal resistance
    n = the factor by which the original meter scale is to be multiplied

Quite often the internal resistance of a particular meter will be unknown. This is usually the case when the meter is purchased at a flea market or is obtained from a commercial piece of equipment. Unfortunately, the internal resistance of the meter can not be directly measured with a VOM or VTVM without risk of

injury to the meter movement. In most cases, the currents in the measuring equipment are high enough to damage the delicate meter movement.

Fig. 2 illustrates a method that can be used to safely determine the internal resistance of a meter. A calibrated meter capable of measuring the same current as the unknown meter is required. The system works as follows: S1 is placed in the open position and R2 is set for maximum resistance. A supply of constant voltage is connected to the terminals + and − (a battery will work fine) and R2 is adjusted so that the unknown meter reads exactly full scale. Note the current shown on M2. Close S1 and alternately adjust R1 and R2 so that the unknown meter (M1) reads exactly half scale and the known meter (M2) reads the same value as in the step above. At this point the current in the circuit is divided in half. Half of the current flows through M1 and half through R1. To determine the internal resistance of the meter simply open S1 and read the resistance of R1 with a VTVM, VOM or digital volt-ohmmeter.

The values for R1 and R2 will depend on the meter sensitivity and the voltage of the supply. The maximum resistance value for R1 should be approximately twice the expected internal resistance of the meter. For highly sensitive meters (100μA and less) 1000 ohms should be adequate. For less sensitive meters 100 ohms should suffice.

The value for minimum resistance at R2 can be calculated using Ohm's Law. For example, if the meter is a 0-1 mA type and the supply is a 1.5-volt battery, the minimum resistance required at R2 will be

$$R_2 = \frac{1.5}{0.001}$$

$$R_{2(min)} = 1500 \text{ ohms}$$

In practice a 2000- or 2500-ohm potentiometer would be used.

Making Shunts

Homemade shunts can be constructed from any of various special kinds of resistance wire, or from ordinary copper wire if no resistance wire is available. The copper wire table in this handbook gives the resistance per 1000 feet (305 m) for various sizes of copper wire. After computing the resistance required, determine the smallest wire size that will carry the full-scale current (250 circular mils per ampere is a satisfactory figure for this purpose). Measure off enough wire to provide the required resistance. A 1- or 2-watt carbon resistor makes an excellent form on which to wind the wire.

The Voltmeter

If a large resistance is connected in *series* with a current-reading meter, as in Fig. 3, the current multiplied by the resistance will be the voltage drop across

the resistance. This is known as a *multiplier*. An instrument used in this way is calibrated in terms of the voltage drop across the multiplier resistor and is called a *voltmeter*.

Sensitivity

Voltmeter sensitivity is usually expressed in *ohms per volt*, meaning that the meter *full-scale* reading multiplied by the sensitivity will give the total resistance of the voltmeter. For example, the resistance of a 1000-ohms-per-volt voltmeter is 1000 times the full-scale calibration voltage, and by Ohm's Law the current required for full-scale deflection is 1 milliampere. A sensitivity of 20,000 ohms per volt, a commonly used value, means that the instrument is a 50-microampere meter.

The higher the resistance of the

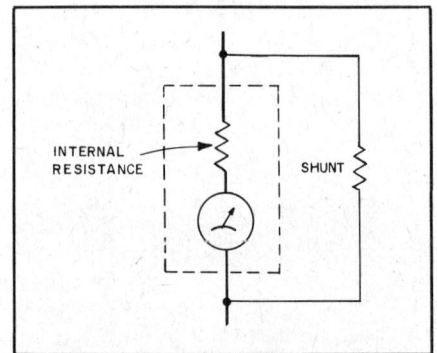

Fig. 1 — Use of a shunt to extend the calibration range of a current-reading instrument.

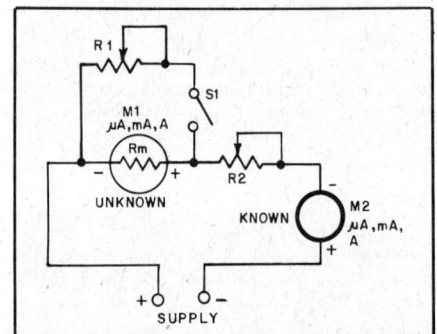

Fig. 2 — A safe method for determining the internal resistance of a meter.

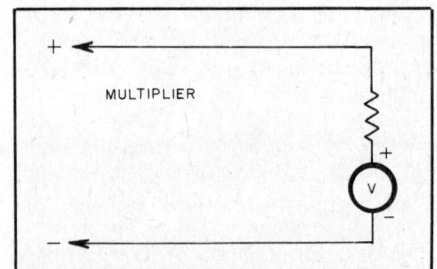

Fig. 3 — A voltmeter is a current-indicating instrument in series with a high resistance, the "multiplier."

voltmeter, the more accurate the measurements, especially in high-resistance circuits. Current flowing through the voltmeter will cause a change in the voltage between the points where the meter is connected, compared with the voltage with the meter absent. This is illustrated in Fig. 4.

Multipliers

The required multiplier resistance is found by dividing the desired full-scale voltage by the current, in amperes, required for full-scale deflection of the

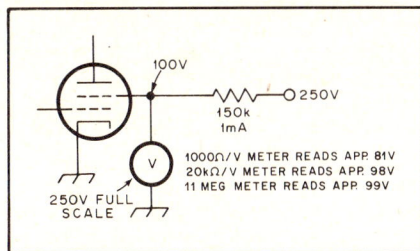

Fig. 4 — Effect of voltmeter resistance on accuracy of readings. It is assumed that the dc resistance of the screen circuit is constant at 100 kilohms. The actual current and voltage without the voltmeter connected are 1 mA and 100 volts. The voltmeter readings will differ because the different types of meters draw different amounts of current through the 150-kilohm resistor.

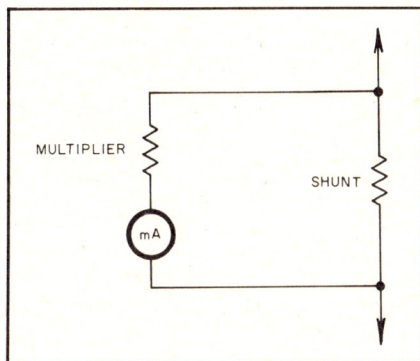

Fig. 5 — Voltmeter method of measuring current. This method permits using relatively large values of resistance in the shunt, standard values of fixed resistors frequently being usable. If the multiplier resistance is 20 (or more) times the shunt resistance, the error in assuming that all the current flows through the shunt will not be of consequence in most practical applications.

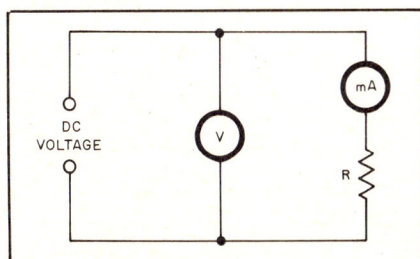

Fig. 6 — Measurement of power requires both current and voltage measurements; once these values are known the power is equal to the product $P = EI$. The same circuit can be used for measurement of an unknown resistance.

meter alone. Strictly, the internal resistance of the meter should be subtracted from the calculated value but this is seldom necessary (except perhaps for very low ranges), since the meter resistance will be negligibly small compared with the multiplier resistance. An exception is when the instrument is already a voltmeter and is provided with an internal multiplier, in which case the multiplier resistance required to extend the range is $R = R_m(n - 1)$
where
  R_m = total resistance of the instrument
  n = factor by which the scale is to be multiplied

For example, if a 1000-ohms-per-volt voltmeter having a calibrated range of 0-10 volts is to be extended to 1000 volts, R_m is $1000 \times 10 = 10{,}000$ ohms, n is $1000/10 = 100$, and $R = 10{,}000 (100 - 1) = 990{,}000$ ohms.

When extending the range of a voltmeter or converting a low-range meter into a voltmeter, the rated accuracy of the instrument is retained only when the multiplier resistance is precise. Precision wire-wound resistors are used in the multipliers of high-quality instruments. These are relatively expensive, but the home constructor can do quite well with 1-percent tolerance composition resistors. They should be "derated" when used for this purpose — that is, the actual power dissipated in the resistor should not be more than 1/4 to 1/2 the rated dissipation — and care should be used to avoid overheating the body of the resistor when soldering to the leads. These precautions will help prevent permanent change in the resistance of the unit.

Ordinary composition resistors are generally furnished in 10- or 5-percent tolerance ratings. If possible errors of this order can be accepted, resistors of this type may be used as multipliers. They should be operated below the rated power dissipation figure, in the interests of long-time stability.

DC Measurement Circuits: the Voltmeter

A current-measuring instrument should have very low resistance compared with the resistance of the circuit being measured; otherwise, inserting the instrument will cause the current to differ from its value with the instrument out of the circuit. The resistance of many circuits in radio equipment is high and the circuit operation is affected little, if at all, by adding as much as a few hundred ohms in series. In such cases the voltmeter method of measuring current, shown in Fig. 5, is frequently convenient. A voltmeter (or low-range milliammeter provided with a multiplier and operating as a voltmeter) having a full-scale voltage range of a few volts is used to measure the voltage drop across a suitable value of resistance acting as a shunt.

The value of shunt resistance must be calculated from the known or estimated maximum current expected in the circuit (allowing a safe margin) and the voltage required for full-scale deflection of the meter with its multiplier.

Power

Power in direct-current circuits is determined by measuring the current and voltage. When these are known, the power is equal to the voltage in volts multiplied by the current in amperes. If the current is measured with a milliammeter, the reading of the instrument must be divided by 1000 to convert it to amperes.

The setup for measuring power is shown in Fig. 6, where R is any dc "load," not necessarily an actual resistor.

Resistance

Obviously, if both voltage and current are measured in a circuit such as that in Fig. 6 the value of resistance R (in case it is unknown) can be calculated from Ohm's Law. For accurate results the internal resistance of the ammeter or milliammeter, mA, should be very low compared with the resistance, R, being measured, since the voltage read by the voltmeter, V, is the voltage across mA and R in series. The instruments and the dc voltage should be chosen so that the readings are in the upper half of the scale, if possible, since the percentage error is less in this region.

The Ohmmeter

Although Fig. 6 suffices for occasional resistance measurements, it is inconvenient when frequent measurements over a wide range of resistance are to be made. The device generally used for this purpose is the *ohmmeter*. This consists fundamentally of a voltmeter (or milliammeter, depending on the circuit used) and a small battery, the meter being calibrated so the value of an unknown resistance can be read directly from the scale. Typical ohmmeter circuits are shown in Fig. 7. In the simplest type, shown in Fig. 7 A, the meter and battery are connected in series with the unknown resistance. If a given deflection is obtained with terminals A-B shorted, inserting the resistance to be measured will cause the meter reading to decrease. When the resistance of the voltmeter is known, the following formula can be applied:

$$R = \frac{eR_m}{E} - R_m$$

where
  R = resistance to be found
  e = voltage applied (A-B shorted)
  E = voltmeter reading with R connected, and
  R_m = resistance of the voltmeter.

The circuit of Fig. 7A is not suited to measuring low values of resistance (below a hundred ohms or so) with a high-resistance voltmeter. For such measurements the circuit of Fig. 7B can be used. The unknown resistance is

$$R = \frac{I_2 R_m}{I_1 - I_2}$$

The formula is based on the assumption that the current in the complete circuit will be essentially constant whether or not the "unknown" terminals are short-circuited. This requires that R1 be very large compared with R_m e.g., 300 ohms for a 1-mA meter having an internal resistance of perhaps 50 ohms. A 3-volt battery would be necessary in this case in order to obtain a full-scale deflection with the "unknown" terminals open. R1 can be an adjustable resistor, to permit setting the open-terminals current to exact full scale.

A third circuit for measuring resistance is shown in Fig. 7C. In this case a high-resistance voltmeter is used to measure the voltage drop across a reference resistor, R2, when the unknown resistor is connected so that current flows through it, R2 and the battery in series. By suitable choice of R2 (low values for low-resistance, high values for high-resistance unknowns) this circuit will give equally good results on all resistance values in the range from one ohm to several megohms, provided that the voltmeter resistance, R_m, is always very high (50 times or more) compared with the resistance of R2. A 20,000-ohm-per-volt instrument (50- A movement) is generally used. Assuming that the current through the voltmeter is negligible compared with the current through R2, the formula for the unknown is

$$R = \frac{eR2}{E} - R2$$

where

R and R2 are as shown in Fig. 7C.

e = the voltmeter reading with A-B open circuited

E = voltmeter reading with R connected.

The "zero adjuster," R_1, is used to set the voltmeter reading exactly to full scale when the meter is calibrated in ohms. A 10,000-ohm variable resistor is suitable with a 20,000-ohms-per-volt meter. The battery voltage is usually 3 volts for ranges up to 100,000 ohms or so and 6 volts for higher ranges.

Bridge Circuits

An important class of measurement circuits is the bridge. A desired result is obtained by balancing the voltages at two different points in the circuit against each other so that there is zero potential difference between them. A voltmeter bridged between the two points will read zero (null) when this balance exists, but will indicate some definite value of voltage when the bridge is not balanced.

Bridge circuits are useful both on direct current and on ac of all frequencies. The majority of amateur applications are at radio frequencies, as shown later in this chapter. However, the principles of bridge operation are most easily introduced in terms of dc, where the bridge takes its simplest form.

The Wheatstone Bridge

The simple resistance bridge, known as the Wheatstone bridge, is shown in Fig. 8. All other bridge circuits — some of which are rather elaborate, especially those designed for ac — derive from this. The four resistors, R1, R2, R3, and R4 shown in A, are known as the bridge *arms*. For the voltmeter reading to be zero, the voltages across R3 and R4 in series must add algebraically to zero; that is, E1 must equal E2. R1R3 and R2R4 form voltage dividers across the dc source, so that if

$$\frac{R3}{R1 + R3} = \frac{R4}{R2 + R4}$$
$$E1 = E3$$

The circuit is customarily drawn as shown at 8B when used for resistance measurement. The equation above can be rewritten

$$R_x = R_s \frac{R2}{R1}$$

to find R_x, the unknown resistance. R1 and R2 are frequently made equal; then the calibrated adjustable resistance (the *standard*), R_s, will have the same value as R_x when R_s is set to show a null on the voltmeter.

Note that the resistance *ratios*, rather than the actual resistance values, determine the voltage balance. However, the values do have important practical effects on the sensitivity and power consumption. The *bridge sensitivity* is the readiness with which the meter responds to small amounts of unbalance about the null point; the "sharper" the null the more accurate the setting of R_s at balance.

The Wheatstone bridge is rarely used by amateurs for resistance measurement, the ohm-meter being the favorite instrument for that purpose. However, it is worthwhile to understand its operation because it is the basis of more complex bridges.

Electronic Voltmeters

It has been pointed out that for many purposes the resistance of a voltmeter must be extremely high in order to avoid "loading" errors caused by the current that necessarily flows through the meter. This tends to cause difficulty in measuring relatively low voltages (under perhaps 1000 volts) because a meter movement of given sensitivity takes a progressively smaller multiplier resistance as the voltage range is lowered.

The voltmeter resistance can be made independent of the voltage range by using vacuum tubes or field-effect transistors as electronic dc amplifiers between the circuit being measured and the actual indicator, which is usually a conventional

meter movement. As the input resistance of the electronic devices is extremely high — hundreds of megohms — they have essentially no loading effect on the circuit to which they are connected. They do, however, require a closed dc path in their input circuits (although this path can have very high resistance) and are limited in the amplitude of voltage that their input circuits can handle. Because of this, the

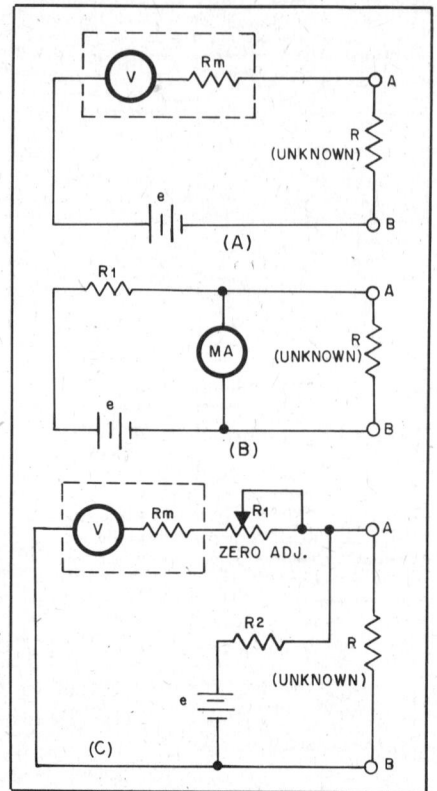

Fig. 7 — Ohmmeter circuits. Values are discussed in the text.

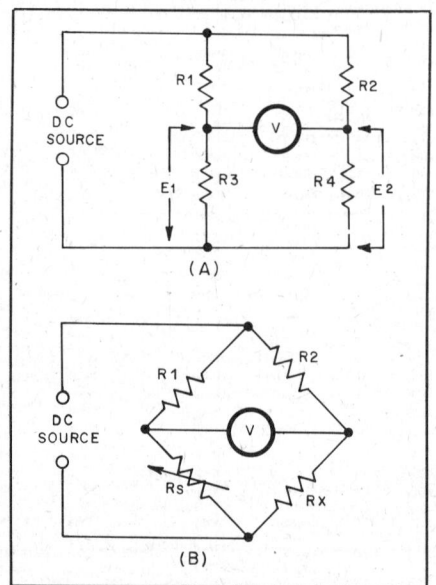

Fig. 8 — The Wheatstone bridge circuit. It is frequently drawn as at (B) for emphasizing its special function.

device actually measures a small voltage across a portion of a high-resistance voltage divider connected to the circuit being measured. Various voltage ranges are obtained by appropriate taps on the voltage divider.

In the design of electronic voltmeters it has become practically standard to use a voltage divider having a resistance of 10 megohms, tapped as required, in series with a 1-megohm resistor incorporated in a *probe* that makes the actual contact with the "hot" side of the circuit under measurement. The total voltmeter resistance, including probe, is therefore 11 megohms. The probe resistor serves to isolate the voltmeter circuit from the "active" circuit.

AN FET VOLT-OHMMETER

The circuit shown in Fig. 10 makes use of two field-effect transistors in a balanced circuit. Since no two active devices have exactly the same characteristics, some means must be incorporated to balance the circuit under static conditions. The ZERO potentiometer does just that since the meter will read exactly "0" when the circuit is balanced. Any imbalance causes the meter

Fig. 9 — Exterior view of the FET volt-ohmmeter.

to deflect, the amount of deflection pro-

portional to the degree of imbalance.

Voltage scales for both ac and dc are 0-0.5, 0-5, 0-50 and 0-500. A series of dividers (R2 through R5) feed a portion of the voltage being measured to the bridge circuitry. A 1-MΩ resistor is used in the tip of the dc probe bringing the total input impedance to approximately 7 megohms. The use of potentiometers in the divider alleviates the need for precision, special-value resistors, thereby reducing the cost of the unit.

Measurements of ac voltage are facilitated by rectifying the ac and reading the resulting dc directly. Two 1N4007 diodes, a 22-MΩ resistor, and a 0.05-μF capacitor form the rectifier circuit. R1 is used to calibrate the instrument for ac measurements.

Resistance measurements are made in ohms using five ranges: 0-50, 0-500, 0-5000, 0-50k and 0-500k. This circuit makes use of a linear ohms-readout system. Conventional VTVMs and VOMs use scales that are cramped on the high end and expanded on the low resistance end. This logarithmic system is impractical for a home-constructed instrument since special meter faces are not generally available. Linear readout of resistance allows the user to read the value

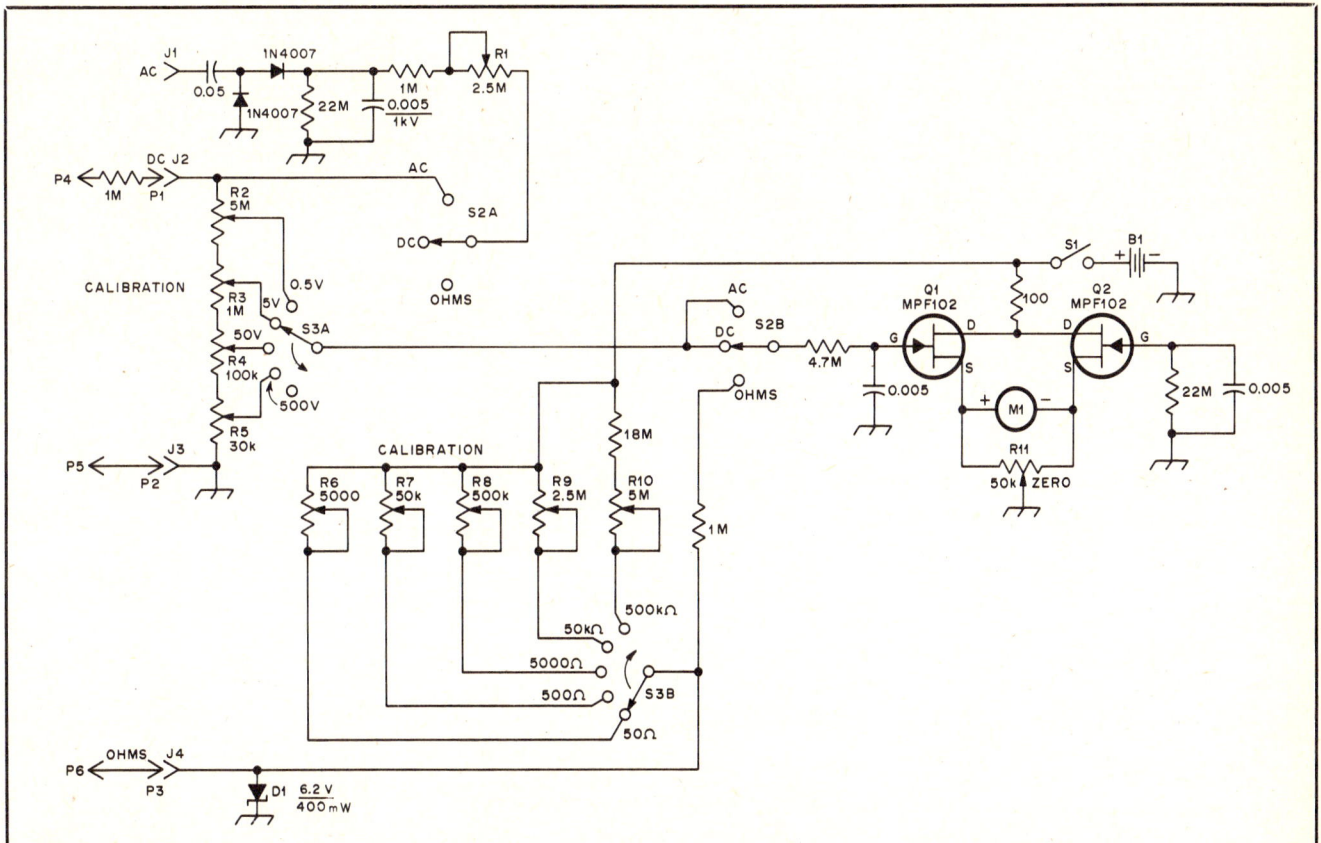

Fig. 10 — Schematic diagram of the FET VOM. All resistors are 1/4-watt carbon types except for the potentiometers. Numbered components not appearing in the parts list are for text callout only. All controls except R11 are for calibration.

BT1 — Battery, 9-volt rectangular.	P1-P3, incl. — Standard tip plugs.	Q1, Q2 — Motorola MPF102; do not substitute.
D1 — Zener diode, 6.2-V.	P4 — Standard test probe. Mount 1-MΩ 1/4-	S1 — Spst toggle.
J1-J4, incl. — Standard tip jacks.	watt resistor inside probe tip.	S2 — 2-pole, 3-position rotary.
M1 — Panel meter, 0-50 μA dc; Radio Shack	P5, P6 — Standard test probe.	S3 — 2-pole, 5-position rotary.

22-051 or equiv.

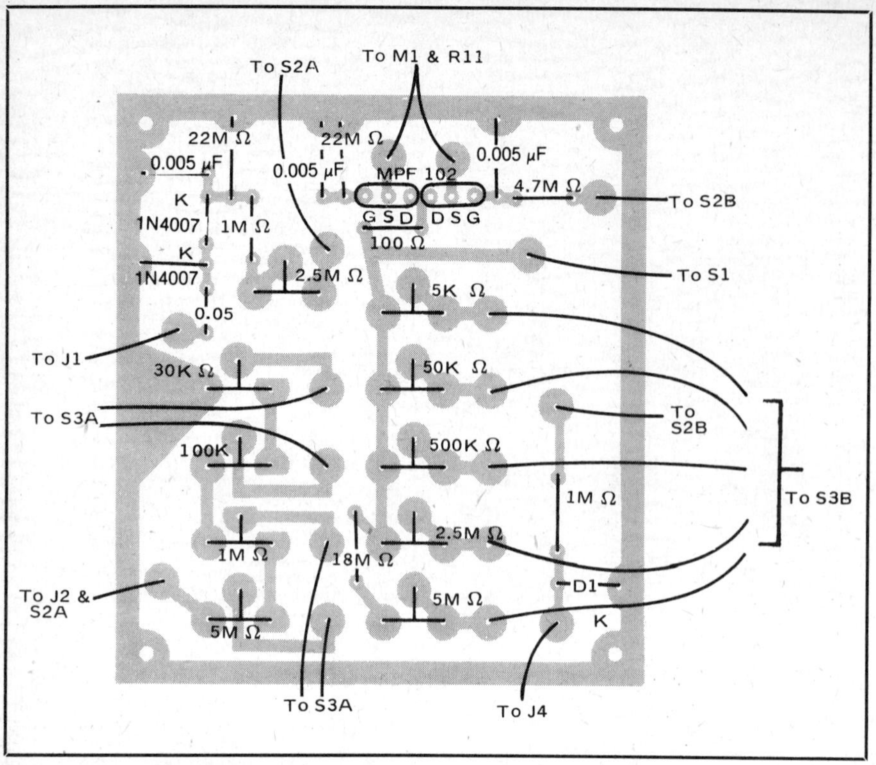

Fig. 11 — Circuit-board layout with parts overlay for the FET VOM. Shaded areas represent unetched copper areas of circuit board. This view is from the foil side of the board.

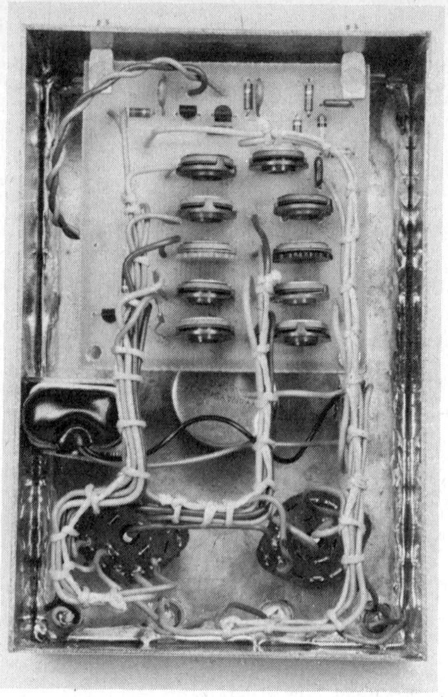

Fig. 12 — The inside of the FET VOM. Leads are dressed with wire-tying twine to provide a neat appearance.

of resistance directly from a standard meter face.

Potentiometers R6 through R10 are used in place of precision, nonstandard-value resistors. Each potentiometer controls the voltage division for its associated range.

Under normal circuit conditions with the instrument placed in the ohms position, the meter will rest gently against the peg, off scale at the high end. When the ohmmeter leads are connected together, the zero potentiometer is adjusted so that the meter indicates zero resistance. Separating the leads causes the pointer to return to its position resting against the high-end peg. D1 is used to limit the voltage fed to the bridge so that the pointer does not slam against the peg.

Construction

The enclosure is made from pieces of double-sided, glass-epoxy circuit board material with the overall dimensions measuring 4 × 6 × 2-1/4 inches (100 × 150 × 60 mm). All seams are soldered together along their entire length to ensure a rigid construction. The battery is held between two pieces of circuit board material soldered to the sides at several locations.

All components other than the meter switches, battery and zero potentiometer are mounted on a circuit board that measures 2-7/8 × 3-1/8 inches (73 × 79 mm). A suitable foil pattern with parts layout is shown in Fig. 11.

The schematic shows a number of connections to ground. In this particular circuit, ground is not the cabinet of the instrument but rather a "floating" ground. By not connecting any of the circuitry to the cabinet there is no chance of having dangerous voltages on the case. *This means that the circuit-board ground foil should not be allowed to contact the cabinet.*

Although the unit shown in the photographs was left natural (tarnished copper) with a clear acrylic coating, there is no reason why the builder should not paint the finished VOM. Treat the copper like any other metal surface when painting. Any type of labeling that suits the builder's fancy may be used. Dry transfer-type labels were used on the unit shown.

Calibration

Adjustment of the completed FET VOM is simple. However, it does require the use of a calibrated meter and a source of variable-voltage dc. The dc ranges should be adjusted first. Connect the calibrated voltmeter in parallel with the FET VOM and attach these connection points to the variable-voltage dc supply. Start with the lowest range (0-0.5) and set the supply voltage for a midscale reading (0.25 volt). Adjust R2 so that the FET VOM reading conforms with the reading on the calibrated meter. Do the same for each of the other ranges using a voltage that will allow the meter to read near midscale. Should 250 volts not be available for the high-range calibration, 50 volts could be used, yielding only a small difference in accuracy. *Care should be taken to touch only the plastic insulation on the potentiometers since potentially dangerous voltages are present in the circuit.*

Ac calibration is somewhat simpler since the basic voltage dividers have already been calibrated. The ac line voltage should be used for calibration, again conforming the reading on the FET VOM with the calibrated meter. R1 is provided for this adjustment.

Calibration of the ohmmeter circuitry is done in a similar manner. A resistor that will allow the meter to read approximately midscale for each range will be required. If the resistors are of the precision variety, a calibrated ohmmeter will not be required. However, if the resistors used for calibration are of five-percent tolerance or greater, it would be wise to use a calibrated meter. For example, a 27-ohm precision resistor could be used for the lowest resistance range. R6 would be adjusted for a reading of exactly 27 ohms on the FET VOM. A 10- or 20-percent tolerance resistor could be used provided a calibrated meter is available. In that case the FET VOM reading should be made the same as the calibrated meter. Simply do the same for the remainder of the resistance ranges. That completes the calibration of the instrument. It is now ready for use in those many applications around your shack.

AC Instruments and Circuits

Although purely electromagnetic instruments which operate directly from alternating current are available, they are seen infrequently in present-day amateur equipment. For one thing, their use is not feasible above power-line frequencies.

Practical instruments for audio and radio frequencies generally use a dc meter movement in conjunction with a rectifier. Voltage measurements suffice for nearly all test purposes. Current, as such, is seldom measured in the af range. When rf current is measured the instrument used is a thermocouple milliammeter or ammeter.

The Thermocouple Meter

In a *thermocouple meter* the alternating current flows through a low-resistance heating element. The power lost in the resistance generates heat that warms a "thermocouple," a junction of certain dissimilar metals which has the property of developing a small dc voltage when heated. This voltage is applied to a dc milliammeter calibrated in suitable ac units. The heater-thermocouple-dc meter combination is usually housed in a regular meter case.

Thermocouple meters can be obtained in ranges from about 100 mA to many amperes. Their useful upper frequency limit is in the neighborhood of 100 MHz. Their principal value in amateur work is in measuring current into a known load resistance for calculating the rf power delivered to the load. A suitable mounting for this is shown in Fig. 13, for use in coaxial lines.

Rectifier Instruments

The response of a rectifier-type meter is proportional (depending on the design) to either the peak amplitude or average amplitude of the rectified ac wave, and never directly responsive to the rms value. The meter therefore cannot be calibrated in rms without preknowledge of the relationship that happens to exist between the "real" reading and the rms value. This relationship, in general, is not known, except in the case of single-frequency ac (a sine wave). Very many practical measurements involve nonsinusoidal wave forms, so it is necessary to know what kind of instrument you have, and what it is actually reading, in order to make measurements intelligently.

Peak and Average with Sine-Wave Rectification

Fig. 14 shows the relative peak and average values in the outputs of half- and full-wave rectifiers (see power-supply chapter for further details). As the positive and negative half cycles of the sine wave have the same shape (A), half-wave rectification of either the positive half (B)

Fig. 13 — Rf ammeter mounted in a Minibox, with connectors for placing the meter in series with a coaxial line. A bakelite-case meter should be used to minimize shunt capacitance (which introduces error) although a metal-case meter can be used if mounted on bakelite sheet with a large cut-out in the case around the rim. The meter can be used for rf power measurements (P = I²R) when connected between the transmitter and a nonreactive load of known resistance.

or the negative half (C) gives exactly the same result. With full-wave rectification (D) the peak is still the same, but the average is doubled, since there are twice as many half cycles per unit of time.

Unsymmetrical Wave Forms

A nonsinusoidal waveform is shown in Fig. 15A. When the positive half cycles of this wave are rectified the peak and average values are as shown at B. If the polarity is reversed and the negative half cycles are rectified, the peak value is different but the average value is unchanged. The fact that the average of the positive side is equal to the average of the negative side is true of *all* ac waveforms, but different waveforms have different averages. Full-wave rectification of such a "lopsided" wave doubles the average value, but the peak reading is always the same as it is with the half cycle that produces the *highest* peak in half-wave rectification.

Effective-Value Calibration

The actual scale calibration of commercially made rectifier-type voltmeters is very often (almost always, in fact) in terms of rms values. For sine waves this is satisfactory, and useful since rms is the standard measure at power-line frequency. It is also useful for many rf applications where the waveform is often closely sinusoidal. But in other cases, particularly in the af range, the error may

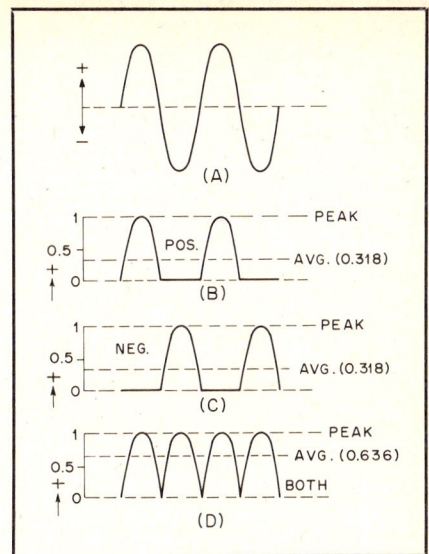

Fig. 14 — Sine-wave alternating current or voltage (A), with half-wave rectification of the positive half cycle (B) and negative half cycle (C). D — full-wave rectification. Average values are shown with relation to a peak value of 1.

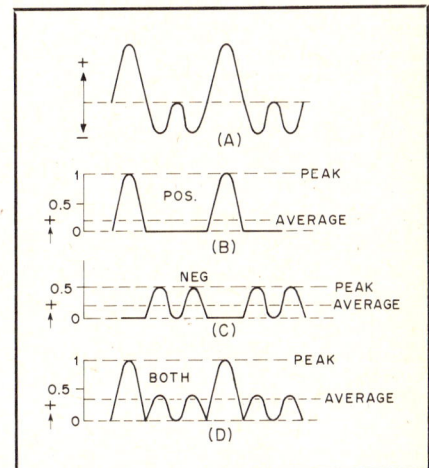

Fig. 15 — Same as Fig. 14 for an unsymmetrical waveform. The peak values are different with positive and negative half-cycle rectification.

be considerable when the waveform is not pure.

Turn-Over

From Fig. 15 it is apparent that the calibration of an average-reading meter will be the same whether the positive or negative sides are rectified. A half-wave *peak*-reading instrument, however, will indicate different values when its connections to the circuit are reversed (*turn-over effect*). Very often readings are taken both ways, in which case the sum of the two is the *peak-to-peak* (pk-pk) value, a useful figure in much audio and video work.

Average- and Peak-Reading Circuits

The basic difference between average-

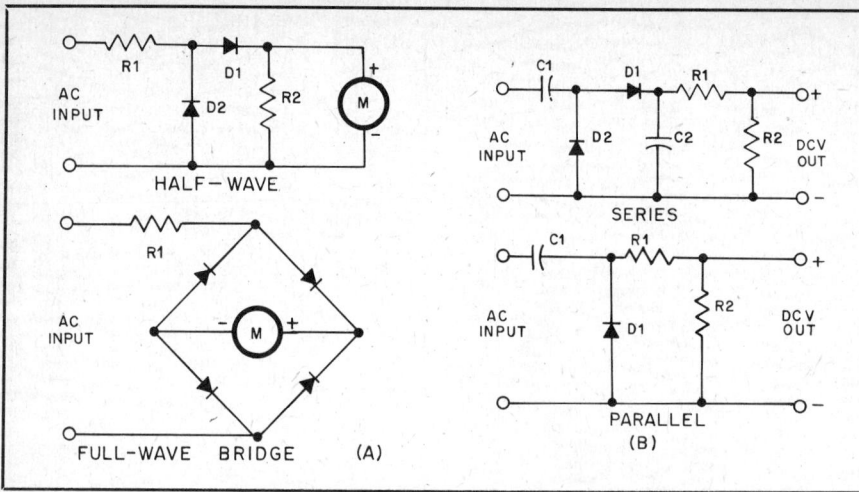

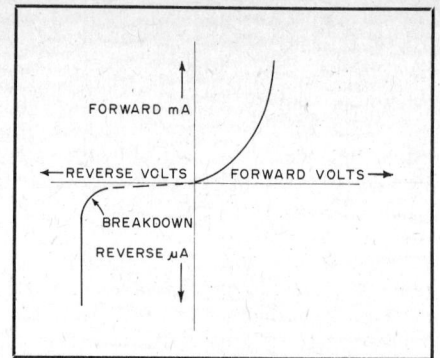

Fig. 17 — Typical semiconductor diode characteristic. Actual current and voltage values vary with the type of diode, but the forward-current curve would be in its steep part with only a volt or so applied. Note change in current scale for reverse current. Breakdown voltage, again depending on diode type, may range from 15 or 20 volts to several hundred.

Fig. 16 — At A, half-wave and full-wave rectification for an instrument intended to operate on average values. At B, half-wave circuits for a peak-reading meter.

and peak-reading rectifier circuits is that in the former the output is not filtered while in the latter a filter capacitor is charged up to the peak value of the output voltage. Fig. 16A shows typical average-reading circuits, one half-wave and the other full-wave. In the absence of dc filtering the meter responds to wave forms such as are shown at B, C and D in Figs. 14 and 15, and since the inertia of the pointer system makes it unable to follow the rapid variations in current, it averages them out mechanically.

In Fig. 16A, D1 actuates the meter; D2 provides a low-resistance dc return in the meter circuit on the negative half cycles. R1 is the voltmeter multiplier resistance. R2 forms a voltage divider with R1 (through D1) which prevents more than a few ac volts from appearing across the rectifier-meter combination. A corresponding resistor can be used across the full-wave bridge circuit.

In these two circuits no provision is made for isolating the meter from any dc voltage that may be on the circuit under measurement. The error caused by this can be avoided by connecting a large capacitance in series with the "hot" lead. The reactance must be low compared with the meter impedance (see next section) in order for the full ac voltage to be applied to the meter circuit. As much as 1 μF may be required at line frequencies with some meters. The capacitor is not usually included in a VOM.

Series and shunt peak-reading circuits are shown in Fig. 16B. Capacitor C1 isolates the rectifier from dc voltage on the circuit under measurement. In the series circuit (which is seldom used) the time constant of the C2-R1-R2 combination must be very large compared with the period of the lowest ac frequency to be measured; similarly with C1-R1-R2 in the shunt circuit. The reason is that the capacitor is charged to the peak value of voltage when the ac wave reaches its maxi-

mum, and then must hold the charge (so it can register on a dc meter) until the next maximum of the same polarity. If the time constant is 20 times the ac period the charge will have decreased by about five percent by the time the next charge occurs. The *average* drop will be smaller, so the error is appreciably less. The error will decrease rapidly with increasing frequency, assuming no change in the circuit values, but will increase at lower frequencies.

In Fig. 16B, R1 and R2 form a voltage divider which reduces the peak dc voltage to 71 percent of its actual value. This converts the peak reading to rms on sine-wave ac. Since the peak-reading circuits are incapable of delivering appreciable current without considerable error, R2 is usually the 11-megohm input resistance of an electronic voltmeter. R1 is therefore approximately 4.7 megohms, making the total resistance approach 16 megohms. A capacitance of 0.05 μF is sufficient for low audio frequencies under these conditions. Much smaller values of capacitance suffice for radio frequencies, obviously.

Voltmeter Impedance

The impedance of the voltmeter at the frequency being measured may have an effect on the accuracy similar to the error caused by the resistance of a dc voltmeter, as discussed earlier. The ac meter acts like a resistance in parallel with a capacitance, and since the capacitive reactance decreases with increasing frequency, the impedance also decreases with frequency. The resistance is subject to some variation with voltage level, particularly at very low voltages (of the order of 10 volts or less) depending upon the sensitivity of the meter movement and the kind of rectifier used.

The ac load resistance represented by a diode rectifier is approximately equal to one-half its dc load resistance. In Fig. 16A the dc load is essentially the meter

resistance, which is generally quite low compared with the multiplier resistance R1, to the total resistance will be about the same as the multiplier resistance. The capacitance depends on the components and construction, test lead length and disposition, and other such factors. In general, it has little or no effect at lower-line and low audio frequencies, but the ordinary VOM loses accuracy at the high audio frequencies and is of little use at rf. For radio frequencies it is necessary to use a rectifier having very low inherent capacitance.

Similar limitations apply to the peak-reading circuits. In the parallel circuit the resistive component of the impedance is smaller than in the series circuit, since the dc load resistance, R1/R2, is directly across the circuit being measured, and is therefore in parallel with the diode ac load resistance. In both peak-reading circuits the effective capacitance may range from 1 or 2 to a few hundred pF. Values of the order of 100 pF are to be expected in electronic voltmeters of customary design and construction.

Linearity

Fig. 17, a typical current/voltage characteristic of a small semiconductor rectifier, indicates that the forward dynamic resistance of the diode is not constant, but rapidly decreases as the forward voltage is increased from zero. The transition from high to low resistance occurs at considerably less than 1 volt, but is in the range of voltage required by the associated dc meter. With an average-reading circuit the current tends to be proportional to the *square* of the applied voltage. This crowds the calibration points at the low end of the meter scale. For most measurement purposes, however, it is far more desirable for the output to be "linear;" that is, for the reading to be *directly* proportional to the applied voltage.

To achieve linearity it is necessary to use a relatively large load resistance for the diode — large enough so that this resistance, rather than the diode's own resistance, will govern the current flow. A linear or equally spaced scale is thus gained at the expense of sensitivity. The amount of resistance needed depends on the type of diode; 5000 to 50,000 ohms usually suffices for a germanium rectifier, depending on the dc meter sensitivity, but several times as much may be needed for silicon. The higher the resistance, the greater the meter sensitivity required; i.e., the basic meter must be a microammeter rather than a low-range milliammeter.

Reverse Current

When voltage is applied in the reverse direction there is a small leakage current in semiconductor diodes. This is equivalent to a resistance connected across the rectifier, allowing current to flow during the half cycle which should be completely nonconducting, and causing an error in the dc meter reading. This "back resistance" is so high as to be practically unimportant with silicon, but may be less than 100 kilohms with germanium.

The practical effect of back resistance is to limit the amount of resistance that can be used in the dc load resistance. This in turn affects the linearity of the meter scale.

The back resistance of vacuum-tube diodes is infinite, for practical purposes.

RF Voltage

Special precautions must be taken to minimize the capacitive component of the voltmeter impedance at radio frequencies. If possible, the rectifier circuit should be installed permanently at the point where the rf voltage to be measured exists, using the shortest possible rf connections. The dc meter can be remotely located, however.

For general rf measurements an *rf probe* is used in conjunction with an electronic voltmeter, substituted for the dc probe mentioned earlier. The circuit of Fig. 19, essentially the peak-reading shunt circuit of Fig. 16B, is generally used. The series resistor, installed in the probe close to the rectifier, prevents rf from being fed through the probe cable to the electronic voltmeter, being helped in this by the cable capacitance. This resistor, in conjunction with the 10-MΩ divider resistance of the electronic voltmeter, also reduces the peak rectified voltage to a dc value equivalent to the rms of the rf signal, to make the rf readings consistent with the regular ac calibration.

Of the diodes readily available to amateurs, the germanium point-contact type is preferred for rf applications. It has low capacitance (of the order of 1 pF) and in the high-back-resistance types the reverse current is not serious. The principal limitation is that its safe reverse

voltage is only about 50-75 volts, which limits the rms applied voltage to 15 or 20 volts, approximately. Diodes can be connected in series to raise the overall rating.

An RF Probe for Electronic Voltmeters

The isolation capacitor, C1, crystal diode, and filter/divider resistor are mounted on a bakelite five-lug terminal strip, as shown in Fig. 21. One end lug should be rotated 90 degrees so that it extends off the end of the strip. All other lugs should be cut off flush with the edge of the strip. Where the inner conductor connects to the terminal lug, unravel the shield three-quarters of an inch, slip a piece of spaghetti over it, and then solder the braid to the ground lug on the terminal strip. Remove the spring from the tube shield, slide it over the cable, and crimp it to the remaining quarter inch of shield braid. Solder both the spring and a 12-inch (305-mm) length of flexible copper braid to the shield.

Next, cut off the pins on a seven-pin miniature shield-base tube socket. Use a socket with a cylindrical center post. Crimp the terminal lug previously bent out at the end of the strip and insert it into the center post of the tube socket from the top. Insert the end of a phone tip or a pointed piece of heavy wire into the bottom of the tube socket center post, and solder the lug and tip to the center post. Insert a half-inch grommet at the top of the tube shield, and slide the shield over the cable and flexible braid down onto the tube socket. The spring should make good contact with the tube shield to insure that the tube shield (probe case) is grounded. Solder an alligator clip to the other end of the flexible braid and mount a phone plug on the free end of the shielded wire.

Mount components close to the terminal strip, to keep lead lengths as short as possible and minimize stray capacitance. Use spaghetti over all wires to prevent accidental shorts.

The phone plug on the probe cable plugs into the dc input jack of the electronic voltmeter and rms voltages are read on the voltmeter's negative dc scale.

The accuracy of the probe is within ±10 percent from 50 kHz to 250 MHz. The approximate input impedance is 6000 ohms shunted by 1.75 pF (at 200 MHz).

RF Power

Power at radio frequencies can be measured by means of an accurately calibrated rf voltmeter connected across the load in which the power is being dissipated. If the load is a known pure resistance the power, by Ohm's Law, is equal to E^2/R, where E is the rms value of the voltage.

The method only indicates *apparent* power if the load is not a pure resistance. The load can be a terminated transmission line tuned, with the aid of bridge circuits

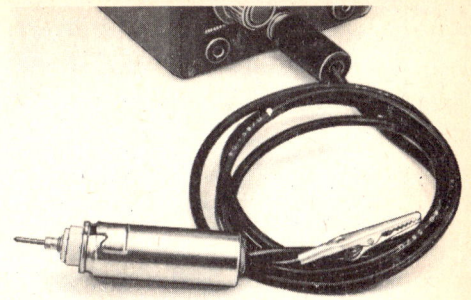

Fig. 18 — Rf probe for use with an electronic voltmeter. The case of the probe is constructed from a seven-pin ceramic tube socket and a 2-1/4-inch (57-mm) tube shield. A half-inch (13-mm) grommet at the top of the tube shield prevents the output lead from chafing. A flexible copper-braid grounding lead and alligator clip provide a low-inductance return path from the test circuit.

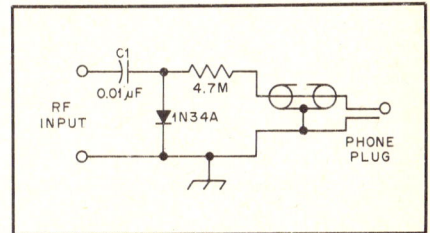

Fig. 19 — The rf probe circuit.

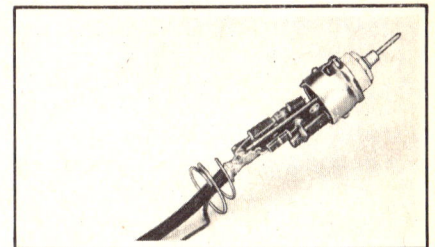

Fig. 20 — Inside the probe. The 1N34A diode, calibrating resistor and input capacitor are mounted tight to the terminal strip with shortest leads possible. Spaghetti tubing is placed on the diode leads to prevent accidental short circuits. The tube-shield spring and flexible-copper grounding lead are soldered to the cable braid (the cable is RG-58/U coax). The tip can be either a phone tip or a short pointed piece of heavy wire.

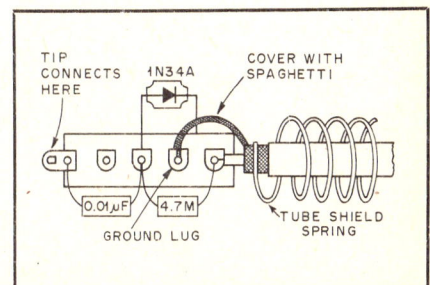

Fig. 21 — Component mounting details.

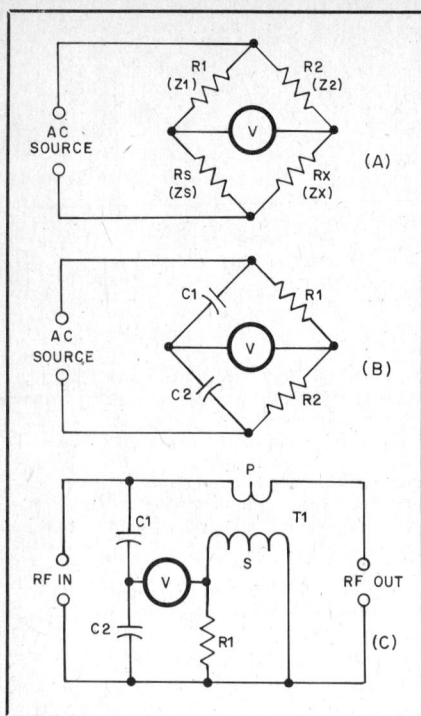

Fig. 22 — (A) Generalized form of bridge circuit for either ac or dc. (B) One form of ac bridge frequently used for rf measurements. (C) SWR bridge for use in transmission lines. This circuit is often calibrated in power rather than voltage.

such as are described in the next section, to act as a known resistance. An alternative load is a "dummy" antenna, a known pure resistance capable of dissipating the rf power safely.

AC Bridges

In its simplest form, the ac bridge is exactly the same as the Wheatstone bridge discussed earlier. However, complex impedances can be substituted for resistances, as suggested by Fig. 22A. The same bridge equation holds if Z is substituted for R in each arm. For the equation to be true, however, *the phase angles as well as the numerical values of the impedances must balance*; otherwise, a true null voltage is impossible to obtain. This means that a bridge with all "pure" arms (pure resistance or reactance) cannot measure complex impedances; a combination of R and X must be present in at least one arm besides the unknown.

The actual circuits of ac bridges take many forms, depending on the type of measurement intended and on the frequency range to be covered. As the frequency is raised, stray effects (unwanted capacitances and inductances, principally) become more pronounced. At radio frequencies special attention must be paid to minimizing them.

Most amateur-built bridges are used for rf measurements, especially SWR measurements on transmission lines. The circuits at B and C, Fig. 22, are favorites for this purpose.

Fig. 22B is useful for measuring both transmission lines and "lumped constant" components. Combinations of resistance and capacitance are often used in one or more arms; this may be required for eliminating the effects of stray capacitance.

Fig. 22C is used only on transmission lines, and only on those lines having the characteristic impedance for which the bridge is designed.

SWR Measurement — the Reflectometer

In measuring standing-wave ratio advantage is taken of the fact that the voltage on transmission line consists of two components traveling in opposite directions. The power going from the transmitter to the load is represented by one voltage (designated "incident" or "forward") and the power reflected from the load is represented by the other. Because the relative amplitudes and phase relationships are definitely established by the line's characteristic impedance, its length and the load impedance in which it is terminated, a bridge circuit can separate the incident and reflected voltages for measurement. This is sufficient for determining the SWR. Bridges designed for this purpose are frequently called *reflectometers*.

Referring to Fig. 22A, if R1 and R2 are made equal, the bridge will be balanced when $R_X = R_S$. This is true whether R_X is an actual resistor or the input resistance of a perfectly matched transmission line, provided R_S is chosen to equal the characteristic impedance of the line. Even if the line is not properly matched, the bridge will still be balanced for power traveling *outward* on the line, since outward-going power sees only the Z_0 of the line until it reaches the load. However, power reflected back from the load does not "see" a bridge circuit, and the reflected voltage registers on the voltmeter. From the known relationship between the incident and reflected voltages the SWR is easily calculated:

$$SWR = \frac{V_0 + V_r}{V_0 - V_r}$$

The "Reflected Power Meter"

Fig. 22C makes use of mutual inductance between the primary and secondary of T1 to establish a balancing circuit. C1 and C2 form a voltage divider in which the voltage across C2 is in the same phase as the voltage at that point on the transmission line. The relative phase of the voltage across R1 is determined by the phase of the *current* in the line. If a pure resistance equal to the design impedance of the bridge is connected to the "RF Out" terminals, the voltages across R1 and C2 will be out of phase and the voltmeter reading will be minimum; if the *amplitudes* of the two voltages are also equal (they are made

so by bridge adjustment) the voltmeter will read zero. Any other value of resistance or impedance connected to the "RF Out" terminals will result in a finite voltmeter reading. When used in a transmission line this reading is proportional to the reflected voltage. To measure the incident voltage the secondary terminals of T1 must be very large compared to the resistance of R1.

Instruments of this type are usually designed for convenient switching between forward and reflected, and are often calibrated to read power in the specified characteristic impedance. As the power calibration is made at an SWR of 1:1, the actual power transmitted to the load must be calculated from the difference between the forward and reflected readings when the SWR is other than 1:1. A reading of 400 watts forward and 300 watts reflected means that 400 W − 300 W = 100 W has been delivered to the load. The SWR can be determined from the VSWR nomograph in chapter 19 or calculated from the formula

$$SWR = \frac{1 + \sqrt{\dfrac{\text{reflected power}}{\text{forward power}}}}{1 - \sqrt{\dfrac{\text{reflected power}}{\text{forward power}}}}$$

$$= \frac{1 + \sqrt{\dfrac{300}{400}}}{1 - \sqrt{\dfrac{300}{400}}} = 13.9$$

Both of these calculations assume that the loss in the transmission line is very small; information on correcting for line loss is given in chapter 19.

Sensitivity vs. Frequency

In all of the circuits in Fig. 22 the sensitivity is independent of the applied frequency, within practical limits. Stray capacitances and couplings generally limit the performance of all three at the high-frequency end of the useful range. Fig. 22A will work right down to dc, but the low-frequency performance of Fig. 22B is degraded when the capacitive reactances become so large that voltmeter impedance becomes low in comparison (in all these bridge circuits, it is assumed that the voltmeter impedance is high compared with the impedance of the bridge arms). In Fig. 22C the performance is limited at low frequencies by the fact that the transformer reactance decreases with frequency, so that eventually the reactance is not very high in comparison with the resistance of R1.

The "Monimatch"

A type of bridge which is quite simple to make, but in which the sensitivity rises directly with frequency, is the *Monimatch* and its various offspring. The circuit cannot be described in terms of lumped constants, as it makes use of the distributed mutual inductance and capacitance between the center conductor of a transmission line and a wire placed parallel to it. The wire is terminated in a resistance approximating the characteristic impedance of the transmission line at one end and feeds a diode rectifier at the other.

Frequency Measurements

The regulations governing amateur operation require that the transmitted signal be maintained inside the limits of certain bands of frequencies.[1] The exact frequency need not be known, so long as it is not outside the limits. On this last point there are no tolerances: It is up to the individual amateur to see that he stays safely "inside."

This is not difficult to do, but requires some simple apparatus and the exercise of some care. The apparatus commonly used is the *frequency-marker generator,* and the method involves use of the station receiver, as in Fig. 23.

The Frequency Marker

The marker generator in its simplest form is a high-stability oscillator generating a series of signals which, when detected in the receiver, mark the exact edges of the amateur assignments. It does this by oscillating at a low frequency that has harmonics falling on the desired frequencies.

U.S. amateur band limits are exact multiples of 25 kHz, whether at the extremes of a band or at points marking the subdivisions between types of emission, license privileges, and so on. A 25-kHz fundamental frequency therefore will produce the desired marker signals if its harmonics at the higher frequencies are strong enough. But since harmonics appear at 25-kHz intervals throughout the spectrum, along with the desired markers, the problem of identifying a *particular* marker arises. This is easily solved if the receiver has a reasonably good calibration. If not, most marker circuits provide for a choice of fundamental outputs of 100 and 50 kHz as well as 25 kHz, so the question can be narrowed down to initial identification of 100-kHz intervals. From these, the desired 25-kHz (or 50-kHz) points can easily be spotted. Coarser frequency intervals are rarely required; there are usually signals available from stations of known frequency, and the 100-kHz points can be counted off from them.

Transmitter Checking

In checking one's own transmitter frequency the signal from the transmitter is first tuned in on the receiver and the dial setting at which it is heard is noted. Then

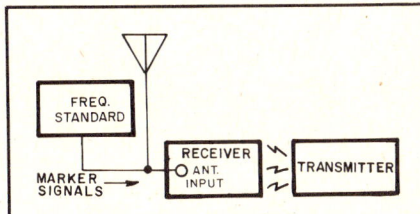

Fig. 23 — Setup for using a frequency standard. It is necessary that the transmitter signal be weak in the receiver — of the same order of strength as the marker signal from the standard. This requirement can usually be met by turning on just the transmitter oscillator, leaving all power off any succeeding stages. In some cases it may also be necessary to disconnect the antenna from the receiver.

the *nearest* marker frequencies above and below the transmitter signal are tuned in and identified. The transmitter frequency is obviously between these two known frequencies.

If the marker frequencies are accurate, this is all that needs to be known — except that the transmitter frequency must not be so close to a band (or subband) edge that sideband frequencies, especially in phone transmission, will extend over the edge.

If the transmitter signal is "inside" a marker at the edge of an assignment, to the extent that there is an audible beat note with the receiver's BFO turned off, normal cw sidebands are safely inside the edge. (This statement does not take into account *abnormal* sidebands such as are caused by clicks and chirps.) For phone the "safety" allowance is usually taken to be about 3 kHz, the nominal width of one sideband. A frequency difference of this order can be estimated by noting the receiver dial settings for the two 25-kHz markers which bracket the signal and dividing 25 by the number of dial divisions between them. This will give the number of kHz per dial division.

Transceivers

The method described above is applicable when the receiver and transmitter are separate pieces of equipment. When a transceiver is used and the transmitting frequency is automatically the same as that to which the receiver is tuned, setting the tuning dial to a spot between two known marker frequencies is all that is required.

The proper dial settings for the markers are those at which, with the BFO on, the signal is tuned to *zero* beat — the spot where the beat disappears as the tuning

makes the beat tone progressively lower. Exact zero beat can be determined by a very slow rise and fall of background noise, caused by a beat of a cycle or less per second.

Frequency-Marker Circuits

The basic frequency-determining element in most amateur frequency markers is a 100-kHz crystal. Although the marker generator should produce harmonics at 25-kHz and 50-kHz intervals, crystals (or other high-stability devices) for frequencies lower than 100 kHz are expensive and difficult to obtain. However, there is really no need for them, since it is easy to divide the basic frequency down to any figure one desires; 50 and 25 kHz require only two successive divisions, each by two. In the division process, the harmonic output of the generator is greatly enhanced, making the generator useful at frequencies well into the vhf range.

Simple Crystal Oscillators

Fig. 24 illustrates two of the simpler circuits. C1 in both circuits is used for exact adjustment of the oscillating frequency to 100 kHz, which is done by using the receiver for comparing one of the oscillator's harmonics with a standard frequency transmitted by WWV, WWVH or a similar station.

Fig. 24A is a field-effect transistor analog of a vacuum-tube circuit. However, it requires a 10-mH coil to operate well, and since the harmonic output is not strong at the higher frequencies the circuit is given principally as an example of a simple transistor arrangement. A much better oscillator is shown at B. This is a cross-connected pair of transistors forming a *multivibrator* of the "free-running" or "asynchronous" type, locked at 100 kHz by using the crystal as one of the coupling elements. While it can use two separate bipolar transistors as shown, it is much simpler to use an integrated-circuit *dual gate,* which will contain all the necessary parts except the crystal and capacitors and is considerably less expensive, as well as more compact, than the separate components.

Frequency Dividers

Electronic division is accomplished by a "bistable" *flip-flop* or cross-coupled circuit which produces one output change for every two impulses applied to its input circuit, thus dividing the applied frequency by two. All division therefore must

[1]These limits depend on the type of emission and class of license held, as well as on international agreements. See the latest edition of *The Radio Amateur's License Manual* for current status.

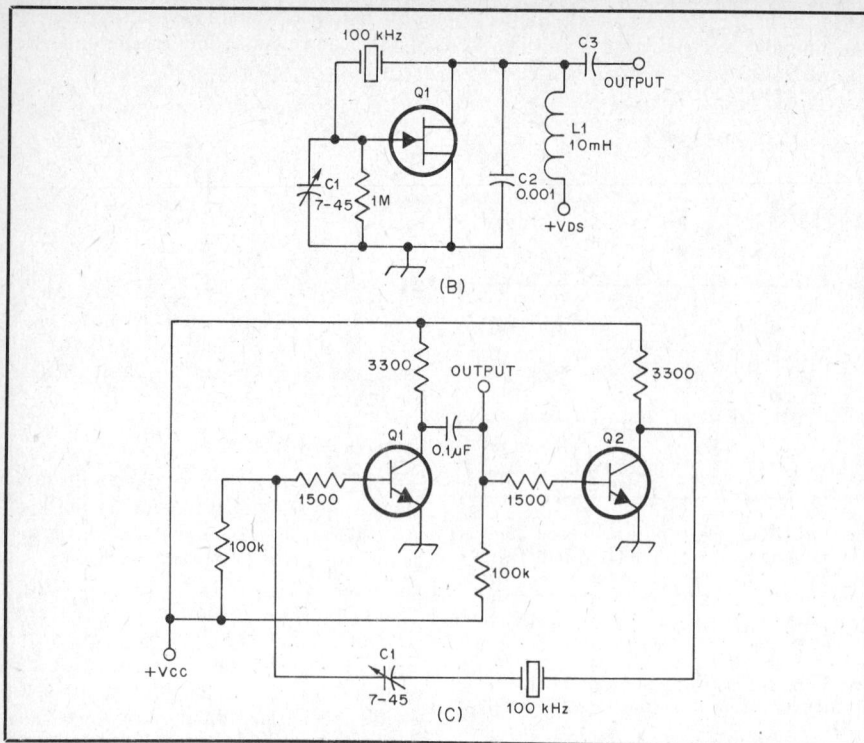

Fig. 24 — Two simple 100-kHz oscillator circuits. B is the most suitable of available transistor circuits (for marker generators) and is recommended where solid-state is to be used. In both circuits C1 is for fine frequency adjustment. The output coupling capacitor, C3 is generally small — 20 to 50 pF — a compromise to avoid loading the oscillator by the receiver antenna input while maintaining adequate coupling for good harmonic strength.

Fig. 25 — The 100, 50 and 25 kHz marker generator is housed in a 4-1/2 × 3 × 2-1/2 inch (114 × 76 × 64 mm) aluminum cabinet. A 9-volt battery makes the instrument completely self-contained.

be in terms of some power of two. In practice this is no handicap since with modern integrated-circuit flip-flops, circuit arrangements can be worked out for division by any desired number.

As flip-flops and gates in integrated circuits come in compatible series — meaning that they work at the same supply voltage and can be directly connected together — a combination of a dual-gate version of Fig. 24B and a dual flip-flop make an attractively simple combination for the marker generator.

There are several different basic types of flip-flops, the variations having to do with methods of driving (dc or pulse operation) and control of the counting function. Information on the operating principles and ratings of a specific type usually can be obtained from the manufacturer. The counting-control functions are not needed in using the flip-flop in a simple marker generator, although they come into play when dividing by some number other than a power of two.

Marker Generator for 100, 50 and 25 kHz

The signal source in the accompanying illustrations will deliver usable calibration markers throughout the hf spectrum. When built into an enclosure and powered from a battery or regulated dc supply, the unit is a test instrument that is very helpful in aligning receivers. Alternatively, the pc assembly can be incorporated into a communications receiver; in this service it

becomes a "crystal calibrator."

Three integrated circuits are used in the generator. A 100-kHz crystal oscillator is designed around a dual JFET-input operational amplifier. Crystals in this frequency range show considerable variation in their characteristics as compared to hf ones. A characteristic they all share, however, is their high expense — for this reason builders often purchase "bargain" crystals, sometimes with disappointing results. When the crystal characteristics (resonance mode, load capacitance and equivalent series resistance) are known, it's a fairly simple matter to design an appropriate oscillator circuit. Instruments of this type usually have oscillators made from one or two digital gates. These circuits are attractive for their simplicity, but they sometimes fail to oscillate with some crystals. The op-amp circuit in this generator is designed to start and oscillate reliably with practically any crystal having an equivalent series resistance up to 20 kilohms. Oscillation is in the series mode, regardless of the cut of the crystal. The start-up time depends on the crystal activity — 100 milliseconds is typical. The second op-amp section serves as a buffer and comparator to provide a waveform suitable for driving the frequency divider stages. A CMOS dual D flip-flop IC provides two divide-by-two stages for the 50- and 25-kHz outputs. Signal routing and divisor selection is handled by a quad NOR gate and a diode matrix. A single-

wafer rotary switch and a trimmer capacitor are the only controls. The complete schematic diagram is drawn in Fig. 26.

Construction

A component-placement guide and etching pattern are given in Figs. 27 and 28. The layout is not critical, but the rotor and/or adjustment screw of the trimmer capacitor should be grounded so that contact with a screwdriver won't affect the frequency. Similarly, the crystal should be positioned so that a screwdriver won't come too near it during frequency adjustment. Fig. 27 shows a successful arrangement. Give due respect to the CMOS ICs during assembly — they can be damaged by static charges. Keep them in their protective material and don't insert them until the other components have been installed. The output coupling capacitor isn't critical — the value shown allows fairly constant-amplitude harmonics up to 30 MHz into a 50-ohm load. If the unit is installed in a receiver, a small twisted-wire "gimmick" capacitor should be used. This capacitor should be adjusted for minimal loading of the input circuit, consistent with adequate marker strength.

Adjustment and Operation

The unit can be set to precisely 100 kHz by zero-beating a harmonic against WWV or a known broadcast station. A small fixed-value capacitor can be shunted across the trimmer if necessary. Use a silver mica or NPO ceramic component for this purpose. The dissipation in the crystal and op amp is minimal, so the generator is stable with time. As a test instrument, the marker generator should be set before beginning an alignment job. As a crystal calibrator, it should be located away from the heat-generating components of the receiver. After the receiver has reached operating temperature, the marker frequency should be checked against WWV from time to time. The generator can be powered from any 9- to 12-volt dc source. The frequency varies with the applied voltage, so a Zener diode regulator should be used where the voltage varies, such as in mobile service.

Other Methods of Frequency Checking

The simplest possible frequency-measuring device is a parallel LC circuit, tunable over a desired frequency range of having its tuning dial calibrated in terms of frequency. It can be used only for checking circuits in which at least a small amount of rf power is present, because the energy required to give a detectable indication is not available in the LC circuit itself; it has to be extracted from the circuit being measured; hence the name *absorption frequency meter*. It will be observed that what is actually measured is the frequency of the rf *energy, not* the frequency to which the circuit in which the energy is present may be tuned.

The measurement accuracy of such an instrument is low, compared with the accuracy of a marker generator, because the Q of a practicable LC circuit is not high enough to make precise reading of the dial possible. Also, any two circuits coupled together react on each others' tuning. (This can be minimized by using the loosest coupling that will give an adequate indication.)

The absorption frequency meter has one useful advantage over the marker generator — it will respond *only* to the frequency to which it is tuned, or to a band of frequencies very close to it. Thus there is no harmonic ambiguity, as there sometimes is when using a marker generator.

Absorption Circuit

A typical absorption frequency-meter circuit is shown in Fig. 29. In addition to the adjustable tuned circuit, L1-C1, it includes a pickup coil, L2, wound over L1, a high-frequency semiconductor diode, D1, and a microammeter or low-range (usually not more than 0-1 mA) milliammeter. A phone jack is included so the device can be used for listening to the signal.

The sensitivity of the frequency meter depends on the sensitivity of the dc meter movement and the size of L2 in relation to L1. There is an optimum size for this coil which has to be found by experiment. An alternative is to make the rectifier connection to an adjustable tap on L1, in which case there is an optimum tap point. In general, the rectifier coupling should be a little *below* (that is, less tight) the point that gives maximum response, since this will make the indications sharper.

Calibration

The absorption frequency meter must be calibrated by taking a series of readings on various frequencies from circuits carrying rf power, the frequency of the rf

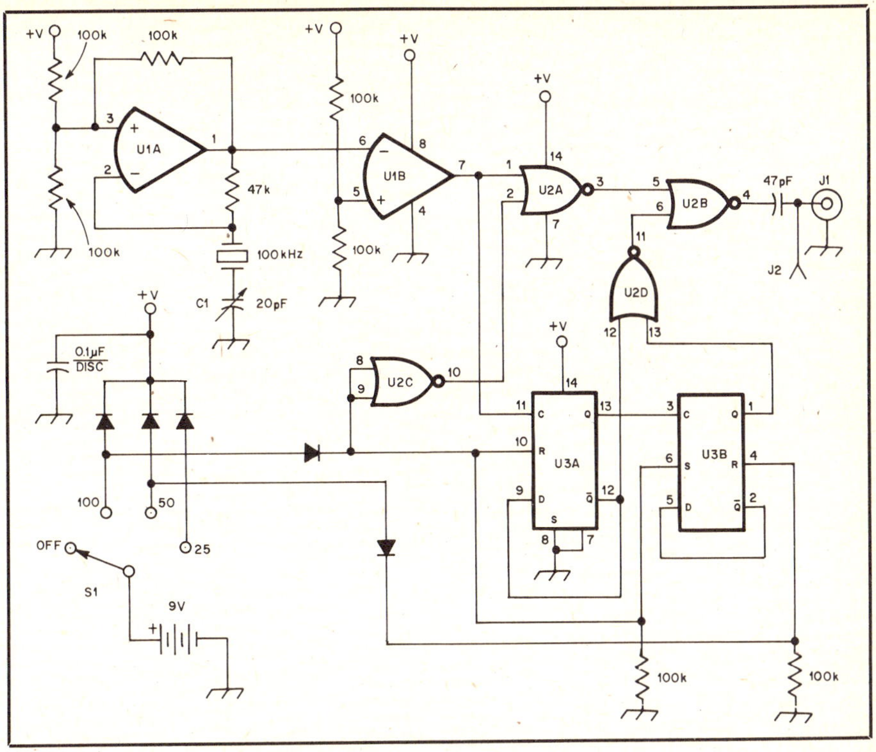

Fig. 26 — Schematic diagram of the 100, 50 and 25 kHz marker generator. Resistors are 1/4-watt composition types. Diodes are 1N914 or similar switching types.

C1 — 20-pF trimmer, Johnson 189-508-5 or equiv.
J1 — Coaxial connector, builder's choice.
J2 — Miniature binding post.
S1 — Rotary, 1 pole, 4 position.
U1 — LF-353N.
U2 — 4001.
U3 — 4013.

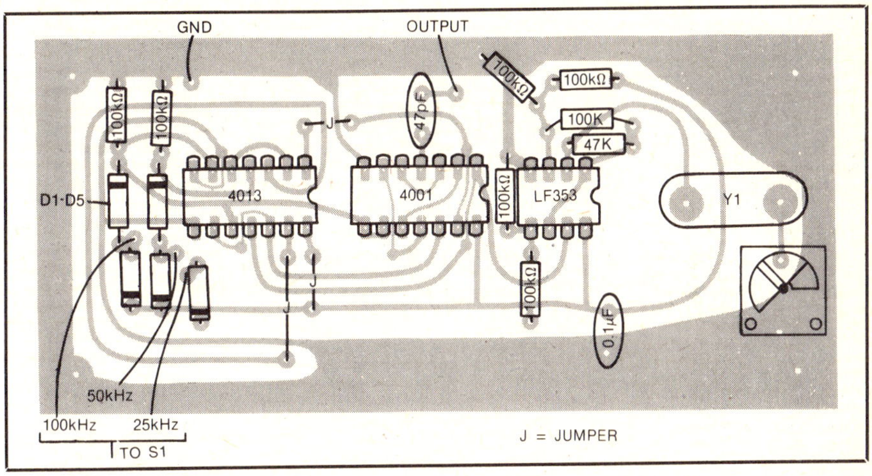

Fig. 27 — Parts-placement guide for the 100, 50 and 25 kHz marker generator. The component side of the board is shown. Shaded areas represent an X-ray view of the copper foil.

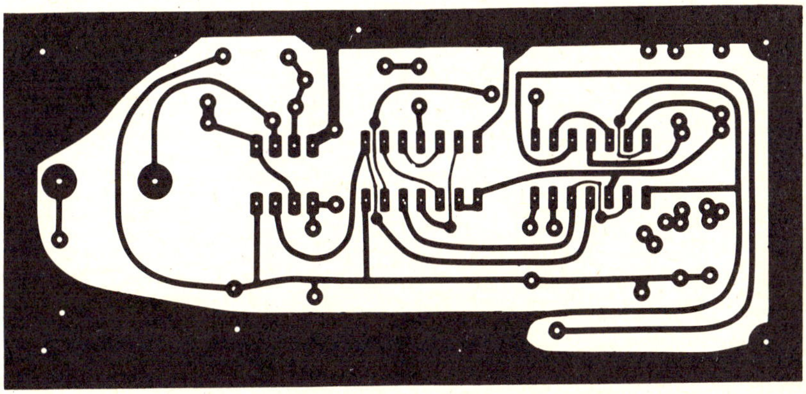

Fig. 28 — Etching pattern for the 100, 50 and 25 kHz marker generator pc board. Black areas indicate copper.

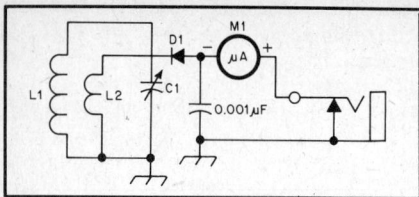

Fig. 29 — Absorption frequency-meter circuit. The closed-circuit phone jack may be omitted if listening is not wanted, in which case the positive terminal of M1 goes to common ground.

Fig. 30 — Exterior view of the 600-MHz frequency counter. The last digit is not blanked in order to provide an on-off indicator.

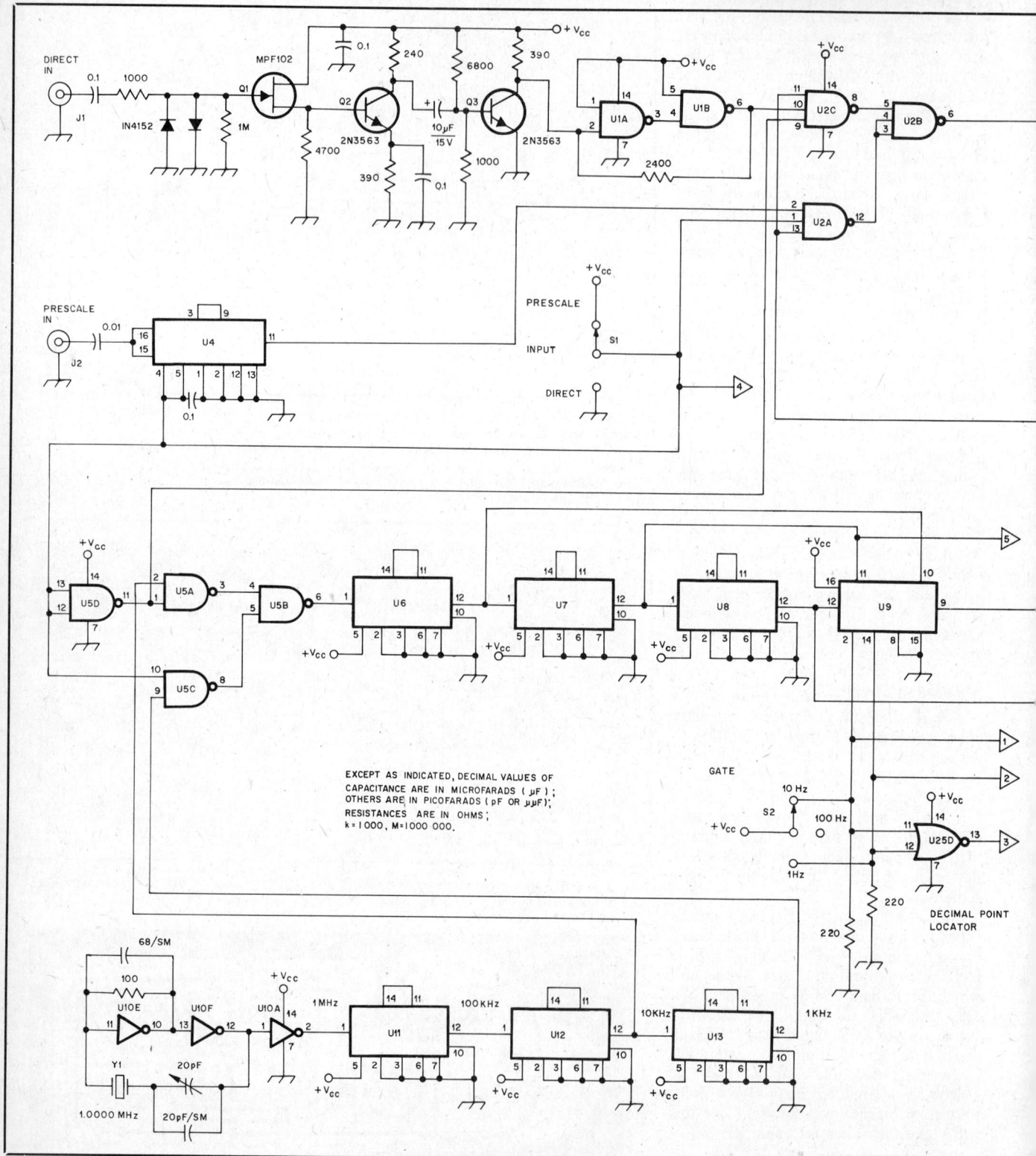

EXCEPT AS INDICATED, DECIMAL VALUES OF CAPACITANCE ARE IN MICROFARADS (µF); OTHERS ARE IN PICOFARADS (pF OR µµF); RESISTANCES ARE IN OHMS; k=1000, M=1000 000.

Fig. 31 — Schematic diagram of the 600 MHz counter. All resistors are 1/4-watt, 5-percent types. Nonpolarized capacitors are miniature ceramic unless noted otherwise. Polarized capacitors are tantalum types.

J1, J2 — Coaxial connector, BNC type.
S1 — Toggle, spdt.
S2 — Toggle, spdt, center off.

U1, U5 — 74S00.
U2, U15 — 74S10.
U3 — 74LS196.

U4 — 11C90.
U6, U7, U8, U11, U12, U13 — 74LS90.

energy first being determined by some other means such as a marker generator and receiver. The setting of the dial that gives the highest meter indication is the calibration point for that frequency. This point should be determined by tuning through it with loose coupling to the circuit being measured.

Frequency Standards

The difference between a marker generator and a *frequency standard* is that in the latter special pains are taken to make the oscillator frequency as stable as possible in the face of variations in temperature, humidity, line voltage, and other factors which could cause a small change in frequency.

While there are no definite criteria that distinguish the two in this respect, a circuit designated as a "standard" for

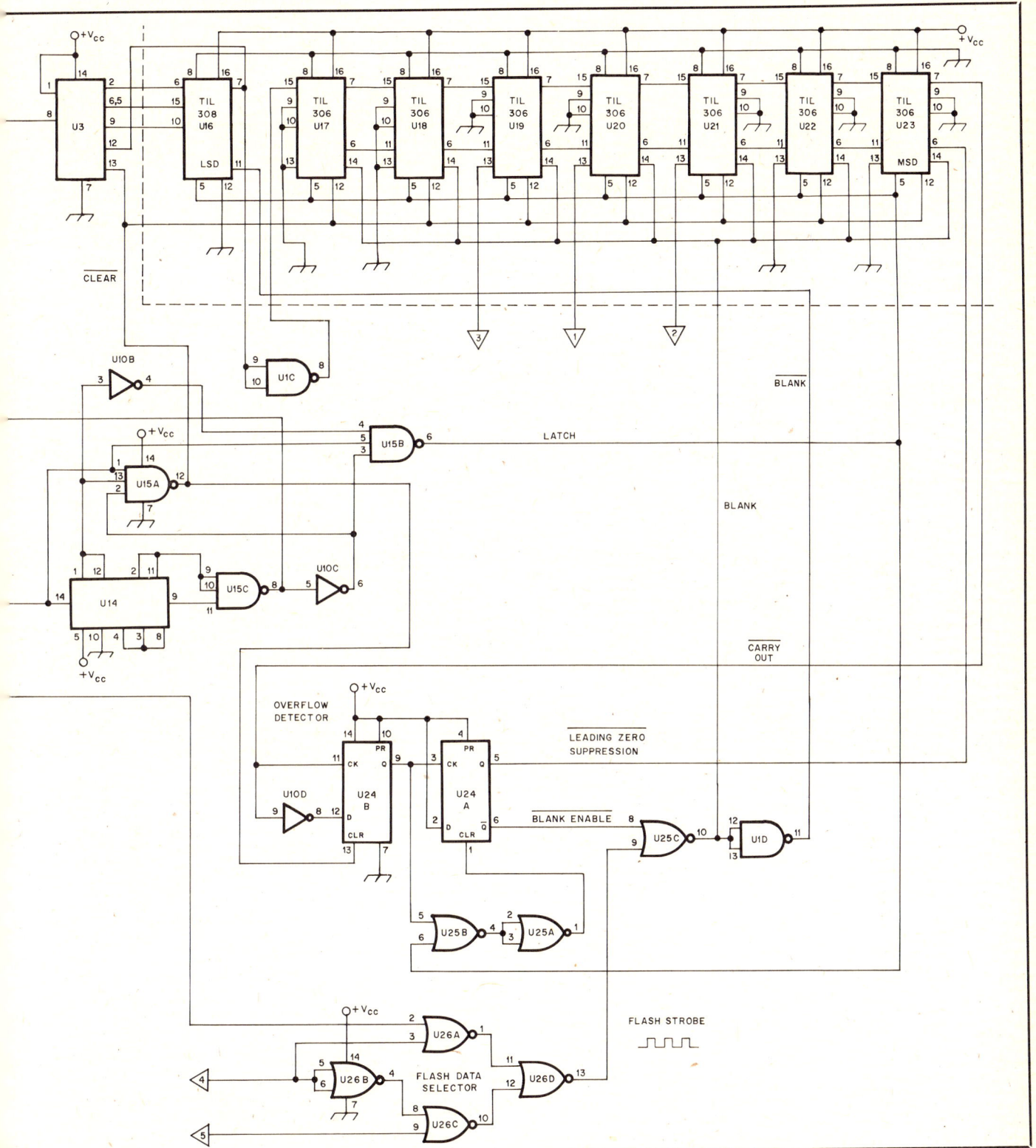

U9 — 74LS153.
U10 — 7404.

U14 — 74LS93.
U16 — TIL308.
U17-U23, incl. — TIL306.

U24 — 74LS74.
U25, U26 — 74LS02.
Y1 — 1.000-MHz crystal.

amateur purposes should be capable of maintaining frequency within at least a few parts per million under normal variations in ambient conditions, without adjustment. A simple marker generator using a 100-kHz crystal can be expected to have frequency variations 10 times (or more) greater under similar conditions. It can of course be adjusted to exact frequency at any time the WWV (or equivalent) signal is available.

The design considerations of high-precision frequency standards are outside the scope of this chapter, but information is available from time to time in periodicals.

Frequency Counters

One of the most accurate means of measuring frequency is the frequency counter. This instrument is capable of displaying numerically the frequency of the signal supplied to its input. For example, if an oscillator operating at 8.244 MHz is connected to the counter input, 8.244 would be displayed. At present, there are counters that are usable well up into the GHz range. Most counters that are to be used at high frequencies make use of a prescaler ahead of a basic low-frequency counter. Basically, the prescaler divides the high-frequency signal by 10, 100, 1000 or some other amount so that the low-frequency counter can display the operating frequency.

The accuracy of the counter depends on an internal crystal reference. The more accurate the crystal reference, the more accurate will be the readings. Crystals for frequency counters are manufactured to close tolerances. Most counters have a trimmer capacitor so that the crystal can be set exactly on frequency. A crystal frequency of 1 MHz has become more or less standard. The 10th harmonic of the crystal can be compared to the 10-MHz signal of WWV or WWVH and adjusted for zero beat.

A 600-MHZ FREQUENCY COUNTER

A frequency counter is one of the most versatile instruments a radio amateur can own. To align or troubleshoot a synthesized transceiver, a counter is practically mandatory. The unit described here should satisfy the most demanding amateur requirements. While it can't be considered low in cost, it is inexpensive compared to commercially manufactured units of similar sophistication. A cash expenditure in the neighborhood of $100 (1980 prices) will secure the parts.

Two input connectors are provided. The one marked "direct" has a frequency range of 20 Hz to over 40 MHz, with an ultimate resolution of 1 Hz. The sensitivity on this range is 100 mV. For frequencies up to 600 MHz, the "prescale" input is used. On this input the sensitivity is 100 mV, and the resolution is 10 Hz. Leading-zero suppression, automatic decimal point

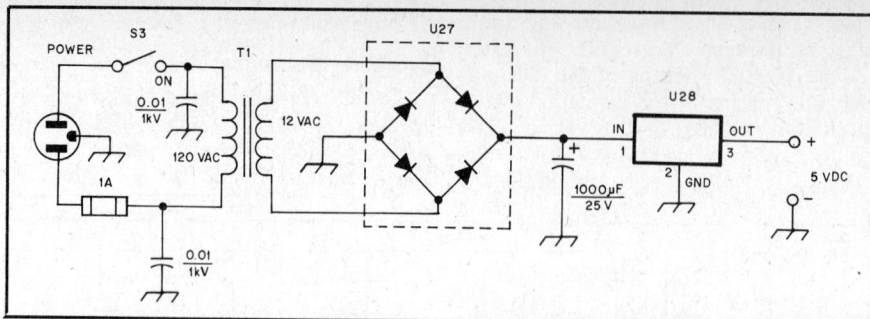

Fig. 32 — Schematic diagram of the supply used to power the 600-MHz frequency counter. The 0.01-μF capacitors are disc ceramic and the 1000-μF unit is an electrolytic type.
S3 — Toggle, spst.
T1 — Transformer, 117-volt primary, 12.6-volt c.t. secondary, c.t. not used in this circuit. Stancor P-8384 or equiv.
U28 — 5-volt regulator, LM-309K or equiv.
U27 — Bridge rectifier assembly, 5-ampere, 50 PRV or greater.

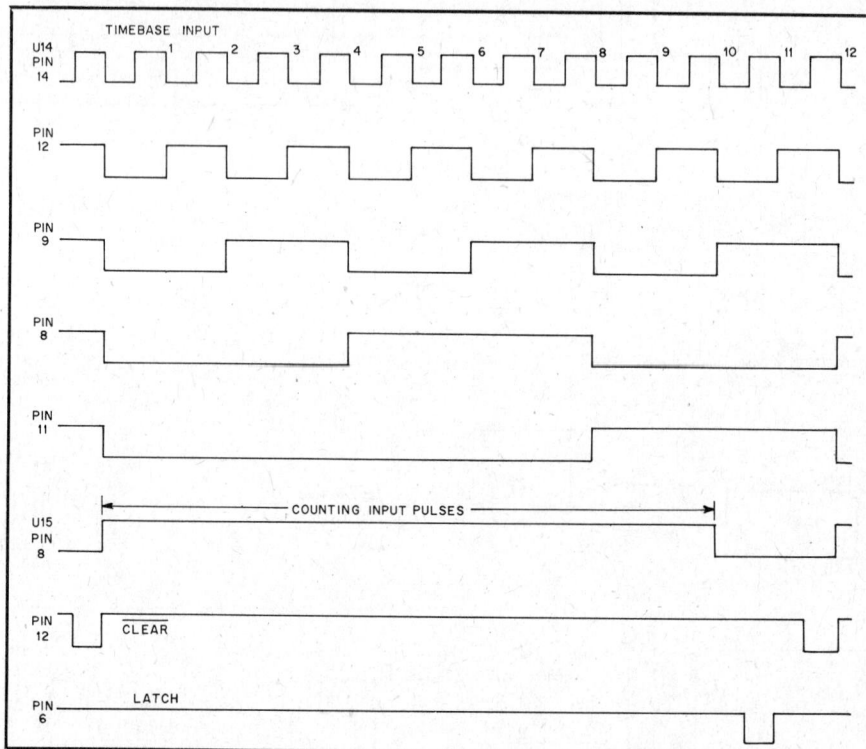

Fig. 33 — Timing diagram for the 600-MHz counter as it should appear at key points in the circuit.

placement and positive overflow indication minimize readout ambiguity and enhance the usefulness of the instrument.

Technical Details

Fig. 31 is the schematic diagram. The counter is designed almost entirely with TTL integrated circuits. At the heart of the circuit is a 1-MHz crystal-controlled time base or clock oscillator. No special temperature compensation is used, but the stability is adequate for the usual indoor environment. If operation over a wide temperature range is contemplated, the builder should consider an oven or TCXO. A string of decade counters, U11 to U13 and U6 to U8, divides the clock frequency down to 100, 10 or 1 Hz, depending on the frequency range and resolution selected. The divider chain is interrupted by U5, which gates a 1-kHz

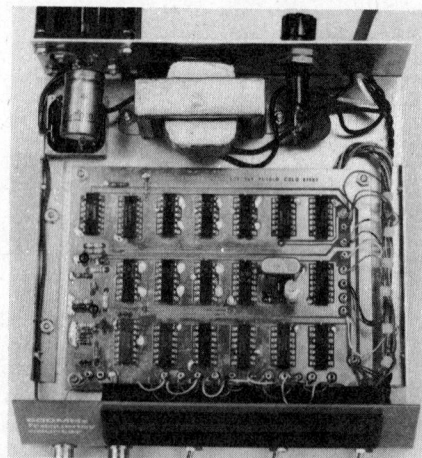

Fig. 34 — Interior view of the 600 MHz frequency counter. Cable ties are used on the wiring harness for a neat appearance. A large heat sink is visible at the upper-left corner of the counter.

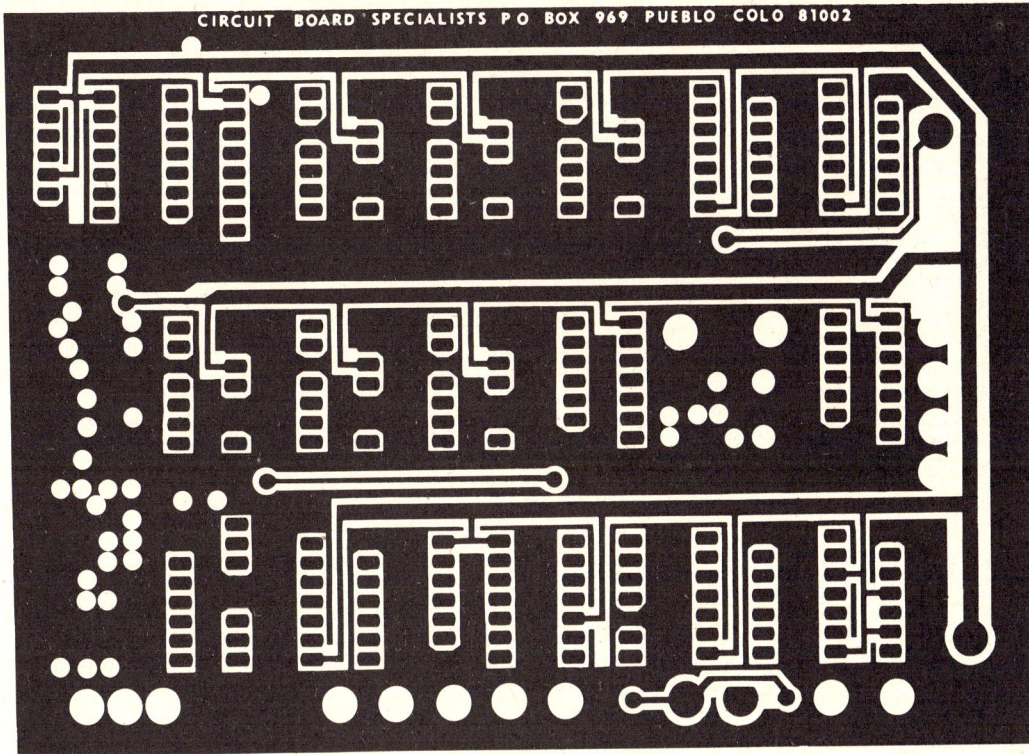

(A)

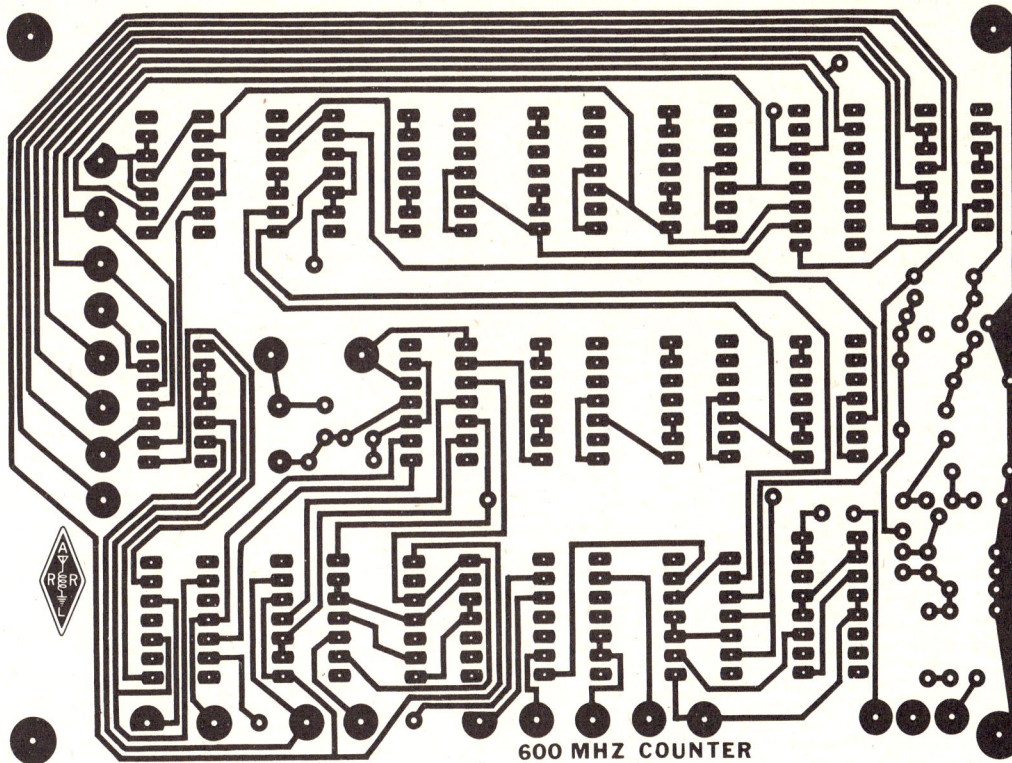

600 MHZ COUNTER

(B)

Fig. 35 — Full-scale etching pattern for the 600-MHz counter pc board. The component side of the board is shown in A, and the bottom side in B.

signal to U6 when the prescale input is selected via S1, or 10 kHz for the direct input. The final clock frequency is determined by the gating interval, which is controlled by U9 in conjunction with S2. S2

(with U25D) also positions the decimal point in the display.

U17 to U23 are counters with integral displays. Use of these LSI devices simplifies the design and reduces the

number of ICs and connections. Each counter/display IC has provisions for leading-zero suppression and complete blanking in addition to a decimal point. Leading-zero suppression makes the display easier to read by eliminating superfluous figures while preventing unnecessary energy consumption. The complete blanking feature is used in the overflow indicator circuit. The counting and display functions are separated in the LSD to allow greater speed.

The timing diagram is given in Fig. 33. The count sequence begins when the CLEAR line (output of U15A) goes LOW for one clock interval. This CLEAR pulse resets all the counters to zero in preparation for receiving input signals. U10, U14 and U15 perform the "bookkeeping" chores. Input pulses are gated into the counter chain through U2 when the output of U15C is HIGH. The input gate is enabled for 10 clock pulses. The eleventh clock pulse triggers the LATCH line (output of U15B) LOW for one clock interval, which causes the number of input pulses accumulated in the counter during the gate period (in other words, the input frequency) to be displayed. When the LATCH line returns to its HIGH state, the display remains fixed throughout the next gate period.

The counter can display readings up to 99999999. (The decimal point position depends on the range and resolution selected.) If the input frequency increased by two counts per gate interval, the proper reading would be 100000001. However, the instrument can only display the eight least significant digits, so the leading "1" would be lost. Since the first seven digits in the register are zeros, the leading-zero suppression circuitry within the counter/display ICs would normally cause all the digits except the last "1" to be blanked. Thus, one could be misled into thinking he is measuring a very low frequency when the actual value is eight orders of magnitude higher! The positive overflow indicator circuitry overcomes this ambiguity. Under ordinary conditions the Q output of U24B is LOW because it is periodically reset by the CLEAR line. The LOW output of U24B enables U25A and B so that U24A is periodically reset by the LATCH line. With U24A in the reset state the leading-zero suppression function operates normally and the blanking function is locked out.

If the display overflows, U23 delivers a CARRY OUT pulse to U10D and U24B, which sets the data input of U24B HIGH. The end of the CARRY OUT pulse (LOW to HIGH transition) transfers the HIGH data to the output of U24B. (The propagation delay of U10D prevents loss of data before it can be transferred by the rising edge of the clock.) The rising edge of the U24B output clocks the U24A Q output HIGH, (which disables the leading-zero suppression) and the U24A Q output LOW (which enables U25C, the blanking gate.) The HIGH U24B output also locks out U25A and B, preventing U24A from being reset by the LATCH signal. U26 selects an output from U8 or U9 (whichever happens to be toggling at the proper frequency) and flashes the digits, including all leading zeros, at a 10 Hz rate. The display will continue to flash until the input frequency no longer overflows the display. There is no delay between an input overflow and a flashing display. Because the readout displays the number of pulses accumulated by the register during the previous gate interval, the digits may begin to flash before the reading changes to leading zeros.

A simple signal conditioner is used at the "direct" input. Q1, a source follower, presents a 1 MΩ input resistance to signals having amplitudes less than 1.2 V pk-pk. Above this level the input resistance drops

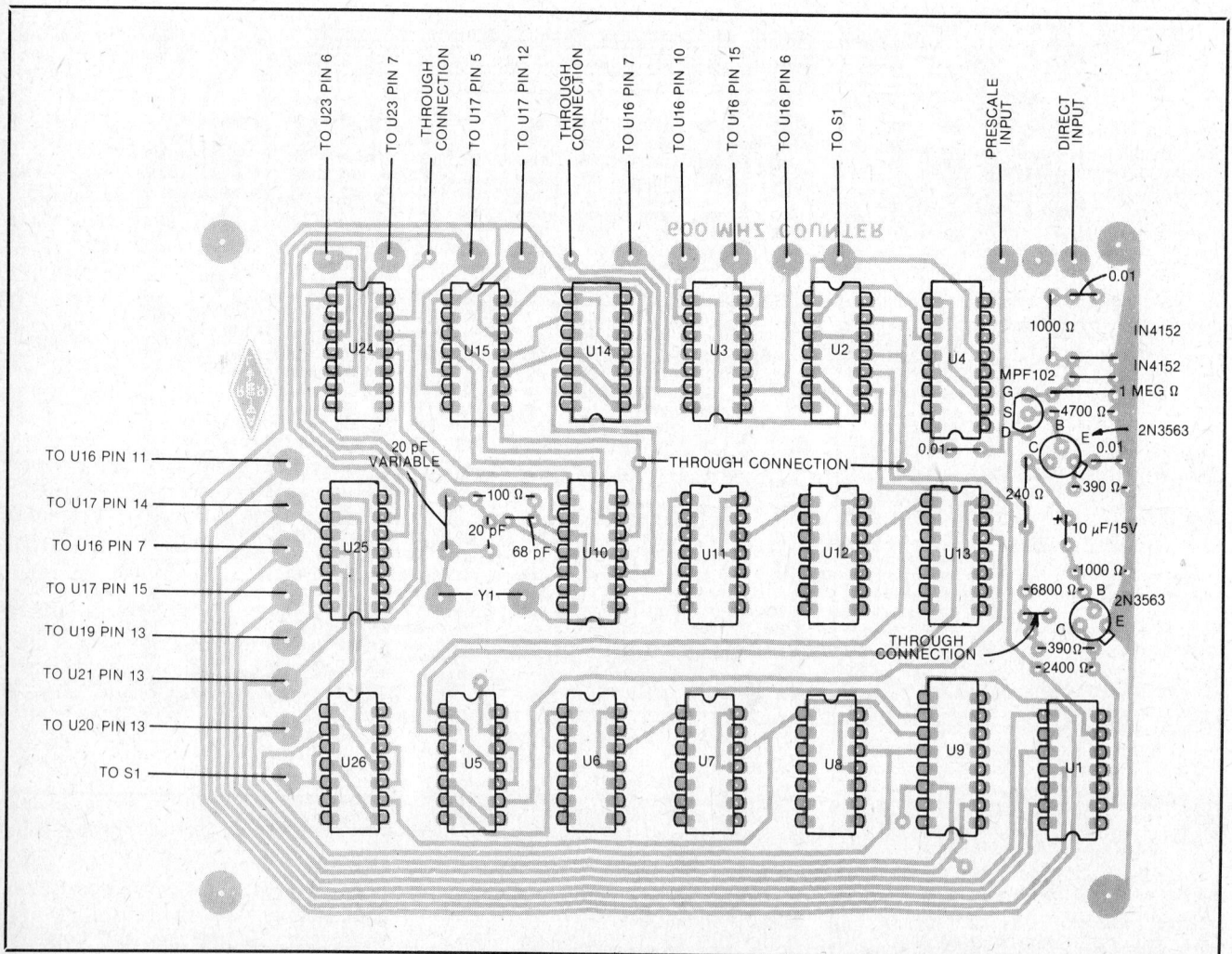

Fig. 36 — Parts-placement diagram for the 600-MHz counter pc board as seen from the component side of the board.

to about 1-kΩ, determined by the current-limiting resistor for the silicon protective diodes. A two-stage RC-coupled amplifier consisting of Q2 and Q3 provides broadband gain and wave squaring. U1A and B convert the signal to TTL levels compatible with the remainder of the circuit. The positive feedback around U1A and B causes Schmitt trigger action, which discourages oscillation. Any oscillations on the logic transitions would be counted as input pulses, causing erroneous readings.

U4 divides the "prescale" input frequency by 10. This IC is an ECL divider with a built-in ECL-to-TTL translator. The input is internally biased, so no external conditioning circuitry is necessary except the coupling capacitor.

Construction

All of the integrated circuits (with the exception of the readouts), prescaler and the input signal-conditioning circuit are contained on a double-sided circuit board that measures 4 × 5-1/2 inches (102 × 140 mm). The bottom pattern of the board contains most of the connections between ICs. The power bus and a few IC interconnections are on the top side of the board. Large amounts of foil were left on the top of the board to provide a ground plane. This appears to be effective since no bypass capacitors on the power bus were required for glitch-free counter operation. Circuit-board etching patterns and parts-placement guide information is given in Figs. 35 and 36. Cambion terminals are used for each of the connection points to the board. This provides for easy removal of the board should that be necessary.

The readouts are contained in an IEE-Atlas 1750 bezel assembly that is attached to the front panel. Each of the readouts plugs into this assembly and a red filter, supplied with the bezel, mounts over them. Wire wrap pins are provided for connection to the readouts. Connection from the display section to the circuit board is accomplished by means of a wiring harness as can be seen in the photograph of the counter interior. Tie wraps impart a professional appearance.

The power supply components are located against the rear wall of the counter. A simple power supply, shown in Fig. 32 provides the required 5-volts dc. A hefty heat sink is used on the regulator and the two remain only perceptibly warm even after long periods of use. It is a good idea to use as large a heat sink as is practical so that the chassis does not become unreasonably warm. Chassis heating may affect the frequency stability of the time base oscillator.

Three front-panel switches control the on/off power function, resolution and the selection of either the direct or prescale input. Separate BNC-type connectors are provided for the direct and prescale input. Two connectors are convenient if the builder desires to install an amplifier ahead of the prescaler.

The front panel escutcheon was made from 3M Scotchcal material. This involves a photographic process similar to circuit board production. These materials can be obtained from many artist supply houses. Alternatively, the panel can be labeled with press-on lettering.

Adjustment

Adjustment is limited to that of the time base oscillator. This can be accomplished by receiving WWV or a similar frequency-standard station. Couple a small amount of power from the time base oscillator, along with the WWV signal into the station receiver. A small coupling loop of wire connected to the receiver by a length of miniature coaxial cable should work fine. Place the loop in the vicinity of the oscillator but do not allow it to touch any of the circuit components, as a slight frequency "pulling" may occur. While listening to the two signals adjust the oscillator trimmer for zero beat. That completes the adjustment procedure.

Other Instruments and Measurements

Many measurements require a source of ac power of adjustable frequency (and sometimes adjustable amplitude as well) in addition to what is already available from the transmitter to receiver. Rf and af test oscillators, for example, provide signals for purposes such as receiver alignment, testing of phone transmitters, and so on. Another valuable adjunct to the station is the oscilloscope, especially useful for checking phone modulation.

RF Oscillators for Circuit Alignment

Receiver testing and alignment, covered in an earlier chapter, uses equipment common to ordinary radio service work. Inexpensive rf signal generators are available, both complete and in kit form. However, any source of signal that is weak enough to avoid overloading the receiver usually will serve for alignment work. The frequency marker generator is a satisfactory signal source. In addition, its frequencies, although not continuously adjustable, are known far more precisely, since the usual signal-generator calibration is not highly accurate. For rough work the dip meter described in the next section will serve.

A DUAL-GATE MOSFET DIP METER

The dip meter reverses the absorption-wave-meter procedure in that *it* supplies the rf power by incorporating a tunable oscillator from which the circuit being checked absorbs energy when this circuit and the oscillator are tuned to the same frequency and coupled together. In the vacuum-tube version the energy absorption causes a decrease or "dip" in the

Fig. 37 — Exterior view of the dual-gate MOSFET dip meter with plug-in coils.

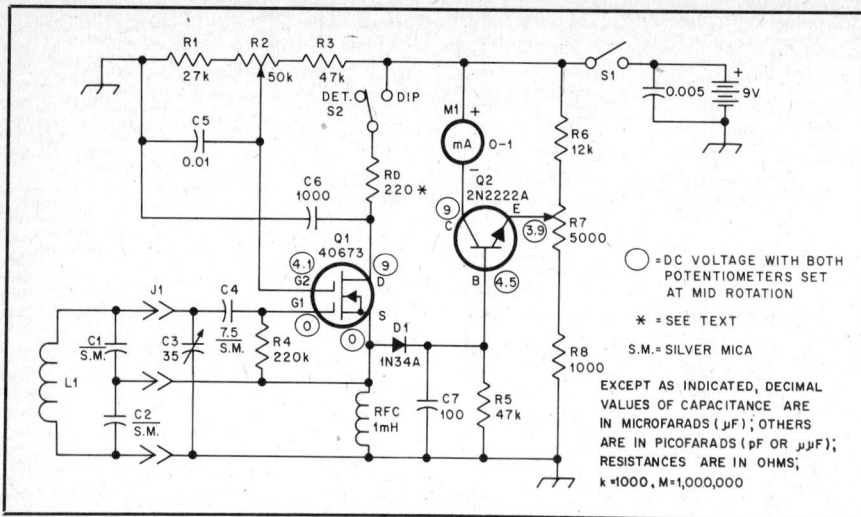

Fig. 38 — Schematic diagram of the dual-gate MOSFET dip meter. All resistors are 1/2-watt composition type. Capacitors are disc ceramic unless noted otherwise.
C1, C2 — See Table 1.
C3 — Variable capacitor, 35 pF, Millen 20035 or equiv.
D1 — 1N34A or equiv.
J1 — Socket, Amphenol type S4.
L1 — See Table 1 for values. All coils wound on Millen 45004 coil forms.

M1 — Edgewise panel meter, 0-1 mA, Calectro D1-905 or equiv.
Q1 — Dual-gate MOSFET, RCA 40673.
Q2 — Npn transistor, 2N2222A.
R2 — Potentiometer, 50 kΩ.
R7 — Potentiometer, 5000 ohms.
S1 — Spst on-off switch mounted on R7.

Table 1
Coil-Winding Information

Freq. Range MHz	C1 pF	C2 pF	L1 Turns
2.3-4	15	15	71-1/2
3.4-5.1	33	10	39-1/2
4.8-8	10	33	25-1/2
7.9-13	10	33	14-1/2
12.8-21.2	10	33	6-1/2
21-34	10	33	4-1/2
34-60	10	33	2-1/2
60-110	10	33	*
90-200	not used	not used	**

*denotes a 1-1/2-turn coil of no. 18 enam. wire wound on a 1/2-inch (13-mm) form spaced 1/8 inch (3 mm) between turns. It should be placed so that the coil is near the top of the coil form.
**denotes a hairpin loop made from flashing copper, 3/8-inch (9.5-mm) wide × 1-7/8-inch (89 mm) total length.
All other coils are wound with no. 24 enam. wire.

oscillator's rectified grid current, measured by a dc microammeter.

Described here and shown in Figs. 37 through 40 is a simple-to-build dip meter that covers 2.3 to 200 MHz. By opening switch S2 the circuit will function as a wavemeter, eliminating the need for two separate test instruments. The layout is not especially critical; however, you should try to keep the leads from the coil socket to the remainder of the circuitry as short as possible. This will help prevent unwanted resonances in the higher frequency ranges. Such parasitic resonances can cause false dips and erratic operation.

Circuit Details

The circuit shown in Fig. 38 is a grounded-drain Colpitts oscillator employing an RCA n-channel, dual-gate MOSFET. The oscillation level, detected by a diode and amplified by a 2N2222A transistor, is displayed on a 0- to 1-mA meter. Transconductance of the MOSFET, and hence the output signal, is controlled by potentiometer R2 and reaches a maximum of 10-volts pk-pk at the source when V_{G2} (voltage from gate 2 to source) is set to +5 volts. The meter is adjusted for the desired deflection by R7. R8 must be selected according to the meter used and should be 1000 ohms for a 1-mA meter movement. Frequency of oscillation depends on C1, C2, C3 and L1, and may reach 250 MHz or so when L1 is reduced to a hairpin.

Higher frequencies may be obtainable by using a uhf D-MOSFET, such as a Signetics SD300, or by placing C3 and L1 in series in a Clapp-oscillator configuration. The circuit is designed to operate from a 12-volt supply, but it also works fine with a 9-volt transistor-radio type of battery if the drain resistor (R_D) is shorted. In either case the unit draws approximately 20 mA.

Construction

Most of the components that comprise the oscillator and meter-driver circuits are mounted on a circuit board that measures approximately 1-1/4 × 2-1/2 (32 × 63 mm) inches. The foil pattern is shown in Fig. 39. A Minibox that measures 5-1/2 × 3 × 2-1/2 inches (140 × 76 × 63 mm) contains the circuit board, variable capacitor, meter, controls and four-pin coil socket. Nine plug-in coils are used to cover the frequency range from 2.3 to 200 MHz. The coils are wound on Millen 45004 coil forms to which L brackets are mounted for the dial scale. Winding information is given in Table 1. Epoxy cement holds the aluminum brackets to the forms. The use of nine separate coils instead of five or six greatly expands the calibration scales so more accurate frequency measurements may be made. To reduce the fast tuning rate of the variable capacitor, a reduction vernier is used. It

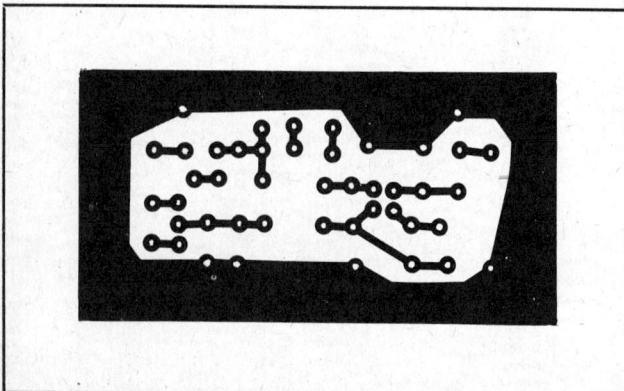

Fig. 39 — Full-scale circuit-board pattern for the MOSFET dip meter — foil side of board. Black areas represent unetched copper.

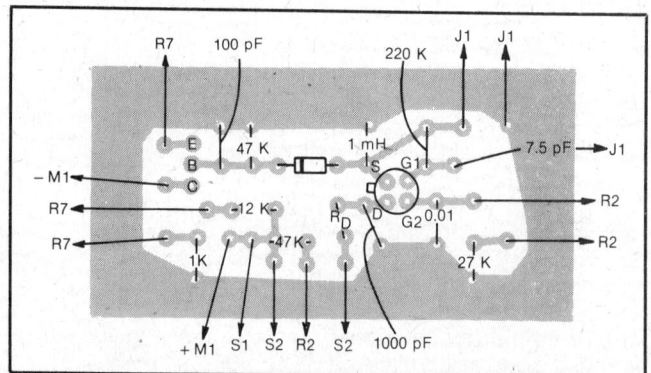

Fig. 40 — Parts-placement guide for the MOSFET dip meter shown from the component side of the board. Gray areas represent an X-ray view of the unetched copper.

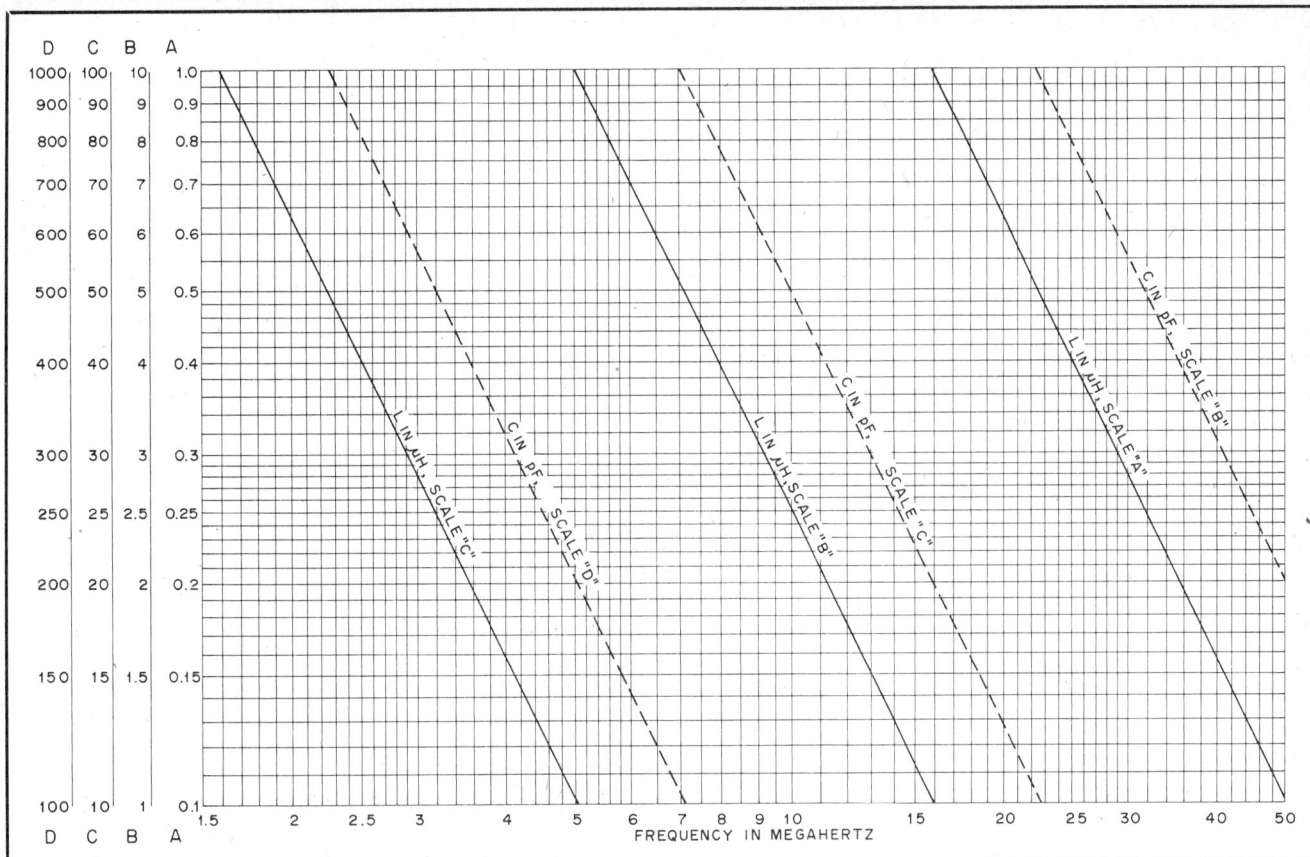

Fig. 41 — Chart for determining unknown values of L and C in the range of 0.1 to 100 μH and 2 to 1000 pF, using standards of 100 pF and 5 μH.

was removed from a Japanese vernier dial assembly. An aluminum bracket supports the variable capacitor inside the box. A rectangular piece of thin Plexiglas is used for the dial. A thin line is scribed down the center of the dial and is colored with a permanent-marking felt pen.

Alignment

A general-coverage receiver or another dip meter (calibrated) will be required to align the instrument. Plug in the appropriate coil for the range to be calibrated and turn the power switch to the ON position and advance R7 to approximately one-third scale. If a receiver is being used to calibrate the instrument, tune it to the lowest frequency covered by the particular coil in use. With the coil of the dip meter in close proximity to the receiver antenna terminal and the variable capacitor fully meshed, the dip-meter oscillator should be heard somewhere close to that frequency. Start by marking this frequency on the paper of thin cardboard dial attached to the plate. Next tune the receiver higher in frequency (approximately 100 kHz on the lower range coils and 1 MHz on the higher frequency ranges) and mark this frequency on the dial. Continue this procedure until the complete range of the particular coil has been marked. Do the same for each of the other coils. If another dip meter is used for the calibration process, it should be

placed in the DETECTOR mode and used in a similar fashion as that of the receiver outlined above.

Operating the Dip Meter

The dip meter will check only resonant circuits, since nonresonant circuits or components will not absorb energy at a specific frequency. The circuit may be either lumped or linear (a transmission-line type circuit) provided only that it has enough Q to give sufficient coupling to the dip-meter coil for detectable absorption of rf energy. Generally the coupling is principally inductive, although at times there may be sufficient capacitive coupling between the meter and a circuit point that is at relatively high potential with respect to ground to permit a reading. For inductive coupling, maximum energy absorption will occur when the meter is coupled to a coil (the same coupling rules which apply to any two coils are operative here) in the tuned circuit being checked, or to a high-current point in a linear circuit.

Because of distributed capacitance (and sometimes inductance) most circuits resonant at the lower amateur frequencies will show quasi-linear-type resonances at or close to the vhf region. A vhf dip meter will uncover these, often with beneficial results since such "parasitic" resonances can cause unwanted responses at harmonics of the intended frequency, or be

responsible for parasitic oscillations in amplifiers. Caution must be used in checking transmission lines or antennas — and, especially, combinations of antenna and line — on this account, because these linear circuits have well-defined series of harmonic responses, based on the lowest resonant frequency, which may lead to false conclusions respecting the behavior of the system.

Measurements with the dip meter are essentially frequency measurements, and for best accuracy the coupling between the meter and circuit under checking must be as loose as will allow a perceptible dip. In this respect the dip meter is similar to the absorption wavemeter.

Measuring Inductance and Capacitance with the Dip Meter

With a carefully calibrated dip meter, properly operated inductance and capacitance in the values ordinarily used for the 1.5-50 MHz range can be measured with ample accuracy for practical work. The method requires two accessories: an inductance "standard" of known value, and a capacitance standard also known with reasonable accuracy. Values of 100 pF for the capacitance and 5 μH for the inductance are convenient. The chart of Fig. 41 is based on these values.

The L and C standards can be quite ordinary components. A small silver-mica capacitor is satisfactory for the

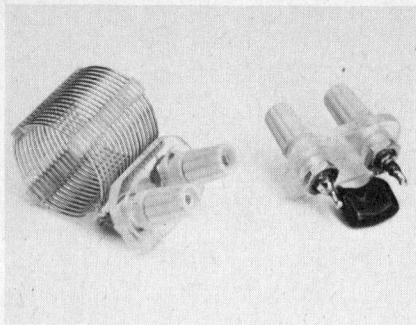

Fig. 42 — A convenient mounting, using bind-ingpost plates, for L and C standards made from commercially available parts. The capacitor is a 100-pF silver mica unit, mounted so the lead length is as nearly zero as possible. The inductance standard, 5 μH, is 17 turns of coil stock, 1-inch (25-mm) diameter, 16 turns per inch.

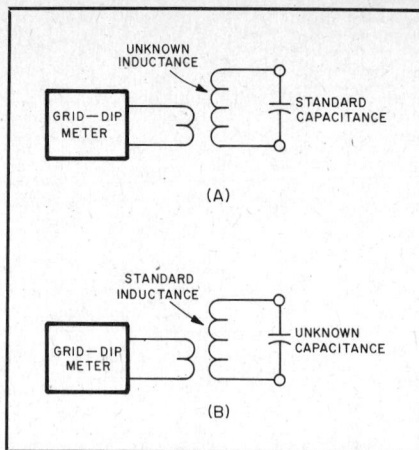

Fig. 43 — Setups for measuring inductance and capacitance with the dip meter.

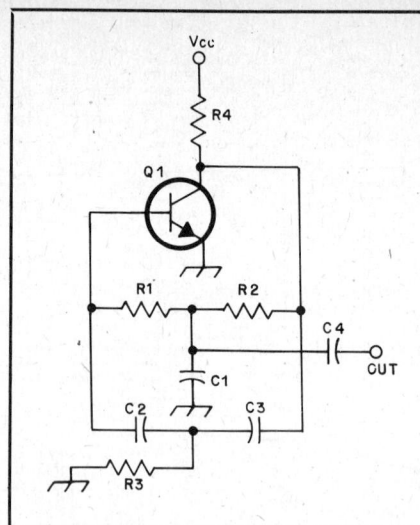

Fig. 44 — Twin-T audio oscillator circuit. Representative values for R1-R2 and C1 range from 18 kΩ and 0.05 μF for 750 Hz to 15 kΩ and 0.02 μF for 1800 Hz. For the same frequency range, R3 and C2-C3 vary from 1800 ohms and 0.02 μF to 1500 ohms and 0.01 μF. R4 should be approximately 3300 ohms. C4, the output coupling capacitor, can be 0.05 μF for high-impedance loads.

capacitance, since the customary tolerance is ±5 percent. The inductance standard can be cut from commercial machine-wound coil stock; if none is available, a homemade equivalent in diameter, turn spacing, and number of turns can be substituted. The inductance will be 5 μH within amply close tolerances if the specifications in Fig. 42 are followed closely. In any case, the inductance can easily be adjusted to the proper value; it should resonate with the 100-pF capacitor at 7100 kHz.

The setup for measuring an unknown is shown in Fig. 43. Inductance is measured with the unknown connected to the standard capacitance. Couple the dip meter to the coil and adjust the meter for the dip, using the loosest possible coupling that will give a usable indication. Similar procedure is followed for capacitance measurement, except that the unknown is connected to the standard inductance. Values are read off the chart for the frequency indicated by the dip meter.

Coefficient of Coupling

The same equipment can be used for measurement of the coefficient of coupling between two coils. This simply requires two measurements of inductance (of *one* of the coils) with the coupled coil first open-circuited and then short-circuited. Connect the 100-pF standard capacitor to one coil and measure the inductance with the terminals of the second coil open. Then short the terminals of the second coil and again measure the inductance of the first. The coefficient of coupling is given by

$$k = \sqrt{1 - \frac{L_2}{L_1}}$$

where

- k = coefficient of coupling
- L1 = inductance of first coil with terminals of second coil open

L2 = inductance of first coil with terminals of second coil shortened

Audio-Frequency Oscillators

Tests requiring an audio-frequency signal generally call for one that is a reasonably good sine wave, and best oscillator circuits for this are RC-coupled, operating as close to a Class A amplifier as possible. Variable frequency covering the entire audio range is needed for determining frequency response of audio amplifiers, but this is a relatively unimportant type of test in amateur equipment. The variable-frequency af signal generator is best purchased complete; kits are readily available at prices that compare very favorably with the cost of parts.

For most phone-transmitter testing, and for simple trouble shooting in amplifiers, an oscillator generating one or two frequencies with good wave form is adequate. A "two-tone" (dual) oscillator is particularly useful for testing sideband transmitters, and adjusting them for on-the-air use.

The circuit of a simple RC oscillator useful for general test purposes is given in Fig. 44. This "Twin-T" arrangement gives a waveform that is satisfactory for most purposes, and by choice of circuit constants the oscillator can be operated at any frequency in the usual audio range. R1, R2 and C1 form a low-pass type network, while C2C3R3 is high-pass. As the phase shifts are opposite, there is only one frequency at which the total phase shift from collector to base is 180 degrees, and oscillation will occur at this frequency. Optimum operation results when C1 is approximately twice the capacitance of C2 or C3, and R3 has a resistance about 0.1 that of R1 or R2 (C2 = C3 and R1 = R2). Output is taken across C1, where the harmonic distortion is least. A relatively high-impedance load should be used — 0.1 megohm or more.

A small-signal af transistor is suitable for Q1. Either npn or pnp types can be

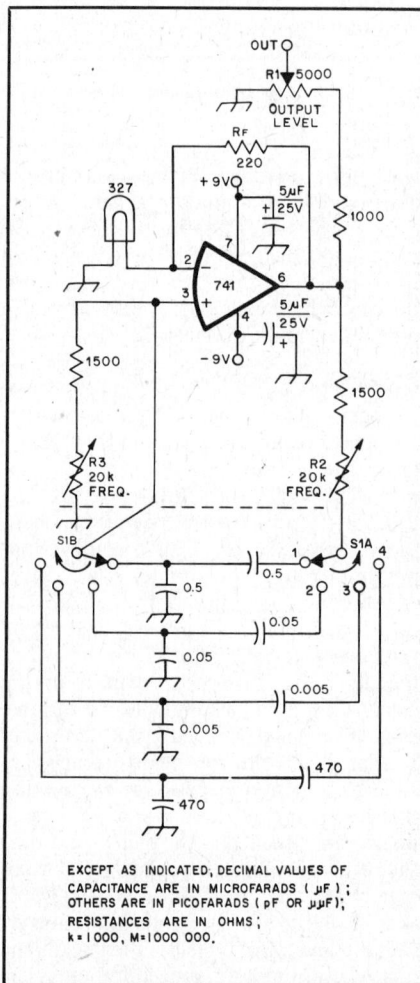

Fig. 45 — A simple audio oscillator that provides a selectable frequency range. R2 and R3 control the frequency and R1 varies the output level.

used, with due regard for supply polarity. R4, the collector load resistor, must be large enough for normal amplification, and may be varied somewhat to adjust the operating conditions for best waveform.

A WIDE-RANGE AUDIO OSCILLATOR

A wide-range audio oscillator that will provide a moderate output level can be built from a single 741 operational amplifier (Fig. 45). Power is supplied by two 9-volt batteries, from which the circuit draws 4 mA. The frequency range is selectable from 15 Hz to 150 kHz, although a 1.5-to 15-Hz range can be included with the addition of two 5-μF non-polarized capacitors and an extra switch position. Distortion is approximately one percent. The output level under a light load (10 k ohms) is 4 to 5 volts. This can be increased by using higher battery voltages, up to a maximum of plus and minus 18 volts, with a corresponding adjustment of R_f.

Pin connections shown are for the TO-5 case. If another package configuration is used, the pin connections may be different. R_f (220 ohms) is trimmed for an output level about five percent below clipping. This should be done for the temperature at which the oscillator will normally operate, as the lamp is sensitive to ambient temperature. Note that the output of this oscillator is direct coupled. If you are connecting this unit into circuits where dc voltage is present, use a coupling capacitor. As with any solid-state equipment, be cautious around plate circuits of tube-type equipment, as the voltage spike caused by charging a coupling capacitor may destroy the IC. This unit was originally described by Schultz in November 1974 *QST*.

A TWO-TONE AUDIO GENERATOR

The audio frequency generator shown in Figs. 46 and 47 makes a very convenient signal source for testing the linearity of a single-sideband transmitter. To be suitable for transmitter evaluation, a generator of this type must produce two nonharmonically related tones of equal amplitude. The level of harmonic and intermodulation distortion must be sufficiently low so as not to confuse the measurement. The frequencies used in this generator are 700 and 1900 Hz, both well inside the normal audio passband of an ssb transmitter. Spectral analysis and practical application with many different transmitters has shown this generator to meet all of the requirements mentioned above. While designed specifically for transmitter testing, it is also useful anytime a fixed-frequency, low-level audio tone is needed. Details on distortion measurement and the two-tone test can be found in chapter 12.

Circuit Details

Each of the two tones is generated by a

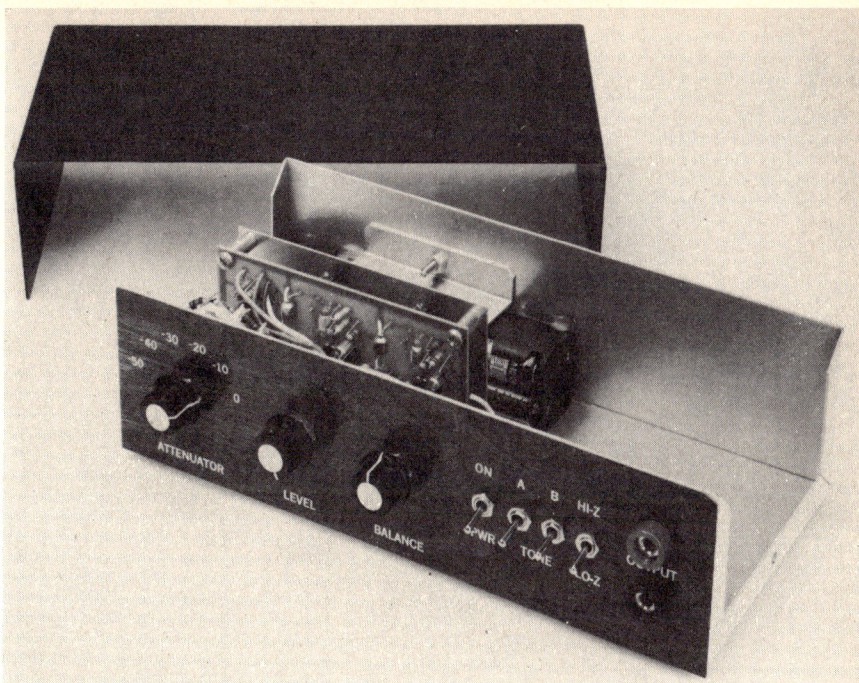

Fig. 46 — Exterior view of the two-tone audio generator.

separate Wein bridge oscillator, U1B and U2B. The oscillators are followed by RC active low-pass filters, U1A and U2A. Because the filters require nonstandard capacitor values, provisions have been made on the circuit board for placing two capacitors in parallel in those cases where standard values cannot be used. The oscillator and filter capacitors should be polystyrene or Mylar film types if available. The two tones are combined at the summing junction of op amp U3A. This amplifier has a variable resistor, R4, in its feedback loop which serves as the output LEVEL control. While R4 varies both tones together, R3, the BALANCE control, allows the level of tone A to be changed without affecting the level of tone B. This is necessary because some transmitters do not have equal audio response at both frequencies. Following the summing amplifier is a step attenuator, S3, which controls the output level in 10-dB steps. The use of two output level controls, R4 and S3, allows the output to cover a wide range and still be easy to set to a specific level. The remaining op amp, U3B, is connected as a voltage follower and serves to buffer the output while providing a high impedance load for the step attenuator. Either a high- or a low-output impedance can be selected by S4. The values shown are suitable for most transmitters using either high or low impedance microphones.

Construction and Adjustment

Component layout and wiring are not critical and any type of construction can be used with good results. For those who

wish to use a printed circuit board, the layout shown in Fig. 48 is recommended. Parts placement for this layout is shown in Fig. 49. Because the generator will normally be used near a transmitter, it should be enclosed in some type of metal case for shielding. Battery power was chosen to reduce the possibility of rf entering the unit through the ac line. With careful shielding and filtering, the builder should be able to use an ac power supply in place of the batteries.

The only adjustment required before use is the setting of the oscillator feedback trimmers, R1 and R2. These should be set so that the output of each oscillator, measured at pin 7 of U1 and U2, is about 0.5 volts rms. A *VTVM* or oscilloscope can be used for this measurement. If neither of these is available, the feedback should be adjusted to the minimum level that allows the oscillators to start reliably and stabilize quickly. When the oscillators are first turned on, they take a few seconds before they will have a stable output amplitude. This is caused by the lamps, DS1 and DS2, used in the oscillator feedback circuit. This is normal and should cause no difficulty. The connection to the transmitter should be through a shielded cable.

RESISTORS AT RADIO FREQUENCIES

Measuring equipment, in some part of its circuit, often requires essentially pure resistance — that is, resistance exhibiting only negligible reactive effects on the frequencies at which measurement is intended. Of the resistors available to

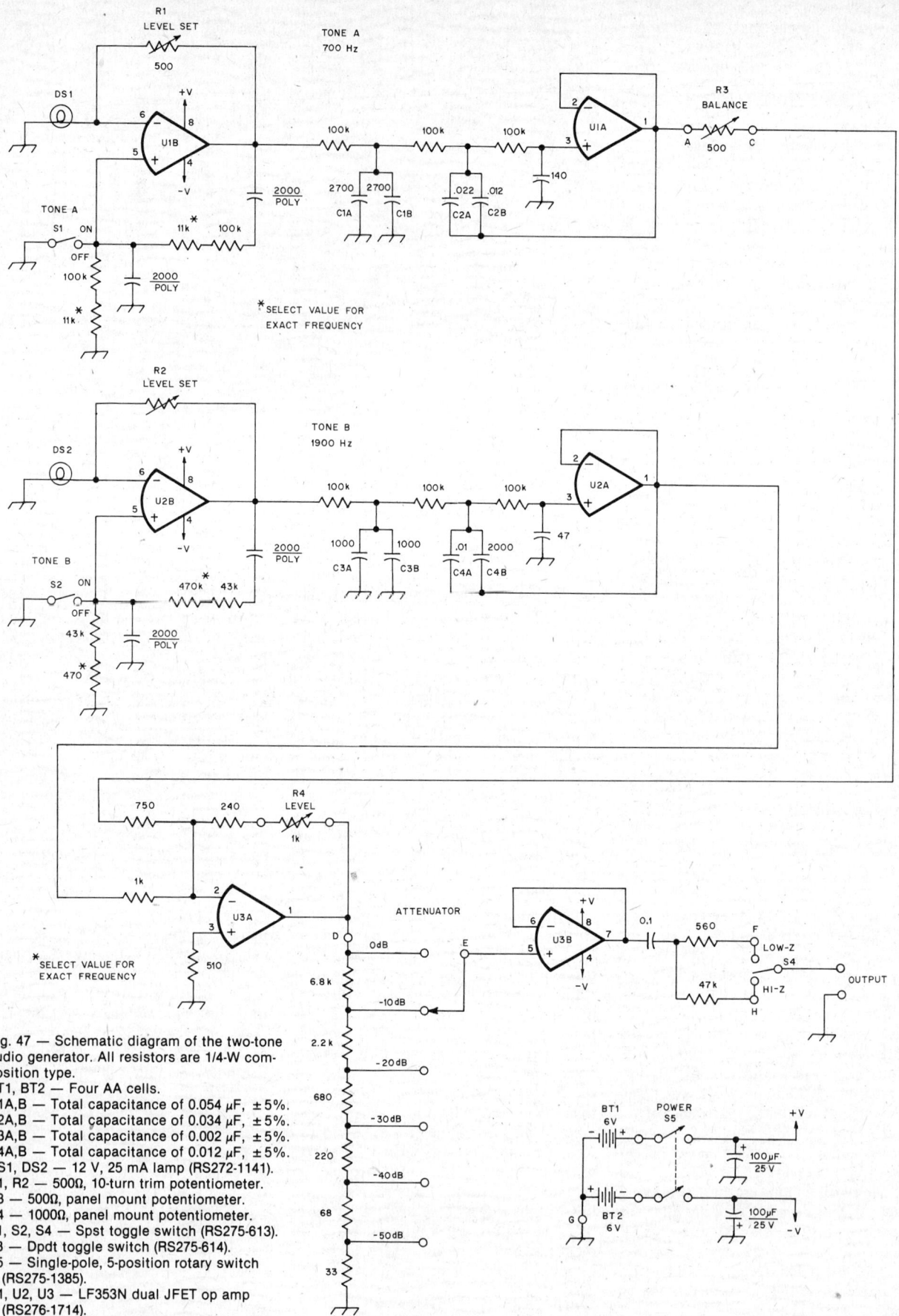

Fig. 47 — Schematic diagram of the two-tone audio generator. All resistors are 1/4-W composition type.

BT1, BT2 — Four AA cells.
C1A,B — Total capacitance of 0.054 µF, ±5%.
C2A,B — Total capacitance of 0.034 µF, ±5%.
C3A,B — Total capacitance of 0.002 µF, ±5%.
C4A,B — Total capacitance of 0.012 µF, ±5%.
DS1, DS2 — 12 V, 25 mA lamp (RS272-1141).
R1, R2 — 500Ω, 10-turn trim potentiometer.
R3 — 500Ω, panel mount potentiometer.
R4 — 1000Ω, panel mount potentiometer.
S1, S2, S4 — Spst toggle switch (RS275-613).
S3 — Dpdt toggle switch (RS275-614).
S5 — Single-pole, 5-position rotary switch
   (RS275-1385).
U1, U2, U3 — LF353N dual JFET op amp
   (RS276-1714).

amateurs, this requirement is met only by small composition (carbon) resistors. The inductance of wire-wound resistors makes them useless for amateur frequencies.

The reactances to be considered arise from the inherent inductance of the resistor itself and its leads, and from small stray capacitances from one part of the resistor to another and to surrounding conductors. Although both the inductance and capacitance are small, their reactances become increasingly important as the frequency is raised. Small composition resistors, properly mounted, show negligible capacitive reactance up to 100 MHz or so in resistance values up to a few hundred ohms; similarly, the inductive reactance is negligible in values *higher* than a few hundred ohms. The optimum resistance region in this respect is in the 50 to 200-ohm range, approximately.

Proper mounting includes reducing lead length as much as possible, and keeping the resistor separated from other resistors and conductors. Care must also be taken in some applications to ensure that the resistor, with its associated components, does not form a closed loop into which a voltage could be induced magnetically.

So installed, the resistance is essentially pure. In composition resistors the skin effect is very small, and the rf resistance up to vhf is very nearly the same as the dc resistance.

Dummy Antennas

A *dummy antenna* is simply a resistor that, in impedance characteristics, can be substituted for an antenna or transmission line for test purposes. It permits leisurely transmitter testing without radiating a signal. (The amateur regulations strictly limit the amount of "on-the-air" testing that may be done.) It is useful in testing receivers, in that electrically it resembles an antenna, but does not pick up external noise and signals, a desirable feature in some tests.

For transmitter tests the dummy antenna must be capable of dissipating safely the entire power output of the transmitter. Since for most testing it is desirable that the dummy simulate a perfectly matched transmission line, it should be a pure resistance, usually of approximately 52 or 73 ohms. This is a severe limitation in home construction, because nonreactive resistors of more than a few watts rated safe dissipation are very difficult to obtain. (There are, however, dummy antenna kits available that can handle up to a kilowatt.)

For receiver and minipower transmitter testing an excellent dummy antenna can be made by installing a 51- or 75-ohm composition resistor in a PL-259 fitting as shown in Fig. 50. Sizes from one half to two watts are satisfactory. The disc at the end helps reduce lead inductance and completes the shielding. Dummy antennas

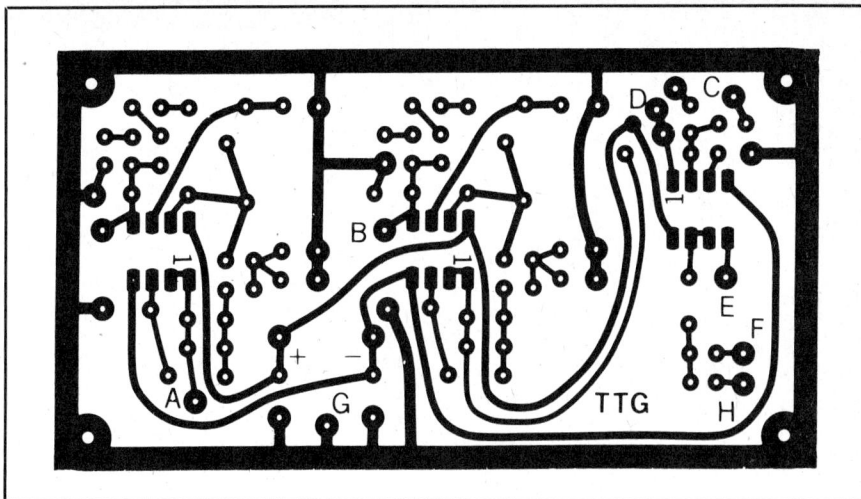

Fig. 48 — Full-scale pc-board etching pattern for the two-tone audio generator. Foil-side view with black areas representing unetched copper. To aid in wiring those components that are not mounted on the pc board, some pc-board pads have been labeled with letters corresponding to points marked on the diagram in Fig. 47.

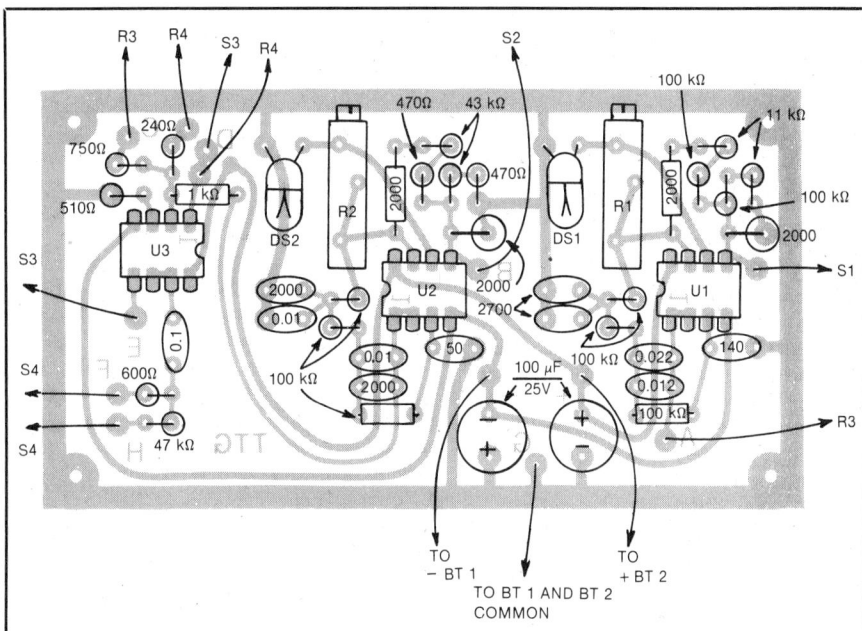

Fig. 49 — Parts-placement diagram for the two-tone audio generator, shown from the component-side of the board.

made in this way have good characteristics through the vhf bands as well as at all lower frequencies.

Increasing Power Ratings

More power can be handled by using a number of 2-watt resistors in parallel, or series-parallel, but at the expense of introducing some reactance. Nevertheless, if some departure from the ideal impedance characteristics can be tolerated this is a practical method for getting increased dissipations. The principal problem is stray inductance which can be minimized by mounting the resistors on flat copper strips or sheets, as suggested in Fig. 51.

The power rating on resistors is a *continuous* rating in free air. In practice,

the maximum power dissipated can be increased in proportion to the reduction in duty cycle. Thus with keying, which has a duty cycle of about one half, the rating

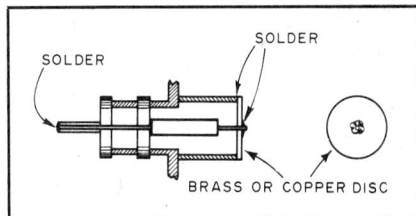

Fig. 50 — Dummy antenna made by mounting a composition resistor in a PL-259 coaxial plug. Only the inner portion of the plug is shown; the cap screws on after the assembly is completed.

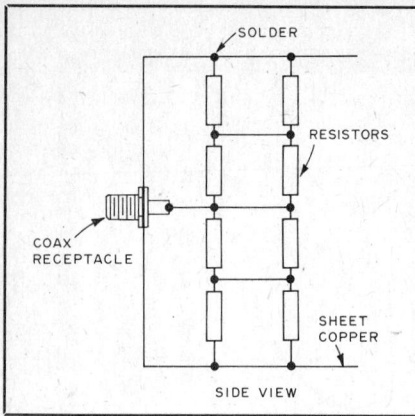

Fig. 51 — Using resistors in series-parallel to increase the power rating of a small dummy antenna. Mounted in this way on pieces of flat copper, inductance is reduced to a minimum. Eight 100-ohm 2-watt composition resistors in two groups, each four resistors in parallel, can be connected in series to form a 50-ohm dummy. The open construction shown permits free air circulation.

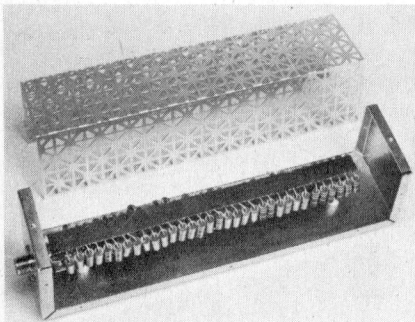

Fig. 52 — 100-watt dummy antenna made up of 66 2-watt carbon resistors.

can be doubled. With sideband the duty cycle is usually not over about one-third. The best way of judging is to feel the resistors occasionally (with power off); if too hot to touch, they may be dissipating more power than they are rated for.

The dummy load shown in Fig. 52 was constructed to test the feasibility of using a large number of 2-watt resistors to form a high-power load. This example uses 66 3300-ohm carbon resistors. The load will handle power levels up to 100 watts for reasonable lengths of time, and the SWR is less than 1.3 to 1 at frequencies up to 30 MHz. Because variations in construction may affect the stray reactances, it is recommended that the builder follow the general layout shown in the photograph. Complete details can be found in January 1981 *QST*.

The Oscilloscope

The electrostatically deflected cathode-ray tube, with appropriate associated equipment, is capable of displaying both low- and radio-frequency signals on its fluorescent screen, in a form which lends itself to ready interpretation. (In contrast, the magnetically deflected television pic-

ture tube is not at all suitable for measurement purposes.) In the usual display presentation, the fluorescent spot moves across the screen horizontally at some known rate (*horizontal deflection* or *horizontal sweep*) and simultaneously is moved vertically by the signal voltage being examined (*vertical deflection*). Because of the retentivity of the screen and the eye, a rapidly deflected spot appears as a continuous line. Thus a varying signal voltage causes a *pattern* to appear on the screen.

Conventionally, oscilloscope circuits are designed so that in vertical deflection the spot moves upward as the signal voltage becomes more positive with respect to ground, and vice versa (there are exceptions, however). Also, the horizontal deflection is such that with an ac sweep voltage — the simplest form — positive is to the right; with a *linear sweep* — one which moves the spot at a uniform rate across the screen and then at the end of its travel snaps it back very quickly to the starting point — *time* progresses to the right.

Most cathode-ray tubes for oscilloscope work require a deflection amplitude of about 50 volts per inch. For displaying small signals, therefore, considerable amplification is needed. Also, special circuits have to be used for linear deflection. The design of amplifiers and linear deflection circuits is complicated, and extensive texts are available. For checking modulation of transmitters, a principal amateur use of the scope, quite simple circuits suffice. A 60-Hz voltage from the power line makes a satisfactory horizontal sweep, and the voltage required for vertical deflection can easily be obtained from transmitter rf circuits without amplification.

For general measurement purposes amplifiers and linear deflection circuits are needed. The most economical and satisfactory way to obtain a scope having these features is to assemble one of the many kits available.

When using any oscilloscope, care must be taken not to damage the screen of the CRT. Too high beam intensity, a stationary focused spot, or even a stationary pattern left on the screen can cause overheating that may desensitize, burn or actually pierce the internal coating of phosphor.

Simple Oscilloscope Circuit

Fig. 53 is an oscilloscope circuit that has all the essentials for modulation monitoring: controls for centering, focusing, and adjusting the brightness of the fluorescent spot; voltage dividers to supply proper electrode potentials to the cathode-ray tube; and means for coupling the vertical and horizontal signals to the deflection plates.

The circuit can be used with electrostatic-deflection tubes from two to five inches in face diameter, with voltages up

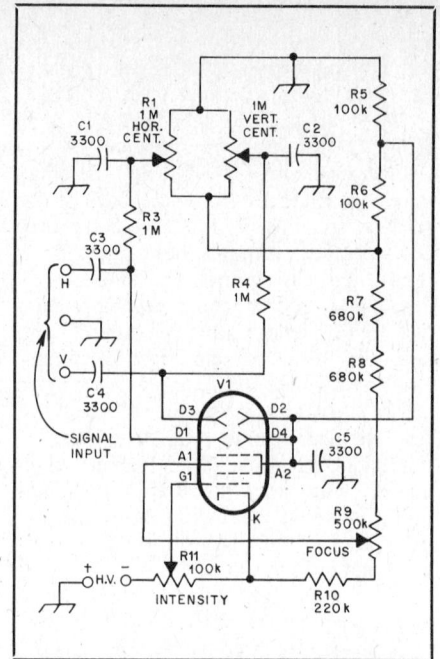

Fig. 53 — Oscilloscope circuit for modulation monitoring. Constants are for 1500- to 2500-volt high-voltage supply. For 1000 to 1500 volts, omit R8 and connect the bottom end of R7 to the top end of R9.
C1-C5, incl. — 1000-volt disc ceramic.
R1, R2, R9, R11 — Volume-control type, linear taper. R9 and R11 must be well insulated from chassis.
R3, R4, R5, R6, R10 — 1/2 watt.
R7, R8 — 1 watt.
V1 — Electrostatic-deflection cathode-ray tube, 2- to 5-inch (51- to 127-mm). Base connections and heater ratings vary with type chosen.

to 2500. Either set of deflecting electrodes (D1-D2, or D3-D4) may be used for either horizontal or vertical deflection, depending on how the tube is mounted.

In Fig. 53, the centering controls are not too high above electrical ground, so they do not need special insulation. However, the focusing and intensity controls are at a high voltage above ground and therefore should be carefully insulated. Insulated couplings or extension shafts should be used.

The tube should be protected from stray magnetic fields, either by enclosing it in an iron or steel box or by using one of the special CR tube shields available. If the heater transformer (or other transformer) is mounted in the same cabinet, care must be used to place it so the stray field around it does not deflect the spot. The spot cannot be focused to a fine point when influenced by a transformer field. The heater transformer must be well insulated, and one side of the heater should be connected to the cathode. The high-voltage dc can be taken from the transmitter plate supply; the current required is negligible.

Methods for connecting the oscilloscope to a transmitter for checking or monitoring modulation are given in earlier chapters.

Quasi-Linear Sweep

For wave-envelope patterns that require a fairly linear horizontal sweep, Fig. 54 shows a method of using the substantially linear portion of the 60-Hz sine wave — the "center" portion where the wave goes through zero and reverses polarity. A 60-Hz transformer with a center-tapped secondary winding is required. The voltage should be sufficient to deflect the spot well off the screen on both sides — 250 to 350 volts, usually. With such "over-deflection" the sweep is fairly linear, but it is as bright on retrace as on left-to-right. To blank it in one direction, it is necessary to couple the ac to the no. 1 grid of the CR tube as shown.

Lissajous Figures

When sinusoidal ac voltages are applied to both sets of deflecting plates in the oscilloscope the resultant pattern depends on the relative amplitudes, frequencies and phases of the two voltages. If the ratio between the two frequencies is constant and can be expressed in integers, a stationary pattern will be produced.

The stationary patterns obtained in this way are called *Lissajous figures*. Examples of some of the simpler Lissajous figures are given in Fig. 55. The frequency ratio is found by counting the number of loops along two adjacent edges. Thus in the second figure on the left there are three loops along a horizontal edge and only one along the vertical, so the ratio of the vertical frequency to the horizontal frequency is 3:1. Similarly, in the bottom figure there are four loops along the horizontal edge and three along the vertical edge, giving a ratio of 4:3. Assuming that the known frequency is applied to the horizontal plates, the unknown frequency is

$$f2 = \frac{n2}{n1} \, f1$$

where

f1 = known frequency applied to horizontal plates
f2 = unknown frequency applied to vertical plates
n1 = number of loops along a vertical edge
n2 = number of loops along a horizontal edge.

An important application of Lissajous figures is in the calibration of audio-frequency signal generators. For very low frequencies the 60-Hz power-line frequency is held accurately enough to be used as a standard in most localities. The medium audio-frequency range can be covered by comparison with the 440- and 600-Hz modulation on the WWV transmissions. It is possible to calibrate over a 10:1 range, both upward and downward from each of the latter frequencies and

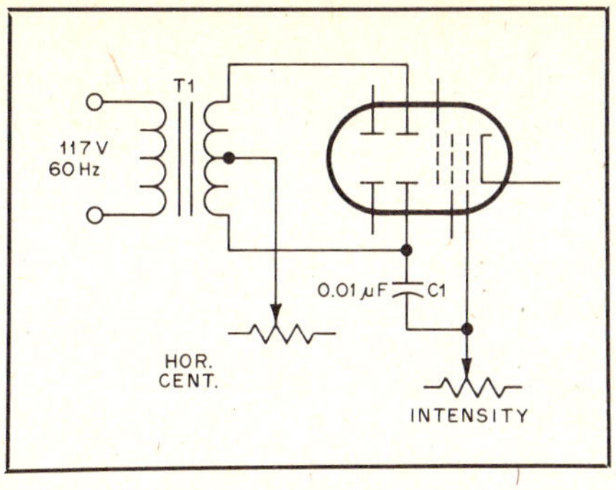

Fig. 54 — A quasi-linear time base for an oscilloscope can be obtained from the "center" portion of a sine-wave. Coupling the ac to the grid gives intensity modulation that blanks the retrace.
C1 — Ceramic capacitor of adequate voltage rating.
T1 — 250- to 350-volt center-tapped secondary. If voltage is too high, use dropping resistor in primary side.

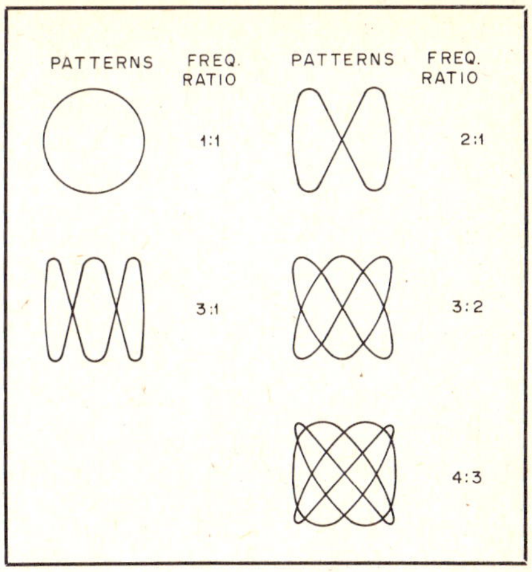

Fig. 55 — Lissajous figures and corresponding frequency ratios for a 90-degree phase relationship between the voltages applied to the two sets of deflecting plates.

thus cover the audio range useful for voice communication.

An oscilloscope having both horizontal and vertical amplifiers is desirable, since it is convenient to have a means for adjusting the voltages applied to the deflection plates to secure a suitable pattern size.

A Tester for FET and Bipolar Transistors

The circuit shown is intended solely as a tester for npn and pnp transistors, junction FETs and dual-gate MOSETs. This equipment is not for use in checking audio or high-power rf transistors.

The circuit of Fig. 57 is an oscillator which is wired so that it will test various small-signal transistors by switching the battery polarity and bias voltage. A crystal for the upper range of the hf spectrum is wired into the circuit permanently, but could be installed in a crystal socket if the builder so desires. A 20-MHz crystal was chosen for this model. Any hf crystal cut for fundamental mode operation can be used.

When testing FETs the bias switch, S3, is placed in the FET position, thus removing R2 from the circuit. However, when testing bipolar transistors the switch position must be changed to BIPOL so that forward bias can be applied to the base of the bipolar transistor under test. R1 is always in the circuit, and serves as a gate-leak resistor for FETs being evaluated. It becomes part of the bias network when bipolars are under test. C1 is used for feedback in combination with the internal capacitances of the transistors being checked. Its value may have to be changed experimentally if crystals for lower frequencies are utilized in the circuit. Generally speaking, the lower the crystal frequency, the greater the amount of capacitance needed to assure oscillation. Use only that amount necessary to provide quick starting of the oscillator.

Fig. 56 — Exterior view of the transistor/FET tester.

Components R3 and R4 are used as a voltage divider to provide bias for dual-gate MOSFETs. C2 is kept small in value to minimize loading of the oscillator by the low-impedance voltage doubler, D1 and D2. Rectified rf from the oscillator is monitored on M1. Meter deflection is regulated manually by means of control R5. S1 is used to select the desired supply voltage polarity — negative ground for testing n-channel FETs and npn bipolars, and a positive ground when working with p-channel and pnp devices.

When testing MOSFETs which are not gate protected (3N140 for one), make certain that the transistor leads are shorted together until the device is seated

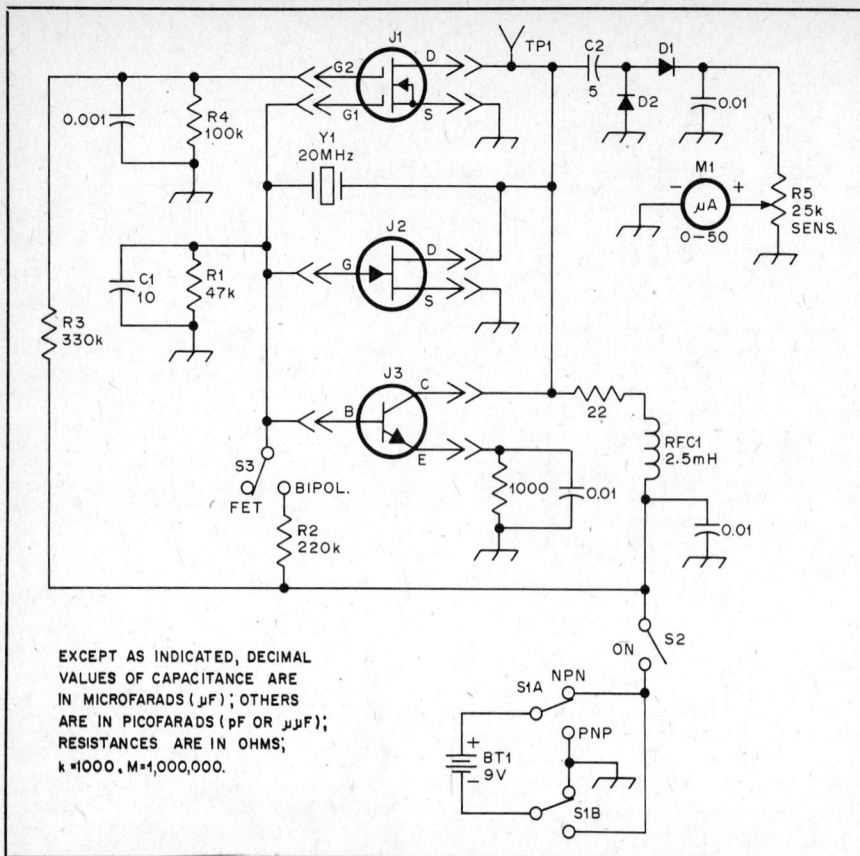

Fig. 57 — Schematic diagram of the transistor tester. Capacitors are disc ceramic or mica. Resistors are 1/2- or 1/4-watt composition except for R5. Numbered components not appearing in parts list are so designated for text discussion.

BT1 — Small 9-V transistor-radio battery.
D1, D2 — 1N34A germanium diode or equiv.
J1 — Four-terminal transistor socket.
M1 — Microammeter. Calectro D1-910 used here.
R5 — 25 kΩ linear-taper composition control

with switch.
RFC1 — 2.5-mH rf choke.
S1 — Two-pole double-throw miniature toggle.
S2 — Part of R5.
S3 — Spst miniature toggle.
Y1 — Surplus crystal (see text).

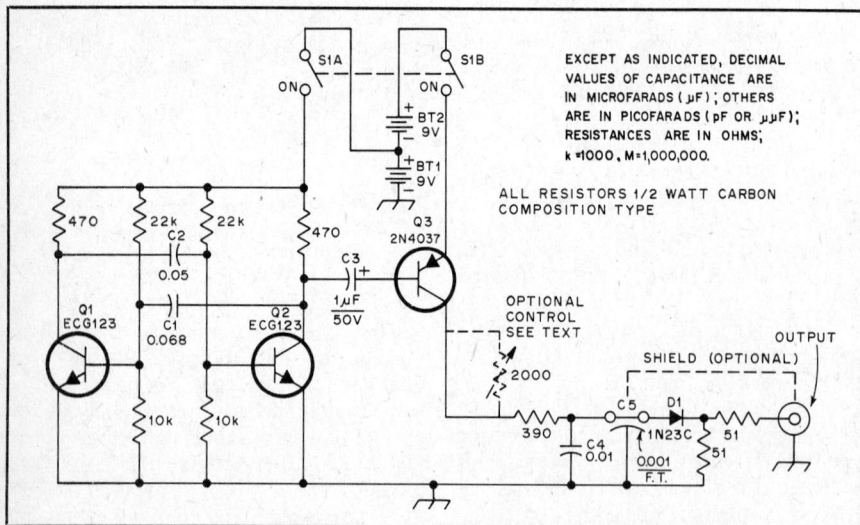

Fig. 58 — Schematic diagram of the gated noise source.
B1, B2 — 9-volt battery, Eveready 216 or equiv.
C5 — 0.001-μF feedthrough capacitor, Sprague BH-340.

S1 — Double-pole, single-throw miniature toggle.

in the test socket. Static charges on one's hands can be sufficiently great to damage the insulation within the transistor. Use a single strand of wire from some no. 22 or 24 stranded hookup wire, wrapping it two or three times around the pigtails of the FET as close to the transistor body as possible. After the FET is plugged into the socket, unwrap the wire and perform the tests. (It's not a bad idea to have an earth

ground connected to the case of the tester when checking unprotected FETs.) Put the shorting wire back on the FET leads before removing the unit from the tester.

The meter indication is significant in checking any type of transistor. If the device is open, shorted, or extremely leaky, no oscillation will take place, and the meter will not deflect. The higher the meter reading, the greater the vigor of the transistor at the operating frequency. High meter readings suggest that the transistor is made for vhf or uhf service, and that its beta is medium to high. Lower readings may indicate that the transistor is designed for hf use, or that it has very low gain. Transistors that are known to be good but will not cause the circuit to oscillate are most likely made for low-frequency or audio applications.

A Gated Noise Source

This circuit (Fig. 58) provides a simple low-cost method to optimize a converter or receiver for best noise figure. The simplicity of this system makes effective tune-up possible without a lot of test equipment.

Numerous articles have described units where noise-figure tests may be made. With the exception of certain thermal-limited diodes (5722, for example), an absolute value of noise figure is not obtainable with these units; this device is no exception.

Anyone using a classic noise-figure meter soon learns that the tune-up of a system is a cut-and-try procedure where an adjustment is made and its influence is observed by calibrating the system. Then the excess-noise source is applied and the effect evaluated. This is basically an after-the-fact method of testing after an adjustment is made, and is consequently time consuming.

The gated noise source doesn't require a special detector or any detector at all, other than your ear. By turning the noise source on and off at an audio rate, the ratio of noise contributed by the system to noise of the system *plus* excess noise appears as an audio note. The louder the note, the greater the differential in levels and hence the greater the influence of the excess noise or the better the noise figure.

If greater precision is desired than subjectively listening to the signal, an oscilloscope may be used. Hook the scope vertical input to any point in the audio system of the receiver, such as the speaker terminals. Adjust the scope for a display of several multiples of the train of square pulses. Proceed by adjusting the device(s) being tested for greatest vertical deflection.

The result of an adjustment is instantly visible as an increase or decrease in the recovered audio. This method of noise evaluation is by no means new. Most modern automatic noise-figure meters turn the excess-noise source on and off and then, through rather sophisticated

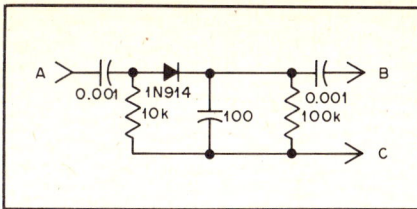

Fig. 59 — A simple detector which can be used when aligning ssb and fm receivers (see text for details).

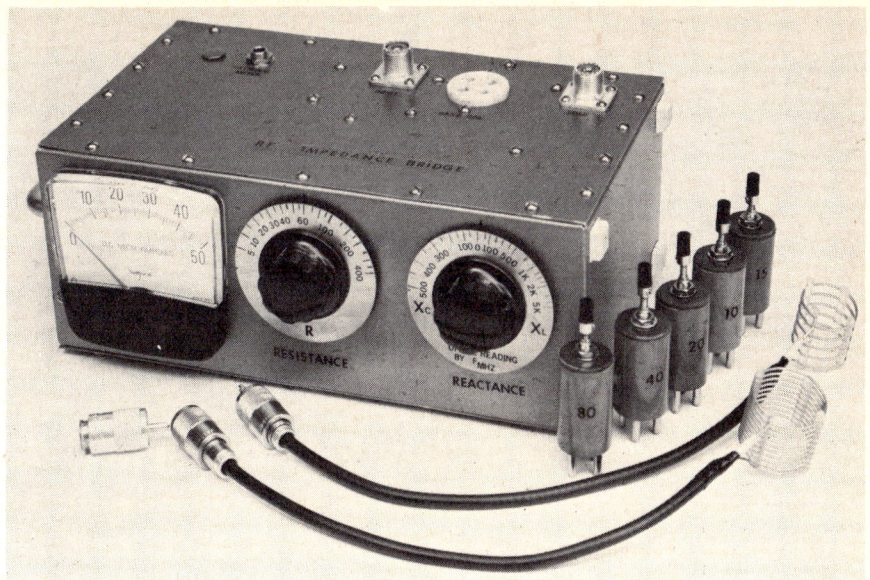

Fig. 60 — An RCL bridge for measuring unknown values of complex impedances. A plug-in coil is used for each frequency band. The bridge operates at an rf input level of about 5 volts; pickup-link assemblies for use with a grid-dip oscillator are shown. Before measurements are made, the bridge must be balanced with a nonreactive load connected at its measurement terminals. This load consists of a resistor mounted inside a coaxial plug, shown in front of the instrument at the left. The aluminum box measures 4-1/4 × 10-3/4 × 6-1/8 inches (108 × 273 × 156 mm) and is fitted with a carrying handle on the left end and self-sticking rubber feet on the right end and bottom. Dials are Millen no. 10009 with skirts reversed and calibrations added.

methods, evaluate the results. This technique is sometimes called "Y"-parameter testing.

While the method and circuit described here are not exceptional, they represent a fresh approach to noise evaluation. This approach does not require long-term integrating detectors and tedious "twice-power" measurements which, without absolute calibration, can result in no more than simply optimizing the system.

In some cases the available noise generated by this unit may be too great. The output may be reduced by inserting attenuators between the generator output and the device under test or by adding a 2000-ohm potentiometer at the point marked in the circuit (see Fig. 58). The use of an attenuator is preferred because it reduces the apparent output VSWR of the generator by increasing the return loss. If a control is used it must be returned to its minimum insertion-loss position when starting a test or no signal may be heard.

This circuit uses readily available junk-box parts and may be easily duplicated. The lead placement in and around the diode itself should follow good vhf practices with short leads and direct placement.

Theory of Operation

Q1 and Q2 are used in a cross-coupled multivibrator circuit, operating at approximately 700 Hz. The value of C1 is greater than C2 to cause the duty cycle to favor the conduction of Q2 slightly. When Q2 conducts, the pulse is coupled to Q3 via C3, turning on Q3 and causing current flow through R7, CR1 and R8.

The diode generates broadband noise which is passed through R9 to the output. R7, C4 and C5 form a low-pass filter to prevent high-order harmonics of the switching pulses from appearing in the output.

The influence of stray rf signals entering the device under test through the generator may be minimized by shielding the components shown. A simple box may be built by using pc-board scraps. For best match, this source should be connected directly to the input of the device under test; therefore, the unit is equipped with a male connector. This matching becomes a greater consideration as the frequency of

interest increases.

Addendum

The gated noise source was first developed in November 1975. Subsequently, some interesting things have been learned regarding its application. Some contemporary receivers and transceivers cannot be operated in the a-m mode, and consequently the noise source seems not to operate. The detection of noise is the process by which the noise source operates; therefore, it will not work through an fm detector, nor will it work through a product detector since one of the terms of the detection (the noise) is not coherent.

The "scope" jack on most receivers is loosely coupled to the i-f amplifier, preceding the detector. A wide-band scope connected to this point will show the train of pulses and eliminate the need for aural detection. The alignment of the later i-f stages of a system should have the least impact on the noise performance, and maximum signal response will always occur at the same setting. With this thought in mind, the simple prototype detector of Fig. 59 will generally work for aural a-m detection. Connect point A to the last i-f amplifier plate or collector. Connect point B to the audio amplifier, at or near the volume control and ground point C. With this arrangement the normal detector output is turned down with the volume control, and the temporary detector provides a-m detection.

The gated noise source has been used

for literally hundreds of applications and has proved to be a powerful yet simple addition to the test bench. While no guarantee of duplication may be made, these units develop approximately 18 dB of excess noise in the region of 50-300 MHz. This unit was originally described by Hartsen in January 1977 *QST*.

RF IMPEDANCE BRIDGE FOR COAX LINES

The bridge shown in Figs. 60 through 63 may be used to measure unknown complex impedances at frequencies below 30 MHz. Measured values are of equivalent series form, $R + jX$. The useful range of the instrument is from about 5 to 400 ohms if the unknown load is purely resistive, or 10 to 150 ohms resistive component in the presence of reactance. The reactance range is from 0 to approximately 100 ohms for either inductive or capacitive loads. Although the instrument cannot indicate impedances with the accuracy of a laboratory type of bridge, its readings are quite adequate for the measurement and adjustment of antenna systems for amateur use, including the taking of line lengths into account with a Smith chart or Smith transmission-line calculator.

The bridge incorporates a differential capacitor, C1, to obtain an adjustable ratio for measurement of the resistive component of the load. The capacitor consists of two identical sections on the same frame, arranged so that when the shaft is rotated to increase the capacitance

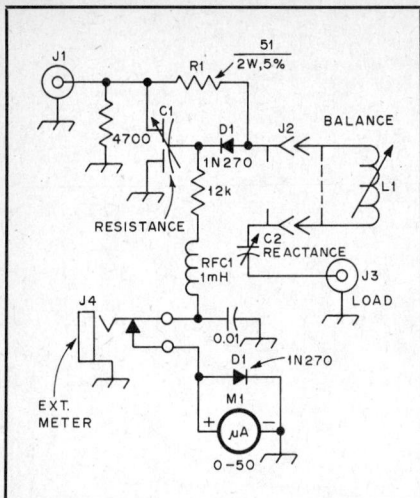

Fig. 61 — Schematic diagram of the impedance bridge. Capacitance is in microfarads; resistances are in ohms. Resistors are 1/2-W, 10-percent tolerance unless otherwise indicated.

C1 — Differential capacitor, 5.6 to 150 pF per section, Jackson Bros. C709 (see text).
C2 — 17.5-327 pF with straight-line capacitance characteristic, Hammarlund RMC-325-S.
D1, D2 — Germanium diode, high back resistance.
J1, J3 — Coaxial connectors, chassis type.
J2 — To mate plug of L1, ceramic.
J4 — Phone jack, disconnecting type.
L1 — See text and Table 1.
M1 — 0-50 µA dc, Simpson Model 1223 Bold-Vue, Cat. No. 15560 or equiv.
R1 — For text reference.
RFC1 — Subminiature rf choke, Miller 70F103A1 or equiv.

Fig. 62 — All components except the meter are mounted on the top of the box. C1 is visible inside the shield at the left, with C2 at the right and J2 mounted between them. J1 is hidden beneath C1 in this view; a part of J3 may be seen in the lower right corner of the box. Components for the dc metering circuit are mounted on a tie-point strip which is affixed to the shield wall for C1; all other components are interconnected with very short leads. The 4700-ohm input resistor is connected across J1.

of one section, the capacitance of the other section decreases. The capacitor is adjusted for a null reading on M1, and its settings are calibrated in terms of resistance at J3 so the unknown value can be read off the calibration. A coil-and-capacitor combination is used to determine the amount and type of reactance, inductive or capacitive. L1 and C2 in the bridge circuit are connected in series with the load. The instrument is initially balanced at the frequency of measurement with a purely resistive load connected at J3, so that the reactances of L1 and of C2 at its midsetting are equal. Thus, these reactances cancel each other in this arm of the bridge. With an unknown complex-impedance load then connected at J3, the setting of C2 is varied either to increase or decrease the capacitive reactance, as required, to cancel any reactance present in the load. If the load is inductive more reactance is needed from C2 to obtain a balance, indicated by a null on M1, with less reactance needed from C2 if the load is capacitive. The settings of C2 are calibrated in terms of the value and type of reactance at J3. Because of the relationship of capacitive reactance to frequency, the calibration for the dial of C2 is valid at only one frequency. It is therefore convenient to calibrate this dial

Table 2
Coil Data for RF Impedance Bridge

Band	Nominal Inductance Range (µH)	Frequency Coverage (MHz)	Coil Type or Data
80	6.5 - 13.8	3.2 - 4.8	28 turns no. 30 enam. wire close-wound on Miller form 42A000CBI.
40	2.0 - 4.4	5.8 - 8.5	Miller 42A336CBI or 16 turns no. 22 enam. wire close-wound on Miller form 42A000CBI.
20	0.6 - 1.1	11.5 - 16.6	8 turns no. 18 enam. wire close-wound on Miller form 42A000CBI.
15	0.3 - 0.48	18.5 - 23.5	4-1/2 turns no. 18 enam. wire close-wound on Miller form 42A000CBI.
10	0.18 - 0.28	25.8 - 32.0	3 turns no. 16 or 18 enam. or tinned bus wire spaced over 1/4-inch (6.3-mm) winding length on Miller form 42A000CBI.

for equivalent reactances at 1 MHz, as shown in Fig. 63. Frequency corrections may then be made simply by dividing the reactance dial reading by the measurement frequency in megahertz.

Construction

In any rf-bridge type of instrument, the leads must be kept as short as possible to reduce stray reactances. Placement of component parts, while not critical, must be such that lead lengths greater than about 1/2 inch (13 mm) (except in the dc metering circuit) are avoided. Shorter leads are desirable, especially for R1, the

"standard" resistor for the bridge. In the unit photographed, the body of this resistor just fits between the terminals of C1 and J2 where it is connected. C1 should be enclosed in shield and connections made with leads passing through holes drilled through the shield wall. The frames of both variable capacitors, C1 and C2, must be insulated from the chassis, with insulated couplings used on the shafts. C2 is mounted on 1-inch (25 mm) ceramic insulating pillars.

Band-switching arrangements for L1 complicate the construction and contribute to stray reactances in the bridge

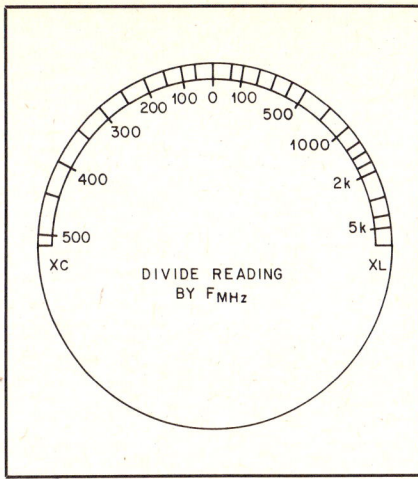

Fig. 63 — Calibration scale for the reactance dial associated with C2. See text.

circuit. For these reasons plug-in coils are used at L1, one coil for each band over which the instrument is used. The coils must be adjustable, to permit initial balancing of the bridge with C2 set at the zero-reactance calibration point. Coil data are given in Table 2. Millen 45004 coil forms with the coils supported inside provide a convenient method of constructing these slug-tuned plug-in coils. A phenolic washer cut to the proper diameter is epoxied to the top or open end of each form, giving a rigid support for mounting of the coil by its bushing. Small knobs for 1/8-inch (3.2-mm) shafts, threaded with a no. 6-32 tap, are screwed onto the coil slug-tuning screws to permit ease of adjustment without a tuning tool. Knobs with setscrews should be used to prevent slipping. A ceramic socket to mate with the pins of the coil form is used for J2.

The capacitor used for the resistance dial, C1, in the original unit (and shown in Fig. 62) was a Millen 28801. This capacitor is no longer available, but the Jackson Brothers capacitor specified in the parts list is a suitable substitute. Jackson Brothers capacitors are sold by Leeds Radio. (See parts supplier list in chapter 17.)

Calibration

The resistance dial of the bridge may be calibrated by using a number of 1/2- or 1-watt 5-percent-tolerance composition resistors of different values in the 5- to 400-ohm range as loads. For this calibration, the appropriate frequency coil must be inserted at J2 and its inductance adjusted for the best null reading on the meter when C2 is set with its plates half meshed. For each test resistor, C1 is then adjusted for a null reading. Alternate adjustment of L1 and C1 should be made for a complete null. The leads between the test resistor and J3 should be as short as possible, and the calibration preferably

should be done in the 3.5-MHz band where stray inductance and capacitance will have the least effect.

If the constructional layout of the bridge closely follows that shown in the photographs, the calibration scale of Fig. 63 may be used for the reactance dial. This calibration was obtained by connecting various reactances, measured on a laboratory bridge, in series with a 47-ohm 1-W resistor connected at J3. The scale is applied so that maximum capacitive reactance is indicated with C2 fully meshed. If it is desired to obtain an individual calibration for C2, known values of inductance and capacitance may be used in series with a fixed resistor of the same approximate value as R1. For this calibration it is *very important* to keep the leads to the test components as short as possible, and calibration should be performed in the 3.5-MHz range to minimize the effects of stray reactances. Begin the calibration by setting C2 at half mesh, marking this point as 0 ohms reactance. With a purely resistive load connected at J3, adjust L1 and C1 for the best null on M1. From this point on during calibration, do not adjust L1 except to rebalance the bridge for a new calibration frequency. The ohmic value of the known reactance for the frequency of calibration is multiplied by the frequency in MHz to obtain the calibration value for the dial.

Using the Impedance Bridge

This instrument is a low-input-power device, and is *not* of the type to be excited from a transmitter or left in the antenna line during station operation. Sufficient sensitivity for all measurements results when a 5-V rms rf signal is applied at J1. This amount of voltage can be delivered by most grid-dip oscillators. In no case should the power applied to J1 exceed 1 watt or calibration inaccuracy may result from a permanent change in the value of R1. The input impedance of the bridge at J1 is low, in the order of 50 to 100 ohms, so it is convenient to excite the bridge through a length of 52- or 75-ohm line such as RG-58A/U or RG-59/U. If a grid-dip oscillator is used, a link coupling arrangement to the oscillator coil may be used. Fig. 60 shows two pick-up link assemblies. The larger coil, 10 turns of 1-1/4-inch-dia stock with turns spaced at 8 turns per inch, is used for the 80-, 40- and 20-meter bands. The smaller coil, 5 turns of 1-inch (25-mm) dia stock with turns spaced at 4 turns per inch, is used for the 15- and 10-meter bands. Coupling to the oscillator should be as light as possible, while obtaining sufficient sensitivity, to prevent severe "pulling" of the oscillator frequency.

Before measurements are made, it is necessary to balance the bridge. Set the reactance dial at zero and adjust L1 and C1 for a null with a nonreactive load connected at J3. The bridge must be rebal-

Fig. 64 — Front view of the 2-30 MHz rf watt-meter/VSWR indicator.

anced after any appreciable change is made in the measurement frequency. A 51-ohm, 1-W resistor mounted inside a PL-259 plug, as shown in Fig. 60, makes a load which is essentially nonreactive. After the bridge is balanced, connect the unknown load to J3, and alternately adjust C1 and C2 for the best null.

The calibration of the reactance dial is shown in Fig. 63. The measurement range for capacitive loads may be extended by "zeroing" the reactance dial at some value other than 0. For example, if the bridge is initially balanced with the reactance dial set at 500 in the X_L range, the 0 dial indication is now equivalent to an X_C reading of 500, and the total range of measurement for X_C has been extended to 1000.

A VSWR INDICATOR AND POWER METER FOR 2-30 MHz

The wattmeter/VSWR indicator illustrated in Figs. 64-67 will allow the user to measure rf power and VSWR in the hf frequency range. This unit makes use of two meters which serve as an aid for adjusting Transmatches by making it possible to simultaneously monitor forward and reflected power. A dual-meter power indicator is much more convenient than a single-meter type that must be switched manually for a FORWARD or REFLECTED reading. One other feature of this meter is its ability to indicate transmitter peak power during an ssb transmission. This feature can be selected with the panel control labelled NORM and PEAK.

The Circuit

The circuit for the wattmeter/VSWR indicator is shown in Fig. 65 and is relatively straightforward. The sampling section is made up of a toroid coil through which a short length of RG-8/U coaxial cable is passed. This circuit is shown in Fig. 66. Symmetry is maintained in order to prevent measurement inaccuracies at the higher frequencies. The braid of the

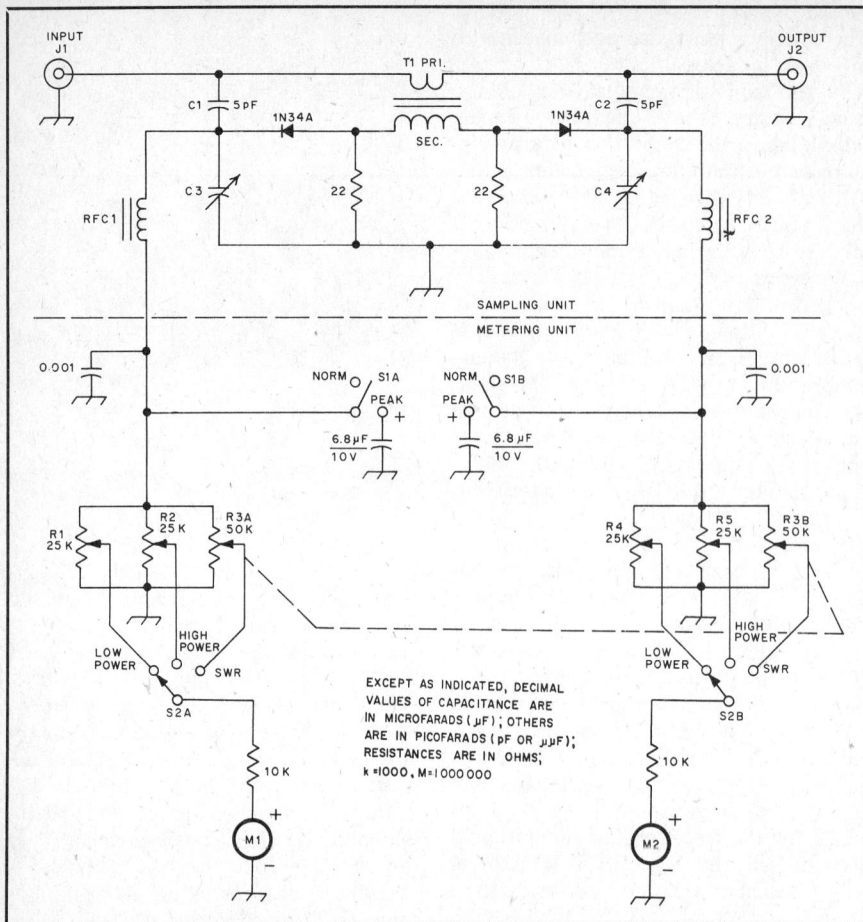

Fig. 65 — Schematic diagram of the wattmeter/VSWR indicator. Parts designations called out in the diagram, but not appearing in the parts list, are for text reference only.

C1, C2 — 5 pF, silver mica.
C3, C4 — 170-780 pF trimmer capacitor. Elmenco 469 or equiv.
J1, J2 — Coaxial connectors. Builder's choice.
M1, M2 — 50 µA dc meter.
R3 — Dual 50-kΩ potentiometer, panel mount.

RFC1, RFC2 — 4 turns no. 22 enameled wire on a 3/8-in. OD ferrite bead (950µ).
S1 — Rotary switch, two pole, two position.
S2 — Rotary switch, two pole, three position.
T1 — Primary: see text; secondary: 40 turns no. 22 enameled wire on T80-2 core.

Fig. 66 — This is a photograph of the sampling unit with the shield removed. The braid of the coaxial cable is grounded at only one end.

Fig. 67 — Interior view of the wattmeter. Cable ties are used to provide a neat appearance. Three interconnecting wires are run from the sampling unit to the meter circuit.

Table 3
Power Calibration Chart

Wattmeter reading	Watts (low-power scale)	Watts (high-power scale)
14	20	200
20	40	400
25	60	600
30	80	800
35	100	1000
38	120	1200
41	140	1400
44	160	1600
47	180	1800
50	200	2000

Table 4
SWR Calibration Chart

Reflected reading	VSWR
4	1.5 : 1
9	2.0 : 1
14	2.5 : 1
18	3.0 : 1

RG-8/U cable is grounded at only one end (either end may be grounded) thereby providing an effective electrostatic shield over the primary of T1.

The lower capacitors in the voltage dividers (C1/C3 and C2/C4) are made variable since this will allow for easy adjustment, especially if the rotors of these capacitors are connected to ground. This will eliminate the "hand capacitance" effects during the adjustment procedure.

The value of 6.8 µF for the peak detector capacitors is not especially critical. Any values in the 5- to 10-µF range should work fine. Both capacitors should be of the same value. The low- and high-power potentiometers are circuit-board types and will be adjusted and left at that setting. The SWR CAL potentiometer is a dual 50-kΩ, panel-mount type.

Construction

The unit is housed in a homemade aluminum enclosure that measures 5-1/4 × 7 × 3 inches (133 × 178 × 76 mm). The only critical portion of the circuitry is that of the sampling unit, which must be shielded. Additionally, all leads within the sampling circuit should be kept as short as possible. The shield is made from scraps of single-sided, printed-circuit board material cut to size and soldered along the edges. Solder lugs attached to the shield are bolted to the case at several locations thus providing a good ground connection.

The remainder of the circuit is not critical. Once the FORWARD and REFLECTED leads leave the shielded enclosure they carry only dc voltages and can be made any length. The builder may wish to locate the sampling unit in a separate enclosure from the meter circuit. This will allow the sampling unit to be placed at a more convenient location in the shack. Only three leads are needed between the sampling unit and the meter circuitry (the FORWARD and REFLECTED leads and a ground lead).

Adjustment

For initial adjustment connect a transmitter to the input and a 50-ohm, non-reactive load to the output. Set S2 to the SWR position and R3 for maximum sensitivity. Gradually increase the transmitter power until the FORWARD meter reads full scale. Adjust C4 for a null on the REFLECTED meter. Next, reverse the input and output connections and adjust C3 for a null as indicated on the FORWARD meter. This completes alignment of the sampling unit.

In order to calibrate the power scales, an accurate wattmeter or rf ammeter is required. Alignment is a simple matter of adjusting potentiometers R1, R2, R4 and R5 to make the meter readings conform with the readings obtained with the calibrated wattmeter or ammeter. This

Fig. 68 — Exterior view of the noise bridge. The unit is finished in red enamel. Press-on lettering is used for the calibration marks.

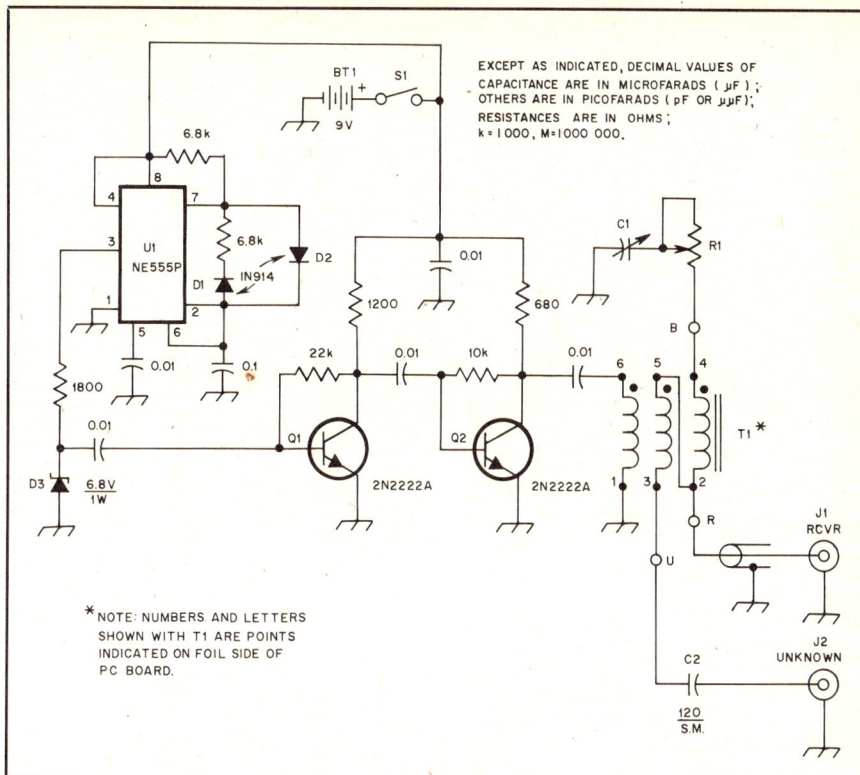

Fig. 69 — Schematic diagram of the noise bridge. Resistors are 1/4-watt composition types. Capacitors are miniature ceramic units unless indicated otherwise. Component designations indicated in the schematic but not called out in the parts list are for text and parts-placement reference only.

BT1 — 9-volt battery, NEDA 1604A or equiv.
C1 — Variable, 250 pF maximum. Use a good grade of capacitor.
C2 — Approximately 1/2 of C1 value. Selection may be necessary — see text.
J1, J2 — Coaxial connector, BNC type.
R1 — Linear, 250 ohm, AB type. Use a good grade of resistor.
S1 — Toggle, spst.
T1 — Broadband transformer, 8-trifilar turns of no. 26 enameled wire or an Amidon FT-37-43 toroid core.
U1 — Timer, NE555 or equiv.

unit was calibrated for a full-scale reading of 200 watts on the low-power range and 2000 watts on the high-power range. There is no reason why other power levels could not be used as full-scale readings.

Table 3 is a power calibration chart for the wattmeter described here. Table 4 is an SWR calibration chart. The values given here assume a full-scale reading for the FORWARD meter.

A NOISE BRIDGE FOR 160 TO 10 METERS

The noise bridge, sometimes referred to as an antenna (RX) noise bridge, is an instrument that will allow the user to measure the impedance of an antenna or other electrical circuits. The unit described here, designed for use in the 160-through 10-meter range, provides adequate accuracy for most measurements. Battery operation and small physical size make this unit ideal for remote-location use. Tone modulation is applied to the wide-band noise generator as an aid for obtaining a null indication. A detector, such as the station receiver, is required for operation of the unit.

The Circuit

The noise bridge consists of two parts — the noise generator and the bridge circuitry. See Fig. 69. A 6.8-volt Zener diode serves as the noise source. U1 generates an approximate 50-per-cent duty cycle, 1000-Hz, square wave signal which is applied to the cathode of the Zener diode. The 1000-Hz modulation appears on the noise signal and provides a useful null detection enhancement effect. The broadband-noise signal is amplified by Q1, Q2 and associated components to a level which produces an approximate S-9 signal in the receiver. Slightly more noise is available at the lower end of the frequency range, as no frequency compensation is applied to the amplifier. Roughly 20 mA of current is drawn from the 9-volt battery, thus assuring long battery life — as long as the power is switched off after use!

The bridge portion of the circuit consists of T1, C1, C2 and R1. T1 is a trifilar-wound transformer with one of the windings used to couple noise energy into the bridge circuit. The remaining two windings are arranged so that each one is in an arm of the bridge. C1 and R1 complete one arm and the UNKNOWN circuit along with C2 comprise the remainder of the bridge. The terminal labeled RCVR is for connection to the detector.

Construction

The noise bridge is contained in a homemade aluminum enclosure that measures 5 × 2 3/8 × 3-3/4 inches (127 × 60 × 95 mm). Many of the circuit components are mounted on a circuit board that is fastened to the rear wall of the cabinet. The circuit-board layout is such that the lead lengths to the board from the bridge and coaxial connectors are at a minimum. Etching pattern and parts-placement-guide information for the circuit board are shown in Figs. 71 and 72.

Care must be taken when mounting the potentiometer. For accurate readings the potentiometer must be well insulated from ground. In the unit shown this was accomplished by mounting the control to a piece of Plexiglas, which in turn was fastened to the chassis with a piece of

Fig. 70 — Interior view of the noise bridge. Note that the potentiometer must be isolated from ground.

aluminum angle stock. Additionally, a 1/4-inch (6.4-mm), control-shaft coupling and a length of phenolic rod were used to further isolate the control from ground where the shaft passes through the front panel. A high-quality potentiometer is a

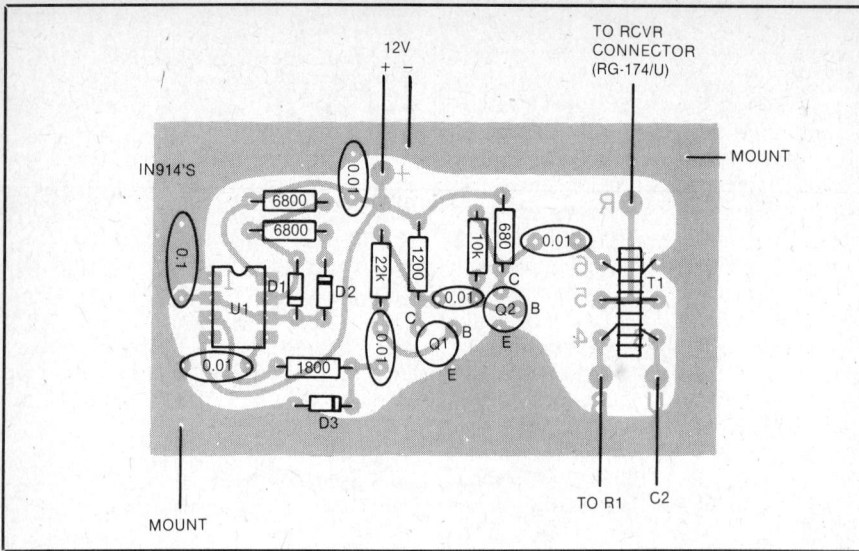

Fig. 71 — Parts-placement guide for the noise bridge as viewed from the component side of the board.

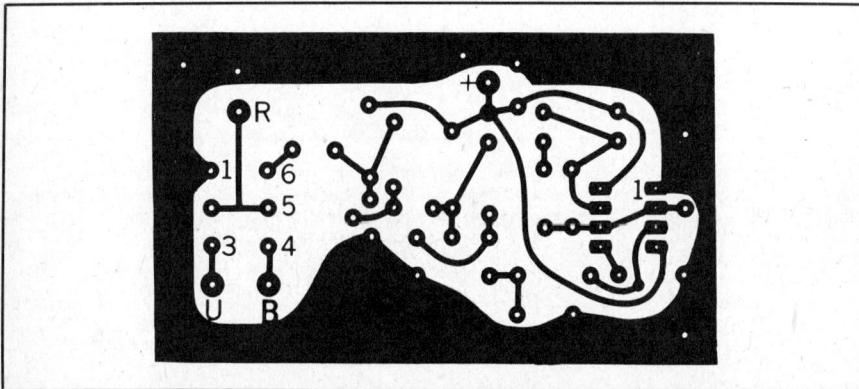

Fig. 72 — Etching pattern for the noise bridge pc board. This is the pattern for the bottom side of the board. The top side of the board is a complete ground plane with a small amount of copper removed from around the component holes. Mounting holes are located in two corners of the board.

must if good measurement results are to be obtained.

Mounting the variable capacitor is not a problem since the rotor is grounded. As with the potentiometer, a good grade of capacitor is important. If you must cut corners to save money, look elsewhere in the circuit. Two BNC-type female coaxial fittings are provided on the rear panel for connection to a detector (receiver) and to the UNKNOWN circuit. There is no reason why other types of connectors can't be used. One should avoid the use of plastic insulated phono connectors, however, as these might influence the accuracy at the higher frequencies. As can be seen from the photograph, a length of miniature coaxial cable (RG-174/U) is used between the RCVR connector and the appropriate circuit board foils. Also, C2 has one lead attached to the circuit board and the other connected directly to the UNKNOWN circuit connector.

Calibration and Use

Calibration of the bridge is straightfor-

ward and requires no special instruments. A receiver tuned to any portion of the 15-meter band is connected to the RCVR terminal of the bridge. The power is switched on and a broadband noise with a 1000-Hz note should be heard in the receiver. Calibration of the resistance dial should be performed first. This is accomplished by inserting small composition resistors of appropriate values across the UNKNOWN connector of the bridge. The resistors should have the shortest lead lengths possible in order to mate with the connector. Start with 25 ohms of resistance (this may be made up of series or parallel connected units). Adjust the capacitance and resistance dials for a null of the signal as heard in the receiver. Place a calibration mark on the front panel at that location of the resistance dial. Remove the 25-ohm resistor and insert a 50-ohm resistor, 100-ohm unit and so on until the dial is completely calibrated.

The capacitance dial is calibrated in a similar manner. Initially, this dial is set so that the plates of C1 are exactly half

meshed. A 50-ohm resistor is connected to the UNKNOWN terminal and the resistance control is adjusted for a null. Next, the reactance dial is adjusted for a null and its position is noted. If this setting is significantly different than the half-meshed position the value of C2 will need to be changed. Unit-to-unit value variations of 120-pF capacitors may be sufficient to provide a suitable unit. Alternatively, other values can be connected in series or parallel and tried in place of the 120-pF capacitor. The idea is to have the capacitance dial null as close as possible to the half-meshed position of C1.

Once the final value of C2 has been determined and the appropriate component installed in the circuit, the bridge should be adjusted for a null. The 0 capacitance point can be marked on the face of the unit. The next step is to place a 20-pF capacitor in series with the 50-ohm load resistor. Use a good grade of capacitor, such as a silver-mica type and keep the leads as short as possible. Null with the capacitance dial and make a calibration mark at that point. Remove the 20-pF capacitor and insert a 40-pF unit in series with the 50-ohm resistor. Again null the bridge and make a calibration mark for 40 pF. Continue on in a similar manner until that half of the dial is completely calibrated.

To calibrate the other half of the scale, the same capacitors may be used. This time they must be placed temporarily in parallel with C2. Connect the 50-ohm resistor to the UNKNOWN terminal and the 20-pF capacitor in parallel with C2. Null the bridge and place a calibration mark on the panel. Remove the 20-pF unit and temporarily install the 40-pF capacitor. Again null the bridge and make a calibration mark at that point. Continue this procedure until the capacitance dial is completely calibrated. It should be pointed out that the exact resistance and capacitance values used for calibration can be determined by the builder. If resistance values of 20, 40, 60, 80, 100 ohms and so on are more in line with the builder's needs, the scale may be calibrated in those terms. The same is true for the capacitance dial. The accuracy of the bridge is determined by the components that are used in the calibration process.

Operation

The resistance dial is calibrated directly in ohms, but the capacitance dial is calibrated in terms of pF or capacitance. The + C half of the dial indicates that the load is capacitive and the − C portion is for inductive loads. To find the reactance of the load, the dial setting must be applied to the standard capacitive reactance formula:

$$X = \frac{1}{2\pi f C}$$

The result will be a capacitive reactance for readings in the +C area of the dial and inductive reactance for −C portions.

When using the bridge remember that the instrument measures the impedance of loads as connected at the UNKNOWN terminal. This means that the actual load to be measured must be directly at the connector rather than being attached to the bridge by a length of coaxial cable. Even a short length of cable will transform the load impedance to some other value. Unless the electrical length of line is known and taken into account, it is necessary to place the bridge at the load. An exception to this would be if the antenna were to be matched to the characteristic impedance of the cable. In this case the bridge controls may be preset for 50-ohms resistance and 0-pF capacitance. With the bridge placed at any point along the coaxial line, the load (antenna) may be adjusted until a null is obtained. If the length of line is known to be an even multiple of a half-wavelength at the frequency of interest, the readings obtained from the bridge will be accurate. Consult the Transmission Line chapter of this volume or the ARRL *Antenna Book* for more information on impedance-measurement techniques.

Interpreting the Readings

A couple of words on how to interpret the measurements may be in order. For example, assume that the impedance of a 40-meter inverted-V antenna fed with a half-wavelength of cable was measured. The antenna had been cut for roughly the center of the band (7.150 MHz) and the bridge was nulled with the aid of a receiver tuned to that frequency. The results were 45-ohms resistive and 70 picofarads of capacitance. The 45-ohm resistance reading is close to 50 ohms as would be expected for this type of antenna. The capacitive reactance calculates to be 318 ohms from the equation:

$$X = \frac{1}{2\pi(7.15 \times 10^6)(70 \times 10^{(-12)})}$$

$$= 318\ \Omega$$

When an antenna is adjusted for resonance the capacitive or inductance reactance will be zero and the antenna in question is a long way from that mark. Since the antenna looks capacitive it is too short and wire should be added to each side of the antenna. An approximation of how much wire to add can be made by tuning the receiver higher in frequency until a point is reached where the bridge nulls with the capacitance dial at zero. The percentage difference between this new frequency and the desired frequency indicates the approximate amount that the antenna should be lengthened. The same system will work if the antenna has been cut too long. In this case the capacitance dial would have nulled in the −C region,

indicating an inductive reactance. This procedure will work for most any antenna.

AN OPERATING IMPEDANCE BRIDGE FOR 160 TO 6 METERS

The instrument described in this section is an rf bridge combined with a directional coupler. Loose coupling (approximately 100 dB) allows the load impedance to be characterized at high power levels. Another advantage of this scheme is relative immunity to erroneous readings caused by external fields, such as those generated by local broadcast stations. An operating impedance bridge (OIB) is particularly useful for analyzing impedances that vary with the applied power. Many amateur power amplifiers exhibit this characteristic. The OIB principle is patented by Delta Electronics, Inc., but that company has granted amateurs permission to duplicate this unit for private use. The bridge pictured in Fig. 73, built by Robert Luetzow, K9ZLU, is an improved and simplified version of his OIB published in November 1979 QST. Significant features of this instrument include a reactance measurement range of 0 to 350 ohms (inductive or capacitive) at 10 MHz, a resistance range of 0 to 300 ohms, a useful frequency range of at least 1 to 54 MHz, and the ability to withstand any legal amateur power level. The OIB has very little insertion effect, so it can be permanently installed in a transmission line for continuous monitoring. The total cost of components (all new) for this project was under $40 in 1980.

Design Information

The standing-wave pattern on the "through" transmission line of the directional coupler appears at reduced amplitude on the "pickup" line. If the coupler is symmetrical, the voltage and current on the pickup line exhibit the same phase relationship as those on the through line. "Box constant" tabs allow the coupling coefficient at each end of the line to be adjusted for symmetry. Setting the reactance and resistance in the bridge circuit equal to the load values (at the OIB output connector) causes the voltage at the midpoint of the pickup line to null. One end of the pickup line has a fixed "standard" capacitor. The variable capacitor is adjusted so that it, in combination with the reactance coupled from the load, balances the reactance of the standard. Since capacitive and inductive reactances of equal magnitude simply cause opposite phase shifts, the bridge can measure either type of reactance by interchanging the fixed and variable capacitors.

Because of the low coupling coefficient, the signal must be amplified to provide a useful meter indication. Q1 serves as a broadband preamplifier. The rf is rectified by D1, which is followed by a dc

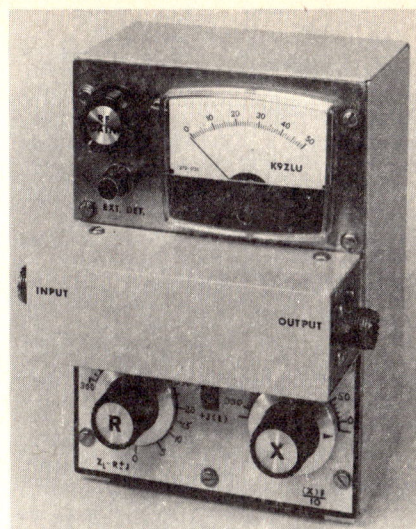

Fig. 73 — The OIB is enclosed in a 7 × 5 × 3-in. (178 × 127 × 76-mm) aluminum chassis. Tinnerman Speed Nuts and machine screws are used to secure the front panel to the chassis flange. Rectangular holes with cover plates in the rear panel permit alignment of the unit while it is fastened to the chassis.

amplifier, U1. An agc loop (augmented by a manual gain control) from the op-amp output to gate 2 or Q1 allows sensitive low power null indications while keeping the readings on the meter scale during high-power operation. This null-detector circuit functions well with as little as two watts of drive. The amplifier has no selectivity, so the driving signal must be spectrally pure if true sharp nulls are to be obtained. Fig. 74 contains the complete schematic diagram for the operating impedance bridge. More comprehensive OIB theory is presented in the QST article and by Wright in "Unique Bridge Measures Antenna Operating Impedance," *Electronics*, February 23, 1963.

Construction

Three major assemblies comprise the OIB: The directional coupler or "pickup box," the impedance bridge and the null detector. Copper-clad pc board is the principal building material. All of the components are fastened to the front panel, which is a double thickness of 1/16 inch (1.5 mm) single-clad pc board measuring 5 × 7 inches (127 × 178 mm). The two pieces are bonded with "five-minute" epoxy. Rough the unclad surfaces with sandpaper before joining them.

A view of the OIB, with the pickup box elements exposed, is presented in Fig. 75. The components are soldered directly to the front panel. Line drawings for the pickup box are given in Fig. 76. Careful work on this assembly will be rewarded by an accurate, well-performing instrument. The copper and brass stock can be obtained at a hobby or model-airplane shop.

The rear of the panel can be seen in

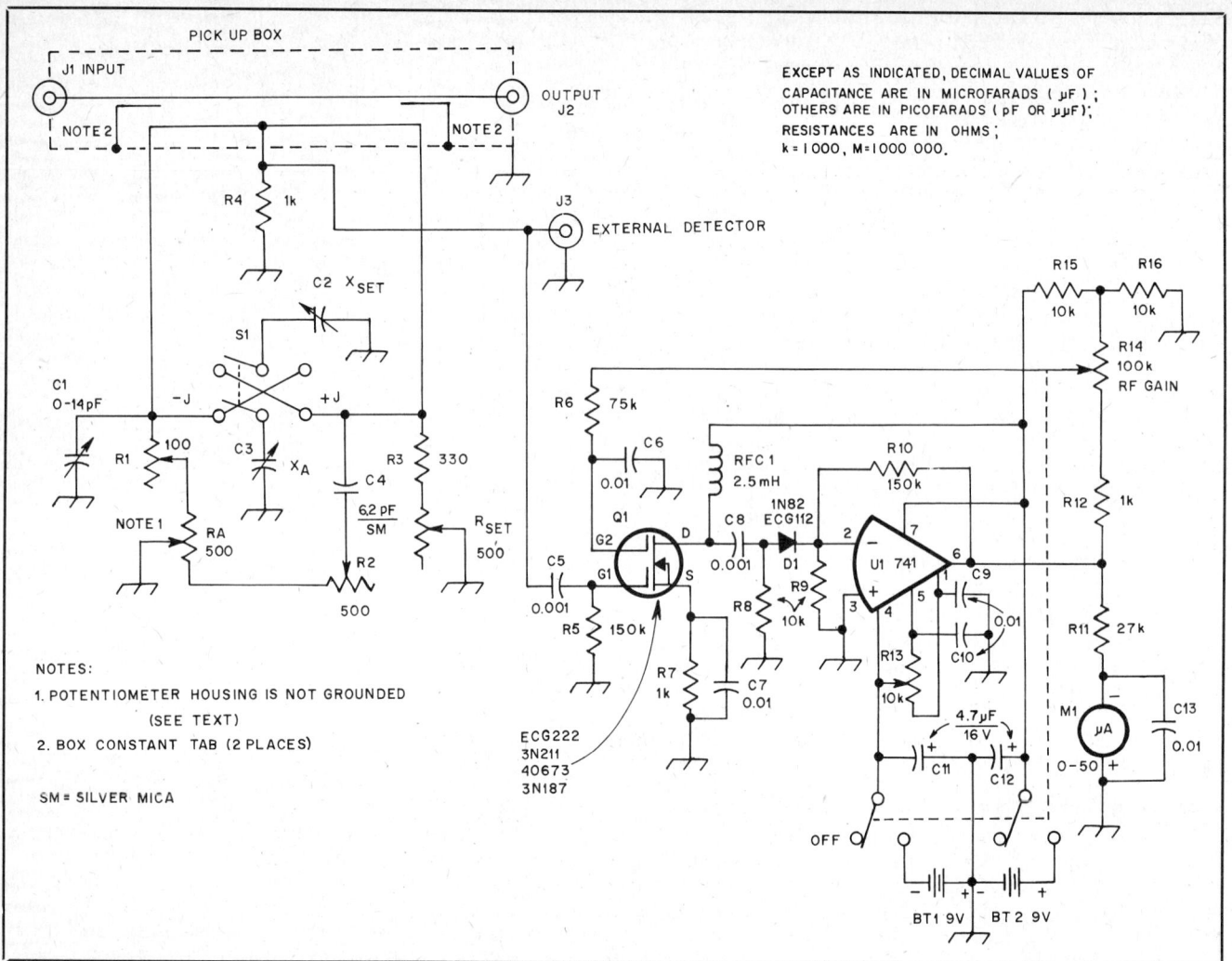

Fig. 74 — Schematic diagram of the operating impedance bridge.

PICK UP BOX

J1 INPUT
OUTPUT J2
NOTE 2
NOTE 2

EXCEPT AS INDICATED, DECIMAL VALUES OF
CAPACITANCE ARE IN MICROFARADS (μF);
OTHERS ARE IN PICOFARADS (pF OR μμF);
RESISTANCES ARE IN OHMS;
k = 1000, M = 1000 000.

R4 1k
J3 EXTERNAL DETECTOR
C2 X_SET
S1
C1 0-14 pF
−J   +J
R1 100
C3   X_A
C4 6.2 pF SM
R3 330
R_SET 500
NOTE 1
RA 500
R2 500

R6 75k
C6 0.01
RFC 1 2.5 mH
R10 150k
1N82 ECG112
C8 0.001
Q1
G2   D
G1   S
C5 0.001
R5 150k
R8
D1
R9 10k
U1 741
C9
C10 0.01
R13 10k
C7 0.01
R7 1k

R15 10k   R16 10k
R14 100k RF GAIN
R12 1k
R11 27k
C13 0.01
M1 μA 0-50

4.7 μF 16 V
C11   C12
OFF
BT1 9V   BT2 9V

ECG222
3N211
40673
3N187

NOTES:
1. POTENTIOMETER HOUSING IS NOT GROUNDED (SEE TEXT)
2. BOX CONSTANT TAB (2 PLACES)

SM = SILVER MICA

C1 — 0 to 14-pF air trimmer.
C2 — 6 to 60-pF mica trimmer.
C3 — 365-pF broadcast-radio tuning capacitor (Calectro A1-227 or equiv.) with 6 rotor plates removed.
J1, J2 — Uhf female coaxial connector — SO-239.
J3 — BNC or RCA phono (female) connector, as available.
M1 — 0 to 50 μA dc microammeter (Radio Shack 270-1751).

R_A — Hot-molded carbon potentiometer, 500 Ω, 2 W, log taper.
R1, R2, R_SET, R13 — Tc-mount trimmer potentiometers.
R14 — Potentiometer, 100 kΩ with dpdt switch (Radio Shack 271-216).

Fig. 77. Connections from the pickup line to the bridge and detector circuits are made via solder lugs under the pickup line supporting screws. These screws are insulated from the front panel with nylon shoulder washers. To retain simplicity with high performance, the aesthetics of the bridge circuit must be compromised in favor of minimum stray capacitance. For this reason, an open-air construction technique is employed here. Short, stiff leads prevent mechanical (and attendant electrical) instability. The R1, R2, R_A and R_set control should be good quality hot-molded or ceramic-metal potentiometers to ensure low contact noise. The C_set capacitor is a mica trimmer, and the X_A capacitor is a broadcast-band-receiver replacement unit having semicircular plates. Six plates must be removed. If other plate profiles are used, the dial escutcheon must be calibrated empirically. The R_A potentiometer housing is floated above ground to reduce the stray

Fig. 75 — The OIB with the pickup-box cover removed. Brass 6-32 nuts soldered into the pickup box side flanges allow the cover to be attached with machine screws.

capacitance. Mounting the component on an insulating plate, spaced from the panel, accomplishes this objective. To avoid hand-capacitance effects, the control shaft must be insulated. Epoxy was used to secure the insulated shaft to the potentiometer.

Figs. 78 and 79 are the parts-placement guide and etching pattern for the null detector pc board. The board can be attached to the front panel with lugs soldered to the ground foil.

Calibration and Operation

The first step in calibrating the OIB is to zero the null-detector circuit with no signal applied. Adjust R13 to set the meter to zero. Connect a dummy load to the output port. Next, apply some rf power and balance the bridge with the X_A capacitor set to approximately 20 pF, which corresponds to the zero mark on the reactance dial. A reduced dial escutcheon is shown in Fig. 80. A full-scale

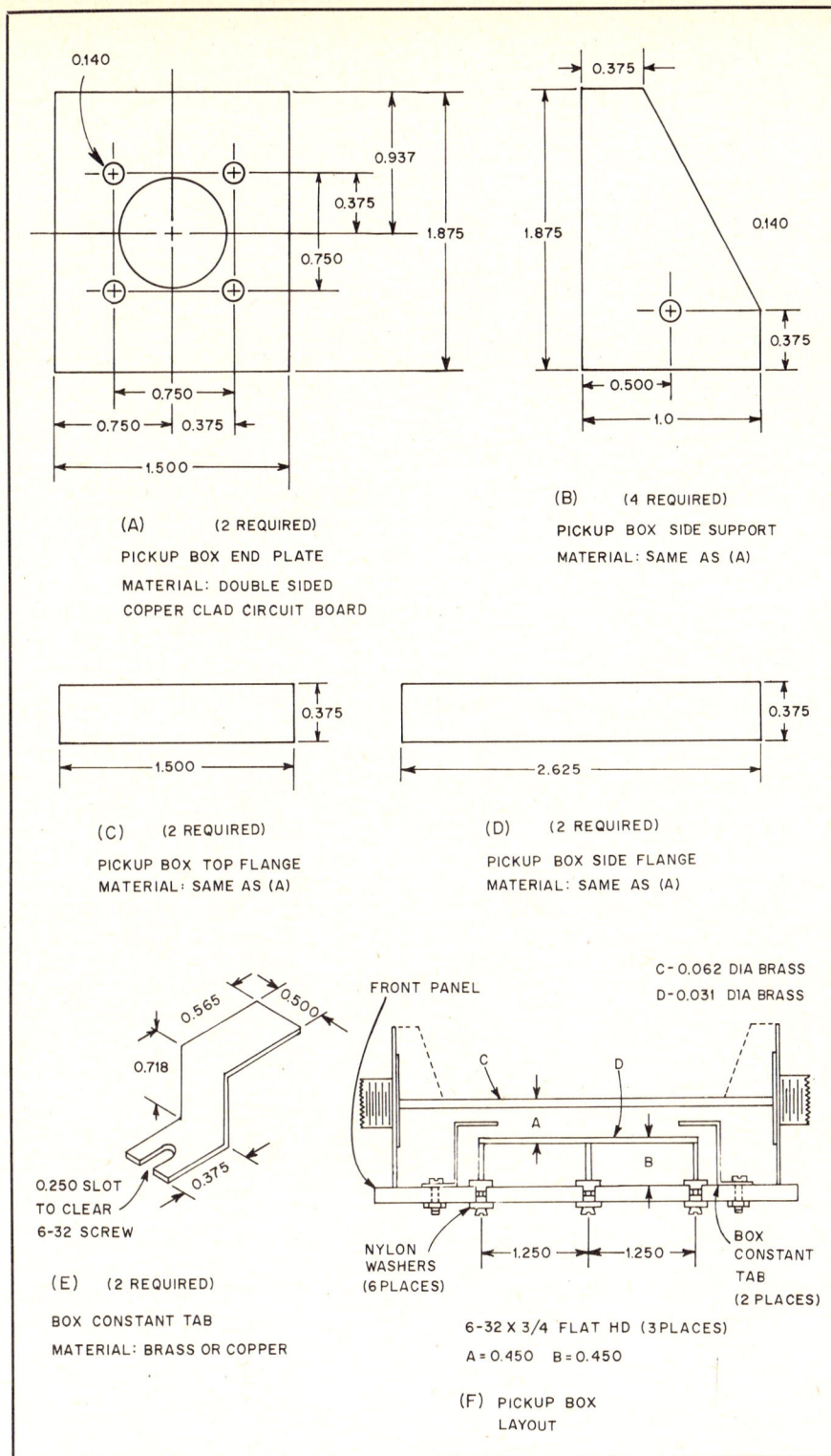

(A)   (2 REQUIRED)
PICKUP BOX END PLATE
MATERIAL: DOUBLE SIDED
COPPER CLAD CIRCUIT BOARD

(B)   (4 REQUIRED)
PICKUP BOX SIDE SUPPORT
MATERIAL: SAME AS (A)

(C)   (2 REQUIRED)
PICKUP BOX TOP FLANGE
MATERIAL: SAME AS (A)

(D)   (2 REQUIRED)
PICKUP BOX SIDE FLANGE
MATERIAL: SAME AS (A)

(E)   (2 REQUIRED)
BOX CONSTANT TAB
MATERIAL: BRASS OR COPPER

C - 0.062 DIA BRASS
D - 0.031 DIA BRASS

FRONT PANEL

NYLON WASHERS (6 PLACES)

BOX CONSTANT TAB (2 PLACES)

6-32 X 3/4 FLAT HD (3 PLACES)
A = 0.450   B = 0.450

(F) PICKUP BOX LAYOUT

Fig. 76 — Mechanical drawings for the OIB pickup box. Dimensions are in inches (mm = in. × 25.4).

Fig. 77 — Rear view of the OIB front-panel assembly. Mechanical rigidity and symmetry in the bridge circuit is essential to proper performance.

made using the station transmitter to supply power. Non-zero reactances are calibrated at 10 MHz. A well-filtered crystal oscillator as the signal source, and a receiver as the indicator provides a good calibration set-up.

After the zero adjustment is completed, connect a capacitive reactance of 200 ohms to the bridge output and adjust the box-constant tabs over the ends of the pickup line so the zero and 200-ohms settings match the dial calibration. Both tabs must be adjusted. If the tabs aren't located symmetrically with respect to the transmission lines, the instrument will display unequal readings for inductive and capacitive reactances of equal magnitude. A capacitive reactance other than 200 ohms can be used, but remember that the capacitances of the connectors and leads must be included. A good check on the proper setting of the tabs is to obtain a null, transpose the input and output connections and set the L/C switch to inductance. The instrument should stay nulled.

The resistance dial is easier to calibrate. First, adjust the R_A potentiometer so that about 85 percent of its resistance is between R1 and the wiper of R_A. Terminate the pickup box with a shorted connector, and with R1 set to midrange, adjust the R_{set} control for a null indication. This null indication locates the zero mark for the resistance dial. Now adjust the R control to 200 ohms (as indicated by the dial), and with a 200-ohm resistive load terminating the pickup box, adjust R1 for null indication. The X_A control will need to be readjusted to balance the load capacitance. R2 is used to readjust the zero setting of the R dial at higher frequencies.

The controls will have to be readjusted several times to secure accurate readings from the R and X dials. There is a little interaction between the R and X calibrating controls, but the box constant tabs are the only controls that greatly affect both the resistance and reactance settings. After the calibration is completed, the box-constant tabs should be soldered in place.

template that can be pasted on the front panel is available from ARRL. The bridge is in a balanced condition when a resistive load of 20 to 50 ohms is measured, and the L/C switch can be toggled without disturbing the null reading. This null condition is found by adjusting the X_{set} capacitor and C1 while toggling the L/C switch. When the bridge is balanced this

way, the zero-reactance reading is valid only for the particular connector or leads used to connect the load. If the connecting arrangement is changed, the bridge must be rebalanced. Also, when measuring a high resistive load, the capacitive reactance is much more apparent. A reactance reading of zero is frequency-independent, so this adjustment can be

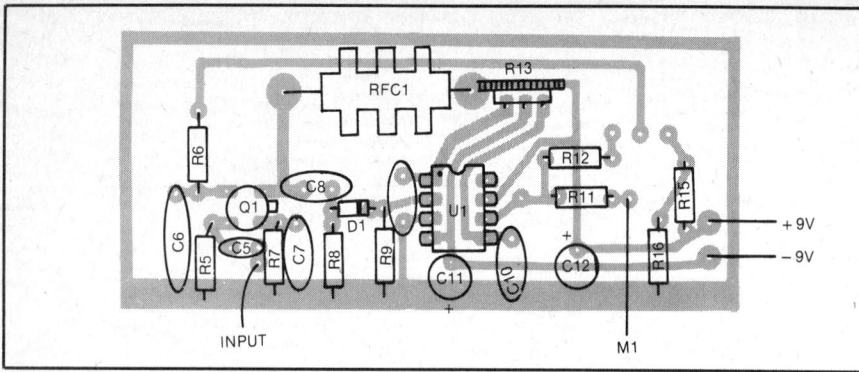

Fig. 78 — Parts-placement guide for the OIB null detector pc board. The component side is shown with an X-ray view of the copper foil. R10 is mounted with short leads on the foil side.

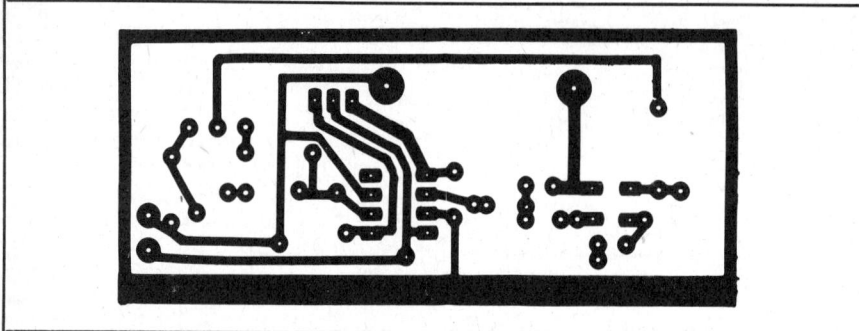

Fig. 81 — Exterior view of the signal generators suitable for use in receiver performance measurements. Each uses an OE-10 oscillator available from International Crystal Mfg. Co., Inc.

Fig. 79 — Full-scale etching pattern for the OIB null detector pc board. The foil side is shown; black represents copper.

Fig. 82 — Interior view of one of the signal generators. To the right of the oscillator module is the 7-element Chebyshev low-pass filter network. Notice the use of the feed-through capacitor with additional low-frequency bypassing for the dc lead.

Since the OIB is calibrated at 10 MHz, a correction must be made to the reactance dial reading when operating at other frequencies. All one need do is express the operating frequency as a multiple of 10 MHz and multiply the dial reading by that number.

When a length of cable is used between the OIB and the load, the cable will act as an impedance transformer unless it is perfectly matched to the load. Therefore, graphical or analytical methods must be used to determine the actual load impedance. Don't forget to consider the velocity factor of the cable. As the operating frequency is increased, the effect of the pickup box becomes greater. One must add half the length of the pickup box to the total length of the transmission line used between the bridge and the load.

When this operating impedance bridge is used in an rf power transmission line, high voltages may appear on the unit if the coax shield is broken, disconnected or improperly grounded to the total transmitting system. Antenna currents on the shield, caused by an unbalanced condition at the load, are another possible source of high voltage on the case. Always be alert to these situations to avoid rf burns.

SIGNAL GENERATORS FOR RECEIVER TESTING

Here are two simple signal generators that can be used for receiver performance measurements. Since many receiver tests can be carried out at a fixed frequency or frequencies, two International Crystal Mfg. Co., Inc. OE-10 oscillators make simple yet effective generators. The units described here were designed for 14.040 and 14.060 MHz and each provides an output power of −5 dBm. A 7-pole Chebyshev low-pass filter is contained in each oscillator assembly since the harmonics of the oscillator alone are down only 10 to 30 dB up through the 9th harmonic. This magnitude of harmonic energy will cause significant errors when measuring receiver intermodulation performance. A spectral photograph of the output of the oscillator and 7-element low-pass filter is shown in Fig. 84.

The oscillator assemblies are housed in chassis made from double-sided circuit-board material. The circuit-board panels

are soldered along each seam to construct an "rf-tight" enclosure. This reduces the amount of leakage from the box, which is important when measuring the receiver noise floor. A feedthrough type of capacitor with additional low-frequency bypassing is used to feed dc to the oscillator. This is shown in the schematic diagram and accompanying photographs. Reactance values for the filter are shown in Table 5 so that filters may be constructed for any frequency range. The nearest standard-value capacitor is sufficient.

Since the output of these generators is at a fixed level, it is assumed the user will use a step attenuator to lower the output level to appropriate values. Constructional information on step attenuators can be found elsewhere in this chapter.

A Hybrid Combiner for Signal Generators

Many receiver performance measurements require two signal generators to be attached simultaneously to a receiver. A combiner that isolates the two signal generators is necessary to keep one generator from being frequency- or phase-modulated by the other. The combiners described here provide 40 to 50 dB of isolation between ports while attenuating the desired signal paths (each input to output) by 6 dB. A second feature of these combiners is that of maintaining the 50-ohm impedance of the system — very important if accurate measurements are to be made.

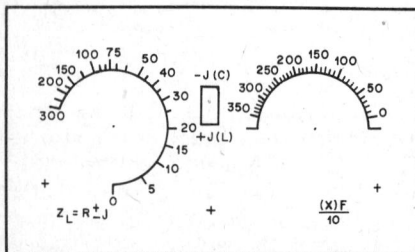

Fig. 80 — Reduced dial escutcheon for the OIB. Send s.a.s.e. to ARRL for full-scale template.

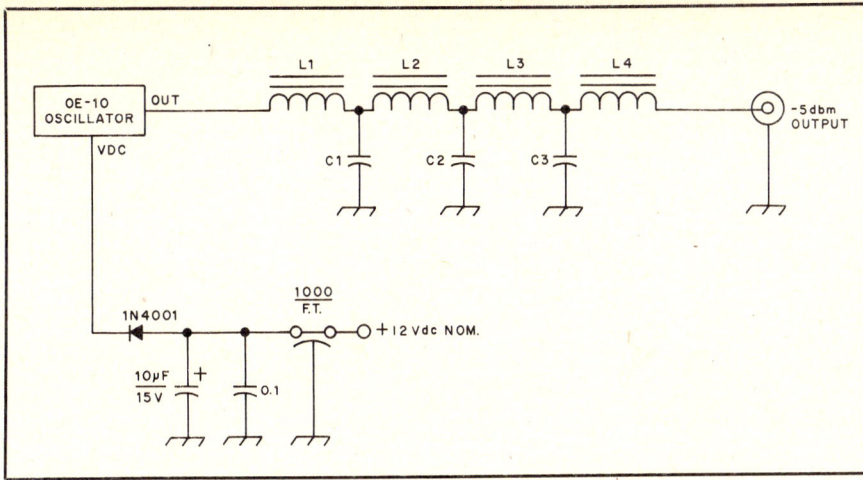

Fig. 83 — Schematic diagram of the signal generator assembly. Reactance values for the filter components are given. From this information, filters can be built for any frequency range which the user may need.

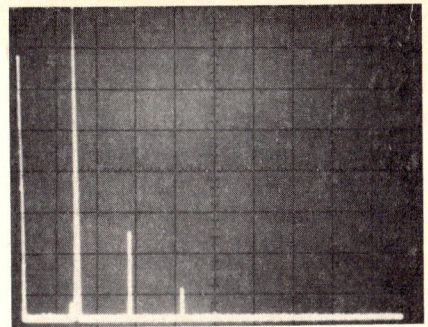

Fig. 84 — Spectral display of the output from one of the signal generators. Each vertical division represents 10 dB and each horizontal division is 10 MHz. The second harmonic is 55 dB below the main signal and the third harmonic is some 68 dB down. Higher-order harmonics are not visible in this photograph.

Fig. 85 — Exterior view of two hybrid combiners. The one on the left is designed to cover the 1 to 50 MHz range; the one on the right 50 to 500 MHz.

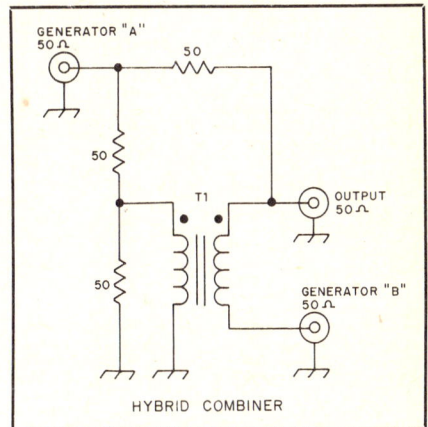

Fig. 86 — Schematic diagram of the hybrid combiners. For the 1 to 50 MHz model, T1 is 10 turns no. 30 bifilar wound on an FT-23-72 ferrite core. For the 50 to 500 MHz model, T1 consists of 10 turns no. 30 bilfilar wire wound on an FT-23-63 ferrite core. Keep all leads as short as possible when constructing these units.

The combiners are constructed in small boxes made from double-sized, circuit-board material. Each piece is soldered to the adjacent one along the entire length of the seam. This makes for an "rf-tight" enclosure. BNC coaxial fittings are used on the units shown. However, any type of coaxial connector can be used. Leads must be kept as short as possible and precision resistors (or matched units from the junk box) should be used. The circuit diagram for the combiners is shown in Fig. 86.

A Low-Power Step Attenuator

Described here is a simple low-power step attenuator suitable for receiver front-end protection, and as a calibrated attenuator for receiver performance evaluation. This attenuator uses double-pole, double-throw toggle switches to select different amounts of attenuation. Coaxial fittings are used at each end of the attenuator.

This unit provides 0-147 dB of attenuation in 1-dB steps. Careful attention must be paid to circuit layout, with good shielding between each of the attenuator sections being essential. A suitable enclosure can be made from double-sided, printed-circuit board material with individual compartments for each section. The resistances shown in Fig. 87 are the nearest standard values to those resistances appearing in Tables 6 and 7. Although some of the values are a few ohms off, they should be more than acceptable for amateur work.

Measuring Receiver Performance

Comparing the performance of one receiver to another is difficult at best. The features of one receiver may outweigh a second, even though its performance under strong-signal conditions is not as good as it should be. Although the final decision on which receiver to own will more than likely be based on personal preference, there are ways to compare receiver performance characteristics. The most important parameters are noise floor, intermodulation distortion, blocking (gain compression) and cross modulation.

The general test setup is shown in Fig. 88. Two calibrated signal generators are required, along with a hybrid combiner, a step attenuator and an ac voltmeter. A

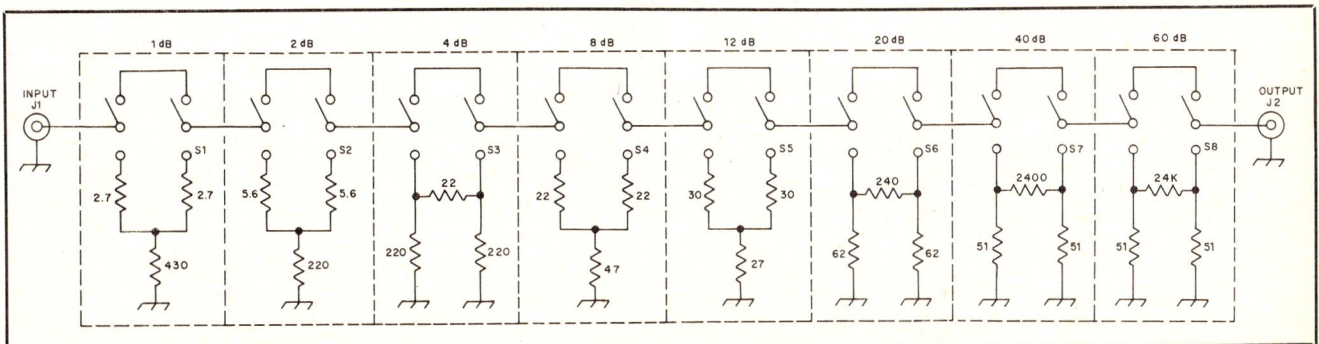

Fig. 87 — Schematic diagram of the 0 to 147 dB step attenuator. Resistance values for other amounts of attenuation are given in Tables 6 and 7. S1-S8 are dpdt toggle switches.

Table 5
Filter Reactance Values

Inductance and capacitance values normalized to 1 MHz for the 7-element Chebyshev low-pass filter. Values for 0.1 and 1 dB ripple are given. For filters at other cut-off frequencies, simply divide the normalized values by the desired frequency in MHz.

	L1	L2	L3	L4	C1	C2	C3
(0.1 dB ripple)	9.4	16.68	16.68	9.4	4529	5008	4529
(1 dB ripple)	17.24	24.62	24.62	17.24	252.7	266.8	252.7

Inductance values are in μH and capacitance values are in pF.

Table 6
Pi-Network Resistive Attenuator

dB Atten.	R1 (Ohms)	R2 (Ohms)
1	870.0	5.8
2	436.0	11.6
3	292.0	17.6
4	221.0	23.8
5	178.6	30.4
6	150.5	37.3
7	130.7	44.8
8	116.0	52.8
9	105.0	61.6
10	96.2	71.2
11	89.2	81.6
12	83.5	93.2
13	78.8	106.0
14	74.9	120.3
15	71.6	136.1
16	68.8	153.8
17	66.4	173.4
18	64.4	195.4
19	62.6	220.0
20	61.0	247.5
21	59.7	278.2
22	58.6	312.7
23	57.6	351.9
24	56.7	394.6
25	56.0	443.1
30	53.2	789.7
35	51.8	1405.4
40	51.0	2500.0
45	50.5	4446.0
50	50.3	7905.6
55	50.2	14,058.0
60	50.1	25,000.0

Table 7
T-Network Resistive Attenuator

dB Atten.	R1 (Ohms)	R2 (Ohms)
1	2.9	433.3
2	5.7	215.2
3	8.5	141.9
4	11.3	104.8
5	14.0	82.2
6	16.6	66.9
7	19.0	55.8
8	21.5	47.3
9	23.8	40.6
10	26.0	35.0
11	28.0	30.6
12	30.0	26.8
13	31.7	23.5
14	33.3	20.8
15	35.0	18.4
16	36.3	16.2
17	37.6	14.4
18	38.8	12.8
19	40.0	11.4
20	41.0	10.0
21	41.8	9.0
22	42.6	8.0
23	43.4	7.1
24	44.0	6.3
25	44.7	5.6
30	47.0	3.2
35	48.2	1.8
40	49.0	1.0
45	49.4	0.56
50	49.7	0.32
55	49.8	0.18
60	49.9	0.10

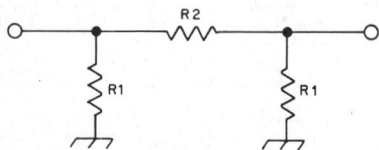

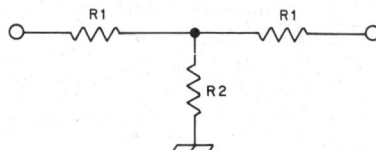

hybrid combiner is essentially a unit with three ports. The device is used to combine the signals from a pair of generators. This box has the characteristic that signals applied at ports 1 or 2 appear at port 3, and are attenuated by 6 dB. However, a signal from port 1 is attenuated 30 or 40 dB when sampled at port 2. Similarly, signals applied at port 2 are isolated from port 1 some 30 to 40 dB. The isolating properties of the box prevent one signal generator from being frequency- or phase-modulated by the other. A second feature of a hybrid combiner is that a 50-ohm impedance level is maintained throughout the system. A commercial example of a hybrid coupler of this kind is an HP-8721A.

The signal generators used in the test setup must be calibrated accurately in dBm or microvolts. The generators should have extremely low leakage. That is, when the output of the generator is disconnected, no signal should be detected at the operating frequency with a sensitive receiver. Ideally, at least one of the signal generators should be capable of amplitude modulation. A suitable lab-quality piece would be the HP-8640B.

While most signal generators are calibrated in terms of microvolts, the real concern is not with the voltage from the generator but with the power available. The fundamental unit of power is the watt. However, the unit which is used for most low-level rf work is the milliwatt,

and power is often specified in dB with respect to one milliwatt (dBm). Hence, a 0 dBm would be one milliwatt. The dBM level, in a 50 ohm load, can be calculated with the aid of the following equation

$$dBm = 10 \, Log_{10} \, [20(V_{RMS})^2]$$

where dBm is the power with respect to one milliwatt and V is the rms voltage available at the output of the signal generator.

The convenience of a logarithmic power unit like the dBm becomes apparent when signals are amplified or attenuated. For example, a −107-dBm signal that is applied to an amplifier with a gain of 20 dB will result in an output of −107 dBm + 20 dB, or −87 dBm. Similarly, a −107-dBm signal which is applied to an attenuator with a loss of 10 dB will result in an output of −107 dBm − 10 dB, or −117 dBm.

Noise-Floor Measurement

A generator that is tuned to the same frequency as the receiver is used for this test. Output from the generator is increased until the ac voltmeter at the audio-output jack of the receiver shows a 3-dB increase. This measurement indicates the minimum discernable signal (MDS) which could be detected with the receiver. This level is defined as that which will produce the same audio-output power as the internally generated receiver noise. Hence, the term "noise floor." As a hypothetical example, say the output of the signal generator is −127 dBm, the loss through the combiner is fixed at 6 dB and the step attenuator is set to 4 dB. The equivalent noise floor can then be calculated as follows:

$$Noise \, floor = -127 \, dBm - 6 \, dB - 4 \, dB$$
$$= -137 \, dBm$$

where noise floor is the power available at the receiver antenna terminal, the −6 dB is the loss through the coupler and −4 dB is the loss through the attenuator. Refer to Fig. 89.

Blocking

This measurement concerns *gain compression*. Both signal generators are used. One is set for a weak signal of roughly −110 dBm and the receiver is tuned to this frequency. The other generator is set to a frequency 20 kHz away and is increased in amplitude until the receiver output drops by 1 dB, as measured with the ac voltmeter. A blocking measurement is indicative of the signal level that can be tolerated at the receiver antenna terminal before desensitization will occur.

As an example, say that the output of the generator is −27 dBm, the loss through the combiner is fixed at 6 dB and there is 0-dB attenuation through the attenuator (effectively switched out of the

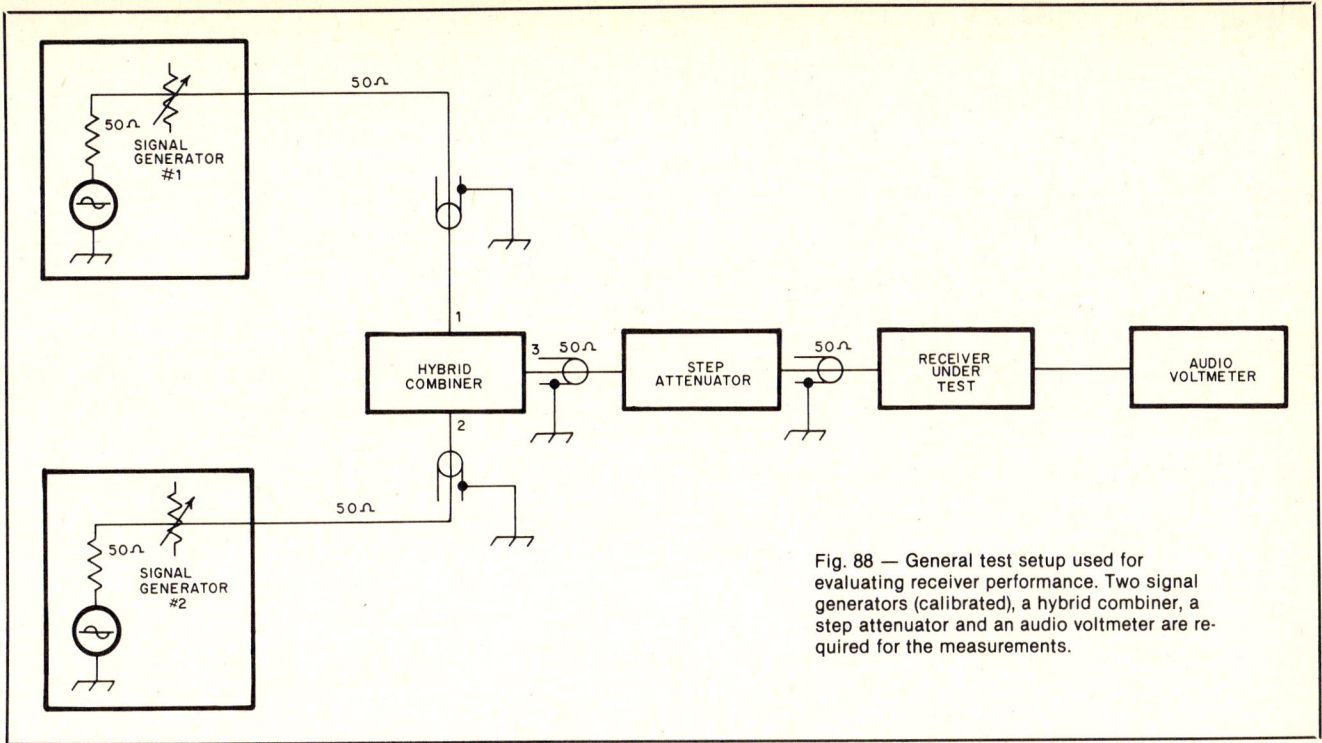

Fig. 88 — General test setup used for evaluating receiver performance. Two signal generators (calibrated), a hybrid combiner, a step attenuator and an audio voltmeter are required for the measurements.

line). See Fig. 90. The signal level at the receiver terminal that will cause gain compression is calculated as follows:

Blocking level = − 27 dBm − 6 dB
= − 33 dBm

This can be expressed as *dynamic range* when this level is referenced to the receiver noise floor that was calculated earlier. This term can be called "receiver blocking dynamic range." Calculate as follows:

Blocking dynamic range
= noise floor − blocking level
= − 137 dBm − (− 33 dBm)
= − 104 dB

This value is usually taken in terms of absolute value and would be referred to as 104 dB.

Two-Tone IMD Test

This figure is one of the most significant parameters that can be specified for a receiver. It is a measure of the range of signals that can be tolerated while producing essentially no undesired spurious responses. It is generally a conservative evaluation for other effects, such as blocking, which will occur only for signals well outside the IMD dynamic range of the receiver.

Two signals of equal level spaced 20 kHz apart are injected into the input of the receiver. Call the frequencies F1 and F2. The so-called third-order intermodulation-distortion products will appear at frequencies of $(2F_1-F_2)$ and $(2F_2-F_1)$. Assume that the two input frequencies are 14.040 and 14.060 MHz. The third-order

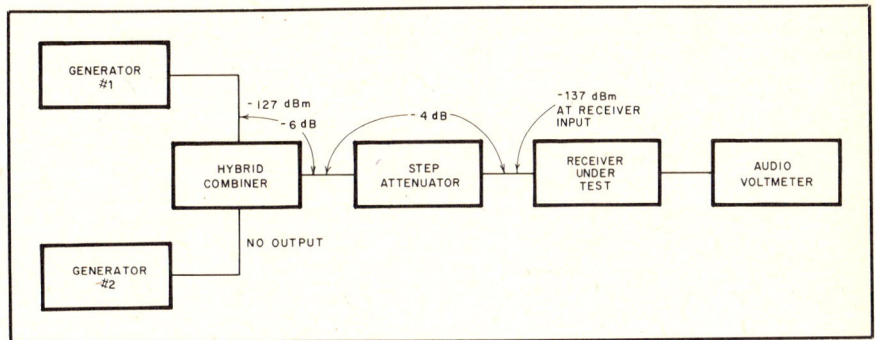

Fig. 89 — General test setup for measuring receiver noise floor. Signal levels for a hypothetical measurement are indicated. See text for a detailed discussion.

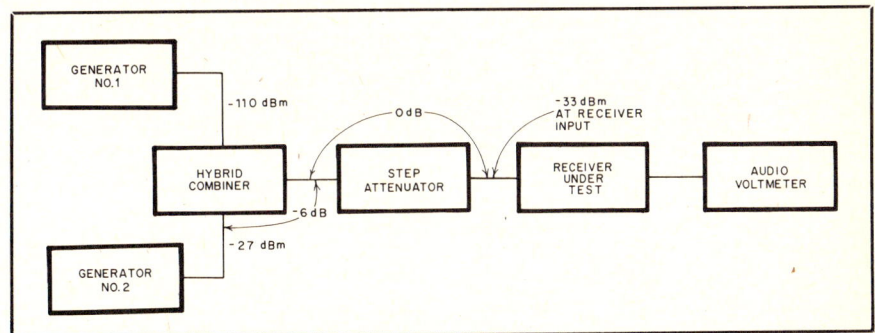

Fig. 90 — Test setup for measuring receiver blocking performance. Again, signal levels for a hypothetical measurement are included on the drawing.

products will be at 14.020 and 14.080 MHz.

The step attenuator will be useful in this experiment. Adjust the two generators for an output of − 10 dBm each at frequencies spaced 20 kHz. Tune the receiver to either of the third-order IMD products.

Adjust the step attenuator until the IMD product produces an output 3 dB above the noise level as read on the ac voltmeter.

For an example, say the output of the generator is − 10 dBm, the loss through the combiner is 6 dB and the amount of attenuation used is 40 dB. See Fig. 91. The

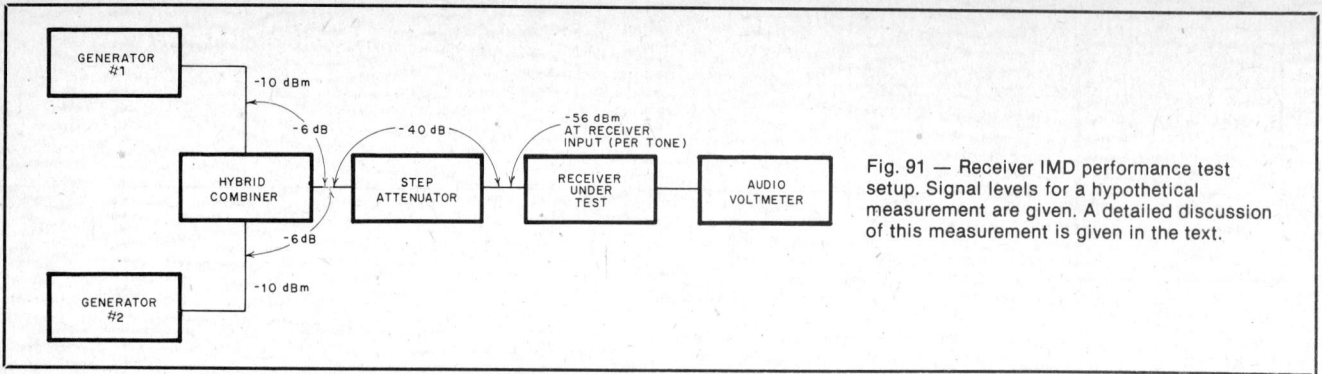

Fig. 91 — Receiver IMD performance test setup. Signal levels for a hypothetical measurement are given. A detailed discussion of this measurement is given in the text.

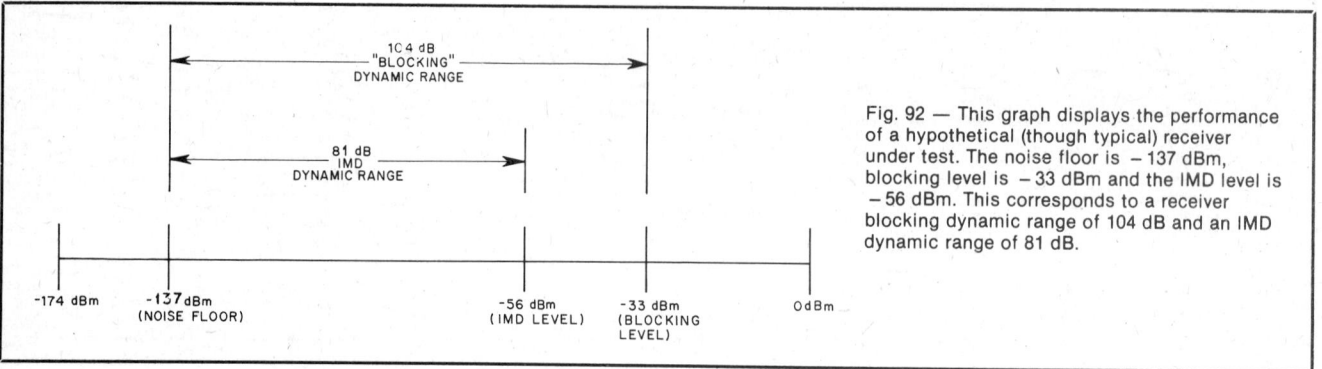

Fig. 92 — This graph displays the performance of a hypothetical (though typical) receiver under test. The noise floor is − 137 dBm, blocking level is − 33 dBm and the IMD level is − 56 dBm. This corresponds to a receiver blocking dynamic range of 104 dB and an IMD dynamic range of 81 dB.

signal level at the receiver antenna terminal that just begins to cause IMD problems is calculated as:

IMD level = − 10 dBm − 6 dB − 40 dB
= − 56 dBm

This can be expressed as a dynamic range when this level is referenced to the noise floor. This term is referred to as "IMD dynamic range" and can be calculated

IMD dynamic range
= noise floor − IM level
= − 137 dBm − (− 56 dBm)
= − 81 dB

Therefore, the IMD dynamic range of this receiver would be 81 dB.

Evaluating the Data

Thus far a fair amount of data has been gathered with no mention of what the numbers really mean. It is somewhat easier to understand exactly what is hap-

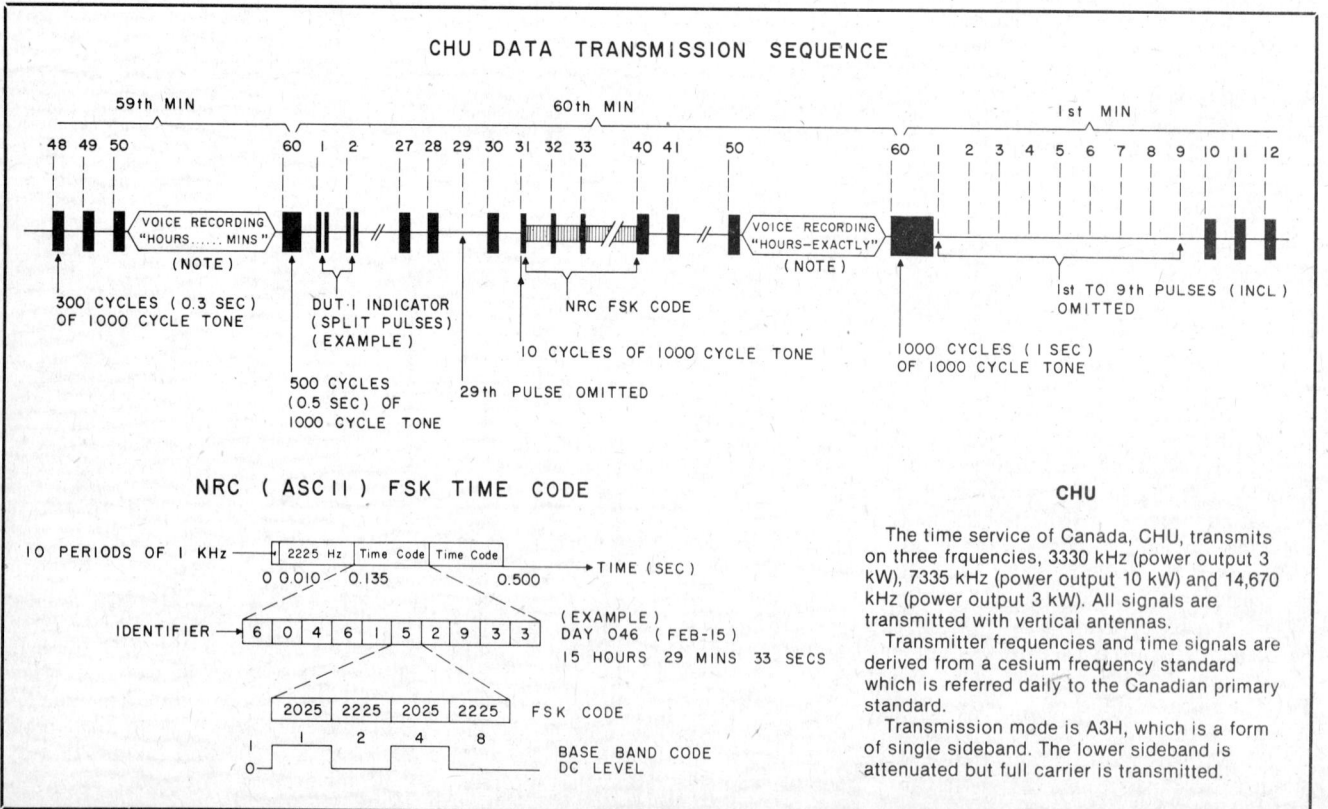

CHU DATA TRANSMISSION SEQUENCE

NRC (ASCII) FSK TIME CODE

CHU

The time service of Canada, CHU, transmits on three frquencies, 3330 kHz (power output 3 kW), 7335 kHz (power output 10 kW) and 14,670 kHz (power output 3 kW). All signals are transmitted with vertical antennas.

Transmitter frequencies and time signals are derived from a cesium frequency standard which is referred daily to the Canadian primary standard.

Transmission mode is A3H, which is a form of single sideband. The lower sideband is attenuated but full carrier is transmitted.

pening by arranging the data in a form something like that in Fig. 92. The base line is just a power line with a very small level of power at the left and a high level (0 dBm) at the right.

The noise floor of the hypothetical receiver is drawn in at −137 dBm, the IMD level (the level at which signals will begin to create spurious responses) at −56 dBm and the blocking level (the level at which signals will begin to desense the receiver) at −33 dBm. As can be seen, the IMD dynamic range is some 23 dB smaller than the blocking dynamic range. This means that IMD products will be heard across the band long before the receiver will begin to desense — some 23 dB sooner.

The figures for the hypothetical receiver represent those which would be expected from a typical communications receiver on the market today. It is interesting to note that it is possible for the home constructor to build a receiver that will outperform commercially available units (even the high-priced ones).

WWV BROADCAST FORMAT

VIA TELEPHONE (303) 499-7111
(NOT A TOLL-FREE NUMBER)

● BEGINNING OF EACH HOUR IS IDENTIFIED BY 0.8-SECOND LONG, 1500-Hz TONE.
● BEGINNING OF EACH MINUTE IS IDENTIFIED BY 0.8-SECOND LONG, 1000-Hz TONE.
● THE 29th & 59th SECOND PULSE OF EACH MINUTE IS OMITTED.

WWVH BROADCAST FORMAT

VIA TELEPHONE (808) 335-4363
(NOT A TOLL-FREE NUMBER)

● BEGINNING OF EACH HOUR IS IDENTIFIED BY 0.8-SECOND LONG, 1500-Hz TONE.
● BEGINNING OF EACH MINUTE IS IDENTIFIED BY 0.8-SECOND LONG, 1200-Hz TONE.
● THE 29th & 59th SECOND PULSE OF EACH MINUTE IS OMITTED.

Standard Frequencies and Time Signals

The National Bureau of Standards maintains two radio transmitting stations, WWV at Ft. Collins, CO, and WWVH near Kekaha, Dauai, HI, for broadcasting standard radio frequencies of high accuracy. WWV and WWVH broadcasts are on 2.5, 5, 10 and 15 MHz. The broadcasts of both stations are continuous, night and day. Standard audio frequencies of 440, 500 and 600 Hz on each radio carrier frequency by WWV and WWVH. The duration of each tone is approximately 45 seconds. A 500-Hz tone is broadcast during even alternate minutes unless voice announcements or silent periods are scheduled. A 440-Hz tone is broadcast beginning one minute after the hour by WWVH and two minutes after the hour by WWV. The 440-Hz tone period is omitted during the first hour of the UTC day.

Transmitted frequencies from the two stations are accurate to ±1 part in 10. Atomic frequency standards are used to maintain this accuracy. Voice announcements of the time, in English, are given every minute. WWV utilizes a male voice, and WWVH features a female voice to distinguish between the two stations. WWV time and frequency broadcasts can be heard by telephone also. The number to call is 303-499-7111, Boulder, CO.

All official announcements are made by voice. Time announcements are in UTC (Universal Coordinated Time). One-second markers are transmitted throughout all programs except that the 29th and 59th markers of each minute are omitted. Detailed information on hourly broadcast schedules is given in the accompanying format chart. Complete information on the services can be found in NBS Special Publication 432, *NBS Frequency and Time Dissemination Services*, available for 60 cents from the Superintendent of Documents, U.S. Government Printing Office, Washington, DC 20402.

Geophysical Alerts

"Geoalerts" are broadcast in voice during the 18th minute of each hour from WWV. The messages are changed each day at 0400 UT with provisions to schedule immediate alerts of outstanding occurring events. Geoalerts tell of geophysical events affecting radio propagation, stratospheric warming and related events.

Chapter 17

Construction Practices and Data Tables

While a better job can be done with a variety of tools, by taking a little care it is possible to turn out a fine piece of equipment with only a few common hand tools. A list of tools which are indispensable in the construction of electronic equipment is found on this page. To convert English dimensions in the list to millimeters, multiply inches × 25.4. With these tools it should be possible to prepare panels and metal chassis for assembly and wiring. It is an excellent idea for the amateur who builds gear to add to his supply of tools from time to time as finances permit.

Recommended Tools and Materials

Long-nose pliers, 6- and 4-inch
Diagonal cutters, 6- and 4-inch
Combination pliers, 6-inch
Screwdriver, 6- to 7-inch, 1/4-inch blade
Screwdriver, 4- to 5-inch, 1/8-inch blade
Phillips screwdriver, 6- to 7-inch
Phillips screwdriver, 3- to 4-inch
Long-shank screwdriver with holding clip on blade
Scratch awl or scriber for marking metal
Combination square, 12-inch, for layout work
Hand drill, 1/4-inch chuck or larger
Soldering pencil, 30-watt, 1/8-inch tip
Soldering iron, 200-watt, 5/8-inch tip
Hacksaw and 12-inch blades
Hand nibbling tool, for chassis-hole cutting
Hammer, ball-peen 1-lb. head
Heavy-duty jackknife
File set, flat, round, half-round, and triangular. Large and miniature types recommended.
High-speed drill bits, no. 60 through 3/8-inch diameter.
Set of "Spintite" socket wrenches for hex nuts
Adjustable wrenches, 6- and 10-inch
Machine-screw taps, 4-40 through 10-32 thread
Socket punches, 1/2", 5/8", 3/4", 1-1/8", 1-1/4", and 1-1/2"

Tapered reamer, T-handle, 1/2-inch maximum width
Bench vise, 4-inch jaws or larger
Medium-weight machine oil
Tin shears, 10-inch size
Motor-driven emery wheel for grinding
Solder, *rosin core only*
Contact cleaner, liquid or spray can
Duco cement or equivalent
Electrical tape, vinyl plastic

Radio-supply houses, mail-order retail stores and most hardware stores carry the various tools required for building or servicing amateur radio equipment. While power tools (electric drill or drill press, grinding wheel, etc.) are very useful and

will save a lot of time, they are not essential.

Twist Drills

Twist drills are made of either high-speed steel or carbon steel. The latter type is more common and will usually be supplied unless specific request is made for high-speed drills. The carbon drill will suffice for most ordinary equipment construction work and costs less than the high-speed type.

While twist drills are available in a number of sizes, those listed in bold type in Table 1 will be most commonly used in construction of amateur equipment. It is

Table 1

Numbered Drill Sizes

No.	Diameter (Mils)	Will Clear Screw	Drilled for Tapping from Steel or Brass	No.	Diameter (Mils)	Will Clear Screw	Drilled for Tapping from Steel or Brass
1	228.0	12-24	-	**28**	**140.0**	6-32	-
2	221.0	-	-	**29**	**136.0**	-	8-32
3	213.0	-	14-24	30	128.5	-	-
4	209.0	12-20	-	31	120.0	-	-
5	205.0	-	-	32	116.0	-	-
6	204.0	-	-	**33**	**113.0**	4-40	-
7	201.0	-	-	34	111.0	-	-
8	199.0	-	-	35	110.0	-	-
9	196.0	-	-	**36**	**106.5**	-	6-32
10	193.5	-	-	37	104.0	-	-
11	**191.0**	10-24, 10-32	-	38	101.5	-	-
12	189.0	-	-	**39**	**099.5**	3-48	-
13	185.0	-	-	40	098.0	-	-
14	182.0	-	-	41	096.0	-	-
15	180.0	-	-	**42**	**093.5**	-	-
16	177.0	-	12-24	43	089.0	-	4-40
17	173.0	-	-	**44**	**086.0**	2-56	-
18	169.5	-	-	45	082.0	-	-
19	**166.0**	8-32	12-20	46	081.0	-	-
20	161.0	-	-	**47**	**078.5**	-	3-48
21	**159.0**	-	10-32	48	076.0	-	-
22	157.0	-	-	49	073.0	-	-
23	154.0	-	-	**50**	**070.0**	-	2-56
24	152.0	-	-	51	067.0	-	-
25	149.5	-	10-24	52	063.5	-	-
26	147.0	-	-	53	059.5	-	-
27	144.0	-	-	54	055.0	-	-

Fig. 1 — A compact assembly of commonly available items, this soldering station sanitizes the electronics assembly process. Miniature toggle switches are used because of the minimal force required to manipulate them. The force required to operate standard-size switches could destabilize the unit.

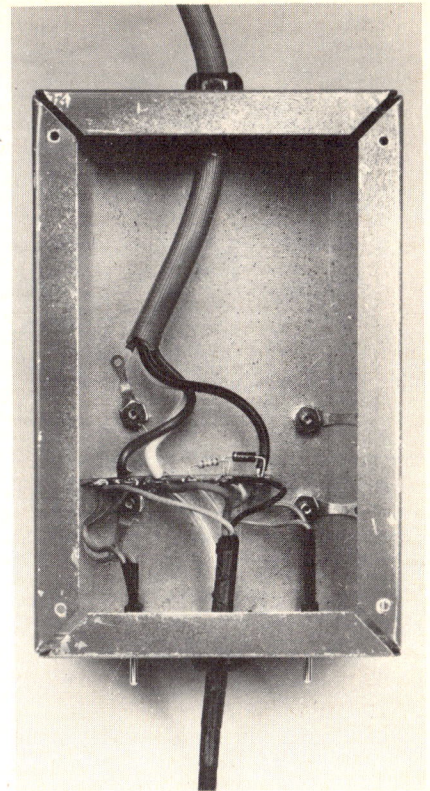

Fig. 2 — View of the chassis underside with the bottom plate removed. No. 24 hookup wire is adequate for all connections. Use sleeving wherever the possibility of a short circuit exists. The diode may be installed in either direction.

usually desirable to purchase several of each of the commonly used sizes rather than a standard set, most of which will be used infrequently, if at all.

Although Table 1 lists drills down to no. 54, the series extends to no. 80. No. 68 and no. 70 are useful for drilling printed-circuit boards for component leads.

Care of Tools

The proper care of tools is not only a matter of pride to a good worker. He also recognizes the energy saved and the annoyance avoided by possessing a full kit of well-kept, sharp-edged tools.

Drills should be sharpened at frequent intervals so that grinding is kept at a minimum each time. This makes it easier to maintain the rather critical surface angles required for best cutting with least wear. Occasional oilstoning of the cutting edges of a drill or reamer will extend the time between grindings.

The soldering iron can be kept in good condition by keeping the tip well tinned with solder and not allowing it to run at full voltage for long periods when it is not being used. After each period of use, the tip should be removed and cleaned of any scale which may have accumulated. An oxidized tip may be cleaned by dipping it in sal ammoniac (ammonium chloride) while hot and then wiping it clean with a rag. If a copper tip becomes pitted it

should be filed until smooth and bright, and then tinned immediately by dipping it in solder. Most modern soldering iron tips are iron-clad and cannot be filed.

Useful Materials

Small stocks of various miscellaneous materials will be required in constructing radio apparatus. Most of these are available from hardware or radio-supply stores. A representative list follows:

Sheet aluminum, solid and perforated, 16 or 18 gauge, for brackets and shielding.

1/2 × 1/2-inch (12 × 13-mm) aluminum angle stock.

1/4-inch (6-mm) diameter round brass or aluminum rod for shaft extensions.

Machine screws: Round-head and flat head, with nuts to fit. Most useful sizes: 4-40, 6-32 and 8-32, in lengths from 1/4-inch (6-mm) to 1-1/2 inches (38 mm). (Nickel-plated iron will be found satisfactory except in strong rf fields, where brass should be used.)

Bakelite, Lucite, polystyrene and copper-clad pc-board scraps.

Soldering lugs, panel bearings, rubber grommets, terminal-lug wiring strips, varnished-cambric insulating tubing, heat-shrinkable tubing.

Shielded and unshielded wire.

Tinned bare wire, nos. 22, 14 and 12.

Machine screws, nuts, washers, soldering lugs, etc., are most reasonably pur-

chased in quantities of a gross. Many of the radio-supply stores sell small quantities and assortments that come in handy.

A Deluxe Soldering Station

The simple device shown on this page can enhance the versatility of longevity of a soldering iron as well as make electronic assembly more convenient. Fig. 1 depicts the obvious convenience features — a protective heat sink and cage, and a tip-cleaning sponge rigidly attached to a sturdy base for efficient one-handed operation. Inside the chassis are some electrical refinements that justify the sophisticated name "soldering station."

Soldering iron tips and heating elements last longer if operated at lower-than-maximum temperature when idling. Many solder connections can be made satisfactorily with reduced heat, and some small semiconductor devices *require* lower temperatures to avoid junction damage. In the unit described here temperature reduction is accomplished by halving the duty cycle of the applied ac voltage. D1 in Fig. 3 conducts only when the "hot" ac line is positive with respect to neutral. If the diode were reversed, the soldering iron would be heated only on the negative half cycles, but the result would be the same. (This is one of the rare applications of rectifier diodes where the polarity is *not* important.) With current flowing only in one

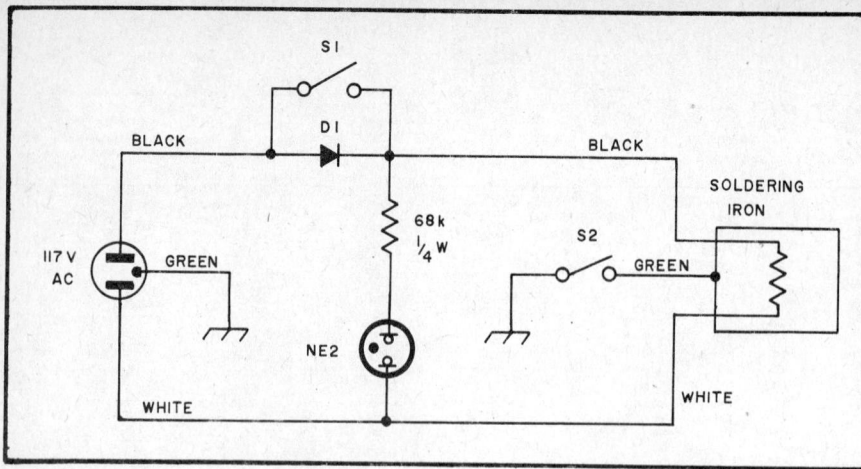

Fig. 3 — Schematic diagram of the soldering station. D1 is a silicon power rectifier, 1-A, 400-PRV. S1 and S2 are miniature spst toggle switches rated 3 A at 125 V. This circuit is satisfactory for use with irons having power ratings up to 100 W.

direction, only one electrode of the neon bulb will appear to glow. Closing S1 short-circuits the diode and applies full power to the soldering iron, igniting both bulb electrodes brightly.

CMOS ICs are prone to damage by static charges, so they should be soldered with an iron having a grounded tip. This requirement is fulfilled by most irons having 3-wire power cords. Unfortunately, a grounded tip precludes using the iron on a live circuit. If the potentials are low (less than about 25 volts) and the operator is thoroughly familiar with the circuit, a soldering iron may facilitate experimentation or trouble-shooting. This technique should be used only when complete safety is assured. The simplest way to increase the usefulness of a 3-wire soldering iron is to install a switch in the ground lead. S2 in Fig. 3 serves this purpose. Before clipping the cord on your soldering iron, be certain that the tip is common to the ground prong and isolated from the other ac prongs.

The base for the unit is a 2 × 6 × 4-inch (51 × 152 × 102 mm) (HWD) aluminum chassis (Bud AC-431 or equivalent). An Ungar model 8000 soldering iron holder fits neatly on the chassis top. The holder has two mounting holes in each foot. A sponge tray nests between the feet and the cage. In this model a sardine tin is used for the sponge tray, although a suitable watertight enclosure can also be fabricated from strips of copper-clad circuit-board material. The tray and iron holder are secured to the chassis by 6-32 × 1/2-inch pan head machine screws and nuts, with flat washers under the screw heads (sponge tray) and lock washers under the nuts (chassis underside). One of these nuts fastens a 6-lug tie point strip to the chassis bottom. Use the soldering iron holder base as a template for drilling the chassis and sponge tray. The floor of the sponge tray must be sealed around the screw heads to prevent moisture from leaking

into the electrical components below the chassis. RTV compound was used for this purpose in the unit pictured.

Purchase a separate 3-wire cord for the power input. Merely splicing the soldering station into the existing soldering iron cord will shorten the operating radius of the iron and make it awkward to use. Heyco bushings were used to anchor both cords in the unit described, but if these aren't available, grommets and cable clamps will work as well. Knotting the cords inside the chassis is a simple expedient that sometimes provides adequate strain relief.

The underchassis assembly is shown in Fig. 2. The neon bulb is forced through a 3/16-inch (5-mm) ID grommet. The leads are sleeved to prevent short circuits. If you mount the bulb in a fixture or socket, use a clear lens to ensure that the electrodes are distinctly visible. Fit a cover to the bottom of the chassis to prevent accidental contact with the live ac wiring. Stick-on rubber feet will ensure a skid-free unit that won't mar your work surface.

The total cost of this project with all new parts (including sardines) was $28. One could trim that figure considerably with even a modest junkbox. The soldering iron, an Ungar model 127, represents half of the investment.

Chassis Working

With a few essential tools and proper procedure, building radio gear on a metal chassis is a relatively simple matter. Aluminum is preferred to steel, not only because it is a superior shielding material, but because it is much easier to work and provides good chassis contacts.

The placement of components on the chassis is shown quite clearly in the photographs in this *Handbook*. Aside from certain essential dimensions, which usually are given in the text, exact duplication is not necessary.

Much trouble and energy can be saved by spending sufficient time in planning

the job. When all details are worked out beforehand the actual construction is greatly simplified.

Cover the top of the chassis with a piece of wrapping paper, or, preferably, cross-section paper, folding the edges down over the sides of the chassis and fastening with adhesive tape. Then assemble the parts to be mounted on top of the chassis and move them about until a satisfactory arrangement has been found, keeping in mind any parts which are to be mounted underneath, so interference in mounting can be avoided. Place capacitors and other parts with shafts extending through the panel first, and arrange them so that the controls will form the desired pattern of the panel. Be sure to line up the shafts squarely with the chassis front. Locate any partition shields and panel brackets next, and then the tube sockets and any other parts, marking the mounting-hole centers of each accurately on the paper. Watch out for capacitors whose shafts are off center and do not line up with the mounting holes. Do not forget to mark the centers of socket holes and holes for wiring leads. The small holes for socket-mounting screws are best located and center-punched, using the socket itself as a template, after the main center hole has been cut.

By means of the square, lines indicating accurately the centers of shafts should be extended to the chassis front and marked on the panel at the chassis line, the panel being fastened on temporarily. The hole centers may then be punched in the chassis with the center punch. After drilling, the parts which require mounting underneath may be located and the mounting holes drilled, making sure by trial that no interferences exist with parts mounted on top. Mounting holes along the front edge of the chassis should be transferred to the panel by once again fastening the panel to the chassis and marking it from the rear.

Next, mount on the chassis the capacitors and any other parts with shafts extending to the panel, and measure accurately the height of the center of each shaft above the chassis, as illustrated in Fig. 4. The horizontal displacement of

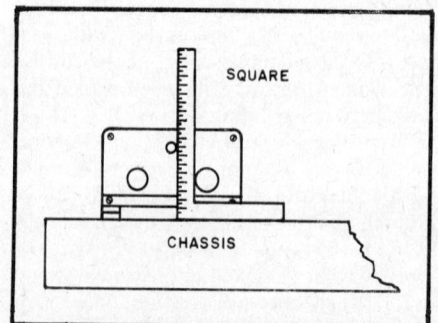

Fig. 4 — Method of measuring the heights of capacitor shafts. If the square is adjustable, the end of the scale should be set flush with the face of the head.

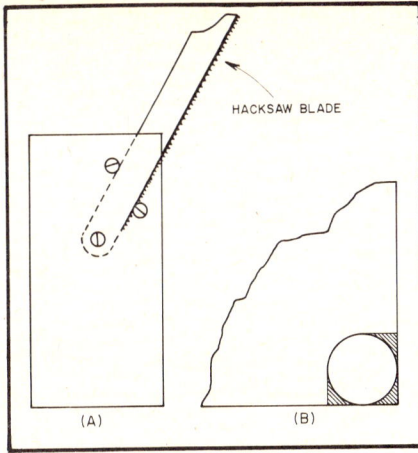

Fig. 5 — To cut rectangular holes in a chassis corner, holes may be filed out as shown in the shaded portion of B, making it possible to start the hacksaw blade along the cutting line. A shows how a single-ended handle may be constructed for a hacksaw blade.

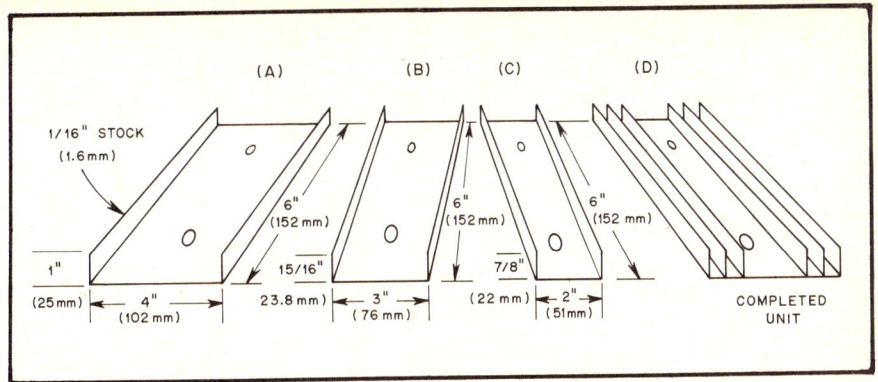

Fig. 6 — Details for forming channel-type heat sinks.

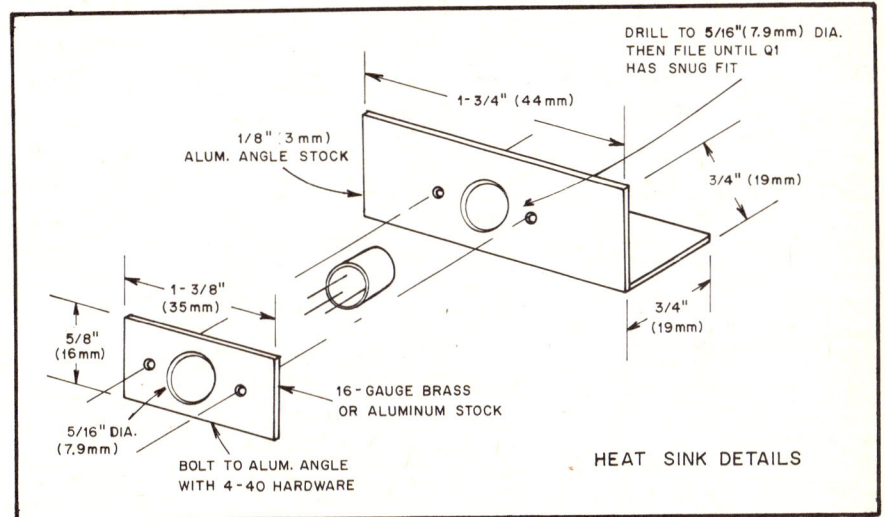

Fig. 7 — Layout and assembly details of another homemade heat sink. The completed assembly can be insulated from the main chassis of the transmitter by using insulating washers.

shafts having already been marked on the chassis line on the panel, the vertical displacement can be measured from this line. The shaft centers may now be marked on the back of the panel, and the holes drilled. Holes for any other panel equipment coming above the chassis line may then be marked and drilled, and the remainder of the apparatus mounted. Holes for terminals and other parts of the rear edge of the chassis should be marked and drilled at the same time that they are done for the top.

Drilling and Cutting Holes

When drilling holes in metal with a hand drill it is important that the centers first be located with a center punch, so that the drill point will not "walk" away from the center when starting the hole. When the drill starts to break through, special care must be used. Often it is an advantage to shift a two-speed drill to low gear at this point. Holes more than 1/4-inch (6-mm) in diameter should be started with a smaller drill and reamed out with the larger drill.

The chuck on the usual type of hand drill is limited to 1/4-inch (6-mm) drills. The 1/4-inch (6-mm) hole may be filed out to larger diameters with round files. Another method possible with limited tools is to drill a series of small holes with the hand drill along the inside of the circumference of the large hole, placing the holes as close together as possible. The center may then be knocked out with a cold chisel and the edges smoothed with a file. Taper reamers which fit into the carpenter's brace will make the job easier. A large rat-tail file clamped in the brace makes a very good reamer for holes up to the diameter of the file.

For socket holes and other large holes in an aluminum chassis, socket-hole punches should be used. They require first drilling a guide hole to pass the bolt that is

turned to squeeze the punch through the chassis. The threads of the bolt should be oiled occasionally.

Large holes in steel panels or chassis are best cut with an adjustable circle cutter. Occasional application of machine oil in the cutting groove will help. The cutter first should be tried out on a block of wood, to make sure that it is set for the right diameter.

The burrs or rough edges which usually result after drilling or cutting holes may be removed with a file, or sometimes more conveniently with a sharp knife or chisel. It is a good idea to keep an old wood chisel sharpened and available for this purpose.

Rectangular Holes

Square or rectangular holes may be cut out by making a row of small holes as previously described, but is more easily done by drilling a 1/2-inch (13-mm) hole inside each corner, as illustrated in Fig. 5, and using these holes for starting and turning the hacksaw. The socket-hole punch and the square punches which are now available also may be of considerable assistance in cutting out large openings.

Semiconductor Heat Sinks

Homemade heat sinks can be fashioned from brass, copper or aluminum stock by employing ordinary workshop tools. The dimensions of the heat sink will depend upon the type of transistor used and the amount of heat that must be conducted away from the body of the semiconductor.

Fig. 6 shows the order of progression for forming a large heat sink from aluminum or brass channels of near-equal height and depth. The width is lessened in parts B and C so that each channel will fit into the preceding one as shown in the completed model at D. The three pieces are bolted together with 8-32 screws and nuts. Dimensions given are for illustrative purposes only.

Heat sinks for smaller transistors can be fabricated as shown in Fig. 8. Select a drill bit that is one size smaller than the diameter of the transistor case and form the heat sink from 1/16-inch (1.6-mm) thick brass, copper or aluminum stock as shown in steps A, B and C. Form the stock around the drill bit by compressing it in a vise (A). The completed heat sink is press-fitted over the body of the semicon-

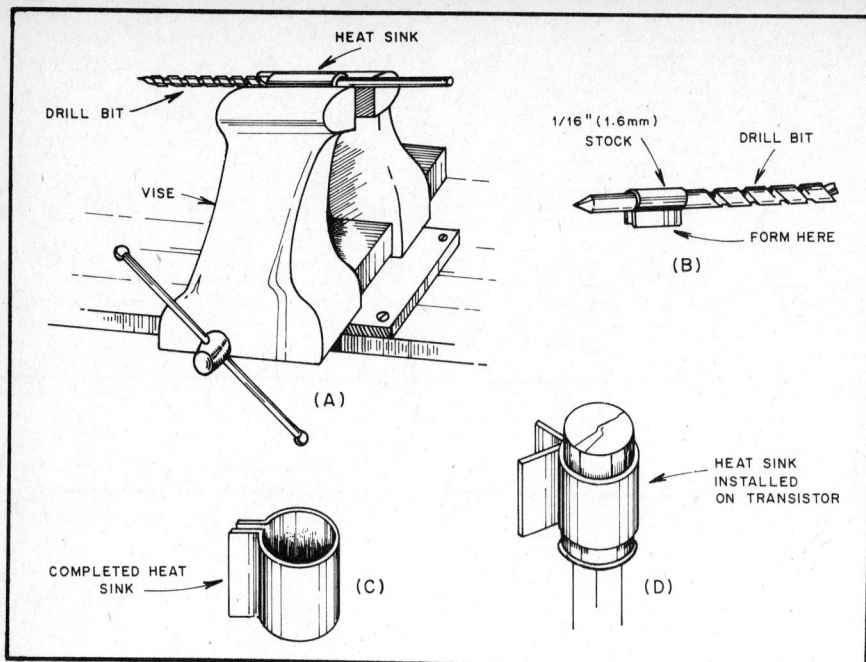

Fig. 8 — Steps used in constructing heat sinks for small transistors.

Table 2

Standard Metal Gauges

Gauge No.	American or BS[1]	U.S. Standard[2]	Birmingham or Stubs[3]	Gauge No.	American or BS[1]	U.S. Standard[2]	Birmingham or Stubs[3]
1	0.2893	0.28125	0.300	24	0.02010	0.025	0.022
2	0.2576	0.265625	0.284	25	0.01790	0.021875	0.020
3	0.2294	0.25	0.259	26	0.01594	0.01875	0.018
4	0.2043	0.234375	0.238	27	0.01420	0.0171875	0.016
5	0.1819	0.21875	0.220	28	0.01264	0.015625	0.014
6	0.1620	0.203125	0.203	29	0.01126	0.0140625	0.013
7	0.1443	0.1875	0.180	30	0.01003	0.0125	0.012
8	0.1285	0.171875	0.165	31	0.008928	0.0109375	0.010
9	0.1144	0.15625	0.148	32	0.007950	0.01015625	0.009
10	0.1019	0.140625	0.134	33	0.007080	0.009375	0.008
11	0.09074	0.125	0.120	34	0.006350	0.00859375	0.007
12	0.08081	0.109375	0.109	35	0.005615	0.0078125	0.005
13	0.07196	0.09375	0.095	36	0.005000	0.00703125	0.004
14	0.06408	0.078125	0.083	37	0.004453	0.006640626	.
15	0.05707	0.0703125	0.072	38	0.003965	0.00625	.
16	0.05082	0.0625	0.065	39	0.003531	.	.
17	0.04526	0.05625	0.058	40	0.003145	.	.
18	0.04030	0.05	0.049				
19	0.03589	0.04375	0.042				
20	0.03196	0.0375	0.035				
21	0.02846	0.034375	0.032				
22	0.02535	0.03125	0.028				
23	0.02257	0.028125	0.025				

[1]Used for aluminum, copper, brass and nonferrous alloy sheets, wire and rods.
[2]Used for iron, steel, nickel and ferrous alloy sheets, wire and rods.
[3]Used for seamless tubes; also by some manufacturers for copper and brass.

ductor as illustrated at D. The larger the heat sink area, the greater will be the amount of heat conducted away from the transistor body. In some applications, the heat sinks shown in Fig. 8 may be two or three inches in height (power transistor stages).

Another technique for making heat sinks for TO-5 type transistors and larger models is shown in Fig. 7. This style of heat sink will dissipate considerably more heat than will the type shown in Fig. 8. The main body of the sink is fashioned from a piece of 1/8-inch (3-mm) thick aluminum angle bracket — available from most hardware stores. A hole is bored in the angle stock to allow the transistor case to fit *snugly* into it. The transistor is held in place by a small metal plate whose center hole is slightly smaller in diameter than the case of the transistor. Details are given in Fig. 7.

A thin coating of silicone grease, available from most electronics supply houses, can be applied between the case of the transistor and the part of the heat sink with which it comes in contact. The silicone grease will aid the transfer of heat from the transistor to the sink. This practice can be applied to all models shown here. In the example given in Fig. 6, the grease should be applied between the three channels before they are bolted together, as well as between the transistor and the channel it contacts.

Construction Notes

If a control shaft must be extended or insulated, a flexible shaft coupling with adequate insulation should be used. Satisfactory support for the shaft extension, as well as electrical contact for safety, can be provided by means of a metal panel bearing made for the purpose. These can be obtained singly for use with existing shafts, or they can be bought with a captive extension shaft included. In either case the panel bearing gives a "solid" feel to the control. The use of fiber washers between ceramic insulation and metal brackets, screws or nuts will prevent the ceramic parts from breaking.

Cutting and Bending Sheet Metal

If a metal sheet is too large to be cut conveniently with a hacksaw, it may be marked with scratches as deep as possible along the line of the cut on both sides of the sheet, and then clamped in a vise and worked back and forth until the sheet breaks at the line. Do not carry the bending too far until the break begins to weaken; otherwise the edge of the sheet may become bent. A pair of iron bars or pieces of heavy angle stock, as long or longer than the width of the sheet, to hold it in the vise, will make the job easier. C clamps may be used to keep the bars from spreading at the ends. The rough edges may be smoothed with a file or by placing a large piece of emery cloth or sandpaper on a flat surface and running the edge of the metal back and forth over the sheet. Bends may be made similarly.

Today much of the tedium of sheet metal work can be relieved by using copper-clad printed-circuit board material wherever possible. Copper-clad stock is manufactured with phenolic, G-10 fiberglass and Teflon base materials in thicknesses up to 1/8 inch (3 mm). While it is manufactured in large sheets for industrial use, some hobby electronics stores and surplus outlets market usable scraps at reasonable prices. Pc-board stock is easily cut with a small hacksaw. Because the nonmetallic base material isn't malleable, it can't be bent in the usual way. However, corners are easily formed by holding two pieces at right angles and soldering the seam. Excellent rf-tight enclosures can be fabricated in this manner. Many projects in this *Handbook* were constructed using this technique. If mechanical rigidity is required of a large copper-clad surface, stiffening ribs may be soldered at right angles to the sheet.

Finishing Aluminum

Aluminum chassis, panels and parts

may be given a sheen finish by treating them in a caustic bath. An enameled or plastic container, such as a dishpan or infant's bathtub, should be used for the solution. Dissolve ordinary household lye in cold water in a proportion of one-quarter to one-half can of lye per gallon of water. The stronger solution will do the job more rapidly. Stir the solution with a stick of wood until the lye crystals are completely dissolved. Be very careful to avoid skin contact with the solution. It is also harmful to clothing. Sufficient solution should be prepared to cover the piece completely. When the aluminum is immersed, a very pronounced bubbling takes place and ventilation should be provided to disperse the escaping gas. A half hour to two hours in the solution should be sufficient, depending upon the strength of the solution and the desired surface.

Remove the aluminum from the solution with sticks and rinse thoroughly in cold water while swabbing with a rag to remove the black deposit. When dry, finish by spraying on a light coat of clear lacquer.

Raw aluminum can be prepared for painting by abrading the surface with medium-grade sandpaper, making certain the strokes are applied in the same direction (not circular or random). This process will create tiny grooves on the otherwise smooth surface. As a result, paint or lacquer will adhere well. Before painting, wash the abraded aluminum with soap and hot water, dry thoroughly: Avoid touching the prepared surface before painting it.

Soldering

The secret of good soldering is to use the right amount of heat. Too little heat will produce a "cold-soldered joint"; too much may injure a component. The iron and the solder should be applied simultaneously to the joint. Keep the iron clean by brushing the hot tip with a paper towel or a moist sponge, as illustrated in the soldering station described earlier in this chapter. Always use rosin-core solder; never acid-core. Solders have different melting points, depending upon the ratio of tin to lead. A 50-50 solder melts at 425° F (218° C), while 60-40 melts at 371° F (188° C). When it is desirable to protect from excessive heat the components being soldered, the 60-40 solder is preferable to the 50-50. (A less-common solder, 63-37, melts at 361° F or 182° C.)

When soldering transistors, crystal diodes or small resistors, the lead should be gripped with a pair of pliers up close to the unit so that the heat will be conducted away. Overheating of a transistor or diode while soldering can cause permanent damage. Also, mechanical stress will have a similar effect. Therefore, a small unit should be mounted so that there is no appreciable mechanical strain on the leads.

Trouble is sometimes experienced in soldering to the pins of coil forms or male cable plugs. It helps if the pins are first cleaned on the inside with a suitable twist drill and then tinned by flowing rosin-core solder into them. Immediately clear the surplus solder from each hot pin by a whipping motion or by blowing through the pin from the inside of the form or plug. Before inserting the wire in the pin, file the nickel plate from the tip. After soldering, round the solder tip off with a file.

When soldering to the pins of polystyrene coil forms, hold the pin to be soldered with a pair of heavy pliers to form a "heat sink" and insure that the pin does not heat enough in the coil form to loosen and become misaligned.

Some connections carrying very high current can't be made with ordinary tin-lead solder because the heat generated by the joint resistance would melt the solder. Automotive starter brushes and uhf transmitter tank circuits are two cases in which this situation can occur. Silver solder prevents this condition in two ways: It melts at a significantly higher temperature than tin-lead solder (about 600° F or 315° C) and generates less heat because of its superior conductivity. A propane torch may be necessary for large silver soldering jobs. The special flux used with silver solder releases toxic fumes, so follow the manufacturer's instructions carefully and work only in a well-ventilated area.

Wiring

The wire used in connecting amateur equipment should be selected by considering both the maximum current it will be called upon to handle and the voltage its insulation must stand without breakdown. Also, from the consideration of TVI, the power wiring of all transmitters should be done with wire that has a braided shielding cover. Receiver and audio circuits may also require the use of shielded wire at some points for stability or the elimination of hum.

No. 20 stranded wire is commonly used for most receiver wiring (except for the high-frequency circuits) where the current does not exceed 2 or 3 amperes. For higher-current heater circuits, no. 18 is available. Wire with cellulose acetate insulation is good for voltages up to about 500. For higher voltages, Teflon-insulated or other special HV wire should be used. Inexpensive wire strippers that make the removal of insulation from hookup wire an easy job are available on the market.

When power leads have several branches in the chassis, it is convenient to use fiber-insulated multiple tie points as anchorages or junction points. Strips of this type are also useful as insulated supports for resistors, rf chokes and capacitors. Exposed points of high-voltage wiring should be held to a minimum; those which cannot be avoided should be made as inaccessible as possible to accidental contact or short-circuit.

Where shielded wire is called for and capacitance to ground is not a factor, Belden type 8885 shielded grid wire may be used. If capacitance must be minimized, it may be necessary to use a piece of car-radio low-capacitance lead-in wire or coaxial cable.

For wiring high-frequency circuits, rigid wire is often used. Bare soft-drawn tinned wire, size 22 to 12 (depending on mechanical requirements) is suitable. Kinks can be removed by stretching a piece of 10 or 15 feet (3 or 4.5 m) long and then cutting it into short lengths that can be handled conveniently. Rf wiring should be run directly from point to point with a minimum of sharp bends and the wire kept well spaced from the chassis or other grounded metal surfaces. Where the wiring must pass through the chassis or a partition, a clearance hole should be cut and lined with a rubber grommet. In case insulation becomes necessary, varnished cambric tubing (spaghetti) can be slipped over the wire.

In transmitters where the peak voltage does not exceed 2500, the shielded grid wire mentioned above should be satisfactory for power circuits. For higher voltages, Belden type 8656, Birnbach type 1820, or shielded ignition cable can be used. In the case of filament circuits carrying heavy current, it may be necessary to use no. 10 or 12 bare or enameled wire, slipped through spaghetti, and then covered with copper braid pulled tightly over the spaghetti. If the shielding is simply slid back over the insulation and solder flowed into the end of the braid, the braid usually will stay in place without the necessity for cutting it back or binding it in place. The braid should be cleaned first so that solder will take with a minimum of heat. Rf wiring in transmitters usually follows the method described above for receivers, with due respect to the voltages involved.

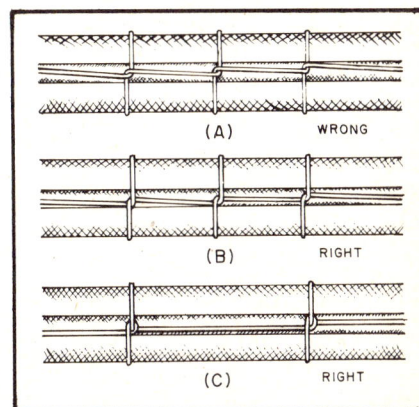

Fig. 9 — Methods of lacing cables. The method shown at C is more secure, but takes more time than the method of B. The latter is usually adequate for most amateur requirements.

Where power or control leads run together for more than a few inches, they will present a better appearance when bound together in a single cable. The correct technique is illustrated in Fig. 9; both plastic and waxed-linen lacing cords are available. Plastic cable clamps are available to hold the laced cable.

To give a "commercial look" to the wiring of any unit, run any cabled leads along the edge of the chassis. If this isn't possible, the cabled leads should then run parallel to an edge of the chassis. Further, the generous use of tie points mounted parallel to an edge of the chassis, for the support of one or both ends of a resistor or fixed capacitor, will add to the appearance of the finished unit. In a similar manner, "dress" the small components so that they are parallel to the panel or sides of the chassis.

Winding Coils

Close-wound coils are readily wound on the specified form by anchoring one end of the length of wire (in a vise or to a doorknob) and the other end to the coil form. Straighten any kinks in the wire and then pull to keep the wire under slight tension. Wind the coil to the required number of turns while walking toward the anchor, always maintaining a slight tension on the wire.

To space-wind the coil, wind the coil simultaneously with a suitable spacing medium (heavy thread, string or wire) in the manner described above. When the winding is complete, secure the end of the coil to the coil-form terminal and then carefully unwind the spacing material. If the coil is wound under suitable tension, the spacing material can be easily removed without disturbing the winding. Finish the

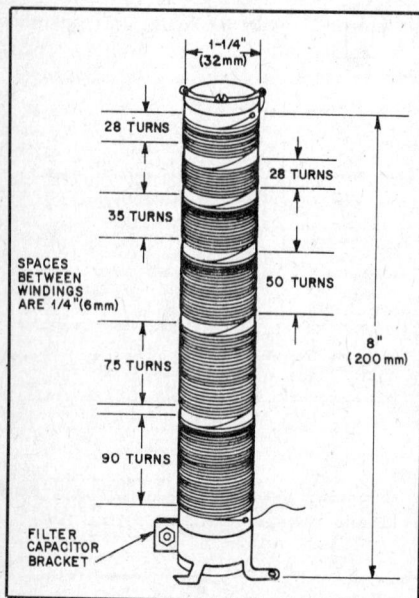

Fig. 11 — Bifilar filament choke wound on a ferrite rod. Heat-shrink tubing will help anchor the winding. If enameled wire is used the form should be insulated before winding.

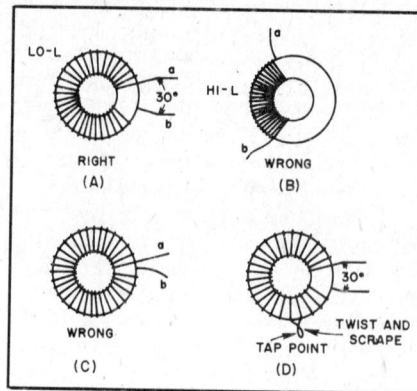

Fig. 12 — The suggested winding method for a single-layer toroid as shown at A. A 30° gap is recommended (see text). Wrong methods are shown at B and C. At D is a method for placing a tap on the coil.

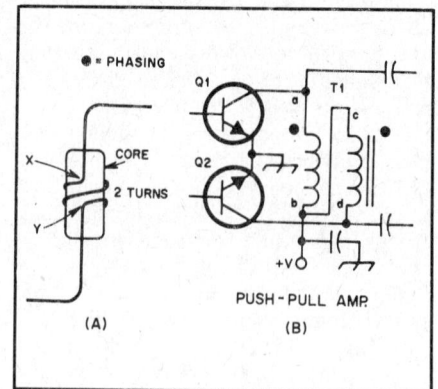

Fig. 13 — The view at A shows how the turns on a toroid should be counted. The large black dots in the diagram at B are used to indicate the polarity of the windings (phasing).

space-wound coil by judicious applications of Duco cement to hold the turns in place.

The "cold" end of a coil is the end at or close to chassis or ground potential. Coupling links should be wound on the cold end of a coil to minimize capacitive coupling.

Rf chokes must often present a high impedance over a broad frequency range. This requirement calls for the avoidance of series resonances within the range. Such resonances can be avoided in single-layer solenoids by separating the winding into progressively shorter sections. A practical choke suitable for hf amateur service at plate impedances up to 5 kΩ and currents up to 600 mA is shown in Fig. 10.

Another way to build a broad-band choke is to wind a small number of turns on a high-permeability ferrite rod (such as used for antennas in some portable radios). The magnetic core supplies a large inductance with a small winding. Keeping the number of turns small reduces the distributed capacitance and raises the self-resonant frequency of the choke. Ferrite chokes are best suited to low-impedance applications. A bifilar (this term is explained in the following paragraphs) filament choke for grounded-grid kilowatt hf amplifiers appears in Fig. 11.

Toroidal inductors and transformers are specified for many projects in this *Handbook*. The advantages of this type of winding include compactness and a self-shielding property. June 1979 *QST* contains a comprehensive treatment of the

theoretical and practical aspects of toroids. Figs. 12 and 13 illustrate the proper way to wind and count the turns on a toroidal core.

A *bifilar* winding is one which has two identical lengths of wire, which when placed on the core result in the same number of turns for each wire. The two wires can be put on the core side by side at the same time, just as if a single winding were being applied. An easier and more popular method is to twist the two wires (8 to 15 times per inch or 1-1/2 to 3 mm per "twist" will suffice), then wind the pair on the core. The wires can be twisted handily by placing one end of the length of two wires in a bench vise. The remaining wire-pair ends are tightened into the chuck of a small hand drill, and the twisting is done.

A *trifilar* winding has three wires and a *quadrifilar* winding has four. The procedure for preparation and winding is otherwise the same as for a bifilar winding. Fig. 14 shows a bifilar type of

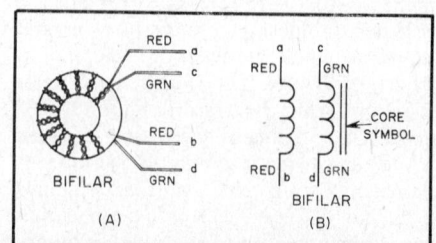

Fig. 10 — Pictorial diagram of a heavy-duty plate choke for the hf bands. No. 26 enameled wire is used for the windings. Low-loss material should be used for the form.

Fig. 14 — Schematic and pictorial presentation of a bifilar-wound toroidal transformer.

toroid in schematic and pictorial form. The wires have been twisted together prior to placing them on the core. It is helpful, though by no means essential, to use wires of different color when multifilar-winding a core. The more wires used, the more perplexing it is to identify the end of the windings correctly once the core has been wound. There are various colors of enamel insulation available, but it is not easy for amateurs to find this wire locally or in small-quantity lots. This problem can be solved by taking lengths of wire (enameled magnet wire), cleaning them to remove dirt and grease, then spray-painting them. Ordinary aerosol-can spray enamel works fine. Spray lacquer is not as satisfactory because it is brittle when dry and tends to flake off the wire.

The winding sense of a multifilar toroidal transformer is important in most circuits. Fig. 13B illustrates this principle. The black dots (called phasing dots) at the top of the T1 windings indicate polarity. That is, points *a* and *c* are both start or finish ends of their respective windings. In this example, points *a* and *d* are of opposite phase (180° phase difference) to provide push-pull voltage feed to Q1 and Q2.

Circuit-Board Fabrication

Modern-day builders prefer the neatness and miniaturization made possible by the use of etched or printed circuit boards. There are additional benefits to be realized from the use of circuit boards: Low lead inductances, excellent physical stability of the components and interconnecting leads, and good repeatability of the basic layout of a given project. The latter attribute makes the use of circuit boards ideal for group projects.

Planning and Layout

The constructor should first plan the physical layout of the circuit by sketching a pictorial diagram on paper, drawing it to scale. Once this has been done, the interconnecting leads can be inked in to represent the copper strips that will remain on the etched board. The Vector Company sells layout paper for this purpose. It is marked with the same patterns that are used on their perforated boards.

After the basic etched-circuit design has been completed the designer should go over the proposed layout several times to insure against errors. When the foregoing has been done, the pattern can be painted on the copper surface of the board to be etched. Etch-resistant solutions are available from commercial suppliers and can be selected from their catalogs. Some builders prefer to use India ink for this purpose. Perhaps the most readily available material for use in etch-resist applications is ordinary exterior enamel paint. The portions of the board to be retained are covered with a layer of paint, applied with an artist's brush, duplicating the pattern that was drawn on the layout

paper. The job can be made a bit easier by tracing over the original layout with a ballpoint pen and carbon paper while the pattern is taped to the copper side of the unetched circuit board. The carbon paper is placed between the pattern and the circuit board. After the paint has been applied, it should be allowed to dry for at least 24 hours prior to the etching process. The Vector Company produces a rub-on transfer material that can also be used as etch-resist when laying out circuit-board patterns. Thin strips of ordinary masking tape, cut to size and firmly applied, serve nicely as etch-resist material too.

When making "one-shot" pc boards it is convenient to cover the copper surface with masking tape, transfer the circuit pattern by means of carbon paper, then cut out and remove the sections of masking tape where the copper is to be etched away. An X-acto hobby knife is excellent for the purpose. Masking tape, securely applied, serves as a fine etch-resist material.

Many magazine articles feature printed-circuit layouts. The more-complex patterns (those containing ICs and high component densities) are difficult to duplicate accurately by hand. A photographic process is the most efficient way to transfer a layout from a magazine page to a circuit board. A Thermofax transparency-producing machine (most schools have these) will copy the circuit on a clear plastic sheet for use as a negative. Pressing this negative against a photosensitive copper-clad board with a piece of glass and exposing the assembly to sunlight for about 90 seconds will deactivate the etchant resist on the exposed part of the board. The portion of the copper that is shielded from the light by the negative will resist etching. This process is described in detail by Taylor, W4POS, in August 1979 *QST*. Photosensitive pc-board material is manufactured by Kepro Company.

The Etching Process

Almost any strong acid bath will serve as an etchant, but the two chemical preparations recommended here are the safest to use. A bath can be prepared by mixing one part ammonium persulphate crystals with two parts clear water. A normal quantity of working solution for most amateur radio applications is composed of one cup of crystals and two cups of water. To this mixture add 1/4 teaspoon of mercuric chloride crystals. The latter serves as an activator for the bath. Ready-made etchant kits which use these chemicals are available from Vector. Complete kits which contain circuit boards, etchant powders, etch-resist transfers, layout paper, and plastic etchant bags are also available from Vector at moderate prices.

Another chemical bath that works satisfactorily for copper etching is made up from one part ferric chloride crystals and two parts water. No activator is re-

Fig. 15 — A homemade stand for processing etched-circuit boards. The heat lamp maintains the etchant-bath temperature between 90 and 115° F (32 and 46° C) and is mounted on an adjustable arm. The tray for the bath is raised and lowered at one end by the action of a motor-driven eccentric disc, providing the necessary agitation of the chemical solution. A darkroom thermometer monitors the temperature of the bath.

quired with this bath. Ready-made solutions (one-pint and one-gallon sizes) are available through some mail-order houses at low cost. They are manufactured by Kepro Company and carry stock numbers E-1PT and E-1G, respectively.

Etchant solutions become exhausted after a certain amount of copper has been processed. Therefore, it is wise to keep a quantity of the bath on hand if frequent use is anticipated. With either chemical bath, the working solution should be maintained at a temperature between 90 and 115° F (32 and 46° C). A heat lamp can be directed toward the bath during the etching period, its distance set to maintain the required temperature. A darkroom thermometer is handy for monitoring the temperature of the bath.

While the circuit board is immersed in the solution, it should be agitated continuously to permit uniform reaction to the chemicals. This action will also speed up the etching process somewhat. Normally, the circuit board should be placed in the bath with the copper side facing down, toward the bottom of the tray. The tray should be non-metallic preferably a Pyrex dish or a photographic darkroom tray.

The photograph, Fig. 15, shows a homemade etching stand made up from a heat lamp, some lumber, and an 8-rpm motor. An eccentric disc has been mounted on the motor shaft and butts against the bottom of the etchant tray. As the motor turns, the eccentric disc raises and lowers one end of the tray, thus providing continuous agitation of the solution. The heat lamp is mounted on an adjustable, slotted wooden arm. Its height above the solution tray is adjusted to provide the desired bath temperature. Because the etching process takes between 15 minutes and one hour — dependent upon the strength and temperature of the bath — such an accessory is convenient.

After the etching process is completed, the board is removed from the tray and washed thoroughly with fresh, clear water. The etch-resist material can then be rubbed off by applying a few brisk strokes with medium-grade steel wool. WARNING: *Always use rubber gloves when working with etchant powders and solutions. Should the acid bath come in contact with the body, immediately wash the affected area with clear water. Protect the eyes when using acid baths.*

Alternative Construction Methods

Some would-be builders express revulsion at the prospect of pc-board fabrication. The distaste for chemical processes should not deter a person, however, for several alternatives exist.

Practically all designs are "breadboarded" before being committed to a printed circuit. The fact that these prototypes work proves that etched circuit boards aren't an absolute necessity. Where a ground plane is required, Teflon terminals pushed through holes in a copper-clad board allow neat and rigid component mounting. High-value resistors with one end soldered to the ground plane can also be used for standoff terminals.

Low- and medium-speed digital circuits are often assembled on a *wire wrap* board. The IC sockets have long pins around which small solid wires are wrapped. An electric or pneumatic "gun" is used to make the connections in industry, but a manual wrapping tool can be used when time is not of the essence. Toltec Corporation, listed in Table 11, sells wire-wrapping supplies in small quantities to amateurs.

Radio Shack and the Vector Company produce a variety of breadboarding fixtures that can also be used in permanent assemblies. The deluxe models feature several power and ground buses, as well as IC hole patterns.

Table 3

Approximate Series-Resonant Frequencies of Disc Ceramic Bypass Capacitors

Capacitance	Freq.[1]	Freq.[2]
0.01 µF	13 MHz	15 MHz
0.0047	18	22
0.002	31	38
0.001	46	55
0.0005	65	80
0.0001	135	165

[1]Total lead length of 1 inch (25-mm)
[2]Total lead length of 1/2-inch (13-mm)

Table 4

Resistor-Capacitor Color Code

Color	Significant Figure	Decimal Multiplier	Tolerance (%)	Voltage Rating*
Black	0	1		
Brown	1	10	1*	100
Red	2	100	2*	200
Orange	3	1,000	3*	300
Yellow	4	10,000	4*	400
Green	5	100,000	5*	500
Blue	6	1,000,000	6*	600
Violet	7	10,000,000	7*	700
Gray	8	100,000,000	8*	800
White	9	1,000,000,000	9*	900
Gold	-	0.1	5	1000
Silver	-	0.01	10	2000
No color	-		20	500

*Applies to capacitors only.

A construction technique that is practically indistinguishable from true printed circuitry is excavating a copper-clad board with a hand-held grinding tool, such as the Moto-tool manufactured by the Dremel Company. The simpler circuits can be cut out of the board with an X-acto knife.

Perhaps the least complicated approach to circuit-board fabrication is the use of unclad perforated board into which a number of push-in terminals have been installed. The perforated board can be obtained with one of many hole patterns, dependent upon the needs of the builder. Perforated terminal boards are manufactured by several companies. Their products are available from most mail-order houses.

Once the builder plots the layout of his circuit on paper, push-in terminals can be installed in the "perf" board to match the layout which was done on paper. The terminals serve as tie points and provide secure mounting-post anchors for the various components. Selected terminals can be wired together to provide ground and B + lines. Although this technique is the most basic of the methods, it is entirely practical.

Component Values

Values of composition resistors and small capacitors (mica and ceramic) are specified throughout this *Handbook* in terms of "preferred values." In the preferred-number system, all values represent (approximately) a constant-percentage increase over the next lower value. The base of the system is the number 10. Only two significant figures are used.

"Tolerance" means that a variation of plus or minus the percentage given is considered satisfactory. For example, the actual resistance of a "4700-ohm" 20-percent resistor can lie anywhere between 3700 and 5600 ohms, approximately. The permissible variation in the same resistance value with 5-percent tolerance would be in the range from 4500 to 4900 ohms, approximately.

In the component specifications in this

Handbook, it is to be understood that when no tolerance is specified the *largest* tolerance available in that value will be satisfactory.

Values that do not easily fit into the preferred-number system (such as 500, 25,000) can be substituted. It is obvious, for example, that a 5000-ohm resistor falls well within the tolerance range of the 4700-ohm 20-percent resistor used in the example above. It would not, however, be usable if the tolerance were specified as 5 percent.

Color Codes

Standardized color codes are used to mark values on small components such as composition resistors and mica capacitors, and to identify leads from transformers and other large components. The resistor-capacitor number color code is given in Table 4.

Fixed-Value Capacitors

The methods of marking "postage-stamp" mica capacitors, molded paper capacitors and tubular ceramic capacitors are shown in Fig. 16.

Capacitors made to American War Standards (AWS) or Joint Army-Navy (JAN) specifications are marked with the six-dot code shown at the top. Practically all surplus capacitors are in this category.

The three-dot EIA code is used for capacitors having a rating of 500 volts and ± 20 percent tolerance only; other ratings and tolerances are covered by the six-dot EIA code.

Example: A capacitor with a six-dot code has the following markings: Top row, left to right, black, yellow, violet; bottom row, right to left, brown, silver, red. Since the first color in the top row is black (significant figure zero) this is the AWS code and the capacitor has mica dielectric. The significant figures are 4 and 7, the decimal multiplier 10 (brown, at right of second row), so the capacitance is 470 pF. The tolerance is ± 10 percent. The final color, the characteristic, deals with temperature coefficients and methods of testing (see Table 6).

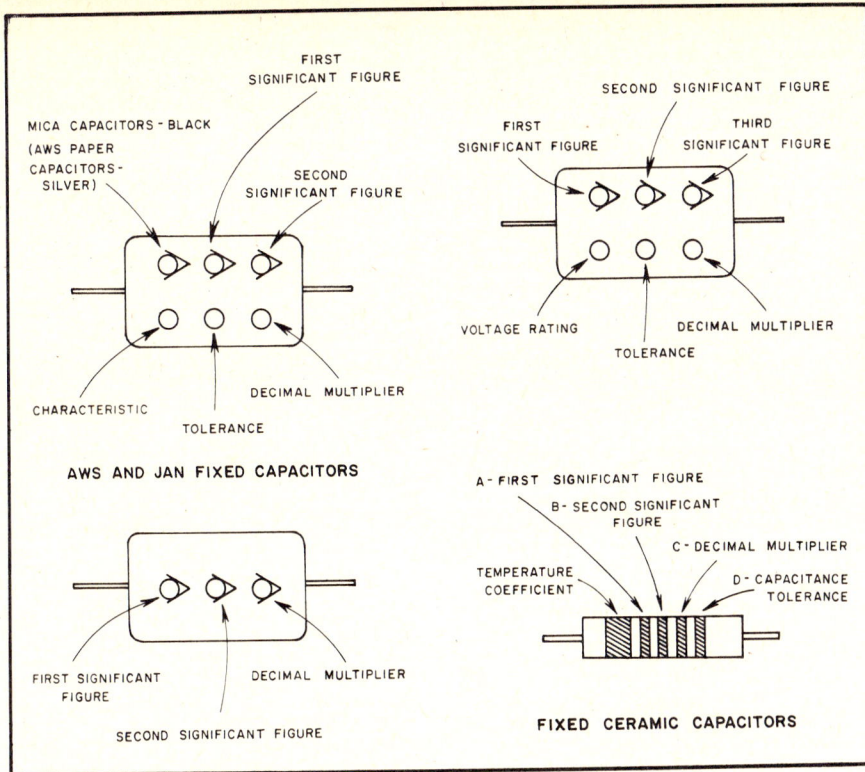

Fig. 16 — Color coding of fixed mica, molded paper and tubular ceramic capacitors. The color code for mica and molded paper capacitors is given in Table 4. Table 5 gives the color code for tubular ceramic capacitors.

Table 5

Color Code for Ceramic Capacitors

Color	Significant Figure	Decimal Multiplier	Capacitance Tolerance More than 10 pF (in %)	Capacitance Tolerance Less than 10 pF (in pF)	Temp. Coeff. ppm/ deg. C.
Black	0	1	± 20	2.0	0
Brown	1	10	± 1		− 30
Red	2	100	± 2		− 80
Orange	3	1000			−150
Yellow	4				−220
Green	5				−330
Blue	6		± 5	0.5	−470
Violet	7				−750
Gray	8	0.01		0.25	30
White	9	0.1	± 10	1.0	500

Table 6

Capacitor Characteristic Code

Color Sixth Dot	Temperature Coefficient ppm/deg. C.	Capacitance Drift
Black	± 1000	± 5% + 1 pF
Brown	± 500	± 3% + 1 pF
Red	± 200	± 0.5%
Orange	± 100	± 0.3%
Yellow	− 20 to + 100	± 0.1% + 0.1 pF
Green	0 to + 70	± 0.05% + 0.1 pF

A capacitor with a three-dot code has the following colors, left to right: brown, black, red. The significant figures are 1, 0 (10) and the multiplier is 100. The capacitance is therefore 100 pF.

A capacitor with a six-dot code has the following markings: Top row, left to right, brown, black, black; bottom row, right to left, black, gold, blue. Since the first color in the top row is neither black nor silver, this is the EIA code. The significant figures are 1, 0, 0 (100) and the decimal multiplier is 1 (black). The capacitance is therefore 100 pF. The gold dot shows that the tolerance is ± 5 percent and the blue dot indicates 600-volt rating.

Ceramic Capacitors

Conventional markings for ceramic capacitors are shown in the lower drawing of Fig. 16. The colors have the meanings indicated in Table 4. In practice, dots may be used instead of the *narrow* bands indicated in Fig. 16.

Example: A ceramic capacitor has the following markings: Broad band, violet; narrow bands or dots, green, brown, black, green. The significant figures are 5, 1 (51) and the decimal multiplier is 1, so the capacitance is 51 pF. The temperature coefficient is − 750 parts per million per degree celsius, as given by the broad band, the capacitance tolerance is ± 5 percent.

Fixed-Value Composition Resistors

Composition resistors (including small wire-wound units molded in cases identical with the composition type) are color-coded as shown in Fig. 17. Colored bands are used on resistors having axial leads; on radial-lead resistors the colors are placed as shown in the drawing. When bands are used for color coding the body color has no significance.

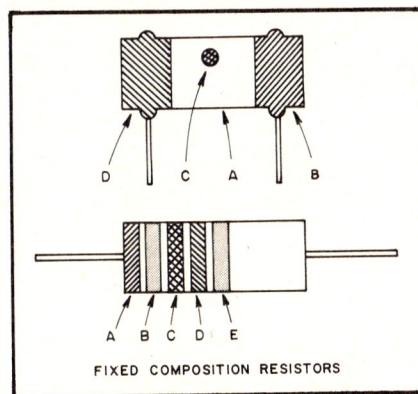

Fig. 17 — Color coding for fixed composition resistors. The color code is given in Table 4. The colored areas have the following significance:
A — First significant figure of resistance in ohms.
B — Second significant figure.
C — Decimal multiplier.
D — Resistance tolerance in percent. If no color is shown the tolerance is ± 20%.
E — Relative percent change in value per 1000 hours of operation; Brown, 1%; Red, 0.1%; Orange, 0.01%; Yellow, 0.001%.

Examples: A resistor of the type shown in the lower drawing of Fig. 17 has the following color bands: A, red; B, red; C, orange; D, no color. The significant figures are 2, 2 (22) and the decimal multiplier is 1000. The value of resistance is therefore 22,000 ohms and the tolerance is ± 20 percent.

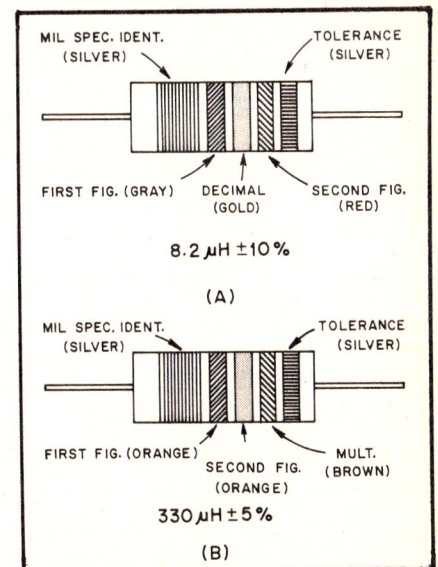

Fig. 18 — Color coding for tubular encapsulated rf chokes. At A, an example of the coding for an 8.2-µH choke is given. At B, the color bands for a 330-µH inductor are illustrated. The color code is given in Table 4.

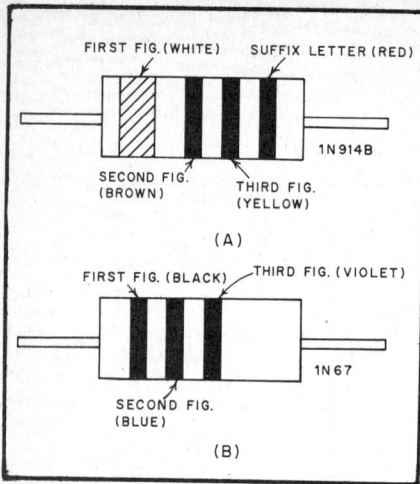

Fig. 19 — Color coding for semiconductor diodes. At A, the cathode is identified by the double-width first band. At B, the bands are grouped toward the cathode. Two-figure designations are signified by a black first band. The color code is given in Table 4. The suffix-letter code is: A — brown, B — red, C — orange, D — yellow, E — green, F — blue. The 1N prefix is understood.

A resistor of the type shown in the upper drawing of Fig. 17 has the following colors: Body (A), blue; end (B), gray; dot, red; end (D), gold. The significant figures are 6, 8 (68) and the decimal multiplier is 100, so the resistance is 6800 ohms. The tolerance is ± 5 percent.

I-F Transformers

Blue — plate lead.
Red — B + lead.
Green — grid (or diode) lead.
Black — grid (or diode) return.

Note: If the secondary of the i-f transformer is center-tapped, the second diode plate lead is green-and-black striped, and black is used for the center-tap lead.

Audio Transformers

Blue — plate (finish) lead of primary
Red — B + lead (this applies whether the primary is plain or center-tapped).
Brown — plate (start) lead on center-tapped primaries. (Blue may be used for this lead if polarity is not important.)
Green — grid (finish) lead to secondary.
Black — grid return (this applies whether the secondary is plain or center-tapped).
Yellow — grid (start) lead on center-tapped secondaries. (Green may be used for this lead if polarity is not important.)

Note: These markings apply also to line-to-grid and tube-to-line transformers.

Power Transformers

1) Primary Leads: black
    If tapped:
        Common: black
        Tap: black and yellow striped

Table 7

Metric Multiplier Prefixes

Multiples and submultiples of fundamental units (e.g., ampere, farad, gram, meter, watt) may be indicated by the following prefixes.

Prefix	Abbreviation	Multiplier
tera	T	10^{12}
giga	G	10^{9}
mega	M	10^{6}
kilo	k	10^{3}
hecto	h	10^{2}
deci	d	10^{-1}
centi	c	10^{-2}
milli	m	10^{-3}
micro	μ	10^{-6}
nano	n	10^{-9}
pico	p	10^{-12}

Finish: black and red striped
2) High-Voltage Place Winding: red
    Center-Tap: red and yellow striped
3) Rectifier Filament Winding: yellow
    Center-Tap: yellow and blue striped
4) Filament Winding no. 1: green
    Center-Tap: green and yellow striped
5) Filament Winding no. 2: brown
    Center-Tap: brown and yellow striped
6) Filament Winding no. 3: slate
    Center-Tap: slate and yellow striped

Finding Parts

No chapter on construction would be

83-58FCP

1. Strip cable — *don't nick braid, dielectric or conductor.* Slide ferrule, then coupling ring on cable. Flare braid slightly by rotating conductor and dielectric in circular motion.

2. Slide body on dielectric, barb going under braid until flange is against outer jacket. Braid will fan out against body flange.

3. Slide nut over body. Grasp cable with hand and push ferrule over barb until braid is captured between ferrule and body flange. Squeeze crimp tip only of center contact with pliers; alternate-solder tip.

83-1SP PLUG (PL-259)

1. Strip cable, *don't nick braid, dielectric or conductor.* Tin exposed braid and conductor. Slide coupling ring on cable.

2. Screw body on cable. Solder braid through solder holes. Solder conductor to center contact.

3. Screw coupling ring on body.

83-1SP PLUG WITH ADAPTERS

1. Strip jacket. *Don't nick braid.* Slide coupling ring and adapter on cable. Note — use 83-168 adapter for RG-58/U and 83-185 for RG-59/U.

2. Fan braid slightly, fold back over adapter and trim to 3/8''. Strip dielectric and tin exposed conductor. *Don't nick conductor.*

3. Screw body on adapter. Follow 2 and 3 under 83-1SP plug.

Fig. 20 — Cable stripping dimensions and assembly instructions for several popular coaxial cable connectors. This material courtesy of Amphenol* Electronic Components, RF Division, Bunker Ramo Corp. (Dimensions on this drawing are in English inches. Multiply inches × 25.4 to obtain mm).

Table 8

Pilot-Lamp Data

Lamp No.	Bead Color	Base (Miniature)	Bulb Type	Rating Volts	Amp.
40	Brown	Screw	T-3 1/4	6-8	0.15
40A[1]	Brown	Bayonet	T-3 1/4	6-8	0.15
41	White	Screw	T-3 1/4	2.5	0.5
42	Green	Screw	T-3 1/4	3.2	**
43	White	Bayonet	T-3 1/4	2.5	0.5
44	Blue	Bayonet	T-3 1/4	6-8	0.25
45	*	Bayonet	T-3 1/4	3.2	**
46[2]	Blue	Screw	T-3 1/4	6-8	0.25
47[1]	Brown	Bayonet	T-3 1/4	6-9	0.15
48	Pink	Screw	T-3 1/4	2.0	0.06
49[3]	Pink	Bayonet	T-3 1/4	2.0	0.06
49A[3]	White	Bayonet	T-3 1/4	2.1	0.12
50	White	Screw	G-3 1/2	6-8	0.2
51[2]	White	Bayonet	G-3 1/2	6-8	0.2
53	-	Bayonet	G-3 1/2	14.4	0.12
55	White	Bayonet	G-4 1/2	6-8	0.4
292[5]	White	Screw	T-3 1/4	2.9	0.17
292A[5]	White	Bayonet	T-3 1/4	2.9	0.17
1455	Brown	Screw	G-5	18.0	0.25
1455A	Brown	Bayonet	G-5	18.0	0.25
1487	-	Screw	T-3 1/4	12-16	0.20
1488	-	Bayonet	T-3 1/4	14	0.15
1813	-	Bayonet	T-3 1/4	14.4	0.10
1815	-	Bayonet	T-3 1/4	12-16	0.20

[1]40A and 47 are interchangeable.
[2]Have frosted bulbs.
[3]49 and 49A are interchangeable.
[4]Replace with no. 48.
[5]Use in 2.5-volt sets where regular bulb burns out too frequently.
*White in G.E. and Sylvania; green in National Union, Raytheon and Tung-Sol.
**0.35 in G.E. and Sylvania; 0.5 in National Union, Raytheon and Tung-Sol.

Table 9

Frequency-Spectrum Reference Chart of nonamateur channel assignments and other frequency data.

Frequency (kHz)

15.734264 ± 0.000044: TV hor. scan freq.
17.8(0.5)[1]: NAA Cutler, Maine
18.6(0.5)[1]: NPG/NLK Jim Creek, Washington
21.4(0.5)[1]: NSS Annapolis, Maryland
24.0(0.5)[1]: NBA Balboa, Panama, Canal Zone
26.1(0.5)[1]: NPM, Hawaii
60.0(0.5)[1,2]: WWVB Ft. Collins, Colorado
85: Receiver i-f (command set of "Q5er")
100.0(0.5)[1]: Loran C (regional)
179: WGU20 CD Station, East Coast, Bc of WX and time (a-m)
285-325: Marine RDF bnad. Two cw tones 1020-Hz apart
285-405: Aero RDF; aero WX (a-m) 325-405.
415-490: Marine (cw)
455: Receiver i-f/mech. filters (Collins)
535-1605: bc (a-m), 107 chans. every 10 kHz from 540 (carrier)

Frequency (MHz)

1.8-2.0: Loran A (pulse transmission)
2.5 (0.5)[1,2]: WWV, Ft. Collins, Colorado, WWVH Kekaha, Hawaii
3.33 (50)[1,2]: CHU, Ottawa, Canada
3.395: Transceiver i-f (Heath, Kenwood)
3.579545 ± 10[-5]: TV chrominance subcarrier
5.0 (0.5)[1]: WWV, WWVH
5.645: receiver i-f (Drake)
7.335 (50)[1,2]: CHU
9.0: Xtal filters (KVG)
10.0 (0.5)[1]: WWV, WWVH
10.7: Receiver i-f (fm bc)
14.67 (50)[1,2]: CHU

15.0 (0.5)[1,2]: WWV
26.965-26.985: Citizens Band chan. 1-3 (10-kHz sep.)
27.005-27.035: CB chan. 4-7
27.055-27.085: CB chan. 8-11
27.105-27.135: CB chan. 12-15
27.155-27.185: CB chan. 16-19
27.205-27.225: CB chan. 20-22
27.235-27.255: CB chan. 24, 25, 23
27.265-27.405: CB chan. 26-40
41.25: TV sound carrier (location in receiver i-f)
42.17: TV color subcarrier (location in receiver i-f)
45.75: TV picture carrier (location in receiver i-f)
54-72: TV chan. 2-4. (Three 6-MHz chans. starting from 54)
72, 75: RC chans.
76-88: TV chan. 5-6
88.1-107.9: Bc (fm) 100 chan. from 88.1 (carrier) in 200-kHz steps
120-130: Aero, RDF WX
137.5, 137.62: WX Sat. (A4). Ref. W1AW Bul. for orbital data
162.4: Marine WX bc (fm, regional)
174-216: TV chan. 7-13
470-890: TV chan. 14-83 (70 chan. 6-MHz wide)

[1]Standard-frequency transmission figure in brackets is error in parts 10[10] (*Electronics Engineers' Handbook*, McGraw Hill, pp. 1-48).

[2]Standard time station. A3 transmissions include time, weather and propagation on WWV/WWVH. A3 time transmission on CHU (English/French). WWVB has no A3; info in BCD format generated by reducing carrier by 10-dB (binary 0).

BNC CONNECTORS (STANDARD CLAMP)

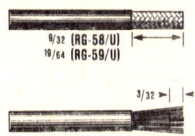

1. Strip jacket. Fray braid and strip dielectric. *Don't nick braid or conductor.* Tin conductor.

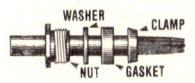

2. Taper braid. Slide nut, washer, gasket and clamp over braid. Clamp inner shoulder should fit squarely against end of jacket.

3. With clamp in place, comb out braid, fold back smooth as shown. Trim 3/32" from end.

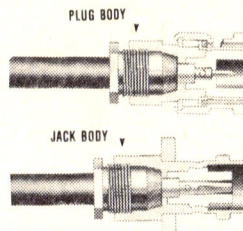

4. Solder contact on conductor through solder hole. Contact should butt against dielectric. Remove excess solder from outside of contact. Avoid excess heat to prevent swollen dielectric which would interfere with connector body.

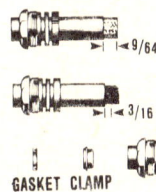

5. Push assembly into body. Screw nut into body with wrench until tight. *Don't rotate body on cable to tighten.*

BNC CONNECTORS (IMPROVED CLAMP)

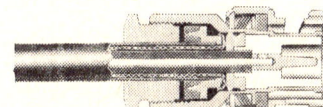

1. Follow 1, 2, 3 and 4 in BNC connectors (standard clamp) except as noted. Strip cable as shown. Slide gasket on cable *with groove facing clamp.* Slide clamp on cable *with sharp edge facing gasket.* Clamp *should* cut gasket to seal properly.

complete without information on where to buy parts. Amateurs, on a dwarfed scale, must function as purchasing agents in these perplexing times. A properly equipped buyer maintains as complete a catalog file as possible. Many of the companies listed in Table 11 will provide free catalogs upon written request. Others may charge a small fee for catalogs. Mail ordering, especially for those distant from metropolitan areas, is today's means to the desired end when collecting component parts for an amateur project. Prices are, to some extent, competitive. A wise buyer will study the catalogs and select his merchandise accordingly.

Delays in shipment can be lessened by avoiding the use of personal checks when ordering, especially for those distant from metropolitan areas. Personal checks often take a week to clear, thereby causing frustrating delays in the order reaching you.

Table 11 is updated with each new edition of this *Handbook*. Suppliers wishing to be listed in the table are urged to contact the editors.

Table 10

Copper-Wire Table

Wire Size A. W. G. (B&S)	Diam. in Mils [1]	Circular Mil Area	Turns per Enamel	Linear Inch S.C.E.	(25.4-mm)[2] D.C.C.	Cont.-duty current [3] single wire in open air	Cont.-duty current [3] wires or cables in conduits or bundles	Feet per Pound (0.45 kg) Bare	Ohms per 1000 ft. 25° C	Current Carrying Capacity [4] at 700 C.M. per Amp.	Diam. in mm.	Nearest British S.W.G. No.
1	289.3	83690	—	—	—	—	—	3.947	.1264	119.6	7.348	1
2	257.6	66370	—	—	—	—	—	4.977	.1593	94.8	6.544	3
3	229.4	52640	—	—	—	—	—	6.276	.2009	75.2	5.827	4
4	204.3	41740	—	—	—	—	—	7.914	.2533	59.6	5.189	5
5	181.9	33100	—	—	—	—	—	9.980	.3195	47.3	4.621	7
6	162.0	26250	—	—	—	—	—	12.58	.4028	37.5	4.115	8
7	144.3	20820	—	—	—	—	—	15.87	.5080	29.7	3.665	9
8	128.5	16510	7.6	—	7.1	73	46	20.01	.6405	23.6	3.264	10
9	114.4	13090	8.6	—	7.8	—	—	25.23	.8077	18.7	2.906	11
10	101.9	10380	9.6	9.1	8.9	55	33	31.82	1.018	14.8	2.588	12
11	90.7	8234	10.7	—	9.8	—	—	40.12	1.284	11.8	2.305	13
12	80.8	6530	12.0	11.3	10.9	41	23	50.59	1.619	9.33	2.053	14
13	72.0	5178	13.5	—	12.8	—	—	63.80	2.042	7.40	1.828	15
14	64.1	4107	15.0	14.0	13.8	32	17	80.44	2.575	5.87	1.628	16
15	57.1	3257	16.8	—	14.7	—	—	101.4	3.247	4.65	1.450	17
16	50.8	2583	18.9	17.3	16.4	22	13	127.9	4.094	3.69	1.291	18
17	45.3	2048	21.2	—	18.1	—	—	161.3	5.163	2.93	1.150	18
18	40.3	1624	23.6	21.2	19.8	16	10	203.4	6.510	2.32	1.024	19
19	35.9	1288	26.4	—	21.8	—	—	256.5	8.210	1.84	.912	20
20	32.0	1022	29.4	25.8	23.8	11	7.5	323.4	10.35	1.46	.812	21
21	28.5	810	33.1	—	26.0	—	—	407.8	13.05	1.16	.723	22
22	25.3	642	37.0	31.3	30.0	—	5	514.2	16.46	.918	.644	23
23	22.6	510	41.3	—	37.6	—	—	648.4	20.76	.728	.573	24
24	20.1	404	46.3	37.6	35.6	—	—	817.7	26.17	.577	.511	25
25	17.9	320	51.7	—	38.6	—	—	1031	33.00	.458	.455	26
26	15.9	254	58.0	46.1	41.8	—	—	1300	41.62	.363	.405	27
27	14.2	202	64.9	—	45.0	—	—	1639	52.48	.288	.361	29
28	12.6	160	72.7	54.6	48.5	—	—	2067	66.17	.228	.321	30
29	11.3	127	81.6	—	51.8	—	—	2607	83.44	.181	.286	31
30	10.0	101	90.5	64.1	55.5	—	—	3287	105.2	.144	.255	33
31	8.9	80	101	—	59.2	—	—	4145	132.7	.114	.227	34
32	8.0	63	113	74.1	61.6	—	—	5227	167.3	.090	.202	36
33	7.1	50	127	—	66.3	—	—	6591	211.0	.072	.180	37
34	6.3	40	143	86.2	70.0	—	—	8310	266.0	.057	.160	38
35	5.6	32	158	—	73.5	—	—	10480	335	.045	.143	38-39
36	5.0	25	175	103.1	77.0	—	—	13210	423	.036	.127	39-40
37	4.5	20	198	—	80.3	—	—	16660	533	.028	.113	41
38	4.0	16	224	116.3	83.6	—	—	21010	673	.022	.101	42
39	3.5	12	248	—	86.6	—	—	26500	848	.018	.090	43
40	3.1	10	282	131.6	89.7	—	—	33410	1070	.014	.080	44

[1] A mil is 0.001 inch.
[2] Figures given are approximate only; insulation thickness varies with manufacturer.
[3] Max. wire temp. of 212° F (100° C) and max. ambient temp. of 135° F (57° C).
[4] 700 circular mils per ampere is a satisfactory design figure for small transformers, but values from 500 to 1000 c.m. are commonly used.

Table 11
ARRL Parts Supplier List

A, E, I, L, M *35¢ stamp **$10	Adva Electronics Box 4181 Woodside, CA 94062		Hammond Mfg. Co., Inc. (U.S.) 1690 Walden Ave. Buffalo, NY 14225	J	Piezo Technology, Inc. P.O. Box 7859 Orlando, FL 32854
A, C, D, E, F	Alaska Microwave Labs 4335 E. 5th St. Anchorage, AK 99504	L	Harrison Radio 20 Smith St. Farmingdale, NY 11713	E, M, N, P *free	Poly Paks Box 942 Lynnfield, MA 01940
B, C	Amidon Associates 12033 Otsego Street N. Hollywood, CA 91607	A, B, I, K, M, N, T	Herbach and Rademan, Inc. 401 E. Erie Ave. Philadelphia, PA 19134	A, B, D, E, F, G, H, J, K, L *25¢	Radiokit P.O. Box 411 Greenville, NH 03048
N, O	Atlantic Surplus Sales (facsimile equipment) 3730 Nautilus Ave. Brooklyn, NY 11224	L *25¢ **$1	HI, Inc. (25¢ in coin for manual list) Box 864 Council Bluffs, IA 51502	E	Semiconductors Surplus 2822 N. 32nd St. Unit 1 Phoenix, AZ 85008
A, E, I, T, U	ATV Research 13th & Broadway Dakota City, NE 68731	C, E, H, I, L, M, N, U **$10	Hobbyworld 19511 Business Center Dr. Northridge, CA 91324	D, J	Sentry Mfg. Co. Crystal Park Chickasha, OK 73108
A, C, D, E, H, I, L	Azotic Industries 2293 N. Clybourn Chicago, IL 60614	A, B, C, D, E, F, G, H, I, J, K, L, M, N, O, P, T, U, V	Integrated Circuits Unlimited, Inc. 7895 Clairemont Mesa Blvd. San Diego, CA 92111	D, J	Sherwood Engineering, Inc. 1268 S. Ogden St. Denver, CO 80210
A, B, D, F, G, H	Barker and Williamson, Inc. 10 Canal St. Bristol, PA 19007	A, B, C, E, H, I, K, L, M, T, U, W **$10	Jameco Electronics 1355 Shoreway Rd. Belmont, CA 92111	F, V *30¢ stamp	Skylane Products 406 Bon Air Ave. Temple Terrace, FL 33617
U	Byte-Me Computer Shop 327 Captain's Walk New London, CT 06320	D	JAN Crystals 2400 Crystal Dr. P.O. Box 06017 Ft. Meyers, FL 33906	**$5	Small Parts, Inc. (mechanical components and metal stock) P.O. Box 381736 Miami, FL 33138
B	Caddell Coil Corp. (coils for ARRL projects) 35 Main St. Poultney, VT 05764	F	Jug Wire Co. (Surplus Dept.) 2234 36th St. Woolsey, NY 11105	A, D, J	Spectrum International P.O. Box 1084 Concord, MA 01742
A, H, L *free **$10	Cambridge Thermionic Corp. 445 Concord Ave. Cambridge, MA 02138	A, E, F, H, M, N	Marlin P. Jones and Assoc. P.O. Box 12685 Lake Park, FL 33403	M, N *$4	Star Tronics P.O. Box 683 McMinnville, OR 97128
H	Caywood Electronics (Millen Capacitors) 67 Maplewood St., P.O. Box U Malden, MA 02148	C	Kepro Circuit Systems, Inc. 630 Axminister Dr. Fenton, MO 63026	B, M, O, U **$10	Teleprinter Corp. of America 550G Springfield Ave. Berkeley Heights, NJ 07922
A, C, W	Circuit Board Specialists (circuit boards for ARRL projects, kits) P.O. Box 969 Pueblo, CO 81002	A, E, M, U	Key Electronics P.O. Box 3506 Schenectady, NY 12303	K	Ten-Tec, Inc. Highway 411, E. Sevierville, TN 37862
A, B, D, H, L *stamp	D and V Radio Parts 12805 W. Sarle Freeland, MI 48623	F **$10	Kirk Electronics 73 Ferry Rd. Chester, CT 06412	F, V	Texas Towers 1108 Summit Ave. Plano, TX 75074
D, I, M	Peter W. Dahl 4007 Fort Blvd. El Paso, TX 79930	D, H, G, L, N	Leeds Radio 57 Warren St. New York, NY 10007	B, O *s.a.s.e.	Typetronics Box 8873 Ft. Lauderdale, FL 33310
D, E, F, H, I, L, M, N, U **$7	Diamondback Electronics Co. P.O. Box 12095 Sarasota, FL 33578	K, L	MFJ Enterprises P.O. Box 494 Mississippi State, MS 39762	R	Western Nebraska Electronics Rte. 1 — Box 1 Potter, NE 69156
W	Dynaclad Industries P.O. Box 296 Meadowlands, PA 15347	A, B, D, E, H, M, N, U	MHZ Electronics 2111 W. Camelback Rd. Phoenix, AZ 85015	E, G, D	Workman Electronic Products, Inc (will refer customer to nearest dealer) Box 3828 Sarasota, FL 33578
L *$10	Electro Sonic, Inc. 1100 Gordon Baker Rd. Willowdale, ON M2H 3B3	A, G, L	Millen Components Div. of E.I. & S. Corp. 42 Pleasant St. Stoneham, MA 02180		
C, D, E, G, H, I, M	Etco Electronics North Country Shopping Center Rte. 9 North Plattsburgh, NY 12901	A, B, H, J	J. W. Miller Div., Bell Industries 19070 Reyes Ave. Compton, CA 90224		

Chart Coding

A — New Components
B — Toroids and Ferrites
C — Etched Circuit Board Materials
D — Transmitting and Receiving Materials
E — Solid State Devices
F — Antenna Hardware
G — Dials and Knobs
H — Variable Capacitors
I — Transformers
J — I-f Filters
K — Cabinets and Boxes
L — General Supplier
M — Surplus Parts
N — Surplus Assemblies
O — RTTY Equipment & Parts
P — Surplus FM Gear and Parts
R — Service of Collins Equipment
T — Amateur TV Cameras & Components
U — Microcomputer Peripheral Equipment
V — Towers
W — Ready-made Printed Circuit Boards
*Catalog price
**Minimum Order

J	Fox-Tango Corp. (modification kits for amateur equipment) Box 15944 W. Palm Beach, FL 33406	A, B, D, E, F, G, H, K, M *s.a.s.e. **$15	Milo Associates Box 2323 Indianapolis, IN 46206
D, F, L	Gregory Electronics Corp. 249 Route 46 Saddle Brook, NJ 07662	E, F, G, H, L *50¢	Modern Radio Laboratories P.O. Box 1477-Q Garden Grove, CA 92642
A, E, H, U, W	Hamilton Avnet 2111 W. Walnut Hill Lane Irving, TX 75062	A, C, E, F, G, H, I, K, M, N	Olson Electronics 260 S. Forge St. Akron, OH 44327
I, K	Hammond Mfg., Ltd. 394 Edinburg Rd. N. Guelph, ON N1H 1E5 Canada	B	Palomar Engineers Box 455 Escondido, CA 92025
		T, U, W	P. C. Electronics 2522 Paxson Arcadia, CA 91006
		L	C. M. Peterson Co., Ltd. 220 Adelaide St. N. London, ON N6E 3H4

To the best of our knowledge the suppliers shown are willing to sell components to amateurs in small quantities by mail. This listing does not necessarily indicate that these firms have the approval of ARRL.

Chapter 18

Wave Propagation

Though great advances have been made in recent years in understanding the many modes of propagation of radio waves, variables affecting long-distance communication are very complex, and not entirely predictable. Amateur attempts to schedule operating time and frequencies for optimum results may not always succeed, but familiarity with the nature of radio propagation can reduce the margin of failure and add greatly to one's enjoyment of the pursuit of any kind of DX.

The sun, ultimate source of life and energy on earth, influences all radio communication beyond the local range. Conditions vary with such obvious sun-related earthly cycles as time of day and season of the year. Since these differ for appreciable changes in latitude and longitude, almost every communications circuit is unique in some respects. There are also short- and long-term solar cycles which influence propagation in less obvious ways. Furthermore, the state of the sun at a given moment is critical to long-distance communication, so it is understandable that propagation forecasting is still a rather inexact science.

With every part of the radio spectrum open to our use differing in its response to solar phenomena, amateurs have been, and still are, in a position to contribute to advancement of the art, both by accident and by careful investigation.

SOLAR PHENOMENA

Man's interest in the sun is older than recorded history. Records of sunspot observations translatable into modern terms go back nearly 300 years. Current observations are statistically "smoothed" to maintain a continuous record, in the form of the *Zurich sunspot number*, on which propagation predictions are based.

A useful modern indication of overall solar activity is the solar flux index. A 2800-MHz measurement made at 1700 UT daily in Ottawa is transmitted hourly by WWV. Because it is essentially current information, directly related to the sunspot number (see Fig. 1) and more immediately useful, it tends to displace the latter as a means of predicting propagation conditions.[1]

Sunspot Cycles

Even before their correlation with radio propagation variations was well-known, the periodic rise and fall of sunspot numbers had been studied for many years. These cycles average roughly 11 years in length, but have been as short as 9 and as long as 13 years. The highs and lows of the cycles also vary greatly. Cycle 19 peaked in 1958 with a sunspot number of over 200. Cycle 20, of nearer average intensity, reached 120 in 1969. By contrast, one of the lowest, Cycle 14, peaked at only 60 in 1907. Several cycle lows have not reached zero levels on the Zurich scale for

[1]Tilton, "The DXer's Crystal Ball," *QST*, June, August and September 1975. The WWV bulletin form changed in 1976, but basic principles apply.

any appreciable period, while others have had several months of little or no activity.

Sunspot cycles do not have a sine-wave shape. The rise is shorter than the decay, but neither is clearly defined. October 1974 had a solar flux range of 73 to 144. June 1976, the last month of Cycle 20, had several quiet-sun days (solar flux 66), but April and August had readings in the 80s. November 1979, peak month of Cycle 21, had 383 (equalling the Cycle 19 record high) and 154, within 18 days.

Solar Radiation

Insofar as it affects most radio propagation, solar radiation is of two principal kinds: ultraviolet light and charged particles. The first travels at just under 300,000,000 meters (186,000 miles)

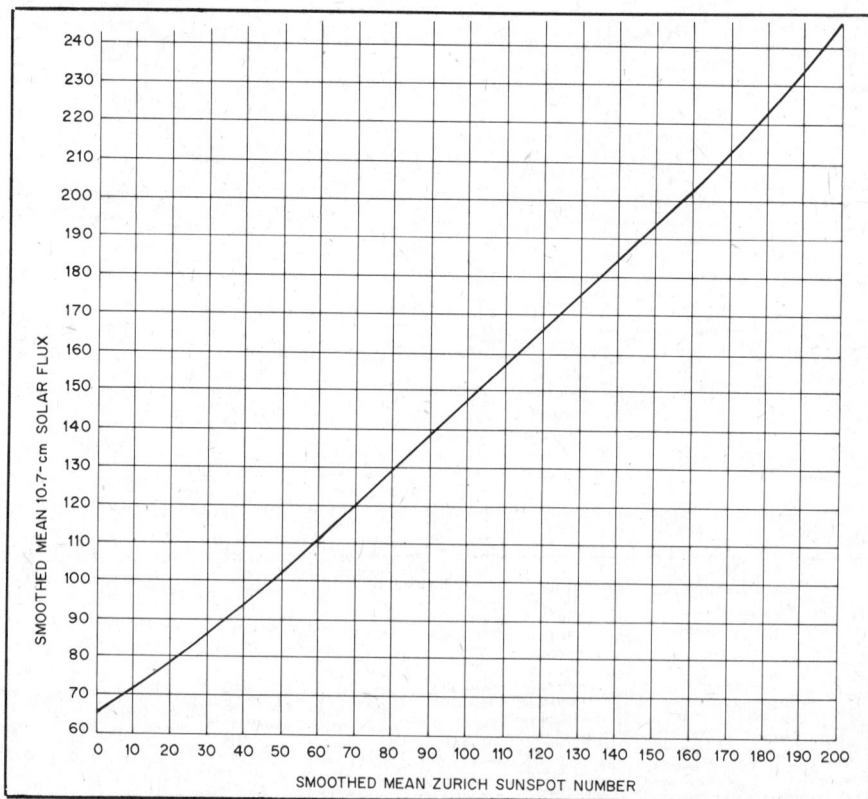

Fig. 1 — Relationship between the smoothed mean Sunspot Number and the 2800-MHz solar flux. In the low months of a solar cycle flux values run between about 66 and 86. Intermediate years may see 85 to 150. The peak years of Cycle 21 have brought readings between 140 and 380, through 1979 and the first half of 1980.

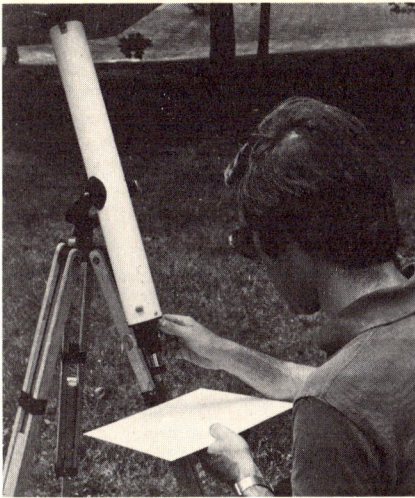

Fig. 2 — A simple sunspot projection system, demonstrated by ARRL staffer AC1Y. The cardboard baffle at the top provides a shaded area for viewing the sun's image. *Never look at the sun with the naked eye, binoculars or telescope, except through a known safe solar filter. See page 18-9 for additional cautions.*

per second, as does all electromagnetic radiation, so UV effects on wave propagation develop simultaneously with increases in observed solar noise, approximately eight minutes after the actual solar event. Particle radiation moves more slowly, and by varying routes, so it may take up to 40 hours to affect radio propagation. Its principal effects are high absorption of radio energy and the production of auroras, both visual and the radio variety.

Variations in the level of solar radiation can be gradual, as with the passage of some sunspot groups and other long-lived activity centers across the solar disk, or sudden, as with solar flares. An important clue for anticipating variations in solar radiation levels and radio propagation changes resulting from them is the rotational period of the sun, approximately 27 days. Sudden events (flares) may be short-lived, but active areas capable of influencing radio propagation may recur at four-week intervals for four or five solar rotations. Evidence of the "27-day cycle" is most marked during years of low solar activity.

Solar activity can be observed quite easily. Simple projection of the sun's image, as in Fig. 2, is most useful in the low years of the "11-year" cycle. In times of high activity the visible evidence may be difficult to sort out, unless observations are made daily and the results are recorded with care. Enough definition for our purposes is possible with the simplest telescopes. Low-cost instruments, 10- to 30-power, are adequate. A principal requirement is provision for mounting on a tripod having a pan-tilt head.[2]

Adjust the aiming to give a circular shadow of the scope body, then move the scope slowly until a bright spot appears on the projection surface. Put a baffle on the scope to enlarge the shaded area and adjust the focus to give a sharp-edged image of the solar disc. If there are any sunspots you will see them now. Draw a rough sketch of what you see, every time an observation is made, and keep it with your record of propagation observations.

Spots move across the image from left to right, on the projected image, as it is viewed with the sun at the observer's back. The line of movement is parallel to the solar equator. Not all activity capable of affecting propagation can be seen, but any spots seen have significance. Active areas may develop before spots are visible and may persist after spots associated with them are gone, but once identified by date they are likely to recur about 27 days later, emphasizing the worth of detailed records.

Vhf or uhf arrays capable of movement in elevation as well as azimuth are useful for solar noise monitoring. With a good system, the "quiet sun" can be "heard" at a low level.[3] Bursts that can be many dB higher indicate the start of a major event, such as a solar flare capable of producing an hf blackout and possibly vhf auroral propagation.

TYPES OF PROPAGATION

Depending on the means of propagation, radio waves can be classified as *ionospheric, tropospheric,* or *ground waves.* The ionospheric or *skywave* is that main portion of the total radiation leaving the antenna at angles somewhat above the horizontal. Except for the reflecting qualities of the ionosphere, it would be lost in space. The tropospheric wave is that portion of the radiation kept close to the earth's surface as the result of bending in the lower atmosphere. The ground wave is that portion of the radiation directly affected by the surface of the earth. It has two components, an earth-guided *surface wave,* and the *space wave.* The latter is the resultant of two components, direct and ground-reflected. The terms "tropospheric wave" and "ground wave" are often used interchangeably, though this is not strictly correct.

THE IONOSPHERE

Long-distance communication and much over shorter distances, on frequencies below 30 MHz, is the result of bending of the wave in the *ionosphere,* a region be-

tween about 60 and 200 miles above the earth's surface where free ions and electrons exist in sufficient quantity to affect the direction of wave travel.

Ionization of the upper atmosphere is attributed to ultraviolet radiation from the sun. The result is not a single region, but several layers of varying densities at various heights surrounding the earth. Each layer has a central region of relatively dense ionization that tapers off both above and below.

Ionospheric Layers

The lowest useful region of the ionosphere is called the E layer. Its average height of maximum ionization is about 70 miles. The atmosphere here is still dense enough so that ions and electrons set free by solar radiation do not have to travel far before they meet and recombine to form neutral particles: The layer can maintain its ability to bend radio waves only when continuously in sunlight. Ionization is thus greatest around local noon, and it practically disappears after sundown.

In the daylight hours there is a still lower area called the D region where ionization is proportional to the height of the sun. Wave energy in the two lowest frequency amateur bands, 1.8 and 3.5 MHz, is almost completely absorbed by this layer. Only the highest-angle radiation passes through it and is reflected to earth by the E layer. Communication on these bands in daylight is thus limited to short distances, as the lower-angle radiation needed for longer distances travels farther in the D region and is absorbed.

The region of ionization mainly responsible for long-distance communication is called the F layer. At its altitude, about 175 miles at night, the air is so thin that recombination takes place very slowly. Ionization decreases slowly after sundown, reaching a minimum just before sunrise. The obvious effect of this change is the early disappearance of long-distance signals on the highest frequency that was usable that day, followed by loss of communication on progressively lower frequencies during the night. In the daytime the F layer splits into two parts, F_1 and F_2, having heights of about 140 and 200 miles, respectively. They merge again at sunset.

Scattered patches of relatively dense ionization develop seasonally at E-layer height. Such *sporadic E* is most prevalent in the equatorial regions, but it is common in the temperate latitudes in late spring and early summer, and to a lesser degree in early winter. Its effects become confused with those of other ionization on the lower amateur frequencies, but they stand out above 21 MHz, especially in the low-activity years of the solar cycle, when other forms of DX are not consistently available.

Duration of openings decreases and the length of skip increases with progressively higher frequencies. Skip distance is com-

---

[2]Projection of the sun and interpretation of results are discussed in reference 1, and in *QST,* December 1974, p. 83; January 1975, p. 84 and October 1976, p. 11. A black-box viewing device (Tomcik, K4OU) for sun projection is shown in July 1964, *QST.* (Photocopy from ARRL, 75 cents and stamped envelope.)

[3]Bray and Kirchner, "Antenna Patterns from the Sun," *QST,* July 1960. Wilson, "432-MHz Solar Patrol," *QST,* August 1967.

monly a few hundred miles on 21 or 28 MHz, but multiple hop propagation can extend the range to 2500 miles or more. June and July are the peak months in the northern hemisphere. E_s propagation is most common in midmorning and early evening, but may extend almost around the clock at times. The highest frequency for E_s is not known, but the number of opportunities for using the mode drops off rapidly between the amateur 50- and 144-MHz bands, whereas 28 and 50 MHz are quite similar.

The greater the intensity of ionization in a layer, the more the wavepath is bent. The bending also depends on wavelength; the longer the wave the more its path is modified for a given degree of ionization. Thus, for a given level of solar radiation, ionospheric communication is available for a longer period of time on the lower-frequency amateur bands than on those near the upper limit of hf spectrum. The intensity and character of solar radiation are subject to many short-term and long-term variables, the former still predictable with only partial success.

Absorption

In traveling through the ionosphere, a radio wave gives up some of its energy by setting the ionized particles in motion. When moving particles collide with others, this energy is lost. Such *absorption* is greater at lower frequencies. It also increases with the intensity of ionization, and with the density of the atmosphere. This leads to a propagation factor often not fully appreciated: *Signal levels and quality tend to be best when the operating frequency is near the maximum that is reflected back to earth at the time.*

Virtual Height

An ionospheric layer is a region of considerable depth, but for practical purposes it is convenient to think of it as having finite height, from which a simple reflection would give the same effects (observed from the ground) as result from the gradual bending that actually takes place. It is given several names, such as *group height, equivalent height,* and *virtual height.*

The virtual height of an ionospheric layer for various frequencies and vertical incidence is determined with a variable-frequency sounding device that directs pulses of energy vertically and measures the time required for the round-trip path shown in Fig. 3. As the frequency rises, a point is reached where no energy is returned vertically. This is known as the *critical frequency* for the layer under consideration. A representation of a typical *ionogram* is shown in Fig. 4.[4] In this sounding the virtual height for 3.5 to 4 MHz was 400 km. Because the ionogram is a graphical presentation of wave travel time, double-hop propagation appears as an 800-km return for the same frequency. The critical frequency was just over 5 MHz on this occasion. Such a clear F-layer ionogram is possible only under magnetically quiet conditions, and at night, when little or no *E-* and *D*-layer ionization is present.

Effects of the Earth's Magnetic Field

The ionosphere has been discussed thus far in terms of simple bending, or refraction, a concept useful for some

[4]Davies, "Ionospheric Radio Propagation," NBS Monograph 80, out of print. Available in some technical libraries.

explanatory purposes. But an understanding of long-distance propagation must take the earth's magnetic field into account. Because of it, the ionosphere is a birefringent medium (double refracting) which breaks up plane-polarized waves into what are known as the *ordinary* and *extraordinary* waves, f_oF_2 and f_xF_2 in the ionogram. This helps to explain the dispersal of plane polarization encountered in most ionospheric communication.[5]

Sudden marked increases in solar radiation, such as with solar flares, trigger instantaneous effects in the *F, E* and *D* regions; slightly delayed effects, mainly in the polar areas; and geomagnetic effects, delayed up to 40 hours.

Onset of the *D*-region absorption is usually sudden, lasting a few minutes to several hours, leading to use of the term SID (sudden ionospheric disturbance). Shortwave fadeouts (SWFs) and SIDs exhibit wide variations in intensity, duration and number of events, all tending to be greater in periods of high solar activity.

Radiation Angle and Skip Distance

The lower the angle above the horizon at which a wave leaves the antenna, the less refraction in the ionosphere or troposphere is required to bring it back, or to maintain useful signal levels in the case of tropospheric bending. This results in the emphasis on low radiation angles in the pursuit of DX, on the hf or vhf bands.

Some of the effects of radiation angle are illustrated in Fig. 3. The high-angle wave at the left is bent only slightly in the ionosphere, and so goes through it. The wave at the somewhat lower angle is just capable of being returned by the ionosphere. In daylight it might be returned via the *E* layer. Its area of return from the *F* layer, R2, is closer to the transmitting point, T, than is that of the lowest-angle wave. If R2 is at the shortest distance where returned energy is usable, the area between R1 and the outer reaches of the ground wave, near the transmitter, is called the *skip zone*. The distance between R2 and T is called the *skip distance*. The distances to both R1 and R2 depend on the ionization density, the radiation angle at T, and the frequency in use. The maximum distance for single-hop propagation via the *F* layer is about 2500 miles (4000 kilometers). The maximum *E*-layer single hop is about 1250 miles (2000 kilometers).

The maximum usable frequency (muf) for *E*-layer communication is about three times the critical frequency for vertical return, as in Fig. 3. For *F*-layer propagation it is about five times.

Multiple-Hop Propagation

On its return to earth, the ionospherical-

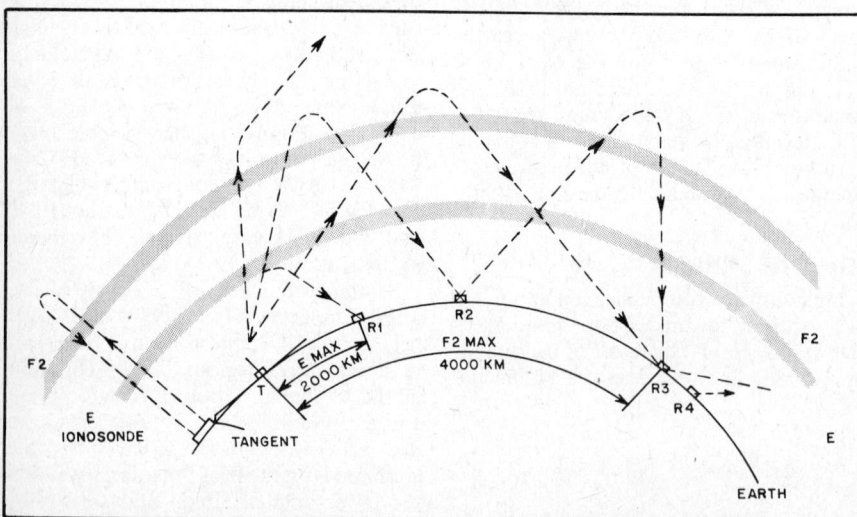

Fig. 3 — Three types of ionospheric propagation. Sounder, left measures virtual height and critical frequency of F_2 layer. Transmitter T is shown radiating at three different angles. Highest passes through the ionosphere after slight refraction. Lower-angle wave is returned to earth by the E layer, if frequency is low enough, at a maximum distance of 2000 km. The F-layer reflection returns at a maximum distance of about 4000 km, depending on the radiation angle. It is shown traversing a second path (double hop) from R2 to R4, the latter beyond single-hop range. The lowest-angle wave reaches the maximum practical single-hop distance at R3.

[5]See reference 4.

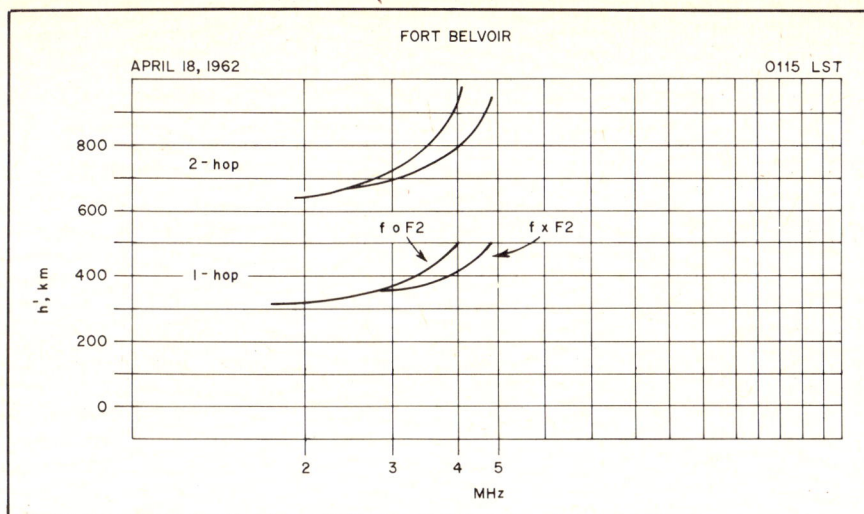

Fig. 4 — F-layer ionogram taken at night during magnetically quiet conditions. The traces show the breaking up into ordinary and extraordinary waves. Because it required twice the travel time, the double-hop return appears as having come from twice the height of the single-hop.

ly progagated wave can be reflected upward near R1 or R2, travel again to the ionosphere, and be refracted to earth. This process can be repeated several times under ideal propagation conditions, leading to communication beyond halfway around the world. Ordinarily ionospheric absorption and ground-reflection losses exact tolls in signal level and quality, so multiple-hop propagation usually yields lower signal levels and more distorted modulation than single-hop. This is not always the case, and under ideal conditions even long-way-around communication is possible with good signals. There is evidence to support the theory that signals for such communications, rather than hopping, may be ducted through the ionosphere for a good part of the distance.

Fading

Two or more parts of the wave may follow different paths, causing phase differences between wave components at the receiving end. Total field strength may be greater or smaller than that of one component. Fluctuating signal levels also result from the changing nature of the wave path, as in the case of moving air-mass boundaries, in tropospheric propagation on the higher frequencies. Changes in signal level, lumped under the term *fading*, arise from a variety of phenomena; some natural, some man-made. Reflections from aircraft, and ionospheric "holes" produced by the exhaust from large rocket engines, are in the latter category.

Under some circumstances the wave path may vary with very small changes in frequency, so that modulation sidebands arrive at the receiver out of phase, causing distortion that may be mild or severe. Called *selective fading*, this problem increases with signal bandwidth. Double-sideband a-m signals suffer much more

than single-sideband signals with suppressed carrier do.

THE SCATTER MODES

Much long-distance propagation can be described in terms of discrete reflection, though the analogy is never precise since true reflection would be possible only with perfect mirrors, and in a vacuum. All electromagnetic wave propagation is subject to scattering influences which alter idealized patterns to a great degree. The earth's atmosphere and ionospheric layers are scattering media, as are most objects that intervene in the wave path as it leaves the earth. Strong returns are thought of as reflections and weaker ones as scattering, but both influences prevail. Scatter modes have become useful tools in many kinds of communication.

Forward Scatter

We describe a skip zone as if there were no signal heard between the end of useful ground-wave range and the points R1 or R2 of Fig. 3, but actually the transmitted signal can be detected over much of the skip zone, with sufficiently sensitive devices and methods. A small portion of the transmitted energy is scattered back to earth in several ways, depending on the frequency in use.

Tropospheric scatter extends the local communications range to an increasing degree with frequency, above about 20 MHz, becoming most useful in the vhf range. *Ionospheric scatter*, mostly from the height of the E region, is most marked at frequencies up to about 60 or 70 MHz. Vhf tropospheric scatter is usable within the limits of amateur power levels and antenna techniques, out to nearly 500 miles. Ionospheric forward scatter is discernible in the skip zone at distances up to 1200 miles or so.

A major component of ionospheric scatter is that contributed by short-lived

columns of ionization formed around meteors entering the earth's atmosphere. This can be anything from very short bursts of little communications value to sustained periods of usable signal level, lasting up to a minute or more. Meteor scatter is most common in the early morning hours, and it can be an interesting adjunct to amateur communication at 21 MHz and higher, especially in periods of low solar activity. It is at its best during major meteor showers.[6]

Backscatter

A complex form of scatter is readily observed when working near the maximum usable frequency for the F layer at the time. The transmitted wave is refracted back to earth at some distant point, which may be an ocean area or a land mass where there is no use of the frequency in question at the time. A small part of the energy is scattered back to the skip zone of the transmitter via the ionospheric route.

Backscatter signals are generally rather weak, and subject to some distortion from multipath effects. But with optimum equipment they are usable at distances from just beyond the reliable local range out to several hundred miles. Under ideal conditions backscatter communication is possible over 3000 miles or more, though the term "sidescatter" is more descriptive of what probably happens on such long paths.

The scatter modes contribute to the usefulness of the higher parts of the DX spectrum, especially during periods of low solar activity when the normal ionospheric modes are less often available.

MF AND HF PROPAGATION

The 1.8-MHz band offers reliable communication over distances up to at least 50 miles during daylight. On winter nights ranges up to several thousand miles are possible.

The 3.5-MHz band is seldom usable beyond 200 miles in daylight, but long distances are not unusual at night, especially in years of low solar activity. Atmospheric noise tends to be high in the summer months on both 3.5 and 1.8 MHz.

The 7-MHz band has characteristics similar to 3.5 MHz, except that much greater distances are possible in daylight, and more often at night. In winter dawn and dusk periods it is possible to work the other side of the world, as signals follow the darkness path.

The 14-MHz band is the most widely used DX band. In the peak years of the solar cycle it is open to distant parts of the world almost continuously. During low solar activity it is open mainly in the

[6]Bain, "VHF Propagation by Meteor Trail Ionization," *QST*, May 1974. Table of major meteor showers, *Radio Amateur's VHF Manual*, Ch. 2.

daylight hours, and is especially good in the dawn and dusk periods. There is almost always a skip zone on this band.

The 21-MHz band shows highly variable propagation depending on the level of solar activity. During sunspot maxima it is useful for long-distance work almost around the clock. At intermediate levels it is mainly a daylight DX band. In the low years it is useful for transequatorial paths much of the year, but is open less often to the high latitudes. Sporadic-E skip is common in early summer and midwinter.

The 28-MHz band is excellent for DX communication in the peak solar-cycle years, but mostly in the daylight hours. The open time is shorter in the intermediate years, and is more confined to low-latitude and transequatorial paths as solar activity drops off. For about two years near the solar minimum, F-layer openings tend to be infrequent, and largely on north-south paths, with very long skip.

Sporadic-E propagation keeps things interesting in the period from late April through early August on this band, and on 21 MHz, providing single-hop communication out to 1300 miles or so, and multiple-hop to 2600 miles. Effects discussed in the following section on vhf propagation also show up in this band, though tropospheric bending is less than on 50 MHz.

THE WORLD ABOVE 50 MHz

50 to 54 MHz

This borderline region has some of the characteristics of both higher and lower frequencies. Just about every form of wave propagation is found occasionally in the 50-MHz band, which has contributed greatly to its popularity. Its utility for service-area communication should not be overlooked. In the absence of any favorable condition, the well-equipped 50-MHz station should be able to work regularly over a radius of 75 to 100 miles or more, depending on terrain and antenna size and height.

Changing weather patterns extend coverage to 300 miles or more at times, mainly in the warmer months. Sporadic-E skip provides seasonal openings for work over 400 to 2500 miles in seasons centered on the longest and shortest days of the year. Auroral effects afford vhf operators in the temperate latitudes an intriguing form of DX up to about 1300 miles. During the peak of "11-year" sunspot cycle 50-MHz DX of worldwide proportions may be workable by reflections of waves by the ionospheric F_2 layer. Various weak-signal scatter modes round out the 50-MHz propagation fare.

144 to 148 MHz

Ionospheric effects are greatly reduced at 144 MHz. F-layer propagation is unknown. Sporadic-E skip is rare and much more limited in duration and coverage

than on 50 MHz. Auroral propagation is quite similar to that on 50 MHz, except that signals tend to be somewhat weaker and more distorted at 144. Tropospheric propagation improves with increasing frequency. It has been responsible for 144-MHz work over distances up to 2500 miles, and 500-mile contacts are fairly common in the warmer months. Reliable range on 144 is slightly less than on 50 under minimum conditions.

220 MHz and Higher

Ionospheric propagation of the sorts discussed above is virtually unknown above about 200 MHz. Auroral communication is possible on 220 and 420 MHz, but probably not on higher frequencies, with amateur power levels. Tropospheric bending is very marked, and may be better on 432 than on 144 MHz, for example. Communication has been carried on over paths far beyond line of sight, on all amateur frequencies up through 10,000 MHz. Under minimum conditions, signal levels drop off slightly with each higher band.

VHF/UHF PROPAGATION MODES

Known means by which vhf signals are propagated beyond the horizon are described below.

F_2-Layer Reflection

Most communication on lower frequencies is by reflection of the wave in the F region, highest of the ionized layers. Its density varies with solar activity, the maximum usable frequency (muf) being highest in peak years of the sunspot cycle. The F_2-layer muf is believed to have reached an all-time high, near 70 MHz, in the fall of 1958 (Cycle 19). Wherever the band was in use in late 1979, the best months of Cycle 21, there was phenomenal worldwide 50-MHz DX, and an muf of at least 65 MHz on the North Atlantic path. Cycle 20 brought marginal 50-MHz DX in 1968 to 1970, but less than Cycle 18, 1946 to 1950, and Cycle 21, not yet complete.

The muf for F_2-layer propagation follows daily, monthly and seasonal cycles, all related to conditions on the sun, as with the hf bands. Frequency checks will show if the muf is rising or falling, and the times and directions for which it is highest. Two-way work has been done over about 1800 to 12,500 miles; even greater, if daylight routes around the earth the long way are included.

The TE Mode

Also associated with high solar activity is a transequatorial mode, having an muf somewhat higher than the F_2. This is observed most often between points up to 2500 miles north and south of the geomagnetic equator, mainly in late afternoon or early evening.[7] Work done by amateurs, beginning in 1976, has shown that the TE mode works in the 144-MHz band,

and even as high as 432 MHz, under optimum conditions.

Sporadic-E Skip

Patchy ionization of the E region of the ionosphere often propagates 28- and 50-MHz signals over 400 to 1300 miles or more. Often called "short skip," this is most common in May, June and July, with a shorter season around year end. Seasons are reversed in the southern hemisphere. E skip can occur at any time or season, but is most likely in mid-morning or early evening. Multiple-hop effects may extend the range to 2500 miles or more.

E_s propagation has been observed in the 144-MHz band, and on TV channels up to about 200 MHz. Minimum skip distance is greater, and duration of openings much shorter, on 144 MHz than on 50. Reception of strong E_s signals from under 300 miles on 50 MHz indicates some possibility of skip propagation on 144, probably to 800 miles or more.

Aurora Effect

High-frequency communication may be wiped out or seriously impaired by absorption in the ionosphere, during disturbances associated with high solar activity and variations in the earth's magnetic field. If this occurs at night in clear weather, there may be a visible aurora; but the condition also develops in daylight, usually in late afternoon. Weak wavery signals in the 3.5- and 7-MHz bands are good indicators.

Vhf waves can be returned to earth from the auroral region, but the varying intensity of the aurora and its porosity as a propagation medium impart a multipath distortion to the signal, which garbles or even destroys any modulation. Distortion increases with signal frequency and varies, often quite quickly, with the nature of the aurora. Single-sideband is preferred to modes requiring more bandwidth. The most effective mode is cw, which may be the only reliable communications method at 144 MHz and higher, during most auroras.

Propagation is generally from the north, but probing with a directional array is recommended. Maximum range is about 1300 miles, though 50-MHz signals are heard occasionally over greater distances, usually with little or no auroral distortion.

How often auroral communication is possible is related to the *geomagnetic* latitude of participating stations, auroras being most frequent in northeastern USA and adjacent areas of Canada. They are rare below about 32° N in the Southeast and about 38 to 40° N in the Southwest. The highest frequency for auroral returns depends on equipment and antennas, but auroral communication has been achieved up to at least 432 MHz.

[7] *World Radio and TV Handbook,* Billboard Publications, London W1V 1PG, England; 2160 Patterson St., Cincinnati, OH 45124. Gilfer Associates, Box 239, Park Ridge, NJ 07656.

Tropospheric Bending

An easily anticipated extension of normal vhf coverage results from abrupt changes in the refractive index of the atmosphere, at boundaries between air masses of differing temperature and humidity characteristics. Such warm-dry over cool-moist boundaries often lie along the southern and western edges of stable slow-moving areas of fair weather and high barometric presure. Tropospheric bending can increase signal levels from within the normal working range, or bring in more distant stations, not normally heard.

A condition known as *ducting* or *trapping* may simulate propagation within a waveguide, causing vhf waves to follow earth curvature for hundreds or even thousands of miles. Ducting incidence increases with frequency. It is rare on 50 MHz, fairly common on 144, and more so on higher frequencies. It occurs most often in temperate or low latitudes. It was the medium for the W6NLZ-KH6UK work on 144, 220 and 432 MHz, over a 2540-mile path. Gulf Coast states see it often, the Atlantic Seaboard, Great Lakes and Mississippi Valley areas occasionally, usually in September and October.

Many local conditions contribute to tropospheric bending. Convection in coastal areas in warm weather; rapid cooling of the earth after a hot day, with upper air cooling more slowly; warming of air aloft with the summer sunrise; subsidence of cool moist air into valleys on calm summer evenings — these familiar situations create upper-air conditions which can extend normal vhf coverage.

The alert vhf enthusiast soon learns to correlate various weather signs and propagation patterns. Temperature and barometric-pressure trends, changing cloud formations, wind direction, visibility and other natural indicators can give him clues as to what is in store in the way of tropospheric propagation.

The 50-MHz band is more responsive to weather effects than 28, and 144 MHz is much more active than 50. This trend continues into the microwave region, as evidenced by tropospheric records on all our bands, up to and including work over many long paths on 10,000 MHz.

The Scatter Modes

Though they provide signal levels too low for routine communication, several scatter modes attract the advanced vhf operator.

Tropospheric scatter offers marginal communication up to 500 miles or so, almost regardless of conditions and frequency, when optimum equipment and methods are used.

Ionospheric scatter is useful mainly on 50 MHz, where it usually is a composite of meteor bursts and a weak residual scatter signal. The latter may be heard only when optimum conditions prevail. The best

distances are 600 to 1200 miles.

Backscatter, common on lower frequencies, is observed on 50 MHz during ionospheric propagation, mainly of the F_2 variety. Conditions for 50-MHz backscatter are similar to those for the hf bands, detailed earlier in this chapter.

Scatter from meteor trails in the *E* region can cause signal enhancement, or isolated bursts of signal from a station not otherwise heard. Strength and duration of meteor bursts decrease with increasing signal frequency, but the mode is popular for marginal communication in the 50- and 144-MHz bands. It has been used on 220 MHz, and, more marginally, on 432 MHz.

Random meteor bursts can be heard by cooperating vhf stations at any time or season, but early-morning hours are preferred. Major meteor showers (August Perseids and December Geminids) provide frequent bursts. Some other showers have various periods, and may show phenomenal burst counts in peak years. Distances are similar to other *E*-layer communication.

All scatter communication requires good equipment and optimum operating methods. The narrow-band modes are superior to wide-band systems.

PROPAGATION PREDICTION

Information on the prediction of maximum usable frequencies (muf) and optimum working frequencies for F-layer propagation was formerly available from the U.S. Government Printing Office. The material took several forms, as methods developed for military communications use were adapted to worldwide civilian needs. Though the service was terminated in 1975, the basic methods are still of interest. A full description may be found in *QST* for March, 1972.[8] The government information is available in some technical libraries. The propagation charts which appear regularly in *QST* are computer-derived from information similar to that described in reference 8.

Other means are available to amateurs who wish to make their own predictions, both short- and long-term. An appreciable amount of observing and record-keeping time is involved at first, but the work can be streamlined with practice. Many amateurs who try it find the task almost as interesting as any operational success that may result from it. Properly organized, data collection and propagation prediction can become an ideal group project.

Getting Started

Because most factors have well-defined cyclical trends, the first step in propagation prediction is to become familiar with the rhythm of these trends for the geographical location and season under

[8]Hall, "High-Frequency Propagation Estimations for the Radio Amateur," *QST*, March 1972.

consideration. This job is made easier if we understand the causes of the ups and downs, so familiarity with basic information given earlier in this chapter is helpful.

What frequencies are "open," and where the cutoff in ionospheric propagation lies in the spectrum can be determined quite readily by tuning upward in frequency with a general-coverage receiver, until ionospherically propagated signals are no longer heard. The muf for the day and the times that a given frequency band opens or closes can be found in this way. A daily log will show if conditions are improving or deteriorating.

Listening in the amateur bands and on immediately adjacent frequencies may be the only way to do this, if the receiver is the amateur-bands-only variety. Most DX bands are narrower in other parts of the world than in the Americas, so there is no lack of round-the-clock occupancy by other services, ordinarily. Most receivers also cover somewhat more than the actual amateur assignments, at their widest, so some commercial and governmental signals can be found close by our band edges. A worldwide listing of stations, by frequency, is useful in identifying signals for propagation monitoring purpose. Don't overlook W1AW; frequencies and schedule are listed in every *QST*.

Ability to tune to 5 MHz and multiples thereof, to receive the standard time-and-frequency stations now operating in many parts of the world, is a great aid. See Table 1. Most such stations operate continuously, with appreciable power and omnidirectional antennas. WWV and WWVH are excellent indicators, at any suitable distance from Colorado or Hawaii. Their signal behavior can tell the experienced observer at least as much about propagation — at the moment — as does the content of their propagation bulletins. Many receivers can be made to tune some of these frequencies by detuning their front-panel tracking controls. See September 1975 *QST*, page 23, for suggestions. Simple crystal-controlled converters for the standard frequencies offer another possibility (June 1976 *QST*, page 25).

Recurring Phenomena

Because the sun is responsible for all radio-propagation variables, their rhythmic qualities are related to time, season and other sun-earth factors. Some are obvious. Others, particularly the rotational period of the sun, about 27.5 days, show best in long-term chart records kept on a monthly or four-week basis. Recurrence data are used in nearly all prediction work done presently, and the data can yield fair accuracy.

If the muf is high and conditions are generally good for several days, a similar condition is likely to prevail four weeks later, when the same area of the sun will be in view from the earth. Ionospheric disturbances also generally follow the

Table 1

Some time and frequency stations useful for propagation monitoring.

Call	Frequency (kHz)	Location
WWV	2500, 5000, 10,000, 15,000, 20,000	Ft. Collins, Colorado
WWVH	Same as WWV, but no 20,000	Kekaha, Kauai, Hawaii
CHU	3330, 7335, 14,670	Ottawa, Ontario, Canada
RID	5004, 10,004, 15,004	Irkutsk, USSR*
RWM	4996, 9996, 14,996	Novosibirsk, USSR*
ZUO	2500, 5000	Pretoria, South Africa
VNG	7500	Lyndhurst, Australia
BPV	5000, 10,000, 15,000	Shanghai, China
JJY	2500, 5000, 10,000, 15,000	Tokyo, Japan
LOL	5000, 10,000, 15,000	Buenos Aires, Argentina

*Call, from international table, may not check with actual reception. Locations and frequencies appear to be as given.

27-day cycle, though there may be marked differences in level from one period to the next.

Some solar-activity centers are short-lived, lasting less than a full rotation. Others go on and on, recognizable from their propagation effects for a year or more. Recurring phenomena are more apparent in the low-activity years of the solar cycle, most of them being far enough apart to be clearly identifiable. In April and May, 1976, for example, there were three well-separated areas affecting radio propagation. All were of "the old cycle." There were also three new-cycle areas seen briefly, but with no recognizable radio-propagation influence. The WWV propagation bulletins described will be seen to show recurring effects, if their content is charted for extended periods.

WWV Propagation Bulletins

The National Bureau of Standards radio station WWV, Ft. Collins, CO, transmits hourly bulletins on solar activity and the condition of the earth's magnetic field. This information is essentially current, and thus invaluable to any student of radio propagation. Putting it into simple chart form and using it in conjunction with visual observation of the sun provides an excellent base for propagation understanding and prediction. The WWV (and WWVH) signals themselves are also very useful in gathering propagation data, as the stations are on the air continuously, using constant power levels and omnidirectional antennas. It should be stated here that WWVH (Kekaha, HI) does not carry the bulletin service, but its signals provide valuable propagation evidence. Other time-and-frequency stations, some of which are listed in Table 1, can be used similarly.

In order of their presentation, the WWV bulletins give the solar flux and A-index for the previous Universal Time (UT) day; the Boulder K-index (changed every 3 hours); the state of solar activity and the condition of the earth's magnetic field for the previous 24 hours, and the same factors predicted for the coming 24 hours. The bulletin starts after the time

announcement at 18 minutes after each hour.

The *solar flux* is a measure of the sun's radiation at 2800 MHz, taken daily at 1700 UTC in Ottawa. Similar measurements are made on many different frequencies at other observatories. The 2800-MHz flux is given because this value correlates well with the ionization density of the ionospheric *F* region, with the Zurich sunspot number (see Fig. 1), and with the maximum usable frequency for long-distance communication.

A solar flux value of 66 represents "quiet sun" conditions. It will be heard only in the years near the minimum of the solar cycle, most recently 1975 and 1976. At such times any flux variation is worthy of note, as it will produce an observable change in long-distance propagation in the hf range. Large increases in flux values come from large new active areas on the sun, which will be apparent to the regular viewer of the sun. The short-term effects of flares do not appear in the solar flux values, unless they are very large and numerous. Flares are mentioned in the bulletins, if they are major events capable of affecting radio propagation.

A steady rise or fall in solar flux will

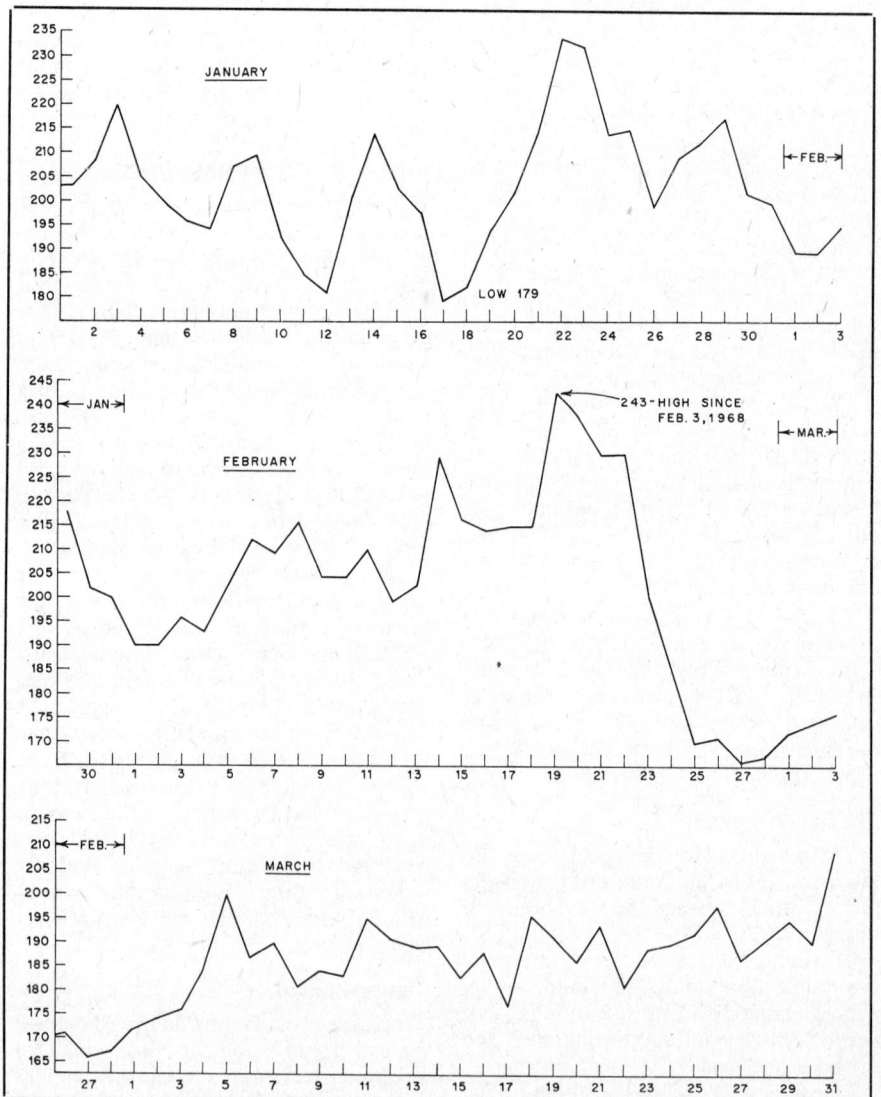

Fig. 5 — Graphs of 2800-MHz solar flux values for the first three months of 1979, approaching the peak of Cycle 21. The same period in 1976, near the bottom of Cycle 20, had values between 68 and 85. The dates are arranged so that days one solar rotation (27.5 days, average) apart line up vertically. Recurring solar phenomena are clearly apparent.

show clearly in radio propagation effects, and also in the observer's view of sunspot activity. Trends either way are important to the propagation student. They often run for several days, during which the associated changes in muf, and in the duration and geographical distribution of openings on frequencies above about 15 MHz, are easily observed. Flux readings of 80 or higher will make the 21- and 28-MHz bands come alive, even near the bottom of the solar cycle.

In the intermediate years of the cycle, as in 1977 to 1979, flux values tend to range between about 100 and 200, the latter being high enough to make even the 50-MHz band worth watching for world-wide communication on the more favored paths. When values above 200 come, the 21- and 28-MHz bands are open almost around the clock during the cooler months of the year. Sustained periods above 250, particularly in spring and fall, will bring widespread 50-MHz openings.

The *A-index* is a 24-hour figure for the activity of the geomagnetic field, on a scale of 0 to 400 or so, though values of 100 or more are rare. A quiet field (A indices of 10 or lower) is characteristic of the best propagation conditions. Absorption of wave energy is low at such times, so signals are strong on long paths. This effect is most noticeable on circuits crossing the higher latitudes, where very low K values must prevail or propagation will be very poor or nonexistent. The effect of geomagnetic activity is very slight on low-latitude paths or any circuit crossing the equator. In fact, it is observed that north-south or transequatorial propagation may *improve* briefly at the onset of a period of high A (and K) indices, especially on the highest frequency that is usable at all.

The *K-index* is similar to the A-index, except that it is as near to a current figure as can be given in bulletin form. It is also given on a different mathematical scale, in order to make short-term changes more apparent. The information given by WWV is for Boulder, Colorado. It is likely to be higher for Boston, and lower for San Francisco or New Orleans, both the latter cities having much lower geomagnetic latitude than either Boulder or Boston, while Boston's is higher than Boulder's.

The K-index is given for three-hour periods beginning at 0000, 0300, 0600, 0900, 1200, 1500, 1800 and 2100 UT. It represents the conditions during the last three hours just before the bulletin's time of issue. Thus, it is close to a *now* statement of a factor of vital importance to any user of the hf radio spectrum. It may interest vhf operators as well, when the values go above 3. The *trend* is important — a rising trend means degraded hf propagation; values of 4 and up may mean auroral conditions on the vhf bands.

Solar Activity, as the term is used in the final portion of the WWV bulletins,

relates to fast-changing conditions that affect propagation adversely. Activity is given as *very low, low, moderate, high* or *very high*.

Geomagnetic field conditions are termed *quiet, unsettled* or *active*. These relate roughly to K indices of 0 to 1 for *quiet*, 1 to 3 for *unsettled*, and 4 or higher for *active*. These three words rather aptly describe the propagation conditions associated with them and with the K indices. Often any K value of 4 or higher will be associated with a "geomagnetic storm," usually described as "minor" or "major." Either is bad news for the amateur interested in high-latitude hf communication.

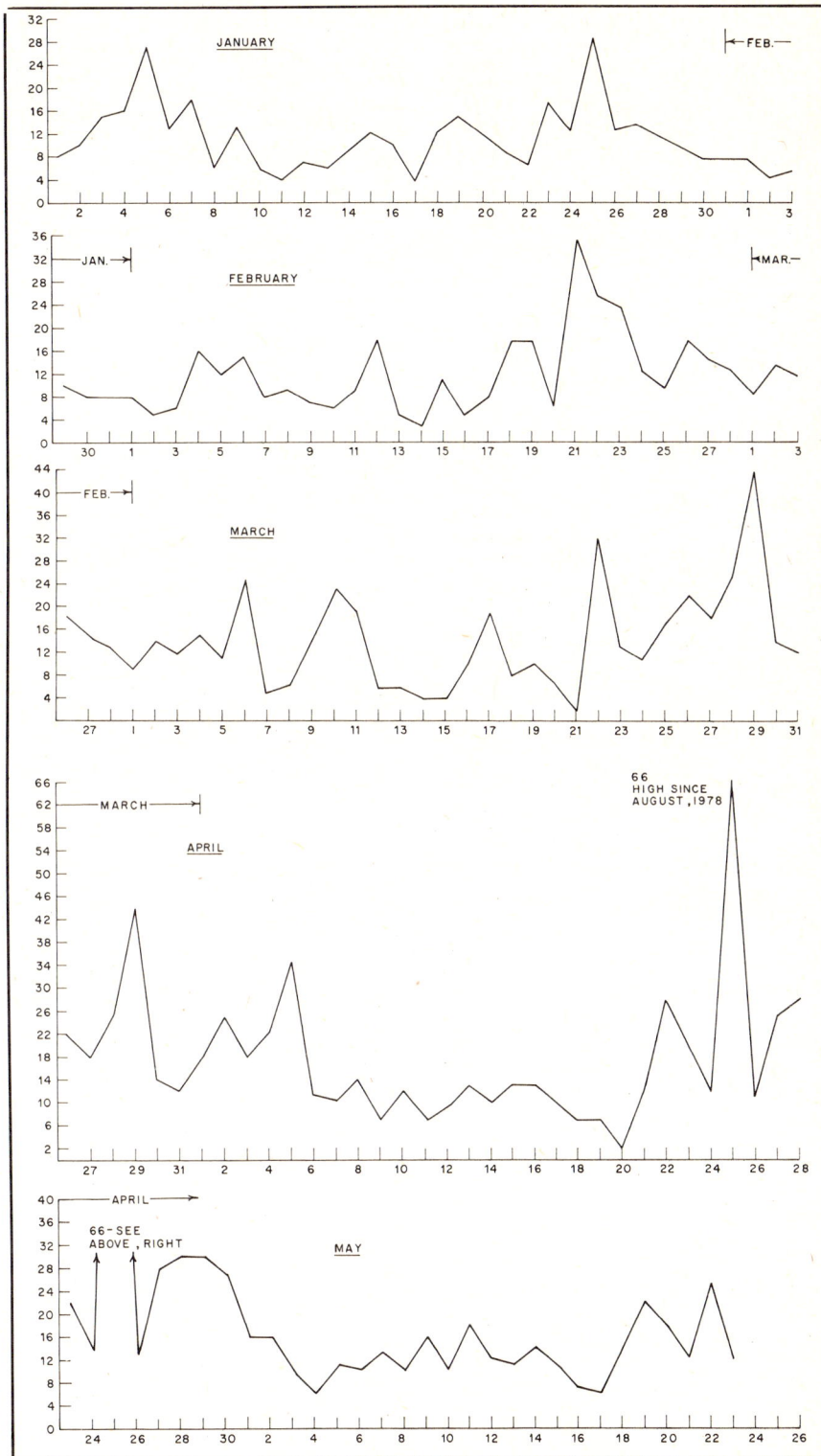

Fig 6 — A-index information transmitted by WWV for the first four months of 1979, arranged to show effects of recurring solar phenomena, as in Fig. 5. Recurrence dates vary because of the varying travel time of charged particles from the sun that cause geomagnetic disturbances on Earth. A-indices of 10 and lower go with good hf propagation. Peaks above 20 nearly always are accompanied by high signal absorption. Auroral effects are common in high latitudes at these times.

"Major" may include a total or near-total interruption of all communication on the lower hf bands. It is likely to be accompanied by auroral conditions on the vhf bands, at least in the higher latitudes.

K values are used to derive the A-index for the whole UT day. The manner in which this is done is rather involved. It is spelled out in detail in reference 1. Explanation of the bulletin data in greater detail is also given, though the form of the bulletins is now different from that described in the three 1975 *QST* articles. Information on all WWV services can be had by writing the National Bureau of Standards, Boulder, CO 80302.

Keeping Records

In a group project, or for the individual observer who has the time for it, charting all WWV information is very useful. A less time-consuming effort, since it requires logging only a single bulletin each day, is a chart record of the solar flux and A-index only. Depending on the time available, this record can take many forms. In the interest of readability here, Figs. 5 and 6 are separate records of the solar flux and A-index. Both are plotted in four-week periods, lined up vertically to show recurrent effects resulting from the solar rotational period.

For his own purposes the author developed a compact form containing much more detail than could be reproduced clearly. Graph paper having five blocks to the inch is cut to just over 54 blocks high and 73 blocks wide. This format was chosen to fit a standard binder 11-1/2 × 15-1/2 inches (292 × 394 mm) in size. Two blocks per day are used horizontally, the record beginning a day or two before the first of the month, and running over up to three days at the end, for better continuity in the total information. Solar flux values are plotted in the upper portion of the sheet, using two squares for each five flux units. The A and K indices share the lower portion, using different scales. The A-index is plotted at two units per block and the K-index at one unit per block. Using a different color for each item helps to keep them sorted out visually.

Much more information can be added. Brief propagation notes, drawings of the sun for the more significant dates, and a record of the solar flux and A-index values for the corresponding day four weeks before are commonly used. If the record for the previous solar rotation is inked in lightly (solar flux and A-index only) when the chart is first set up, it will make anticipation of recurring effects much easier. It should be emphasized that solar rotation time is by no means the whole story. Clearly there are occasional surprises which are not explained by this factor alone. Activity centers die out and new ones are born, seemingly without reason. Prediction of these seeming anomalies presents a challenge not yet met

Fig. 7 — Projection viewing of the sun's image with a 4-inch (102-mm) reflector-type telescope. White-paper viewing surface is cemented in the bottom of a black-sprayed cardboard box.

Fig. 8 — *Direct viewing of the sun should be done only with a telescope equipped with an aperture filter known to be safe for this purpose.* Telescope is a Celestron 5 with the maker's solar filter, which passes 0.01 percent of impinging light. A brimmed hat shading the observer's eyes from direct rays of the sun helps to improve visual acuity.

fully by anyone, including professionals. It is a wide-open field.

Advanced Solar Observation

Regular viewing of the sun should be a part of any major propagation-prediction effort. Even simple projection with a low-cost telescope, as shown in Fig. 2, is well worthwhile, if it is done regularly and drawings are made to record what is seen. Improvements in technique need not be costly. A desirable first step is a light-exclusion box. An example, shown in use in Fig. 7, uses a corrugated-paper box sprayed dull black, inside and out, with a white viewing surface cemented in the bottom. The observer should position himself so as to see as little light other than that of the solar image as possible. A wide-brim hat is useful in this. For a complete black-box viewing system that can be built from plywood and simple optical components, see reference 2.

The small refractor telescope shown in Fig. 2 is a low-priced model. Scopes in this general range work well, especially with the light-exclusion methods described above. Many radio amateurs are also interested in astronomy, and thus may

W1AW Propagation Bulletins

For those who may not have time or interest to do their own predicting, short-term predictions of propagation conditions are carried daily on the Headquarters station, W1AW. (See schedule in April and October *QST*, or write ARRL for a copy.) These bulletins are designed to supplement the propagation charts given in *QST*, and to provide the best available information on the conditions to be expected in the next few days after the bulletin date. Normally the text is changed weekly, but new information is included whenever it becomes available.

The bulletin contents are made up from a combination of daily monitoring and charting of all WWV information, observation of the sun with advanced equipment, and frequent monitoring of the amateur hf and vhf bands. Interesting propagation events of recent days are reported, in order to put new information in proper perspective.

already have much better instruments available. Moderately priced 2- or 3-inch (51- or 76-mm) refractors give beautiful detail with light-exclusion viewing. Telescopes with more than 4-inch (102-mm) aperture are not recommended for projection, as they may develop enough internal heating to damage optical components. Large telescopes also introduce mechanical stability problems, unless mounted on permanent supports or tripods of exceptional sturdiness.

Better definition is available with larger telescopes when they are used for direct viewing, as shown in Fig. 8. **This requires a safe solar filter,** mounted over the telescope aperture. **Never look at the sun directly, with the naked eye, binoculars or any telescope, except through a solar filter known to be safe. Such a filter should pass no more than 0.01 percent of the impinging light, and it should be reasonably uniform across the whole solar spectrum. Do not use eyepiece filters. Be sure that the aperture filter is mounted firmly so that it cannot come off while the telescope is in use. Never look at the sun without it.**

Visual acuity is very important. Even people who think they have satisfactory eyesight, with or without glasses, may find that keener eyes will see much solar surface detail that they miss. Have a younger helper, if you are middle aged or older.

Interpreting What You See and Hear

In viewing the sun with a celestial telescope equipped with a star diagonal and a vertical eyepiece, one sees the solar disk with the east limb on the right and the west limb on the left. This is the opposite of the view obtained with the setup of Fig. 2, but is more natural since it simulates a map. Visible solar activity moves across the disk from right to left, on a line parallel with the solar equator. The apparent position of the equator varies with the time of day and the position of the viewer, but it can be determined readily if drawings are made during each observation. Knowing the position of the equator is important in identifying activity as belonging to the old or new cycle, in times of transition. Old-cycle spots move near the equator. New-cycle activity appears some 30° above or below.

In good projection, or with properly safeguarded direct viewing, bright patches may be seen, especially near the east or west limbs. Known as plage, faculae, or flocculi, these patches identify active areas that may or may not include visible spots. When seen on the east limb, they may be advance notice of spots due in another day or two. They serve as warning of propagation changes several days away, and their appearance may coincide with the start of a steady rise in solar flux and in the muf as well. Faculae may identify new activity in which spots will appear four weeks later, or they may be the residue of declining activity that contained spots last time around. They can be a vital part of visual records, and their significance will increase as records accumulate.

In their first or last day on the east or west limb, respectively, sizeable spots or groups usually show as fine lines on or close to the edge of the image. Some detail will begin to show on the second day of new or recurring activity, and sketches should be made as accurately as possible. Note any changes in additional sketches, marked with date and time. Changes in appearance and growth or decay are significant indicators, becoming more so on consecutive rounds of long-lived activity centers.

Increasing size and number of spots will be reflected in a rise in solar flux on the WWV bulletins, particularly the one for 0000 UT and in rising F-layer muf. Sudden large growth, or a major breakup of a large spot or group, may show radio effects at once — a rise in muf and perhaps a considerable increase in noise level. The latter is more obvious when using a directive array that can be aimed at the sun.

The noise burst and visible change will almost certainly be accompanied by particle radiation increase, the radio effects of which will be increased absorption of hf signal energy, and possibly auroral conditions on 28 MHz or the vhf bands, one to four days later. (Rising K-index on WWV, possibly without warning on previous bulletins.)

Slower growth, barely distinguishable from day to day, will be accompanied by rising solar-flux numbers, probably a

point or two daily, and a gradual improvement in hf conditions that will last as long as the K-index remains low. A rise in muf will be apparent at such times, and propagation will remain good on all frequencies for several days, barring sudden solar change which is always a possibility.

If, on the first attempt at solar viewing, one sees sizeable spots or groups, it is well to remember that these may represent activity in a declining phase. If so, they may move across the disk with only minor apparent change. Keep watch though — the area could be brought back to active state again by forces not yet fully understood. This is why long-term predictions are doomed to occasional abject failure and why short-term prediction, using all the tools available, is such an exciting and useful pursuit.

SUNSPOT NUMBERS AND THE SOLAR FLUX

Information about solar cycles is usually based on the statistically smoothed Zurich Sunspot Number. Because sunspot records have been kept for more than 200 years in essentially this form, Zurich numbers are still widely used, especially for study of long-term trends in solar activity. The 2800-MHz solar flux (see "Solar Phenomena," at the beginning of this chapter) is much more useful for short-term propagation study and prediction.

Daily and smoothed Zurich numbers were compiled from observations made all over the world. The Solar Division of the American Association of Variable Star Observers (AAVSO) has been a major contributor to this program since 1944. As this text was in preparation it seemed likely that the Zurich effort would be terminated in 1980. If it is, the work will be continued by the AAVSO, with financial and technical support from the National Oceanic and Atmospheric Administration (NOAA), Boulder, Colorado.

Interpreting Sunspot Numbers

"Sunspot Number" is not number of sunspots. Often called the "Wolf Number," after its Swiss originator, the relative sunspot number, R, is derived from

$$R = k(10G + S),$$

where G is the number of sunspot *groups*, and S is the number of sunspots, both

regardless of size or position on the sun. (Both size and position are important considerations in assessing the effects of changing solar activity on radio propagation.) The constant, k, is in effect a "rating" of the individual observer, based on his equipment and skill, and the quality of his observing site. Though this system is cumbersome, and vague at best, it is the only statistical link with solar history prior to 1947, when the solar flux program began. Its continuance on the same basis is thus of some considerable value.

Provisional American daily sunspot numbers, $R_A 1$, are issued weekly by AAVSO. This program is handled by Casper Hossfield, KA2DKD, chairman of the Solar Division. The smoothed sunspot number, often shown graphically in solar records and referred to in Fig. 1, is prepared as the cycle progresses, but it is tentative until after the cycle is over. Daily sunspot number comparisons with the solar flux may not follow the curve of Fig. 1, though long-term averages should do so.

Using Solar Flux Data

The daily solar flux reading at 2800 MHz, available hourly on WWV, is more closely related to radio conditions than is the sunspot number. It correlates well with the ionization density of the F_2 layer, and thus with the maximum usable frequency for long-distance communication.

Solar flux values range from about 66 (quiet sun, near the bottom of the solar cycle) to highs of over 300, reached in the peak months of only the more-active cycles. There were four days over 300 in November 1979, and readings in the 280s in May 1980.

The significance of rising or falling solar flux values differs with the season and with the latitude of the stations involved. In the temperate latitudes the muf is highest, for a given solar flux value, in the cool months of the year. There is much less seasonal effect in the equatorial regions. Transequatorial paths also show little seasonal change. In general, a solar flux of 80 or so, even in the lowest years of the cycle, can bring brief flurries of DX on 28 MHz, in the northern hemisphere fall. Values over 200 work similarly for 50-MHz F-layer activity, if they come between late October and early December. A sustained period of high flux values has more significance in this respect than a single peak day.

Chapter 19

Transmission Lines

Transmission lines, and the theory behind them, play an important role in many phases of radio communication. This is because the basic principles involved can be applied to a wide variety of problems. Types of transmission lines include simple two-conductor configurations such as the familiar coaxial cable and TV parallel-wire line. Such lines are useful from power frequencies to well up into the microwave region and form perhaps the most important class. The waveguide is representative of a second type. Here, the conductor configuration is rather complex and ordinary concepts such as voltage and current tend to become obscure. As a consequence, various parameters are expressed in terms of the electric and magnetic fields associated with the line. Finally, the propagation of electromagnetic energy through space itself is closely related to similar phenomena in wave guides and transmission lines. In fact, the only significant physical difference is that the power density in a wave propagated in space decreases with increasing distance while it is possible to transmit power over long distances with conventional lines with little attenuation. This is because power flow is essentially confined to one dimension in the latter case while the three-dimensional aspect of space does not permit such confinement.

Transmission Lines and Circuits

A transmission line differs from an ordinary circuit in one very important aspect. Delay effects associated with the finite propagation time of electromagnetic energy are often neglected in network design since the dimensions involved are normally small compared to the wavelength of any frequencies present in the circuit. This is not true in transmission-line considerations. The finite propagation time becomes a factor of paramount importance. This can be illustrated with the aid of Fig. 1. A transmission line separates a source at point g from a load at point a by a distance ℓ. If the line is uniform (same conductor shape at any cross section along the line), only two parameters are required to express the line properties completely. These are the phase velocity, v_p and the characteristic impedance, Z_o. If the line can be considered lossless as well, Z_o becomes a pure resistance, R_o.

Assume that a very short burst of power is emitted from the source. This is represented by the vertical line at the left of the series of lines in Fig. 2. As the pulse voltage appears across the load Z_a, all the energy may be absorbed or part of it may be reflected in much the same manner energy in a wave in water is reflected as the wave hits a steep breakwater or the end of a container. This reflected wave is represented by the second line in the series and the arrow above indicates the direction of travel. As the latter wave reaches the source, the process is again repeated with either all of the energy being absorbed or partially reflected.

The back-and-forth cycle is actually an infinite one but after a few reflections, the intensity of the wave becomes very small. If, instead of a short pulse, a continuous voltage is applied to the terminals of a transmission line, the voltage at any point along the line will consist of a sum of voltages of the composite of waves traveling toward the right and a composite of waves traveling toward the left. The total sum of the waves traveling toward the right is called the *forward wave* or incident wave while the one traveling toward the left is called the *reflected wave*. Provided certain conditions concerning Z_a are met, there will be a net flow of energy from the source to the load, with a fraction of the energy being stored in the "standing" waves on the line. This phenomenon is identical to the case of a coupled resonator with ordinary circuit elements. Sections of transmission line are often used for this purpose, especially in the vhf/uhf region. The duplexer found in many vhf repeaters is a common example.

Line Factors and Equations

Since transmission lines are usually connected between lumped or discrete circuitry, it is convenient to be able to express the input impedance of a line in terms of the output or load impedance. A line treated this way is then similar to a filter or matching network with a given load impedance. One caution should be kept in mind in applying such relations and that is the manner in which the source and load are connected to the line can be important. There are always some "parasitic" effects arising from connectors and post-connector circuit configuration that may cause the line to "see" a different impedance than if measurements were made at the load terminals directly. This is indicated by the abrupt change in line dimensions at points a and g in Fig. 1. Even though the short line connecting the generator to the main transmission line (and the one connecting the load to the line) might have the same characteristic

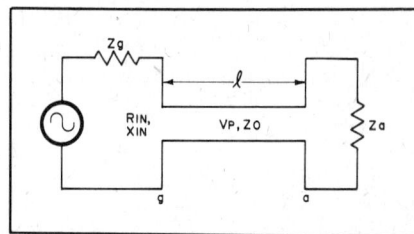

Fig. 1 — Source and load connected by means of a transmission line.

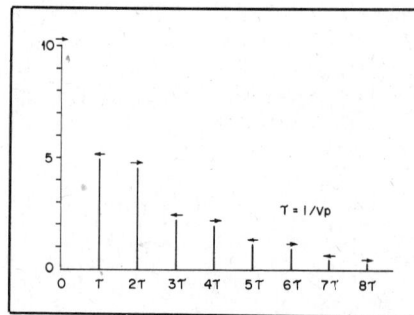

Fig. 2 — Magnitudes of components for forward and reverse traveling waves of a short pulse on a transmission line.

impedance, if the sizes are different a mismatch will still occur. Normally, this effect can be neglected at hf but becomes important as the frequency of operation is extended into the vhf region and above.

In referring to the previous example shown in Fig. 2, the ratio of the voltage in the reflected wave to that of the voltage in the incident wave is defined as the voltage reflection coefficient designated by the Greek letter, Γ, or by ρ. The relation between the output resistance, R_a, the output reactance, X_a, the line impedance, Z_0, and the *magnitude* of the reflection coefficient is

$$\Gamma = \sqrt{\frac{(R_a - R_o)^2 + X_a^2}{(R_o + R_a)^2 + X_a^2}}$$

Note that if R_a is equal to R_o, and if X_a is 0 the reflection coefficient is 0, which represents "matched" conditions. All the energy in the incident wave is transferred to the load. In effect, it was as if there were an infinite line of characteristic impedance Z_o connected at a. On the other hand, if R_a is 0, regardless of the value of X_a the reflection coefficient is 1.0. This means all the power is reflected in much the same manner as radiant energy is reflected from a mirror.

If there are no reflections from the load, the voltage distribution along the line is constant or "flat" while if reflections exist, a standing-wave pattern will result. The ratio of the maximum voltage on the line to the minimum value (provided the line is longer than a quarter wavelength) is defined as the voltage standing-wave ratio (VSWR). The VSWR is related to the reflection coefficient by

$$VSWR = \frac{1 + \Gamma}{1 - \Gamma}$$

This latter definition is a more general one, valid for any line length. Quite often, the actual load impedance is unknown. An alternate way of expressing the reflection coefficient is

$$\Gamma = \sqrt{\frac{P_r}{P_f}}$$

where

P_r = the power in the reflected wave
and
P_f = the power in the forward wave.

The parameters are relatively easy to measure with power meters available commercially or with homemade designs. However, it is obvious there can be no other power sources at the load if the foregoing definition is to hold. For instance, the reflection coefficient of the generator in the example shown in Fig. 2 is 0.9. This value could have been obtained by substituting the generator resistance and reactance into a previous formula for reflection coefficient, but not by measurement if the source were activated.

Fortunately, it is possible to determine the input resistance and reactance of a terminated line if the load resistance and reactance are known, along with the line length and characteristic impedance. (With actual lines, the physical length must be divided by the velocity factor of the cable which gives the value of l in the following formula.) The equations are

$$r_{in} = \frac{r_a(1 + \tan^2\beta l)}{(1 - x_a\tan\beta l)^2 + (r_a\tan\beta l)^2}$$

$$x_{in} =$$

$$\frac{x_a(1 - \tan^2\beta l) + (1 - r_a^2 - x_a^2)\tan\beta l}{(1 - x_a\tan\beta l)^2 + (r_a\tan\beta l)^2}$$

for a 1-ohm line. Equations are often "normalized" this way in order to make universal tables or plots that cover a wide range of values. If characteristic impedances (Z_o) other than 1.0 are to be used, the following set of conversions apply where R_a and X_a are the load resistance and reactance and R_{in} and X_{in} represent the resistance and reactance at the input end of the line.

$$r_a = \frac{R_a}{Z_o} \qquad\qquad R_{in} = Z_o r_{in}$$

$$x_a = \frac{X_a}{Z_o} \qquad\qquad X_{in} = Z_o X_{in}$$

In order to determine the value of the tangent function, either the line length in meters or feet, along with the frequency in MHz, can be substituted into the following expressions:

$$\beta l_{(degrees)} = 1.2f_{(MHz)} \times l_{(meters)}$$
$$\beta l_{(degrees)} = 0.367f_{(MHz)} \times l_{(feet)}$$

Since the foregoing transmission-line equations are somewhat awkward to work with, various plots have been devised that permit a graphical solution. However, with modern programmable calculators, even those in a moderate price class, it takes approximately four seconds to solve both equations. The plots shown in Fig. 3A and Fig. 3B were computed in this manner. The curves are for r_{in} and x_{in} for various values of r_a (x_a equal to 0) and line length in degrees. Note that 90 degrees appears to be a "critical" value and represents a line length of a quarter wavelength. As this value is approached, the transmission-line equations can be approximated by the formulas:

$$r_{in} \approx \frac{r_a}{r_a^2 + x_a^2}$$

$$x_{in} = \frac{-x_a}{r_a^2 + x_a^2}$$

If x_a is zero, the formula for a quarter-wavelength transformer is obtained:

$$R_{in} = Z_o^2/R_a$$

Quite often, it is mistakenly assumed that power reflected from a load represents power "lost" in some way. This is only true if there is considerable loss in the line itself and the power is dissipated on the way back to the source. On the other hand, the quarter-wavelength transformer is an example where reflections on a lossless line can actually be used to advantage in matching a load impedance that is different from the source impedance.

If the terminating resistance is zero, the input resistance is also zero. In effect, the line and load act as a pure reactance which is given by the formula:

$$x_{in} = \frac{x_a + \tan\beta l}{1 - x_a\tan\beta l}$$

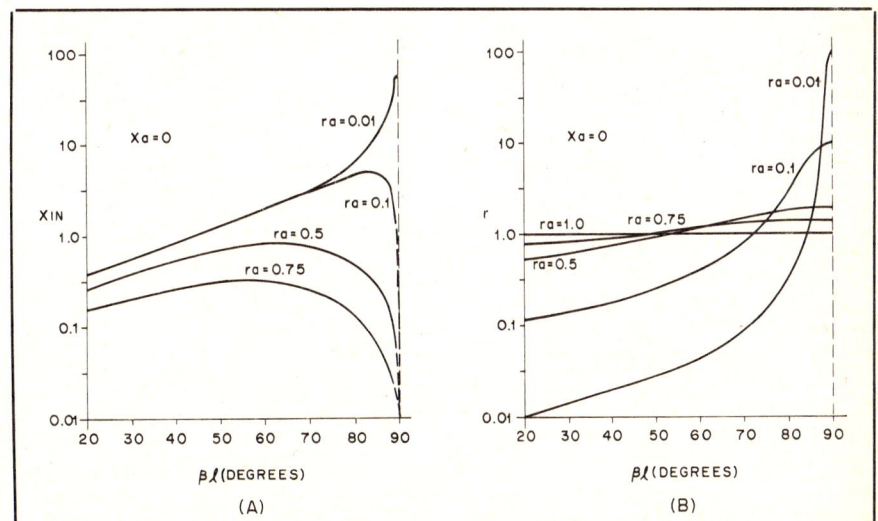

Fig. 3 — Normalized input reactance and resistance vs. line length for various values of r_a ($x_a = 0$).

The special cases in which the terminating reactance is either zero or infinity are given by the respective formulas

$$x_{in} = \tan \beta l$$
$$x_{in} = \cot \beta l$$

A short length of line with a short circuit as a terminating load appears as an inductor while an open-circuited line appears as a capacitance.

Waveguides

A waveguide is a conducting tube through which energy is transmitted in the form of electromagnetic waves. The tube is not considered as carrying a current in the same sense that the wires of a two-conductor line do, but rather as a *boundary* which confines the waves to the enclosed space. Skin effect prevents any electromagnetic effects from being evident outside the guide. The energy is injected at one end, either through capacitive or inductive coupling or by radiation, and is received at the other end. The waveguide then merely confines the energy of the fields, which are propagated through it to the receiving end by means of reflections against its inner walls.

Analysis of waveguide operation is based on the assumption that the guide material is a perfect conductor of electricity. Typical distributions of electric and magnetic fields in a rectangular guide are shown in Fig. 4. It will be observed that the intensity of the electric field is greatest (as indicated by closer spacing of the lines of force) at the center along the x dimension, Fig. 4 (B), diminishing to zero at the end walls. The latter is a necessary condition, since the existence of any electric field parallel to the walls at the surface would cause an infinite current to flow in a perfect conductor. This represents an impossible situation.

Modes of Propagation

Fig. 4 represents a relatively simple distribution of the electric and magnetic fields. There is in general an infinite number of ways in which the fields can arrange themselves in a guide so long as there is no upper limit to the frequency to be transmitted. Each field configuration is called a mode. All modes may be separated into two general groups. One group, designated *TM* (transverse magnetic), has the magnetic field entirely transverse to the direction of propagation, but has a component of electric field in that direction. The other type, designated *TE* (transverse electric) has the electric field entirely transverse, but has a component of magnetic field in the direction of propagation. *TM* waves are sometimes called *E* waves, and *TE* waves are sometimes called *H* waves, but the *TM* and *TE* designations are preferred.

The particular mode of transmission is identified by the group letters followed by two subscript numerals; for example,

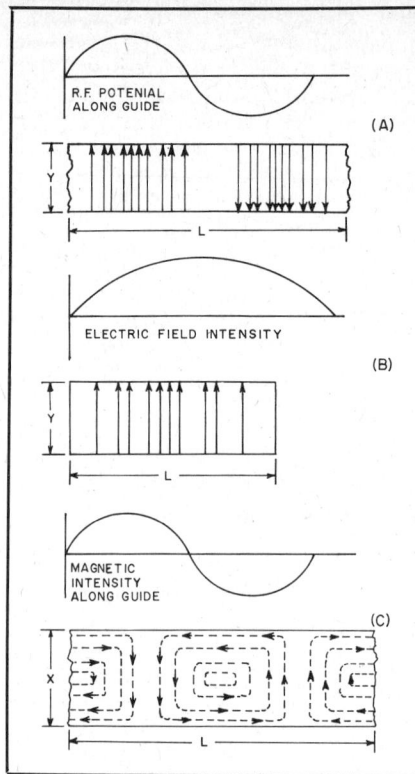

Fig. 4 — Field distribution in a rectangular waveguide. The $TE_{1,0}$ mode of propagation is depicted.

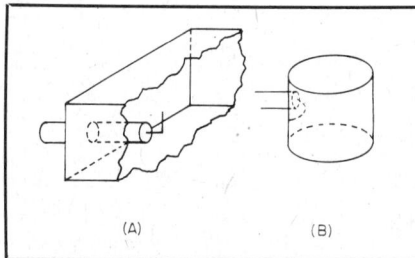

Fig. 5 — Coupling to waveguide and resonators.

$TE_{1,0}$, $TM_{1,1}$, etc. The number of possible modes increases with frequency for a given size of guide. There is only one possible mode (called the dominant mode) for the lowest frequency that can be transmitted. The dominant mode is the one generally used in practical work.

Waveguide Dimensions

In the rectangular guide the critical dimension is x in Fig. 4; this dimension must be more than one-half wavelength at the lowest frequency to be transmitted. In practice, the y dimension usually is made about equal to $1/2\,x$ to avoid the possibility of operation at other than the dominant mode.

Other cross-sectional shapes than the rectangle can be used, the most important being the circular pipe. Much the same considerations apply as in the rectangular case.

Wavelength formulas for rectangular and circular guides are given in the following table, where x is the width of a rectangular guide and r is the radius of a circular guide. All figures are in terms of the dominant mode.

	Rectangular	Circular
Cut-off wavelength	2x	3.41r
Longest wavelength transmitted with little attenuation	1.6x	3.2r
Shortest wavelength before next mode becomes possible	1.1x	2.8r

Coupling to Waveguides

Energy may be introduced into or extracted from a waveguide or resonator by means of either the electric or magnetic field. The energy transfer frequently is through a coaxial line, two methods for coupling to which are shown in Fig. 5. The probe shown at A is simply a short extension of the inner conductor of the coaxial line, so oriented that it is parallel to the electric lines of force. The loop shown at B is arranged so that it encloses some of the magnetic lines of force. The point at which maximum coupling will be secured depends upon the particular mode of propagation in the guide or cavity; the coupling will be maximum when the coupling device is in the most intense field.

Coupling can be varied by turning the probe or loop through a 90-degree angle. When the probe is perpendicular to the electric lines the coupling will be minimum; similarly, when the plane of the loop is parallel to the magnetic lines the coupling will have its minimum value.

Evolution of a Waveguide

Suppose an open-wire line is used to convey rf energy from a generator to a load. If the line has any appreciable length it must be supported mechanically. The line must be well insulated from the supports if high losses are to be avoided. Since high-quality insulators are difficult to realize at microwave frequencies, the logical alternative is to support the transmission line with quarter-wavelength stubs, shorted at the far end. The open end of such a stub presents an infinite impedance to the transmission line, provided the shorted stub is non-reactive. However, the shorting link has finite length and, therefore, some inductance. This inductance can be nullified by making the rf current flow on the surface of a plate rather than a thin wire. If the plate is large enough, it will prevent the magnetic lines of force from encircling the rf current.

Infinitely many of these quarter-wave stubs may be connected in parallel without affecting the standing waves of voltage and current. The transmission line may be supported from the top as well as the bottom, and when infinitely many supports are added, they form the walls of a waveguide at its cutoff frequency. Fig. 6 illustrates how a rectangular waveguide

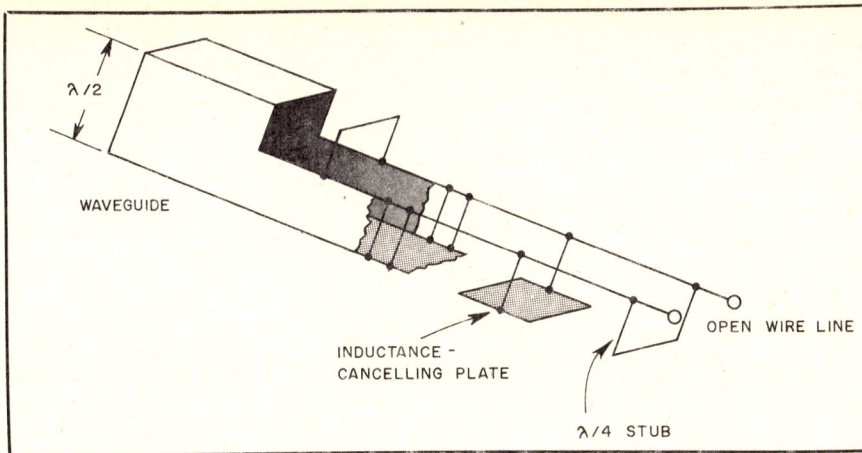

Fig. 6 — At its cutoff frequency a rectangular waveguide can be analyzed as a parallel two-conductor transmission line supported from top and bottom by infinitely many quarter-wavelength stubs.

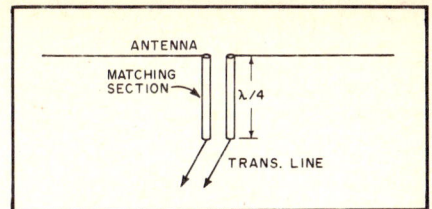

Fig. 7 — "Q" matching section, a quarter-wave impedance transformer.

$$Z = \sqrt{Z_1 Z_o}$$

where

Z_1 = antenna impedance
Z_o = characteristic impedance of the line to which it is to be matched.

Example: To match a 600-ohm line to an antenna presenting a 72-ohm load, the quarter-wave matching section would require a characteristic impedance of

$$\sqrt{72 \times 600} = \sqrt{43,200} = 208 \text{ ohms}$$

The spacing between conductors and the conductor size determines the characteristic impedance of the transmission line. As an example, for the 208-ohm transmission line required above, the line could be made from 1/2-inch (13-mm) diameter tubing spaced 1.5 inches (38-mm) between conductors.

The length of the quarter-wave matching section may be calculated from

$$\text{Length (feet)} = \frac{246V}{f}$$

$$\text{Length (meters)} = \frac{75V}{f}$$

where

V = velocity factor
f = frequency in MHz

Example: A quarter-wave transformer of RG-11/U is to be used at 28.7 MHz. From Table 1 of this chapter, V = 0.66.

$$\text{Length} = \frac{246 \times 0.66}{28.7} = 5.65 \text{ feet}$$

$$= 5 \text{ feet } 8 \text{ inches (1.72 m)}$$

The antenna must be resonant at the operating frequency. Setting the antenna length by formula is amply accurate with single-wire antennas, but in other systems, particularly close-spaced arrays, the antenna should be adjusted to resonance before the matching section is connected.

When the antenna input impedance is not known accurately, it is advisable to construct the matching section so that the spacing between conductors can be changed. The spacing then may be adjusted to give the lowest possible SWR on the transmission line.

Folded Dipoles

A half-wave antenna element can be made to match various line impedances if it is split into two or more parallel conductors with the transmission line attached at

---

evolves from a two-wire parallel transmission line. This simplified analysis also shows why the cutoff dimension is a half wavelength.

While the operation of waveguides is usually described in terms of fields, current flows on the inside walls, just as fields exist between the conductors of a two-wire transmission line. At the waveguide cutoff frequency, the current is concentrated in the center of the walls, and disperses toward the floor and ceiling as the frequency increases.

Matching the Antenna to the Line

The load for a transmission line may be any device capable of dissipating rf energy. When lines are used for transmitting applications the most common type of load is an antenna. When a transmission line is connected between an antenna and a receiver, the receiver input circuit (not the antenna) is the load, because the power taken from a passing wave is delivered to the receiver.

Whatever the application, the conditions existing at the load, and *only* the load, determine the standing-wave ratio on the line. If the load is purely resistive and equal in value to the characteristic impedance of the line, there will be no standing waves. In case the load is not purely resistive, and/or is not equal to the line Z_o, there will be standing waves. No adjustments that can be made at the input end of the line can change the SWR, nor is it affected by changing the line length.

Only in a few special cases is the load inherently of the proper value to match a practicable transmission line. In all other cases it is necessary either to operate with a mismatch and accept the SWR that results, or else to take steps to bring about a proper match between the line and load by means of transformers or similar devices. Impedance-matching transformers may take a variety of physical forms, depending on the circumstances.

Note that it is essential, if the SWR is to be made as low as possible, that the load at the point of connection to the transmission line be purely resistive. In general, this requires that the load be tuned to resonance. If the load itself is not resonant at the operating frequency the tuning sometimes can be accomplished in the matching system.

The Antenna as a Load

Every antenna system, no matter what its physical form, will have a definite value of impedance at the point where the line is to be connected. The problem is to transform this *antenna input impedance* to the proper value to match the line. In this respect there is no one "best" type of line for a particular antenna system, because it is possible to transform impedances in any desired ratio. Consequently, any type of line may be used with any type of antenna. There are frequently reasons other than impedance matching that dictate the use of one type of line in preference to another, such as ease of installation, inherent loss in the line, and so on, but these are not considered in this section.

Although the input impedance of an antenna system is seldom known very accurately, it is often possible to make a reasonably close estimate of its value.

Matching circuits can be built using ordinary coils and capacitors, but are not used very extensively because they must be supported at the antenna and must be weatherproofed. The systems to be described use *linear transformers*.

The Quarter-Wave Transformer or "Q" Section

As mentioned previously, a quarter-wave transmission line may be used as an impedance transformer. Knowing the antenna impedance and the characteristic impedance of the line to be matched, the characteristic impedance of a matching section such as is shown in Fig. 7 is

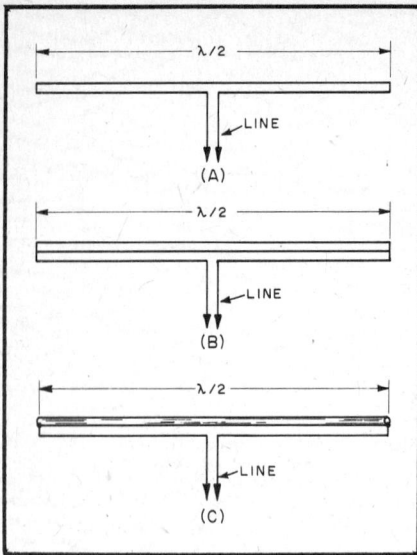

Fig. 8 — The folded dipole, a method for using the antenna element itself to provide an impedance transformation.

the center of only one of them. Various forms of such "folded dipoles" are shown in Fig. 8. Currents in all conductors are in phase in a folded dipole, and since the conductor spacing is small the folded dipole is equivalent in radiating properties to an ordinary single-conductor dipole. However, the current flowing into the input terminals of the antenna from the line is the current in one conductor only, and the entire power from the line is delivered at this value of current. This is equivalent to saying that the input impedance of the antenna has been raised by splitting it up into two or more conductors.

The ratio by which the input impedance of the antenna is stepped up depends not only on the number of conductors in the folded dipole but also on their relative diameters, since the distribution of current between conductors is a function of their diameters. (When one conductor is larger than the other, as in Fig. 8, the larger one carries the greater current.) The ratio also depends, in general, on the spacing between the conductors, as shown by the graphs of Figs. 9 and 10. An important special case is the two-conductor dipole with conductors of equal diameter; as a simple antenna, not a part of a directive array, it has an input impedance close enough to 300 ohms to afford a good match to 300-ohm twin-lead.

The required ratio of conductor diameters to give a desired impedance ratio using two conductors may be obtained from Fig. 9. Similar information for a three-conductor dipole is given in Fig. 10. This graph applies if all three conductors are in the same plane. The two conductors not connected to the transmission line must be equally spaced from the fed conductor, and must have equal diameters. The fed conductor may have a different diameter, however. The unequal-conductor method has been

found particularly useful in matching to low-impedance antennas such as directive arrays using close-spaced parasitic elements.

The length of the antenna element should be such as to be approximately self-resonant at the median operating frequency. The length is usually not highly critical, because a folded dipole tends to have the characteristics of a "thick" antenna and thus has a relatively broad frequency-response curve.

"T" and "Gamma Matching" Sections

The method of matching shown in Fig. 11 is based on the fact that the impedance between any two points along a resonant antenna is resistive, and has a value which depends on the spacing between the two points. It is therefore possible to choose a pair of points between which the impedance will have the right value to match a transmission line. In practice, the line cannot be connected directly at these points because the distance between them is much greater than the conductor spacing of a practical transmission line. The T arrangement in Fig. 11 overcomes this difficulty by using a second conductor paralleling the antenna to form a matching section to which the line may be connected.

The T is particularly suited to use with a parallel-conductor line, in which case the two points along the antenna should be equidistant from the center so that electrical balance is maintained. The operation of this system is somewhat complex. Each T conductor (y in the drawing) forms with the antenna conductor opposite it a short section of transmission line. Each of these transmission-line sections can be considered to be terminated in the impedance that exists at the point of connection to the antenna. Thus the part of the antenna between the two points carries a transmission-line current in addition to the normal antenna current. The two transmission-line matching sections are in series, as seen by the main transmission line.

If the antenna by itself is resonant at the operating frequency, its impedance will be purely resistive. In such case the matching-section lines are terminated in a resistive load. However, since these sections are shorter than a quarter wavelength, their input impedance — i.e., the impedance seen by the main transmission line looking into the matching-section terminals — will be reactive as well as resistive. This prevents a perfect match to the main transmission line, since its load must be a pure resistance for perfect matching. The reactive component of the input impedance must be tuned out before a proper match can be secured.

One way to do this is to detune the antenna just enough, by changing its length, to cause reactance of the opposite kind to be reflected to the input terminals

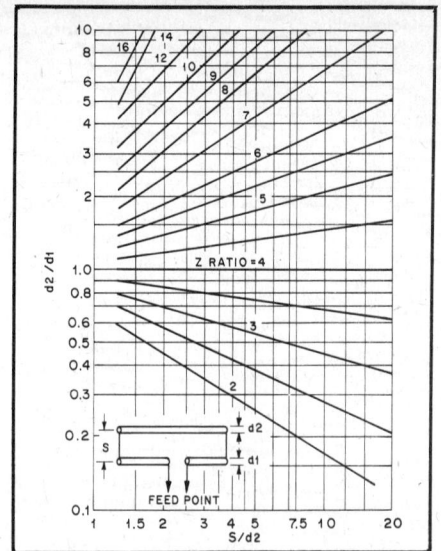

Fig. 9 — Impedance transformation ratio, two-conductor folded dipole. The dimensions d1, d2 and s are shown on the inset drawing. Curves show the ratio of the impedance (resistive) seen by the transmission line to the radiation resistance of the resonant antenna system.

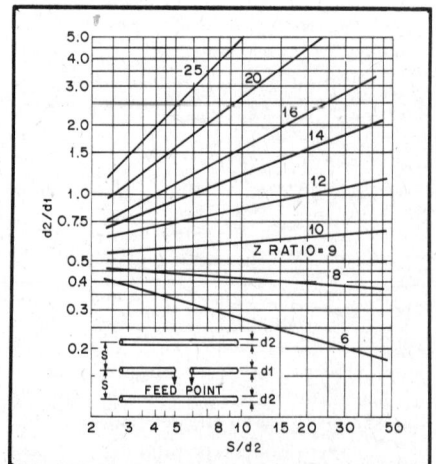

Fig. 10 — Impedance transformation ratio, three-conductor folded dipole. The dimensions d1, d2 and s are shown on the inset drawing. Curves show the ratio of the impedance (resistive) seen by the transmission line to the radiation resistance of the resonant antenna system.

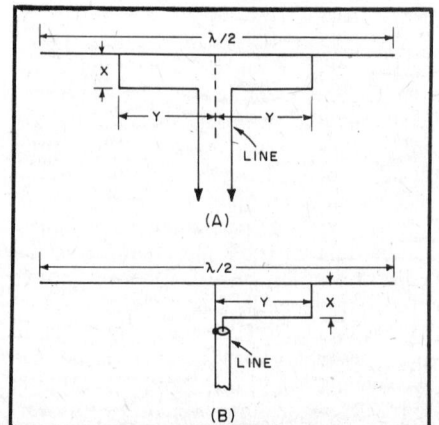

Fig. 11 — The "T" match and "gamma" match.

of the matching section, thus cancelling the reactance introduced by the latter. Another method, which is considerably easier to adjust, is to insert a variable capacitor in series with the matching section where it connects to the transmission line, as shown in chapter 20. The capacitor must be protected from the weather.

The method of adjustment commonly used is to cut the antenna for approximate resonance and then make the spacing x some value that is convenient constructionally. The distance y is then adjusted, while maintaining symmetry with respect to the center, until the SWR on the transmission line is as low as possible. If the SWR is not below 2:1 after this adjustment, the antenna length should be changed slightly and the matching section taps adjusted again. This procedure may be continued until the SWR is as close to 1:1 as possible.

When the series-capacitor method of reactance compensation is used, the antenna should be the proper length to be resonant at the operating frequency. Trial positions of the matching-section taps are then taken, each time adjusting the capacitor for minimum SWR, until the standing waves on the transmission line are brought down to the lowest possible value.

The unbalanced ("gamma") arrangement in Fig. 11 is similar in principle to the T, but is adapted for use with single coax line. The method of adjustment is the same.

Balancing Devices

An antenna with open ends, of which the half-wave type is an example, is inherently a balanced radiator. When opened at the center and fed with a parallel-conductor line, this balance is maintained throughout the system, so long as the causes of unbalance discussed in this chapter are avoided.

If the antenna is fed at the center through a coaxial line, as indicated in Fig. 12, this balance is upset because one side of the radiator is connected to the shield while the other is connected to the inner conductor. On the side connected to the shield, a current can flow down over the *outside* of the coaxial line. The fields thus set up cannot be canceled by the fields from the inner conductor because the fields *inside* the line cannot escape through the shielding afforded by the outer conductor. Hence these "antenna" currents flowing on the outside of the line will be responsible for radiation.

Linear Baluns

Line radiation can be prevented by a number of devices whose purpose is to detune or decouple the line for "antenna" currents and thus greatly reduce their amplitude. Such devices generally are known as *baluns* (a contraction for

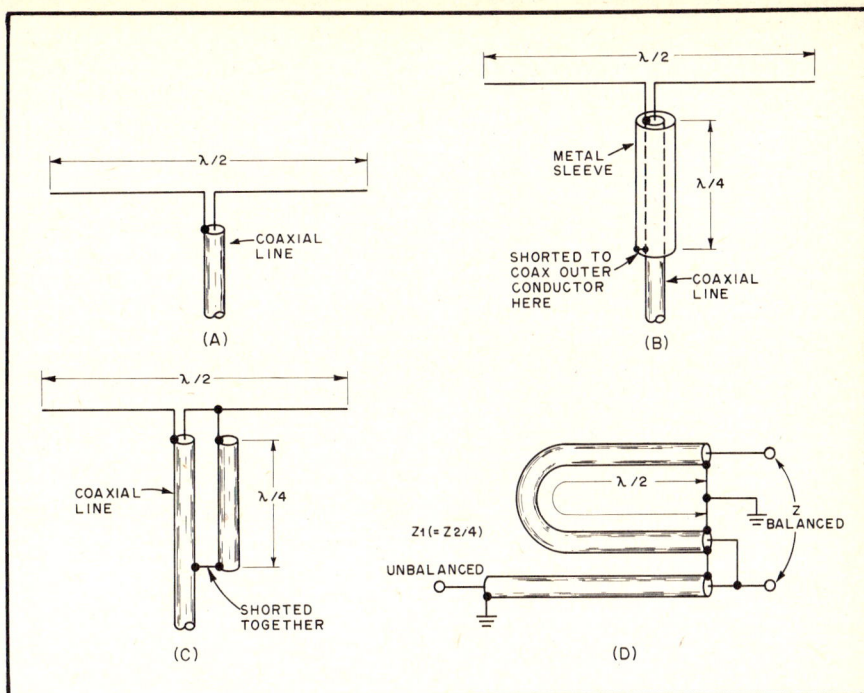

Fig. 12 — Radiator with coaxial feed (A) and methods of preventing unbalanced currents from flowing on the outside of the transmission line (B and C). The half-wave phasing section shown at D is used for coupling between an unbalanced and a balanced circuit when a 4:1 impedance ratio is desired or can be accepted.

"balanced to unbalanced"). Fig. 12B shows once such arrangement, known as a *bazooka*, which uses a sleeve over the transmission line to form, with the outside of the outer line conductor, a shorted quarter-wave line section. As described earlier in this chapter, the impedance looking into the open end of such a section is very high, so that the end of the outer conductor of the coaxial line is effectively isolated from the part of the line below the sleeve. The length is an *electrical* quarter wave, and may be physically shorter if the insulation between the sleeve and the line is other than air. The bazooka has no effect on the impedance relationships between the antenna and the coaxial line.

Another method that gives an equivalent effect is shown at C. Since the voltages at the antenna terminals are equal and opposite (with reference to ground), equal and opposite currents flow on the surfaces of the line and second conductor. Beyond the shorting point, in the direction of the transmitter, these currents combine to cancel out. The balancing section "looks like" an open circuit to the antenna, since it is a quarter-wave parallel-conductor line shorted at the far end, and thus has no effect on the normal antenna operation. However, this is not essential to the line-balancing function of the device, and baluns of this type are sometimes made shorter than a quarter wavelength to provide the shunt inductive reactance required in certain matching systems.

Fig. 12D shows a third balun, in which equal and opposite voltages, balanced to ground, are taken from the inner conductors of the main transmission line and half-wave phasing section. Since the voltages at the balanced end are in series while the voltages at the unbalanced end are in parallel, there is a 4:1 step-down in impedance from the balanced to the unbalanced side. This arrangement is useful for coupling between a 300-ohm balanced line and a 75-ohm coaxial line, for example.

Other Loads and Balancing Devices

The most important practical load for a transmission line is an antenna which in most cases, will be "balanced" — that is, symmetrically constructed with respect to the feed point. Aside from considerations of matching the actual impedance of the antenna at the feed point to the characteristic impedance of the line (if such matching is attempted) a balanced antenna should be fed through a balanced transmission line in order to preserve symmetry with respect to ground. This will avoid difficulties with unbalanced currents on the line and consequent undesirable radiation from the transmission line itself.

If, as is often the case, the antenna is to be fed through coaxial line, (which is inherently unbalanced) some method should be used for connecting the line to the antenna without upsetting the symmetry of the antenna itself. This requires a circuit that will isolate the balanced load from the unbalanced line while providing efficient power transfer. Devices for doing this are called *baluns*. The types used between the antenna and transmission line are generally linear, consisting of transmission-line sections.

The need for baluns also arises in coupling a transmitter to a balanced transmission line, since the output circuits of most transmitters have one side grounded. (This type of output circuit is desirable for a number of reasons, including TVI reduction.) The most flexible type of balun for this purpose is the inductively coupled matching network described in a subsequent section in this chapter. This combines impedance matching with balanced-to-unbalanced operation, but has the disadvantage that it uses resonant circuits and thus can work over only a limited band of frequencies without readjustment. However, if a fixed impedance ratio in the balun can be tolerated, the coil balun described below can be used without adjustment over a frequency range of about 10:1 — 3 to 30 MHz, for example.

Coil Baluns

The type of balun known as the "coil balun" is based on the principles of linear-transmission-line balun as shown in the upper drawing of Fig. 13. Two transmission lines of equal length having a characteristic impedance (Z_o) are connected in series at one end and in parallel at the other. At the series-connected end the lines are balanced to ground and will match an impedance equal to $2Z_o$. At the parallel-connected end the lines will be matched by an impedance equal to $Z_o/2$. One side may be connected to ground at the parallel-connected end, provided the two lines have a length such that, considering each line as a single wire, the balanced end is effectively decoupled from the parallel-connected end. This requires a length that is an odd multiple of π wavelength.

A definite line length is required only for decoupling purposes, and so long as there is adequate decoupling the system will act as a 4:1 impedance transformer regardless of line length. If each line is wound into a coil, as in the lower drawing, the inductances so formed will act as choke coils and will tend to isolate the series-connected end from any ground connection that may be placed on the parallel-connected end. Balun coils made in this way will operate over a wide frequency range, since the choke inductance is not critical. The lower frequency limit is where the coils are no longer effective in isolating one from the other; the length of line in each coil should be about equal to a quarter wave-length at the lowest frequency to be used.

The principal application of such coils is in going from a 300-ohm balanced line to a 75-ohm coaxial line. This requires

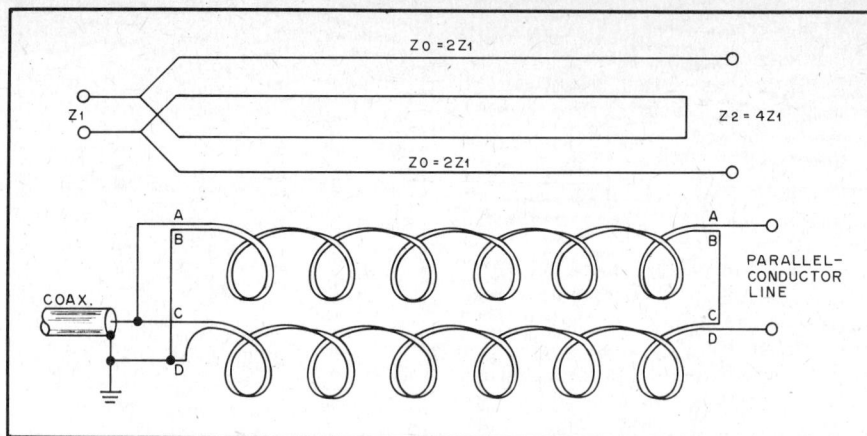

Fig. 13 — Baluns for matching between push-pull and single-ended circuits. The impedance ratio is 4:1 from the push-pull side to the unbalanced side. Coiling the lines (lower drawing) increases the frequency range over which satisfactory operation is obtained.

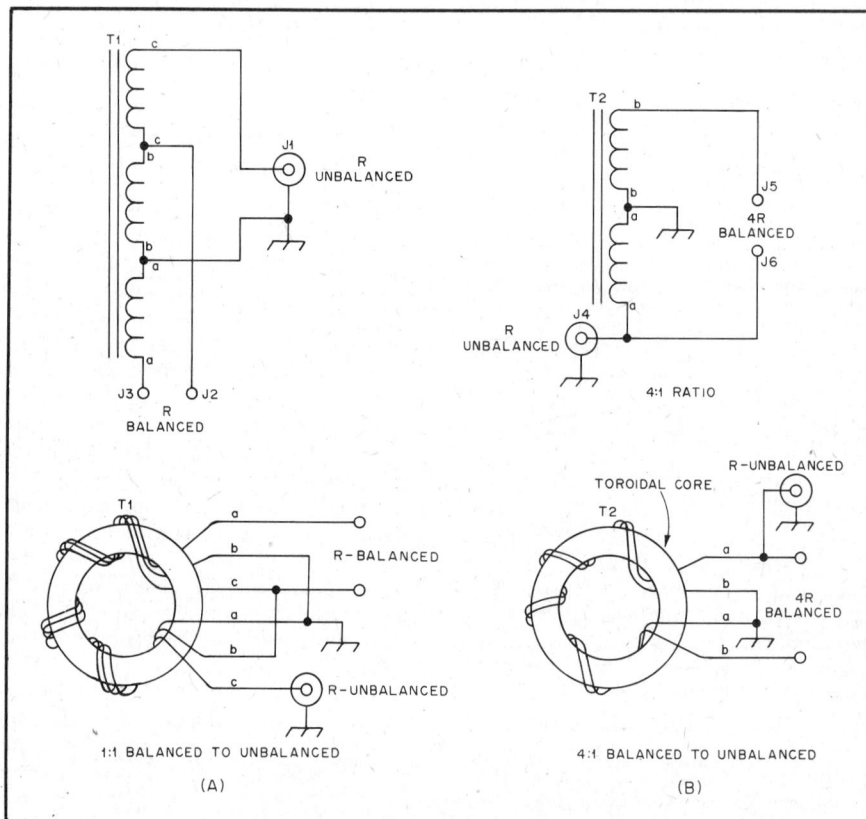

Fig. 14 — Schematic and pictorial representations of the balun transformers. T1 and T2 are wound on CF-123 toroid cores (see footnote 4 and the text). J1 and J4 are SO-239-type coax connectors, or similar. J2, J3, J5, and J6 are steatite feedthrough bushings. The windings are labeled a, b and c to show the relationship between the pictorial and schematic illustrations.

that the Z_o of the lines forming the coils be 150 ohms.

A balun of this type is simply a fixed-ratio transformer, when matched. It cannot compensate for inaccurate matching elsewhere in the system. With a "300-ohm" line on the balanced end, for example, a 75-ohm coax cable will not be matched unless the 300-ohm line actually is terminated in a 300-ohm load.

Two Broadband Toroidal Baluns

Air-wound balun transformers are some-

what bulky when designed for operation in the 1.8- to 30-MHz range. A more compact broadband transformer can be realized by using toroidal ferrite core material as the foundation for bifilar-wound coil balun transformers. Two such baluns are described here.

In Fig. 14 at A, a 1:1 ratio balanced-to-unbalanced-line transformer is shown. This transformer is useful in converting a 50-ohm balanced line condition to one that is 50 ohms, unbalanced. Similarly, the transformer will work between

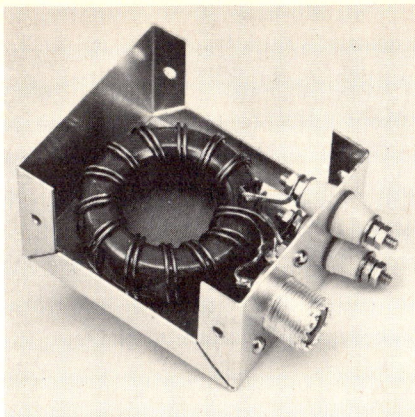

Fig. 15 — Layout of a kilowatt 4:1 toroidal balun transformer. Phenolic insulating board is mounted between the transformer and the Minibox wall to prevent short-circuiting. The board is held in place with epoxy cement. Cement is also used to secure the transformer to the board. For outdoor use, the Minibox cover can be installed, then sealed against the weather by applying epoxy cement along the seams of the box.

balanced and unbalanced 75-ohm impedances. A 4:1 ratio transformer is illustrated in Fig. 14 at B. This balun is useful for converting a 200-ohm balanced condition to one that is 50 ohms, unbalanced. In a like manner, the transformer can be used between a balanced 300-ohm point and a 75-ohm unbalanced line. Both balun transformers will handle 1000 watts of rf power and are designed to operate from 1.8 through 60 MHz.

Low-loss high-frequency ferrite core material is used for T1 and T2. [1,2] The cores are made from Q-2 material and cost approximately $5.50 in single-lot quantity. They are 0.5 inches (13-mm) thick, have an OD of 2.4 inches (61-mm) and the ID is 1.4 inches (36-mm) The permeability rating of the cores is 40. A packaged 1-kilowatt balun kit, with winding instructions for 1:1 or 4:1 impedance transformation ratios, is available, but uses a core of slightly different dimensions. [3]

Winding Information

The transformer shown in Fig. 14 at A has a trifilar winding consisting of 10 turns of No. 14 Formvar-insulated copper wire. A 10-turn bifilar winding of the same type of wire is used for the balun of Fig. 14 at B. If the cores have rough edges, they should be carefully sanded until smooth enough to prevent damage to the wire's Formvar insulation. The windings should be spaced around the entire core as shown in Fig. 15. Insulation

[1] Available in single-lot quantity from Permag Corp., 88-06 Van Wyck Expy., Jamaica, NY 11418

[2] Toroid cores are also available from Ferroxcube Corp. of America, 5083 Kings Hwy., Saugerties, NY 12477.

[3] Amidon Associates, 12033 Otsego St., North Hollywood, CA 91601.

can be used between the core material and the windings to increase the breakdown voltage of the balun.

A 50- to 75-Ohm Broadband Transformer

Shown in Figs. 16 through 18 is a simple 50- to 75-ohm or 75- to 50-ohm transformer that is suitable for operation in the 2- to 30-MHz frequency range. A pair of these transformers is ideal for using 75-ohm CATV hardline in a 50-ohm system. In this application one transformer is used at each end of the cable run. At the antenna one transformer raises the 50-ohm impedance of the antenna to 75 ohms, thereby presenting a match to the 75-ohm cable. At the station end a transformer is used to step the 75-ohm line impedance down to 50 ohms.

The schematic diagram of the transformer is shown in Fig. 16, and the winding details are given in Fig. 17. C1 and C2 are compensating capacitors; the values shown were determined through swept return-loss measurements using a spectrum analyzer and tracking generator. The transformer consists of a trifilar winding of no. 14 enameled copper wire wound over an FT-200-61 (Q1 material) or equivalent core. As shown in Fig. 17, one winding has only half the number of turns of the other two. Care must be taken when connecting the loose ends so that the proper phasing of the turns is maintained. Improper phasing will become apparent when power is applied to the transformer.

If the core has sharp edges it is a good idea either to sand the edges until they are relatively smooth or wrap the core with tape. The one shown in the photograph was wrapped with ordinary vinyl electrical tape, although glass-cloth insulating tape would be better. The idea is to prevent chafing of the wire insulation.

Construction

The easiest way to construct the transformer is wind the three lengths of wire on the core at the same time. Different color wires will aid in identifying the ends of the windings. After all three windings are securely in place, the appropriate winding may be unwound three turns as shown in the diagram. This wire is the 75-ohm connection point. Connections at the 50-ohm end are a bit tricky, but if the information in Fig. 17 is followed carefully no problems should be encountered. Use the shortest connections possible, as long leads will degrade the high-frequency performance.

The balun is housed in a homemade aluminum enclosure measuring 3-1/2 × 3-3/4 × 1-1/4 inches (89 × 95 × 32 mm). Any commercial cabinet of similar dimensions will work fine. In the unit shown in the photograph, several "blobs" of silicone seal (RTV) were used to hold the core in position. Alternatively, a piece of phenolic insulating material may be used between the core and the aluminum

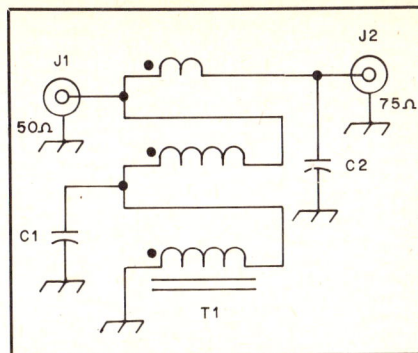

Fig. 16 — Schematic diagram of the 50- to 75-ohm transformer described in the text. C1 and C2 are compensating capacitors.
C1 — 100 pF, silver mica.
C2 — 10 pF, silver mica.
J1, J2 — Coaxial connectors, builder's choice.
T1 — Transformer, 6 trifilar turns no. 14 enameled copper wire on an FT-200-61 (Q1 material, $\mu_i = 125$) core. One winding has one-half the number of turns of the other two.

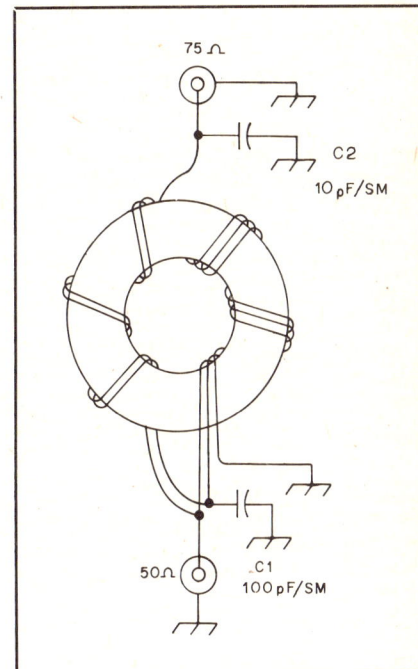

Fig. 17 — Pictorial drawing of the 50- to 75-ohm transformer showing details of the windings.

Fig. 18 — This is a photograph of a pair of the 50- to 75-ohm transformers. The units are identical.

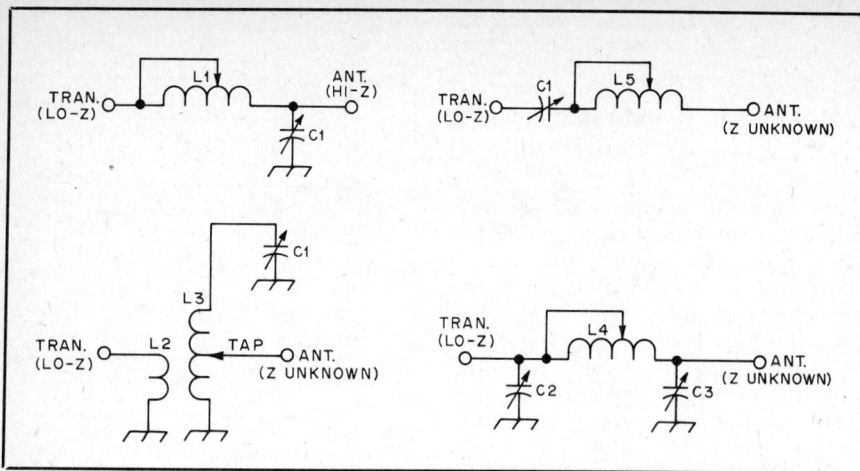

Fig. 19 — Networks for matching a low-Z transmitter output to random-length end-fed wire antennas.

Checkout

Checkout of the completed transformer or transformers is quite simple. If a 75-ohm dummy load is available connect it to the 75-ohm terminal of the transformer. Connect a transmitter and VSWR indicator (50 ohm) to the 50-ohm terminal of the transformer. Apply power (on each of the hf bands) and measure the VSWR looking into the transformer. Readings should be well under 1.3 to 1 on each of the bands. If a 75-ohm load is not available and two transformers have been constructed they may be checked out simultaneously as follows. Connect the 75-ohm terminals of both transformers together, either directly through a coaxial adaptor or through a length of 75-ohm cable. Attach a 50-ohm load to one of the 50-ohm terminals and connect a transmitter and VSWR indicator (50 ohm) to the remaining 50-ohm terminal. Apply power as outlined above and record the measurements. Readings should be under 1.3 to 1.

The transformers were checked in the ARRL laboratory under various mismatched conditions at the 1500-watt power level. No spurious signals (indicative of core saturation) could be found while viewing the lf, hf and vhf frequency range with a spectrum analyzer. A key-down, 1500-watt signal produced no noticeable core heating and only a slight increase in the temperature of the windings.

Using the Transformers

For indoor applications, the transformers can be assembled open style, without benefit of a protective enclosure. For outdoor installations, such as at the antenna feed point, the balun should be encapsulated in epoxy resin or mounted in a suitable weatherproof enclosure. A Minibox, sealed against moisture, works nicely for the latter.

Nonradiating Loads

Typical examples of nonradiating loads for a transmission line are the grid circuit of a power amplifier (considered in the chapter on transmitters), the input circuit of a receiver, and another transmission line. This last case includes the "antenna tuner" — a misnomer because it is actually a device for coupling a transmission line to the transmitter. Because of its importance in amateur installations, the antenna coupler is considered separately in a later part of this chapter.

Coupling to a Receiver

A good match between an antenna and its transmission line does not guarantee a low standing-wave ratio on the line when the antenna system is used for receiving. The SWR is determined wholly by what the line "sees" at the receiver's antenna-input terminals. For minimum SWR the receiver input circuit must be matched to the line. The rated input impedance of a receiver is a nominal value that varies over a considerable range with frequency. Most hf receivers are sensitive enough that exact matching is not necessary. The most desirable condition is when the receiver is matched to the Z_o and the line in turn is matched to the antenna. This transfers maximum power from the antenna to the receiver with the least transmission line loss.

Coupling to Random-Length Antennas

Several impedance-matching schemes are shown in Fig. 19, permitting random-length wires to be matched to normal low-Z transmitter outputs. The circuit used will depend upon the length of the antenna wire and its impedance at the desired operating frequency. Ordinarily, one of the four methods shown will provide a suitable impedance match to an end-fed random wire, but the configuration will have to be determined experimentally. For

operation between 3.5 and 30 MHz, C1 can be a 200-pF type with suitable plate spacing for the power level in use. C2 and C3 should be 500-pF units to allow for flexibility in matching. L1, L4 and L5 should be tapped or rotary inductors with sufficient L for the operating frequency. L3 can be tapped Miniductor coil with ample turns for the band being used. An SWR bridge should be used as a match indicator.

Coupling the Transmitter to the Line

The type of coupling system that will be needed to transfer power adequately from the final rf amplifier to the transmission line depends almost entirely on the input impedance of the line. As shown earlier in this chapter, the input impedance is determined by the standing-wave ratio and the line length. The simplest case is that where the line is terminated in its characteristic impedance so that the SWR is 1:1 and the input impedance is equal to the Z_o of the line, regardless of line length.

Coupling systems that will deliver power into a flat line are readily designed. For all practical purposes the line can be considered to be flat if the SWR is no greater than about 1.5:1. That is, a coupling system designed to work into a pure resistance equal to the line Z_o will have enough leeway to take care of the small variations in input impedance that will occur when the line length is changed, if the SWR is higher than 1:1 but no greater than 1.5:1.

Current practice in transmitter design is to provide an output circuit that will work into such a line, usually a coaxial line of 50 to 75 ohms characteristic impedance. The design of such output circuits is discussed in the chapter on high-frequency transmitters. If the input impedance of the transmission line that is to be connected to the transmitter differs appreciably from the impedance value that the transmitter output circuit is designed to operate, an impedance-matching network must be inserted between the transmitter and the line input terminals.

Impedance-Matching Circuits for Transmission Lines

As shown earlier in this chapter, the input impedance of a line that is operating with a high standing-wave ratio can vary over quite wide limits. The simplest type of circuit that will match such a range of impedances to 50 to 75 ohms is a simple series- or parallel-tuned circuit, approximately resonant at the operating frequency. If the load presented by the line at the operating frequency is low (below a few hundred ohms), a series-tuned circuit should be used. When the load is higher than this, the parallel-tuned circuit is easier to use.

Typical simple circuits for coupling between the transmitter with 50- to 75-ohm coaxial-line output and a balanced transmission line are shown in Fig. 20. The inductor L1 should have a reactance of about 60 ohms when adjustable inductive coupling is used (Figs. 20A and 20B). When a variable series capacitor is used, L1 should have a reactance of about 120 ohms. The variable capacitor, C1, should have a reactance at maximum capacitance of about 100 ohms.

On the secondary side, L_s and C_s should be capable of being tuned to resonance at about 80 percent of the operating frequency. In the series-tuned circuits, for a given low-impedance load, looser coupling can be used between L1 and L_s as the L_s-to-C_s ratio is increased. In the parallel-tuned circuits, for a given high-impedance load looser coupling can be used between L1 and Lp as the C_p-to-L_p ratio is increased. The constants are not critical; the rules of thumb are mentioned to assist in correcting a marginal condition where sufficient transmitter loading cannot be obtained.

Coupling to a coaxial line that has a high SWR, and that consequently may present a transmitter with a load it cannot couple to, is done with an unbalanced version of the series-tuned circuit, as shown in Fig. 21. The rule given above for coupling ease and L_s-to-C_s ratio applies to these circuits as well.

The most satisfactory way to set up initially any of the circuits of Fig. 20 or 21 is to connect a coaxial SWR bridge in the line to the transmitter, as shown in Fig. 21. The "Monimatch" type of bridge, which can handle the full transmitter power and may be left in the line for continuous monitoring, is excellent for this purpose. However, a simple resistance bridge is perfectly adequate, requiring only that the transmitter output be reduced to a very low value so that the bridge will not be overloaded. To adjust the circuit, make a trial setting of the coupling (coil spacing in Figs. 20A and B and 21A, C1 setting in others) and adjust C_s or C_p for minimum SWR as indicated by the bridge. If the SWR is not close to practically 1:1, readjust the coupling and return C_s or C_p, continuing this procedure until the SWR is practically 1:1. The settings may then be logged for future reference.

In the series-tuned circuits of Figs. 20A and 20C, the two capacitors should be set at similar settings. The "$2C_s$" indicates that a balanced series-tuned coupler requires twice the capacitance in each of two capacitors as does an unbalanced series-tuned circuit, all other things being equal.

It is possible to use circuits of this type without initially setting them up with an SWR bridge. In such a case it is a matter of cut-and-try until adequate power transfer between the amplifier and main transmission line is secured. However, this method frequently results in a high SWR in the link, with consequent power loss, "hot spots" in the coaxial cable, and

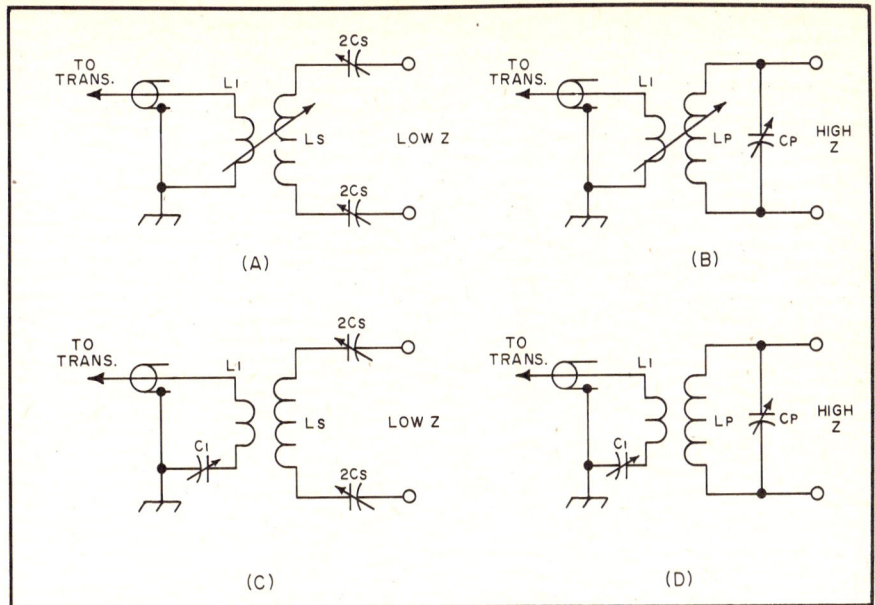

Fig. 20 — Simple circuits for coupling a transmitter to a balanced line that presents a load different than the transmitter output impedance. (A) and (B) respectively are series- and parallel-tuned circuits using variable inductive coupling between coils, and (C) and (D) are similar but use fixed inductive coupling and a variable series capacitor, C1. A series-tuned circuit works well with a low-impedance load; the parallel circuit is better with high-impedance loads (several hundred ohms or more).

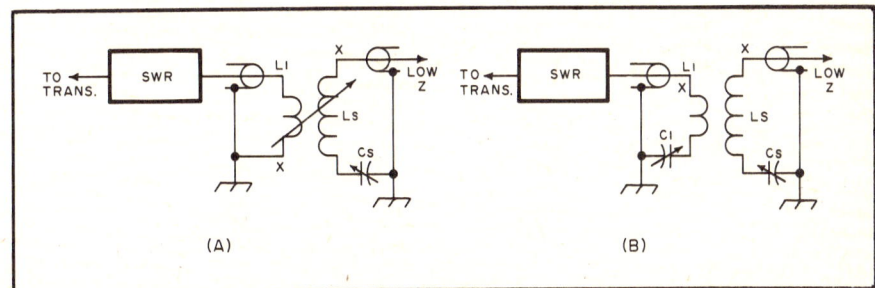

Fig. 21 — Coupling from a transmitter designed for 50- to 75-ohm output to a coaxial line with a 3 or 4:1 SWR is readily accomplished with these circuits. Essential difference between the circuits is (A) adjustable inductive coupling and (B) fixed inductive coupling with variable series capacitor. In either case the circuit can be adjusted to give a 1:1 SWR on the meter in the line to the transmitter. The coil ends marked "x" should be adjacent, for minimum capacitive coupling.

tuning that is critical with frequency. The bridge method is simple and gives the optimum operating conditions quickly and with certainty.

A Transmatch for Balanced or Unbalanced Lines

Most modern transmitters are designed to operate into loads of approximately 50 ohms. Solid-state transmitters produce progressively lower output power as the SWR on the transmission line increases, owing to the built-in SWR protection circuits. Therefore, it is useful to employ a matching network between the transmitter and the antenna feeder when antennas with complex impedances are used. One example of this need can be seen in the case of an 80-meter, coax-fed dipole antenna which has been cut for resonance at, say, 3.6 MHz. If this antenna were used in the 75-meter phone band, the SWR would be fairly high. A Transmatch

could be used to give the transmitter a 50-ohm load, even though a significant mismatch was present at the antenna feed point. It is important to remember that the Transmatch will *not* correct the actual SWR condition; it only conceals it as far as the transmitter is concerned. A Transmatch is useful also when using a single-wire antenna for multiband use. By means of a balun at the Transmatch output it is possible to operate the transmitter into a balanced transmission line, such as a 300- or 600-ohm feed system of the type that would be used with a multiband tuned dipole, V beam or rhombic antenna.

A secondary benefit can be realized from Transmatches of certain varieties: The matching network can, if it has a *bandpass* response, attenuate harmonics from the transmitter. The amount of attenuation is dependent upon the loaded Q (Q_L) of the network after the impedance

Fig. 22 — Exterior view of the SPC Transmatch. Radio Shack vernier drives are used for adjusting the tuning capacitors. A James Millen turns-counter drive is coupled to the rotary inductor. Green paint and green Dymo tape labels are used for panel decor. The cover is plain aluminum with a lightly grooved finish (sandpapered) which has been coated with clear lacquer. An aluminum foot holds the Transmatch at an easy access angle.

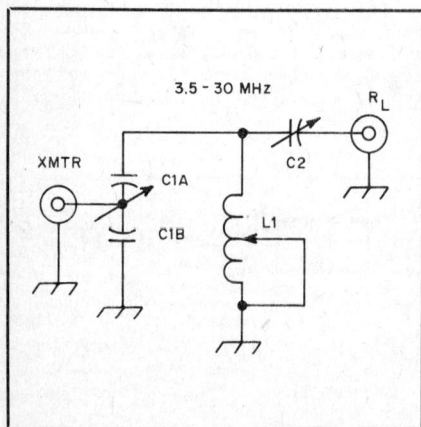

Fig. 23 — Circuit of the Ultimate Transmatch showing the network which can degenerate to a high-pass network under some conditions of transformation (see text).

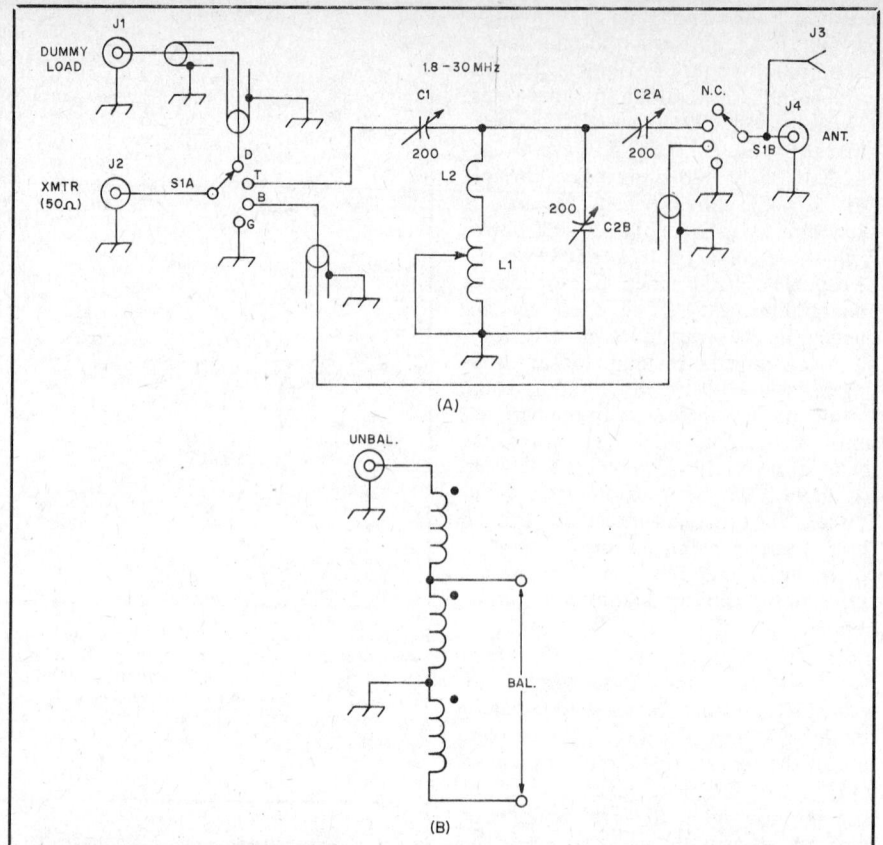

Fig. 24 — Schematic diagram of the SPC circuit. Capacitance is in pF.

C1 — 200-pF transmitting variable with plate spacing of 0.075 inch (2 mm) or greater. J. W. Miller Co. no. 2150 used here.

C2 — Dual-section variable, 200 pF per section. Same plate spacing as C1. J. W. Miller Co. no. 2151 used here. (Catalog no. 79, J. W. Miller Co., 19070 Reyes Ave., Compton, CA 90224.)

J1, J2, J4 — SO-239 style coaxial connector. J4 should have high-dielectric insulation if high-Z single-wire antennas are used at J3. Teflon insulation is recommended.

J3 — Ceramic feedthrough bushing.

L1 — Rotary inductor, 25 μH min. inductance. E. F. Johnson 229-203 or equiv.

L2 — Three turns no. 8 copper wire, 1 inch (25 mm) ID × 1-1/2 inches (38 mm) long.

S1 — Large ceramic rotary wafer switch with heavy contacts. Two-pole, 4-position type. Surplus Centralab JV-9033 or equiv., two positions unused.

Z1 — Balun transformer. 12 turns no. 12 Formvar wire, trifilar, close-wound on 1-inch (25-mm) OD phenolic or PVC-tubing form.

has been matched. The higher the QL the greater the attenuation. Some Transmatches, such as the "Ultimate Transmatch" of Fig. 23, can exhibit a high-pass response (undesirable), depending upon the transformation ratio they are adjusted to accommodate. In a worst-case condition the attenuation of harmonic currents may be as low as 3 to 6 dB. Under different conditions of impedance transformation (better) the attenuation can be as great as 20 to 25 dB.

The "SPC Transmatch" described here was developed to correct for the sometimes poor harmonic attenuation of the network in the Ultimate Transmatch. The SPC (series-parallel capacitance) circuit maintains a bandpass response under load conditions of less than 25 ohms to more than 1000 ohms (from a 50-ohm transmitter). This is because a substantial amount of capacitance is always in parallel with the rotary inductor (C2B and L1 of Fig. 24). In comparison with the "Ultimate" circuit of Fig. 23, it can be

seen that at high load impedances, the Ultimate Transmatch will have minimal effective output capacitance in shunt with the inductor, giving rise to a high-pass response.

Another advantage of the SPC Transmatch is its greater frequency range with the same component values used in the Ultimate Transmatch. The circuit of Fig. 24 operates from 1.8 to 30 MHz with the values shown. Only 3/4 of the available inductance of L1 is needed on 160 meters.

The notable difference in outward performance over the circuit in Fig. 23 is somewhat sharper tuning. This is because of the increased network Q. This is especially prominent at 40, 80 and 160 meters. For this reason there are vernier-drive dials on C1 and C2. They are also useful in logging the dial settings for changing bands or antennas.

Spectrographs of an Ultimate Transmatch and the SPC Transmatch bandpass characteristics are shown in

Fig. 26. The example at A shows the second harmonic down only 14 dB from the fundamental when looking into 1000 ohms with the Ultimate. The display at B of Fig. 26 shows the SPC response at 1000 ohms with the second harmonic down 28 dB from the fundamental energy. The response at A resembles a high-pass characteristic.

Construction

Figs. 22 and 25 show the structural details of the Transmatch. The cabinet is homemade from 16-gauge aluminum sheeting. L brackets are affixed to the right and left sides of the lower part of the cabinet to permit attachment of the U-shaped cover.

The conductors which join the components should be of heavy-gauge material to minimize stray inductance and heating. Wide strips of flashing copper are suitable for the conductor straps. The center conductor and insulation from RG-59/U polyfoam coaxial cable is used in

Fig. 25 — Interior view of the W1FB SPC Transmatch. L2 is mounted on the rear wall by means of two ceramic standoff insulators. C1 is on the left and C2 is at the right. The coaxial connectors, ground post and J3 are on the lower part of the rear panel.

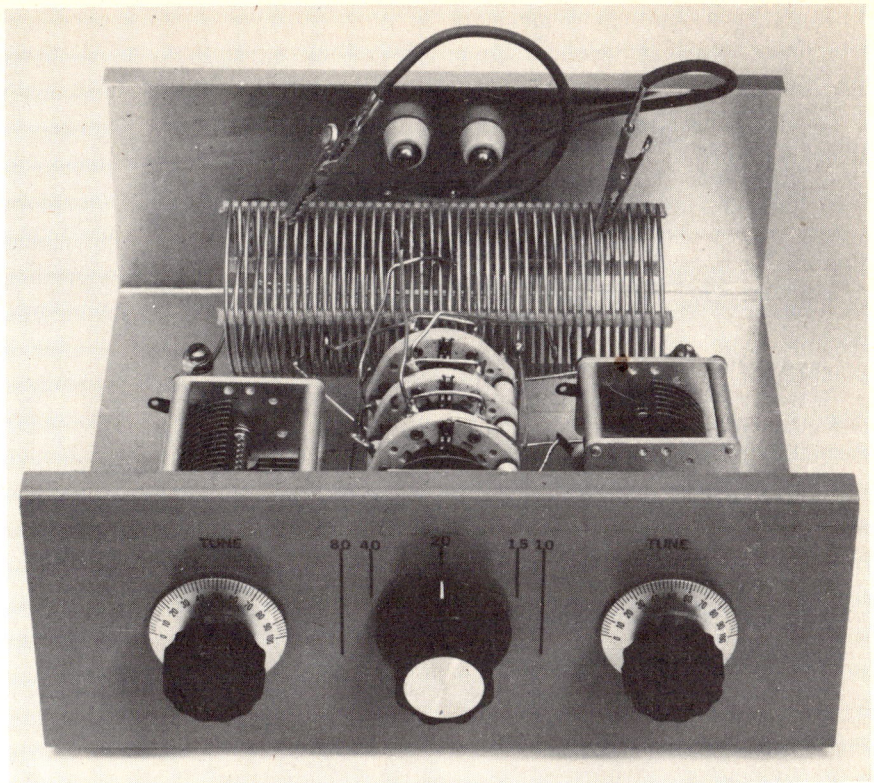

Fig. 27 — Exterior view of the band-switched link coupler. Alligator clips are used to select the proper tap positions of the coil.

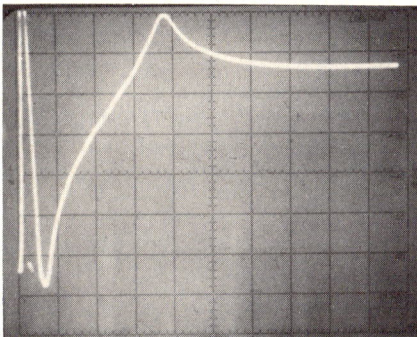

(A)

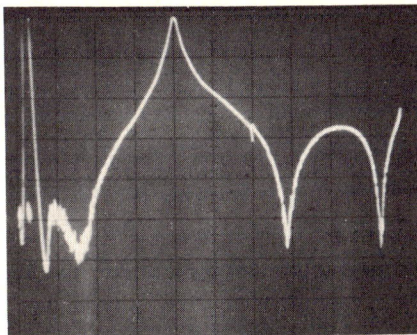

(B)

Fig. 26 — Spectrographs of the response characteristics of the Ultimate Transmatch (A) and the SPC Transmatch (B) looking into a 1000-ohm termination from a 50-ohm signal source. Greater harmonic attenuation is obtained with the SPC Transmatch (see text). The scale divisions are 2 MHz horizontal and 10 dB vertical. The fundamental frequency is 8 MHz.

this model for the wiring between the switch and the related components. The insulation is sufficient to prevent breakdown and arcing at 2 kW PEP input to the transmitter.

All leads should be kept as short as possible to help prevent degradation of the circuit Q. The stators of C1 and C2 should face toward the cabinet cover to minimize the stray capacitance between the capacitor plates and the bottom of the cabinet (important at the upper end of the Transmatch frequency range). Insulated ceramic shaft couplings are used between the vernier drives and C1 and C2, since the rotors of both capacitors are "floating" in this circuit. C1 and C2 are supported above the bottom plate on steatite cone insulators. S1 is attached to the rear apron of the cabinet by means of two metal standoff posts.

Operation

The SPC Transmatch is designed to handle the output from transmitters which operate up to 2 kW PEP. L2 has been added to improve the circuit Q at 10 and 15 meters. However, it may be omitted from the circuit if the rotary inductor (L1) has a tapered pitch at the minimum-inductance end. It may be necessary to omit L2 if the stray wiring inductance of the builder's version is high. Otherwise, it may be impossible to obtain a matched condition at 28 MHz with certain loads.

An SWR indicator is used between the transmitter and the Transmatch to show when a matched condition is achieved. The builder may want to integrate an SWR meter in the Transmatch circuit between J2 and the arm of S1A (Fig.

24A). If this is done there should be room for an edgewise panel meter above the vernier drive for C2.

Initial transmitter tuning should be done with a dummy load connected to J1, and with S1 in the D position. This will prevent interference which could otherwise occur if tuning is done "on the air." After the transmitter is properly tuned into the dummy load, unkey the transmitter and switch S1 to T (Transmatch). Never "hot switch" a Transmatch, as this can damage both transmitter and Transmatch. Set C1 and C2 at midrange. With a few watts of rf, adjust L1 for a decrease in reflected power. Then adjust C1 and C2 alternately for the lowest possible SWR. If the SWR cannot be reduced to 1:1, adjust L1 slightly and repeat the procedure. Finally, increase the transmitter power to maximum and touch up the Transmatch controls if necessary. When tuning, keep your transmissions brief and identify your station.

The air-wound balun of Fig. 24B can be used outboard from the Transmatch if a low-impedance balanced feeder is contemplated. Ferrite or powdered-iron core material is not used in the interest of avoiding TVI and harmonics which can result from core saturation.

The B position of S1 permits switched-through operation when the Transmatch is not needed. The G position is used for grounding the antenna system, as necessary; a quality earth ground should be attached at all times to the Transmatch chassis.

Final Comments

Surplus coils and capacitors are okay in this circuit. L1 should have at least 25 uH of inductance, and the tuning capacitors need to have 150 pF or more of capacitance per section. Insertion loss through this Transmatch was measured at less than 0.5 dB at 600 watts of rf power on 7 MHz.

A Link-Coupled Matching Network

Link coupling offers many advantages over other types of systems where a direct connection between the transmitter and antenna is required. This is particularly true on 80 meters, where commercial broadcast stations often induce sufficient voltage to cause either rectification or front-end overload. Transceivers and receivers that show this tendency can usually be cured by using only magnetic coupling between the transceiver and antenna system. There is no direct connection and better isolation results along with the inherent band-pass characteristics of magnetically coupled tuned circuits.

Although link coupling can be used with either single-ended or balanced antenna systems, its most common application is with balanced feed. The model shown here is designed for 80- through 10-meter operation.

The Circuit

The circuit shown in Fig. 28 and the accompanying photographs is that of a band-switched link coupler. L2 is the link and C1 is used to adjust the coupling. S1b selects the proper amount of link inductance for each band. L1 and L3 are located on each side of the link and are the coils to which the antenna is connected. Alligator clips are used to connect the antenna to the coil because antennas of different impedances must be connected at different points (taps) along the coil. Also, with many antennas it will be necessary to change taps for different bands of operation. C2 tunes L1 and L3 to resonance at the operating frequency.

Switch sections S1a and S1b select the amount of inductance necessary for each of the hf bands. The inductance of each of the coils has been optimized for antennas in the impedance range of roughly 20 - 600 ohms. Antennas that exhibit impedances well outside this range may require that some of the fixed connections to L1 and L3 be changed. Should this be necessary remember that the L1 and L3 sections must be kept symmetrical — the same number of turns on each coil.

Construction

The unit is housed in a homemade aluminum enclosure that measures 9 × 8

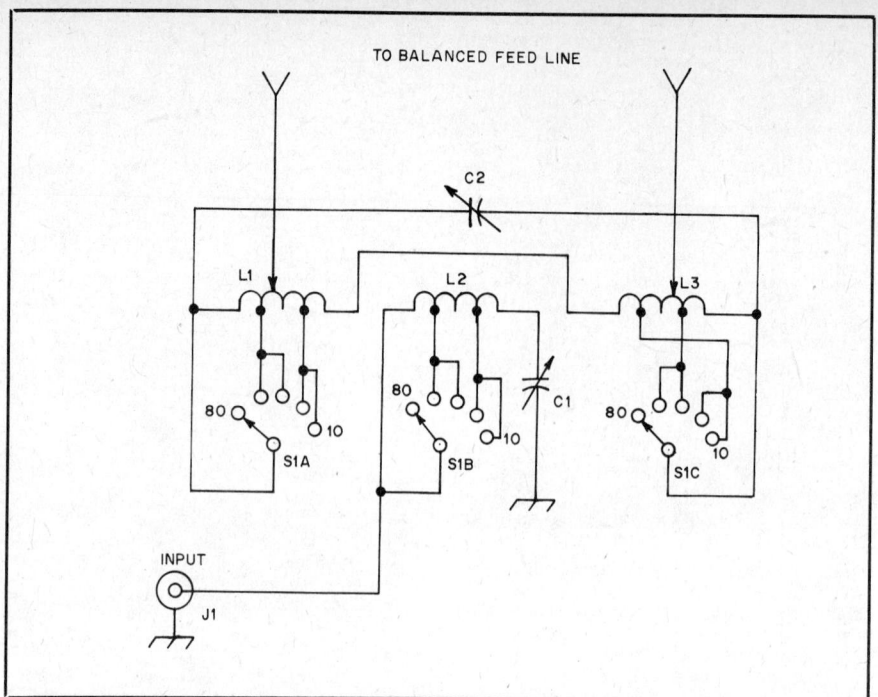

Fig. 28 — Schematic diagram of the link coupler. The connections marked as "to balanced feed line" are steatite feedthrough insulators. The arrows on the other ends of these connections are alligator clips.

C1 — 350 pF maximum, 0.0435 inch plate spacing or greater.
C2 — 100 pF maximum, 0.0435 inch plate spacing or greater.
J1 — Coaxial connector.
L1, L2, L3 — B&W 3026 Miniductor stock, 2-inch diameter, 8 turns per inch, no. 14 wire.

Coils assembly consists of 48 turns. L1 and L3 are each 17 turns tapped at 8 and 11 turns from outside ends. L2 is 14 turns tapped at 8 and 12 turns from C1 end. See text for additional details.
S1 — 3-pole, 5-position ceramic rotary switch.

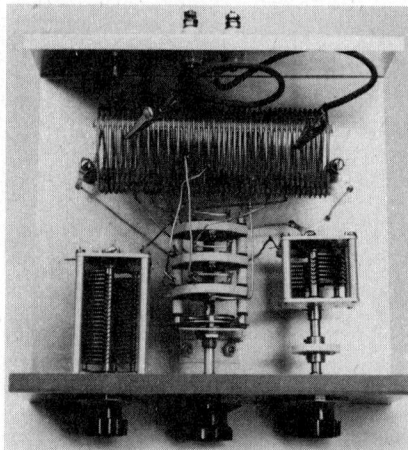

Fig. 29 — Interior view of the coupler showing the basic positions of the major components. Component placement is not critical, but the unit should be laid out for minimum lead lengths.

× 3 1/2 inches (229 × 203 × 89 mm). Any cabinet with similar dimensions that will accommodate the components may be used. L1, L2 and L3 are a one-piece assembly of B&W 3026 Miniductor stock. The individual coils are separated from each other by cutting two of the turns at the appropriate spots along the length of the coil. Then, the inner ends of the outer sections are joined by a short wire that is run through the center of L2. Position the wire so that it will not come into contact

with L2. Each of the fixed tap points on L1, L2 and L3 is located and lengths of hookup wire are attached. The coil is mounted in the enclosure and the connections between the coil and the bandswitch are made. Every other turn of L1 and L3 are pressed in toward the center of the coil to facilitate connection of the alligator clips.

As can be seen from the schematic, C2 must be isolated from ground. This can be accomplished by mounting the capacitor on steatite cones or other suitable insulating material. Make sure that the hole through the front panel for the shaft of C2 is large enough so the shaft does not come into contact with the chassis.

Tuneup

The transmitter should be connected to the input of the Transmatch through some sort of instrument that will indicate SWR. S1 is set to the band of operation and the balanced line is connected to the insulators on the rear panel of the coupler. The alligator clips are attached to the mid points of coils L1 and L3 and power is applied. Adjust C1 and C2 for minimum reflected power. If a good match is not obtained, move the antenna tap points either closer to the ends or center of the coils. Again apply power, tune C1 and C2 until the best possible match is obtained. Continue moving the antenna taps until a 1-to-1 match is obtained.

Table 1

Characteristics of Commonly Used Transmission Lines

Type of line	Z_o Ohms	Vel %	pF per foot	OD	Diel. Material	Max Operating Volts (RMS)
RG-8/U	52.0	66	29.5	.405	PE	4,000
RG-8/U Foam	50.0	80	25.4	.405	Foam PE	1,500
RG-8A/U	52.0	66	29.5	.405	PE	5,000
RG-9/U	51.0	66	30.0	.420	PE	4,000
RG-9A/U	51.0	66	30.0	.420	PE	4,000
RG-9B/U	50.0	66	30.8	.420	PE	5,000
RG-11/U	75.0	66	20.6	.405	PE	4,000
RG-11/U Foam	75.0	80	16.9	.405	Foam PE	1,600
RG-11A/U	75.0	66	20.6	.405	PE	5,000
RG-12/U	75.0	66	20.6	.475	PE	4,000
RG-12A/U	75.0	66	20.6	.475	PE	5,000
RG-17/U	52.0	66	29.5	.870	PE	11,000
RG-17A/U	52.0	66	29.5	.870	PE	11,000
RG-55/U	53.5	66	28.5	.216	PE	1,900
RG-55A/U	50.0	66	30.8	.216	PE	1,900
RG-55B/U	53.5	66	28.5	.216	PE	1,900
RG-58/U	53.5	66	28.5	.195	PE	1,900
RG-58/U Foam	53.5	79	28.5	.195	Foam PE	600
RG-58A/U	53.5	66	28.5	.195	PE	1,900
RG-58B/U	53.5	66	28.5	.195	PE	1,900
RG-58C/U	50.0	66	30.8	.195	PE	1,900
RG-59/U	73.0	66	21.0	.242	PE	2,300
RG-59/U Foam	75.0	79	16.9	.242	Foam PE	800
RG-59A/U	73.0	66	21.0	.242	PE	2,300
RG-62/U	93.0	86	13.5	.242	Air Space PE	750
RG-62/U Foam	95.0	79	13.4	.242	Foam PE	700
RG-62A/U	93.0	86	13.5	.242	Air Space PE	750
RG-62B/U	93.0	86	13.5	.242	Air Space PE	750
RG-133A/U	95.0	66	16.2	.405	PE	4,000
RG-141/U	50.0	70	29.4	.190	PTFE	1,900
RG-141A/U	50.0	70	29.4	.190	PTFE	1,900
RG-142/U	50.0	70	29.4	.206	PTFE	1,900
RG-142A/U	50.0	70	29.4	.206	PTFE	1,900
RG-142B/U	50.0	70	29.4	.195	PTFE	1,900
RG-174/U	50.0	66	30.8	.1	PE	1,500
RG-213/U	50.0	66	30.8	.405	PE	5,000
RG-215/U	50.0	66	30.8	.475	PE	5,000
RG-216/U	75.0	66	20.6	.425	PE	5,000
Aluminum Jacket Foam Dielectric						
1/2 inch	50.0	81	25.0	.5		2,500
3/4 inch	50.0	81	25.0	.75		4,000
7/8 inch	50.0	81	25.0	.875		4,500
1/2 inch	75.0	81	16.7	.5		2,500
3/4 inch	75.0	81	16.7	.75		3,500
7/8 inch	75.0	81	16.7	.875		4,000
Open wire	—	97	—	—		—
75-ohm transmitting twin lead	75.0	67	19.0	—		—
300-ohm twin lead	300.0	82	5.8	—		—
300-ohm tubular	300.0	80	4.6	—		—
Open wire, TV type						
1/2 inch	300.0	95	—	—		—
1 inch	450.0	95	—	—		—

Dielectric Designation	Name	Temperature Limits
PE	Polyethylene	−65° to +80° C
Foam PE	Foamed Polyethylene	−65° to +80° C
PTFE	Polytetrafluoroethylene (Teflon)	−250° to +250° C

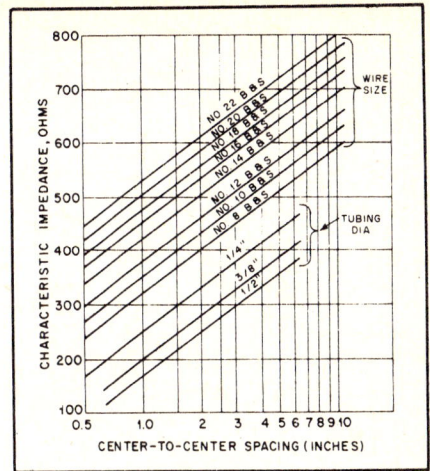

Fig. 30 — Chart showing the characteristic impedance of spaced-conductor parallel transmission lines with air dielectric. Tubing sizes given for outside diameters.

sion lines used by amateurs. Open-wire line has a velocity factor of essential unity because it lacks a substantial amount of solid insulating material. Conversely, molded 300-ohm TV line has a velocity factor of 0.80 to 0.82. The higher cost of the larger coaxial lines is often worth the expenditure in terms of reduced feeder losses.

Amateurs can construct their own parallel transmission lines by following the chart contained in Fig. 30. When using wire conductors it is an easy matter to fabricate open-wire feed lines. Spacers made of high-dielectric material need to be affixed to the conductors at appropriate distances apart to maintain the spacing between the wires (constant impedance) and to prevent shorting of the conductors.

Characteristic Impedance

The characteristic impedance of an air-insulated parallel-conductor line, neglecting the effect of the insulating spacers, is given by

$$Z_o = 276 \log \frac{b}{a}$$

where

Z_o = Characteristic impedance
b = Center-to-center distance between conductors
a = Radius of conductor (in same units as b)

The characteristic impedance of an air-insulated coaxial line is given by the formula

$$Z_o = 138 \log \frac{b}{a}$$

where

Z_o = Characteristic impedance
b = Inside diameter of outer conductors
a = Outside diameter of inner conductor (in same units as b)

It does not matter what units are used for

The circuit described here is intended for power levels up to roughly 200 watts. Balance was checked by means of two rf ammeters, one in each leg of the feed line. Results showed the balance to be well within 1 dB.

Transmission Line Characteristics

Each type of line has a characteristic *velocity factor,* owing to the insulating-material properties. The velocity factor must be taken into account when cutting a transmission line to a specific part of a wavelength — such as with a quarter-wavelength transformer. For example, if RG-8A/U were employed to make a quarter-wavelength line at 3.5 MHz, the line dimension should be 246 f(MHz) × 0.66. Thus, the line would be 46.4 feet (14.14 m) long instead of the free-space length of 70.3 feet (21.4 m). Table 1 shows various velocity factors for the transmis-

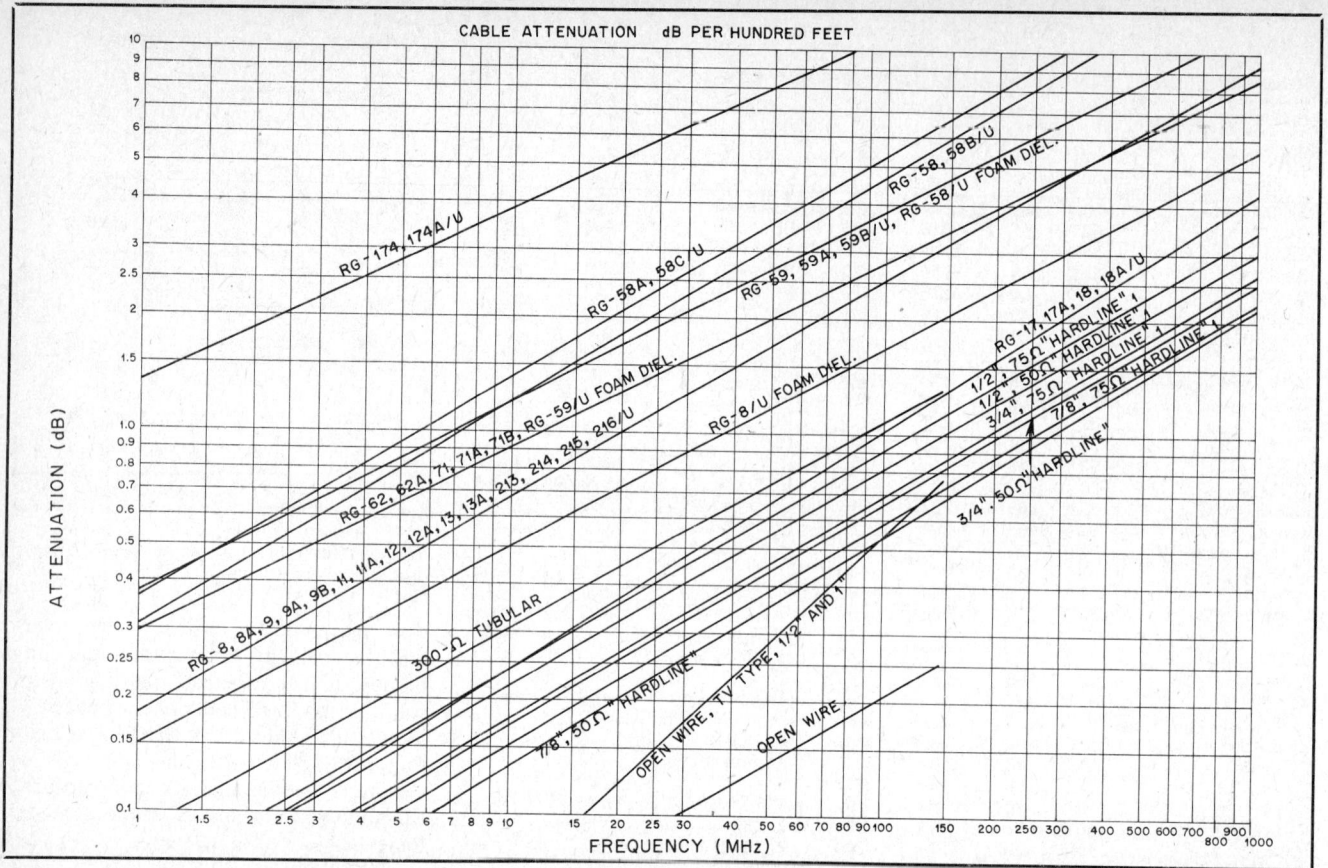

Fig. 31 — This graph displays the attenuation in dB per 100-foot lengths of many popular transmission lines. The vertical axis represents attenuation and horizontal axis is frequency, 1 — 1000 MHz.

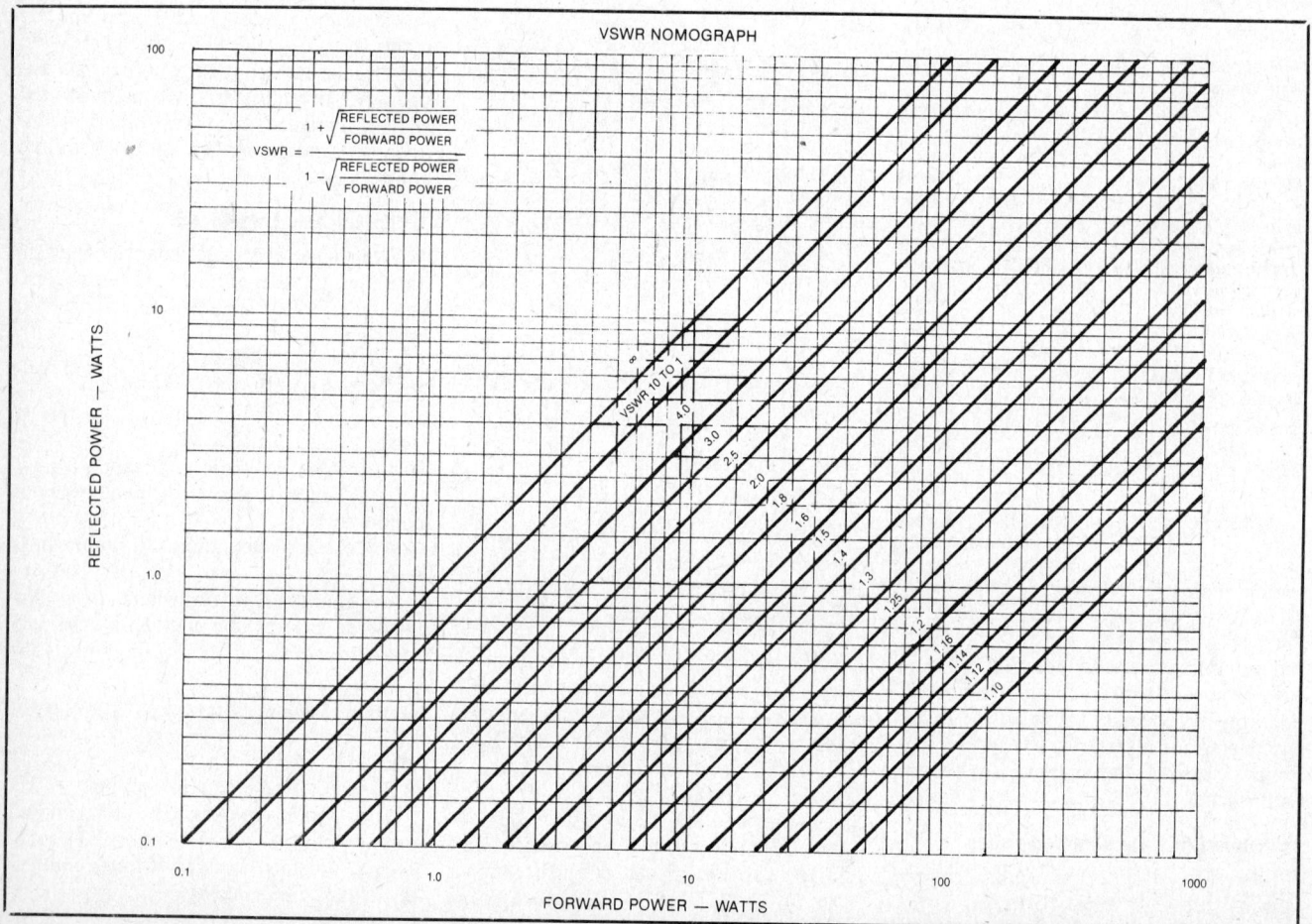

$$VSWR = \frac{1 + \sqrt{\dfrac{REFLECTED\ POWER}{FORWARD\ POWER}}}{1 - \sqrt{\dfrac{REFLECTED\ POWER}{FORWARD\ POWER}}}$$

Fig. 32 — VSWR as a function of forward and reflected power.

a and b so long as they are the *same* units. Both quantities may be measured in centimeters, inches, etc.

VSWR Nomograph and VSWR/Loss Chart

The graph displayed in Fig. 32 can be used to determine the VSWR on a transmission line when a forward and reflected power measurement are known. As an example, suppose a forward power measurement is taken as 100 watts and the reflected power measurement is 11 watts. The 100-watt line on the horizontal axis is located and the 11-watt line on the vertical axis is noted. The intersection of these two lines on the graph is at the 2 to 1 VSWR line. Therefore the VSWR for this set of conditions is 2 to 1.

The graph at Fig. 33 provides a convenient means of determining total losses if the VSWR at either the input or the load is known and if the loss in the line without standing waves is known. (This latter factor may be obtained from Table 1 of this chapter or the manufacturer's literature). Conversely, if the VSWR values at the input to the line and at the load are measured with a reliable instrument, the total line loss and the loss of the line without standing waves (matched loss) may be determined from the graph.

The horizontal axis of the graph is calibrated in values representing the VSWR at the load, while the vertical axis represents total loss of the line in decibels. The curves that are predominantly vertical (dotted lines) in the body of the graph represent the VSWR value at the line input, and the curves that are predominantly horizontal in the lower portion of the graph (solid lines) represent the matched-line loss. Interpolation of values may be made between curves, and the curves are interrelated so that each set or family may be considered as another "axis" of the overall graph.

A couple of examples best illustrate use of the graph. Suppose 100 feet of RG-8 feed line connects a 15-meter transmitter and its load. If the VSWR as measured at the load is 3:1, what is the total line loss and what is the VSWR at the line input? First, it is necessary to determine the matched loss of the length of the coaxial cable. Table 1 indicates that the line has a loss of 1.0 dB per hundred feet at 21 MHz. Proceed by running a finger along the scale at the bottom of the graph and locate the value of 3, representing the 3:1 VSWR at the load. Follow the vertical "3" line up until it intersects with the "1" solid-line curve which represents 1 dB of matched-line loss. The calibration scale at the left indicated that the total line loss under these conditions is 1.5 dB. At this same intersection, by interpolating

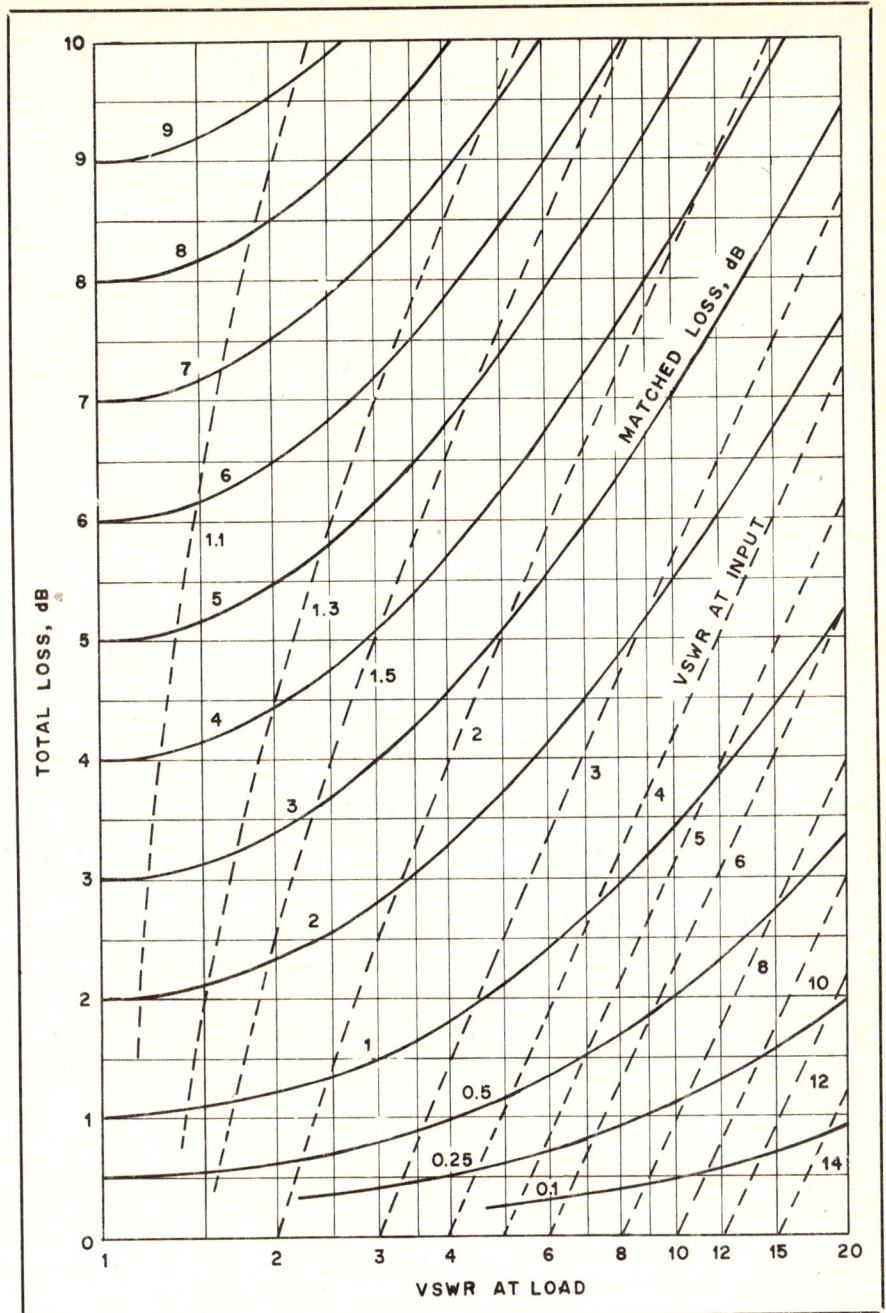

Fig. 33 — Transmission loss as a function of source and load VSWR. See text for applications.

between the dotted-line curves, it may be seen that the VSWR at the line input is approximately 2.3:1.

As another example, assume the use of a line with a matched loss of 3 dB and that the measured VSWR at the line input is 2:1. What is the total line loss and the VSWR at the load? In this case begin as before, by running a finger along the scale at the bottom of the graph until the value of 2 is located (it will be the same at the line input if the line is lossless, or has 0 dB of matched loss). This time, however, proceed by following the "2" dotted line up and slightly to the right, as this dotted

line represents a 2:1 VSWR at the line input. Follow this line until it intersects with the 3 solid curve, representing the matched loss of the line. From this intersection it may be read (scale at left) that the total line loss is just a bit more than 5 dB, and that the VSWR at the load (scale at the bottom) is 5:1. Had the attenuation values not been known in this example but the VSWR at both the load and line input were known, it would have been possible to determine the matched loss and the total loss from the intersection of the 5 VSWR-at-load vertical line and the 2 VSWR-at-input curve.

Chapter 20

Antennas for High Frequency

An antenna *system* is comprised of all the components which are used between the transmitter or receiver and the actual radiator. Therefore, such items as the antenna proper, transmission line, matching transformers, baluns and Transmatch qualify as parts of an antenna system.

Only the antenna does the radiating in a well-designed system. It is noteworthy that any type of feed line can be utilized with a given antenna, provided a suitable matching device is used to ensure a low standing-wave ratio (SWR) between the feed line and the antenna, and again between the feed line and the transmitter and/or the receiver. Some antennas possess a characteristic impedance at the feed point close to that of certain transmission lines. For example, a half-wavelength center-fed dipole, placed a correct height above ground, will have a feed impedance of approximately 75 ohms. In such a case it is practical to use 75-ohm coaxial or balanced line to feed the antenna. But few amateur half-wavelength dipoles actually exhibit a 75-ohm impedance. This is because at the lower end of the high-frequency spectrum the typical height above ground is rarely more than 1/4 wavelength. The 75-ohm characteristic is most likely to be realized in a practical installation when the horizontal dipole is approximately one-half, three-quarters or one wavelength above ground. At other heights the feed impedance may vary from roughly 58 to 95 ohms (see Fig. 1). This general principle applies to nearly all antennas which are erected horizontally above an earth ground. Furthermore, the precise conductivity of the earth at one location may differ markedly at another site, and this phenomenon has a direct effect on the electrical height of the radiator above ground. A curve (dotted line) is included in Fig. 1 to demonstrate the radiation resistance of a vertical dipole at various feed-point heights above ground.

The Antenna Choice

Paramount among the factors to consider when selecting an antenna is the matter of available space. Those who live in urban areas have frequent need to accept a compromise type of antenna for the hf bands because the city lot won't accommodate full-size wire dipoles, end-fed systems or high supporting structures. Other constrictions are imposed by the amount of money available for an antenna system (inclusive of supporting hardware), the number of amateur bands to be worked, and local zoning ordinances. Finally, the operational objective comes into play: To dedicate one's self to DXing, or settle for a general type of operation that will yield short- and long-haul QSOs during periods of good propagation. Because of the foregoing influences, it is impossible to suggest one type of antenna system over another. Perhaps a general rule of thumb might be to erect the biggest and best antenna collection that space and finances will allow. If a modest system is the order of the day, then use whatever is practical and accept a tradeoff between elaboration and performance. Practically any radiator will enable the operator to make good contacts under some conditions of propagation, assuming that the radiator is able to accept power and radiate it at some useful angle respective to earth.

In general, the height of the antenna above ground is the most critical factor at the higher end of the hf spectrum — 20, 15 and 10 meters. This is because the antenna should be clear of conductive objects such as power lines, phone wires, gutters and the like, plus high enough to have a low radiation angle. This is not nearly as important at 160, 80 and 40 meters, but it is still recommended that lower-frequency antennas be well away from conductive objects and as high above ground as possible in the interest of good performance. The exception is a ground-mounted vertical antenna. Ground-plane verticals, however, should be installed as high above ground as possible so that their performance will not be degraded by conductive objects.

Antenna Polarization

Most hf-band antennas are either vertically or horizontally polarized, although circular polarization is possible, just as it is at vhf and uhf. The polarization is determined by the position of the radiating element or wire with respect to earth. Thus, a radiator that is parallel to earth radiates horizontally, while an antenna at a right angle to earth (vertical) radiates a vertical wave. If a wire antenna is *slanted* above earth, it radiates waves which are between vertical and horizontal in nature. During line-of-sight communications, maximum signal strength will be realized when the antennas at both ends of the circuit have the same polarity. Cross polarization results in many decibels of signal reduction. However, during propagation via the ionosphere (sky wave) it is not essential to have the same polarization as the station at the opposite end of the circuit. This is because the radiated wave is bent and tumbled considerably during its travel from the atmospheric layer from which it is refracted. At the far end of the communications path the wave may be horizontal, vertical or somewhere in between at a given instant. On *multihop* transmissions, where the signal is refracted more than once from the atmosphere, and similarly reflected from the earth's surface during its travel (skip), considerable polarization shift will occur. Therefore, the main consideration for a good DX antenna is a low angle of

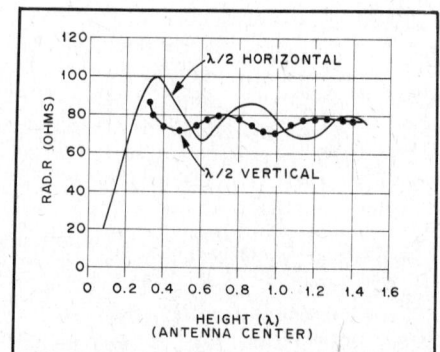

Fig. 1 — Curves showing the radiation resistance of vertical and horizontal half-wavelength dipoles at various heights above ground.

radiation rather than the polarization. It should be said, however, that most DX antennas for hf work are horizontally polarized. The major exception is the ground-plane vertical and phased vertical arrays.

Impedance

The impedance at a given point in the antenna is determined by the ratio of the voltage to the current at that point. For example, if there were 100 rf volts and 1.4 amperes of current at a specified point in an antenna, the impedance would be approximately 71 ohms. The impedance is significant with regard to matching the feeder to the feed point: Maximum power transfer takes place under a perfectly matched condition. As the mismatch increases so does the reflected power. If the feed line is not too lossy or long, good performance can be had at hf when the standing-wave ratio (SWR) is 3:1 or less. When feeder loss is very low — such as with open-wire transmission line — much higher SWR is not particularly detrimental to performance provided the transmitter is able to lead into the mismatched condition satisfactorily. In this regard, a *Transmatch* (matching network between the transmitter and the feed line) is often employed to "disguise" the mismatch condition, thereby enabling the operator to load the transmitter to its full rated power.

Antenna impedance can be either resistive or complex. This will depend upon whether or not the antenna is resonant at the operating frequency. Many operators believe that a mismatch, however small, is a serious matter, and that their signals won't be heard well even if the SWR is as low as 1.3:1. This unfortunate fallacy has cost many man-hours and dollars among some amateur groups as individuals attempted to obtain a "perfect" match: A perfect match, however ideal the concept may be, is not necessary. The significance of a perfect match becomes more pronounced at vhf and higher, where feeder losses are a major problem.

Antenna Bandwidth

The bandwidth of an antenna refers generally to the range of frequencies over which the antenna can be used to obtain good performance. The bandwidth is usually referenced to some SWR value, such as "The 2:1 SWR bandwidth is 3.5 to 3.8 MHz." Some more specific bandwidth terms are used also, such as the *gain bandwidth* and the *front-to-back ratio bandwidth*. The gain bandwidth is significant because the higher the antenna gain the narrower the gain bandwidth will be, for a given gain-bandwidth product.

For the most part, the lower the operating frequency of a given antenna design, the narrower the bandwidth. This

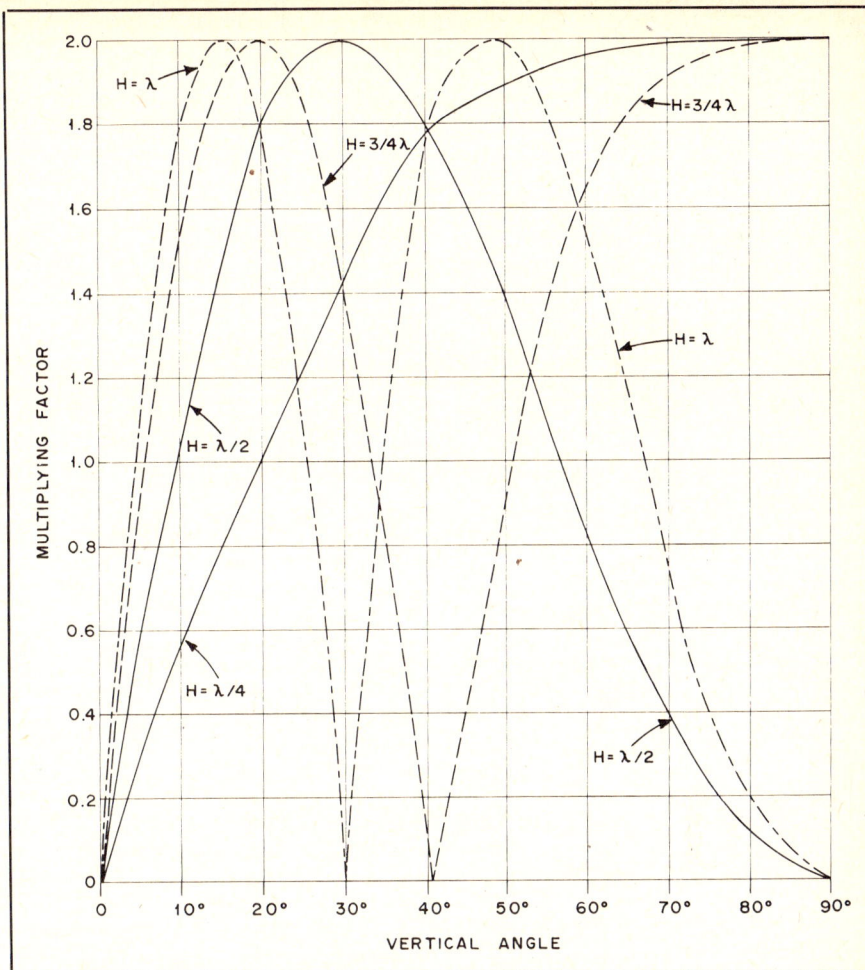

Fig. 2 — Effect of ground on the radiation of horizontal antennas at vertical angles for four antenna heights. These data are based on perfectly conducting ground.

follows the rule where the bandwidth of a resonant circuit doubles as the frequency of operation is increased one octave (doubled) assuming the Q is the same for each case. Therefore, it is often difficult to achieve sufficient bandwidth to cover all of the 160- and 80-meter bands with a dipole antenna cut for each of those bands. The situation can be aided by applying broadbanding techniques, such as fanning the far ends of a dipole to simulate a conical type of dipole.

Radiation Angle

The vertical angle of maximum radiation is of primary importance, especially at the higher frequencies. It is advantageous, therefore, to erect the antenna at a height that will take advantage of ground reflection in such a way as to reinforce the space radiation at the most desirable angle. Since low angles usually are most effective, this generally means that the antenna should be high — at least one-half wavelength at 14 MHz, and preferably three-quarters or one wavelength, and at least one wavelength, and preferably higher, at 28 MHz. The physical height required for a given height in wavelengths decreases as the frequency

is increased, so that good heights are not impracticable; a half wavelength at 14 MHz is only 35 feet, approximately, while the same height represents a full wavelength at 28 MHz. At 7 MHz and lower frequencies the higher radiation angles are effective, so that again a useful antenna height is not difficult to attain. But, greater height is important at 7 MHz and lower when it is desired to work DX consistently. Heights between 35 and 70 feet are suitable for the upper bands, the higher figures being preferable. It is well to remember that most simple horizontally polarized antennas do not exhibit the directivity they are capable of unless they are one-half wavelength above ground, or greater, at their operating frequency. Therefore, with dipole-type antennas it is not important to choose a favored broadside direction unless the antenna is *at least one-half wavelength above ground*.

Imperfect Ground

Fig. 2 is based on ground having perfect conductivity, whereas the earth is not a perfect conductor. The principal effect of actual ground is to make the curves inaccurate at the lowest angles; appreciable high-frequency radiation at

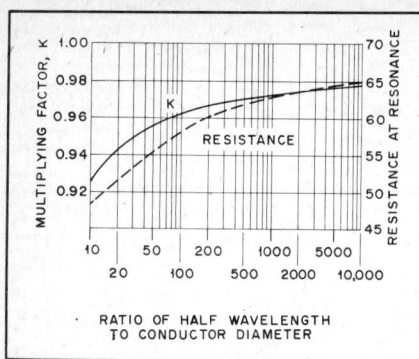

Fig. 3 — Effect of antenna diameter on length for half-wavelength resonance, shown as a multiplying factor, K, to be applied to the free-space, half-wavelength equation (Eq. 1). The effect of conductor diameter on the center impedance is shown also.

angles smaller than a few degrees is practically impossible to obtain over horizontal ground. Above 15 degrees, however, the curves are accurate enough for all practical purposes, and may be taken as indicative of the result to be expected at angles between 5 and 15 degrees.

The effective ground plane — that is, the plane from which ground reflections can be considered to take place — seldom is the actual surface of the ground but is a few feet below it, depending upon the characteristics of the soil.

Current and Voltage Distribution

When power is fed to an antenna, the current and voltage vary along its length. The current is maximum (*loop*) at the center and nearly zero (*node*) at the ends, while the opposite is true of the rf voltage. The current does not actually reach zero at the current nodes, because of the end effect; similarly, the voltage is not zero at its node because of the resistance of the antenna, which consists of both the rf resistance of the wire (*ohmic resistance*) and the *radiation resistance*. The radiation resistance is an *equivalent* resistance, a convenient conception to indicate the radiation properties of an antenna. The radiation resistance is the equivalent resistance that would dissipate the power the antenna radiates, with a current flowing in it equal to the antenna current at a current loop (maximum). The ohmic resistance of a half-wavelength antenna is ordinarily small enough, compared with the radiation resistance, to be neglected for all practical purposes.

Conductor Size

The impedance of the antenna also depends upon the diameter of the conductor in relation to the wavelength, as indicated in Fig. 3. If the diameter of the conductor is increased, the capacitance per unit length increases and the inductance per unit length decreases. Since the radiation resistance is affected

relatively little, the decreased L/C ratio causes the Q of the antenna to decrease, so that the resonance curve becomes less sharp. Hence, the antenna is capable of working over a wide frequency range. This effect is greater as the diameter is increased, and is a property of some importance at the very high frequencies where the wavelength is small.

The Half-Wave Wavelength Antenna

A fundamental form of antenna is a single wire whose length is approximately equal to half the transmitting wavelength. It is the unit from which many more-complex forms of antennas are constructed. It is known as a *dipole antenna*.

The length of a half-wave in free space is

$$\text{Length (ft)} = \frac{492}{f(\text{MHz})} \qquad \text{1a}$$

$$\text{Length (m)} = \frac{150}{f(\text{MHz})} \qquad \text{1b}$$

The actual length of a half-wavelength antenna will not be exactly equal to the half-wavelength in space, but depends upon the thickness of the conductor in relation to the wavelength as shown in Fig. 3, where K is a factor that must be multiplied by the half wavelength in free space to obtain the resonant antenna length. An additional shortening effect occurs with wire antennas supported by insulators at the ends because of the capacitance added to the system by the insulators (end effect). The following formula is sufficiently accurate for wire antennas for frequencies up to 30 MHz.

Length of half-wave antenna (ft) =

$$\frac{492 \times 0.95}{f(\text{MHz})} = \frac{468}{f(\text{MHz})} \qquad \text{2a}$$

$$\frac{150 \times 0.95}{f(\text{MHz})} = \frac{143}{f(\text{MHz})} \qquad \text{2b}$$

Example:
A half-wave antenna for 7150 kHz (7.15 MHz) is $\frac{468}{7.15}$ = 65.45 ft, or 65 ft 5 in. (19.9 m).

Above 30 MHz the following formulas should be used, particularly for antennas constructed from rod or tubing. K is taken from Fig. 3.

Length of half-wave antenna (ft) =

$$\frac{492 \times K}{f(\text{MHz})} \qquad \text{3a}$$

$$\text{Length (in.)} = \frac{5905 \times K}{f(\text{MHz})} \qquad \text{3b}$$

$$\text{Length (m)} = \frac{150 \times K}{f(\text{MHz})} \qquad \text{3c}$$

$$\text{Length (mm)} = \frac{150,000 \times K}{f(\text{MHz})} \qquad \text{3d}$$

Example: Find the length of a half-wavelength antenna at 28.7 MHz, if the antenna is made of 1/2-inch (12.7 mm) diameter tubing. At 28.7 MHz, a half-wavelength in space is

$$\frac{492}{28.7} = 17.14 \text{ ft} \ (5.3\text{m})$$

from Equation 1. The ratio of half wavelength to conductor diameter (changing wavelength to inches) is

$$\frac{(17.14 \times 12)}{0.5} = 411$$

From Fig. 3, K = 0.97 for this ratio. The length of the antenna, from Equation 3a is

$$\frac{(492 \times 0.97)}{28.7} = 16.63 \text{ ft} \ (5.06 \text{ m})$$

or 16 feet 7-1/2 inches. The answer is obtained directly in inches by substitution of Equation 3b.

$$\frac{(5905 \times 0.97)}{28.7} = 199.6 \text{ inches} \ (5.06 \text{ m})$$

The length of a half-wavelength antenna is affected also by the proximity of the dipole ends to nearby conductive and semiconductive objects. In practice, it is often necessary after cutting the antenna to the computed length to do some experimental "pruning" of the wire, shortening it in increments to obtain a low SWR. This can be done by applying rf power through an SWR indicator and observing the reflected-power reading. When the lowest SWR is obtained for the desired part of an amateur band, it does not necessarily follow that the antenna is resonant at that frequency. However, a matched condition will have been secured — the basic objective in preparing an antenna for use.

Radiation Characteristics

The classic radiation pattern of a dipole antenna is most intense perpendicular to the wire. A figure-8 pattern (Fig. 4) can be assumed off the broad side of the antenna (bidirectional pattern) if the dipole is 1/2 wavelength or greater above earth and not degraded by nearby conductive objects. This assumption is based also on a symmetrical feed system. In practice, a coaxial feed line will distort this pattern slightly. Minimum horizontal radiation occurs off the ends of the dipole. The foregoing relates to a half-wavelength antenna which is parallel to the earth. However, if the dipole is erected vertically, uniform radiation in all compass directions (a doughnut pattern if it could be viewed from above the antenna) will result.

One of the greatest errors made by some beginners is to assume that a dipole

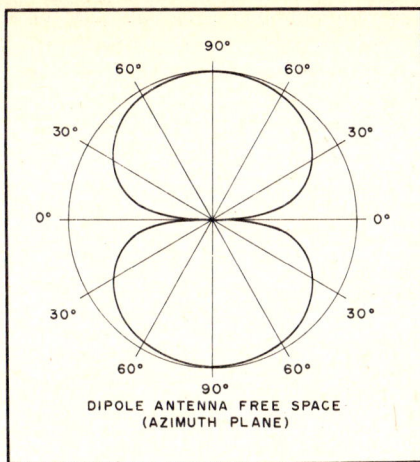

Fig. 4 — Azimuth-plane response of a half-wavelength dipole in free space.

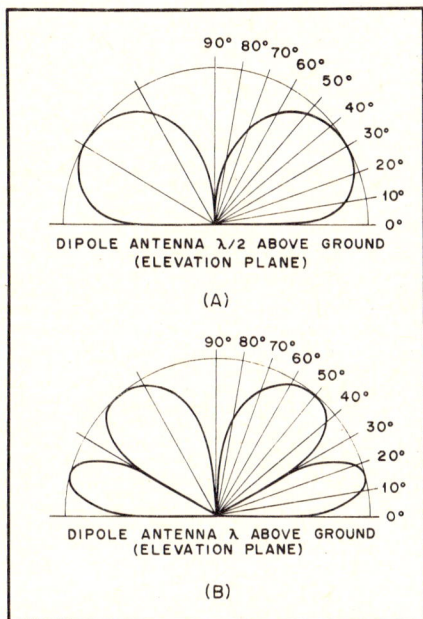

Fig. 5 — Elevation-plane response of a dipole antenna placed 1/2-wavelength above a perfectly conducting ground (A), and the pattern for the same antenna (B) when it is raised to one-wavelength above ground.

antenna will exhibit a "broadside pattern" at any height above ground. As the antenna is brought closer to ground, the radiation pattern deteriorates until the antenna is, for the most part, an omnidirectional radiator of high-angle waves. Many are tempted to use any convenient height, such as 20 or 30 feet above ground for an 80-meter dipole, only to learn that the system is effective in all directions over a relatively short distance (out to 500 or 1000 miles under good conditions). It can be seen from this that height above ground is important for a host of reasons. Fig. 5A and B illustrate clearly the advantage gained from antenna height. The radiation angle of Fig. 5A is roughly 35 degrees, whereas at a height of one wavelength (Fig. 5B) the lobes split and the lower ones provide a good

DX-communications angle of approximately 15 degrees. The higher-angle lobes (50°) are useful for short-haul communications and compare favorably in practice with the lobe angle seen in Fig. 5A. At heights appreciably lower than 1/2 wavelength, the lobe angle becomes higher, and eventually the two lobes converge to create a discrete "ball of radiation" which has a very high-angle nature (poor for long-distance communications).

Feed Methods for Half-Wavelength Antennas

Most amateur single-wire dipole antennas (half wavelength) have a feed impedance between 50 and 75 ohms, depending on the installation. Therefore, standard coaxial cable is suitable for most installations. The smaller types of cable (RG-58/U and RG-59/U) are satisfactory for power levels up to a few hundred watts if the SWR of the system is low. For high-power stations it is recommended that the larger cables be employed (RG-8/U or RG-11/U). These cables can be connected at the center of the antenna, as shown in Fig. 6. A plastic insulating block is used as a central reinforcement for the cable and the dipole wires. The coax shield braid connects to one leg of the dipole and the center conductor is soldered to the remaining leg. The exposed end of the cable should be sealed against dirt and moisture to prevent degradation of the transmission line.

Symmetrical feed can be achieved by inserting a 1:1 balun transformer at the dipole feed point. If one is not used, it is unlikely that the slight pattern skew resulting from nonsymmetrical feed will be noticed. The effects of unbalanced feed are most significant in beam antennas at vhf and higher. The narrower the beam pattern the more annoying the condition will be.

The characteristic impedance of a dipole antenna can be increased by using a *two-wire* or *folded* dipole of the type seen in Fig. 7. This antenna offers a good match to 300-ohm feed line. In fact, the dipole itself can be fashioned from a length of 300-ohm TV ribbon. Alternatively, two pieces of wire can be used to form the equivalent of the TV-line dipole. If this is done it will be necessary to locate insulating spacers every few feet along the length of the dipole to keep the wires spaced apart uniformly and to prevent short circuiting. Open-wire TV "ladder line" is excellent for use in a 300-ohm folded-dipole antenna, both for the radiator and the feed line. Feeder losses with this type of construction will be very low as opposed to molded TV twin-lead.

A dipole antenna can be used as an "all-band" radiator by using tuned open-wire feed line. This principle is seen in Fig. 8A. In this example the dipole is cut to a

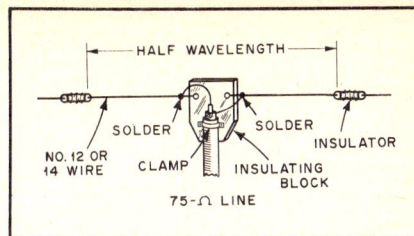

Fig. 6 — Method for affixing the feed line to the center of a dipole antenna. A plastic block is used as a center insulator. The coaxial cable is held in place by means of a metal clamp.

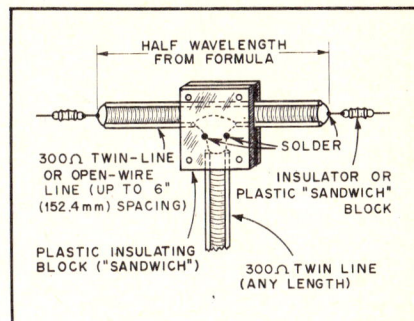

Fig. 7 — Construction details for a folded-dipole antenna. TV ribbon line is used as the dipole and feed line. Two pieces of plastic form an insulator/sandwich at the center to hold the conductor junction secure.

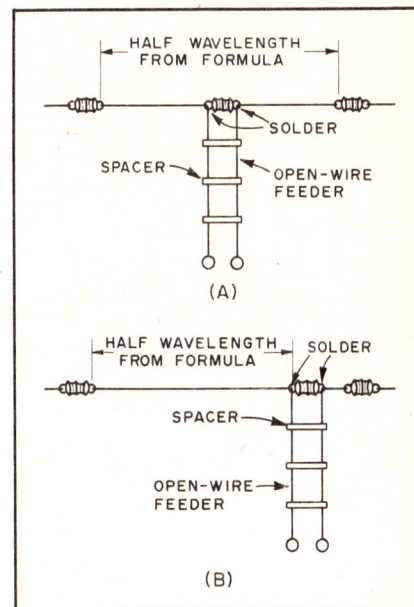

Fig. 8 — Center-fed Zepp antenna (A) and an end-fed Zepp at B.

half wavelength for the lowest desired amateur band. It is operated on its harmonics when used for the other chosen amateur bands. A typical antenna of this type might be utilized from 80 through 10 meters. This style of radiator is known by some amateurs as the "center-fed Zepp." An end-fed version (end-fed Zepp) is shown in Fig. 8 at B. The latter is not quite as desirable as the center-fed version

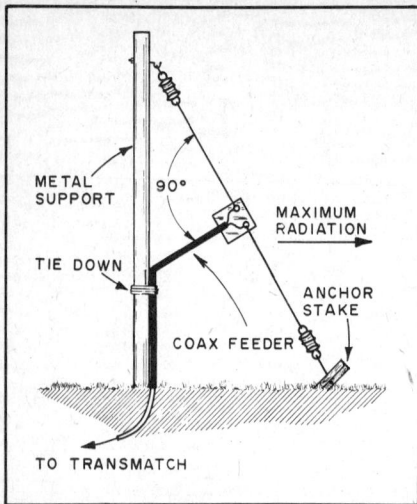

Fig. 9 — Example of a sloping half-wavelength dipole. Maximum directivity is as shown provided a metal mast is used. If a tree or nonconducting mast is employed the pattern will be similar to that of a vertical ground-plane antenna.

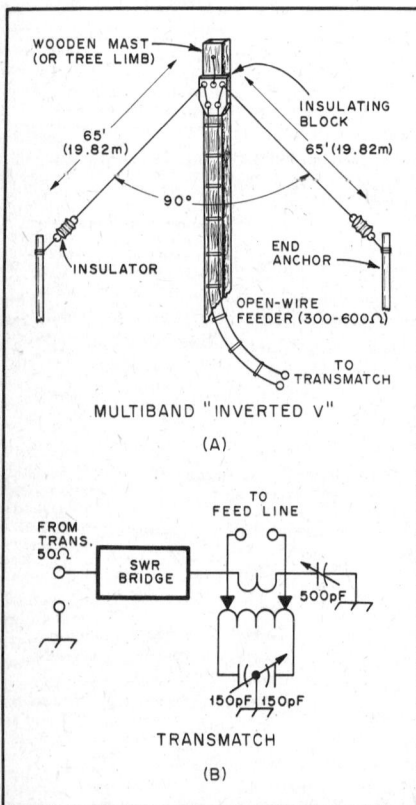

Fig. 10 — Details for an inverted-V dipole which can be used for multiband hf operation (A). A Transmatch is seen at B. It is suitable for matching the antenna to the transmitter over a wide frequency range.

because the feed system is not symmetrical. This can cause feeder radiation and a distortion of the antenna radiation pattern. Both types of Zepp antenna require a matching network (Transmatch) at the transmitter end of the line to convert the feeder impedance to 50 ohms, and to change the balanced condition to an

unbalanced one. Although the feed line may be anything from 200 ohms up to 600 ohms (not critical), losses will be insignificant when open-wire line is used. This is true despite the variations in dipole feedpoint impedance from band to band. The feed impedance will be high at even harmonics and will be low at the lowest operating frequency and at odd harmonics thereof. For example, if the dipole is cut for 40 meters, the feed impedance will be low on 40 and 15 meters, but it will be high at 20 and 10 meters.

When using any dipole antenna it is recommended that the feed line be routed away from the antenna at a right angle for as great a distance as possible. This will help prevent current unbalance in the line caused by rf pickup from the dipole. A right-angle departure of 1/4 wavelength or greater is suggested.

Under some circumstances it may be necessary to experiment with the length of the open-wire feeders when using an all-band Zepp. This is because at some operating frequencies the line may present an "awkward" impedance to the Transmatch, making it impossible to obtain a suitable load condition for the transmitter. This will depend essentially on the capability of the Transmatch being used.

Dipole Variations

The physical application of dipoles can be varied to obtain radiation properties which differ from those of the more conventional "horizontal dipole." Furthermore, the nature and amount of property at the installation site will often dictate certain departures from the conventional when erecting a dipole.

A sloping dipole can be useful for DX work because of the low angle of radiation which results when it is erected as shown in Fig. 9. The higher the feed point is above ground the lower the radiation angle will be. However, excellent results can be had when the ground end of the antenna is only a few feet above the earth. Maximum directivity is off the sloped front of the antenna, as shown. This characteristic is obtained when the "sloper" is supported by a metal mast or tower: The metal structure tends to act as a reflector, actually providing a slight amount of gain. Some amateurs install four slopers for a given amateur band, spaced equidistantly around the tower. A feed-line switching system is used to obtain directivity in the chosen direction.

Another popular type of dipole antenna is the so-called inverted V, or drooping doublet. An inverted V is shown in Fig. 10. Newcomers to amateur radio are frequently led to regard this antenna as a panacea, but there is nothing magical or superior about an inverted-V dipole. The main attributes are that it radiates more or less omnidirectionally at typical heights above ground, requires only one supporting structure, and can be used for

single or multiband use. As is true of a horizontal dipole, the higher the inverted V is mounted above ground (feed point), the better it will perform respective to long-distance communications.

Best results seem to be obtained when the enclosed angle of the inverted V is between 90 and 120 degrees. At angles less than 90 degrees considerable cancellation is likely, resulting in reduced antenna performance. At angles greater than 120 degrees, the antenna begins to function as a horizontal dipole.

When the ends of the inverted V are relatively close to the earth, pruning of the dipole legs may be necessary to compensate for the capacitive effect to ground. Thus, if the dipole is cut to length by means of the standard half-wavelength dipole equation, it may be too long as an inverted V. Incremental trimming of each end of the dipole can be done while using an SWR indicator. This will show when the lowest SWR is obtained. A 50-ohm feed line offers a good match for inverted Vs when single-band operation is desired. For multiband use, the inverted V can be fed by means of open-wire line and matched to the transmitter by means of a Transmatch.

The most ideal supporting structure for an inverted V is a wooden or other nonconductive one. This type of support will have the least effect on antenna pattern and performance. When a metal structure is used, the mast of tower is in the field of the antenna and will affect the radiation pattern considerably. The effect is similar to that of the sloping dipole, where maximum directivity is off the sloping front of the antenna (Fig. 9). In the case of an inverted V, two prominent lobes result — off two sides of the tower, as if two sloping dipoles were being used.

Inverted Vs are most effective when used for a single band — the one for which the dipole has been cut. As the operating frequency is increased, the radiation angle deteriorates: Results with DX are often mediocre. Good results can be had when using an inverted V for two-band operation, however. An 80-meter inverted V can be fed with open-wire line and used also on 40 meters as two half waves in phase. Although ARRL tests have not been made to verify the effect, some gain seems to be realized when operating the inverted V on its second harmonic.

Tests performed with an inverted V on 80 meters showed that with an apex height of 60 feet (18 m) and an enclosed angle of 90 degrees, no directivity could be observed during ground-wave signal evaluation. The test stations were 10 miles apart during daylight, and the inverted V was rotated through the various compass headings while the signal level was monitored at the far end of the communications circuit. The supporting structure was a 60-foot (18 m) tower in an open

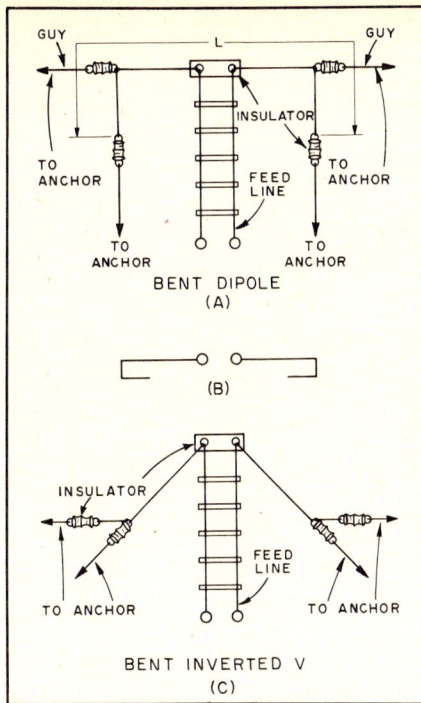

Fig. 11 — When limited space is available for a dipole antenna the ends can be bent downward (A), or back on the radiator as shown at B. The inverted V at C can be erected with the voltage ends bent parallel with the ground when the available supporting structure is not high enough to permit an enclosed angle of approximately 90 degrees.

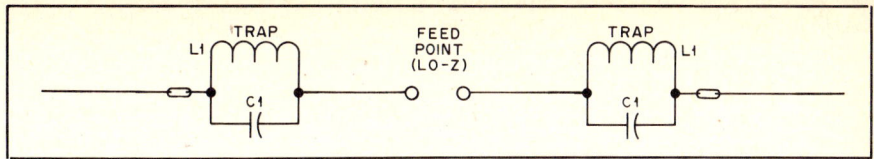

Fig. 12 — Example of a trap dipole antenna. L1 and C1 can be tuned to the desired frequency by means of a dip meter before they are installed in the antenna.

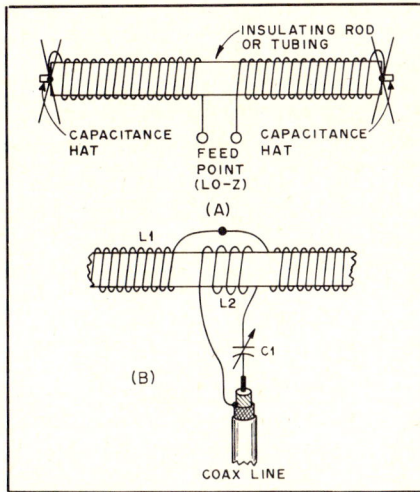

Fig. 13 — A helically wound dipole is illustrated at A. As shown, the radiation resistance will be very low and will require a broadband matching transformer. The coupling method seen at B is more satisfactory for providing a matched condition.

20-acre field. This verified the omni-directional property mentioned earlier.

Bent Dipoles and Trap Dipoles

When there is insufficient real estate to permit the erection of a full-size horizontal or inverted-V dipole, certain compromises are possible in the interest of getting an antenna installed. The voltage ends (far ends) of a dipole can be bent downward toward earth to effect resonance, and the performance will not be reduced markedly. Fig. 11 illustrates the technique under discussion. At A the dipole ends are bent downward and secured to anchors by means of guy line. Some pattern distortion will result from bending the ends. The dipole ends can also be bent back over the wire halves of the antenna, as seen at B in Fig. 11. This causes some signal cancellation (more severe than with the system of Fig. 11A), so it is not a preferred technique.

Fig. 11C demonstrates a bending technique for inverted Vs when the available supporting mast or tree is too short to permit normal installation. The ends of the dipole are guyed off by means of insulators and wires, as shown. Alternatively, but not preferred, is the fold-back method at B in Fig. 11.

All of the shortening systems highlighted in Fig. 11 will have an effect on the overall length of the dipole. Therefore, some cutting and testing will be necessary to ensure a low SWR in the favored part of the amateur band for which the antenna is built. If open-wire feeders and a Transmatch are used, the dipole length will not be a critical factor, provided it is close to the length required for a fully extended half-wavelength dipole. Pruning will be required if single-band operation with coaxial feed line is planned.

Trap dipoles offer one solution to multiband operation with a shortened radiator. The concept is seen in Fig. 12. In this example the dipole is structured for two-band use. Assuming in this case that the antenna is made for operation on 80 and 40 meters, the overall radiator (inclusive of the traps) must be resonant at the center of the chosen section of the 80-meter band. The traps add loading to the dipole, so the length from the feed point to the far end of each leg will be somewhat shorter than normal. During 40-meter operation the traps present a high impedance to the signal and "divorce" the wires beyond the traps. Therefore, the wire length from the feed point to each trap is approximately what it would be if the dipole were cut for just 40 meters, with no traps in the line. This principle can be extended for additional bands, using a new set of traps for each additional band. Since there is considerable interaction between the various segments of a multiband trap dipole, considerable experimentation with the

wire lengths between the traps and beyond will be necessary.

The trap capacitors should be high-voltage and-current units (C1). Transmitting mica capacitors offer good performance. Transmitting ceramic capacitors are usable, but change value with extreme changes of temperature. Therefore, they are more suitable for use in regions where the climate is fairly constant throughout the year. The coils (L1) should be of reasonably heavy wire gauge to minimize I^2R losses. The X_L and X_C values in the traps are not critical. Generally the reactance can be on the order of 100 to 300 ohms. The traps are checked for resonance before they are installed in the antenna system. This can be accomplished by means of a dip meter and a calibrated receiver. Weatherproofing should be added to the traps as a measure against detuning and damage from ice snow and dirt.

Helically Wound Dipoles

The overall length of a half-wavelength dipole can be reduced considerably by employing helically-wound elements. Fig. 13A shows the general form taken with this type of antenna. A length of insulating rod or tubing (fiberglass or phenolic) is used to contain the wire turns of the dipole. The material should be of high dielectric quality. Varnished bamboo has been used successfully by some in lieu of the more expensive materials. A hardwood pole from a lumber yard can be used after being coated one or more times with exterior spar varnish.

To minimize losses, the wire used should be of the largest diameter practical. The turns can be close wound or spaced apart with little difference in performance. The ends of the helical dipole should contain capacitance hats (disks or wire spokes preferred) of the largest size practical. The hats will lower the Q of the antenna and broaden its response. If no disks are used, extremely high rf voltage can appear at the ends of the antenna. At medium power levels and higher the insulating material can burn when no hats are used. The voltage effect is similar to that of a Tesla coil.

The feed impedance of helical dipoles or verticals is quite low. Therefore, it may be necessary to employ some form of matching network to interface the antenna with 50-ohm coaxial cable. A broadband, variable-impedance transformer is convenient for determining the turns ratio of the final transformer used. The feed

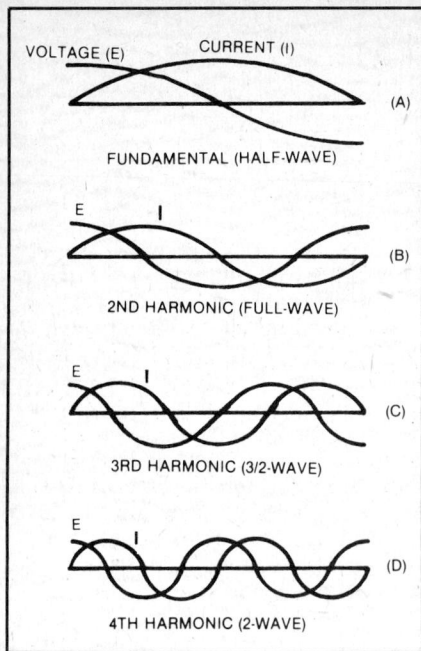

Fig. 14 — Standing-wave current and voltage distribution along an antenna when it is operated at various harmonics of its fundamental resonant frequency.

method shown at B of Fig. 13 can be used to secure a matched condition. L2 is wound over L1, or between the two halves of L1, as illustrated. C1 is adjusted for an SWR of 1 at the center of the desired operating range. The bandwidth of this type of antenna is quite narrow. A 40-meter version with an 18-foot (5.5-m) overall length exhibited a 2:1 SWR bandwidth of 50 kHz. The capacitance hats on that model were merely 18-inch (457-mm) lengths (spikes) of no. 8 copperweld wire. Greater bandwidth would result with larger capacitance hats.

To obtain half-wavelength performance it is necessary to wind approximately one wavelength of wire on the tubing. Final pruning can be accomplished while observing an SWR indicator placed in the transmission line. Proximity to nearby conductive objects and the earth will have a significant effect on the resonance of the antenna. Ideally, final adjustments should be made with the antenna situated where it will be during use. Marine spar varnish should be painted on the elements after all tuning is finished. This will protect the antenna from the weather and will lock the turns in place so that detuning will not occur later on.

A reasonably linear current and voltage distribution will result when using a helically wound dipole or vertical. The same is not true of center-, mid- or end-loaded (lumped inductance) dipoles. The efficiency of this antenna will be somewhat less than a full-size dipole. The performance will degrade as the helixes are made shorter. Despite the gain-length

tradeoff, this antenna is capable of good performance when there is no room for a full-size dipole.

LONG-WIRE ANTENNAS

An antenna will be resonant so long as an integral number of standing waves of current and voltage can exist along its length; in other words, so long as its length is some integral multiple of a half wavelength. When the antenna is more than one wavelength long it usually is called a long-wire antenna, or a harmonic antenna.

Current and Voltage Distribution

Fig. 14 shows the current and voltage distribution along a wire operating at its fundamental frequency (where its length is equal to a half wavelength) and at its second, third and fourth harmonics. For example, if the fundamental frequency of the antenna is 7 MHz, the current and voltage distribution will be as shown at A. The same antenna excited at 14 MHz would have current and voltage distribution as shown at B. At 21 MHz, the third harmonic of 7 MHz, the current and voltage distribution would be as in C; and at 28 MHz, the fourth harmonic, as in D. The number of the harmonic is the number of half waves contained in the antenna at the particular operating frequency.

The polarity of current or voltage in each standing wave is opposite to that in the adjacent standing waves. This is shown in the figure by drawing the current and voltage curves successively above and below the antenna (taken as a zero reference line), to indicate that the polarity reverses when the current or voltage goes through zero. Currents flowing in the same direction are *in phase;* in opposite directions, *out of phase.*

It is evident that one antenna may be used for harmonically-related frequencies, such as the various amateur bands. The long-wire or harmonic antenna is the basis of multiband operation with one antenna.

Physical Lengths

The length of a long-wire antenna is not an exact multiple of that of a half-wave antenna because the end effects operate only on the end sections of the antenna; in other parts of the wire these effects are absent, and the wire length is approximately that of an equivalent portion of the wave in space. The formula for the length of a long-wire antenna, therefore, is

$$Length \text{ (feet)} = \frac{492 \ (N \ - \ 0.05)}{\text{Freq. (MHz)}}$$

where N is the number of *half*-waves on the antenna.

Example: An antenna 4 half-waves long at 14.2 MHz would be

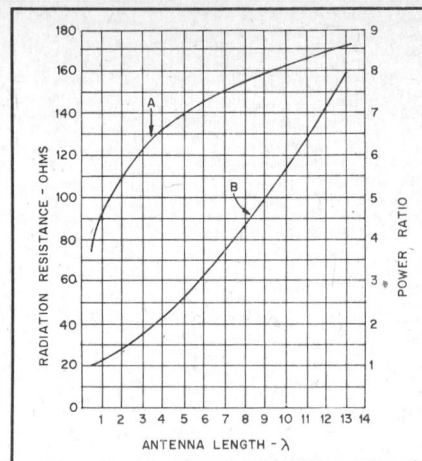

Fig. 15 — Curve A shows variation in radiation resistance with antenna length. Curve B shows power in lobes of maximum radiation for long-wire antennas as a ratio to the maximum radiation for a half-wave antenna.

$$\frac{492 \ (4 \ - \ 0.05)}{14.2} = \frac{492 \ (3.95)}{14.2}$$

= 136.7 feet, or 136 feet 8 inches.

It is apparent that an antenna cut as a halfwave for a given frequency will be slightly off resonance at exactly twice that frequency (the second harmonic), because of the decreased influence of the end effects when the antenna is more than one-half wavelength long. The effect is not very important, except for a possible unbalance in the feeder system and consequent radiation from the feedline. If the antenna is fed in the exact center, no unbalance will occur at any frequency, but end-fed systems will show an unbalance on all but one frequency in each harmonic range.

Impedance and Power Gain

The radiation resistance as measured at a current loop becomes higher as the antenna length is increased. Also, a long-wire antenna radiates more power in its most favorable direction than does a half-wave antenna in its most favorable direction. This power gain is secured at the expense of radiation in other directions. Fig. 15 shows how the radiation resistance and the power in the lobe of maximum radiation vary with the antenna length.

Directional Characteristics

As the wire is made longer in terms of the number of half wavelengths, the directional effects change. Instead of the "doughnut" pattern of the half-wave antenna, the directional characteristic splits up into "lobes" which make various angles with the wire. In general, as the length of the wire is increased the direction in which maximum radiation occurs tends to approach the line of the antenna itself.

Directional characteristics for antennas one wavelength, three half-wavelengths, and two wavelengths long are given in Figs. 16, 17 and 18, for three vertical angles of radiation. Note that, as the wire length increases, the radiation along the line of the antenna becomes more pronounced. Still longer antennas can be considered to have practically "end-on" directional characteristics, even at the lower radiation angles.

When a long-wire antenna is fed at one end or at the current loop closest to that end, the radiation is most pronounced from the long section. This unidirectional pattern can be accentuated by terminating the far end in a resistance to ground. The load resistor will dissipate energy that would ordinarily be radiated toward the feed point. Depending on the pattern symmetry of the unterminated antenna, this resistor must handle up to half the power delivered to the feed point. The exact resistance must be determined empirically, but the voltage-to-current ratio at a current node is a good starting value. A quarter-wavelength wire beyond the resistor can serve as a pseudo ground for the system. Low-angle radiation from a long wire can be enhanced by sloping the wire down toward the favored direction.

Methods of Feeding

In a long-wire antenna, the currents in adjacent half-wave sections must be out of phase, as shown in Fig. 14. The feeder system must not upset this phase relationship. This is satisfied by feeding the antenna at either end or at any current loop. A two-wire feeder cannot be inserted at a current node, however, because this invariably brings the currents in two adjacent half-wave sections in phase. A long wire antenna is usually made a half wavelength at the lowest frequency and fed at the end.

LONG-WIRE DIRECTIVE ARRAYS

Two long wires can be combined in the form of a horizontal "V", in the form of a horizontal rhombus, or in parallel, to provide a long-wire directive array. In the "V" and rhombic antennas the main lobes reinforce along a line bisecting the acute angle between the wires; in the parallel antenna the reinforcement is along the line of the lobe. This reinforcement provides both gain and directivity along the line, since the lobes in other directions tend to cancel. When the proper configuration for a given length and height above ground is used, the power gain depends upon the length (in wavelengths) of the wires.

Rhombic and "V" antennas are normally bidirectional along the bisector line mentioned above. They can be made unidirectional by terminating the ends of the wires away from the feed point in the proper value of resistance. When properly terminated, "V" and rhombic antennas

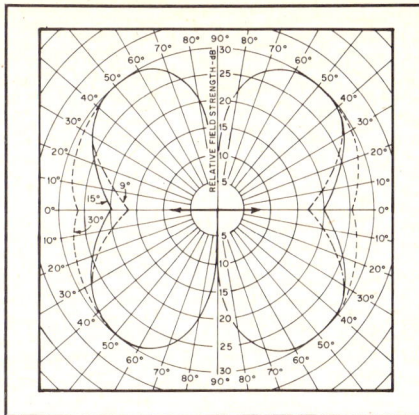

Fig. 16 — Horizontal patterns of radiation from a full-wave antenna. The solid line shows the pattern for a vertical angle of 15 degrees; dotted lines show deviation from the 15-degree pattern at 9 and 30 degrees. All three patterns are drawn to the same relative scale; actual amplitudes will depend upon the height of the antenna.

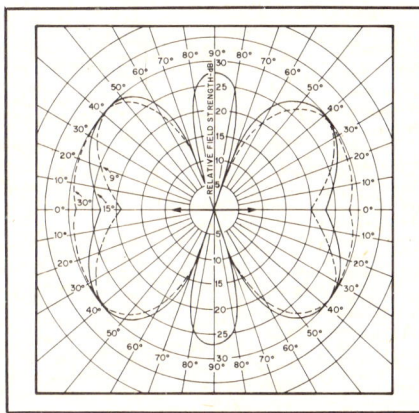

Fig. 17 — Horizontal patterns of radiation from an antenna three half-waves long. The solid line shows the pattern for a vertical angle of 15 degrees; dotted lines show deviation from the 15-degree pattern at 9 and 30 degrees. Minor lobes coincide for all three angles.

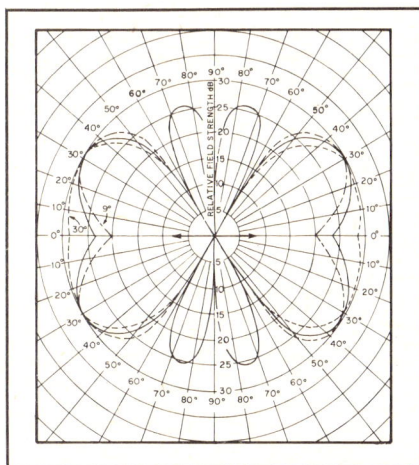

Fig. 18 — Horizontal patterns of radiation from an antenna two wavelengths long. The solid line shows the pattern for a vertical angle of 15 degrees; dotted lines show deviation from the 15-degree pattern at 9 and 30 degrees. The minor lobes coincide for all three angles.

of sufficient length work well over a three-to-one or four-to-one frequency range and hence are useful for multiband operation.

Antenna gains of the order of 10 to 15 dB can be obtained with properly constructed long-wire arrays. However, the pattern is rather sharp with gains of this order, and rhombic and "V" beams are not used by amateurs as commonly as they were, having been displaced by the rotatable multi-element Yagi beam. Further information on these antennas can be found in *The ARRL Antenna Book*.

BEAMS WITH DRIVEN ELEMENTS

By combining individual half-wave antennas into an array with suitable spacing between the antennas (called elements) and feeding power to them simultaneously, it is possible to make the radiation from the elements add up along a single direction and form a beam. In other directions the radiation tends to cancel, so a power gain is obtained in one direction at the expense of radiation in other directions. There are several methods of arranging the elements. If they are strung end to end, so that all lie on the same straight line, the elements are said to be collinear. If they are parallel and all lying in the same plane, the elements are said to be broadside when the phase of the current is the same in all, and end-fire when the currents are not in phase.

Collinear Arrays

Simple forms of collinear arrays, with the current distribution, are shown in Fig. 19. The two-element array at A is popularly known as a Franklin array, "two half-waves in phase" or a double Zepp antenna. It will be recognized as simply a center-fed dipole operated at its second harmonic.

By extending the antenna, as at B, the additional gain of an extended double Zepp antenna can be obtained. Carrying the length beyond that shown will result in an "X"-shaped pattern that no longer has the maximum radiation at right angles to the wire.

Collinear arrays may be mounted either horizontally or vertically. Horizontal mounting gives increased azimuthal directivity, while the vertical directivity remains the same as for a single element at the same height. Vertical mounting gives the same horizontal pattern as a single element, but improves the low-angle radiation.

Broadside Arrays

Parallel antenna elements with currents in phase may be combined as shown in Fig. 20 to form a broadside array, so named because the direction of maximum radiation is broadside to the plane containing the antennas. Again the gain and directivity depend upon the spacing of the elements.

Broadside arrays may be suspended

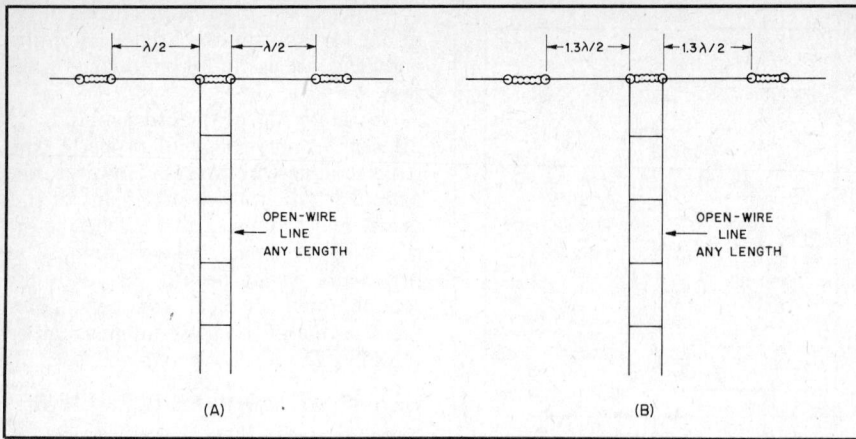

Fig. 19 — Collinear antennas in phase. The system at A is known as "two half waves in phase" and has a gain of 1.8 dB over a half-wave antenna. By lengthening the antenna slightly, as in B, the gain can be increased to 3 dB. Maximum radiation is at right angles to the antenna. The antenna at A is sometimes called a "double Zepp" antenna, and that of B is known as an "extended double Zepp."

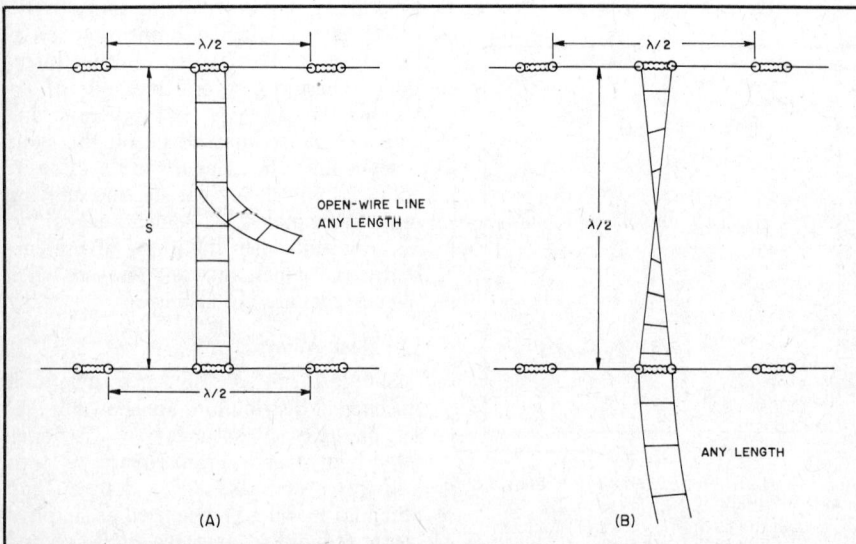

Fig. 20 — Simple broadside array using horizontal elements. By making the spacing S equal to 3/8 wavelength, the antenna at A can be used at the corresponding frequency and up to twice that frequency. Thus when designed for 14 MHz it can also be used on 18, 21, 25 and 28 MHz. The antenna at B can be used on only the design band. This array is bidirectional, with maximum radiation "broadside" or perpendicular to the antenna plane (perpendicularly through this page). Gain varies with the spacing S, running from 2-1/2 to almost 5 dB. (See Fig. 22)

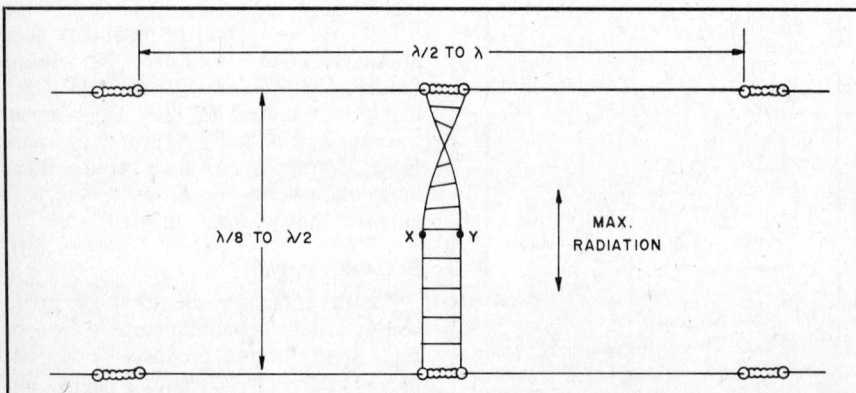

Fig. 21 — Top view of a horizontal end-fire array. The system is fed with an open-wire line at X and Y; the line can be of any length. Feed points X and Y are equidistant from the two insulators, and the feed line should drop down vertically from the antenna. The gain of the system will vary with the spacing, as shown in Fig. 22, and is a maximum at 1/8 wavelength. By using a length of 33 feet and a spacing of 8 feet, the antenna will work on 20, 17, 15, 12 and 10 meters.

either with the elements all vertical or with them horizontal and one above the other *(stacked)*. In the former case the horizontal pattern becomes quite sharp, while the vertical pattern is the same as that of one element alone. If the array is suspended horizontally, the horizontal pattern is equivalent to that of one element while the vertical pattern is sharpened, giving low-angle radiation.

Broadside arrays may be fed either by tuned open-wire lines or through quarter-wave matching sections and flat lines. In Fig. 20B, note the "crossing over" of the phasing section, which is necessary to bring the elements into proper phase relationship.

End-Fire Arrays

Fig. 21 shows a pair of parallel half-wave elements with currents out of phase. This is known as an end-fire array because it radiates best along the plane of the antennas, as shown. The end-fire principle was first demonstrated by John Kraus, W8JK, and 2-element arrays of this type are often called "8JK" antennas.

The end-fire array may be used either vertically or horizontally (elements at the same height), and is well adapted to amateur work because it gives maximum gain with relatively close element spacing. Fig. 22 shows how the gain varies with spacing. End-fire elements may be combined with additional collinear and broadside elements to give a further increase in gain and directivity.

Either tuned or untuned lines may be used with this type of array. Untuned lines preferably are matched to the antenna through a quarter-wave matching section or phasing stub.

Unidirectional End-Fire Arrays

Two parallel elements spaced 1/4 wavelength apart and fed equal currents 90 degrees out of phase will have a directional pattern in the plane of the array at right angles to the elements. The maximum radiation is in the direction from the element in which the current lags. In the opposite direction the fields from the two elements cancel.

One way in which the 90-degree phase difference can be obtained is shown in Fig. 23. Each element must be matched to its transmission line, the two lines being of the same type except that one is an electrical quarter wavelength longer than the other. The length *L* can be any convenient value. Open quarter-wave matching sections could be used instead. The two transmission lines are connected in parallel at the transmitter coupling circuit.

When the currents in the elements are neither in phase nor 180 degrees out of phase the radiation resistances of the elements are not equal. This complicates the problem of feeding equal currents to the elements. If the currents are not equal one or more minor lobes will appear in the pattern and decrease the front-to-back ratio. The adjustment process is likely to be tedious and requires field-strength measurements in order to get the best performance.

More than two elements can be used in unidirectional end-fire array. The requirement for unidirectivity is that there must be a progressive phase shift in the element currents equal to the spacing, in electrical degrees, between the elements, and the amplitudes of the currents in the various elements also must be properly related. This requires "binomial" current distribution — i.e., the ratios of the currents in the elements must be proportional to the coefficients of the binomial series. In the case of three elements, this requires that the current in the center element be twice that in the two outside elements, for 90-degree (quarter-wave) spacing and element current phasing. This antenna has an overall length of 1/2 wavelength.

Combined Arrays

Broadside, collinear and end-fire arrays may be combined to give both horizontal and vertical directivity, as well as additional gain. The lower angle of radiation resulting from stacking elements in the vertical plane is desirable at the higher frequencies. In general, doubling the number of elements in an array by stacking will raise the gain from 2 to 4 db.

Although arrays can be fed at one end as in Fig. 20 B, it is not especially desirable in the case of large arrays. Better distribution of energy between elements, and hence better overall performance will result when the feeders are attached as nearly as possible to the center of the array.

A four-element array, known as the "lazy-H" antenna, has been quite frequently used. This arrangement is shown, with the feed point indicated, in Fig. 24. (Compare with Fig. 20B). For best results, the bottom section should be at least a half wavelength above ground.

It will usually suffice to make the length of each element that given by the dipole formula. The phasing line between the parallel elements should be of open-wire construction, and its length can be calculated from:

Length of half-wave line (feet)

$$= \frac{480}{\text{Freq. (MHz)}}$$

Example: A half-wavelength phasing line for 28.8 MHz would be

$$\frac{480}{28.8} = 16.66 \text{ feet} = 16 \text{ feet 8 inches.}$$

The spacing between elements can be made equal to the length of the phasing line. No special adjustments of line or element length or spacing are needed, provided the formulas are followed closely.

The Vertical Antenna

One of the more popular amateur

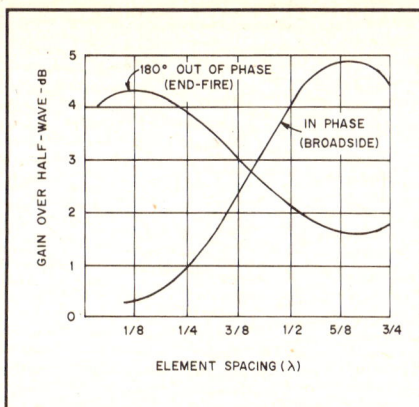

Fig. 22 — Gain vs. spacing for two parallel half-wave elements combined as either broadside or end-fire arrays.

antennas is the vertical type. With this style of antenna it is possible to obtain low-angle radiation for ground-wave and DX work. Additionally, the space occupied by vertical antennas is relatively small, making them ideal for city-lot property and apartment buildings. The principal limitation in performance is the omnidirectional pattern. This means that QRM can't be nulled out from the directions which are not of interest at a given period. The exception is, of course, when phased arrays of vertical elements are used. Despite the limitation of a single vertical element with a ground screen or radial system, cost versus performance is an incentive that inspires many antenna builders.

For use on the lower-frequency amateur bands — notably 160 and 80 meters — it is not always practical to erect a full-size vertical. In such instances it is satisfactory to accept a shorter radiating element and employ some form of *loading* to obtain an electrical length of one's choice. Most constructors design a system that contains a 1/4-wavelength driven element. However, good results and lower radiation angles are sometimes realized when using a 3/8- or 1/2-wavelength vertical. At the lower amateur frequencies the larger verticals become prohibitive, especially in urban areas where zoning ordinances may exist, and where limited acreage may rule out the installation of guy-wire systems.

Fig. 25 provides curves for physical height of verticals in wavelength versus radiation resistance and reactance. The plots are based on perfectly conducting ground, a condition which is seldom realized in practical installations. It can be seen that the shorter the radiator the lower the radiation resistance, with 6 ohms being typical for a 0.1-wavelength antenna. The lower the radiation resistance the more the antenna efficiency becomes dependent on ground conductivity. Also, the bandwidth decreases markedly as the length is reduced toward the left of the scale in Fig. 25. Difficulty is also experienced in developing a suitable match-

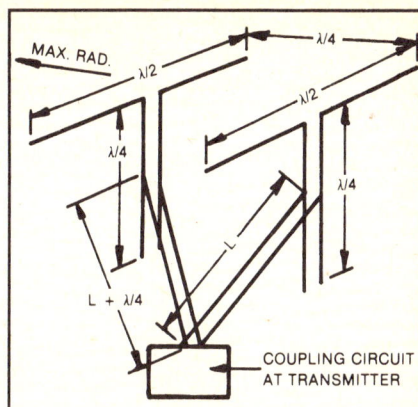

Fig. 23 — Unidirectional two-element end-fire array and method of obtaining 90-degree phasing.

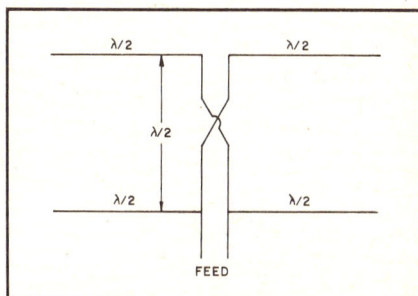

Fig. 24 — A four-element combination broadside collinear array, popularly known as the "lazy-H" antenna. A closed quarter-wave stub may be used at the feed point to match into an untuned transmission line, or tuned feeders may be attached at the point indicated. The gain over a half-wave antenna is 5 to 6 dB.

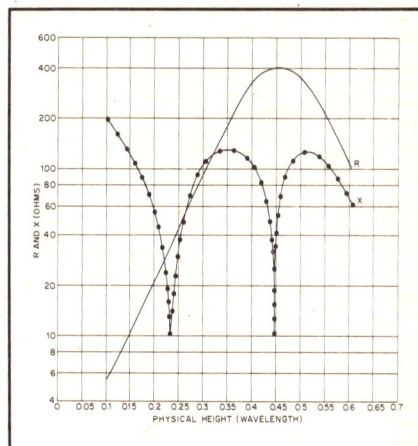

Fig. 25 — Radiation resistance (solid curve) and reactance (dotted curve) of vertical antennas as a function of physical height.

ing network when the radiation resistance is very low.

Illustrations of some vertical-antenna radiation patterns are given in Fig. 26. The example at A is for a quarter-wavelength radiator over a theoretically ideal ground. The dashed lines show the current distribution, inclusive of the *image* portion below ground. The image can

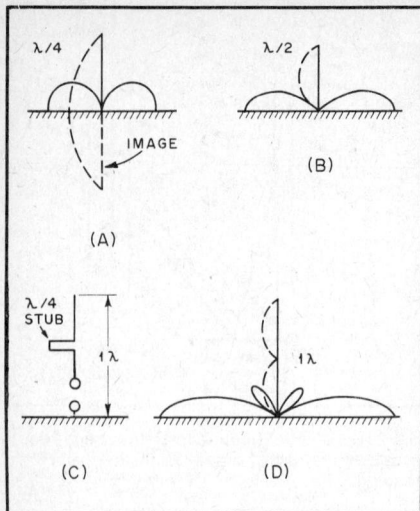

Fig. 26 — Elevation plane responses for a quarter-wavelength vertical antenna (A), a 1/2-wavelength type (B) and two half waves in phase (C and D). It can be seen that the examples at B and D provide lower radiation angles than the version at A.

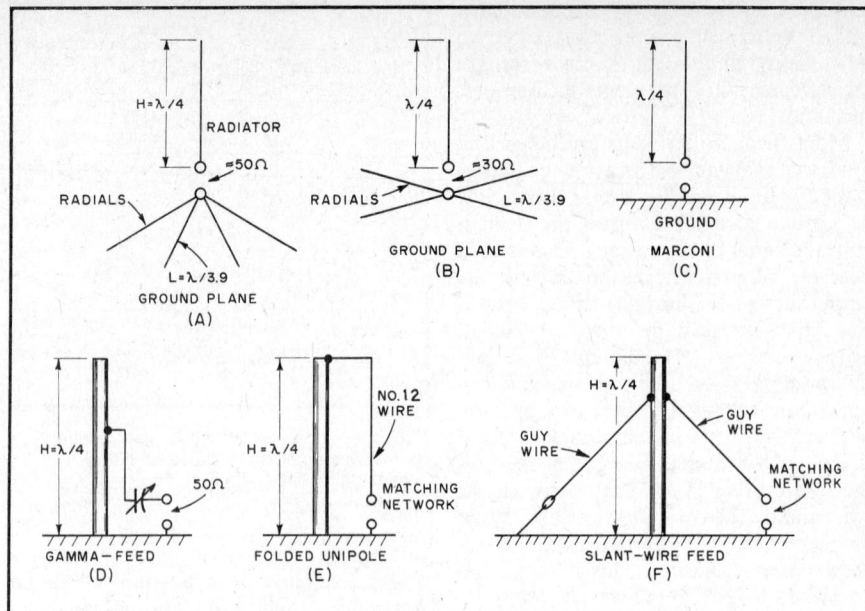

Fig. 27 — Various types of vertical antennas.

be equated to one half of a dipole antenna, with the vertical radiator representing the remaining dipole half. The illustration at B characterizes the pattern of a half-wavelength vertical. It can be seen that the radiation angle is somewhat lower than that of the quarter-wavelength version at A. The lower angles enhance the DX capability of the antenna. Two half wavelengths in phase are shown in Fig. 26 at C and D. From a practical point of view, few amateurs could erect such an antenna unless it was built for use on the higher hf bands, such as 20, 15 or 10 meters. The very low radiation angle is excellent for DXing, however.

Full-Size Vertical Antennas

When it is practical to erect a full-size vertical antenna, the forms shown in Fig. 27 are worthy of consideration. The example at A is the well-known *vertical ground plane*. The ground system consists of four or more above-ground radial wires against which the driven element is worked. The driven element length in feet is derived from the standard equation

$$L \text{ (feet)} = \frac{234}{f(\text{MHz})}$$

$$L \text{ (meters)} = \frac{71.3}{f(\text{MHz})}$$

The radial wires are slightly longer, approximately $\lambda/3.9$ yielding the dimension in feet. It has been established generally that with four equidistant radial wires drooped at approximately 45 degrees (Fig. 27A) the feed impedance is roughly 50 ohms. When the radials are at right angles to the radiator (Fig. 27B) the feed impedance approaches 30 ohms. The major advantage in this type of vertical antenna over a ground-mounted type is

that the system can be elevated well above nearby conductive objects (power lines, trees, buildings, etc.). When drooping radials are utilized they can be used as guy wires for the mast which supports the antenna. The coaxial cable shield braid is connected to the radials and the center conductor is common to the driven element.

The Marconi antenna seen in Fig. 27 at C is the classic form taken by a ground-mounted vertical. It can be grounded at the base and shunt fed, or it can be isolated from ground, as shown, and series fed. This antenna depends upon an effective ground system for efficient performance. The subject of ground screens is treated later in this section. If a perfect ground were located below the antenna, the feed impedance would be near 30 ohms. In a practical case, owing to imperfect ground, the impedance is more apt to be in the vicinity of 50 to 75 ohms.

A gamma-feed system for a grounded 1/4-wavelength vertical is presented in Fig. 27D. Some rules of thumb for arriving at workable gamma-arm and capacitor dimensions are to make the rod length 0.04 to 0.05 wavelength, its diameter 1/3 to 1/2 that of the driven element and the center-to-center spacing between the gamma arm and the driven element roughly 0.007 wavelength. The capacitance of C1 at a 50-ohm matched condition will be some 7 pF per meter. The absolute value at C1 will depend upon whether the vertical is resonant and the precise value of the radiation resistance. Generally, best results can be had when the radiator is approximately three percent shorter than the resonant length. Amateur antenna towers lend themselves well to use as shunt fed verticals, even though an hf-band beam antenna may be mounted on the tower. The overall system should

be close to resonance at the desired operating frequency if gamma feed is to be used. The hf-band beam will contribute somewhat to top loading of the tower. The natural resonance of such a system can be checked by dropping a no. 12 or 14 wire down from the top of the tower (making it common to the tower top) to form a folded unipole (Fig. 27E). A four- or five-turn link can be inserted between the lower end of the drop wire and the ground system, then a dip meter inserted in the link to observe the resonant frequency. If the tower is equipped with guy wires, the latter should be broken up with strain insulators to prevent unwanted loading of the vertical. In such cases where the tower and beam antenna are not able to provide 1/4-wavelength resonance, portions of the top guy wires can be used as top-loading capacitance. It will be necessary to experiment with the guy-wire lengths (using the dip-meter technique) while determining the proper dimensions.

A folded-unipole type of vertical is depicted at E of Fig. 27. This system has the advantage of increased feed-point impedance. Furthermore, a Transmatch can be connected between the bottom of the drop wire and the ground system to permit operation on more than one band. For example, if the tower is resonant on 80 meters, it can be used as shown on 160 and 40 meters with reasonable results, even though it is not electrically long enough on 160. The drop wire need not be a specific distance from the tower, but spacings between 12 and 30 inches are suggested.

The method of feed seen at Fig. 27F is commonly referred to as "slant-wire feed." The guy wires and the tower combine to provide quarter-wave resonance. A matching network is placed

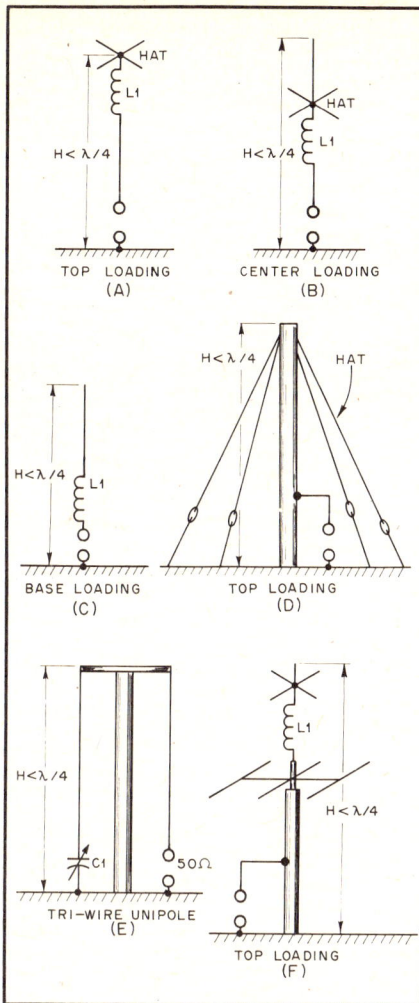

Fig. 28 — Vertical antennas that are less than one quarter wavelength in height.

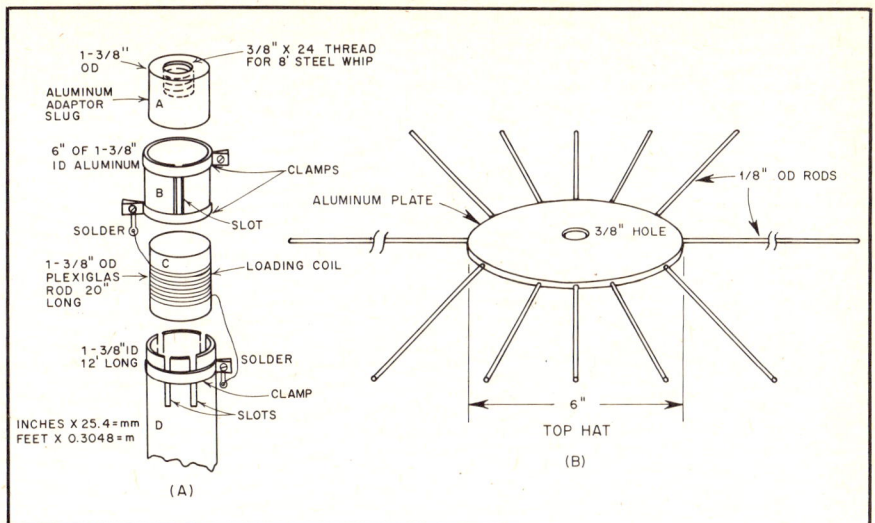

Fig. 29 — At A are the details for the tubing section of the loading assembly. Illustration B shows the top hat and its spokes. The longer the spokes, the better.

between the lower end of one guy wire and ground and adjusted for an SWR of 1. It does not matter at which level on the tower the guy wires are connected, assuming that the Transmatch is capable of effecting a match to 50 ohms.

Physically Short Verticals

A group of short vertical radiators is presented in Fig. 28. Illustrations A and B are for top and center loading. A capacitance hat is shown in each example. It should be as large as practical to increase the radiation resistance of the antenna and improve the bandwidth. The wire in the loading coil is chosen for the largest gauge consistent with ease of winding and coil-form size. The larger wire diameters will reduce the I²R losses in the system. The coil-form material should be of the medium- or high-voltage breakdown resistance dielectric type. Phenolic or fiberglass tubing is entirely adequate.

A base-loaded vertical is shown at C of Fig. 28. Since this is the least effective method of loading in terms of antenna performance, it should be used only as a last choice. The primary limitation is that

the current portion of the vertical exists in the coil rather than the driven element. With center loading the portion of the antenna below the coil carries current, and with the top-loading version the entire vertical element carries current. Since the current part of the antenna is responsible for most of the radiating, base loading is the least effective of the three methods. The radiation resistance of the coil-loaded antennas shown is usually less than 16 ohms.

A method for using guy wires to top load a short vertical is illustrated in Fig. 28 at D. This system works well with gamma feed. The loading wires are trimmed to provide an electrical quarter wavelength for the overall system. This method of loading will result in a higher radiation resistance and greater band-width than the systems shown in Fig. 28 at A, B and C. If an hf-band or vhf array is atop the tower, it will simply contribute to the top loading.

A tri-wire unipole is shown at E of Fig. 28. Two no. 8 drop wires are connected to the top of the tower and brought to ground level. The wires can be spaced any convenient distance from the tower — normally 12 to 30 inches (0.3 to 0.76 m) from one side. C1 is adjusted for an SWR of 1. This type of vertical has a fairly narrow bandwidth, but because C1 can be motor-driven and controlled from the operating position, QSYing is accomplished easily. This technique will not be suitable for matching to 50-ohm line unless the tower is less than an electrical quarter wavelength.

A different method for top loading is shown at F of Fig. 28. W9UCW described this system in December 1974 QST as "The Minooka Special." An extension is used at the top of the tower to effect an electrical quarter-wavelength vertical. L1 is a loading coil with sufficient inductance to provide antenna resonance. This type

of antenna lends itself nicely to operation on 160 meters. L1 and the pipe extension above the hf-band beam can be tuned at ground level against the ground system. It should be made resonant approximately 100 kHz higher than the desired operating frequency for use on 160 meters. After it is in place on the tower, the overall system resonance will drop some 100 kHz.

A method for effecting the top-loading of Fig. 28, illustration F, is shown in the drawing of Fig. 29. Pipe section D is mated with the mast above the hf-band beam antenna. A loading coil is wound on solid Plexiglas rod or phenolic rod (item C), then clamped inside the collet (B). An aluminum slug (part A) is clamped inside item B. The top of part A is bored and threaded for a 3/8 inch × 24 thread stud. This will permit a standard 8-foot (2.4 m) stainless-steel mobile whip to be threaded into item A above the loading coil. The capacitance hat (Fig. 29, illustration B) can be made from a 1/4-inch (6.3-mm) thick brass or aluminum plate. It may be round or square. Lengths of 1/8-inch (3-mm) brazing rod can be threaded for a 6-32 format to permit the rods to be screwed into the edge of the aluminum plate. The plate contains a row of holes along its perimeter, each having been tapped for a 6-32 thread. The capacitance hat is affixed to item A by means of the 8-foot whip antenna. The whip will increase the effective height of the vertical antenna.

Cables and Control Wires on Towers

Most vertical antennas of the type shown in Fig. 28 consist of towers and hf or vhf beam antennas. The rotator control wires and the coaxial feeders to the top of the tower will not affect antenna performance adversely. In fact, they become a part of the composite antenna. To prevent unwanted rf currents from following the wires into the shack, simply dress them close to the tower legs and bring

GUY ○—————○——● GUY
|← L →|
L+H=λ/4
H
○
FLAT-TOP T
(A)

GUY ←————○——○————→ GUY
|← L →|
H λ/4
○
INVERTED L
(B)

≈50Ω
○
L λ/4
GUY
HALF SLOPER
(C)

L1 - TOWER ∠ ≈ 45°
L2 - TOWER ∠ ≈ 30°
$L1, L2 ≈ 1.045 \left(\frac{234}{f(MHz)}\right)$ FT.

HF BEAM
L1
≈ 90°
L2
GUY
GUY
(D)
GROUND SYSTEM

0°
315° 45°
L1
270° L2 90°
225° 135°
180°
(E)

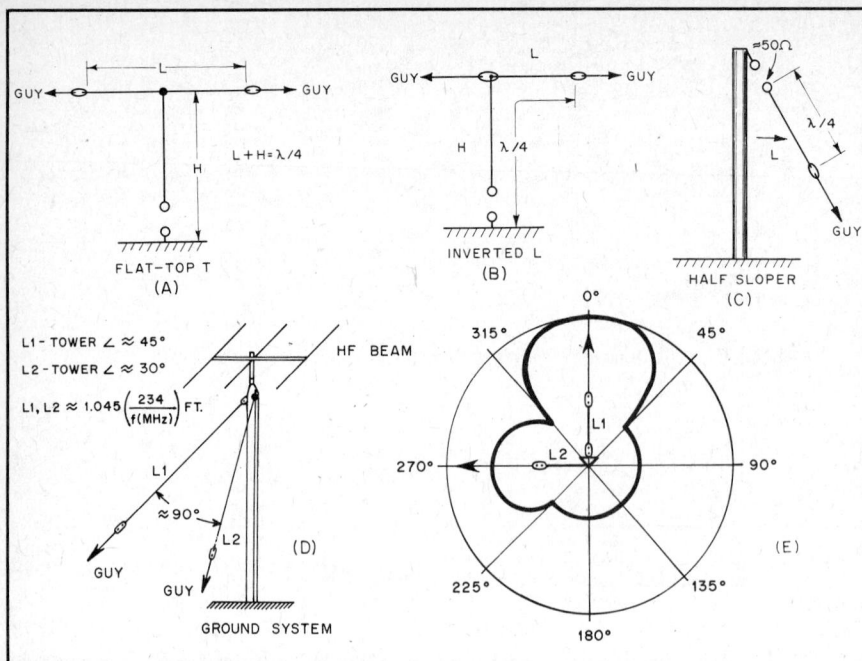

Fig. 30 — Some variations in vertical antennas which offer excellent performance.

them to ground level. This decouples the wires at rf. The wires should then be routed along the earth surface (or buried underground) to the operating position. It is not necessary to use bypass capacitors or rf chokes in the rotator control leads if this is done, even when maximum legal power is employed.

Variations in Verticals

A number of configurations qualify for use as vertical antennas even though the radiators are fashioned from lengths of wire. Fig. 30 at A shows a flat-top T vertical. Dimension H should be as tall as possible for best results. The horizontal section, L, is adjusted to a length which provides resonance. Maximum radiation is polarized vertically despite the horizontal top-loading wire. A variation of the T antenna is depicted at B of Fig. 30. This antenna is commonly referred to as an "inverted L." Again, vertical member H should be as long as possible. L is added to provide an electrical quarter wavelength overall. Some amateurs believe that a 3/8-wavelength version of this antenna is more effective, since the current portion of the wire is elevated higher above ground than is the case with a quarter-wavelength wire.

Half-Sloper Antenna

A basic half-sloper antenna is shown at C of Fig. 31. The wire portion represents one half of a dipole or inverted-V. The feed point is between the tower leg and the upper end of the slope wire, with the coax shield braid being connected to the tower. The feed-point impedance is generally between 30 and 60 ohms, depending on the length of the wire, the tower height and the enclosed angle between the slope wire

and the tower. Polarization is vertical and there is directivity off the sloper side of the tower. Gain in the favored direction is on the order of 3 dB, depending on the quality of the ground below the tower and antenna. The tower constitutes a portion of this antenna and will have a voltage maximum somewhere between the feed point and ground. The radiation angle appears to be between 20 and 30 degrees, making the half-sloper useful for close-in work as well as for DXing. A VSWR as low as 1.5:1 can be obtained (using 50-ohm coax feed line with the system of Fig. 30 at C by experimenting with the slope-wire length and the enclosed angle. This assumes that there are no guy wires that are common to the tower. If guys are used they should be insulated from the tower and broken into nonresonant lengths for the band of operation.

Most amateurs are likely to have towers that support hf-band Yagis or quad beams. The conductors situated above the sloper feed point have a marked effect on the tuning of the sloper system. The beam antenna becomes a portion of the overall tower/sloper system. Although the presence of the beam seems to have no appreciable effect on the radiation pattern of the sloper, it does greatly affect the VSWR. If a low VSWR can't be obtained for a given amateur antenna installation, the system illustrated in Fig. 30 at D can be applied. The sloper is erected in the usual manner, but a second wire (L2) is added. It is connected to the tower at the antenna feed point, but insulated at the low end. L2 plus the tower constitute the missing half of the dipole. This half is effectively a fanned element (similar to a conical element) and tends to increase the overall bandwidth of the sloper system

over that which can be obtained with the version at C of Fig. 30. The "compensating wire" (L2) is spaced approximately 90° from L1, and is roughly 30° away from the side of the tower. The VSWR can be brought to a low value (usually 1:1) by varying the L2/tower and L1/L2 enclosed angles. The degree of these angles will be dependent upon the tower height, the beam atop the tower and the ground condition below the system.

Fig. 30E shows the relative radiation pattern of the antenna in Fig. 30D. The maximum radiation is off slope wire L1, with a minor lobe occurring off the compensating wire, L2. The pattern shown was obtained with a field-strength meter placed at the ends of 2-wavelength radials. Without L2 in the system, a single prominent lobe prevails. However, the half-sloper is otherwise (and for all practical purposes) omnidirectional. Users may want to orient the pattern in some favored DX direction.

Tests indicate this antenna to be effective on its harmonics for DX and local work. If harmonic operation is planned, the 50-ohm coaxial feeder should be replaced with open-wire line, which is coupled to the transmitter by means of a Transmatch. With open-wire feed, the L2 compensating wire of Fig. 30D will not be necessary. Best performance will be had when the base of the tower is well grounded. Buried radials are highly recommended as part of the ground system. Operation on 160 meters can be had by feeding the low end of L2.

The bandwidth of the system in Fig. 30 will be approximately 50 kHz on 160 meters, 100 kHz on 80 meters, 200 kHz on 40 meters, and so on, between the 2:1 VSWR points.

Ground Systems

The importance of an effective ground system for vertical antennas cannot be emphasized too strongly. However, is it not always possible to install a radial network that approaches the ideal. It might be said that "a poor ground is better than no ground at all," and therefore the amateur should experiment with whatever is physically possible rather than exclude vertical antennas from his or her construction plans. It is often possible to obtain excellent DX results with practically no ground system at all. Although the matter of ground systems could be debated almost endlessly, some practical rules of thumb are in order for those wishing to erect vertical antennas. Generally, if the physical height of the vertical is an eighth wavelength, the radial wires should be of the same length and dispersed uniformly from the base of the tower. In this example approximately 60 radial wires will suffice. The conductor size of the radials is not especially significant. Wire gauges from no. 4 to no. 20 have been used successfully by amateurs.

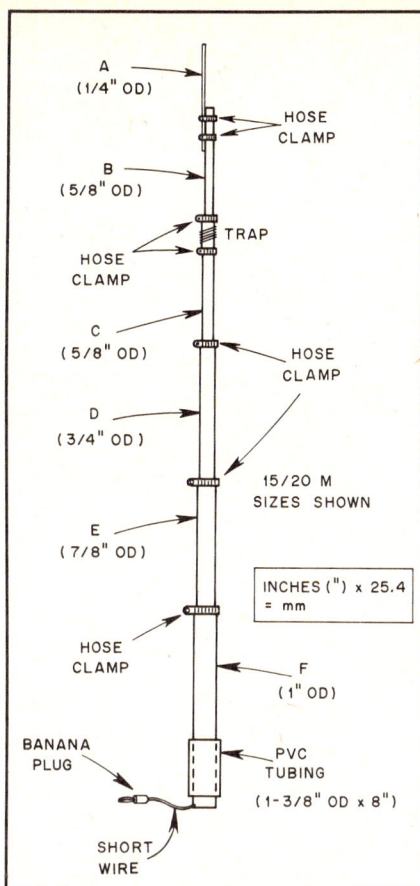

Fig. 31 — Details of the two-band trap vertical, which telescopes to 39 inches when dismantled. Stainless-steel hose clamps are used to hold the tubing sections together and to affix the trap to the tubing. A short length of flexible wire and a banana plug are connected to the base of the antenna for joining the antenna to the coax connector of Fig. 33.

Table 1

Tubing-Section Lengths for 2-Band Vertical

	Band (MHz)	A	B	C	D	E	F	C1 (pF)[1]	L1 (approx. µH)
Tubing Length (Inches)	21/28	25	16	25	25	25	33	18	1.70
	14/21	38	33	37	37	37	33	25	2.25
	10/14[1]	42	42	54	54	54	49	39	3.25
Tubing Length at Resonance (approx. inches)[3]	21/28	20	16	21.5	21.5	21.5	33	—	—
	14/21	33	33	33	33	33	33	—	—
	10/14[1]	37	37	49.6	49.6	49.6	49	—	—

mm = in. × 25.4 dimensions.
[1] New WARC-79 band.
[2] See text.
[3] Midband Dimensions X_{C1}, X_{L1} ≈ 300 Ω.

Copper wire is preferred, but where soil acid or alkali is not high in level, aluminum wire can be used. The wires can be bare or insulated, and they can be laid on the earth surface or buried a few inches below ground. The insulated wires will have greater longevity by virtue of reduced corrosion and dissolution from soil chemicals.

The longer the vertical antenna the fewer and shorter the radials need be. For example, a vertical which is 1/4 wavelength high will provide good field strength with 16 to 18 radial wires, and the wires need be only as long as the vertical is high. If time and expense are not a prime consideration, the amateur should bury as much ground wire as possible. Some operators have literally *miles* of wire buried radially beneath their vertical antennas.

When property dimensions do not allow a classic installation of equally spaced radial wires, they can be placed in the ground wherever space will permit. They may run away from the antenna in only one or two compass directions. Results will still exceed those of when no ground system is used.

A single ground rod, or a group of them bonded together, is seldom as effective as a collection of random-length radial wires. In some instances a group of short radial wires can be used in combination with ground rods driven into the soil near the base of the antenna. The power-company ground can be tied in also, and if a metal fence skirts the property it can also be used as part of the ground system. A good rule is to use anything that will serve as a ground when developing a radial-ground system.

All radial wires must be connected together at the base of the vertical antenna. The electrical bond needs to be of low resistance. Best results will be obtained when the wires are soldered together at the junction point. When a grounded vertical is used, the ground wires should be affixed securely to the base of the driven element. A lawn-edging tool is excellent for cutting slits in the soil when laying radial wires.

Trap Verticals

Although a full-size, single-band antenna is more effective than a lumped-constant one, there is justification for using trap types of multiband antennas. The concept is especially useful to operators who have limited antenna space on their property. Multiband "Compromise" antennas are also appealing to persons who engage in portable operation and are unwilling to transport large amounts of antenna hardware to the field.

The trap vertical antenna operates in much the same manner as a trap dipole or trap-style Yagi. The notable difference is that the vertical is one half of a dipole. The radial system (in-ground or above ground) functions as a ground plane for the antenna, and represents the missing half of the dipole. Therefore, the more effective the ground system the better the antenna performance.

Trap verticals are adjusted as quarter-wavelength radiators. The portion of the antenna below the trap is adjusted as a quarter-wavelength radiator at the highest proposed operating frequency, i.e., a 20/15-meter trap vertical would be a resonant quarter wavelength at 15-meters from the feed point to the bottom of the trap. The trap and that portion of the antenna above the trap (plus the 15-meter section below the trap) constitute the complete antenna during 20-meter operation. But, because the trap is in the circuit the overall physical length of the vertical antenna will be slightly less than that of a single-band, full-size 20-meter vertical.

Traps

The trap functions as the name implies: It traps the 15-meter energy and confines it to the part of the antenna below the trap. During 20-meter operation it allows the rf energy to reach all of the antenna. Therefore, the trap should in this example be tuned as a parallel resonant circuit to 21 MHz. At this frequency it "divorces" the top section of the vertical from the lower section because it presents a high-impedance (barrier) at 21 MHz. Generally, the trap inductor and capacitor have a reactance of 100 to 300 ohms. Within that range it is not critical.

The trap is built and adjusted separately from the antenna. It should be resonated at the center of the portion of the band to be operated. Thus, if one's favorite part of the 15-meter band is between 21,000 and 21,100 kHz, the trap would be tuned to 21,050 kHz.

Resonance is checked by using a dip meter and beating the dipper signal against a calibrated receiver. Once the trap is adjusted, it can be installed in the antenna, and no further adjustment will be required. It is easy, however, to be misled after the system is assembled: Attempts to check the trap with a dip meter will suggest that the trap has moved much lower in frequency (approximately 5 MHz lower in a 20/15-meter vertical). This is because the trap has become absorbed into the overall antenna, and the resultant resonance is that of the total antenna. Ignore this phenomenon.

Multiband operation for three or four bands is entirely practical by using the appropriate number of traps and tubing sec-

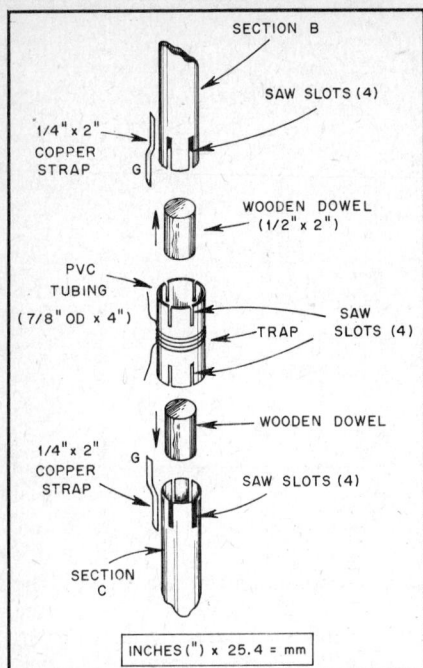

Fig. 32 — Close-up details of the trap construction and how it connects to tubing sections B and C of Fig. 31.

A Practical Two-Band Vertical

Fig. 31 contains a pictorial view of a two-band trap vertical (20/15 meters) which can be collapsed to 39 inches (991 mm) for easy transportation on holidays, DXpeditions or camping trips. All of the tubing sections except B telescope together to make a compact package. The trap and base plate will be separate from the remainder of the antenna during storage or transport.

If portability is not a requirement, a single section of aluminum tubing can be used below the trap, although two sections (telescoping) are recommended to facilitate adjustment of the 15-meter portion of the system. Similarly, two telescoping tubes can be used above the trap (as shown) to permit adjustment for 20-meter operation.

Table 1 contains data on the starting lengths of the tubing sections, plus approximate dimensions for resonance on a variety of band pairings. Final adjustment is done for the lowest VSWR attainable in the chosen part of each band (resonance). The adjustment must be done while the antenna is mounted for use with the ground system in place.

Fig. 32 shows the details of a simple trap with the tubing sections keyed to the nomenclature of Fig. 31. The ID of the PVC tubing is too small to accommodate the 1/2-inch OD (13-mm) tubing. Therefore, a hacksaw is used to cut four slots at the ends of sections B and C so they will compress and fit into the PVC tubing. The wooden dowel plugs permit a tight bond when the hose clamps are compressed over the ends of the PVC tubing. Innovative builders can find other methods for mounting the trap in the antenna.

Copper straps G are slid into the PVC tubing to provide electrical contact with tubing sections B and C. The straps are bent (as shown) so they will fit under the hose clamps that affix the trap to the tubing. The trap capacitor and coil leads are soldered to the copper straps.

Silicone grease is applied in a thin layer between the copper straps and the tubing sections to prevent undue oxidation. Similarly, the grease is applied to the mating surfaces of all of the tubing sections.

If coax cable is used for the trap capacitor, it can be taped (after soldering) to the upper tubing section (B) of Fig. 31. If a fixed-value transmitting capacitor is used, it should be located at the trap coil by means of stiff wire leads, which are soldered to copper tabs G. If a fixed-value capacitor is used, the trap can be brought to resonance by adjusting the coil turns. The exposed ends of the coax capacitor should be weather-sealed with non-corrosive RTV compound.

Ground System

If the antenna is mounted at ground

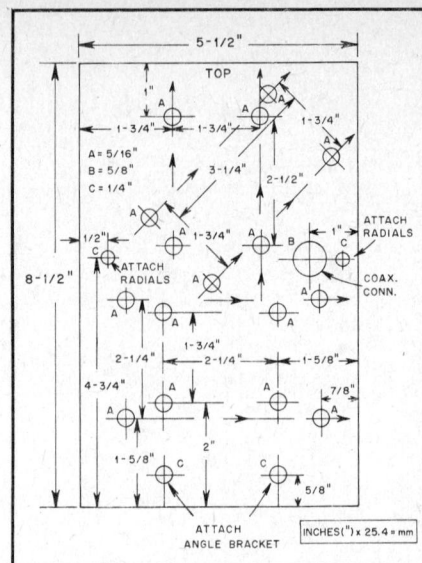

Fig. 33 — Layout details for a universal mounting plate. The hole sizes and spacing will depend on the type of U bolts used (see text).

level, section F of Fig. 31 should be as close to ground as possible. At least 20 radials are recommended. They need not be longer than 20 feet, and should be buried from 2 to 4 inches below the surface of the soil or lawn.

If an above-ground installation is planned, a minimum of two radial wires should be used for each band. It is suggested that four radials be used for each band. The wires are cut to one-quarter wavelength for each of the bands, although some amateurs prefer to make them 5 percent longer than one-quarter wavelength to increase the antenna band-width slightly.

The slope of the above-ground radials can be changed to help provide a match to 50-ohm feed line. The greater the angle between the vertical element and the radial wires, the higher the feed impedance. The feed impedance will be approximately 30 ohms when the radials are at right angles to the vertical element. If such an installation is contemplated, a 1.6:1 broadband toroidal transformer can be used at the feed point to effect a matched condition.

Universal Mounting Plate

Fig. 33 illustrates a mounting plate which will satisfy a host of conditions one might encounter when operating from an unfamiliar place. The hole size and spacing will depend on the U bolts or muffler clamps used with the antenna. The lower set of holes (except the bottom-most two) permit using a supporting mast that is either vertical or horizontal. The holes in the top half of the plate permit the antenna to be mounted vertically or at a 45° angle. Hole B is for a female-to-female bulkhead connector. The feed line attaches to one side of the fitting and the banana plug of Fig. 31 connects to the

tions. The construction and adjustment procedure is the same, irrespective of the number of bands covered. The highest-frequency section is always closest to the feed end of the antenna, and the lowest-frequency section is always the farthest from the feed point. As the operating frequency is progressively lowered, more traps and more tubing sections become a functional part of the antenna.

The trap should be weather-proofed to prevent moisture from detuning it. Several coatings of high-dielectric compound, such as Polystyrene Q Dope, are effective. Alternatively, a protective sleeve of heat-shrink tubing can be applied to the coil after completion. The coil form for the trap should be of high-dielectric quality and be rugged enough to sustain stress during periods of wind.

The trap capacitor must be capable of withstanding the rf voltage developed across it. The amount of voltage present will depend on the operating power of the transmitter. Fixed-value ceramic transmitting capacitors are suitable for most power levels if they are rated at 5000 to 10,000 volts. A length of RG-58 or RG-59 coax cable can be used successfully up to 200 watts. (Check to see how many pF per foot your cable is before cutting it for the trap.) RG-8 or RG-11 cable is recommended for the trap capacitor at powers in excess of 200 watts. The advantage of using coax cable is that it can be trimmed easily to adjust the trap capacitance.

Large-diameter copper magnet wire is suggested for the trap coil. The heavier the wire gauge the lower the trap losses and the higher the Q. The larger wire sizes will reduce coil heating.

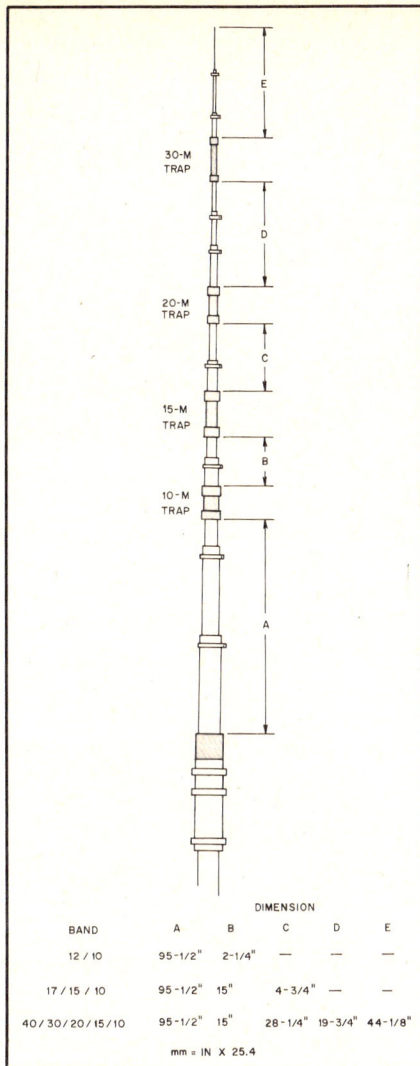

Fig. 34 — Modified dimensions for the ATV-series Cushcraft vertical antennas for some frequency combinations that include the WARC bands. The 30-meter trap inductor consists of 20 turns of no. 16 enameled wire close-wound on a 5/8-in. dia. Plexiglas rod. The capacitor is a 29-3/4-in. length of RG-58/U cable.

BAND	DIMENSION				
	A	B	C	D	E
12 / 10	95-1/2"	2-1/4"	—	—	—
17 / 15 / 10	95-1/2"	15"	4-3/4"	—	—
40 / 30 / 20 / 15 / 10	95-1/2"	15"	28-1/4"	19-3/4"	44-1/8"

mm = IN X 25.4

center hole of the opposite end of the connector. This permits easy disconnection when disassembling the antenna. The radials are bolted to the two holes marked C, at the left and right center of the plate. The two holes (C) at the bottom of the plate are for bolting an iron or aluminum angle stock to the plate. A second angle-stock piece is cut to the same size as the first and is used with the mounting hardware when it is convenient to clamp the mounting plate to a porch railing, window sill and so on. A pair of large C clamps can be used for this mounting technique.

The plate is made from 1/4-inch (6-mm) aluminum or iron stock. Brass or copper material could be used equally well, if available.

Adapting Commercial Trap Verticals to the WARC Bands

The frequency coverage of a multiband

vertical antenna can be modified simply by altering the lengths of the tubing sections and/or adding a trap. Several companies manufacture trap verticals covering 40, 20, 15 and 10 meters. Many amateurs roof-mount these antennas, because an effective ground radial system isn't practical, to keep children away from the antenna, or to clear metal-frame buildings. On the three highest bands, the tubing and radial lengths are convenient for rooftop installations, but 40 meters sometimes presents problems. Prudence dictates erecting an antenna with the assumption that it will fall down. When the antenna falls, it and the radial system must clear any nearby power lines. Where this consideration rules out 40-meter operation, careful measurement may show that 30-meter dimensions will allow adequate safety. The antenna is resonated by pruning the tubing above the 20-meter trap and installing tuned radials.

Several new frequency combinations are possible. The simpler ones, 12/10, 17/15/10, and 40/30/20/15/10 meters, are shown in Fig. 34 applied to the popular ATV series of trap verticals manufactured by Cushcraft. Operation in the 30-meter band requires an additional trap — use Fig. 32 as a guide for constructing this component.

A 360° Steerable Vertical Phased Array for 7 MHz

The original design information for the array presented here appeared in April 1976 *QST* in an article written by Atchley, Stinehelfer and White. That article featured a system designed for use in the 3.5-to 4.0-MHz range.

The configuration shown here makes use of four 1/4 wavelength vertical elements in a square, with quarter-wave spacing between adjacent elements, as shown, with its predicted pattern, in Fig. 35. All elements are fed with equal amplitudes, the rear element at 0°, the two side elements at −90° and the lead element at −180°. The beam is transmitted along the diagonal from the rear to the lead element. Gain due to horizontal beam formation alone is approximately 5.3 dB. Front-to-back ratio is on the order of 25 dB. Front-to-side ratio is 12 dB at 90° either side, increasing to much higher levels at 135° either side. Since most of the vertical energy is concentrated at low angles, as much as 4 dB of gain additional to the predicted 5.3 dB in the horizontal plane can be achieved in theory, with perfectly conducting ground. With a good radial system and less-than-perfect ground one can expect an approximate additional gain of 2 dB, or a total gain just over 7 dB for the system described.

A computer program calculation predicts the half-power beamwidth to be 97°. A suitable switching matrix is used to direct the beam pattern to four different

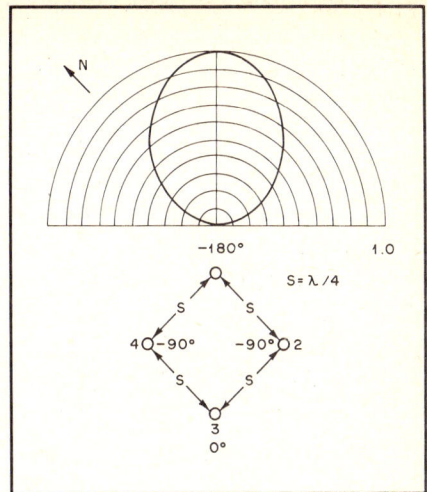

Fig. 35 — Polar plot of relative power, and planar view of the four-element diamond array showing the pattern obtained with no dc voltage on the switching relays, as in Fig. 36. Minor lobes are too far down to show on this scale.

quadrants, with only a slight loss of forward gain at the crossover points, and virtually no deterioration of the front-to-back and front-to-side suppression.

RF Power Dividers

Good power splitters are essential to the operation of phased arrays. The system described here makes use of 2-way Wilkinson Power Divider.[1] Those readers versed in microwave technology are no doubt familiar with these devices. As shown in Fig. 36, power from the transmitter is fed through 50-ohm line of any length to a T connection, feeding two quarter-wavelength lines, W1 and W2. The two inner conductors of the 70-ohm lines are connected through a 100-ohm noninductive resistance, R1. This type of divider gives an equal power split, matches to 50-ohm loads at the two outputs, and has the unique property that any energy returning to the two outputs out of phase, due to mismatches or mutuals, is absorbed in the 100-ohm resistor.

The 90-degree phase-delay cable (DL) used in one side of each of the three power-splitting hybrids serves to assure that equal power reflections from the antennas are absorbed. Theoretically the resistors absorb none of the forward power. This technique provides approximately 30 dB of isolation from one output terminal to the other, to unwanted energy. The Wilkinson Power Divider, when used to feed phased arrays, reduces the problems associated with element interaction due to mutual impedances between elements.

Feeding, Switching and Phasing

The four-element system shown here

[1]Wilkinson, "An N-Way Hybrid Power Divider," IRE *Transactions on Microwave Theory and Techniques*, January 1960.

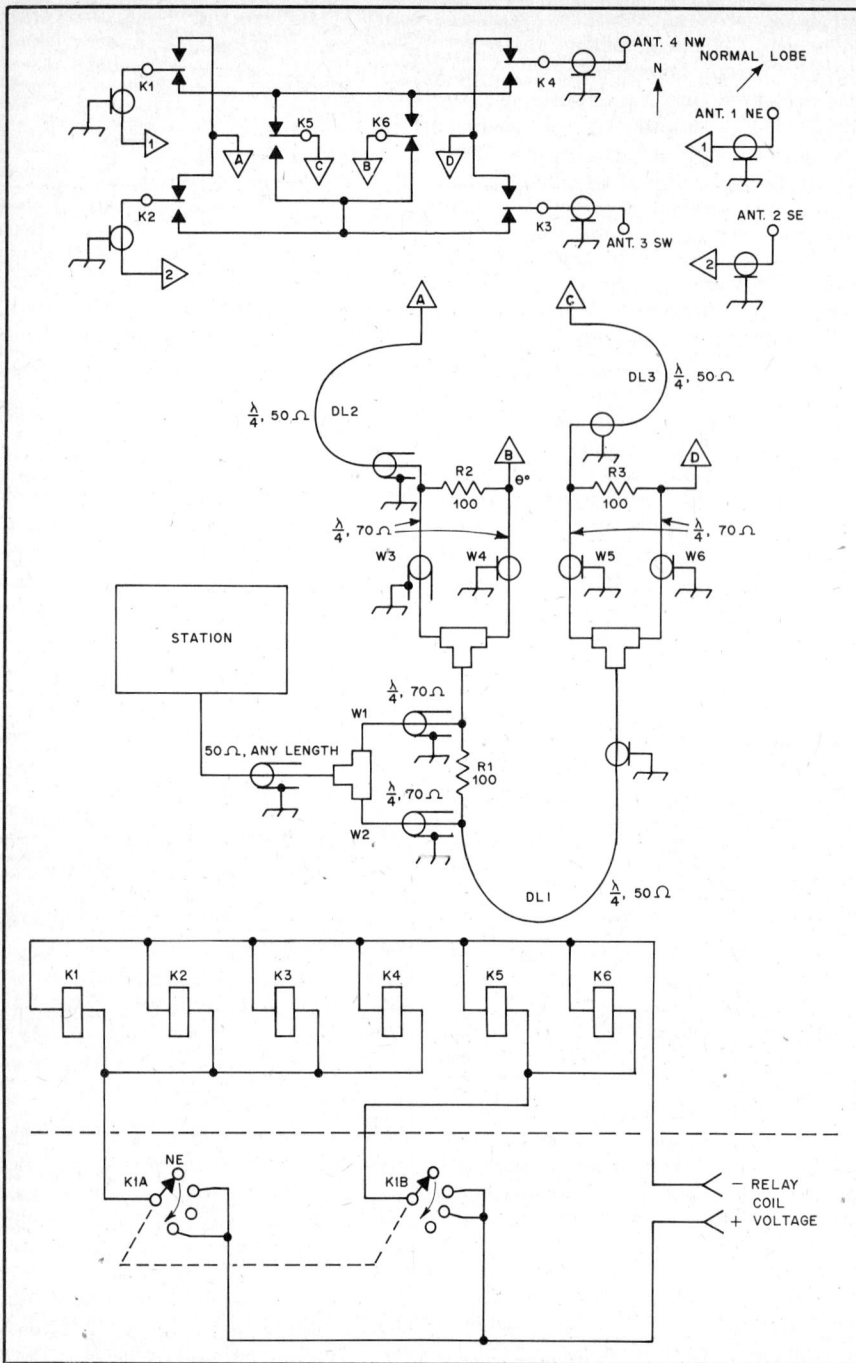

Fig. 37 — Close-up view of the relay and resistor assembly.

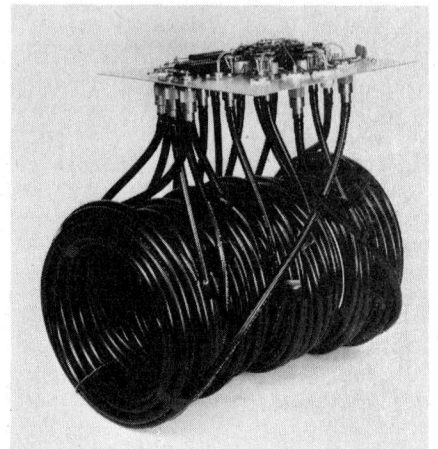

Fig. 38 — Photograph of the assembled phase system.

Fig. 36 — Schematic diagram of the phased array. Relays K1 through K6 are three-pole, double-throw units with all three sections of each relay connected in parallel. Each of the 100-ohm resistances are comprised of two 200-ohm noninductive 50-watt resistors manufactured by Nytronics. The quarter-wavelength lines should be adjusted for 7.1 MHz with the aid of a noise bridge, RX bridge or similar device. Calculated lengths of line for the velocity factors of 0.81 and 0.66 are 28.1 feet and 22.9 feet respectively. (Feet × 0.3048 = M.)

used three Wilkinson 2-way power dividers, as shown in Fig. 36. Proper element phasing is accomplished with three 90° sections of RG-8/U cable; DL1, DL2 and DL3. The four outputs are at 0°, −90°, −90° and −180°, respectively in phase relationship, with power from the transmitter divided into four equal parts. Special attention should be paid to preserving symmetry throughout the system, in the hybrids, phase shifters, rf switching, feeds and antenna placement, in

order to have the array perform uniformly as it is switched between the four headings.

The switching was done with six relays as shown in Fig. 36 and the accompanying photograph (Fig. 37). With no voltage applied to the relay coils, the arms are in the positions shown in Fig. 36, giving northeasterly directivity — a convenient heading for operation from New England. The three Wilkinson power divider resistor networks along with the switching

relays are mounted on a piece of sheet aluminum measuring 10 × 12 inches (254 × 305 mm). Each of the three phasing lines and quarter-wavelength lines associated with the power dividers connects to a coaxial fitting mounted to the piece of sheet aluminum. The entire phasing/switching system can be transported or worked on as a single unit. See Fig. 38. Basic layout of this system is shown in Figs. 37 and 39. Measured phase errors due to unequal lead lengths in the switching system are less than 2°. The builder need not follow the layout shown here — generally speaking, if the leads are kept short and direct the system should function according to specifications listed earlier.

The relays used in this system were surplus 3-pole double-throw units. The three sections were connected in parallel to assure that the relays would handle the transmitter power. A three-wire control system is used to steer the array. Four different commands are required for four different directions as shown in Fig. 36.

An aluminum chassis measuring 10 × 12 × 3 inches (254 × 305 × 76 mm) placed over the piece of aluminum on the

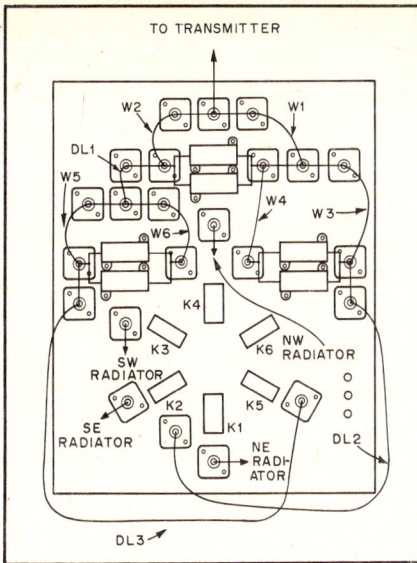

Fig. 39 — This drawing shows where each of the phasing and delay lines is connected to the divider/switching system.

resistor/relay side protects these components from the weather. Silicone seal is used to fill any holes or slots in the chassis.

Radiators and Radials

Relays K1 through K4 (Fig. 36) connect to their respective radiators through equal lengths of 50-ohm line. The radiators are constructed from 6061-T6 aluminum tubing which is available in 12 foot (3.7 m) lengths. Three pieces are required for each radiator: one each 1-1/4 inches (32 mm), 1-1/8 inches (28 mm), and 1 inch (25 mm) diameter. Constructional details are shown in Fig. 40. Base insulators are made from schedule 40 PVC pipe measuring 1-3/8 inch (35-mm) diameter. The 1-1/4-inch aluminum tubing does not fit securely inside the PVC pipe necessitating the use of shims between the aluminum tubing and PVC. One simple way of solving this problem is to cut thin strips (about 1/2 inch or 13 mm) of sheet aluminum and wind them over the aluminum tubing, sliding both the tubing and shim material into the PVC pipe.

The radiators are supported at the base by 4-foot (1.2-m) long pieces of 1-inch galvanized water pipe driven approximately 3 feet into the ground. Care should be taken to ensure that the pipe is kept truly vertical when inserted into the ground. Although the radiators are self-supporting, it is highly recommended that at least one set of guy wires be used for each radiator. Moderate winds have no difficulty bending the radiators if guys are not used. With the system described here, one set of guys, located at the 24-foot (7.3 m) point is used. Heavy nylon cord was used for this purpose.

Though 120 radials is considered to be the optimum number, only 40 per element are used with this system. All are 35 feet

(10.66 m) long, no. 15 aluminum fence wire, lying directly on the ground. The radials at the center of the array, in addition to providing considerable symmetry of the mutuals, allow a higher packing density, reducing ground losses.

Testing the System

Before power is applied to the array a quick resistance check should be made. One person should go to each element and short the input, while another person watches an ohmmeter placed across the main line at the station end. Make sure that a very low dc resistance is measured. Then, with the array in the normal position, that shown in the schematic, make sure that the resistance across the input is high. If it is low, check for moisture in the cables or connectors or for other leakage resistance. Silicone grease in the connectors is a good moisture preventive measure. It is recommended that these resistance checks be repeated periodically to be sure that all is well.

When the array is ready for use, go easy at first, as any reflected energy will be dissipated in the 100-ohm resistances. If they become hot, there is either a problem with the system or better matching of the elements will be necessary. Each 100-ohm resistance is made of two Sage 200-ohm, 50-watt resistors connected in parallel.

Using the System

When receiving in the "search" mode, one hand tunes the receiver while the other operates the lobe selector switch, to see which position "listens" best. Large rotary arrays typically take 45 to 60 seconds to rotate 360°, which tends to discourage frequent directional checks. With the phased array a complete scan takes but a few seconds. The high front and side rejection eliminates most of the interference from signals in unwanted directions, and in transmitting the clean pattern helps to prevent interference to stations off the lobe of the beam that might be on the same frequency.

One useful by-product of this system has been the reduced atmospheric noise pickup from unwanted directions. In particular, when listening toward Europe, atmospheric noise coming from electrical storms in the southwest is greatly reduced, improving the signal-to-noise ratio on signals arriving from the favored direction.

YAGI AND QUAD DIRECTIVE ANTENNAS

Most of the antennas described earlier in this chapter have unity gain or just slightly more. The notable exception is the phased vertical array. For the purpose of obtaining gain and directivity it is convenient to use the *Yagi-Uda* or *cubical-quad* types of hf-band beam antennas. The former is commonly called a "Yagi" and the latter is referred to as a

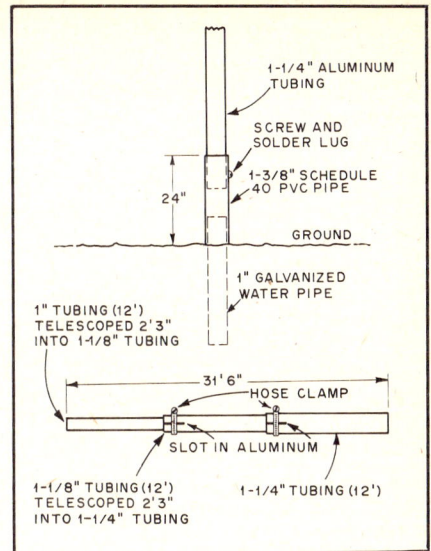

Fig. 40 — Details of element construction and mounting method. The water pipe should be driven into the ground so that approximately one foot of the pipe extends above ground level. Feet × 0.3048 = m and inches × 25.4 = mm.

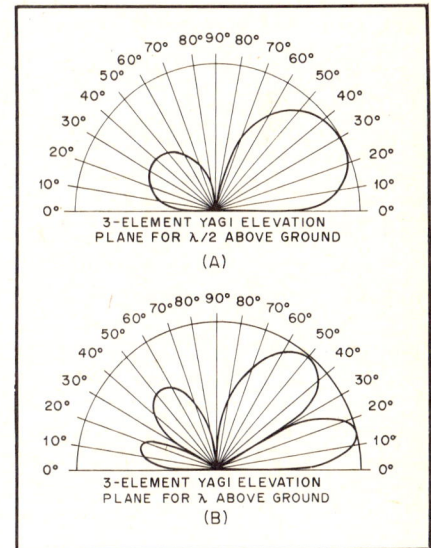

Fig. 41 — Elevation-plane response of a three-element Yagi placed 1/2 wavelength above a perfect ground (A) and the same antenna spaced one wavelength above ground (B).

"quad" in the amateur vernacular.

Most operators prefer to erect these antennas for horizontal polarization, but they can be used as vertically polarized arrays as well by merely rotating the feed point 90 degrees. In effect, the beam antenna is turned on its side for vertical polarity. The number of elements employed will depend on the gain desired and the capability of the supporting structure to contain the array safely. Many amateurs obtain satisfactory results with only two elements in a beam antenna, while others have several elements operating for a single amateur band. Regardless of the number of elements used, the height-above-ground rule shown

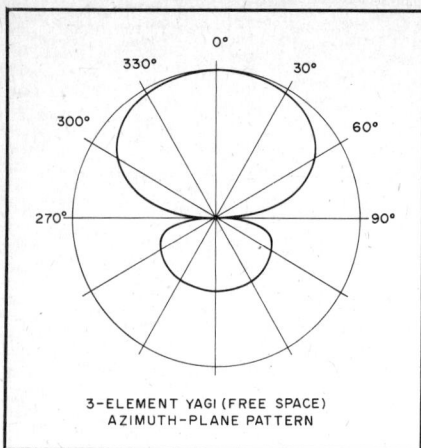

Fig. 42 — Azimuth-plane pattern of a three element Yagi in free space.

earlier for dipole antennas remains valid with respect to the angle of radiation. This is demonstrated in Fig. 41 at A and B where a comparison of radiation characteristics is given for a three-element Yagi at one-half and one wavelength above a perfectly conducting ground. It can be seen that the higher antenna has a lobe that is more favorable for DX work (roughly 15 degrees) than the larger lobe of Fig. 41 (approximately 30°). The pattern at B shows that some useful high-angle radiation exists also, and the higher lobe is suitable for short-skip contacts when propagation conditions dictate the need. A free-space azimuth pattern for the same antenna is provided in Fig. 42. The back-lobe pattern reveals that most of the power is concentrated in the forward lobe. The power difference dictates the front-to-back ratio in dB. It is infrequent that two three-element Yagis with different element spacings will yield the same lobe patterns. The data in Fig. 42 are given only for illustrative purposes.

Parasitic Excitation

In most of these arrangements the additional elements receive power by induction or radiation from the driven element, generally called the "antenna," and reradiate it in the proper phase relationship to achieve the desired effect. These elements are called *parasitic* elements, as contrasted to the driven elements which receive power directly from the transmitter through the transmission line.

The parasitic element is called a *director* when it reinforces radiation on a line pointing to it from the antenna, and a *reflector* when the reverse is the case. Whether the parasitic element is a director or reflector depends upon the parasitic-element tuning, which usually is adjusted by changing its length.

Gain vs. Spacing

The gain of an antenna with parasitic elements varies with the spacing and

tuning of the elements and thus for any given spacing there is a tuning condition that will give maximum gain at this spacing. The maximum front-to-back ratio seldom, if ever, occurs at the same condition that gives maximum forward gain. The impedance of the driven element also varies with the tuning and spacing, and thus the antenna system must be tuned to its final condition before the match between the line and the antenna can be completed. However, the tuning and matching may interlock to some extent, and it is usually necessary to run through the adjustments several times to insure that the best possible tuning has been obtained.

Two-Element Beams

A two-element beam is useful where space or other considerations prevent the use of the larger structure required for a three-element beam. The general practice is to tune the parasitic element as a reflector and space it about 0.15 wavelength from the driven element, although some successful antennas have been built with 0.1-wavelentgh spacing and director tuning. Gain vs. element spacing for a two-element antenna is given in Fig. 43 for the special case where the parasitic element is resonant. It is indicative of the performance to be expected under maximum-gain tuning conditions.

Three-Element Beams

A theoretical investigation of the three-element case (director, driven element and reflector) has indicated a maximum gain of slightly more than 7 dB. A number of experimental investigations have shown that the optimum spacing between the driven element and reflector is in the region of 0.15 to 0.25 wavelength, with 0.2 wavelength representing probably the best overall choice. With 0.2-wavelength reflector spacing, Fig. 44 shows the gain variation with director spacing is not especially critical and that the overall length of the array (boom length in the case of a rotatable antenna) can be anywhere between 0.35 and 0.45 wavelength with no appreciable difference in gain.

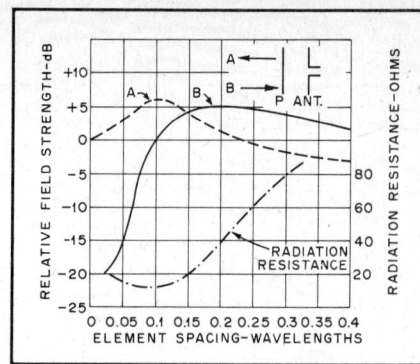

Fig. 43 — Gain vs. element spacing for an antenna and one parasitic element. The reference point, 0 dB, is the field strength from a half-wave antenna alone. The greatest gain is in the direction A at spacings of less than 0.14 wavelength, and in direction B at greater spacings. The front-to-back ratio is the difference in dB between curves A and B. Variation in radiation resistance of the driven element is also shown. These curves are for a self-resonant parasitic element. At most spacings the gain as a reflector can be increased by slight lengthening of the parasitic element; the gain as a director can be increased by shortening. This also improves the front-to-back ratio.

Wide spacing of both elements is desirable not only because it results in high gain but also because adjustment of tuning or element length is less critical and the input resistance of the driven element is higher than with close spacing. The latter feature improves the efficiency of the antenna and makes a greater bandwidth possible. However, a total antenna length, director to reflector, of more than 0.3 wavelength at frequencies of the order of 14 MHz introduces considerable difficulty from a constructional standpoint. Lengths of 0.25 to 0.3 wavelength are therefore frequently used for this band, even though they are less than optimum.

In general, the antenna gain drops off less rapidly when the reflector length is increased beyond the optimum value than it does for a corresponding decrease below the optimum value. The opposite is true of a director. It is therefore advisable to err, if necessary, on the long side for a reflector and on the short side for a director. This also tends to make the

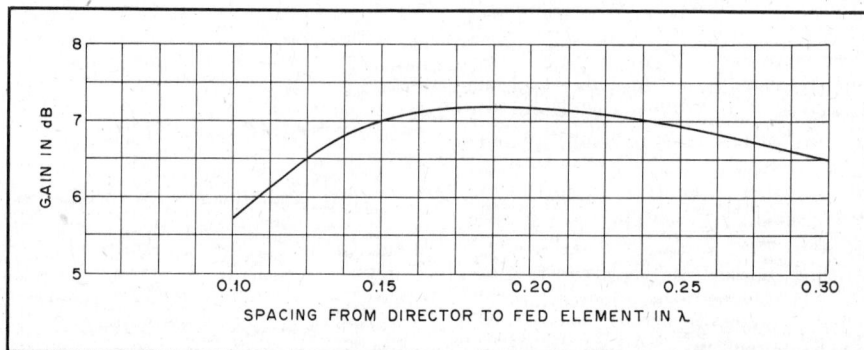

Fig. 44 — Gain of three-element Yagi versus director spacing, the reflector being fixed at 0.2 wavelength.

Table 2
Table 2
Element Lengths for 20, 15 and 10 Meters, Phone and CW

Freq. (kHz)	Driven Element		Reflector		First Director		Second Director	
	A	B	A	B	A	B	A	B
14,050	33' 5-3/8"	33' 8"	35' 2-1/2"	35' 5-1/4'	31' 9-3/8"	31' 11-5/8"	31' 1-1/4"	31' 3-5/8"
14,250	32' 11-3/4"	33' 2-1/4"	34' 8-1/2"	34' 11-1/4"	31' 4"	31' 6-3/8"	30' 8"	30' 10-1/2"
21,050	22' 4"	22' 5-5/8"	23' 6"	23' 7-3/4"	21' 2-1/2"	21' 4"	20' 9-1/8"	20' 10-7/8"
21,300	22' 3/4"	22' 2-3/8"	23' 2-5/8"	23' 4-1/2"	20' 11-1/2"	21' 1"	20' 6-1/4"	20' 7-3/4"
28,050	16' 9"	16' 10-1/4"	17' 7-5/8"	17' 8-7/8"	15' 11"	16'	15' 7"	15' 9-1/2"
28,600	16' 5-1/4"	16' 6-3/8"	17' 3-1/2"	17' 4-3/4"	15' 7-1/4"	15' 8-1/2"	15' 3-3/8"	15' 4-1/2"

A
| 0.2 | 0.2 | 0.2 |

B
| 0.15 | 0.15 | 0.15 |

These lengths are for 0.2- or 0.15-wavelength element spacing.

To convert ft to meters multiply ft × 0.3048.
Convert in. to mm by multiplying in. × 25.4.

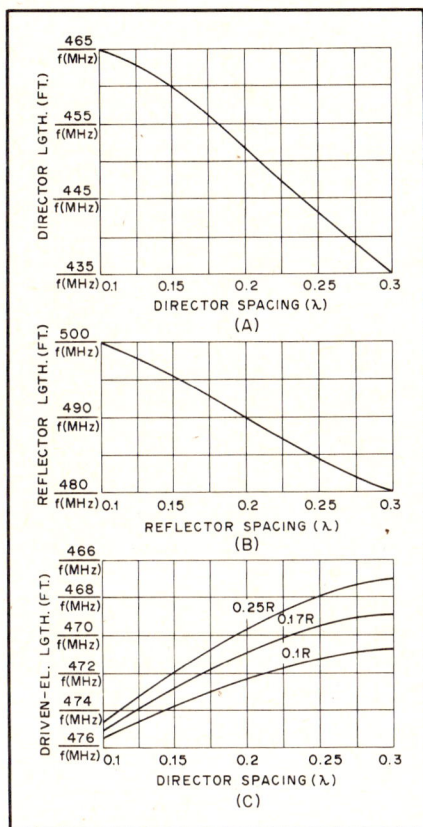

Fig. 45 — Element lengths for a three-element beam. These lengths will hold closely for tubing elements supported at or near the center.

antenna performance less dependent on the exact frequency at which it is operated, because an increase above the design frequency has the same effect as increasing the length of both parasitic elements, while a decrease in frequency has the same effect as shortening both elements. By making the director slightly short and the reflector slightly long, there will be a greater spread between the upper and lower frequencies at which the gain starts to show a rapid decrease.

When the over all length has been decided upon, the element lengths can be found by referring to Fig. 45. The lengths determined by these charts will vary slightly in actual practice with the element diameter and the method of supporting the elements. The tuning of a beam should always be checked after installation. However, the lengths obtained by the use of the charts will be close to correct in practically all cases, and they can be used without checking if the beam is difficult to access.

In order to make it even easier for the Yagi builder, Table 2 can be used to determine the element lengths needed. Both cw and phone lengths are included for the three bands, 20, 15 and 10 meters. The 0.2 wavelength spacing will provide greater bandwidth than the 0.15 spacing. Antenna gain is essentially the same with either spacing. The element lengths given will be the same whether the beam has two, three or four elements. It is recommended that "plumber's delight" type construction be used where all the elements are mounted directly on and grounded to the boom. This puts the entire array at dc ground potential, affording better lightning protection. A gamma section can be used for matching the feed line to the array.

Tuning Adjustments

The preferable method for checking the beam is by means of a field-strength meter or the S meter of a communications receiver, used in conjunction with a dipole antenna located at least 10 wavelengths away and as high as or higher than the beam that is being checked. A few watts of power fed into the antenna will give a useful signal at the observation point, and the power input to the transmitter (and hence the antenna) should be held constant for all the readings.

Preliminary matching adjustments can be done on the ground. The beam should be set up so the reflector element rests on earth, with the remaining elements in a vertical configuration. In other words, the beam should be aimed straight up. The matching system is then adjusted for 1:1 SWR between the feed line and driven element. When the antenna is raised into its operating height, only slight touch-up of the matching network will be required.

A great deal has been printed about the need for tuning the elements of a Yagi-type beam. However, experience has shown that lengths given in Fig. 45 and Table 2 are close enough to the desired length that no further tuning should be required. This is true for Yagi arrays made from metal tubing. However, in the case of quad antennas, made from wire, the reflectors and directors *should* be tuned with the antenna in its operating location. The reason is that it is practically impossible to cut and install wire to the *exact* dimensions required for maximum gain or front-to-back.

Simple Systems: the Rotary Beam

Two- and three-element systems are popular for rotary-beam antennas, where the entire antenna system is rotated, to permit its gain and directivity to be utilized for any compass direction. They may be mounted either horizontally (with the plane containing the elements parallel to the earth) or vertically.

A four-element beam will give still more gain than a three-element one, provided the support is sufficient for about 0.2 wavelength spacing between elements. The tuning for maximum gain involves many variables, and complete gain and tuning data are not available.

The elements in close-spaced (less than 1/4-wavelength element spacing) arrays preferably should be made of tubing of 1/2 to 1 inch (13 to 25-mm) diameter. A conductor of large diameter not only has less ohmic resistance but also has lower Q; both these factors are important in close-spaced arrays because the impedance of the driven element usually is quite low compared to that of a simple dipole antenna. With three- and four-element close-spaced arrays the radiation resistance of the driven element may be so

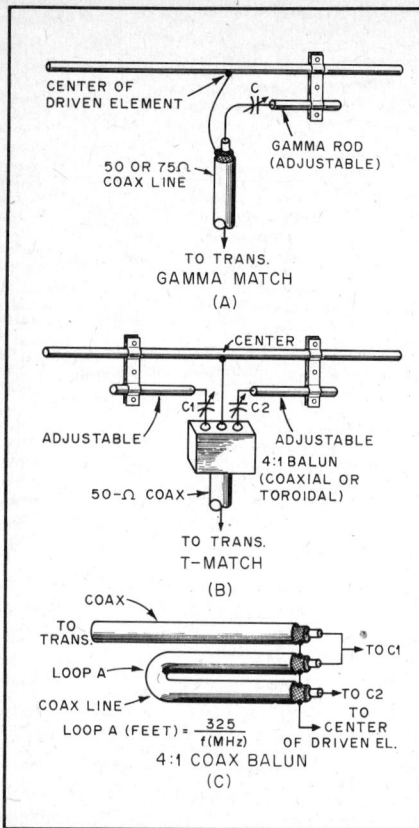

Fig. 46 — Illustrations of gamma and T-matching systems. At A, the gamma rod is adjusted along with C until the lowest SWR is obtained. A T-match is shown at B. It is the same as two gamma-match rods. The rods and C1 and C2 are adjusted alternately for a 1:1 SWR. A coaxial 4:1 balun transformer is shown at C. A toroidal balun can be used in place of the coax model shown. The toroidal version has a broader frequency range than the coaxial one. The T-match is adjusted for 200 ohms and the balun steps this *balanced* value down to 50 ohms, *unbalanced*. Or, the T-match can be set for 300 ohms, and the balun used to step this down to 75 ohms, unbalanced. Dimensions for the gamma and T-match rods are not given by formula. Their lengths and spacing will depend upon the tubing size used, and the spacing of the parasitic elements of the beam. Capacitors C, C1 and C2 can be 140 pF for 14-MHz beams. Somewhat less capacitance will be needed at 21 and 28 MHz.

low that ohmic losses in the conductor can consume an appreciable fraction of the power.

Feeding the Rotary Beam

Any of the usual methods of feed (described later under "Matching the Antenna to the Line") can be applied to the driven element of a rotary beam. The popular choices for feeding a beam are the gamma match with series capacitor and the T match with series capacitors and a half-wavelength phasing section, as shown in Fig. 46. These methods are preferred over any others because they permit adjustment of the matching and the use of coaxial-line feed. The variable capacitors can be housed in small plastic cups for

weatherproofing; receiving types with close spacing can be used at powers up to a few hundred watts. Maximum capacitance required is usually 140 pF at 14 MHz and proportionately less at the higher frequencies.

If physically possible, it is better to adjust the matching device after the antenna has been installed at its ultimate height, since a match made with the antenna near the ground may not hold for the same antenna in the air.

Sharpness of Resonance

Peak performance of a multielement parasitic array depends upon proper phasing or tuning of the elements, which can be exact for one frequency only. In the case of close-spaced arrays, which because of the low radiation resistance, usually are quite sharp-tuning, the frequency range over which optimum results can be secured is only of the order of one or two percent of the resonant frequency, or up to about 500 kHz at 28 MHz. However, the antenna can be made to work satisfactorily over a wider frequency range by adjusting the director or directors to give maximum gain at the *highest* frequency to be covered, and by adjusting the reflector to give optimum gain at the lowest frequency. This sacrifices some gain at all frequencies, but maintains more uniform gain over a wider frequency range.

The use of large-diameter conductors will broaden the response curve of an array because the larger diameter lowers the Q. This causes the reactances of the elements to change rather slowly with frequency, with the result that the tuning stays near the optimum over a considerably wider frequency range than is the case with wire conductors.

Delta Loops and Quad Beams

One of the more effective DX arrays is called the "cubical quad" or, simply, "quad" antenna. It consists of two or more square loops of wire, each supported by a bamboo or fiberglass cross-arm assembly. The loops are a quarter wavelength per side (full wavelength overall) one loop being driven, and the other serving as a parasitic element — usually a reflector. A variation of the quad is called the delta loop. The electrical properties of both antennas are the same, generally speaking, though some operators report better DX results with the delta loop. Both antennas are shown in Fig. 47. They differ mainly in their physical properties, one being of plumber's delight construction, while the other uses insulating support members. One or more directors can be added to either antenna if additional gain and directivity is desired, though most operators use the two-element arrangement.

It is possible to interlace quads or

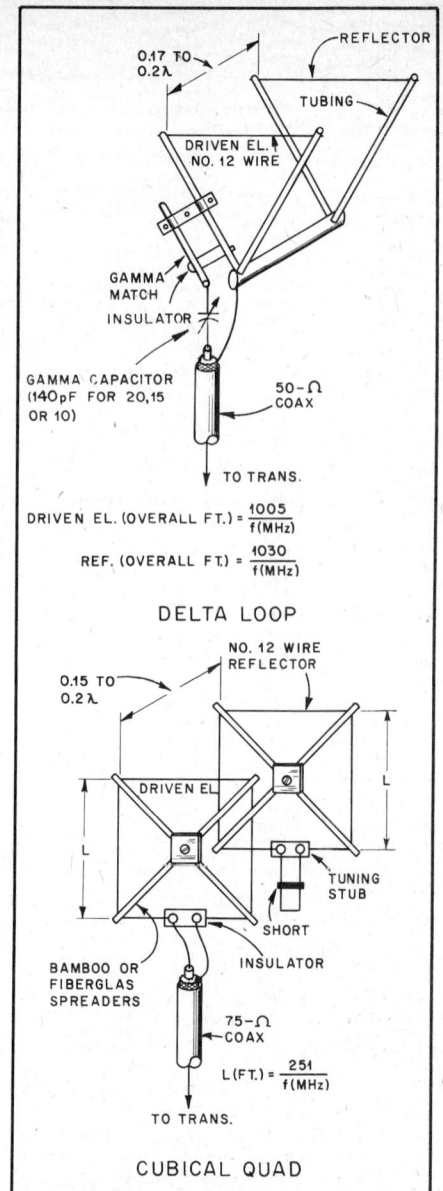

DRIVEN EL. (OVERALL FT.) $= \dfrac{1005}{f(MHz)}$

REF. (OVERALL FT.) $= \dfrac{1030}{f(MHz)}$

DELTA LOOP

$$L (FT.) = \dfrac{251}{f(MHz)}$$

CUBICAL QUAD

Fig. 47 — Information on building a quad of a delta-loop antenna. The antennas are electrically similar, but the delta-loop uses "plumber's delight" construction.

"deltas" for two or more bands, but if this is done the formulas given in Fig. 47 may have to be changed slightly to compensate for the proximity effect of the second antenna. For quads the length of the full-wave loop can be computed from

Full-wave loop (ft) $= \dfrac{1005}{f(MHz)}$

Full-wave loop (m) $= \dfrac{306}{f(MHz)}$

If multiple arrays are used, each antenna should be tuned up separately for maximum forward gain as noted on a field-strength meter. The reflector stub on the quad should be adjusted for the foregoing condition. The delta loop gamma match should be adjusted for a 1:1 SWR. No reflector tuning is needed. The delta loop antenna has a broader frequency response than the quad, and holds at an SWR of 1.5:1 or better across the band it is cut for.

The resonance of the quad antenna can be found by checking the frequency at which the lowest SWR occurs. The element length (driven element) can be adjusted for resonance in the most-used portion of the band by lengthening or shortening it.

A two-element quad or delta loop antenna compares favorably with a three-element Yagi array in terms of gain (see *QST*, May, 1963 and January, 1969, for additional information). The quad and delta-loop antennas perform very well at 50 and 144 MHz. A discussion of radiation patterns and gain, quads vs. Yagis, was presented by Lindsay in May 1968 *QST*.

An Optimum-Gain Two-Band Yagi Array

If optimum performance is desired from a Yagi, the dual four-element array shown in Fig. 48 will be of interest. This antenna consists of four elements on 15 meters interlaced with the same number for 10. Wide spacing is used, providing excellent gain and good bandwidth on both bands. Each driven element is fed separately with 50-ohm coax; gamma-matching systems are employed. If desired, a single feed line can be run to the array and then switched by a remotely controlled relay.

The element lengths shown in Fig. 48 are for the phone portions of the band, centered at 21,300 and 28,600 kHz. If desired, the element lengths can be changed for cw operation, using the dimensions given in Table 2. The spacing of the elements will remain the same for both phone and cw.

Construction Details

The elements are supported by commercially made U-bolt assemblies. Muffler clamps also make excellent element supports. The boom-to-mast support (Fig. 49) is also a manufactured item that is designed to hold a 2-inch (51-mm) diameter boom and that can be used with mast sizes up to 2-1/2 inches (63.5 mm) in diameter. Another feature of this device is that it permits the beam to be tilted after it is mounted in place on the tower, providing access to the elements if they need to be adjusted once the beam has been mounted on the tower.

A Small Yagi for 40 Meters

A 7-MHz antenna for most amateur installations consists of a half-wavelength dipole attached between two convenient supports and fed power at the center with coaxial cable. When antenna gain is a requirement on this frequency, the dimensions of the system can become overwhelming. A full-size three-element Yagi typically would have 68-foot (20.7-m)

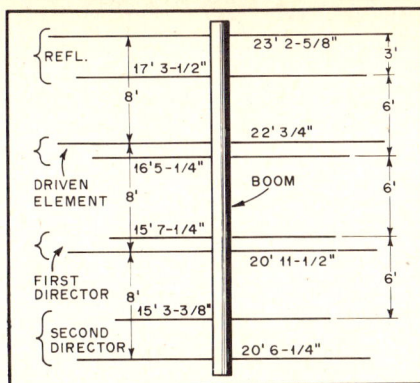

Fig. 48 — The element lengths shown are for the phone sections of the bands. Table 2 provides the dimensions for cw frequencies.

Fig. 49 — The boom-to-mast fixture that holds the two 12-foot boom sections together. The unit is made by Hy-Gain Electronics.

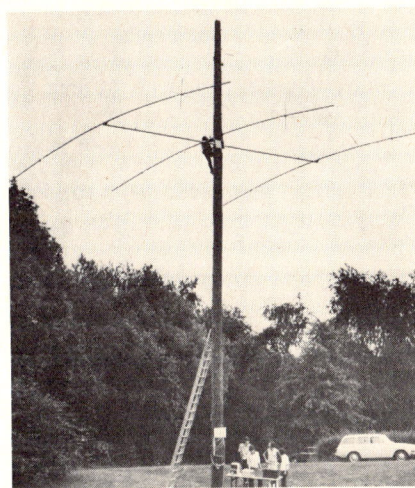

Fig. 50 — The shortened 40-meter Yagi beam closely approximates the size of a standard 20-meter Yagi. It is shown on a 60-foot (1.8-m) telephone pole.

elements and a 36-foot (10.9-m) boom. Accordingly, half-size elements present some distinct mechanical as well as economical advantages. Reducing the spacing between elements is not recommended since it would severely restrict the bandwidth of operation and make the

Table 3
Materials for Two-Band Yagi

Quantity	Length (ft)	Diameter (in.)	Reynolds No.
2	8	1	9A
4	8	3/4	8A
1	8	1-1/4	10A
1	6	7/8	4231

2 U-bolts, TV antenna to mast type, 1 variable capacitor, 150 pF maximum, any type, 1 plastic freezer container, approximately 5 × 5 × 5 inches, to house gamma capacitor.

Gamma rod, 3/8- to 1/2-inch diameter aluminum tubing, 36 inches long. (Aluminum curtain rod or similar.)

Ft × 0.3048 = m. In. × 25.4 = mm.

tuning critical. This array (Fig. 50) features good directivity and reasonable gain, yet the mechanical design allows the use of a "normal" heavy-duty rotator and a conventional tower support. Element loading is accomplished by lumped inductance and capacitance hats along the 38-foot (11.6-m) elements. This design concept can be applied on any of the amateur hf bands.

Construction

The system described here uses standard sizes and lengths of aluminum tubing available through most aluminum suppliers. For best mechanical and electrical performance, 6061-T6 alloy should be used. All three elements are the same length; the tuning of the inductor is slightly different on each element, however. The two parasitic elements are grounded at the center with the associated boom-to-element hardware. A helical hairpin match is used to provide a proper match to the split and insulated driven element. Two sections of steel angle stock are used to reinforce the driven-element mounting plate since the Plexiglas center insulating material is not rigid and element sag might otherwise result. The parasitic element center sections are continuous sections of aluminum tubing and additional support is not needed here. Fig. 51 and 52 show the details clearly.

The inductors for each element are wound on 1-1/8-inch (28-mm) diameter solid Plexiglas cast rod. Each end of the coil is secured in place with a solder lug and the Plexiglas is held in position with an automotive compression clamp. The total number or turns needed to resonate the elements correctly is given in Fig. 54. The capacitance hats consist of 1/2-inch (13-mm) tubing 3 feet (0.9 m) long (two pieces used) attached to the element directly next to the coil on each parasitic element and 2 inches (51 mm) away from the coil for the driven element. Complete details are given in Figs. 53 and 54.

The boom is constructed from three sections of aluminum tubing which measures 2-1/2 inches (63.5 mm) diameter and

Fig. 51 — An aluminum plate and four automotive muffler clamps are used to affix the parasitic beam elements to the boom.

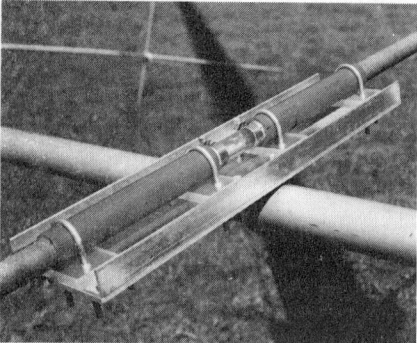

Fig. 52 — The driven element of the antenna is insulated from the boom by means of PVC tubing, as shown.

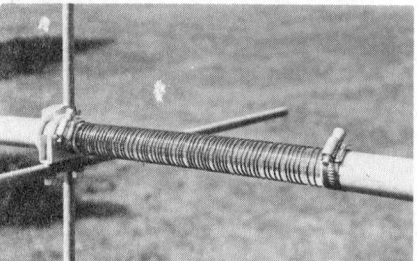

Fig. 53 — Each loading coil is wound on Plexiglas rod. The capacitance hats for the parasitic elements are mounted next to the coil, as shown here. The hose clamps compress the tubing against the Plexiglas rod. Each capacitance hat consists of two sections of tubing and associated muffler clamps.

original ones to assure mechanical security.

The helical hairpin details are given in Fig. 55. Quarter-inch copper tubing is formed into seven turns approximately 4 inches long and 2-1/4 inches (102 mm) ID.

Tuning and Matching

The builder is encouraged to carefully follow the dimensions given in Fig. 54. Tuning the elements with the aid of a dip oscillator has proved to be somewhat unreliable and accordingly, no resonant frequencies will be given.

The hairpin matching system may not resemble the usual form but its operation and adjustment are essentially the same. For a detailed explanation of this network see *The ARRL Antenna Book*, 13th edition. The driven element resonant frequency required for the hairpin match is determined by the placement of the capacitance hats with respect to the ends of the coils. Sliding the capacitance hats away from the ends of the coils increases the resonant frequency (capacitive reactance) of the element to cancel the effect of the hairpin inductive reactance. The model shown here had capacitance hats mounted 2-1/2 inches (63.5 mm) out from the ends of the coils (on the driven element only). An SWR indicator or wattmeter should be installed in series with the feed line *at the antenna*. The hairpin coil may be spread or compressed with an insulated tool (or by hand if power is removed!) to provide minimum reflected power at 7.050 MHz. The builder should not necessarily strive for a perfect match by changing the position of the capacitance hats since this may reduce the bandwidth of the matching system. An SWR of less than 2:1 was achieved across the entire 40-meter band with the antenna mounted atop an 80-foot tower.

The tuning of the array can be checked by making front-to-back ratio measurements across the band. With the di-

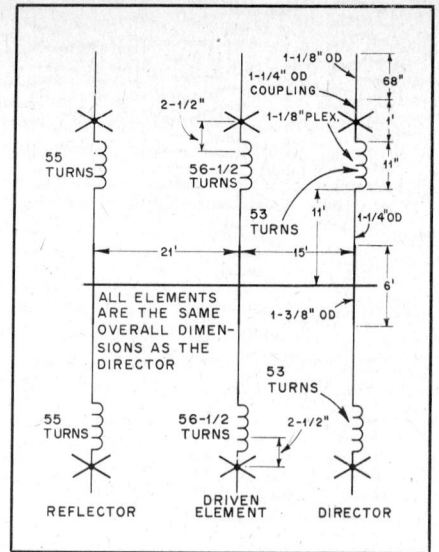

Fig. 54 — Mechanical details and dimensions for the 40-meter Yagi. Each of the elements uses the same dimensions; the difference is only the number of turns on the inductors and the placement of the capacitance hats. Feet × 0.3048 = m. Inches × 25.4 = mm. See the text for more details.

mensions given here, the best figures of front-to-back (approximately 25 to 30 dB) should be noticed in the cw portion of the band. Should the builder suspect the tuning is incorrect or if the antenna is mounted at some height greatly different than 80 feet (24.3 m) retuning of the elements may be necessary.

A Three-Band Quad Antenna System

Quads have been popular with amateurs during the past few decades because of their light weight, relatively small turning radius, and their unique ability to provide good DX performance when mounted close to the earth. A two-element three-band quad, for instance, with the elements mounted only 35 feet

12 feet (3.65 m) long. These pieces are joined together with inner tubes made from 2-1/4-inch (57 mm) stock shimmed with aluminum flashing. Long strips, approximately one inch wide, are wound on the inner tubing before it is placed inside the boom sections. A pair of 3/8 × 3-1/2 inch (9.5 × 89 mm) steel bolts are placed at right angles to each other at every connection point to secure the boom. Caution: Do not overtighten the bolts since this will distort the tubing making it impossible to pull apart sections, should the need arise. It is much better to install locking nuts over the

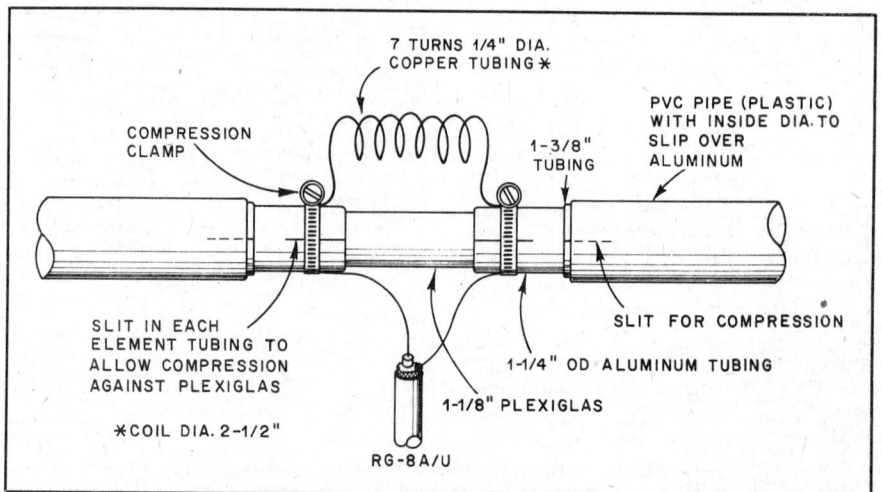

Fig. 55 — Details of the hairpin matching network on the driven element. See Fig. 54 for metric conversion.

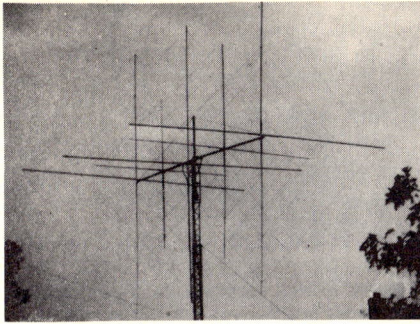

Fig. 56 — The assembled and installed three-band cubical-quad beam antenna.

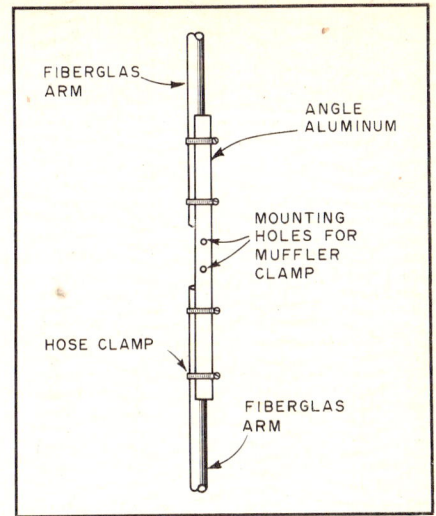

Fig. 58 — Details of one of two assemblies for a spreader frame. The two assemblies are jointed to form an × with a muffler clamp mounted at the position shown.

parasitic loops is closed (ends soldered together) and requires no tuning. All of the loop sizes are listed in Table 4 and are designed for a center frequency of 14.1, 21.1 and 28.3 MHz. Since quad antennas are rather broad-tuning devices excellent performance is achieved in both cw and ssb band segments of each band (with the possible exception of the very high end of 10 meters). Changing the dimensions to favor a frequency 200 kHz higher in each band to create a "phone" antenna is not necessary.

One question which comes up quite often is whether to mount the loops in a diamond or a square configuration. In other words, should one spreader be horizontal to the earth, or should the wire be horizontal to the ground (spreaders mounted in the fashion of an X)? From the electrical point of view, it is probably a trade-off. While the square configuration has its lowest point higher above ground than a diamond version (which may lower the angle of radiation slightly), the top is also lower than that of a diamond shaped array. Some authorities indicate that separation of the current points in the diamond system gives slightly more gain than is possible with a square layout. It should be pointed out, however, that there

(10.7 m) above the ground, will give good performance in situations where a triband Yagi will not. Fig. 56 shows a large quad antenna which can be used as a basis for design for either smaller or larger arrays.

Five sets of element spreaders are used to support the three-element 20-meter, four-element 15-meter, and five-element 10-meter wire-loop system. The spacing between elements has been chosen to provide optimum performance consistent with boom length and mechanical construction. See Fig. 57. Each of the

never has been any substantial proof in favor of one or the other, electrically.

Spreader supports (sometimes called spiders) are available from many different manufacturers. If the builder is keeping the cost at a minimum, he should consider building his own. The expense is about half that of a commercially manufactured equivalent and, according to some authorities, the homemade arm supports described below are less likely to rotate on the boom as a result of wind pressure.

A 3-foot (0.9-m) long section of 1-inch (25-mm)-per-side steel angle stock is used to interconnect the pairs of spreader arms. The steel is drilled at the center to accept a muffler clamp of sufficient size to clamp the assembly to the boom. The fiberglass is attached to the steel angle stock with automotive hose clamps, two per pole. Each quad-loop spreader frame consists of two assemblies of the type shown in Fig. 58.

Beam-Antenna Elements

Most Yagi antennas are made from sections of aluminum tubing which has been extruded or drawn. Compromise beams have been fashioned from less-expensive materials such as electrical conduit (iron) or bamboo poles wrapped with conductive tape or aluminum foil. The iron conduit is heavy, a poor conductor and is subject to rust. Therefore, it is best suited to experimental antennas or emergency use. Similarly, bamboo with conducting material affixed to it will deteriorate rapidly when exposed to the natural elements for a period of time. For the foregoing reasons it is wise to use aluminum tubing for Yagi elements and booms.

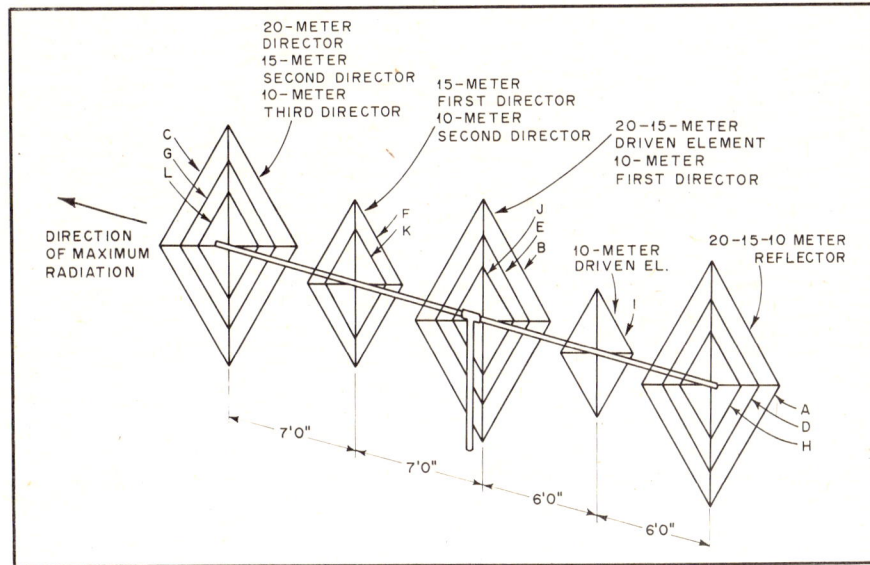

Fig. 57 — Dimensions of the three-band cubical quad. See Table 4 for the dimensions of the lettered wires. Note: Feet × 0.3048 = meters.

Table 4
Three-Band Quad Loop Dimensions

Band	Reflector	Driven Element	First Director	Second Director	Third Director
20 meters	(A) 72'8''	(B) 71'3''	(C) 69'6''	—	—
15 meters	(D) 48'6 1/2''	(E) 47'7 1/2''	(F) 46'5''	(G) 46'5''	—
10 meters	(H) 36' 2 1/2''	(I) 35'6''	(J) 34'7''	(K) 34'7''	(L) 34'7''

Letters indicate loops identified in Fig. 54. Feet × 0.3048 =' m. Inches × 25.4 = mm.

Chapter 21

VHF and UHF Antennas

Improving an antenna system is one of the most productive moves open to the vhf enthusiast. It can increase transmitting range, improve reception, reduce interference problems, and bring other practical benefits. The work itself is by no means the least attractive part of the job. Even with high-gain antennas, experimentation is greatly simplified at vhf and uhf because an array is a workable size, and much can be learned about the nature and adjustment of antennas. No large investment in test equipment is necessary.

Whether we buy or build our antennas, we soon find that there is no one "best" design for all purposes. Selecting the antenna best suited to our needs involves much more than scanning gain figures and prices in a manufacturer's catalog. The first step should be to establish priorities.

OBJECTIVES: GAIN

Shaping the pattern of an antenna to concentrate radiated energy, or received-signal pickup, in some directions at the expense of others is the only way to develop gain. This is best explained by starting with the hypothetical *isotropic antenna*, which would radiate equally in all directions. A point source of light illuminating the inside of a globe uniformly, from its center, is a visual analogy. No practical antenna can do this, so all antennas have "gain over isotropic" *(dBi)*. A half-wave dipole in free space has 2.1 dBi. If we can plot the radiation pattern of an antenna in all planes, we can compute its gain, so quoting it with respect to isotropic is a logical base for agreement and understanding. It is rarely possible to erect a half-wave antenna that has anything approaching a free-space pattern, and this fact is responsible for much of the confusion about true antenna gain.

Radiation patterns can be controlled in various ways. One is to use two or more driven elements, fed in phase. Such *collinear* arrays provide gain without markedly sharpening the frequency response, compared to that of a single element. More gain per element, but with a sacrifice in frequency coverage, is obtained by placing *parasitic* elements, longer and shorter than the driven one, in the plane of the first element, but not driven from the feedline. The reflector and directors of a *Yagi* array are highly frequency sensitive and such an antenna is at its best over frequency changes of less than one percent of the operating frequency.

Frequency Response

Ability to work over an entire vhf band may be important in some types of work. The response of an antenna element can be broadened somewhat by increasing the conductor diameter, and by tapering it to something approximating a cigar shape, but this is done mainly with simple antennas. More practically, wide frequency coverage may be a reason to select a collinear array, rather than a Yagi. On the other hand, the growing tendency to channelize operations in small segments of our bands tends to place broad frequency coverage low on the priority list of most vhf stations.

Radiation Pattern

Antenna radiation can be made omnidirectional, bidirectional, practically unidirectional, or anything between these conditions. A vhf net operator may find an omnidirectional system almost a necessity, but it may be a poor choice otherwise. Noise pickup and other interference problems tend to be greater with such antennas, and those having some gain are especially bad in these respects. Maximum gain and low radiation angle are usually prime interests of the weak-signal DX aspirant. A clean pattern, with lowest possible pickup and radiation off the sides and back, may be important in high-activity areas, or where the noise level is high.

Height Gain

In general, the higher the better in vhf antenna installations. If raising the antenna clears its view over nearby obstructions, it may make dramatic improvements in coverage. Within reason, greater height is almost always worth its cost, but height gain must be balanced against increased transmission-line loss. The latter is considerable, and it increases with frequency. The best available line may be none too good, if the run is long in terms of wavelength. Give line-loss information, shown in table form in chapter 19, close scrutiny in any antenna planning.

Physical Size

A given antenna design for 432 MHz will have the same gain as one for 144 MHz, but being only one-third the size it will intercept only one-third as much energy in receiving. Thus, to be equal in communication effectiveness, the 432-MHz array should be at least equal in *size* to the 144-MHz one, which will require roughly three times as many elements. With all the extra difficulties involved in going higher in frequency, it is well to be on the big side in building an antenna for the higher bands.

DESIGN FACTORS

Having sorted out objectives in a general way, we face decisions on specifics, such as polarization, type of transmission line, matching methods and mechanical design.

Polarization

Whether to position the antenna elements vertically or horizontally has been a moot point since early vhf pioneering. Tests show little evidence on which to set up a uniform polarization policy. On long paths there is no consistent advantage, either way. Shorter paths tend to yield higher signal levels with horizontal in some kinds of terrain. Man-made noise, especially ignition interference, tends to be lower with horizontal. Verticals are markedly simpler to use in omnidirectional systems, and in mobile work.

Early vhf communication was largely vertical, but horizontal gained favor when directional arrays became widely used. The major trend to fm and repeaters, particularly in the 144-MHz band, has tipped the balance in favor of verticals in mobile work and for repeaters. Horizontal predominates in other communication on 50 MHz and higher frequencies. It is well to check in advance in any new area in which you expect to operate, however, as some localities still use vertical almost exclusively. A circuit loss of 20-dB or more can be expected with cross-polarization.

Transmission Lines

There are two main categories of transmission lines: balanced and unbalanced. The former include open-wire lines separated by insulating spreaders, and twin-lead, in which the wires are embedded in solid or foamed insulation. Line losses result from ohmic resistance, radiation from the line and deficiencies in the insulation. Large conductors, closely spaced in terms of wavelength, and using a minimum of insulation, make the best balanced lines. Impedances are mainly 300 to 500 ohms. Balanced lines are best in straight runs. If bends are unavoidable, the angles should be as obtuse as possible. Care should be taken to prevent one wire from coming closer to metal objects than the other. Wire spacing should be less than 1/20 wavelength.

Properly built, open-wire line can operate with very low loss in vhf and even uhf installations. A total line loss under 2 dB per hundred feet at 432 MHz is readily obtained. A line made of no. 12 wire, spaced 3/4 inch (19 mm) or less with Teflon spreaders, and running essentially straight from antenna to station, can be better than anything but the most expensive coax, at a fraction of the cost. This assumes the use of baluns to match into and out of the line, with a short length of quality coax for the moving section from the top of the tower to the antenna. A similar 144-MHz setup could have a line loss under 1 dB.

Small coax such as RG-58/U or -59/U should never be used in vhf work if the run is more than a few feet. Half-inch (13-mm) lines (RG-8 or -11) work fairly well at 50 MHz, and are acceptable for 144-MHz runs of 50 feet or less. If these lines have foam rather than solid insulation they are about 30 percent better. Aluminum-jacket lines with large inner conductors and foam insulation are well worth their cost. They are readily waterproofed, and can last almost indefinitely. Beware of any "bargains" in coax for vhf or uhf uses. Lost transmitter power can be made up to some extent by increasing power, but once lost, a weak signal can never be recovered in the receiver.

Effects of weather should not be ignored. A well-constructed open-wire line works well in nearly any weather, and it stands up well. Twin-lead is almost useless in heavy rain, wet snow

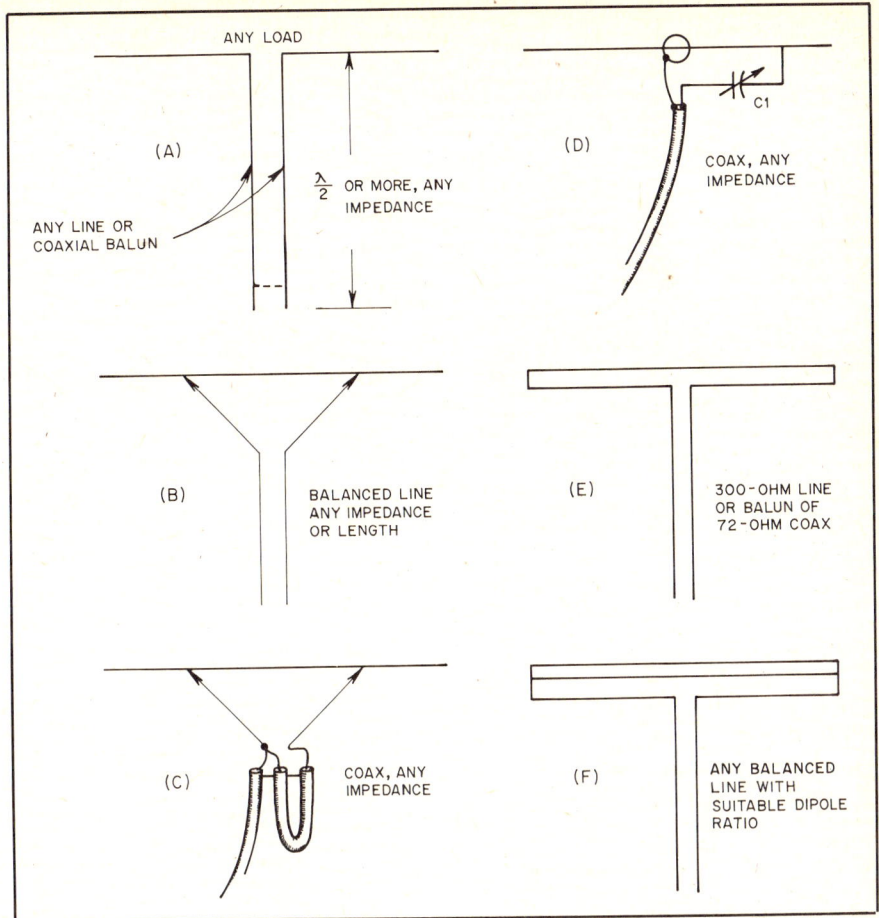

Fig. 1 — Matching methods commonly used in vhf antennas. The universal stub, A, combines tuning and matching. The adjustable short on the stub, and the points of connection of the transmission line, are adjusted for minimum reflected power in the line. In the delta match, B and C, the line is fanned out to tap on the dipole at the point of best impedance match. Impedances need not be known in A, B and C. The gamma-match, D, is for direct connection of coax. C1 tunes out inductance in the arm. Folded dipole of uniform conductor size, E, steps up antenna impedance by a factor of four. Using a larger conductor in the unbroken portion of the folded dipole, E, gives higher orders of impedance transformation.

or icing. The best grades of coax are impervious to weather. They can be run underground, fastened to metal towers without insulation, or bent into any convenient position, with no adverse effects on performance.

Impedance Matching

Theory and practice in impedance matching are given in detail in earlier chapters, and theory, at least, is the same for frequencies above 50 MHz. Practice may be similar, but physical size can be a major modifying factor in choice of methods. Only the matching devices used in practical construction examples later in this chapter will be discussed in detail here. This should not rule out consideration of other methods, however, and a reading of relevant portions of chapters 19 and 20 is recommended.

Universal Stub

As its name implies, the double-adjustment stub of Fig. 1A is useful for many matching purposes. The stub length is varied to resonate the system, and the transmission line and stub impedances are equal. In practice this involves moving both the sliding short and the point of line connection for zero reflected power, as indicated on an SWR bridge connected in the line.

The universal stub allows for tuning out any small reactance present in the driven part of the system. It permits matching the antenna to the line without knowledge of the actual impedances involved. The position of the short yielding the best match gives some indication of amount of reactance present. With little or no reactive component to be tuned out, the stub will be approximately a half-wavelength from load to short.

The stub should be stiff bare wire or rod, spaced no more than 1/20 wavelength. Preferably it should be mounted rigidly, on insulators. Once the position of the short is determined, the center of the short can be grounded, if desired, and the portion of the stub no longer needed can be removed.

It is not necessary that the stub be connected directly to the driven

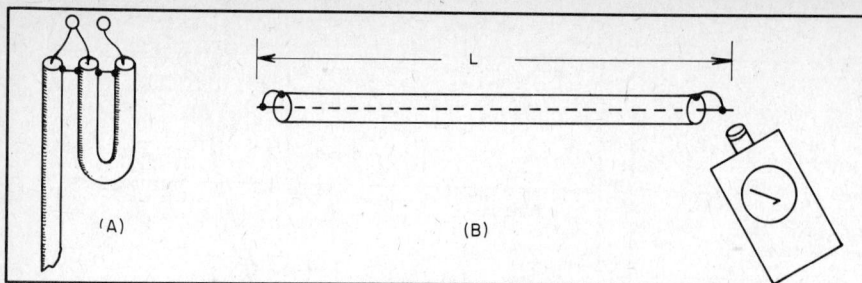

Fig. 2 — Conversion from unbalanced coax to a balanced load can be done with a half-wave coaxial balun, A. Electrical length of the looped section should be checked with a dip-meter, with ends shorted, B. The half-wave balun gives a 4:1 impedance step up.

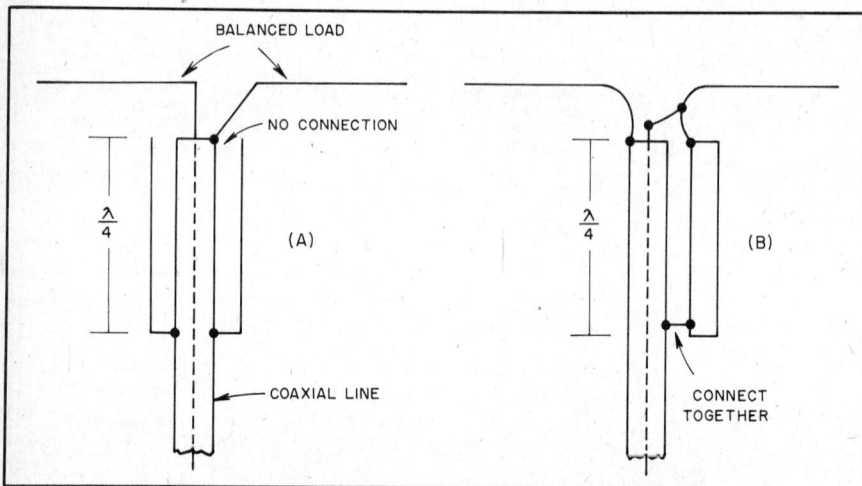

Fig. 3 — The balun conversion function, with no impedance change, is accomplished with quarter-wave lines, open at the top and connected to the coax outer conductor at the bottom. The coaxial sleeve, A, is preferred.

element. It can be made part of an open-wire line, as a device to match into or out of the line with coax. It can be connected to the lower end of a delta match, or placed at the feedpoint of a phased array. Examples of these uses are given later.

Delta Match

Probably the first impedance match was made when the ends of an open line were fanned out and tapped onto a half-wave antenna at the point of most efficient power transfer, as in Fig. 1B. Both the side length and the points of connection either side of the center of the element must be adjusted for minimum reflected power in the line, but as with the universal stub, the impedances need not be known. The delta makes no provision for tuning out reactance, so the universal stub is often used as a termination for it, to this end.

Once thought to be inferior for vhf applications because of its tendency to radiate if improperly adjusted, the delta has come back to favor, now that we have good methods for measuring the effects of matching. It is very handy for phasing multiple-bay arrays with open lines, and its dimensions in this use are not particularly critical. It should be

checked out carefully in applications like that of Fig. 1C, having no tuning device.

Gamma Match

An application of the same principle to direct connection of coax is the gamma match, Fig. 1D. There being no rf voltage at the center of a half-wave dipole, the outer conductor of the coax is connected to the element at this point, which may also be the junction with a metallic or wooden boom. The inner conductor, carrying the rf current, is tapped out on the element at the matching point. Inductance of the arm is tuned out by means of C1, resulting in electrical balance. Both the point of contact with the element and the setting of the capacitor are adjusted for zero reflected power, with a bridge connected in the coaxial line.

The capacitor can be made variable temporarily, then replaced with a suitable fixed unit when the required capacitance value is found, or C1 can be mounted in a waterproof box. Maximum should be about 100 pF for 50 MHz and 35 to 50 pF for 144. The capacitor and arm can be combined in one coaxial assembly, with the arm connecting to the driven element by

means of a sliding clamp, and the inner end of the arm sliding inside a sleeve connected to the inner conductor of the coax. A commercially supplied assembly of this type is used in a 50-MHz array described later, or one can be constructed from concentric pieces of tubing, insulated by plastic sleeving. Rf voltage across the capacitor is low, once the match is adjusted properly, so with a good dielectric, insulation presents no great problem, if the initial adjustment is made with low power level. A clean, permanent high-conductivity bond between arm and element is important, as the rf current is high at this point.

Folded Dipole

The impedance of a half-wave antenna broken at its center is 72 ohms. If a single conductor of uniform size is folded to make a half-wave dipole as shown in Fig. 1E, the impedance is stepped up four times. Such a folded dipole can thus be fed directly with 300-ohm line with no appreciable mismatch. Coaxial line of 70 to 75 ohms impedance may also be used if a 4:1 balun is added. (See balun information presented later in this chapter.) Higher impedance step up can be obtained if the unbroken portion is made larger in cross-section than the fed portion, as in Fig. 1F. For design information, see chapter 19.

Baluns and Transmatches

Conversion from balanced loads to unbalanced lines, or vice versa, can be performed with electrical circuits, or their equivalents made of coaxial line. A balun made from flexible coax is shown in Fig. 2A. The looped portion is an electrical half-wavelength. The physical length depends on the propagation factor of the line used, so it is well to check its resonant frequency, as shown at B. The two ends are shorted, and the loop at one end is coupled to a dip-meter coil. This type of balun gives an impedance stepup of 4:1 in impedance, 50 to 200 ohms, or 75 to 300 ohms, typically.

Coaxial baluns giving 1:1 impedance transfer are shown in Fig. 3. The coaxial sleeve, open at the top and connected to the outer conductor of the line at the lower end (A) is the preferred type. A conductor of approximately the same size as the line is used with the outer conductor to form a quarter-wave stub, in B. Another piece of coax, using only the outer conductor, will serve this purpose. Both baluns are intended to present an infinite impedance to any rf current that might otherwise tend to flow on the outer conductor of the coax.

The functions of the balun and the impedance transformer can be handled by various tuned circuits. Such a device, commonly called an antenna coupler or Transmatch, can provide a wide range of impedance transformations. Additional selectivity inherent to the Transmatch can reduce RFI problems.

The VHF, UHF Yagi

The small size of vhf and, especially, uhf arrays opens up a wide range of construction possibilities. Finding components is becoming difficult for home constructors of ham gear, but it should not hold back antenna work. Radio and TV distributors have many useful antenna parts and materials. Hardware stores, metals suppliers, lumber yards, welding-supply and plumbing-supply houses and even junkyards should not be overlooked. With a little imagination, the possibilities are endless.

Boom Materials

Wood is very useful in antenna work. It is available in a great variety of shapes and sizes. Rug poles of wood or bamboo make fine booms. Round wood stock (doweling) is found in many hardware stores in sizes suitable for small arrays. Square or rectangular boom and frame materials can be cut to order in most lumber yards if they are not available from the racks in suitable sizes.

There is no rf voltage at the center of a half-wave dipole or parasitic element, so no insulation is required in mounting elements that are centered in the support, whether the latter is wood or metal. Wood is good for the framework of multibay arrays for the higher bands, as it keeps down the amount of metal in the active area of the array.

Wood used for antenna construction should be well-seasoned and free of knots or damage. Available materials vary, depending on local sources. Your lumber dealer can help you better than anyone else in choosing suitable materials. Joining wood members at right angles is often done advantageously with gusset plates. These can be of thin outdoor-grade plywood or Masonite. Round materials can be handled in ways similar to those used with metal components, with U clamps and with other hardware.

Metal booms have a small "shorting effect" on elements that run through them. With materials sizes commonly employed, this is not more than one percent of the element length, and may not be noticeable in many applications. It is just perceptible with 1/2-inch (13-mm) tubing booms used on 432 MHz, for example. Formula lengths can be used as given, if the matching is adjusted in the frequency range one expects to use. The center frequency of an all-metal array will tend to be 0.5 to 1 percent higher than a similar system built of wooden supporting members.

Element Materials

Antennas for 50 MHz need not have

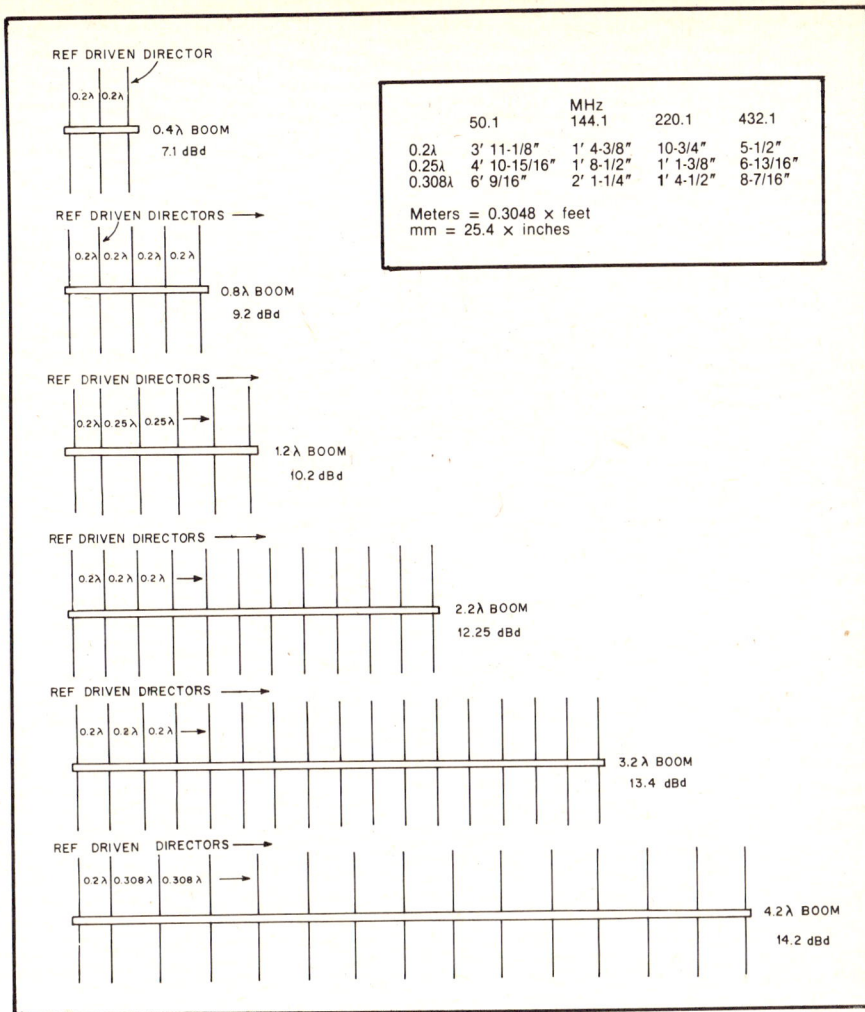

Fig. 4 — Element spacing for the various arrays, in terms of boom wavelength.

elements larger than 1/2-inch diameter, though up to 1 inch (25 mm) is used occasionally. At 144 and 220 MHz the elements are usually 1/8 to 1/4 inch (3 to 6 mm) in diameter. For 420, elements as small as 1/16 inch (1.6 mm) in diameter work well, if made of stiff rod. Aluminum welding rod, 3/32 to 1/8 (2.4 to 3 mm) inch in diameter is fine for 420-MHz arrays, and 1/8 inch or larger is good for the 220 band. Aluminum rod or hard-drawn wire works well at 144 MHz. Very strong elements can be made with stiff-rod inserts in hollow tubing. If the latter is slotted, and tightened down with a small clamp, the element lengths can be adjusted experimentally with ease.

Sizes recommended above are usable with formula dimensions given in Table 1. Larger diameters broaden frequency response; smaller ones sharpen it. Much smaller diameters than those recommended will require longer elements, especially in 50-MHz arrays.

Element and Boom Dimensions

Tables 1 through 4 list element and boom dimensions for several Yagi configurations for operation on 50, 144, 220

and 432 MHz. These figures are based on information contained in the *National Bureau of Standards Technical Note 688* which offers element dimensions for maximum-gain Yagi arrays as well as other types of antennas. The original information provides various element and boom diameters. The information shown in the Tables represents a highly condensed set of antenna designs, however, making use of standard and readily available material. Element and boom diameters have been chosen so as to produce lightweight, yet very rugged, antennas.

Since these antennas are designed for maximum forward gain, the front-to-back pattern ratios may be a bit lower than those for some other designs. Ratios on the order of 15 to 25 dB are common for these antennas and should be more than adequate for most installations. Additionally, the patterns are quite clean, with the side lobes well suppressed. The driven-element lengths for the antennas represent good starting-point dimensions. The type of feed system used on the array may require longer or shorter lengths, as appropriate. Full details of the various

Table 1
NBS 50.1-MHz Yagi Dimensions

meters = 0.3048 × feet
mm = 25.4 × inches

Boom Length	Boom Diameter	Element Diameter	Insulated Elements	Ref.	Driven	Dir. 1	Dir. 2	Dir. 3	Dir. 4	Dir. 5	Dir. 7	Dir. 8	Dir. 9	Dir. 10
7' 10"(0.4 λ)	1-1/4"	1/2"	YES	9' 7"	9' 1-3/4"	9' 5/8"								
			NO	9' 7-3/4"	9' 1-3/4"	9' 1-3/8"	8' 8-3/8"	8' 9-1/8"	8' 9-1/8"					
15' 8-1/2"(0.8 λ)	2"	3/4"	YES	9' 6-1/2"	9' 1-3/4"	8' 9-1/8"	8' 8-7/8"	8' 9-1/8"	8' 10-1/4"					
			NO	9' 7"	9' 1-3/4"	8' 9-5/8"	8' 7-3/4"	8' 7-3/4"	8' 3-1/2"					
23' 6-7/8"(1.2λ)	2"	3/4"	YES	9' 6-1/2"	9' 1-3/4"	8' 9-1/8"	8' 8-7/8"	8' 8-7/8"	8' 4-5/8"	8' 1-3/4"	8' 1-3/4"	8' 1-3/4"	8' 3-1/2"	8' 5-3/8"
			YES	9' 7-3/4"	9' 1-3/4"	8' 10-1/4"	8' 7"	8' 5-3/8"		8' 3"	8' 3"	8' 3"	8' 4-5/8"	8' 6-1/2"
39' 3-3/8"(2.2λ)	2"	3/4"	YES	9' 6-1/2"	9' 1-3/4"	8' 9-7/8"	8' 8-7/8"	8' 5-3/8"					8' 3-1/2"	8' 5-3/8"
			NO	9' 7-3/4"	9' 1-3/4"	8' 11"	8' 8-1/8"	8' 6-1/2"					8' 4-5/8"	8' 6-1/2"

Table 2
NBS 144.1-MHz Yagi Dimensions

meters = 0.3048 × feet
mm = 25.4 × inches

Boom Length	Boom Diameter	Element Diameter	Insulated Elements	Ref.	Driven*	Dir. 1	Dir. 2	Dir. 3	Dir. 4	Dir. 5	Dir. 6	Dir. 7	Dir. 8	Dir. 9	Dir. 10	Dir. 11	Dir. 12	Dir. 13	Dir. 14	Dir. 15
5' 5-9/16"(0.8λ)	1"	3/16"	YES	3'4"	3'2-3/16"	3'7/8"	3'11/16"	3'7/8"												
			NO	3'4-5/8"	2'1	3'1-1/2"	3'1-3/8"	3'1-1/2"	3'7/8"											
8' 2-5/16"(1.2λ)	1"	3/16"	YES	3'4"		3'7/8"	3'7/16"	3'7/16"	3'1-1/8"											
			NO	3'4-5/8"		3'1-1/2"	3'1-1/8"	3'1-1/2"	3'1-1/2"											
15' 1/4"(2.2λ)	1 1/4"	3/16"	YES	3'4"		3'1-1/8"	2'11-13/16"	2'11-1/4"	2'10-9/16"	2'10-9/16"	2'10-9/16"	2'10-9/16"	2'10-9/16"	2'10-5/16"	2'10-5/16"	2'10-5/16"	2'10-5/16"	2'10-5/16"	2'10-5/16"	2'10-5/16"
			NO	3'4-13/16"		3'1-15/16"	3'5/8"	3'	2'11-3/8"	2'11-3/8"	2'11-3/8"	2'11-3/8"	2'11-3/8"	3'	3'-5/8"	2'11-3/8"	2'11-3/8"	2'11-3/8"	2'11-3/8"
21'10-1/16"(3.2λ)	1 1/2"	3/16"	YES	3'4"		3'7/8"	3'9/16"	2'11-3/4"	3'	3'	3'1-5/8"	2'11-3/8"	2'11-3/8"	2'10-5/16"	2'10-5/16"	2'10-9/16"	2'10-9/16"	2'10-9/16"	2'10-9/16"	
			NO	3'5-1/16"		3'1-15/16"	3'1-3/8"	3'3/16"	3'3/16"	3'11-1/8"	3'1-5/8"	2'11-3/8"	2'11-3/8"	2'11-3/8"	2'11-3/8"	2'11-5/8"	2'11-5/8"	2'11-3/8"	2'11-3/8"	
28'8-1/8"(4.2λ)	1 1/2"	3/16"	YES	3'3-3/8"		3'9/16"	3'1-3/8"	3'3/8"	3'11/16"	3'9/16"	3'3/16"	2'11-1/2"	2'11-5/8"	2'11-5/8"	2'11-5/8"					
			NO	3'4-1/2"		3'1-5/8"	3'1-5/8"	3'1-7/16"	3'11/16"	3'3/16"	3'3/16"	3'1-7/8"	2'11-5/8"	2'11-5/8"	2'11-5/8"					

Table 3
NBS 220.1-MHz Yagi Dimensions

meters = 0.3048 × feet
mm = 25.4 × inches

Boom Length	Boom Diameter	Element Diameter	Insulated Elements	Ref.	Driven*	Dir. 1	Dir. 2	Dir. 3	Dir. 4	Dir. 5	Dir. 6	Dir. 7	Dir. 8	Dir. 9	Dir. 10	Dir. 11	Dir. 12	Dir. 13	Dir. 14	Dir. 15
3'6-15/16"(0.8λ)	1"	3/16"	YES	2'2-1/16"	2'1	1'11-13/16"	1'11-11/16"	1'11-13/16"	1'11-3/16"											
			NO	2'2-3/4"		2'1/2"	2'3/8"	2'1/2"	1'11/32"											
5'4-3/8"(1.2λ)	1"	3/16"	YES	2'2-1/16"		1'11-13/16"	1'11-9/16"	1'11-9/16"	1'11-3/16"											
			NO	2'2-3/4"		2'1/2"	2'1/4"	2'1/4"	2'1/2"											
9'10"(2.2λ)	1"	3/16"	YES	2'2-1/16"		2'1/16"	1'11-5/16"	1'10-15/16"	1'10-1/2"	1'10-1/8"	1'10-1/8"	1'10-1/8"	1'10-1/8"	1'10-1/2"	1'10-15/16"	1'10-1/8"	1'10-13/16"	1'10-13/16"	1'10-13/16"	1'10-13/16"
			NO	2'2-3/4"		2'3/4"	2'1/16"	1'11-5/8"	1'10-7/8"	1'10-7/8"	1'10-7/8"	1'10-7/8"	1'10-7/8"	1'10-1/4"	1'11-5/8"	1'10-13/16"	1'10-13/16"	1'10-1/8"	1'10-1/8"	1'10-13/16"
14'3-11/16"(3.2λ)	1 1/4"	3/16"	YES	2'2-1/16"		1'11-13/16"	2'1/16"	1'11-5/8"	1'11-1/4"	1'11-1/4"	1'11"	1'10-13/16"	1'10-13/16"	1'9-7/8"	1'9-7/8"	1'10-13/16"	1'10-13/16"	1'10-1/8"	1'9-7/8"	
			NO	2'3"		2'3/4"	2'7/16"	1'11-7/8"	1'11-7/16"	1'11-1/4"	1'11"	1'10-1/8"	1'10-13/16"	1'10-13/16"	1'10-13/16"	1'10-1/8"	1'10-1/8"	1'10-1/8"	1'10-1/8"	
18'9-5/16"(4.2λ)	1 1/2"	3/16"	YES	2'1-11/16"		1'11-5/8"	2'1/16"	1'11-7/8"	1'10-3/4"	1'11-5/32"	1'10-1/2"	1'10-5/16"	1'10-13/16"	1'10-5/16"	1'10-1/8"	1'10-5/16"	1'11-3/16"	1'10-13/16"	1'10-13/16"	
			NO	2'2-3/4"		2'11/16"	2'11/16"	2'1/2"	2'	1'11-13/16"	1'11-9/16"	1'11-3/8"	1'11-3/16"	1'11-3/16"	1'11-3/16"	1'11-3/16"	1'11-5/8"	1'11-3/16"	1'11-3/16"	

Table 4
NBS 432.1-MHz Yagi Dimensions

meters = 0.3048 × feet
mm = 25.4 × inches

Boom Length	Boom Diameter	Element Diameter	Insulated Elements	Ref.	Driven*	Dir. 1	Dir. 2	Dir. 3	Dir. 4	Dir. 5	Dir. 6	Dir. 7	Dir. 8	Dir. 9	Dir. 10	Dir. 11	Dir. 12	Dir. 13	Dir. 14	Dir. 15
2'8-13/16"(1.2λ)	1"	3/16"	YES	1'1-3/16"	1'23/32"	11-13/16"	11-5/8"	11-5/8"	11-13/16"	10-13/16"										
			NO	1'1-15/16"		1'17/32"	1'11/32"	1'11/32"	1'17/32"	11-17/32"										
5'1/8"(2.2λ)	1"	3/16"	YES	1'1-3/16"		11-29/32"	11-7/16"	11-1/4"	11"	10-29/32"	10-13/16"	10-13/16"	10-13/16"	11"	11-1/4"	10-11/16"	10-11/16"	10-11/16"	10-11/16"	10-11/16"
			NO	1'1-15/16"		1'21/32"	1'3/16"	1'	11-3/4"	11-5/8"	11"	11-1/16"	11-17/32"	11-3/4"	1'	11-13/32"	11-13/32"	11-13/32"	11-13/32"	11-13/32"
7'3-15/32"(3.2λ)	1"	3/16"	YES	1'1-3/32"		1'27/32"	11-5/8"	11-1/4"	11"	11-5/8"	10-13/16"	10-11/16"	10-11/16"	10-11/16"	10-11/16"	10-13/16"	10-13/16"	10-13/16"	10-13/16"	10-13/16"
			NO	1'1-15/16"		1'9/16"	1'11/32"	11-1/4"	11"	11-5/32"	11-5/32"	10-29/32"	10-13/16"	10-13/16"	11-1/4"	10-13/16"	10-13/16"	10-13/16"	10-13/16"	10-13/16"
9'6-25/32"(4.2λ)	1"	3/16"	YES	1'1"		11-22/32"	1'11/32"	11-19/32"	11-3/4"	11-5/32"	11"	10-29/32"	10-13/16"	10-13/16"	10-13/16"	10-13/16"	11-17/32"	11-17/32"	11-13/32"	11-13/32"
			NO	1'1-3/4"		1'7/16"	1'7/16"	1'11/32"	1'	11-7/8"	11-3/4"	11-5/8"	11-17/32"	11-17/32"	11-17/32"	11-17/32"	11-17/32"	11-17/32"	11-17/32"	

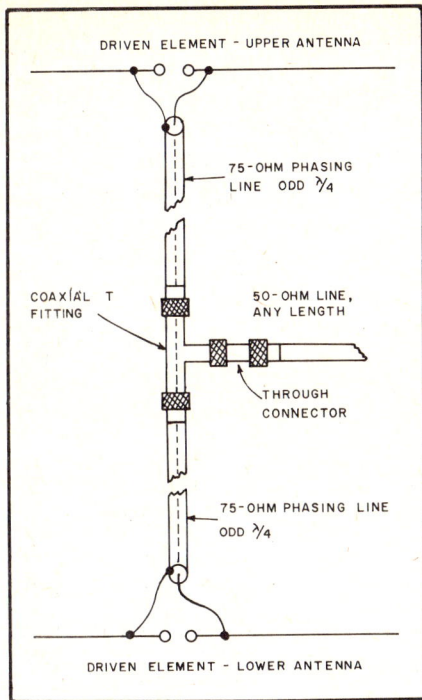

Fig. 5 — A method for feeding a stacked Yagi array.

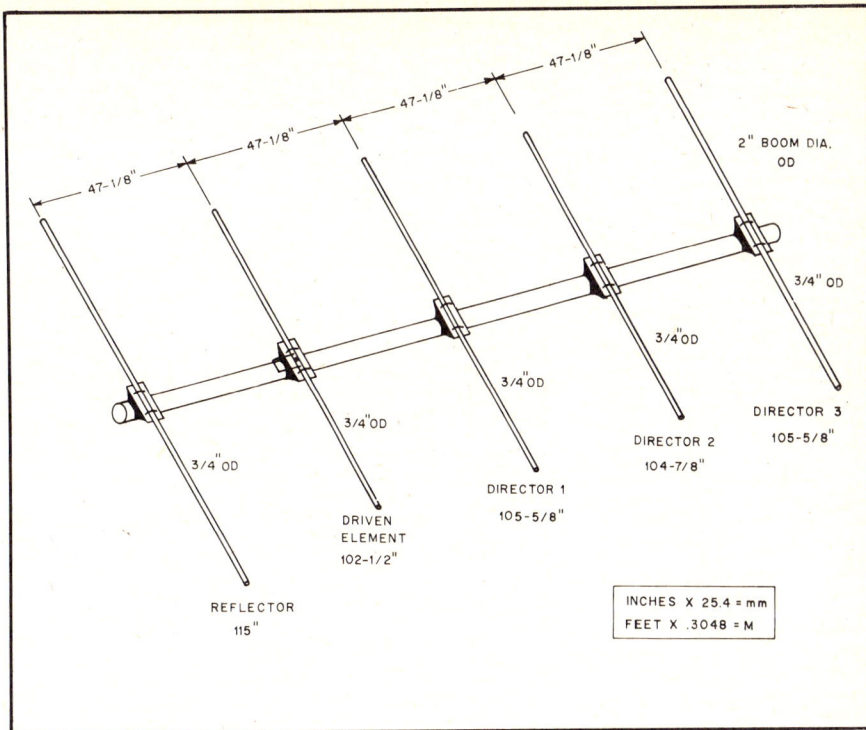

Fig. 7 — Dimensional drawing of the 5-element, 50-MHz Yagi antenna described in the text.

methods for feeding a Yagi array are given in the hf antennas chapter of this volume, and also in the *ARRL Antenna Book*. Generally speaking, a balanced feed system is preferred in order to prevent pattern skewing and the possibility of unwanted side lobes, which can occur with an unbalanced feed system.

Element spacing for the various arrays is presented in Fig. 4 in terms of the wavelength of the boom, as noted in the first column of Tables 1 through 4. The 0.4-, 0.8-, 2.2- and 3.2-wavelength boom antennas have equally spaced elements for both reflector and directors; 1.2- and 4.2-wavelength boom antennas have different reflector and director spacings. As all of the antenna parameters are inter-related, changes in element diameter, boom diameter and element spacing will

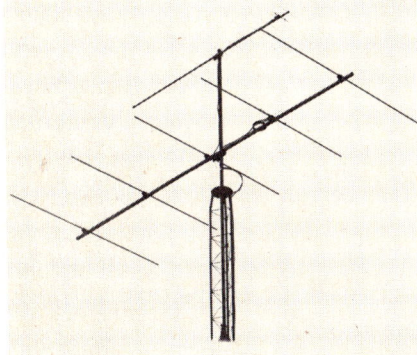

Fig. 6 — Photograph of the 5-element, 50-MHz Yagi. A 432-MHz beam is mounted above.

require that a new design be worked out. As the information presented in the *NBS Technical Note 688* is straightforward, the serious antenna experimenter should have no difficulty designing antennas with different dimensions from those presented in the Tables. For antennas with the same element and boom diameters, but for different frequencies within the band, standard scaling techniques may be applied.

Stacking Yagis

Where suitable provision can be made for supporting them, two Yagis mounted one above the other and fed in phase may be preferable to one long Yagi having the same theoretical or measured gain. The pair will require a much smaller turning space for the same gain, and their lower radiation angle can provide interesting results. On long ionospheric paths a stacked pair occasionally may show an *apparent* gain much greater than the 2 to 3 dB that can be measured locally as the gain due to stacking.

Optimum spacing for Yagis of five elements or more is one wavelength, but this may be too much for many builders of 50-MHz antennas to handle. Worthwhile results can be obtained with as little as one half-wavelength (10 feet/3 m), and 5/8 wavelength (12 feet/3.66 m) is markedly better. The difference between 12 and 20 feet may not be worth the added structural problems involved in the wider spacing, at 50 MHz, at least. The closer spacings give lower measured gain, but the antenna patterns are cleaner than will be obtained with one-wavelength spacing. The extra gain with wider spacings is

usually the objective on 144 MHz and higher bands, where the structural problems are not severe.

One method for feeding two 50-ohm antennas, as might be used in a stacked Yagi array, is shown in Fig. 5. The transmission lines from each antenna to the common feed point must be equal in length and an odd multiple of a quarter wavelength. This line acts as an impedance transformer and raises the feed impedance of each antenna to 100 ohms. When the two antennas are connected in parallel at the coaxial "T" fitting, the resulting impedance is close to 50 ohms.

A 5-Element Yagi for 50 MHz

The antenna described here was designed from information contained in Table 1 and Fig. 4. This antenna has a theoretical gain of 9.2 dB over a half-wavelength dipole and should exhibit a front-to-back ratio of roughly 18 dB. The pattern is quite clean, with side lobes well suppressed. A hairpin matching system is used, and if the dimensions are followed closely no adjustment should be necessary. The completed antenna is rugged, yet lightweight, and should be easy to install on any tower or mast.

Mechanical Details

Constructional details of the antenna are given in Figs. 7 and 8. The boom of the antenna is 17 feet (5.18 m) long and is made from a single piece of 2-inch (50.8 mm) aluminum irrigation tubing that has a wall thickness of 0.047 inch (1.2 mm). Irrigation tubing is normally supplied in 20 foot (6.1 m) sections so several feet

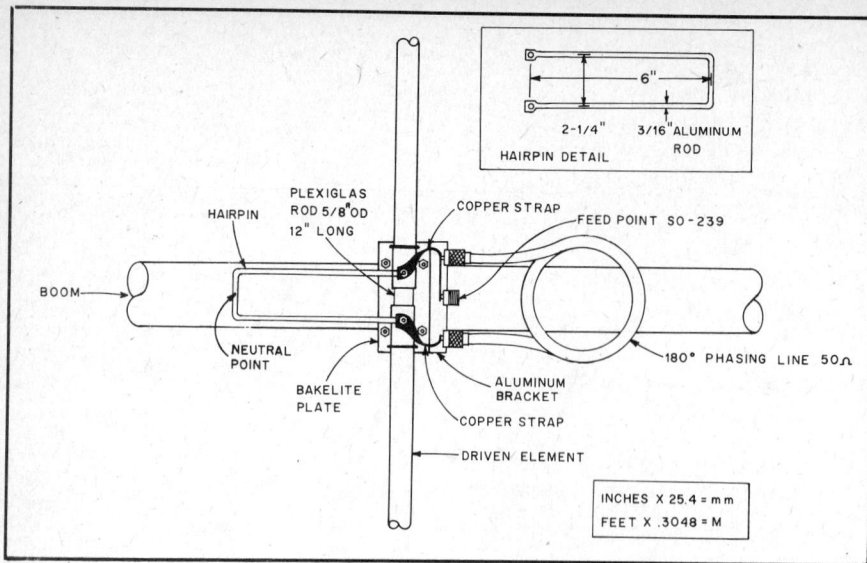

Fig. 8 — Detailed drawing of the feed system used with the 50-MHz Yagi. Phasing-line lengths are: for cable with 0.80 velocity factor — 7′ 10-3/8″ (2.4 m); for cable with 0.66 velocity factor — 6′ 5-3/4″ (1.97 m).

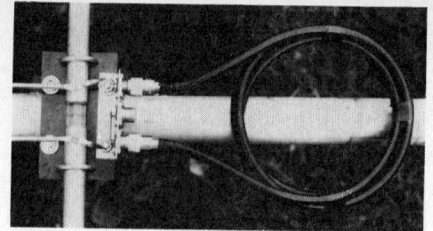

Fig. 9 — Closeup of the driven element and feed system. The phasing line is coiled and taped to the boom.

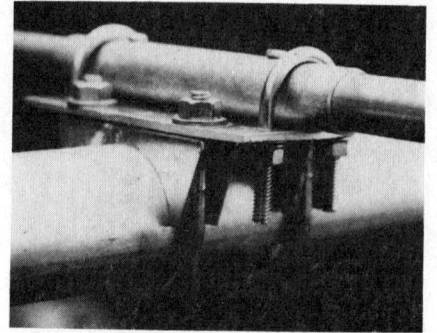

Fig. 10 — Photograph of the element-to-boom clamp. U-bolts are used to hold the element to the plate, and 2-inch (50.8 mm) plated muffler clamps hold the plates to the boom.

may be removed from the length.

Elements are constructed from 3/4-inch (19 mm) OD aluminum tubing of the 6061-T6 variety, with a wall thickness of 0.058 inch (1.5 mm). Each element, with the exception of the driven element, is made from a single length of tubing. The driven element is split in the center and insulated from the boom to provide a balanced feed system. The reflector and directors have short lengths of 7/8-inch (22.2 mm) aluminum tubing telescoped over the center of the elements for reinforcement purposes. Boom-to-element clamps were fashioned from 3/16-inch (4.8 mm) thick aluminum-plate stock, as shown in the photographs. Two muffler clamps hold each plate to the boom, and two U-bolts affix each element to the plate. Exact dimensions of the plate are not critical, but should be great enough to accommodate the two muffler clamps and two U-bolts. The element and clamp structure may seem to be a bit over-engineered. However, the antenna is designed to withstand the severe weather conditions common to New England. This antenna has withstood many wind storms and several ice storms, and no maintenance has been required.

The driven element is mounted to the boom on a Bakelite plate of similar dimension to the reflector and director element-to-boom plates. A piece of 5/8-inch (15.9 mm) Plexiglas rod, 12 inches in length, is inserted into each half of the driven element. The Plexiglas piece allows the use of a single clamp on each side of the element and also seals the center of the elements against moisture. Self-tapping screws are used for connection to the driven element. A length of 1/4-inch (6.4 mm) polypropylene rope is inserted into each element, and end caps

are placed on the elements. The rope damps element vibrations, which could lead to element or hardware fatigue.

Feed System

Details of the feed systems are shown in Figs. 8 and 9. A bracket fashioned from a piece of scrap aluminum is used to mount the three SO-239 connectors to the driven element plate. A half-wavelength phasing line connects the two element halves, providing the necessary 180-degree phase difference between them. The "hairpin" is connected directly across the element halves. It should be noted that the exact center of the hairpin is electrically neutral and may be fastened to the boom or allowed to hang. Phasing-line lengths are: for cable with 0.80 velocity factor — 7′ 10-3/8″ (2.4 m); for cable with 0.66 velocity factor — 6′ 5-3/4″ (1.97 m). It will be noted that the driven element is the shortest element in this array. While this may seem a bit unusual it is necessary with the hairpin matching system.

A 15-Element Yagi for 432 MHz

This 432-MHz Yagi antenna was designed using the information shown in Table 4 and Fig. 4. The theoretical gain for this antenna is 14.2 dBd, with a front-to-back pattern ratio of approximately 22 dB. The pattern is very sharp and quite clean, as would be expected from a well-tuned array of this size. Four of these antennas in a "box" or "H" array would serve well for terrestrial work, while eight would make a respectable EME system.

Mechanical Details

Dimensions for the antenna are given in Figs. 12 and 13. The boom of the antenna is made from a length of 1-inch (25.4 mm) aluminum tubing. Each of the elements is

mounted through the boom, and only the driven element is insulated. Auveco 8715 external retaining rings secure each of the parasitic elements in place. These rings are available at most hardware supply houses. Consult your local telephone company Yellow Pages for a hardware dealer near you.

The driven element is insulated from the boom by a length of 1/2-inch (12.7-mm) Teflon rod. The length of rod is drilled to accept the 1/4 inch (6.4-mm) thick aluminum tubing driven element. A press fit was used to secure the Teflon piece in the boom of the antenna. An exact fit can be achieved by drilling the hole slightly undersized and enlarging the hole with a hand reamer, a small amount at a time. Should the hole turn out to be oversized, a small amount of RTV (silicone seal) can be used to secure the Teflon in place. Although it wasn't tried with this antenna, it should be possible to use a driven element that is not insulated from the boom. Small changes in the position of the matching rods and/or clamps might be necessary.

Details of the feed system are shown in Fig. 13. This is a form of the "T" match, where the driven element is shortened from its resonant length to provide the necessary capacitance to tune out the reactance of the matching rods. With this system no variable capacitors are required, as in the more conventional T-match systems used at hf. This is a definite plus in terms of antenna endurance in harsh-weather environments.

The center pin of the UG-58A/U N-type connector attaches to one of the matching rods. A half wavelength of 50-ohm, foam-dielectric cable is used to provide the 180-degree phase shift from one half of the element to the other. An alternative to the large and cumbersome cable used here would be the miniature copper Hardline with Teflon dielectric material, such as RG-401.

Each of the matching rods is secured to two threaded steatite standoffs at the center of the antenna. These standoffs provide tie points for the ends of the phasing lines, the center pin of the coaxial connector as well as the ends of the matching rods. Solder lugs are used for each of the connections for easy assembly or disassembly. The clamps that connect the matching rods to the driven element are constructed from pieces of aluminum measuring 1/4 × 1/2 × 1-5/16 inches (6.4 × 12.7 × 33.3 mm). These pieces are drilled and slotted so that when the screws are tightened the pieces will compress slightly to provide a snug fit. Alternatively, simple clamps could be fashioned from strips of aluminum.

Adjustment

If the dimensions given in the drawings are followed closely, little adjustment should be necessary. If adjustment is necessary, as indicated by an SWR greater than 1.5 to 1, move the clamps a short distance along the driven element. Keep in mind that the clamps should be located equidistant from the center of the boom.

THE VHF QUAGI

First described by K6YNB in April 1977 *QST*, the quagi has become a very popular antenna for use on 144 MHz and above. The long-boom quagi was presented by K6YNB/N6NB in February 1978 *QST*.

How to Build a Quagi

There are a few tricks to quagi building.

Fig. 11 — The completed 15-element Yagi for 432 MHz, ready for installation atop the tower.

The designer mass produced as many as 16 in one day. Table 6 gives the dimensions for various frequencies.

The boom is *wood* or any other nonconductor (e.g., fiberglass). If a metal boom is used, a new design and new element lengths will be required. Many vhf antenna builders go wrong by failing to follow this rule: If the original uses a metal boom, use the same size and shape metal boom when you duplicate it. If it calls for a wood boom, use a nonconductor. Many amateurs dislike wood booms, but in a salt-air environment they outlast aluminum (and surely cost less). Varnish the boom for added protection.

The 2-meter version is usually built on a 14-foot (4.3-m), 1 × 3 inch (20 × 60 mm)

boom, with the boom cut down to taper it to one inch at both ends. Clear pine is best because of its light weight, but construction grade Douglas fir works well. At 220 MHz, the boom is under 10 feet (3 m) long and most builders use 1 × 2 (20 × 40 mm) or (preferably) 3/4 by 1-1/4-inch (19 × 32 mm) pine molding stock. On 432 MHz the boom must be 1/2-inch (13-mm) thick or less. Most builders use strips of 1/2-inch (13-mm) exterior plywood for 432.

The quad elements are supported at the current maxima (the top and bottom, the latter beside the feed point) with Plexiglas or small strips of wood. The quad elements are made of no. 12 copper wire, commonly used in house wiring. Some

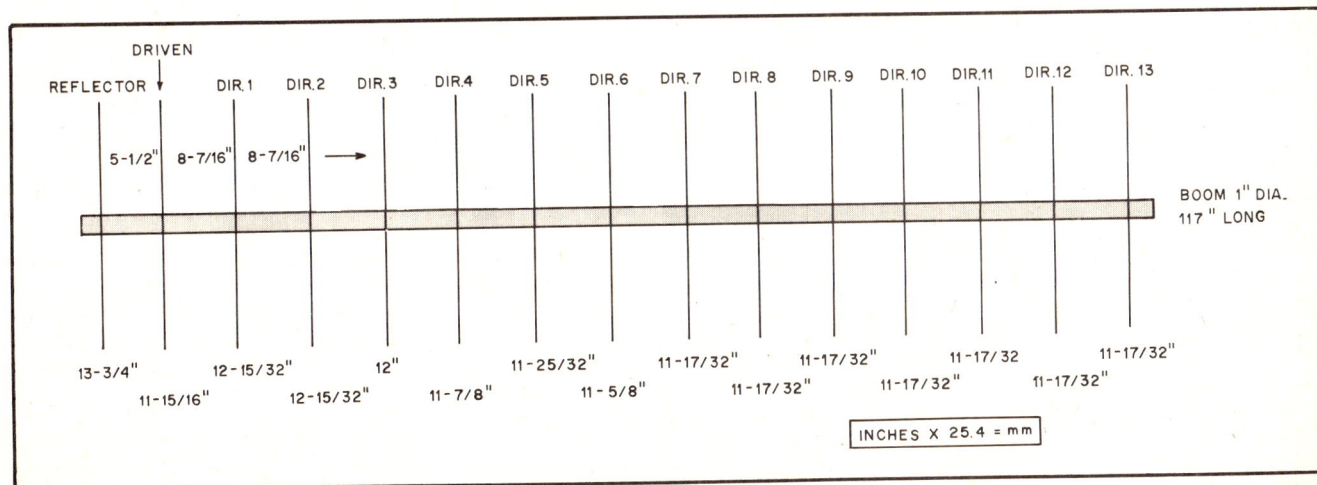

Fig. 12 — Dimensional drawing of the 15-element Yagi for 432 MHz. The balance point of the antenna is between the fifth and sixth director. The antenna was designed from the information presented in Table 4 and Fig. 4 of this chapter.

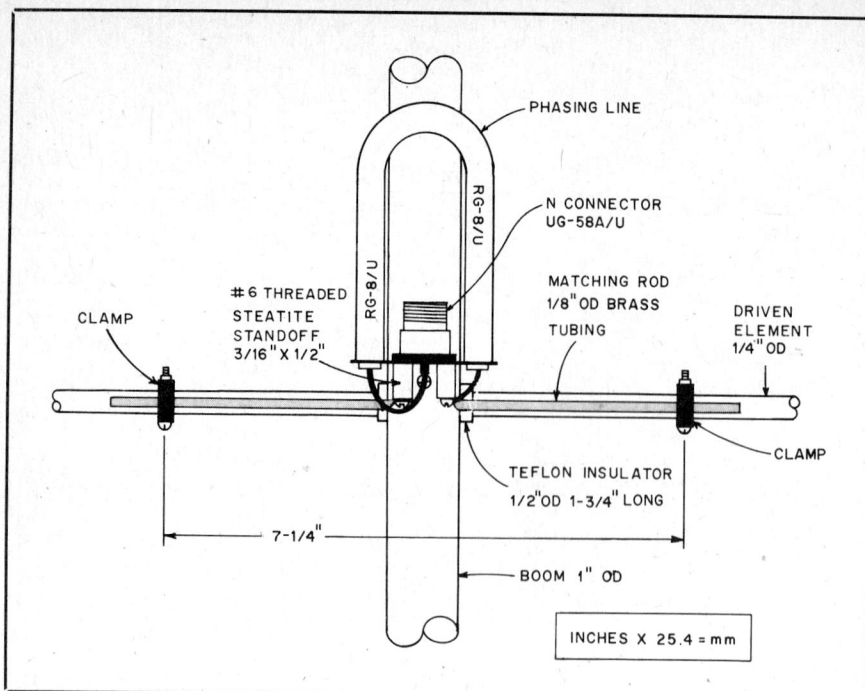

Fig. 13 — Detailed drawing of the feed system used with the Yagi. A small copper plate is attached to the coaxial-connector plate assembly for connection of the phasing line braid. As indicated, the braid is soldered to the plate. A coat of clear lacquer or enamel is recommended for waterproofing the feed system.

Fig. 14 — A close-up view of the feed method used on a 432-MHz quagi. This arrangement produces an excellent SWR and an actual measured gain in excess of 13 dB over an isotropic antenna with a 4-foot 10-inch (1.5 m) boom! The same basic arrangement is used on lower frequencies, but wood may be substituted for the Plexiglas spreaders. The boom is 1/2-inch (13 mm) exterior plywood.

Table 5

Dimensions, Eight-Element Quagi

Element Lengths	144.5 MHz	147 MHz	222 MHz	432 MHz	446 MHz
Reflector (all no. 12 TW wire, closed)	86-5/8" (loop)	85"	56-3/8"	28"	27-1/8"
Driven element (no. 12 TW, fed at bottom)	82" (loop)	80"	53.5"	26-5/8"	25-7/8"
Directors	35-15/16" to 35" in 3/16" steps	35-5/16" to 34-3/8" in 3/16" steps	23-3/8" to 22-3/4" in 1/8" steps	11-3/4" to 11-7/16" in 1/16" steps	11-3/8" to 11" in 1/16" steps
Spacing					
R-DE	21"	20-1/2"	13-5/8"	7"	6.8"
DE-D1	15-3/4"	15-3/8"	10-1/4"	5-1/4"	5.1"
D1-D2	33"	32-1/2"	21-1/2"	11"	10.7"
D2-D3	17-1/2"	17-1/8"	11-3/8"	5.85"	5.68"
D3-D4	26.1"	25-5/8"	17"	8.73"	8.46"
D4-D5	26.1"	25-5/8"	17"	8.73"	8.46"
D5-D6	26.1"	25-5/8"	17"	8.73"	8.46"
Stacking Distance Between Bays	11'	10'10"	7'1-1/2"	3'7"	3'5-5/8"

Inches × 25.4 = mm   Feet × 0.3048 = m

builders may elect to use no. 10 wire on 144 MHz and no. 14 wire on 432 MHz, although this will change the resonant frequency slightly. Solder a type-N connector (an SO-239 is often used at 2 meters) at the midpoint of the driven element bottom side, and close the reflector loop.

The directors are mounted through the boom. They can be made of almost any metal rod or wire of about 1/8-inch (3 mm) diameter. Welding rod or aluminum clothesline wire will work well if straight. (The designer used 1/8-inch stainless-steel rod secured from an aircraft surplus store.)

A TV-type U-bolt mounts the antenna on a mast. The author uses a single machine screw, washers and nut to secure the spreaders to the boom so the antenna can be quickly "flattened" for travel. In permanent installations two screws are recommended.

Construction Reminders

Here are a couple of hints based on the experiences of some who have built the quagi. First, remember that at 432 MHz even a 1/8-inch measuring error will deteriorate performance. Cut the loops and elements as carefully as possible. No precision tools are needed but be careful about accuracy. Also, make sure to get the elements in the right order. The longest director goes closest to the driven element.

Finally, remember that a balanced antenna is being fed with an unbalanced line. Every balun the designer tried introduced more losses than the feed imbalance problem. Some builders have tightly coiled several turns of the feed line near the feed point to limit radiation further down the line. In any case, the feed line should be kept at right angles to the antenna. Run it from the driven element directly to the supporting mast and then up or down perpendicularly for best results.

COLLINEAR ANTENNAS

Information given thus far is mainly on parasitic arrays, but the collinear antenna has much to recommend it. Inherently broad in frequency response, it is a logical choice where coverage of an entire band is wanted. This tolerance also makes a collinear easy to build and adjust for any vhf application, and the use of many driven elements is popular in very large phased arrays, such as may be required for moonbounce (EME) communication.

Large Collinear Arrays

Bidirectional curtain arrays of four, six, and eight half-waves in phase are shown in Fig. 15. Usually reflector elements are added, normally at about 0.2 wavelength in back of each driven element, for more

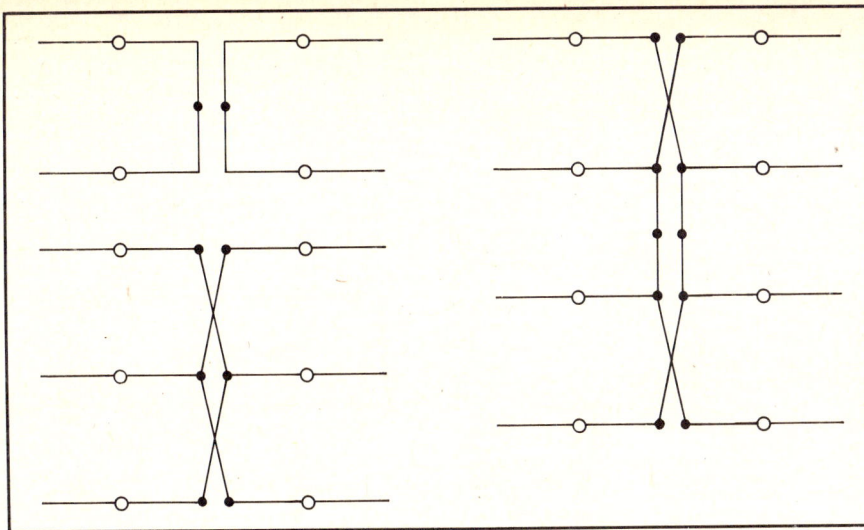

Fig. 15 — Element arrangements for 8-, 12- and 16-element collinear arrays. Parasitic reflectors omitted here for clarity, are 5 percent longer and 0.2 wavelength in back of the driven elements. Feed points are indicated by black dots. Open circles are recommended support points. The elements can run through wood or metal booms, without insulation, if supported at their centers in this way. Insulators at the element ends (points of high rf voltage) tend to detune and unbalance the system.

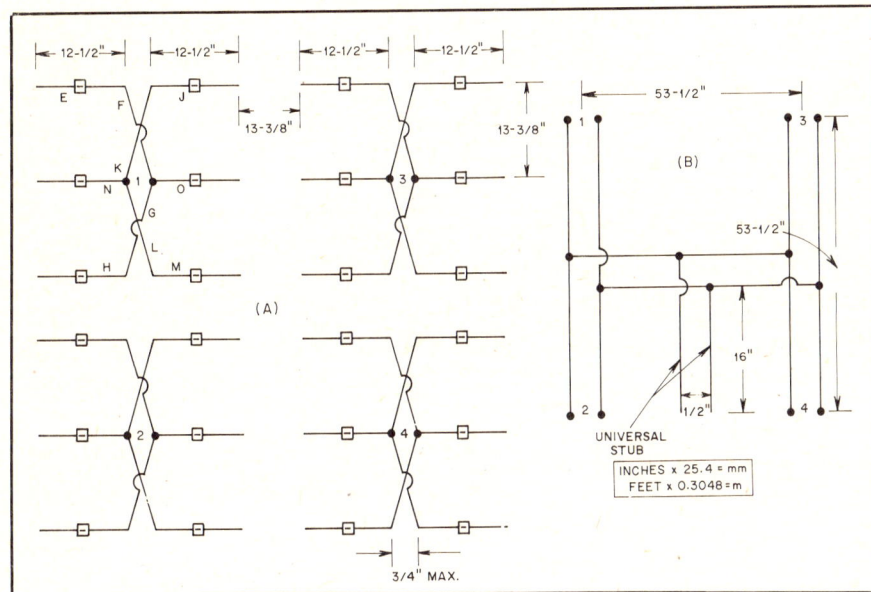

Fig. 16 — Large collinear arrays should be fed as sets of no more than eight driven elements each, interconnected by phasing lines. This 48-element array for 432 MHz (A) is treated as if it were four 12-element collinears. Reflector elements are omitted for clarity. Phasing harness is shown at B.

Table 6

432-MHz, 15-Element, Long-Boom Quagi Construction Data

Element Lengths — Inches		Interelement Spacing — Inches	
R — 28″ loop	D7 — 11-3/8	R-DE — 7	D6-D7 — 12
DE — 26-5/8″ loop	D8 — 11-5/16	DE-D1 — 5-1/4	D7-D8 — 12
D1 — 11-3/4	D9 — 11-5/16	D1-D2 — 11	D8-D9 — 11-1/4
D2 — 11-11/16	D10 — 11-1/4	D2-D3 — 5-7/8	D9-D10 — 11-1/2
D3 — 11-5/8	D11 — 11-3/16	D3-D4 — 8-3/4	D10-D11 — 9-3/16
D4 — 11-9/16	D12 — 11-1/8	D4-D5 — 8-3/4	D11-D12 — 12-3/8
D5 — 11-1/2	D13 — 11-1/16	D5-D6 — 8-3/4	D12-D13 — 13-3/4
D6 — 11-7/16			

Boom — 1 × 2-inch × 12-ft Douglas fir, tapered to 5/8 inch at both ends.
Driven element — No. 12 TW copper-wire loop in square configuration, fed at center bottom with type N connector and 52-ohm coax.
Reflector — No. 12 TW copper-wire loop, closed at bottom.
Directors — 1/8-inch rod passing through boom.

gain and a unidirectional pattern. Such parasitic elements are omitted from the sketch in the interest of clarity.

When parasitic elements are added, the feed impedance is low enough for direct connection to open line or twin-lead, connected at the points indicated by black dots. With coaxial line and a balun, it is suggested that the universal stub match, Fig. 1A, be used at the feedpoint. All elements should be mounted at their electrical centers, as indicated by open circles in Fig. 15. The framework can be metal or insulating material, with equally good results. The metal supporting structure is entirely in back of the plane of the reflector elements. Sheet-metal clamps can be cut from scraps of aluminum to make this kind of assembly, which is very light in weight and rugged as well. Collinear elements should always be mounted at their centers, where rf voltage is zero — never at their ends, where the voltage is high and insulation losses and detuning can be very harmful.

Collinear arrays of 32, 48, 64 and even 128 elements can be made to give outstanding performance. Any collinear should be fed at the center of the system, for balanced current distribution. This is very important in large arrays, which are treated as sets of six or eight driven elements each, and fed through a balanced harness, each section of which is a resonant length, usually of open-wire line. A 48-element collinear array for 432 MHz, Fig. 16, illustrates this principle.

A reflecting plane, which may be sheet metal, wire mesh, or even closely spaced elements of tubing or wire, can be used in place of parasitic reflectors. To be effective, the plane reflector must extend on all sides to at least a quarter-wavelength beyond the area occupied by the driven elements. The plane reflector provides high front-to-back ratio, a clean pattern, and somewhat more gain than parasitic elements, but large physical size rules it out for amateur use below 420 MHz. An interesting space-saving possibility lies in using a single plane reflector with elements for two different bands mounted on opposite sides. Reflector spacing from the driven element is not critical. About 0.2 wavelength is common.

CIRCULAR POLARIZATION

Polarization is described as "horizontal" or "vertical," but these terms have no meaning once the reference of the earth's surface is lost. Many propagation factors can cause polarization change: reflection or refraction, passage through magnetic fields (Faraday rotation) and satellite rolling, for example. Polarization of vhf waves is often random, so an antenna capable of accepting any polarization is useful. Circular polarization, generated with helical antennas or with crossed elements fed 90 degrees out of phase, has this quality.

The circularly polarized wave, in effect, threads its way through space, and it can be left- or right-hand polarized. These polarization "senses" are mutually exclusive, but either will respond to any plane polarization. A wave generated with right-hand polarization comes back with left-hand, when reflected from the moon, a fact to be borne in mind in setting up EME circuits. Stations communicating on direct paths should have the same polarization sense.

Both senses can be generated with crossed dipoles, with the aid of a switchable phasing harness. With helical arrays, both senses are provided with two antennas, wound in opposite directions.

Helical Antenna for 432 MHz

The eight turn helix of Fig. 17 is designed for 432 MHz, with left-hand polarization. It is made form 213 inches (5.4 m) of aluminum clothesline wire, including 6 inches (152 mm) that is used for cutting back to adjust the feed impedance.

Each turn is one wavelength long, and the pitch is about 0.25 wavelength. Turns are stapled to the wooden supports, which should be waterproofed with liquid fiberglass or exterior varnish. The reflecting screen is one wavelength square, with a type-N coaxial fitting soldered at its center, for connection of the required coaxial Q section.

The nominal impedance of a helical antenna is 140 ohms, calling for an 84-ohm matching section to match to a 50-ohm line. This can be approximated with copper tubing of 0.4-inch (10-mm) inside diameter, with no. 10 inner conductor, both 6-1/2 inches (165 mm) long. With the antenna and transformer connected, apply power and trim the outer end of the helix until the reflected power approaches zero.

The support arms are made from sections of 1 × 1 inch (25 × 25 mm) wood and are each 60 inches (1.52 m) long. The spacing between them is 8-1/4 inches (210 mm), outer dimension. The screen of the antenna in Fig. 17 is tacked to the support arms for temporary use. A wooden framework for the screen would provide a more rugged antenna structure. The theoretical gain of an eight-turn helical is approximately 14 decibels. Where both right- and left-hand circularity is desired, two antennas can be mounted on a common framework, a few wavelengths apart, and wound for opposite senses.

VHF AND UHF PARABEAMS

Fig. 18 shows two types of "Parabeam" Yagis for use at vhf and uhf. This style of gain antenna was developed in the UK by J-Beam, Ltd. to offer high gain and wide bandwidth. The design is suitable for use at 144, 220, 432 and 1296 MHz. The Parabeam utilizes a "skeleton slot" radiator and reflector (original J-Beam format), but employs conven-

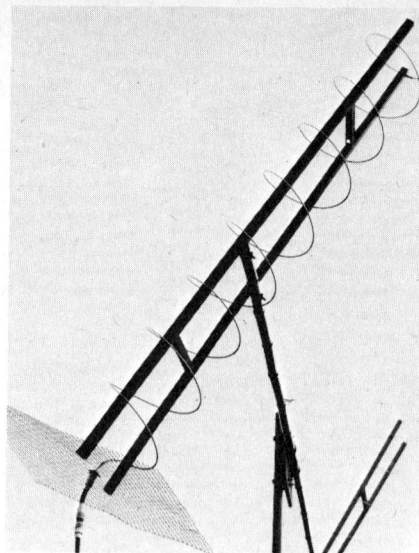

Fig. 17 — An eight-turn 432-MHz helical array, wound from aluminum clothesline wire. Left-hand polarization is shown. Each turn is one wavelength, with a pitch of 0.25 wavelength. Feed is with 50-ohm coax, through an 84-ohm Q section.

tional Yagi directors of the parasitic type.

The 2-meter Parabeam of Fig. 18A has claimed gain of 15 dB and a half-power horizontal beam width of 24°, according to information contained in the RSGB *Radio Communication Handbook*. (This has not been proved by the ARRL.)

Illustration B of Fig. 18 is for a 432-MHz Parabeam for which a gain of 17 dB is stated along with a half-power beam width of 28°. The reflector of the 432-MHz version consists of a pair of half-wavelength elements joined to form a full-wavelength loop. Bandwidth for this and the 2-meter version is on the order of 20 percent of frequency between the 2:1 VSWR points on the curve.

Details of the skeleton-slot driven element are seen in Fig. 19. It can be observed that a forward pitch of 11° is provided. According to the designers this is necessary to obtain optimum "launching" into the parasitic directors. The feed

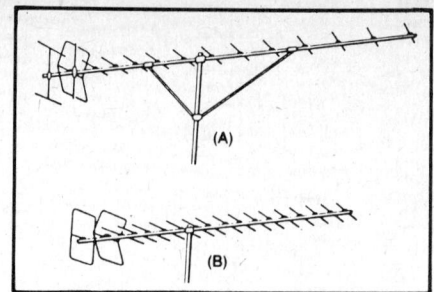

Fig. 18 — Illustration of a 2-meter Parabeam (A) and a 432-MHz version (B).

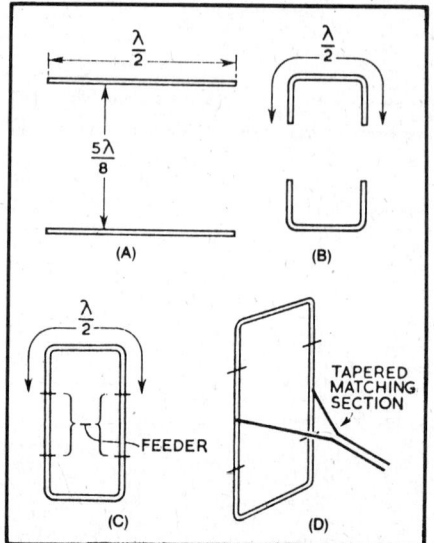

Fig. 19 — Developmental progression of the driven element of a Parabeam Yagi.

impedance of these Yagis is on the order of 280 to 300 ohms. A 4:1 balun can be used to provide a match to 75-Ω transmission line.

Dimensional details for a 432-MHz version of the Parabeam are given in Fig. 20. There are 18 elements used in the system, and stacking can be done if greater gain and aperture are desired. Good results should be possible by scaling this design to 144, 220 or 1296 MHz. However ex-

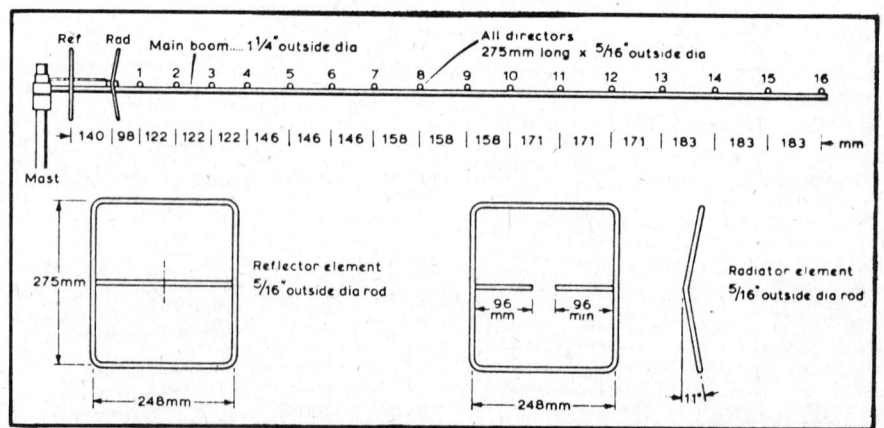

Fig. 20 — RSGB version of an 18-element Parabeam for 432 MHz. To convert mm to inches use in. = mm x 0.03937.

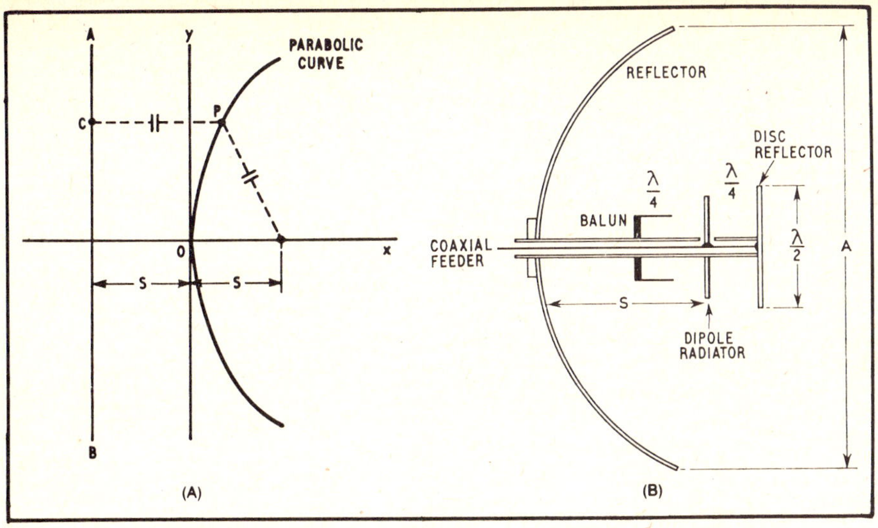

Fig. 21 — Details of the parabolic curve. $Y^2 = 4SX$. This curve (illustration A) is the locus of points which are equidistant from a fixed point, the focus (F), and a fixed line (AB) which is called the "directrix". Hence, FP = PC. The focus (F) contains the coordinates S and O. At B is a practical design which shows the feed arrangement of a parabolic antenna. The dipole must be at the focal point of the parabola. Distance S is made adjustable to provide maximum gain.

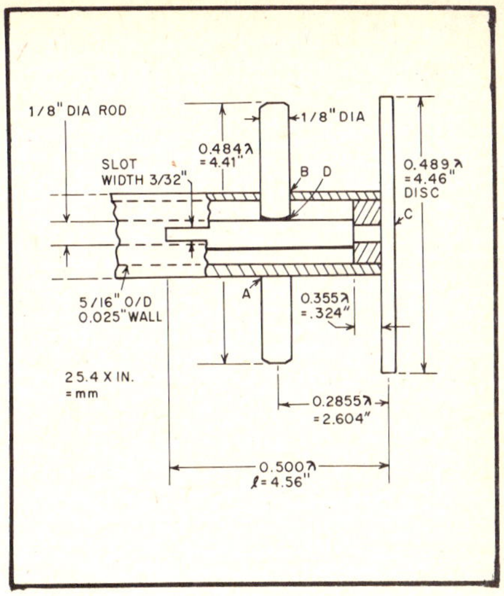

Fig. 22 — Symmetrical feed system for a 1296-MHz parabolic antenna. One half of the dipole is soldered in place at A. The remaining dipole half is threaded and screwed into the center conductor at point D. It is soldered to the outer conductor at B.

perimental adjustment of the element spacing should be carried out to ensure optimum forward gain from the scaled versions.

Parabolic Antennas

When an antenna is located at the focus point of a parabolic reflector it is possible to obtain considerable gain. Furthermore, the beam width of the radiated energy will be very narrow, provided all of the energy from the driven element is directed toward the focal point of the reflector.

When the rf energy is applied so that uniform illumination of the paraboloid is effected, the angular width of the radiated wave will be based on the diameter of the parabolic reflector, A. A close approximation can be obtained from

$$\varnothing = \frac{58}{A}$$

where A is expressed in wavelengths. Fig. 21A shows the fundamental shape of a parabolic curve. At B of Fig. 21 is a detailed sketch of the system, showing dimension A.

Wire mesh of close spacing can be used through 1296 MHz with good results. A wooden support form may be fashioned to provide the proper parabolic shape by plotting a curve (Fig. 21A) from

$$y^2 = 4SX$$

as shown in the figure.

The gain will be a function of the dish diameter and the illumination of it from the feed point. In a classic design there will be no radiation from the feed system. Such "spillover" will cause a loss of gain.

Fig. 22 shows details for a 1296-MHz symmetrical excitation system for a parabolic reflector. These dimensions were given in the RSGB *Radio Communication Handbook,* 5th Edition, Vol. 2. Practical details for constructing this type of antenna were written by R. Knadle, K2RIW, in "Twelve-Foot Stressed Parabolic Dish," *QST*, August 1972. The same author described modern uhf-antenna test procedures in, "UHF Antenna Ratiometry," *QST*, February 1976.

Chapter 22

Operating a Station

Although this Handbook serves primarily the technical phases of Amateur Radio — and indeed radio/electronics in general — as far as hamming is concerned, only about half the game deals with how you handle the soldering iron. The other half is concerned with what you do with your radio equipment once you get it operating. This is the subject of this chapter.

It's a pretty big subject, one which we can't hope to cover in these few pages, so at the very start we urge that you obtain the two principal operating publications put out by ARRL. They're both free. *Operating an Amateur Radio Station* covers basic and intermediate operating, details all the League's awards, contests and other operating activities, the complete ARRL operating organization and how it works, abbreviations and a lot more. The *Public Service Communications Manual* goes into even further detail on ARRL's important operating organization, comprising the Amateur Radio Emergency Service (ARES) and the National Traffic System (NTS).

The ARRL Operating Manual is the definitive source of information on all phases of on-the-air activity for radio amateurs. Newcomer and veteran alike will profit from in-depth treatment of everything from DXing to computers. This publication is available at your local dealer or direct from ARRL Hq. for $5 ($5.50 foreign).

Operating Standards

Amateur Radio through the years has developed a number of operating standards and procedures. Some of these are borrowed from other services, such as commercial or military. Some have been coined by amateurs for our particular use, as a part of "ham jargon." Still others have been innovated by the League to fulfill a need. All of them together make up a "standard operating procedure" for amateurs differing, at least in part, from the procedures used in any other radio service. We hams are a service; we aren't just a bunch of licensed operators randomly pursuing a "hobby." We have standard procedures that are recommended by ARRL but based on our particular needs. If we all use different procedures, we will have difficulty, at times, in communicating with each other. If we all use the same procedures, even though the ARRL recommendation may not be our personal preference, we will communicate effectively. And since ARRL is ultimately controlled by its members, any procedure you don't like can be changed if members share your opinion.

So let's all use the same standards. Let's be a single, organized, operating *service*.

Establishing Contact

If you are looking for a contact with *anyone*, then you may want to call "CQ." Before your signal uses valuable spectrum space, however, listen to see if the frequency appears to be in use. Follow that up with, "Is the frequency in use?" on voice, or QRL? on cw. You don't need a long CQ — just a 3 × 3. If still no answer, it means one of two things: Either no one heard you or no one listening wants to contact you. Try it again, another 3 × 3. If still no answer, that's enough. Move to another (clear) frequency before you try it again. A CQ call on cw might go like this: CQ CQ CQ DE AB1U AB1U AB1U K. On voice, "CQ CQ, this is AB1U, Alpha Bravo One Uniform, go ahead."

If you call CQ, it means you are willing to talk to anyone. If you want to be fussy don't call CQ. Find someone you would like to talk to and call *that person*. In this case, you can observe the same principle; zero beat his frequency and give him a short call, such as: K1MM K1MM K1MM DE VE3GT VE3GT $\overline{AR}$, or even shorter. On voice, "K1MM K1MM from VE3GT, Victor Echo Three Golf Tango, over."

Notice the "ending signals." These aren't just happenstance; each one of them means something. On cw, for example, K at the end of a transmission means "anyone go ahead," while $\overline{AR}$ means "I have just called another station and want

This neat "hide-away" station of W9OBF is very active in contest and DX activities.

only him to reply." On voice, "go" or "go ahead" refers to anybody listening, while "over" refers to a specific station. And so on. There is a complete list of ending prosigns and prowords. Use them properly, even if others you hear do not.

Now, about the QSO itself. You are introducing yourself to a brand-new aquaintance. Don't bore him. The usual procedure is to give him a signal report (see RST system), and tell him your name and location. After that, tell him the things about yourself that you would like to know about him, but make your transmission short. Start each transmission with your own call, i.e., DE K1MEM or "This is K1MEM," and end it with both calls, i.e., W6R1P DE (or from) K1MEM. Most voice contacts use "VOX" or "push to talk," which makes it possible to talk back and forth rapidly, much like a telephone conversation. When you do this, identification of each transmission is unnecessary. But be sure your station identification is completed at least every 10 minutes. At the end of the contact identify both stations; that is, each station must identify not only itself but the station with which contact is being made.

Repeater operating uses somewhat different procedures because of special circumstances, but we'll discuss that a little later.

The ARRL QSL Bureaus

Only one thing is left to do before you can consider the contact complete — confirm it with a "QSL" card. The QSL is considered the final courtesy of a contact and is an Amateur Radio tradition. Most amateurs have printed cards, some personalized (at greater cost), some using standard setups provided by printers for the purpose. Whatever form you adopt, be sure your QSL card shows very clearly the *correct* call of the station contacted, the date (including year) and UTC time, the band on which contact was made and the mode. Most awards based on QSL cards require at least these essentials, along with, of course, your street address, city, state or province, and country. Some awards include county as well. Other interesting data might include some details of your equipment, antennas, former calls held, class of license, signal report and any friendly comments.

If everybody waited to receive a card before sending one, there would be no QSLing. Admittedly, printing and postage are expensive and if you are very active on the air, QSLing every contact may seem like a needless expense. If you want a QSL from your contact, send him your card; and as a matter of common courtesy, send a card to everyone who sends you one.

Receiving DX QSLs

Within the U.S. and Canada, the ARRL DX QSL Bureau System is made

Table 1
Some Facts About Time Conversion

The chart below has been arranged to show time zones used by most amateurs in the North American continent and Universal Coordinated Time, used universally as a standard. The advantage of UTC is that it is the universally understood reference throughout the world. ARRL recommends that all amateur logging be done in UTC.

All times shown are in 24-hour time for convenience. To convert to 12-hour time; for times between 0000 and 0059, change the first two ciphers to 12, insert a colon and add A.M.; for times between 1200 and 1259, insert a colon and add P.M.; for times between 1300 and 2400, subtract 12, insert a colon and add P.M.

Time zone letters may be used to identify the kind of time being used. For example, UTC is designated by the letter Z, EDT/AST by the letter Q, CDT/EST by R, MDT/CST by S, PDT/MST by T, PST by U; thus, 1200R would indicate noon in the CDT/EST zone, which would convert to 1700 UTC or 1700 Z.

In converting from one time to another, be sure the day or date corresponds to the new time. That is, 2100R (EST) on January 1 would be 0200Z (UTC) on January 2; similarly, 0400Z on January 2 would be 2000U (PST) on January 1.

A good method is to use UTC (Z) for *all* amateur logging, schedule-making, QSLing and other amateur work. Confusion, with all the different time zones, is inevitable. Leave your clock on UTC.

The Canadian Maritime provinces and Puerto Rico use AST (Q) time, or ADST (P) time. Most of Alaska and Hawaii use W time (+ 10 to UTC).

UTC	EDT/AST	CDT/EST	MDT/CST	PDT/MST	PST
0000*	2000	1900	1800	1700	1600
0100	2100	2000	1900	1800	1700
0200	2200	2100	2000	1900	1800
0300	2300	2200	2100	2000	1900
0400	0000*	2300	2200	2100	2000
0500	0100	0000*	2300	2200	2100
0600	0200	0100	0000*	2300	2200
0700	0300	0200	0100	0000*	2300
0800	0400	0300	0200	0100	0000*
0900	0500	0400	0300	0200	0100
1000	0600	0500	0400	0300	0200
1100	0700	0600	0500	0400	0300
1200	0800	0700	0600	0500	0400
1300	0900	0800	0700	0600	0500
1400	1000	0900	0800	0700	0600
1500	1100	1000	0900	0800	0700
1600	1200	1100	1000	0900	0800
1700	1300	1200	1100	1000	0900
1800	1400	1300	1200	1100	1000
1900	1500	1400	1300	1200	1100
2000	1600	1500	1400	1300	1200
2100	1700	1600	1500	1400	1300
2200	1800	1700	1600	1500	1400
2300	1900	1800	1700	1600	1500
2400	2000	1900	1800	1700	1600

Time changes one hour with each change of 15° in longitude. The five time zones in the U.S. proper and Canada roughly follow these lines.
*0000 and 2400 are interchangeable. 2400 is associated with the date of the day ending, 0000 with the day just starting.

up of 22 call area bureaus. Most of the cards from DX bureaus go directly to the individual bureaus.

At the individual bureaus, the incoming cards are sorted by the first letter of the suffix. This sorting divides the work load into portions that can be handled by a single individual.

To claim your cards, send a 5 × 7-1/2-inch self-addressed stamped envelope to the bureau serving your district. Addresses for the U.S. and Canadian bureaus are shown in *QST* in the "QSL Corner" column.

These envelopes should have your call sign printed neatly in the upper left corner of the envelope to assist the sorter of your cards. Some bureaus will sell envelopes or postage credits as well as handling s.a.s.e.s. The bureau will provide the proper-size envelope and affix appropriate postage upon prepayment of a certain fee. The exact arrangement of

your area bureau can be obtained by sending your bureau an s.a.s.e. with your inquiry.

Since many of the DX stations use the bureau system, this area bureau can be very important to someone who works DX. But, it is a complex volunteer arrangement requiring good cooperation on the part of the DXers to function properly.

Sending Your DX QSLs

Each month, every member of the ARRL (except family and sightless members) is mailed a copy of *QST*. The address label on the wrapper of *QST* is the member's "ticket" for use of the Overseas QSL Service. Twelve times per year, an ARRL member may send any number of QSL cards for amateurs overseas. With each mailing the member must include the address label from the

Table 2
ARRL Communications Procedures

Voice	Code	Situation
Go ahead	K	Used after calling CQ, or at the end of a transmission, to indicate any station is invited to transmit.
Over	$\overline{AR}$	Used after a call to a specific station, before the contact has been established.
——	$\overline{KN}$	Used at the end of any transmission when only the specific station contacted is invited to answer.
Stand by or wait	$\overline{AS}$	A temporary interruption of the contact.
Roger	R	Indicates a transmission has been received correctly and in full.
Clear	$\overline{SK}$	End of contact. $\overline{SK}$ is sent before the final identification.
Leaving the air or closing station	CL	Indicates that a station is going off the air, and will not listen for or answer any further calls. CL is sent after the final identification.

current copy of *QST* and $1 (check or money order) for each pound or part of a pound of cards; 155 typical cards weigh one pound. QSLs must be presorted by prefix. Nothing but the cards, address label and money may be included in the package. Wrap the package securely and address it to ARRL-Membership Overseas QSL Service, 225 Main St., Newington, CT 06111.

"Family" members of ARRL, to whom only one copy of *QST* is sent, may send cards in the same package but must include $1 for each member sending cards and indicate that the *QST* address label includes a "family membership."

Sightless members, who do not receive a copy of *QST*, need only include $1 with a note indicating that the cards are from a sightless member. Associate (unlicensed) members may use the Overseas QSL Service to send SWL reports to overseas *amateur* stations. No cards will be sent to individual QSL managers.

Additional information is available from ARRL. Send an s.a.s.e. and request the QSL Bureau reprint.

Your Station Record — Logging

"Official" logging requirements have

been eased in recent years, but an accurate, complete and neat log book should be a matter of personal pride. It can also be a strong form of protection for you against possible claims by others of intentional interference or against troubles caused by unauthorized use of your call by "bootleggers."

A log should be convenient to use — a bound one is best so pages will not be lost. The ARRL log is designed to make it as easy as possible for you to comply legally with FCC requirements. Your log becomes a written diary of your amateur operation and should include everything that will be of interest to you in years to come. FCC requires you to retain your log for only a year, but most amateurs retain theirs indefinitely as an historical record.

The Voice Modes

The use of proper procedure to get best results is very important. Voice operators *say* what they want to have understood, while cw operators have to spell it out or abbreviate. Since on phone the speed of transmission is generally between 150 and 200 words per minute, the matter of readability and understandability is critical to good communications. The

good voice operator uses operating habits that are beyond reproach.

Phone Operating Practices

Listen with care. It is very natural to answer the loudest station that calls, but with a little digging, if need be, answer the *best* signal instead. Not all amateurs can run a kilowatt, but there is no reason every amateur cannot have a signal of the highest quailty. Do not reward the operator who cranks up the transmitter gain and splatters by answering him if another station is calling.

Use VOX or push-to-talk. If you use VOX, don't defeat its purpose by saying "ahhh" to keep the relay closed. If you use push-to-talk, let go of the mic button every so often to make sure you are not "doubling" with the other station. Don't be a monologuist.

Listen before transmitting. Make sure the frequency isn't being used before you come barging into it. If you don't hear any station on the frequency, make this announcement: "Is the frequency in use? W1CKK." If it is still clear, you are ready to make your call.

Interpose your call frequently. In distinct, measured tones say your call often. Use approved phonetics if your call sign is hard to understand or conditions are poor. Remember you can be cited for improper identification if it cannot be understood.

Keep microphone gain (MIC GAIN) constant. Don't "ride" the microphone gain. Try to speak in an even amplitude the same distance from the microphone, keeping the gain down to eliminate room noise. Follow the manufacturers' instructions for use of the microphone; some require close-talking, while some need to be turned at an angle to the speaker's mouth.

The speed of radiotelephone transmission (with perfect accuracy) depends

	FIXED					VARIABLE						
DATE	FREQ.	MODE	POWER	TIME	STATION WORKED	REPORT SENT	REC'D	TIME OFF	QTH	COMMENTS NAME	QSL VIA	QSL S	R
28 JUL	146.52	FM	10	0430	WA1CCR				Wallingford	Eric	New converter works!	
3 OCT	7.0	A1	150	2319	WA6VEF	001	322	CONTRA COS	CALIFORNIA QSO PARTY			
				22	N6OJ	002	157	SONO				
				24	K6NA	003	331	SD				
				31	N6OP/M	004	117	CALAV				
9 OCT	28.6	A3	1 KW	0301	JA1OCA	59	57		Tokyo	Isao	BURO	✓
	21	A1		1545	EA9GD	559	579		Melilla	Jose	Box 348	✓ ✓
				56	6Ø0DX	599	599		Somalia		I2YAE	✓
5 NOV	3.8	A3	150	0030	W9NA	59+	59+	0117	Wausau, WI	Reno		
9 NOV	21	A1	10	1642	G4BUE	339	449	1657		1 watt!		

Labels: See inside front cover. — Input in Watts. — UTC recommended. — RST. See back inside cover. — This column may also be used for contest exchange info. received.

The ARRL Universal Log is adaptable for all types of operating — ragchewing, contesting, DXing. References are to pages in the ARRL Log.

Table 3
International Telecommunication Union
Phonetics

A — Alpha	J — Juliet	S — Sierra
B — Bravo	K — Kilo	T — Tango
C — Charlie	L — Lima	U — Uniform
D — Delta	M — Mike	V — Victor
E — Echo	N — November	W — Whiskey
F — Foxtrot	O — Oscar	X — X-Ray
G — Golf	P — Papa	Y — Yankee
H — Hotel	Q — Quebec	Z — Zulu
I — India	R — Romeo	

almost entirely on the skill of the two operators concerned. One must use a rate of speech allowing perfect understanding as well as permitting the receiving operator to record the information.

Voice Operating Hints

1) Listen before calling.

2) Make short calls with breaks to listen. Avoid long CQs; do not answer overlong CQs.

3) Use push-to-talk or voice control (VOX). Give data concisely in first transmissions.

4) Make reports honest. Use definitions of strength and readability for reference. Make your reports informative and useful. Honest reports and *full* word descriptions of signals save operators from FCC trouble.

5) Limit transmission length. Two minutes or less will convey much information. When three or more stations converse in roundtables, brevity is essential.

6) Display sportsmanship and courtesy. Bands are congested . . . make transmissions meaningful . . . give others a break.

7) Check transmitter adjustment . . . avoid splatter. On ssb, check carrier balance carefully. Do not transmit when moving VFO frequency. Complete testing before busy hours — use a dummy load!

Repeater Operating

A repeater is a device that receives a signal on one frequency and simultaneously transmits ("repeats") the received signal on another frequency. Often located atop a tall building or high mountain, repeaters greatly extend the operating coverage of amateurs using mobile and hand-held tranceivers.

To use a repeater you must have a transceiver with the capability of transmitting on the repeater's *input frequency* (the frequency that the repeater listens on) and receiving on the repeater's *output frequency* (the frequency the repeater transmits on). This capability can be acquired by installing the correct crystals in your transceiver or, if you have a synthesized rig, by dialing the correct frequency and selecting the proper *offset* (frequency difference between input and output).

When you have the frequency capability, all that you need do is key the

microphone button and you will turn on ("access") the repeater. Some repeaters have limited access requiring the transmission of a subaudible tone, series of tones or bursts in order to gain access. Most repeaters briefly transmit a carrier after a user has stopped transmitting to inform the user that he is actually accessing a repeater.

After acquiring the ability to access a repeater you should become acquainted with the operating practices that are inherent to this unique mode of Amateur Radio:

1) Monitor the repeater to become familiar with any peculiarities in its operation.

2) To initiate a contact simply indicate that you are on frequency. Various geographical areas have different practices on making yourself heard, but, generally, "This is W1XZ monitoring" will suffice. One practice that is usually looked upon with disfavor throughout the U.S. and Canada is calling CQ on a repeater.

3) Identify legally; you must transmit your call sign at the beginning and end of each contact and every 10 minutes in between. At the end you must also identify the station you were in QSO with. It is illegal to key a repeater without identification.

4) Pause between transmissions. This allows other hams to use the repeater (someone may have an emergency). On most repeaters a pause is necessary to reset the timer.

5) Keep transmissions short and thoughtful. Your monologue may prevent someone with an emergency from using the repeater. If your monologue is long enough, you may *time-out* the repeater. Your transmissions are being heard by many listeners including nonhams with "public service band" monitors and scanners; don't give a bad impression of our hobby.

6) Use simplex whenever possible. If you can complete your QSO on a direct frequency, there is no need to tie up the repeater and prevent others from using it.

7) Use the minimum amount of power necessary to maintain communciations. This FCC regulation minimizes the possibility of accessing distant repeaters on the same frequency.

8) Don't break into a contact unless you have something to add. Interrupting is no more polite on the air than it is in person.

9) Many repeaters are equipped with autopatch facilities which, when properly accessed, connect the repeater to the telephone system to provide a public service. The FCC forbids using an autopatch for anything that could be construed as business communications. Nor should an autopatch be used to avoid a toll call. Do not use an autopatch where regular telephone service is available. Abuses of

This W1AW 120-foot tower holds phased 4-element 20-meter Yagis at 60 and 120 feet and a 3-element 40-meter Yagi at 90 feet. The 5-element 10-meter Yagi in the background is on a 60-foot tower.

autopatch privileges may lead to their loss.

10) All repeaters are assembled and maintained at considerable expense and inconvenience. Usually an individual or a group is responsible and it behooves those who are regular users of a repeater to support the efforts of keeping the repeater on the air.

A directory listing all registered repeaters is available from the ARRL.

CW Operating

If you spend your entire Amateur Radio career on phone, once you have mastered enough cw to pass the necessary tests, you are missing out on at least 50 percent of the fun of hamming. Mastering the art of cw communication is 10 times easier than learning to talk, and you did this when you were two years old. All it takes is some basic learning principles, then practice, practice and more practice. This is not drudgery as you might think, because you can combine learning with listening to actual signals on the band, and even with operating, since the Novice class license requires only five wpm.

But listening is the best way to go, in the beginning. You don't need a license for this. Once you have learned the basic sounds of code, you will soon start recognizing common words — the, and, CQ, DE (from), etc. Practicing by copying calls is an excellent means of getting prelicense practice. You can even get good

Table 5
The R-S-T System

Readability

1 — Unreadable.
2 — Barely readable, occasional words distinguishable.
3 — Readable with considerable difficulty.
4 — Readable with practically no difficulty.
5 — Perfectly readable.

Signal Strength

1 — Faint signals, barely perceptible.
2 — Very weak signals.
3 — Weak signals.
4 — Fair signals.
5 — Fairly good signals.
6 — Good signals.
7 — Moderately strong signals.
8 — Strong signals.
9 — Extremely strong signals.

Tone

1 — Sixty cycle ac or less, very rough and broad.
2 — Very rough ac, very harsh and broad.
3 — Rough ac tone, rectified but not filtered.
4 — Rough note, some trace of filtering.
5 — Filtered rectified ac but strongly ripple-modulated.
6 — Filtered tone, definite trace of ripple modulation.
7 — Near pure tone, trace of ripple modulation.
8 — Near perfect tone, slight trace of modulation.
9 — Perfect tone, no trace of ripple or modulation of any kind.

If the signal has the characteristic steadiness of crystal control, add the letter X to the RST report. If there is a chirp, the letter C may be added to so indicate. Similarly for a click, add K. See FCC Regulations 97.73, purity of emissions. The above reporting system is used on both cw and voice, leaving out the "tone" report on voice.

sending practice by pretending to call the CQing station on your code practice oscillator. Sending practice is important. Start with a simple "straight" key: the time to "graduate" to an electronic keyer is later.

There are many pitfalls to developing into a finished cw operator, not the least of which is the acquisition of bad habits. Many of these come from mimicking your peers or elders, some of whom themselves are the victims of bad cw habits. Don't let them rub off on you. Most of these unfavorable traits of beginner cw operators are called the "Novice accent," which identify you as a beginner.

Probably the worst of these is carelessness about spacing. Your early cw training should have taught you that spacing length is just as important as dit and dah length. One way to improve your spacing is to practice sending in step with W1AW using a code-practice oscillator (but not on the air!). If you can send "in step" with W1AW, your sending is perfect. The source of each W1AW code practice text is sent several times during

the transmission to enable checking your copy.

On phone it is unnecessary and therefore improper to use jargon and abbreviations, but on cw abbreviations are a necessity. Without them, it takes a long time to say what you want to say, especially at beginner speeds. Most of the abbreviations we hams use have developed within the fraternity; some of them are borrowed from or are carryovers from old-time telegraphy abbreviations. Which is which doesn't matter; it is all ham radio to us. Learn to use ham cw abbreviations liberally and you will get much more said in much less time. When you reach a high-proficiency level, you will find that cw is almost as fast as talking, thus all but eliminating one of its principal disadvantages while still retaining its many advantages.

Despite the fact that learning cw is easier than learning to talk, nearly everybody can talk but few people can communicate by cw. Thus, there is considerable pride of accomplishment in cw operating. Regardless of your level of proficiency, chances are that you still have something to learn about cw operating, additional proficiency goals you can achieve. ARRL offers a series of awards in this field, starting at 10 wpm and progressing all the way through 40 wpm, in five-wpm increments. You will find more details elsewhere in this chapter and full details in *Operating An Amateur Radio Station*.

Copying cw and comprehending it are not the same thing. The word "copy" implies something written, so "if you don't put it down, that ain't copying." On the other hand, for conversational cw purposes, copying really isn't necessary and can be quite cumbersome. Most beginners on cw learn by copying everything down, and some find this habit hard to break as they achieve higher levels of proficiency. However, it does involve a "translation" process that can and should be eliminated for conversational purposes. Sooner or later, in order to realize the full value of cw communication, you must learn just to listen to it, as you do the spoken word, rather than to "copy" it. It should not be necessary to translate the cw into written copy and then translate the written copy into intelligence by reading it. Eventually, the sound of code should directly trigger your consciousness just as the spoken word does, and then the "copy" and understanding functions are reversed; that is, you understand it first, then you copy it. If written copy is unnecessary, the function stops at understanding it.

Cw is not just something that has been imposed on us to make passing the amateur test a little more difficult. It is an entirely different method of communication, and a long way from obsolete. Learn it well and you will enjoy Amateur Radio much more.

SSTV Procedures

The popularity of slow-scan television (SSTV) continues growing with more and more of the newer amateurs getting on this mode. To operate SSTV, first listen around the calling frequencies, then either respond on voice to a station calling "CQ SSTV" or call CQ SSTV yourself after first ascertaining that the frequency is not in use. Call CQ on voice in the normal manner adding "SSTV" to your call: "CQ SSTV CQ SSTV CQ SSTV this is K1WJ K1WJ K1WJ over." Once contact on voice is established, an exchange of video is undertaken. You should identify your video as you start your transmission and as you end transmission of video as follows: "K1KI this is K1WJ, video follows. SSTV pictures sent. OK, how do you like that one? K1KI this is K1WJ over."

SSTV operators should carefully monitor their station audio to ensure that they are not overmodulating (allowing the SSTV signal to exceed normal voice bandwidth limitations). It is possible to store properly modulated signals on ordinary cassette tape, so that you can play back some rare DX QSO, or a nice series of shack pictures to visitors in your shack when the band is dead, and you're not on the air.

RTTY Operating Procedures

Radioteletype (RTTY) operation involves specialized communications techniques and practices. One should be familiar with the unique aspects of this mode before getting on the air and communicating via the "green keys."

Either the Baudot or ASCII code may be used for radioteletype communications. Baudot is a five-level code and is usually transmitted at 60 or 100 wpm. ASCII, a code of communications in the computer world, is a seven-level code and may be transmitted at maximum speeds ranging from 110 baud to 19.6 kilobaud, depending on the operating frequency.

A *baud* is a unit of signaling rate, and is derived from the shortest mark or space interval. This signaling interval is called a unit pulse, and a transmission rate of one unit pulse per second is one baud. A *bit* (contraction of binary digit) is the smallest single unit of information in a binary system. Each ASCII alphanumeric character is composed of seven bits. Optional bits for error detection and timing may be added to each character. There is a common misconception that bit rate (information transfer rate) and baud (signaling rate) are equivalent. This is true only if the transmission is a continuous stream of alternating marks and spaces. The proper relationship between information rate and signaling rate is expressed by the equation: bits per second = bauds × no. of bits per signaling interval.

ASCII is a recent addition to the world

of Amateur Radio. It is expected that hams who are also computer hobbyists will use ASCII to relay computer programs and set up computer communication networks. The most common application on the hf bands at the present time is a start-stop Teletype using 110 bps and 170 Hz shift. On vhf and higher frequencies, high speed "packet" lends itself to computerized networks. There is some limited use of Bell 103 AFSK modems using 200 Hz shift. Since this is a rapidly changing field, watch *QST* for latest developments.

Although the regulations permit RTTY (Baudot) transmissions at 60, 67, 75 and 100 words per minute (wpm), the majority of RTTY operators are set up for 60 wpm. With the influx of computer-geared equipment into the mode, however, 100 wpm is increasing in popularity.

By gentlemen's agreement, RTTY activity is centered in certain portions of each hf band. On 80 meters almost all of the activity is between 3.600 and 3.630 MHz. On 20 through 10 meters, the activity is between the first 80 and 100 kHz of each band. For example, on 20 meters, 14.080 to 14.100 MHz is the RTTY hotbed.

Along with the growth of 2-meter activity, there has been a parallel growth of RTTY on this band. A number of repeaters are devoted to RTTY; these often operate on the 146.10/.70 pair of frequencies. Two-meter simplex frequencies are also popular spots for RTTY.

The regulations requiring station identification also apply to RTTY transmissions. Identification in RTTY does not fulfill this requirement; identification must be in cw (or phone). If you have the capability to punch and send messages with paper tape, be sure that these automated transmissions do not extend beyond 10 minutes without proper identification.

When transmitting via RTTY, you are in control of the receiving station's copy format and it is imperative that you transmit the carriage return (CR), line feed (LF) and letters (LTRS) characters when necessary. Two CRS, one LF and two LTRS characters should be sent at the end of each line. CR and LTRS are repeated to ensure their reception through any radio interference that may be present. LF is sent only once in order to conserve the receiving station's paper supply. Add an extra LF at the end of each transmission to separate your message from the next message being sent (or received).

Some RTTY equipment automatically downshifts when a space is sent; a LTRS need not be sent to clear the FIGS to permit printing of letter characters. Even though *your* gear may have this capability, everyone is not as lucky; you should send a LTRS after every transmission of figures to assure that the receiving station, whether or not it is equipped with

downshift on space, will receive perfect copy and not receive RTTY hieroglyphics.

These RTTY procedures may seem strange to the uninitiated, but after a few sessions at the keyboard you will appreciate their usage and learn to perform them automatically.

Working DX

Most amateurs at one time or another make "working DX" a major aim. As in every other phase of amateur work, there are right and wrong ways to go about getting best results in working foreign stations. This section will outline a few of them.

The ham who has trouble raising DX stations readily may find that poor transmitter efficiency is not the reason. He may find that his sending is poor, his call ill-timed, or his judgment in error. Working DX requires the know-how that comes with experience. If you just call CQ DX you may get a call from a foreign station, but it isn't likely to be a "rare one." On the other hand, unless you are experienced enough to know that conditions are right, your receiver is sensitive and selective enough, and your transmitter and antenna properly tuned and oriented, you may get no calls at all and succeed only in causing some unnecessary QRM.

The call CQ DX means slightly different things to amateurs on different bands:

a) On vhf, CQ DX is a general call ordinarily used only when the band is open, under favorable "skip" conditions. For vhf work, such a call is used for looking for new states and countries, also for distances beyond the customary "line-of-sight" range on most vhf bands.

b) CQ DX on our 7-, 14-, 21- and 28-MHz bands may be taken to mean "general call to any foreign station." The term "foreign station" usually refers to any station on a different continent. If you do call CQ DX, remember that it implies you will answer any DX who calls. If you don't mean "general call to any DX station," then listen and call the station you do want.

Codes and Ethics

One of the most effective ways to work DX is to know the operating habits of the DX stations sought, and to abide by the procedures they use. Know when and where to call, and for how long, and when to remain silent while waiting your chance. DXing has certain understood codes of ethics and procedures that will make this popular amateur pursuit more fun for everybody if everybody follows them. One of the sad things about DXing is to listen to some of the abuse that goes on, mostly by stations on "this" side, as they trample on each other trying to raise their quarry. DX stations have been known to go off the air in disgust at some of the tactics.

An overview of the W1AW station. The main operating console on the left houses the control equipment for simultaneous code practice/bulletin transmissions on eight bands. Two visitor operating positions are on the right.

The modern RTTY position at W1AW. Note the neat order of equipment. This permits the operator to select all RTTY converter functions while tuning the transceiver. The proximity of the keyboard to transmit controls and the visibility of the equipment operating parameters (plate current and frequency) make for convenient operating. The required cw identification is accomplished by changing mode on the HAL keyboard at the end of transmissions. Many stations have an fsk identification technique that uses a very small shifting network across the keyboard contacts.

If W and VE stations will use the procedure in the "DX Operating Code" detailed elsewhere on these pages, we can all make a good impression on the air. ARRL has also recommended some operating procedures for DX stations aimed at controlling some of the thoughtless practices used by W/VE amateurs. A copy of these recommendations (CD-215) can be obtained free of charge from ARRL Headquarters.

Snagging the Rare Ones

Once in a while a CQ DX will result in snagging a rare DX contact, if you're lucky. This seldom happens, however; usually, what you have to do is listen — and listen — and then listen some more. You gotta hear 'em before you can work 'em! If everybody transmits, nobody is going to hear anything. Be a snooper. Usually, unless you are lucky enough to be among the first to hear him, a rare DX station will be found under a pileup, with stations swarming all over him like worker bees over a queen. The bedlam will subside when the DX station is transmitting (although some stations keep right on calling him), and you can hear him. Don't immediately join the pack; be a little cagey. Listen a while, get an idea of his habits, find out where he is listening (if not zero on himself), bide your time, and wait your chance.

Make your calls short, snappy and distinct. No need to repeat his call (he knows it very well; all he needs to know is that you are calling him), but send your own call a couple of times. Try to find a time when few stations are calling him and he is not transmitting; then get in there! With experience, you'll learn all kinds of tricks, some of them clever, some just plain dirty. You'll have no trouble discerning which is which. Learn to use the clever ones, and shun the dirty ones. More than you think depends on the impression we make on our foreign friends!

Choosing Your Band

If it does nothing else in furthering your education, striving to work DX will certainly teach you a few things about propagation. You will find that four principal factors determine propagation characteristics: (1) the frequency of the band on which you do your operating, (2) the time of day or night, (3) the season of the year, and (4) the sunspot cycle. The proper choice of band depends pretty much on the other three factors. For example, the 3.5- to 4.0-MHz band at high noon in the summertime at the "node" part of the sunspot cycle is the poorest possible choice, while the same band at midnight during the wintertime at the "null" part of the cycle might produce some very exciting DX. Similarly, you will learn by experience when to operate on which band for the best DX by juggling the above factors using both long-range and other indications of band conditions. WWV transmissions can also be helpful in indicating both current and immediate-forecast band conditions.

On some bands, such as 10 and 6 meters, beacons have been established to give an indication of band openings. Listen between 28.2 and 28.3 MHz on 10 meters and around 50.110 MHz on 6 meters. Commercial stations near ham-band edges are also a fair indication of openings. But remember that many of these run many times the maximum amateur power, and consequently may be heard well before skip improves to the point necessary to sustain amateur communications.

Conditions in the transmission medium often make it possible for the signals from low-powered transmitters to be received at great distances. In general, the higher the frequency band, the less important power considerations become, for occasional

DX work. This accounts in part for the relative popularity of the 14-, 21- and 28-MHz bands among amateurs who like to work DX.

DX Century Club Award

The DXCC is one of the most popular and sought-after awards in all of Amateur Radio, and among the more difficult to acquire. Its issuance is carefully supervised at ARRL headquarters by three staff members.

To obtain DXCC, an amateur must make two-way contact with 100 "countries" on the ARRL DXCC List. Written confirmations are required for proof of contact. These must show clearly your call sign, date, time, frequency and mode. Such confirmations must be sent to ARRL headquarters, where each one is carefully scrutinized to make sure it actually confirms a contact with the applying amateur, that it was not altered or tampered with, and that the "country" claimed is actually on the ARRL list. Further safeguards are applied to maintain the high standards of this award. A handsome, king-size certificate and "DXCC" lapel pin are sent to each amateur qualifying.

The term "country" is an arbitrary one, not necessarily agreeing with the dictionary definition. For DXCC purposes, many bodies of land not having independent status politically are classified as countries. For example, Alaska and Hawaii, states of the U.S., are considered separate "countries" because of their distance from the mainland. There are over 300 such designations on the ARRL list. Once a basic DXCC is issued, the certificate can be endorsed, by sticker, for additional countries by sending the additional cards to Headquarters for checking.

Separate DXCC awards are available for mixed modes, all phone, all cw, RTTY, 160 meters and satellite.

Before applying, familiarize yourself with full information. Application forms (CD-164) and the ARRL DXCC List (detailing rules) may be obtained from Headquarters for a stamped, addressed envelope.

Five-Band DXCC

Entirely separate from DXCC, ARRL also offers a Five-Band DXCC (5BDXCC) Award for those amateurs who submit written proof of having made two-way contact with 100 or more countries on each of five amateur bands since January 1, 1969.

For a copy of the complete rules, drop a line to ARRL Headquarters, 225 Main St., Newington, CT 06111.

DX Operating Code (for W/VE Amateurs)

The points below, if observed by all W/VE amateurs, will go a long way

toward making DX more enjoyable for everybody.

1) Call DX only after he calls CQ, QRZ?, signs $\overline{SK}$, or phone equivalent thereof. thereof.

2) Do not call a DX station:

a) on the frequency of the station he is working until you are sure the QSO is over. This is indicated by the ending signal $\overline{SK}$ on cw and any indication that the operator is listening, on phone.

b) because you hear someone else calling him.

c) when he signs $\overline{KN}$, $\overline{AR}$, CL or phone equivalents.

d) after he calls a directional CQ, unless of course you are in the right direction or area.

3) Keep within *your* frequency-band limits. Many DX stations are permitted outside U.S. band segments but you are not.

4) Observe calling instructions given by DX stations. "10U" means call 10 kHz up from his frequency, "15D" means 15 kHz down, etc.

5) Give honest reports. Many foreign stations depend on W and VE reports for adjustment of station and equipment.

6) Keep your signal clean. Key clicks, chirps, hum or splatter give you a bad reputation and may get you a citation from FCC or DOC.

7) Listen for and call the station you want. Calling CQ DX is not the best assurance that the rare DX will reply.

8) When there are several W or VE stations waiting to work a DX station, avoid asking him to "listen for a friend." Let your friend take his chances with the rest. Also avoid engaging DX stations in rag chews against their wishes.

WAC Award

The ever-popular Worked All Continents Award (WAC), sponsored by the International Amateur Radio Union (IARU), can be yours by simply submitting proof of contact with each of the six continents. Two-way confirmed contacts must be made with amateurs in each of six continental areas of the world: Africa, Asia, Europe, North America, Oceania and South America. Confirmations submitted with all cw or mixed cw and phone will receive the basic award. Special endorsements include 1.8 MHz, 3.5 MHz, 50 MHz, 144 MHz, 432 MHz, RTTY,

SSTV, ssb and phone. A unique version of the award is available for working all continents on each of five or six bands after January 1, 1974. Amateurs residing in the USA or it possessions and Canadian amateurs may obtain full details about the WAC Awards from ARRL headquarters. QSLs (not copies) must be sent to ARRL headquarters for checking. ARRL membership is required of all W/VE applicants. Applicants in other countries must send QSLs to their IARU amateur society, the latter of which will certify their eligibility to the IARU headquarters society (ARRL) for issuance of the award on behalf of the Union. Applicants in countries not belonging to the IARU may send their QSLs direct to ARRL headquarters for checking. IARU member-society addresses will be sent, upon request, to those amateurs who do not reside within continental limits of the USA, its possessions or Canada.

Awards

League-sponsored operating activities have useful objectives and provide much enjoyment for members of the fraternity. Achievement in Amateur Radio is also recognized by various awards offered by ARRL and detailed below. Basic rules require that sufficient funds be included with all submissions of cards to ensure their safe return. A basic fee for return postage is included with each award application. Applicants in the U.S., its possessions and Canada must be ARRL members to participate in the WAS and DXCC programs. DX stations are exempted from this requirement.

WAS Award

"WAS" means Worked All States. This award is universal and may be obtained by any amateur who has worked each of the U.S. states and submits original proof of contacts to ARRL headquarters for examination. Contacts may be made over any period of time on any or all of the amateur bands.

Special endorsements are also available for WAS on cw, ssb, phone, all on one band, and so on — confirmations must clearly state that contact took place under circumstance of desired endorsement. QSLs must be accompanied by sufficient postage for their safe return. Please send an s.a.s.e to the ARRL Communications Department before mailing QSLs, and request a copy of the rules and application, which will include a guideline for postage fees for return of cards.

5BWAS Award

A handsome, specially engraved plaque will be issued to all amateurs who submit original proof of contact with all of the 50 states on each of five amateur bands, made after January 1, 1970 (only contacts made after that day will be eligible). Rules require applicants in the U.S., its possessions and Canada to be full ARRL members. Standard WAS rules apply. Write to ARRL for the application, full rules and QSL postage fee return.

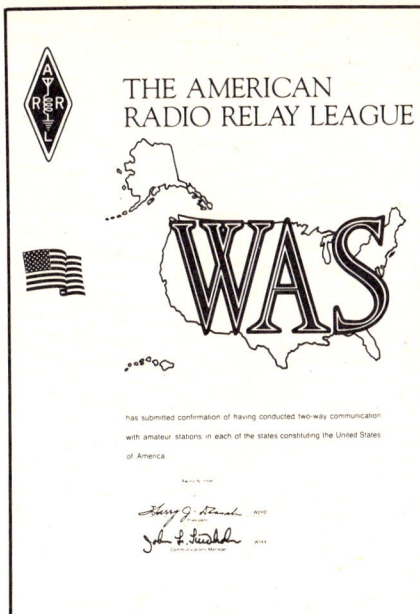

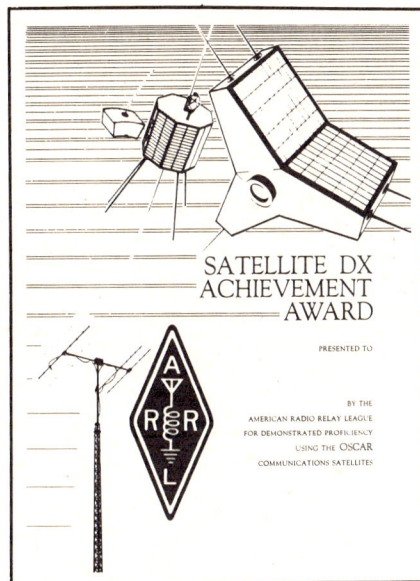

6-Meter "600 Club"

The 6-Meter "600 Club" award counts contacts made on 6 meters on or after January 1, 1977. A total of 600 points is required. Scoring is based on the sum of QSOs (times 2), ARRL sections (times 6) and countries (times 25). Some of the scoring areas are complex so be sure to send in for a form before you apply. Applicants must include sufficient postage for safe return of QSLs. Applicants in the U.S., its possessions and Canada must be full members to apply.

Satellite "1000" Award

This award, known also as the Satellite Achievement Award, is gaining in popularity as more amateurs are mastering the art of using the OSCAR com-

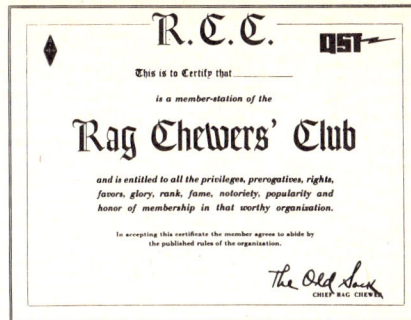

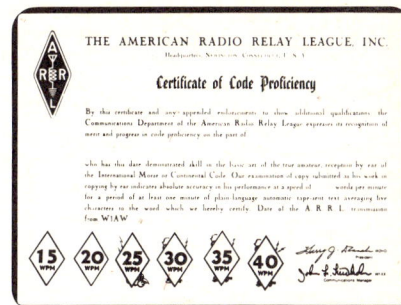

munications satellites. Contacts count if made on or after December 15, 1972. Only one contact per station is accepted, regardless of mode. Each contact with a new station counts 10 points, with a new country 50 points and with a new continent 250 points. To qualify for this award, amass 1000 points. An s.a.s.e. to the League will provide you with complete rules and an application form.

A-1 Operator Club

The A-1 Operator Club should include in its ranks every good operator. To become a member one must be nominated by two persons who already belong. General keying (not speed) or voice technique, procedure, copying ability, judgment and courtesy all count in rating candidates under the club rules. These are detailed at length in the booklet *Operating an Amateur Radio Station*. Aim to make yourself a fine operator and one of these days you will be pleasantly surprised when your mailman arrives at your QTH with your certificate of membership in the A-1 Operator Club.

Old-Timers Club

If you held an Amateur Radio license

20 or more years ago and are licensed at the present time, you are eligible to become a member of the Old Timers Club. Lapses in activity during intervening years are permitted. An s.a.s.e. (legal-size, at least 10 × 4 inches) will expedite your certificate.

Rag Chewers Club

Your first contact as a licensed amateur may very well earn your first award. The Rag Chewers Club is designed to encourage friendly contacts and discourage the "contest" type of QSO with nothing more than an exchange of calls, signal reports and so on. It furthers fraternalism through Amateur Radio.

Membership certificates are awarded to amateurs who report a fraternal-type contact with another amateur lasting a half hour or longer. This does not mean a half hour spent trying to work a rare DX station, but a solid half hour of pleasant "visiting" with another amateur, discussing subjects of mutual interest. If nominating someone for RCC, please send the information to the nominee who will (in turn) apply to Headquarters for membership. Or if you know *you* qualify for the RCC, just report the conversation to ARRL (c/o RCC) and back will come your member certificate. A legal-size envelope (at least 10 × 4 inches) is appreciated when requesting this award.

Code Proficiency Award

Many hams can follow the general idea of a contact "by ear," but when pressed to "write it down" they "muff" the copy. The Code Proficiency Award permits each amateur to prove himself as a proficient operator, and sets up a system of awards for step-by-step gains in copying proficiency. It enables every amateur to check his code proficiency, to better that proficiency, and to receive a certification of his receiving speed.

This program is a lot of fun. The League will award a certificate to any interested individual who demonstrates that he can copy perfectly for at least one minute, plain-language Continental code at 10, 15, 20, 25, 30, 35 or 40 words per minute, as transmitted twice monthly from W1AW and once a month from W6OWP. Neither an amateur license nor ARRL membership is required to participate.

As part of the ARRL Code Proficiency program W1AW transmits plain-language practice material several times daily at speeds from 5 to 35 wpm, occasionally in reverse order. All amateurs are invited to use these transmissions to increase their code-copying ability. Nonamateurs are invited to utilize the lower speeds, 5, 7-1/2 and 10 wpm, which are transmitted for the benefit of persons studying the code in preparation for the amateur license examination. Check the W1AW material earlier in this chapter and/or refer to any

Youngsters like KA8HXX, age 12, find Amateur Radio a most enjoyable hobby.

issue of *QST* for details.

Contesting

Contesting is to Amateur Radio what the Olympic Games are to worldwide amateur athletic competition: a showcase to display talent and learned skills, as well as a stimulus for further achievement through competition. Increased operating skills and greater station efficiency are the predominant end results of Amateur Radio contesting, whether the operator is a serious contender or a casual participant.

Don't believe it? Tune across the band, any band, and listen for the most efficient operators. Chances are better than ever that they are avid contesters or at least have contesting as one of their favorite Amateur Radio activities. How can one tell who is a contester just by listening to a particular operator's style? It is easier to tell who is *not* interested in contesting by listening. The contester is *not* likely to be the one, who (while thousands on the frequency are gnashing their teeth in anger) asks the operator of the rarest DXpedition in two decades what the weather is like in "Lower Slobbovia." The contester is *not* likely to be the operator who, when working a much-sought-after station on one of the many award nets, punctuates his repeating of the needed exchange 37 times with a long series of "uhhs, duhs" and assorted other noises for the lack of anything better to say. The contest operator knows from experience that conciseness and brevity are aids in efficient and courteous operating.

The contest operator is also likely to have one of the better signals on the band — not necessarily the most elaborate station equipment, but a signal enhanced by the most efficient use of station components available. Contest operation encourages optimization of station and operator efficiency.

The ARRL contest program is so diverse that it holds appeal for almost every operator — the beginning contester and the old hand, the newest Novice and oldest Extra-Classer, "Top Band" buff

and microwave enthusiast.

A thumbnail sketch of many of the contests sponsored by the ARRL follows. Complete entry rules and details appear in *QST*, usually the month before the contest occurs.

January

CD (Communications Department) Party. Ten hours of cw, 10 hours of phone, where CD appointees only work each other in a QSO party format.

VHF Sweepstakes. Premier vhf operating event. All bands, 50 MHz and up. ARRL affiliated-club competition, based on members' aggregate total scores.

February

Novice Roundup. Competition geared for the beginning (Novice and Technician) amateur. Increase code speed through operating, work stations needed for WAS, and other achievements. Awards for ARRL Section winners. Fun for all.

International DX Contest, CW. W/VE amateurs work the rest of the world for individual section, country and ARRL affiliated club honors; single-band, QRP, and multi-operator categories also, with many plaques awarded.

March

International DX Contest, Phone.

April

"Open" CD Party. Open to all ARRL members, as well as CD appointees. Same format as "closed" parties in January and October. Exchange appointment and/or membership status (i.e. member, life member, charter life member) and ARRL Section.

EME Contest. First weekend of two for moonbounce enthusiasts. Listeners' reports as well as active participants entries are welcomed.

May

EME Contest. Second part of same contest begun in April.

June

VHF QSO Party. One of two vhf QSO parties. This one (and the September party) lends itself to multioperator expedition operation. Use all bands above 50 MHz. ARRL sections and DXCC countries are scoring multipliers.

Field Day. The *number one* operating event of the year. More than 20,000 participants take to the fields to operate some 1700 emergency stations for informal competition, a score listing in *QST* and an all-around good time. Don't miss this one.

July

IARU Radiosport Championship. Worldwide competition. Everybody works everybody else for continental, country and ARRL section honors. Vary-

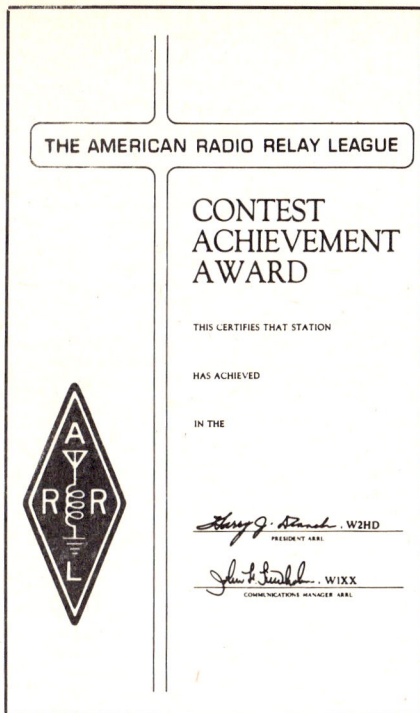

Section leaders earn this handsome certificate, with embossed ARRL emblem, in many ARRL-sponsored contests.

The Wellesley (Massachusetts) Amateur Radio Society coordinated this most successful Amateur Radio exhibit and message-handling service at a local shopping mall. (*WA1RGA photo*)

ing point scale, ITU zones are scoring multipliers. Some of those hard-to-work DXCC countries turn out for this contest.

August

UHF Contest. Similar to the vhf contests, but utilizes the 220 MHz and above bands. Scoring multipliers are determined by the longitude/latitude coordinates of the participants. The uhf bands come alive for this contest weekend.

September

VHF QSO Party. Second of two vhf QSO parties (see June also).

October

CD Party. Same as the January CD Party. For CD appointees only.

November

Sweepstakes. The most prestigious domestic contest. Two weekends (actually separate contests and separate listings); one weekend for phone and one weekend for cw. Twenty-four hour time limit on each mode. W and VE operators work each other. ARRL sections are the scoring multipliers. Awards for both high and low power ARRL section winners. An ARRL-affiliated club competition highlights the Sweepstakes activity.

December

160-Meter Contest. A gathering of "top band" enthusiasts. W/VE types work each other and DX stations for contest credit.

10-Meter Contest. The last (but certainly not least) one on the year's contest calendar. A 10-meter operator's dream come true as 28 MHz springs to life and everyone, worldwide, tries to work everyone else for top scorer (in country, continent and ARRL section) honors.

That's the ARRL Contest Program in a nutshell. Of course, more detailed rules and descriptions of the award structure (certificates and plaques awarded to designated top scorers) are announced in *QST* for each of these events. The monthly "Contest Corral" column of *QST* also details the entry rules for many contests other than those sponsored directly by ARRL, including the very popular state QSO parties and most other major contests.

Public Service

Tens of thousands of U.S. and Canadian amateurs are involved with public service. Where do you fit in? The emergency preparedness and third-party traffic-handling facets of Amateur Radio beckon. There's a place for every ham in the League's Amateur Radio Emergency Service (ARES), an emergency-preparedness group of approximately 60,000 amateurs who have signed up voluntarily to keep Amateur Radio in the forefront of public service operating. The National Traffic System (NTS) functions as a message-handling network operating 365 days a year for the systematic handling of third-party traffic.

Also recognized by ARRL as a part of the organized public service effort are the Radio Amateur Civil Emergency Service (RACES), a part of the Amateur Service serving civil defense under a separate subpart of the amateur regulations; the Military Affiliate Radio System (MARS) sponsored by the armed forces to provide military communications training for amateurs; and the numerous amateur groups organized into nets or monitoring

services by individuals, clubs or other amateur entities for public service. The detailed workings of the League's emergency and traffic programs are covered briefly herein and in more depth in the *Public Service Communications Manual* and *Operating an Amateur Radio Station*, available free from ARRL Hq. (please provide an s.a.s.e.). Additional information can be found in the *ARRL Operating Manual*, a for-sale publication available at your local radio dealer or direct from Hq.

ARES and NTS —
How it Applies to You

As a member of the local ARES group, you'll be training to provide communications at the city or county level. Each group is headed by an ARRL Emergency Coordinator. Most ARES activities are centered on 2-meter fm, so it's advantageous to have your own emergency-powered vhf gear. However, you really don't need *any* equipment to join; it's the training and practice that are most important. All you really need is an interest in serving your community through Amateur Radio and participation in periodic tests as time permits. What kinds of tests? Well, they run the gamut from serious simulated emergencies to providing communications for parades and walkathons, or conducting a message-handling service at a shopping center during the Christmas season. Many hams have trained with the National Weather Service to become tornado and storm spotters. All these activities exist, so that when a flood or an ice storm disrupts the community, experienced hams will know exactly what to do.

Becoming involved is as simple as requesting a registration card (CD-98) from the ARRL Communications Department and filling it out. These cards are turned over to the local EC, who registers you in the local ARES organization. Should your community not have an EC, why not volunteer yourself? You qualify if you are a licensed amateur of Technician class or higher, an ARRL member, have a sincere interest in public service and a willingness to put in the time and effort to fulfill the appointment. If this sounds like you, contact your Section Communications Manager. His/her name, address and telephone number appear on page 8 of each issue of *QST*.

The bulk of recent localized emergency communications has been handled on vhf. Much has involved repeater operation. The reason is simple. Repeaters can be accessed with low-cost and lightweight equipment, hand-held or mobile. Best of all, they provide clear, reliable communications up to 100 miles or so. Many repeaters have emergency-power capabilities as well, making them the mainstay of any widespread emergency.

What's the National Traffic System all

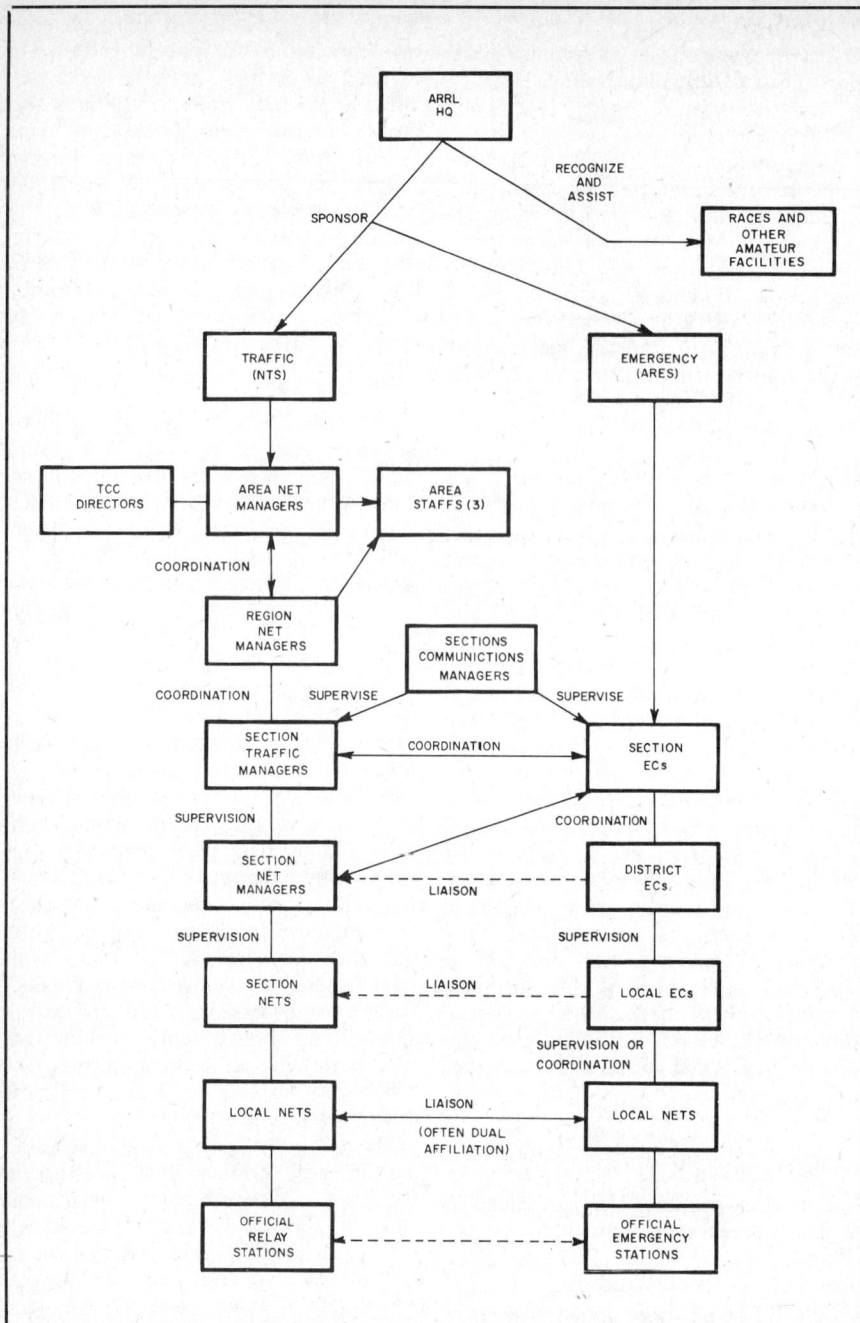

Phone on the hour: 1835, 3990, 7290, 14,290, 21,390, 28,590, 50,190, 147,555 kHz. *RTTY 15 minutes past the hour:* 3625, 7095, 14,095, 21,095, 28,095, 147,555 kHz. *Cw on the half hour:* 1835, 3580, 7080, 14,080, 21,080, 28,080, 50,080, 147,555 kHz.

Table 5
Operating Aids for Public Service

The ARRL Communications Department makes available the following free operating aids for public service communications:

Operating an Amateur Radio Station
Public Service Communications Manual CD-235
Net Directory   CD-50
ARRL numbered radiograms   CD-3
Sample emergency plan   CD-27
ARES registration form   CD-98
Amateur message form   CD-218
Emergency Reference Information   CD-255

This entire Public Service Package can be obtained by sending a large (9 × 12-inch) envelope with postage for six ounces, first class.

for it by keeping your station and emergency power supply (if you have one) in good working order. Participate in the annual nationwide Simulated Emergency Test, contests and Field Day.

Register your station with your local EC. During an emergency, report to him at once and follow his suggestions.

Monitor your local emergency net frequency, *but don't transmit* unless you are specifically requested to or are certain you can be of assistance.

Copy special W1AW bulletins for latest developments.

Use your receiver more, your transmitter less. Interference can be intense during an emergency.

After an emergency, tell your EC or net manager of your activities, so he can submit a timely report to ARRL hq. Each month, *QST* chronicles Amateur Radio's emergency communications efforts.

The Amateur Radio Service has been a vital part of emergency communications for more than 50 years, whether it be relaying medical traffic into an earthquake-ravaged village in South America, answering a "Mayday" from a ship in the Pacific Ocean, or finding out if a neighbor's relative survived a blizzard in the Midwest.

Why not become part of it!

W1AW: ARRL Hq. Station

The Maxim Memorial Station, W1AW, is dedicated to service to the amateur fraternity. It is adjacent to the Headquarters offices and is operated by the Headquarters Operators Club. Operating hours are 7:30 A.M. to 1 A.M. Monday through Friday and 3:30 P.M. to 1 A.M. Saturday and Sunday. The station is open

about, then? NTS serves a dual purpose: The rapid movement of long-haul traffic from origin to destination and the training of amateur operators in the handling of formal radiogram traffic in efficient directed nets. A rundown of the NTS schedule of nets and functions can be found in the *Public Service Communications Manual.* In the overview, however, NTS can be visualized as somewhat of a pony express of the airwaves, with assigned amateurs carrying traffic to and from the next higher (or lower) level in the system. NTS operations are concentrated mainly on the high frequencies (hf), but local nets on 2 meters have become more and more popular as the ideal place to distribute traffic for delivery after the long journey.

Most ARRL sections have section nets on 80 meters, both phone and cw. You can find the traffic and emergency nets that service your area in the annual ARRL *Net Directory* (free with an s.a.s.e.). Directed net procedures, especially on cw, take a little getting used to, but if you consult the League's reference material and do a little monitoring beforehand, you shouldn't have much problem. In fact, the many slow-speed nets that meet on the Novice frequencies are a boon to learning the simple procedures.

Each day, hams enjoy the challenge of these activities. Why not check into a net and check it out!

What You Should Do

Before an emergency occurs, prepare

to visitors at all times it is in operation. If you wish to operate W1AW while visiting, the period between 1 and 4 P.M. Monday through Friday is available. Be sure to bring a copy of your FCC license with you if you plan to operate.

W1AW Code Practice

Code practice is sent on approximately 1.835, 3.58, 7.08, 14.08, 21.08, 28.08, 50.08 and 147.555 MHz. Texts are from recent issues of *QST* and checking references are sent several times in each session. For practice purposes, the order of words in each line of text may be reversed during the 5-15 wpm transmissions. Two code proficiency qualifying runs are sent each month; dates are in the "Contest Corral" section of *QST*. Here is the code-practice schedule:

EST/EDST   *PST/PDST*

Speeds
5, 7-1/2, 10, 13, 15

9 A.M. MWF	6 A.M. MWF
7 P.M. MWF	4 P.M MWF
4&10 P.M. TTHSSU	1&7 P.M. TTHSSU

Speeds
35, 30, 25, 20, 15, 13, 10

9 A.M. TTH	6 A.M. TTH
7 P.M. TTHSSU	4 P.M. TTHSSU
4&10 P.M. MWF	1&7 P.M. MWF

Bulletins

Cw bulletins at 18 wpm are sent daily at 5, 8 and 11 P.M. EST/EDST, and Monday through Friday at 10 A.M. Frequencies are the same as those used for code practice.

Teletype bulletins, at 60 wpm with 170-Hz shift, are daily at 6, 9 and 12 P.M. EST/EDST and Monday through Friday at 11 A.M. Each Teletype bulletin is followed by a repeat on 110-baud ASCII. Frequencies are 3.625, 7.095, 14.095, 21.095, 28.095 and 147.555 MHz.

Voice bulletins are daily at 9:30 P.M. and 12:30 A.M. EST/EDST on 1.835, 3.990, 7.290, 14.290, 21.390, 28.590, 50.19 and 147.555 MHz.

A complete W1AW schedule is available from ARRL headquarters for an s.a.s.e.

General Operation

W1AW is equipped for operation on all bands from 1.8 to 144 MHz, and for RTTY, SSTV and satellite communications as well as for cw, ssb and nbfm.

AMATEUR MESSAGE FORM

Every message originated and handled should contain the following component parts in the order given.

I   PREAMBLE
   a. Number (begin with 1 each month or year)
   b. Precedence (R, W, P or EMERGENCY)
   c. Handling Instructions (optional, see text)
   d. Station of Origin (first amateur handler)
   e. Check (number of words/groups in text only)
   f. Place of Origin (not necessarily location of station of origin)
   g. Time Filed (optional with originating station)
   h. Date (must agree with date of time filed)
II ADDRESS (as complete as possible, include zip code and telephone number)
III TEXT (limit to 25 words or less, if possible)
IV SIGNATURE

CW MESSAGE EXAMPLE

I   NR 1 R HXG W1AW 8 NEWINGTON CONN 1830Z July 1
     a   b c   d   e   f   g   h

II DONALD R SMITH AA
    164 EAST SIXTH AVE AA
    NORTH RIVER CITY MO 00789 AA
    733-3968 BT

III HAPPY BIRTHDAY X SEE YOU   SOON X LOVE BT

IV   DIANA AR

CW: Note that X, when used in the text as punctuation, counts as a word. The prosign AA separates the parts of the address, BT separates the address from the text and the text from the signature. AR marks end of message; this is followed by B if there is another message to follow, by N if this is the only or last message. It is customary to copy the preamble, parts of the address, text and signature on separate line.

RTTY: Same as cw procedure above, except (1) use extra space between parts of address, instead of AA; (2) omit cw procedure sign BT to separate text from address and signature, using line spaces instead; (3) add a CFM line under the signature, consisting of all names, numerals and unusual words in the message in the order transmitted.

PHONE: In general, use *prowords* in place of procedural signals or *prosigns*. The above message on phone would go something like this: *"Message Follows* number one, routine HX Golf, W1AW, eight, Newington, Connecticut, one eight thuhree zero zulu, July one, Donald *Initial* R Smith, *Figures* one six fower, East Sixth Avenue, North River City, Missouri zero zero seven eight nine, *Telephone* sev-ven thuhree thuhree, thuhree niyen six eight. *Break* Happy Birthday X-ray see you soon X-ray love *Break* Diana, *End of Message, Over."* Speak in measured tones, emphasizing *every* syllable. Spell out phonetically all difficult or unusual words, but do *not* spell out common ones.

PRECEDENCES

The precedence will follow the message number. For example, on cw 207R or 207 EMERGENCY. On phone, "Two Zero Seven, Routine (or Emergency)."

EMERGENCY — Any message having life and death urgency to any person or group of persons, which is transmitted by Amateur Radio in the absence of regular commercial facilities. This includes official messages of welfare agencies during emergencies requesting supplies, materials or instructions vital to relief of stricken populace in emergency areas. During normal times, it will be *very rare.* On cw, this designation will *always* be spelled out. When in doubt, do *not* use it.

PRIORITY — Important messages having a specific time limit. Official messages not covered in the Emergency category. Press dispatches and other emergency-related traffic not of the utmost urgency. Notification of death or injury in a disaster area, personal or official. Use the abbreviation P on cw.

WELFARE — A message that is either a) an inquiry as to the health and welfare of an individual in the disaster area or b) an advisory or reply from the disaster area that indicates all is well should carry this precedence, which is abbreviated W on cw. These messages are handled *after* Emergency and Priority traffic but before Routine.

ROUTINE — Most traffic normal times will bear this designation. In disaster situations, traffic labeled Routine (R on cw) should be handled *last*, or not at all when circuits are busy with Emergency, Priority or Welfare traffic.

Handling Instructions

HXA — (Followed by number.) Collect landline delivery authorized by addressee within......miles. (If no number, authorization is unlimited.)
HXB — (Followed by number.) Cancel message if not delivered within.....hours of filing time; service originating station.
HXC — Report date and time of delivery (TOD) to originating station.
HXD — Report to originating station the identity of station from which received, plus date and time. Report identity of station to which relayed, plus date and time, or if delivered report date, time and method of delivery.
HXE — Delivering station get reply from addressee, originate message back.
HXF — (Followed by number.) Hold delivery until......(date).
HXG — Delivery by mail or landline toll call not required. If toll or other expense involved, cancel message and service originating station.

This prosign (when used) will be inserted in the message preamble before the station of origin, thus(NR 207 R HXA50 W1AW 12.....(etc.). If more than one HX prosign is used, they can be combined if no numbers are to be inserted, otherwise the HX should be repeated, thus: NR 207 R HXAC W1AW.....(etc.), but: NR 207 R HXA50 HXC W1AW....(etc.); On phone, use phonetics for the letter or letters following the HX, to insure accuracy.

CD 218 4/80   ARRL, 225 Main St., Newington, CT 06111

Chapter 23

Vacuum Tubes and Semiconductors

Receiving types of vacuum tubes are being phased out by the industry because the demand for new designs in which tubes are used has diminished to practically zero. Furthermore, the cost of new tubes has risen to impractical levels for amateurs who wish to build a piece of equipment in which they are used. Therefore, this and future editions of the *Handbook*, will list only the base-connection diagrams for receiving tubes. Electrical specifications can be obtained from past editions of the *Handbook*, or from the tube manuals of those manufacturers who still fabricate tubes.

Complete data is given herein for transmitting types of tubes, along with the base diagrams. It is unlikely that semiconductors will replace tubes entirely for high-power operation in the near future, hence the retention of this information.

The tabular data for transmitting tubes lists recommended typical operating conditions under the ICAS (Intermittent Commercial and Amateur Service) ratings of the tube manufacturers. Other values of plate voltage, plate current, screen voltage and screen current can be used provided none of these parameters exceeds the maximum values listed.

This edition includes a table of some of the more common television horizontal deflection tubes. These are suitable for amateur transmitting applications. The table contains rf data for some types, as well as base diagrams not shown in the first section.

Semiconductors

Selected semiconductors are listed in this section of the book. Since there are literally thousands of small-signal transistors available today, only those that are relatively common are specified here. The range of types given in tabular form are capable of serving the needs of most radio amateurs who design and build their own equipment. A complete listing of the available transistors would consume a book the size of this publication. For this reason it would be impractical to offer such a complete set of tables.

Similarly, a representative list of power FETs and rf-power bipolar transistors is provided in these pages. Devices of various power and frequency ratings are highlighted in this part of the book, thereby offering the reader a wide assortment of choices for his or her design work. Base or case diagrams are included where applicable. Further data on these transistors can be obtained from the manufacturers. Most semiconductor fabricators publish complete manuals which provide in-depth electrical profiles of their solid-state devices. Individual data sheets for each type of transistor are also available from the manufacturers.

A brief table of IC operational amplifiers appears at the end of the semiconductor section. The types listed are becoming widely available from hobby and surplus outlets.

Index to Receiving-Tube Types and Bases

*Base diagrams appear later in this chapter.

Type	Base	Type	Base	Type	Base	Type	Base	Type	Base	Type	Base
00-A	4D	1C3	5CF	1LC5	7AO	2B4	5A	3A3	8EZ	3FP7	14B
01-A	4D	1C5GT	6X	1LC6	7AK	2B6	7J	3A4	7BB	3FP7A	14J
0A2	5BO	1C6	6L	1LD5	6AX	2B7	7D	3A5	7BC	3GP1-4-5-11	11A
0A3A	4AJ	1C7G	7Z	1LE3	4AA	2B22	Fig. 22	3A8GT	8AS	3GP1A	11N
0A4G	4V	1C21	4V	1LF3	4AA	2B25	3T	3ACP1-7-11	14J	3GP4A	11N
0A5	Fig. 19	1D5GP	5Y	1LG5	7AO	2BP1-11	12E	3AP1-4	7AN	3JP1-12	14J
0B2	5BO	1D5GT	5R	1LH4	5AG	2C4	5AS	3AP1A	7CE	3JP1A-11A	14J
0B3	4AJ	1D7G	7Z	1LN5	7AO	2C21	7BH	3AQP1	12E	3KP1-4-11	11M
0C2	5BO	1D8GT	8AJ	1N5GT	5Y	2C22	4AM	3B4	7CY	3LE4	6BA
0C3A	4AJ	1DN5	6BW	1N6G	7AM	2C25	4D	3B5GT	7AP	3LF4	6BB
0D3A	4AJ	1E3	9BG	1P5GT	5Y	2C26A	4BB	3B7	7BE	3MP1	12F
0G3	5BO	1E4G	5S	1Q5GT	6AF	2C34	Fig. 70	3B24*	Fig. 49	3Q4	7BA
0Y4	4BU	1E5GP	5Y	1R4	4AH	2C36	Fig. 21	3B25	4P	3Q5GT	7AP
0Z4	4R,	1E7G	8C	1R5	7AT	2C37	Fig. 21	3B26	Fig. 18	3RP1-4	12E
0Z4A	4R	1EP1-2-11	11V	1S4	7AV	2C39*		3B27	4P	3RP1A	12E
0Z4G	4R	1F4	5K	1S5	6AU	2C40*	Fig. 11	3B28*	4P	3S4	7BA
1.	4G	1F5G	6X	1SA6GT	6CA	2C43*	Fig. 11	3BP1-4-11	14A	3SP1-4-7	12E
1A3	5AP	1F6	6W	1SB6GT	6CB	2C51	8CJ	3BP1A	14G	3UP1	12F
1A4P	4M	1F7G	7AD	1T4	6AR	2C52	8BD	3C4	6BX	3V4	6BX
1A4T	4K	1G3-GT/1B3-GT	3C	1T5GT	6X	2D21	7BN	3C5GT	7AQ	3WP1-2-11	12T
1A5GT	6X	1G4GT	5S	1U4	6A	2E5	6R	3C6	7BW	3-25A3	3G
1A6	6L	1G5G	6X	1U5	6BR	2E22*	5J	3C22	Fig. 17	3-50A4	3G
1A7GT	7Z	1G6GT	7AB	1U6	7DW	2E24	7CL	3C23	3G	3-50D4	2D
1AB5	5BF	1H4G	5S	1-V	4G	2E25*	5BJ	3C24*	2D	3-50G2	2D
1AB6	7DH	1H5GT	5Z	1V2	9U	2E26*	7CK	3CP1	11C	3-75A3	2D
1AC6	7DH	1H6G	7AA	1W4	5BZ	2E30	7CQ	3CX100A5*		3-100A2*	2D
1AE4	6AR	1J3	3C	1X2	9Y	2EA5	7EW	3D6	6BB	3-400Z*	Fig. 3
1AF4	6AR	1J5G	6X	1X2A	9Y	2EN5	7FL	3D23	Fig. 30	3-500Z*	Fig. 3
1AF5	6AU	1J6GT	7AB	1X2B	9Y	2G5	6R	3D24*	Fig. 75	3-1000Z*	Fig. 3
1AH5	6AU	1K3	3C	1Y2	4P	2S/4S	5D	3DK6	7CM	4A6G	8L
1AJ4	6AR	1K3	3C	1Z2	7CB	2V2	8FV	3DX3	Fig. 24	4C32	2N
1AX2	9Y	1L4	6AR	2A3	4D	2V3G	4Y	3E5	6BX	4C34	2N
1B3GT	3C	1L6	7DC	2A4G	5S	2W3	4X	3E6	7CJ	4C36	Fig. 31
1B4	4M	1LA4	5AD	2A5	6B	2X2-A	4AB	3E22	8BY	4CX250	Fig. 75
1B5	6M	1LA6	7AK	2A6	6G	2Y2	4AB	3E29*	7BP	4CX300A*	
1B7GT	7Z	1LB4	5AD	2A7	7C	2Z2	4B	3EA5	7EW	4CX1000A*	
1B8GT	8AW	1LB6	8AX	2AP1A	11L	3A2	9DT	3EP1	11N	4D21*	5BK

Type	Base	Type	Base	Type	Base	Type	Base	Type	Base	Type	Base
4D22*	Fig. 26	6AK7	8Y	6BX8	9AJ	6EZ8	9KA	6MQ8	9AE	7E6	8W
4D23	5BK	6AK8	9E	6BY5G*	6CN	6F4*	7BR	6MU8	9AE	7E7	8AE
4D32*	Fig. 27	6AL3	9CB	6BY6	7CH	6F4	7BR	6N4	7CA	7EP4	11N
4DK6	7CM	6AL5	6BT	6BY7	9AQ	6F5	5M	6N5	6R	7EV6	7AC
4E27*	7BM	6AL6G	6AM	6BY8	9FN	6F6	7S	6N6G	7AU	7EY6	7AC
4E27A*	7BM	6AL7GT	8CH	6BZ6	7CM	6F7	7E	6N7GT	8B	7F7	8AC
4EW6	7CM	6AM4	9BX	6BZ7	9AJ	6F8G	8G	6N7GT*	8B	7F8	8BW
4X150A*	Fig. 75	6AM5	6CH	6BZ8	9AJ	6FD7	9HF	6N8	9T	7G7	8V
4X150D*	Fig. 75	6AM6	7DB	6C4	6BG	6FG5	7GA	6P5GT	6Q	7G8	8BV
4X150G*		6AM8A	9CY	6C4*	6BG	6FG7	9GF	6P7G	7U	7GP4	14G
4X250B*	Fig. 75	6AN4	7DK	6C5	6Q	6FH5	7FP	6P8G	8K	7H7	8V
6-65A*	Fig. 25	6AN5	7BD	6C6	6F	6FH6	6AM	6Q4	9S	7J7	8BL
4-125A*	5BK	6AN6	7BJ	6C7	7G	6FJ7	12BM	6Q5G	6Q	7JP1-4-7	14R
4-250A*	5BK	6AN7	9Q	6C8G	8G	6FM7	12BJ	6Q6G	6Y	7K7	8BF
4-400A*	5BK	6AN8A	9DA	6C10	12BQ	6FM8	9KR	6Q7	7V	7L7	8V
4-100A/8166*		6AQ4	7DT	6CA4*	9M	6FQ5A	7FP	6Q11	12BY	7N7	8AC
5A6	9L	6AQ5A	7BZ	6CA5	7CV	6FS5	7GA	6R4	9R	7Q7	8AL
5ABP1-7-11	14J	6AQ6	7BT	6CB5A	8GD	6FV6	7FQ	6R6G	6AW	7R7	8AE
5ADP1-7-11	14J	6AQ7GT	8CK	6CB6A	7CM	6FV8A	9FA	6R7	7V	7S7	8BL
5AMP1	14U	6AQ8	9AJ	6CD6GA	5BT	6FW8	9AJ	6R8	9E	7T7	8V
5AP1-4	11A	6AR5	6CC	6CE5	7BD	6FY5	7FN	6S4A	9AC	7V7	8V
5AQP1	14G	6AR6	6BQ	6CF6	7CM	6G5	6R	6S6GT	5AK	7VP1	14R
5AS4A*	5T	6AR7GT	7DE	6CG6	7BK	6G8G	7S	6S7	7R	7W7	8BJ
5AT4*	5L	6AR8	9DP	6CG7	9AJ	6GC5	9EU	6S8GT	8CB	7X6	7AJ
5ATP1-11	14V	6AS5	7CV	6CG8A	9GF	6GE5	12BJ	6SA7GT	8R	7X7	8BZ
5AU4*	5T	6AS6	7CM	6CH6	9BA	6GF5	12BJ	6SB7Y	8R	7Y4	5AB
5AW4*	5T	6AS7GA	8BD	6CH7	9EW	6GH8A	9AE	6SC7	8S	7Z4	5AB
5AX4GT	5T	6AS8	9DS	6CH8	9F	6GJ5	9NM	6SD7GT	8N	8BP4	14G
5AZ4	5T	6AT6	7BT	6CJ6	9AS	6GJ8	9AE	6SE7GT	8N	9BM5	7BZ
5BC3	9NT	6AT8A	9DW	6CK4	8JB	6GK5	7FP	6SF5	6AB	9BW6	9AM
5BP1	11A	6AU4GT	4CG	6CK6	9AR	6GK6	9GK	6SF7	7AZ	9NP1	6BN
5BP1A	11N	6AU5GT	6CK	6CL5	8GD	6GM6	7CM	6SG7	8BK	10	4D
5BP7A	11N	6AU6A	7BK	6CL6	9BV	6GM8	9DE	6SH7	8BK	10EB8	9DX
5BP1-11	14B	6AU7	9A	6CL8A	9FX	6GN8	9DX	6SH7L	8BK	10GP4	14G
5BP1A	14J	6AU8A	9DX	6CM6	9CK	6GS8	9LW	6SJ7	8N	10HP4	14G
5CP1B-11B	14J	6AV4*	5BS	6CM7	9ES	6GT5	9NZ	6SJ7Y*	8N	10Y	4D
5CP7A	14J	6AV5GA	6CK	6CM8	9FZ	6GU5	7GA	6SK7	8N	11/12	4F
5CP11A	14J	6AV5GT	6CK	6CN7	9EN	6GV8	9LY	6SL7GT	8BD	12A4	9AG
5CP12	14J	6AV6	7BT	6CQ6	7DB	6GW6	6AM	6SN7GTA	8BD	12A5	7F
5D22*	5BK	6AV11	12BY	6CQ8	9GE	6GW8	9MB	6SN7GTB	8BD	12A6	7S
5DJ4	8KS	6AW7GT	8CQ	6CR6	7EA	6GY8	9MB	6SQ7GT	8Q	12A7	7K
5EA8	9AE	6AW8A	9DX	6CR8	9GJ	6GZ5	7CV	6SR7	8Q	12A8GT	8A
5FV8	9FA	6AX4GT	4CG	6CS5	9CK	6H4GT	5AF	6SS7	8N	12AB5	9EU
5GP1	11A	6AX5GT*	6S	6CS6	7CH	6H5	6R	6ST7	8Q	12AB6	7CC
5HP1-4	11A	6AX6G	7Q	6CS7	F9E	6H6	7Q	6SU7GTY	8BD	12AC6	7BK
5BP1A	11N	6AX7	9A	6CS8	9FZ	6H8G	8E	6SV7	7AZ	12AD6	7CH
5JP1A-4A	11S	6AX8	9AE	6CU5	7CV	6HA5	76M	6SZ7	8Q	12AD7	9A
5LP1A-4A	11T	6AX8	9ED	6CU6	6AM	6HB5	12BJ	6T4	7DK	12AE6A	7BT
5MP1-11	7AN	6B4G	5S	6CU8	9GM	6HB6	9PU	6T5	6R	12AE7	9A
5NP1-4	11A	6B5	6AS	6CW4	12AQ	6HB7	9QA	6T6GM	6Z	12AF6	7BK
5R4GY*	5T	6B6G	7V	6CW5	9CV	6HF5	12FB	6T7	7V	12AG6	7CH
5R4GYA*	5T	6B7	7D	6CX7	9FC	6HF8	9DX	6T8	9E	12AH7GT	8BE
5RP1A-4A	14P	6B8	8E	6CX8	9DX	6HG8	7BZ	6T8A	9E	12AJ6	7BT
5SP1-4	14K	6B10	12BF	6CY5	7EW	6HJ8	9CY	6T9	12FM	12AL5	6BT
5T4	5T	6BA6	7BK	6CY7	9EF	6HK5	7GM	6U3	9BM	12AL8	9GS
5U4G*	5T	6BA7	8CT	6CZ5	9HN	6HM5	76M	6U4GT	4CG	12AQ5	7BZ
5U4GA-GB*	5T	6BA8A	9DX	6D4	5AY	6HQ5	7GM	6U5	6R	12AT6	7BT
5UP1-11	12E	6BC4	9DR	6D6	6F	6HS5	7BK	6U6GT	7S	12AT7	9A
5V3A*	5T	6BC5	7BD	6D7	7H	6HS8	9LW	6U7G	7R	12AU6	7BK
5V4GA*	5L	6BC7	9AX	6D8G	8A	6HZ6	7EN	6U8	9AE	12AU7A*	9A
5VP7	11N	6BC8	9AJ	6D10	12BQ	6HZ8	9DX	6U8A	9AE	12AV5GA	6CK
5W4GT	5T	6BD4A	Fig. 80	6DA4A	4CQ	6J4	7BQ	6V3	9BD	12AV6	7BT
5X3	4C	6BD5GT	6CK	6DB5	9GR	6J5	6Q	6V3A	9BD	12AV7	9A
5X4G	5Q	6BD6	7BK	6DB6	7CM	6J6A	7BF	6V4*	9M	12AW6	7CM
5XP1	14P	6BD7	9Z	6DC6	7CM	6J6A*	7BF	6V5GT	6AO	12AW7	7CM
5XP1A-11A	14P	6BE6	7CH	6DE4*	4CG	6J7	7R	6V6GTA	7S	12AX4GT	4CG
5Y3-G-GT*	5T	6BE7	9AA	6DE6	7CM	6J8G	8H	6V7G	7V	12AC4GTA	4CG
5Y4-G-GT	5Q	6BE8A	9EG	6DE7	9HF	6J11	12BW	6V8	9AH	12AX7A	9A
5Z3*	4C	6BF5	7BZ	6DG8GT	7S	6JB6	9QL	6W4GT	4CG	12AV7*	9A
5Z4*	5L	6BF6	7BT	6DJ8	9AJ	6JC6A	9PM	6W5G	6S	12AZ7A	9A
5-125B*	7BM	6BG6GA	5BT	6DK6	7CM	6JC8	9PA	6W6GT	7S	12B4	9AG
5-500A*		6BH5	9AZ	6DN6	5BT	6JD6	9PM	6W7G	7R	12B4A	9AG
6A3	4D	6BH6	7CM	6DN7	8BD	6JE6A	9QL	6X4/6063*	7CF	12B6M	6Y
6A4	5B	6BH8	9DX	6DQ5	8JC	6JF6	9QL	6X5GT*	6S	12B7	8V
6A5GT	6T	6BJ5	6CH	6DQ6B	6AM	6JH6	7CM	6X6G	7AL	12B7ML	8V
6A6	7B	6BJ6A	7CM	6DR7	9HF	6JH8	9DP	6X8	9AK	12B8GT	8T
6A7	7C	6BJ7	9AX	6DS5	7BZ	6JK8	9AJ	6X8A	9AK	12BA6	7BK
6A8	8A	6BJ8	9ER	6DT5	9HN	6JN8	9FA	6Y3G	4AC	12BA7	8CT
6AB4	5CE	6BK5	9BQ	6DT6	7EN	6JT8	9DX	6Y5	6J	12BD6	7BK
6AB5	6R	6BK6	7BT	6DT8	9DE	6JU6	9DX	6Y6GA	7S	12BE6	7CH
6AB6G	7AU	6BK7A	9AJ	6DV4	12EA	6K5GT	5U	6Y6GT	7S	12BF6	7BT
6AB7	8N	6BK7B	9AJ	6DW5	9CK	6K6GT	7S	6Z3*	4G	12BH7	9A
6AB8	9AT	6BL7GTA	8BD	6DX4	7DK	6K7	7R	6Z4	5D	12BH7A	9A
6AC5GT	6Q	6BL8	9DC	6DX8	9HX	6K8	8K	6Z5	6K	12BK5	9BQ
6AC6G	7AU	6BM5	7BZ	6DZ4	7DK	6K11	12BY	6Z7G	8B	12BK6	7BT
6AC7	8N	6BN4A	7EG	6DZ7	8JP	6KD6	12GW	6ZY5G	6S	12BL6	7BK
6AD5G	6Q	6BN6	7DF	6E5	6R	6KD8	9AE	7A4	5AC	12BN6	7DF
6AD6G	7AG	6BN7	9AJ	6E6	7B	6KE8	9DC	7A5	6AA	12BQ6GA	6AM
6AD7G	8AY	6BN8	9ER	6E7	7H	6KM6	9QL	7A6	7AJ	12BQ6BT	6AM
6AD8	9T	6BQ5	9CV	6E8G	8O	6KR8	9DX	7A7	8V	12BQ6GTB	6AM
6AE5G	6Q	6BQ6GTB/6CU6	6AM	6EA5	7EW	6KT6	9PM	7A8	8U	12BR7A	9CF
6AE6G	7AH	6BQ7A	9AJ	6EA7	8BD	6KV8	9QP	7AB7	8BO	12BT6	7BT
6AE7GT	7AX	6BR7	9BC	6EA8	9AE	6KY6	9GK	7AD7	8V	12BU6	7BT
6AE8	8DU	6BR8A	9FA	6EB5	6BT	6KZ8	9FZ	7AF7	8AC	12BW4	9DJ
6AF3	9CB	6BS5	9BK	6EB8	9DX	6L4	7BR	7AG7	8V	12BV7	9BF
6AF4A	7DK	6BS7	9BB	6EF6	7S	6L5G	6Q	7AH7	8V	12BX6	9AQ
6AF5G	6Q	6BS8	9AJ	6EH5	7CV	6L6GA	7S	7AJ7	8V	12BY7	9BF
6AF6G	7AG	6BT6	7BT	6EH7	9AQ	6L6GB	7S	7AK7	8V	12BY7A	9BF
6AF7G	8AG	6BT8	9FE	6EH8	9JG	6L6GX	7S	7B4	5AC	12BZ6	7CM
6AG5	7BD	6BU5	8FP	6EJ7	9AQ	6L7	7T	7B5	6AE	12BZ7	9A
6AG6G	7S	6BU6	7BT	6ER5	7FN	6LF6	12GW	7B6	8W	12C5	7CV
6AG7	8Y	6BU8	9FG	6ES5	7FP	6LJ8	9GF	7B7	8V	12C8	8E
6AH4GT	8EL	6BV7	9BU	6ES8	9DE	6LM8	9AE	7B8	8X	12CA5	7CV
6AH5G	6AP	6BV8	9FJ	6EU7	9LS	6LQ6	9QL	7C4	4AH	12CM6	9CK
6AH6	7BK	6BW4*	9DJ	6EU8	9JF	6LY8	9DX	7C5	6AA	12CN5	7CV
6AH7GT	8BE	6BW6	9AM	6EV6	7CM	6M5	9N	7C6	8W	12CR6	7EA
6AJ4	9BX	6BW7	9AQ	6EW7	9HF	6M6G	7S	7C7	8V	12CS5	9CK
6AJ5	7BD	6BW8	9HK	6EX6	5BT	6M7G	7R	7D7	8AR	12CS7	7CH
6AJ7	8N	6BW11	12HD	6EY6	7AC	6M8GT	8AU	7E5	8BN	12CT8	9DA
6AJ8	9CA	6BX4*	5BS	6EZ5	7AC	6M11	12CA			12CU5	7CV
6AK5	7BD	6BX6	9AQ								
6AK6	7BK	6BX7GT	8BD								

Type	Base	Type	Base	Type	Base	Type	Base	Type	Base	Type	Base
12CU6	6A	14Z3	4G	48	6A	312-E	Fig. 44	1221	6F	5844	7BF
12CX6	7BK	15	5F	49	5C	327-A	Fig. 50	1223	7R	5845	5CA
12DB5	9GR	15A6	9AR	50	4D	327-B	Fig. 50	1229	4K	5847	9X
12DE8	Fig. 81	15E	Fig. 51	50A5	6AA	342-B	4E	1230	4D	5852	6S
12DF5	9BS	16A5	9BL	50AX6G	7Q	356-A	Fig. 55	1231	8V	5857	9AB
12DF7	9A	17Z3	9CB	50B5	7BZ	361-A	4E	1232	8V	5866*	Fig. 3
12DK7	9HZ	18	6B	50BK5	9BQ	376-A	4E	1265	4AJ	5869*	Fig. 3
12DL8	9HR	18FW6A	7CC	50C5	7CV	408A	7BD	1266	4AJ	5871	7AC
12DM7	9A	18FX6A	7CH	50C6G	7S	417-A	9V	1273	8V	5876	Fig. 21
12DQ6A	6AM	18FY6A	7BT	50C6GA	7S	482-B	4D	1274	6S	5879	9AD
12DQ7	9BF	19	6C	50DC4*	5BQ	483	4D	1275	4C	5881	7AC
12DS7	9JU	19CL8A	9FX	50EH5	7CV	485	5A	1276	4D	5890	12J
12DT5	9HN	19X3	9BM	50FK5	7CV	527	Fig. 53	1284	8V	5893*	Fig. 21
12DT6	7EN	19Y3	9BM	50L6GT	7S	559	Fig. 10	1291	7BE	5894A*	Fig. 7
12DT7	9A	20	4D	50T	2D	572B*	3G	1293	4AA	5910	6AR
12DT8	9DE	20AP1-4	12A	50X6	7AJ	575-A	4AT	1294	4AH	5915	7CH
12DU7	9JX	20J8GT	8H	50Y6GT	7Q	592	Fig. 28	1299	6BB	5920	7BF
12DV7	9JY	21A6	9AS	50Y7GT	8AN	705-A	Fig. 45	1602	4D	5933*	5AW
12DV8	9HR	21A7	8AR	50Z6G	7Q	717-A	8BK	1603	6F	5961	8R
12DW5	9CK	21EX6	5BT	50Z7G	8AN	756	4D	1608	4D	5962	2AG
12DW7	9A	22	4K	51	5E	800	2D	1609	5B	5963	9A
12DW8	9JC	24-A	5E	52	5C	801A/801	4D	1610	Fig. 62	5964	7BF
12DY8	9JD	24-G	2D	53	7B	802	6BM	1611	7S	5965	9A
12DZ6	7BK	24XH	Fig. 1	53A	Fig. 53	803*	5J	1612	7T	5993*	Fig. 35
12E5GT	6Q	25A6	7S	55	6G	804	Fig. 61	1613	7S	5998	8BD
12EA6	7BK	25A7GT	8F	56	5A	805	3N	1614	7AC	6005	7BZ
12EC8	9FA	25AC5GT	6Q	56AS	5A	806	2N	1616	4P	6023	9CD
12ED5	7CV	25AV5GA	6CK	57	6F	807	5AW	1619	Fig. 74	6026	Fig. 16
12EF6	7S	25AV5GT	6CK	57AS	6F	807W*	5AW	1620	7R	6028	7BD
12EG6	7CH	25AX4GT	4CG	58	6F	808	2D	1621	7S	6045	7BF
12EK6	7BK	25B5	6D	58AS	6F	809	3G	1622	7AC	6046	7AC
12EL6	7FB	25B6G	7S	59	7A	810*	2N	1623*	3G	6057	9A
12EM6	9HV	25B8GT	8T	70A7GT	8AB	811*	3G	1624	Fig. 66	6058	6BT
12EN6	7S	25BK5	9BQ	70L7GT	8AA	811A*	3G	1625*	5AZ	6059	9BC
12F5GT	5M	25BQ6GA	6AM	71-A	4D	812*	3G	1626	6Q	6060	9A
12F8	9FH	25BQ6GT	6AM	72	4P	812A*	3G	1627	2N	6061	9AM
12FK6	7BT	25BQ6GTB	6AM	73	4Y	812H	3G	1628	Fig. 54	6062	9K
12FM6	7BT	25C5	7CV	75	6G	813	5BA	1629	6RA	6063*	7CF
12FP7	14E	25C6G	7AC	75TH	2D	814*	Fig. 54	1631	7AC	6064	7DB
12FQ8	9KT	25C6GA	7S	75TL	2D	815	8BY	1633	8BD	6065	7DB
12FR8	9KU	25CA5	7CV	76	5A	816	4P	1634	8S	6066	7BT
12FT6	7BT	25CD6G	5BT	77	6F	822	3N	1635	8B	6067	9A
12FX5	7CV	25CD6GA	5BT	78	6F	822S	2N	1641	Fig. 52	6072	9A
12FX8A	9KV	25CD6GB	5BT	79	6H	826	7BO	1642	7BH	6073	5BO
12G4	6BG	25CU6	6AM	80	4C	828	5JP	1644	Fig. 4	6074	5BO
12G7G	7V	25D8GT	8AF	81	4B	829	7BP	1654	2Z	6080	8BD
12G8	9CZ	25DN6	5BT	82	4C	829A	7BP	1802P1-11	11A	6082	8BD
12GA6	7CH	25DQ6	6AM	83	4C	820B*	7BP	1805P1-4	11A	6083	Fig. 5
12GE5	12BJ	25EC6	5BT	83-V	4AD	830	4D	1806P1	11N	6084	9BJ
12GJ5	9NM	25EH5	7CV	84/6Z4	5D	830B	3G	1851	7R	6085	9A
12GW6	6AM	25F5	7CV	85	6G	831	Fig. 40	1852	8N	6086	9BK
12GN7	9BF	25L6GT	7S	85AS	6G	832	7BP	1853	8N	6087	5L
12GP7	14S	25N6G	7W	89	6F	832A*	3G	2002	Fig. 1	6101	7BF
12H4	7DW	25S	6M	90C1	4D	833A*	Fig. 41	2005	Fig. 1	6132	9BA
12H6	7Q	25SA7GT	8AD	99	4D	834	2D	2050	8BA	6135	6BG
12HP7	11J	25T	3G	100TH*	2D	835	4E	2051	8BA	6136	7BK
12J5GT	6Q	25W4GT	4CG	100TL*	2D	836*	4P	2523N/128A	5A	6137	8N
12J7GT	7R	25W6GT	7S	111H	2D	837	6BM	4604	7CL	6140	9BY
12J8	9GC	25X6GT	7Q	112-A	4D	838	4E	4652	Fig. 51	6141	9BZ
12K5	7EK	25Y4GT	5AA	117L7GT	8AO	840	5J	5514	4BO	6146	7CK
12K7GT	7R	25Y5	6E	117M7GT	8AO	841	4D	5516	7CL	6146A*	7CK
12K8	8K	25Z3	4G	117N7GT*	8AV	841A	3G	5517	5BU	6146B*	7CK
12L6GT	7S	25Z4	5AA	117P7GT*	8AV	841SW	3G	5556	4D	6155*	5BK
12L8GT	8BU	25Z5*	6E	117Z3*	4CB	843	5A	5562	Fig. 30	6156*	5BK
12Q7GT	7V	25Z6	7Q	117Z4GT*	5AA	844	5AW	5590	7BD	6157	Fig. 36
12R5	7CV	26	4D	117Z6GT*	7Q	849	Fig. 39	5591	7BD	6158	9A
12S8GT	8CB	26A6	7BK	128AS	5A	850	Fig. 47	5608	7BD	6159B*	7CK
12SA7	8R	26A7GT	8BU	150T	2N	852	2D	5608A	7B	6173	Fig. 34
12SC7	8S	26BK6	7BT	152TH	4BC	860	Fig. 58	5610	6CG	6186	7BD
12SF5	6AB	26C6	7BT	152TL	4BC	861	Fig. 42	5618	7CU	6197	9BV
12SF7	7AZ	26CG6	7BK	175A*	Fig. 78	864	4D	5651	5BO	6201	9A
12SG7	8BK	26D6	7CH	182-B	4D	865	Fig. 57	5654	7BD	6211	9A
12SH7	8BK	26Z5W	9BS	183	4D	866	4P	5656	9F	6216	Fig. 37
12SJ7	8N	27	5A	203-A	4E	866A-AX*	4P	5662	Fig. 79	6218	9CG
12SK7	8N	28Z5	5AB	203-H	3N	866B*	4P	5663	6CE	6227	9BA
12SL7GT	8BD	30	4D	204-A	Fig. 39	866jr*	4B	5670	8CJ	6252*	Fig. 7
12SN7GT	8BD	31	4D	205-D	4D	871	4P	5675*	Fig. 21	6263	
12SN7GTA	8BD	32	4K	211	4E	872A/872*	4AT	5679	7CX	6264	
12SQ7	8Q	32ET5	7CV	212-E	Fig. 43	874	4S	5686	9G	6265	7CM
12SR7	8Q	32L7GT	8Z	217-A	4AT	878	4P	5687	9H	6287	9CT
12SW7	8BD	33	5K	217-C	4AT	879	4AB	5690	Fig. 38	6308	8EX
12SX7	8BD	34	4M	227-A	Fig. 53	884	6Q	5691	8BD	6336A	8BD
12SY7	8R	34GD5	7CV	241-B	Fig. 44	885	5A	5692	8BD	6350	9CZ
12U7	9A	35/51	5E	242-A	4E	902A	8CD	5693	8N	6354	Fig. 12
12V6GT	7S	35A5	6AA	242-B	4E	906P1-11	7AN	5694	8CS	6360*	Fig. 13
12W6GT	7S	35B5	7BZ	242-C	4E	908A	7CE	5696	7BN	6374	9BW
12X4*	5BS	35C5	7CV	249-B	Fig. 29	909	5BP	5722	5CB	6386	8CJ
12Z3	4G	35L6GT	7S	250TH*	2N	910	7AN	5725	7CM	6417*	9K
12Z5	7L	35T	3G	250TL*	2N	911	7AN	5726	6BT	6443	9BW
14A4	5AC	35TG	2D	254	2N	914A	6BF	5727	7BN	6485	7BK
14A5	6AA	35W4*	5BQ	254-A	Fig. 57	930B	3G	5731	5BC	6524	Fig. 76
14A7	8V	35Y4	5AL	254-B	Fig. 57	938	4E	5749	7BK	6550	7S
14AF7	8AC	35Z3	4Z	261-A	4E	950	5K	5750	7CH	6627	5BO
14AP1-4	12A	35Z4GT*	5AA	270-A	Fig. 39	951	4M	5751	9A	6660	7CC
14B6	8W	35Z5G*	6AD	276-A	4E	954	5BB	5755	9J	6661	7CM
14B8	8X	35Z6G	7Q	282-A	Fig. 57	955	5BC	5763*	9K	6662	7CM
14C5	6AA	36	5E	284-B	3N	955	5BC	5764	Fig. 21	6663	6BT
14C7	8V	36AM3*	5BQ	284-D	4E	956	5BB	5765	Fig. 21	6664	5CE
14E6	8W	37	5A	295-A	4E	957	5BD	5766	See 2C37	6669	7BZ
14E7	8AE	38	5F	300T	2N	958	5BD	5767	See 2C37	6676	7CM
14F7	8AC	39/44	5F	303-A	4E	958A	5BD	5768	Fig. 21	6677	9BV
14F8	8BW	40	4D	304-A	Fig. 39	959	5BE	5794	9A	6678	9AE
14H7	8V	40Z5GT	6AD	304-B	2D	967	3G	5812	7CQ	6679	9A
14J7	8BL	41	6B	304TH*	4BC	975A	4AT	5814	9A	6680	9A
14N7	8AC	42	6B	304TL*	4BC	1003	4R	5823	4CK	6681	9A
14Q7	8AL	43	6B	305-A	Fig. 59	1005	5AQ	5824	7S	6816*	Fig. 77
14R7	8AE	44	4D	306-A	Fig. 63	1006	4C	5825	4P	6829	9A
14S7	8BL	45	5C	307-A	Fig. 61	1201	8BN	5839	6S	6850*	Fig. 76
14V7	8V	45Z3	5AM	308-B	Fig. 43	1203	4AH	5842	9V	6883*	7CK
14W7	8BJ	45Z5GT	6AD	310	4D	1204	8BO			6884*	Fig. 77
14X7	8BZ	46	5C	311	4E	1206	8BV			6887	6BT
14Y4	5AB	47	5B	311CH	Fig. 32	1218A	7DK			6893*	7CK
				312-A	Fig. 68						

E.I.A. Vacuum-Tube Base Diagrams

Socket connections correspond to the base designations given in the column headed "Base" in the classified tube-data tables. Bottom views are shown throughout. Terminal designations are as follows:

A — Anode	B — Deflecting Plate	IS — Internal Shield	RC — Ray-Control Electrode
B — Beam	F — Filament	K — Cathode	Ref — Reflector
BP — Bayonet Pin	FE — Focus Elect.	NC — No Connection	S — Shell
BS — Base Sleeve	G — Grid	P — Plate (Anode)	TA — Target
C — Ext. Coating	H — Heater	P_1 — Starter-Anode	U — Unit
CL — Collector	IC — Internal Con.	P_{BF} — Beam Plates	● — Gas-Type Tube

Alphabetical subscripts D, P, T and HX indicate, respectively, diode unit, pentode unit, triode unit or hexode unit in multi-unit types. Subscript CT indicates filament or heater tap.

Generally when the No. 1 pin of a metal-type tube, with the exception of all triodes, is shown connected to the shell, the No. 1 pin in the glass (G or GT) equivalent is connected to an internal shield.

*On 12AQ, 12AS and 12CT: index = large lug; ● = pin cut off

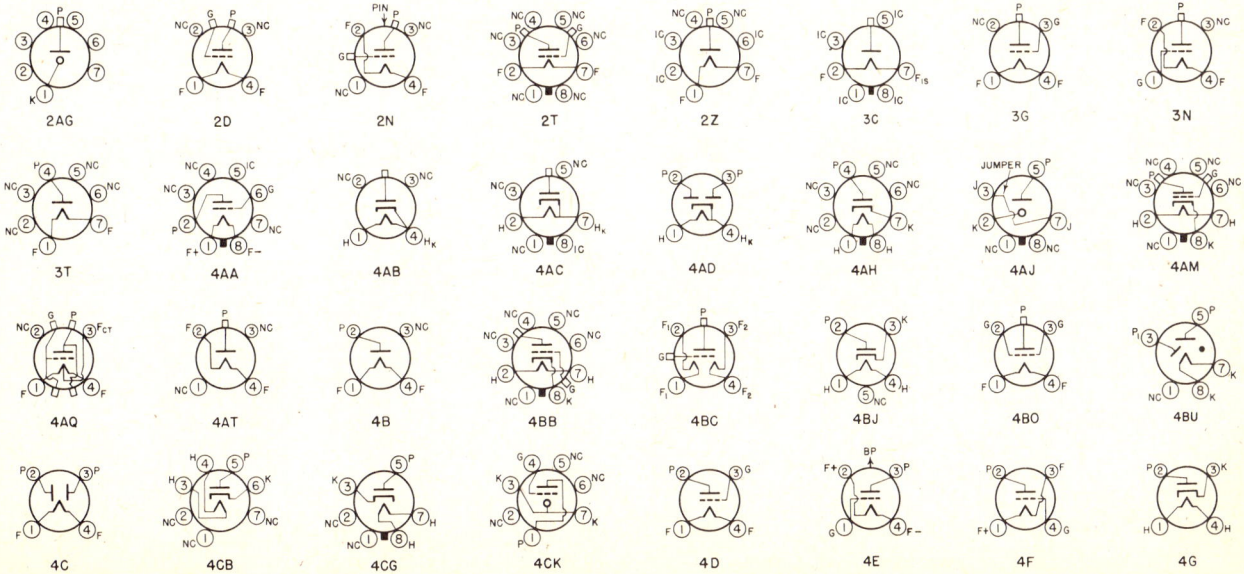

2AG 2D 2N 2T 2Z 3C 3G 3N

3T 4AA 4AB 4AC 4AD 4AH 4AJ 4AM

4AQ 4AT 4B 4BB 4BC 4BJ 4BO 4BU

4C 4CB 4CG 4CK 4D 4E 4F 4G

Tube Base Diagrams

Bottom views are shown. Terminal designations on sockets are given on page 23-4.

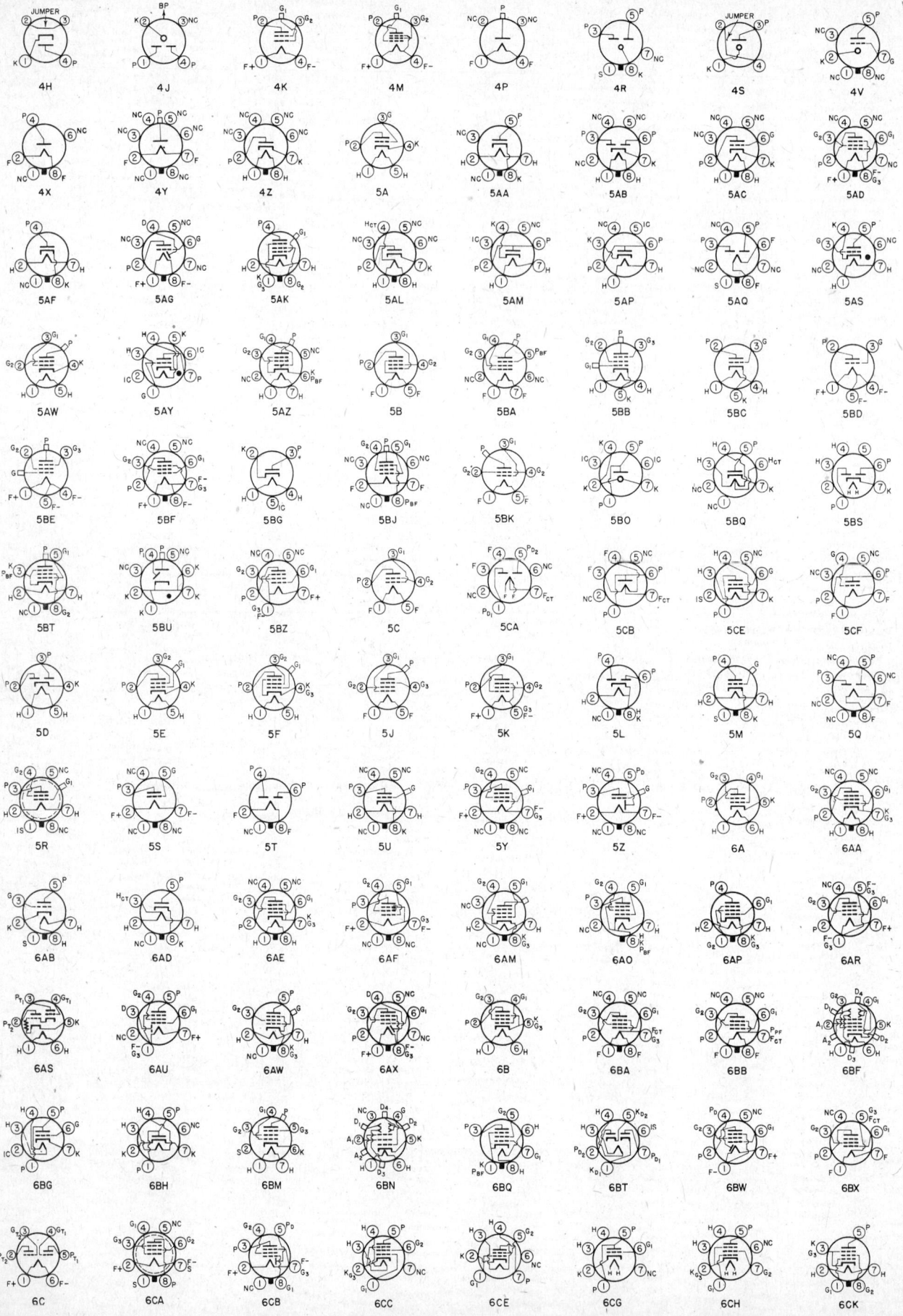

4H 4J 4K 4M 4P 4R 4S 4V

4X 4Y 4Z 5A 5AA 5AB 5AC 5AD

5AF 5AG 5AK 5AL 5AM 5AP 5AQ 5AS

5AW 5AY 5AZ 5B 5BA 5BB 5BC 5BD

5BE 5BF 5BG 5BJ 5BK 5BO 5BQ 5BS

5BT 5BU 5BZ 5C 5CA 5CB 5CE 5CF

5D 5E 5F 5J 5K 5L 5M 5Q

5R 5S 5T 5U 5Y 5Z 6A 6AA

6AB 6AD 6AE 6AF 6AM 6AO 6AP 6AR

6AS 6AU 6AW 6AX 6B 6BA 6BB 6BF

6BG 6BH 6BM 6BN 6BQ 6BT 6BW 6BX

6C 6CA 6CB 6CC 6CE 6CG 6CH 6CK

6CN 6D 6E 6F 6G 6H 6J 6K

6L 6M 6Q 6R 6RA 6S 6T 6W

6X 6Y 6Z 7A 7AA 7AB 7AC 7AD

7AG 7AH 7AJ 7AK 7AL 7AM 7AN 7AO

7AP 7AQ 7AT 7AU 7AV 7AX 7AZ 7B

7BA 7BB 7BC 7BD 7BE 7BF 7BH 7BJ

7BK 7BM 7BN 7BO 7BP 7BQ 7BR 7BS

7BT 7BW 7BZ 7C 7CA 7CB 7CC 7CE

7CF 7CH 7CJ 7CK 7CL 7CM 7CQ 7CU

7CV 7CX 7CY 7D 7DB 7DC 7DE 7DF

7DH 7DK 7DT 7DW 7E 7EA 7EG 7EK

7EN 7EW 7F 7FB 7FL 7FN 7FP 7FQ

Tube Base Diagrams (cont.)

Bottom views are shown. Terminal designations on sockets are given on page 23-4.

7G 7GA 7GK 7GM 7H 7J 7K 7L

7Q 7R 7S 7T 7U 7V 7W 7Z

8A 8AA 8AB 8AC 8AE 8AF 8AG 8AJ

8AL 8AN 8AO 8AR 8AS 8AU 8AV 8AW

8AX 8AY 8B 8BA 8BD 8BE 8BF 8BJ

8BK 8BL 8BN 8BO 8BS 8BU 8BV 8BW

8BY 8BZ 8C 8CB 8CD 8CH 8CJ 8CK

8CQ 8CS 8CT 8DU 8E 8EL 8EX 8EZ

8F 8FP 8FV 8G 8GD 8GS 8H 8HY

8JB 8JC 8JP 8K 8KQ 8KS 8L 8N

8O 8Q 8R 8S 8T 8U 8V 8W

8X 8Y 8Z 9A 9AA 9AB 9AC 9AD

Tube Base Diagrams (cont.)

Bottom views are shown. Terminal designations on sockets are given on page 23-4.

 9AE
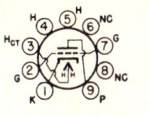 9AG
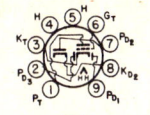 9AH
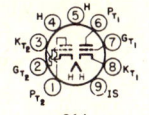 9AJ
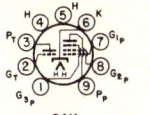 9AK
 9AM
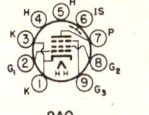 9AQ
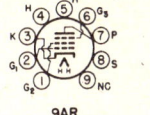 9AR

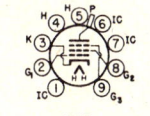 9AS
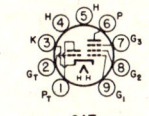 9AT
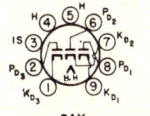 9AX
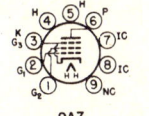 9AZ
 9BA
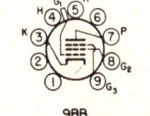 9BB
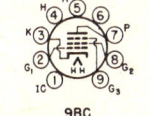 9BC
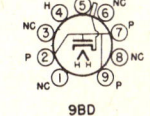 9BD

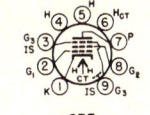 9BF
 9BG
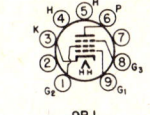 9BJ
 9BK
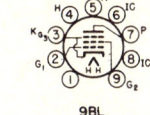 9BL
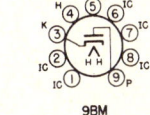 9BM
 9BP
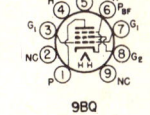 9BQ

 9BS
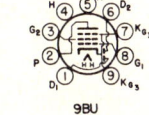 9BU
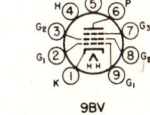 9BV
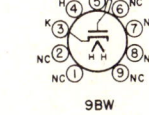 9BW
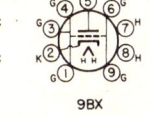 9BX
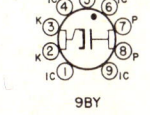 9BY
 9BZ
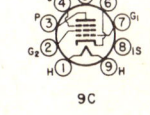 9C

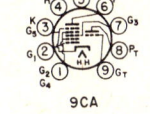 9CA
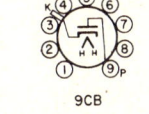 9CB
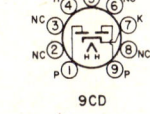 9CD
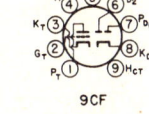 9CF
 9CG
 9CK
 9CT
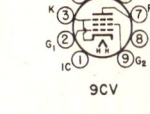 9CV

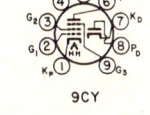 9CY
 9CZ
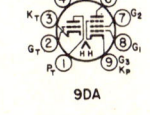 9DA
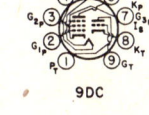 9DC
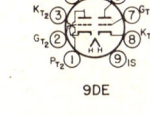 9DE
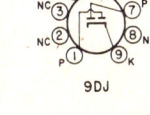 9DJ
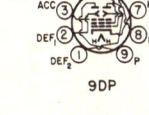 9DP
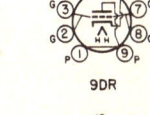 9DR

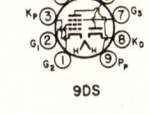 9DS
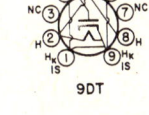 9DT
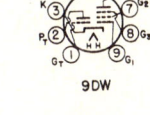 9DW
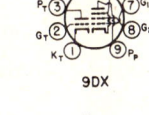 9DX
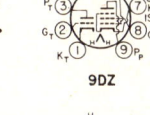 9DZ
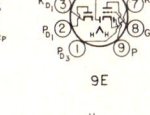 9E
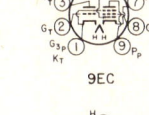 9EC
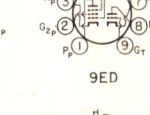 9ED

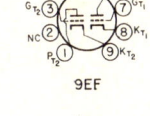 9EF
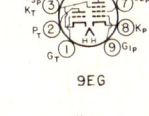 9EG
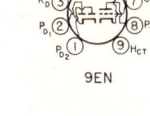 9EN
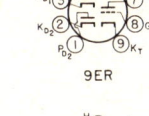 9ER
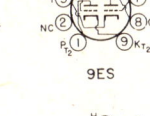 9ES
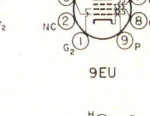 9EU
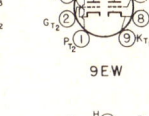 9EW
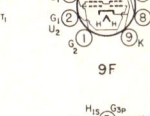 9F

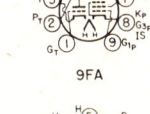 9FA
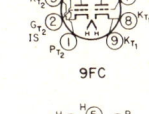 9FC
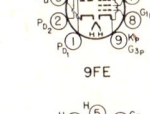 9FE
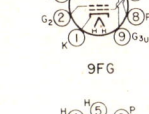 9FG
 9FH
 9FJ
 9FN
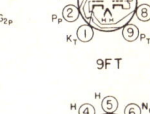 9FT

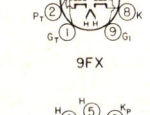 9FX
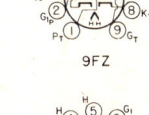 9FZ
 9G
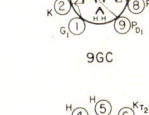 9GC
 9GE
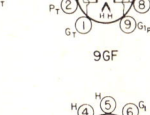 9GF
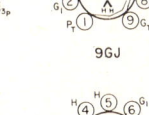 9GJ
 9GK

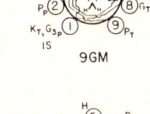 9GM
 9GR
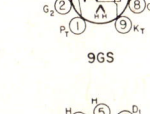 9GS
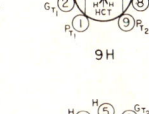 9H
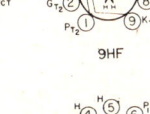 9HF
 9HK
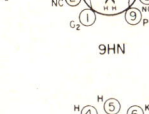 9HN
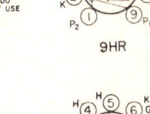 9HR

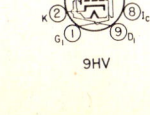 9HV
9HX
9HZ
9J
9JC
9JD
9JF
9JG

Vacuum Tubes and Semiconductors 23-8

Tube Base Diagrams (cont.)

Bottom views are shown. Terminal designations on sockets are given on page 23-4.

9JU 9JX 9JY 9K 9KA 9KR 9KS 9KT

9KU 9L 9LK 9LS 9KV 9LW 9LY 9M

9MS 9N 9NM 9NT 9NZ 9PA 9PB 9PM

9PU 9PX 9Q 9QA 9QL 9QP 9R 9S

9T 9U 9V 9X 9Y 9Z 11A 11B

11C 11J 11L 11M 11N 11S 11T 11V

12A 12AQ 12AS 12BF 12BJ 12BM 12BQ 12BW

12BY 12CA 12CT 12E 12EA 12EU 12F 12FB

12FM 12GW 12HD 12J 12T 14A 14B 14E

14G 14J 14K 14P 14R 14S 14U 14V

FIG. 1 FIG. 2 FIG. 3 FIG. 4 FIG. 5 FIG. 6 FIG. 7 FIG. 8

FIG. 9 FIG. 10 FIG. 11 FIG. 12 FIG. 13 FIG. 14 FIG. 15 FIG. 16

FIG. 17 FIG. 18 FIG. 19 FIG. 20 FIG. 21 FIG. 22 FIG. 23 FIG. 24

Tube Base Diagrams (cont.)

Bottom views are shown. Terminal designations on sockets are given on page 23-4.

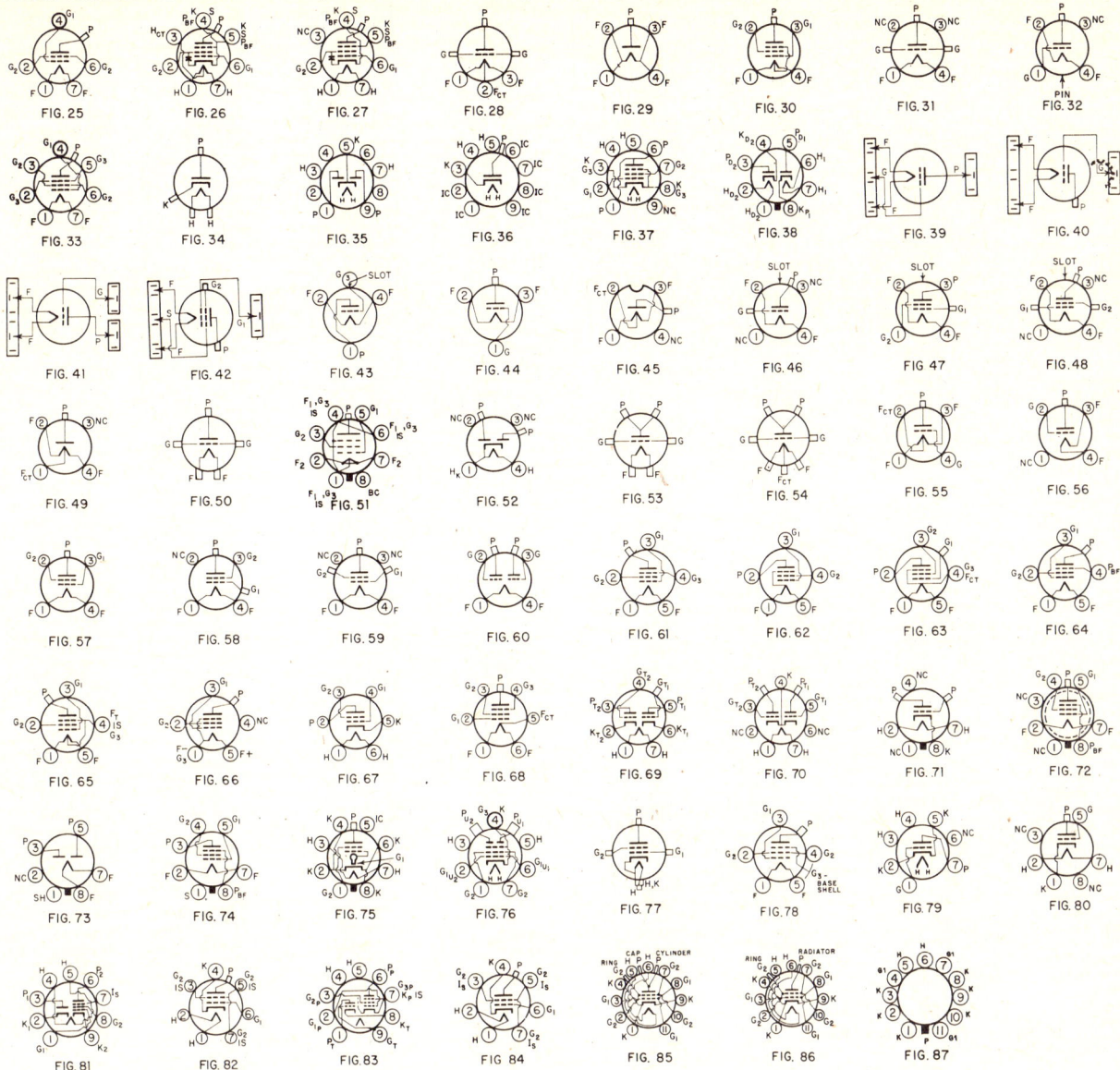

FIG. 25 FIG. 26 FIG. 27 FIG. 28 FIG. 29 FIG. 30 FIG. 31 FIG. 32

FIG. 33 FIG. 34 FIG. 35 FIG. 36 FIG. 37 FIG. 38 FIG. 39 FIG. 40

FIG. 41 FIG. 42 FIG. 43 FIG. 44 FIG. 45 FIG. 46 FIG 47 FIG. 48

FIG. 49 FIG. 50 FIG. 51 FIG. 52 FIG. 53 FIG. 54 FIG. 55 FIG. 56

FIG. 57 FIG. 58 FIG. 59 FIG. 60 FIG. 61 FIG. 62 FIG. 63 FIG. 64

FIG. 65 FIG. 66 FIG. 67 FIG. 68 FIG. 69 FIG. 70 FIG. 71 FIG. 72

FIG. 73 FIG. 74 FIG. 75 FIG. 76 FIG. 77 FIG. 78 FIG. 79 FIG. 80

FIG. 81 FIG. 82 FIG. 83 FIG 84 FIG. 85 FIG. 86 FIG. 87

Table 1 — Triode Transmitting Tubes

Type	Maximum Ratings						Cathode		Capacitances			Base	Typical Operation							
	Plate Dissi-pation Watts	Plate Voltage	Plate Current mA	DC Grid Current mA	Freq. MHz. Full Ratings	Amplification Factor	Volts	Amperes	C_{in} pF	C_{gp} pF	C_{out} pF		Class of Service	Plate Voltage	Grid Voltage	Plate Current mA	DC Grid Current mA	Approx. Driving Power Watts	P-to-P Load Ohms	Approx. Output Power Watts
6J6A [2]	1.5	300	30	16	250	32	6.3	0.45	2.2	1.6	0.4	7BF	C-T	150	−10	30	1.6	0.035	—	3.5
6F4	2.0	150	20	8.0	500	17	6.3	0.225	2.0	1.9	0.6	7BR	C-T-O	150	−15 550* 2000[4]	20	7.5	0.2	—	1.8
12AU7A[2]	2.76[6]	350	12[6]	3.5[6]	54	18	6.3	0.3	1.5	1.5	0.5	9A	C-T-O	350	−100	24	7	—	—	6.0
6C4	5.0	350	25	8.0	54	18	6.3	0.15	1.8	1.6	1.3	6BG	C-T-O	300	−27	25	7.0	0.35	—	5.5
5675	5	165	30	8	3000	20	6.3	0.135	2.3	1.3	0.09	Fig. 21	G-G-O	120	−8	25	4	—	—	0.05
6N7GT[2]	5.5[6]	350	30[6]	5.0[6]	10	35	6.3	0.8	—	—	—	8B	C-T-O	350	−100	60	10	—	—	14.5
2C40	6.5	500	25	—	500	36	6.3	0.75	2.1	1.3	0.05	Fig. 11	C-T-O	250	−5	20	0.3	—	—	0.075
5893	8.0	400	40	13	1000	27	6.0	0.33	2.5	1.75	0.07	Fig. 21	C-T	350	−33	35	13	2.4	—	6.5
													C-P	300	−45	30	12	2.0	—	6.5
2C43	12	500	40	—	1250	48	6.3	0.9	2.9	1.7	0.05	Fig. 11	C-T-O	470	—	38[7]	—	—	—	9[7]
3C24	25	2000	75	7[13]	60	24	6.3	3.0	1.7	1.6	0.2	2D	C-T	2000	−130	63	18	4	—	100
	17	1600	60										C-P	1600	−170	53	11	3.1	—	68
	25	2000	75										AB₂[7]	1250	−42	24/130	270[9]	3.4[8]	21.4K	112
1623	30	1000	100	25	60	20	6.3	2.5	5.7	6.7	0.9	3G	C-T-O	1000	−90	100	20	3.1	—	75
													C-P	750	−125	100	20	4.0	—	55
													B[7]	1000	−40	30/200	230[9]	4.2[8]	12K	145

Table 1 — Triode Transmitting Tubes — Continued

Type	Maximum Ratings — Plate Dissipation Watts	Plate Voltage	Plate Current mA	DC Grid Current mA	Freq. MHz Full Ratings	Amplification Factor	Cathode — Volts	Amperes	Capacitances — Cin pF	Cgp pF	Cout pF	Base	Class of Service	Typical Operation — Plate Voltage	Grid Voltage	Plate Current mA	DC Grid Current mA	Approx. Driving Power Watts	P-to-P Load Ohms	Approx. Output Power Watts
811-A	65	1500	175	50	60	160	6.3	4.0	5.9	5.6	0.7	3G	C·T	1500	-70	173	40	7.1	—	200
													C·P	1250	-120	140	45	10.0	—	135
													G·G·B	1250	0	27/175	28	12	—	165
													AB1	1250	0	27/175	13	3.0	—	155
812-A	65	1500	175	35	60	29	6.3	4.0	5.4	5.5	0.77	3G	C·T	1500	-120	173	30	6.5	—	190
													C·P	1250	-115	140	35	7.6	—	130
													B7	1500	-48	28/310	270[9]	5.0	13.2K	340
100TH	100	3000	225	60	40	40	5.0	6.3	2.9	2.0	0.4	2D	C·T / C·P	3000	-200	165	51	18	—	400
													B7	3000	-65	40/215	335[9]	5.0[8]	31K	650
3-100A2 / 100TL	100	3000	225	50	40	14	5.0	6.3	2.3	2.0	0.4	2D	C·T / C·P	3000	-400	165	30	20	—	400
													G·M·A	3000	-560	60	2.0	7.0	—	90
													B7	3000	-185	40/215	640[9]	6.0[8]	30K	450
3CX100A5[15]	100	1000	125[14]	50	2500	100	6.0	1.05	7.0	2.15	0.035	—	G·G·A	800	-20	80	30	6	—	27
	70	600	100[14]										C·P	600	-15	75	40	6	—	18
2C39	100	1000	60	40	500	100	6.3	1.1	6.5	1.95	0.03	—	G·I·C	600	-35	60	40	5.0	—	20
													C·T·O	900	-40	90	30	—	—	40
													C·P	600	-150	100[14]	50	—	—	—
AX9900/5866[15]	135	2500	200	40	150	25	6.3	5.4	5.8	5.5	0.1	Fig. 3	C·T	2500	-200	200	40	16	—	390
													C·P	2000	-225	127	40	16	—	204
													B7	2500	-90	80/330	350[9]	14[8]	15.68K	560
572B/T160L	160	2750	275	—	—	170	6.3	4.0	—	—	—	3G	C·T	1650	-70	165	32	6	—	205
													G·G·B7	2400	-2.0	90/500	—	100	—	600
810	175	2500	300	75	30	36	10	4.5	8.7	4.8	12	2N	C·T	2500	-180	300	60	19	—	575
													C·P	2000	-350	250	70	35	—	380
													G·M·A	2250	-140	100	2.0	4	—	75
													B7	2250	-60	70/450	380[9]	13[8]	11.6K	725
8873	200	2200	250	—	500	160	6.3	3.2	19.5	7.0	0.03	Fig. 87	AB2	2000	—	22/500	98[8]	27[8]	—	505
250TH	250	4000	350	40[13]	40	37	5.0	10.5	4.6	2.9	0.5	2N	C·T·O	2000	-100	357	94	29	—	464
													C·T·O	3000	-150	333	90	32	—	750
													C·P	2000	-160	250	60	22	—	335
													C·P	2500	-180	225	45	17	—	400
													C·P	3000	-200	200	38	14	—	435
													AB2[7]	1500	0	220/700	460[9]	46[8]	4.2K	630
250TL	250	4000	350	35[13]	40	14	5.0	10.5	3.7	3.0	0.7	2N	C·T·O	2000	-200	350	45	22	—	455
													C·T·O	3000	-350	335	45	29	—	750
													C·P	2000	-520	250	29	24	—	335
													C·P	2500	-520	225	20	16	—	400
													C·P	3000	-520	200	14	11	—	435
													AB2[7]	1500	-40	200/700	780[9]	38[8]	3.8K	580
PL-6569	250	4000	300	120	30	45	5.0	14.5	7.6	3.7	0.1	Fig. 3	G·G·A	2500	-70	300	85	75[11]	—	555
														3000	-95	300	110	85[11]	—	710
														3500	-110	285	90	85[11]	—	805
														4000	-120	250	50	70[11]	—	820
8875	300	2200	250	—	500	160	6.3	3.2	19.5	7.0	0.03	—	AB2	2000	—	22/500	98[8]	27[8]	—	505
304TH	300	3000	900	60[13]	40	20	5.0 / 10	25 / 12.5	13.5	10.2	0.7	4BC	C·T·O	1500	-125	665	115	25	—	700
														2000	-200	600	125	39	—	900
													C·P	1500	-200	420	55	18	—	500
														2000	-300	440	60	26	—	680
														3000	-350	400	60	29	—	800
													AB2[7]	1500	-65	1065[8]	330[9]	25[8]	2.84K	1000
304TL	300	3000	900	50[13]	40	12	5.0 / 10	25 / 12.5	12.1	8.6	0.8	4BC	C·T·O	1500	-250	665	90	33	—	700
														2000	-300	600	85	36	—	900
													C·P	2000	-500	250	30	18	—	410
														2000	-500	500	75	52	—	810
														2500	-525	200	18	11	—	425
														2500	-550	400	50	36	—	830
													AB1[7]	1500	-118	270/572	236[9]	0	2.54K	256
														2500	-230	160/483	460[9]	0	8.5K	610
													AB2[7]	1500	-118	1140[8]	490[9]	39[8]	2.75K	1100
833A	350 / 450[15]	3300 / 4000[15]	500	100 / 20[15]	30 / 20[15]	35	10	10	12.3	6.3	8.5	Fig. 41	C·T·O	2250	-125	445	85	23	—	780
														3000	-160	335	70	20	—	800
													C·P	2500	-300	335	75	30	—	635
														3000	-240	335	70	26	—	800
													B7	3000	-70	100/750	400[9]	20[8]	9.5K	1650
8874	400	2200	250	—	500	160	6.3	3.2	19.5	7.0	0.03	—	AB2	2000	—	22/500	98[8]	27[8]	—	505
3-400Z	400	3000	400	—	110	200	5	14.5	7.4	4.1	0.07	Fig. 3	G·G·B	3000	0	100/333	120	32	—	655
PL-6580	400	4000[15]	350	120	—	45	5.0	14.5	7.6	3.9	0.1	5BK	G·G·A	4000	-110	350	92	105[11]	—	1080
														2500	-70	350	95	85	—	660
8163	400	3000	400	20[13]	30	350	5.0	14.1	8.0	5.0	0.3	Fig. 3	G·G·B	2500	0	72/400	140	35	—	640
3-500Z	500	4000	400	—	110	160	5	14.5	7.4	4.1	0.07	Fig. 3	G·G·B	3000		370	115	30	5K	750
													C·T	3500	-75	300	115	22	—	850
3-1000Z	1000	3000	800	—	110	200	7.5	21.3	17	6.9	0.12	Fig. 3	G·G·B	3000	0	180/670	300	65	—	1360
8877	1500	4000	1000	—	250	200	5.0	10	42	10	0.1		AB2	2500	-8.2	1000	—	57	—	1520

* Cathode resistor in ohms.

[1] KEY TO CLASS-OF-SERVICE ABBREVIATIONS

A1 = Class-A1 af modulator.
AB1 = Class-AB1 push-pull af modulator.
AB2 = Class-AB2 push-pull af modulator.
B = Class-B push-pull af modulator.
C·M = Frequency multiplier.
C·P = Class-C plate-modulated telephone.
C·T = Class-C telegraph.
C·T·O = Class-C amplifier-osc.
G·G·A = Grounded-grid class-A amp.
G·G·B = Grounded-grid class-B amp. (Single Tone).

G·G·O = Grounded-grid osc.
G·I·C = Grid-isolation circuit.
G·M·A = Grid-modulated amp.

[2] Twin triode. Values, except interelectrode capacitances, are for both sections in push-pull.
[3] Output at 112 MHz.
[4] Grid leak resistor in ohms.
[5] Peak values.
[6] Per section.
[7] Values are for two tubes in push-pull.
[8] Max. signal value.

[9] Peak af grid-to-grid volts.
[10] Plate-pulsed 1000-MHz. osc.
[11] Includes bias loss, grid dissipation, and feed-through power.
[12] 1000-MHz. cw osc.
[13] Max. grid dissipation in watts.
[14] Max. cathode current in mA.
[15] Forced-air cooling required.
[16] Plate-pulsed 3300-MHz. osc.
[17] 1900-MHz. cw osc.
[18] No Class-B data available.

Table 2 — Tetrode and Pentode Transmitting Tubes

Type	Maximum Ratings					Cathode		Capacitances			Base	Typical Operation										
	Plate Dissipation Watts	Plate Voltage	Screen Dissipation Watts	Screen Voltage	Freq. MHz. Full Ratings	Volts	Amperes	C_{in} pF	C_{gp} pF	C_{out} pF		Class of Service[14]	Plate Voltage	Screen Voltage	Suppressor Voltage	Grid Voltage	Plate Current mA	Screen Current mA	Grid Current mA	Approx. Driving Power Watts	P-to-P Load Ohms	Approx. Output Power Watts
8203	1.8	400	—	—	250	6.3	0.16	4.2	2.2	1.6	12AQ	C·P/C·T	155	—	—	14/2700[1]	21	—	5	0.4	—	1.55
6939[3]	7.5	275	3	200	500	6.3 / 12.6	0.75 / 0.375	6.6	0.15	1.55	Fig. 13	C·T	200	200	—	−20	60	13	2	1.0	—	7.5
												C·P	180	180	—	−20	55	11.5	1.7	1.0	—	6
												C·M	200	190	—	68K[1]	46	10	2.2	0.9	—	—
7551 7558	12	300	2	250	175	12.6 / 6.3	0.38 / 0.8	10	0.15	5.5	9LK	C·T	300	250	—	−55	80	5.1	1.6	1.5	—	10
												C·P	250	250	—	−75	70	3.0	2.3	1.0	—	7.5
5763 6417	13.5	350	2	250	50	6.3 / 12.6	0.75 / 0.375	9.5	0.3	4.5	9K	C·T	350	250	—	−28.5	48.5	6.2	1.6	0.1	—	12
												C·P	300	250	—	−42.5	50	6	2.4	0.15	—	10
												C·M[2]	300	250	—	−75	40	4	1	0.6	—	2.1
												C·M[4]	300	235	—	−100	35	5	1	0.6	—	1.3
2E26 6893	13.5	600	2.5	200	125	6.3 / 12.6	0.8 / 0.4	12.5	0.2	7	7CK	C·T	600	185	—	−45	66	10	3	0.17	—	27
												C·P	500	180	—	−50	54	9	2.5	0.15	—	18
												AB₁	500	200	—	−25	9/45	10[7]	0	0	—	15
6360[3]	14	300	2	200	200	6.3 / 12.6	0.82 / 0.41	6.2	0.1	2.6	Fig. 13	C·T	300	200	—	−45	100	3	3	0.2	—	18.5
												C·P	200	100	—	15K[1]	86	3.1	3.3	0.2	—	9.8
												C·M[11]	300	150	—	−100	65	3.5	3.8	0.45	—	4.8
												AB₂	300	200	—	−21.5	30/100	1/11.4	64[8]	0.04	6.5K	17.5
2E25	15	450	4	250	125	6	0.8	8.5	0.15	6.7	5BJ	C·T·O	450	250	—	−45	75	15	3	0.4	—	24
												C·P	400	200	—	−45	60	12	3	0.4	—	16
												AB₂[6]	450	250	—	−30	44/150	10/40	3	0.9[7]	6K	40
832A[3]	15	750	5	250	200	6.3 / 12.6	1.6 / 0.8	8	0.07	3.8	7BP	C·T	750	200	—	−65	48	15	2.8	0.19	—	26
												C·P	600	200	—	−65	36	16	2.6	0.16	—	17
6252/ AX9910[3]	20	750	4	300	300	6.3 / 12.6	1.3 / 0.65	6.5	—	2.5	Fig. 7	C·T	600	250	—	−60	140	14	4	2.0	—	—
												C·P	500	250	—	−80	100	12	3	4.0	—	—
												B	500	250	—	−26	25/73	0.7/16	52[8]	—	20K	23.5
1614	25	450	3.5	300	80	10	0.9	10	0.4	12.5	7AC	C·T	450	250	—	−45	100	8	2	0.15	—	31
												C·P	375	250	—	−50	93	7	2	0.15	—	24.5
												AB₁[6]	530	340	—	−36	60/160	20[7]	—	—	7.2K	50
815[3]	25	500	4	200	125	6.3 / 12.6	1.6 / 0.8	13.3	0.2	8.5	8BY	C·T·O	500	200	—	−45	150	17	2.5	0.13	—	56
												AB₂	500	125	—	−15	22/150	32[7]	—	0.36[7]	8K	54
6146 6146A	25	750	3	250	60	6.3	1.25	13	0.24	8.5	7CK	C·T	500	170	—	−66	135	9	2.5	0.2	—	48
												C·T	750	160	—	−62	120	11	3.1	0.2	—	70
8032 6883						12.6	0.585					C·T[12]	400	190	—	−54	150	10.4	2.2	3.0	—	35
												C·P	400	150	—	−87	112	7.8	3.4	0.4	—	32
												C·P	600	150	—	−87	112	7.8	3.4	0.4	—	52
												AB₂[6]	600	190	—	−48	28/270	1.2/20	2[7]	0.3	5K	113
6159B						26.5	0.3					AB₂[6]	750	165	—	−46	22/240	0.3/20	2.6[7]	0.4	7.4K	131
												AB₁[6]	750	195	—	−50	23/220	1/26	100[8]	0	8K	120
6524[3] 6850	25	600	—	300	100	6.3 / 12.6	1.25 / 0.625	7	0.11	3.4	Fig. 76	C·T	600	200	—	−44	120	8	3.7	0.2	—	56
												C·P	500	200	—	−61	100	7	2.5	0.2	—	40
												AB₂	500	200	—	−26	20/116	0.1/10	2.6	0.1	11.1K	40
807 807W 5933	30	750	3.5	300	60	6.3	0.9	12	0.2	7	5AW	C·T	750	250	—	−45	100	6	3.5	0.22	—	50
												C·P	600	275	—	−90	100	6.5	4	0.4	—	42.5
												AB₁	750	300	—	−35	15/70	3/8	75[8]	0	—	72
1625						12.6	0.45				5AZ	B[10]	750	—	—	0	15/240	—	555[8]	5.3[7]	6.65K	120
2E22	30	750	10	250	—	6.3	1.5	13	0.2	8	5J	C·T·O	750	250	22.5	−60	100	16	6	0.55	—	53
6146B/ 8298A	35	750	3	250	60	6.3	1.125	13	0.22	8.5	7CK	C·T	750	200	—	−77	160	10	2.7	0.3	—	85
												C·P	600	175	—	−92	140	9.5	3.4	0.5	—	62
												AB₁	750	200	—	−48	25/125	6.3	—	—	3.6K	61
AX-9903[3] 5894A	40	600	7	250	250	6.3 / 12.6	1.8 / 0.9	6.7	0.08	2.1	Fig. 7	C·T	600	250	—	−80	200	16	2	0.2	—	80
829B[3] 3E29[3]	40	750	7	240	200	6.3 / 12.6	2.25 / 1.125	14.5	0.12	7	7BP	C·T	500	200	—	−45	240	32	12	0.7	—	83
												C·P	425	200	—	−60	212	35	11	0.8	—	63
												B	500	200	—	−18	27/230	—	56[8]	0.39	4.8K	76
3D24	45	2000	10	400	125	6.3	3	6.5	0.2	2.4	Fig. 75	C·T·O	2000	375	—	−300	90	20	10	4.0	—	140
													1500	375	—	−300	90	22	10	4.0	—	105
4D22	50	750	14	350	60	12.6 / 25.2	1.6 / 0.8	28	0.27	13	Fig. 26	C·T	750	300	—	−100	240	26	12	1.5	—	135
													600	300	—	−100	215	30	10	1.25	—	100
4D32						6.3	3.75				Fig. 27	C·P	600	—	—	−100	220	28	10	1.25	—	100
													550	—	—	−100	175	17	6	0.6	—	70
												AB₂[6]	600	250	—	−25	100/365	26[7]	70[8]	0.45[7]	3K	125
8117[3]	60	750	7	300	175	6.3 / 12.6	1.8 / 0.9	11.8	3.7	0.09	Fig. 7	AB₁	600	250	—	−32.5	60/212	1.9/25	—	—	1410	76
814	65	1500	10	300	30	10	3.25	13.5	0.1	13.5	Fig. 64	C·T	1500	300	—	−90	150	24	10	1.5	—	160
												C·P	1250	300	—	−150	145	20	10	3.2	—	130
4-65A	65	3000	10	600	150	6	3.5	8	0.08	2.1	Fig. 25	C·T·O	1500	250	—	−85	150	40	18	3.2	—	165
													3000	250	—	−100	115	22	10	1.7	—	280
												C·P	1500	250	—	−125	120	40	16	3.5	—	140
													2500	250	—	−135	110	25	12	2.6	—	230
												AB₁	2500	400	—	−85	15/66	3[7]	—	—	—	100
7854[3]	68	1000	8	300	175	6.3 / 12.6	1.8 / 0.9	6.7	2.1	0.09	Fig. 7	C·T	750	260	—	−75	240	12.7	5.5	3.5	—	123
												C·P	600	225	—	−75	200	7.8	5.5	3.5	—	85
4E27/ 8001	75	4000	30	750	75	5	7.5	12	0.06	6.5	7BM	C·T	2000	500	60	−200	150	11	6	1.4	—	230
												C·P	1800	400	60	−130	135	11	8	1.7	—	178
PL-177A	75	2000	10	600	175	6	3.2	7.5	0.06	4.2	Fig. 14	C·T·C·P	2000	400	0	−125	150	12	5	0.8	—	220
													1000	400	0	−105	150	16	5	0.7	—	100
												AB₁	2000	600	0	−115	25/175	0/7	0	0	—	210
7270 7271	80	1350	—	425	175	6.3 / 13.5	3.1 / 1.25	8	0.4	0.14	Fig. 84	C·T	850	400	—	−100	275	15	8	10	—	135
												AB₁	665	400	—	−119	220	15	6	10	—	85
8072	100	2200	8	400	500	13.5	1.3	16	0.13	0.011	Fig. 85	C·T·O	700	200	—	−30	300	10	20	5	—	85

Table 2 — Tetrode and Pentode Transmitting Tubes — Continued

Type	Plate Dissipation Watts	Plate Voltage	Screen Dissipation Watts	Screen Voltage	Freq. MHz. Full Ratings	Volts	Amperes	Cin pF	Cgp pF	Cout pF	Base	Class of Service[14]	Plate Voltage	Screen Voltage	Suppressor Voltage	Grid Voltage	Plate Current mA	Screen Current mA	Grid Current mA	Approx. Driving Power Watts	P-to-P Load Ohms	Approx. Output Power Watts
6816[9] 6884	115	1000	4.5	300	400	6.3	2.1	14	0.085	0.015	Fig. 77	C·T·O	900	300	—	−30	170	1	10	3	—	80
												C·P	700	250	—	−50	130	10	10	3	—	45
						26.5	0.52					AB1[6]	850	300	—	−15	80/200	0/20	30[8]	0	7K	80
												AB2[6]	850	300	—	−15	80/335	0/25	46[8]	0.3	3.96K	140
813[13]	125	2500	20	800	30	10	5	16.3	0.25	14	5BA	C·T·O	1250	300	0	−75	180	35	12	1.7	—	170
												C·T·O	2250	400	0	−155	220	40	15	4	—	375
												AB1	2500	750	0	−95	25/145	27[7]	0	0	—	245
												AB2[6]	2000	750	0	−90	40/315	1.5/58	230[8]	0.17	16K	455
												AB2[6]	2500	750	0	−95	35/360	1.2/55	235[8]	0.35[7]	17K	650
4-125A 4D21 6155	125	3000	20	600	120	5	6.5	10.8	0.07	3.1	5BK	C·T·O	2000	350	—	−100	200	50	12	2.8	—	275
												C·T·O	3000	350	—	−150	167	30	9	2.5	—	375
												AB2[6]	2500	350	—	−43	93/260	0/6	178[8]	1.0[7]	22K	400
												AB1[6]	2500	600	—	−96	50/232	0.3/8.5	192[8]	0	20.3K	330
												GG	2000	0	—	0	10/105[17]	30[17]	55[17]	16[17]	10.5K	145
4E27A/ 5-125B	125	4000	20	750	75	5	7.5	10.5	0.08	4.7	7BM	C·	3000	500	60	−200	167	5	6	1.6	—	375
													1000	750	0	−170	160	21	3	0.6	—	115
803	125	2000	30	600	20	10	5	17.5	0.15	29	5J	C·T	2000	500	40	−90	160	45	12	2	—	210
												C·P	1600	400	100	−80	150	45	25	5	—	155
7094	125	2000	20	400	60	6.3	3.2	9.0	0.5	1.8	Fig. 82	C·T	1500	400	—	−100	330	20	5	4	—	340
												C·P	1200	400	—	−130	275	20	5	5	—	240
												AB1	2000	400	—	−65	30/200	35[7]	60[8]	0	12K	250
4X150A 4X150G[15]	150[9]	2000	12	400	500	6	2.6	15.5	0.03	4.5	Fig. 75	C·T·O	1250	250	—	−90	200	20	10	0.8	—	195
												C·P	1000	250	—	−105	200	20	15	2	—	140
						2.5	6.25	27	0.035	4.5	—	AB2[6]	1250	300	—	−44	475[7]	0/65	100[8]	0.15[7]	5.6K	425
8121	150	2200	8	400	500	13.5	1.3	16	0.13	0.011	Fig. 5	C·T·O	1000	200	—	−30	300	10	30	5	—	165
4-250A 5D22 6156	250[9]	4000	35	600	110	5	14.5	12.7	0.12	4.5	5BK	C·T·O	2500	500	—	−150	300	60	9	1.7	—	575
												C·T·O	3000	500	—	+180	345	60	10	2.6	—	800
												C·P	2500	400	—	−200	200	30	9	2.2	—	375
												C·P	3000	400	—	−310	225	30	9	3.2	—	510
												AB2[6]	2000	300	—	−48	510[7]	0/26	198[8]	5.5[7]	8K	650
												AB1[6]	2500	600	—	−110	430[7]	0.3/13	180[8]	0	11.4K	625
4X250B	250[9]	2000	12	400	175	6	2.1	18.5	0.04	4.7	Fig. 75	C·T·O	2000	250	—	−90	250	25	27	2.8	—	410
												C·P	1500	250	—	−100	200	25	17	2.1	—	250
												AB1[6]	2000	350	—	−50	500[7]	30[7]	100[8]	0	8.26K	650
7034/[9] 4X150A	250	2000	12	300	150	6	2.6	16	0.03	4.4	Fig. 75	C·T·O	2000	250	—	−88	250	24	8	2.5	—	370
												C·P	1600	250	—	−118	200	23	5	3	—	230
7035/ 4X150D	250	2000	12	400		26.5	0.58				Fig. 75	AB2[6]	2000	300	—	−50	100/500	0/36	106[8]	0.2	8.1K	630
												AB1[6]	2000	300	—	−50	100/470	0/36	100[8]	0	8.76K	580
4CX-300A	300[9]	2000	12	400	500	6	2.75	29.5	0.04	4.8	—	C·T	2000	250	—	−90	250	25	27	2.8	—	410
												C·P	1500	250	—	−100	200	25	17	2.1	—	250
												AB1[6]	2000	350	—	−50	500[7]	30[7]	100[8]	0	8.26K	650
175A	400	4000	25	600		5	14.5	15.1	0.06	9.8	Fig. 86	C·T·C·P	4000	600	0	−200	350	29	6	1.4	—	960
												C·T·C·P	2500	600	0	−180	350	40	7	1.6	—	600
												AB1	2500	750	—	−143	100/350	1/35	0	0	—	570
4-400A	400[9]	4000	35	600	110	5	14.5	12.5	0.12	4.7	5BK	C·T·C·P	4000	300	—	−170	270	22.5	10	10	—	720
												GG	2500	0	—	0	80/270[17]	55[17]	100[17]	38[17]	4.0K	325
												AB1	2500	750	—	−130	95/317	0/14	0	0	—	425
8122	400	2200	8	400	500	13.5	1.3	16	0.13	0.011	Fig. 86	C·T·O	2000	200	—	−30	300	5	30	5	—	300
5-500A	500	4000	35	600	30	10	10.2	19	0.10	12	—	C·T	3000	500	0	−220	432	65	35	12	—	805
												C·T	3100	470	0	−310	260	50	15	6	—	580
												AB1	3000	750	0	−112	320	26	—	—	—	612
8166/ 4-1000A	1000	6000	75	1000	—	7.5	21	27.2	.24	7.6	—	C·T	3000	500	—	−150	700	146	38	11	—	1430
												C·P	3000	500	—	−200	600	145	36	12	—	1390
												AB2	4000	500	—	−60	300/1200	0/95	—	11	7K	3000
												GG	3000	0	—	0	100/700[17]	105[17]	170[17]	130[17]	2.5K	1475
4CX1000A	1000	3000	12	400	400	6	12.5	35	.005	12	—	AB1[6]	2000	325	—	−55	500/2000	−4/60	—	—	2.8K	2160
												AB1[6]	2500	325	—	−55	500/2000	−4/60	—	—	3.1K	2920
													3000	325	—	−55	500/1800	−4/60	—	—	3.85K	3360
8295/ 172	1000	3000	30	600	—	6	8.2	38	.09	18	—	C·T	2000	500	35	−175	850	42	10	1.9	—	1155
												C·T	2500	500	35	−200	840	40	10	2.1	—	1440
												C·T	3000	500	35	−200	820	42	10	2.1	—	1770
												AB1	2000	500	35	−110	200/800	12/43	110[8]	—	2.65K	1040
												AB1	2500	500	35	−110	200/800	11/40	115[8]	—	3.5K	1260
												AB1	3000	500	35	−115	220/800	11/39	115[8]	—	4.6K	1590

[1] Grid-resistor.
[2] Doubler to 175 MHz.
[3] Dual tube. Values for both sections, in push-pull. Interelectrode capacitances, however, are for each section.
[4] Tripler to 175 MHz.
[5] Filament limited to intermittent operation.
[6] Values are for two tubes.
[7] Max.-signal value.
[8] Peak grid-to-grid volts.
[9] Forced-air cooling required.
[10] Two tubes triode connected, G_2 to G_1 through 20K Ω. Input to G_2.
[11] Tripler to 200 MHz.
[12] Typical Operation at 175 MHz.
[13] ± 1.5 volts.

[14] KEY TO CLASS-OF-SERVICE ABBREVIATIONS
AB1 = Class-AB1.
AB2 = Class-AB2.
B = Class-B push-pull af modulator.
C·M = Frequency multiplier.
C·P = Class-C plate-modulated telephone.
C·T = Class-C telegraph.
C·T·O = Class-C amplifier-osc.
GG = Grounded-grid (grid and screen connected together).
[15] No Class B data available.
[16] HK257B 120 MHz. full rating.
[17] Single tone.

Table 3
TV Deflection Tubes

Type	Plate Dissipation Watts	Screen Dissipation Watts	Transconductance Micromhos	Heater (6.3V) Amperes	Cin pF	Cgp pF	Cout pF	Base	Class of Service	Plate Voltage	Screen Voltage	Grid Voltage	Plate Current mA	Screen Current mA	Grid Current mA	Approx. Driving Power Watts	Approx. Output Power Watts
																RF Operation (Up to 30 MHz)	
6DQ5	24	3.2	10.5k	2.5	23	0.5	11	8JC									
6DQ6B	18	3.6	7.3k	1.2	15	0.5	7	6AM	C	400	200	−40	100	12	1.5	0.1	25
6FH6	17	3.6	6k	1.2	33	0.4	8	6AM									
6GC6	17.5	4.5	6.6k	1.2	15	0.55	7	8JX									
6GJ5	17.5	3.5	7.1k	1.2	15	0.26	6.5	9NM	C	500	200	−75	180	15	5	0.43	63
									AB1	500	200	−43	85	4			35
6HF5	28	5.5	11.3k	2.25	24	0.56	10	12FB	C	500	140	−85	232	12.5	8	0.76	77
									AB1	500	140	−46	133	4.5			58
6JB6	17.5	3.5	7.1k	1.2	15	0.2	6	9QL	C	500	200	−75	180	13.3	5	0.43	63
									AB1	500	200	−42	85	4.2			35
6JE6C	30	5	10.5k	2.5	24.3		14.5	9QL	C	500	125	−85	222	17	8	0.82	76
									AB1	500	125	−44	110	3.9			47
6JG6A	17	3.5	10k	1.6	22	0.7	9	9QU	C	450	150	−80	202	20	8	0.75	63
									AB1	450	150	−35	98	4.5			38
6JM6	17.5	3.5	7.3k	1.2	16	0.6	7	12FJ	C	500	200	−75	190	13.7	4	0.32	61
									AB1	500	200	−42	85	4.4			37
6JN6	17.5	3.5	7.3k	1.2	16	0.34	7	12FK									
6JS6C	30	5.5		2.25	24	0.7	10	12FY	GG	800	0	−11	150			12.5	82
6KD6	33	5	14k	2.85	40	0.8	16	12GW									
6LB6	30	5	13.4k	2.25	33	0.4	18	12GJ									
6LG6	28	5	11.5k	2	25	0.8	13	12HL									
6LQ6	30	5	9.6k	2.5	22	0.46	11	9QL									
6MH6	38.5	7	14k	2.65	40	1.0	20	12GW									

Note: For AB1 operation, input data is average
2-tone value. Output power is PEP.

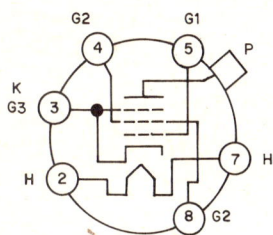

8JX

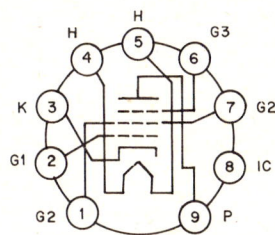

9QU

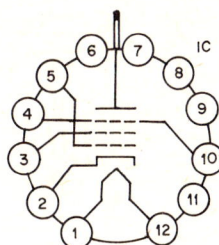

12FJ

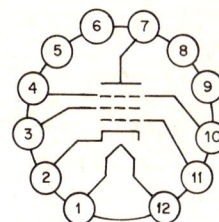

12FK

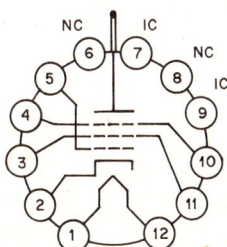

12FY

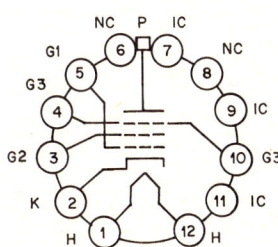

12GJ

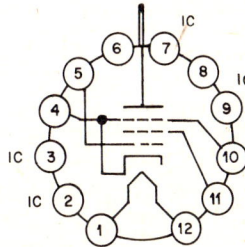

12HL

Table 4

RF Small-Signal and Power Transistors
High Frequency, Low Voltage Amplifier Transistors

The transistors listed in this table are specified for operation in RF Power amplifiers and are listed by specific application at a given test frequency. Modulation type is given in each application heading. *Courtesy Motorola Semiconductor Prod. Inc.*

Device Type	P_{in} Input Power Watts	P_{out} Output Power Watts	G_{PE} Power Gain dB Min	V_{CC} Supply Voltage Volts	Package
2-30 MHz, SSB Transistors					
MRF476	0.1	3.0 PEP	15	12.5	TO-220
2N6367	0.36	9.0 PEP	14	12.5	211-07
MRF475	1.2	12 PEP	10	13.6	TO-220
MRF432*	0.125	12.5 PEP	20	12.5	211-07
MRF433*	0.125	12.5 PEP	20	12.5	211-07
MRF406	1.25	20 PEP	12	12.5	211-07
MRF460	2.5	40 PEP	12	12.5	211-10
MRF421	10	100 PEP	10	12.5	211-08

*PNP/NPN Complements for Complementary Symmetry Driver. For Matched Pairs, Order MK433.

Device Type	P_{in} Input Power Watts	P_{out} Output Power Watts	G_{PE} Power Gain dB Min	V_{CC} Supply Voltage Volts	Package
2N6370	0.62	10 PEP	12	28	211-07
MRF432	0.125	12.5 PEP	20	12.5	211-07
MRF433	0.125	12.5 PEP	20	12.5	211-07
2N5070	1.25	25 PEP	13	28	TO-60
MRF401	1.25	25 PEP	13	28	145A-09
MRF427A	1.56	25 PEP	12	50	145A-10
2N5941	2.0	40 PEP	13	28	211-07
MRF463	2.53	80 PEP	15	28	211-08
MRF464	2.53	80 PEP	15	28	211-11
MRF464A	2.53	80 PEP	15	28	145A-10
MRF422	15	150 PEP	10	28	211-08
MRF428	7.5	150 PEP	13	50	211-08
MRF428A	7.5	150 PEP	13	50	307-01
14-30 MHz Amateur Transistors					
MRF8003	0.025	0.5	13	12.5	TO-39
MRF8004	0.35	3.5	10	12.5	TO-39
MRF449	0.30	30	10	13.6	211-07
MRF449A	0.30	30	10	13.6	145A-09
MRF450	4.0	50	11	13.6	211-07
MRF450A	4.0	50	11	13.6	145A-09
MRF453	4.8	60	11	12.5	211-10
MRF453A	4.8	60	11	12.5	145A-10
MRF455	4.8	60	11	12.5	211-07
MRF455A	4.8	60	11	12.5	145A-09
MRF454	5.0	80	12	12.5	211-11
MRF454A	5.0	80	12	12.5	145A-10
130-175 MHz VHF FM Transistors					
MRF604	0.1	1.0	10	12.5	TO-46
2N4427	0.1	1.0	10	12	TO-39
MRF607	0.1	1.75	12.5	12.5	TO-39
2N6255	0.5	3.0	7.8	12.5	TO-39
2N5589	0.44	3.0	8.2	13.6	144B-05
MRF237*	0.25	4.0	12	12.5	TO-39
2N6080	0.25	4.0	12	12.5	145A-09
2N5590	3.0	10	5.2	13.6	145A-09
MRF212	1.25	10	9.0	12.5	145A-09
2N6081	3.3	15	6.3	12.5	145A-09
MRF221	3.3	15	6.3	12.5	211-07
MRF215**	0.33	20	8.2	12.5	316.01
2N5591	9.0	25	4.4	13.6	145A-09
2N6082	6.0	25	6.2	12.5	145A-09
MRF222	6.0	25	6.2	12.5	211-01
2N6083	8.1	30	5.7	12.5	145A-09
MRF223	8.1	30	5.7	12.5	211-07
2N6084	14.3	40	4.5	12.5	145A-09
MRF224	14.3	40	4.5	12.5	211-07

Device Type	P_{in} Input Power Watts	P_{out} Output Power Watts	G_{PE} Power Gain dB Min	V_{CC} Supply Voltage Volts	Package
MRF216**	8.5	40	6.7	12.5	316-01
MRF243**	12.0	60	7.0	12.5	316-01
MRF245**	18.2	80	6.4	12.5	316-01

*Grounded Emitter TO-39 Package.
**Controlled "Q" Transistor.

Device Type	P_{in} Input Power Watts	P_{out} Output Power Watts	G_{PE} Power Gain dB Min	V_{CC} Supply Voltage Volts	Package
220 MHz FM Transistors					
MRF207	0.15	1.0	8.2	12.5	TO-39
MRF225	0.18	1.5	9.0	12.5	TO-39
MRF227*	0.13	3.0	13.5	12.5	TO-39
MRF208	0.1	10	10	12.5	145A-09
MRF226	1.6	13	9.0	12.5	145A-09
MRF209	9.1	25	4.4	12.5	145A-09

*Grounded Emitter TO-39 Package.

Device Type	P_{in} Input Power Watts	P_{out} Output Power Watts	G_{PE} Power Gain dB Min	V_{CC} Supply Voltage Volts	Package
407-512 MHz, UHF FM Transistors					
2N6256	0.05	0.5	10	12.5	249-05
MRF626	0.05	0.5	10	12.5	305-01
MRF627	0.05	0.5	10	12.5	305A-01
MRF628	0.05	0.5	10	12.5	249-05
MRF515	0.12	0.75	8.0	12.5	TO-39
2N3948	0.25	1.0	6.0	13.6	TO-39
2N5644	0.20	1.0	7.0	12.5	145A-09
MRF629*	0.32	2.0	8.0	12.5	TO-39
2N5944	0.25	2.0	9.0	12.5	244-04
2N5945	0.64	4.0	8.0	12.5	244-04
2N5946	2.5	10	6.0	12.5	244-04
MRF641**	3.75	15	6.0	12.5	316.01
MRF644**	5.9	25	6.2	12.5	316-01
MRF646**	13.3	40	4.8	12.5	316-01
MRF648**	22	60	4.4	12.5	316-01

*Grounded Emitter TO-39 Package. Case 79-03
**Controlled "Q" Transistor.

Device Type	P_{in} Input Power Watts	P_{out} Output Power Watts	G_{PE} Power Gain dB Min	V_{CC} Supply Voltage Volts	Package
106-175 MHz, VHF A-M Transistors					
2N3866	0.1	1.0	10	28	TO-39
2N3553	0.25	2.5	10	28	TO-39
2N5641	1.0	7.0	8.4	28	144-05
2N5642	3.0	20	8.2	28	145A-09
MRF314	3.0	30	10	28	211-07
MRF314A	3.0	30	10	28	145A-09
2N5643	6.9	40	7.6	28	145A-09
MRF315	5.7	45	9.0	28	211-07
MRF315A	5.7	45	9.0	28	145A-09
MRF316*	8.0	80	10	28	316-01
MRF317*	12.5	100	9.0	28	316-01

*Controlled "Q" Transistor.

Device Type	P_{in} Input Power Watts	P_{out} Output Power Watts	G_{PE} Power Gain dB Min	V_{CC} Supply Voltage Volts	Package
806-947 MHz, UHF FM Transistors					
MRF816	0.075	0.75	10	12.5	249.05
MRF838	0.22	1.0	6.5	12.5	305A-01
MRF838A	0.22	1.0	6.5	12.5	305A-01
MRF817	0.59	2.5	6.2	13.6	244-04
MRF840	1.1	8.0	8.0	12.5	319-01
MRF842	5.0	20	6.0	12.5	319-01
MRF844	8.1	30	5.7	12.5	319-01
MRF846	10	40	6.0	12.5	319-01

Table 5

UHF and Microwave Oscillators

The transistors listed below are for uhf and microwave-oscillator applications as initial signal sources or as output stages of low-power transmitters. Devices are listed in order of increasing test frequency.

Device Type	Test Conditions f MHz	V_{CC} Volts	P_{out} mW Min	f_T MHz Min	Package	Device Type	Test Conditions F MHz	V_{CC} Volts	P_{out} mW Min	f_T MHz Min	Package
2N5179	500	10	20	900	TO-72	2N5108	1680	20	300	1200	TO-39
2N2857	500	10	30	1000	TO-72	MRF905	1680	20	500	2200	TO-46
2N3839	500	6.0	30	1000	TO-72	2N3866	400	15	1000	500	TO-39
MM8009	1680	20	200	1000	TO-39						

Table 6

Low-Noise Transistors

The low-noise devices listed are produced with carefully controlled $r_{b'}$ and f_T to optimize device noise performance. Devices listed in the matrix are classified according to noise figure performance versus frequency.

NF dB	Frequency MHz 60	100	200	450	1000	2000	Polarity
1.5	2N5829	2N5829					PNP
	2N5031	2N5031	MRF904				NPN
2.0	2N4957	2N4957	2N5829				PNP
	2N5032	2N5032	2N5031	MRF904	MRF901		NPN
2.5	2N4958	2N4958	2N4957	2N5829			PNP
	2N5032	2N5032	2N5032	2N5031	MRF901		NPN
					2N6603		NPN
3.0	2N4959	2N4959	2N4958	2N4957	2N5829		PNP
	2N2857	2N2857	2N5032	3N5032	MRF901	MRF902	NPN
					2N6604		NPN
3.5	2N4959	2N4959	2N4959	2N4958	2N4957		PNP
	2N5179	2N5179	2N2857	3N5032	2N5031	MRF901	NPN
4.0	2N4959	2N4959	2N4959	2N4959	2N4958		PNP
	2N5179	2N5179	2N5179	2N2857	2N5031		NPN
4.5	2N4959	2N4959	2N4959	2N4959	2N4959		PNP
	2N5179	2N5179	2N5179	2N2857	2N5032		NPN

Table 7

Small-Signal RF Transistors

High-Speed Switches

The transistors listed below are for use as high-frequency current-mode switches. They are also suitable for rf amplifier and oscillator applications. The devices are listed in ascending order of collector current.

Device Type	Test Conditions I_C/V_{CE} mA/Volts	f_T MHz Min.	$r_{b'}C_c$ Max	Package
MD4957	2.0/10	1000	20	TO-78
2N3959	10/10	1300	25	TO-18
2N3960	10/10	1600	40	TO-18
2N5835	10/6.0	2500	5.0**	TO-72
MM4049*	20/5.0	4000	15	TO-72
MRF914	20/10	4500**	—	TO-72
2N5842	25/4.0	1700	40	TO-72
2N5841	25/4.0	2200	25	TO-72
MRF531	50/25	500	—	TO-39
MRF532*	50/25	500	—	TO-39
2N5583*	50/10	1000	8.0**	TO-39
2N5836	50/6.0	2000	6.0**	TO-46
2N5837	100/3.0	1700	6.0**	TO-46

*PNP
**Typ.

VHF and UHF Class A Linear Transistors

The devices listed below are excellent for Class A linear applications. The devices are listed according to increasing current-gain (f_T).

Device Type	Nominal Test Conditions V_{CE}/I_C Volts/mA	f_T MHz Min.	Noise Figure Max/Freq. MHz	Distortion Specifications 2nd Order IMD	3rd Order IMD	Output Level dbMV	Package
MRF501	6/5	600	4.5*/200				TO-72
MRF502	6/5	800	4.0*/200				TO-72
2N5179	6/5	900	4.5/200				TO-72
BFY90	5/2	1000	5.0/500				TO-72
BFW92	5/2	1000	4.0/500				302A-01
2N6305	5/10	1200	5.5/450				TO-72
BFX89	5/25	1200	6.5/500				TO-72
2N5109	15/50	1200	3.0*/200				TO-39
2N5943	15/50	1200	3.4/200	−50		+50	TO-39
2N6304	5/10	1400	4.5/450				TO-72
MRF511	20/80	1500	7.3*/200	−50	−65	+50	144D-04
MRF517	15/60	2200	7.5/300	−60	−72	+45	TO-39
BFR90	10/14	5000*	2.4*/500				302A-01
BFR91	5/35	5000*	1.9*/500				302A-01
BFR96	10/50	5000*	3.3*/500				302A-01

*Typ.

Table 8
Package Information

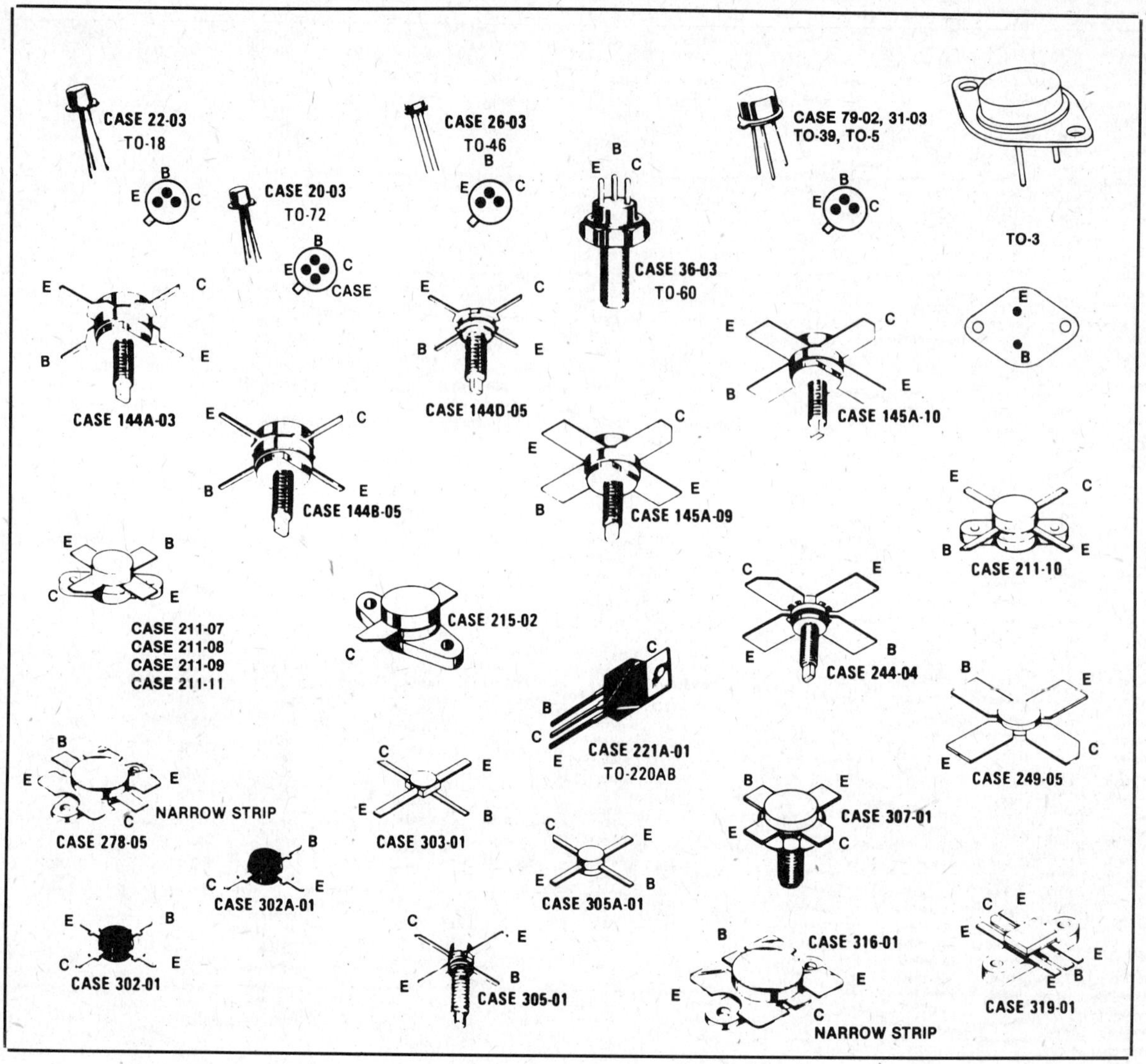

CASE 22-03 TO-18

CASE 20-03 TO-72

CASE 26-03 TO-46

CASE 36-03 TO-60

CASE 79-02, 31-03 TO-39, TO-5

TO-3

CASE 144A-03

CASE 144D-05

CASE 145A-10

CASE 144B-05

CASE 145A-09

CASE 211-10

CASE 211-07
CASE 211-08
CASE 211-09
CASE 211-11

CASE 215-02

CASE 244-04

CASE 221A-01 TO-220AB

CASE 249-05

CASE 278-05 NARROW STRIP

CASE 303-01

CASE 307-01

CASE 302A-01

CASE 305A-01

CASE 316-01

CASE 302-01

CASE 305-01

NARROW STRIP

CASE 319-01

Table 10
Power FETs

Device No.	Type	Max. Diss. (W)	Max. V_{DS} (volts)	Max. I_D (A)	G_{fs} μmhos (typ.)	Input C C_{iss} (pF)	Output C C_{oss}	Approx. Upper Freq. (MHz)	Case Type	Base Conn. Mfgr.	Applications (general)
DV1202S	N-Chan. (No Zener)	10	50	0.5	100k	14	20	500	.380 SOE	1 / S	RF pwr. amp., osc.
DV1202W	N-Chan. (No Zener)	10	50	0.5	100k	14	20	500	C-220	5 / S	RF pwr. amp., osc.
DV1205S	N-Chan. (No Zener)	20	50	1	200k	26	38	500	.380 SOE	1 / S	RF pwr. amp., osc.
DV1205W	N-Chan. (No Zener)	20	50	1	200k	26	38	500	C-220	5 / S	RF pwr. amp., osc.
2SK133	N-Chan.	100	120	7	1M	600	350	1	TO-3	6 / H	AF pwr. amp., switch (complement to 2SJ48)
2SK134	N-Chan.	100	140	7	1M	600	350	1	TO-3	6 / H	AF pwr. amp., switch (complement to 2SJ49)
2SK135	N-Chan.	100	160	7	1M	600	350	1	TO-3	6 / H	AF pwr. amp., switch (complement to 2SJ50)
2SJ48	P-Chan.	100	120	7	1M	900	400	1	TO-3	6 / H	AF pwr. amp., switch (complement to 2SK133)
2SJ49	P-Chan.	100	140	7	1M	900	400	1	TO-3	6 / H	AF pwr. amp., switch (complement 2SK134)
2SJ50	P-Chan.	100	160	7	1M	900	400	1	TO-3	6 / H	AF pwr. amp., switch (complement 2SK135)

Table 9
Suggested Small-Signal FETs

Device No.	Type	Max. Diss. (mW)	Max. V_{DS} (volts)	$V_{GS(off)}$ volts	Min. gfs (μmhos)	Input C (pF)	Max. I_D (mA)*	Upper Freq. (MHz)	Noise Figure (typ.)	Case Type	Base Conn.	Mfgr. (see code)	Applications (general)
2N4416	N-JFET	300	30	−6	4500	4	15	450	400 MHz 4 dB	TO-72	1	S, M	vhf/uhf/rf amp., mix., osc.
2N5484	N-JFET	310	25	−3	2500	5	30	200	200 MHz 4dB	TO-92	2	M	vhf/uhf amp., mix., osc.
2N5485	N-JFET	310	25	−4	3500	5	30	400	400 MHz 4 dB	TO-92	2	S	vhf/uhf rf amp., mix., osc.
3N200	N-Dual-Gate MOSFET	330	20	−6	10,000	4-8.5	50	500	400 MHz 4.5 dB	TO-72	3	R	vhf/uhf rf amp., mix., osc.
3N202	N-Dual-Gate MOSFET	360	25	−5	8000	6	50	200	200 MHz 4.5 dB	TO-72	3	S	vhf amp., mixer
MPF102	N-JFET	310	25	−8	2000	4.5	20	200		TO-92	2	N, M	hf/vhf amp., mix., osc.
MPF106/ 2N5484	N-JFET	310	25	−6	2500	5	30	400	200 MHz 4 dB	TO-92	2	N, M	hf/uhf/uhf amp., mix., osc.
40673	N-Dual-Gate MOSFET	330	20	−4	12,000	6	50	400	200 MHz 6 dB	TO-72	3	R	hf/vhf/uhf amp., mix., osc.
U300	P-JFET	300	−40	+10	8000	20	−50	—	40 nV/√Hz	TO-18	4	S	General-purpose amp.
U304	P-JFET	350	−30	+10		27	−50	—	—	TO-18	4	S	analog switch, chopper
U310	N-JFET	500	30	−6	10,000	2.5	60	450	450 MHz 3.2 dB	TO-52	5	S	common-gate vhf/uhf amp., osc., mix.
U350	N-JFET Quad	1W	25	−6	9000	5	60	100	100 MHz 7 dB	TO-99	6	S	matched JFET doubly bal. mix.
U431	N-JFET Dual	300	25	−6	10,000	5	30	100		TO-99	7	S	matched JFET cascode amp. and bal. mix.

*25°C   M = Motorola.   N = National Semiconductor.   R = RCA.   S = Siliconix Inc.   D = Drain.   S = Source.   G = Gate.

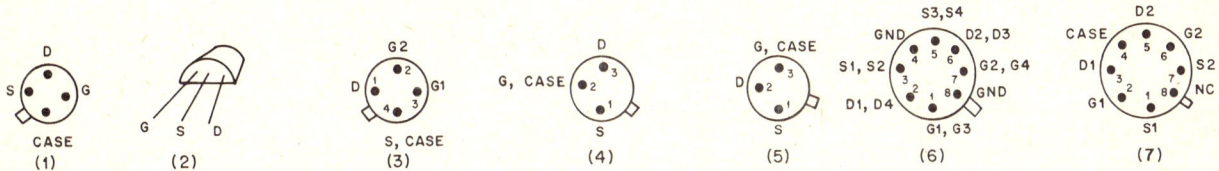

Table 10, Continued
Power FETs

Device No.	Type	Max. Diss. (W)*	Max. V_{DS} (Volts)	Max. I_D (A)	Gfs μmhos (typ.)	Input C C_{iss} (pF)	Output C C_{OSS} (pF)	Approx. Upper Freq. (MHz)	Case Type	Base Conn. Mfgr.	Applications (general)
VMP4	N-Chan. (No Zener)	25	60	2	170,000	32	4.8	200	.380-SOE	1/S	VHF pwr. amp., rcvr front end (rf amp., mixer).
VN10KM	N-Chan. (Zenered)	1	60	0.5	100,000	48	16	—	TO-92	2/S	High-speed line driver, relay driver, LED stroke driver
VN64GA	N-Chan. (No Zener)	80	60	12.5	150,000	700	325	30	TO-3	3/S	Linear amp., power-supply switch, motor control
VN66AF	N-Chan. (Zenered)	15	60	2	150,000	50	50	—	TO-202	4/S	High-speed switch, hf linear amp., audio amp. line driver.
VN66AK	N-Chan. (No Zener)	8.3	60	2	250,000	33	6	100	TO-39	5/S	RF pwr. amp., high-current analog switching.
VN67AJ	N-Chan. (No Zener)	25	60	2	250,000	33	7	100	TO-3	3/S	RF pwr. amp., high-current switching.
VN89AA	N-Chan. (Zenered)	25	80	2	250,000	50	10	100	TO-3	3/S	High-speed switching, hf linear amps., line drivers.
IRF100	N-Chan. (No Zener)	125	80	16	300,000	900	25	—	TO-3	3/IR	High-speed switching, audio amps., motor control, inverters.
IRF101	N-Chan. (No Zener)	125	60	16	300,000	900	25	—	TO-3	3/IR	Same as IRF100

*25°C (case)   S = Siliconix Inc.   IR = International Rectifier.   H = Hitachi   Zenered units have built-in 15-V Zener diode from gate to source.

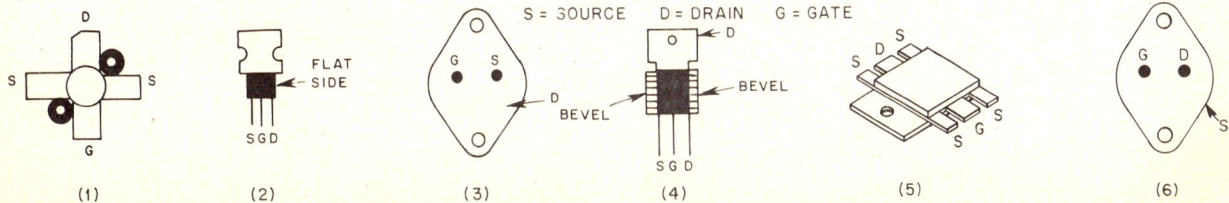

Table 11

Integrated Circuit Operational Amplifiers

Device Type	Fabrication Technology	Freq. Comp.	Max.* Supply Voltage	Min. Input Resistance	Max. Input Offset Voltage	Min. DC Open-loop Gain (dB)	Min. Output Current (mA)	Min. Small-sig. Band-width (MHz)	Min. Slew Rate (V/μS)	Notes
301	bipolar	ext	36	0.5M	7.5m	88	5	1	10	Bandwidth extendable with spec. comp.
324	bipolar	int	32		7m	100	10	1		Quad — Single-supply applications
353	bipolar/JFET	int	36	1T	2m			4	13	Dual — low noise
358	bipolar	int	32		2m	100	10	1		Dual — Single-supply applications
709	bipolar	ext	36	50k	7.5m	84	5	0.3	0.15	
741	bipolar	int	36	0.3M	6m	88	5	0.4	0.2	
741S	bipolar	int	36	0.3M	6m	86	5	1	3	Improved 741 for audio applications
747	bipolar	int	36	0.3M	6m	88	5	0.4	0.2	Dual 741
1456	bipolar	int	36	1M	12m	88	5	0.5	1	
1458	bipolar	int	36	0.15M	10m	84	4	0.5	0.3	Dual
1458S	bipolar	int	36	0.3M	6m	86	5	0.5	3	Improved 1458 for audio applications
3140	bipolar/MOSFET	int	36	1.5T	2m	86	1	3.7	9	Strobable output
5534	bipolar	int	44	30k	5m	100	38	10	13	Low noise — can swing 20 V Pk-pk across 600Ω
5556	bipolar	int								Same as type 1456
5558	bipolar	int								Same as type 1458
TL084	bipolar/JFET	int	36		15m				13	Quad — High-performance audio applications

* From V + to V − terminals

Table 12

IC Op Amp Base Diagrams (Top view — not to scale)

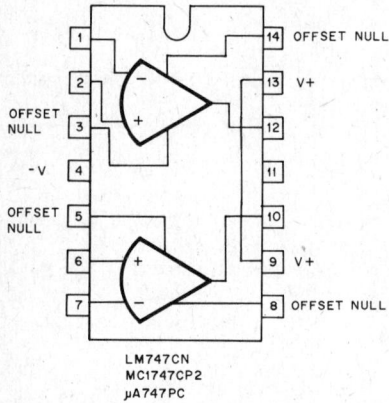

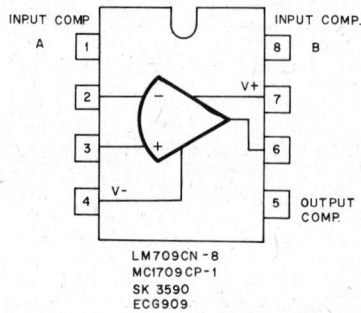

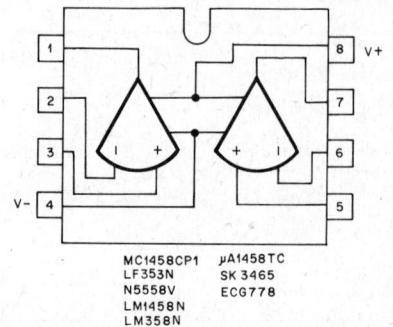

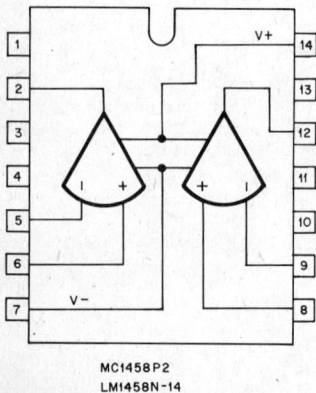

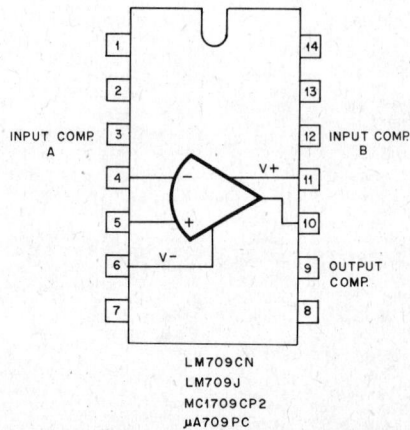

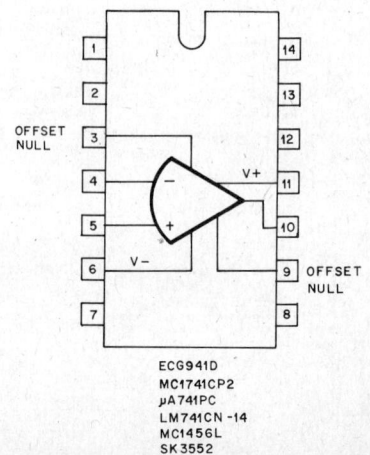

LM747CN
MC1747CP2
µA747PC

LM709CN-8
MC1709CP-1
SK 3590
ECG909

MC1458CP1
LF353N
N5558V
LM1458N
LM358N
µA1458TC
SK 3465
ECG778

MC1458P2
LM1458N-14

LM709CN
LM709J
MC1709CP2
µA709PC

ECG941D
MC1741CP2
µA741PC
LM741CN-14
MC1456L
SK3552

Table 12 (Cont.)

INPUT COMP. A   1

INPUT COMP B   8

V+   7

2

−

+

6

3

V−   5   OUTPUT COMP

4   CASE

MC1709G   SK3551
LM709H   µA709HC

OFFSET NULL   8

1

V+   7

2

−

+

6

3

V−   5   OFFSET NULL

4

LF356H   µAF356HC
LM741CH   MC1456G
MC1741G   SK3553
µA741HC

FREQ. COMP.
OFFSET NULL   8

OFFSET NULL   1

V+   7

2

−

+

6

3

V−   5   FREQ. COMP.

NE5534T

1   14

2   −   13   −

3   +   12   +

4   V+   11   V−

5   +   10   +

6   −   9   −

7   8

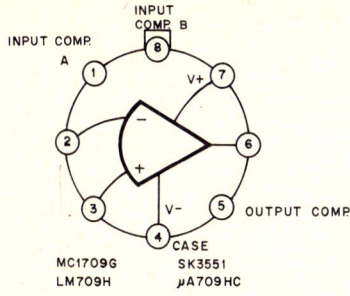

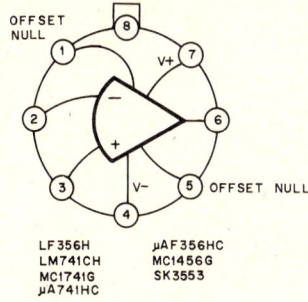

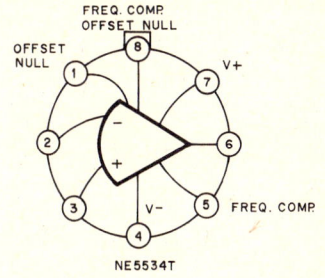

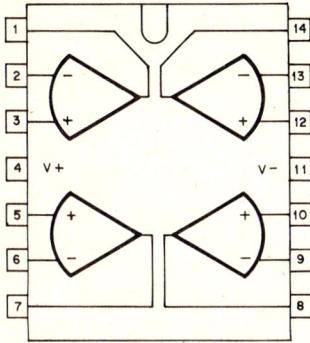

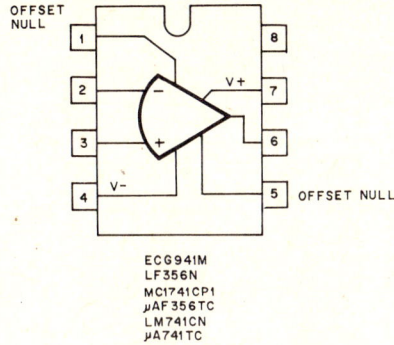

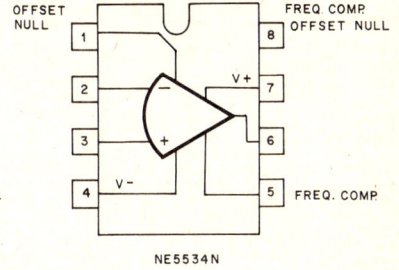

µA324PC
µA348PC
LM324N
LM348N
TL084CN

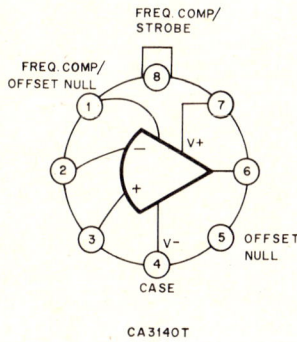

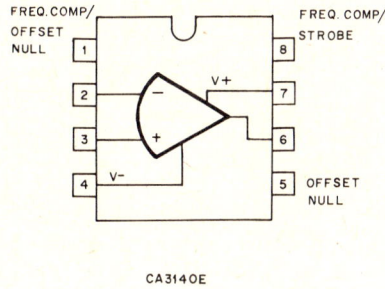

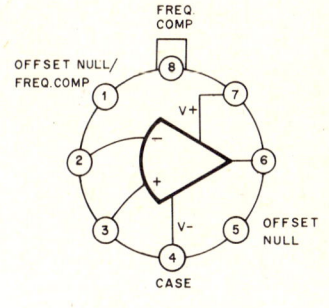

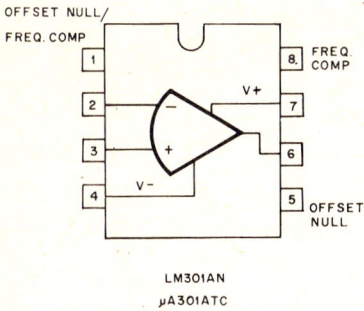

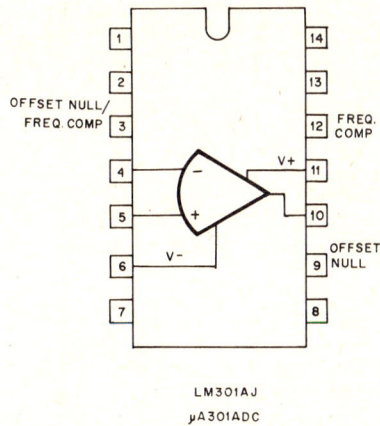

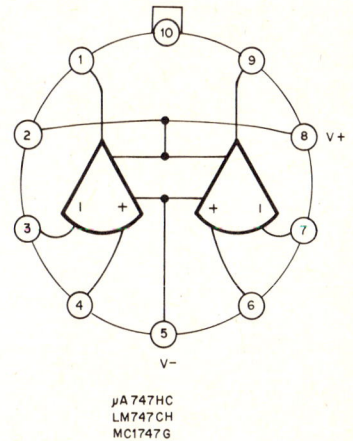

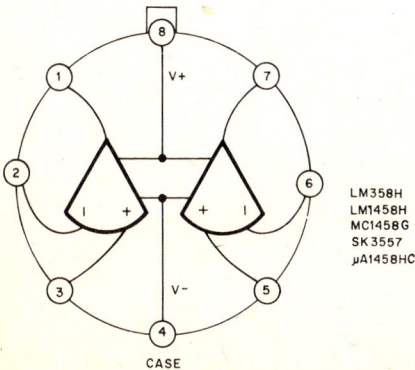

OFFSET NULL   8

1

2   −   7   V+

3   +   6

4   V−   5   OFFSET NULL

ECG941M
LF356N
MC1741CP1
µAF356TC
LM741CN
µA741TC

OFFSET NULL   8

FREQ. COMP.
OFFSET NULL

1

2   −   7   V+

3   +   6

4   V−   5   FREQ. COMP.

NE5534N

FREQ. COMP/
STROBE   8

FREQ. COMP/
OFFSET NULL   1

7

2   −

+   6

3   V+

4   V−   5   OFFSET NULL

CASE

CA3140T

FREQ. COMP/
OFFSET NULL   1   8   FREQ. COMP/
STROBE

2   −   7   V+

3   +   6

4   V−   5   OFFSET NULL

CA3140E

FREQ.
COMP   8

OFFSET NULL/
FREQ. COMP   1

7

2   −

+   6

3   V+

4   V−   5   OFFSET NULL

CASE

LM301AH
µA301AHC

OFFSET NULL/
FREQ. COMP   1   8   FREQ. COMP

2   −   7   V+

3   +   6

4   V−   5   OFFSET NULL

LM301AN
µA301ATC

1   14

2   13

OFFSET NULL/
FREQ. COMP   3   12   FREQ. COMP

4   −   11   V+

5   +   10

6   V−   9   OFFSET NULL

7   8

LM301AJ
µA301ADC

10

1   9

2   8   V+

3   −   +   7

4   6

5

V−

µA747HC
LM747CH
MC1747G

8

1   7

V+

2   6

−   +

3   5

V−

4   CASE

LM358H
LM1458H
MC1458G
SK3557
µA1458HC

MANUFACTURER'S PREFIXES

MC	MOTOROLA
LF, LM	NATIONAL
N, S	SIGNETICS
ECG	SYLVANIA
CA, SK	RCA
µA	FAIRCHILD
TL	TEXAS INSTRUMENTS